SIXTH EDITION

Introduction to Heat Transfer

THEODORE L. BERGMAN

Department of Mechanical Engineering
University of Connecticut

ADRIENNE S. LAVINE

Mechanical and Aerospace Engineering Department
University of California, Los Angeles

FRANK P. INCROPERA

College of Engineering
University of Notre Dame

DAVID P. DEWITT

School of Mechanical Engineering
Purdue University

JOHN WILEY & SONS, INC.

VICE PRESIDENT & PUBLISHER	Don Fowley
EXECUTIVE EDITOR	Linda Ratts
EDITORIAL ASSISTANT	Renata Marchione
MARKETING MANAGER	Christopher Ruel
PRODUCTION MANAGER	Dorothy Sinclair
PRODUCTION EDITOR	Sandra Dumas
DESIGNER	Wendy Lai
EXECUTIVE MEDIA EDITOR	Thomas Kulesa
PRODUCTION MANAGEMENT SERVICES	MPS Ltd.

This book was typeset in 10.5/12 Times Roman by MPS Limited, a Macmillan Company and printed and bound by R. R. Donnelley (Jefferson City). The cover was printed by R. R. Donnelley (Jefferson City).

Founded in 1807, John Wiley & Sons, Inc. has been a valued source of knowledge and understanding for more than 200 years, helping people around the world meet their needs and fulfill their aspirations. Our company is built on a foundation of principles that include responsibility to the communities we serve and where we live and work. In 2008, we launched a Corporate Citizenship Initiative, a global effort to address the environmental, social, economic, and ethical challenges we face in our business. Among the issues we are addressing are carbon impact, paper specifications and procurement, ethical conduct within our business and among our vendors, and community and charitable support. For more information, please visit our website: www.wiley.com/go/citizenship.

The paper in this book was manufactured by a mill whose forest management programs include sustained yield harvesting of its timberlands. Sustained yield-harvesting principles ensure that the number of trees cut each year does not exceed the amount of new growth.

This book is printed on acid-free paper. 

Copyright © 2011, 2007, 2002 by John Wiley & Sons, Inc. All rights reserved.

No part of this publication may be reproduced, stored in a retrieval system or transmitted in any form or by any means, electronic, mechanical, photocopying, recording, scanning or otherwise, except as permitted under Sections 107 or 108 of the 1976 United States Copyright Act, without either the prior written permission of the Publisher or authorization through payment of the appropriate per-copy fee to the Copyright Clearance Center, 222 Rosewood Drive, Danvers, MA 01923, (978) 750-8400, fax (978) 646-8600. Requests to the Publisher for permission should be addressed to the Permissions Department, John Wiley & Sons, Inc., 111 River Street, Hoboken, NJ 07030-5774, (201) 748-6011, fax (201) 748-6008.

Evaluation copies are provided to qualified academics and professionals for review purposes only, for use in their courses during the next academic year. These copies are licensed and may not be sold or transferred to a third party. Upon completion of the review period, please return the evaluation copy to Wiley. Return instructions and a free of charge return shipping label are available at www.wiley.com/go/returnlabel. If you have chosen to adopt this textbook for use in your course, please accept this book as your complimentary desk copy. Outside of the United States, please contact your local representative.

ISBN 13 978-0470-50196-2

Printed in the United States of America

10 9 8 7 6 5 4 3 2

Preface

In the Preface to the previous edition, we posed questions regarding trends in engineering education and practice, and whether the discipline of heat transfer would remain relevant. After weighing various arguments, we concluded that the future of engineering was bright and that heat transfer would remain a vital and enabling discipline across a range of emerging technologies including but not limited to information technology, biotechnology, pharmacology, and alternative energy generation.

Since we drew these conclusions, many changes have occurred in both engineering education and engineering practice. Driving factors have been a contracting global economy, coupled with technological and environmental challenges associated with energy production and energy conversion. The impact of a weak global economy on higher education has been sobering. Colleges and universities around the world are being forced to set priorities and answer tough questions as to which educational programs are crucial, and which are not. *Was our previous assessment of the future of engineering, including the relevance of heat transfer, too optimistic?*

Faced with economic realities, many colleges and universities have set clear priorities. In recognition of its value and relevance to society, investment in engineering education has, in many cases, *increased.* Pedagogically, there is renewed emphasis on the fundamental principles that are the foundation for *lifelong learning*. The important and sometimes dominant role of heat transfer in many applications, particularly in conventional as well as in alternative energy generation and concomitant environmental effects, has reaffirmed its relevance. We believe our previous conclusions were correct: The future of engineering is bright, and heat transfer is a topic that is crucial to address a broad array of technological and environmental challenges.

In preparing this edition, we have sought to incorporate recent heat transfer research at a level that is appropriate for an undergraduate student. We have strived to include new examples and problems that motivate students with interesting applications, but whose solutions are based firmly on fundamental principles. We have remained true to the pedagogical approach of previous editions by retaining a rigorous and systematic methodology for problem solving. We have attempted to continue the tradition of providing a text that will serve as a valuable, everyday resource for students and practicing engineers throughout their careers.

Approach and Organization

Previous editions of the text have adhered to four learning objectives:

1. The student should internalize the meaning of the terminology and physical principles associated with heat transfer.
2. The student should be able to delineate pertinent transport phenomena for any process or system involving heat transfer.
3. The student should be able to use requisite inputs for computing heat transfer rates and/or material temperatures.
4. The student should be able to develop representative models of real processes and systems and draw conclusions concerning process/system design or performance from the attendant analysis.

Moreover, as in previous editions, specific learning objectives for each chapter are clarified, as are means by which achievement of the objectives may be assessed. The summary of each chapter highlights key terminology and concepts developed in the chapter and poses questions designed to test and enhance student comprehension.

It is recommended that problems involving complex models and/or exploratory, what-if, and parameter sensitivity considerations be addressed using a computational equation-solving package. To this end, the *Interactive Heat Transfer* (*IHT*) package available in previous editions has been updated. Specifically, a simplified user interface now delineates between the basic and advanced features of the software. It has been our experience that most students and instructors will use primarily the basic features of *IHT*. By clearly identifying which features are advanced, we believe students will be motivated to use *IHT* on a daily basis. A second software package, *Finite Element Heat Transfer* (FEHT), developed by F-Chart Software (Madison, Wisconsin), provides enhanced capabilities for solving two-dimensional conduction heat transfer problems.

To encourage use of *IHT*, a *Quickstart User's Guide* has been installed in the software. Students and instructors can become familiar with the basic features of *IHT* in approximately one hour. It has been our experience that once students have read the Quickstart guide, they will use *IHT* heavily, even in courses other than heat transfer. Students report that *IHT* significantly reduces the time spent on the mechanics of lengthy problem solutions, reduces errors, and allows more attention to be paid to substantive aspects of the solution. Graphical output can be generated for homework solutions, reports, and papers.

As in previous editions, some homework problems require a computer-based solution. Other problems include both a hand calculation and an extension that is computer based. The latter approach is time-tested and promotes the habit of checking a computer-generated solution with a hand calculation. Once validated in this manner, the computer solution can be utilized to conduct parametric calculations. Problems involving both hand- and computer-generated solutions are identified by enclosing the exploratory part in a red rectangle, as, for example, [(b)], [(c)], or [(d)]. This feature also allows instructors who wish to limit their assignments of computer-based problems to benefit from the richness of these problems without assigning their computer-based parts. Solutions to problems for which the number is highlighted (for example, [1.19]) are entirely computer based.

What's New in the Sixth Edition

Chapter-by-Chapter Content Changes In the previous edition, *Chapter 1 Introduction* was modified to emphasize the relevance of heat transfer in various contemporary applications. Responding to today's challenges involving energy production and its environmental impact, an expanded discussion of the efficiency of energy conversion and the production of greenhouse gases has been added. Chapter 1 has also been modified to embellish the complementary nature of heat transfer and thermodynamics. The existing treatment of the first law of thermodynamics is augmented with a new section on the relationship between heat transfer and the second law of thermodynamics as well as the efficiency of heat engines. Indeed, the influence of heat transfer on the efficiency of energy conversion is a recurring theme throughout this edition.

The coverage of micro- and nanoscale effects in *Chapter 2 Introduction to Conduction* has been updated, reflecting recent advances. For example, the description of the thermophysical properties of composite materials is enhanced, with a new discussion of nanofluids. *Chapter 3 One-Dimensional, Steady-State Conduction* has undergone extensive revision and includes new material on conduction in porous media, thermoelectric power generation, and micro- as well as nanoscale systems. Inclusion of these new topics follows recent fundamental discoveries and is presented through the use of the thermal resistance network concept. Hence the power and utility of the resistance network approach is further emphasized in this edition.

Chapter 4 Two-Dimensional, Steady-State Conduction has been reduced in length. Today, systems of linear, algebraic equations are readily solved using standard computer software or even handheld calculators. Hence the focus of the shortened chapter is on the application of heat transfer principles to derive the systems of algebraic equations to be solved and on the discussion and interpretation of results. The discussion of Gauss–Seidel iteration has been moved to an appendix for instructors wishing to cover that material.

Chapter 5 Transient Conduction was substantially modified in the previous edition and has been augmented in this edition with a streamlined presentation of the lumped-capacitance method.

Chapter 6 Introduction to Convection includes clarification of how temperature-dependent properties should be evaluated when calculating the convection heat transfer coefficient. The fundamental aspects of compressible flow are introduced to provide the reader with guidelines regarding the limits of applicability of the treatment of convection in the text.

Chapter 7 External Flow has been updated and reduced in length. Specifically, presentation of the similarity solution for flow over a flat plate has been simplified. New results for flow over noncircular cylinders have been added, replacing the correlations of previous editions. The discussion of flow across banks of tubes has been shortened, eliminating redundancy without sacrificing content.

Chapter 8 Internal Flow entry length correlations have been updated, and the discussion of micro- and nanoscale convection has been modified and linked to the content of Chapter 3.

Changes to *Chapter 9 Free Convection* include a new correlation for free convection from flat plates, replacing a correlation from previous editions. The discussion of boundary layer effects has been modified.

Aspects of condensation included in *Chapter 10 Boiling and Condensation* have been updated to incorporate recent advances in, for example, external condensation on finned tubes. The effects of surface tension and the presence of noncondensable gases in modifying condensation phenomena and heat transfer rates are elucidated. The coverage of forced convection condensation and related enhancement techniques has been expanded, again reflecting advances reported in the recent literature.

The content of *Chapter 11 Heat Exchangers* is experiencing a resurgence in interest due to the critical role such devices play in conventional and alternative energy generation technologies. A new section illustrates the applicability of heat exchanger analysis to heat sink design and materials processing. Much of the coverage of compact heat exchangers included in the previous edition was limited to a specific heat exchanger. Although general coverage of compact heat exchangers has been retained, the discussion that is limited to the specific heat exchanger has been relegated to supplemental material, where it is available to instructors who wish to cover this topic in greater depth.

The concepts of emissive power, irradiation, radiosity, and net radiative flux are now introduced early in *Chapter 12 Radiation: Processes and Properties,* allowing early assignment of end-of-chapter problems dealing with surface energy balances and properties, as well as radiation detection. The coverage of environmental radiation has undergone substantial revision, with the inclusion of separate discussions of solar radiation, the atmospheric radiation balance, and terrestrial solar irradiation. Concern for the potential impact of anthropogenic activity on the temperature of the earth is addressed and related to the concepts of the chapter.

Much of the modification to *Chapter 13 Radiation Exchange Between Surfaces* emphasizes the difference between geometrical surfaces and radiative surfaces, a key concept that is often difficult for students to appreciate. Increased coverage of radiation exchange between multiple blackbody surfaces, included in older editions of the text, has been returned to Chapter 13. In doing so, radiation exchange between differentially small surfaces is briefly introduced and used to illustrate the limitations of the analysis techniques included in Chapter 13.

Problem Sets Approximately 225 new end-of-chapter problems have been developed for this edition. An effort has been made to include new problems that (*a*) are amenable to short solutions or (*b*) involve finite-difference solutions. A significant number of solutions to existing end-of-chapter problems have been modified due to the inclusion of the new convection correlations in this edition.

Classroom Coverage

The content of the text has evolved over many years in response to a variety of factors. Some factors are obvious, such as the development of powerful, yet inexpensive calculators and software. There is also the need to be sensitive to the diversity of users of the text, both in terms of (*a*) the broad background and research interests of instructors and (*b*) the wide range of missions associated with the departments and institutions at which the text is used. Regardless of these and other factors, it is important that the four previously identified learning objectives be achieved.

Mindful of the broad diversity of users, the authors' intent is *not* to assemble a text whose content is to be covered, in entirety, during a single semester- or quarter-long course. Rather, the text includes both (*a*) fundamental material that we believe must be covered and (*b*) optional material that instructors can use to address specific interests or that can be covered in a second, intermediate heat transfer course. To assist instructors in preparing a syllabus for a *first course in heat transfer*, we have several recommendations.

Chapter 1 Introduction sets the stage for any course in heat transfer. It explains the linkage between heat transfer and thermodynamics, and it reveals the relevance and richness of the subject. It should be covered in its entirety. Much of the content of *Chapter 2 Introduction to Conduction* is critical in a first course, especially Section 2.1 The Conduction Rate Equation, Section 2.3 The Heat Diffusion Equation, and Section 2.4 Boundary and Initial Conditions. It is recommended that Chapter 2 be covered in its entirety.

Chapter 3 One-Dimensional, Steady-State Conduction includes a substantial amount of optional material from which instructors can *pick-and-choose* or defer to a subsequent, intermediate heat transfer course. The optional material includes Section 3.1.5 Porous Media, Section 3.7 The Bioheat Equation, Section 3.8 Thermoelectric Power Generation, and Section 3.9 Micro- and Nanoscale Conduction. Because the content of these sections is not interlinked, instructors may elect to cover any or all of the optional material.

The content of *Chapter 4 Two-Dimensional, Steady-State Conduction* is important because both (*a*) fundamental concepts and (*b*) powerful and practical solution techniques are presented. We recommend that all of Chapter 4 be covered in any introductory heat transfer course.

The optional material in *Chapter 5 Transient Conduction* is Section 5.9 Periodic Heating. Also, some instructors do not feel compelled to cover Section 5.10 Finite-Difference Methods in an introductory course, especially if time is short.

The content of *Chapter 6 Introduction to Convection* is often difficult for students to absorb. However, Chapter 6 introduces fundamental concepts and lays the foundation for the subsequent convection chapters. It is recommended that all of Chapter 6 be covered in an introductory course.

Chapter 7 External Flow introduces several important concepts and presents convection correlations that students will utilize throughout the remainder of the text and in subsequent professional practice. Sections 7.1 through 7.5 should be included in any first course in heat transfer. However, the content of Section 7.6 Flow Across Banks of Tubes, Section 7.7 Impinging Jets, and Section 7.8 Packed Beds is optional. Since the content of these sections is not interlinked, instructors may select from any of the optional topics.

Likewise, *Chapter 8 Internal Flow* includes matter that is used throughout the remainder of the text and by practicing engineers. However, Section 8.7 Heat Transfer Enhancement, and Section 8.8 Flow in Small Channels may be viewed as optional.

Buoyancy-induced flow and heat transfer is covered in *Chapter 9 Free Convection.* Because free convection thermal resistances are typically large, they are often the dominant resistance in many thermal systems and govern overall heat transfer rates. Therefore, most of Chapter 9 should be covered in a first course in heat transfer. Optional material includes Section 9.7 Free Convection Within Parallel Plate Channels and Section 9.9 Combined Free and Forced Convection. In contrast to resistances associated with free convection, thermal resistances corresponding to liquid-vapor phase change are typically small, and they can sometimes be neglected. Nonetheless, the content of *Chapter 10 Boiling and Condensation* that should be covered in a first heat transfer course includes Sections 10.1 through 10.4, Sections 10.6 through 10.8, and Section 10.11. Section 10.5 Forced Convection Boiling may be material appropriate for an intermediate heat transfer course. Similarly, Section 10.9 Film Condensation on Radial Systems and Section 10.10 Condensation in Horizontal Tubes may be either covered as time permits or included in a subsequent heat transfer course.

We recommend that all of *Chapter 11 Heat Exchangers* be covered in a first heat transfer course.

A distinguishing feature of the text, from its inception, is the in-depth coverage of radiation heat transfer in *Chapter 12 Radiation: Processes and Properties*. The content of the chapter is perhaps more relevant today than ever, with applications ranging from advanced manufacturing, to radiation detection and monitoring, to environmental issues related to global climate change. Although Chapter 12 has been reorganized to accommodate instructors who may wish to skip ahead to Chapter 13 after Section 12.4, we encourage instructors to cover Chapter 12 in its entirety.

Chapter 13 Radiation Exchange Between Surfaces may be covered as time permits or in an intermediate heat transfer course.

Acknowledgments

We wish to acknowledge and thank many of our colleagues in the heat transfer community. In particular, we would like to express our appreciation to Diana Borca-Tasciuc of the Rensselaer Polytechnic Institute and David Cahill of the University of Illinois Urbana-Champaign for their assistance in developing the periodic heating material of Chapter 5. We thank John Abraham of the University of St. Thomas for recommendations that have led to an improved treatment of flow over noncircular tubes in Chapter 7. We are very grateful to Ken Smith, Clark Colton, and William Dalzell of the Massachusetts Institute of Technology for the stimulating and detailed discussion of thermal entry effects in Chapter 8. We acknowledge Amir Faghri of the University of Connecticut for his advice regarding the treatment of condensation in Chapter 10. We extend our gratitude to Ralph Grief of the University of California, Berkeley for his many constructive suggestions pertaining to material throughout the text. Finally, we wish to thank the many students, instructors, and practicing engineers from around the globe who have offered countless interesting, valuable, and stimulating suggestions.

In closing, we are deeply grateful to our spouses and children, Tricia, Nate, Tico, Greg, Elias, Jacob, Andrea, Terri, Donna, and Shaunna for their endless love and patience. We extend appreciation to Tricia Bergman who expertly processed solutions for the end-of-chapter problems.

Theodore L. Bergman (tberg@engr.uconn.edu)
Storrs, Connecticut

Adrienne S. Lavine (lavine@seas.ucla.edu)
Los Angeles, California

Frank P. Incropera (fpi@nd.edu)
Notre Dame, Indiana

Supplemental and Web Site Material

The companion web site for the texts is www.wiley.com/college/bergman. By selecting one of the two texts and clicking on the "student companion site" link, students may access the **Answers to Selected Exercises** and the **Supplemental Sections** of the text. Supplemental Sections are identified throughout the text with the icon shown in the margin to the left.

Material available *for instructors only* may also be found by selecting one of the two texts at www.wiley.com/college/bergman and clicking on the "instructor companion site" link. The available content includes the **Solutions Manual**, **PowerPoint Slides** that can be used by instructors for lectures, and **Electronic Versions** of figures from the text for those wishing to prepare their own materials for electronic classroom presentation. *The* Instructor Solutions Manual *is copyrighted material for use only by instructors who are requiring the text for their course.*[1]

Interactive Heat Transfer 4.0/FEHT is available either with the text or as a separate purchase. As described by the authors in the *Approach and Organization*, this simple-to-use software tool provides modeling and computational features useful in solving many problems in the text, and it enables rapid what-if and exploratory analysis of many types of problems. Instructors interested in using this tool in their course can download the software from the book's web site at www.wiley.com/college/bergman. Students can download the software by registering on the student companion site; for details, see the registration card provided in this book. The software is also available as a stand-alone purchase at the web site. Any questions can be directed to your local Wiley representative.

This mouse icon identifies Supplemental Sections and is used throughout the text.

[1]Excerpts from the Solutions Manual may be reproduced by instructors for distribution on a not-for-profit basis for testing or instructional purposes only to students enrolled in courses for which the textbook has been adopted. *Any other reproduction or translation of the contents of the Solutions Manual beyond that permitted by Sections 107 or 108 of the 1976 United States Copyright Act without the permission of the copyright owner is unlawful.*

Contents

CHAPTER 4 *Two-Dimensional, Steady-State Conduction* 229

CHAPTER 5 *Transient Conduction* 279

CHAPTER 9 *Free Convection* 561

CHAPTER 10 *Boiling and Condensation* 619

Symbols

A	area, m^2
A_b	area of prime (unfinned) surface, m^2
A_c	cross-sectional area, m^2
A_p	fin profile area, m^2
A_r	nozzle area ratio
a	acceleration, m/s^2; speed of sound, m/s
Bi	Biot number
Bo	Bond number
C_D	drag coefficient
C_f	friction coefficient
C_t	thermal capacitance, J/K
Co	Confinement number
c	specific heat, J/kg · K; speed of light, m/s
c_p	specific heat at constant pressure, J/kg · K
c_v	specific heat at constant volume, J/kg · K
D	diameter, m
D_b	bubble diameter, m
D_h	hydraulic diameter, m
d	diameter of gas molecule, nm
E	thermal plus mechanical energy, J; electric potential, V; emissive power, W/m^2
E^{tot}	total energy, J
Ec	Eckert number
$\dot{E}_g$	rate of energy generation, W
$\dot{E}_{\text{in}}$	rate of energy transfer into a control volume, W
$\dot{E}_{\text{out}}$	rate of energy transfer out of control volume, W
$\dot{E}_{\text{st}}$	rate of increase of energy stored within a control volume, W
e	thermal internal energy per unit mass, J/kg; surface roughness, m
F	force, N; fraction of blackbody radiation in a wavelength band; view factor
Fo	Fourier number
Fr	Froude number
f	friction factor; similarity variable
G	irradiation, W/m^2; mass velocity, kg/s · m^2
Gr	Grashof number
Gz	Graetz number
g	gravitational acceleration, m/s^2
H	nozzle height, m; Henry's constant, bars
h	convection heat transfer coefficient, W/m^2 · K; Planck's constant, J · s
h_{fg}	latent heat of vaporization, J/kg
h'_{fg}	modified heat of vaporization, J/kg
h_{sf}	latent heat of fusion, J/kg
h_{rad}	radiation heat transfer coefficient, W/m^2 · K
I	electric current, A; radiation intensity, W/m^2 · sr
i	electric current density, A/m^2; enthalpy per unit mass, J/kg
J	radiosity, W/m^2
Ja	Jakob number
j_H	Colburn *j* factor for heat transfer
k	thermal conductivity, W/m · K
k_B	Boltzmann's constant, J/K
L	length, m
M	mass, kg
$\dot{M}_{\text{in}}$	rate at which mass enters a control volume, kg/s
$\dot{M}_{\text{out}}$	rate at which mass leaves a control volume, kg/s
$\dot{M}_{\text{st}}$	rate of increase of mass stored within a control volume, kg/s
$\mathcal{M}_i$	molecular weight of species *i*, kg/kmol
Ma	Mach number
m	mass, kg
$\dot{m}$	mass flow rate, kg/s
N	integer number
N_L, N_T	number of tubes in longitudinal and transverse directions
Nu	Nusselt number

NTU	number of transfer units
$\mathcal{N}$	Avogadro's number
P	power, W; perimeter, m
P_L, P_T	dimensionless longitudinal and transverse pitch of a tube bank
Pe	Peclet number
Pr	Prandtl number
p	pressure, N/m^2
Q	energy transfer, J
q	heat transfer rate, W
$\dot{q}$	rate of energy generation per unit volume, W/m^3
q'	heat transfer rate per unit length, W/m
q''	heat flux, W/m^2
q^*	dimensionless conduction heat rate
R	cylinder radius, m; gas constant, J/kg · K
$\mathcal{R}$	universal gas constant, J/kmol · K
Ra	Rayleigh number
Re	Reynolds number
R_e	electric resistance, Ω
R_f	fouling factor, $m^2 \cdot K/W$
$R_{m,n}$	residual for the *m, n* nodal point
R_t	thermal resistance, K/W
$R_{t,c}$	thermal contact resistance, K/W
$R_{t,f}$	fin thermal resistance, K/W
$R_{t,o}$	thermal resistance of fin array, K/W
r_o	cylinder or sphere radius, m
r, ϕ, z	cylindrical coordinates
r, θ, ϕ	spherical coordinates
S	shape factor for two-dimensional conduction, m; nozzle pitch, m; plate spacing, m; Seebeck coefficient, V/K
S_c	solar constant, W/m^2
S_D, S_L, S_T	diagonal, longitudinal, and transverse pitch of a tube bank, m
St	Stanton number
T	temperature, K
t	time, s
U	overall heat transfer coefficient, $W/m^2 \cdot K$; internal energy, J
u, v, w	mass average fluid velocity components, m/s
V	volume, m^3; fluid velocity, m/s
v	specific volume, m^3/kg
W	width of a slot nozzle, m
$\dot{W}$	rate at which work is performed, W
We	Weber number
X	vapor quality
X_{tt}	Martinelli parameter
X, Y, Z	components of the body force per unit volume, N/m^3
x, y, z	rectangular coordinates, m
x_c	critical location for transition to turbulence, m
$x_{fd,h}$	hydrodynamic entry length, m
$x_{fd,t}$	thermal entry length, m
Z	thermoelectric material property, K^{-1}

Greek Letters

α	thermal diffusivity, m^2/s; accommodation coefficient; absorptivity
β	volumetric thermal expansion coefficient, K^{-1}
Γ	mass flow rate per unit width in film condensation, kg/s · m
γ	ratio of specific heats
δ	hydrodynamic boundary layer thickness, m
δ_p	thermal penetration depth, m
δ_t	thermal boundary layer thickness, m
ε	emissivity; porosity; heat exchanger effectiveness
ε_f	fin effectiveness
η	thermodynamic efficiency; similarity variable
η_f	fin efficiency
η_o	overall efficiency of fin array
θ	zenith angle, rad; temperature difference, K
κ	absorption coefficient, m^{-1}
λ	wavelength, μm
λ_{mfp}	mean free path length, nm
μ	viscosity, kg/s · m
ν	kinematic viscosity, m^2/s; frequency of radiation, s^{-1}
ρ	mass density, kg/m^3; reflectivity
ρ_e	electric resistivity, Ω/m
σ	Stefan–Boltzmann constant, $W/m^2 \cdot K^4$; electrical conductivity, 1/Ω · m; normal viscous stress, N/m^2; surface tension, N/m
Φ	viscous dissipation function, s^{-2}
φ	volume fraction
ϕ	azimuthal angle, rad
ψ	stream function, m^2/s
τ	shear stress, N/m^2; transmissivity
ω	solid angle, sr; perfusion rate, s^{-1}

Subscripts

abs	absorbed
am	arithmetic mean
atm	atmospheric
b	base of an extended surface; blackbody
C	Carnot
c	cross-sectional; cold fluid; critical
cr	critical insulation thickness
cond	conduction
conv	convection
CF	counterflow
D	diameter; drag
e	excess; emission; electron
f	fluid properties; fin conditions; saturated liquid conditions
fc	forced convection
fd	fully developed conditions
g	saturated vapor conditions
H	heat transfer conditions
h	hydrodynamic; hot fluid; helical
i	inner surface of an annulus; initial condition; tube inlet condition; incident radiation
L	based on characteristic length
l	saturated liquid conditions
lat	latent energy
lm	log mean condition

m	mean value over a tube cross section
max	maximum
o	center or midplane condition; tube outlet condition; outer
p	momentum
ph	phonon
R	reradiating surface
r, ref	reflected radiation
rad	radiation
S	solar conditions
s	surface conditions; solid properties; saturated solid conditions
sat	saturated conditions
sens	sensible energy
sky	sky conditions
ss	steady state
sur	surroundings
t	thermal
tr	transmitted
v	saturated vapor conditions
x	local conditions on a surface
λ	spectral
∞	free stream conditions

Superscripts

*	dimensionless quantity

Overbar

$\overline{\ }$	surface average conditions; time mean

CHAPTER 1

Introduction

From the study of thermodynamics, you have learned that energy can be transferred by interactions of a system with its surroundings. These interactions are called work and heat. However, thermodynamics deals with the end states of the process during which an interaction occurs and provides no information concerning the nature of the interaction or the time rate at which it occurs. The objective of this text is to extend thermodynamic analysis through the study of the *modes* of heat transfer and through the development of relations to calculate heat transfer *rates.*

In this chapter we lay the foundation for much of the material treated in the text. We do so by raising several questions: *What is heat transfer? How is heat transferred? Why is it important?* One objective is to develop an appreciation for the fundamental concepts and principles that underlie heat transfer processes. A second objective is to illustrate the manner in which a knowledge of heat transfer may be used with the first law of thermodynamics (*conservation of energy*) to solve problems relevant to technology and society.

1.1 What and How?

A simple, yet general, definition provides sufficient response to the question: What is heat transfer?

Heat transfer (or heat) is thermal energy in transit due to a spatial temperature difference.

Whenever a temperature difference exists in a medium or between media, heat transfer must occur.

As shown in Figure 1.1, we refer to different types of heat transfer processes as *modes.* When a temperature gradient exists in a stationary medium, which may be a solid or a fluid, we use the term *conduction* to refer to the heat transfer that will occur across the medium. In contrast, the term *convection* refers to heat transfer that will occur between a surface and a moving fluid when they are at different temperatures. The third mode of heat transfer is termed *thermal radiation.* All surfaces of finite temperature emit energy in the form of electromagnetic waves. Hence, in the absence of an intervening medium, there is net heat transfer by radiation between two surfaces at different temperatures.

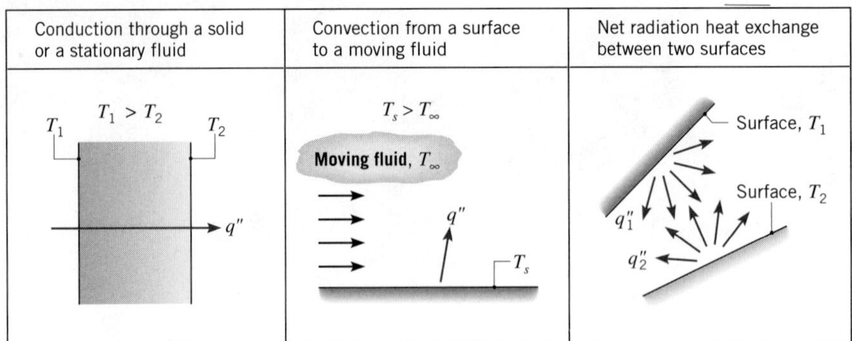

FIGURE 1.1 Conduction, convection, and radiation heat transfer modes.

1.2 Physical Origins and Rate Equations

As engineers, it is important that we understand the *physical mechanisms* which underlie the heat transfer modes and that we be able to use the rate equations that quantify the amount of energy being transferred per unit time.

1.2.1 Conduction

At mention of the word *conduction*, we should immediately conjure up concepts of *atomic* and *molecular activity* because processes at these levels sustain this mode of heat transfer. Conduction may be viewed as the transfer of energy from the more energetic to the less energetic particles of a substance due to interactions between the particles.

The physical mechanism of conduction is most easily explained by considering a gas and using ideas familiar from your thermodynamics background. Consider a gas in which a temperature gradient exists, and assume that there is *no bulk, or macroscopic, motion.* The gas may occupy the space between two surfaces that are maintained at different temperatures, as shown in Figure 1.2. We associate the temperature at any point with the energy of gas molecules in proximity to the point. This energy is related to the random translational motion, as well as to the internal rotational and vibrational motions, of the molecules.

Higher temperatures are associated with higher molecular energies. When neighboring molecules collide, as they are constantly doing, a transfer of energy from the more energetic to the less energetic molecules must occur. In the presence of a temperature gradient, energy transfer by conduction must then occur in the direction of decreasing temperature. This would be true even in the absence of collisions, as is evident from Figure 1.2. The hypothetical plane at x_o is constantly being crossed by molecules from above and below due to their *random* motion. However, molecules from above are associated with a higher temperature than those from below, in which case there must be a *net* transfer of energy in the positive x-direction. Collisions between molecules enhance this energy transfer. We may speak of the net transfer of energy by random molecular motion as a *diffusion* of energy.

The situation is much the same in liquids, although the molecules are more closely spaced and the molecular interactions are stronger and more frequent. Similarly, in a solid, conduction may be attributed to atomic activity in the form of lattice vibrations. The modern

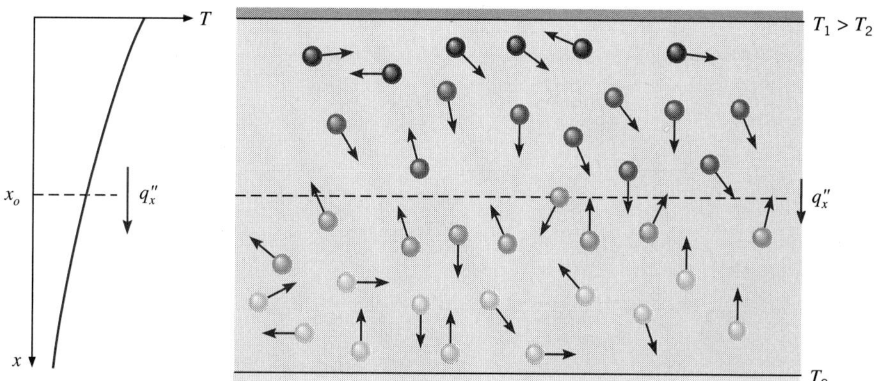

FIGURE 1.2 Association of conduction heat transfer with diffusion of energy due to molecular activity.

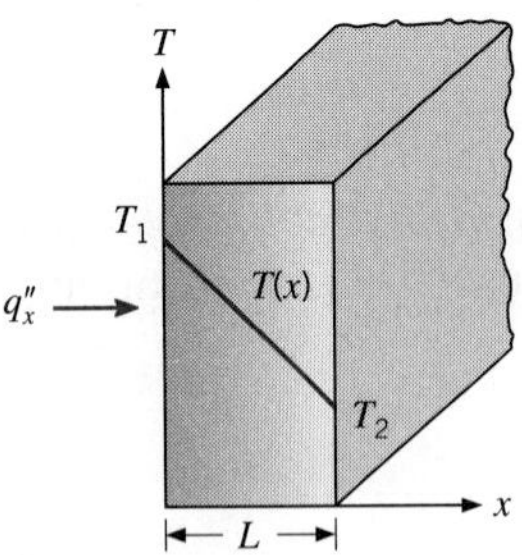

FIGURE 1.3 One-dimensional heat transfer by conduction (diffusion of energy).

view is to ascribe the energy transfer to *lattice waves* induced by atomic motion. In an electrical nonconductor, the energy transfer is exclusively via these lattice waves; in a conductor, it is also due to the translational motion of the free electrons. We treat the important properties associated with conduction phenomena in Chapter 2 and in Appendix A.

Examples of conduction heat transfer are legion. The exposed end of a metal spoon suddenly immersed in a cup of hot coffee is eventually warmed due to the conduction of energy through the spoon. On a winter day, there is significant energy loss from a heated room to the outside air. This loss is principally due to conduction heat transfer through the wall that separates the room air from the outside air.

Heat transfer processes can be quantified in terms of appropriate *rate equations.* These equations may be used to compute the amount of energy being transferred per unit time. For heat conduction, the rate equation is known as *Fourier's law.* For the one-dimensional plane wall shown in Figure 1.3, having a temperature distribution $T(x)$, the rate equation is expressed as

$$q''_x = -k\frac{dT}{dx} \tag{1.1}$$

The *heat flux* q''_x (W/m^2) is the heat transfer rate in the x-direction *per* unit area *perpendicular* to the direction of transfer, and it is proportional to the *temperature gradient, dT/dx,* in this direction. The parameter k is a *transport* property known as the *thermal conductivity* (W/m·K) and is a characteristic of the wall material. The minus sign is a consequence of the fact that heat is transferred in the direction of decreasing temperature. Under the *steady-state conditions* shown in Figure 1.3, where the temperature distribution is *linear,* the temperature gradient may be expressed as

$$\frac{dT}{dx} = \frac{T_2 - T_1}{L}$$

and the heat flux is then

$$q''_x = -k\frac{T_2 - T_1}{L}$$

or

$$q''_x = k\frac{T_1 - T_2}{L} = k\frac{\Delta T}{L} \tag{1.2}$$

Note that this equation provides a *heat flux,* that is, the rate of heat transfer per *unit area.* The *heat rate* by conduction, q_x (W), through a plane wall of area A is then the product of the flux and the area, $q_x = q''_x \cdot A$.

IHT* **EXAMPLE 1.1**

The wall of an industrial furnace is constructed from 0.15-m-thick fireclay brick having a thermal conductivity of 1.7 W/m · K. Measurements made during steady-state operation reveal temperatures of 1400 and 1150 K at the inner and outer surfaces, respectively. What is the rate of heat loss through a wall that is 0.5 m × 1.2 m on a side?

SOLUTION

Known: Steady-state conditions with prescribed wall thickness, area, thermal conductivity, and surface temperatures.

Find: Wall heat loss.

Schematic:

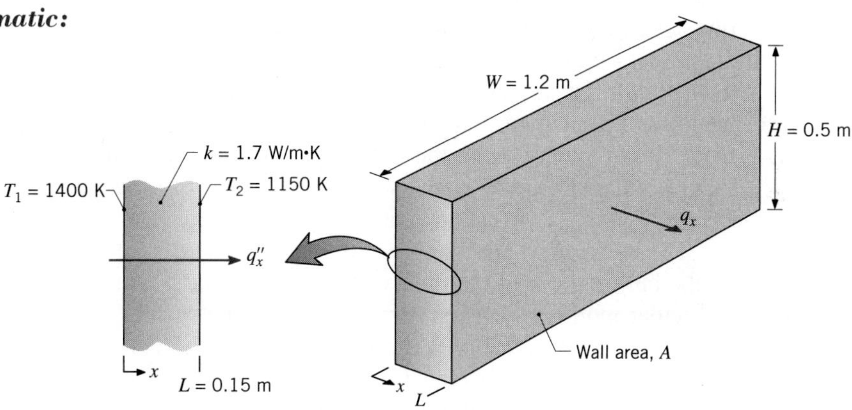

Assumptions:

1. Steady-state conditions.
2. One-dimensional conduction through the wall.
3. Constant thermal conductivity.

Analysis: Since heat transfer through the wall is by conduction, the heat flux may be determined from Fourier's law. Using Equation 1.2, we have

$$q''_x = k\,\frac{\Delta T}{L} = 1.7\ \text{W/m}\cdot\text{K} \times \frac{250\ \text{K}}{0.15\ \text{m}} = 2833\ \text{W/m}^2$$

The heat flux represents the rate of heat transfer through a section of unit area, and it is uniform (invariant) across the surface of the wall. The heat loss through the wall of area $A = H \times W$ is then

$$q_x = (HW)\,q''_x = (0.5\ \text{m} \times 1.2\ \text{m})\,2833\ \text{W/m}^2 = 1700\ \text{W} \qquad \triangleleft$$

Comments: Note the direction of heat flow and the distinction between heat flux and heat rate.

*This icon identifies examples that are available in tutorial form in the *Interactive Heat Transfer (IHT)* software that accompanies the text. Each tutorial is brief and illustrates a basic function of the software. *IHT* can be used to solve simultaneous equations, perform parameter sensitivity studies, and graph the results. Use of *IHT* will reduce the time spent solving more complex end-of-chapter problems.

1.2.2 Convection

The convection heat transfer *mode* is comprised of *two mechanisms.* In addition to energy transfer due to *random molecular motion* (*diffusion*), energy is also transferred by the *bulk*, or *macroscopic, motion* of the fluid. This fluid motion is associated with the fact that, at any instant, large numbers of molecules are moving collectively or as aggregates. Such motion, in the presence of a temperature gradient, contributes to heat transfer. Because the molecules in the aggregate retain their random motion, the total heat transfer is then due to a superposition of energy transport by the random motion of the molecules and by the bulk motion of the fluid. The term *convection* is customarily used when referring to this cumulative transport, and the term *advection* refers to transport due to bulk fluid motion.

We are especially interested in convection heat transfer, which occurs between a fluid in motion and a bounding surface when the two are at different temperatures. Consider fluid flow over the heated surface of Figure 1.4. A consequence of the fluid–surface interaction is the development of a region in the fluid through which the velocity varies from zero at the surface to a finite value u_∞ associated with the flow. This region of the fluid is known as the *hydrodynamic*, or *velocity, boundary layer.* Moreover, if the surface and flow temperatures differ, there will be a region of the fluid through which the temperature varies from T_s at $y = 0$ to T_∞ in the outer flow. This region, called the *thermal boundary layer*, may be smaller, larger, or the same size as that through which the velocity varies. In any case, if $T_s > T_\infty$, convection heat transfer will occur from the surface to the outer flow.

The convection heat transfer mode is sustained both by random molecular motion and by the bulk motion of the fluid within the boundary layer. The contribution due to random molecular motion (diffusion) dominates near the surface where the fluid velocity is low. In fact, at the interface between the surface and the fluid ($y = 0$), the fluid velocity is zero, and heat is transferred by this mechanism only. The contribution due to bulk fluid motion originates from the fact that the boundary layer *grows* as the flow progresses in the x-direction. In effect, the heat that is conducted into this layer is swept downstream and is eventually transferred to the fluid outside the boundary layer. Appreciation of boundary layer phenomena is essential to understanding convection heat transfer. For this reason, the discipline of fluid mechanics will play a vital role in our later analysis of convection.

Convection heat transfer may be classified according to the nature of the flow. We speak of *forced convection* when the flow is caused by external means, such as by a fan, a pump, or atmospheric winds. As an example, consider the use of a fan to provide forced convection air cooling of hot electrical components on a stack of printed circuit boards (Figure 1.5*a*). In contrast, for *free* (or *natural*) *convection*, the flow is induced by buoyancy forces, which are due to density differences caused by temperature variations in the fluid. An example is the free convection heat transfer that occurs from hot components on a vertical array of circuit

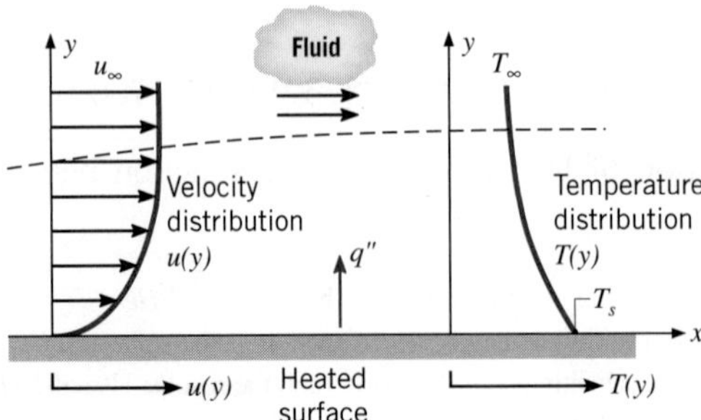

FIGURE 1.4 Boundary layer development in convection heat transfer.

boards in air (Figure 1.5*b*). Air that makes contact with the components experiences an increase in temperature and hence a reduction in density. Since it is now lighter than the surrounding air, buoyancy forces induce a vertical motion for which warm air ascending from the boards is replaced by an inflow of cooler ambient air.

While we have presumed *pure* forced convection in Figure 1.5*a* and *pure* natural convection in Figure 1.5*b*, conditions corresponding to *mixed (combined) forced* and *natural convection* may exist. For example, if velocities associated with the flow of Figure 1.5*a* are small and/or buoyancy forces are large, a secondary flow that is comparable to the imposed forced flow could be induced. In this case, the buoyancy-induced flow would be normal to the forced flow and could have a significant effect on convection heat transfer from the components. In Figure 1.5*b*, mixed convection would result if a fan were used to force air upward between the circuit boards, thereby assisting the buoyancy flow, or downward, thereby opposing the buoyancy flow.

We have described the convection heat transfer mode as energy transfer occurring within a fluid due to the combined effects of conduction and bulk fluid motion. Typically, the energy that is being transferred is the *sensible*, or internal thermal, energy of the fluid. However, for some convection processes, there is, in addition, *latent* heat exchange. This latent heat exchange is generally associated with a phase change between the liquid and vapor states of the fluid. Two special cases of interest in this text are *boiling* and *condensation.* For example, convection heat transfer results from fluid motion induced by vapor bubbles generated at the bottom of a pan of boiling water (Figure 1.5*c*) or by the condensation of water vapor on the outer surface of a cold water pipe (Figure 1.5*d*).

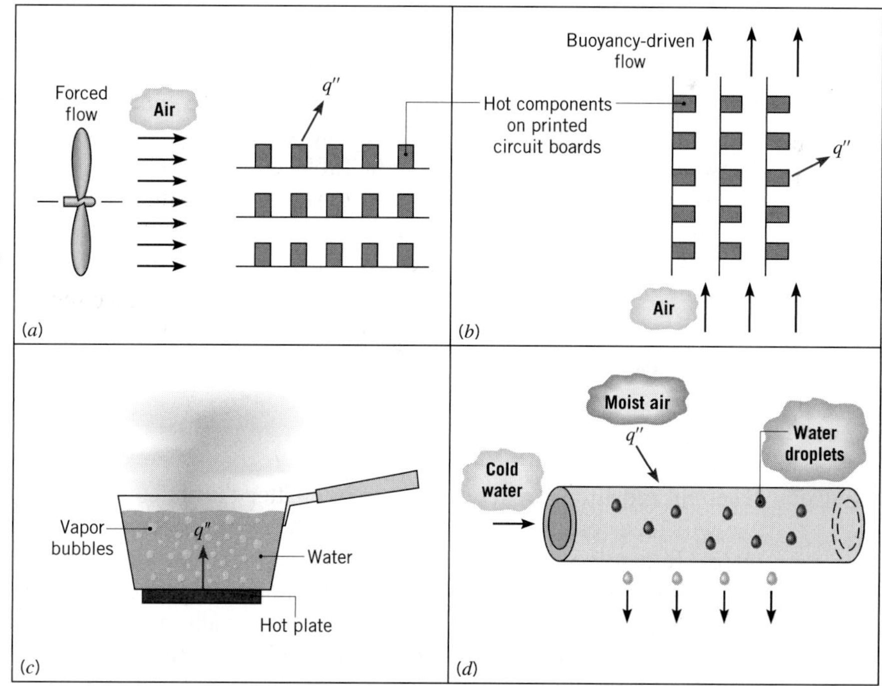

Figure 1.5 Convection heat transfer processes. (*a*) Forced convection. (*b*) Natural convection. (*c*) Boiling. (*d*) Condensation.

TABLE 1.1 Typical values of the convection heat transfer coefficient

Process	h $(W/m^2 \cdot K)$
Free convection	
Gases	2–25
Liquids	50–1000
Forced convection	
Gases	25–250
Liquids	100–20,000
Convection with phase change	
Boiling or condensation	2500–100,000

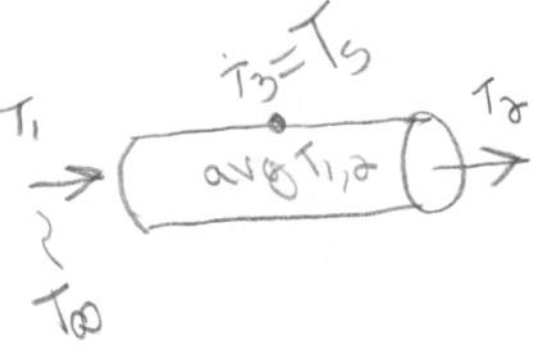

Regardless of the nature of the convection heat transfer process, the appropriate rate equation is of the form

$$q'' = h(T_s - T_\infty) \tag{1.3a}$$

where q'', the convective *heat flux* (W/m^2), is proportional to the difference between the surface and fluid temperatures, T_s and T_∞, respectively. This expression is known as *Newton's law of cooling*, and the parameter h $(W/m^2 \cdot K)$ is termed the *convection heat transfer coefficient.* This coefficient depends on conditions in the boundary layer, which are influenced by surface geometry, the nature of the fluid motion, and an assortment of fluid thermodynamic and transport properties.

Any study of convection ultimately reduces to a study of the means by which h may be determined. Although consideration of these means is deferred to Chapter 6, convection heat transfer will frequently appear as a boundary condition in the solution of conduction problems (Chapters 2 through 5). In the solution of such problems we presume h to be known, using typical values given in Table 1.1.

When Equation 1.3a is used, the convection heat flux is presumed to be *positive* if heat is transferred *from* the surface ($T_s > T_\infty$) and *negative* if heat is transferred *to* the surface ($T_\infty > T_s$). However, nothing precludes us from expressing Newton's law of cooling as

$$q'' = h(T_\infty - T_s) \tag{1.3b}$$

in which case heat transfer is positive if it is to the surface.

1.2.3 Radiation

Thermal radiation is energy *emitted* by matter that is at a nonzero temperature. Although we will focus on radiation from solid surfaces, emission may also occur from liquids and gases. Regardless of the form of matter, the emission may be attributed to changes in the electron configurations of the constituent atoms or molecules. The energy of the radiation field is transported by electromagnetic waves (or alternatively, photons). While the transfer of energy by conduction or convection requires the presence of a material medium, radiation does not. In fact, radiation transfer occurs most efficiently in a vacuum.

Consider radiation transfer processes for the surface of Figure 1.6*a*. Radiation that is *emitted* by the surface originates from the thermal energy of matter bounded by the surface, and the rate at which energy is released per unit area (W/m^2) is termed the surface *emissive power, E.* There is an upper limit to the emissive power, which is prescribed by the *Stefan–Boltzmann law*

$$E_b = \sigma T_s^4 \tag{1.4}$$

where T_s is the *absolute temperature* (K) of the surface and σ is the *Stefan–Boltzmann constant* ($\sigma = 5.67 \times 10^{-8}$ W/m$^2 \cdot$ K^4). Such a surface is called an ideal radiator or *blackbody.*

The heat flux emitted by a real surface is less than that of a blackbody at the same temperature and is given by

$$E = \varepsilon \sigma T_s^4 \tag{1.5}$$

where ε is a radiative property of the surface termed the *emissivity.* With values in the range $0 \leq \varepsilon \leq 1$, this property provides a measure of how efficiently a surface emits energy relative to a blackbody. It depends strongly on the surface material and finish, and representative values are provided in Appendix A.

Radiation may also be *incident* on a surface from its surroundings. The radiation may originate from a special source, such as the sun, or from other surfaces to which the surface of interest is exposed. Irrespective of the source(s), we designate the rate at which all such radiation is incident on a unit area of the surface as the *irradiation G* (Figure 1.6*a*).

A portion, or all, of the irradiation may be *absorbed* by the surface, thereby increasing the thermal energy of the material. The rate at which radiant energy is absorbed per unit surface area may be evaluated from knowledge of a surface radiative property termed the *absorptivity* α. That is,

$$G_{abs} = \alpha G \tag{1.6}$$

where $0 \leq \alpha \leq 1$. If $\alpha < 1$ and the surface is *opaque*, portions of the irradiation are *reflected.* If the surface is *semitransparent*, portions of the irradiation may also be *transmitted.* However, whereas absorbed and emitted radiation increase and reduce, respectively, the thermal energy of matter, reflected and transmitted radiation have no effect on this energy. Note that the value of α depends on the nature of the irradiation, as well as on the surface itself. For example, the absorptivity of a surface to solar radiation may differ from its absorptivity to radiation emitted by the walls of a furnace.

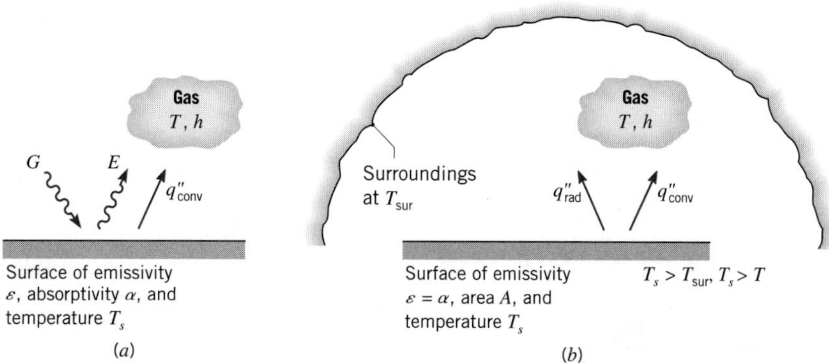

FIGURE 1.6 Radiation exchange: (*a*) at a surface and (*b*) between a surface and large surroundings.

In many engineering problems (a notable exception being problems involving solar radiation or radiation from other very high temperature sources), liquids can be considered opaque to radiation heat transfer, and gases can be considered transparent to it. Solids can be opaque (as is the case for metals) or *semitransparent* (as is the case for thin sheets of some polymers and some semiconducting materials).

A special case that occurs frequently involves radiation exchange between a small surface at T_s and a much larger, isothermal surface that completely surrounds the smaller one (Figure 1.6*b*). The *surroundings* could, for example, be the walls of a room or a furnace whose temperature T_{sur} differs from that of an enclosed surface ($T_{sur} \neq T_s$). We will show in Chapter 12 that, for such a condition, the irradiation may be approximated by emission from a blackbody at T_{sur}, in which case $G = \sigma T_{sur}^4$. If the surface is assumed to be one for which $\alpha = \varepsilon$ (a *gray surface*), the *net* rate of radiation heat transfer *from* the surface, expressed per unit area of the surface, is

$$q''_{rad} = \frac{q}{A} = \varepsilon E_b(T_s) - \alpha G = \varepsilon\sigma(T_s^4 - T_{sur}^4) \qquad (1.7)$$

This expression provides the difference between thermal energy that is released due to radiation emission and that gained due to radiation absorption.

For many applications, it is convenient to express the net radiation heat exchange in the form

$$q_{rad} = h_r A(T_s - T_{sur}) \qquad (1.8)$$

where, from Equation 1.7, the *radiation heat transfer coefficient h_r is*

$$h_r \equiv \varepsilon\sigma(T_s + T_{sur})(T_s^2 + T_{sur}^2) \qquad (1.9)$$

Here we have modeled the radiation mode in a manner similar to convection. In this sense we have *linearized* the radiation rate equation, making the heat rate proportional to a temperature difference rather than to the difference between two temperatures to the fourth power. Note, however, that h_r depends strongly on temperature, whereas the temperature dependence of the convection heat transfer coefficient h is generally weak.

The surfaces of Figure 1.6 may also simultaneously transfer heat by convection to an adjoining gas. For the conditions of Figure 1.6*b*, the total rate of heat transfer *from* the surface is then

$$q = q_{conv} + q_{rad} = hA(T_s - T_\infty) + \varepsilon A\sigma(T_s^4 - T_{sur}^4) \qquad (1.10)$$

IHT

Example 1.2

An uninsulated steam pipe passes through a room in which the air and walls are at 25°C. The outside diameter of the pipe is 70 mm, and its surface temperature and emissivity are 200°C and 0.8, respectively. What are the surface emissive power and irradiation? If the coefficient associated with free convection heat transfer from the surface to the air is 15 W/m$^2 \cdot$K, what is the rate of heat loss from the surface per unit length of pipe?

Solution

Known: Uninsulated pipe of prescribed diameter, emissivity, and surface temperature in a room with fixed wall and air temperatures.

Find:

1. Surface emissive power and irradiation.
2. Pipe heat loss per unit length, q'.

Schematic:

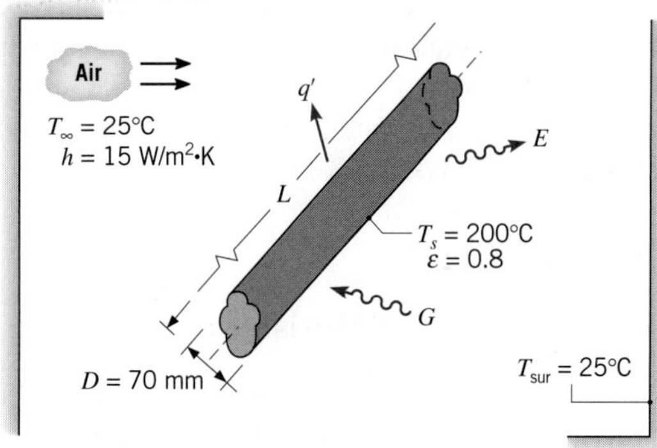

Assumptions:

1. Steady-state conditions.
2. Radiation exchange between the pipe and the room is between a small surface and a much larger enclosure.
3. The surface emissivity and absorptivity are equal.

Analysis:

1. The surface emissive power may be evaluated from Equation 1.5, while the irradiation corresponds to $G = \sigma T_{\text{sur}}^4$. Hence

$$E = \varepsilon \sigma T_s^4 = 0.8(5.67 \times 10^{-8}\ \text{W/m}^2 \cdot \text{K}^4)(473\ \text{K})^4 = 2270\ \text{W/m}^2 \quad \triangleleft$$

$$G = \sigma T_{\text{sur}}^4 = 5.67 \times 10^{-8}\ \text{W/m}^2 \cdot \text{K}^4\ (298\ \text{K})^4 = 447\ \text{W/m}^2 \quad \triangleleft$$

2. Heat loss from the pipe is by convection to the room air and by radiation exchange with the walls. Hence, $q = q_{\text{conv}} + q_{\text{rad}}$ and from Equation 1.10, with $A = \pi DL$,

$$q = h(\pi DL)(T_s - T_\infty) + \varepsilon(\pi DL)\sigma(T_s^4 - T_{\text{sur}}^4)$$

The heat loss per unit length of pipe is then

$$q' = \frac{q}{L} = 15\ \text{W/m}^2 \cdot \text{K}(\pi \times 0.07\ \text{m})(200 - 25)°\text{C}$$

$$+\ 0.8(\pi \times 0.07\ \text{m})\ 5.67 \times 10^{-8}\ \text{W/m}^2 \cdot \text{K}^4\ (473^4 - 298^4)\ \text{K}^4$$

$$q' = 577\ \text{W/m} + 421\ \text{W/m} = 998\ \text{W/m} \quad \triangleleft$$

Comments:

1. Note that temperature may be expressed in units of °C or K when evaluating the temperature difference for a convection (or conduction) heat transfer rate. However, temperature must be expressed in kelvins (K) when evaluating a radiation transfer rate.

2. The net rate of radiation heat transfer from the pipe may be expressed as

$$q'_{\text{rad}} = \pi D \,(E - \alpha G)$$

$$q'_{\text{rad}} = \pi \times 0.07 \text{ m } (2270 - 0.8 \times 447) \text{ W/m}^2 = 421 \text{ W/m}$$

3. In this situation, the radiation and convection heat transfer rates are comparable because T_s is large compared to T_{sur} and the coefficient associated with free convection is small. For more moderate values of T_s and the larger values of h associated with forced convection, the effect of radiation may often be neglected. The radiation heat transfer coefficient may be computed from Equation 1.9. For the conditions of this problem, its value is $h_r = 11 \text{ W/m}^2 \cdot \text{K}$.

1.2.4 The Thermal Resistance Concept

The three modes of heat transfer were introduced in the preceding sections. As is evident from Equations 1.2, 1.3, and 1.8, the heat transfer rate can be expressed in the form

$$q = q''A = \frac{\Delta T}{R_t} \tag{1.11}$$

where ΔT is a relevant temperature difference and A is the area normal to the direction of heat transfer. The quantity R_t is called a *thermal resistance* and takes different forms for the three different modes of heat transfer. For example, Equation 1.2 may be multiplied by the area A and rewritten as $q_x = \Delta T/R_{t,c}$, where $R_{t,c} = L/kA$ is a thermal resistance associated with conduction, having the units K/W. The thermal resistance concept will be considered in detail in Chapter 3 and will be seen to have great utility in solving complex heat transfer problems.

1.3 *Relationship to Thermodynamics*

The subjects of heat transfer and thermodynamics are highly complementary and interrelated, but they also have fundamental differences. If you have taken a thermodynamics course, you are aware that heat exchange plays a vital role in the first and second laws of thermodynamics because it is one of the primary mechanisms for energy transfer between a system and its surroundings. While thermodynamics may be used to determine the *amount* of energy required in the form of heat for a system to pass from one state to another, it considers neither the mechanisms that provide for heat exchange nor the methods that exist for computing the *rate* of heat exchange. The discipline of heat transfer specifically seeks to quantify the rate at which heat is exchanged through the rate equations expressed, for example, by Equations 1.2, 1.3, and 1.7. Indeed, heat transfer principles often enable the engineer to implement the concepts of thermodynamics. For example, the actual size of a power plant to be constructed cannot be determined from thermodynamics alone; the principles of heat transfer must also be invoked at the design stage.

The remainder of this section considers the relationship of heat transfer to thermodynamics. Since the *first law* of thermodynamics (the *law of conservation of energy*) provides a useful, often essential, starting point for the solution of heat transfer problems, Section 1.3.1 will provide a development of the general formulations of the first law. The ideal

(Carnot) efficiency of a *heat engine*, as determined by the *second law* of thermodynamics will be reviewed in Section 1.3.2. It will be shown that a realistic description of the heat transfer between a heat engine and its surroundings *further limits* the actual efficiency of a heat engine.

1.3.1 Relationship to the First Law of Thermodynamics (Conservation of Energy)

At its heart, the first law of thermodynamics is simply a statement that the total energy of a system is conserved, and therefore the only way that the amount of energy in a system can change is if energy crosses its boundaries. The first law also addresses the ways in which energy can cross the boundaries of a system. For a closed system (a region of fixed mass), there are only two ways: heat transfer through the boundaries and work done on or by the system. This leads to the following statement of the first law for a closed system, which is familiar if you have taken a course in thermodynamics:

$$\Delta E_{st}^{tot} = Q - W \tag{1.12a}$$

where ΔE_{st}^{tot} is the change in the total energy stored in the system, Q is the *net* heat transferred to the system, and W is the *net* work done by the system. This is schematically illustrated in Figure 1.7*a*.

The first law can also be applied to a *control volume* (or *open system*), a region of space bounded by a *control surface* through which mass may pass. Mass entering and leaving the control volume carries energy with it; this process, termed *energy advection*, adds a third way in which energy can cross the boundaries of a control volume. To summarize, the first law of thermodynamics can be very simply stated as follows for both a control volume and a closed system.

First Law of Thermodynamics over a Time Interval (Δt)

The increase in the amount of energy stored in a control volume must equal the amount of energy that enters the control volume, minus the amount of energy that leaves the control volume.

In applying this principle, it is recognized that energy can enter and leave the control volume due to heat transfer through the boundaries, work done on or by the control volume, and energy advection.

The first law of thermodynamics addresses *total* energy, which consists of kinetic and potential energies (together known as mechanical energy) and internal energy. Internal energy can be further subdivided into thermal energy (which will be defined more carefully later)

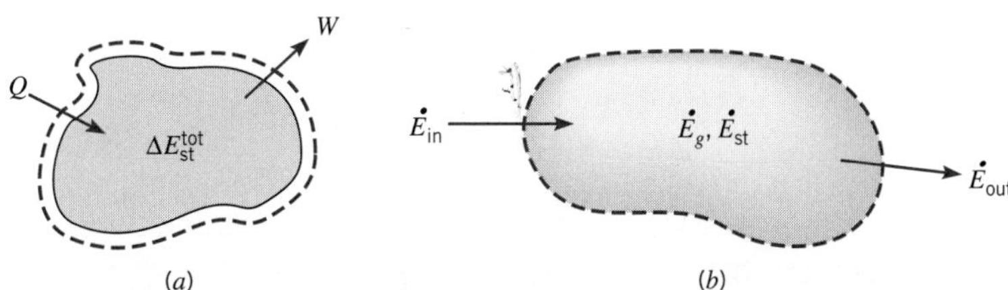

FIGURE 1.7 Conservation of energy: (*a*) for a closed system over a time interval and (*b*) for a control volume at an instant.

and other forms of internal energy, such as chemical and nuclear energy. For the study of heat transfer, we wish to focus attention on the thermal and mechanical forms of energy. We must recognize that the sum of thermal and mechanical energy is *not* conserved, because conversion can occur between other forms of energy and thermal or mechanical energy. For example, if a chemical reaction occurs that decreases the amount of chemical energy in the system, it will result in an increase in the thermal energy of the system. If an electric motor operates within the system, it will cause conversion from electrical to mechanical energy. We can think of such energy conversions as resulting in *thermal or mechanical energy generation* (which can be either positive or negative). So a statement of the first law that is well suited for heat transfer analysis is:

Thermal and Mechanical Energy Equation over a Time Interval (Δt)

The increase in the amount of thermal and mechanical energy stored in the control volume must equal the amount of thermal and mechanical energy that enters the control volume, minus the amount of thermal and mechanical energy that leaves the control volume, plus the amount of thermal and mechanical energy that is generated within the control volume.

This expression applies over a *time interval* Δt, and all the energy terms are measured in joules. Since the first law must be satisfied at each and every *instant* of time t, we can also formulate the law on a *rate basis*. That is, at any instant, there must be a balance between all *energy rates*, as measured in joules per second (W). In words, this is expressed as follows:

Thermal and Mechanical Energy Equation at an Instant (t)

The rate of increase of thermal and mechanical energy stored in the control volume must equal the rate at which thermal and mechanical energy enters the control volume, minus the rate at which thermal and mechanical energy leaves the control volume, plus the rate at which thermal and mechanical energy is generated within the control volume.

If the inflow and generation of thermal and mechanical energy exceed the outflow, the amount of thermal and mechanical energy stored (accumulated) in the control volume must increase. If the converse is true, thermal and mechanical energy storage must decrease. If the inflow and generation equal the outflow, a *steady-state* condition must prevail such that there will be no change in the amount of thermal and mechanical energy stored in the control volume.

We will now define symbols for each of the energy terms so that the boxed statements can be rewritten as equations. We let E stand for the sum of thermal and mechanical energy (in contrast to the symbol E^{tot} for total energy). Using the subscript *st* to denote energy stored in the control volume, the change in thermal and mechanical energy stored over the time interval Δt is then ΔE_{st}. The subscripts *in* and *out* refer to energy entering and leaving the control volume. Finally, thermal and mechanical energy generation is given the symbol E_g. Thus, the first boxed statement can be written as:

$$\Delta E_{st} = E_{in} - E_{out} + E_g \tag{1.12b}$$

Next, using a dot over a term to indicate a rate, the second boxed statement becomes:

$$\dot{E}_{st} \equiv \frac{dE_{st}}{dt} = \dot{E}_{in} - \dot{E}_{out} + \dot{E}_g \tag{1.12c}$$

This expression is illustrated schematically in Figure 1.7*b*.

Equations 1.12b,c provide important and, in some cases, essential tools for solving heat transfer problems. Every application of the first law must begin with the identification of an appropriate control volume and its control surface, to which an analysis is subsequently applied. The first step is to indicate the control surface by drawing a dashed line. The second step is to decide whether to perform the analysis for a time interval Δt (Equation 1.12b) or on a rate basis (Equation 1.12c). This choice depends on the objective of the solution and on how information is given in the problem. The next step is to identify the energy terms that are relevant in the problem you are solving. To develop your confidence in taking this last step, the remainder of this section is devoted to clarifying the following energy terms:

- Stored thermal and mechanical energy, E_{st}.
- Thermal and mechanical energy generation, E_g.
- Thermal and mechanical energy transport across the control surfaces, that is, the inflow and outflow terms, E_{in} and E_{out}.

In the statement of the first law (Equation 1.12a), the total energy, E^{tot}, consists of kinetic energy (KE $= \frac{1}{2}mV^2$, where m and V are mass and velocity, respectively), potential energy (PE $= mgz$, where g is the gravitational acceleration and z is the vertical coordinate), and *internal energy* (U). Mechanical energy is defined as the sum of kinetic and potential energy. Most often in heat transfer problems, the changes in kinetic and potential energy are small and can be neglected. The internal energy consists of a *sensible component*, which accounts for the translational, rotational, and/or vibrational motion of the atoms/molecules comprising the matter; a *latent component*, which relates to intermolecular forces influencing phase change between solid, liquid, and vapor states; a *chemical component*, which accounts for energy stored in the chemical bonds between atoms; and a *nuclear component*, which accounts for the binding forces in the nucleus.

For the study of heat transfer, we focus attention on the sensible and latent components of the internal energy (U_{sens} and U_{lat}, respectively), which are together referred to as *thermal energy*, U_t. The sensible energy is the portion that we associate mainly with changes in temperature (although it can also depend on pressure). The latent energy is the component we associate with changes in phase. For example, if the material in the control volume changes from solid to liquid (*melting*) or from liquid to vapor (*vaporization, evaporation, boiling*), the latent energy increases. Conversely, if the phase change is from vapor to liquid (*condensation*) or from liquid to solid (*solidification, freezing*), the latent energy decreases. Obviously, if no phase change is occurring, there is no change in latent energy, and this term can be neglected.

Based on this discussion, the *stored thermal and mechanical energy* is given by $E_{\text{st}} =$ KE + PE + U_t, where $U_t = U_{\text{sens}} + U_{\text{lat}}$. In many problems, the only relevant energy term will be the sensible energy, that is, $E_{\text{st}} = U_{\text{sens}}$.

The *energy generation term* is associated with conversion from some other form of internal energy (chemical, electrical, electromagnetic, or nuclear) to thermal or mechanical energy. It is a *volumetric phenomenon*. That is, it occurs within the control volume and is generally proportional to the magnitude of this volume. For example, an exothermic chemical reaction may be occurring, converting chemical energy to thermal energy. The net effect is an increase in the thermal energy of the matter within the control volume. Another source of thermal energy is the conversion from electrical energy that occurs due to resistance heating when an electric current is passed through a conductor. That is, if an electric current I passes through a resistance R in the control volume, electrical energy is dissipated at a rate I^2R, which corresponds to the rate at which thermal energy is generated (released) within the volume. In all applications of interest in this text, if chemical, electrical, or nuclear effects exist, they are treated as sources (or *sinks*, which correspond to negative sources) of thermal or mechanical energy and hence are included in the generation terms of Equations 1.12b,c.

The inflow and outflow terms are *surface phenomena*. That is, they are associated exclusively with processes occurring at the control surface and are generally proportional to the surface area. As discussed previously, the energy inflow and outflow terms include heat transfer (which can be by conduction, convection, and/or radiation) and work interactions occurring at the system boundaries (e.g., due to displacement of a boundary, a rotating shaft, and/or electromagnetic effects). For cases in which mass crosses the control volume boundary (e.g., for situations involving fluid flow), the inflow and outflow terms also include energy (thermal and mechanical) that is advected (carried) by mass entering and leaving the control volume. For instance, if the mass flow rate entering through the boundary is $\dot{m}$, then the rate at which thermal and mechanical energy enters with the flow is $\dot{m}\,(u_t + \frac{1}{2}V^2 + gz)$, where u_t is the thermal energy per unit mass.

When the first law is applied to a control volume with fluid crossing its boundary, it is customary to divide the work term into two contributions. The first contribution, termed *flow work*, is associated with work done by pressure forces moving fluid through the boundary. For a *unit mass*, the amount of work is equivalent to the product of the pressure and the specific volume of the fluid (pv). The symbol $\dot{W}$ is traditionally used for the rate at which the remaining work (not including flow work) is perfomed. If operation is under steady-state conditions ($dE_{st}/dt = 0$) and if there is no thermal or mechanical energy generation, Equation 1.12c reduces to the following form of the steady-flow energy equation (see Figure 1.8), which will be familiar if you have taken a thermodynamics course:

$$\dot{m}(u_t + pv + \tfrac{1}{2}V^2 + gz)_{\text{in}} - \dot{m}(u_t + pv + \tfrac{1}{2}V^2 + gz)_{\text{out}} + q - \dot{W} = 0 \tag{1.12d}$$

Terms within the parentheses are expressed for a unit mass of fluid at the inflow and outflow locations. When multiplied by the mass flow rate $\dot{m}$, they yield the rate at which the corresponding form of the energy (thermal, flow work, kinetic, and potential) enters or leaves the control volume. The sum of thermal energy and flow work per unit mass may be replaced by the enthalpy per unit mass, $i = u_t + pv$.

In most open system applications of interest in this text, changes in latent energy between the inflow and outflow conditions of Equation 1.12d may be neglected, so the thermal energy reduces to only the sensible component. If the fluid is approximated as an *ideal gas* with *constant specific heats*, the difference in enthalpies (per unit mass) between the inlet and outlet flows may then be expressed as $(i_{\text{in}} - i_{\text{out}}) = c_p(T_{\text{in}} - T_{\text{out}})$, where c_p is

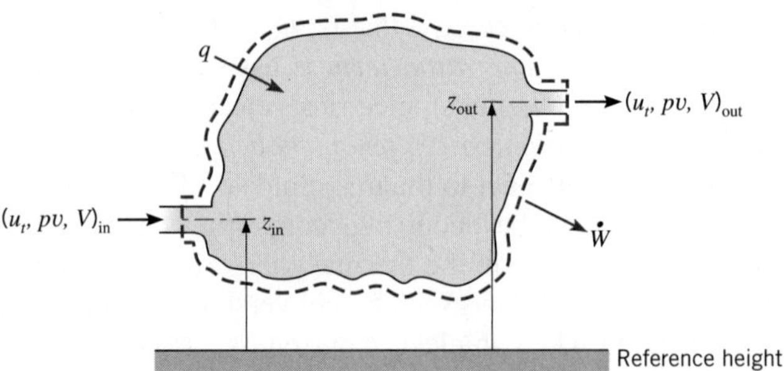

FIGURE 1.8 Conservation of energy for a steady-flow, open system.

the specific heat at constant pressure and T_{in} and T_{out} are the inlet and outlet temperatures, respectively. If the fluid is an *incompressible liquid*, its specific heats at constant pressure and volume are equal, $c_p = c_v \equiv c$, and for Equation 1.12d the change in sensible energy (per unit mass) reduces to $(u_{t,\text{in}} - u_{t,\text{out}}) = c(T_{\text{in}} - T_{\text{out}})$. Unless the pressure drop is extremely large, the difference in flow work terms, $(pv)_{\text{in}} - (pv)_{\text{out}}$, is negligible for a liquid.

Having already assumed steady-state conditions, no changes in latent energy, and no thermal or mechanical energy generation, there are at least four cases in which further assumptions can be made to reduce Equation 1.12d to the *simplified steady-flow thermal energy equation*:

$$q = \dot{m}c_p(T_{\text{out}} - T_{\text{in}}) \tag{1.12e}$$

The right-hand side of Equation 1.12e represents the net rate of outflow of enthalpy (thermal energy plus flow work) for an ideal gas or of thermal energy for an incompressible liquid.

The first two cases for which Equation 1.12e holds can readily be verified by examining Equation 1.12d. They are:

1. An ideal gas with negligible kinetic and potential energy changes and negligible work (other than flow work).
2. An incompressible liquid with negligible kinetic and potential energy changes and negligible work, *including* flow work. As noted in the preceding discussion, flow work is negligible for an incompressible liquid provided the pressure variation is not too great.

The second pair of cases cannot be directly derived from Equation 1.12d but require further knowledge of how mechanical energy is converted into thermal energy. These cases are:

3. An ideal gas with negligible viscous dissipation and negligible pressure variation.
4. An incompressible liquid with negligible viscous dissipation.

Viscous dissipation is the conversion from mechanical energy to thermal energy associated with viscous forces acting in a fluid. It is important only in cases involving high-speed flow and/or highly viscous fluid. Since so many engineering applications satisfy one or more of the preceding four conditions, Equation 1.12e is commonly used for the analysis of heat transfer in moving fluids. It will be used in Chapter 8 in the study of convection heat transfer in internal flow.

The *mass flow rate* $\dot{m}$ of the fluid may be expressed as $\dot{m} = \rho V A_c$, where ρ is the fluid density and A_c is the cross-sectional area of the channel through which the fluid flows. The *volumetric flow rate* is simply $\dot{\forall} = VA_c = \dot{m}/\rho$.

Example 1.3

The blades of a wind turbine turn a large shaft at a relatively slow speed. The rotational speed is increased by a gearbox that has an efficiency of $\eta_{\text{gb}} = 0.93$. In turn, the gearbox output shaft drives an electric generator with an efficiency of $\eta_{\text{gen}} = 0.95$. The cylindrical *nacelle*, which houses the gearbox, generator, and associated equipment, is of length $L = 6$ m and diameter $D = 3$ m. If the turbine produces $P = 2.5$ MW of electrical power, and the air and surroundings temperatures are $T_\infty = 25°\text{C}$ and $T_{\text{sur}} = 20°\text{C}$, respectively, determine the minimum possible operating temperature inside the nacelle. The emissivity of the nacelle is $\varepsilon = 0.83$, and the convective heat transfer coefficient is $h = 35\ \text{W/m}^2 \cdot \text{K}$. The surface of the nacelle that is adjacent to the blade hub can be considered to be adiabatic, and solar irradiation may be neglected.

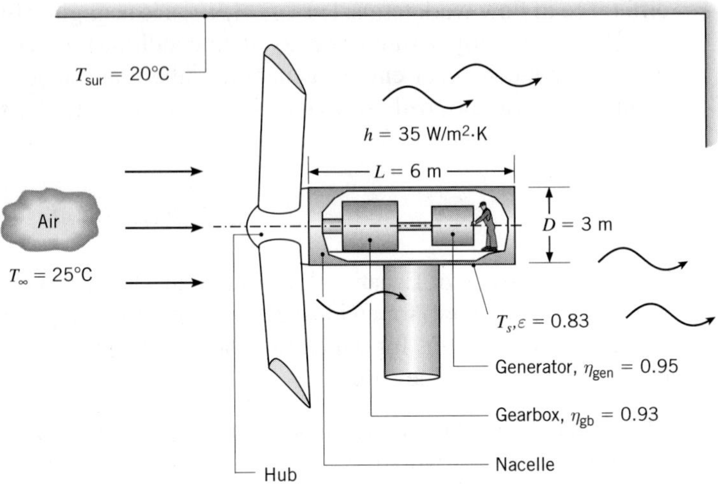

SOLUTION

Known: Electrical power produced by a wind turbine. Gearbox and generator efficiencies, dimensions and emissivity of the nacelle, ambient and surrounding temperatures, and heat transfer coefficient.

Find: Minimum possible temperature inside the enclosed nacelle.

Schematic:

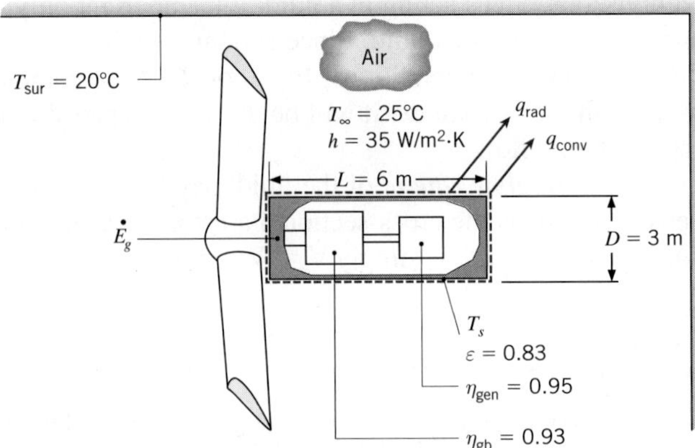

Assumptions:

1. Steady-state conditions.
2. Large surroundings.
3. Surface of the nacelle that is adjacent to the hub is adiabatic.

Analysis: The nacelle temperature represents the minimum possible temperature inside the nacelle, and the first law of thermodynamics may be used to determine this temperature. The first step is to perform an energy balance on the nacelle to determine the rate of heat transfer from the nacelle to the air and surroundings under steady-state conditions. This step can be accomplished using either conservation of *total* energy or conservation of *thermal and mechanical* energy; we will compare these two approaches.

Conservation of Total Energy The first of the three boxed statements of the first law in Section 1.3 can be converted to a rate basis and expressed in equation form as follows:

$$\frac{dE_{\text{st}}^{\text{tot}}}{dt} = \dot{E}_{\text{in}}^{\text{tot}} - \dot{E}_{\text{out}}^{\text{tot}} \tag{1}$$

Under steady-state conditions, this reduces to $\dot{E}_{\text{in}}^{\text{tot}} - \dot{E}_{\text{out}}^{\text{tot}} = 0$. The $\dot{E}_{\text{in}}^{\text{tot}}$ term corresponds to the mechanical work entering the nacelle $\dot{W}$, and the $\dot{E}_{\text{out}}^{\text{tot}}$ term includes the electrical power output P and the rate of heat transfer leaving the nacelle q. Thus

$$\dot{W} - P - q = 0 \tag{2}$$

Conservation of Thermal and Mechanical Energy Alternatively, we can express conservation of thermal and mechanical energy, starting with Equation 1.12c. Under steady-state conditions, this reduces to

$$\dot{E}_{\text{in}} - \dot{E}_{\text{out}} + \dot{E}_g = 0 \tag{3}$$

Here, $\dot{E}_{\text{in}}$ once again corresponds to the mechanical work $\dot{W}$. However, $\dot{E}_{\text{out}}$ now includes *only* the rate of heat transfer leaving the nacelle q. It does *not* include the electrical power, since E represents only the thermal and mechanical forms of energy. The electrical power appears in the generation term, because mechanical energy is converted to electrical energy in the generator, giving rise to a negative source of mechanical energy. That is, $\dot{E}_g = -P$. Thus, Equation (3) becomes

$$\dot{W} - q - P = 0 \tag{4}$$

which is equivalent to Equation (2), as it must be. Regardless of the manner in which the first law of thermodynamics is applied, the following expression for the rate of heat transfer evolves:

$$q = \dot{W} - P \tag{5}$$

The mechanical work and electrical power are related by the efficiencies of the gearbox and generator,

$$P = \dot{W}\eta_{\text{gb}}\eta_{\text{gen}} \tag{6}$$

Equation (5) can therefore be written as

$$q = P\left(\frac{1}{\eta_{\text{gb}}\eta_{\text{gen}}} - 1\right) = 2.5 \times 10^6\ \text{W} \times \left(\frac{1}{0.93 \times 0.95} - 1\right) = 0.33 \times 10^6\ \text{W} \tag{7}$$

Application of the Rate Equations Heat transfer is due to convection and radiation from the exterior surface of the nacelle, governed by Equations 1.3a and 1.7, respectively. Thus

$$q = q_{\text{rad}} + q_{\text{conv}} = A[q''_{\text{rad}} + q''_{\text{conv}}]$$

$$= \left[\pi DL + \frac{\pi D^2}{4}\right][\varepsilon\sigma(T_s^4 - T_{\text{sur}}^4) + h(T_s - T_\infty)] = 0.33 \times 10^6 \text{ W}$$

or

$$\left[\pi \times 3 \text{ m} \times 6 \text{ m} + \frac{\pi \times (3 \text{ m})^2}{4}\right]$$

$$\times \left[0.83 \times 5.67 \times 10^{-8} \text{ W/m}^2 \cdot \text{K}^4 \, (T_s^4 - (273 + 20)^4)\text{K}^4\right.$$

$$\left. + 35 \text{ W/m}^2 \cdot \text{K} \, (T_s - (273 + 25)\text{K})\right] = 0.33 \times 10^6 \text{ W}$$

The preceding equation does not have a closed-form solution, but the surface temperature can be easily determined by trial and error or by using a software package such as the *Interactive Heat Transfer* (*IHT*) software accompanying your text. Doing so yields

$$T_s = 416 \text{ K} = 143°\text{C}$$

We know that the temperature inside the nacelle must be greater than the exterior surface temperature of the nacelle T_s, because the heat generated within the nacelle must be transferred from the interior of the nacelle to its surface, and from the surface to the air and surroundings. Therefore, T_s represents the minimum possible temperature inside the enclosed nacelle. ◁

Comments:

1. The temperature inside the nacelle is very high. This would preclude, for example, performance of routine maintenance by a worker, as illustrated in the problem statement. Thermal management approaches involving fans or blowers must be employed to reduce the temperature to an acceptable level.
2. Improvements in the efficiencies of either the gearbox or the generator would not only provide more electrical power, but would also reduce the size and cost of the thermal management hardware. As such, improved efficiencies would increase revenue generated by the wind turbine and decrease both its capital and operating costs.
3. The heat transfer coefficient would not be a steady value but would vary periodically as the blades sweep past the nacelle. Therefore, the value of the heat transfer coefficient represents a *time-averaged* quantity.

IHT

EXAMPLE 1.4

A long conducting rod of diameter D and electrical resistance per unit length R'_e is initially in thermal equilibrium with the ambient air and its surroundings. This equilibrium is disturbed when an electrical current I is passed through the rod. Develop an equation that could be used to compute the variation of the rod temperature with time during the passage of the current.

Solution

Known: Temperature of a rod of prescribed diameter and electrical resistance changes with time due to passage of an electrical current.

Find: Equation that governs temperature change with time for the rod.

Schematic:

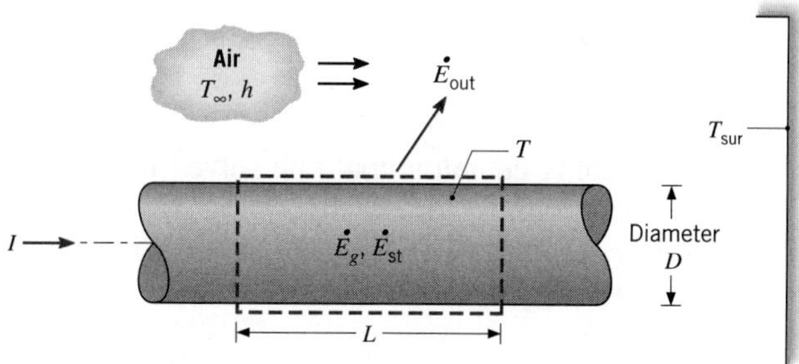

Assumptions:

1. At any time t, the temperature of the rod is uniform.
2. Constant properties (ρ, c, $\varepsilon = \alpha$).
3. Radiation exchange between the outer surface of the rod and the surroundings is between a small surface and a large enclosure.

Analysis: The first law of thermodynamics may often be used to determine an unknown temperature. In this case, there is no mechanical energy component. So relevant terms include heat transfer by convection and radiation from the surface, thermal energy generation due to ohmic heating within the conductor, and a change in thermal energy storage. Since we wish to determine the rate of change of the temperature, the first law should be applied at an instant of time. Hence, applying Equation 1.12c to a control volume of length L about the rod, it follows that

$$\dot{E}_g - \dot{E}_{out} = \dot{E}_{st}$$

where thermal energy generation is due to the electric resistance heating,

$$\dot{E}_g = I^2 R'_e L$$

Heating occurs uniformly within the control volume and could also be expressed in terms of a volumetric heat generation rate $\dot{q}$(W/m^3). The generation rate for the entire control volume is then $\dot{E}_g = \dot{q}V$, where $\dot{q} = I^2 R'_e/(\pi D^2/4)$. Energy outflow is due to convection and net radiation from the surface, Equations 1.3a and 1.7, respectively,

$$\dot{E}_{out} = h(\pi DL)(T - T_\infty) + \varepsilon\sigma(\pi DL)(T^4 - T_{sur}^4)$$

and the change in energy storage is due to the temperature change,

$$\dot{E}_{st} = \frac{dU_t}{dt} = \frac{d}{dt}(\rho VcT)$$

The term $\dot{E}_{st}$ is associated with the rate of change in the internal thermal energy of the rod, where ρ and c are the mass density and the specific heat, respectively, of the rod material, and V is the volume of the rod, $V = (\pi D^2/4)L$. Substituting the rate equations into the energy balance, it follows that

$$I^2R_e'L - h(\pi DL)(T - T_\infty) - \varepsilon\sigma(\pi DL)(T^4 - T_{sur}^4) = \rho c\left(\frac{\pi D^2}{4}\right)L\frac{dT}{dt}$$

Hence

$$\frac{dT}{dt} = \frac{I^2R_e' - \pi Dh(T - T_\infty) - \pi D\varepsilon\sigma(T^4 - T_{sur}^4)}{\rho c(\pi D^2/4)} \quad \triangleleft$$

Comments:

1. The preceding equation could be solved for the time dependence of the rod temperature by integrating numerically. A steady-state condition would eventually be reached for which $dT/dt = 0$. The rod temperature is then determined by an algebraic equation of the form

$$\pi Dh(T - T_\infty) + \pi D\varepsilon\sigma(T^4 - T_{sur}^4) = I^2R_e'$$

2. For fixed environmental conditions (h, T_∞, T_{sur}), as well as a rod of fixed geometry (D) and properties (ε, R_e'), the steady-state temperature depends on the rate of thermal energy generation and hence on the value of the electric current. Consider an uninsulated copper wire ($D = 1$ mm, $\varepsilon = 0.8$, $R_e' = 0.4\ \Omega$/m) in a relatively large enclosure ($T_{sur} = 300$ K) through which cooling air is circulated ($h = 100\ \text{W/m}^2\cdot\text{K}$, $T_\infty = 300$ K). Substituting these values into the foregoing equation, the rod temperature has been computed for operating currents in the range $0 \le I \le 10$ A, and the following results were obtained:

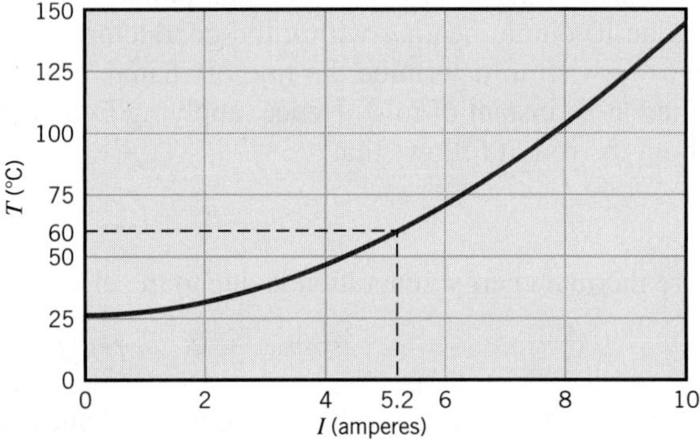

3. If a maximum operating temperature of $T = 60°\text{C}$ is prescribed for safety reasons, the current should not exceed 5.2 A. At this temperature, heat transfer by radiation (0.6 W/m) is much less than heat transfer by convection (10.4 W/m). Hence, if one wished to operate at a larger current while maintaining the rod temperature within the safety limit, the convection coefficient would have to be increased by increasing the velocity of the circulating air. For $h = 250\ \text{W/m}^2\cdot\text{K}$, the maximum allowable current could be increased to 8.1 A.

4. The *IHT* software is especially useful for solving equations, such as the energy balance in Comment 1, and generating the graphical results of Comment 2.

IHT | **EXAMPLE 1.5**

A hydrogen-air Proton Exchange Membrane (PEM) fuel cell is illustrated below. It consists of an *electrolytic membrane* sandwiched between porous *cathode* and *anode* materials, forming a very thin, three-layer *membrane electrode assembly* (MEA). At the anode, protons and electrons are generated ($2H_2 \rightarrow 4H^+ + 4e^-$); at the cathode, the protons and electrons recombine to form water ($O_2 + 4e^- + 4H^+ \rightarrow 2H_2O$). The overall reaction is then $2H_2 + O_2 \rightarrow 2H_2O$. The dual role of the electrolytic membrane is to transfer hydrogen ions and serve as a barrier to electron transfer, forcing the electrons to the electrical load that is external to the fuel cell.

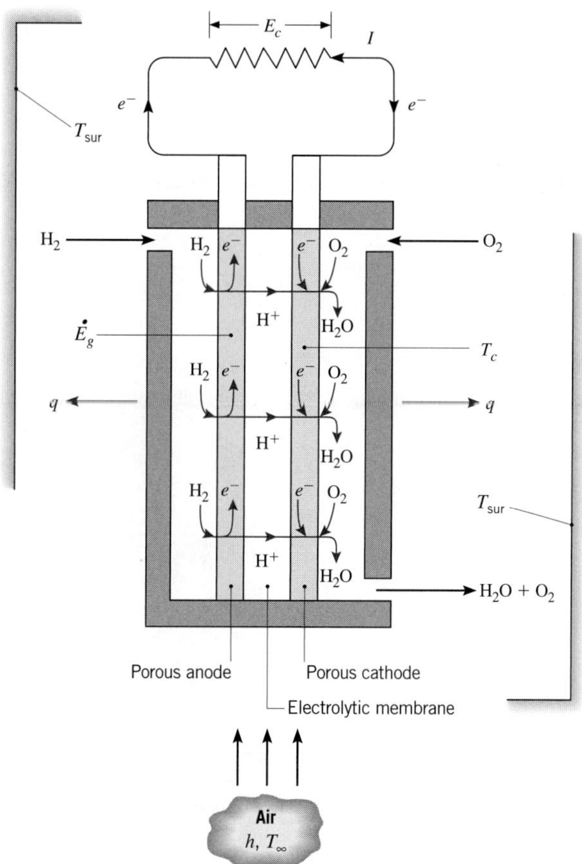

The membrane must operate in a moist state in order to conduct ions. However, the presence of liquid water in the cathode material may block the oxygen from reaching the cathode reaction sites, resulting in the failure of the fuel cell. Therefore, it is critical to control the temperature of the fuel cell, T_c, so that the cathode side contains saturated water vapor.

For a given set of H_2 and air inlet flow rates and use of a 50 mm × 50 mm MEA, the fuel cell generates $P = I \cdot E_c = 9$ W of electrical power. Saturated vapor conditions exist in the fuel cell, corresponding to $T_c = T_{sat} = 56.4°C$. The overall electrochemical reaction is exothermic, and the corresponding thermal generation rate of $\dot{E}_g = 11.25$ W must be removed from the fuel cell by convection and radiation. The ambient and surrounding temperatures are $T_\infty = T_{sur} = 25°C$, and the relationship between the cooling air velocity and the convection heat transfer coefficient h is

$$h = 10.9\ \text{W} \cdot \text{s}^{0.8}/\text{m}^{2.8} \cdot \text{K} \times V^{0.8}$$

where V has units of m/s. The exterior surface of the fuel cell has an emissivity of $\varepsilon = 0.88$. Determine the value of the cooling air velocity needed to maintain steady-state operating conditions. Assume the edges of the fuel cell are well insulated.

SOLUTION

Known: Ambient and surrounding temperatures, fuel cell output voltage and electrical current, heat generated by the overall electrochemical reaction, and the desired fuel cell operating temperature.

Find: The required cooling air velocity V needed to maintain steady-state operation at $T_c \approx 56.4°C$.

Schematic:

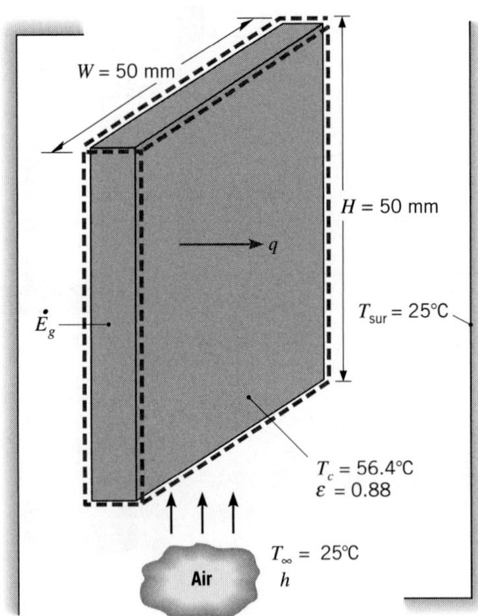

Assumptions:

1. Steady-state conditions.
2. Negligible temperature variations within the fuel cell.
3. Fuel cell is placed in large surroundings.
4. Edges of the fuel cell are well insulated.
5. Negligible energy entering or leaving the control volume due to gas or liquid flows.

Analysis: To determine the required cooling air velocity, we must first perform an energy balance on the fuel cell. Noting that there is no mechanical energy component, we see that $\dot{E}_{in} = 0$ and $\dot{E}_{out} = \dot{E}_g$. This yields

$$q_{conv} + q_{rad} = \dot{E}_g = 11.25 \text{ W}$$

where

$$\begin{aligned} q_{rad} &= \varepsilon A \sigma (T_c^4 - T_{sur}^4) \\ &= 0.88 \times (2 \times 0.05 \text{ m} \times 0.05 \text{ m}) \times 5.67 \times 10^{-8} \text{ W/m}^2 \cdot \text{K}^4 \times (329.4^4 - 298^4) \text{ K}^4 \\ &= 0.97 \text{ W} \end{aligned}$$

Therefore, we may find

$$\begin{aligned} q_{conv} &= 11.25 \text{ W} - 0.97 \text{ W} = 10.28 \text{ W} \\ &= hA(T_c - T_\infty) \\ &= 10.9 \text{ W} \cdot \text{s}^{0.8}/\text{m}^{2.8} \cdot \text{K} \times V^{0.8} A(T_c - T_\infty) \end{aligned}$$

which may be rearranged to yield

$$V = \left[\frac{10.28 \text{ W}}{10.9 \text{ W} \cdot \text{s}^{0.8}/\text{m}^{2.8} \cdot \text{K} \times (2 \times 0.05 \text{ m} \times 0.05 \text{ m}) \times (56.4 - 25)\,^\circ\text{C}} \right]^{1.25}$$

$$V = 9.4 \text{ m/s} \quad \triangleleft$$

Comments:

1. Temperature and humidity of the MEA will vary from location to location within the fuel cell. Prediction of the *local* conditions within the fuel cell would require a more detailed analysis.
2. The required cooling air velocity is quite high. Decreased cooling velocities could be used if heat transfer enhancement devices were added to the exterior of the fuel cell.
3. The convective heat rate is significantly greater than the radiation heat rate.
4. The chemical energy (20.25 W) of the hydrogen and oxygen is converted to electrical (9 W) and thermal (11.25 W) energy. This fuel cell operates at a conversion efficiency of (9 W)/(20.25 W) × 100 = 44%.

IHT | **EXAMPLE 1.6**

Large PEM fuel cells, such as those used in automotive applications, often require internal cooling using pure liquid water to maintain their temperature at a desired level (see Example 1.5). In cold climates, the cooling water must be drained from the fuel cell to an adjoining container when the automobile is turned off so that harmful freezing does not occur within the fuel cell. Consider a mass M of ice that was frozen while the automobile was not being operated. The ice is at the fusion temperature ($T_f = 0°\text{C}$) and is enclosed in a cubical container of width W on a side. The container wall is of thickness L and thermal conductivity k. If the outer surface of the wall is heated to a temperature $T_1 > T_f$ to melt the ice, obtain an expression for the time needed to melt the entire mass of ice and, in turn, deliver cooling water to, and energize, the fuel cell.

SOLUTION

Known: Mass and temperature of ice. Dimensions, thermal conductivity, and outer surface temperature of containing wall.

Find: Expression for time needed to melt the ice.

Schematic:

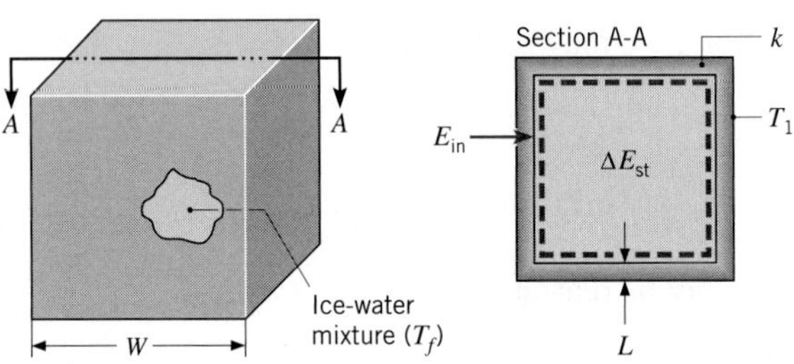

Assumptions:

1. Inner surface of wall is at T_f throughout the process.
2. Constant properties.
3. Steady-state, one-dimensional conduction through each wall.
4. Conduction area of one wall may be approximated as W^2 ($L \ll W$).

Analysis: Since we must determine the melting time t_m, the first law should be applied over the time interval $\Delta t = t_m$. Hence, applying Equation 1.12b to a control volume about the ice–water mixture, it follows that

$$E_{\text{in}} = \Delta E_{\text{st}} = \Delta U_{\text{lat}}$$

where the increase in energy stored within the control volume is due exclusively to the change in latent energy associated with conversion from the solid to liquid state. Heat is transferred to the ice by means of conduction through the container wall. Since the temperature difference across the wall is assumed to remain at $(T_1 - T_f)$ throughout the melting process, the wall conduction rate is constant

$$q_{\text{cond}} = k(6W^2)\frac{T_1 - T_f}{L}$$

and the amount of energy inflow is

$$E_{\text{in}} = \left[k(6W^2)\frac{T_1 - T_f}{L}\right]t_m$$

The amount of energy required to effect such a phase change per unit mass of solid is termed the *latent heat of fusion* h_{sf}. Hence the increase in energy storage is

$$\Delta E_{\text{st}} = Mh_{sf}$$

By substituting into the first law expression, it follows that

$$t_m = \frac{Mh_{sf}L}{6W^2k(T_1 - T_f)} \qquad \triangleleft$$

Comments:

1. Several complications would arise if the ice were initially subcooled. The storage term would have to include the change in sensible (internal thermal) energy required to take the ice from the subcooled to the fusion temperature. During this process, temperature gradients would develop in the ice.
2. Consider a cavity of width $W = 100$ mm on a side, wall thickness $L = 5$ mm, and thermal conductivity $k = 0.05$ W/m · K. The mass of the ice in the cavity is

$$M = \rho_s(W - 2L)^3 = 920 \text{ kg/m}^3 \times (0.100 - 0.01)^3 \text{ m}^3 = 0.67 \text{ kg}$$

If the outer surface temperature is $T_1 = 30°\text{C}$, the time required to melt the ice is

$$t_m = \frac{0.67 \text{ kg} \times 334{,}000 \text{ J/kg} \times 0.005 \text{ m}}{6(0.100 \text{ m})^2 \times 0.05 \text{ W/m} \cdot \text{K } (30 - 0)°\text{C}} = 12{,}430 \text{ s} = 207 \text{ min}$$

The density and latent heat of fusion of the ice are $\rho_s = 920$ kg/m^3 and $h_{sf} = 334$ kJ/kg, respectively.
3. Note that the units of K and °C cancel each other in the foregoing expression for t_m. Such cancellation occurs frequently in heat transfer analysis and is due to both units appearing in the context of a *temperature difference.*

The Surface Energy Balance We will frequently have occasion to apply the conservation of energy requirement at the surface of a medium. In this special case, the control surfaces are located on either side of the physical boundary and enclose no mass or volume (see Figure 1.9). Accordingly, the generation and storage terms of the conservation

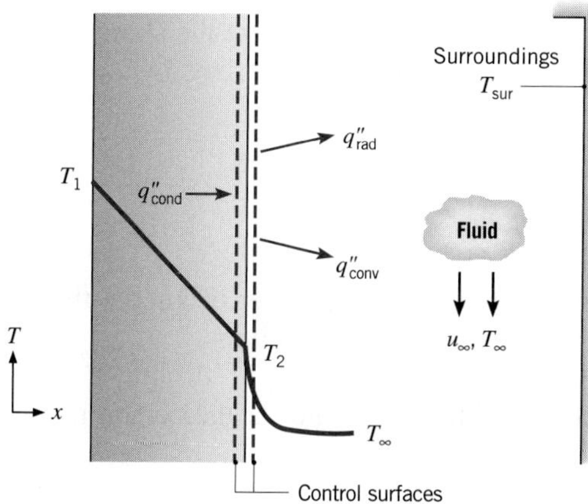

FIGURE 1.9 The energy balance for conservation of energy at the surface of a medium.

expression, Equation 1.12c, are no longer relevant, and it is necessary to deal only with surface phenomena. For this case, the conservation requirement becomes

$$\dot{E}_{\text{in}} - \dot{E}_{\text{out}} = 0 \tag{1.13}$$

Even though energy generation may be occurring in the medium, the process would not affect the energy balance at the control surface. Moreover, this conservation requirement holds for both *steady-state* and *transient* conditions.

In Figure 1.9, three heat transfer terms are shown for the control surface. On a unit area basis, they are conduction from the medium *to* the control surface (q''_{cond}), convection *from* the surface to a fluid (q''_{conv}), and net radiation exchange from the surface to the surroundings (q''_{rad}). The energy balance then takes the form.

$$q''_{\text{cond}} - q''_{\text{conv}} - q''_{\text{rad}} = 0 \tag{1.14}$$

and we can express each of the terms using the appropriate rate equations, Equations 1.2, 1.3a, and 1.7.

IHT

EXAMPLE 1.7

Humans are able to control their heat production rate and heat loss rate to maintain a nearly constant core temperature of $T_c = 37°\text{C}$ under a wide range of environmental conditions. This process is called *thermoregulation*. From the perspective of calculating heat transfer between a human body and its surroundings, we focus on a layer of skin and fat, with its outer surface exposed to the environment and its inner surface at a temperature slightly less than the core temperature, $T_i = 35°\text{C} = 308\ \text{K}$. Consider a person with a skin/fat layer of thickness $L = 3\ \text{mm}$ and effective thermal conductivity $k = 0.3\ \text{W/m}\cdot\text{K}$. The person has a surface area $A = 1.8\ \text{m}^2$ and is dressed in a bathing suit. The emissivity of the skin is $\varepsilon = 0.95$.

1. When the person is in still air at $T_\infty = 297\ \text{K}$, what is the skin surface temperature and rate of heat loss to the environment? Convection heat transfer to the air is characterized by a free convection coefficient of $h = 2\ \text{W/m}^2\cdot\text{K}$.

2. When the person is in water at $T_\infty = 297\ \text{K}$, what is the skin surface temperature and heat loss rate? Heat transfer to the water is characterized by a convection coefficient of $h = 200\ \text{W/m}^2\cdot\text{K}$.

SOLUTION

Known: Inner surface temperature of a skin/fat layer of known thickness, thermal conductivity, emissivity, and surface area. Ambient conditions.

Find: Skin surface temperature and heat loss rate for the person in air and the person in water.

Schematic:

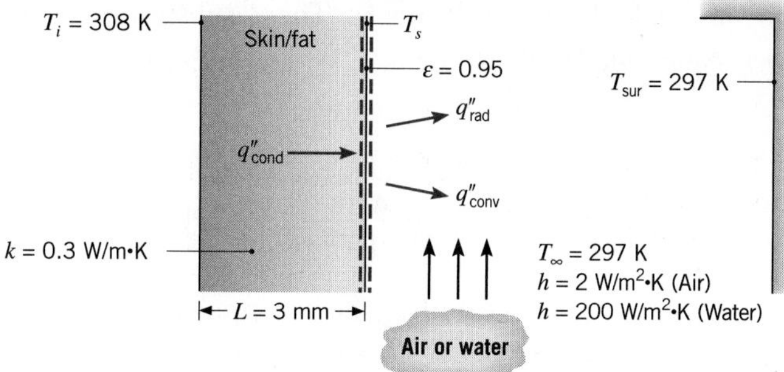

Assumptions:

1. Steady-state conditions.
2. One-dimensional heat transfer by conduction through the skin/fat layer.
3. Thermal conductivity is uniform.
4. Radiation exchange between the skin surface and the surroundings is between a small surface and a large enclosure at the air temperature.
5. Liquid water is opaque to thermal radiation.
6. Bathing suit has no effect on heat loss from body.
7. Solar radiation is negligible.
8. Body is completely immersed in water in part 2.

Analysis:

1. The skin surface temperature may be obtained by performing an energy balance at the skin surface. From Equation 1.13,

$$\dot{E}_{\text{in}} - \dot{E}_{\text{out}} = 0$$

It follows that, on a unit area basis,

$$q''_{\text{cond}} - q''_{\text{conv}} - q''_{\text{rad}} = 0$$

or, rearranging and substituting from Equations 1.2, 1.3a, and 1.7,

$$k\frac{T_i - T_s}{L} = h(T_s - T_\infty) + \varepsilon\sigma(T_s^4 - T_{\text{sur}}^4)$$

The only unknown is T_s, but we cannot solve for it explicitly because of the fourth-power dependence of the radiation term. Therefore, we must solve the equation iteratively, which can be done by hand or by using *IHT* or some other equation solver. To expedite a hand solution, we write the radiation heat flux in terms of the radiation heat transfer coefficient, using Equations 1.8 and 1.9:

$$k\frac{T_i - T_s}{L} = h(T_s - T_\infty) + h_r(T_s - T_{\text{sur}})$$

Solving for T_s, with $T_{\text{sur}} = T_\infty$, we have

$$T_s = \frac{\dfrac{kT_i}{L} + (h + h_r)T_\infty}{\dfrac{k}{L} + (h + h_r)}$$

We estimate h_r using Equation 1.9 with a guessed value of $T_s = 305$ K and $T_\infty = 297$ K, to yield $h_r = 5.9\ \text{W/m}^2 \cdot \text{K}$. Then, substituting numerical values into the preceding equation, we find

$$T_s = \frac{\dfrac{0.3\ \text{W/m} \cdot \text{K} \times 308\ \text{K}}{3 \times 10^{-3}\ \text{m}} + (2 + 5.9)\ \text{W/m}^2 \cdot \text{K} \times 297\ \text{K}}{\dfrac{0.3\ \text{W/m} \cdot \text{K}}{3 \times 10^{-3}\ \text{m}} + (2 + 5.9)\ \text{W/m}^2 \cdot \text{K}} = 307.2\ \text{K}$$

With this new value of T_s, we can recalculate h_r and T_s, which are unchanged. Thus the skin temperature is 307.2 K ≅ 34°C. ◁

The rate of heat loss can be found by evaluating the conduction through the skin/fat layer:

$$q_s = kA\frac{T_i - T_s}{L} = 0.3\ \text{W/m} \cdot \text{K} \times 1.8\ \text{m}^2 \times \frac{(308 - 307.2)\ \text{K}}{3 \times 10^{-3}\ \text{m}} = 146\ \text{W} \quad \triangleleft$$

2. Since liquid water is opaque to thermal radiation, heat loss from the skin surface is by convection only. Using the previous expression with $h_r = 0$, we find

$$T_s = \frac{\dfrac{0.3\ \text{W/m} \cdot \text{K} \times 308\ \text{K}}{3 \times 10^{-3}\ \text{m}} + 200\ \text{W/m}^2 \cdot \text{K} \times 297\ \text{K}}{\dfrac{0.3\ \text{W/m} \cdot \text{K}}{3 \times 10^{-3}\ \text{m}} + 200\ \text{W/m}^2 \cdot \text{K}} = 300.7\ \text{K} \quad \triangleleft$$

and

$$q_s = kA\frac{T_i - T_s}{L} = 0.3\ \text{W/m} \cdot \text{K} \times 1.8\ \text{m}^2 \times \frac{(308 - 300.7)\ \text{K}}{3 \times 10^{-3}\ \text{m}} = 1320\ \text{W} \quad \triangleleft$$

Comments:

1. When using energy balances involving radiation exchange, the temperatures appearing in the radiation terms must be expressed in kelvins, and it is good practice to use kelvins in all terms to avoid confusion.
2. In part 1, heat losses due to convection and radiation are 37 W and 109 W, respectively. Thus, it would not have been reasonable to neglect radiation. Care must be taken to include radiation when the heat transfer coefficient is small (as it often is for natural convection to a gas), even if the problem statement does not give any indication of its importance.
3. A typical rate of metabolic heat generation is 100 W. If the person stayed in the water too long, the core body temperature would begin to fall. The large heat loss in water is due to the higher heat transfer coefficient, which in turn is due to the much larger thermal conductivity of water compared to air.
4. The skin temperature of 34°C in part 1 is comfortable, but the skin temperature of 28°C in part 2 is uncomfortably cold.

Application of the Conservation Laws: Methodology In addition to being familiar with the transport rate equations described in Section 1.2, the heat transfer analyst must be able to work with the energy conservation requirements of Equations 1.12 and 1.13. The application of these balances is simplified if a few basic rules are followed.

1. The appropriate control volume must be defined, with the control surfaces represented by a dashed line or lines.
2. The appropriate time basis must be identified.
3. The relevant energy processes must be identified, and each process should be shown on the control volume by an appropriately labeled arrow.
4. The conservation equation must then be written, and appropriate rate expressions must be substituted for the relevant terms in the equation.

Note that the energy conservation requirement may be applied to a *finite* control volume or a *differential* (infinitesimal) control volume. In the first case, the resulting expression governs overall system behavior. In the second case, a differential equation is obtained that can be solved for conditions at each point in the system. Differential control volumes are introduced in Chapter 2, and both types of control volumes are used extensively throughout the text.

1.3.2 Relationship to the Second Law of Thermodynamics and the Efficiency of Heat Engines

In this section, we are interested in the efficiency of heat engines. The discussion builds on your knowledge of thermodynamics and shows how heat transfer plays a crucial role in managing and promoting the efficiency of a broad range of energy conversion devices. Recall that a heat engine is any device that operates continuously or cyclically and that converts heat to work. Examples include internal combustion engines, power plants, and thermoelectric devices (to be discussed in Section 3.8). Improving the efficiency of heat engines is a subject of extreme importance; for example, more efficient combustion engines consume less fuel to produce a given amount of work and reduce the corresponding emissions of pollutants and carbon dioxide. More efficient thermoelectric devices can generate more electricity from waste heat. Regardless of the energy conversion device, its size, weight, and cost can all be reduced through improvements in its energy conversion efficiency.

The second law of thermodynamics is often invoked when efficiency is of concern and can be expressed in a variety of different but equivalent ways. The *Kelvin–Planck statement* is particularly relevant to the operation of heat engines [1]. It states:

> *It is impossible for any system to operate in a thermodynamic cycle and deliver a net amount of work to its surroundings while receiving energy by heat transfer from a single thermal reservoir.*

Recall that a thermodynamic cycle is a process for which the initial and final states of the system are identical. Consequently, the energy stored in the system does not change between the initial and final states, and the first law of thermodynamics (Equation 1.12a) reduces to $W = Q$.

A consequence of the Kelvin–Planck statement is that a heat engine must exchange heat with two (or more) reservoirs, gaining thermal energy from the higher-temperature reservoir and rejecting thermal energy to the lower-temperature reservoir. Thus, converting all of the input heat to work is impossible, and $W = Q_{in} - Q_{out}$, where Q_{in} and Q_{out} are both defined to be positive. That is, Q_{in} is the heat transferred from the high temperature source to the heat engine, and Q_{out} is the heat transferred from the heat engine to the low temperature sink.

The efficiency of a heat engine is defined as the fraction of heat transferred into the system that is converted to work, namely

$$\eta \equiv \frac{W}{Q_{in}} = \frac{Q_{in} - Q_{out}}{Q_{in}} = 1 - \frac{Q_{out}}{Q_{in}} \tag{1.15}$$

The second law also tells us that, for a *reversible* process, the ratio Q_{out}/Q_{in} is equal to the ratio of the absolute temperatures of the respective reservoirs [1]. Thus, the efficiency of a heat engine undergoing a reversible process, called the *Carnot efficiency* η_C, is given by

$$\eta_C = 1 - \frac{T_c}{T_h} \tag{1.16}$$

where T_c and T_h are the absolute temperatures of the low- and high-temperature reservoirs, respectively. The Carnot efficiency is the maximum possible efficiency that any heat engine can achieve operating between those two temperatures. Any *real* heat engine, which will necessarily undergo an irreversible process, will have a lower efficiency.

From our knowledge of thermodynamics, we know that, for heat transfer to take place reversibly, it must occur through an infinitesimal temperature difference between the reservoir and heat engine. However, from our newly acquired knowledge of heat transfer mechanisms, as embodied, for example, in Equations 1.2, 1.3, and 1.7, we now realize that, for heat transfer to occur, there *must* be a nonzero temperature difference between the reservoir and the heat engine. This reality introduces irreversibility and reduces the efficiency.

With the concepts of the preceding paragraph in mind, we now consider a more realistic model of a heat engine [2–5] in which heat is transferred into the engine through a thermal resistance $R_{t,h}$, while heat is extracted from the engine through a second thermal resistance $R_{t,c}$ (Figure 1.10). The subscripts h and c refer to the hot and cold sides of the heat engine, respectively. As discussed in Section 1.2.4, these thermal resistances are associated with heat transfer between the heat engine and the reservoirs across a nonzero temperature difference, by way of the mechanisms of conduction, convection, and/or radiation. For example, the resistances could represent conduction through the walls separating the heat engine from the two reservoirs. Note that the reservoir temperatures are still T_h and T_c but that the temperatures seen by the heat engine are $T_{h,i} < T_h$ and $T_{c,i} > T_c$, as shown in the diagram. The heat engine is still assumed to be *internally* reversible, and its efficiency is still the Carnot efficiency. However,

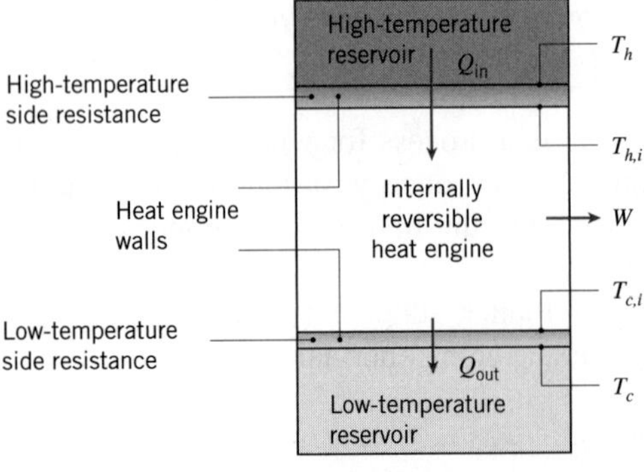

FIGURE 1.10 Internally reversible heat engine exchanging heat with high- and low-temperature reservoirs through thermal resistances.

the Carnot efficiency is *now based on the internal temperatures* $T_{h,i}$ and $T_{c,i}$. Therefore, a modified efficiency that accounts for realistic (irreversible) heat transfer processes η_m is

$$\eta_m = 1 - \frac{Q_{\text{out}}}{Q_{\text{in}}} = 1 - \frac{q_{\text{out}}}{q_{\text{in}}} = 1 - \frac{T_{c,i}}{T_{h,i}} \tag{1.17}$$

where the ratio of heat *flows* over a time interval, $Q_{\text{out}}/Q_{\text{in}}$, has been replaced by the corresponding ratio of heat *rates*, $q_{\text{out}}/q_{\text{in}}$. This replacement is based on applying energy conservation at an instant in time,[1] as discussed in Section 1.3.1. Utilizing the definition of a thermal resistance, the heat transfer rates into and out of the heat engine are given by

$$q_{\text{in}} = (T_h - T_{h,i})/R_{t,h} \tag{1.18a}$$

$$q_{\text{out}} = (T_{c,i} - T_c)/R_{t,c} \tag{1.18b}$$

Equations 1.18 can be solved for the internal temperatures, to yield

$$T_{h,i} = T_h - q_{\text{in}}R_{t,h} \tag{1.19a}$$

$$T_{c,i} = T_c + q_{\text{out}}R_{t,c} = T_c + q_{\text{in}}(1 - \eta_m)R_{t,c} \tag{1.19b}$$

In Equation 1.19b, q_{out} has been related to q_{in} and η_m, using Equation 1.17. The more realistic, modified efficiency can then be expressed as

$$\eta_m = 1 - \frac{T_{c,i}}{T_{h,i}} = 1 - \frac{T_c + q_{\text{in}}(1 - \eta_m)R_{t,c}}{T_h - q_{\text{in}}R_{t,h}} \tag{1.20}$$

Solving for η_m results in

$$\eta_m = 1 - \frac{T_c}{T_h - q_{\text{in}}R_{\text{tot}}} \tag{1.21}$$

where $R_{\text{tot}} = R_{t,h} + R_{t,c}$. It is readily evident that $\eta_m = \eta_C$ only if the thermal resistances $R_{t,h}$ and $R_{t,c}$ could somehow be made infinitesimally small (or if $q_{\text{in}} = 0$). For realistic (nonzero) values of R_{tot}, $\eta_m < \eta_C$, and η_m further deteriorates as either R_{tot} or q_{in} increases. As an extreme case, note that $\eta_m = 0$ when $T_h = T_c + q_{\text{in}}R_{\text{tot}}$, meaning that no power could be produced even though the Carnot efficiency, as expressed in Equation 1.16, is nonzero.

In addition to the efficiency, another important parameter to consider is the power output of the heat engine, given by

$$\dot{W} = q_{\text{in}}\eta_m = q_{\text{in}}\left[1 - \frac{T_c}{T_h - q_{\text{in}}R_{\text{tot}}}\right] \tag{1.22}$$

It has already been noted in our discussion of Equation 1.21 that the efficiency is equal to the maximum Carnot efficiency ($\eta_m = \eta_C$) if $q_{\text{in}} = 0$. However, under these circumstances

[1]The heat engine is assumed to undergo a continuous, steady-flow process, so that all heat and work processes are occurring simultaneously, and the corresponding terms would be expressed in watts (W). For a heat engine undergoing a cyclic process with sequential heat and work processes occurring over different time intervals, we would need to introduce the time intervals for each process, and each term would be expressed in joules (J).

the power output $\dot{W}$ is zero according to Equation 1.22. To increase $\dot{W}$, q_{in} must be increased at the expense of decreased efficiency. In any real application, a balance must be struck between maximizing the efficiency and maximizing the power output. If provision of the heat input is inexpensive (for example, if waste heat is converted to power), a case could be made for sacrificing efficiency to maximize power output. In contrast, if fuel is expensive or emissions are detrimental (such as for a conventional fossil fuel power plant), the efficiency of the energy conversion may be as or more important than the power output. In any case, heat transfer and thermodyamic principles should be used to determine the actual efficiency and power output of a heat engine.

Although we have limited our discussion of the second law to heat engines, the preceding analysis shows how the principles of thermodynamics and heat transfer can be combined to address significant problems of contemporary interest.

Example 1.8

In a large steam power plant, the combustion of coal provides a heat rate of $q_{in} = 2500$ MW at a flame temperature of $T_h = 1000$ K. Heat is rejected from the plant to a river flowing at $T_c = 300$ K. Heat is transferred from the combustion products to the exterior of large tubes in the boiler by way of radiation and convection, through the boiler tubes by conduction, and then from the interior tube surface to the working fluid (water) by convection. On the cold side, heat is extracted from the power plant by condensation of steam on the exterior condenser tube surfaces, through the condenser tube walls by conduction, and from the interior of the condenser tubes to the river water by convection. Hot and cold side thermal resistances account for the combined effects of conduction, convection, and radiation and, under *design conditions*, they are $R_{t,h} = 8 \times 10^{-8}$ K/W and $R_{t,c} = 2 \times 10^{-8}$ K/W, respectively.

1. Determine the efficiency and power output of the power plant, accounting for heat transfer effects to and from the cold and hot reservoirs. Treat the power plant as an internally reversible heat engine.
2. Over time, coal slag will accumulate on the combustion side of the boiler tubes. This *fouling process* increases the hot side resistance to $R_{t,h} = 9 \times 10^{-8}$ K/W. Concurrently, biological matter can accumulate on the river water side of the condenser tubes, increasing the cold side resistance to $R_{t,c} = 2.2 \times 10^{-8}$ K/W. Find the efficiency and power output of the plant under fouled conditions.

Solution

Known: Source and sink temperatures and heat input rate for an internally reversible heat engine. Thermal resistances separating heat engine from source and sink under clean and fouled conditions.

Find:

1. Efficiency and power output for clean conditions.
2. Efficiency and power output under fouled conditions.

Schematic:

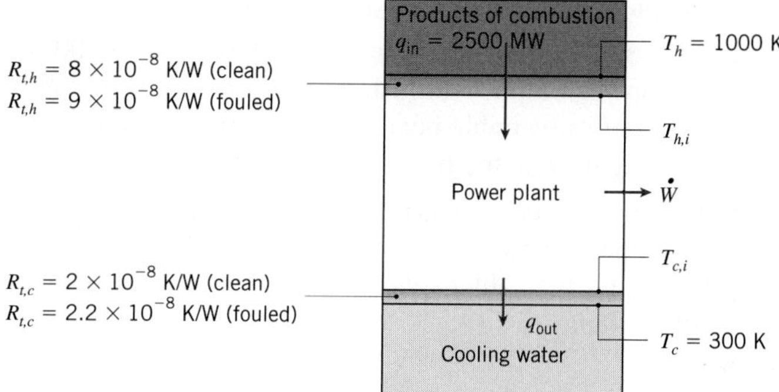

Assumptions:

1. Steady-state conditions.
2. Power plant behaves as an internally reversible heat engine, so its efficiency is the modified efficiency.

Analysis:

1. The modified efficiency of the internally reversible power plant, considering realistic heat transfer effects on the hot and cold side of the power plant, is given by Equation 1.21:

$$\eta_m = 1 - \frac{T_c}{T_h - q_{\text{in}} R_{\text{tot}}}$$

where, for clean conditions

$$R_{\text{tot}} = R_{t,h} + R_{t,c} = 8 \times 10^{-8}\ \text{K/W} + 2 \times 10^{-8}\ \text{K/W} = 1.0 \times 10^{-7}\ \text{K/W}$$

Thus

$$\eta_m = 1 - \frac{T_c}{T_h - q_{\text{in}} R_{\text{tot}}} = 1 - \frac{300\ \text{K}}{1000\ \text{K} - 2500 \times 10^6\ \text{W} \times 1.0 \times 10^{-7}\ \text{K/W}} = 0.60 = 60\% \quad \triangleleft$$

The power output is given by

$$\dot{W} = q_{\text{in}} \eta_m = 2500\ \text{MW} \times 0.60 = 1500\ \text{MW} \quad \triangleleft$$

2. Under fouled conditions, the preceding calculations are repeated to find

$$\eta_m = 0.583 = 58.3\% \text{ and } \dot{W} = 1460\ \text{MW} \quad \triangleleft$$

Comments:

1. The actual efficiency and power output of a power plant operating between these temperatures would be much less than the foregoing values, since there would be other irreversibilities internal to the power plant. Even if these irreversibilities

were considered in a more comprehensive analysis, fouling effects would still reduce the plant efficiency and power output.

2. The Carnot efficiency is $\eta_C = 1 - T_c/T_h = 1 - 300\text{ K}/1000\text{ K} = 70\%$. The corresponding power output would be $\dot{W} = q_{in}\eta_C = 2500\text{ MW} \times 0.70 = 1750\text{ MW}$. Thus, if the effect of irreversible heat transfer from and to the hot and cold reservoirs, respectively, were neglected, the power output of the plant would be significantly overpredicted.
3. Fouling reduces the power output of the plant by $\Delta P = 40$ MW. If the plant owner sells the electricity at a price of \$0.08/kW · h, the daily lost revenue associated with operating the fouled plant would be $C = 40{,}000\text{ kW} \times \$0.08/\text{kW}\cdot\text{h} \times 24\text{ h/day} =$ \$76,800/day.

1.4 Units and Dimensions

The physical quantities of heat transfer are specified in terms of *dimensions,* which are measured in terms of units. Four *basic* dimensions are required for the development of heat transfer: length (L), mass (M), time (t), and temperature (T). All other physical quantities of interest may be related to these four basic dimensions.

In the United States, dimensions have been customarily measured in terms of the *English system of units*, for which the *base units* are:

Dimension		Unit
Length (L)	→	foot (ft)
Mass (M)	→	pound mass (lb_m)
Time (t)	→	second (s)
Temperature (T)	→	degree Fahrenheit (°F)

The units required to specify other physical quantities may then be inferred from this group.

In recent years, there has been a strong trend toward the global usage of a standard set of units. In 1960, the SI (Système International d'Unités) system of units was defined by the Eleventh General Conference on Weights and Measures and recommended as a worldwide standard. In response to this trend, the American Society of Mechanical Engineers (ASME) has required the use of SI units in all of its publications since 1974. For this reason and because SI units are operationally more convenient than the English system, the SI system is used for calculations of this text. However, because for some time to come, engineers might also have to work with results expressed in the English system, you should be able to convert from one system to the other. For your convenience, conversion factors are provided on the inside back cover of the text.

The SI *base* units required for this text are summarized in Table 1.2. With regard to these units, note that 1 mol is the amount of substance that has as many atoms or molecules as there are atoms in 12 g of carbon-12 (^{12}C); this is the gram-mole (mol). Although the mole has been recommended as the unit quantity of matter for the SI system, it is more consistent to work with the kilogram-mol (kmol, kg-mol). One kmol is simply the amount of substance that has as many atoms or molecules as there are atoms in 12 kg of ^{12}C. As long as the use is consistent within a given problem, no difficulties arise in using either mol or kmol. The molecular weight of a substance is the mass associated with a mole or kilogram-mole. For oxygen, as an example, the molecular weight $\mathcal{M}$ is 16 g/mol or 16 kg/kmol.

Although the SI unit of temperature is the kelvin, use of the Celsius temperature scale remains widespread. Zero on the Celsius scale (0°C) is equivalent to 273.15 K on the thermodynamic scale,[2] in which case

$$T(\text{K}) = T(^\circ\text{C}) + 273.15$$

However, temperature *differences* are equivalent for the two scales and may be denoted as °C or K. Also, although the SI unit of time is the second, other units of time (minute, hour, and day) are so common that their use with the SI system is generally accepted.

The SI units comprise a coherent form of the metric system. That is, all remaining units may be derived from the base units using formulas that do not involve any numerical factors. *Derived* units for selected quantities are listed in Table 1.3. Note that force is measured in newtons, where a 1-N force will accelerate a 1-kg mass at 1 m/s^2. Hence 1 N = 1 kg · m/s^2. The unit of pressure (N/m^2) is often referred to as the pascal. In the SI system, there is one unit of energy (thermal, mechanical, or electrical) called the joule (J); 1 J = 1 N · m. The unit for energy rate, or power, is then J/s, where one joule per second is equivalent to one watt (1 J/s = 1 W). Since working with extremely large or small numbers is frequently necessary, a set of standard prefixes has been introduced to simplify matters (Table 1.4). For example, 1 megawatt (MW) = 10^6 W, and 1 micrometer (μm) = 10^{-6} m.

TABLE 1.2 SI base and supplementary units

Quantity and Symbol	Unit and Symbol
Length (L)	meter (m)
Mass (M)	kilogram (kg)
Amount of substance	mole (mol)
Time (t)	second (s)
Electric current (I)	ampere (A)
Thermodynamic temperature (T)	kelvin (K)
Plane angle[a] (θ)	radian (rad)
Solid angle[a] (ω)	steradian (sr)

[a]Supplementary unit.

TABLE 1.3 SI derived units for selected quantities

Quantity	Name and Symbol	Formula	Expression in SI Base Units
Force	newton (N)	m · kg/s^2	m · kg/s^2
Pressure and stress	pascal (Pa)	N/m^2	kg/m · s^2
Energy	joule (J)	N · m	m^2 · kg/s^2
Power	watt (W)	J/s	m^2 · kg/s^3

[2]The degree symbol is retained for designating the Celsius temperature (°C) to avoid confusion with the use of C for the unit of electrical charge (coulomb).

TABLE 1.4 Multiplying prefixes

Prefix	Abbreviation	Multiplier
femto	f	10^{-15}
pico	p	10^{-12}
nano	n	10^{-9}
micro	μ	10^{-6}
milli	m	10^{-3}
centi	c	10^{-2}
hecto	h	10^{2}
kilo	k	10^{3}
mega	M	10^{6}
giga	G	10^{9}
tera	T	10^{12}
peta	P	10^{15}
exa	E	10^{18}

1.5 *Analysis of Heat Transfer Problems: Methodology*

A major objective of this text is to prepare you to solve engineering problems that involve heat transfer processes. To this end, numerous problems are provided at the end of each chapter. In working these problems you will gain a deeper appreciation for the fundamentals of the subject, and you will gain confidence in your ability to apply these fundamentals to the solution of engineering problems.

In solving problems, we advocate the use of a systematic procedure characterized by a prescribed format. We consistently employ this procedure in our examples, and we require our students to use it in their problem solutions. It consists of the following steps:

1. *Known:* After carefully reading the problem, state briefly and concisely what is known about the problem. Do not repeat the problem statement.
2. *Find:* State briefly and concisely what must be found.
3. *Schematic:* Draw a schematic of the physical system. If application of the conservation laws is anticipated, represent the required control surface or surfaces by dashed lines on the schematic. Identify relevant heat transfer processes by appropriately labeled arrows on the schematic.
4. *Assumptions:* List all pertinent simplifying assumptions.
5. *Properties:* Compile property values needed for subsequent calculations and identify the source from which they are obtained.
6. *Analysis:* Begin your analysis by applying appropriate conservation laws, and introduce rate equations as needed. Develop the analysis as completely as possible before substituting numerical values. Perform the calculations needed to obtain the desired results.
7. *Comments:* Discuss your results. Such a discussion may include a summary of key conclusions, a critique of the original assumptions, and an inference of trends obtained by performing additional *what-if* and *parameter sensitivity* calculations.

The importance of following steps 1 through 4 should not be underestimated. They provide a useful guide to thinking about a problem before effecting its solution. In step 7, we hope you will take the initiative to gain additional insights by performing calculations that may be computer based. The software accompanying this text provides a suitable tool for effecting such calculations.

IHT

EXAMPLE 1.9

The coating on a plate is cured by exposure to an infrared lamp providing a uniform irradiation of 2000 W/m^2. It absorbs 80% of the irradiation and has an emissivity of 0.50. It is also exposed to an airflow and large surroundings for which temperatures are 20°C and 30°C, respectively.

1. If the convection coefficient between the plate and the ambient air is 15 W/m$^2 \cdot$K, what is the cure temperature of the plate?
2. The final characteristics of the coating, including wear and durability, are known to depend on the temperature at which curing occurs. An airflow system is able to control the air velocity, and hence the convection coefficient, on the cured surface, but the process engineer needs to know how the temperature depends on the convection coefficient. Provide the desired information by computing and plotting the surface temperature as a function of h for $2 \leq h \leq 200$ W/m$^2 \cdot$K. What value of h would provide a cure temperature of 50°C?

SOLUTION

Known: Coating with prescribed radiation properties is cured by irradiation from an infrared lamp. Heat transfer from the coating is by convection to ambient air and radiation exchange with the surroundings.

Find:

1. Cure temperature for $h = 15$ W/m$^2 \cdot$K.
2. Effect of airflow on the cure temperature for $2 \leq h \leq 200$ W/m$^2 \cdot$K. Value of h for which the cure temperature is 50°C.

Schematic:

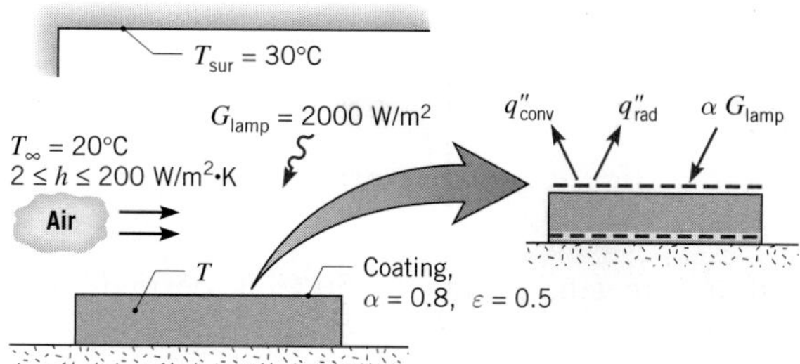

Assumptions:

1. Steady-state conditions.
2. Negligible heat loss from back surface of plate.
3. Plate is small object in large surroundings, and coating has an absorptivity of $\alpha_{sur} = \varepsilon = 0.5$ with respect to irradiation from the surroundings.

Analysis:

1. Since the process corresponds to steady-state conditions and there is no heat transfer at the back surface, the plate must be isothermal ($T_s = T$). Hence the desired temperature may be determined by placing a control surface about the exposed surface and applying Equation 1.13 or by placing the control surface about the entire plate and applying Equation 1.12c. Adopting the latter approach and recognizing that there is no energy generation ($\dot{E}_g = 0$), Equation 1.12c reduces to

$$\dot{E}_{in} - \dot{E}_{out} = 0$$

where $\dot{E}_{st} = 0$ for steady-state conditions. With energy inflow due to absorption of the lamp irradiation by the coating and outflow due to convection and net radiation transfer to the surroundings, it follows that

$$(\alpha G)_{lamp} - q''_{conv} - q''_{rad} = 0$$

Substituting from Equations 1.3a and 1.7, we obtain

$$(\alpha G)_{lamp} - h(T - T_\infty) - \varepsilon\sigma(T^4 - T^4_{sur}) = 0$$

Substituting numerical values

$$0.8 \times 2000\ \text{W/m}^2 - 15\ \text{W/m}^2 \cdot \text{K}\ (T - 293)\ \text{K}$$
$$- 0.5 \times 5.67 \times 10^{-8}\ \text{W/m}^2 \cdot \text{K}^4\ (T^4 - 303^4)\ \text{K}^4 = 0$$

and solving by trial-and-error, we obtain

$$T = 377\ \text{K} = 104°\text{C} \quad \triangleleft$$

2. Solving the foregoing energy balance for selected values of h in the prescribed range and plotting the results, we obtain

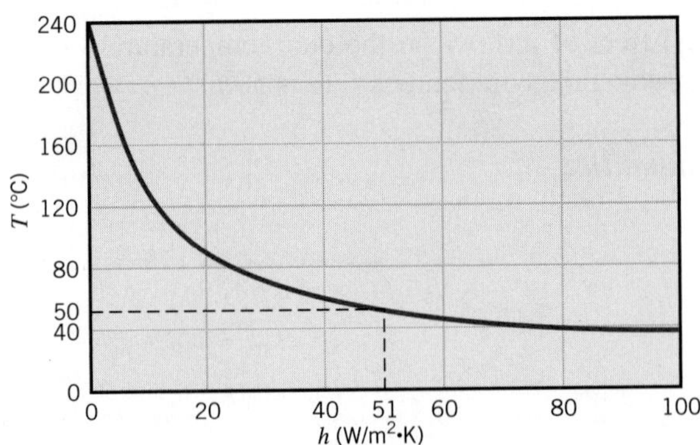

If a cure temperature of 50°C is desired, the airflow must provide a convection coefficient of

$$h(T = 50°\text{C}) = 51.0\ \text{W/m}^2 \cdot \text{K} \quad \triangleleft$$

Comments:

1. The coating (plate) temperature may be reduced by decreasing T_∞ and T_{sur}, as well as by increasing the air velocity and hence the convection coefficient.
2. The relative contributions of convection and radiation to heat transfer from the plate vary greatly with h. For $h = 2\ \text{W/m}^2\cdot\text{K}$, $T = 204°\text{C}$ and radiation dominates ($q''_{rad} \approx 1232\ \text{W/m}^2$, $q''_{conv} \approx 368\ \text{W/m}^2$). Conversely, for $h = 200\ \text{W/m}^2\cdot\text{K}$, $T = 28°\text{C}$ and convection dominates ($q''_{conv} \approx 1606\ \text{W/m}^2$, $q''_{rad} \approx -6\ \text{W/m}^2$). In fact, for this condition the plate temperature is slightly less than that of the surroundings and net radiation exchange is *to* the plate.

1.6 *Relevance of Heat Transfer*

We will devote much time to acquiring an understanding of heat transfer effects and to developing the skills needed to predict heat transfer rates and temperatures that evolve in certain situations. What is the value of this knowledge? To what problems may it be applied? A few examples will serve to illustrate the rich breadth of applications in which heat transfer plays a critical role.

The challenge of providing sufficient amounts of energy for humankind is well known. Adequate supplies of energy are needed not only to fuel industrial productivity, but also to supply safe drinking water and food for much of the world's population and to provide the sanitation necessary to control life-threatening diseases.

To appreciate the role heat transfer plays in the energy challenge, consider a flow chart that represents energy use in the United States, as shown in Figure 1.11*a*. Currently, about 58% of the nearly 110 EJ of energy that is consumed annually in the United States is wasted in the form of heat. Nearly 70% of the energy used to generate electricity is lost in the form of heat. The transportation sector, which relies almost exclusively on petroleum-based fuels, utilizes only 21.5% of the energy it consumes; the remaining 78.5% is released in the form of heat. Although the industrial and residential/commercial use of energy is relatively more efficient, opportunities for *energy conservation* abound. Creative thermal engineering, utilizing the tools of thermodynamics *and* heat transfer, can lead to new ways to (1) increase the efficiency by which energy is *generated* and *converted*, (2) reduce energy *losses*, and (3) *harvest* a large portion of the waste heat.

As evident in Figure 1.11*a*, *fossil fuels* (petroleum, natural gas, and coal) dominate the energy portfolio in many countries, such as the United States. The combustion of fossil fuels produces massive amounts of carbon dioxide; the amount of CO_2 released in the United States on an annual basis due to combustion is currently 5.99 Eg (5.99×10^{15} kg). As more CO_2 is pumped into the atmosphere, mechanisms of radiation heat transfer *within* the atmosphere are modified, resulting in potential changes in global temperatures. In a country like the United States, electricity generation and transportation are responsible for nearly 75% of the total CO_2 released into the atmosphere due to energy use (Figure 1.11*b*).

What are some of the ways engineers are applying the principles of heat transfer to address issues of energy and environmental *sustainability*?

The efficiency of a *gas turbine engine* can be significantly increased by increasing its operating temperature. Today, the temperatures of the combustion gases inside these

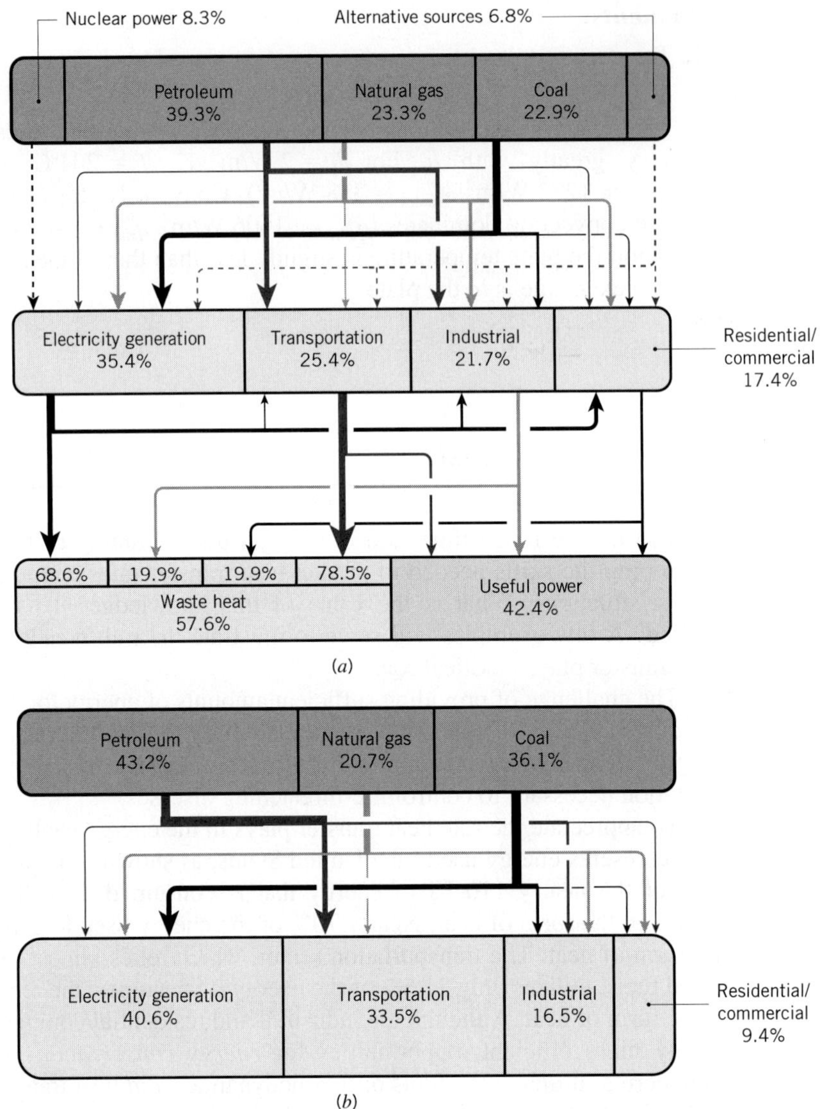

FIGURE 1.11 Flow charts for energy consumption and associated CO_2 emissions in the United States in 2007. (*a*) Energy production and consumption. (*b*) Carbon dioxide by source of fossil fuel and end-use application. Arrow widths represent relative magnitudes of the flow streams. (Credit: U.S. Department of Energy and the Lawrence Livermore National Laboratory.)

engines far exceed the melting point of the exotic alloys used to manufacture the turbine blades and vanes. Safe operation is typically achieved by three means. First, relatively cool gases are injected through small holes at the leading edge of a turbine blade (Figure 1.12). These gases hug the blade as they are carried downstream and help insulate the blade from the hot combustion gases. Second, thin layers of a very low thermal conductivity, ceramic *thermal barrier coating* are applied to the blades and vanes to provide an extra layer of insulation. These coatings are produced by spraying molten ceramic powders onto the engine components using extremely high temperature sources such as plasma spray guns

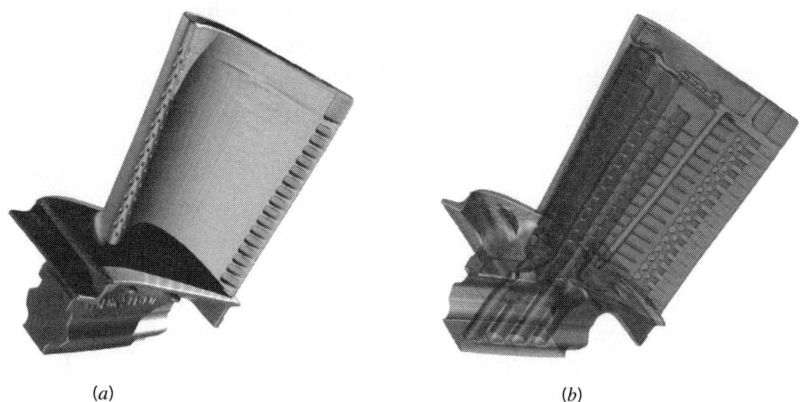

(*a*) (*b*)

FIGURE 1.12 Gas turbine blade. (*a*) External view showing holes for injection of cooling gases. (*b*) X ray view showing internal cooling passages. (Credit: Images courtesy of FarField Technology, Ltd., Christchurch, New Zealand.)

that can operate in excess of 10,000 kelvins. Third, the blades and vanes are designed with intricate, internal cooling passages, all carefully configured by the heat transfer engineer to allow the gas turbine engine to operate under such extreme conditions.

Alternative sources constitute a small fraction of the energy portfolio of many nations, as illustrated in the flow chart of Figure 1.11*a* for the United States. The intermittent nature of the power generated by sources such as the wind and solar irradiation limits their widespread utilization, and creative ways to *store* excess energy for use during low-power generation periods are urgently needed. Emerging energy conversion devices such as *fuel cells* could be used to (1) combine excess electricity that is generated during the day (in a solar power station, for example) with liquid water to produce hydrogen, and (2) subsequently convert the stored hydrogen at night by recombining it with oxygen to produce electricity and water. Roadblocks hindering the widespread use of hydrogen fuel cells are their size, weight, and limited durability. As with the gas turbine engine, the efficiency of a fuel cell increases with temperature, but high operating temperatures and large temperature gradients can cause the delicate polymeric materials within a hydrogen fuel cell to fail.

More challenging is the fact that water exists inside any hydrogen fuel cell. If this water should freeze, the polymeric materials within the fuel cell would be destroyed, and the fuel cell would cease operation. Because of the necessity to utilize very pure water in a hydrogen fuel cell, common remedies such as antifreeze cannot be used. What heat transfer mechanisms must be controlled to avoid freezing of pure water within a fuel cell located at a wind farm or solar energy station in a cold climate? How might your developing knowledge of internal forced convection, evaporation, or condensation be applied to control the operating temperatures and enhance the durability of a fuel cell, in turn promoting more widespread use of solar and wind power?

Due to the *information technology* revolution of the last two decades, strong industrial productivity growth has brought an improved quality of life worldwide. Many information technology breakthroughs have been enabled by advances in heat transfer engineering that have ensured the precise control of temperatures of systems ranging in size from nanoscale integrated circuits, to microscale storage media including compact discs, to large data centers filled with heat-generating equipment. As electronic devices become faster and incorporate greater functionality, they generate more thermal energy. Simultaneously, the devices have become smaller. Inevitably, heat fluxes (W/m^2) and volumetric energy generation rates (W/m^3) keep increasing, but the operating temperatures of the devices must be held to reasonably low values to ensure their reliability.

For *personal computers*, cooling fins (also known as *heat sinks*) are fabricated of a high thermal conductivity material (usually aluminum) and attached to the microprocessors to reduce their operating temperatures, as shown in Figure 1.13. Small fans are used to induce forced convection over the fins. The cumulative energy that is consumed worldwide, just to (1) power the small fans that provide the airflow over the fins and (2) manufacture the heat sinks for personal computers, is estimated to be over 10^9 kW·h per year [6]. How might your knowledge of conduction, convection, and radiation be used to, for example, eliminate the fan and minimize the size of the heat sink?

Further improvements in microprocessor technology are currently limited by our ability to cool these tiny devices. Policy makers have voiced concern about our ability to continually reduce the cost of computing and, in turn as a society, continue the growth in productivity that has marked the last 30 years, specifically citing the need to enhance heat transfer in electronics cooling [7]. How might your knowledge of heat transfer help ensure continued industrial productivity into the future?

Heat transfer is important not only in engineered systems but also in nature. Temperature regulates and triggers biological responses in all living systems and ultimately marks the boundary between sickness and health. Two common examples include *hypothermia*, which results from excessive cooling of the human body, and *heat stroke*, which is triggered in warm, humid environments. Both are deadly, and both are associated with core temperatures of the body exceeding physiological limits. Both are directly linked to the convection, radiation, and evaporation processes occurring at the surface of the body, the transport of heat within the body, and the metabolic energy generated volumetrically within the body.

Recent advances in *biomedical engineering*, such as laser surgery, have been enabled by successfully applying fundamental heat transfer principles [8, 9]. While increased temperatures resulting from contact with hot objects may cause thermal *burns*, beneficial *hyperthermal treatments* are used to purposely destroy, for example, cancerous lesions. In a

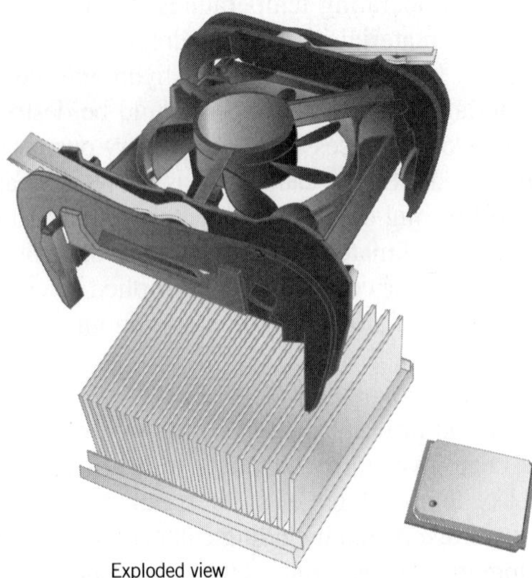

FIGURE 1.13 A finned heat sink and fan assembly (left) and microprocessor (right).

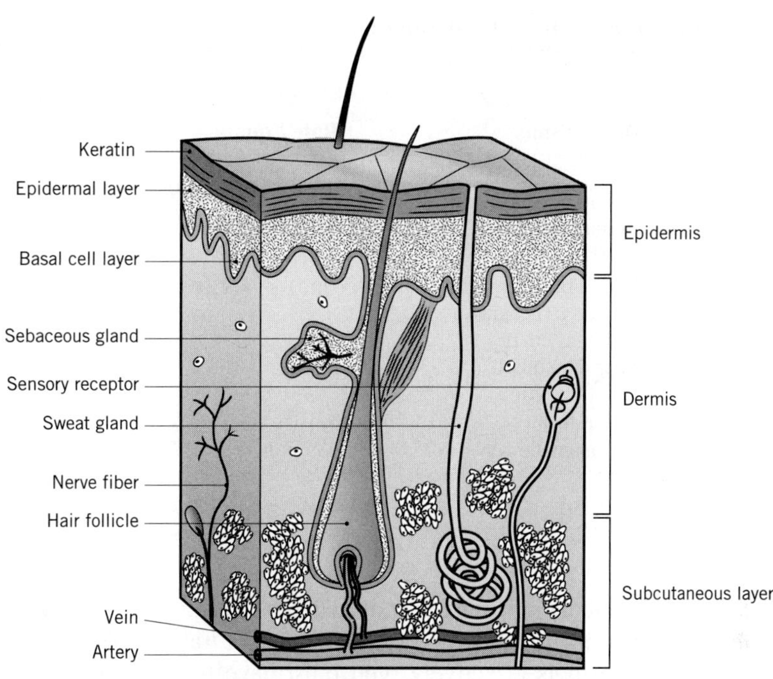

FIGURE 1.14 Morphology of human skin.

similar manner, very low temperatures might induce *frostbite*, but purposeful localized freezing can selectively destroy diseased tissue during *cryosurgery*. Many medical therapies and devices therefore operate by destructively heating or cooling diseased tissue, while leaving the surrounding healthy tissue unaffected.

The ability to design many medical devices and to develop the appropriate protocol for their use hinges on the engineer's ability to predict and control the distribution of temperatures during thermal treatment and the distribution of chemical species in chemotherapies. The treatment of mammalian tissue is made complicated by its morphology, as shown in Figure 1.14. The flow of blood within the venular and capillary structure of a thermally treated area affects heat transfer through advection processes. Larger veins and arteries, which commonly exist in pairs throughout the body, carry blood at different temperatures and advect thermal energy at different rates. Therefore, the veins and arteries exist in a *counterflow heat exchange* arrangement with warm, arteriolar blood exchanging thermal energy with the cooler, venular blood through the intervening solid tissue. Networks of smaller capillaries can also affect local temperatures by *perfusing* blood through the treated area.

In subsequent chapters, example and homework problems will deal with the analysis of these and many other *thermal systems*.

1.7 *Summary*

Although much of the material of this chapter will be discussed in greater detail, you should now have a reasonable overview of heat transfer. You should be aware of the

TABLE 1.5 Summary of heat transfer processes

Mode	Mechanism(s)	Rate Equation	Equation Number	Transport Property or Coefficient
Conduction	Diffusion of energy due to random molecular motion	$q''_x(\text{W/m}^2) = -k\dfrac{dT}{dx}$	(1.1)	k (W/m·K)
Convection	Diffusion of energy due to random molecular motion plus energy transfer due to bulk motion (advection)	$q''(\text{W/m}^2) = h(T_s - T_\infty)$	(1.3a)	h (W/m^2·K)
Radiation	Energy transfer by electromagnetic waves	$q''(\text{W/m}^2) = \varepsilon\sigma(T_s^4 - T_{sur}^4)$ or $q(\text{W}) = h_r A(T_s - T_{sur})$	(1.7) (1.8)	ε h_r (W/m^2·K)

several modes of transfer and their physical origins. You will be devoting much time to acquiring the tools needed to calculate heat transfer phenomena. However, before you can use these tools effectively, you must have the intuition to determine what is happening physically. Specifically, given a physical situation, you must be able to identify the relevant transport phenomena; the importance of developing this facility must not be underestimated. The example and problems at the end of this chapter will launch you on the road to developing this intuition.

You should also appreciate the significance of the rate equations and feel comfortable in using them to compute transport rates. These equations, summarized in Table 1.5, *should be committed to memory.* You must also recognize the importance of the conservation laws and the need to carefully identify control volumes. With the rate equations, the conservation laws may be used to solve numerous heat transfer problems.

Lastly, you should have begun to acquire an appreciation for the terminology and physical concepts that underpin the subject of heat transfer. Test your understanding of the important terms and concepts introduced in this chapter by addressing the following questions:

- What are the *physical mechanisms* associated with heat transfer by *conduction*, *convection*, and *radiation*?
- What is the driving potential for heat transfer? What are analogs to this potential and to heat transfer itself for the transport of electric charge?
- What is the difference between a heat *flux* and a heat *rate*? What are their units?
- What is a *temperature gradient*? What are its units? What is the relationship of heat flow to a temperature gradient?
- What is the *thermal conductivity*? What are its units? What role does it play in heat transfer?
- What is *Fourier's law*? Can you write the equation from memory?
- If heat transfer by conduction through a medium occurs under *steady-state* conditions, will the temperature at a particular instant vary with location in the medium? Will the temperature at a particular location vary with time?

- What is the difference between *natural convection* and *forced convection*?
- What conditions are necessary for the development of a *hydrodynamic boundary layer*? A *thermal boundary layer*? What varies across a hydrodynamic boundary layer? Across a thermal boundary layer?
- If convection heat transfer for flow of a liquid or a vapor is not characterized by liquid/vapor phase change, what is the nature of the energy being transferred? What is it if there is such a phase change?
- What is *Newton's law of cooling*? Can you write the equation from memory?
- What role is played by the *convection heat transfer coefficient* in Newton's law of cooling? What are its units?
- What effect does convection heat transfer from or to a surface have on the solid bounded by the surface?
- What is predicted by the Stefan–Boltzmann law, and what unit of temperature must be used with the law? Can you write the equation from memory?
- What is the *emissivity*, and what role does it play in characterizing radiation transfer at a surface?
- What is *irradiation*? What are its units?
- What two outcomes characterize the response of an *opaque* surface to incident radiation? Which outcome affects the thermal energy of the medium bounded by the surface and how? What property characterizes this outcome?
- What conditions are associated with use of the *radiation heat transfer coefficient*?
- Can you write the equation used to express net radiation exchange between a small isothermal surface and a large isothermal enclosure?
- Consider the surface of a solid that is at an elevated temperature and exposed to cooler surroundings. By what mode(s) is heat transferred from the surface if (1) it is in intimate (perfect) contact with another solid, (2) it is exposed to the flow of a liquid, (3) it is exposed to the flow of a gas, and (4) it is in an evacuated chamber?
- What is the inherent difference between the application of conservation of energy over a *time interval* and at an *instant of time*?
- What is *thermal energy storage*? How does it differ from *thermal energy generation*? What role do the terms play in a surface energy balance?

EXAMPLE 1.10

A closed container filled with hot coffee is in a room whose air and walls are at a fixed temperature. Identify all heat transfer processes that contribute to the cooling of the coffee. Comment on features that would contribute to a superior container design.

SOLUTION

Known: Hot coffee is separated from its cooler surroundings by a plastic flask, an air space, and a plastic cover.

Find: Relevant heat transfer processes.

Schematic:

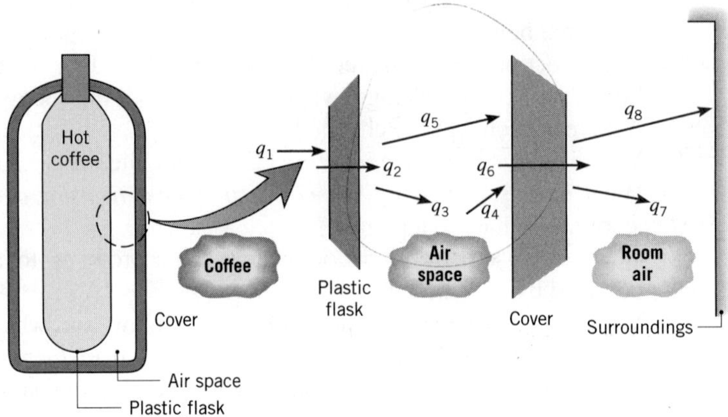

Pathways for energy transfer from the coffee are as follows:

q_1: free convection from the coffee to the flask.

q_2: conduction through the flask.

q_3: free convection from the flask to the air.

q_4: free convection from the air to the cover.

q_5: net radiation exchange between the outer surface of the flask and the inner surface of the cover.

q_6: conduction through the cover.

q_7: free convection from the cover to the room air.

q_8: net radiation exchange between the outer surface of the cover and the surroundings.

Comments: Design improvements are associated with (1) use of aluminized (low-emissivity) surfaces for the flask and cover to reduce net radiation, and (2) evacuating the air space or using a filler material to retard free convection.

References

1. Moran, M. J., and H. N. Shapiro, *Fundamentals of Engineering Thermodynamics*, Wiley, Hoboken, NJ, 2004.
2. Curzon, F. L., and B. Ahlborn, *American J. Physics*, **43**, 22, 1975.
3. Novikov, I. I., *J. Nuclear Energy II*, **7**, 125, 1958.
4. Callen, H. B., *Thermodynamics and an Introduction to Thermostatistics*, Wiley, Hoboken, NJ, 1985.
5. Bejan, A., *American J. Physics*, **64**, 1054, 1996.
6. Bar-Cohen, A., and I. Madhusudan, *IEEE Trans. Components and Packaging Tech.*, **25**, 584, 2002.
7. Miller, R., *Business Week*, November 11, 2004.
8. Diller, K. R., and T. P. Ryan, *J. Heat Transfer*, **120**, 810, 1998.
9. Datta, A.K., *Biological and Bioenvironmental Heat and Mass Transfer*, Marcel Dekker, New York, 2002.

Problems

Conduction

1.1 The thermal conductivity of a sheet of rigid, extruded insulation is reported to be $k = 0.029$ W/m·K. The measured temperature difference across a 20-mm-thick sheet of the material is $T_1 - T_2 = 10°C$.

(a) What is the heat flux through a 2 m × 2 m sheet of the insulation?

(b) What is the rate of heat transfer through the sheet of insulation?

1.2 The heat flux that is applied to the left face of a plane wall is $q'' = 20$ W/m^2. The wall is of thickness $L = 10$ mm and of thermal conductivity $k = 12$ W/m·K. If the surface temperatures of the wall are measured to be 50°C on the left side and 30°C on the right side, do steady-state conditions exist?

1.3 A concrete wall, which has a surface area of 20 m^2 and is 0.30 m thick, separates conditioned room air from ambient air. The temperature of the inner surface of the wall is maintained at 25°C, and the thermal conductivity of the concrete is 1 W/m·K.

(a) Determine the heat loss through the wall for outer surface temperatures ranging from −15°C to 38°C, which correspond to winter and summer extremes, respectively. Display your results graphically.

(b) On your graph, also plot the heat loss as a function of the outer surface temperature for wall materials having thermal conductivities of 0.75 and 1.25 W/m·K. Explain the family of curves you have obtained.

1.4 The concrete slab of a basement is 11 m long, 8 m wide, and 0.20 m thick. During the winter, temperatures are nominally 17°C and 10°C at the top and bottom surfaces, respectively. If the concrete has a thermal conductivity of 1.4 W/m·K, what is the rate of heat loss through the slab? If the basement is heated by a gas furnace operating at an efficiency of $\eta_f = 0.90$ and natural gas is priced at C_g = \$0.02/MJ, what is the daily cost of the heat loss?

1.5 Consider Figure 1.3. The heat flux in the x-direction is $q''_x = 10$ W/m^2, the thermal conductivity and wall thickness are $k = 2.3$ W/m·K and $L = 20$ mm, respectively, and steady-state conditions exist. Determine the value of the temperature gradient in units of K/m. What is the value of the temperature gradient in units of °C/m?

1.6 The heat flux through a wood slab 50 mm thick, whose inner and outer surface temperatures are 40 and 20°C, respectively, has been determined to be 40 W/m^2. What is the thermal conductivity of the wood?

1.7 The inner and outer surface temperatures of a glass window 5 mm thick are 15 and 5°C. What is the heat loss through a 1 m × 3 m window? The thermal conductivity of glass is 1.4 W/m·K.

1.8 A thermodynamic analysis of a proposed Brayton cycle gas turbine yields $P = 5$ MW of net power production. The compressor, at an average temperature of $T_c = 400°C$, is driven by the turbine at an average temperature of $T_h = 1000°C$ by way of an L = 1-m-long, d = 70-mm-diameter shaft of thermal conductivity $k = 40$ W/m·K.

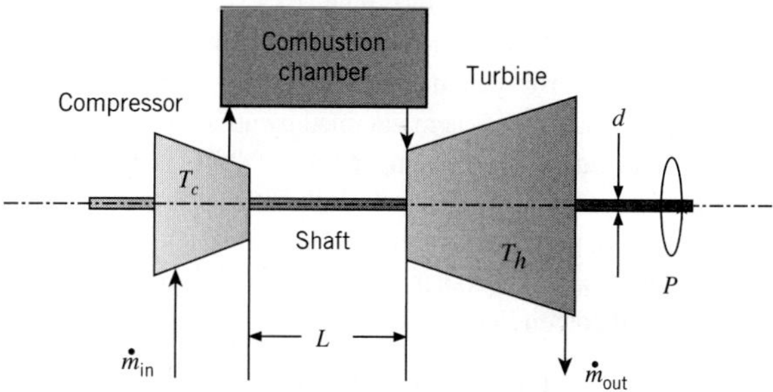

(a) Compare the steady-state conduction rate through the shaft connecting the hot turbine to the warm compressor to the net power predicted by the thermodynamics-based analysis.

(b) A research team proposes to scale down the gas turbine of part (a), keeping all dimensions in the same proportions. The team assumes that the same hot and cold temperatures exist as in part (a) and that the net power output of the gas turbine is proportional to the overall volume of the device. Plot the ratio of the conduction through the shaft to the net power output of the turbine over the range 0.005 m ≤ L ≤ 1 m. Is a scaled-down device with L = 0.005 m feasible?

1.9 A glass window of width W = 1 m and height H = 2 m is 5 mm thick and has a thermal conductivity of k_g = 1.4 W/m·K. If the inner and outer surface temperatures of the glass are 15°C and −20°C, respectively, on a cold winter day, what is the rate of heat loss through the glass? To reduce heat loss through windows, it is customary to use a double pane construction in which adjoining panes are separated by an air space. If the spacing is 10 mm and the glass surfaces in contact with the air have temperatures of 10°C and −15°C, what is the rate of heat loss from a 1 m × 2 m window? The thermal conductivity of air is k_a = 0.024 W/m·K.

1.10 A freezer compartment consists of a cubical cavity that is 2 m on a side. Assume the bottom to be perfectly insulated. What is the minimum thickness of styrofoam insulation ($k = 0.030$ W/m·K) that must be applied to the top and side walls to ensure a heat load of less than 500 W, when the inner and outer surfaces are −10 and 35°C?

1.11 The heat flux that is applied to one face of a plane wall is $q'' = 20$ W/m^2. The opposite face is exposed to air at temperature 30°C, with a convection heat transfer coefficient of 20 W/m^2·K. The surface temperature of the wall exposed to air is measured and found to be 50°C. Do steady-state conditions exist? If not, is the temperature of the wall increasing or decreasing with time?

1.12 An inexpensive food and beverage container is fabricated from 25-mm-thick polystyrene ($k = 0.023$ W/m·K) and has interior dimensions of 0.8 m × 0.6 m × 0.6 m. Under conditions for which an inner surface temperature of approximately 2°C is maintained by an ice-water mixture and an outer surface temperature of 20°C is maintained by the ambient, what is the heat flux through the container wall? Assuming negligible heat gain through the 0.8 m × 0.6 m base of the cooler, what is the total heat load for the prescribed conditions?

1.13 What is the thickness required of a masonry wall having thermal conductivity 0.75 W/m·K if the heat rate is to be 80% of the heat rate through a composite structural wall having a thermal conductivity of 0.25 W/m·K and a thickness of 100 mm? Both walls are subjected to the same surface temperature difference.

1.14 A wall is made from an inhomogeneous (nonuniform) material for which the thermal conductivity varies through the thickness according to $k = ax + b$, where a and b are constants. The heat flux is known to be constant. Determine expressions for the temperature gradient and the temperature distribution when the surface at $x = 0$ is at temperature T_1.

1.15 The 5-mm-thick bottom of a 200-mm-diameter pan may be made from aluminum ($k = 240$ W/m·K) or copper ($k = 390$ W/m·K). When used to boil water, the surface of the bottom exposed to the water is nominally at 110°C. If heat is transferred from the stove to the pan at a rate of 600 W, what is the temperature of the surface in contact with the stove for each of the two materials?

1.16 A square silicon chip ($k = 150$ W/m·K) is of width $w = 5$ mm on a side and of thickness $t = 1$ mm. The chip is mounted in a substrate such that its side and back surfaces are insulated, while the front surface is exposed to a coolant. If 4 W are being dissipated in circuits mounted to the back surface of the chip, what is the steady-state temperature difference between back and front surfaces?

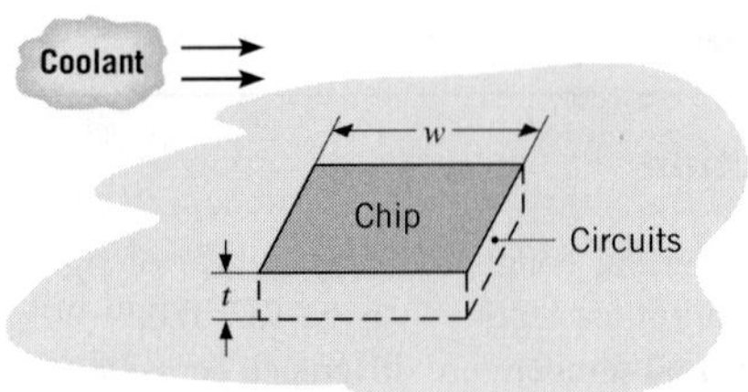

Convection

1.17 For a boiling process such as shown in Figure 1.5c, the ambient temperature T_∞ in Newton's law of cooling is replaced by the saturation temperature of the fluid T_{sat}. Consider a situation where the heat flux from the hot plate is $q'' = 20 \times 10^5$ W/m^2. If the fluid is water at atmospheric pressure and the convection heat transfer coefficient is $h_w = 20 \times 10^3$ W/m^2·K, determine the upper surface temperature of the plate, $T_{s,w}$. In an effort to minimize the surface temperature, a technician proposes replacing the water with a dielectric fluid whose saturation temperature is $T_{sat,d} = 52$°C. If the heat transfer coefficient associated with the dielectric fluid is $h_d = 3 \times 10^3$ W/m^2·K, will the technician's plan work?

1.18 You've experienced convection cooling if you've ever extended your hand out the window of a moving vehicle or into a flowing water stream. With the surface of your hand at a temperature of 30°C, determine the convection heat flux for (a) a vehicle speed of 35 km/h in air at −5°C with a convection coefficient of 40 W/m^2·K and (b) a velocity of 0.2 m/s in a water stream at 10°C with a convection coefficient of 900 W/m^2·K. Which condition would *feel* colder? Contrast these results with a heat loss of approximately 30 W/m^2 under normal room conditions.

1.19 Air at 40°C flows over a long, 25-mm-diameter cylinder with an embedded electrical heater. In a series of tests, measurements were made of the power per unit length, P', required to maintain the cylinder surface temperature at 300°C for different free stream velocities V of the air. The results are as follows:

Air velocity, V (m/s)	1	2	4	8	12
Power, P' (W/m)	450	658	983	1507	1963

(a) Determine the convection coefficient for each velocity, and display your results graphically.

(b) Assuming the dependence of the convection coefficient on the velocity to be of the form $h = CV^n$, determine the parameters C and n from the results of part (a).

1.20 A wall has inner and outer surface temperatures of 16 and 6°C, respectively. The interior and exterior air temperatures are 20 and 5°C, respectively. The inner and outer convection heat transfer coefficients are 5 and 20 W/m$^2 \cdot$K, respectively. Calculate the heat flux from the interior air to the wall, from the wall to the exterior air, and from the wall to the interior air. Is the wall under steady-state conditions?

1.21 An electric resistance heater is embedded in a long cylinder of diameter 30 mm. When water with a temperature of 25°C and velocity of 1 m/s flows crosswise over the cylinder, the power per unit length required to maintain the surface at a uniform temperature of 90°C is 28 kW/m. When air, also at 25°C, but with a velocity of 10 m/s is flowing, the power per unit length required to maintain the same surface temperature is 400 W/m. Calculate and compare the convection coefficients for the flows of water and air.

1.22 The free convection heat transfer coefficient on a thin hot vertical plate suspended in still air can be determined from observations of the change in plate temperature with time as it cools. Assuming the plate is isothermal and radiation exchange with its surroundings is negligible, evaluate the convection coefficient at the instant of time when the plate temperature is 225°C and the change in plate temperature with time (dT/dt) is -0.022 K/s. The ambient air temperature is 25°C and the plate measures 0.3×0.3 m with a mass of 3.75 kg and a specific heat of 2770 J/kg $\cdot$ K.

1.23 A transmission case measures $W = 0.30$ m on a side and receives a power input of $P_i = 150$ hp from the engine.

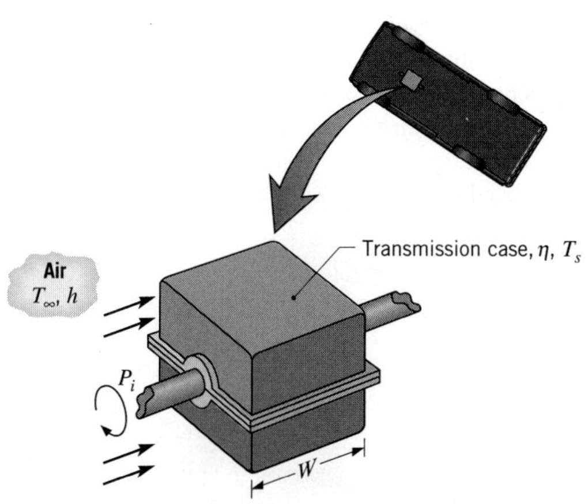

If the transmission efficiency is $\eta = 0.93$ and airflow over the case corresponds to $T_\infty = 30$°C and $h = 200$ W/m$^2 \cdot$K, what is the surface temperature of the transmission?

1.24 A cartridge electrical heater is shaped as a cylinder of length $L = 200$ mm and outer diameter $D = 20$ mm. Under normal operating conditions, the heater dissipates 2 kW while submerged in a water flow that is at 20°C and provides a convection heat transfer coefficient of $h = 5000$ W/m$^2 \cdot$K. Neglecting heat transfer from the ends of the heater, determine its surface temperature T_s. If the water flow is inadvertently terminated while the heater continues to operate, the heater surface is exposed to air that is also at 20°C but for which $h = 50$ W/m$^2 \cdot$K. What is the corresponding surface temperature? What are the consequences of such an event?

1.25 A common procedure for measuring the velocity of an airstream involves the insertion of an electrically heated wire (called a *hot-wire anemometer*) into the airflow, with the axis of the wire oriented perpendicular to the flow direction. The electrical energy dissipated in the wire is assumed to be transferred to the air by forced convection. Hence, for a prescribed electrical power, the temperature of the wire depends on the convection coefficient, which, in turn, depends on the velocity of the air. Consider a wire of length $L = 20$ mm and diameter $D = 0.5$ mm, for which a calibration of the form $V = 6.25 \times 10^{-5} h^2$ has been determined. The velocity V and the convection coefficient h have units of m/s and W/m$^2 \cdot$K, respectively. In an application involving air at a temperature of $T_\infty = 25$°C, the surface temperature of the anemometer is maintained at $T_s = 75$°C with a voltage drop of 5 V and an electric current of 0.1 A. What is the velocity of the air?

1.26 A square isothermal chip is of width $w = 5$ mm on a side and is mounted in a substrate such that its side and back surfaces are well insulated; the front surface is exposed to the flow of a coolant at $T_\infty = 15$°C. From reliability considerations, the chip temperature must not exceed $T = 85$°C.

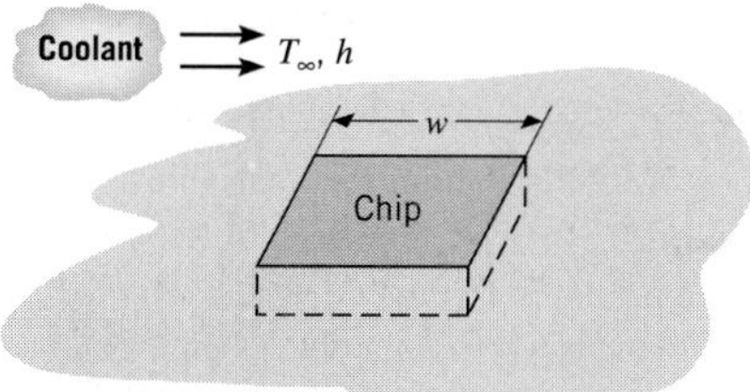

If the coolant is air and the corresponding convection coefficient is $h = 200$ W/m$^2 \cdot$K, what is the maximum allowable chip power? If the coolant is a dielectric liquid for which $h = 3000$ W/m$^2 \cdot$K, what is the maximum allowable power?

1.27 The temperature controller for a clothes dryer consists of a bimetallic switch mounted on an electrical heater attached to a wall-mounted insulation pad.

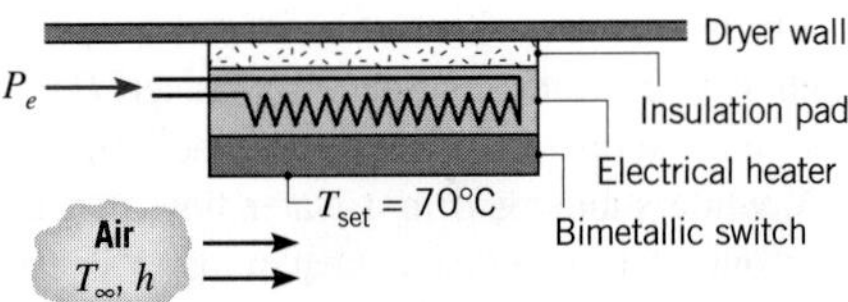

The switch is set to open at 70°C, the maximum dryer air temperature. To operate the dryer at a lower air temperature, sufficient power is supplied to the heater such that the switch reaches 70°C (T_{set}) when the air temperature T is less than T_{set}. If the convection heat transfer coefficient between the air and the exposed switch surface of 30 mm^2 is 25 W/m$^2 \cdot$K, how much heater power P_e is required when the desired dryer air temperature is T_∞ = 50°C?

Radiation

1.28 An overhead 25-m-long, uninsulated industrial steam pipe of 100-mm diameter is routed through a building whose walls and air are at 25°C. Pressurized steam maintains a pipe surface temperature of 150°C, and the coefficient associated with natural convection is $h = 10$ W/m$^2 \cdot$K. The surface emissivity is $\varepsilon = 0.8$.

(a) What is the rate of heat loss from the steam line?

(b) If the steam is generated in a gas-fired boiler operating at an efficiency of $\eta_f = 0.90$ and natural gas is priced at C_g = \$0.02 per MJ, what is the annual cost of heat loss from the line?

1.29 Under conditions for which the same room temperature is maintained by a heating or cooling system, it is not uncommon for a person to feel chilled in the winter but comfortable in the summer. Provide a plausible explanation for this situation (with supporting calculations) by considering a room whose air temperature is maintained at 20°C throughout the year, while the walls of the room are nominally at 27°C and 14°C in the summer and winter, respectively. The exposed surface of a person in the room may be assumed to be at a temperature of 32°C throughout the year and to have an emissivity of 0.90. The coefficient associated with heat transfer by natural convection between the person and the room air is approximately 2 W/m$^2 \cdot$K.

1.30 A spherical interplanetary probe of 0.5-m diameter contains electronics that dissipate 150 W. If the probe surface has an emissivity of 0.8 and the probe does not receive radiation from other surfaces, as, for example, from the sun, what is its surface temperature?

1.31 An instrumentation package has a spherical outer surface of diameter D = 100 mm and emissivity $\varepsilon = 0.25$. The package is placed in a large space simulation chamber whose walls are maintained at 77 K. If operation of the electronic components is restricted to the temperature range $40 \leq T \leq 85$°C, what is the range of acceptable power dissipation for the package? Display your results graphically, showing also the effect of variations in the emissivity by considering values of 0.20 and 0.30.

1.32 Consider the conditions of Problem 1.22. However, now the plate is in a vacuum with a surrounding temperature of 25°C. What is the emissivity of the plate? What is the rate at which radiation is emitted by the surface?

1.33 If $T_s \approx T_{sur}$ in Equation 1.9, the radiation heat transfer coefficient may be approximated as

$$h_{r,a} = 4\varepsilon\sigma\overline{T}^3$$

where $\overline{T} \equiv (T_s + T_{sur})/2$. We wish to assess the validity of this approximation by comparing values of h_r and $h_{r,a}$ for the following conditions. In each case, represent your results graphically and comment on the validity of the approximation.

(a) Consider a surface of either polished aluminum (ε = 0.05) or black paint (ε = 0.9), whose temperature may exceed that of the surroundings (T_{sur} = 25°C) by 10 to 100°C. Also compare your results with values of the coefficient associated with free convection in air ($T_\infty = T_{sur}$), where $h(\text{W/m}^2 \cdot \text{K}) = 0.98\,\Delta T^{1/3}$.

(b) Consider initial conditions associated with placing a workpiece at T_s = 25°C in a large furnace whose wall temperature may be varied over the range $100 \leq T_{sur} \leq 1000$°C. According to the surface finish or coating, its emissivity may assume values of 0.05, 0.2, and 0.9. For each emissivity, plot the relative error, $(h_r - h_{r,a})/h_r$, as a function of the furnace temperature.

1.34 A vacuum system, as used in sputtering electrically conducting thin films on microcircuits, is comprised of a baseplate maintained by an electrical heater at 300 K and a shroud within the enclosure maintained at 77 K by a liquid-nitrogen coolant loop. The circular baseplate, insulated on the lower side, is 0.3 m in diameter and has an emissivity of 0.25.

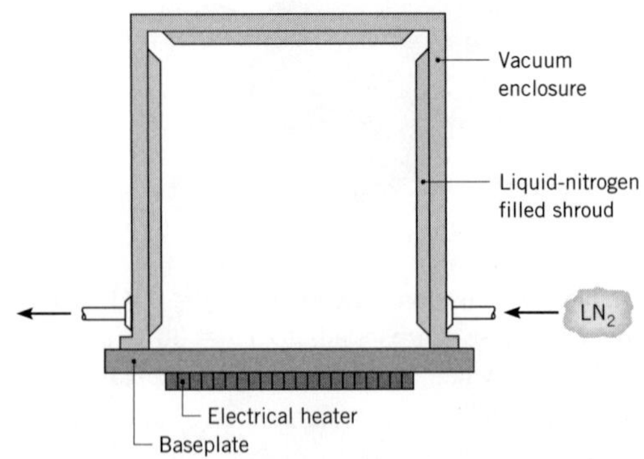

(a) How much electrical power must be provided to the baseplate heater?

(b) At what rate must liquid nitrogen be supplied to the shroud if its heat of vaporization is 125 kJ/kg?

(c) To reduce the liquid nitrogen consumption, it is proposed to bond a thin sheet of aluminum foil ($\varepsilon = 0.09$) to the baseplate. Will this have the desired effect?

Relationship to Thermodynamics

1.35 An electrical resistor is connected to a battery, as shown schematically. After a brief transient, the resistor assumes a nearly uniform, steady-state temperature of 95°C, while the battery and lead wires remain at the ambient temperature of 25°C. Neglect the electrical resistance of the lead wires.

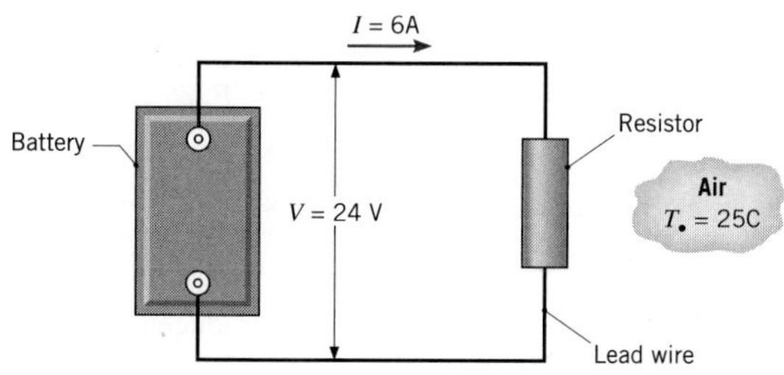

(a) Consider the resistor as a system about which a control surface is placed and Equation 1.12c is applied. Determine the corresponding values of $\dot{E}_{in}(W)$, $\dot{E}_g(W)$, $\dot{E}_{out}(W)$, and $\dot{E}_{st}(W)$. If a control surface is placed about the entire system, what are the values of $\dot{E}_{in}$, $\dot{E}_g$, $\dot{E}_{out}$, and $\dot{E}_{st}$?

(b) If electrical energy is dissipated uniformly within the resistor, which is a cylinder of diameter $D = 60$ mm and length $L = 250$ mm, what is the volumetric heat generation rate, $\dot{q}$ (W/m^3)?

(c) Neglecting radiation from the resistor, what is the convection coefficient?

1.36 Pressurized water ($p_{in} = 10$ bar, $T_{in} = 110°C$) enters the bottom of an $L = 10$-m-long vertical tube of diameter $D = 100$ mm at a mass flow rate of $\dot{m} = 1.5$ kg/s. The tube is located inside a combustion chamber, resulting in heat transfer to the tube. Superheated steam exits the top of the tube at $p_{out} = 7$ bar, $T_{out} = 600°C$. Determine the change in the rate at which the following quantities enter and exit the tube: (a) the combined thermal and flow work, (b) the mechanical energy, and (c) the total energy of the water. Also, (d) determine the heat transfer rate, q. *Hint*: Relevant properties may be obtained from a thermodynamics text.

1.37 Consider the tube and inlet conditions of Problem 1.36. Heat transfer at a rate of $q = 3.89$ MW is delivered to the tube. For an exit pressure of $p = 8$ bar, determine (a) the temperature of the water at the outlet as well as the change in (b) combined thermal and flow work, (c) mechanical energy, and (d) total energy of the water from the inlet to the outlet of the tube. *Hint:* As a first estimate, neglect the change in mechanical energy in solving part (a). Relevant properties may be obtained from a thermodynamics text.

1.38 An internally reversible refrigerator has a modified coefficient of performance accounting for realistic heat transfer processes of

$$\text{COP}_m = \frac{q_{in}}{\dot{W}} = \frac{q_{in}}{q_{out} - q_{in}} = \frac{T_{c,i}}{T_{h,i} - T_{c,i}}$$

where q_{in} is the refrigerator cooling rate, q_{out} is the heat rejection rate, and $\dot{W}$ is the power input. Show that COP_m can be expressed in terms of the reservoir temperatures T_c and T_h, the cold and hot thermal resistances $R_{t,c}$ and $R_{t,h}$, and q_{in}, as

$$\text{COP}_m = \frac{T_c - q_{in}R_{tot}}{T_h - T_c + q_{in}R_{tot}}$$

where $R_{tot} = R_{t,c} + R_{t,h}$. Also, show that the power input may be expressed as

$$\dot{W} = q_{in}\frac{T_h - T_c + q_{in}R_{tot}}{T_c - q_{in}R_{tot}}$$

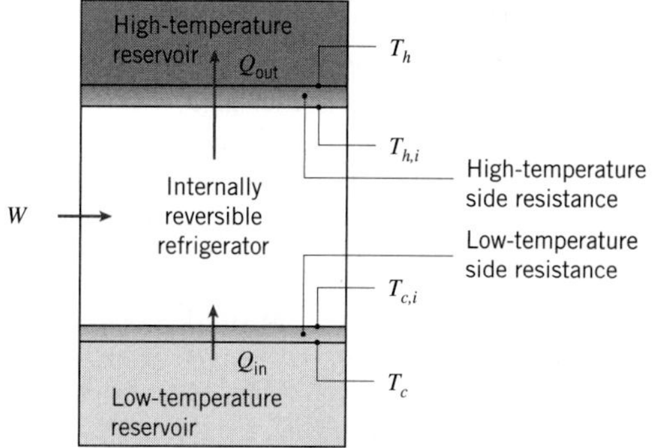

1.39 A household refrigerator operates with cold- and hot-temperature reservoirs of $T_c = 5°C$ and $T_h = 25°C$, respectively. When new, the cold and hot side resistances are $R_{c,n} = 0.05$ K/W and $R_{h,n} = 0.04$ K/W, respectively. Over time, dust accumulates on the refrigerator's condenser coil, which is located behind the refrigerator, increasing the hot side resistance to $R_{h,d} = 0.1$ K/W. It is desired to have a refrigerator cooling rate of $q_{in} = 750$ W. Using the results of Problem 1.38, determine the modified coefficient of performance and the required power input $\dot{W}$ under (a) clean and (b) dusty coil conditions.

Energy Balance and Multimode Effects

1.40 Chips of width $L = 15$ mm on a side are mounted to a substrate that is installed in an enclosure whose walls and air are maintained at a temperature of $T_{sur} = 25°C$. The chips have an emissivity of $\varepsilon = 0.60$ and a maximum allowable temperature of $T_s = 85°C$.

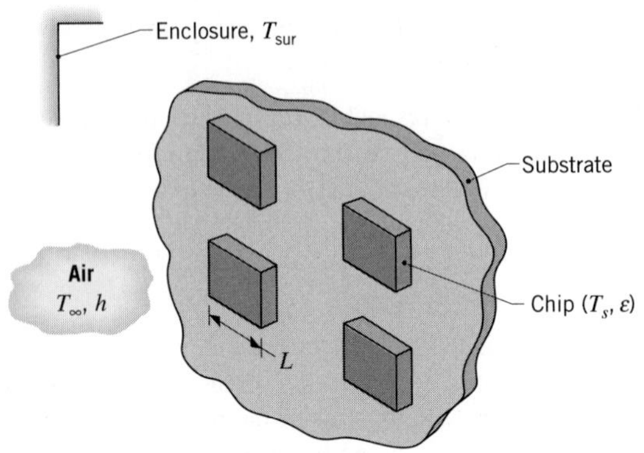

(a) If heat is rejected from the chips by radiation and natural convection, what is the maximum operating power of each chip? The convection coefficient depends on the chip-to-air temperature difference and may be approximated as $h = C(T_s - T_\infty)^{1/4}$, where $C = 4.2\ \text{W/m}^2 \cdot \text{K}^{5/4}$.

(b) If a fan is used to maintain airflow through the enclosure and heat transfer is by forced convection, with $h = 250\ \text{W/m}^2 \cdot \text{K}$, what is the maximum operating power?

1.41 Consider the transmission case of Problem 1.23, but now allow for radiation exchange with the ground/chassis, which may be approximated as large surroundings at $T_{sur} = 30°C$. If the emissivity of the case is $\varepsilon = 0.80$, what is the surface temperature?

1.42 One method for growing thin silicon sheets for photovoltaic solar panels is to pass two thin strings of high melting temperature material upward through a bath of molten silicon. The silicon solidifies on the strings near the surface of the molten pool, and the solid silicon sheet is pulled slowly upward out of the pool. The silicon is replenished by supplying the molten pool with solid silicon powder. Consider a silicon sheet that is $W_{si} = 85$ mm wide and $t_{si} = 150\ \mu$m thick that is pulled at a velocity of $V_{si} = 20$ mm/min. The silicon is melted by supplying electric power to the cylindrical growth chamber of height $H = 350$ mm and diameter $D = 300$ mm. The exposed surfaces of the growth chamber are at $T_s = 320$ K, the corresponding convection coefficient at the exposed surface is $h = 8\ \text{W/m}^2 \cdot \text{K}$, and the surface is characterized by an emissivity of $\varepsilon_s = 0.9$. The solid silicon powder is at $T_{si,i} = 298$ K, and the solid silicon sheet exits the chamber at $T_{si,o} = 420$ K. Both the surroundings and ambient temperatures are $T_\infty = T_{sur} = 298$ K.

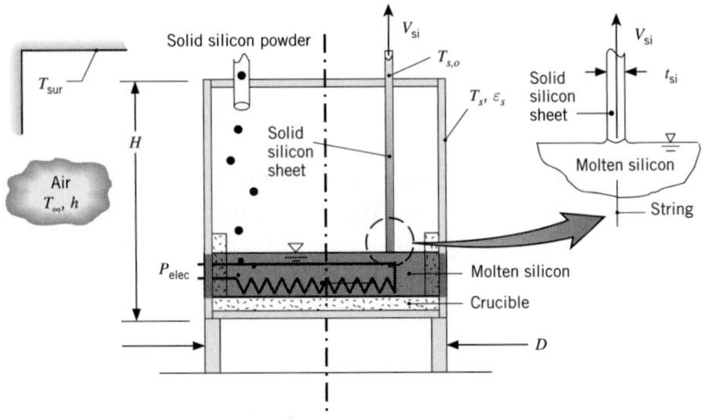

(a) Determine the electric power, P_{elec}, needed to operate the system at steady state.

(b) If the photovoltaic panel absorbs a time-averaged solar flux of $q''_{sol} = 180\ \text{W/m}^2$ and the panel has a conversion efficiency (the ratio of solar power absorbed to electric power produced) of $\eta = 0.20$, how long must the solar panel be operated to produce enough electric energy to offset the electric energy that was consumed in its manufacture?

1.43 Heat is transferred by radiation and convection between the inner surface of the nacelle of the wind turbine of Example 1.3 and the outer surfaces of the gearbox and generator. The convection heat flux associated with the gearbox and the generator may be described by $q''_{conv,gb} = h(T_{gb} - T_\infty)$ and $q''_{conv,gen} = h(T_{gen} - T_\infty)$, respectively, where the ambient temperature $T_\infty \approx T_s$ (which is the nacelle temperature) and $h = 40\ \text{W/m}^2 \cdot \text{K}$. The outer surfaces of both the gearbox and the generator are characterized by an emissivity of $\varepsilon = 0.9$. If the surface areas of the gearbox and generator are $A_{gb} = 6\ \text{m}^2$ and $A_{gen} = 4\ \text{m}^2$, respectively, determine their surface temperatures.

1.44 Radioactive wastes are packed in a long, thin-walled cylindrical container. The wastes generate thermal energy nonuniformly according to the relation $\dot{q} = \dot{q}_o[1 - (r/r_o)^2]$, where $\dot{q}$ is the local rate of energy generation per unit volume, $\dot{q}_o$ is a constant, and r_o is the radius of the container. Steady-state conditions are maintained by submerging the container in a liquid that is at T_∞ and provides a uniform convection coefficient h.

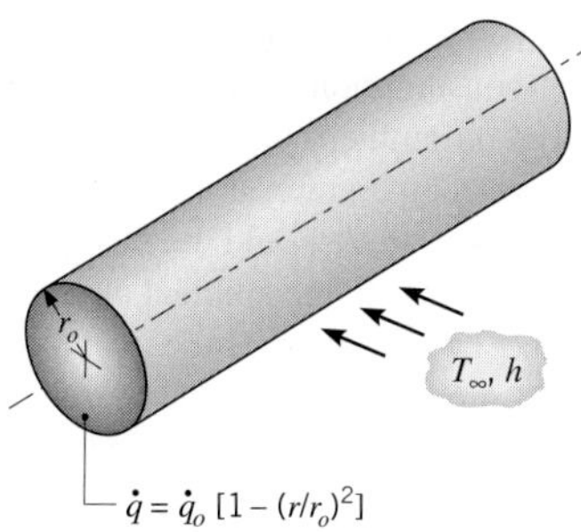

Obtain an expression for the total rate at which energy is generated in a unit length of the container. Use this result to obtain an expression for the temperature T_s of the container wall.

1.45 An aluminum plate 4 mm thick is mounted in a horizontal position, and its bottom surface is well insulated. A special, thin coating is applied to the top surface such that it absorbs 80% of any incident solar radiation, while having an emissivity of 0.25. The density ρ and specific heat c of aluminum are known to be 2700 kg/m^3 and 900 J/kg · K, respectively.

(a) Consider conditions for which the plate is at a temperature of 25°C and its top surface is suddenly exposed to ambient air at T_∞ = 20°C and to solar radiation that provides an incident flux of 900 W/m^2. The convection heat transfer coefficient between the surface and the air is h = 20 W/m^2·K. What is the initial rate of change of the plate temperature?

(b) What will be the equilibrium temperature of the plate when steady-state conditions are reached?

(c) The surface radiative properties depend on the specific nature of the applied coating. Compute and plot the steady-state temperature as a function of the emissivity for $0.05 \le \varepsilon \le 1$, with all other conditions remaining as prescribed. Repeat your calculations for values of α_S = 0.5 and 1.0, and plot the results with those obtained for α_S = 0.8. If the intent is to maximize the plate temperature, what is the most desirable combination of the plate emissivity and its absorptivity to solar radiation?

1.46 A blood warmer is to be used during the transfusion of blood to a patient. This device is to heat blood taken from the blood bank at 10°C to 37°C at a flow rate of 200 ml/min. The blood passes through tubing of length 2 m, with a rectangular cross section 6.4 mm × 1.6 mm At what rate must heat be added to the blood to accomplish the required temperature increase? If the fluid originates from a large tank with nearly zero velocity and flows vertically downward for its 2-m length, estimate the magnitudes of kinetic and potential energy changes. Assume the blood's properties are similar to those of water.

1.47 Consider a carton of milk that is refrigerated at a temperature of T_m = 5°C. The kitchen temperature on a hot summer day is T_∞ = 30°C. If the four sides of the carton are of height and width L = 200 mm and w = 100 mm, respectively, determine the heat transferred to the milk carton as it sits on the kitchen counter for durations of t = 10 s, 60 s, and 300 s before it is returned to the refrigerator. The convection coefficient associated with natural convection on the sides of the carton is h = 10 W/m^2·K. The surface emissivity is 0.90. Assume the milk carton temperature remains at 5°C during the process. Your parents have taught you the importance of refrigerating certain foods from the food safety perspective. Comment on the importance of quickly returning the milk carton to the refrigerator from an energy conservation point of view.

1.48 The energy consumption associated with a home water heater has two components: (i) the energy that must be supplied to bring the temperature of groundwater to the heater storage temperature, as it is introduced to replace hot water that has been used; (ii) the energy needed to compensate for heat losses incurred while the water is stored at the prescribed temperature. In this problem, we will evaluate the first of these components for a family of four, whose daily hot water consumption is approximately 100 gal. If groundwater is available at 15°C, what is the annual energy consumption associated with heating the water to a storage temperature of 55°C? For a unit electrical power cost of \$0.18/kW·h, what is the annual cost associated with supplying hot water by means of (a) electric resistance heating or (b) a heat pump having a COP of 3.

1.49 Liquid oxygen, which has a boiling point of 90 K and a latent heat of vaporization of 214 kJ/kg, is stored in a spherical container whose outer surface is of 500-mm diameter and at a temperature of −10°C. The container is housed in a laboratory whose air and walls are at 25°C.

(a) If the surface emissivity is 0.20 and the heat transfer coefficient associated with free convection at the outer surface of the container is 10 W/m^2·K, what is the rate, in kg/s, at which oxygen vapor must be vented from the system?

(b) Moisture in the ambient air will result in frost formation on the container, causing the surface emissivity to increase. Assuming the surface temperature and convection coefficient to remain at −10°C and

10 W/m$^2 \cdot$K, respectively, compute the oxygen evaporation rate (kg/s) as a function of surface emissivity over the range $0.2 \leq \varepsilon \leq 0.94$.

1.50 The emissivity of galvanized steel sheet, a common roofing material, is $\varepsilon = 0.13$ at temperatures around 300 K, while its absorptivity for solar irradiation is $\alpha_S = 0.65$. Would the neighborhood cat be comfortable walking on a roof constructed of the material on a day when $G_S = 750$ W/m^2, $T_\infty = 16°$C, and $h = 7$ W/m$^2 \cdot$K? Assume the bottom surface of the steel is insulated.

1.51 Three electric resistance heaters of length $L = 250$ mm and diameter $D = 25$ mm are submerged in a 10-gal tank of water, which is initially at 295 K. The water may be assumed to have a density and specific heat of $\rho = 990$ kg/m^3 and $c = 4180$ J/kg$\cdot$K.

(a) If the heaters are activated, each dissipating $q_1 = 500$ W, estimate the time required to bring the water to a temperature of 335 K.

(b) If the natural convection coefficient is given by an expression of the form $h = 370\ (T_s - T)^{1/3}$, where T_s and T are temperatures of the heater surface and water, respectively, what is the temperature of each heater shortly after activation and just before deactivation? Units of h and $(T_s - T)$ are W/m$^2 \cdot$K and K, respectively.

(c) If the heaters are inadvertently activated when the tank is empty, the natural convection coefficient associated with heat transfer to the ambient air at $T_\infty = 300$ K may be approximated as $h = 0.70\ (T_s - T_\infty)^{1/3}$. If the temperature of the tank walls is also 300 K and the emissivity of the heater surface is $\varepsilon = 0.85$, what is the surface temperature of each heater under steady-state conditions?

1.52 A hair dryer may be idealized as a circular duct through which a small fan draws ambient air and within which the air is heated as it flows over a coiled electric resistance wire.

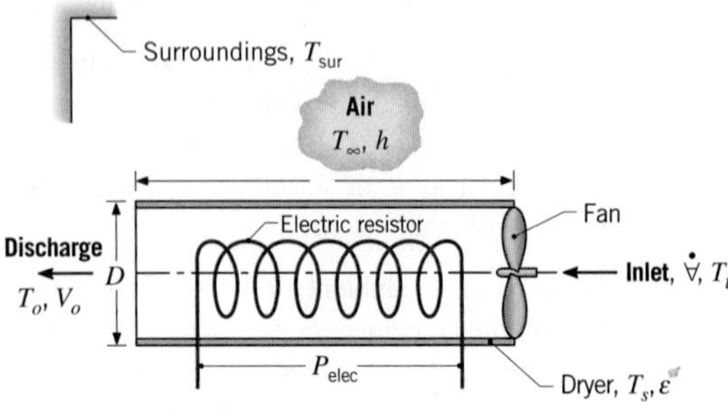

(a) If a dryer is designed to operate with an electric power consumption of $P_{elec} = 500$ W and to heat air from an ambient temperature of $T_i = 20°$C to a discharge temperature of $T_o = 45°$C, at what volumetric flow rate $\dot{\forall}$ should the fan operate? Heat loss from the casing to the ambient air and the surroundings may be neglected. If the duct has a diameter of $D = 70$ mm, what is the discharge velocity V_o of the air? The density and specific heat of the air may be approximated as $\rho = 1.10$ kg/m^3 and $c_p = 1007$ J/kg$\cdot$K, respectively.

(b) Consider a dryer duct length of $L = 150$ mm and a surface emissivity of $\varepsilon = 0.8$. If the coefficient associated with heat transfer by natural convection from the casing to the ambient air is $h = 4$ W/m$^2 \cdot$K and the temperature of the air and the surroundings is $T_\infty = T_{sur} = 20°$C, confirm that the heat loss from the casing is, in fact, negligible. The casing may be assumed to have an average surface temperature of $T_s = 40°$C.

1.53 In one stage of an annealing process, 304 stainless steel sheet is taken from 300 K to 1250 K as it passes through an electrically heated oven at a speed of $V_s = 10$ mm/s. The sheet thickness and width are $t_s = 8$ mm and $W_s = 2$ m, respectively, while the height, width, and length of the oven are $H_o = 2$ m, $W_o = 2.4$ m, and $L_o = 25$ m, respectively. The top and four sides of the oven are exposed to ambient air and large surroundings, each at 300 K, and the corresponding surface temperature, convection coefficient, and emissivity are $T_s = 350$ K, $h = 10$ W/m$^2\cdot$K, and $\varepsilon_s = 0.8$. The bottom surface of the oven is also at 350 K and rests on a 0.5-m-thick concrete pad whose base is at 300 K. Estimate the required electric power input, P_{elec}, to the oven.

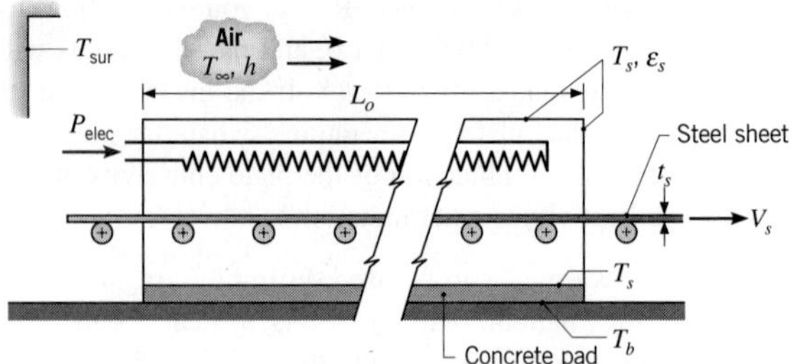

1.54 Convection ovens operate on the principle of inducing forced convection inside the oven chamber with a fan. A *small* cake is to be baked in an oven when the convection feature is disabled. For this situation, the free convection coefficient associated with the cake and its pan is $h_{fr} = 3\ \text{W/m}^2\cdot\text{K}$. The oven air and wall are at temperatures $T_\infty = T_{sur} = 180°\text{C}$. Determine the heat flux delivered to the cake pan and cake batter when they are initially inserted into the oven and are at a temperature of $T_i = 24°\text{C}$. If the convection feature is activated, the forced convection heat transfer coefficient is $h_{fo} = 27\ \text{W/m}^2\cdot\text{K}$. What is the heat flux at the batter or pan surface when the oven is operated in the convection mode? Assume a value of 0.97 for the emissivity of the cake batter and pan.

1.55 Annealing, an important step in semiconductor materials processing, can be accomplished by rapidly heating the silicon wafer to a high temperature for a short period of time. The schematic shows a method involving the use of a hot plate operating at an elevated temperature T_h. The wafer, initially at a temperature of $T_{w,i}$, is suddenly positioned at a gap separation distance L from the hot plate. The purpose of the analysis is to compare the heat fluxes by conduction through the gas within the gap and by radiation exchange between the hot plate and the cool wafer. The initial time rate of change in the temperature of the wafer, $(dT_w/dt)_i$, is also of interest. Approximating the surfaces of the hot plate and the wafer as blackbodies and assuming their diameter D to be much larger than the spacing L, the radiative heat flux may be expressed as $q''_{rad} = \sigma(T_h^4 - T_w^4)$. The silicon wafer has a thickness of $d = 0.78$ mm, a density of 2700 kg/m^3, and a specific heat of 875 J/kg·K. The thermal conductivity of the gas in the gap is 0.0436 W/m·K.

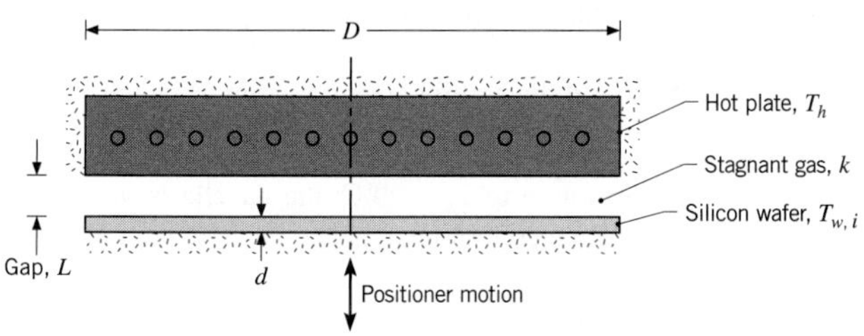

(a) For $T_h = 600°\text{C}$ and $T_{w,i} = 20°\text{C}$, calculate the radiative heat flux and the heat flux by conduction across a gap distance of $L = 0.2$ mm. Also determine the value of $(dT_w/dt)_i$, resulting from each of the heating modes.

(b) For gap distances of 0.2, 0.5, and 1.0 mm, determine the heat fluxes and temperature-time change as a function of the hot plate temperature for $300 \le T_h \le 1300°\text{C}$. Display your results graphically. Comment on the relative importance of the two heat transfer modes and the effect of the gap distance on the heating process. Under what conditions could a wafer be heated to 900°C in less than 10 s?

1.56 In the thermal processing of semiconductor materials, annealing is accomplished by heating a silicon wafer according to a temperature-time recipe and then maintaining a fixed elevated temperature for a prescribed period of time. For the process tool arrangement shown as follows, the wafer is in an evacuated chamber whose walls are maintained at 27°C and within which heating lamps maintain a radiant flux q''_s at its upper surface. The wafer is 0.78 mm thick, has a thermal conductivity of 30 W/m·K, and an emissivity that equals its absorptivity to the radiant flux ($\varepsilon = \alpha_l = 0.65$). For $q''_s = 3.0 \times 10^5\ \text{W/m}^2$, the temperature on its lower surface is measured by a radiation thermometer and found to have a value of $T_{w,l} = 997°\text{C}$.

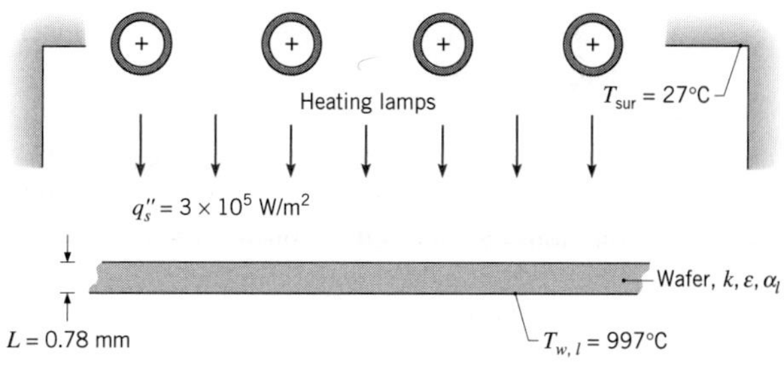

To avoid warping the wafer and inducing slip planes in the crystal structure, the temperature difference across the thickness of the wafer must be less than 2°C. Is this condition being met?

1.57 A furnace for processing semiconductor materials is formed by a silicon carbide chamber that is zone-heated on the top section and cooled on the lower section. With the elevator in the lowest position, a robot arm inserts the silicon wafer on the mounting pins. In a production operation, the wafer is rapidly moved toward the hot zone to achieve the temperature-time history required for the process recipe. In this position, the top and bottom surfaces of the wafer exchange radiation with the hot and cool zones, respectively, of the chamber. The zone temperatures are $T_h = 1500$ K and $T_c = 330$ K, and the emissivity and thickness of the wafer are $\varepsilon = 0.65$ and $d = 0.78$ mm, respectively. With the ambient gas at $T_\infty = 700$ K, convection coefficients at the upper and lower surfaces of the wafer are 8 and 4 W/m^2·K, respectively. The silicon wafer has a density of 2700 kg/m^3 and a specific heat of 875 J/kg·K.

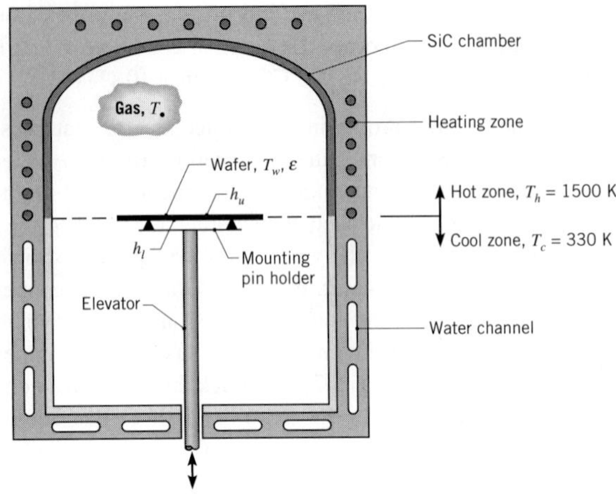

(a) For an initial condition corresponding to a wafer temperature of $T_{w,i} = 300$ K and the position of the wafer shown schematically, determine the corresponding time rate of change of the wafer temperature, $(dT_w/dt)_i$.

(b) Determine the steady-state temperature reached by the wafer if it remains in this position. How significant is convection heat transfer for this situation? Sketch how you would expect the wafer temperature to vary as a function of vertical distance.

1.58 Single fuel cells such as the one of Example 1.5 can be scaled up by arranging them into a *fuel cell stack.* A stack consists of multiple electrolytic membranes that are sandwiched between electrically conducting *bipolar plates.* Air and hydrogen are fed to each membrane through *flow channels* within each bipolar plate, as shown in the sketch. With this stack arrangement, the individual fuel cells are connected in series, electrically, producing a stack voltage of $E_{stack} = N \times E_c$, where E_c is the voltage produced across each membrane and N is the number of membranes in the stack. The electrical current is the same for each membrane. The cell voltage, E_c, as well as the cell efficiency, increases with temperature (the air and hydrogen fed to the stack are humidified to allow operation at temperatures greater than in Example 1.5), but the membranes will fail at temperatures exceeding $T \approx 85°C$. Consider $L \times w$ membranes, where $L = w = 100$ mm, of thickness $t_m = 0.43$ mm, that each produce $E_c = 0.6$ V at $I = 60$ A, and $\dot{E}_{c,g} = 45$ W of thermal energy when operating at $T = 80°C$. The external surfaces of the stack are exposed to air at $T_\infty = 25°C$ and surroundings at $T_{sur} = 30°C$, with $\varepsilon = 0.88$ and $h = 150\ W/m^2 \cdot K$.

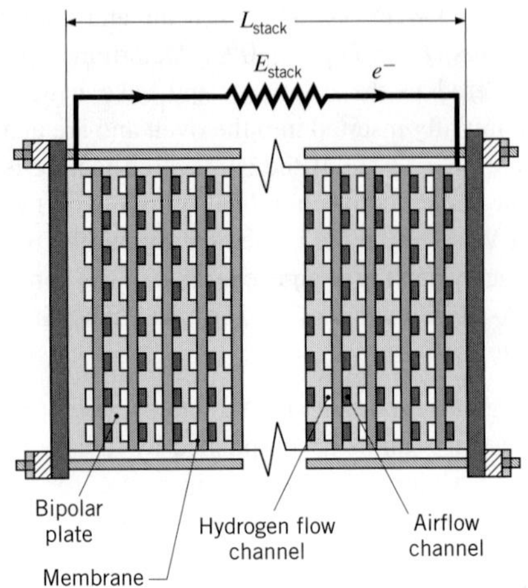

(a) Find the electrical power produced by a stack that is $L_{stack} = 200$ mm long, for bipolar plate thickness in the range 1 mm $< t_{bp} <$ 10 mm. Determine the total thermal energy generated by the stack.

(b) Calculate the surface temperature and explain whether the stack needs to be internally heated or cooled to operate at the optimal internal temperature of 80°C for various bipolar plate thicknesses.

(c) Identify how the internal stack operating temperature might be lowered or raised for a given bipolar plate thickness, and discuss design changes that would promote a more uniform temperature distribution within the stack. How would changes in the external air and surroundings temperature affect your answer? Which membrane in the stack is most likely to fail due to high operating temperature?

1.59 Consider the wind turbine of Example 1.3. To reduce the nacelle temperature to $T_s = 30°C$, the nacelle is vented and a fan is installed to force ambient air into and out of the nacelle enclosure. What is the minimum mass flow rate of air required if the air temperature increases to the nacelle surface temperature before exiting the nacelle? The specific heat of air is 1007 J/kg·K.

1.60 Consider the conducting rod of Example 1.4 under steady-state conditions. As suggested in Comment 3, the temperature of the rod may be controlled by varying the speed of airflow over the rod, which, in turn, alters the convection heat transfer coefficient. To consider the effect of the convection coefficient, generate plots of T versus I for values of $h = 50$, 100, and 250 $W/m^2 \cdot K$. Would variations in the surface emissivity have a significant effect on the rod temperature?

1.61 A long bus bar (cylindrical rod used for making electrical connections) of diameter D is installed in a large conduit having a surface temperature of 30°C and in which the ambient air temperature is $T_\infty =$ 30°C. The electrical resistivity, $\rho_e(\mu\Omega \cdot \text{m})$, of the bar material is a function of temperature, $\rho_{e,o} = \rho_e [1 + \alpha (T - T_o)]$, where $\rho_{e,o} = 0.0171\ \mu\Omega \cdot \text{m}$, $T_o =$ 25°C, and $\alpha = 0.00396\ \text{K}^{-1}$. The bar experiences free convection in the ambient air, and the convection coefficient depends on the bar diameter, as well as on the difference between the surface and ambient temperatures. The governing relation is of the form, $h = CD^{-0.25}\ (T - T_\infty)^{0.25}$, where $C = 1.21\ \text{W} \cdot \text{m}^{-1.75} \cdot \text{K}^{-1.25}$. The emissivity of the bar surface is $\varepsilon = 0.85$.

(a) Recognizing that the electrical resistance per unit length of the bar is $R'_e = \rho_e/A_c$, where A_c is its cross-sectional area, calculate the current-carrying capacity of a 20-mm-diameter bus bar if its temperature is not to exceed 65°C. Compare the relative importance of heat transfer by free convection and radiation exchange.

(b) To assess the trade-off between current-carrying capacity, operating temperature, and bar diameter, for diameters of 10, 20, and 40 mm, plot the bar temperature T as a function of current for the range $100 \leq I \leq 5000$ A. Also plot the ratio of the heat transfer by convection to the total heat transfer.

1.62 A small sphere of reference-grade iron with a specific heat of 447 J/kg·K and a mass of 0.515 kg is suddenly immersed in a water–ice mixture. Fine thermocouple wires suspend the sphere, and the temperature is observed to change from 15 to 14°C in 6.35 s. The experiment is repeated with a metallic sphere of the same diameter, but of unknown composition with a mass of 1.263 kg. If the same observed temperature change occurs in 4.59 s, what is the specific heat of the unknown material?

1.63 A 50 mm × 45 mm × 20 mm cell phone charger has a surface temperature of $T_s = 33$°C when plugged into an electrical wall outlet but not in use. The surface of the charger is of emissivity $\varepsilon = 0.92$ and is subject to a free convection heat transfer coefficient of $h = 4.5\ \text{W/m}^2 \cdot \text{K}$. The room air and wall temperatures are $T_\infty = 22$°C and $T_{\text{sur}} = 20$°C, respectively. If electricity costs $C = \$0.18/\text{kW} \cdot \text{h}$, determine the daily cost of leaving the charger plugged in when not in use.

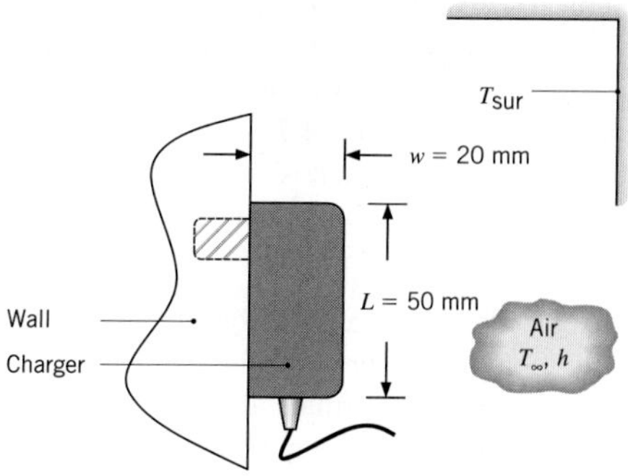

1.64 A spherical, stainless steel (AISI 302) canister is used to store reacting chemicals that provide for a uniform heat flux q''_i to its inner surface. The canister is suddenly submerged in a liquid bath of temperature $T_\infty < T_i$, where T_i is the initial temperature of the canister wall.

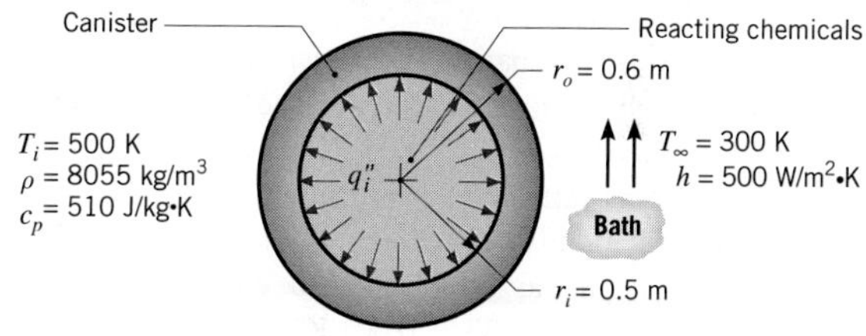

(a) Assuming negligible temperature gradients in the canister wall and a constant heat flux q''_i, develop an equation that governs the variation of the wall temperature with time during the transient process. What is the initial rate of change of the wall temperature if $q''_i = 10^5\ \text{W/m}^2$?

(b) What is the steady-state temperature of the wall?

(c) The convection coefficient depends on the velocity associated with fluid flow over the canister and whether the wall temperature is large enough to induce boiling in the liquid. Compute and plot the steady-state temperature as a function of h for the range $100 \leq h \leq 10{,}000\ \text{W/m}^2 \cdot \text{K}$. Is there a value of h below which operation would be unacceptable?

1.65 A freezer compartment is covered with a 2-mm-thick layer of frost at the time it malfunctions. If the compartment is in ambient air at 20°C and a coefficient of $h = 2\ \text{W/m}^2 \cdot \text{K}$ characterizes heat transfer by natural convection from the exposed surface of the layer, estimate the time required to completely melt the frost. The frost may be assumed to have a mass density of 700 kg/m^3 and a latent heat of fusion of 334 kJ/kg.

1.66 A vertical slab of Wood's metal is joined to a substrate on one surface and is melted as it is uniformly irradiated by a laser source on the opposite surface. The metal is initially at its fusion temperature of $T_f = 72°C$, and the melt runs off by gravity as soon as it is formed. The absorptivity of the metal to the laser radiation is $\alpha_l = 0.4$, and its latent heat of fusion is $h_{sf} = 33$ kJ/kg.

(a) Neglecting heat transfer from the irradiated surface by convection or radiation exchange with the surroundings, determine the instantaneous rate of melting in kg/s·m² if the laser irradiation is 5 kW/m². How much material is removed if irradiation is maintained for a period of 2 s?

(b) Allowing for convection to ambient air, with $T_\infty = 20°C$ and $h = 15$ W/m²·K, and radiation exchange with large surroundings ($\varepsilon = 0.4$, $T_{sur} = 20°C$), determine the instantaneous rate of melting during irradiation.

1.67 A photovoltaic panel of dimension 2 m × 4 m is installed on the roof of a home. The panel is irradiated with a solar flux of $G_S = 700$ W/m², oriented normal to the top panel surface. The absorptivity of the panel to the solar irradiation is $\alpha_S = 0.83$, and the efficiency of conversion of the absorbed flux to electrical power is $\eta = P/\alpha_S G_S A = 0.553 - 0.001\ \text{K}^{-1} T_p$, where T_p is the panel temperature expressed in kelvins and A is the solar panel area. Determine the electrical power generated for (a) a still summer day, in which $T_{sur} = T_\infty = 35°C$, $h = 10$ W/m²·K, and (b) a breezy winter day, for which $T_{sur} = T_\infty = -15°C$, $h = 30$ W/m²·K. The panel emissivity is $\varepsilon = 0.90$.

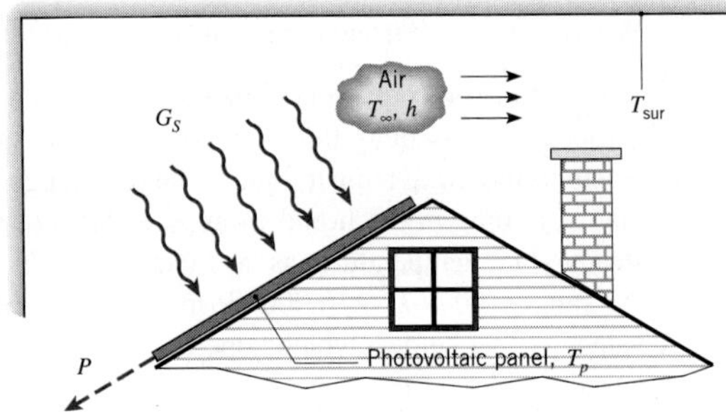

1.68 Following the hot vacuum forming of a paper-pulp mixture, the product, an egg carton, is transported on a conveyor for 18 s toward the entrance of a gas-fired oven where it is dried to a desired final water content. Very little water evaporates during the travel time. So, to increase the productivity of the line, it is proposed that a bank of infrared radiation heaters, which provide a uniform radiant flux of 5000 W/m², be installed over the conveyor. The carton has an exposed area of 0.0625 m² and a mass of 0.220 kg, 75% of which is water after the forming process.

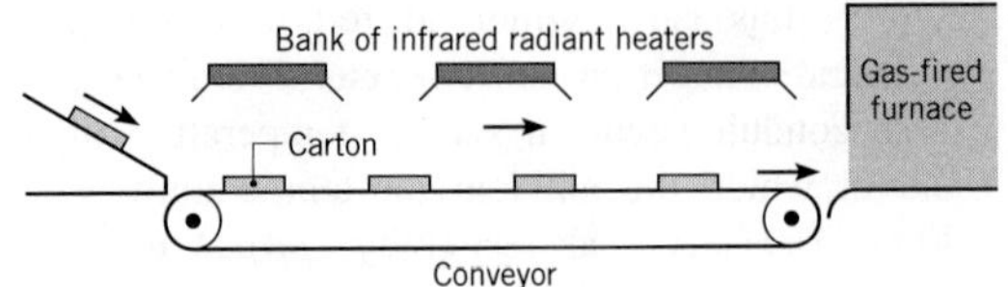

The chief engineer of your plant will approve the purchase of the heaters if they can reduce the water content by 10% of the total mass. Would you recommend the purchase? Assume the heat of vaporization of water is $h_{fg} = 2400$ kJ/kg.

1.69 Electronic power devices are mounted to a heat sink having an exposed surface area of 0.045 m² and an emissivity of 0.80. When the devices dissipate a total power of 20 W and the air and surroundings are at 27°C, the average sink temperature is 42°C. What average temperature will the heat sink reach when the devices dissipate 30 W for the same environmental condition?

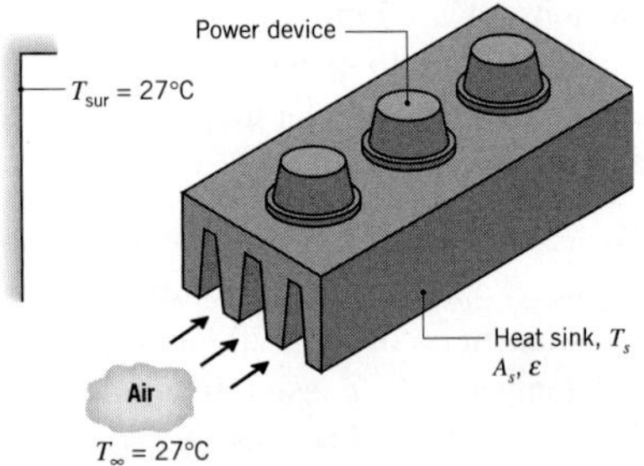

1.70 A computer consists of an array of five printed circuit boards (PCBs), each dissipating $P_b = 20$ W of power. Cooling of the electronic components on a board is provided by the forced flow of air, equally distributed in passages formed by adjoining boards, and the convection coefficient associated with heat transfer from the components to the air is approximately $h = 200$ W/m²·K. Air enters the computer console at a temperature of $T_i = 20°C$, and flow is driven by a fan whose power consumption is $P_f = 25$ W.

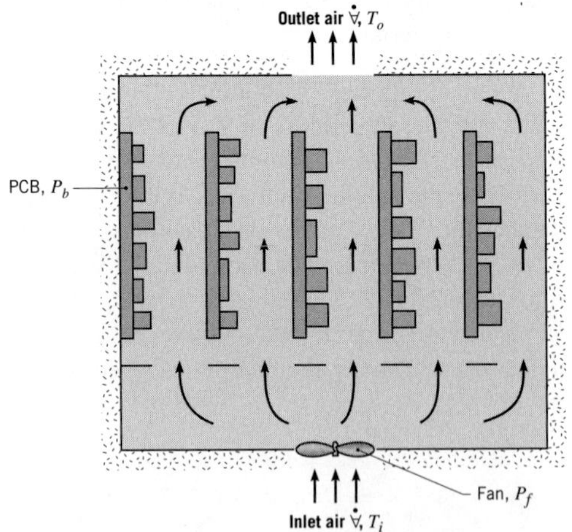

(a) If the temperature rise of the airflow, $(T_o - T_i)$, is not to exceed 15°C, what is the minimum allowable volumetric flow rate $\dot{V}$ of the air? The density and specific heat of the air may be approximated as $\rho = 1.161$ kg/m^3 and $c_p = 1007$ J/kg·K, respectively.

(b) The component that is most susceptible to thermal failure dissipates 1 W/cm^2 of surface area. To minimize the potential for thermal failure, where should the component be installed on a PCB? What is its surface temperature at this location?

1.71 Consider a surface-mount type transistor on a circuit board whose temperature is maintained at 35°C. Air at 20°C flows over the upper surface of dimensions 4 mm × 8 mm with a convection coefficient of 50 W/m^2·K. Three wire leads, each of cross section 1 mm × 0.25 mm and length 4 mm, conduct heat from the case to the circuit board. The gap between the case and the board is 0.2 mm.

(a) Assuming the case is isothermal and neglecting radiation, estimate the case temperature when 150 mW is dissipated by the transistor and (i) stagnant air or (ii) a conductive paste fills the gap. The thermal conductivities of the wire leads, air, and conductive paste are 25, 0.0263, and 0.12 W/m·K, respectively.

(b) Using the conductive paste to fill the gap, we wish to determine the extent to which increased heat dissipation may be accommodated, subject to the constraint that the case temperature not exceed 40°C. Options include increasing the air speed to achieve a larger convection coefficient h and/or changing the lead wire material to one of larger thermal conductivity. Independently considering leads fabricated from materials with thermal conductivities of 200 and 400 W/m·K, compute and plot the maximum allowable heat dissipation for variations in h over the range $50 \le h \le 250$ W/m^2·K.

1.72 The roof of a car in a parking lot absorbs a solar radiant flux of 800 W/m^2, and the underside is perfectly insulated. The convection coefficient between the roof and the ambient air is 12 W/m^2·K.

(a) Neglecting radiation exchange with the surroundings, calculate the temperature of the roof under steady-state conditions if the ambient air temperature is 20°C.

(b) For the same ambient air temperature, calculate the temperature of the roof if its surface emissivity is 0.8.

(c) The convection coefficient depends on airflow conditions over the roof, increasing with increasing air speed. Compute and plot the roof temperature as a function of h for $2 \le h \le 200$ W/m^2·K.

1.73 Consider the conditions of Problem 1.22, but the surroundings temperature is 25°C and radiation exchange with the surroundings is not negligible. If the convection coefficient is 6.4 W/m^2·K and the emissivity of the plate is $\varepsilon = 0.42$, determine the time rate of change of the plate temperature, dT/dt, when the plate temperature is 225°C. Evaluate the heat loss by convection and the heat loss by radiation.

1.74 Most of the energy we consume as food is converted to thermal energy in the process of performing all our bodily functions and is ultimately lost as heat from our bodies. Consider a person who consumes 2100 kcal per day (note that what are commonly referred to as food calories are actually kilocalories), of which 2000 kcal is converted to thermal energy. (The remaining 100 kcal is used to do work on the environment.) The person has a surface area of 1.8 m^2 and is dressed in a bathing suit.

(a) The person is in a room at 20°C, with a convection heat transfer coefficient of 3 W/m^2·K. At this air temperature, the person is not perspiring much. Estimate the person's average skin temperature.

(b) If the temperature of the environment were 33°C, what rate of perspiration would be needed to maintain a comfortable skin temperature of 33°C?

1.75 Consider Problem 1.1.

(a) If the exposed cold surface of the insulation is at $T_2 = 20$°C, what is the value of the convection heat transfer coefficient on the cold side of the insulation if the surroundings temperature is $T_{sur} = 320$ K, the ambient temperature is $T_\infty = 5$°C, and the emissivity is $\varepsilon = 0.95$? Express your results in units of W/m^2·K and W/m^2·°C.

(b) Using the convective heat transfer coefficient you calculated in part (a), determine the surface temperature, T_2, as the emissivity of the surface is varied over the range $0.05 \le \varepsilon \le 0.95$. The hot wall temperature of the insulation remains fixed at $T_1 = 30$°C. Display your results graphically.

1.76 The wall of an oven used to cure plastic parts is of thickness $L = 0.05$ m and is exposed to large surroundings and air at its outer surface. The air and the surroundings are at 300 K.

(a) If the temperature of the outer surface is 400 K and its convection coefficient and emissivity are

$h = 20$ W/m$^2 \cdot$K and $\varepsilon = 0.8$, respectively, what is the temperature of the inner surface if the wall has a thermal conductivity of $k = 0.7$ W/m$^2 \cdot$K?

(b) Consider conditions for which the temperature of the inner surface is maintained at 600 K, while the air and large surroundings to which the outer surface is exposed are maintained at 300 K. Explore the effects of variations in k, h, and ε on (i) the temperature of the outer surface, (ii) the heat flux through the wall, and (iii) the heat fluxes associated with convection and radiation heat transfer from the outer surface. Specifically, compute and plot the foregoing dependent variables for parametric variations about baseline conditions of $k = 10$ W/m$\cdot$K, $h = 20$ W/m$^2 \cdot$K, and $\varepsilon = 0.5$. The suggested ranges of the independent variables are $0.1 \leq k \leq 400$ W/m$\cdot$K, $2 \leq h \leq 200$ W/m$^2 \cdot$K, and $0.05 \leq \varepsilon \leq 1$. Discuss the physical implications of your results. Under what conditions will the temperature of the outer surface be less than 45°C, which is a reasonable upper limit to avoid burn injuries if contact is made?

1.77 An experiment to determine the convection coefficient associated with airflow over the surface of a thick stainless steel casting involves the insertion of thermocouples into the casting at distances of 10 and 20 mm from the surface along a hypothetical line normal to the surface. The steel has a thermal conductivity of 15 W/m$\cdot$K. If the thermocouples measure temperatures of 50 and 40°C in the steel when the air temperature is 100°C, what is the convection coefficient?

1.78 A thin electrical heating element provides a uniform heat flux q''_o to the outer surface of a duct through which airflows. The duct wall has a thickness of 10 mm and a thermal conductivity of 20 W/m$\cdot$K.

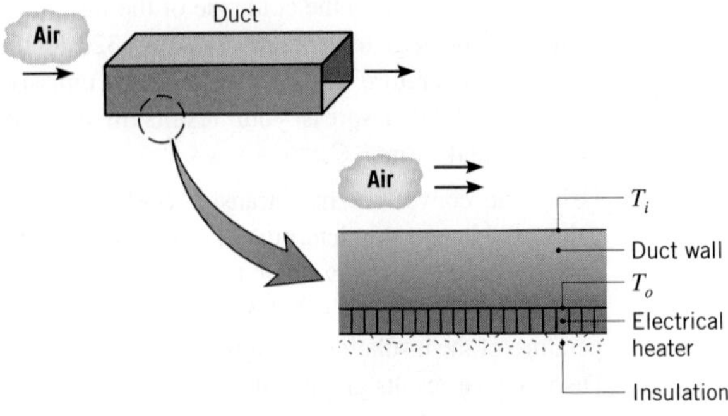

(a) At a particular location, the air temperature is 30°C and the convection heat transfer coefficient between the air and inner surface of the duct is 100 W/m$^2 \cdot$K. What heat flux q''_o is required to maintain the inner surface of the duct at $T_i = 85$°C?

(b) For the conditions of part (a), what is the temperature (T_o) of the duct surface next to the heater?

(c) With $T_i = 85$°C, compute and plot q''_o and T_o as a function of the air-side convection coefficient h for the range $10 \leq h \leq 200$ W/m$^2 \cdot$K. Briefly discuss your results.

1.79 A rectangular forced air heating duct is suspended from the ceiling of a basement whose air and walls are at a temperature of $T_\infty = T_{sur} = 5$°C. The duct is 15 m long, and its cross section is 350 mm × 200 mm.

(a) For an uninsulated duct whose average surface temperature is 50°C, estimate the rate of heat loss from the duct. The surface emissivity and convection coefficient are approximately 0.5 and 4 W/m$^2 \cdot$K, respectively.

(b) If heated air enters the duct at 58°C and a velocity of 4 m/s and the heat loss corresponds to the result of part (a), what is the outlet temperature? The density and specific heat of the air may be assumed to be $\rho = 1.10$ kg/m^3 and $c_p = 1008$ J/kg$\cdot$K, respectively.

1.80 Consider the steam pipe of Example 1.2. The facilities manager wants you to recommend methods for reducing the heat loss to the room, and two options are proposed. The first option would restrict air movement around the outer surface of the pipe and thereby reduce the convection coefficient by a factor of two. The second option would coat the outer surface of the pipe with a low emissivity ($\varepsilon = 0.4$) paint.

(a) Which of the foregoing options would you recommend?

(b) To prepare for a presentation of your recommendation to management, generate a graph of the heat loss q' as a function of the convection coefficient for $2 \leq h \leq 20$ W/m$^2 \cdot$K and emissivities of 0.2, 0.4, and 0.8. Comment on the relative efficacy of reducing heat losses associated with convection and radiation.

1.81 During its manufacture, plate glass at 600°C is cooled by passing air over its surface such that the convection heat transfer coefficient is $h = 5$ W/m$^2 \cdot$K. To prevent cracking, it is known that the temperature gradient must not exceed 15°C/mm at any point in the glass during the cooling process. If the thermal conductivity of the glass is 1.4 W/m$\cdot$K and its surface emissivity is 0.8, what is the lowest temperature of the air that can initially be used for the cooling? Assume that the temperature of the air equals that of the surroundings.

1.82 The curing process of Example 1.9 involves exposure of the plate to irradiation from an infrared lamp and attendant cooling by convection and radiation exchange with the surroundings. Alternatively, in lieu of the lamp, heating may be achieved by inserting the plate in an oven whose walls (the surroundings) are maintained at an elevated temperature.

(a) Consider conditions for which the oven walls are at 200°C, airflow over the plate is characterized by $T_\infty = 20°C$ and $h = 15\ W/m^2 \cdot K$, and the coating has an emissivity of $\varepsilon = 0.5$. What is the temperature of the plate?

(b) For ambient air temperatures of 20, 40, and 60°C, determine the plate temperature as a function of the oven wall temperature over the range from 150 to 250°C. Plot your results, and identify conditions for which acceptable curing temperatures between 100 and 110°C may be maintained.

1.83 The diameter and surface emissivity of an electrically heated plate are $D = 300$ mm and $\varepsilon = 0.80$, respectively.

(a) Estimate the power needed to maintain a surface temperature of 200°C in a room for which the air and the walls are at 25°C. The coefficient characterizing heat transfer by natural convection depends on the surface temperature and, in units of $W/m^2 \cdot K$, may be approximated by an expression of the form $h = 0.80(T_s - T_\infty)^{1/3}$.

(b) Assess the effect of surface temperature on the power requirement, as well as on the relative contributions of convection and radiation to heat transfer from the surface.

1.84 Bus bars proposed for use in a power transmission station have a rectangular cross section of height $H = 600$ mm and width $W = 200$ mm. The electrical resistivity, $\rho_e(\mu\Omega \cdot m)$, of the bar material is a function of temperature, $\rho_e = \rho_{e,o}[1 + \alpha(T - T_o)]$, where $\rho_{e,o} = 0.0828\ \mu\Omega \cdot m$, $T_o = 25°C$, and $\alpha = 0.0040\ K^{-1}$. The emissivity of the bar's painted surface is 0.8, and the temperature of the surroundings is 30°C. The convection coefficient between the bar and the ambient air at 30°C is 10 $W/m^2 \cdot K$.

(a) Assuming the bar has a uniform temperature T, calculate the steady-state temperature when a current of 60,000 A passes through the bar.

(b) Compute and plot the steady-state temperature of the bar as a function of the convection coefficient for $10 \leq h \leq 100\ W/m^2 \cdot K$. What minimum convection coefficient is required to maintain a safe-operating temperature below 120°C? Will increasing the emissivity significantly affect this result?

1.85 A solar flux of 700 W/m^2 is incident on a flat-plate solar collector used to heat water. The area of the collector is 3 m^2, and 90% of the solar radiation passes through the cover glass and is absorbed by the absorber plate. The remaining 10% is reflected away from the collector. Water flows through the tube passages on the back side of the absorber plate and is heated from an inlet temperature T_i to an outlet temperature T_o. The cover glass, operating at a temperature of 30°C, has an emissivity of 0.94 and experiences radiation exchange with the sky at −10°C. The convection coefficient between the cover glass and the ambient air at 25°C is 10 $W/m^2 \cdot K$.

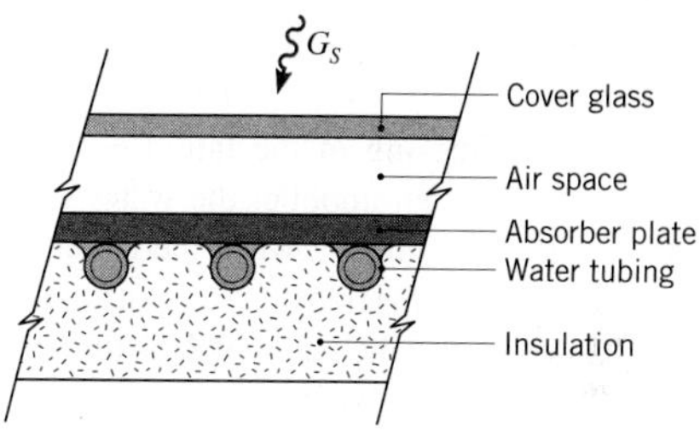

(a) Perform an overall energy balance on the collector to obtain an expression for the rate at which useful heat is collected per unit area of the collector, q''_u. Determine the value of q''_u.

(b) Calculate the temperature rise of the water, $T_o - T_i$, if the flow rate is 0.01 kg/s. Assume the specific heat of the water to be 4179 J/kg·K.

(c) The collector efficiency η is defined as the ratio of the useful heat collected to the rate at which solar energy is incident on the collector. What is the value of η?

Process Identification

1.86 In analyzing the performance of a thermal system, the engineer must be able to identify the relevant heat transfer processes. Only then can the system behavior be properly quantified. For the following systems, identify the pertinent processes, designating them by appropriately labeled arrows on a sketch of the system. Answer additional questions that appear in the problem statement.

(a) Identify the heat transfer processes that determine the temperature of an asphalt pavement on a summer day. Write an energy balance for the surface of the pavement.

(b) Microwave radiation is known to be transmitted by plastics, glass, and ceramics but to be absorbed by materials having polar molecules such as water. Water molecules exposed to microwave radiation align and reverse alignment with the microwave radiation at frequencies up to $10^9\ s^{-1}$, causing heat to be generated. Contrast cooking in a microwave oven with cooking in a conventional radiant or convection oven. In each case, what is the physical mechanism responsible for heating the food? Which oven has the greater energy utilization efficiency? Why? Microwave heating is being considered for drying clothes. How would the operation of a microwave clothes dryer differ from a conventional dryer? Which is likely to have the greater energy utilization efficiency? Why?

(c) To prevent freezing of the liquid water inside the fuel cell of an automobile, the water is drained to an onboard storage tank when the automobile is not in use. (The water is transferred from the tank back to the fuel cell when the automobile is turned on.) Consider a fuel cell–powered automobile that is parked outside on a very cold evening with $T_\infty = -20°C$. The storage tank is initially empty at $T_{i,t} = -20°C$, when liquid water, at atmospheric pressure and temperature $T_{i,w} = 50°C$, is introduced into the tank. The tank has a wall thickness t_t and is blanketed with insulation of thickness t_{ins}. Identify the heat transfer processes that will promote freezing of the water. Will the likelihood of freezing change as the insulation thickness is modified? Will the likelihood of freezing depend on the tank wall's thickness and material? Would freezing of the water be more likely if plastic (low thermal conductivity) or stainless steel (moderate thermal conductivity) tubing is used to transfer the water to and from the tank? Is there an optimal tank shape that would minimize the probability of the water freezing? Would freezing be more likely or less likely to occur if a thin sheet of aluminum foil (high thermal conductivity, low emissivity) is applied to the outside of the insulation?

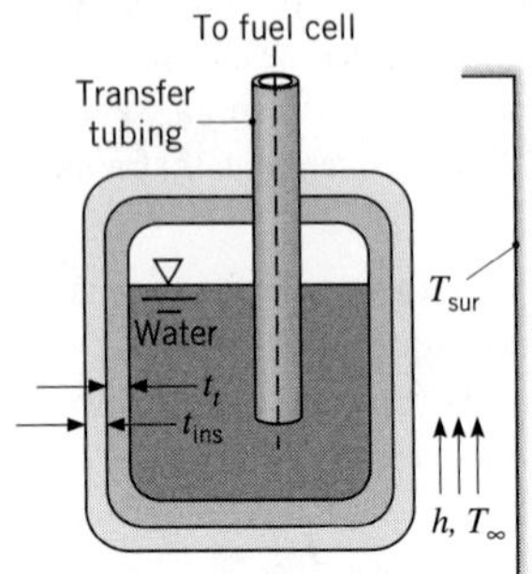

(d) Your grandmother is concerned about reducing her winter heating bills. Her strategy is to loosely fit rigid polystyrene sheets of insulation over her double-pane windows right after the first freezing weather arrives in the autumn. Identify the relevant heat transfer processes on a cold winter night when the foamed insulation sheet is placed (i) on the inner surface and (ii) on the outer surface of her window. To avoid condensation damage, which configuration is preferred? Condensation on the window pane does not occur when the foamed insulation is not in place.

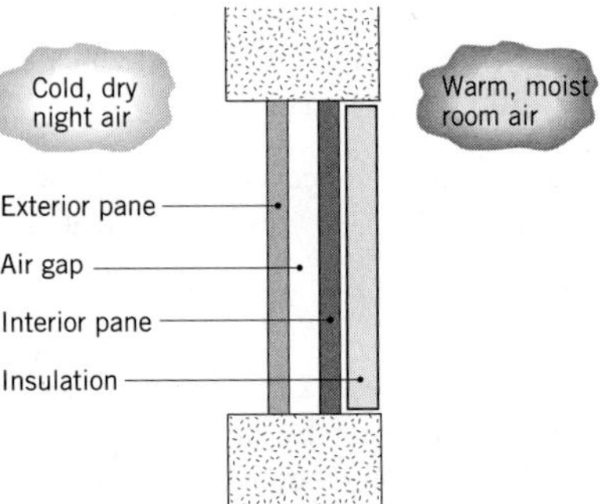

Insulation on inner surface

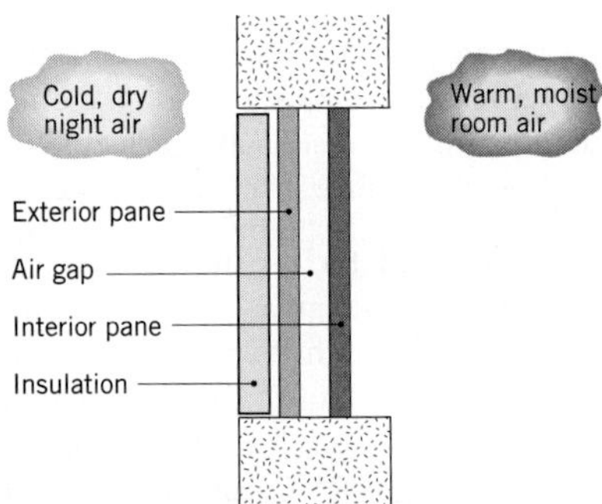

Insulation on outer surface

(e) There is considerable interest in developing building materials with improved insulating qualities. The development of such materials would do much to enhance energy conservation by reducing space heating requirements. It has been suggested that superior structural and insulating qualities could be obtained by using the composite shown. The material consists of a honeycomb, with cells of square cross section, sandwiched between solid slabs. The cells are filled with air, and the slabs, as well as the honeycomb matrix, are fabricated from plastics of low thermal conductivity. For heat transfer normal to the slabs, identify all heat transfer processes pertinent to the performance of the composite. Suggest ways in which this performance could be enhanced.

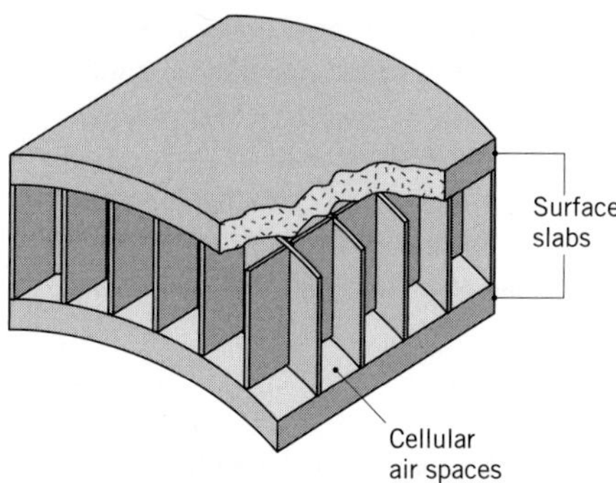

(f) A thermocouple junction (bead) is used to measure the temperature of a hot gas stream flowing through a channel by inserting the junction into the mainstream of the gas. The surface of the channel is cooled such that its temperature is well below that of the gas. Identify the heat transfer processes associated with the junction surface. Will the junction sense a temperature that is less than, equal to, or greater than the gas temperature? A radiation shield is a small, open-ended tube that encloses the thermocouple junction, yet allows for passage of the gas through the tube. How does use of such a shield improve the accuracy of the temperature measurement?

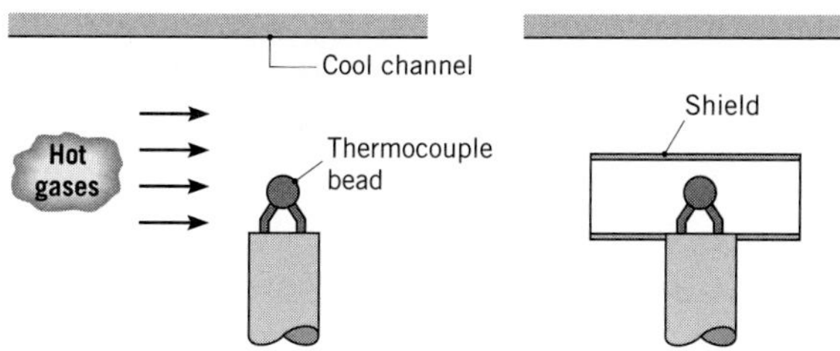

(g) A double-glazed, glass fire screen is inserted between a wood-burning fireplace and the interior of a room. The screen consists of two vertical glass plates that are separated by a space through which room air may flow (the space is open at the top and bottom). Identify the heat transfer processes associated with the fire screen.

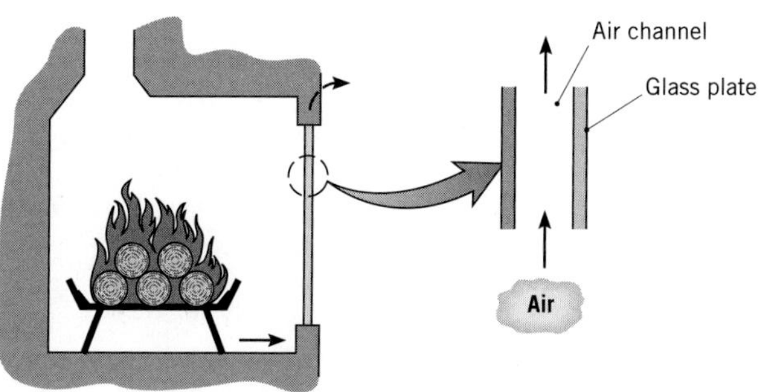

(h) A thermocouple junction is used to measure the temperature of a solid material. The junction is inserted into a small circular hole and is held in place by epoxy. Identify the heat transfer processes associated with the junction. Will the junction sense a temperature less than, equal to, or greater than the solid temperature? How will the thermal conductivity of the epoxy affect the junction temperature?

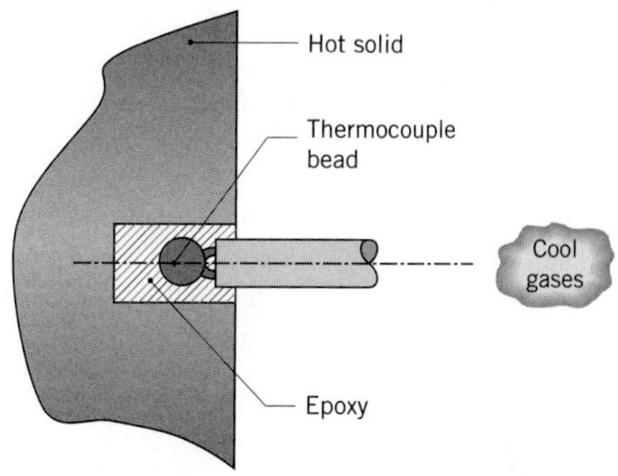

1.87 In considering the following problems involving heat transfer in the natural environment (outdoors), recognize that solar radiation is comprised of long and short wavelength components. If this radiation is incident on a *semi-transparent medium*, such as water or glass, two things will happen to the nonreflected portion of the radiation. The long wavelength component will be absorbed at the surface of the medium, whereas the short wavelength component will be transmitted by the surface.

(a) The number of panes in a window can strongly influence the heat loss from a heated room to the outside ambient air. Compare the single- and double-paned units shown by identifying relevant heat transfer processes for each case.

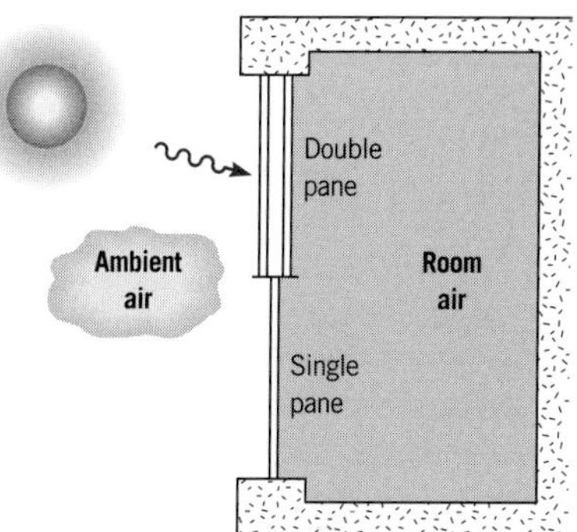

(b) In a typical flat-plate solar collector, energy is collected by a working fluid that is circulated through tubes that are in good contact with the back face of an absorber plate. The back face is insulated from the surroundings, and the absorber plate receives solar radiation on its front face, which is typically covered by one or more transparent plates. Identify the relevant heat transfer processes, first for the absorber plate with no cover plate and then for the absorber plate with a single cover plate.

(c) The solar energy collector design shown in the schematic has been used for agricultural applications. Air is blown through a long duct whose cross section is in the form of an equilateral triangle. One side of the triangle is comprised of a double-paned, semitransparent cover; the other two sides are constructed from aluminum sheets painted flat black on the inside and covered on the outside with a layer of styrofoam insulation. During sunny periods, air entering the system is heated for delivery to either a greenhouse, grain drying unit, or storage system.

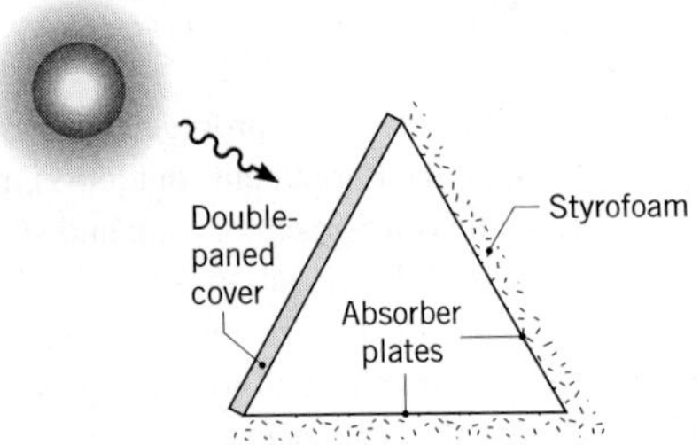

Identify all heat transfer processes associated with the cover plates, the absorber plate(s), and the air.

(d) Evacuated-tube solar collectors are capable of improved performance relative to flat-plate collectors. The design consists of an inner tube enclosed in an outer tube that is transparent to solar radiation. The annular space between the tubes is evacuated. The outer, opaque surface of the inner tube absorbs solar radiation, and a working fluid is passed through the tube to collect the solar energy. The collector design generally consists of a row of such tubes arranged in front of a reflecting panel. Identify all heat transfer processes relevant to the performance of this device.

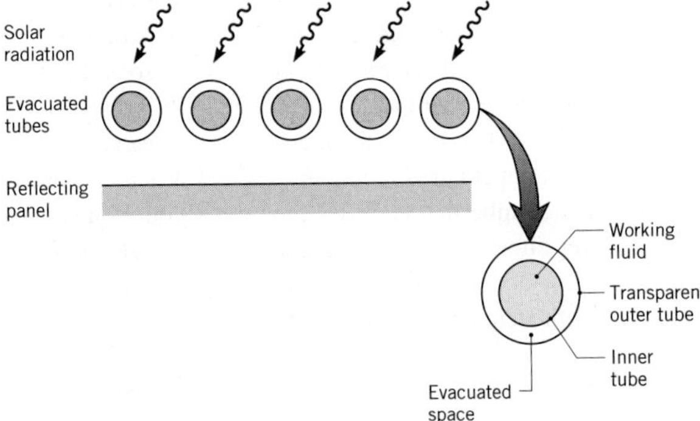

CHAPTER 2

Introduction to Conduction

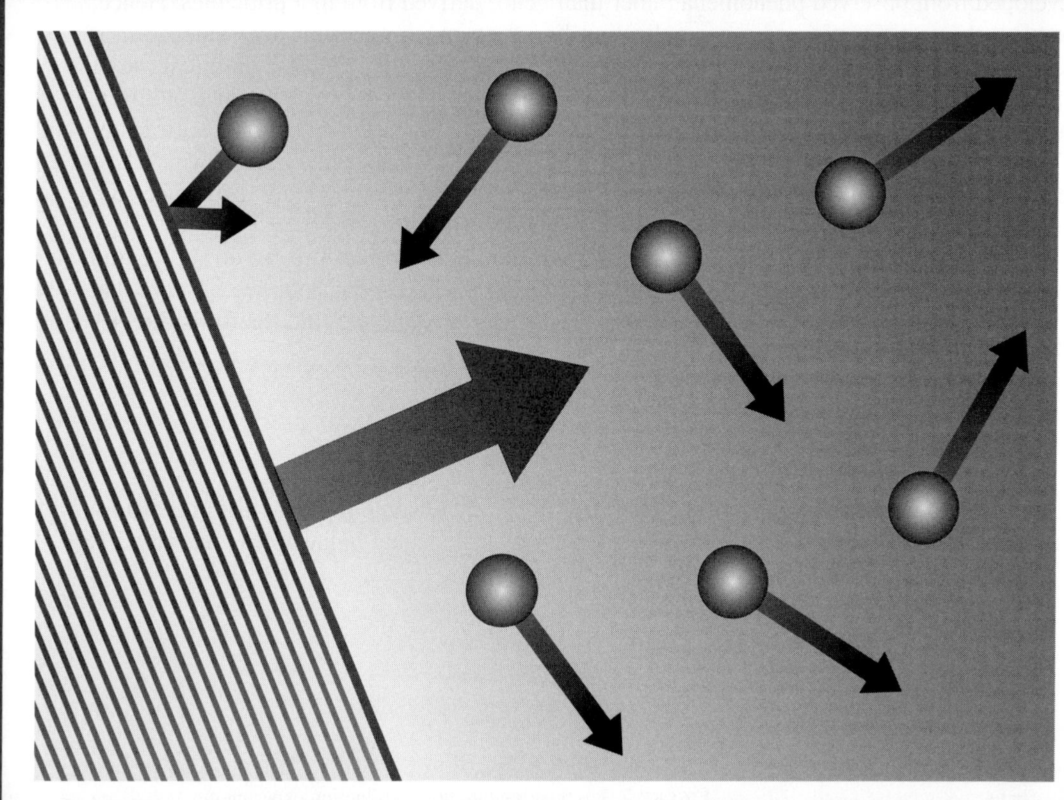

Recall that *conduction* is the transport of energy in a medium due to a temperature gradient, and the physical mechanism is one of random atomic or molecular activity. In Chapter 1 we learned that conduction heat transfer is governed by *Fourier's law* and that use of the law to determine the heat flux depends on knowledge of the manner in which temperature varies within the medium (the *temperature distribution*). By way of introduction, we restricted our attention to simplified conditions (one-dimensional, steady-state conduction in a plane wall). However, Fourier's law is applicable to transient, multidimensional conduction in complex geometries.

The objectives of this chapter are twofold. First, we wish to develop a deeper understanding of Fourier's law. What are its origins? What form does it take for different geometries? How does its proportionality constant (the *thermal conductivity*) depend on the physical nature of the medium? Our second objective is to develop, from basic principles, the general equation, termed the *heat equation*, which governs the temperature distribution in a medium. The solution to this equation provides knowledge of the temperature distribution, which may then be used with Fourier's law to determine the heat flux.

2.1 The Conduction Rate Equation

Although the conduction rate equation, Fourier's law, was introduced in Section 1.2, it is now appropriate to consider its origin. Fourier's law is *phenomenological*; that is, it is developed from observed phenomena rather than being derived from first principles. Hence, we view the rate equation as a generalization based on much experimental evidence. For example, consider the steady-state conduction experiment of Figure 2.1. A cylindrical rod of known material is insulated on its lateral surface, while its end faces are maintained at different temperatures, with $T_1 > T_2$. The temperature difference causes conduction heat transfer in the positive x-direction. We are able to measure the heat transfer rate q_x, and we seek to determine how q_x depends on the following variables: ΔT, the temperature difference; Δx, the rod length; and A, the cross-sectional area.

We might imagine first holding ΔT and Δx constant and varying A. If we do so, we find that q_x is directly proportional to A. Similarly, holding ΔT and A constant, we observe that q_x varies inversely with Δx. Finally, holding A and Δx constant, we find that q_x is directly proportional to ΔT. The collective effect is then

$$q_x \propto A \frac{\Delta T}{\Delta x}$$

In changing the material (e.g., from a metal to a plastic), we would find that this proportionality remains valid. However, we would also find that, for equal values of A, Δx, and ΔT,

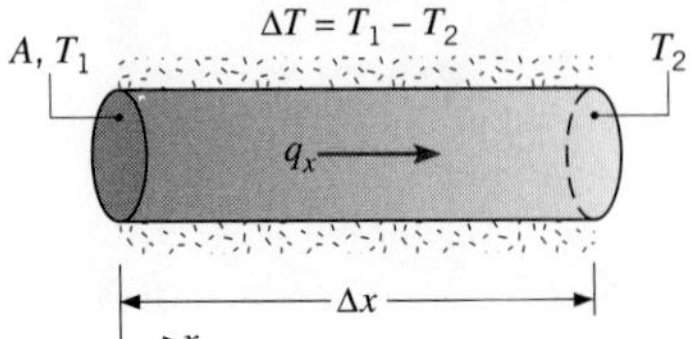

FIGURE 2.1 Steady-state heat conduction experiment.

the value of q_x would be smaller for the plastic than for the metal. This suggests that the proportionality may be converted to an equality by introducing a coefficient that is a measure of the material behavior. Hence, we write

$$q_x = kA\frac{\Delta T}{\Delta x}$$

where k, the *thermal conductivity* (W/m·K) is an important *property* of the material. Evaluating this expression in the limit as $\Delta x \rightarrow 0$, we obtain for the heat *rate*

$$q_x = -kA\frac{dT}{dx} \tag{2.1}$$

or for the heat *flux*

$$q''_x = \frac{q_x}{A} = -k\frac{dT}{dx} \tag{2.2}$$

Recall that the minus sign is necessary because heat is always transferred in the direction of decreasing temperature.

Fourier's law, as written in Equation 2.2, implies that the heat flux is a directional quantity. In particular, the direction of q''_x is *normal* to the cross-sectional area A. Or, more generally, the direction of heat flow will always be normal to a surface of constant temperature, called an *isothermal* surface. Figure 2.2 illustrates the direction of heat flow q''_x in a plane wall for which the *temperature gradient* dT/dx is negative. From Equation 2.2, it follows that q''_x is positive. Note that the isothermal surfaces are planes normal to the x-direction.

Recognizing that the heat flux is a vector quantity, we can write a more general statement of the conduction rate equation (*Fourier's law*) as follows:

$$\mathbf{q}'' = -k\nabla T = -k\left(\mathbf{i}\frac{\partial T}{\partial x} + \mathbf{j}\frac{\partial T}{\partial y} + \mathbf{k}\frac{\partial T}{\partial z}\right) \tag{2.3}$$

where ∇ is the three-dimensional del operator and $T(x, y, z)$ is the scalar temperature field. It is implicit in Equation 2.3 that the heat flux vector is in a direction perpendicular to the isothermal surfaces. An alternative form of Fourier's law is therefore

$$\mathbf{q}'' = q''_n\mathbf{n} = -k\frac{\partial T}{\partial n}\mathbf{n} \tag{2.4}$$

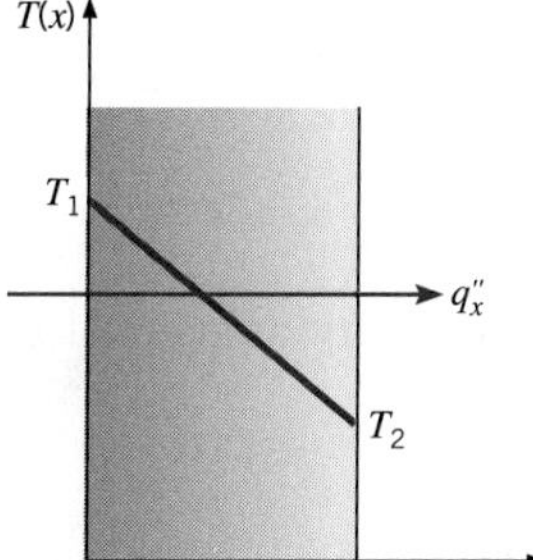

FIGURE 2.2 The relationship between coordinate system, heat flow direction, and temperature gradient in one dimension.

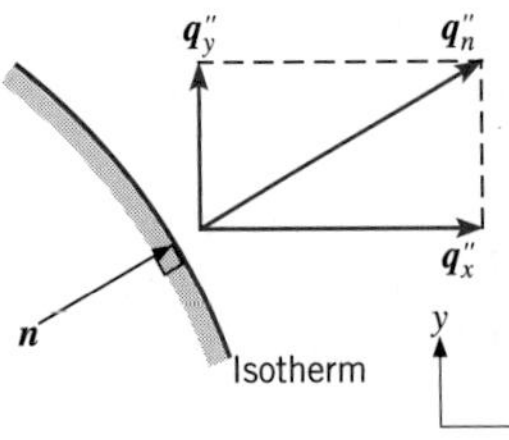

FIGURE 2.3 The heat flux vector normal to an isotherm in a two-dimensional coordinate system.

where q''_n is the heat flux in a direction n, which is normal to an *isotherm*, and $\mathbf{n}$ is the unit normal vector in that direction. This is illustrated for the two-dimensional case in Figure 2.3. The heat transfer is sustained by a temperature gradient along $\mathbf{n}$. Note also that the heat flux vector can be resolved into components such that, in Cartesian coordinates, the general expression for $\mathbf{q}''$ is

$$\mathbf{q}'' = \mathbf{i}q''_x + \mathbf{j}q''_y + \mathbf{k}q''_z \tag{2.5}$$

where, from Equation 2.3, it follows that

$$q''_x = -k\frac{\partial T}{\partial x} \qquad q''_y = -k\frac{\partial T}{\partial y} \qquad q''_z = -k\frac{\partial T}{\partial z} \tag{2.6}$$

Each of these expressions relates the heat flux *across a surface* to the temperature gradient in a direction perpendicular to the surface. It is also implicit in Equation 2.3 that the medium in which the conduction occurs is *isotropic*. For such a medium, the value of the thermal conductivity is independent of the coordinate direction.

Fourier's law is the cornerstone of conduction heat transfer, and its key features are summarized as follows. It is *not* an expression that may be derived from first principles; it is instead a generalization based on experimental evidence. It is an expression that *defines* an important material property, the thermal conductivity. In addition, Fourier's law is a vector expression indicating that the heat flux is normal to an isotherm and in the direction of decreasing temperature. Finally, note that Fourier's law applies for all matter, regardless of its state (solid, liquid, or gas).

2.2 The Thermal Properties of Matter

To use Fourier's law, the thermal conductivity of the material must be known. This property, which is referred to as a *transport property,* provides an indication of the rate at which energy is transferred by the diffusion process. It depends on the physical structure of matter, atomic and molecular, which is related to the state of the matter. In this section we consider various forms of matter, identifying important aspects of their behavior and presenting typical property values.

2.2.1 Thermal Conductivity

From Fourier's law, Equation 2.6, the thermal conductivity associated with conduction in the x-direction is defined as

$$k_x \equiv -\frac{q''_x}{(\partial T/\partial x)}$$

Similar definitions are associated with thermal conductivities in the y- and z-directions (k_y, k_z), but for an isotropic material the thermal conductivity is independent of the direction of transfer, $k_x = k_y = k_z \equiv k$.

From the foregoing equation, it follows that, for a prescribed temperature gradient, the conduction heat flux increases with increasing thermal conductivity. In general, the thermal conductivity of a solid is larger than that of a liquid, which is larger than that of a gas. As illustrated in Figure 2.4, the thermal conductivity of a solid may be more than four orders of magnitude larger than that of a gas. This trend is due largely to differences in intermolecular spacing for the two states.

The Solid State In the modern view of materials, a solid may be comprised of free electrons and atoms bound in a periodic arrangement called the lattice. Accordingly, transport of thermal energy may be due to two effects: the migration of free electrons and lattice vibrational waves. When viewed as a particle-like phenomenon, the lattice vibration quanta are termed *phonons*. In pure metals, the electron contribution to conduction heat transfer dominates, whereas in nonconductors and semiconductors, the phonon contribution is dominant.

Kinetic theory yields the following expression for the thermal conductivity [1]:

$$k = \frac{1}{3} C \bar{c} \lambda_{\text{mfp}} \tag{2.7}$$

For conducting materials such as metals, $C \equiv C_e$ is the electron specific heat per unit volume, $\bar{c}$ is the mean electron velocity, and $\lambda_{\text{mfp}} \equiv \lambda_e$ is the electron mean free path, which is defined as the average distance traveled by an electron before it collides with either an imperfection in the material or with a phonon. In nonconducting solids, $C \equiv C_{\text{ph}}$ is the phonon specific heat, $\bar{c}$ is the average speed of sound, and $\lambda_{\text{mfp}} \equiv \lambda_{\text{ph}}$ is the phonon mean free path, which again is determined by collisions with imperfections or other phonons. In all cases, the thermal conductivity increases as the mean free path of the *energy carriers* (electrons or phonons) is increased.

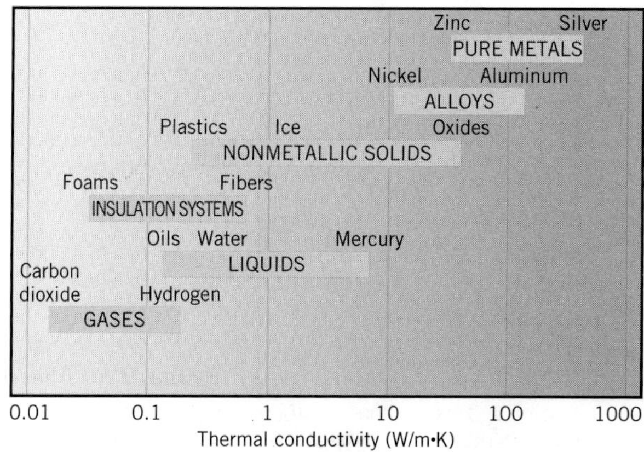

FIGURE 2.4 Range of thermal conductivity for various states of matter at normal temperatures and pressure.

When electrons and phonons carry thermal energy leading to conduction heat transfer in a solid, the thermal conductivity may be expressed as

$$k = k_e + k_{\text{ph}} \tag{2.8}$$

To a first approximation, k_e is inversely proportional to the electrical resistivity, ρ_e. For pure metals, which are of low ρ_e, k_e is much larger than k_{ph}. In contrast, for alloys, which are of substantially larger ρ_e, the contribution of k_{ph} to k is no longer negligible. For nonmetallic solids, k is determined primarily by k_{ph}, which increases as the frequency of interactions between the atoms and the lattice decreases. The regularity of the lattice arrangement has an important effect on k_{ph}, with crystalline (well-ordered) materials like quartz having a higher thermal conductivity than amorphous materials like glass. In fact, for crystalline, nonmetallic solids such as diamond and beryllium oxide, k_{ph} can be quite large, exceeding values of k associated with good conductors, such as aluminum.

The temperature dependence of k is shown in Figure 2.5 for representative metallic and nonmetallic solids. Values for selected materials of technical importance are also provided in Table A.1 (metallic solids) and Tables A.2 and A.3 (nonmetallic solids). More detailed treatments of thermal conductivity are available in the literature [2].

The Solid State: Micro- and Nanoscale Effects In the preceding discussion, the *bulk* thermal conductivity is described, and the thermal conductivity values listed in Tables A.1 through A.3 are appropriate for use when the physical dimensions of the material of interest are relatively large. This is the case in many commonplace engineering problems. However, in several

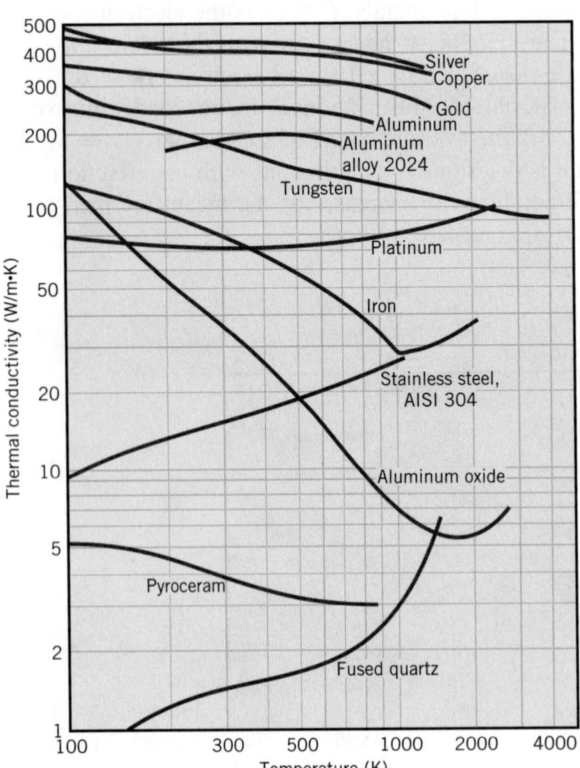

Figure 2.5 The temperature dependence of the thermal conductivity of selected solids.

areas of technology, such as microelectronics, the material's characteristic dimensions can be on the order of micrometers or nanometers, in which case care must be taken to account for the possible modifications of k that can occur as the physical dimensions become small.

Cross sections of *films* of the same material having thicknesses L_1 and L_2 are shown in Figure 2.6. Electrons or phonons that are associated with conduction of thermal energy are also shown qualitatively. Note that the physical boundaries of the film act to *scatter* the energy carriers and *redirect* their propagation. For large L/λ_{mfp}[1] (Figure 2.6*a*), the effect of the boundaries on reducing the *average* energy carrier path length is minor, and conduction heat transfer occurs as described for bulk materials. However, as the film becomes thinner, the physical boundaries of the material can decrease the average *net* distance traveled by the energy carriers, as shown in Figure 2.6*b*. Moreover, electrons and phonons moving in the thin x-direction (representing conduction in the x-direction) are affected by the boundaries to a more significant degree than energy carriers moving in the y-direction. As such, for films characterized by small L/λ_{mfp}, we find that $k_x < k_y < k$, where k is the bulk thermal conductivity of the film material.

For $L/\lambda_{mfp} \geq 1$, the predicted values of k_x and k_y may be estimated to within 20% from the following expression [1]:

$$\frac{k_x}{k} = 1 - \frac{\lambda_{mfp}}{3L} \tag{2.9a}$$

$$\frac{k_y}{k} = 1 - \frac{2\lambda_{mfp}}{3\pi L} \tag{2.9b}$$

Equations 2.9a,b reveal that the values of k_x and k_y are within approximately 5% of the bulk thermal conductivity if $L/\lambda_{mfp} > 7$ (for k_x) and $L/\lambda_{mfp} > 4.5$ (for k_y). Values of the mean free path as well as critical film thicknesses below which microscale effects must be considered, L_{crit}, are included in Table 2.1 for several materials at $T \approx 300\,K$. For films with $\lambda_{mfp} < L < L_{crit}$, k_x and k_y are reduced from the bulk value as indicated in Equations 2.9a,b.

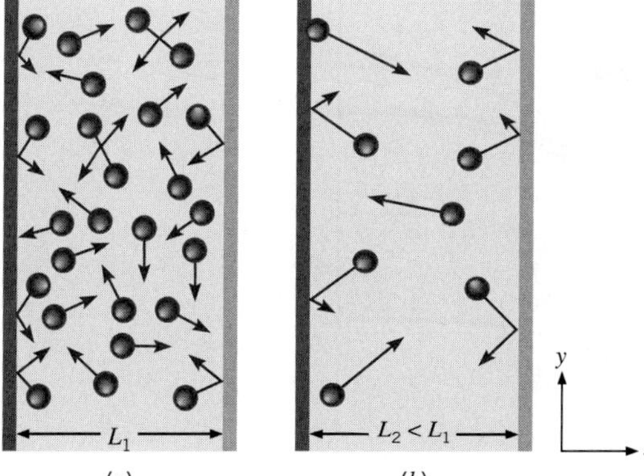

FIGURE 2.6 Electron or phonon trajectories in (*a*) a relatively thick film and (*b*) a relatively thin film with boundary effects.

[1]The quantity λ_{mfp}/L is a dimensionless parameter known as the Knudsen number. Large Knudsen numbers (small L/λ_{mfp}) suggest potentially significant nano- or microscale effects.

No general guidelines exist for predicting values of the thermal conductivities for $L/\lambda_{mfp} < 1$. Note that, in solids, the value of λ_{mfp} decreases as the temperature increases.

In addition to scattering from physical boundaries, as in the case of Figure 2.6*b*, energy carriers may be redirected by chemical *dopants* embedded within a material or by *grain boundaries* that separate individual clusters of material in otherwise homogeneous matter. *Nanostructured materials* are chemically identical to their conventional counterparts but are processed to provide very small grain sizes. This feature impacts heat transfer by increasing the scattering and reflection of energy carriers at grain boundaries.

Measured values of the thermal conductivity of a bulk, nanostructured yttria-stabilized zirconia material are shown in Figure 2.7. This particular ceramic is widely used for insulation purposes in high-temperature combustion devices. Conduction is dominated by phonon transfer, and the mean free path of the phonon energy carriers is, from Table 2.1, $\lambda_{mfp} = 25$ nm at 300 K. As the grain sizes are reduced to characteristic dimensions less than 25 nm (and more grain boundaries are introduced in the material per unit volume), significant reduction of the thermal conductivity occurs. Extrapolation of the results of Figure 2.7 to higher temperatures is not recommended, since the mean free path decreases with increasing temperature ($\lambda_{mfp} \approx 4$ nm at $T \approx 1525$ K) *and* grains of the material may coalesce, merge, and enlarge at elevated temperatures. Therefore, L/λ_{mfp} becomes larger at high temperatures, and

TABLE 2.1 Mean free path and critical film thickness for various materials at $T \approx 300$ K [3,4]

Material	λ_{mfp} (nm)	$L_{crit,x}$ (nm)	$L_{crit,y}$ (nm)
Aluminum oxide	5.08	36	22
Diamond (IIa)	315	2200	1400
Gallium arsenide	23	160	100
Gold	31	220	140
Silicon	43	290	180
Silicon dioxide	0.6	4	3
Yttria-stabilized zirconia	25	170	110

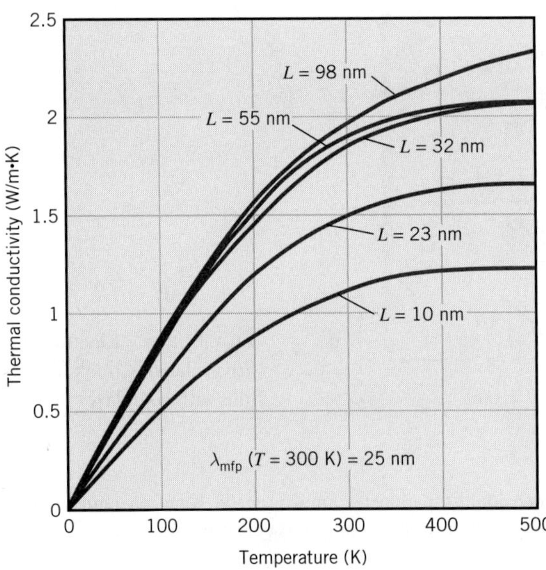

FIGURE 2.7 Measured thermal conductivity of yttria-stabilized zirconia as a function of temperature and mean grain size, L [3].

reduction of k due to nanoscale effects becomes less pronounced. Research on heat transfer in nanostructured materials continues to reveal novel ways engineers can manipulate the nanostructure to reduce or increase thermal conductivity [5]. Potentially important consequences include applications such as gas turbine engine technology [6], microelectronics [7], and renewable energy [8].

The Fluid State The fluid state includes both liquids and gases. Because the intermolecular spacing is much larger and the motion of the molecules is more random for the fluid state than for the solid state, thermal energy transport is less effective. The thermal conductivity of gases and liquids is therefore generally smaller than that of solids.

The effect of temperature, pressure, and chemical species on the thermal conductivity of a gas may be explained in terms of the kinetic theory of gases [9]. From this theory it is known that the thermal conductivity is directly proportional to the density of the gas, the mean molecular speed $\bar{c}$, and the mean free path λ_{mfp}, which is the average distance traveled by an energy carrier (a molecule) before experiencing a collision.

$$k \approx \frac{1}{3} c_v \rho \bar{c} \lambda_{\text{mfp}} \tag{2.10}$$

For an ideal gas, the mean free path may be expressed as

$$\lambda_{\text{mfp}} = \frac{k_B T}{\sqrt{2}\pi d^2 p} \tag{2.11}$$

where k_B is Boltzmann's constant, $k_B = 1.381 \times 10^{-23}$ J/K, d is the diameter of the gas molecule, representative values of which are included in Figure 2.8, and p is the pressure.

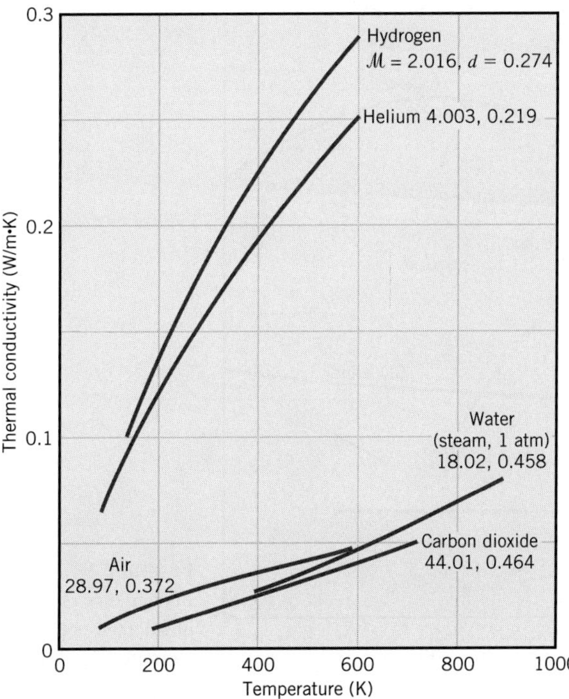

FIGURE 2.8 The temperature dependence of the thermal conductivity of selected gases at normal pressures. Molecular diameters (d) are in nm [10]. Molecular weights ($\mathcal{M}$) of the gases are also shown.

As expected, the mean free path is small for high pressure or low temperature, which causes densely packed molecules. The mean free path also depends on the diameter of the molecule, with larger molecules more likely to experience collisions than small molecules; in the limiting case of an infinitesimally small molecule, the molecules cannot collide, resulting in an infinite mean free path. The mean molecular speed, $\bar{c}$, can be determined from the kinetic theory of gases, and Equation 2.10 may ultimately be expressed as

$$k = \frac{9\gamma - 5}{4} \frac{c_v}{\pi d^2} \sqrt{\frac{\mathcal{M} k_B T}{\mathcal{N} \pi}} \qquad (2.12)$$

where the parameter γ is the ratio of specific heats, $\gamma \equiv c_p/c_v$, and $\mathcal{N}$ is Avogadro's number, $\mathcal{N} = 6.022 \times 10^{23}$ molecules per mol. Equation 2.12 can be used to estimate the thermal conductivity of gas, although more accurate models have been developed [10].

It is important to note that the thermal conductivity is independent of pressure except in extreme cases as, for example, when conditions approach that of a perfect vacuum. Therefore, the assumption that k is independent of gas pressure for large volumes of gas is appropriate for the pressures of interest in this text. Accordingly, although the values of k presented in Table A.4 pertain to atmospheric pressure or the saturation pressure corresponding to the prescribed temperature, they may be used over a much wider pressure range.

Molecular conditions associated with the liquid state are more difficult to describe, and physical mechanisms for explaining the thermal conductivity are not well understood [11]. The thermal conductivity of nonmetallic liquids generally decreases with increasing temperature. As shown in Figure 2.9, water, glycerine, and engine oil are notable exceptions. The thermal conductivity of liquids is usually insensitive to pressure except near the critical point. Also, thermal conductivity generally decreases with increasing molecular weight. Values of

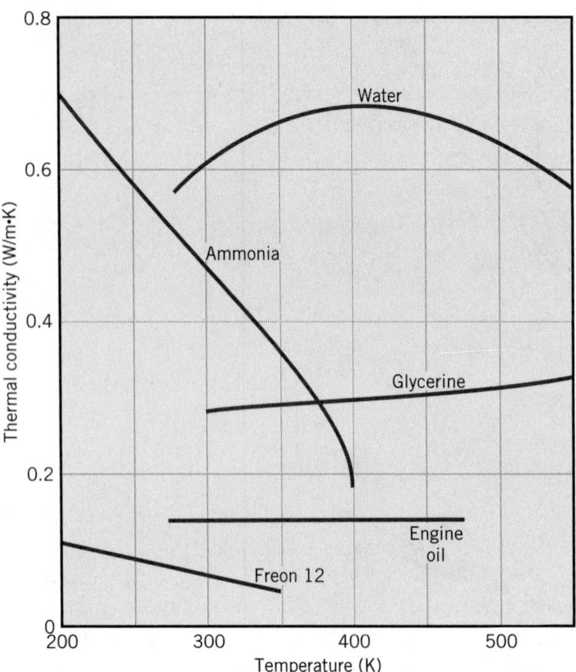

FIGURE 2.9 The temperature dependence of the thermal conductivity of selected nonmetallic liquids under saturated conditions.

the thermal conductivity are often tabulated as a function of temperature for the saturated state of the liquid. Tables A.5 and A.6 present such data for several common liquids.

Liquid metals are commonly used in high heat flux applications, such as occur in nuclear power plants. The thermal conductivity of such liquids is given in Table A.7. Note that the values are much larger than those of the nonmetallic liquids [12].

The Fluid State: Micro- and Nanoscale Effects As for the solid state, the bulk thermal conductivity of a fluid may be modified when the characteristic dimension of the system becomes small, in particular for small values of L/λ_{mfp}. Similar to the situation of a thin solid film shown in Figure 2.6*b*, the molecular mean free path is restricted when a fluid is constrained by a small physical dimension, affecting conduction across a thin fluid layer.

Mixtures of fluids and solids can also be formulated to tailor the transport properties of the resulting *suspension*. For example, *nanofluids* are *base liquids* that are seeded with nanometer-sized solid particles. Their very small size allows the solid particles to remain suspended within the base liquid for a long time. From the heat transfer perspective, a nanofluid exploits the high thermal conductivity that is characteristic of most solids, as is evident in Figure 2.5, to boost the relatively low thermal conductivity of base liquids, typical values of which are shown in Figure 2.9. Typical nanofluids involve liquid water seeded with nominally spherical nanoparticles of Al_2O_3 or CuO.

Insulation Systems Thermal insulations consist of low thermal conductivity materials combined to achieve an even lower system thermal conductivity. In conventional *fiber-*, *powder-*, and *flake*-type insulations, the solid material is finely dispersed throughout an air space. Such systems are characterized by an *effective thermal conductivity*, which depends on the thermal conductivity and surface radiative properties of the solid material, as well as the nature and volumetric fraction of the air or void space. A special parameter of the system is its bulk density (solid mass/total volume), which depends strongly on the manner in which the material is packed.

If small voids or hollow spaces are formed by bonding or fusing portions of the solid material, a rigid matrix is created. When these spaces are sealed from each other, the system is referred to as a *cellular* insulation. Examples of such rigid insulations are *foamed* systems, particularly those made from plastic and glass materials. *Reflective* insulations are composed of multilayered, parallel, thin sheets or foils of high reflectivity, which are spaced to reflect radiant energy back to its source. The spacing between the foils is designed to restrict the motion of air, and in high-performance insulations, the space is evacuated. In all types of insulation, evacuation of the air in the void space will reduce the effective thermal conductivity of the system.

Heat transfer through any of these insulation systems may include several modes: conduction through the solid materials; conduction or convection through the air in the void spaces; and radiation exchange between the surfaces of the solid matrix. The effective thermal conductivity accounts for all of these processes, and values for selected insulation systems are summarized in Table A.3. Additional background information and data are available in the literature [13, 14].

As with thin films, micro- and nanoscale effects can influence the effective thermal conductivity of insulating materials. The value of k for a nanostructured silica aerogel material that is composed of approximately 5% by volume solid material and 95% by volume air that is trapped within pores of $L \approx 20$ nm is shown in Figure 2.10. Note that at $T \approx 300$ K, the mean free path for air at atmospheric pressure is approximately 80 nm. As the gas pressure is reduced, λ_{mfp} would increase for an unconfined gas, but the molecular

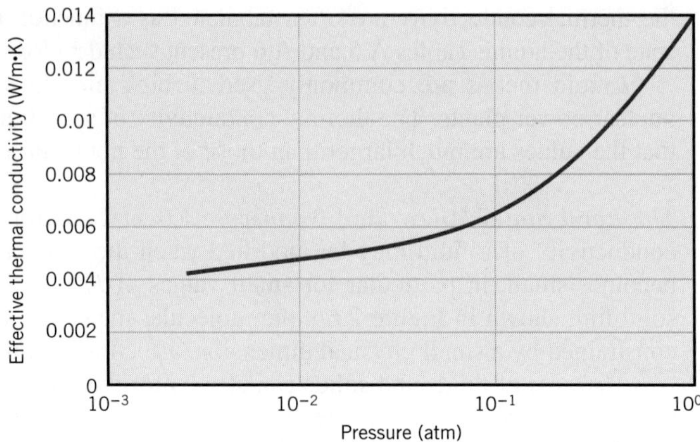

FIGURE 2.10 Measured thermal conductivity of carbon-doped silica aerogel as a function of pressure at $T \approx 300\,\text{K}$ [15].

motion of the trapped air is restricted by the walls of the small pores and k is reduced to extremely small values relative to the thermal conductivities of conventional matter reported in Figure 2.4.

2.2.2 Other Relevant Properties

In our analysis of heat transfer problems, it will be necessary to use several properties of matter. These properties are generally referred to as *thermophysical* properties and include two distinct categories, *transport* and *thermodynamic* properties. The transport properties include the diffusion rate coefficients such as k, the thermal conductivity (for heat transfer), and ν, the kinematic viscosity (for momentum transfer). Thermodynamic properties, on the other hand, pertain to the equilibrium state of a system. Density (ρ) and specific heat (c_p) are two such properties used extensively in thermodynamic analysis. The product ρc_p ($\text{J/m}^3 \cdot \text{K}$), commonly termed the *volumetric heat capacity*, measures the ability of a material to store thermal energy. Because substances of large density are typically characterized by small specific heats, many solids and liquids, which are very good energy storage media, have comparable heat capacities ($\rho c_p > 1\,\text{MJ/m}^3 \cdot \text{K}$). Because of their very small densities, however, gases are poorly suited for thermal energy storage ($\rho c_p \approx 1\,\text{kJ/m}^3 \cdot \text{K}$). Densities and specific heats are provided in the tables of Appendix A for a wide range of solids, liquids, and gases.

In heat transfer analysis, the ratio of the thermal conductivity to the heat capacity is an important property termed the *thermal diffusivity* α, which has units of m^2/s:

$$\alpha = \frac{k}{\rho c_p}$$

It measures the ability of a material to conduct thermal energy relative to its ability to store thermal energy. Materials of large α will respond quickly to changes in their thermal environment, whereas materials of small α will respond more sluggishly, taking longer to reach a new equilibrium condition.

The accuracy of engineering calculations depends on the accuracy with which the thermophysical properties are known [16–18]. Numerous examples could be cited of flaws in equipment and process design or failure to meet performance specifications that were attributable to misinformation associated with the selection of key property values used in the initial system analysis. Selection of reliable property data is an integral part of any careful engineering analysis. The casual use of data that have not been well characterized or evaluated, as may be found in some literature or handbooks, is to be avoided. Recommended data values for many thermophysical properties can be obtained from Reference 19. This reference, available in most institutional libraries, was prepared by the Thermophysical Properties Research Center (TPRC) at Purdue University.

IHT

EXAMPLE 2.1

The thermal diffusivity α is the controlling transport property for transient conduction. Using appropriate values of k, ρ, and c_p from Appendix A, calculate α for the following materials at the prescribed temperatures: pure aluminum, 300 and 700 K; silicon carbide, 1000 K; paraffin, 300 K.

SOLUTION

Known: Definition of the thermal diffusivity α.

Find: Numerical values of α for selected materials and temperatures.

Properties: Table A.1, pure aluminum (300 K):

$$\left.\begin{aligned} \rho &= 2702 \text{ kg/m}^3 \\ c_p &= 903 \text{ J/kg}\cdot\text{K} \\ k &= 237 \text{ W/m}\cdot\text{K} \end{aligned}\right\} \alpha = \frac{k}{\rho c_p} = \frac{237 \text{ W/m}\cdot\text{K}}{2702 \text{ kg/m}^3 \times 903 \text{ J/kg}\cdot\text{K}}$$

$$= 97.1 \times 10^{-6} \text{ m}^2\text{/s} \qquad \triangleleft$$

Table A.1, pure aluminum (700 K):

$$\begin{aligned} \rho &= 2702 \text{ kg/m}^3 && \text{at 300 K} \\ c_p &= 1090 \text{ J/kg}\cdot\text{K} && \text{at 700 K (by linear interpolation)} \\ k &= 225 \text{ W/m}\cdot\text{K} && \text{at 700 K (by linear interpolation)} \end{aligned}$$

Hence

$$\alpha = \frac{k}{\rho c_p} = \frac{225 \text{ W/m}\cdot\text{K}}{2702 \text{ kg/m}^3 \times 1090 \text{ J/kg}\cdot\text{K}} = 76 \times 10^{-6} \text{ m}^2\text{/s} \qquad \triangleleft$$

Table A.2, silicon carbide (1000 K):

$$\left.\begin{aligned} \rho &= 3160 \text{ kg/m}^3 && \text{at 300 K} \\ c_p &= 1195 \text{ J/kg}\cdot\text{K} && \text{at 1000 K} \\ k &= 87 \text{ W/m}\cdot\text{K} && \text{at 1000 K} \end{aligned}\right\} \alpha = \frac{87 \text{ W/m}\cdot\text{K}}{3160 \text{ kg/m}^3 \times 1195 \text{ J/kg}\cdot\text{K}}$$

$$= 23 \times 10^{-6} \text{ m}^2\text{/s} \qquad \triangleleft$$

Table A.3, paraffin (300 K):

$$\left.\begin{aligned} \rho &= 900\ \text{kg/m}^3 \\ c_p &= 2890\ \text{J/kg}\cdot\text{K} \\ k &= 0.24\ \text{W/m}\cdot\text{K} \end{aligned}\right\} \alpha = \frac{k}{\rho c_p} = \frac{0.24\ \text{W/m}\cdot\text{K}}{900\ \text{kg/m}^3 \times 2890\ \text{J/kg}\cdot\text{K}}$$

$$= 9.2 \times 10^{-8}\ \text{m}^2/\text{s} \qquad \triangleleft$$

Comments:

1. Note the temperature dependence of the thermophysical properties of aluminum and silicon carbide. For example, for silicon carbide, $\alpha(1000\ \text{K}) \approx 0.1 \times \alpha(300\ \text{K})$; hence properties of this material have a strong temperature dependence.
2. The physical interpretation of α is that it provides a measure of heat transport (k) relative to energy storage (ρc_p). In general, metallic solids have higher α, whereas nonmetallics (e.g., paraffin) have lower values of α.
3. Linear interpolation of property values is generally acceptable for engineering calculations.
4. Use of the low-temperature (300 K) density at higher temperatures ignores thermal expansion effects but is also acceptable for engineering calculations.
5. The *IHT* software provides a library of thermophysical properties for selected solids, liquids, and gases that can be accessed from the toolbar button, *Properties*. See Example 2.1 in *IHT*.

EXAMPLE 2.2

The bulk thermal conductivity of a nanofluid containing uniformly dispersed, noncontacting spherical nanoparticles may be approximated by

$$k_{\text{nf}} = \left[\frac{k_p + 2k_{\text{bf}} + 2\varphi(k_p - k_{\text{bf}})}{k_p + 2k_{\text{bf}} - \varphi(k_p - k_{\text{bf}})}\right] k_{\text{bf}}$$

where φ is the volume fraction of the nanoparticles, and k_{bf}, k_p, and k_{nf} are the thermal conductivities of the base fluid, particle, and nanofluid, respectively. Likewise, the dynamic viscosity may be approximated as [20]

$$\mu_{\text{nf}} = \mu_{\text{bf}}(1 + 2.5\varphi)$$

Determine the values of k_{nf}, ρ_{nf}, $c_{p,\text{nf}}$, μ_{nf}, and α_{nf} for a mixture of water and Al_2O_3 nanoparticles at a temperature of $T = 300$ K and a particle volume fraction of $\varphi = 0.05$. The thermophysical properties of the particle are $k_p = 36.0\ \text{W/m}\cdot\text{K}$, $\rho_p = 3970\ \text{kg/m}^3$, and $c_{p,p} = 0.765\ \text{kJ/kg}\cdot\text{K}$.

SOLUTION

Known: Expressions for the bulk thermal conductivity and viscosity of a nanofluid with spherical nanoparticles. Nanoparticle properties.

Find: Values of the nanofluid thermal conductivity, density, specific heat, dynamic viscosity, and thermal diffusivity.

Schematic:

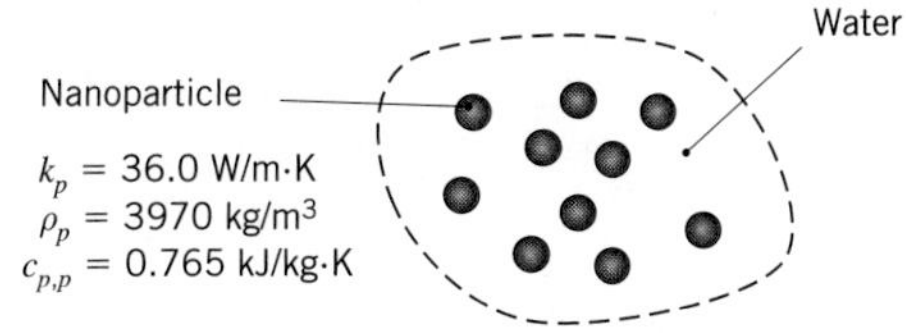

Assumptions:

1. Constant properties.
2. Density and specific heat are not affected by nanoscale phenomena.
3. Isothermal conditions.

Properties: Table A.6 ($T = 300\,\text{K}$): Water; $k_{\text{bf}} = 0.613\,\text{W/m}\cdot\text{K}$, $\rho_{\text{bf}} = 997\,\text{kg/m}^3$, $c_{p,\text{bf}} = 4.179\,\text{kJ/kg}\cdot\text{K}$, $\mu_{\text{bf}} = 855 \times 10^{-6}\,\text{N}\cdot\text{s/m}^2$.

Analysis: From the problem statement,

$$k_{\text{nf}} = \left[\frac{k_p + 2k_{\text{bf}} + 2\varphi(k_p - k_{\text{bf}})}{k_p + 2k_{\text{bf}} - \varphi(k_p - k_{\text{bf}})}\right] k_{\text{bf}}$$

$$= \left[\frac{36.0\ \text{W/m}\cdot\text{K} + 2 \times 0.613\ \text{W/m}\cdot\text{K} + 2 \times 0.05(36.0 - 0.613)\ \text{W/m}\cdot\text{K}}{36.0\ \text{W/m}\cdot\text{K} + 2 \times 0.613\ \text{W/m}\cdot\text{K} - 0.05(36.0 - 0.613)\ \text{W/m}\cdot\text{K}}\right]$$

$$\times 0.613\ \text{W/m}\cdot\text{K}$$

$$= 0.705\ \text{W/m}\cdot\text{K} \qquad \triangleleft$$

Consider the control volume shown in the schematic to be of total volume V. Then the conservation of mass principle yields

$$\rho_{\text{nf}} V = \rho_{\text{bf}} V(1 - \varphi) + \rho_p V \varphi$$

or, after dividing by the volume V,

$$\rho_{\text{nf}} = 997\ \text{kg/m}^3 \times (1 - 0.05) + 3970\ \text{kg/m}^3 \times 0.05 = 1146\ \text{kg/m}^3 \qquad \triangleleft$$

Similarly, the conservation of energy principle yields,

$$\rho_{\text{nf}} V c_{p,\text{nf}} T = \rho_{\text{bf}} V(1 - \varphi) c_{p,\text{bf}} T + \rho_p V \varphi c_{p,p} T$$

Dividing by the volume V, temperature T, and nanofluid density ρ_{nf} yields

$$c_{p,\text{nf}} = \frac{\rho_{\text{bf}} c_{p,\text{bf}}(1 - \varphi) + \rho_p c_{p,p} \varphi}{\rho_{\text{nf}}}$$

$$= \frac{997\ \text{kg/m}^3 \times 4.179\ \text{kJ/kg}\cdot\text{K} \times (1 - 0.05) + 3970\ \text{kg/m}^3 \times 0.765\ \text{kJ/kg}\cdot\text{K} \times (0.05)}{1146\ \text{kg/m}^3}$$

$$= 3.587\ \text{kJ/kg}\cdot\text{K} \qquad \triangleleft$$

From the problem statement, the dynamic viscosity of the nanofluid is

$$\mu_{nf} = 855 \times 10^{-6}\,\text{N}\cdot\text{s/m}^2 \times (1 + 2.5 \times 0.05) = 962 \times 10^{-6}\,\text{N}\cdot\text{s/m}^2 \qquad \triangleleft$$

The nanofluid's thermal diffusivity is

$$\alpha_{nf} = \frac{k_{nf}}{\rho_{nf} c_{p,nf}} = \frac{0.705\ \text{W/m}\cdot\text{K}}{1146\ \text{kg/m}^3 \times 3587\ \text{J/kg}\cdot\text{K}} = 171 \times 10^{-9}\ \text{m}^2/\text{s} \qquad \triangleleft$$

Comments:

1. Ratios of the properties of the nanofluid to the properties of water are as follows.

$$\frac{k_{nf}}{k_{bf}} = \frac{0.705\ \text{W/m}\cdot\text{K}}{0.613\ \text{W/m}\cdot\text{K}} = 1.150 \qquad \frac{\rho_{nf}}{\rho_{bf}} = \frac{1146\ \text{kg/m}^3}{997\ \text{kg/m}^3} = 1.149$$

$$\frac{c_{p,nf}}{c_{p,bf}} = \frac{3587\ \text{J/kg}\cdot\text{K}}{4179\ \text{J/kg}\cdot\text{K}} = 0.858 \qquad \frac{\mu_{nf}}{\mu_{bf}} = \frac{962 \times 10^{-6}\,\text{N}\cdot\text{s/m}^2}{855 \times 10^{-6}\,\text{N}\cdot\text{s/m}^2} = 1.130$$

$$\frac{\alpha_{nf}}{\alpha_{bf}} = \frac{171 \times 10^{-9}\,\text{m}^2/\text{s}}{147 \times 10^{-9}\,\text{m}^2/\text{s}} = 1.166$$

The relatively large thermal conductivity and thermal diffusivity of the nanofluid enhance heat transfer rates in some applications. However, all of the thermophysical properties are affected by the addition of the nanoparticles, and, as will become evident in Chapters 6 through 9, properties such as the viscosity and specific heat are adversely affected. This condition can degrade thermal performance when the use of nanofluids involves convection heat transfer.

2. The expression for the nanofluid's thermal conductivity (and viscosity) is limited to dilute mixtures of noncontacting, spherical particles. In some cases, the particles do not remain separated but can *agglomerate* into long chains, providing effective paths for heat conduction through the fluid and larger bulk thermal conductivities. Hence, the expression for the thermal conductivity represents the *minimum* possible enhancement of the thermal conductivity by spherical nanoparticles. An expression for the maximum possible *isotropic* thermal conductivity of a nanofluid, corresponding to agglomeration of the spherical particles, is available [21], as are expressions for dilute suspensions of nonspherical particles [22]. Note that these expressions can also be applied to nanostructured *composite materials* consisting of a particulate phase interspersed within a host binding medium, as will be discussed in more detail in Chapter 3.
3. The nanofluid's density and specific heat are determined by applying the principles of mass and energy conservation, respectively. As such, these properties do not depend on the manner in which the nanoparticles are dispersed within the base liquid.

2.3 *The Heat Diffusion Equation*

A major objective in a conduction analysis is to determine the *temperature field* in a medium resulting from conditions imposed on its boundaries. That is, we wish to know the *temperature distribution*, which represents how temperature varies with position in the medium. Once this distribution is known, the conduction heat flux at any point in the medium or on its surface may be computed from Fourier's law. Other important quantities of interest may also be determined. For a solid, knowledge of the temperature distribution could be used to ascertain structural integrity through determination of thermal stresses, expansions, and deflections. The temperature distribution could also be used to optimize the thickness of an insulating material or to determine the compatibility of special coatings or adhesives used with the material.

We now consider the manner in which the temperature distribution can be determined. The approach follows the methodology described in Section 1.3.1 of applying the energy conservation requirement. In this case, we define a *differential control volume*, identify the relevant energy transfer processes, and introduce the appropriate rate equations. The result is a differential equation whose solution, for prescribed boundary conditions, provides the temperature distribution in the medium.

Consider a homogeneous medium within which there is no bulk motion (advection) and the temperature distribution $T(x, y, z)$ is expressed in Cartesian coordinates. Following the methodology of applying conservation of energy (Section 1.3.1), we first define an infinitesimally small (differential) control volume, $dx \cdot dy \cdot dz$, as shown in Figure 2.11. Choosing to formulate the first law at an instant of time, the second step is to consider the energy processes that are relevant to this control volume. In the absence of motion (or with uniform motion), there are no changes in mechanical energy and no work being done on the system. Only thermal forms of energy need be considered. Specifically, if there are temperature gradients, conduction heat transfer will occur across each of the control surfaces. The conduction heat rates perpendicular to each of the control surfaces at the x-, y-, and z-coordinate locations are indicated by the terms q_x, q_y, and q_z, respectively. The conduction heat rates at the opposite surfaces can then be expressed as a Taylor series expansion where, neglecting higher-order terms,

$$q_{x+dx} = q_x + \frac{\partial q_x}{\partial x} dx \tag{2.13a}$$

$$q_{y+dy} = q_y + \frac{\partial q_y}{\partial y} dy \tag{2.13b}$$

$$q_{z+dz} = q_z + \frac{\partial q_z}{\partial z} dz \tag{2.13c}$$

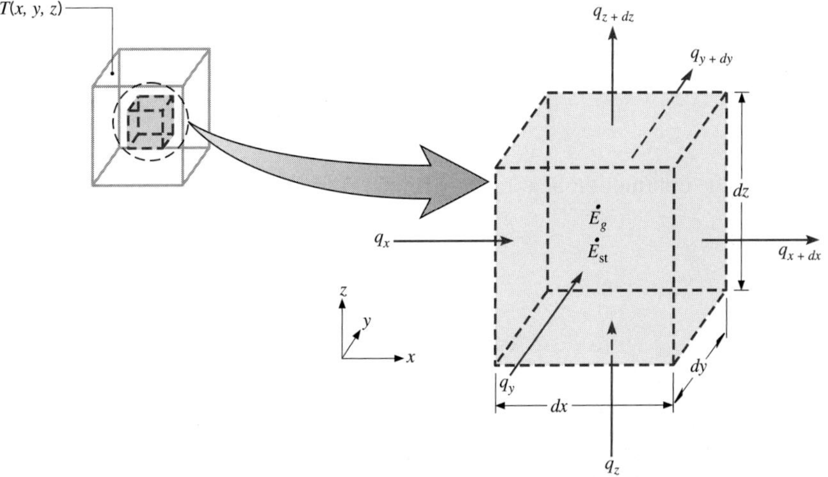

FIGURE 2.11 Differential control volume, $dx\,dy\,dz$, for conduction analysis in Cartesian coordinates.

In words, Equation 2.13a simply states that the x-component of the heat transfer rate at $x + dx$ is equal to the value of this component at x plus the amount by which it changes with respect to x times dx.

Within the medium there may also be an *energy source* term associated with the rate of thermal energy generation. This term is represented as

$$\dot{E}_g = \dot{q}\, dx\, dy\, dz \tag{2.14}$$

where $\dot{q}$ is the rate at which energy is generated per unit volume of the medium (W/m^3). In addition, changes may occur in the amount of the internal thermal energy stored by the material in the control volume. If the material is not experiencing a change in phase, latent energy effects are not pertinent, and the *energy storage* term may be expressed as

$$\dot{E}_{st} = \rho c_p \frac{\partial T}{\partial t}\, dx\, dy\, dz \tag{2.15}$$

where $\rho c_p\, \partial T/\partial t$ is the time rate of change of the sensible (thermal) energy of the medium per unit volume.

Once again it is important to note that the terms $\dot{E}_g$ and $\dot{E}_{st}$ represent different physical processes. The energy generation term $\dot{E}_g$ is a manifestation of some energy conversion process involving thermal energy on one hand and some other form of energy, such as chemical, electrical, or nuclear, on the other. The term is positive (a *source*) if thermal energy is being generated in the material at the expense of some other energy form; it is negative (a *sink*) if thermal energy is being consumed. In contrast, the energy storage term $\dot{E}_{st}$ refers to the rate of change of thermal energy stored by the matter.

The last step in the methodology outlined in Section 1.3.1 is to express conservation of energy using the foregoing rate equations. On a *rate* basis, the general form of the conservation of energy requirement is

$$\dot{E}_{in} + \dot{E}_g - \dot{E}_{out} = \dot{E}_{st} \tag{1.12c}$$

Hence, recognizing that the conduction rates constitute the energy inflow $\dot{E}_{in}$ and outflow $\dot{E}_{out}$, and substituting Equations 2.14 and 2.15, we obtain

$$q_x + q_y + q_z + \dot{q}\, dx\, dy\, dz - q_{x+dx} - q_{y+dy} - q_{z+dz} = \rho c_p \frac{\partial T}{\partial t}\, dx\, dy\, dz \tag{2.16}$$

Substituting from Equations 2.13, it follows that

$$-\frac{\partial q_x}{\partial x}\, dx - \frac{\partial q_y}{\partial y}\, dy - \frac{\partial q_z}{\partial z}\, dz + \dot{q}\, dx\, dy\, dz = \rho c_p \frac{\partial T}{\partial t}\, dx\, dy\, dz \tag{2.17}$$

The conduction heat rates in an isotropic material may be evaluated from Fourier's law,

$$q_x = -k\, dy\, dz \frac{\partial T}{\partial x} \tag{2.18a}$$

$$q_y = -k\, dx\, dz \frac{\partial T}{\partial y} \tag{2.18b}$$

$$q_z = -k\, dx\, dy \frac{\partial T}{\partial z} \tag{2.18c}$$

where each heat flux component of Equation 2.6 has been multiplied by the appropriate control surface (differential) area to obtain the heat transfer rate. Substituting

Equations 2.18 into Equation 2.17 and dividing out the dimensions of the control volume ($dx\, dy\, dz$), we obtain

$$\frac{\partial}{\partial x}\left(k\frac{\partial T}{\partial x}\right)+\frac{\partial}{\partial y}\left(k\frac{\partial T}{\partial y}\right)+\frac{\partial}{\partial z}\left(k\frac{\partial T}{\partial z}\right)+\dot{q}=\rho c_p\frac{\partial T}{\partial t} \tag{2.19}$$

Equation 2.19 is the general form, in Cartesian coordinates, of the *heat diffusion equation.* This equation, often referred to as the *heat equation*, provides the basic tool for heat conduction analysis. From its solution, we can obtain the temperature distribution $T(x, y, z)$ as a function of time. The apparent complexity of this expression should not obscure the fact that it describes an important physical condition, that is, conservation of energy. You should have a clear understanding of the physical significance of each term appearing in the equation. For example, the term $\partial(k\partial T/\partial x)/\partial x$ is related to the *net* conduction heat flux *into* the control volume for the x-coordinate direction. That is, multiplying by dx,

$$\frac{\partial}{\partial x}\left(k\frac{\partial T}{\partial x}\right)dx = q''_x - q''_{x+dx} \tag{2.20}$$

with similar expressions applying for the fluxes in the y- and z-directions. In words, the heat equation, Equation 2.19, therefore states that *at any point in the medium the net rate of energy transfer by conduction into a unit volume plus the volumetric rate of thermal energy generation must equal the rate of change of thermal energy stored within the volume.*

It is often possible to work with simplified versions of Equation 2.19. For example, if the thermal conductivity is constant, the heat equation is

$$\frac{\partial^2 T}{\partial x^2}+\frac{\partial^2 T}{\partial y^2}+\frac{\partial^2 T}{\partial z^2}+\frac{\dot{q}}{k}=\frac{1}{\alpha}\frac{\partial T}{\partial t} \tag{2.21}$$

where $\alpha = k/\rho c_p$ is the *thermal diffusivity.* Additional simplifications of the general form of the heat equation are often possible. For example, under *steady-state* conditions, there can be no change in the amount of energy storage; hence Equation 2.19 reduces to

$$\frac{\partial}{\partial x}\left(k\frac{\partial T}{\partial x}\right)+\frac{\partial}{\partial y}\left(k\frac{\partial T}{\partial y}\right)+\frac{\partial}{\partial z}\left(k\frac{\partial T}{\partial z}\right)+\dot{q}=0 \tag{2.22}$$

Moreover, if the heat transfer is *one-dimensional* (e.g., in the x-direction) and there is *no energy generation*, Equation 2.22 reduces to

$$\frac{d}{dx}\left(k\frac{dT}{dx}\right)=0 \tag{2.23}$$

The important implication of this result is that, *under steady-state, one-dimensional conditions with no energy generation*, the heat flux is a constant in the direction of transfer ($dq''_x/dx = 0$).

The heat equation may also be expressed in cylindrical and spherical coordinates. The differential control volumes for these two coordinate systems are shown in Figures 2.12 and 2.13.

$\frac{d^2T}{dx^2} = -\dot{q}/k$ → 1 dim steady state

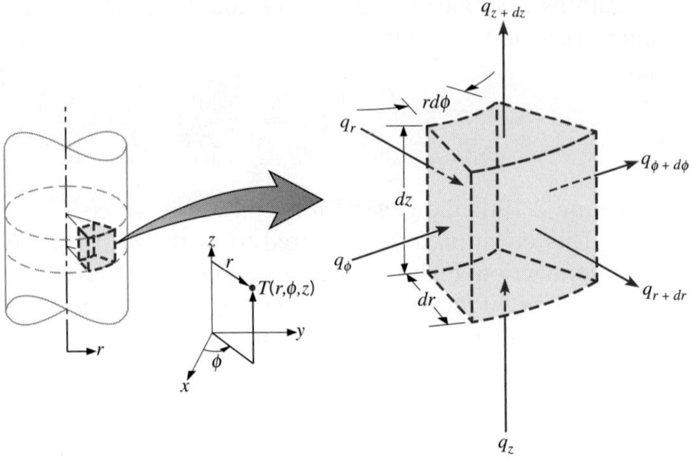

FIGURE 2.12 Differential control volume, $dr \cdot r\,d\phi \cdot dz$, for conduction analysis in cylindrical coordinates (r, ϕ, z).

Cylindrical Coordinates When the del operator ∇ of Equation 2.3 is expressed in cylindrical coordinates, the general form of the heat flux vector and hence of Fourier's law is

$$\mathbf{q}'' = -k\nabla T = -k\left(\mathbf{i}\frac{\partial T}{\partial r} + \mathbf{j}\frac{1}{r}\frac{\partial T}{\partial \phi} + \mathbf{k}\frac{\partial T}{\partial z}\right) \tag{2.24}$$

where

$$q''_r = -k\frac{\partial T}{\partial r} \qquad q''_\phi = -\frac{k}{r}\frac{\partial T}{\partial \phi} \qquad q''_z = -k\frac{\partial T}{\partial z} \tag{2.25}$$

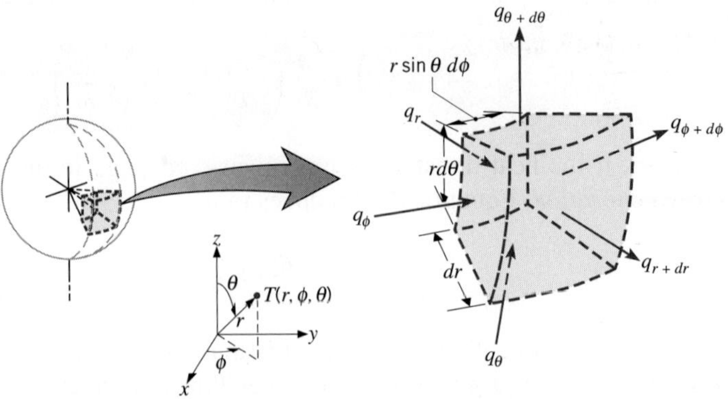

FIGURE 2.13 Differential control volume, $dr \cdot r\sin\theta\,d\phi \cdot r\,d\theta$, for conduction analysis in spherical coordinates (r, ϕ, θ).

are heat flux components in the radial, circumferential, and axial directions, respectively. Applying an energy balance to the differential control volume of Figure 2.12, the following general form of the heat equation is obtained:

$$\frac{1}{r}\frac{\partial}{\partial r}\left(kr\frac{\partial T}{\partial r}\right)+\frac{1}{r^2}\frac{\partial}{\partial \phi}\left(k\frac{\partial T}{\partial \phi}\right)+\frac{\partial}{\partial z}\left(k\frac{\partial T}{\partial z}\right)+\dot{q}=\rho c_p\frac{\partial T}{\partial t} \tag{2.26}$$

Spherical Coordinates In spherical coordinates, the general form of the heat flux vector and Fourier's law is

$$\boldsymbol{q}''=-k\nabla T=-k\left(\boldsymbol{i}\frac{\partial T}{\partial r}+\boldsymbol{j}\frac{1}{r}\frac{\partial T}{\partial \theta}+\boldsymbol{k}\frac{1}{r\sin\theta}\frac{\partial T}{\partial \phi}\right) \tag{2.27}$$

where

$$q''_r=-k\frac{\partial T}{\partial r}\qquad q''_\theta=-\frac{k}{r}\frac{\partial T}{\partial \theta}\qquad q''_\phi=-\frac{k}{r\sin\theta}\frac{\partial T}{\partial \phi} \tag{2.28}$$

are heat flux components in the radial, polar, and azimuthal directions, respectively. Applying an energy balance to the differential control volume of Figure 2.13, the following general form of the heat equation is obtained:

$$\frac{1}{r^2}\frac{\partial}{\partial r}\left(kr^2\frac{\partial T}{\partial r}\right)+\frac{1}{r^2\sin^2\theta}\frac{\partial}{\partial \phi}\left(k\frac{\partial T}{\partial \phi}\right)+\frac{1}{r^2\sin\theta}\frac{\partial}{\partial \theta}\left(k\sin\theta\frac{\partial T}{\partial \theta}\right)+\dot{q}=\rho c_p\frac{\partial T}{\partial t} \tag{2.29}$$

You should attempt to derive Equation 2.26 or 2.29 to gain experience in applying conservation principles to differential control volumes (see Problems 2.35 and 2.36). Note that the temperature gradient in Fourier's law must have units of K/m. Hence, when evaluating the gradient for an angular coordinate, it must be expressed in terms of the differential change in arc *length*. For example, the heat flux component in the circumferential direction of a cylindrical coordinate system is $q''_\phi=-(k/r)(\partial T/\partial \phi)$, *not* $q''_\phi=-k(\partial T/\partial \phi)$.

EXAMPLE 2.3

The temperature distribution across a wall 1 m thick at a certain instant of time is given as

$$T(x)=a+bx+cx^2$$

where T is in degrees Celsius and x is in meters, while $a=900°\text{C}$, $b=-300°\text{C/m}$, and $c=-50°\text{C/m}^2$. A uniform heat generation, $\dot{q}=1000\ \text{W/m}^3$, is present in the wall of area $10\ \text{m}^2$ having the properties $\rho=1600\ \text{kg/m}^3$, $k=40\ \text{W/m}\cdot\text{K}$, and $c_p=4\ \text{kJ/kg}\cdot\text{K}$.

1. Determine the rate of heat transfer entering the wall ($x = 0$) and leaving the wall ($x = 1$ m).
2. Determine the rate of change of energy storage in the wall.
3. Determine the time rate of temperature change at $x = 0$, 0.25, and 0.5 m.

SOLUTION

Known: Temperature distribution $T(x)$ at an instant of time t in a one-dimensional wall with uniform heat generation.

Find:

1. Heat rates entering, q_{in} ($x = 0$), and leaving, q_{out} ($x = 1$ m), the wall.
2. Rate of change of energy storage in the wall, $\dot{E}_{st}$.
3. Time rate of temperature change at $x = 0$, 0.25, and 0.5 m.

Schematic:

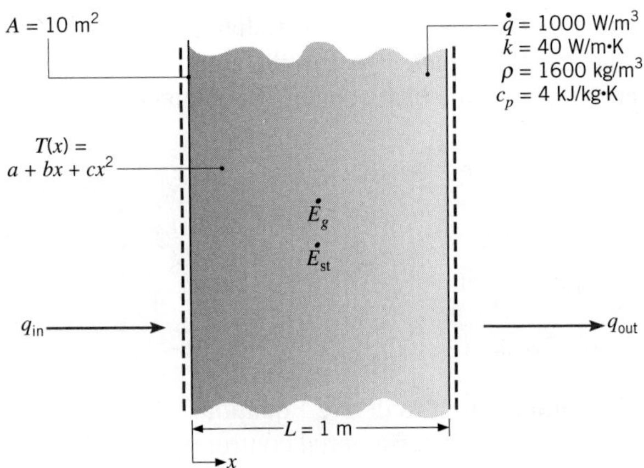

Assumptions:

1. One-dimensional conduction in the x-direction.
2. Isotropic medium with constant properties.
3. Uniform internal heat generation, $\dot{q}$ (W/m^3).

Analysis:

1. Recall that once the temperature distribution is known for a medium, it is a simple matter to determine the conduction heat transfer rate at any point in the medium or at its surfaces by using Fourier's law. Hence the desired heat rates may be determined by using the prescribed temperature distribution with Equation 2.1. Accordingly,

$$q_{in} = q_x(0) = -kA\left.\frac{\partial T}{\partial x}\right|_{x=0} = -kA(b + 2cx)_{x=0}$$

$$q_{in} = -bkA = 300°\text{C/m} \times 40\ \text{W/m}\cdot\text{K} \times 10\ \text{m}^2 = 120\ \text{kW} \qquad \triangleleft$$

Similarly,

$$q_{\text{out}} = q_x(L) = -kA\frac{\partial T}{\partial x}\bigg|_{x=L} = -kA(b + 2cx)_{x=L}$$

$$q_{\text{out}} = -(b + 2cL)kA = -[-300°\text{C/m}$$

$$+\,2(-50°\text{C/m}^2) \times 1\ \text{m}] \times 40\ \text{W/m}\cdot\text{K} \times 10\ \text{m}^2 = 160\ \text{kW} \qquad \triangleleft$$

2. The rate of change of energy storage in the wall $\dot{E}_{st}$ may be determined by applying an overall energy balance to the wall. Using Equation 1.12c for a control volume about the wall,

$$\dot{E}_{\text{in}} + \dot{E}_g - \dot{E}_{\text{out}} = \dot{E}_{\text{st}}$$

where $\dot{E}_g = \dot{q}AL$, it follows that

$$\dot{E}_{\text{st}} = \dot{E}_{\text{in}} + \dot{E}_g - \dot{E}_{\text{out}} = q_{\text{in}} + \dot{q}AL - q_{\text{out}}$$

$$\dot{E}_{\text{st}} = 120\ \text{kW} + 1000\ \text{W/m}^3 \times 10\ \text{m}^2 \times 1\ \text{m} - 160\ \text{kW}$$

$$\dot{E}_{\text{st}} = -30\ \text{kW} \qquad \triangleleft$$

3. The time rate of change of the temperature at any point in the medium may be determined from the heat equation, Equation 2.21, rewritten as

$$\frac{\partial T}{\partial t} = \frac{k}{\rho c_p}\frac{\partial^2 T}{\partial x^2} + \frac{\dot{q}}{\rho c_p}$$

From the prescribed temperature distribution, it follows that

$$\frac{\partial^2 T}{\partial x^2} = \frac{\partial}{\partial x}\left(\frac{\partial T}{\partial x}\right)$$

$$= \frac{\partial}{\partial x}(b + 2cx) = 2c = 2(-50°\text{C/m}^2) = -100°\text{C/m}^2$$

Note that this derivative is independent of position in the medium. Hence the time rate of temperature change is also independent of position and is given by

$$\frac{\partial T}{\partial t} = \frac{40\ \text{W/m}\cdot\text{K}}{1600\ \text{kg/m}^3 \times 4\ \text{kJ/kg}\cdot\text{K}} \times (-100°\text{C/m}^2)$$

$$+\,\frac{1000\ \text{W/m}^3}{1600\ \text{kg/m}^3 \times 4\ \text{kJ/kg}\cdot\text{K}}$$

$$\frac{\partial T}{\partial t} = -6.25 \times 10^{-4}°\text{C/s} + 1.56 \times 10^{-4}°\text{C/s}$$

$$= -4.69 \times 10^{-4}°\text{C/s} \qquad \triangleleft$$

Comments:

1. From this result, it is evident that the temperature at every point within the wall is decreasing with time.
2. Fourier's law can always be used to compute the conduction heat rate from knowledge of the temperature distribution, even for unsteady conditions with internal heat generation.

Microscale Effects For most practical situations, the heat diffusion equations generated in this text may be used with confidence. However, these equations are based on Fourier's law, which does not account for the finite speed at which thermal information is propagated within the medium by the various energy carriers. The consequences of the finite propagation speed may be neglected if the heat transfer events of interest occur over a sufficiently long time scale, Δt, such that

$$\frac{\lambda_{\text{mfp}}}{\bar{c}\Delta t} \ll 1 \tag{2.30}$$

The heat diffusion equations of this text are likewise invalid for problems where boundary scattering must be explicitly considered. For example, the temperature distribution *within* the thin film of Figure 2.6*b* cannot be determined by applying the foregoing heat diffusion equations. Additional discussion of micro- and nanoscale heat transfer applications and analysis methods is available in the literature [1, 5, 10, 23].

2.4 Boundary and Initial Conditions

To determine the temperature distribution in a medium, it is necessary to solve the appropriate form of the heat equation. However, such a solution depends on the physical conditions existing at the *boundaries* of the medium and, if the situation is time dependent, on conditions existing in the medium at some *initial time.* With regard to the *boundary conditions,* there are several common possibilities that are simply expressed in mathematical form. Because the heat equation is second order in the spatial coordinates, two boundary conditions must be expressed for each coordinate needed to describe the system. Because the equation is first order in time, however, only one condition, termed the *initial condition*, must be specified.

Three kinds of boundary conditions commonly encountered in heat transfer are summarized in Table 2.2. The conditions are specified at the surface $x = 0$ for a one-dimensional system. Heat transfer is in the positive x-direction with the temperature distribution, which may be time dependent, designated as $T(x, t)$. The first condition corresponds to a situation for which the surface is maintained at a fixed temperature T_s. It is commonly termed a *Dirichlet condition*, or a boundary condition of the *first kind.* It is closely approximated, for example, when the surface is in contact with a melting solid or a boiling liquid. In both cases, there is heat transfer at the surface, while the surface remains at the temperature of the phase change process. The second condition corresponds to the existence of a fixed or constant heat flux q''_s at the surface. This heat flux is related to the temperature gradient at the surface by Fourier's law, Equation 2.6, which may be expressed as

$$q''_x(0) = -k\frac{\partial T}{\partial x}\bigg|_{x=0} = q''_s$$

TABLE 2.2 Boundary conditions for the heat diffusion equation at the surface ($x = 0$)

1. Constant surface temperature

$$T(0, t) = T_s \qquad (2.31)$$

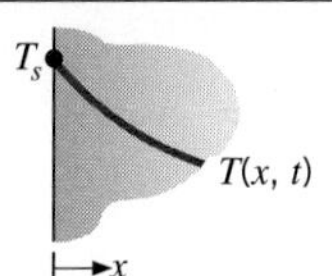

2. Constant surface heat flux

   (a) Finite heat flux

$$-k \frac{\partial T}{\partial x}\bigg|_{x=0} = q_s'' \qquad (2.32)$$

q_s'' — $T(x, t)$ — x

   (b) Adiabatic or insulated surface

$$\frac{\partial T}{\partial x}\bigg|_{x=0} = 0 \qquad (2.33)$$

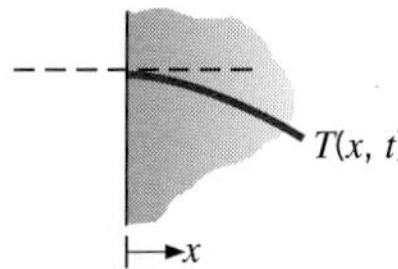

3. Convection surface condition

$$-k \frac{\partial T}{\partial x}\bigg|_{x=0} = h[T_\infty - T(0, t)] \qquad (2.34)$$

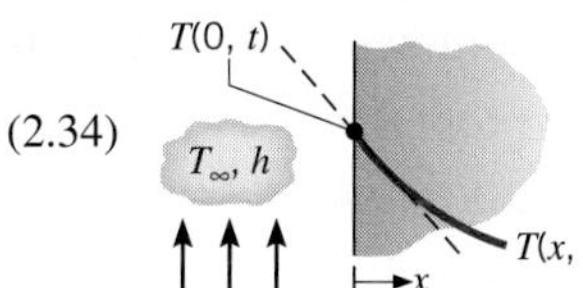

It is termed a *Neumann condition*, or a boundary condition of the *second kind*, and may be realized by bonding a thin film electric heater to the surface. A special case of this condition corresponds to the *perfectly insulated*, or *adiabatic*, surface for which $\partial T/\partial x|_{x=0} = 0$. The boundary condition of the *third kind* corresponds to the existence of convection heating (or cooling) at the surface and is obtained from the surface energy balance discussed in Section 1.3.1.

EXAMPLE 2.4

A long copper bar of rectangular cross section, whose width w is much greater than its thickness L, is maintained in contact with a heat sink at its lower surface, and the temperature throughout the bar is approximately equal to that of the sink, T_o. Suddenly, an electric current is passed through the bar and an airstream of temperature T_∞ is passed over the top surface, while the bottom surface continues to be maintained at T_o. Obtain the differential equation and the boundary and initial conditions that could be solved to determine the temperature as a function of position and time in the bar.

SOLUTION

Known: Copper bar initially in thermal equilibrium with a heat sink is suddenly heated by passage of an electric current.

Find: Differential equation and boundary and initial conditions needed to determine temperature as a function of position and time within the bar.

Schematic:

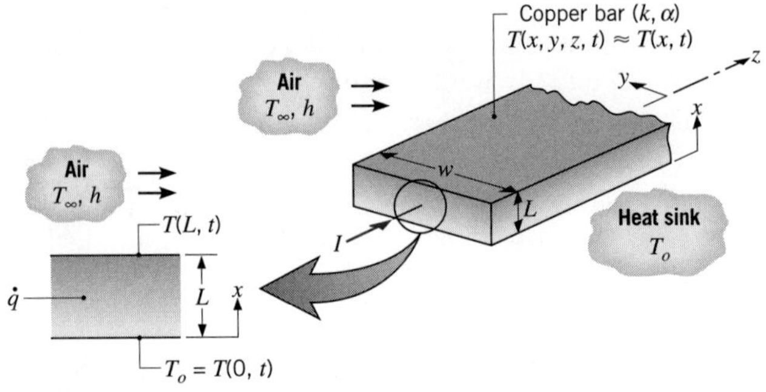

Assumptions:

1. Since the bar is long and $w \gg L$, end and side effects are negligible and heat transfer within the bar is primarily one dimensional in the x-direction.
2. Uniform volumetric heat generation, $\dot{q}$.
3. Constant properties.

Analysis: The temperature distribution is governed by the heat equation (Equation 2.19), which, for the one-dimensional and constant property conditions of the present problem, reduces to

$$\frac{\partial^2 T}{\partial x^2} + \frac{\dot{q}}{k} = \frac{1}{\alpha}\frac{\partial T}{\partial t} \qquad (1) \quad \triangleleft$$

where the temperature is a function of position and time, $T(x, t)$. Since this differential equation is second order in the spatial coordinate x and first order in time t, there must be two boundary conditions for the x-direction and one condition, termed the initial condition, for time. The boundary condition at the bottom surface corresponds to case 1 of Table 2.2. In particular, since the temperature of this surface is maintained at a value, T_o, which is fixed with time, it follows that

$$T(0, t) = T_o \qquad (2) \quad \triangleleft$$

The convection surface condition, case 3 of Table 2.2, is appropriate for the top surface. Hence

$$-k\frac{\partial T}{\partial x}\bigg|_{x=L} = h[T(L, t) - T_\infty] \qquad (3) \quad \triangleleft$$

The initial condition is inferred from recognition that, before the change in conditions, the bar is at a uniform temperature T_o. Hence

$$T(x, 0) = T_o \qquad (4) \quad \triangleleft$$

If T_o, T_∞, $\dot{q}$, and h are known, Equations 1 through 4 may be solved to obtain the time-varying temperature distribution $T(x, t)$ following imposition of the electric current.

Comments:

1. The *heat sink* at $x = 0$ could be maintained by exposing the surface to an ice bath or by attaching it to a *cold plate.* A cold plate contains coolant channels machined in a solid of large thermal conductivity (usually copper). By circulating a liquid (usually water) through the channels, the plate and hence the surface to which it is attached may be maintained at a nearly uniform temperature.
2. The temperature of the top surface $T(L, t)$ will change with time. This temperature is an unknown and may be obtained after finding $T(x, t)$.
3. We may use our physical intuition to sketch temperature distributions in the bar at selected times from the beginning to the end of the transient process. If we assume that $T_\infty > T_o$ and that the electric current is sufficiently large to heat the bar to temperatures in excess of T_∞, the following distributions would correspond to the initial condition ($t \leq 0$), the final (steady-state) condition ($t \to \infty$), and two intermediate times.

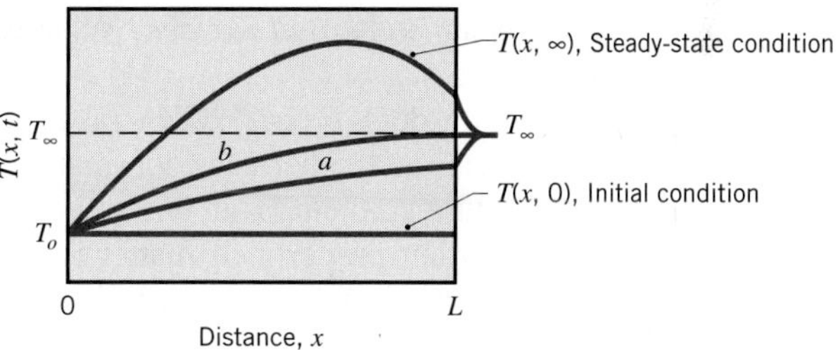

Note how the distributions comply with the initial and boundary conditions. What is a special feature of the distribution labeled (b)?

4. Our intuition may also be used to infer the manner in which the heat flux varies with time at the surfaces ($x = 0, L$) of the bar. On $q''_x - t$ coordinates, the transient variations are as follows.

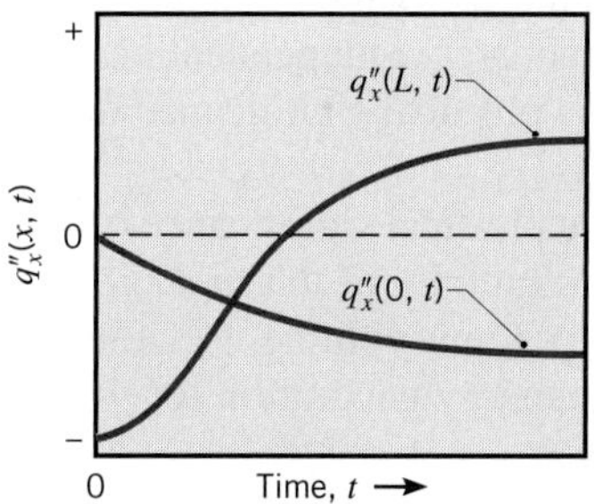

Convince yourself that the foregoing variations are consistent with the temperature distributions of Comment 3. For $t \to \infty$, how are $q''_x(0)$ and $q''_x(L)$ related to the volumetric rate of energy generation?

2.5 Summary

Despite the relative brevity of this chapter, its importance must not be underestimated. Understanding the conduction rate equation, Fourier's law, is essential. You must be cognizant of the importance of thermophysical properties; over time, you will develop a sense of the magnitudes of the properties of many real materials. Likewise, you must recognize that the heat equation is derived by applying the conservation of energy principle to a differential control volume and that it is used to determine temperature distributions within matter. From knowledge of the distribution, Fourier's law can be used to determine the corresponding conduction heat rates. A firm grasp of the various types of thermal boundary conditions that are used in conjunction with the heat equation is vital. Indeed, Chapter 2 is the foundation on which Chapters 3 through 5 are based, and you are encouraged to revisit this chapter often. You may test your understanding of various concepts by addressing the following questions.

- In the general formulation of *Fourier's law* (applicable to any geometry), what are the vector and scalar quantities? Why is there a minus sign on the right-hand side of the equation?
- What is an *isothermal surface*? What can be said about the heat flux at any location on this surface?
- What form does *Fourier's law* take for each of the orthogonal directions of Cartesian, cylindrical, and spherical coordinate systems? In each case, what are the units of the temperature gradient? Can you write each equation from memory?
- An important property of matter is defined by *Fourier's law*. What is it? What is its physical significance? What are its units?
- What is an *isotropic* material?
- Why is the thermal conductivity of a solid generally larger than that of a liquid? Why is the thermal conductivity of a liquid larger than that of a gas?
- Why is the thermal conductivity of an electrically conducting solid generally larger than that of a nonconductor? Why are materials such as beryllium oxide, diamond, and silicon carbide (see Table A.2) exceptions to this rule?
- Is the *effective thermal conductivity* of an insulation system a true manifestation of the efficacy with which heat is transferred through the system by conduction alone?
- Why does the thermal conductivity of a gas increase with increasing temperature? Why is it approximately independent of pressure?
- What is the physical significance of the *thermal diffusivity*? How is it defined and what are its units?
- What is the physical significance of each term appearing in the *heat equation*?
- Cite some examples of *thermal energy generation*. If the rate at which thermal energy is generated per unit volume, $\dot{q}$, varies with location in a medium of volume V, how can the rate of energy generation for the entire medium, $\dot{E}_g$, be determined from knowledge of $\dot{q}(x, y, z)$?
- For a chemically reacting medium, what kind of reaction provides a *source* of thermal energy ($\dot{q} > 0$)? What kind of reaction provides a *sink* for thermal energy ($\dot{q} < 0$)?
- To solve the *heat equation* for the temperature distribution in a medium, *boundary conditions* must be prescribed at the surfaces of the medium. What physical conditions are commonly suitable for this purpose?

References

1. Flik, M. I., B.-I. Choi, and K. E. Goodson, *J. Heat Transfer*, **114**, 666, 1992.
2. Klemens, P. G., "Theory of the Thermal Conductivity of Solids," in R. P. Tye, Ed., *Thermal Conductivity*, Vol. 1, Academic Press, London, 1969.
3. Yang, H.-S., G.-R. Bai, L. J. Thompson, and J. A. Eastman, *Acta Materialia*, **50**, 2309, 2002.
4. Chen, G., *J. Heat Transfer*, **118**, 539, 1996.
5. Carey, V. P., G. Chen, C. Grigoropoulos, M. Kaviany, and A. Majumdar, *Nano. and Micro. Thermophys. Engng*. **12**, 1, 2008.
6. Padture, N. P., M. Gell, and E. H. Jordan, *Science*, **296**, 280, 2002.
7. Schelling, P. K., L. Shi, and K. E. Goodson, *Mat. Today*, **8**, 30, 2005.
8. Baxter, J., Z. Bian, G. Chen, D. Danielson, M. S. Dresselhaus, A. G. Federov, T. S. Fisher, C. W. Jones, E. Maginn, W. Kortshagen, A. Manthiram, A. Nozik, D. R. Rolison, T. Sands, L. Shi, D. Sholl, and Y. Wu, *Energy and Environ. Sci.*, **2**, 559, 2009.
9. Vincenti, W. G., and C. H. Kruger Jr., *Introduction to Physical Gas Dynamics*, Wiley, New York, 1986.
10. Zhang, Z. M., *Nano/Microscale Heat Transfer*, McGraw-Hill, New York, 2007.
11. McLaughlin, E., "Theory of the Thermal Conductivity of Fluids," in R. P. Tye, Ed., *Thermal Conductivity*, Vol. 2, Academic Press, London, 1969.
12. Foust, O. J., Ed., "Sodium Chemistry and Physical Properties," in *Sodium-NaK Engineering Handbook*, Vol. 1, Gordon & Breach, New York, 1972.
13. Mallory, J. F., *Thermal Insulation*, Reinhold Book Corp., New York, 1969.
14. American Society of Heating, Refrigeration and Air Conditioning Engineers, *Handbook of Fundamentals*, Chapters 23–25 and 31, ASHRAE, New York, 2001.
15. Zeng, S. Q., A. Hunt, and R. Greif, *J. Heat Transfer*, **117**, 1055, 1995.
16. Sengers, J. V., and M. Klein, Eds., *The Technical Importance of Accurate Thermophysical Property Information*, National Bureau of Standards Technical Note No. 590, 1980.
17. Najjar, M. S., K. J. Bell, and R. N. Maddox, *Heat Transfer Eng.*, **2**, 27, 1981.
18. Hanley, H. J. M., and M. E. Baltatu, *Mech. Eng.*, **105**, 68, 1983.
19. Touloukian, Y. S., and C. Y. Ho, Eds., *Thermophysical Properties of Matter, The TPRC Data Series* (13 volumes on thermophysical properties: thermal conductivity, specific heat, thermal radiative, thermal diffusivity, and thermal linear expansion), Plenum Press, New York, 1970 through 1977.
20. Chow, T. S., *Phys. Rev. E*, **48**, 1977, 1993.
21. Keblinski, P., R. Prasher, and J. Eapen, *J. Nanopart. Res.*, **10**, 1089, 2008.
22. Hamilton, R. L., and O. K. Crosser, *I&EC Fundam.* **1**, 187, 1962.
23. Cahill, D. G., W. K. Ford, K. E. Goodson, G. D. Mahan, A. Majumdar, H. J. Maris, R. Merlin, and S. R. Phillpot, *App. Phys. Rev.*, **93**, 793, 2003.

Problems

Fourier's Law

2.1 Assume steady-state, one-dimensional heat conduction through the axisymmetric shape shown below.

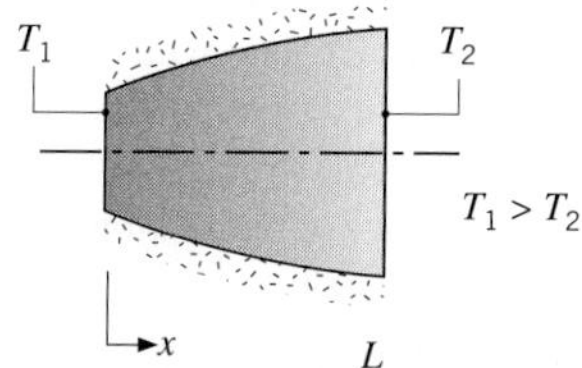

Assuming constant properties and no internal heat generation, sketch the temperature distribution on $T - x$ coordinates. Briefly explain the shape of your curve.

2.2 Assume steady-state, one-dimensional conduction in the axisymmetric object below, which is insulated around its perimeter.

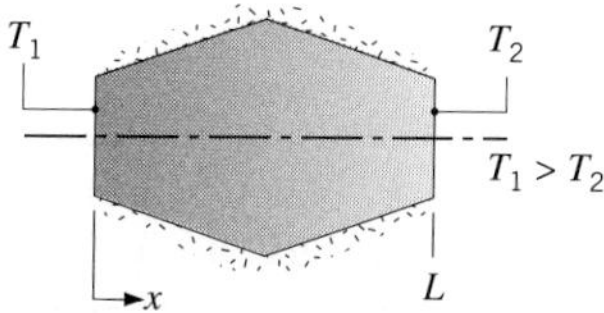

If the properties remain constant and no internal heat generation occurs, sketch the heat flux distribution, $q''_x(x)$, and the temperature distribution, $T(x)$. Explain the shapes of your curves. How do your curves depend on the thermal conductivity of the material?

2.3 A hot water pipe with outside radius r_1 has a temperature T_1. A thick insulation, applied to reduce the heat loss, has an outer radius r_2 and temperature T_2. On $T - r$ coordinates, sketch the temperature distribution in the insulation for one-dimensional, steady-state heat transfer with constant properties. Give a brief explanation, justifying the shape of your curve.

2.4 A spherical shell with inner radius r_1 and outer radius r_2 has surface temperatures T_1 and T_2, respectively, where $T_1 > T_2$. Sketch the temperature distribution on $T - r$ coordinates assuming steady-state, one-dimensional conduction with constant properties. Briefly justify the shape of your curve.

2.5 Assume steady-state, one-dimensional heat conduction through the symmetric shape shown.

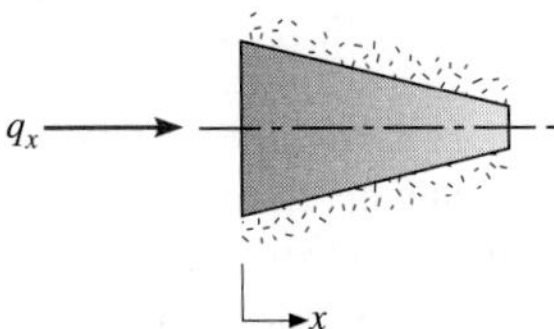

Assuming that there is no internal heat generation, derive an expression for the thermal conductivity $k(x)$ for these conditions: $A(x) = (1 - x)$, $T(x) = 300(1 - 2x - x^3)$, and $q = 6000$ W, where A is in square meters, T in kelvins, and x in meters.

2.6 A composite rod consists of two different materials, A and B, each of length $0.5L$.

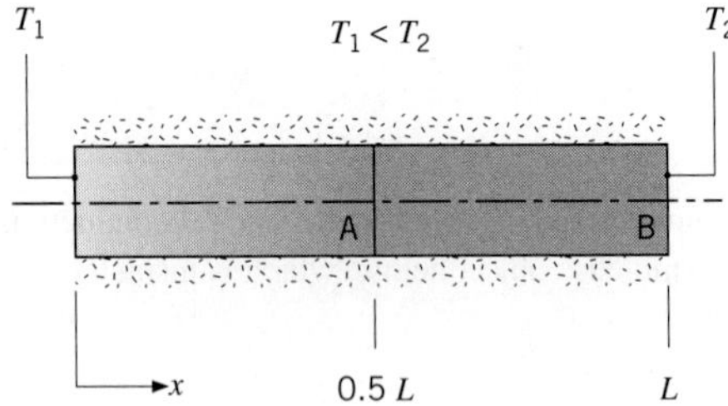

The thermal conductivity of Material A is half that of Material B, that is, $k_A/k_B = 0.5$. Sketch the steady-state temperature and heat flux distributions, $T(x)$ and q''_x, respectively. Assume constant properties and no internal heat generation in either material.

2.7 A solid, truncated cone serves as a support for a system that maintains the top (truncated) face of the cone at a temperature T_1, while the base of the cone is at a temperature $T_2 < T_1$.

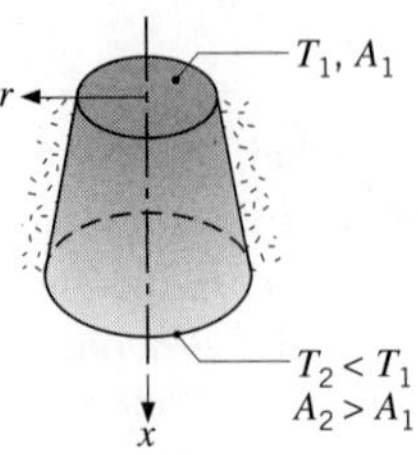

The thermal conductivity of the solid depends on temperature according to the relation $k = k_0 - aT$, where a is a positive constant, and the sides of the cone are well insulated. Do the following quantities increase, decrease, or remain the same with increasing x: the heat transfer rate q_x, the heat flux q''_x, the thermal conductivity k, and the temperature gradient dT/dx?

2.8 To determine the effect of the temperature dependence of the thermal conductivity on the temperature distribution in a solid, consider a material for which this dependence may be represented as

$$k = k_o + aT$$

where k_o is a positive constant and a is a coefficient that may be positive or negative. Sketch the steady-state temperature distribution associated with heat transfer in a plane wall for three cases corresponding to $a > 0$, $a = 0$, and $a < 0$.

2.9 A young engineer is asked to design a thermal protection barrier for a sensitive electronic device that might be exposed to irradiation from a high-powered infrared laser. Having learned as a student that a low thermal conductivity material provides good insulating characteristics, the engineer specifies use of a nanostructured aerogel, characterized by a thermal conductivity of $k_a = 0.005$ W/m·K, for the protective barrier. The engineer's boss questions the wisdom of selecting the aerogel *because* it has a low thermal conductivity. Consider the sudden laser irradiation of (a) pure aluminum, (b) glass, and (c) aerogel. The laser provides irradiation of $G = 10 \times 10^6$ W/m^2. The absorptivities of the materials are $\alpha = 0.2$, 0.9, and 0.8 for the aluminum, glass, and aerogel, respectively, and the initial temperature of the barrier is $T_i = 300$ K. Explain why the boss is concerned. *Hint:* All materials experience thermal expansion (or contraction), and local stresses that develop within a material are, to a first approximation, proportional to the local temperature gradient.

2.10 A one-dimensional plane wall of thickness $2L = 100$ mm experiences uniform thermal energy generation of $\dot{q} = 1000$ W/m^3 and is convectively cooled at $x = \pm 50$ mm by an ambient fluid characterized by $T_\infty = 20°$C. If the steady-state temperature distribution within the wall is $T(x) = a(L^2 - x^2) + b$ where $a = 10°C/m^2$ and $b = 30°C$, what is the thermal conductivity of the wall? What is the value of the convection heat transfer coefficient, h?

2.11 Consider steady-state conditions for one-dimensional conduction in a plane wall having a thermal conductivity $k = 50\,W/m \cdot K$ and a thickness $L = 0.25\,m$, with no internal heat generation.

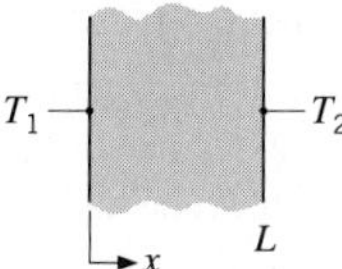

Determine the heat flux and the unknown quantity for each case and sketch the temperature distribution, indicating the direction of the heat flux.

Case	T_1(°C)	T_2(°C)	dT/dx (K/m)
1	50	−20	
2	−30	−10	
3	70		160
4		40	−80
5		30	200

2.12 Consider a plane wall 100 mm thick and of thermal conductivity $100\,W/m \cdot K$. Steady-state conditions are known to exist with $T_1 = 400\,K$ and $T_2 = 600\,K$. Determine the heat flux q''_x and the temperature gradient dT/dx for the coordinate systems shown.

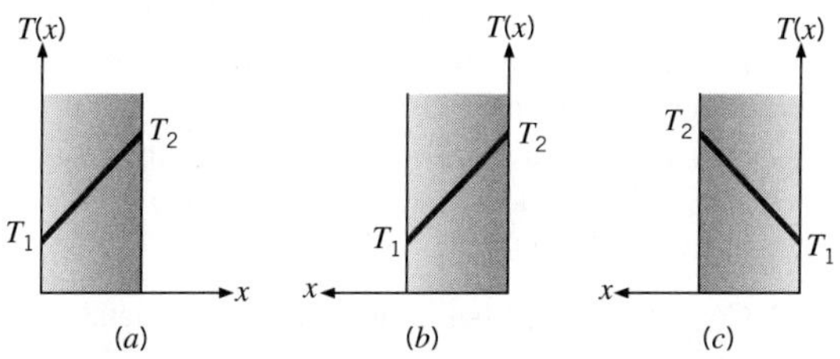

2.13 A cylinder of radius r_o, length L, and thermal conductivity k is immersed in a fluid of convection coefficient h and unknown temperature T_∞. At a certain instant the temperature distribution in the cylinder is $T(r) = a + br^2$, where a and b are constants. Obtain expressions for the heat transfer rate at r_o and the fluid temperature.

2.14 In the two-dimensional body illustrated, the gradient at surface A is found to be $\partial T/\partial y = 30\,K/m$. What are $\partial T/\partial y$ and $\partial T/\partial x$ at surface B?

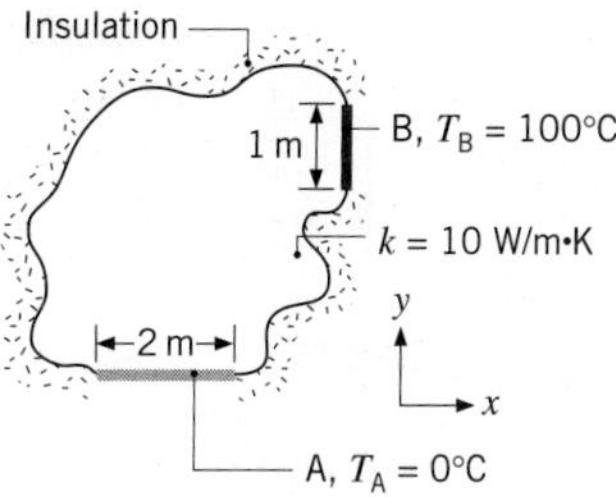

2.15 Consider the geometry of Problem 2.14 for the case where the thermal conductivity varies with temperature as $k = k_o + aT$, where $k_o = 10\,W/m \cdot K$, $a = -10^{-3}\,W/m \cdot K^2$, and T is in kelvins. The gradient at surface B is $\partial T/\partial x = 30\,K/m$. What is $\partial T/\partial y$ at surface A?

2.16 Steady-state, one-dimensional conduction occurs in a rod of constant thermal conductivity k and variable cross-sectional area $A_x(x) = A_o e^{ax}$, where A_o and a are constants. The lateral surface of the rod is well insulated.

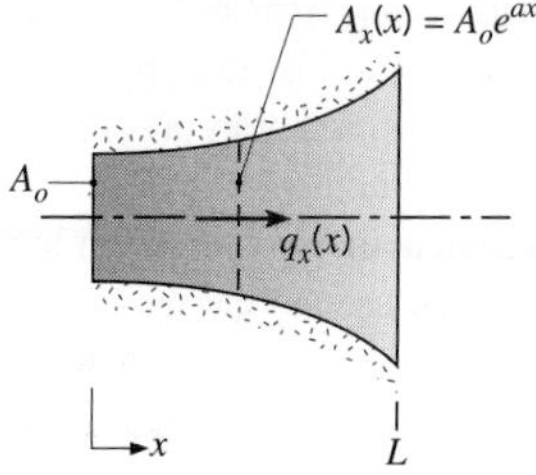

(a) Write an expression for the conduction heat rate, $q_x(x)$. Use this expression to determine the temperature distribution $T(x)$ and qualitatively sketch the distribution for $T(0) > T(L)$.

(b) Now consider conditions for which thermal energy is generated in the rod at a volumetric rate $\dot{q} = \dot{q}_o \exp(-ax)$, where $\dot{q}_o$ is a constant. Obtain an expression for $q_x(x)$ when the left face $(x = 0)$ is well insulated.

Thermophysical Properties

2.17 An apparatus for measuring thermal conductivity employs an electrical heater sandwiched between two identical samples of diameter 30 mm and length 60 mm, which are pressed between plates maintained at a uniform temperature $T_o = 77°C$ by a circulating fluid. A conducting grease is placed between all the surfaces to ensure good thermal contact. Differential thermocouples are imbedded in the samples with a spacing of 15 mm. The lateral sides of the samples are insulated to ensure one-dimensional heat transfer through the samples.

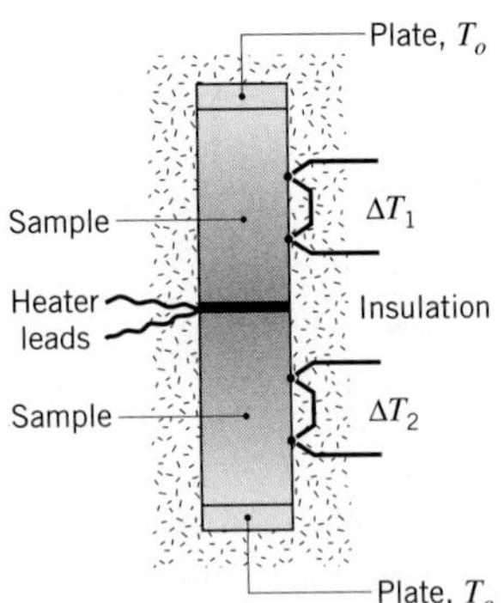

(a) With two samples of SS316 in the apparatus, the heater draws 0.353 A at 100 V, and the differential thermocouples indicate $\Delta T_1 = \Delta T_2 = 25.0°\text{C}$. What is the thermal conductivity of the stainless steel sample material? What is the average temperature of the samples? Compare your result with the thermal conductivity value reported for this material in Table A.1.

(b) By mistake, an Armco iron sample is placed in the lower position of the apparatus with one of the SS316 samples from part (a) in the upper portion. For this situation, the heater draws 0.601 A at 100 V, and the differential thermocouples indicate $\Delta T_1 = \Delta T_2 = 15.0°\text{C}$. What are the thermal conductivity and average temperature of the Armco iron sample?

(c) What is the advantage in constructing the apparatus with two identical samples sandwiching the heater rather than with a single heater–sample combination? When would heat leakage out of the lateral surfaces of the samples become significant? Under what conditions would you expect $\Delta T_1 \neq \Delta T_2$?

2.18 An engineer desires to measure the thermal conductivity of an aerogel material. It is expected that the aerogel will have an extremely small thermal conductivity.

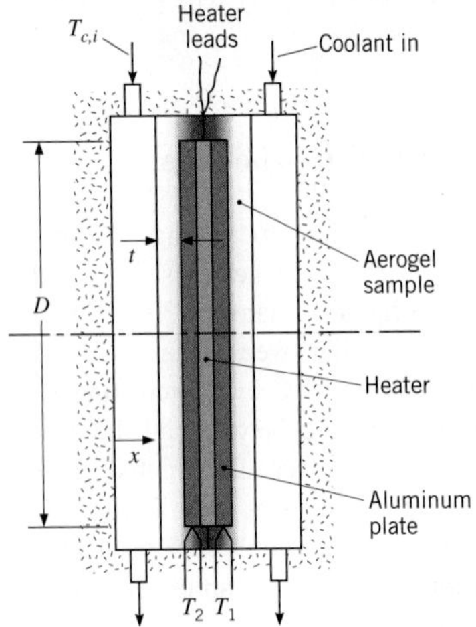

(a) Explain why the apparatus of Problem 2.17 cannot be used to obtain an accurate measurement of the aerogel's thermal conductivity.

(b) The engineer designs a new apparatus for which an electric heater of diameter $D = 150\text{ mm}$ is sandwiched between two thin plates of aluminum. The steady-state temperatures of the 5-mm-thick aluminum plates, T_1 and T_2, are measured with thermocouples. Aerogel sheets of thickness $t = 5\text{ mm}$ are placed outside the aluminum plates, while a coolant with an inlet temperature of $T_{c,i} = 25°\text{C}$ maintains the exterior surfaces of the aerogel at a low temperature. The circular aerogel sheets are formed so that they encase the heater and aluminum sheets, providing insulation to minimize radial heat losses. At steady state, $T_1 = T_2 = 55°\text{C}$, and the heater draws 125 mA at 10 V. Determine the value of the aerogel thermal conductivity k_a.

(c) Calculate the temperature difference across the thickness of the 5-mm-thick aluminum plates. Comment on whether it is important to know the axial locations at which the temperatures of the aluminum plates are measured.

(d) If liquid water is used as the coolant with a total flow rate of $\dot{m} = 1\text{ kg/min}$ (0.5 kg/min for each of the two streams), calculate the outlet temperature of the water, $T_{c,o}$.

2.19 Consider a 300 mm $\times$ 300 mm window in an aircraft. For a temperature difference of 80°C from the inner to the outer surface of the window, calculate the heat loss through L = 10-mm-thick polycarbonate, soda lime glass, and aerogel windows, respectively. The thermal conductivities of the aerogel and polycarbonate are $k_{ag} = 0.014\text{ W/m}\cdot\text{K}$ and $k_{pc} = 0.21\text{ W/m}\cdot\text{K}$, respectively. Evaluate the thermal conductivity of the soda lime glass at 300 K. If the aircraft has 130 windows and the cost to heat the cabin air is \$1/kW·h, compare the costs associated with the heat loss through the windows for an 8-hour intercontinental flight.

2.20 Consider a small but known volume of metal that has a large thermal conductivity.

(a) Since the thermal conductivity is large, spatial temperature gradients that develop within the metal in response to mild heating are small. Neglecting spatial temperature gradients, derive a differential equation that could be solved for the temperature of the metal versus time $T(t)$ if the metal is subjected to a fixed surface heat rate q supplied by an electric heater.

(b) A student proposes to identify the unknown metal by comparing measured and predicted thermal responses. Once a match is made, relevant thermophysical properties might be determined, and, in turn, the metal may be identified by comparison to published property data. Will this approach work? Consider aluminum, gold, and silver as the candidate metals.

2.21 Use *IHT* to perform the following tasks.

(a) Graph the thermal conductivity of pure copper, 2024 aluminum, and AISI 302 stainless steel over the temperature range $300 \leq T \leq 600$ K. Include all data on a single graph, and comment on the trends you observe.

(b) Graph the thermal conductivity of helium and air over the temperature range $300 \leq T \leq 800$ K. Include the data on a single graph, and comment on the trends you observe.

(c) Graph the kinematic viscosity of engine oil, ethylene glycol, and liquid water over the temperature range $300 \leq T \leq 360$ K. Include all data on a single graph, and comment on the trends you observe.

(d) Graph the thermal conductivity of a water-Al_2O_3 nanofluid at $T = 300$ K over the volume fraction range $0 \leq \varphi \leq 0.08$. See Example 2.2.

2.22 Calculate the thermal conductivity of air, hydrogen, and carbon dioxide at 300 K, assuming ideal gas behavior. Compare your calculated values to values from Table A.4.

2.23 A method for determining the thermal conductivity k and the specific heat c_p of a material is illustrated in the sketch. Initially the two identical samples of diameter $D = 60$ mm and thickness $L = 10$ mm and the thin heater are at a uniform temperature of $T_i = 23.00°C$, while surrounded by an insulating powder. Suddenly the heater is energized to provide a uniform heat flux q''_o on each of the sample interfaces, and the heat flux is maintained constant for a period of time, Δt_o. A short time after sudden heating is initiated, the temperature at this interface T_o is related to the heat flux as

$$T_o(t) - T_i = 2q''_o \left(\frac{t}{\pi \rho c_p k} \right)^{1/2}$$

For a particular test run, the electrical heater dissipates 15.0 W for a period of $\Delta t_o = 120$ s, and the temperature at the interface is $T_o(30\text{ s}) = 24.57°C$ after 30 s of heating. A long time after the heater is deenergized, $t \gg \Delta t_0$, the samples reach the uniform temperature of $T_o(\infty) = 33.50°C$. The density of the sample materials, determined by measurement of volume and mass, is $\rho = 3965\text{ kg/m}^3$.

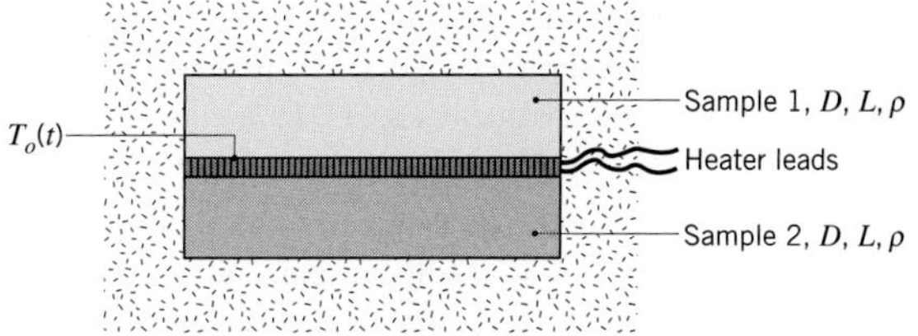

Determine the specific heat and thermal conductivity of the test material. By looking at values of the thermophysical properties in Table A.1 or A.2, identify the test sample material.

2.24 Compare and contrast the heat capacity ρc_p of common brick, plain carbon steel, engine oil, water, and soil. Which material provides the greatest amount of thermal energy storage per unit volume? Which material would you expect to have the lowest cost per unit heat capacity? Evaluate properties at 300 K.

2.25 A cylindrical rod of stainless steel is insulated on its exterior surface except for the ends. The steady-state temperature distribution is $T(x) = a - bx/L$, where $a = 305$ K and $b = 10$ K. The diameter and length of the rod are $D = 20$ mm and $L = 100$ mm, respectively. Determine the heat flux along the rod, q''_x. *Hint:* The mass of the rod is $M = 0.248$ kg.

The Heat Equation

2.26 At a given instant of time, the temperature distribution within an infinite homogeneous body is given by the function

$$T(x, y, z) = x^2 - 2y^2 + z^2 - xy + 2yz$$

Assuming constant properties and no internal heat generation, determine the regions where the temperature changes with time.

2.27 A pan is used to boil water by placing it on a stove, from which heat is transferred at a fixed rate q_o. There are two stages to the process. In Stage 1, the water is taken from its initial (room) temperature T_i to the boiling point, as heat is transferred from the pan by natural convection. During this stage, a constant value of the convection coefficient h may be assumed, while the bulk temperature of the water increases with time, $T_\infty = T_\infty(t)$. In Stage 2, the water has come to a boil, and its temperature remains at a fixed value, $T_\infty = T_b$, as heating continues. Consider a pan bottom of thickness L and diameter D, with a coordinate system corresponding to $x = 0$ and $x = L$ for the surfaces in contact with the stove and water, respectively.

(a) Write the form of the heat equation and the boundary/initial conditions that determine the variation of temperature with position and time, $T(x, t)$, in the pan bottom during Stage 1. Express your result in terms of the parameters q_o, D, L, h, and T_∞, as well as appropriate properties of the pan material.

(b) During Stage 2, the surface of the pan in contact with the water is at a fixed temperature, $T(L, t) = T_L > T_b$. Write the form of the heat equation and boundary conditions that determine the temperature distribution $T(x)$ in the pan bottom. Express your result in terms of the parameters q_o, D, L, and T_L, as well as appropriate properties of the pan material.

2.28 Uniform internal heat generation at $\dot{q} = 5 \times 10^7\ \text{W/m}^3$ is occurring in a cylindrical nuclear reactor fuel rod of 50-mm diameter, and under steady-state conditions the temperature distribution is of the form $T(r) = a + br^2$, where T is in degrees Celsius and r is in meters, while $a = 800°\text{C}$ and $b = -4.167 \times 10^5\ °\text{C/m}^2$. The fuel rod properties are $k = 30\ \text{W/m}\cdot\text{K}$, $\rho = 1100\ \text{kg/m}^3$, and $c_p = 800\ \text{J/kg}\cdot\text{K}$.

(a) What is the rate of heat transfer per unit length of the rod at $r = 0$ (the centerline) and at $r = 25$ mm (the surface)?

(b) If the reactor power level is suddenly increased to $\dot{q}_2 = 10^8\ \text{W/m}^3$, what is the initial time rate of temperature change at $r = 0$ and $r = 25$ mm?

2.29 Consider a one-dimensional plane wall with constant properties and uniform internal generation $\dot{q}$. The left face is insulated, and the right face is held at a uniform temperature.

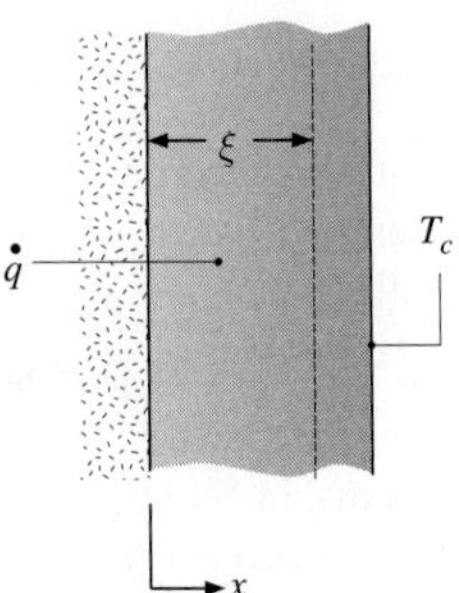

(a) Using the appropriate form of the heat equation, derive an expression for the x-dependence of the steady-state heat flux $q''(x)$.

(b) Using a finite volume spanning the range $0 \le x \le \xi$, derive an expression for $q''(\xi)$ and compare the expression to your result for part (a).

2.30 The steady-state temperature distribution in a one-dimensional wall of thermal conductivity 50 W/m·K and thickness 50 mm is observed to be $T(°\text{C}) = a + bx^2$, where $a = 200°\text{C}$, $b = -2000°\text{C/m}^2$, and x is in meters.

(a) What is the heat generation rate $\dot{q}$ in the wall?

(b) Determine the heat fluxes at the two wall faces. In what manner are these heat fluxes related to the heat generation rate?

2.31 The temperature distribution across a wall 0.3 m thick at a certain instant of time is $T(x) = a + bx + cx^2$, where T is in degrees Celsius and x is in meters, $a = 200°\text{C}$, $b = -200°\text{C/m}$, and $c = 30°\text{C/m}^2$. The wall has a thermal conductivity of 1 W/m·K.

(a) On a unit surface area basis, determine the rate of heat transfer into and out of the wall and the rate of change of energy stored by the wall.

(b) If the cold surface is exposed to a fluid at 100°C, what is the convection coefficient?

2.32 A plane wall of thickness $2L = 40$ mm and thermal conductivity $k = 5\ \text{W/m}\cdot\text{K}$ experiences uniform volumetric heat generation at a rate $\dot{q}$, while convection heat transfer occurs at both of its surfaces ($x = -L, +L$), each of which is exposed to a fluid of temperature $T_\infty = 20°\text{C}$. Under steady-state conditions, the temperature distribution in the wall is of the form $T(x) = a + bx + cx^2$ where $a = 82.0°\text{C}$, $b = -210°\text{C/m}$, $c = -2 \times 10^4 °\text{C/m}^2$, and x is in meters. The origin of the x-coordinate is at the midplane of the wall.

(a) Sketch the temperature distribution and identify significant physical features.

(b) What is the volumetric rate of heat generation $\dot{q}$ in the wall?

(c) Determine the surface heat fluxes, $q''_x(-L)$ and $q''_x(+L)$. How are these fluxes related to the heat generation rate?

(d) What are the convection coefficients for the surfaces at $x = -L$ and $x = +L$?

(e) Obtain an expression for the heat flux distribution $q''_x(x)$. Is the heat flux zero at any location? Explain any significant features of the distribution.

(f) If the source of the heat generation is suddenly deactivated ($\dot{q} = 0$), what is the rate of change of energy stored in the wall at this instant?

(g) What temperature will the wall eventually reach with $\dot{q} = 0$? How much energy must be removed by the fluid per unit area of the wall (J/m²) to reach this state? The density and specific heat of the wall material are 2600 kg/m³ and 800 J/kg·K, respectively.

2.33 Temperature distributions within a series of one-dimensional plane walls at an initial time, at steady state, and at several intermediate times are as shown.

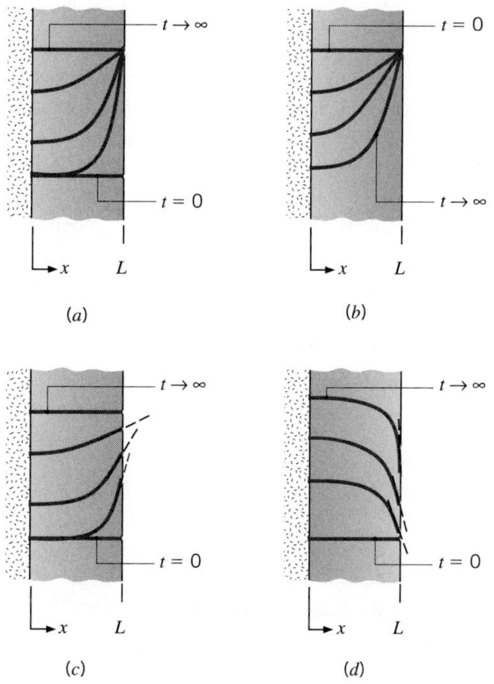

For each case, write the appropriate form of the heat diffusion equation. Also write the equations for the initial condition and the boundary conditions that are applied at $x = 0$ and $x = L$. If volumetric generation occurs, it is uniform throughout the wall. The properties are constant.

2.34 One-dimensional, steady-state conduction with uniform internal energy generation occurs in a plane wall with a thickness of 50 mm and a constant thermal conductivity of 5 W/m·K. For these conditions, the temperature distribution has the form $T(x) = a + bx + cx^2$. The surface at $x = 0$ has a temperature of $T(0) \equiv T_o = 120°\text{C}$ and experiences convection with a fluid for which $T_\infty = 20°\text{C}$ and $h = 500\ \text{W/m}^2\cdot\text{K}$. The surface at $x = L$ is well insulated.

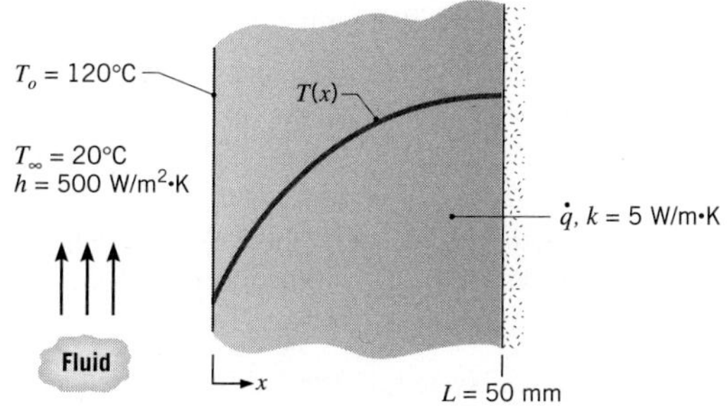

(a) Applying an overall energy balance to the wall, calculate the volumetric energy generation rate $\dot{q}$.

(b) Determine the coefficients a, b, and c by applying the boundary conditions to the prescribed temperature distribution. Use the results to calculate and plot the temperature distribution.

(c) Consider conditions for which the convection coefficient is halved, but the volumetric energy generation rate remains unchanged. Determine the new values of a, b, and c, and use the results to plot the temperature distribution. *Hint:* recognize that $T(0)$ is no longer 120°C.

(d) Under conditions for which the volumetric energy generation rate is doubled, and the convection coefficient remains unchanged ($h = 500\ \text{W/m}^2\cdot\text{K}$), determine the new values of a, b, and c and plot the corresponding temperature distribution. Referring to the results of parts (b), (c), and (d) as Cases 1, 2, and 3, respectively, compare the temperature distributions for the three cases and discuss the effects of h and $\dot{q}$ on the distributions.

2.35 Derive the heat diffusion equation, Equation 2.26, for cylindrical coordinates beginning with the differential control volume shown in Figure 2.12.

2.36 Derive the heat diffusion equation, Equation 2.29, for spherical coordinates beginning with the differential control volume shown in Figure 2.13.

2.37 The steady-state temperature distribution in a semi-transparent material of thermal conductivity k and thickness L exposed to laser irradiation is of the form

$$T(x) = -\frac{A}{ka^2}e^{-ax} + Bx + C$$

where A, a, B, and C are known constants. For this situation, radiation absorption in the material is manifested by a distributed heat generation term, $\dot{q}(x)$.

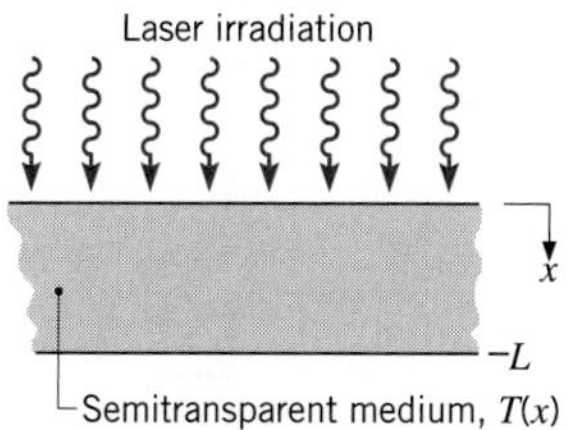

(a) Obtain expressions for the conduction heat fluxes at the front and rear surfaces.

(b) Derive an expression for $\dot{q}(x)$.

(c) Derive an expression for the rate at which radiation is absorbed in the entire material, per unit surface area. Express your result in terms of the known constants for the temperature distribution, the thermal conductivity of the material, and its thickness.

2.38 One-dimensional, steady-state conduction with no energy generation is occurring in a cylindrical shell of inner radius r_1 and outer radius r_2. Under what condition is the linear temperature distribution shown possible?

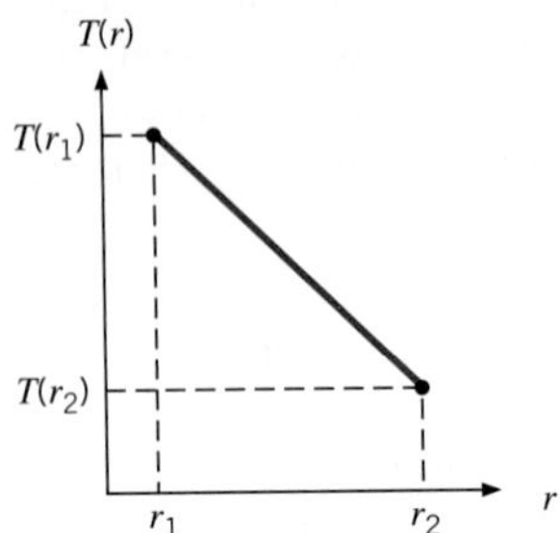

2.39 One-dimensional, steady-state conduction with no energy generation is occurring in a spherical shell of inner radius r_1 and outer radius r_2. Under what condition is the linear temperature distribution shown in Problem 2.38 possible?

2.40 The steady-state temperature distribution in a one-dimensional wall of thermal conductivity k and thickness L is of the form $T = ax^3 + bx^2 + cx + d$. Derive expressions for the heat generation rate per unit volume in the wall and the heat fluxes at the two wall faces ($x = 0, L$).

2.41 One-dimensional, steady-state conduction with no energy generation is occurring in a plane wall of constant thermal conductivity.

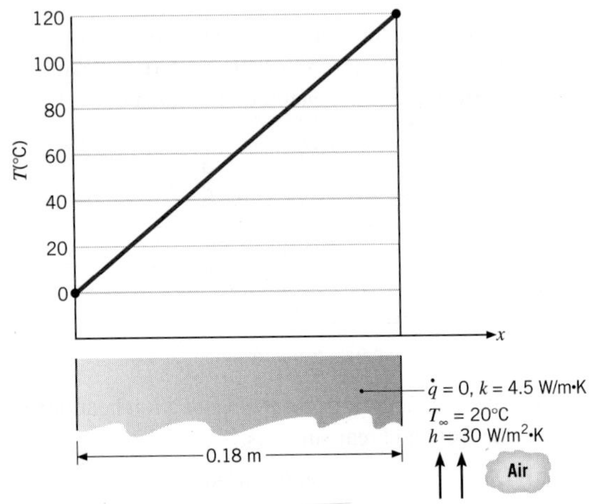

(a) Is the prescribed temperature distribution possible? Briefly explain your reasoning.

(b) With the temperature at $x = 0$ and the fluid temperature fixed at $T(0) = 0°C$ and $T_\infty = 20°C$, respectively, compute and plot the temperature at $x = L$, $T(L)$, as a function of h for $10 \leq h \leq 100\ \mathrm{W/m^2 \cdot K}$. Briefly explain your results.

2.42 A plane layer of coal of thickness $L = 1$ m experiences uniform volumetric generation at a rate of $\dot{q} = 20\ \mathrm{W/m^3}$ due to slow oxidation of the coal particles. Averaged over a daily period, the top surface of the layer transfers heat by convection to ambient air for which $h = 5\ \mathrm{W/m^2 \cdot K}$ and $T_\infty = 25°C$, while receiving solar irradiation in the amount $G_S = 400\ \mathrm{W/m^2}$. Irradiation from the atmosphere may be neglected. The solar absorptivity and emissivity of the surface are each $\alpha_S = \varepsilon = 0.95$.

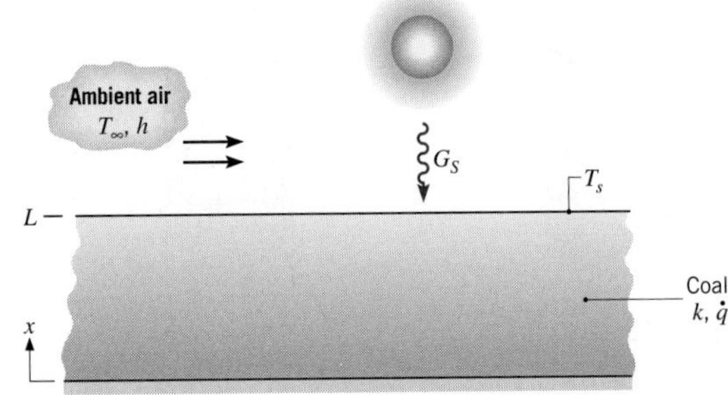

(a) Write the steady-state form of the heat diffusion equation for the layer of coal. Verify that this equation is satisfied by a temperature distribution of the form

$$T(x) = T_s + \frac{\dot{q}L^2}{2k}\left(1 - \frac{x^2}{L^2}\right)$$

From this distribution, what can you say about conditions at the bottom surface ($x = 0$)? Sketch the temperature distribution and label key features.

(b) Obtain an expression for the rate of heat transfer by conduction per unit area at $x = L$. Applying an energy balance to a control surface about the top surface of the layer, obtain an expression for T_s. Evaluate T_s and $T(0)$ for the prescribed conditions.

(c) Daily average values of G_S and h depend on a number of factors, such as time of year, cloud cover, and wind conditions. For $h = 5\ \mathrm{W/m^2 \cdot K}$, compute and plot T_S and $T(0)$ as a function of G_S for $50 \leq G_S \leq 500\ \mathrm{W/m^2}$. For $G_S = 400\ \mathrm{W/m^2}$, compute and plot T_S and $T(0)$ as a function of h for $5 \leq h \leq 50\ \mathrm{W/m^2 \cdot K}$.

2.43 The cylindrical system illustrated has negligible variation of temperature in the r- and z-directions. Assume that $\Delta r = r_o - r_i$ is small compared to r_i, and denote the length in the z-direction, normal to the page, as L.

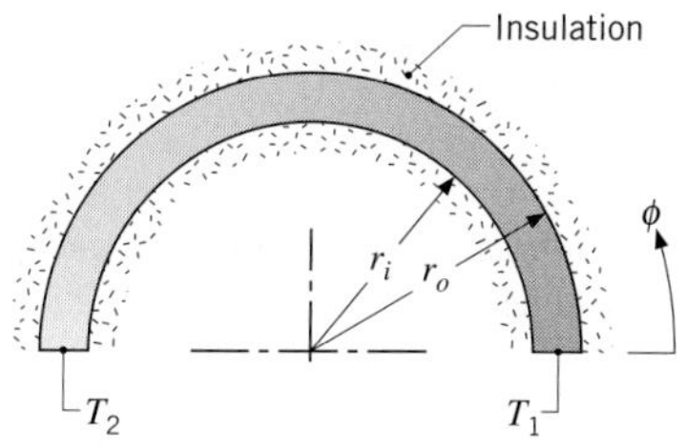

(a) Beginning with a properly defined control volume and considering energy generation and storage effects, derive the differential equation that prescribes the variation in temperature with the angular coordinate ϕ. Compare your result with Equation 2.26.

(b) For steady-state conditions with no internal heat generation and constant properties, determine the temperature distribution $T(\phi)$ in terms of the constants T_1, T_2, r_i, and r_o. Is this distribution linear in ϕ?

(c) For the conditions of part (b) write the expression for the heat rate q_ϕ.

2.44 Beginning with a differential control volume in the form of a cylindrical shell, derive the heat diffusion equation for a one-dimensional, cylindrical, radial coordinate system with internal heat generation. Compare your result with Equation 2.26.

2.45 Beginning with a differential control volume in the form of a spherical shell, derive the heat diffusion equation for a one-dimensional, spherical, radial coordinate system with internal heat generation. Compare your result with Equation 2.29.

2.46 A steam pipe is wrapped with insulation of inner and outer radii r_i and r_o, respectively. At a particular instant the temperature distribution in the insulation is known to be of the form

$$T(r) = C_1 \ln\left(\frac{r}{r_o}\right) + C_2$$

Are conditions steady-state or transient? How do the heat flux and heat rate vary with radius?

2.47 For a long circular tube of inner and outer radii r_1 and r_2, respectively, uniform temperatures T_1 and T_2 are maintained at the inner and outer surfaces, while thermal energy generation is occurring within the tube wall ($r_1 < r < r_2$). Consider steady-state conditions for which $T_1 < T_2$. Is it possible to maintain a *linear* radial temperature distribution in the wall? If so, what special conditions must exist?

2.48 Passage of an electric current through a long conducting rod of radius r_i and thermal conductivity k_r results in uniform volumetric heating at a rate of $\dot{q}$. The conducting rod is wrapped in an electrically nonconducting cladding material of outer radius r_o and thermal conductivity k_c, and convection cooling is provided by an adjoining fluid.

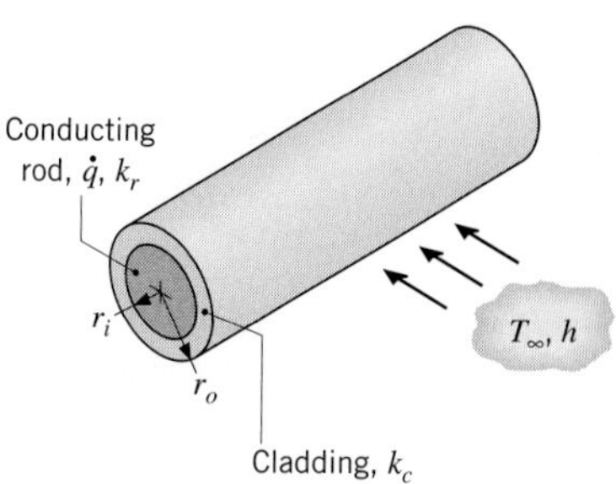

For steady-state conditions, write appropriate forms of the heat equations for the rod and cladding. Express appropriate boundary conditions for the solution of these equations.

2.49 Two-dimensional, steady-state conduction occurs in a hollow cylindrical solid of thermal conductivity $k = 16$ W/m·K, outer radius $r_o = 1$ m and overall length $2z_o = 5$ m, where the origin of the coordinate system is located at the midpoint of the center line. The inner surface of the cylinder is insulated, and the temperature distribution within the cylinder has the form $T(r, z) = a + br^2 + c\ln r + dz^2$, where $a = -20°C$, $b = 150°C/m^2$, $c = -12°C$, $d = -300°C/m^2$ and r and z are in meters.

(a) Determine the inner radius r_i of the cylinder.

(b) Obtain an expression for the volumetric rate of heat generation, $\dot{q}(W/m^3)$.

(c) Determine the axial distribution of the heat flux at the outer surface, $q''_r(r_o, z)$. What is the heat rate at the outer surface? Is it into or out of the cylinder?

(d) Determine the radial distribution of the heat flux at the end faces of the cylinder, $q''_r(r, +z_o)$ and $q''_r(r, -z_o)$. What are the corresponding heat rates? Are they into or out of the cylinder?

(e) Verify that your results are consistent with an overall energy balance on the cylinder.

2.50 An electric cable of radius r_1 and thermal conductivity k_c is enclosed by an insulating sleeve whose outer surface is of radius r_2 and experiences convection heat transfer and radiation exchange with the adjoining air and large surroundings, respectively. When electric current passes through the cable, thermal energy is generated within the cable at a volumetric rate $\dot{q}$.

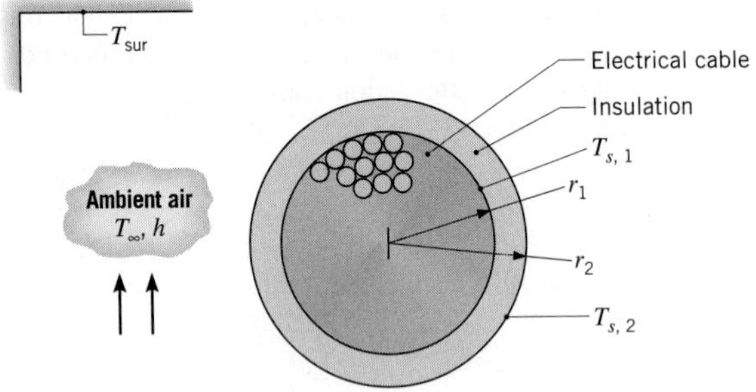

(a) Write the steady-state forms of the heat diffusion equation for the insulation and the cable. Verify that these equations are satisfied by the following temperature distributions:

$$\text{Insulation: } T(r) = T_{s,2} + (T_{s,1} - T_{s,2})\frac{\ln(r/r_2)}{\ln(r_1/r_2)}$$

$$\text{Cable: } T(r) = T_{s,1} + \frac{\dot{q}r_1^2}{4k_c}\left(1 - \frac{r^2}{r_1^2}\right)$$

Sketch the temperature distribution, $T(r)$, in the cable and the sleeve, labeling key features.

(b) Applying Fourier's law, show that the rate of conduction heat transfer per unit length through the sleeve may be expressed as

$$q'_r = \frac{2\pi k_s(T_{s,1} - T_{s,2})}{\ln(r_2/r_1)}$$

Applying an energy balance to a control surface placed around the cable, obtain an alternative expression for q'_r, expressing your result in terms of $\dot{q}$ and r_1.

(c) Applying an energy balance to a control surface placed around the outer surface of the sleeve, obtain an expression from which $T_{s,2}$ may be determined as a function of $\dot{q}$, r_1, h, T_∞, ε, and T_{sur}.

(d) Consider conditions for which 250 A are passing through a cable having an electric resistance per unit length of $R'_e = 0.005\ \Omega/\text{m}$, a radius of $r_1 = 15$ mm, and a thermal conductivity of $k_c = 200\ \text{W/m}\cdot\text{K}$. For $k_s = 15\ \text{W/m}\cdot\text{K}$, $r_2 = 15.5$ mm, $h = 25\ \text{W/m}^2\cdot\text{K}$, $\varepsilon = 0.9$, $T_\infty = 25°\text{C}$, and $T_{sur} = 35°\text{C}$, evaluate the surface temperatures, $T_{s,1}$ and $T_{s,2}$, as well as the temperature T_o at the centerline of the cable.

(e) With all other conditions remaining the same, compute and plot T_o, $T_{s,1}$, and $T_{s,2}$ as a function of r_2 for $15.5 \le r_2 \le 20$ mm.

2.51 A spherical shell of inner and outer radii r_i and r_o, respectively, contains heat-dissipating components, and at a particular instant the temperature distribution in the shell is known to be of the form

$$T(r) = \frac{C_1}{r} + C_2$$

Are conditions steady-state or transient? How do the heat flux and heat rate vary with radius?

2.52 A chemically reacting mixture is stored in a thin-walled spherical container of radius $r_1 = 200$ mm, and the exothermic reaction generates heat at a uniform, but temperature-dependent volumetric rate of $\dot{q} = \dot{q}_o \exp(-A/T_o)$, where $\dot{q}_o = 5000\ \text{W/m}^3$, $A = 75$ K, and T_o is the mixture temperature in kelvins. The vessel is enclosed by an insulating material of outer radius r_2, thermal conductivity k, and emissivity ε. The outer surface of the insulation experiences convection heat transfer and net radiation exchange with the adjoining air and large surroundings, respectively.

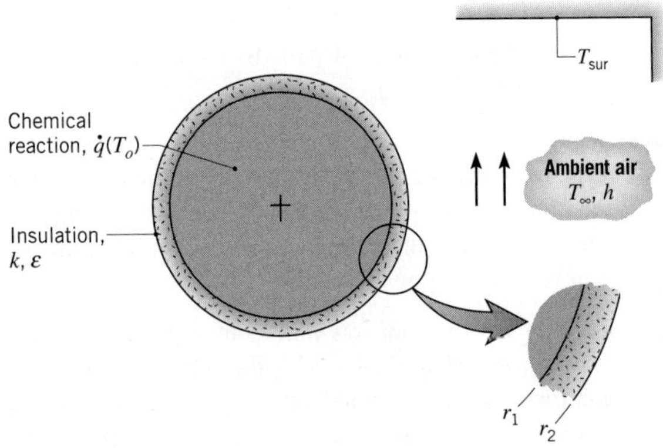

(a) Write the steady-state form of the heat diffusion equation for the insulation. Verify that this equation is satisfied by the temperature distribution

$$T(r) = T_{s,1} - (T_{s,1} - T_{s,2})\left[\frac{1 - (r_1/r)}{1 - (r_1/r_2)}\right]$$

Sketch the temperature distribution, $T(r)$, labeling key features.

(b) Applying Fourier's law, show that the rate of heat transfer by conduction through the insulation may be expressed as

$$q_r = \frac{4\pi k(T_{s,1} - T_{s,2})}{(1/r_1) - (1/r_2)}$$

Applying an energy balance to a control surface about the container, obtain an alternative expression for q_r, expressing your result in terms of $\dot{q}$ and r_1.

(c) Applying an energy balance to a control surface placed around the outer surface of the insulation, obtain an expression from which $T_{s,2}$ may be determined as a function of $\dot{q}$, r_1, h, T_∞, ε, and T_{sur}.

(d) The process engineer wishes to maintain a reactor temperature of $T_o = T(r_1) = 95°C$ under conditions for which $k = 0.05\ W/m \cdot K$, $r_2 = 208$ mm, $h = 5\ W/m^2 \cdot K$, $\varepsilon = 0.9$, $T_\infty = 25°C$, and $T_{sur} = 35°C$. What is the actual reactor temperature and the outer surface temperature $T_{s,2}$ of the insulation?

(e) Compute and plot the variation of $T_{s,2}$ with r_2 for $201 \leq r_2 \leq 210$ mm. The engineer is concerned about potential burn injuries to personnel who may come into contact with the exposed surface of the insulation. Is increasing the insulation thickness a practical solution to maintaining $T_{s,2} \leq 45°C$? What other parameter could be varied to reduce $T_{s,2}$?

Graphical Representations

2.53 A thin electrical heater dissipating 4000 W/m^2 is sandwiched between two 25-mm-thick plates whose exposed surfaces experience convection with a fluid for which $T_\infty = 20°C$ and $h = 400\ W/m^2 \cdot K$. The thermophysical properties of the plate material are $\rho = 2500\ kg/m^3$, $c = 700\ J/kg \cdot K$, and $k = 5\ W/m \cdot K$.

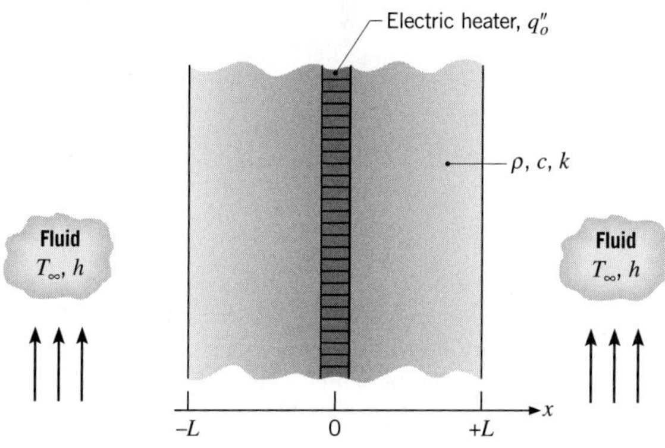

(a) On $T - x$ coordinates, sketch the steady-state temperature distribution for $-L \leq x \leq +L$. Calculate values of the temperatures at the surfaces, $x = \pm L$, and the midpoint, $x = 0$. Label this distribution as Case 1, and explain its salient features.

(b) Consider conditions for which there is a loss of coolant and existence of a nearly adiabatic condition on the $x = +L$ surface. On the $T - x$ coordinates used for part (a), sketch the corresponding steady-state temperature distribution and indicate the temperatures at $x = 0, \pm L$. Label the distribution as Case 2, and explain its key features.

(c) With the system operating as described in part (b), the surface $x = -L$ also experiences a sudden loss of coolant. This dangerous situation goes undetected for 15 min, at which time the power to the heater is deactivated. Assuming no heat losses from the surfaces of the plates, what is the eventual ($t \rightarrow \infty$), uniform, steady-state temperature distribution in the plates? Show this distribution as Case 3 on your sketch, and explain its key features. *Hint*: Apply the conservation of energy requirement on a time-interval basis, Eq. 1.12b, for the initial and final conditions corresponding to Case 2 and Case 3, respectively.

(d) On $T - t$ coordinates, sketch the temperature history at the plate locations $x = 0, \pm L$ during the transient period between the distributions for Cases 2 and 3. Where and when will the temperature in the system achieve a maximum value?

2.54 The one-dimensional system of mass M with constant properties and no internal heat generation shown in the figure is initially at a uniform temperature T_i. The electrical heater is suddenly energized, providing a uniform heat flux q''_o at the surface $x = 0$. The boundaries at $x = L$ and elsewhere are perfectly insulated.

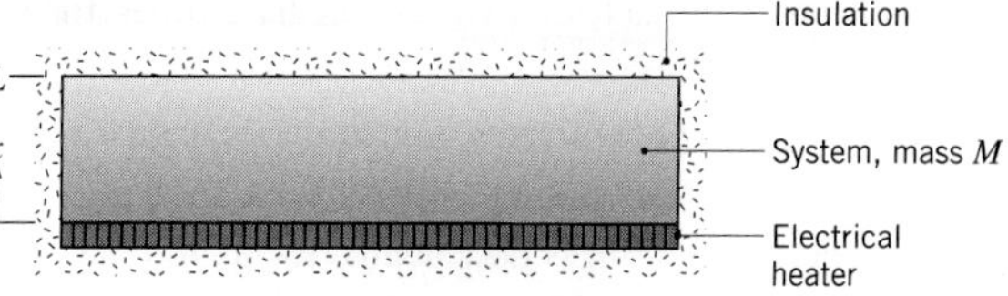

(a) Write the differential equation, and identify the boundary and initial conditions that could be used to determine the temperature as a function of position and time in the system.

(b) On $T - x$ coordinates, sketch the temperature distributions for the initial condition ($t \leq 0$) and for several times after the heater is energized. Will a steady-state temperature distribution ever be reached?

(c) On $q''_x - t$ coordinates, sketch the heat flux $q''_x(x, t)$ at the planes $x = 0$, $x = L/2$, and $x = L$ as a function of time.

(d) After a period of time t_e has elapsed, the heater power is switched off. Assuming that the insulation is perfect, the system will eventually reach a final uniform temperature T_f. Derive an expression that can be used to determine T_f as a function of the parameters q''_o, t_e, T_i, and the system characteristics M, c_p, and A_s (the heater surface area).

2.55 Consider a one-dimensional plane wall of thickness $2L$. The surface at $x = -L$ is subjected to convective conditions characterized by $T_{\infty,1}$, h_1, while the surface at $x = +L$ is subjected to conditions $T_{\infty,2}$, h_2. The initial temperature of the wall is $T_o = (T_{\infty,1} + T_{\infty,2})/2$ where $T_{\infty,1} > T_{\infty,2}$.

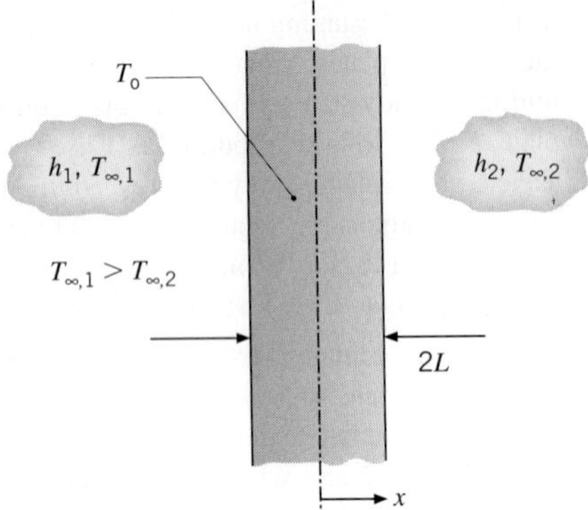

(a) Write the differential equation, and identify the boundary and initial conditions that could be used to determine the temperature distribution $T(x, t)$ as a function of position and time.

(b) On $T - x$ coordinates, sketch the temperature distributions for the initial condition, the steady-state condition, and for two intermediate times for the case $h_1 = h_2$.

(c) On $q''_x - t$ coordinates, sketch the heat flux $q''_x(x, t)$ at the planes $x = 0, -L$, and $+L$.

(d) The value of h_1 is now doubled with all other conditions being identical as in parts (a) through (c). On $T - x$ coordinates drawn to the same scale as used in part (b), sketch the temperature distributions for the initial condition, the steady-state condition, and for two intermediate times. Compare the sketch to that of part (b).

(e) Using the doubled value of h_1, sketch the heat flux $q''_x(x, t)$ at the planes $x = 0, -L$, and $+L$ on the same plot you prepared for part (c). Compare the two responses.

2.56 A large plate of thickness $2L$ is at a uniform temperature of $T_i = 200°C$, when it is suddenly quenched by dipping it in a liquid bath of temperature $T_\infty = 20°C$. Heat transfer to the liquid is characterized by the convection coefficient h.

(a) If $x = 0$ corresponds to the midplane of the wall, on $T - x$ coordinates, sketch the temperature distributions for the following conditions: initial condition ($t \leq 0$), steady-state condition ($t \rightarrow \infty$), and two intermediate times.

(b) On $q''_x - t$ coordinates, sketch the variation with time of the heat flux at $x = L$.

(c) If $h = 100\ \text{W/m}^2 \cdot \text{K}$, what is the heat flux at $x = L$ and $t = 0$? If the wall has a thermal conductivity of $k = 50\ \text{W/m} \cdot \text{K}$ what is the corresponding temperature gradient at $x = L$?

(d) Consider a plate of thickness $2L = 20$ mm with a density of $\rho = 2770\ \text{kg/m}^3$ and a specific heat $c_p = 875\ \text{J/kg} \cdot \text{K}$. By performing an energy balance on the plate, determine the amount of energy per unit surface area of the plate (J/m²) that is transferred to the bath over the time required to reach steady-state conditions.

(e) From other considerations, it is known that, during the quenching process, the heat flux at $x = +L$ and $x = -L$ decays exponentially with time according to the relation, $q''_x = A \exp(-Bt)$, where t is in seconds, $A = 1.80 \times 10^4\ \text{W/m}^2$, and $B = 4.126 \times 10^{-3}\ \text{s}^{-1}$. Use this information to determine the energy per unit surface area of the plate that is transferred to the fluid during the quenching process.

2.57 The plane wall with constant properties and no internal heat generation shown in the figure is initially at a uniform temperature T_i. Suddenly the surface at $x = L$ is heated by a fluid at T_∞ having a convection heat transfer coefficient h. The boundary at $x = 0$ is perfectly insulated.

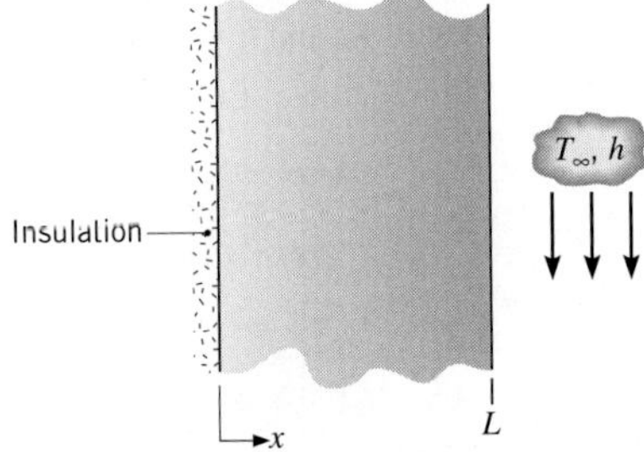

(a) Write the differential equation, and identify the boundary and initial conditions that could be used to determine the temperature as a function of position and time in the wall.

(b) On $T - x$ coordinates, sketch the temperature distributions for the following conditions: initial condition ($t \leq 0$), steady-state condition ($t \rightarrow \infty$), and two intermediate times.

(c) On $q''_x - t$ coordinates, sketch the heat flux at the locations $x = 0$, $x = L$. That is, show qualitatively how $q''_x(0, t)$ and $q''_x(L, t)$ vary with time.

(d) Write an expression for the total energy transferred to the wall per unit volume of the wall (J/m³).

2.58 Consider the steady-state temperature distributions within a composite wall composed of Material A and Material B for the two cases shown. There is no internal generation, and the conduction process is one-dimensional.

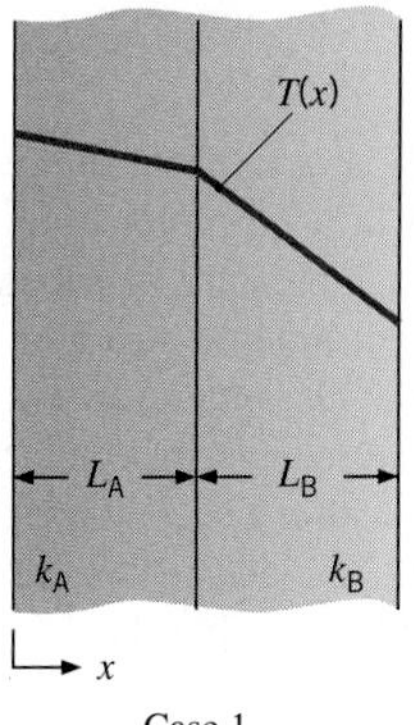

Case 1

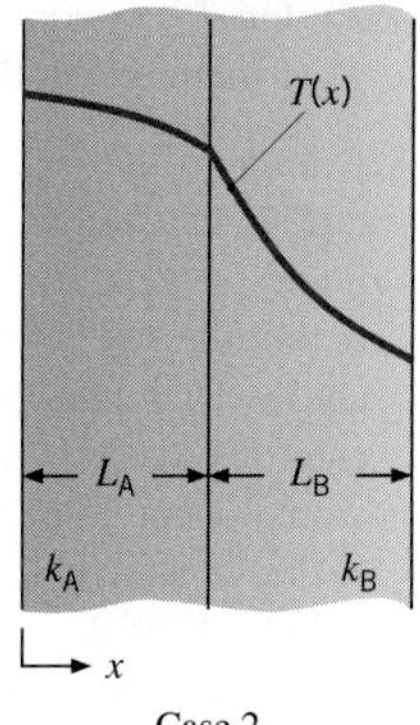

Case 2

Answer the following questions for each case. Which material has the higher thermal conductivity? Does the thermal conductivity vary significantly with temperature? If so, how? Describe the heat flux distribution $q''_x(x)$ through the composite wall. If the thickness and thermal conductivity of each material were both doubled and the boundary temperatures remained the same, what would be the effect on the heat flux distribution?

Case 1. Linear temperature distributions exist in both materials, as shown.
Case 2. Nonlinear temperature distributions exist in both materials, as shown.

2.59 A plane wall has constant properties, no internal heat generation, and is initially at a uniform temperature T_i. Suddenly, the surface at $x = L$ is heated by a fluid at T_∞ having a convection coefficient h. At the same instant, the electrical heater is energized, providing a constant heat flux q''_o at $x = 0$.

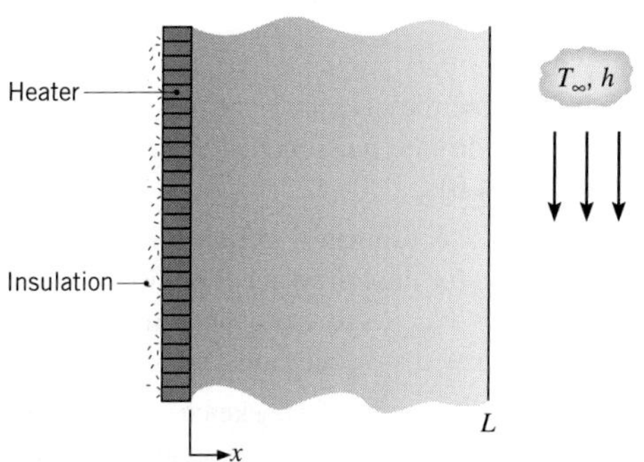

(a) On $T - x$ coordinates, sketch the temperature distributions for the following conditions: initial condition ($t \leq 0$), steady-state condition ($t \rightarrow \infty$), and for two intermediate times.

(b) On $q''_x - x$ coordinates, sketch the heat flux corresponding to the four temperature distributions of part (a).

(c) On $q''_x - t$ coordinates, sketch the heat flux at the locations $x = 0$ and $x = L$. That is, show qualitatively how $q''_x(0, t)$ and $q''_x(L, t)$ vary with time.

(d) Derive an expression for the steady-state temperature at the heater surface, $T(0, \infty)$, in terms of q''_o, T_∞, k, h, and L.

2.60 A plane wall with constant properties is initially at a uniform temperature T_o. Suddenly, the surface at $x = L$ is exposed to a convection process with a fluid at $T_\infty (> T_o)$ having a convection coefficient h. Also, suddenly the wall experiences a uniform internal volumetric heating $\dot{q}$ that is sufficiently large to induce a maximum steady-state temperature within the wall, which exceeds that of the fluid. The boundary at $x = 0$ remains at T_o.

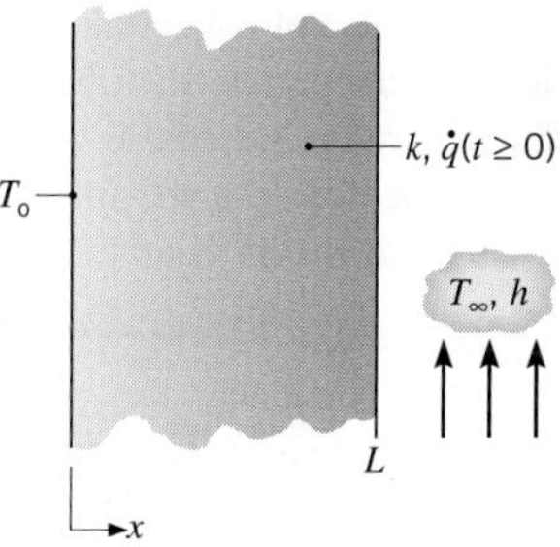

(a) On $T - x$ coordinates, sketch the temperature distributions for the following conditions: initial condition ($t \leq 0$), steady-state condition ($t \rightarrow \infty$), and for two intermediate times. Show also the distribution for the special condition when there is no heat flow at the $x = L$ boundary.

(b) On $q''_x - t$ coordinates, sketch the heat flux for the locations $x = 0$ and $x = L$, that is, $q''_x(0, t)$ and $q''_x(L, t)$, respectively.

2.61 Consider the conditions associated with Problem 2.60, but now with a convection process for which $T_\infty < T_o$.

(a) On $T - x$ coordinates, sketch the temperature distributions for the following conditions: initial condition ($t \leq 0$), steady-state condition ($t \rightarrow \infty$), and for two intermediate times. Identify key features of the distributions, especially the location of the maximum temperature and the temperature gradient at $x = L$.

(b) On $q''_x - t$ coordinates, sketch the heat flux for the locations $x = 0$ and $x = L$, that is, $q''_x(0, t)$ and $q''_x(L, t)$, respectively. Identify key features of the flux histories.

1650-1710 Uniform prep GMC
1710-1745 Guidons GMC
1650-1745 ADTD

2.62 Consider the steady-state temperature distribution within a composite wall composed of Materials A and B.

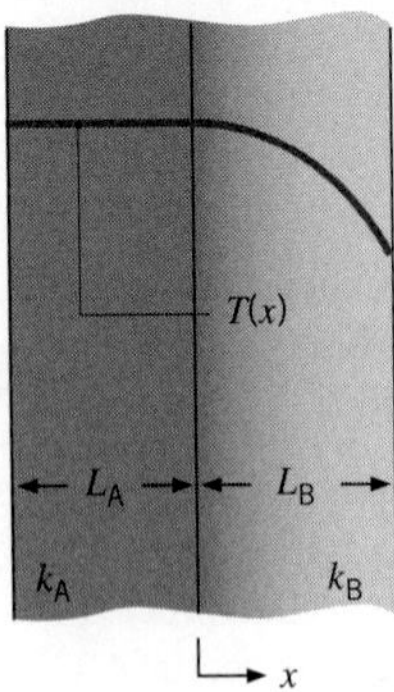

The conduction process is one-dimensional. Within which material does uniform volumetric generation occur? What is the boundary condition at $x = -L_A$? How would the temperature distribution change if the thermal conductivity of Material A were doubled? How would the temperature distribution change if the thermal conductivity of Material B were doubled? Does a contact resistance exist at the interface between the two materials? Sketch the heat flux distribution $q''_x(x)$ through the composite wall.

2.63 A spherical particle of radius r_1 experiences uniform thermal generation at a rate of $\dot{q}$. The particle is encapsulated by a spherical shell of outside radius r_2 that is cooled by ambient air. The thermal conductivities of the particle and shell are k_1 and k_2, respectively, where $k_1 = 2k_2$.

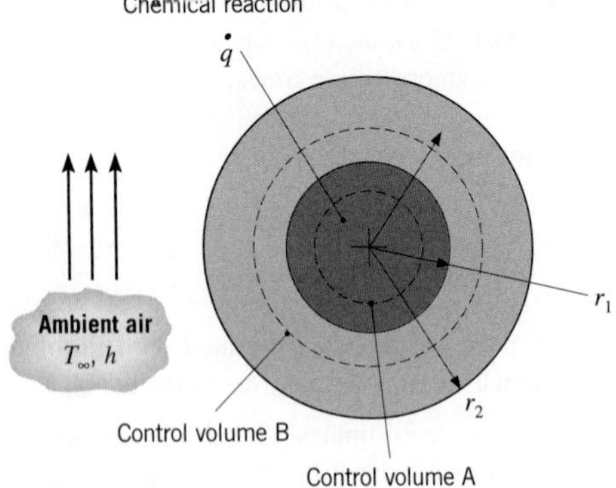

(a) By applying the conservation of energy principle to spherical control volume A, which is placed at an arbitrary location within the sphere, determine a relationship between the temperature gradient dT/dr and the local radius r, for $0 \le r \le r_1$.

(b) By applying the conservation of energy principle to spherical control volume B, which is placed at an arbitrary location within the spherical shell, determine a relationship between the temperature gradient dT/dr and the local radius r, for $r_1 \le r \le r_2$.

(c) On $T - r$ coordinates, sketch the temperature distribution over the range $0 \le r \le r_2$.

2.64 A *long* cylindrical rod, initially at a uniform temperature T_i, is suddenly immersed in a *large* container of liquid at $T_\infty < T_i$. Sketch the temperature distribution within the rod, $T(r)$, at the initial time, at steady state, and at two intermediate times. *On the same graph*, carefully sketch the temperature distributions that would occur at the same times within a second rod that is the same size as the first rod. The densities and specific heats of the two rods are identical, but the thermal conductivity of the second rod is very large. Which rod will approach steady-state conditions sooner? Write the appropriate boundary conditions that would be applied at $r = 0$ and $r = D/2$ for either rod.

2.65 A plane wall of thickness $L = 0.1$ m experiences uniform volumetric heating at a rate $\dot{q}$. One surface of the wall ($x = 0$) is insulated, and the other surface is exposed to a fluid at $T_\infty = 20°C$, with convection heat transfer characterized by $h = 1000\ W/m^2 \cdot K$. Initially, the temperature distribution in the wall is $T(x, 0) = a + bx^2$, where $a = 300°C$, $b = -1.0 \times 10^4 °C/m^2$, and x is in meters. Suddenly, the volumetric heat generation is deactivated ($\dot{q} = 0$ for $t \ge 0$), while convection heat transfer continues to occur at $x = L$. The properties of the wall are $\rho = 7000\ kg/m^3$, $c_p = 450\ J/kg \cdot K$, and $k = 90\ W/m \cdot K$.

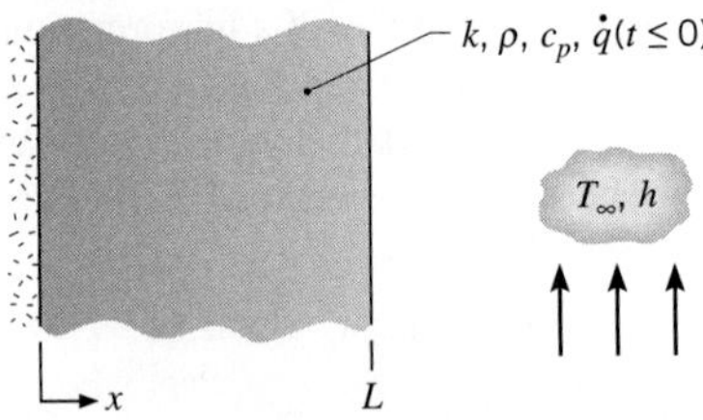

(a) Determine the magnitude of the volumetric energy generation rate $\dot{q}$ associated with the initial condition ($t < 0$).

(b) On $T - x$ coordinates, sketch the temperature distribution for the following conditions: initial condition ($t < 0$), steady-state condition ($t \to \infty$), and two intermediate conditions.

(c) On $q''_x - t$ coordinates, sketch the variation with time of the heat flux at the boundary exposed to the convection process, $q''_x(L, t)$. Calculate the corresponding value of the heat flux at $t = 0$, $q''_x(L, 0)$.

(d) Calculate the amount of energy removed from the wall per unit area (J/m²) by the fluid stream as the wall cools from its initial to steady-state condition.

2.66 A plane wall that is insulated on one side ($x = 0$) is initially at a uniform temperature T_i, when its exposed surface at $x = L$ is suddenly raised to a temperature T_s.

(a) Verify that the following equation satisfies the heat equation and boundary conditions:

$$\frac{T(x, t) - T_s}{T_i - T_s} = C_1 \exp\left(-\frac{\pi^2}{4}\frac{\alpha t}{L^2}\right)\cos\left(\frac{\pi}{2}\frac{x}{L}\right)$$

where C_1 is a constant and α is the thermal diffusivity.

(b) Obtain expressions for the heat flux at $x = 0$ and $x = L$.

(c) Sketch the temperature distribution $T(x)$ at $t = 0$, at $t \rightarrow \infty$, and at an intermediate time. Sketch the variation with time of the heat flux at $x = L$, $q''_L(t)$.

(d) What effect does α have on the thermal response of the material to a change in surface temperature?

2.67 A composite one-dimensional plane wall is of overall thickness $2L$. Material A spans the domain $-L \leq x < 0$ and experiences an *exothermic* chemical reaction leading to a uniform volumetric generation rate of $\dot{q}_A$. Material B spans the domain $0 \leq x \leq L$ and undergoes an *endothermic* chemical reaction corresponding to a uniform volumetric generation rate of $\dot{q}_B = -\dot{q}_A$. The surfaces at $x = \pm L$ are insulated. Sketch the steady-state temperature and heat flux distributions $T(x)$ and $q''_x(x)$, respectively, over the domain $-L \leq x \leq L$ for $k_A = k_B$, $k_A = 0.5k_B$, and $k_A = 2k_B$. Point out the important features of the distributions you have drawn. If $\dot{q}_B = -2\dot{q}_A$, can you sketch the steady-state temperature distribution?

2.68 Typically, air is heated in a hair dryer by blowing it across a coiled wire through which an electric current is passed. Thermal energy is generated by electric resistance heating within the wire and is transferred by convection from the surface of the wire to the air. Consider conditions for which the wire is initially at room temperature, T_i, and resistance heating is concurrently initiated with airflow at $t = 0$.

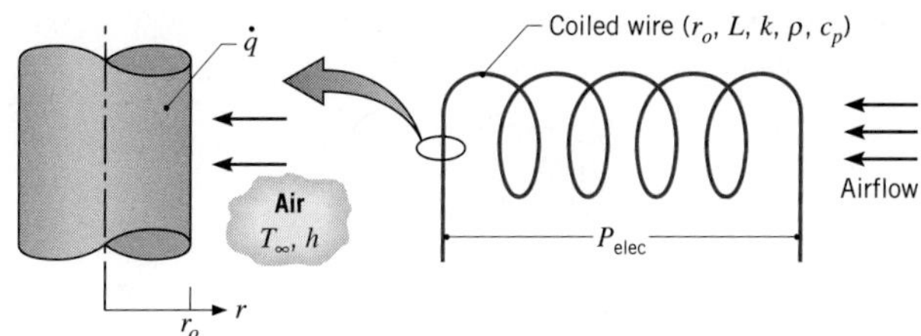

(a) For a wire radius r_o, an air temperature T_∞, and a convection coefficient h, write the form of the heat equation and the boundary/initial conditions that govern the transient thermal response, $T(r, t)$, of the wire.

(b) If the length and radius of the wire are 500 mm and 1 mm, respectively, what is the volumetric rate of thermal energy generation for a power consumption of $P_{elec} = 500$ W? What is the convection heat flux under steady-state conditions?

(c) On $T - r$ coordinates, sketch the temperature distributions for the following conditions: initial condition ($t \leq 0$), steady-state condition ($t \rightarrow \infty$), and for two intermediate times.

(d) On $q''_r - t$ coordinates, sketch the variation of the heat flux with time for locations at $r = 0$ and $r = r_o$.

2.69 The steady-state temperature distribution in a composite plane wall of three different materials, each of constant thermal conductivity, is shown.

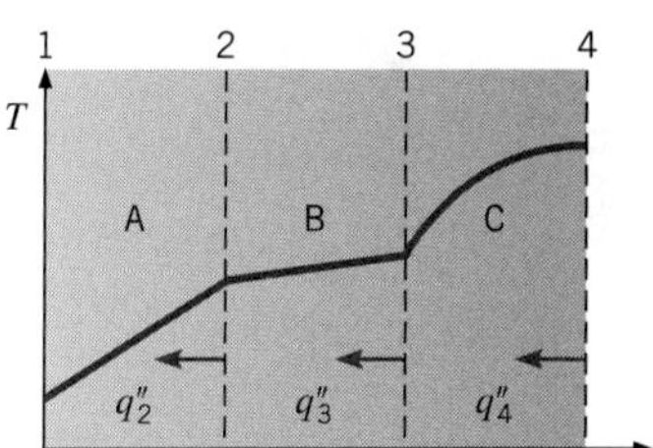

(a) Comment on the relative magnitudes of q''_2 and q''_3, and of q''_3 and q''_4.

(b) Comment on the relative magnitudes of k_A and k_B, and of k_B and k_C.

(c) Sketch the heat flux as a function of x.

CHAPTER 3

One-Dimensional, Steady-State Conduction

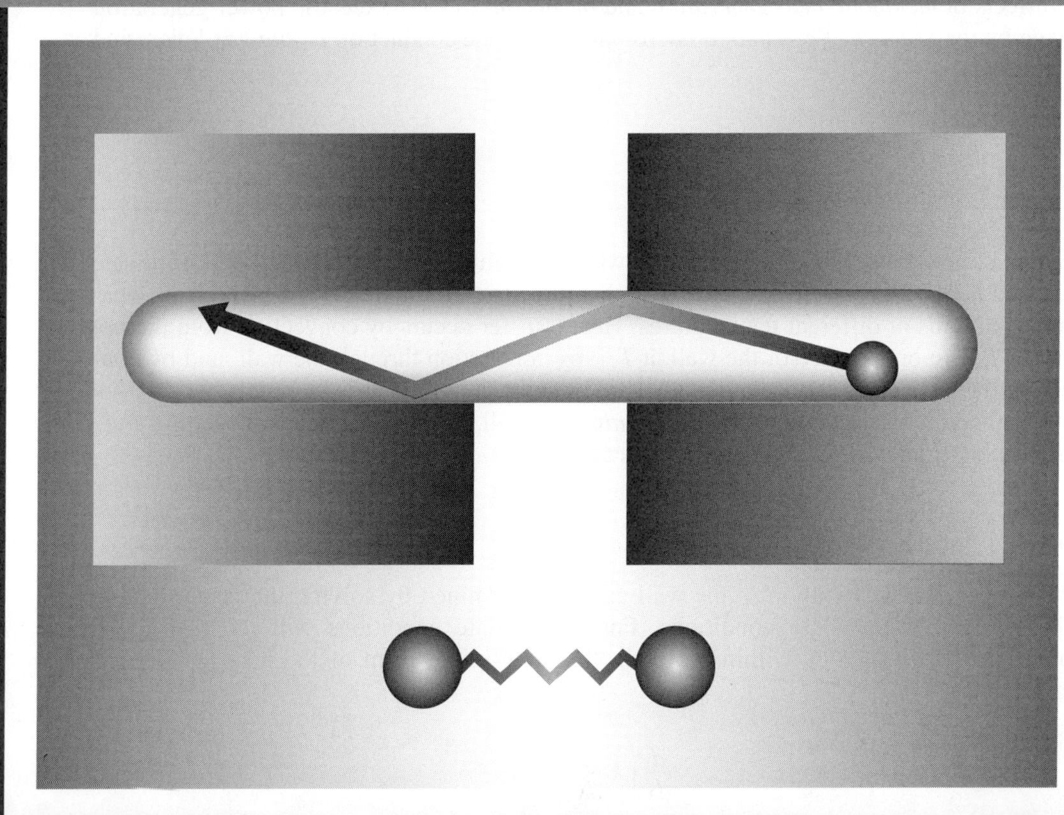

In this chapter we treat situations for which heat is transferred by diffusion under one-dimensional, steady-state conditions. The term *one-dimensional* refers to the fact that only one coordinate is needed to describe the spatial variation of the dependent variables. Hence, in a one-dimensional system, temperature gradients exist along only a single coordinate direction, and heat transfer occurs exclusively in that direction. The system is characterized by *steady-state* conditions if the temperature at each point is independent of time. Despite their inherent simplicity, one-dimensional, steady-state models may be used to accurately represent numerous engineering systems.

We begin our consideration of one-dimensional, steady-state conduction by discussing heat transfer with no internal generation of thermal energy (Sections 3.1 through 3.4). The objective is to determine expressions for the temperature distribution and heat transfer rate in common (planar, cylindrical, and spherical) geometries. For such geometries, an additional objective is to introduce the concept of *thermal resistance* and to show how *thermal circuits* may be used to model heat flow, much as electrical circuits are used for current flow. The effect of internal heat generation is treated in Section 3.5, and again our objective is to obtain expressions for determining temperature distributions and heat transfer rates. In Section 3.6, we consider the special case of one-dimensional, steady-state conduction for *extended surfaces*. In their most common form, these surfaces are termed *fins* and are used to *enhance* heat transfer by convection to an adjoining fluid. In addition to determining related temperature distributions and heat rates, our objective is to introduce *performance parameters* that may be used to determine their efficacy. Finally, in Sections 3.7 through 3.9 we apply heat transfer and thermal resistance concepts to the human body, including the effects of *metabolic heat generation* and *perfusion*; to thermoelectric power generation driven by the *Seebeck effect*; and to micro- and nanoscale conduction in *thin gas layers* and *thin solid films*.

3.1 The Plane Wall

For one-dimensional conduction in a plane wall, temperature is a function of the x-coordinate only and heat is transferred exclusively in this direction. In Figure 3.1a, a plane wall separates two fluids of different temperatures. Heat transfer occurs by convection from the hot fluid at $T_{\infty,1}$ to one surface of the wall at $T_{s,1}$, by conduction through the wall, and by convection from the other surface of the wall at $T_{s,2}$ to the cold fluid at $T_{\infty,2}$.

We begin by considering conditions *within* the wall. We first determine the temperature distribution, from which we can then obtain the conduction heat transfer rate.

3.1.1 Temperature Distribution

The temperature distribution in the wall can be determined by solving the heat equation with the proper boundary conditions. For steady-state conditions with no distributed source or sink of energy within the wall, the appropriate form of the heat equation is Equation 2.23

$$\frac{d}{dx}\left(k\frac{dT}{dx}\right) = 0 \tag{3.1}$$

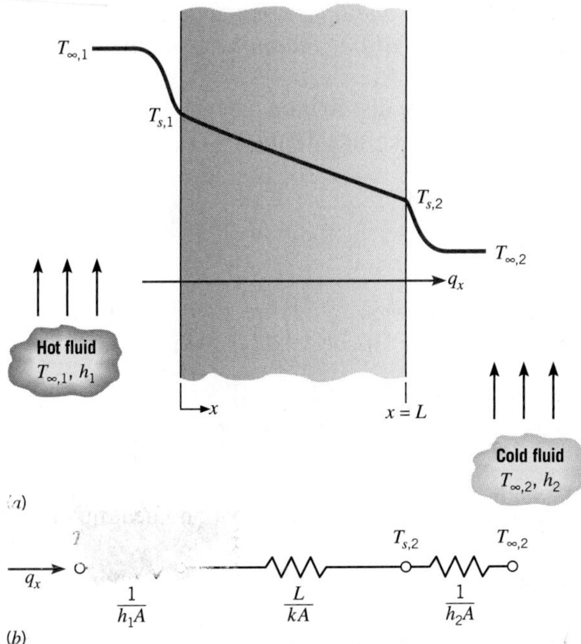

FIGURE 3.1 Heat transfer through a plane wall. (*a*) Temperature distribution. (*b*) Equivalent thermal circuit.

Hence, from Equation 2.2, it follows that, for *one-dimensional, steady-state conduction in a plane wall with no heat generation, the heat flux is a constant, independent of x.* If the thermal conductivity of the wall material is assumed to be constant, the equation may be integrated twice to obtain the *general solution*

$$T(x) = C_1x + C_2 \tag{3.2}$$

To obtain the constants of integration, C_1 and C_2, boundary conditions must be introduced. We choose to apply conditions of the first kind at $x = 0$ and $x = L$, in which case

$$T(0) = T_{s,1} \quad \text{and} \quad T(L) = T_{s,2}$$

Applying the condition at $x = 0$ to the general solution, it follows that

$$T_{s,1} = C_2$$

Similarly, at $x = L$,

$$T_{s,2} = C_1L + C_2 = C_1L + T_{s,1}$$

in which case

$$\frac{T_{s,2} - T_{s,1}}{L} = C_1$$

Substituting into the general solution, the temperature distribution is then

$$T(x) = (T_{s,2} - T_{s,1})\frac{x}{L} + T_{s,1} \tag{3.3}$$

From this result it is evident that, *for one-dimensional, steady-state conduction in a plane wall with no heat generation and constant thermal conductivity, the temperature varies linearly with x.*

Now that we have the temperature distribution, we may use Fourier's law, Equation 2.1, to determine the conduction heat transfer rate. That is,

$$q_x = -kA\frac{dT}{dx} = \frac{kA}{L}(T_{s,1} - T_{s,2}) \tag{3.4}$$

Note that A is the area of the wall *normal* to the direction of heat transfer and, for the plane wall, it is a constant independent of x. The heat flux is then

$$q''_x = \frac{q_x}{A} = \frac{k}{L}(T_{s,1} - T_{s,2}) \tag{3.5}$$

Equations 3.4 and 3.5 indicate that both the heat rate q_x and heat flux q''_x are constants, independent of x.

In the foregoing paragraphs we have used the *standard approach* to solving conduction problems. That is, the general solution for the temperature distribution is first obtained by solving the appropriate form of the heat equation. The boundary conditions are then applied to obtain the particular solution, which is used with Fourier's law to determine the heat transfer rate. Note that we have opted to prescribe surface temperatures at $x = 0$ and $x = L$ as boundary conditions, even though it is the fluid temperatures, not the surface temperatures, that are typically known. However, since adjoining fluid and surface temperatures are easily related through a surface energy balance (see Section 1.3.1), it is a simple matter to express Equations 3.3 through 3.5 in terms of fluid, rather than surface, temperatures. Alternatively, equivalent results could be obtained directly by using the surface energy balances as boundary conditions of the third kind in evaluating the constants of Equation 3.2 (see Problem 3.1).

3.1.2 Thermal Resistance

At this point we note that, for the special case of one-dimensional heat transfer with no internal energy generation and with constant properties, a very important concept is suggested by Equation 3.4. In particular, an analogy exists between the diffusion of heat and electrical charge. Just as an electrical resistance is associated with the conduction of electricity, a thermal resistance may be associated with the conduction of heat. Defining resistance as the ratio of a driving potential to the corresponding transfer rate, it follows from Equation 3.4 that the *thermal resistance for conduction* in a plane wall is

$$R_{t,\text{cond}} \equiv \frac{T_{s,1} - T_{s,2}}{q_x} = \frac{L}{kA} \tag{3.6}$$

Similarly, for electrical conduction in the same system, Ohm's law provides an electrical resistance of the form

$$R_e = \frac{E_{s,1} - E_{s,2}}{I} = \frac{L}{\sigma A} \tag{3.7}$$

The analogy between Equations 3.6 and 3.7 is obvious. A thermal resistance may also be associated with heat transfer by convection at a surface. From Newton's law of cooling,

$$q = hA(T_s - T_\infty) \tag{3.8}$$

The *thermal resistance for convection* is then

$$R_{t,\text{conv}} \equiv \frac{T_s - T_\infty}{q} = \frac{1}{hA} \tag{3.9}$$

Circuit representations provide a useful tool for both conceptualizing and quantifying heat transfer problems. The *equivalent thermal circuit* for the plane wall with convection surface conditions is shown in Figure 3.1*b*. The heat transfer rate may be determined from separate consideration of each element in the network. Since q_x is constant throughout the network, it follows that

$$q_x = \frac{T_{\infty,1} - T_{s,1}}{1/h_1A} = \frac{T_{s,1} - T_{s,2}}{L/kA} = \frac{T_{s,2} - T_{\infty,2}}{1/h_2A} \tag{3.10}$$

In terms of the *overall temperature difference*, $T_{\infty,1} - T_{\infty,2}$, and the *total thermal resistance*, R_{tot}, the heat transfer rate may also be expressed as

$$q_x = \frac{T_{\infty,1} - T_{\infty,2}}{R_{\text{tot}}} \tag{3.11}$$

Because the conduction and convection resistances are in series and may be summed, it follows that

$$R_{\text{tot}} = \frac{1}{h_1A} + \frac{L}{kA} + \frac{1}{h_2A} \tag{3.12}$$

Radiation exchange between the surface and surroundings may also be important if the convection heat transfer coefficient is small (as it often is for natural convection in a gas). A *thermal resistance for radiation* may be defined by reference to Equation 1.8:

$$R_{t,\text{rad}} = \frac{T_s - T_{\text{sur}}}{q_{\text{rad}}} = \frac{1}{h_rA} \tag{3.13}$$

For radiation between a surface and *large surroundings*, h_r is determined from Equation 1.9. Surface radiation and convection resistances act in parallel, and if $T_\infty = T_{\text{sur}}$, they may be combined to obtain a single, effective surface resistance.

3.1.3 The Composite Wall

Equivalent thermal circuits may also be used for more complex systems, such as *composite walls.* Such walls may involve any number of series and parallel thermal resistances due to layers of different materials. Consider the series composite wall of Figure 3.2. The one-dimensional heat transfer rate for this system may be expressed as

$$q_x = \frac{T_{\infty,1} - T_{\infty,4}}{\Sigma R_t} \tag{3.14}$$

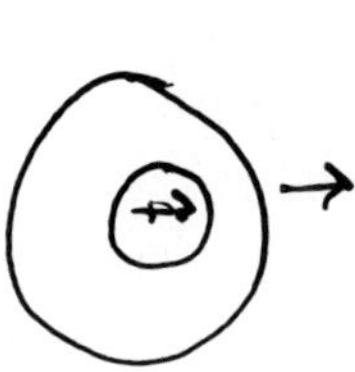

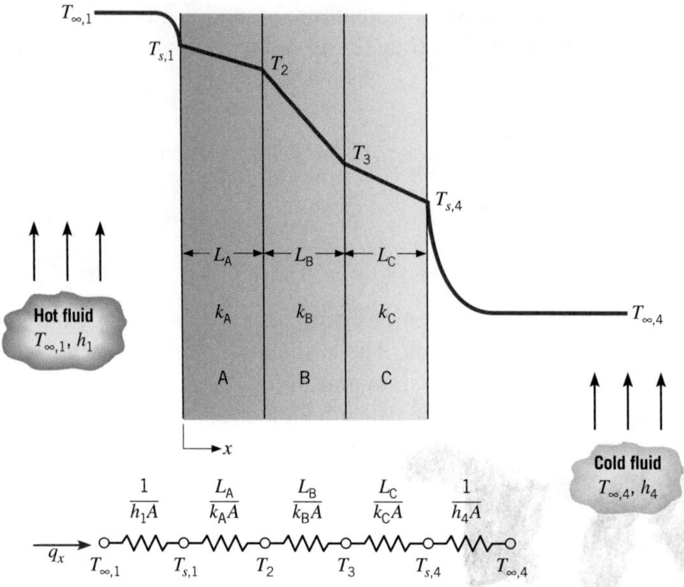

FIGURE 3.2 Equivalent thermal circuit for a series composite wall.

where $T_{\infty,1} - T_{\infty,4}$ is the *overall* temperature difference, and the summation includes all thermal resistances. Hence

$$q_x = \frac{T_{\infty,1} - T_{\infty,4}}{[(1/h_1A) + (L_A/k_AA) + (L_B/k_BA) + (L_C/k_CA) + (1/h_4A)]} \tag{3.15}$$

Alternatively, the heat transfer rate can be related to the temperature difference and resistance associated with each element. For example,

$$q_x = \frac{T_{\infty,1} - T_{s,1}}{(1/h_1A)} = \frac{T_{s,1} - T_2}{(L_A/k_AA)} = \frac{T_2 - T_3}{(L_B/k_BA)} = \cdots \tag{3.16}$$

With composite systems, it is often convenient to work with an *overall heat transfer coefficient U*, which is defined by an expression analogous to Newton's law of cooling. Accordingly,

$$q_x \equiv UA\,\Delta T \tag{3.17}$$

where ΔT is the overall temperature difference. The overall heat transfer coefficient is related to the total thermal resistance, and from Equations 3.14 and 3.17 we see that $UA = 1/R_{tot}$. Hence, for the composite wall of Figure 3.2,

$$U = \frac{1}{R_{tot}A} = \frac{1}{[(1/h_1) + (L_A/k_A) + (L_B/k_B) + (L_C/k_C) + (1/h_4)]} \tag{3.18}$$

In general, we may write

$$R_{tot} = \sum R_t = \frac{\Delta T}{q} = \frac{1}{UA} \tag{3.19}$$

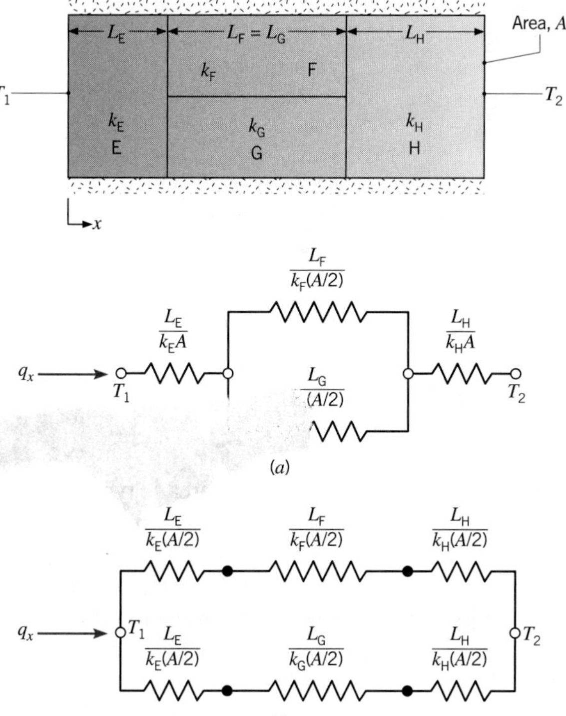

FIGURE 3.3 Equivalent thermal circuits for a series–parallel composite wall.

Composite walls may also be characterized by series–parallel configurations, such as that shown in Figure 3.3. Although the heat flow is now multidimensional, it is often reasonable to assume one-dimensional conditions. Subject to this assumption, two different thermal circuits may be used. For case (*a*) it is presumed that surfaces normal to the *x*-direction are isothermal, whereas for case (*b*) it is assumed that surfaces parallel to the *x*-direction are adiabatic. Different results are obtained for R_{tot}, and the corresponding values of q bracket the actual heat transfer rate. These differences increase with increasing $|k_F - k_G|$, as multidimensional effects become more significant.

3.1.4 Contact Resistance

Although neglected until now, it is important to recognize that, in composite systems, the temperature drop across the interface between materials may be appreciable. This temperature change is attributed to what is known as the *thermal contact resistance*, $R_{t,c}$. The effect is shown in Figure 3.4, and for a unit area of the interface, the resistance is defined as

$$R''_{t,c} = \frac{T_A - T_B}{q''_x} \tag{3.20}$$

The existence of a finite contact resistance is due principally to surface roughness effects. Contact spots are interspersed with gaps that are, in most instances, air filled. Heat transfer is therefore due to conduction across the actual contact area and to conduction and/or radiation across the gaps. The contact resistance may be viewed as two parallel resistances: that due to

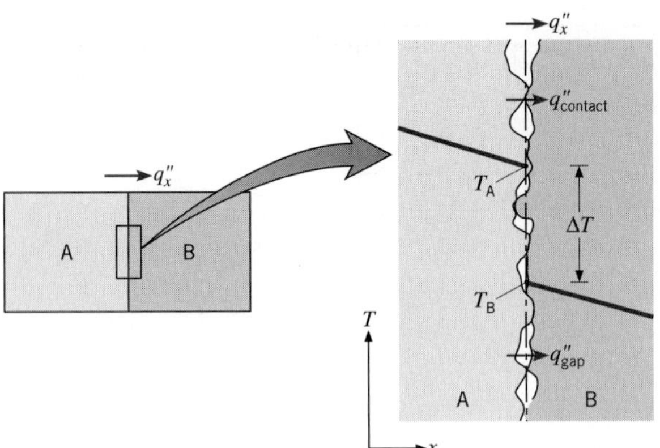

FIGURE 3.4 Temperature drop due to thermal contact resistance.

the contact spots and that due to the gaps. The contact area is typically small, and, especially for rough surfaces, the major contribution to the resistance is made by the gaps.

For solids whose thermal conductivities exceed that of the interfacial fluid, the contact resistance may be reduced by increasing the area of the contact spots. Such an increase may be effected by increasing the joint pressure and/or by reducing the roughness of the mating surfaces. The contact resistance may also be reduced by selecting an interfacial fluid of large thermal conductivity. In this respect, no fluid (an evacuated interface) eliminates conduction across the gap, thereby increasing the contact resistance. Likewise, if the characteristic gap width L becomes small (as, for example, in the case of very smooth surfaces in contact), L/λ_{mfp} can approach values for which the thermal conductivity of the interfacial gas is reduced by microscale effects, as discussed in Section 2.2.

Although theories have been developed for predicting $R''_{t,c}$, the most reliable results are those that have been obtained experimentally. The effect of loading on metallic interfaces can be seen in Table 3.1*a*, which presents an approximate range of thermal resistances under vacuum conditions. The effect of interfacial fluid on the thermal resistance of an aluminum interface is shown in Table 3.1*b*.

Contrary to the results of Table 3.1, many applications involve contact between dissimilar solids and/or a wide range of possible interstitial (filler) materials (Table 3.2). Any interstitial substance that fills the gap between contacting surfaces and whose thermal conductivity exceeds that of air will decrease the contact resistance. Two classes of materials that are well suited for this purpose are soft metals and thermal greases. The metals, which include

TABLE 3.1 Thermal contact resistance for (*a*) metallic interfaces under vacuum conditions and (*b*) aluminum interface (10-μm surface roughness, 10^5 N/m^2) with different interfacial fluids [1]

Thermal Resistance, $R''_{t,c} \times 10^4$ (m$^2 \cdot$K/W)				
(*a*) Vacuum Interface			**(*b*) Interfacial Fluid**	
Contact pressure	100 kN/m^2	10,000 kN/m^2	Air	2.75
Stainless steel	6–25	0.7–4.0	Helium	1.05
Copper	1–10	0.1–0.5	Hydrogen	0.720
Magnesium	1.5–3.5	0.2–0.4	Silicone oil	0.525
Aluminum	1.5–5.0	0.2–0.4	Glycerine	0.265

TABLE 3.2 Thermal resistance of representative solid/solid interfaces

Interface	$R''_{t,c} \times 10^4$ ($m^2 \cdot K/W$)	Source
Silicon chip/lapped aluminum in air (27–500 kN/m^2)	0.3–0.6	[2]
Aluminum/aluminum with indium foil filler (~100 kN/m^2)	~0.07	[1, 3]
Stainless/stainless with indium foil filler (~3500 kN/m^2)	~0.04	[1, 3]
Aluminum/aluminum with metallic (Pb) coating	0.01–0.1	[4]
[illegible]/aluminum with Dow Corning [illegible] (~100 kN/m^2)	~0.07	[1, 3]
Stainless/stainless with Dow Corning 340 grease (~3500 kN/m^2)	~0.04	[1, 3]
Silicon chip/aluminum with 0.02-mm epoxy	0.2–0.9	[5]
Brass/brass with 15-μm tin solder	0.025–0.14	[6]

indium, lead, tin, and silver, may be inserted as a thin foil or applied as a thin coating to one of the parent materials. Silicon-based thermal greases are attractive on the basis of their ability to completely fill the interstices with a material whose thermal conductivity is as much as 50 times that of air.

Unlike the foregoing interfaces, which are not permanent, many interfaces involve permanently bonded joints. The joint could be formed from an epoxy, a soft solder rich in lead, or a hard solder such as a gold/tin alloy. Due to interface resistances between the parent and bonding materials, the actual thermal resistance of the joint exceeds the theoretical value (L/k) computed from the thickness L and thermal conductivity k of the joint material. The thermal resistance of epoxied and soldered joints is also adversely affected by voids and cracks, which may form during manufacture or as a result of thermal cycling during normal operation.

Comprehensive reviews of thermal contact resistance results and models are provided by Snaith et al. [3], Madhusudana and Fletcher [7], and Yovanovich [8].

3.1.5 Porous Media

In many applications, heat transfer occurs within *porous media* that are combinations of a stationary solid and a fluid. When the fluid is *either* a gas *or* a liquid, the resulting porous medium is said to be *saturated.* In contrast, all three phases coexist in an *unsaturated* porous medium. Examples of porous media include beds of powder with a fluid occupying the interstitial regions between individual granules, as well as the insulation systems and nanofluids of Section 2.2.1. A saturated porous medium that consists of a stationary solid phase through which a fluid flows is referred to as a *packed bed* and is discussed in Section 7.8.

Consider a saturated porous medium that is subjected to surface temperatures T_1 at $x = 0$ and T_2 at $x = L$, as shown in Figure 3.5*a*. After steady-state conditions are reached and if $T_1 > T_2$, the heat rate may be expressed as

$$q_x = \frac{k_{\text{eff}} A}{L} (T_1 - T_2) \tag{3.21}$$

where k_{eff} is an effective thermal conductivity. Equation 3.21 is valid if fluid motion, as well as radiation heat transfer *within* the medium, are negligible. The effective thermal conductivity varies with the porosity or void fraction of the medium ε which is defined as the volume of fluid relative to the total volume (solid and fluid). In addition, k_{eff} depends on the thermal conductivities of each of the phases and, in this discussion, it is assumed that $k_s > k_f$. The detailed solid phase geometry, for example the size distribution and packing arrangement of individual powder particles, also affects the value of k_{eff}. Contact resistances that might evolve at interfaces between adjacent solid particles can impact the value of k_{eff}. As discussed in Section 2.2.1, nanoscale phenomena might also influence the effective thermal conductivity. Hence, prediction of k_{eff} can be difficult and, in general, requires detailed knowledge of parameters that might not be readily available.

Despite the complexity of the situation, the value of the effective thermal conductivity may be bracketed by considering the composite walls of Figures 3.5*b* and 3.5*c*. In Figure 3.5*b*, the medium is modeled as an equivalent, series composite wall consisting of a fluid region of length εL and a solid region of length $(1-\varepsilon)L$. Applying Equations 3.17 and 3.18 to this model for which there is no convection ($h_1 = h_2 = 0$) and only two conduction terms, it follows that

$$q_x = \frac{A\,\Delta T}{(1-\varepsilon)L/k_s + \varepsilon L/k_f} \tag{3.22}$$

Equating this result to Equation 3.21, we then obtain

$$k_{\text{eff,min}} = \frac{1}{(1-\varepsilon)/k_s + \varepsilon/k_f} \tag{3.23}$$

Alternatively, the medium of Figure 3.5*a* could be described by the equivalent, parallel composite wall consisting of a fluid region of width εw and a solid region of width $(1-\varepsilon)w$, as shown in Figure 3.5*c*. Combining Equation 3.21 with an expression for the equivalent resistance of two resistors in parallel gives

$$k_{\text{eff,max}} = \varepsilon k_f + (1-\varepsilon)k_s \tag{3.24}$$

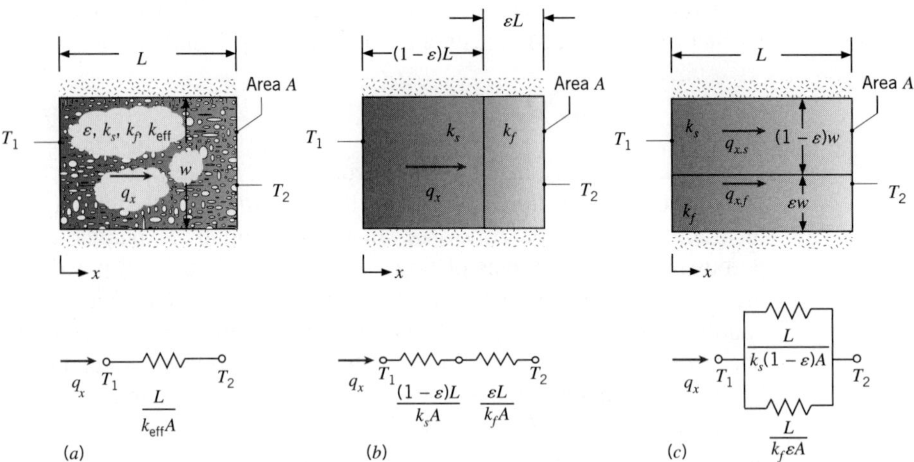

FIGURE 3.5 A porous medium. (*a*) The medium and its properties. (*b*) Series thermal resistance representation. (*c*) Parallel resistance representation.

While Equations 3.23 and 3.24 provide the minimum and maximum possible values of k_{eff}, more accurate expressions have been derived for specific composite systems within which nanoscale effects are negligible. Maxwell [9] derived an expression for the effective electrical conductivity of a solid matrix interspersed with uniformly distributed, noncontacting spherical inclusions. Noting the analogy between Equations 3.6 and 3.7, Maxwell's result may be used to determine the effective thermal conductivity of a saturated porous medium consisting of an interconnected solid phase within which a dilute distribution of spherical fluid regions exists, resulting in an expression of the form [10]

$$k_{\text{eff}} = \left[\frac{k_f + 2k_s - 2\varepsilon(k_s - k_f)}{k_f + 2k_s + \varepsilon(k_s - k_f)}\right]k_s \tag{3.25}$$

Equation 3.25 is valid for relatively small porosities ($\varepsilon \lesssim 0.25$) as shown schematically in Figure 3.5*a* [11]. It is equi[illegible]he expression introduced in Example 2.2 for a fluid that contains a dilute mixture [illegible] particles, but with reversal of the fluid and solid.

When analyzing conduct[illegible]thin porous media, it is important to consider the potential directional dependence of the effective thermal conductivity. For example, the media represented in Figure 3.5*b* or Figure 3.5*c* would *not* be characterized by isotropic properties, since the effective thermal conductivity in the x-direction is clearly different from values of k_{eff} in the vertical direction. Hence, although Equations 3.23 and 3.24 can be used to bracket the actual value of the effective thermal conductivity, they will generally overpredict the possible range of k_{eff} for isotropic media. For isotropic media, expressions have been developed to determine the minimum and maximum possible effective thermal conductivities based solely on knowledge of the porosity and the thermal conductivities of the solid and fluid. Specifically, the *maximum* possible value of k_{eff} in an isotropic porous medium is given by Equation 3.25, which corresponds to an interconnected, high thermal conductivity solid phase. The *minimum* possible value of k_{eff} for an isotropic medium corresponds to the case where the fluid phase forms long, randomly oriented fingers within the medium [12]. Additional information regarding conduction in saturated porous media is available [13].

IHT

EXAMPLE 3.1

In Example 1.7, we calculated the heat loss rate from a human body in air and water environments. Now we consider the same conditions except that the surroundings (air or water) are at 10°C. To reduce the heat loss rate, the person wears special sporting gear (snow suit and wet suit) made from a nanostructured silica aerogel insulation with an extremely low thermal conductivity of 0.014 W/m · K. The emissivity of the outer surface of the snow and wet suits is 0.95. What thickness of aerogel insulation is needed to reduce the heat loss rate to 100 W (a typical metabolic heat generation rate) in air and water? What are the resulting skin temperatures?

SOLUTION

Known: Inner surface temperature of a skin/fat layer of known thickness, thermal conductivity, and surface area. Thermal conductivity and emissivity of snow and wet suits. Ambient conditions.

Find: Insulation thickness needed to reduce heat loss rate to 100 W and corresponding skin temperature.

Schematic:

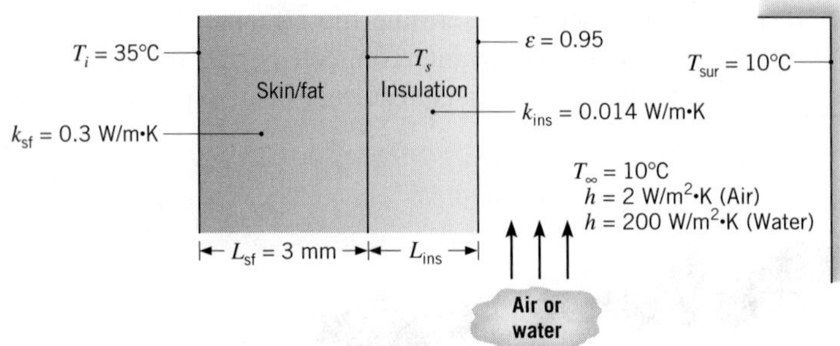

Assumptions:

1. Steady-state conditions.
2. One-dimensional heat transfer by conduction through the skin/fat and insulation layers.
3. Contact resistance is negligible.
4. Thermal conductivities are uniform.
5. Radiation exchange between the skin surface and the surroundings is between a small surface and a large enclosure at the air temperature.
6. Liquid water is opaque to thermal radiation.
7. Solar radiation is negligible.
8. Body is completely immersed in water in part 2.

Analysis: The thermal circuit can be constructed by recognizing that resistance to heat flow is associated with conduction through the skin/fat and insulation layers and convection and radiation at the outer surface. Accordingly, the circuit and the resistances are of the following form (with $h_r = 0$ for water):

$q \rightarrow T_i$ — $\frac{L_{sf}}{k_{sf}A}$ — T_s — $\frac{L_{ins}}{k_{ins}A}$ — [parallel: $\frac{1}{h_r A}$ — T_{sur}; $\frac{1}{hA}$ — T_∞], $T_{sur} = T_\infty$

The total thermal resistance needed to achieve the desired heat loss rate is found from Equation 3.19,

$$R_{tot} = \frac{T_i - T_\infty}{q} = \frac{(35 - 10)\text{ K}}{100\text{ W}} = 0.25\text{ K/W}$$

The total thermal resistance between the inside of the skin/fat layer and the cold surroundings includes conduction resistances for the skin/fat and insulation layers and an effective resistance associated with convection and radiation, which act in parallel. Hence,

$$R_{\text{tot}} = \frac{L_{\text{sf}}}{k_{\text{sf}}A} + \frac{L_{\text{ins}}}{k_{\text{ins}}A} + \left(\frac{1}{1/hA} + \frac{1}{1/h_rA}\right)^{-1} = \frac{1}{A}\left(\frac{L_{\text{sf}}}{k_{\text{sf}}} + \frac{L_{\text{ins}}}{k_{\text{ins}}} + \frac{1}{h + h_r}\right)$$

This equation can be solved for the insulation thickness.

Air

The radiation heat transfer coefficient is approximated as having the same value as in Example 1.7: $h_r = 5.9\ \text{W/m}^2\cdot\text{K}$.

$$L_{\text{ins}} = k_{\text{ins}}\left[AR_{\text{tot}} - \frac{L_{\text{sf}}}{k_{\text{sf}}} - \frac{1}{h + h_r}\right]$$

$$= 0.014\ \text{W/m}\cdot\text{K}\left[1.8\ \text{m}^2 \times 0.25\ \text{K/W} - \frac{3 \times 10^{-3}\ \text{m}}{0.3\ \text{W/m}\cdot\text{K}} - \frac{1}{(2 + 5.9)\ \text{W/m}^2\cdot\text{K}}\right]$$

$$= 0.0044\ \text{m} = 4.4\ \text{mm} \qquad \triangleleft$$

Water

$$L_{\text{ins}} = k_{\text{ins}}\left[AR_{\text{tot}} - \frac{L_{\text{sf}}}{k_{\text{sf}}} - \frac{1}{h}\right]$$

$$= 0.014\ \text{W/m}\cdot\text{K}\left[1.8\ \text{m}^2 \times 0.25\ \text{K/W} - \frac{3 \times 10^{-3}\ \text{m}}{0.3\ \text{W/m}\cdot\text{K}} - \frac{1}{200\ \text{W/m}^2\cdot\text{K}}\right]$$

$$= 0.0061\ \text{m} = 6.1\ \text{mm} \qquad \triangleleft$$

These required thicknesses of insulation material can easily be incorporated into the snow and wet suits.

The skin temperature can be calculated by considering conduction through the skin/fat layer:

$$q = \frac{k_{\text{sf}}A(T_i - T_s)}{L_{\text{sf}}}$$

or solving for T_s,

$$T_s = T_i - \frac{qL_{\text{sf}}}{k_{\text{sf}}A} = 35^\circ\text{C} - \frac{100\ \text{W} \times 3 \times 10^{-3}\ \text{m}}{0.3\ \text{W/m}\cdot\text{K} \times 1.8\ \text{m}^2} = 34.4^\circ\text{C} \qquad \triangleleft$$

The skin temperature is the same in both cases because the heat loss rate and skin/fat properties are the same.

Comments:

1. The nanostructured silica aerogel is a porous material that is only about 5% solid. Its thermal conductivity is less than the thermal conductivity of the gas that fills its pores. As explained in Section 2.2, the reason for this seemingly impossible result is that the pore size is only around 20 nm, which reduces the mean free path of the gas and hence decreases its thermal conductivity.

2. By reducing the heat loss rate to 100 W, a person could remain in the cold environments indefinitely without becoming chilled. The skin temperature of 34.4°C would feel comfortable.

3. In the water case, the thermal resistance of the insulation dominates and all other resistances can be neglected.

4. The convection heat transfer coefficient associated with the air depends on the wind conditions, and it can vary over a broad range. As it changes, so will the outer surface temperature of the insulation layer. Since the radiation heat transfer coefficient depends on this temperature, it will also vary. We can perform a more complete analysis that takes this into account. The radiation heat transfer coefficient is given by Equation 1.9:

$$h_r = \varepsilon\sigma(T_{s,o} + T_{\text{sur}})(T_{s,o}^2 + T_{\text{sur}}^2) \tag{1}$$

Here $T_{s,o}$ is the outer surface temperature of the insulation layer, which can be calculated from

$$T_{s,o} = T_i - q\left[\frac{L_{\text{sf}}}{k_{\text{sf}}A} + \frac{L_{\text{ins}}}{k_{\text{ins}}A}\right] \tag{2}$$

Since this depends on the insulation thickness, we also need the previous equation for L_{ins}:

$$L_{\text{ins}} = k_{\text{ins}}\left(AR_{\text{tot}} - \frac{L_{\text{sf}}}{k_{\text{sf}}} - \frac{1}{h + h_r}\right) \tag{3}$$

With all other values known, these three equations can be solved for the required insulation thickness. Using all the values from above, these equations have been solved for values of h in the range $0 \leq h \leq 100\ \text{W/m}^2 \cdot \text{K}$, and the results are represented graphically.

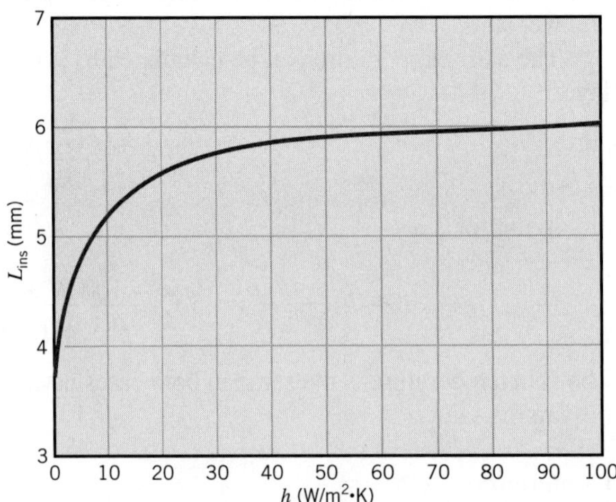

Increasing h reduces the corresponding convection resistance, which then requires additional insulation to maintain the heat transfer rate at 100 W. Once the heat transfer coefficient exceeds approximately $60\ \text{W/m}^2 \cdot \text{K}$, the convection resistance is negligible and further increases in h have little effect on the required insulation thickness.

The outer surface temperature and radiation heat transfer coefficient can also be calculated. As h increases from 0 to 100 W/m$^2 \cdot$K, $T_{s,o}$ decreases from 294 to 284 K, while h_r decreases from 5.2 to 4.9 W/m$^2 \cdot$K. The initial estimate of $h_r = 5.9$ W/m$^2 \cdot$K was not highly accurate. Using this more complete model of the radiation heat transfer, with $h = 2$ W/m$^2 \cdot$K, the radiation heat transfer coefficient is 5.1 W/m$^2 \cdot$K, and the required insulation thickness is 4.2 mm, close to the value calculated in the first part of the problem.

5. See Example 3.1 in *IHT*. This problem can also be solved using the thermal resistance network builder, *Models/Resistance Networks,* available in *IHT*.

EXAMPLE 3.2

A thin silicon chip and an 8-mm-thick aluminum substrate are separated by a 0.02-mm-thick epoxy joint. The chip and substrate are each 10 mm on a side, and their exposed surfaces are cooled by air, which is at a temperature of 25°C and provides a convection coefficient of 100 W/m$^2 \cdot$K. If the chip dissipates 10^4 W/m^2 under normal conditions, will it operate below a maximum allowable temperature of 85°C?

SOLUTION

Known: Dimensions, heat dissipation, and maximum allowable temperature of a silicon chip. Thickness of aluminum substrate and epoxy joint. Convection conditions at exposed chip and substrate surfaces.

Find: Whether maximum allowable temperature is exceeded.

Schematic:

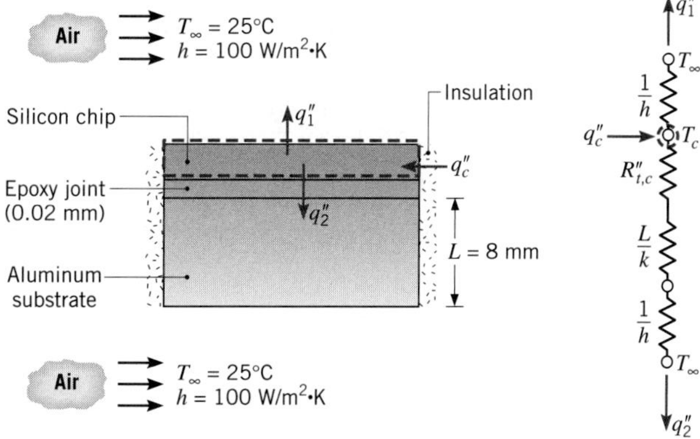

Assumptions:

1. Steady-state conditions.
2. One-dimensional conduction (negligible heat transfer from sides of composite).
3. Negligible chip thermal resistance (an isothermal chip).
4. Constant properties.
5. Negligible radiation exchange with surroundings.

Properties: Table A.1, pure aluminum ($T \sim 350$ K): $k = 239$ W/m·K.

Analysis: Heat dissipated in the chip is transferred to the air directly from the exposed surface and indirectly through the joint and substrate. Performing an energy balance on a control surface about the chip, it follows that, on the basis of a unit surface area,

$$q''_c = q''_1 + q''_2$$

or

$$q''_c = \frac{T_c - T_\infty}{(1/h)} + \frac{T_c - T_\infty}{R''_{t,c} + (L/k) + (1/h)}$$

To conservatively estimate T_c, the maximum possible value of $R''_{t,c} = 0.9 \times 10^{-4}$ m^2·K/W is obtained from Table 3.2. Hence

$$T_c = T_\infty + q''_c\left[h + \frac{1}{R''_{t,c} + (L/k) + (1/h)}\right]^{-1}$$

or

$$T_c = 25°\text{C} + 10^4 \text{ W/m}^2$$

$$\times\left[100 + \frac{1}{(0.9 + 0.33 + 100) \times 10^{-4}}\right]^{-1} \text{m}^2\cdot\text{K/W}$$

$$T_c = 25°\text{C} + 50.3°\text{C} = 75.3°\text{C}$$ ◁

Hence the chip will operate below its maximum allowable temperature.

Comments:

1. The joint and substrate thermal resistances are much less than the convection resistance. The joint resistance would have to increase to the unrealistically large value of 50×10^{-4} m^2·K/W, before the maximum allowable chip temperature would be exceeded.
2. The allowable power dissipation may be increased by increasing the convection coefficients, either by increasing the air velocity and/or by replacing the air with a more effective heat transfer fluid. Exploring this option for $100 \leq h \leq 2000$ W/m^2·K with $T_c = 85°$C, the following results are obtained.

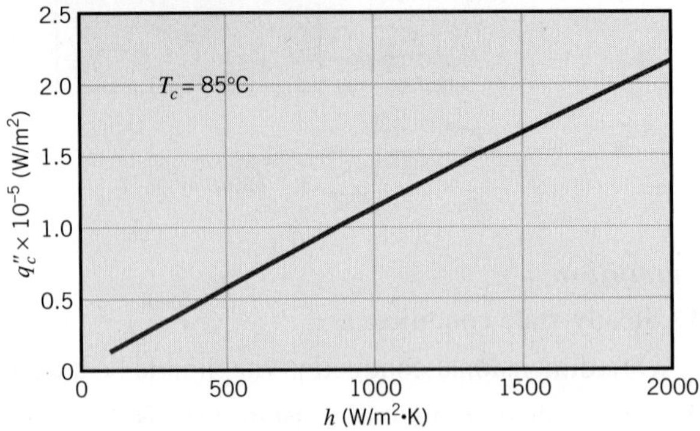

As $h \to \infty$, $q''_2 \to 0$ and virtually all of the chip power is transferred directly to the fluid stream.

3. As calculated, the *difference* between the air temperature ($T_\infty = 25°C$) and the chip temperature ($T_c = 75.3°C$) is 50.3 K. Keep in mind that this is a temperature *difference* and therefore is the same as 50.3°C.

4. Consider conditions for which airflow over the chip (upper) or substrate (lower) surface ceases due to a blockage in the air supply channel. If heat transfer from either surface is negligible, what are the resulting chip temperatures for $q''_c = 10^4\ W/m^2$? [Answer, 126°C or 125°C]

EXAMPLE 3.3

A photovoltaic panel consists of (top to bottom) a 3-mm-thick ceria-doped glass ($k_g = 1.4$ W/m·K), a 0.1-mm-thick optical grade adhesive ($k_a = 145$ W/m·K), a *very thin* layer of silicon within which solar energy is converted to electrical energy, a 0.1-mm-thick solder layer ($k_{sdr} = 50$ W/m·K), and a 2-mm-thick aluminum nitride substrate ($k_{an} = 120$ W/m·K). The solar-to-electrical conversion efficiency within the silicon layer η decreases with increasing silicon temperature, T_{si}, and is described by the expression $\eta = a - bT_{si}$, where $a = 0.553$ and $b = 0.001\ K^{-1}$. The temperature T is expressed in kelvins over the range $300\ K \leq T_{si} \leq 525\ K$. Of the incident solar irradiation, $G_S = 700\ W/m^2$, 7% is reflected from the top surface of the glass, 10% is absorbed at the top surface of the glass, and 83% is transmitted to and absorbed within the silicon layer. Part of the solar irradiation absorbed in the silicon is converted to thermal energy, and the remainder is converted to electrical energy. The glass has an emissivity of $\varepsilon = 0.90$, and the bottom as well as the sides of the panel are insulated. Determine the electric power P produced by an $L = 1$-m-long, $w = 0.1$-m-wide solar panel for conditions characterized by $h = 35\ W/m^2 \cdot K$ and $T_\infty = T_{sur} = 20°C$.

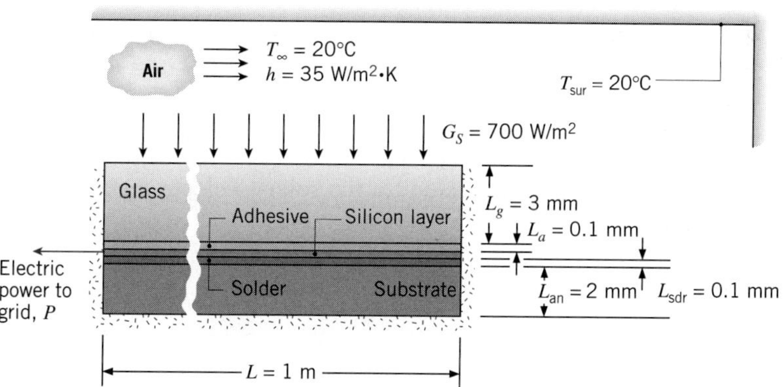

SOLUTION

Known: Dimensions and materials of a photovoltaic solar panel. Material properties, solar irradiation, convection coefficient and ambient temperature, emissivity of top panel surface and surroundings temperature. Partitioning of the solar irradiation, and expression for the solar-to-electrical conversion efficiency.

Find: Electric power produced by the photovoltaic panel.

Schematic:

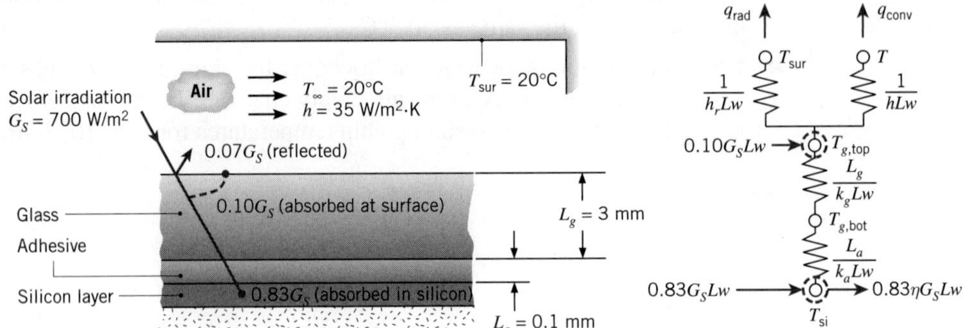

Assumptions:

1. Steady-state conditions.
2. One-dimensional heat transfer.
3. Constant properties.
4. Negligible thermal contact resistances.
5. Negligible temperature differences within the silicon layer.

Analysis: Recognize that there is no heat transfer to the bottom insulated surface of the solar panel. Hence, the solder layer and aluminum nitride substrate do not affect the solution, and all of the solar energy absorbed by the panel must ultimately leave the panel in the form of radiation and convection heat transfer from the top surface of the glass, and electric power to the grid, $P = \eta 0.83\, G_S L w$. Performing an energy balance on the node associated with the silicon layer yields

$$0.83\, G_S L w - \eta 0.83\, G_S L w = \frac{T_{si} - T_{g,top}}{\dfrac{L_a}{k_a L w} + \dfrac{L_g}{k_g L w}}$$

Substituting the expression for the solar-to-electrical conversion efficiency and simplifying leads to

$$0.83\, G_S(1 - a + bT_{si}) = \frac{T_{si} - T_{g,top}}{\dfrac{L_a}{k_a} + \dfrac{L_g}{k_g}} \qquad (1)$$

Performing a second energy balance on the node associated with the top surface of the glass gives

$$0.83\, G_S L w(1 - \eta) + 0.1\, G_S L w = hLw(T_{g,top} - T_\infty) + \varepsilon\sigma L w(T_{g,top}^4 - T_{sur}^4)$$

Substituting the expression for the solar-to-electrical conversion efficiency into the preceding equation and simplifying provides

$$0.83\, G_S(1 - a + bT_{si}) + 0.1\, G_S = h(T_{g,top} - T_\infty) + \varepsilon\sigma(T_{g,top}^4 - T_{sur}^4) \qquad (2)$$

Finally, substituting known values into Equations 1 and 2 and solving simultaneously yields $T_{si} = 307\ \text{K} = 34°\text{C}$, providing a solar-to-electrical conversion efficiency of $\eta = 0.553 - 0.001\ \text{K}^{-1} \times 307\ \text{K} = 0.247$. Hence, the power produced by the photovoltaic panel is

$$P = \eta 0.83\, G_S L w = 0.247 \times 0.83 \times 700\ \text{W/m}^2 \times 1\ \text{m} \times 0.1\ \text{m} = 14.3\ \text{W} \qquad \triangleleft$$

Comments:

1. The correct application of the conservation of energy requirement is crucial to determining the silicon temperature and the electric power. Note that solar energy is converted to *both* thermal and electrical energy, and the thermal circuit is used to quantify *only* the thermal energy transfer.
2. Because of the thermally insulated boundary condition, it is not necessary to include the solder or substrate layers in the analysis. This is because there is no conduction through these materials and, from Fourier's law, there can be no temperature gradients within these materials. At steady state, $T_{sdr} = T_{an} = T_{si}$.
3. As the convection coefficient increases, the temperature of the silicon decreases. This leads to a higher solar-to-electrical conversion efficiency and increased electric power output. Similarly, higher silicon temperatures and less power production are associated with smaller convection coefficients. For example, $P = 13.6\ \text{W}$ and $14.6\ \text{W}$ for $h = 15\ \text{W/m}^2 \cdot \text{K}$ and $55\ \text{W/m}^2 \cdot \text{K}$, respectively.
4. The cost of a photovoltaic system can be reduced significantly by *concentrating* the solar energy onto the relatively expensive photovoltaic panel using inexpensive focusing mirrors or lenses. However, good thermal management then becomes even more important. For example, if the irradiation supplied to the panel were increased to $G_S = 7000\ \text{W/m}^2$ through concentration, the conversion efficiency drops to $\eta = 0.160$ as the silicon temperature increases to $T_{si} = 119°\text{C}$, even for $h = 55\ \text{W/m}^2 \cdot \text{K}$. A key to reducing the cost of photovoltaic power generation is developing innovative cooling technologies for use in concentrating photovoltaic systems.
5. The simultaneous solution of Equations 1 and 2 may be achieved by using *IHT*, another commercial code, or a handheld calculator. A trial-and-error solution could also be obtained, but with considerable effort. Equations 1 and 2 could be combined to write a single transcendental expression for the silicon temperature, but the equation must still be solved numerically or by trial-and-error.

EXAMPLE 3.4

The thermal conductivity of a $D = 14$-nm-diameter carbon nanotube is measured with an instrument that is fabricated of a wafer of silicon nitride at a temperature of $T_\infty = 300\ \text{K}$. The 20-μm-long nanotube rests on two 0.5-μm-thick, $10\ \mu\text{m} \times 10\ \mu\text{m}$ square islands that are separated by a distance $s = 5\ \mu\text{m}$. A thin layer of platinum is used as an electrical resistor on the *heated island* (at temperature T_h) to dissipate $q = 11.3\ \mu\text{W}$ of electrical power. On the *sensing island*, a similar layer of platinum is used to determine its temperature, T_s. The platinum's electrical resistance, $R(T_s) = E/I$, is found by measuring the voltage drop and electrical current across the platinum layer. The temperature of the sensing island, T_s, is then determined from the relationship of the platinum electrical resistance to its temperature.

Each island is suspended by two $L_{sn} = 250$-μm-long silicon nitride beams that are $w_{sn} = 3\ \mu$m wide and $t_{sn} = 0.5\ \mu$m thick. A platinum line of width $w_{pt} = 1\ \mu$m and thickness $t_{pt} = 0.2\ \mu$m is deposited within each silicon nitride beam to power the heated island or to detect the voltage drop associated with the determination of T_s. The entire experiment is performed in a vacuum with $T_{sur} = 300$ K and at steady state, $T_s = 308.4$ K. Estimate the thermal conductivity of the carbon nanotube.

Solution

Known: Dimensions, heat dissipated at the heated island, and temperatures of the sensing island and surrounding silicon nitride wafer.

Find: The thermal conductivity of the carbon nanotube.

Schematic:

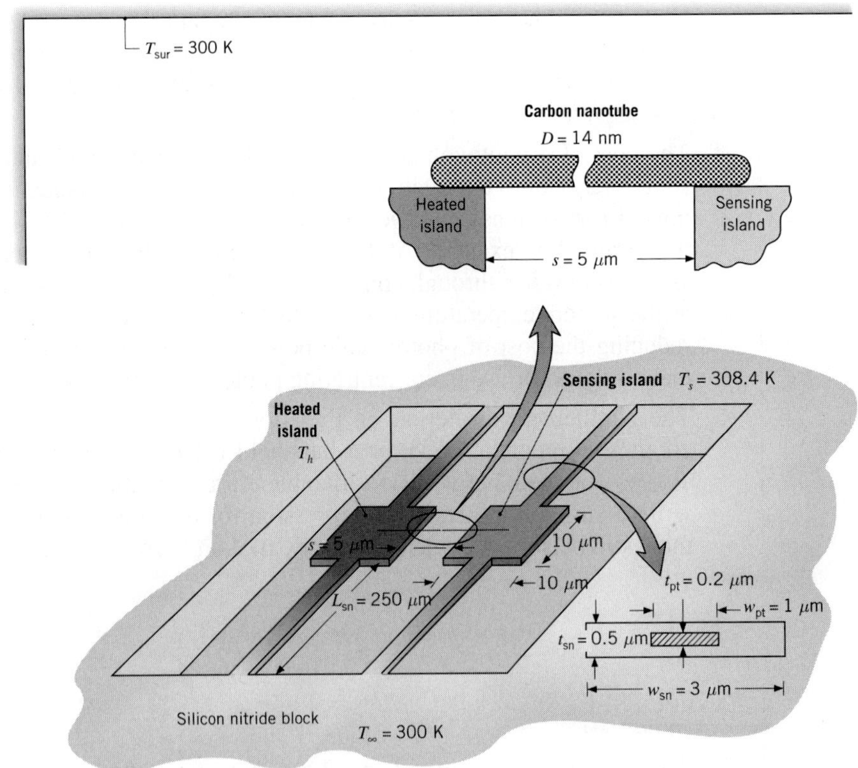

Assumptions:

1. Steady-state conditions.
2. One-dimensional heat transfer.
3. The heated and sensing islands are isothermal.
4. Radiation exchange between the surfaces and the surroundings is negligible.
5. Negligible convection losses.

6. Ohmic heating in the platinum signal lines is negligible.
7. Constant properties.
8. Negligible contact resistance between the nanotube and the islands.

Properties: Table A.1, platinum (325 K, assumed): $k_{pt} = 71.6$ W/m·K. Table A.2, silicon nitride (325 K, assumed): $k_{sn} = 15.5$ W/m·K.

Analysis: Energy that is dissipated at the heated island is transferred to the silicon nitride block through the support beams of the heated island, the carbon nanotube, and subsequently through the support beams of the sensing island. Therefore, the thermal circuit may be constructed as follows

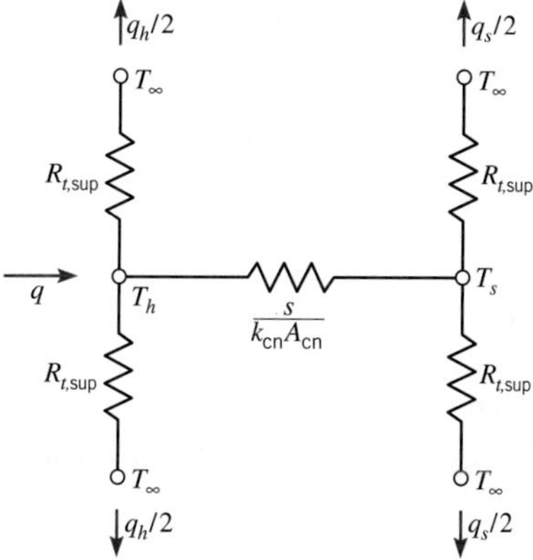

where each supporting beam provides a thermal resistance $R_{t,sup}$ that is composed of a resistance due to the silicon nitride (sn) in parallel with a resistance due to the platinum (pt) line.

The cross-sectional areas of the materials in the support beams are

$$A_{pt} = w_{pt}t_{pt} = (1 \times 10^{-6}\,\text{m}) \times (0.2 \times 10^{-6}\,\text{m}) = 2 \times 10^{-13}\,\text{m}^2$$

$$A_{sn} = w_{sn}t_{sn} - A_{pt} = (3 \times 10^{-6}\,\text{m}) \times (0.5 \times 10^{-6}\,\text{m}) - 2 \times 10^{-13}\,\text{m}^2 = 1.3 \times 10^{-12}\,\text{m}^2$$

while the cross-sectional area of the carbon nanotube is

$$A_{cn} = \pi D^2/4 = \pi(14 \times 10^{-9}\,\text{m})^2/4 = 1.54 \times 10^{-16}\,\text{m}^2$$

The thermal resistance of each support is

$$R_{t,sup} = \left[\frac{k_{pt}A_{pt}}{L_{pt}} + \frac{k_{sn}A_{sn}}{L_{sn}}\right]^{-1}$$

$$= \left[\frac{71.6\,\text{W/m}\cdot\text{K} \times 2 \times 10^{-13}\,\text{m}^2}{250 \times 10^{-6}\,\text{m}} + \frac{15.5\,\text{W/m}\cdot\text{K} \times 1.3 \times 10^{-12}\,\text{m}^2}{250 \times 10^{-6}\,\text{m}}\right]^{-1}$$

$$= 7.25 \times 10^6\,\text{K/W}$$

The combined heat loss through both sensing island supports is

$$q_s = 2(T_s - T_\infty)/R_{t,sup} = 2 \times (308.4\,\text{K} - 300\,\text{K})/(7.25 \times 10^6\,\text{K/W})$$

$$= 2.32 \times 10^{-6}\,\text{W} = 2.32\,\mu\text{W}$$

It follows that

$$q_h = q - q_s = 11.3\ \mu\text{W} - 2.32\ \mu\text{W} = 8.98\ \mu\text{W}$$

and T_h attains a value of

$$T_h = T_\infty + \frac{1}{2} q_h R_{t,\text{sup}} = 300\ \text{K} + \frac{8.98 \times 10^{-6}\ \text{W} \times 7.25 \times 10^{6}\ \text{K/W}}{2} = 332.6\ \text{K}$$

For the portion of the thermal circuit connecting T_h and T_s,

$$q_s = \frac{T_h - T_s}{s/(k_{\text{cn}} A_{\text{cn}})}$$

from which

$$k_{\text{cn}} = \frac{q_s s}{A_{\text{cn}}(T_h - T_s)} = \frac{2.32 \times 10^{-6}\ \text{W} \times 5 \times 10^{-6}\ \text{m}}{1.54 \times 10^{-16}\ \text{m}^2 \times (332.6\ \text{K} - 308.4\ \text{K})}$$

$$k_{\text{cn}} = 3113\ \text{W/m} \cdot \text{K} \qquad \triangleleft$$

Comments:

1. The measured thermal conductivity is extremely large, as evident by comparing its value to the thermal conductivities of pure metals shown in Figure 2.4. Carbon nanotubes might be used to *dope* otherwise low thermal conductivity materials to improve heat transfer.
2. Contact resistances between the carbon nanotube and the heated and sensing islands were neglected because little is known about such resistances at the nanoscale. However, *if* a contact resistance were included in the analysis, the measured thermal conductivity of the carbon nanotube would be even higher than the predicted value.
3. The significance of radiation heat transfer may be estimated by approximating the heated island as a blackbody radiating to T_{sur} from both its top and bottom surfaces. Hence, $q_{\text{rad},b} \approx 5.67 \times 10^{-8}\ \text{W/m}^2 \cdot \text{K}^4 \times 2 \times (10 \times 10^{-6}\ \text{m})^2 \times (332.6^4 - 300^4)\text{K}^4 = 4.7 \times 10^{-8}\ \text{W} = 0.047\ \mu\text{W}$, and radiation is negligible.

3.2 An Alternative Conduction Analysis

The conduction analysis of Section 3.1 was performed using the *standard approach.* That is, the heat equation was solved to obtain the temperature distribution, Equation 3.3, and Fourier's law was then applied to obtain the heat transfer rate, Equation 3.4. However, an alternative approach may be used for the conditions presently of interest. Considering conduction in the system of Figure 3.6, we recognize that, for *steady-state conditions* with *no heat generation* and *no heat loss from the sides*, the heat transfer rate q_x must be a constant independent of x. That is, for any differential element dx, $q_x = q_{x+dx}$. This condition is, of course, a consequence of the energy conservation requirement, and it must apply even if the area varies with position $A(x)$ and the thermal conductivity varies with temperature $k(T)$. Moreover, even though the temperature distribution may be two-dimensional, varying with x and y, it is often reasonable to neglect the y-variation and to assume a *one-dimensional* distribution in x.

For the above conditions it is possible to work exclusively with Fourier's law when performing a conduction analysis. In particular, since the conduction rate is a *constant*, the

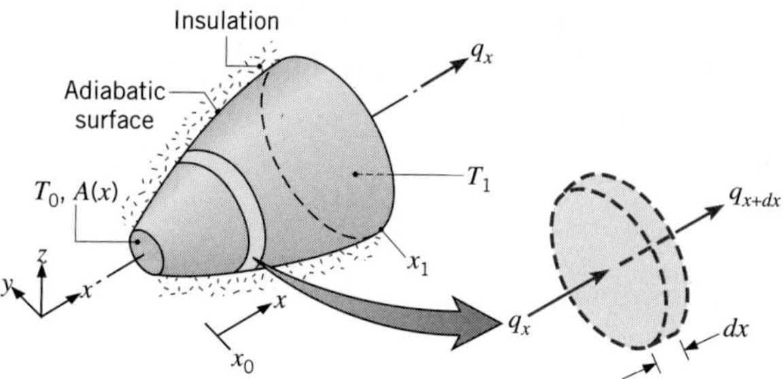

FIGURE 3.6 System with a constant conduction heat transfer rate.

rate equation may be *integrated*, even though neither the rate nor the temperature distribution is known. Consider Fourier's law, Equation 2.1, which may be applied to the system of Figure 3.6. Although we may have no knowledge of the value of q_x or the form of $T(x)$, we do know that q_x is a constant. Hence we may express Fourier's law in the integral form

$$q_x \int_{x_0}^{x} \frac{dx}{A(x)} = -\int_{T_0}^{T} k(T)\,dT \tag{3.26}$$

The cross-sectional area may be a known function of x, and the material thermal conductivity may vary with temperature in a known manner. If the integration is performed from a point x_0 at which the temperature T_0 is known, the resulting equation provides the functional form of $T(x)$. Moreover, if the temperature $T = T_1$ at some $x = x_1$ is also known, integration between x_0 and x_1 provides an expression from which q_x may be computed. Note that, if the area A is uniform and k is independent of temperature, Equation 3.26 reduces to

$$\frac{q_x\,\Delta x}{A} = -k\,\Delta T \tag{3.27}$$

where $\Delta x = x_1 - x_0$ and $\Delta T = T_1 - T_0$.

We frequently elect to solve diffusion problems by working with integrated forms of the diffusion rate equations. However, the limiting conditions for which this may be done should be firmly fixed in our minds: *steady-state* and *one-dimensional* transfer with *no heat generation*.

EXAMPLE 3.5

The diagram shows a conical section fabricated from pyroceram. It is of circular cross section with the diameter $D = ax$, where $a = 0.25$. The small end is at $x_1 = 50$ mm and the large end at $x_2 = 250$ mm. The end temperatures are $T_1 = 400$ K and $T_2 = 600$ K, while the lateral surface is well insulated.

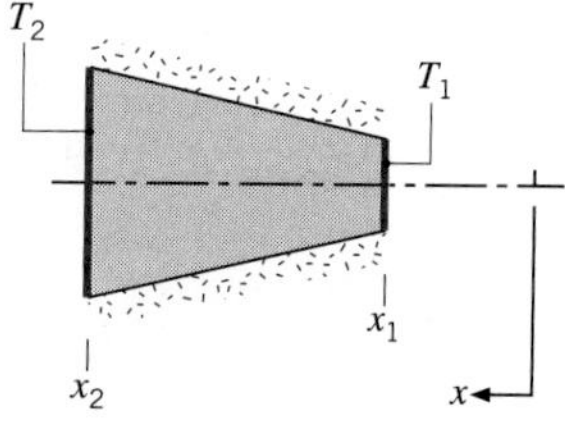

1. Derive an expression for the temperature distribution $T(x)$ in symbolic form, assuming one-dimensional conditions. Sketch the temperature distribution.
2. Calculate the heat rate q_x through the cone.

SOLUTION

Known: Conduction in a circular conical section having a diameter $D = ax$, where $a = 0.25$.

Find:

1. Temperature distribution $T(x)$.
2. Heat transfer rate q_x.

Schematic:

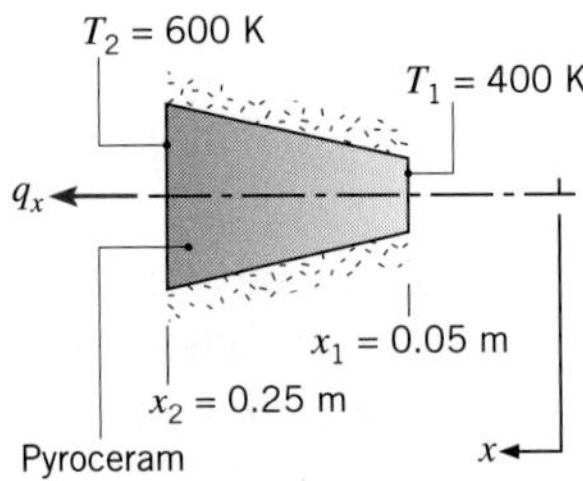

Assumptions:

1. Steady-state conditions.
2. One-dimensional conduction in the x-direction.
3. No internal heat generation.
4. Constant properties.

Properties: Table A.2, pyroceram (500 K): $k = 3.46$ W/m·K.

Analysis:

1. Since heat conduction occurs under steady-state, one-dimensional conditions with no internal heat generation, the heat transfer rate q_x is a constant independent of x. Accordingly, Fourier's law, Equation 2.1, may be used to determine the temperature distribution

$$q_x = -kA\frac{dT}{dx}$$

where $A = \pi D^2/4 = \pi a^2 x^2/4$. Separating variables,

$$\frac{4q_x dx}{\pi a^2 x^2} = -k dT$$

Integrating from x_1 to any x within the cone, and recalling that q_x and k are constants, it follows that

$$\frac{4q_x}{\pi a^2}\int_{x_1}^{x}\frac{dx}{x^2} = -k\int_{T_1}^{T} dT$$

Hence

$$\frac{4q_x}{\pi a^2}\left(-\frac{1}{x}+\frac{1}{x_1}\right)=-k(T-T_1)$$

or solving for T

$$T(x)=T_1-\frac{4q_x}{\pi a^2 k}\left(\frac{1}{x_1}-\frac{1}{x}\right)$$

Although q_x is a constant, it is as yet an unknown. However, it may be determined by evaluating the above expression at $x = x_2$, where $T(x_2) = T_2$. Hence

$$T_2=T_1-\frac{4q_x}{\pi a^2 k}\left(\frac{1}{x_1}-\frac{1}{x_2}\right)$$

and solving for q_x

$$q_x=\frac{\pi a^2 k(T_1-T_2)}{4[(1/x_1)-(1/x_2)]}$$

Substituting for q_x into the expression for $T(x)$, the temperature distribution becomes

$$T(x)=T_1+(T_1-T_2)\left[\frac{(1/x)-(1/x_1)}{(1/x_1)-(1/x_2)}\right] \quad \triangleleft$$

From this result, temperature may be calculated as a function of x and the distribution is as shown.

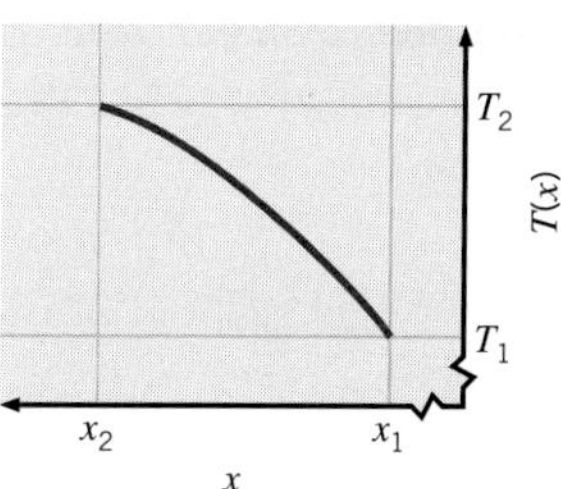

Note that, since $dT/dx = -4q_x/k\pi a^2 x^2$ from Fourier's law, it follows that the temperature gradient and heat flux decrease with increasing x.

2. Substituting numerical values into the foregoing result for the heat transfer rate, it follows that

$$q_x=\frac{\pi(0.25)^2\times 3.46\ \text{W/m}\cdot\text{K}\,(400-600)\ \text{K}}{4\,(1/0.05\ \text{m}-1/0.25\ \text{m})}=-2.12\ \text{W} \quad \triangleleft$$

Comments: When the parameter a increases, the cross-sectional area changes more rapidly with distance, causing the one-dimensional assumption to become less appropriate.

3.3 *Radial Systems*

Cylindrical and spherical systems often experience temperature gradients in the radial direction only and may therefore be treated as one-dimensional. Moreover, under steady-state conditions with no heat generation, such systems may be analyzed by using the *standard* method, which begins with the appropriate form of the heat equation, or the *alternative* method, which begins with the appropriate form of Fourier's law. In this section, the cylindrical system is analyzed by means of the standard method and the spherical system by means of the alternative method.

3.3.1 The Cylinder

A common example is the hollow cylinder whose inner and outer surfaces are exposed to fluids at different temperatures (Figure 3.7). For steady-state conditions with no heat generation, the appropriate form of the heat equation, Equation 2.26, is

$$\frac{1}{r}\frac{d}{dr}\left(kr\frac{dT}{dr}\right) = 0 \tag{3.28}$$

where, for the moment, k is treated as a variable. The physical significance of this result becomes evident if we also consider the appropriate form of Fourier's law. The rate at which energy is conducted across any cylindrical surface in the solid may be expressed as

$$q_r = -kA\frac{dT}{dr} = -k(2\pi rL)\frac{dT}{dr} \tag{3.29}$$

where $A = 2\pi rL$ is the area normal to the direction of heat transfer. Since Equation 3.28 dictates that the quantity $kr(dT/dr)$ is independent of r, it follows from Equation 3.29 that the conduction *heat transfer rate* q_r (*not* the heat flux q''_r) is a *constant in the radial direction.*

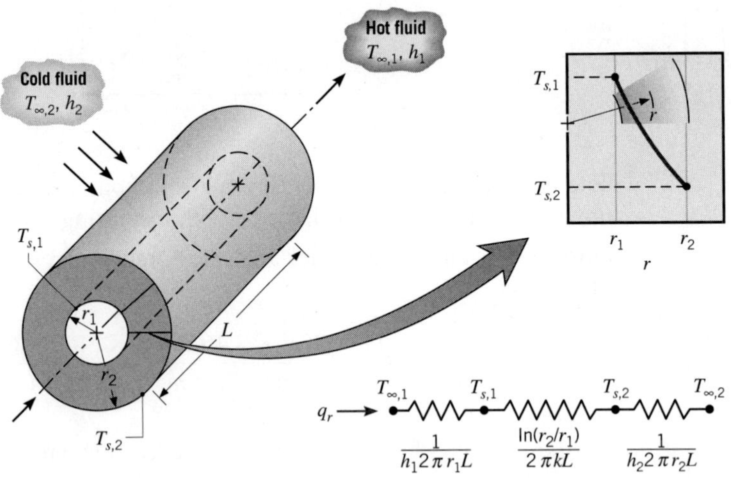

FIGURE 3.7 Hollow cylinder with convective surface conditions.

We may determine the temperature distribution in the cylinder by solving Equation 3.28 and applying appropriate boundary conditions. Assuming the value of k to be constant, Equation 3.28 may be integrated twice to obtain the general solution

$$T(r) = C_1 \ln r + C_2 \tag{3.30}$$

To obtain the constants of integration C_1 and C_2, we introduce the following boundary conditions:

$$T(r_1) = T_{s,1} \quad \text{and} \quad T(r_2) = T_{s,2}$$

Applying these conditions to the general solution, we then obtain

$$T_{s,1} = C_1 \ln r_1 + C_2 \quad \text{and} \quad T_{s,2} = C_1 \ln r_2 + C_2$$

Solving for C_1 and C_2 and substituting into the general solution, we then obtain

$$T(r) = \frac{T_{s,1} - T_{s,2}}{\ln(r_1/r_2)} \ln\left(\frac{r}{r_2}\right) + T_{s,2} \tag{3.31}$$

Note that the temperature distribution associated with radial conduction through a cylindrical wall is logarithmic, not linear, as it is for the plane wall under the same conditions. The logarithmic distribution is sketched in the inset of Figure 3.7.

If the temperature distribution, Equation 3.31, is now used with Fourier's law, Equation 3.29, we obtain the following expression for the heat transfer rate:

$$q_r = \frac{2\pi Lk(T_{s,1} - T_{s,2})}{\ln(r_2/r_1)} \tag{3.32}$$

From this result it is evident that, for radial conduction in a cylindrical wall, the thermal resistance is of the form

$$R_{t,\text{cond}} = \frac{\ln(r_2/r_1)}{2\pi Lk} \tag{3.33}$$

This resistance is shown in the series circuit of Figure 3.7. Note that since the value of q_r is independent of r, the foregoing result could have been obtained by using the alternative method, that is, by integrating Equation 3.29.

Consider now the composite system of Figure 3.8. Recalling how we treated the composite plane wall and neglecting the interfacial contact resistances, the heat transfer rate may be expressed as

$$q_r = \frac{T_{\infty,1} - T_{\infty,4}}{\dfrac{1}{2\pi r_1 L h_1} + \dfrac{\ln(r_2/r_1)}{2\pi k_A L} + \dfrac{\ln(r_3/r_2)}{2\pi k_B L} + \dfrac{\ln(r_4/r_3)}{2\pi k_C L} + \dfrac{1}{2\pi r_4 L h_4}} \tag{3.34}$$

The foregoing result may also be expressed in terms of an overall heat transfer coefficient. That is,

$$q_r = \frac{T_{\infty,1} - T_{\infty,4}}{R_{\text{tot}}} = UA(T_{\infty,1} - T_{\infty,4}) \tag{3.35}$$

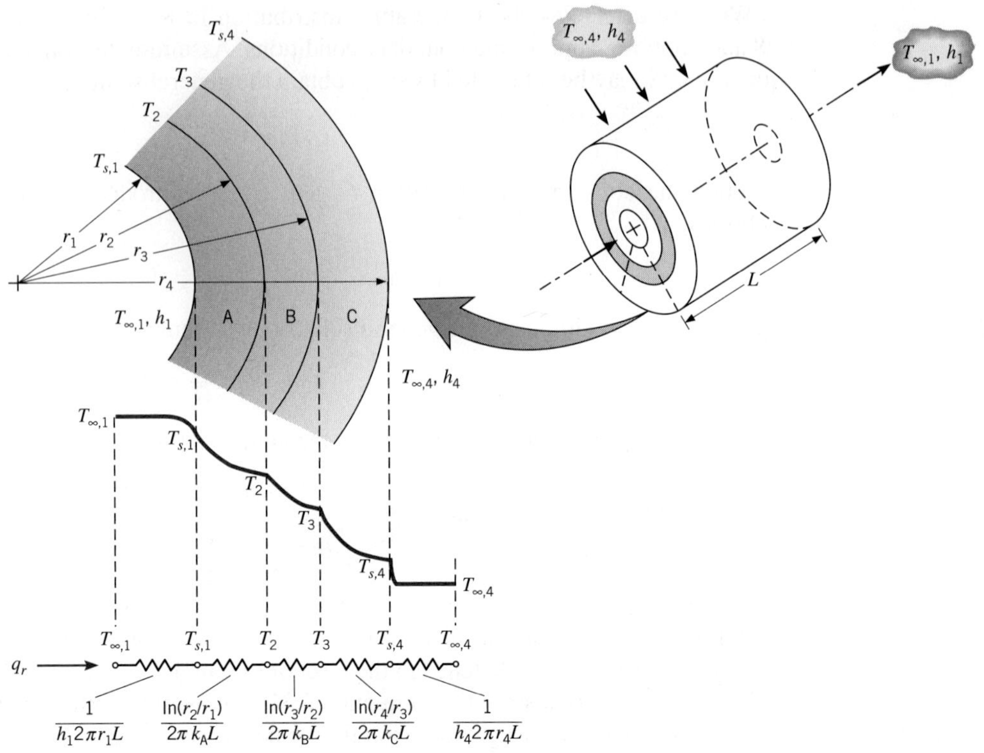

FIGURE 3.8 Temperature distribution for a composite cylindrical wall.

If U is defined in terms of the inside area, $A_1 = 2\pi r_1 L$, Equations 3.34 and 3.35 may be equated to yield

$$U_1 = \frac{1}{\frac{1}{h_1} + \frac{r_1}{k_A}\ln\frac{r_2}{r_1} + \frac{r_1}{k_B}\ln\frac{r_3}{r_2} + \frac{r_1}{k_C}\ln\frac{r_4}{r_3} + \frac{r_1}{r_4}\frac{1}{h_4}} \tag{3.36}$$

This definition is *arbitrary,* and the overall coefficient may also be defined in terms of A_4 or any of the intermediate areas. Note that

$$U_1A_1 = U_2A_2 = U_3A_3 = U_4A_4 = (\Sigma R_t)^{-1} \tag{3.37}$$

and the specific forms of U_2, U_3, and U_4 may be inferred from Equations 3.34 and 3.35.

EXAMPLE 3.6

The possible existence of an optimum insulation thickness for radial systems is suggested by the presence of competing effects associated with an increase in this thickness. In particular, although the conduction resistance increases with the addition of insulation, the convection resistance decreases due to increasing outer surface area. Hence there may exist an insulation thickness that minimizes heat loss by maximizing the total resistance to heat transfer. Resolve this issue by considering the following system.

1. A thin-walled copper tube of radius r_i is used to transport a low-temperature refrigerant and is at a temperature T_i that is less than that of the ambient air at T_∞ around the tube. Is there an optimum thickness associated with application of insulation to the tube?
2. Confirm the above result by computing the total thermal resistance per unit length of tube for a 10-mm-diameter tube having the following insulation thicknesses: 0, 2, 5, 10, 20, and 40 mm. The insulation is composed of cellular glass, and the outer surface convection coefficient is 5 W/m$^2 \cdot$K.

SOLUTION

Known: Radius r_i and temperature T_i of a thin-walled copper tube to be insulated from the ambient air.

Find:

1. Whether there exists an optimum insulation thickness that minimizes the heat transfer rate.
2. Thermal resistance associated with using cellular glass insulation of varying thickness.

Schematic:

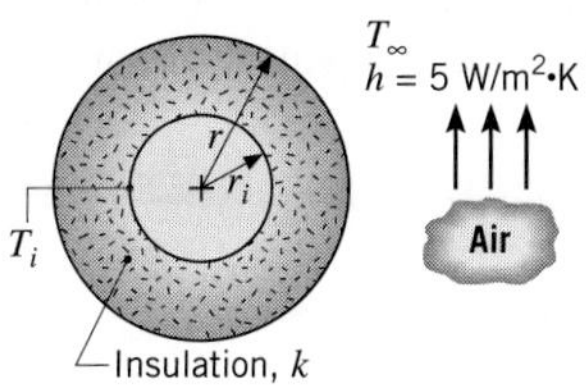

Assumptions:

1. Steady-state conditions.
2. One-dimensional heat transfer in the radial (cylindrical) direction.
3. Negligible tube wall thermal resistance.
4. Constant properties for insulation.
5. Negligible radiation exchange between insulation outer surface and surroundings.

Properties: Table A.3, cellular glass (285 K, assumed): $k = 0.055$ W/m$\cdot$K.

Analysis:

1. The resistance to heat transfer between the refrigerant and the air is dominated by conduction in the insulation and convection in the air. The thermal circuit is therefore

$q' \leftarrow$ T_i —— $\frac{\ln(r/r_i)}{2\pi k}$ —— $\frac{1}{2\pi r h}$ —— T_∞

where the conduction and convection resistances per unit length follow from Equations 3.33 and 3.9, respectively. The total thermal resistance per unit length of tube is then

$$R'_{\text{tot}} = \frac{\ln(r/r_i)}{2\pi k} + \frac{1}{2\pi r h}$$

where the rate of heat transfer per unit length of tube is

$$q' = \frac{T_\infty - T_i}{R'_{\text{tot}}}$$

An optimum insulation thickness would be associated with the value of r that minimized q' or maximized R'_{tot}. Such a value could be obtained from the requirement that

$$\frac{dR'_{\text{tot}}}{dr} = 0$$

Hence

$$\frac{1}{2\pi kr} - \frac{1}{2\pi r^2 h} = 0$$

or

$$r = \frac{k}{h}$$

To determine whether the foregoing result maximizes or minimizes the total resistance, the second derivative must be evaluated. Hence

$$\frac{d^2R'_{\text{tot}}}{dr^2} = -\frac{1}{2\pi kr^2} + \frac{1}{\pi r^3 h}$$

or, at $r = k/h$,

$$\frac{d^2R'_{\text{tot}}}{dr^2} = \frac{1}{\pi(k/h)^2}\left(\frac{1}{k} - \frac{1}{2k}\right) = \frac{1}{2\pi k^3/h^2} > 0$$

Since this result is always positive, it follows that $r = k/h$ is the insulation radius for which the total resistance is a minimum, not a maximum. Hence an *optimum* insulation thickness *does not exist.*

From the above result it makes more sense to think in terms of a *critical insulation radius*

$$r_{\text{cr}} \equiv \frac{k}{h}$$

which maximizes heat transfer, that is, below which q' increases with increasing r and above which q' decreases with increasing r.

2. With $h = 5\ \text{W/m}^2 \cdot \text{K}$ and $k = 0.055\ \text{W/m} \cdot \text{K}$, the critical radius is

$$r_{\text{cr}} = \frac{0.055\ \text{W/m} \cdot \text{K}}{5\ \text{W/m}^2 \cdot \text{K}} = 0.011\ \text{m}$$

Hence $r_{\text{cr}} > r_i$ and heat transfer will increase with the addition of insulation up to a thickness of

$$r_{\text{cr}} - r_i = (0.011 - 0.005)\ \text{m} = 0.006\ \text{m}$$

The thermal resistances corresponding to the prescribed insulation thicknesses may be calculated and are plotted as follows:

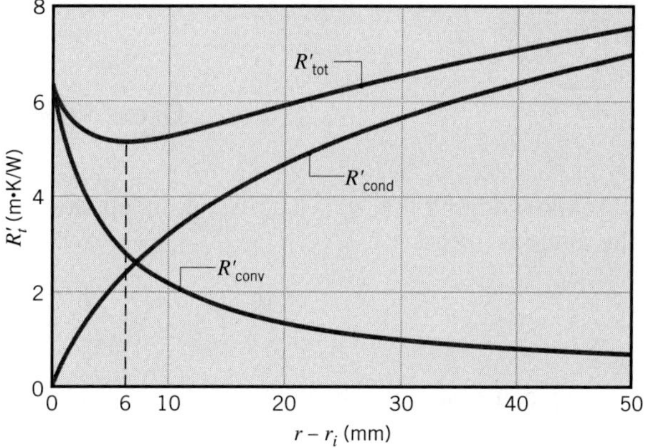

Comments:

1. The effect of the critical radius is revealed by the fact that, even for 20 mm of insulation, the total resistance is not as large as the value for no insulation.
2. If $r_i < r_{cr}$, as it is in this case, the total resistance decreases and the heat rate therefore increases with the addition of insulation. This trend continues until the outer radius of the insulation corresponds to the critical radius. The trend is desirable for electrical current flow through a wire, since the addition of electrical insulation would aid in transferring heat dissipated in the wire to the surroundings. Conversely, if $r_i > r_{cr}$, any addition of insulation would increase the total resistance and therefore decrease the heat loss. This behavior would be desirable for steam flow through a pipe, where insulation is added to reduce heat loss to the surroundings.
3. For radial systems, the problem of reducing the total resistance through the application of insulation exists only for small diameter wires or tubes and for small convection coefficients, such that $r_{cr} > r_i$. For a typical insulation ($k \approx 0.03$ W/m·K) and free convection in air ($h \approx 10$ W/m^2·K), $r_{cr} = (k/h) \approx 0.003$ m. Such a small value tells us that, normally, $r_i > r_{cr}$ and we need not be concerned with the effects of a critical radius.
4. The existence of a critical radius requires that the heat transfer area change in the direction of transfer, as for radial conduction in a cylinder (or a sphere). In a plane wall the area perpendicular to the direction of heat flow is constant and there is no critical insulation thickness (the total resistance always increases with increasing insulation thickness).

3.3.2 The Sphere

Now consider applying the alternative method to analyzing conduction in the hollow sphere of Figure 3.9. For the differential control volume of the figure, energy conservation requires that $q_r = q_{r+dr}$ for steady-state, one-dimensional conditions with no heat generation. The appropriate form of Fourier's law is

$$q_r = -kA\frac{dT}{dr} = -k(4\pi r^2)\frac{dT}{dr} \tag{3.38}$$

where $A = 4\pi r^2$ is the area normal to the direction of heat transfer.

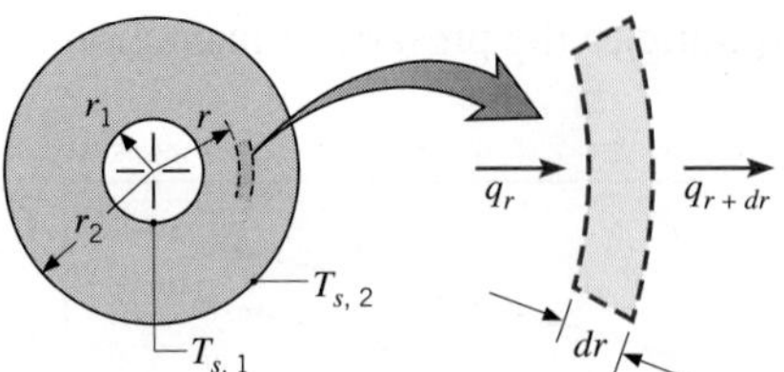

FIGURE 3.9 Conduction in a spherical shell.

Acknowledging that q_r is a constant, independent of r, Equation 3.38 may be expressed in the integral form

$$\frac{q_r}{4\pi}\int_{r_1}^{r_2}\frac{dr}{r^2} = -\int_{T_{s,1}}^{T_{s,2}} k(T)\,dT \tag{3.39}$$

Assuming constant k, we then obtain

$$q_r = \frac{4\pi k(T_{s,1} - T_{s,2})}{(1/r_1) - (1/r_2)} \tag{3.40}$$

Remembering that the thermal resistance is defined as the temperature difference divided by the heat transfer rate, we obtain

$$R_{t,\text{cond}} = \frac{1}{4\pi k}\left(\frac{1}{r_1} - \frac{1}{r_2}\right) \tag{3.41}$$

Note that the temperature distribution and Equations 3.40 and 3.41 could have been obtained by using the standard approach, which begins with the appropriate form of the heat equation.

Spherical composites may be treated in much the same way as composite walls and cylinders, where appropriate forms of the total resistance and overall heat transfer coefficient may be determined.

3.4 *Summary of One-Dimensional Conduction Results*

Many important problems are characterized by one-dimensional, steady-state conduction in plane, cylindrical, or spherical walls without thermal energy generation. Key results for these three geometries are summarized in Table 3.3, where ΔT refers to the temperature difference, $T_{s,1} - T_{s,2}$, between the inner and outer surfaces identified in Figures 3.1, 3.7, and 3.9. In each case, beginning with the heat equation, you should be able to derive the corresponding expressions for the temperature distribution, heat flux, heat rate, and thermal resistance.

3.5 *Conduction with Thermal Energy Generation*

In the preceding section we considered conduction problems for which the temperature distribution in a medium was determined solely by conditions at the boundaries of the medium. We now want to consider the additional effect on the temperature distribution of processes that may be occurring *within* the medium. In particular, we wish to consider situations for which thermal energy is being *generated* due to *conversion* from some other energy form.

TABLE 3.3 One-dimensional, steady-state solutions to the heat equation with no generation

	Plane Wall	Cylindrical Wall[a]	Spherical Wall[a]
Heat equation	$\frac{d^2T}{dx^2} = 0$	$\frac{1}{r}\frac{d}{dr}\left(r\frac{dT}{dr}\right) = 0$	$\frac{1}{r^2}\frac{d}{dr}\left(r^2\frac{dT}{dr}\right) = 0$
Temperature distribution	$T_{s,1} - \Delta T\frac{x}{L}$	$T_{s,2} + \Delta T\frac{\ln(r/r_2)}{\ln(r_1/r_2)}$	$T_{s,1} - \Delta T\left[\frac{1-(r_1/r)}{1-(r_1/r_2)}\right]$
Heat flux (q'')	$k\frac{\Delta T}{L}$	$\frac{k\,\Delta T}{r\ln(r_2/r_1)}$	$\frac{k\,\Delta T}{r^2[(1/r_1)-(1/r_2)]}$
Heat rate (q)	$kA\frac{\Delta T}{L}$	$\frac{2\pi Lk\,\Delta T}{\ln(r_2/r_1)}$	$\frac{4\pi k\,\Delta T}{(1/r_1)-(1/r_2)}$
Thermal resistance ($R_{t,\mathrm{cond}}$)	$\frac{L}{kA}$	$\frac{\ln(r_2/r_1)}{2\pi Lk}$	$\frac{(1/r_1)-(1/r_2)}{4\pi k}$

[a]The critical radius of insulation is $r_{cr} = k/h$ for the cylinder and $r_{cr} = 2k/h$ for the sphere.

A common thermal energy generation process involves the conversion from *electrical to thermal energy* in a current-carrying medium (*Ohmic*, or *resistance*, or *Joule heating*). The rate at which energy is generated by passing a current I through a medium of electrical resistance R_e is

$$\dot{E}_g = I^2R_e \tag{3.42}$$

If this power generation (W) occurs uniformly throughout the medium of volume V, the volumetric generation rate (W/m^3) is then

$$\dot{q} \equiv \frac{\dot{E}_g}{V} = \frac{I^2R_e}{V} \tag{3.43}$$

Energy generation may also occur as a result of the deceleration and absorption of neutrons in the fuel element of a nuclear reactor or exothermic chemical reactions occurring within a medium. Endothermic reactions would, of course, have the inverse effect (a thermal energy sink) of converting thermal energy to chemical bonding energy. Finally, a conversion from electromagnetic to thermal energy may occur due to the absorption of radiation within the medium. The process occurs, for example, when gamma rays are absorbed in external nuclear reactor components (cladding, thermal shields, pressure vessels, etc.) or when visible radiation is absorbed in a semitransparent medium. Remember not to confuse energy generation with energy storage (Section 1.3.1).

3.5.1 The Plane Wall

Consider the plane wall of Figure 3.10*a*, in which there is *uniform* energy generation per unit volume ($\dot{q}$ is constant) and the surfaces are maintained at $T_{s,1}$ and $T_{s,2}$. For constant thermal conductivity k, the appropriate form of the heat equation, Equation 2.22, is

$$\frac{d^2T}{dx^2} + \frac{\dot{q}}{k} = 0 \tag{3.44}$$

The general solution is

$$T = -\frac{\dot{q}}{2k}x^2 + C_1 x + C_2 \tag{3.45}$$

where C_1 and C_2 are the constants of integration. For the prescribed boundary conditions,

$$T(-L) = T_{s,1} \quad \text{and} \quad T(L) = T_{s,2}$$

The constants may be evaluated and are of the form

$$C_1 = \frac{T_{s,2} - T_{s,1}}{2L} \quad \text{and} \quad C_2 = \frac{\dot{q}}{2k}L^2 + \frac{T_{s,1} + T_{s,2}}{2}$$

in which case the temperature distribution is

$$T(x) = \frac{\dot{q}L^2}{2k}\left(1 - \frac{x^2}{L^2}\right) + \frac{T_{s,2} - T_{s,1}}{2}\frac{x}{L} + \frac{T_{s,1} + T_{s,2}}{2} \tag{3.46}$$

The heat flux at any point in the wall may, of course, be determined by using Equation 3.46 with Fourier's law. Note, however, that *with generation the heat flux is no longer independent of x.*

The preceding result simplifies when both surfaces are maintained at a common temperature, $T_{s,1} = T_{s,2} \equiv T_s$. The temperature distribution is then *symmetrical* about the midplane, Figure 3.10*b,* and is given by

$$T(x) = \frac{\dot{q}L^2}{2k}\left(1 - \frac{x^2}{L^2}\right) + T_s \tag{3.47}$$

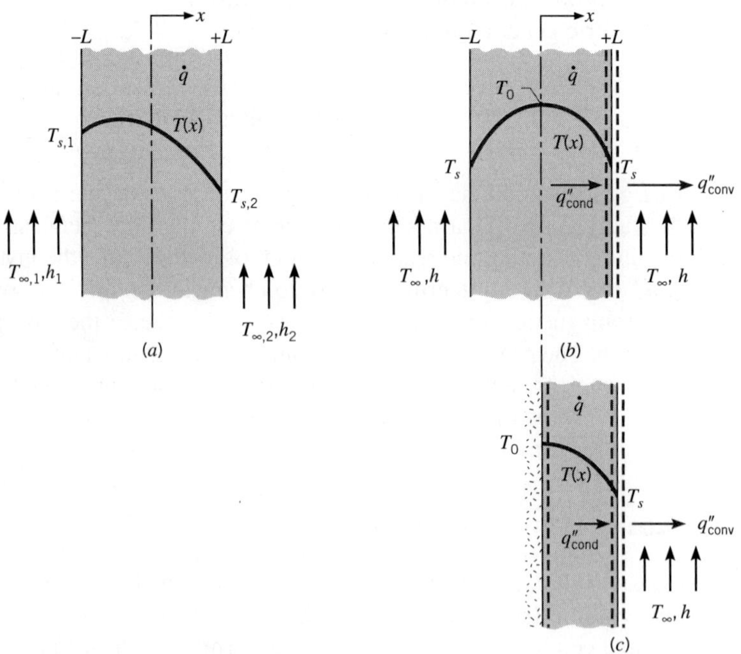

FIGURE 3.10 Conduction in a plane wall with uniform heat generation. (*a*) Asymmetrical boundary conditions. (*b*) Symmetrical boundary conditions. (*c*) Adiabatic surface at midplane.

The maximum temperature exists at the midplane

$$T(0) \equiv T_0 = \frac{\dot{q}L^2}{2k} + T_s \tag{3.48}$$

in which case the temperature distribution, Equation 3.47, may be expressed as

$$\frac{T(x) - T_0}{T_s - T_0} = \left(\frac{x}{L}\right)^2 \tag{3.49}$$

It is important to note that at the plane of symmetry in Figure 3.10*b*, the temperature gradient is zero, $(dT/dx)_{x=0} = 0$. Accordingly, there is no heat transfer across this plane, and it may be represented by the *adiabatic* surface shown in Figure 3.10*c*. One implication of this result is that Equation 3.47 also applies to plane walls that are perfectly insulated on one side ($x = 0$) and maintained at a fixed temperature T_s on the other side ($x = L$).

To use the foregoing results, the surface temperature(s) T_s must be known. However, a common situation is one for which it is the temperature of an adjoining fluid, T_∞, and not T_s, which is known. It then becomes necessary to relate T_s to T_∞. This relation may be developed by applying a surface energy balance. Consider the surface at $x = L$ for the symmetrical plane wall (Figure 3.10*b*) or the insulated plane wall (Figure 3.10*c*). Neglecting radiation and substituting the appropriate rate equations, the energy balance given by Equation 1.13 reduces to

$$-k\left.\frac{dT}{dx}\right|_{x=L} = h(T_s - T_\infty) \tag{3.50}$$

Substituting from Equation 3.47 to obtain the temperature gradient at $x = L$, it follows that

$$T_s = T_\infty + \frac{\dot{q}L}{h} \tag{3.51}$$

Hence T_s may be computed from knowledge of T_∞, $\dot{q}$, L, and h.

Equation 3.51 may also be obtained by applying an *overall* energy balance to the plane wall of Figure 3.10*b* or 3.10*c*. For example, relative to a control surface about the wall of Figure 3.10*c*, the rate at which energy is generated within the wall must be balanced by the rate at which energy leaves via convection at the boundary. Equation 1.12c reduces to

$$\dot{E}_g = \dot{E}_{\text{out}} \tag{3.52}$$

or, for a unit surface area,

$$\dot{q}L = h(T_s - T_\infty) \tag{3.53}$$

Solving for T_s, Equation 3.51 is obtained.

Equation 3.51 may be combined with Equation 3.47 to eliminate T_s from the temperature distribution, which is then expressed in terms of the known quantities $\dot{q}$, L, k, h, and T_∞. The same result may be obtained directly by using Equation 3.50 as a boundary condition to evaluate the constants of integration appearing in Equation 3.45.

EXAMPLE 3.7

A plane wall is a composite of two materials, A and B. The wall of material A has uniform heat generation $\dot{q} = 1.5 \times 10^6\ \text{W/m}^3$, $k_A = 75\ \text{W/m}\cdot\text{K}$, and thickness $L_A = 50\ \text{mm}$. The wall material B has no generation with $k_B = 150$ W/m·K and thickness $L_B = 20$ mm. The inner surface of material A is well insulated, while the outer surface of material B is cooled by a water stream with $T_\infty = 30°$C and $h = 1000$ W/m^2·K.

1. Sketch the temperature distribution that exists in the composite under steady-state conditions.
2. Determine the temperature T_0 of the insulated surface and the temperature T_2 of the cooled surface.

SOLUTION

Known: Plane wall of material A with internal heat generation is insulated on one side and bounded by a second wall of material B, which is without heat generation and is subjected to convection cooling.

Find:

1. Sketch of steady-state temperature distribution in the composite.
2. Inner and outer surface temperatures of the composite.

Schematic:

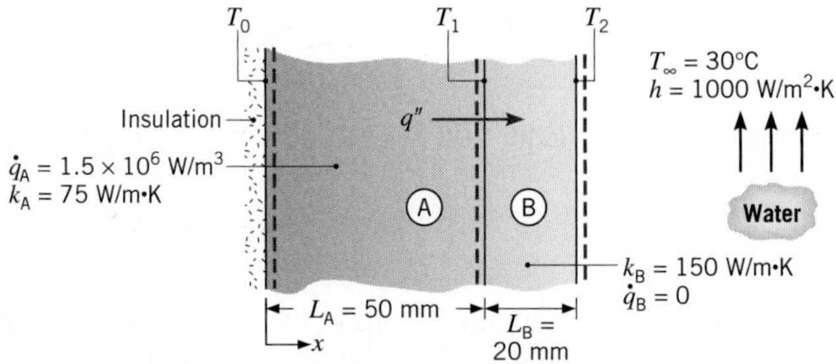

Assumptions:

1. Steady-state conditions.
2. One-dimensional conduction in x-direction.
3. Negligible contact resistance between walls.
4. Inner surface of A adiabatic.
5. Constant properties for materials A and B.

Analysis:

1. From the prescribed physical conditions, the temperature distribution in the composite is known to have the following features, as shown:
   (a) Parabolic in material A.
   (b) Zero slope at insulated boundary.
   (c) Linear in material B.
   (d) Slope change $= k_B/k_A = 2$ at interface.

The temperature distribution in the water is characterized by

(e) Large gradients near the surface.

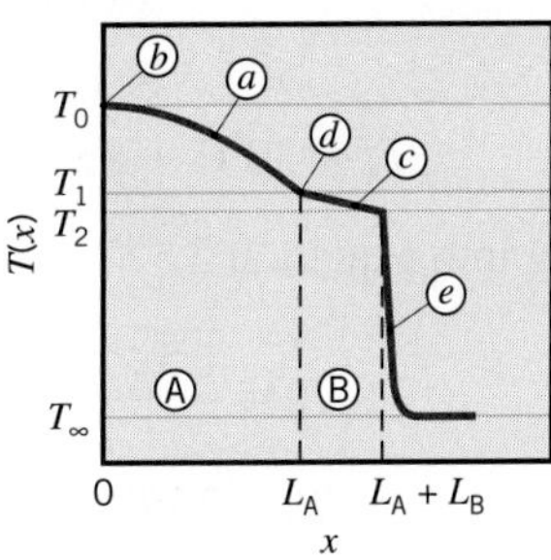

2. The outer surface temperature T_2 may be obtained by performing an energy balance on a control volume about material B. Since there is no generation in this material, it follows that, for steady-state conditions and a unit surface area, the heat flux into the material at $x = L_A$ must equal the heat flux from the material due to convection at $x = L_A + L_B$. Hence

$$q'' = h(T_2 - T_\infty) \tag{1}$$

The heat flux q'' may be determined by performing a second energy balance on a control volume about material A. In particular, since the surface at $x = 0$ is adiabatic, there is no inflow and the rate at which energy is generated must equal the outflow. Accordingly, for a unit surface area,

$$\dot{q}L_A = q'' \tag{2}$$

Combining Equations 1 and 2, the outer surface temperature is

$$T_2 = T_\infty + \frac{\dot{q}L_A}{h}$$

$$T_2 = 30°\text{C} + \frac{1.5 \times 10^6 \text{ W/m}^3 \times 0.05 \text{ m}}{1000 \text{ W/m}^2 \cdot \text{K}} = 105°\text{C} \qquad \triangleleft$$

From Equation 3.48 the temperature at the insulated surface is

$$T_0 = \frac{\dot{q}L_A^2}{2k_A} + T_1 \tag{3}$$

where T_1 may be obtained from the following thermal circuit:

$$q'' \longrightarrow \; T_1 \; R''_{\text{cond, B}} \; T_2 \; R''_{\text{conv}} \; T_\infty$$

That is,

$$T_1 = T_\infty + (R''_{\text{cond,B}} + R''_{\text{conv}})\, q''$$

where the resistances for a unit surface area are

$$R''_{\text{cond, B}} = \frac{L_B}{k_B} \qquad R''_{\text{conv}} = \frac{1}{h}$$

Hence,

$$T_1 = 30°\text{C} + \left(\frac{0.02\ \text{m}}{150\ \text{W/m}\cdot\text{K}} + \frac{1}{1000\ \text{W/m}^2\cdot\text{K}}\right)$$
$$\times 1.5 \times 10^6\ \text{W/m}^3 \times 0.05\ \text{m}$$
$$T_1 = 30°\text{C} + 85°\text{C} = 115°\text{C}$$

Substituting into Equation 3,

$$T_0 = \frac{1.5 \times 10^6\ \text{W/m}^3 (0.05\ \text{m})^2}{2 \times 75\ \text{W/m}\cdot\text{K}} + 115°\text{C}$$
$$T_0 = 25°\text{C} + 115°\text{C} = 140°\text{C} \qquad \triangleleft$$

Comments:

1. Material A, having heat generation, cannot be represented by a thermal circuit element.
2. Since the resistance to heat transfer by convection is significantly larger than that due to conduction in material B, $R''_{conv}/R''_{cond} = 7.5$, the surface-to-fluid temperature difference is much larger than the temperature drop across material B, $(T_2 - T_\infty)/(T_1 - T_2) = 7.5$. This result is consistent with the temperature distribution plotted in part 1.
3. The surface and interface temperatures (T_0, T_1, and T_2) depend on the generation rate $\dot{q}$, the thermal conductivities k_A and k_B, and the convection coefficient h. Each material will have a maximum allowable operating temperature, which must not be exceeded if thermal failure of the system is to be avoided. We explore the effect of one of these parameters by computing and plotting temperature distributions for values of $h = 200$ and $1000\ \text{W/m}^2\cdot\text{K}$, which would be representative of air and liquid cooling, respectively.

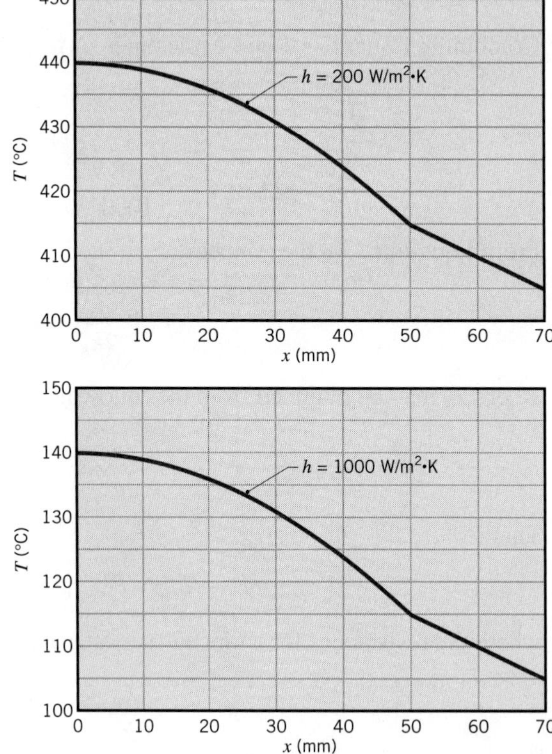

For $h = 200\ \text{W/m}^2\cdot\text{K}$, there is a significant increase in temperature throughout the system and, depending on the selection of materials, thermal failure could be a problem. Note the slight discontinuity in the temperature gradient, dT/dx, at $x = 50$ mm. What is the physical basis for this discontinuity? We have assumed negligible contact resistance at this location. What would be the effect of such a resistance on the temperature distribution throughout the system? Sketch a representative distribution. What would be the effect on the temperature distribution of an increase in $\dot{q}$, k_A, or k_B? Qualitatively sketch the effect of such changes on the temperature distribution.

4. This example is solved in the *Advanced* section of *IHT*.

3.5.2 Radial Systems

Heat generation may occur in a variety of radial geometries. Consider the long, solid cylinder of Figure 3.11, which could represent a current-carrying wire or a fuel element in a nuclear reactor. For steady-state conditions, the rate at which heat is generated within the cylinder must equal the rate at which heat is convected from the surface of the cylinder to a moving fluid. This condition allows the surface temperature to be maintained at a fixed value of T_s.

To determine the temperature distribution in the cylinder, we begin with the appropriate form of the heat equation. For constant thermal conductivity k, Equation 2.26 reduces to

$$\frac{1}{r}\frac{d}{dr}\left(r\frac{dT}{dr}\right) + \frac{\dot{q}}{k} = 0 \tag{3.54}$$

Separating variables and assuming uniform generation, this expression may be integrated to obtain

$$r\frac{dT}{dr} = -\frac{\dot{q}}{2k}r^2 + C_1 \tag{3.55}$$

Repeating the procedure, the general solution for the temperature distribution becomes

$$T(r) = -\frac{\dot{q}}{4k}r^2 + C_1 \ln r + C_2 \tag{3.56}$$

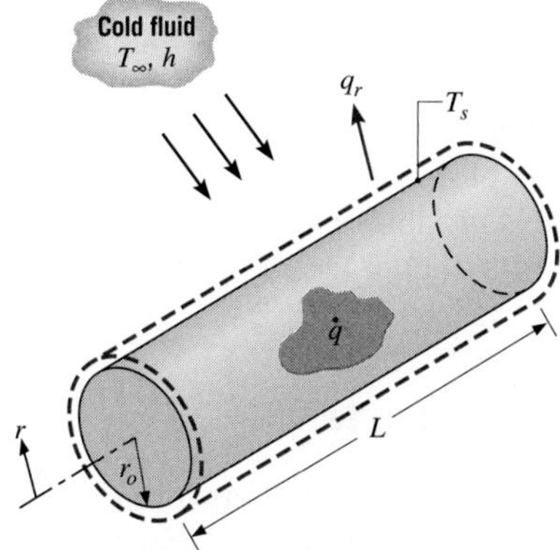

FIGURE 3.11 Conduction in a solid cylinder with uniform heat generation.

To obtain the constants of integration C_1 and C_2, we apply the boundary conditions

$$\left.\frac{dT}{dr}\right|_{r=0} = 0 \quad \text{and} \quad T(r_0) = T_s$$

The first condition results from the symmetry of the situation. That is, for the solid cylinder the centerline is a line of symmetry for the temperature distribution and the temperature gradient must be zero. Recall that similar conditions existed at the midplane of a wall having symmetrical boundary conditions (Figure 3.10*b*). From the symmetry condition at $r = 0$ and Equation 3.55, it is evident that $C_1 = 0$. Using the surface boundary condition at $r = r_o$ with Equation 3.56, we then obtain

$$C_2 = T_s + \frac{\dot{q}}{4k} r_o^2 \tag{3.57}$$

The temperature distribution is therefore

$$T(r) = \frac{\dot{q} r_o^2}{4k}\left(1 - \frac{r^2}{r_o^2}\right) + T_s \tag{3.58}$$

Evaluating Equation 3.58 at the centerline and dividing the result into Equation 3.58, we obtain the temperature distribution in nondimensional form,

$$\frac{T(r) - T_s}{T_o - T_s} = 1 - \left(\frac{r}{r_o}\right)^2 \tag{3.59}$$

where T_o is the centerline temperature. The heat rate at any radius in the cylinder may, of course, be evaluated by using Equation 3.58 with Fourier's law.

To relate the surface temperature, T_s, to the temperature of the cold fluid T_∞, either a surface energy balance or an overall energy balance may be used. Choosing the second approach, we obtain

$$\dot{q}(\pi r_o^2 L) = h(2\pi r_o L)(T_s - T_\infty)$$

or

$$T_s = T_\infty + \frac{\dot{q} r_o}{2h} \tag{3.60}$$

3.5.3 Tabulated Solutions

Appendix C provides a convenient and systematic procedure for treating the different combinations of surface conditions that may be applied to one-dimensional planar and radial (cylindrical and spherical) geometries with uniform thermal energy generation. From the tabulated results of this appendix, it is a simple matter to obtain distributions of the temperature, heat flux, and heat rate for boundary conditions of the *second kind* (a uniform surface heat flux) and the *third kind* (a surface heat flux that is proportional to a convection coefficient *h* or the overall heat transfer coefficient *U*). You are encouraged to become familiar with the contents of the appendix.

3.5.4 Application of Resistance Concepts

We conclude our discussion of heat generation effects with a word of caution. In particular, when such effects are present, the heat transfer rate is not a constant, independent of the spatial coordinate. Accordingly, it would be *incorrect* to use the conduction resistance concepts and the related heat rate equations developed in Sections 3.1 and 3.3.

IHT

EXAMPLE 3.8

Consider a long solid tube, insulated at the outer radius r_2 and cooled at the inner radius r_1, with uniform heat generation $\dot{q}$ (W/m^3) within the solid.

1. Obtain the general solution for the temperature distribution in the tube.
2. In a practical application a limit would be placed on the maximum temperature that is permissible at the insulated surface ($r = r_2$). Specifying this limit as $T_{s,2}$, identify appropriate boundary conditions that could be used to determine the arbitrary constants appearing in the general solution. Determine these constants and the corresponding form of the temperature distribution.
3. Determine the heat removal rate per unit length of tube.
4. If the coolant is available at a temperature T_∞, obtain an expression for the convection coefficient that would have to be maintained at the inner surface to allow for operation at prescribed values of $T_{s,2}$ and $\dot{q}$.

SOLUTION

Known: Solid tube with uniform heat generation is insulated at the outer surface and cooled at the inner surface.

Find:

1. General solution for the temperature distribution $T(r)$.
2. Appropriate boundary conditions and the corresponding form of the temperature distribution.
3. Heat removal rate for specified maximum temperature.
4. Corresponding required convection coefficient at the inner surface.

Schematic:

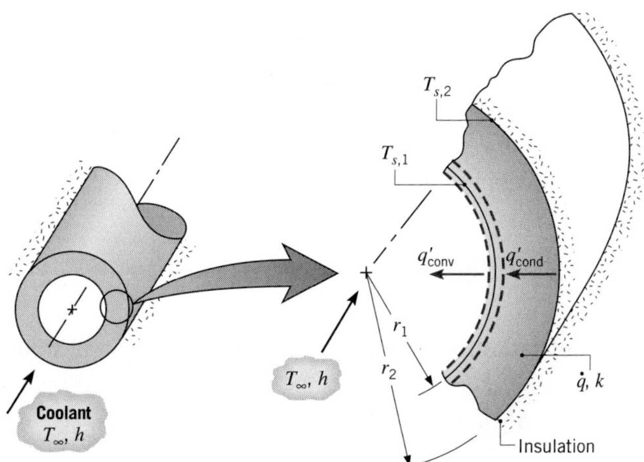

Assumptions:

1. Steady-state conditions.
2. One-dimensional radial conduction.
3. Constant properties.
4. Uniform volumetric heat generation.
5. Outer surface adiabatic.

Analysis:

1. To determine $T(r)$, the appropriate form of the heat equation, Equation 2.26, must be solved. For the prescribed conditions, this expression reduces to Equation 3.54, and the general solution is given by Equation 3.56. Hence, this solution applies in a cylindrical shell, as well as in a solid cylinder (Figure 3.11).
2. Two boundary conditions are needed to evaluate C_1 and C_2, and in this problem it is appropriate to specify both conditions at r_2. Invoking the prescribed temperature limit,

$$T(r_2) = T_{s,2} \tag{1}$$

and applying Fourier's law, Equation 3.29, at the adiabatic outer surface

$$\left.\frac{dT}{dr}\right|_{r_2} = 0 \tag{2}$$

Using Equations 3.56 and 1, it follows that

$$T_{s,2} = -\frac{\dot{q}}{4k} r_2^2 + C_1 \ln r_2 + C_2 \tag{3}$$

Similarly, from Equations 3.55 and 2

$$0 = -\frac{\dot{q}}{2k} r_2^2 + C_1 \tag{4}$$

Hence, from Equation 4,

$$C_1 = \frac{\dot{q}}{2k} r_2^2 \tag{5}$$

and from Equation 3

$$C_2 = T_{s,2} + \frac{\dot{q}}{4k} r_2^2 - \frac{\dot{q}}{2k} r_2^2 \ln r_2 \tag{6}$$

Substituting Equations 5 and 6 into the general solution, Equation 3.56, it follows that

$$T(r) = T_{s,2} + \frac{\dot{q}}{4k} (r_2^2 - r^2) - \frac{\dot{q}}{2k} r_2^2 \ln \frac{r_2}{r} \tag{7}$$

3. The heat removal rate may be determined by obtaining the conduction rate at r_1 or by evaluating the total generation rate for the tube. From Fourier's law

$$q'_r = -k2\pi r \frac{dT}{dr}$$

Hence, substituting from Equation 7 and evaluating the result at r_1,

$$q'_r(r_1) = -k2\pi r_1\left(-\frac{\dot{q}}{2k} r_1 + \frac{\dot{q}}{2k}\frac{r_2^2}{r_1}\right) = -\pi\dot{q}(r_2^2 - r_1^2) \tag{8}$$

Alternatively, because the tube is insulated at r_2, the rate at which heat is generated in the tube must equal the rate of removal at r_1. That is, for a control volume about the tube, the energy conservation requirement, Equation 1.12c, reduces to $\dot{E}_g - \dot{E}_{out} = 0$, where $\dot{E}_g = \dot{q}\pi(r_2^2 - r_1^2)L$ and $\dot{E}_{out} = q'_{cond} L = -q'_r(r_1)L$. Hence

$$q'_r(r_1) = -\pi\dot{q}(r_2^2 - r_1^2) \tag{9}$$

4. Applying the energy conservation requirement, Equation 1.13, to the inner surface, it follows that

$$q'_{cond} = q'_{conv}$$

or

$$\pi\dot{q}(r_2^2 - r_1^2) = h2\pi r_1(T_{s,1} - T_\infty)$$

Hence

$$h = \frac{\dot{q}(r_2^2 - r_1^2)}{2r_1(T_{s,1} - T_\infty)} \tag{10}$$

where $T_{s,1}$ may be obtained by evaluating Equation 7 at $r = r_1$.

Comments:

1. Note that, through application of Fourier's law in part 3, the sign on $q'_r(r_1)$ was found to be negative, Equation 8, implying that heat flow is in the negative r-direction. However, in applying the energy balance, we acknowledged that heat flow was *out* of the wall. Hence we expressed q'_{cond} as $-q'_r(r_1)$ and we expressed q'_{conv} in terms of $(T_{s,1} - T_\infty)$, rather than $(T_\infty - T_{s,1})$.
2. Results of the foregoing analysis may be used to determine the convection coefficient required to maintain the maximum tube temperature $T_{s,2}$ below a prescribed value. Consider a tube of thermal conductivity $k = 5\ \text{W/m}\cdot\text{K}$ and inner and outer radii of $r_1 = 20$ mm and $r_2 = 25$ mm, respectively, with a maximum allowable temperature of $T_{s,2} = 350°\text{C}$. The tube experiences heat generation at a rate of $\dot{q} = 5 \times 10^6\ \text{W/m}^3$, and the coolant is at a temperature of $T_\infty = 80°\text{C}$. Obtaining $T(r_1) = T_{s,1} = 336.5°\text{C}$ from Equation 7 and substituting into Equation 10, the required convection coefficient is found to be $h = 110\ \text{W/m}^2\cdot\text{K}$. Using the *IHT Workspace*, parametric calculations may be performed to determine the effects of the convection coefficient and the generation rate on the maximum tube temperature, and results are plotted as a function of h for three values of $\dot{q}$.

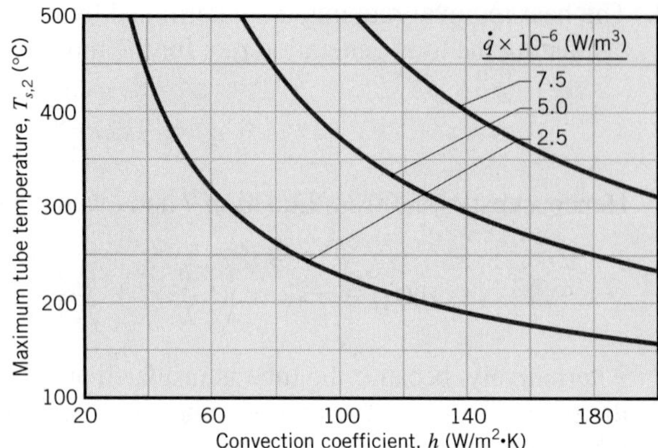

For each generation rate, the minimum value of h needed to maintain $T_{s,2} \leq 350°C$ may be determined from the figure.

3. The temperature distribution, Equation 7, may also be obtained by using the results of Appendix C. Applying a surface energy balance at $r = r_1$, with $q(r) = -\dot{q}\pi(r_2^2 - r_1^2)L$, $(T_{s,2} - T_{s,1})$ may be determined from Equation C.8 and the result substituted into Equation C.2 to eliminate $T_{s,1}$ and obtain the desired expression.

3.6 *Heat Transfer from Extended Surfaces*

The term *extended surface* is commonly used to depict an important special case involving heat transfer by conduction within a solid and heat transfer by convection (and/or radiation) from the boundaries of the solid. Until now, we have considered heat transfer from the boundaries of a solid to be in the same direction as heat transfer by conduction in the solid. In contrast, for an extended surface, the direction of heat transfer from the boundaries is perpendicular to the principal direction of heat transfer in the solid.

Consider a strut that connects two walls at different temperatures and across which there is fluid flow (Figure 3.12). With $T_1 > T_2$, temperature gradients in the x-direction sustain heat transfer by conduction in the strut. However, with $T_1 > T_2 > T_\infty$, there is concurrent heat

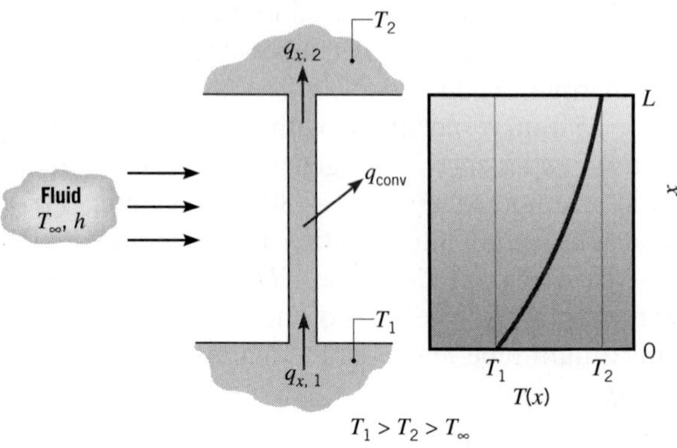

FIGURE 3.12 Combined conduction and convection in a structural element.

transfer by convection to the fluid, causing q_x, and hence the magnitude of the temperature gradient, $|dT/dx|$, to decrease with increasing x.

Although there are many different situations that involve such combined conduction–convection effects, the most frequent application is one in which an extended surface is used specifically to *enhance* heat transfer between a solid and an adjoining fluid. Such an extended surface is termed a *fin.*

Consider the plane wall of Figure 3.13*a* . If T_s is fixed, there are two ways in which the heat transfer rate may be increased. The convection coefficient h could be increased by increasing the fluid velocity, and/or the fluid temperature T_∞ could be reduced. However, there are many situations for which increasing h to the maximum possible value is either insufficient to obtain the desired heat transfer rate or the associated costs are prohibitive. Such costs are related to the blower or pump power requirements needed to increase h through increased fluid motion. Moreover, the second option of reducing T_∞ is often impractical. Examining Figure 3.13*b* , however, we see that there exists a third option. That is, the heat transfer rate may be increased by increasing the surface area across which the convection occurs. This may be done by employing *fins* that *extend* from the wall into the surrounding fluid. The thermal conductivity of the fin material can have a strong effect on the temperature distribution along the fin and therefore influences the degree to which the heat transfer rate is enhanced. Ideally, the fin material should have a large thermal conductivity to minimize temperature variations from its base to its tip. In the limit of infinite thermal conductivity, the entire fin would be at the temperature of the base surface, thereby providing the maximum possible heat transfer enhancement.

Examples of fin applications are easy to find. Consider the arrangement for cooling engine heads on motorcycles and lawn mowers or for cooling electric power transformers. Consider also the tubes with attached fins used to promote heat exchange between air and the working fluid of an air conditioner. Two common finned-tube arrangements are shown in Figure 3.14.

Different fin configurations are illustrated in Figure 3.15. A *straight fin* is any extended surface that is attached to a *plane wall.* It may be of uniform cross-sectional area, or its cross-sectional area may vary with the distance x from the wall. An *annular fin* is one that is circumferentially attached to a cylinder, and its cross section varies with radius from the wall of the cylinder. The foregoing fin types have rectangular cross sections, whose area may be expressed as a product of the fin thickness t and the width w for straight fins or the circumference $2\pi r$ for annular fins. In contrast a *pin fin,* or *spine,* is an extended surface of circular cross section. Pin fins may also be of uniform or nonuniform cross section. In any

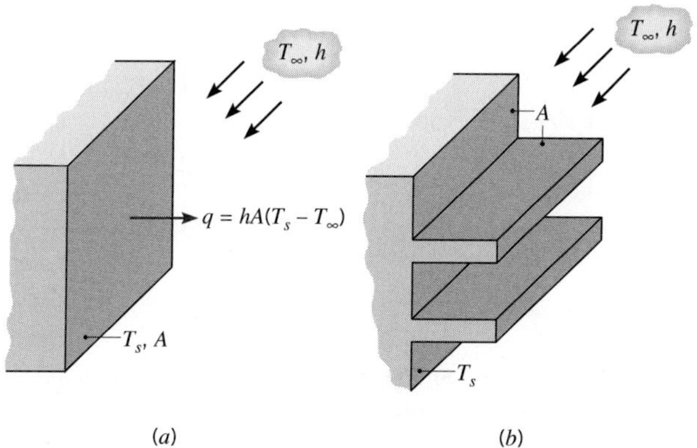

FIGURE 3.13 Use of fins to enhance heat transfer from a plane wall. (*a*) Bare surface. (*b*) Finned surface.

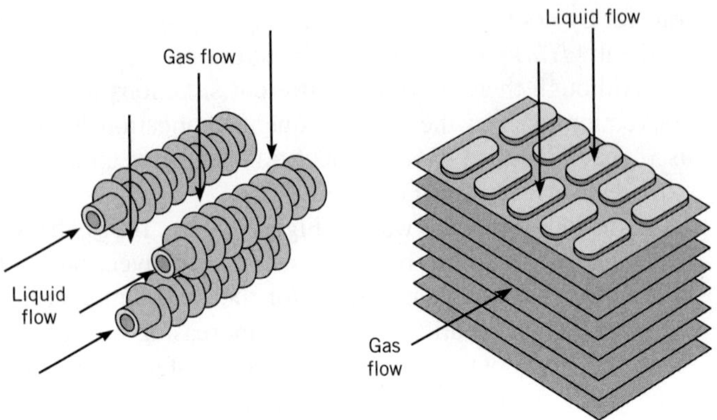

FIGURE 3.14 Schematic of typical finned-tube heat exchangers.

application, selection of a particular fin configuration may depend on space, weight, manufacturing, and cost considerations, as well as on the extent to which the fins reduce the surface convection coefficient and increase the pressure drop associated with flow over the fins.

3.6.1 A General Conduction Analysis

As engineers we are primarily interested in knowing the extent to which particular extended surfaces or fin arrangements could improve heat transfer from a surface to the surrounding fluid. To determine the heat transfer rate associated with a fin, we must first obtain the temperature distribution along the fin. As we have done for previous systems, we begin by performing an energy balance on an appropriate differential element. Consider the extended surface of Figure 3.16. The analysis is simplified if certain assumptions are made. We choose to assume one-dimensional conditions in the longitudinal (x-) direction, even though conduction within the fin is actually two-dimensional. The rate at which energy is convected to the fluid from any point on the fin surface must be balanced by the net rate at which energy reaches that point due to conduction in the transverse (y-, z-) direction. However, in practice the fin is thin, and temperature changes in the transverse

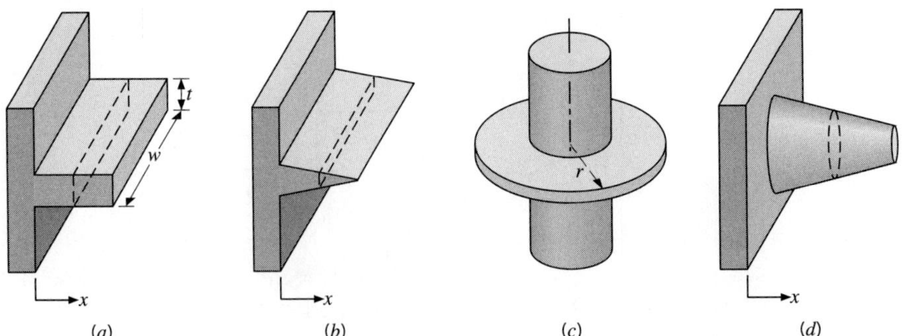

FIGURE 3.15 Fin configurations. (*a*) Straight fin of uniform cross section. (*b*) Straight fin of nonuniform cross section. (*c*) Annular fin. (*d*) Pin fin.

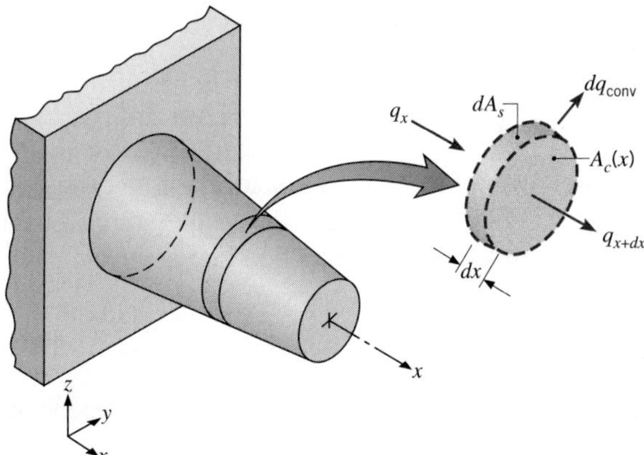

FIGURE 3.16 Energy balance for an extended surface.

direction within the fin are small compared with the temperature difference between the fin and the environment. Hence, we may assume that the temperature is uniform across the fin thickness, that is, it is only a function of x. We will consider steady-state conditions and also assume that the thermal conductivity is constant, that radiation from the surface is negligible, that heat generation effects are absent, and that the convection heat transfer coefficient h is uniform over the surface.

Applying the conservation of energy requirement, Equation 1.12c, to the differential element of Figure 3.16, we obtain

$$q_x = q_{x+dx} + dq_{\text{conv}} \tag{3.61}$$

From Fourier's law we know that

$$\boxed{q_x = -kA_c \frac{dT}{dx}} \tag{3.62}$$

where A_c is the *cross-sectional* area, which may vary with x. Since the conduction heat rate at $x + dx$ may be expressed as

$$q_{x+dx} = q_x + \frac{dq_x}{dx} dx \tag{3.63}$$

it follows that

$$q_{x+dx} = -kA_c \frac{dT}{dx} - k \frac{d}{dx}\left(A_c \frac{dT}{dx}\right) dx \tag{3.64}$$

The convection heat transfer rate may be expressed as

$$dq_{\text{conv}} = h dA_s (T - T_\infty) \tag{3.65}$$

where dA_s is the *surface* area of the differential element. Substituting the foregoing rate equations into the energy balance, Equation 3.61, we obtain

$$\frac{d}{dx}\left(A_c \frac{dT}{dx}\right) - \frac{h}{k} \frac{dA_s}{dx} (T - T_\infty) = 0$$

or

$$\frac{d^2T}{dx^2} + \left(\frac{1}{A_c}\frac{dA_c}{dx}\right)\frac{dT}{dx} - \left(\frac{1}{A_c}\frac{h}{k}\frac{dA_s}{dx}\right)(T - T_\infty) = 0 \tag{3.66}$$

This result provides a general form of the energy equation for an extended surface. Its solution for appropriate boundary conditions provides the temperature distribution, which may be used with Equation 3.62 to calculate the conduction rate at any x.

3.6.2 Fins of Uniform Cross-Sectional Area

To solve Equation 3.66 it is necessary to be more specific about the geometry. We begin with the simplest case of straight rectangular and pin fins of uniform cross section (Figure 3.17). Each fin is attached to a base surface of temperature $T(0) = T_b$ and extends into a fluid of temperature T_∞.

For the prescribed fins, A_c is a constant and $A_s = Px$, where A_s is the surface area measured from the base to x and P is the fin perimeter. Accordingly, with $dA_c/dx = 0$ and $dA_s/dx = P$, Equation 3.66 reduces to

$$\frac{d^2T}{dx^2} - \frac{hP}{kA_c}(T - T_\infty) = 0 \tag{3.67}$$

To simplify the form of this equation, we transform the dependent variable by defining an *excess temperature* θ as

$$\theta(x) \equiv T(x) - T_\infty \tag{3.68}$$

where, since T_∞ is a constant, $d\theta/dx = dT/dx$. Substituting Equation 3.68 into Equation 3.67, we then obtain

$$\frac{d^2\theta}{dx^2} - m^2\theta = 0 \tag{3.69}$$

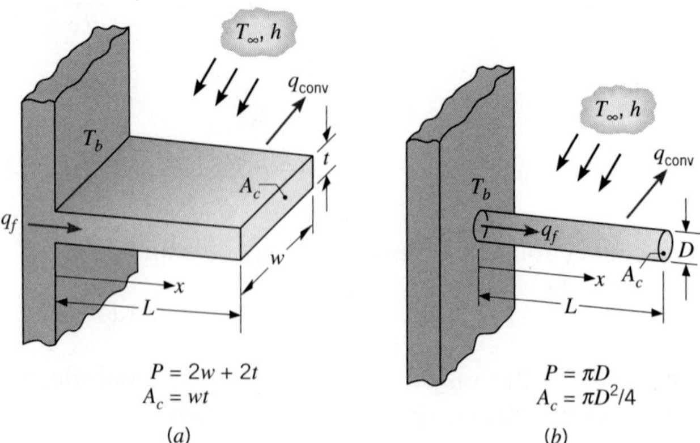

FIGURE 3.17 Straight fins of uniform cross section. (*a*) Rectangular fin. (*b*) Pin fin.

where

$$m^2 \equiv \frac{hP}{kA_c} \tag{3.70}$$

Equation 3.69 is a linear, homogeneous, second-order differential equation with constant coefficients. Its general solution is of the form

$$\theta(x) = C_1 e^{mx} + C_2 e^{-mx} \tag{3.71}$$

By substitution it may readily be verified that Equation 3.71 is indeed a solution to Equation 3.69.

To evaluate the constants C_1 and C_2 of Equation 3.71, it is necessary to specify appropriate boundary conditions. One such condition may be specified in terms of the temperature at the *base* of the fin ($x = 0$)

$$\theta(0) = T_b - T_\infty \equiv \theta_b \tag{3.72}$$

The second condition, specified at the fin tip ($x = L$), may correspond to one of four different physical situations.

The first condition, Case A, considers convection heat transfer from the fin tip. Applying an energy balance to a control surface about this tip (Figure 3.18), we obtain

$$hA_c[T(L) - T_\infty] = -kA_c \frac{dT}{dx}\bigg|_{x=L}$$

or

$$h\theta(L) = -k \frac{d\theta}{dx}\bigg|_{x=L} \tag{3.73}$$

That is, the rate at which energy is transferred to the fluid by convection from the tip must equal the rate at which energy reaches the tip by conduction through the fin. Substituting Equation 3.71 into Equations 3.72 and 3.73, we obtain, respectively,

$$\theta_b = C_1 + C_2 \tag{3.74}$$

and

$$h(C_1 e^{mL} + C_2 e^{-mL}) = km(C_2 e^{-mL} - C_1 e^{mL})$$

Solving for C_1 and C_2, it may be shown, after some manipulation, that

$$\frac{\theta}{\theta_b} = \frac{\cosh m(L - x) + (h/mk) \sinh m(L - x)}{\cosh mL + (h/mk) \sinh mL} \tag{3.75}$$

The form of this temperature distribution is shown schematically in Figure 3.18. Note that the magnitude of the temperature gradient decreases with increasing x. This trend is a consequence of the reduction in the conduction heat transfer $q_x(x)$ with increasing x due to continuous convection losses from the fin surface.

We are particularly interested in the amount of heat transferred from the entire fin. From Figure 3.18, it is evident that the fin heat transfer rate q_f may be evaluated in two

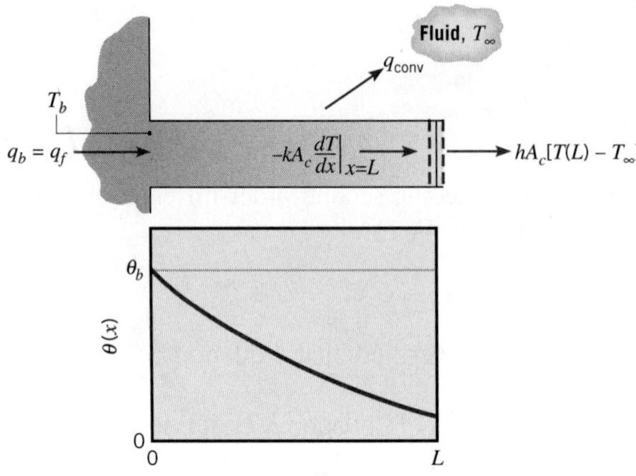

FIGURE 3.18 Conduction and convection in a fin of uniform cross section.

alternative ways, both of which involve use of the temperature distribution. The simpler procedure, and the one that we will use, involves applying Fourier's law at the fin base. That is,

$$q_f = q_b = -kA_c \frac{dT}{dx}\bigg|_{x=0} = -kA_c \frac{d\theta}{dx}\bigg|_{x=0} \tag{3.76}$$

Hence, knowing the temperature distribution, $\theta(x)$, q_f may be evaluated, giving

$$q_f = \sqrt{hPkA_c}\,\theta_b \frac{\sinh mL + (h/mk)\cosh mL}{\cosh mL + (h/mk)\sinh mL} \tag{3.77}$$

However, conservation of energy dictates that the rate at which heat is transferred by convection from the fin must equal the rate at which it is conducted through the base of the fin. Accordingly, the alternative formulation for q_f is

$$q_f = \int_{A_f} h[T(x) - T_\infty]\, dA_s$$

$$q_f = \int_{A_f} h\theta(x)\, dA_s \tag{3.78}$$

where A_f is the *total,* including the tip, *fin surface area.* Substitution of Equation 3.75 into Equation 3.78 would yield Equation 3.77.

The second tip condition, Case B, corresponds to the assumption that the convective heat loss from the fin tip is negligible, in which case the tip may be treated as adiabatic and

$$\frac{d\theta}{dx}\bigg|_{x=L} = 0 \tag{3.79}$$

Substituting from Equation 3.71 and dividing by m, we then obtain

$$C_1 e^{mL} - C_2 e^{-mL} = 0$$

Using this expression with Equation 3.74 to solve for C_1 and C_2 and substituting the results into Equation 3.71, we obtain

$$\frac{\theta}{\theta_b} = \frac{\cosh m(L - x)}{\cosh mL} \tag{3.80}$$

Using this temperature distribution with Equation 3.76, the fin heat transfer rate is then

$$q_f = \sqrt{hPkA_c}\,\theta_b \tanh mL \tag{3.81}$$

In the same manner, we can obtain the fin temperature distribution and heat transfer rate for Case C, where the temperature is prescribed at the fin tip. That is, the second boundary condition is $\theta(L) = \theta_L$, and the resulting expressions are of the form

$$\frac{\theta}{\theta_b} = \frac{(\theta_L/\theta_b)\sinh mx + \sinh m(L - x)}{\sinh mL} \tag{3.82}$$

$$q_f = \sqrt{hPkA_c}\,\theta_b \frac{\cosh mL - \theta_L/\theta_b}{\sinh mL} \tag{3.83}$$

The *very long fin,* Case D, is an interesting extension of these results. In particular, as $L \rightarrow \infty$, $\theta_L \rightarrow 0$ and it is easily verified that

$$\frac{\theta}{\theta_b} = e^{-mx} \tag{3.84}$$

$$q_f = \sqrt{hPkA_c}\,\theta_b \tag{3.85}$$

The foregoing results are summarized in Table 3.4. A table of hyperbolic functions is provided in Appendix B.1.

TABLE 3.4 Temperature distribution and heat loss for fins of uniform cross section

Case	Tip Condition ($x = L$)	Temperature Distribution θ/θ_b	Fin Heat Transfer Rate q_f
A	Convection heat transfer: $h\theta(L) = -kd\theta/dx\rvert_{x=L}$	$\dfrac{\cosh m(L-x) + (h/mk)\sinh m(L-x)}{\cosh mL + (h/mk)\sinh mL}$ (3.75)	$M\dfrac{\sinh mL + (h/mk)\cosh mL}{\cosh mL + (h/mk)\sinh mL}$ (3.77)
B	Adiabatic: $d\theta/dx\rvert_{x=L} = 0$	$\dfrac{\cosh m(L-x)}{\cosh mL}$ (3.80)	$M \tanh mL$ (3.81)
C	Prescribed temperature: $\theta(L) = \theta_L$	$\dfrac{(\theta_L/\theta_b)\sinh mx + \sinh m(L-x)}{\sinh mL}$ (3.82)	$M\dfrac{(\cosh mL - \theta_L/\theta_b)}{\sinh mL}$ (3.83)
D	Infinite fin ($L \rightarrow \infty$): $\theta(L) = 0$	e^{-mx} (3.84)	M (3.85)

$\theta \equiv T - T_\infty$ $\quad m^2 \equiv hP/kA_c$

$\theta_b = \theta(0) = T_b - T_\infty$ $\quad M \equiv \sqrt{hPkA_c}\,\theta_b$

Example 3.9

A very long rod 5 mm in diameter has one end maintained at 100°C. The surface of the rod is exposed to ambient air at 25°C with a convection heat transfer coefficient of 100 W/m$^2 \cdot$K.

1. Determine the temperature distributions along rods constructed from pure copper, 2024 aluminum alloy, and type AISI 316 stainless steel. What are the corresponding heat losses from the rods?
2. Estimate how long the rods must be for the assumption of *infinite length* to yield an accurate estimate of the heat loss.

Solution

Known: A long circular rod exposed to ambient air.

Find:

1. Temperature distribution and heat loss when rod is fabricated from copper, an aluminum alloy, or stainless steel.
2. How long rods must be to assume infinite length.

Schematic:

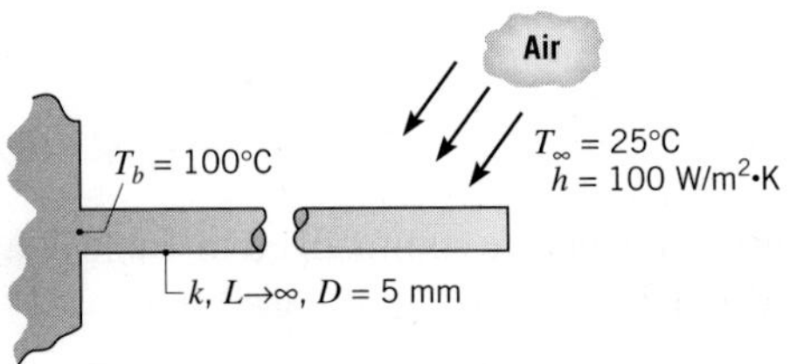

Assumptions:

1. Steady-state conditions.
2. One-dimensional conduction along the rod.
3. Constant properties.
4. Negligible radiation exchange with surroundings.
5. Uniform heat transfer coefficient.
6. Infinitely long rod.

Properties: Table A.1, copper [$T = (T_b + T_\infty)/2 = 62.5°\text{C} \approx 335$ K]: $k = 398$ W/m$\cdot$K. Table A.1, 2024 aluminum (335 K): $k = 180$ W/m$\cdot$K. Table A.1, stainless steel, AISI 316 (335 K): $k = 14$ W/m$\cdot$K.

Analysis:

1. Subject to the assumption of an infinitely long fin, the temperature distributions are determined from Equation 3.84, which may be expressed as

$$T = T_\infty + (T_b - T_\infty)e^{-mx}$$

where $m = (hP/kA_c)^{1/2} = (4h/kD)^{1/2}$. Substituting for h and D, as well as for the thermal conductivities of copper, the aluminum alloy, and the stainless steel, respectively, the values of m are 14.2, 21.2, and 75.6 m^{-1}. The temperature distributions may then be computed and plotted as follows:

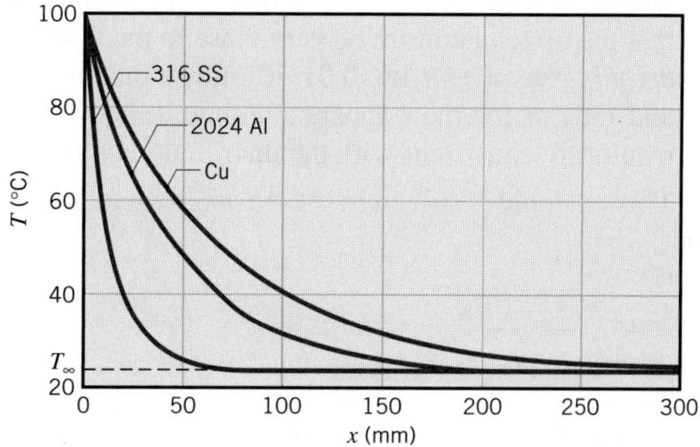

From these distributions, it is evident that there is little additional heat transfer associated with extending the length of the rod much beyond 50, 200, and 300 mm, respectively, for the stainless steel, the aluminum alloy, and the copper.

From Equation 3.85, the heat loss is

$$q_f = \sqrt{hPkA_c}\,\theta_b$$

Hence for copper,

$$q_f = \left[100\ \text{W/m}^2\cdot\text{K} \times \pi \times 0.005\ \text{m} \times 398\ \text{W/m}\cdot\text{K} \times \frac{\pi}{4}(0.005\ \text{m})^2\right]^{1/2} (100 - 25)^\circ\text{C}$$

$$= 8.3\ \text{W} \quad \triangleleft$$

Similarly, for the aluminum alloy and stainless steel, respectively, the heat rates are $q_f = 5.6$ W and 1.6 W.

2. Since there is no heat loss from the tip of an infinitely long rod, an estimate of the validity of this approximation may be made by comparing Equations 3.81 and 3.85. To a satisfactory approximation, the expressions provide equivalent results if tanh $mL \geq 0.99$ or $mL \geq 2.65$. Hence a rod may be assumed to be infinitely long if

$$L \geq L_\infty \equiv \frac{2.65}{m} = 2.65\left(\frac{kA_c}{hP}\right)^{1/2}$$

For copper,

$$L_\infty = 2.65\left[\frac{398\ \text{W/m}\cdot\text{K} \times (\pi/4)(0.005\ \text{m})^2}{100\ \text{W/m}^2\cdot\text{K} \times \pi(0.005\ \text{m})}\right]^{1/2} = 0.19\ \text{m} \quad \triangleleft$$

Results for the aluminum alloy and stainless steel are $L_\infty = 0.13$ m and $L_\infty = 0.04$ m, respectively.

Comments:

1. The foregoing results suggest that the fin heat transfer rate may accurately be predicted from the infinite fin approximation if $mL \geq 2.65$. However, if the infinite fin approximation is to accurately predict the temperature distribution $T(x)$, a larger value of mL would be required. This value may be inferred from Equation 3.84 and the requirement that the tip temperature be very close to the fluid temperature. Hence, if we require that $\theta(L)/\theta_b = \exp(-mL) < 0.01$, it follows that $mL > 4.6$, in which case $L_\infty \approx 0.33$, 0.23, and 0.07 m for the copper, aluminum alloy, and stainless steel, respectively. These results are consistent with the distributions plotted in part 1.
2. This example is solved in the *Advanced* section of *IHT*.

3.6.3 Fin Performance

Recall that fins are used to increase the heat transfer from a surface by increasing the effective surface area. However, the fin itself represents a conduction resistance to heat transfer from the original surface. For this reason, there is no assurance that the heat transfer rate will be increased through the use of fins. An assessment of this matter may be made by evaluating the *fin effectiveness* ε_f. It is defined as the *ratio of the fin heat transfer rate to the heat transfer rate that would exist without the fin.* Therefore

$$\varepsilon_f = \frac{q_f}{hA_{c,b}\theta_b} \tag{3.86}$$

where $A_{c,b}$ is the fin cross-sectional area at the base. In any rational design the value of ε_f should be as large as possible, and in general, the use of fins may rarely be justified unless $\varepsilon_f \gtrsim 2$.

Subject to any one of the four tip conditions that have been considered, the effectiveness for a fin of uniform cross section may be obtained by dividing the appropriate expression for q_f in Table 3.4 by $hA_{c,b}\theta_b$. Although the installation of fins will alter the surface convection coefficient, this effect is commonly neglected. Hence, assuming the convection coefficient of the finned surface to be equivalent to that of the unfinned base, it follows that, for the infinite fin approximation (Case D), the result is

$$\varepsilon_f = \left(\frac{kP}{hA_c}\right)^{1/2} \tag{3.87}$$

Several important trends may be inferred from this result. Obviously, fin effectiveness is enhanced by the choice of a material of high thermal conductivity. Aluminum alloys and copper come to mind. However, although copper is superior from the standpoint of thermal conductivity, aluminum alloys are the more common choice because of additional benefits related to lower cost and weight. Fin effectiveness is also enhanced by increasing the ratio of the perimeter to the cross-sectional area. For this reason, the use of *thin,* but closely spaced fins, is preferred, with the proviso that the fin gap not be reduced to a value for which flow between the fins is severely impeded, thereby reducing the convection coefficient.

Equation 3.87 also suggests that the use of fins can be better justified under conditions for which the convection coefficient h is small. Hence from Table 1.1 it is evident that the need for fins is stronger when the fluid is a gas rather than a liquid and when the surface heat transfer is by *free* convection. If fins are to be used on a surface separating a gas and a liquid, they are generally placed on the gas side, which is the side of lower convection coefficient. A common example is the tubing in an automobile radiator. Fins are applied to the outer tube surface, over which there is flow of ambient air (small h), and not to the inner surface, through which there is flow of water (large h). Note that, if $\varepsilon_f > 2$ is used as a criterion to justify the implementation of fins, Equation 3.87 yields the requirement that $(kP/hA_c) > 4$.

Equation 3.87 provides an upper limit to ε_f, which is reached as L approaches infinity. However, it is certainly not necessary to use very long fins to achieve near maximum heat transfer enhancement. As seen in Example 3.9, 99% of the maximum possible fin heat transfer rate is achieved for $mL = 2.65$. Hence, it would make no sense to extend the fins beyond $L = 2.65/m$.

Fin performance may also be quantified in terms of a thermal resistance. Treating the difference between the base and fluid temperatures as the driving potential, a *fin resistance* may be defined as

$$R_{t,f} = \frac{\theta_b}{q_f} \tag{3.88}$$

This result is extremely useful, particularly when representing a finned surface by a thermal circuit. Note that, according to the fin tip condition, an appropriate expression for q_f may be obtained from Table 3.4.

Dividing Equation 3.88 into the expression for the thermal resistance due to convection at the exposed base,

$$R_{t,b} = \frac{1}{hA_{c,b}} \tag{3.89}$$

and substituting from Equation 3.86, it follows that

$$\varepsilon_f = \frac{R_{t,b}}{R_{t,f}} \tag{3.90}$$

Hence the fin effectiveness may be interpreted as a ratio of thermal resistances, and to increase ε_f it is necessary to reduce the conduction/convection resistance of the fin. If the fin is to enhance heat transfer, its resistance must not exceed that of the exposed base.

Another measure of fin thermal performance is provided by the *fin efficiency* η_f. The maximum driving potential for convection is the temperature difference between the base ($x = 0$) and the fluid, $\theta_b = T_b - T_\infty$. Hence the maximum rate at which a fin could dissipate energy is the rate that would exist *if* the entire fin surface were at the base temperature. However, since any fin is characterized by a finite conduction resistance, a temperature gradient must exist along the fin and the preceding condition is an idealization. A logical definition of fin efficiency is therefore

$$\eta_f \equiv \frac{q_f}{q_{\text{max}}} = \frac{q_f}{hA_f\theta_b} \tag{3.91}$$

where A_f is the surface area of the fin. For a straight fin of uniform cross section and an adiabatic tip, Equations 3.81 and 3.91 yield

$$\eta_f = \frac{M \tanh mL}{hPL\theta_b} = \frac{\tanh mL}{mL} \tag{3.92}$$

Referring to Table B.1, this result tells us that η_f approaches its maximum and minimum values of 1 and 0, respectively, as L approaches 0 and ∞.

In lieu of the somewhat cumbersome expression for heat transfer from a straight rectangular fin with an active tip, Equation 3.77, it has been shown that approximate, yet accurate, predictions may be obtained by using the adiabatic tip result, Equation 3.81, with a corrected fin length of the form $L_c = L + (t/2)$ for a rectangular fin and $L_c = L + (D/4)$ for a pin fin [14]. The correction is based on assuming equivalence between heat transfer from the actual fin with tip convection and heat transfer from a longer, hypothetical fin with an adiabatic tip. Hence, with tip convection, the fin heat rate may be approximated as

$$q_f = M \tanh mL_c \tag{3.93}$$

and the corresponding efficiency as

$$\eta_f = \frac{\tanh mL_c}{mL_c} \tag{3.94}$$

Errors associated with the approximation are negligible if (ht/k) or $(hD/2k) \lesssim 0.0625$ [15].

If the width of a rectangular fin is much larger than its thickness, $w \gg t$, the perimeter may be approximated as $P = 2w$, and

$$mL_c = \left(\frac{hP}{kA_c}\right)^{1/2} L_c = \left(\frac{2h}{kt}\right)^{1/2} L_c$$

Multiplying numerator and denominator by $L_c^{1/2}$ and introducing a corrected fin profile area, $A_p = L_c t$, it follows that

$$mL_c = \left(\frac{2h}{kA_p}\right)^{1/2} L_c^{3/2} \tag{3.95}$$

Hence, as shown in Figures 3.19 and 3.20, the efficiency of a rectangular fin with tip convection may be represented as a function of $L_c^{3/2}(h/kA_p)^{1/2}$.

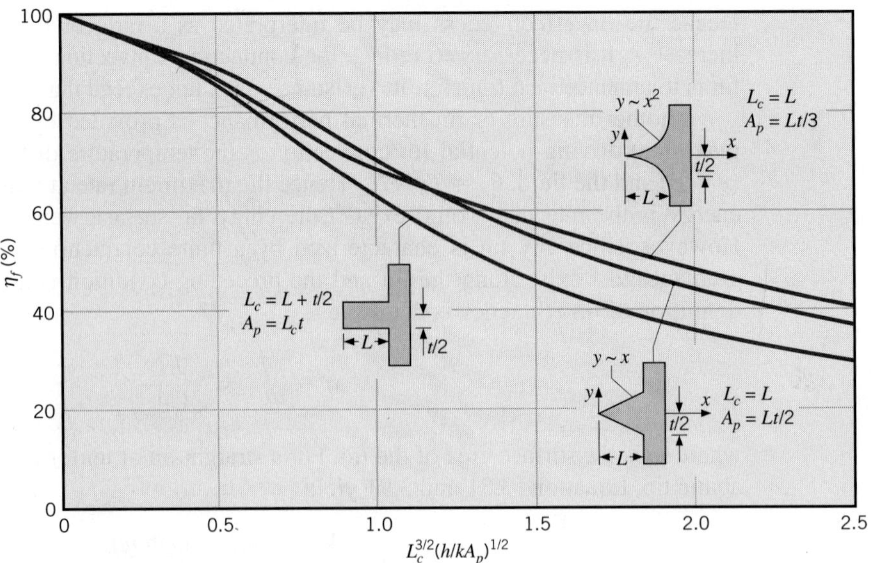

FIGURE 3.19 Efficiency of straight fins (rectangular, triangular, and parabolic profiles).

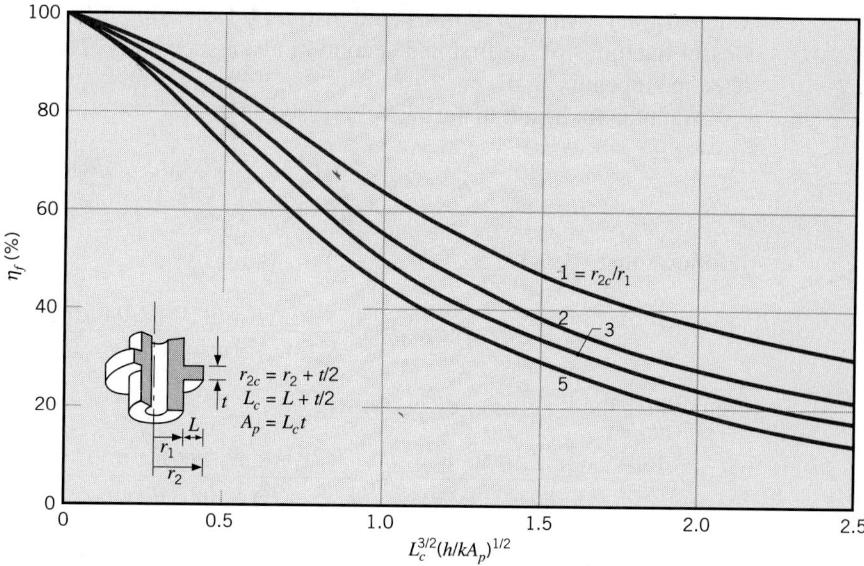

FIGURE 3.20 Efficiency of annular fins of rectangular profile.

3.6.4 Fins of Nonuniform Cross-Sectional Area

Analysis of fin thermal behavior becomes more complex if the fin is of nonuniform cross section. For such cases the second term of Equation 3.66 must be retained, and the solutions are no longer in the form of simple exponential or hyperbolic functions. As a special case, consider the annular fin shown in the inset of Figure 3.20. Although the fin thickness is uniform (t is independent of r), the cross-sectional area, $A_c = 2\pi rt$, varies with r. Replacing x by r in Equation 3.66 and expressing the surface area as $A_s = 2\pi(r^2 - r_1^2)$, the general form of the fin equation reduces to

$$\frac{d^2T}{dr^2} + \frac{1}{r}\frac{dT}{dr} - \frac{2h}{kt}(T - T_\infty) = 0$$

or, with $m^2 \equiv 2h/kt$ and $\theta \equiv T - T_\infty$,

$$\frac{d^2\theta}{dr^2} + \frac{1}{r}\frac{d\theta}{dr} - m^2\theta = 0$$

The foregoing expression is a *modified Bessel equation* of order zero, and its general solution is of the form

$$\theta(r) = C_1 I_0(mr) + C_2 K_0(mr)$$

where I_0 and K_0 are modified, zero-order Bessel functions of the first and second kinds, respectively. If the temperature at the base of the fin is prescribed, $\theta(r_1) = \theta_b$, and an adiabatic tip is presumed, $d\theta/dr|_{r_2} = 0$, C_1 and C_2 may be evaluated to yield a temperature distribution of the form

$$\frac{\theta}{\theta_b} = \frac{I_0(mr)K_1(mr_2) + K_0(mr)I_1(mr_2)}{I_0(mr_1)K_1(mr_2) + K_0(mr_1)I_1(mr_2)}$$

where $I_1(mr) = d[I_0(mr)]/d(mr)$ and $K_1(mr) = -d[K_0(mr)]/d(mr)$ are modified, first-order Bessel functions of the first and second kinds, respectively. The Bessel functions are tabulated in Appendix B.

With the fin heat transfer rate expressed as

$$q_f = -kA_{c,b}\left.\frac{dT}{dr}\right|_{r=r_1} = -k(2\pi r_1 t)\left.\frac{d\theta}{dr}\right|_{r=r_1}$$

it follows that

$$q_f = 2\pi k r_1 t \theta_b m \frac{K_1(mr_1)I_1(mr_2) - I_1(mr_1)K_1(mr_2)}{K_0(mr_1)I_1(mr_2) + I_0(mr_1)K_1(mr_2)}$$

from which the fin efficiency becomes

$$\eta_f = \frac{q_f}{h2\pi(r_2^2 - r_1^2)\theta_b} = \frac{2r_1}{m(r_2^2 - r_1^2)} \frac{K_1(mr_1)I_1(mr_2) - I_1(mr_1)K_1(mr_2)}{K_0(mr_1)I_1(mr_2) + I_0(mr_1)K_1(mr_2)} \tag{3.96}$$

This result may be applied for an active (convecting) tip, if the tip radius r_2 is replaced by a corrected radius of the form $r_{2c} = r_2 + (t/2)$. Results are represented graphically in Figure 3.20.

Knowledge of the thermal efficiency of a fin may be used to evaluate the fin resistance, where, from Equations 3.88 and 3.91, it follows that

$$R_{t,f} = \frac{1}{hA_f\eta_f} \tag{3.97}$$

Expressions for the efficiency and surface area of several common fin geometries are summarized in Table 3.5. Although results for the fins of uniform thickness or diameter

TABLE 3.5 Efficiency of common fin shapes

Straight Fins

Rectangular[a]

$A_f = 2wL_c$

$L_c = L + (t/2)$

$A_p = tL$

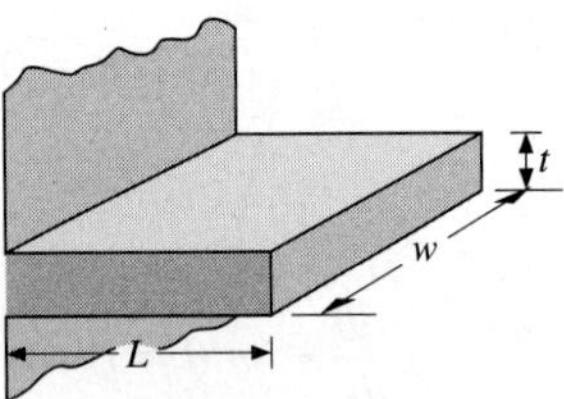

$$\eta_f = \frac{\tanh mL_c}{mL_c} \tag{3.94}$$

Triangular[a]

$A_f = 2w[L^2 + (t/2)^2]^{1/2}$

$A_p = (t/2)L$

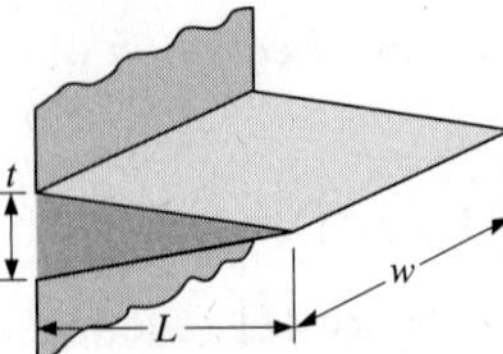

$$\eta_f = \frac{1}{mL}\frac{I_1(2mL)}{I_0(2mL)} \tag{3.98}$$

Parabolic[a]

$A_f = w[C_1L + (L^2/t)\ln(t/L + C_1)]$

$C_1 = [1 + (t/L)^2]^{1/2}$

$A_p = (t/3)L$

$y = (t/2)(1 - x/L)^2$

$$\eta_f = \frac{2}{[4(mL)^2 + 1]^{1/2} + 1} \tag{3.99}$$

TABLE 3.5 Continued

Annular Fin *Rectangular*[a] $A_f = 2\pi\,(r_{2c}^2 - r_1^2)$ $r_{2c} = r_2 + (t/2)$ $V = \pi\,(r_2^2 - r_1^2)t$	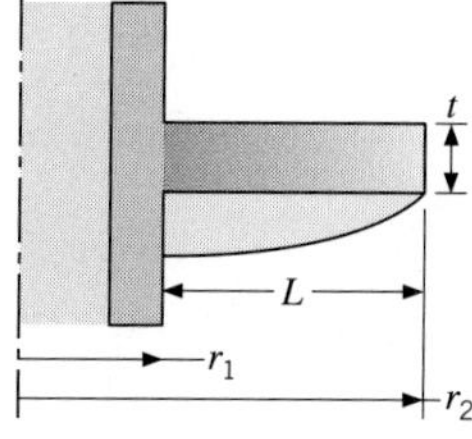	$\eta_f = C_2 \dfrac{K_1(mr_1)I_1(mr_{2c}) - I_1(mr_1)K_1(mr_{2c})}{I_0(mr_1)K_1(mr_{2c}) + K_0(mr_1)I_1(mr_{2c})}$ $C_2 = \dfrac{(2r_1/m)}{(r_{2c}^2 - r_1^2)}$	(3.96)
Pin Fins *Rectangular*[b] $A_f = \pi D L_c$ $L_c = L + (D/4)$ $V = (\pi D^2/4)L$	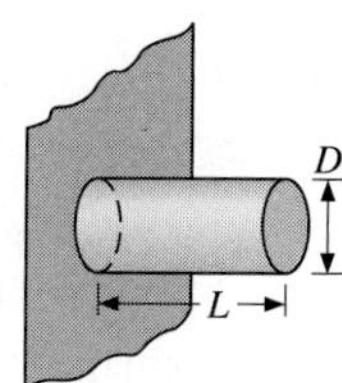	$\eta_f = \dfrac{\tanh mL_c}{mL_c}$	(3.100)
Triangular[b] $A_f = \dfrac{\pi D}{2}[L^2 + (D/2)^2]^{1/2}$ $V = (\pi/12)D^2L$	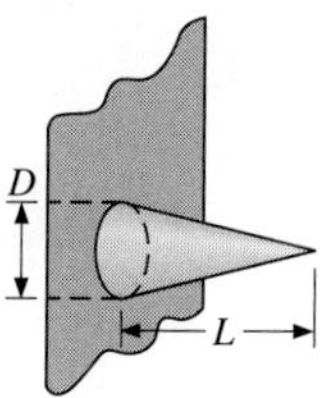	$\eta_f = \dfrac{2}{mL}\dfrac{I_2(2mL)}{I_1(2mL)}$	(3.101)
Parabolic[b] $A_f = \dfrac{\pi L^3}{8D}\{C_3C_4 - \dfrac{L}{2D}\ln[(2DC_4/L) + C_3]\}$ $C_3 = 1 + 2(D/L)^2$ $C_4 = [1 + (D/L)^2]^{1/2}$ $V = (\pi/20)D^2L$	D $y = (D/2)(1 - x/L)^2$ L x	$\eta_f = \dfrac{2}{[4/9(mL)^2 + 1]^{1/2} + 1}$	(3.102)

[a] $m = (2h/kt)^{1/2}$.
[b] $m = (4h/kD)^{1/2}$.

were obtained by assuming an adiabatic tip, the effects of convection may be treated by using a corrected length (Equations 3.94 and 3.100) or radius (Equation 3.96). The triangular and parabolic fins are of nonuniform thickness that reduces to zero at the fin tip.

Expressions for the profile area, A_p, or the volume, V, of a fin are also provided in Table 3.5. The volume of a straight fin is simply the product of its width and profile area, $V = wA_p$.

Fin design is often motivated by a desire to minimize the fin material and/or related manufacturing costs required to achieve a prescribed cooling effectiveness. Hence, a straight *triangular* fin is attractive because, for equivalent heat transfer, it requires much less volume (fin material) than a rectangular profile. In this regard, heat dissipation per unit volume, $(q/V)_f$, is largest for a *parabolic* profile. However, since $(q/V)_f$ for the parabolic profile is only slightly larger than that for a triangular profile, its use can rarely be justified in view of its larger manufacturing costs. The *annular* fin of rectangular profile is commonly used to enhance heat transfer to or from circular tubes.

3.6.5 Overall Surface Efficiency

In contrast to the fin efficiency η_f, which characterizes the performance of a single fin, the *overall surface efficiency* η_o characterizes an *array* of fins and the base surface to which they are attached. Representative arrays are shown in Figure 3.21, where S designates the fin pitch. In each case the overall efficiency is defined as

$$\eta_o = \frac{q_t}{q_{\max}} = \frac{q_t}{hA_t\theta_b} \tag{3.103}$$

where q_t is the total heat rate from the surface area A_t associated with both the fins and the exposed portion of the base (often termed the *prime* surface). If there are N fins in the array, each of surface area A_f, and the area of the prime surface is designated as A_b, the total surface area is

$$A_t = NA_f + A_b \tag{3.104}$$

The maximum possible heat rate would result if the entire fin surface, as well as the exposed base, were maintained at T_b.

The total rate of heat transfer by convection from the fins and the prime (unfinned) surface may be expressed as

$$q_t = N\eta_f hA_f\theta_b + hA_b\theta_b \tag{3.105}$$

where the convection coefficient h is assumed to be equivalent for the finned and prime surfaces and η_f is the efficiency of a single fin. Hence

$$q_t = h[N\eta_f A_f + (A_t - NA_f)]\theta_b = hA_t\left[1 - \frac{NA_f}{A_t}(1 - \eta_f)\right]\theta_b \tag{3.106}$$

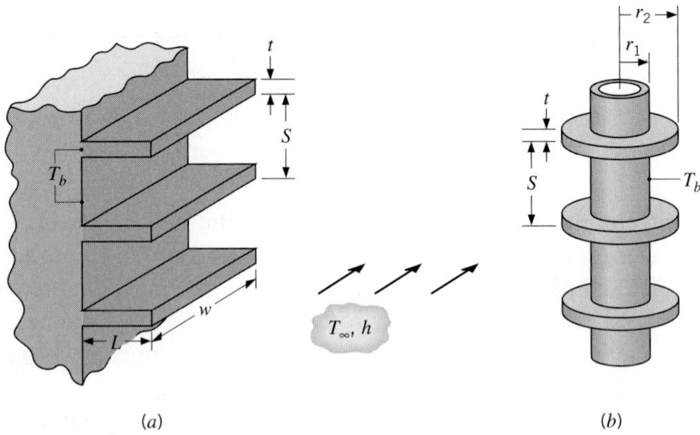

FIGURE 3.21 Representative fin arrays. (*a*) Rectangular fins. (*b*) Annular fins.

Substituting Equation (3.106) into (3.103), it follows that

$$\eta_o = 1 - \frac{NA_f}{A_t}(1 - \eta_f) \tag{3.107}$$

From knowledge of η_o, Equation 3.103 may be used to calculate the total heat rate for a fin array.

Recalling the definition of the fin thermal resistance, Equation 3.88, Equation 3.103 may be used to infer an expression for the thermal resistance of a fin array. That is,

$$R_{t,o} = \frac{\theta_b}{q_t} = \frac{1}{\eta_o h A_t} \tag{3.108}$$

where $R_{t,o}$ is an effective resistance that accounts for parallel heat flow paths by conduction/convection in the fins and by convection from the prime surface. Figure 3.22 illustrates the thermal circuits corresponding to the parallel paths and their representation in terms of an effective resistance.

If fins are machined as an integral part of the wall from which they extend (Figure 3.22*a*), there is no contact resistance at their base. However, more commonly, fins are manufactured separately and are attached to the wall by a metallurgical or adhesive joint. Alternatively, the attachment may involve a *press fit,* for which the fins are forced into slots machined on the wall material. In such cases (Figure 3.22*b*), there is a thermal contact resistance $R_{t,c}$, which

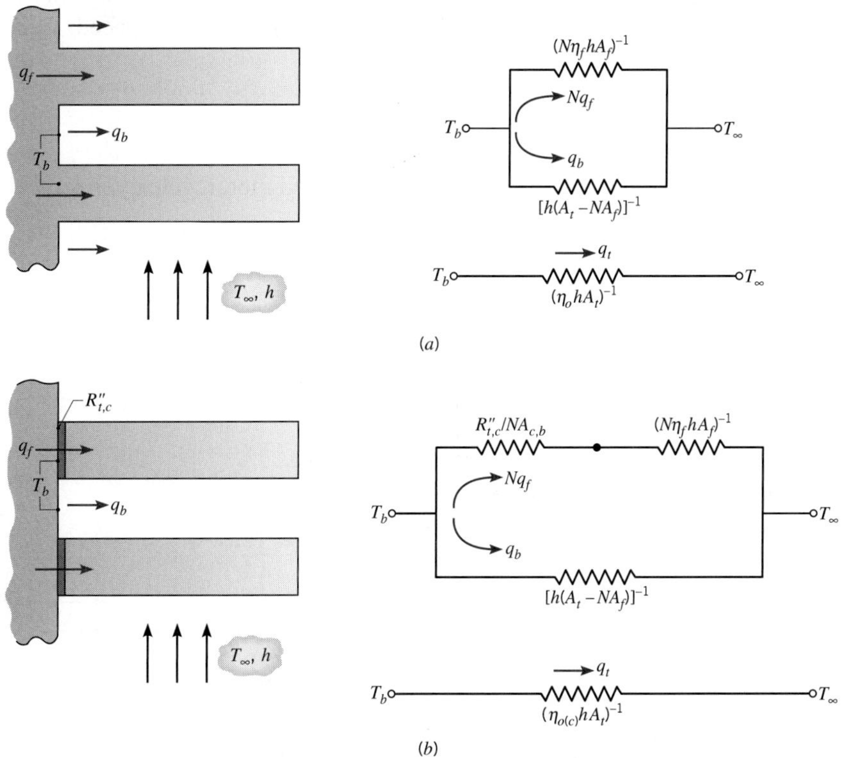

FIGURE 3.22 Fin array and thermal circuit. (*a*) Fins that are integral with the base. (*b*) Fins that are attached to the base.

may adversely influence overall thermal performance. An effective circuit resistance may again be obtained, where, with the contact resistance,

$$R_{t,o(c)} = \frac{\theta_b}{q_t} = \frac{1}{\eta_{o(c)} h A_t} \tag{3.109}$$

It is readily shown that the corresponding overall surface efficiency is

$$\eta_{o(c)} = 1 - \frac{NA_f}{A_t}\left(1 - \frac{\eta_f}{C_1}\right) \tag{3.110a}$$

where

$$C_1 = 1 + \eta_f h A_f (R''_{t,c}/A_{c,b}) \tag{3.110b}$$

In manufacturing, care must be taken to render $R_{t,c} \ll R_{t,f}$.

EXAMPLE 3.10

The engine cylinder of a motorcycle is constructed of 2024-T6 aluminum alloy and is of height $H = 0.15$ m and outside diameter $D = 50$ mm. Under typical operating conditions the outer surface of the cylinder is at a temperature of 500 K and is exposed to ambient air at 300 K, with a convection coefficient of 50 W/m$^2 \cdot$K. Annular fins are integrally cast with the cylinder to increase heat transfer to the surroundings. Consider five such fins, which are of thickness $t = 6$ mm, length $L = 20$ mm, and equally spaced. What is the increase in heat transfer due to use of the fins?

SOLUTION

Known: Operating conditions of a finned motorcycle cylinder.

Find: Increase in heat transfer associated with using fins.

Schematic:

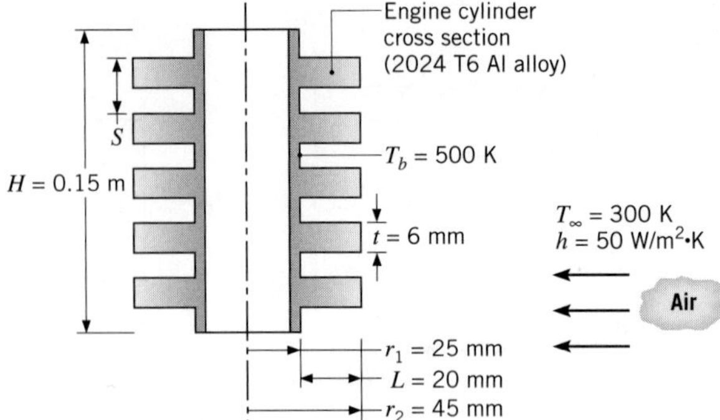

Assumptions:

1. Steady-state conditions.
2. One-dimensional radial conduction in fins.
3. Constant properties.

4. Negligible radiation exchange with surroundings.
5. Uniform convection coefficient over outer surface (with or without fins).

Properties: Table A.1, 2024-T6 aluminum ($T = 400$ K): $k = 186$ W/m·K.

Analysis: With the fins in place, the heat transfer rate is given by Equation 3.106

$$q_t = hA_t\left[1 - \frac{NA_f}{A_t}(1 - \eta_f)\right]\theta_b$$

where $A_f = 2\pi(r_{2c}^2 - r_1^2) = 2\pi[(0.048 \text{ m})^2 - (0.025 \text{ m})^2] = 0.0105 \text{ m}^2$ and, from Equation 3.104, $A_t = NA_f + 2\pi r_1(H - Nt) = 0.0527 \text{ m}^2 + 2\pi(0.025 \text{ m})\,[0.15 \text{ m} - 0.03 \text{ m}] = 0.0716 \text{ m}^2$. With $r_{2c}/r_1 = 1.92$, $L_c = 0.023$ m, $A_p = 1.380 \times 10^{-4} \text{ m}^2$, we obtain $L_c^{3/2}(h/kA_p)^{1/2} = 0.15$. Hence, from Figure 3.20, the fin efficiency is $\eta_f \approx 0.95$. With the fins, the total heat transfer rate is then

$$q_t = 50 \text{ W/m}^2\cdot\text{K} \times 0.0716 \text{ m}^2\left[1 - \frac{0.0527 \text{ m}^2}{0.0716 \text{ m}^2}(0.05)\right]200 \text{ K} = 690 \text{ W}$$

Without the fins, the convection heat transfer rate would be

$$q_{\text{wo}} = h(2\pi r_1 H)\theta_b = 50 \text{ W/m}^2\cdot\text{K}(2\pi \times 0.025 \text{ m} \times 0.15 \text{ m})200 \text{ K} = 236 \text{ W}$$

Hence

$$\Delta q = q_t - q_{\text{wo}} = 454 \text{ W}$$ ◁

Comments:

1. Although the fins significantly increase heat transfer from the cylinder, considerable improvement could still be obtained by increasing the number of fins. We assess this possibility by computing q_t as a function of N, first by fixing the fin thickness at $t = 6$ mm and increasing the number of fins by reducing the spacing between fins. Prescribing a fin clearance of 2 mm at each end of the array and a minimum fin gap of 4 mm, the maximum allowable number of fins is $N = H/S = 0.15 \text{ m}/(0.004 + 0.006)$ m $= 15$. The parametric calculations yield the following variation of q_t with N:

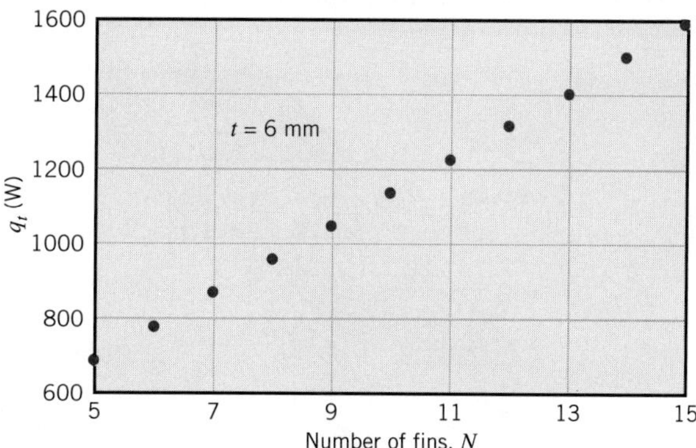

The number of fins could also be increased by reducing the fin thickness. If the fin gap is fixed at $(S - t) = 4$ mm and manufacturing constraints dictate a minimum allowable fin thickness of 2 mm, up to $N = 25$ fins may be accommodated. In this case the parametric calculations yield

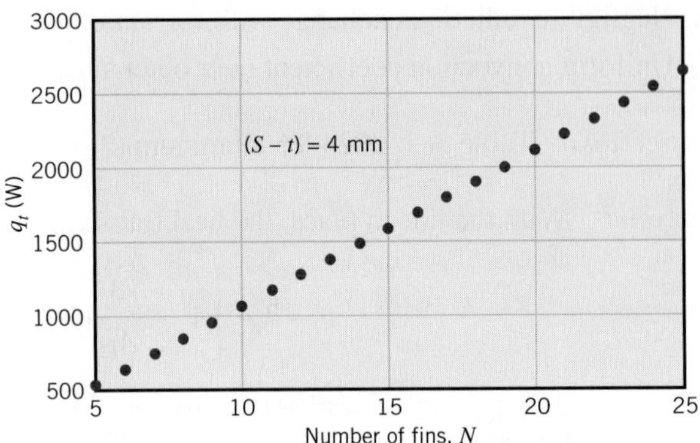

The foregoing calculations are based on the assumption that h is not affected by a reduction in the fin gap. The assumption is reasonable as long as there is no interaction between boundary layers that develop on the opposing surfaces of adjoining fins. Note that, since $NA_f \gg 2\pi r_1(H - Nt)$ for the prescribed conditions, q_t increases nearly linearly with increasing N.

2. The *Models/Extended Surfaces* option in the *Advanced* section of *IHT* provides ready-to-solve models for straight, pin, and annular fins, as well as for fin arrays. The models include the efficiency relations of Figures 3.19 and 3.20 and Table 3.5.

EXAMPLE 3.11

In Example 1.5, we saw that to generate an electrical power of $P = 9$ W, the temperature of the PEM fuel cell had to be maintained at $T_c \approx 56.4°C$, which required removal of 11.25 W from the fuel cell and a cooling air velocity of $V = 9.4$ m/s for $T_\infty = 25°C$. To provide these convective conditions, the fuel cell is centered in a 50 mm $\times$ 26 mm rectangular duct, with 10-mm gaps between the exterior of the 50 mm $\times$ 50 mm $\times$ 6 mm fuel cell and the top and bottom of the well-insulated duct wall. A small fan, powered by the fuel cell, is used to circulate the cooling air. Inspection of a particular fan vendor's data sheets suggests that the ratio of the fan power consumption to the fan's volumetric flow rate is $P_f/\dot{\forall}_f = C = 1000\ \text{W/(m}^3\text{/s)}$ for the range $10^{-4} \leq \dot{\forall}_f \leq 10^{-2}\ \text{m}^3\text{/s}$.

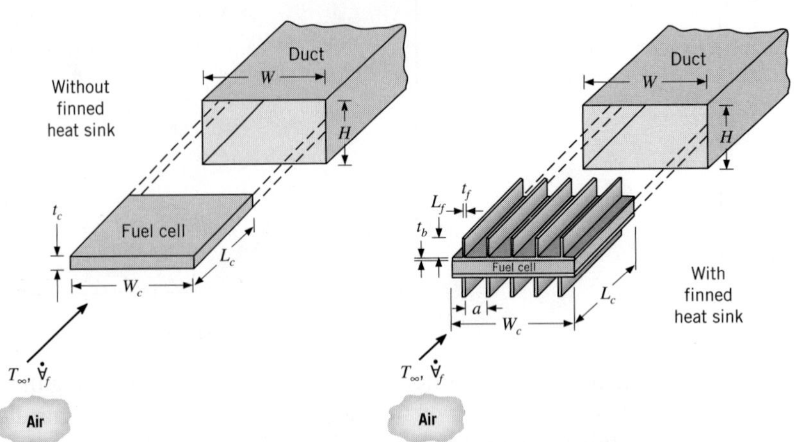

1. Determine the *net* electric power produced by the fuel cell–fan system, $P_{net} = P - P_f$.
2. Consider the effect of attaching an aluminum ($k = 200\ W/m \cdot K$) *finned heat sink*, of identical top and bottom sections, onto the fuel cell body. The contact joint has a thermal resistance of $R''_{t,c} = 10^{-3}\ m^2 \cdot K/W$, and the base of the heat sink is of thickness $t_b = 2$ mm. Each of the N rectangular fins is of length $L_f = 8$ mm and thickness $t_f = 1$ mm, and spans the entire length of the fuel cell, $L_c = 50$ mm. With the heat sink in place, radiation losses are negligible and the convective heat transfer coefficient may be related to the size and geometry of a typical air channel by an expression of the form $h = 1.78\ k_{air}\ (L_f + a)/(L_f \cdot a)$, where a is the distance between fins. Draw an equivalent thermal circuit for part 2 and determine the total number of fins needed to reduce the fan power consumption to half of the value found in part 1.

Solution

Known: Dimensions of a fuel cell and finned heat sink, fuel cell operating temperature, rate of thermal energy generation, power production. Relationship between power consumed by a cooling fan and the fan airflow rate. Relationship between the convection coefficient and the air channel dimensions.

Find:

1. The net power produced by the fuel cell–fan system when there is no heat sink.
2. The number of fins needed to reduce the fan power consumption found in part 1 by 50%.

Schematic:

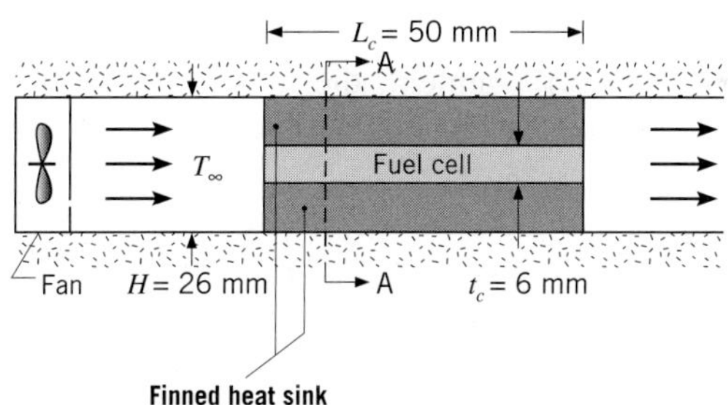

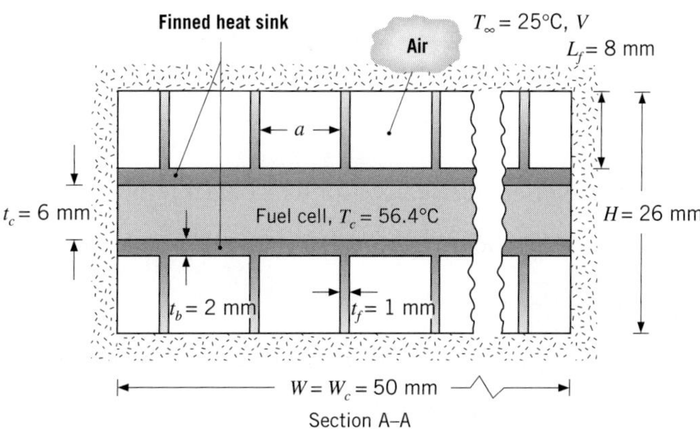

Section A–A

Assumptions:

1. Steady-state conditions.
2. Negligible heat transfer from the edges of the fuel cell, as well as from the front and back faces of the finned heat sink.
3. One-dimensional heat transfer through the heat sink.
4. Adiabatic fin tips.
5. Constant properties.
6. Negligible radiation when the heat sink is in place.

Properties: Table A.4. air ($\bar{T} = 300$ K): $k_{air} = 0.0263$ W/m·K, $c_p = 1007$ J/kg·K, $\rho = 1.1614$ kg/m^3.

Analysis:

1. The volumetric flow rate of cooling air is $\dot{\forall}_f = VA_c$, where $A_c = W(H - t_c)$ is the cross-sectional area of the flow region between the duct walls and the unfinned fuel cell. Therefore,

$$\dot{\forall}_f = V[W(H - t_c)] = 9.4 \text{ m/s} \times [0.05 \text{ m} \times (0.026 \text{ m} - 0.006 \text{ m})]$$

$$= 9.4 \times 10^{-3} \text{ m}^3\text{/s}$$

and

$$P_{net} = P - C\dot{\forall}_f = 9.0 \text{ W} - 1000 \text{ W/(m}^3\text{/s)} \times 9.4 \times 10^{-3} \text{ m}^3\text{/s} = -0.4 \text{ W} \quad \triangleleft$$

With this arrangement, the fan consumes more power than is generated by the fuel cell, and the system cannot produce net power.

2. To reduce the fan power consumption by 50%, the volumetric flow rate of air must be reduced to $\dot{\forall}_f = 4.7 \times 10^{-3}$ m^3/s. The thermal circuit includes resistances for the contact joint, conduction through the base of the finned heat sink, and resistances for the exposed base of the finned side of the heat sink, as well as the fins.

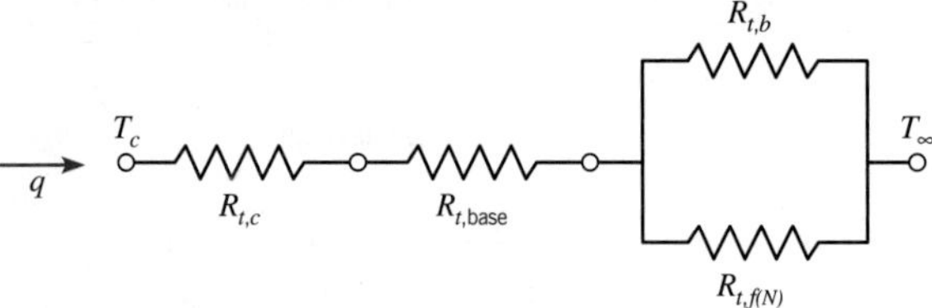

The thermal resistances for the contact joint and the base are

$$R_{t,c} = R''_{t,c}/2L_cW_c = (10^{-3} \text{ m}^2\cdot\text{K/W})/(2 \times 0.05 \text{ m} \times 0.05 \text{ m}) = 0.2 \text{ K/W}$$

and

$$R_{t,\text{base}} = t_b/(2kL_cW_c) = (0.002 \text{ m})/(2 \times 200 \text{ W/m}\cdot\text{K} \times 0.05 \text{ m} \times 0.05 \text{ m})$$

$$= 0.002 \text{ K/W}$$

where the factors of two account for the two sides of the heat sink assembly. For the portion of the base exposed to the cooling air, the thermal resistance is

$$R_{t,b} = 1/[h\,(2W_c - Nt_f)L_c] = 1/[h \times (2 \times 0.05\text{ m} - N \times 0.001\text{ m}) \times 0.05\text{ m}]$$

which cannot be evaluated until the total number of fins on both sides, N, and h are determined.

For a single fin, $R_{t,f} = \theta_b/q_f$, where, from Table 3.4 for a fin with an insulated fin tip, $R_{t,f} = (hPkA_c)^{-1/2}/\tanh(mL_f)$. In our case, $P = 2(L_c + t_f) = 2 \times (0.05\text{ m} + 0.001\text{ m}) = 0.102\text{ m}$, $A_c = L_c t_f = 0.05\text{ m} \times 0.001\text{ m} = 0.00005\text{ m}^2$, and

$$m = \sqrt{hP/kA_c} = [h \times 0.102\text{ m}/(200\text{ W/m}\cdot\text{K} \times 0.00005\text{ m}^2)]^{1/2}$$

Hence,

$$R_{t,f} = \frac{(h \times 0.102\text{ m} \times 200\text{ W/m}\cdot\text{K} \times 0.00005\text{ m}^2)^{-1/2}}{\tanh(m \times 0.008\text{ m})}$$

and for N fins, $R_{t,f(N)} = R_{t,f}/N$. As for $R_{t,b}$, $R_{t,f}$ cannot be evaluated until h and N are determined. Also, h depends on a, the distance between fins, which in turn depends on N, according to $a = (2W_c - Nt_f)/N = (2 \times 0.05\text{ m} - N \times 0.001\text{ m})/N$. Thus, specification of N will make it possible to calculate all resistances. From the thermal resistance network, the total thermal resistance is $R_{\text{tot}} = R_{t,c} + R_{t,\text{base}} + R_{\text{equiv}}$, where $R_{\text{equiv}} = [R_{t,b}^{-1} + R_{t,f(N)}^{-1}]^{-1}$.

The equivalent fin resistance, R_{equiv}, corresponding to the desired fuel cell temperature is found from the expression

$$q = \frac{T_c - T_\infty}{R_{\text{tot}}} = \frac{T_c - T_\infty}{R_{t,c} + R_{t,\text{base}} + R_{\text{equiv}}}$$

in which case,

$$\begin{aligned} R_{\text{equiv}} &= \frac{T_c - T_\infty}{q} - (R_{t,c} + R_{t,\text{base}}) \\ &= (56.4°\text{C} - 25°\text{C})/11.25\text{ W} - (0.2 + 0.002)\text{ K/W} = 2.59\text{ K/W} \end{aligned}$$

For $N = 22$, the following values of the various parameters are obtained: $a = 0.0035$ m, $h = 19.1\text{ W/m}^2\cdot\text{K}$, $m = 13.9\text{ m}^{-1}$, $R_{t,f(N)} = 2.94$ K/W, $R_{t,b} = 13.5$ K/W, $R_{\text{equiv}} = 2.41$ K/W, and $R_{\text{tot}} = 2.61$ K/W, resulting in a fuel cell temperature of 54.4°C. Fuel cell temperatures associated with $N = 20$ and $N = 24$ fins are $T_c = 58.9°\text{C}$ and 50.7°C, respectively.

The actual fuel cell temperature is closest to the desired value when $N = 22$. Therefore, a total of 22 fins, 11 on top and 11 on the bottom, should be specified, resulting in

$$P_{\text{net}} = P - P_f = 9.0\text{ W} - 4.7\text{ W} = 4.3\text{ W} \qquad \triangleleft$$

Comments:

1. The performance of the fuel cell–fan system is enhanced significantly by combining the finned heat sink with the fuel cell. Good thermal management can transform an impractical proposal into a viable concept.
2. The temperature of the cooling air increases as heat is transferred from the fuel cell. The temperature of the air leaving the finned heat sink may be calculated from an overall energy balance on the airflow, which yields $T_o = T_i + q/(\rho c_p \dot{\forall}_f)$. For part 1, $T_o = 25°\text{C} + 10.28\text{ W}/(1.1614\text{ kg/m}^3 \times 1007\text{ J/kg}\cdot\text{K} \times 9.4 \times 10^{-3}\text{ m}^3/\text{s}) = 25.9°\text{C}$. For part 2, the

outlet air temperature is $T_o = 27.0°C$. Hence, the operating temperature of the fuel cell will be slightly higher than predicted under the assumption that the cooling air temperature is constant at 25°C and will be closer to the desired value.

3. For the conditions in part 2, the convection heat transfer coefficient does not vary with the air velocity. The insensitivity of the value of h to the fluid velocity occurs frequently in cases where the flow is confined within passages of small cross-sectional area, as will be discussed in detail in Chapter 8. The fin's influence on increasing or reducing the value of h relative to that of an unfinned surface should be taken into account in critical applications.

4. A more detailed analysis of the system would involve prediction of the pressure drop associated with the fan-induced flow of air through the gaps between the fins.

5. The adiabatic fin tip assumption is valid since the duct wall is well insulated.

3.7 *The Bioheat Equation*

The topic of heat transfer within the human body is becoming increasingly important as new medical treatments are developed that involve extreme temperatures [16] and as we explore more adverse environments, such as the Arctic, underwater, or space. There are two main phenomena that make heat transfer in living tissues more complex than in conventional engineering materials: metabolic heat generation and the exchange of thermal energy between flowing blood and the surrounding tissue. Pennes [17] introduced a modification to the heat equation, now known as the Pennes or bioheat equation, to account for these effects. The bioheat equation is known to have limitations, but it continues to be a useful tool for understanding heat transfer in living tissues. In this section, we present a simplified version of the bioheat equation for the case of steady-state, one-dimensional heat transfer.

Both the metabolic heat generation and exchange of thermal energy with the blood can be viewed as effects of thermal energy generation. Therefore, we can rewrite Equation 3.44 to account for these two heat sources as

$$\frac{d^2T}{dx^2} + \frac{\dot{q}_m + \dot{q}_p}{k} = 0 \tag{3.111}$$

where $\dot{q}_m$ and $\dot{q}_p$ are the *metabolic* and *perfusion* heat source terms, respectively. The perfusion term accounts for energy exchange between the blood and the tissue and is an energy source or sink according to whether heat transfer is from or to the blood, respectively. The thermal conductivity has been assumed constant in writing Equation 3.111.

Pennes proposed an expression for the perfusion term by assuming that within any small volume of tissue, the blood flowing in the small capillaries enters at an arterial temperature, T_a, and exits at the local tissue temperature, T. The rate at which heat is gained by the tissue is the rate at which heat is lost from the blood. If the perfusion rate is ω (m^3/s of volumetric blood flow per m^3 of tissue), the heat lost from the blood can be calculated from Equation 1.12e, or on a unit volume basis,

$$\dot{q}_p = \omega \rho_b c_b (T_a - T) \tag{3.112}$$

where ρ_b and c_b are the blood density and specific heat, respectively. Note that $\omega \rho_b$ is the blood mass flow rate per unit volume of tissue.

Substituting Equation 3.112 into Equation 3.111, we find

$$\frac{d^2T}{dx^2} + \frac{\dot{q}_m + \omega\rho_b c_b(T_a - T)}{k} = 0 \tag{3.113}$$

Drawing on our experience with extended surfaces, it is convenient to define an excess temperature of the form $\theta \equiv T - T_a - \dot{q}_m/\omega\rho_b c_b$. Then, if we assume that T_a, $\dot{q}_m$, ω, and the blood properties are all constant, Equation 3.113 can be rewritten as

$$\frac{d^2\theta}{dx^2} - \tilde{m}^2\theta = 0 \tag{3.114}$$

where $\tilde{m}^2 = \omega\rho_b c_b/k$. This equation is identical in form to Equation 3.69. Depending on the form of the boundary conditions, it may therefore be possible to use the results of Table 3.4 to estimate the temperature distribution within the living tissue.

Example 3.12

In Example 1.7, the temperature at the inner surface of the skin/fat layer was given as 35°C. In reality, this temperature depends on the existing heat transfer conditions, including phenomena occurring farther inside the body. Consider a region of muscle with a skin/fat layer over it. At a depth of $L_m = 30$ mm into the muscle, the temperature can be assumed to be at the core body temperature of $T_c = 37°\text{C}$. The muscle thermal conductivity is $k_m = 0.5$ W/m·K. The metabolic heat generation rate within the muscle is $\dot{q}_m = 700$ W/m^3. The perfusion rate is $\omega = 0.0005$ s^{-1}; the blood density and specific heat are $\rho_b = 1000$ kg/m^3 and $c_b = 3600$ J/kg·K, respectively, and the arterial blood temperature T_a is the same as the core body temperature. The thickness, emissivity, and thermal conductivity of the skin/fat layer are as given in Example 1.7; perfusion and metabolic heat generation within this layer can be neglected. We wish to predict the heat loss rate from the body and the temperature at the inner surface of the skin/fat layer for air and water environments of Example 1.7.

Solution

Known: Dimensions and thermal conductivities of a muscle layer and a skin/fat layer. Skin emissivity and surface area. Metabolic heat generation rate and perfusion rate within the muscle layer. Core body and arterial temperatures. Blood density and specific heat. Ambient conditions.

Find: Heat loss rate from body and temperature at inner surface of the skin/fat layer.

Schematic:

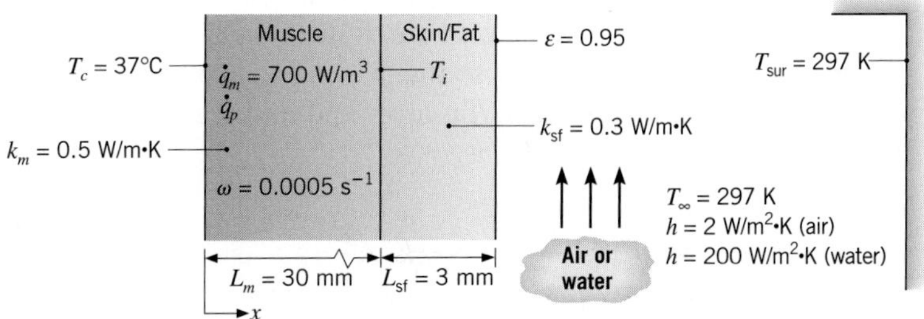

Assumptions:

1. Steady-state conditions.
2. One-dimensional heat transfer through the muscle and skin/fat layers.
3. Metabolic heat generation rate, perfusion rate, arterial temperature, blood properties, and thermal conductivities are all uniform.
4. Radiation heat transfer coefficient is known from Example 1.7.
5. Solar irradiation is negligible.

Analysis: We will combine an analysis of the muscle layer with a treatment of heat transfer through the skin/fat layer and into the environment. The rate of heat transfer through the skin/fat layer and into the environment can be expressed in terms of a total resistance, R_{tot}, as

$$q = \frac{T_i - T_\infty}{R_{\text{tot}}} \tag{1}$$

As in Example 3.1 and for exposure of the skin to the air, R_{tot} accounts for conduction through the skin/fat layer in series with heat transfer by convection and radiation, which act in parallel with each other. Thus,

$$R_{\text{tot}} = \frac{L_{\text{sf}}}{k_{\text{sf}}A} + \left(\frac{1}{1/hA} + \frac{1}{1/h_r A}\right)^{-1} = \frac{1}{A}\left(\frac{L_{\text{sf}}}{k_{\text{sf}}} + \frac{1}{h + h_r}\right)$$

Using the values from Example 1.7 for air,

$$R_{\text{tot}} = \frac{1}{1.8\ \text{m}^2}\left(\frac{0.003\ \text{m}}{0.3\ \text{W/m}\cdot\text{K}} + \frac{1}{(2 + 5.9)\ \text{W/m}^2\cdot\text{K}}\right) = 0.076\ \text{K/W}$$

For water, with $h_r = 0$ and $h = 200\ \text{W/m}^2\cdot\text{K}$, $R_{\text{tot}} = 0.0083\ \text{W/m}^2\cdot\text{K}$.

Heat transfer in the muscle layer is governed by Equation 3.114. The boundary conditions are specified in terms of the temperatures, T_c and T_i, where T_i is, as yet, unknown. In terms of the excess temperature θ, the boundary conditions are then

$$\theta(0) = T_c - T_a - \frac{\dot{q}_m}{\omega \rho_b c_b} = \theta_c \quad \text{and} \quad \theta(L_m) = T_i - T_a - \frac{\dot{q}_m}{\omega \rho_b c_b} = \theta_i$$

Since we have two boundary conditions involving prescribed temperatures, the solution for θ is given by case C of Table 3.4,

$$\frac{\theta}{\theta_c} = \frac{(\theta_i/\theta_c)\sinh \tilde{m}x + \sinh \tilde{m}(L_m - x)}{\sinh \tilde{m}L_m}$$

The value of q_f given in Table 3.4 would correspond to the heat transfer rate at $x = 0$, but this is not of particular interest here. Rather, we seek the rate at which heat leaves the muscle and enters the skin/fat layer so that we can equate this quantity with the rate at which heat is transferred through the skin/fat layer and into the environment. Therefore, we calculate the heat transfer rate at $x = L_m$ as

$$q\bigg|_{x=L_m} = -k_m A \frac{dT}{dx}\bigg|_{x=L_m} = -k_m A \frac{d\theta}{dx}\bigg|_{x=L_m} = -k_m A \tilde{m} \theta_c \frac{(\theta_i/\theta_c)\cosh \tilde{m}L_m - 1}{\sinh \tilde{m}L_m} \tag{2}$$

Combining Equations 1 and 2 yields

$$-k_m A\tilde{m}\theta_c \frac{(\theta_i/\theta_c)\cosh \tilde{m}L_m - 1}{\sinh \tilde{m}L_m} = \frac{T_i - T_\infty}{R_{\text{tot}}}$$

This expression can be solved for T_i, recalling that T_i also appears in θ_i.

$$T_i = \frac{T_\infty \sinh \tilde{m}L_m + k_m A\tilde{m}R_{\text{tot}}\left[\theta_c + \left(T_a + \dfrac{\dot{q}_m}{\omega\rho_b c_b}\right)\cosh \tilde{m}L_m\right]}{\sinh \tilde{m}L_m + k_m A\tilde{m}R_{\text{tot}}\cosh \tilde{m}L_m}$$

where

$$\tilde{m} = \sqrt{\omega\rho_b c_b/k_m} = [0.0005\ \text{s}^{-1} \times 1000\ \text{kg/m}^3 \times 3600\ \text{J/kg}\cdot\text{K}/0.5\ \text{W/m}\cdot\text{K}]^{1/2}$$
$$= 60\ \text{m}^{-1}$$

$$\sinh(\tilde{m}L_m) = \sinh(60\ \text{m}^{-1} \times 0.03\ \text{m}) = 2.94$$

and

$$\cosh(\tilde{m}L_m) = \cosh(60\ \text{m}^{-1} \times 0.03\ \text{m}) = 3.11$$

$$\theta_c = T_c - T_a - \frac{\dot{q}_m}{\omega\rho_b c_b} = -\frac{\dot{q}_m}{\omega\rho_b c_b} = -\frac{700\ \text{W/m}^3}{0.0005\ \text{s}^{-1} \times 1000\ \text{kg/m}^3 \times 3600\ \text{J/kg}\cdot\text{K}}$$
$$= -0.389\ \text{K}$$

The excess temperature can be expressed in kelvins or degrees Celsius, since it is a temperature difference.

Thus, for air:

$$T_i = \frac{\{24°\text{C} \times 2.94 + 0.5\ \text{W/m}\cdot\text{K} \times 1.8\ \text{m}^2 \times 60\ \text{m}^{-1} \times 0.076\ \text{K/W}[-0.389°\text{C} + (37°\text{C} + 0.389°\text{C}) \times 3.11]\}}{2.94 + 0.5\ \text{W/m}\cdot\text{K} \times 1.8\ \text{m}^2 \times 60\ \text{m}^{-1} \times 0.076\ \text{K/W} \times 3.11} = 34.8°\text{C} \quad \triangleleft$$

This result agrees well with the value of 35°C that was assumed for Example 1.7. Next we can find the heat loss rate:

$$q = \frac{T_i - T_\infty}{R_{\text{tot}}} = \frac{34.8°\text{C} - 24°\text{C}}{0.076\ \text{K/W}} = 142\ \text{W} \quad \triangleleft$$

Again this agrees well with the previous result. Repeating the calculation for water, we find

$$T_i = 28.2°\text{C} \quad \triangleleft$$

$$q = 514\ \text{W} \quad \triangleleft$$

Here the calculation of Example 1.7 was not accurate because it incorrectly assumed that the inside of the skin/fat layer would be at 35°C. Furthermore, the skin temperature in this case would be only 25.4°C based on this more complete calculation.

Comments:

1. In reality, our bodies adjust in many ways to the thermal environment. For example, if we are too cold, we will shiver, which increases our metabolic heat generation rate. If we are too warm, the perfusion rate near the skin surface will increase, locally raising the skin temperature to increase heat loss to the environment.

2. Measuring the true thermal conductivity of living tissue is very challenging, first because of the necessity of making *invasive* measurements in a living being, and second because it is difficult to experimentally separate the effects of heat conduction and perfusion. It is easier to measure an effective thermal conductivity that would account for the combined contributions of conduction and perfusion. However, this effective conductivity value necessarily depends on the perfusion rate, which in turn varies with the thermal environment and physical condition of the specimen.
3. The calculations can be repeated for a range of values of the perfusion rate, and the dependence of the heat loss rate on the perfusion rate is illustrated below. The effect is stronger for the case of the water environment, because the muscle temperature is lower and therefore the effect of perfusion by the warm arterial blood is more pronounced.

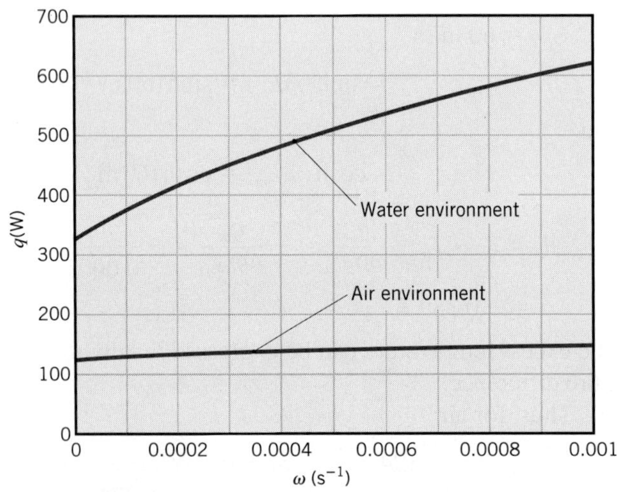

3.8 *Thermoelectric Power Generation*

As noted in Section 1.6, approximately 60% of the energy consumed globally is wasted in the form of low-grade heat. As such, an opportunity exists to *harvest* this energy stream and convert some of it to useful power. One approach involves *thermoelectric power generation*, which operates on a fundamental principle termed the *Seebeck effect* that states when a temperature gradient is established within a material, a corresponding voltage gradient is induced. The *Seebeck coefficient S* is a material property representing the proportionality between voltage and temperature gradients and, accordingly, has units of volts/K. For a constant property material experiencing one-dimensional conduction, as illustrated in Figure 3.23*a*,

$$(E_1 - E_2) = S(T_1 - T_2) \tag{3.115}$$

Electrically conducting materials can exhibit either positive or negative values of the Seebeck coefficient, depending on how they scatter electrons. The Seebeck coefficient is very small in metals, but can be relatively large in some semiconducting materials.

If the material of Figure 3.23*a* is installed in an electric circuit, the voltage difference induced by the Seebeck effect can drive an electric current *I*, and electric power can be

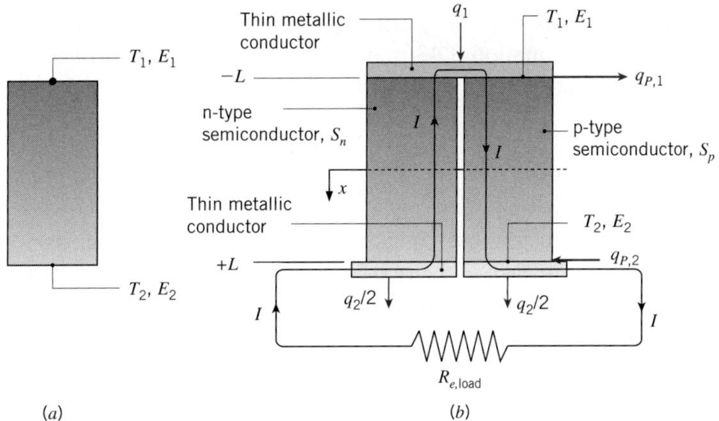

FIGURE 3.23 Thermoelectric phenomena. (*a*) The Seebeck effect. (*b*) A simplified thermoelectric circuit consisting of one pair ($N = 1$) of semiconducting pellets.

generated from waste heat that induces a temperature difference across the material. A simplified *thermoelectric circuit*, consisting of two *pellets* of semiconducting material, is shown in Figure 3.23*b*. By blending minute amounts of a secondary element into the pellet material, the *direction* of the current induced by the Seebeck effect can be manipulated. The resulting *p*- and *n-type* semiconductors, which are characterized by positive and negative Seebeck coefficients, respectively, can be arranged as shown in the figure. Heat is supplied to the top and lost from the bottom of the assembly, and thin metallic conductors connect the semiconductors to an external load represented by the electrical resistance, $R_{e,\text{load}}$. Ultimately, the amount of electric power that is produced is governed by the heat transfer rates to and from the pair of semiconducting pellets shown in Figure 3.23*b*.

In addition to inducing an electric current I, thermoelectric effects also induce the generation or absorption of heat *at the interface* between two dissimilar materials. This heat source or heat sink phenomenon is known as the *Peltier effect*, and the amount of heat absorbed q_P is related to the Seebeck coefficients of the adjoining materials by an equation of the form

$$q_P = I(S_p - S_n)T = IS_{p\text{-}n}T \tag{3.116}$$

where the individual Seebeck coefficients in the preceding expression, S_p and S_n, correspond to the p- and n-type semiconductors, and the differential Seebeck coefficient is $S_{p\text{-}n} \equiv S_p - S_n$. Temperature is expressed in kelvins in Equation 3.116. The heat absorption is positive (generation is negative) when the electric current flows from the n-type to the p-type semiconductor. Hence, in Figure 3.23*b*, Peltier heat absorption occurs at the warm interface between the semiconducting pellets and the upper, thin metallic conductor, while Peltier heat generation occurs at the cool interface between the pellets and the lower conductor.

When $T_1 > T_2$, the heat transfer rates to and from the device, q_1 and q_2, respectively, may be found by solving the appropriate form of the energy equation. For steady-state, one-dimensional conduction within the assembly of Figure 3.23*b* the analysis proceeds as follows.

Assuming the thin metallic connectors are of relatively high thermal and electrical conductivity, Ohmic dissipation occurs exclusively within the semiconducting pellets, each of which has a cross-sectional area $A_{c,s}$. The thermal resistances of the metallic conductors are assumed to be negligible, as is heat transfer within any gas trapped between the semiconducting pellets. Recognizing that the electrical resistance of each of the two pellets may be expressed as $R_{e,s} = \rho_{e,s}(2L)/A_{c,s}$ where $\rho_{e,s}$ is the electrical resistivity of the semiconducting material, Equation 3.43 may be used to find the uniform volumetric generation rate within each pellet

$$\dot{q} = \frac{I^2 \rho_{e,s}}{A_{c,s}^2} \tag{3.117}$$

Assuming negligible contact resistances and identical, as well as constant, thermophysical properties in each of the two pellets (with the exception being $S_p = -S_n$), Equation C.7 may be used to write expressions for the heat conduction out of and into the semiconducting material

$$q(x = L) = 2A_{c,s}\left[\frac{k_s}{2L}(T_1 - T_2) + \frac{I^2 \rho_{e,s} L}{A_{c,s}^2}\right] \tag{3.118a}$$

$$q(x = -L) = 2A_{c,s}\left[\frac{k_s}{2L}(T_1 - T_2) - \frac{I^2 \rho_{e,s} L}{A_{c,s}^2}\right] \tag{3.118b}$$

The factor of 2 outside the brackets accounts for heat transfer in *both* pellets and, as evident, $q(x = L) > q(x = -L)$.

Because of the Peltier effect, q_1 and q_2 are *not* equal to the heat transfer rates into and out of the pellets as expressed in Equations 3.118a,b. Incorporating Equation 3.116 in an energy balance for a control surface about the interface between the thin metallic conductor and the semiconductor material at $x = -L$ yields

$$q_1 = q(x = -L) + q_{P,1} = q(x = -L) + IS_{p\text{-}n}T_1 \tag{3.119}$$

Similarly at $x = L$,

$$q_2 = q(x = L) - IS_{n\text{-}p}T_2 = q(x = L) + IS_{p\text{-}n}T_2 \tag{3.120}$$

Combining Equations 3.118b and 3.119 yields

$$q_1 = \frac{A_{c,s}k_s}{L}(T_1 - T_2) + IS_{p\text{-}n}T_1 - 2\frac{I^2 \rho_{e,s} L}{A_{c,s}} \tag{3.121}$$

Similarly, combining Equations 3.118a and 3.120 gives

$$q_2 = \frac{A_{c,s}k_s}{L}(T_1 - T_2) + IS_{p\text{-}n}T_2 + 2\frac{I^2 \rho_{e,s} L}{A_{c,s}} \tag{3.122}$$

From an overall energy balance on the thermoelectric device, the electric power produced by the Seebeck effect is

$$P = q_1 - q_2 \tag{3.123}$$

Substituting Equations 3.121 and 3.122 into this expression yields

$$P = IS_{p\text{-}n}(T_1 - T_2) - 4\frac{I^2 \rho_{e,s} L}{A_{c,s}} = IS_{p\text{-}n}(T_1 - T_2) - I^2 R_{e,\mathrm{tot}} \tag{3.124}$$

where $R_{e,\mathrm{tot}} = 2R_{e,s}$.

The voltage difference induced by the Seebeck effect is relatively small for a single pair of semiconducting pellets. To amplify the voltage difference, thermoelectric *modules* are fabricated, as shown schematically in Figure 3.24*a* where $N \gg 1$ pairs of semiconducting pellets are wired in series. Thin layers of a dielectric material, usually a ceramic, sandwich the module to provide structural rigidity and electrical insulation from the surroundings. Assuming the

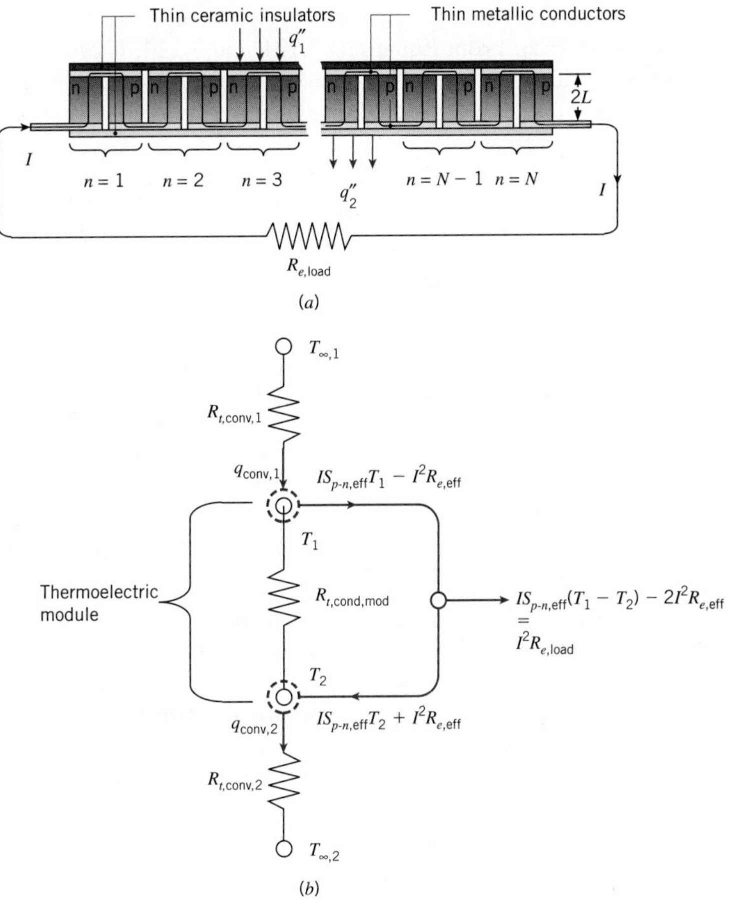

FIGURE 3.24 Thermoelectric module. (*a*) Cross-section of a module consisting of N semiconductor pairs. (*b*) Equivalent thermal circuit for a convectively heated and cooled module.

thermal resistances of the thin ceramic layers are negligible, q_1, q_2, and the total module electric power, P_N, can be written by modifying Equations 3.121, 3.122, 3.124 as

$$q_1 = \frac{1}{R_{t,\text{cond,mod}}}(T_1 - T_2) + IS_{p\text{-}n,\text{eff}}T_1 - I^2R_{e,\text{eff}} \tag{3.125}$$

$$q_2 = \frac{1}{R_{t,\text{cond,mod}}}(T_1 - T_2) + IS_{p\text{-}n,\text{eff}}T_2 + I^2R_{e,\text{eff}} \tag{3.126}$$

$$P_N = q_1 - q_2 = IS_{p\text{-}n,\text{eff}}(T_1 - T_2) - 2I^2R_{e,\text{eff}} \tag{3.127}$$

where $S_{p\text{-}n,\text{eff}} = NS_{p\text{-}n}$, and $R_{e,\text{eff}} = NR_{e,s}$ are the *effective* Seebeck coefficient and the total internal electrical resistance of the module while $R_{t,\text{cond,mod}} = L/NA_sk_s$ is the conduction resistance associated with the module's p-n semiconductor matrix. An equivalent thermal circuit for a convectively heated and cooled thermoelectric module is shown in Figure 3.24*b*. If heating or cooling were to be applied by radiation or conduction, the resistance network outside of the thermoelectric module portion of the circuit would be modified accordingly.

Returning to the single thermoelectric circuit of Figure 3.23*b*, the efficiency is defined as $\eta_{TE} \equiv P/q_1$. From Equations 3.121 and 3.124, it can be seen that efficiency depends on the electrical current in a complex manner. However, the efficiency can be maximized by adjusting the current through changes in the load resistance. The resulting maximum efficiency is given as [18]

$$\eta_{TE} = \left(1 - \frac{T_2}{T_1}\right) \frac{\sqrt{1 + Z\overline{T}} - 1}{\sqrt{1 + Z\overline{T} + T_2/T_1}} \qquad (3.128)$$

where $\overline{T} = (T_1 + T_2)/2$, $S \equiv S_p = -S_n$, and

$$Z = \frac{S^2}{\rho_{e,s} k_s} \qquad (3.129)$$

Since the efficiency increases with increasing $Z\overline{T}$, $Z\overline{T}$ may be seen as a dimensionless *figure of merit* associated with thermoelectric generation [19]. As $Z\overline{T} \rightarrow \infty$, $\eta_{TE} \rightarrow (1 - T_2/T_1) = (1 - T_c/T_h) \equiv \eta_C$ where η_C is the Carnot efficiency. As discussed in Section 1.3.2, the Carnot efficiency and, in turn, the thermoelectric efficiency cannot be determined until the appropriate hot and cold temperatures are calculated from a heat transfer analysis.

Because $Z\overline{T}$ is defined in terms of interrelated electrical and thermal conductivities, extensive research is being conducted to tailor the properties of the semiconducting pellets, primarily by manipulating the nanostructure of the material so as to independently control phonon and electron motion and, in turn, the thermal and electrical conductivities of the material. Currently, $Z\overline{T}$ values of approximately unity at room temperature are readily achieved. Finally, we note that thermoelectric modules can be operated in reverse; supplying electric power *to* the module allows one to control the heat transfer rates to or from the outer ceramic surfaces. Such *thermoelectric chillers* or *thermoelectric heaters* are used in a wide variety of applications. A comprehensive discussion of one-dimensional, steady-state heat transfer modeling associated with thermoelectric heating and cooling modules is available [20].

EXAMPLE 3.13

An array of $M = 48$ thermoelectric modules is installed on the exhaust of a sports car. Each module has an effective Seebeck coefficient of $S_{p\text{-}n,\text{eff}} = 0.1435$ V/K, and an internal electrical resistance of $R_{e,\text{eff}} = 4\ \Omega$. In addition, each module is of width and length $W = 54$ mm and contains $N = 100$ pairs of semiconducting pellets. Each pellet has an overall length of $2L = 5$ mm and cross-sectional area $A_{c,s} = 1.2 \times 10^{-5}\ \text{m}^2$ and is characterized by a thermal conductivity of $k_s = 1.2$ W/m·K. The hot side of each module is exposed to exhaust gases at $T_{\infty,1} = 550°\text{C}$ with $h_1 = 40\ \text{W/m}^2\cdot\text{K}$, while the opposite side of each module is cooled by pressurized water at $T_{\infty,2} = 105°\text{C}$ with $h_2 = 500\ \text{W/m}^2\cdot\text{K}$. If the modules are wired in series, and the load resistance is $R_{e,\text{load}} = 400\ \Omega$, what is the electric power harvested from the hot exhaust gases?

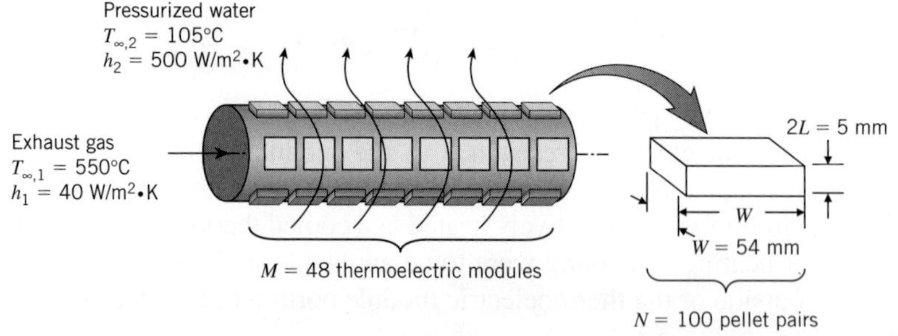

SOLUTION

Known: Thermoelectric module properties and dimensions, number of semiconductor pairs in each module, and number of modules in the array. Temperature of exhaust gas and pressurized water, as well as convection coefficients at the hot and cold module surfaces. Modules are wired in series, and the electrical resistance of the load is known.

Find: Power produced by the module array.

Schematic:

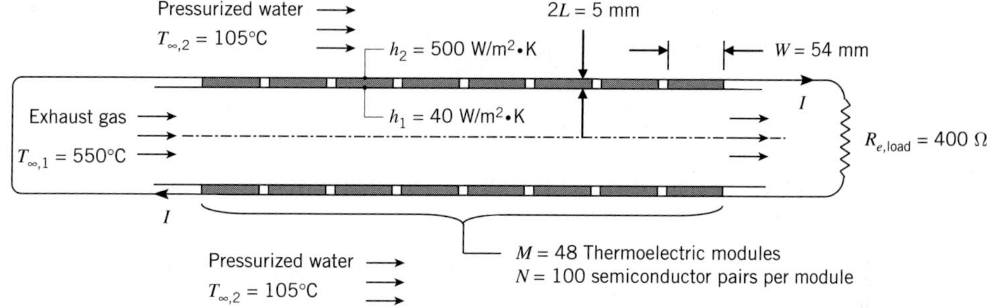

Assumptions:

1. Steady-state conditions.
2. One-dimensional heat transfer.
3. Constant properties.
4. Negligible electrical and thermal contact resistances.
5. Negligible radiation exchange and negligible heat transfer within the gas inside the modules.
6. Negligible conduction resistance posed by the metallic contacts and ceramic insulators of the modules.

Analysis: We begin by analyzing a single module. The conduction resistance of each module's semiconductor array is

$$R_{t,\text{cond,mod}} = \frac{L}{NA_{c,s}k_s} = \frac{2.5 \times 10^{-3}\text{ m}}{100 \times 1.2 \times 10^{-5}\text{ m}^2 \times 1.2\text{ W/m}\cdot\text{K}} = 1.736\text{ K/W}$$

From Equation 3.125,

$$q_1 = \frac{1}{R_{t,\text{cond,mod}}}(T_1 - T_2) + IS_{p\text{-}n,\text{eff}}T_1 - I^2R_{e,\text{eff}} = \frac{(T_1 - T_2)}{1.736\text{ K/W}}$$

$$+ I \times 0.1435\text{ V/K} \times T_1 - I^2 \times 4\ \Omega \tag{1}$$

while from Equation 3.126,

$$q_2 = \frac{1}{R_{t,\text{cond,mod}}}(T_1 - T_2) + IS_{p\text{-}n,\text{eff}}T_2 + I^2R_{e,\text{eff}} = \frac{(T_1 - T_2)}{1.736\text{ K/W}}$$

$$+ I \times 0.1435\text{ V/K} \times T_2 + I^2 \times 4\ \Omega \tag{2}$$

At the hot surface, Newton's law of cooling may be written as

$$q_1 = h_1 W^2(T_{\infty,1} - T_1) = 40\ \text{W/m}^2\cdot\text{K} \times (0.054\ \text{m})^2 \times [(550 + 273)\ \text{K} - T_1] \quad (3)$$

whereas at the cool surface,

$$q_2 = h_2 W^2(T_2 - T_{\infty,2}) = 500\ \text{W/m}^2\cdot\text{K} \times (0.054\ \text{m})^2 \times [T_2 - (105 + 273)\ \text{K}] \quad (4)$$

Four equations have been written that include five unknowns, q_1, q_2, T_1, T_2, and I. An additional equation is obtained from the electrical circuit. With the modules wired in series, the total electric power produced by all $M = 48$ modules is equal to the electric power dissipated in the load resistance. Equation 3.127 yields

$$P_{\text{tot}} = MP_N = M[IS_{p\text{-}n,\text{eff}}(T_1 - T_2) - 2I^2R_{e,\text{eff}}] = 48[I \times 0.1435\ \text{V/K} \times (T_1 - T_2) - 2I^2 \times 4\ \Omega] \quad (5)$$

Since the electric power produced by the thermoelectric module is dissipated in the electrical load, it follows that

$$P_{\text{tot}} = I^2R_{\text{load}} = I^2 \times 400\ \Omega \quad (6)$$

Equations 1 through 6 may be solved simultaneously, yielding $P_{\text{tot}} = 46.9$ W. ◁

Comments:

1. Equations 1 through 5 can be readily written by inspecting the equivalent thermal circuit of Figure 3.24*b*.
2. The module surface temperatures are $T_1 = 173°\text{C}$ and $T_2 = 134°\text{C}$, respectively. If these surface temperatures were specified in the problem statement, the electric power could be obtained directly from Equations 5 and 6. In any practical design of a thermoelectric generator, however, a heat transfer analysis must be conducted to determine the power generated.
3. Power generation is very sensitive to the convection heat transfer resistances. For $h_1 = h_2 \rightarrow \infty$, $P_{\text{tot}} = 5900$ W. To reduce the thermal resistance between the module and fluid streams, finned heat sinks are often used to increase the temperature difference across the modules and, in turn, increase their power output. Good thermal management and design are crucial to maximizing the power generation.
4. Harvesting the thermal energy contained in the exhaust with thermoelectrics can eliminate the need for an alternator, resulting in an increase in the net power produced by the engine, a reduction in the automobile's weight, and an increase in gas mileage of up to 10%.
5. Thermoelectric modules, operating in the heating mode, can be embedded in car seats and powered by thermoelectric exhaust harvesters, reducing energy costs associated with heating the entire passenger cabin. The seat modules can also be operated in the cooling mode, potentially eliminating the need for vapor compression air conditioning. Common refrigerants, such as R134a, are harmful greenhouse gases, and are emitted into the atmosphere by leakage through seals and connections, and by catastrophic leaks due to collisions. Replacing automobile vapor compression air conditioners with personalized thermoelectric seat coolers can eliminate the equivalent of 45 million metric tons of CO_2 released into the atmosphere every year in the United States alone.

3.9 Micro- and Nanoscale Conduction

We conclude the discussion of one-dimensional, steady-state conduction by considering situations for which the physical dimensions are on the order of, or smaller than, the mean free path of the energy carriers, leading to potentially important nano- or microscale effects.

3.9.1 Conduction Through Thin Gas Layers

Figure 3.25 shows instantaneous trajectories of gas molecules between two isothermal, solid surfaces separated by a distance L. As discussed in Section 1.2.1, even in the absence of *bulk* fluid motion individual molecules continually impinge on the two solid boundaries that are held at uniform surface temperatures $T_{s,1}$ and $T_{s,2}$, respectively. The molecules also collide with each other, exchanging energy *within* the gaseous medium. When the thickness of the gas layer is large, $L = L_1$ (Figure 3.25*a*), a particular gas molecule will collide more frequently with other gas molecules than with either of the solid boundaries. Alternatively, for a very thin gas layer, $L = L_2 \ll L_1$ (Figure 3.25*b*), the probability of a molecule striking either of the solid boundaries is high relative to the likelihood of it colliding with another molecule.

The energy content of a gas molecule is associated with its translational, rotational, and vibrational kinetic energies. It is this molecular-scale kinetic energy that ultimately defines the temperature of the gas, and collisions between individual molecules determine the value of the thermal conductivity, as discussed in Section 2.2.1. However, the manner in which a gas molecule is reflected or scattered from the solid walls also affects its level of kinetic energy and, in turn, its temperature. Hence, wall–molecule collisions can become important in determining the heat rate, q_x, as L/λ_{mfp} becomes small.

The collision with and subsequent scattering of an individual gas molecule from a solid wall can be described by a *thermal accommodation coefficient*, α_t,

$$\alpha_t = \frac{T_i - T_{sc}}{T_i - T_s} \tag{3.130}$$

where T_i is the effective molecule temperature just prior to striking the solid surface, T_{sc} is the temperature of the molecule immediately after it is scattered or reflected by the surface, and T_s is the surface temperature. When the temperature of the scattered molecule is identical to the wall temperature, $\alpha_t = 1$. Alternatively, if $T_{sc} = T_i$, the molecule's kinetic energy and temperature are unaffected by a collision with the wall and $\alpha_t = 0$.

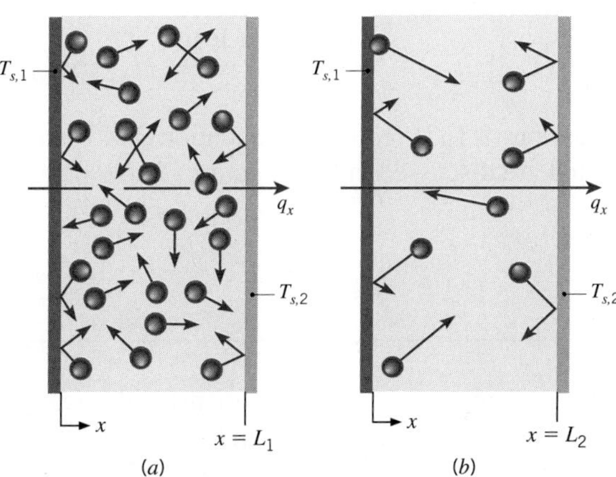

FIGURE 3.25 Molecule trajectories in (*a*) a relatively thick gas layer and (*b*) a relatively thin gas layer. Molecules collide with each other, and with the two solid walls.

For one-dimensional conduction within an ideal gas contained between two surfaces held at temperatures $T_{s,1}$ and $T_{s,2} < T_{s,1}$, the heat rate through the gas layer may be expressed as [21]

$$q = \frac{T_{s,1} - T_{s,2}}{(R_{t,m-m} + R_{t,m-s})} \tag{3.131}$$

where, at the molecular level, the thermal resistances are associated with molecule–molecule and molecule-surface collisions

$$R_{t,m-m} = \frac{L}{kA} \quad \text{and} \quad R_{t,m-s} = \frac{\lambda_{\text{mfp}}}{kA}\left[\frac{2-\alpha_t}{\alpha_t}\right]\left[\frac{9\gamma - 5}{\gamma + 1}\right] \tag{3.132a,b}$$

In the preceding expression, $\gamma \equiv c_p/c_v$ is the specific heat ratio of the ideal gas. The two solids are assumed to be the same material with equal values of α_t, and the temperature difference is assumed to be small relative to the cold wall, $(T_{s,1} - T_{s,2})/T_{s,2} \ll 1$. Equations 3.132a,b may be combined to yield

$$\frac{R_{t,m-s}}{R_{t,m-m}} = \frac{\lambda_{\text{mfp}}}{L}\left[\frac{2-\alpha_t}{\alpha_t}\right]\left[\frac{9\gamma - 5}{\gamma + 1}\right]$$

from which it is evident that $R_{t,m-s}$ may be neglected if L/λ_{mfp} is large and $\alpha_t \neq 0$. In this case, Equation 3.131 reduces to Equation 3.6. However, $R_{t,m-s}$ can be significant if L/λ_{mfp} is small. From Equation 2.11 the mean free path increases as the gas pressure is decreased. Hence, $R_{t,m-s}$ increases with decreasing gas pressure, and the heat rate can be pressure dependent when L/λ_{mfp} is small. Values of α_t for specific gas and surface combinations range from 0.87 to 0.97 for air–aluminum and air–steel, but can be less than 0.02 when helium interacts with clean metallic surfaces [21]. Equations 3.131, 3.132a,b may be applied to situations for which $L/\lambda_{\text{mfp}} \gtrsim 0.1$. For air at atmospheric pressure, this corresponds to $L \gtrsim 10$ nm.

3.9.2 Conduction Through Thin Solid Films

One-dimensional conduction across or along thin solid films was discussed in Section 2.2.1 in terms of the thermal conductivities k_x and k_y. The heat transfer rate across a thin solid film may be approximated by combining Equation 2.9a with Equation 3.5, yielding

$$q_x = \frac{k_x A}{L}(T_{s,1} - T_{s,2}) = \frac{k[1 - \lambda_{\text{mfp}}/(3L)]A}{L}(T_{s,1} - T_{s,2}) \tag{3.133}$$

When L/λ_{mfp} is large, Equation (3.133) reduces to Equation 3.4. Many alternative expressions for k_x are available and are discussed in the literature [21].

3.10 *Summary*

Despite its inherent mathematical simplicity, one-dimensional, steady-state heat transfer occurs in numerous engineering applications. Although one-dimensional, steady-state conditions may not apply exactly, the assumptions may often be made to obtain results of reasonable accuracy. You should therefore be thoroughly familiar with the means by which such problems are treated. In particular, you should be comfortable with the use of equivalent thermal circuits and with the expressions for the conduction resistances that pertain to each of the three common geometries. You should also be familiar with how the heat equation and Fourier's law may be used to obtain temperature distributions and the corresponding fluxes. The implications of an internally distributed source of energy should also be clearly understood. In addition, you should appreciate the important role that extended surfaces can play in the design of thermal systems and should have the facility to effect design and performance calculations for such surfaces. Finally, you should understand how the preceding concepts can be applied to analyze heat transfer in the human body, thermoelectric power generation, and micro- and nanoscale conduction.

You may test your understanding of this chapter's key concepts by addressing the following questions.

- Under what conditions may it be said that the *heat flux* is a constant, independent of the direction of heat flow? For each of these conditions, use physical considerations to convince yourself that the heat flux would not be independent of direction if the condition were not satisfied.
- For one-dimensional, steady-state conduction in a cylindrical or spherical shell without heat generation, is the radial heat flux independent of radius? Is the radial heat rate independent of radius?
- For one-dimensional, steady-state conduction without heat generation, what is the shape of the temperature distribution in a *plane wall*? In a *cylindrical shell*? In a *spherical shell*?
- What is the *thermal resistance*? How is it defined? What are its units?
- For conduction across a *plane wall*, can you write the expression for the thermal resistance from memory? Similarly, can you write expressions for the thermal resistance associated with conduction across *cylindrical* and *spherical* shells? From memory, can you express the thermal resistances associated with convection from a surface and net radiation exchange between the surface and large surroundings?
- What is the physical basis for existence of a *critical insulation radius*? How do the thermal conductivity and the convection coefficient affect its value?
- How is the conduction resistance of a solid affected by its thermal conductivity? How is the convection resistance at a surface affected by the convection coefficient? How is the radiation resistance affected by the surface emissivity?
- If heat is transferred from a surface by convection and radiation, how are the corresponding thermal resistances represented in a circuit?
- Consider steady-state conduction through a plane wall separating fluids of different temperatures, $T_{\infty,i}$ and $T_{\infty,o}$, adjoining the inner and outer surfaces, respectively. If the convection coefficient at the outer surface is five times larger than that at the inner surface, $h_o = 5h_i$, what can you say about relative proximity of the corresponding surface temperatures, $T_{s,o}$ and $T_{s,i}$, to their adjoining fluid temperatures?
- Can a thermal conduction resistance be applied to a *solid* cylinder or sphere?
- What is a *contact resistance*? How is it defined? What are its units for an interface of prescribed area? What are they for a unit area?
- How is the contact resistance affected by the roughness of adjoining surfaces?
- If the air in the contact region between two surfaces is replaced by helium, how is the thermal contact resistance affected? How is it affected if the region is evacuated?
- What is the *overall heat transfer coefficient*? How is it defined, and how is it related to the *total thermal resistance*? What are its units?
- In a solid circular cylinder experiencing uniform volumetric heating and convection heat transfer from its surface, how does the heat flux vary with radius? How does the heat rate vary with radius?

- In a solid sphere experiencing uniform volumetric heating and convection heat transfer from its surface, how does the heat flux vary with radius? How does the heat rate vary with radius?
- Is it possible to achieve steady-state conditions in a solid cylinder or sphere that is experiencing heat generation and whose surface is perfectly insulated? Explain.
- Can a material experiencing heat generation be represented by a thermal resistance and included in a circuit analysis? If so, why? If not, why not?
- What is the physical mechanism associated with cooking in a microwave oven? How do conditions differ from a conventional (convection or radiant) oven?
- If radiation is incident on the surface of a semitransparent medium and is absorbed as it propagates through the medium, will the corresponding volumetric rate of heat generation $\dot{q}$ be distributed uniformly in the medium? If not, how will $\dot{q}$ vary with distance from the surface?
- In what way is a plane wall that is of thickness $2L$ and experiences uniform volumetric heating and equivalent convection conditions at both surfaces similar to a plane wall that is of thickness L and experiences the same volumetric heating and convection conditions at one surface but whose opposite surface is well insulated?
- What purpose is served by attaching *fins* to a surface?
- In the derivation of the general form of the energy equation for an extended surface, why is the assumption of one-dimensional conduction an approximation? Under what conditions is it a good approximation?
- Consider a straight fin of uniform cross section (Figure 3.15*a*). For an x-location in the fin, sketch the temperature distribution in the transverse (y-) direction, placing the origin of the coordinate at the midplane of the fin ($-t/2 \leq y \leq t/2$). What is the form of a *surface* energy balance applied at the location $(x, t/2)$?
- What is the *fin effectiveness*? What is its range of possible values? Under what conditions are fins most effective?
- What is the *fin efficiency*? What is its range of possible values? Under what conditions will the efficiency be large?
- What is the *fin resistance*? What are its units?
- How are the effectiveness, efficiency, and thermal resistance of a fin affected if its thermal conductivity is increased? If the convection coefficient is increased? If the length of the fin is increased? If the thickness (or diameter) of the fin is increased?
- Heat is transferred from hot water flowing through a tube to air flowing over the tube. To enhance the rate of heat transfer, should fins be installed on the tube interior or exterior surface?
- A fin may be manufactured as an integral part of a surface by using a casting or extrusion process, or it may be separately brazed or adhered to the surface. From thermal considerations, which option is preferred?
- Describe the physical origins of the two heat source terms in the bioheat equation. Under what conditions is the perfusion term a heat sink?
- How do heat sinks increase the electric power generated by a thermoelectric device?
- Under what conditions do thermal resistances associated with molecule–wall interactions become important?

References

1. Fried, E., "Thermal Conduction Contribution to Heat Transfer at Contacts," in R. P. Tye, Ed., *Thermal Conductivity,* Vol. 2, Academic Press, London, 1969.
2. Eid, J. C., and V. W. Antonetti, "Small Scale Thermal Contact Resistance of Aluminum Against Silicon," in C. L. Tien, V. P. Carey, and J. K. Ferrel, Eds., *Heat Transfer—1986,* Vol. 2, Hemisphere, New York, 1986, pp. 659–664.
3. Snaith, B., P. W. O'Callaghan, and S. D. Probert, *Appl. Energy,* **16,** 175, 1984.
4. Yovanovich, M. M., "Theory and Application of Constriction and Spreading Resistance Concepts for Microelectronic Thermal Management," Presented at the International Symposium on Cooling Technology for Electronic Equipment, Honolulu, 1987.
5. Peterson, G. P., and L. S. Fletcher, "Thermal Contact Resistance of Silicon Chip Bonding Materials," Proceedings of the International Symposium on Cooling Technology for Electronic Equipment, Honolulu, 1987, pp. 438–448.
6. Yovanovich, M. M., and M. Tuarze, *AIAA J. Spacecraft Rockets,* **6**, 1013, 1969.
7. Madhusudana, C. V., and L. S. Fletcher, *AIAA J.,* **24**, 510, 1986.
8. Yovanovich, M. M., "Recent Developments in Thermal Contact, Gap and Joint Conductance Theories and Experiment," in C. L. Tien, V. P. Carey, and J. K. Ferrel, Eds., *Heat Transfer—1986*, Vol. 1, Hemisphere, New York, 1986, pp. 35–45.
9. Maxwell, J. C., *A Treatise on Electricity and Magnetism*, 3rd ed., Oxford University Press, Oxford, 1892.
10. Hamilton, R. L., and O. K. Crosser, *I&EC Fund.* **1**, 187, 1962.
11. Jeffrey, D. J., *Proc. Roy. Soc. A*, **335**, 355, 1973.
12. Hashin Z., and S. Shtrikman, *J. Appl. Phys*., **33**, 3125, 1962.
13. Aichlmayr, H. T., and F. A. Kulacki, "The Effective Thermal Conductivity of Saturated Porous Media," in J. P. Hartnett, A. Bar-Cohen, and Y. I Cho, Eds., *Advances in Heat Transfer*, Vol. 39, Academic Press, London, 2006.
14. Harper, D. R., and W. B. Brown, "Mathematical Equations for Heat Conduction in the Fins of Air Cooled Engines," NACA Report No. 158, 1922.
15. Schneider, P. J., *Conduction Heat Transfer*, Addison-Wesley, Reading, MA, 1957.
16. Diller, K. R., and T. P. Ryan, *J. Heat Transfer*, **120**, 810, 1998.
17. Pennes, H. H., *J. Applied Physiology*, **85**, 5, 1998.
18. Goldsmid, H. J., "Conversion Efficiency and Figure-of-Merit," in D. M. Rowe, Ed., *CRC Handbook of Thermoelectrics*, Chap. 3, CRC Press, Boca Raton, 1995.
19. Majumdar, A., *Science*, **303**, 777, 2004.
20. Hodes, M., *IEEE Trans. Com. Pack. Tech*., **28**, 218, 2005.
21. Zhang, Z. M., *Nano/Microscale Heat Transfer*, McGraw-Hill, New York, 2007.

Problems

Plane and Composite Walls

3.1 Consider the plane wall of Figure 3.1, separating hot and cold fluids at temperatures $T_{\infty,1}$ and $T_{\infty,2}$, respectively. Using surface energy balances as boundary conditions at $x = 0$ and $x = L$ (see Equation 2.34), obtain the temperature distribution within the wall and the heat flux in terms of $T_{\infty,1}$, $T_{\infty,2}$, h_1, h_2, k, and L.

3.2 A new building to be located in a cold climate is being designed with a basement that has an $L = 200$-mm-thick wall. Inner and outer basement wall temperatures are $T_i = 20°\text{C}$ and $T_o = 0°\text{C}$, respectively. The architect can specify the wall material to be either aerated concrete block with $k_{ac} = 0.15\ \text{W/m}\cdot\text{K}$, or stone mix concrete. To reduce the conduction heat flux through the stone mix wall to a level equivalent to that of the aerated concrete wall, what thickness of extruded polystyrene sheet must be applied onto the inner surface of the stone mix concrete wall? Floor dimensions of the basement are $20\ \text{m} \times 30\ \text{m}$, and the expected rental rate is $\$50/\text{m}^2$/month. What is the yearly cost, in terms of lost rental income, if the stone mix concrete wall with polystyrene insulation is specified?

3.3 The rear window of an automobile is defogged by passing warm air over its inner surface.

(a) If the warm air is at $T_{\infty,i} = 40°\text{C}$ and the corresponding convection coefficient is $h_i = 30\ \text{W/m}^2\cdot\text{K}$, what are the inner and outer surface temperatures of 4-mm-thick window glass, if the outside ambient air temperature is $T_{\infty,o} = -10°\text{C}$ and the associated convection coefficient is $h_o = 65\ \text{W/m}^2\cdot\text{K}$?

(b) In practice $T_{\infty,o}$ and h_o vary according to weather conditions and car speed. For values of $h_o = 2$, 65, and $100\ \text{W/m}^2\cdot\text{K}$, compute and plot the inner and outer surface temperatures as a function of $T_{\infty,o}$ for $-30 \leq T_{\infty,o} \leq 0°\text{C}$.

3.4 The rear window of an automobile is defogged by attaching a thin, transparent, film-type heating element to its inner surface. By electrically heating this element, a uniform heat flux may be established at the inner surface.

(a) For 4-mm-thick window glass, determine the electrical power required per unit window area to maintain an inner surface temperature of 15°C when the interior air temperature and convection coefficient are $T_{\infty,i} = 25°C$ and $h_i = 10\ W/m^2 \cdot K$, while the exterior (ambient) air temperature and convection coefficient are $T_{\infty,o} = -10°C$ and $h_o = 65\ W/m^2 \cdot K$.

(b) In practice $T_{\infty,o}$ and h_o vary according to weather conditions and car speed. For values of $h_o = 2$, 20, 65, and $100\ W/m^2 \cdot K$, determine and plot the electrical power requirement as a function of $T_{\infty,o}$ for $-30 \leq T_{\infty,o} \leq 0°C$. From your results, what can you conclude about the need for heater operation at low values of h_o? How is this conclusion affected by the value of $T_{\infty,o}$? If $h \propto V^n$, where V is the vehicle speed and n is a positive exponent, how does the vehicle speed affect the need for heater operation?

3.5 A dormitory at a large university, built 50 years ago, has exterior walls constructed of L_s = 25-mm-thick sheathing with a thermal conductivity of $k_s = 0.1\ W/m \cdot K$. To reduce heat losses in the winter, the university decides to encapsulate the entire dormitory by applying an L_i = 25-mm-thick layer of extruded insulation characterized by $k_i = 0.029\ W/m \cdot K$ to the exterior of the original sheathing. The extruded insulation is, in turn, covered with an L_g = 5-mm-thick architectural glass with $k_g = 1.4\ W/m \cdot K$. Determine the heat flux through the original and retrofitted walls when the interior and exterior air temperatures are $T_{\infty,i} = 22°C$ and $T_{\infty,o} = -20°C$, respectively. The inner and outer convection heat transfer coefficients are $h_i = 5\ W/m^2 \cdot K$ and $h_o = 25\ W/m^2 \cdot K$, respectively.

3.6 In a manufacturing process, a transparent film is being bonded to a substrate as shown in the sketch. To cure the bond at a temperature T_0, a radiant source is used to provide a heat flux q''_0 (W/m^2), all of which is absorbed at the bonded surface. The back of the substrate is maintained at T_1 while the free surface of the film is exposed to air at T_∞ and a convection heat transfer coefficient h.

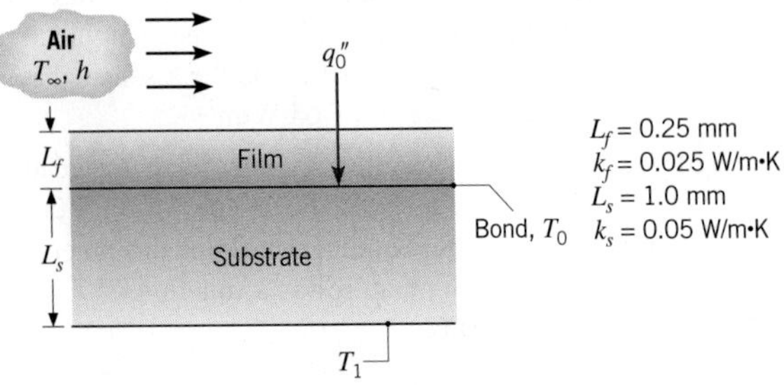

(a) Show the thermal circuit representing the steady-state heat transfer situation. Be sure to label *all* elements, nodes, and heat rates. Leave in symbolic form.

(b) Assume the following conditions: $T_\infty = 20°C$, $h = 50\ W/m^2 \cdot K$, and $T_1 = 30°C$. Calculate the heat flux q''_0 that is required to maintain the bonded surface at $T_0 = 60°C$.

(c) Compute and plot the required heat flux as a function of the film thickness for $0 \leq L_f \leq 1$ mm.

(d) If the film is not transparent and all of the radiant heat flux is absorbed at its upper surface, determine the heat flux required to achieve bonding. Plot your results as a function of L_f for $0 \leq L_f \leq 1$ mm.

3.7 The walls of a refrigerator are typically constructed by sandwiching a layer of insulation between sheet metal panels. Consider a wall made from fiberglass insulation of thermal conductivity $k_i = 0.046\ W/m \cdot K$ and thickness $L_i = 50$ mm and steel panels, each of thermal conductivity $k_p = 60\ W/m \cdot K$ and thickness $L_p = 3$ mm. If the wall separates refrigerated air at $T_{\infty,i} = 4°C$ from ambient air at $T_{\infty,o} = 25°C$, what is the heat gain per unit surface area? Coefficients associated with natural convection at the inner and outer surfaces may be approximated as $h_i = h_o = 5\ W/m^2 \cdot K$.

3.8 A t = 10-mm-thick horizontal layer of water has a top surface temperature of $T_c = -4°C$ and a bottom surface temperature of $T_h = 2°C$. Determine the location of the solid–liquid interface at steady state.

3.9 A technique for measuring convection heat transfer coefficients involves bonding one surface of a thin metallic foil to an insulating material and exposing the other surface to the fluid flow conditions of interest.

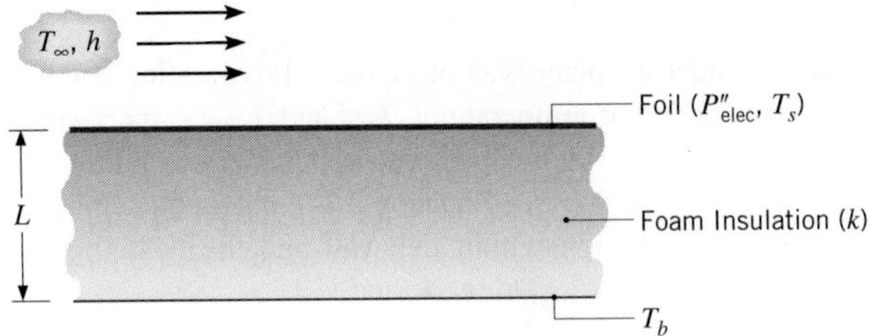

By passing an electric current through the foil, heat is dissipated uniformly within the foil and the corresponding flux, P''_{elec}, may be inferred from related voltage and current measurements. If the insulation thickness L and thermal conductivity k are known and the fluid, foil, and insulation temperatures (T_∞, T_s, T_b) are measured, the convection coefficient may be determined. Consider conditions for which $T_\infty = T_b = 25°C$, $P''_{elec} = 2000\ W/m^2$, $L = 10$ mm, and $k = 0.040\ W/m \cdot K$.

(a) With water flow over the surface, the foil temperature measurement yields $T_s = 27°C$. Determine the convection coefficient. What error would be incurred by assuming all of the dissipated power to be transferred to the water by convection?

(b) If, instead, air flows over the surface and the temperature measurement yields $T_s = 125°C$, what is the convection coefficient? The foil has an emissivity of 0.15 and is exposed to large surroundings at 25°C. What error would be incurred by assuming all of the dissipated power to be transferred to the air by convection?

(c) Typically, heat flux gages are operated at a fixed temperature (T_s), in which case the power dissipation provides a direct measure of the convection coefficient. For $T_s = 27°C$, plot P''_{elec} as a function of h_o for $10 \le h_o \le 1000$ W/m²·K. What effect does h_o have on the error associated with neglecting conduction through the insulation?

3.10 The *wind chill*, which is experienced on a cold, windy day, is related to increased heat transfer from exposed human skin to the surrounding atmosphere. Consider a layer of fatty tissue that is 3 mm thick and whose interior surface is maintained at a temperature of 36°C. On a calm day the convection heat transfer coefficient at the outer surface is 25 W/m²·K, but with 30 km/h winds it reaches 65 W/m²·K. In both cases the ambient air temperature is −15°C.

(a) What is the ratio of the heat loss per unit area from the skin for the calm day to that for the windy day?

(b) What will be the skin outer surface temperature for the calm day? For the windy day?

(c) What temperature would the air have to assume on the calm day to produce the same heat loss occurring with the air temperature at −15°C on the windy day?

3.11 Determine the thermal conductivity of the carbon nanotube of Example 3.4 when the heating island temperature is measured to be $T_h = 332.6$ K, without evaluating the thermal resistances of the supports. The conditions are the same as in the example.

3.12 A thermopane window consists of two pieces of glass 7 mm thick that enclose an air space 7 mm thick. The window separates room air at 20°C from outside ambient air at −10°C. The convection coefficient associated with the inner (room-side) surface is 10 W/m²·K.

(a) If the convection coefficient associated with the outer (ambient) air is $h_o = 80$ W/m²·K, what is the heat loss through a window that is 0.8 m long by 0.5 m wide? Neglect radiation, and assume the air enclosed between the panes to be stagnant.

(b) Compute and plot the effect of h_o on the heat loss for $10 \le h_o \le 100$ W/m²·K. Repeat this calculation for a triple-pane construction in which a third pane and a second air space of equivalent thickness are added.

3.13 A house has a composite wall of wood, fiberglass insulation, and plaster board, as indicated in the sketch. On a cold winter day, the convection heat transfer coefficients are $h_o = 60$ W/m²·K and $h_i = 30$ W/m²·K. The total wall surface area is 350 m².

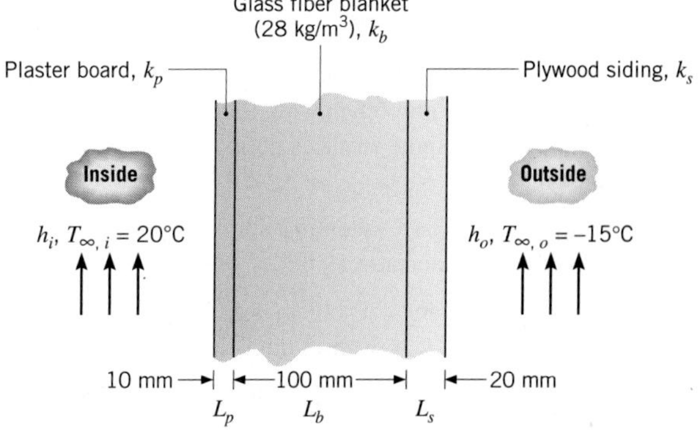

(a) Determine a symbolic expression for the total thermal resistance of the wall, including inside and outside convection effects for the prescribed conditions.

(b) Determine the total heat loss through the wall.

(c) If the wind were blowing violently, raising h_o to 300 W/m²·K, determine the percentage increase in the heat loss.

(d) What is the controlling resistance that determines the amount of heat flow through the wall?

3.14 Consider the composite wall of Problem 3.13 under conditions for which the inside air is still characterized by $T_{\infty,i} = 20°C$ and $h_i = 30$ W/m²·K. However, use the more realistic conditions for which the outside air is characterized by a diurnal (time) varying temperature of the form

$$T_{\infty,o}(\text{K}) = 273 + 5\sin\left(\frac{2\pi}{24}t\right) \qquad 0 \le t \le 12\text{ h}$$

$$T_{\infty,o}(\text{K}) = 273 + 11\sin\left(\frac{2\pi}{24}t\right) \qquad 12 \le t \le 24\text{ h}$$

with $h_o = 60$ W/m²·K. Assuming quasi-steady conditions for which changes in energy storage within the wall may be neglected, estimate the daily heat loss through the wall if its total surface area is 200 m².

3.15 Consider a composite wall that includes an 8-mm-thick hardwood siding, 40-mm by 130-mm hardwood studs on 0.65-m centers with glass fiber insulation (paper faced, 28 kg/m^3), and a 12-mm layer of gypsum (vermiculite) wall board.

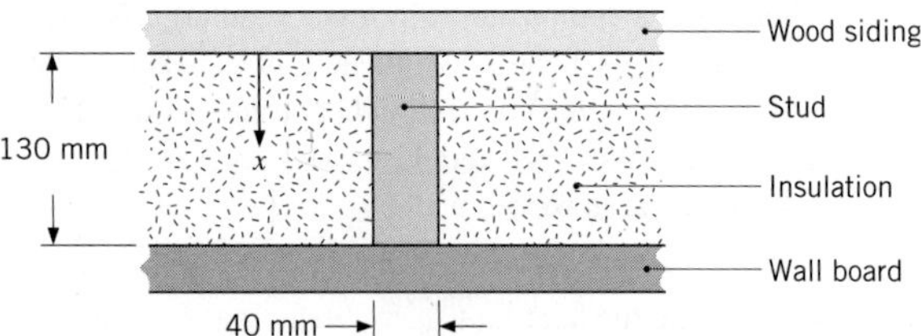

What is the thermal resistance associated with a wall that is 2.5 m high by 6.5 m wide (having 10 studs, each 2.5 m high)? Assume surfaces normal to the x-direction are isothermal.

3.16 Work Problem 3.15 assuming surfaces parallel to the x-direction are adiabatic.

3.17 Consider the oven of Problem 1.54. The walls of the oven consist of L = 30-mm-thick layers of insulation characterized by k_{ins} = 0.03 W/m·K that are sandwiched between two *thin* layers of sheet metal. The exterior surface of the oven is exposed to air at 23°C with h_{ext} = 2 W/m^2·K. The interior oven air temperature is 180°C. Neglecting radiation heat transfer, determine the steady-state heat flux through the oven walls when the convection mode is disabled and the free convection coefficient at the inner oven surface is h_{fr} = 3 W/m^2·K. Determine the heat flux through the oven walls when the convection mode is activated, in which case the forced convection coefficient at the inner oven surface is h_{fo} = 27 W/m^2·K. Does operation of the oven in its convection mode result in significantly increased heat losses from the oven to the kitchen? Would your conclusion change if radiation were included in your analysis?

3.18 The composite wall of an oven consists of three materials, two of which are of known thermal conductivity, k_A = 20 W/m·K and k_C = 50 W/m·K, and known thickness, L_A = 0.30 m and L_C = 0.15 m. The third material, B, which is sandwiched between materials A and C, is of known thickness, L_B = 0.15 m, but unknown thermal conductivity k_B.

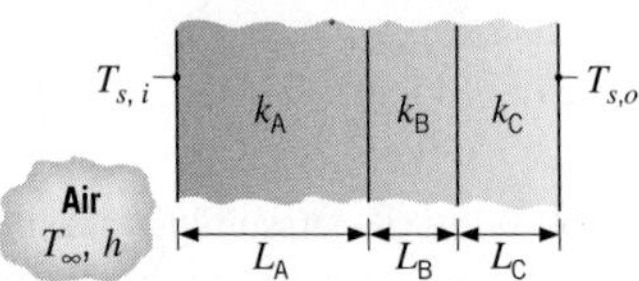

Under steady-state operating conditions, measurements reveal an outer surface temperature of $T_{s,o}$ = 20°C, an inner surface temperature of $T_{s,i}$ = 600°C, and an oven air temperature of T_∞ = 800°C. The inside convection coefficient h is known to be 25 W/m^2·K. What is the value of k_B?

3.19 The wall of a drying oven is constructed by sandwiching an insulation material of thermal conductivity k = 0.05 W/m·K between thin metal sheets. The oven air is at $T_{\infty,i}$ = 300°C, and the corresponding convection coefficient is h_i = 30 W/m^2·K. The inner wall surface absorbs a radiant flux of q''_{rad} = 100 W/m^2 from hotter objects within the oven. The room air is at $T_{\infty,o}$ = 25°C, and the overall coefficient for convection and radiation from the outer surface is h_o = 10 W/m^2·K.

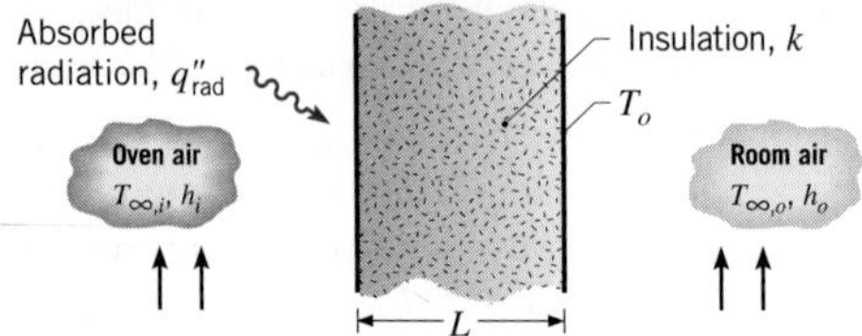

(a) Draw the thermal circuit for the wall and label all temperatures, heat rates, and thermal resistances.

(b) What insulation thickness L is required to maintain the outer wall surface at a *safe-to-touch* temperature of T_o = 40°C?

3.20 The t = 4-mm-thick glass windows of an automobile have a surface area of A = 2.6 m^2. The outside temperature is $T_{\infty,o}$ = 32°C while the passenger compartment is maintained at $T_{\infty,i}$ = 22°C. The convection heat transfer coefficient on the exterior window surface is h_o = 90 W/m^2·K. Determine the heat gain through the windows when the interior convection heat transfer coefficient is h_i = 15 W/m^2·K. By controlling the airflow in the passenger compartment the interior heat transfer coefficient can be reduced to h_i = 5 W/m^2·K without sacrificing passenger comfort. Determine the heat gain through the window for the reduced inside heat transfer coefficient.

3.21 The thermal characteristics of a small, dormitory refrigerator are determined by performing two separate experiments, each with the door closed and the refrigerator placed in ambient air at T_∞ = 25°C. In one case, an electric heater is suspended in the refrigerator cavity, while the refrigerator is unplugged. With the heater dissipating 20 W, a steady-state temperature of 90°C is recorded within the cavity. With the heater removed and the refrigerator now in operation, the second experiment involves maintaining a steady-state cavity temperature of 5°C for a fixed time interval and recording the electrical energy required to operate the refrigerator. In such an experiment for which steady operation is maintained over a 12-hour period, the input electrical energy is 125,000 J. Determine the refrigerator's coefficient of performance (COP).

3.22 In the design of buildings, energy conservation requirements dictate that the exterior surface area, A_s, be minimized. This requirement implies that, for a desired floor space, there may be optimum values associated with the number of floors and horizontal dimensions of the building. Consider a design for which the total floor space, A_f, and the vertical distance between floors, H_f, are prescribed.

(a) If the building has a square cross section of width W on a side, obtain an expression for the value of W that would minimize heat loss to the surroundings. Heat loss may be assumed to occur from the four vertical side walls and from a flat roof. Express your result in terms of A_f and H_f.

(b) If $A_f = 32{,}768\ \text{m}^2$ and $H_f = 4$ m, for what values of W and N_f (the number of floors) is the heat loss minimized? If the average overall heat transfer coefficient is $U = 1\ \text{W/m}^2 \cdot \text{K}$ and the difference between the inside and ambient air temperatures is 25°C, what is the corresponding heat loss? What is the percentage reduction in heat loss compared with a building for $N_f = 2$?

3.23 When raised to very high temperatures, many conventional liquid fuels dissociate into hydrogen and other components. Thus the advantage of a *solid oxide fuel cell* is that such a device can internally *reform* readily available liquid fuels into hydrogen that can then be used to produce electrical power in a manner similar to Example 1.5. Consider a portable solid oxide fuel cell, operating at a temperature of $T_{fc} = 800°\text{C}$. The fuel cell is housed within a cylindrical canister of diameter $D =$ 75 mm and length $L = 120$ mm. The outer surface of the canister is insulated with a low-thermal-conductivity material. For a particular application, it is desired that the *thermal signature* of the canister be small, to avoid its detection by infrared sensors. The degree to which the canister can be detected with an infrared sensor may be estimated by equating the radiation heat flux emitted from the exterior surface of the canister (Equation 1.5; $E_s = \varepsilon_s \sigma T_s^4$) to the heat flux emitted from an *equivalent* black surface, ($E_b = \sigma T_b^4$). If the equivalent black surface temperature T_b is near the surroundings temperature, the thermal signature of the canister is too small to be detected—the canister is indistinguishable from the surroundings.

(a) Determine the required thickness of insulation to be applied to the cylindrical wall of the canister to ensure that the canister does not become highly visible to an infrared sensor (i.e., $T_b - T_{sur} < 5$ K). Consider cases where (i) the outer surface is covered with a very thin layer of dirt ($\varepsilon_s = 0.90$) and (ii) the outer surface is comprised of a very thin polished aluminum sheet ($\varepsilon_s = 0.08$). Calculate the required thicknesses for two types of insulating material, calcium silicate ($k = 0.09\ \text{W/m} \cdot \text{K}$) and aerogel ($k = 0.006\ \text{W/m} \cdot \text{K}$). The temperatures of the surroundings and the ambient are $T_{sur} = 300$ K and $T_\infty = 298$ K, respectively. The outer surface is characterized by a convective heat transfer coefficient of $h = 12\ \text{W/m}^2 \cdot \text{K}$.

(b) Calculate the outer surface temperature of the canister for the four cases (high and low thermal conductivity; high and low surface emissivity).

(c) Calculate the heat loss from the cylindrical walls of the canister for the four cases.

3.24 A firefighter's protective clothing, referred to as a *turnout coat*, is typically constructed as an ensemble of three layers separated by air gaps, as shown schematically.

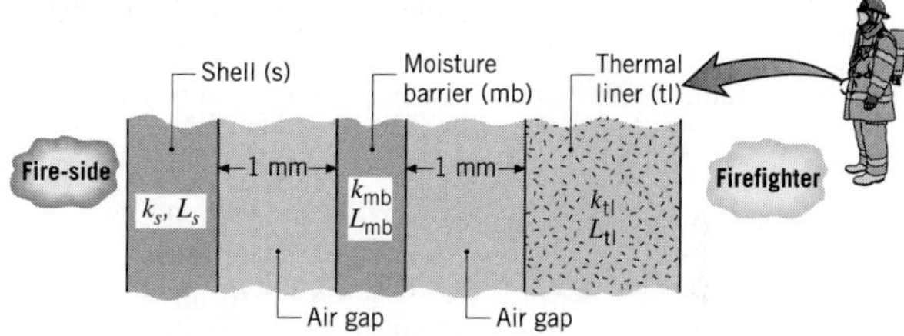

Representative dimensions and thermal conductivities for the layers are as follows.

Layer	*Thickness* (mm)	*k* (W/m · K)
Shell (s)	0.8	0.047
Moisture barrier (mb)	0.55	0.012
Thermal liner (tl)	3.5	0.038

The air gaps between the layers are 1 mm thick, and heat is transferred by conduction and radiation exchange through the stagnant air. The linearized radiation coefficient for a gap may be approximated as, $h_{rad} = \sigma(T_1 + T_2)(T_1^2 + T_2^2) \approx 4\sigma T_{avg}^3$, where T_{avg} represents the average temperature of the surfaces comprising the gap, and the radiation flux across the gap may be expressed as $q''_{rad} = h_{rad}(T_1 - T_2)$.

(a) Represent the turnout coat by a thermal circuit, labeling all the thermal resistances. Calculate and tabulate the thermal resistances per unit area ($\text{m}^2 \cdot$ K/W) for each of the layers, as well as for the conduction and radiation processes in the gaps. Assume that a value of $T_{avg} = 470$ K may be used to approximate the radiation resistance of both gaps. Comment on the relative magnitudes of the resistances.

(b) For a *pre-flash-over* fire environment in which firefighters often work, the typical radiant heat flux on the fire-side of the turnout coat is $0.25\ \text{W/cm}^2$.

What is the outer surface temperature of the turnout coat if the inner surface temperature is 66°C, a condition that would result in burn injury?

3.25 A particular thermal system involves three objects of fixed shape with conduction resistances of $R_1 = 1$ K/W, $R_2 = 2$ K/W and $R_3 = 4$ K/W, respectively. An objective is to minimize the total thermal resistance R_{tot} associated with a combination of R_1, R_2, and R_3. The chief engineer is willing to invest limited funds to specify an alternative material for just one of the three objects; the alternative material will have a thermal conductivity that is twice its nominal value. Which object (1, 2, or 3) should be fabricated of the higher thermal conductivity material to most significantly decrease R_{tot}? *Hint*: Consider two cases, one for which the three thermal resistances are arranged in series, and the second for which the three resistances are arranged in parallel.

Contact Resistance

3.26 A composite wall separates combustion gases at 2600°C from a liquid coolant at 100°C, with gas- and liquid-side convection coefficients of 50 and 1000 $W/m^2 \cdot K$. The wall is composed of a 10-mm-thick layer of beryllium oxide on the gas side and a 20-mm-thick slab of stainless steel (AISI 304) on the liquid side. The contact resistance between the oxide and the steel is 0.05 $m^2 \cdot K/W$. What is the heat loss per unit surface area of the composite? Sketch the temperature distribution from the gas to the liquid.

3.27 Approximately 10^6 discrete electrical components can be placed on a single integrated circuit (chip), with electrical heat dissipation as high as 30,000 W/m^2. The chip, which is very thin, is exposed to a dielectric liquid at its outer surface, with $h_o = 1000$ $W/m^2 \cdot K$ and $T_{\infty,o} = 20°C$, and is joined to a circuit board at its inner surface. The thermal contact resistance between the chip and the board is 10^{-4} $m^2 \cdot K/W$, and the board thickness and thermal conductivity are $L_b = 5$ mm and $k_b = 1$ $W/m \cdot K$, respectively. The other surface of the board is exposed to ambient air for which $h_i = 40$ $W/m^2 \cdot K$ and $T_{\infty,i} = 20°C$.

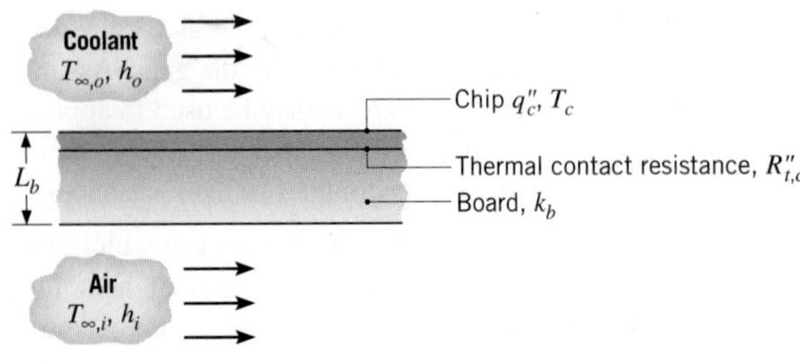

(a) Sketch the equivalent thermal circuit corresponding to steady-state conditions. In variable form, label appropriate resistances, temperatures, and heat fluxes.

(b) Under steady-state conditions for which the chip heat dissipation is $q''_c = 30{,}000$ W/m^2, what is the chip temperature?

(c) The maximum allowable heat flux, $q''_{c,m}$, is determined by the constraint that the chip temperature must not exceed 85°C. Determine $q''_{c,m}$ for the foregoing conditions. If air is used in lieu of the dielectric liquid, the convection coefficient is reduced by approximately an order of magnitude. What is the value of $q''_{c,m}$ for $h_o = 100$ $W/m^2 \cdot K$? With air cooling, can significant improvements be realized by using an aluminum oxide circuit board and/or by using a conductive paste at the chip/board interface for which $R''_{t,c} = 10^{-5}$ $m^2 \cdot K/W$?

3.28 Two stainless steel plates 10 mm thick are subjected to a contact pressure of 1 bar under vacuum conditions for which there is an overall temperature drop of 100°C across the plates. What is the heat flux through the plates? What is the temperature drop across the contact plane?

3.29 Consider a plane composite wall that is composed of two materials of thermal conductivities $k_A = 0.1$ $W/m \cdot K$ and $k_B = 0.04$ $W/m \cdot K$ and thicknesses $L_A = 10$ mm and $L_B = 20$ mm. The contact resistance at the interface between the two materials is known to be 0.30 $m^2 \cdot K/W$. Material A adjoins a fluid at 200°C for which $h = 10$ $W/m^2 \cdot K$, and material B adjoins a fluid at 40°C for which $h = 20$ $W/m^2 \cdot K$.

(a) What is the rate of heat transfer through a wall that is 2 m high by 2.5 m wide?

(b) Sketch the temperature distribution.

3.30 The performance of gas turbine engines may be improved by increasing the tolerance of the turbine blades to hot gases emerging from the combustor. One approach to achieving high operating temperatures involves application of a *thermal barrier coating* (TBC) to the exterior surface of a blade, while passing cooling air through the blade. Typically, the blade is made from a high-temperature superalloy, such as Inconel ($k \approx 25$ $W/m \cdot K$), while a ceramic, such as zirconia ($k \approx 1.3$ $W/m \cdot K$), is used as a TBC.

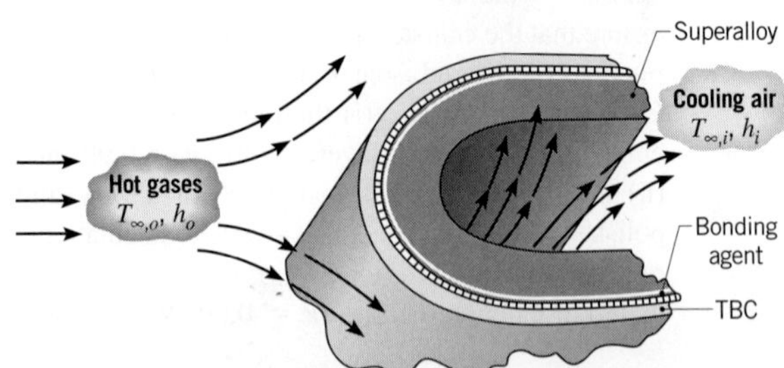

Consider conditions for which hot gases at $T_{\infty,o} =$ 1700 K and cooling air at $T_{\infty,i} = 400$ K provide outer and inner surface convection coefficients of $h_o =$ 1000 W/m$^2 \cdot$K and $h_i = 500$ W/m$^2 \cdot$ K, respectively. If a 0.5-mm-thick zirconia TBC is attached to a 5-mm-thick Inconel blade wall by means of a metallic bonding agent, which provides an interfacial thermal resistance of $R''_{t,c} = 10^{-4}$ m$^2 \cdot$K/W, can the Inconel be maintained at a temperature that is below its maximum allowable value of 1250 K? Radiation effects may be neglected, and the turbine blade may be approximated as a plane wall. Plot the temperature distribution with and without the TBC. Are there any limits to the thickness of the TBC?

3.31 A commercial grade cubical freezer, 3 m on a side, has a composite wall consisting of an exterior sheet of 6.35-mm-thick plain carbon steel, an intermediate layer of 100-mm-thick cork insulation, and an inner sheet of 6.35-mm-thick aluminum alloy (2024). Adhesive interfaces between the insulation and the metallic strips are each characterized by a thermal contact resistance of $R''_{t,c} = 2.5 \times 10^{-4}$ m$^2 \cdot$ K/W. What is the steady-state cooling load that must be maintained by the refrigerator under conditions for which the outer and inner surface temperatures are 22°C and −6°C, respectively?

3.32 Physicists have determined the theoretical value of the thermal conductivity of a carbon nanotube to be $k_{cn,T} =$ 5000 W/m$\cdot$K.

(a) Assuming the actual thermal conductivity of the carbon nanotube is the same as its theoretical value, find the thermal contact resistance, $R_{t,c}$, that exists between the carbon nanotube and the top surfaces of the heated and sensing islands in Example 3.4 .

(b) Using the value of the thermal contact resistance calculated in part (a), plot the fraction of the total resistance between the heated and sensing islands that is due to the thermal contact resistances for island separation distances of 5 μm $\leq s \leq$ 20 μm.

3.33 Consider a power transistor encapsulated in an aluminum case that is attached at its base to a square aluminum plate of thermal conductivity $k = 240$ W/m$\cdot$K, thickness $L = 6$ mm, and width $W = 20$ mm. The case is joined to the plate by screws that maintain a contact pressure of 1 bar, and the back surface of the plate transfers heat by natural convection and radiation to ambient air and large surroundings at $T_\infty = T_{sur} =$ 25°C. The surface has an emissivity of $\varepsilon = 0.9$, and the convection coefficient is $h = 4$ W/m$^2\cdot$K. The case is completely enclosed such that heat transfer may be assumed to occur exclusively through the base plate.

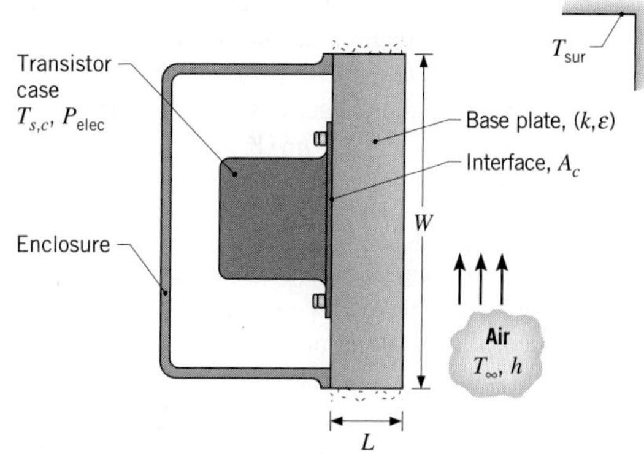

(a) If the air-filled aluminum-to-aluminum interface is characterized by an area of $A_c = 2 \times 10^{-4}$ m^2 and a roughness of 10 μm, what is the maximum allowable power dissipation if the surface temperature of the case, $T_{s,c}$, is not to exceed 85°C?

(b) The convection coefficient may be increased by subjecting the plate surface to a forced flow of air. Explore the effect of increasing the coefficient over the range $4 \leq h \leq 200$ W/m$^2 \cdot$ K.

Porous Media

3.34 *Ring-porous* woods, such as oak, are characterized by grains. The dark grains consist of very low-density material that forms early in the springtime. The surrounding lighter-colored wood is composed of high-density material that forms slowly throughout most of the growing season.

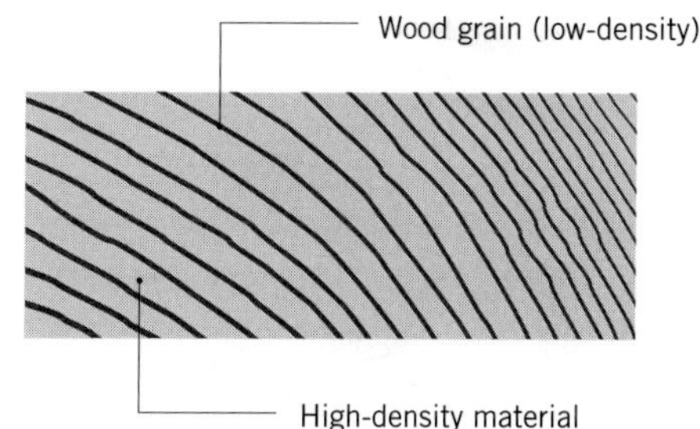

Assuming the low-density material is highly porous and the oak is dry, determine the fraction of the oak cross-section that appears as being grained. *Hint*: Assume the thermal conductivity parallel to the grains is the same as the radial conductivity of Table A.3.

3.35 A batt of glass fiber insulation is of density $\rho =$ 28 kg/m^3. Determine the maximum and minimum possible values of the effective thermal conductivity of the insulation at $T = 300$ K, and compare with the value reported in Table A.3.

3.36 Air usually constitutes up to half of the volume of commercial ice creams and takes the form of small spherical bubbles interspersed within a matrix of frozen matter. The thermal conductivity of ice cream that contains no air is $k_{na} = 1.1$ W/m·K at $T = -20°C$. Determine the thermal conductivity of commercial ice cream characterized by $\varepsilon = 0.20$, also at $T = -20°C$.

3.37 Determine the density, specific heat, and thermal conductivity of a lightweight aggregate concrete that is composed of 65% stone mix concrete and 35% air by volume. Evaluate properties at $T = 300$ K.

3.38 A one-dimensional plane wall of thickness L is constructed of a solid material with a linear, nonuniform porosity distribution described by $\varepsilon(x) = \varepsilon_{max}(x/L)$. Plot the steady-state temperature distribution, $T(x)$, for $k_s = 10$ W/m·K, $k_f = 0.1$ W/m·K, $L = 1$ m, $\varepsilon_{max} = 0.25$, $T(x = 0) = 30°C$ and $q''_x = 100$ W/m^2 using the expression for the minimum effective thermal conductivity of a porous medium, the expression for the maximum effective thermal conductivity of a porous medium, Maxwell's expression, and for the case where $k_{eff}(x) = k_s$.

Alternative Conduction Analysis

3.39 The diagram shows a conical section fabricated from pure aluminum. It is of circular cross section having diameter $D = ax^{1/2}$, where $a = 0.5$ m$^{1/2}$. The small end is located at $x_1 = 25$ mm and the large end at $x_2 = 125$ mm. The end temperatures are $T_1 = 600$ K and $T_2 = 400$ K, while the lateral surface is well insulated.

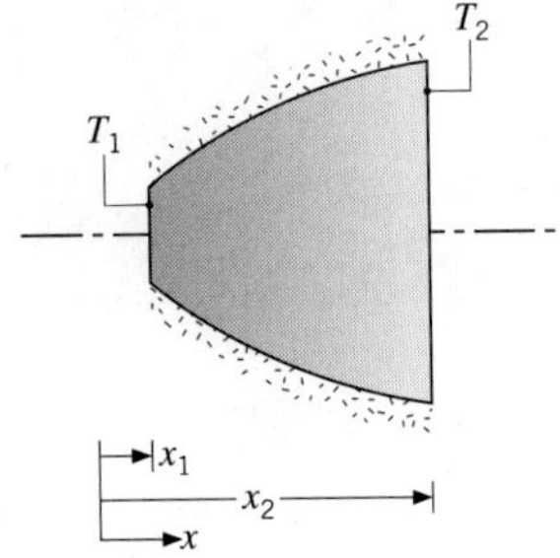

(a) Derive an expression for the temperature distribution $T(x)$ in symbolic form, assuming one-dimensional conditions. Sketch the temperature distribution.

(b) Calculate the heat rate q_x.

3.40 A truncated solid cone is of circular cross section, and its diameter is related to the axial coordinate by an expression of the form $D = ax^{3/2}$, where $a = 1.0$ m$^{-1/2}$.

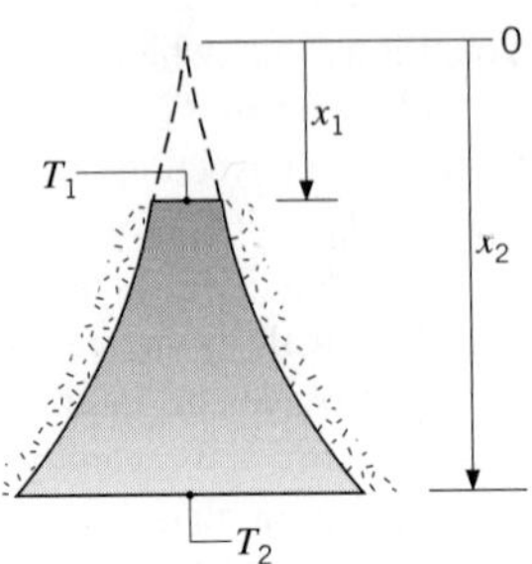

The sides are well insulated, while the top surface of the cone at x_1 is maintained at T_1 and the bottom surface at x_2 is maintained at T_2.

(a) Obtain an expression for the temperature distribution $T(x)$.

(b) What is the rate of heat transfer across the cone if it is constructed of pure aluminum with $x_1 = 0.075$ m, $T_1 = 100°C$, $x_2 = 0.225$ m, and $T_2 = 20°C$?

3.41 From Figure 2.5 it is evident that, over a wide temperature range, the temperature dependence of the thermal conductivity of many solids may be approximated by a linear expression of the form $k = k_o + aT$, where k_o is a positive constant and a is a coefficient that may be positive or negative. Obtain an expression for the heat flux across a plane wall whose inner and outer surfaces are maintained at T_0 and T_1, respectively. Sketch the forms of the temperature distribution corresponding to $a > 0$, $a = 0$, and $a < 0$.

3.42 Consider a tube wall of inner and outer radii r_i and r_o, whose temperatures are maintained at T_i and T_o, respectively. The thermal conductivity of the cylinder is temperature dependent and may be represented by an expression of the form $k = k_o(1 + aT)$, where k_o and a are constants. Obtain an expression for the heat transfer per unit length of the tube. What is the thermal resistance of the tube wall?

3.43 Measurements show that steady-state conduction through a plane wall without heat generation produced a convex temperature distribution such that the midpoint temperature was ΔT_o higher than expected for a linear temperature distribution.

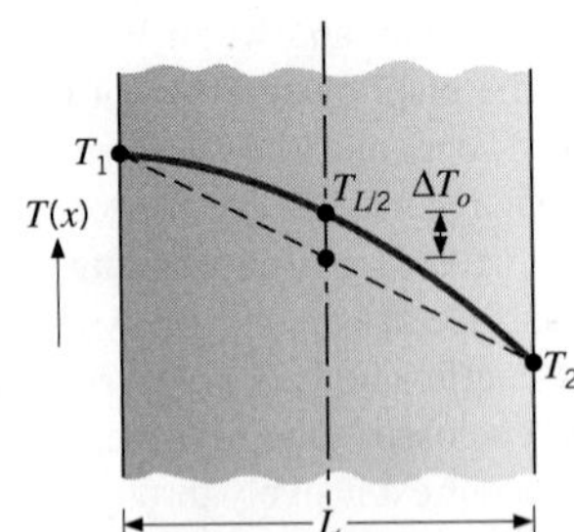

Assuming that the thermal conductivity has a linear dependence on temperature, $k = k_o(1 + \alpha T)$, where α is a constant, develop a relationship to evaluate α in terms of ΔT_o, T_1, and T_2.

3.44 A device used to measure the surface temperature of an object to within a spatial resolution of approximately 50 nm is shown in the schematic. It consists of an extremely sharp-tipped stylus and an extremely small cantilever that is scanned across the surface. The probe tip is of circular cross section and is fabricated of polycrystalline silicon dioxide. The ambient temperature is measured at the pivoted end of the cantilever as T_∞ = 25°C, and the device is equipped with a sensor to measure the temperature at the upper end of the sharp tip, T_{sen}. The thermal resistance between the sensing probe and the pivoted end is $R_t = 5 \times 10^6$ K/W.

(a) Determine the thermal resistance between the surface temperature and the sensing temperature.

(b) If the sensing temperature is T_{sen} = 28.5°C, determine the surface temperature.

Hint: Although nanoscale heat transfer effects may be important, assume that the conduction occurring in the air adjacent to the probe tip can be described by Fourier's law and the thermal conductivity found in Table A.4.

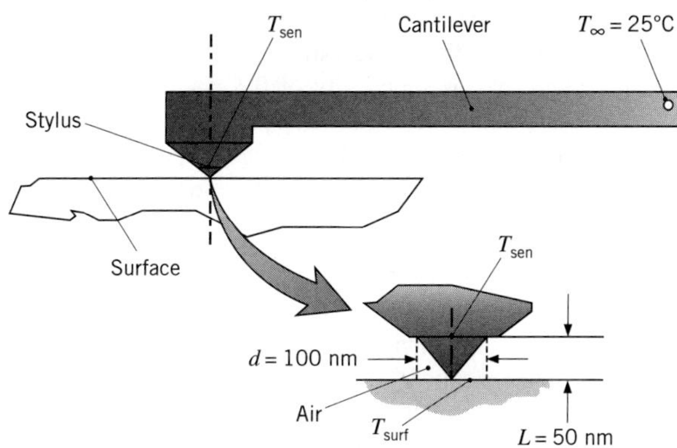

Cylindrical Wall

3.45 A steam pipe of 0.12-m outside diameter is insulated with a layer of calcium silicate.

(a) If the insulation is 20 mm thick and its inner and outer surfaces are maintained at $T_{s,1}$ = 800 K and $T_{s,2}$ = 490 K, respectively, what is the heat loss per unit length (q') of the pipe?

(b) We wish to explore the effect of insulation thickness on the heat loss q' and outer surface temperature $T_{s,2}$, with the inner surface temperature fixed at $T_{s,1}$ = 800 K. The outer surface is exposed to an airflow (T_∞ = 25°C) that maintains a convection coefficient of h = 25 W/m$^2\cdot$K and to large surroundings for which $T_{sur} = T_\infty$ = 25°C. The surface emissivity of calcium silicate is approximately 0.8. Compute and plot the temperature distribution in the insulation as a function of the dimensionless radial coordinate, $(r - r_1)/(r_2 - r_1)$, where r_1 = 0.06 m and r_2 is a variable (0.06 $< r_2 \leq$ 0.20 m). Compute and plot the heat loss as a function of the insulation thickness for $0 \leq (r_2 - r_1) \leq 0.14$ m.

3.46 Consider the water heater described in Problem 1.48. We now wish to determine the energy needed to compensate for heat losses incurred while the water is stored at the prescribed temperature of 55°C. The cylindrical storage tank (with flat ends) has a capacity of 100 gal, and foamed urethane is used to insulate the side and end walls from ambient air at an annual average temperature of 20°C. The resistance to heat transfer is dominated by conduction in the insulation and by free convection in the air, for which $h \approx 2$ W/m$^2\cdot$K. If electric resistance heating is used to compensate for the losses and the cost of electric power is \$0.18/kWh, specify tank and insulation dimensions for which the annual cost associated with the heat losses is less than \$50.

3.47 To maximize production and minimize pumping costs, crude oil is heated to reduce its viscosity during transportation from a production field.

(a) Consider a *pipe-in-pipe* configuration consisting of concentric steel tubes with an intervening insulating material. The inner tube is used to transport warm crude oil through cold ocean water. The inner steel pipe (k_s = 35 W/m$\cdot$K) has an inside diameter of $D_{i,1}$ = 150 mm and wall thickness t_i = 10 mm while the outer steel pipe has an inside diameter of $D_{i,2}$ = 250 mm and wall thickness $t_o = t_i$. Determine the maximum allowable crude oil temperature to ensure the polyurethane foam insulation (k_p = 0.075 W/m$\cdot$K) between the two pipes does not exceed its maximum service temperature of $T_{p,\max}$ = 70°C. The ocean water is at $T_{\infty,o}$ = –5°C and provides an external convection heat transfer coefficient of h_o = 500 W/m$^2\cdot$K. The convection coefficient associated with the flowing crude oil is h_i = 450 W/m$^2\cdot$K.

(b) It is proposed to enhance the performance of the pipe-in-pipe device by replacing a thin (t_a = 5 mm) section of polyurethane located at the outside of the inner pipe with an aerogel insulation material (k_a = 0.012 W/m$\cdot$K). Determine the maximum allowable crude oil temperature to ensure maximum polyurethane temperatures are below $T_{p,\max}$ = 70°C.

3.48 A thin electrical heater is wrapped around the outer surface of a long cylindrical tube whose inner surface is maintained at a temperature of 5°C. The tube wall has inner and outer radii of 25 and 75 mm, respectively, and a thermal conductivity of 10 W/m·K. The thermal contact resistance between the heater and the outer surface of the tube (per unit length of the tube) is $R'_{t,c} =$ 0.01 m·K/W. The outer surface of the heater is exposed to a fluid with $T_\infty = -10$°C and a convection coefficient of $h =$ 100 W/m^2·K. Determine the heater power per unit length of tube required to maintain the heater at $T_o = 25$°C.

3.49 In Problem 3.48, the electrical power required to maintain the heater at $T_o = 25$°C depends on the thermal conductivity of the wall material k, the thermal contact resistance $R'_{t,c}$ and the convection coefficient h. Compute and plot the separate effect of changes in k ($1 \leq k \leq 200$ W/m·K), $R'_{t,c}$ ($0 \leq R'_{t,c} \leq 0.1$ m·K/W), and h ($10 \leq h \leq 1000$ W/m^2·K) on the total heater power requirement, as well as the rate of heat transfer to the inner surface of the tube and to the fluid.

3.50 A stainless steel (AISI 304) tube used to transport a chilled pharmaceutical has an inner diameter of 36 mm and a wall thickness of 2 mm. The pharmaceutical and ambient air are at temperatures of 6°C and 23°C, respectively, while the corresponding inner and outer convection coefficients are 400 W/m^2·K and 6 W/m^2·K, respectively.

(a) What is the heat gain per unit tube length?

(b) What is the heat gain per unit length if a 10-mm-thick layer of calcium silicate insulation ($k_{ins} =$ 0.050 W/m·K) is applied to the tube?

3.51 Superheated steam at 575°C is routed from a boiler to the turbine of an electric power plant through steel tubes ($k = 35$ W/m·K) of 300-mm inner diameter and 30-mm wall thickness. To reduce heat loss to the surroundings and to maintain a *safe-to-touch* outer surface temperature, a layer of calcium silicate insulation ($k = 0.10$ W/m·K) is applied to the tubes, while degradation of the insulation is reduced by wrapping it in a thin sheet of aluminum having an emissivity of $\varepsilon = 0.20$. The air and wall temperatures of the power plant are 27°C.

(a) Assuming that the inner surface temperature of a steel tube corresponds to that of the steam and the convection coefficient outside the aluminum sheet is 6 W/m^2·K, what is the minimum insulation thickness needed to ensure that the temperature of the aluminum does not exceed 50°C? What is the corresponding heat loss per meter of tube length?

(b) Explore the effect of the insulation thickness on the temperature of the aluminum and the heat loss per unit tube length.

3.52 A thin electrical heater is inserted between a long circular rod and a concentric tube with inner and outer radii of 20 and 40 mm. The rod (A) has a thermal conductivity of $k_A = 0.15$ W/m·K, while the tube (B) has a thermal conductivity of $k_B = 1.5$ W/m·K and its outer surface is subjected to convection with a fluid of temperature $T_\infty = -15$°C and heat transfer coefficient 50 W/m^2·K. The thermal contact resistance between the cylinder surfaces and the heater is negligible.

(a) Determine the electrical power per unit length of the cylinders (W/m) that is required to maintain the outer surface of cylinder B at 5°C.

(b) What is the temperature at the center of cylinder A?

3.53 A wire of diameter $D = 2$ mm and uniform temperature T has an electrical resistance of 0.01 Ω/m and a current flow of 20 A.

(a) What is the rate at which heat is dissipated per unit length of wire? What is the heat dissipation per unit volume within the wire?

(b) If the wire is not insulated and is in ambient air and large surroundings for which $T_\infty = T_{sur} = 20$°C, what is the temperature T of the wire? The wire has an emissivity of 0.3, and the coefficient associated with heat transfer by natural convection may be approximated by an expression of the form, $h = C[(T - T_\infty)/D]^{1/4}$, where $C = 1.25$ W/m$^{7/4}$·K$^{5/4}$.

(c) If the wire is coated with plastic insulation of 2-mm thickness and a thermal conductivity of 0.25 W/m·K, what are the inner and outer surface temperatures of the insulation? The insulation has an emissivity of 0.9, and the convection coefficient is given by the expression of part (b). Explore the effect of the insulation thickness on the surface temperatures.

3.54 A 2-mm-diameter electrical wire is insulated by a 2-mm-thick rubberized sheath ($k = 0.13$ W/m·K), and the wire/sheath interface is characterized by a thermal contact resistance of $R''_{t,c} = 3 \times 10^{-4}$ m^2·K/W. The convection heat transfer coefficient at the outer surface of the sheath is 10 W/m^2·K, and the temperature of the ambient air is 20°C. If the temperature of the insulation may not exceed 50°C, what is the maximum allowable electrical power that may be dissipated per unit length of the conductor? What is the critical radius of the insulation?

3.55 Electric current flows through a long rod generating thermal energy at a uniform volumetric rate of $\dot{q} = 2 \times 10^6$ W/m^3. The rod is concentric with a hollow ceramic cylinder, creating an enclosure that is filled with air.

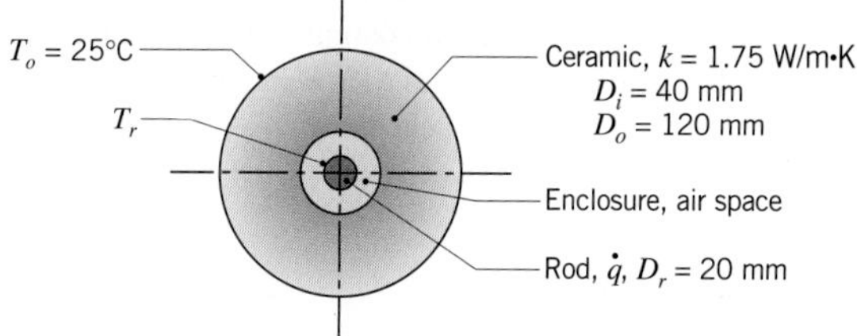

The thermal resistance per unit length due to radiation between the enclosure surfaces is $R'_{\text{rad}} = 0.30\ \text{m} \cdot \text{K/W}$, and the coefficient associated with free convection in the enclosure is $h = 20\ \text{W/m}^2 \cdot \text{K}$.

(a) Construct a thermal circuit that can be used to calculate the surface temperature of the rod, T_r. Label all temperatures, heat rates, and thermal resistances, and evaluate each thermal resistance.

(b) Calculate the surface temperature of the rod for the prescribed conditions.

3.56 The evaporator section of a refrigeration unit consists of thin-walled, 10-mm-diameter tubes through which refrigerant passes at a temperature of $-18°$C. Air is cooled as it flows over the tubes, maintaining a surface convection coefficient of $100\ \text{W/m}^2 \cdot \text{K}$, and is subsequently routed to the refrigerator compartment.

(a) For the foregoing conditions and an air temperature of $-3°$C, what is the rate at which heat is extracted from the air per unit tube length?

(b) If the refrigerator's defrost unit malfunctions, frost will slowly accumulate on the outer tube surface. Assess the effect of frost formation on the cooling capacity of a tube for frost layer thicknesses in the range $0 \leq \delta \leq 4$ mm. Frost may be assumed to have a thermal conductivity of $0.4\ \text{W/m} \cdot \text{K}$.

(c) The refrigerator is disconnected after the defrost unit malfunctions and a 2-mm-thick layer of frost has formed. If the tubes are in ambient air for which $T_\infty = 20°$C and natural convection maintains a convection coefficient of $2\ \text{W/m}^2 \cdot \text{K}$, how long will it take for the frost to melt? The frost may be assumed to have a mass density of $700\ \text{kg/m}^3$ and a latent heat of fusion of 334 kJ/kg.

3.57 A composite cylindrical wall is composed of two materials of thermal conductivity k_{A} and k_{B}, which are separated by a very thin, electric resistance heater for which interfacial contact resistances are negligible.

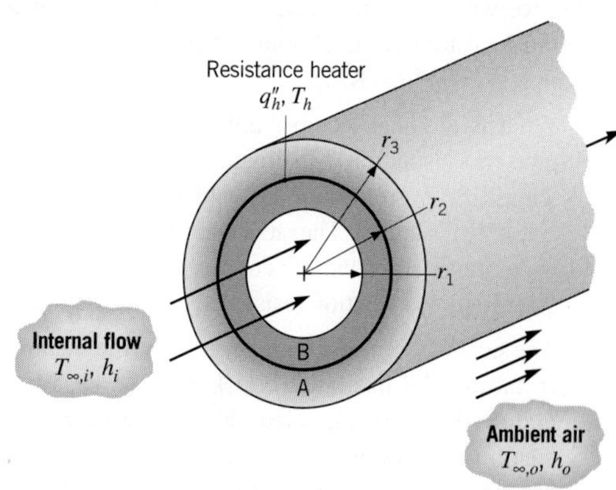

Liquid pumped through the tube is at a temperature $T_{\infty,i}$ and provides a convection coefficient h_i at the inner surface of the composite. The outer surface is exposed to ambient air, which is at $T_{\infty,o}$ and provides a convection coefficient of h_o. Under steady-state conditions, a uniform heat flux of q''_h is dissipated by the heater.

(a) Sketch the equivalent thermal circuit of the system and express all resistances in terms of relevant variables.

(b) Obtain an expression that may be used to determine the heater temperature, T_h.

(c) Obtain an expression for the ratio of heat flows to the outer and inner fluids, q'_o/q'_i. How might the variables of the problem be adjusted to minimize this ratio?

3.58 An electrical current of 700 A flows through a stainless steel cable having a diameter of 5 mm and an electrical resistance of $6 \times 10^{-4}\ \Omega$/m (i.e., per meter of cable length). The cable is in an environment having a temperature of 30°C, and the total coefficient associated with convection and radiation between the cable and the environment is approximately $25\ \text{W/m}^2 \cdot \text{K}$.

(a) If the cable is bare, what is its surface temperature?

(b) If a very thin coating of electrical insulation is applied to the cable, with a contact resistance of $0.02\ \text{m}^2 \cdot \text{K/W}$, what are the insulation and cable surface temperatures?

(c) There is some concern about the ability of the insulation to withstand elevated temperatures. What thickness of this insulation ($k = 0.5\ \text{W/m} \cdot \text{K}$) will yield the lowest value of the maximum insulation temperature? What is the value of the maximum temperature when this thickness is used?

3.59 A 0.20-m-diameter, thin-walled steel pipe is used to transport saturated steam at a pressure of 20 bars in a room for which the air temperature is 25°C and the convection heat transfer coefficient at the outer surface of the pipe is 20 $W/m^2 \cdot K$.

(a) What is the heat loss per unit length from the bare pipe (no insulation)? Estimate the heat loss per unit length if a 50-mm-thick layer of insulation (magnesia, 85%) is added. The steel and magnesia may each be assumed to have an emissivity of 0.8, and the steam-side convection resistance may be neglected.

(b) The costs associated with generating the steam and installing the insulation are known to be $\$4/10^9$ J and \$100/m of pipe length, respectively. If the steam line is to operate 7500 h/yr, how many years are needed to pay back the initial investment in insulation?

3.60 An uninsulated, thin-walled pipe of 100-mm diameter is used to transport water to equipment that operates outdoors and uses the water as a coolant. During particularly harsh winter conditions, the pipe wall achieves a temperature of –15°C and a cylindrical layer of ice forms on the inner surface of the wall. If the mean water temperature is 3°C and a convection coefficient of 2000 $W/m^2 \cdot K$ is maintained at the inner surface of the ice, which is at 0°C, what is the thickness of the ice layer?

3.61 Steam flowing through a long, thin-walled pipe maintains the pipe wall at a uniform temperature of 500 K. The pipe is covered with an insulation blanket comprised of two different materials, A and B.

The interface between the two materials may be assumed to have an infinite contact resistance, and the entire outer surface is exposed to air for which T_∞ = 300 K and h = 25 $W/m^2 \cdot K$.

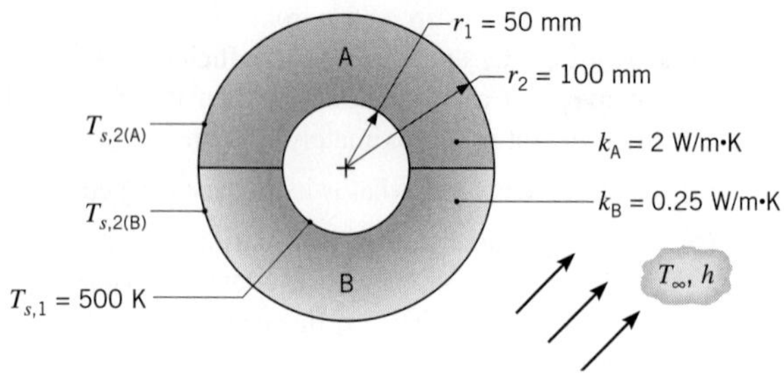

(a) Sketch the thermal circuit of the system. Label (using the preceding symbols) all pertinent nodes and resistances.

(b) For the prescribed conditions, what is the total heat loss from the pipe? What are the outer surface temperatures $T_{s,2(A)}$ and $T_{s,2(B)}$?

3.62 A bakelite coating is to be used with a 10-mm-diameter conducting rod, whose surface is maintained at 200°C by passage of an electrical current. The rod is in a fluid at 25°C, and the convection coefficient is 140 $W/m^2 \cdot K$. What is the critical radius associated with the coating? What is the heat transfer rate per unit length for the bare rod and for the rod with a coating of bakelite that corresponds to the critical radius? How much bakelite should be added to reduce the heat transfer associated with the bare rod by 25%?

Spherical Wall

3.63 A storage tank consists of a cylindrical section that has a length and inner diameter of L = 2 m and D_i = 1 m, respectively, and two hemispherical end sections. The tank is constructed from 20-mm-thick glass (Pyrex) and is exposed to ambient air for which the temperature is 300 K and the convection coefficient is 10 $W/m^2 \cdot K$. The tank is used to store heated oil, which maintains the inner surface at a temperature of 400 K. Determine the electrical power that must be supplied to a heater submerged in the oil if the prescribed conditions are to be maintained. Radiation effects may be neglected, and the Pyrex may be assumed to have a thermal conductivity of 1.4 $W/m \cdot K$.

3.64 Consider the liquid oxygen storage system and the laboratory environmental conditions of Problem 1.49. To reduce oxygen loss due to vaporization, an insulating layer should be applied to the outer surface of the container. Consider using a laminated aluminum foil/glass mat insulation, for which the thermal conductivity and surface emissivity are k = 0.00016 $W/m \cdot K$ and ε = 0.20, respectively.

(a) If the container is covered with a 10-mm-thick layer of insulation, what is the percentage reduction in oxygen loss relative to the uncovered container?

(b) Compute and plot the oxygen evaporation rate (kg/s) as a function of the insulation thickness t for $0 \leq t \leq 50$ mm.

3.65 A spherical Pyrex glass shell has inside and outside diameters of D_1 = 0.1 m and D_2 = 0.2 m, respectively. The inner surface is at $T_{s,1}$ = 100°C while the outer surface is at $T_{s,2}$ = 45°C.

(a) Determine the temperature at the midpoint of the shell thickness, $T(r_m = 0.075 \text{ m})$.

(b) For the same surface temperatures and dimensions as in part (a), show how the midpoint temperature would change if the shell material were aluminum.

3.66 In Example 3.6, an expression was derived for the critical insulation radius of an insulated, cylindrical tube.

Derive the expression that would be appropriate for an insulated sphere.

3.67 A hollow aluminum sphere, with an electrical heater in the center, is used in tests to determine the thermal conductivity of insulating materials. The inner and outer radii of the sphere are 0.15 and 0.18 m, respectively, and testing is done under steady-state conditions with the inner surface of the aluminum maintained at 250°C. In a particular test, a spherical shell of insulation is cast on the outer surface of the sphere to a thickness of 0.12 m. The system is in a room for which the air temperature is 20°C and the convection coefficient at the outer surface of the insulation is $30 \text{ W/m}^2 \cdot \text{K}$. If 80 W are dissipated by the heater under steady-state conditions, what is the thermal conductivity of the insulation?

3.68 A spherical tank for storing liquid oxygen on the space shuttle is to be made from stainless steel of 0.80-m outer diameter and 5-mm wall thickness. The boiling point and latent heat of vaporization of liquid oxygen are 90 K and 213 kJ/kg, respectively. The tank is to be installed in a large compartment whose temperature is to be maintained at 240 K. Design a thermal insulation system that will maintain oxygen losses due to boiling below 1 kg/day.

3.69 A spherical, cryosurgical probe may be imbedded in diseased tissue for the purpose of freezing, and thereby destroying, the tissue. Consider a probe of 3-mm diameter whose surface is maintained at −30°C when imbedded in tissue that is at 37°C. A spherical layer of frozen tissue forms around the probe, with a temperature of 0°C existing at the phase front (interface) between the frozen and normal tissue. If the thermal conductivity of frozen tissue is approximately $1.5 \text{ W/m} \cdot \text{K}$ and heat transfer at the phase front may be characterized by an effective convection coefficient of $50 \text{ W/m}^2 \cdot \text{K}$, what is the thickness of the layer of frozen tissue (assuming negligible perfusion)?

3.70 A spherical vessel used as a reactor for producing pharmaceuticals has a 10-mm-thick stainless steel wall ($k = 17 \text{ W/m} \cdot \text{K}$) and an inner diameter of 1 m. The exterior surface of the vessel is exposed to ambient air ($T_\infty = 25°\text{C}$) for which a convection coefficient of $6 \text{ W/m}^2 \cdot \text{K}$ may be assumed.

(a) During steady-state operation, an inner surface temperature of 50°C is maintained by energy generated within the reactor. What is the heat loss from the vessel?

(b) If a 20-mm-thick layer of fiberglass insulation ($k = 0.040 \text{ W/m} \cdot \text{K}$) is applied to the exterior of the vessel and the rate of thermal energy generation is unchanged, what is the inner surface temperature of the vessel?

3.71 The wall of a spherical tank of 1-m diameter contains an exothermic chemical reaction and is at 200°C when the ambient air temperature is 25°C. What thickness of urethane foam is required to reduce the exterior temperature to 40°C, assuming the convection coefficient is $20 \text{ W/m}^2 \cdot \text{K}$ for both situations? What is the percentage reduction in heat rate achieved by using the insulation?

3.72 A composite spherical shell of inner radius $r_1 = 0.25$ m is constructed from lead of outer radius $r_2 = 0.30$ m and AISI 302 stainless steel of outer radius $r_3 = 0.31$ m. The cavity is filled with radioactive wastes that generate heat at a rate of $\dot{q} = 5 \times 10^5 \text{ W/m}^3$. It is proposed to submerge the container in oceanic waters that are at a temperature of $T_\infty = 10°\text{C}$ and provide a uniform convection coefficient of $h = 500 \text{ W/m}^2 \cdot \text{K}$ at the outer surface of the container. Are there any problems associated with this proposal?

3.73 The energy transferred from the anterior chamber of the eye through the cornea varies considerably depending on whether a contact lens is worn. Treat the eye as a spherical system and assume the system to be at steady state. The convection coefficient h_o is unchanged with and without the contact lens in place. The cornea and the lens cover one-third of the spherical surface area.

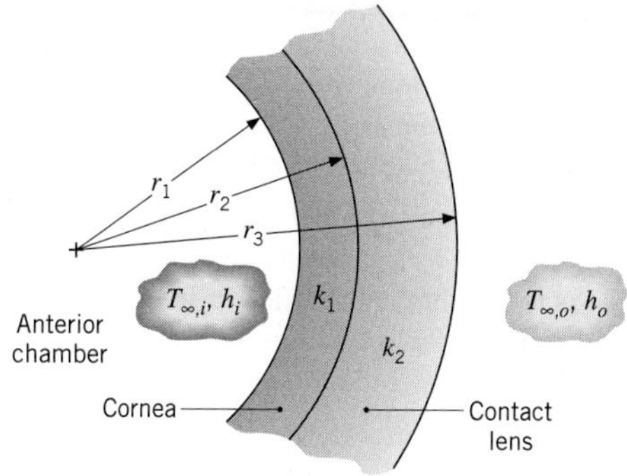

Values of the parameters representing this situation are as follows:

$r_1 = 10.2$ mm	$r_2 = 12.7$ mm
$r_3 = 16.5$ mm	$T_{\infty,o} = 21°\text{C}$
$T_{\infty,i} = 37°\text{C}$	$k_2 = 0.80 \text{ W/m} \cdot \text{K}$
$k_1 = 0.35 \text{ W/m} \cdot \text{K}$	$h_o = 6 \text{ W/m}^2 \cdot \text{K}$
$h_i = 12 \text{ W/m}^2 \cdot \text{K}$	

(a) Construct the thermal circuits, labeling all potentials and flows for the systems excluding the contact lens and including the contact lens. Write resistance elements in terms of appropriate parameters.

(b) Determine the heat loss from the anterior chamber with and without the contact lens in place.

(c) Discuss the implication of your results.

3.74 The outer surface of a hollow sphere of radius r_2 is subjected to a uniform heat flux q''_2. The inner surface at r_1 is held at a constant temperature $T_{s,1}$.

(a) Develop an expression for the temperature distribution $T(r)$ in the sphere wall in terms of q''_2, $T_{s,1}$, r_1, r_2, and the thermal conductivity of the wall material k.

(b) If the inner and outer tube radii are $r_1 = 50$ mm and $r_2 = 100$ mm, what heat flux q''_2 is required to maintain the outer surface at $T_{s,2} = 50°C$, while the inner surface is at $T_{s,1} = 20°C$? The thermal conductivity of the wall material is $k = 10$ W/m·K.

3.75 A spherical shell of inner and outer radii r_i and r_o, respectively, is filled with a heat-generating material that provides for a uniform volumetric generation rate (W/m^3) of $\dot{q}$. The outer surface of the shell is exposed to a fluid having a temperature T_∞ and a convection coefficient h. Obtain an expression for the steady-state temperature distribution $T(r)$ in the shell, expressing your result in terms of r_i, r_o, $\dot{q}$, h, T_∞, and the thermal conductivity k of the shell material.

3.76 A spherical tank of 3-m diameter contains a liquified-petroleum gas at −60°C. Insulation with a thermal conductivity of 0.06 W/m·K and thickness 250 mm is applied to the tank to reduce the heat gain.

(a) Determine the radial position in the insulation layer at which the temperature is 0°C when the ambient air temperature is 20°C and the convection coefficient on the outer surface is 6 W/m^2·K.

(b) If the insulation is pervious to moisture from the atmospheric air, what conclusions can you reach about the formation of ice in the insulation? What effect will ice formation have on heat gain to the LP gas? How could this situation be avoided?

3.77 A transistor, which may be approximated as a hemispherical heat source of radius $r_o = 0.1$ mm, is embedded in a large silicon substrate ($k = 125$ W/m·K) and dissipates heat at a rate q. All boundaries of the silicon are maintained at an ambient temperature of $T_\infty = 27°C$, except for the top surface, which is well insulated.

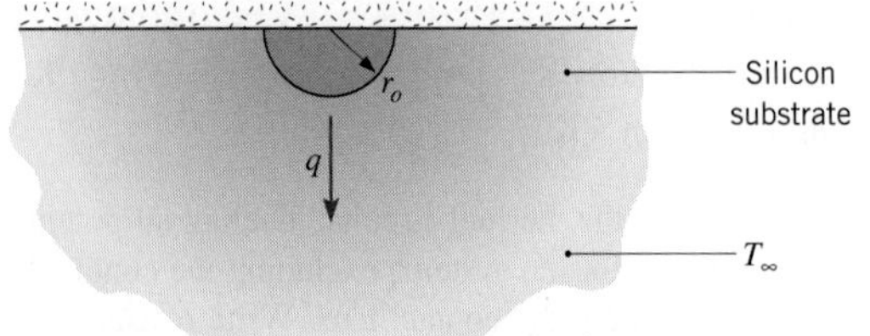

Obtain a general expression for the substrate temperature distribution and evaluate the surface temperature of the heat source for $q = 4$ W.

3.78 One modality for destroying malignant tissue involves imbedding a small spherical heat source of radius r_o within the tissue and maintaining local temperatures above a critical value T_c for an extended period. Tissue that is well removed from the source may be assumed to remain at normal body temperature ($T_b = 37°C$). Obtain a general expression for the radial temperature distribution in the tissue under steady-state conditions for which heat is dissipated at a rate q. If $r_o = 0.5$ mm, what heat rate must be supplied to maintain a tissue temperature of $T \geq T_c = 42°C$ in the domain $0.5 \leq r \leq 5$ mm? The tissue thermal conductivity is approximately 0.5 W/m · K. Assume negligible perfusion.

Conduction with Thermal Energy Generation

3.79 The air *inside* a chamber at $T_{\infty,i} = 50°C$ is heated convectively with $h_i = 20$ W/m^2 · K by a 200-mm-thick wall having a thermal conductivity of 4 W/m·K and a uniform heat generation of 1000 W/m^3. To prevent any heat generated within the wall from being lost to the *outside* of the chamber at $T_{\infty,o} = 25°C$ with $h_o = 5$ W/m^2 · K, a very thin electrical strip heater is placed on the outer wall to provide a uniform heat flux, q''_o.

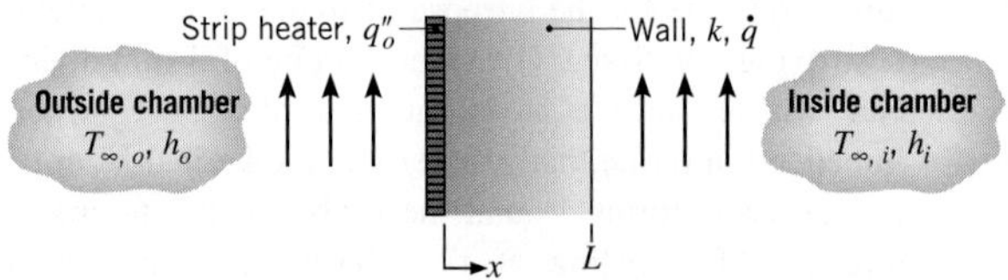

(a) Sketch the temperature distribution in the wall on $T - x$ coordinates for the condition where no heat generated within the wall is lost to the *outside* of the chamber.

(b) What are the temperatures at the wall boundaries, $T(0)$ and $T(L)$, for the conditions of part (a)?

(c) Determine the value of q''_o that must be supplied by the strip heater so that all heat generated within the wall is transferred to the *inside* of the chamber.

(d) If the heat generation in the wall were switched off while the heat flux to the strip heater remained constant, what would be the steady-state temperature, $T(0)$, of the outer wall surface?

3.80 Consider cylindrical and spherical shells with inner and outer surfaces at r_1 and r_2 maintained at uniform temperatures $T_{s,1}$ and $T_{s,2}$, respectively. If there is uniform heat generation within the shells, obtain expressions for the steady-state, one-dimensional radial distributions of the temperature, heat flux, and heat rate. Contrast your results with those summarized in Appendix C.

3.81 A plane wall of thickness 0.1 m and thermal conductivity 25 W/m·K having uniform volumetric heat generation of 0.3 MW/m^3 is insulated on one side, while the other side is exposed to a fluid at 92°C. The convection heat transfer coefficient between the wall and the fluid is 500 W/m^2·K. Determine the maximum temperature in the wall.

3.82 Large, cylindrical bales of hay used to feed livestock in the winter months are $D = 2$ m in diameter and are stored end-to-end in long rows. Microbial energy generation occurs in the hay and can be excessive if the farmer bales the hay in a too-wet condition. Assuming the thermal conductivity of baled hay to be $k = 0.04$ W/m·K, determine the maximum steady-state hay temperature for dry hay ($\dot{q} = 1$ W/m^3), moist hay ($\dot{q} = 10$ W/m^3), and wet hay ($\dot{q} = 100$ W/m^3). Ambient conditions are $T_\infty = 0$°C and $h = 25$ W/m^2·K.

3.83 Consider the cylindrical bales of hay in Problem 3.82. It is proposed to utilize the microbial energy generation associated with wet hay to heat water. Consider a 30-mm diameter, thin-walled tube inserted lengthwise through the middle of a cylindrical bale. The tube carries water at $T_{\infty,i} = 20$°C with $h_i = 200$ W/m^2·K.

(a) Determine the steady-state heat transfer to the water per unit length of tube.

(b) Plot the radial temperature distribution in the hay, $T(r)$.

(c) Plot the heat transfer to the water per unit length of tube for bale diameters of $0.2 \text{ m} \le D \le 2$ m.

3.84 Consider one-dimensional conduction in a plane composite wall. The outer surfaces are exposed to a fluid at 25°C and a convection heat transfer coefficient of 1000 W/m^2·K. The middle wall B experiences uniform heat generation $\dot{q}_B$, while there is no generation in walls A and C. The temperatures at the interfaces are $T_1 = 261$°C and $T_2 = 211$°C.

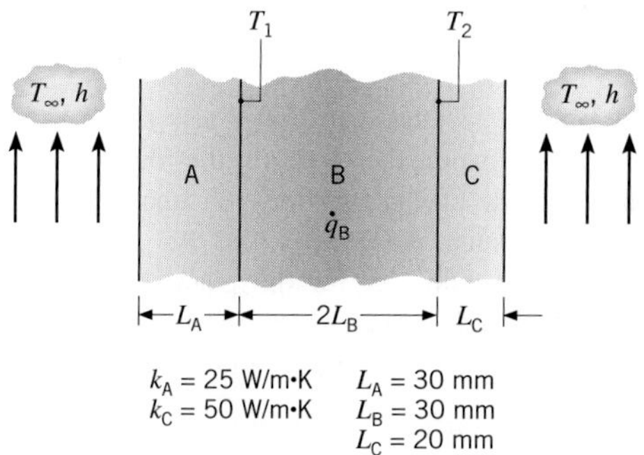

(a) Assuming negligible contact resistance at the interfaces, determine the volumetric heat generation $\dot{q}_B$ and the thermal conductivity k_B.

(b) Plot the temperature distribution, showing its important features.

(c) Consider conditions corresponding to a *loss of coolant* at the exposed surface of material A ($h = 0$). Determine T_1 and T_2 and plot the temperature distribution throughout the system.

3.85 Consider a plane composite wall that is composed of three materials (materials A, B, and C are arranged left to right) of thermal conductivities $k_A = 0.24$ W/m·K, $k_B = 0.13$ W/m·K, and $k_C = 0.50$ W/m·K. The thicknesses of the three sections of the wall are $L_A = 20$ mm, $L_B = 13$ mm, and $L_C = 20$ mm. A contact resistance of $R''_{t,c} = 10^{-2}$ m^2·K/W exists at the interface between materials A and B, as well as at the interface between materials B and C. The left face of the composite wall is insulated, while the right face is exposed to convective conditions characterized by $h = 10$ W/m^2·K, $T_\infty = 20$°C. For Case 1, thermal energy is generated within material A at the rate $\dot{q}_A = 5000$ W/m^3. For Case 2, thermal energy is generated within material C at the rate $\dot{q}_C = 5000$ W/m^3.

(a) Determine the maximum temperature within the composite wall under steady-state conditions for Case 1.

(b) Sketch the steady-state temperature distribution on $T - x$ coordinates for Case 1.

(c) Sketch the steady-state temperature distribution for Case 2 on the same $T - x$ coordinates used for Case 1.

3.86 An air heater may be fabricated by coiling Nichrome wire and passing air in cross flow over the wire. Consider a heater fabricated from wire of diameter $D = 1$ mm, electrical resistivity $\rho_e = 10^{-6}$ Ω·m, thermal conductivity $k = 25$ W/m·K, and emissivity $\varepsilon = 0.20$. The heater is designed to deliver air at a temperature of $T_\infty = 50$°C under flow conditions that provide a convection coefficient of $h = 250$ W/m^2·K for the wire. The temperature of the housing that encloses the wire and through which the air flows is $T_{sur} = 50$°C.

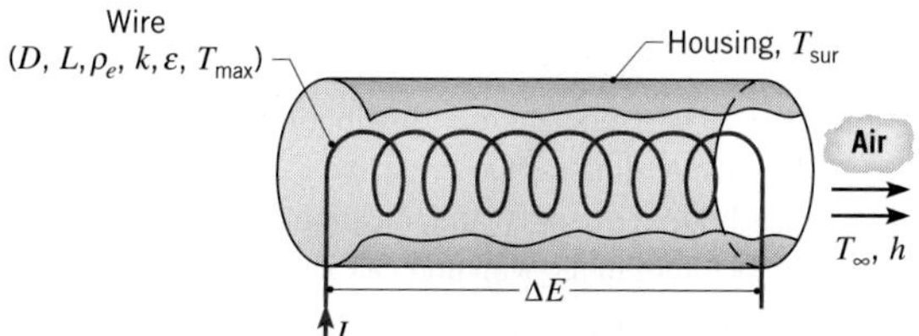

If the maximum allowable temperature of the wire is $T_{max} = 1200$°C, what is the maximum allowable electric current I? If the maximum available voltage is $\Delta E = 110$ V, what is the corresponding length L of wire that may be used in the heater and the power rating of the heater? *Hint:* In your solution, assume negligible temperature variations within the wire, but after obtaining the desired results, assess the validity of this assumption.

3.87 Consider the composite wall of Example 3.7. In the Comments section, temperature distributions in the wall were determined assuming negligible contact resistance between materials A and B. Compute and plot the temperature distributions if the thermal contact resistance is $R''_{t,c} = 10^{-4}\ \text{m}^2 \cdot \text{K/W}$.

3.88 Consider uniform thermal energy generation inside a one-dimensional plane wall of thickness L with one surface held at $T_{s,1}$ and the other surface insulated.

(a) Find an expression for the conduction heat flux to the cold surface and the temperature of the hot surface $T_{s,2}$, expressing your results in terms of k, $\dot{q}$, L, and $T_{s,1}$.

(b) Compare the heat flux found in part (a) with the heat flux associated with a plane wall without energy generation whose surface temperatures are $T_{s,1}$ and $T_{s,2}$.

3.89 A plane wall of thickness $2L$ and thermal conductivity k experiences a uniform volumetric generation rate $\dot{q}$. As shown in the sketch for Case 1, the surface at $x = -L$ is perfectly insulated, while the other surface is maintained at a uniform, constant temperature T_o. For Case 2, a very thin dielectric strip is inserted at the midpoint of the wall ($x = 0$) in order to electrically isolate the two sections, A and B. The thermal resistance of the strip is $R''_t = 0.0005\ \text{m}^2 \cdot \text{K/W}$. The parameters associated with the wall are $k = 50\ \text{W/m} \cdot \text{K}$, $L = 20\ \text{mm}$, $\dot{q} = 5 \times 10^6\ \text{W/m}^3$, and $T_o = 50°\text{C}$.

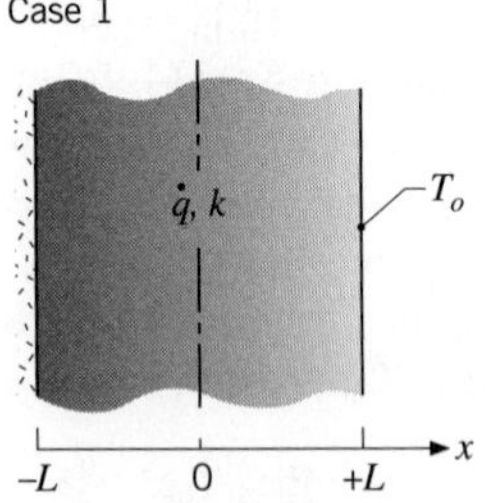

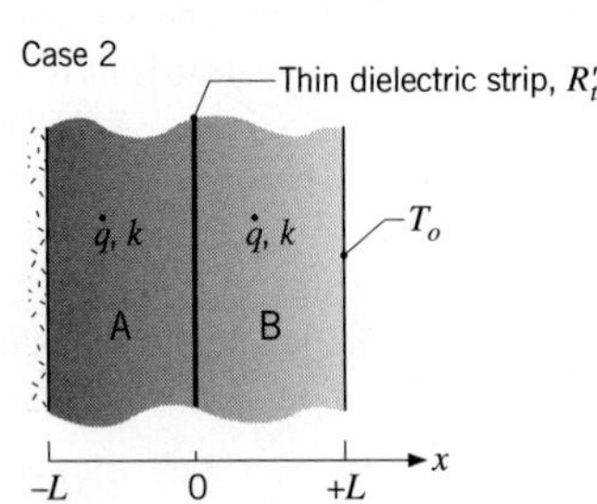

(a) Sketch the temperature distribution for Case 1 on $T - x$ coordinates. Describe the key features of this distribution. Identify the location of the maximum temperature in the wall and calculate this temperature.

(b) Sketch the temperature distribution for Case 2 on the same $T - x$ coordinates. Describe the key features of this distribution.

(c) What is the temperature difference between the two walls at $x = 0$ for Case 2?

(d) What is the location of the maximum temperature in the composite wall of Case 2? Calculate this temperature.

3.90 A nuclear fuel element of thickness $2L$ is covered with a steel cladding of thickness b. Heat generated within the nuclear fuel at a rate $\dot{q}$ is removed by a fluid at T_∞, which adjoins one surface and is characterized by a convection coefficient h. The other surface is well insulated, and the fuel and steel have thermal conductivities of k_f and k_s, respectively.

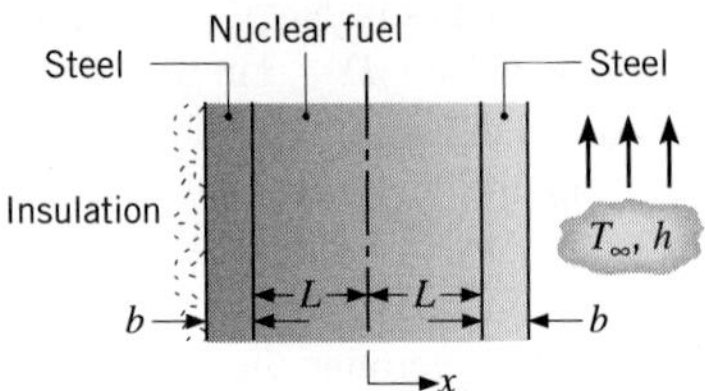

(a) Obtain an equation for the temperature distribution $T(x)$ in the nuclear fuel. Express your results in terms of $\dot{q}$, k_f, L, b, k_s, h, and T_∞.

(b) Sketch the temperature distribution $T(x)$ for the entire system.

3.91 Consider the clad fuel element of Problem 3.90.

(a) Using appropriate relations from Tables C.1 and C.2, obtain an expression for the temperature distribution $T(x)$ in the fuel element. For $k_f = 60\ \text{W/m} \cdot \text{K}$, $L = 15\ \text{mm}$, $b = 3\ \text{mm}$, $k_s = 15\ \text{W/m} \cdot \text{K}$, $h = 10{,}000\ \text{W/m}^2 \cdot \text{K}$, and $T_\infty = 200°\text{C}$, what are the largest and smallest temperatures in the fuel element if heat is generated uniformly at a volumetric rate of $\dot{q} = 2 \times 10^7\ \text{W/m}^3$? What are the corresponding locations?

(b) If the insulation is removed and equivalent convection conditions are maintained at each surface, what is the corresponding form of the temperature distribution in the fuel element? For the conditions of part (a), what are the largest and smallest temperatures in the fuel? What are the corresponding locations?

(c) For the conditions of parts (a) and (b), plot the temperature distributions in the fuel element.

3.92 In Problem 3.79 the strip heater acts to *guard* against heat losses from the wall to the outside, and the required heat flux q''_o depends on chamber operating conditions such as $\dot{q}$ and $T_{\infty,i}$. As a first step in designing a controller for the guard heater, compute and plot q''_o and $T(0)$ as a function of $\dot{q}$ for $200 \leq \dot{q} \leq 2000\ \text{W/m}^3$ and $T_{\infty,i} = 30$, 50, and 70°C.

3.93 The exposed surface ($x = 0$) of a plane wall of thermal conductivity k is subjected to microwave radiation that causes volumetric heating to vary as

$$\dot{q}(x) = \dot{q}_o \left(1 - \frac{x}{L}\right)$$

where $\dot{q}_o$ (W/m^3) is a constant. The boundary at $x = L$ is perfectly insulated, while the exposed surface is maintained at a constant temperature T_o. Determine the temperature distribution $T(x)$ in terms of x, L, k, $\dot{q}_o$, and T_o.

3.94 A quartz window of thickness L serves as a viewing port in a furnace used for annealing steel. The inner surface ($x = 0$) of the window is irradiated with a uniform heat flux q''_o due to emission from hot gases in the furnace. A fraction, β, of this radiation may be assumed to be absorbed at the inner surface, while the remaining radiation is partially absorbed as it passes through the quartz. The volumetric heat generation due to this absorption may be described by an expression of the form

$$\dot{q}(x) = (1 - \beta)q''_o\alpha e^{-\alpha x}$$

where α is the absorption coefficient of the quartz. Convection heat transfer occurs from the outer surface ($x = L$) of the window to ambient air at T_∞ and is characterized by the convection coefficient h. Convection and radiation emission from the inner surface may be neglected, along with radiation emission from the outer surface. Determine the temperature distribution in the quartz, expressing your result in terms of the foregoing parameters.

3.95 For the conditions described in Problem 1.44, determine the temperature distribution, $T(r)$, in the container, expressing your result in terms of $\dot{q}_o$, r_o, T_∞, h, and the thermal conductivity k of the radioactive wastes.

3.96 A cylindrical shell of inner and outer radii, r_i and r_o, respectively, is filled with a heat-generating material that provides a uniform volumetric generation rate (W/m^3) of $\dot{q}$. The inner surface is insulated, while the outer surface of the shell is exposed to a fluid at T_∞ and a convection coefficient h.

(a) Obtain an expression for the steady-state temperature distribution $T(r)$ in the shell, expressing your result in terms of r_i, r_o, $\dot{q}$, h, T_∞, and the thermal conductivity k of the shell material.

(b) Determine an expression for the heat rate, $q'(r_o)$, at the outer radius of the shell in terms of $\dot{q}$ and shell dimensions.

3.97 The cross section of a long cylindrical fuel element in a nuclear reactor is shown. Energy generation occurs uniformly in the thorium fuel rod, which is of diameter $D = 25$ mm and is wrapped in a thin aluminum cladding.

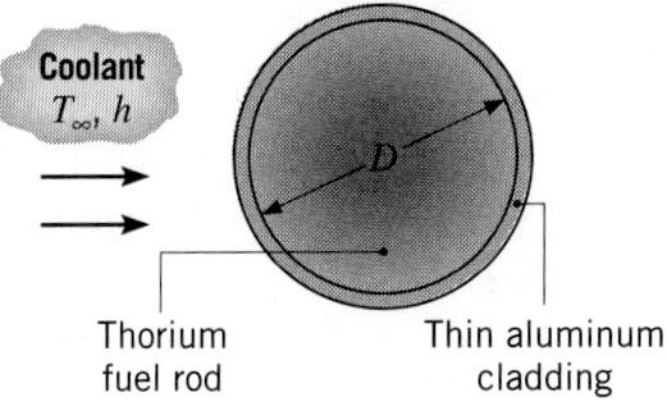

(a) It is proposed that, under steady-state conditions, the system operates with a generation rate of $\dot{q} = 7 \times 10^8$ W/m^3 and cooling system characteristics of $T_\infty = 95$°C and $h = 7000$ W/m$^2 \cdot$ K. Is this proposal satisfactory?

[(b)] Explore the effect of variations in $\dot{q}$ and h by plotting temperature distributions $T(r)$ for a range of parameter values. Suggest an envelope of acceptable operating conditions.

3.98 A nuclear reactor fuel element consists of a solid cylindrical pin of radius r_1 and thermal conductivity k_f. The fuel pin is in good contact with a cladding material of outer radius r_2 and thermal conductivity k_c. Consider steady-state conditions for which uniform heat generation occurs within the fuel at a volumetric rate $\dot{q}$ and the outer surface of the cladding is exposed to a coolant that is characterized by a temperature T_∞ and a convection coefficient h.

(a) Obtain equations for the temperature distributions $T_f(r)$ and $T_c(r)$ in the fuel and cladding, respectively. Express your results exclusively in terms of the foregoing variables.

(b) Consider a uranium oxide fuel pin for which $k_f = 2$ W/m $\cdot$ K and $r_1 = 6$ mm and cladding for which $k_c = 25$ W/m$\cdot$K and $r_2 = 9$ mm. If $\dot{q} = 2 \times 10^8$ W/m^3, $h = 2000$ W/m$^2\cdot$K, and $T_\infty = 300$ K, what is the maximum temperature in the fuel element?

[(c)] Compute and plot the temperature distribution, $T(r)$, for values of $h = 2000$, 5000, and 10,000 W/m$^2\cdot$K. If the operator wishes to maintain the centerline temperature of the fuel element below 1000 K, can she do so by adjusting the coolant flow and hence the value of h?

3.99 Consider the configuration of Example 3.8, where uniform volumetric heating within a stainless steel tube is induced by an electric current and heat is transferred by convection to air flowing through the tube. The tube wall has inner and outer radii of $r_1 = 25$ mm and $r_2 = 35$ mm, a thermal conductivity of $k = 15$ W/m$\cdot$K, an electrical resistivity of $\rho_e = 0.7 \times 10^{-6}$ $\Omega\cdot$m, and a maximum allowable operating temperature of 1400 K.

(a) Assuming the outer tube surface to be perfectly insulated and the airflow to be characterized by a temperature and convection coefficient of $T_{\infty,1} = 400$ K and $h_1 = 100$ W/m$^2\cdot$K, determine the maximum allowable electric current I.

[(b)] Compute and plot the radial temperature distribution in the tube wall for the electric current of part (a) and three values of h_1 (100, 500, and 1000 W/m$^2\cdot$K). For each value of h_1, determine the rate of heat transfer to the air per unit length of tube.

(c) In practice, even the best of insulating materials would be unable to maintain adiabatic conditions at the outer tube surface. Consider use of a refractory insulating material of thermal conductivity $k = 1.0$ W/m · K and neglect radiation exchange at its outer surface. For $h_1 = 100$ W/m^2 · K and the maximum allowable current determined in part (a), compute and plot the temperature distribution in the *composite* wall for two values of the insulation thickness ($\delta = 25$ and 50 mm). The outer surface of the insulation is exposed to room air for which $T_{\infty,2} = 300$ K and $h_2 = 25$ W/m^2·K. For each insulation thickness, determine the rate of heat transfer per unit tube length to the inner airflow and the ambient air.

3.100 A high-temperature, gas-cooled nuclear reactor consists of a composite cylindrical wall for which a thorium fuel element ($k \approx 57$ W/m · K) is encased in graphite ($k \approx 3$ W/m·K) and gaseous helium flows through an annular coolant channel. Consider conditions for which the helium temperature is $T_\infty = 600$ K and the convection coefficient at the outer surface of the graphite is $h = 2000$ W/m^2·K.

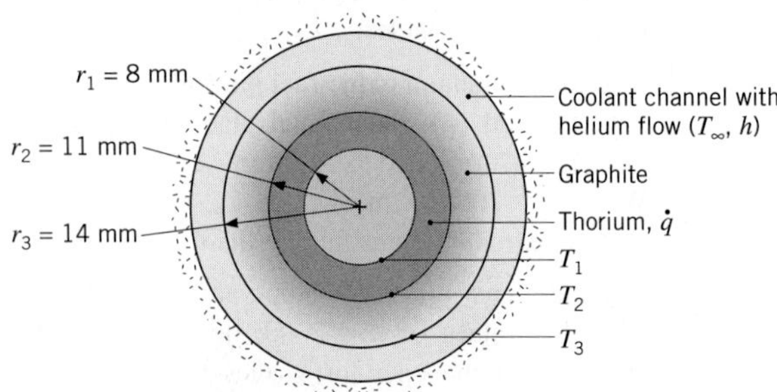

(a) If thermal energy is uniformly generated in the fuel element at a rate $\dot{q} = 10^8$ W/m^3, what are the temperatures T_1 and T_2 at the inner and outer surfaces, respectively, of the fuel element?

(b) Compute and plot the temperature distribution in the composite wall for selected values of $\dot{q}$. What is the maximum allowable value of $\dot{q}$?

3.101 A long cylindrical rod of diameter 200 mm with thermal conductivity of 0.5 W/m·K experiences uniform volumetric heat generation of 24,000 W/m^3. The rod is encapsulated by a circular sleeve having an outer diameter of 400 mm and a thermal conductivity of 4 W/m · K. The outer surface of the sleeve is exposed to cross flow of air at 27°C with a convection coefficient of 25 W/m^2·K.

(a) Find the temperature at the interface between the rod and sleeve and on the outer surface.

(b) What is the temperature at the center of the rod?

3.102 A radioactive material of thermal conductivity k is cast as a solid sphere of radius r_o and placed in a liquid bath for which the temperature T_∞ and convection coefficient h are known. Heat is uniformly generated within the solid at a volumetric rate of $\dot{q}$. Obtain the steady-state radial temperature distribution in the solid, expressing your result in terms of r_o, $\dot{q}$, k, h, and T_∞.

3.103 Radioactive wastes are packed in a thin-walled spherical container. The wastes generate thermal energy nonuniformly according to the relation $\dot{q} = \dot{q}_o[1 - (r/r_o)^2]$ where $\dot{q}$ is the local rate of energy generation per unit volume, $\dot{q}$ is a constant, and r_o is the radius of the container. Steady-state conditions are maintained by submerging the container in a liquid that is at T_∞ and provides a uniform convection coefficient h.

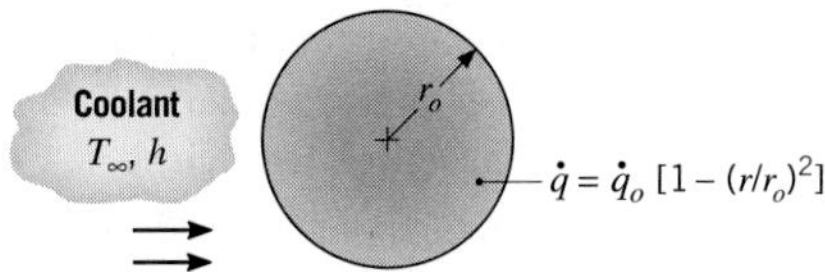

Determine the temperature distribution, $T(r)$, in the container. Express your result in terms of $\dot{q}_o$, r_o, T_∞, h, and the thermal conductivity k of the radioactive wastes.

3.104 Radioactive wastes ($k_{rw} = 20$ W/m·K) are stored in a spherical, stainless steel ($k_{ss} = 15$ W/m·K) container of inner and outer radii equal to $r_i = 0.5$ m and $r_o = 0.6$ m. Heat is generated volumetrically within the wastes at a uniform rate of $\dot{q} = 10^5$ W/m^3, and the outer surface of the container is exposed to a water flow for which $h = 1000$ W/m^2·K and $T_\infty = 25$°C.

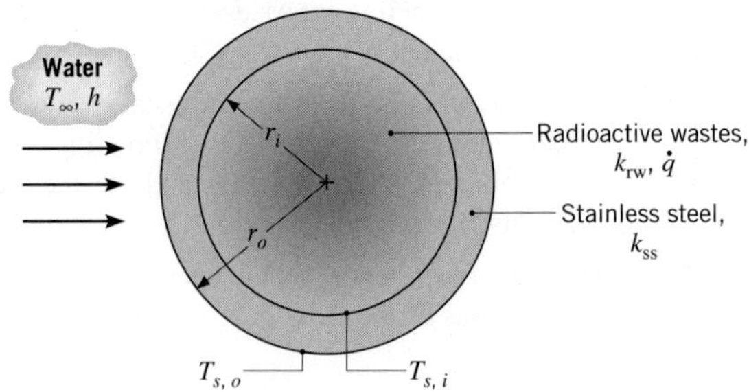

(a) Evaluate the steady-state outer surface temperature, $T_{s,o}$.

(b) Evaluate the steady-state inner surface temperature, $T_{s,i}$.

(c) Obtain an expression for the temperature distribution, $T(r)$, in the radioactive wastes. Express your result in terms of r_i, $T_{s,i}$, k_{rw}, and $\dot{q}$. Evaluate the temperature at $r = 0$.

(d) A proposed extension of the foregoing design involves storing waste materials having the same thermal conductivity but twice the heat generation ($\dot{q} = 2 \times 10^5$ W/m^3) in a stainless steel container of equivalent inner radius ($r_i = 0.5$ m). Safety considerations dictate that the maximum system temperature not exceed 475°C and that the container wall thickness be no less than $t = 0.04$ m and preferably at or close to the original design ($t = 0.1$ m). Assess the effect of varying the outside convection coefficient to a maximum achievable value of $h = 5000$ W/m$^2 \cdot$K (by increasing the water velocity) and the container wall thickness. Is the proposed extension feasible? If so, recommend suitable operating and design conditions for h and t, respectively.

3.105 Unique characteristics of biologically active materials such as fruits, vegetables, and other products require special care in handling. Following harvest and separation from producing plants, glucose is catabolized to produce carbon dioxide, water vapor, and heat, with attendant internal energy generation. Consider a carton of apples, each of 80-mm diameter, which is ventilated with air at 5°C and a velocity of 0.5 m/s. The corresponding value of the heat transfer coefficient is 7.5 W/m$^2 \cdot$K. Within each apple thermal energy is uniformly generated at a total rate of 4000 J/kg$\cdot$day. The density and thermal conductivity of the apple are 840 kg/m^3 and 0.5 W/m$\cdot$K, respectively.

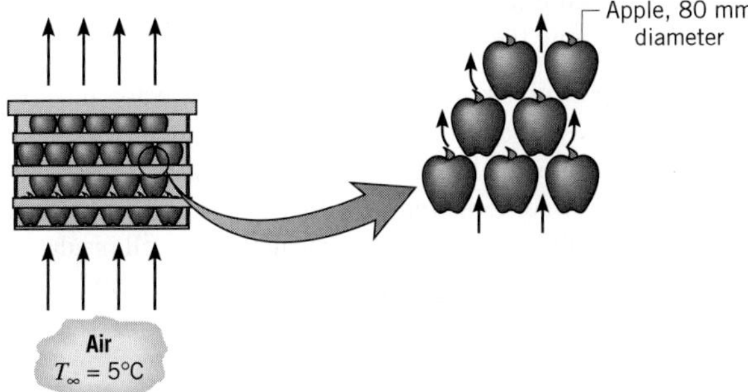

(a) Determine the apple center and surface temperatures.

(b) For the stacked arrangement of apples within the crate, the convection coefficient depends on the velocity as $h = C_1 V^{0.425}$, where $C_1 = 10.1$ W/m$^2 \cdot$K$\cdot$(m/s)$^{0.425}$. Compute and plot the center and surface temperatures as a function of the air velocity for $0.1 \leq V \leq 1$ m/s.

3.106 Consider the plane wall, long cylinder, and sphere shown schematically, each with the same characteristic length a, thermal conductivity k, and uniform volumetric energy generation rate $\dot{q}$.

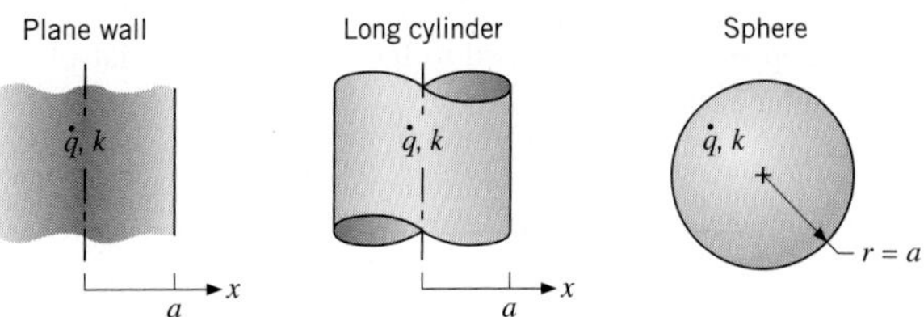

(a) On the same graph, plot the steady-state dimensionless temperature, $[T(x \text{ or } r) - T(a)]/[(\dot{q}a^2)/2k]$, versus the dimensionless characteristic length, x/a or r/a, for each shape.

(b) Which shape has the smallest temperature difference between the center and the surface? Explain this behavior by comparing the ratio of the volume-to-surface area.

(c) Which shape would be preferred for use as a nuclear fuel element? Explain why.

Extended Surfaces

3.107 The radiation heat gage shown in the diagram is made from constantan metal foil, which is coated black and is in the form of a circular disk of radius R and thickness t. The gage is located in an evacuated enclosure. The incident radiation flux absorbed by the foil, q_i'', diffuses toward the outer circumference and into the larger copper ring, which acts as a heat sink at the constant temperature $T(R)$. Two copper lead wires are attached to the center of the foil and to the ring to complete a thermocouple circuit that allows for measurement of the temperature difference between the foil center and the foil edge, $\Delta T = T(0) - T(R)$.

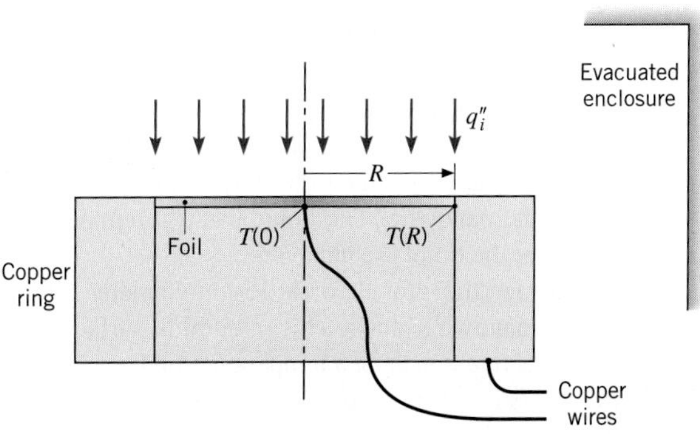

Obtain the differential equation that determines $T(r)$, the temperature distribution in the foil, under steady-state conditions. Solve this equation to obtain an expression relating ΔT to q_i''. You may neglect radiation exchange between the foil and its surroundings.

3.108 Copper tubing is joined to the absorber of a flat-plate solar collector as shown.

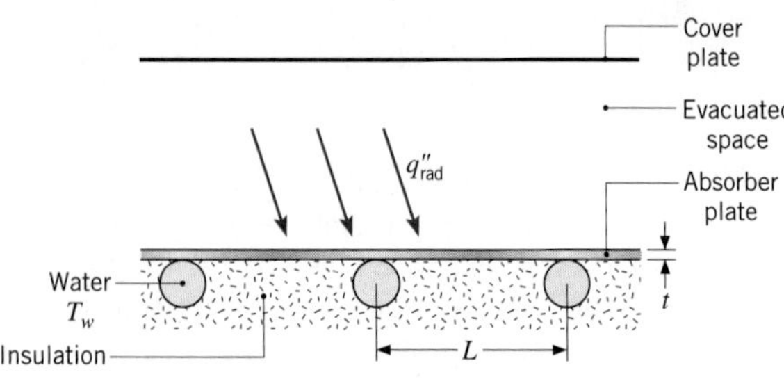

The aluminum alloy (2024-T6) absorber plate is 6 mm thick and well insulated on its bottom. The top surface of the plate is separated from a transparent cover plate by an evacuated space. The tubes are spaced a distance L of 0.20 m from each other, and water is circulated through the tubes to remove the collected energy. The water may be assumed to be at a uniform temperature of $T_w = 60°C$. Under steady-state operating conditions for which the *net* radiation heat flux to the surface is $q''_{rad} = 800\ W/m^2$, what is the maximum temperature on the plate and the heat transfer rate per unit length of tube? Note that q''_{rad} represents the net effect of solar radiation absorption by the absorber plate and radiation exchange between the absorber and cover plates. You may assume the temperature of the absorber plate directly above a tube to be equal to that of the water.

3.109 One method that is used to grow nanowires (nanotubes with solid cores) is to initially deposit a small droplet of a liquid catalyst onto a flat surface. The surface and catalyst are heated and simultaneously exposed to a higher-temperature, low-pressure gas that contains a mixture of chemical species from which the nanowire is to be formed. The catalytic liquid *slowly* absorbs the species from the gas through its top surface and converts these to a solid material that is deposited onto the underlying liquid-solid interface, resulting in construction of the nanowire. The liquid catalyst remains suspended at the tip of the nanowire.

Consider the growth of a 15-nm-diameter silicon carbide nanowire onto a silicon carbide surface. The surface is maintained at a temperature of $T_s = 2400$ K, and the particular liquid catalyst that is used must be maintained in the range $2400\ K \le T_c \le 3000$ K to perform its function. Determine the maximum length of a nanowire that may be grown for conditions characterized by $h = 10^5\ W/m^2 \cdot K$ and $T_\infty = 8000$ K. Assume properties of the nanowire are the same as for bulk silicon carbide.

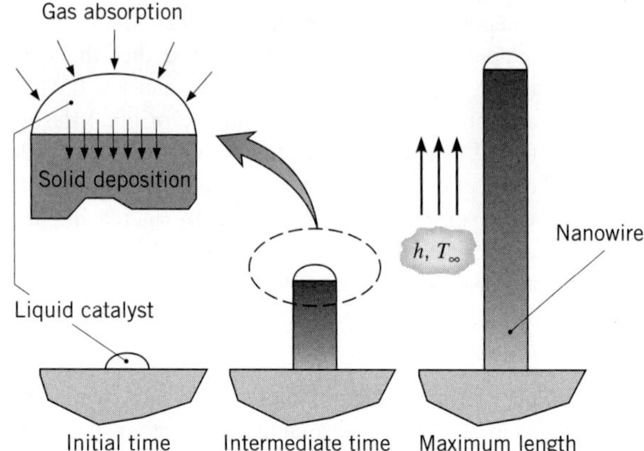

3.110 Consider the manufacture of photovoltaic silicon, as described in Problem 1.42. The thin sheet of silicon is pulled from the pool of molten material *very slowly* and is subjected to an ambient temperature of $T_\infty = 527°C$ within the growth chamber. A convection coefficient of $h = 7.5\ W/m^2 \cdot K$ is associated with the exposed surfaces of the silicon sheet when it is inside the growth chamber. Calculate the maximum allowable velocity of the silicon sheet V_{si}. The latent heat of fusion for silicon is $h_{sf} = 1.8 \times 10^6$ J/kg. It can be assumed that the thermal energy released due to solidification is removed by conduction along the sheet.

3.111 Copper tubing is joined to a solar collector plate of thickness t, and the working fluid maintains the temperature of the plate above the tubes at T_o. There is a uniform net radiation heat flux q''_{rad} to the top surface of the plate, while the bottom surface is well insulated. The top surface is also exposed to a fluid at T_∞ that provides for a uniform convection coefficient h.

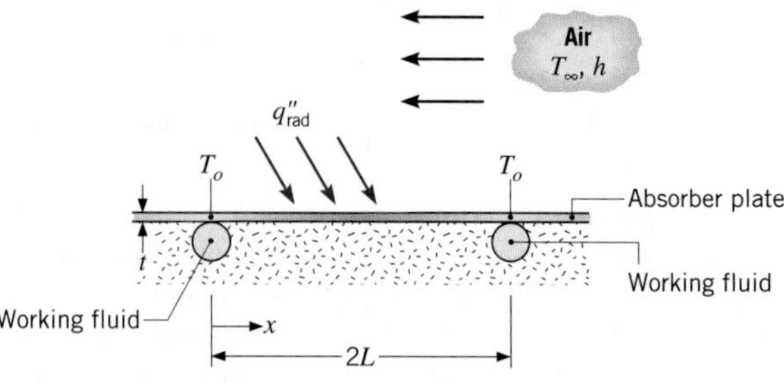

(a) Derive the differential equation that governs the temperature distribution $T(x)$ in the plate.

(b) Obtain a solution to the differential equation for appropriate boundary conditions.

3.112 A thin flat plate of length L, thickness t, and width $W \gg L$ is thermally joined to two large heat sinks that are maintained at a temperature T_o. The bottom of the plate is well insulated, while the net heat flux to the top surface of the plate is known to have a uniform value of q''_o.

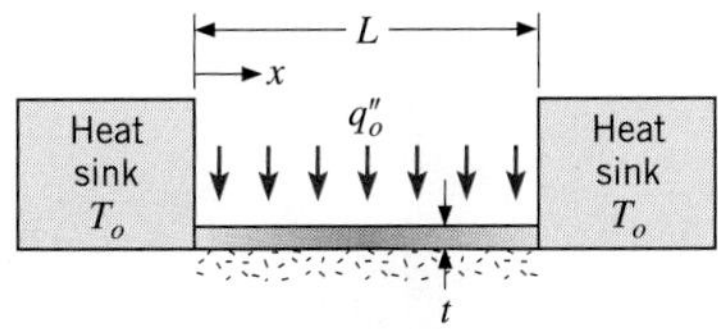

(a) Derive the differential equation that determines the steady-state temperature distribution $T(x)$ in the plate.

(b) Solve the foregoing equation for the temperature distribution, and obtain an expression for the rate of heat transfer from the plate to the heat sinks.

3.113 Consider the flat plate of Problem 3.112, but with the heat sinks at different temperatures, $T(0) = T_o$ and $T(L) = T_L$, and with the bottom surface no longer insulated. Convection heat transfer is now allowed to occur between this surface and a fluid at T_∞, with a convection coefficient h.

(a) Derive the differential equation that determines the steady-state temperature distribution $T(x)$ in the plate.

(b) Solve the foregoing equation for the temperature distribution, and obtain an expression for the rate of heat transfer from the plate to the heat sinks.

(c) For $q''_o = 20{,}000\ \text{W/m}^2$, $T_o = 100°\text{C}$, $T_L = 35°\text{C}$, $T_\infty = 25°\text{C}$, $k = 25\ \text{W/m}\cdot\text{K}$, $h = 50\ \text{W/m}^2\cdot\text{K}$, $L = 100$ mm, $t = 5$ mm, and a plate width of $W = 30$ mm, plot the temperature distribution and determine the sink heat rates, $q_x(0)$ and $q_x(L)$. On the same graph, plot three additional temperature distributions corresponding to changes in the following parameters, with the remaining parameters unchanged: (i) $q''_o = 30{,}000\ \text{W/m}^2$, (ii) $h = 200\ \text{W/m}^2\cdot\text{K}$, and (iii) the value of q''_o for which $q_x(0) = 0$ when $h = 200\ \text{W/m}^2\cdot\text{K}$.

3.114 The temperature of a flowing gas is to be measured with a thermocouple junction and wire stretched between two legs of a *sting*, a wind tunnel test fixture. The junction is formed by butt-welding two wires of different material, as shown in the schematic. For wires of diameter $D = 125\ \mu\text{m}$ and a convection coefficient of $h = 700\ \text{W/m}^2\cdot\text{K}$, determine the minimum separation distance between the two legs of the sting, $L = L_1 + L_2$, to ensure that the sting temperature does not influence the junction temperature and, in turn, invalidate the gas temperature measurement. Consider two different types of thermocouple junctions consisting of (i) copper and constantan wires and (ii) chromel and alumel wires. Evaluate the thermal conductivity of copper and constantan at $T = 300$ K. Use $k_{\text{Ch}} = 19\ \text{W/m}\cdot\text{K}$ and $k_{\text{Al}} = 29\ \text{W/m}\cdot\text{K}$ for the thermal conductivities of the chromel and alumel wires, respectively.

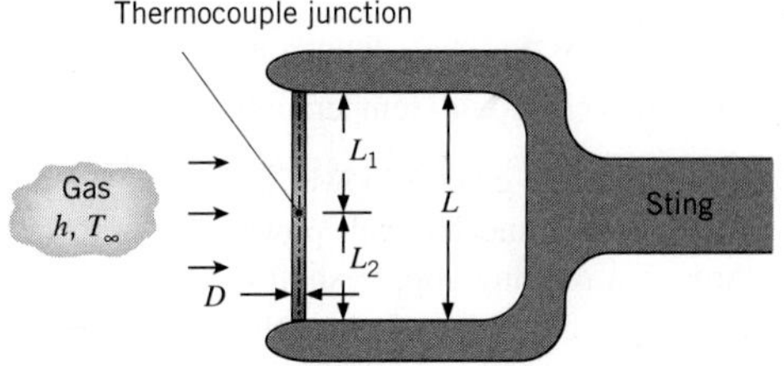

3.115 A bonding operation utilizes a laser to provide a constant heat flux, q''_o, across the top surface of a thin adhesive-backed, plastic film to be affixed to a metal strip as shown in the sketch. The metal strip has a thickness $d = 1.25$ mm, and its width is large relative to that of the film. The thermophysical properties of the strip are $\rho = 7850\ \text{kg/m}^3$, $c_p = 435\ \text{J/kg}\cdot\text{K}$, and $k = 60\ \text{W/m}\cdot\text{K}$. The thermal resistance of the plastic film of width $w_1 = 40$ mm is negligible. The upper and lower surfaces of the strip (including the plastic film) experience convection with air at 25°C and a convection coefficient of $10\ \text{W/m}^2\cdot\text{K}$. The strip and film are very long in the direction normal to the page. Assume the edges of the metal strip are at the air temperature (T_∞).

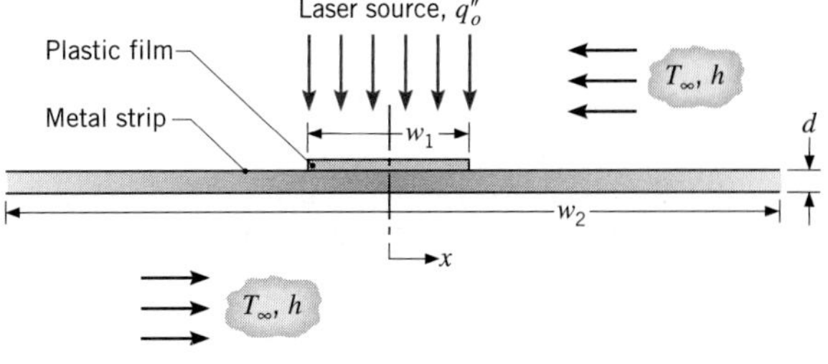

(a) Derive an expression for the temperature distribution in the portion of the steel strip with the plastic film ($-w_1/2 \le x \le +w_1/2$).

(b) If the heat flux provided by the laser is 10,000 W/m², determine the temperature of the plastic film at the center ($x = 0$) and its edges ($x = \pm w_1/2$).

(c) Plot the temperature distribution for the entire strip and point out its special features.

3.116 A thin metallic wire of thermal conductivity k, diameter D, and length $2L$ is annealed by passing an electrical current through the wire to induce a uniform volumetric heat generation $\dot{q}$. The ambient air around the wire is at a temperature T_∞, while the ends of the wire at $x = \pm L$ are also maintained at T_∞. Heat transfer from the wire to the air is characterized by the convection coefficient h. Obtain expressions for the following:

(a) The steady-state temperature distribution $T(x)$ along the wire,

(b) The maximum wire temperature.

(c) The average wire temperature.

3.117 A motor draws electric power P_{elec} from a supply line and delivers mechanical power P_{mech} to a pump through a rotating copper shaft of thermal conductivity k_s, length L, and diameter D. The motor is mounted on a square pad of width W, thickness t, and thermal conductivity k_p. The surface of the housing exposed to ambient air at T_∞ is of area A_h, and the corresponding convection coefficient is h_h. Opposite ends of the shaft are at temperatures of T_h and T_∞, and heat transfer from the shaft to the ambient air is characterized by the convection coefficient h_s. The base of the pad is at T_∞.

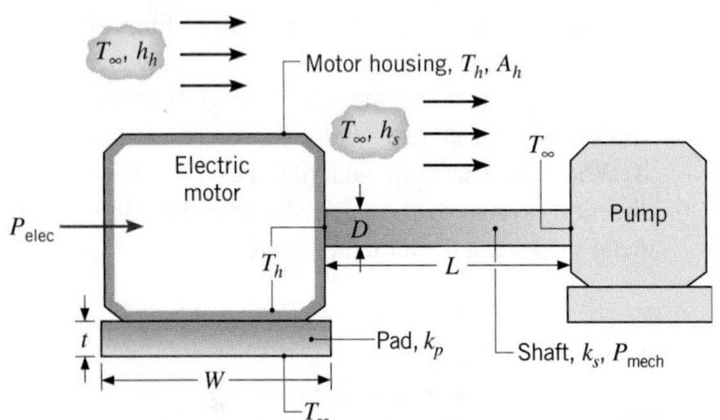

(a) Expressing your result in terms of P_{elec}, P_{mech}, k_s, L, D, W, t, k_p, A_h, h_h, and h_s, obtain an expression for $(T_h - T_\infty)$.

(b) What is the value of T_h if $P_{elec} = 25$ kW, $P_{mech} = 15$ kW, $k_s = 400$ W/m·K, $L = 0.5$ m, $D = 0.05$ m, $W = 0.7$ m, $t = 0.05$ m, $k_p = 0.5$ W/m·K, $A_h = 2$ m^2, $h_h = 10$ W/m^2·K, $h_s = 300$ W/m^2·K, and $T_\infty = 25°$C?

3.118 Consider the fuel cell stack of Problem 1.58. The $t = 0.42$-mm-thick membranes have a nominal thermal conductivity of $k = 0.79$ W/m·K that can be increased to $k_{eff,x} = 15.1$ W/m·K by loading 10%, by volume, carbon nanotubes into the catalyst layers. The membrane experiences uniform volumetric energy generation at a rate of $\dot{q} = 10 \times 10^6$ W/m^3. Air at $T_a = 80°$C provides a convection coefficient of $h_a = 35$ W/m^2·K on one side of the membrane, while hydrogen at $T_h = 80°$C, $h_h = 235$ W/m^2·K flows on the opposite side of the membrane. The flow channels are $2L = 3$ mm wide. The membrane is clamped between bipolar plates, each of which is at a temperature $T_{bp} = 80°$C.

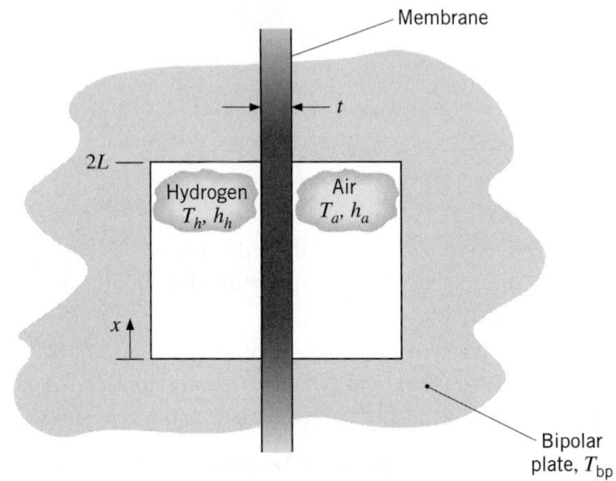

(a) Derive the differential equation that governs the temperature distribution $T(x)$ in the membrane.

(b) Obtain a solution to the differential equation, assuming the membrane is at the bipolar plate temperature at $x = 0$ and $x = 2L$.

(c) Plot the temperature distribution $T(x)$ from $x = 0$ to $x = L$ for carbon nanotube loadings of 0% and 10% by volume. Comment on the ability of the carbon nanotubes to keep the membrane below its softening temperature of 85°C.

3.119 Consider a rod of diameter D, thermal conductivity k, and length $2L$ that is perfectly insulated over one portion of its length, $-L \le x \le 0$, and experiences convection with a fluid (T_∞, h) over the other portion, $0 \le x \le +L$. One end is maintained at T_1, while the other is separated from a heat sink at T_3 by an interfacial thermal contact resistance $R''_{t,c}$.

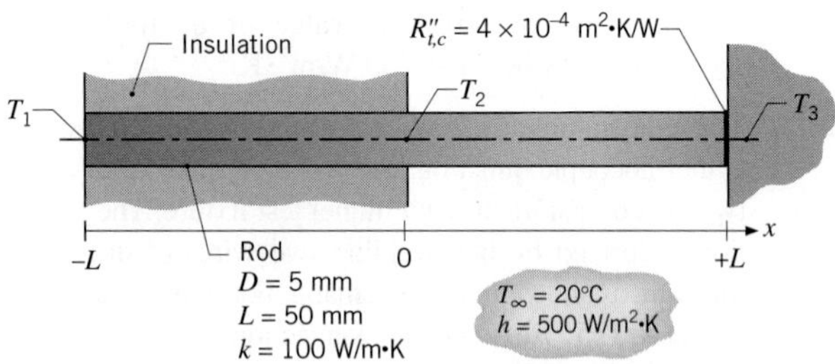

(a) Sketch the temperature distribution on $T - x$ coordinates and identify its key features. Assume that $T_1 > T_3 > T_\infty$.

(b) Derive an expression for the midpoint temperature T_2 in terms of the thermal and geometric parameters of the system.

(c) For $T_1 = 200°C$, $T_3 = 100°C$, and the conditions provided in the schematic, calculate T_2 and plot the temperature distribution. Describe key features of the distribution and compare it to your sketch of part (a).

3.120 A carbon nanotube is suspended across a trench of width $s = 5\ \mu m$ that separates two islands, each at $T_\infty = 300$ K. A focused laser beam irradiates the nanotube at a distance ξ from the left island, delivering $q = 10\ \mu W$ of energy to the nanotube. The nanotube temperature is measured at the midpoint of the trench using a point probe. The measured nanotube temperature is $T_1 = 324.5$ K for $\xi_1 = 1.5\ \mu m$ and $T_2 = 326.4$ K for $\xi_2 = 3.5\ \mu m$.

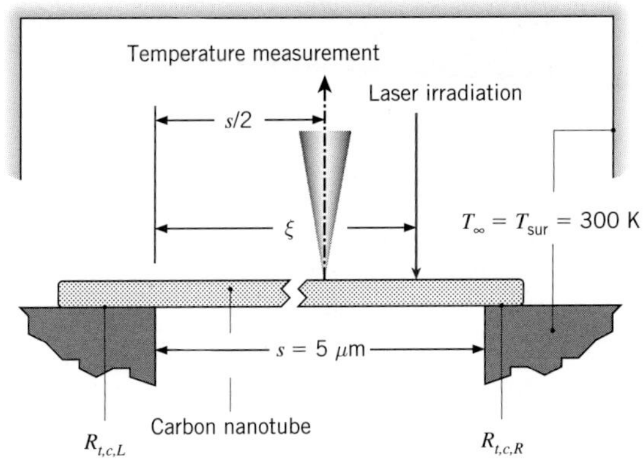

Determine the two contact resistances, $R_{t,c,L}$ and $R_{t,c,R}$ at the left and right ends of the nanotube, respectively. The experiment is performed in a vacuum with $T_{sur} = 300$ K. The nanotube thermal conductivity and diameter are $k_{cn} = 3100$ W/m · K and $D = 14$ nm, respectively.

3.121 A probe of overall length $L = 200$ mm and diameter $D = 12.5$ mm is inserted through a duct wall such that a portion of its length, referred to as the immersion length L_i, is in contact with the water stream whose temperature, $T_{\infty,i}$, is to be determined. The convection coefficients over the immersion and ambient-exposed lengths are $h_i = 1100$ W/m^2·K and $h_o = 10$ W/m^2·K, respectively. The probe has a thermal conductivity of 177 W/m·K and is in poor thermal contact with the duct wall.

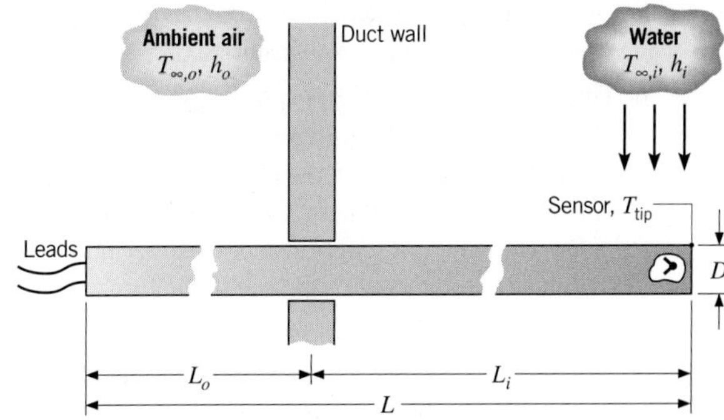

(a) Derive an expression for evaluating the measurement error, $\Delta T_{err} = T_{tip} - T_{\infty,i}$, which is the difference between the tip temperature, T_{tip}, and the water temperature, $T_{\infty,i}$. *Hint:* Define a coordinate system with the origin at the duct wall and treat the probe as two fins extending inward and outward from the duct, but having the same base temperature. Use Case A results from Table 3.4.

(b) With the water and ambient air temperatures at 80 and 20°C, respectively, calculate the measurement error, ΔT_{err}, as a function of immersion length for the conditions $L_i/L = 0.225$, 0.425, and 0.625.

(c) Compute and plot the effects of probe thermal conductivity and water velocity (h_i) on the measurement error.

3.122 A rod of diameter $D = 25$ mm and thermal conductivity $k = 60$ W/m·K protrudes normally from a furnace wall that is at $T_w = 200°C$ and is covered by insulation of thickness $L_{ins} = 200$ mm. The rod is welded to the furnace wall and is used as a hanger for supporting instrumentation cables. To avoid damaging the cables, the temperature of the rod at its exposed surface, T_o, must be maintained below a specified operating limit of $T_{max} = 100°C$. The ambient air temperature is $T_\infty = 25°C$, and the convection coefficient is $h = 15$ W/m^2·K.

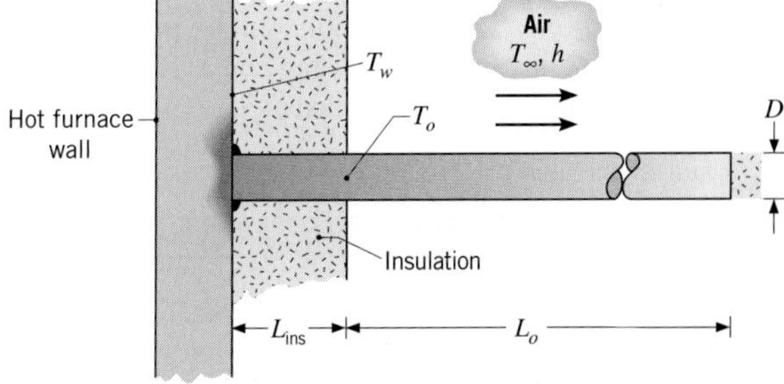

(a) Derive an expression for the exposed surface temperature T_o as a function of the prescribed thermal and geometrical parameters. The rod has an exposed length L_o, and its tip is well insulated.

(b) Will a rod with $L_o = 200$ mm meet the specified operating limit? If not, what design parameters would you change? Consider another material, increasing the thickness of the insulation, and increasing the rod length. Also, consider how you might attach the base of the rod to the furnace wall as a means to reduce T_o.

3.123 A metal rod of length $2L$, diameter D, and thermal conductivity k is inserted into a perfectly insulating wall, exposing one-half of its length to an airstream that is of temperature T_∞ and provides a convection coefficient h at the surface of the rod. An electromagnetic field induces volumetric energy generation at a uniform rate $\dot{q}$ within the *embedded* portion of the rod.

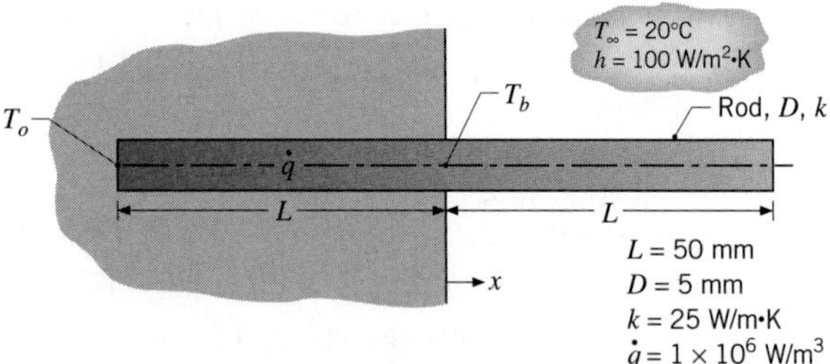

(a) Derive an expression for the steady-state temperature T_b at the base of the exposed half of the rod. The exposed region may be approximated as a very long fin.

(b) Derive an expression for the steady-state temperature T_o at the end of the embedded half of the rod.

(c) Using numerical values provided in the schematic, plot the temperature distribution in the rod and describe key features of the distribution. Does the rod behave as a very long fin?

3.124 A very long rod of 5-mm diameter and uniform thermal conductivity $k = 25$ W/m·K is subjected to a heat treatment process. The center, 30-mm-long portion of the rod within the induction heating coil experiences uniform volumetric heat generation of 7.5×10^6 W/m³.

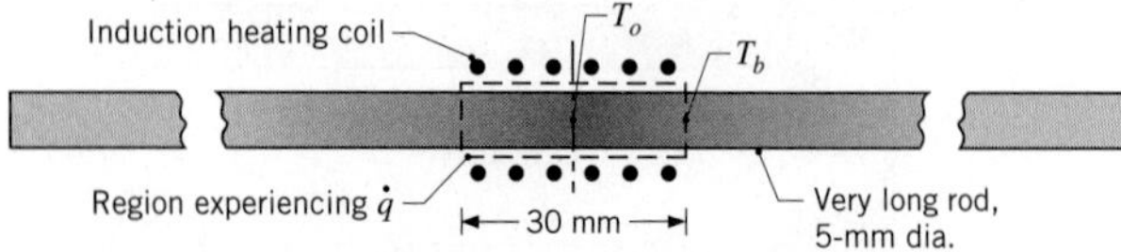

The unheated portions of the rod, which protrude from the heating coil on either side, experience convection with the ambient air at $T_\infty = 20$°C and $h = 10$ W/m²·K. Assume that there is no convection from the surface of the rod within the coil.

(a) Calculate the steady-state temperature T_o of the rod at the midpoint of the heated portion in the coil.

(b) Calculate the temperature of the rod T_b at the edge of the heated portion.

3.125 From Problem 1.71, consider the wire leads connecting the transistor to the circuit board. The leads are of thermal conductivity k, thickness t, width w, and length L. One end of a lead is maintained at a temperature T_c corresponding to the transistor case, while the other end assumes the temperature T_b of the circuit board. During steady-state operation, current flow through the leads provides for uniform volumetric heating in the amount $\dot{q}$, while there is convection cooling to air that is at T_∞ and maintains a convection coefficient h.

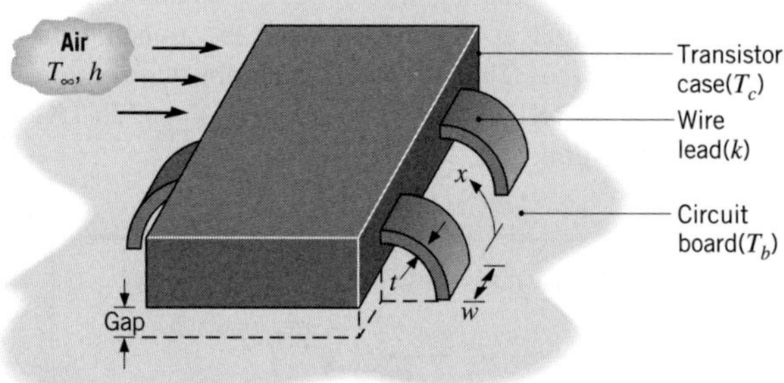

(a) Derive an equation from which the temperature distribution in a wire lead may be determined. List all pertinent assumptions.

(b) Determine the temperature distribution in a wire lead, expressing your results in terms of the prescribed variables.

3.126 Turbine blades mounted to a rotating disc in a gas turbine engine are exposed to a gas stream that is at $T_\infty = 1200$°C and maintains a convection coefficient of $h = 250$ W/m²·K over the blade.

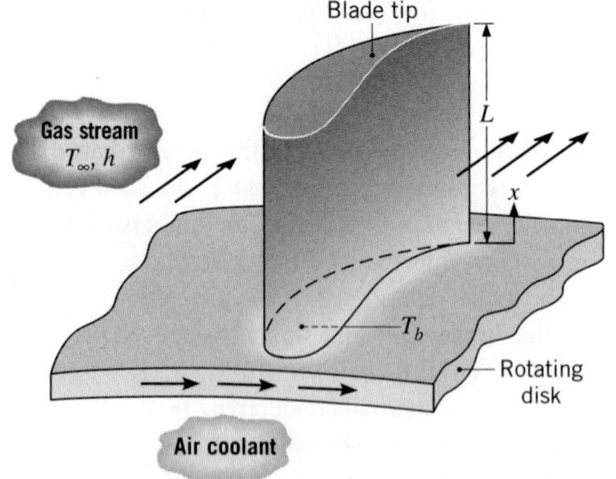

The blades, which are fabricated from Inconel, $k \approx 20$ W/m · K, have a length of $L = 50$ mm. The blade profile has a uniform cross-sectional area of $A_c = 6 \times 10^{-4}$ m^2 and a perimeter of $P = 110$ mm. A proposed blade-cooling scheme, which involves routing air through the supporting disc, is able to maintain the base of each blade at a temperature of $T_b = 300°$C.

(a) If the maximum allowable blade temperature is 1050°C and the blade tip may be assumed to be adiabatic, is the proposed cooling scheme satisfactory?

(b) For the proposed cooling scheme, what is the rate at which heat is transferred from each blade to the coolant?

3.127 In a test to determine the friction coefficient μ associated with a disk brake, one disk and its shaft are rotated at a constant angular velocity ω, while an equivalent disk/shaft assembly is stationary. Each disk has an outer radius of $r_2 = 180$ mm, a shaft radius of $r_1 = 20$ mm, a thickness of $t = 12$ mm, and a thermal conductivity of $k = 15$ W/m·K. A known force F is applied to the system, and the corresponding torque τ required to maintain rotation is measured. The disk contact pressure may be assumed to be uniform (i.e., independent of location on the interface), and the disks may be assumed to be well insulated from the surroundings.

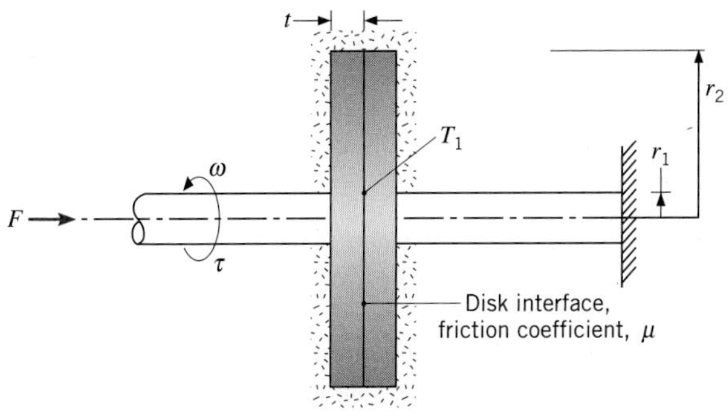

(a) Obtain an expression that may be used to evaluate μ from known quantities.

(b) For the region $r_1 \leq r \leq r_2$, determine the radial temperature distribution $T(r)$ in the disk, where $T(r_1) = T_1$ is presumed to be known.

(c) Consider test conditions for which $F = 200$ N, $\omega = 40$ rad/s, $\tau = 8$ N·m, and $T_1 = 80°$C. Evaluate the friction coefficient and the maximum disk temperature.

3.128 Consider an extended surface of rectangular cross section with heat flow in the longitudinal direction.

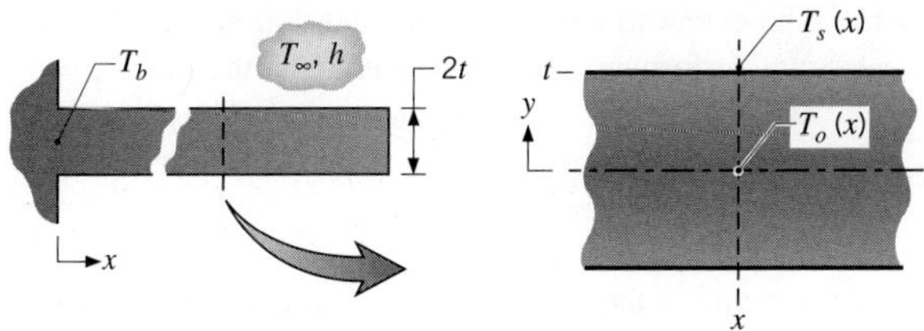

In this problem we seek to determine conditions for which the transverse (y-direction) temperature difference within the extended surface is negligible compared to the temperature difference between the surface and the environment, such that the one-dimensional analysis of Section 3.6.1 is valid.

(a) Assume that the transverse temperature distribution is parabolic and of the form

$$\frac{T(y) - T_o(x)}{T_s(x) - T_o(x)} = \left(\frac{y}{t}\right)^2$$

where $T_s(x)$ is the surface temperature and $T_o(x)$ is the centerline temperature at any x-location. Using Fourier's law, write an expression for the conduction heat flux at the surface, $q''_y(t)$, in terms of T_s and T_o.

(b) Write an expression for the convection heat flux at the surface for the x-location. Equating the two expressions for the heat flux by conduction and convection, identify the parameter that determines the ratio $(T_o - T_s)/(T_s - T_\infty)$.

(c) From the foregoing analysis, develop a criterion for establishing the validity of the one-dimensional assumption used to model an extended surface.

Simple Fins

3.129 A long, circular aluminum rod is attached at one end to a heated wall and transfers heat by convection to a cold fluid.

(a) If the diameter of the rod is tripled, by how much would the rate of heat removal change?

(b) If a copper rod of the same diameter is used in place of the aluminum, by how much would the rate of heat removal change?

3.130 A brass rod 100 mm long and 5 mm in diameter extends horizontally from a casting at 200°C. The rod is in an air environment with $T_\infty = 20°$C and $h = 30$ W/m^2 · K. What is the temperature of the rod 25, 50, and 100 mm from the casting?

3.131 The extent to which the tip condition affects the thermal performance of a fin depends on the fin geometry and thermal conductivity, as well as the convection coefficient. Consider an alloyed aluminum (k = 180 W/m·K) rectangular fin of length L = 10 mm, thickness t = 1 mm, and width $w \gg t$. The base temperature of the fin is T_b = 100°C, and the fin is exposed to a fluid of temperature T_∞ = 25°C.

(a) Assuming a uniform convection coefficient of h = 100 W/m^2 · K over the entire fin surface, determine the fin heat transfer rate per unit width q'_f, efficiency η_f, effectiveness ε_f, thermal resistance per unit width $R'_{t,f}$, and the tip temperature $T(L)$ for Cases A and B of Table 3.4. Contrast your results with those based on an *infinite fin* approximation.

(b) Explore the effect of variations in the convection coefficient on the heat rate for $10 < h < 1000$ W/m^2·K. Also consider the effect of such variations for a stainless steel fin (k = 15 W/m·K).

3.132 A pin fin of uniform, cross-sectional area is fabricated of an aluminum alloy (k = 160 W/m·K). The fin diameter is D = 4 mm, and the fin is exposed to convective conditions characterized by h = 220 W/m^2·K. It is reported that the fin efficiency is η_f = 0.65. Determine the fin length L and the fin effectiveness ε_f. Account for tip convection.

3.133 The extent to which the tip condition affects the thermal performance of a fin depends on the fin geometry and thermal conductivity, as well as the convection coefficient. Consider an alloyed aluminum (k = 180 W/m·K) rectangular fin whose base temperature is T_b = 100°C. The fin is exposed to a fluid of temperature T_∞ = 25°C, and a uniform convection coefficient of h = 100 W/m^2·K may be assumed for the fin surface.

(a) For a fin of length L = 10 mm, thickness t = 1 mm, and width $w \gg t$, determine the fin heat transfer rate per unit width q'_f, efficiency η_f, effectiveness ε_f, thermal resistance per unit width $R'_{t,f}$, and tip temperature $T(L)$ for Cases A and B of Table 3.4. Contrast your results with those based on an *infinite fin* approximation.

(b) Explore the effect of variations in L on the heat rate for $3 < L < 50$ mm. Also consider the effect of such variations for a stainless steel fin (k = 15 W/m·K).

3.134 A straight fin fabricated from 2024 aluminum alloy (k = 185 W/m·K) has a base thickness of t = 3 mm and a length of L = 15 mm. Its base temperature is T_b = 100°C, and it is exposed to a fluid for which T_∞ = 20°C and h = 50 W/m^2·K. For the foregoing conditions and a fin of unit width, compare the fin heat rate, efficiency, and volume for rectangular, triangular, and parabolic profiles.

3.135 Triangular and parabolic straight fins are subjected to the same thermal conditions as the rectangular straight fin of Problem 3.134.

(a) Determine the length of a triangular fin of unit width and base thickness t = 3 mm that will provide the same fin heat rate as the straight rectangular fin. Determine the ratio of the mass of the triangular straight fin to that of the rectangular straight fin.

(b) Repeat part (a) for a parabolic straight fin.

3.136 Two long copper rods of diameter D = 10 mm are soldered together end to end, with solder having a melting point of 650°C. The rods are in air at 25°C with a convection coefficient of 10 W/m^2·K. What is the minimum power input needed to effect the soldering?

3.137 Circular copper rods of diameter D = 1 mm and length L = 25 mm are used to enhance heat transfer from a surface that is maintained at $T_{s,1}$ = 100°C. One end of the rod is attached to this surface (at x = 0), while the other end (x = 25 mm) is joined to a second surface, which is maintained at $T_{s,2}$ = 0°C. Air flowing between the surfaces (and over the rods) is also at a temperature of T_∞ = 0°C, and a convection coefficient of h = 100 W/m^2·K is maintained.

(a) What is the rate of heat transfer by convection from a single copper rod to the air?

(b) What is the total rate of heat transfer from a 1 m × 1 m section of the surface at 100°C, if a bundle of the rods is installed on 4-mm centers?

3.138 During the initial stages of the growth of the nanowire of Problem 3.109, a slight perturbation of the liquid catalyst droplet can cause it to be suspended on the top of the nanowire in an off-center position. The resulting nonuniform deposition of solid at the solid-liquid interface can be manipulated to form engineered shapes such as a *nanospring*, that is characterized by a spring radius r, spring pitch s, overall chord length L_c (length running along the spring), and end-to-end length L, as shown in the sketch. Consider a silicon carbide nanospring of diameter D = 15 nm, r = 30 nm, s = 25 nm, and L_c = 425 nm. From experiments, it is known that the average spring pitch $\bar{s}$ varies with average temperature $\bar{T}$ by the relation $d\bar{s}/d\bar{T}$ = 0.1 nm/K. Using this information, a student suggests that a *nanoactuator* can be constructed by connecting one end of the nanospring to a small heater and raising the temperature of that end of the nano spring above its initial value. Calculate the actuation distance ΔL for conditions where h = 10^6 W/m^2·K, $T_\infty = T_i$ = 25°C, with a base temperature of $T_b = 50°C$. If the base temperature can be controlled to within 1°C, calculate the accuracy to which the actuation distance can be controlled. *Hint:* Assume the spring radius does not change when the spring is heated. The overall spring length may be approximated by the formula,

$$L = \frac{\bar{s}}{2\pi} \frac{L_c}{\sqrt{r^2 + (\bar{s}/2\pi)^2}}$$

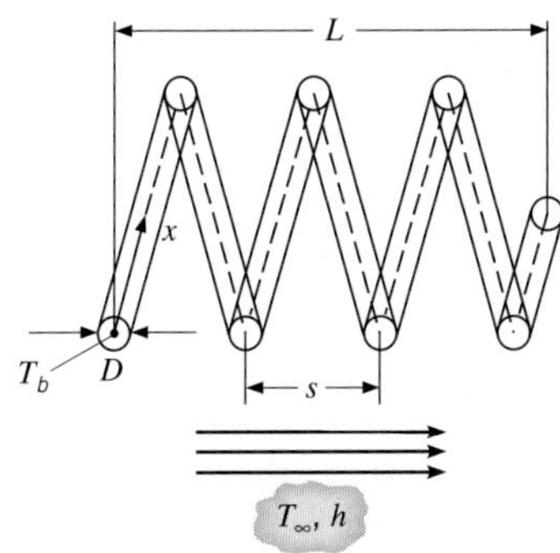

3.139 Consider two long, slender rods of the same diameter but different materials. One end of each rod is attached to a base surface maintained at 100°C, while the surfaces of the rods are exposed to ambient air at 20°C. By traversing the length of each rod with a thermocouple, it was observed that the temperatures of the rods were equal at the positions $x_A = 0.15$ m and $x_B = 0.075$ m, where x is measured from the base surface. If the thermal conductivity of rod A is known to be $k_A = 70$ W/m · K, determine the value of k_B for rod B.

3.140 A 40-mm-long, 2-mm-diameter pin fin is fabricated of an aluminum alloy ($k = 140$ W/m · K).

(a) Determine the fin heat transfer rate for $T_b = 50°C$, $T_\infty = 25°C$, $h = 1000$ W/m^2 · K, and an adiabatic tip condition.

(b) An engineer suggests that by holding the fin tip at a low temperature, the fin heat transfer rate can be increased. For $T(x = L) = 0°C$, determine the new fin heat transfer rate. Other conditions are as in part (a).

(c) Plot the temperature distribution, $T(x)$, over the range $0 \leq x \leq L$ for the adiabatic tip case and the prescribed tip temperature case. Also show the ambient temperature in your graph. Discuss relevant features of the temperature distribution.

(d) Plot the fin heat transfer rate over the range $0 \leq h \leq 1000$ W/m^2·K for the adiabatic tip case and the prescribed tip temperature case. For the prescribed tip temperature case, what would the calculated fin heat transfer rate be if Equation 3.78 were used to determine q_f rather than Equation 3.76?

3.141 An experimental arrangement for measuring the thermal conductivity of solid materials involves the use of two long rods that are equivalent in every respect, except that one is fabricated from a standard material of known thermal conductivity k_A while the other is fabricated from the material whose thermal conductivity k_B is desired. Both rods are attached at one end to a heat source of fixed temperature T_b, are exposed to a fluid of temperature T_∞, and are instrumented with thermocouples to measure the temperature at a fixed distance x_1 from the heat source. If the standard material is aluminum, with $k_A = 200$ W/m · K, and measurements reveal values of $T_A = 75°C$ and $T_B = 60°C$ at x_1 for $T_b = 100°C$ and $T_\infty = 25°C$, what is the thermal conductivity k_B of the test material?

Fin Systems and Arrays

3.142 Finned passages are frequently formed between parallel plates to enhance convection heat transfer in compact heat exchanger cores. An important application is in electronic equipment cooling, where one or more air-cooled stacks are placed between heat-dissipating electrical components. Consider a single stack of rectangular fins of length L and thickness t, with convection conditions corresponding to h and T_∞.

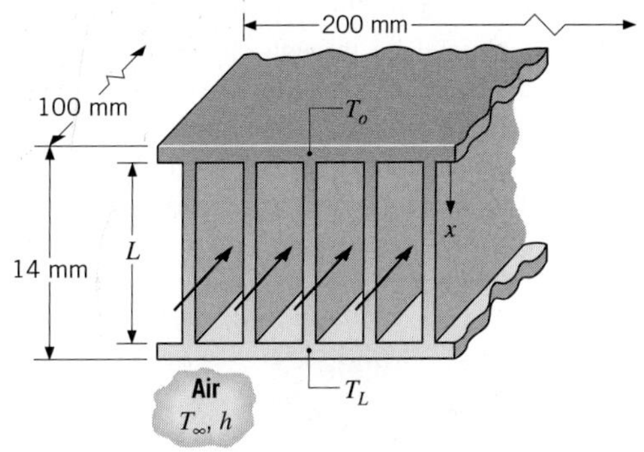

(a) Obtain expressions for the fin heat transfer rates, $q_{f,o}$ and $q_{f,L}$, in terms of the base temperatures, T_o and T_L.

(b) In a specific application, a stack that is 200 mm wide and 100 mm deep contains 50 fins, each of length $L = 12$ mm. The entire stack is made from aluminum, which is everywhere 1.0 mm thick. If temperature limitations associated with electrical components joined to opposite plates dictate maximum allowable plate temperatures of $T_o = 400$ K and $T_L = 350$ K, what are the corresponding maximum power dissipations if $h = 150$ W/m$^2\cdot$K and $T_\infty = 300$ K?

3.143 The fin array of Problem 3.142 is commonly found in *compact heat exchangers,* whose function is to provide a large surface area per unit volume in transferring heat from one fluid to another. Consider conditions for which the second fluid maintains equivalent temperatures at the parallel plates, $T_o = T_L$, thereby establishing symmetry about the midplane of the fin array. The heat exchanger is 1 m long in the direction of the flow of air (first fluid) and 1 m wide in a direction normal to both the airflow and the fin surfaces. The length of the fin passages between adjoining parallel plates is $L = 8$ mm, whereas the fin thermal conductivity and convection coefficient are $k = 200$ W/m$\cdot$K (aluminum) and $h = 150$ W/m$^2\cdot$K, respectively.

(a) If the fin thickness and pitch are $t = 1$ mm and $S = 4$ mm, respectively, what is the value of the thermal resistance $R_{t,o}$ for a one-half section of the fin array?

(b) Subject to the constraints that the fin thickness and pitch may not be less than 0.5 and 3 mm, respectively, assess the effect of changes in t and S.

3.144 An isothermal silicon chip of width $W = 20$ mm on a side is soldered to an aluminum heat sink ($k =$ 180 W/m$\cdot$K) of equivalent width. The heat sink has a base thickness of $L_b = 3$ mm and an array of rectangular fins, each of length $L_f = 15$ mm. Airflow at $T_\infty =$ 20°C is maintained through channels formed by the fins and a cover plate, and for a convection coefficient of $h = 100$ W/m$^2\cdot$K, a minimum fin spacing of 1.8 mm is dictated by limitations on the flow pressure drop. The solder joint has a thermal resistance of $R''_{t,c} = 2 \times 10^{-6}$ m$^2\cdot$K/W.

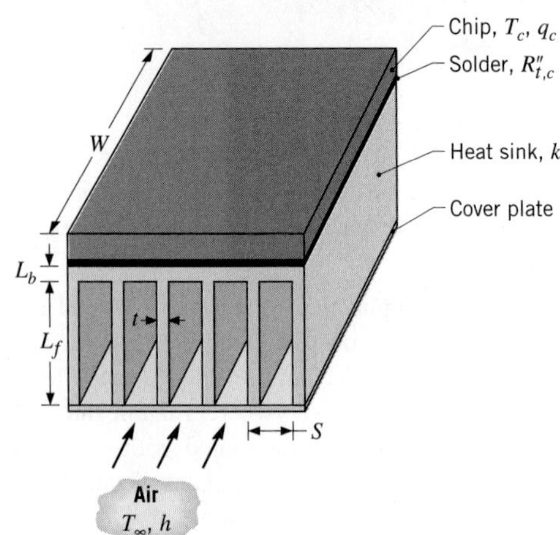

(a) Consider limitations for which the array has $N = 11$ fins, in which case values of the fin thickness $t = 0.182$ mm and pitch $S = 1.982$ mm are obtained from the requirements that $W = (N - 1)S + t$ and $S - t = 1.8$ mm. If the maximum allowable chip temperature is $T_c = 85$°C, what is the corresponding value of the chip power q_c? An adiabatic fin tip condition may be assumed, and airflow along the outer surfaces of the heat sink may be assumed to provide a convection coefficient equivalent to that associated with airflow through the channels.

(b) With $(S - t)$ and h fixed at 1.8 mm and 100 W/m$^2\cdot$K, respectively, explore the effect of increasing the fin thickness by reducing the number of fins. With $N = 11$ and $S - t$ fixed at 1.8 mm, but relaxation of the constraint on the pressure drop, explore the effect of increasing the airflow, and hence the convection coefficient.

3.145 As seen in Problem 3.109, silicon carbide nanowires of diameter $D = 15$ nm can be grown onto a solid silicon carbide surface by carefully depositing droplets of catalyst liquid onto a flat silicon carbide substrate. Silicon carbide nanowires grow upward from the deposited drops, and if the drops are deposited in a pattern, an array of nanowire fins can be grown, forming a silicon carbide *nano-heat sink.* Consider finned and unfinned electronics packages in which an extremely small, 10 μm $\times$ 10 μm electronics device is sandwiched between two $d =$ 100-nm-thick silicon carbide sheets. In both cases, the coolant is a dielectric liquid at 20°C. A heat transfer coefficient of $h = 1 \times 10^5$ W/m$^2\cdot$K exists on the top and bottom of the unfinned package and on all surfaces of the exposed silicon carbide fins, which are each $L =$ 300 nm long. Each nano-heat sink includes a 200 $\times$ 200 array of nanofins. Determine the maximum allowable heat rate that can be generated by the electronic device so that its temperature is maintained at $T_t < 85$°C for the unfinned and finned packages.

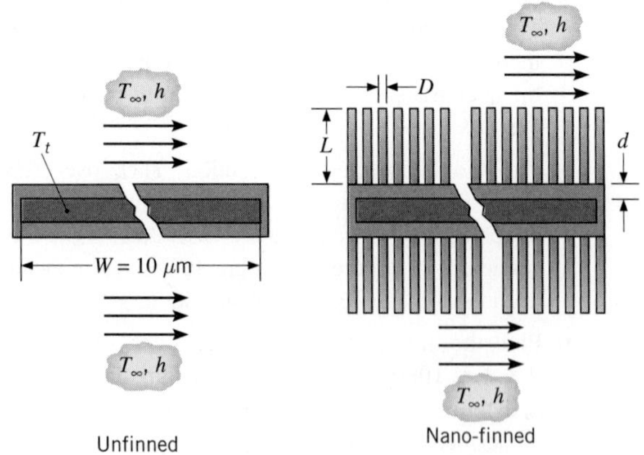

3.146 As more and more components are placed on a single integrated circuit (chip), the amount of heat that is dissipated continues to increase. However, this increase is limited by the maximum allowable chip operating temperature, which is approximately 75°C. To maximize heat dissipation, it is proposed that a 4 × 4 array of copper pin fins be metallurgically joined to the outer surface of a square chip that is 12.7 mm on a side.

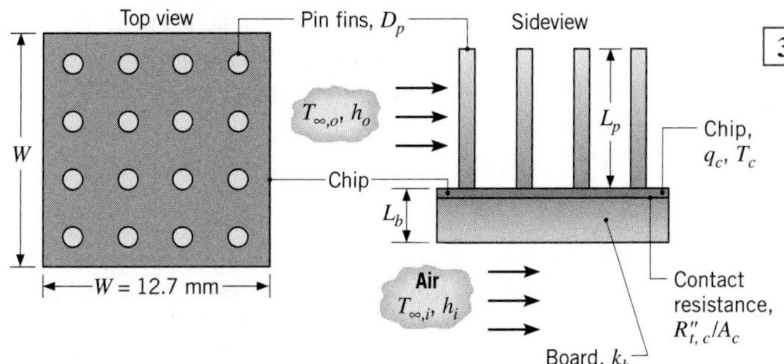

(a) Sketch the equivalent thermal circuit for the pin–chip–board assembly, assuming one-dimensional, steady-state conditions and negligible contact resistance between the pins and the chip. In variable form, label appropriate resistances, temperatures, and heat rates.

(b) For the conditions prescribed in Problem 3.27, what is the maximum rate at which heat can be dissipated in the chip when the pins are in place? That is, what is the value of q_c for T_c = 75°C? The pin diameter and length are D_p = 1.5 mm and L_p = 15 mm.

3.147 A homeowner's wood stove is equipped with a top burner for cooking. The D = 200-mm-diameter burner is fabricated of cast iron (k = 65 W/m·K). The bottom (combustion) side of the burner has 8 straight fins of uniform cross section, arranged as shown in the sketch. A very thin ceramic coating (ε = 0.95) is applied to all surfaces of the burner. The top of the burner is exposed to room conditions ($T_{sur,t} = T_{\infty,t}$ = 20°C, h_t = 40 W/m^2·K), while the bottom of the burner is exposed to combustion conditions ($T_{sur,b} = T_{\infty,b}$ = 450°C, h_b = 50 W/m^2·K). Compare the top surface temperature of the finned burner to that which would exist for a burner without fins. *Hint*: Use the same expression for radiation heat transfer to the bottom of the finned burner as for the burner with no fins.

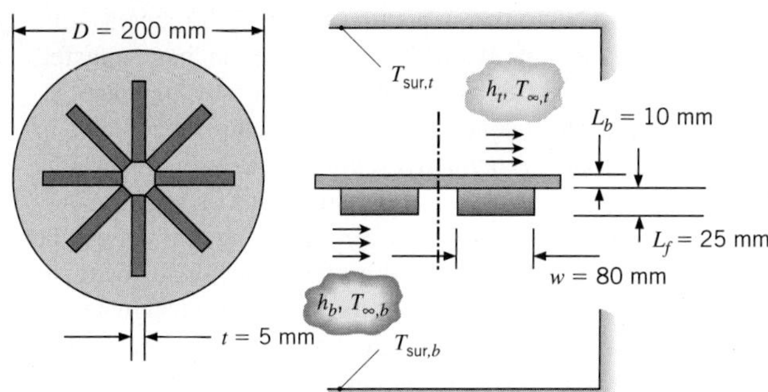

3.148 In Problem 3.146, the prescribed value of h_o = 1000 W/m^2·K is large and characteristic of liquid cooling. In practice it would be far more preferable to use air cooling, for which a reasonable upper limit to the convection coefficient would be h_o = 250 W/m^2·K. Assess the effect of changes in the pin fin geometry on the chip heat rate if the remaining conditions of Problem 3.146, including a maximum allowable chip temperature of 75°C, remain in effect. Parametric variations that may be considered include the total number of pins N in the square array, the pin diameter D_p, and the pin length L_p. However, the product $N^{1/2}D_p$ should not exceed 9 mm to ensure adequate airflow passage through the array. Recommend a design that enhances chip cooling.

3.149 Water is heated by submerging 50-mm-diameter, thin-walled copper tubes in a tank and passing hot combustion gases (T_g = 750 K) through the tubes. To enhance heat transfer to the water, four straight fins of uniform cross section, which form a cross, are inserted in each tube. The fins are 5 mm thick and are also made of copper (k = 400 W/m·K).

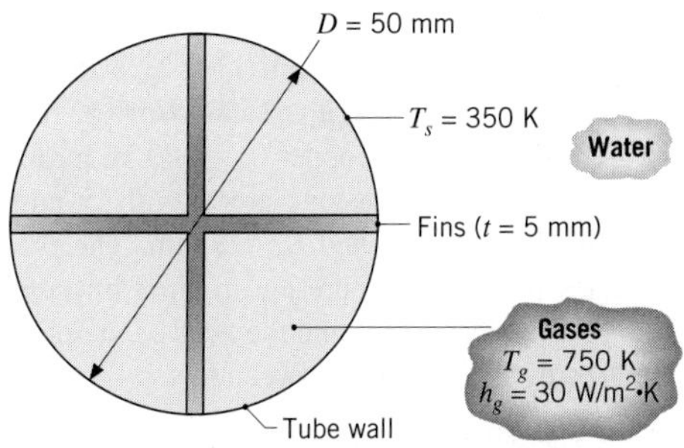

If the tube surface temperature is T_s = 350 K and the gas-side convection coefficient is h_g = 30 W/m^2·K, what is the rate of heat transfer to the water per meter of pipe length?

3.150 As a means of enhancing heat transfer from high-performance logic chips, it is common to attach a *heat sink* to the chip surface in order to increase the surface area available for convection heat transfer. Because of the ease with which it may be manufactured (by taking orthogonal sawcuts in a block of material), an attractive option is to use a heat sink consisting of an array of square fins of width w on a side. The spacing between adjoining fins would be determined by the width of the sawblade, with the sum of this spacing and the fin width designated as the fin pitch S. The method by which the heat sink is joined to the chip would determine the interfacial contact resistance, $R''_{t,c}$.

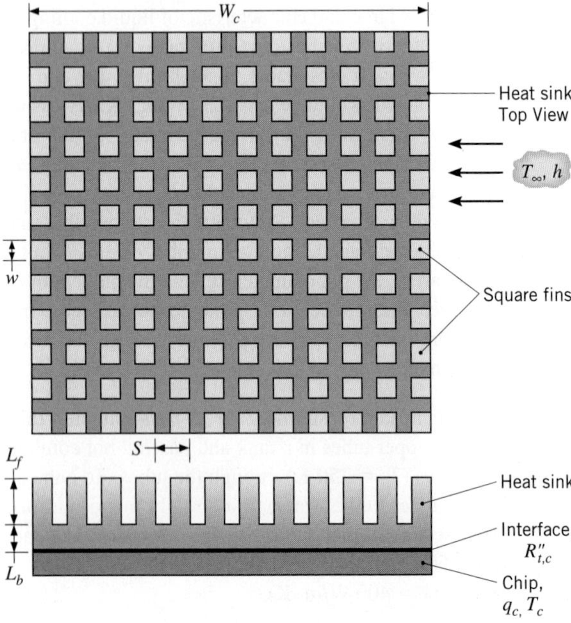

Consider a square chip of width $W_c = 16$ mm and conditions for which cooling is provided by a dielectric liquid with $T_\infty = 25°C$ and $h = 1500$ W/m²·K. The heat sink is fabricated from copper ($k = 400$ W/m·K), and its characteristic dimensions are $w = 0.25$ mm, $S = 0.50$ mm, $L_f = 6$ mm, and $L_b = 3$ mm. The prescribed values of w and S represent minima imposed by manufacturing constraints and the need to maintain adequate flow in the passages between fins.

(a) If a metallurgical joint provides a contact resistance of $R''_{t,c} = 5 \times 10^{-6}$ m²·K/W and the maximum allowable chip temperature is 85°C, what is the maximum allowable chip power dissipation q_c? Assume all of the heat to be transferred through the heat sink.

(b) It may be possible to increase the heat dissipation by increasing w, subject to the constraint that $(S - w) \geq 0.25$ mm, and/or increasing L_f (subject to manufacturing constraints that $L_f \leq 10$ mm). Assess the effect of such changes.

3.151 Because of the large number of devices in today's PC chips, finned heat sinks are often used to maintain the chip at an acceptable operating temperature. Two fin designs are to be evaluated, both of which have base (unfinned) area dimensions of 53 mm × 57 mm. The fins are of square cross section and fabricated from an extruded aluminum alloy with a thermal conductivity of 175 W/m · K. Cooling air may be supplied at 25°C, and the maximum allowable chip temperature is 75°C. Other features of the design and operating conditions are tabulated.

	Fin Dimensions			
Design	Cross Section $w \times w$ (mm)	Length L (mm)	Number of Fins in Array	Convection Coefficient (W/m²·K)
A	3 × 3	30	6 × 9	125
B	1 × 1	7	14 × 17	375

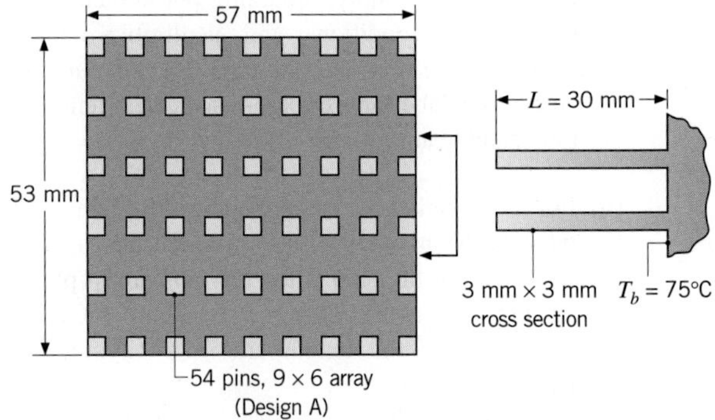

Determine which fin arrangement is superior. In your analysis, calculate the heat rate, efficiency, and effectiveness of a single fin, as well as the total heat rate and overall efficiency of the array. Since real estate inside the computer enclosure is important, compare the total heat rate per unit volume for the two designs.

3.152 Consider design B of Problem 3.151. Over time, dust can collect in the fine grooves that separate the fins. Consider the buildup of a dust layer of thickness L_d, as shown in the sketch. Calculate and plot the total heat rate for design B for dust layers in the range $0 \leq L_d \leq 5$ mm. The thermal conductivity of the dust can be taken as $k_d = 0.032$ W/m·K. Include the effects of convection from the fin tip.

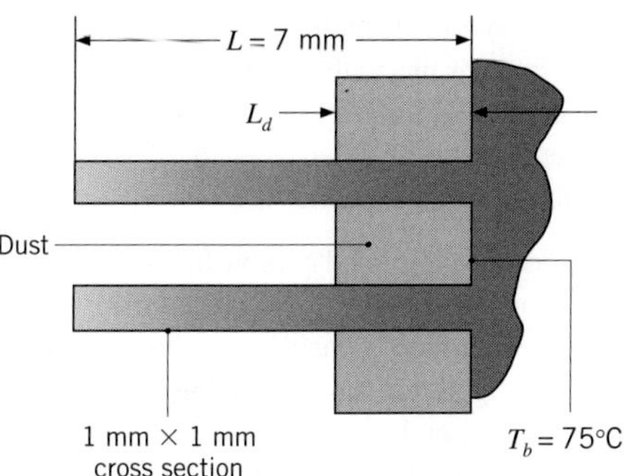

3.153 A long rod of 20-mm diameter and a thermal conductivity of 1.5 W/m·K has a uniform internal volumetric thermal energy generation of 10^6 W/m^3. The rod is covered with an electrically insulating sleeve of 2-mm thickness and thermal conductivity of 0.5 W/m·K. A spider with 12 ribs and dimensions as shown in the sketch has a thermal conductivity of 175 W/m·K, and is used to support the rod and to maintain concentricity with an 80-mm-diameter tube. Air at $T_\infty = 25°C$ passes over the spider surface, and the convection coefficient is 20 W/m^2·K. The outer surface of the tube is well insulated.

We wish to increase volumetric heating within the rod, while not allowing its centerline temperature to exceed 100°C. Determine the impact of the following changes, which may be effected independently or concurrently: (i) increasing the air speed and hence the convection coefficient; (ii) changing the number and/or thickness of the ribs; and (iii) using an electrically nonconducting sleeve material of larger thermal conductivity (e.g., amorphous carbon or quartz). Recommend a realistic configuration that yields a significant increase in $\dot{q}$.

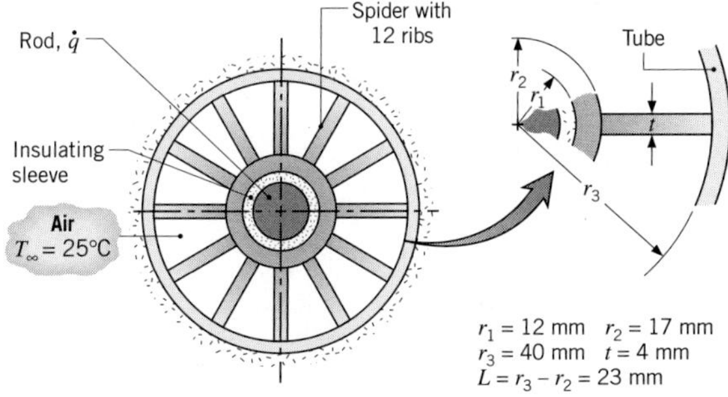

3.154 An air heater consists of a steel tube (k = 20 W/m·K), with inner and outer radii of $r_1 = 13$ mm and $r_2 = 16$ mm, respectively, and eight integrally machined longitudinal fins, each of thickness $t = 3$ mm. The fins extend to a concentric tube, which is of radius $r_3 = 40$ mm and insulated on its outer surface. Water at a temperature $T_{\infty,i} = 90°C$ flows through the inner tube, while air at $T_{\infty,o} = 25°C$ flows through the annular region formed by the larger concentric tube.

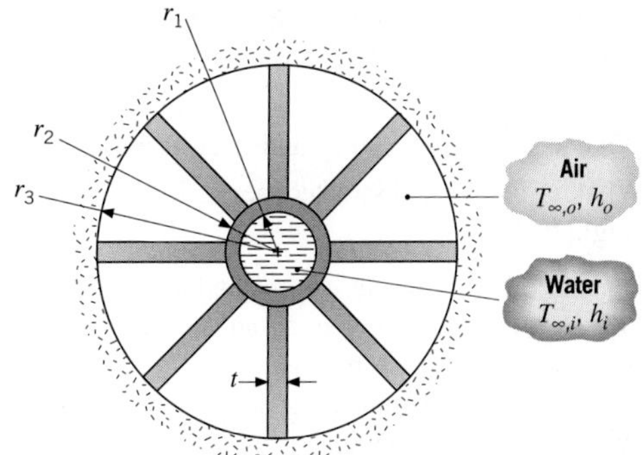

(a) Sketch the equivalent thermal circuit of the heater and relate each thermal resistance to appropriate system parameters.

(b) If h_i = 5000 W/m^2·K and h_o = 200 W/m^2·K, what is the heat rate per unit length?

(c) Assess the effect of increasing the number of fins N and/or the fin thickness t on the heat rate, subject to the constraint that $Nt < 50$ mm.

3.155 Determine the percentage increase in heat transfer associated with attaching aluminum fins of rectangular profile to a plane wall. The fins are 50 mm long, 0.5 mm thick, and are equally spaced at a distance of 4 mm (250 fins/m). The convection coefficient associated with the bare wall is 40 W/m^2·K, while that resulting from attachment of the fins is 30 W/m^2·K.

3.156 Heat is uniformly generated at the rate of 2×10^5 W/m^3 in a wall of thermal conductivity 25 W/m·K and thickness 60 mm. The wall is exposed to convection on both sides, with different heat transfer coefficients and temperatures as shown. There are straight rectangular fins on the right-hand side of the wall, with dimensions as shown and thermal conductivity of 250 W/m·K. What is the maximum temperature that will occur in the wall?

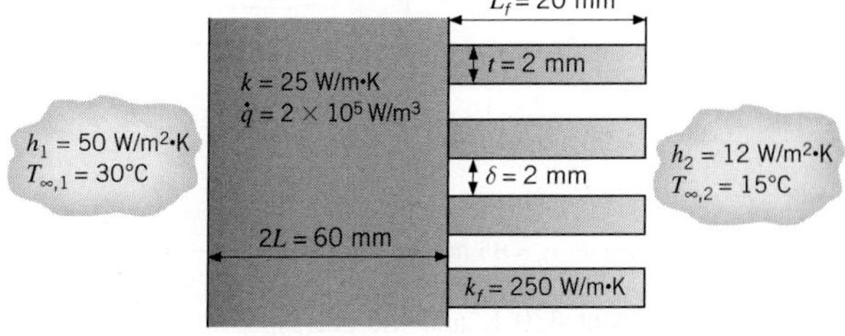

3.157 Aluminum fins of triangular profile are attached to a plane wall whose surface temperature is 250°C. The fin base thickness is 2 mm, and its length is 6 mm. The system is in ambient air at a temperature of 20°C, and the surface convection coefficient is 40 W/m$^2\cdot$K.

(a) What are the fin efficiency and effectiveness?

(b) What is the heat dissipated per unit width by a single fin?

3.158 An annular aluminum fin of rectangular profile is attached to a circular tube having an outside diameter of 25 mm and a surface temperature of 250°C. The fin is 1 mm thick and 10 mm long, and the temperature and the convection coefficient associated with the adjoining fluid are 25°C and 25 W/m$^2\cdot$K, respectively.

(a) What is the heat loss per fin?

(b) If 200 such fins are spaced at 5-mm increments along the tube length, what is the heat loss per meter of tube length?

3.159 Annular aluminum fins of rectangular profile are attached to a circular tube having an outside diameter of 50 mm and an outer surface temperature of 200°C. The fins are 4 mm thick and 15 mm long. The system is in ambient air at a temperature of 20°C, and the surface convection coefficient is 40 W/m$^2\cdot$K.

(a) What are the fin efficiency and effectiveness?

(b) If there are 125 such fins per meter of tube length, what is the rate of heat transfer per unit length of tube?

3.160 It is proposed to air-cool the cylinders of a combustion chamber by joining an aluminum casing with annular fins ($k = 240$ W/m$\cdot$K) to the cylinder wall ($k = 50$ W/m$\cdot$K).

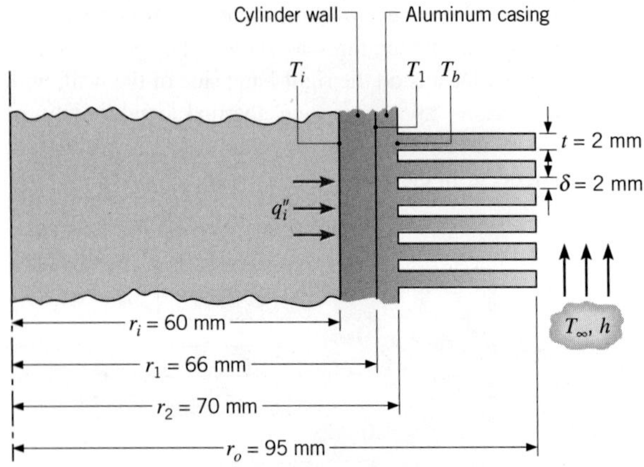

The air is at 320 K and the corresponding convection coefficient is 100 W/m$^2\cdot$K. Although heating at the inner surface is periodic, it is reasonable to assume steady-state conditions with a time-averaged heat flux of $q_i'' = 10^5$ W/m^2. Assuming negligible contact resistance between the wall and the casing, determine the wall inner temperature T_i, the interface temperature T_1, and the fin base temperature T_b. Determine these temperatures if the interface contact resistance is $R''_{t,c} = 10^{-4}$ m$^2\cdot$K/W.

3.161 Consider the air-cooled combustion cylinder of Problem 3.160, but instead of imposing a uniform heat flux at the inner surface, consider conditions for which the time-averaged temperature of the combustion gases is $T_g = 1100$ K and the corresponding convection coefficient is $h_g = 150$ W/m$^2\cdot$K. All other conditions, including the cylinder/casing contact resistance, remain the same. Determine the heat rate per unit length of cylinder (W/m), as well as the cylinder inner temperature T_i, the interface temperatures $T_{1,i}$ and $T_{1,o}$, and the fin base temperature T_b. Subject to the constraint that the fin gap is fixed at $\delta = 2$ mm, assess the effect of increasing the fin thickness at the expense of reducing the number of fins.

3.162 Heat transfer from a transistor may be enhanced by inserting it in an aluminum sleeve ($k = 200$ W/m$\cdot$K) having 12 integrally machined longitudinal fins on its outer surface. The transistor radius and height are $r_1 = 2.5$ mm and $H = 4$ mm, respectively, while the fins are of length $L = r_3 - r_2 = 8$ mm and uniform thickness $t = 0.8$ mm. The thickness of the sleeve base is $r_2 - r_1 = 1$ mm, and the contact resistance of the sleeve-transistor interface is $R''_{t,c} = 0.6 \times 10^{-3}$ m$^2\cdot$K/W. Air at $T_\infty = 20$°C flows over the fin surface, providing an approximately uniform convection coeffficient of $h = 30$ W/m$^2\cdot$K.

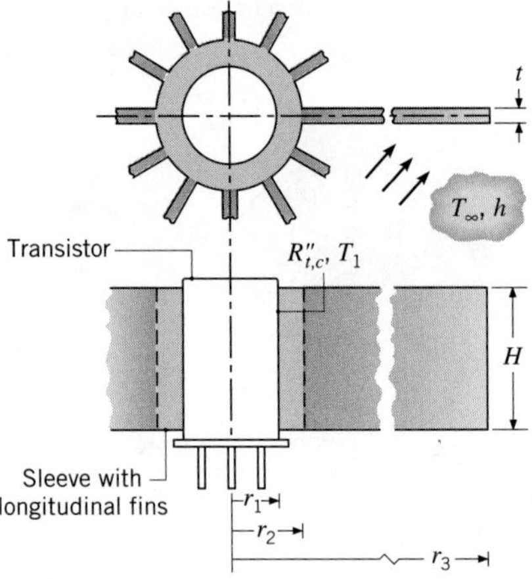

(a) When the transistor case temperature is 80°C, what is the rate of heat transfer from the sleeve?

(b) Identify all of the measures that could be taken to improve design and/or operating conditions, such that heat dissipation may be increased while still maintaining a case temperature of 80°C. In words, assess the relative merits of each measure. Choose what you believe to be the three most promising measures, and numerically assess the effect of corresponding changes in design and/or operating conditions on thermal performance.

3.163 Consider the conditions of Problem 3.149 but now allow for a tube wall thickness of 5 mm (inner and outer diameters of 50 and 60 mm), a fin-to-tube thermal contact resistance of 10^{-4} m$^2\cdot$K/W, and the fact that the water temperature, $T_w = 350$ K, is known, not the tube surface temperature. The water-side convection coefficient is $h_w = 2000$ W/m$^2\cdot$K. Determine the rate of heat transfer per unit tube length (W/m) to the water. What would be the separate effect of each of the following design changes on the heat rate: (i) elimination of the contact resistance; (ii) increasing the number of fins from four to eight; and (iii) changing the tube wall and fin material from copper to AISI 304 stainless steel ($k = 20$ W/m$\cdot$K)?

3.164 A scheme for concurrently heating separate water and air streams involves passing them through and over an array of tubes, respectively, while the tube wall is heated electrically. To enhance gas-side heat transfer, annular fins of rectangular profile are attached to the outer tube surface. Attachment is facilitated with a dielectric adhesive that electrically isolates the fins from the current-carrying tube wall.

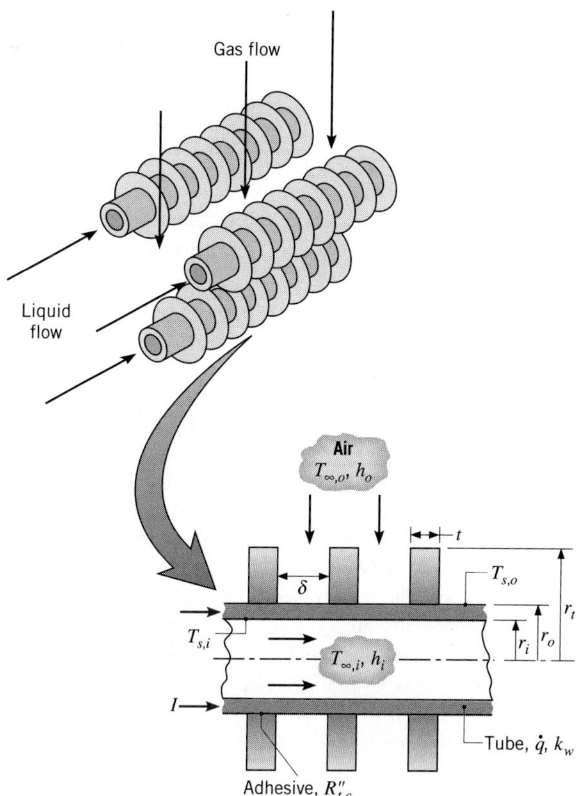

(a) Assuming uniform volumetric heat generation within the tube wall, obtain expressions for the heat rate per unit tube length (W/m) at the inner (r_i) and outer (r_o) surfaces of the wall. Express your results in terms of the tube inner and outer surface temperatures, $T_{s,i}$ and $T_{s,o}$, and other pertinent parameters.

(b) Obtain expressions that could be used to determine $T_{s,i}$ and $T_{s,o}$ in terms of parameters associated with the water- and air-side conditions.

(c) Consider conditions for which the water and air are at $T_{\infty,i} = T_{\infty,o} = 300$ K, with corresponding convection coefficients of $h_i = 2000$ W/m$^2\cdot$K and $h_o = 100$ W/m$^2\cdot$K. Heat is uniformly dissipated in a stainless steel tube ($k_w = 15$ W/m$\cdot$K), having inner and outer radii of $r_i = 25$ mm and $r_o = 30$ mm, and aluminum fins ($t = \delta = 2$ mm, $r_t = 55$ mm) are attached to the outer surface, with $R''_{t,c} = 10^{-4}$ m$^2\cdot$K/W. Determine the heat rates and temperatures at the inner and outer surfaces as a function of the rate of volumetric heating $\dot{q}$. The upper limit to $\dot{q}$ will be determined by the constraints that $T_{s,i}$ not exceed the boiling point of water (100°C) and $T_{s,o}$ not exceed the decomposition temperature of the adhesive (250°C).

The Bioheat Equation

3.165 Consider the conditions of Example 3.12, except that the person is now exercising (in the air environment), which increases the metabolic heat generation rate by a factor of 8, to 5600 W/m^3. At what rate would the person have to perspire (in liters/s) to maintain the same skin temperature as in that example?

3.166 Consider the conditions of Example 3.12 for an air environment, except now the air and surroundings temperatures are both 15°C. Humans respond to cold by shivering, which increases the metabolic heat generation rate. What would the metabolic heat generation rate (per unit volume) have to be to maintain a comfortable skin temperature of 33°C under these conditions?

3.167 Consider heat transfer in a forearm, which can be approximated as a cylinder of muscle of radius 50 mm (neglecting the presence of bones), with an outer layer of skin and fat of thickness 3 mm. There is metabolic heat generation and perfusion within the muscle. The metabolic heat generation rate, perfusion rate, arterial temperature, and properties of blood, muscle, and skin/fat layer are identical to those in Example 3.12.

The environment and surroundings are the same as for the air environment in Example 3.12.

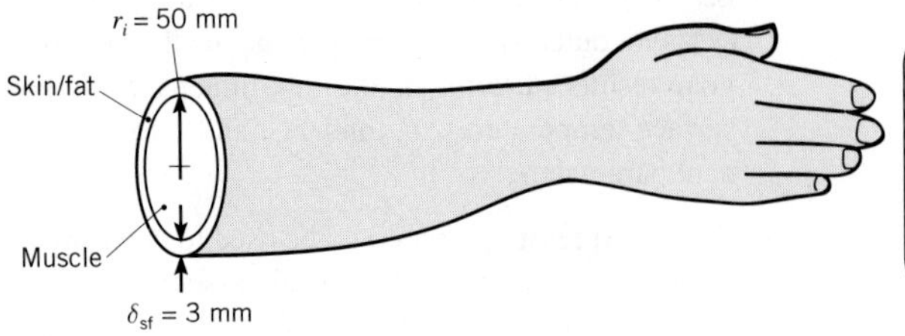

(a) Write the bioheat transfer equation in radial coordinates. Write the boundary conditions that express symmetry at the centerline of the forearm and specified temperature at the outer surface of the muscle. Solve the differential equation and apply the boundary conditions to find an expression for the temperature distribution. Note that the derivatives of the modified Bessel functions are given in Section 3.6.4.

(b) Equate the heat flux at the outer surface of the muscle to the heat flux through the skin/fat layer and into the environment to determine the temperature at the outer surface of the muscle.

(c) Find the maximum forearm temperature.

Thermoelectric Power Generation

3.168 For one of the $M = 48$ modules of Example 3.13, determine a variety of different efficiency values concerning the conversion of waste heat to electrical energy.

(a) Determine the thermodynamic efficiency, $\eta_{\text{therm}} \equiv P_{M=1}/q_1$.

(b) Determine the figure of merit $Z\overline{T}$ for one module, and the thermoelectric efficiency, η_{TE} using Equation 3.128.

(c) Determine the Carnot efficiency, $\eta_{\text{Carnot}} = 1 - T_2/T_1$.

(d) Determine both the thermoelectric efficiency and the Carnot efficiency for the case where $h_1 = h_2 \rightarrow \infty$.

(e) The energy conversion efficiency of thermoelectric devices is commonly reported by evaluating Equation 3.128, but with $T_{\infty,1}$ and $T_{\infty,2}$ used instead of T_1 and T_2, respectively. Determine the value of η_{TE} based on the inappropriate use of $T_{\infty,1}$ and $T_{\infty,2}$, and compare with your answers for parts (b) and (d).

3.169 One of the thermoelectric modules of Example 3.13 is installed between a hot gas at $T_{\infty,1} = 450°C$ and a cold gas at $T_{\infty,2} = 20°C$. The convection coefficient associated with the flowing gases is $h = h_1 = h_2 = 80\ \text{W/m}^2\cdot\text{K}$ while the electrical resistance of the load is $R_{e,\text{load}} = 4\ \Omega$.

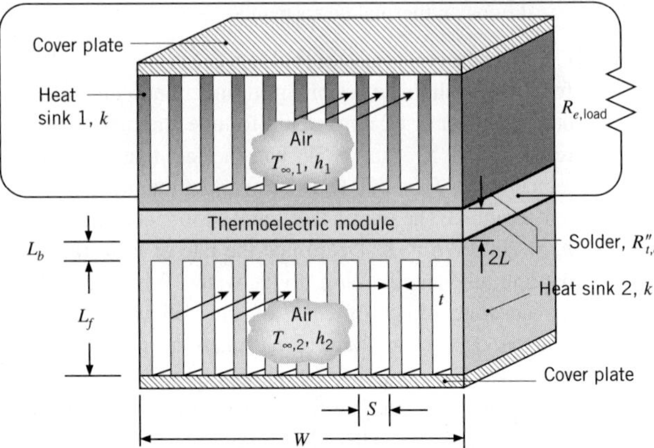

(a) Sketch the equivalent thermal circuit and determine the electric power generated by the module for the situation where the hot and cold gases provide convective heating and cooling directly to the module (no heat sinks).

(b) Two heat sinks ($k = 180\ \text{W/m}\cdot\text{K}$; see sketch), each with a base thickness of $L_b = 4$ mm and fin length $L_f = 20$ mm, are soldered to the upper and lower sides of the module. The fin spacing is 3 mm, while the solder joints each have a thermal resistance of $R''_{t,c} = 2.5 \times 10^{-6}\ \text{m}^2\cdot\text{K/W}$. Each heat sink has $N = 11$ fins, so that $t = 2.182$ mm and $S = 5.182$ mm, as determined from the requirements that $W = (N - 1)S + t$ and $S - t = 3$ mm. Sketch the equivalent thermal circuit and determine the electric power generated by the module. Compare the electric power generated to your answer for part (a). Assume adiabatic fin tips and convection coefficients that are the same as in part (a).

3.170 Thermoelectric modules have been used to generate electric power by tapping the heat generated by wood stoves. Consider the installation of the thermoelectric module of Example 3.13 on a vertical surface of a wood stove that has a surface temperature of $T_s = 375°C$. A thermal contact resistance of $R''_{t,c} = 5 \times 10^{-6}\ \text{m}^2\cdot\text{K/W}$ exists at the interface between the stove and the thermoelectric module, while the room air and walls are at $T_\infty = T_{\text{sur}} = 25°C$. The exposed surface of the thermoelectric module has an emissivity of $\varepsilon = 0.90$ and is subjected to a convection coefficient of $h = 15\ \text{W/m}^2\cdot\text{K}$. Sketch the equivalent thermal circuit and determine the electric power generated by the module. The load electrical resistance is $R_{e,\text{load}} = 3\ \Omega$.

3.171 The electric power generator for an orbiting satellite is composed of a long, cylindrical uranium heat source that is housed within an enclosure of square cross section. The only way for heat that is generated by the uranium to leave the enclosure is through four rows of the thermoelectric modules of Example 3.13. The thermoelectric modules generate electric power and also radiate heat into deep space characterized by $T_{\text{sur}} = 4$ K. Consider the situation for which there are 20 modules in each row for a total of $M = 4 \times 20 = 80$ modules. The modules are wired in series with an electrical load of $R_{e,\text{load}} = 250\ \Omega$, and have an emissivity of $\varepsilon = 0.93$. Determine the electric power generated for $\dot{E}_g = 1$, 10, and 100 kW. Also determine the surface temperatures of the modules for the three thermal energy generation rates.

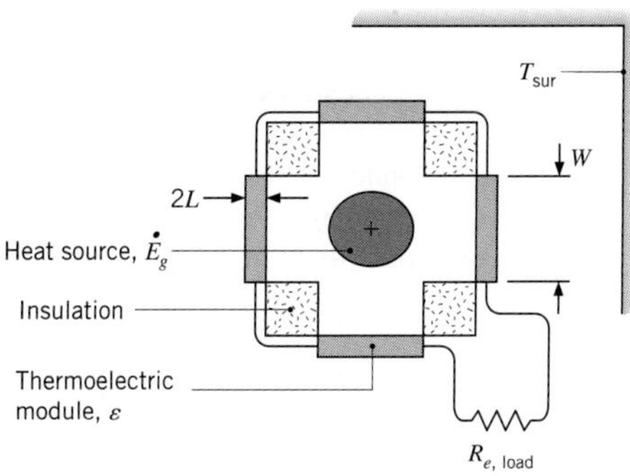

3.172 Rows of the thermoelectric modules of Example 3.13 are attached to the flat absorber plate of Problem 3.108. The rows of modules are separated by $L_{\text{sep}} = 0.5$ m and the backs of the modules are cooled by water at a temperature of $T_w = 40°\text{C}$, with $h = 45\ \text{W/m}^2 \cdot \text{K}$.

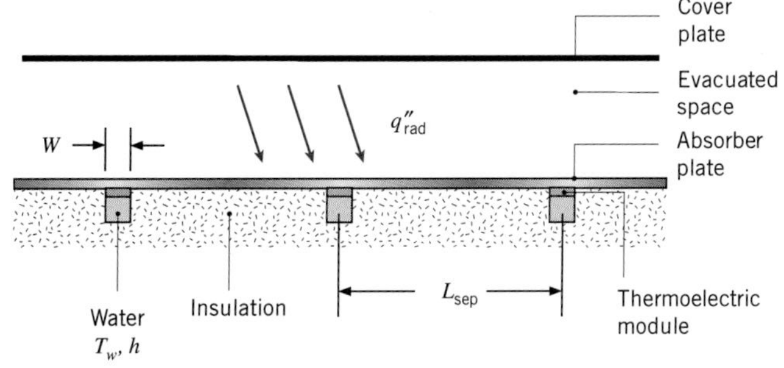

Determine the electric power produced by one row of thermoelectric modules connected in series electrically with a load resistance of 60 Ω. Calculate the heat transfer rate to the flowing water. Assume rows of 20 immediately adjacent modules, with the lengths of both the module rows and water tubing to be $L_{\text{row}} = 20W$ where $W = 54$ mm is the module dimension taken from Example 3.13. Neglect thermal contact resistances and the temperature drop across the tube wall, and assume that the high thermal conductivity tube wall creates a uniform temperature around the tube perimeter. Because of the thermal resistance provided by the thermoelectric modules, it is no longer appropriate to assume that the temperature of the absorber plate directly above a tube is equal to that of the water.

Micro- and Nanoscale Conduction

3.173 Determine the conduction heat transfer through an air layer held between two 10 mm × 10 mm parallel aluminum plates. The plates are at temperatures $T_{s,1} = 305$ K and $T_{s,2} = 295$ K, respectively, and the air is at atmospheric pressure. Determine the conduction heat rate for plate spacings of $L = 1$ mm, $L = 1\ \mu$m, and $L = 10$ nm. Assume a thermal accommodation coefficient of $\alpha_t = 0.92$.

3.174 Determine the parallel plate separation distance L, above which the thermal resistance associated with molecule-surface collisions $R_{t,m-s}$ is less than 1% of the resistance associated with molecule–molecule collisions, $R_{t,m-m}$ for (i) air between steel plates with $\alpha_t = 0.92$ and (ii) helium between clean aluminum plates with $\alpha_t = 0.02$. The gases are at atmospheric pressure, and the temperature is $T = 300$ K.

3.175 Determine the conduction heat flux through various plane layers that are subjected to boundary temperatures of $T_{s,1} = 301$ K and $T_{s,2} = 299$ K at atmospheric pressure. *Hint*: Do not account for micro- or nanoscale effects within the solid, and assume the thermal accommodation coefficient for an aluminum–air interface is $\alpha_t = 0.92$.

(a) Case A: The plane layer is aluminum. Determine the heat flux q''_x for $L_{\text{tot}} = 600\ \mu$m and $L_{\text{tot}} = 600$ nm.

(b) Case B: Conduction occurs through an air layer. Determine the heat flux q''_x for $L_{\text{tot}} = 600\ \mu$m and $L_{\text{tot}} = 600$ nm.

(c) Case C: The composite wall is composed of air held between two aluminum sheets. Determine the heat flux q''_x for $L_{\text{tot}} = 600\ \mu$m (with aluminum sheet thicknesses of $\delta = 40\ \mu$m) and $L_{\text{tot}} = 600$ nm (with aluminum sheet thicknesses of $\delta = 40$ nm).

(d) Case D: The composite wall is composed of 7 air layers interspersed between 8 aluminum sheets.

Determine the heat flux q''_x for $L_{tot} = 600\ \mu m$ (with aluminum sheet and air layer thicknesses of $\delta = 40\ \mu m$) and $L_{tot} = 600$ nm (with aluminum sheet and air layer thicknesses of $\delta = 40$ nm).

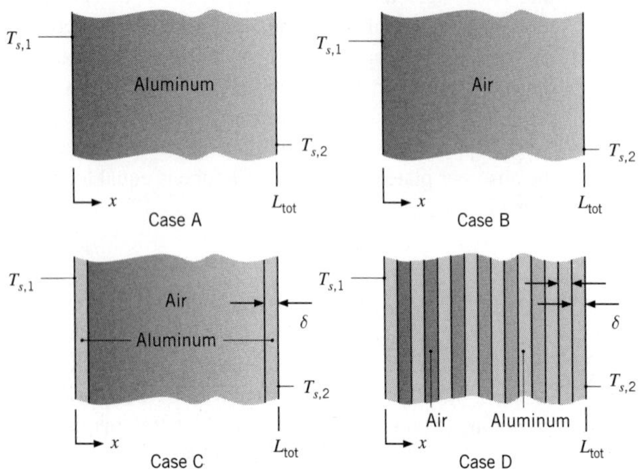

3.176 The Knudsen number, $Kn = \lambda_{mfp}/L$, is a dimensionless parameter used to describe potential micro- or nanoscale effects. Derive an expression for the ratio of the thermal resistance due to molecule–surface collisions to the thermal resistance associated with molecule–molecule collisions, $R_{t,m-s}/R_{t,m-m}$, in terms of the Knudsen number, the thermal accommodation coefficient α_t, and the ratio of specific heats γ, for an ideal gas. Plot the critical Knudsen number, Kn_{crit}, that is associated with $R_{t,m-s}/R_{t,m-m} = 0.01$ versus α_t, for $\gamma = 1.4$ and 1.67 (corresponding to air and helium, respectively).

3.177 A nanolaminated material is fabricated with an atomic layer deposition process, resulting in a series of stacked, alternating layers of tungsten and aluminum oxide, each layer being $\delta = 0.5$ nm thick. Each tungsten–aluminum oxide interface is associated with a thermal resistance of $R''_{t,i} = 3.85 \times 10^{-9}\ m^2 \cdot K/W$. The theoretical values of the thermal conductivities of the *thin* aluminum oxide and tungsten layers are $k_A = 1.65$ W/m·K and $k_T = 6.10$ W/m·K, respectively. The properties are evaluated at $T = 300$ K.

(a) Determine the effective thermal conductivity of the nanolaminated material. Compare the value of the effective thermal conductivity to the bulk thermal conductivities of aluminum oxide and tungsten, given in Tables A.1 and A.2.

(b) Determine the effective thermal conductivity of the nanolaminated material assuming that the thermal conductivities of the tungsten and aluminum oxide layers are equal to their bulk values.

3.178 Gold is commonly used in semiconductor packaging to form interconnections (also known as interconnects) that carry electrical signals between different devices in the package. In addition to being a good electrical conductor, gold interconnects are also effective at protecting the heat-generating devices to which they are attached by conducting thermal energy away from the devices to surrounding, cooler regions. Consider a thin film of gold that has a cross section of 60 nm × 250 nm.

(a) For an applied temperature difference of 20°C, determine the energy conducted along a 1-μm-long, thin-film interconnect. Evaluate properties at 300 K.

(b) Plot the lengthwise (in the 1-μm direction) and spanwise (in the thinnest direction) thermal conductivities of the gold film as a function of the film thickness L for $30 \leq L \leq 140$ nm.

CHAPTER 4

Two-Dimensional, Steady-State Conduction

To this point, we have restricted our attention to conduction problems in which the temperature gradient is significant for only one coordinate direction. However, in many cases such problems are grossly oversimplified if a one-dimensional treatment is used, and it is necessary to account for multidimensional effects. In this chapter, we consider several techniques for treating two-dimensional systems under steady-state conditions.

We begin our consideration of two-dimensional, steady-state conduction by briefly reviewing alternative approaches to determining temperatures and heat rates (Section 4.1). The approaches range from *exact solutions*, which may be obtained for idealized conditions, to *approximate methods* of varying complexity and accuracy. In Section 4.2 we consider some of the mathematical issues associated with obtaining an exact solution. In Section 4.3, we present compilations of existing exact solutions for a variety of simple geometries. Our objective in Sections 4.4 and 4.5 is to show how, with the aid of a computer, *numerical* (*finite-difference* or *finite-element*) *methods* may be used to accurately predict temperatures and heat rates within the medium and at its boundaries.

4.1 Alternative Approaches

Consider a long, prismatic solid in which there is two-dimensional heat conduction (Figure 4.1). With two surfaces insulated and the other surfaces maintained at different temperatures, $T_1 > T_2$, heat transfer by conduction occurs from surface 1 to 2. According to Fourier's law, Equation 2.3 or 2.4, the local heat flux in the solid is a vector that is everywhere perpendicular to lines of constant temperature (*isotherms*). The directions of the heat flux vector are represented by the *heat flow lines* of Figure 4.1, and the vector itself is the resultant of heat flux components in the x- and y-directions. These components are determined by Equation 2.6. Since the heat flow lines are, by definition, in the direction of heat flow, *no heat can be conducted across a heat flow line*, and they are therefore sometimes referred to as *adiabats*. Conversely, adiabatic surfaces (or symmetry lines) are heat flow lines.

Recall that, in any conduction analysis, there exist two major objectives. The first objective is to determine the temperature distribution in the medium, which, for the present problem, necessitates determining $T(x, y)$. This objective is achieved by solving the appropriate form of the heat equation. For two-dimensional, steady-state conditions with no generation and constant thermal conductivity, this form is, from Equation 2.22,

$$\frac{\partial^2 T}{\partial x^2} + \frac{\partial^2 T}{\partial y^2} = 0 \tag{4.1}$$

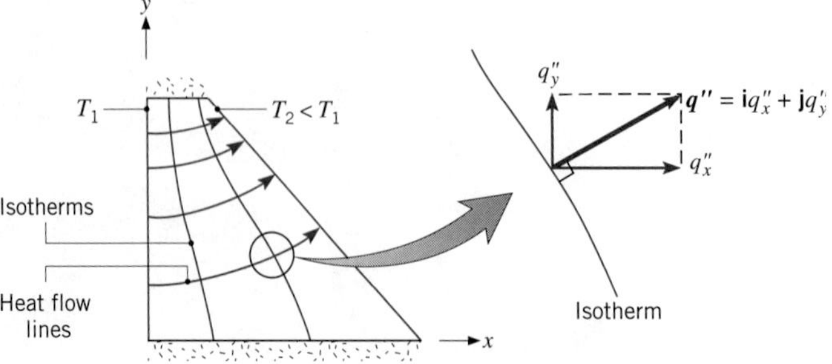

FIGURE 4.1 Two-dimensional conduction.

If Equation 4.1 can be solved for $T(x, y)$, it is then a simple matter to satisfy the second major objective, which is to determine the heat flux components q''_x and q''_y by applying the rate equations (2.6). Methods for solving Equation 4.1 include the use of *analytical, graphical,* and *numerical* (*finite-difference, finite-element,* or *boundary-element*) approaches.

The analytical method involves effecting an exact mathematical solution to Equation 4.1. The problem is more difficult than those considered in Chapter 3, since it now involves a partial, rather than an ordinary, differential equation. Although several techniques are available for solving such equations, the solutions typically involve complicated mathematical series and functions and may be obtained for only a restricted set of simple geometries and boundary conditions [1–5]. Nevertheless, the solutions are of value, since the dependent variable T is determined as a continuous function of the independent variables (x, y). Hence the solution could be used to compute the temperature at *any* point of interest in the medium. To illustrate the nature and importance of analytical techniques, an exact solution to Equation 4.1 is obtained in Section 4.2 by the method of *separation of variables.* Conduction shape factors and dimensionless conduction heat rates (Section 4.3) are compilations of existing solutions for geometries that are commonly encountered in engineering practice.

In contrast to the analytical methods, which provide *exact* results at *any* point, graphical and numerical methods can provide only *approximate* results at *discrete* points. Although superseded by computer solutions based on numerical procedures, the graphical, or flux-plotting, method may be used to obtain a quick estimate of the temperature distribution. Its use is restricted to two-dimensional problems involving adiabatic and isothermal boundaries. The method is based on the fact that isotherms must be perpendicular to heat flow lines, as noted in Figure 4.1. Unlike the analytical or graphical approaches, numerical methods (Sections 4.4 and 4.5) may be used to obtain accurate results for complex, two- or three-dimensional geometries involving a wide variety of boundary conditions.

4.2 The Method of Separation of Variables

To appreciate how the method of separation of variables may be used to solve two-dimensional conduction problems, we consider the system of Figure 4.2. Three sides of a thin rectangular plate or a long rectangular rod are maintained at a constant temperature T_1, while the fourth side is maintained at a constant temperature $T_2 \neq T_1$. Assuming negligible heat transfer from the surfaces of the plate or the ends of the rod, temperature gradients normal to the x–y plane may be neglected ($\partial^2 T/\partial z^2 \approx 0$) and conduction heat transfer is primarily in the x- and y-directions.

We are interested in the temperature distribution $T(x, y)$, but to simplify the solution we introduce the transformation

$$\theta \equiv \frac{T - T_1}{T_2 - T_1} \tag{4.2}$$

Substituting Equation 4.2 into Equation 4.1, the transformed differential equation is then

$$\frac{\partial^2 \theta}{\partial x^2} + \frac{\partial^2 \theta}{\partial y^2} = 0 \tag{4.3}$$

The graphical method is described, and its use is demonstrated, in Section 4S.1.

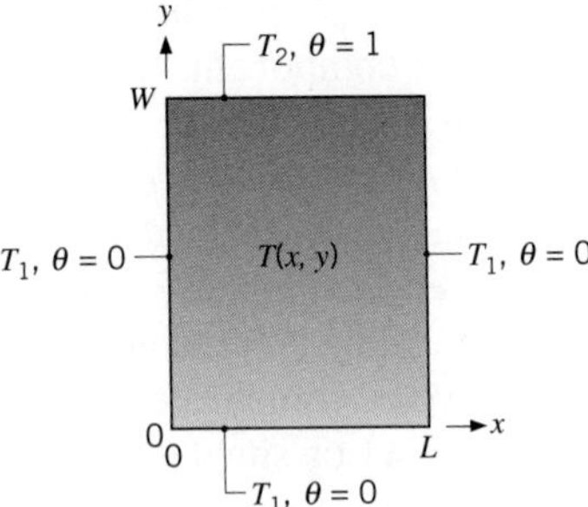

FIGURE 4.2 Two-dimensional conduction in a thin rectangular plate or a long rectangular rod.

Since the equation is second order in both x and y, two boundary conditions are needed for each of the coordinates. They are

$$\theta(0, y) = 0 \quad \text{and} \quad \theta(x, 0) = 0$$
$$\theta(L, y) = 0 \quad \text{and} \quad \theta(x, W) = 1$$

Note that, through the transformation of Equation 4.2, three of the four boundary conditions are now homogeneous and the value of θ is restricted to the range from 0 to 1.

We now apply the separation of variables technique by assuming that the desired solution can be expressed as the product of two functions, one of which depends only on x while the other depends only on y. That is, we assume the existence of a solution of the form

$$\theta(x, y) = X(x) \cdot Y(y) \tag{4.4}$$

Substituting into Equation 4.3 and dividing by XY, we obtain

$$-\frac{1}{X}\frac{d^2X}{dx^2} = \frac{1}{Y}\frac{d^2Y}{dy^2} \tag{4.5}$$

and it is evident that the differential equation is, in fact, separable. That is, the left-hand side of the equation depends only on x and the right-hand side depends only on y. Hence the equality can apply in general (for any x or y) only if both sides are equal to the same constant. Identifying this, as yet unknown, *separation constant* as λ^2, we then have

$$\frac{d^2X}{dx^2} + \lambda^2 X = 0 \tag{4.6}$$

$$\frac{d^2Y}{dy^2} - \lambda^2 Y = 0 \tag{4.7}$$

and the partial differential equation has been reduced to two ordinary differential equations. Note that the designation of λ^2 as a positive constant was not arbitrary. If a negative value were selected or a value of $\lambda^2 = 0$ was chosen, it is readily shown (Problem 4.1) that it would be impossible to obtain a solution that satisfies the prescribed boundary conditions.

The general solutions to Equations 4.6 and 4.7 are, respectively,

$$X = C_1 \cos \lambda x + C_2 \sin \lambda x$$
$$Y = C_3 e^{-\lambda y} + C_4 e^{+\lambda y}$$

in which case the general form of the two-dimensional solution is

$$\theta = (C_1 \cos \lambda x + C_2 \sin \lambda x)(C_3 e^{-\lambda y} + C_4 e^{\lambda y}) \tag{4.8}$$

Applying the condition that $\theta(0, y) = 0$, it is evident that $C_1 = 0$. In addition from the requirement that $\theta(x, 0) = 0$, we obtain

$$C_2 \sin \lambda x (C_3 + C_4) = 0$$

which may only be satisfied if $C_3 = -C_4$. Although the requirement could also be satisfied by having $C_2 = 0$, this would result in $\theta(x, y) = 0$, which does not satisfy the boundary condition $\theta(x, W) = 1$. If we now invoke the requirement that $\theta(L, y) = 0$, we obtain

$$C_2 C_4 \sin \lambda L (e^{\lambda y} - e^{-\lambda y}) = 0$$

The only way in which this condition may be satisfied (and still have a nonzero solution) is by requiring that λ assume discrete values for which $\sin \lambda L = 0$. These values must then be of the form

$$\lambda = \frac{n\pi}{L} \qquad n = 1, 2, 3, \ldots \tag{4.9}$$

where the integer $n = 0$ is precluded, since it implies $\theta(x, y) = 0$. The desired solution may now be expressed as

$$\theta = C_2 C_4 \sin \frac{n\pi x}{L} (e^{n\pi y/L} - e^{-n\pi y/L}) \tag{4.10}$$

Combining constants and acknowledging that the new constant may depend on n, we obtain

$$\theta(x, y) = C_n \sin \frac{n\pi x}{L} \sinh \frac{n\pi y}{L}$$

where we have also used the fact that $(e^{n\pi y/L} - e^{-n\pi y/L}) = 2 \sinh (n\pi y/L)$. In this form we have really obtained an infinite number of solutions that satisfy the differential equation and boundary conditions. However, since the problem is linear, a more general solution may be obtained from a superposition of the form

$$\theta(x, y) = \sum_{n=1}^{\infty} C_n \sin \frac{n\pi x}{L} \sinh \frac{n\pi y}{L} \tag{4.11}$$

To determine C_n we now apply the remaining boundary condition, which is of the form

$$\theta(x, W) = 1 = \sum_{n=1}^{\infty} C_n \sin \frac{n\pi x}{L} \sinh \frac{n\pi W}{L} \tag{4.12}$$

Although Equation 4.12 would seem to be an extremely complicated relation for evaluating C_n, a standard method is available. It involves writing an infinite series expansion in terms of *orthogonal functions*. An infinite set of functions $g_1(x), g_2(x), \ldots, g_n(x), \ldots$ is said to be orthogonal in the domain $a \leq x \leq b$ if

$$\int_a^b g_m(x) g_n(x)\, dx = 0 \qquad m \neq n \tag{4.13}$$

Many functions exhibit orthogonality, including the trigonometric functions $\sin(n\pi x/L)$ and $\cos(n\pi x/L)$ for $0 \le x \le L$. Their utility in the present problem rests with the fact that any function $f(x)$ may be expressed in terms of an infinite series of orthogonal functions

$$f(x) = \sum_{n=1}^{\infty} A_n g_n(x) \tag{4.14}$$

The form of the coefficients A_n in this series may be determined by multiplying each side of the equation by $g_m(x)$ and integrating between the limits a and b.

$$\int_a^b f(x) g_m(x)\, dx = \int_a^b g_m(x) \sum_{n=1}^{\infty} A_n g_n(x)\, dx \tag{4.15}$$

However, from Equation 4.13 it is evident that all but one of the terms on the right-hand side of Equation 4.15 must be zero, leaving us with

$$\int_a^b f(x) g_m(x)\, dx = A_m \int_a^b g_m^2(x)\, dx$$

Hence, solving for A_m, and recognizing that this holds for any A_n by switching m to n:

$$A_n = \frac{\displaystyle\int_a^b f(x) g_n(x)\, dx}{\displaystyle\int_a^b g_n^2(x)\, dx} \tag{4.16}$$

The properties of orthogonal functions may be used to solve Equation 4.12 for C_n by formulating an infinite series for the appropriate form of $f(x)$. From Equation 4.14 it is evident that we should choose $f(x) = 1$ and the orthogonal function $g_n(x) = \sin(n\pi x/L)$. Substituting into Equation 4.16 we obtain

$$A_n = \frac{\displaystyle\int_0^L \sin\frac{n\pi x}{L}\, dx}{\displaystyle\int_0^L \sin^2\frac{n\pi x}{L}\, dx} = \frac{2}{\pi}\,\frac{(-1)^{n+1} + 1}{n}$$

Hence from Equation 4.14, we have

$$1 = \sum_{n=1}^{\infty} \frac{2}{\pi}\,\frac{(-1)^{n+1} + 1}{n} \sin\frac{n\pi x}{L} \tag{4.17}$$

which is simply the expansion of unity in a Fourier series. Comparing Equations 4.12 and 4.17 we obtain

$$C_n = \frac{2[(-1)^{n+1} + 1]}{n\pi \sinh(n\pi W/L)} \qquad n = 1, 2, 3, \ldots \tag{4.18}$$

Substituting Equation 4.18 into Equation 4.11, we then obtain for the final solution

$$\theta(x, y) = \frac{2}{\pi} \sum_{n=1}^{\infty} \frac{(-1)^{n+1} + 1}{n} \sin\frac{n\pi x}{L}\, \frac{\sinh(n\pi y/L)}{\sinh(n\pi W/L)} \tag{4.19}$$

$$T(x,y) = \frac{2}{\pi} \sum_{n=1}^{\infty} T_1 \frac{(-1)^{n+1} + 1}{n} \sin(n\pi x)\, \frac{\sinh(n\pi y)}{\sinh(n\pi)}$$

$$\frac{1-\cos(n\pi)}{n}$$

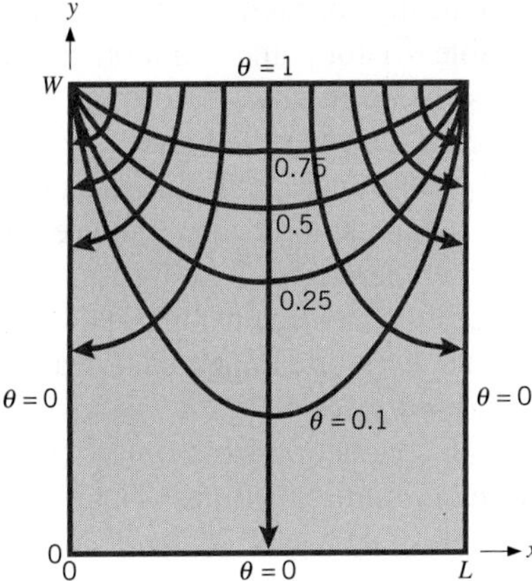

FIGURE 4.3 Isotherms and heat flow lines for two-dimensional conduction in a rectangular plate.

Equation 4.19 is a convergent series, from which the value of θ may be computed for any x and y. Representative results are shown in the form of isotherms for a schematic of the rectangular plate (Figure 4.3). The temperature T corresponding to a value of θ may be obtained from Equation 4.2, and components of the heat flux may be determined by using Equation 4.19 with Equation 2.6. The heat flux components determine the heat flow lines, which are shown in the figure. We note that the temperature distribution is symmetric about $x = L/2$, with $\partial T/\partial x = 0$ at that location. Hence, from Equation 2.6, we know the symmetry plane at $x = L/2$ is adiabatic and therefore is a heat flow line. However, note that the discontinuities prescribed at the upper corners of the plate are physically untenable. In reality, large temperature gradients could be maintained in proximity to the corners, but discontinuities could not exist.

Exact solutions have been obtained for a variety of other geometries and boundary conditions, including cylindrical and spherical systems. Such solutions are presented in specialized books on conduction heat transfer [1–5].

4.3 The Conduction Shape Factor and the Dimensionless Conduction Heat Rate

In general, finding analytical solutions to the two- or three-dimensional heat equation is time-consuming and, in many cases, not possible. Therefore, a different approach is often taken. For example, in many instances, two- or three-dimensional conduction problems may be rapidly solved by utilizing *existing* solutions to the heat diffusion equation. These solutions are reported in terms of a *shape factor* S or a steady-state *dimensionless conduction heat rate,* q^*_{ss}. The shape factor is defined such that

$$q = Sk\Delta T_{1-2} \tag{4.20}$$

where ΔT_{1-2} is the temperature difference between boundaries, as shown in, for example, Figure 4.2. It also follows that a two-dimensional conduction resistance may be expressed as

$$R_{t,\text{cond(2D)}} = \frac{1}{Sk} \tag{4.21}$$

Shape factors have been obtained analytically for numerous two- and three-dimensional systems, and results are summarized in Table 4.1 for some common configurations. Results are also available for other configurations [6–9]. In cases 1 through 8 and case 11, two-dimensional conduction is presumed to occur between the boundaries that are maintained at uniform temperatures, with $\Delta T_{1-2} = T_1 - T_2$. In case 9, three-dimensional conduction exists in the corner region, while in case 10 conduction occurs between an isothermal disk (T_1) and a semi-infinite medium of uniform temperature (T_2) at locations well removed from the disk. Shape factors may also be defined for one-dimensional geometries, and from the results of Table 3.3, it follows that for plane, cylindrical, and spherical walls, respectively, the shape factors are A/L, $2\pi L/\ln(r_2/r_1)$, and $4\pi r_1 r_2/(r_2 - r_1)$.

Cases 12 through 15 are associated with conduction from objects held at an isothermal temperature (T_1) that are embedded within an infinite medium of uniform temperature (T_2)

Shape factors for two-dimensional geometries may also be estimated with the graphical method that is described in Section 4S.1.

TABLE 4.1 Conduction shape factors and dimensionless conduction heat rates for selected systems.

(a) Shape factors [$q = Sk(T_1 - T_2)$]

System	Schematic	Restrictions	Shape Factor
Case 1 Isothermal sphere buried in a semi-infinite medium	T_2, z, T_1, D	$z > D/2$	$\dfrac{2\pi D}{1 - D/4z}$
Case 2 Horizontal isothermal cylinder of length L buried in a semi-infinite medium	T_2, z, L, T_1, D	$L \gg D$	$\dfrac{2\pi L}{\cosh^{-1}(2z/D)}$
		$L \gg D$ $z > 3D/2$	$\dfrac{2\pi L}{\ln(4z/D)}$
Case 3 Vertical cylinder in a semi-infinite medium	T_2, T_1, L, D	$L \gg D$	$\dfrac{2\pi L}{\ln(4L/D)}$
Case 4 Conduction between two cylinders of length L in infinite medium	T_1, D_1, D_2, T_2, w	$L \gg D_1, D_2$ $L \gg w$	$\dfrac{2\pi L}{\cosh^{-1}\left(\dfrac{4w^2 - D_1^2 - D_2^2}{2D_1D_2}\right)}$

TABLE 4.1 *Continued*

System	Schematic	Restrictions	Shape Factor
Case 5 Horizontal circular cylinder of length L midway between parallel planes of equal length and infinite width	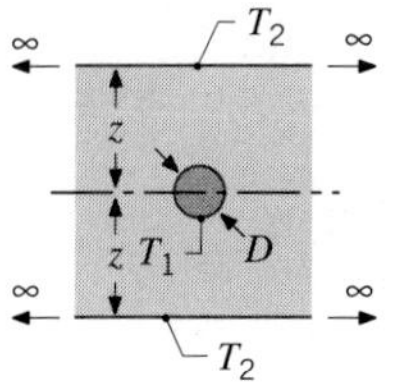	$z \gg D/2$ $L \gg z$	$\dfrac{2\pi L}{\ln(8z/\pi D)}$
Case 6 Circular cylinder of length L centered in a square solid of equal length	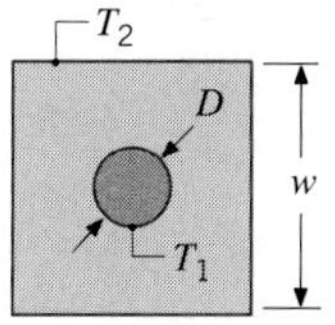	$w > D$ $L \gg w$	$\dfrac{2\pi L}{\ln(1.08\,w/D)}$
Case 7 Eccentric circular cylinder of length L in a cylinder of equal length	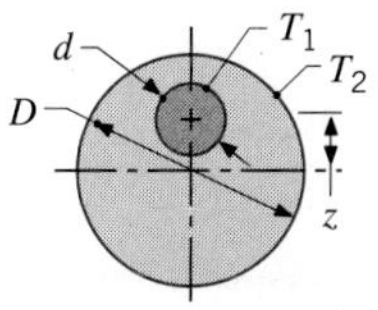	$D > d$ $L \gg D$	$\dfrac{2\pi L}{\cosh^{-1}\left(\dfrac{D^2 + d^2 - 4z^2}{2Dd}\right)}$
Case 8 Conduction through the edge of adjoining walls	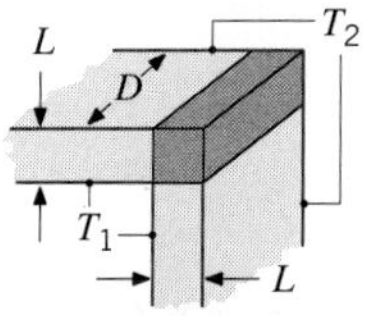	$D > 5L$	$0.54D$
Case 9 Conduction through corner of three walls with a temperature difference ΔT_{1-2} across the walls	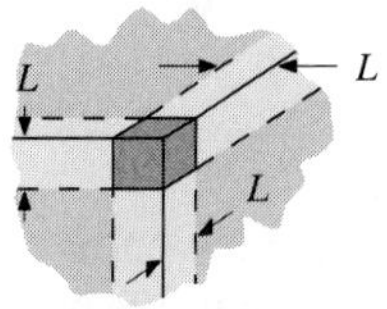	$L \ll$ length and width of wall	$0.15L$
Case 10 Disk of diameter D and temperature T_1 on a semi-infinite medium of thermal conductivity k and temperature T_2	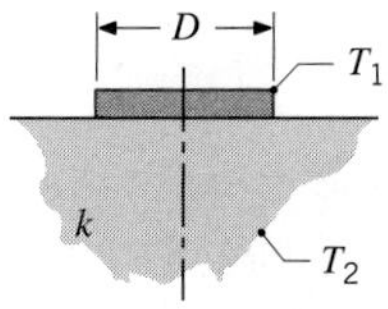	None	$2D$
Case 11 Square channel of length L		$\dfrac{W}{w} < 1.4$	$\dfrac{2\pi L}{0.785\ln(W/w)}$
		$\dfrac{W}{w} > 1.4$	$\dfrac{2\pi L}{0.930\ln(W/w) - 0.050}$
		$L \gg W$	

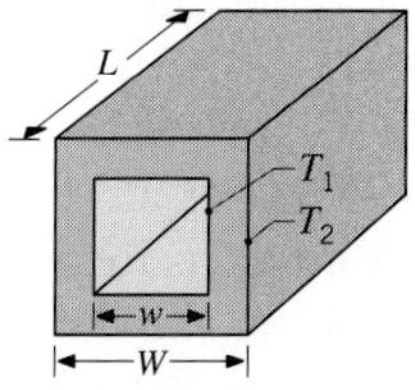

TABLE 4.1 *Continued*

(*b*) Dimensionless conduction heat rates [$q = q^*_{ss} kA_s(T_1 - T_2)/L_c$; $L_c \equiv (A_s/4\pi)^{1/2}$]

System	Schematic	Active Area, A_s	q^*_{ss}
Case 12 Isothermal sphere of diameter D and temperature T_1 in an infinite medium of temperature T_2	T_1, D, T_2	πD^2	1
Case 13 Infinitely thin, isothermal disk of diameter D and temperature T_1 in an infinite medium of temperature T_2	T_1, D, T_2	$\frac{\pi D^2}{2}$	$\frac{2\sqrt{2}}{\pi} = 0.900$
Case 14 Infinitely thin rectangle of length L, width w, and temperature T_1 in an infinite medium of temperature T_2	L, w, T_1, T_2	$2wL$	0.932
Case 15 Cuboid shape of height d with a square footprint of width D and temperature T_1 in an infinite medium of temperature T_2	D, d, T_1, T_2	$2D^2 + 4Dd$	d/D — q^*_{ss}: 0.1 — 0.943 1.0 — 0.956 2.0 — 0.961 10 — 1.111

at locations removed from the object. For these infinite medium cases, useful results may be obtained by defining a *characteristic length*

$$L_c \equiv (A_s/4\pi)^{1/2} \tag{4.22}$$

where A_s is the surface area of the object. Conduction heat transfer rates from the object to the infinite medium may then be reported in terms of a *dimensionless conduction heat rate* [10]

$$q^*_{ss} \equiv qL_c/kA_s(T_1 - T_2) \tag{4.23}$$

From Table 4.1, it is evident that the values of q^*_{ss}, which have been obtained analytically and numerically, are similar for a wide range of geometrical configurations. As a consequence of this similarity, values of q^*_{ss} may be *estimated* for configurations that are *similar* to those for which q^*_{ss} is known. For example, dimensionless conduction heat rates from cuboid shapes (case 15) over the range $0.1 \leq d/D \leq 10$ may be closely approximated by interpolating the values of q^*_{ss} reported in Table 4.1. Additional procedures that may be exploited to estimate values of q^*_{ss} for other geometries are explained in [10]. Note that results for q^*_{ss} in Table 4.1*b* may be converted to expressions for S listed in Table 4.1*a*. For example, the shape factor of case 10 may be derived from the dimensionless conduction heat rate of case 13 (recognizing that the infinite medium can be viewed as two adjacent semi-infinite media).

The shape factors and dimensionless conduction heat rates reported in Table 4.1 are associated with objects that are held at uniform temperatures. For uniform heat flux conditions, the object's temperature is no longer uniform but varies spatially with the coolest temperatures located near the periphery of the heated object. Hence, the temperature difference that is used to define S or q^*_{ss} is replaced by a temperature difference involving the *spatially averaged* surface temperature of the object $(\overline{T}_1 - T_2)$ or by the difference between the *maximum* surface temperature of the heated object and the far field temperature of the surrounding medium, $(T_{1,\max} - T_2)$. For the *uniformly heated* geometry of case 10 (a disk of diameter D in contact with a semi-infinite medium of thermal conductivity k and temperature T_2), the values of S are $3\pi^2D/16$ and $\pi D/2$ for temperature differences based on the average and maximum disk temperatures, respectively.

EXAMPLE 4.1

A metallic electrical wire of diameter $d = 5$ mm is to be coated with insulation of thermal conductivity $k = 0.35$ W/m·K. It is expected that, for the typical installation, the coated wire will be exposed to conditions for which the total coefficient associated with convection and radiation is $h = 15$ W/m^2·K. To minimize the temperature rise of the wire due to ohmic heating, the insulation thickness is specified so that the *critical insulation radius* is achieved (see Example 3.6). During the wire coating process, however, the insulation thickness sometimes varies around the periphery of the wire, resulting in eccentricity of the wire relative to the coating. Determine the change in the thermal resistance of the insulation due to an eccentricity that is 50% of the critical insulation thickness.

SOLUTION

Known: Wire diameter, convective conditions, and insulation thermal conductivity.

Find: Thermal resistance of the wire coating associated with peripheral variations in the coating thickness.

Schematic:

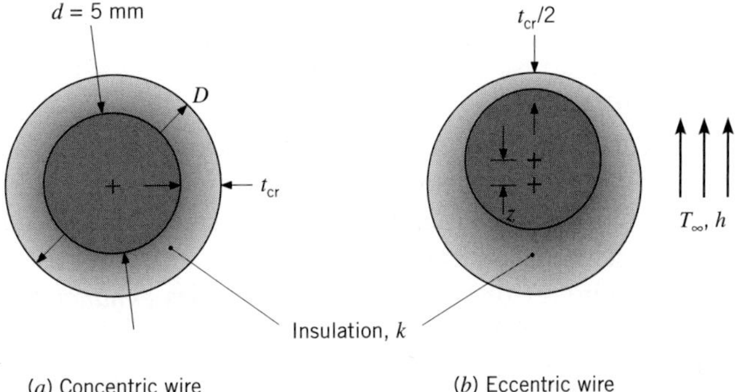

Assumptions:

1. Steady-state conditions.
2. Two-dimensional conduction.

3. Constant properties.
4. Both the exterior and interior surfaces of the coating are at uniform temperatures.

Analysis: From Example 3.6, the critical insulation radius is

$$r_{cr} = \frac{k}{h} = \frac{0.35 \text{ W/m} \cdot \text{K}}{15 \text{ W/m}^2 \cdot \text{K}} = 0.023 \text{ m} = 23 \text{ mm}$$

Therefore, the critical insulation thickness is

$$t_{cr} = r_{cr} - d/2 = 0.023 \text{ m} - \frac{0.005 \text{ m}}{2} = 0.021 \text{ m} = 21 \text{ mm}$$

The thermal resistance of the coating associated with the concentric wire may be evaluated using Equation 3.33 and is

$$R'_{t,\text{cond}} = \frac{\ln[r_{cr}/(d/2)]}{2\pi k} = \frac{\ln[0.023 \text{ m}/(0.005 \text{ m}/2)]}{2\pi(0.35 \text{ W/m} \cdot \text{K})} = 1.0 \text{ m} \cdot \text{K/W}$$

For the eccentric wire, the thermal resistance of the insulation may be evaluated using case 7 of Table 4.1, where the eccentricity is $z = 0.5 \times t_{cr} = 0.5 \times 0.021 \text{ m} = 0.010 \text{ m}$

$$R'_{t,\text{cond(2D)}} = \frac{1}{Sk} = \frac{\cosh^{-1}\left(\dfrac{D^2 + d^2 - 4z^2}{2Dd}\right)}{2\pi k}$$

$$= \frac{\cosh^{-1}\left(\dfrac{(2 \times 0.023 \text{ m})^2 + (0.005 \text{ m})^2 - 4(0.010 \text{ m})^2}{2 \times (2 \times 0.023 \text{ m}) \times 0.005 \text{ m}}\right)}{2\pi \times 0.35 \text{ W/m} \cdot \text{K}}$$

$$= 0.91 \text{ m} \cdot \text{K/W}$$

Therefore, the reduction in the thermal resistance of the insulation is 0.10 m·K/W, or 10%. ◁

Comments:

1. Reduction in the local insulation thickness leads to a smaller local thermal resistance of the insulation. Conversely, locations associated with thicker coatings have increased local thermal resistances. These effects offset each other, but not exactly; the maximum resistance is associated with the concentric wire case. For this application, eccentricity of the wire relative to the coating provides *enhanced* thermal performance relative to the concentric wire case.
2. The interior surface of the coating will be at nearly uniform temperature if the thermal conductivity of the wire is high relative to that of the insulation. Such is the case for metallic wire. However, the exterior surface temperature of the coating will not be perfectly uniform due to the variation in the local insulation thickness.

4.4 Finite-Difference Equations

As discussed in Sections 4.1 and 4.2, analytical methods may be used, in certain cases, to effect exact mathematical solutions to steady, two-dimensional conduction problems. These solutions have been generated for an assortment of simple geometries and boundary conditions and are well documented in the literature [1–5]. However, more often than not, two-dimensional problems involve geometries and/or boundary conditions that preclude such solutions. In these cases, the best alternative is often one that uses a *numerical* technique such as the *finite-difference, finite-element,* or *boundary-element* method. Another strength of numerical methods is that they can be readily extended to three-dimensional problems. Because of its ease of application, the finite-difference method is well suited for an introductory treatment of numerical techniques.

4.4.1 The Nodal Network

In contrast to an analytical solution, which allows for temperature determination at *any* point of interest in a medium, a numerical solution enables determination of the temperature at only *discrete* points. The first step in any numerical analysis must therefore be to select these points. Referring to Figure 4.4, this may be done by subdividing the medium of interest into a number of small regions and assigning to each a reference point that is at its center.

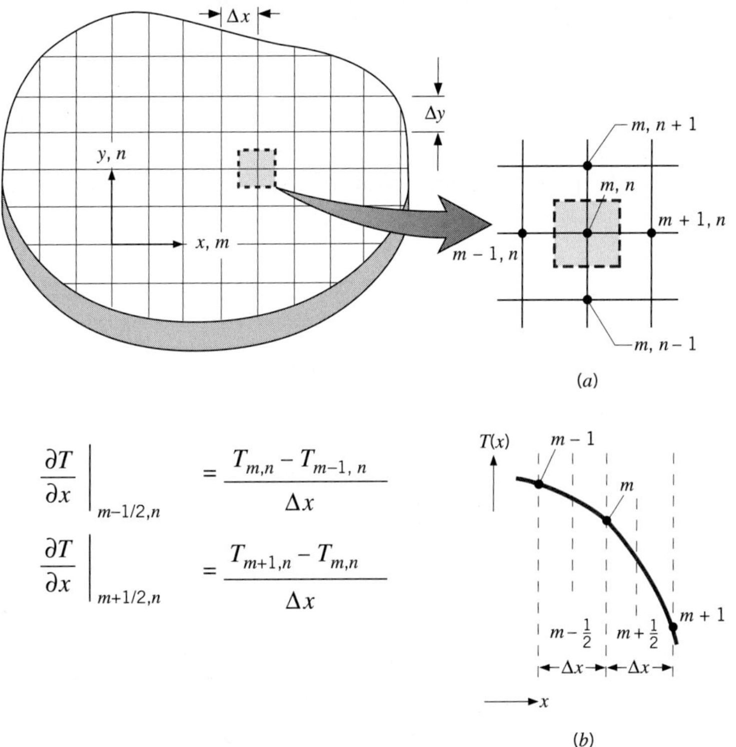

FIGURE 4.4 Two-dimensional conduction. (*a*) Nodal network. (*b*) Finite-difference approximation.

The reference point is frequently termed a *nodal point* (or simply a *node*), and the aggregate of points is termed a *nodal network, grid,* or *mesh.* The nodal points are designated by a numbering scheme that, for a two-dimensional system, may take the form shown in Figure 4.4*a*. The x and y locations are designated by the m and n indices, respectively.

Each node represents a certain region, and its temperature is a measure of the *average* temperature of the region. For example, the temperature of the node (m, n) of Figure 4.4*a* may be viewed as the average temperature of the surrounding shaded area. The selection of nodal points is rarely arbitrary, depending often on matters such as geometric convenience and the desired accuracy. The numerical accuracy of the calculations depends strongly on the number of designated nodal points. If this number is large (a *fine mesh*), accurate solutions can be obtained.

4.4.2 Finite-Difference Form of the Heat Equation

Determination of the temperature distribution numerically dictates that an appropriate conservation equation be written for *each* of the nodal points of unknown temperature. The resulting set of equations may then be solved simultaneously for the temperature at each node. For *any interior* node of a two-dimensional system with no generation and uniform thermal conductivity, the *exact* form of the energy conservation requirement is given by the heat equation, Equation 4.1. However, if the system is characterized in terms of a nodal network, it is necessary to work with an *approximate,* or *finite-difference,* form of this equation.

A finite-difference equation that is suitable for the interior nodes of a two-dimensional system may be inferred directly from Equation 4.1. Consider the second derivative, $\partial^2T/\partial x^2$. From Figure 4.4*b*, the value of this derivative at the (m, n) nodal point may be approximated as

$$\left.\frac{\partial^2 T}{\partial x^2}\right|_{m,n} \approx \frac{\partial T/\partial x|_{m+1/2,n} - \partial T/\partial x|_{m-1/2,n}}{\Delta x} \tag{4.24}$$

The temperature gradients may in turn be expressed as a function of the nodal temperatures. That is,

$$\left.\frac{\partial T}{\partial x}\right|_{m+1/2,n} \approx \frac{T_{m+1,n} - T_{m,n}}{\Delta x} \tag{4.25}$$

$$\left.\frac{\partial T}{\partial x}\right|_{m-1/2,n} \approx \frac{T_{m,n} - T_{m-1,n}}{\Delta x} \tag{4.26}$$

Substituting Equations 4.25 and 4.26 into 4.24, we obtain

$$\left.\frac{\partial^2 T}{\partial x^2}\right|_{m,n} \approx \frac{T_{m+1,n} + T_{m-1,n} - 2T_{m,n}}{(\Delta x)^2} \tag{4.27}$$

Proceeding in a similar fashion, it is readily shown that

$$\left.\frac{\partial^2 T}{\partial y^2}\right|_{m,n} \approx \frac{\partial T/\partial y|_{m,n+1/2} - \partial T/\partial y|_{m,n-1/2}}{\Delta y}$$

$$\approx \frac{T_{m,n+1} + T_{m,n-1} - 2T_{m,n}}{(\Delta y)^2} \tag{4.28}$$

Using a network for which $\Delta x = \Delta y$ and substituting Equations 4.27 and 4.28 into Equation 4.1, we obtain

$$T_{m,n+1} + T_{m,n-1} + T_{m+1,n} + T_{m-1,n} - 4T_{m,n} = 0 \tag{4.29}$$

Hence for the (m, n) node, the heat equation, which is an *exact differential equation,* is reduced to an *approximate algebraic equation.* This approximate, *finite-difference form of the heat equation* may be applied to any interior node that is equidistant from its four neighboring nodes. It requires simply that the temperature of an interior node be equal to the average of the temperatures of the four neighboring nodes.

4.4.3 The Energy Balance Method

In many cases, it is desirable to develop the finite-difference equations by an alternative method called the *energy balance method.* As will become evident, this approach enables one to analyze many different phenomena such as problems involving multiple materials, embedded heat sources, or exposed surfaces that do not align with an axis of the coordinate system. In the energy balance method, the finite-difference equation for a node is obtained by applying conservation of energy to a control volume about the nodal region. Since the actual direction of heat flow (into or out of the node) is often unknown, it is convenient to formulate the energy balance by *assuming* that *all* the heat flow is *into the node.* Such a condition is, of course, impossible, but if the rate equations are expressed in a manner consistent with this assumption, the correct form of the finite-difference equation is obtained. For steady-state conditions with generation, the appropriate form of Equation 1.12c is then

$$\dot{E}_{\text{in}} + \dot{E}_g = 0 \tag{4.30}$$

Consider applying Equation 4.30 to a control volume about the interior node (m, n) of Figure 4.5. For two-dimensional conditions, energy exchange is influenced by conduction between (m, n) and its four adjoining nodes, as well as by generation. Hence Equation 4.30 reduces to

$$\sum_{i=1}^{4} q_{(i)\rightarrow(m,n)} + \dot{q}(\Delta x \cdot \Delta y \cdot 1) = 0$$

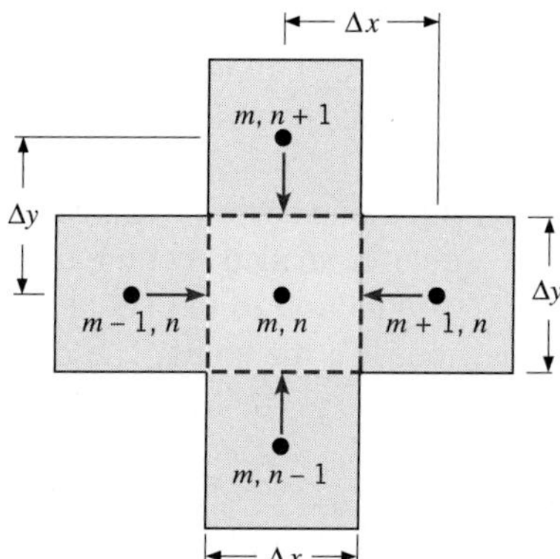

FIGURE 4.5 Conduction to an interior node from its adjoining nodes.

where i refers to the neighboring nodes, $q_{(i)\rightarrow(m,n)}$ is the conduction rate between nodes, and unit depth is assumed. To evaluate the conduction rate terms, we *assume* that conduction transfer occurs exclusively through *lanes* that are oriented in either the x- or y-direction. Simplified forms of Fourier's law may therefore be used. For example, the rate at which energy is transferred by conduction from node $(m-1, n)$ to (m, n) may be expressed as

$$q_{(m-1,n)\rightarrow(m,n)} = k(\Delta y \cdot 1)\frac{T_{m-1,n} - T_{m,n}}{\Delta x} \tag{4.31}$$

The quantity $(\Delta y \cdot 1)$ is the heat transfer area, and the term $(T_{m-1,n} - T_{m,n})/\Delta x$ is the finite-difference approximation to the temperature gradient at the boundary between the two nodes. The remaining conduction rates may be expressed as

$$q_{(m+1,n)\rightarrow(m,n)} = k(\Delta y \cdot 1)\frac{T_{m+1,n} - T_{m,n}}{\Delta x} \tag{4.32}$$

$$q_{(m,n+1)\rightarrow(m,n)} = k(\Delta x \cdot 1)\frac{T_{m,n+1} - T_{m,n}}{\Delta y} \tag{4.33}$$

$$q_{(m,n-1)\rightarrow(m,n)} = k(\Delta x \cdot 1)\frac{T_{m,n-1} - T_{m,n}}{\Delta y} \tag{4.34}$$

Note that, in evaluating each conduction rate, we have subtracted the temperature of the (m, n) node from the temperature of its adjoining node. This convention is necessitated by the assumption of heat flow into (m, n) and is consistent with the direction of the arrows shown in Figure 4.5. Substituting Equations 4.31 through 4.34 into the energy balance and remembering that $\Delta x = \Delta y$, it follows that the finite-difference equation for an interior node with generation is

$$T_{m,n+1} + T_{m,n-1} + T_{m+1,n} + T_{m-1,n} + \frac{\dot{q}(\Delta x)^2}{k} - 4T_{m,n} = 0 \tag{4.35}$$

If there is no internally distributed source of energy ($\dot{q} = 0$), this expression reduces to Equation 4.29.

It is important to note that a finite-difference equation is needed for each nodal point at which the temperature is unknown. However, it is not always possible to classify all such points as interior and hence to use Equation 4.29 or 4.35. For example, the temperature may be unknown at an insulated surface or at a surface that is exposed to convective conditions. For points on such surfaces, the finite-difference equation must be obtained by applying the energy balance method.

To further illustrate this method, consider the node corresponding to the internal corner of Figure 4.6. This node represents the three-quarter shaded section and exchanges energy by convection with an adjoining fluid at T_∞. Conduction to the nodal region (m, n) occurs along four different lanes from neighboring nodes in the solid. The conduction heat rates q_{cond} may be expressed as

$$q_{(m-1,n)\rightarrow(m,n)} = k(\Delta y \cdot 1)\frac{T_{m-1,n} - T_{m,n}}{\Delta x} \tag{4.36}$$

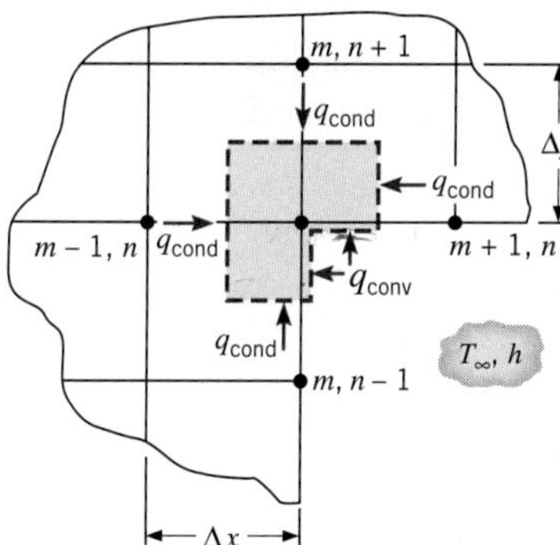

FIGURE 4.6 Formulation of the finite-difference equation for an internal corner of a solid with surface convection.

$$q_{(m,n+1)\rightarrow(m,n)} = k(\Delta x \cdot 1)\,\frac{T_{m,n+1} - T_{m,n}}{\Delta y} \tag{4.37}$$

$$q_{(m+1,n)\rightarrow(m,n)} = k\left(\frac{\Delta y}{2} \cdot 1\right)\frac{T_{m+1,n} - T_{m,n}}{\Delta x} \tag{4.38}$$

$$q_{(m,n-1)\rightarrow(m,n)} = k\left(\frac{\Delta x}{2} \cdot 1\right)\frac{T_{m,n-1} - T_{m,n}}{\Delta y} \tag{4.39}$$

Note that the areas for conduction from nodal regions $(m - 1, n)$ and $(m, n + 1)$ are proportional to Δy and Δx, respectively, whereas conduction from $(m + 1, n)$ and $(m, n - 1)$ occurs along lanes of width $\Delta y/2$ and $\Delta x/2$, respectively.

Conditions in the nodal region (m, n) are also influenced by convective exchange with the fluid, and this exchange may be viewed as occurring along half-lanes in the x- and y-directions. The total convection rate q_{conv} may be expressed as

$$q_{(\infty)\rightarrow(m,n)} = h\left(\frac{\Delta x}{2} \cdot 1\right)(T_\infty - T_{m,n}) + h\left(\frac{\Delta y}{2} \cdot 1\right)(T_\infty - T_{m,n}) \tag{4.40}$$

Implicit in this expression is the assumption that the exposed surfaces of the corner are at a uniform temperature corresponding to the nodal temperature $T_{m,n}$. This assumption is consistent with the concept that the entire nodal region is characterized by a single temperature, which represents an average of the actual temperature distribution in the region. In the absence of transient, three-dimensional, and generation effects, conservation of energy, Equation 4.30, requires that the sum of Equations 4.36 through 4.40 be zero. Summing these equations and rearranging, we therefore obtain

$$T_{m-1,n} + T_{m,n+1} + \frac{1}{2}(T_{m+1,n} + T_{m,n-1}) + \frac{h\Delta x}{k}T_\infty - \left(3 + \frac{h\Delta x}{k}\right)T_{m,n} = 0 \tag{4.41}$$

where again the mesh is such that $\Delta x = \Delta y$.

Nodal energy balance equations pertinent to several common geometries for situations where there is no internal energy generation are presented in Table 4.2.

TABLE 4.2 Summary of nodal finite-difference equations

Configuration	Finite-Difference Equation for $\Delta x = \Delta y$	
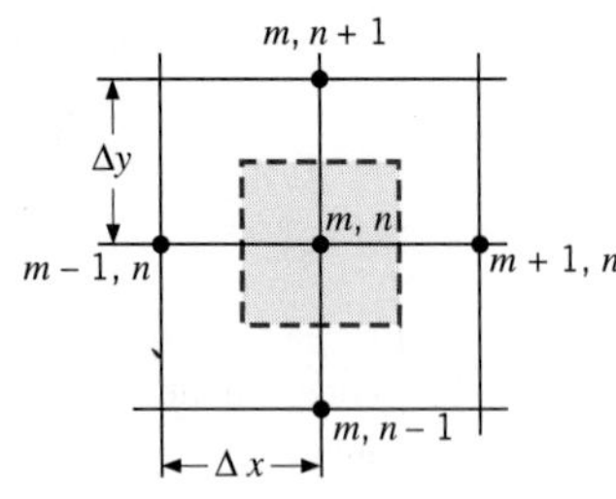	$T_{m,n+1} + T_{m,n-1} + T_{m+1,n} + T_{m-1,n} - 4T_{m,n} = 0$ **Case 1.** Interior node	(4.29)
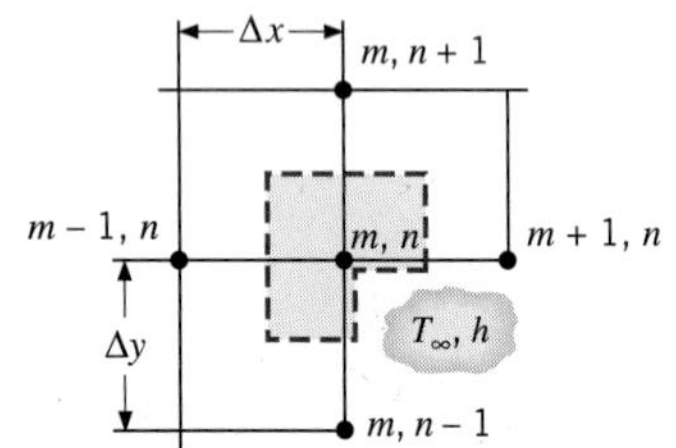	$2(T_{m-1,n} + T_{m,n+1}) + (T_{m+1,n} + T_{m,n-1}) + 2\frac{h\,\Delta x}{k}T_\infty - 2\left(3 + \frac{h\,\Delta x}{k}\right)T_{m,n} = 0$ **Case 2.** Node at an internal corner with convection	(4.41)
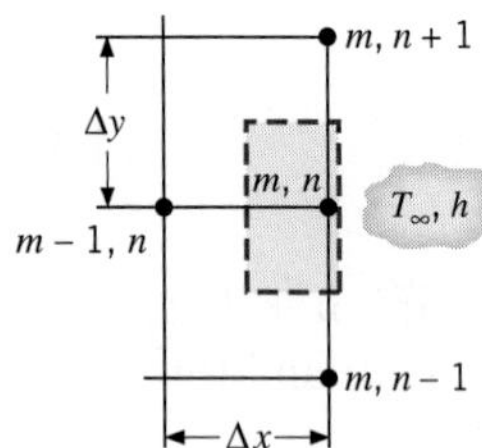	$(2T_{m-1,n} + T_{m,n+1} + T_{m,n-1}) + \frac{2h\,\Delta x}{k}T_\infty - 2\left(\frac{h\,\Delta x}{k} + 2\right)T_{m,n} = 0$ **Case 3.** Node at a plane surface with convection	(4.42)[a]
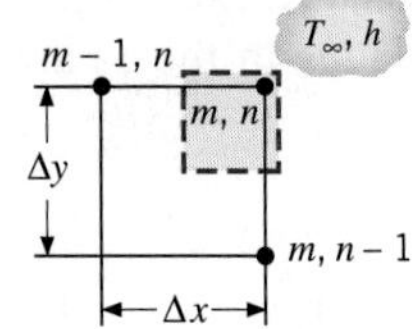	$(T_{m,n-1} + T_{m-1,n}) + 2\frac{h\,\Delta x}{k}T_\infty - 2\left(\frac{h\,\Delta x}{k} + 1\right)T_{m,n} = 0$ **Case 4.** Node at an external corner with convection	(4.43)
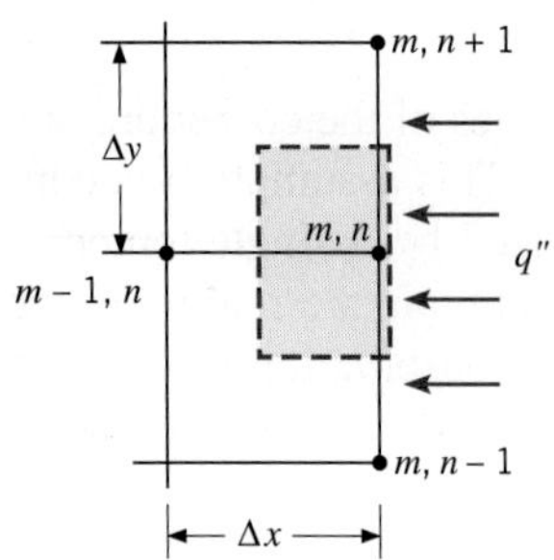	$(2T_{m-1,n} + T_{m,n+1} + T_{m,n-1}) + \frac{2q''\,\Delta x}{k} - 4T_{m,n} = 0$ **Case 5.** Node at a plane surface with uniform heat flux	(4.44)[b]

[a,b]To obtain the finite-difference equation for an adiabatic surface (or surface of symmetry), simply set h or q'' equal to zero.

EXAMPLE 4.2

Using the energy balance method, derive the finite-difference equation for the (m, n) nodal point located on a plane, insulated surface of a medium with uniform heat generation.

Solution

Known: Network of nodal points adjoining an insulated surface.

Find: Finite-difference equation for the surface nodal point.

Schematic:

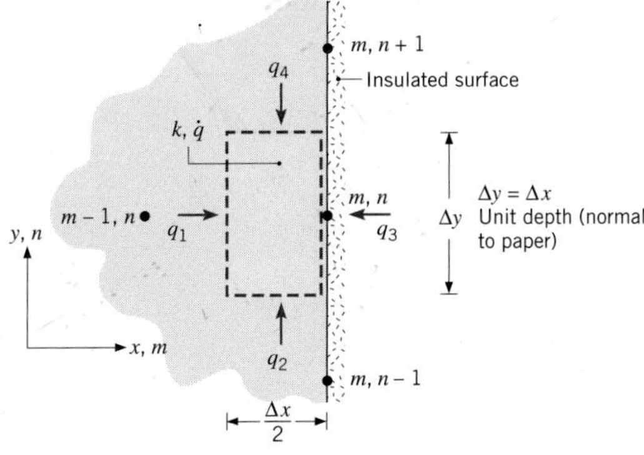

Assumptions:

1. Steady-state conditions.
2. Two-dimensional conduction.
3. Constant properties.
4. Uniform internal heat generation.

Analysis: Applying the energy conservation requirement, Equation 4.30, to the control surface about the region $(\Delta x/2 \cdot \Delta y \cdot 1)$ associated with the (m, n) node, it follows that, with volumetric heat generation at a rate $\dot{q}$,

$$q_1 + q_2 + q_3 + q_4 + \dot{q}\left(\frac{\Delta x}{2} \cdot \Delta y \cdot 1\right) = 0$$

where

$$q_1 = k(\Delta y \cdot 1)\frac{T_{m-1,n} - T_{m,n}}{\Delta x}$$

$$q_2 = k\left(\frac{\Delta x}{2} \cdot 1\right)\frac{T_{m,n-1} - T_{m,n}}{\Delta y}$$

$$q_3 = 0$$

$$q_4 = k\left(\frac{\Delta x}{2} \cdot 1\right)\frac{T_{m,n+1} - T_{m,n}}{\Delta y}$$

Substituting into the energy balance and dividing by $k/2$, it follows that

$$2T_{m-1,n} + T_{m,n-1} + T_{m,n+1} - 4T_{m,n} + \frac{\dot{q}(\Delta x \cdot \Delta y)}{k} = 0 \qquad \triangleleft$$

Comments:

1. The same result could be obtained by using the symmetry condition, $T_{m+1,n} = T_{m-1,n}$, with the finite-difference equation (Equation 4.35) for an interior nodal point.

If $\dot{q} = 0$, the desired result could also be obtained by setting $h = 0$ in Equation 4.42 (Table 4.2).

2. As an application of the foregoing finite-difference equation, consider the following two-dimensional system within which thermal energy is uniformly generated at an unknown rate $\dot{q}$. The thermal conductivity of the solid is known, as are convection conditions at one of the surfaces. In addition, temperatures have been measured at locations corresponding to the nodal points of a finite-difference mesh.

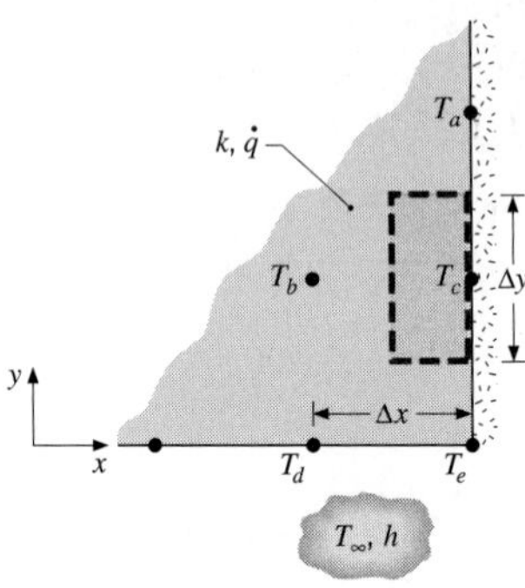

$T_a = 235.9°C$ $T_b = 227.6°C$

$T_c = 230.9°C$ $T_d = 220.1°C$

$T_e = 222.4°C$ $T_\infty = 200.0°C$

$h = 50 \text{ W/m}^2 \cdot \text{K}$ $k = 1 \text{ W/m} \cdot \text{K}$

$\Delta x = 10 \text{ mm}$ $\Delta y = 10 \text{ mm}$

The generation rate can be determined by applying the finite-difference equation to node c.

$$2T_b + T_e + T_a - 4T_c + \frac{\dot{q}(\Delta x \cdot \Delta y)}{k} = 0$$

$$(2 \times 227.6 + 222.4 + 235.9 - 4 \times 230.9)°\text{C} + \frac{\dot{q}(0.01 \text{ m})^2}{1 \text{ W/m} \cdot \text{K}} = 0$$

$$\dot{q} = 1.01 \times 10^5 \text{ W/m}^3$$

From the prescribed thermal conditions and knowledge of $\dot{q}$, we can also determine whether the conservation of energy requirement is satisfied for node e. Applying an energy balance to a control volume about this node, it follows that

$$q_1 + q_2 + q_3 + q_4 + \dot{q}(\Delta x/2 \cdot \Delta y/2 \cdot 1) = 0$$

$$k(\Delta x/2 \cdot 1)\frac{T_c - T_e}{\Delta y} + 0 + h(\Delta x/2 \cdot 1)(T_\infty - T_e) + k(\Delta y/2 \cdot 1)\frac{T_d - T_e}{\Delta x}$$
$$+ \dot{q}(\Delta x/2 \cdot \Delta y/2 \cdot 1) = 0$$

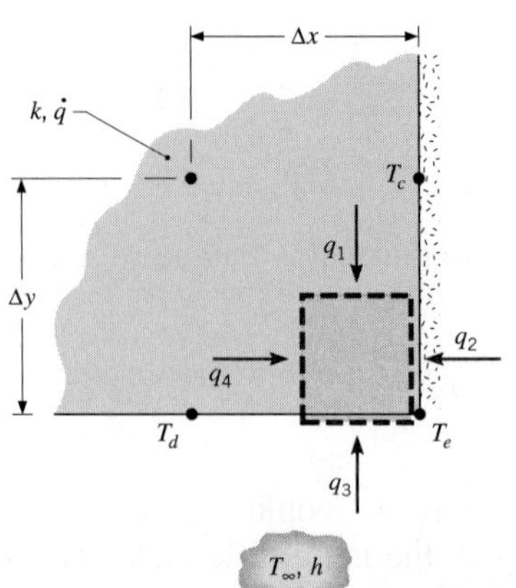

If the energy balance is satisfied, the left-hand side of this equation will be identically equal to zero. Substituting values, we obtain

$$1\ \text{W/m}\cdot\text{K}(0.005\ \text{m}^2)\frac{(230.9 - 222.4)^\circ\text{C}}{0.010\ \text{m}}$$

$$+\ 0 + 50\ \text{W/m}^2\cdot\text{K}(0.005\ \text{m}^2)\,(200 - 222.4)^\circ\text{C}$$

$$+\ 1\ \text{W/m}\cdot\text{K}(0.005\ \text{m}^2)\frac{(220.1 - 222.4)^\circ\text{C}}{0.010\ \text{m}} + 1.01 \times 10^5\ \text{W/m}^3(0.005)^2\ \text{m}^3 = 0(?)$$

$$4.250\ \text{W} + 0 - 5.600\ \text{W} - 1.150\ \text{W} + 2.525\ \text{W} = 0(?)$$

$$0.025\ \text{W} \approx 0$$

The inability to precisely satisfy the energy balance is attributable to temperature measurement errors, the approximations employed in developing the finite-difference equations, and the use of a relatively coarse mesh.

It is useful to note that heat rates between adjoining nodes may also be formulated in terms of the corresponding thermal resistances. Referring, for example, to Figure 4.6, the rate of heat transfer by conduction from node $(m - 1, n)$ to (m, n) may be expressed as

$$q_{(m-1,n)\rightarrow(m,n)} = \frac{T_{m-1,n} - T_{m,n}}{R_{t,\text{cond}}} = \frac{T_{m-1,n} - T_{m,n}}{\Delta x/k\,(\Delta y \cdot 1)}$$

yielding a result that is equivalent to that of Equation 4.36. Similarly, the rate of heat transfer by convection to (m, n) may be expressed as

$$q_{(\infty)\rightarrow(m,n)} = \frac{T_\infty - T_{m,n}}{R_{t,\text{conv}}} = \frac{T_\infty - T_{m,n}}{\{h[(\Delta x/2)\cdot 1 + (\Delta y/2)\cdot 1]\}^{-1}}$$

which is equivalent to Equation 4.40.

As an example of the utility of resistance concepts, consider an interface that separates two dissimilar materials and is characterized by a thermal contact resistance $R''_{t,c}$ (Figure 4.7). The rate of heat transfer from node (m, n) to $(m, n - 1)$ may be expressed as

$$q_{(m,n)\rightarrow(m,n-1)} = \frac{T_{m,n} - T_{m,n-1}}{R_{\text{tot}}} \tag{4.45}$$

where, for a unit depth,

$$R_{\text{tot}} = \frac{\Delta y/2}{k_{\text{A}}(\Delta x \cdot 1)} + \frac{R''_{t,c}}{\Delta x \cdot 1} + \frac{\Delta y/2}{k_{\text{B}}(\Delta x \cdot 1)} \tag{4.46}$$

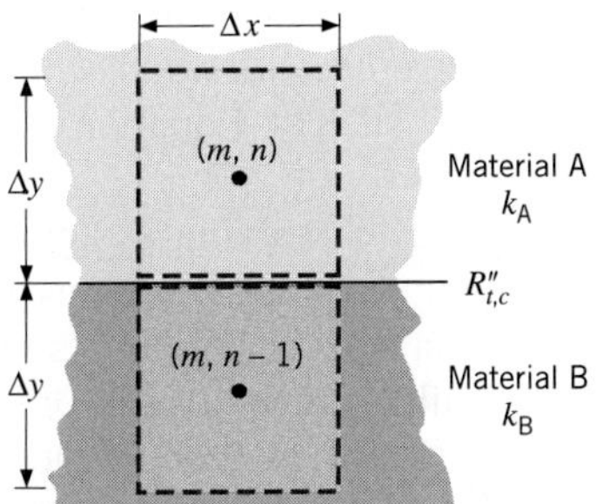

FIGURE 4.7 Conduction between adjoining, dissimilar materials with an interface contact resistance.

4.5 *Solving the Finite-Difference Equations*

Once the nodal network has been established and an appropriate finite-difference equation has been written for each node, the temperature distribution may be determined. The problem reduces to one of solving a system of linear, algebraic equations. In this section, we formulate the system of linear, algebraic equations as a matrix equation and briefly discuss its solution by the matrix inversion method. We also present some considerations for verifying the accuracy of the solution.

4.5.1 Formulation as a Matrix Equation

Consider a system of N finite-difference equations corresponding to N unknown temperatures. Identifying the nodes by a single integer subscript, rather than by the double subscript (m, n), the procedure for performing a matrix inversion begins by expressing the equations as

$$\begin{aligned} a_{11}T_1 + a_{12}T_2 + a_{13}T_3 + \cdots + a_{1N}T_N &= C_1 \\ a_{21}T_1 + a_{22}T_2 + a_{23}T_3 + \cdots + a_{2N}T_N &= C_2 \\ \vdots \quad\quad \vdots \quad\quad \vdots \quad\quad \vdots \quad\quad \vdots \quad &\quad \vdots \\ a_{N1}T_1 + a_{N2}T_2 + a_{N3}T_3 + \cdots + a_{NN}T_N &= C_N \end{aligned} \tag{4.47}$$

where the quantities $a_{11}, a_{12}, \ldots, C_1, \ldots$ are known coefficients and constants involving quantities such as Δx, k, h, and T_∞. Using matrix notation, these equations may be expressed as

$$[A][T] = [C] \tag{4.48}$$

where

$$A \equiv \begin{bmatrix} a_{11} & a_{12} & \cdots & a_{1N} \\ a_{21} & a_{22} & \cdots & a_{2N} \\ \vdots & \vdots & & \vdots \\ a_{N1} & a_{N2} & \cdots & a_{NN} \end{bmatrix}, \quad T \equiv \begin{bmatrix} T_1 \\ T_2 \\ \vdots \\ T_N \end{bmatrix}, \quad C \equiv \begin{bmatrix} C_1 \\ C_2 \\ \vdots \\ C_N \end{bmatrix}$$

The *coefficient matrix* $[A]$ is square $(N \times N)$, and its *elements* are designated by a double subscript notation, for which the first and second subscripts refer to rows and columns, respectively. The matrices $[T]$ and $[C]$ have a single column and are known as *column vectors*. Typically, they are termed the *solution* and *right-hand side vectors,* respectively. If the matrix multiplication implied by the left-hand side of Equation 4.48 is performed, Equations 4.47 are obtained.

Numerous mathematical methods are available for solving systems of linear, algebraic equations [11, 12], and many computational software programs have the built-in capability to solve Equation 4.48 for the solution vector $[T]$. For small matrices, the solution can be found using a programmable calculator or by hand. One method suitable for hand or computer calculation is the Gauss–Seidel method, which is presented in Appendix D.

4.5.2 Verifying the Accuracy of the Solution

It is good practice to verify that a numerical solution has been correctly formulated by performing an energy balance on a control surface surrounding all nodal regions whose temperatures have been evaluated. The temperatures should be substituted into the energy balance equation, and if the balance is not satisfied to a high degree of precision, the finite-difference equations should be checked for errors.

Even when the finite-difference equations have been properly formulated and solved, the results may still represent a coarse approximation to the actual temperature field. This behavior is a consequence of the finite spacings (Δx, Δy) between nodes and of finite-difference approximations, such as $k(\Delta y \cdot 1)(T_{m-1,n} - T_{m,n})/\Delta x$, to Fourier's law of conduction, $-k(dy \cdot 1)\partial T/\partial x$. The finite-difference approximations become more accurate as the nodal network is refined (Δx and Δy are reduced). Hence, if accurate results are desired, grid studies should be performed, whereby results obtained for a fine grid are compared with those obtained for a coarse grid. One could, for example, reduce Δx and Δy by a factor of 2, thereby increasing the number of nodes and finite-difference equations by a factor of 4. If the agreement is unsatisfactory, further grid refinements could be made until the computed temperatures no longer depend significantly on the choice of Δx and Δy. Such *grid-independent* results would provide an accurate solution to the physical problem.

Another option for validating a numerical solution involves comparing results with those obtained from an exact solution. For example, a finite-difference solution of the physical problem described in Figure 4.2 could be compared with the exact solution given by Equation 4.19. However, this option is limited by the fact that we seldom seek numerical solutions to problems for which there exist exact solutions. Nevertheless, if we seek a numerical solution to a complex problem for which there is no exact solution, it is often useful to test our finite-difference procedures by applying them to a simpler version of the problem.

IHT **EXAMPLE 4.3**

A major objective in advancing gas turbine engine technologies is to increase the temperature limit associated with operation of the gas turbine blades. This limit determines the permissible turbine gas inlet temperature, which, in turn, strongly influences overall system performance. In addition to fabricating turbine blades from special, high-temperature, high-strength superalloys, it is common to use internal cooling by machining flow channels within the blades and routing air through the channels. We wish to assess the effect of such a scheme by approximating the blade as a rectangular solid in which rectangular channels are machined. The blade, which has a thermal conductivity of $k = 25$ W/m·K, is 6 mm thick, and each channel has a 2 mm $\times$ 6 mm rectangular cross section, with a 4-mm spacing between adjoining channels.

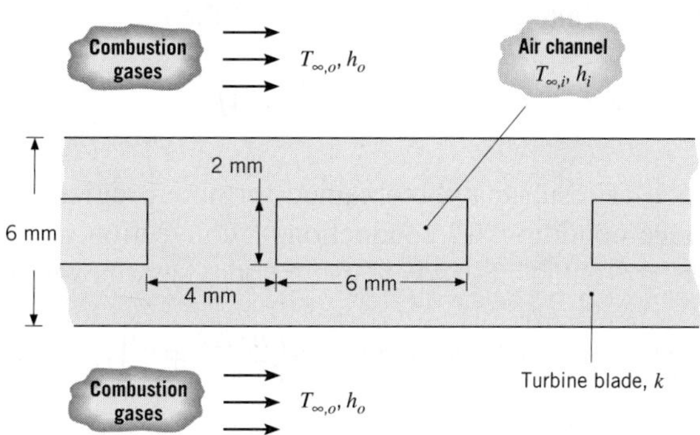

Under operating conditions for which $h_o = 1000\ \text{W/m}^2\cdot\text{K}$, $T_{\infty,o} = 1700\ \text{K}$, $h_i = 200\ \text{W/m}^2\cdot\text{K}$, and $T_{\infty,i} = 400\ \text{K}$, determine the temperature field in the turbine blade and the rate of heat transfer per unit length to the channel. At what location is the temperature a maximum?

SOLUTION

Known: Dimensions and operating conditions for a gas turbine blade with embedded channels.

Find: Temperature field in the blade, including a location of maximum temperature. Rate of heat transfer per unit length to the channel.

Schematic:

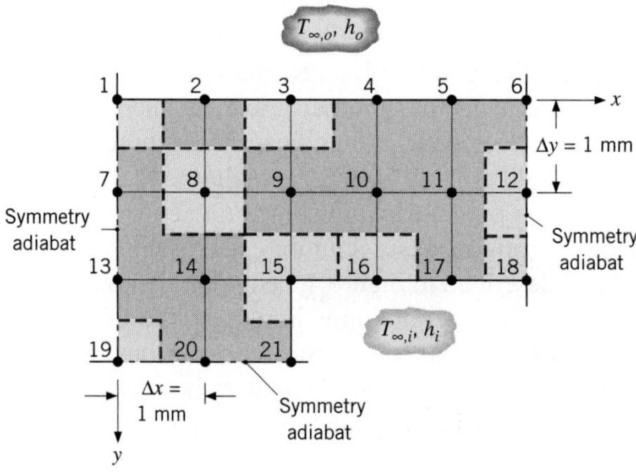

Assumptions:

1. Steady-state, two-dimensional conduction.
2. Constant properties.

Analysis: Adopting a grid space of $\Delta x = \Delta y = 1$ mm and identifying the three lines of symmetry, the foregoing nodal network is constructed. The corresponding finite-difference equations may be obtained by applying the energy balance method to nodes 1, 6, 18, 19, and 21 and by using the results of Table 4.2 for the remaining nodes.

Heat transfer to node 1 occurs by conduction from nodes 2 and 7, as well as by convection from the outer fluid. Since there is no heat transfer from the region beyond the symmetry adiabat, application of an energy balance to the one-quarter section associated with node 1 yields a finite-difference equation of the form

$$\text{Node 1:} \quad T_2 + T_7 - \left(2 + \frac{h_o \Delta x}{k}\right) T_1 = -\frac{h_o \Delta x}{k} T_{\infty,o}$$

A similar result may be obtained for nodal region 6, which is characterized by equivalent surface conditions (2 conduction, 1 convection, 1 adiabatic). Nodes 2 to 5 correspond to case 3 of Table 4.2, and choosing node 3 as an example, it follows that

$$\text{Node 3:} \quad T_2 + T_4 + 2T_9 - 2\left(\frac{h_o \Delta x}{k} + 2\right) T_3 = -\frac{2h_o \Delta x}{k} T_{\infty,o}$$

Nodes 7, 12, 13, and 20 correspond to case 5 of Table 4.2, with $q'' = 0$, and choosing node 12 as an example, it follows that

$$\text{Node 12:} \quad T_6 + 2T_{11} + T_{18} - 4T_{12} = 0$$

Nodes 8 to 11 and 14 are interior nodes (case 1), in which case the finite-difference equation for node 8 is

$$\text{Node 8:} \quad T_2 + T_7 + T_9 + T_{14} - 4T_8 = 0$$

Node 15 is an internal corner (case 2) for which

$$\text{Node 15:} \quad 2T_9 + 2T_{14} + T_{16} + T_{21} - 2\left(3 + \frac{h_i \Delta x}{k}\right)T_{15} = -2\frac{h_i \Delta x}{k} T_{\infty,i}$$

while nodes 16 and 17 are situated on a plane surface with convection (case 3):

$$\text{Node 16:} \quad 2T_{10} + T_{15} + T_{17} - 2\left(\frac{h_i \Delta x}{k} + 2\right)T_{16} = -\frac{2h_i \Delta x}{k} T_{\infty,i}$$

In each case, heat transfer to nodal regions 18 and 21 is characterized by conduction from two adjoining nodes and convection from the internal flow, with no heat transfer occurring from an adjoining adiabat. Performing an energy balance for nodal region 18, it follows that

$$\text{Node 18:} \quad T_{12} + T_{17} - \left(2 + \frac{h_i \Delta x}{k}\right) T_{18} = -\frac{h_i \Delta x}{k} T_{\infty,i}$$

The last special case corresponds to nodal region 19, which has two adiabatic surfaces and experiences heat transfer by conduction across the other two surfaces.

$$\text{Node 19:} \quad T_{13} + T_{20} - 2T_{19} = 0$$

The equations for nodes 1 through 21 may be solved simultaneously using *IHT*, another commercial code, or a handheld calculator. The following results are obtained:

T_1	T_2	T_3	T_4	T_5	T_6
1526.0 K	1525.3 K	1523.6 K	1521.9 K	1520.8 K	1520.5 K
T_7	T_8	T_9	T_{10}	T_{11}	T_{12}
1519.7 K	1518.8 K	1516.5 K	1514.5 K	1513.3 K	1512.9 K
T_{13}	T_{14}	T_{15}	T_{16}	T_{17}	T_{18}
1515.1 K	1513.7 K	1509.2 K	1506.4 K	1505.0 K	1504.5 K
T_{19}	T_{20}	T_{21}			
1513.4 K	1511.7 K	1506.0 K			

The temperature field may also be represented in the form of isotherms, and four such lines of constant temperature are shown schematically. Also shown are heat flux lines that have been carefully drawn so that they are everywhere perpendicular to the isotherms and coincident with the symmetry adiabat. The surfaces that are exposed to the combustion gases and air are not isothermal, and therefore the heat flow lines are not perpendicular to these boundaries.

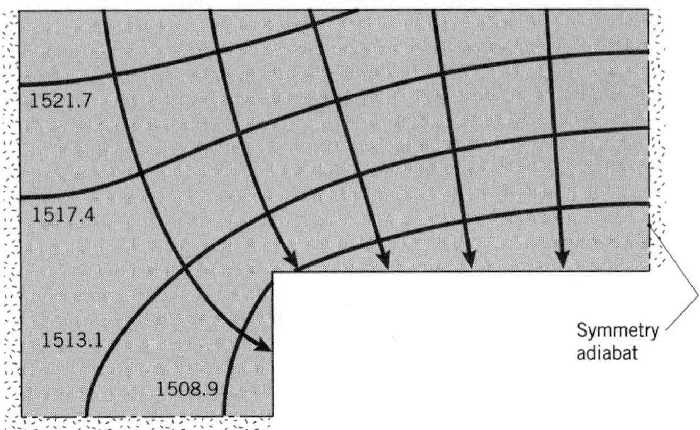

As expected, the maximum temperature exists at the location farthest removed from the coolant, which corresponds to node 1. Temperatures along the surface of the turbine blade exposed to the combustion gases are of particular interest. The finite-difference predictions are plotted below (with straight lines connecting the nodal temperatures).

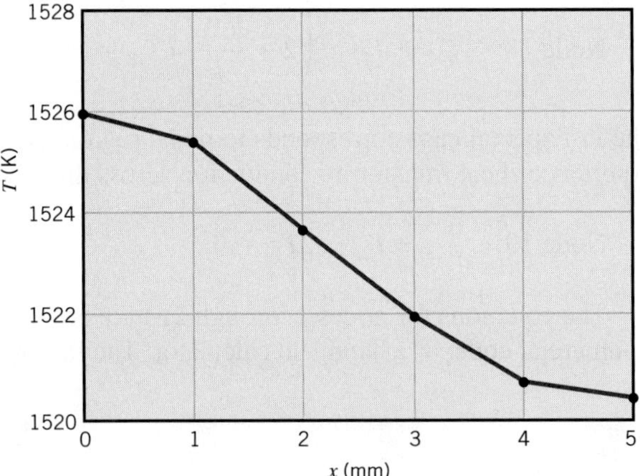

The rate of heat transfer per unit length of channel may be calculated in two ways. Based on heat transfer from the blade to the air, it is

$$q' = 4h_i[(\Delta y/2)(T_{21} - T_{\infty,i}) + (\Delta y/2 + \Delta x/2)(T_{15} - T_{\infty,i}) + (\Delta x)(T_{16} - T_{\infty,i}) + \Delta x(T_{17} - T_{\infty,i}) + (\Delta x/2)(T_{18} - T_{\infty,i})]$$

Alternatively, based on heat transfer from the combustion gases to the blade, it is

$$q' = 4h_o[(\Delta x/2)(T_{\infty,o} - T_1) + (\Delta x)(T_{\infty,o} - T_2) + (\Delta x)(T_{\infty,o} - T_3) + (\Delta x)(T_{\infty,o} - T_4) + (\Delta x)(T_{\infty,o} - T_5) + (\Delta x/2)(T_{\infty,o} - T_6)]$$

where the factor of 4 originates from the symmetry conditions. In both cases, we obtain

$$q' = 3540.6 \text{ W/m} \qquad \triangleleft$$

Comments:

1. In matrix notation, following Equation 4.48, the equations for nodes 1 through 21 are of the form $[A][T] = [C]$, where

$$[A] = \begin{bmatrix}
-a & 1 & 0 & 0 & 0 & 0 & 1 & 0 & 0 & 0 & 0 & 0 & 0 & 0 & 0 & 0 & 0 & 0 & 0 & 0 & 0 \\
1 & -b & 1 & 0 & 0 & 0 & 0 & 2 & 0 & 0 & 0 & 0 & 0 & 0 & 0 & 0 & 0 & 0 & 0 & 0 & 0 \\
0 & 1 & -b & 1 & 0 & 0 & 0 & 0 & 2 & 0 & 0 & 0 & 0 & 0 & 0 & 0 & 0 & 0 & 0 & 0 & 0 \\
0 & 0 & 1 & -b & 1 & 0 & 0 & 0 & 0 & 2 & 0 & 0 & 0 & 0 & 0 & 0 & 0 & 0 & 0 & 0 & 0 \\
0 & 0 & 0 & 1 & -b & 1 & 0 & 0 & 0 & 0 & 2 & 0 & 0 & 0 & 0 & 0 & 0 & 0 & 0 & 0 & 0 \\
0 & 0 & 0 & 0 & 1 & -a & 0 & 0 & 0 & 0 & 0 & 1 & 0 & 0 & 0 & 0 & 0 & 0 & 0 & 0 & 0 \\
1 & 0 & 0 & 0 & 0 & 0 & -4 & 2 & 0 & 0 & 0 & 0 & 1 & 0 & 0 & 0 & 0 & 0 & 0 & 0 & 0 \\
0 & 1 & 0 & 0 & 0 & 0 & 1 & -4 & 1 & 0 & 0 & 0 & 0 & 1 & 0 & 0 & 0 & 0 & 0 & 0 & 0 \\
0 & 0 & 1 & 0 & 0 & 0 & 0 & 1 & -4 & 1 & 0 & 0 & 0 & 0 & 1 & 0 & 0 & 0 & 0 & 0 & 0 \\
0 & 0 & 0 & 1 & 0 & 0 & 0 & 0 & 1 & -4 & 1 & 0 & 0 & 0 & 0 & 1 & 0 & 0 & 0 & 0 & 0 \\
0 & 0 & 0 & 0 & 1 & 0 & 0 & 0 & 0 & 1 & -4 & 1 & 0 & 0 & 0 & 0 & 1 & 0 & 0 & 0 & 0 \\
0 & 0 & 0 & 0 & 0 & 1 & 0 & 0 & 0 & 0 & 2 & -4 & 0 & 0 & 0 & 0 & 0 & 1 & 0 & 0 & 0 \\
0 & 0 & 0 & 0 & 0 & 0 & 1 & 0 & 0 & 0 & 0 & 0 & -4 & 2 & 0 & 0 & 0 & 0 & 1 & 0 & 0 \\
0 & 0 & 0 & 0 & 0 & 0 & 0 & 1 & 0 & 0 & 0 & 0 & 1 & -4 & 1 & 0 & 0 & 0 & 0 & 1 & 0 \\
0 & 0 & 0 & 0 & 0 & 0 & 0 & 0 & 2 & 0 & 0 & 0 & 0 & 2 & -c & 1 & 0 & 0 & 0 & 0 & 1 \\
0 & 0 & 0 & 0 & 0 & 0 & 0 & 0 & 0 & 2 & 0 & 0 & 0 & 0 & 1 & -d & 1 & 0 & 0 & 0 & 0 \\
0 & 0 & 0 & 0 & 0 & 0 & 0 & 0 & 0 & 0 & 2 & 0 & 0 & 0 & 0 & 1 & -d & 1 & 0 & 0 & 0 \\
0 & 0 & 0 & 0 & 0 & 0 & 0 & 0 & 0 & 0 & 0 & 1 & 0 & 0 & 0 & 0 & 1 & -e & 0 & 0 & 0 \\
0 & 0 & 0 & 0 & 0 & 0 & 0 & 0 & 0 & 0 & 0 & 0 & 1 & 0 & 0 & 0 & 0 & 0 & -2 & 1 & 0 \\
0 & 0 & 0 & 0 & 0 & 0 & 0 & 0 & 0 & 0 & 0 & 0 & 0 & 2 & 0 & 0 & 0 & 0 & 1 & -4 & 1 \\
0 & 0 & 0 & 0 & 0 & 0 & 0 & 0 & 0 & 0 & 0 & 0 & 0 & 0 & 1 & 0 & 0 & 0 & 0 & 1 & -e
\end{bmatrix} \qquad [C] = \begin{bmatrix} -f \\ -2f \\ -2f \\ -2f \\ -2f \\ -f \\ 0 \\ 0 \\ 0 \\ 0 \\ 0 \\ 0 \\ 0 \\ 0 \\ -2g \\ -2g \\ -2g \\ -g \\ 0 \\ 0 \\ -g \end{bmatrix}$$

   With $h_o \Delta x/k = 0.04$ and $h_i \Delta x/k = 0.008$, the following coefficients in the equations can be calculated: $a = 2.04$, $b = 4.08$, $c = 6.016$, $d = 4.016$, $e = 2.008$, $f = 68$, and $g = 3.2$. By framing the equations as a matrix equation, standard tools for solving matrix equations may be used.

2. To ensure that no errors have been made in formulating and solving the finite-difference equations, the calculated temperatures should be used to verify that conservation of energy is satisfied for a control surface surrounding all nodal regions. This check has already been performed, since it was shown that the heat transfer rate from the combustion gases to the blade is equal to that from the blade to the air.

3. The accuracy of the finite-difference solution may be improved by refining the grid. If, for example, we halve the grid spacing ($\Delta x = \Delta y = 0.5$ mm), thereby increasing the number of unknown nodal temperatures to 65, we obtain the following results for selected temperatures and the heat rate:

$$\begin{aligned}
&T_1 = 1525.9 \text{ K}, \quad T_6 = 1520.5 \text{ K}, \quad T_{15} = 1509.2 \text{ K}, \\
&T_{18} = 1504.5 \text{ K}, \quad T_{19} = 1513.5 \text{ K}, \quad T_{21} = 1505.7 \text{ K}, \\
&q' = 3539.9 \text{ W/m}
\end{aligned}$$

   Agreement between the two sets of results is excellent. Of course, use of the finer mesh increases setup and computation time, and in many cases the results obtained from a coarse grid are satisfactory. Selection of the appropriate grid is a judgment that the engineer must make.

4. In the gas turbine industry, there is great interest in adopting measures that reduce blade temperatures. Such measures could include use of a different alloy of larger thermal conductivity and/or increasing coolant flow through the channel, thereby increasing h_i. Using the finite-difference solution with $\Delta x = \Delta y = 1$ mm, the following results are obtained for parametric variations of k and h_i:

k (W/m·K)	h_i (W/m^2·K)	T_1 (K)	q' (W/m)
25	200	1526.0	3540.6
50	200	1523.4	3563.3
25	1000	1154.5	11,095.5
50	1000	1138.9	11,320.7

Why do increases in k and h_i reduce temperature in the blade? Why is the effect of the change in h_i more significant than that of k?

5. Note that, because the exterior surface of the blade is at an extremely high temperature, radiation losses to its surroundings may be significant. In the finite-difference analysis, such effects could be considered by linearizing the radiation rate equation (see Equations 1.8 and 1.9) and treating radiation in the same manner as convection. However, because the radiation coefficient h_r depends on the surface temperature, an iterative finite-difference solution would be necessary to ensure that the resulting surface temperatures correspond to the temperatures at which h_r is evaluated at each nodal point.

6. See Example 4.3 in *IHT*. This problem can also be solved using *Tools, Finite-Difference Equations* in the *Advanced* section of *IHT*.

7. A second software package accompanying this text, *Finite-Element Heat Transfer (FEHT)*, may also be used to solve one- and two-dimensional forms of the heat equation. This example is provided as a solved model in *FEHT* and may be accessed through *Examples* on the *Toolbar*.

4.6 Summary

The primary objective of this chapter was to develop an appreciation for the nature of a two-dimensional conduction problem and the methods that are available for its solution. When confronted with a two-dimensional problem, one should first determine whether an exact solution is known. This may be done by examining some of the excellent references in which exact solutions to the heat equation are obtained [1–5]. One may also want to determine whether the shape factor or dimensionless conduction heat rate is known for the system of interest [6–10]. However, often, conditions are such that the use of a shape factor, dimensionless conduction heat rate, or an exact solution is not possible, and it is necessary to use a finite-difference or finite-element solution. You should therefore appreciate the inherent nature of the *discretization process* and know how to formulate and solve the finite-difference equations for the discrete points of a nodal network. You may test your understanding of related concepts by addressing the following questions.

- What is an *isotherm*? What is a *heat flow line*? How are the two lines related geometrically?
- What is an *adiabat*? How is it related to a line of symmetry? How is it intersected by an isotherm?
- What parameters characterize the effect of geometry on the relationship between the heat rate and the overall temperature difference for steady conduction in a two-dimensional system? How are these parameters related to the conduction resistance?
- What is represented by the temperature of a *nodal point*, and how does the accuracy of a nodal temperature depend on prescription of the *nodal network*?

References

1. Schneider, P. J., *Conduction Heat Transfer,* Addison-Wesley, Reading, MA, 1955.
2. Carslaw, H. S., and J. C. Jaeger, *Conduction of Heat in Solids*, Oxford University Press, London, 1959.
3. Özisik, M. N., *Heat Conduction*, Wiley Interscience, New York, 1980.
4. Kakac, S., and Y. Yener, *Heat Conduction,* Hemisphere Publishing, New York, 1985.
5. Poulikakos, D., *Conduction Heat Transfer,* Prentice-Hall, Englewood Cliffs, NJ, 1994.
6. Sunderland, J. E., and K. R. Johnson, *Trans. ASHRAE,* **10**, 237–241, 1964.
7. Kutateladze, S. S., *Fundamentals of Heat Transfer,* Academic Press, New York, 1963.
8. General Electric Co. (Corporate Research and Development), *Heat Transfer Data Book,* Section 502, General Electric Company, Schenectady, NY, 1973.
9. Hahne, E., and U. Grigull, *Int. J. Heat Mass Transfer,* **18**, 751–767, 1975.
10. Yovanovich, M. M., in W. M. Rohsenow, J. P. Hartnett, and Y. I. Cho, Eds., *Handbook of Heat Transfer*, McGraw-Hill, New York, 1998, pp. 3.1–3.73.
11. Gerald, C. F., and P. O. Wheatley, *Applied Numerical Analysis*, Pearson Education, Upper Saddle River, NJ, 1998.
12. Hoffman, J. D., *Numerical Methods for Engineers and Scientists,* McGraw-Hill, New York, 1992.

Problems

Exact Solutions

4.1 In the method of separation of variables (Section 4.2) for two-dimensional, steady-state conduction, the separation constant λ^2 in Equations 4.6 and 4.7 must be a positive constant. Show that a negative or zero value of λ^2 will result in solutions that cannot satisfy the prescribed boundary conditions.

4.2 A two-dimensional rectangular plate is subjected to prescribed boundary conditions. Using the results of the exact solution for the heat equation presented in Section 4.2, calculate the temperature at the midpoint (1, 0.5) by considering the first five nonzero terms of the infinite series that must be evaluated. Assess the error resulting from using only the first three terms of the infinite series. Plot the temperature distributions $T(x, 0.5)$ and $T(1.0, y)$.

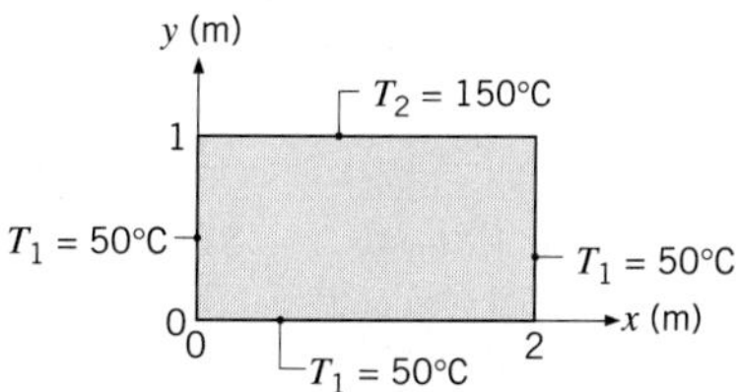

4.3 Consider the two-dimensional rectangular plate of Problem 4.2 having a thermal conductivity of 50 W/m·K. Beginning with the exact solution for the temperature distribution, derive an expression for the heat transfer rate per unit thickness from the plate along the lower surface ($0 \le x \le 2$, $y = 0$). Evaluate the heat rate considering the first five nonzero terms of the infinite series.

4.4 A two-dimensional rectangular plate is subjected to the boundary conditions shown. Derive an expression for the steady-state temperature distribution $T(x, y)$.

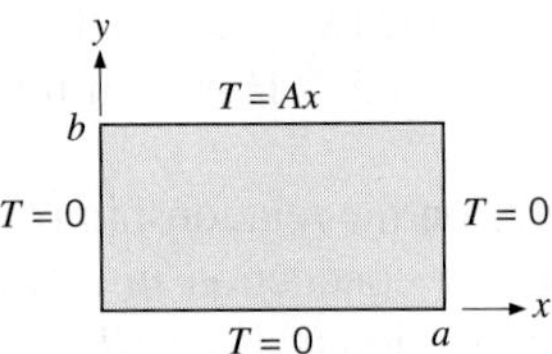

4.5 A two-dimensional rectangular plate is subjected to prescribed temperature boundary conditions on three sides and a uniform heat flux *into* the plate at the top surface. Using the general approach of Section 4.2, derive an expression for the temperature distribution in the plate.

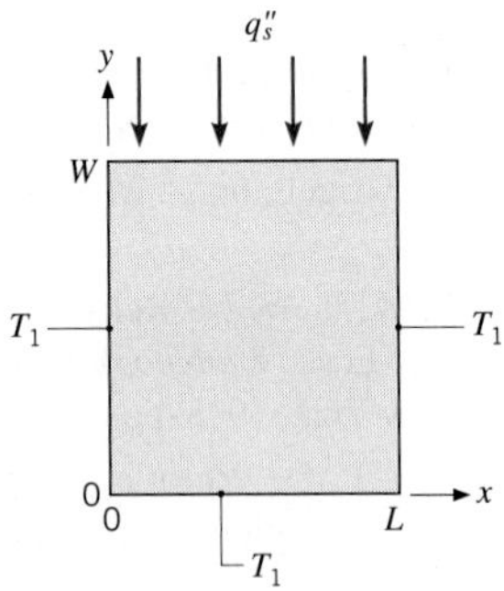

Shape Factors and Dimensionless Conduction Heat Rates

4.6 Using the thermal resistance relations developed in Chapter 3, determine shape factor expressions for the following geometries:

(a) Plane wall, cylindrical shell, and spherical shell.

(b) Isothermal sphere of diameter D buried in an infinite medium.

4.7 Free convection heat transfer is sometimes quantified by writing Equation 4.20 as $q_{conv} = Sk_{eff}\,\Delta T_{1-2}$, where k_{eff} is an *effective* thermal conductivity. The ratio k_{eff}/k is greater than unity because of fluid motion driven by buoyancy forces, as represented by the dashed streamlines.

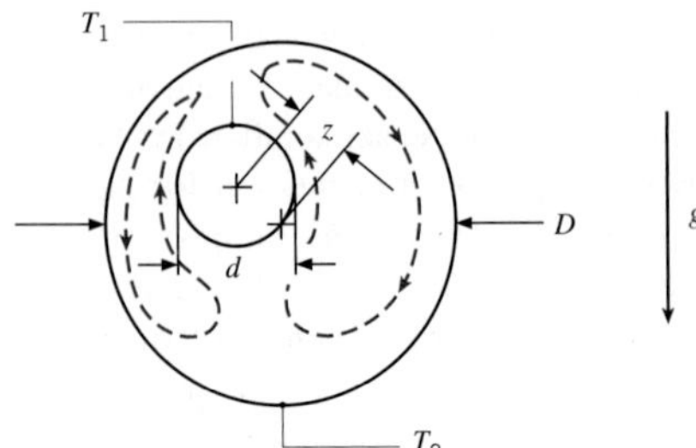

An experiment for the configuration shown yields a heat transfer rate per unit length of $q'_{conv} = 110$ W/m for surface temperatures of $T_1 = 53°C$ and $T_2 = 15°C$, respectively. For inner and outer cylinders of diameters $d = 20$ mm and $D = 60$ mm, and an eccentricity factor of $z = 10$ mm, determine the value of k_{eff}. The actual thermal conductivity of the fluid is $k = 0.255$ W/m·K.

4.8 Consider Problem 4.5 for the case where the plate is of square cross section, $W = L$.

(a) Derive an expression for the shape factor, S_{max}, associated with the *maximum* top surface temperature, such that $q = S_{max}\, k\, (T_{2,max} - T_1)$ where $T_{2,max}$ is the maximum temperature along $y = W$.

(b) Derive an expression for the shape factor, S_{avg}, associated with the *average* top surface temperature, $q = S_{avg} k(\overline{T}_2 - T_1)$ where $\overline{T}_2$ is the average temperature along $y = W$.

(c) Evaluate the shape factors that can be used to determine the maximum and average temperatures along $y = W$. Evaluate the maximum and average temperatures for $T_1 = 0°C$, $L = W = 10$ mm, $k = 20$ W/m·K, and $q''_s = 1000$ W/m^2.

4.9 Radioactive wastes are temporarily stored in a spherical container, the center of which is buried a distance of 10 m below the earth's surface. The outside diameter of the container is 2 m, and 500 W of heat are released as a result of radioactive decay. If the soil surface temperature is 20°C, what is the outside surface temperature of the container under steady-state conditions? On a sketch of the soil–container system drawn to scale, show representative isotherms and heat flow lines in the soil.

4.10 Based on the dimensionless conduction heat rates for cases 12–15 in Table 4.1*b*, find shape factors for the following objects having temperature T_1, located at the surface of a semi-infinite medium having temperature T_2. The surface of the semi-infinite medium is adiabatic.

(a) A buried hemisphere, flush with the surface.

(b) A disk on the surface. Compare your result to Table 4.1*a*, case 10.

(c) A square on the surface.

(d) A buried cube, flush with the surface.

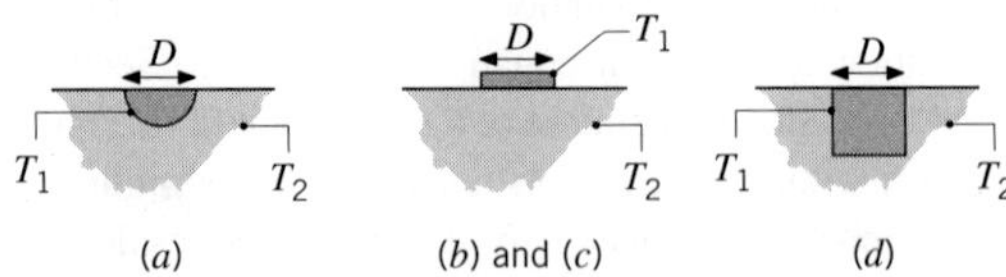

4.11 Determine the heat transfer rate between two particles of diameter $D = 100\ \mu$m and temperatures $T_1 = 300.1$ K and $T_2 = 299.9$ K, respectively. The particles are in contact and are surrounded by air.

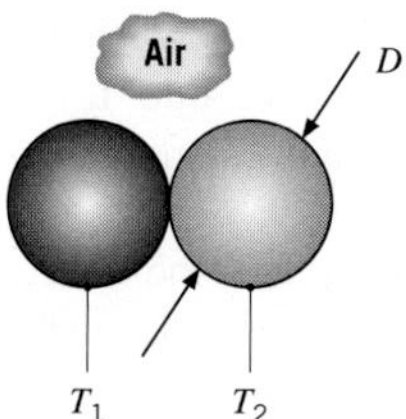

4.12 A two-dimensional object is subjected to isothermal conditions at its left and right surfaces, as shown in the schematic. Both diagonal surfaces are adiabatic and the depth of the object is $L = 100$ mm.

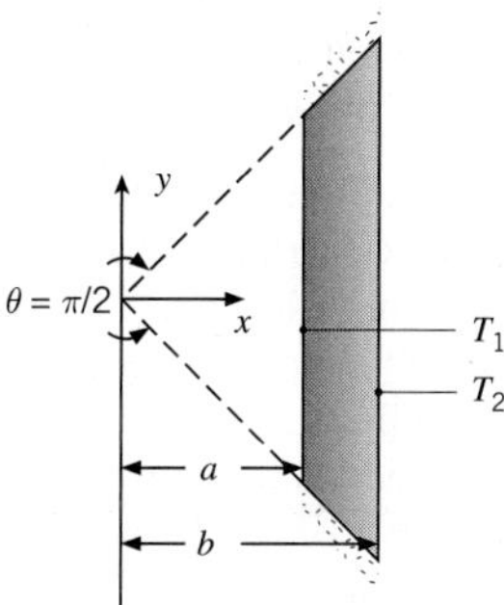

(a) Determine the two-dimensional shape factor for the object for $a = 10$ mm, $b = 12$ mm.

(b) Determine the two-dimensional shape factor for the object for $a = 10$ mm, $b = 15$ mm.

(c) Use the alternative conduction analysis of Section 3.2 to estimate the shape factor for parts (a) and (b). Compare the values of the approximate shape factors of the alternative conduction analysis to the two-dimensional shape factors of parts (a) and (b).

(d) For $T_1 = 100°C$ and $T_2 = 60°C$, determine the heat transfer rate per unit depth for $k = 15$ W/m·K for parts (a) and (b).

4.13 An electrical heater 100 mm long and 5 mm in diameter is inserted into a hole drilled normal to the surface of a large block of material having a thermal conductivity of 5 W/m·K. Estimate the temperature reached by the heater when dissipating 50 W with the surface of the block at a temperature of 25°C.

4.14 Two parallel pipelines spaced 0.5 m apart are buried in soil having a thermal conductivity of 0.5 W/m·K. The pipes have outer diameters of 100 and 75 mm with surface temperatures of 175°C and 5°C, respectively. Estimate the heat transfer rate per unit length between the two pipelines.

4.15 A small water droplet of diameter $D = 100$ μm and temperature $T_{mp} = 0°C$ falls on a nonwetting metal surface that is at temperature $T_s = -15°C$. Determine how long it will take for the droplet to freeze completely. The latent heat of fusion is $h_{sf} = 334$ kJ/kg.

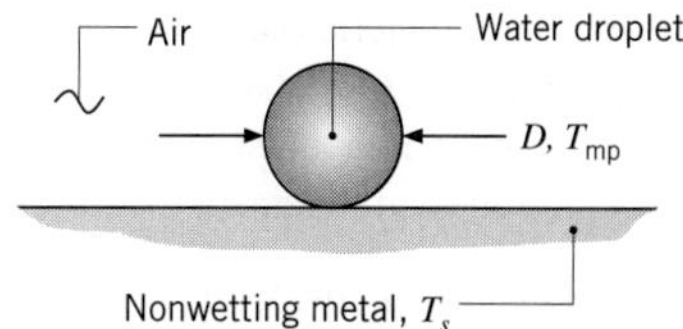

4.16 A tube of diameter 50 mm having a surface temperature of 85°C is embedded in the center plane of a concrete slab 0.1 m thick with upper and lower surfaces at 20°C. Using the appropriate tabulated relation for this configuration, find the shape factor. Determine the heat transfer rate per unit length of the tube.

4.17 Pressurized steam at 450 K flows through a long, thin-walled pipe of 0.5-m diameter. The pipe is enclosed in a concrete casing that is of square cross section and 1.5 m on a side. The axis of the pipe is centered in the casing, and the outer surfaces of the casing are maintained at 300 K. What is the heat loss per unit length of pipe?

4.18 The temperature distribution in laser-irradiated materials is determined by the power, size, and shape of the laser beam, along with the properties of the material being irradiated. The beam shape is typically Gaussian, and the local beam irradiation flux (often referred to as the laser *fluence*) is

$$q''(x, y) = q''(x = y = 0)\exp(-x/r_b)^2 \exp(-y/r_b)^2$$

The x- and y-coordinates determine the location of interest on the surface of the irradiated material. Consider the case where the center of the beam is located at $x = y = r = 0$. The beam is characterized by a radius r_b, defined as the radial location where the local fluence is $q''(r_b) = q''(r = 0)/e \approx 0.368q''(r = 0)$.

A shape factor for Gaussian heating is $S = 2\pi^{1/2}r_b$, where S is defined in terms of $T_{1,max} - T_2$ [Nissin, Y. I., A. Lietoila, R. G. Gold, and J. F. Gibbons, *J. Appl. Phys.*, **51**, 274, 1980]. Calculate the maximum steady-state surface temperature associated with irradiation of a material of thermal conductivity $k = 27$ W/m·K and absorptivity $\alpha = 0.45$ by a Gaussian beam with $r_b = 0.1$ mm and power $P = 1$ W. Compare your result with the maximum temperature that would occur if the irradiation was from a circular beam of the same diameter and power, but characterized by a uniform fluence (a *flat* beam). Also calculate the average temperature of the irradiated surface for the uniform fluence case. The temperature far from the irradiated spot is $T_2 = 25°C$.

4.19 Hot water at 85°C flows through a thin-walled copper tube of 30-mm diameter. The tube is enclosed by an eccentric cylindrical shell that is maintained at 35°C and has a diameter of 120 mm. The eccentricity, defined as the separation between the centers of the tube and shell, is 20 mm. The space between the tube and shell is filled with an insulating material having a thermal conductivity of 0.05 W/m·K. Calculate the heat loss per unit length of the tube, and compare the result with the heat loss for a concentric arrangement.

4.20 A furnace of cubical shape, with external dimensions of 0.35 m, is constructed from a refractory brick (fireclay). If the wall thickness is 50 mm, the inner surface temperature is 600°C, and the outer surface temperature is 75°C, calculate the heat loss from the furnace.

4.21 Laser beams are used to thermally process materials in a wide range of applications. Often, the beam is scanned along the surface of the material in a desired pattern. Consider the laser heating process of Problem 4.18, except now the laser beam scans the material at a scanning velocity of U. A dimensionless maximum surface temperature can be well correlated by an expression of the form [Nissin, Y. I., A. Lietoila, R. G. Gold, and J. F. Gibbons, *J. Appl. Phys.*, **51**, 274, 1980]

$$\frac{T_{1,\max,U=0} - T_2}{T_{1,\max,U\neq 0} - T_2} = 1 + 0.301Pe - 0.0108Pe^2$$

for the range $0 < Pe < 10$, where Pe is a dimensionless velocity known as the Peclet number. For this problem, $Pe = Ur_b/\sqrt{2}\alpha$ where α is the thermal diffusivity of the material. The maximum material temperature does not occur directly below the laser beam, but at a lag distance δ behind the center of the moving beam. The dimensionless lag distance can be correlated to Pe by [Sheng, I. C., and Y. Chen, *J. Thermal Stresses*, **14**, 129, 1991]

$$\frac{\delta U}{\alpha} = 0.944Pe^{1.55}$$

(a) For the laser beam size and shape and material of Problem 4.18, determine the laser power required to achieve $T_{1,\max} = 200°\text{C}$ for $U = 2$ m/s. The density and specific heat of the material are $\rho = 2000$ kg/m^3 and $c = 800$ J/kg·K, respectively.

(b) Determine the lag distance δ associated with $U = 2$ m/s.

(c) Plot the required laser power to achieve $T_{\max,1} = 200°\text{C}$ for $0 \leq U \leq 2$ m/s.

Shape Factors with Thermal Circuits

4.22 A double-glazed window consists of two sheets of glass separated by an $L = 0.2$-mm-thick gap. The gap is evacuated, eliminating conduction and convection across the gap. Small cylindrical pillars, each $L = 0.2$ mm long and $D = 0.15$ mm in diameter, are inserted between the glass sheets to ensure that the glass does not break due to stresses imposed by the pressure difference across each glass sheet. A contact resistance of $R''_{t,c} = 1.5 \times 10^{-6}\ \text{m}^2\cdot\text{K/W}$ exists between the pillar and the sheet. For nominal glass temperatures of $T_1 = 20°\text{C}$ and $T_2 = -10°\text{C}$, determine the conduction heat transfer through an individual stainless steel pillar.

4.23 A pipeline, used for the transport of crude oil, is buried in the earth such that its centerline is a distance of 1.5 m below the surface. The pipe has an outer diameter of 0.5 m and is insulated with a layer of cellular glass 100 mm thick. What is the heat loss per unit length of pipe when heated oil at 120°C flows through the pipe and the surface of the earth is at a temperature of 0°C?

4.24 A long power transmission cable is buried at a depth (ground-to-cable-centerline distance) of 2 m. The cable is encased in a thin-walled pipe of 0.1-m diameter, and, to render the cable *superconducting* (with essentially zero power dissipation), the space between the cable and pipe is filled with liquid nitrogen at 77 K. If the pipe is covered with a superinsulator ($k_i = 0.005$ W/m·K) of 0.05-m thickness and the surface of the earth ($k_g = 1.2$ W/m·K) is at 300 K, what is the cooling load (W/m) that must be maintained by a cryogenic refrigerator per unit pipe length?

4.25 A small device is used to measure the surface temperature of an object. A thermocouple bead of diameter $D = 120\ \mu$m is positioned a distance $z = 100\ \mu$m from the surface of interest. The two thermocouple wires, each of diameter $d = 25\ \mu$m and length $L = 300\ \mu$m, are held by a large manipulator that is at a temperature of $T_m = 23°\text{C}$.

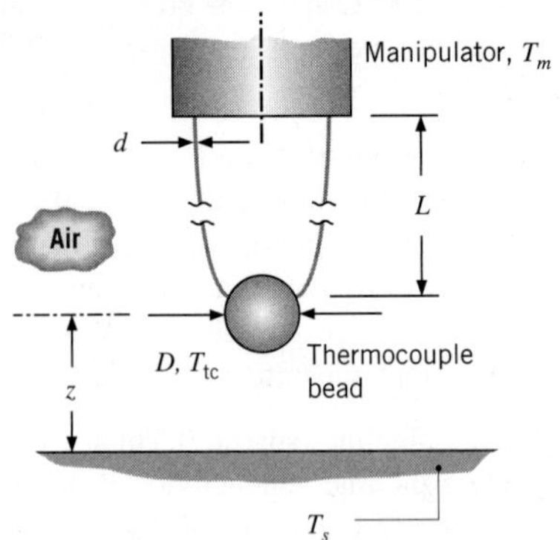

If the thermocouple registers a temperature of $T_{tc} = 29°\text{C}$, what is the surface temperature? The thermal conductivities of the chromel and alumel thermocouple wires are $k_{Ch} = 19\ \text{W/m}\cdot\text{K}$ and $k_{Al} = 29\ \text{W/m}\cdot\text{K}$, respectively. You may neglect radiation and convection effects.

4.26 A cubical glass melting furnace has exterior dimensions of width $W = 5$ m on a side and is constructed from refractory brick of thickness $L = 0.35$ m and thermal conductivity $k = 1.4\ \text{W/m}\cdot\text{K}$. The sides and top of the furnace are exposed to ambient air at 25°C, with free convection characterized by an average coefficient of $h = 5\ \text{W/m}^2\cdot\text{K}$. The bottom of the furnace rests on a framed platform for which much of the surface is exposed to the ambient air, and a convection coefficient of $h = 5\ \text{W/m}^2\cdot\text{K}$ may be assumed as a first approximation. Under operating conditions for which combustion gases maintain the inner surfaces of the furnace at 1100°C, what is the heat loss from the furnace?

4.27 A hot fluid passes through circular channels of a cast iron platen (A) of thickness $L_A = 30$ mm which is in poor contact with the cover plates (B) of thickness $L_B = 7.5$ mm. The channels are of diameter $D = 15$ mm with a centerline spacing of $L_o = 60$ mm. The thermal conductivities of the materials are $k_A = 20\ \text{W/m}\cdot\text{K}$ and $k_B = 75\ \text{W/m}\cdot\text{K}$, while the contact resistance between the two materials is $R''_{t,c} = 2.0 \times 10^{-4}\ \text{m}^2\cdot\text{K/W}$. The hot fluid is at $T_i = 150°\text{C}$, and the convection coefficient is $1000\ \text{W/m}^2\cdot\text{K}$. The cover plate is exposed to ambient air at $T_\infty = 25°\text{C}$ with a convection coefficient of $200\ \text{W/m}^2\cdot\text{K}$. The shape factor between one channel and the platen top and bottom surfaces is 4.25.

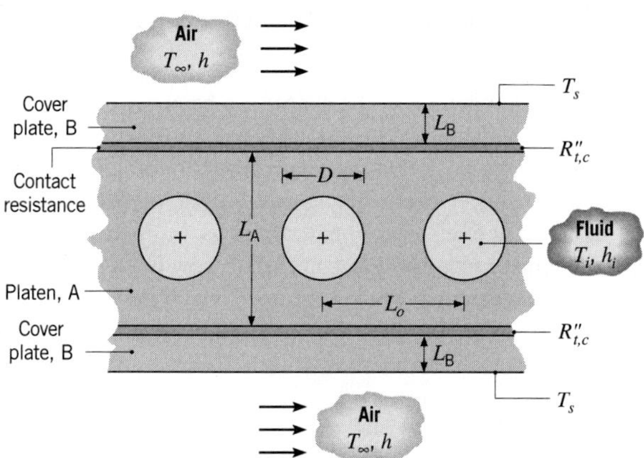

(a) Determine the heat rate from a single channel per unit length of the platen normal to the page, q'_i.

(b) Determine the outer surface temperature of the cover plate, T_s.

(c) Comment on the effects that changing the centerline spacing will have on q'_i and T_s. How would insulating the lower surface affect q'_i and T_s?

4.28 An aluminum heat sink ($k = 240\ \text{W/m}\cdot\text{K}$), used to cool an array of electronic chips, consists of a square channel of inner width $w = 25$ mm, through which liquid flow may be assumed to maintain a uniform surface temperature of $T_1 = 20°\text{C}$. The outer width and length of the channel are $W = 40$ mm and $L = 160$ mm, respectively.

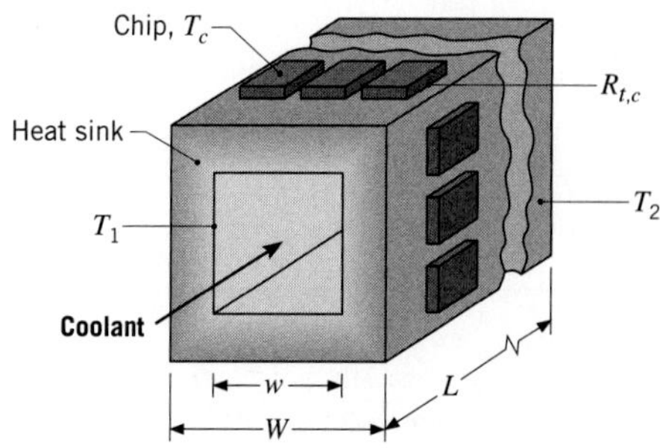

If $N = 120$ chips attached to the outer surfaces of the heat sink maintain an approximately uniform surface temperature of $T_2 = 50°\text{C}$ and all of the heat dissipated by the chips is assumed to be transferred to the coolant, what is the heat dissipation per chip? If the contact resistance between each chip and the heat sink is $R_{t,c} = 0.2$ K/W, what is the chip temperature?

4.29 Hot water is transported from a cogeneration power station to commercial and industrial users through steel pipes of diameter $D = 150$ mm, with each pipe centered in concrete ($k = 1.4\ \text{W/m}\cdot\text{K}$) of square cross section ($w = 300$ mm). The outer surfaces of the concrete are exposed to ambient air for which $T_\infty = 0°\text{C}$ and $h = 25\ \text{W/m}^2\cdot\text{K}$.

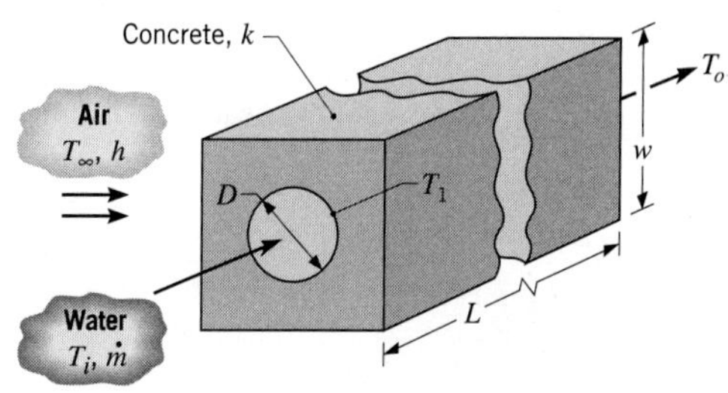

(a) If the inlet temperature of water flowing through the pipe is $T_i = 90°\text{C}$, what is the heat loss per unit length of pipe in proximity to the inlet? The temperature of the pipe T_1 may be assumed to be that of the inlet water.

(b) If the difference between the inlet and outlet temperatures of water flowing through a 100-m-long pipe is not to exceed 5°C, estimate the minimum allowable flow rate $\dot{m}$. A value of $c = 4207$ J/kg·K may be used for the specific heat of the water.

4.30 A long constantan wire of 1-mm diameter is butt welded to the surface of a large copper block, forming a thermocouple junction. The wire behaves as a fin, permitting heat to flow from the surface, thereby depressing the sensing junction temperature T_j below that of the block T_o.

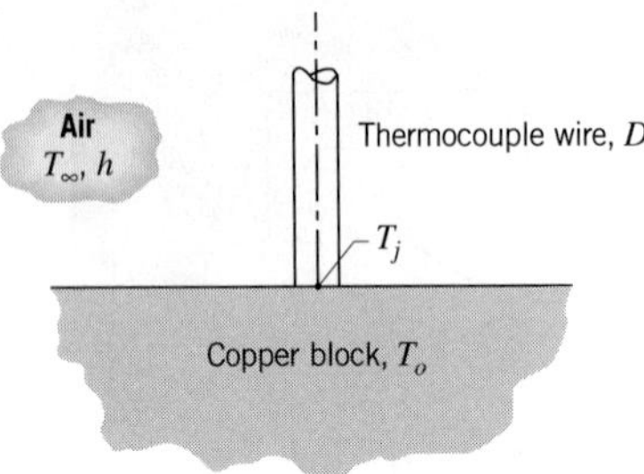

(a) If the wire is in air at 25°C with a convection coefficient of 10 W/m^2·K, estimate the measurement error $(T_j - T_o)$ for the thermocouple when the block is at 125°C.

(b) For convection coefficients of 5, 10, and 25 W/m^2·K, plot the measurement error as a function of the thermal conductivity of the block material over the range 15 to 400 W/m·K. Under what circumstances is it advantageous to use smaller diameter wire?

4.31 A hole of diameter $D = 0.25$ m is drilled through the center of a solid block of square cross section with $w = 1$ m on a side. The hole is drilled along the length, $l = 2$ m, of the block, which has a thermal conductivity of $k = 150$ W/m·K. The four outer surfaces are exposed to ambient air, with $T_{\infty,2} = 25$°C and $h_2 = 4$ W/m^2·K, while hot oil flowing through the hole is characterized by $T_{\infty,1} = 300$°C and $h_1 = 50$ W/m^2·K. Determine the corresponding heat rate and surface temperatures.

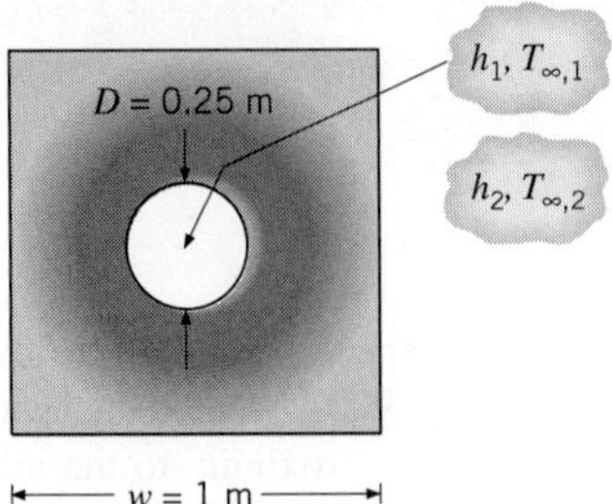

4.32 In Chapter 3 we assumed that, whenever fins are attached to a base material, the base temperature is unchanged. What in fact happens is that, if the temperature of the base material exceeds the fluid temperature, attachment of a fin depresses the junction temperature T_j below the original temperature of the base, and heat flow from the base material to the fin is two-dimensional.

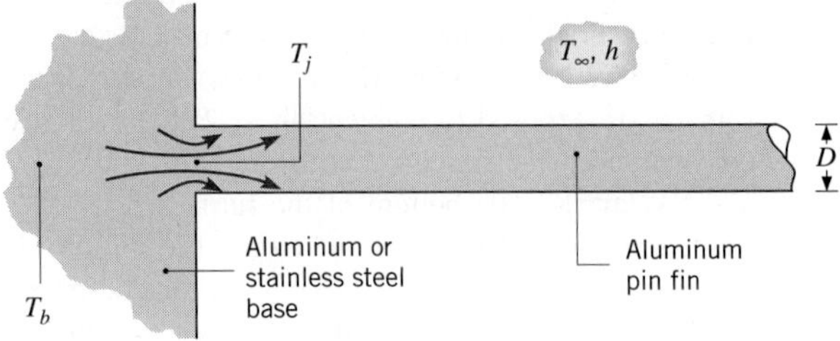

Consider conditions for which a long aluminum pin fin of diameter $D = 5$ mm is attached to a base material whose temperature far from the junction is maintained at $T_b = 100$°C. Fin convection conditions correspond to $h = 50$ W/m^2·K and $T_\infty = 25$°C.

(a) What are the fin heat rate and junction temperature when the base material is (i) aluminum (k = 240 W/m·K) and (ii) stainless steel (k = 15 W/m·K)?

(b) Repeat the foregoing calculations if a thermal contact resistance of $R''_{t,j} = 3 \times 10^{-5}$ m^2·K/W is associated with the method of joining the pin fin to the base material.

(c) Considering the thermal contact resistance, plot the heat rate as a function of the convection coefficient over the range $10 \leq h \leq 100$ W/m^2·K for each of the two materials.

4.33 An igloo is built in the shape of a hemisphere, with an inner radius of 1.8 m and walls of compacted snow that are 0.5 m thick. On the inside of the igloo, the surface heat transfer coefficient is 6 W/m^2·K; on the outside, under normal wind conditions, it is 15 W/m^2·K. The thermal conductivity of compacted snow is 0.15 W/m·K. The temperature of the ice cap on which the igloo sits is −20°C and has the same thermal conductivity as the compacted snow.

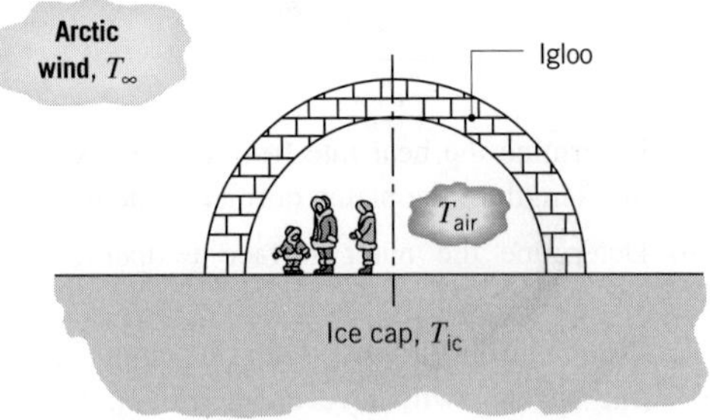

(a) Assuming that the occupants' body heat provides a continuous source of 320 W within the igloo, calculate the inside air temperature when the outside air temperature is $T_\infty = -40°C$. Be sure to consider heat losses through the floor of the igloo.

(b) Using the thermal circuit of part (a), perform a parameter sensitivity analysis to determine which variables have a significant effect on the inside air temperature. For instance, for very high wind conditions, the outside convection coefficient could double or even triple. Does it make sense to construct the igloo with walls half or twice as thick?

4.34 Consider the thin integrated circuit (chip) of Problem 3.150. Instead of attaching the heat sink to the chip surface, an engineer suggests that sufficient cooling might be achieved by mounting the top of the chip onto a large copper ($k = 400$ W/m·K) surface that is located nearby. The metallurgical joint between the chip and the substrate provides a contact resistance of $R''_{t,c} = 5 \times 10^{-6}$ m²·K/W, and the maximum allowable chip temperature is 85°C. If the large substrate temperature is $T_2 = 25°C$ at locations far from the chip, what is the maximum allowable chip power dissipation q_c?

4.35 An electronic device, in the form of a disk 20 mm in diameter, dissipates 100 W when mounted flush on a large aluminum alloy (2024) block whose temperature is maintained at 27°C. The mounting arrangement is such that a contact resistance of $R''_{t,c} = 5 \times 10^{-5}$ m²·K/W exists at the interface between the device and the block.

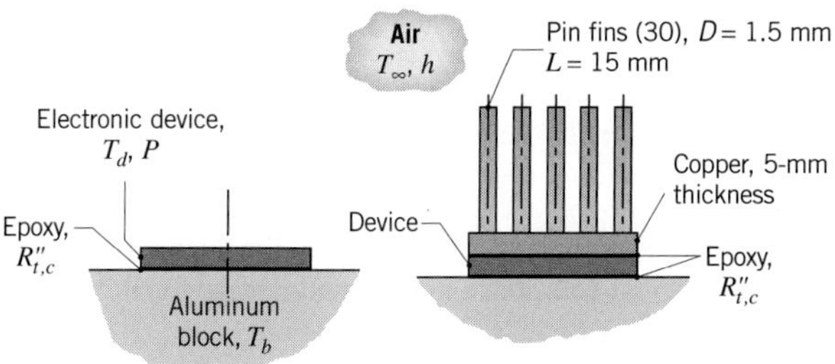

(a) Calculate the temperature the device will reach, assuming that all the power generated by the device must be transferred by conduction to the block.

(b) To operate the device at a higher power level, a circuit designer proposes to attach a finned heat sink to the top of the device. The pin fins and base material are fabricated from copper ($k = 400$ W/m·K) and are exposed to an airstream at 27°C for which the convection coefficient is 1000 W/m²·K. For the device temperature computed in part (a), what is the permissible operating power?

4.36 The elemental unit of an air heater consists of a long circular rod of diameter D, which is encapsulated by a finned sleeve and in which thermal energy is generated by ohmic heating. The N fins of thickness t and length L are integrally fabricated with the square sleeve of width w. Under steady-state operating conditions, the rate of thermal energy generation corresponds to the rate of heat transfer to airflow over the sleeve.

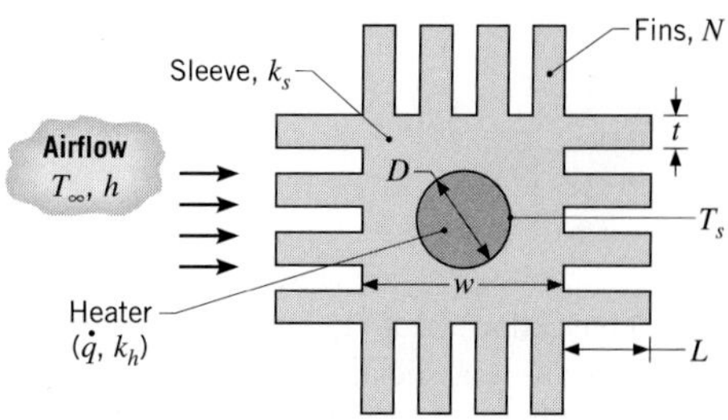

(a) Under conditions for which a uniform surface temperature T_s is maintained around the circumference of the heater and the temperature T_∞ and convection coefficient h of the airflow are known, obtain an expression for the rate of heat transfer per unit length to the air. Evaluate the heat rate for $T_s = 300°C$, $D = 20$ mm, an aluminum sleeve ($k_s = 240$ W/m·K), $w = 40$ mm, $N = 16$, $t = 4$ mm, $L = 20$ mm, $T_\infty = 50°C$, and $h = 500$ W/m²·K.

(b) For the foregoing heat rate and a copper heater of thermal conductivity $k_h = 400$ W/m·K, what is the required volumetric heat generation within the heater and its corresponding centerline temperature?

(c) With all other quantities unchanged, explore the effect of variations in the fin parameters (N, L, t) on the heat rate, subject to the constraint that the fin thickness and the spacing between fins cannot be less than 2 mm.

4.37 For a small heat source attached to a large substrate, the *spreading* resistance associated with multidimensional conduction in the substrate may be approximated by the expression [Yovanovich, M. M., and V. W. Antonetti, in *Adv. Thermal Modeling Elec. Comp. and Systems*, Vol. 1, A. Bar-Cohen and A. D. Kraus, Eds., Hemisphere, NY, 79–128, 1988]

$$R_{t(\mathrm{sp})} = \frac{1 - 1.410\,A_r + 0.344\,A_r^3 + 0.043\,A_r^5 + 0.034\,A_r^7}{4k_{\mathrm{sub}}A_{s,h}^{1/2}}$$

where $A_r = A_{s,h}/A_{s,\mathrm{sub}}$ is the ratio of the heat source area to the substrate area. Consider application of the expression to an in-line array of square chips of width $L_h = 5$ mm on a side and pitch $S_h = 10$ mm. The interface between the chips and a large substrate of thermal conductivity $k_{sub} = 80$ W/m·K is characterized by a thermal contact resistance of $R''_{t,c} = 0.5 \times 10^{-4}$ m²·K/W.

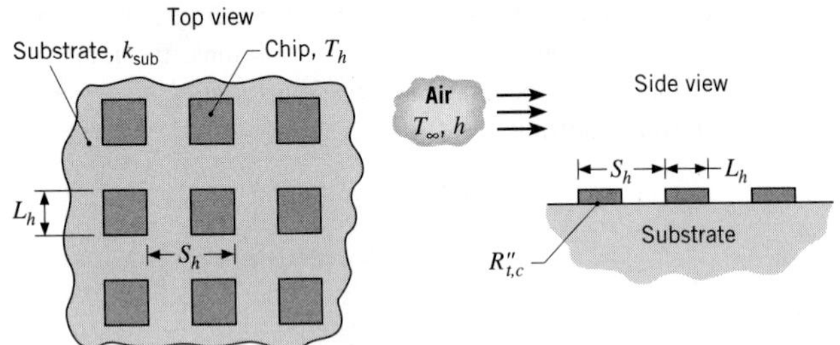

If a convection heat transfer coefficient of $h = 100$ W/m²·K is associated with airflow ($T_\infty = 15°C$) over the chips and substrate, what is the maximum allowable chip power dissipation if the chip temperature is not to exceed $T_h = 85°C$?

Finite-Difference Equations: Derivations

4.38 Consider nodal configuration 2 of Table 4.2. Derive the finite-difference equations under steady-state conditions for the following situations.

(a) The horizontal boundary of the internal corner is perfectly insulated and the vertical boundary is subjected to the convection process (T_∞, h).

(b) Both boundaries of the internal corner are perfectly insulated. How does this result compare with Equation 4.41?

4.39 Consider nodal configuration 3 of Table 4.2. Derive the finite-difference equations under steady-state conditions for the following situations.

(a) The boundary is insulated. Explain how Equation 4.42 can be modified to agree with your result.

(b) The boundary is subjected to a constant heat flux.

4.40 Consider nodal configuration 4 of Table 4.2. Derive the finite-difference equations under steady-state conditions for the following situations.

(a) The upper boundary of the external corner is perfectly insulated and the side boundary is subjected to the convection process (T_∞, h).

(b) Both boundaries of the external corner are perfectly insulated. How does this result compare with Equation 4.43?

4.41 One of the strengths of numerical methods is their ability to handle complex boundary conditions. In the sketch, the boundary condition changes from specified heat flux q''_s (into the domain) to convection, at the location of the node (m, n). Write the steady-state, two-dimensional finite difference equation at this node.

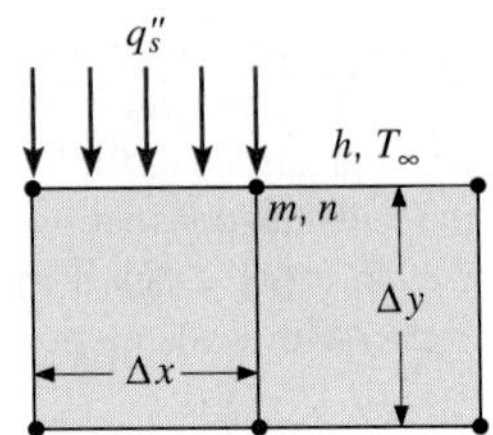

4.42 Determine expressions for $q_{(m-1,n)\to(m,n)}$, $q_{(m+1,n)\to(m,n)}$, $q_{(m,n+1)\to(m,n)}$ and $q_{(m,n-1)\to(m,n)}$ for conduction associated with a control volume that spans two different materials. There is no contact resistance at the interface between the materials. The control volumes are L units long into the page. Write the finite difference equation under steady-state conditions for node (m, n).

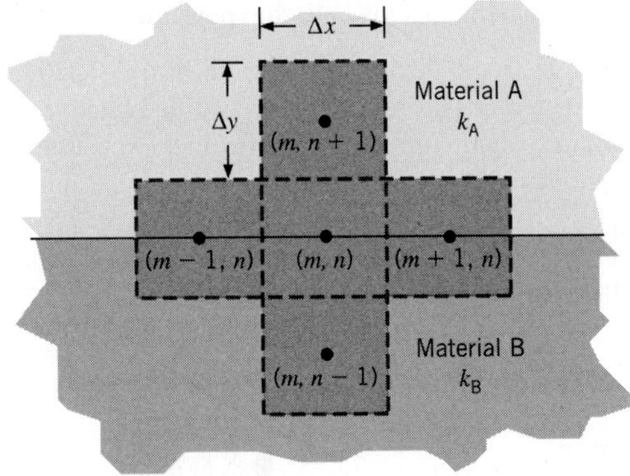

4.43 Consider heat transfer in a one-dimensional (radial) cylindrical coordinate system under steady-state conditions with volumetric heat generation.

(a) Derive the finite-difference equation for any interior node m.

(b) Derive the finite-difference equation for the node n located at the external boundary subjected to the convection process (T_∞, h).

4.44 In a two-dimensional cylindrical configuration, the radial (Δr) and angular ($\Delta\phi$) spacings of the nodes are uniform. The boundary at $r = r_i$ is of uniform temperature T_i. The boundaries in the radial direction are adiabatic (insulated) and exposed to surface convection (T_∞, h), as illustrated. Derive the finite-difference equations for (i) node 2, (ii) node 3, and (iii) node 1.

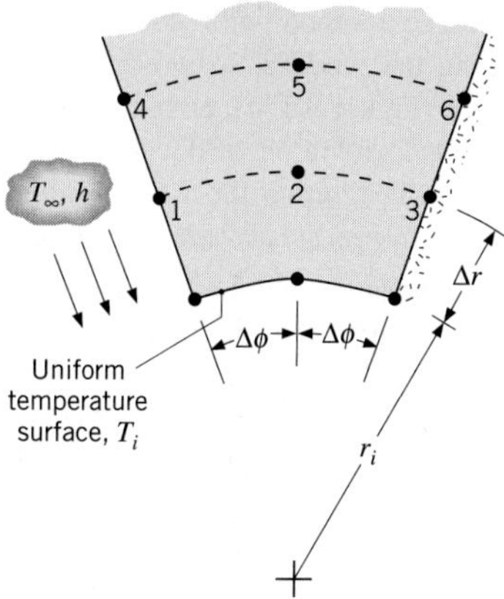

4.45 Upper and lower surfaces of a bus bar are convectively cooled by air at T_∞, with $h_u \neq h_l$. The sides are cooled by maintaining contact with heat sinks at T_o, through a thermal contact resistance of $R''_{t,c}$. The bar is of thermal conductivity k, and its width is twice its thickness L.

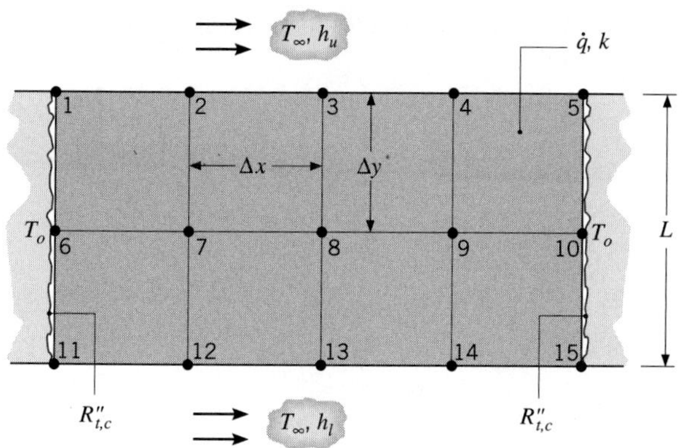

Consider steady-state conditions for which heat is uniformly generated at a volumetric rate $\dot{q}$ due to passage of an electric current. Using the energy balance method, derive finite-difference equations for nodes 1 and 13.

4.46 Derive the nodal finite-difference equations for the following configurations.

(a) Node (m, n) on a diagonal boundary subjected to convection with a fluid at T_∞ and a heat transfer coefficient h. Assume $\Delta x = \Delta y$.

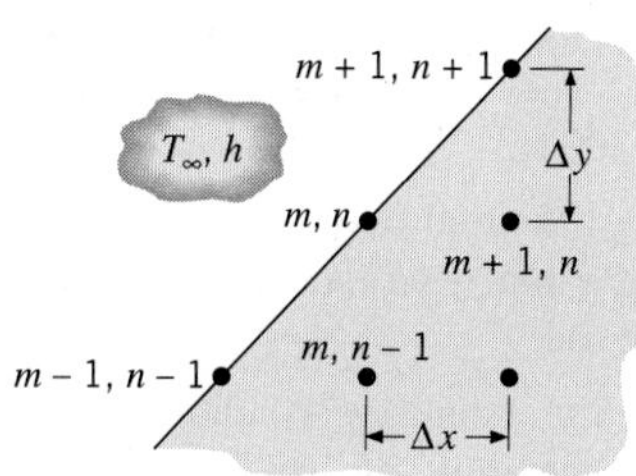

(b) Node (m, n) at the tip of a cutting tool with the upper surface exposed to a constant heat flux q''_o, and the diagonal surface exposed to a convection cooling process with the fluid at T_∞ and a heat transfer coefficient h. Assume $\Delta x = \Delta y$.

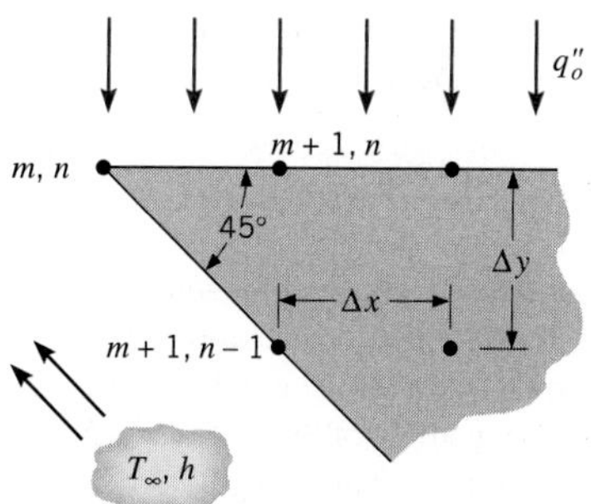

4.47 Consider the nodal point 0 located on the boundary between materials of thermal conductivity k_A and k_B.

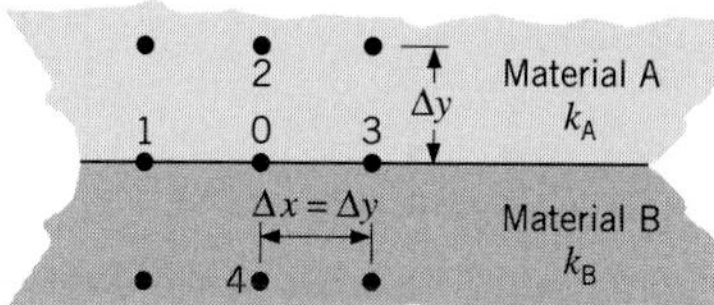

Derive the finite-difference equation, assuming no internal generation.

4.48 Consider the two-dimensional grid ($\Delta x = \Delta y$) representing steady-state conditions with no internal volumetric generation for a system with thermal conductivity k. One of the boundaries is maintained at a constant temperature T_s while the others are adiabatic.

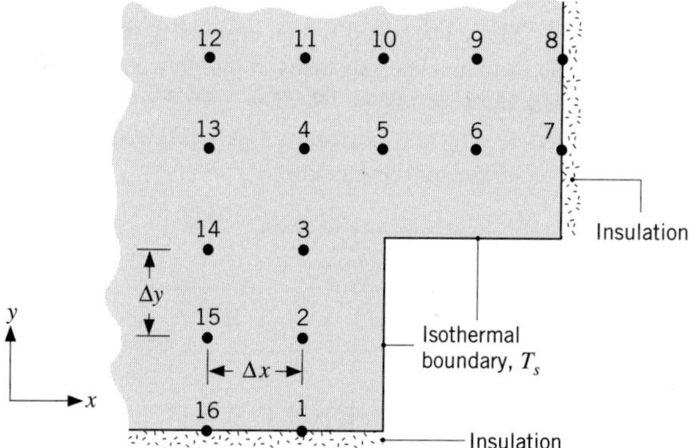

Derive an expression for the heat rate per unit length normal to the page crossing the isothermal boundary (T_s).

4.49 Consider a one-dimensional fin of uniform cross-sectional area, insulated at its tip, $x = L$. (See Table 3.4, case B). The temperature at the base of the fin T_b and of the adjoining fluid T_∞, as well as the heat transfer coefficient h and the thermal conductivity k, are known.

(a) Derive the finite-difference equation for any interior node m.

(b) Derive the finite-difference equation for a node n located at the insulated tip.

Finite-Difference Equations: Analysis

4.50 Consider the network for a two-dimensional system without internal volumetric generation having nodal temperatures shown below. If the grid spacing is 125 mm and the thermal conductivity of the material is 50 W/m·K, calculate the heat rate per unit length normal to the page from the isothermal surface (T_s).

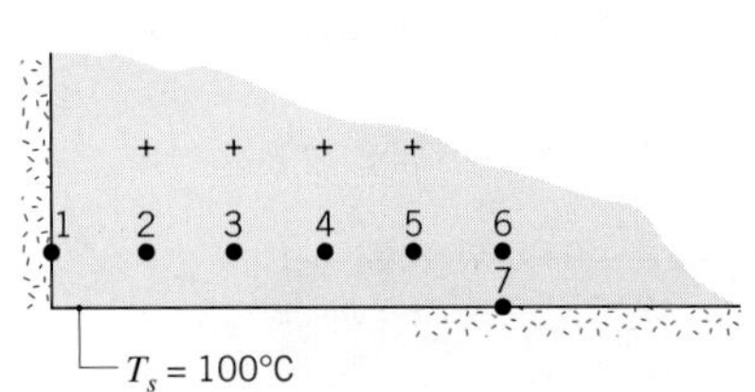

Node	T_i (°C)
1	120.55
2	120.64
3	121.29
4	123.89
5	134.57
6	150.49
7	147.14

4.51 An ancient myth describes how a wooden ship was destroyed by soldiers who reflected sunlight from their polished bronze shields onto its hull, setting the ship ablaze. To test the validity of the myth, a group of college students are given mirrors and they reflect sunlight onto a 100 mm × 100 mm area of a t = 10-mm-thick plywood mockup characterized by k = 0.8 W/m·K. The bottom of the mockup is in water at T_w = 20°C, while the air temperature is T_∞ = 25°C. The surroundings are at T_{sur} = 23°C. The wood has an emissivity of ε = 0.90; both the front and back surfaces of the plywood are characterized by h = 5 W/m²·K. The absorbed irradiation from the N students' mirrors is $G_{S,N}$ = 70,000 W/m² on the front surface of the mockup.

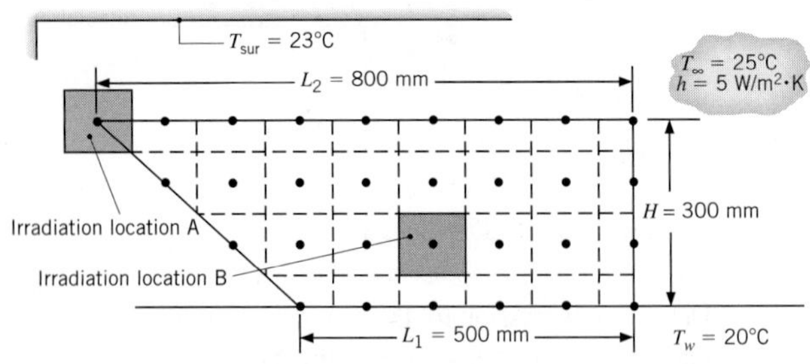

(a) A debate ensues concerning where the beam should be focused, location A or location B. Using a finite difference method with $\Delta x = \Delta y$ = 100 mm and treating the wood as a two-dimensional extended surface (Figure 3.17*a*), enlighten the students as to whether location A or location B will be more effective in igniting the wood by determining the maximum local steady-state temperature.

(b) Some students wonder whether the same technique can be used to melt a stainless steel hull. Repeat part (a) considering a stainless steel mockup of the same dimensions with k = 15 W/m·K and ε = 0.2. The value of the absorbed irradiation is the same as in part (a).

4.52 Consider the square channel shown in the sketch operating under steady-state conditions. The inner surface of the channel is at a uniform temperature of 600 K, while the outer surface is exposed to convection with a fluid at 300 K and a convection coefficient of 50 W/m²·K. From a symmetrical element of the channel, a two-dimensional grid has been constructed and the nodes labeled. The temperatures for nodes 1, 3, 6, 8, and 9 are identified.

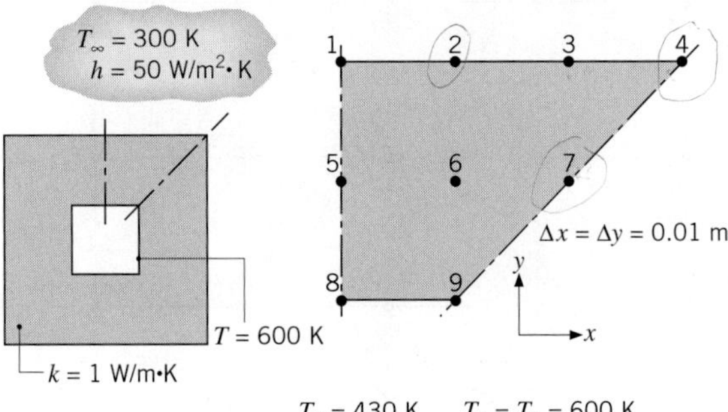

(a) Beginning with properly defined control volumes, derive the finite-difference equations for nodes 2, 4, and 7 and determine the temperatures T_2, T_4, and T_7 (K).

(b) Calculate the heat loss per unit length from the channel.

4.53 A long conducting rod of rectangular cross section (20 mm × 30 mm) and thermal conductivity k = 20 W/m·K experiences uniform heat generation at a rate $\dot{q} = 5 \times 10^7$ W/m³, while its surfaces are maintained at 300 K.

(a) Using a finite-difference method with a grid spacing of 5 mm, determine the temperature distribution in the rod.

(b) With the boundary conditions unchanged, what heat generation rate will cause the midpoint temperature to reach 600 K?

4.54 A flue passing hot exhaust gases has a square cross section, 300 mm to a side. The walls are constructed of refractory brick 150 mm thick with a thermal conductivity of 0.85 W/m · K. Calculate the heat loss from the flue per unit length when the interior and exterior surfaces are maintained at 350 and 25°C, respectively. Use a grid spacing of 75 mm.

4.55 Steady-state temperatures (K) at three nodal points of a long rectangular rod are as shown. The rod experiences a uniform volumetric generation rate of 5×10^7 W/m^3 and has a thermal conductivity of 20 W/m·K. Two of its sides are maintained at a constant temperature of 300 K, while the others are insulated.

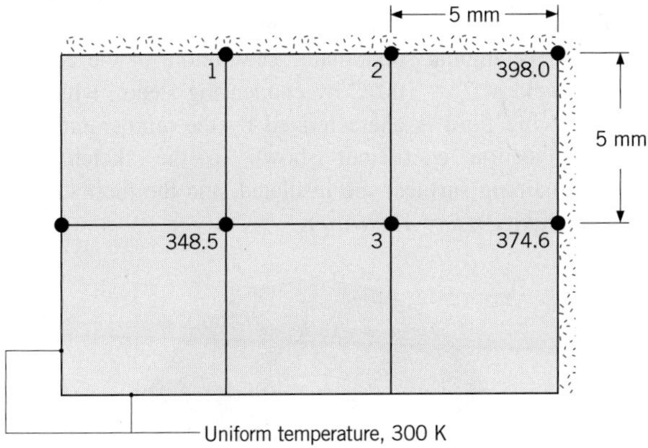

(a) Determine the temperatures at nodes 1, 2, and 3.

(b) Calculate the heat transfer rate per unit length (W/m) from the rod using the nodal temperatures. Compare this result with the heat rate calculated from knowledge of the volumetric generation rate and the rod dimensions.

4.56 Functionally graded materials are intentionally fabricated to establish a spatial distribution of properties in the final product. Consider an $L \times L$ two-dimensional object with L = 20 mm. The thermal conductivity distribution of the functionally graded material is $k(x)$ = 20 W/m·K + (7070 W/m$^{5/2}$·K) $x^{3/2}$. Two sets of boundary conditions, denoted as cases 1 and 2, are applied.

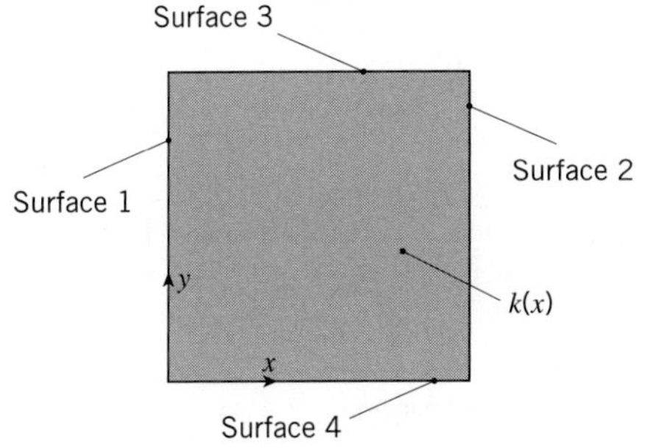

Case	Surface	Boundary Condition
1	1	T = 100°C
—	2	T = 50°C
—	3	Adiabatic
—	4	Adiabatic
2	1	Adiabatic
—	2	Adiabatic
—	3	T = 50°C
—	4	T = 100°C

(a) Determine the spatially averaged value of the thermal conductivity $\bar{k}$. Use this value to estimate the heat rate per unit length for cases 1 and 2.

(b) Using a grid spacing of 2 mm, determine the heat rate per unit depth for case 1. Compare your result to the estimated value calculated in part (a).

(c) Using a grid spacing of 2 mm, determine the heat rate per unit depth for case 2. Compare your result to the estimated value calculated in part (a).

4.57 Steady-state temperatures at selected nodal points of the symmetrical section of a flow channel are known to be T_2 = 95.47°C, T_3 = 117.3°C, T_5 = 79.79°C, T_6 = 77.29°C, T_8 = 87.28°C, and T_{10} = 77.65°C. The wall experiences uniform volumetric heat generation of $\dot{q} = 10^6$ W/m^3 and has a thermal conductivity of k = 10 W/m·K. The inner and outer surfaces of the channel experience convection with fluid temperatures of $T_{\infty,i}$ = 50°C and $T_{\infty,o}$ = 25°C and convection coefficients of h_i = 500 W/m^2·K and h_o = 250 W/m^2·K.

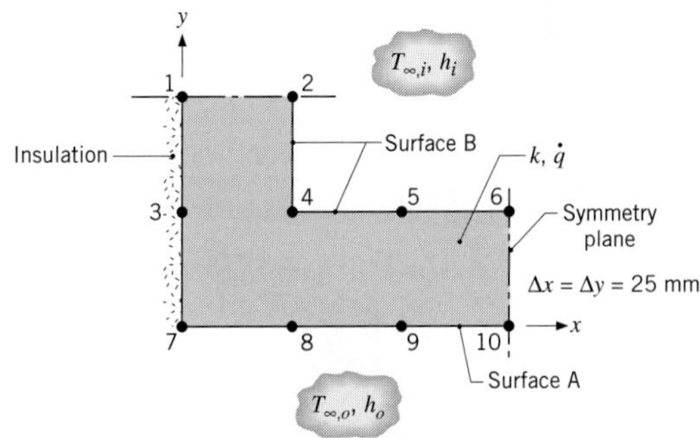

(a) Determine the temperatures at nodes 1, 4, 7, and 9.

(b) Calculate the heat rate per unit length (W/m) from the outer surface A to the adjacent fluid.

(c) Calculate the heat rate per unit length from the inner fluid to surface B.

(d) Verify that your results are consistent with an overall energy balance on the channel section.

4.58 Consider an aluminum heat sink (k = 240 W/m·K), such as that shown schematically in Problem 4.28. The inner and outer widths of the square channel are $w = 20$ mm and $W = 40$ mm, respectively, and an outer surface temperature of $T_s = 50°C$ is maintained by the array of electronic chips. In this case, it is not the inner surface temperature that is known, but conditions (T_∞, h) associated with coolant flow through the channel, and we wish to determine the rate of heat transfer to the coolant per unit length of channel. For this purpose, consider a symmetrical section of the channel and a two-dimensional grid with $\Delta x = \Delta y = 5$ mm.

(a) For $T_\infty = 20°C$ and $h = 5000\ \text{W/m}^2\cdot\text{K}$, determine the unknown temperatures, $T_1, \ldots, T_7$, and the rate of heat transfer per unit length of channel, q'.

(b) Assess the effect of variations in h on the unknown temperatures and the heat rate.

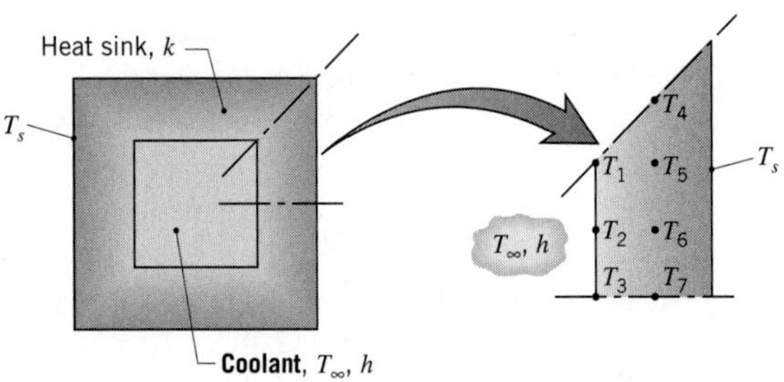

4.59 Conduction within relatively complex geometries can sometimes be evaluated using the finite-difference methods of this text that are applied to *subdomains* and *patched* together. Consider the two-dimensional domain formed by rectangular and cylindrical subdomains patched at the common, dashed control surface. Note that, along the dashed control surface, temperatures in the two subdomains are identical and local conduction heat fluxes to the cylindrical subdomain are identical to local conduction heat fluxes from the rectangular subdomain.

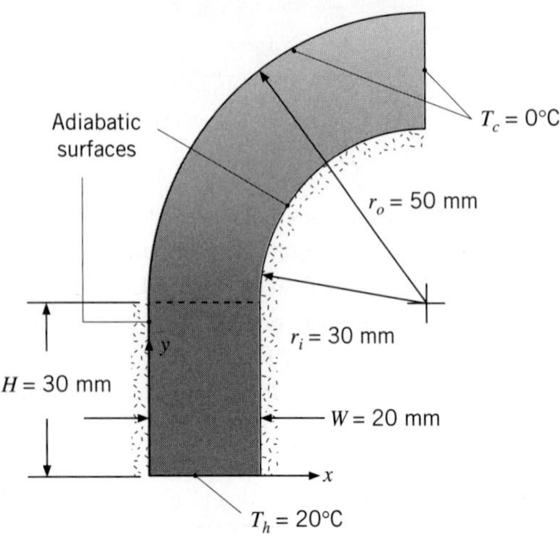

Calculate the heat transfer per unit depth into the page, q', using $\Delta x = \Delta y = \Delta r = 10$ mm and $\Delta\phi = \pi/8$. The base of the rectangular subdomain is held at $T_h = 20°C$, while the vertical surface of the cylindrical subdomain and the surface at outer radius r_o are at $T_c = 0°C$. The remaining surfaces are adiabatic, and the thermal conductivity is $k = 10\ \text{W/m}\cdot\text{K}$.

4.60 Consider the two-dimensional tube of a noncircular cross section formed by rectangular and semicylindrical subdomains patched at the common dashed control surfaces in a manner similar to that described in Problem 4.59. Note that, along the dashed control surfaces, temperatures in the two subdomains are identical and local conduction heat fluxes to the semicylindrical subdomain are identical to local conduction heat fluxes from the rectangular subdomain. The bottom of the domain is held at $T_s = 100°C$ by condensing steam, while the flowing fluid is characterized by the temperature and convection coefficient shown in the sketch. The remaining surfaces are insulated, and the thermal conductivity is $k = 15\ \text{W/m}\cdot\text{K}$.

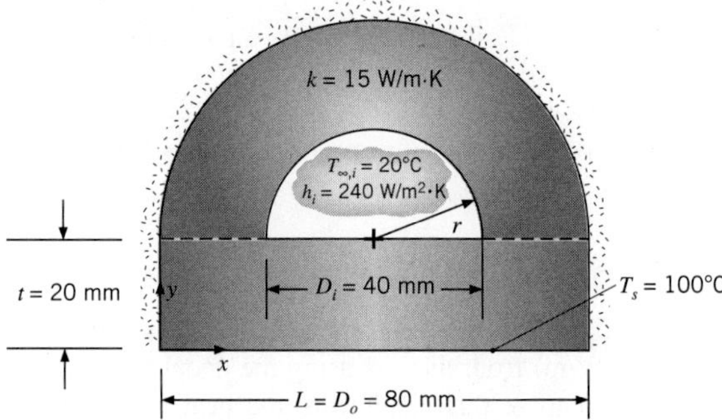

Find the heat transfer rate per unit length of tube, q', using $\Delta x = \Delta y = \Delta r = 10$ mm and $\Delta\phi = \pi/8$. *Hint*: Take advantage of the symmetry of the problem by considering only half of the entire domain.

4.61 The steady-state temperatures (°C) associated with selected nodal points of a two-dimensional system having a thermal conductivity of $1.5\ \text{W/m}\cdot\text{K}$ are shown on the accompanying grid.

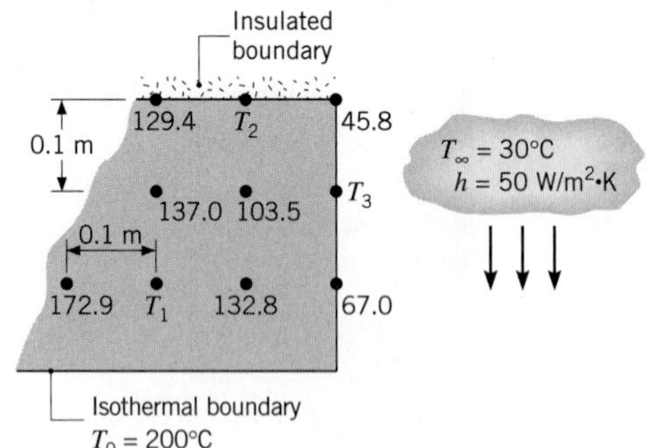

(a) Determine the temperatures at nodes 1, 2, and 3.

(b) Calculate the heat transfer rate per unit thickness normal to the page from the system to the fluid.

4.62 A steady-state, finite-difference analysis has been performed on a cylindrical fin with a diameter of 12 mm and a thermal conductivity of 15 W/m·K. The convection process is characterized by a fluid temperature of 25°C and a heat transfer coefficient of 25 W/m²·K.

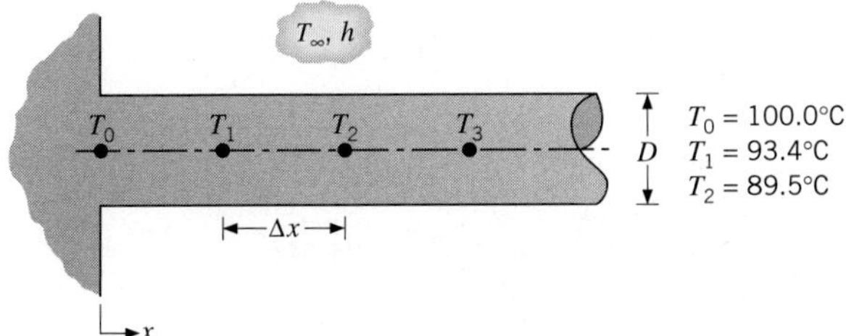

(a) The temperatures for the first three nodes, separated by a spatial increment of x = 10 mm, are given in the sketch. Determine the fin heat rate.

(b) Determine the temperature at node 3, T_3.

4.63 Consider the two-dimensional domain shown. All surfaces are insulated except for the isothermal surfaces at $x = 0$ and L.

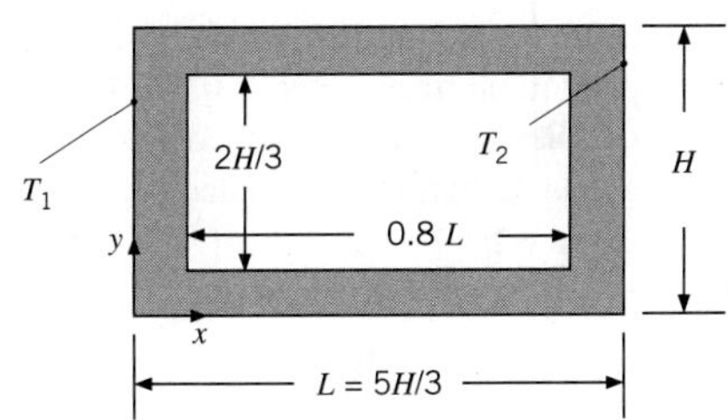

(a) Use a one-dimensional analysis to estimate the shape factor S.

(b) Estimate the shape factor using a finite difference analysis with $\Delta x = \Delta y = 0.05L$. Compare your answer with that of part (a), and explain the difference between the two solutions.

4.64 Consider two-dimensional, steady-state conduction in a square cross section with prescribed surface temperatures.

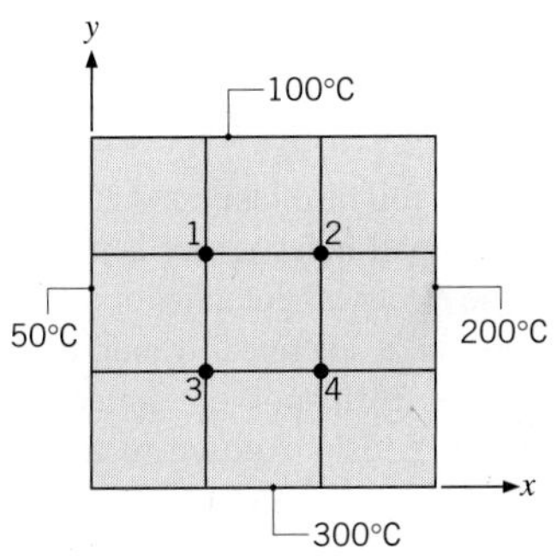

(a) Determine the temperatures at nodes 1, 2, 3, and 4. Estimate the midpoint temperature.

(b) Reducing the mesh size by a factor of 2, determine the corresponding nodal temperatures. Compare your results with those from the coarser grid.

(c) From the results for the finer grid, plot the 75, 150, and 250°C isotherms.

4.65 Consider a long bar of square cross section (0.8 m to the side) and of thermal conductivity 2 W/m·K. Three of these sides are maintained at a uniform temperature of 300°C. The fourth side is exposed to a fluid at 100°C for which the convection heat transfer coefficient is 10 W/m²·K.

(a) Using an appropriate numerical technique with a grid spacing of 0.2 m, determine the midpoint temperature and heat transfer rate between the bar and the fluid per unit length of the bar.

(b) Reducing the grid spacing by a factor of 2, determine the midpoint temperature and heat transfer rate. Plot the corresponding temperature distribution across the surface exposed to the fluid. Also, plot the 200 and 250°C isotherms.

4.66 Consider a two-dimensional, straight triangular fin of length L = 50 mm and base thickness t = 20 mm. The thermal conductivity of the fin is k = 25 W/m·K. The base temperature is T_b = 50°C, and the fin is exposed to convection conditions characterized by h = 50 W/m²·K, T_∞ = 20°C. Using a finite difference mesh with Δx = 10 mm and Δy = 2 mm, and taking advantage of symmetry, determine the fin efficiency, η_f. Compare your value of the fin efficiency with that reported in Figure 3.19.

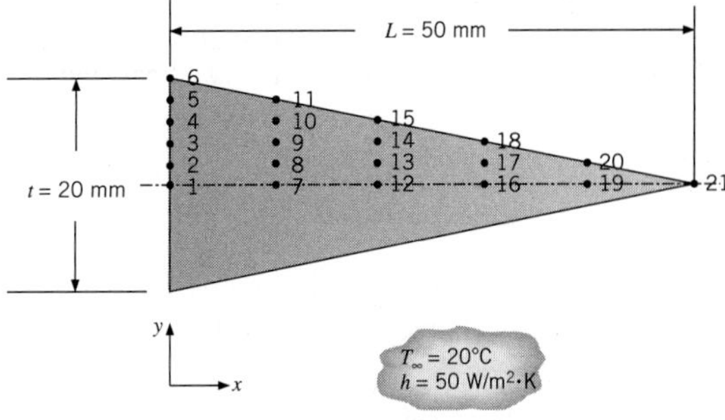

4.67 A common arrangement for heating a large surface area is to move warm air through rectangular ducts below the surface. The ducts are square and located midway between the top and bottom surfaces that are exposed to room air and insulated, respectively.

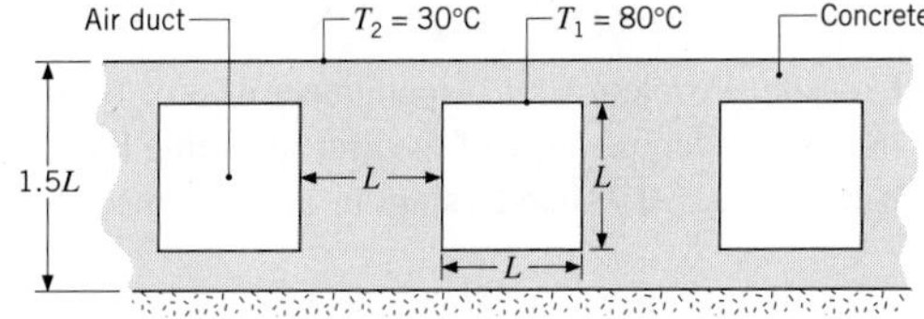

For the condition when the floor and duct temperatures are 30 and 80°C, respectively, and the thermal conductivity of concrete is 1.4 W/m·K, calculate the heat rate from each duct, per unit length of duct. Use a grid spacing with $\Delta x = 2\,\Delta y$, where $\Delta y = 0.125L$ and $L = 150$ mm.

4.68 Consider the gas turbine cooling scheme of Example 4.3. In Problem 3.23, advantages associated with applying a *thermal barrier coating* (*TBC*) to the exterior surface of a turbine blade are described. If a 0.5-mm-thick zirconia coating ($k = 1.3$ W/m·K, $R''_{t,c} = 10^{-4}$ m²·K/W) is applied to the outer surface of the air-cooled blade, determine the temperature field in the blade for the operating conditions of Example 4.3.

4.69 A long, solid cylinder of diameter $D = 25$ mm is formed of an insulating core that is covered with a *very thin* ($t = 50\ \mu$m), highly polished metal sheathing of thermal conductivity $k = 25$ W/m·K. Electric current flows through the stainless steel from one end of the cylinder to the other, inducing uniform volumetric heating within the sheathing of $\dot{q} = 5 \times 10^6$ W/m³. As will become evident in Chapter 6, values of the convection coefficient between the surface and air for this situation are spatially nonuniform, and for the airstream conditions of the experiment, the convection heat transfer coefficient varies with the angle θ as $h(\theta) = 26 + 0.637\theta - 8.92\theta^2$ for $0 \le \theta < \pi/2$ and $h(\theta) = 5$ for $\pi/2 \le \theta \le \pi$.

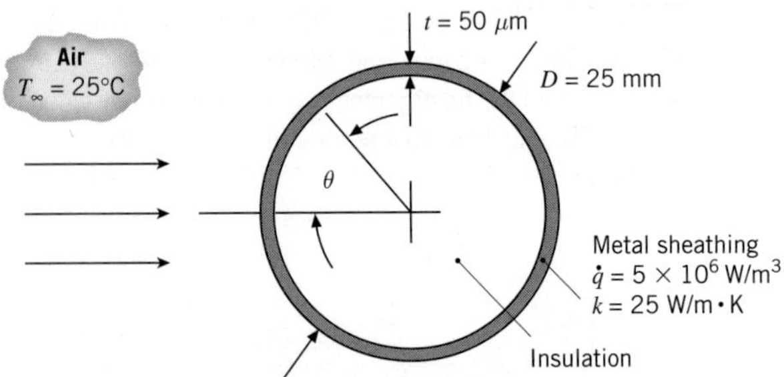

(a) Neglecting conduction in the θ-direction within the stainless steel, plot the temperature distribution $T(\theta)$ for $0 \le \theta \le \pi$ for $T_\infty = 25$°C.

(b) Accounting for θ-direction conduction in the stainless steel, determine temperatures in the stainless steel at increments of $\Delta\theta = \pi/20$ for $0 \le \theta \le \pi$. Compare the temperature distribution with that of part (a).

Hint: The temperature distribution is symmetrical about the horizontal centerline of the cylinder.

4.70 Consider Problem 4.69. An engineer desires to measure the surface temperature of the thin sheathing by painting it black ($\varepsilon = 0.98$) and using an infrared measurement device to nonintrusively determine the surface temperature distribution. Predict the temperature distribution of the painted surface, accounting for radiation heat transfer with large surroundings at $T_{sur} = 25$°C.

4.71 Consider using the experimental methodology of Problem 4.70 to determine the convection coefficient distribution about an airfoil of complex shape.

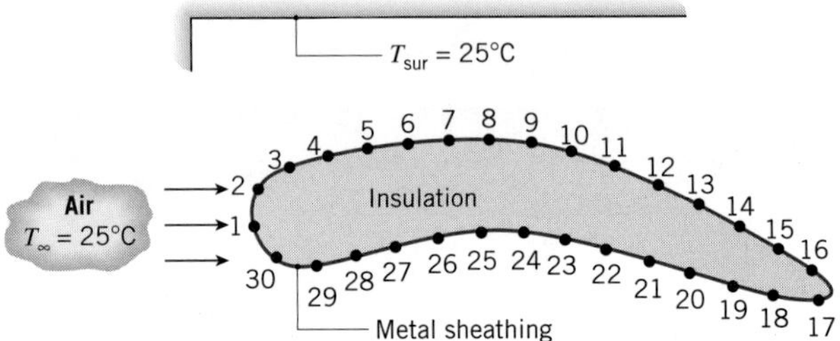

Accounting for conduction in the metal sheathing and radiation losses to the large surroundings, determine the convection heat transfer coefficients at the locations shown. The surface locations at which the temperatures are measured are spaced 2 mm apart. The thickness of the metal sheathing is $t = 20\ \mu$m, the volumetric generation rate is $\dot{q} = 20 \times 10^6$ W/m³, the sheathing's thermal conductivity is $k = 25$ W/m·K, and the emissivity of the painted surface is $\varepsilon = 0.98$. Compare your results to cases where (i) both conduction along the sheathing and radiation are neglected, and (ii) when only radiation is neglected.

Location	Temperature (°C)	Location	Temperature (°C)	Location	Temperature (°C)
1	27.77	11	34.29	21	31.13
2	27.67	12	36.78	22	30.64
3	27.71	13	39.29	23	30.60
4	27.83	14	41.51	24	30.77
5	28.06	15	42.68	25	31.16
6	28.47	16	42.84	26	31.52
7	28.98	17	41.29	27	31.85
8	29.67	18	37.89	28	31.51
9	30.66	19	34.51	29	29.91
10	32.18	20	32.36	30	28.42

4.72 A thin metallic foil of thickness 0.25 mm with a pattern of extremely small holes serves as an acceleration grid to control the electrical potential of an ion beam. Such a grid is used in a chemical vapor deposition (CVD) process for the fabrication of semiconductors. The top surface of the grid is exposed to a uniform heat flux caused by absorption of the ion beam, $q''_s = 600\ \text{W/m}^2$. The edges of the foil are thermally coupled to water-cooled sinks maintained at 300 K. The upper and lower surfaces of the foil experience radiation exchange with the vacuum enclosure walls maintained at 300 K. The effective thermal conductivity of the foil material is 40 W/m·K, and its emissivity is 0.45.

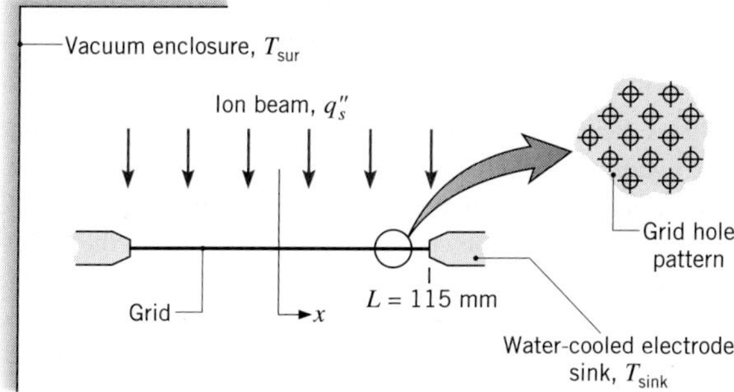

Assuming one-dimensional conduction and using a finite-difference method representing the grid by 10 nodes in the x-direction, estimate the temperature distribution for the grid. *Hint:* For each node requiring an energy balance, use the linearized form of the radiation rate equation, Equation 1.8, with the radiation coefficient h_r, Equation 1.9, evaluated for each node.

4.73 A long bar of rectangular cross section, 0.4 m × 0.6 m on a side and having a thermal conductivity of 1.5 W/m·K, is subjected to the boundary conditions shown.

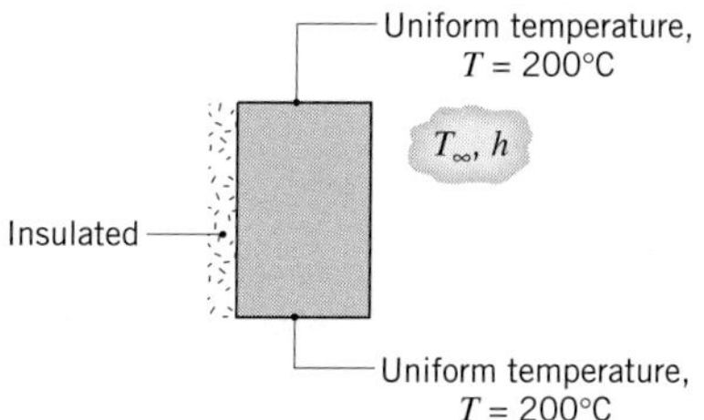

Two of the sides are maintained at a uniform temperature of 200°C. One of the sides is adiabatic, and the remaining side is subjected to a convection process with $T_\infty = 30°\text{C}$ and $h = 50\ \text{W/m}^2\cdot\text{K}$. Using an appropriate numerical technique with a grid spacing of 0.1 m, determine the temperature distribution in the bar and the heat transfer rate between the bar and the fluid per unit length of the bar.

4.74 The top surface of a plate, including its grooves, is maintained at a uniform temperature of $T_1 = 200°\text{C}$. The lower surface is at $T_2 = 20°\text{C}$, the thermal conductivity is 15 W/m·K, and the groove spacing is 0.16 m.

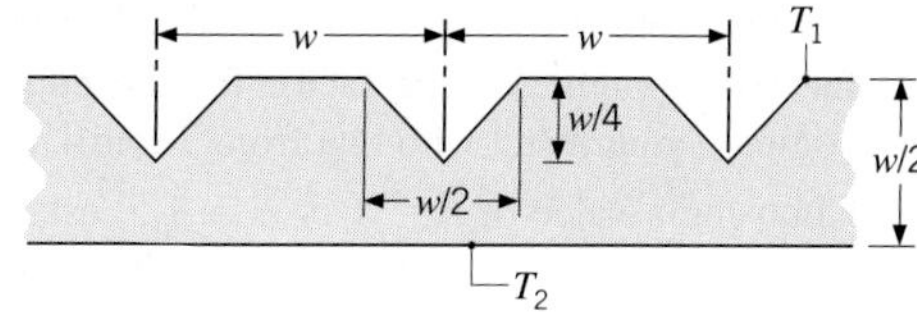

(a) Using a finite-difference method with a mesh size of $\Delta x = \Delta y = 40$ mm, calculate the unknown nodal temperatures and the heat transfer rate per width of groove spacing (w) and per unit length normal to the page.

(b) With a mesh size of $\Delta x = \Delta y = 10$ mm, repeat the foregoing calculations, determining the temperature field and the heat rate. Also, consider conditions for which the bottom surface is not at a uniform temperature T_2 but is exposed to a fluid at $T_\infty = 20°\text{C}$. With $\Delta x = \Delta y = 10$ mm, determine the temperature field and heat rate for values of h = 5, 200, and 1000 W/m²·K, as well as for $h \to \infty$.

4.75 Refer to the two-dimensional rectangular plate of Problem 4.2. Using an appropriate numerical method with $\Delta x = \Delta y = 0.25$ m, determine the temperature at the midpoint (1, 0.5).

4.76 The shape factor for conduction through the edge of adjoining walls for which $D > L/5$, where D and L are the wall depth and thickness, respectively, is shown in Table 4.1. The two-dimensional symmetrical element of the edge, which is represented by inset (a), is bounded by the diagonal symmetry adiabat and a section of the wall thickness over which the temperature distribution is assumed to be linear between T_1 and T_2.

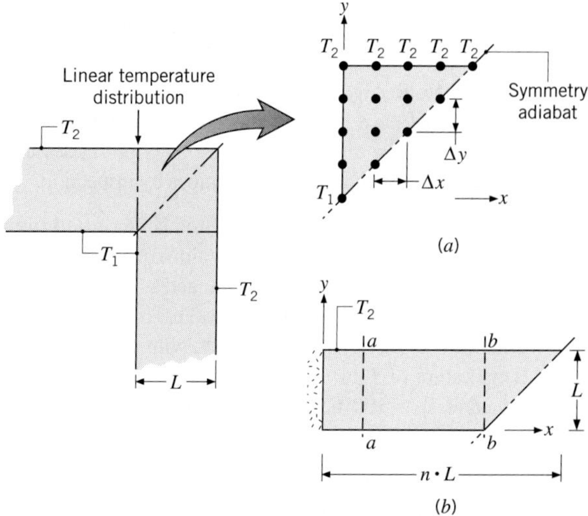

(a) Using the nodal network of inset (a) with L = 40 mm, determine the temperature distribution in the element for $T_1 = 100°\text{C}$ and $T_2 = 0°\text{C}$. Evaluate the heat rate per unit depth ($D = 1$ m) if $k = 1$ W/m·K. Determine the corresponding shape factor for the edge, and compare your result with that from Table 4.1.

(b) Choosing a value of $n = 1$ or $n = 1.5$, establish a nodal network for the trapezoid of inset (*b*) and determine the corresponding temperature field. Assess the validity of assuming linear temperature distributions across sections *a–a* and *b–b*.

4.77 The diagonal of a long triangular bar is well insulated, while sides of equivalent length are maintained at uniform temperatures T_a and T_b.

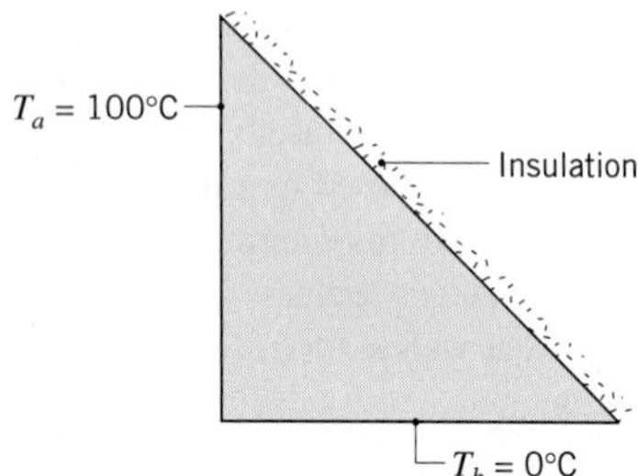

(a) Establish a nodal network consisting of five nodes along each of the sides. For one of the nodes on the diagonal surface, define a suitable control volume and derive the corresponding finite-difference equation. Using this form for the diagonal nodes and appropriate equations for the interior nodes, find the temperature distribution for the bar. On a scale drawing of the shape, show the 25, 50, and 75°C isotherms.

(b) An alternate and simpler procedure to obtain the finite-difference equations for the diagonal nodes follows from recognizing that the insulated diagonal surface is a symmetry plane. Consider a square 5×5 nodal network, and represent its diagonal as a symmetry line. Recognize which nodes on either side of the diagonal have identical temperatures. Show that you can treat the diagonal nodes as "interior" nodes and write the finite-difference equations by inspection.

4.78 A straight fin of uniform cross section is fabricated from a material of thermal conductivity 50 W/m·K, thickness $w = 6$ mm, and length $L = 48$ mm, and it is very long in the direction normal to the page. The convection heat transfer coefficient is 500 W/m^2·K with an ambient air temperature of $T_\infty = 30°C$. The base of the fin is maintained at $T_b = 100°C$, while the fin tip is well insulated.

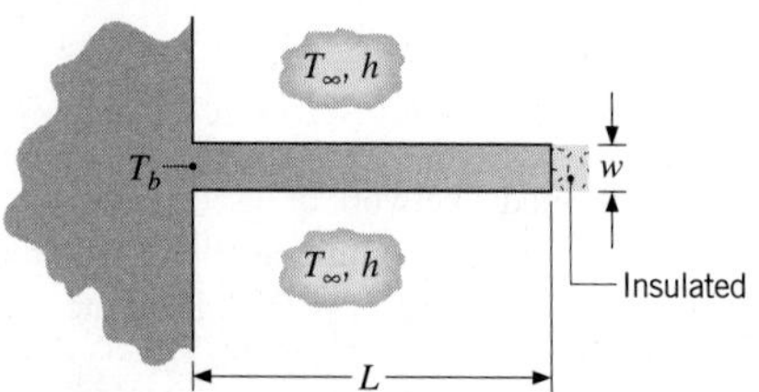

(a) Using a finite-difference method with a space increment of 4 mm, estimate the temperature distribution within the fin. Is the assumption of one-dimensional heat transfer reasonable for this fin?

(b) Estimate the fin heat transfer rate per unit length normal to the page. Compare your result with the one-dimensional, analytical solution, Equation 3.81.

(c) Using the finite-difference mesh of part (a), compute and plot the fin temperature distribution for values of $h = 10$, 100, 500, and 1000 W/m^2·K. Determine and plot the fin heat transfer rate as a function of h.

4.79 A rod of 10-mm diameter and 250-mm length has one end maintained at 100°C. The surface of the rod experiences free convection with the ambient air at 25°C and a convection coefficient that depends on the difference between the temperature of the surface and the ambient air. Specifically, the coefficient is prescribed by a correlation of the form, $h_{fc} = 2.89[0.6 + 0.624(T - T_\infty)^{1/6}]^2$, where the units are h_{fc} (W/m^2·K) and T (K). The surface of the rod has an emissivity $\varepsilon = 0.2$ and experiences radiation exchange with the surroundings at $T_{sur} = 25°C$. The fin tip also experiences free convection and radiation exchange.

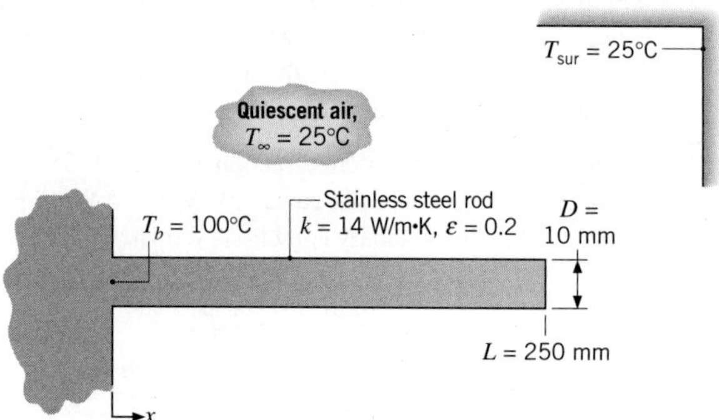

Assuming one-dimensional conduction and using a finite-difference method representing the fin by five nodes, estimate the temperature distribution for the fin. Determine also the fin heat rate and the relative contributions of free convection and radiation exchange. *Hint:* For each node requiring an energy balance, use the linearized form of the radiation rate equation, Equation 1.8, with the radiation coefficient h_r, Equation 1.9, evaluated for each node. Similarly, for the convection rate equation associated with each node, the free convection coefficient h_{fc} must be evaluated for each node.

4.80 A simplified representation for cooling in very large-scale integration (VLSI) of microelectronics is shown in the sketch. A silicon chip is mounted in a dielectric substrate, and one surface of the system is convectively cooled, while the remaining surfaces are well insulated from the surroundings. The problem is rendered two-dimensional by assuming the system to be very long in the direction perpendicular to the paper. Under steady-state operation, electric power dissipation in the chip provides for uniform volumetric heating at a rate of $\dot{q}$. However, the heating rate is limited by restrictions on the maximum temperature that the chip is allowed to achieve.

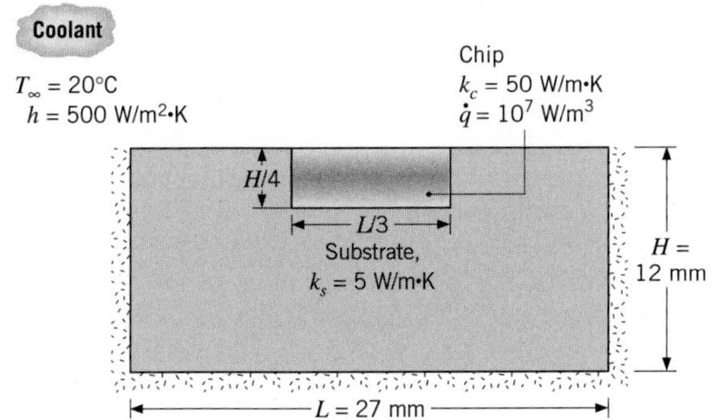

For the conditions shown on the sketch, will the maximum temperature in the chip exceed 85°C, the maximum allowable operating temperature set by industry standards? A grid spacing of 3 mm is suggested.

4.81 A heat sink for cooling computer chips is fabricated from copper ($k_s = 400$ W/m·K), with machined microchannels passing a cooling fluid for which $T = 25°C$ and $h = 30{,}000$ W/m^2·K. The lower side of the sink experiences no heat removal, and a preliminary heat sink design calls for dimensions of $a = b = w_s = w_f = 200\ \mu m$. A symmetrical element of the heat path from the chip to the fluid is shown in the inset.

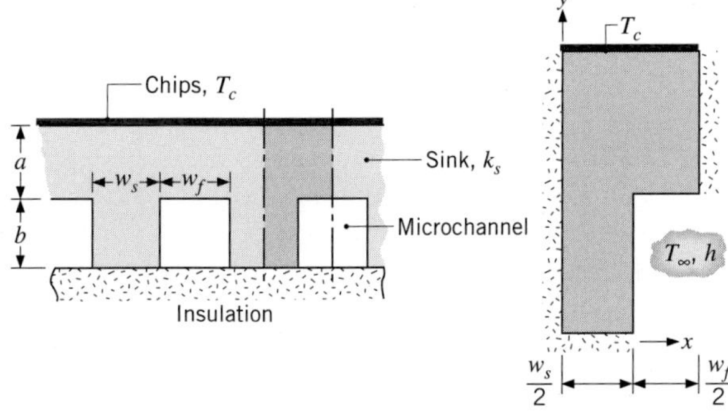

(a) Using the symmetrical element with a square nodal network of $\Delta x = \Delta y = 100\ \mu m$, determine the corresponding temperature field and the heat rate q' to the coolant per unit channel length (W/m) for a maximum allowable chip temperature $T_{c,\max} = 75°C$. Estimate the corresponding thermal resistance between the chip surface and the fluid, $R'_{t,c-f}$(m·K/W). What is the maximum allowable heat dissipation for a chip that measures 10 mm × 10 mm on a side?

(b) The grid spacing used in the foregoing finite-difference solution is coarse, resulting in poor precision for the temperature distribution and heat removal rate. Investigate the effect of grid spacing by considering spatial increments of 50 and 25 μm.

(c) Consistent with the requirement that $a + b = 400\ \mu m$, can the heat sink dimensions be altered in a manner that reduces the overall thermal resistance?

4.82 A plate ($k = 10$ W/m·K) is stiffened by a series of longitudinal ribs having a rectangular cross section with length $L = 8$ mm and width $w = 4$ mm. The base of the plate is maintained at a uniform temperature $T_b = 45°C$, while the rib surfaces are exposed to air at a temperature of $T_\infty = 25°C$ and a convection coefficient of $h = 600$ W/m^2·K.

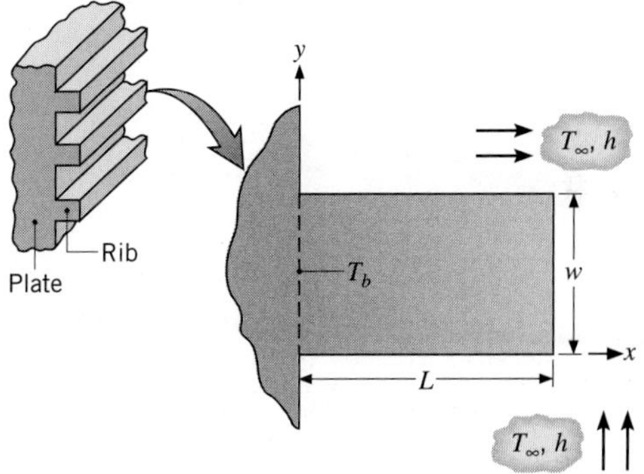

(a) Using a finite-difference method with $\Delta x = \Delta y = 2$ mm and a total of 5 × 3 nodal points and regions, estimate the temperature distribution and the heat rate from the base. Compare these results with those obtained by assuming that heat transfer in the rib is one-dimensional, thereby approximating the behavior of a fin.

(b) The grid spacing used in the foregoing finite-difference solution is coarse, resulting in poor precision for estimates of temperatures and the heat rate. Investigate the effect of grid refinement by reducing the nodal spacing to $\Delta x = \Delta y = 1$ mm (a 9 × 3 grid) considering symmetry of the center line.

(c) Investigate the nature of two-dimensional conduction in the rib and determine a criterion for which the one-dimensional approximation is reasonable. Do so by extending your finite-difference analysis to determine the heat rate from the base as a function of the length of the rib for the range $1.5 \le L/w \le 10$, keeping the length L constant. Compare your results with those determined by approximating the rib as a fin.

4.83 The bottom half of an I-beam providing support for a furnace roof extends into the heating zone. The web is well insulated, while the flange surfaces experience convection with hot gases at $T_\infty = 400°C$ and a convection coefficient of $h = 150\ W/m^2 \cdot K$. Consider the symmetrical element of the flange region (inset a), assuming that the temperature distribution across the web is uniform at $T_w = 100°C$. The beam thermal conductivity is $10\ W/m \cdot K$, and its dimensions are $w_f = 80\ mm$, $w_w = 30\ mm$, and $L = 30\ mm$.

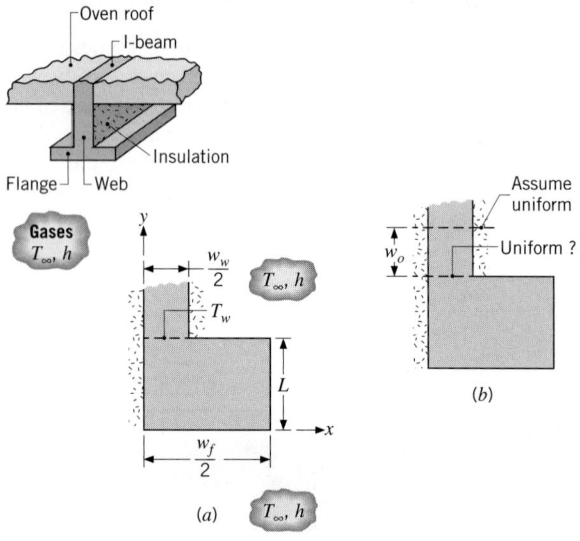

(a) Calculate the heat transfer rate per unit length to the beam using a 5 × 4 nodal network.

(b) Is it reasonable to assume that the temperature distribution across the web–flange interface is uniform? Consider the L-shaped domain of inset (b) and use a fine grid to obtain the temperature distribution across the web–flange interface. Make the distance $w_o \geq w_w/2$.

4.84 A long bar of rectangular cross section is 60 mm × 90 mm on a side and has a thermal conductivity of $1\ W/m \cdot K$. One surface is exposed to a convection process with air at 100°C and a convection coefficient of $100\ W/m^2 \cdot K$, while the remaining surfaces are maintained at 50°C.

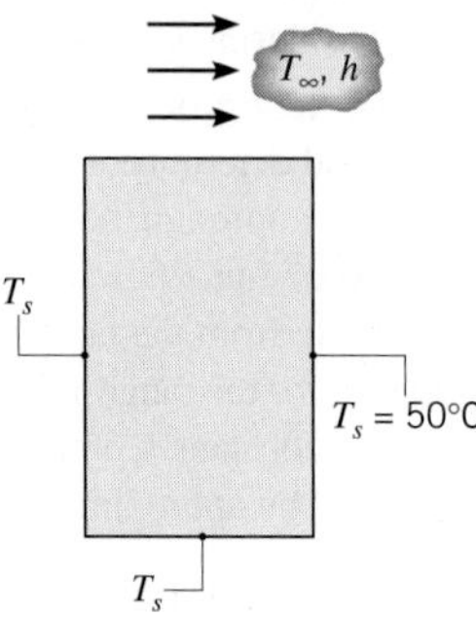

(a) Using a grid spacing of 30 mm and the Gauss-Seidel iteration method, determine the nodal temperatures and the heat rate per unit length normal to the page into the bar from the air.

(b) Determine the effect of grid spacing on the temperature field and heat rate. Specifically, consider a grid spacing of 15 mm. For this grid, explore the effect of changes in h on the temperature field and the isotherms.

4.85 A long trapezoidal bar is subjected to uniform temperatures on two surfaces, while the remaining surfaces are well insulated. If the thermal conductivity of the material is $20\ W/m \cdot K$, estimate the heat transfer rate per unit length of the bar using a finite-difference method. Use the Gauss–Seidel method of solution with a space increment of 10 mm.

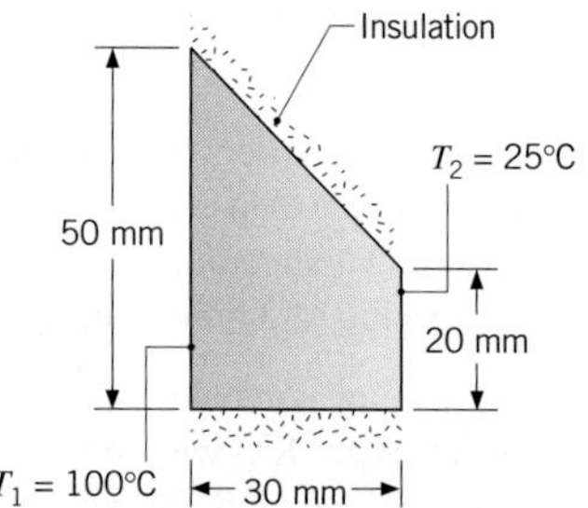

4.86 Small-diameter electrical heating elements dissipating 50 W/m (length normal to the sketch) are used to heat a ceramic plate of thermal conductivity $2\ W/m \cdot K$. The upper surface of the plate is exposed to ambient air at 30°C with a convection coefficient of $100\ W/m^2 \cdot K$, while the lower surface is well insulated.

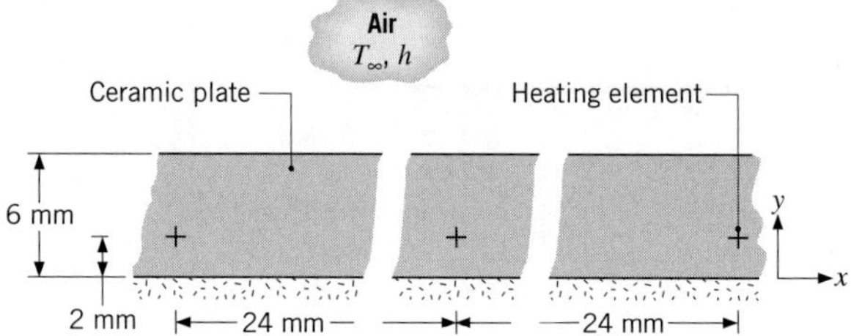

(a) Using the Gauss–Seidel method with a grid spacing of $\Delta x = 6\ mm$ and $\Delta y = 2\ mm$, obtain the temperature distribution within the plate.

(b) Using the calculated nodal temperatures, sketch four isotherms to illustrate the temperature distribution in the plate.

(c) Calculate the heat loss by convection from the plate to the fluid. Compare this value with the element dissipation rate.

(d) What advantage, if any, is there in not making $\Delta x = \Delta y$ for this situation?

(e) With $\Delta x = \Delta y = 2$ mm, calculate the temperature field within the plate and the rate of heat transfer from the plate. Under no circumstances may the temperature at any location in the plate exceed 400°C. Would this limit be exceeded if the airflow were terminated and heat transfer to the air were by natural convection with $h = 10\ \text{W/m}^2\cdot\text{K}$?

Special Applications: Finite Element Analysis

4.87 A straight fin of uniform cross section is fabricated from a material of thermal conductivity $k = 5\ \text{W/m}\cdot\text{K}$, thickness $w = 20$ mm, and length $L = 200$ mm. The fin is very long in the direction normal to the page. The base of the fin is maintained at $T_b = 200°\text{C}$, and the tip condition allows for convection (case A of Table 3.4), with $h = 500\ \text{W/m}^2\cdot\text{K}$ and $T_\infty = 25°\text{C}$.

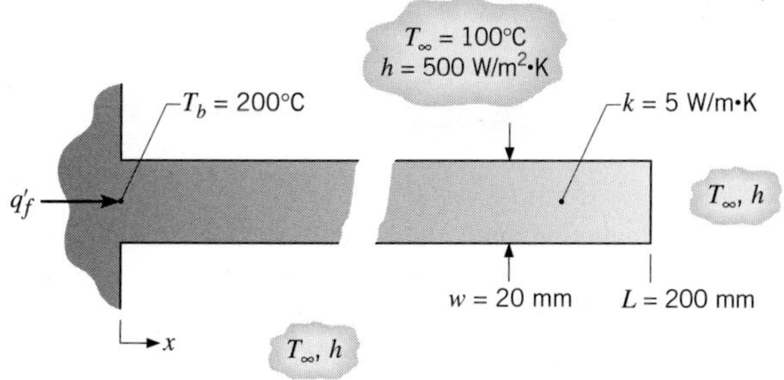

(a) Assuming one-dimensional heat transfer in the fin, calculate the fin heat rate, q'_f (W/m), and the tip temperature T_L. Calculate the Biot number for the fin to determine whether the one-dimensional assumption is valid.

(b) Using the finite-element method of *FEHT*, perform a two-dimensional analysis on the fin to determine the fin heat rate and tip temperature. Compare your results with those from the one-dimensional, analytical solution of part (a). Use the *View/Temperature Contours* option to display isotherms, and discuss key features of the corresponding temperature field and heat flow pattern. *Hint*: In drawing the outline of the fin, take advantage of symmetry. Use a fine mesh near the base and a coarser mesh near the tip. Why?

(c) Validate your *FEHT* model by comparing predictions with the analytical solution for a fin with thermal conductivities of $k = 50\ \text{W/m}\cdot\text{K}$ and $500\ \text{W/m}\cdot\text{K}$. Is the one-dimensional heat transfer assumption valid for these conditions?

4.88 Consider the long rectangular bar of Problem 4.84 with the prescribed boundary conditions.

(a) Using the finite-element method of *FEHT*, determine the temperature distribution. Use the *View/Temperature Contours* command to represent the isotherms. Identify significant features of the distribution.

(b) Using the *View/Heat Flows* command, calculate the heat rate per unit length (W/m) from the bar to the airstream.

(c) Explore the effect on the heat rate of increasing the convection coefficient by factors of two and three. Explain why the change in the heat rate is not proportional to the change in the convection coefficient.

4.89 Consider the long rectangular rod of Problem 4.53, which experiences uniform heat generation while its surfaces are maintained at a fixed temperature.

(a) Using the finite-element method of *FEHT*, determine the temperature distribution. Use the *View/Temperature Contours* command to represent the isotherms. Identify significant features of the distribution.

(b) With the boundary conditions unchanged, what heat generation rate will cause the midpoint temperature to reach 600 K?

4.90 Consider the symmetrical section of the flow channel of Problem 4.57, with the prescribed values of $\dot{q}$, k, $T_{\infty,i}$, $T_{\infty,o}$, h_i, and h_o. Use the finite-element method of *FEHT* to obtain the following results.

(a) Determine the temperature distribution in the symmetrical section, and use the *View/Temperature Contours* command to represent the isotherms. Identify significant features of the temperature distribution, including the hottest and coolest regions and the region with the steepest gradients. Describe the heat flow field.

(b) Using the *View/Heat Flows* command, calculate the heat rate per unit length (W/m) from the outer surface A to the adjacent fluid.

(c) Calculate the heat rate per unit length from the inner fluid to surface B.

(d) Verify that your results are consistent with an overall energy balance on the channel section.

4.91 The hot-film heat flux gage shown schematically may be used to determine the convection coefficient of an adjoining fluid stream by measuring the electric power dissipation per unit area, P''_e (W/m²), and the average surface temperature, $T_{s,f}$, of the film. The power dissipated in the film is transferred directly to the fluid by convection, as well as by conduction into the substrate.

If substrate conduction is negligible, the gage measurements can be used to determine the convection coefficient without application of a correction factor. Your assignment is to perform a two-dimensional, steady-state conduction analysis to estimate the fraction of the power dissipation that is conducted into a 2-mm-thick quartz substrate of width $W = 40$ mm and thermal conductivity $k = 1.4$ W/m·K. The thin, hot-film gage has a width of $w = 4$ mm and operates at a uniform power dissipation of 5000 W/m^2. Consider cases for which the fluid temperature is 25°C and the convection coefficient is 500, 1000, and 2000 W/m^2·K.

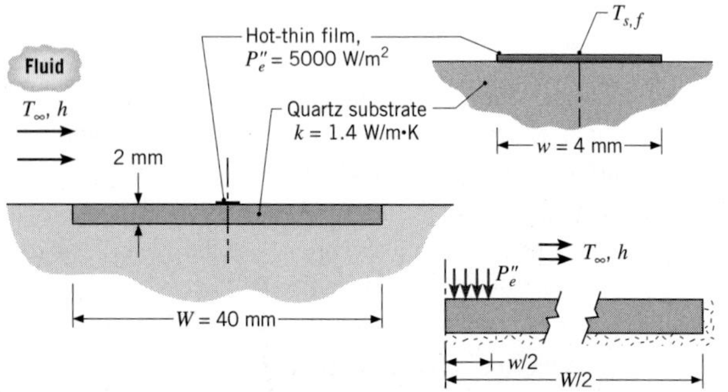

Use the finite-element method of *FEHT* to analyze a symmetrical half-section of the gage and the quartz substrate. Assume that the lower and end surfaces of the substrate are perfectly insulated, while the upper surface experiences convection with the fluid.

(a) Determine the temperature distribution and the conduction heat rate into the region below the hot film for the three values of h. Calculate the fractions of electric power dissipation represented by these rates. *Hint*: Use the *View/Heat Flow* command to find the heat rate across the boundary elements.

(b) Use the *View/Temperature Contours* command to view the isotherms and heat flow patterns. Describe the heat flow paths, and comment on features of the gage design that influence the paths. What limitations on applicability of the gage have been revealed by your analysis?

4.92 Consider the system of Problem 4.54. The interior surface is exposed to hot gases at 350°C with a convection coefficient of 100 W/m^2·K, while the exterior surface experiences convection with air at 25°C and a convection coefficient of 5 W/m^2·K.

(a) Using a grid spacing of 75 mm, calculate the temperature field within the system and determine the heat loss per unit length by convection from the outer surface of the flue to the air. Compare this result with the heat gained by convection from the hot gases to the air.

(b) Determine the effect of grid spacing on the temperature field and heat loss per unit length to the air. Specifically, consider a grid spacing of 25 mm and plot appropriately spaced isotherms on a schematic of the system. Explore the effect of changes in the convection coefficients on the temperature field and heat loss.

4.93 Electronic devices dissipating electrical power can be cooled by conduction to a heat sink. The lower surface of the sink is cooled, and the spacing of the devices w_s, the width of the device w_d, and the thickness L and thermal conductivity k of the heat sink material each affect the thermal resistance between the device and the cooled surface. The function of the heat sink is to *spread* the heat dissipated in the device throughout the sink material.

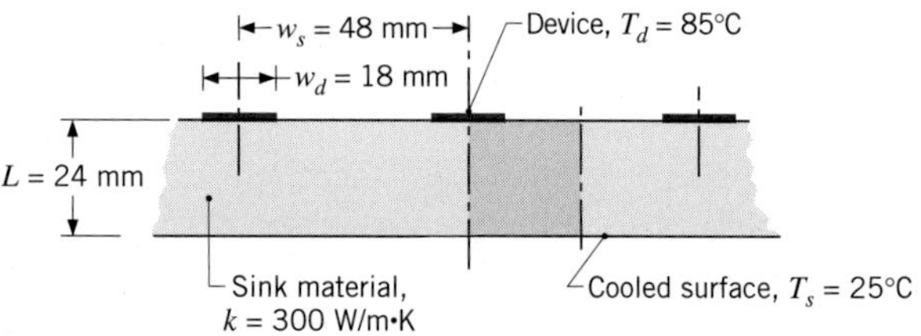

(a) Beginning with the shaded symmetrical element, use a coarse (5 × 5) nodal network to estimate the thermal resistance per unit depth between the device and lower surface of the sink, $R'_{t,d-s}$ (m·K/W). How does this value compare with thermal resistances based on the assumption of one-dimensional conduction in rectangular domains of (i) width w_d and length L and (ii) width w_s and length L?

(b) Using nodal networks with grid spacings three and five times smaller than that in part (a), determine the effect of grid size on the precision of the thermal resistance calculation.

(c) Using the finer nodal network developed for part (b), determine the effect of device width on the thermal resistance. Specifically, keeping w_s and L fixed, find the thermal resistance for values of w_d/w_s = 0.175, 0.275, 0.375, and 0.475.

4.94 Consider one-dimensional conduction in a plane composite wall. The exposed surfaces of materials A and B are maintained at $T_1 = 600$ K and $T_2 = 300$ K, respectively. Material A, of thickness $L_a = 20$ mm, has a temperature-dependent thermal conductivity of $k_a = k_o[1 + \alpha(T - T_o)]$, where $k_o = 4.4$ W/m·K, $\alpha = 0.008\ \text{K}^{-1}$, $T_o = 300$ K, and T is in kelvins. Material B is of thickness $L_b = 5$ mm and has a thermal conductivity of $k_b = 1$ W/m·K.

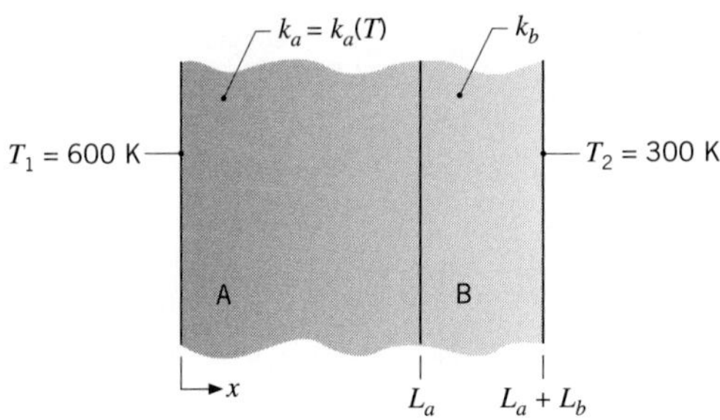

(a) Calculate the heat flux through the composite wall by assuming material A to have a uniform thermal conductivity evaluated at the average temperature of the section.

(b) Using a space increment of 1 mm, obtain the finite-difference equations for the internal nodes and calculate the heat flux considering the temperature-dependent thermal conductivity for Material A. If the *IHT* software is employed, call-up functions from *Tools/Finite-Difference Equations* may be used to obtain the nodal equations. Compare your result with that obtained in part (a).

(c) As an alternative to the finite-difference method of part (b), use the finite-element method of *FEHT* to calculate the heat flux, and compare the result with that from part (a). *Hint*: In the *Specify/Material Properties* box, properties may be entered as a function of temperature (T), the space coordinates (x, y), or time (t). See the *Help* section for more details.

4.95 A platen of thermal conductivity $k = 15\,\text{W/m}\cdot\text{K}$ is heated by flow of a hot fluid through channels of width $L = 20$ mm, with $T_{\infty,i} = 200°\text{C}$ and $h_i = 500\,\text{W/m}^2\cdot\text{K}$. The upper surface of the platen is used to heat a process fluid at $T_{\infty,o} = 25°\text{C}$ with a convection coefficient of $h_o = 250\,\text{W/m}^2\cdot\text{K}$. The lower surface of the platen is insulated. To heat the process fluid uniformly, the temperature of the platen's upper surface must be uniform to within 5°C. Use a finite-difference method, such as that of *IHT*, or the finite-element method of *FEHT* to obtain the following results.

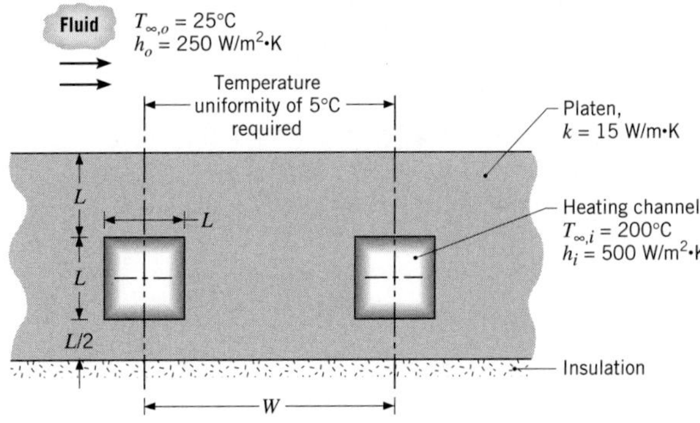

(a) Determine the maximum allowable spacing W between the channel centerlines that will satisfy the specified temperature uniformity requirement.

(b) What is the corresponding heat rate per unit length from a flow channel?

4.96 Consider the cooling arrangement for the very large-scale integration (VLSI) chip of Problem 4.93. Use the finite-element method of *FEHT* to obtain the following results.

(a) Determine the temperature distribution in the chip-substrate system. Will the maximum temperature exceed 85°C?

(b) Using the *FEHT* model developed for part (a), determine the volumetric heating rate that yields a maximum temperature of 85°C.

(c) What effect would reducing the substrate thickness have on the maximum operating temperature? For a volumetric generation rate of $\dot{q} = 10^7\,\text{W/m}^3$, reduce the thickness of the substrate from 12 to 6 mm, keeping all other dimensions unchanged. What is the maximum system temperature for these conditions? What fraction of the chip power generation is removed by convection directly from the chip surface?

CHAPTER 5

Transient Conduction

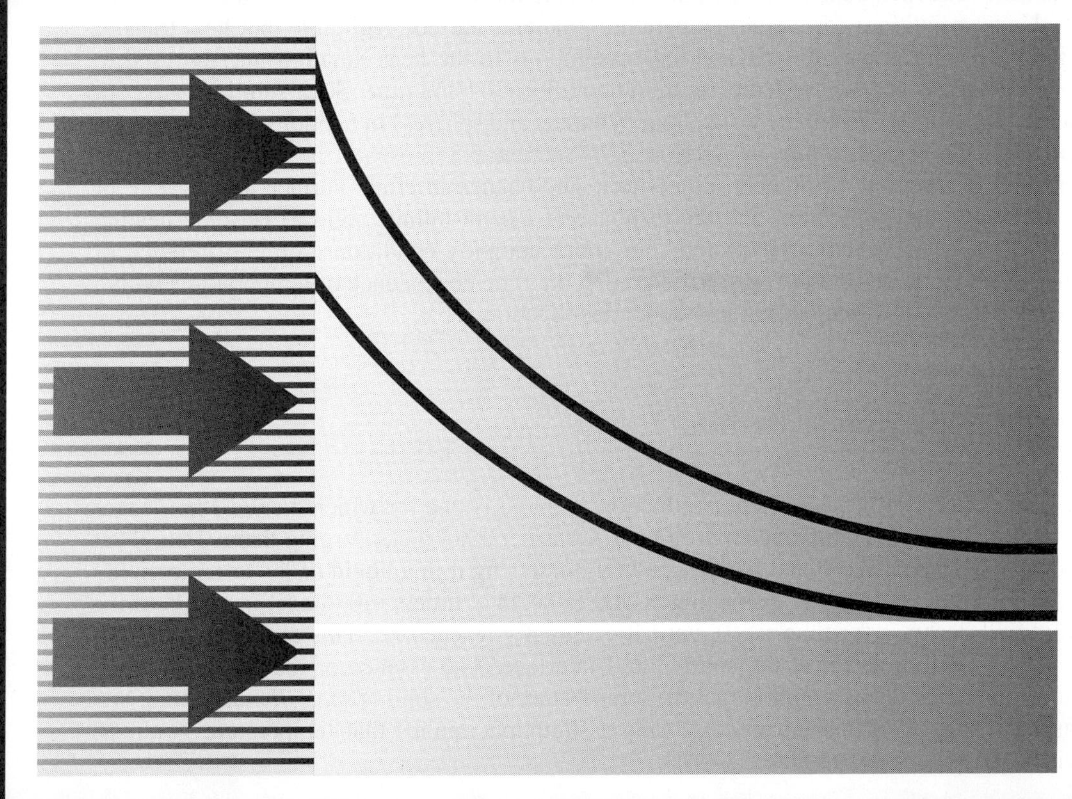

In our treatment of conduction we have gradually considered more complicated conditions. We began with the simple case of one-dimensional, steady-state conduction with no internal generation, and we subsequently considered more realistic situations involving multidimensional and generation effects. However, we have not yet considered situations for which conditions change with time.

We now recognize that many heat transfer problems are time dependent. Such *unsteady,* or *transient,* problems typically arise when the boundary conditions of a system are changed. For example, if the surface temperature of a system is altered, the temperature at each point in the system will also begin to change. The changes will continue to occur until a *steady-state* temperature distribution is reached. Consider a hot metal billet that is removed from a furnace and exposed to a cool airstream. Energy is transferred by convection and radiation from its surface to the surroundings. Energy transfer by conduction also occurs from the interior of the metal to the surface, and the temperature at each point in the billet decreases until a steady-state condition is reached. The final properties of the metal will depend significantly on the time-temperature history that results from heat transfer. Controlling the heat transfer is one key to fabricating new materials with enhanced properties.

Our objective in this chapter is to develop procedures for determining the time dependence of the temperature distribution within a solid during a transient process, as well as for determining heat transfer between the solid and its surroundings. The nature of the procedure depends on assumptions that may be made for the process. If, for example, temperature gradients within the solid may be neglected, a comparatively simple approach, termed the *lumped capacitance method,* may be used to determine the variation of temperature with time. The method is developed in Sections 5.1 through 5.3.

Under conditions for which temperature gradients are not negligible, but heat transfer within the solid is one-dimensional, exact solutions to the heat equation may be used to compute the dependence of temperature on both location and time. Such solutions are considered for *finite solids* (plane walls, long cylinders and spheres) in Sections 5.4 through 5.6 and for *semi-infinite solids* in Section 5.7. Section 5.8 presents the transient thermal response of a variety of objects subject to a step change in either surface temperature or surface heat flux. In Section 5.9, the response of a semi-infinite solid to periodic heating conditions at its surface is explored. For more complex conditions, finite-difference or finite-element methods must be used to predict the time dependence of temperatures within the solid, as well as heat rates at its boundaries (Section 5.10).

5.1 The Lumped Capacitance Method

A simple, yet common, transient conduction problem is one for which a solid experiences a sudden change in its thermal environment. Consider a hot metal forging that is initially at a uniform temperature T_i and is quenched by immersing it in a liquid of lower temperature $T_\infty < T_i$ (Figure 5.1). If the quenching is said to begin at time $t = 0$, the temperature of the solid will decrease for time $t > 0$, until it eventually reaches T_∞. This reduction is due to convection heat transfer at the solid–liquid interface. The essence of the lumped capacitance method is the assumption that the temperature of the solid is *spatially uniform* at any instant during the transient process. This assumption implies that temperature gradients within the solid are negligible.

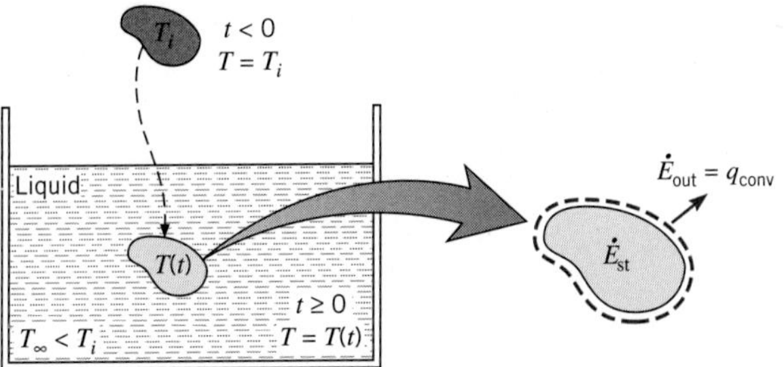

FIGURE 5.1 Cooling of a hot metal forging.

From Fourier's law, heat conduction in the absence of a temperature gradient implies the existence of infinite thermal conductivity. Such a condition is clearly impossible. However, the condition is closely approximated if the resistance to conduction within the solid is small compared with the resistance to heat transfer between the solid and its surroundings. For now we assume that this is, in fact, the case.

In neglecting temperature gradients within the solid, we can no longer consider the problem from within the framework of the heat equation, since the heat equation is a differential equation governing the spatial temperature distribution within the solid. Instead, the transient temperature response is determined by formulating an overall energy balance on the entire solid. This balance must relate the rate of heat loss at the surface to the rate of change of the internal energy. Applying Equation 1.12c to the control volume of Figure 5.1, this requirement takes the form

$$-\dot{E}_{out} = \dot{E}_{st} \tag{5.1}$$

or

$$-hA_s(T - T_\infty) = \rho Vc \frac{dT}{dt} \tag{5.2}$$

Introducing the temperature difference

$$\theta \equiv T - T_\infty \tag{5.3}$$

and recognizing that $(d\theta/dt) = (dT/dt)$ if T_∞ is constant, it follows that

$$\frac{\rho Vc}{hA_s}\frac{d\theta}{dt} = -\theta$$

Separating variables and integrating from the initial condition, for which $t = 0$ and $T(0) = T_i$, we then obtain

$$\frac{\rho Vc}{hA_s}\int_{\theta_i}^{\theta}\frac{d\theta}{\theta} = -\int_0^t dt$$

where

$$\theta_i \equiv T_i - T_\infty \tag{5.4}$$

Evaluating the integrals, it follows that

$$\frac{\rho Vc}{hA_s} \ln \frac{\theta_i}{\theta} = t \tag{5.5}$$

or

$$\frac{\theta}{\theta_i} = \frac{T - T_\infty}{T_i - T_\infty} = \exp\left[-\left(\frac{hA_s}{\rho Vc} \right) t \right] \tag{5.6}$$

Equation 5.5 may be used to determine the time required for the solid to reach some temperature T, or, conversely, Equation 5.6 may be used to compute the temperature reached by the solid at some time t.

The foregoing results indicate that the difference between the solid and fluid temperatures must decay exponentially to zero as t approaches infinity. This behavior is shown in Figure 5.2. From Equation 5.6 it is also evident that the quantity $(\rho Vc/hA_s)$ may be interpreted as a *thermal time constant* expressed as

$$\tau_t = \left(\frac{1}{hA_s} \right)(\rho Vc) = R_t C_t \tag{5.7}$$

where, from Equation 3.9, R_t is the resistance to convection heat transfer and C_t is the *lumped thermal capacitance* of the solid. Any increase in R_t or C_t will cause a solid to respond more slowly to changes in its thermal environment. This behavior is analogous to the voltage decay that occurs when a capacitor is discharged through a resistor in an electrical RC circuit.

To determine the total energy transfer Q occurring up to some time t, we simply write

$$Q = \int_0^t q\, dt = hA_s \int_0^t \theta\, dt$$

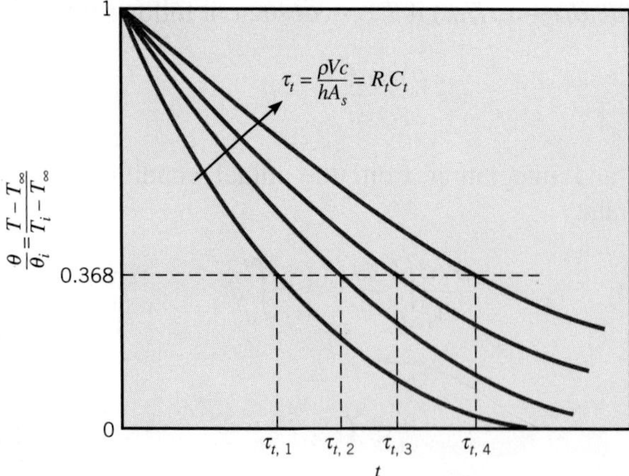

FIGURE 5.2 Transient temperature response of lumped capacitance solids for different thermal time constants τ_t.

Substituting for θ from Equation 5.6 and integrating, we obtain

$$Q = (\rho Vc)\theta_i\left[1 - \exp\left(-\frac{t}{\tau_t}\right)\right] \tag{5.8a}$$

The quantity Q is, of course, related to the change in the internal energy of the solid, and from Equation 1.12b

$$-Q = \Delta E_{st} \tag{5.8b}$$

For quenching, Q is positive and the solid experiences a decrease in energy. Equations 5.5, 5.6, and 5.8a also apply to situations where the solid is heated ($\theta < 0$), in which case Q is negative and the internal energy of the solid increases.

5.2 *Validity of the Lumped Capacitance Method*

From the foregoing results it is easy to see why there is a strong preference for using the lumped capacitance method. It is certainly the simplest and most convenient method that can be used to solve transient heating and cooling problems. Hence it is important to determine under what conditions it may be used with reasonable accuracy.

To develop a suitable criterion consider steady-state conduction through the plane wall of area A (Figure 5.3). Although we are assuming steady-state conditions, the following criterion is readily extended to transient processes. One surface is maintained at a temperature $T_{s,1}$ and the other surface is exposed to a fluid of temperature $T_\infty < T_{s,1}$. The temperature of this surface will be some intermediate value $T_{s,2}$, for which $T_\infty < T_{s,2} < T_{s,1}$. Hence under steady-state conditions the surface energy balance, Equation 1.13, reduces to

$$\frac{kA}{L}(T_{s,1} - T_{s,2}) = hA(T_{s,2} - T_\infty)$$

where k is the thermal conductivity of the solid. Rearranging, we then obtain

$$\frac{T_{s,1} - T_{s,2}}{T_{s,2} - T_\infty} = \frac{(L/kA)}{(1/hA)} = \frac{R_{t,\text{cond}}}{R_{t,\text{conv}}} = \frac{hL}{k} \equiv Bi \tag{5.9}$$

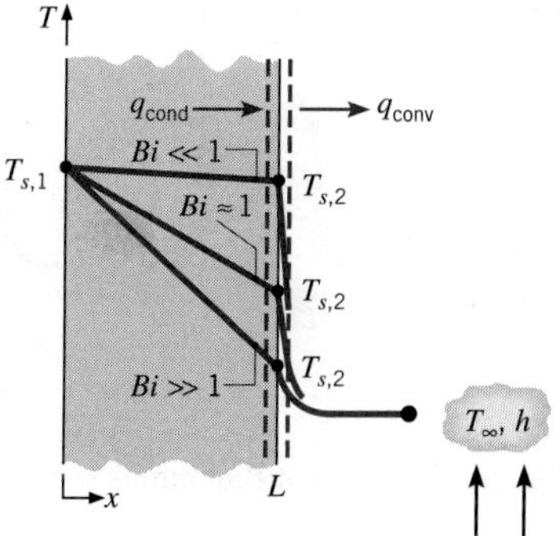

FIGURE 5.3 Effect of Biot number on steady-state temperature distribution in a plane wall with surface convection.

The quantity (hL/k) appearing in Equation 5.9 is a *dimensionless parameter.* It is termed the *Biot number,* and it plays a fundamental role in conduction problems that involve surface convection effects. According to Equation 5.9 and as illustrated in Figure 5.3, the Biot number provides a measure of the temperature drop in the solid relative to the temperature difference between the solid's surface and the fluid. From Equation 5.9, it is also evident that the Biot number may be interpreted as a ratio of thermal resistances. *In particular, if* $Bi \ll 1$, *the resistance to conduction within the solid is much less than the resistance to convection across the fluid boundary layer. Hence, the assumption of a uniform temperature distribution within the solid is reasonable if the Biot number is small.*

Although we have discussed the Biot number in the context of steady-state conditions, we are reconsidering this parameter because of its significance to transient conduction problems. Consider the plane wall of Figure 5.4, which is initially at a uniform temperature T_i and experiences convection cooling when it is immersed in a fluid of $T_\infty < T_i$. The problem may be treated as one-dimensional in x, and we are interested in the temperature variation with position and time, $T(x, t)$. This variation is a strong function of the Biot number, and three conditions are shown in Figure 5.4. Again, for $Bi \ll 1$ the temperature gradients in the solid are small and the assumption of a uniform temperature distribution, $T(x, t) \approx T(t)$ is reasonable. Virtually all the temperature difference is between the solid and the fluid, and the solid temperature remains nearly uniform as it decreases to T_∞. For moderate to large values of the Biot number, however, the temperature gradients within the solid are significant. Hence $T = T(x, t)$. Note that for $Bi \gg 1$, the temperature difference across the solid is much larger than that between the surface and the fluid.

We conclude this section by emphasizing the importance of the lumped capacitance method. Its inherent simplicity renders it the preferred method for solving transient heating and cooling problems. Hence, when confronted with such a problem, *the very first thing that one should do is calculate the Biot number.* If the following condition is satisfied

$$Bi = \frac{hL_c}{k} < 0.1 \tag{5.10}$$

the error associated with using the lumped capacitance method is small. For convenience, it is customary to define the *characteristic length* of Equation 5.10 as the ratio of the solid's

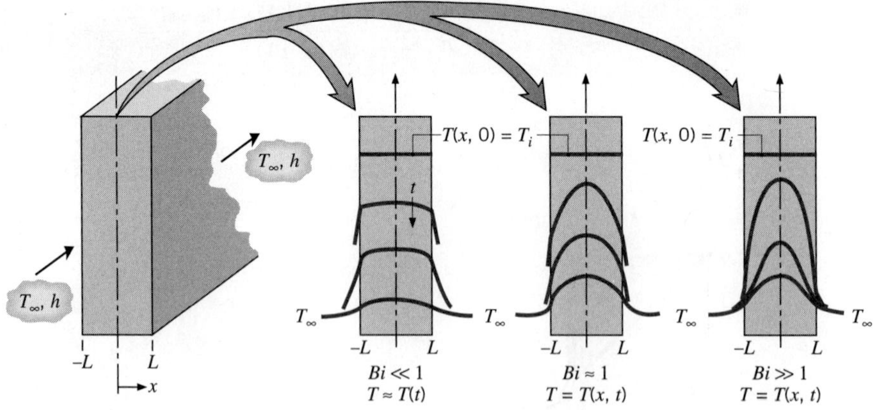

FIGURE 5.4 Transient temperature distributions for different Biot numbers in a plane wall symmetrically cooled by convection.

volume to surface area $L_c \equiv V/A_s$. Such a definition facilitates calculation of L_c for solids of complicated shape and reduces to the half-thickness L for a plane wall of thickness $2L$ (Figure 5.4), to $r_o/2$ for a long cylinder, and to $r_o/3$ for a sphere. However, if one wishes to implement the criterion in a conservative fashion, L_c should be associated with the length scale corresponding to the maximum spatial temperature difference. Accordingly, for a symmetrically heated (or cooled) plane wall of thickness $2L$, L_c would remain equal to the half-thickness L. However, for a long cylinder or sphere, L_c would equal the actual radius r_o, rather than $r_o/2$ or $r_o/3$.

Finally, we note that, with $L_c \equiv V/A_s$, the exponent of Equation 5.6 may be expressed as

$$\frac{hA_s t}{\rho V c} = \frac{ht}{\rho c L_c} = \frac{hL_c}{k}\frac{k}{\rho c}\frac{t}{L_c^2} = \frac{hL_c}{k}\frac{\alpha t}{L_c^2}$$

or

$$\frac{hA_s t}{\rho V c} = Bi \cdot Fo \tag{5.11}$$

where

$$Fo \equiv \frac{\alpha t}{L_c^2} \tag{5.12}$$

is termed the Fourier number. It is a *dimensionless time*, which, with the Biot number, characterizes transient conduction problems. Substituting Equation 5.11 into 5.6, we obtain

$$\frac{\theta}{\theta_i} = \frac{T - T_\infty}{T_i - T_\infty} = \exp(-Bi \cdot Fo) \tag{5.13}$$

Example 5.1

A thermocouple junction, which may be approximated as a sphere, is to be used for temperature measurement in a gas stream. The convection coefficient between the junction surface and the gas is $h = 400\ \text{W/m}^2 \cdot \text{K}$, and the junction thermophysical properties are $k = 20\ \text{W/m} \cdot \text{K}$, $c = 400\ \text{J/kg} \cdot \text{K}$, and $\rho = 8500\ \text{kg/m}^3$. Determine the junction diameter needed for the thermocouple to have a time constant of 1 s. If the junction is at 25°C and is placed in a gas stream that is at 200°C, how long will it take for the junction to reach 199°C?

Solution

Known: Thermophysical properties of thermocouple junction used to measure temperature of a gas stream.

Find:

1. Junction diameter needed for a time constant of 1 s.
2. Time required to reach 199°C in gas stream at 200°C.

Schematic:

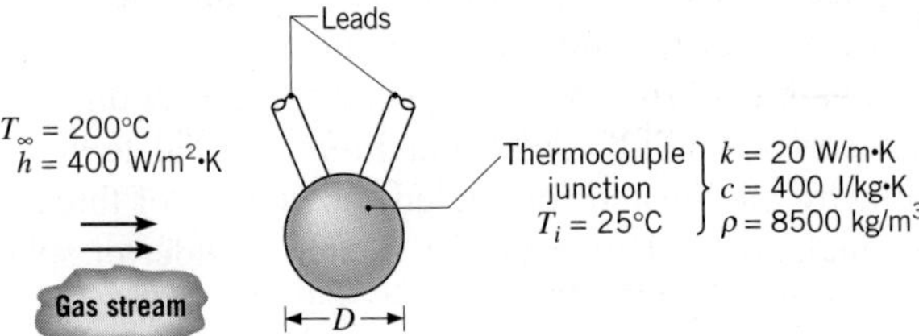

Assumptions:

1. Temperature of junction is uniform at any instant.
2. Radiation exchange with the surroundings is negligible.
3. Losses by conduction through the leads are negligible.
4. Constant properties.

Analysis:

1. Because the junction diameter is unknown, it is not possible to begin the solution by determining whether the criterion for using the lumped capacitance method, Equation 5.10, is satisfied. However, a reasonable approach is to use the method to find the diameter and to then determine whether the criterion is satisfied. From Equation 5.7 and the fact that $A_s = \pi D^2$ and $V = \pi D^3/6$ for a sphere, it follows that

$$\tau_t = \frac{1}{h\pi D^2} \times \frac{\rho \pi D^3}{6} c$$

Rearranging and substituting numerical values,

$$D = \frac{6h\tau_t}{\rho c} = \frac{6 \times 400 \text{ W/m}^2 \cdot \text{K} \times 1 \text{ s}}{8500 \text{ kg/m}^3 \times 400 \text{ J/kg} \cdot \text{K}} = 7.06 \times 10^{-4} \text{ m} \qquad \triangleleft$$

With $L_c = r_o/3$ it then follows from Equation 5.10 that

$$Bi = \frac{h(r_o/3)}{k} = \frac{400 \text{ W/m}^2 \cdot \text{K} \times 3.53 \times 10^{-4} \text{ m}}{3 \times 20 \text{ W/m} \cdot \text{K}} = 2.35 \times 10^{-3}$$

Accordingly, Equation 5.10 is satisfied (for $L_c = r_o$, as well as for $L_c = r_o/3$) and the lumped capacitance method may be used to an excellent approximation.

2. From Equation 5.5 the time required for the junction to reach $T = 199°\text{C}$ is

$$t = \frac{\rho(\pi D^3/6)c}{h(\pi D^2)} \ln \frac{T_i - T_\infty}{T - T_\infty} = \frac{\rho D c}{6h} \ln \frac{T_i - T_\infty}{T - T_\infty}$$

$$t = \frac{8500 \text{ kg/m}^3 \times 7.06 \times 10^{-4} \text{ m} \times 400 \text{ J/kg} \cdot \text{K}}{6 \times 400 \text{ W/m}^2 \cdot \text{K}} \ln \frac{25 - 200}{199 - 200}$$

$$t = 5.2 \text{ s} \approx 5\tau_t \qquad \triangleleft$$

Comments: Heat transfer due to radiation exchange between the junction and the surroundings and conduction through the leads would affect the time response of the junction and would, in fact, yield an equilibrium temperature that differs from T_∞.

5.3 General Lumped Capacitance Analysis

Although transient conduction in a solid is commonly initiated by convection heat transfer to or from an adjoining fluid, other processes may induce transient thermal conditions within the solid. For example, a solid may be separated from large surroundings by a gas or vacuum. If the temperatures of the solid and surroundings differ, radiation exchange could cause the internal thermal energy, and hence the temperature, of the solid to change. Temperature changes could also be induced by applying a heat flux at a portion, or all, of the surface or by initiating thermal energy generation within the solid. Surface heating could, for example, be applied by attaching a film or sheet electrical heater to the surface, while thermal energy could be generated by passing an electrical current through the solid.

Figure 5.5 depicts the general situation for which thermal conditions within a solid may be influenced simultaneously by convection, radiation, an applied surface heat flux, and internal energy generation. It is presumed that, initially ($t = 0$), the temperature of the solid T_i differs from that of the fluid T_∞, and the surroundings T_{sur}, and that both surface and volumetric heating (q''_s and $\dot{q}$) are initiated. The imposed heat flux q''_s and the convection–radiation heat transfer occur at mutually exclusive portions of the surface, $A_{s(h)}$ and $A_{s(c,r)}$, respectively, and convection–radiation transfer is presumed to be *from* the surface. Moreover, although convection and radiation have been prescribed for the same surface, the surfaces may, in fact, differ ($A_{s,c} \neq A_{s,r}$). Applying conservation of energy at any instant t, it follows from Equation 1.12c that

$$q''_s A_{s,h} + \dot{E}_g - (q''_{conv} + q''_{rad})A_{s(c,r)} = \rho Vc\frac{dT}{dt} \tag{5.14}$$

or, from Equations 1.3a and 1.7,

$$q''_s A_{s,h} + \dot{E}_g - [h(T - T_\infty) + \varepsilon\sigma(T^4 - T^4_{sur})]A_{s(c,r)} = \rho Vc\frac{dT}{dt} \tag{5.15}$$

Equation 5.15 is a nonlinear, first-order, nonhomogeneous, ordinary differential equation that cannot be integrated to obtain an exact solution.[1] However, exact solutions may be obtained for simplified versions of the equation.

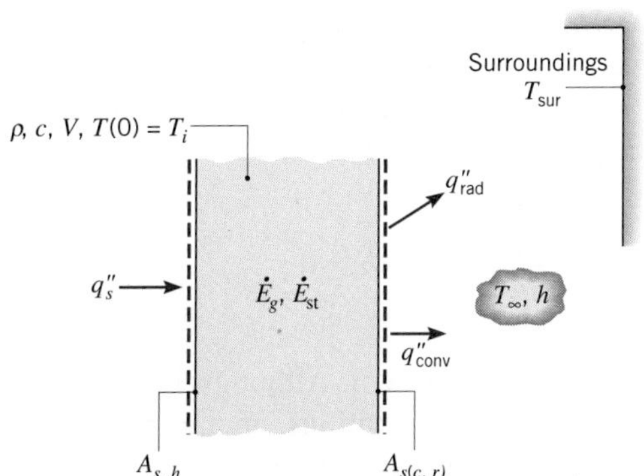

FIGURE 5.5 Control surface for general lumped capacitance analysis.

[1]An approximate, finite-difference solution may be obtained by *discretizing* the time derivative (Section 5.10) and *marching* the solution out in time.

5.3.1 Radiation Only

If there is no imposed heat flux or generation and convection is either nonexistent (a vacuum) or negligible relative to radiation, Equation 5.15 reduces to

$$\rho V c \frac{dT}{dt} = -\varepsilon A_{s,r}\sigma(T^4 - T_{sur}^4) \tag{5.16}$$

Separating variables and integrating from the initial condition to any time t, it follows that

$$\frac{\varepsilon A_{s,r}\sigma}{\rho V c}\int_0^t dt = \int_{T_i}^{T} \frac{dT}{T_{sur}^4 - T^4} \tag{5.17}$$

Evaluating both integrals and rearranging, the time required to reach the temperature T becomes

$$t = \frac{\rho V c}{4\varepsilon A_{s,r}\sigma T_{sur}^3}\left\{\ln\left|\frac{T_{sur}+T}{T_{sur}-T}\right| - \ln\left|\frac{T_{sur}+T_i}{T_{sur}-T_i}\right| + 2\left[\tan^{-1}\left(\frac{T}{T_{sur}}\right) - \tan^{-1}\left(\frac{T_i}{T_{sur}}\right)\right]\right\} \tag{5.18}$$

This expression cannot be used to evaluate T explicitly in terms of t, T_i, and T_{sur}, nor does it readily reduce to the limiting result for $T_{sur} = 0$ (radiation to deep space). However, returning to Equation 5.17, its solution for $T_{sur} = 0$ yields

$$t = \frac{\rho V c}{3\varepsilon A_{s,r}\sigma}\left(\frac{1}{T^3} - \frac{1}{T_i^3}\right) \tag{5.19}$$

5.3.2 Negligible Radiation

An exact solution to Equation 5.15 may also be obtained if radiation may be neglected and all quantities (other than T, of course) are independent of time. Introducing a temperature difference $\theta \equiv T - T_\infty$, where $d\theta/dt = dT/dt$, Equation 5.15 reduces to a linear, first-order, nonhomogeneous differential equation of the form

$$\frac{d\theta}{dt} + a\theta - b = 0 \tag{5.20}$$

where $a \equiv (hA_{s,c}/\rho V c)$ and $b \equiv [(q_s'' A_{s,h} + \dot{E}_g)/\rho V c]$. Although Equation 5.20 may be solved by summing its homogeneous and particular solutions, an alternative approach is to eliminate the nonhomogeneity by introducing the transformation

$$\theta' \equiv \theta - \frac{b}{a} \tag{5.21}$$

Recognizing that $d\theta'/dt = d\theta/dt$, Equation 5.21 may be substituted into (5.20) to yield

$$\frac{d\theta'}{dt} + a\theta' = 0 \tag{5.22}$$

Separating variables and integrating from 0 to t (θ'_i to θ'), it follows that

$$\frac{\theta'}{\theta'_i} = \exp(-at) \tag{5.23}$$

or substituting for θ' and θ,

$$\frac{T - T_\infty - (b/a)}{T_i - T_\infty - (b/a)} = \exp(-at) \tag{5.24}$$

Hence

$$\frac{T - T_\infty}{T_i - T_\infty} = \exp(-at) + \frac{b/a}{T_i - T_\infty}[1 - \exp(-at)] \tag{5.25}$$

As it must, Equation 5.25 reduces to Equation 5.6 when $b = 0$ and yields $T = T_i$ at $t = 0$. As $t \to \infty$, Equation 5.25 reduces to $(T - T_\infty) = (b/a)$, which could also be obtained by performing an energy balance on the control surface of Figure 5.5 for steady-state conditions.

5.3.3 Convection Only with Variable Convection Coefficient

In some cases, such as those involving free convection or boiling, the convection coefficient h varies with the temperature difference between the object and the fluid. In these situations, the convection coefficient can often be approximated with an expression of the form

$$h = C(T - T_\infty)^n \tag{5.26}$$

where n is a constant and the parameter C has units of $W/m^2 \cdot K^{(1+n)}$. If radiation, surface heating, and volumetric generation are negligible, Equation 5.15 may be written as

$$-C(T - T_\infty)^n A_{s,c}(T - T_\infty) = -CA_{s,c}(T - T_\infty)^{1+n} = \rho Vc\frac{dT}{dt} \tag{5.27}$$

Substituting θ and $d\theta/dt = dT/dt$ into the preceding expression, separating variables and integrating yields

$$\frac{\theta}{\theta_i} = \left[\frac{nCA_{s,c}\theta_i^n}{\rho Vc}t + 1\right]^{-1/n} \tag{5.28}$$

It can be shown that Equation 5.28 reduces to Equation 5.6 if the heat transfer coefficient is independent of temperature, $n = 0$.

5.3.4 Additional Considerations

In some cases the ambient or surroundings temperature may vary with time. For example, if the container of Figure 5.1 is insulated and of finite volume, the liquid temperature will increase as the metal forging is cooled. An analytical solution for the time-varying solid (and liquid) temperature is presented in Example 11.8. As evident in Examples 5.2 through 5.4, the heat equation can be solved numerically for a wide variety of situations involving variable properties or time-varying boundary conditions, internal energy generation rates, or surface heating or cooling.

IHT

EXAMPLE 5.2

Consider the thermocouple and convection conditions of Example 5.1, but now allow for radiation exchange with the walls of a duct that encloses the gas stream. If the duct walls are at 400°C and the emissivity of the thermocouple bead is 0.9, calculate the steady-state temperature of the junction. Also, determine the time for the junction temperature to increase from an initial condition of 25°C to a temperature that is within 1°C of its steady-state value.

SOLUTION

Known: Thermophysical properties and diameter of the thermocouple junction used to measure temperature of a gas stream passing through a duct with hot walls.

Find:

1. Steady-state temperature of the junction.
2. Time required for the thermocouple to reach a temperature that is within 1°C of its steady-state value.

Schematic:

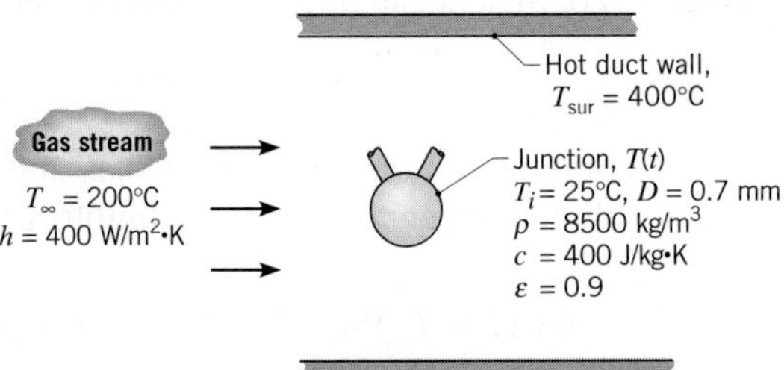

Assumptions: Same as Example 5.1, but radiation transfer is no longer treated as negligible and is approximated as exchange between a small surface and large surroundings.

Analysis:

1. For steady-state conditions, the energy balance on the thermocouple junction has the form

$$\dot{E}_{\text{in}} - \dot{E}_{\text{out}} = 0$$

Recognizing that net radiation to the junction must be balanced by convection from the junction to the gas, the energy balance may be expressed as

$$[\varepsilon\sigma(T_{\text{sur}}^4 - T^4) - h(T - T_\infty)]A_s = 0$$

Substituting numerical values, we obtain

$$T = 218.7°\text{C} \quad \triangleleft$$

2. The temperature-time history, $T(t)$, for the junction, initially at $T(0) = T_i = 25°\text{C}$, follows from the energy balance for transient conditions,

$$\dot{E}_{\text{in}} - \dot{E}_{\text{out}} = \dot{E}_{\text{st}}$$

From Equation 5.15, the energy balance may be expressed as

$$-[h(T - T_\infty) + \varepsilon\sigma(T^4 - T_{\text{sur}}^4)]A_s = \rho Vc\frac{dT}{dt}$$

The solution to this first-order differential equation can be obtained by numerical integration, giving the result, $T(4.9\text{ s}) = 217.7°\text{C}$. Hence, the time required to reach a temperature that is within 1°C of the steady-state value is

$$t = 4.9\text{ s.} \quad \triangleleft$$

Comments:

1. The effect of radiation exchange with the hot duct walls is to increase the junction temperature, such that the thermocouple indicates an erroneous gas stream temperature that exceeds the actual temperature by 18.7°C. The time required to reach a temperature that is within 1°C of the steady-state value is slightly less than the result of Example 5.1, which only considers convection heat transfer. Why is this so?
2. The response of the thermocouple and the indicated gas stream temperature depend on the velocity of the gas stream, which in turn affects the magnitude of the convection coefficient. Temperature–time histories for the thermocouple junction are shown in the following graph for values of h = 200, 400, and 800 $\text{W/m}^2 \cdot \text{K}$.

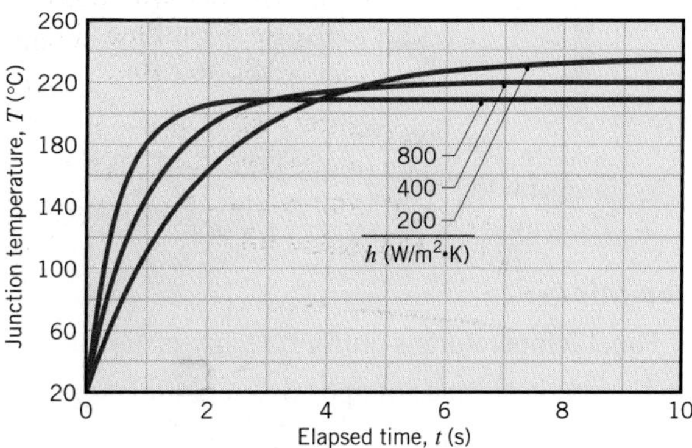

The effect of increasing the convection coefficient is to cause the junction to indicate a temperature closer to that of the gas stream. Further, the effect is to reduce the time required for the junction to reach the near-steady-state condition. What physical explanation can you give for these results?

3. The *IHT* software includes an integral function, Der(T, t), that can be used to represent the temperature–time derivative and to integrate first-order differential equations.

EXAMPLE 5.3

A 3-mm-thick panel of aluminum alloy ($k = 177$ W/m·K, $c = 875$ J/kg·K, and $\rho = 2770$ kg/m^3) is finished on both sides with an epoxy coating that must be cured at or above $T_c = 150°C$ for at least 5 min. The production line for the curing operation involves two steps: (1) heating in a large oven with air at $T_{\infty,o} = 175°C$ and a convection coefficient of $h_o = 40$ W/m^2·K, and (2) cooling in a large chamber with air at $T_{\infty,c} = 25°C$ and a convection coefficient of $h_c = 10$ W/m^2·K. The heating portion of the process is conducted over a time interval t_e, which exceeds the time t_c required to reach 150°C by 5 min ($t_e = t_c + 300$ s). The coating has an emissivity of $\varepsilon = 0.8$, and the temperatures of the oven and chamber walls are 175 and 25°C, respectively. If the panel is placed in the oven at an initial temperature of 25°C and removed from the chamber at a *safe-to-touch* temperature of 37°C, what is the total elapsed time for the two-step curing operation?

SOLUTION

Known: Operating conditions for a two-step heating/cooling process in which a coated aluminum panel is maintained at or above a temperature of 150°C for at least 5 min.

Find: Total time t_t required for the two-step process.

Schematic:

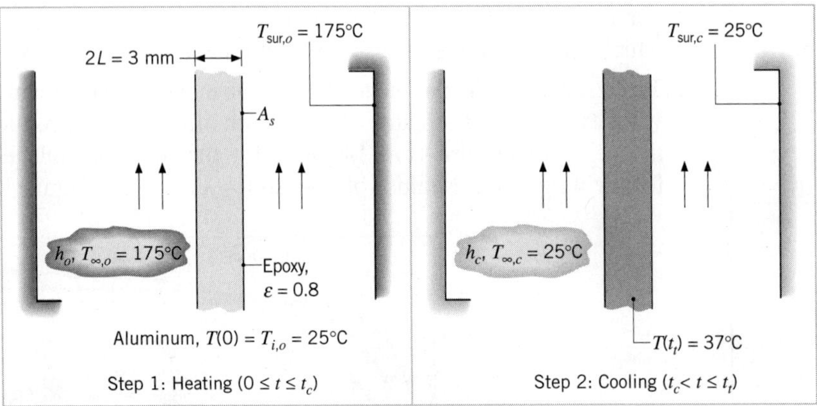

Assumptions:

1. Panel temperature is uniform at any instant.
2. Thermal resistance of epoxy is negligible.
3. Constant properties.

Analysis: To assess the validity of the lumped capacitance approximation, we begin by calculating Biot numbers for the heating and cooling processes.

$$Bi_h = \frac{h_o L}{k} = \frac{(40\ \text{W/m}^2\cdot\text{K})(0.0015\ \text{m})}{177\ \text{W/m}\cdot\text{K}} = 3.4 \times 10^{-4}$$

$$Bi_c = \frac{h_c L}{k} = \frac{(10\ \text{W/m}^2\cdot\text{K})(0.0015\ \text{m})}{177\ \text{W/m}\cdot\text{K}} = 8.5 \times 10^{-5}$$

Hence the lumped capacitance approximation is excellent.

To determine whether radiation exchange between the panel and its surroundings should be considered, the radiation heat transfer coefficient is determined from Equation 1.9. A representative value of h_r for the heating process is associated with the cure condition, in which case

$$\begin{aligned} h_{r,o} &= \varepsilon\sigma(T_c + T_{\text{sur},o})(T_c^2 + T_{\text{sur},o}^2) \\ &= 0.8 \times 5.67 \times 10^{-8}\ \text{W/m}^2\cdot\text{K}^4(423 + 448)\text{K}(423^2 + 448^2)\text{K}^2 \\ &= 15\ \text{W/m}^2\cdot\text{K} \end{aligned}$$

Using $T_c = 150°\text{C}$ with $T_{\text{sur},c} = 25°\text{C}$ for the cooling process, we also obtain $h_{r,c} = 8.8\ \text{W/m}^2\cdot\text{K}$. Since the values of $h_{r,o}$ and $h_{r,c}$ are comparable to those of h_o and h_c, respectively, radiation effects must be considered.

With $V = 2LA_s$ and $A_{s,c} = A_{s,r} = 2A_s$, Equation 5.15 may be expressed as

$$-[h(T - T_\infty) + \varepsilon\sigma(T^4 - T_{\text{sur}}^4)] = \rho c L \frac{dT}{dt}$$

Selecting a suitable time increment, Δt, the equation may be integrated numerically to obtain the panel temperature at $t = \Delta t$, $2\Delta t$, $3\Delta t$, and so on. Selecting $\Delta t = 10$ s, calculations for the heating process are extended to $t_e = t_c + 300$ s, which is 5 min beyond the time required for the panel to reach $T_c = 150°\text{C}$. At t_e the cooling process is initiated and continued until the panel temperature reaches 37°C at $t = t_t$. The integration was performed using *IHT*, and results of the calculations are plotted as follows:

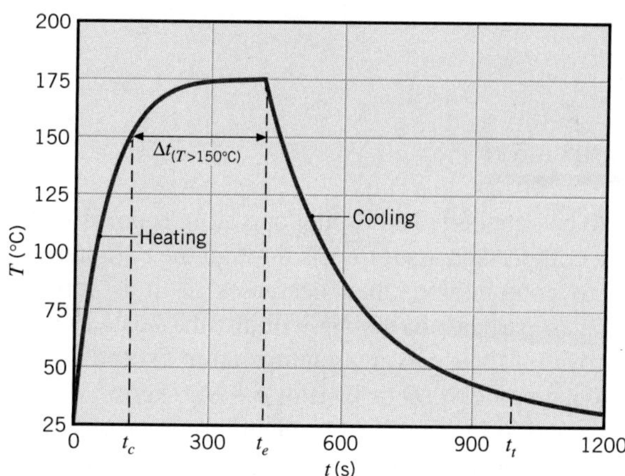

The total time for the two-step process is

$$t_t = 989\ \text{s} \quad \triangleleft$$

with intermediate times of $t_c = 124$ s and $t_e = 424$ s.

Comments:

1. The duration of the two-step process may be reduced by increasing the convection coefficients and/or by reducing the period of extended heating. The second option is made possible by the fact that, during a portion of the cooling period, the panel

temperature remains above 150°C. Hence, to satisfy the cure requirement, it is not necessary to extend heating for as much as 5 min from $t = t_c$. If the convection coefficients are increased to $h_o = h_c = 100\ \text{W/m}^2 \cdot \text{K}$ and an extended heating period of 300 s is maintained, the numerical integration yields $t_c = 58$ s and $t_t = 445$ s. The corresponding time interval over which the panel temperature exceeds 150°C is $\Delta t_{(T>150^\circ\text{C})} = 306$ s ($58\ \text{s} \leq t \leq 364\ \text{s}$). If the extended heating period is reduced to 294 s, the numerical integration yields $t_c = 58$ s, $t_t = 439$ s, and $\Delta t_{(T>150^\circ\text{C})} = 300$ s. Hence the total process time is reduced, while the curing requirement is still satisfied.

2. Generally, the accuracy of a numerical integration improves with decreasing Δt, but at the expense of increased computation time. In this case, however, results obtained for $\Delta t = 1$ s are virtually identical to those obtained for $\Delta t = 10$ s, indicating that the larger time interval is sufficient to accurately depict the temperature history.
3. The complete solution for this example is provided as a ready-to-solve model in the *Advanced* section of *IHT*, using *Models, Lumped Capacitance*. The model can be used to check the results of Comment 1 or to independently explore modifications of the cure process.
4. If the Biot numbers were not small, it would be inappropriate to apply the lumped capacitance method. For moderate or large Biot numbers, temperatures near the solid's centerline would continue to increase for some time after the conclusion of heating, as thermal energy near the solid's surface propagates inward. The temperatures near the centerline would subsequently reach a maximum and would then decrease to the steady-state value. Correlations for the maximum temperature experienced at the panel's centerline, along with the time at which these maximum temperatures are reached, have been correlated for a broad range of Bi_h and Bi_c values [1].

EXAMPLE 5.4

Air to be supplied to a hospital operating room is first purified by forcing it through a single-stage compressor. As it travels through the compressor, the air temperature initially increases due to compression, then decreases as it is returned to atmospheric pressure. Harmful pathogen particles in the air will also be heated and subsequently cooled, and they will be destroyed if their maximum temperature exceeds a *lethal* temperature T_d. Consider spherical pathogen particles ($D = 10\ \mu\text{m}$, $\rho = 900\ \text{kg/m}^3$, $c = 1100\ \text{J/kg} \cdot \text{K}$, and $k = 0.2\ \text{W/m} \cdot \text{K}$) that are dispersed in unpurified air. During the process, the air temperature may be described by an expression of the form $T_\infty(t) = 125^\circ\text{C} - 100^\circ\text{C} \cdot \cos(2\pi t/t_p)$, where t_p is the process time associated with flow through the compressor. If $t_p = 0.004$ s, and the initial and lethal pathogen temperatures are $T_i = 25^\circ\text{C}$ and $T_d = 220^\circ\text{C}$, respectively, will the pathogens be destroyed? The value of the convection heat transfer coefficient associated with the pathogen particles is $h = 4600\ \text{W/m}^2 \cdot \text{K}$.

SOLUTION

Known: Air temperature versus time, convection heat transfer coefficient, pathogen geometry, size, and properties.

Find: Whether the pathogens are destroyed for $t_p = 0.004$ s.

Schematic:

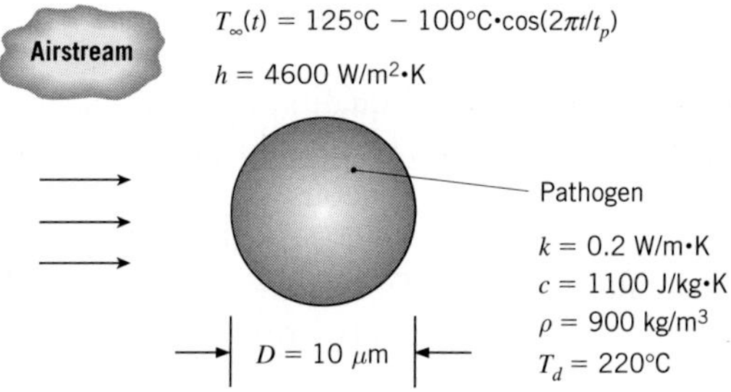

Assumptions:

1. Constant properties.
2. Negligible radiation.

Analysis: The Biot number associated with a spherical pathogen particle is

$$Bi = \frac{h(D/6)}{k} = \frac{4600\ \text{W/m}^2\cdot\text{K} \times (10 \times 10^{-6}\ \text{m}/6)}{0.2\ \text{W/m}\cdot\text{K}} = 0.038$$

Hence, the lumped capacitance approximation is valid and we may apply Equation 5.2.

$$\frac{dT}{dt} = -\frac{hA_s}{\rho Vc}[T - T_\infty(t)] = -\frac{6h}{\rho cD}[T - 125°\text{C} + 100°\text{C}\cdot\cos(2\pi t/t_p)] \quad (1)$$

The solution to this first-order differential equation may be obtained analytically, or by numerical integration.

Numerical Integration A numerical solution of Equation 1 may be obtained by specifying the initial particle temperature, T_i, and using *IHT* or an equivalent numerical solver to integrate the equation. The plot of the numerical solution follows.

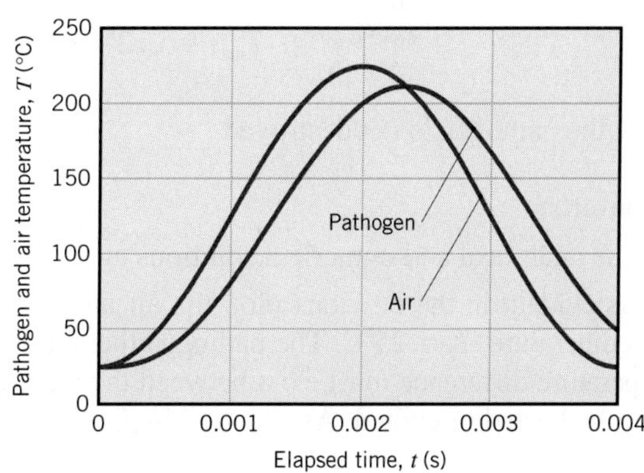

Inspection of the predicted pathogen temperatures yields

$$T_{max} = 212°C < 220°C$$

Hence, the pathogen is not destroyed. ◁

Analytical Solution Equation 1 is a linear nonhomogeneous differential equation, therefore the solution can be found as the sum of a homogeneous and a particular solution, $T = T_h + T_p$. The homogeneous part, T_h, corresponds to the homogeneous differential equation, $dT_h/dt = -(6h/\rho cD)T_h$, which has the familiar solution, $T_h = C_0 \exp(-6ht/\rho cD)$. The particular solution, T_p, can then be found using the method of undetermined coefficients; for a nonhomogeneous term that includes a cosine function and a constant term, the particular solution is assumed to be of the form $T_p = C_1 \cos(2\pi t/t_p) + C_2 \sin(2\pi t/t_p) + C_3$. Substituting this expression into Equation 1 yields values for the coefficients, resulting in

$$T_p = 125°C - 100°C \times A\left[\cos\left(\frac{2\pi t}{t_p}\right) + \frac{2\pi\rho cD}{6ht_p}\sin\left(\frac{2\pi t}{t_p}\right)\right] \tag{2}$$

where

$$A = \frac{(6h/\rho cD)^2}{(6h/\rho cD)^2 + (2\pi/t_p)^2}$$

The initial condition, $T(0) = T_i$, is then applied to the complete solution, $T = T_h + T_p$, to yield $C_0 = 100°C(A - 1)$. Thus, the particle temperature is

$$T(t) = 125°C + 100°C\left\{(A-1)\exp\left(-\frac{6ht}{\rho cD}\right) - A\left[\cos\left(\frac{2\pi t}{t_p}\right) + \frac{2\pi\rho cD}{6ht_p}\sin\left(\frac{2\pi t}{t_p}\right)\right]\right\} \tag{3}$$

To find the maximum pathogen temperature, we could differentiate Equation 3 and set the result equal to zero. This yields a lengthy, implicit equation for the critical time t_{crit} at which the maximum temperature is reached. The maximum temperature may then be found by substituting $t = t_{crit}$ into Equation 3. Alternatively, Equation 3 can be plotted or $T(t)$ may be tabulated to find

$$T_{max} = 212°C < 220°C$$

Hence, the pathogen is not destroyed. ◁

Comments:

1. The analytical and numerical solutions agree, as they must.
2. As evident in the previous plot, the air and pathogen particles initially have the same temperature, $T_i = 25°C$. The pathogen thermal response lags that of the air since a temperature difference must exist between the air and the particle in order for the pathogen to be heated or cooled. As required by Equation 1 and as evident in the plot, the maximum particle temperature is reached when there is no temperature difference between the air and the pathogen.

3. The maximum pathogen temperature may be increased by extending the duration of the process. For a process time of $t_p = 0.008$ s, the air and pathogen particle temperatures are as follows.

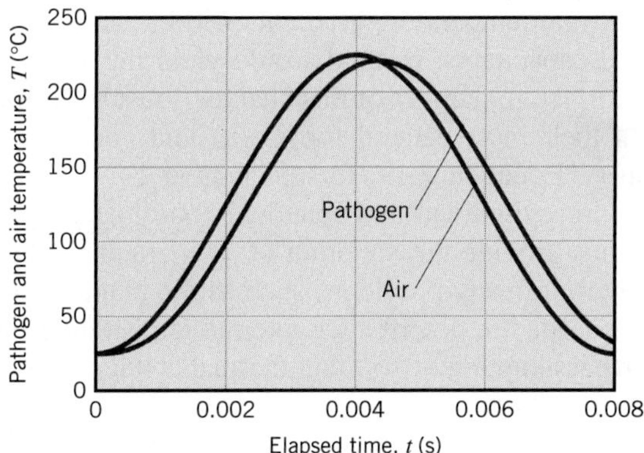

The maximum particle temperature is now $T_{\max} = 221°\text{C} > T_d = 220°\text{C}$, and the pathogen would be killed. However, because the duration of the cycle is twice as long as originally specified, approximately half of the air could be supplied to the operating room compared to the $t_p = 0.004$ s case. A trade-off exists between the amount of air that can be delivered to the operating room and its purity.

4. The maximum possible radiation heat transfer coefficient may be calculated based on the extreme temperatures of the problem and by assuming a particle emissivity of unity. Hence,

$$h_{r,\max} = \sigma(T_{\max} + T_{\min})(T_{\max}^2 + T_{\min}^2)$$

$$= 5.67 \times 10^{-8}\ \text{W/m}^2 \cdot \text{K}^4 \times (498 + 298)\text{K} \times (498^2 + 298^2)\text{K}^2 = 15.2\ \text{W/m}^2 \cdot \text{K}$$

Since $h_{r,\max} \ll h$, radiation heat transfer is negligible.

5. The Der(T, t) function of the *IHT* software was used to generate the numerical solution for this problem. See Comment 3 of Example 5.2. If one is familiar with a numerical solver such as *IHT*, it is often much faster to obtain a numerical solution than an analytical solution, as is the case in this example. Moreover, if one seeks maximum or minimum values of the dependent variable or variables, such as the pathogen temperature in this example, it is often faster to determine the maxima or minima by inspection, rather than with an analytical solution. However, analytical solutions often explicitly show parameter dependencies and can provide insights that numerical solutions might obscure.

6. A time increment of $\Delta t = 0.00001$ s was used to generate the numerical solutions. Generally, the accuracy of a numerical integration improves with decreasing Δt, but at the expense of increased computation time. For this example, results for $\Delta t = 0.000005$ s are virtually identical to those obtained for the larger time increment, indicating that either increment is sufficient to accurately depict the temperature history and to determine the maximum particle temperature.

7. Assumption of instantaneous pathogen death at the lethal temperature is an approximation. Pathogen destruction also depends on the duration of exposure to the high temperatures [2].

5.4 Spatial Effects

Situations frequently arise for which the Biot number is not small, and we must cope with the fact that temperature gradients within the medium are no longer negligible. Use of the lumped capacitance method would yield incorrect results, so alternative approaches, presented in the remainder of this chapter, must be utilized.

In their most general form, transient conduction problems are described by the heat equation, Equation 2.19, for rectangular coordinates or Equations 2.26 and 2.29, respectively, for cylindrical and spherical coordinates. The solutions to these partial differential equations provide the variation of temperature with both time and the spatial coordinates. However, in many problems, such as the plane wall of Figure 5.4, only one spatial coordinate is needed to describe the internal temperature distribution. With no internal generation and the assumption of constant thermal conductivity, Equation 2.19 then reduces to

$$\frac{\partial^2 T}{\partial x^2} = \frac{1}{\alpha}\frac{\partial T}{\partial t} \tag{5.29}$$

To solve Equation 5.29 for the temperature distribution $T(x, t)$, it is necessary to specify an *initial* condition and two *boundary conditions.* For the typical transient conduction problem of Figure 5.4, the initial condition is

$$T(x, 0) = T_i \tag{5.30}$$

and the boundary conditions are

$$\left.\frac{\partial T}{\partial x}\right|_{x=0} = 0 \tag{5.31}$$

and

$$-k\left.\frac{\partial T}{\partial x}\right|_{x=L} = h[T(L, t) - T_\infty] \tag{5.32}$$

Equation 5.30 presumes a uniform temperature distribution at time $t = 0$; Equation 5.31 reflects the *symmetry requirement* for the midplane of the wall; and Equation 5.32 describes the surface condition experienced for time $t > 0$. From Equations 5.29 through 5.32, it is evident that, in addition to depending on x and t, temperatures in the wall also depend on a number of physical parameters. In particular

$$T = T(x, t, T_i, T_\infty, L, k, \alpha, h) \tag{5.33}$$

The foregoing problem may be solved analytically or numerically. These methods will be considered in subsequent sections, but first it is important to note the advantages that may be obtained by *nondimensionalizing* the governing equations. This may be done by arranging the relevant variables into suitable *groups.* Consider the dependent variable T. If the temperature difference $\theta \equiv T - T_\infty$ is divided by the *maximum possible temperature difference* $\theta_i \equiv T_i - T_\infty$, a dimensionless form of the dependent variable may be defined as

$$\theta^* \equiv \frac{\theta}{\theta_i} = \frac{T - T_\infty}{T_i - T_\infty} \tag{5.34}$$

Accordingly, θ^* must lie in the range $0 \le \theta^* \le 1$. A dimensionless spatial coordinate may be defined as

$$x^* \equiv \frac{x}{L} \tag{5.35}$$

where L is the half-thickness of the plane wall, and a dimensionless time may be defined as

$$t^* \equiv \frac{\alpha t}{L^2} \equiv Fo \tag{5.36}$$

where t^* is equivalent to the dimensionless *Fourier number,* Equation 5.12.

Substituting the definitions of Equations 5.34 through 5.36 into Equations 5.29 through 5.32, the heat equation becomes

$$\frac{\partial^2 \theta^*}{\partial x^{*2}} = \frac{\partial \theta^*}{\partial Fo} \tag{5.37}$$

and the initial and boundary conditions become

$$\theta^*(x^*, 0) = 1 \tag{5.38}$$

$$\left.\frac{\partial \theta^*}{\partial x^*}\right|_{x^*=0} = 0 \tag{5.39}$$

and

$$\left.\frac{\partial \theta^*}{\partial x^*}\right|_{x^*=1} = -Bi\, \theta^*(1, t^*) \tag{5.40}$$

where the *Biot number* is $Bi \equiv hL/k$. In dimensionless form the functional dependence may now be expressed as

$$\theta^* = f(x^*, Fo, Bi) \tag{5.41}$$

Recall that a similar functional dependence, without the x^* variation, was obtained for the lumped capacitance method, as shown in Equation 5.13.

Comparing Equations 5.33 and 5.41, the considerable advantage associated with casting the problem in dimensionless form becomes apparent. Equation 5.41 implies that *for a prescribed geometry, the transient temperature distribution is a universal function of x^*, Fo, and Bi.* That is, the *dimensionless solution* has a prescribed form that does not depend on the particular value of T_i, T_∞, L, k, α, or h. Since this generalization greatly simplifies the presentation and utilization of transient solutions, the dimensionless variables are used extensively in subsequent sections.

5.5 *The Plane Wall with Convection*

Exact, analytical solutions to transient conduction problems have been obtained for many simplified geometries and boundary conditions and are well documented [3–6]. Several mathematical techniques, including the method of separation of variables (Section 4.2), may be used for this purpose, and typically the solution for the dimensionless temperature distribution, Equation 5.41, is in the form of an infinite series. However, except for very small values of the Fourier number, this series may be approximated by a single term, considerably simplifying its evaluation.

5.5.1 Exact Solution

Consider the *plane wall* of thickness $2L$ (Figure 5.6*a*). If the thickness is small relative to the width and height of the wall, it is reasonable to assume that conduction occurs exclusively in the x-direction. If the wall is initially at a uniform temperature, $T(x, 0) = T_i$, and is suddenly immersed in a fluid of $T_\infty \neq T_i$, the resulting temperatures may be obtained by solving Equation 5.37 subject to the conditions of Equations 5.38 through 5.40. Since the convection conditions for the surfaces at $x^* = \pm 1$ are the same, the temperature distribution at any instant must be symmetrical about the midplane ($x^* = 0$). An exact solution to this problem is of the form [4]

$$\theta^* = \sum_{n=1}^{\infty} C_n \exp(-\zeta_n^2 Fo) \cos(\zeta_n x^*) \tag{5.42a}$$

where $Fo = \alpha t/L^2$, the coefficient C_n is

$$C_n = \frac{4 \sin \zeta_n}{2\zeta_n + \sin(2\zeta_n)} \tag{5.42b}$$

and the discrete values of ζ_n (*eigenvalues*) are positive roots of the transcendental equation

$$\zeta_n \tan \zeta_n = Bi \tag{5.42c}$$

The first four roots of this equation are given in Appendix B.3. The exact solution given by Equation 5.42a is valid for any time, $0 \leq Fo \leq \infty$.

5.5.2 Approximate Solution

It can be shown (Problem 5.43) that for values of $Fo > 0.2$, the infinite series solution, Equation 5.42a, can be approximated by the first term of the series, $n = 1$. Invoking this approximation, the dimensionless form of the temperature distribution becomes

$$\theta^* = C_1 \exp(-\zeta_1^2 Fo) \cos(\zeta_1 x^*) \tag{5.43a}$$

or

$$\theta^* = \theta_o^* \cos(\zeta_1 x^*) \tag{5.43b}$$

where $\theta_o^* \equiv (T_o - T_\infty)/(T_i - T_\infty)$ represents the midplane ($x^* = 0$) temperature

$$\theta_o^* = C_1 \exp(-\zeta_1^2 Fo) \tag{5.44}$$

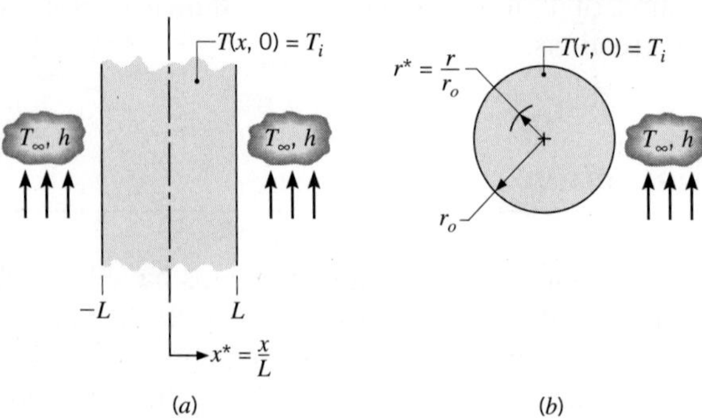

FIGURE 5.6 One-dimensional systems with an initial uniform temperature subjected to sudden convection conditions: (*a*) Plane wall. (*b*) Infinite cylinder or sphere.

Graphical representations of the one-term approximations are presented in Section 5S.1.

An important implication of Equation 5.43b is that *the time dependence of the temperature at any location within the wall is the same as that of the midplane temperature.* The coefficients C_1 and ζ_1 are evaluated from Equations 5.42b and 5.42c, respectively, and are given in Table 5.1 for a range of Biot numbers.

TABLE 5.1 Coefficients used in the one-term approximation to the series solutions for transient one-dimensional conduction

	Plane Wall		Infinite Cylinder		Sphere	
Bi^a	ζ_1 (rad)	C_1	ζ_1 (rad)	C_1	ζ_1 (rad)	C_1
0.01	0.0998	1.0017	0.1412	1.0025	0.1730	1.0030
0.02	0.1410	1.0033	0.1995	1.0050	0.2445	1.0060
0.03	0.1723	1.0049	0.2440	1.0075	0.2991	1.0090
0.04	0.1987	1.0066	0.2814	1.0099	0.3450	1.0120
0.05	0.2218	1.0082	0.3143	1.0124	0.3854	1.0149
0.06	0.2425	1.0098	0.3438	1.0148	0.4217	1.0179
0.07	0.2615	1.0114	0.3709	1.0173	0.4551	1.0209
0.08	0.2791	1.0130	0.3960	1.0197	0.4860	1.0239
0.09	0.2956	1.0145	0.4195	1.0222	0.5150	1.0268
0.10	0.3111	1.0161	0.4417	1.0246	0.5423	1.0298
0.15	0.3779	1.0237	0.5376	1.0365	0.6609	1.0445
0.20	0.4328	1.0311	0.6170	1.0483	0.7593	1.0592
0.25	0.4801	1.0382	0.6856	1.0598	0.8447	1.0737
0.30	0.5218	1.0450	0.7465	1.0712	0.9208	1.0880
0.4	0.5932	1.0580	0.8516	1.0932	1.0528	1.1164
0.5	0.6533	1.0701	0.9408	1.1143	1.1656	1.1441
0.6	0.7051	1.0814	1.0184	1.1345	1.2644	1.1713
0.7	0.7506	1.0919	1.0873	1.1539	1.3525	1.1978
0.8	0.7910	1.1016	1.1490	1.1724	1.4320	1.2236
0.9	0.8274	1.1107	1.2048	1.1902	1.5044	1.2488
1.0	0.8603	1.1191	1.2558	1.2071	1.5708	1.2732
2.0	1.0769	1.1785	1.5994	1.3384	2.0288	1.4793
3.0	1.1925	1.2102	1.7887	1.4191	2.2889	1.6227
4.0	1.2646	1.2287	1.9081	1.4698	2.4556	1.7202
5.0	1.3138	1.2402	1.9898	1.5029	2.5704	1.7870
6.0	1.3496	1.2479	2.0490	1.5253	2.6537	1.8338
7.0	1.3766	1.2532	2.0937	1.5411	2.7165	1.8673
8.0	1.3978	1.2570	2.1286	1.5526	1.7654	1.8920
9.0	1.4149	1.2598	2.1566	1.5611	2.8044	1.9106
10.0	1.4289	1.2620	2.1795	1.5677	2.8363	1.9249
20.0	1.4961	1.2699	2.2881	1.5919	2.9857	1.9781
30.0	1.5202	1.2717	2.3261	1.5973	3.0372	1.9898
40.0	1.5325	1.2723	2.3455	1.5993	3.0632	1.9942
50.0	1.5400	1.2727	2.3572	1.6002	3.0788	1.9962
100.0	1.5552	1.2731	2.3809	1.6015	3.1102	1.9990
∞	1.5708	1.2733	2.4050	1.6018	3.1415	2.0000

[a] $Bi = hL/k$ for the plane wall and hr_o/k for the infinite cylinder and sphere. See Figure 5.6.

5.5.3 Total Energy Transfer

In many situations it is useful to know the total energy that has left (or entered) the wall up to any time t in the transient process. The conservation of energy requirement, Equation 1.12b, may be applied for the time interval bounded by the initial condition ($t = 0$) and any time $t > 0$

$$E_{\text{in}} - E_{\text{out}} = \Delta E_{\text{st}} \tag{5.45}$$

Equating the energy transferred from the wall Q to E_{out} and setting $E_{\text{in}} = 0$ and $\Delta E_{\text{st}} = E(t) - E(0)$, it follows that

$$Q = -[E(t) - E(0)] \tag{5.46a}$$

or

$$Q = -\int \rho c[T(x, t) - T_i]dV \tag{5.46b}$$

where the integration is performed over the volume of the wall. It is convenient to nondimensionalize this result by introducing the quantity

$$Q_o = \rho c V(T_i - T_\infty) \tag{5.47}$$

which may be interpreted as the initial internal energy of the wall relative to the fluid temperature. It is also the *maximum* amount of energy transfer that could occur if the process were continued to time $t = \infty$. Hence, assuming constant properties, the ratio of the total energy transferred from the wall over the time interval t to the maximum possible transfer is

$$\frac{Q}{Q_o} = \int \frac{-[T(x, t) - T_i]}{T_i - T_\infty} \frac{dV}{V} = \frac{1}{V}\int (1 - \theta^*)dV \tag{5.48}$$

Employing the approximate form of the temperature distribution for the plane wall, Equation 5.43b, the integration prescribed by Equation 5.48 can be performed to obtain

$$\frac{Q}{Q_o} = 1 - \frac{\sin \zeta_1}{\zeta_1}\theta_o^* \tag{5.49}$$

where θ_o^* can be determined from Equation 5.44, using Table 5.1 for values of the coefficients C_1 and ζ_1.

5.5.4 Additional Considerations

Because the mathematical problem is precisely the same, the foregoing results may also be applied to a plane wall of thickness L that is insulated on one side ($x^* = 0$) and experiences convective transport on the other side ($x^* = +1$). This equivalence is a consequence of the fact that, regardless of whether a symmetrical or an adiabatic requirement is prescribed at $x^* = 0$, the boundary condition is of the form $\partial\theta^*/\partial x^* = 0$.

Also note that the foregoing results may be used to determine the transient response of a plane wall to a sudden change in *surface* temperature. The process is equivalent to having an infinite convection coefficient, in which case the Biot number is infinite ($Bi = \infty$) and the fluid temperature T_∞ is replaced by the prescribed surface temperature T_s.

5.6 Radial Systems with Convection

For an infinite cylinder or sphere of radius r_o (Figure 5.6*b*), which is at an initial uniform temperature and experiences a change in convective conditions, results similar to those of Section 5.5 may be developed. That is, an exact series solution may be obtained for the time dependence of the radial temperature distribution, and a one-term approximation may be used for most conditions. The infinite cylinder is an idealization that permits the assumption of one-dimensional conduction in the radial direction. It is a reasonable approximation for cylinders having $L/r_o \gtrsim 10$.

5.6.1 Exact Solutions

For a uniform initial temperature and convective boundary conditions, the exact solutions [4], applicable at any time ($Fo > 0$), are as follows.

Infinite Cylinder In dimensionless form, the temperature is

$$\theta^* = \sum_{n=1}^{\infty} C_n \exp(-\zeta_n^2 Fo) J_0(\zeta_n r^*) \tag{5.50a}$$

where $Fo = \alpha t / r_o^2$,

$$C_n = \frac{2}{\zeta_n} \frac{J_1(\zeta_n)}{J_0^2(\zeta_n) + J_1^2(\zeta_n)} \tag{5.50b}$$

and the discrete values of ζ_n are positive roots of the transcendental equation

$$\zeta_n \frac{J_1(\zeta_n)}{J_0(\zeta_n)} = Bi \tag{5.50c}$$

where $Bi = hr_o/k$. The quantities J_1 and J_0 are Bessel functions of the first kind, and their values are tabulated in Appendix B.4. Roots of the transcendental equation (5.50c) are tabulated by Schneider [4].

Sphere Similarly, for the sphere

$$\theta^* = \sum_{n=1}^{\infty} C_n \exp(-\zeta_n^2 Fo) \frac{1}{\zeta_n r^*} \sin(\zeta_n r^*) \tag{5.51a}$$

where $Fo = \alpha t / r_o^2$,

$$C_n = \frac{4[\sin(\zeta_n) - \zeta_n \cos(\zeta_n)]}{2\zeta_n - \sin(2\zeta_n)} \tag{5.51b}$$

and the discrete values of ζ_n are positive roots of the transcendental equation

$$1 - \zeta_n \cot \zeta_n = Bi \tag{5.51c}$$

where $Bi = hr_o/k$. Roots of the transcendental equation are tabulated by Schneider [4].

5.6.2 Approximate Solutions

For the infinite cylinder and sphere, the foregoing series solutions can again be approximated by a single term, $n = 1$, for $Fo > 0.2$. Hence, as for the case of the plane wall, the time dependence of the temperature at any location within the radial system is the same as that of the centerline or centerpoint.

Infinite Cylinder The one-term approximation to Equation 5.50a is

$$\theta^* = C_1 \exp(-\zeta_1^2 Fo) J_0(\zeta_1 r^*) \tag{5.52a}$$

or

$$\theta^* = \theta_o^* J_0(\zeta_1 r^*) \tag{5.52b}$$

where θ_o^* represents the centerline temperature and is of the form

$$\theta_o^* = C_1 \exp(-\zeta_1^2 Fo) \tag{5.52c}$$

Values of the coefficients C_1 and ζ_1 have been determined and are listed in Table 5.1 for a range of Biot numbers.

Sphere From Equation 5.51a, the one-term approximation is

$$\theta^* = C_1 \exp(-\zeta_1^2 Fo) \frac{1}{\zeta_1 r^*} \sin(\zeta_1 r^*) \tag{5.53a}$$

or

$$\theta^* = \theta_o^* \frac{1}{\zeta_1 r^*} \sin(\zeta_1 r^*) \tag{5.53b}$$

where θ_o^* represents the center temperature and is of the form

$$\theta_o^* = C_1 \exp(-\zeta_1^2 Fo) \tag{5.53c}$$

Values of the coefficients C_1 and ζ_1 have been determined and are listed in Table 5.1 for a range of Biot numbers.

5.6.3 Total Energy Transfer

As in Section 5.5.3, an energy balance may be performed to determine the total energy transfer from the infinite cylinder or sphere over the time interval $\Delta t = t$. Substituting from

---

Graphical representations of the one-term approximations are presented in Section 5S.1.

the approximate solutions, Equations 5.52b and 5.53b, and introducing Q_o from Equation 5.47, the results are as follows.

Infinite Cylinder

$$\frac{Q}{Q_o} = 1 - \frac{2\theta_o^*}{\zeta_1} J_1(\zeta_1) \tag{5.54}$$

Sphere

$$\frac{Q}{Q_o} = 1 - \frac{3\theta_o^*}{\zeta_1^3} [\sin(\zeta_1) - \zeta_1 \cos(\zeta_1)] \tag{5.55}$$

Values of the center temperature θ_o^* are determined from Equation 5.52c or 5.53c, using the coefficients of Table 5.1 for the appropriate system.

5.6.4 Additional Considerations

As for the plane wall, the foregoing results may be used to predict the transient response of long cylinders and spheres subjected to a sudden change in *surface* temperature. Namely, an infinite Biot number would be prescribed, and the fluid temperature T_∞ would be replaced by the constant surface temperature T_s.

EXAMPLE 5.5

Consider a steel pipeline (AISI 1010) that is 1 m in diameter and has a wall thickness of 40 mm. The pipe is heavily insulated on the outside, and, before the initiation of flow, the walls of the pipe are at a uniform temperature of −20°C. With the initiation of flow, hot oil at 60°C is pumped through the pipe, creating a convective condition corresponding to $h = 500 \text{ W/m}^2 \cdot \text{K}$ at the inner surface of the pipe.

1. What are the appropriate Biot and Fourier numbers 8 min after the initiation of flow?
2. At $t = 8$ min, what is the temperature of the exterior pipe surface covered by the insulation?
3. What is the heat flux $q''(\text{W/m}^2)$ to the pipe from the oil at $t = 8$ min?
4. How much energy per meter of pipe length has been transferred from the oil to the pipe at $t = 8$ min?

SOLUTION

Known: Wall subjected to sudden change in convective surface condition.

Find:

1. Biot and Fourier numbers after 8 min.
2. Temperature of exterior pipe surface after 8 min.
3. Heat flux to the wall at 8 min.
4. Energy transferred to pipe per unit length after 8 min.

Schematic:

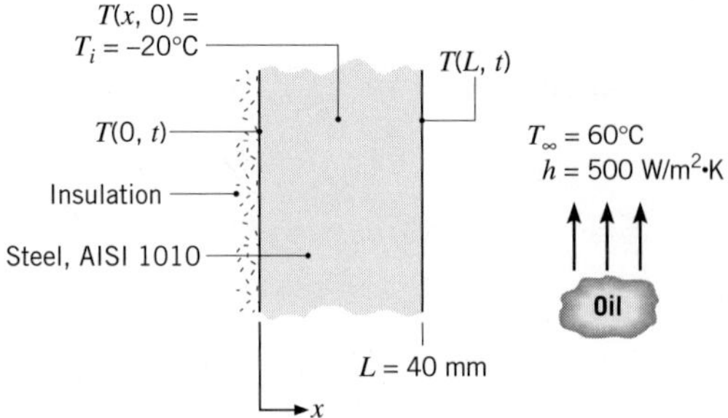

Assumptions:

1. Pipe wall can be approximated as plane wall, since thickness is much less than diameter.
2. Constant properties.
3. Outer surface of pipe is adiabatic.

Properties: Table A.1, steel type AISI 1010 $[T = (-20 + 60)°C/2 \approx 300$ K]: $\rho = 7832$ kg/m^3, $c = 434$ J/kg·K, $k = 63.9$ W/m·K, $\alpha = 18.8 \times 10^{-6}$ m^2/s.

Analysis:

1. At $t = 8$ min, the Biot and Fourier numbers are computed from Equations 5.10 and 5.12, respectively, with $L_c = L$. Hence

$$Bi = \frac{hL}{k} = \frac{500 \text{ W/m}^2 \cdot \text{K} \times 0.04 \text{ m}}{63.9 \text{ W/m} \cdot \text{K}} = 0.313 \qquad \triangleleft$$

$$Fo = \frac{\alpha t}{L^2} = \frac{18.8 \times 10^{-6} \text{ m}^2\text{/s} \times 8 \text{ min} \times 60 \text{ s/min}}{(0.04 \text{ m})^2} = 5.64 \qquad \triangleleft$$

2. With $Bi = 0.313$, use of the lumped capacitance method is inappropriate. However, since $Fo > 0.2$ and transient conditions in the insulated pipe wall of thickness L correspond to those in a plane wall of thickness $2L$ experiencing the same surface condition, the desired results may be obtained from the one-term approximation for a plane wall. The midplane temperature can be determined from Equation 5.44

$$\theta_o^* = \frac{T_o - T_\infty}{T_i - T_\infty} = C_1 \exp(-\zeta_1^2 Fo)$$

where, with $Bi = 0.313$, $C_1 = 1.047$ and $\zeta_1 = 0.531$ rad from Table 5.1. With $Fo = 5.64$,

$$\theta_o^* = 1.047 \exp[-(0.531 \text{ rad})^2 \times 5.64] = 0.214$$

Hence after 8 min, the temperature of the exterior pipe surface, which corresponds to the midplane temperature of a plane wall, is

$$T(0, 8 \text{ min}) = T_\infty + \theta_o^*(T_i - T_\infty) = 60°\text{C} + 0.214(-20 - 60)°\text{C} = 42.9°\text{C} \qquad \triangleleft$$

3. Heat transfer to the inner surface at $x = L$ is by convection, and at any time t the heat flux may be obtained from Newton's law of cooling. Hence at $t = 480$ s,

$$q''_x(L, 480\text{ s}) \equiv q''_L = h[T(L, 480\text{ s}) - T_\infty]$$

Using the one-term approximation for the surface temperature, Equation 5.43b with $x^* = 1$ has the form

$$\theta^* = \theta_o^* \cos(\zeta_1)$$

$$T(L, t) = T_\infty + (T_i - T_\infty)\theta_o^* \cos(\zeta_1)$$

$$T(L, 8\text{ min}) = 60°\text{C} + (-20 - 60)°\text{C} \times 0.214 \times \cos(0.531\text{ rad})$$

$$T(L, 8\text{ min}) = 45.2°\text{C}$$

The heat flux at $t = 8$ min is then

$$q''_L = 500\text{ W/m}^2\cdot\text{K}\,(45.2 - 60)°\text{C} = -7400\text{ W/m}^2 \qquad \triangleleft$$

4. The energy transfer to the pipe wall over the 8-min interval may be obtained from Equations 5.47 and 5.49. With

$$\frac{Q}{Q_o} = 1 - \frac{\sin(\zeta_1)}{\zeta_1}\theta_o^*$$

$$\frac{Q}{Q_o} = 1 - \frac{\sin(0.531\text{ rad})}{0.531\text{ rad}} \times 0.214 = 0.80$$

it follows that

$$Q = 0.80\,\rho c V(T_i - T_\infty)$$

or with a volume per unit pipe length of $V' = \pi DL$,

$$Q' = 0.80\,\rho c \pi DL(T_i - T_\infty)$$

$$Q' = 0.80 \times 7832\text{ kg/m}^3 \times 434\text{ J/kg}\cdot\text{K}$$

$$\times\, \pi \times 1\text{ m} \times 0.04\text{ m}\,(-20 - 60)°\text{C}$$

$$Q' = -2.73 \times 10^7\text{ J/m} \qquad \triangleleft$$

Comments:

1. The minus sign associated with q'' and Q' simply implies that the direction of heat transfer is from the oil to the pipe (into the pipe wall).
2. The solution for this example is provided as a ready-to-solve model in the *Advanced* section of *IHT*, which uses the *Models, Transient Conduction, Plane Wall* option. Since the *IHT* model uses a multiple-term approximation to the series solution, the results are more accurate than those obtained from the foregoing one-term approximation. *IHT Models* for *Transient Conduction* are also provided for the radial systems treated in Section 5.6.

EXAMPLE 5.6

A new process for treatment of a special material is to be evaluated. The material, a sphere of radius $r_o = 5$ mm, is initially in equilibrium at 400°C in a furnace. It is suddenly removed from the furnace and subjected to a two-step cooling process.

Step 1 Cooling in air at 20°C for a period of time t_a until the center temperature reaches a critical value, $T_a(0, t_a) = 335°C$. For this situation, the convection heat transfer coefficient is $h_a = 10\ \text{W/m}^2\cdot\text{K}$.

After the sphere has reached this critical temperature, the second step is initiated.

Step 2 Cooling in a well-stirred water bath at 20°C, with a convection heat transfer coefficient of $h_w = 6000\ \text{W/m}^2\cdot\text{K}$.

The thermophysical properties of the material are $\rho = 3000\ \text{kg/m}^3$, $k = 20\ \text{W/m}\cdot\text{K}$, $c = 1000\ \text{J/kg}\cdot\text{K}$, and $\alpha = 6.66 \times 10^{-6}\ \text{m}^2/\text{s}$.

1. Calculate the time t_a required for step 1 of the cooling process to be completed.
2. Calculate the time t_w required during step 2 of the process for the center of the sphere to cool from 335°C (the condition at the completion of step 1) to 50°C.

SOLUTION

Known: Temperature requirements for cooling a sphere.

Find:

1. Time t_a required to accomplish desired cooling in air.
2. Time t_w required to complete cooling in water bath.

Schematic:

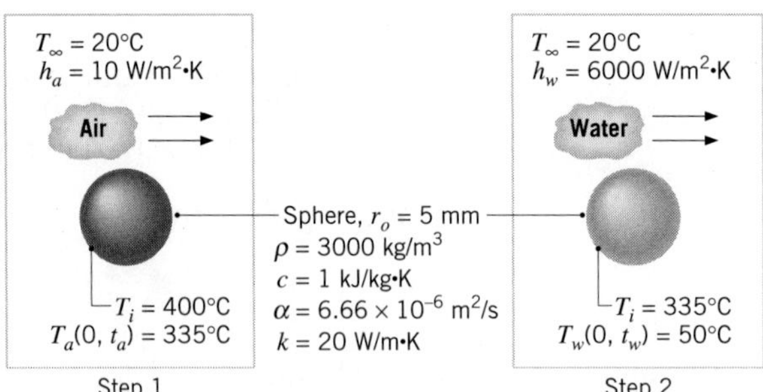

Assumptions:

1. One-dimensional conduction in r.
2. Constant properties.

Analysis:

1. To determine whether the lumped capacitance method can be used, the Biot number is calculated. From Equation 5.10, with $L_c = r_o/3$,

$$Bi = \frac{h_a r_o}{3k} = \frac{10\ \text{W/m}^2 \cdot \text{K} \times 0.005\ \text{m}}{3 \times 20\ \text{W/m} \cdot \text{K}} = 8.33 \times 10^{-4}$$

Accordingly, the lumped capacitance method may be used, and the temperature is nearly uniform throughout the sphere. From Equation 5.5 it follows that

$$t_a = \frac{\rho V c}{h_a A_s} \ln \frac{\theta_i}{\theta_a} = \frac{\rho r_o c}{3h_a} \ln \frac{T_i - T_\infty}{T_a - T_\infty}$$

where $V = (4/3)\pi r_o^3$ and $A_s = 4\pi r_o^2$. Hence

$$t_a = \frac{3000\ \text{kg/m}^3 \times 0.005\ \text{m} \times 1000\ \text{J/kg} \cdot \text{K}}{3 \times 10\ \text{W/m}^2 \cdot \text{K}} \ln \frac{400 - 20}{335 - 20} = 94\ \text{s} \qquad \triangleleft$$

2. To determine whether the lumped capacitance method may also be used for the second step of the cooling process, the Biot number is again calculated. In this case

$$Bi = \frac{h_w r_o}{3k} = \frac{6000\ \text{W/m}^2 \cdot \text{K} \times 0.005\ \text{m}}{3 \times 20\ \text{W/m} \cdot \text{K}} = 0.50$$

and the lumped capacitance method is not appropriate. However, to an excellent approximation, the temperature of the sphere is uniform at $t = t_a$ and the one-term approximation may be used for the calculations. The time t_w at which the center temperature reaches 50°C, that is, $T(0, t_w) = 50°\text{C}$, can be obtained by rearranging Equation 5.53c

$$Fo = -\frac{1}{\zeta_1^2} \ln \left[\frac{\theta_o^*}{C_1} \right] = -\frac{1}{\zeta_1^2} \ln \left[\frac{1}{C_1} \times \frac{T(0, t_w) - T_\infty}{T_i - T_\infty} \right]$$

where $t_w = Fo\, r_o^2/\alpha$. With the Biot number now defined as

$$Bi = \frac{h_w r_o}{k} = \frac{6000\ \text{W/m}^2 \cdot \text{K} \times 0.005\ \text{m}}{20\ \text{W/m} \cdot \text{K}} = 1.50$$

Table 5.1 yields $C_1 = 1.376$ and $\zeta_1 = 1.800$ rad. It follows that

$$Fo = -\frac{1}{(1.800\ \text{rad})^2} \ln \left[\frac{1}{1.376} \times \frac{(50 - 20)°\text{C}}{(335 - 20)°\text{C}} \right] = 0.82$$

and

$$t_w = Fo \frac{r_o^2}{\alpha} = 0.82 \frac{(0.005\ \text{m})^2}{6.66 \times 10^{-6}\ \text{m}^2/\text{s}} = 3.1\,\text{s} \qquad \triangleleft$$

Note that, with $Fo = 0.82$, use of the one-term approximation is justified.

Comments:

1. If the temperature distribution in the sphere at the conclusion of step 1 were not uniform, the one-term approximation could not be used for the calculations of step 2.
2. The surface temperature of the sphere at the conclusion of step 2 may be obtained from Equation 5.53b. With $\theta_o^* = 0.095$ and $r^* = 1$,

$$\theta^*(r_o) = \frac{T(r_o) - T_\infty}{T_i - T_\infty} = \frac{0.095}{1.800 \text{ rad}} \sin(1.800 \text{ rad}) = 0.0514$$

and

$$T(r_o) = 20°\text{C} + 0.0514(335 - 20)°\text{C} = 36°\text{C}$$

The infinite series, Equation 5.51a, and its one-term approximation, Equation 5.53b, may be used to compute the temperature at any location in the sphere and at any time $t > t_a$. For $(t - t_a) < 0.2(0.005 \text{ m})^2/6.66 \times 10^{-6} \text{ m}^2/\text{s} = 0.75 \text{ s}$, a sufficient number of terms must be retained to ensure convergence of the series. For $(t - t_a) > 0.75$ s, satisfactory convergence is provided by the one-term approximation. Computing and plotting the temperature histories for $r = 0$ and $r = r_o$, we obtain the following results for $0 \le (t - t_a) \le 5$ s:

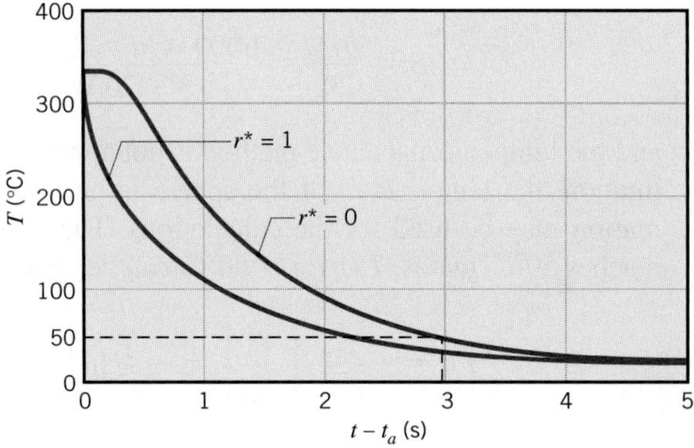

3. The *IHT Models*, *Transient Conduction*, *Sphere* option could be used to analyze the cooling processes experienced by the sphere in air and water, steps 1 and 2. The *IHT Models*, *Lumped Capacitance* option may only be used to analyze the air-cooling process, step 1.

5.7 *The Semi-Infinite Solid*

An important simple geometry for which analytical solutions may be obtained is the *semi-infinite solid.* Since, in principle, such a solid extends to infinity in all but one direction, it is characterized by a single identifiable surface (Figure 5.7). If a sudden change of conditions is imposed at this surface, transient, one-dimensional conduction will occur within the

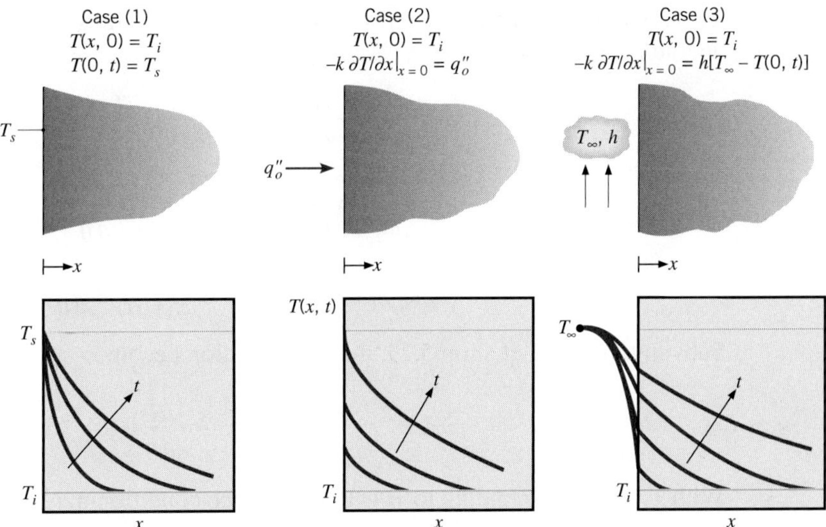

FIGURE 5.7 Transient temperature distributions in a semi-infinite solid for three surface conditions: constant surface temperature, constant surface heat flux, and surface convection.

solid. The semi-infinite solid provides a *useful idealization* for many practical problems. It may be used to determine transient heat transfer near the surface of the earth or to approximate the transient response of a finite solid, such as a thick slab. For this second situation the approximation would be reasonable for the early portion of the transient, during which temperatures in the slab interior (well removed from the surface) are essentially uninfluenced by the change in surface conditions. These early portions of the transient might correspond to very small Fourier numbers, and the approximate solutions of Sections 5.5 and 5.6 would not be valid. Although the exact solutions of the preceding sections could be used to determine the temperature distributions, many terms might be required to evaluate the infinite series expressions. The following semi-infinite solid solutions often eliminate the need to evaluate the cumbersome infinite series exact solutions at small Fo. It will be shown that a plane wall of thickness $2L$ can be accurately approximated as a semi-infinite solid for $Fo = \alpha t/L^2 \lesssim 0.2$.

The heat equation for transient conduction in a semi-infinite solid is given by Equation 5.29. The initial condition is prescribed by Equation 5.30, and the interior boundary condition is of the form

$$T(x \rightarrow \infty, t) = T_i \tag{5.56}$$

Closed-form solutions have been obtained for three important surface conditions, instantaneously applied at $t = 0$ [3, 4]. These conditions are shown in Figure 5.7. They include application of a constant surface temperature $T_s \neq T_i$, application of a constant surface heat flux q''_o, and exposure of the surface to a fluid characterized by $T_\infty \neq T_i$ and the convection coefficient h.

The solution for case 1 may be obtained by recognizing the existence of a *similarity variable* η, through which the heat equation may be transformed from a partial differential equation, involving two independent variables (x and t), to an ordinary differential equation expressed in terms of the single similarity variable. To confirm that such a requirement is satisfied by $\eta \equiv x/(4\alpha t)^{1/2}$, we first transform the pertinent differential operators, such that

$$\frac{\partial T}{\partial x} = \frac{dT}{d\eta}\frac{\partial \eta}{\partial x} = \frac{1}{(4\alpha t)^{1/2}}\frac{dT}{d\eta}$$

$$\frac{\partial^2 T}{\partial x^2} = \frac{d}{d\eta}\left[\frac{\partial T}{\partial x}\right]\frac{\partial \eta}{\partial x} = \frac{1}{4\alpha t}\frac{d^2 T}{d\eta^2}$$

$$\frac{\partial T}{\partial t} = \frac{dT}{d\eta}\frac{\partial \eta}{\partial t} = -\frac{x}{2t(4\alpha t)^{1/2}}\frac{dT}{d\eta}$$

Substituting into Equation 5.29, the heat equation becomes

$$\frac{d^2T}{d\eta^2} = -2\eta\frac{dT}{d\eta} \tag{5.57}$$

With $x = 0$ corresponding to $\eta = 0$, the surface condition may be expressed as

$$T(\eta = 0) = T_s \tag{5.58}$$

and with $x \rightarrow \infty$, as well as $t = 0$, corresponding to $\eta \rightarrow \infty$, both the initial condition and the interior boundary condition correspond to the single requirement that

$$T(\eta \rightarrow \infty) = T_i \tag{5.59}$$

Since the transformed heat equation and the initial/boundary conditions are independent of x and t, $\eta \equiv x/(4\alpha t)^{1/2}$ is, indeed, a similarity variable. Its existence implies that, irrespective of the values of x and t, the temperature may be represented as a unique function of η.

The specific form of the temperature dependence, $T(\eta)$, may be obtained by separating variables in Equation 5.57, such that

$$\frac{d(dT/d\eta)}{(dT/d\eta)} = -2\eta\, d\eta$$

Integrating, it follows that

$$\ln(dT/d\eta) = -\eta^2 + C_1'$$

or

$$\frac{dT}{d\eta} = C_1 \exp(-\eta^2)$$

Integrating a second time, we obtain

$$T = C_1\int_0^{\eta} \exp(-u^2)\, du + C_2$$

where u is a dummy variable. Applying the boundary condition at $\eta = 0$, Equation 5.58, it follows that $C_2 = T_s$ and

$$T = C_1\int_0^{\eta} \exp(-u^2)\, du + T_s$$

From the second boundary condition, Equation 5.59, we obtain

$$T_i = C_1 \int_0^\infty \exp(-u^2)\, du + T_s$$

or, evaluating the definite integral,

$$C_1 = \frac{2(T_i - T_s)}{\pi^{1/2}}$$

Hence the temperature distribution may be expressed as

$$\frac{T - T_s}{T_i - T_s} = (2/\pi^{1/2}) \int_0^\eta \exp(-u^2)\, du \equiv \operatorname{erf} \eta \tag{5.60}$$

where the *Gaussian error function,* erf η, is a standard mathematical function that is tabulated in Appendix B. Note that erf η asymptotically approaches unity as η becomes infinite. Thus, at any nonzero time, temperatures everywhere are predicted to have changed from T_i (become closer to T_s). The infinite speed at which boundary-condition information propagates into the semi-infinite solid is physically unrealistic, but this limitation of Fourier's law is not important except at extremely small time scales, as discussed in Section 2.3. The surface heat flux may be obtained by applying Fourier's law at $x = 0$, in which case

$$q_s'' = -k \frac{\partial T}{\partial x}\bigg|_{x=0} = -k(T_i - T_s) \frac{d(\operatorname{erf} \eta)}{d\eta} \frac{\partial \eta}{\partial x}\bigg|_{\eta=0}$$

$$q_s'' = k(T_s - T_i)(2/\pi^{1/2}) \exp(-\eta^2)(4\alpha t)^{-1/2}\big|_{\eta=0}$$

$$q_s'' = \frac{k(T_s - T_i)}{(\pi \alpha t)^{1/2}} \tag{5.61}$$

Analytical solutions may also be obtained for the case 2 and case 3 surface conditions, and results for all three cases are summarized as follows.

Case 1 Constant Surface Temperature: $T(0, t) = T_s$

$$\frac{T(x, t) - T_s}{T_i - T_s} = \operatorname{erf}\left(\frac{x}{2\sqrt{\alpha t}}\right) \tag{5.60}$$

$$q_s''(t) = \frac{k(T_s - T_i)}{\sqrt{\pi \alpha t}} \tag{5.61}$$

Case 2 Constant Surface Heat Flux: $q_s'' = q_o''$

$$T(x, t) - T_i = \frac{2q_o''(\alpha t/\pi)^{1/2}}{k} \exp\left(\frac{-x^2}{4\alpha t}\right) - \frac{q_o'' x}{k} \operatorname{erfc}\left(\frac{x}{2\sqrt{\alpha t}}\right) \tag{5.62}$$

Case 3 Surface Convection: $-k\left.\frac{\partial T}{\partial x}\right|_{x=0} = h[T_\infty - T(0, t)]$

$$\frac{T(x, t) - T_i}{T_\infty - T_i} = \operatorname{erfc}\left(\frac{x}{2\sqrt{\alpha t}}\right) - \left[\exp\left(\frac{hx}{k} + \frac{h^2\alpha t}{k^2}\right)\right]\left[\operatorname{erfc}\left(\frac{x}{2\sqrt{\alpha t}} + \frac{h\sqrt{\alpha t}}{k}\right)\right] \tag{5.63}$$

The *complementary error function,* erfc w, is defined as erfc $w \equiv 1 - \operatorname{erf} w$.

Temperature histories for the three cases are shown in Figure 5.7, and distinguishing features should be noted. With a step change in the surface temperature, case 1, temperatures within the medium monotonically approach T_s with increasing t, while the magnitude of the surface temperature gradient, and hence the surface heat flux, decreases as $t^{-1/2}$. A *thermal penetration depth* δ_p can be defined as the depth to which significant temperature effects propagate within a medium. For example, defining δ_p as the x-location at which $(T - T_s)/(T_i - T_s) = 0.90$, Equation 5.60 results in $\delta_p = 2.3\sqrt{\alpha t}$.[2] Hence, the penetration depth increases as $t^{1/2}$ and is larger for materials with higher thermal diffusivity. For a fixed surface heat flux (case 2), Equation 5.62 reveals that $T(0, t) = T_s(t)$ increases monotonically as $t^{1/2}$. For surface convection (case 3), the surface temperature and temperatures within the medium approach the fluid temperature T_∞ with increasing time. As T_s approaches T_∞, there is, of course, a reduction in the surface heat flux, $q''_s(t) = h[T_\infty - T_s(t)]$. Specific temperature histories computed from Equation 5.63 are plotted in Figure 5.8. The result corresponding to $h = \infty$ is equivalent to that associated with a sudden change in surface temperature, case 1. That is, for $h = \infty$, the surface instantaneously achieves the imposed fluid temperature ($T_s = T_\infty$), and with the second term on the right-hand side of Equation 5.63 reducing to zero, the result is equivalent to Equation 5.60.

An interesting permutation of case 1 occurs when two semi-infinite solids, initially at uniform temperatures $T_{A,i}$ and $T_{B,i}$, are placed in contact at their free surfaces (Figure 5.9).

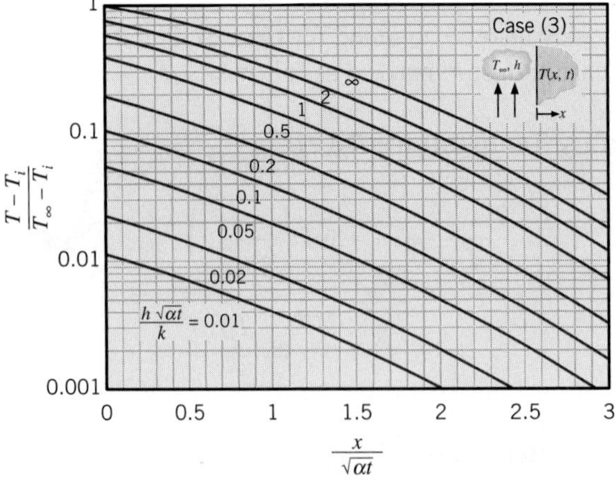

FIGURE 5.8 Temperature histories in a semi-infinite solid with surface convection.

[2]To apply the semi-infinite approximation to a plane wall of thickness $2L$, it is necessary that $\delta_p < L$. Substituting $\delta_p = L$ into the expression for the thermal penetration depth yields $Fo = 0.19 \approx 0.2$. Hence, a plane wall of thickness $2L$ can be accurately approximated as a semi-infinite solid for $Fo = \alpha t/L^2 \lesssim 0.2$. This restriction will also be demonstrated in Section 5.8.

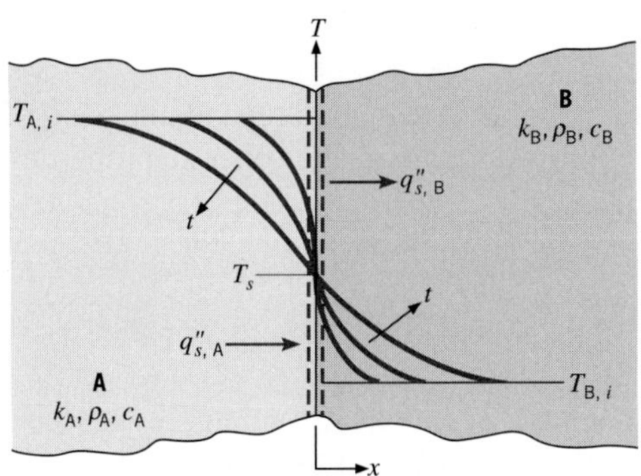

FIGURE 5.9 Interfacial contact between two semi-infinite solids at different initial temperatures.

If the contact resistance is negligible, the requirement of thermal equilibrium dictates that, at the instant of contact ($t = 0$), both surfaces must assume the same temperature T_s, for which $T_{B,i} < T_s < T_{A,i}$. Since T_s does not change with increasing time, it follows that the transient thermal response and the surface heat flux of each of the solids are determined by Equations 5.60 and 5.61, respectively.

The equilibrium surface temperature of Figure 5.9 may be determined from a surface energy balance, which requires that

$$q''_{s,A} = q''_{s,B} \tag{5.64}$$

Substituting from Equation 5.61 for $q''_{s,A}$ and $q''_{s,B}$ and recognizing that the x-coordinate of Figure 5.9 requires a sign change for $q''_{s,A}$, it follows that

$$\frac{-k_A(T_s - T_{A,i})}{(\pi\alpha_A t)^{1/2}} = \frac{k_B(T_s - T_{B,i})}{(\pi\alpha_B t)^{1/2}} \tag{5.65}$$

or, solving for T_s,

$$T_s = \frac{(k\rho c)_A^{1/2} T_{A,i} + (k\rho c)_B^{1/2} T_{B,i}}{(k\rho c)_A^{1/2} + (k\rho c)_B^{1/2}} \tag{5.66}$$

Hence the quantity $m \equiv (k\rho c)^{1/2}$ is a weighting factor that determines whether T_s will more closely approach $T_{A,i}$ ($m_A > m_B$) or $T_{B,i}$ ($m_B > m_A$).

EXAMPLE 5.7

On a hot and sunny day, the concrete deck surrounding a swimming pool is at a temperature of $T_d = 55°C$. A swimmer walks across the dry deck to the pool. The soles of the swimmer's dry feet are characterized by an $L_{sf} = 3$-mm-thick skin/fat layer of thermal conductivity $k_{sf} = 0.3$ W/m·K. Consider two types of concrete decking; (i) a dense stone mix and (ii) a lightweight aggregate characterized by density, specific heat, and thermal conductivity of $\rho_{lw} = 1495$ kg/m^3, $c_{p,lw} = 880$ J/kg·K, and $k_{lw} = 0.28$ W/m·K, respectively. The density and specific heat of the skin/fat layer may be approximated to be those of liquid water, and the skin/fat layer is at an initial temperature of $T_{sf,i} = 37°C$. What is the temperature of the bottom of the swimmer's feet after an elapsed time of $t = 1$ s?

SOLUTION

Known: Concrete temperature, initial foot temperature, and thickness of skin/fat layer on the sole of the foot. Skin/fat and lightweight aggregate concrete properties.

Find: The temperature of the bottom of the swimmer's feet after 1 s.

Schematic:

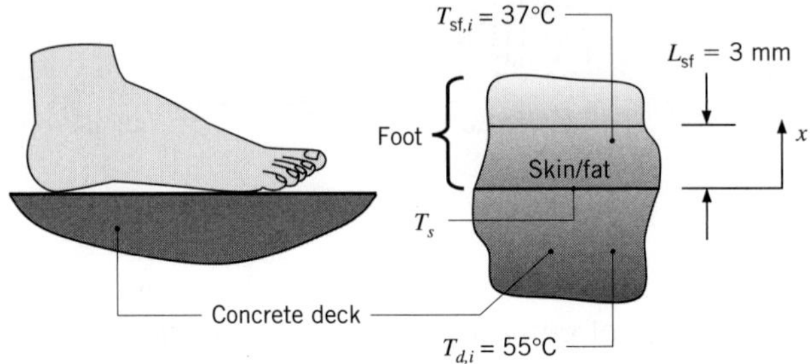

Assumptions:

1. One-dimensional conduction in the x-direction.
2. Constant and uniform properties.
3. Negligible contact resistance.

Properties: Table A.3 stone mix concrete (T = 300 K): ρ_{sm} = 2300 kg/m^3, k_{sm} = 1.4 W/m·K, c_{sm} = 880 J/kg·K. Table A.6 water (T = 310 K): ρ_{sf} = 993 kg/m^3, c_{sf} = 4178 J/kg·K.

Analysis: If the skin/fat layer and the deck are both semi-infinite media, from Equation 5.66 the surface temperature T_s is constant when the swimmer's foot is in contact with the deck. For the lightweight aggregate concrete decking, the thermal penetration depth at t = 1 s is

$$\delta_{p,\mathrm{lw}} = 2.3\sqrt{\alpha_{\mathrm{lw}} t} = 2.3\sqrt{\frac{k_{\mathrm{lw}} t}{\rho_{\mathrm{lw}} c_{\mathrm{lw}}}} = 2.3\sqrt{\frac{0.28\ \mathrm{W/m\cdot K} \times 1\mathrm{s}}{1495\ \mathrm{kg/m^3} \times 880\ \mathrm{J/kg\cdot K}}}$$

$$= 1.06 \times 10^{-3}\,\mathrm{m} = 1.06\ \mathrm{mm}$$

Since the thermal penetration depth is relatively small, it is reasonable to assume that the lightweight aggregate deck behaves as a semi-infinite medium. Similarly, the thermal penetration depth in the stone mix concrete is $\delta_{p,\mathrm{sm}}$ = 1.91 mm, and the thermal penetration depth associated with the skin/fat layer of the foot is $\delta_{p,\mathrm{sf}}$ = 0.62 mm. Hence, it is reasonable to assume that the stone mix concrete deck responds as a semi-infinite medium, and, since $\delta_{p,\mathrm{sf}} < L_{\mathrm{sf}}$, it is also correct to assume that the skin/fat layer behaves as a semi-infinite medium. Therefore, Equation 5.66 may be used to determine the surface temperature of the swimmer's foot for exposure to the two types of concrete decking. For the lightweight aggregate,

$$T_{s,\mathrm{lw}} = \frac{(k\rho c)_{\mathrm{lw}}^{1/2}\, T_{d,i} + (k\rho c)_{\mathrm{sf}}^{1/2}\, T_{\mathrm{sf},i}}{(k\rho c)_{\mathrm{lw}}^{1/2} + (k\rho c)_{\mathrm{sf}}^{1/2}}$$

$$= \frac{\begin{bmatrix}(0.28\ \mathrm{W/m\cdot K} \times 1495\ \mathrm{kg/m^3} \times 880\ \mathrm{J/kg\cdot K})^{1/2} \times 55°\mathrm{C} \\ + (0.3\ \mathrm{W/m\cdot K} \times 993\ \mathrm{kg/m^3} \times 4178\ \mathrm{J/kg\cdot K})^{1/2} \times 37°\mathrm{C}\end{bmatrix}}{\begin{bmatrix}(0.28\ \mathrm{W/m\cdot K} \times 1495\ \mathrm{kg/m^3} \times 880\ \mathrm{J/kg\cdot K})^{1/2} \\ + (0.3\ \mathrm{W/m\cdot K} \times 993\ \mathrm{kg/m^3} \times 4178\ \mathrm{J/kg\cdot K})^{1/2}\end{bmatrix}} = 43.3°\mathrm{C} \quad \triangleleft$$

Repeating the calculation for the stone mix concrete gives $T_{s,\mathrm{sm}} = 47.8°\mathrm{C}$. ◁

Comments:

1. The lightweight aggregate concrete feels cooler to the swimmer, relative to the stone mix concrete. Specifically, the temperature rise from the initial skin/fat temperature that is associated with the stone mix concrete is $\Delta T_{\mathrm{sm}} = T_{\mathrm{sm}} - T_{\mathrm{sf},i} = 47.8°\mathrm{C} - 37°\mathrm{C} = 10.8°\mathrm{C}$, whereas the temperature rise associated with the lightweight aggregate is $\Delta T_{\mathrm{lw}} = T_{\mathrm{lw}} - T_{\mathrm{sf},i} = 43.3°\mathrm{C} - 37°\mathrm{C} = 6.3°\mathrm{C}$.
2. The thermal penetration depths associated with an exposure time of $t = 1$ s are small. Stones and air pockets within the concrete may be of the same size as the thermal penetration depth, making the uniform property assumption somewhat questionable. The predicted foot temperatures should be viewed as representative values.

5.8 *Objects with Constant Surface Temperatures or Surface Heat Fluxes*

In Sections 5.5 and 5.6, the transient thermal response of plane walls, cylinders, and spheres to an applied convection boundary condition was considered in detail. It was pointed out that the solutions in those sections may be used for cases involving a step change in surface temperature by allowing the Biot number to be infinite. In Section 5.7, the response of a semi-infinite solid to a step change in surface temperature, or to an applied constant heat flux, was determined. This section will conclude our discussion of transient heat transfer in one-dimensional objects experiencing constant surface temperature or constant surface heat flux boundary conditions. A variety of approximate solutions are presented.

5.8.1 Constant Temperature Boundary Conditions

In the following discussion, the transient thermal response of objects to a step change in surface temperature is considered.

Semi-Infinite Solid Insight into the thermal response of objects to an applied constant temperature boundary condition may be obtained by casting the heat flux in Equation 5.61 into the nondimensional form

$$q^* \equiv \frac{q_s'' L_c}{k(T_s - T_i)} \tag{5.67}$$

where L_c is a *characteristic length* and q^* is the *dimensionless conduction heat rate* that was introduced in Section 4.3. Substituting Equation 5.67 into Equation 5.61 yields

$$q^* = \frac{1}{\sqrt{\pi Fo}} \tag{5.68}$$

where the Fourier number is defined as $Fo \equiv \alpha t/L_c^2$. Note that the value of q_s'' is independent of the choice of the characteristic length, as it must be for a semi-infinite solid. Equation 5.68 is plotted in Figure 5.10*a*, and since $q^* \propto Fo^{-1/2}$, the slope of the line is $-1/2$ on the log-log plot.

Interior Heat Transfer: Plane Wall, Cylinder, and Sphere Results for heat transfer to the *interior* of a plane wall, cylinder, and sphere are also shown in Figure 5.10*a*. These results are generated by using Fourier's law in conjunction with Equations 5.42, 5.50, and 5.51 for $Bi \rightarrow \infty$. As in Sections 5.5 and 5.6, the characteristic length is $L_c = L$ or r_o for a plane wall of thickness $2L$ or a cylinder (or sphere) of radius r_o, respectively. For each geometry, q^* initially follows the semi-infinite solid solution but at some point decreases rapidly as the objects approach their equilibrium temperature and $q_s''(t \rightarrow \infty) \rightarrow 0$. The value of q^* is expected to decrease more rapidly for geometries that possess large surface area to volume ratios, and this trend is evident in Figure 5.10*a*.

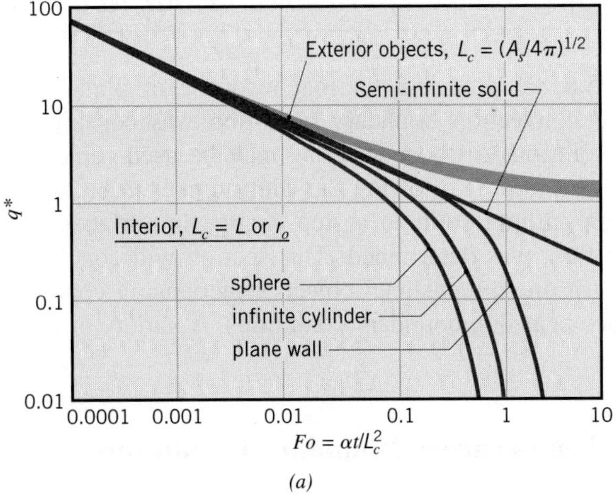

(*a*)

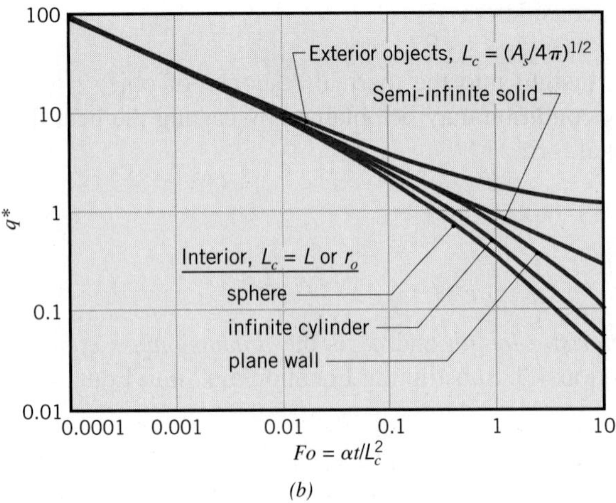

(*b*)

FIGURE 5.10 Transient dimensionless conduction heat rates for a variety of geometries. (*a*) Constant surface temperature. Results for the geometries of Table 4.1 lie within the shaded region and are from Yovanovich [7]. (*b*) Constant surface heat flux.

Exterior Heat Transfer: Various Geometries Additional results are shown in Figure 5.10*a* for objects that are embedded in an exterior (surrounding) medium of infinite extent. The infinite medium is initially at temperature T_i, and the surface temperature of the object is suddenly changed to T_s. For the exterior cases, L_c is the characteristic length used in Section 4.3, namely $L_c = (A_s/4\pi)^{1/2}$. For the sphere in a surrounding infinite medium, the exact solution for $q^*(Fo)$ is [7]

$$q^* = \frac{1}{\sqrt{\pi Fo}} + 1 \tag{5.69}$$

As seen in the figure, for all of the *exterior cases* q^* closely mimics that of the sphere when the appropriate length scale is used in its definition, regardless of the object's shape. Moreover, in a manner consistent with the interior cases, q^* initially follows the semi-infinite solid solution. In contrast to the interior cases, q^* eventually reaches the nonzero, steady-state value of q^*_{ss} that is listed in Table 4.1. Note that q''_s in Equation 5.67 is the *average* surface heat flux for geometries that have nonuniform surface heat flux.

As seen in Figure 5.10*a*, *all* of the thermal responses collapse to that of the semi-infinite solid for *early* times, that is, for Fo less than approximately 10^{-3}. This remarkable consistency reflects the fact that temperature variations are confined to thin layers adjacent to the surface of any object at early times, regardless of whether internal or external heat transfer is of interest. At early times, therefore, Equations 5.60 and 5.61 may be used to predict the temperatures and heat transfer rates within the thin regions adjacent to the boundaries of any object. For example, predicted local heat fluxes and local dimensionless temperatures using the semi-infinite solid solutions are within approximately 5% of the predictions obtained from the exact solutions for the interior and exterior heat transfer cases involving spheres when $Fo \leq 10^{-3}$.

5.8.2 Constant Heat Flux Boundary Conditions

When a constant surface heat flux is applied to an object, the resulting surface temperature history is often of interest. In this case, the heat flux in the numerator of Equation 5.67 is now constant, and the temperature difference in the denominator, $T_s - T_i$, increases with time.

Semi-Infinite Solid In the case of a semi-infinite solid, the surface temperature history can be found by evaluating Equation 5.62 at $x = 0$, which may be rearranged and combined with Equation 5.67 to yield

$$q^* = \frac{1}{2}\sqrt{\frac{\pi}{Fo}} \tag{5.70}$$

As for the constant temperature case, $q^* \propto Fo^{-1/2}$, but with a different coefficient. Equation 5.70 is presented in Figure 5.10*b*.

Interior Heat Transfer: Plane Wall, Cylinder, and Sphere A second set of results is shown in Figure 5.10*b* for the *interior* cases of the plane wall, cylinder, and sphere. As for the constant surface temperature results of Figure 5.10*a*, q^* initially follows the semi-infinite solid solution and subsequently decreases more rapidly, with the decrease occurring first for the sphere, then the cylinder, and finally the plane wall. Compared to the constant surface temperature case, the rate at which q^* decreases is not as dramatic, since steady-state conditions are never reached; the surface temperature must continue to increase with time. At *late* times (large Fo), the surface temperature increases linearly with time, yielding $q^* \propto Fo^{-1}$, with a slope of -1 on the log-log plot.

Exterior Heat Transfer: Various Geometries Results for heat transfer between a sphere and an exterior infinite medium are also presented in Figure 5.10*b*. The exact solution for the embedded sphere is

$$q^* = [1 - \exp(Fo)\,\text{erfc}(Fo^{1/2})]^{-1} \tag{5.71}$$

As in the constant surface temperature case of Figure 5.10*a*, this solution approaches steady-state conditions, with $q''_{ss} = 1$. For objects of other shapes that are embedded within an infinite medium, q^* would follow the semi-infinite solid solution at small Fo. At larger Fo, q^* must asymptotically approach the value of q''_{ss} given in Table 4.1 where T_s in Equation 5.67 is the *average* surface temperature for geometries that have nonuniform surface temperatures.

5.8.3 Approximate Solutions

Simple expressions have been developed for $q^*(Fo)$ [8]. These expressions may be used to approximate all the results included in Figure 5.10 over the *entire range of Fo*. These expressions are listed in Table 5.2, along with the corresponding exact solutions. Table 5.2*a* is for the constant surface temperature case, while Table 5.2*b* is for the constant surface heat flux situation. For each of the geometries listed in the left-hand column, the tables provide the length scale to be used in the definition of both Fo and q^*, the exact solution for $q^*(Fo)$, the approximation solutions for early times ($Fo < 0.2$) and late times ($Fo \geq 0.2$), and the maximum percentage error associated with use of the approximations (which occurs at $Fo \approx 0.2$ for all results except the external sphere with constant heat flux).

EXAMPLE 5.8

Derive an expression for the ratio of the total energy transferred from the isothermal surfaces of a plane wall to the interior of the plane wall, Q/Q_o, that is valid for $Fo < 0.2$. Express your results in terms of the Fourier number Fo.

SOLUTION

Known: Plane wall with constant surface temperatures.

Find: Expression for Q/Q_o as a function of $Fo = \alpha t/L^2$.

Schematic:

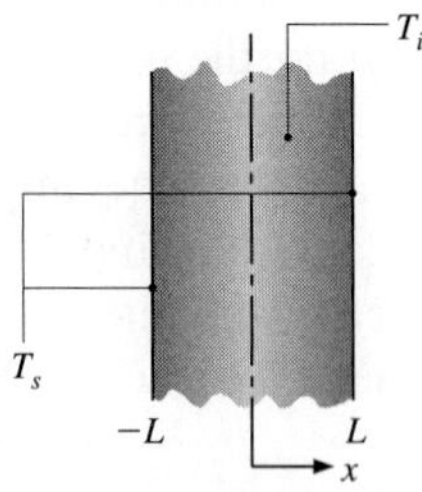

TABLE 5.2a Summary of transient heat transfer results for constant surface temperature cases[a] [8]

Geometry	Length Scale, L_c	$q^*(Fo)$ Exact Solutions	$q^*(Fo)$ Approximate Solutions $Fo < 0.2$	$q^*(Fo)$ Approximate Solutions $Fo \geq 0.2$	Maximum Error (%)
Semi-infinite	L (arbitrary)	$\frac{1}{\sqrt{\pi Fo}}$	Use exact solution.		None
Interior Cases					
Plane wall of thickness $2L$	L	$2\sum_{n=1}^{\infty}\exp(-\zeta_n^2 Fo) \quad \zeta_n = (n - \frac{1}{2})\pi$	$\frac{1}{\sqrt{\pi Fo}}$	$2\exp(-\zeta_1^2 Fo) \quad \zeta_1 = \pi/2$	1.7
Infinite cylinder	r_o	$2\sum_{n=1}^{\infty}\exp(-\zeta_n^2 Fo) \quad J_0(\zeta_n) = 0$	$\frac{1}{\sqrt{\pi Fo}} - 0.50 - 0.65\,Fo$	$2\exp(-\zeta_1^2 Fo) \quad \zeta_1 = 2.4050$	0.8
Sphere	r_o	$2\sum_{n=1}^{\infty}\exp(-\zeta_n^2 Fo) \quad \zeta_n = n\pi$	$\frac{1}{\sqrt{\pi Fo}} - 1$	$2\exp(-\zeta_1^2 Fo) \quad \zeta_1 = \pi$	6.3
Exterior Cases					
Sphere	r_o	$\frac{1}{\sqrt{\pi Fo}} + 1$	Use exact solution.		None
Various shapes (Table 4.1, cases 12–15)	$(A_s/4\pi)^{1/2}$	None	$\frac{1}{\sqrt{\pi Fo}} + q^*_{ss}, \quad q^*_{ss}$ from Table 4.1		7.1

[a] $q^* \equiv q_s'' L_c/k(T_s - T_i)$ and $Fo \equiv \alpha t/L_c^2$, where L_c is the length scale given in the table, T_s is the object surface temperature, and T_i is (a) the initial object temperature for the interior cases and (b) the temperature of the infinite medium for the exterior cases.

TABLE 5.2b Summary of transient heat transfer results for constant surface heat flux cases[a] [8]

Geometry	Length Scale, L_c	$q^*(Fo)$ Exact Solutions	Approximate Solutions $Fo < 0.2$	Approximate Solutions $Fo \geq 0.2$	Maximum Error (%)
Semi-infinite	L (arbitrary)	$\frac{1}{2}\sqrt{\frac{\pi}{Fo}}$	Use exact solution.		None
Interior Cases					
Plane wall of thickness $2L$	L	$\left[Fo + \frac{1}{3} - 2\sum_{n=1}^{\infty}\frac{\exp(-\zeta_n^2 Fo)}{\zeta_n^2}\right]^{-1}$ $\zeta_n = n\pi$	$\frac{1}{2}\sqrt{\frac{\pi}{Fo}}$	$\left[Fo + \frac{1}{3}\right]^{-1}$	5.3
Infinite cylinder	r_o	$\left[2Fo + \frac{1}{4} - 2\sum_{n=1}^{\infty}\frac{\exp(-\zeta_n^2 Fo)}{\zeta_n^2}\right]^{-1}$ $J_1(\zeta_n) = 0$	$\frac{1}{2}\sqrt{\frac{\pi}{Fo}} - \frac{\pi}{8}$	$\left[2Fo + \frac{1}{4}\right]^{-1}$	2.1
Sphere	r_o	$\left[3Fo + \frac{1}{5} - 2\sum_{n=1}^{\infty}\frac{\exp(-\zeta_n^2 Fo)}{\zeta_n^2}\right]^{-1}$ $\tan(\zeta_n) = \zeta_n$	$\frac{1}{2}\sqrt{\frac{\pi}{Fo}} - \frac{\pi}{4}$	$\left[3Fo + \frac{1}{5}\right]^{-1}$	4.5
Exterior Cases					
Sphere	r_o	$[1 - \exp(Fo)\mathrm{erfc}(Fo^{1/2})]^{-1}$	$\frac{1}{2}\sqrt{\frac{\pi}{Fo}} + \frac{\pi}{4}$	$\frac{0.77}{\sqrt{Fo}} + 1$	3.2
Various shapes (Table 4.1, cases 12–15)	$(A_s/4\pi)^{1/2}$	None	$\frac{1}{2}\sqrt{\frac{\pi}{Fo}} + \frac{\pi}{4}$	$\frac{0.77}{\sqrt{Fo}} + q_{ss}^*$	Unknown

[a] $q^* \equiv q_s'' L_c / k(T_s - T_i)$ and $Fo \equiv \alpha t / L_c^2$, where L_c is the length scale given in the table, T_s is the object surface temperature, and T_i is (a) the initial object temperature for the interior cases and (b) the temperature of the infinite medium for the exterior cases.

Assumptions:

1. One-dimensional conduction.
2. Constant properties.
3. Validity of the approximate solution of Table 5.2*a*.

Analysis: From Table 5.2*a* for a plane wall of thickness $2L$ and $Fo < 0.2$,

$$q^* = \frac{q_s''L}{k(T_s - T_i)} = \frac{1}{\sqrt{\pi Fo}} \quad \text{where} \quad Fo = \frac{\alpha t}{L^2}$$

Combining the preceding equations yields

$$q_s'' = \frac{k(T_s - T_i)}{\sqrt{\pi\alpha t}}$$

Recognizing that Q is the accumulated heat that has entered the wall up to time t, we can write

$$\frac{Q}{Q_o} = \frac{\int_{t=0}^{t} q_s'' dt}{L\rho c(T_s - T_i)} = \frac{\alpha}{L\sqrt{\pi\alpha}} \int_{t=0}^{t} t^{-1/2} dt = \frac{2}{\sqrt{\pi}}\sqrt{Fo} \qquad \triangleleft$$

Comments:

1. The exact solution for Q/Q_o at small Fourier number involves many terms that would need to be evaluated in the infinite series expression. Use of the approximate solution simplifies the evaluation of Q/Q_o considerably.
2. At $Fo = 0.2$, $Q/Q_o \approx 0.5$. Approximately half of the total possible change in thermal energy of the plane wall occurs during $Fo \leq 0.2$.
3. Although the Fourier number may be viewed as a dimensionless time, it has an important physical interpretation for problems involving heat transfer by conduction through a solid concurrent with thermal energy storage in the solid. Specifically, as suggested by the solution, the Fourier number provides a measure of the amount of energy stored in the solid at any time.

EXAMPLE 5.9

A proposed cancer treatment utilizes small, composite *nanoshells* whose size and composition are carefully specified so that the particles efficiently absorb laser irradiation at particular wavelengths [9]. Prior to treatment, antibodies are attached to the nanoscale particles. The doped particles are then injected into the patient's bloodstream and are distributed throughout the body. The antibodies are attracted to malignant sites, and therefore carry and adhere the nanoshells only to cancerous tissue. After the particles have come to rest within the tumor, a laser beam penetrates through the tissue between the skin and the cancer, is absorbed by the nanoshells, and, in turn, heats and destroys the cancerous tissues.

Consider an approximately spherical tumor of diameter $D_t = 3$ mm that is uniformly infiltrated with nanoshells that are highly absorptive of incident radiation from a laser located outside the patient's body.

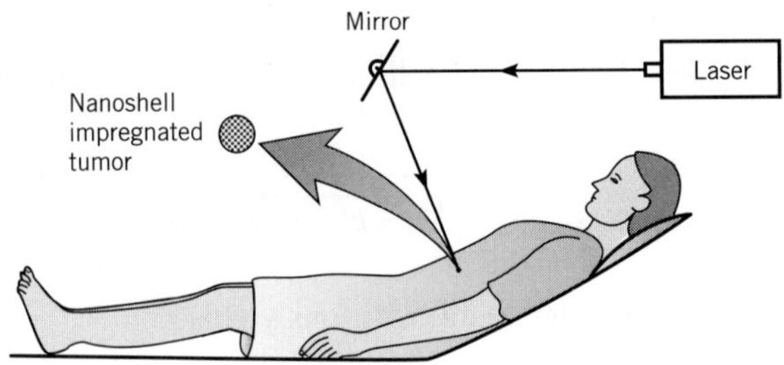

1. Estimate the heat transfer rate from the tumor to the surrounding healthy tissue for a steady-state treatment temperature of $T_{t,ss} = 55°C$ at the surface of the tumor. The thermal conductivity of healthy tissue is approximately $k = 0.5$ W/m·K, and the body temperature is $T_b = 37°C$.
2. Find the laser power necessary to sustain the tumor surface temperature at $T_{t,ss} = 55°C$ if the tumor is located $d = 20$ mm beneath the surface of the skin, and the laser heat flux decays exponentially, $q_l''(x) = q_{l,o}''(1 - \rho)\, e^{-\kappa x}$, between the surface of the body and the tumor. In the preceding expression, $q_{l,o}''$ is the laser heat flux outside the body, $\rho = 0.05$ is the reflectivity of the skin surface, and $\kappa = 0.02\ \text{mm}^{-1}$ is the *extinction coefficient* of the tissue between the tumor and the surface of the skin. The laser beam has a diameter of $D_l = 5$ mm.
3. Neglecting heat transfer to the surrounding tissue, estimate the time at which the tumor temperature is within 3°C of $T_{t,ss} = 55°C$ for the laser power found in part 2. Assume the tissue's density and specific heat are that of water.
4. Neglecting the thermal mass of the tumor but accounting for heat transfer to the surrounding tissue, estimate the time needed for the surface temperature of the tumor to reach $T_t = 52°C$.

SOLUTION

Known: Size of a small sphere; thermal conductivity, reflectivity, and extinction coefficient of tissue; depth of sphere below the surface of the skin.

Find:

1. Heat transferred from the tumor to maintain its surface temperature at $T_{t,ss} = 55°C$.
2. Laser power needed to sustain the tumor surface temperature at $T_{t,ss} = 55°C$.
3. Time for the tumor to reach $T_t = 52°C$ when heat transfer to the surrounding tissue is neglected.
4. Time for the tumor to reach $T_t = 52°C$ when heat transfer to the surrounding tissue is considered and the thermal mass of the tumor is neglected.

Schematic:

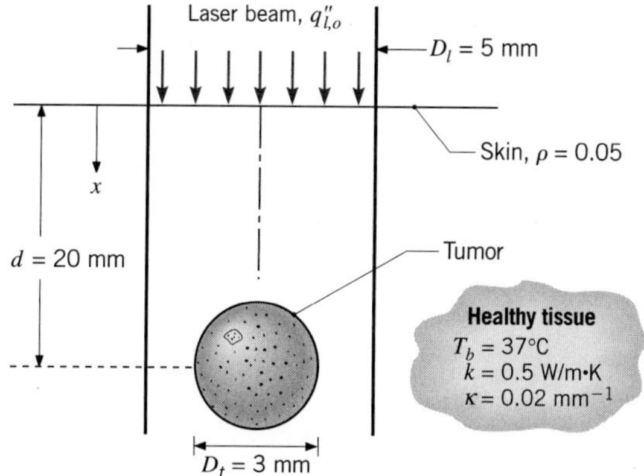

Assumptions:

1. One-dimensional conduction in the radial direction.
2. Constant properties.
3. Healthy tissue can be treated as an infinite medium.
4. The treated tumor absorbs all irradiation incident from the laser.
5. Lumped capacitance behavior for the tumor.
6. Neglect potential nanoscale heat transfer effects.
7. Neglect the effect of perfusion.

Properties: Table A.6, water (320 K, assumed): $\rho = v_f^{-1} = 989.1\ \text{kg/m}^3$, $c_p = 4180\ \text{J/kg}\cdot\text{K}$.

Analysis:

1. The steady-state heat loss from the spherical tumor may be determined by evaluating the dimensionless heat rate from the expression for case 12 of Table 4.1:

$$q = 2\pi k D_t(T_{t,\text{ss}} - T_b) = 2 \times \pi \times 0.5\ \text{W/m}\cdot\text{K} \times 3 \times 10^{-3}\ \text{m} \times (55 - 37)°\text{C}$$
$$= 0.170\ \text{W} \qquad \triangleleft$$

2. The laser irradiation will be absorbed over the projected area of the tumor, $\pi D_t^2/4$. To determine the laser power corresponding to $q = 0.170$ W, we first write an energy balance for the sphere. For a control surface about the sphere, the energy absorbed from the laser irradiation is offset by heat conduction to the healthy tissue, $q = 0.170\ \text{W} \approx q_l''(x = d)\pi D_t^2/4$, where, $q_l''(x = d) = q_{l,o}''(1 - \rho)e^{-\kappa d}$ and the laser power is $P_l = q_{l,o}''\pi D_l^2/4$. Hence,

$$P_l = qD_l^2 e^{\kappa d}/[(1 - \rho)D_t^2]$$
$$= 0.170\ \text{W} \times (5 \times 10^{-3}\ \text{m})^2 \times e^{(0.02\ \text{mm}^{-1} \times 20\ \text{mm})}/[(1 - 0.05) \times (3 \times 10^{-3}\ \text{m})^2]$$
$$= 0.74\ \text{W} \qquad \triangleleft$$

3. The general lumped capacitance energy balance, Equation 5.14, may be written

$$q_l''(x = d)\pi D_t^2/4 = q = \rho V c_p \frac{dT}{dt}$$

Separating variables and integrating between appropriate limits,

$$\frac{q}{\rho V c}\int_{t=0}^{t} dt = \int_{T_b}^{T_t} dT$$

yields

$$t = \frac{\rho V c_p}{q}(T_t - T_b) = \frac{989.1\ \text{kg/m}^3 \times (\pi/6) \times (3 \times 10^{-3}\ \text{m})^3 \times 4180\ \text{J/kg}\cdot\text{K}}{0.170\ \text{W}} \times (52°\text{C} - 37°\text{C})$$

or

$$t = 5.16\ \text{s} \qquad \triangleleft$$

4. Using Equation 5.71,

$$q/2\pi k D_t(T_t - T_b) = q^* = [1 - \exp(Fo)\text{erfc}(Fo^{1/2})]^{-1}$$

which may be solved by trial-and-error to yield $Fo = 10.3 = 4\alpha t/D_t^2$. Then, with $\alpha = k/\rho c_p = 0.50\ \text{W/m}\cdot\text{K}/(989.1\ \text{kg/m}^3 \times 4180\ \text{J/kg}\cdot\text{K}) = 1.21 \times 10^{-7}\ \text{m}^2\text{/s}$, we find

$$t = FoD_t^2/4\alpha = 10.3 \times (3 \times 10^{-3}\ \text{m})^2/(4 \times 1.21 \times 10^{-7}\ \text{m}^2\text{/s}) = 192\ \text{s} \qquad \triangleleft$$

Comments:

1. The analysis does not account for blood perfusion. The flow of blood would lead to advection of warmed fluid away from the tumor (and relatively cool blood to the vicinity of the tumor), increasing the power needed to reach the desired treatment temperature.
2. The laser power needed to treat various-sized tumors, calculated as in parts 1 and 2 of the problem solution, is shown below. Note that as the tumor becomes smaller, a higher-powered laser is needed, which may seem counterintuitive. The power required to heat the tumor, which is the same as the heat loss calculated in part 1, increases in direct proportion to the diameter, as might be expected. However, since the laser power flux remains constant, a smaller tumor cannot absorb as much energy (the energy absorbed has a D_t^2 dependence). Less of the overall laser power is utilized to heat the tumor, and the required laser power increases for smaller tumors.

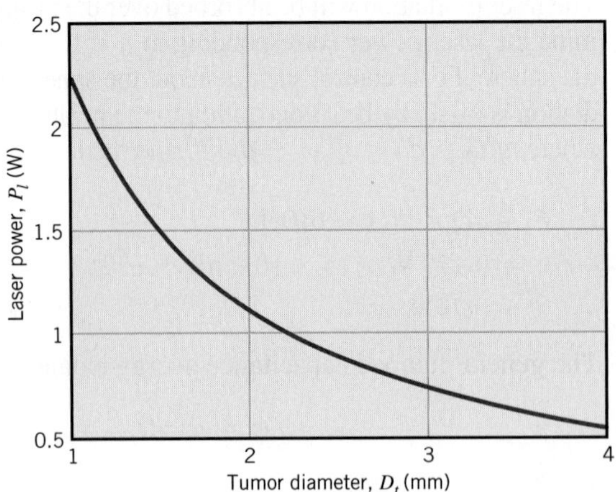

3. To determine the *actual* time needed for the tumor temperature to approach steady-state conditions, a numerical solution of the heat diffusion equation applied to the surrounding tissue, *coupled* with a solution for the temperature history within the tumor, would be required. However, we see that significantly more time is needed for the surrounding tissue to reach steady-state conditions than to increase the temperature of the isolated spherical tumor. This is due to the fact that higher temperatures propagate into a large volume when heating of the surrounding tissue is considered, while in contrast the thermal mass of the tumor is limited by the tumor's size. Hence, the actual time to heat *both* the tumor and the surrounding tissue will be slightly greater than 192 s.
4. Since temperatures are likely to increase at a considerable distance from the tumor, the assumption that the surroundings are of infinite size would need to be checked by inspecting results of the proposed numerical solution described in Comment 3.

5.9 *Periodic Heating*

In the preceding discussion of transient heat transfer, we have considered objects that experience constant surface temperature or constant surface heat flux boundary conditions. In many practical applications the boundary conditions are not constant, and analytical solutions have been obtained for situations where the conditions vary with time. One situation involving nonconstant boundary conditions is periodic heating, which describes various applications, such as thermal processing of materials using pulsed lasers, and occurs naturally in situations such as those involving the collection of solar energy.

Consider, for example, the semi-infinite solid of Figure 5.11*a*. For a surface temperature history described by $T(0, t) = T_i + \Delta T \sin \omega t$, the solution of Equation 5.29 subject to the interior boundary condition given by Equation 5.56 is

$$\frac{T(x, t) - T_i}{\Delta T} = \exp[-x\sqrt{\omega/2\alpha}]\sin[\omega t - x\sqrt{\omega/2\alpha}] \tag{5.72}$$

This solution applies after sufficient time has passed to yield a *quasi*-steady state for which all temperatures fluctuate periodically about a time-invariant mean value. At locations in the solid, the fluctuations have a time lag relative to the surface temperature.

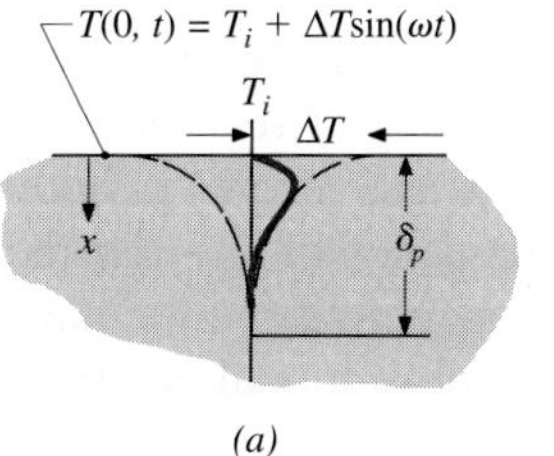

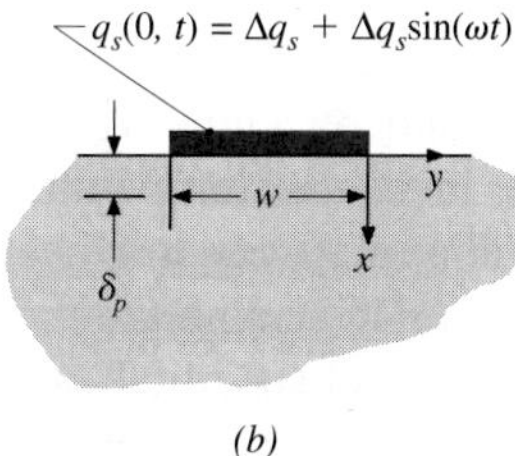

FIGURE 5.11 Schematic of (*a*) a periodically heated, one-dimensional semi-infinite solid and (*b*) a periodically heated strip attached to a semi-infinite solid.

In addition, the amplitude of the fluctuations within the material decays exponentially with distance from the surface. Consistent with the earlier definition of the thermal penetration depth, δ_p can be defined as the x-location at which the amplitude of the temperature fluctuation is reduced by approximately 90% relative to that of the surface. This results in $\delta_p = 4\sqrt{\alpha/\omega}$. The heat flux at the surface may be determined by applying Fourier's law at $x = 0$, yielding

$$q_s''(t) = k\Delta T\sqrt{\omega/\alpha}\,\sin(\omega t + \pi/4) \tag{5.73}$$

Equation 5.73 reveals that the surface heat flux is periodic, with a time-averaged value of zero.

Periodic heating can also occur in two- or three-dimensional arrangements, as shown in Figure 5.11*b*. Recall that for this geometry, a steady state can be attained with constant heating of the strip placed upon a semi-infinite solid (Table 4.1, case 13). In a similar manner, a quasi-steady state may be achieved when sinusoidal heating ($q_s = \Delta q_s + \Delta q_s \sin \omega t$) is applied to the strip. Again, a quasi-steady state is achieved for which all temperatures fluctuate about a time-invariant mean value.

The solution of the two-dimensional, transient heat diffusion equation for the two-dimensional configuration shown in Figure 5.11*b* has been obtained, and the relationship between the amplitude of the applied sinusoidal heating and the amplitude of the temperature response of the heated strip can be approximated as [10]

$$\Delta T \approx \frac{\Delta q_s}{L\pi k}\left[-\frac{1}{2}\ln(\omega/2) - \ln(w^2/4\alpha) + C_1\right] = \frac{\Delta q_s}{L\pi k}\left[-\frac{1}{2}\ln(\omega/2) + C_2\right] \tag{5.74}$$

where the constant C_1 depends on the thermal contact resistance at the interface between the heated strip and the underlying material. Note that the amplitude of the temperature fluctuation, ΔT, corresponds to the spatially averaged temperature of the rectangular strip of length L and width w. The heat flux from the strip to the semi-infinite medium is assumed to be spatially uniform. The approximation is valid for $L \gg w$. For the system of Figure 5.11*b*, the thermal penetration depth is smaller than that of Figure 5.11*a* because of the lateral spreading of thermal energy and is $\delta_p \approx \sqrt{\alpha/\omega}$.

EXAMPLE 5.10

A nanostructured dielectric material has been fabricated, and the following method is used to measure its thermal conductivity. A long metal strip 3000 angstroms thick, $w = 100\ \mu\text{m}$ wide, and $L = 3.5$ mm long is deposited by a photolithography technique on the top surface of a $d = 300$-μm-thick sample of the new material. The strip is heated periodically by an electric current supplied through two connector pads. The heating rate is $q_s(t) = \Delta q_s + \Delta q_s \sin(\omega t)$, where Δq_s is 3.5 mW. The instantaneous, spatially averaged temperature of the metal strip is found experimentally by measuring the time variation of its electrical resistance, $R(t) = E(t)/I(t)$, and by knowing how the electrical resistance of the metal varies with temperature. The measured temperature of the metal strip is periodic; it has an amplitude of $\Delta T = 1.37$ K at a relatively low heating frequency of $\omega = 2\pi$ rad/s and 0.71 K at a frequency of 200π rad/s. Determine the thermal conductivity of the nanostructured dielectric material. The density and specific heats of the conventional version of the material are 3100 kg/m^3 and 820 J/kg·K, respectively.

SOLUTION

Known: Dimensions of a thin metal strip, the frequency and amplitude of the electric power dissipated within the strip, the amplitude of the induced oscillating strip temperature, and the thickness of the underlying nanostructured material.

Find: The thermal conductivity of the nanostructured material.

Schematic:

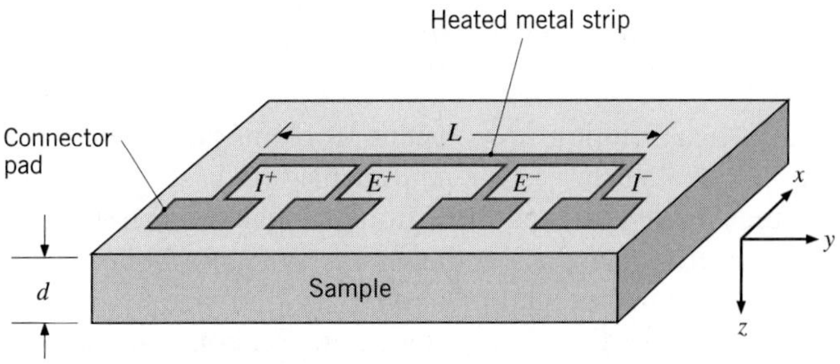

Assumptions:

1. Two-dimensional transient conduction in the x- and z-directions.
2. Constant properties.
3. Negligible radiation and convection losses from the metal strip and top surface of the sample.
4. The nanostructured material sample is a semi-infinite solid.
5. Uniform heat flux at the interface between the heated strip and the nanostructured material.

Analysis: Substitution of $\Delta T = 1.37$ K at $\omega = 2\pi$ rad/s and $\Delta T = 0.71$ K at $\omega = 200\pi$ rad/s into Equation 5.74 results in two equations that may be solved simultaneously to yield

$$C_2 = 5.35 \qquad k = 1.11 \text{ W/m} \cdot \text{K} \qquad \triangleleft$$

The thermal diffusivity is $\alpha = 4.37 \times 10^{-7}$ m^2/s, while the thermal penetration depths are estimated by $\delta_p \approx \sqrt{\alpha/\omega}$, resulting in $\delta_p = 260$ μm and $\delta_p = 26$ μm at $\omega = 2\pi$ rad/s and $\omega = 200\pi$ rad/s, respectively.

Comments:

1. The foregoing experimental technique, which is widely used to measure the thermal conductivity of microscale devices and nanostructured materials, is referred to as the *3 ω method* [10].
2. Because this technique is based on measurement of a temperature that fluctuates about a mean value that is approximately the same as the temperature of the surroundings, the measured value of k is relatively insensitive to radiation heat transfer losses from the top of the metal strip. Likewise, the technique is insensitive to thermal contact resistances that may exist at the interface between the sensing strip and the underlying material since these effects cancel when measurements are made at two different excitation frequencies [10].

3. The specific heat and density are not strongly dependent on the nanostructure of most solids, and properties of conventional material may be used.
4. The thermal penetration depth is less than the sample thickness. Therefore, treating the sample as a semi-infinite solid is a valid approach. Thinner samples could be used if higher heating frequencies were employed.

5.10 *Finite-Difference Methods*

Analytical solutions to transient problems are restricted to simple geometries and boundary conditions, such as the one-dimensional cases considered in the preceding sections. For some simple two- and three-dimensional geometries, analytical solutions are still possible. However, in many cases the geometry and/or boundary conditions preclude the use of analytical techniques, and recourse must be made to *finite-difference* (or *finite-element*) methods. Such methods, introduced in Section 4.4 for steady-state conditions, are readily extended to transient problems. In this section we consider *explicit* and *implicit* forms of finite-difference solutions to transient conduction problems.

5.10.1 Discretization of the Heat Equation: The Explicit Method

Once again consider the two-dimensional system of Figure 4.4. Under transient conditions with constant properties and no internal generation, the appropriate form of the heat equation, Equation 2.21, is

$$\frac{1}{\alpha}\frac{\partial T}{\partial t} = \frac{\partial^2 T}{\partial x^2} + \frac{\partial^2 T}{\partial y^2} \tag{5.75}$$

To obtain the finite-difference form of this equation, we may use the *central-difference* approximations to the spatial derivatives prescribed by Equations 4.27 and 4.28. Once again the m and n subscripts may be used to designate the x- and y-locations of *discrete nodal points.* However, in addition to being discretized in space, the problem must be discretized in time. The integer p is introduced for this purpose, where

$$t = p\Delta t \tag{5.76}$$

and the finite-difference approximation to the time derivative in Equation 5.75 is expressed as

$$\left.\frac{\partial T}{\partial t}\right|_{m,n} \approx \frac{T_{m,n}^{p+1} - T_{m,n}^{p}}{\Delta t} \tag{5.77}$$

The superscript p is used to denote the time dependence of T, and the time derivative is expressed in terms of the difference in temperatures associated with the *new* $(p + 1)$ and *previous* (p) times. Hence calculations must be performed at successive times separated by the interval Δt, and just as a finite-difference solution restricts temperature determination to discrete points in space, it also restricts it to discrete points in time.

Analytical solutions for some simple two- and three-dimensional geometries are found in Section 5S.2.

If Equation 5.77 is substituted into Equation 5.75, the nature of the finite-difference solution will depend on the specific time at which temperatures are evaluated in the finite-difference approximations to the spatial derivatives. In the *explicit method* of solution, these temperatures are evaluated at the *previous* (p) time. Hence Equation 5.77 is considered to be a *forward-difference* approximation to the time derivative. Evaluating terms on the right-hand side of Equations 4.27 and 4.28 at p and substituting into Equation 5.75, the explicit form of the finite-difference equation for the interior node (m, n) is

$$\frac{1}{\alpha}\frac{T_{m,n}^{p+1}-T_{m,n}^{p}}{\Delta t}=\frac{T_{m+1,n}^{p}+T_{m-1,n}^{p}-2T_{m,n}^{p}}{(\Delta x)^2}+\frac{T_{m,n+1}^{p}+T_{m,n-1}^{p}-2T_{m,n}^{p}}{(\Delta y)^2} \tag{5.78}$$

Solving for the nodal temperature at the new ($p + 1$) time and assuming that $\Delta x = \Delta y$, it follows that

$$T_{m,n}^{p+1}=Fo(T_{m+1,n}^{p}+T_{m-1,n}^{p}+T_{m,n+1}^{p}+T_{m,n-1}^{p})+(1-4Fo)T_{m,n}^{p} \tag{5.79}$$

where Fo is a finite-difference form of the Fourier number

$$Fo=\frac{\alpha\,\Delta t}{(\Delta x)^2} \tag{5.80}$$

This approach can easily be extended to one- or three-dimensional systems. If the system is one-dimensional in x, the explicit form of the finite-difference equation for an interior node m reduces to

$$T_{m}^{p+1}=Fo(T_{m+1}^{p}+T_{m-1}^{p})+(1-2Fo)T_{m}^{p} \tag{5.81}$$

Equations 5.79 and 5.81 are *explicit* because *unknown* nodal temperatures for the new time are determined exclusively by *known* nodal temperatures at the previous time. Hence calculation of the unknown temperatures is straightforward. Since the temperature of each interior node is known at $t = 0$ ($p = 0$) from prescribed initial conditions, the calculations begin at $t = \Delta t$ ($p = 1$), where Equation 5.79 or 5.81 is applied to each interior node to determine its temperature. With temperatures known for $t = \Delta t$, the appropriate finite-difference equation is then applied at each node to determine its temperature at $t = 2\,\Delta t$ ($p = 2$). In this way, the transient temperature distribution is obtained by *marching out in time,* using intervals of Δt.

The accuracy of the finite-difference solution may be improved by decreasing the values of Δx and Δt. Of course, the number of interior nodal points that must be considered increases with decreasing Δx, and the number of time intervals required to carry the solution to a prescribed final time increases with decreasing Δt. Hence the computation time increases with decreasing Δx and Δt. The choice of Δx is typically based on a compromise between accuracy and computational requirements. Once this selection has been made, however, the value of Δt may not be chosen independently. It is, instead, determined by *stability* requirements.

An undesirable feature of the explicit method is that it is not unconditionally *stable.* In a transient problem, the solution for the nodal temperatures should continuously approach final (steady-state) values with increasing time. However, with the explicit method, this solution may be characterized by numerically induced oscillations, which are physically impossible. The oscillations may become *unstable,* causing the solution to diverge from the actual steady-state conditions. To prevent such erroneous results, the prescribed value of Δt must be maintained below a certain limit, which depends on Δx and other parameters of the system. This dependence is termed a *stability criterion,* which may be obtained mathematically or demonstrated from a thermodynamic argument (see Problem 5.108). For the problems of interest in this text, *the criterion is determined by requiring that the coefficient associated with the node of interest at the previous time is greater than or equal to zero.*

In general, this is done by collecting all terms involving $T^p_{m,n}$ to obtain the form of the coefficient. This result is then used to obtain a limiting relation involving Fo, from which the maximum allowable value of Δt may be determined. For example, with Equations 5.79 and 5.81 already expressed in the desired form, it follows that the stability criterion for a one-dimensional interior node is $(1 - 2Fo) \geq 0$, or

$$Fo \leq \frac{1}{2} \tag{5.82}$$

and for a two-dimensional node, it is $(1 - 4Fo) \geq 0$, or

$$Fo \leq \frac{1}{4} \tag{5.83}$$

For prescribed values of Δx and α, these criteria may be used to determine upper limits to the value of Δt.

Equations 5.79 and 5.81 may also be derived by applying the energy balance method of Section 4.4.3 to a control volume about the interior node. Accounting for changes in thermal energy storage, a general form of the energy balance equation may be expressed as

$$\dot{E}_{\text{in}} + \dot{E}_g = \dot{E}_{\text{st}} \tag{5.84}$$

In the interest of adopting a consistent methodology, it is again assumed that all heat flow is *into* the node.

To illustrate application of Equation 5.84, consider the surface node of the one-dimensional system shown in Figure 5.12. To more accurately determine thermal conditions near the surface, this node has been assigned a thickness that is one-half that of the interior nodes. Assuming convection transfer from an adjoining fluid and no generation, it follows from Equation 5.84 that

$$hA(T_\infty - T_0^p) + \frac{kA}{\Delta x}(T_1^p - T_0^p) = \rho c A \frac{\Delta x}{2} \frac{T_0^{p+1} - T_0^p}{\Delta t}$$

or, solving for the surface temperature at $t + \Delta t$,

$$T_0^{p+1} = \frac{2h\,\Delta t}{\rho c\,\Delta x}(T_\infty - T_0^p) + \frac{2\alpha\,\Delta t}{\Delta x^2}(T_1^p - T_0^p) + T_0^p$$

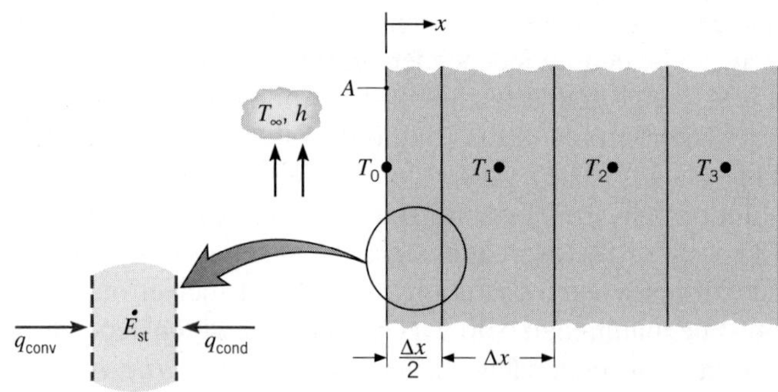

FIGURE 5.12 Surface node with convection and one-dimensional transient conduction.

Recognizing that $(2h\Delta t/\rho c\Delta x) = 2(h\Delta x/k)(\alpha\Delta t/\Delta x^2) = 2\,Bi\,Fo$ and grouping terms involving T_0^p, it follows that

$$T_0^{p+1} = 2Fo(T_1^p + Bi\,T_\infty) + (1 - 2Fo - 2Bi\,Fo)T_0^p \tag{5.85}$$

The finite-difference form of the Biot number is

$$Bi = \frac{h\,\Delta x}{k} \tag{5.86}$$

Recalling the procedure for determining the stability criterion, we require that the coefficient for T_0^p be greater than or equal to zero. Hence

$$1 - 2Fo - 2Bi\,Fo \geq 0$$

or

$$Fo(1 + Bi) \leq \frac{1}{2} \tag{5.87}$$

Since the complete finite-difference solution requires the use of Equation 5.81 for the interior nodes, as well as Equation 5.85 for the surface node, Equation 5.87 must be contrasted with Equation 5.82 to determine which requirement is more stringent. Since $Bi \geq 0$, it is apparent that the limiting value of Fo for Equation 5.87 is less than that for Equation 5.82. To ensure stability for all nodes, Equation 5.87 should therefore be used to select the maximum allowable value of Fo, and hence Δt, to be used in the calculations.

Forms of the explicit finite-difference equation for several common geometries are presented in Table 5.3*a*. Each equation may be derived by applying the energy balance method to a control volume about the corresponding node. To develop confidence in your ability to apply this method, you should attempt to verify at least one of these equations.

Example 5.11

A fuel element of a nuclear reactor is in the shape of a plane wall of thickness $2L = 20$ mm and is convectively cooled at both surfaces, with $h = 1100\ \text{W/m}^2\cdot\text{K}$ and $T_\infty = 250°\text{C}$. At normal operating power, heat is generated uniformly within the element at a volumetric rate of $\dot{q}_1 = 10^7\ \text{W/m}^3$. A departure from the steady-state conditions associated with normal operation will occur if there is a change in the generation rate. Consider a sudden change to $\dot{q}_2 = 2 \times 10^7\ \text{W/m}^3$, and use the explicit finite-difference method to determine the fuel element temperature distribution after 1.5 s. The fuel element thermal properties are $k = 30\ \text{W/m}\cdot\text{K}$ and $\alpha = 5 \times 10^{-6}\ \text{m}^2\text{/s}$.

Solution

Known: Conditions associated with heat generation in a rectangular fuel element with surface cooling.

Find: Temperature distribution 1.5 s after a change in operating power.

TABLE 5.3 Transient, two-dimensional finite-difference equations ($\Delta x = \Delta y$)

Configuration	(a) Explicit Method: Finite-Difference Equation	(a) Explicit Method: Stability Criterion	(b) Implicit Method
	$T^{p+1}_{m,n} = Fo(T^p_{m+1,n} + T^p_{m-1,n} + T^p_{m,n+1} + T^p_{m,n-1}) + (1 - 4Fo)T^p_{m,n}$ (5.79) 1. Interior node	$Fo \leq \frac{1}{4}$ (5.83)	$(1 + 4Fo)T^{p+1}_{m,n} - Fo(T^{p+1}_{m+1,n} + T^{p+1}_{m-1,n} + T^{p+1}_{m,n+1} + T^{p+1}_{m,n-1}) = T^p_{m,n}$ (5.95)
	$T^{p+1}_{m,n} = \frac{2}{3}Fo(T^p_{m+1,n} + 2T^p_{m-1,n} + 2T^p_{m,n+1} + T^p_{m,n-1} + 2Bi\, T_\infty) + (1 - 4Fo - \frac{4}{3}Bi\, Fo)T^p_{m,n}$ (5.88) 2. Node at interior corner with convection	$Fo(3 + Bi) \leq \frac{3}{4}$ (5.89)	$(1 + 4Fo(1 + \frac{1}{3}Bi))T^{p+1}_{m,n} - \frac{2}{3}Fo \cdot (T^{p+1}_{m+1,n} + 2T^{p+1}_{m-1,n} + 2T^{p+1}_{m,n+1} + T^{p+1}_{m,n-1}) = T^p_{m,n} + \frac{4}{3}Bi\, Fo\, T_\infty$ (5.98)
	$T^{p+1}_{m,n} = Fo(2T^p_{m-1,n} + T^p_{m,n+1} + T^p_{m,n-1} + 2Bi\, T_\infty) + (1 - 4Fo - 2Bi\, Fo)T^p_{m,n}$ (5.90) 3. Node at plane surface with convection[a]	$Fo(2 + Bi) \leq \frac{1}{2}$ (5.91)	$(1 + 2Fo(2 + Bi))T^{p+1}_{m,n} - Fo(2T^{p+1}_{m-1,n} + T^{p+1}_{m,n+1} + T^{p+1}_{m,n-1}) = T^p_{m,n} + 2Bi\, Fo\, T_\infty$ (5.99)
	$T^{p+1}_{m,n} = 2Fo(T^p_{m-1,n} + T^p_{m,n-1} + 2Bi\, T_\infty) + (1 - 4Fo - 4Bi\, Fo)T^p_{m,n}$ (5.92) 4. Node at exterior corner with convection	$Fo(1 + Bi) \leq \frac{1}{4}$ (5.93)	$(1 + 4Fo(1 + Bi))T^{p+1}_{m,n} - 2Fo(T^{p+1}_{m-1,n} + T^{p+1}_{m,n-1}) = T^p_{m,n} + 4Bi\, Fo\, T_\infty$ (5.100)

[a] To obtain the finite-difference equation and/or stability criterion for an adiabatic surface (or surface of symmetry), simply set *Bi* equal to zero.

Schematic:

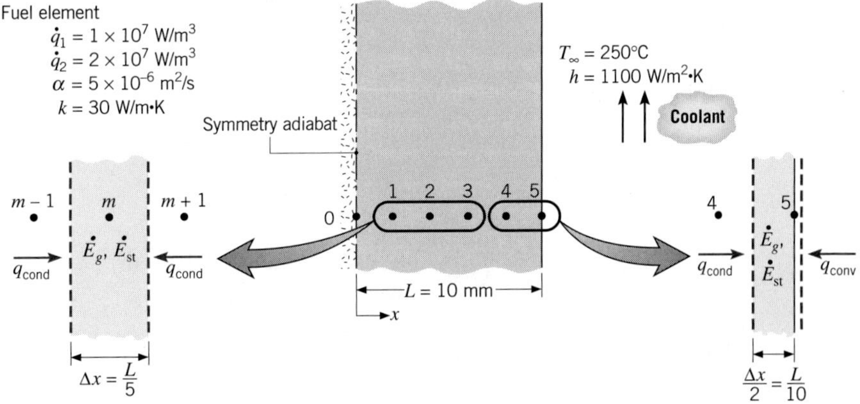

Assumptions:

1. One-dimensional conduction in x.
2. Uniform generation.
3. Constant properties.

Analysis: A numerical solution will be obtained using a space increment of $\Delta x = 2$ mm. Since there is symmetry about the midplane, the nodal network yields six unknown nodal temperatures. Using the energy balance method, Equation 5.84, an explicit finite-difference equation may be derived for any interior node m.

$$kA\frac{T_{m-1}^p - T_m^p}{\Delta x} + kA\frac{T_{m+1}^p - T_m^p}{\Delta x} + \dot{q}A\,\Delta x = \rho A\,\Delta x\,c\frac{T_m^{p+1} - T_m^p}{\Delta t}$$

Solving for T_m^{p+1} and rearranging,

$$T_m^{p+1} = Fo\left[T_{m-1}^p + T_{m+1}^p + \frac{\dot{q}(\Delta x)^2}{k}\right] + (1 - 2Fo)T_m^p \qquad (1)$$

This equation may be used for node 0, with $T_{m-1}^p = T_{m+1}^p$, as well as for nodes 1, 2, 3, and 4. Applying energy conservation to a control volume about node 5,

$$hA(T_\infty - T_5^p) + kA\frac{T_4^p - T_5^p}{\Delta x} + \dot{q}A\frac{\Delta x}{2} = \rho A\frac{\Delta x}{2}c\frac{T_5^{p+1} - T_5^p}{\Delta t}$$

or

$$T_5^{p+1} = 2Fo\left[T_4^p + Bi\,T_\infty + \frac{\dot{q}(\Delta x)^2}{2k}\right] + (1 - 2Fo - 2Bi\,Fo)T_5^p \qquad (2)$$

Since the most restrictive stability criterion is associated with Equation 2, we select Fo from the requirement that

$$Fo(1 + Bi) \le \frac{1}{2}$$

Hence, with

$$Bi = \frac{h\,\Delta x}{k} = \frac{1100\ \text{W/m}^2\cdot\text{K}\,(0.002\ \text{m})}{30\ \text{W/m}\cdot\text{K}} = 0.0733$$

it follows that

$$Fo \leq 0.466$$

or

$$\Delta t = \frac{Fo(\Delta x)^2}{\alpha} \leq \frac{0.466(2\times10^{-3}\ \text{m})^2}{5\times10^{-6}\ \text{m}^2/\text{s}} \leq 0.373\ \text{s}$$

To be well within the stability limit, we select $\Delta t = 0.3$ s, which corresponds to

$$Fo = \frac{5\times10^{-6}\ \text{m}^2/\text{s}(0.3\ \text{s})}{(2\times10^{-3}\ \text{m})^2} = 0.375$$

Substituting numerical values, including $\dot{q} = \dot{q}_2 = 2\times10^7\ \text{W/m}^3$, the nodal equations become

$$T_0^{p+1} = 0.375(2T_1^p + 2.67) + 0.250T_0^p$$

$$T_1^{p+1} = 0.375(T_0^p + T_2^p + 2.67) + 0.250T_1^p$$

$$T_2^{p+1} = 0.375(T_1^p + T_3^p + 2.67) + 0.250T_2^p$$

$$T_3^{p+1} = 0.375(T_2^p + T_4^p + 2.67) + 0.250T_3^p$$

$$T_4^{p+1} = 0.375(T_3^p + T_5^p + 2.67) + 0.250T_4^p$$

$$T_5^{p+1} = 0.750(T_4^p + 19.67) + 0.195T_5^p$$

To begin the marching solution, the initial temperature distribution must be known. This distribution is given by Equation 3.47, with $\dot{q} = \dot{q}_1$. Obtaining $T_s = T_5$ from Equation 3.51,

$$T_5 = T_\infty + \frac{\dot{q}L}{h} = 250°\text{C} + \frac{10^7\ \text{W/m}^3\times0.01\ \text{m}}{1100\ \text{W/m}^2\cdot\text{K}} = 340.91°\text{C}$$

it follows that

$$T(x) = 16.67\left(1 - \frac{x^2}{L^2}\right) + 340.91°\text{C}$$

Computed temperatures for the nodal points of interest are shown in the first row of the accompanying table.

Using the finite-difference equations, the nodal temperatures may be sequentially calculated with a time increment of 0.3 s until the desired final time is reached. The results are illustrated in rows 2 through 6 of the table and may be contrasted with the new steady-state condition (row 7), which was obtained by using Equations 3.47 and 3.51 with $\dot{q} = \dot{q}_2$:

Tabulated Nodal Temperatures

p	t(s)	T_0	T_1	T_2	T_3	T_4	T_5
0	0	357.58	356.91	354.91	351.58	346.91	340.91
1	0.3	358.08	357.41	355.41	352.08	347.41	341.41
2	0.6	358.58	357.91	355.91	352.58	347.91	341.88
3	0.9	359.08	358.41	356.41	353.08	348.41	342.35
4	1.2	359.58	358.91	356.91	353.58	348.89	342.82
5	1.5	360.08	359.41	357.41	354.07	349.37	343.27
∞	∞	465.15	463.82	459.82	453.15	443.82	431.82

Comments:

1. It is evident that, at 1.5 s, the wall is in the early stages of the transient process and that many additional calculations would have to be made to reach steady-state conditions with the finite-difference solution. The computation time could be reduced slightly by using the maximum allowable time increment ($\Delta t = 0.373$ s), but with some loss of accuracy. In the interest of maximizing accuracy, the time interval should be reduced until the computed results become independent of further reductions in Δt.

   Extending the finite-difference solution, the time required to achieve the new steady-state condition may be determined, with temperature histories computed for the midplane (0) and surface (5) nodes having the following forms:

   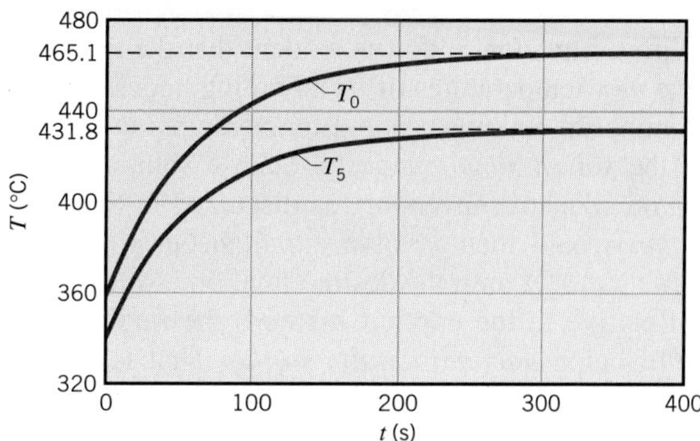

   With steady-state temperatures of $T_0 = 465.15$°C and $T_5 = 431.82$°C, it is evident that the new equilibrium condition is reached within 250 s of the step change in operating power.

2. This problem can be solved using *Tools, Finite-Difference Equations, One-Dimensional, Transient* in the *Advanced* section of *IHT*. The problem may also be solved using *Finite-Element Heat Transfer* (*FEHT*).

5.10.2 Discretization of the Heat Equation: The Implicit Method

In the *explicit* finite-difference scheme, the temperature of any node at $t + \Delta t$ may be calculated from knowledge of temperatures at the same and neighboring nodes for the *preceding time t*. Hence determination of a nodal temperature at some time is *independent* of temperatures at other nodes for the *same time.* Although the method offers computational convenience, it suffers from limitations on the selection of Δt. For a given space increment, the time interval must be compatible with stability requirements. Frequently, this dictates the use of extremely small values of Δt, and a very large number of time intervals may be necessary to obtain a solution.

A reduction in the amount of computation time may often be realized by employing an *implicit,* rather than explicit, finite-difference scheme. The implicit form of a finite-difference equation may be derived by using Equation 5.77 to approximate the time derivative, while evaluating all other temperatures at the *new* $(p + 1)$ time, instead of the previous (p) time. Equation 5.77 is then considered to provide a *backward-difference* approximation to the time derivative. In contrast to Equation 5.78, the implicit form of the finite-difference equation for the interior node of a two-dimensional system is then

$$\frac{1}{\alpha}\frac{T_{m,n}^{p+1} - T_{m,n}^{p}}{\Delta t} = \frac{T_{m+1,n}^{p+1} + T_{m-1,n}^{p+1} - 2T_{m,n}^{p+1}}{(\Delta x)^2} + \frac{T_{m,n+1}^{p+1} + T_{m,n-1}^{p+1} - 2T_{m,n}^{p+1}}{(\Delta y)^2} \tag{5.94}$$

Rearranging and assuming $\Delta x = \Delta y$, it follows that

$$(1 + 4Fo)T_{m,n}^{p+1} - Fo(T_{m+1,n}^{p+1} + T_{m-1,n}^{p+1} + T_{m,n+1}^{p+1} + T_{m,n-1}^{p+1}) = T_{m,n}^{p} \tag{5.95}$$

From Equation 5.95 it is evident that the *new* temperature of the (m, n) node depends on the *new* temperatures of its adjoining nodes, which are, in general, unknown. Hence, to determine the unknown nodal temperatures at $t + \Delta t$, the corresponding nodal equations must be *solved simultaneously.* Such a solution may be effected by using Gauss–Seidel iteration or matrix inversion, as discussed in Section 4.5 and Appendix D. The *marching solution* would then involve simultaneously solving the nodal equations at each time $t = \Delta t, 2\Delta t, \ldots,$ until the desired final time was reached.

Relative to the explicit method, the implicit formulation has the important advantage of being *unconditionally stable.* That is, the solution remains stable for all space and time intervals, in which case there are no restrictions on Δx and Δt. Since larger values of Δt may therefore be used with an implicit method, computation times may often be reduced, with little loss of accuracy. Nevertheless, to maximize accuracy, Δt should be sufficiently small to ensure that the results are independent of further reductions in its value.

The implicit form of a finite-difference equation may also be derived from the energy balance method. For the surface node of Figure 5.12, it is readily shown that

$$(1 + 2Fo + 2Fo\,Bi)T_0^{p+1} - 2Fo\,T_1^{p+1} = 2Fo\,Bi\,T_\infty + T_0^p \tag{5.96}$$

For any interior node of Figure 5.12, it may also be shown that

$$(1 + 2Fo)T_m^{p+1} - Fo\,(T_{m-1}^{p+1} + T_{m+1}^{p+1}) = T_m^p \tag{5.97}$$

Forms of the implicit finite-difference equation for other common geometries are presented in Table 5.3*b*. Each equation may be derived by applying the energy balance method.

IHT

EXAMPLE 5.12

A thick slab of copper initially at a uniform temperature of 20°C is suddenly exposed to radiation at one surface such that the net heat flux is maintained at a constant value of 3×10^5 W/m^2. Using the explicit and implicit finite-difference techniques with a space increment of $\Delta x = 75$ mm, determine the temperature at the irradiated surface and at an interior point that is 150 mm from the surface after 2 min have elapsed. Compare the results with those obtained from an appropriate analytical solution.

SOLUTION

Known: Thick slab of copper, initially at a uniform temperature, is subjected to a constant net heat flux at one surface.

Find:

1. Using the explicit finite-difference method, determine temperatures at the surface and 150 mm from the surface after an elapsed time of 2 min.
2. Repeat the calculations using the implicit finite-difference method.
3. Determine the same temperatures analytically.

Schematic:

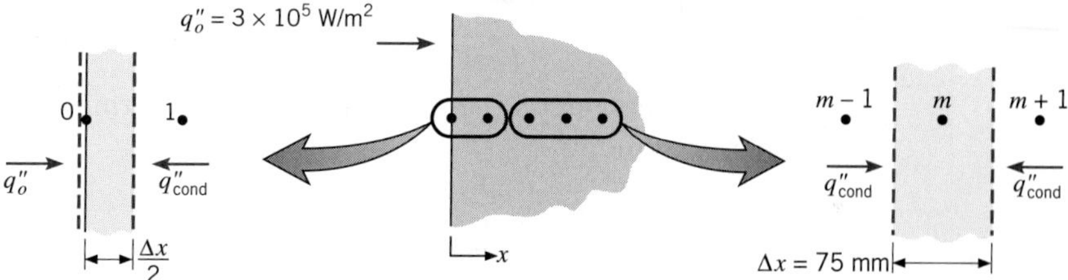

Assumptions:

1. One-dimensional conduction in x.
2. For the analytical solution, the thick slab may be approximated as a semi-infinite medium with constant surface heat flux. For the finite-difference solutions, implementation of the boundary condition $T(x \to \infty) = T_i$ will be discussed below in this example.
3. Constant properties.

Properties: Table A.1, copper (300 K): $k = 401$ W/m·K, $\alpha = 117 \times 10^{-6}$ m^2/s.

Analysis:

1. An explicit form of the finite-difference equation for the surface node may be obtained by applying an energy balance to a control volume about the node.

$$q_o''A + kA\frac{T_1^p - T_0^p}{\Delta x} = \rho A \frac{\Delta x}{2} c \frac{T_0^{p+1} - T_0^p}{\Delta t}$$

or

$$T_0^{p+1} = 2Fo\left(\frac{q_0''\Delta x}{k} + T_1^p\right) + (1 - 2Fo)T_0^p$$

The finite-difference equation for any interior node is given by Equation 5.81. Both the surface and interior nodes are governed by the stability criterion

$$Fo \leq \frac{1}{2}$$

Noting that the finite-difference equations are simplified by choosing the maximum allowable value of Fo, we select $Fo = \frac{1}{2}$. Hence

$$\Delta t = Fo\frac{(\Delta x)^2}{\alpha} = \frac{1}{2}\frac{(0.075\text{ m})^2}{117 \times 10^{-6}\text{ m}^2/\text{s}} = 24\text{ s}$$

With

$$\frac{q_o''\Delta x}{k} = \frac{3 \times 10^5\text{ W/m}^2\,(0.075\text{ m})}{401\text{ W/m}\cdot\text{K}} = 56.1^\circ\text{C}$$

the finite-difference equations become

$$T_0^{p+1} = 56.1^\circ\text{C} + T_1^p \qquad \text{and} \qquad T_m^{p+1} = \frac{T_{m+1}^p + T_{m-1}^p}{2}$$

for the surface and interior nodes, respectively. Performing the calculations, the results are tabulated as follows:

Explicit Finite-Difference Solution for $Fo = \frac{1}{2}$

p	$t(s)$	T_0	T_1	T_2	T_3	T_4
0	0	20	20	20	20	20
1	24	76.1	20	20	20	20
2	48	76.1	48.1	20	20	20
3	72	104.2	48.1	34.0	20	20
4	96	104.2	69.1	34.0	27.0	20
5	120	125.2	69.1	48.1	27.0	23.5

After 2 min, the surface temperature and the desired interior temperature are $T_0 = 125.2^\circ\text{C}$ and $T_2 = 48.1^\circ\text{C}$.

It can be seen from the explicit finite-difference solution that, with each successive time step, one more nodal temperature changes from its initial condition. For this reason, it is not necessary to formally implement the second boundary condition $T(x \rightarrow \infty) = T$. Also note that calculation of identical temperatures at successive times for the same node is an idiosyncrasy of using the maximum allowable value of Fo with the explicit finite-difference technique. The actual physical condition is, of course, one in which the temperature changes continuously with time. The idiosyncrasy is diminished and the accuracy of the calculations is improved by reducing the value of Fo.

To determine the extent to which the accuracy may be improved by reducing *Fo*, let us redo the calculations for $Fo = \frac{1}{4}$ ($\Delta t = 12$ s). The finite-difference equations are then of the form

$$T_0^{p+1} = \frac{1}{2}(56.1°\text{C} + T_1^p) + \frac{1}{2}T_0^p$$

$$T_m^{p+1} = \frac{1}{4}(T_{m+1}^p + T_{m-1}^p) + \frac{1}{2}T_m^p$$

and the results of the calculations are tabulated as follows:

Explicit Finite-Difference Solution for $Fo = \frac{1}{4}$

p	t(s)	T_0	T_1	T_2	T_3	T_4	T_5	T_6	T_7	T_8
0	0	20	20	20	20	20	20	20	20	20
1	12	48.1	20	20	20	20	20	20	20	20
2	24	62.1	27.0	20	20	20	20	20	20	20
3	36	72.6	34.0	21.8	20	20	20	20	20	20
4	48	81.4	40.6	24.4	20.4	20	20	20	20	20
5	60	89.0	46.7	27.5	21.3	20.1	20	20	20	20
6	72	95.9	52.5	30.7	22.5	20.4	20.0	20	20	20
7	84	102.3	57.9	34.1	24.1	20.8	20.1	20.0	20	20
8	96	108.1	63.1	37.6	25.8	21.5	20.3	20.0	20.0	20
9	108	113.6	67.9	41.0	27.6	22.2	20.5	20.1	20.0	20.0
10	120	118.8	72.6	44.4	29.6	23.2	20.8	20.2	20.0	20.0

After 2 min, the desired temperatures are $T_0 = 118.8°\text{C}$ and $T_2 = 44.4°\text{C}$. Comparing the above results with those obtained for $Fo = \frac{1}{2}$, it is clear that by reducing *Fo* we have diminished the problem of recurring temperatures. We have also predicted greater thermal penetration (to node 6 instead of node 3). An assessment of the improvement in accuracy will be given later, by comparison with an exact solution. In the absence of an exact solution, the value of *Fo* could be successively reduced until the results became essentially independent of *Fo*.

2. Performing an energy balance on a control volume about the surface node, the implicit form of the finite-difference equation is

$$q_o'' + k\frac{T_1^{p+1} - T_0^{p+1}}{\Delta x} = \rho\frac{\Delta x}{2}c\frac{T_0^{p+1} - T_0^p}{\Delta t}$$

or

$$(1 + 2Fo)T_0^{p+1} - 2FoT_1^{p+1} = \frac{2\alpha q_o''\Delta t}{k\,\Delta x} + T_0^p$$

Arbitrarily choosing $Fo = \frac{1}{2}$ ($\Delta t = 24$ s), it follows that

$$2T_0^{p+1} - T_1^{p+1} = 56.1 + T_0^p$$

From Equation 5.97, the finite-difference equation for any interior node is then of the form

$$-T_{m-1}^{p+1} + 4T_m^{p+1} - T_{m+1}^{p+1} = 2T_m^p$$

In contrast to the explicit method, the implicit method requires the simultaneous solution of the nodal equations for all nodes at time $p + 1$. Hence, the number of nodes under consideration must be limited to some finite number, and a boundary condition must be applied at the last node. The number of nodes may be limited to those that are affected significantly by the change in boundary condition for the time of interest. From the results of the explicit method, it is evident that we are safe in choosing nine nodes corresponding to $T_0, T_1, \ldots, T_8$. We are thereby assuming that, at $t = 120$ s, there has been no change in T_9, and the boundary condition is implemented numerically as $T_9 = 20°C$.

We now have a set of nine equations that must be solved simultaneously for each time increment. We can express the equations in the form $[A][T] = [C]$, where

$$[A] = \begin{bmatrix} 2 & -1 & 0 & 0 & 0 & 0 & 0 & 0 & 0 \\ -1 & 4 & -1 & 0 & 0 & 0 & 0 & 0 & 0 \\ 0 & -1 & 4 & -1 & 0 & 0 & 0 & 0 & 0 \\ 0 & 0 & -1 & 4 & -1 & 0 & 0 & 0 & 0 \\ 0 & 0 & 0 & -1 & 4 & -1 & 0 & 0 & 0 \\ 0 & 0 & 0 & 0 & -1 & 4 & -1 & 0 & 0 \\ 0 & 0 & 0 & 0 & 0 & -1 & 4 & -1 & 0 \\ 0 & 0 & 0 & 0 & 0 & 0 & -1 & 4 & -1 \\ 0 & 0 & 0 & 0 & 0 & 0 & 0 & -1 & 4 \end{bmatrix}$$

$$[C] = \begin{bmatrix} 56.1 + T_0^p \\ 2T_1^p \\ 2T_2^p \\ 2T_3^p \\ 2T_4^p \\ 2T_5^p \\ 2T_6^p \\ 2T_7^p \\ 2T_8^p + T_9^{p+1} \end{bmatrix}$$

Note that numerical values for the components of $[C]$ are determined from previous values of the nodal temperatures. Note also how the finite-difference equation for node 8 appears in matrices $[A]$ and $[C]$, with $T_9^{p+1} = 20°C$, as indicated previously.

A table of nodal temperatures may be compiled, beginning with the first row ($p = 0$) corresponding to the prescribed initial condition. To obtain nodal temperatures for subsequent times, the matrix equation must be solved. At each time step $p + 1$, $[C]$ is updated using the previous time step (p) values. The process is carried out five times to determine the nodal temperatures at 120 s. The desired temperatures are $T_0 = 114.7°C$ and $T_2 = 44.2°C$.

Implicit Finite-Difference Solution for $Fo = \frac{1}{2}$

p	t(s)	T_0	T_1	T_2	T_3	T_4	T_5	T_6	T_7	T_8
0	0	20.0	20.0	20.0	20.0	20.0	20.0	20.0	20.0	20.0
1	24	52.4	28.7	22.3	20.6	20.2	20.0	20.0	20.0	20.0
2	48	74.0	39.5	26.6	22.1	20.7	20.2	20.1	20.0	20.0
3	72	90.2	50.3	32.0	24.4	21.6	20.6	20.2	20.1	20.0
4	96	103.4	60.5	38.0	27.4	22.9	21.1	20.4	20.2	20.1
5	120	114.7	70.0	44.2	30.9	24.7	21.9	20.8	20.3	20.1 ◁

3. Approximating the slab as a semi-infinite medium, the appropriate analytical expression is given by Equation 5.62, which may be applied to any point in the slab.

$$T(x, t) - T_i = \frac{2q_o''(\alpha t/\pi)^{1/2}}{k} \exp\left(-\frac{x^2}{4\alpha t}\right) - \frac{q_o''x}{k} \operatorname{erfc}\left(\frac{x}{2\sqrt{\alpha t}}\right)$$

At the surface, this expression yields

$$T(0, 120\text{ s}) - 20°\text{C} = \frac{2 \times 3 \times 10^5 \text{ W/m}^2}{401 \text{ W/m} \cdot \text{K}} (117 \times 10^{-6} \text{ m}^2\text{/s} \times 120 \text{ s}/\pi)^{1/2}$$

or

$$T(0, 120\text{ s}) = 120.0°\text{C} \quad ◁$$

At the interior point ($x = 0.15$ m)

$$\begin{aligned} T(0.15\text{ m}, 120\text{ s}) - 20°\text{C} = {} & \frac{2 \times 3 \times 10^5 \text{ W/m}^2}{401 \text{ W/m} \cdot \text{K}} \\ & \times (117 \times 10^{-6} \text{ m}^2\text{/s} \times 120 \text{ s}/\pi)^{1/2} \\ & \times \exp\left[-\frac{(0.15 \text{ m})^2}{4 \times 117 \times 10^{-6} \text{ m}^2\text{/s} \times 120 \text{ s}}\right] \\ & - \frac{3 \times 10^5 \text{ W/m}^2 \times 0.15 \text{ m}}{401 \text{ W/m} \cdot \text{K}} \\ & \times \left[1 - \operatorname{erf}\left(\frac{0.15 \text{ m}}{2\sqrt{117 \times 10^{-6} \text{ m}^2\text{/s} \times 120 \text{ s}}}\right)\right] \end{aligned}$$

$$T(0.15\text{ m}, 120\text{ s}) = 45.4°\text{C} \quad ◁$$

Comments:

1. Comparing the exact results with those obtained from the three approximate solutions, it is clear that the explicit method with $Fo = \frac{1}{4}$ provides the most accurate predictions.

Method	**$T_0 = T(0, 120\text{ s})$**	**$T_2 = T(0.15\text{ m}, 120\text{ s})$**
Explicit ($Fo = \frac{1}{2}$)	125.2	48.1
Explicit ($Fo = \frac{1}{4}$)	118.8	44.4
Implicit ($Fo = \frac{1}{2}$)	114.7	44.2
Exact	120.0	45.4

This is not unexpected, since the corresponding value of Δt is 50% smaller than that used in the other two methods. Although computations are simplified by using the maximum allowable value of Fo in the explicit method, the accuracy of the results is seldom satisfactory.

2. The accuracy of the foregoing calculations is adversely affected by the coarse grid ($\Delta x = 75$ mm), as well as by the large time steps ($\Delta t = 24$ s, 12 s). Applying the implicit method with $\Delta x = 18.75$ mm and $\Delta t = 6$ s ($Fo = 2.0$), the solution yields $T_0 = T(0, 120\text{ s}) = 119.2°\text{C}$ and $T_2 = T(0.15\text{ m}, 120\text{ s}) = 45.3°\text{C}$, both of which are in good agreement with the exact solution. Complete temperature distributions may be plotted at any of the discrete times, and results obtained at $t = 60$ and 120 s are as follows:

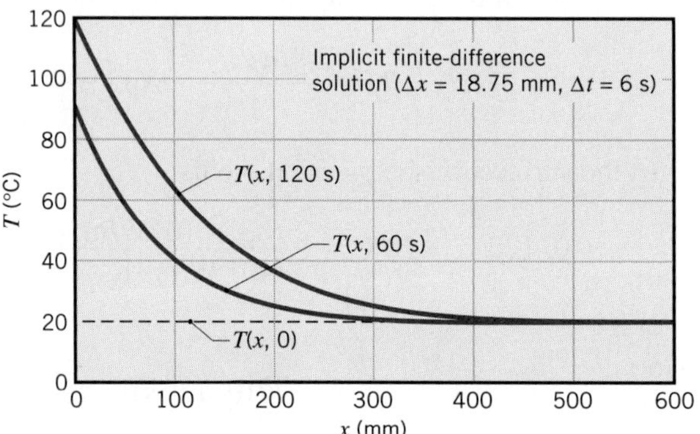

Note that, at $t = 120$ s, the assumption of a semi-infinite medium would remain valid if the thickness of the slab exceeded approximately 500 mm.

3. Note that the coefficient matrix $[A]$ is *tridiagonal.* That is, all elements are zero except those that are on, or to either side of, the main diagonal. Tridiagonal matrices are associated with one-dimensional conduction problems.

4. A more general radiative heating condition would be one in which the surface is suddenly exposed to large surroundings at an elevated temperature T_{sur} (Problem 5.126). The net rate at which radiation is transferred to the surface may then be calculated from Equation 1.7. Allowing for convection heat transfer to the surface, application of conservation of energy to the surface node yields an explicit finite-difference equation of the form

$$\varepsilon\sigma[T_{sur}^4 - (T_0^p)^4] + h(T_\infty - T_0^p) + k\frac{T_1^p - T_0^p}{\Delta x} = \rho\frac{\Delta x}{2}c\frac{T_0^{p+1} - T_0^p}{\Delta t}$$

Use of this finite-difference equation in a numerical solution is complicated by the fact that it is *nonlinear.* However, the equation may be *linearized* by introducing the radiation heat transfer coefficient h_r defined by Equation 1.9, and the finite-difference equation is

$$h_r^p(T_{sur} - T_0^p) + h(T_\infty - T_0^p) + k\frac{T_1^p - T_0^p}{\Delta x} = \rho\frac{\Delta x}{2}c\frac{T_0^{p+1} - T_0^p}{\Delta t}$$

The solution may proceed in the usual manner, although the effect of a radiative Biot number ($Bi_r \equiv h_r\,\Delta x/k$) must be included in the stability criterion, and the value of h_r must be updated at each step in the calculations. If the implicit method is used, h_r is calculated at $p + 1$, in which case an iterative calculation must be made at each time step.

5. This problem can be solved using *Tools, Finite-Difference Equations, One-Dimensional, Transient* in the *Advanced* section of *IHT*. This example is also included in *FEHT* as a solved model accessed through the *Toolbar* menu, *Examples*. The input screen summarizes key pre- and postprocessing steps, as well as results for nodal spacings of 1 and 0.125 mm. As an exercise, press *Run* to solve for the nodal temperatures, and in the *View* menu, select *Temperature Contours* to represent the temperature field in the form of isotherms.

5.11 *Summary*

Transient conduction occurs in numerous engineering applications and may be treated using different methods. There is certainly much to be said for simplicity, in which case, when confronted with a transient problem, the first thing you should do is calculate the Biot number. If this number is much less than unity, you may use the lumped capacitance method to obtain accurate results with minimal computational requirements. However, if the Biot number is not much less than unity, spatial effects must be considered, and some other method must be used. Analytical results are available in convenient graphical and equation form for the plane wall, the infinite cylinder, the sphere, and the semi-infinite solid. You should know when and how to use these results. If geometrical complexities and/or the form of the boundary conditions preclude their use, recourse must be made to an approximate numerical technique, such as the finite-difference method.

You may test your understanding of key concepts by addressing the following questions:

- Under what conditions may the *lumped capacitance method* be used to predict the transient response of a solid to a change in its thermal environment?
- What is the physical interpretation of the *Biot number*?
- Is the lumped capacitance method of analysis likely to be more applicable for a hot solid being cooled by forced convection in air or in water? By forced convection in air or natural convection in air?
- Is the lumped capacitance method of analysis likely to be more applicable for cooling of a hot solid made of copper or aluminum? For silicon nitride or glass?
- What parameters determine the *time constant* associated with the transient thermal response of a lumped capacitance solid? Is this response accelerated or decelerated by an increase in the convection coefficient? By an increase in the density or specific heat of the solid?
- For one-dimensional, transient conduction in a plane wall, a long cylinder, or a sphere with surface convection, what dimensionless parameters may be used to simplify the representation of thermal conditions? How are these parameters defined?
- Why is the semi-infinite solution applicable to any geometry at early times?
- What is the physical interpretation of the *Fourier number*?
- What requirement must be satisfied for use of a *one-term approximation* to determine the transient thermal response of a plane wall, a long cylinder, or a sphere experiencing one-dimensional conduction due to a change in surface conditions? At what stage of a transient process is the requirement not satisfied?
- What does transient heating or cooling of a plane wall with equivalent convection conditions at opposite surfaces have in common with a plane wall heated or cooled by convection at one surface and well insulated at the other surface?

- How may a one-term approximation be used to determine the transient thermal response of a plane wall, long cylinder, or sphere subjected to a sudden change in surface temperature?
- For one-dimensional, transient conduction, what is implied by the idealization of a *semi-infinite* solid? Under what conditions may the idealization be applied to a plane wall?
- What differentiates an *explicit*, finite-difference solution to a transient conduction problem from an *implicit* solution?
- What is meant by characterization of the implicit finite-difference method *as unconditionally stable*? What constraint is placed on the explicit method to ensure a stable solution?

References

1. Bergman, T. L., *J. Heat Transfer,* **130**, 094503, 2008.
2. Peleg, M., *Food Res. Int.,* **33**, 531–538, 2000.
3. Carslaw, H. S., and J. C. Jaeger, *Conduction of Heat in Solids,* 2nd ed., Oxford University Press, London, 1986.
4. Schneider, P. J., *Conduction Heat Transfer,* Addison-Wesley, Reading, MA, 1957.
5. Kakac, S., and Y. Yener, *Heat Conduction,* Taylor & Francis, Washington, DC, 1993.
6. Poulikakos, D., *Conduction Heat Transfer,* Prentice-Hall, Englewood Cliffs, NJ, 1994.
7. Yovanovich, M. M., "Conduction and Thermal Contact Resistances (Conductances)," in W. M. Rohsenow, J. P. Hartnett, and Y. I. Cho, Eds., *Handbook of Heat Transfer,* McGraw-Hill, New York, 1998, pp. 3.1–3.73.
8. Lavine, A. S., and T. L. Bergman, *J. Heat Transfer,* **130**, 101302, 2008.
9. Hirsch, L. R., R. J. Stafford, J. A. Bankson, S. R. Sershen, B. Rivera, R. E. Price, J. D. Hazle, N. J. Halas, and J. L. West, *Proc. Nat. Acad. Sciences of the U.S.*, **100**, 13549–13554, 2003.
10. Cahill, D. G., *Rev. Sci. Instrum.*, **61**, 802–808, 1990.

Problems

Qualitative Considerations

5.1 Consider a thin electrical heater attached to a plate and backed by insulation. Initially, the heater and plate are at the temperature of the ambient air, T_∞. Suddenly, the power to the heater is activated, yielding a constant heat flux q''_o (W/m^2) at the inner surface of the plate.

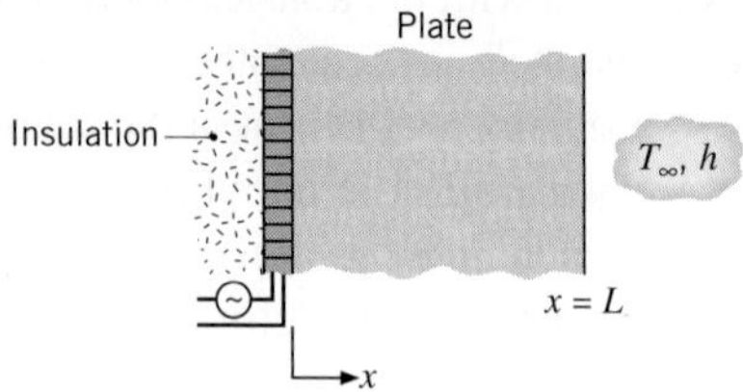

(a) Sketch and label, on $T - x$ coordinates, the temperature distributions: initial, steady-state, and at two intermediate times.

(b) Sketch the heat flux at the outer surface $q''_x(L, t)$ as a function of time.

5.2 The inner surface of a plane wall is insulated while the outer surface is exposed to an airstream at T_∞. The wall is at a uniform temperature corresponding to that of the airstream. Suddenly, a radiation heat source is switched on, applying a uniform flux q''_o to the outer surface.

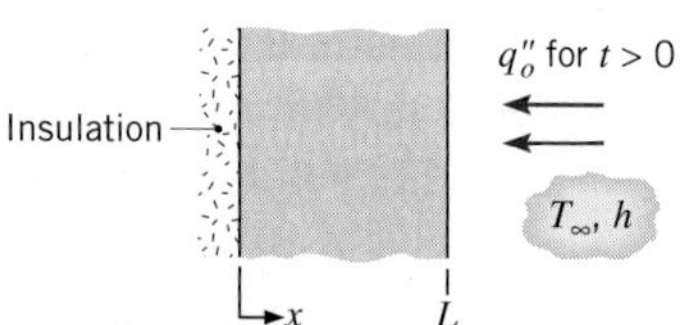

(a) Sketch and label, on $T - x$ coordinates, the temperature distributions: initial, steady-state, and at two intermediate times.

(b) Sketch the heat flux at the outer surface $q''_x(L, t)$ as a function of time.

5.3 A microwave oven operates on the principle that application of a high-frequency field causes electrically polarized molecules in food to oscillate. The net effect is a nearly *uniform generation* of thermal energy within the food. Consider the process of cooking a slab of beef of thickness $2L$ in a microwave oven and compare it with cooking in a conventional oven, where *each side* of the slab is *heated by radiation.* In each case the meat is to be heated from 0°C to a *minimum* temperature of 90°C. Base your comparison on a sketch of the temperature distribution at selected times for each of the cooking processes. In particular, consider the time t_0 at which heating is initiated, a time t_1 during the heating process, the time t_2 corresponding to the conclusion of heating, and a time t_3 well into the subsequent cooling process.

5.4 A plate of thickness $2L$, surface area A_s, mass M, and specific heat c_p, initially at a uniform temperature T_i, is suddenly heated on both surfaces by a convection process (T_∞, h) for a period of time t_o, following which the plate is insulated. Assume that the midplane temperature does not reach T_∞ within this period of time.

(a) Assuming $Bi \gg 1$ for the heating process, sketch and label, on $T-x$ coordinates, the following temperature distributions: initial, steady-state ($t \rightarrow \infty$), $T(x, t_o)$, and at two intermediate times between $t = t_o$ and $t \rightarrow \infty$.

(b) Sketch and label, on $T-t$ coordinates, the midplane and exposed surface temperature distributions.

(c) Repeat parts (a) and (b) assuming $Bi \ll 1$ for the plate.

(d) Derive an expression for the steady-state temperature $T(x, \infty) = T_f$, leaving your result in terms of plate parameters (M, c_p), thermal conditions (T_i, T_∞, h), the surface temperature $T(L, t)$, and the heating time t_o.

Lumped Capacitance Method

5.5 For each of the following cases, determine an appropriate characteristic length L_c and the corresponding Biot number Bi that is associated with the transient thermal response of the solid object. State whether the lumped capacitance approximation is valid. If temperature information is not provided, evaluate properties at $T = 300$ K.

(a) A toroidal shape of diameter $D = 50$ mm and cross-sectional area $A_c = 5\ \text{mm}^2$ is of thermal conductivity $k = 2.3$ W/m·K. The surface of the torus is exposed to a coolant corresponding to a convection coefficient of $h = 50\ \text{W/m}^2\cdot\text{K}$.

(b) A long, hot AISI 304 stainless steel bar of rectangular cross section has dimensions $w = 3$ mm, $W = 5$ mm, and $L = 100$ mm. The bar is subjected to a coolant that provides a heat transfer coefficient of $h = 15\ \text{W/m}^2\cdot\text{K}$ at all exposed surfaces.

(c) A long extruded aluminum (Alloy 2024) tube of inner and outer dimensions $w = 20$ mm and $W = 24$ mm, respectively, is suddenly submerged in water, resulting in a convection coefficient of $h = 37\ \text{W/m}^2\cdot\text{K}$ at the four exterior tube surfaces. The tube is plugged at both ends, trapping stagnant air inside the tube.

(d) An $L = 300$-mm-long solid stainless steel rod of diameter $D = 13$ mm and mass $M = 0.328$ kg is exposed to a convection coefficient of $h = 30\ \text{W/m}^2\cdot\text{K}$.

(e) A solid sphere of diameter $D = 12$ mm and thermal conductivity $k = 120$ W/m·K is suspended in a large vacuum oven with internal wall temperatures of $T_{sur} = 20$°C. The initial sphere temperature is $T_i = 100$°C, and its emissivity is $\varepsilon = 0.73$.

(f) A long cylindrical rod of diameter $D = 20$ mm, density $\rho = 2300\ \text{kg/m}^3$, specific heat $c_p = 1750$ J/kg·K, and thermal conductivity $k = 16$ W/m·K is suddenly exposed to convective conditions with $T_\infty = 20$°C. The rod is initially at a uniform temperature of $T_i = 200$°C and reaches a spatially averaged temperature of $T = 100$°C at $t = 225$ s.

(g) Repeat part (f) but now consider a rod diameter of $D = 200$ mm.

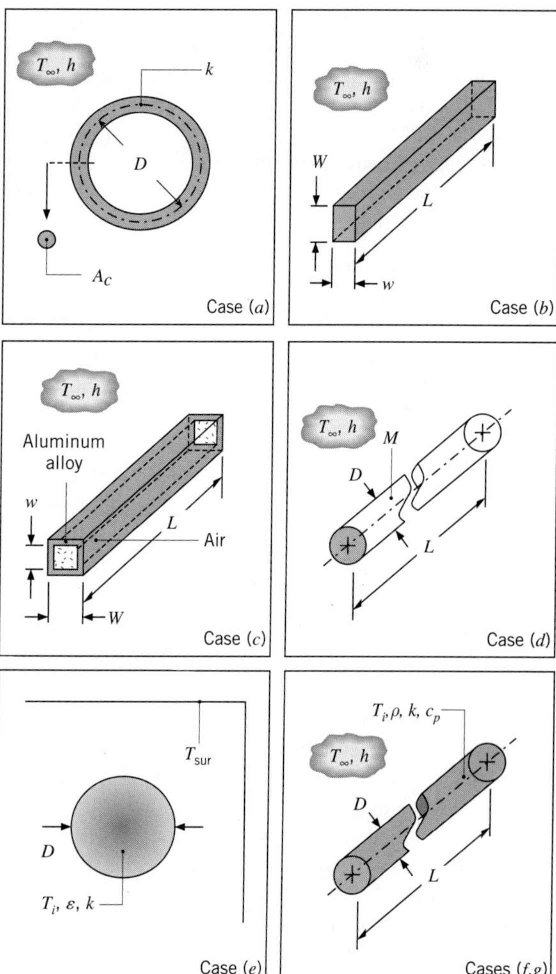

5.6 Steel balls 12 mm in diameter are annealed by heating to 1150 K and then slowly cooling to 400 K in an air environment for which $T_\infty = 325$ K and $h = 20\ \text{W/m}^2 \cdot \text{K}$. Assuming the properties of the steel to be $k = 40\ \text{W/m} \cdot \text{K}$, $\rho = 7800\ \text{kg/m}^3$, and $c = 600\ \text{J/kg} \cdot \text{K}$, estimate the time required for the cooling process.

5.7 Consider the steel balls of Problem 5.6, except now the air temperature increases with time as $T_\infty(t) = 325\ \text{K} + at$, where $a = 0.1875$ K/s.

(a) Sketch the ball temperature versus time for $0 \le t \le 1$ h. Also show the ambient temperature, T_∞, in your graph. Explain special features of the ball temperature behavior.

(b) Find an expression for the ball temperature as a function of time $T(t)$, and plot the ball temperature for $0 \le t \le 1$ h. Was your sketch correct?

5.8 The heat transfer coefficient for air flowing over a sphere is to be determined by observing the temperature–time history of a sphere fabricated from pure copper. The sphere, which is 12.7 mm in diameter, is at 66°C before it is inserted into an airstream having a temperature of 27°C. A thermocouple on the outer surface of the sphere indicates 55°C 69 s after the sphere is inserted into the airstream. Assume and then justify that the sphere behaves as a spacewise isothermal object and calculate the heat transfer coefficient.

5.9 A solid steel sphere (AISI 1010), 300 mm in diameter, is coated with a dielectric material layer of thickness 2 mm and thermal conductivity $0.04\ \text{W/m} \cdot \text{K}$. The coated sphere is initially at a uniform temperature of 500°C and is suddenly quenched in a large oil bath for which $T_\infty = 100°\text{C}$ and $h = 3300\ \text{W/m}^2 \cdot \text{K}$. Estimate the time required for the coated sphere temperature to reach 140°C. *Hint:* Neglect the effect of energy storage in the dielectric material, since its thermal capacitance (ρcV) is small compared to that of the steel sphere.

5.10 A flaked cereal is of thickness $2L = 1.2$ mm. The density, specific heat, and thermal conductivity of the flake are $\rho = 700\ \text{kg/m}^3$, $c_p = 2400\ \text{J/kg} \cdot \text{K}$, and $k = 0.34\ \text{W/m} \cdot \text{K}$, respectively. The product is to be baked by increasing its temperature from $T_i = 20°\text{C}$ to $T_f = 220°\text{C}$ in a convection oven, through which the product is carried on a conveyor. If the oven is $L_o = 3$ m long and the convection heat transfer coefficient at the product surface and oven air temperature are $h = 55\ \text{W/m}^2 \cdot \text{K}$ and $T_\infty = 300°\text{C}$, respectively, determine the required conveyor velocity, V. An engineer suggests that if the flake thickness is reduced to $2L = 1.0$ mm the conveyor velocity can be increased, resulting in higher productivity. Determine the required conveyor velocity for the thinner flake.

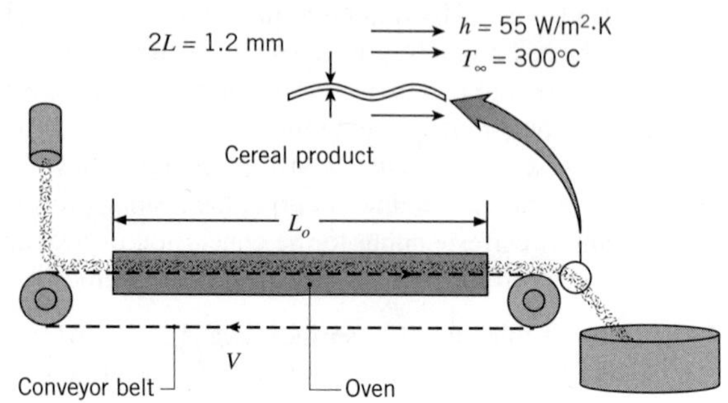

5.11 The base plate of an iron has a thickness of $L = 7$ mm and is made from an aluminum alloy ($\rho = 2800\ \text{kg/m}^3$, $c = 900\ \text{J/kg} \cdot \text{K}$, $k = 180\ \text{W/m} \cdot \text{K}$, $\varepsilon = 0.80$). An electric resistance heater is attached to the inner surface of the plate, while the outer surface is exposed to ambient air and large surroundings at $T_\infty = T_{sur} = 25°\text{C}$. The areas of both the inner and outer surfaces are $A_s = 0.040\ \text{m}^2$.

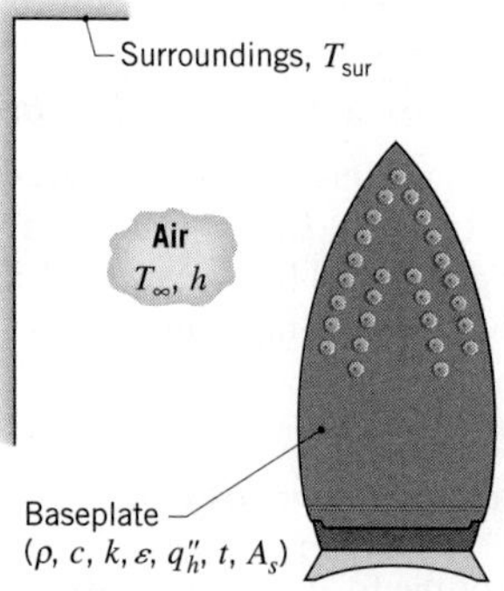

If an approximately uniform heat flux of $q''_h = 1.25 \times 10^4\ \text{W/m}^2$ is applied to the inner surface of the base plate and the convection coefficient at the outer surface is $h = 10\ \text{W/m}^2 \cdot \text{K}$, estimate the time required for the plate to reach a temperature of 135°C. *Hint:* Numerical integration is suggested in order to solve the problem.

5.12 Thermal energy storage systems commonly involve a *packed bed* of solid spheres, through which a hot gas flows if the system is being charged, or a cold gas if it is being discharged. In a charging process, heat transfer from the hot gas increases thermal energy stored within the colder spheres; during discharge, the stored energy decreases as heat is transferred from the warmer spheres to the cooler gas.

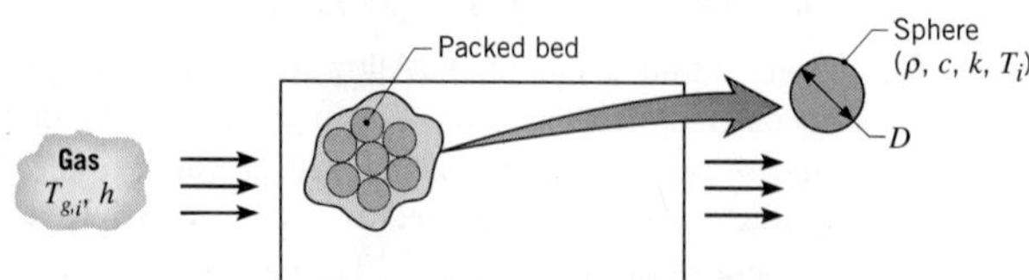

Consider a packed bed of 75-mm-diameter aluminum spheres ($\rho = 2700\ \text{kg/m}^3$, $c = 950\ \text{J/kg}\cdot\text{K}$, $k = 240\ \text{W/m}\cdot\text{K}$) and a charging process for which gas enters the storage unit at a temperature of $T_{g,i} = 300°\text{C}$. If the initial temperature of the spheres is $T_i = 25°\text{C}$ and the convection coefficient is $h = 75\ \text{W/m}^2\cdot\text{K}$, how long does it take a sphere near the inlet of the system to accumulate 90% of the maximum possible thermal energy? What is the corresponding temperature at the center of the sphere? Is there any advantage to using copper instead of aluminum?

5.13 A tool used for fabricating semiconductor devices consists of a chuck (thick metallic, cylindrical disk) onto which a very thin silicon wafer ($\rho = 2700\ \text{kg/m}^3$, $c = 875\ \text{J/kg}\cdot\text{K}$, $k = 177\ \text{W/m}\cdot\text{K}$) is placed by a robotic arm. Once in position, an electric field in the chuck is energized, creating an electrostatic force that holds the wafer firmly to the chuck. To ensure a reproducible thermal contact resistance between the chuck and the wafer from cycle to cycle, pressurized helium gas is introduced at the center of the chuck and flows (very slowly) radially outward between the asperities of the interface region.

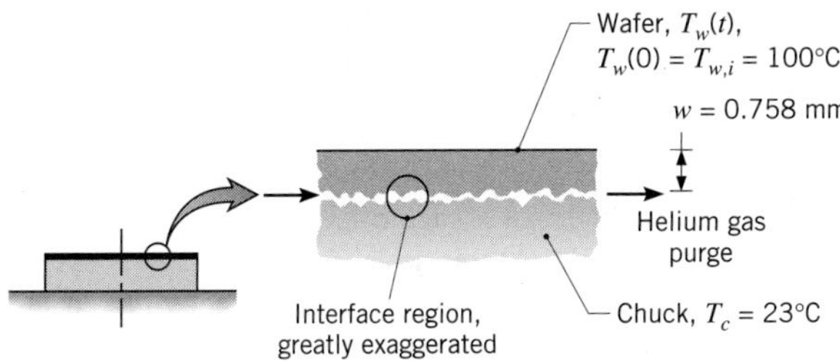

An experiment has been performed under conditions for which the wafer, initially at a uniform temperature $T_{w,i} = 100°\text{C}$, is suddenly placed on the chuck, which is at a uniform and constant temperature $T_c = 23°\text{C}$. With the wafer in place, the electrostatic force and the helium gas flow are applied. After 15 s, the temperature of the wafer is determined to be 33°C. What is the thermal contact resistance $R''_{t,c}$ ($\text{m}^2\cdot\text{K/W}$) between the wafer and chuck? Will the value of $R''_{t,c}$ increase, decrease, or remain the same if air, instead of helium, is used as the purge gas?

5.14 A copper sheet of thickness $2L = 2$ mm has an initial temperature of $T_i = 118°\text{C}$. It is suddenly quenched in liquid water, resulting in boiling at its two surfaces. For boiling, Newton's law of cooling is expressed as $q'' = h(T_s - T_{\text{sat}})$, where T_s is the solid surface temperature and T_{sat} is the saturation temperature of the fluid (in this case $T_{\text{sat}} = 100°\text{C}$). The convection heat transfer coefficient may be expressed as $h = 1010\ \text{W/m}^2\cdot\text{K}^3(T - T_{\text{sat}})^2$. Determine the time needed for the sheet to reach a temperature of $T = 102°\text{C}$. Plot the copper temperature versus time for $0 \le t \le 0.5$ s. On the same graph, plot the copper temperature history assuming the heat transfer coefficient is constant, evaluated at the average copper temperature $\overline{T} = 110°\text{C}$. Assume lumped capacitance behavior.

5.15 Carbon steel (AISI 1010) shafts of 0.1-m diameter are heat treated in a gas-fired furnace whose gases are at 1200 K and provide a convection coefficient of $100\ \text{W/m}^2\cdot\text{K}$. If the shafts enter the furnace at 300 K, how long must they remain in the furnace to achieve a centerline temperature of 800 K?

5.16 A thermal energy storage unit consists of a large rectangular channel, which is well insulated on its outer surface and encloses alternating layers of the storage material and the flow passage.

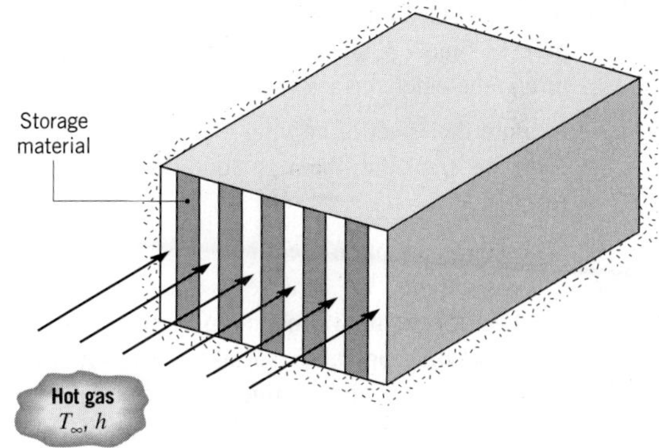

Each layer of the storage material is an aluminum slab of width $W = 0.05$ m, which is at an initial temperature of 25°C. Consider conditions for which the storage unit is charged by passing a hot gas through the passages, with the gas temperature and the convection coefficient assumed to have constant values of $T_\infty = 600°\text{C}$ and $h = 100\ \text{W/m}^2\cdot\text{K}$ throughout the channel. How long will it take to achieve 75% of the maximum possible energy storage? What is the temperature of the aluminum at this time?

5.17 Small spherical particles of diameter $D = 50\ \mu\text{m}$ contain a fluorescent material that, when irradiated with white light, emits at a wavelength corresponding to the material's temperature. Hence the color of the particle varies with its temperature. Because the small particles are *neutrally buoyant* in liquid water, a researcher wishes to use them to measure instantaneous local water temperatures in a turbulent flow by observing their emitted color. If the particles are characterized by a density, specific heat, and thermal conductivity of $\rho = 999\ \text{kg/m}^3$, $k = 1.2\ \text{W/m}\cdot\text{K}$, and $c_p = 1200\ \text{J/kg}\cdot\text{K}$, respectively, determine the time constant of the particles. *Hint*: Since the particles travel with the flow, heat transfer between the particle and the fluid occurs by conduction. Assume lumped capacitance behavior.

5.18 A spherical vessel used as a reactor for producing pharmaceuticals has a 5-mm-thick stainless steel wall ($k = 17$ W/m·K) and an inner diameter of $D_i = 1.0$ m. During production, the vessel is filled with reactants for which $\rho = 1100$ kg/m^3 and $c = 2400$ J/kg·K, while exothermic reactions release energy at a volumetric rate of $\dot{q} = 10^4$ W/m^3. As first approximations, the reactants may be assumed to be well stirred and the thermal capacitance of the vessel may be neglected.

(a) The exterior surface of the vessel is exposed to ambient air ($T_\infty = 25°$C) for which a convection coefficient of $h = 6$ W/m^2·K may be assumed. If the initial temperature of the reactants is 25°C, what is the temperature of the reactants after 5 h of process time? What is the corresponding temperature at the outer surface of the vessel?

(b) Explore the effect of varying the convection coefficient on transient thermal conditions within the reactor.

5.19 Batch processes are often used in chemical and pharmaceutical operations to achieve a desired chemical composition for the final product and typically involve a transient heating operation to take the product from room temperature to the desired process temperature. Consider a situation for which a chemical of density $\rho = 1200$ kg/m^3 and specific heat $c = 2200$ J/kg·K occupies a volume of $V = 2.25$ m^3 in an insulated vessel. The chemical is to be heated from room temperature, $T_i = 300$ K, to a process temperature of $T = 450$ K by passing saturated steam at $T_h = 500$ K through a coiled, thin-walled, 20-mm-diameter tube in the vessel. Steam condensation within the tube maintains an interior convection coefficient of $h_i = 10{,}000$ W/m^2·K, while the highly agitated liquid in the stirred vessel maintains an outside convection coefficient of $h_o = 2000$ W/m^2·K.

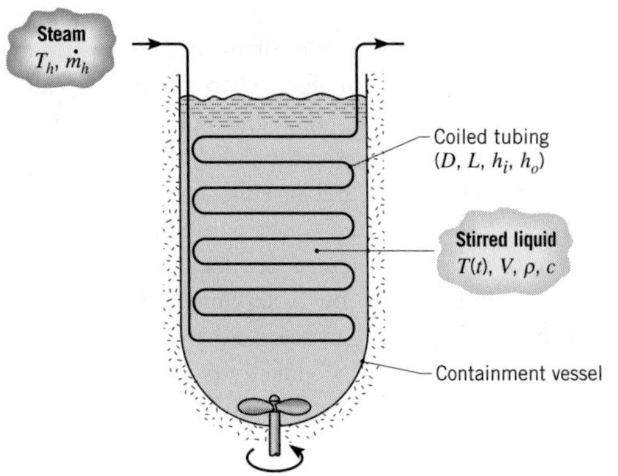

If the chemical is to be heated from 300 to 450 K in 60 min, what is the required length L of the submerged tubing?

5.20 An electronic device, such as a power transistor mounted on a finned heat sink, can be modeled as a spatially isothermal object with internal heat generation and an external convection resistance.

(a) Consider such a system of mass M, specific heat c, and surface area A_s, which is initially in equilibrium with the environment at T_∞. Suddenly, the electronic device is energized such that a constant heat generation $\dot{E}_g$(W) occurs. Show that the temperature response of the device is

$$\frac{\theta}{\theta_i} = \exp\left(-\frac{t}{RC}\right)$$

where $\theta \equiv T - T(\infty)$ and $T(\infty)$ is the steady-state temperature corresponding to $t \rightarrow \infty$; $\theta_i = T_i - T(\infty)$; T_i = initial temperature of device; R = thermal resistance $1/\bar{h}A_s$; and C = thermal capacitance Mc.

(b) An electronic device, which generates 60 W of heat, is mounted on an aluminum heat sink weighing 0.31 kg and reaches a temperature of 100°C in ambient air at 20°C under steady-state conditions. If the device is initially at 20°C, what temperature will it reach 5 min after the power is switched on?

5.21 *Molecular electronics* is an emerging field associated with computing and data storage utilizing energy transfer at the molecular scale. At this scale, thermal energy is associated exclusively with the vibration of molecular chains. The primary resistance to energy transfer in these proposed devices is the contact resistance at metal-molecule interfaces. To measure the contact resistance, individual molecules are *self-assembled* in a regular pattern onto a very thin gold substrate. The substrate is suddenly heated by a short pulse of laser irradiation, simultaneously transferring thermal energy to the molecules. The molecules vibrate rapidly in their "hot" state, and their vibrational intensity can be measured by detecting the randomness of the electric field produced by the molecule tips, as indicated by the dashed, circular lines in the schematic.

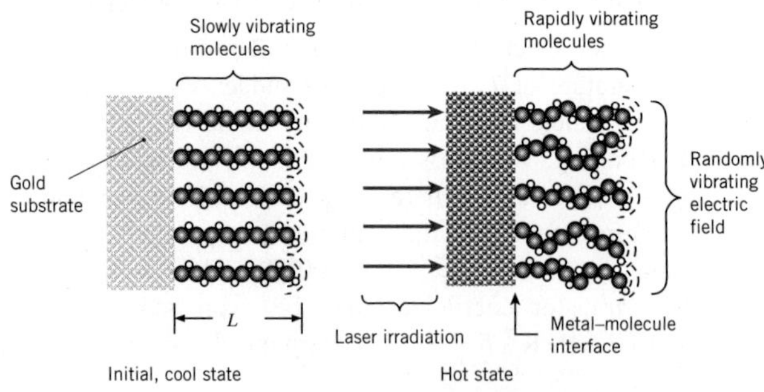

Molecules that are of density $\rho = 180\,\text{kg/m}^3$ and specific heat $c_p = 3000\,\text{J/kg}\cdot\text{K}$ have an initial, relaxed length of $L = 2\,\text{nm}$. The intensity of the molecular vibration increases exponentially from an initial value of I_i to a steady-state value of $I_{ss} > I_i$ with the time constant associated with the exponential response being $\tau_I = 5\,\text{ps}$. Assuming the intensity of the molecular vibration represents temperature on the molecular scale and that each molecule can be viewed as a cylinder of initial length L and cross-sectional area A_c, determine the thermal contact resistance, $R''_{t,c}$, at the metal–molecule interface.

5.22 A plane wall of a furnace is fabricated from plain carbon steel ($k = 60\,\text{W/m}\cdot\text{K}$, $\rho = 7850\,\text{kg/m}^3$, $c = 430\,\text{J/kg}\cdot\text{K}$) and is of thickness $L = 10\,\text{mm}$. To protect it from the corrosive effects of the furnace combustion gases, one surface of the wall is coated with a thin ceramic film that, for a unit surface area, has a thermal resistance of $R''_{t,f} = 0.01\,\text{m}^2\cdot\text{K/W}$. The opposite surface is well insulated from the surroundings.

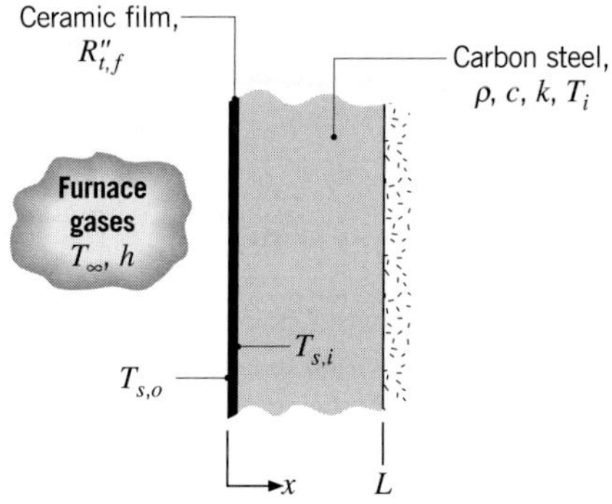

At furnace start-up the wall is at an initial temperature of $T_i = 300\,\text{K}$, and combustion gases at $T_\infty = 1300\,\text{K}$ enter the furnace, providing a convection coefficient of $h = 25\,\text{W/m}^2\cdot\text{K}$ at the ceramic film. Assuming the film to have negligible thermal capacitance, how long will it take for the inner surface of the steel to achieve a temperature of $T_{s,i} = 1200\,\text{K}$? What is the temperature $T_{s,o}$ of the exposed surface of the ceramic film at this time?

5.23 A steel strip of thickness $\delta = 12\,\text{mm}$ is annealed by passing it through a large furnace whose walls are maintained at a temperature T_w corresponding to that of combustion gases flowing through the furnace ($T_w = T_\infty$). The strip, whose density, specific heat, thermal conductivity, and emissivity are $\rho = 7900\,\text{kg/m}^3$, $c_p = 640\,\text{J/kg}\cdot\text{K}$, $k = 30\,\text{W/m}\cdot\text{K}$, and $\varepsilon = 0.7$, respectively, is to be heated from 300°C to 600°C.

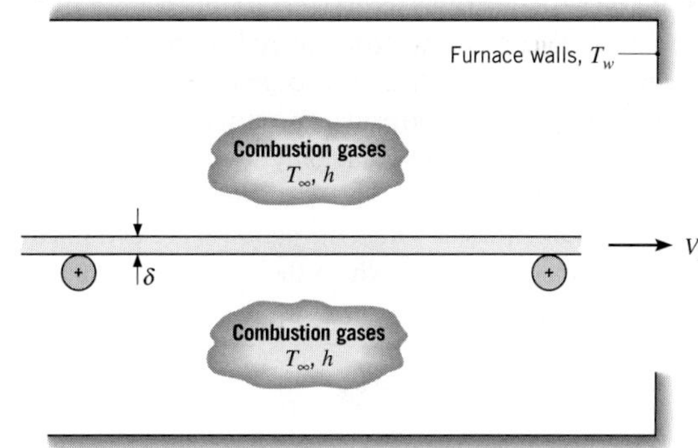

(a) For a uniform convection coefficient of $h = 100\,\text{W/m}^2\cdot\text{K}$ and $T_w = T_\infty = 700°\text{C}$, determine the time required to heat the strip. If the strip is moving at 0.5 m/s, how long must the furnace be?

(b) The annealing process may be accelerated (the strip speed increased) by increasing the environmental temperatures. For the furnace length obtained in part (a), determine the strip speed for $T_w = T_\infty = 850°\text{C}$ and $T_w = T_\infty = 1000°\text{C}$. For each set of environmental temperatures (700, 850, and 1000°C), plot the strip temperature as a function of time over the range $25°\text{C} \le T \le 600°\text{C}$. Over this range, also plot the radiation heat transfer coefficient, h_r, as a function of time.

5.24 In a material processing experiment conducted aboard the space shuttle, a coated niobium sphere of 10-mm diameter is removed from a furnace at 900°C and cooled to a temperature of 300°C. Although properties of the niobium vary over this temperature range, constant values may be assumed to a reasonable approximation, with $\rho = 8600\,\text{kg/m}^3$, $c = 290\,\text{J/kg}\cdot\text{K}$, and $k = 63\,\text{W/m}\cdot\text{K}$.

(a) If cooling is implemented in a large evacuated chamber whose walls are at 25°C, determine the time required to reach the final temperature if the coating is polished and has an emissivity of $\varepsilon = 0.1$. How long would it take if the coating is oxidized and $\varepsilon = 0.6$?

(b) To reduce the time required for cooling, consideration is given to immersion of the sphere in an inert gas stream for which $T_\infty = 25°\text{C}$ and $h = 200\,\text{W/m}^2\cdot\text{K}$. Neglecting radiation, what is the time required for cooling?

(c) Considering the effect of both radiation and convection, what is the time required for cooling if $h = 200\,\text{W/m}^2\cdot\text{K}$ and $\varepsilon = 0.6$? Explore the effect on the cooling time of independently varying h and ε.

5.25 Plasma spray-coating processes are often used to provide surface protection for materials exposed to hostile environments, which induce degradation through factors such as wear, corrosion, or outright thermal failure. *Ceramic* coatings are commonly used for this purpose. By injecting ceramic powder through the nozzle (anode) of a plasma torch, the particles are entrained by the plasma jet, within which they are then accelerated and heated.

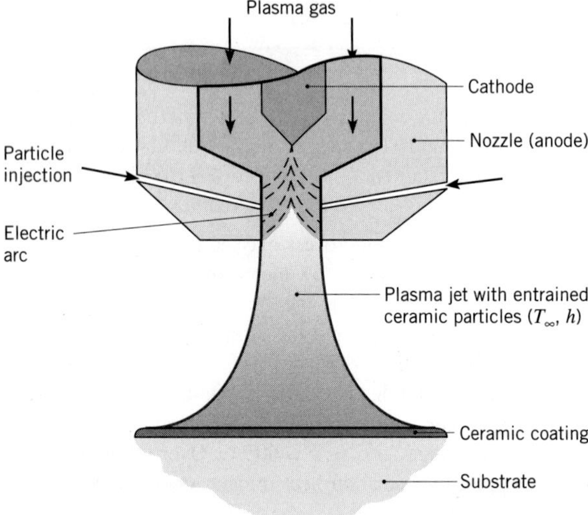

During their *time-in-flight,* the ceramic particles must be heated to their melting point and experience complete conversion to the liquid state. The coating is formed as the molten droplets impinge (*splat*) on the substrate material and experience rapid solidification. Consider conditions for which spherical alumina (Al_2O_3) particles of diameter $D_p = 50\ \mu\text{m}$, density $\rho_p = 3970\ \text{kg/m}^3$, thermal conductivity $k_p = 10.5\ \text{W/m}\cdot\text{K}$, and specific heat $c_p = 1560\ \text{J/kg}\cdot\text{K}$ are injected into an arc plasma, which is at $T_\infty = 10{,}000\ \text{K}$ and provides a coefficient of $h = 30{,}000\ \text{W/m}^2\cdot\text{K}$ for convective heating of the particles. The melting point and latent heat of fusion of alumina are $T_{\text{mp}} = 2318\ \text{K}$ and $h_{sf} = 3577\ \text{kJ/kg}$, respectively.

(a) Neglecting radiation, obtain an expression for the time-in-flight, t_{i-f}, required to heat a particle from its initial temperature T_i to its melting point T_{mp}, and, once at the melting point, for the particle to experience complete melting. Evaluate t_{i-f} for $T_i = 300\ \text{K}$ and the prescribed heating conditions.

(b) Assuming alumina to have an emissivity of $\varepsilon_p = 0.4$ and the particles to exchange radiation with large surroundings at $T_{\text{sur}} = 300\ \text{K}$, assess the validity of neglecting radiation.

5.26 The plasma spray-coating process of Problem 5.25 can be used to produce *nanostructured* ceramic coatings. Such coatings are characterized by low thermal conductivity, which is desirable in applications where the coating serves to protect the substrate from hot gases such as in a gas turbine engine. One method to produce a nanostructured coating involves spraying spherical particles, each of which is composed of agglomerated Al_2O_3 nanoscale granules. To form the coating, particles of diameter $D_p = 50\ \mu\text{m}$ must be *partially* molten when they strike the surface, with the liquid Al_2O_3 providing a means to adhere the ceramic material to the surface, and the unmelted Al_2O_3 providing the many grain boundaries that give the coating its low thermal conductivity. The boundaries between individual granules scatter phonons and reduce the thermal conductivity of the ceramic particle to $k_p = 5\ \text{W/m}\cdot\text{K}$. The density of the porous particle is reduced to $\rho = 3800\ \text{kg/m}^3$. All other properties and conditions are as specified in Problem 5.25.

(a) Determine the *time-in-flight* corresponding to 30% of the particle mass being melted.

(b) Determine the *time-in-flight* corresponding to the particle being 70% melted.

(c) If the particle is traveling at a velocity $V = 35\ \text{m/s}$, determine the *standoff distances* between the nozzle and the substrate associated with your answers in parts (a) and (b).

5.27 A chip that is of length $L = 5\ \text{mm}$ on a side and thickness $t = 1\ \text{mm}$ is encased in a ceramic substrate, and its exposed surface is convectively cooled by a dielectric liquid for which $h = 150\ \text{W/m}^2\cdot\text{K}$ and $T_\infty = 20°\text{C}$.

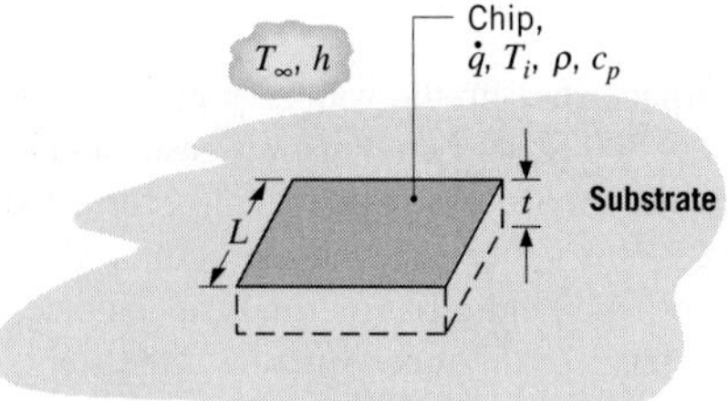

In the off-mode the chip is in thermal equilibrium with the coolant ($T_i = T_\infty$). When the chip is energized, however, its temperature increases until a new steady state is established. For purposes of analysis, the energized chip is characterized by uniform volumetric heating with $\dot{q} = 9 \times 10^6\ \text{W/m}^3$. Assuming an infinite contact resistance between the chip and substrate and negligible conduction resistance within the chip, determine the steady-state chip temperature T_f. Following activation of the chip, how long does it take to come within 1°C of this temperature? The chip density and specific heat are $\rho = 2000\ \text{kg/m}^3$ and $c = 700\ \text{J/kg}\cdot\text{K}$, respectively.

5.28 Consider the conditions of Problem 5.27. In addition to treating heat transfer by convection directly from the chip to the coolant, a more realistic analysis would account for indirect transfer from the chip to the substrate and then from the substrate to the coolant. The total thermal resistance associated with this indirect route includes contributions due to the chip–substrate interface (a contact resistance), multidimensional conduction in the substrate, and convection from the surface of the substrate to the coolant. If this total thermal resistance is $R_t = 200\ \text{K/W}$, what is the steady-state chip temperature T_f? Following activation of the chip, how long does it take to come within 1°C of this temperature?

5.29 A long wire of diameter $D = 1$ mm is submerged in an oil bath of temperature $T_\infty = 25°\text{C}$. The wire has an electrical resistance per unit length of $R'_e = 0.01\ \Omega/\text{m}$. If a current of $I = 100$ A flows through the wire and the convection coefficient is $h = 500\ \text{W/m}^2\cdot\text{K}$, what is the steady-state temperature of the wire? From the time the current is applied, how long does it take for the wire to reach a temperature that is within 1°C of the steady-state value? The properties of the wire are $\rho = 8000\ \text{kg/m}^3$, $c = 500\ \text{J/kg}\cdot\text{K}$, and $k = 20\ \text{W/m}\cdot\text{K}$.

5.30 Consider the system of Problem 5.1 where the temperature of the plate is spacewise isothermal during the transient process.

(a) Obtain an expression for the temperature of the plate as a function of time $T(t)$ in terms of q''_o, T_∞, h, L, and the plate properties ρ and c.

(b) Determine the thermal time constant and the steady-state temperature for a 12-mm-thick plate of pure copper when $T_\infty = 27°\text{C}$, $h = 50\ \text{W/m}^2\cdot\text{K}$, and $q''_o = 5000\ \text{W/m}^2$. Estimate the time required to reach steady-state conditions.

(c) For the conditions of part (b), as well as for $h = 100$ and $200\ \text{W/m}^2\cdot\text{K}$, compute and plot the corresponding temperature histories of the plate for $0 \le t \le 2500$ s.

5.31 Shape memory alloys (SMAs) are metals that undergo a change in crystalline structure within a relatively narrow temperature range. A phase transformation from *martensite* to *austenite* can induce relatively large changes in the overall dimensions of the SMA. Hence, SMAs can be employed as mechanical actuators. Consider an SMA rod that is initially $D_i = 2$ mm in diameter, $L_i = 40$ mm long, and at a uniform temperature of $T_i = 320$ K. The specific heat of the SMA varies significantly with changes in the crystalline phase, hence c varies with the temperature of the material. For a particular SMA, this relationship is well described by $c = 500\ \text{J/kg}\cdot\text{K} + 3630\ \text{J/kg}\cdot\text{K} \times e^{(-0.808\ \text{K}^{-1} \times |T - 336\text{K}|)}$. The density and thermal conductivity of the SMA material are $\rho = 8900\ \text{kg/m}^3$ and $k = 23\ \text{W/m}\cdot\text{K}$, respectively.

The SMA rod is exposed to a hot gas characterized by $T_\infty = 350$ K, $h = 250\ \text{W/m}^2\cdot\text{K}$. Plot the rod temperature versus time for $0 \le t \le 60$ s for the cases of variable and constant ($c = 500\ \text{J/kg}\cdot\text{K}$) specific heats. Determine the time needed for the rod temperature to experience 95% of its maximum temperature change. *Hint:* Neglect the change in the dimensions of the SMA rod when calculating the thermal response of the rod.

5.32 Before being injected into a furnace, pulverized coal is preheated by passing it through a cylindrical tube whose surface is maintained at $T_{sur} = 1000°\text{C}$. The coal pellets are suspended in an airflow and are known to move with a speed of 3 m/s. If the pellets may be approximated as spheres of 1-mm diameter and it may be assumed that they are heated by radiation transfer from the tube surface, how long must the tube be to heat coal entering at 25°C to a temperature of 600°C? Is the use of the lumped capacitance method justified?

5.33 As noted in Problem 5.3, microwave ovens operate by rapidly aligning and reversing water molecules within the food, resulting in volumetric energy generation and, in turn, cooking of the food. When the food is initially frozen, however, the water molecules do not readily oscillate in response to the microwaves, and the volumetric generation rates are between one and two orders of magnitude lower than if the water were in liquid form. (Microwave power that is not absorbed in the food is reflected back to the microwave generator, where it must be dissipated in the form of heat to prevent damage to the generator.)

(a) Consider a frozen, 1-kg spherical piece of ground beef at an initial temperature of $T_i = -20°\text{C}$ placed in a microwave oven with $T_\infty = 30°\text{C}$ and $h = 15\ \text{W/m}^2\cdot\text{K}$. Determine how long it will take the beef to reach a uniform temperature of $T = 0°\text{C}$, with all the water in the form of ice. Assume the properties of the beef are the same as ice, and assume 3% of the oven power ($P = 1$ kW total) is absorbed in the food.

(b) After all the ice is converted to liquid, determine how long it will take to heat the beef to $T_f = 80°\text{C}$ if 95% of the oven power is absorbed in the food. Assume the properties of the beef are the same as liquid water.

(c) When thawing food in microwave ovens, one may observe that some of the food may still be frozen while other parts of the food are overcooked.

Explain why this occurs. Explain why most microwave ovens have thaw cycles that are associated with very low oven powers.

5.34 A metal sphere of diameter D, which is at a uniform temperature T_i, is suddenly removed from a furnace and suspended from a fine wire in a large room with air at a uniform temperature T_∞ and the surrounding walls at a temperature T_{sur}.

(a) Neglecting heat transfer by radiation, obtain an expression for the time required to cool the sphere to some temperature T.

(b) Neglecting heat transfer by convection, obtain an expression for the time required to cool the sphere to the temperature T.

(c) How would you go about determining the time required for the sphere to cool to the temperature T if both convection and radiation are of the same order of magnitude?

(d) Consider an anodized aluminum sphere ($\varepsilon = 0.75$) 50 mm in diameter, which is at an initial temperature of $T_i = 800$ K. Both the air and surroundings are at 300 K, and the convection coefficient is 10 W/m$^2 \cdot$K. For the conditions of parts (a), (b), and (c), determine the time required for the sphere to cool to 400 K. Plot the corresponding temperature histories. Repeat the calculations for a polished aluminum sphere ($\varepsilon = 0.1$).

5.35 A horizontal structure consists of an $L_A = 10$-mm-thick layer of copper and an $L_B = 10$-mm-thick layer of aluminum. The bottom surface of the composite structure receives a heat flux of $q'' = 100$ kW/m^2, while the top surface is exposed to convective conditions characterized by $h = 40$ W/m$^2 \cdot$K, $T_\infty = 25$°C. The initial temperature of both materials is $T_{i,A} = T_{i,B} = 25$°C, and a contact resistance of $R''_{t,c} = 400 \times 10^{-6}$ m$^2 \cdot$K/W exists at the interface between the two materials.

(a) Determine the times at which the copper and aluminum each reach a temperature of $T_f = 90$°C. The copper layer is on the bottom.

(b) Repeat part (a) with the copper layer on the top.

Hint: Modify Equation 5.15 to include a term associated with heat transfer across the contact resistance. Apply the modified form of Equation 5.15 to each of the two slabs. See Comment 3 of Example 5.2.

5.36 As permanent space stations increase in size, there is an attendant increase in the amount of electrical power they dissipate. To keep station compartment temperatures from exceeding prescribed limits, it is necessary to transfer the dissipated heat to space. A novel heat rejection scheme that has been proposed for this purpose is termed a Liquid Droplet Radiator (LDR). The heat is first transferred to a high vacuum oil, which is then injected into outer space as a stream of small droplets. The stream is allowed to traverse a distance L, over which it cools by radiating energy to outer space at absolute zero temperature. The droplets are then collected and routed back to the space station.

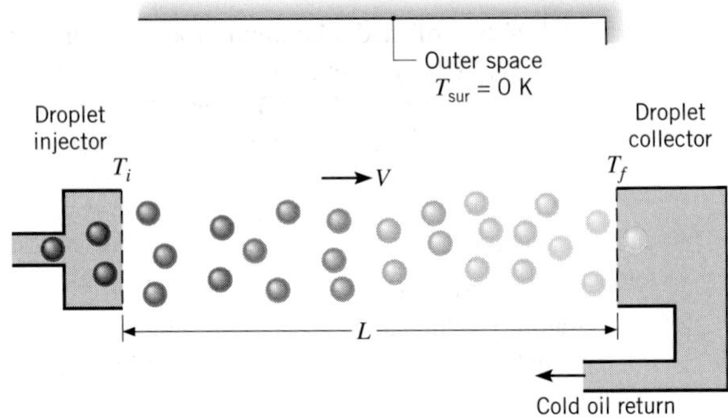

Consider conditions for which droplets of emissivity $\varepsilon = 0.95$ and diameter $D = 0.5$ mm are injected at a temperature of $T_i = 500$ K and a velocity of $V = 0.1$ m/s. Properties of the oil are $\rho = 885$ kg/m^3, $c = 1900$ J/kg$\cdot$K, and $k = 0.145$ W/m$\cdot$K. Assuming each drop to radiate to deep space at $T_{sur} = 0$ K, determine the distance L required for the droplets to impact the collector at a final temperature of $T_f = 300$ K. What is the amount of thermal energy rejected by each droplet?

5.37 Thin film coatings characterized by high resistance to abrasion and fracture may be formed by using microscale composite particles in a plasma spraying process. A spherical particle typically consists of a *ceramic core*, such as tungsten carbide (WC), and a *metallic shell*, such as cobalt (Co). The ceramic provides the thin film coating with its desired hardness at elevated temperatures, while the metal serves to coalesce the particles on the coated surface and to inhibit crack formation. In the plasma spraying process, the particles are injected into a plasma gas jet that heats them to a temperature above the melting point of the metallic casing and melts the casing before the particles impact the surface.

Consider spherical particles comprised of a WC core of diameter $D_i = 16$ μm, which is encased in a Co shell of outer diameter $D_o = 20$ μm. If the particles flow in a plasma gas at $T_\infty = 10{,}000$ K and the coefficient associated with convection from the gas to the particles is $h = 20{,}000$ W/m$^2 \cdot$K, how long does it take to heat the particles from an initial temperature of $T_i = 300$ K to the melting point of cobalt, $T_{mp} = 1770$ K? The density and specific heat of WC (the core of the particle) are $\rho_c = 16{,}000$ kg/m^3 and

$c_c = 300$ J/kg·K, while the corresponding values for Co (the outer shell) are $\rho_s = 8900$ kg/m^3 and $c_s = 750$ J/kg·K. Once having reached the melting point, how much additional time is required to completely melt the cobalt if its latent heat of fusion is $h_{sf} = 2.59 \times 10^5$ J/kg? You may use the lumped capacitance method of analysis and neglect radiation exchange between the particle and its surroundings.

5.38 A long, highly polished aluminum rod of diameter $D = 35$ mm is hung horizontally in a large room. The initial rod temperature is $T_i = 90°C$, and the room air is $T_\infty = 20°C$. At time $t_1 = 1250$ s, the rod temperature is $T_1 = 65°C$, and, at time $t_2 = 6700$ s, the rod temperature is $T_2 = 30°C$. Determine the values of the constants C and n that appear in Equation 5.26. Plot the rod temperature versus time for $0 \leq t \leq 10{,}000$ s. On the same graph, plot the rod temperature versus time for a constant value of the convection heat transfer coefficient, evaluated at a rod temperature of $\overline{T} = (T_i + T_\infty)/2$. For all cases, evaluate properties at $\overline{T} = (T_i + T_\infty)/2$.

5.39 Thermal stress testing is a common procedure used to assess the reliability of an electronic package. Typically, thermal stresses are induced in soldered or wired connections to reveal mechanisms that could cause failure and must therefore be corrected before the product is released. As an example of the procedure, consider an array of silicon chips ($\rho_{ch} = 2300$ kg/m^3, $c_{ch} = 710$ J/kg·K) joined to an alumina substrate ($\rho_{sb} = 4000$ kg/m^3, $c_{sb} = 770$ J/kg·K) by solder balls ($\rho_{sd} = 11{,}000$ kg/m^3, $c_{sd} = 130$ J/kg·K). Each chip of width L_{ch} and thickness t_{ch} is joined to a unit substrate section of width L_{sb} and thickness t_{sb} by solder balls of diameter D.

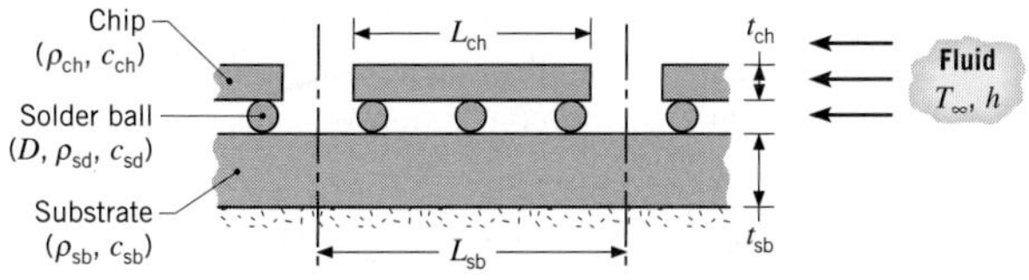

A thermal stress test begins by subjecting the multichip module, which is initially at room temperature, to a hot fluid stream and subsequently cooling the module by exposing it to a cold fluid stream. The process is repeated for a prescribed number of cycles to assess the integrity of the soldered connections.

(a) As a first approximation, assume that there is negligible heat transfer between the components (chip/solder/substrate) of the module and that the thermal response of each component may be determined from a lumped capacitance analysis involving the same convection coefficient h. Assuming no reduction in surface area due to contact between a solder ball and the chip or substrate, obtain expressions for the thermal time constant of each component. Heat transfer is to all surfaces of a chip, but to only the top surface of the substrate. Evaluate the three time constants for $L_{ch} = 15$ mm, $t_{ch} = 2$ mm, $L_{sb} = 25$ mm, $t_{sb} = 10$ mm, $D = 2$ mm, and a value of $h = 50$ W/m^2·K, which is characteristic of an airstream. Compute and plot the temperature histories of the three components for the heating portion of a cycle, with $T_i = 20°C$ and $T_\infty = 80°C$. At what time does each component experience 99% of its maximum possible temperature rise, that is, $(T - T_i)/(T_\infty - T_i) = 0.99$? If the maximum stress on a solder ball corresponds to the maximum difference between its temperature and that of the chip or substrate, when will this maximum occur?

(b) To reduce the time required to complete a stress test, a dielectric liquid could be used in lieu of air to provide a larger convection coefficient of $h = 200$ W/m^2·K. What is the corresponding savings in time for each component to achieve 99% of its maximum possible temperature rise?

5.40 The objective of this problem is to develop thermal models for estimating the steady-state temperature and the transient temperature history of the electrical transformer shown.

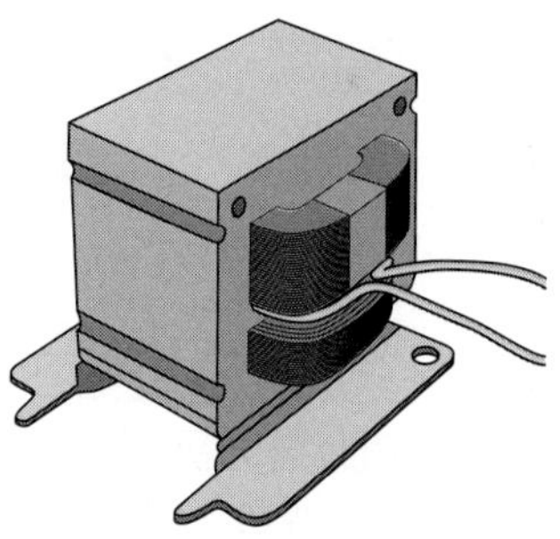

The external transformer geometry is approximately cubical, with a length of 32 mm to a side. The combined mass of the iron and copper in the transformer is 0.28 kg, and its weighted-average specific heat is 400 J/kg·K. The transformer dissipates 4.0 W and is operating in ambient air at $T_\infty = 20°C$, with a convection coefficient of 10 W/m^2·K. List and justify the assumptions made in your analysis, and discuss limitations of the models.

(a) Beginning with a properly defined control volume, develop a model for estimating the steady-state temperature of the transformer, $T(\infty)$. Evaluate $T(\infty)$ for the prescribed operating conditions.

(b) Develop a model for estimating the thermal response (temperature history) of the transformer if it is initially at a temperature of $T_i = T_\infty$ and power is suddenly applied. Determine the time required for the transformer to come within 5°C of its steady-state operating temperature.

5.41 In thermomechanical data storage, a processing head, consisting of M heated cantilevers, is used to write data onto an underlying polymer storage medium. Electrical resistance heaters are microfabricated onto each cantilever, which continually travel over the surface of the medium. The resistance heaters are turned on and off by controlling electrical current to each cantilever. As a cantilever goes through a complete heating and cooling cycle, the underlying polymer is softened, and one bit of data is written in the form of a *surface pit* in the polymer. A track of individual data bits (pits), each separated by approximately 50 nm, can be fabricated. Multiple tracks of bits, also separated by approximately 50 nm, are subsequently fabricated into the surface of the storage medium. Consider a single cantilever that is fabricated primarily of silicon with a mass of 50×10^{-18} kg and a surface area of 600×10^{-15} m^2. The cantilever is initially at $T_i = T_\infty = 300$ K, and the heat transfer coefficient between the cantilever and the ambient is 200×10^3 W/m$^2 \cdot$K.

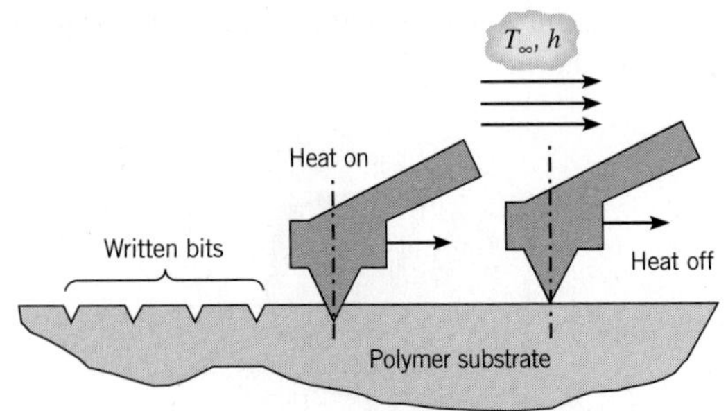

(a) Determine the ohmic heating required to raise the cantilever temperature to $T = 1000$ K within a heating time of $t_h = 1$ μs. *Hint*: See Problem 5.20.

(b) Find the time required to cool the cantilever from 1000 K to 400 K (t_c) and the thermal processing time required for one complete heating and cooling cycle, $t_p = t_h + t_c$.

(c) Determine how many bits (N) can be written onto a 1 mm × 1 mm polymer storage medium. If $M = 100$ cantilevers are ganged onto a single processing head, determine the total thermal processing time needed to write the data.

5.42 The melting of water initially at the fusion temperature, $T_f = 0$°C, was considered in Example 1.6. Freezing of water often occurs at 0°C. However, pure liquids that undergo a cooling process can remain in a *supercooled* liquid state well below their equilibrium freezing temperature, T_f, particularly when the liquid is not in contact with any solid material. Droplets of liquid water in the atmosphere have a supercooled freezing temperature, $T_{f,sc}$, that can be well correlated to the droplet diameter by the expression $T_{f,sc} = -28 + 0.87 \ln(D_p)$ in the diameter range $10^{-7} < D_p < 10^{-2}$ m, where $T_{f,sc}$ has units of degrees Celsius and D_p is expressed in units of meters. For a droplet of diameter $D = 50$ μm and initial temperature $T_i = 10$°C subject to ambient conditions of $T_\infty = -40$°C and $h = 900$ W/m$^2 \cdot$K, compare the time needed to completely solidify the droplet for case A, when the droplet solidifies at $T_f = 0$°C, and case B, when the droplet starts to freeze at $T_{f,sc}$. Sketch the temperature histories from the initial time to the time when the droplets are completely solid. *Hint:* When the droplet reaches $T_{f,sc}$ in case B, rapid solidification occurs during which the latent energy released by the freezing water is absorbed by the remaining liquid in the drop. As soon as any ice is formed within the droplet, the remaining liquid is in contact with a solid (the ice) and the freezing temperature immediately shifts from $T_{f,sc}$ to $T_f = 0$°C.

One-Dimensional Conduction: The Plane Wall

5.43 Consider the series solution, Equation 5.42, for the plane wall with convection. Calculate midplane ($x^* = 0$) and surface ($x^* = 1$) temperatures θ^* for $Fo = 0.1$ and 1, using $Bi = 0.1$, 1, and 10. Consider only the first four eigenvalues. Based on these results, discuss the validity of the approximate solutions, Equations 5.43 and 5.44.

5.44 Consider the one-dimensional wall shown in the sketch, which is initially at a uniform temperature T_i and is suddenly subjected to the convection boundary condition with a fluid at T_∞.

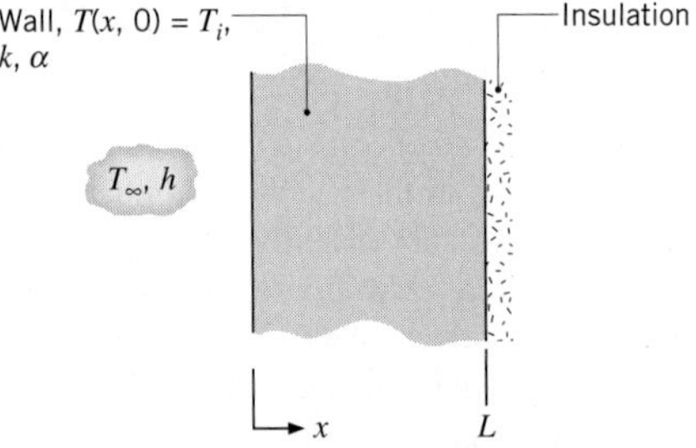

For a particular wall, case 1, the temperature at $x = L_1$ after $t_1 = 100$ s is $T_1(L_1, t_1) = 315$°C. Another wall, case 2, has different thickness and thermal conditions as shown.

Case	L (m)	α (m²/s)	k (W/m · K)	T_i (°C)	T_∞ (°C)	h (W/m² · K)
1	0.10	15×10^{-6}	50	300	400	200
2	0.40	25×10^{-6}	100	30	20	100

How long will it take for the second wall to reach 28.5°C at the position $x = L_2$? Use as the basis for analysis, the dimensionless functional dependence for the transient temperature distribution expressed in Equation 5.41.

5.45 Copper-coated, epoxy-filled fiberglass circuit boards are treated by heating a stack of them under high pressure, as shown in the sketch. The purpose of the pressing–heating operation is to cure the epoxy that bonds the fiberglass sheets, imparting stiffness to the boards. The stack, referred to as a *book,* is comprised of 10 *boards* and 11 *pressing plates,* which prevent epoxy from flowing between the boards and impart a smooth finish to the cured boards. In order to perform simplified thermal analyses, it is reasonable to approximate the book as having an effective thermal conductivity (k) and an effective thermal capacitance (ρc_p). Calculate the effective properties if each of the boards and plates has a thickness of 2.36 mm and the following thermophysical properties: board (b) $\rho_b = 1000\ \text{kg/m}^3$, $c_{p,b} = 1500\ \text{J/kg}\cdot\text{K}$, $k_b = 0.30\ \text{W/m}\cdot\text{K}$; plate ($p$) $\rho_p = 8000\ \text{kg/m}^3$, $c_{p,p} = 480\ \text{J/kg}\cdot\text{K}$, $k_p = 12\ \text{W/m}\cdot\text{K}$.

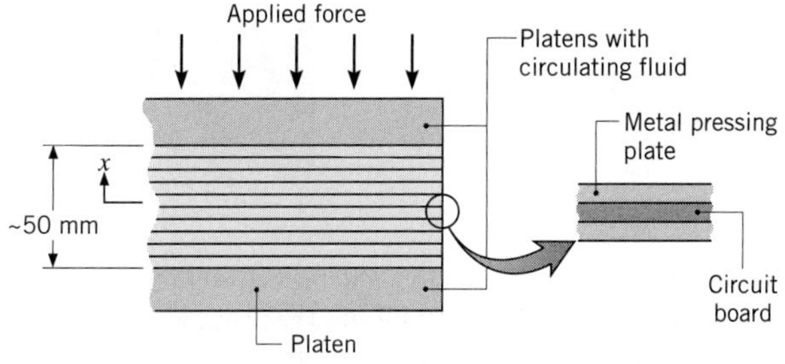

5.46 Circuit boards are treated by heating a stack of them under high pressure, as illustrated in Problem 5.45. The platens at the top and bottom of the stack are maintained at a uniform temperature by a circulating fluid. The purpose of the pressing–heating operation is to cure the epoxy, which bonds the fiberglass sheets, and impart stiffness to the boards. The cure condition is achieved when the epoxy has been maintained at or above 170°C for at least 5 min. The effective thermophysical properties of the stack or *book* (boards and metal pressing plates) are $k = 0.613\ \text{W/m}\cdot\text{K}$ and $\rho c_p = 2.73 \times 10^6\ \text{J/m}^3\cdot\text{K}$.

(a) If the book is initially at 15°C and, following application of pressure, the platens are suddenly brought to a uniform temperature of 190°C, calculate the elapsed time t_e required for the midplane of the book to reach the cure temperature of 170°C.

(b) If, at this instant of time, $t = t_e$, the platen temperature were reduced suddenly to 15°C, how much energy would have to be removed from the book by the coolant circulating in the platen, in order to return the stack to its initial uniform temperature?

5.47 A constant-property, one-dimensional plane slab of width $2L$, initially at a uniform temperature, is heated convectively with $Bi = 1$.

(a) At a dimensionless time of Fo_1, heating is suddenly stopped, and the slab of material is quickly covered with insulation. Sketch the dimensionless surface and midplane temperatures of the slab as a function of dimensionless time over the range $0 < Fo < \infty$. By changing the duration of heating to Fo_2, the *steady-state* midplane temperature can be set equal to the midplane temperature at Fo_1. Is the value of Fo_2 equal to, greater than, or less than Fo_1? *Hint*: Assume both Fo_1 and Fo_2 are greater than 0.2.

(b) Letting $Fo_2 = Fo_1 + \Delta Fo$, derive an analytical expression for ΔFo, and evaluate ΔFo for the conditions of part (a).

(c) Evaluate ΔFo for $Bi = 0.01$, 0.1, 10, 100, and ∞ when Fo_1 and Fo_2 are both greater than 0.2.

5.48 Referring to the semiconductor processing tool of Problem 5.13, it is desired at some point in the manufacturing cycle to cool the chuck, which is made of aluminum alloy 2024. The proposed cooling scheme passes air at 20°C between the air-supply head and the chuck surface.

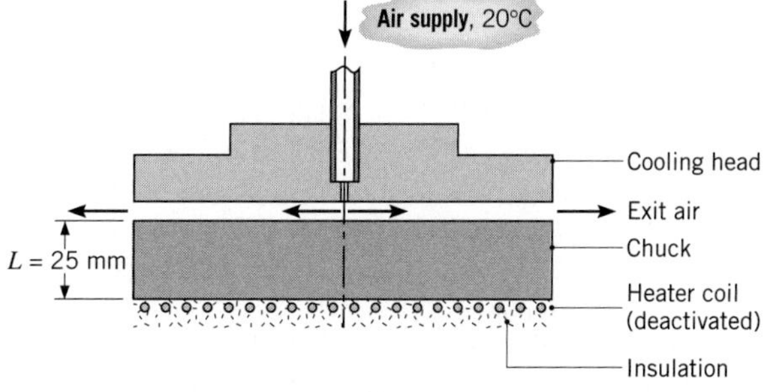

(a) If the chuck is initially at a uniform temperature of 100°C, calculate the time required for its lower surface to reach 25°C, assuming a uniform convection coefficient of $50\ \text{W/m}^2\cdot\text{K}$ at the head–chuck interface.

[(b)] Generate a plot of the time-to-cool as a function of the convection coefficient for the range

$10 \leq h \leq 2000\ \mathrm{W/m^2 \cdot K}$. If the lower limit represents a free convection condition without any head present, comment on the effectiveness of the head design as a method for cooling the chuck.

5.49 Annealing is a process by which steel is reheated and then cooled to make it less brittle. Consider the reheat stage for a 100-mm-thick steel plate ($\rho = 7830\ \mathrm{kg/m^3}$, $c = 550\ \mathrm{J/kg \cdot K}$, $k = 48\ \mathrm{W/m \cdot K}$), which is initially at a uniform temperature of $T_i = 200°\mathrm{C}$ and is to be heated to a minimum temperature of 550°C. Heating is effected in a gas-fired furnace, where products of combustion at $T_\infty = 800°\mathrm{C}$ maintain a convection coefficient of $h = 250\ \mathrm{W/m^2 \cdot K}$ on both surfaces of the plate. How long should the plate be left in the furnace?

5.50 Consider an acrylic sheet of thickness $L = 5$ mm that is used to coat a hot, isothermal metal substrate at $T_h = 300°\mathrm{C}$. The properties of the acrylic are $\rho = 1990\ \mathrm{kg/m^3}$, $c = 1470\ \mathrm{J/kg \cdot K}$, and $k = 0.21\ \mathrm{W/m \cdot K}$. Neglecting the thermal contact resistance between the acrylic and the metal substrate, determine how long it will take for the insulated back side of the acrylic to reach its softening temperature, $T_{soft} = 90°\mathrm{C}$. The initial acrylic temperature is $T_i = 20°\mathrm{C}$.

5.51 The 150-mm-thick wall of a gas-fired furnace is constructed of fireclay brick ($k = 1.5\ \mathrm{W/m \cdot K}$, $\rho = 2600\ \mathrm{kg/m^3}$, $c_p = 1000\ \mathrm{J/kg \cdot K}$) and is well insulated at its outer surface. The wall is at a uniform initial temperature of 20°C, when the burners are fired and the inner surface is exposed to products of combustion for which $T_\infty = 950°\mathrm{C}$ and $h = 100\ \mathrm{W/m^2 \cdot K}$.

(a) How long does it take for the outer surface of the wall to reach a temperature of 750°C?

(b) Plot the temperature distribution in the wall at the foregoing time, as well as at several intermediate times.

5.52 Steel is sequentially heated and cooled (*annealed*) to relieve stresses and to make it less brittle. Consider a 100-mm-thick plate ($k = 45\ \mathrm{W/m \cdot K}$, $\rho = 7800\ \mathrm{kg/m^3}$, $c_p = 500\ \mathrm{J/kg \cdot K}$) that is initially at a uniform temperature of 300°C and is heated (on both sides) in a gas-fired furnace for which $T_\infty = 700°\mathrm{C}$ and $h = 500\ \mathrm{W/m^2 \cdot K}$. How long will it take for a minimum temperature of 550°C to be reached in the plate?

5.53 Stone mix concrete slabs are used to absorb thermal energy from flowing air that is carried from a large concentrating solar collector. The slabs are heated during the day and release their heat to cooler air at night. If the daytime airflow is characterized by a temperature and convection heat transfer coefficient of $T_\infty = 200°\mathrm{C}$ and $h = 35\ \mathrm{W/m^2 \cdot K}$, respectively, determine the slab thickness $2L$ required to transfer a total amount of energy such that $Q/Q_o = 0.90$ over a $t = 8$-h period. The initial concrete temperature is $T_i = 40°\mathrm{C}$.

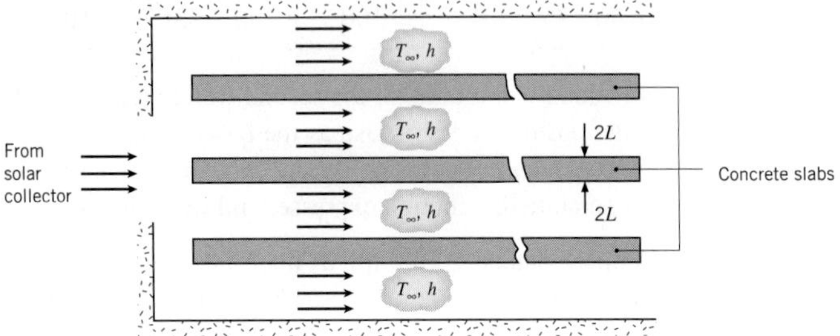

5.54 A plate of thickness $2L = 25$ mm at a temperature of 600°C is removed from a hot pressing operation and must be cooled rapidly to achieve the required physical properties. The process engineer plans to use air jets to control the rate of cooling, but she is uncertain whether it is necessary to cool both sides (case 1) or only one side (case 2) of the plate. The concern is not just for the *time-to-cool*, but also for the maximum temperature difference within the plate. If this temperature difference is too large, the plate can experience significant warping.

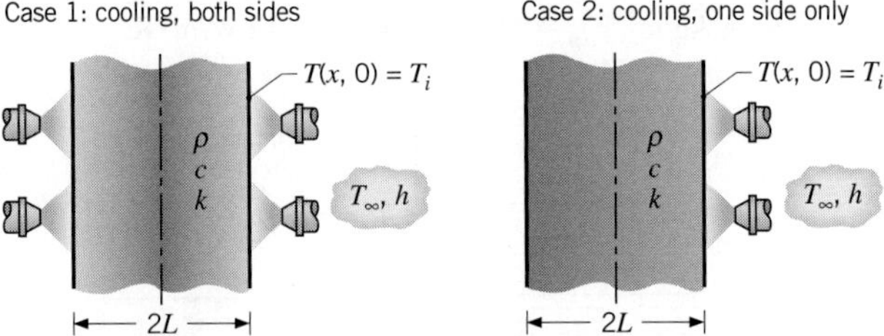

The air supply is at 25°C, and the convection coefficient on the surface is $400\ \mathrm{W/m^2 \cdot K}$. The thermophysical properties of the plate are $\rho = 3000\ \mathrm{kg/m^3}$, $c = 750\ \mathrm{J/kg \cdot K}$, and $k = 15\ \mathrm{W/m \cdot K}$.

(a) Using the *IHT* software, calculate and plot on one graph the temperature histories for cases 1 and 2 for a 500-s cooling period. Compare the times required for the maximum temperature in the plate to reach 100°C. Assume no heat loss from the unexposed surface of case 2.

(b) For both cases, calculate and plot on one graph the variation with time of the maximum temperature difference in the plate. Comment on the relative magnitudes of the temperature gradients within the plate as a function of time.

5.55 During transient operation, the steel nozzle of a rocket engine must not exceed a maximum allowable operating temperature of 1500 K when exposed to combustion gases characterized by a temperature of 2300 K and a convection coefficient of 5000 W/m^2·K. To extend the duration of engine operation, it is proposed that a ceramic *thermal barrier coating* (k = 10 W/m·K, $\alpha = 6 \times 10^{-6}$ m^2/s) be applied to the interior surface of the nozzle.

(a) If the ceramic coating is 10 mm thick and at an initial temperature of 300 K, obtain a conservative estimate of the maximum allowable duration of engine operation. The nozzle radius is much larger than the combined wall and coating thickness.

(b) Compute and plot the inner and outer surface temperatures of the coating as a function of time for $0 \leq t \leq 150$ s. Repeat the calculations for a coating thickness of 40 mm.

5.56 Two plates of the same material and thickness L are at different initial temperatures $T_{i,1}$ and $T_{i,2}$, where $T_{i,2} > T_{i,1}$. Their faces are suddenly brought into contact. The external surfaces of the two plates are insulated.

(a) Let a dimensionless temperature be defined as $T^*(Fo) \equiv (T - T_{i,1})/(T_{i,2} - T_{i,1})$. Neglecting the thermal contact resistance at the interface between the plates, what are the steady-state dimensionless temperatures of each of the two plates, $T^*_{ss,1}$ and $T^*_{ss,2}$? What is the dimensionless interface temperature T^*_{int} at any time?

(b) An effective overall heat transfer coefficient between the two plates can be defined based on the instantaneous, spatially averaged dimensionless plate temperatures, $U^*_{eff} \equiv q^*/(\bar{T}^*_2 - \bar{T}^*_1)$. Noting that a dimensionless heat transfer rate to or from either of the two plates may be expressed as $q^* = d(Q/Q_o)/dFo$, determine an expression for U^*_{eff} for $Fo > 0.2$.

5.57 In a tempering process, glass plate, which is initially at a uniform temperature T_i, is cooled by suddenly reducing the temperature of both surfaces to T_s. The plate is 20 mm thick, and the glass has a thermal diffusivity of 6×10^{-7} m^2/s.

(a) How long will it take for the midplane temperature to achieve 50% of its maximum possible temperature reduction?

(b) If $(T_i - T_s)$ = 300°C, what is the maximum temperature gradient in the glass at the time calculated in part (a)?

5.58 The strength and stability of tires may be enhanced by heating both sides of the rubber (k = 0.14 W/m·K, $\alpha = 6.35 \times 10^{-8}$ m^2/s) in a steam chamber for which T_∞ = 200°C. In the heating process, a 20-mm-thick rubber wall (assumed to be untreaded) is taken from an initial temperature of 25°C to a midplane temperature of 150°C.

(a) If steam flow over the tire surfaces maintains a convection coefficient of h = 200 W/m^2·K, how long will it take to achieve the desired midplane temperature?

(b) To accelerate the heating process, it is recommended that the steam flow be made sufficiently vigorous to maintain the tire surfaces at 200°C throughout the process. Compute and plot the midplane and surface temperatures for this case, as well as for the conditions of part (a).

5.59 A plastic coating is applied to wood panels by first depositing molten polymer on a panel and then cooling the surface of the polymer by subjecting it to airflow at 25°C. As first approximations, the heat of reaction associated with solidification of the polymer may be neglected and the polymer/wood interface may be assumed to be adiabatic.

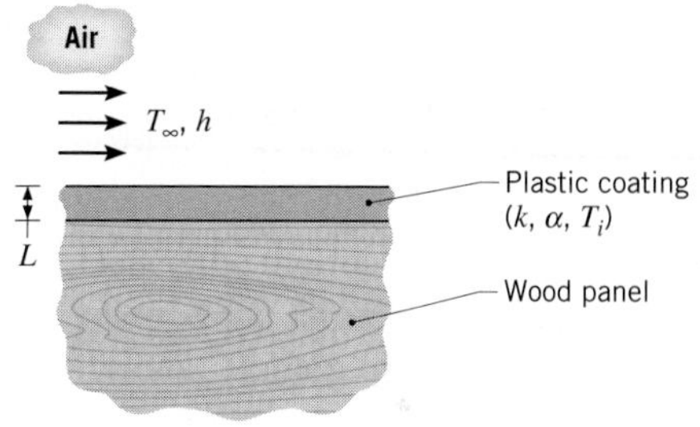

If the thickness of the coating is L = 2 mm and it has an initial uniform temperature of T_i = 200°C, how long will it take for the surface to achieve a *safe-to-touch* temperature of 42°C if the convection coefficient is h = 200 W/m^2·K? What is the corresponding value of the interface temperature? The thermal conductivity and diffusivity of the plastic are k = 0.25 W/m·K and $\alpha = 1.20 \times 10^{-7}$ m^2/s, respectively.

One-Dimensional Conduction: The Long Cylinder

5.60 A long rod of 60-mm diameter and thermophysical properties ρ = 8000 kg/m^3, c = 500 J/kg·K, and k = 50 W/m·K is initially at a uniform temperature and is heated in a forced convection furnace maintained at 750 K. The convection coefficient is estimated to be 1000 W/m^2·K.

(a) What is the centerline temperature of the rod when the surface temperature is 550 K?

(b) In a heat-treating process, the centerline temperature of the rod must be increased from $T_i = 300$ K to $T = 500$ K. Compute and plot the centerline temperature histories for $h = 100$, 500, and 1000 W/m$^2 \cdot$K. In each case the calculation may be terminated when $T = 500$ K.

5.61 A long cylinder of 30-mm diameter, initially at a uniform temperature of 1000 K, is suddenly quenched in a large, constant-temperature oil bath at 350 K. The cylinder properties are $k = 1.7$ W/m·K, $c = 1600$ J/kg·K, and $\rho = 400$ kg/m^3, while the convection coefficient is 50 W/m$^2 \cdot$K.

(a) Calculate the time required for the surface of the cylinder to reach 500 K.

(b) Compute and plot the surface temperature history for $0 \le t \le 300$ s. If the oil were agitated, providing a convection coefficient of 250 W/m$^2 \cdot$K, how would the temperature history change?

5.62 Work Problem 5.47 for a cylinder of radius r_o and length $L = 20\ r_o$.

5.63 A long pyroceram rod of diameter 20 mm is clad with a very thin metallic tube for mechanical protection. The bonding between the rod and the tube has a thermal contact resistance of $R'_{t,c} = 0.12$ m·K/W.

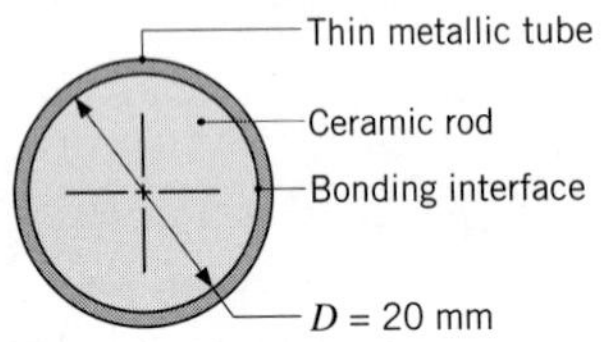

(a) If the rod is initially at a uniform temperature of 900 K and is suddenly cooled by exposure to an airstream for which $T_\infty = 300$ K and $h = 100$ W/m$^2 \cdot$K, at what time will the centerline reach 600 K?

(b) Cooling may be accelerated by increasing the airspeed and hence the convection coefficient. For values of $h = 100$, 500, and 1000 W/m$^2 \cdot$K, compute and plot the centerline and surface temperatures of the pyroceram as a function of time for $0 \le t \le 300$ s. Comment on the implications of achieving enhanced cooling solely by increasing h.

5.64 A long rod 40 mm in diameter, fabricated from sapphire (aluminum oxide) and initially at a uniform temperature of 800 K, is suddenly cooled by a fluid at 300 K having a heat transfer coefficient of 1600 W/m$^2 \cdot$K. After 35 s, the rod is wrapped in insulation and experiences no heat losses. What will be the temperature of the rod after a long period of time?

5.65 A cylindrical stone mix concrete beam of diameter $D = 0.5$ m initially at $T_i = 20$°C is exposed to hot gases at $T_\infty = 500$°C. The convection coefficient is $h = 10$ W/m$^2 \cdot$K.

(a) Determine the centerline temperature of the beam after an exposure time of $t = 6$ h.

(b) Determine the centerline temperature of a second beam that is of the same size and exposed to the same conditions as in part (a) but fabricated of lightweight aggregate concrete with density $\rho = 1495$ kg/m^3, thermal conductivity $k = 0.789$ W/m·K, and specific heat $c_p = 880$ J/kg·K.

5.66 A long plastic rod of 30-mm diameter ($k = 0.3$ W/m·K and $\rho c_p = 1040$ kJ/m$^3 \cdot$K) is uniformly heated in an oven as preparation for a pressing operation. For best results, the temperature in the rod should not be less than 200°C. To what uniform temperature should the rod be heated in the oven if, for the worst case, the rod sits on a conveyor for 3 min while exposed to convection cooling with ambient air at 25°C and with a convection coefficient of 8 W/m$^2 \cdot$K? A further condition for good results is a maximum–minimum temperature difference of less than 10°C. Is this condition satisfied? If not, what could you do to satisfy it?

5.67 As part of a heat treatment process, cylindrical, 304 stainless steel rods of 100-mm diameter are cooled from an initial temperature of 500°C by suspending them in an oil bath at 30°C. If a convection coefficient of 500 W/m$^2 \cdot$K is maintained by circulation of the oil, how long does it take for the centerline of a rod to reach a temperature of 50°C, at which point it is withdrawn from the bath? If 10 rods of length $L = 1$ m are processed per hour, what is the nominal rate at which energy must be extracted from the bath (the cooling load)?

5.68 In a manufacturing process, long rods of different diameters are at a uniform temperature of 400°C in a curing oven, from which they are removed and cooled by forced convection in air at 25°C. One of the line operators has observed that it takes 280 s for a 40-mm-diameter rod to cool to a *safe-to-handle* temperature of 60°C. For an equivalent convection coefficient, how long will it take for an 80-mm-diameter rod to cool to the same temperature? The thermophysical properties of the rod are $\rho = 2500$ kg/m^3, $c = 900$ J/kg·K, and $k = 15$ W/m·K. Comment on your result. Did you anticipate this outcome?

5.69 The density and specific heat of a particular material are known ($\rho = 1200$ kg/m^3, $c_p = 1250$ J/kg·K), but its thermal conductivity is unknown. To determine the thermal conductivity, a long cylindrical specimen of diameter $D = 40$ mm is machined, and a thermocouple is inserted through a small hole drilled along the centerline.

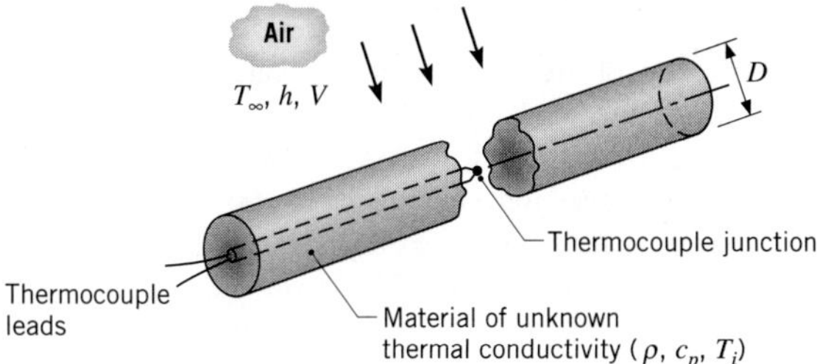

The thermal conductivity is determined by performing an experiment in which the specimen is heated to a uniform temperature of $T_i = 100°C$ and then cooled by passing air at $T_\infty = 25°C$ in cross flow over the cylinder. For the prescribed air velocity, the convection coefficient is $h = 55\ W/m^2 \cdot K$.

(a) If a centerline temperature of $T(0, t) = 40°C$ is recorded after $t = 1136$ s of cooling, verify that the material has a thermal conductivity of $k = 0.30\ W/m \cdot K$.

(b) For air in cross flow over the cylinder, the prescribed value of $h = 55\ W/m^2 \cdot K$ corresponds to a velocity of $V = 6.8$ m/s. If $h = CV^{0.618}$, where the constant C has units of $W \cdot s^{0.618}/m^{2.618} \cdot K$, how does the centerline temperature at $t = 1136$ s vary with velocity for $3 \leq V \leq 20$ m/s? Determine the centerline temperature histories for $0 \leq t \leq 1500$ s and velocities of 3, 10, and 20 m/s.

5.70 In Section 5.2 we noted that the value of the Biot number significantly influences the nature of the temperature distribution in a solid during a transient conduction process. Reinforce your understanding of this important concept by using the *IHT* model for one-dimensional transient conduction to determine radial temperature distributions in a 30-mm-diameter, stainless steel rod ($k = 15\ W/m \cdot K$, $\rho = 8000\ kg/m^3$, $c_p = 475\ J/kg \cdot K$), as it is cooled from an initial uniform temperature of 325°C by a fluid at 25°C. For the following values of the convection coefficient and the designated times, determine the radial temperature distribution: $h = 100\ W/m^2 \cdot K$ ($t = 0, 100, 500$ s); $h = 1000\ W/m^2 \cdot K$ ($t = 0, 10, 50$ s); $h = 5000\ W/m^2 \cdot K$ ($t = 0, 1, 5, 25$ s). Prepare a separate graph for each convection coefficient, on which temperature is plotted as a function of dimensionless radius at the designated times.

One-Dimensional Conduction: The Sphere

5.71 In heat treating to harden steel ball bearings ($c = 500\ J/kg \cdot K$, $\rho = 7800\ kg/m^3$, $k = 50\ W/m \cdot K$), it is desirable to increase the surface temperature for a short time without significantly warming the interior of the ball. This type of heating can be accomplished by sudden immersion of the ball in a molten salt bath with $T_\infty = 1300$ K and $h = 5000\ W/m^2 \cdot K$. Assume that any location within the ball whose temperature exceeds 1000 K will be hardened. Estimate the time required to harden the outer millimeter of a ball of diameter 20 mm, if its initial temperature is 300 K.

5.72 A cold air chamber is proposed for quenching steel ball bearings of diameter $D = 0.2$ m and initial temperature $T_i = 400°C$. Air in the chamber is maintained at $-15°C$ by a refrigeration system, and the steel balls pass through the chamber on a conveyor belt. Optimum bearing production requires that 70% of the initial thermal energy content of the ball above $-15°C$ be removed. Radiation effects may be neglected, and the convection heat transfer coefficient within the chamber is $1000\ W/m^2 \cdot K$. Estimate the residence time of the balls within the chamber, and recommend a drive velocity of the conveyor. The following properties may be used for the steel: $k = 50\ W/m \cdot K$, $\alpha = 2 \times 10^{-5}\ m^2/s$, and $c = 450\ J/kg \cdot K$.

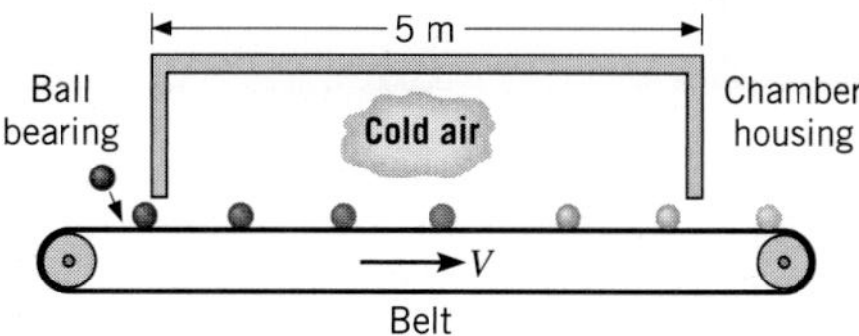

5.73 A soda lime glass sphere of diameter $D_1 = 25$ mm is encased in a bakelite spherical shell of thickness $L = 10$ mm. The composite sphere is initially at a uniform temperature, $T_i = 40°C$, and is exposed to a fluid at $T_\infty = 10°C$ with $h = 30\ W/m^2 \cdot K$. Determine the center temperature of the glass at $t = 200$ s. Neglect the thermal contact resistance at the interface between the two materials.

5.74 Stainless steel (AISI 304) ball bearings, which have uniformly been heated to 850°C, are hardened by quenching them in an oil bath that is maintained at 40°C. The ball diameter is 20 mm, and the convection coefficient associated with the oil bath is $1000\ W/m^2 \cdot K$.

(a) If quenching is to occur until the surface temperature of the balls reaches 100°C, how long must the balls be kept in the oil? What is the center temperature at the conclusion of the cooling period?

(b) If 10,000 balls are to be quenched per hour, what is the rate at which energy must be removed by the oil bath cooling system in order to maintain its temperature at 40°C?

5.75 A sphere 30 mm in diameter initially at 800 K is quenched in a large bath having a constant temperature of 320 K with a convection heat transfer coefficient of 75 W/m$^2 \cdot$K. The thermophysical properties of the sphere material are: $\rho = 400$ kg/m^3, $c = 1600$ J/kg$\cdot$K, and $k = 1.7$ W/m$\cdot$K.

(a) Show, in a qualitative manner on $T - t$ coordinates, the temperatures at the center and at the surface of the sphere as a function of time.

(b) Calculate the time required for the surface of the sphere to reach 415 K.

(c) Determine the heat flux (W/m^2) at the outer surface of the sphere at the time determined in part (b).

(d) Determine the energy (J) that has been lost by the sphere during the process of cooling to the surface temperature of 415 K.

(e) At the time determined by part (b), the sphere is quickly removed from the bath and covered with perfect insulation, such that there is no heat loss from the surface of the sphere. What will be the temperature of the sphere after a long period of time has elapsed?

(f) Compute and plot the center and surface temperature histories over the period $0 \leq t \leq 150$ s. What effect does an increase in the convection coefficient to $h = 200$ W/m$^2 \cdot$K have on the foregoing temperature histories? For $h = 75$ and 200 W/m$^2 \cdot$K, compute and plot the surface heat flux as a function of time for $0 \leq t \leq 150$ s.

5.76 Work Problem 5.47 for the case of a sphere of radius r_o.

5.77 Spheres A and B are initially at 800 K, and they are simultaneously quenched in large constant temperature baths, each having a temperature of 320 K. The following parameters are associated with each of the spheres and their cooling processes.

	Sphere A	Sphere B
Diameter (mm)	300	30
Density (kg/m^3)	1600	400
Specific heat (kJ/kg $\cdot$ K)	0.400	1.60
Thermal conductivity (W/m $\cdot$ K)	170	1.70
Convection coefficient (W/m^2 $\cdot$ K)	5	50

(a) Show in a qualitative manner, on $T - t$ coordinates, the temperatures at the center and at the surface for each sphere as a function of time. Briefly explain the reasoning by which you determine the relative positions of the curves.

(b) Calculate the time required for the surface of each sphere to reach 415 K.

(c) Determine the energy that has been gained by each of the baths during the process of the spheres cooling to 415 K.

5.78 Spheres of 40-mm diameter heated to a uniform temperature of 400°C are suddenly removed from the oven and placed in a forced-air bath operating at 25°C with a convection coefficient of 300 W/m$^2 \cdot$K on the sphere surfaces. The thermophysical properties of the sphere material are $\rho = 3000$ kg/m^3, $c = 850$ J/kg$\cdot$K, and $k = 15$ W/m$\cdot$K.

(a) How long must the spheres remain in the air bath for 80% of the thermal energy to be removed?

(b) The spheres are then placed in a packing carton that prevents further heat transfer to the environment. What uniform temperature will the spheres eventually reach?

5.79 To determine which parts of a spider's brain are triggered into neural activity in response to various optical stimuli, researchers at the University of Massachusetts Amherst desire to examine the brain as it is shown images that might evoke emotions such as fear or hunger. Consider a spider at $T_i = 20$°C that is shown a frightful scene and is then immediately immersed in liquid nitrogen at $T_\infty = 77$ K. The brain is subsequently dissected in its frozen state and analyzed to determine which parts of the brain reacted to the stimulus. Using your knowledge of heat transfer, determine how much time elapses before the spider's brain begins to freeze. Assume the brain is a sphere of diameter $D_b = 1$ mm, centrally located in the spider's cephalothorax, which may be approximated as a spherical shell of diameter $D_c = 3$ mm. The brain and cephalothorax properties correspond to those of liquid water. Neglect the effects of the latent heat of fusion and assume the heat transfer coefficient is $h = 100$ W/m$^2 \cdot$K.

5.80 Consider the packed bed operating conditions of Problem 5.12, but with Pyrex ($\rho = 2225$ kg/m^3, $c = 835$ J/kg$\cdot$K, $k = 1.4$ W/m$\cdot$K) used instead of aluminum. How long does it take a sphere near the inlet of the system to accumulate 90% of the maximum possible thermal energy? What is the corresponding temperature at the center of the sphere?

5.81 The convection coefficient for flow over a solid sphere may be determined by submerging the sphere, which is initially at 25°C, into the flow, which is at 75°C, and measuring its surface temperature at some time during the transient heating process.

(a) If the sphere has a diameter of 0.1 m, a thermal conductivity of 15 W/m$\cdot$K, and a thermal diffusivity of 10^{-5} m^2/s, at what time will a surface temperature of 60°C be recorded if the convection coefficient is 300 W/m^2·K?

(b) Assess the effect of thermal diffusivity on the thermal response of the material by computing center and surface temperature histories for $\alpha = 10^{-6}$, 10^{-5}, and 10^{-4} m^2/s. Plot your results for the period $0 \leq t \leq 300$ s. In a similar manner, assess the effect of thermal conductivity by considering values of $k = 1.5$, 15, and 150 W/m·K.

5.82 Consider the sphere of Example 5.6, which is initially at a uniform temperature when it is suddenly removed from the furnace and subjected to a two-step cooling process. Use the *Transient Conduction, Sphere* model of *IHT* to obtain the following solutions.

(a) For step 1, calculate the time required for the center temperature to reach $T(0, t) = 335$°C, while cooling in air at 20°C with a convection coefficient of 10 W/m^2·K. What is the Biot number for this cooling process? Do you expect radial temperature gradients to be appreciable? Compare your results to those of the example.

(b) For step 2, calculate the time required for the center temperature to reach $T(0, t) = 50$°C, while cooling in a water bath at 20°C with a convection coefficient of 6000 W/m^2·K.

(c) For the step 2 cooling process, calculate and plot the temperature histories, $T(r, t)$, for the center and surface of the sphere. Identify and explain key features of the histories. When do you expect the temperature gradients in the sphere to be the largest?

Semi-Infinite Media

5.83 Two large blocks of different materials, such as copper and concrete, have been sitting in a room (23°C) for a very long time. Which of the two blocks, if either, will feel colder to the touch? Assume the blocks to be semi-infinite solids and your hand to be at a temperature of 37°C.

5.84 A plane wall of thickness 0.6 m ($L = 0.3$ m) is made of steel ($k = 30$ W/m·K, $\rho = 7900$ kg/m^3, $c = 640$ J/kg·K). It is initially at a uniform temperature and is then exposed to air on both surfaces. Consider two different convection conditions: natural convection, characterized by $h = 10$ W/m^2·K, and forced convection, with $h = 100$ W/m^2·K. You are to calculate the surface temperature at three different times—$t = 2.5$ min, 25 min, and 250 min—for a total of six different cases.

(a) For each of these six cases, calculate the nondimensional surface temperature, $\theta_s^* = (T_s - T_\infty)/(T_i - T_\infty)$, using four different methods: exact solution, first-term-of-the-series solution, lumped capacitance, and semi-infinite solid. Present your results in a table.

(b) Briefly explain the conditions for which (i) the first-term solution is a good approximation to the exact solution, (ii) the lumped capacitance solution is a good approximation, (iii) the semi-infinite solid solution is a good approximation.

5.85 Asphalt pavement may achieve temperatures as high as 50°C on a hot summer day. Assume that such a temperature exists throughout the pavement, when suddenly a rainstorm reduces the surface temperature to 20°C. Calculate the total amount of energy (J/m^2) that will be transferred from the asphalt over a 30-min period in which the surface is maintained at 20°C.

5.86 A thick steel slab ($\rho = 7800$ kg/m^3, $c = 480$ J/kg·K, $k = 50$ W/m·K) is initially at 300°C and is cooled by water jets impinging on one of its surfaces. The temperature of the water is 25°C, and the jets maintain an extremely large, approximately uniform convection coefficient at the surface. Assuming that the surface is maintained at the temperature of the water throughout the cooling, how long will it take for the temperature to reach 50°C at a distance of 25 mm from the surface?

5.87 A tile-iron consists of a massive plate maintained at 150°C by an embedded electrical heater. The iron is placed in contact with a tile to soften the adhesive, allowing the tile to be easily lifted from the subflooring. The adhesive will soften sufficiently if heated above 50°C for at least 2 min, but its temperature should not exceed 120°C to avoid deterioration of the adhesive. Assume the tile and subfloor to have an initial temperature of 25°C and to have equivalent thermophysical properties of $k = 0.15$ W/m·K and $\rho c_p = 1.5 \times 10^6$ J/m^3·K.

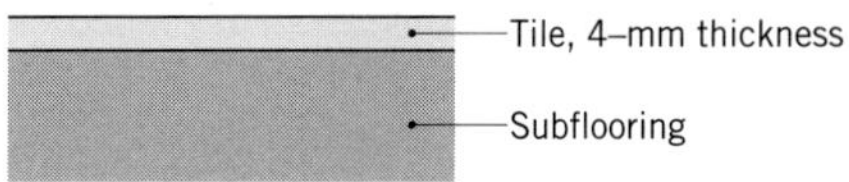

(a) How long will it take a worker using the tile-iron to lift a tile? Will the adhesive temperature exceed 120°C?

(b) If the tile-iron has a square surface area 254 mm to the side, how much energy has been removed from it during the time it has taken to lift the tile?

5.88 A simple procedure for measuring surface convection heat transfer coefficients involves coating the surface with a thin layer of material having a precise melting point temperature. The surface is then heated and, by determining the time required for melting to occur, the convection coefficient is determined. The following experimental arrangement uses the procedure to determine the convection coefficient for gas flow normal to a surface. Specifically, a long copper rod is encased in a super insulator of very low thermal conductivity, and a very thin coating is applied to its exposed surface.

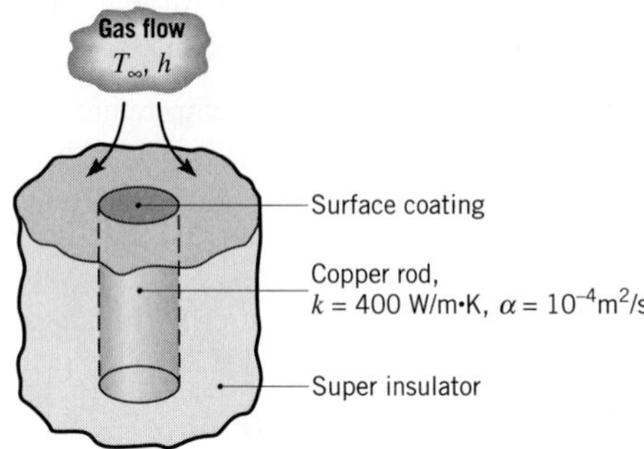

If the rod is initially at 25°C and gas flow for which $h = 200\ \text{W/m}^2\cdot\text{K}$ and $T_\infty = 300°\text{C}$ is initiated, what is the melting point temperature of the coating if melting is observed to occur at $t = 400$ s?

5.89 An insurance company has hired you as a consultant to improve their understanding of burn injuries. They are especially interested in injuries induced when a portion of a worker's body comes into contact with machinery that is at elevated temperatures in the range of 50 to 100°C. Their medical consultant informs them that irreversible thermal injury (cell death) will occur in any living tissue that is maintained at $T \geq 48°\text{C}$ for a duration $\Delta t \geq 10$ s. They want information concerning the extent of irreversible tissue damage (as measured by distance from the skin surface) as a function of the machinery temperature and the time during which contact is made between the skin and the machinery. Assume that living tissue has a normal temperature of 37°C, is isotropic, and has constant properties equivalent to those of liquid water.

(a) To assess the seriousness of the problem, compute locations in the tissue at which the temperature will reach 48°C after 10 s of exposure to machinery at 50°C and 100°C.

(b) For a machinery temperature of 100°C and $0 \leq t \leq 30$ s, compute and plot temperature histories at tissue locations of 0.5, 1, and 2 mm from the skin.

5.90 A procedure for determining the thermal conductivity of a solid material involves embedding a thermocouple in a thick slab of the solid and measuring the response to a prescribed change in temperature at one surface. Consider an arrangement for which the thermocouple is embedded 10 mm from a surface that is suddenly brought to a temperature of 100°C by exposure to boiling water. If the initial temperature of the slab was 30°C and the thermocouple measures a temperature of 65°C, 2 min after the surface is brought to 100°C, what is its thermal conductivity? The density and specific heat of the solid are known to be 2200 kg/m³ and 700 J/kg·K.

5.91 A very thick slab with thermal diffusivity $5.6 \times 10^{-6}\ \text{m}^2/\text{s}$ and thermal conductivity 20 W/m·K is initially at a uniform temperature of 325°C. Suddenly, the surface is exposed to a coolant at 15°C for which the convection heat transfer coefficient is 100 W/m²·K.

(a) Determine temperatures at the surface and at a depth of 45 mm after 3 min have elapsed.

(b) Compute and plot temperature histories ($0 \leq t \leq 300$ s) at $x = 0$ and $x = 45$ mm for the following parametric variations: (i) $\alpha = 5.6 \times 10^{-7}$, 5.6×10^{-6}, and $5.6 \times 10^{-5}\ \text{m}^2/\text{s}$; and (ii) $k = 2$, 20, and 200 W/m·K.

5.92 A thick oak wall, initially at 25°C, is suddenly exposed to combustion products for which $T_\infty = 800°\text{C}$ and $h = 20\ \text{W/m}^2\cdot\text{K}$.

(a) Determine the time of exposure required for the surface to reach the ignition temperature of 400°C.

(b) Plot the temperature distribution $T(x)$ in the medium at $t = 325$ s. The distribution should extend to a location for which $T \approx 25°\text{C}$.

5.93 Standards for firewalls may be based on their thermal response to a prescribed radiant heat flux. Consider a 0.25-m-thick concrete wall ($\rho = 2300\ \text{kg/m}^3$, $c = 880\ \text{J/kg}\cdot\text{K}$, $k = 1.4\ \text{W/m}\cdot\text{K}$), which is at an initial temperature of $T_i = 25°\text{C}$ and irradiated at one surface by lamps that provide a uniform heat flux of $q_s'' = 10^4\ \text{W/m}^2$. The absorptivity of the surface to the irradiation is $\alpha_s = 1.0$. If building code requirements dictate that the temperatures of the irradiated and back surfaces must not exceed 325°C and 25°C, respectively, after 30 min of heating, will the requirements be met?

5.94 It is well known that, although two materials are at the same temperature, one may feel cooler to the touch than the other. Consider thick plates of copper and glass, each at an initial temperature of 300 K. Assuming your finger to be at an initial temperature of 310 K and to have thermophysical properties of $\rho = 1000\ \text{kg/m}^3$, $c = 4180\ \text{J/kg}\cdot\text{K}$, and $k = 0.625\ \text{W/m}\cdot\text{K}$, determine whether the copper or the glass will feel cooler to the touch.

5.95 Two stainless steel plates ($\rho = 8000\ \text{kg/m}^3$, $c = 500$ J/kg·K, $k = 15\ \text{W/m}\cdot\text{K}$), each 20 mm thick and insulated on one surface, are initially at 400 and 300 K when they are pressed together at their uninsulated surfaces. What is the temperature of the insulated surface of the hot plate after 1 min has elapsed?

5.96 Special coatings are often formed by depositing thin layers of a molten material on a solid substrate. Solidification begins at the substrate surface and proceeds until the thickness S of the solid layer becomes equal to the thickness δ of the deposit.

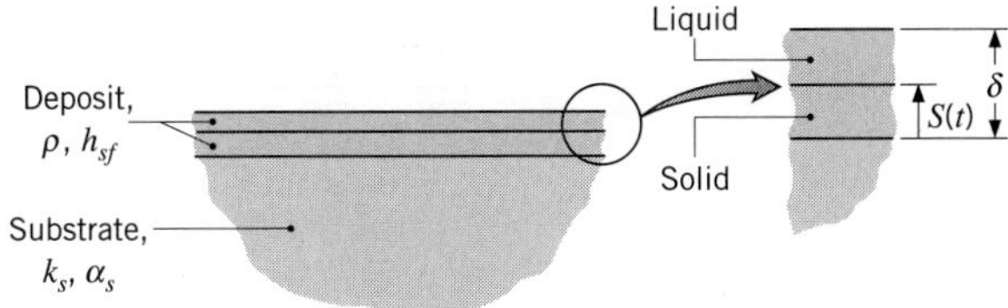

(a) Consider conditions for which molten material at its fusion temperature T_f is deposited on a *large* substrate that is at an initial uniform temperature T_i. With $S = 0$ at $t = 0$, develop an expression for estimating the time t_d required to completely solidify the deposit if it remains at T_f throughout the solidification process. Express your result in terms of the substrate thermal conductivity and thermal diffusivity (k_s, α_s), the density and latent heat of fusion of the deposit (ρ, h_{sf}), the deposit thickness δ, and the relevant temperatures (T_f, T_i).

(b) The plasma spray deposition process of Problem 5.25 is used to apply a thin ($\delta = 2$ mm) alumina coating on a thick tungsten substrate. The substrate has a uniform initial temperature of $T_i = 300$ K, and its thermal conductivity and thermal diffusivity may be approximated as $k_s = 120$ W/m·K and $\alpha_s = 4.0 \times 10^{-5}$ m²/s, respectively. The density and latent heat of fusion of the alumina are $\rho = 3970$ kg/m³ and $h_{sf} = 3577$ kJ/kg, respectively, and the alumina solidifies at its fusion temperature ($T_f = 2318$ K). Assuming that the molten layer is instantaneously deposited on the substrate, estimate the time required for the deposit to solidify.

5.97 When a molten metal is cast in a mold that is a poor conductor, the dominant resistance to heat flow is within the mold wall. Consider conditions for which a liquid metal is solidifying in a thick-walled mold of thermal conductivity k_w and thermal diffusivity α_w. The density and latent heat of fusion of the metal are designated as ρ and h_{sf}, respectively, and in both its molten and solid states, the thermal conductivity of the metal is very much larger than that of the mold.

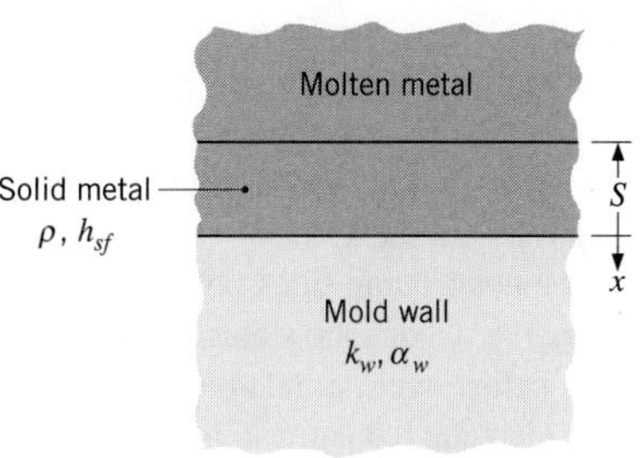

Just before the start of solidification ($S = 0$), the mold wall is everywhere at an initial uniform temperature T_i and the molten metal is everywhere at its fusion (melting point) temperature of T_f. Following the start of solidification, there is conduction heat transfer into the mold wall and the thickness of the solidified metal S increases with time t.

(a) Sketch the one-dimensional temperature distribution, $T(x)$, in the mold wall and the metal at $t = 0$ and at two subsequent times during the solidification. Clearly indicate any underlying assumptions.

(b) Obtain a relation for the variation of the solid layer thickness S with time t, expressing your result in terms of appropriate parameters of the system.

5.98 Joints of high quality can be formed by friction welding. Consider the friction welding of two 40-mm-diameter Inconel rods. The bottom rod is stationary, while the top rod is forced into a back-and-forth linear motion characterized by an instantaneous horizontal displacement, $d(t) = a \cos(\omega t)$ where $a = 2$ mm and $\omega = 1000$ rad/s. The coefficient of sliding friction between the two pieces is $\mu = 0.3$. Determine the compressive force that must be applied to heat the joint to the Inconel melting point within $t = 3$ s, starting from an initial temperature of 20°C. *Hint*: The frequency of the motion and resulting heat rate are very high. The temperature response can be approximated as if the heating rate were constant in time, equal to its average value.

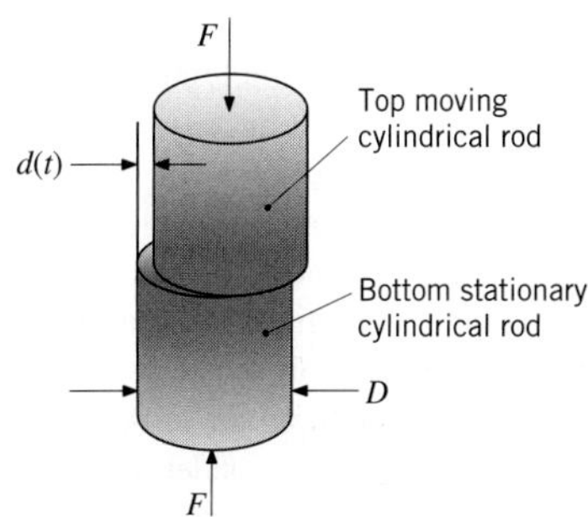

Objects with Constant Surface Temperatures or Surface Heat Fluxes and Periodic Heating

5.99 A rewritable optical disc (DVD) is formed by sandwiching a 15-nm-thick binary compound storage material between two 1-mm-thick polycarbonate sheets. Data are *written* to the opaque storage medium by irradiating it from below with a relatively high-powered laser beam of diameter 0.4 μm and power 1 mW, resulting in rapid heating of the compound material (the polycarbonate is transparent to the laser irradiation). If the temperature of the storage medium exceeds 900 K, a noncrystalline, amorphous material forms at the heated spot when the laser irradiation is curtailed and the spot is allowed to cool rapidly. The resulting spots of amorphous material have a different reflectivity from the surrounding crystalline material, so they can subsequently be *read* by irradiating them with a second, low-power laser and detecting the changes in laser radiation transmitted through the entire DVD thickness. Determine the irradiation (write) time needed to raise the storage medium temperature from an initial value of 300 K to 1000 K. The absorptivity of the storage medium is 0.8. The polycarbonate properties are $\rho = 1200 \text{ kg/m}^3$, $k = 0.21 \text{ W/m} \cdot \text{K}$, and $c_p = 1260 \text{ J/kg} \cdot \text{K}$.

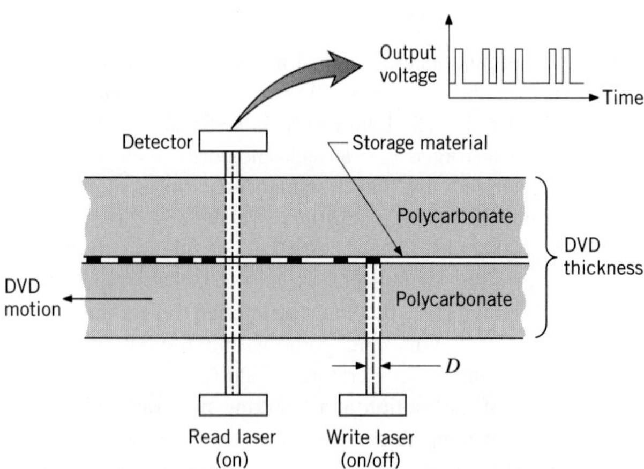

5.100 Ground source heat pumps operate by using the soil, rather than ambient air, as the heat source (or sink) for heating (or cooling) a building. A liquid transfers energy from (to) the soil by way of buried plastic tubing. The tubing is at a depth for which annual variations in the temperature of the soil are much less than those of the ambient air. For example, at a location such as South Bend, Indiana, deep-ground temperatures may remain at approximately 11°C, while annual excursions in the ambient air temperature may range from –25°C to +37°C. Consider the tubing to be laid out in a *closely spaced* serpentine arrangement.

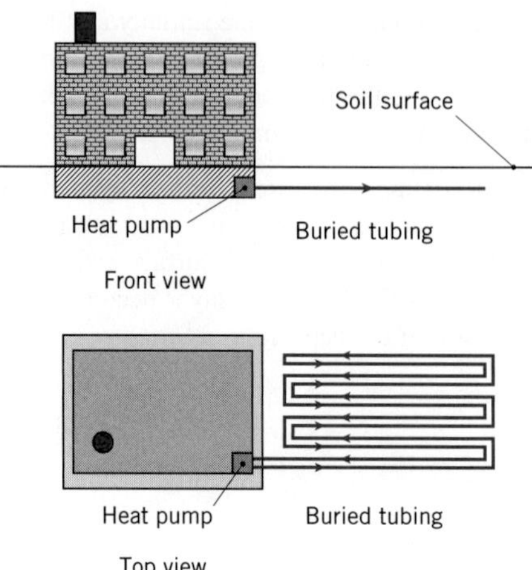

To what depth should the tubing be buried so that the soil can be viewed as an infinite medium at constant temperature over a 12-month period? Account for the periodic cooling (heating) of the soil due to both annual changes in ambient conditions and variations in heat pump operation from the winter heating to the summer cooling mode.

5.101 To enable cooking a wider range of foods in microwave ovens, thin, metallic packaging materials have been developed that will readily absorb microwave energy. As the packaging material is heated by the microwaves, conduction simultaneously occurs from the hot packaging material to the cold food. Consider the spherical piece of frozen ground beef of Problem 5.33 that is now wrapped in the thin microwave-absorbing packaging material. Determine the time needed for the beef that is immediately adjacent to the packaging material to reach $T = 0°\text{C}$ if 50% of the oven power ($P = 1$ kW total) is absorbed in the packaging material.

5.102 Derive an expression for the ratio of the total energy transferred from the isothermal surface of an infinite cylinder to the interior of the cylinder, Q/Q_o, that is valid for $Fo < 0.2$. Express your results in terms of the Fourier number Fo.

5.103 The structural components of modern aircraft are commonly fabricated of high-performance composite materials. These materials are fabricated by impregnating mats of extremely strong fibers that are held within a form with an epoxy or thermoplastic liquid. After the liquid cures or cools, the resulting component is of extremely high strength and low weight. Periodically, these components must be inspected to ensure that the fiber mats and bonding material do not become delaminated and, in turn, the component loses its airworthiness. One inspection method involves application of a uniform, constant radiation heat flux to the surface being inspected. The thermal response of the surface is measured with an infrared imaging system, which captures the emission from the surface and converts it to a color-coded map of the surface temperature distribution. Consider the case where a uniform flux of $5\ \text{kW/m}^2$ is applied to the top skin of an airplane wing initially at 20°C. The opposite side of the 15-mm-thick skin is adjacent to stagnant air and can be treated as well insulated. The density and specific heat of the skin material are $1200\ \text{kg/m}^3$ and $1200\ \text{J/kg}\cdot\text{K}$, respectively. The effective thermal conductivity of the intact skin material is $k_1 = 1.6\ \text{W/m}\cdot\text{K}$. Contact resistances develop internal to the structure as a result of delamination between the fiber mats and the bonding material, leading to a reduced effective thermal conductivity of $k_2 = 1.1\ \text{W/m}\cdot\text{K}$. Determine the surface temperature of the component after 10 and 100 s of irradiation for (i) an area where the material is structurally intact and (ii) an adjacent area where delamination has occurred within the wing.

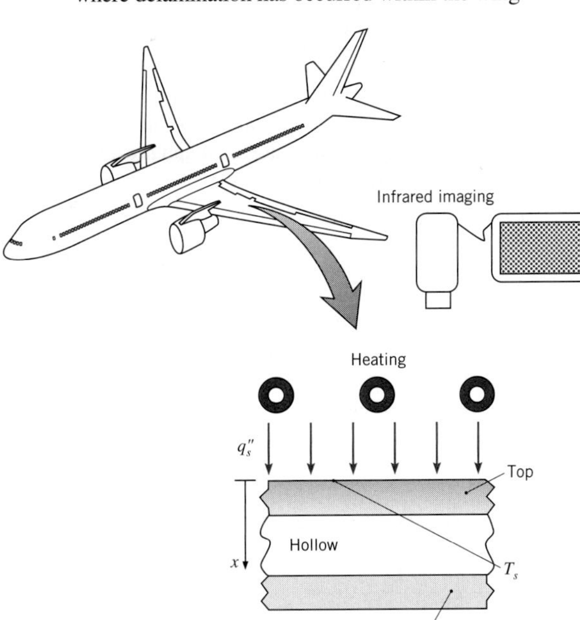

5.104 Consider the plane wall of thickness $2L$, the infinite cylinder of radius r_o, and the sphere of radius r_o. Each configuration is subjected to a constant surface heat flux q''_s. Using the approximate solutions of Table 5.2*b* for $Fo \geq 0.2$, derive expressions for each of the three geometries for the quantity $(T_{s,\text{act}} - T_i)/(T_{s,\text{lc}} - T_i)$. In this expression, $T_{s,\text{act}}$ is the actual surface temperature as determined by the relations of Table 5.2b, and $T_{s,\text{lc}}$ is the temperature associated with lumped capacitance behavior. Determine criteria associated with $(T_{s,\text{act}} - T_i)/\ (T_{s,\text{lc}} - T_i) \leq 1.1$, that is, determine when the lumped capacitance approximation is accurate to within 10%.

5.105 Problem 4.9 addressed radioactive wastes stored underground in a spherical container. Because of uncertainty in the thermal properties of the soil, it is desired to measure the steady-state temperature using a test container (identical to the real container) that is equipped with internal electrical heaters. Estimate how long it will take the test container to come within 10°C of its steady-state value, assuming it is buried very far underground. Use the soil properties from Table A.3 in your analysis.

5.106 Derive an expression for the ratio of the total energy transferred from the isothermal surface of a sphere to the interior of the sphere Q/Q_o that is valid for $Fo < 0.2$. Express your result in terms of the Fourier number, Fo.

5.107 Consider the experimental measurement of Example 5.10. It is desired to measure the thermal conductivity of an extremely thin sample of the same nanostructured material having the same length and width. To minimize experimental uncertainty, the experimenter wishes to keep the amplitude of the temperature response, ΔT, above a value of 0.1°C. What is the minimum sample thickness that can be measured? Assume the properties of the thin sample and the magnitude of the applied heating rate are the same as those measured and used in Example 5.10.

Finite-Difference Equations: Derivations

5.108 The stability criterion for the explicit method requires that the coefficient of the T_m^p term of the one-dimensional, finite-difference equation be zero or positive. Consider the situation for which the temperatures at the two neighboring nodes (T_{m-1}^p, T_{m+1}^p) are 100°C while the center node (T_m^p) is at 50°C. Show that for values of $Fo > \frac{1}{2}$ the finite-difference equation will predict a value of T_m^{p+1} that violates the second law of thermodynamics.

5.109 A thin rod of diameter D is initially in equilibrium with its surroundings, a large vacuum enclosure at temperature T_{sur}. Suddenly an electrical current I (A) is passed through the rod having an electrical resistivity ρ_e and emissivity ε. Other pertinent thermophysical properties are identified in the sketch. Derive the transient, finite-difference equation for node m.

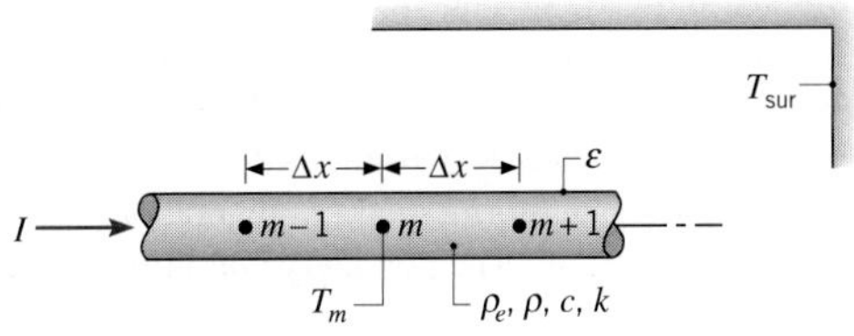

5.110 A one-dimensional slab of thickness $2L$ is initially at a uniform temperature T_i. Suddenly, electric current is passed through the slab causing uniform volumetric heating $\dot{q}$ (W/m^3). At the same time, both outer surfaces ($x = \pm L$) are subjected to a convection process at T_∞ with a heat transfer coefficient h.

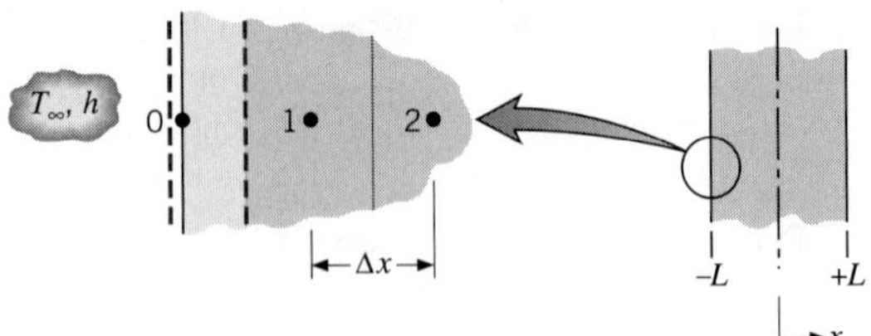

Write the finite-difference equation expressing conservation of energy for node 0 located on the outer surface at $x = -L$. Rearrange your equation and identify any important dimensionless coefficients.

5.111 Consider Problem 5.9 except now the combined volume of the oil bath and the sphere is $V_{tot} = 1\ m^3$. The oil bath is well mixed and well insulated.

(a) Assuming the quenching liquid's properties are that of engine oil at 380 K, determine the steady-state temperature of the sphere.

(b) Derive explicit finite difference expressions for the sphere and oil bath temperatures as a function of time using a single node each for the sphere and oil bath. Determine any stability requirements that might limit the size of the time step Δt.

(c) Evaluate the sphere and oil bath temperatures after one time step using the explicit expressions of part (b) and time steps of 1000, 10,000, and 20,000 s.

(d) Using an implicit formulation with $\Delta t = 100$ s, determine the time needed for the coated sphere to reach 140°C. Compare your answer to the time associated with a large, well-insulated oil bath. Plot the sphere and oil temperatures as a function of time over the interval $0\ h \leq t \leq 15\ h$. *Hint:* See Comment 3 of Example 5.2.

5.112 A plane wall ($\rho = 4000\ kg/m^3$, $c_p = 500\ J/kg \cdot K$, $k = 10\ W/m \cdot K$) of thickness $L = 20$ mm initially has a linear, steady-state temperature distribution with boundaries maintained at $T_1 = 0$°C and $T_2 = 100$°C. Suddenly, an electric current is passed through the wall, causing uniform energy generation at a rate $\dot{q} = 2 \times 10^7\ W/m^3$. The boundary conditions T_1 and T_2 remain fixed.

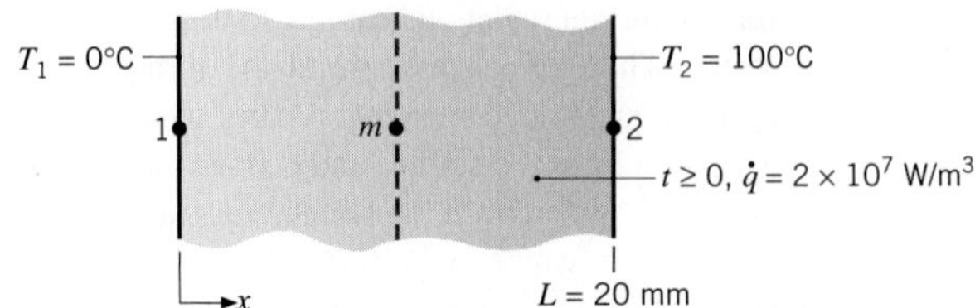

(a) On $T - x$ coordinates, sketch temperature distributions for the following cases: (i) initial condition ($t \leq 0$); (ii) steady-state conditions ($t \rightarrow \infty$), assuming that the maximum temperature in the wall exceeds T_2; and (iii) for two intermediate times. Label all important features of the distributions.

(b) For the system of three nodal points shown schematically (1, m, 2), define an appropriate control volume for node m and, identifying all relevant processes, derive the corresponding finite-difference equation using either the *explicit* or *implicit* method.

(c) With a time increment of $\Delta t = 5$ s, use the finite-difference method to obtain values of T_m for the first 45 s of elapsed time. Determine the corresponding heat fluxes at the boundaries, that is, q''_x (0, 45 s) and q''_x (20 mm, 45 s).

(d) To determine the effect of mesh size, repeat your analysis using grids of 5 and 11 nodal points ($\Delta x = 5.0$ and 2.0 mm, respectively).

5.113 A round solid cylinder made of a plastic material ($\alpha = 6 \times 10^{-7}\ m^2/s$) is initially at a uniform temperature of 20°C and is well insulated along its lateral surface and at one end. At time $t = 0$, heat is applied to the left boundary causing T_0 to increase linearly with time at a rate of 1°C/s.

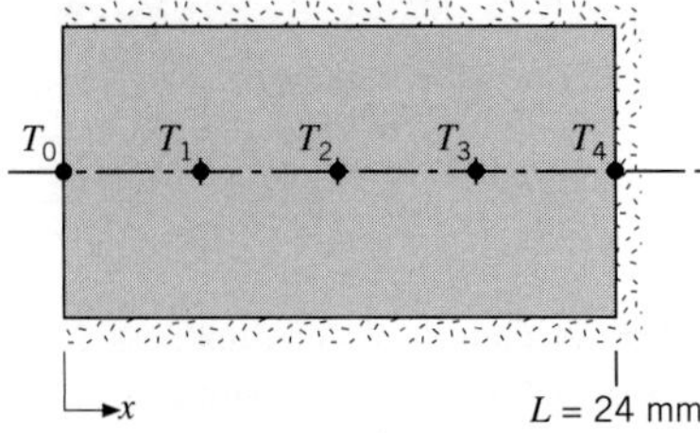

(a) Using the explicit method with $Fo = \frac{1}{2}$, derive the finite-difference equations for nodes 1, 2, 3, and 4.

(b) Format a table with headings of p, t(s), and the nodal temperatures T_0 to T_4. Determine the surface temperature T_0 when $T_4 = 35$°C.

5.114 Derive the explicit finite-difference equation for an interior node for three-dimensional transient conduction. Also determine the stability criterion. Assume constant properties and equal grid spacing in all three directions.

5.115 Derive the transient, two-dimensional finite-difference equation for the temperature at nodal point 0 located on the boundary between two different materials.

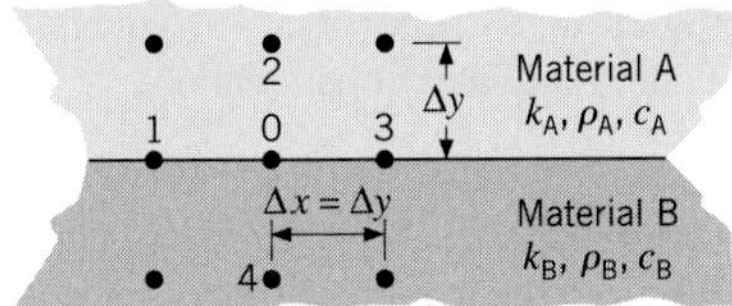

Finite-Difference Solutions: One-Dimensional Systems

5.116 A wall 0.12 m thick having a thermal diffusivity of 1.5×10^{-6} m²/s is initially at a uniform temperature of 85°C. Suddenly one face is lowered to a temperature of 20°C, while the other face is perfectly insulated.

(a) Using the explicit finite-difference technique with space and time increments of 30 mm and 300 s, respectively, determine the temperature distribution at $t = 45$ min.

(b) With $\Delta x = 30$ mm and $\Delta t = 300$ s, compute $T(x, t)$ for $0 \leq t \leq t_{ss}$, where t_{ss} is the time required for the temperature at each nodal point to reach a value that is within 1°C of the steady-state temperature. Repeat the foregoing calculations for $\Delta t = 75$ s. For each value of Δt, plot temperature histories for each face and the midplane.

5.117 A molded plastic product ($\rho = 1200$ kg/m³, $c = 1500$ J/kg·K, $k = 0.30$ W/m·K) is cooled by exposing one surface to an array of air jets, while the opposite surface is well insulated. The product may be approximated as a slab of thickness $L = 60$ mm, which is initially at a uniform temperature of $T_i = 80$°C. The air jets are at a temperature of $T_\infty = 20$°C and provide a uniform convection coefficient of $h = 100$ W/m²·K at the cooled surface.

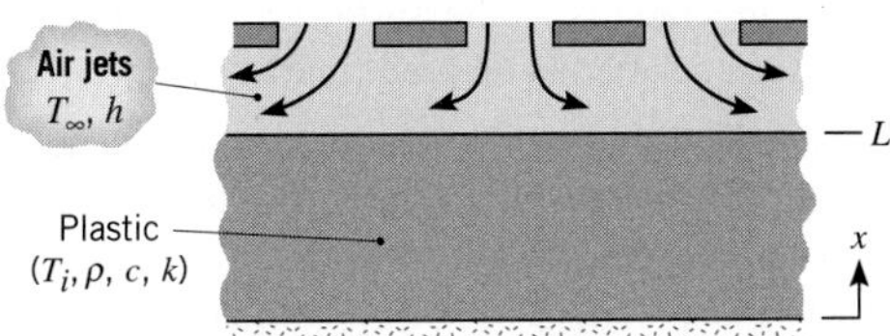

Using a finite-difference solution with a space increment of $\Delta x = 6$ mm, determine temperatures at the cooled and insulated surfaces after 1 h of exposure to the gas jets.

5.118 Consider a one-dimensional plane wall at a uniform initial temperature T_i. The wall is 10 mm thick, and has a thermal diffusivity of $\alpha = 6 \times 10^{-7}$ m²/s. The left face is insulated, and suddenly the right face is lowered to a temperature $T_{s,r}$.

(a) Using the implicit finite-difference technique with $\Delta x = 2$ mm and $\Delta t = 2$ s, determine how long it will take for the temperature at the left face $T_{s,l}$ to achieve 50% of its maximum possible temperature reduction.

(b) At the time determined in part (a), the right face is suddenly returned to the initial temperature. Determine how long it will take for the temperature at the left face to recover to a 20% temperature reduction, that is, $T_i - T_{s,l} = 0.2(T_i - T_{s,r})$.

5.119 The plane wall of Problem 2.60 ($k = 50$ W/m·K, $\alpha = 1.5 \times 10^{-6}$ m²/s) has a thickness of $L = 40$ mm and an initial uniform temperature of $T_o = 25$°C. Suddenly, the boundary at $x = L$ experiences heating by a fluid for which $T_\infty = 50$°C and $h = 1000$ W/m²·K, while heat is uniformly generated within the wall at $\dot{q} = 1 \times 10^7$ W/m³. The boundary at $x = 0$ remains at T_o.

(a) With $\Delta x = 4$ mm and $\Delta t = 1$ s, plot temperature distributions in the wall for (i) the initial condition, (ii) the steady-state condition, and (iii) two intermediate times.

(b) On $q''_x - t$ coordinates, plot the heat flux at $x = 0$ and $x = L$. At what elapsed time is there zero heat flux at $x = L$?

5.120 Consider the fuel element of Example 5.11. Initially, the element is at a uniform temperature of 250°C with no heat generation. Suddenly, the element is inserted into the reactor core, causing a uniform volumetric heat generation rate of $\dot{q} = 10^8$ W/m³. The surfaces are convectively cooled with $T_\infty = 250$°C and $h = 1100$ W/m²·K. Using the explicit method with a space increment of 2 mm, determine the temperature distribution 1.5 s after the element is inserted into the core.

5.121 Consider two plates, A and B, that are each initially isothermal and each of thickness $L = 5$ mm. The faces of the plates are suddenly brought into contact in a joining process. Material A is acrylic, initially at $T_{i,A} = 20$°C with $\rho_A = 1990$ kg/m³, $c_A = 1470$ J/kg·K, and $k_A = 0.21$ W/m·K. Material B is steel initially at $T_{i,B} = 300$°C with $\rho_B = 7800$ kg/m³, $c_B = 500$ J/kg·K, and $k_B = 45$ W/m·K. The external (back) surfaces of the acrylic and steel are insulated. Neglecting the thermal contact resistance between the plates, determine how long it will take for the external surface of the acrylic to reach its softening temperature, $T_{soft} = 90$°C. Plot the acrylic's external surface temperature as well as the average temperatures of both materials over the time span $0 \leq t \leq 300$ s. Use 20 equally spaced nodal points.

5.122 Consider the fuel element of Example 5.11, which operates at a uniform volumetric generation rate of $\dot{q} = 10^7$ W/m^3, until the generation rate suddenly changes to $\dot{q} = 2 \times 10^7$ W/m^3. Use the *Finite-Difference Equations, One-Dimensional, Transient* conduction model builder of *IHT* to obtain the implicit form of the finite-difference equations for the 6 nodes, with $\Delta x = 2$ mm, as shown in the example.

(a) Calculate the temperature distribution 1.5 s after the change in operating power, and compare your results with those tabulated in the example.

(b) Use the *Explore* and *Graph* options of *IHT* to calculate and plot temperature histories at the midplane (00) and surface (05) nodes for $0 \leq t \leq 400$ s. What are the steady-state temperatures, and approximately how long does it take to reach the new equilibrium condition after the step change in operating power?

5.123 In a thin-slab, continuous casting process, molten steel leaves a mold with a thin solid shell, and the molten material solidifies as the slab is quenched by water jets en route to a section of rollers. Once fully solidified, the slab continues to cool as it is brought to an acceptable handling temperature. It is this portion of the process that is of interest.

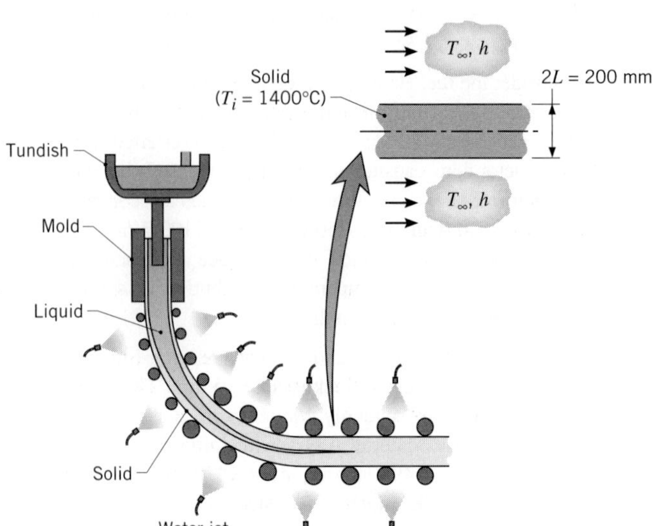

Consider a 200-mm-thick solid slab of steel ($\rho = 7800$ kg/m^3, $c = 700$ J/kg·K, $k = 30$ W/m·K), initially at a uniform temperature of $T_i = 1400$°C. The slab is cooled at its top and bottom surfaces by water jets ($T_\infty = 50$°C), which maintain an approximately uniform convection coefficient of $h = 5000$ W/m^2·K at both surfaces. Using a finite-difference solution with a space increment of $\Delta x = 1$ mm, determine the time required to cool the surface of the slab to 200°C. What is the corresponding temperature at the midplane of the slab? If the slab moves at a speed of $V = 15$ mm/s, what is the required length of the cooling section?

5.124 Determine the temperature distribution at $t = 30$ min for the conditions of Problem 5.116.

(a) Use an explicit finite-difference technique with a time increment of 600 s and a space increment of 30 mm.

(b) Use the implicit method of the *IHT Finite-Difference Equation* Tool Pad for *One-Dimensional Transient Conduction.*

5.125 A very thick plate with thermal diffusivity 5.6×10^{-6} m^2/s and thermal conductivity 20 W/m·K is initially at a uniform temperature of 325°C. Suddenly, the surface is exposed to a coolant at 15°C for which the convection heat transfer coefficient is 100 W/m^2·K. Using the finite-difference method with a space increment of $\Delta x = 15$ mm and a time increment of 18 s, determine temperatures at the surface and at a depth of 45 mm after 3 min have elapsed.

5.126 Referring to Example 5.12, Comment 4, consider a sudden exposure of the surface to large surroundings at an elevated temperature (T_{sur}) and to convection (T_∞, h).

(a) Derive the explicit, finite-difference equation for the surface node in terms of *Fo*, *Bi*, and Bi_r.

(b) Obtain the stability criterion for the surface node. Does this criterion change with time? Is the criterion more restrictive than that for an interior node?

(c) A thick slab of material ($k = 1.5$ W/m·K, $\alpha = 7 \times 10^{-7}$ m^2/s, $\varepsilon = 0.9$), initially at a uniform temperature of 27°C, is suddenly exposed to large surroundings at 1000 K. Neglecting convection and using a space increment of 10 mm, determine temperatures at the surface and 30 mm from the surface after an elapsed time of 1 min.

5.127 A constant-property, one-dimensional plane wall of width $2L$, at an initial uniform temperature T_i, is heated convectively (both surfaces) with an ambient fluid at $T_\infty = T_{\infty,1}$, $h = h_1$. At a later instant in time, $t = t_1$, heating is curtailed, and convective cooling is initiated. Cooling conditions are characterized by $T_\infty = T_{\infty,2} = T_i$, $h = h_2$.

(a) Write the heat equation as well as the initial and boundary conditions in their dimensionless form for the heating phase (Phase 1). Express the equations in terms of the dimensionless quantities θ^*, x^*, Bi_1, and Fo, where Bi_1 is expressed in terms of h_1.

(b) Write the heat equation as well as the initial and boundary conditions in their dimensionless form for the cooling phase (Phase 2). Express the equations in terms of the dimensionless quantities θ^*, x^*, Bi_2, Fo_1, and Fo where Fo_1 is the dimensionless time associated with t_1, and Bi_2 is expressed in terms of h_2. To be consistent with part (a), express the dimensionless temperature in terms of $T_\infty = T_{\infty,1}$.

(c) Consider a case for which $Bi_1 = 10$, $Bi_2 = 1$, and $Fo_1 = 0.1$. Using a finite-difference method with $\Delta x^* = 0.1$ and $\Delta Fo = 0.001$, determine the transient thermal response of the surface ($x^* = 1$), midplane ($x^* = 0$), and quarter-plane ($x^* = 0.5$) of the slab. Plot these three dimensionless temperatures as a function of dimensionless time over the range $0 \leq Fo \leq 0.5$.

(d) Determine the minimum dimensionless temperature at the midplane of the wall, and the dimensionless time at which this minimum temperature is achieved.

5.128 Consider the thick slab of copper in Example 5.12, which is initially at a uniform temperature of 20°C and is suddenly exposed to a net radiant flux of 3×10^5 W/m². Use the *Finite-Difference Equations/One-Dimensional/Transient* conduction model builder of *IHT* to obtain the implicit form of the finite-difference equations for the interior nodes. In your analysis, use a space increment of $\Delta x = 37.5$ mm with a total of 17 nodes (00–16), and a time increment of $\Delta t = 1.2$ s. For the surface node 00, use the finite-difference equation derived in Section 2 of the Example.

(a) Calculate the 00 and 04 nodal temperatures at $t = 120$ s, that is, $T(0, 120\text{ s})$ and $T(0.15\text{ m}, 120\text{ s})$, and compare the results with those given in Comment 1 for the exact solution. Will a time increment of 0.12 s provide more accurate results?

(b) Plot temperature histories for $x = 0$, 150, and 600 mm, and explain key features of your results.

5.129 In Section 5.5, the one-term approximation to the series solution for the temperature distribution was developed for a plane wall of thickness $2L$ that is initially at a uniform temperature and suddenly subjected to convection heat transfer. If $Bi < 0.1$, the wall can be approximated as isothermal and represented as a lumped capacitance (Equation 5.7). For the conditions shown schematically, we wish to compare predictions based on the one-term approximation, the lumped capacitance method, and a finite-difference solution.

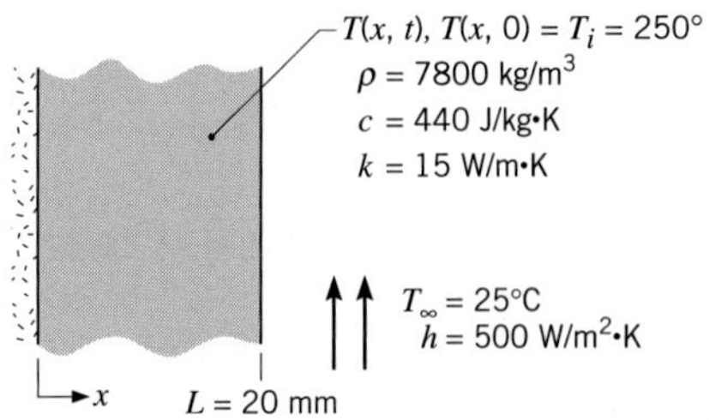

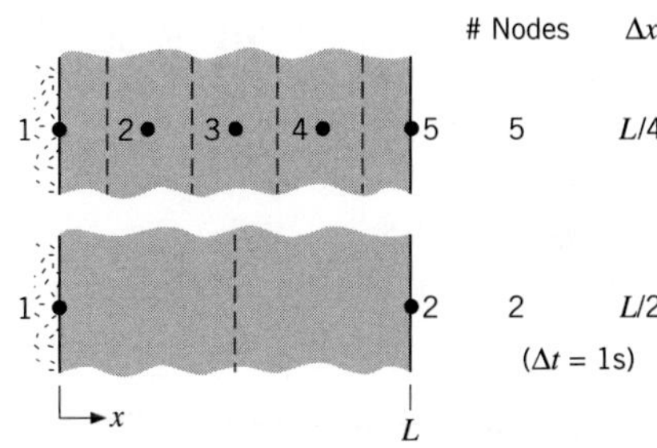

(a) Determine the midplane, $T(0, t)$, and surface, $T(L, t)$, temperatures at $t = 100$, 200, and 500 s using the one-term approximation to the series solution, Equation 5.43, What is the Biot number for the system?

(b) Treating the wall as a lumped capacitance, calculate the temperatures at $t = 50$, 100, 200, and 500 s. Did you expect these results to compare favorably with those from part (a)? Why are the temperatures considerably higher?

(c) Consider the 2- and 5-node networks shown schematically. Write the implicit form of the finite-difference equations for each network, and determine the temperature distributions for $t = 50$, 100, 200, and 500 s using a time increment of $\Delta t = 1$ s. You may use *IHT* to solve the finite-difference equations by representing the rate of change of the nodal temperatures by the intrinsic function, Der(T, t). Prepare a table summarizing the results of parts (a), (b), and (c). Comment on the relative differences of the predicted temperatures. *Hint*: See the *Solver/Intrinsic Functions* section of *IHT/Help* or the *IHT Examples* menu (Example 5.2) for guidance on using the Der(T, t) function.

5.130 Steel-reinforced concrete pillars are used in the construction of large buildings. Structural failure can occur at high temperatures due to a fire because of softening of the metal core. Consider a 200-mm-thick composite pillar consisting of a central steel core (50 mm thick) sandwiched between two 75-mm-thick concrete walls. The pillar is at a uniform initial temperature of $T_i = 27°C$ and is suddenly exposed to combustion products at $T_\infty = 900°C$, $h = 40\ W/m^2 \cdot K$ on both exposed surfaces. The surroundings temperature is also 900°C.

(a) Using an implicit finite difference method with $\Delta x = 10$ mm and $\Delta t = 100$ s, determine the temperature of the exposed concrete surface and the center of the steel plate at $t = 10{,}000$ s. Steel properties are: $k_s = 55\ W/m \cdot K$, $\rho_s = 7850\ kg/m^3$, and $c_s = 450\ J/kg \cdot K$. Concrete properties are: $k_c = 1.4\ W/m \cdot K$, $\rho_c = 2300\ kg/m^3$, $c_c = 880\ J/kg \cdot K$, and $\varepsilon = 0.90$. Plot the maximum and minimum concrete temperatures along with the maximum and minimum steel temperatures over the duration $0 \leq t \leq 10{,}000$ s.

(b) Repeat part (a) but account for a thermal contact resistance of $R''_{t,c} = 0.20\ m^2 \cdot K/W$ at the concrete-steel interface.

(c) At $t = 10{,}000$ s, the fire is extinguished, and the surroundings and ambient temperatures return to $T_\infty = T_{sur} = 27°C$. Using the same convection heat transfer coefficient and emissivity as in parts (a) and (b), determine the maximum steel temperature and the *critical time* at which the maximum steel temperature occurs for cases with and without the contact resistance. Plot the concrete surface temperature, the concrete temperature adjacent to the steel, and the steel temperatures over the duration $10{,}000 \leq t \leq 20{,}000$ s.

5.131 Consider the bonding operation described in Problem 3.115, which was analyzed under steady-state conditions. In this case, however, the laser will be used to heat the film for a prescribed period of time, creating the transient heating situation shown in the sketch.

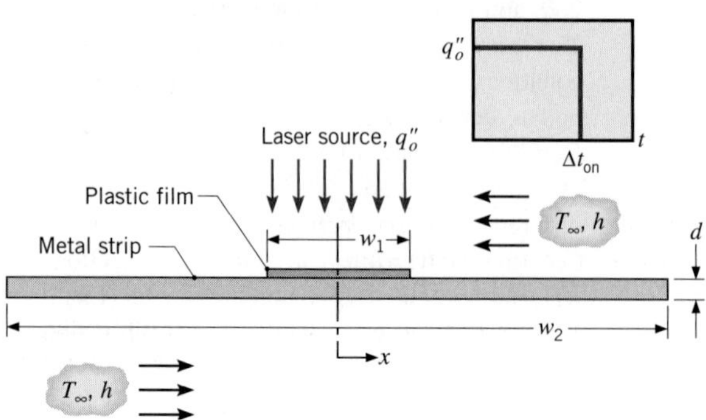

The strip is initially at 25°C and the laser provides a uniform flux of 85,000 W/m² over a time interval of $\Delta t_{on} = 10$ s. The system dimensions and thermophysical properties remain the same, but the convection coefficient to the ambient air at 25°C is now $100\ W/m^2 \cdot K$ and $w_1 = 44$ mm.

Using an implicit finite-difference method with $\Delta x = 4$ mm and $\Delta t = 1$ s, obtain temperature histories for $0 \leq t \leq 30$ s at the center and film edge, $T(0, t)$ and $T(w_1/2, t)$, respectively, to determine if the adhesive is satisfactorily cured above 90°C for 10 s and if its degradation temperature of 200°C is exceeded.

5.132 One end of a stainless steel (AISI 316) rod of diameter 10 mm and length 0.16 m is inserted into a fixture maintained at 200°C. The rod, covered with an insulating sleeve, reaches a uniform temperature throughout its length. When the sleeve is removed, the rod is subjected to ambient air at 25°C such that the convection heat transfer coefficient is $30\ W/m^2 \cdot K$.

(a) Using the explicit finite-difference technique with a space increment of $\Delta x = 0.016$ m, estimate the time required for the midlength of the rod to reach 100°C.

(b) With $\Delta x = 0.016$ m and $\Delta t = 10$ s, compute $T(x, t)$ for $0 \leq t \leq t_1$, where t_1 is the time required for the midlength of the rod to reach 50°C. Plot the temperature distribution for $t = 0$, 200 s, 400 s, and t_1.

5.133 A tantalum rod of diameter 3 mm and length 120 mm is supported by two electrodes within a large vacuum enclosure. Initially the rod is in equilibrium with the electrodes and its surroundings, which are maintained at 300 K. Suddenly, an electrical current, $I = 80$ A, is passed through the rod. Assume the emissivity of the rod is 0.1 and the electrical resistivity is $95 \times 10^{-8}\ \Omega \cdot m$. Use Table A.1 to obtain the other thermophysical properties required in your solution. Use a finite-difference method with a space increment of 10 mm.

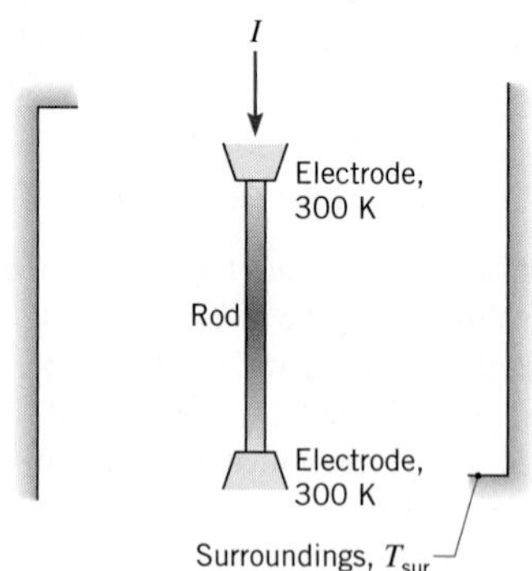

(a) Estimate the time required for the midlength of the rod to reach 1000 K.

(b) Determine the steady-state temperature distribution and estimate approximately how long it will take to reach this condition.

5.134 A support rod ($k = 15$ W/m·K, $\alpha = 4.0 \times 10^{-6}$ m²/s) of diameter $D = 15$ mm and length $L = 100$ mm spans a channel whose walls are maintained at a temperature of $T_b = 300$ K. Suddenly, the rod is exposed to a cross flow of hot gases for which $T_\infty = 600$ K and $h = 75$ W/m²·K. The channel walls are cooled and remain at 300 K.

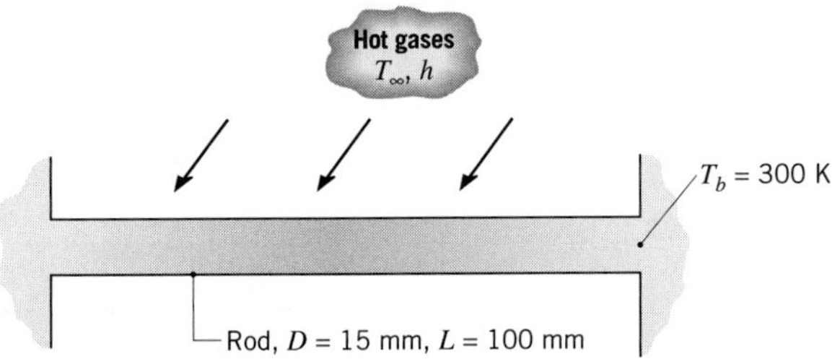

(a) Using an appropriate numerical technique, determine the thermal response of the rod to the convective heating. Plot the midspan temperature as a function of elapsed time. Using an appropriate analytical model of the rod, determine the steady-state temperature distribution, and compare the result with that obtained numerically for very long elapsed times.

(b) After the rod has reached steady-state conditions, the flow of hot gases is suddenly terminated, and the rod cools by free convection to ambient air at $T_\infty = 300$ K and by radiation exchange with large surroundings at $T_{sur} = 300$ K. The free convection coefficient can be expressed as h (W/m²·K) $= C\,\Delta T^n$, where $C = 4.4$ W/m²·K$^{1.188}$ and $n = 0.188$. The emissivity of the rod is 0.5. Determine the subsequent thermal response of the rod. Plot the midspan temperature as a function of cooling time, and determine the time required for the rod to reach a *safe-to-touch* temperature of 315 K.

5.135 Consider the acceleration-grid foil ($k = 40$ W/m·K, $\alpha = 3 \times 10^{-5}$ m²/s, $\varepsilon = 0.45$) of Problem 4.72. Develop an implicit, finite-difference model of the foil, which can be used for the following purposes.

(a) Assuming the foil to be at a uniform temperature of 300 K when the ion beam source is activated, obtain a plot of the midspan temperature–time history. At what elapsed time does this point on the foil reach a temperature within 1 K of the steady-state value?

(b) The foil is operating under steady-state conditions when, suddenly, the ion beam is deactivated. Obtain a plot of the subsequent midspan temperature–time history. How long does it take for the hottest point on the foil to cool to 315 K, a *safe-to-touch* condition?

5.136 Circuit boards are treated by heating a stack of them under high pressure as illustrated in Problem 5.45 and described further in Problem 5.46. A finite-difference method of solution is sought with two additional considerations. First, the book is to be treated as having distributed, rather than lumped, characteristics, by using a grid spacing of $\Delta x = 2.36$ mm with nodes at the center of the individual circuit board or plate. Second, rather than bringing the platens to 190°C in one sudden change, the heating schedule $T_p(t)$ shown in the sketch is to be used to minimize excessive thermal stresses induced by rapidly changing thermal gradients in the vicinity of the platens.

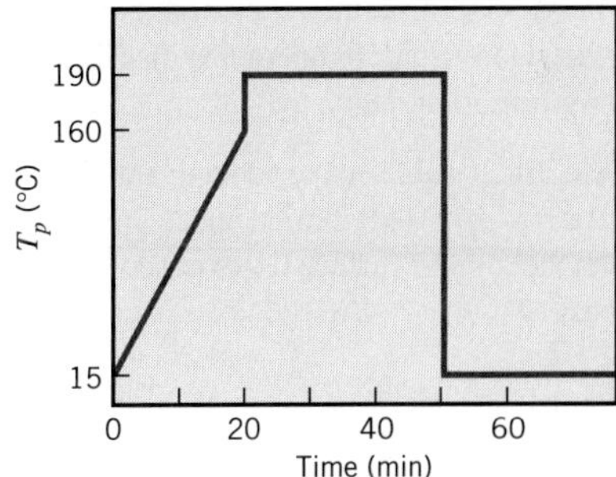

(a) Using a time increment of $\Delta t = 60$ s and the implicit method, find the temperature history of the midplane of the book and determine whether curing will occur (170°C for 5 min).

(b) Following the reduction of the platen temperatures to 15°C ($t = 50$ min), how long will it take for the midplane of the book to reach 37°C, a safe temperature at which the operator can begin unloading the press?

(c) Validate your program code by using the heating schedule of a sudden change of platen temperature from 15 to 190°C and compare results with those from an appropriate analytical solution (see Problem 5.46).

Finite-Difference Equations: Cylindrical Coordinates

5.137 A thin circular disk is subjected to induction heating from a coil, the effect of which is to provide a uniform heat generation within a ring section as shown.

Convection occurs at the upper surface, while the lower surface is well insulated.

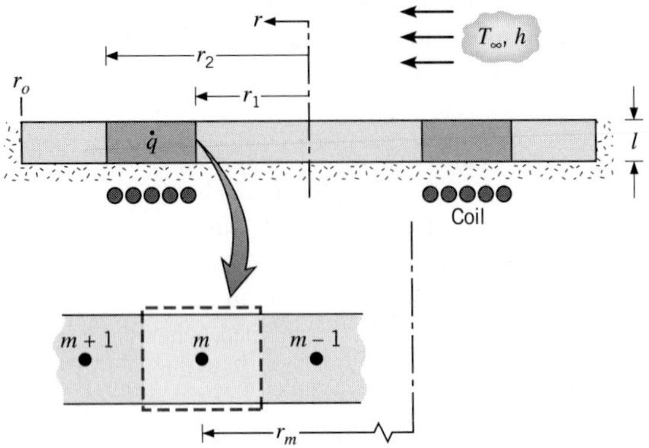

(a) Derive the transient, finite-difference equation for node m, which is within the region subjected to induction heating.

(b) On $T - r$ coordinates sketch, in qualitative manner, the steady-state temperature distribution, identifying important features.

5.138 An electrical cable, experiencing uniform volumetric generation $\dot{q}$, is half buried in an insulating material while the upper surface is exposed to a convection process (T_∞, h).

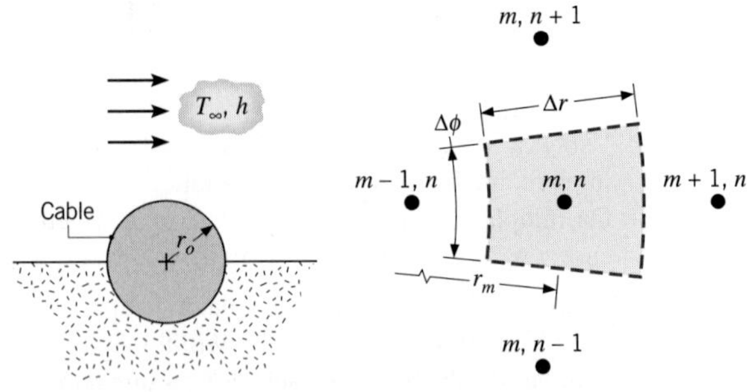

(a) Derive the explicit, finite-difference equations for an interior node (m, n), the center node ($m = 0$), and the outer surface nodes (M, n) for the convection and insulated boundaries.

(b) Obtain the stability criterion for each of the finite-difference equations. Identify the most restrictive criterion.

Finite-Difference Solutions: Two-Dimensional Systems

5.139 Two very long (in the direction normal to the page) bars having the prescribed initial temperature distributions are to be soldered together. At time $t = 0$, the $m = 3$ face of the copper (pure) bar contacts the $m = 4$ face of the steel (AISI 1010) bar. The solder and flux act as an interfacial layer of negligible thickness and effective contact resistance $R''_{t,c} = 2 \times 10^{-5}\ \text{m}^2 \cdot \text{K/W}$.

Initial Temperatures (K)

n/m	1	2	3	4	5	6
1	700	700	700	1000	900	800
2	700	700	700	1000	900	800
3	700	700	700	1000	900	800

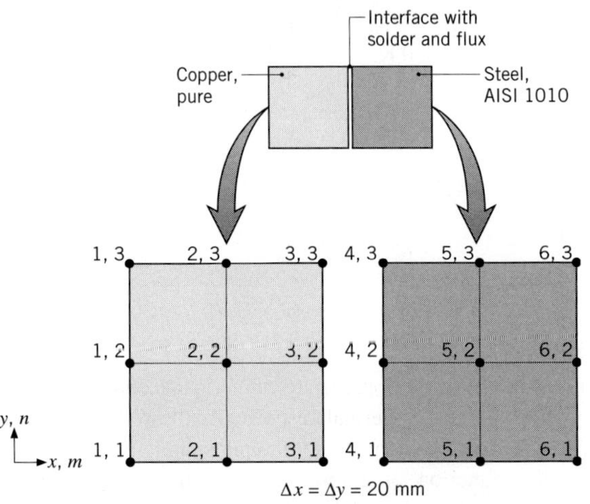

(a) Derive the explicit, finite-difference equation in terms of Fo and $Bi_c = \Delta x / k R''_{t,c}$ for $T_{4,2}$ and determine the corresponding stability criterion.

(b) Using $Fo = 0.01$, determine $T_{4,2}$ one time step after contact is made. What is Δt? Is the stability criterion satisfied?

5.140 Consider the system of Problem 4.92. Initially with no flue gases flowing, the walls ($\alpha = 5.5 \times 10^{-7}\ \text{m}^2/\text{s}$) are at a uniform temperature of 25°C. Using the implicit, finite-difference method with a time increment of 1 h, find the temperature distribution in the wall 5, 10, 50, and 100 h after introduction of the flue gases.

5.141 Consider the system of Problem 4.86. Initially, the ceramic plate ($\alpha = 1.5 \times 10^{-6}\ \text{m}^2/\text{s}$) is at a uniform temperature of 30°C, and suddenly the electrical heating elements are energized. Using the implicit, finite-difference method, estimate the time required for the difference between the surface and initial temperatures to reach 95% of the difference for steady-state conditions. Use a time increment of 2 s.

Special Applications: Finite Element Analysis

5.142 Consider the fuel element of Example 5.11, which operates at a uniform volumetric generation rate of $\dot{q}_1 = 10^7\ \text{W/m}^3$ until the generation rate suddenly changes to $\dot{q}_2 = 2 \times 10^7$ W/m^3. Use the finite-element software *FEHT* to obtain the following solutions.

(a) Calculate the temperature distribution 1.5 s after the change in operating power and compare your results with those tabulated in the example. *Hint*: First determine the steady-state temperature distribution for $\dot{q}_1$, which represents the initial condition for the transient temperature distribution after the step change in power to $\dot{q}_2$. Next, in the *Setup* menu, click on *Transient*: in the *Specify/Internal Generation* box, change the value to $\dot{q}_2$; and in the *Run* command, click on *Continue* (not *Calculate*). See the *Run* menu in the *FEHT* Help section for background information on the *Continue* option.

(b) Use your *FEHT* model to plot temperature histories at the midplane and surface for $0 \le t \le 400$ s. What are the steady-state temperatures, and approximately how long does it take to reach the new equilibrium condition after the step change in operating power?

5.143 Consider the thick slab of copper in Example 5.12, which is initially at a uniform temperature of 20°C and is suddenly exposed to large surroundings at 1000°C (instead of a prescribed heat flux).

(a) For a surface emissivity of 0.94, calculate the temperatures $T(0, 120\text{ s})$ and $T(0.15\text{ m}, 120\text{ s})$ using the finite-element software *FEHT*. *Hint*: In the *Convection Coefficient* box of the *Specify/Boundary Conditions* menu of *FEHT*, enter the linearized radiation coefficient (see Equation 1.9) for the surface ($x = 0$). Enter the temperature of the surroundings in the *Fluid Temperature* box. See also the *Help* section on *Entering Equations*. Click on *Setup/Temperatures in K* to enter all temperatures in kelvins.

(b) Plot the temperature histories for $x = 0$, 150, and 600 mm, and explain key features of your results.

5.144 Consider the composite wall of Problem 2.53. In part (d), you are asked to sketch the temperature histories at $x = 0, L$ during the transient period between cases 2 and 3. Calculate and plot these histories using the finite-element method of *FEHT*, the finite-difference method of *IHT* (with $\Delta x = 5$ mm and $\Delta t = 1.2$ s), and/or an alternative procedure of your choice. If you use more than one method, compare the respective results. Note that, in using *FEHT* or *IHT*, a look-up table must be created for prescribing the variation of the heater flux with time (see the appropriate *Help* section for guidance).

5.145 Common transmission failures result from the glazing of clutch surfaces by deposition of oil oxidation and decomposition products. Both the oxidation and decomposition processes depend on temperature histories of the surfaces. Because it is difficult to measure these surface temperatures during operation, it is useful to develop models to predict clutch-interface thermal behavior. The relative velocity between mating clutch plates, from the initial engagement to the zero-sliding (*lock-up*) condition, generates heat that is transferred to the plates. The relative velocity decreases at a constant rate during this period, producing a heat flux that is initially very large and decreases linearly with time, until lock-up occurs. Accordingly, $q''_f = q''_o = [1 - (t/t_{lu})]$, where $q''_o = 1.6 \times 10^7$ W/m^2 and $t_{lu} = 100$ ms is the lock-up time. The plates have an initial uniform temperature of $T_i = 40$°C, when the prescribed frictional heat flux is suddenly applied to the surfaces. The reaction plate is fabricated from steel, while the composite plate has a thinner steel center section bonded to low-conductivity friction material layers. The thermophysical properties are $\rho_s = 7800$ kg/m^3, $c_s = 500$ J/kg · K, and $k_s = 40$ W/m · K for the steel and $\rho_{fm} = 1150$ kg/m^3, $c_{fm} = 1650$ J/kg · K, and $k_{fm} = 4$ W/m · K for the friction material.

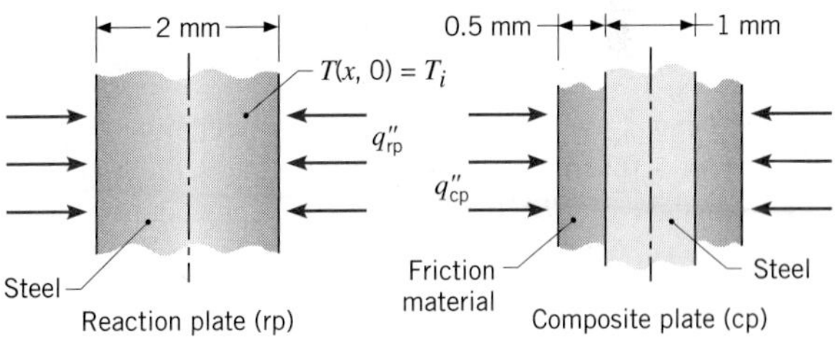

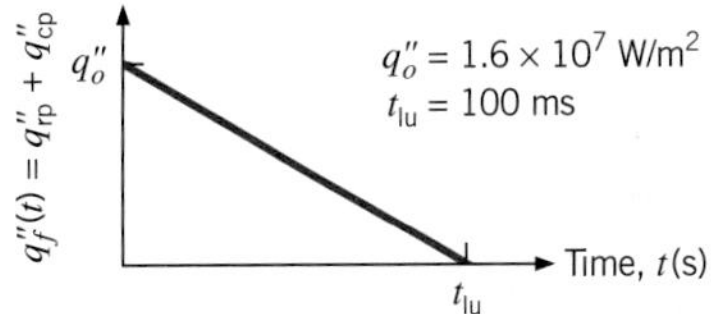

(a) On $T - t$ coordinates, sketch the temperature history at the midplane of the reaction plate, at the interface between the clutch pair, and at the midplane of the composite plate. Identify key features.

(b) Perform an energy balance on the clutch pair over the time interval $\Delta t = t_{lu}$ to determine the steady-state temperature resulting from clutch engagement. Assume negligible heat transfer from the plates to the surroundings.

(c) Compute and plot the three temperature histories of interest using the finite-element method of *FEHT* or the finite-difference method of *IHT* (with $\Delta x = 0.1$ mm and $\Delta t = 1$ ms). Calculate and plot the frictional heat fluxes to the reaction and composite plates, q''_{rp} and q''_{cp}, respectively, as a function of time. Comment on features of the temperature and heat flux histories. Validate your model by comparing predictions with the results from part (b). *Note*: Use of both *FEHT* and *IHT* requires creation of a look-up data table for prescribing the heat flux as a function of time.

5.146 A process mixture at 200°C flows at a rate of 207 kg/min onto a conveyor belt of 3-mm thickness, 1-m width, and 30-m length traveling with a velocity of 36 m/min. The underside of the belt is cooled by a water spray at a temperature of 30°C, and the convection coefficient is $3000\ \text{W/m}^2\cdot\text{K}$. The thermophysical properties of the process mixture are $\rho_m = 960\ \text{kg/m}^3$, $c_m = 1700\ \text{J/kg}\cdot\text{K}$, and $k_m = 1.5\ \text{W/m}\cdot\text{K}$, while the properties for the conveyor (metallic) belt are $\rho_b = 8000\ \text{kg/m}^3$, $c_b = 460\ \text{J/kg}\cdot\text{K}$, and $k_b = 15\ \text{W/m}\cdot\text{K}$.

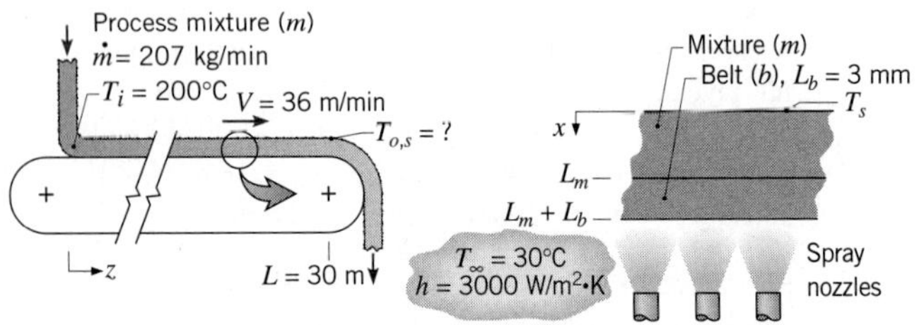

Using the finite-difference method of *IHT* ($\Delta x = 0.5$ mm, $\Delta t = 0.05$ s), the finite-element method of *FEHT*, or a numerical procedure of your choice, calculate the surface temperature of the mixture at the end of the conveyor belt $T_{o,s}$. Assume negligible heat transfer to the ambient air by convection or by radiation to the surroundings.

5.147 In a manufacturing process, stainless steel cylinders (AISI 304) initially at 600 K are quenched by submersion in an oil bath maintained at 300 K with $h = 500\ \text{W/m}^2\cdot\text{K}$. Each cylinder is of length $2L = 60$ mm and diameter $D = 80$ mm. Use the *ready-to-solve* model in the *Examples* menu of *FEHT* to obtain the following solutions.

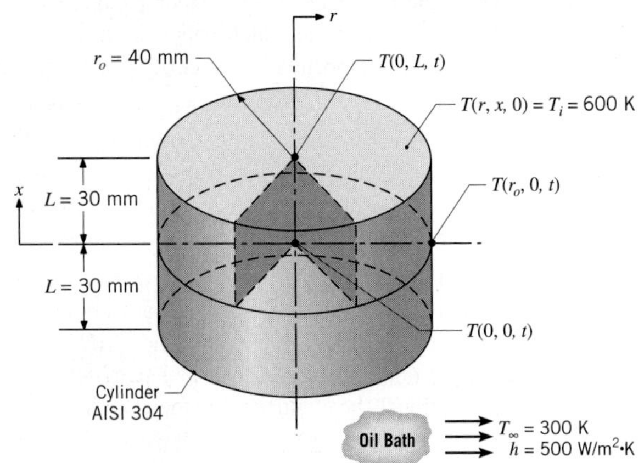

(a) Calculate the temperatures, $T(r, x, t)$, after 3 min at the cylinder center, $T(0, 0, 3\ \text{min})$, at the center of a circular face, $T(0, L, 3\ \text{min})$, and the midheight of the side, $T(r_o, 0, 3\ \text{min})$.

(b) Plot the temperature history at the center, $T(0, 0, t)$, and at the midheight of the side, $T(r_o, 0, t)$, for $0 \le t \le 10$ min using the *View/Temperatures vs. Time* command. Comment on the gradients occurring at these locations and what effect they might have on phase transformations and thermal stresses.

(c) Having solved the model for a total integration time of 10 min in part (b), now use the *View/Temperature Contours* command with the shaded band option for the isotherm contours. Select the *From Start to Stop* time option, and view the temperature contours as the cylinder cools during the quench process. Describe the major features of the cooling process revealed by this display. Use other options of this command to create a 10-isotherm temperature distribution for $t = 3$ min.

(d) For the location of part (a), calculate the temperatures after 3 min if the convection coefficient is doubled ($h = 1000\ \text{W/m}^2\cdot\text{K}$). Also, for convection coefficients of 500 and $1000\ \text{W/m}^2\cdot\text{K}$, determine how long the cylinder needs to remain in the oil bath to achieve a *safe-to-touch* surface temperature of 316 K. Tabulate and comment on the results of parts (a) and (d).

5.148 The operations manager for a metals processing plant anticipates the need to repair a large furnace and has come to you for an estimate of the time required for the furnace interior to cool to a safe working temperature. The furnace is cubical with a 16-m interior dimension and 1-m thick walls for which $\rho = 2600\ \text{kg/m}^3$, $c = 960\ \text{J/kg}\cdot\text{K}$, and $k = 1\ \text{W/m}\cdot\text{K}$. The operating temperature of the furnace is 900°C, and the outer surface experiences convection with ambient air at 25°C and a convection coefficient of $20\ \text{W/m}^2\cdot\text{K}$.

(a) Use a numerical procedure to estimate the time required for the inner surface of the furnace to cool to a safe working temperature of 35°C. *Hint*: Consider a two-dimensional cross section of the furnace, and perform your analysis on the smallest symmetrical section.

(b) Anxious to reduce the furnace downtime, the operations manager also wants to know what effect circulating ambient air through the furnace would have on the cool-down period. Assume equivalent convection conditions for the inner and outer surfaces.

CHAPTER 6

Introduction to Convection

Thus far we have focused on heat transfer by conduction and have considered convection only to the extent that it provides a possible boundary condition for conduction problems. In Section 1.2.2 we used the term *convection* to describe energy transfer between a surface and a fluid moving over the surface. Convection includes energy transfer by both the bulk fluid motion (advection) and the random motion of fluid molecules (conduction or diffusion).

In our treatment of convection, we have two major objectives. In addition to obtaining an understanding of the physical mechanisms that underlie convection transfer, we wish to develop the means to perform convection transfer calculations. This chapter and the material of Appendix E are devoted primarily to achieving the former objective. Physical origins are discussed, and relevant dimensionless parameters, as well as important analogies, are developed.

With conceptual foundations established, subsequent chapters are used to develop useful tools for quantifying convection effects. Chapters 7 and 8 present methods for computing the coefficients associated with forced convection in external and internal flow configurations, respectively. Chapter 9 describes methods for determining these coefficients in free convection, and Chapter 10 considers the problem of convection with phase change (boiling and condensation). Chapter 11 develops methods for designing and evaluating the performance of heat exchangers, devices that are widely used in engineering practice to effect heat transfer between fluids.

Accordingly, we begin by developing our understanding of the nature of convection.

6.1 *The Convection Boundary Layers*

The concept of boundary layers is central to the understanding of convection heat transfer between a surface and a fluid flowing past it. In this section, velocity and thermal boundary layers are described, and their relationships to the friction coefficient and convection heat transfer coefficient are introduced.

6.1.1 The Velocity Boundary Layer

To introduce the concept of a boundary layer, consider flow over the flat plate of Figure 6.1. When fluid particles make contact with the surface, their velocity is reduced significantly relative to the fluid velocity upstream of the plate, and for most situations it is valid to assume that the particle velocity is zero at the wall.[1] These particles then act to retard the motion of particles in the adjoining fluid layer, which act to retard the motion of particles in the next layer, and so on until, at a distance $y = \delta$ from the surface, the effect becomes negligible. This retardation of fluid motion is associated with *shear stresses* τ acting in planes that are parallel to the fluid velocity (Figure 6.1). With increasing distance y from the surface, the x velocity component of the fluid, u, must then increase until it approaches the free stream value u_∞. The subscript ∞ is used to designate conditions in the *free stream* outside the boundary layer.

[1]This is an approximation of the situation discussed in Section 3.9, wherein fluid molecules or particles continually collide with and are reflected from the surface. The momentum of an individual fluid particle will change in response to its collision with the surface. This effect may be described by *momentum accommodation coefficients*, as will be discussed in Section 8.8. In this chapter, we assume that nano- and microscale effects are not important, in which case the assumption of zero fluid velocity at the wall is valid.

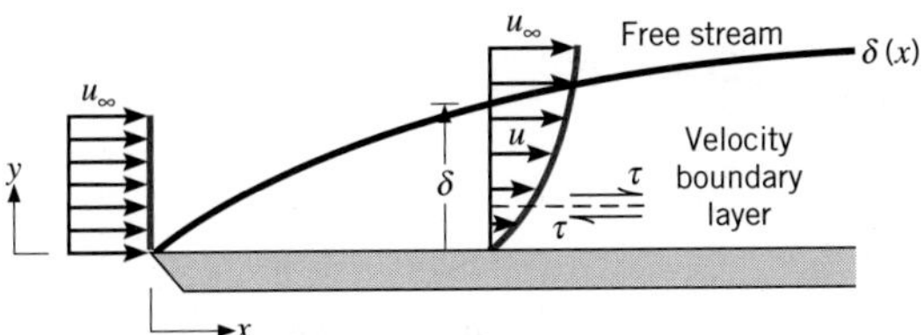

Figure 6.1 Velocity boundary layer development on a flat plate.

The quantity δ is termed the *boundary layer thickness*, and it is typically defined as the value of y for which $u = 0.99u_\infty$. The *boundary layer velocity profile* refers to the manner in which u varies with y through the boundary layer. Accordingly, the fluid flow is characterized by two distinct regions, a thin fluid layer (the boundary layer) in which velocity gradients and shear stresses are large and a region outside the boundary layer in which velocity gradients and shear stresses are negligible. With increasing distance from the leading edge, the effects of viscosity penetrate farther into the free stream and the boundary layer grows (δ increases with x).

Because it pertains to the fluid velocity, the foregoing boundary layer may be referred to more specifically as the *velocity boundary layer*. It develops whenever there is fluid flow over a surface, and it is of fundamental importance to problems involving convection transport. In fluid mechanics its significance to the engineer stems from its relation to the surface shear stress τ_s, and hence to surface frictional effects. For external flows it provides the basis for determining the local *friction coefficient*

$$C_f \equiv \frac{\tau_s}{\rho u_\infty^2/2} \tag{6.1}$$

a key dimensionless parameter from which the surface frictional drag may be determined. Assuming a *Newtonian fluid*, the surface shear stress may be evaluated from knowledge of the velocity gradient at the surface

$$\tau_s = \mu \left.\frac{\partial u}{\partial y}\right|_{y=0} \tag{6.2}$$

where μ is a fluid property known as the *dynamic viscosity*. In a velocity boundary layer, the velocity gradient at the surface depends on the distance x from the leading edge of the plate. Therefore, the surface shear stress and friction coefficient also depend on x.

6.1.2 The Thermal Boundary Layer

Just as a velocity boundary layer develops when there is fluid flow over a surface, a *thermal boundary layer* must develop if the fluid free stream and surface temperatures differ. Consider flow over an isothermal flat plate (Figure 6.2). At the leading edge the *temperature profile* is uniform, with $T(y) = T_\infty$. However, fluid particles that come into contact with the plate achieve thermal equilibrium at the plate's surface temperature.[2] In turn, these particles exchange energy with those in the adjoining fluid layer, and temperature gradients develop

[2]Micro- and nanoscale effects are assumed to be negligible in this chapter. Hence, the thermal accommodation coefficient of Section 3.9 attains a value of unity, in which case the fluid particles achieve thermal equilibrium with the surface of the plate. Micro- and nanoscale effects will be discussed in Section 8.8.

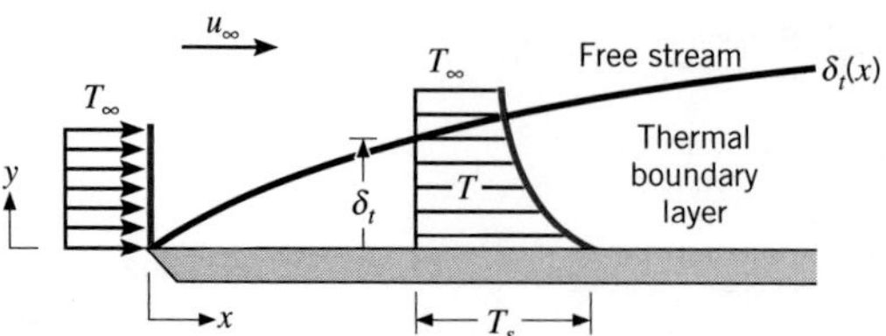

FIGURE 6.2 Thermal boundary layer development on an isothermal flat plate.

in the fluid. The region of the fluid in which these temperature gradients exist is the thermal boundary layer, and its thickness δ_t is typically defined as the value of y for which the ratio $[(T_s - T)/(T_s - T_\infty)] = 0.99$. With increasing distance from the leading edge, the effects of heat transfer penetrate farther into the free stream and the thermal boundary layer grows.

The relation between conditions in this boundary layer and the convection heat transfer coefficient may readily be demonstrated. At any distance x from the leading edge, the *local* surface heat flux may be obtained by applying Fourier's law to the *fluid* at $y = 0$. That is,

$$q_s'' = -k_f \frac{\partial T}{\partial y}\bigg|_{y=0} \tag{6.3}$$

The subscript s has been used to emphasize that this is the surface heat flux, but it will be dropped in later sections. This expression is appropriate because, *at the surface, there is no fluid motion and energy transfer occurs only by conduction.* Recalling Newton's law of cooling, we see that

$$q_s'' = h(T_s - T_\infty) \tag{6.4}$$

and combining this with Equation 6.3, we obtain

$$h = \frac{-k_f\, \partial T/\partial y|_{y=0}}{T_s - T_\infty} \tag{6.5}$$

Hence, conditions in the thermal boundary layer, which strongly influence the wall temperature gradient $\partial T/\partial y|_{y=0}$, determine the rate of heat transfer across the boundary layer. Since $(T_s - T_\infty)$ is a constant, independent of x, while δ_t increases with increasing x, temperature gradients in the boundary layer must decrease with increasing x. Accordingly, the magnitude of $\partial T/\partial y|_{y=0}$ decreases with increasing x, and it follows that q_s'' and h decrease with increasing x.

6.1.3 Significance of the Boundary Layers

For flow over any surface, there will always exist a velocity boundary layer and hence surface friction. Likewise, a thermal boundary layer, and hence convection heat transfer, will always exist if the surface and free stream temperatures differ. The velocity boundary layer is of extent $\delta(x)$ and is characterized by the presence of velocity gradients and shear stresses. The thermal boundary layer is of extent $\delta_t(x)$ and is characterized by temperature gradients and heat transfer. When both boundary layers are present, they rarely grow at the same rate, and the values of δ and δ_t at a given location are not the same.

For the engineer, the principal manifestations of the velocity and thermal boundary layers are, respectively, *surface friction* and *convection heat transfer*. The key boundary layer parameters are then the *friction coefficient* C_f and the *heat transfer convection coefficient* h. We now turn our attention to examining these key parameters, which are central to the analysis of convection heat transfer problems.

6.2 Local and Average Convection Coefficients

6.2.1 Heat Transfer

Consider the conditions of Figure 6.3*a*. A fluid of velocity V and temperature T_∞ flows over a surface of arbitrary shape and of area A_s. The surface is presumed to be at a uniform temperature, T_s, and if $T_s \neq T_\infty$, we know that convection heat transfer will occur. From Section 6.1.2, we also know that the surface heat flux and convection heat transfer coefficient both vary along the surface. The *total heat transfer rate* q may be obtained by integrating the local flux over the entire surface. That is,

$$q = \int_{A_s} q'' dA_s \tag{6.6}$$

or, from Equation 6.4,

$$q = (T_s - T_\infty) \int_{A_s} h dA_s \tag{6.7}$$

Defining an *average convection coefficient* $\bar{h}$ for the entire surface, the total heat transfer rate may also be expressed as

$$q = \bar{h} A_s (T_s - T_\infty) \tag{6.8}$$

Equating Equations 6.7 and 6.8, it follows that the average and local convection coefficients are related by an expression of the form

$$\bar{h} = \frac{1}{A_s} \int_{A_s} h dA_s \tag{6.9}$$

Note that for the special case of flow over a flat plate (Figure 6.3*b*), h varies only with the distance x from the leading edge and Equation 6.9 reduces to

$$\bar{h} = \frac{1}{L} \int_0^L h dx \tag{6.10}$$

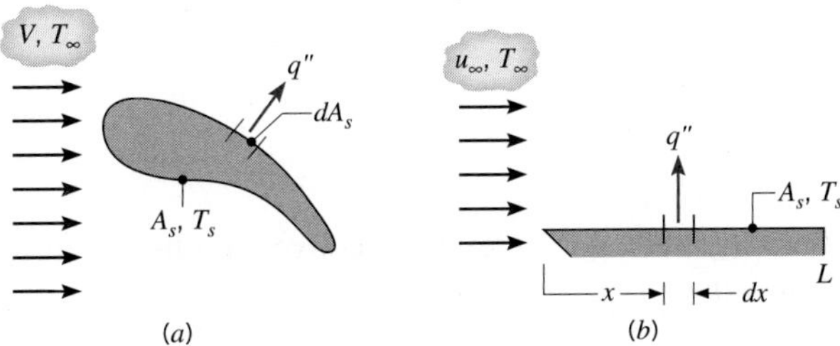

FIGURE 6.3 Local and total convection heat transfer. (*a*) Surface of arbitrary shape. (*b*) Flat plate.

6.2.2 The Problem of Convection

The local flux and/or the total transfer rate are of paramount importance in any convection problem. These quantities may be determined from the rate equations, Equations 6.4 and 6.8, which depend on knowledge of the local (h) and average ($\overline{h}$) convection coefficients. It is for this reason that determination of these coefficients is viewed as *the problem of convection.* However, the problem is not a simple one, for in addition to depending on numerous *fluid properties* such as density, viscosity, thermal conductivity, and specific heat, the coefficients depend on the *surface geometry* and the *flow conditions.* This multiplicity of independent variables is attributable to the dependence of convection transfer on the boundary layers that develop on the surface.

EXAMPLE 6.1

Experimental results for the local heat transfer coefficient h_x for flow over a flat plate with an extremely rough surface were found to fit the relation

$$h_x(x) = ax^{-0.1}$$

where a is a coefficient (W/m$^{1.9}\cdot$K) and x (m) is the distance from the leading edge of the plate.

1. Develop an expression for the ratio of the average heat transfer coefficient $\overline{h}_x$ for a plate of length x to the local heat transfer coefficient h_x at x.
2. Plot the variation of h_x and $\overline{h}_x$ as a function of x.

SOLUTION

Known: Variation of the local heat transfer coefficient, $h_x(x)$.

Find:

1. The ratio of the average heat transfer coefficient $\overline{h}(x)$ to the local value $h_x(x)$.
2. Plot of the variation of h_x and $\overline{h}_x$ with x.

Schematic:

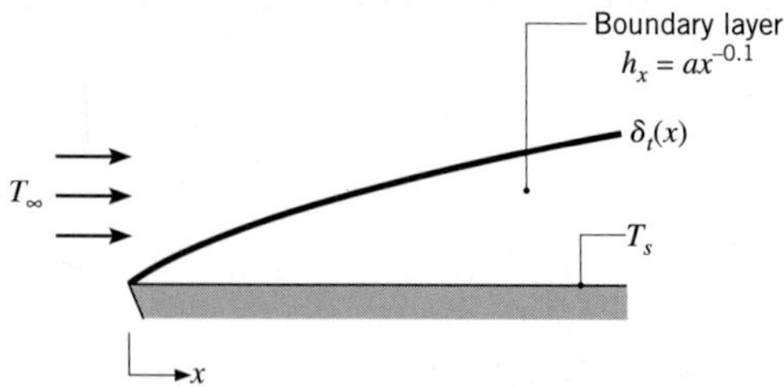

Analysis:

1. From Equation 6.10 the average value of the convection heat transfer coefficient over the region from 0 to x is

$$\overline{h}_x = \overline{h}_x(x) = \frac{1}{x}\int_0^x h_x(x)\,dx$$

Substituting the expression for the local heat transfer coefficient

$$h_x(x) = ax^{-0.1}$$

and integrating, we obtain

$$\bar{h}_x = \frac{1}{x}\int_0^x ax^{-0.1}\,dx = \frac{a}{x}\int_0^x x^{-0.1}\,dx = \frac{a}{x}\left(\frac{x^{+0.9}}{0.9}\right) = 1.11ax^{-0.1}$$

or

$$\bar{h}_x = 1.11h_x \quad \triangleleft$$

2. The variation of h_x and $\bar{h}_x$ with x is as follows:

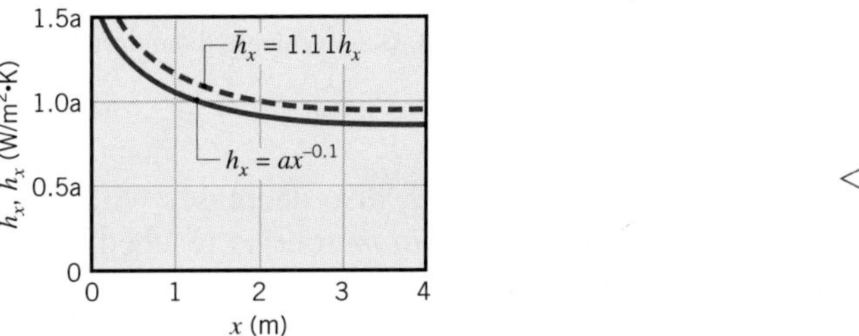

$\triangleleft$

Comments: Boundary layer development causes both the local and average coefficients to decrease with increasing distance from the leading edge. The average coefficient up to x must therefore exceed the local value at x.

6.3 *Laminar and Turbulent Flow*

In the discussion of convection so far, we have not addressed the significance of the *flow conditions*. An essential step in the treatment of any convection problem is to determine whether the boundary layer is *laminar* or *turbulent*. Surface friction and the convection transfer rate depend strongly on which of these conditions exists.

6.3.1 Laminar and Turbulent Velocity Boundary Layers

Boundary layer *development* on a flat plate is illustrated in Figure 6.4. In many cases, laminar and turbulent flow conditions both occur, with the laminar section preceding the turbulent section. For either condition, the fluid motion is characterized by velocity components in the x- and y-directions. Fluid motion away from the surface is necessitated by the slowing of the fluid near the wall as the boundary layer grows in the x-direction. Figure 6.4 shows that there are sharp differences between laminar and turbulent flow conditions, as described in the following paragraphs.

In the laminar boundary layer, the fluid flow is highly ordered and it is possible to identify streamlines along which fluid particles move. From Section 6.1.1 we know that the boundary layer thickness grows and that velocity gradients at $y = 0$ decrease in the streamwise (increasing x) direction. From Equation 6.2, we see that the local surface

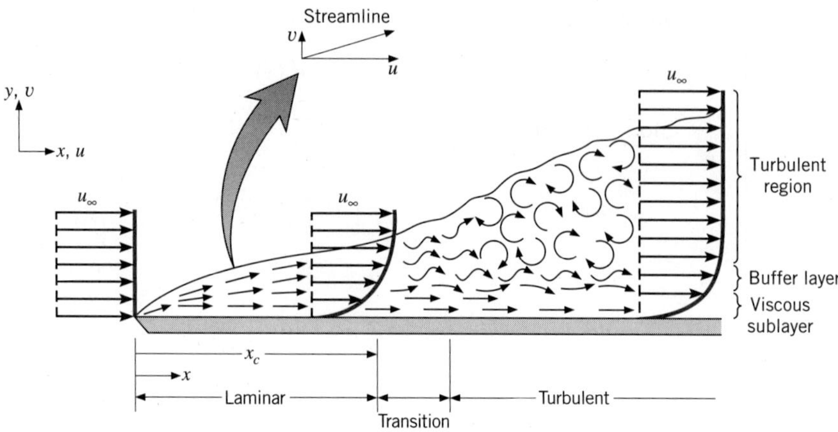

FIGURE 6.4 Velocity boundary layer development on a flat plate.

shear stress τ_s also decreases with increasing x. The highly ordered behavior continues until a *transition* zone is reached, across which a conversion from laminar to turbulent conditions occurs. Conditions within the transition zone change with time, with the flow sometimes exhibiting laminar behavior and sometimes exhibiting the characteristics of turbulent flow.

Flow in the fully turbulent boundary layer is, in general, highly irregular and is characterized by random, three-dimensional motion of relatively large parcels of fluid. Mixing within the boundary layer carries high-speed fluid toward the solid surface and transfers slower-moving fluid farther into the free stream. Much of the mixing is promoted by streamwise vortices called *streaks* that are generated intermittently near the flat plate, where they rapidly grow and decay. Recent analytical and experimental studies have suggested that these and other *coherent structures* within the turbulent flow can travel in *waves* at velocities that can exceed u_∞, interact nonlinearly, and spawn the chaotic conditions that characterize turbulent flow [1].

As a result of the interactions that lead to chaotic flow conditions, velocity and pressure fluctuations occur at any point within the turbulent boundary layer. Three different regions may be delineated within the turbulent boundary layer as a function of distance from the surface. We may speak of a *viscous sublayer* in which transport is dominated by diffusion and the velocity profile is nearly linear. There is an adjoining *buffer layer* in which diffusion and turbulent mixing are comparable, and there is a *turbulent zone* in which transport is dominated by turbulent mixing. A comparison of the laminar and turbulent boundary layer profiles for the x-component of the velocity, provided in Figure 6.5, shows that the turbulent velocity profile is relatively flat due to the mixing that occurs within the buffer layer and turbulent region, giving rise to large velocity gradients within the viscous sublayer. Hence, τ_s is generally larger in the turbulent portion of the boundary layer of Figure 6.4 than in the laminar portion.

The transition from laminar to turbulent flow is ultimately due to *triggering mechanisms*, such as the interaction of unsteady flow structures that develop naturally within the fluid or small disturbances that exist within many typical boundary layers. These disturbances may originate from fluctuations in the free stream, or they may be induced by surface roughness or minute surface vibrations. The onset of turbulence depends on whether

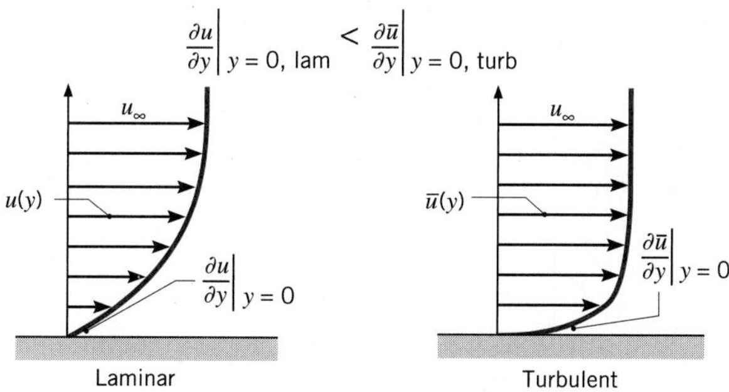

FIGURE 6.5 Comparison of laminar and turbulent velocity boundary layer profiles for the same free stream velocity.[3]

the triggering mechanisms are amplified or attenuated in the direction of fluid flow, which in turn depends on a dimensionless grouping of parameters called the *Reynolds number*,

$$Re_x = \frac{\rho u_\infty x}{\mu} \tag{6.11}$$

where, for a flat plate, the characteristic length is x, the distance from the leading edge. It will be shown later that the Reynolds number represents the ratio of the inertia to viscous forces. If the Reynolds number is small, inertia forces are insignificant relative to viscous forces. The disturbances are then dissipated, and the flow remains laminar. For a large Reynolds number, however, the inertia forces can be sufficient to amplify the triggering mechanisms, and a transition to turbulence occurs.

In determining whether the boundary layer is laminar or turbulent, it is frequently reasonable to assume that transition begins at some location x_c, as shown in Figure 6.4. This location is determined by the *critical* Reynolds number, $Re_{x,c}$. For flow over a flat plate, $Re_{x,c}$ is known to vary from approximately 10^5 to 3×10^6, depending on surface roughness and the turbulence level of the free stream. A representative value of

$$Re_{x,c} \equiv \frac{\rho u_\infty x_c}{\mu} = 5 \times 10^5 \tag{6.12}$$

is often assumed for boundary layer calculations and, unless otherwise noted, is used for the calculations of this text that involve a flat plate.

6.3.2 Laminar and Turbulent Thermal Boundary Layers

Since the velocity distribution determines the advective component of thermal energy transfer within the boundary layer, the nature of the flow also has a profound effect on the convective heat transfer rate. Similar to the laminar velocity boundary layer, the thermal boundary layer grows in the streamwise (increasing x) direction, temperature gradients in the fluid at $y = 0$ decrease in the streamwise direction, and, from Equation 6.5, the heat transfer coefficient also decreases with increasing x.

Just as it induces large velocity gradients at $y = 0$, as shown in Figure 6.5, turbulent mixing promotes large temperature gradients adjacent to the solid surface as well as a corresponding increase in the heat transfer coefficient across the transition region. These effects are illustrated in Figure 6.6 for the velocity boundary layer thickness δ and the local convection heat transfer coefficient h. Because turbulence induces mixing, which in turn reduces

[3]Since velocity fluctuates with time in turbulent flow, the time-averaged velocity, $\bar{u}$, is plotted in Figure 6.5.

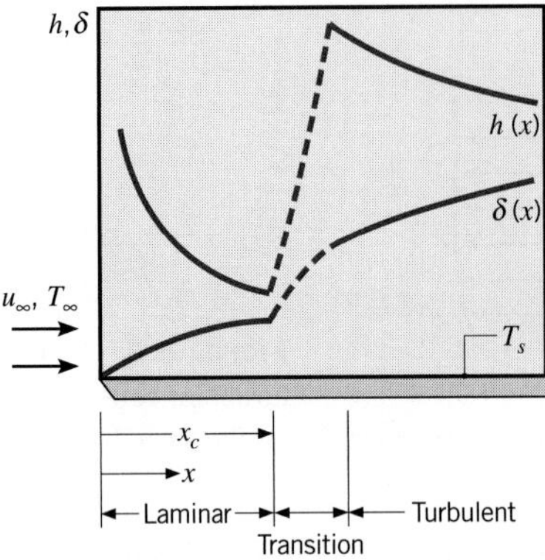

FIGURE 6.6 Variation of velocity boundary layer thickness δ and the local heat transfer coefficient h for flow over an isothermal flat plate.

the importance of conduction in determining the thermal boundary layer thickness, *differences* in the thicknesses of the velocity and thermal boundary layers tend to be much smaller in turbulent flow than in laminar flow. As is evident in Equation 6.12, the presence of heat transfer can affect the location of the transition from laminar to turbulent flow x_c since the density and dynamic viscosity of the fluid can depend on the temperature.

EXAMPLE 6.2

Water flows at a velocity $u_\infty = 1$ m/s over a flat plate of length $L = 0.6$ m. Consider two cases, one for which the water temperature is approximately 300 K and the other for an approximate water temperature of 350 K. In the laminar and turbulent regions, experimental measurements show that the local convection coefficients are well described by

$$h_{\text{lam}}(x) = C_{\text{lam}}x^{-0.5} \qquad h_{\text{turb}}(x) = C_{\text{turb}}x^{-0.2}$$

where x has units of m. At 300 K,

$$C_{\text{lam},300} = 395 \text{ W/m}^{1.5} \cdot \text{K} \qquad C_{\text{turb},300} = 2330 \text{ W/m}^{1.8} \cdot \text{K}$$

while at 350 K,

$$C_{\text{lam},350} = 477 \text{ W/m}^{1.5} \cdot \text{K} \qquad C_{\text{turb},350} = 3600 \text{ W/m}^{1.8} \cdot \text{K}$$

As is evident, the constant C depends on the nature of the flow as well as the water temperature because of the thermal dependence of various properties of the fluid.

Determine the average convection coefficient, $\bar{h}$, over the entire plate for the two water temperatures.

SOLUTION

Known: Water flow over a flat plate, expressions for the dependence of the local convection coefficient with distance from the plate's leading edge x, and approximate temperature of the water.

Find: Average convection coefficient, $\bar{h}$.

Schematic:

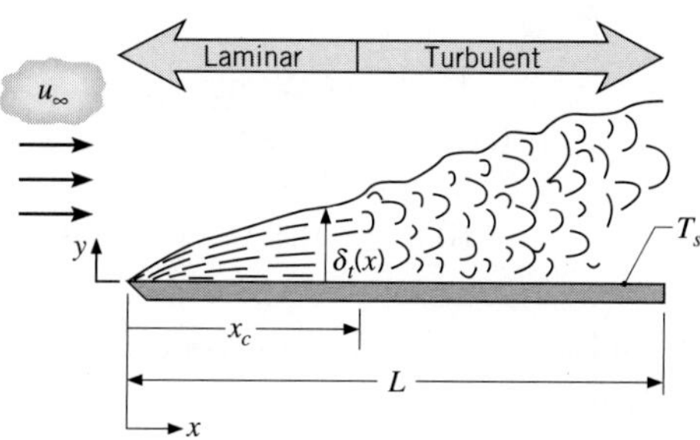

Assumptions:

1. Steady-state conditions.
2. Transition occurs at a critical Reynolds number of $Re_{x,c} = 5 \times 10^5$.

Properties: Table A.6, water ($\overline{T} \approx 300$ K): $\rho = v_f^{-1} = 997$ kg/m^3, $\mu = 855 \times 10^{-6}$ N·s/m^2. Table A.6 ($\overline{T} \approx 350$ K): $\rho = v_f^{-1} = 974$ kg/m^3, $\mu = 365 \times 10^{-6}$ N·s/m^2.

Analysis: The local convection coefficient is highly dependent on whether laminar or turbulent conditions exist. Therefore, we first determine the extent to which these conditions exist by finding the location where transition occurs, x_c. From Equation 6.12, we know that at 300 K,

$$x_c = \frac{Re_{x,c}\mu}{\rho u_\infty} = \frac{5 \times 10^5 \times 855 \times 10^{-6}\ \text{N}\cdot\text{s/m}^2}{997\ \text{kg/m}^3 \times 1\ \text{m/s}} = 0.43\ \text{m}$$

while at 350 K,

$$x_c = \frac{Re_{x,c}\mu}{\rho u_\infty} = \frac{5 \times 10^5 \times 365 \times 10^{-6}\ \text{N}\cdot\text{s/m}^2}{974\ \text{kg/m}^3 \times 1\ \text{m/s}} = 0.19\ \text{m}$$

From Equation 6.10 we know that

$$\overline{h} = \frac{1}{L}\int_0^L h dx = \frac{1}{L}\left[\int_0^{x_c} h_{\text{lam}}\, dx + \int_{x_c}^L h_{\text{turb}}\, dx\right]$$

or

$$\overline{h} = \frac{1}{L}\left[\frac{C_{\text{lam}}}{0.5}x^{0.5}\bigg|_0^{x_c} + \frac{C_{\text{turb}}}{0.8}x^{0.8}\bigg|_{x_c}^{L}\right]$$

At 300 K,

$$\overline{h} = \frac{1}{0.6\ \text{m}}\left[\frac{395\ \text{W/m}^{1.5}\cdot\text{K}}{0.5} \times (0.43^{0.5})\ \text{m}^{0.5} + \frac{2330\ \text{W/m}^{1.8}\cdot\text{K}}{0.8} \times (0.6^{0.8} - 0.43^{0.8})\ \text{m}^{0.8}\right] = 1620\ \text{W/m}^2\cdot\text{K} \quad \triangleleft$$

while at 350 K,

$$\bar{h} = \frac{1}{0.6\ \text{m}}\left[\frac{477\ \text{W/m}^{1.5}\cdot\text{K}}{0.5} \times (0.19^{0.5})\ \text{m}^{0.5} + \frac{3600\ \text{W/m}^{1.8}\cdot\text{K}}{0.8}\right.$$

$$\left. \times\ (0.6^{0.8} - 0.19^{0.8})\ \text{m}^{0.8}\right] = 3710\ \text{W/m}^2\cdot\text{K} \qquad \triangleleft$$

The local and average convection coefficient distributions for the plate are shown in the following figure.

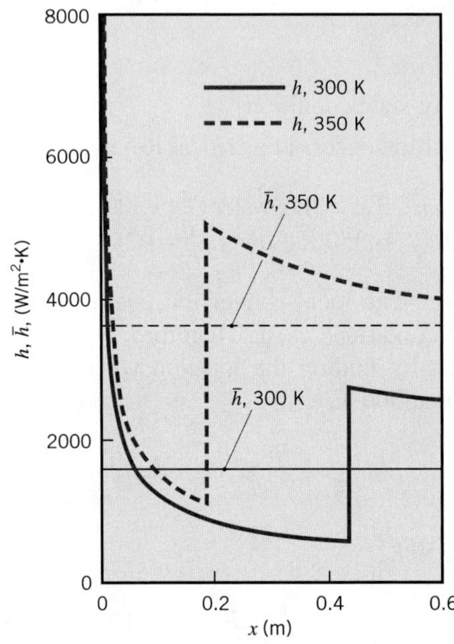

Comments:

1. The average convection coefficient at $T \approx 350$ K is over twice as large as the value at $T \approx 300$ K. This strong temperature dependence is due primarily to the shift of x_c that is associated with the smaller viscosity of the water at the higher temperature. Careful consideration of the temperature dependence of fluid properties is *crucial* when performing a convection heat transfer analysis.
2. Spatial variations in the local convection coefficient are significant. The largest local convection coefficients occur at the leading edge of the flat plate, where the laminar thermal boundary layer is extremely thin, and just downstream of x_c, where the turbulent boundary layer is thinnest.

6.4 *The Boundary Layer Equations*

We can improve our understanding of the physical effects that determine boundary layer behavior and further illustrate its relevance to convection transport by considering the equations that govern boundary layer conditions, such as those illustrated in Figure 6.7.

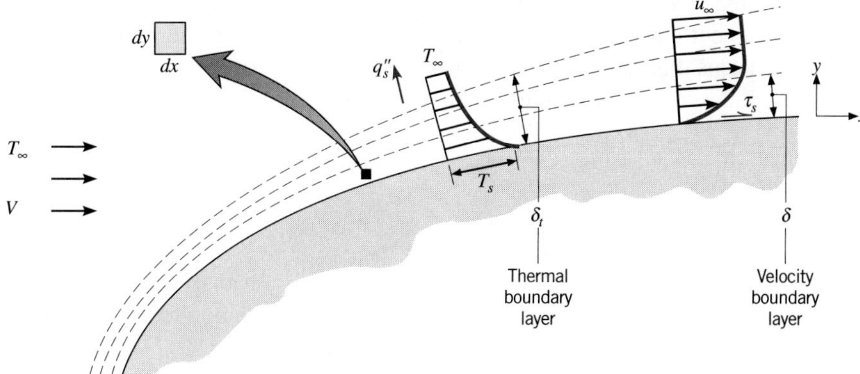

FIGURE 6.7 Development of the velocity and thermal boundary layers for an arbitrary surface.

As discussed in Section 6.1, the velocity boundary layer results from the difference between the free stream velocity and the zero velocity at the wall, while the thermal boundary layer results from a difference between the free stream and surface temperatures. Illustration of the relative thicknesses ($\delta_t > \delta_c$) in Figure 6.7 is arbitrary, for the moment, and the factors that influence relative boundary layer development are discussed later in this chapter.

Our objective in this section is to examine the differential equations that govern the velocity and temperature fields that are applicable to boundary layer flow with heat transfer. Section 6.4.1 presents the laminar boundary layer equations, and Appendix F gives the corresponding equations for turbulent conditions. In Section 6.5, these equations will be used to determine important dimensionless parameters associated with convection that will be used extensively in subsequent chapters.

6.4.1 Boundary Layer Equations for Laminar Flow

Motion of a fluid in which there are coexisting velocity and temperature gradients must comply with several *fundamental laws of nature*. In particular, at each point in the fluid, *conservation of mass* and *energy*, as well as *Newton's second law of motion*, must be satisfied. Equations representing these requirements are derived by applying the laws to a differential control volume situated in the flow. The resulting equations, in Cartesian coordinates, for the *steady, two-dimensional flow* of an *incompressible fluid* with *constant properties* are given in Appendix E. These equations serve as starting points for our analysis of laminar boundary layers. Note that turbulent flows are inherently unsteady, and the equations governing them are presented in Appendix F.

We begin by restricting attention to applications for which *body forces are negligible* ($X = Y = 0$ in Equations E.2 and E.3) and there is *no thermal energy generation in the fluid* ($\dot{q} = 0$ in Equation E.4). Additional simplifications may be made by invoking approximations pertinent to conditions in the velocity and thermal boundary layers. The boundary layer thicknesses are typically very small relative to the size of the object upon which they form, and the x-direction velocity and temperature must change from their surface to their

These equations are derived in Section 6S.1.

free stream values over these very small distances. Therefore, gradients normal to the object's surface are much larger than those along the surface. As a result, we can neglect terms that represent x-direction diffusion of momentum and thermal energy relative to their y-direction counterparts. That is [2, 3]:

$$\frac{\partial^2 u}{\partial x^2} \ll \frac{\partial^2 u}{\partial y^2} \qquad \frac{\partial^2 T}{\partial x^2} \ll \frac{\partial^2 T}{\partial y^2} \tag{6.13}$$

By neglecting the x-direction terms, we are assuming the net shear stress and conduction heat flux in the x-direction to be negligible. Furthermore, because the boundary layer is so thin, the x-direction pressure gradient within the boundary layer can be approximated as the free stream pressure gradient:

$$\frac{\partial p}{\partial x} \approx \frac{dp_\infty}{dx} \tag{6.14}$$

The form of $p_\infty(x)$ depends on the surface geometry and may be obtained from a separate consideration of flow conditions in the free stream where shear stresses are negligible [4]. Hence, the pressure gradient may be treated as a known quantity.

With the foregoing simplifications and approximations, the overall continuity equation is unchanged from Equation E.1:

$$\frac{\partial u}{\partial x} + \frac{\partial v}{\partial y} = 0 \tag{6.15}$$

This equation is an outgrowth of applying conservation of mass to the differential, $dx \cdot dy \cdot 1$ control volume shown in Figure 6.7. The two terms represent the *net* outflow (outflow minus inflow) of mass in the x- and y-directions, the sum of which must be zero for steady flow.

The x-momentum equation (Equation E.2) reduces to:

$$u\frac{\partial u}{\partial x} + v\frac{\partial u}{\partial y} = -\frac{1}{\rho}\frac{dp_\infty}{dx} + \nu\frac{\partial^2 u}{\partial y^2} \tag{6.16}$$

This equation results from application of Newton's second law of motion in the x-direction to the $dx \cdot dy \cdot 1$ differential control volume in the fluid. The left-hand side represents the net rate at which x-momentum leaves the control volume due to fluid motion across its boundaries. The first term on the right-hand side represents the net pressure force, and the second term represents the net force due to viscous shear stresses.

The energy equation (Equation E.4) reduces to

$$u\frac{\partial T}{\partial x} + v\frac{\partial T}{\partial y} = \alpha\frac{\partial^2 T}{\partial y^2} + \frac{\nu}{c_p}\left(\frac{\partial u}{\partial y}\right)^2 \tag{6.17}$$

This equation results from application of conservation of energy to the $dx \cdot dy \cdot 1$ differential control volume in the flowing fluid. Terms on the left-hand side account for the net rate at which thermal energy leaves the control volume due to bulk fluid motion (advection). The first term on the right-hand side accounts for the net inflow of thermal energy due to y-direction conduction. The last term on the right-hand side is what remains of the viscous dissipation, Equation E.5, when it is acknowledged that, in a boundary layer, the velocity component in the direction along the surface, u, is much larger than that normal to the surface, v, and gradients normal to the surface are much larger than those along the surface. In many situations this term may be neglected relative to those that account for advection and conduction. However, aerodynamic heating that accompanies high-speed (especially supersonic) flight is a noteworthy situation in which this term is important.

After specifying appropriate boundary conditions, Equations 6.15 through 6.17 may be solved to determine the spatial variations of u, v, and T in the different laminar boundary layers. For incompressible, constant property flow, Equations 6.15 and 6.16 are *uncoupled* from Equation 6.17. That is, Equations 6.15 and 6.16 may be solved for the *velocity field*, $u(x, y)$ and $v(x, y)$, without consideration of Equation 6.17. From knowledge of $u(x, y)$, the velocity gradient $(\partial u/\partial y)|_{y=0}$ could then be evaluated, and the wall shear stress could be obtained from Equation 6.2. In contrast, through the appearance of u and v in Equation 6.17, the temperature is *coupled* to the velocity field. Hence $u(x, y)$ and $v(x, y)$ must be known before Equation 6.17 may be solved for $T(x, y)$. Once $T(x, y)$ has been obtained from such solutions, the convection heat transfer coefficient may be determined from Equation 6.5. It follows that this coefficient depends strongly on the velocity field.

Because boundary layer solutions generally involve mathematics beyond the scope of this text, our treatment of such solutions will be restricted to the analysis of laminar parallel flow over a flat plate (Section 7.2 and Appendix G). However, other analytical solutions are discussed in advanced texts on convection [5–7], and detailed boundary layer solutions may be obtained by using numerical (finite-difference or finite-element) techniques [8]. It is also essential to recognize that a wide array of situations of engineering relevance involve turbulent convective heat transfer, which is both mathematically and physically more complex than laminar convection. The boundary layer equations for turbulent flow are included in Appendix F.

It is important to stress that we have not developed the laminar boundary layer equations primarily for the purpose of obtaining solutions. Rather, we have been motivated mainly by two other considerations. One motivation has been to obtain an appreciation for the physical processes that occur in boundary layers. These processes affect wall friction and energy transfer in the boundary layers. A second important motivation arises from the fact that the equations may be used to identify key *boundary layer similarity parameters*, as well as important *analogies* between *momentum* and *heat* transfer that have numerous practical applications. The laminar governing equations will be used for this purpose in Sections 6.5 through 6.7, but the same key parameters and analogies hold true for turbulent conditions as well.

6.4.2 Compressible Flow

The equations of the foregoing section and Appendix E are restricted to incompressible flows, that is, for cases where the fluid density can be treated as constant.[4] Flows in which the fluids experience significant density changes as a result of *pressure variations* associated with the fluid motion are deemed to be compressible. The treatment of convection heat transfer associated with *compressible flow* is beyond the scope of this text. Although liquids may nearly

[4]Chapter 9 addresses flows that arise due to the variation of density with temperature. These *free convection* flows can nearly always be treated as if the fluid is incompressible but with an extra term in the momentum equation to account for buoyancy forces.

always be treated as incompressible, density variations in flowing gases should be considered when the velocity approaches or exceeds the speed of sound. Specifically, a gradual transition from incompressible to compressible flow in gases occurs at a critical Mach number of $Ma_c \approx 0.3$, where $Ma \equiv V/a$ and V and a are the gas velocity and speed of sound, respectively [9, 10]. For an ideal gas, $a = \sqrt{\gamma RT}$ where γ is the ratio of specific heats, $\gamma \equiv c_p/c_v$, R is the gas constant, and the temperature is expressed in kelvins. As an example, for air at $T = 300$ K and $p = 1$ atm, we may assume ideal gas behavior. The gas constant is $R \equiv \mathcal{R}/\mathcal{M} =$ 8315 J/kmol·K/28.7 kg/kmol = 287 J/kg·K and $c_v \equiv c_p - R =$ 1007 J/kg·K − 287 J/kg·K = 720 J/kg·K. The ratio of specific heats is therefore $\gamma = c_p/c_v =$ 1007 J/kg·K/720 J/kg·K = 1.4, and the speed of sound is $a = \sqrt{1.4 \times 287 \text{ J/kg}\cdot\text{K} \times 300 \text{ K}} = 347$ m/s. Hence air flowing at 300 K must be treated as being compressible if $V > 0.3 \times 347$ m/s $\cong 100$ m/s.

Since the material in Chapters 6 through 9 is restricted to incompressible or *low-speed* flow, it is important to confirm that compressibility effects are not important when utilizing the material to solve a convection heat transfer problem.[5]

6.5 Boundary Layer Similarity: The Normalized Boundary Layer Equations

If we examine Equations 6.16 and 6.17 we note a strong similarity. In fact, if the pressure gradient appearing in Equation 6.16 and the viscous dissipation term of Equation 6.17 are negligible, the two equations are of the same form. *Both equations are characterized by advection terms on the left-hand side and a diffusion term on the right-hand side.* This situation describes *low-speed, forced convection flows,* which are found in many engineering applications. Implications of this similarity may be developed in a rational manner by first *nondimensionalizing* the governing equations.

6.5.1 Boundary Layer Similarity Parameters

The boundary layer equations are *normalized* by first defining dimensionless independent variables of the forms

$$x^* \equiv \frac{x}{L} \qquad \text{and} \qquad y^* \equiv \frac{y}{L} \tag{6.18}$$

where L is a *characteristic length* for the surface of interest (e.g., the length of a flat plate). Moreover, dependent dimensionless variables may also be defined as

$$u^* \equiv \frac{u}{V} \qquad \text{and} \qquad v^* \equiv \frac{v}{V} \tag{6.19}$$

where V is the velocity upstream of the surface (Figure 6.7), and as

$$T^* \equiv \frac{T - T_s}{T_\infty - T_s} \tag{6.20}$$

[5]Turbulence and compressibility often coincide, since large velocities can lead to large Reynolds and Mach numbers. It can be shown (Problem 6.26) that, for sufficiently small geometries, any flow that is turbulent is also compressible.

The dimensional variables may then be written in terms of the new dimensionless variables (for example, from Equation 6.18 $x \equiv x^*L$ and $y \equiv y^*L$) and the resulting expressions for x, y, u, v, and T may be substituted into Equations 6.16 and 6.17 to obtain the dimensionless forms of the conservation equations shown in Table 6.1. Note that viscous dissipation has been neglected and that $p^* \equiv (p_\infty/\rho V^2)$ is a dimensionless pressure. The y-direction boundary conditions required to solve the equations are also shown in the table.

By normalizing the boundary layer equations, two very important dimensionless *similarity parameters* evolve and are introduced in Table 6.1. They are the Reynolds number, Re_L and Prandtl number, Pr. Such similarity parameters are important because they allow us to apply results obtained for a surface experiencing one set of convective conditions to *geometrically similar* surfaces experiencing entirely different conditions. These conditions may vary, for example, with the fluid, the fluid velocity as described by the free stream value V, and/or the size of the surface as described by the characteristic length L. As long as the similarity parameters *and* dimensionless boundary conditions are the same for two sets of conditions, the solutions of the differential equations of Table 6.1 for the nondimensional velocity and temperature will be *identical*. This concept will be amplified in the remainder of this section.

6.5.2 Functional Form of the Solutions

Equations 6.21 through 6.26 in Table 6.1 are extremely useful from the standpoint of suggesting how important boundary layer results may be simplified and generalized. The momentum equation (6.21) suggests that, although conditions in the velocity boundary layer depend on the fluid properties ρ and μ, the velocity V, and the length scale L, this dependence may be simplified by grouping these variables in the form of the Reynolds number. We therefore anticipate that the solution to Equation 6.21 will be of the functional form

$$u^* = f\left(x^*, y^*, Re_L, \frac{dp^*}{dx^*}\right) \tag{6.27}$$

Since the pressure distribution $p^*(x^*)$ depends on the surface geometry and may be obtained independently by considering flow conditions in the free stream, the appearance of dp^*/dx^* in Equation 6.27 represents the influence of geometry on the velocity distribution.

From Equations 6.2, 6.18, and 6.19, the shear stress at the surface, $y^* = 0$, may be expressed as

$$\tau_s = \mu \left.\frac{\partial u}{\partial y}\right|_{y=0} = \left(\frac{\mu V}{L}\right) \left.\frac{\partial u^*}{\partial y^*}\right|_{y^*=0}$$

and from Equations 6.1 and 6.25, it follows that the friction coefficient is

Friction Coefficient:

$$C_f = \frac{\tau_s}{\rho V^2/2} = \frac{2}{Re_L} \left.\frac{\partial u^*}{\partial y^*}\right|_{y^*=0} \tag{6.28}$$

From Equation 6.27 we also know that

$$\left.\frac{\partial u^*}{\partial y^*}\right|_{y^*=0} = f\left(x^*, Re_L, \frac{dp^*}{dx^*}\right)$$

TABLE 6.1 The boundary layer equations and their y-direction boundary conditions in nondimensional form

Boundary Layer	Conservation Equation	Boundary Conditions		Similarity Parameter(s)
		Wall	Free Stream	
Velocity	$u^*\frac{\partial u^*}{\partial x^*} + v^*\frac{\partial u^*}{\partial y^*} = -\frac{dp^*}{dx^*} + \frac{1}{Re_L}\frac{\partial^2 u^*}{\partial y^{*2}}$ (6.21)	$u^*(x^*, 0) = 0$	$u^*(x^*, \infty) = \frac{u_\infty(x^*)}{V}$ (6.23)	$Re_L = \frac{VL}{\nu}$ (6.25)
Thermal	$u^*\frac{\partial T^*}{\partial x^*} + v^*\frac{\partial T^*}{\partial y^*} = \frac{1}{Re_L Pr}\frac{\partial^2 T^*}{\partial y^{*2}}$ (6.22)	$T^*(x^*, 0) = 0$	$T^*(x^*, \infty) = 1$ (6.24)	$Re_L, Pr = \frac{\nu}{\alpha}$ (6.26)

Convection Similarity Parameters Re_L and Pr

Hence, *for a prescribed geometry* Equation 6.28 may be expressed as

$$C_f = \frac{2}{Re_L} f(x^*, Re_L) \tag{6.29}$$

The significance of this result should not be overlooked. Equation 6.29 states that the friction coefficient, a dimensionless parameter of considerable importance to the engineer, may be expressed exclusively in terms of a dimensionless space coordinate and the Reynolds number. Hence, for a prescribed geometry we expect the function that relates C_f to x^* and Re_L to be *universally* applicable. That is, we expect it to apply to different fluids and over a wide range of values for V and L.

Similar results may be obtained for the heat transfer coefficient. Intuitively, we might anticipate that h depends on the fluid properties (k, c_p, μ, and ρ), the fluid velocity V, the length scale L, and the surface geometry. However, Equation 6.22 suggests the manner in which this dependence may be simplified. In particular, the solution to this equation may be expressed in the form

$$T^* = f\left(x^*, y^*, Re_L, Pr, \frac{dp^*}{dx^*}\right) \tag{6.30}$$

where the dependence on dp^*/dx^* originates from the influence of the geometry on the fluid motion (u^* and v^*), which, in turn, affects the thermal conditions. Once again the term dp^*/dx^* represents the effect of surface geometry. From the definition of the convection coefficient, Equation 6.5, and the dimensionless variables, Equations 6.18 and 6.20, we also obtain

$$h = -\frac{k_f (T_\infty - T_s)}{L (T_s - T_\infty)} \frac{\partial T^*}{\partial y^*}\bigg|_{y^*=0} = +\frac{k_f}{L}\frac{\partial T^*}{\partial y^*}\bigg|_{y^*=0}$$

This expression suggests defining an important dependent dimensionless parameter termed the Nusselt number.

Nusselt Number:

$$Nu \equiv \frac{hL}{k_f} = +\frac{\partial T^*}{\partial y^*}\bigg|_{y^*=0} \tag{6.31}$$

This parameter is equal to the dimensionless temperature gradient at the surface, and it provides a measure of the convection heat transfer occurring at the surface. From Equation 6.30 it follows that, *for a prescribed geometry,*

$$Nu = f(x^*, Re_L, Pr) \tag{6.32}$$

The Nusselt number is to the thermal boundary layer what the friction coefficient is to the velocity boundary layer. Equation 6.32 implies that for a given geometry, the Nusselt number must be some *universal function* of x^*, Re_L, and Pr. If this function were known, it could be used to compute the value of Nu for different fluids and for different values of V and L. From knowledge of Nu, the local convection coefficient h may be found and the *local* heat flux may then be computed from Equation 6.4. Moreover, since the *average* heat transfer coefficient is obtained by integrating over the surface of the body, it must be independent of the spatial variable x^*. Hence the functional dependence of the *average* Nusselt number is

$$\overline{Nu} = \frac{\bar{h}L}{k_f} = f(Re_L, Pr) \tag{6.33}$$

From the foregoing development we have obtained the relevant dimensionless parameters for low-speed, forced-convection boundary layers. We have done so by nondimensionalizing the differential equations that describe the physical processes occurring within the boundary layers. An alternative approach could have involved the use of dimensional analysis in the form of the Buckingham pi theorem [4]. However, the success of that method depends on one's ability to select, largely from intuition, the various parameters that influence a problem. For example, knowing beforehand that $\bar{h} = f(k, c_p, \rho, \mu, V, L)$, one could use the Buckingham pi theorem to obtain Equation 6.33. However, having begun with the differential form of the conservation equations, we have eliminated the guesswork and have established the similarity parameters in a rigorous fashion.

The importance of an expression such as Equation 6.33 should be fully appreciated. It states that values of the average heat transfer coefficient $\bar{h}$, whether obtained theoretically, experimentally, or numerically, can be completely represented in terms of only three dimensionless groups, instead of the original seven dimensional parameters. The convenience and power afforded by such simplification will become evident in Chapters 7 through 10. Moreover, once the form of the functional dependence of Equation 6.33 has been obtained for a particular surface geometry, let us say from laboratory measurements, it is known to be *universally* applicable. By this we mean that it may be applied for different fluids, velocities, and length scales, as long as the assumptions implicit in the originating boundary layer equations remain valid (e.g., negligible viscous dissipation and body forces).

Example 6.3

Experimental tests using air as the working fluid are conducted on a portion of the turbine blade shown in the sketch. The heat flux to the blade at a particular point (x^*) on the surface is measured to be $q'' = 95{,}000\ \text{W/m}^2$. To maintain a steady-state surface temperature of 800°C, heat transferred to the blade is removed by circulating a coolant inside the blade.

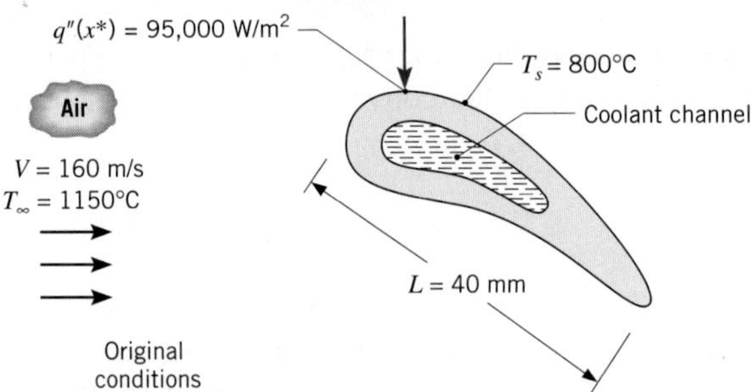

1. Determine the heat flux to the blade at x^* if its temperature is reduced to $T_{s,1} = 700°\text{C}$ by increasing the coolant flow.
2. Determine the heat flux at the same dimensionless location x^* for a similar turbine blade having a chord length of $L = 80$ mm, when the blade operates in an airflow at $T_\infty = 1150°\text{C}$ and $V = 80$ m/s, with $T_s = 800°\text{C}$.

SOLUTION

Known: Operating conditions of an internally cooled turbine blade.

Find:

1. Heat flux to the blade at a point x^* when the surface temperature is reduced.
2. Heat flux at the same dimensionless location to a larger turbine blade of the same shape with reduced air velocity.

Schematic:

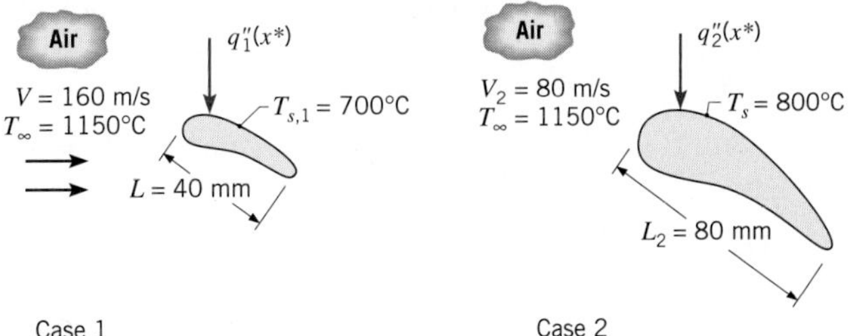

Assumptions:

1. Steady-state, incompressible flow.
2. Constant air properties.

Analysis:

1. When the surface temperature is 800°C, the local convection heat transfer coefficient between the surface and the air at x^* can be obtained from Newton's law of cooling:

$$q'' = h(T_\infty - T_s)$$

Thus,

$$h = \frac{q''}{(T_\infty - T_s)}$$

We proceed without calculating the value for now. From Equation 6.32, it follows that, for the prescribed geometry,

$$Nu = \frac{hL}{k} = f(x^*, Re_L, Pr)$$

Hence, since there is no change in x^*, Re_L, or Pr associated with a change in T, for constant properties, the local Nusselt number is unchanged. Moreover, since L and k are unchanged, the local convection coefficient remains the same. Thus, when the surface temperature is reduced to 700°C, the heat flux may be obtained from Newton's law of cooling, using the same local convection coefficient:

$$q_1'' = h(T_\infty - T_{s,1}) = \frac{q''}{(T_\infty - T_s)}(T_\infty - T_{s,1}) = \frac{95{,}000\ \text{W/m}^2}{(1150 - 800)°\text{C}}(1150 - 700)°\text{C}$$

$$= 122{,}000\ \text{W/m}^2 \qquad \triangleleft$$

2. To determine the heat flux at x^* associated with the larger blade and the reduced airflow (case 2), we first note that, although L has increased by a factor of 2, the velocity has decreased by the same factor and the Reynolds number has not changed. That is,

$$Re_{L,2} = \frac{V_2 L_2}{\nu} = \frac{VL}{\nu} = Re_L$$

Accordingly, since x^* and Pr are also unchanged, the local Nusselt number remains the same.

$$Nu_2 = Nu$$

Because the characteristic length is different, however, the local convection coefficient changes, where

$$\frac{h_2 L_2}{k} = \frac{hL}{k} \quad \text{or} \quad h_2 = h\frac{L}{L_2} = \frac{q''}{(T_\infty - T_s)}\frac{L}{L_2}$$

The heat flux at x^* is then

$$q''_2 = h_2(T_\infty - T_s) = q''\frac{(T_\infty - T_s)}{(T_\infty - T_s)}\frac{L}{L_2}$$

$$q''_2 = 95{,}000 \text{ W/m}^2 \times \frac{0.04 \text{ m}}{0.08 \text{ m}} = 47{,}500 \text{ W/m}^2 \qquad \triangleleft$$

Comments:

1. If the Reynolds numbers for the two situations of part 2 differed, that is, $Re_{L,2} \neq Re_L$, the local heat flux q''_2 could be obtained only if the particular functional dependence of Equation 6.32 were known. Such forms are provided for many different shapes in subsequent chapters.
2. Air temperatures in the boundary layer range from the blade surface temperature T_s to the ambient value T_∞. Hence, as will become evident in Section 7.1 representative air properties could be evaluated at arithmetic mean or *film* temperatures of $T_{f,1} = (T_{s,1} + T_\infty)/2 = (700°\text{C} + 1150°\text{C})/2 = 925°\text{C}$ and $T_{f,2} = (800°\text{C} + 1150°\text{C})/2 = 975°\text{C}$, respectively. Based on properties corresponding to these film temperatures, the Reynolds (and Nusselt) numbers for the two cases would be slightly different. However, the difference would not be large enough to significantly change the calculated value of the local heat flux for case 2.
3. At $T = 1150°\text{C} = 1423 \text{ K}$, $c_v \equiv c_p - R = 1167 \text{ J/kg}\cdot\text{K} - 287 \text{ J/kg}\cdot\text{K} = 880 \text{ J/kg}\cdot\text{K}$, and the specific heat ratio is $\gamma = c_p/c_v = 1167 \text{ J/kg}\cdot\text{K}/880 \text{ J/kg}\cdot\text{K} = 1.33$. Assuming the air behaves as an ideal gas, the speed of sound in the air is $a = \sqrt{\gamma RT} = \sqrt{1.33 \times 287 \text{ J/kg}\cdot\text{K} \times 1423 \text{ K}} = 736 \text{ m/s}$. Therefore $Ma = V/a = 0.22$ and 0.11 for cases 1 and 2, respectively. Hence the flow is incompressible in both cases. If the flow were to be compressible for either case, the Nusselt number would also depend on the Mach number, and the two cases would not be similar.

EXAMPLE 6.4

Consider convective cooling of a two-dimensional streamlined strut of characteristic length $L_{H_2} = 40$ mm. The strut is exposed to hydrogen flowing at $p_{H_2} = 2$ atm, $V_{H_2} = 8.1$ m/s, and $T_{\infty,H_2} = -30°C$. Of interest is the value of the average heat transfer coefficient $\bar{h}_{H_2}$, when the surface temperature is $T_{s,H_2} = -15°C$. Rather than conducting expensive experiments involving pressurized hydrogen, an engineer proposes to take advantage of similarity by performing wind tunnel experiments using air at atmospheric pressure with $T_{\infty,\text{Air}} = 23°C$. A geometrically similar strut of characteristic length $L_{\text{Air}} = 60$ mm and perimeter $P = 150$ mm is placed in the wind tunnel. Measurements reveal a surface temperature of $T_{s,\text{Air}} = 30°C$ when the heat loss per unit object length (into the page) is $q'_{\text{Air}} = 50$ W/m. Determine the required air velocity in the wind tunnel experiment V_{Air} and the average convective heat transfer coefficient in the hydrogen $\bar{h}_{H_2}$.

SOLUTION

Known: Flow across a strut. Hydrogen pressure, velocity, and temperature. Air temperature and pressure, as well as heat loss per unit length. Surface temperatures of the strut in hydrogen and in air.

Find: Air velocity and average convective heat transfer coefficient for the strut that is exposed to hydrogen.

Schematic:

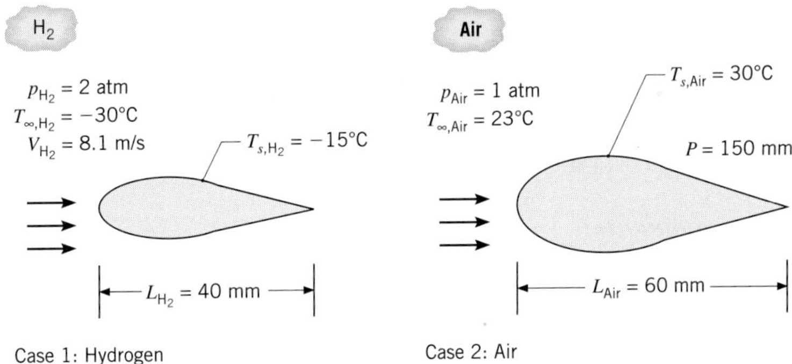

Assumptions:

1. Steady-state, incompressible boundary layer behavior.
2. Ideal gas behavior.
3. Constant properties.
4. Negligible viscous dissipation.

Properties: Table A.4, air ($p = 1$ atm, $T_f = (23°C + 30°C)/2 = 26.5°C \approx 300$ K): $Pr = 0.707$, $\nu = 15.89 \times 10^{-6}$ m²/s, $k = 26.3 \times 10^{-3}$ W/m·K. Table A.4 hydrogen ($p = 1$ atm, $T_f = -22.5°C \approx 250$ K): $Pr = 0.707$, $\nu = 81.4 \times 10^{-6}$ m²/s, $k = 157 \times 10^{-3}$ W/m·K.

The properties k, Pr, c_p, and μ may be assumed to be independent of pressure to an excellent approximation. However, for a gas, the kinematic viscosity $\nu = \mu/\rho$ will vary with pressure through its dependence on density. From the ideal gas law, $\rho = p/RT$, it follows that the ratio of kinematic viscosities for a gas at the same temperature but at different pressures, p_1 and p_2, is $(\nu_1/\nu_2) = (p_2/p_1)$. Hence, the kinematic viscosity of hydrogen at 250 K and 2 atm is $\nu_{H_2} = 81.4 \times 10^{-6}\ \text{m}^2/\text{s} \times 1\ \text{atm}/2\ \text{atm} = 40.7 \times 10^{-6}\ \text{m}^2/\text{s}$. Since Pr is independent of pressure, $Pr_{H_2}(p = 2\ \text{atm}, T_f = -22.5°\text{C}) = Pr_{\text{Air}}\ (p = 1\ \text{atm}, T_f = 26.5°\text{C}) = 0.707$.

Analysis: From Equation 6.33, we know that the average Nusselt numbers are related to the Reynolds and Prandtl numbers by the functional dependence

$$\overline{Nu}_{H_2} = \frac{\bar{h}_{H_2} L_{H_2}}{k_{H_2}} = f(Re_{L,H_2}, Pr_{H_2}) \quad \text{and} \quad \overline{Nu}_{\text{Air}} = \frac{\bar{h}_{\text{Air}} L_{\text{Air}}}{k_{\text{Air}}} = f(Re_{L,\text{Air}}, Pr_{\text{Air}})$$

Since $Pr_{H_2} = Pr_{\text{Air}}$, similarity exists if $Re_{L,\text{Air}} = Re_{L,H_2}$, in which case the average Nusselt numbers for the air and hydrogen will be identical, $\overline{Nu}_{\text{Air}} = \overline{Nu}_{H_2}$. Equating the Reynolds numbers for the hydrogen and air yields the expression

$$V_{\text{Air}} = \frac{Re_{L,\text{Air}} \nu_{\text{Air}}}{L_{\text{Air}}} = \frac{Re_{L,H_2} \nu_{\text{Air}}}{L_{\text{Air}}} = \frac{V_{H_2} L_{H_2} \nu_{\text{Air}}}{\nu_{H_2} L_{\text{Air}}}$$

$$= \frac{8.1\ \text{m/s} \times 0.04\ \text{m} \times 15.89 \times 10^{-6}\ \text{m}^2/\text{s}}{40.7 \times 10^{-6}\ \text{m}^2/\text{s} \times 0.06\ \text{m}} = 2.10\ \text{m/s} \qquad \triangleleft$$

With $Re_{L,\text{Air}} = Re_{L,H_2}$ and $Pr_{\text{Air}} = Pr_{H_2}$, we may equate the Nusselt numbers for the hydrogen and air, and incorporate Newton's law of cooling. Doing so gives

$$\bar{h}_{H_2} = \bar{h}_{\text{Air}} \frac{L_{\text{Air}} k_{H_2}}{L_{H_2} k_{\text{Air}}} = \frac{q'_{\text{Air}}}{P(T_{s,\text{Air}} - T_{\infty,\text{Air}})} \times \frac{L_{\text{Air}} k_{H_2}}{L_{H_2} k_{\text{Air}}}$$

$$= \frac{50\ \text{W/m}}{150 \times 10^{-3}\ \text{m} \times (30 - 23)°\text{C}} \times \frac{0.06\ \text{m} \times 0.157\ \text{W/m} \cdot \text{K}}{0.04\ \text{m} \times 0.0263\ \text{W/m} \cdot \text{K}} = 426\ \text{W/m}^2 \cdot \text{K} \qquad \triangleleft$$

Comments:

1. The fluid properties are evaluated at the arithmetic mean of the free stream and surface temperatures. As will become evident in Section 7.1, the temperature dependence of fluid properties is often accounted for by evaluating properties at the film temperature, $T_f = (T_s + T_\infty)/2$.
2. Experiments involving pressurized hydrogen can be relatively expensive because care must be taken to prevent leakage of this small-molecule, flammable gas.

6.6 *Physical Interpretation of the Dimensionless Parameters*

All of the foregoing dimensionless parameters have physical interpretations that relate to conditions in the flow, not only for boundary layers but also for other flow types, such as the internal flows we will see in Chapter 8. Consider the *Reynolds number* Re_L

(Equation 6.25), which may be interpreted as the *ratio of inertia to viscous forces* in a region of characteristic dimension L. Inertia forces are associated with an increase in the momentum of a moving fluid. From Equation 6.16, it is evident that these forces (per unit mass) are of the form $u\partial u/\partial x$, in which case an order-of-magnitude approximation gives $F_I \approx V^2/L$. Similarly, the net shear force (per unit mass) is found on the right-hand side of Equation 6.16 as $\nu(\partial^2 u/\partial y^2)$ and may be approximated as $F_s \approx \nu V/L^2$. Therefore, the ratio of forces is

$$\frac{F_I}{F_s} \approx \frac{\rho V^2/L}{\mu V/L^2} = \frac{\rho VL}{\mu} = Re_L$$

We therefore expect inertia forces to dominate for large values of Re_L and viscous forces to dominate for small values of Re_L.

There are several important implications of this result. Recall from Section 6.3.1 that the Reynolds number determines the existence of laminar or turbulent flow. We should also expect the magnitude of the Reynolds number to influence the velocity boundary layer thickness δ. With increasing Re_L at a fixed location on a surface, we expect viscous forces to become less influential relative to inertia forces. Hence the effects of viscosity do not penetrate as far into the free stream, and the value of δ diminishes.

The *Prandtl number* is defined as the ratio of the kinematic viscosity, also referred to as the momentum diffusivity, ν, to the thermal diffusivity α. It is therefore a fluid property. The Prandtl number provides a *measure of the relative effectiveness of momentum and energy transport by diffusion in the velocity and thermal boundary layers*, respectively. From Table A.4 we see that the Prandtl number of gases is near unity, in which case energy and momentum transfer by diffusion are comparable. In a liquid metal (Table A.7), $Pr \ll 1$ and the energy diffusion rate greatly exceeds the momentum diffusion rate. The opposite is true for oils (Table A.5), for which $Pr \gg 1$. From this interpretation it follows that the value of Pr strongly influences the relative growth of the velocity and thermal boundary layers. In fact for laminar boundary layers (in which transport by diffusion is *not* overshadowed by turbulent mixing), it is reasonable to expect that

$$\frac{\delta}{\delta_t} \approx Pr^n \tag{6.34}$$

where n is a positive exponent. Hence for a gas $\delta_t \approx \delta$; for a liquid metal $\delta_t \gg \delta$; for an oil $\delta_t \ll \delta$.

Table 6.2 lists the dimensionless groups that appear frequently in heat transfer. The list includes groups already considered, as well as those yet to be introduced for special conditions. As a new group is confronted, its definition and interpretation should be committed to memory. Note that the *Grashof number* provides a measure of the ratio of buoyancy forces to viscous forces in the velocity boundary layer. Its role in free convection (Chapter 9) is much the same as that of the Reynolds number in forced convection. The *Eckert number* provides a measure of the kinetic energy of the flow relative to the enthalpy difference across the thermal boundary layer. It plays an important role in high-speed flows for which viscous dissipation is significant. Note also that, although similar in form, the Nusselt and Biot numbers differ in both definition and interpretation. Whereas the Nusselt number is defined in terms of the thermal conductivity of the fluid, the Biot number is based on the solid thermal conductivity, Equation 5.9.

TABLE 6.2 Selected dimensionless groups of heat transfer

Group	Definition	Interpretation
Biot number (Bi)	$\frac{hL}{k_s}$	Ratio of the internal thermal resistance of a solid to the boundary layer thermal resistance
Bond number (Bo)	$\frac{g(\rho_l - \rho_v)L^2}{\sigma}$	Ratio of gravitational and surface tension forces
Coefficient of friction (C_f)	$\frac{\tau_s}{\rho V^2/2}$	Dimensionless surface shear stress
Eckert number (Ec)	$\frac{V^2}{c_p(T_s - T_\infty)}$	Kinetic energy of the flow relative to the boundary layer enthalpy difference
Fourier number (Fo)	$\frac{\alpha t}{L^2}$	Ratio of the heat conduction rate to the rate of thermal energy storage in a solid. Dimensionless time
Friction factor (f)	$\frac{\Delta p}{(L/D)(\rho u_m^2/2)}$	Dimensionless pressure drop for internal flow
Grashof number (Gr_L)	$\frac{g\beta(T_s - T_\infty)L^3}{\nu^2}$	Measure of the ratio of buoyancy forces to viscous forces
Colburn j factor (j_H)	$St\,Pr^{2/3}$	Dimensionless heat transfer coefficient
Jakob number (Ja)	$\frac{c_p(T_s - T_{sat})}{h_{fg}}$	Ratio of sensible to latent energy absorbed during liquid–vapor phase change
Mach number (Ma)	$\frac{V}{a}$	Ratio of velocity to speed of sound
Nusselt number (Nu_L)	$\frac{hL}{k_f}$	Ratio of convection to pure conduction heat transfer
Peclet number (Pe_L)	$\frac{VL}{\alpha} = Re_L\,Pr$	Ratio of advection to conduction heat transfer rates
Prandtl number (Pr)	$\frac{c_p\mu}{k} = \frac{\nu}{\alpha}$	Ratio of the momentum and thermal diffusivities
Reynolds number (Re_L)	$\frac{VL}{\nu}$	Ratio of the inertia and viscous forces
Stanton number (St)	$\frac{h}{\rho V c_p} = \frac{Nu_L}{Re_L\,Pr}$	Modified Nusselt number
Weber number (We)	$\frac{\rho V^2 L}{\sigma}$	Ratio of inertia to surface tension forces

6.7 *Momentum and Heat Transfer (Reynolds) Analogy*

As engineers, our interest in boundary layer behavior is directed principally toward the dimensionless parameters C_f and Nu. From knowledge of these parameters, we may compute the wall shear stress and the convection heat transfer rate. It is therefore understandable that an expression that relates C_f and Nu can be a useful tool in convection analysis. Such an expression is available in the form of a *boundary layer analogy*.

If two or more processes are governed by dimensionless equations of the same form, the processes are said to be *analogous*. From Table 6.1, we note that, if $dp^*/dx^* = 0$ and $Pr = 1$, Equations 6.21 and 6.22 are of precisely the same form. Each equation is composed of analogous advection and diffusion terms, and the coefficient of the diffusion term is the reciprocal Reynolds number. Moreover, since $u_\infty = V$ if $dp^*/dx^* = 0$, the boundary conditions, Equations 6.23 and 6.24, are equivalent. With governing differential equations and boundary conditions of the same form, it follows that dimensionless relations that govern velocity boundary layer behavior must be the same as those that govern the thermal boundary layer. Hence the boundary layer velocity and temperature profiles must be of the same functional form.

Recalling the discussion of Section 6.5.2, features of which are summarized in Table 6.3, the momentum and heat transfer analogy may be obtained. From the foregoing paragraph, it follows that the functional form of Equation 6.27 must be the same as that of Equation 6.30. From Equations 6.28 and 6.31 it also follows that the dimensionless velocity and temperature gradients at the surface, and therefore the values of C_f and Nu, are analogous. That is, the functional form of Equation 6.29 is the same as that of Equation 6.32. Accordingly, friction and heat transfer relations for a particular geometry are interchangeable. We conclude that

$$C_f \frac{Re_L}{2} = Nu \tag{6.35}$$

Replacing Nu by the *Stanton number* (St),

$$St \equiv \frac{h}{\rho V c_p} = \frac{Nu}{Re\,Pr} \tag{6.36}$$

the analogy takes the form

$$\frac{C_f}{2} = St \tag{6.37}$$

Equation 6.37 is known as the *Reynolds analogy,* and it relates the key engineering parameters of the velocity and thermal boundary layers. If the velocity parameter is known, the analogy may be used to obtain the heat transfer parameter, and vice versa. However, there are restrictions associated with using this result. In addition to relying on

TABLE 6.3 Functional relations pertinent to the Reynolds analogy

Fluid Flow		Heat Transfer	
$u^* = f\left(x^*, y^*, Re_L, \frac{dp^*}{dx^*}\right)$	(6.27)	$T^* = f\left(x^*, y^*, Re_L, Pr, \frac{dp^*}{dx^*}\right)$	(6.30)
$C_f = \frac{2}{Re_L}\frac{\partial u^*}{\partial y^*}\Big\vert_{y^*=0}$	(6.28)	$Nu = \frac{hL}{k} = +\frac{\partial T^*}{\partial y^*}\Big\vert_{y^*=0}$	(6.31)
$C_f = \frac{2}{Re_L} f(x^*, Re_L)$	(6.29)	$Nu = f(x^*, Re_L, Pr)$	(6.32)
		$\overline{Nu} = f(Re_L, Pr)$	(6.33)

the validity of the boundary layer approximations, the accuracy of Equation 6.37 depends on having $Pr \approx 1$ and $dp^*/dx^* \approx 0$. However, it has been shown that the analogy may be applied over a wide range of Pr, if a Prandtl number correction is added. In particular the *modified Reynolds,* or *Chilton–Colburn, analogy* [11, 12], has the form

$$\frac{C_f}{2} = St\,Pr^{2/3} \equiv j_H \qquad 0.6 < Pr < 60 \tag{6.38}$$

where j_H is the *Colburn j factor* for heat transfer. For laminar flow, Equation 6.38 is only appropriate when $dp^*/dx^* \approx 0$, but in turbulent flow, conditions are less sensitive to the effect of pressure gradients and the equation remains approximately valid. If the analogy is applicable at every point on a surface, it may be applied to the surface average coefficients.

6.8 Summary

In this chapter we have considered several fundamental issues related to convection transport phenomena. In the process, however, you should not lose sight of what remains *the problem of convection.* Our primary objective is still one of developing means to determine the convection coefficient h. Although this coefficient may be obtained by solving the boundary layer equations, it is only for simple flow situations that such solutions are readily effected. The more practical approach frequently involves calculating h from empirical relations of the form given by Equation 6.32. The particular form of these equations is obtained by *correlating* measured convection heat transfer results in terms of appropriate dimensionless groups. It is this approach that is emphasized in the chapters that follow.

To test your comprehension of the material, you should challenge yourself with appropriate questions.

- What is the difference between a *local* convection heat transfer coefficient and an *average* coefficient? What are their units?
- What are the forms of *Newton's law of cooling* for a *heat flux* and a *heat rate*?
- What are the *velocity* and *thermal boundary layers*? Under what conditions do they develop?
- What quantities change with location in a *velocity boundary layer*? A *thermal boundary layer*?
- Recognizing that convection heat transfer is strongly influenced by conditions associated with fluid flow over a surface, how is it that we may determine the convection heat flux by applying Fourier's law to the fluid at the surface?
- Do we expect heat transfer to change with transition from a laminar to a turbulent boundary layer? If so, how?
- What laws of nature are embodied by the *convection transfer equations*?
- What physical processes are represented by the terms of the x-momentum equation (6.16)? By the energy equation (6.17)?
- What special approximations may be made for conditions within *thin* velocity and thermal boundary layers?
- What is the *film* temperature and how is it used?
- How is the *Reynolds number* defined? What is its physical interpretation? What role is played by the *critical Reynolds number*?

- What is the definition of the *Prandtl number*? How does its value affect relative growth of the velocity and thermal boundary layers for laminar flow over a surface? What are representative room-temperature values of the Prandtl number for a liquid metal, a gas, water, and an oil?
- What is the *coefficient of friction*? The *Nusselt number*? For flow over a prescribed geometry, what are the independent parameters that determine local and average values of these quantities?
- Under what conditions may velocity and thermal boundary layers be termed *analogous*? What is the physical basis of analogous behavior?
- What important boundary layer parameters are linked by the *Reynolds analogy*?
- What physical features distinguish a turbulent flow from a laminar flow?

References

1. Hof, B., C. W. H. van Doorne, J. Westerweel, F. T. M. Nieuwstadt, H. Faisst, B. Eckhardt, H. Wedin, R. R. Kerswell, and F. Waleffe, *Science*, **305**, 1594, 2004.
2. Schlichting, H., and K. Gersten, *Boundary Layer Theory*, 8th ed., Spinger-Verlag, New York, 1999.
3. Bird, R. B., W. E. Stewart, and E. N. Lightfoot, *Transport Phenomena*, 2nd ed., Wiley, New York, 2002.
4. Fox, R. W., A. T. McDonald, and P. J. Pritchard, *Introduction to Fluid Mechanics*, 6th ed., Wiley, Hoboken, NJ, 2003.
5. Kays, W. M., M. E. Crawford, and B. Weigand, *Convective Heat and Mass Transfer*, 4th ed., McGraw-Hill Higher Education, Boston, 2005.
6. Burmeister, L. C., *Convective Heat Transfer,* 2nd ed., Wiley, New York, 1993.
7. Kaviany, M., *Principles of Convective Heat Transfer*, Springer-Verlag, New York, 1994.
8. Patankar, S. V., *Numerical Heat Transfer and Fluid Flow,* Hemisphere Publishing, New York, 1980.
9. Oosthuizen, P. H., and W. E. Carscallen, *Compressible Fluid Flow*, McGraw-Hill, New York, 1997.
10. John, J. E. A., and T. G. Keith, *Gas Dynamics*, 3rd ed., Pearson Prentice Hall, Upper Saddle River, NJ, 2006.
11. Colburn, A. P., *Trans. Am. Inst. Chem. Eng.,* **29**, 174, 1933.
12. Chilton, T. H., and A. P. Colburn, *Ind. Eng. Chem.,* **26,** 1183, 1934.

Problems

Boundary Layer Profiles

6.1 The temperature distribution within a laminar thermal boundary layer associated with flow over an isothermal flat plate is shown in the sketch. The temperature distribution shown is located at $x = x_2$.

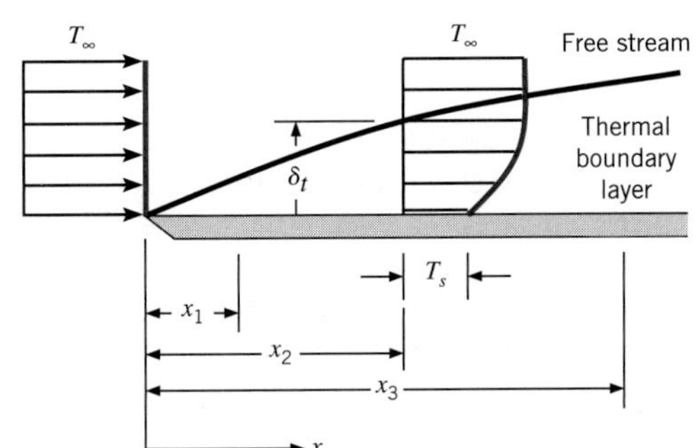

(a) Is the plate being heated or cooled by the fluid?

(b) Carefully sketch the temperature distributions at $x = x_1$ and $x = x_3$. Based on your sketch, at which of the three x-locations is the local heat flux largest? At which location is the local heat flux smallest?

(c) As the free stream velocity increases, the velocity and thermal boundary layers both become thinner. Carefully sketch the temperature distributions at $x = x_2$ for (i) a low free stream velocity and (ii) a high free stream velocity. Based on your sketch, which velocity condition will induce the larger local convective heat flux?

6.2 In flow over a surface, velocity and temperature profiles are of the forms

$$u(y) = Ay + By^2 - Cy^3 \quad \text{and}$$
$$T(y) = D + Ey + Fy^2 - Gy^3$$

where the coefficients A through G are constants. Obtain expressions for the friction coefficient C_f and the convection coefficient h in terms of u_∞, T_∞, and appropriate profile coefficients and fluid properties.

6.3 In a particular application involving airflow over a heated surface, the boundary layer temperature distribution may be approximated as

$$\frac{T - T_s}{T_\infty - T_s} = 1 - \exp\left(-Pr\frac{u_\infty y}{\nu}\right)$$

where y is the distance normal to the surface and the Prandtl number, $Pr = c_p\mu/k = 0.7$, is a dimensionless fluid property. If $T_\infty = 400$ K, $T_s = 300$ K, and $u_\infty/\nu = 5000\ \text{m}^{-1}$, what is the surface heat flux?

6.4 Water at a temperature of $T_\infty = 25°\text{C}$ flows over one of the surfaces of a steel wall (AISI 1010) whose temperature is $T_{s,1} = 40°\text{C}$. The wall is 0.35 m thick, and its other surface temperature is $T_{s,2} = 100°\text{C}$. For steady-state conditions what is the convection coefficient associated with the water flow? What is the temperature gradient in the wall and in the water that is in contact with the wall? Sketch the temperature distribution in the wall and in the adjoining water.

Heat Transfer Coefficients

6.5 For laminar flow over a flat plate, the local heat transfer coefficient h_x is known to vary as $x^{-1/2}$, where x is the distance from the leading edge ($x = 0$) of the plate. What is the ratio of the average coefficient between the leading edge and some location x on the plate to the local coefficient at x?

6.6 A flat plate is of planar dimension 1 m × 0.75 m. For parallel laminar flow over the plate, calculate the ratio of the average heat transfer coefficients over the entire plate, $\bar{h}_{L,1}/\bar{h}_{L,2}$, for two cases. In Case 1, flow is in the short direction ($L = 0.75$ m); in Case 2, flow is in the long direction ($L = 1$ m). Which orientation will result in the larger heat transfer rate? See Problem 6.5.

6.7 Parallel flow of atmospheric air over a flat plate of length $L = 3$ m is disrupted by an array of stationary rods placed in the flow path over the plate.

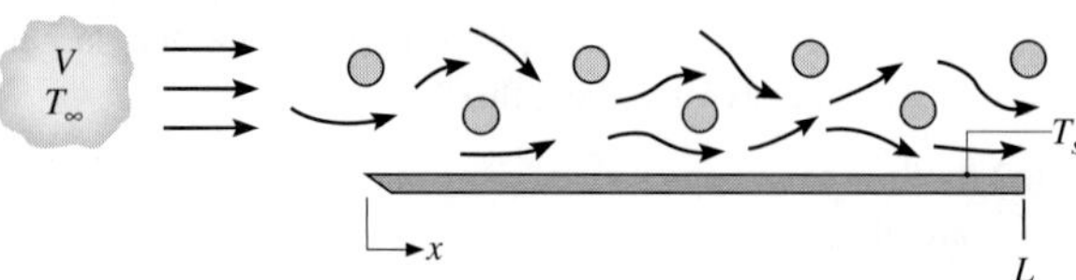

Laboratory measurements of the local convection coefficient at the surface of the plate are made for a prescribed value of V and $T_s > T_\infty$. The results are correlated by an expression of the form $h_x = 0.7 + 13.6x - 3.4x^2$, where h_x has units of W/m²·K and x is in meters. Evaluate the average convection coefficient $\bar{h}_L$ for the entire plate and the ratio $\bar{h}_L/h_L$ at the trailing edge.

6.8 For laminar free convection from a heated vertical surface, the local convection coefficient may be expressed as $h_x = Cx^{-1/4}$, where h_x is the coefficient at a distance x from the leading edge of the surface and the quantity C, which depends on the fluid properties, is independent of x. Obtain an expression for the ratio $\bar{h}_x/h_x$, where $\bar{h}_x$ is the average coefficient between the leading edge ($x = 0$) and the x-location. Sketch the variation of h_x and $\bar{h}_x$ with x.

6.9 A circular, hot gas jet at T_∞ is directed normal to a circular plate that has radius r_o and is maintained at a uniform temperature T_s. Gas flow over the plate is axisymmetric, causing the local convection coefficient to have a radial dependence of the form $h(r) = a + br^n$, where a, b, and n are constants. Determine the rate of heat transfer to the plate, expressing your result in terms of T_∞, T_s, r_o, a, b, and n.

6.10 Experiments have been conducted to determine local heat transfer coefficients for flow perpendicular to a long, isothermal bar of rectangular cross section. The bar is of width c parallel to the flow, and height d normal to the flow. For Reynolds numbers in the range $10^4 \le Re_d \le 5 \times 10^4$, the *face-averaged* Nusselt numbers are well correlated by an expression of the form

$$\overline{Nu}_d = \bar{h}d/k = C\,Re_d^m\,Pr^{1/3}$$

The values of C and m for the front face, side faces, and back face of the rectangular rod are found to be the following:

Face	c/d	C	m
Front	$0.33 \le c/d \le 1.33$	0.674	1/2
Side	0.33	0.153	2/3
Side	1.33	0.107	2/3
Back	0.33	0.174	2/3
Back	1.33	0.153	2/3

Determine the value of the average heat transfer coefficient for the entire exposed surface (that is, averaged over all four faces) of a c = 40-mm-wide, d = 30-mm-tall rectangular rod. The rod is exposed to air in cross flow at $V = 10$ m/s, $T_\infty = 300$ K. Provide a plausible explanation of the relative values of the face-averaged heat transfer coefficients on the front, side, and back faces.

6.11 A concentrating solar collector consists of a parabolic reflector and a collector tube of diameter D, through which flows a working fluid that is heated with concentrated solar irradiation. Throughout the day, the reflector is slowly repositioned to track the sun. For wind conditions characterized by a steady, horizontal flow normal to the tube axis, the local heat transfer coefficient on the tube surface varies, as shown in the schematic for various reflector positions.

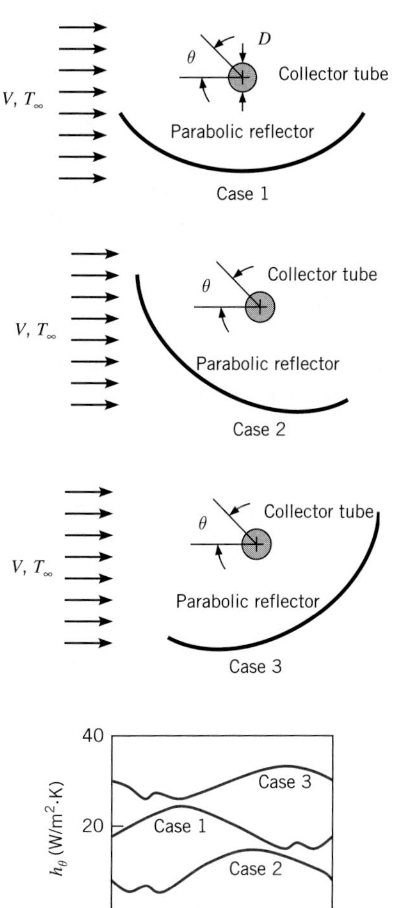

(a) Estimate the value of the average heat transfer coefficient over the entire collector tube surface for each of the three cases.

(b) Assuming the tube receives the same amount of solar irradiation in each case, which case would have the highest collector efficiency?

6.12 Air at a free stream temperature of $T_\infty = 20°C$ is in parallel flow over a flat plate of length $L = 5$ m and temperature $T_s = 90°C$. However, obstacles placed in the flow intensify mixing with increasing distance x from the leading edge, and the spatial variation of temperatures measured in the boundary layer is correlated by an expression of the form $T(°C) = 20 + 70 \exp(-600xy)$, where x and y are in meters. Determine and plot the manner in which the local convection coefficient h varies with x. Evaluate the average convection coefficient $\bar{h}$ for the plate.

6.13 The heat transfer rate per unit width (normal to the page) from a longitudinal section, $x_2 - x_1$, can be expressed as $q'_{12} = \bar{h}_{12}(x_2 - x_1)(T_s - T_\infty)$, where $\bar{h}_{12}$ is the average coefficient for the section of length $(x_2 - x_1)$. Consider laminar flow over a flat plate with a uniform temperature T_s. The spatial variation of the local convection coefficient is of the form $h_x = Cx^{-1/2}$, where C is a constant.

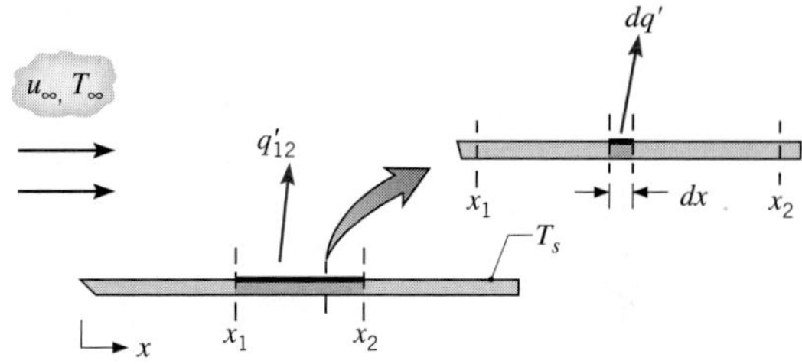

(a) Beginning with the convection rate equation in the form $dq' = h_x\, dx(T_s - T_\infty)$, derive an expression for $\bar{h}_{12}$ in terms of C, x_1, and x_2.

(b) Derive an expression for $\bar{h}_{12}$ in terms of x_1, x_2, and the average coefficients $\bar{h}_1$ and $\bar{h}_2$, corresponding to lengths x_1 and x_2, respectively.

6.14 Experiments to determine the local convection heat transfer coefficient for uniform flow normal to a heated circular disk have yielded a radial Nusselt number distribution of the form

$$Nu_D = \frac{h(r)D}{k} = Nu_o\left[1 + a\left(\frac{r}{r_o}\right)^n\right]$$

where both n and a are positive. The Nusselt number at the stagnation point is correlated in terms of the Reynolds ($Re_D = VD/\nu$) and Prandtl numbers

$$Nu_o = \frac{h(r = 0)D}{k} = 0.814\, Re_D^{1/2}\, Pr^{0.36}$$

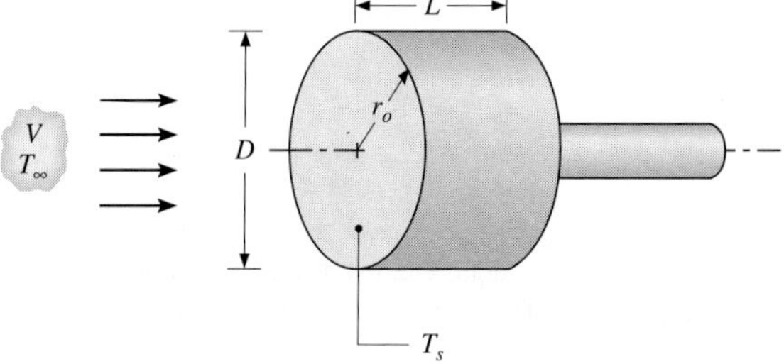

Obtain an expression for the average Nusselt number, $\overline{Nu}_D = \bar{h}D/k$, corresponding to heat transfer from an isothermal disk. Typically, boundary layer development from a stagnation point yields a decaying convection coefficient with increasing distance from the stagnation point. Provide a plausible explanation for why the opposite trend is observed for the disk.

6.15 An experimental procedure for validating results of Problem 6.14 involves preheating a copper disk to an initial elevated temperature T_i and recording its temperature history $T(t)$ as it is subsequently cooled by the impinging flow to a final temperature T_f. The measured temperature decay may then be compared with predictions based on the correlation for $\overline{Nu}_D$. Assume that values of $a = 0.30$ and $n = 2$ are associated with the correlation.

Consider experimental conditions for which a disk of diameter $D = 50$ mm and length $L = 25$ mm is preheated to $T_i = 1000$ K and cooled to $T_f = 400$ K by an impinging airflow at $T_\infty = 300$ K. The cooled surface of the disk has an emissivity of $\varepsilon = 0.8$ and is exposed to large, isothermal surroundings for which $T_{sur} = T_\infty$. The remaining surfaces of the disk are well insulated, and heat transfer through the supporting rod may be neglected. Using results from Problem 6.14, compute and plot temperature histories corresponding to air velocities of $V = 4$, 20, and 50 m/s. Constant properties may be assumed for the copper ($\rho = 8933$ kg/m^3, $c_p = 425$ J/kg·K, $k = 386$ W/m·K) and air ($\nu = 38.8 \times 10^{-6}$ m^2/s, $k = 0.0407$ W/m·K, $Pr = 0.684$).

6.16 If laminar flow is induced at the surface of a disk due to rotation about its axis, the local convection coefficient is known to be a constant, $h = C$, independent of radius. Consider conditions for which a disk of radius $r_o = 100$ mm is rotating in stagnant air at $T_\infty = 20°$C and a value of $C = 20$ W/m^2·K is maintained.

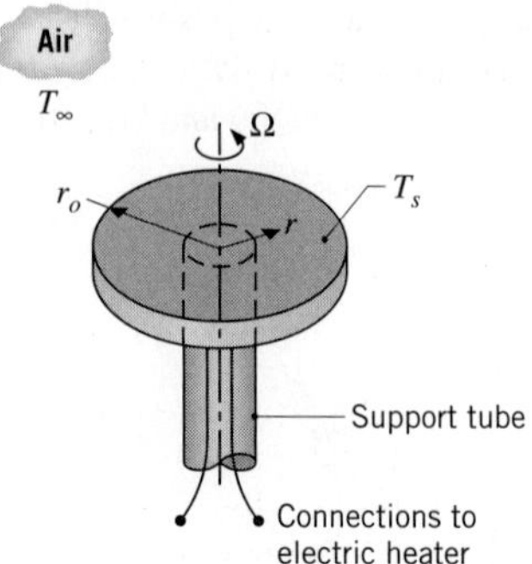

If an embedded electric heater maintains a surface temperature of $T_s = 50°$C, what is the local heat flux at the top surface of the disk? What is the total electric power requirement? What can you say about the nature of boundary layer development on the disk?

6.17 Consider the rotating disk of Problem 6.16. A disk-shaped, stationary plate is placed a short distance away from the rotating disk, forming a gap of width g. The stationary plate and ambient air are at $T_\infty = 20°$C. If the flow is laminar and the gap-to-radius ratio, $G = g/r_o$, is small, the local radial Nusselt number distribution is of the form

$$Nu_r = \frac{h(r)r}{k} = 70(1 + e^{-140G})\, Re_{r_o}^{-0.456}\, Re_r^{0.478}$$

where $Re_r = \Omega r^2/\nu$ [Pelle J., and S. Harmand, *Exp. Thermal Fluid Science*, **31**, 165, 2007]. Determine the value of the average Nusselt number, $\overline{Nu}_D = \bar{h}D/k$ where $D = 2r_o$. If the rotating disk temperature is $T_s = 50°$C, what is the total heat flux from the disk's top surface for $g = 1$ mm, $\Omega = 150$ rad/s? What is the total electric power requirement? What can you say about the nature of the flow between the disks?

Boundary Layer Transition

6.18 Consider airflow over a flat plate of length $L = 1$ m under conditions for which transition occurs at $x_c = 0.5$ m based on the critical Reynolds number, $Re_{x,c} = 5 \times 10^5$.

(a) Evaluating the thermophysical properties of air at 350 K, determine the air velocity.

(b) In the laminar and turbulent regions, the local convection coefficients are, respectively,

$$h_{lam}(x) = C_{lam}\, x^{-0.5} \text{ and } h_{turb} = C_{turb}\, x^{-0.2}$$

where, at $T = 350$ K, $C_{lam} = 8.845$ W/m$^{3/2}$·K, $C_{turb} = 49.75$ W/m$^{1.8}$·K, and x has units of m. Develop an expression for the average convection coefficient, $\bar{h}_{lam}(x)$, as a function of distance from the leading edge, x, for the laminar region, $0 \leq x \leq x_c$.

(c) Develop an expression for the average convection coefficient, $\bar{h}_{turb}(x)$, as a function of distance from the leading edge, x, for the turbulent region, $x_c \leq x \leq L$.

(d) On the same coordinates, plot the local and average convection coefficients, h_x and $\bar{h}_x$, respectively, as a function of x for $0 \leq x \leq L$.

6.19 A fan that can provide air speeds up to 50 m/s is to be used in a low-speed wind tunnel with atmospheric air at 25°C. If one wishes to use the wind tunnel to study flat-plate boundary layer behavior up to Reynolds numbers of $Re_x = 10^8$, what is the minimum plate length that should be used? At what distance from the leading edge would transition occur if the critical Reynolds number were $Re_{x,c} = 5 \times 10^5$?

6.20 Consider the flow conditions of Example 6.2 for two situations, one in which the flow is completely laminar, and the second for flow that is tripped to turbulence at the leading edge of the plate. Determine whether there is a plate length L for which the average convection coefficient for laminar flow is the same as the average convection coefficient for turbulent flow. Assume a water temperature of 300 K.

6.21 Assuming a transition Reynolds number of 5×10^5, determine the distance from the leading edge of a flat plate at which transition will occur for each of the following fluids when $u_\infty = 1$ m/s: atmospheric air, engine oil, and mercury. In each case, calculate the transition location for fluid temperatures of 27°C and 77°C.

6.22 To a good approximation, the dynamic viscosity μ, the thermal conductivity k, and the specific heat c_p are independent of pressure. In what manner do the kinematic viscosity ν and thermal diffusivity α vary with pressure for an incompressible liquid and an ideal gas? Determine α of air at 350 K for pressures of 1, 5, and 10 atm. Assuming a transition Reynolds number of 5×10^5, determine the distance from the leading edge of a flat plate at which transition will occur for air at 350 K at pressures of 1, 5, and 10 atm with $u_\infty = 2$ m/s.

6.23 For the situation described in Example 6.2, the boundary layer can be *tripped* into a turbulent state by applying roughness to the surface of the flat plate at a particular x-location. Hence the location where transition occurs, x_c, can be moved upstream relative to the transition location associated with the smooth plate of the example. Calculate and plot the average convection coefficient over the entire plate $\overline{h}$ for roughness applied over the range $0 \le x_r \le L$. What values of x_r provide the minimum and maximum values of $\overline{h}$? Assume the water temperature is 300 K.

Similarity and Dimensionless Parameters

6.24 Consider a laminar boundary layer developing over a flat plate. The flow is incompressible.

(a) Substitute Equations 6.18 and 6.19 into Equation 6.23 to determine the boundary conditions in dimensional form associated with flow over a flat plate of length L.

(b) Substitute Equations 6.18, 6.19, as well as the definition of Re_L into Equation 6.21, and compare the resulting expression with Equation 6.16. Note that for a flat plate, $dp/dx = 0$ and $u_\infty = V$.

6.25 Consider a laminar boundary layer developing over an isothermal flat plate. The flow is incompressible, and viscous dissipation is negligible.

(a) Substitute Equations 6.18 and 6.20 into Equation 6.24 to determine the thermal boundary conditions in dimensional form associated with flow over a flat plate of length L and temperature T_s.

(b) Substitute Equations 6.18, 6.19, and 6.20, as well as the definitions of Re_L and Pr, into Equation 6.22, and compare the resulting dimensional expression with Equation 6.17.

6.26 Experiments have shown that the transition from laminar to turbulent conditions for flow normal to the axis of a long cylinder occurs at a critical Reynolds number of $Re_{D,c} \approx 2 \times 10^5$, where D is the cylinder diameter. Moreover, the transition from incompressible to compressible flow occurs at a critical Mach number of $Ma_c \approx 0.3$. For air at a pressure of $p = 1$ atm and temperature $T = 27°C$, determine the critical cylinder diameter D_c below which, if the flow is turbulent, compressibility effects are likely to be important.

6.27 An object of irregular shape has a characteristic length of $L = 1$ m and is maintained at a uniform surface temperature of $T_s = 400$ K. When placed in atmospheric air at a temperature of $T_\infty = 300$ K and moving with a velocity of $V = 100$ m/s, the average heat flux from the surface to the air is 20,000 W/m^2. If a second object of the same shape, but with a characteristic length of $L = 5$ m, is maintained at a surface temperature of $T_s = 400$ K and is placed in atmospheric air at $T_\infty = 300$ K, what will the value of the average convection coefficient be if the air velocity is $V = 20$ m/s?

6.28 Experiments have shown that, for airflow at $T_\infty = 35°C$ and $V_1 = 100$ m/s, the rate of heat transfer from a turbine blade of characteristic length $L_1 = 0.15$ m and surface temperature $T_{s,1} = 300°C$ is $q_1 = 1500$ W. What would be the heat transfer rate from a second turbine blade of characteristic length $L_2 = 0.3$ m operating at $T_{s,2} = 400°C$ in airflow of $T_\infty = 35°C$ and $V_2 = 50$ m/s? The surface area of the blade may be assumed to be directly proportional to its characteristic length.

6.29 Experimental measurements of the convection heat transfer coefficient for a square bar in cross flow yielded the following values:

$\overline{h}_1 = 50$ W/m$^2 \cdot$K when $V_1 = 20$ m/s

$\overline{h}_2 = 40$ W/m$^2 \cdot$K when $V_2 = 15$ m/s

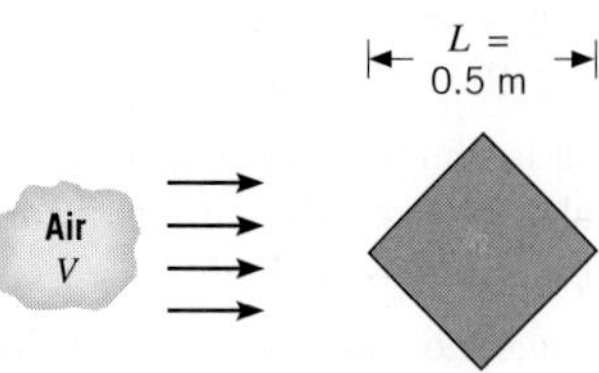

Assume that the functional form of the Nusselt number is $\overline{Nu} = C\,Re^m\,Pr^n$, where C, m, and n are constants.

(a) What will be the convection heat transfer coefficient for a similar bar with $L = 1$ m when $V = 15$ m/s?

(b) What will be the convection heat transfer coefficient for a similar bar with $L = 1$ m when $V = 30$ m/s?

(c) Would your results be the same if the side of the bar, rather than its diagonal, were used as the characteristic length?

6.30 To assess the efficacy of different liquids for cooling an object of given size and shape by forced convection, it is convenient to introduce a *figure of merit*, F_F, which combines the influence of all pertinent fluid properties on the convection coefficient. If the Nusselt number is governed by an expression of the form, $\overline{Nu}_L \sim Re_L^m\,Pr^n$, obtain the corresponding relationship between F_F and the fluid properties. For representative values of $m = 0.80$ and $n = 0.33$, calculate values of F_F for air ($k = 0.026$ W/m·K, $\nu = 1.6 \times 10^{-5}$ m²/s, $Pr = 0.71$), water ($k = 0.600$ W/m·K, $\nu = 10^{-6}$ m²/s, $Pr = 5.0$), and a dielectric liquid ($k = 0.064$ W/m·K, $\nu = 10^{-6}$ m²/s, $Pr = 25$). Which fluid is the most effective cooling agent?

6.31 Gases are often used instead of liquids to cool electronics in avionics applications because of weight considerations. The cooling systems are often *closed* so that coolants other than air may be used. Gases with high figures of merit (see Problem 6.30) are desired. For representative values of $m = 0.85$ and $n = 0.33$ in the expression of Problem 6.30, determine the figures of merit for air, pure helium, pure xenon ($k = 0.006$ W/m·K, $\mu = 24.14 \times 10^{-6}$ N·s/m²), and an ideal He-Xe mixture containing 0.75 mole fraction of helium ($k = 0.0713$ W/m·K, $\mu = 25.95 \times 10^{-6}$ N·s/m²). Evaluate properties at 300 K and atmospheric pressure. For monatomic gases such as helium and xenon and their mixtures, the specific heat at constant pressure is well described by the relation $c_p = (5/2)\mathcal{R}/\mathcal{M}$.

6.32 Experimental results for heat transfer over a flat plate with an extremely rough surface were found to be correlated by an expression of the form

$$Nu_x = 0.04\,Re_x^{0.9}\,Pr^{1/3}$$

where Nu_x is the local value of the Nusselt number at a position x measured from the leading edge of the plate. Obtain an expression for the ratio of the average heat transfer coefficient $\bar{h}_x$ to the local coefficient h_x.

6.33 Consider conditions for which a fluid with a free stream velocity of $V = 1$ m/s flows over a surface with a characteristic length of $L = 1$ m, providing an average convection heat transfer coefficient of $\bar{h} = 100$ Wm²·K. Calculate the dimensionless parameters $\overline{Nu}_L$, Re_L, Pr, and $\bar{j}_H$ for the following fluids: air, engine oil, mercury, and water. Assume the fluids to be at 300 K.

6.34 Consider the nanofluid of Example 2.2.

(a) Calculate the Prandtl numbers of the base fluid and nanofluid, using information provided in the example problem.

(b) For a geometry of fixed characteristic dimension L, and a fixed characteristic velocity V, determine the ratio of the Reynolds numbers associated with the two fluids, Re_{nf}/Re_{bf}. Calculate the ratio of the average Nusselt numbers, $\overline{Nu}_{L,nf}/\overline{Nu}_{L,bf}$, that is associated with identical average heat transfer coefficients for the two fluids, $\bar{h}_{nf} = \bar{h}_{bf}$.

(c) The functional dependence of the average Nusselt number on the Reynolds and Prandtl numbers for a broad array of various geometries may be expressed in the general form

$$\overline{Nu}_L = \bar{h}L/k = C\,Re^m\,Pr^{1/3}$$

where C and m are constants whose values depend on the geometry from or to which convection heat transfer occurs. Under most conditions the value of m is positive. For positive m, is it possible for the base fluid to provide greater convection heat transfer rates than the nanofluid, for conditions involving a fixed geometry, the same characteristic velocities, and identical surface and ambient temperatures?

6.35 For flow over a flat plate of length L, the local heat transfer coefficient h_x is known to vary as $x^{-1/2}$, where x is the distance from the leading edge of the plate. What is the ratio of the average Nusselt number for the entire plate ($\overline{Nu}_L$) to the local Nusselt number at $x = L$ (Nu_L)?

6.36 For laminar boundary layer flow over a flat plate with air at 20°C and 1 atm, the thermal boundary layer thickness δ_t is approximately 13% larger than the velocity boundary layer thickness δ. Determine the ratio δ/δ_t if the fluid is ethylene glycol under the same flow conditions.

6.37 Sketch the variation of the velocity and thermal boundary layer thicknesses with distance from the leading edge of a flat plate for the laminar flow of air, water, engine oil, and mercury. For each case assume a mean fluid temperature of 300 K.

6.38 Consider parallel flow over a flat plate for air at 300 K and engine oil at 380 K. The free stream velocity is $u_\infty = 2$ m/s. The temperature difference between the surface and the free stream is the same in both cases, with $T_s > T_\infty$.

(a) Determine the location where transition to turbulence occurs, x_c, for both fluids.

(b) For laminar flow over a flat plate, the velocity boundary layer thickness is given by

$$\frac{\delta}{x} = \frac{5}{\sqrt{Re_x}}$$

Calculate and plot the velocity boundary layer thickness δ over the range $0 \le x \le x_c$ for each fluid.

(c) Calculate and plot the thermal boundary layer thickness δ_t for the two fluids over the same range of x used in part (b). At an x-location where both fluids experience laminar flow conditions, explain which fluid has the largest temperature gradient at the plate surface, $-\partial T/\partial y|_{y=0}$. Which fluid is associated with the largest local Nusselt number Nu? Which fluid is associated with the largest local heat transfer coefficient h?

6.39 Forced air at $T_\infty = 25°\text{C}$ and $V = 10$ m/s is used to cool electronic elements on a circuit board. One such element is a chip, 4 mm × 4 mm, located 120 mm from the leading edge of the board. Experiments have revealed that flow over the board is disturbed by the elements and that convection heat transfer is correlated by an expression of the form

$$Nu_x = 0.04\, Re_x^{0.85}\, Pr^{1/3}$$

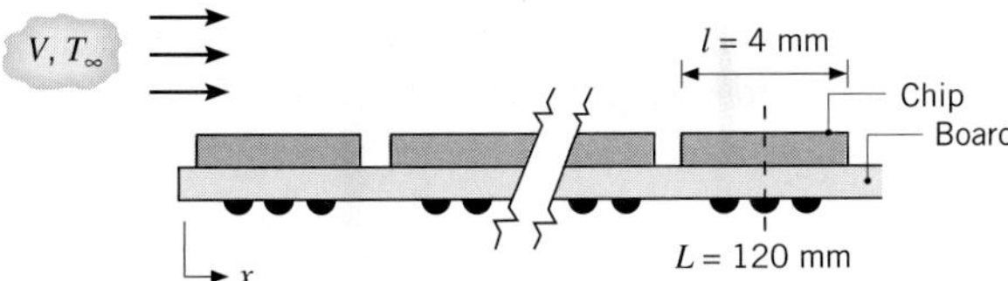

Estimate the surface temperature of the chip if it is dissipating 30 mW.

6.40 Consider the electronic elements that are cooled by forced convection in Problem 6.39. The cooling system is designed and tested at sea level ($p \approx 1$ atm), but the circuit board is sold to a customer in Mexico City, with an elevation of 2250 m and atmospheric pressure of 76.5 kPa.

(a) Estimate the surface temperature of the chip located 120 mm from the leading edge of the board when the board is operated in Mexico City. The dependence of various thermophysical properties on pressure is noted in Problem 6.22.

(b) It is desirable for the chip operating temperature to be independent of the location of the customer. What air velocity is required for operation in Mexico City if the chip temperature is to be the same as at sea level?

6.41 Consider the chip on the circuit board of Problem 6.39. To ensure reliable operation over extended periods, the chip temperature should not exceed 85°C. Assuming the availability of forced air at $T_\infty = 25°\text{C}$ and applicability of the prescribed heat transfer correlation, compute and plot the maximum allowable chip power dissipation P_c as a function of air velocity for $1 \le V \le 25$ m/s. If the chip surface has an emissivity of 0.80 and the board is mounted in a large enclosure whose walls are at 25°C, what is the effect of radiation on the $P_c - V$ plot?

6.42 A major contributor to product defects in electronic modules relates to stresses induced during thermal cycling (intermittent heating and cooling). For example, in circuit cards having active and passive components with materials of different thermal expansion coefficients, thermal stresses are the principal source of failure in component joints, such as soldered and wired connections. Although concern is generally for fatigue failure resulting from numerous excursions during the life of a product, it is possible to identify defective joints by performing accelerated thermal stress tests before the product is released to the customer. In such cases, it is important to achieve rapid thermal cycling to minimize disruptions to production schedules.

A manufacturer of circuit cards wishes to develop an apparatus for imposing rapid thermal transients on the cards by subjecting them to forced convection characterized by a relation of the form $\overline{Nu}_L = C\, Re_L^m\, Pr^n$, where $m = 0.8$ and $n = 0.33$. However, he does not know whether to use air ($k = 0.026$ W/m·K, $\nu = 1.6 \times 10^{-5}$ m²/s, $Pr = 0.71$) or a dielectric liquid ($k = 0.064$ W/m·K, $\nu = 10^{-6}$ m²/s, $Pr = 25$) as the working fluid. Assuming equivalent air and liquid velocities and validity of the lumped capacitance model for the components, obtain a quantitative estimate of the ratio of the thermal time constants for the two fluids. What fluid provides the faster thermal response?

6.43 The defroster of an automobile functions by discharging warm air on the inner surface of the windshield. To prevent condensation of water vapor on the surface, the temperature of the air and the surface convection coefficient ($T_{\infty,i}$, $\overline{h}_i$) must be large enough to maintain a surface temperature $T_{s,i}$ that is at least as high as the dewpoint ($T_{s,i} \ge T_{dp}$).

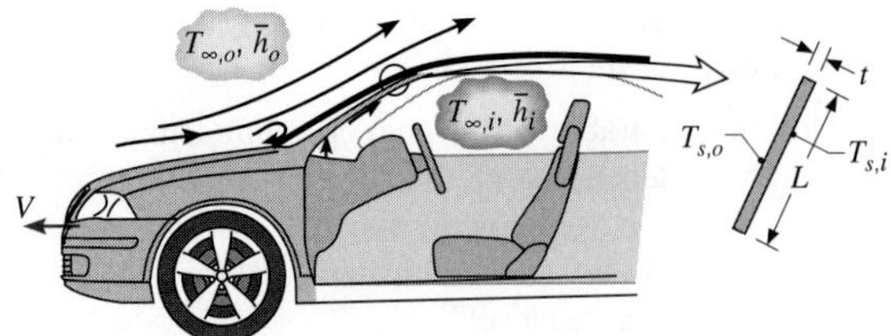

Consider a windshield of length $L = 800$ mm and thickness $t = 6$ mm and driving conditions for which the vehicle moves at a velocity of $V = 70$ mph in ambient air at $T_{\infty,o} = -15°C$. From laboratory experiments performed on a model of the vehicle, the average convection coefficient on the outer surface of the windshield is known to be correlated by an expression of the form $\overline{Nu}_L = 0.030\, Re_L^{0.8}\, Pr^{1/3}$, where $Re_L \equiv VL/\nu$. Air properties may be approximated as $k = 0.023$ W/m·K, $\nu = 12.5 \times 10^{-6}$ m²/s, and $Pr = 0.71$. If $T_{dp} = 10°C$ and $T_{\infty,i} = 50°C$, what is the smallest value of $\bar{h}_i$ required to prevent condensation on the inner surface?

6.44 A microscale detector monitors a steady flow (T_∞ = 27°C, $V = 10$ m/s) of air for the possible presence of small, hazardous particulate matter that may be suspended in the room. The sensor is heated to a slightly higher temperature to induce a chemical reaction associated with certain substances of interest that might impinge on the sensor's active surface. The active surface produces an electric current if such surface reactions occur; the electric current is then sent to an alarm. To maximize the sensor head's surface area and, in turn, the probability of capturing and detecting a particle, the sensor head is designed with a very complex shape. The value of the average heat transfer coefficient associated with the heated sensor must be known so that the required electrical power to the sensor can be determined.

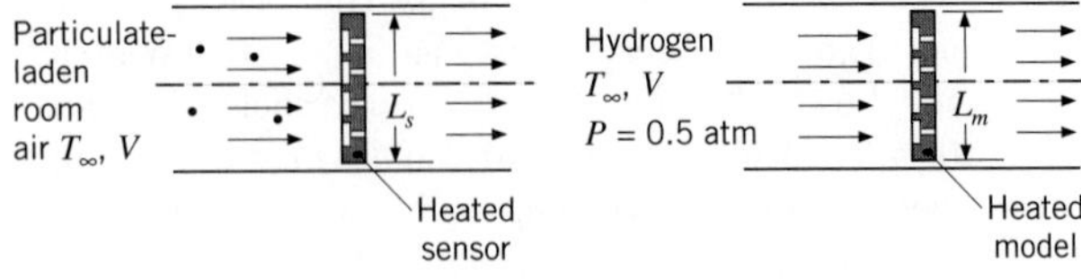

Consider a sensor with a characteristic dimension of $L_s = 80\ \mu$m. A scale model of the sensor is placed in a recirculating (closed) wind tunnel using hydrogen as the working fluid. If the wind tunnel operates at a hydrogen absolute pressure of 0.5 atm and velocity of $V = 0.5$ m/s, find the required hydrogen temperature and characteristic dimension of the scale model, L_m.

Reynolds Analogy

6.45 A thin, flat plate that is 0.2 m × 0.2 m on a side is oriented parallel to an atmospheric airstream having a velocity of 40 m/s. The air is at a temperature of $T_\infty = 20°C$, while the plate is maintained at $T_s = 120°C$. The air flows over the top and bottom surfaces of the plate, and measurement of the drag force reveals a value of 0.075 N. What is the rate of heat transfer from both sides of the plate to the air?

6.46 Atmospheric air is in parallel flow ($u_\infty = 15$ m/s, $T_\infty = 15°C$) over a flat heater surface that is to be maintained at a temperature of 140°C. The heater surface area is 0.25 m², and the airflow is known to induce a drag force of 0.25 N on the heater. What is the electrical power needed to maintain the prescribed surface temperature?

6.47 Determine the drag force imparted to the top surface of the flat plate of Example 6.2 for water temperatures of 300 K and 350 K. Assume the plate dimension in the z-direction is $W = 1$ m.

6.48 For flow over a flat plate with an extremely rough surface, convection heat transfer effects are known to be correlated by the expression of Problem 6.32. For airflow at 50 m/s, what is the surface shear stress at $x = 1$ m from the leading edge of the plate? Assume the air to be at a temperature of 300 K.

6.49 A thin, flat plate that is 0.2 m × 0.2 m on a side with rough top and bottom surfaces is placed in a wind tunnel so that its surfaces are parallel to an atmospheric airstream having a velocity of 30 m/s. The air is at a temperature of $T_\infty = 20°C$ while the plate is maintained at $T_s = 80°C$. The plate is rotated 45° about its center point, as shown in the schematic. Air flows over the top and bottom surfaces of the plate, and measurement of the heat transfer rate is 2000 W. What is the drag force on the plate?

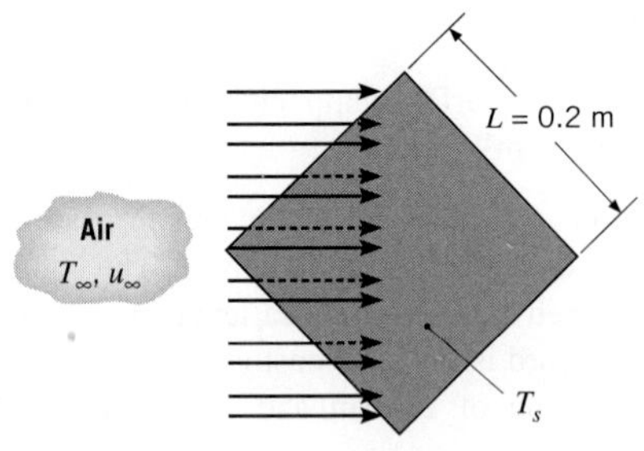

Top view of thin, flat plate

6.50 As a means of preventing ice formation on the wings of a small, private aircraft, it is proposed that electric resistance heating elements be installed within the wings. To determine representative power requirements, consider nominal flight conditions for which the plane moves at 100 m/s in air that is at a temperature of −23°C. If the characteristic length of the airfoil is $L = 2$ m and wind tunnel measurements indicate an average friction coefficient of $\overline{C}_f = 0.0025$ for the nominal conditions, what is the average heat flux needed to maintain a surface temperature of $T_s = 5°C$?

6.51 A circuit board with a dense distribution of integrated circuits (ICs) and dimensions of 120 mm × 120 mm on a side is cooled by the parallel flow of atmospheric air with a velocity of 2 m/s.

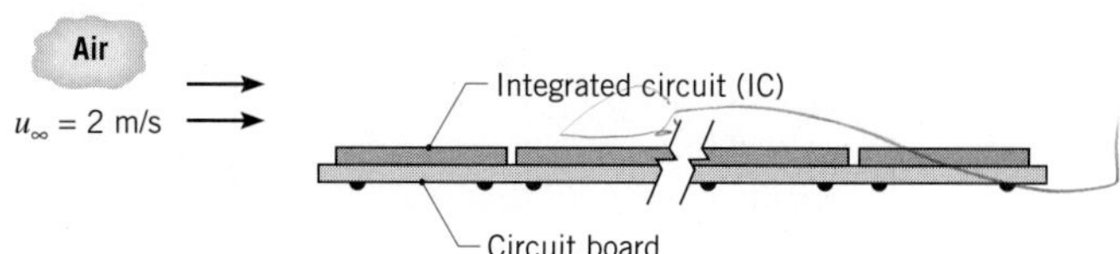

From wind tunnel tests under the same flow conditions, the average frictional shear stress on the upper surface is determined to be 0.0625 N/m^2. What is the allowable power dissipation from the upper surface of the board if the average surface temperature of the ICs must not exceed the ambient air temperature by more than 25°C? Evaluate the thermophysical properties of air at 300 K.

CHAPTER 7
External Flow

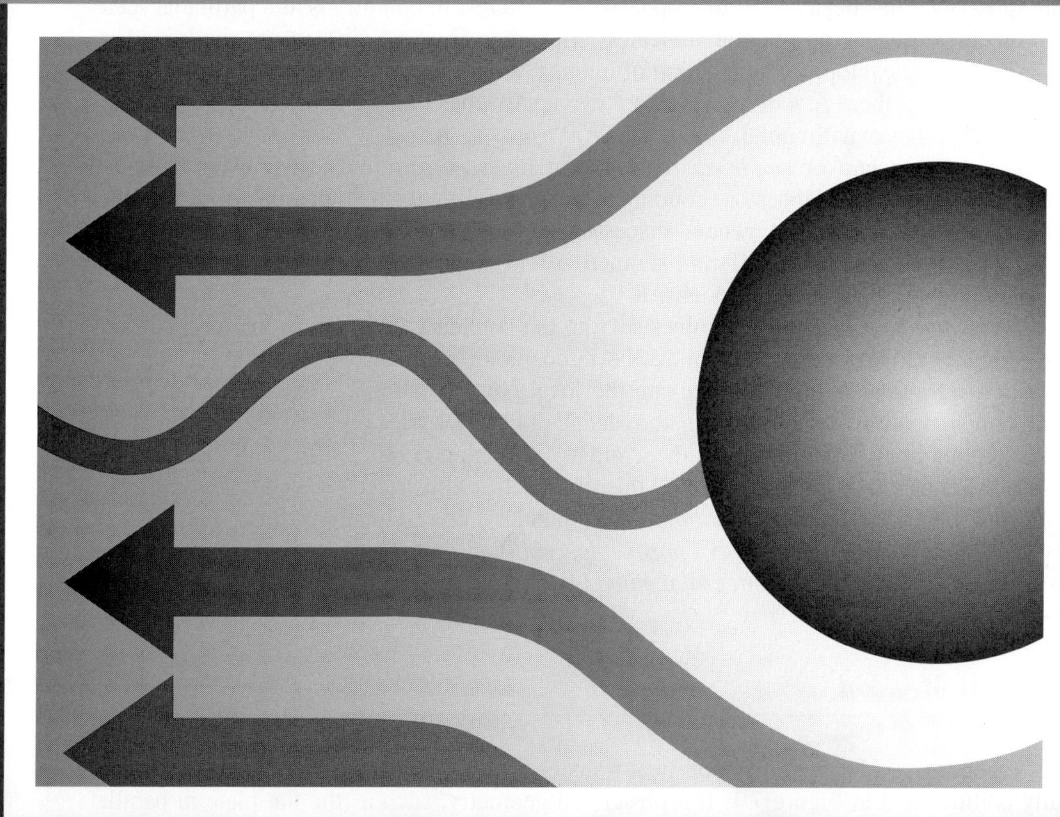

In this chapter we focus on the problem of computing heat transfer rates to or from a surface in *external flow*. In such a flow boundary layers develop freely, without constraints imposed by adjacent surfaces. Accordingly, there will always exist a region of the flow outside the boundary layer in which velocity and/or temperature gradients are negligible. Examples include fluid motion over a flat plate (inclined or parallel to the free stream velocity) and flow over curved surfaces such as a sphere, cylinder, airfoil, or turbine blade.

For the moment we confine our attention to problems of *low-speed, forced convection* with *no phase change* occurring within the fluid. In addition, we will not consider potential micro- or nanoscale effects within the fluid, as described in Section 2.2, in this chapter. In *forced* convection, the relative motion between the fluid and the surface is maintained by external means, such as a fan or a pump, and not by buoyancy forces due to temperature gradients in the fluid (*natural* convection). *Internal flows, natural convection,* and *convection with phase change* are treated in Chapters 8, 9, and 10, respectively.

Our primary objective is to determine convection coefficients for different flow geometries. In particular, we wish to obtain specific forms of the functions that represent these coefficients. By nondimensionalizing the boundary layer equations in Chapter 6, we found that the local and average convection coefficients may be correlated by equations of the form

$$Nu_x = f(x^*, Re_x, Pr) \tag{6.32}$$

$$\overline{Nu}_x = f(Re_x, Pr) \tag{6.33}$$

The subscript x has been added to emphasize our interest in conditions at a particular location on the surface. The overbar indicates an average from $x^* = 0$, where the boundary layer begins to develop, to the location of interest. Recall that *the problem of convection* is one of obtaining these functions. There are two approaches that we could take, one theoretical and the other experimental.

The *experimental* or *empirical approach* involves performing heat transfer measurements under controlled laboratory conditions and correlating the data in terms of appropriate dimensionless parameters. A general discussion of the approach is provided in Section 7.1. It has been applied to many different geometries and flow conditions, and important results are presented in Sections 7.2 through 7.8.

The *theoretical approach* involves solving the boundary layer equations for a particular geometry. For example, obtaining the temperature profile T^* from such a solution, Equation 6.31 may be used to evaluate the local Nusselt number Nu_x, and therefore the local convection coefficient h_x. With knowledge of how h_x varies over the surface, Equation 6.9 may then be used to determine the average convection coefficient $\overline{h}_x$, and therefore the Nusselt number $\overline{Nu}_x$. In Section 7.2.1 this approach is illustrated by using the *similarity method* to obtain an *exact solution* of the boundary layer equations for a flat plate in parallel, laminar flow [1–3]. An *approximate solution* to the same problem is obtained in Appendix G by using the *integral method* [4].

7.1 The Empirical Method

The manner in which a convection heat transfer correlation may be obtained experimentally is illustrated in Figure 7.1. If a prescribed geometry, such as the flat plate in parallel

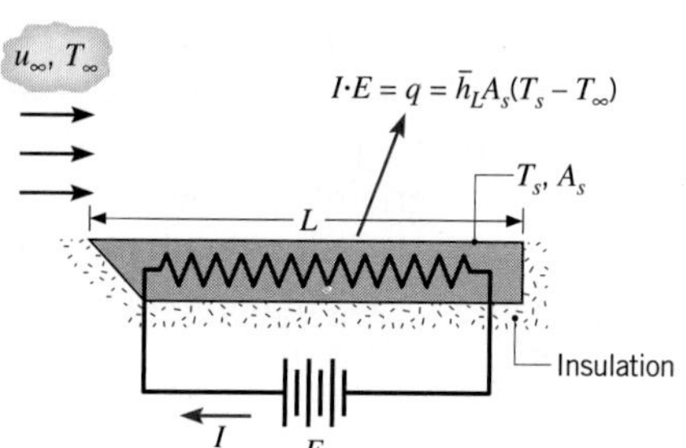

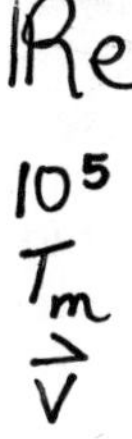

FIGURE 7.1 Experiment for measuring the average convection heat transfer coefficient $\bar{h}_L$.

flow, is heated electrically to maintain $T_s > T_\infty$, convection heat transfer occurs from the surface to the fluid. It would be a simple matter to measure T_s and T_∞, as well as the electrical power, $E \cdot I$, which is equal to the total heat transfer rate q. The convection coefficient $\bar{h}_L$, which is an average associated with the entire plate, could then be computed from Newton's law of cooling, Equation 6.8. Moreover, from knowledge of the characteristic length L and the fluid properties, the Nusselt, Reynolds, and Prandtl numbers could be computed from their definitions, Equations 6.33, 6.25, and 6.26, respectively.

The foregoing procedure could be repeated for a variety of test conditions. We could vary the velocity u_∞ and the plate length L, as well as the nature of the fluid, using, for example, air, water, and engine oil, which have substantially different Prandtl numbers. We would then be left with many different values of the Nusselt number corresponding to a wide range of Reynolds and Prandtl numbers, and the results could be plotted on a *log–log* scale, as shown in Figure 7.2*a*. Each symbol represents a unique set of test conditions. As is often the case, the results associated with a given fluid, and hence a fixed Prandtl number, fall close to a straight line, indicating a power law dependence of the Nusselt number on the Reynolds number. Considering all the fluids, the data may then be represented by an algebraic expression of the form

$$\overline{Nu}_L = C\, Re_L^m\, Pr^n \tag{7.1}$$

Since the values of C, m, and n are often independent of the nature of the fluid, the family of straight lines corresponding to different Prandtl numbers can be collapsed to a single line by plotting the results in terms of the ratio, $\overline{Nu}_L/Pr^n$, as shown in Figure 7.2*b*.

Because Equation 7.1 is inferred from experimental measurements, it is termed an *empirical correlation.* The specific values of the coefficient C and the exponents m and n vary with the nature of the surface geometry and the type of flow.

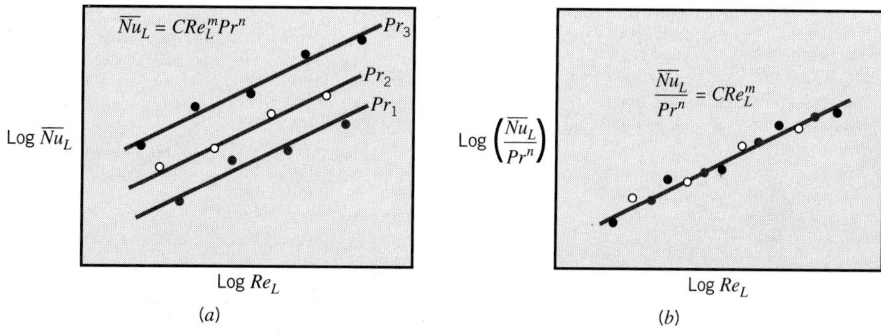

FIGURE 7.2 Dimensionless representation of convection heat transfer measurements.

We will use expressions of the form given by Equation 7.1 for many special cases, and it is important to note that the assumption of *constant fluid properties* is often implicit in the results. However, we know that the fluid properties vary with temperature across the boundary layer and that this variation can certainly influence the heat transfer rate. This influence may be handled in one of two ways. In one method, Equation 7.1 is used with all properties evaluated at a mean boundary layer temperature T_f, termed the *film temperature*.

$$T_f \equiv \frac{T_s + T_\infty}{2} \tag{7.2}$$

The alternate method is to evaluate all properties at T_∞ and to multiply the right-hand side of Equation 7.1 by an additional parameter to account for the property variations. The parameter is commonly of the form $(Pr_\infty/Pr_s)^r$ or $(\mu_\infty/\mu_s)^r$, where the subscripts ∞ and s designate evaluation of the properties at the free stream and surface temperatures, respectively. Both methods are used in the results that follow.

7.2 The Flat Plate in Parallel Flow

Despite its simplicity, parallel flow over a flat plate (Figure 7.3) occurs in numerous engineering applications. As discussed in Section 6.3, laminar boundary layer development begins at the leading edge ($x = 0$) and transition to turbulence may occur at a downstream location (x_c) for which a critical Reynolds number $Re_{x,c}$ is achieved. We begin by considering conditions in the laminar boundary layer. Specifically, we will analytically determine the velocity and temperature distributions that are shown qualitatively in Figures 6.1 and 6.2, respectively. From knowledge of these distributions, we will then determine expressions for the local and average friction coefficients and Nusselt numbers.

7.2.1 Laminar Flow over an Isothermal Plate: A Similarity Solution

The major convection parameters may be obtained by solving the appropriate form of the boundary layer equations. Assuming *steady, incompressible, laminar* flow with *constant fluid properties* and *negligible viscous dissipation* and recognizing that $dp/dx = 0$, the boundary layer equations (6.15, 6.16, and 6.17) reduce to

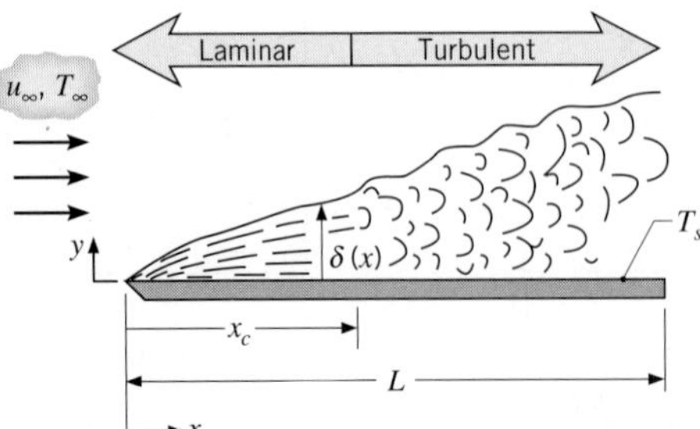

FIGURE 7.3 The flat plate in parallel flow.

Continuity:

$$\frac{\partial u}{\partial x} + \frac{\partial v}{\partial y} = 0 \tag{7.3}$$

Momentum:

$$u\frac{\partial u}{\partial x} + v\frac{\partial u}{\partial y} = \nu\frac{\partial^2 u}{\partial y^2} \tag{7.4}$$

Energy:

$$u\frac{\partial T}{\partial x} + v\frac{\partial T}{\partial y} = \alpha\frac{\partial^2 T}{\partial y^2} \tag{7.5}$$

Solution of these equations is simplified by the fact that for constant properties, conditions in the velocity (hydrodynamic) boundary layer are independent of temperature. Hence we may begin by solving the hydrodynamic problem, Equations 7.3 and 7.4, to the exclusion of Equation 7.5. Once the hydrodynamic problem has been solved, a solution to Equation 7.5, which depends on u and v, may be obtained.

Hydrodynamic Solution The hydrodynamic solution follows the method of Blasius [1, 2]. The first step is to define a stream function $\psi(x, y)$, such that

$$u \equiv \frac{\partial \psi}{\partial y} \quad \text{and} \quad v \equiv -\frac{\partial \psi}{\partial x} \tag{7.6}$$

Equation 7.3 is then automatically satisfied and hence is no longer needed. New dependent and independent variables, f and η, respectively, are then defined such that

$$f(\eta) \equiv \frac{\psi}{u_\infty \sqrt{\nu x / u_\infty}} \tag{7.7}$$

$$\eta \equiv y\sqrt{u_\infty / \nu x} \tag{7.8}$$

As we will find, use of these variables simplifies matters by reducing the partial differential equation, Equation 7.4, to an ordinary differential equation.

The Blasius solution is termed a *similarity solution,* and η is a *similarity variable.* This terminology is used because, despite growth of the boundary layer with distance x from the leading edge, the velocity profile u/u_∞ remains *geometrically similar.* This similarity is of the functional form

$$\frac{u}{u_\infty} = \phi\left(\frac{y}{\delta}\right)$$

where δ is the boundary layer thickness. We will find from the Blasius solution that δ varies as $(\nu x/u_\infty)^{1/2}$; thus, it follows that

$$\frac{u}{u_\infty} = \phi(\eta) \tag{7.9}$$

Hence the velocity profile is uniquely determined by the similarity variable η, which depends on both x and y.

From Equations 7.6 through 7.8 we obtain

$$u = \frac{\partial \psi}{\partial y} = \frac{\partial \psi}{\partial \eta}\frac{\partial \eta}{\partial y} = u_\infty \sqrt{\frac{\nu x}{u_\infty}}\frac{df}{d\eta}\sqrt{\frac{u_\infty}{\nu x}} = u_\infty \frac{df}{d\eta} \tag{7.10}$$

and

$$v = -\frac{\partial \psi}{\partial x} = -\left(u_\infty \sqrt{\frac{\nu x}{u_\infty}}\frac{\partial f}{\partial x} + \frac{u_\infty}{2}\sqrt{\frac{\nu}{u_\infty x}} f\right)$$

$$v = \frac{1}{2}\sqrt{\frac{\nu u_\infty}{x}}\left(\eta \frac{df}{d\eta} - f\right) \tag{7.11}$$

By differentiating the velocity components, it may also be shown that

$$\frac{\partial u}{\partial x} = -\frac{u_\infty}{2x}\eta \frac{d^2 f}{d\eta^2} \tag{7.12}$$

$$\frac{\partial u}{\partial y} = u_\infty \sqrt{\frac{u_\infty}{\nu x}}\frac{d^2 f}{d\eta^2} \tag{7.13}$$

$$\frac{\partial^2 u}{\partial y^2} = \frac{u_\infty^2}{\nu x}\frac{d^3 f}{d\eta^3} \tag{7.14}$$

Substituting these expressions into Equation 7.4, we then obtain

$$2\frac{d^3 f}{d\eta^3} + f\frac{d^2 f}{d\eta^2} = 0 \tag{7.15}$$

Hence the hydrodynamic boundary layer problem is reduced to one of solving a nonlinear, third-order ordinary differential equation. The appropriate boundary conditions are

$$u(x, 0) = v(x, 0) = 0 \quad \text{and} \quad u(x, \infty) = u_\infty$$

or, in terms of the similarity variables,

$$\left.\frac{df}{d\eta}\right|_{\eta=0} = f(0) = 0 \quad \text{and} \quad \left.\frac{df}{d\eta}\right|_{\eta\to\infty} = 1 \tag{7.16}$$

The solution to Equation 7.15, subject to the conditions of Equation 7.16, may be obtained by a series expansion [2] or by numerical integration [3]. Selected results are presented in Table 7.1, from which useful information may be extracted. The x-component velocity distribution from the third column of the table is plotted in Figure 7.4a. We also note that, to a good approximation, $(u/u_\infty) = 0.99$ for $\eta = 5.0$. Defining the boundary layer thickness δ as that value of y for which $(u/u_\infty) = 0.99$, it follows from Equation 7.8 that

$$\delta = \frac{5.0}{\sqrt{u_\infty/\nu x}} = \frac{5x}{\sqrt{Re_x}} \tag{7.17}$$

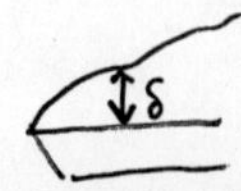

TABLE 7.1 Flat plate laminar boundary layer functions [3]

$\eta = y\sqrt{\dfrac{u_\infty}{\nu x}}$	f	$\dfrac{df}{d\eta} = \dfrac{u}{u_\infty}$	$\dfrac{d^2f}{d\eta^2}$
0	0	0	0.332
0.4	0.027	0.133	0.331
0.8	0.106	0.265	0.327
1.2	0.238	0.394	0.317
1.6	0.420	0.517	0.297
2.0	0.650	0.630	0.267
2.4	0.922	0.729	0.228
2.8	1.231	0.812	0.184
3.2	1.569	0.876	0.139
3.6	1.930	0.923	0.098
4.0	2.306	0.956	0.064
4.4	2.692	0.976	0.039
4.8	3.085	0.988	0.022
5.2	3.482	0.994	0.011
5.6	3.880	0.997	0.005
6.0	4.280	0.999	0.002
6.4	4.679	1.000	0.001
6.8	5.079	1.000	0.000

From Equation 7.17 it is clear that δ increases with increasing x and ν but decreases with increasing u_∞ (the larger the free stream velocity, the *thinner* the boundary layer). In addition, from Equation 7.13 the wall shear stress may be expressed as

$$\tau_s = \mu \left.\frac{\partial u}{\partial y}\right|_{y=0} = \mu u_\infty \sqrt{u_\infty/\nu x} \left.\frac{d^2 f}{d\eta^2}\right|_{\eta=0}$$

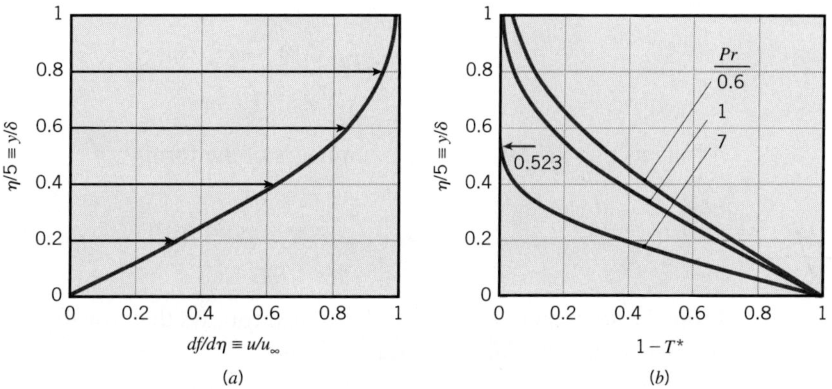

FIGURE 7.4 Similarity solution for laminar flow over an isothermal plate. (*a*) The *x*-component of the velocity. (*b*) Temperature distributions for Pr = 0.6, 1, and 7.

Hence from Table 7.1

$$\tau_s = 0.332 u_\infty \sqrt{\rho \mu u_\infty / x}$$

The *local* friction coefficient is then

$$C_{f,x} \equiv \frac{\tau_{s,x}}{\rho u_\infty^2/2} = 0.664\, Re_x^{-1/2} \tag{7.18}$$

Heat Transfer Solution From knowledge of conditions in the velocity boundary layer, the energy equation may now be solved. We begin by introducing the dimensionless temperature $T^* \equiv [(T - T_s)/(T_\infty - T_s)]$ and assume a similarity solution of the form $T^* = T^*(\eta)$. Making the necessary substitutions, Equation 7.5 reduces to

$$\frac{d^2T^*}{d\eta^2} + \frac{Pr}{2} f \frac{dT^*}{d\eta} = 0 \tag{7.19}$$

Note the dependence of the thermal solution on hydrodynamic conditions through appearance of the variable f in Equation 7.19. The appropriate boundary conditions are

$$T^*(0) = 0 \quad \text{and} \quad T^*(\infty) = 1 \tag{7.20}$$

Subject to the conditions of Equation 7.20, Equation 7.19 may be solved by numerical integration for different values of the Prandtl number; representative temperature distributions for $Pr = 0.6$, 1, and 7 are shown in Figure 7.4*b*. Thermal effects penetrate farther into the velocity boundary layer with decreasing Prandtl number and transcend the velocity boundary layer for $Pr < 1$. One important consequence of this solution is that, for $Pr \gtrsim 0.6$, results for the surface temperature gradient $dT^*/d\eta|_{\eta=0}$ may be correlated by the following relation:

$$\left.\frac{dT^*}{d\eta}\right|_{\eta=0} = 0.332\, Pr^{1/3}$$

Expressing the local convection coefficient as

$$h_x = \frac{q_s''}{T_s - T_\infty} = -\frac{T_\infty - T_s}{T_s - T_\infty} k \left.\frac{\partial T^*}{\partial y}\right|_{y=0}$$

$$h_x = k\left(\frac{u_\infty}{\nu x}\right)^{1/2} \left.\frac{dT^*}{d\eta}\right|_{\eta=0}$$

it follows that the *local* Nusselt number is of the form

$$Nu_x \equiv \frac{h_x x}{k} = 0.332\, Re_x^{1/2}\, Pr^{1/3} \qquad Pr \gtrsim 0.6 \tag{7.21}$$

From the solution to Equation 7.19, it also follows that, for $Pr \gtrsim 0.6$, the ratio of the velocity to thermal boundary layer thickness is

$$\frac{\delta}{\delta_t} \approx Pr^{1/3} \tag{7.22}$$

where δ is given by Equation 7.17. For example, for $Pr = 7$, $\delta/\delta_t = 1.91$ ($\delta_t/\delta = 0.523$), as shown in Figure 7.4*b*.

The foregoing results may be used to compute important *laminar* boundary layer parameters for $0 < x < x_c$, where x_c is the distance from the leading edge at which transition begins. Equations 7.18 and 7.21 imply that $\tau_{s,x}$ and h_x are, in principle, infinite at the leading edge and decrease as $x^{-1/2}$ in the flow direction. Equation 7.22 also implies that, for values of Pr close to unity, which is the case for most gases, the two boundary layers experience nearly identical growth.

Average Boundary Layer Parameters From the foregoing local results, average boundary layer parameters may be determined. With the average friction coefficient defined as

$$\overline{C}_{f,x} \equiv \frac{\overline{\tau}_{s,x}}{\rho u_\infty^2/2} \tag{7.23}$$

where

$$\overline{\tau}_{s,x} \equiv \frac{1}{x}\int_0^x \tau_{s,x}\, dx$$

the form of $\tau_{s,x}$ may be substituted from Equation 7.18 and the integration performed to obtain

$$\overline{C}_{f,x} = 1.328\, Re_x^{-1/2} \tag{7.24}$$

Moreover, from Equations 6.10 and 7.21, the *average* heat transfer coefficient for laminar flow is

$$\overline{h}_x = \frac{1}{x}\int_0^x h_x\, dx = 0.332\left(\frac{k}{x}\right)Pr^{1/3}\left(\frac{u_\infty}{\nu}\right)^{1/2}\int_0^x \frac{dx}{x^{1/2}}$$

Integrating and substituting from Equation 7.21, it follows that $\overline{h}_x = 2h_x$. Hence

$$\overline{Nu}_x \equiv \frac{\overline{h}_x x}{k} = 0.664\, Re_x^{1/2}\, Pr^{1/3} \qquad Pr \gtrsim 0.6 \tag{7.25}$$

If the flow is laminar over the entire surface, the subscript x may be replaced by L, and Equations 7.24 and 7.25 may be used to predict average conditions for the entire surface.

From the foregoing expressions we see that, for laminar flow over a flat plate, the *average* friction and convection coefficients from the leading edge to a point x on the surface are *twice* the *local* coefficients at that point. We also note that, in using these expressions, the effect of variable properties can be treated by evaluating all properties at the *film temperature,* Equation 7.2.

Liquid Metals For fluids of small Prandtl number, namely *liquid metals,* Equation 7.21 does not apply. However, for this case the thermal boundary layer development is much more rapid than that of the velocity boundary layer ($\delta_t \gg \delta$), and it is reasonable to assume uniform velocity ($u = u_\infty$) throughout the thermal boundary layer. From a solution to the thermal boundary layer equation based on this assumption [5], it may then be shown that

$$Nu_x = 0.564\, Pe_x^{1/2} \qquad Pr \lesssim 0.05, \quad Pe_x \gtrsim 100 \tag{7.26}$$

where $Pe_x \equiv Re_x\, Pr$ is the *Peclet number* (Table 6.2). Despite the corrosive and reactive nature of liquid metals, their unique properties (low melting point and vapor pressure, as well as high thermal capacity and conductivity) render them attractive as coolants in applications requiring high heat transfer rates.

A single correlating equation, which applies for all Prandtl numbers, has been recommended by Churchill and Ozoe [6]. For laminar flow over an isothermal plate, the local convection coefficient may be obtained from

$$Nu_x = \frac{0.3387\, Re_x^{1/2}\, Pr^{1/3}}{[1 + (0.0468/Pr)^{2/3}]^{1/4}} \qquad Pe_x \gtrsim 100 \qquad (7.27)$$

with $\overline{Nu}_x = 2Nu_x$.

7.2.2 Turbulent Flow over an Isothermal Plate

It is not possible to obtain exact analytical solutions for turbulent boundary layers, which are inherently unsteady. From experiment [2] it is known that, for turbulent flows with Reynolds numbers up to approximately 10^8, the *local* friction coefficient is correlated to within 15% accuracy by an expression of the form

$$C_{f,x} = 0.0592\, Re_x^{-1/5} \qquad Re_{x,c} \lesssim Re_x \lesssim 10^8 \qquad (7.28)$$

Moreover, it is known that, to a reasonable approximation, the velocity boundary layer thickness may be expressed as

$$\delta = 0.37x\, Re_x^{-1/5} \qquad (7.29)$$

Comparing these results with those for the laminar boundary layer, Equations 7.17 and 7.18, we see that turbulent boundary layer growth is much more rapid (δ varies as $x^{4/5}$ in contrast to $x^{1/2}$ for laminar flow) and that the decay in the friction coefficient is more gradual ($x^{-1/5}$ versus $x^{-1/2}$). For turbulent flow, boundary layer development is influenced strongly by random fluctuations in the fluid and not by molecular diffusion. Hence relative boundary layer growth does not depend on the value of Pr, and Equation 7.29 may be used to obtain the thermal, as well as the velocity, boundary layer thicknesses. That is, for turbulent flow, $\delta \approx \delta_t$.

Using Equation 7.28 with the modified Reynolds, or Chilton–Colburn, analogy, Equation 6.38, the *local* Nusselt number for turbulent flow is

$$Nu_x = St\, Re_x\, Pr = 0.0296\, Re_x^{4/5}\, Pr^{1/3} \qquad 0.6 \lesssim Pr \lesssim 60 \qquad (7.30)$$

Enhanced mixing causes the turbulent boundary layer to grow more rapidly than the laminar boundary layer and to have larger friction and convection coefficients.

Expressions for the average coefficients may now be determined. However, since the turbulent boundary layer is generally preceded by a laminar boundary layer, we first consider *mixed* boundary layer conditions.

7.2.3 Mixed Boundary Layer Conditions

For laminar flow over the entire plate, Equations 7.24 and 7.25 may be used to compute the average coefficients. Moreover, if transition occurs toward the rear of the plate, for example, in the range $0.95 \lesssim (x_c/L) \lesssim 1$, these equations may be used to compute the average coefficients to a reasonable approximation. However, when transition occurs sufficiently upstream of the trailing edge, $(x_c/L) \lesssim 0.95$, the surface average coefficients will be influenced by conditions in both the laminar and turbulent boundary layers.

In the mixed boundary layer situation (Figure 7.3), Equation 6.10 may be used to obtain the average convection heat transfer coefficient for the entire plate. Integrating over the laminar region ($0 \leq x \leq x_c$) and then over the turbulent region ($x_c < x \leq L$), this equation may be expressed as

$$\bar{h}_L = \frac{1}{L}\left(\int_0^{x_c} h_{\text{lam}}\, dx + \int_{x_c}^{L} h_{\text{turb}}\, dx\right)$$

where it is assumed that transition occurs abruptly at $x = x_c$. Substituting from Equations 7.21 and 7.30 for h_{lam} and h_{turb}, respectively, we obtain

$$\bar{h}_L = \left(\frac{k}{L}\right)\left[0.332\left(\frac{u_\infty}{\nu}\right)^{1/2}\int_0^{x_c}\frac{dx}{x^{1/2}} + 0.0296\left(\frac{u_\infty}{\nu}\right)^{4/5}\int_{x_c}^{L}\frac{dx}{x^{1/5}}\right]Pr^{1/3}$$

Integrating, we then obtain

$$\overline{Nu}_L = (0.037\, Re_L^{4/5} - A)\, Pr^{1/3} \tag{7.31}$$

$$\begin{bmatrix} 0.6 \lesssim Pr \lesssim 60 \\ Re_{x,c} \lesssim Re_L \lesssim 10^8 \end{bmatrix}$$

where the bracketed relations indicate the range of applicability and the constant A is determined by the value of the critical Reynolds number, $Re_{x,c}$. That is,

$$A = 0.037\, Re_{x,c}^{4/5} - 0.664\, Re_{x,c}^{1/2} \tag{7.32}$$

Similarly, the average friction coefficient may be found using the expression

$$\bar{C}_{f,L} = \frac{1}{L}\left(\int_0^{x_c} C_{f,x,\text{lam}}\, dx + \int_{x_c}^{L} C_{f,x,\text{turb}}\, dx\right)$$

Substituting expressions for $C_{f,x,\text{lam}}$ and $C_{f,x,\text{turb}}$ from Equations 7.18 and 7.28, respectively, and carrying out the integration provides an expression of the form

$$\bar{C}_{f,L} = 0.074\, Re_L^{-1/5} - \frac{2A}{Re_L} \tag{7.33}$$

$$[Re_{x,c} \lesssim Re_L \lesssim 10^8]$$

For a completely turbulent boundary layer ($Re_{x,c} = 0$), $A = 0$. Such a condition may be realized by *tripping* the boundary layer at the leading edge, using a fine wire or some other turbulence promoter. For a transition Reynolds number of $Re_{x,c} = 5 \times 10^5$, $A = 871$.

All of the foregoing correlations require evaluation of the fluid properties at the film temperature, Equation 7.2.

7.2.4 Unheated Starting Length

All the foregoing Nusselt number expressions are restricted to situations for which the surface temperature T_s is uniform. A common exception involves existence of an *unheated starting length* ($T_s = T_\infty$) upstream of a heated section ($T_s \neq T_\infty$). As shown in Figure 7.5, velocity boundary layer growth begins at $x = 0$, while thermal boundary layer development begins at $x = \xi$. Hence there is no heat transfer for $0 \leq x \leq \xi$. Through use of an integral boundary layer solution [5], it is known that, for laminar flow,

$$Nu_x = \frac{Nu_x|_{\xi=0}}{[1 - (\xi/x)^{3/4}]^{1/3}} \tag{7.34}$$

where $Nu_x|_{\xi=0}$, is given by Equation 7.21. In both Nu_x and $Nu_x|_{\xi=0}$, the characteristic length x is measured from the leading edge of the unheated starting length. It has also been found that, for turbulent flow,

$$Nu_x = \frac{Nu_x|_{\xi=0}}{[1 - (\xi/x)^{9/10}]^{1/9}} \tag{7.35}$$

where $Nu_x|_{\xi=0}$ is given by Equation 7.30.

By using Equation 6.10 with local convection coefficients given by the foregoing relations, expressions may be obtained for the *average Nusselt number* of an isothermal plate with an unheated starting length [7]. For a plate of total length L, with laminar *or* turbulent flow over the entire surface, the expressions are of the form

$$\overline{Nu}_L = \overline{Nu}_L|_{\xi=0} \frac{L}{L - \xi}[1 - (\xi/L)^{(p+1)/(p+2)}]^{p/(p+1)} \tag{7.36}$$

where $p = 2$ for laminar flow and $p = 8$ for turbulent flow. The quantity $\overline{Nu}_L|_{\xi=0}$ is the average Nusselt number for a plate of length L when heating starts at the leading edge of the plate. For laminar flow, it can be obtained from Equation 7.25 (with x replaced by L); for turbulent flow it is given by Equation 7.31 with $A = 0$ (assuming turbulent flow over the entire surface). Note that $\overline{Nu}_L$ is equal to $\overline{h}L/k$, where $\overline{h}$ is averaged over the heated portion of the plate only, which is of length $(L - \xi)$. The corresponding value of $\overline{h}_L$ must therefore be multiplied by the area of the heated section to determine the total heat rate from the plate.

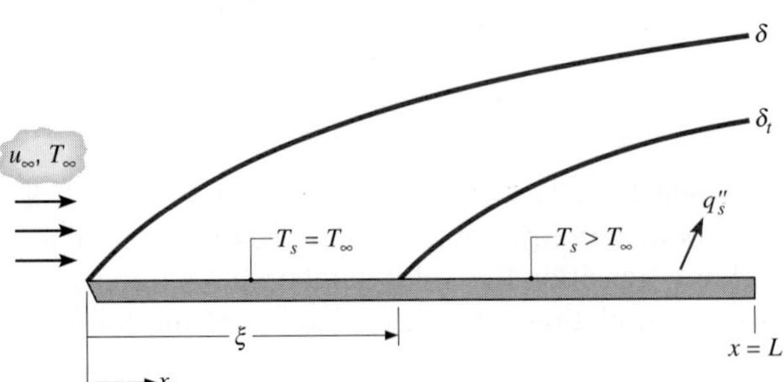

FIGURE 7.5 Flat plate in parallel flow with unheated starting length.

7.2.5 Flat Plates with Constant Heat Flux Conditions

It is also possible to have a uniform surface heat flux, rather than a uniform temperature, imposed at the plate. For laminar flow, it may be shown that [5]

$$Nu_x = 0.453\, Re_x^{1/2}\, Pr^{1/3} \qquad Pr \gtrsim 0.6 \tag{7.37}$$

while for turbulent flow

$$Nu_x = 0.0308\, Re_x^{4/5}\, Pr^{1/3} \qquad 0.6 \lesssim Pr \lesssim 60 \tag{7.38}$$

Hence the Nusselt number is 36% and 4% larger than the constant surface temperature result for laminar and turbulent flow, respectively. Correction for the effect of an unheated starting length may be made by using Equations 7.37 and 7.38 with Equations 7.34 and 7.35, respectively. If the heat flux is known, the convection coefficient may be used to determine the local surface temperature

$$T_s(x) = T_\infty + \frac{q_s''}{h_x} \tag{7.39}$$

Since the total heat rate is readily determined from the product of the uniform flux and the surface area, $q = q_s''A_s$, it is not necessary to introduce an average convection coefficient for the purpose of determining q. However, one may still wish to determine an *average surface temperature* from an expression of the form

$$\overline{(T_s - T_\infty)} = \frac{1}{L}\int_0^L (T_s - T_\infty)dx = \frac{q_s''}{L}\int_0^L \frac{x}{k\,Nu_x}dx$$

where Nu_x is obtained from the appropriate convection correlation. Substituting from Equation 7.37, it follows that

$$\overline{(T_s - T_\infty)} = \frac{q_s''L}{k\overline{Nu}_L} \tag{7.40}$$

where

$$\overline{Nu}_L = 0.680\, Re_L^{1/2}\, Pr^{1/3} \tag{7.41}$$

This result is only 2% larger than that obtained by evaluating Equation 7.25 at $x = L$. Differences are even smaller for turbulent flow, suggesting that any of the $\overline{Nu}_L$ results obtained for a uniform surface temperature may be used with Equation 7.40 to evaluate $\overline{(T_s - T_\infty)}$. Expressions for the average temperature of a plate that is subjected to a uniform heat flux downstream of an unheated starting section have been obtained by Ameel [7].

7.2.6 Limitations on Use of Convection Coefficients

Although the equations of this section are suitable for most engineering calculations, in practice they rarely provide exact values for the convection coefficients. Conditions vary according to free stream turbulence and surface roughness, and errors as large as 25% may be incurred by using the expressions. A detailed description of free stream turbulence effects is provided by Blair [8].

7.3 Methodology for a Convection Calculation

Although we have only discussed correlations for parallel flow over a flat plate, selection and application of a convection correlation for *any flow situation* are facilitated by following a few simple rules.

1. *Become immediately cognizant of the flow geometry.* For example, does the problem involve flow over a flat plate, a sphere, or a cylinder? The specific form of the convection correlation depends, of course, on the geometry.
2. *Specify the appropriate reference temperature and evaluate the pertinent fluid properties at that temperature.* For moderate boundary layer temperature differences, the film temperature, Equation 7.2, may be used for this purpose. However, we will consider correlations that require property evaluation at the free stream temperature and include a property ratio to account for the nonconstant property effect.
3. *Calculate the Reynolds number.* Boundary layer conditions are strongly influenced by this parameter. If the geometry is a flat plate in parallel flow, determine whether the flow is laminar or turbulent.
4. *Decide whether a local or surface average coefficient is required.* Recall that for constant surface temperature, the local coefficient is used to determine the flux at a particular point on the surface, whereas the average coefficient determines the transfer rate for the entire surface.
5. *Select the appropriate correlation.*

IHT **EXAMPLE 7.1**

Air at a pressure of 6 kN/m^2 and a temperature of 300°C flows with a velocity of 10 m/s over a flat plate 0.5 m long. Estimate the cooling rate per unit width of the plate needed to maintain it at a surface temperature of 27°C.

SOLUTION

Known: Airflow over an isothermal flat plate.

Find: Cooling rate per unit width of the plate, q' (W/m).

Schematic:

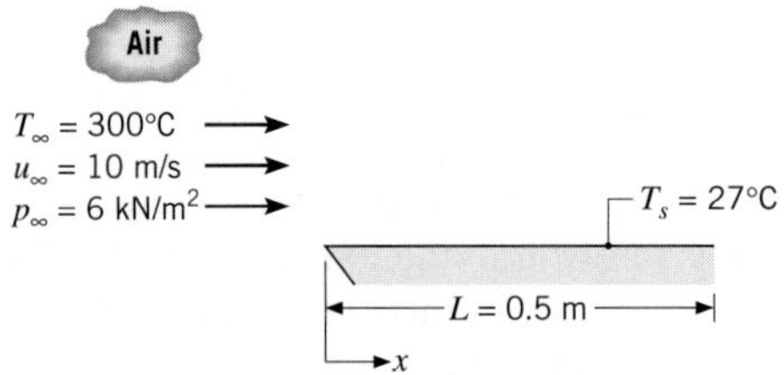

Assumptions:

1. Steady-state, incompressible flow conditions.
2. Negligible radiation effects.

Properties: Table A.4, air ($T_f = 437$ K, $p = 1$ atm): $\nu = 30.84 \times 10^{-6}$ m²/s, $k = 36.4 \times 10^{-3}$ W/m·K, $Pr = 0.687$. As noted in Example 6.4, the properties k, Pr, c_p, and μ may be assumed to be independent of pressure. However, for an ideal gas, the kinematic viscosity is inversely proportional to pressure. Hence the kinematic viscosity of air at 437 K and $p_\infty = 6 \times 10^3$ N/m² is

$$\nu = 30.84 \times 10^{-6}\,\text{m}^2/\text{s} \times \frac{1.0133 \times 10^5\,\text{N/m}^2}{6 \times 10^3\,\text{N/m}^2} = 5.21 \times 10^{-4}\,\text{m}^2/\text{s}$$

Analysis: For a plate of unit width, it follows from Newton's law of cooling that the rate of convection heat transfer *to* the plate is

$$q' = \overline{h}L(T_\infty - T_s)$$

To determine the appropriate convection correlation for computing $\overline{h}$, the Reynolds number must first be determined

$$Re_L = \frac{u_\infty L}{\nu} = \frac{10\,\text{m/s} \times 0.5\,\text{m}}{5.21 \times 10^{-4}\,\text{m}^2/\text{s}} = 9597$$

Hence the flow is laminar over the entire plate, and the appropriate correlation is given by Equation 7.25.

$$\overline{Nu}_L = 0.664\, Re_L^{1/2}\, Pr^{1/3} = 0.664(9597)^{1/2}(0.687)^{1/3} = 57.4$$

The average convection coefficient is then

$$\overline{h} = \frac{\overline{Nu}_L k}{L} = \frac{57.4 \times 0.0364\,\text{W/m}\cdot\text{K}}{0.5\,\text{m}} = 4.18\,\text{W/m}^2\cdot\text{K}$$

and the required cooling rate per unit width of plate is

$$q' = 4.18\,\text{W/m}^2\cdot\text{K} \times 0.5\,\text{m}\,(300 - 27)^\circ\text{C} = 570\,\text{W/m} \quad \triangleleft$$

Comments:

1. The results of Table A.4 apply to gases at atmospheric pressure.
2. Example 7.1 in *IHT* demonstrates how to use the *Correlations* and *Properties* tools, which can facilitate performing convection calculations.

EXAMPLE 7.2

A flat plate of width $w = 1$ m is maintained at a uniform surface temperature, $T_s = 230^\circ$C, by using independently controlled, electrical strip heaters, each of which is 50 mm long. If atmospheric air at 25°C flows over the plate at a velocity of 60 m/s, at what heater is the electrical input a maximum? What is the value of this input?

SOLUTION

Known: Airflow over a flat plate with segmented heaters.

Find: Maximum heater power requirement.

Schematic:

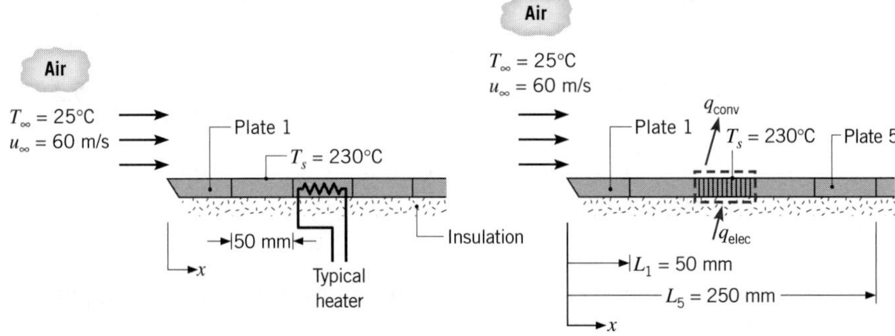

Assumptions:

1. Steady-state, incompressible flow conditions.
2. Negligible radiation effects.
3. Bottom surface of plate adiabatic.

Properties: Table A.4, air ($T_f = 400$ K, $p = 1$ atm): $\nu = 26.41 \times 10^{-6}$ m^2/s, $k =$ 0.0338 W/m·K, $Pr = 0.690$.

Analysis: The location of the heater requiring the maximum electrical power may be determined by first finding the point of boundary layer transition. The Reynolds number based on the length L_1 of the first heater is

$$Re_1 = \frac{u_\infty L_1}{\nu} = \frac{60 \text{ m/s} \times 0.05 \text{ m}}{26.41 \times 10^{-6} \text{ m}^2/\text{s}} = 1.14 \times 10^5$$

If the transition Reynolds number is assumed to be $Re_{x,c} = 5 \times 10^5$, it follows that transition will occur on the fifth heater, or more specifically at

$$x_c = \frac{\nu}{u_\infty} Re_{x,c} = \frac{26.41 \times 10^{-6} \text{ m}^2/\text{s}}{60 \text{ m/s}} 5 \times 10^5 = 0.22 \text{ m}$$

The heater requiring the maximum electrical power is that for which the average convection coefficient is largest. Knowing how the local convection coefficient varies with distance from the leading edge, we conclude that there are three possibilities:

1. Heater 1, since it corresponds to the largest local, laminar convection coefficient.
2. Heater 5, since it corresponds to the largest local, turbulent convection coefficient.
3. Heater 6, since turbulent conditions exist over the entire heater.

For each of these heaters, conservation of energy requires that

$$q_{\text{elec}} = q_{\text{conv}}$$

For the first heater,

$$q_{\text{conv},1} = \overline{h}_1 L_1 w(T_s - T_\infty)$$

where $\overline{h}_1$ is determined from Equation 7.25,

$$\overline{Nu}_1 = 0.664\, Re_1^{1/2}\, Pr^{1/3} = 0.664(1.14 \times 10^5)^{1/2}(0.69)^{1/3} = 198$$

Hence

$$\overline{h}_1 = \frac{\overline{Nu}_1 k}{L_1} = \frac{198 \times 0.0338 \text{ W/m}\cdot\text{K}}{0.05 \text{ m}} = 134 \text{ W/m}^2\cdot\text{K}$$

and

$$q_{\text{conv},1} = 134 \text{ W/m}^2\cdot\text{K}(0.05 \times 1)\text{ m}^2(230 - 25)^\circ\text{C} = 1370 \text{ W}$$

The power requirement for the fifth heater may be obtained by subtracting the total heat loss associated with the first four heaters from that associated with the first five heaters. Accordingly,

$$q_{\text{conv},5} = \overline{h}_{1-5} L_5 w(T_s - T_\infty) - \overline{h}_{1-4} L_4 w(T_s - T_\infty)$$

$$q_{\text{conv},5} = (\overline{h}_{1-5} L_5 - \overline{h}_{1-4} L_4) w(T_s - T_\infty)$$

The value of $\overline{h}_{1-4}$ may be obtained from Equation 7.25, where

$$\overline{Nu}_4 = 0.664\, Re_4^{1/2}\, Pr^{1/3}$$

With $Re_4 = 4\, Re_1 = 4.56 \times 10^5$,

$$\overline{Nu}_4 = 0.664(4.56 \times 10^5)^{1/2}(0.69)^{1/3} = 396$$

Hence

$$\overline{h}_{1-4} = \frac{\overline{Nu}_4 k}{L_4} = \frac{396 \times 0.0338 \text{ W/m}\cdot\text{K}}{0.2 \text{ m}} = 67 \text{ W/m}^2\cdot\text{K}$$

In contrast, the fifth heater is characterized by mixed boundary layer conditions, and $\overline{h}_{1-5}$ must be obtained from Equation 7.31, with $A = 871$. With $Re_5 = 5\, Re_1 = 5.70 \times 10^5$,

$$\overline{Nu}_5 = (0.037\, Re_5^{4/5} - 871) Pr^{1/3}$$

$$\overline{Nu}_5 = [0.037(5.70 \times 10^5)^{4/5} - 871](0.69)^{1/3} = 546$$

Hence

$$\overline{h}_{1-5} = \frac{\overline{Nu}_5 k}{L_5} = \frac{546 \times 0.0338 \text{ W/m}\cdot\text{K}}{0.25 \text{ m}} = 74 \text{ W/m}^2\cdot\text{K}$$

The rate of heat transfer from the fifth heater is then

$$q_{conv,5} = (74\ \text{W/m}^2\cdot\text{K} \times 0.25\ \text{m} - 67\ \text{W/m}^2\cdot\text{K} \times 0.20\ \text{m})$$
$$\times 1\ \text{m}(230 - 25)°\text{C}$$
$$q_{conv,5} = 1050\ \text{W}$$

Similarly, the power requirement for the sixth heater may be obtained by subtracting the total heat loss associated with the first five heaters from that associated with the first six heaters. Hence

$$q_{conv,6} = (\bar{h}_{1-6} L_6 - \bar{h}_{1-5} L_5) w (T_s - T_\infty)$$

where $\bar{h}_{1-6}$ may be obtained from Equation 7.31. With $Re_6 = 6\ Re_1 = 6.84 \times 10^5$,

$$\overline{Nu}_6 = [0.037(6.84 \times 10^5)^{4/5} - 871](0.69)^{1/3} = 753$$

Hence

$$\bar{h}_{1-6} = \frac{\overline{Nu}_6 k}{L_6} = \frac{753 \times 0.0338\ \text{W/m}\cdot\text{K}}{0.30\ \text{m}} = 85\ \text{W/m}^2\cdot\text{K}$$

and

$$q_{conv,6} = (85\ \text{W/m}^2\cdot\text{K} \times 0.30\ \text{m} - 74\ \text{W/m}^2\cdot\text{K} \times 0.25\ \text{m})$$
$$\times\ 1\ \text{m}\ (230 - 25)°\text{C}$$
$$q_{conv,6} = 1440\ \text{W} \qquad \triangleleft$$

Hence $q_{conv,6} > q_{conv,1} > q_{conv,5}$, and the sixth plate has the largest power requirement.

Comments:

1. An alternative, less accurate method of finding the convection heat transfer rate from a particular plate involves estimating an average local convection coefficient for the surface. For example, Equation 7.30 could be used to evaluate the local convection coefficient at the midpoint of the sixth plate. With $x = 0.275$ m, $Re_x = 6.25 \times 10^5$, $Nu_x = 1130$, and $h_x = 139\ \text{W/m}^2\cdot\text{K}$, the convection heat transfer rate from the sixth plate is

$$q_{conv,6} = h_x(L_6 - L_5)w(T_s - T_\infty)$$
$$q_{conv,6} = 139\ \text{W/m}^2\cdot\text{K}\ (0.30 - 0.25)\ \text{m} \times 1\ \text{m}\ (230 - 25)°\text{C} = 1430\ \text{W}$$

   This procedure must be used with great caution and only when the variation of the local convection coefficient with distance is gradual, such as in turbulent flow. It could lead to significant error when used for a surface that experiences transition.

2. The variation of the local convection coefficient along the flat plate may be determined from Equations 7.21 and 7.30 for laminar and turbulent flow, respectively, and the results are represented by the solid curves of the following schematic:

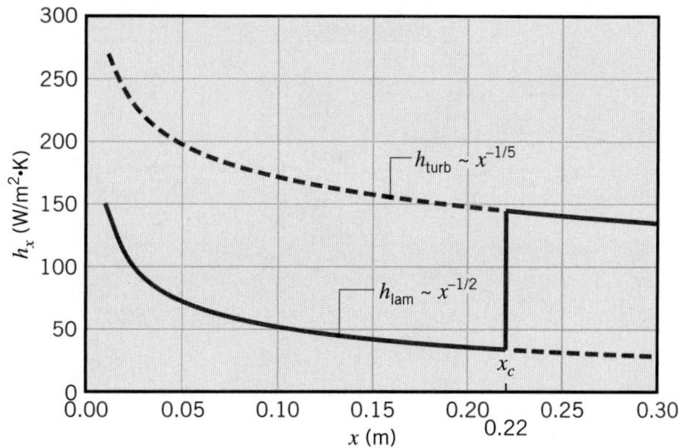

The $x^{-1/2}$ decay of the laminar convection coefficient is presumed to conclude abruptly at $x_c = 0.22$ m, where transition yields more than a fourfold increase in the local convection coefficient. For $x > x_c$, the decay in the convection coefficient is more gradual ($x^{-1/5}$). The dashed lines represent extensions of the distributions, which would apply if the value of x_c were shifted. For example, if the free stream turbulence were to increase and/or the surface were to be roughened, $Re_{x,c}$ would decrease. The smaller value of x_c would cause the laminar and turbulent distributions, respectively, to extend over smaller and larger portions of the plate. A similar effect may be achieved by increasing u_∞. In this case larger values of h_x would be associated with the laminar and turbulent distributions ($h_{\text{lam}} \sim u_\infty^{1/2}$, $h_{\text{turb}} \sim u_\infty^{4/5}$).

3. This example is solved in the *Advanced* section of *IHT*.

7.4 *The Cylinder in Cross Flow*

7.4.1 Flow Considerations

Another common external flow involves fluid motion normal to the axis of a circular cylinder. As shown in Figure 7.6, the free stream fluid is brought to rest at the *forward stagnation point,* with an accompanying rise in pressure. From this point, the pressure decreases with increasing x, the streamline coordinate, and the boundary layer develops under the influence of a *favorable pressure gradient* ($dp/dx < 0$). However, the pressure must

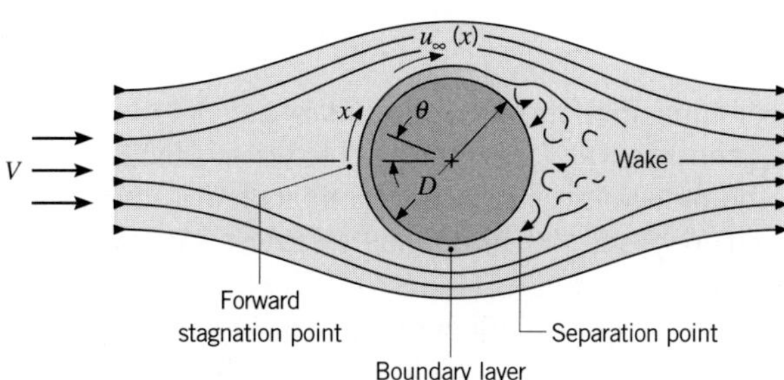

FIGURE 7.6 Boundary layer formation and separation on a circular cylinder in cross flow.

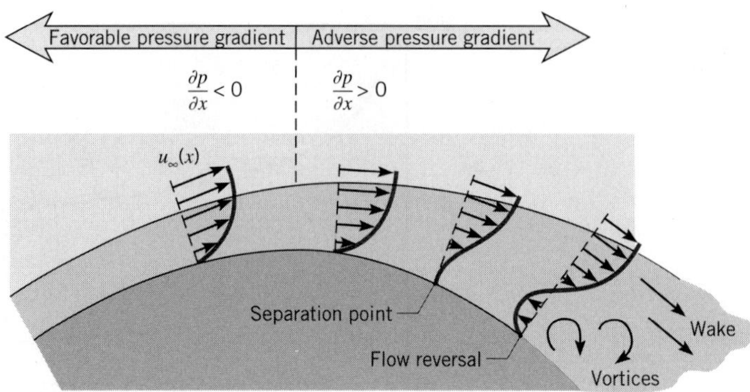

FIGURE 7.7 Velocity profile associated with separation on a circular cylinder in cross flow.

eventually reach a minimum, and toward the rear of the cylinder further boundary layer development occurs in the presence of an *adverse pressure gradient* ($dp/dx > 0$).

In Figure 7.6, the distinction between the upstream velocity V and the free stream velocity u_∞ should be noted. Unlike conditions for the flat plate in parallel flow, these velocities differ, with u_∞ now depending on the distance x from the stagnation point. From Euler's equation for an inviscid flow [9], $u_\infty(x)$ must exhibit behavior opposite to that of $p(x)$. That is, from $u_\infty = 0$ at the stagnation point, the fluid accelerates because of the favorable pressure gradient ($du_\infty/dx > 0$ when $dp/dx < 0$), reaches a maximum velocity when $dp/dx = 0$, and decelerates because of the adverse pressure gradient ($du_\infty/dx < 0$ when $dp/dx > 0$). As the fluid decelerates, the velocity gradient at the surface, $\partial u/\partial y|_{y=0}$, eventually becomes zero (Figure 7.7). At this location, termed the *separation point,* fluid near the surface lacks sufficient momentum to overcome the pressure gradient, and continued downstream movement is impossible. Since the oncoming fluid also precludes flow back upstream, *boundary layer separation* must occur. This is a condition for which the boundary layer detaches from the surface, and a *wake* is formed in the downstream region. Flow in this region is characterized by vortex formation and is highly irregular. The *separation point* is the location for which $\partial u/\partial y|_{y=0} = 0$. An excellent review of flow conditions in the wake of a circular cylinder is provided by Coutanceau and Defaye [10].

The occurrence of *boundary layer transition,* which depends on the Reynolds number, strongly influences the position of the separation point. For the circular cylinder the characteristic length is the diameter, and the Reynolds number is defined as

$$Re_D \equiv \frac{\rho VD}{\mu} = \frac{VD}{\nu}$$

Since the momentum of fluid in a turbulent boundary layer is larger than in the laminar boundary layer, it is reasonable to expect transition to delay the occurrence of separation. If $Re_D \lesssim 2 \times 10^5$, the boundary layer remains laminar, and separation occurs at $\theta \approx 80°$ (Figure 7.8). However, if $Re_D \gtrsim 2 \times 10^5$, boundary layer transition occurs, and separation is delayed to $\theta \approx 140°$.

The foregoing processes strongly influence the drag force, F_D, acting on the cylinder. This force has two components, one of which is due to the boundary layer surface shear

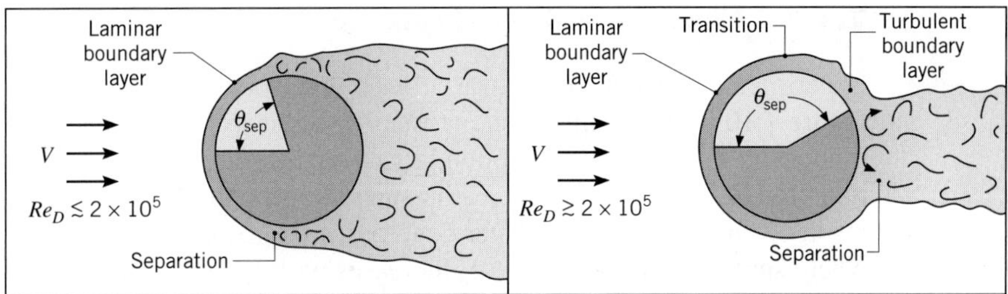

FIGURE 7.8 The effect of turbulence on separation.

stress (*friction drag*). The other component is due to a pressure differential in the flow direction resulting from formation of the wake (*form,* or *pressure, drag*). A dimensionless *drag coefficient* C_D may be defined as

$$C_D \equiv \frac{F_D}{A_f(\rho V^2/2)} \tag{7.42}$$

where A_f is the cylinder frontal area (the area projected perpendicular to the free stream velocity). The drag coefficient is a function of the Reynolds number and results are presented in Figure 7.9. For $Re_D \lesssim 2$ separation effects are negligible, and conditions are dominated by friction drag. However, with increasing Reynolds number, the effect of separation, and therefore form drag, becomes more important. The large reduction in C_D that occurs for $Re_D \gtrsim 2 \times 10^5$ is due to boundary layer transition, which delays separation, thereby reducing the extent of the wake region and the magnitude of the form drag.

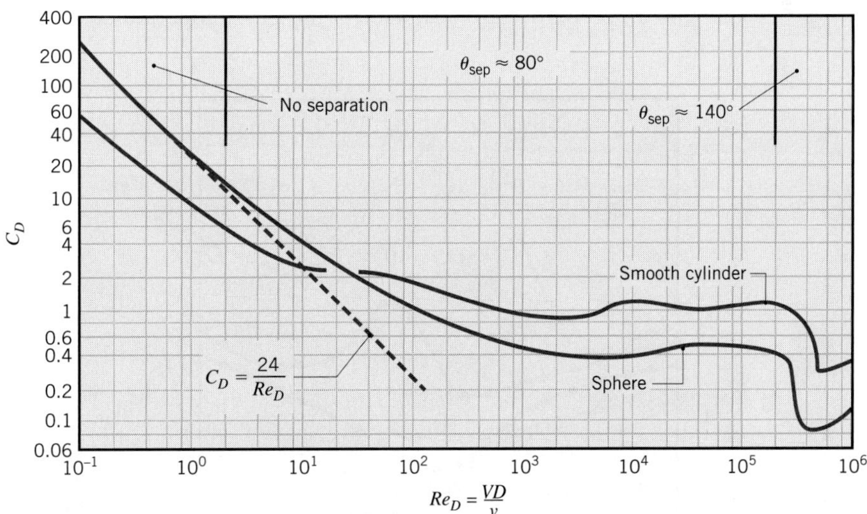

FIGURE 7.9 Drag coefficients for a smooth circular cylinder in cross flow and for a sphere. Boundary layer separation angles are for a cylinder. Adapted from [2].

7.4.2 Convection Heat Transfer

Experimental results for the variation of the local Nusselt number with θ are shown in Figure 7.10 for the cylinder in a cross flow of air. Not unexpectedly, the results are strongly influenced by the nature of boundary layer development on the surface. Consider conditions for $Re_D \lesssim 10^5$. Starting at the stagnation point, Nu_θ decreases with increasing θ as a result of laminar boundary layer development. However, a minimum is reached at $\theta \approx 80°$, where separation occurs and Nu_θ increases with θ due to mixing associated with vortex formation in the wake. In contrast, for $Re_D \gtrsim 10^5$ the variation of Nu_θ with θ is characterized by two minima. The decline in Nu_θ from the value at the stagnation point is again due to laminar boundary layer development, but the sharp increase that occurs between 80° and 100° is now due to boundary layer transition to turbulence. With further development of the turbulent boundary layer, Nu_θ again begins to decline. Eventually separation occurs ($\theta \approx 140°$), and Nu_θ increases as a result of mixing in the wake region. The increase in Nu_θ with increasing Re_D is due to a corresponding reduction in the boundary layer thickness.

Correlations may be obtained for the local Nusselt number, and at the forward stagnation point for $Pr \gtrsim 0.6$, boundary layer analysis [5] yields an expression of the following form, which is most accurate at low Reynolds number:

$$Nu_D(\theta = 0) = 1.15\, Re_D^{1/2}\, Pr^{1/3} \tag{7.43}$$

However, from the standpoint of engineering calculations, we are more interested in overall average conditions. An empirical correlation due to Hilpert [11] that has been modified to account for fluids of various Prandtl numbers,

$$\overline{Nu}_D \equiv \frac{\bar{h}D}{k} = C\, Re_D^m\, Pr^{1/3} \tag{7.44}$$

is widely used for $Pr \gtrsim 0.7$, where the constants C and m are listed in Table 7.2. Equation 7.44 may also be used for flow over cylinders of noncircular cross section, with the characteristic length D and the constants obtained from Table 7.3. In working with Equations 7.43 and 7.44 all properties are evaluated at the film temperature.

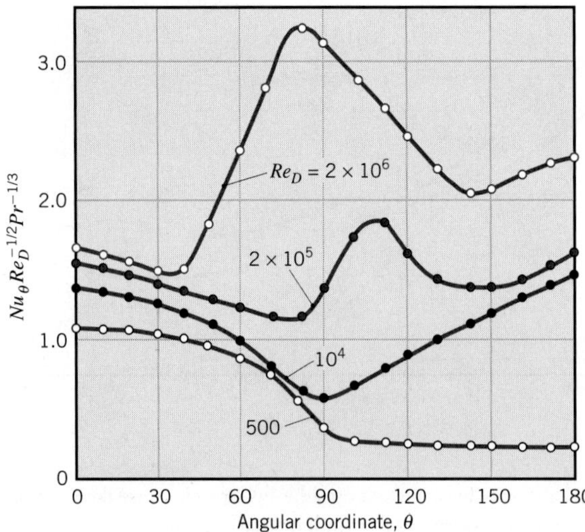

FIGURE 7.10 Local Nusselt number for airflow normal to a circular cylinder. (Adapted with permission from Zukauskas, A., "Convective Heat Transfer in Cross Flow," in S. Kakac, R. K. Shah, and W. Aung, Eds., *Handbook of Single-Phase Convective Heat Transfer*, Wiley, New York, 1987.)

TABLE 7.2 Constants of Equation 7.44 for the circular cylinder in cross flow [11, 12]

Re_D	C	m
0.4–4	0.989	0.330
4–40	0.911	0.385
40–4000	0.683	0.466
4000–40,000	0.193	0.618
40,000–400,000	0.027	0.805

Other correlations have been suggested for the circular cylinder in cross flow [15, 16, 17]. The correlation due to Zukauskas [16] is of the form

$$\overline{Nu}_D = C\, Re_D^m\, Pr^n \left(\frac{Pr}{Pr_s}\right)^{1/4} \tag{7.45}$$

$$\left[\begin{array}{c} 0.7 \lesssim Pr \lesssim 500 \\ 1 \lesssim Re_D \lesssim 10^6 \end{array}\right]$$

where all properties are evaluated at T_∞, except Pr_s, which is evaluated at T_s. Values of C and m are listed in Table 7.4. If $Pr \lesssim 10$, $n = 0.37$; if $Pr \gtrsim 10$, $n = 0.36$. Churchill and Bernstein [17] have proposed a single comprehensive equation that covers the entire range

TABLE 7.3 Constants of Equation 7.44 for noncircular cylinders in cross flow of a gas [13, 14][a]

Geometry		Re_D	C	m
Square				
$V \rightarrow$ ◇ D		6000–60,000	0.304	0.59
$V \rightarrow$ □ D		5000–60,000	0.158	0.66
Hexagon				
$V \rightarrow$ ⬡ D		5200–20,400	0.164	0.638
		20,400–105,000	0.039	0.78
$V \rightarrow$ ⬢ D		4500–90,700	0.150	0.638
Thin plate perpendicular to flow				
$V \rightarrow$ ▯ D	Front	10,000–50,000	0.667	0.500
	Back	7000–80,000	0.191	0.667

[a]These tabular values are based on the recommendations of Sparrow et al. [14] for air, with extension to other fluids through the $Pr^{1/3}$ dependence of Equation 7.44. A Prandtl number of $Pr = 0.7$ was assumed for the experimental results for air that are described in [14].

TABLE 7.4 Constants of Equation 7.45 for the circular cylinder in cross flow [17]

Re_D	C	m
1–40	0.75	0.4
40–1000	0.51	0.5
10^3–2×10^5	0.26	0.6
2×10^5–10^6	0.076	0.7

of Re_D for which data are available, as well as a wide range of Pr. The equation is recommended for all $Re_D\,Pr \gtrsim 0.2$ and has the form

$$\overline{Nu}_D = 0.3 + \frac{0.62\,Re_D^{1/2}\,Pr^{1/3}}{[1 + (0.4/Pr)^{2/3}]^{1/4}}\left[1 + \left(\frac{Re_D}{282{,}000}\right)^{5/8}\right]^{4/5} \tag{7.46}$$

where all properties are evaluated at the film temperature.

Again we caution the reader not to view any of the foregoing correlations as sacrosanct. Each correlation is reasonable over a certain range of conditions, but for most engineering calculations one should not expect accuracy to much better than 20%. Because they are based on more recent results encompassing a wide range of conditions, Equations 7.45 and 7.46 are generally used for the calculations of this text. Detailed reviews of the many correlations that have been developed for the circular cylinder have been provided by Sparrow et al. [14] as well as Morgan [18].

EXAMPLE 7.3

Experiments have been conducted on a metallic cylinder 12.7 mm in diameter and 94 mm long. The cylinder is heated internally by an electrical heater and is subjected to a cross flow of air in a low-speed wind tunnel. Under a specific set of operating conditions for which the upstream air velocity and temperature were maintained at $V = 10$ m/s and 26.2°C, respectively, the heater power dissipation was measured to be $P = 46$ W, while the average cylinder surface temperature was determined to be $T_s = 128.4$°C. It is estimated that 15% of the power dissipation is lost through the cumulative effect of surface radiation and conduction through the endpieces.

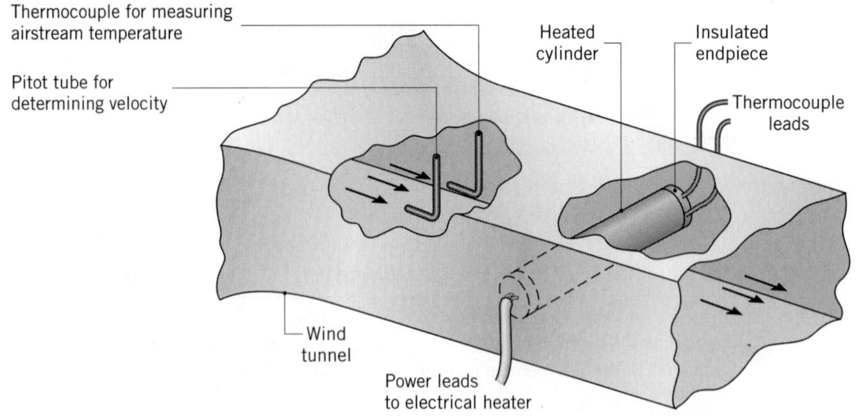

1. Determine the convection heat transfer coefficient from the experimental observations.
2. Compare the experimental result with the convection coefficient computed from an appropriate correlation.

SOLUTION

Known: Operating conditions for a heated cylinder.

Find:

1. Convection coefficient associated with the operating conditions.
2. Convection coefficient from an appropriate correlation.

Schematic:

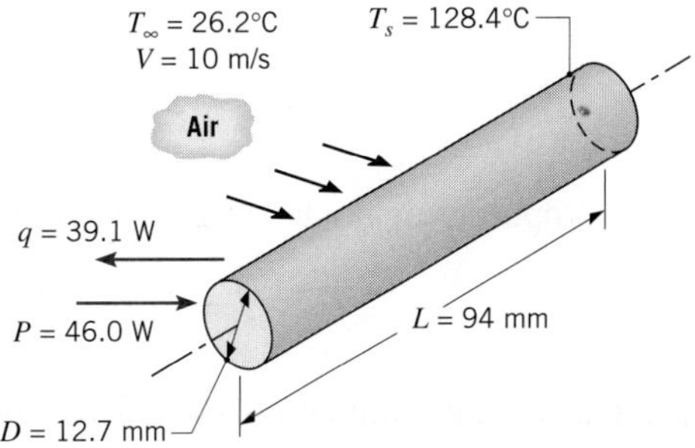

Assumptions:

1. Steady-state, incompressible flow conditions.
2. Uniform cylinder surface temperature.

Properties: Table A.4, air ($T_\infty = 26.2°C \approx 300$ K): $\nu = 15.89 \times 10^{-6}$ m²/s, $k = 26.3 \times 10^{-3}$ W/m·K, $Pr = 0.707$. Table A.4, air ($T_f \approx 350$ K): $\nu = 20.92 \times 10^{-6}$ m²/s, $k = 30 \times 10^{-3}$ W/m·K, $Pr = 0.700$. Table A.4, air ($T_s = 128.4°C = 401$ K): $Pr = 0.690$.

Analysis:

1. The convection heat transfer coefficient may be determined from the data by using Newton's law of cooling. That is,

$$\bar{h} = \frac{q}{A(T_s - T_\infty)}$$

With $q = 0.85P$ and $A = \pi DL$, it follows that

$$\bar{h} = \frac{0.85 \times 46 \text{ W}}{\pi \times 0.0127 \text{ m} \times 0.094 \text{ m} (128.4 - 26.2)°\text{C}} = 102 \text{ W/m}^2 \cdot \text{K} \quad \triangleleft$$

2. Working with the Zukauskas relation, Equation 7.45,

$$\overline{Nu}_D = C\, Re_D^m\, Pr^n \left(\frac{Pr}{Pr_s}\right)^{1/4}$$

all properties, except Pr_s, are evaluated at T_∞. Accordingly,

$$Re_D = \frac{VD}{\nu} = \frac{10 \text{ m/s} \times 0.0127 \text{ m}}{15.89 \times 10^{-6} \text{ m}^2\text{/s}} = 7992$$

Hence, from Table 7.4, $C = 0.26$ and $m = 0.6$. Also, since $Pr < 10$, $n = 0.37$. It follows that

$$\overline{Nu}_D = 0.26(7992)^{0.6}(0.707)^{0.37}\left(\frac{0.707}{0.690}\right)^{0.25} = 50.5$$

$$\bar{h} = \overline{Nu}_D \frac{k}{D} = 50.5 \frac{0.0263 \text{ W/m} \cdot \text{K}}{0.0127 \text{ m}} = 105 \text{ W/m}^2 \cdot \text{K} \quad \triangleleft$$

Comments:

1. Using the Churchill relation, Equation 7.46,

$$\overline{Nu}_D = 0.3 + \frac{0.62\, Re_D^{1/2}\, Pr^{1/3}}{[1 + (0.4/Pr)^{2/3}]^{1/4}}\left[1 + \left(\frac{Re_D}{282{,}000}\right)^{5/8}\right]^{4/5}$$

With all properties evaluated at T_f, $Pr = 0.70$ and

$$Re_D = \frac{VD}{\nu} = \frac{10 \text{ m/s} \times 0.0127 \text{ m}}{20.92 \times 10^{-6} \text{ m}^2\text{/s}} = 6071$$

Hence the Nusselt number and the convection coefficient are

$$\overline{Nu}_D = 0.3 + \frac{0.62(6071)^{1/2}(0.70)^{1/3}}{[1 + (0.4/0.70)^{2/3}]^{1/4}}\left[1 + \left(\frac{6071}{282{,}000}\right)^{5/8}\right]^{4/5} = 40.6$$

$$\bar{h} = \overline{Nu}_D \frac{k}{D} = 40.6 \frac{0.030 \text{ W/m} \cdot \text{K}}{0.0127 \text{ m}} = 96.0 \text{ W/m}^2 \cdot \text{K}$$

Alternatively, from the Hilpert correlation, Equation 7.44,

$$\overline{Nu}_D = C\, Re_D^m\, Pr^{1/3}$$

With all properties evaluated at the film temperature, $Re_D = 6071$ and $Pr = 0.70$. Hence, from Table 7.2, $C = 0.193$ and $m = 0.618$. The Nusselt number and the convection coefficient are then

$$\overline{Nu}_D = 0.193(6071)^{0.618}(0.700)^{0.333} = 37.3$$

$$\bar{h} = \overline{Nu}_D \frac{k}{D} = 37.3 \frac{0.030 \text{ W/m} \cdot \text{K}}{0.0127 \text{ m}} = 88 \text{ W/m}^2 \cdot \text{K}$$

2. Uncertainties associated with measuring the air velocity, estimating the heat loss from cylinder ends, and averaging the cylinder surface temperature, which varies axially and circumferentially, render the experimental result accurate to no better than 15%. Accordingly, calculations based on each of the three correlations are within the experimental uncertainty of the measured result.
3. Recognize the importance of using the proper temperature when evaluating fluid properties.

EXAMPLE 7.4

Because the molecular weight of hydrogen is very small, storing significant amounts in its gaseous form requires very large, high-pressure containers. In situations where use of such high-pressure storage is not feasible, such as in automotive applications, the H_2 is typically stored by *adsorbing* it into a metal hydride powder. The hydrogen is subsequently *desorbed* as needed, by heating the metal hydride throughout its volume.

Gaseous, desorbed hydrogen is present within the interstitial regions of the powder at a pressure that depends on the metal hydride temperature as

$$p_{H_2} = \exp(-3550/T + 12.9)$$

where p_{H_2} is the hydrogen pressure in atmospheres and T is the metal hydride temperature in kelvins. The desorption process is an *endothermic* chemical reaction corresponding to a thermal generation rate expressed as

$$\dot{E}_g = -\dot{m}_{H_2} \times (29.5 \times 10^3 \text{ kJ/kg})$$

where $\dot{m}_{H_2}$ is the hydrogen desorption rate (kg/s). Thermal energy must be supplied to the metal hydride in order to maintain a sufficiently high operating temperature. The operating temperature is determined by the requirement that the hydrogen pressure remain above 1 atm so that hydrogen can be removed from the metal hydride.

At a steady-state cruising speed of $V = 25$ m/s, a fuel cell–powered automobile consumes $\dot{m}_{H_2} = 1.35 \times 10^{-4}$ kg/s of hydrogen, which is supplied from a cylindrical, stainless steel canister with inside diameter $D_i = 0.1$ m, length $L = 0.8$ m, and wall thickness $t = 0.5$ mm. The canister, which is loaded with metal hydride powder, is installed in the vehicle so that it is subject to air in cross flow at $V = 25$ m/s, $T_\infty = 23°$C. Determine how much additional heating, beyond that supplied by convection from the warm air, should be supplied to the canister so that $p_{H_2} > p_{fc} = 1$ atm.

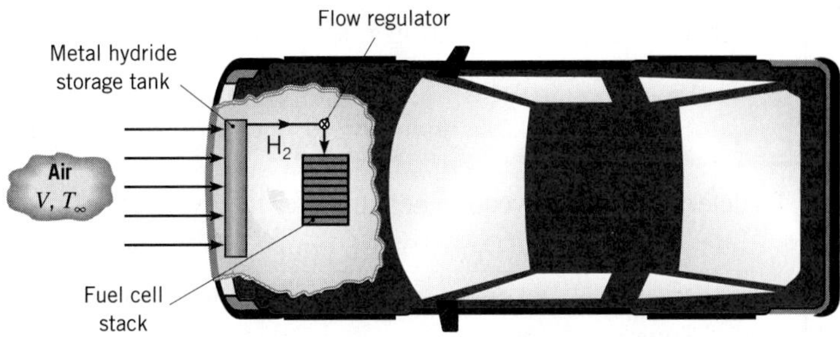

SOLUTION

Known: Size of a hydrogen storage canister, hydrogen desorption rate, required hydrogen operating pressure, velocity, and temperature of air in cross flow.

Find: The convective heat transfer to the canister and the additional heating needed to sustain $p_{H_2} > p_{fc}$.

Schematic:

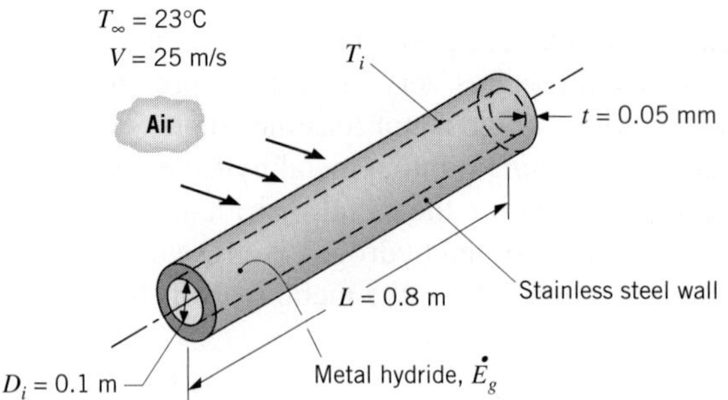

Assumptions:

1. Steady-state, incompressible flow conditions.
2. Uniform cylinder surface temperature.
3. Negligible heat gain through the ends of the cylinder.
4. Uniform metal hydride temperature.
5. Negligible contact resistance between the canister wall and the metal hydride.

Properties: Table A.4, air ($T_f \approx 285$ K): $\nu = 14.56 \times 10^{-6}$ m²/s, $k = 25.2 \times 10^{-3}$ W/m·K, $Pr = 0.712$. Table A.1, AISI 316 stainless steel ($T_{ss} \approx 300$ K): $k_{ss} = 13.4$ W/m·K.

Analysis: We begin by finding the minimum allowable operating temperature of the metal hydride, T_{min}, corresponding to $p_{H_2,min} = 1$ atm. The relationship between the operating temperature and pressure may be rearranged to yield

$$T_{min} = \frac{-3550}{\ln(p_{H_2,min}) - 12.9} = \frac{-3550}{\ln(1) - 12.9} = 275.2 \text{ K}$$

The thermal energy generation rate associated with the desorption of hydrogen from the metal hydride at the required flow rate is

$$\dot{E}_g = -(1.35 \times 10^{-4} \text{ kg/s}) \times (29.5 \times 10^6 \text{ J/kg}) = -3982 \text{ W}$$

To determine the convective heat transfer rate, we begin by calculating the Reynolds number:

$$Re_D = \frac{V(D_i + 2t)}{\nu} = \frac{23 \text{ m/s} \times (0.1 \text{ m} + 2 \times 0.005 \text{ m})}{14.56 \times 10^{-6} \text{ m}^2\text{/s}} = 173{,}760$$

Use of Equation 7.46

$$\overline{Nu}_D = 0.3 + \frac{0.62\, Re_D^{1/2}\, Pr^{1/3}}{[1 + (0.4/Pr)^{2/3}]^{1/4}}\left[1 + \left(\frac{Re_D}{282{,}000}\right)^{5/8}\right]^{4/5}$$

yields

$$\overline{Nu}_D = 0.3 + \frac{0.62(173{,}760)^{1/2}(0.712)^{1/3}}{[1 + (0.4/0.712)^{2/3}]^{1/4}}\left[1 + \left(\frac{173{,}760}{282{,}000}\right)^{5/8}\right]^{4/5} = 315.8$$

Therefore, the average convection heat transfer coefficient is

$$\bar{h} = \overline{Nu}_D \frac{k}{(D_i + 2t)} = 315.8 \times \frac{25.3 \times 10^{-3}\ \text{W/m}\cdot\text{K}}{(0.1\ \text{m} + 2 \times 0.005\ \text{m})} = 72.6\ \text{W/m}^2\cdot\text{K}$$

Simplifying Equation 3.34, we find

$$q_{\text{conv}} = \frac{T_\infty - T_i}{\dfrac{1}{\pi L(D_i + 2t)\bar{h}} + \dfrac{\ln[(D_i + 2t)/D_i]}{2\pi k_{\text{ss}} L}}$$

or, substituting values,

$$q_{\text{conv}} = \frac{296\ \text{K} - 275.2\ \text{K}}{\dfrac{1}{\pi(0.8\ \text{m})(0.1\ \text{m} + 2 \times 0.005\ \text{m})(72.6\ \text{W/m}^2\cdot\text{K})} + \dfrac{\ln[(0.1\ \text{m} + 2 \times 0.005\ \text{m})/0.1\ \text{m}]}{2\pi(13.4\ \text{W/m}\cdot\text{K})(0.8\ \text{m})}}$$
$$= 406\ \text{W}$$

The additional thermal energy, q_{add}, that must be supplied to the canister to maintain the steady-state operating temperature may be found from an energy balance, $q_{\text{add}} + q_{\text{conv}} + \dot{E}_g = 0$. Therefore,

$$q_{\text{add}} = -q_{\text{conv}} - \dot{E}_g = -406\ \text{W} + 3982\ \text{W} = 3576\ \text{W} \qquad \triangleleft$$

Comments:

1. Additional heating will occur due to radiation, conduction from the canister mounting hardware and fuel lines, and possibly condensation of water vapor on the cool canister. Waste heat from the fuel cell (see Example 3.11) might also be used as a source of thermal energy for the hydrogen storage canister.
2. The thermal resistances associated with conduction in the canister wall and convection are 0.0014 K/W and 0.053 K/W, respectively. The convection resistance dominates and can be reduced by adding fins to the exterior of the canister.
3. The amount of additional heating that is required will increase if the automobile is operated at a higher speed, since the hydrogen consumption scales as V^3, while the convective heat transfer coefficient increases as $V^{0.7}$ to $V^{0.8}$. Additional heating is also needed when the automobile is operated in a cooler climate.

7.5 The Sphere

Boundary layer effects associated with flow over a sphere are much like those for the circular cylinder, with transition and separation playing prominent roles. Results for the drag coefficient, which is defined by Equation 7.42, are presented in Figure 7.9. In the limit of very small Reynolds numbers (*creeping flow*), the coefficient is inversely proportional to the Reynolds number and the specific relation is termed *Stokes' law*

$$C_D = \frac{24}{Re_D} \qquad Re_D \lesssim 0.5 \tag{7.47}$$

Numerous heat transfer correlations have been proposed, and Whitaker [15] recommends an expression of the form

$$\overline{Nu}_D = 2 + (0.4\, Re_D^{1/2} + 0.06\, Re_D^{2/3})Pr^{0.4}\left(\frac{\mu}{\mu_s}\right)^{1/4} \tag{7.48}$$

$$\begin{bmatrix} 0.71 \lesssim Pr \lesssim 380 \\ 3.5 \lesssim Re_D \lesssim 7.6 \times 10^4 \\ 1.0 \lesssim (\mu/\mu_s) \lesssim 3.2 \end{bmatrix}$$

All properties except μ_s are evaluated at T_∞. A special case of convection heat transfer from spheres relates to transport from freely falling liquid drops, and the correlation of Ranz and Marshall [19] is often used

$$\overline{Nu}_D = 2 + 0.6\, Re_D^{1/2}\, Pr^{1/3} \tag{7.49}$$

In the limit $Re_D \rightarrow 0$, Equations 7.48 and 7.49 reduce to $\overline{Nu}_D = 2$, which corresponds to heat transfer by conduction from a spherical surface to a stationary, infinite medium around the surface, as may be derived from Case 1 of Table 4.1.

EXAMPLE 7.5

Electrical circuitry is written onto a photovoltaic panel by depositing a stream of small ($D = 55\ \mu\text{m}$) droplets of electrically conducting ink from a thermal inkjet printer. The drops are at an initial temperature of $T_i = 200°\text{C}$, and it is desirable for them to strike the panel at a temperature of $T_{\text{final}} = 50°\text{C}$. The quiescent air and surroundings are at $T_\infty = T_{\text{sur}} = 25°\text{C}$, and the drops are ejected from the print head at their terminal velocity. Determine the required standoff distance L between the printer and the photovoltaic panel. The properties of the electrically conducting ink drop are $\rho_d = 2400\ \text{kg/m}^3$, $c_d = 800\ \text{J/kg}\cdot\text{K}$, and $k_d = 5.0\ \text{W/m}\cdot\text{K}$.

SOLUTION

Known: Droplet size and properties, initial and desired final droplet temperature. Droplet injected at its terminal velocity.

Find: Required standoff distance between the printer and the photovoltaic panel.

Schematic:

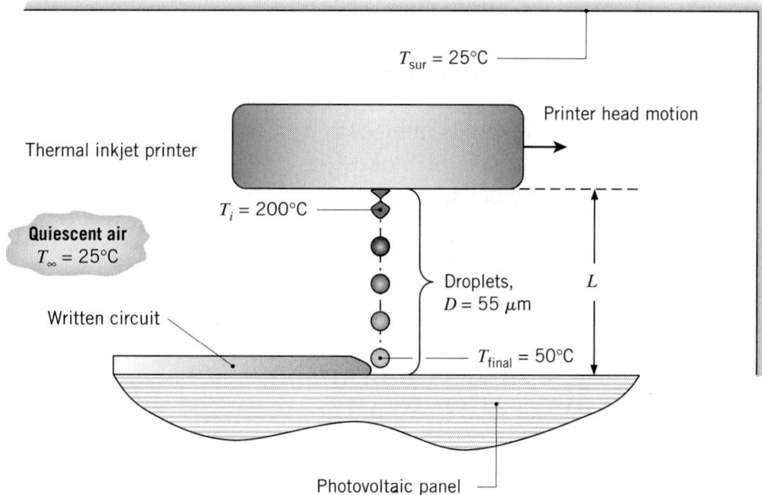

Assumptions:

1. The Nusselt number is approximated by the Ranz and Marshall correlation for a falling droplet, Equation 7.49.
2. Constant air properties evaluated at 25°C.
3. Negligible radiation effects.
4. Negligible temperature variation within the droplets (lumped capacitance approximation).
5. Drag coefficient is determined by Stokes' law.

Properties: Table A.4, air (T_f = 75°C): $\rho = 1.002$ kg/m³, $\nu = 20.72 \times 10^{-6}$ m²/s. Table A.4, air (T_∞ = 25°C): $\nu = 15.71 \times 10^{-6}$ m²/s, $k = 0.0261$ W/m·K, $Pr = 0.708$.

Analysis: Since the droplets travel at their terminal velocities, the net force on each drop must be zero. Hence the weight of the drop is offset by the buoyancy force associated with the displaced air and the drag force:

$$\rho_d g\left(\pi\frac{D^3}{6}\right) = \rho g\left(\pi\frac{D^3}{6}\right) + C_D\left(\frac{\pi D^2}{4}\right)\left(\rho\frac{V^2}{2}\right) \quad (1)$$

where Equation 7.42 has been used to express the drag force F_D. Since the droplets are small, we anticipate that the Reynolds number will also be small. If this is the case, Stokes' law, Equation 7.47 may be used to express the drag coefficient as

$$C_D = \frac{24}{Re_D} = \frac{24\nu}{VD} \quad (2)$$

Substituting Equation (2) into Equation (1) and solving for the velocity,

$$V = \frac{gD^2}{18\nu\rho}(\rho_d - \rho) = \frac{9.8\ \text{m/s}^2 \times (55 \times 10^{-6}\ \text{m})^2}{18 \times 20.72 \times 10^{-6}\ \text{m}^2/\text{s} \times 1.002\ \text{kg/m}^3} \times (2400 - 1.002)\text{kg/m}^3$$

$$= 0.190\ \text{m/s} = 190\ \text{mm/s}$$

Therefore, the Reynolds number is $Re_D = VD/\nu = 0.190\ \text{m/s} \times 55 \times 10^{-6}\ \text{m}/20.72 \times 10^{-6}\ \text{m}^2/\text{s} = 0.506$, and use of Stokes' law is appropriate. The Nusselt number and convection coefficient can be calculated from the Ranz and Marshall correlation, Equation 7.49, using properties evaluated at the free stream temperature (see Table 7.7):

$$\overline{Nu}_D = 2 + 0.6\, Re_D^{1/2}\, Pr^{1/3} = 2 + 0.6 \times \left(\frac{0.190\ \text{m/s} \times 55 \times 10^{-6}\ \text{m}}{15.71 \times 10^{-6}\ \text{m}^2/\text{s}}\right)^{1/2} \times 0.708^{1/3} = 2.44$$

$$\bar{h} = \frac{\overline{Nu}_D k}{D} = \frac{2.44 \times 0.0261\ \text{W/m} \cdot \text{K}}{55 \times 10^{-6}\ \text{m}} = 1160\ \text{W/m}^2 \cdot \text{K}$$

Applying the lumped capacitance method, Equation 5.5, the required *time-of-flight* is then

$$t = \frac{\rho_d V c_d}{\bar{h} A_s} \ln\left(\frac{\theta_i}{\theta_{\text{final}}}\right) = \frac{\rho_d c_d D}{6\bar{h}} \ln\left(\frac{T_i - T_\infty}{T_{\text{final}} - T_\infty}\right)$$

$$= \frac{2400\ \text{kg/m}^3 \times 800\ \text{J/kg} \cdot \text{K} \times 55 \times 10^{-6}\ \text{m}}{6 \times 1160\ \text{W/m}^2 \cdot \text{K}} \ln\left(\frac{(200 - 25)^\circ\text{C}}{(50 - 25)^\circ\text{C}}\right)$$

$$= 0.030\ \text{s}$$

and the standoff distance is

$$L = Vt = 0.190\ \text{m/s} \times 0.030\ \text{s} = 0.0056\ \text{m} = 5.6\ \text{mm} \quad \triangleleft$$

Comments:

1. The validity of the lumped capacitance method may be determined by calculating the Biot number. Applying Equation 5.10 in the conservative fashion with $L_c = D/2$,

$$Bi = \frac{\bar{h}(D/2)}{k_p} = \left(\frac{1160\ \text{W/m}^2 \cdot \text{K} \times 55 \times 10^{-6}\ \text{m}}{2}\right) / 5.0\ \text{W/m} \cdot \text{K} = 0.006 < 0.1$$

   and the criterion is satisfied.

2. Use of Equation 7.47, Stokes' law, to describe the Reynolds number dependence of the drag coefficient is valid since $Re_D \lesssim 0.5$. For larger particles, Figure 7.9 would need to be consulted to determine the relationship between C_D and Re_D.
3. If the particles were not injected at their terminal velocity, they would either accelerate or decelerate during flight, complicating the analysis.
4. Assuming blackbody behavior and using the maximum (initial) temperature of the particle, $T_s = 473$ K, the maximum radiation heat transfer coefficient is $h_r = \sigma(T_s + T_{\text{sur}})(T_s^2 + T_{\text{sur}}^2) = 5.67 \times 10^{-8}\ \text{W/m}^2 \cdot \text{K}^4 \times (473\ \text{K} + 298\ \text{K}) \times [(473\ \text{K})^2 + (298\ \text{K})^2] = 13.7\ \text{W/m}^2 \cdot \text{K}$. Since $h_r \ll \bar{h}$, radiation heat transfer is negligible.

7.6 Flow Across Banks of Tubes

Heat transfer to or from a bank (or bundle) of tubes in cross flow is relevant to numerous industrial applications, such as steam generation in a boiler or air cooling in the coil of an air conditioner. The geometric arrangement is shown schematically in Figure 7.11. Typically, one fluid moves over the tubes, while a second fluid at a different temperature passes through the tubes. In this section we are specifically interested in the convection heat transfer associated with cross flow over the tubes.

The tube rows of a bank can be either *aligned* or *staggered* in the direction of the fluid velocity V (Figure 7.12). The configuration is characterized by the tube diameter D and

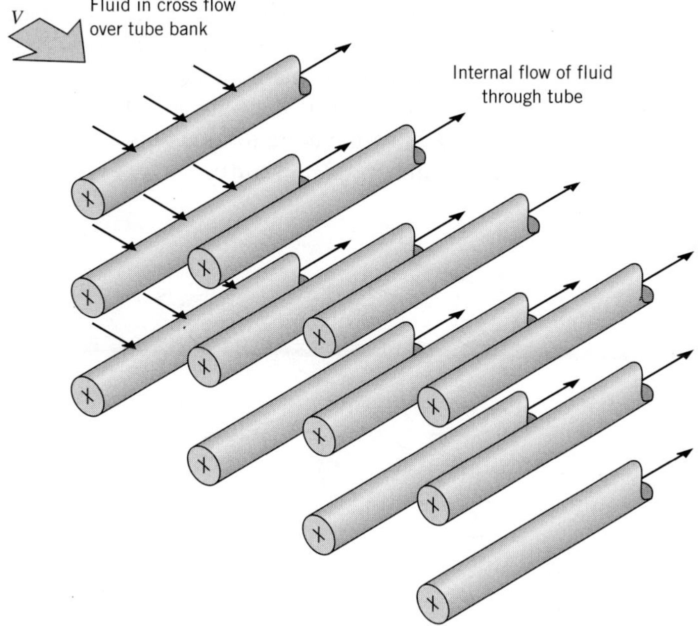

FIGURE 7.11 Schematic of a tube bank in cross flow.

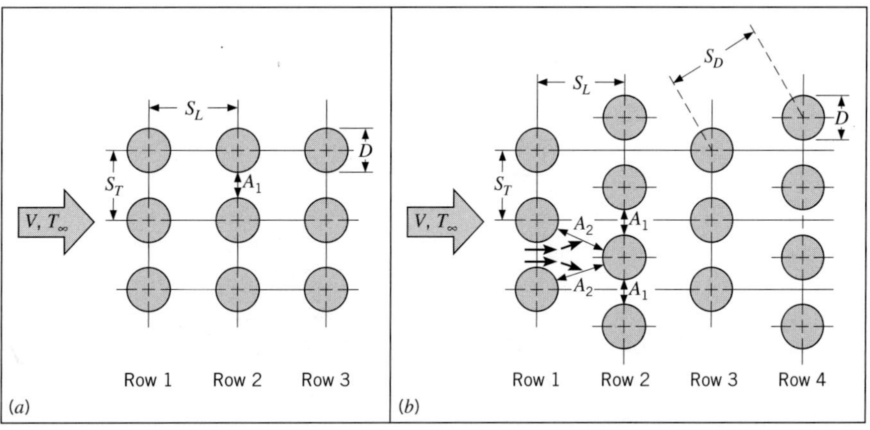

FIGURE 7.12 Tube arrangements in a bank. (*a*) Aligned. (*b*) Staggered.

by the *transverse pitch* S_T and *longitudinal pitch* S_L measured between tube centers. Flow conditions within the bank are dominated by boundary layer separation effects and by wake interactions, which in turn influence convection heat transfer.

Flow around the tubes in the first row of a tube bank is similar to that for a single (isolated) cylinder in cross flow. Correspondingly, the heat transfer coefficient for a tube in the first row is approximately equal to that for a single tube in cross flow. For downstream rows, flow conditions depend strongly on the tube bank arrangement (Figure 7.13). Aligned tubes beyond the first row reside in the wakes of upstream tubes, and for moderate values of S_L convection coefficients associated with downstream rows are enhanced by mixing, or turbulation, of the flow. Typically, the convection coefficient of a row increases with increasing row number until approximately the fifth row, after which there is little change in flow conditions and hence in the convection coefficient. For large S_L, the influence of upstream rows decreases, and heat transfer in the downstream rows is not enhanced. For this reason, operation of aligned tube banks with $S_T/S_L < 0.7$ is undesirable. For the staggered tube array, the path of the main flow is more tortuous, and mixing of the cross-flowing fluid is increased relative to the aligned tube arrangement. In general, heat transfer enhancement is favored by the more tortuous flow of a staggered arrangement, particularly for small Reynolds numbers ($Re_D \lesssim 100$).

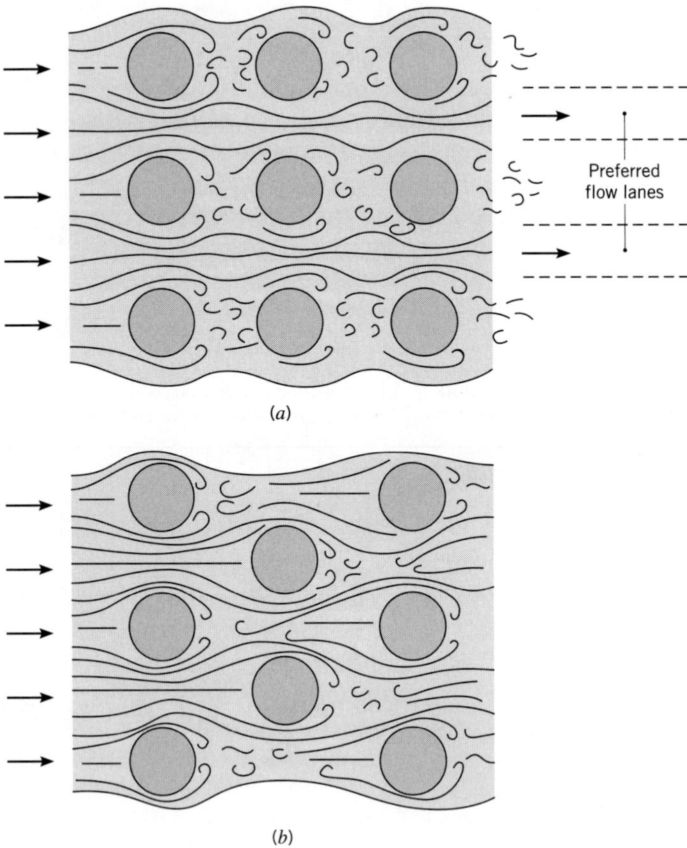

FIGURE 7.13 Flow conditions for (*a*) aligned and (*b*) staggered tubes.

Typically, we wish to know the *average* heat transfer coefficient for the *entire* tube bank. Zukauskas [16] has proposed a correlation of the form

$$\overline{Nu}_D = C_1 \, Re_{D,\max}^m \, Pr^{0.36} \left(\frac{Pr}{Pr_s}\right)^{1/4} \tag{7.50}$$

$$\left[\begin{array}{l} N_L \geq 20 \\ 0.7 \lesssim Pr \lesssim 500 \\ 10 \lesssim Re_{D,\max} \lesssim 2 \times 10^6 \end{array}\right]$$

where N_L is the number of tube rows, all properties except Pr_s are evaluated at the arithmetic mean of the fluid inlet ($T_i = T_\infty$) and outlet (T_o) temperatures, and the constants C_1 and m are listed in Table 7.5. The need to evaluate fluid properties at the arithmetic mean of the inlet and outlet temperatures is dictated by the fact that the fluid temperature will decrease or increase, respectively, due to heat transfer to or from the tubes. If the change of the mean fluid temperature, $|T_i - T_o|$, is large, significant error could result from the evaluation of the properties at the inlet temperature.

If there are 20 or fewer rows of tubes, $N_L \leq 20$, the average heat transfer coefficient is typically reduced, and a correction factor may be applied such that

$$\overline{Nu}_D\Big|_{(N_L<20)} = C_2 \overline{Nu}_D\Big|_{(N_L \geq 20)} \tag{7.51}$$

where C_2 is given in Table 7.6.

The Reynolds number $Re_{D,\max}$ for the foregoing correlation is based on the *maximum fluid velocity* occurring within the tube bank, $Re_{D,\max} \equiv \rho V_{\max} D/\mu$. For the aligned arrangement, $V_{\max}$ occurs at the transverse plane A_1 of Figure 7.12*a*, and from the mass conservation requirement for an incompressible fluid

$$V_{\max} = \frac{S_T}{S_T - D} V \tag{7.52}$$

TABLE 7.5 Constants of Equation 7.50 for the tube bank in cross flow [16]

Configuration	$Re_{D,\max}$	C_1	m
Aligned	10–10^2	0.80	0.40
Staggered	10–10^2	0.90	0.40
Aligned	10^2–10^3	Approximate as a single (isolated) cylinder	
Staggered	10^2–10^3	Approximate as a single (isolated) cylinder	
Aligned ($S_T/S_L > 0.7$)[a]	10^3–2×10^5	0.27	0.63
Staggered ($S_T/S_L < 2$)	10^3–2×10^5	$0.35(S_T/S_L)^{1/5}$	0.60
Staggered ($S_T/S_L > 2$)	10^3–2×10^5	0.40	0.60
Aligned	2×10^5–2×10^6	0.021	0.84
Staggered	2×10^5–2×10^6	0.022	0.84

[a]For $S_T/S_L < 0.7$, heat transfer is inefficient and aligned tubes should not be used.

TABLE 7.6 Correction factor C_2 of Equation 7.51 for $N_L < 20$ ($Re_{D,\max} \gtrsim 10^3$) [16]

N_L	1	2	3	4	5	7	10	13	16
Aligned	0.70	0.80	0.86	0.90	0.92	0.95	0.97	0.98	0.99
Staggered	0.64	0.76	0.84	0.89	0.92	0.95	0.97	0.98	0.99

For the staggered configuration, the maximum velocity may occur at either the transverse plane A_1 or the diagonal plane A_2 of Figure 7.12*b*. It will occur at A_2 if the rows are spaced such that

$$2(S_D - D) < (S_T - D)$$

The factor of 2 results from the bifurcation experienced by the fluid moving from the A_1 to the A_2 planes. Hence $V_{\max}$ occurs at A_2 if

$$S_D = \left[S_L^2 + \left(\frac{S_T}{2}\right)^2\right]^{1/2} < \frac{S_T + D}{2}$$

in which case it is given by

$$V_{\max} = \frac{S_T}{2(S_D - D)} V \tag{7.53}$$

If $V_{\max}$ occurs at A_1 for the staggered configuration, it may again be computed from Equation 7.52.

Since the fluid may experience a large change in temperature as it moves through the tube bank, the heat transfer rate could be significantly overpredicted by using $\Delta T = T_s - T_\infty$ as the temperature difference in Newton's law of cooling. As the fluid moves through the bank, its temperature approaches T_s and $|\Delta T|$ decreases. In Chapter 8, the appropriate form of ΔT is shown to be a *log-mean temperature difference,*

$$\Delta T_{\text{lm}} = \frac{(T_s - T_i) - (T_s - T_o)}{\ln\left(\dfrac{T_s - T_i}{T_s - T_o}\right)} \tag{7.54}$$

where T_i and T_o are temperatures of the fluid as it enters and leaves the bank, respectively. The outlet temperature, which is needed to determine ΔT_{lm}, may be estimated from

$$\frac{T_s - T_o}{T_s - T_i} = \exp\left(-\frac{\pi D N \bar{h}}{\rho V N_T S_T c_p}\right) \tag{7.55}$$

where N is the total number of tubes in the bank and N_T is the number of tubes in each row. Once ΔT_{lm} is known, the heat transfer rate per unit length of the tubes may be computed from

$$q' = N(\bar{h}\pi D\Delta T_{lm}) \tag{7.56}$$

Additional results, obtained for specific values of S_T/D and S_L/D, are reported by Zukauskas [16] and Grimison [20]. The results of Grimison are restricted to air as the cross-flowing fluid, and predicted values of the average Nusselt numbers generated by the correlations of the two references agree to within approximately 15% over a broad range of $Re_{D,max}$.

We close by recognizing that there is generally as much interest in the pressure drop associated with flow across a tube bank as in the overall heat transfer rate. The power required to move the fluid across the bank is often a major operating expense and is directly proportional to the pressure drop, which may be expressed as [16]

$$\Delta p = N_L\chi\left(\frac{\rho V_{max}^2}{2}\right)f \tag{7.57}$$

The friction factor f and the correction factor χ are plotted in Figures 7.14 and 7.15. Figure 7.14 pertains to a square, in-line tube arrangement for which the dimensionless longitudinal and transverse pitches, $P_L \equiv S_L/D$ and $P_T \equiv S_T/D$, respectively, are equal. The correction factor χ, plotted in the inset, is used to apply the results to other in-line arrangements. Similarly, Figure 7.15 applies to a staggered arrangement of tubes in the form of an equilateral triangle ($S_T = S_D$), and the correction factor enables extension of the results to other

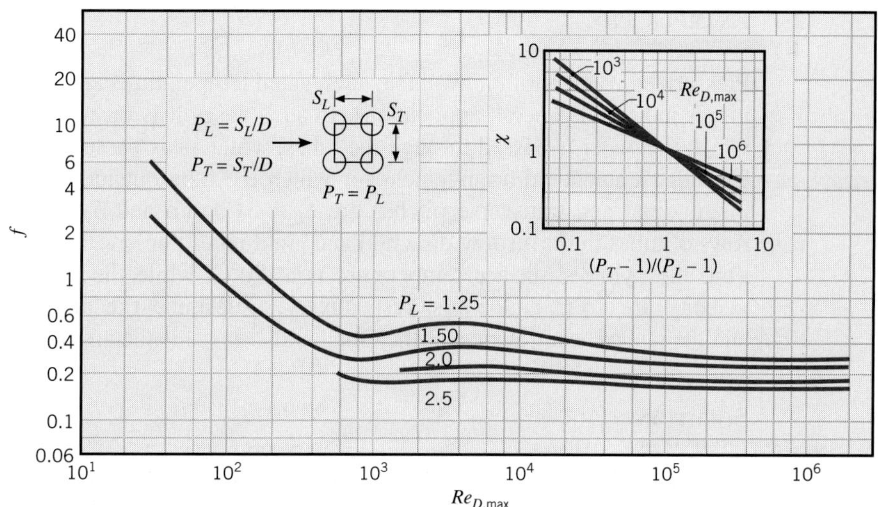

FIGURE 7.14 Friction factor f and correction factor χ for Equation 7.57. In-line tube bundle arrangement [16]. (Used with permission.)

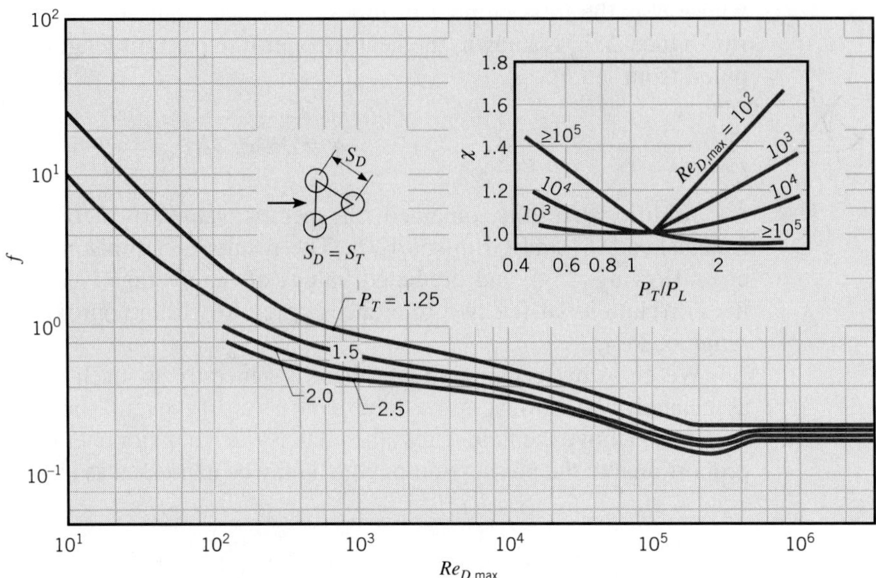

FIGURE 7.15 Friction factor f and correction factor χ for Equation 7.57. Staggered tube bundle arrangement [16]. (Used with permission.)

staggered arrangements. Note that the Reynolds number appearing in Figures 7.14 and 7.15 is based on the maximum fluid velocity V_{max}.

EXAMPLE 7.6

Pressurized water is often available at elevated temperatures and may be used for space heating or industrial process applications. In such cases it is customary to use a tube bundle in which the water is passed through the tubes, while air is passed in cross flow over the tubes. Consider a staggered arrangement for which the tube outside diameter is 16.4 mm and the longitudinal and transverse pitches are $S_L = 34.3$ mm and $S_T = 31.3$ mm. There are seven rows of tubes in the airflow direction and eight tubes per row. Under typical operating conditions the cylinder surface temperature is at 70°C, while the air upstream temperature and velocity are 15°C and 6 m/s, respectively. Determine the air-side convection coefficient and the rate of heat transfer for the tube bundle. What is the air-side pressure drop?

SOLUTION

Known: Geometry and operating conditions of a tube bank.

Find:

1. Air-side convection coefficient and heat rate.
2. Pressure drop.

Schematic:

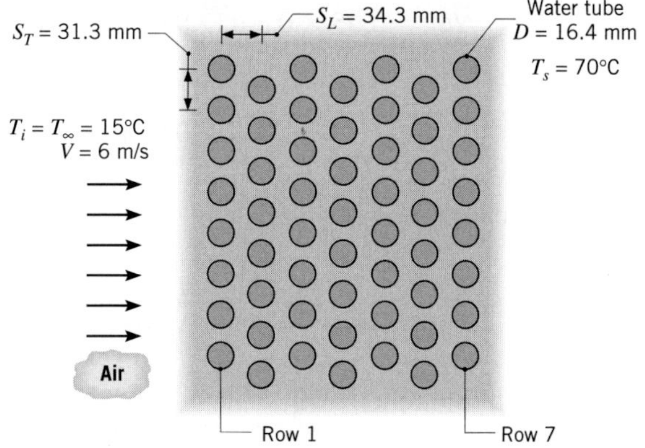

Assumptions:

1. Steady-state, incompressible flow conditions.
2. Negligible radiation effects.
3. Negligible effect of change in air temperature across tube bank on air properties.

Properties: Table A.4, air ($T_\infty = 15°C$): $\rho = 1.217\ \text{kg/m}^3$, $c_p = 1007\ \text{J/kg}\cdot\text{K}$, $\nu = 14.82 \times 10^{-6}\ \text{m}^2/\text{s}$, $k = 0.0253\ \text{W/m}\cdot\text{K}$, $Pr = 0.710$. Table A.4, air ($T_s = 70°C$): $Pr = 0.701$. Table A.4, air ($T_f = 43°C$): $\nu = 17.4 \times 10^{-6}\ \text{m}^2/\text{s}$, $k = 0.0274\ \text{W/m}\cdot\text{K}$, $Pr = 0.705$.

Analysis:

1. From Equations 7.50 and 7.51, the air-side Nusselt number is

$$\overline{Nu}_D = C_2 C_1 Re_{D,\max}^m Pr^{0.36}\left(\frac{Pr}{Pr_s}\right)^{1/4}$$

Since $S_D = [S_L^2 + (S_T/2)^2]^{1/2} = 37.7$ mm is greater than $(S_T + D)/2$, the maximum velocity occurs on the transverse plane, A_1, of Figure 7.12. Hence from Equation 7.52

$$V_{\max} = \frac{S_T}{S_T - D}V = \frac{31.3\ \text{mm}}{(31.3 - 16.4)\ \text{mm}} 6\ \text{m/s} = 12.6\ \text{m/s}$$

With

$$Re_{D,\max} = \frac{V_{\max}D}{\nu} = \frac{12.6\ \text{m/s} \times 0.0164\ \text{m}}{14.82 \times 10^{-6}\ \text{m}^2/\text{s}} = 13{,}943$$

and

$$\frac{S_T}{S_L} = \frac{31.3\ \text{mm}}{34.3\ \text{mm}} = 0.91 < 2$$

it follows from Tables 7.5 and 7.6 that

$$C_1 = 0.35\left(\frac{S_T}{S_L}\right)^{1/5} = 0.34, \quad m = 0.60, \quad \text{and} \quad C_2 = 0.95$$

Hence

$$\overline{Nu}_D = 0.95 \times 0.34(13{,}943)^{0.60}(0.71)^{0.36}\left(\frac{0.710}{0.701}\right)^{0.25} = 87.9$$

and

$$\bar{h} = \overline{Nu}_D \frac{k}{D} = 87.9 \times \frac{0.0253 \text{ W/m} \cdot \text{K}}{0.0164 \text{ m}} = 135.6 \text{ W/m}^2 \cdot \text{K} \qquad \triangleleft$$

From Equation 7.55

$$T_s - T_o = (T_s - T_i) \exp\left(-\frac{\pi D N \bar{h}}{\rho V N_T S_T c_p}\right)$$

$$T_s - T_o = (55°\text{C}) \exp\left(-\frac{\pi(0.0164 \text{ m})\, 56\, (135.6 \text{ W/m}^2 \cdot \text{K})}{1.217 \text{ kg/m}^3\, (6 \text{ m/s})\, 8\, (0.0313 \text{ m})\, 1007 \text{ J/kg} \cdot \text{K}}\right)$$

$$T_s - T_o = 44.5°\text{C}$$

Hence from Equations 7.54 and 7.56

$$\Delta T_{\text{lm}} = \frac{(T_s - T_i) - (T_s - T_o)}{\ln\left(\frac{T_s - T_i}{T_s - T_o}\right)} = \frac{(55 - 44.5)°\text{C}}{\ln\left(\frac{55}{44.5}\right)} = 49.6°\text{C}$$

and

$$q' = N(\bar{h}\pi D \Delta T_{\text{lm}}) = 56\pi \times 135.6 \text{ W/m}^2 \cdot \text{K} \times 0.0164 \text{ m} \times 49.6°\text{C}$$

$$q' = 19.4 \text{ kW/m} \qquad \triangleleft$$

2. The pressure drop may be obtained from Equation 7.57.

$$\Delta p = N_L \chi \left(\frac{\rho V_{\max}^2}{2}\right) f$$

With $Re_{D,\max} = 13{,}943$, $P_T = (S_T/D) = 1.91$, $P_L = (S_L/D) = 2.09$, and $(P_T/P_L) = 0.91$, it follows from Figure 7.15 that $\chi \approx 1.04$ and $f \approx 0.35$. Hence with $N_L = 7$

$$\Delta p = 7 \times 1.04 \left[\frac{1.217 \text{ kg/m}^3 (12.6 \text{ m/s})^2}{2}\right] 0.35$$

$$\Delta p = 246 \text{ N/m}^2 = 2.46 \times 10^{-3} \text{ bars} \qquad \triangleleft$$

Comments:

1. Had $\Delta T_i \equiv T_s - T_i$ been used in lieu of ΔT_{lm} in Equation 7.56, the heat rate would have been overpredicted by 11%.
2. Since the air temperature is predicted to increase by only 10.5°C, evaluation of the air properties at $T_i = 15°\text{C}$ is a reasonable approximation. However, if improved accuracy is desired, the calculations could be repeated with the properties reevaluated at $(T_i + T_o)/2 = 20.25°\text{C}$. An exception pertains to the density ρ in the exponential term of Equation 7.55. As it appears in the denominator of this term, ρ is matched with the inlet velocity to provide a product (ρV) that is linked to the mass flow rate of air entering the tube bank. Hence, in this term, ρ should be evaluated at T_i.

3. The air outlet temperature and heat rate may be increased by increasing the number of tube rows, and for a fixed number of rows, they may be varied by adjusting the air velocity. For $5 \le N_L \le 25$ and $V = 6$ m/s, parametric calculations based on Equations 7.50, 7.51, and 7.54 through 7.56 yield the following results:

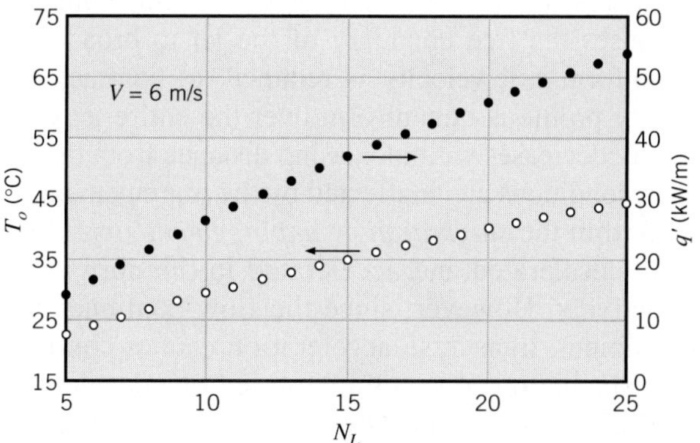

The air outlet temperature would asymptotically approach the surface temperature with increasing N_L, at which point the heat rate approaches a constant value and there is no advantage to adding more tube rows. Note that Δp increases linearly with increasing N_L. For $N_L = 25$ and $1 \le V \le 20$ m/s, we obtain

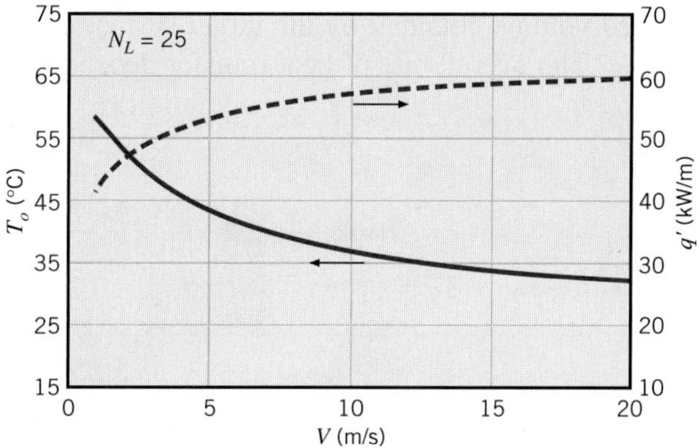

Although the heat rate increases with increasing V, the air outlet temperature decreases, approaching T_i as $V \rightarrow \infty$.

7.7 Impinging Jets

A single gas jet or an array of such jets, impinging normally on a surface, may be used to achieve enhanced coefficients for convective heating, cooling, or drying. Applications include tempering of glass plate, annealing of metal sheets, drying of textile and paper products, cooling of heated components in gas turbine engines, and deicing of aircraft systems.

7.7.1 Hydrodynamic and Geometric Considerations

As shown in Figure 7.16, gas jets are typically discharged into a quiescent ambient from a round nozzle of diameter D or a slot (rectangular) nozzle of width W. Typically, the jet is turbulent and, at the nozzle exit, is characterized by a uniform velocity profile. However, with increasing distance from the exit, momentum exchange between the jet and the ambient causes the free boundary of the jet to broaden and the *potential core,* within which the uniform exit velocity is retained, to contract. Downstream of the potential core the velocity profile is nonuniform over the entire jet cross section and the maximum (center) velocity decreases with increasing distance from the nozzle exit. The region of the flow over which conditions are unaffected by the impingement (target) surface is termed the *free jet.*

Within the *stagnation* or *impingement zone,* flow is influenced by the target surface and is decelerated and accelerated in the normal (z) and transverse (r or x) directions, respectively. However, since the flow continues to entrain zero momentum fluid from the ambient, transverse acceleration cannot continue indefinitely and accelerating flow in the stagnation zone is transformed to a decelerating *wall jet.* Hence, with increasing r or x, velocity components parallel to the surface increase from a value of zero to some maximum and subsequently decay to zero. Velocity profiles within the wall jet are characterized by zero velocity at both the impingement and free surfaces. If $T_s \neq T_e$, convection heat transfer occurs in both the stagnation and wall jet regions.

Many impingement heat transfer schemes involve an array of jets, as, for example, the array of slot jets shown in Figure 7.17. In addition to flow from each nozzle exhibiting free jet, stagnation, and wall jet regions, secondary stagnation zones result from the interaction of adjoining wall jets. In many such schemes the jets are discharged into a restricted volume bounded by the target surface and the nozzle plate from which the jets originate. The overall rate of heat transfer depends strongly on the manner in which *spent*

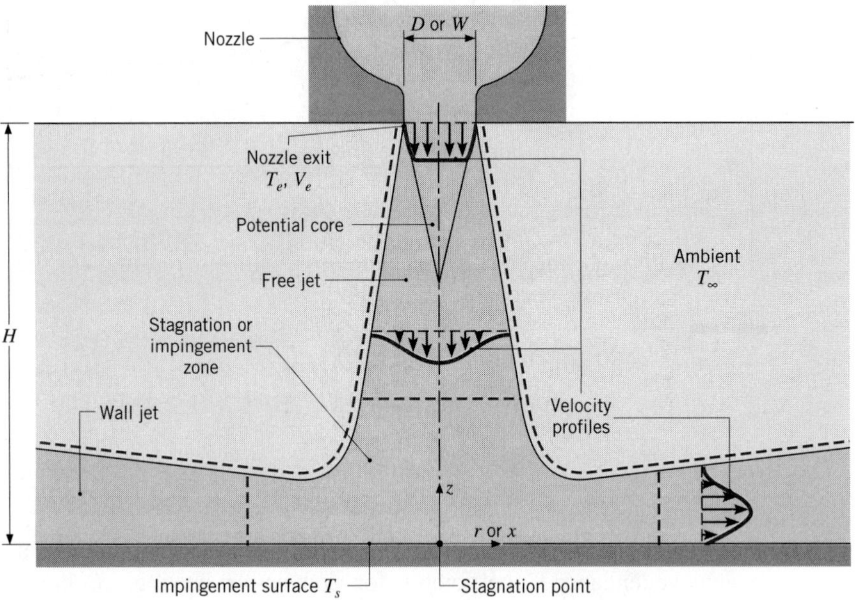

FIGURE 7.16 Surface impingement of a single round or slot gas jet.

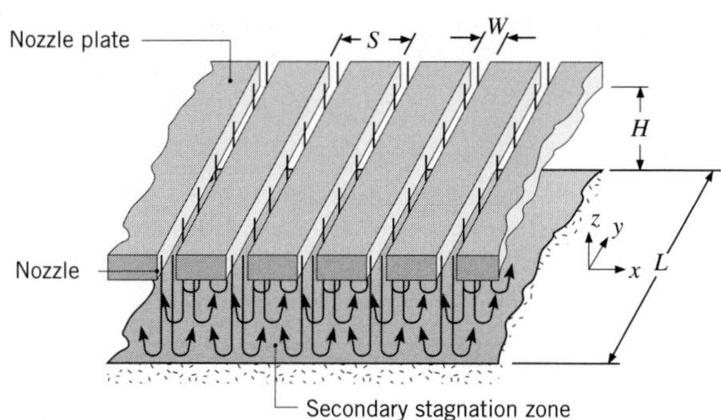

FIGURE 7.17 Surface impingement of an array of slot jets.

gas, whose temperature is between values associated with the nozzle exit and the impingement surface, is vented from the system. For the configuration of Figure 7.17, spent gas cannot flow upward between the nozzles but must instead flow symmetrically in the $\pm y$-directions. As the temperature of the gas increases with increasing $|y|$, the local surface-to-gas temperature difference decreases, causing a reduction in local convection fluxes. A preferable situation is one for which the space between adjoining nozzles is open to the ambient, thereby permitting continuous upflow and direct discharge of the spent gas.

Plan (top) views of single round and slot nozzles, as well as regular arrays of round and slot nozzles, are shown in Figure 7.18. For the isolated nozzles (Figures 7.18*a*, *d*), local and average convection coefficients are associated with any $r > 0$ and $x > 0$. For the arrays,

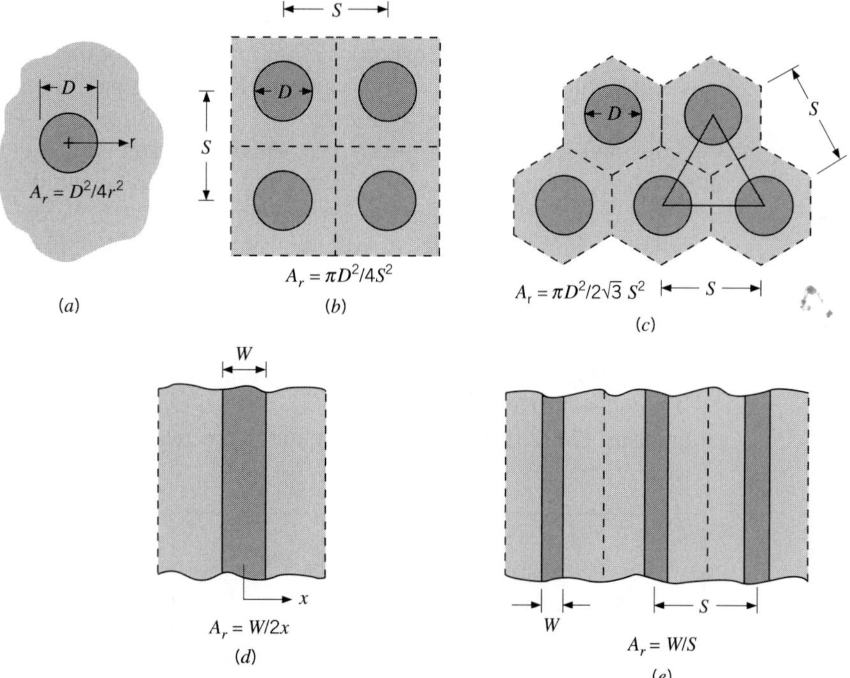

FIGURE 7.18 Plan view of pertinent geometrical features for (*a*) single round jet, (*b*) in-line array of round jets, (*c*) staggered array of round jets, (*d*) single slot jet, and (*e*) array of slot jets.

with discharge of the spent gas in the vertical (z) direction, symmetry dictates equivalent local and average values for each of the unit cells delineated by dashed lines. For a large number of square-in-line (Figure 7.18*b*) or equilaterally staggered (Figure 7.18*c*) round jets, the unit cells correspond to a square or hexagon, respectively. A pertinent geometric parameter is the relative nozzle area, which is defined as the ratio of the nozzle exit cross-sectional area to the surface area of the cell ($A_r \equiv A_{c,e}/A_{\text{cell}}$). In each case, S represents the pitch of the array.

7.7.2 Convection Heat Transfer

In the results that follow, it is presumed that the gas jet exits its nozzle with a uniform velocity V_e and temperature T_e. Thermal equilibrium with the ambient is presumed ($T_e = T_\infty$), while convection heat transfer may occur at an impingement surface of uniform temperature ($T_s \neq T_e$). Newton's law of cooling is then

$$q'' = h(T_s - T_e) \tag{7.58}$$

Conditions are presumed to be uninfluenced by the level of turbulence at the nozzle exit, and the surface is presumed to be stationary. However, this requirement may be relaxed for surface velocities which are much less than the jet impact velocity.

An extensive review of available convection coefficient data for impinging gas jets has been performed by Martin [21], and for a single round or slot nozzle, distributions of the *local* Nusselt number have the characteristic forms shown in Figure 7.19. The characteristic length is the *hydraulic diameter* of the nozzle, which is defined as four times its cross-sectional area divided by its wetted perimeter ($D_h \equiv 4A_{c,e}/P$). Hence the characteristic length is the diameter of a round nozzle, and assuming $L \gg W$, it is twice the width of a slot nozzle. It follows that $Nu = hD/k$ for a round nozzle and $Nu = h(2W/k)$ for a slot nozzle. For large nozzle-to-plate separations, Figure 7.19*a*, the distribution is characterized by a bell-shaped curve for which Nu monotonically decays from a maximum value at the *stagnation point*, r/D or $x/2W$ equal to zero.

For small separations, Figure 7.19*b*, the distribution is characterized by a second maximum, whose value increases with increasing jet Reynolds number and may exceed that of the first maximum. The threshold separation of $H/D \approx 5$, below which there is a second maximum, is loosely associated with the length of the potential core (Figure 7.16). Appearance of the second maximum is attributed to a sharp rise in the turbulence level which accompanies the transition from an accelerating stagnation region flow to a decelerating wall jet [21]. Additional maxima have been observed and attributed to the formation of vortices in the stagnation zone, as well as transition to a turbulent wall jet [22].

Secondary maxima in Nu are also associated with the interaction of adjoining wall jets for an array [21, 23]. However, distributions are two-dimensional, exhibiting, for example,

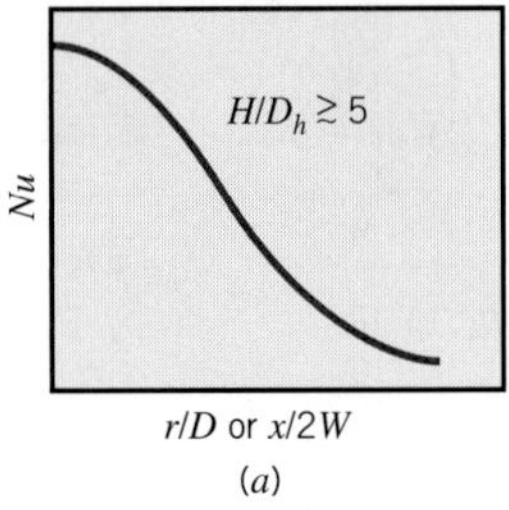

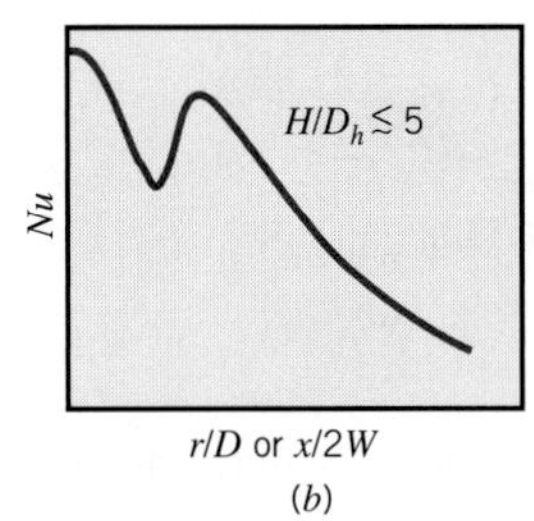

FIGURE 7.19 Distribution of local Nusselt number associated with a single round or slot nozzle for (*a*) large and (*b*) small relative nozzle-to-plate spacings.

variations with both x and y for the slot jet array of Figure 7.17. Variations with x could be expected to yield maxima at the jet centerline and halfway between adjoining jets, while constraint of the exhaust flow to the $\pm y$-direction would induce acceleration with increasing $|y|$ and hence a monotonically increasing Nu with $|y|$. However, variations with y decrease with increasing cross-sectional area of the outflow and may be neglected if $S \times H \gtrsim W \times L$ [21].

Average Nusselt numbers may be obtained by integrating local results over the appropriate surface area. The resulting correlations are reported in the form

$$\overline{Nu} = f(Re, Pr, A_r, H/D_h) \tag{7.59}$$

where

$$\overline{Nu} \equiv \frac{\bar{h}D_h}{k} \tag{7.60}$$

$$Re = \frac{V_e D_h}{\nu} \tag{7.61}$$

and $D_h = D$ (round nozzle) or $D_h = 2W$ (slot nozzle).

Round Nozzles Having assessed data from several sources, Martin [21] recommends the following correlation for a *single round nozzle* ($A_r = D^2/4r^2$)

$$\frac{\overline{Nu}}{Pr^{0.42}} = G\left(A_r, \frac{H}{D}\right)[2\,Re^{1/2}(1 + 0.005\,Re^{0.55})^{1/2}] \tag{7.62}$$

where

$$G = 2A_r^{1/2}\,\frac{1 - 2.2A_r^{1/2}}{1 + 0.2(H/D - 6)A_r^{1/2}} \tag{7.63}$$

The ranges of validity are

$$\begin{bmatrix} 2000 \lesssim Re \lesssim 400{,}000 \\ 2 \lesssim H/D \lesssim 12 \\ 0.004 \lesssim A_r \lesssim 0.04 \end{bmatrix}$$

For $A_r \gtrsim 0.04$, results for $\overline{Nu}$ are available in graphical form [21].

For an *array of round nozzles* ($A_r = \pi D^2/4S^2$ or $\pi D^2/2\sqrt{3}S^2$ for in-line and staggered arrays, respectively),

$$\frac{\overline{Nu}}{Pr^{0.42}} = 0.5\,K\left(A_r, \frac{H}{D}\right)G\left(A_r, \frac{H}{D}\right)Re^{2/3} \tag{7.64}$$

where

$$K = \left[1 + \left(\frac{H/D}{0.6/A_r^{1/2}}\right)^6\right]^{-0.05} \tag{7.65}$$

and G is the single nozzle function given by Equation 7.63. The function K accounts for the fact that, for $H/D \gtrsim 0.6/A_r^{1/2}$, the average Nusselt number for the array decays more rapidly with increasing H/D than that for the single nozzle. The correlation is valid over the ranges

$$\begin{bmatrix} 2000 \lesssim Re \lesssim 100{,}000 \\ 2 \lesssim H/D \lesssim 12 \\ 0.004 \lesssim A_r \lesssim 0.04 \end{bmatrix}$$

Slot Nozzles For a *single slot nozzle* ($A_r = W/2x$), the recommended correlation is

$$\frac{\overline{Nu}}{Pr^{0.42}} = \frac{3.06}{0.5/A_r + H/W + 2.78} Re^m \tag{7.66}$$

where

$$m = 0.695 - \left[\left(\frac{1}{4A_r}\right) + \left(\frac{H}{2W}\right)^{1.33} + 3.06\right]^{-1} \tag{7.67}$$

and the ranges of validity are

$$\begin{bmatrix} 3000 \lesssim Re \lesssim 90{,}000 \\ 2 \lesssim H/W \lesssim 10 \\ 0.025 \lesssim A_r \lesssim 0.125 \end{bmatrix}$$

As a *first approximation*, Equation 7.66 may be used for $A_r \gtrsim 0.125$, yielding predictions for the stagnation point ($x = 0$, $A_r \rightarrow \infty$) that are within 40% of measured values.

For an *array of slot nozzles* ($A_r = W/S$), the recommended correlation is

$$\frac{\overline{Nu}}{Pr^{0.42}} = \frac{2}{3} A_{r,o}^{3/4} \left(\frac{2\,Re}{A_r/A_{r,o} + A_{r,o}/A_r}\right)^{2/3} \tag{7.68}$$

where

$$A_{r,o} = \left[60 + 4\left(\frac{H}{2W} - 2\right)^2\right]^{-1/2} \tag{7.69}$$

The correlation pertains to conditions for which the outflow of spent gas is restricted to the $\pm y$-directions of Figure 7.17 and the outflow area is large enough to satisfy the requirement $(S \times H)/(W \times L) \gtrsim 1$. Additional restrictions are

$$\begin{bmatrix} 1500 \lesssim Re \lesssim 40{,}000 \\ 2 \lesssim H/W \lesssim 80 \\ 0.008 \lesssim A_r \lesssim 2.5A_{r,o} \end{bmatrix}$$

An *optimal* arrangement of nozzles would be one for which the values of H, S, and D_h yielded the largest value of $\overline{Nu}$ for a prescribed total gas flow rate per unit surface area of the target. For fixed H and for arrays of both round and slot nozzles, optimal values of D_h and S have been found to be [21]

$$D_{h,\text{op}} \approx 0.2H \tag{7.70}$$

$$S_{\text{op}} \approx 1.4H \tag{7.71}$$

The optimum value of $(D_h/H)^{-1} \approx 5$ coincides approximately with the length of the potential core. Beyond the potential core, the midline jet velocity decays, causing an attendant reduction in convection coefficients.

Application of these equations should be restricted to conditions for which they were developed. For example, in their present form, the correlations may not be used if the jets emanate from sharp-edged orifices instead of bell-shaped nozzles. The orifice jet is strongly affected by a flow contraction phenomenon that alters convection heat transfer [21, 22]. Conditions are also influenced by differences between the jet exit and ambient temperatures ($T_e \neq T_\infty$). The exit temperature is then an inappropriate temperature in Newton's law of cooling, Equation 7.58, and should be replaced by what is commonly termed the recovery, or adiabatic wall, temperature [24, 25]. Finally, care should be taken in situations involving small nozzle diameters or narrow slot widths. For these situations, high jet velocities are necessary for Reynolds numbers to be within the range of application of Equations 7.62, 7.64, 7.66, or 7.68. When the Mach number based on the jet velocity exceeds 0.3 ($V_e/a \gtrsim 0.3$), compressibility effects may be significant [26], invalidating use of the correlations of this section. Additional correlations for liquid jet impingement, flame jet impingement, and impingement on nonflat surfaces are available [27–29].

7.8 *Packed Beds*

Gas flow through a *packed bed* of solid particles (Figure 7.20) is relevant to many industrial processes, which include the transfer and storage of thermal energy, heterogeneous catalytic reactions, and drying. The term *packed bed* refers to a condition for which the position of the particles is *fixed.* In contrast, a *fluidized bed* is one for which the particles are in motion due to advection with the fluid.

For a packed bed a large amount of heat transfer surface area can be obtained in a small volume, and the irregular flow that exists in the voids of the bed enhances transport through mixing. Many correlations that have been developed for different particle shapes, sizes, and packing densities are described in the literature [30, 31]. One such correlation, which has been recommended for gas flow in a bed of spheres, is of the form

$$\varepsilon \bar{j}_H = 2.06\, Re_D^{-0.575} \qquad \left[\begin{array}{l} Pr \approx 0.7 \\ 90 \lesssim Re_D \lesssim 4000 \end{array}\right] \tag{7.72}$$

where $\bar{j}_H$ is the Colburn j factor defined by Equation 6.38. The Reynolds number $Re_D = VD/\nu$ is defined in terms of the sphere diameter and the upstream velocity V that would exist in the empty channel without the packing. The quantity ε is the *porosity,* or *void fraction,* of the bed (volume of void space per unit volume of bed), and its value typically ranges from 0.30 to 0.50. The correlation may be applied to packing materials other than spheres by multiplying the right-hand side by an appropriate correction factor. For a bed of uniformly sized cylinders, with length-to-diameter ratio of 1, the factor is 0.79; for a bed of cubes it is 0.71.

In using Equation 7.72, properties should be evaluated at the arithmetic mean of the fluid temperatures entering and leaving the bed. If the particles are at a uniform temperature T_s, the heat transfer rate for the bed may be computed from

$$q = \bar{h} A_{p,t}\, \Delta T_{\text{lm}} \tag{7.73}$$

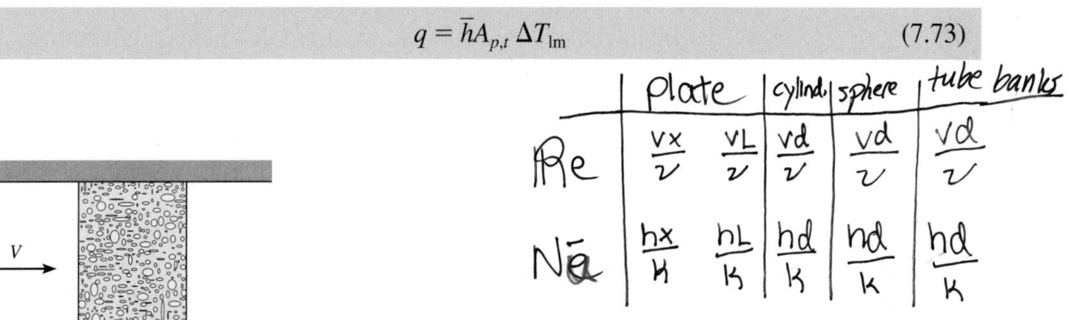

FIGURE 7.20 Gas flow through a packed bed of solid particles.

where $A_{p,t}$ is the total surface area of the particles and ΔT_{lm} is the log-mean temperature difference defined by Equation 7.54. The outlet temperature, which is needed to compute ΔT_{lm}, may be estimated from

$$\frac{T_s - T_o}{T_s - T_i} = \exp\left(-\frac{\overline{h}A_{p,t}}{\rho V A_{c,b} c_p}\right) \tag{7.74}$$

where ρ and V are the inlet density and velocity, respectively, and $A_{c,b}$ is the bed (channel) cross-sectional area.

7.9 Summary

In this chapter we have considered *forced convection* heat transfer for an important class of problems involving *external flow* at low-to-moderate speeds. Consideration was given to several common geometries, for which convection coefficients depend on the nature of boundary layer development. You should test your understanding of related concepts by addressing the following questions.

- What is an *external flow*?
- What is an empirical heat transfer coefficient?
- What are the inherent dimensionless parameters for forced convection?
- How does the velocity boundary layer thickness vary with distance from the leading edge for laminar flow over a flat plate? For turbulent flow? What determines the relative velocity and thermal boundary layer thicknesses for laminar flow? For turbulent flow?
- How does the *local* convection heat transfer coefficient vary with distance from the leading edge for *laminar flow* over a flat plate? For *turbulent flow*? For flow in which *transition* to turbulence occurs on the plate?
- How is local heat transfer from the surface of a flat plate affected by the existence of an *unheated starting length*?
- What are the manifestations of *boundary layer separation* from the surface of a circular cylinder in cross flow? How is separation influenced by whether the upstream flow is laminar or turbulent?
- How is variation of the local convection coefficient on the surface of a circular cylinder in cross flow affected by boundary layer separation? By boundary layer transition? Where do local maxima and minima in the convection coefficient occur on the surface?
- How does the average convection coefficient of a tube vary with its location in a tube bank?
- For jet impingement on a surface, what are distinguishing features of the *free jet*? The *potential core*? The *impingement zone*? The *wall jet*?
- At what location on the surface of an impinging jet will a maximum in the convection coefficient always exist? Under what conditions will there be a secondary maximum?
- For an *array* of impinging jets, how are flow and heat transfer affected by the manner in which *spent fluid* is discharged from the system?
- What is the difference between a *packed bed* and a *fluidized bed* of solid particles?
- What is the *film temperature*?
- What temperature difference must be used when computing the total rate of heat transfer from a bank of tubes or a packed bed?

In this chapter we have also compiled convection correlations that may be used to estimate convection transfer rates for a variety of external flow conditions. For simple surface geometries these results may be derived from a boundary layer analysis, but in most cases they are obtained from generalizations based on experiment. You should know when and how to use the various expressions, and you should be familiar with the general methodology of a convection calculation. To facilitate their use, the correlations are summarized in Table 7.7.

TABLE 7.7 Summary of convection heat transfer correlations for external flow[a]

Correlation		Geometry	Conditions[b]
$\delta = 5x\, Re_x^{-1/2}$	(7.17)	Flat plate	Laminar, T_f
$C_{f,x} = 0.664\, Re_x^{-1/2}$	(7.18)	Flat plate	Laminar, local, T_f
$Nu_x = 0.332\, Re_x^{1/2}\, Pr^{1/3}$	(7.21)	Flat plate	Laminar, local, T_f, $Pr \gtrsim 0.6$
$\delta_t = \delta\, Pr^{-1/3}$	(7.22)	Flat plate	Laminar, T_f
$\overline{C}_{f,x} = 1.328\, Re_x^{-1/2}$	(7.24)	Flat plate	Laminar, average, T_f
$\overline{Nu}_x = 0.664\, Re_x^{1/2}\, Pr^{1/3}$	(7.25)	Flat plate	Laminar, average, T_f, $Pr \gtrsim 0.6$
$Nu_x = 0.564\, Pe_x^{1/2}$	(7.26)	Flat plate	Laminar, local, T_f, $Pr \lesssim 0.05$, $Pe_x \gtrsim 100$
$C_{f,x} = 0.0592\, Re_x^{-1/5}$	(7.28)	Flat plate	Turbulent, local, T_f, $Re_x \lesssim 10^8$
$\delta = 0.37x\, Re_x^{-1/5}$	(7.29)	Flat plate	Turbulent, T_f, $Re_x \lesssim 10^8$
$Nu_x = 0.0296\, Re_x^{4/5}\, Pr^{1/3}$	(7.30)	Flat plate	Turbulent, local, T_f, $Re_x \lesssim 10^8$, $0.6 \lesssim Pr \lesssim 60$
$\overline{C}_{f,L} = 0.074\, Re_L^{-1/5} - 1742\, Re_L^{-1}$	(7.33)	Flat plate	Mixed, average, T_f, $Re_{x,c} = 5 \times 10^5$, $Re_L \lesssim 10^8$
$\overline{Nu}_L = (0.037\, Re_L^{4/5} - 871)Pr^{1/3}$	(7.31)	Flat plate	Mixed, average, T_f, $Re_{x,c} = 5 \times 10^5$, $Re_L \lesssim 10^8$, $0.6 \lesssim Pr \lesssim 60$
$\overline{Nu}_D = C\, Re_D^m\, Pr^{1/3}$ (Table 7.2)	(7.44)	Cylinder	Average, T_f, $0.4 \lesssim Re_D \lesssim 4 \times 10^5$, $Pr \gtrsim 0.7$
$\overline{Nu}_D = C\, Re_D^m\, Pr^n(Pr/Pr_s)^{1/4}$ (Table 7.4)	(7.45)	Cylinder	Average, T_∞, $1 \lesssim Re_D \lesssim 10^6$, $0.7 \lesssim Pr \lesssim 500$
$\overline{Nu}_D = 0.3 + [0.62\, Re_D^{1/2}\, Pr^{1/3} \times [1 + (0.4/Pr)^{2/3}]^{-1/4}] \times [1 + (Re_D/282{,}000)^{5/8}]^{4/5}$	(7.46)	Cylinder	Average, T_f, $Re_D\, Pr \gtrsim 0.2$
$\overline{Nu}_D = 2 + (0.4\, Re_D^{1/2} + 0.06\, Re_D^{2/3})Pr^{0.4} \times (\mu/\mu_s)^{1/4}$	(7.48)	Sphere	Average, T_∞, $3.5 \lesssim Re_D \lesssim 7.6 \times 10^4$, $0.71 \lesssim Pr \lesssim 380$, $1.0 \lesssim (\mu/\mu_s) \lesssim 3.2$
$\overline{Nu}_D = 2 + 0.6\, Re_D^{1/2}\, Pr^{1/3}$	(7.49)	Falling drop	Average, T_∞

TABLE 7.7 *(Continued)*

Correlation		Geometry	Conditions[b]
$\overline{Nu}_D = C_1 C_2 Re_{D,\max}^m Pr^{0.36}(Pr/Pr_s)^{1/2}$ (Tables 7.5, 7.6)	(7.50), (7.51)	Tube bank[c]	Average, $\overline{T}$, $10 \lesssim Re_D \lesssim 2 \times 10^6$, $0.7 \lesssim Pr \lesssim 500$
Single round nozzle	(7.62)	Impinging jet	Average, T_f, $2000 \lesssim Re \lesssim 4 \times 10^5$, $2 \lesssim (H/D) \lesssim 12$, $2.5 \lesssim (r/D) \lesssim 7.5$
Single slot nozzle	(7.66)	Impinging jet	Average, T_f, $3000 \lesssim Re \lesssim 9 \times 10^4$, $2 \lesssim (H/W) \lesssim 10$, $4 \lesssim (x/W) \lesssim 20$
Array of round nozzles	(7.64)	Impinging jet	Average, T_f, $2000 \lesssim Re \lesssim 10^5$, $2 \lesssim (H/D) \lesssim 12$, $0.004 \lesssim A_r \lesssim 0.04$
Array of slot nozzles	(7.68)	Impinging jet	Average, T_f, $1500 \lesssim Re \lesssim 4 \times 10^4$, $2 \lesssim (H/W) \lesssim 80$, $0.008 \lesssim A_r \lesssim 2.5A_{r,o}$
$\varepsilon \bar{j}_H = 2.06\, Re_D^{-0.575}$	(7.72)	Packed bed of spheres[c]	Average, $\overline{T}$, $90 \lesssim Re_D \lesssim 4000$, $Pr \approx 0.7$

[a]Correlations in this table pertain to isothermal surfaces; for special cases involving an unheated starting length or a uniform surface heat flux, see Section 7.2.4 or 7.2.5.
[b]The temperature listed under "Conditions" is the temperature at which properties should be evaluated.
[c]For tube banks and packed beds, properties are evaluated at the average fluid temperature, $\overline{T} = (T_i + T_o)/2$.

References

1. Blasius, H., *Z. Math. Phys.*, **56**, 1, 1908. English translation in National Advisory Committee for Aeronautics Technical Memo No. 1256.
2. Schlichting, H., and K. Gertsen, *Boundary Layer Theory*, Springer, New York, 2000.
3. Howarth, L., *Proc. R. Soc. Lond., Ser. A*, **164**, 547, 1938.
4. Pohlhausen, E., *Z. Angew. Math. Mech.*, **1**, 115, 1921.
5. Kays, W. M., M. E. Crawford, and B. Weigand, *Convective Heat and Mass Transfer*, 4th ed. McGraw-Hill Higher Education, Boston, 2005.
6. Churchill, S. W., and H. Ozoe, *J. Heat Transfer*, **95**, 78, 1973.
7. Ameel, T. A., *Int. Comm. Heat Mass Transfer*, **24**, 1113, 1997.
8. Blair, M. F., *J. Heat Transfer*, **105**, 33 and 41, 1983.
9. Fox, R. W., A. T. McDonald, and P. J. Pritchard, *Introduction to Fluid Mechanics,* 6th ed., Wiley, New York, 2003.
10. Coutanceau, M., and J.-R. Defaye, *Appl. Mech. Rev.*, **44**, 255, 1991.
11. Hilpert, R., *Forsch. Geb. Ingenieurwes.*, **4**, 215, 1933.
12. Knudsen, J. D., and D. L. Katz, *Fluid Dynamics and Heat Transfer*, McGraw-Hill, New York, 1958.
13. Jakob, M., *Heat Transfer*, Vol. 1, Wiley, New York, 1949.
14. Sparrow, E. M., J. P. Abraham, and J. C. K. Tong, *Int. J. Heat Mass Transfer*, **47**, 5285, 2004.
15. Whitaker, S., *AIChE J.*, **18**, 361, 1972.
16. Zukauskas, A., "Heat Transfer from Tubes in Cross Flow," in J. P. Hartnett and T. F. Irvine, Jr., Eds., *Advances in Heat Transfer*, Vol. 8, Academic Press, New York, 1972.
17. Churchill, S. W., and M. Bernstein, *J. Heat Transfer*, **99**, 300, 1977.
18. Morgan, V. T., "The Overall Convective Heat Transfer from Smooth Circular Cylinders," in T. F. Irvine, Jr. and J. P. Hartnett, Eds., *Advances in Heat Transfer*, Vol. 11, Academic Press, New York, 1975.
19. Ranz, W., and W. Marshall, *Chem. Eng. Prog.*, **48**, 141, 1952.
20. Grimison, E. D., *Trans. ASME*, **59**, 583, 1937.
21. Martin, H., "Heat and Mass Transfer between Impinging Gas Jets and Solid Surfaces," in J. P. Hartnett and T. F. Irvine, Jr., Eds., *Advances in Heat Transfer*, Vol. 13, Academic Press, New York, 1977.
22. Popiel, Cz. O., and L. Bogusiawski, "Mass or Heat Transfer in Impinging Single Round Jets Emitted by a Bell-Shaped Nozzle and Sharp-Ended Orifice," in C. L. Tien,

V. P. Carey, and J. K. Ferrell, Eds., *Heat Transfer 1986*, Vol. 3, Hemisphere Publishing, New York, 1986.

23. Goldstein, R. J., and J. F. Timmers, *Int. J. Heat Mass Transfer*, **25**, 1857, 1982.
24. Hollworth, B. R., and L. R. Gero, *J. Heat Transfer*, **107**, 910, 1985.
25. Goldstein, R. J., A. I. Behbahani, and K. K. Heppelman, *Int. J. Heat Mass Transfer*, **29**, 1227, 1986.
26. Pence, D. V., P. A. Boeschoten, and J. A. Liburdy, *J. Heat Transfer*, **125**, 447, 2003.
27. Webb, B. W., and C.-F. Ma, in J. P. Hartnett, T. F. Irvine, Y. I. Cho, and G. A. Greene, Eds., *Advances in Heat Transfer*, Vol. 26, Academic Press, New York, 1995.
28. Baukal, C. E., and B. Gebhart, *Int. J. Heat and Fluid Flow*, **4**, 386, 1996.
29. Chander, S., and A. Ray, *Energy Conversion and Management*, **46**, 2803, 2005.
30. Bird, R. B., W. E. Stewart, and E. N. Lightfoot, *Transport Phenomena*, 2nd ed. Wiley, New York, 2002.
31. Jakob, M., *Heat Transfer*, Vol. 2, Wiley, New York, 1957.

Problems

Flat Plate in Parallel Flow

7.1 Consider the following fluids at a film temperature of 300 K in parallel flow over a flat plate with velocity of 1 m/s: atmospheric air, water, engine oil, and mercury.

(a) For each fluid, determine the velocity and thermal boundary layer thicknesses at a distance of 40 mm from the leading edge.

(b) For each of the prescribed fluids and on the same coordinates, plot the boundary layer thicknesses as a function of distance from the leading edge to a plate length of 40 mm.

7.2 Engine oil at 100°C and a velocity of 0.1 m/s flows over both surfaces of a 1-m-long flat plate maintained at 20°C. Determine:

(a) The velocity and thermal boundary layer thicknesses at the trailing edge.

(b) The local heat flux and surface shear stress at the trailing edge.

(c) The total drag force and heat transfer per unit width of the plate.

(d) Plot the boundary layer thicknesses and local values of the surface shear stress, convection coefficient, and heat flux as a function of x for $0 \leq x \leq 1$ m.

7.3 Consider steady, parallel flow of atmospheric air over a flat plate. The air has a temperature and free stream velocity of 300 K and 25 m/s.

(a) Evaluate the boundary layer thickness at distances of $x = 1$, 10, and 100 mm from the leading edge. If a second plate were installed parallel to and at a distance of 3 mm from the first plate, what is the distance from the leading edge at which boundary layer merger would occur?

(b) Evaluate the surface shear stress and the y-velocity component at the outer edge of the boundary layer for the single plate at $x = 1$, 10, and 100 mm.

(c) Comment on the validity of the boundary layer approximations.

7.4 Consider a liquid metal ($Pr \ll 1$), with free stream conditions u_∞ and T_∞, in parallel flow over an isothermal flat plate at T_s. Assuming that $u = u_\infty$ throughout the thermal boundary layer, write the corresponding form of the boundary layer energy equation. Applying appropriate initial ($x = 0$) and boundary conditions, solve this equation for the boundary layer temperature field, $T(x, y)$. Use the result to obtain an expression for the local Nusselt number Nu_x. *Hint:* This problem is analogous to one-dimensional heat transfer in a semi-infinite medium with a sudden change in surface temperature.

7.5 Consider the velocity boundary layer profile for flow over a flat plate to be of the form $u = C_1 + C_2 y$. Applying appropriate boundary conditions, obtain an expression for the velocity profile in terms of the boundary layer thickness δ and the free stream velocity u_∞. Using the integral form of the boundary layer momentum equation (Appendix G), obtain expressions for the boundary layer thickness and the local friction coefficient, expressing your result in terms of the local Reynolds number. Compare your results with those obtained from the exact solution (Section 7.2.1) and the integral solution with a cubic profile (Appendix G).

7.6 Consider a steady, turbulent boundary layer on an isothermal flat plate of temperature T_s. The boundary layer is "tripped" at the leading edge $x = 0$ by a fine wire. Assume constant physical properties and velocity and temperature profiles of the form

$$\frac{u}{u_\infty} = \left(\frac{y}{\delta}\right)^{1/7} \quad \text{and} \quad \frac{T - T_\infty}{T_s - T_\infty} = 1 - \left(\frac{y}{\delta_t}\right)^{1/7}$$

(a) From experiment it is known that the surface shear stress is related to the boundary layer thickness by an expression of the form

$$\tau_s = 0.0228\,\rho u_\infty^2 \left(\frac{u_\infty \delta}{\nu}\right)^{-1/4}$$

Beginning with the momentum integral equation (Appendix G), show that

$$\delta/x = 0.376\, Re_x^{-1/5}.$$

Determine the average friction coefficient $\overline{C}_{f,x}$.

(b) Beginning with the energy integral equation, obtain an expression for the local Nusselt number Nu_x and use this result to evaluate the average Nusselt number $\overline{Nu}_x$.

7.7 Consider flow over a flat plate for which it is desired to determine the average heat transfer coefficient over the short span x_1 to x_2, $\overline{h}_{1-2}$, where $(x_2 - x_1) \ll L$.

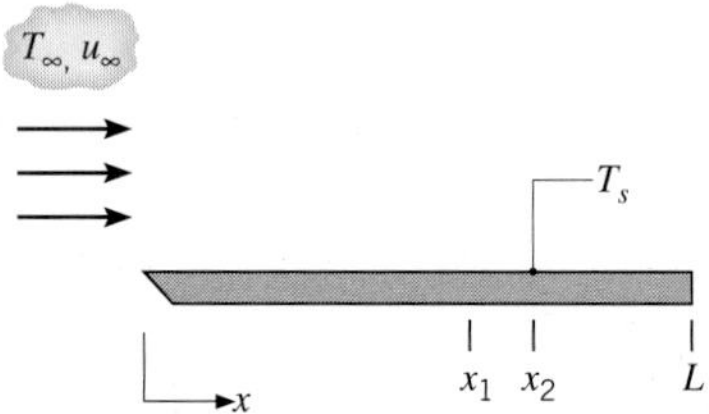

Provide three different expressions that can be used to evaluate $\overline{h}_{1-2}$ in terms of (a) the local coefficient at $x = (x_1 + x_2)/2$, (b) the local coefficients at x_1 and x_2, and (c) the average coefficients at x_1 and x_2. Indicate which of the expressions is approximate. Considering whether the flow is laminar, turbulent, or mixed, indicate when it is appropriate or inappropriate to use each of the equations.

7.8 A flat plate of width 1 m is maintained at a uniform surface temperature of $T_s = 150°C$ by using independently controlled, heat-generating rectangular modules of thickness $a = 10$ mm and length $b = 50$ mm. Each module is insulated from its neighbors, as well as on its back side. Atmospheric air at 25°C flows over the plate at a velocity of 30 m/s. The thermophysical properties of the module are $k = 5.2$ W/m·K, $c_p = 320$ J/kg·K, and $\rho = 2300$ kg/m³.

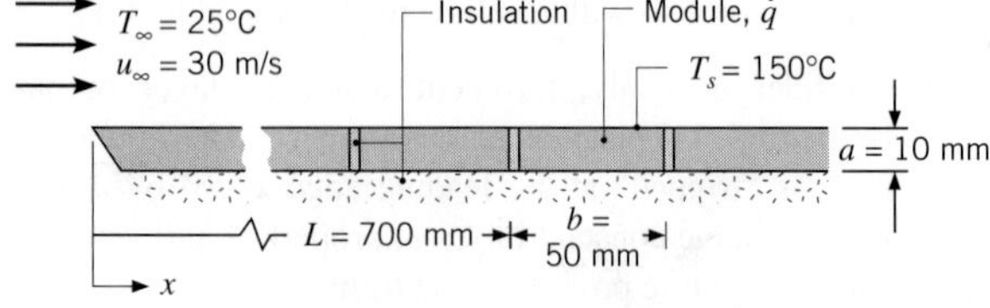

(a) Find the required power generation, $\dot{q}$ (W/m³), in a module positioned at a distance 700 mm from the leading edge.

(b) Find the maximum temperature T_{max} in the heat-generating module.

7.9 An electric air heater consists of a horizontal array of thin metal strips that are each 10 mm long in the direction of an airstream that is in parallel flow over the top of the strips. Each strip is 0.2 m wide, and 25 strips are arranged side by side, forming a continuous and smooth surface over which the air flows at 2 m/s. During operation, each strip is maintained at 500°C and the air is at 25°C.

(a) What is the rate of convection heat transfer from the first strip? The fifth strip? The tenth strip? All the strips?

(b) For air velocities of 2, 5, and 10 m/s, determine the convection heat rates for all the locations of part (a). Represent your results in tabular or bar graph form.

(c) Repeat part (b), but under conditions for which the flow is fully turbulent over the entire array of strips.

7.10 Consider atmospheric air at 25°C and a velocity of 25 m/s flowing over both surfaces of a 1-m-long flat plate that is maintained at 125°C. Determine the rate of heat transfer per unit width from the plate for values of the critical Reynolds number corresponding to 10^5, 5×10^5, and 10^6.

7.11 Consider laminar, parallel flow past an isothermal flat plate of length L, providing an average heat transfer coefficient of $\overline{h}_L$. If the plate is divided into N smaller plates, each of length $L_N = L/N$, determine an expression for the ratio of the heat transfer coefficient averaged over the N plates to the heat transfer coefficient averaged over the single plate, $\overline{h}_{L,N}/\overline{h}_{L,1}$.

7.12 Repeat Problem 7.11 for the case when the boundary layer is tripped to a turbulent condition at its leading edge.

7.13 Consider a flat plate subject to parallel flow (top and bottom) characterized by $u_\infty = 5$ m/s, $T_\infty = 20°C$.

(a) Determine the average convection heat transfer coefficient, convective heat transfer rate, and drag force associated with an $L = 2$-m-long, $w = 2$-m-wide flat plate for airflow and surface temperatures of $T_s = 50°C$ and 80°C.

(b) Determine the average convection heat transfer coefficient, convective heat transfer rate, and drag force associated with an $L = 0.1$-m-long, $w = 0.1$-m-wide flat plate for water flow and surface temperatures of $T_s = 50°C$ and 80°C.

7.14 Consider water at 27°C in parallel flow over an isothermal, 1-m-long flat plate with a velocity of 2 m/s.

(a) Plot the variation of the local heat transfer coefficient, $h_x(x)$, with distance along the plate for three flow conditions corresponding to transition Reynolds numbers of (i) 5×10^5, (ii) 3×10^5, and (iii) 0 (the flow is fully turbulent).

(b) Plot the variation of the average heat transfer coefficient $\overline{h}_x(x)$ with distance for the three flow conditions of part (a).

(c) What are the average heat transfer coefficients for the entire plate $\overline{h}_L$ for the three flow conditions of part (a)?

7.15 Explain under what conditions the total rate of heat transfer from an isothermal flat plate of dimensions $L \times 2L$ would be the same, independent of whether parallel flow over the plate is directed along the side of length L or $2L$. With a critical Reynolds number of 5×10^5, for what values of Re_L would the total heat transfer be independent of orientation?

7.16 In fuel cell stacks, it is desirable to operate under conditions that promote uniform surface temperatures for the electrolytic membranes. This is especially true in high-temperature fuel cells where the membrane is constructed of a brittle ceramic material. Electrochemical reactions in the electrolytic membranes generate thermal energy, while gases flowing above and below the membranes cool it. The stack designer may specify top and bottom flows that are in the same, opposite, or orthogonal directions. A preliminary study of the effect of the relative flow directions is conducted whereby a 150 mm × 150 mm *thin* sheet of material, producing a uniform heat flux of 100 W/m², is cooled (top and bottom) by air with a free stream temperature and velocity of 25°C and 2 m/s, respectively.

(a) Determine the minimum and maximum local membrane temperatures for top and bottom flows that are in the same, opposite, and orthogonal directions. Which flow configuration minimizes the membrane temperature? *Hint*: For the opposite and orthogonal flow cases, the boundary layers are subject to boundary conditions that are neither uniform temperature nor uniform heat flux. It is, however, reasonable to expect that the resulting temperatures would be *bracketed* by your answers based on the constant heat flux and constant temperature boundary conditions.

(b) Plot the surface temperature distribution $T(x)$ for the cases involving flow in the opposite and same directions. Thermal stresses are undesirable and are related to the spatial temperature gradient along the membrane. Which configuration minimizes spatial temperature gradients?

7.17 Air at a pressure of 1 atm and a temperature of 50°C is in parallel flow over the top surface of a flat plate that is heated to a uniform temperature of 100°C. The plate has a length of 0.20 m (in the flow direction) and a width of 0.10 m. The Reynolds number based on the plate length is 40,000. What is the rate of heat transfer from the plate to the air? If the free stream velocity of the air is doubled and the pressure is increased to 10 atm, what is the rate of heat transfer?

7.18 Consider the photovoltaic solar panel of Example 3.3. The heat transfer coefficient should no longer be taken to be a specified value.

(a) Determine the silicon temperature and the electric power produced by the solar cell for an air velocity of 4 m/s parallel to the long direction, with air and surroundings temperatures of 20°C. The boundary layer is tripped to a turbulent condition at the leading edge of the panel.

(b) Repeat part (a), except now the panel is oriented with its short side parallel to the airflow, that is, $L = 0.1$ m and $w = 1$ m.

(c) Plot the electric power output and the silicon temperature versus air velocity over the range $0 \leq u_m \leq 10$ m/s for the $L = 0.1$ m and $w = 1$ m case.

7.19 Concentration of sunlight onto photovoltaic cells is desired since the concentrating mirrors and lenses are less expensive than the photovoltaic material. Consider the solar photovoltaic cell of Example 3.3. A 100 mm × 100 mm photovoltaic cell is irradiated with concentrated solar energy. Since the concentrating lens is glass, it absorbs 10% of the irradiation instead of the top surface of the solar cell, as in Example 3.3. The remaining irradiation is reflected from the system (7%) or is absorbed in the silicon semiconductor material of the photovoltaic cell (83%). The photovoltaic cell is cooled by air directed parallel to its top and bottom surfaces. The air temperature and velocity are 25°C and 5 m/s, respectively, and the bottom surface is coated with a high-emissivity paint, $\varepsilon_b = 0.95$.

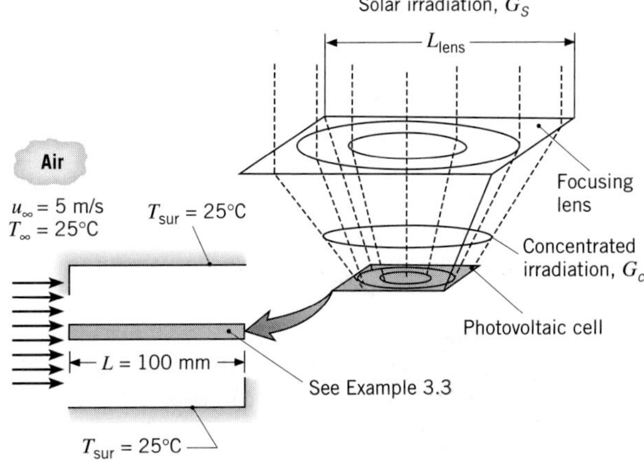

(a) Determine the electric power produced by the photovoltaic cell and the silicon temperature for a square concentrating lens with $L_{lens} = 400$ mm, which focuses the irradiation falling on the lens to the smaller area of the photovoltaic cell. Assume the concentrating lens temperature is 25°C and does not interfere with boundary layer development over the photovoltaic cell's top surface. The top and bottom boundary layers are both tripped to turbulent conditions at the leading edge of the photovoltaic material.

(b) Determine the electric power output of the photovoltaic cell and the silicon temperature over the range 100 mm $\leq L_{lens} \leq$ 600 mm.

7.20 The roof of a refrigerated truck compartment is of composite construction, consisting of a layer of foamed urethane insulation ($t_2 = 50$ mm, $k_i = 0.026$ W/m·K) sandwiched between aluminum alloy panels ($t_1 = 5$ mm, $k_p = 180$ W/m·K). The length and width of the roof are $L = 10$ m and W = 3.5 m, respectively, and the temperature of the inner surface is $T_{s,i} = -10$°C. Consider conditions for which the truck is moving at a speed of $V = 105$ km/h, the air temperature is $T_\infty = 32$°C, and the solar irradiation is $G_S = 750$ W/m^2. Turbulent flow may be assumed over the entire length of the roof.

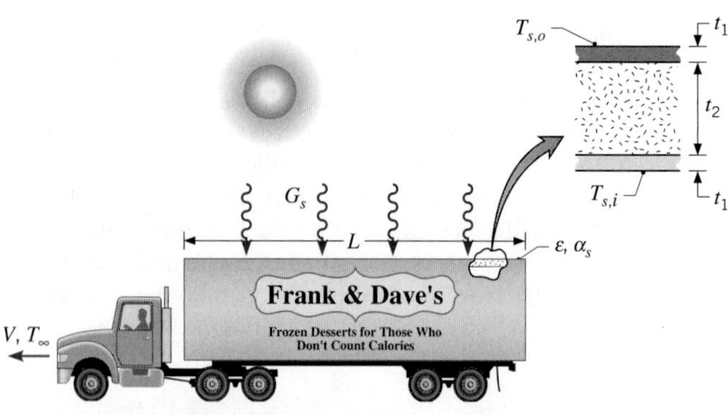

(a) For equivalent values of the solar absorptivity and the emissivity of the outer surface ($\alpha_S = \varepsilon = 0.5$), estimate the average temperature $T_{s,o}$ of the outer surface. What is the corresponding heat load imposed on the refrigeration system?

(b) A special finish ($\alpha_S = 0.15$, $\varepsilon = 0.8$) may be applied to the outer surface. What effect would such an application have on the surface temperature and the heat load?

(c) If, with $\alpha_S = \varepsilon = 0.5$, the roof is not insulated ($t_2 = 0$), what are the corresponding values of the surface temperature and the heat load?

7.21 The top surface of a heated compartment consists of very smooth (A) and highly roughened (B) portions, and the surface is placed in an atmospheric airstream. In the interest of minimizing total convection heat transfer from the surface, which orientation, (1) or (2), is preferred? If $T_s = 100$°C, $T_\infty = 20$°C, and $u_\infty =$ 20 m/s, what is the convection heat transfer from the entire surface for this orientation?

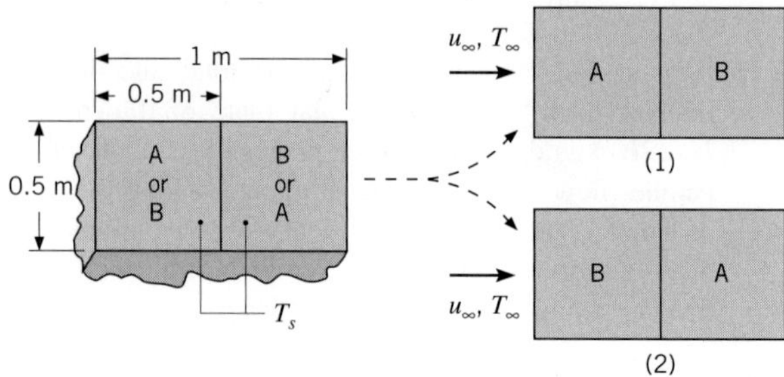

7.22 Calculate the value of the average heat transfer coefficient for the plate of Problem 7.21 when the entire plate is rotated 90° so that half of the leading edge consists of a very smooth portion (A) and the other half consists of a highly roughened portion (B).

7.23 The proposed design for an anemometer to determine the velocity of an airstream in a wind tunnel is comprised of a thin metallic strip whose ends are supported by stiff rods serving as electrodes for passage of current used to heat the strip. A fine-wire thermocouple is attached to the trailing edge of the strip and serves as the sensor for a system that controls the power to maintain the strip at a constant operating temperature for variable airstream velocities. Design conditions pertain to an airstream at $T_\infty = 25$°C and $1 \leq u_\infty \leq 50$ m/s, with a strip temperature of $T_s = 35$°C.

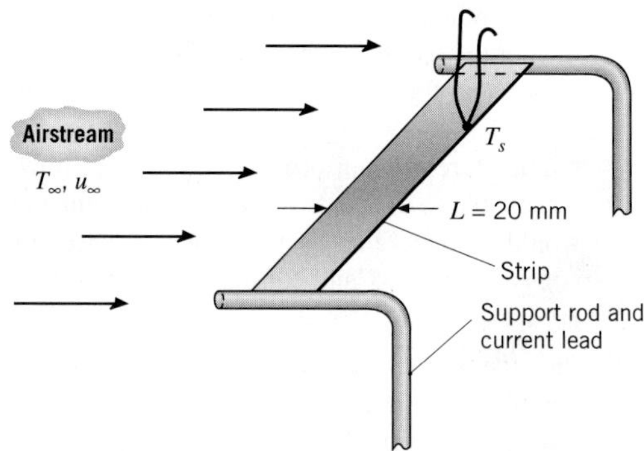

(a) Determine the relationship between the electrical power dissipation per unit width of the strip in the transverse direction, P' (mW/mm), and the airstream velocity. Show this relationship graphically for the specified range of u_∞.

(b) If the accuracy with which the temperature of the operating strip can be measured and maintained constant is ±0.2°C, what is the uncertainty in the airstream velocity?

(c) The proposed design operates in a strip constant-temperature mode for which the airstream velocity is related to the measured power. Consider now an alternative mode wherein the strip is provided with a constant power, say, 30 mW/mm, and the airstream velocity is related to the measured strip temperature T_s. For this mode of operation, show the graphical relationship between the strip temperature and airstream velocity. If the temperature can be measured with an uncertainty of $\pm 0.2°C$, what is the uncertainty in the airstream velocity?

(d) Compare the features associated with each of the anemometer operating modes.

7.24 Steel (AISI 1010) plates of thickness $\delta = 6$ mm and length $L = 1$ m on a side are conveyed from a heat treatment process and are concurrently cooled by atmospheric air of velocity $u_\infty = 10$ m/s and $T_\infty = 20°C$ in parallel flow over the plates.

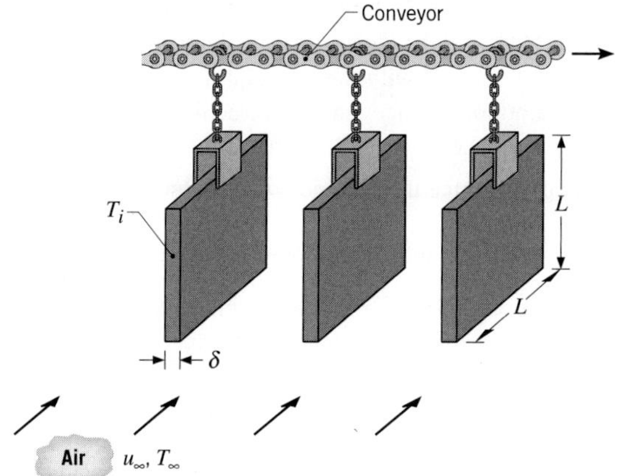

For an initial plate temperature of $T_i = 300°C$, what is the rate of heat transfer from the plate? What is the corresponding rate of change of the plate temperature? The velocity of the air is much larger than that of the plate.

7.25 Consider weather conditions for which the prevailing wind blows past the penthouse tower on a tall building. The tower length in the wind direction is 10 m and there are 10 window panels.

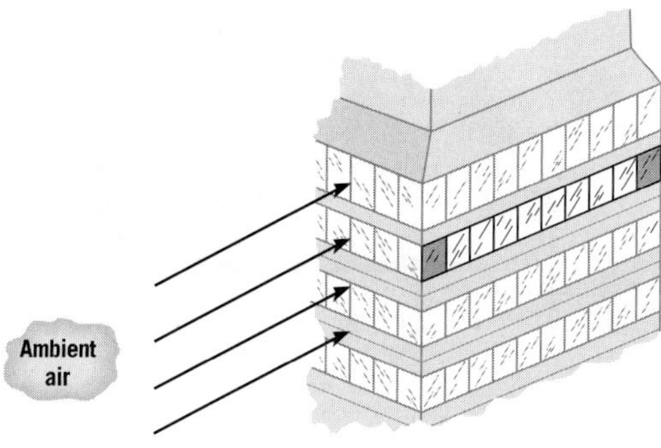

(a) Calculate the average convection coefficient for the first, third, and tenth window panels when the wind speed is 5 m/s. Use a film temperature of 300 K to evaluate the thermophysical properties required of the correlation. Would this be a suitable value of the film temperature for ambient air temperatures in the range $-15 \leq T_\infty \leq 38°C$?

(b) For the first, third, and tenth windows, on one graph, plot the variation of the average convection coefficient with wind speed for the range $5 \leq u_\infty \leq 100$ km/h. Explain the major features of each curve and their relative magnitudes.

7.26 Consider a rectangular fin that is used to cool a motorcycle engine. The fin is 0.15 m long and at a temperature of 250°C, while the motorcycle is moving at 80 km/h in air at 27°C. The air is in parallel flow over both surfaces of the fin, and turbulent flow conditions may be assumed to exist throughout.

(a) What is the rate of heat removal per unit width of the fin?

(b) Generate a plot of the heat removal rate per unit width of the fin for motorcycle speeds ranging from 10 to 100 km/h.

7.27 The Weather Channel reports that it is a hot, muggy day with an air temperature of 90°F, a 10 mph breeze out of the southwest, and bright sunshine with a solar insolation of 400 W/m^2. Consider the wall of a metal building over which the prevailing wind blows. The length of the wall in the wind direction is 10 m, and the emissivity is 0.93. Assume that all the solar irradiation is absorbed, that irradiation from the sky is negligible, and that flow is fully turbulent over the wall. Estimate the average wall temperature.

7.28 In the production of sheet metals or plastics, it is customary to cool the material before it leaves the production process for storage or shipment to the customer. Typically, the process is continuous, with a sheet of thickness δ and width W cooled as it transits the distance L between two rollers at a velocity V. In this problem, we consider cooling of an aluminum alloy (2024-T6) by an airstream moving at a velocity u_∞ in counter flow over the top surface of the sheet. A turbulence promoter is used to provide turbulent boundary layer development over the entire surface.

(a) By applying conservation of energy to a differential control surface of length dx, which either moves with the sheet *or* is stationary and through which the sheet passes, derive a differential equation that governs the temperature distribution along the sheet. Because of the low emissivity of the aluminum, radiation effects may be neglected. Express

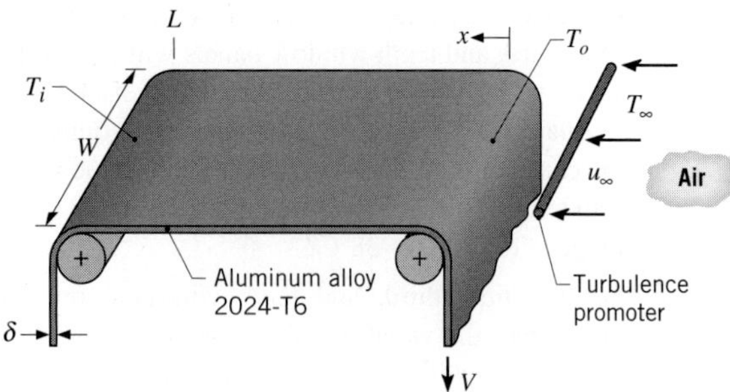

your result in terms of the velocity, thickness, and properties of the sheet (V, δ, ρ, c_p), the local convection coefficient h_x associated with the counter flow, and the air temperature. For a known temperature of the sheet (T_i) at the onset of cooling and a negligible effect of the sheet velocity on boundary layer development, solve the equation to obtain an expression for the outlet temperature T_o.

(b) For $\delta = 2$ mm, $V = 0.10$ m/s, $L = 5$ m, $W = 1$ m, $u_\infty = 20$ m/s, $T_\infty = 20°$C, and $T_i = 300°$C, what is the outlet temperature T_o?

7.29 An array of electronic chips is mounted within a sealed rectangular enclosure, and cooling is implemented by attaching an aluminum heat sink ($k = 180$ W/m·K). The base of the heat sink has dimensions of $w_1 = w_2 = 100$ mm, while the 6 fins are of thickness $t = 10$ mm and pitch $S = 18$ mm. The fin length is $L_f = 50$ mm, and the base of the heat sink has a thickness of $L_b = 10$ mm.

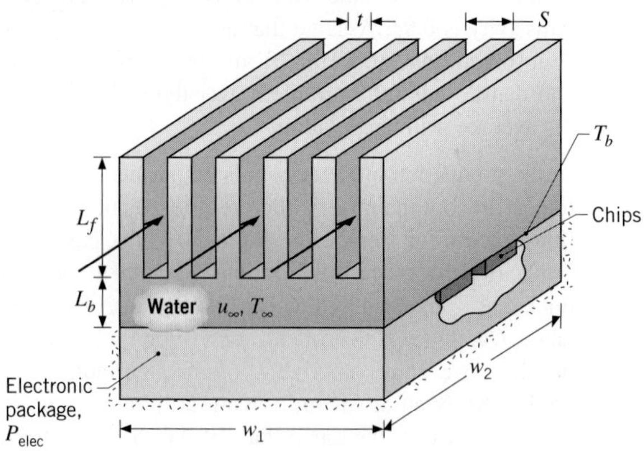

If cooling is implemented by water flow through the heat sink, with $u_\infty = 3$ m/s and $T_\infty = 17°$C, what is the base temperature T_b of the heat sink when power dissipation by the chips is $P_{elec} = 1800$ W? The average convection coefficient for surfaces of the fins and the exposed base may be estimated by assuming parallel flow over a flat plate. Properties of the water may be approximated as $k = 0.62$ W/m·K, $\rho = 995$ kg/m^3, $c_p = 4178$ J/kg·K, $\nu = 7.73 \times 10^{-7}$ m^2/s, and $Pr = 5.2$.

7.30 Consider the concentrating photovoltaic apparatus of Problem 7.19. The apparatus is to be installed in a desert environment, so the space between the concentrating lens and top of the photovoltaic cell is enclosed to protect the cell from sand abrasion in windy conditions. Since convection cooling from the top of the cell is reduced by the enclosure, an engineer proposes to cool the photovoltaic cell by attaching an aluminum heat sink to its bottom surface. The heat sink dimensions and material are the same as those of Problem 7.29. A contact resistance of 0.5×10^{-4} m^2·K/W exists at the photovoltaic cell/heat sink interface and a dielectric liquid ($k = 0.064$ W/m·K, $\rho = 1400$ kg/m^3, $c_p = 1300$ J/kg·K, $\nu = 10^{-6}$ m^2/s, $Pr = 25$) flows between the heat sink fins at $u_\infty = 3$ m/s, $T_\infty = 25°$C.

(a) Determine the electric power produced by the photovoltaic cell and the silicon temperature for a square concentrating lens with $L_{lens} = 400$ mm.

(b) Compare the electric power produced by the photovoltaic cell with the heat sink in place and with the bottom surface cooled directly by the dielectric fluid (i.e., no heat sink) for $L_{lens} = 1.5$ m.

(c) Determine the electric power output and the silicon temperature over the range $100 \text{ mm} \le L_{lens} < 3000$ mm with the aluminum heat sink in place.

7.31 In the production of sheet metals or plastics, it is customary to cool the material before it leaves the production process for storage or shipment to the customer. Typically, the process is continuous, with a sheet of thickness δ and width W cooled as it transits the distance L between two rollers at a velocity V. In this problem, we consider cooling of plain carbon steel by an airstream moving at a velocity u_∞ in cross flow over the top and bottom surfaces of the sheet. A turbulence promoter is used to provide turbulent boundary layer development over the entire surface.

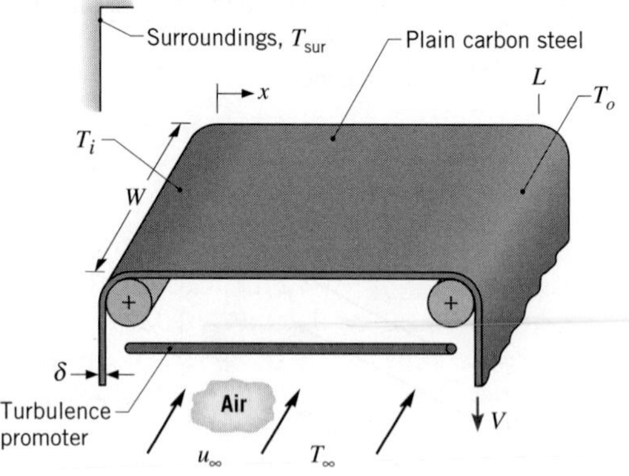

(a) By applying conservation of energy to a differential control surface of length dx, which either moves with the sheet *or* is stationary and through which the sheet passes, and assuming a uniform sheet temperature in the direction of airflow, derive a differential equation that governs the temperature distribution, $T(x)$, along the sheet. Consider the effects of radiation, as well as convection, and express your result in terms of the velocity, thickness, and properties of the sheet (V, δ, ρ, c_p, ε), the average convection coefficient $\overline{h}_W$ associated with the cross flow, and the environmental temperatures (T_∞, T_{sur}).

(b) Neglecting radiation, obtain a closed form solution to the foregoing equation. For $\delta = 3$ mm, $V = 0.10$ m/s, $L = 10$ m, $W = 1$ m, $u_\infty = 20$ m/s, $T_\infty = 20°C$, and a sheet temperature of $T_i = 500°C$ at the onset of cooling, what is the outlet temperature T_o? Assume a negligible effect of the sheet velocity on boundary layer development in the direction of airflow. The density and specific heat of the steel are $\rho = 7850$ kg/m^3 and $c_p = 620$ J/kg·K, while properties of the air may be taken to be $k = 0.044$ W/m·K, $\nu = 4.5 \times 10^{-5}$ m^2/s, $Pr = 0.68$.

(c) Accounting for the effects of radiation, with $\varepsilon = 0.70$ and $T_{sur} = 20°C$, numerically integrate the differential equation derived in part (a) to determine the temperature of the sheet at $L = 10$ m. Explore the effect of V on the temperature distribution along the sheet.

7.32 A steel strip emerges from the hot roll section of a steel mill at a speed of 20 m/s and a temperature of 1200 K. Its length and thickness are $L = 100$ m and $\delta = 0.003$ m, respectively, and its density and specific heat are 7900 kg/m^3 and 640 J/kg·K, respectively.

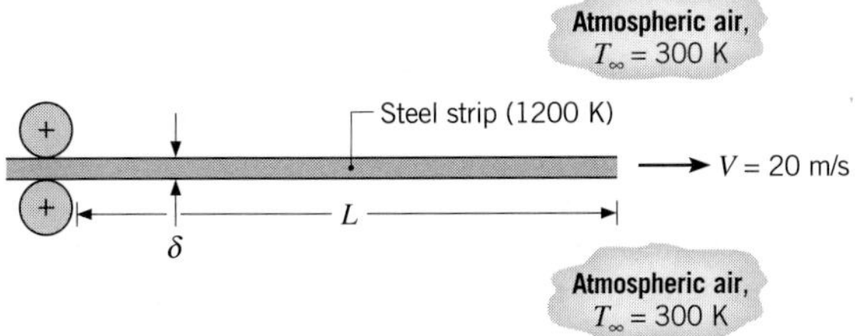

Accounting for heat transfer from the top and bottom surfaces and neglecting radiation and strip conduction effects, determine the time rate of change of the strip temperature at a distance of 1 m from the leading edge and at the trailing edge. Determine the distance from the leading edge at which the minimum cooling rate is achieved.

7.33 In Problem 7.23, an anemometer design was explored, and the assumption was made that the strip temperature was uniform. This is a good assumption when the heat transfer coefficient is low or the strip thermal conductivity high, because then conduction within the strip redistributes the generated heat and makes the strip temperature uniform. However, as the heat transfer coefficient increases or strip thermal conductivity decreases, heat generated at a point in the strip leaves the surface in the vicinity of that point, and the thermal condition is closer to one of uniform surface heat flux.

(a) Develop the calibration equations for both the constant surface temperature and constant heat flux conditions, that is, find the equations that predict the velocity as a function of the power per unit strip width, P'(mW/mm), and the temperature measured at the trailing edge (as in Problem 7.23). Assume laminar flow conditions.

(b) If the true condition is uniform surface heat flux, but the uniform surface temperature calibration is used, what percentage error will be incurred in the velocity determination?

(c) Where could the thermocouple be placed so that the calibration is insensitive to whether the thermal condition is uniform surface temperature or uniform surface heat flux?

7.34 A flat plate of width 1 m and length 0.2 m is maintained at a temperature of 32°C. Ambient fluid at 22°C flows across the top of the plate in parallel flow. Determine the average heat transfer coefficient, the convection heat transfer rate from the top of the plate, and the drag force on the plate for the following:

(a) The fluid is water flowing at a velocity of 0.5 m/s.

(b) The nanofluid of Example 2.2 is flowing at a velocity of 0.5 m/s.

(c) Water is flowing at a velocity of 2.5 m/s.

(d) The nanofluid of Example 2.2 is flowing at a velocity of 2.5 m/s.

7.35 One hundred electrical components, each dissipating 25 W, are attached to one surface of a square (0.2 m × 0.2 m) copper plate, and all the dissipated energy is transferred to water in parallel flow over the opposite surface. A protuberance at the leading edge of the plate acts to *trip* the boundary layer, and the plate itself may be assumed to be isothermal. The water velocity and temperature are $u_\infty = 2$ m/s and $T_\infty = 17°C$, and the water's thermophysical properties may be approximated as $\nu = 0.96 \times 10^{-6}$ m^2/s, $k = 0.620$ W/m·K, and $Pr = 5.2$.

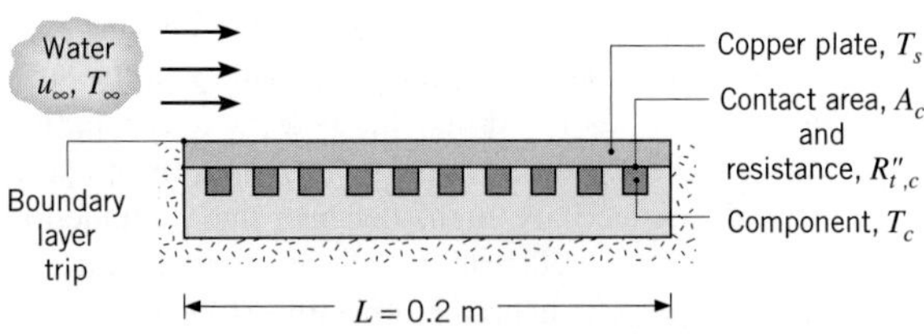

(a) What is the temperature of the copper plate?

(b) If each component has a plate contact surface area of 1 cm^2 and the corresponding contact resistance is $2 \times 10^{-4}\ m^2 \cdot K/W$, what is the component temperature? Neglect the temperature variation across the thickness of the copper plate.

7.36 Air at 27°C with a free stream velocity of 10 m/s is used to cool electronic devices mounted on a printed circuit board. Each device, 4 mm × 4 mm, dissipates 40 mW, which is removed from the top surface. A turbulator is located at the leading edge of the board, causing the boundary layer to be turbulent.

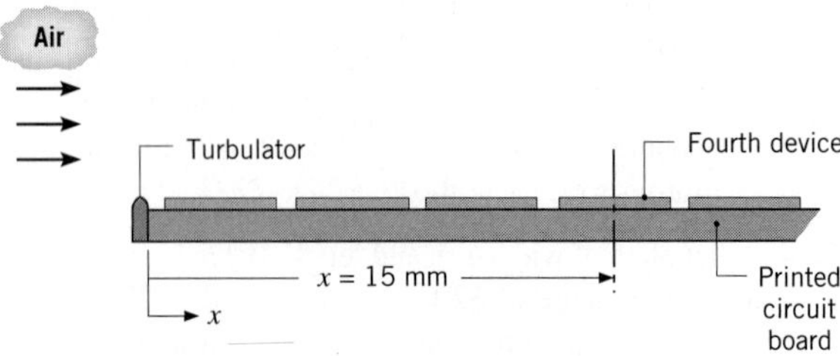

(a) Estimate the surface temperature of the fourth device located 15 mm from the leading edge of the board.

(b) Generate a plot of the surface temperature of the first four devices as a function of the free stream velocity for $5 \leq u_\infty \leq 15$ m/s.

(c) What is the minimum free stream velocity if the surface temperature of the hottest device is not to exceed 80°C?

7.37 The boundary layer associated with parallel flow over an isothermal plate may be tripped at any x-location by using a fine wire that is stretched across the width of the plate. Determine the value of the critical Reynolds number $Re_{x,c,op}$ that is associated with the optimal location of the trip wire from the leading edge that will result in maximum heat transfer from the warm plate to the cool fluid.

7.38 Forced air at 25°C and 10 m/s is used to cool electronic elements mounted on a circuit board. Consider a chip of length 4 mm and width 4 mm located 120 mm from the leading edge. Because the board surface is irregular, the flow is disturbed and the appropriate convection correlation is of the form $Nu_x = 0.04\ Re_x^{0.85}\ Pr^{0.33}$.

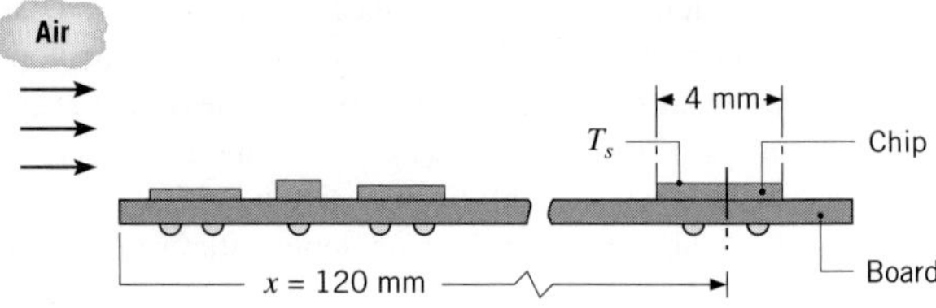

Estimate the surface temperature of the chip, T_s, if its heat dissipation rate is 30 mW.

7.39 Air at atmospheric pressure and a temperature of 25°C is in parallel flow at a velocity of 5 m/s over a 1-m-long flat plate that is heated with a uniform heat flux of 1250 W/m^2. Assume the flow is fully turbulent over the length of the plate.

(a) Calculate the plate surface temperature, $T_s(L)$, and the local convection coefficient, $h_x(L)$, at the trailing edge, $x = L$.

(b) Calculate the average temperature of the plate surface, $\overline{T}_s$.

(c) Plot the variation of the surface temperature, $T_s(x)$, and the convection coefficient, $h_x(x)$, with distance on the same graph. Explain the key features of these distributions.

7.40 Working in groups of two, our students design and perform experiments on forced convection phenomena using the general arrangement shown schematically. The air box consists of two muffin fans, a plenum chamber, and flow straighteners discharging a nearly uniform airstream over the flat *test-plate.* The objectives of one experiment were to measure the heat transfer coefficient and to compare the results with standard convection correlations. The velocity of the airstream was measured using a thermistor-based anemometer, and thermocouples were used to determine the temperatures of the airstream and the test-plate.

With the airstream from the box fully stabilized at T_∞ = 20°C, an aluminum plate was preheated in a convection oven and quickly mounted in the test-plate holder. The subsequent temperature history of the plate was determined from thermocouple measurements, and histories obtained for airstream velocities of 3 and 9 m/s were fitted by the following polynomial:

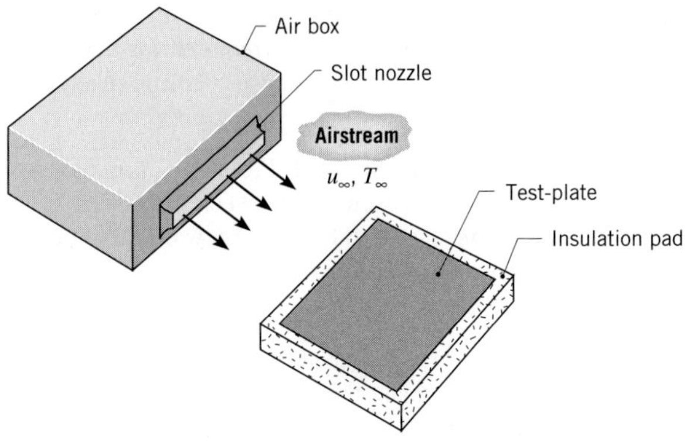

$$T(t) = a + bt + ct^2 + dt^3 + et^4$$

The temperature T and time t have units of °C and s, respectively, and values of the coefficients appropriate for the time interval of the experiments are tabulated as follows:

Velocity (m/s)	**3**	**9**
Elapsed Time (s)	300	160
a (°C)	56.87	57.00
b (°C/s)	−0.1472	−0.2641
c (°C/s²)	3×10^{-4}	9×10^{-4}
d (°C/s³)	-4×10^{-7}	-2×10^{-6}
e (°C/s⁴)	2×10^{-10}	1×10^{-9}

The plate is square, 133 mm to a side, with a thickness of 3.2 mm, and is made from a highly polished aluminum alloy ($\rho = 2770$ kg/m³, $c = 875$ J/kg·K, $k = 177$ W/m·K).

(a) Determine the heat transfer coefficients for the two cases, assuming the plate behaves as a spacewise isothermal object.

(b) Evaluate the coefficients C and m for a correlation of the form

$$\overline{Nu}_L = C\, Re^m\, Pr^{1/3}$$

Compare this result with a standard flat-plate correlation. Comment on the *goodness* of the comparison and explain any differences.

7.41 Consider atmospheric air at $u_\infty = 2$ m/s and $T_\infty = 300$ K in parallel flow over an isothermal flat plate of length $L = 1$ m and temperature $T_s = 350$ K.

(a) Compute the local convection coefficient at the leading and trailing edges of the heated plate with and without an unheated starting length of $\xi = 1$ m.

(b) Compute the average convection coefficient for the plate for the same conditions as part (a).

(c) Plot the variation of the local convection coefficient over the plate with and without an unheated starting length.

7.42 Consider a thin, 50 mm × 50 mm fuel cell similar to that of Example 1.5, with air in parallel flow over its surfaces. Very small-diameter wires are stretched across both sides of the fuel cell at a distance $x = x_c$ from the leading edge in order to trip the flow into turbulent conditions. Using an appropriate correlation from Chapter 7, determine the minimum velocity needed to sustain the fuel cell at $T_c = 77°C$, and the associated location of the wire. The air and large surroundings are at $T_\infty = T_{sur} = 27°C$ and the fuel cell dissipates $\dot{E}_g = 11$ W. The fuel cell emissivity is $\varepsilon = 0.85$.

7.43 The cover plate of a flat-plate solar collector is at 15°C, while ambient air at 10°C is in parallel flow over the plate, with $u_\infty = 2$ m/s.

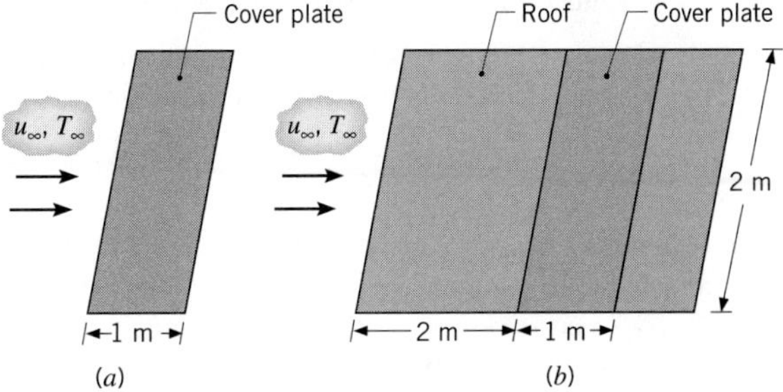

(a) What is the rate of convective heat loss from the plate?

(b) If the plate is installed 2 m from the leading edge of a roof and flush with the roof surface, what is the rate of convective heat loss?

7.44 An array of 10 silicon chips, each of length $L = 10$ mm on a side, is insulated on one surface and cooled on the opposite surface by atmospheric air in parallel flow with $T_\infty = 24°C$ and $u_\infty = 40$ m/s. When in use, the same electrical power is dissipated in each chip, maintaining a uniform heat flux over the entire cooled surface.

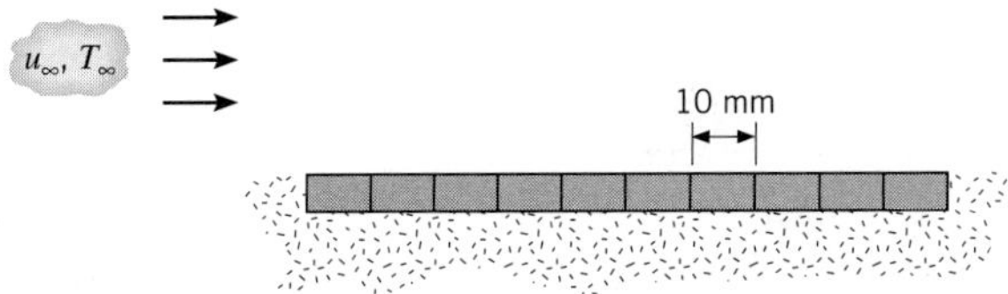

If the temperature of each chip may not exceed 80°C, what is the maximum allowable power per chip? What is the maximum allowable power if a turbulence promoter is used to trip the boundary layer at the leading edge? Would it be preferable to orient the array normal, instead of parallel, to the airflow?

7.45 A square (10 mm × 10 mm) silicon chip is insulated on one side and cooled on the opposite side by atmospheric air in parallel flow at $u_\infty = 20$ m/s and $T_\infty = 24°C$. When in use, electrical power dissipation within the chip maintains a uniform heat flux at the cooled surface. If the chip temperature may not exceed 80°C at any point on its surface, what is the maximum allowable power? What is the maximum allowable power if the chip is flush mounted in a substrate that provides for an unheated starting length of 20 mm?

Cylinder in Cross Flow

7.46 Consider the following fluids, each with a velocity of $V = 5$ m/s and a temperature of $T_\infty = 20°C$, in cross flow over a 10-mm-diameter cylinder maintained at 50°C: atmospheric air, saturated water, and engine oil.

(a) Calculate the rate of heat transfer per unit length, q', using the Churchill–Bernstein correlation.

(b) Generate a plot of q' as a function of fluid velocity for $0.5 \leq V \leq 10$ m/s.

7.47 A circular pipe of 25-mm outside diameter is placed in an airstream at 25°C and 1-atm pressure. The air moves in cross flow over the pipe at 15 m/s, while the outer surface of the pipe is maintained at 100°C. What is the drag force exerted on the pipe per unit length? What is the rate of heat transfer from the pipe per unit length?

7.48 An $L = 1$-m-long vertical copper tube of inner diameter $D_i = 20$ mm and wall thickness $t = 2$ mm contains liquid water at $T_w \approx 0°C$. On a winter day, air at $V = 3$ m/s, $T_\infty = -20°C$ is in cross flow over the tube.

(a) Determine the heat loss per unit mass from the water (W/kg) when the tube is full of water.

(b) Determine the heat loss from the water (W/kg) when the tube is half full.

7.49 A long, cylindrical, electrical heating element of diameter $D = 10$ mm, thermal conductivity $k = 240$ W/m·K, density $\rho = 2700$ kg/m^3, and specific heat $c_p = 900$ J/kg·K is installed in a duct for which air moves in cross flow over the heater at a temperature and velocity of 27°C and 10 m/s, respectively.

(a) Neglecting radiation, estimate the steady-state surface temperature when, per unit length of the heater, electrical energy is being dissipated at a rate of 1000 W/m.

(b) If the heater is activated from an initial temperature of 27°C, estimate the time required for the surface temperature to come within 10°C of its steady-state value.

7.50 Consider the conditions of Problem 7.49, but now allow for radiation exchange between the surface of the heating element ($\varepsilon = 0.8$) and the walls of the duct, which form a large enclosure at 27°C.

(a) Evaluate the steady-state surface temperature.

(b) If the heater is activated from an initial temperature of 27°C, estimate the time required for the surface temperature to come within 10°C of the steady-state value.

(c) To guard against overheating due to unanticipated excursions in the blower output, the heater controller is designed to maintain a fixed surface temperature of 275°C. Determine the power dissipation required to maintain this temperature for air velocities in the range $5 \leq V \leq 10$ m/s.

7.51 Pin fins are to be specified for use in an industrial cooling application. The fins will be subjected to a gas in cross flow at $V = 10$ m/s. The cylindrical fin has a diameter of $D = 15$ mm, and the cross-sectional area is the same for each configuration shown in the sketch.

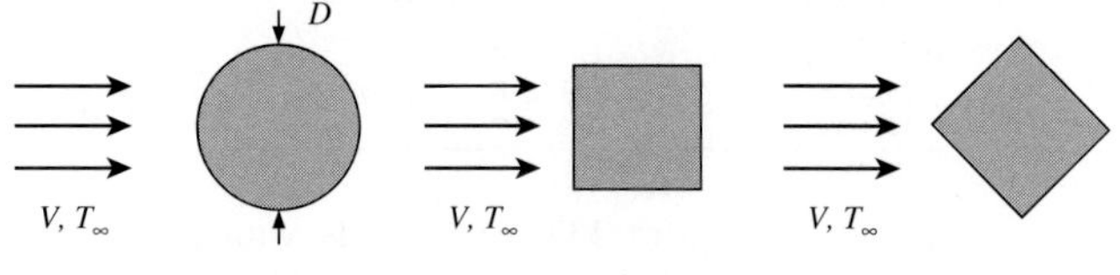

Cross sections of cylindrical and square fins in cross flow

For fins of equal length and therefore equal mass, which fin has the largest heat transfer rate? Assume the gas properties are those of air at $T = 350$ K. *Hint*: Assume the fins can be treated as infinitely long and apply the Hilpert correlation to the fin of circular cross section.

7.52 A pin fin of 10-mm diameter dissipates 30 W by forced convection to air in cross flow with a Reynolds number of 4000. If the diameter of the fin is doubled and all other conditions remain the same, estimate the fin heat rate. Assume the pin to be infinitely long.

7.53 Air at 27°C and a velocity of 5 m/s passes over the small region A_s (20 mm × 20 mm) on a large surface, which is maintained at $T_s = 127°C$. For these conditions, 0.5 W is removed from the surface A_s. To increase the heat removal rate, a stainless steel (AISI 304) pin fin of diameter 5 mm is affixed to A_s, which is assumed to remain at $T_s = 127°C$.

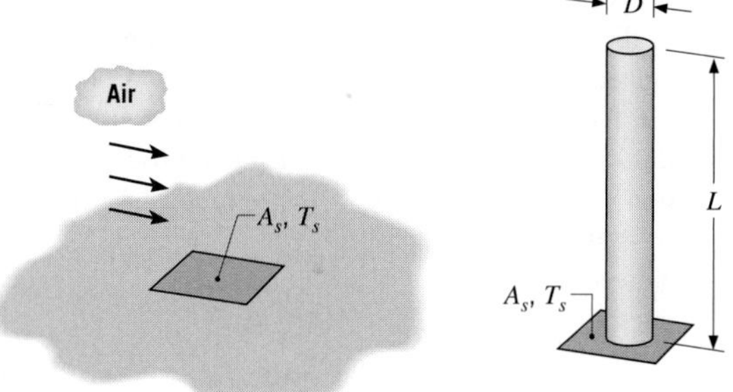

(a) Determine the maximum possible heat removal rate through the fin.

(b) What fin length would provide a close approximation to the heat rate found in part (a)? *Hint*: Refer to Example 3.9.

(c) Determine the fin effectiveness, ε_f.

(d) What is the percentage increase in the heat rate from A_s due to installation of the fin?

7.54 To enhance heat transfer from a silicon chip of width $W = 4$ mm on a side, a copper pin fin is brazed to the surface of the chip. The pin length and diameter are $L = 12$ mm and $D = 2$ mm, respectively, and atmospheric air at $V = 10$ m/s and $T_\infty = 300$ K is in cross flow over the pin. The surface of the chip, and hence the base of the pin, are maintained at a temperature of $T_b = 350$ K.

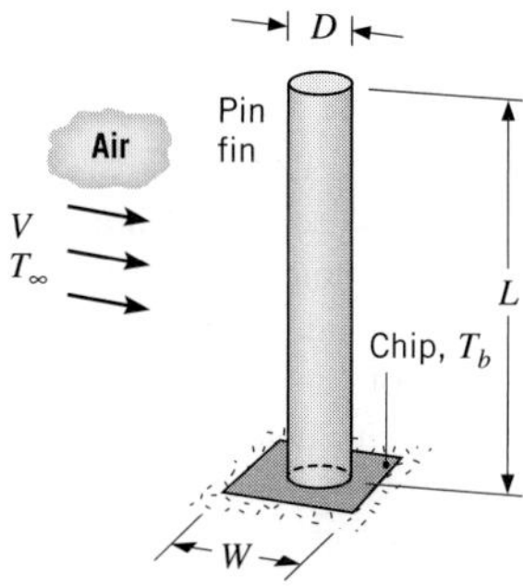

(a) Assuming the chip to have a negligible effect on flow over the pin, what is the average convection coefficient for the surface of the pin?

(b) Neglecting radiation and assuming the convection coefficient at the pin tip to equal that calculated in part (a), determine the pin heat transfer rate.

(c) Neglecting radiation and assuming the convection coefficient at the exposed chip surface to equal that calculated in part (a), determine the total rate of heat transfer from the chip.

(d) Independently determine and plot the effect of increasing velocity ($10 \leq V \leq 40$ m/s) and pin diameter ($2 \leq D \leq 4$ mm) on the total rate of heat transfer from the chip. What is the heat rate for $V = 40$ m/s and $D = 4$ mm?

7.55 Consider the Nichrome wire ($D = 1$ mm, $\rho_e = 10^{-6}$ $\Omega \cdot$m, $k = 25$ W/m·K, $\varepsilon = 0.20$) used to fabricate the air heater of Problem 3.86, but now under conditions for which the convection heat transfer coefficient must be determined.

(a) For atmospheric air at 50°C and a cross-flow velocity of 5 m/s, what are the surface and centerline temperatures of the wire when it carries a current of 25 A and the housing of the heater is also at 50°C?

(b) Explore the effect of variations in the flow velocity and electrical current on the surface and centerline temperatures of the wire.

7.56 Hot water at 50°C is routed from one building in which it is generated to an adjoining building in which it is used for space heating. Transfer between the buildings occurs in a steel pipe ($k = 60$ W/m·K) of 100-mm outside diameter and 8-mm wall thickness. During the winter, representative environmental conditions involve air at $T_\infty = -5$°C and $V = 3$ m/s in cross flow over the pipe.

(a) If the cost of producing the hot water is \$0.10 per kW·h, what is the representative daily cost of heat loss from an uninsulated pipe to the air per meter of pipe length? The convection resistance associated with water flow in the pipe may be neglected.

(b) Determine the savings associated with application of a 10-mm-thick coating of urethane insulation ($k = 0.026$ W/m·K) to the outer surface of the pipe.

7.57 In a manufacturing process, long aluminum rods of square cross section with $d = 25$ mm are cooled from an initial temperature of $T_i = 400$°C. Which configuration in the sketch should be used to minimize the time needed for the rods to reach a *safe-to-handle* temperature of 60°C when exposed to air in cross flow at $V = 8$ m/s, $T_\infty = 30$°C? What is the required cooling time for the preferred configuration? The emissivity of the rods is $\varepsilon = 0.10$ and the surroundings temperature is $T_{sur} = 20$°C.

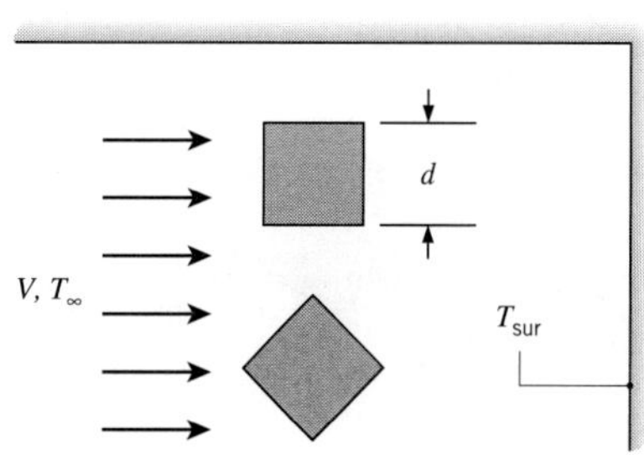

7.58 A fine wire of diameter D is positioned across a passage to determine flow velocity from heat transfer characteristics. Current is passed through the wire to heat it, and the heat is dissipated to the flowing fluid by convection. The resistance of the wire is determined from electrical measurements, and the temperature is known from the resistance.

(a) For a fluid of arbitrary Prandtl number, develop an expression for its velocity in terms of the difference between the temperature of the wire and the free stream temperature of the fluid.

(b) What is the velocity of an airstream at 1 atm and 25°C, if a wire of 0.5-mm diameter achieves a temperature of 40°C while dissipating 35 W/m?

7.59 To determine air velocity changes, it is proposed to measure the electric current required to maintain a platinum wire of 0.5-mm diameter at a constant temperature of 77°C in a stream of air at 27°C.

(a) Assuming Reynolds numbers in the range $40 < Re_D < 1000$, develop a relationship between the wire current and the velocity of the air that is in cross flow over the wire. Use this result to establish a relation between fractional changes in the current, $\Delta I/I$, and the air velocity, $\Delta V/V$.

(b) Calculate the current required when the air velocity is 10 m/s and the electrical resistivity of the platinum wire is $17.1 \times 10^{-5}\ \Omega \cdot \text{m}$.

7.60 Fluid velocities can be measured using hot-film sensors, and a common design is one for which the sensing element forms a thin film about the circumference of a quartz rod. The film is typically comprised of a thin (~100 nm) layer of platinum, whose electrical resistance is proportional to its temperature. Hence, when submerged in a fluid stream, an electric current may be passed through the film to maintain its temperature above that of the fluid. The temperature of the film is controlled by monitoring its electric resistance, and with concurrent measurement of the electric current, the power dissipated in the film may be determined.

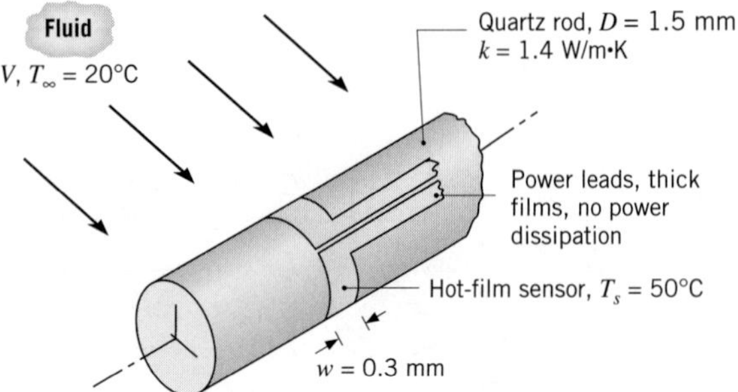

Proper operation is assured only if the heat generated in the film is transferred to the fluid, rather than conducted from the film into the quartz rod. Thermally, the film should therefore be strongly coupled to the fluid and weakly coupled to the quartz rod. This condition is satisfied if the Biot number is very large, $Bi = \overline{h}D/2k \gg 1$, where $\overline{h}$ is the convection coefficient between the fluid and the film and k is the thermal conductivity of the rod.

(a) For the following fluids and velocities, calculate and plot the convection coefficient as a function of velocity: (i) water, $0.5 \le V \le 5$ m/s; (ii) air, $1 \le V \le 20$ m/s.

(b) Comment on the suitability of using this hot-film sensor for the foregoing conditions.

7.61 Consider use of the hot-film sensor described in Problem 7.60 to determine the velocity of water entering the cooling system of an electric power plant from an adjoining lake. The sensor is mounted within an intake pipe, and its controls are set to maintain an average hot-film temperature that is 5°C larger than the fluid temperature ($T_{s,\text{hf}} - T_\infty = 5°\text{C}$).

(a) If an independent measurement of the water temperature yields a value of $T_\infty = 17°\text{C}$, use the Churchill–Bernstein correlation to estimate the velocity of the water under conditions for which the power input to the sensor maintains a heat flux of $q''_{\text{hf}} = 4 \times 10^4\ \text{W/m}^2$ from the film to the water.

(b) If the sensor is exposed to the water for an extended period, its surface will be *fouled* by an accumulation of deposits from the water. Consider conditions for which the deposits form a 0.1-mm-thick shell around the sensor and have a thermal conductivity of $k_d = 2\ \text{W/m} \cdot \text{K}$. For $T_\infty = 17°\text{C}$ and the flow velocity determined in part (a), what heat flux must be supplied to the sensor to maintain its temperature at $T_{s,\text{hf}} = 22°\text{C}$? What is the corresponding error in the velocity measurement? *Note:* Conduction across the deposit may be approximated as that across a plane wall.

7.62 Determine the convection heat loss from both the top and the bottom of a flat plate at $T_s = 80°\text{C}$ with air in parallel flow at $T_\infty = 25°\text{C}$, $u_\infty = 3$ m/s. The plate is $t = 1$ mm thick, $L = 25$ mm long, and of depth $w = 50$ mm. Neglect the heat loss from the edges of the plate. Compare the convection heat loss from the plate to the convection heat loss from an $L_c = 50$-mm-long cylinder of the same volume as that of the plate. The convective conditions associated with the cylinder are the same as those associated with the plate.

7.63 Consider two very long, straight fins of uniform cross section, as shown in Figure 3.17. The rectangular fin has dimensions $t = 1$ mm and $w = 20$ mm. The circular pin fin has the same cross-sectional area as the rectangular fin. Both fins are constructed of aluminum with $k = 237$ W/m·K. In both cases, the base temperature is $T_b = 85°C$. Airflow is directed as shown in the figure, with $T_\infty = 20°C$ and $u_\infty = 5$ m/s.

(a) Calculate the heat loss from each fin. Assume that the heat transfer coefficient on the edges of the rectangular fin is equal to the average value on the upper and lower surfaces.

(b) What diameter cylindrical fin would be needed to provide the same fin heat transfer rate as the rectangular cross-section fin?

7.64 A computer code is being developed to analyze a temperature sensor of 12.5-mm diameter experiencing cross flow of water with a free stream temperature of 80°C and variable velocity. Derive an expression for the convection heat transfer coefficient as a function of the sensor surface temperature T_s for the range $20 < T_s < 80°C$ and for velocities V in the range $0.005 < V < 0.20$ m/s. Use the Zukauskas correlation for the range $40 < Re_D < 1000$ and assume that the Prandtl number of water has a linear temperature dependence.

7.65 A 25-mm-diameter, high-tension line has an electrical resistance of 10^{-4} Ω/m and is transmitting a current of 1000 A.

(a) If ambient air at 10°C and 5 m/s is in cross flow over the line, what is its surface temperature?

(b) If the line may be approximated as a solid copper rod, what is its centerline temperature?

[(c)] Generate a plot that depicts the variation of the surface temperature with air velocity for $1 \leq V \leq 10$ m/s.

7.66 An aluminum transmission line with a diameter of 20 mm has an electrical resistance of $R'_{elec} = 2.636 \times 10^{-4}$ Ω/m and carries a current of 700 A. The line is subjected to frequent and severe cross winds, increasing the probability of contact between adjacent lines, thereby causing sparks and creating a potential fire hazard for nearby vegetation. The remedy is to insulate the line, but with the adverse effect of increasing the conductor operating temperature.

(a) Calculate the conductor temperature when the air temperature is 20°C and the line is subjected to cross flow with a velocity of 10 m/s.

(b) Calculate the conductor temperature for the same conditions, but with a 2-mm-thick insulation having a thermal conductivity of 0.15 W/m·K.

[(c)] Calculate and plot the temperatures of the bare and insulated conductors for wind velocities in the range from 2 to 20 m/s. Comment on features of the curves and the effect of the wind velocity on the conductor temperatures.

7.67 To augment heat transfer between two flowing fluids, it is proposed to insert a 100-mm-long, 5-mm-diameter 2024 aluminum pin fin through the wall separating the two fluids. The pin is inserted to a depth of d into fluid 1. Fluid 1 is air with a mean temperature of 10°C and velocity of 10 m/s. Fluid 2 is air with a mean temperature of 40°C and velocity of 3 m/s.

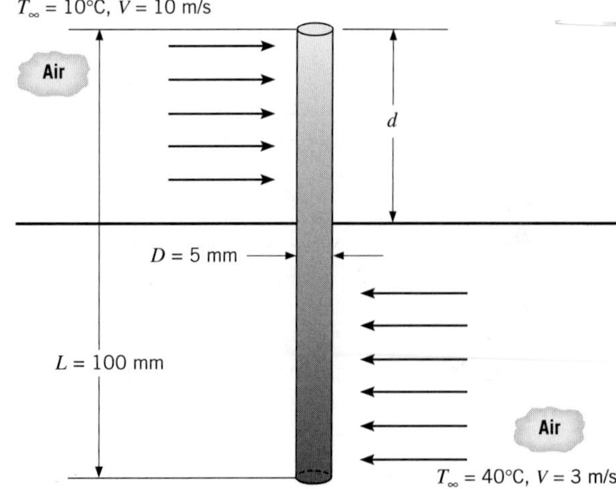

(a) Determine the rate of heat transfer from the warm air to the cool air through the pin fin for $d = 50$ mm.

[(b)] Plot the variation of the heat transfer rate with the insertion distance, d. Does an optimal insertion distance exist?

7.68 An uninsulated steam pipe is used to transport high-temperature steam from one building to another. The pipe is of 0.5-m diameter, has a surface temperature of 150°C, and is exposed to ambient air at −10°C. The air moves in cross flow over the pipe with a velocity of 5 m/s.

(a) What is the heat loss per unit length of pipe?

[(b)] Consider the effect of insulating the pipe with a rigid urethane foam ($k = 0.026$ W/m·K). Evaluate and plot the heat loss as a function of the thickness δ of the insulation layer for $0 \leq \delta \leq 50$ mm.

7.69 A thermocouple is inserted into a hot air duct to measure the air temperature. The thermocouple (T_1) is soldered to the tip of a steel *thermocouple well* of length $L = 0.15$ m and inner and outer diameters of $D_i = 5$ mm and $D_o = 10$ mm. A second thermocouple (T_2) is used to measure the duct wall temperature.

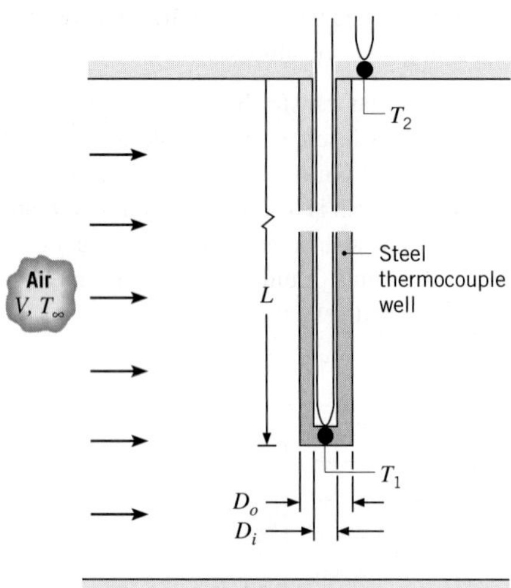

Consider conditions for which the air velocity in the duct is $V = 3$ m/s and the two thermocouples register temperatures of $T_1 = 450$ K and $T_2 = 375$ K. Neglecting radiation, determine the air temperature T_∞. Assume that, for steel, $k = 35$ W/m·K, and, for air, $\rho = 0.774$ kg/m^3, $\mu = 251 \times 10^{-7}$ N·s/m^2, $k = 0.0373$ W/m·K, and $Pr = 0.686$.

7.70 Consider conditions for which a mercury-in-glass thermometer of 4-mm diameter is inserted to a length L through the wall of a duct in which air at 77°C is flowing. If the stem of the thermometer at the duct wall is at the wall temperature $T_w = 15$°C, conduction heat transfer through the glass causes the bulb temperature to be lower than that of the airstream.

(a) Develop a relationship for the *immersion* error, $\Delta T_i = T(L) - T_\infty$, as a function of air velocity, thermometer diameter, and insertion length L.

(b) To what length L must the thermometer be inserted if the immersion error is not to exceed 0.25°C when the air velocity is 10 m/s?

(c) Using the insertion length determined in part (b), calculate and plot the immersion error as a function of air velocity for the range 2 to 20 m/s.

(d) For a given insertion length, will the immersion error increase or decrease if the diameter of the thermometer is increased? Is the immersion error more sensitive to the diameter or air velocity?

7.71 In a manufacturing process, a long, coated plastic rod ($\rho = 2200$ kg/m^3, $c = 800$ J/kg·K, $k = 1$ W/m·K) of diameter $D = 20$ mm is initially at a uniform temperature of 25°C and is suddenly exposed to a cross flow of air at $T_\infty = 350$°C and $V = 50$ m/s.

(a) How long will it take for the surface of the rod to reach 175°C, the temperature above which the special coating will cure?

(b) Generate a plot of the time to reach 175°C as a function of air velocity for $5 \le V \le 50$ m/s.

7.72 In an extrusion process, copper wire emerges from the extruder at a velocity V_e and is cooled by convection heat transfer to air in cross flow over the wire, as well as by radiation to the surroundings.

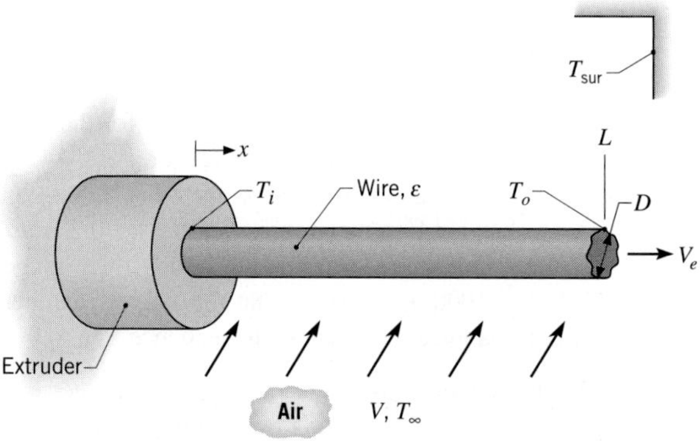

(a) By applying conservation of energy to a differential control surface of length dx, which either moves with the wire *or* is stationary and through which the wire passes, derive a differential equation that governs the temperature distribution, $T(x)$, along the wire. In your derivation, the effect of axial conduction along the wire may be neglected. Express your result in terms of the velocity, diameter, and properties of the wire (V_e, D, ρ, c_p, ε), the convection coefficient associated with the cross flow ($\bar{h}$), and the environmental temperatures (T_∞, T_{sur}).

(b) Neglecting radiation, obtain a closed form solution to the foregoing equation. For $V_e = 0.2$ m/s, $D = 5$ mm, $V = 5$ m/s, $T_\infty = 25$°C, and an initial wire temperature of $T_i = 600$°C, compute the temperature T_o of the wire at $x = L = 5$ m. The density and specific heat of the copper are $\rho = 8900$ kg/m^3 and $c_p = 400$ J/kg·K, while properties of the air may be taken to be $k = 0.037$ W/m·K, $\nu = 3 \times 10^{-5}$ m^2/s, and $Pr = 0.69$.

(c) Accounting for the effects of radiation, with $\varepsilon = 0.55$ and $T_{sur} = 25$°C, numerically integrate the differential equation derived in part (a) to determine the temperature of the wire at $L = 5$ m. Explore the effects of V_e and ε on the temperature distribution along the wire.

7.73 The objective of an experiment performed by our students is to determine the effect of pin fins on the thermal resistance between a flat plate and an airstream. A 25.9-mm-square polished aluminum plate is subjected to an airstream in parallel flow at $T_\infty = 20°C$ and $u_\infty = 6$ m/s. An electrical heating patch is attached to the backside of the plate and dissipates 15.5 W under all conditions. Pin fins of diameter $D = 4.8$ mm and length $L = 25.4$ mm are fabricated from brass and can be firmly attached to the plate at various locations over its surface. Thermocouples are attached to the plate surface and the tip of one of the fins.

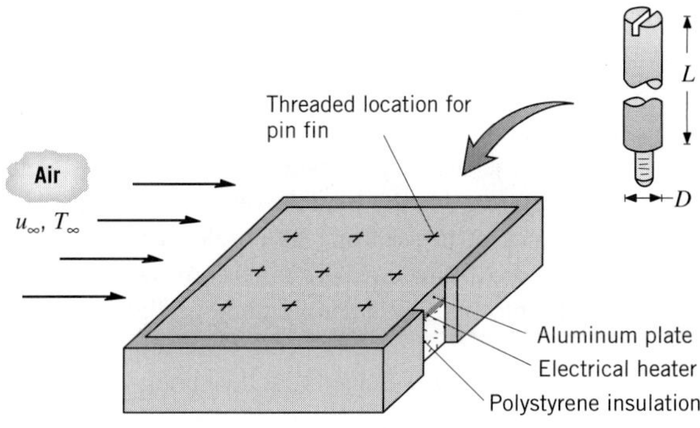

Measured temperatures for five pin-fin configurations are tabulated.

Number of Pin Fins	Temperature (°C) Fin Tip	Plate Base
0	—	70.2
1	40.6	67.4
2	39.5	64.7
5	36.4	57.4
8	34.2	52.1

(a) Using the experimental observations and neglecting the effect of flow interactions between pins, determine the thermal resistance between the plate and the airstream for the five configurations.

(b) Develop a model of the plate–pin fin system and using appropriate convection correlations, predict the thermal resistances for the five configurations. Compare your predictions with the observations and explain any differences.

(c) Use your model to predict the thermal resistances when the airstream velocity is doubled.

Spheres

7.74 Air at 25°C flows over a 10-mm-diameter sphere with a velocity of 25 m/s, while the surface of the sphere is maintained at 75°C.

(a) What is the drag force on the sphere?

(b) What is the rate of heat transfer from the sphere?

(c) Generate a plot of the heat transfer from the sphere as a function of the air velocity for the range 1 to 25 m/s.

7.75 Consider a sphere with a diameter of 20 mm and a surface temperature of 60°C that is immersed in a fluid at a temperature of 30°C and a velocity of 2.5 m/s. Calculate the drag force and the heat rate when the fluid is (a) water and (b) air at atmospheric pressure. Explain why the results for the two fluids are so different.

7.76 Consider the material processing experiment of Problem 5.24, with atmospheric nitrogen used to implement cooling by convection. However, instead of using a prescribed value of the convection coefficient, compute the coefficient from an appropriate correlation.

(a) Neglecting radiation, determine the time required to cool the sphere from 900°C to 300°C if the velocity and temperature of the nitrogen are $V = 5$ m/s and $T_\infty = 25°C$.

(b) Accounting for the effects of both convection and radiation, with $\varepsilon = 0.6$ and $T_{sur} = 25°C$, determine the time required to cool the sphere. Explore the effects of the flow velocity on your result.

7.77 A spherical, underwater instrument pod used to make soundings and to measure conditions in the water has a diameter of 85 mm and dissipates 300 W.

(a) Estimate the surface temperature of the pod when suspended in a bay where the current is 1 m/s and the water temperature is 15°C.

(b) Inadvertently, the pod is hauled out of the water and suspended in ambient air without deactivating the power. Estimate the surface temperature of the pod if the air temperature is 15°C and the wind speed is 3 m/s.

7.78 Worldwide, over a billion solder balls must be manufactured daily for assembling electronics packages. The *uniform droplet spray* method uses a piezoelectric device to vibrate a shaft in a pot of molten solder that, in turn, ejects small droplets of solder through a precision-machined nozzle. As they traverse a collection chamber, the droplets cool and solidify. The collection chamber is flooded with an inert gas such as nitrogen to prevent oxidation of the solder ball surfaces.

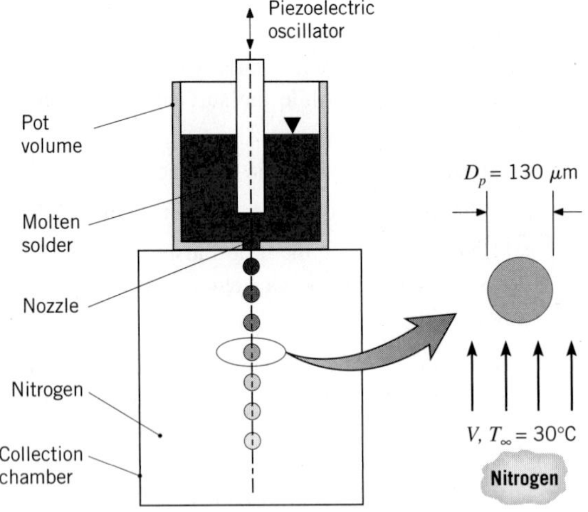

(a) Molten solder droplets of diameter 130 μm are ejected at a velocity of 2 m/s at an initial temperature of 225°C into gaseous nitrogen that is at 30°C and slightly above atmospheric pressure. Determine the terminal velocity of the particles and the distance the particles have traveled when they become completely solidified. The solder properties are $\rho = 8230$ kg/m^3, $c = 240$ J/kg·K, $k = 38$ W/m·K, $h_{sf} = 42$ kJ/kg. The solder's melting temperature is 183°C.

(b) The piezoelectric device oscillates at 1.8 kHz, producing 1800 particles per second. Determine the separation distance between the particles as they traverse the nitrogen gas and the pot volume needed in order to produce the solder balls continuously for one week.

7.79 A spherical workpiece of pure copper with a diameter of 15 mm and an emissivity of 0.5 is suspended in a large furnace with walls at a uniform temperature of 600°C. Air flows over the workpiece at a temperature of 900°C and a velocity of 7.5 m/s.

(a) Determine the steady-state temperature of the workpiece.

(b) Estimate the time required for the workpiece to come within 5°C of the steady-state temperature if it is at an initial, uniform temperature of 25°C.

(c) To decrease the time to heat the workpiece, the air velocity is doubled, with all other conditions remaining the same. Determine the steady-state temperature of the workpiece and the time required for it to come within 5°C of this value. Plot on the same graph the workpiece temperature histories for the two velocities.

7.80 Copper spheres of 20-mm diameter are quenched by being dropped into a tank of water that is maintained at 280 K. The spheres may be assumed to reach the terminal velocity on impact and to drop freely through the water. Estimate the terminal velocity by equating the drag and gravitational forces acting on the sphere. What is the approximate height of the water tank needed to cool the spheres from an initial temperature of 360 K to a center temperature of 320 K?

7.81 For the conditions of Problem 7.80, what are the terminal velocity and the tank height if engine oil at 300 K, rather than water, is used as the coolant?

7.82 Consider the plasma spray coating process of Problem 5.25. In addition to the prescribed conditions, the argon plasma jet is known to have a mean velocity of $V = 400$ m/s, while the *initial* velocity of the injected alumina particles may be approximated as zero. The nozzle exit and the substrate are separated by a distance of $L = 100$ mm, and pertinent properties of the argon plasma may be approximated as $k = 0.671$ W/m·K, $c_p = 1480$ J/kg·K, $\mu = 2.70 \times 10^{-4}$ kg/s·m, and $\nu = 5.6 \times 10^{-3}$ m^2/s.

(a) Assuming the motion of particles entrained by the plasma jet to be governed by Stokes' law, derive expressions for the particle velocity, $V_p(t)$, and its distance of travel from the nozzle exit, $x_p(t)$, as a function of time, t, where $t = 0$ corresponds to particle injection. Evaluate the time-in-flight required for a particle to traverse the separation distance, $x_p = L$, and the velocity V_p at this time.

(b) Assuming an average relative velocity of $\overline{(V - V_p)} = 315$ m/s during the time-of-flight, estimate the convection coefficient associated with heat transfer from the plasma to the particle. Using this coefficient and assuming an initial particle temperature of $T_i = 300$ K, estimate the time-in-flight required to heat a particle to its melting point, T_{mp}, and, once at T_{mp}, for the particle to experience complete melting. Is the prescribed value of L sufficient to ensure complete particle melting before surface impact?

7.83 Highly reflective aluminum coatings may be formed on the surface of a substrate by impacting the surface with molten drops of aluminum. The droplets are discharged from an injector, proceed through an inert gas (helium), and must still be in a molten state at the time of impact.

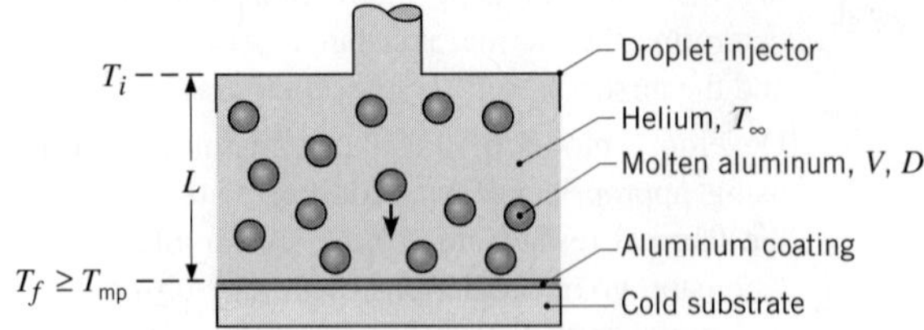

Consider conditions for which droplets with a diameter, velocity, and initial temperature of $D = 500$ μm,

$V = 3$ m/s, and $T_i = 1100$ K, respectively, traverse a stagnant layer of atmospheric helium that is at a temperature of $T_\infty = 300$ K. What is the maximum allowable thickness of the helium layer needed to ensure that the temperature of droplets impacting the substrate is greater than or equal to the melting point of aluminum ($T_f \geq T_{mp} = 933$ K)? Properties of the molten aluminum may be approximated as $\rho = 2500$ kg/m^3, $c = 1200$ J/kg·K, and $k = 200$ W/m·K.

7.84 Tissue engineering involves the development of biological substitutes that restore or improve tissue function. Once manufactured, engineered organs can be implanted and grow within the patient, obviating chronic shortages of natural organs that arise when traditional organ transplant procedures are used. Artificial organ manufacture involves two major steps. First, a porous *scaffold* is fabricated with a specific pore size and pore distribution, as well as overall shape and size. Second, the top surface of the scaffold is seeded with human cells that grow into the pores of the scaffold. The scaffold material is biodegradable and is eventually replaced with healthy tissue. The artificial organ is then ready to be implanted in the patient.

The complex pore shapes, small pore sizes, and unusual organ shapes preclude use of traditional manufacturing methods to fabricate the scaffolds. A method that has been used with success is a *solid freeform fabrication* technique whereby small spherical drops are directed to a substrate. The drops are initially molten and solidify when they impact the room-temperature substrate. By controlling the location of the droplet deposition, complex scaffolds can be built up, one drop at a time. A device similar to that of Problem 7.78 is used to generate uniform, 75-μm-diameter drops at an initial temperature of $T_i = 150$°C. The particles are sent through quiescent air at $T_\infty = 25$°C. The droplet properties are $\rho = 2200$ kg/m^3, $c = 700$ J/kg·K.

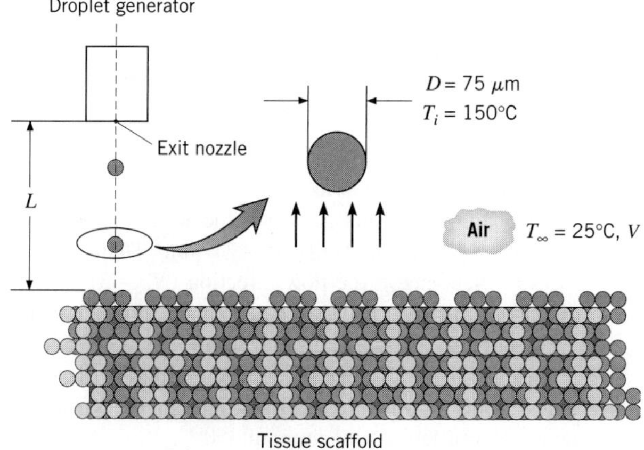

(a) It is desirable for the droplets to exit the nozzle at their terminal velocity. Determine the terminal velocity of the drops.

(b) It is desirable for the droplets to impact the structure at a temperature of $T_2 = 120$°C. What is the required distance between the exit nozzle and the structure, L?

7.85 A spherical thermocouple junction 1.0 mm in diameter is inserted in a combustion chamber to measure the temperature T_∞ of the products of combustion. The hot gases have a velocity of $V = 5$ m/s.

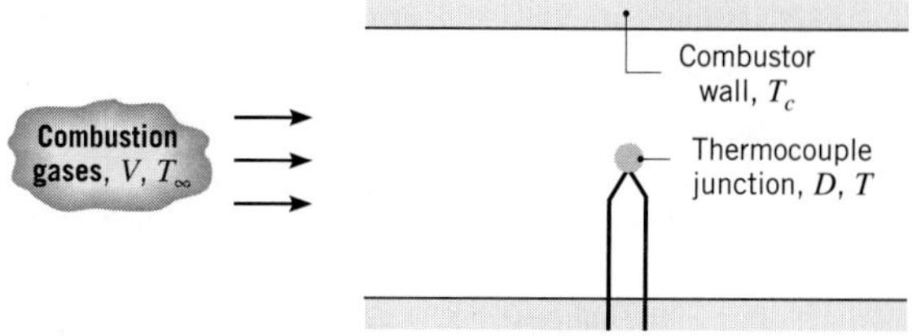

(a) If the thermocouple is at room temperature, T_i, when it is inserted in the chamber, estimate the time required for the temperature difference, $T_\infty - T$, to reach 2% of the initial temperature difference, $T_\infty - T_i$. Neglect radiation and conduction through the leads. Properties of the thermocouple junction are approximated as $k = 100$ W/m·K, $c = 385$ J/kg·K, and $\rho = 8920$ kg/m^3, while those of the combustion gases may be approximated as $k = 0.05$ W/m·K, $\nu = 50 \times 10^{-6}$ m^2/s, and $Pr = 0.69$.

(b) If the thermocouple junction has an emissivity of 0.5 and the cooled walls of the combustor are at $T_c = 400$ K, what is the steady-state temperature of the thermocouple junction if the combustion gases are at 1000 K? Conduction through the lead wires may be neglected.

(c) To determine the influence of the gas velocity on the thermocouple measurement error, compute the steady-state temperature of the thermocouple junction for velocities in the range $1 \leq V \leq 25$ m/s. The emissivity of the junction can be controlled through application of a thin coating. To reduce the measurement error, should the emissivity be increased or decreased? For $V = 5$ m/s, compute the steady-state junction temperature for emissivities in the range $0.1 \leq \varepsilon \leq 1.0$.

7.86 A thermocouple junction is inserted in a large duct to measure the temperature of hot gases flowing through the duct.

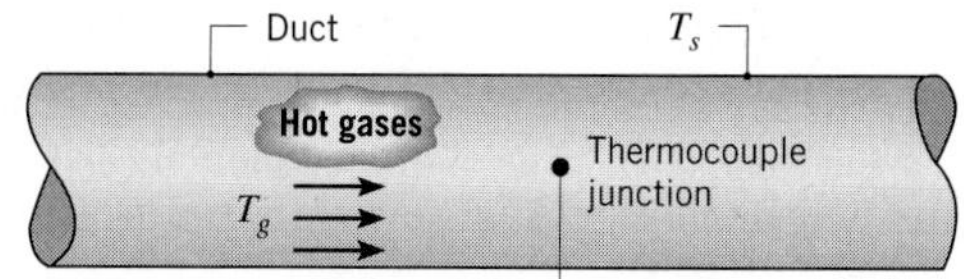

(a) If the duct surface temperature T_s is less than the gas temperature T_g, will the thermocouple sense a temperature that is less than, equal to, or greater than T_g? Justify your answer on the basis of a simple analysis.

(b) A thermocouple junction in the shape of a 2-mm-diameter sphere with a surface emissivity of 0.60 is placed in a gas stream moving at 3 m/s. If the thermocouple senses a temperature of 320°C when the duct surface temperature is 175°C, what is the actual gas temperature? The gas may be assumed to have the properties of air at atmospheric pressure.

(c) How would changes in velocity and emissivity affect the temperature measurement error? Determine the measurement error for velocities in the range $1 \leq V \leq 25$ m/s ($\varepsilon = 0.6$) and for emissivities in the range $0.1 \leq \varepsilon \leq 1.0$ ($V = 3$ m/s).

7.87 Consider temperature measurement in a gas stream using the thermocouple junction described in Problem 7.86 ($D = 2$ mm, $\varepsilon = 0.60$). If the gas velocity and temperature are 3 m/s and 500°C, respectively, what temperature will be indicated by the thermocouple if the duct surface temperature is 200°C? The gas may be assumed to have the properties of atmospheric air. What temperature will be indicated by the thermocouple if the gas pressure is doubled and all other conditions remain the same?

7.88 A silicon chip ($k = 150$ W/m·K, $\rho = 2300$ kg/m^3, $c_p = 700$ J/kg·K), 10 mm on a side and 1 mm thick, is connected to a substrate by solder balls ($k = 40$ W/m·K, $\rho = 10{,}000$ kg/m^3, $c_p = 150$ J/kg·K) of 1-mm diameter, and during an accelerated thermal stress test, the system is exposed to the flow of a dielectric liquid ($k = 0.064$ W/m·K, $\nu = 10^{-6}$ m^2/s, $Pr = 25$). As first approximations, treat the top and bottom surfaces of the chip as flat plates in turbulent, parallel flow and assume the substrate and lower chip surfaces to have a negligible effect on flow over the solder balls. Also assume point contact between the chip and the solder, thereby neglecting heat transfer by conduction between the components.

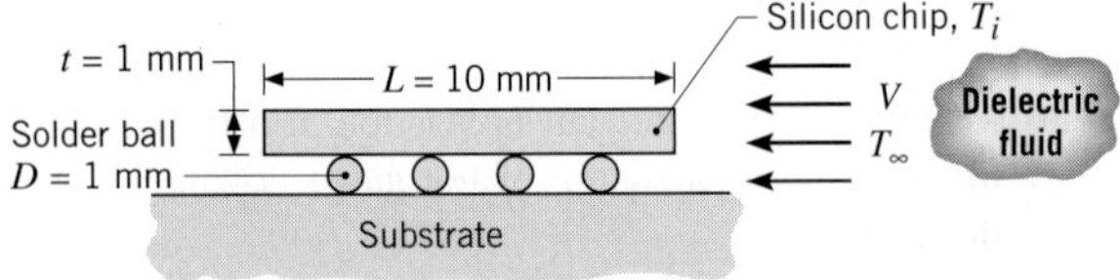

(a) The stress test begins with the components at ambient temperature ($T_i = 20$°C) and proceeds with heating by the fluid at $T_\infty = 80$°C. If the fluid velocity is $V = 0.2$ m/s, estimate the ratio of the time constant of the chip to that of a solder ball. Which component responds more rapidly to the heating process?

(b) The thermal stress acting on the solder joint is proportional to the chip-to-solder temperature difference. What is this temperature difference 0.25 s after the start of heating?

Tube Banks

7.89 Repeat Example 7.6 for a more compact tube bank in which the longitudinal and transverse pitches are $S_L = S_T = 20.5$ mm. All other conditions remain the same.

7.90 A preheater involves the use of condensing steam at 100°C on the inside of a bank of tubes to heat air that enters at 1 atm and 25°C. The air moves at 5 m/s in cross flow over the tubes. Each tube is 1 m long and has an outside diameter of 10 mm. The bank consists of 196 tubes in a square, aligned array for which $S_T = S_L = 15$ mm. What is the total rate of heat transfer to the air? What is the pressure drop associated with the airflow?

7.91 Consider the in-line tube bank of Problem 7.90 ($D = 10$ mm, $L = 1$ m, and $S_T = S_L = 15$ mm), with condensing steam used to heat atmospheric air entering the tube bank at $T_i = 25$°C and $V = 5$ m/s. In this case, however, the desired outlet temperature, not the number of tube rows, is known. What is the minimum value of N_L needed to achieve an outlet temperature of $T_o \geq 75$°C? What is the corresponding pressure drop across the tube bank?

7.92 A tube bank uses an aligned arrangement of 10-mm-diameter tubes with $S_T = S_L = 20$ mm. There are 10 rows of tubes with 50 tubes in each row. Consider an application for which cold water flows through the tubes, maintaining the outer surface temperature at 27°C, while flue gases at 427°C and a velocity of 5 m/s are in cross flow over the tubes. The properties of the flue gas may be approximated as those of atmospheric air at 427°C. What is the total rate of heat transfer per unit length of the tubes in the bank?

7.93 An air duct heater consists of an aligned array of electrical heating elements in which the longitudinal and transverse pitches are $S_L = S_T = 24$ mm. There are 3 rows of elements in the flow direction ($N_L = 3$) and 4 elements per row ($N_T = 4$). Atmospheric air with an upstream velocity of 12 m/s and a temperature of 25°C moves in cross flow over the elements, which have a diameter of 12 mm, a length of 250 mm, and are maintained at a surface temperature of 350°C.

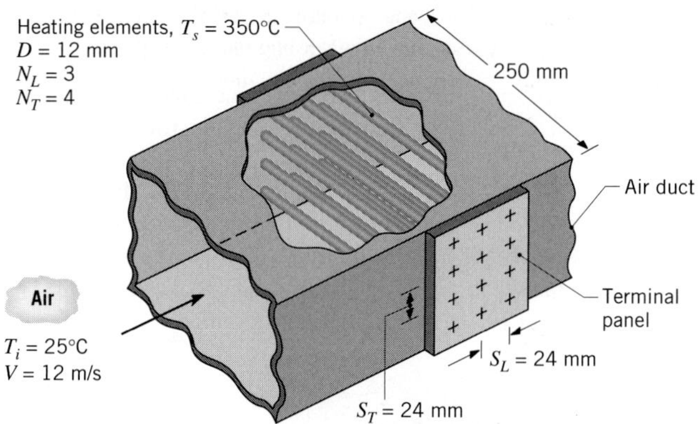

(a) Determine the total heat transfer to the air and the temperature of the air leaving the duct heater.

(b) Determine the pressure drop across the element bank and the fan power requirement.

(c) Compare the average convection coefficient obtained in your analysis with the value for an isolated (single) element. Explain the difference between the results.

(d) What effect would increasing the longitudinal and transverse pitches to 30 mm have on the exit temperature of the air, the total heat rate, and the pressure drop?

7.94 A tube bank uses an aligned arrangement of 30-mm-diameter tubes with $S_T = S_L = 60$ mm and a tube length of 1 m. There are 10 tube rows in the flow direction ($N_L = 10$) and 7 tubes per row ($N_T = 7$). Air with upstream conditions of $T_\infty = 27$°C and $V = 15$ m/s is in cross flow over the tubes, while a tube wall temperature of 100°C is maintained by steam condensation inside the tubes. Determine the temperature of air leaving the tube bank, the pressure drop across the bank, and the fan power requirement.

7.95 Repeat Problem 7.94, but with $N_L = 7$, $N_T = 10$, and $V = 10.5$ m/s.

7.96 Electrical components mounted to each of two isothermal plates are cooled by passing atmospheric air between the plates, and an in-line array of aluminum pin fins is used to enhance heat transfer to the air.

The pins are of diameter $D = 2$ mm, length $L = 100$ mm, and thermal conductivity $k = 240$ W/m·K. The longitudinal and transverse pitches are $S_L = S_T = 4$ mm, with a square array of 625 pins ($N_T = N_L = 25$) mounted to square plates that are each of width $W = 100$ mm on a side. Air enters the pin array with a velocity of 10 m/s and a temperature of 300 K.

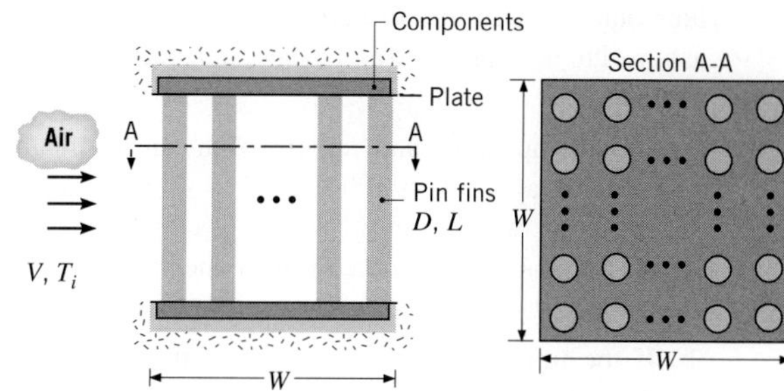

(a) Evaluating air properties at 300 K, estimate the average convection coefficient for the array of pin fins.

(b) Assuming a uniform convection coefficient over all heat transfer surfaces (plates and pins), use the result of part (a) to determine the air outlet temperature and total heat rate when the plates are maintained at 350 K. *Hint:* The air outlet temperature is governed by an exponential relation of the form $[(T_s - T_o)/(T_s - T_i)] = \exp[-(\bar{h}A_t\eta_o)/\dot{m}c_p]$, where $\dot{m} = \rho V L N_T S_T$ is the mass flow rate of air passing through the array, A_t is the total heat transfer surface area (plates and fins), and η_o is the overall surface efficiency defined by Equation 3.107.

7.97 Consider the chip cooling scheme of Problem 3.146, but with an insulated top wall placed at the pin tips to force airflow across the pin array. Air enters the array at 20°C and with a velocity V that may be varied but cannot exceed 10 m/s due to pressure drop considerations. The pin fin geometry, which includes the number of pins in the $N \times N$ square array, as well as the pin diameter D_p and length L_p, may also be varied, subject to the constraint that the product ND_p not exceed 9 mm. Neglecting heat transfer through the board, assess the effect of changes in air velocity, and hence h_o, as well as pin fin geometry, on the air outlet temperature and the chip heat rate, if the remaining conditions of Problems 3.146 and 3.27, including a maximum allowable chip temperature of 75°C, remain in effect. Recommend design and operating conditions for which chip cooling is enhanced. *Hint:* The air outlet temperature is governed by a relation of the form $[(T_s - T_o)/(T_s - T_i)] = \exp[-(\bar{h}A_t\eta_o)/\dot{m}c_p]$, where $\dot{m}$ is the mass flow rate of air passing through the array, A_t is the total heat transfer surface area (chip and pins), and η_o is the overall surface efficiency defined by Equation 3.107.

7.98 An air-cooled steam condenser is operated with air in cross flow over a square, in-line array of 400 tubes ($N_L = N_T = 20$), with an outside tube diameter of 20 mm and longitudinal and transverse pitches of $S_L = 60$ mm and $S_T = 30$ mm, respectively. Saturated steam at a pressure of 2.455 bars enters the tubes, and a uniform tube outer surface temperature of $T_s = 390$ K may be assumed to be maintained as condensation occurs within the tubes.

(a) If the temperature and velocity of the air upstream of the array are $T_i = 300$ K and $V = 4$ m/s, what is the temperature T_o of the air that leaves the array? As a first approximation, evaluate the properties of air at 300 K.

(b) If the tubes are 2 m long, what is the total heat transfer rate for the array? What is the rate at which steam is condensed in kg/s?

(c) Assess the effect of increasing N_L by a factor of 2, while reducing S_L to 30 mm. For this configuration, explore the effect of changes in the air velocity.

Impinging Jets

7.99 Heating and cooling with *miniature* impinging jets has been proposed for numerous applications. For a single round jet, determine the minimum jet diameter for which Equation 7.62 may be applied for air at atmospheric pressure (a) at $T_e = 0°C$ and (b) at $T_e = 500°C$.

7.100 A circular transistor of 10-mm diameter is cooled by impingement of an air jet exiting a 2-mm-diameter round nozzle with a velocity of 20 m/s and a temperature of 15°C. The jet exit and the exposed surface of the transistor are separated by a distance of 10 mm.

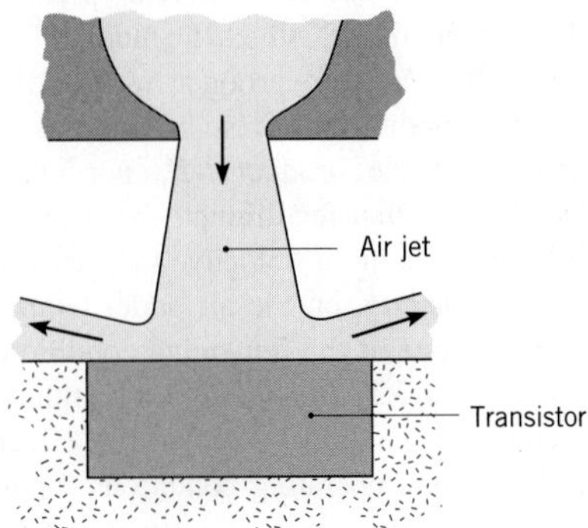

If the transistor is well insulated at all but its exposed surface and the surface temperature is not to exceed 85°C, what is the transistor's maximum allowable operating power?

7.101 A long rectangular plate of AISI 304 stainless steel is initially at 1200 K and is cooled by an array of slot jets (see Figure 7.17). The nozzle width and pitch are $W = 10$ mm and $S = 100$ mm, respectively, and the nozzle-to-plate separation is $H = 200$ mm. The plate thickness and width are $t = 8$ mm and $L = 1$ m, respectively. If air exits the nozzles at a temperature of 400 K and a velocity of 30 m/s, what is the initial cooling rate of the plate?

7.102 A cryogenic probe is used to treat cancerous skin tissue. The probe consists of a single round jet of diameter $D_e = 2$ mm that issues from a nozzle concentrically situated within a larger, enclosed cylindrical tube of outer diameter $D_o = 15$ mm. The wall thickness of the AISI 302 stainless steel probe is $t = 2$ mm, and the separation distance between the nozzle and the inner surface of the probe is $H = 5$ mm.

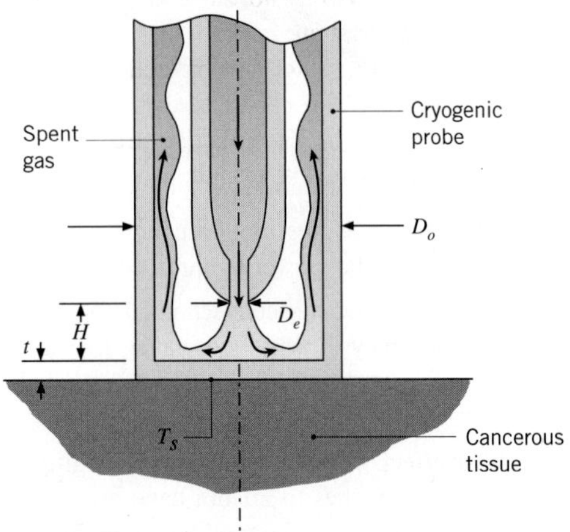

Assuming the cancerous skin tissue to be a semi-infinite medium with $k_c = 0.20$ W/m·K and $T_c = 37°C$ far from the probe location, determine the surface temperature T_s. Neglect the contact resistance between the probe and the tissue. Cold nitrogen exits the jet at $T_e = 100$ K, $V_e = 20$ m/s. *Hint:* Due to the probe walls, the jet is confined and behaves as if it were one in an array such as in Figure 7.18*c*.

7.103 Air at 10 m/s and 15°C is used to cool a square hot molded plastic plate 0.5 m to a side having a surface temperature of 140°C. To increase the throughput of the production process, it is proposed to cool the plate using an array of slotted nozzles with width and pitch of 4 mm and 56 mm, respectively, and a nozzle-to-plate separation of 40 mm. The air exits the nozzle at a temperature of 15°C and a velocity of 10 m/s.

(a) Determine the improvement in cooling rate that can be achieved using the slotted nozzle arrangement in lieu of turbulated air at 10 m/s and 15°C in parallel flow over the plate.

(b) Would the heat rates for both arrangements change significantly if the air velocities were increased by a factor of 2?

(c) What is the air mass rate requirement for the slotted nozzle arrangement?

7.104 Consider Problem 7.103, in which the improvement in performance of slot-jet cooling over parallel-flow cooling was demonstrated. Design an optimal round nozzle array, using the same air jet velocity and temperature, 10 m/s and 15°C, respectively, and compare the cooling rates and supply air requirements. Discuss the features associated with each of the three methods relevant to selecting one for this application of cooling the plastic part.

7.105 Consider the plasma spraying process of Problems 5.25 and 7.82. For a nozzle exit diameter of $D =$ 10 mm and a substrate radius of $r = 25$ mm, *estimate* the rate of heat transfer by convection q_{conv} from the argon plasma to the substrate, if the substrate temperature is maintained at 300 K. Energy transfer to the substrate is also associated with the release of latent heat q_{lat}, which occurs during solidification of the impacted molten droplets. If the mass rate of droplet impingement is $\dot{m}_p = 0.02$ kg/s·m^2, estimate the rate of latent heat release.

7.106 You have been asked to determine the feasibility of using an impinging jet in a soldering operation for electronic assemblies. The schematic illustrates the use of a single, round nozzle to direct high-velocity, hot air to a location where a *surface mount* joint is to be formed.

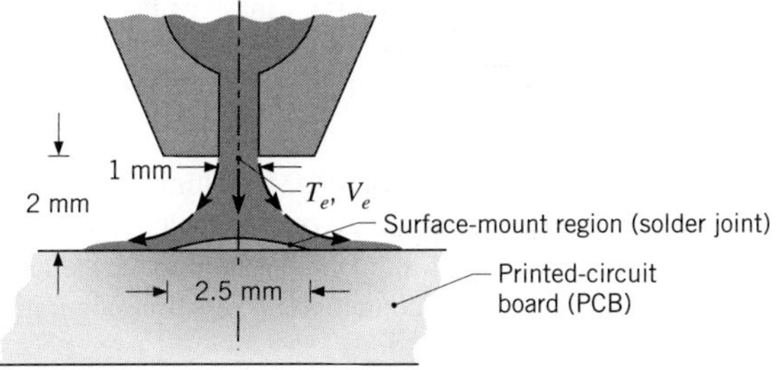

For your study, consider a round nozzle with a diameter of 1 mm located a distance of 2 mm from the region of the surface mount, which has a diameter of 2.5 mm.

(a) For an air jet velocity of 70 m/s and a temperature of 500°C, estimate the average convection coefficient over the area of the surface mount.

(b) Assume that the surface mount region on the printed circuit board (PCB) can be modeled as a semi-infinite medium, which is initially at a uniform temperature of 25°C and suddenly experiences convective heating by the jet. Estimate the time required for the surface to reach 183°C. The thermophysical properties of a typical solder are $\rho = 8333$ kg/m^3, $c_p = 188$ J/kg·K, and $k =$ 51 W/m·K.

(c) For each of three air jet temperatures of 500, 600, and 700°C, calculate and plot the surface temperature as a function of time for $0 \leq t \leq 150$ s. On this plot, identify important temperature limits for the soldering process: the lower limit corresponding to the solder's eutectic temperature, $T_{\text{sol}} =$ 183°C, and the upper limit corresponding to the glass transition temperature, $T_{\text{gl}} = 250$°C, at which the PCB becomes plastic. Comment on the outcome of your study, the appropriateness of the assumptions, and the feasibility of using the jet for a soldering application.

Packed Beds

7.107 Consider the packed bed of aluminum spheres described in Problem 5.12 under conditions for which the bed is charged by hot air with an inlet velocity of $V = 1$ m/s and temperature of $T_{g,i} = 300$°C, but for which the convection coefficient is not prescribed. If the porosity of the bed is $\varepsilon = 0.40$ and the initial temperature of the spheres is $T_i = 25$°C, how long does it take a sphere near the inlet of the bed to accumulate 90% of its maximum possible energy?

7.108 The use of rock pile thermal energy storage systems has been considered for solar energy and industrial process heat applications. A particular system involves a cylindrical container, 2 m long by 1 m in diameter, in which nearly spherical rocks of 0.03-m diameter are packed. The bed has a void space of 0.42, and the density and specific heat of the rock are $\rho = 2300$ kg/m^3 and $c_p = 879$ J/kg·K, respectively. Consider conditions for which atmospheric air is supplied to the rock pile at a steady flow rate of 1 kg/s and a temperature of 90°C. The air flows in the axial direction through the container. If the rock is at a temperature of 25°C, what is the total rate of heat transfer from the air to the rock pile?

7.109 The cylindrical chamber of a *pebble bed nuclear reactor* is of length $L = 10$ m, and diameter $D = 3$ m. The chamber is filled with spherical uranium oxide pellets of core diameter $D_p = 50$ mm. Each pellet generates thermal energy in its core at a rate of $\dot{E}_g$ and is coated with a layer of non-heat-generating graphite, which is of uniform thickness $\delta = 5$ mm, to form a pebble. The uranium oxide and graphite each have a thermal conductivity of 2 W/m·K. The packed bed has a porosity of $\varepsilon = 0.4$. Pressurized helium at 40 bars is used to absorb the thermal energy from the pebbles. The helium enters the packed bed at $T_i = 450$°C with a velocity of 3.2 m/s. The properties of the helium may be assumed to be $c_p = 5193$ J/kg·K, $k = 0.3355$ W/m·K, $\rho = 2.1676$ kg/m^3, $\mu = 4.214 \times 10^{-5}$ kg/s·m, $Pr = 0.654$.

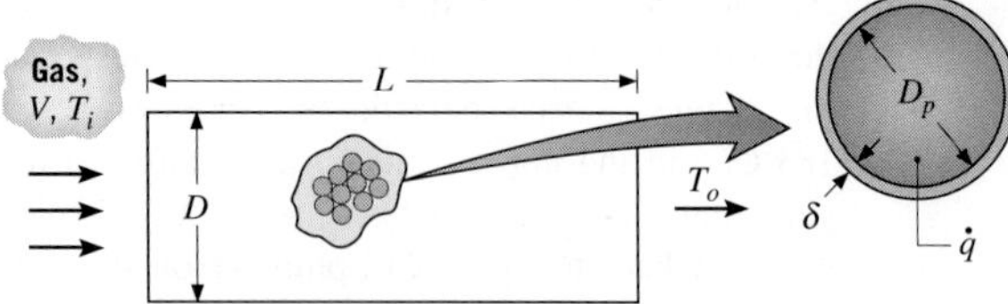

(a) For a desired overall thermal energy transfer rate of $q = 125$ MW, determine the mean outlet temperature of the helium leaving the bed, T_o, and the amount of thermal energy generated by each pellet, $\dot{E}_g$.

(b) The amount of energy generated by the fuel decreases if a maximum operating temperature of approximately 2100°C is exceeded. Determine the maximum internal temperature of the hottest pellet in the packed bed. For Reynolds numbers in the range $4000 \leq Re_D \leq 10{,}000$, Equation 7.72 may be replaced by $\varepsilon \bar{j}_H = 2.876\, Re_D^{-1} + 0.3023\, Re_D^{-0.35}$.

7.110 *Latent heat capsules* consist of a thin-walled spherical shell within which a solid-liquid, phase-change material (PCM) of melting point T_{mp} and latent heat of fusion h_{sf} is enclosed. As shown schematically, the capsules may be packed in a cylindrical vessel through which there is fluid flow. If the PCM is in its solid state and $T_{mp} < T_i$, heat is transferred from the fluid to the capsules and latent energy is stored in the PCM as it melts. Conversely, if the PCM is a liquid and $T_{mp} > T_i$, energy is released from the PCM as it freezes and heat is transferred to the fluid. In either situation, all of the capsules within the packed bed would remain at T_{mp} through much of the phase change process, in which case the fluid outlet temperature would remain at a fixed value T_o.

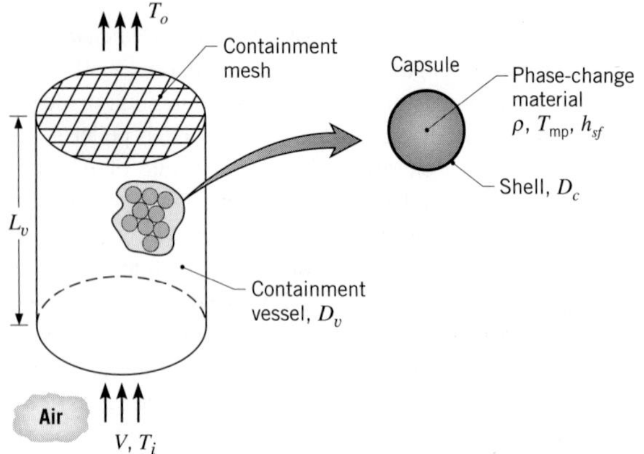

Consider an application for which air at atmospheric pressure is chilled by passing it through a packed bed ($\varepsilon = 0.5$) of capsules ($D_c = 50$ mm) containing an organic compound with a melting point of $T_{mp} = 4$°C. The air enters a cylindrical vessel ($L_v = D_v = 0.40$ m) at $T_i = 25$°C and $V = 1.0$ m/s.

(a) If the PCM in each capsule is in the solid state at T_{mp} as melting occurs within the capsule, what is the outlet temperature of the air? If the density and latent heat of fusion of the PCM are $\rho = 1200$ kg/m^3 and $h_{sf} = 165$ kJ/kg, what is the mass rate (kg/s) at which the PCM is converted from solid to liquid in the vessel?

(b) Explore the effect of the inlet air velocity and capsule diameter on the outlet temperature.

(c) At what location in the vessel will complete melting of the PCM in a capsule first occur? Once complete melting begins to occur, how will the outlet temperature vary with time and what is its asymptotic value?

7.111 The porosity of a packed bed can be decreased by vibrating the containment vessel as the vessel is filled with the particles. The vibration promotes particle settling.

(a) Consider the air chilling process of Problem 7.110a. Determine the outlet air temperature T_o and mass rate at which the PCM is melted for $\varepsilon = 0.30$. Assume the total mass of PCM and the mass flow rate of air are unchanged. The length of the containment vessel L_v is decreased to compensate for the reduced porosity.

(b) Determine T_o and the PCM melting rate for the case where the diameter of the containment vessel D_v is decreased to compensate for the reduced porosity. Which containment vessel configuration is preferred?

7.112 Consider the packed bed ($\varepsilon = 0.5$) of latent heat capsules ($D_c = 50$ mm) described in Problem 7.110, but now for an application in which ambient air is to be heated by passing it through the bed. In this case the capsules contain an organic compound with a melting point of $T_{mp} = 50$°C, and the air enters the vessel ($L_v = D_v = 0.40$ m) at $T_i = 20$°C and $V = 1.0$ m/s.

(a) If the PCM in each capsule is in the liquid state at T_{mp} as solidification occurs within the capsule, what is the outlet temperature of the air? If the density and latent heat of fusion of the PCM are $\rho = 900$ kg/m^3 and $h_{sf} = 200$ kJ/kg, what is the mass rate (kg/s) at which the PCM is converted from liquid to solid in the vessel?

(b) Explore the effect of the inlet air velocity and capsule diameter on the outlet temperature.

(c) At what location in the vessel will complete freezing of the PCM in a capsule first occur? Once complete freezing begins to occur, how will the outlet temperature vary with time and what is its asymptotic value?

7.113 Packed beds of spherical particles can be *sintered* at high temperature to form permeable, rigid foams. A foam sheet of thickness $t = 10$ mm is comprised of sintered bronze spheres, each of diameter $D = 0.6$ mm. The metal foam has a porosity of $\varepsilon = 0.25$, and the foam sheet fills the cross section of an $L = 40$ mm $\times$ $W = 40$ mm wind tunnel. The upper and lower surfaces of the foam are at temperatures $T_s = 80°C$, and the two other foam edges (the front edge shown in the schematic and the corresponding back edge) are insulated. Air flows in the wind tunnel at an upstream temperature and velocity of $T_i = 20°C$ and $V = 10$ m/s, respectively.

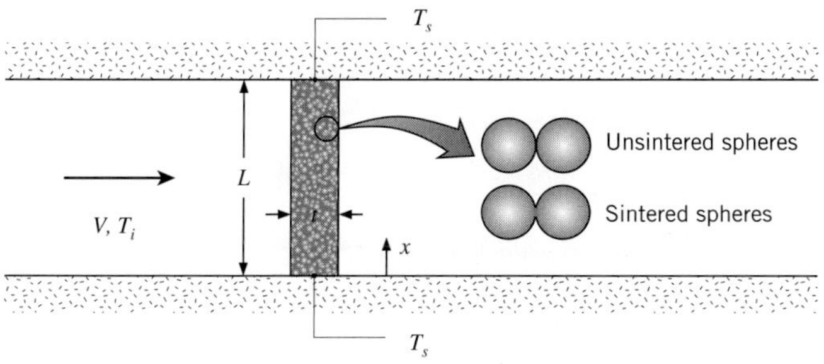

(a) Assuming the foam is at a uniform temperature T_s, *estimate* the convection heat transfer rate to the air. Do you expect the actual heat transfer rate to be equal to, less than, or greater than your estimated value?

(b) Assuming one-dimensional conduction in the x-direction, use an extended surface analysis to *estimate* the heat transfer rate to the air. To do so, show that the *effective* perimeter associated with Equation 3.70 is $P_{\text{eff}} = A_{p,t}/L$. Determine the effective thermal conductivity of the foam k_{eff} by using Equation 3.25. Do you expect the actual heat transfer rate to be equal to, less than, or greater than your estimated value?

CHAPTER 8

Internal Flow

Having acquired the means to compute convection transfer rates for external flow, we now consider the convection transfer problem for *internal flow*. Recall that an external flow is one for which boundary layer development on a surface is allowed to continue without external constraints, as for the flat plate of Figure 6.4. In contrast, an internal flow, such as flow in a pipe, is one for which the fluid is *confined* by a surface. Hence the boundary layer is unable to develop without eventually being constrained. The internal flow configuration represents a convenient geometry for heating and cooling fluids used in chemical processing, environmental control, and energy conversion technologies.

Our objectives are to develop an appreciation for the physical phenomena associated with internal flow and to obtain convection coefficients for flow conditions of practical importance. As in Chapter 7, we will restrict attention to problems of low-speed, forced convection with no phase change occurring in the fluid. We begin by considering velocity (hydrodynamic) effects pertinent to internal flows, focusing on certain unique features of boundary layer development. Thermal boundary layer effects are considered next, and an overall energy balance is applied to determine fluid temperature variations in the flow direction. Finally, correlations for estimating the convection heat transfer coefficient are presented for a variety of internal flow conditions.

8.1 *Hydrodynamic Considerations*

When considering external flow, it is necessary to ask only whether the flow is laminar or turbulent. However, for an internal flow we must also be concerned with the existence of *entrance* and *fully developed* regions.

8.1.1 Flow Conditions

Consider laminar flow in a circular tube of radius r_o (Figure 8.1), where fluid enters the tube with a uniform velocity. We know that when the fluid makes contact with the surface, viscous effects become important, and a boundary layer develops with increasing x. This development occurs at the expense of a shrinking inviscid flow region and concludes with boundary layer merger at the centerline. Following this merger, viscous effects extend over

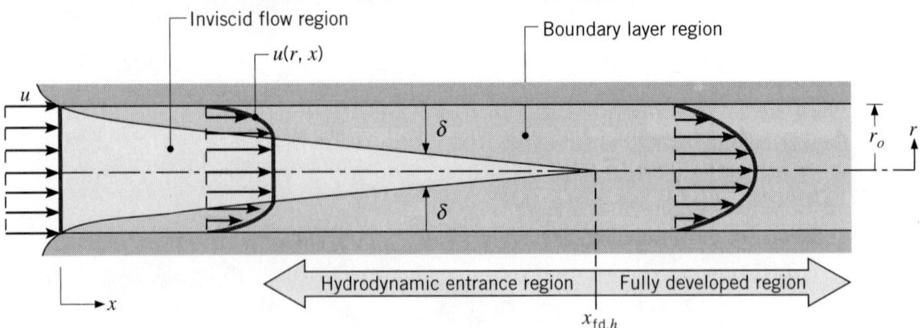

FIGURE 8.1 Laminar, hydrodynamic boundary layer development in a circular tube.

the entire cross section and the velocity profile no longer changes with increasing x. The flow is then said to be *fully developed,* and the distance from the entrance at which this condition is achieved is termed the *hydrodynamic entry length,* $x_{\mathrm{fd},h}$. As shown in Figure 8.1, the *fully developed velocity profile* is parabolic for laminar flow in a circular tube. For turbulent flow, the profile is *flatter* due to turbulent mixing in the radial direction.

When dealing with internal flows, it is important to be cognizant of the extent of the entry region, which depends on whether the flow is laminar or turbulent. The Reynolds number for flow in a circular tube is defined as

$$Re_D \equiv \frac{\rho u_m D}{\mu} = \frac{u_m D}{\nu} \tag{8.1}$$

where u_m is the mean fluid velocity over the tube cross section and D is the tube diameter. In a fully developed flow, the critical Reynolds number corresponding to the *onset* of turbulence is

$$Re_{D,c} \approx 2300 \tag{8.2}$$

although much larger Reynolds numbers ($Re_D \approx 10{,}000$) are needed to achieve fully turbulent conditions. The transition to turbulence is likely to begin in the developing boundary layer of the entrance region.

For laminar flow ($Re_D \lesssim 2300$), the hydrodynamic entry length may be obtained from an expression of the form [1]

$$\left(\frac{x_{\mathrm{fd},h}}{D}\right)_{\mathrm{lam}} \approx 0.05\, Re_D \tag{8.3}$$

This expression is based on the presumption that fluid enters the tube from a rounded converging nozzle and is hence characterized by a nearly uniform velocity profile at the entrance (Figure 8.1). Although there is no satisfactory general expression for the entry length in turbulent flow, we know that it is approximately independent of Reynolds number and that, as a first approximation [2],

$$10 \lesssim \left(\frac{x_{\mathrm{fd},h}}{D}\right)_{\mathrm{turb}} \lesssim 60 \tag{8.4}$$

For the purposes of this text, we shall assume fully developed turbulent flow for $(x/D) > 10$.

8.1.2 The Mean Velocity

Because the velocity varies over the cross section and there is no well-defined free stream, it is necessary to work with a mean velocity u_m when dealing with internal flows. This velocity is defined such that, when multiplied by the fluid density ρ and the cross-sectional area of the tube A_c, it provides the rate of mass flow through the tube. Hence

$$\dot{m} = \rho u_m A_c \tag{8.5}$$

For steady, incompressible flow in a tube of uniform cross-sectional area, $\dot{m}$ and u_m are constants independent of x. From Equations 8.1 and 8.5 it is evident that, for flow in a *circular tube* ($A_c = \pi D^2/4$), the Reynolds number reduces to

$$Re_D = \frac{4\dot{m}}{\pi D \mu} \tag{8.6}$$

Since the mass flow rate may also be expressed as the integral of the mass flux (ρu) over the cross section

$$\dot{m} = \int_{A_c} \rho u(r, x)\, dA_c \tag{8.7}$$

it follows that, for *incompressible* flow in a *circular* tube,

$$u_m = \frac{\int_{A_c} \rho u(r, x)\, dA_c}{\rho A_c} = \frac{2\pi\rho}{\rho \pi r_o^2} \int_0^{r_o} u(r, x) r\, dr = \frac{2}{r_o^2} \int_0^{r_o} u(r, x) r\, dr \tag{8.8}$$

The foregoing expression may be used to determine u_m at any axial location x from knowledge of the velocity profile $u(r)$ at that location.

8.1.3 Velocity Profile in the Fully Developed Region

The form of the velocity profile may readily be determined for the *laminar flow* of an *incompressible, constant property fluid* in the *fully developed region* of a *circular tube.* An important feature of hydrodynamic conditions in the fully developed region is that both the radial velocity component v and the gradient of the axial velocity component ($\partial u/\partial x$) are everywhere zero.

$$v = 0 \quad \text{and} \quad \left(\frac{\partial u}{\partial x}\right) = 0 \tag{8.9}$$

Hence the axial velocity component depends only on r, $u(x, r) = u(r)$.

The radial dependence of the axial velocity may be obtained by solving the appropriate form of the x-momentum equation. This form is determined by first recognizing that, for the conditions of Equation 8.9, the net momentum flux is everywhere zero in the fully developed region. Hence the momentum conservation requirement reduces to a simple balance between shear and pressure forces in the flow. For the annular differential element of Figure 8.2, this force balance may be expressed as

$$\tau_r(2\pi r\, dx) - \left\{\tau_r\,(2\pi r\, dx) + \frac{d}{dr}[\tau_r\,(2\pi r\, dx)]\, dx\right\}$$
$$+\, p(2\pi r\, dr) - \left\{p(2\pi r\, dr) + \frac{d}{dx}[p(2\pi r\, dr)]\, dx\right\} = 0$$

which reduces to

$$-\frac{d}{dr}(r\tau_r) = r\frac{dp}{dx} \tag{8.10}$$

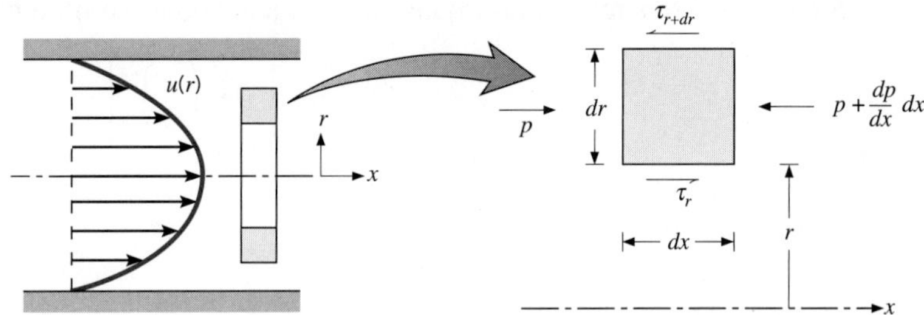

FIGURE 8.2 Force balance on a differential element for laminar, fully developed flow in a circular tube.

With $y = r_o - r$, Newton's law of viscosity, Equation 6S.10, assumes the form

$$\tau_r = -\mu \frac{du}{dr} \tag{8.11}$$

and Equation 8.10 becomes

$$\frac{\mu}{r}\frac{d}{dr}\left(r\frac{du}{dr}\right) = \frac{dp}{dx} \tag{8.12}$$

Since the axial pressure gradient is independent of r, Equation 8.12 may be solved by integrating twice to obtain

$$r\frac{du}{dr} = \frac{1}{\mu}\left(\frac{dp}{dx}\right)\frac{r^2}{2} + C_1$$

and

$$u(r) = \frac{1}{\mu}\left(\frac{dp}{dx}\right)\frac{r^2}{4} + C_1 \ln r + C_2$$

The integration constants may be determined by invoking the boundary conditions

$$u(r_o) = 0 \qquad \text{and} \qquad \left.\frac{\partial u}{\partial r}\right|_{r=0} = 0$$

which, respectively, impose the requirements of zero slip at the tube surface and radial symmetry about the centerline. It is a simple matter to evaluate the constants, and it follows that

$$u(r) = -\frac{1}{4\mu}\left(\frac{dp}{dx}\right) r_o^2 \left[1 - \left(\frac{r}{r_o}\right)^2\right] \tag{8.13}$$

Hence the fully developed velocity profile is *parabolic*, as illustrated in Figure 8.2. Note that the pressure gradient must always be negative.

The foregoing result may be used to determine the mean velocity of the flow. Substituting Equation 8.13 into Equation 8.8 and integrating, we obtain

$$u_m = -\frac{r_o^2}{8\mu}\frac{dp}{dx} \tag{8.14}$$

Substituting this result into Equation 8.13, the velocity profile is then

$$\frac{u(r)}{u_m} = 2\left[1 - \left(\frac{r}{r_o}\right)^2\right] \tag{8.15}$$

Since u_m can be computed from knowledge of the mass flow rate, Equation 8.14 can be used to determine the pressure gradient.

8.1.4 Pressure Gradient and Friction Factor in Fully Developed Flow

The engineer is frequently interested in the pressure drop needed to sustain an internal flow because this parameter determines pump or fan power requirements. To determine the pressure drop, it is convenient to work with the *Moody* (or Darcy) *friction factor,* which is a dimensionless parameter defined as

$$f \equiv \frac{-(dp/dx)D}{\rho u_m^2/2} \tag{8.16}$$

This quantity is not to be confused with the *friction coefficient,* sometimes called the Fanning friction factor, which is defined as

$$C_f \equiv \frac{\tau_s}{\rho u_m^2/2} \tag{8.17}$$

Since $\tau_s = -\mu(du/dr)_{r=r_o}$, it follows from Equation 8.13 that

$$C_f = \frac{f}{4} \tag{8.18}$$

Substituting Equations 8.1 and 8.14 into 8.16, it follows that, for fully developed laminar flow,

$$f = \frac{64}{Re_D} \tag{8.19}$$

For fully developed turbulent flow, the analysis is much more complicated, and we must ultimately rely on experimental results. In addition to depending on the Reynolds number, the friction factor is a function of the tube surface condition and increases with surface roughness e. Measured friction factors covering a wide range of conditions have been correlated by Colebrook [3, 4] and are described by the transcendental expression

$$\frac{1}{\sqrt{f}} = -2.0 \log\left[\frac{e/D}{3.7} + \frac{2.51}{Re_D\sqrt{f}}\right] \tag{8.20}$$

A correlation for the smooth surface condition that encompasses a large Reynolds number range has been developed by Petukhov [5] and is of the form

$$f = (0.790 \ln Re_D - 1.64)^{-2} \qquad 3000 \lesssim Re_D \lesssim 5 \times 10^6 \tag{8.21}$$

Equations 8.19 and 8.20 are plotted in the *Moody diagram* of Figure 8.3.

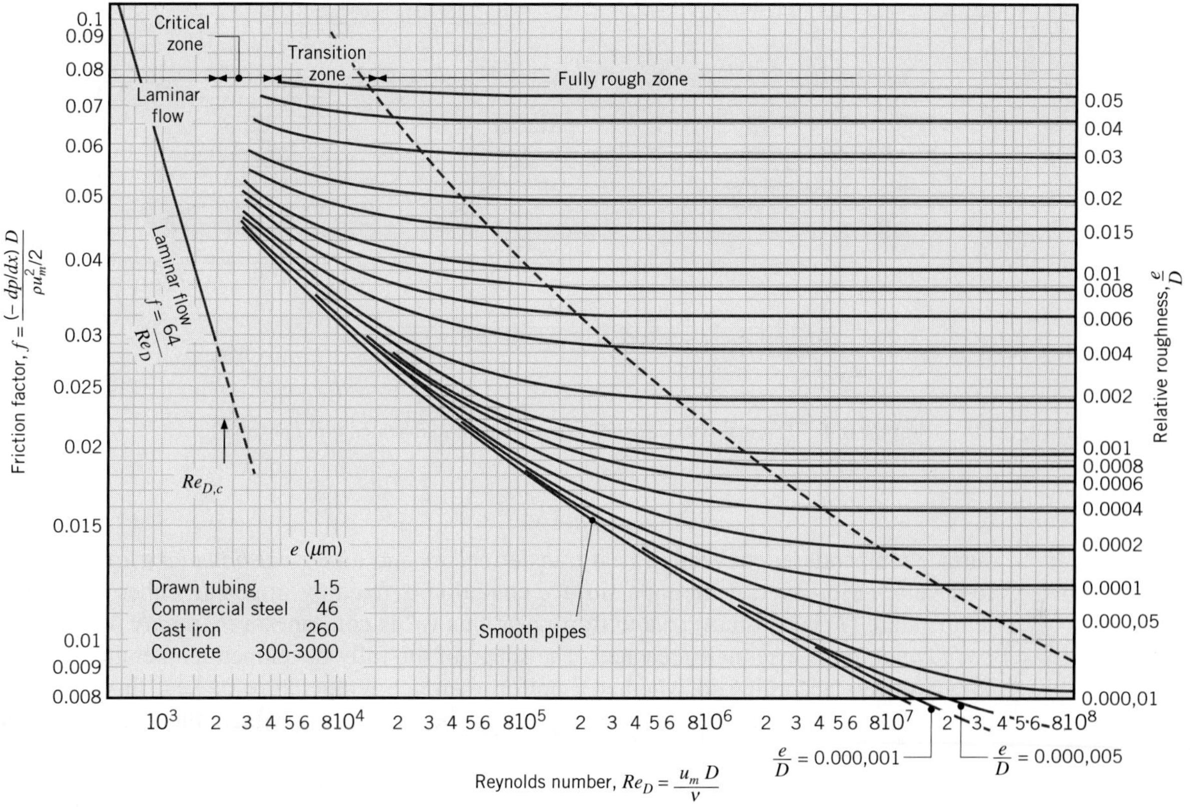

FIGURE 8.3 Friction factor for fully developed flow in a circular tube [6]. Used with permission.

Note that f, hence dp/dx, is a constant in the fully developed region. From Equation 8.16 the pressure drop $\Delta p = p_1 - p_2$ associated with fully developed flow from the axial position x_1 to x_2 may then be expressed as

$$\Delta p = -\int_{p_1}^{p_2} dp = f\frac{\rho u_m^2}{2D}\int_{x_1}^{x_2} dx = f\frac{\rho u_m^2}{2D}(x_2 - x_1) \tag{8.22a}$$

where f is obtained from Figure 8.3 or from Equation 8.19 for laminar flow and from Equation 8.20 or 8.21 for turbulent flow. The pump or fan power required to overcome the resistance to flow associated with this pressure drop may be expressed as

$$P = (\Delta p)\dot{\forall} \tag{8.22b}$$

where the volumetric flow rate $\dot{\forall}$ may, in turn, be expressed as $\dot{\forall} = \dot{m}/\rho$ for an incompressible fluid.

8.2 Thermal Considerations

Having reviewed the fluid mechanics of internal flow, we now consider thermal effects. If fluid enters the tube of Figure 8.4 at a uniform temperature $T(r, 0)$ that is less than the

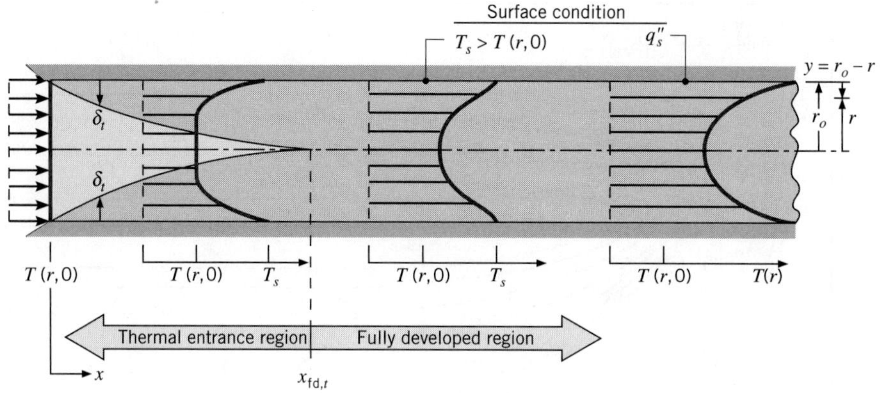

FIGURE 8.4 Thermal boundary layer development in a heated circular tube.

surface temperature, convection heat transfer occurs and a *thermal boundary layer* begins to develop. Moreover, if the tube *surface* condition is fixed by imposing either a uniform temperature (T_s is constant) or a uniform heat flux (q''_s is constant), a *thermally fully developed condition* is eventually reached. The shape of the fully developed temperature profile $T(r, x)$ differs according to whether a uniform surface temperature or heat flux is maintained. For both surface conditions, however, the amount by which fluid temperatures exceed the entrance temperature increases with increasing x.

For laminar flow the *thermal entry length* may be expressed as [2]

$$\left(\frac{x_{\text{fd},t}}{D}\right)_{\text{lam}} \approx 0.05\, Re_D\, Pr \tag{8.23}$$

Comparing Equations 8.3 and 8.23, it is evident that, if $Pr > 1$, the hydrodynamic boundary layer develops more rapidly than the thermal boundary layer ($x_{\text{fd},h} < x_{\text{fd},t}$), while the inverse is true for $Pr < 1$. For large Prandtl number fluids such as oils, $x_{\text{fd},h}$ is much smaller than $x_{\text{fd},t}$ and it is reasonable to assume a fully developed velocity profile throughout the thermal entry region. In contrast, for turbulent flow, conditions are nearly independent of Prandtl number, and to a first approximation, we shall assume $(x_{\text{fd},t}/D) = 10$.

Thermal conditions in the fully developed region are characterized by several interesting and useful features. Before we can consider these features (Section 8.2.3), however, it is necessary to introduce the concept of a mean temperature and the appropriate form of Newton's law of cooling.

8.2.1 The Mean Temperature

Just as the absence of a free stream velocity requires use of a mean velocity to describe an internal flow, the absence of a fixed free stream temperature necessitates using a *mean* (or *bulk*) *temperature*. To provide a definition of the mean temperature, we begin by returning to Equation 1.12e:

$$q = \dot{m} c_p (T_{\text{out}} - T_{\text{in}}) \tag{1.12e}$$

Recall that the terms on the right-hand side represent the thermal energy for an incompressible liquid or the enthalpy (thermal energy plus flow work) for an ideal gas, which is carried by the fluid. In developing this equation, it was implicitly assumed that the temperature was uniform across the inlet and outlet cross-sectional areas. In reality, this is not true if convection heat transfer occurs, and we *define* the mean temperature so that the term $\dot{m}c_pT_m$ is equal to the true rate of thermal energy (or enthalpy) advection integrated over the cross section. This true advection rate may be obtained by integrating the product of mass flux (ρu) and the thermal energy (or enthalpy) per unit mass, c_pT, over the cross section. Therefore, we define T_m from

$$\dot{m}c_pT_m = \int_{A_c} \rho u c_p T dA_c \tag{8.24}$$

or

$$T_m = \frac{\int_{A_c} \rho u c_p T dA_c}{\dot{m}c_p} \tag{8.25}$$

For flow in a circular tube with constant ρ and c_p, it follows from Equations 8.5 and 8.25 that

$$T_m = \frac{2}{u_m r_o^2}\int_0^{r_o} uTr\,dr \tag{8.26}$$

It is important to note that, when multiplied by the mass flow rate and the specific heat, T_m provides the rate at which thermal energy (or enthalpy) is advected with the fluid as it moves along the tube.

8.2.2 Newton's Law of Cooling

The mean temperature T_m is a convenient reference temperature for internal flows, playing much the same role as the free stream temperature T_∞ for external flows. Accordingly, Newton's law of cooling may be expressed as

$$q_s'' = h(T_s - T_m) \tag{8.27}$$

where h is the *local* convection heat transfer coefficient. However, there is an essential difference between T_m and T_∞. Whereas T_∞ is constant in the flow direction, T_m must vary in this direction. That is, dT_m/dx is never zero if heat transfer is occurring. The value of T_m increases with x if heat transfer is from the surface to the fluid ($T_s > T_m$); it decreases with x if the opposite is true ($T_s < T_m$).

8.2.3 Fully Developed Conditions

Since the existence of convection heat transfer between the surface and the fluid dictates that the fluid temperature must continue to change with x, one might legitimately question whether fully developed thermal conditions can ever be reached. The situation is certainly different from the hydrodynamic case, for which $(\partial u/\partial x) = 0$ in the fully developed region. In contrast, if there is heat transfer, (dT_m/dx), as well as $(\partial T/\partial x)$ at any radius r, is not zero.

Accordingly, the temperature profile $T(r)$ is continuously changing with x, and it would seem that a fully developed condition could never be reached. This apparent contradiction may be reconciled by working with a dimensionless form of the temperature, as was done for transient conduction (Chapter 5) and the energy conservation equation (Chapter 6).

Introducing a dimensionless temperature difference of the form $(T_s - T)/(T_s - T_m)$, conditions for which this ratio becomes independent of x are known to exist [2]. That is, although the temperature profile $T(r)$ continues to change with x, the *relative* shape of the profile no longer changes and the flow is said to be *thermally fully developed.* The requirement for such a condition is formally stated as

$$\frac{\partial}{\partial x}\left[\frac{T_s(x) - T(r, x)}{T_s(x) - T_m(x)}\right]_{\text{fd},t} = 0 \qquad (8.28)$$

where T_s is the tube surface temperature, T is the local fluid temperature, and T_m is the mean temperature of the fluid over the cross section of the tube.

The condition given by Equation 8.28 is eventually reached in a tube for which there is either a *uniform surface heat flux* (q''_s is constant) *or a uniform surface temperature* (T_s is constant). These surface conditions arise in many engineering applications. For example, a constant surface heat flux would exist if the tube wall were heated electrically or if the outer surface were uniformly irradiated. In contrast, a constant surface temperature would exist if a phase change (due to boiling or condensation) were occurring at the outer surface. Note that it is impossible to *simultaneously* impose the conditions of constant surface heat flux and constant surface temperature. If q''_s is constant, T_s must vary with x; conversely, if T_s is constant, q''_s must vary with x.

Several important features of thermally developed flow may be inferred from Equation 8.28. Since the temperature ratio is independent of x, the derivative of this ratio with respect to r must also be independent of x. Evaluating this derivative at the tube surface (note that T_s and T_m are constants insofar as differentiation with respect to r is concerned), we then obtain

$$\frac{\partial}{\partial r}\left(\frac{T_s - T}{T_s - T_m}\right)\bigg|_{r=r_o} = \frac{-\partial T/\partial r|_{r=r_o}}{T_s - T_m} \neq f(x)$$

Substituting for $\partial T/\partial r$ from Fourier's law, which, from Figure 8.4, is of the form

$$q''_s = -k\frac{\partial T}{\partial y}\bigg|_{y=0} = k\frac{\partial T}{\partial r}\bigg|_{r=r_o}$$

and for q''_s from Newton's law of cooling, Equation 8.27, we obtain

$$\frac{h}{k} \neq f(x) \qquad (8.29)$$

Hence *in the thermally fully developed flow* of a fluid *with constant properties,* the *local convection coefficient is a constant, independent of x.*

Equation 8.28 is not satisfied in the entrance region, where h varies with x, as shown in Figure 8.5. Because the thermal boundary layer thickness is zero at the tube entrance, the convection coefficient is extremely large at $x = 0$. However, h decays rapidly as the thermal boundary layer develops, until the constant value associated with fully developed conditions is reached.

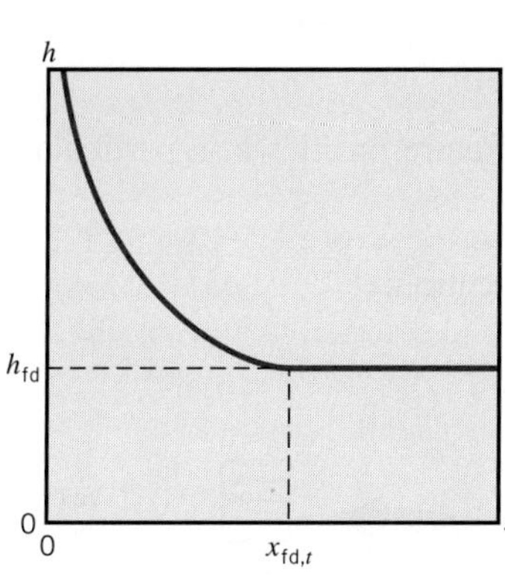

FIGURE 8.5 Axial variation of the convection heat transfer coefficient for flow in a tube.

Additional simplifications are associated with the special case of *uniform surface heat flux.* Since both h and q_s'' are constant in the fully developed region, it follows from Equation 8.27 that

$$\left.\frac{dT_s}{dx}\right|_{\mathrm{fd},t} = \left.\frac{dT_m}{dx}\right|_{\mathrm{fd},t} \qquad q_s'' = \text{constant} \tag{8.30}$$

If we expand Equation 8.28 and solve for $\partial T/\partial x$, it also follows that

$$\left.\frac{\partial T}{\partial x}\right|_{\mathrm{fd},t} = \left.\frac{dT_s}{dx}\right|_{\mathrm{fd},t} - \left.\frac{(T_s - T)}{(T_s - T_m)}\frac{dT_s}{dx}\right|_{\mathrm{fd},t} + \left.\frac{(T_s - T)}{(T_s - T_m)}\frac{dT_m}{dx}\right|_{\mathrm{fd},t} \tag{8.31}$$

Substituting from Equation 8.30, we then obtain

$$\left.\frac{\partial T}{\partial x}\right|_{\mathrm{fd},t} = \left.\frac{dT_m}{dx}\right|_{\mathrm{fd},t} \qquad q_s'' = \text{constant} \tag{8.32}$$

Hence the axial temperature gradient is independent of the radial location. For the case of *constant surface temperature* ($dT_s/dx = 0$), it also follows from Equation 8.31 that

$$\left.\frac{\partial T}{\partial x}\right|_{\mathrm{fd},t} = \left.\frac{(T_s - T)}{(T_s - T_m)}\frac{dT_m}{dx}\right|_{\mathrm{fd},t} \qquad T_s = \text{constant} \tag{8.33}$$

in which case the value of $\partial T/\partial x$ depends on the radial coordinate.

From the foregoing results, it is evident that the mean temperature is a very important variable for internal flows. To describe such flows, its variation with x must be known. This variation may be obtained by applying an *overall energy balance* to the flow, as will be shown in the next section.

EXAMPLE 8.1

For flow of a liquid metal through a circular tube, the velocity and temperature profiles at a particular axial location may be approximated as being uniform and parabolic, respectively. That is, $u(r) = C_1$ and $T(r) - T_s = C_2[1 - (r/r_o)^2]$, where C_1 and C_2 are constants. What is the value of the Nusselt number Nu_D at this location?

SOLUTION

Known: Form of the velocity and temperature profiles at a particular axial location for flow in a circular tube.

Find: Nusselt number at the prescribed location.

Schematic:

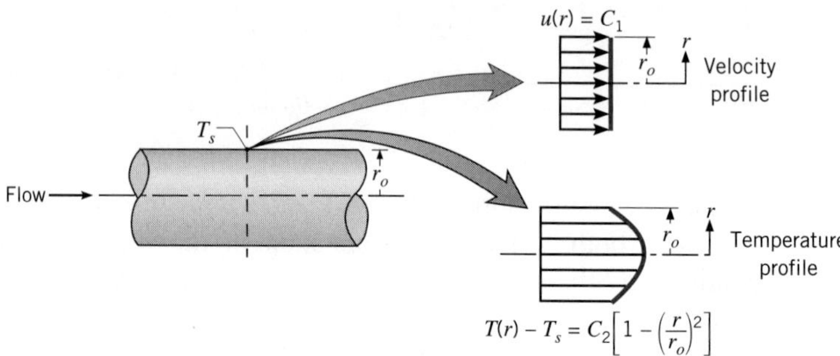

Assumptions: Incompressible, constant property flow.

Analysis: The Nusselt number may be obtained by first determining the convection coefficient, which, from Equation 8.27, is given as

$$h = \frac{q_s''}{T_s - T_m}$$

From Equation 8.26, the mean temperature is

$$T_m = \frac{2}{u_m r_o^2}\int_0^{r_o} uTr\,dr = \frac{2C_1}{u_m r_o^2}\int_0^{r_o}\left\{T_s + C_2\left[1 - \left(\frac{r}{r_o}\right)^2\right]\right\}r\,dr$$

or, since $u_m = C_1$ from Equation 8.8,

$$T_m = \frac{2}{r_o^2}\int_0^{r_o}\left\{T_s + C_2\left[1 - \left(\frac{r}{r_o}\right)^2\right]\right\}r\,dr$$

$$T_m = \frac{2}{r_o^2}\left[T_s\frac{r^2}{2} + C_2\frac{r^2}{2} - \frac{C_2}{4}\frac{r^4}{r_o^2}\right]\Bigg|_0^{r_o}$$

$$T_m = \frac{2}{r_o^2}\left(T_s\frac{r_o^2}{2} + \frac{C_2}{2}r_o^2 - \frac{C_2}{4}r_o^2\right) = T_s + \frac{C_2}{2}$$

The heat flux may be obtained from Fourier's law, in which case

$$q_s'' = k\frac{\partial T}{\partial r}\bigg|_{r=r_o} = -kC_2 2\frac{r}{r_o^2}\bigg|_{r=r_o} = -2C_2\frac{k}{r_o}$$

Hence

$$h = \frac{q_s''}{T_s - T_m} = \frac{-2C_2(k/r_o)}{-C_2/2} = \frac{4k}{r_o}$$

and

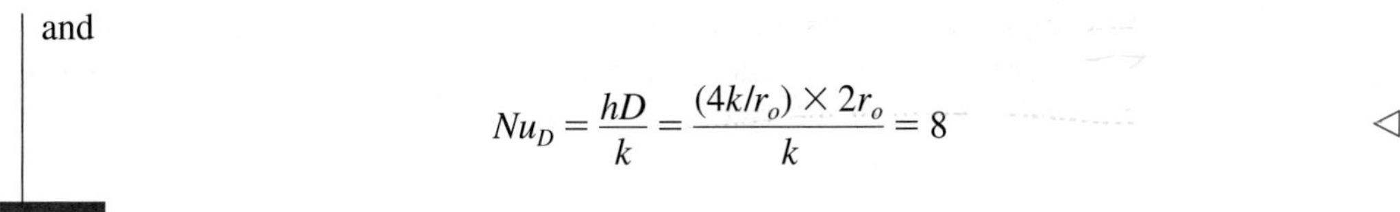

$$Nu_D = \frac{hD}{k} = \frac{(4k/r_o) \times 2r_o}{k} = 8 \quad \triangleleft$$

8.3 The Energy Balance

8.3.1 General Considerations

Because the flow in a tube is completely enclosed, an energy balance may be applied to determine how the mean temperature $T_m(x)$ varies with position along the tube and how the total convection heat transfer q_{conv} is related to the difference in temperatures at the tube inlet and outlet. Consider the tube flow of Figure 8.6. Fluid moves at a constant flow rate $\dot{m}$, and convection heat transfer occurs at the inner surface. Typically, it will be reasonable to make one of the four assumptions in Section 1.3 that leads to the simplified steady-flow thermal energy equation, Equation 1.12e. For example, it is often the case that viscous dissipation is negligible (see Problem 8.10) and that the fluid can be modeled as either an incompressible liquid or an ideal gas with negligible pressure variation. In addition, it is usually reasonable to neglect net heat transfer by conduction in the axial direction, so the heat transfer term in Equation 1.12e includes only q_{conv}. Therefore, Equation 1.12e may be written in the form

$$q_{\text{conv}} = \dot{m}c_p(T_{m,o} - T_{m,i}) \tag{8.34}$$

for a tube of finite length. This simple overall energy balance relates three important thermal variables (q_{conv}, $T_{m,o}$, $T_{m,i}$). *It is a general expression that applies irrespective of the nature of the surface thermal or tube flow conditions.*

Applying Equation 1.12e to the differential control volume of Figure 8.6 and recalling that the mean temperature is defined such that $\dot{m}c_pT_m$ represents the true rate of thermal energy (or enthalpy) advection integrated over the cross section, we obtain

$$dq_{\text{conv}} = \dot{m}c_p[(T_m + dT_m) - T_m] \tag{8.35}$$

or

$$dq_{\text{conv}} = \dot{m}c_p dT_m \tag{8.36}$$

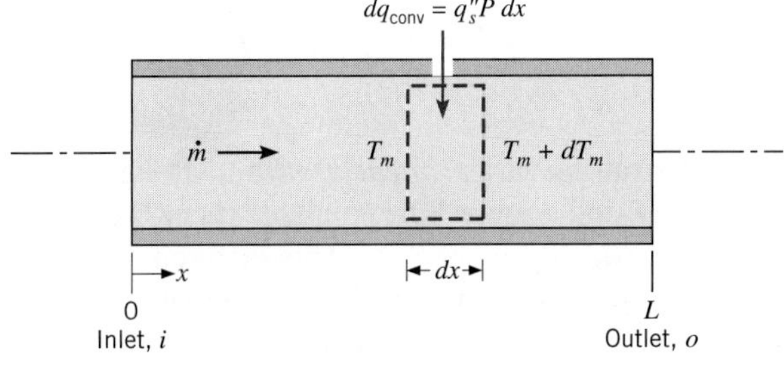

FIGURE 8.6 Control volume for internal flow in a tube.

Equation 8.36 may be cast in a convenient form by expressing the rate of convection heat transfer to the differential element as $dq_{\text{conv}} = q_s'' P\,dx$, where P is the surface perimeter ($P = \pi D$ for a circular tube). Substituting from Equation 8.27, it follows that

$$\frac{dT_m}{dx} = \frac{q_s'' P}{\dot{m} c_p} = \frac{P}{\dot{m} c_p} h(T_s - T_m) \tag{8.37}$$

This expression is an extremely useful result, from which the axial variation of T_m may be determined. If $T_s > T_m$, heat is transferred to the fluid and T_m increases with x; if $T_s < T_m$, the opposite is true.

The manner in which quantities on the right-hand side of Equation 8.37 vary with x should be noted. Although P may vary with x, most commonly it is a constant (a tube of constant cross-sectional area). Hence the quantity $(P/\dot{m} c_p)$ is a constant. In the fully developed region, the convection coefficient h is also constant, although it decreases with x in the entrance region (Figure 8.5). Finally, although T_s may be constant, T_m must always vary with x (except for the trivial case of no heat transfer, $T_s = T_m$).

The solution to Equation 8.37 for $T_m(x)$ depends on the surface thermal condition. Recall that the two special cases of interest are *constant surface heat flux* and *constant surface temperature.* It is common to find one of these conditions existing to a reasonable approximation.

8.3.2 Constant Surface Heat Flux

For constant surface heat flux we first note that it is a simple matter to determine the total heat transfer rate q_{conv}. Since q_s'' is independent of x, it follows that

$$q_{\text{conv}} = q_s''(P \cdot L) \tag{8.38}$$

This expression could be used with Equation 8.34 to determine the fluid temperature change, $T_{m,o} - T_{m,i}$.

For constant q_s'' it also follows that the middle expression in Equation 8.37 is a constant independent of x. Hence

$$\frac{dT_m}{dx} = \frac{q_s'' P}{\dot{m} c_p} \neq f(x) \tag{8.39}$$

Integrating from $x = 0$, it follows that

$$T_m(x) = T_{m,i} + \frac{q_s'' P}{\dot{m} c_p} x \qquad q_s'' = \text{constant} \tag{8.40}$$

Accordingly, the mean temperature varies *linearly* with x along the tube (Figure 8.7*a*). Moreover, from Equation 8.27 and Figure 8.5 we also expect the temperature difference $(T_s - T_m)$ to vary with x, as shown in Figure 8.7*a*. This difference is initially small (due to the large value of h near the entrance) but increases with increasing x due to the decrease in h that occurs as the boundary layer develops. However, in the fully developed region we

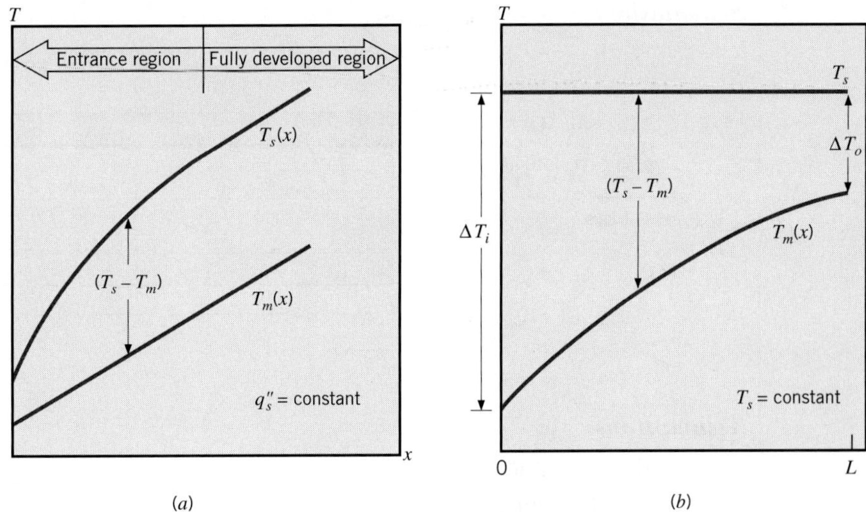

FIGURE 8.7 Axial temperature variations for heat transfer in a tube. (*a*) Constant surface heat flux. (*b*) Constant surface temperature.

know that h is independent of x. Hence from Equation 8.27 it follows that $(T_s - T_m)$ must also be independent of x in this region.

It should be noted that, if the heat flux is not constant but is, instead, a known function of x, Equation 8.37 may still be integrated to obtain the variation of the mean temperature with x. Similarly, the total heat rate may be obtained from the requirement that $q_{\text{conv}} = \int_0^L q_s''(x)P\,dx$.

EXAMPLE 8.2

A system for heating water from an inlet temperature of $T_{m,i} = 20°\text{C}$ to an outlet temperature of $T_{m,o} = 60°\text{C}$ involves passing the water through a thick-walled tube having inner and outer diameters of 20 and 40 mm. The outer surface of the tube is well insulated, and electrical heating within the wall provides for a uniform generation rate of $\dot{q} = 10^6\ \text{W/m}^3$.

1. For a water mass flow rate of $\dot{m} = 0.1$ kg/s, how long must the tube be to achieve the desired outlet temperature?
2. If the inner surface temperature of the tube is $T_s = 70°\text{C}$ at the outlet, what is the local convection heat transfer coefficient at the outlet?

SOLUTION

Known: Internal flow through thick-walled tube having uniform heat generation.

Find:

1. Length of tube needed to achieve the desired outlet temperature.
2. Local convection coefficient at the outlet.

Schematic:

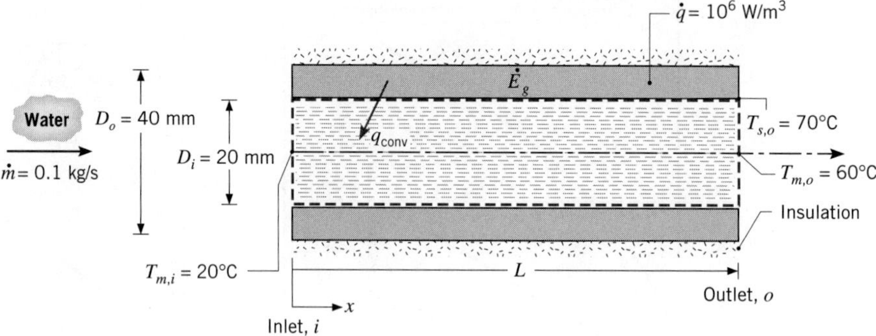

Assumptions:

1. Steady-state conditions.
2. Uniform heat flux.
3. Incompressible liquid and negligible viscous dissipation.
4. Constant properties.
5. Adiabatic outer tube surface.

Properties: Table A.6, water ($\overline{T}_m = 313$ K): $c_p = 4179$ J/kg·K.

Analysis:

1. Since the outer surface of the tube is adiabatic, the rate at which energy is generated within the tube wall must equal the rate at which it is convected to the water.

$$\dot{E}_g = q_{\text{conv}}$$

With

$$\dot{E}_g = \dot{q}\frac{\pi}{4}(D_o^2 - D_i^2)L$$

it follows from Equation 8.34 that

$$\dot{q}\,\frac{\pi}{4}\,(D_o^2 - D_i^2)\,L = \dot{m}c_p(T_{m,o} - T_{m,i})$$

or

$$L = \frac{4\dot{m}c_p}{\pi(D_o^2 - D_i^2)\dot{q}}(T_{m,o} - T_{m,i})$$

$$L = \frac{4 \times 0.1\text{ kg/s} \times 4179\text{ J/kg}\cdot\text{K}}{\pi(0.04^2 - 0.02^2)\text{ m}^2 \times 10^6\text{ W/m}^3}(60 - 20)^\circ\text{C} = 17.7\text{ m} \qquad \triangleleft$$

2. From Newton's law of cooling, Equation 8.27, the local convection coefficient at the tube exit is

$$h_o = \frac{q_s''}{T_{s,o} - T_{m,o}}$$

Assuming that uniform heat generation in the wall provides a constant surface heat flux, with

$$q_s'' = \frac{\dot{E}_g}{\pi D_i L} = \frac{\dot{q}}{4}\frac{D_o^2 - D_i^2}{D_i}$$

$$q_s'' = \frac{10^6 \text{ W/m}^3}{4} \frac{(0.04^2 - 0.02^2) \text{ m}^2}{0.02 \text{ m}} = 1.5 \times 10^4 \text{ W/m}^2$$

it follows that

$$h_o = \frac{1.5 \times 10^4 \text{ W/m}^2}{(70 - 60)°\text{C}} = 1500 \text{ W/m}^2 \cdot \text{K} \qquad \triangleleft$$

Comments:

1. If conditions are fully developed over the entire tube, the local convection coefficient and the temperature difference $(T_s - T_m)$ are independent of x. Hence $h = 1500 \text{ W/m}^2 \cdot \text{K}$ and $(T_s - T_m) = 10°\text{C}$ over the entire tube. The inner surface temperature at the tube inlet is then $T_{s,i} = 30°\text{C}$.
2. The required tube length L could have been computed by applying the expression for $T_m(x)$, Equation 8.40, at $x = L$.

8.3.3 Constant Surface Temperature

Results for the total heat transfer rate and the axial distribution of the mean temperature are entirely different for the *constant surface temperature* condition. Defining ΔT as $T_s - T_m$, Equation 8.37 may be expressed as

$$\frac{dT_m}{dx} = -\frac{d(\Delta T)}{dx} = \frac{P}{\dot{m}c_p} h\, \Delta T$$

Separating variables and integrating from the tube inlet to the outlet,

$$\int_{\Delta T_i}^{\Delta T_o} \frac{d(\Delta T)}{\Delta T} = -\frac{P}{\dot{m}c_p} \int_0^L h\, dx$$

or

$$\ln \frac{\Delta T_o}{\Delta T_i} = -\frac{PL}{\dot{m}c_p} \left(\frac{1}{L} \int_0^L h\, dx \right)$$

From the definition of the average convection heat transfer coefficient, Equation 6.9, it follows that

$$\ln \frac{\Delta T_o}{\Delta T_i} = -\frac{PL}{\dot{m}c_p} \bar{h}_L \qquad T_s = \text{constant} \tag{8.41a}$$

where $\bar{h}_L$, or simply $\bar{h}$, is the average value of h for the entire tube. Rearranging,

$$\frac{\Delta T_o}{\Delta T_i} = \frac{T_s - T_{m,o}}{T_s - T_{m,i}} = \exp\left(-\frac{PL}{\dot{m}c_p}\bar{h}\right) \qquad T_s = \text{constant} \tag{8.41b}$$

Had we integrated from the tube inlet to some axial position x within the tube, we would have obtained the similar, but more general, result that

$$\frac{T_s - T_m(x)}{T_s - T_{m,i}} = \exp\left(-\frac{Px}{\dot{m}c_p}\bar{h}\right) \qquad T_s = \text{constant} \tag{8.42}$$

where $\bar{h}$ is now the average value of h from the tube inlet to x. This result tells us that the temperature difference $(T_s - T_m)$ *decays exponentially* with distance along the tube axis. The axial surface and mean temperature distributions are therefore as shown in Figure 8.7*b*.

Determination of an expression for the total heat transfer rate q_{conv} is complicated by the exponential nature of the temperature decay. Expressing Equation 8.34 in the form

$$q_{\text{conv}} = \dot{m}c_p[(T_s - T_{m,i}) - (T_s - T_{m,o})] = \dot{m}c_p(\Delta T_i - \Delta T_o)$$

and substituting for $\dot{m}c_p$ from Equation 8.41a, we obtain

$$q_{\text{conv}} = \bar{h}A_s\Delta T_{\text{lm}} \qquad T_s = \text{constant} \tag{8.43}$$

where A_s is the tube surface area $(A_s = P \cdot L)$ and ΔT_{lm} is the *log mean temperature difference,*

$$\Delta T_{\text{lm}} \equiv \frac{\Delta T_o - \Delta T_i}{\ln(\Delta T_o/\Delta T_i)} \tag{8.44}$$

Equation 8.43 is a form of Newton's law of cooling for the entire tube, and ΔT_{lm} is the appropriate *average* of the temperature difference over the tube length. The logarithmic nature of this average temperature difference [in contrast, e.g., to an *arithmetic mean temperature* difference of the form $\Delta T_{\text{am}} = (\Delta T_i + \Delta T_o)/2$] is due to the exponential nature of the temperature decay.

Before concluding this section, it is important to note that, in many applications, it is the temperature of an *external* fluid, rather than the tube surface temperature, that is fixed (Figure 8.8). In such cases, it is readily shown that the results of this section may still be used if T_s is replaced by T_∞ (the free stream temperature of the external fluid) and $\bar{h}$ is replaced by $\bar{U}$ (the average overall heat transfer coefficient). For such cases, it follows that

$$\frac{\Delta T_o}{\Delta T_i} = \frac{T_\infty - T_{m,o}}{T_\infty - T_{m,i}} = \exp\left(-\frac{\bar{U}A_s}{\dot{m}c_p}\right) \tag{8.45a}$$

and

$$q = \bar{U}A_s\,\Delta T_{\text{lm}} \tag{8.46a}$$

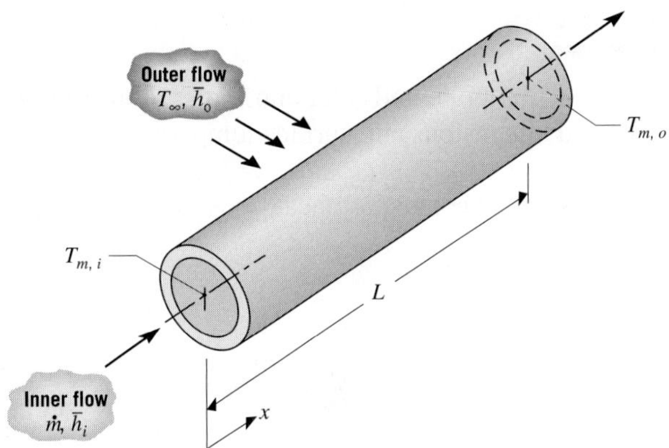

FIGURE 8.8 Heat transfer between fluid flowing over a tube and fluid passing through the tube.

The overall heat transfer coefficient is defined in Section 3.3.1, and for this application it would include contributions due to convection at the tube inner and outer surfaces. For a thick-walled tube of small thermal conductivity, it would also include the effect of conduction across the tube wall. Note that the product $\overline{U}A_s$ yields the same result, irrespective of whether it is defined in terms of the inner ($\overline{U}_iA_{s,i}$) or outer ($\overline{U}_oA_{s,o}$) surface areas of the tube (see Equation 3.37). Also note that $(\overline{U}A_s)^{-1}$ is equivalent to the total thermal resistance between the two fluids, in which case Equations 8.45a and 8.46a may be expressed as

$$\frac{\Delta T_o}{\Delta T_i} = \frac{T_\infty - T_{m,o}}{T_\infty - T_{m,i}} = \exp\left(-\frac{1}{\dot{m}c_pR_{\text{tot}}}\right) \tag{8.45b}$$

and

$$q = \frac{\Delta T_{\text{lm}}}{R_{\text{tot}}} \tag{8.46b}$$

A common variation of the foregoing conditions is one for which the uniform temperature of an *outer* surface, $T_{s,o}$, rather than the free stream temperature of an external fluid, T_∞, is known. In the foregoing equations, T_∞ is then replaced by $T_{s,o}$, and the total resistance embodies the convection resistance associated with the internal flow, as well as the resistance due to conduction between the inner surface of the tube and the surface corresponding to $T_{s,o}$.

EXAMPLE 8.3

Steam condensing on the outer surface of a thin-walled circular tube of diameter $D = 50$ mm and length $L = 6$ m maintains a uniform outer surface temperature of 100°C. Water flows through the tube at a rate of $\dot{m} = 0.25$ kg/s, and its inlet and outlet temperatures are $T_{m,i} = 15$°C and $T_{m,o} = 57$°C. What is the average convection coefficient associated with the water flow?

SOLUTION

Known: Flow rate and inlet and outlet temperatures of water flowing through a tube of prescribed dimensions and surface temperature.

Find: Average convection heat transfer coefficient.

Schematic:

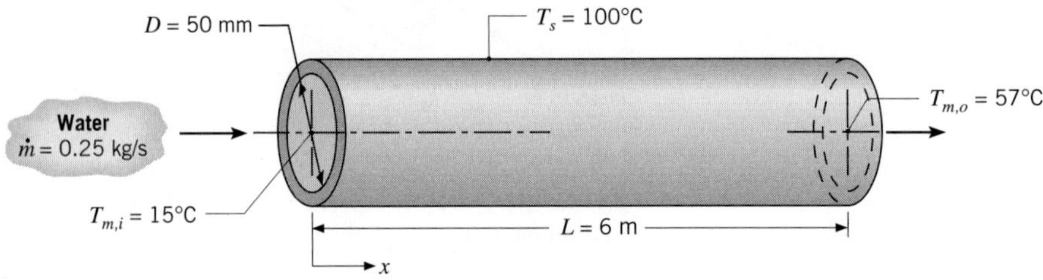

Assumptions:

1. Negligible tube wall conduction resistance.
2. Incompressible liquid and negligible viscous dissipation.
3. Constant properties.

Properties: Table A.6, water ($\overline{T}_m = 36°C$): $c_p = 4178$ J/kg·K.

Analysis: Combining the energy balance, Equation 8.34, with the rate equation, Equation 8.43, the average convection coefficient is given by

$$\overline{h} = \frac{\dot{m}c_p}{\pi DL} \frac{(T_{m,o} - T_{m,i})}{\Delta T_{\text{lm}}}$$

From Equation 8.44

$$\Delta T_{\text{lm}} = \frac{(T_s - T_{m,o}) - (T_s - T_{m,i})}{\ln[(T_s - T_{m,o})/(T_s - T_{m,i})]}$$

$$\Delta T_{\text{lm}} = \frac{(100 - 57) - (100 - 15)}{\ln[(100 - 57)/(100 - 15)]} = 61.6°\text{C}$$

Hence

$$\overline{h} = \frac{0.25 \text{ kg/s} \times 4178 \text{ J/kg} \cdot \text{K}}{\pi \times 0.05 \text{ m} \times 6 \text{ m}} \frac{(57 - 15)°\text{C}}{61.6°\text{C}}$$

or

$$\overline{h} = 755 \text{ W/m}^2 \cdot \text{K}$$ ◁

Comments: If conditions were fully developed over the entire tube, the local convection coefficient would be everywhere equal to 755 W/m^2·K.

8.4 *Laminar Flow in Circular Tubes: Thermal Analysis and Convection Correlations*

To use many of the foregoing results, the convection coefficients must be known. In this section we outline the manner in which such coefficients may be obtained theoretically for laminar flow in a circular tube. In subsequent sections we consider empirical correlations pertinent to turbulent flow in a circular tube, as well as to flows in tubes of noncircular cross section.

8.4.1 The Fully Developed Region

Here, the problem of heat transfer in *laminar flow* of an *incompressible, constant property fluid* in the *fully developed region* of a *circular tube* is treated theoretically. The resulting temperature distribution is used to determine the convection coefficient.

A differential equation governing the temperature distribution is determined by applying the simplified, steady-flow, thermal energy equation, Equation 1.12e [$q = \dot{m}c_p(T_{\text{out}} - T_{\text{in}})$], to the annular differential element of Figure 8.9. If we neglect the effects of net axial conduction, the heat input, q, is due only to conduction through the radial surfaces. Since the radial velocity is zero in the fully developed region, there is no advection of thermal energy through the radial control surfaces, and the only advection is in the axial direction. Thus, Equation 1.12e leads to Equation 8.47, which expresses a balance between radial conduction and axial advection:

$$q_r - q_{r+dr} = (d\dot{m})c_p\left[\left(T + \frac{\partial T}{\partial x}dx\right) - T\right] \tag{8.47a}$$

or

$$(d\dot{m})c_p\frac{\partial T}{\partial x}dx = q_r - \left(q_r + \frac{\partial q_r}{\partial r}dr\right) = -\frac{\partial q_r}{\partial r}dr \tag{8.47b}$$

The differential mass flow rate in the axial direction is $d\dot{m} = \rho u 2\pi r dr$, and the radial heat transfer rate is $q_r = -k(\partial T/\partial r)2\pi r dx$. If we assume constant properties, Equation 8.47b becomes

$$u\frac{\partial T}{\partial x} = \frac{\alpha}{r}\frac{\partial}{\partial r}\left(r\frac{\partial T}{\partial r}\right) \tag{8.48}$$

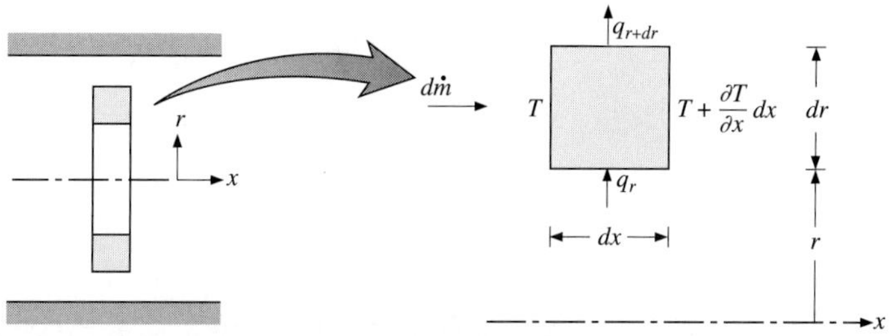

FIGURE 8.9 Thermal energy balance on a differential element for laminar, fully developed flow in a circular tube.

We will now proceed to solve for the temperature distribution for the case of *constant surface heat flux*. In this case, the assumption of negligible net axial conduction is exactly satisfied, that is, $(\partial^2 T/\partial x^2) = 0$. Substituting for the axial temperature gradient from Equation 8.32 and for the axial velocity component, u, from Equation 8.15, the energy equation, Equation 8.48, reduces to

$$\frac{1}{r}\frac{\partial}{\partial r}\left(r\frac{\partial T}{\partial r}\right) = \frac{2u_m}{\alpha}\left(\frac{dT_m}{dx}\right)\left[1 - \left(\frac{r}{r_o}\right)^2\right] \qquad q_s'' = \text{constant} \tag{8.49}$$

where $T_m(x)$ varies linearly with x and $(2u_m/\alpha)(dT_m/dx)$ is a constant. Separating variables and integrating twice, we obtain an expression for the radial temperature distribution:

$$T(r, x) = \frac{2u_m}{\alpha}\left(\frac{dT_m}{dx}\right)\left[\frac{r^2}{4} - \frac{r^4}{16r_o^2}\right] + C_1 \ln r + C_2$$

The constants of integration may be evaluated by applying appropriate boundary conditions. From the requirement that the temperature remain finite at $r = 0$, it follows that $C_1 = 0$. From the requirement that $T(r_o) = T_s$, where T_s varies with x, it also follows that

$$C_2 = T_s(x) - \frac{2u_m}{\alpha}\left(\frac{dT_m}{dx}\right)\left(\frac{3r_o^2}{16}\right)$$

Accordingly, for the fully developed region with constant surface heat flux, the temperature profile is of the form

$$T(r, x) = T_s(x) - \frac{2u_m r_o^2}{\alpha}\left(\frac{dT_m}{dx}\right)\left[\frac{3}{16} + \frac{1}{16}\left(\frac{r}{r_o}\right)^4 - \frac{1}{4}\left(\frac{r}{r_o}\right)^2\right] \tag{8.50}$$

From knowledge of the temperature profile, all other thermal parameters may be determined. For example, if the velocity and temperature profiles, Equations 8.15 and 8.50, respectively, are substituted into Equation 8.26 and the integration over r is performed, the mean temperature is found to be

$$T_m(x) = T_s(x) - \frac{11}{48}\left(\frac{u_m r_o^2}{\alpha}\right)\left(\frac{dT_m}{dx}\right) \tag{8.51}$$

From Equation 8.39, where $P = \pi D$ and $\dot{m} = \rho u_m(\pi D^2/4)$, we then obtain

$$T_m(x) - T_s(x) = -\frac{11}{48}\frac{q_s'' D}{k} \tag{8.52}$$

Combining Newton's law of cooling, Equation 8.27, and Equation 8.52, it follows that

$$h = \frac{48}{11}\left(\frac{k}{D}\right)$$

or

$$Nu_D \equiv \frac{hD}{k} = 4.36 \qquad q_s'' = \text{constant} \tag{8.53}$$

Hence in a *circular tube* characterized by *uniform surface heat flux* and *laminar, fully developed conditions,* the *Nusselt number is a constant,* independent of Re_D, Pr, and axial location.

For *laminar, fully developed conditions* with a *constant surface temperature,* the assumption of negligible axial conduction is often reasonable. Substituting for the velocity profile from Equation 8.15 and for the axial temperature gradient from Equation 8.33, the energy equation becomes

$$\frac{1}{r}\frac{\partial}{\partial r}\left(r\frac{\partial T}{\partial r}\right) = \frac{2u_m}{\alpha}\left(\frac{dT_m}{dx}\right)\left[1-\left(\frac{r}{r_o}\right)^2\right]\frac{T_s - T}{T_s - T_m} \qquad T_s = \text{constant} \tag{8.54}$$

A solution to this equation may be obtained by an iterative procedure, which involves making successive approximations to the temperature profile. The resulting profile is not described by a simple algebraic expression, but the resulting Nusselt number may be shown to be [2]

$$Nu_D = 3.66 \qquad T_s = \text{constant} \tag{8.55}$$

Note that in using Equation 8.53 or 8.55 to determine h, the thermal conductivity should be evaluated at T_m.[1]

EXAMPLE 8.4

In the human body, blood flows from the heart into a series of branching blood vessels having successively smaller diameters. In developing the bioheat equation (Section 3.7), Pennes assumed that blood enters the capillaries (the smallest vessels) at the arterial temperature and exits at the temperature of the surrounding tissue. This problem tests that assumption [7, 8]. The diameters and average blood velocities for three different types of vessels are given in the table. Begin by estimating the length required for the mean blood temperature to approach the tissue temperature, specifically, to satisfy the criterion $(T_t - T_{m,o})/(T_t - T_{m,i}) = 0.05$ for each of these vessels. Heat transfer between the vessel wall and surrounding tissue can be described by an effective heat transfer coefficient, $h_t = k_t/D$, where $k_t = 0.5\ \text{W/m}\cdot\text{K}$.

Vessel	**Diameter, D (mm)**	**Blood Velocity, u_m (mm/s)**
Large artery	3	130
Arteriole	0.02	3
Capillary	0.008	0.7

SOLUTION

Known: Blood vessel diameter and average blood velocity. Tissue thermal conductivity and effective heat transfer coefficient.

Find: Whether the blood enters the capillary vessels at the arterial temperature and leaves at the tissue temperature.

[1] If heat transfer occurs in a liquid that is characterized by a highly temperature-dependent viscosity, as is the case for many oils, experimental results have shown that the Nusselt numbers of Equations 8.53 or 8.55 may be corrected to account for this property variation as described in Section 8.4.3.

Schematic:

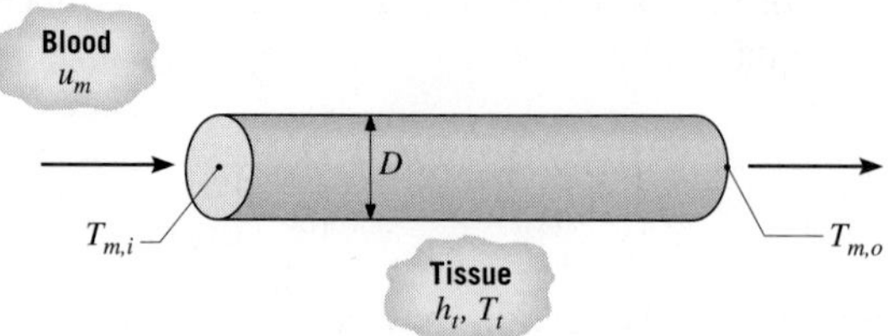

Assumptions:

1. Steady-state conditions.
2. Constant properties.
3. Negligible blood vessel wall thermal resistance.
4. Thermal properties of blood can be approximated by those of water.
5. Blood is incompressible liquid with negligible viscous dissipation.
6. Tissue temperature is fixed.
7. Effects of pulsation of flow are negligible.

Properties: Table A.6, water ($\overline{T}_m = 310$ K) $\rho = v_f^{-1} = 993$ kg/m^3, $c_p = 4178$ J/kg·K, $\mu = 695 \times 10^{-6}$ N·s/m^2, $k = 0.628$ W/m·K, $Pr = 4.62$.

Analysis: Since the tissue temperature T_t is fixed and heat transfer between the blood vessel wall and the tissue can be represented by an effective heat transfer coefficient, Equation 8.45a is applicable, with the "free stream" temperature equal to T_t. This equation can be used to find the length L that satisfies the criterion. However, we must first find $\overline{U}$, which requires knowledge of the heat transfer coefficient for the blood flow, h_b.

For the large artery, the Reynolds number is

$$Re_D = \frac{\rho u_m D}{\mu} = \frac{993 \text{ kg/m}^3 \times 130 \times 10^{-3} \text{ m/s} \times 3 \times 10^{-3} \text{ m}}{695 \times 10^{-6} \text{ N} \cdot \text{s/m}^2} = 557$$

so the flow is laminar. Since the other vessels have smaller diameters and velocities, their flows will also be laminar. We begin by assuming fully developed conditions. Moreover, because the situation is neither one of constant surface temperature nor constant surface heat flux, we will approximate the Nusselt number as $Nu_D \approx 4$, in which case $h_b = 4k_b/D$. Neglecting the thermal resistance of the vessel wall, for the large artery

$$\frac{1}{\overline{U}} = \frac{1}{h_b} + \frac{1}{h_t} = \frac{D}{4k_b} + \frac{D}{k_t} = \frac{3 \times 10^{-3} \text{ m}}{4 \times 0.628 \text{ W/m} \cdot \text{K}} + \frac{3 \times 10^{-3} \text{ m}}{0.5 \text{ W/m} \cdot \text{K}}$$

$$= 7.2 \times 10^{-3} \text{m}^2 \cdot \text{K/W}$$

or

$$\overline{U} = 140 \text{ W/m}^2 \cdot \text{K}$$

The large artery length needed to satisfy the criterion can be found by solving Equation 8.45a, with $\dot{m} = \rho u_m \pi D^2/4$:

$$L = -\frac{\rho u_m D c_p}{4\overline{U}} \ln\left(\frac{T_t - T_{m,o}}{T_t - T_{m,i}}\right)$$

$$= -\frac{993 \text{ kg/m}^3 \times 130 \times 10^{-3} \text{ m/s} \times 3 \times 10^{-3} \text{ m} \times 4178 \text{ J/kg} \cdot \text{K}}{4 \times 140 \text{ W/m}^2 \cdot \text{K}} \ln(0.05)$$

$$= 8.7 \text{ m}$$

Using Equations 8.3 and 8.23:

$$x_{\text{fd},h} = 0.05\, Re_D D = 0.05 \times 557 \times 3 \times 10^{-3} \text{ m} = 0.08 \text{ m}$$

$$x_{\text{fd},t} = x_{\text{fd},h} Pr = 0.08 \text{ m} \times 4.62 = 0.4 \text{ m}$$

Therefore, the flow would become fully developed well within the length of 8.7 m. The calculations can be repeated for the other two cases, and the results are tabulated below.

Vessel	Re_D	$\overline{U}$ (W/m$^2 \cdot$ K)	L (m)	$x_{\text{fd},h}$ (m)	$x_{\text{fd},t}$ (m)
Large artery	557	140	8.7	0.08	0.4
Arteriole	0.086	21,000	8.9×10^{-6}	9×10^{-8}	4×10^{-7}
Capillary	0.0080	52,000	3.3×10^{-7}	3×10^{-9}	1×10^{-8}

The large value of L for the large artery suggests that the temperature remains close to the inlet arterial blood temperature. This is due to its relatively large diameter, which leads to a small overall heat transfer coefficient. In the intermediate arterioles, the blood temperature approaches the tissue temperature within a length on the order of 10 μm. Since arterioles are on the order of millimeters in length, the blood temperature exiting them and entering the capillaries would be approximately equal to the tissue temperature. There could then be no further temperature drop in the capillaries. Thus, it is in the arterioles and slightly larger vessels in which the blood temperature equilibrates to the tissue temperature, not in the small capillaries as Pennes assumed. ◁

Comments:

1. The properties of blood are moderately close to those of water. The property that differs most is the viscosity, as blood is more viscous than water. However, this discrepancy would have no effect on the foregoing results and conclusions. Since the Reynolds number would be even smaller for the higher-viscosity blood, the flow would still be laminar and the heat transfer would be unaffected.
2. Blood cells have dimensions on the order of the capillary diameter. Thus, for the capillaries, an accurate model of blood flow would account for the individual cells surrounded by plasma.
3. Despite the flaw in the assumption employed by Pennes, the bioheat equation is a useful tool in analyzing heat transfer in the human body.

8.4.2 The Entry Region

The results of the preceding section are valid only when *both* the velocity and temperature profiles are fully developed, as determined by the entry length expressions of Equations 8.3 and 8.23. If either or both profiles are not fully developed, the flow is said to be in the *entry region*. The energy equation for the entry region is more complicated than Equation 8.48 because there would be a radial advection term (since $v \neq 0$ in the entry region). In addition, both velocity and temperature now depend on x, as well as r, and the axial temperature gradient $\partial T/\partial x$ may no longer be simplified through Equation 8.32 or 8.33. However, two different entry length solutions have been obtained. The simplest solution is for the *thermal entry length problem,* and it is based on assuming that thermal conditions develop in the presence of a *fully developed velocity profile.* Such a situation would exist if the location at which heat transfer begins were preceded by an *unheated starting length.* It could also be assumed to a reasonable approximation for large Prandtl number fluids, such as oils. Even in the absence of an unheated starting length, velocity boundary layer development would occur far more rapidly than thermal boundary layer development for large Prandtl number fluids, and a thermal entry length approximation could be made. In contrast, the *combined* (thermal and velocity) *entry length problem* corresponds to the case for which the temperature and velocity profiles develop simultaneously. It would never be the case that thermal conditions are fully developed and hydrodynamic conditions are developing. Since the temperature distribution depends on the velocity distribution, as long as the velocity is still changing, thermal conditions cannot be fully developed.

Solutions have been obtained for both the thermal and combined entry length conditions [2], and selected results are shown in Figure 8.10. As evident in Figure 8.10*a*, local

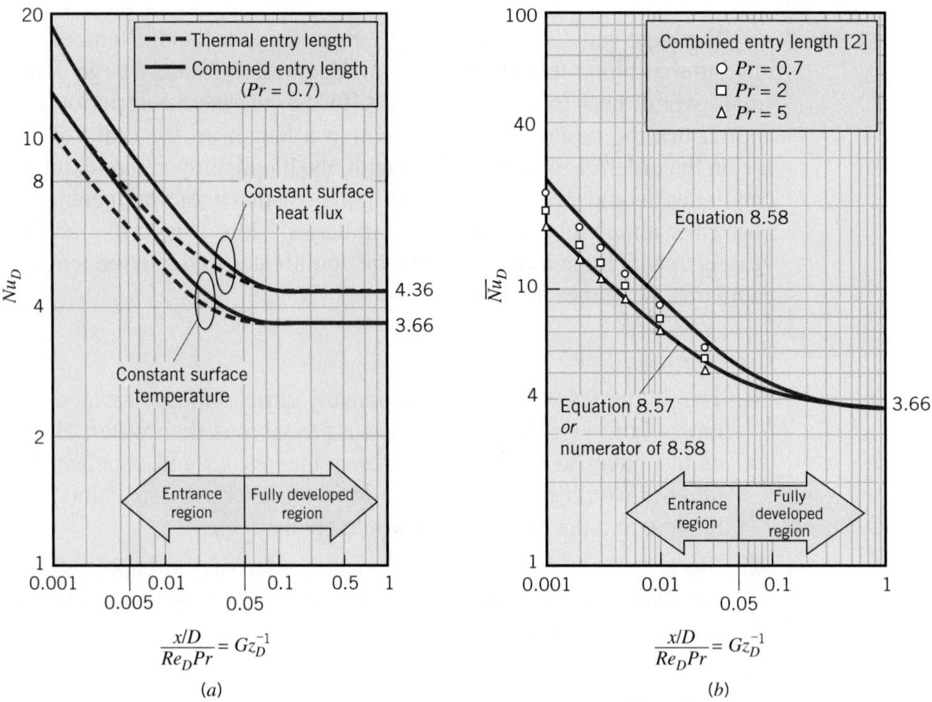

FIGURE 8.10 Results obtained from entry length solutions for laminar flow in a circular tube with constant surface temperature: (*a*) Local Nusselt numbers. (*b*) Average Nusselt numbers.

Nusselt numbers Nu_D are, in principle, infinite at $x = 0$ and decay to their asymptotic (fully developed) values with increasing x. These results are plotted against the dimensionless parameter $x\alpha/(u_m D^2) = x/(D\, Re_D\, Pr)$, which is the reciprocal of the *Graetz* number,

$$Gz_D \equiv (D/x)\, Re_D\, Pr \tag{8.56}$$

The manner in which Nu_D varies with Gz_D^{-1} is independent of Pr for the thermal entry problem, since the fully developed velocity profile, given by Equation 8.13, is independent of the fluid viscosity. In contrast, for the combined entry length problem, results depend on the manner in which the velocity distribution develops, which is highly sensitive to the fluid viscosity. Hence, heat transfer results depend on the Prandtl number for the combined entry length case and are presented in Figure 8.10*a* for $Pr = 0.7$, which is representative of most gases. At any location within the entry region, Nu_D decreases with increasing Pr and approaches the thermal entry length condition as $Pr \rightarrow \infty$. Note that fully developed conditions are reached for $[(x/D)/Re_D\, Pr] \approx 0.05$.

For the *constant surface temperature* condition, it is desirable to know the *average* convection coefficient for use with Equation 8.42 or 8.43. Selection of the appropriate correlation depends on whether a thermal or combined entry length exists.

For the *thermal entry length problem*, Kays [9] presents a correlation attributed to Hausen [10], which is of the form

$$\overline{Nu}_D = 3.66 + \frac{0.0668\, Gz_D}{1 + 0.04\, Gz_D^{2/3}} \tag{8.57}$$

$$\begin{bmatrix} T_s = \text{constant} \\ \text{thermal entry length} \\ or \\ \text{combined entry length with } Pr \gtrsim 5 \end{bmatrix}$$

where $\overline{Nu}_D \equiv \bar{h}D/k$, and $\bar{h}$ is the heat transfer coefficient averaged from the tube inlet to x. Equation 8.57 is applicable to all situations where the velocity profile is fully developed. However, from Figure 8.10*b*, it is apparent that, for $Pr \gtrsim 5$, the thermal entry length approximation is reasonable since it agrees well with the combined entry length solution [2].

For the *combined entry problem*, the Nusselt number depends on the Prandtl and Graetz numbers. Baehr and Stephan [11] recommend a correlation of the form

$$\overline{Nu}_D = \frac{\dfrac{3.66}{\tanh[2.264\, Gz_D^{-1/3} + 1.7\, Gz_D^{-2/3}]} + 0.0499\, Gz_D \tanh(Gz_D^{-1})}{\tanh(2.432\, Pr^{1/6}\, Gz_D^{-1/6})} \tag{8.58}$$

$$\begin{bmatrix} T_s = \text{constant} \\ \text{combined entry length} \\ Pr \gtrsim 0.1 \end{bmatrix}$$

Equation 8.58, evaluated for $Pr = 0.7$, is shown in Figure 8.10*b* and agrees well with the data points obtained by solving the governing equations for the combined entry problem [2]. As $Pr \rightarrow \infty$, the denominator of Equation 8.58 approaches unity. Therefore, the numerator of

Equation 8.58 corresponds to the $Pr \rightarrow \infty$, thermal entry length problem and yields values of $\overline{Nu}_D$ that are within 3% of the Hausen correlation for $0.006 \leq Gz_D^{-1} \leq 1$, also shown in Figure 8.10*b*. All properties appearing in Equations 8.57 and 8.58 should be evaluated at the average mean temperature, $\overline{T}_m = (T_{m,i} + T_{m,o})/2$.

The subject of laminar flow in ducts has been studied extensively, and numerous results are available for a variety of duct cross sections and surface conditions. Representative results have been compiled in a monograph by Shah and London [12] and in an updated review by Shah and Bhatti [13]. Correlations for the combined entry region for non-circular ducts have been developed by Muzychka and Yovanovich [14].

8.4.3 Temperature-Dependent Properties

When differences between the surface temperature T_s and the mean temperature T_m correspond to large fluid property variations, the Nusselt number calculated from Equations 8.53, 8.55, 8.57, or 8.58 can be affected. For gases, this effect is usually small. For liquids, however, the viscosity variation may be particularly important. This is especially true for oils. Viscosity variation changes the radial velocity distribution, which affects the radial temperature distribution and ultimately alters the Nusselt number. Kays et al. [2] recommend applying the following correction factor to the Nusselt number for liquids:

$$\frac{Nu_{D,c}}{Nu_D} = \frac{\overline{Nu}_{D,c}}{\overline{Nu}_D} = \left(\frac{\mu}{\mu_s}\right)^{0.14} \tag{8.59}$$

In this expression, $Nu_{D,c}$ and $\overline{Nu}_{D,c}$ are the corrected Nusselt numbers, while Nu_D and $\overline{Nu}_D$ are Nusselt numbers found from Equations 8.53, 8.55, 8.57, or 8.58. All properties in Equation 8.59 are evaluated at $\overline{T}_m$ except for μ_s, which is evaluated at the surface temperature T_s. This correction factor can be applied to laminar flow of a liquid in a circular tube, regardless of whether the flow is fully developed or in the entry length region. The correction factor may also be applied to tubes of noncircular cross section in the absence of other alternatives [15].

8.5 *Convection Correlations: Turbulent Flow in Circular Tubes*

Since the analysis of turbulent flow conditions is a good deal more involved, greater emphasis is placed on determining empirical correlations. For *fully developed (hydrodynamically and thermally) turbulent flow* in a *smooth circular tube*, the *local* Nusselt number may be obtained from the *Dittus–Boelter equation*[2] [16]:

$$Nu_D = 0.023\, Re_D^{4/5}\, Pr^n \tag{8.60}$$

[2]Although it has become common practice to refer to Equation 8.60 as the *Dittus–Boelter equation*, the original Dittus–Boelter equations are actually of the form

$Nu_D = 0.0243\, Re_D^{4/5}\, Pr^{0.4}$ (Heating)

$Nu_D = 0.0265\, Re_D^{4/5}\, Pr^{0.3}$ (Cooling)

The historical origins of Equation 8.60 are discussed by Winterton [16].

where $n = 0.4$ for heating ($T_s > T_m$) and 0.3 for cooling ($T_s < T_m$). These equations have been confirmed experimentally for the range of conditions

$$\begin{bmatrix} 0.6 \lesssim Pr \lesssim 160 \\ Re_D \gtrsim 10{,}000 \\ \dfrac{L}{D} \gtrsim 10 \end{bmatrix}$$

The equations may be used for small to moderate temperature differences, $T_s - T_m$, with all properties evaluated at T_m. For flows characterized by large property variations, the following equation, due to Sieder and Tate [17], is recommended:

$$Nu_D = 0.027\, Re_D^{4/5}\, Pr^{1/3} \left(\frac{\mu}{\mu_s}\right)^{0.14} \tag{8.61}$$

$$\begin{bmatrix} 0.7 \lesssim Pr \lesssim 16{,}700 \\ Re_D \gtrsim 10{,}000 \\ \dfrac{L}{D} \gtrsim 10 \end{bmatrix}$$

where all properties except μ_s are evaluated at T_m. *To a good approximation, the foregoing correlations may be applied for both the uniform surface temperature and heat flux conditions.*

Although Equations 8.60 and 8.61 are easily applied and are certainly satisfactory for the purposes of this text, errors as large as 25% may result from their use. Such errors may be reduced to less than 10% through the use of more recent, but generally more complex, correlations [5, 18]. One correlation, valid for *smooth tubes* over a large Reynolds number range including the transition region, is provided by Gnielinski [19]:

$$Nu_D = \frac{(f/8)(Re_D - 1000)\, Pr}{1 + 12.7(f/8)^{1/2}(Pr^{2/3} - 1)} \tag{8.62}$$

where the friction factor may be obtained from the Moody diagram or from Equation 8.21. The correlation is valid for $0.5 \lesssim Pr \lesssim 2000$ and $3000 \lesssim Re_D \lesssim 5 \times 10^6$. In using Equation 8.62, which applies for both uniform surface heat flux and temperature, properties should be evaluated at T_m. If temperature differences are large, additional consideration must be given to variable-property effects and available options are reviewed by Kakac [20].

We note that, unless specifically developed for the transition region ($2300 < Re_D < 10^4$), caution should be exercised when applying a turbulent flow correlation for $Re_D < 10^4$. If the correlation was developed for fully turbulent conditions ($Re_D > 10^4$), it may be used as a first approximation at smaller Reynolds numbers, with the understanding that the convection coefficient will be overpredicted. If a higher level of accuracy is desired, the Gnielinski correlation, Equation 8.62, may be used. A comprehensive discussion of heat transfer in the transition region is provided by Ghajar and Tam [21].

We also note that Equations 8.60 through 8.62 pertain to smooth tubes. For turbulent flow in *rough tubes*, the heat transfer coefficient increases with wall roughness, and, as a first approximation, it may be computed by using Equation 8.62 with friction factors obtained from Equation 8.20 or the Moody diagram, Figure 8.3. However, although the general trend is one of increasing h with increasing f, the increase in f is proportionately larger, and when f is approximately four times larger than the corresponding value for a smooth surface, h no longer changes with additional increases in f [22]. Procedures for estimating the effect of wall roughness on convection heat transfer in fully developed turbulent flow are discussed by Bhatti and Shah [18].

Since entry lengths for turbulent flow are typically short, $10 \lesssim (x_{fd}/D) \lesssim 60$, it is often reasonable to assume that the average Nusselt number for the entire tube is equal to the value associated with the fully developed region, $\overline{Nu}_D \approx Nu_{D,fd}$. However, for short tubes $\overline{Nu}_D$ will exceed $Nu_{D,fd}$ and may be calculated from an expression of the form

$$\frac{\overline{Nu}_D}{Nu_{D,fd}} = 1 + \frac{C}{(x/D)^m} \tag{8.63}$$

where C and m depend on the nature of the inlet (e.g., sharp-edged or nozzle) and entry region (thermal or combined), as well as on the Prandtl and Reynolds numbers [2, 18, 23]. Typically, errors of less than 15% are associated with assuming $\overline{Nu}_D = Nu_{D,fd}$ for $(L/D) > 60$. When determining $\overline{Nu}_D$, all fluid properties should be evaluated at the arithmetic average of the mean temperature, $\overline{T}_m \equiv (T_{m,i} + T_{m,o})/2$.

Finally, we note that the foregoing correlations do not apply to *liquid metals.* For fully developed turbulent flow in smooth circular tubes with constant surface heat flux, Skupinski et al. [24] recommend a correlation of the form

$$Nu_D = 4.82 + 0.0185\, Pe_D^{0.827} \qquad q''_s = \text{constant} \tag{8.64}$$

$$\begin{bmatrix} 3 \times 10^{-3} \lesssim Pr \lesssim 5 \times 10^{-2} \\ 3.6 \times 10^{3} \lesssim Re_D \lesssim 9.05 \times 10^{5} \\ 10^{2} \lesssim Pe_D \lesssim 10^{4} \end{bmatrix}$$

Similarly, for constant surface temperature Seban and Shimazaki [25] recommend the following correlation for $Pe_D \gtrsim 100$:

$$Nu_D = 5.0 + 0.025\, Pe_D^{0.8} \qquad T_s = \text{constant} \tag{8.65}$$

Extensive data and additional correlations are available in the literature [26].

Example 8.5

A method to generate electric power from solar irradiation involves concentrating sunlight onto absorber tubes that are placed at the focal points of parabolic reflectors. The absorber tubes carry a liquid *concentrator fluid* that is heated as it flows through the tubes. After it leaves the concentrating field, the fluid enters a heat exchanger, where it transfers thermal energy to the *working fluid* of a Rankine cycle. The cooled concentrator fluid is returned to the concentrator field after it exits the heat exchanger. A power plant consists of many concentrators.

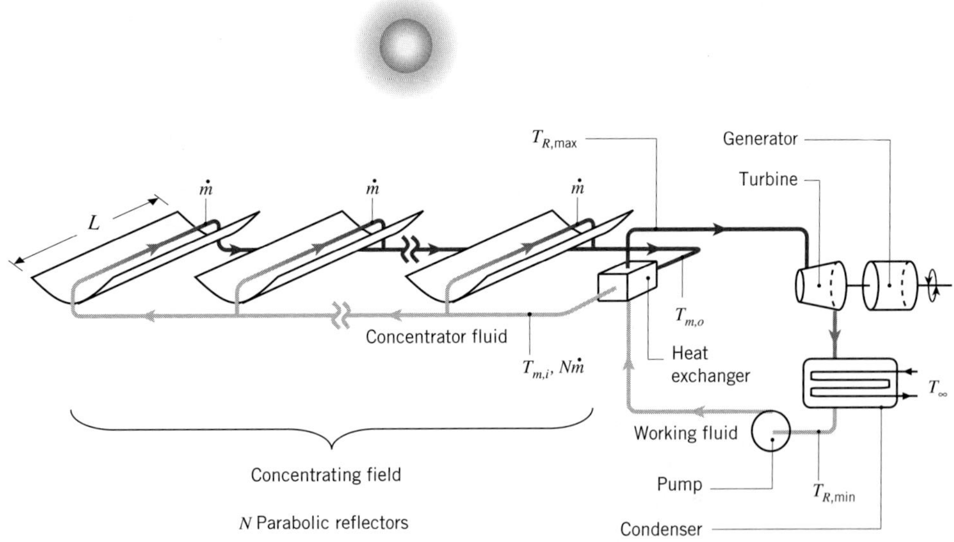

The net effect of a single concentrator-tube arrangement *may be approximated* as one of creating a constant heating condition at the surface of the tube. Consider conditions for which a concentrated heat flux of $q_s'' = 20{,}000\ \text{W/m}^2$, assumed to be uniform over the tube surface, heats a concentrator fluid of density, thermal conductivity, specific heat, and viscosity of $\rho = 700\ \text{kg/m}^3$, $k = 0.078\ \text{W/m}\cdot\text{K}$, $c_p = 2590\ \text{J/kg}\cdot\text{K}$, and $\mu = 0.15 \times 10^{-3}\ \text{N}\cdot\text{s/m}^2$, respectively. The tube diameter is $D = 70$ mm, and the mass flow rate of the fluid in a single concentrator tube is $\dot{m} = 2.5$ kg/s.

1. If the concentrator fluid enters each tube at $T_{m,i} = 400°\text{C}$ and exits at $T_{m,o} = 450°\text{C}$, what is the required concentrator length, L? How much heat q is transferred to the concentrator fluid in a single concentrator-tube arrangement?
2. What is the surface temperature of the tube at the exit of a concentrator, $T_s(L)$?
3. The maximum and minimum temperatures of the entire power plant are the exit temperature of the concentrator fluid $T_{m,o}$ and the ambient temperature T_∞, respectively. If a temperature difference of $\Delta T = T_{m,o} - T_{R,\max} = 20°\text{C}$ occurs across the heat exchanger and a second temperature difference of $\Delta T = T_{R,\min} - T_\infty = 20°\text{C}$ exists across the condenser, where $T_\infty = 20°\text{C}$, determine the minimum number of concentrators N, each of length L, needed to generate $P = 20$ MW of electric power.

SOLUTION

Known: Tube diameter, surface heat flux, fluid mass flow rate and properties. Inlet and outlet mean temperatures of fluid in concentrator. Temperature differences across the heat exchanger and condenser.

Find:

1. Length of concentrator to achieve required temperature increase and the corresponding heat transfer rate q.
2. Tube surface temperature at the end of the concentrator.
3. Minimum number of concentrators needed to generate $P = 20$ MW of electric power.

Schematic:

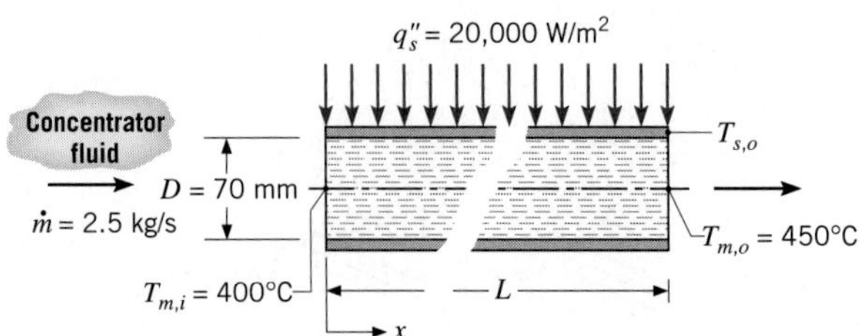

Assumptions:

1. Steady-state conditions.
2. Incompressible liquid with negligible viscous dissipation.
3. Constant properties.
4. Thin tube wall.

Analysis:

1. For constant heat flux conditions, Equation 8.38 may be used with the appropriate energy balance, Equation 8.34, to obtain

$$A_s = \pi DL = \frac{\dot{m}c_p(T_{m,o} - T_{m,i})}{q_s''}$$

$$L = \frac{\dot{m}c_p}{\pi D q_s''}(T_{m,o} - T_{m,i})$$

where L is the length of tube within the concentrator, which is also of length L. Hence

$$L = \frac{2.5\ \text{kg/s} \times 2590\ \text{J/kg}\cdot\text{K}}{\pi \times 0.070\ \text{m} \times 20{,}000\ \text{W/m}^2}(450°\text{C} - 400°\text{C}) = 73.6\ \text{m} \qquad \triangleleft$$

The heat transfer rate is

$$q = q_s''A = q_s''\pi DL = 20{,}000\ \text{W/m}^2 \times \pi \times 0.070\ \text{m} \times 73.6\ \text{m}$$

$$= 0.324 \times 10^6\ \text{W} = 0.324\ \text{MW} \qquad \triangleleft$$

2. The tube surface temperature at the end of the concentrator may be obtained from Newton's law of cooling, Equation 8.27, where

$$T_s(L) = \frac{q_s''}{h} + T_{m,o}$$

To find the local convection coefficient at the tube outlet, the nature of the flow condition must first be established. From Equation 8.6,

$$Re_D = \frac{4\dot{m}}{\pi D \mu} = \frac{4 \times 2.5 \text{ kg/s}}{\pi \times 0.070 \text{ m} \times 0.15 \times 10^{-3} \text{ N} \cdot \text{s/m}^2} = 3.03 \times 10^5$$

Hence the flow is turbulent. The Prandtl number of the concentrator fluid may be determined from

$$Pr = \frac{\nu}{\alpha} = \frac{\mu c_p}{k} = \frac{0.15 \times 10^{-3} \text{ N} \cdot \text{s/m}^2 \times 2590 \text{ J/kg} \cdot \text{K}}{0.078 \text{ W/m} \cdot \text{K}} = 4.98$$

Since $L/D = 73.6 \text{ m}/0.070 \text{ m} = 1050$, we conclude from Equation 8.4 that conditions are fully developed within the tube at the end of the concentrator. The local Nusselt number at $x = L$ is obtained from Equation 8.60

$$Nu_D = 0.023\, Re_D^{4/5}\, Pr^{0.4} = 0.023 \times (3.03 \times 10^5)^{4/5} \times 4.98^{0.4} = 1113$$

from which the local convection heat transfer coefficient is

$$h = \frac{k}{D} Nu_D = \frac{0.078 \text{ W/m} \cdot \text{K}}{0.070 \text{ m}} \times 1113 = 1240 \text{ W/m}^2 \cdot \text{K}$$

The tube surface temperature at the end of the concentrator is

$$T_s(L) = \frac{20{,}000 \text{ W/m}^2}{1240 \text{ W/m}^2 \cdot \text{K}} + 450°\text{C} = 466°\text{C} \quad \triangleleft$$

3. The *minimum* number of concentrators may be determined by first calculating the corresponding minimum amount of thermal energy required to generate $P = 20$ MW of electricity. The maximum possible (Carnot) efficiency is $\eta_C = 1 - T_{R,\min}/T_{R,\max} = 1 - (T_\infty + \Delta T)/(T_{m,o} - \Delta T) = 1 - (293 \text{ K} + 20 \text{ K})/(723 \text{ K} - 20 \text{ K}) = 0.555$. Hence the minimum thermal energy required is

$$q_{\min} = \frac{P}{\eta_C} = \frac{20 \text{ MW}}{0.555} = 36.1 \text{ MW}$$

Correspondingly, the minimum number of concentrators required is

$$N = \frac{q_{\min}}{q} = \frac{36.0 \text{ MW}}{0.324 \text{ MW}} = 111 \quad \triangleleft$$

Comments:

1. If temperature differences within the heat exchanger and the condenser could be eliminated ($\Delta T = 0°\text{C}$), the Carnot efficiency would be $\eta_C = 1 - T_{R,\min}/T_{R,\max} = 1 - T_\infty/T_{m,o} = 1 - (293 \text{ K})/(723 \text{ K}) = 0.595$. This yields $q_{\min} = 33.6$ MW and $N = 104$. Minimizing thermal resistances in the heat exchanger and condenser reduces the number of concentrators required to generate a specified amount of electric power and can reduce the capital cost of the plant.
2. Actual thermal efficiencies are less than the Carnot efficiency, and a nominal value of 38% is associated with parabolic trough *Solar Electric Generating Stations* (SEGS)

operating in Southern California since the mid-1980s. However, the *overall efficiency* of power plants using concentrating solar collectors is typically defined as a ratio of the rate of power generation to the rate at which solar energy is intercepted by the collectors. With a nominal efficiency of 40% for conversion of solar energy to thermal energy, the overall efficiency of the SEGS systems is approximately 15%.

3. A contemporary research challenge is to develop concentrator fluids that do not boil during periods of high solar irradiation and that resist freezing at night. Moreover, development of inexpensive and safe liquids capable of withstanding even higher temperatures will lead to higher maximum Rankine cycle temperatures, $T_{R,\max}$, and, in turn, increased plant efficiency.
4. As implied in Comments 1 and 2, good thermal management and development of new heat transfer fluids can reduce the cost of solar-generated electricity; they are key factors in increasing the amount of electricity generated from the sun.

EXAMPLE 8.6

Hot air flows with a mass rate of $\dot{m} = 0.050$ kg/s through an uninsulated sheet metal duct of diameter $D = 0.15$ m, which is in the crawlspace of a house. The hot air enters at 103°C and, after a distance of $L = 5$ m, cools to 85°C. The heat transfer coefficient between the duct outer surface and the ambient air at $T_\infty = 0°C$ is known to be $h_o = 6\ W/m^2 \cdot K$.

1. Calculate the heat loss (W) from the duct over the length L.
2. Determine the heat flux and the duct surface temperature at $x = L$.

SOLUTION

Known: Hot air flowing in a duct.

Find:

1. Heat loss from the duct over the length L, q(W).
2. Heat flux and surface temperature at $x = L$.

Schematic:

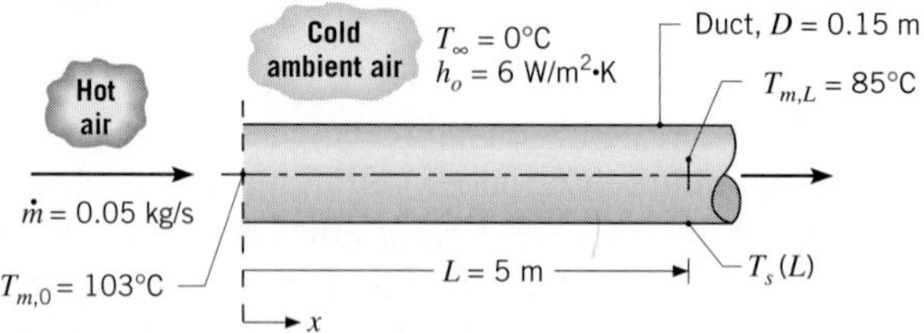

Assumptions:

1. Steady-state conditions.
2. Constant properties.
3. Ideal gas behavior.

4. Negligible viscous dissipation and negligible pressure variations.
5. Negligible duct wall thermal resistance.
6. Uniform convection coefficient at outer surface of duct.
7. Negligible radiation.

Properties: Table A.4, air ($\bar{T}_m = 367$ K): $c_p = 1011$ J/kg·K. Table A.4, air ($T_{m,L} = 358$ K): $k = 0.0306$ W/m·K, $\mu = 211.7 \times 10^{-7}$ N·s/m², $Pr = 0.698$.

Analysis:

1. From the energy balance for the entire tube, Equation 8.34,

$$q = \dot{m}c_p(T_{m,L} - T_{m,0})$$

$$q = 0.05 \text{ kg/s} \times 1011 \text{ J/kg}\cdot\text{K}(85 - 103)°\text{C} = -910 \text{ W} \qquad \triangleleft$$

2. An expression for the heat flux at $x = L$ may be inferred from the resistance network

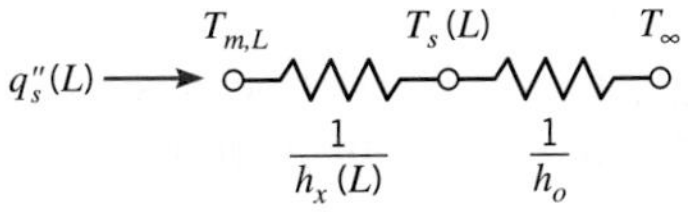

$$q_s''(L) \longrightarrow \underset{T_{m,L}}{\circ} \overset{}{\underset{\frac{1}{h_x(L)}}{\text{—WW—}}} \underset{T_s(L)}{\circ} \underset{\frac{1}{h_o}}{\text{—WW—}} \underset{T_\infty}{\circ}$$

where $h_x(L)$ is the inside convection heat transfer coefficient at $x = L$. Hence

$$q_s''(L) = \frac{T_{m,L} - T_\infty}{1/h_x(L) + 1/h_o}$$

The inside convection coefficient may be obtained from knowledge of the Reynolds number. From Equation 8.6

$$Re_D = \frac{4\dot{m}}{\pi D \mu} = \frac{4 \times 0.05 \text{ kg/s}}{\pi \times 0.15 \text{ m} \times 211.7 \times 10^{-7} \text{ N}\cdot\text{s/m}^2} = 20{,}050$$

Hence the flow is turbulent. Moreover, with $(L/D) = (5/0.15) = 33.3$, it is reasonable to assume fully developed conditions at $x = L$. From Equation 8.60, with $n = 0.3$,

$$Nu_D = \frac{h_x(L)D}{k} = 0.023\, Re_D^{4/5}\, Pr^{0.3} = 0.023(20{,}050)^{4/5}(0.698)^{0.3} = 56.4$$

$$h_x(L) = Nu_D \frac{k}{D} = 56.4 \frac{0.0306 \text{ W/m}\cdot\text{K}}{0.15 \text{ m}} = 11.5 \text{ W/m}^2\cdot\text{K}$$

Therefore

$$q_s''(L) = \frac{(85 - 0)°\text{C}}{(1/11.5 + 1/6.0)\text{m}^2\cdot\text{K/W}} = 335 \text{ W/m}^2$$

Referring back to the network, it also follows that

$$q_s''(L) = \frac{T_{m,L} - T_{s,L}}{1/h_x(L)}$$

in which case

$$T_{s,L} = T_{m,L} - \frac{q_s''(L)}{h_x(L)} = 85°\text{C} - \frac{335 \text{ W/m}^2}{11.5 \text{ W/m}^2\cdot\text{K}} = 55.9°\text{C} \qquad \triangleleft$$

Comments:

1. In using the energy balance of part 1 for the entire tube, properties (in this case, only c_p) are evaluated at $\overline{T}_m = (T_{m,0} + T_{m,L})/2$. However, in using the correlation for a local heat transfer coefficient, Equation 8.60, properties are evaluated at the local mean temperature, $T_{m,L} = 85°\text{C}$.
2. The overall average heat transfer coefficient $\overline{U}$ may be determined from Equation 8.45a, which can be rearranged to yield

$$\overline{U} = -\frac{\dot{m}c_p}{\pi DL}\ln\left[\frac{T_\infty - T_{m,o}}{T_\infty - T_{m,i}}\right] = -\frac{0.05\text{ kg/s} \times 1011\text{ J/kg}\cdot\text{K}}{\pi \times 0.15\text{ m} \times 5\text{ m}}\ln\left[\frac{-85°\text{C}}{-103°\text{C}}\right] = 4.12\text{ W/m}^2\cdot\text{K}$$

   It follows from Assumption 6 that $\overline{h}_o = h_o$ and that $\overline{h}_i = 1/(1/\overline{U} - 1/\overline{h}_o) = 13.2\text{ W/m}^2\cdot\text{K}$. The average inside convection heat transfer coefficient is larger than $h_x(L)$, as expected from Equation 8.63.

3. This problem is characterized neither by constant surface temperature nor by constant surface heat flux. It would therefore be erroneous to presume that the total heat loss from the tube is given by $q_s''(L)\pi DL = 790\text{ W}$. This result is substantially less than the actual heat loss of 910 W because $q_s''(x)$ decreases with increasing x. This decrease in $q_s''(x)$ is due to reductions in both $h_x(x)$ and $[T_m(x) - T_\infty]$ with increasing x.

8.6 *Convection Correlations: Noncircular Tubes and the Concentric Tube Annulus*

Although we have thus far restricted our consideration to internal flows of circular cross section, many engineering applications involve convection transport in *noncircular tubes.* At least to a first approximation, however, many of the circular tube results may be applied by using *an effective diameter* as the characteristic length. It is termed the *hydraulic diameter* and is defined as

$$D_h \equiv \frac{4A_c}{P} \tag{8.66}$$

where A_c and P are the *flow* cross-sectional area and the *wetted perimeter,* respectively. It is this diameter that should be used in calculating parameters such as Re_D and Nu_D.

For turbulent flow, which still occurs if $Re_D \gtrsim 2300$, it is reasonable to use the correlations of Section 8.5 for $Pr \gtrsim 0.7$. However, in a noncircular tube the convection coefficients vary around the periphery, approaching zero in the corners. Hence in using a circular tube correlation, the coefficient is presumed to be an average over the perimeter.

For laminar flow, the use of circular tube correlations is less accurate, particularly with cross sections characterized by sharp corners. For such cases the Nusselt number corresponding to fully developed conditions may be obtained from Table 8.1, which is based on solutions of the differential momentum and energy equations for flow through the different duct cross sections. As for the circular tube, results differ according to the surface thermal

TABLE 8.1 Nusselt numbers and friction factors for fully developed laminar flow in tubes of differing cross section

Cross Section	$\frac{b}{a}$	$Nu_D \equiv \frac{hD_h}{k}$ (Uniform q''_s)	$Nu_D \equiv \frac{hD_h}{k}$ (Uniform T_s)	fRe_{D_h}
Circle	—	4.36	3.66	64
Rectangle (a, b)	1.0	3.61	2.98	57
Rectangle (a, b)	1.43	3.73	3.08	59
Rectangle (a, b)	2.0	4.12	3.39	62
Rectangle (a, b)	3.0	4.79	3.96	69
Rectangle (a, b)	4.0	5.33	4.44	73
Rectangle (a, b)	8.0	6.49	5.60	82
Parallel plates	∞	8.23	7.54	96
Heated / Insulated	∞	5.39	4.86	96
Triangle	—	3.11	2.47	53

Used with permission from [12].

condition. Nusselt numbers tabulated for a uniform surface heat flux presume a constant flux in the axial (flow) direction, but a constant temperature around the perimeter at any cross section. This condition is typical of highly conductive tube wall materials. Results tabulated for a uniform surface temperature apply when the temperature is constant in both the axial and peripheral directions. Care should be taken when comparing the values of the Nusselt numbers associated with different cross-sectional shapes. Specifically, a cross section that is characterized by a larger Nusselt number does not necessarily imply more effective convection heat transfer, since both the hydraulic diameter and the wetted perimeter are cross section–dependent. See Problem 8.87.

Although the foregoing procedures are generally satisfactory, exceptions do exist. Detailed discussions of heat transfer in noncircular tubes are provided in several sources [12, 13, 27].

Many internal flow problems involve heat transfer in a *concentric tube annulus* (Figure 8.11). Fluid passes through the space (annulus) formed by the concentric tubes, and convection heat transfer may occur to or from both the inner and outer tube surfaces. It is possible to independently specify the heat flux or temperature, that is, the thermal

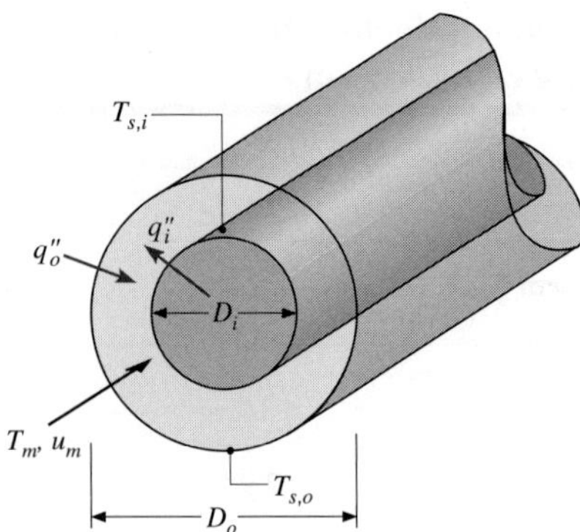

FIGURE 8.11 The concentric tube annulus.

condition, at each of these surfaces. In any case the heat flux from each surface may be computed with expressions of the form

$$q_i'' = h_i(T_{s,i} - T_m) \tag{8.67}$$

$$q_o'' = h_o(T_{s,o} - T_m) \tag{8.68}$$

Note that separate convection coefficients are associated with the inner and outer surfaces. The corresponding Nusselt numbers are of the form

$$Nu_i \equiv \frac{h_i D_h}{k} \tag{8.69}$$

$$Nu_o \equiv \frac{h_o D_h}{k} \tag{8.70}$$

where, from Equation 8.66, the hydraulic diameter D_h is

$$D_h = \frac{4(\pi/4)(D_o^2 - D_i^2)}{\pi D_o + \pi D_i} = D_o - D_i \tag{8.71}$$

For the case of fully developed laminar flow with one surface insulated and the other surface at a constant temperature, Nu_i or Nu_o may be obtained from Table 8.2. Note that in

TABLE 8.2 Nussselt number for fully developed laminar flow in a circular tube annulus with one surface insulated and the other at constant temperature

D_i/D_o	Nu_i	Nu_o	Comments
0	—	3.66	See Equation 8.55
0.05	17.46	4.06	
0.10	11.56	4.11	
0.25	7.37	4.23	
0.50	5.74	4.43	
≈1.00	4.86	4.86	See Table 8.1, $b/a \rightarrow \infty$

Adapted from Lundberg, R.E., W.C. Reynolds, and W.M. Kays, *Heat Transfer with Laminar Flow in Concentric Annuli with Constant and Variable Wall Temperature and Heat Flux,* NASA TN D-1972, 1963.

TABLE 8.3 Influence coefficients for fully developed laminar flow in a circular tube annulus with uniform heat flux maintained at both surfaces

D_i/D_o	Nu_{ii}	Nu_{oo}	θ_i^*	θ_o^*
0	—	4.364[a]	∞	0
0.05	17.81	4.792	2.18	0.0294
0.10	11.91	4.834	1.384	0.0562
0.20	8.499	4.883	0.904	0.1039
0.40	6.583	4.979	0.602	0.1822
0.60	5.912	5.099	0.474	0.2455
0.80	5.58	5.24	0.401	0.298
1.00	5.385	5.385[b]	0.346	0.346

Adapted from Lundberg, R.E., W.C. Reynolds, and W.M. Kays, *Heat Transfer with Laminar Flow in Concentric Annuli with Constant and Variable Wall Temperature and Heat Flux,* NASA TN D-1972, 1963.

[a]See Equation 8.53.
[b]See Table 8.1 for $b/a \rightarrow \infty$ with one surface insulated.

such cases we would be interested only in the convection coefficient associated with the isothermal (nonadiabatic) surface.

If uniform heat flux conditions exist at both surfaces, the Nusselt numbers may be computed from expressions of the form

$$Nu_i = \frac{Nu_{ii}}{1 - (q_o''/q_i'')\theta_i^*} \tag{8.72}$$

$$Nu_o = \frac{Nu_{oo}}{1 - (q_i''/q_o'')\theta_o^*} \tag{8.73}$$

The influence coefficients (Nu_{ii}, Nu_{oo}, θ_i^*, and θ_o^*) appearing in these equations may be obtained from Table 8.3. Note that q_i'' and q_o'' may be positive or negative, depending on whether heat transfer is to or from the fluid, respectively. Moreover, situations may arise for which the values of h_i and h_o are negative. Such results, when used with the sign convention implicit in Equations 8.67 and 8.68, reveal the relative magnitudes of T_s and T_m.

For fully developed turbulent flow, the influence coefficients are a function of the Reynolds and Prandtl numbers [27]. However, to a first approximation the inner and outer convection coefficients may be assumed to be equal, and they may be evaluated by using the hydraulic diameter, Equation 8.71, with the Dittus–Boelter equation, Equation 8.60.

8.7 *Heat Transfer Enhancement*

Several options are available for enhancing heat transfer associated with internal flows. Enhancement may be achieved by increasing the convection coefficient and/or by increasing the convection surface area. For example, h may be increased by introducing surface roughness to enhance turbulence, as, for example, through machining or insertion of a coil-spring wire. The wire insert (Figure 8.12*a*) provides a helical roughness element in contact

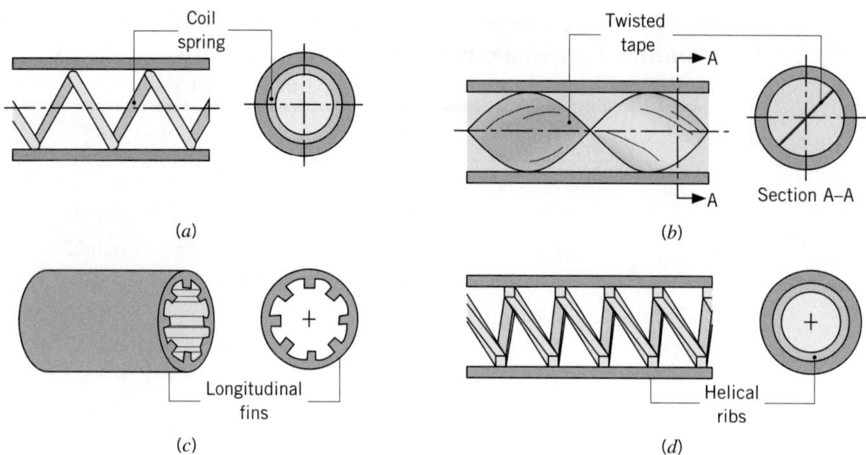

FIGURE 8.12 Internal flow heat transfer enhancement schemes: (*a*) longitudinal section and end view of coil-spring wire insert, (*b*) longitudinal section and cross-sectional view of twisted tape insert, (*c*) cut-away section and end view of longitudinal fins, and (*d*) longitudinal section and end view of helical ribs.

with the tube inner surface. Alternatively, the convection coefficient may be increased by inducing swirl through insertion of a twisted tape (Figure 8.12*b*). The insert consists of a thin strip that is periodically twisted through 360°. Introduction of a tangential velocity component increases the speed of the flow, particularly near the tube wall. The heat transfer area may be increased by manufacturing a tube with a grooved inner surface (Figure 8.12*c*), while both the convection coefficient and area may be increased by using spiral fins or ribs (Figure 8.12*d*). In evaluating any heat transfer enhancement scheme, attention must also be given to the attendant increase in pressure drop and hence fan or pump power requirements. Comprehensive assessments of enhancement options have been published [28–31], and the *Journal of Enhanced Heat Transfer* provides access to recent developments in the field.

By coiling a tube (Figure 8.13), heat transfer may be enhanced without turbulence or additional heat transfer surface area. In this case, centrifugal forces within the fluid induce

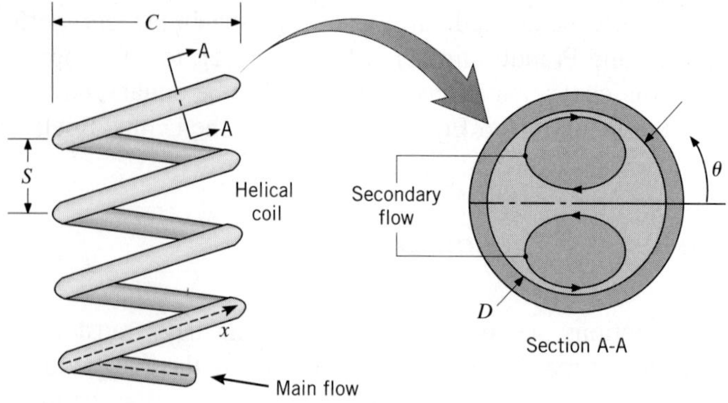

FIGURE 8.13 Schematic of helically coiled tube and secondary flow in enlarged cross-sectional view.

a *secondary flow* consisting of a pair of longitudinal vortices that, in contrast to conditions in a straight tube, can result in highly nonuniform local heat transfer coefficients around the periphery of the tube. Hence, local heat transfer coefficients vary with θ as well as x. If constant heat flux conditions are applied, the mean fluid temperature, $T_m(x)$, may be calculated using the conservation of energy principle, Equation 8.40. For situations where the fluid is heated, maximum fluid temperatures occur at the tube wall, but calculation of the maximum local temperature is not straightforward because of the θ-dependence of the heat transfer coefficient. Therefore, correlations for the *peripherally averaged* Nusselt number are of little use if constant heat flux conditions are applied. In contrast, correlations for the peripherally averaged Nusselt number for constant wall temperature boundary conditions are useful, and the relationships recommended by Shah and Joshi [32] are provided in the next paragraphs.

The secondary flow increases friction losses and heat transfer rates. In addition, the secondary flow decreases entrance lengths and reduces the difference between laminar and turbulent heat transfer rates, relative to the straight tube cases considered previously in this chapter. Pressure drops and heat transfer rates exhibit little dependence on the coil pitch, S. The critical Reynolds number corresponding to the onset of turbulence for the helical tube, $Re_{D,c,h}$, is

$$Re_{D,c,h} = Re_{D,c}[1 + 12(D/C)^{0.5}] \tag{8.74}$$

where $Re_{D,c}$ is given in Equation 8.2 and C is defined in Figure 8.13. Strong secondary flow associated with tightly wound helically coiled tubes delays the transition to turbulence.

For fully developed laminar flow with $C/D \gtrsim 3$, the friction factor is

$$f = \frac{64}{Re_D} \qquad Re_D(D/C)^{1/2} \lesssim 30 \tag{8.19}$$

$$f = \frac{27}{Re_D^{0.725}}(D/C)^{0.1375} \qquad 30 \lesssim Re_D(D/C)^{1/2} \lesssim 300 \tag{8.75a}$$

$$f = \frac{7.2}{Re_D^{0.5}}(D/C)^{0.25} \qquad 300 \lesssim Re_D(D/C)^{1/2} \tag{8.75b}$$

For cases where $C/D \lesssim 3$, recommendations provided in Shah and Joshi [32] should be followed. The heat transfer coefficient for use in Equation 8.27 may be evaluated from a correlation of the form

$$Nu_D = \left[\left(3.66 + \frac{4.343}{a}\right)^3 + 1.158\left(\frac{Re_D(D/C)^{1/2}}{b}\right)^{3/2}\right]^{1/3}\left(\frac{\mu}{\mu_s}\right)^{0.14} \tag{8.76}$$

where

$$a = \left(1 + \frac{927(C/D)}{Re_D^2\,Pr}\right) \qquad \text{and} \qquad b = 1 + \frac{0.477}{Pr} \tag{8.77a,b}$$

$$\begin{bmatrix} 0.005 \lesssim Pr \lesssim 1600 \\ 1 \lesssim Re_D(D/C)^{1/2} \lesssim 1000 \end{bmatrix}$$

Friction factor correlations for turbulent flow are based on limited data. Furthermore, heat transfer augmentation due to the secondary flow is minor when the flow is turbulent and is less than 10% for $C/D \gtrsim 20$. As such, augmentation by using helically coiled tubes is typically employed only for laminar flow situations. In laminar flow, the entrance length is 20% to 50% shorter than that of a straight tube, while the flow becomes fully developed within the first half-turn of the helically coiled tube under turbulent conditions. Therefore, the entrance region may be neglected in most engineering calculations. A compilation of additional correlations is available [33].

When a gas or liquid is heated in a straight tube, a fluid parcel that enters near the centerline of the tube will exit the tube faster, and always be cooler than, a fluid parcel that enters near the tube wall. Hence, the *time-temperature histories* of individual fluid parcels, processed in the same heating tube, can be dramatically different. In addition to augmenting heat transfer, the secondary flow associated with the helically coiled tube serves to *mix* the fluid relative to laminar flow in a straight tube, resulting in similar time-temperature histories for all the fluid parcels. It is for this reason that coiled tubes are routinely used to process and manufacture highly viscous, high value-added fluids, such as pharmaceuticals, cosmetics, and personal care products [33].

8.8 *Flow in Small Channels*

The tubes and channels considered thus far have been characterized by hydraulic diameters of conventional size. However, many technologies involve internal flows with channels of relatively small dimension. An important motivation for developing *microfluidic devices* is readily evident from inspection of Equations 8.53 and 8.55, as well as Tables 8.1 through 8.3, where the heat transfer coefficients are inversely proportional to the hydraulic diameter. That is, as the channel dimensions are decreased, heat transfer coefficients become large [34]. However, care must be taken to consider effects that have not been discussed in the preceding sections of this chapter.

8.8.1 Microscale Convection in Gases ($0.1\ \mu m \lesssim D_h \lesssim 100\ \mu m$)

In most situations, unrealistically high gas velocities are required to achieve $Re_{D,c} = 2300$ in situations involving $D_h \lesssim 100\ \mu m$. Therefore, one is typically not concerned with turbulent microscale convection involving gases.

For very small tubes or channels, the interaction of gas molecules with the tube or channel wall may become important. As discussed in Section 3.9.1, conduction through gas layers can be affected when the characteristic length of the gas volume is of the same magnitude as the mean free path of the gas λ_{mfp}. The same holds true for convection in channels with very small hydraulic diameters. The various correlations presented in the previous sections of this chapter are not expected to apply for gases when $D_h/\lambda_{mfp} \lesssim 100$. For air at atmospheric temperature and pressure, this limit corresponds to $D_h \lesssim 10\ \mu m$.

If the situation shown in Figure 3.25 were to involve a bulk flow in the vertical direction, the manner in which individual gas molecules scatter from the two solid walls would also affect the transfer of momentum throughout the gas and, in turn, the velocity distribution within the gas. Since the gas temperature distribution depends on the gas velocity, a *momentum accommodation coefficient* α_p will influence convection heat transfer rates, as will the thermal accommodation coefficient α_t of Equation 3.130 [35]. Values of the momentum accommodation coefficient are in the range $0 \leq \alpha_p \leq 1$. Specifically, *specular reflection* (where the speed of the molecule is unchanged, and the angle of reflection from the surface is equal to the angle of incidence on the surface) corresponds to $\alpha_p = 0$. On the other hand, *diffuse reflection* (with no preferred angle of reflection) corresponds to $\alpha_p = 1$. Values of α_p for air interacting with most engineering surfaces range from 0.87 to unity, while for nitrogen, argon, or CO_2 in silicon channels $0.75 \lesssim \alpha_p \lesssim 0.85$ [35].

Convection heat transfer in microscale internal gas flow has been analyzed, accounting for both thermal and momentum interactions between the gas molecules and the solid walls. For laminar, fully developed flow in a circular tube of diameter D with a uniform surface heat flux, the Nusselt number may be expressed as [36]

$$Nu_D = \frac{hD}{k} = \frac{48}{11 - 6\zeta + \zeta^2 + 48\Gamma_t} \tag{8.78a}$$

where

$$\zeta = 8\Gamma_p/(1 + 8\Gamma_p) \tag{8.78b}$$

$$\Gamma_p = \frac{2 - \alpha_p}{\alpha_p}\left[\frac{\lambda_{\text{mfp}}}{D}\right] \tag{8.78c}$$

$$\Gamma_t = \frac{2 - \alpha_t}{\alpha_t}\,\frac{2\gamma}{\gamma + 1}\left[\frac{\lambda_{\text{mfp}}}{Pr\,D}\right] \tag{8.78d}$$

The term $\gamma \equiv c_p/c_v$ is the ratio of specific heats of the gas. For large tube diameters ($\lambda_{\text{mfp}}/D \rightarrow 0$), $Nu_D \rightarrow 48/11 = 4.36$, in agreement with Equation 8.53.

Similarly, for laminar, fully developed flow in a channel formed by large plates separated by a spacing a, the Nusselt number for uniform and equal plate heat fluxes is [37]

$$Nu_D = \frac{hD_h}{k} = \frac{140}{17 - 6\zeta + (2/3)\zeta^2 + 70\Gamma_t} \tag{8.79a}$$

where

$$\zeta = 6\Gamma_p/(1 + 6\Gamma_p) \tag{8.79b}$$

Here, Γ_p and Γ_t are defined as in Equations 8.78c,d with $D = D_h$. For infinitely large plates the hydraulic diameter is $D_h = 2a$ and for large plate spacing, $\lambda_{\text{mfp}}/D_h \rightarrow 0$, $Nu_D \rightarrow 140/17 = 8.23$, in agreement with Table 8.1.

The preceding relations may be applied only when the gas flow can be treated as incompressible, that is, when the Mach number is small ($Ma \lesssim 0.3$).

8.8.2 Microscale Convection in Liquids

Experiments have shown that Equations 8.19 and 8.22a may be applied to laminar liquid flows in tubes with diameters as small as 17 μm [38, 39].[3] These equations are expected to be valid

[3]From the discussion of Section 6.3.1, one might anticipate that, since turbulence is characterized by the motion of relatively large parcels of fluid in devices of conventional size, Equations 8.2, 8.20, and 8.21 would not be applicable for flow in microfluidic devices because the volume of the fluid parcel is restricted by the hydraulic diameter of the channel. Nonetheless, careful measurements using various liquids have shown that Equation 8.2 does indeed hold for liquid flow in tubes with diameters at least as small as 17 μm [39].

for most liquids for hydraulic diameters as small as 1 μm [38, 40]. Convection heat transfer in microscale internal flows involving liquids is the subject of ongoing research. The analytical results of Chapters 6 and 8 should be used with caution for liquids when $D_h \lesssim 1\ \mu$m.

8.8.3 Nanoscale Convection ($D_h \lesssim 100$ nm)

As the hydraulic diameter approaches $D_h \approx 0.1\ \mu\text{m} = 100$ nm, molecular interactions must, in general, be accounted for both in the fluid and in the solid wall. Nanoscale convection is an area of current research.

EXAMPLE 8.7

Combinatorial chemistry and biology are used in the pharmaceutical and biotechnology industries to reduce the time and cost associated with producing new drugs. Scientists desire to create large populations of molecules, or *libraries*, that can be subsequently screened en masse. Producing vast libraries increases the probability that novel compounds of significant therapeutic value will be discovered. A crucial variable in producing new compounds is the temperature at which the reactants are processed.

A *microreactor chip* is fabricated by first coating a 1-mm-thick glass microscope slide with a photoresist material. Lines are subsequently etched into the photoresist and a second glass plate is attached to the top of the structure, resulting in multiple parallel channels of rectangular cross section that are $a = 40\ \mu$m deep, $b = 160\ \mu$m wide, and $L = 20$ mm long. The spacing between channels is $s = 40\ \mu$m, so that $N = L/(w + s) = 100$ channels are present within the 20 mm $\times$ 20 mm microreactor. A mixture of two reactants, both initially at $T_{m,i} = 5°$C, is introduced into each channel, and the edges of the chip are maintained at temperatures $T_1 = 125°$C and $T_2 = 25°$C so that the reactants in each channel are subject to a unique processing temperature. Flow is induced through the structure by applying an overall pressure difference of $\Delta p = 500$ kPa. The reactants and the product of reaction have thermophysical properties similar to ethylene glycol. Estimate the time that elapses for the entering reactants to come within 1°C of the desired processing temperature.

SOLUTION

Known: Dimensions and operating conditions for flow of reactants and product of reaction in a microreactor.

Find: Time needed to bring the reactants to within 1°C of the processing temperature.

Schematic:

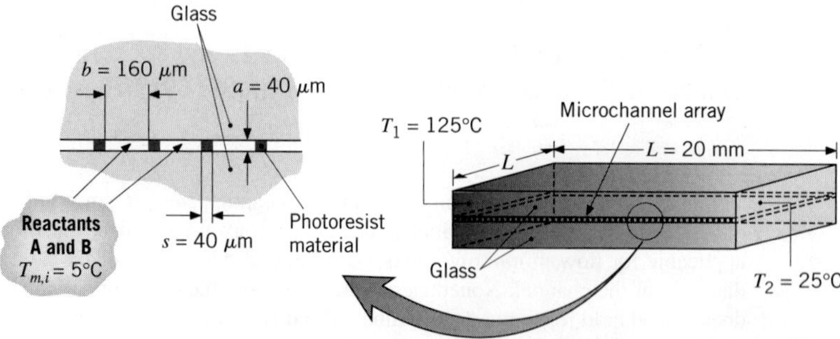

Assumptions:

1. Laminar flow.
2. Linear temperature distribution across the width of the microreactor.
3. Steady-state conditions.
4. Incompressible liquid with constant properties.
5. Negligible viscous dissipation.

Properties: Table A.5, ethylene glycol ($\bar{T}_m = 288$ K): $\rho = 1120.2$ kg/m^3, $c_p = 2359$ J/kg·K, $\mu = 2.82 \times 10^{-2}$ N·s/m^2, $k = 247 \times 10^{-3}$ W/m·K, $Pr = 269$. ($\bar{T}_m = 338$ K): $\rho = 1085$ kg/m^3, $c_p = 2583$ J/kg·K, $\mu = 0.427 \times 10^{-2}$ N·s/m^2, $k = 261 \times 10^{-3}$ W/m·K, $Pr = 45.2$.

Analysis: We will bracket the heat transfer and fluid flow behavior by evaluating the performance at the extreme processing temperatures.

The flow of reactants is induced by the applied pressure difference between the inlet and outlet of the microreactor. Because of the large variation of the viscosity with temperature, we expect the flow rate that is associated with the highest processing temperature to be the largest.

The perimeter of each microchannel is

$$P = 2a + 2b = 2 \times 40 \times 10^{-6}\ \text{m} + 2 \times 160 \times 10^{-6}\ \text{m} = 0.4 \times 10^{-3}\ \text{m}$$

and the hydraulic diameter of each microchannel is found from Equation 8.66 as

$$D_h = \frac{4A_c}{P} = \frac{4ab}{P} = \frac{4 \times 40 \times 10^{-6}\ \text{m} \times 160 \times 10^{-6}\ \text{m}}{0.4 \times 10^{-3}\ \text{m}} = 64 \times 10^{-6}\ \text{m}$$

We begin by assuming a relatively short entrance length, to be verified later, so the flow rate may be estimated by using the friction factor for fully developed conditions. From Table 8.1 for $b/a = 4$, $f = 73/Re_{D_h}$. Substituting this expression into Equation 8.22a, rearranging terms, and using properties at $T = 125°\text{C}$ (in this equation and those following) results in

$$u_m = \frac{2}{73}\frac{D_h^2 \Delta p}{\mu L} = \frac{2}{73} \times \frac{(64 \times 10^{-6}\ \text{m})^2 \times 500 \times 10^3\ \text{N/m}^2}{0.427 \times 10^{-2}\ \text{N}\cdot\text{s/m}^2 \times 20 \times 10^{-3}\ \text{m}} = 0.657\ \text{m/s}$$

Hence, the Reynolds number is

$$Re_{D_h} = \frac{u_m D_h \rho}{\mu} = \frac{0.657\ \text{m/s} \times 64 \times 10^{-6}\ \text{m} \times 1085\ \text{kg/m}^3}{0.427 \times 10^{-2}\ \text{N}\cdot\text{s/m}^2} = 10.7$$

and the flow is deep in the laminar regime. Equation 8.3 may be used to determine the hydrodynamic entrance length, which is

$$x_{\text{fd},h} \approx 0.05\, D_h\, Re_D = 0.05 \times 64 \times 10^{-6}\ \text{m} \times 10.7 = 34.2 \times 10^{-6}\ \text{m}$$

and the thermal entrance length may be obtained from Equation 8.23, yielding

$$x_{\text{fd},t} \approx x_{\text{fd},h}\, Pr = 34.2 \times 10^{-6}\ \text{m} \times 45.2 = 1.55 \times 10^{-3}\ \text{m}$$

Both entrance lengths occupy less than 10% of the total microchannel length, $L = 20$ mm. Therefore, use of fully developed values of f are justified, and the mass flow rate for the $T = 125°\text{C}$ microchannel is

$$\dot{m} = \rho A_c u_m = \rho a b u_m = 1085\ \text{kg/m}^3 \times 40 \times 10^{-6}\ \text{m} \times 160 \times 10^{-6}\ \text{m} \times 0.657\ \text{m/s}$$
$$= 4.56 \times 10^{-6}\ \text{kg/s}$$

Equation 8.42 may now be used to determine the distance from the entrance of the microchannel to the location, x_c, where $T_{m,c} = 124°C$, that is, within 1°C of the surface temperature. The average heat transfer coefficient, $\bar{h}$, is replaced by the fully developed value of the heat transfer coefficient, h, because of the relatively short thermal entrance length. From Table 8.1, we see that for $b/a = 4$, $Nu_D = hD_h/k = 4.44$. Therefore,

$$\bar{h} \approx h = Nu_D \frac{k}{D_h} = 4.44 \times \frac{0.261 \text{ W/m} \cdot \text{K}}{64 \times 10^{-6} \text{ m}} = 1.81 \times 10^4 \text{ W/m}^2 \cdot \text{K}$$

As expected from our discussion of microscale flows, the convection coefficient is very large.

Rearranging Equation 8.42 yields

$$x_c = \frac{\dot{m}c_p}{Ph} \ln\left[\frac{T_s - T_{m,i}}{T_s - T_{m,c}}\right] = \frac{4.56 \times 10^{-6} \text{ kg/s} \times 2583 \text{ J/kg} \cdot \text{K}}{0.4 \times 10^{-3} \text{ m} \times 1.81 \times 10^4 \text{ W/m}^2 \cdot \text{K}} \ln\left[\frac{(125 - 5)°\text{C}}{(125 - 124)°\text{C}}\right]$$

$$= 7.79 \times 10^{-3} \text{ m}$$

Therefore, the time needed for the reactant to reach a mean temperature that is within 1°C of the processing temperature is

$$t_c = x_c/u_m = 7.79 \times 10^{-3} \text{ m}/0.657 \text{ m/s} = 0.012 \text{ s} \qquad \triangleleft$$

Repeating the calculations for the microchannel associated with the smallest processing temperature of 25°C yields $u_m = 0.0995$ m/s, $Re_D = 0.253$, $x_{fd,h} = 8.09 \times 10^{-7}$ m, $x_{fd,t} = 0.218 \times 10^{-3}$ m, $h = 1.71 \times 10^4$ W/m²·K, $x_c = 0.73 \times 10^{-3}$ m, and $t_c = 0.007$ s.

Comments:

1. The total thickness of the glass (2 mm) is 50 times greater than the depth of each microchannel, while the thermal conductivity of the glass, $k_{glass} \approx 1.4$ W/m·K (Table A.3), is 5 times greater than that of the fluid. The presence of such a small amount of fluid is expected to have a negligible effect on the linear temperature distribution that is established across the chip. The temperature difference across the bottom or top surface of each channel is approximately $\Delta T = (T_1 - T_2)b/L = (125 - 25)°\text{C} \times (160 \times 10^{-6} \text{ m})/(20 \times 10^{-3} \text{ m}) = 0.8°\text{C}$.
2. Solving Equation 8.42 over the range $0 \leq x \leq L$ yields the axial variation of the mean temperature for the two extreme processing temperature channels, as shown.

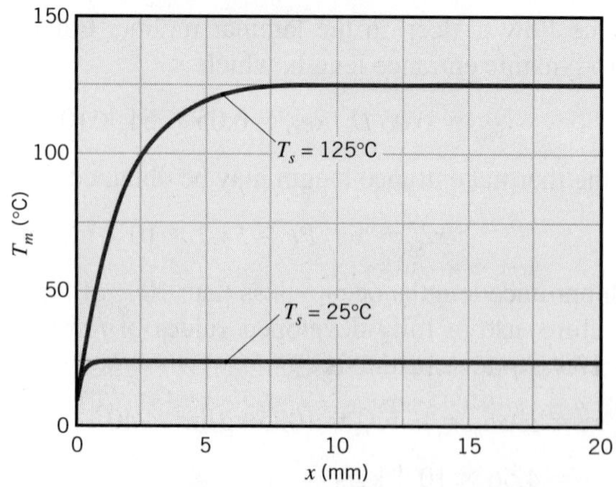

8.9 Summary

In this chapter we have considered forced convection heat transfer for an important class of problems involving *internal flow*. Such flows are encountered in numerous applications, and you should be able to perform engineering calculations that involve an energy balance and appropriate convection correlations. The methodology involves determining whether the flow is laminar or turbulent and establishing the length of the entry region. After deciding whether you are interested in local conditions (at a particular axial location) or in average conditions (for the entire tube), the convection correlation may be selected and used with the appropriate form of the energy balance to solve the problem. A summary of the correlations is provided in Table 8.4.

You should test your understanding of related concepts by addressing the following questions.

- What are the salient features of a *hydrodynamic entry region*? *A thermal entry region*? Are hydrodynamic and thermal entry lengths equivalent? If not, on what do the relative lengths depend?
- What are the salient *hydrodynamic* features of *fully developed flow*? How is the friction factor for fully developed flow affected by wall roughness?
- To what important characteristic of an internal flow is the *mean* or *bulk temperature* linked?
- What are the salient *thermal* features of *fully developed flow*?
- If fluid enters a tube at a uniform temperature and there is heat transfer to or from the surface of the tube, how does the convection coefficient vary with distance along the tube?
- For fluid flow through a tube with a uniform surface heat flux, how does the mean temperature of the fluid vary with distance from the tube entrance in (a) the entrance region and (b) the fully developed region? How does the surface temperature vary with distance in the entrance and fully developed regions?
- For heat transfer to or from a fluid flowing through a tube with a uniform surface temperature, how does the mean temperature of the fluid vary with distance from the entrance? How does the surface heat flux vary with distance from the entrance?
- Why is a *log mean temperature difference*, rather than an arithmetic mean temperature difference, used to calculate the total rate of heat transfer to or from a fluid flowing through a tube with a constant surface temperature?
- What two equations may be used to calculate the total heat rate to a fluid flowing through a tube with a uniform surface heat flux? What two equations may be used to calculate the total heat rate to or from a fluid flowing through a tube with a uniform surface temperature?
- Under what conditions is the Nusselt number associated with internal flow equal to a constant value, independent of Reynolds number and Prandtl number?
- Is the average Nusselt number associated with flow through a tube larger than, equal to, or less than the Nusselt number for fully developed conditions? Why?
- How is the characteristic length defined for a noncircular tube?

Several features that complicate internal flows have not been considered in this chapter. For example, a situation may exist for which there is a prescribed axial variation in T_s or q''_s, rather than uniform surface conditions. Among other things, such a variation would preclude the existence of a fully developed region. There may also exist surface roughness effects, circumferential heat flux or temperature variations, widely varying fluid properties, or transition flow conditions. For a complete discussion of these effects, the literature should be consulted [12, 13, 18, 20, 27].

TABLE 8.4 Summary of convection correlations for flow in a circular tube[a,d]

Correlation		Conditions
$f = 64/Re_D$	(8.19)	Laminar, fully developed
$Nu_D = 4.36$	(8.53)	Laminar, fully developed, uniform q''_s
$Nu_D = 3.66$	(8.55)	Laminar, fully developed, uniform T_s
$\overline{Nu}_D = 3.66 + \dfrac{0.0668\, Gz_D}{1 + 0.04\, Gz_D^{2/3}}$	(8.57)	Laminar, thermal entry (or combined entry with $Pr \gtrsim 5$), uniform $T_s, Gz_D = (D/x)\, Re_D\, Pr$
$\overline{Nu}_D = \dfrac{\dfrac{3.66}{\tanh[2.264\, Gz_D^{-1/3} + 1.7\, Gz_D^{-2/3}]} + 0.0499\, Gz_D \tanh(Gz_D^{-1})}{\tanh(2.432\, Pr^{1/6}\, Gz_D^{-1/6})}$	(8.58)	Laminar, combined entry, $Pr \gtrsim 0.1$, uniform T_s, $Gz_D = (D/x)\, Re_D\, Pr$
$\dfrac{1}{\sqrt{f}} = -2.0 \log\left[\dfrac{e/D}{3.7} + \dfrac{2.51}{Re_D\sqrt{f}}\right]$	(8.20)[b]	Turbulent, fully developed
$f = (0.790 \ln Re_D - 1.64)^{-2}$	(8.21)[b]	Turbulent, fully developed, smooth walls, $3000 \lesssim Re_D \lesssim 5 \times 10^6$
$Nu_D = 0.023\, Re_D^{4/5}\, Pr^n$	(8.60)[c]	Turbulent, fully developed, $0.6 \lesssim Pr \lesssim 160$, $Re_D \gtrsim 10{,}000$, $(L/D) \gtrsim 10$, $n = 0.4$ for $T_s > T_m$ and $n = 0.3$ for $T_s < T_m$
$Nu_D = 0.027\, Re_D^{4/5}\, Pr^{1/3}\left(\dfrac{\mu}{\mu_s}\right)^{0.14}$	(8.61)[c]	Turbulent, fully developed, $0.7 \lesssim Pr \lesssim 16{,}700$, $Re_D \gtrsim 10{,}000$, $L/D \gtrsim 10$
$Nu_D = \dfrac{(f/8)(Re_D - 1000)\, Pr}{1 + 12.7(f/8)^{1/2}(Pr^{2/3} - 1)}$	(8.62)[c]	Turbulent, fully developed, $0.5 \lesssim Pr \lesssim 2000$, $3000 \lesssim Re_D \lesssim 5 \times 10^6$, $(L/D) \gtrsim 10$
$Nu_D = 4.82 + 0.0185(Re_D\, Pr)^{0.827}$	(8.64)	Liquid metals, turbulent, fully developed, uniform q''_s, $3.6 \times 10^3 \lesssim Re_D \lesssim 9.05 \times 10^5$, $3 \times 10^{-3} \lesssim Pr \lesssim 5 \times 10^{-2}$, $10^2 \lesssim Re_D\, Pr \lesssim 10^4$
$Nu_D = 5.0 + 0.025(Re_D\, Pr)^{0.8}$	(8.65)	Liquid metals, turbulent, fully developed, uniform T_s, $Re_D\, Pr \gtrsim 100$

[a]Properties in Equations 8.53, 8.55, 8.60, 8.61, 8.62, 8.64, and 8.65 are based on T_m; properties in Equations 8.19, 8.20, and 8.21 are based on $T_f = (T_s + T_m)/2$; properties in Equations 8.57 and 8.58 are based on $\overline{T}_m = (T_{m,i} + T_{m,o})/2$.
[b]Equation 8.20 pertains to smooth or rough tubes. Equation 8.21 pertains to smooth tubes.
[c]As a first approximation, Equations 8.60, 8.61, or 8.62 may be used to evaluate the average Nusselt number $\overline{Nu}_D$ over the entire tube length, if $(L/D) \gtrsim 10$. The properties should then be evaluated at the average of the mean temperature, $\overline{T}_m = (T_{m,i} + T_{m,o})/2$.
[d]For tubes of noncircular cross section, $Re_D \equiv D_h u_m/\nu$, $D_h \equiv 4A_c/P$, and $u_m = \dot{m}/\rho A_c$. Results for fully developed laminar flow are provided in Table 8.1. For turbulent flow, Equation 8.60 may be used as a first approximation.

References

1. Langhaar, H. L., *J. Appl. Mech.*, **64**, A-55, 1942.
2. Kays, W. M., M. E. Crawford, and B. Weigand, *Convective Heat and Mass Transfer*, 4th ed., McGraw-Hill Higher Education, Boston, 2005.
3. Munson, B. R., D. F. Young, T. H. Okiishi, and W. W. Huebsch, *Fundamentals of Fluid Mechanics*, 6th ed. Wiley, Hoboken, NJ, 2009.
4. Fox, R. W., P. J. Pritchard, and A. T. McDonald, *Introduction to Fluid Mechanics*, 7th ed., Wiley, Hoboken, NJ, 2009.
5. Petukhov, B. S., in T. F. Irvine and J. P. Hartnett, Eds., *Advances in Heat Transfer*, Vol. 6, Academic Press, New York, 1970.
6. Moody, L. F., *Trans. ASME*, **66**, 671, 1944.
7. Chen, M. M., and K. R. Holmes, *Ann. N. Y. Acad. Sci.*, **335**, 137, 1980.
8. Chato, J. C., *J. Biomech. Eng.*, **102**, 110, 1980.
9. Kays, W. M., *Trans. ASME*, **77**, 1265, 1955.
10. Hausen, H., *Z. VDI Beih. Verfahrenstech.*, **4**, 91, 1943.
11. Baehr, H. D., and Stephan, K., *Heat Transfer*, 2nd ed., Springer, Berlin, 2006.
12. Shah, R. K., and A. L. London, *Laminar Flow Forced Convection in Ducts*, Academic Press, New York, 1978.
13. Shah, R. K., and M. S. Bhatti, in S. Kakac, R. K. Shah, and W. Aung, Eds., *Handbook of Single-Phase Convective Heat Transfer*, Chap. 3, Wiley-Interscience, Hoboken, NJ, 1987.
14. Muzychka, Y. S., and Yovanovich, M. M., *J. Heat Transfer*, **126**, 54, 2004.
15. Burmeister, L. C., *Convective Heat Transfer,* 2nd ed., Wiley, Hoboken, NJ, 1993.
16. Winterton, R. H. S., *Int. J. Heat Mass Transfer*, **41**, 809, 1998.
17. Sieder, E. N., and G. E. Tate, *Ind. Eng. Chem.*, **28**, 1429, 1936.
18. Bhatti, M. S., and R. K. Shah, in S. Kakac, R. K. Shah, and W. Aung, Eds., *Handbook of Single-Phase Convective Heat Transfer*, Chap. 4, Wiley-Interscience, Hoboken, NJ, 1987.
19. Gnielinski, V., *Int. Chem. Eng.*, **16**, 359, 1976.
20. Kakac, S., in S. Kakac, R. K. Shah, and W. Aung, Eds., *Handbook of Single-Phase Convective Heat Transfer*, Chap. 18, Wiley-Interscience, Hoboken, NJ, 1987.
21. Ghajar, A. J., and L.-M. Tam, *Exp. Thermal and Fluid Science*, **8**, 79, 1994.
22. Norris, R. H., in A. E. Bergles and R. L. Webb, Eds., *Augmentation of Convective Heat and Mass Transfer*, ASME, New York, 1970.
23. Molki, M., and E. M. Sparrow, *J. Heat Transfer*, **108**, 482, 1986.
24. Skupinski, E. S., J. Tortel, and L. Vautrey, *Int. J. Heat Mass Transfer*, **8**, 937, 1965.
25. Seban, R. A., and T. T. Shimazaki, *Trans. ASME*, **73**, 803, 1951.
26. Reed, C. B., in S. Kakac, R. K. Shah, and W. Aung, Eds., *Handbook of Single-Phase Convective Heat Transfer*, Chap. 8, Wiley-Interscience, Hoboken, NJ, 1987.
27. Kays, W. M., and H. C. Perkins, in W. M. Rohsenow, J. P. Hartnett, and E. N. Ganic, Eds., *Handbook of Heat Transfer, Fundamentals*, Chap. 7, McGraw-Hill, New York, 1985.
28. Bergles, A. E., "Principles of Heat Transfer Augmentation," *Heat Exchangers, Thermal-Hydraulic Fundamentals and Design*, Hemisphere Publishing, New York, 1981, pp. 819–842.
29. Webb, R. L., in S. Kakac, R. K. Shah, and W. Aung, Eds., *Handbook of Single-Phase Convective Heat Transfer*, Chap. 17, Wiley-Interscience, Hoboken, NJ, 1987.
30. Webb, R. L., *Principles of Enhanced Heat Transfer*, Wiley, Hoboken, NJ, 1993.
31. Manglik, R. M., and A. E. Bergles, in J. P. Hartnett, T. F. Irvine, Y. I. Cho, and R. E. Greene, Eds., *Advances in Heat Transfer*, Vol. 36, Academic Press, New York, 2002.
32. Shah, R. K., and S. D. Joshi, in *Handbook of Single-Phase Convective Heat Transfer*, Chap. 5, Wiley-Interscience, Hoboken, NJ, 1987.
33. Vashisth, S., V. Kumar, and K. D. P. Nigam, *Ind. Eng. Chem. Res.*, **47**, 3291, 2008.
34. Jensen, K. F., *Chem. Eng. Sci.*, **56**, 293, 2001.
35. Zhang, Z. M., *Nano/Microscale Heat Transfer*, McGraw-Hill, New York, 2007.
36. Sparrow, E. M., and Lin, S. H., *J. Heat Transfer*, **84**, 363, 1962.
37. Inman, R., *Laminar Slip Flow Heat Transfer in a Parallel Plate Channel or a Round Tube with Uniform Wall Heating*, NASA TN D-2393, 1964.
38. Sharp, K. V., and R. J. Adrian, *Exp. Fluids*, **36**, 741, 2004.
39. Rands, C., B. W. Webb, and D. Maynes, *Int. J. Heat Mass Transfer*, **49**, 2924, 2006.
40. Travis, K. P., B. D. Todd, and D. J. Evans, *Phys. Rev. E*, **55**, 4288, 1997.

Problems

Hydrodynamic Considerations

8.1 Fully developed conditions are known to exist for water flowing through a 25-mm-diameter tube at 0.01 kg/s and 27°C. What is the maximum velocity of the water in the tube? What is the pressure gradient associated with the flow?

8.2 What is the pressure drop associated with water at 27°C flowing with a mean velocity of 0.2 m/s through a 600-m-long cast iron pipe of 0.15-m inside diameter?

8.3 Water at 27°C flows with a mean velocity of 1 m/s through a 1-km-long pipe of 0.25-m inside diameter.

(a) Determine the pressure drop over the pipe length and the corresponding pump power requirement, if the pipe surface is smooth.

(b) If the pipe is made of cast iron and its surface is clean, determine the pressure drop and pump power requirement.

(c) For the smooth pipe condition, generate a plot of pressure drop and pump power requirement for mean velocities in the range from 0.05 to 1.5 m/s.

8.4 An engine oil cooler consists of a bundle of 25 smooth tubes, each of length $L = 2.5$ m and diameter $D = 10$ mm.

(a) If oil at 300 K and a total flow rate of 24 kg/s is in fully developed flow through the tubes, what is the pressure drop and the pump power requirement?

(b) Compute and plot the pressure drop and pump power requirement as a function of flow rate for $10 \leq \dot{m} \leq 30$ kg/s.

8.5 For fully developed laminar flow through a parallel-plate channel, the x-momentum equation has the form

$$\mu\left(\frac{d^2u}{dy^2}\right) = \frac{dp}{dx} = \text{constant}$$

The purpose of this problem is to develop expressions for the velocity distribution and pressure gradient analogous to those for the circular tube in Section 8.1.

(a) Show that the velocity profile, $u(y)$, is parabolic and of the form

$$u(y) = \frac{3}{2}u_m\left[1 - \frac{y^2}{(a/2)^2}\right]$$

where u_m is the mean velocity

$$u_m = -\frac{a^2}{12\mu}\left(\frac{dp}{dx}\right)$$

and $-dp/dx = \Delta p/L$, where Δp is the pressure drop across the channel of length L.

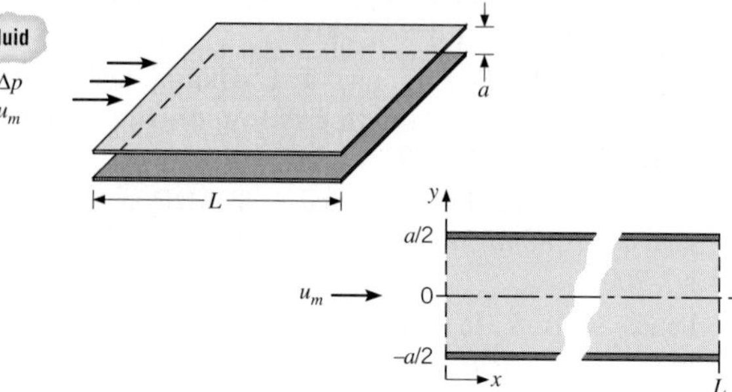

(b) Write an expression defining the friction factor, f, using the hydraulic diameter D_h as the characteristic length. What is the hydraulic diameter for the parallel-plate channel?

(c) The friction factor is estimated from the expression $f = C/Re_{D_h}$, where C depends upon the flow cross section, as shown in Table 8.1. What is the coefficient C for the parallel-plate channel?

(d) Airflow in a parallel-plate channel with a separation of 5 mm and a length of 200 mm experiences a pressure drop of $\Delta p = 3.75$ N/m^2. Calculate the mean velocity and the Reynolds number for air at atmospheric pressure and 300 K. Is the assumption of fully developed flow reasonable for this application? If not, what is the effect on the estimate for u_m?

Thermal Entry Length and Energy Balance Considerations

8.6 Consider pressurized water, engine oil (unused), and NaK (22%/78%) flowing in a 20-mm-diameter tube.

(a) Determine the mean velocity, the hydrodynamic entry length, and the thermal entry length for each of the fluids when the fluid temperature is 366 K and the flow rate is 0.01 kg/s.

(b) Determine the mass flow rate, the hydrodynamic entry length, and the thermal entry length for water and engine oil at 300 and 400 K and a mean velocity of 0.02 m/s.

8.7 Velocity and temperature profiles for laminar flow in a tube of radius $r_o = 10$ mm have the form

$$u(r) = 0.1[1 - (r/r_o)^2]$$
$$T(r) = 344.8 + 75.0(r/r_o)^2 - 18.8(r/r_o)^4$$

with units of m/s and K, respectively. Determine the corresponding value of the mean (or bulk) temperature, T_m, at this axial position.

8.8 At a particular axial station, velocity and temperature profiles for laminar flow in a parallel plate channel have the form

$$u(y) = 0.75[1 - (y/y_o)^2]$$
$$T(y) = 5.0 + 95.66(y/y_o)^2 - 47.83(y/y_o)^4$$

with units of m/s and °C, respectively.

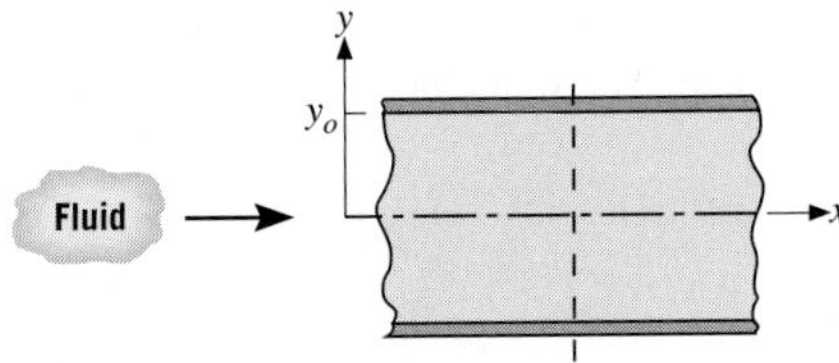

Determine corresponding values of the mean velocity, u_m, and mean (or bulk) temperature, T_m. Plot the velocity and temperature distributions. Do your values of u_m and T_m appear reasonable?

8.9 In Chapter 1, it was stated that for incompressible liquids, flow work could usually be neglected in the steady-flow energy equation (Equation 1.12d). In the trans-Alaska pipeline, the high viscosity of the oil and long distances cause significant pressure drops, and it is reasonable to question whether flow work would be significant. Consider an $L = 100$ km length of pipe of diameter $D = 1.2$ m, with oil flow rate $\dot{m} = 500$ kg/s. The oil properties are $\rho = 900$ kg/m^3, $c_p = 2000$ J/kg·K, $\mu = 0.765$ N·s/m^2. Calculate the pressure drop, the flow work, and the temperature rise caused by the flow work.

8.10 When viscous dissipation is included, Equation 8.48 (multiplied by ρc_p) becomes

$$\rho c_p u \frac{\partial T}{\partial x} = \frac{k}{r}\frac{\partial}{\partial r}\left(r\frac{\partial T}{\partial r}\right) + \mu\left(\frac{du}{dr}\right)^2$$

This problem explores the importance of viscous dissipation. The conditions under consideration are laminar, fully developed flow in a circular pipe, with u given by Equation 8.15.

(a) By integrating the left-hand side over a section of a pipe of length L and radius r_o, show that this term yields the right-hand side of Equation 8.34.

(b) Integrate the viscous dissipation term over the same volume.

(c) Find the temperature rise caused by viscous dissipation by equating the two terms calculated above. Use the same conditions as in Problem 8.9.

8.11 Consider a circular tube of diameter D and length L, with a mass flow rate of $\dot{m}$.

(a) For constant heat flux conditions, derive an expression for the ratio of the temperature difference between the tube wall at the tube exit and the inlet temperature, $T_s(x = L) - T_{m,i}$, to the total heat transfer rate to the fluid q. Express your result in terms of $\dot{m}$, L, the local Nusselt number at the tube exit $Nu_D(x = L)$, and relevant fluid properties.

(b) Repeat part (a) for constant surface temperature conditions. Express your result in terms of $\dot{m}$, L, the average Nusselt number from the tube inlet to the tube exit $\overline{Nu}_D$, and relevant fluid properties.

8.12 Water enters a tube at 27°C with a flow rate of 450 kg/h. The heat transfer from the tube wall to the fluid is given as q_s'(W/m) $= ax$, where the coefficient a is 20 W/m^2 and x (m) is the axial distance from the tube entrance.

(a) Beginning with a properly defined differential control volume in the tube, derive an expression for the temperature distribution $T_m(x)$ of the water.

(b) What is the outlet temperature of the water for a heated section 30 m long?

(c) Sketch the mean fluid temperature, $T_m(x)$, and the tube wall temperature, $T_s(x)$, as a function of distance along the tube for fully developed *and* developing flow conditions.

(d) What value of a uniform wall heat flux, q_s'' (instead of $q_s' = ax$), would provide the same fluid outlet temperature as that determined in part (b)? For this type of heating, sketch the temperature distributions requested in part (c).

8.13 Consider flow in a circular tube. Within the test section length (between 1 and 2) a constant heat flux q_s'' is maintained.

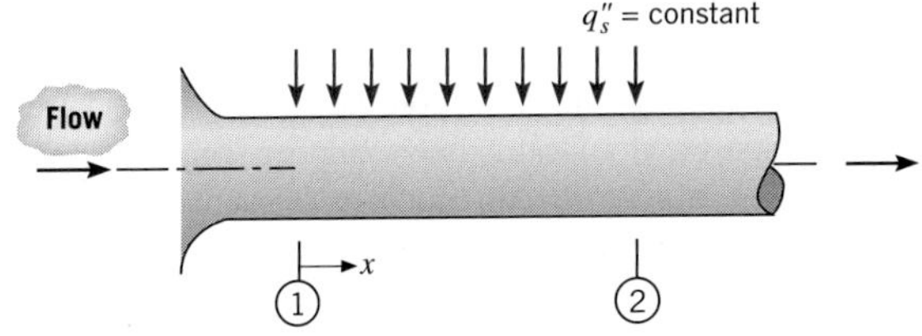

(a) For the following two cases, sketch the surface temperature $T_s(x)$ and the fluid mean temperature $T_m(x)$ as a function of distance along the test section x. In case A, flow is hydrodynamically and thermally fully developed. In case B, flow is not developed.

(b) Assuming that the surface flux q_s'' and the inlet mean temperature $T_{m,1}$ are identical for both cases, will the exit mean temperature $T_{m,2}$ for case A be greater than, equal to, or less than $T_{m,2}$ for case B? Briefly explain why.

8.14 Consider a cylindrical nuclear fuel rod of length L and diameter D that is encased in a concentric tube. Pressurized water flows through the annular region between the rod and the tube at a rate $\dot{m}$, and the outer surface of the tube is well insulated. Heat generation occurs within the fuel rod, and the volumetric generation rate is known to vary sinusoidally with distance along the rod. That is, $\dot{q}(x) = \dot{q}_o \sin(\pi x/L)$, where $\dot{q}_o$(W/m^3) is a constant. A uniform convection coefficient h may be assumed to exist between the surface of the rod and the water.

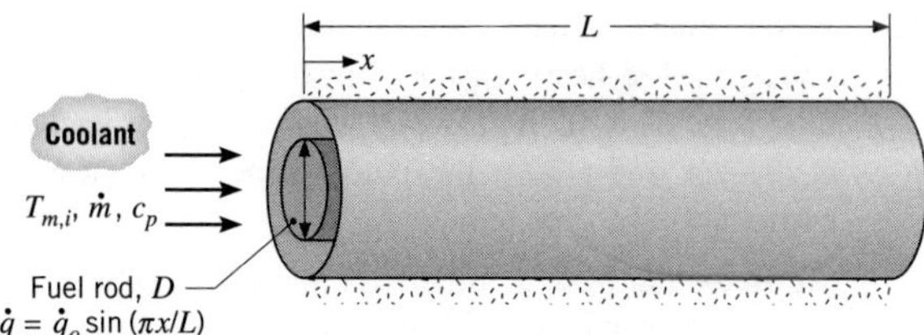

(a) Obtain expressions for the local heat flux $q''(x)$ and the total heat transfer q from the fuel rod to the water.

(b) Obtain an expression for the variation of the mean temperature $T_m(x)$ of the water with distance x along the tube.

(c) Obtain an expression for the variation of the rod surface temperature $T_s(x)$ with distance x along the tube. Develop an expression for the x-location at which this temperature is maximized.

8.15 Consider the laminar thermal boundary layer development near the entrance of the tube shown in Figure 8.4. When the hydrodynamic boundary layer is thin relative to the tube diameter, the inviscid flow region has a uniform velocity that is approximately equal to the mean velocity u_m. Hence the boundary layer development is similar to what would occur for a flat plate.

(a) Beginning with Equation 7.21, derive an expression for the local Nusselt number Nu_D, as a function of the Prandtl number Pr and the inverse Graetz number Gz_D^{-1}. Plot the expression using the coordinates shown in Figure 8.10a for $Pr = 0.7$.

(b) Beginning with Equation 7.25, derive an expression for the average Nusselt number $\overline{Nu}_D$, as a function of the Prandtl number Pr and the inverse Graetz number Gz_D^{-1}. Compare your results with the Nusselt number for the combined entrance length in the limit of small x.

8.16 In a particular application involving fluid flow at a rate $\dot{m}$ through a circular tube of length L and diameter D, the surface heat flux is known to have a sinusoidal variation with x, which is of the form $q_s''(x) = q_{s,m}'' \sin(\pi x/L)$. The maximum flux, $q_{s,m}''$, is a known constant, and the fluid enters the tube at a known temperature, $T_{m,i}$. Assuming the convection coefficient to be constant, how do the mean temperature of the fluid and the surface temperature vary with x?

8.17 A flat-plate solar collector is used to heat atmospheric air flowing through a rectangular channel. The bottom surface of the channel is well insulated, while the top surface is subjected to a uniform heat flux q_o'', which is due to the net effect of solar radiation absorption and heat exchange between the absorber and cover plates.

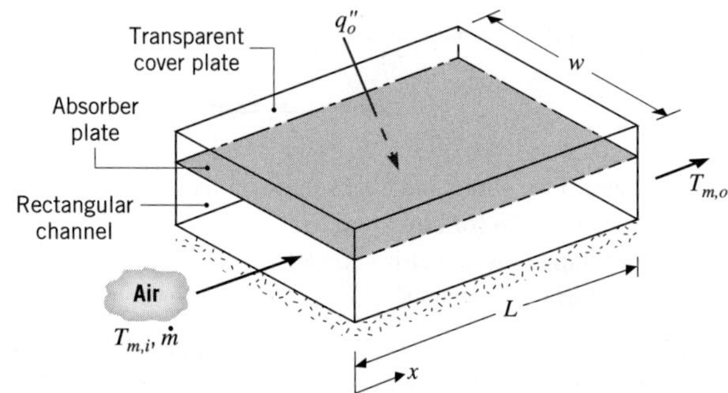

(a) Beginning with an appropriate differential control volume, obtain an equation that could be used to determine the mean air temperature $T_m(x)$ as a function of distance along the channel. Solve this equation to obtain an expression for the mean temperature of the air leaving the collector.

(b) With air inlet conditions of $\dot{m} = 0.1$ kg/s and $T_{m,i} = 40°C$, what is the air outlet temperature if $L = 3$ m, $w = 1$ m, and $q_o'' = 700$ W/m^2? The specific heat of air is $c_p = 1008$ J/kg·K.

8.18 Atmospheric air enters the heated section of a circular tube at a flow rate of 0.005 kg/s and a temperature of 20°C. The tube is of diameter $D = 50$ mm, and fully developed conditions with $h = 25$ W/m^2·K exist over the entire length of $L = 3$ m.

(a) For the case of uniform surface heat flux at $q_s'' = 1000$ W/m^2, determine the total heat transfer rate q and the mean temperature of the air leaving the tube $T_{m,o}$. What is the value of the surface temperature at the tube inlet $T_{s,i}$ and outlet $T_{s,o}$? Sketch the axial variation of T_s and T_m. On the same figure, also sketch (qualitatively) the axial variation of T_s and T_m for the more realistic case in which the local convection coefficient varies with x.

(b) If the surface heat flux varies linearly with x, such that q_s'' (W/m^2) $= 500x$ (m), what are the values of q, $T_{m,o}$, $T_{s,i}$, and $T_{s,o}$? Sketch the axial variation of T_s and T_m. On the same figure, also sketch (qualitatively) the axial variation of T_s and T_m for the more realistic case in which the local convection coefficient varies with x.

(c) For the two heating conditions of parts (a) and (b), plot the mean fluid and surface temperatures, $T_m(x)$ and $T_s(x)$, respectively, as functions of distance along the tube. What effect will a fourfold increase in the convection coefficient have on the temperature distributions?

(d) For each type of heating process, what heat fluxes are required to achieve an air outlet temperature of 125°C? Plot the temperature distributions.

8.19 Fluid enters a tube with a flow rate of 0.015 kg/s and an inlet temperature of 20°C. The tube, which has a length of 6 m and diameter of 15 mm, has a surface temperature of 30°C.

(a) Determine the heat transfer rate to the fluid if it is water.

(b) Determine the heat transfer rate for the nanofluid of Example 2.2.

8.20 Water at 300 K and a flow rate of 5 kg/s enters a black, thin-walled tube, which passes through a large furnace whose walls and air are at a temperature of 700 K. The diameter and length of the tube are 0.25 m and 8 m, respectively. Convection coefficients associated with water flow through the tube and airflow over the tube are 300 W/m$^2 \cdot$ K and 50 W/m$^2 \cdot$K, respectively.

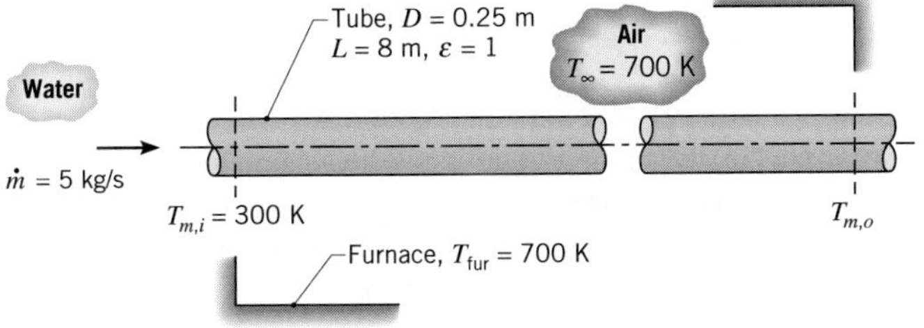

(a) Write an expression for the linearized radiation coefficient corresponding to radiation exchange between the outer surface of the pipe and the furnace walls. Explain how to calculate this coefficient if the surface temperature of the tube is represented by the arithmetic mean of its inlet and outlet values.

(b) Determine the outlet temperature of the water, $T_{m,o}$.

8.21 Slug flow is an idealized tube flow condition for which the velocity is assumed to be uniform over the entire tube cross section. For the case of laminar slug flow with a uniform surface heat flux, determine the form of the fully developed temperature distribution $T(r)$ and the Nusselt number Nu_D.

8.22 Superimposing a control volume that is differential in x on the tube flow conditions of Figure 8.8, derive Equation 8.45a.

8.23 An experimental nuclear core simulation apparatus consists of a long thin-walled metallic tube of diameter D and length L, which is electrically heated to produce the sinusoidal heat flux distribution

$$q_s''(x) = q_o'' \sin\left(\frac{\pi x}{L}\right)$$

where x is the distance measured from the tube inlet. Fluid at an inlet temperature $T_{m,i}$ flows through the tube at a rate of $\dot{m}$. Assuming the flow is turbulent and fully developed over the entire length of the tube, develop expressions for:

(a) the total rate of heat transfer, q, from the tube to the fluid;

(b) the fluid outlet temperature, $T_{m,o}$;

(c) the axial distribution of the wall temperature, $T_s(x)$; and

(d) the magnitude and position of the highest wall temperature.

(e) Consider a 40-mm-diameter tube of 4-m length with a sinusoidal heat flux distribution for which $q_o'' = 10{,}000$ W/m^2. Fluid passing through the tube has a flow rate of 0.025 kg/s, a specific heat of 4180 J/kg$\cdot$K, an entrance temperature of 25°C, and a convection coefficient of 1000 W/m$^2\cdot$K. Plot the mean fluid and surface temperatures as a function of distance along the tube. Identify important features of the distributions. Explore the effect of ±25% changes in the convection coefficient and the heat flux on the distributions.

8.24 Water at 20°C and a flow rate of 0.1 kg/s enters a heated, thin-walled tube with a diameter of 15 mm and length of 2 m. The wall heat flux provided by the heating elements depends on the wall temperature according to the relation

$$q_s''(x) = q_{s,o}''[1 + \alpha(T_s - T_{ref})]$$

where $q_{s,o}'' = 10^4$ W/m^2, $\alpha = 0.2$ K^{-1}, $T_{ref} = 20$°C, and T_s is the wall temperature in °C. Assume fully developed flow and thermal conditions with a convection coefficient of 3000 W/m$^2 \cdot$ K.

(a) Beginning with a properly defined differential control volume in the tube, derive expressions for the variation of the water, $T_m(x)$, and the wall, $T_s(x)$, temperatures as a function of distance from the tube inlet.

(b) Using a numerical integration scheme, calculate and plot the temperature distributions, $T_m(x)$ and $T_s(x)$, on the same graph. Identify and comment on the main features of the distributions. *Hint*: The *IHT* integral function $DER(T_m, x)$ can be used to perform the integration along the length of the tube.

(c) Calculate the total rate of heat transfer to the water.

Heat Transfer Correlations: Circular Tubes

8.25 Engine oil is heated by flowing through a circular tube of diameter $D = 50$ mm and length $L = 25$ m and whose surface is maintained at 150°C.

(a) If the flow rate and inlet temperature of the oil are 0.5 kg/s and 20°C, what is the outlet temperature $T_{m,o}$? What is the total heat transfer rate q for the tube?

(b) For flow rates in the range $0.5 \leq \dot{m} \leq 2.0$ kg/s, compute and plot the variations of $T_{m,o}$ and q with $\dot{m}$. For what flow rate(s) are q and $T_{m,o}$ maximized? Explain your results.

8.26 Engine oil flows through a 25-mm-diameter tube at a rate of 0.5 kg/s. The oil enters the tube at a temperature of 25°C, while the tube surface temperature is maintained at 100°C.

(a) Determine the oil outlet temperature for a 5-m and for a 100-m long tube. For each case, compare the log mean temperature difference to the arithmetic mean temperature difference.

(b) For $5 \leq L \leq 100$ m, compute and plot the average Nusselt number $\overline{Nu}_D$ and the oil outlet temperature as a function of L.

8.27 In the final stages of production, a pharmaceutical is sterilized by heating it from 25 to 75°C as it moves at 0.2 m/s through a straight thin-walled stainless steel tube of 12.7-mm diameter. A uniform heat flux is maintained by an electric resistance heater wrapped around the outer surface of the tube. If the tube is 10 m long, what is the required heat flux? If fluid enters the tube with a fully developed velocity profile and a uniform temperature profile, what is the surface temperature at the tube exit and at a distance of 0.5 m from the entrance? Fluid properties may be approximated as $\rho = 1000$ kg/m^3, $c_p = 4000$ J/kg·K, $\mu = 2 \times 10^{-3}$ kg/s·m, $k = 0.8$ W/m·K, and $Pr = 10$.

8.28 An oil preheater consists of a single tube of 10-mm diameter and 5-m length, with its surface maintained at 175°C by swirling combustion gases. The engine oil (new) enters at 75°C. What flow rate must be supplied to maintain an oil outlet temperature of 100°C? What is the corresponding heat transfer rate?

8.29 Engine oil flows at a rate of 1 kg/s through a 5-mm-diameter straight tube. The oil has an inlet temperature of 45°C and it is desired to heat the oil to a mean temperature of 80°C at the exit of the tube. The surface of the tube is maintained at 150°C. Determine the required length of the tube. *Hint*: Calculate the Reynolds numbers at the entrance and exit of the tube before proceeding with your analysis.

8.30 Air at $p = 1$ atm enters a thin-walled ($D = 5$-mm diameter) long tube ($L = 2$ m) at an inlet temperature of $T_{m,i} = 100$°C. A constant heat flux is applied to the air from the tube surface. The air mass flow rate is $\dot{m} = 135 \times 10^{-6}$ kg/s.

(a) If the tube surface temperature at the exit is $T_{s,o} = 160$°C, determine the heat rate entering the tube. Evaluate properties at $T = 400$ K.

(b) If the tube length of part (a) were reduced to $L = 0.2$ m, how would flow conditions at the tube exit be affected? Would the value of the heat transfer coefficient at the tube exit be greater than, equal to, or smaller than the heat transfer coefficient for part (a)?

(c) If the flow rate of part (a) were increased by a factor of 10, would there be a difference in flow conditions at the tube exit? Would the value of the heat transfer coefficient at the tube exit be greater than, equal to, or smaller than the heat transfer coefficient for part (a)?

8.31 To cool a summer home without using a vapor-compression refrigeration cycle, air is routed through a plastic pipe ($k = 0.15$ W/m·K, $D_i = 0.15$ m, $D_o = 0.17$ m) that is submerged in an adjoining body of water. The water temperature is nominally at $T_\infty = 17$°C, and a convection coefficient of $h_o \approx 1500$ W/m^2·K is maintained at the outer surface of the pipe.

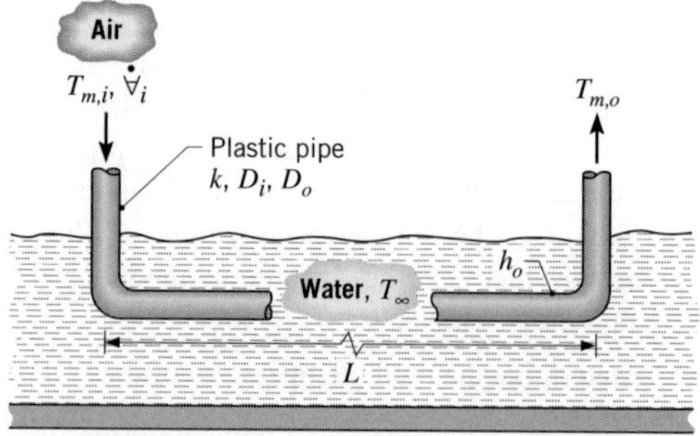

If air from the home enters the pipe at a temperature of $T_{m,i} = 29$°C and a volumetric flow rate of $\dot{\forall}_i = 0.025$ m^3/s, what pipe length L is needed to provide a discharge temperature of $T_{m,o} = 21°C$? What is the fan power required to move the air through this length of pipe if its inner surface is smooth?

8.32 *Batch processes* are often used in chemical and pharmaceutical operations to achieve a desired chemical composition for the final product. Related heat transfer processes are typically transient, involving a liquid of fixed volume that may be heated from room temperature to a desired process temperature, or cooled from the process temperature to room temperature. Consider a batch process for which a pharmaceutical (the cold fluid, c) is poured into an insulated, highly agitated vessel (a *stirred reactor*) and heated by passing a hot fluid (h) through a submerged heat exchanger coil of thin-walled tubing and surface area A_s. The flow rate, $\dot{m}_h$, mean inlet temperature, $T_{h,i}$, and specific heat, $c_{p,h}$, of the hot fluid are known, as are the initial temperature, $T_{c,i} < T_{h,i}$, the volume, V_c, mass density, ρ_c, and specific heat, $c_{v,c}$, of the pharmaceutical. Heat transfer from the hot fluid to the pharmaceutical is governed by an overall heat transfer coefficient U.

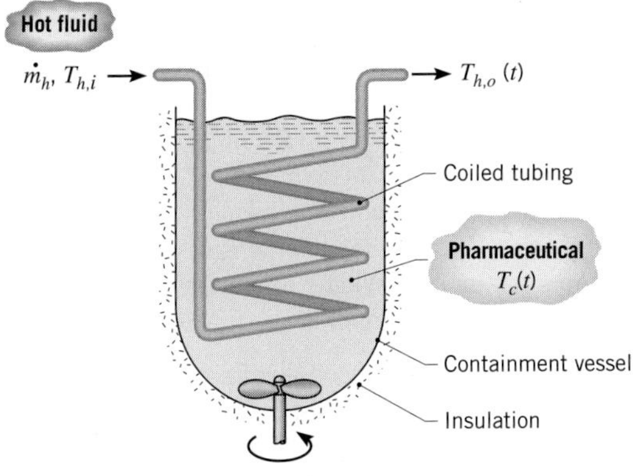

(a) Starting from basic principles, derive expressions that can be used to determine the variation of T_c and $T_{h,o}$ with time during the heating process. *Hint*: Two equations may be written for the rate of heat transfer, $q(t)$, to the pharmaceutical, one based on the log-mean temperature difference and the other on an energy balance for flow of the hot fluid through the tube. Equate these expressions to determine $T_{h,o}(t)$ as a function of $T_c(t)$ and prescribed parameters. Use the expression for $T_{h,o}(t)$ and the energy balance for flow through the tube with an energy balance for a control volume containing the pharmaceutical to obtain an expression for $T_c(t)$.

(b) Consider a pharmaceutical of volume $V_c = 1\ m^3$, density $\rho_c = 1100\ kg/m^3$, specific heat $c_{v,c} = 2000\ J/kg \cdot K$, and an initial temperature of $T_{c,i} = 25°C$. A coiled tube of length $L = 40$ m, diameter $D = 50$ mm, and coil diameter $C = 500$ mm is submerged in the vessel, and hot fluid enters the tubing at $T_{h,i} = 200°C$ and $\dot{m}_h = 2.4$ kg/s. The convection coefficient at the outer surface of the tubing may be approximated as $h_o = 1000\ W/m^2 \cdot K$, and the fluid properties are $c_{p,h} = 2500\ J/kg \cdot K$, $\mu_h = 0.002\ N \cdot s/m^2$, $k_h = 0.260\ W/m \cdot K$, and $Pr_h = 20$. For the foregoing conditions, compute and plot the pharmaceutical temperature T_c and the outlet temperature $T_{h,o}$ as a function of time over the range $0 \le t \le 3600$ s. How long does it take to reach a batch temperature of $T_c = 160°C$? The process operator may control the heating time by varying $\dot{m}_h$. For $1 \le \dot{m}_h \le 5$ kg/s, explore the effect of the flow rate on the time t_c required to reach a value of $T_c = 160°C$.

8.33 The evaporator section of a heat pump is installed in a large tank of water, which is used as a heat source during the winter. As energy is extracted from the water, it begins to freeze, creating an ice/water bath at 0°C, which may be used for air conditioning during the summer. Consider summer cooling conditions for which air is passed through an array of copper tubes, each of inside diameter $D = 50$ mm, submerged in the bath.

(a) If air enters each tube at a mean temperature of $T_{m,i} = 24°C$ and a flow rate of $\dot{m} = 0.01$ kg/s, what tube length L is needed to provide an exit temperature of $T_{m,o} = 14°C$? With 10 tubes passing through a tank of total volume $V = 10\ m^3$, which initially contains 80% ice by volume, how long would it take to completely melt the ice? The density and latent heat of fusion of ice are $920\ kg/m^3$ and 3.34×10^5 J/kg, respectively.

(b) The air outlet temperature may be regulated by adjusting the tube mass flow rate. For the tube length determined in part (a), compute and plot $T_{m,o}$ as a function of $\dot{m}$ for $0.005 \le \dot{m} \le 0.05$ kg/s. If the dwelling cooled by this system requires approximately 0.05 kg/s of air at 16°C, what design and operating conditions should be prescribed for the system?

8.34 A liquid food product is processed in a continuous-flow sterilizer. The liquid enters the sterilizer at a temperature and flow rate of $T_{m,i,h} = 20°C$, $\dot{m} = 1$ kg/s, respectively. A time-at-temperature constraint requires that the product be held at a mean temperature of $T_m = 90°C$ for 10 s to kill bacteria, while a second constraint is that the local product temperature cannot exceed $T_{max} = 230°C$ in order to preserve a pleasing taste. The sterilizer consists of an upstream, $L_h = 5$ m heating section characterized by a uniform heat flux, an intermediate insulated sterilizing section, and a downstream cooling section of length $L_c = 10$ m. The cooling section is composed of an uninsulated tube exposed to a quiescent environment at $T_\infty = 20°C$. The thin-walled tubing is of diameter $D = 40$ mm. Food properties are similar to those of liquid water at $T = 330$ K.

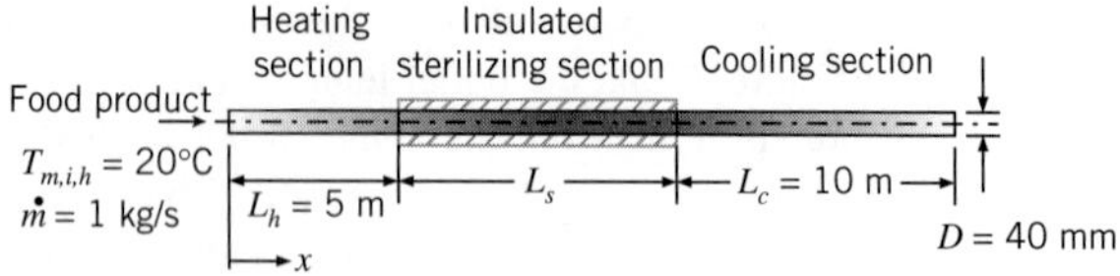

(a) What heat flux is required in the heating section to ensure a maximum mean product temperature of $T_m = 90°C$?

(b) Determine the location and value of the maximum local product temperature. Is the second constraint satisfied?

(c) Determine the minimum length of the sterilizing section needed to satisfy the time-at-temperature constraint.

(d) Sketch the axial distribution of the mean, surface, and centerline temperatures from the inlet of the heating section to the outlet of the cooling section.

8.35 Water flowing at 2 kg/s through a 40-mm-diameter tube is to be heated from 25 to 75°C by maintaining the tube surface temperature at 100°C.

(a) What is the required tube length for these conditions?

(b) To design a water heating system, we wish to consider using tube diameters in the range from 30 to 50 mm. What are the required tube lengths for water flow rates of 1, 2, and 3 kg/s? Represent this design information graphically.

(c) Plot the pressure gradient as a function of tube diameter for the three flow rates. Assume the tube wall is smooth.

8.36 Consider the conditions associated with the hot water pipe of Problem 7.56, but now account for the convection resistance associated with water flow at a mean velocity of $u_m = 0.5$ m/s in the pipe. What is the corresponding daily cost of heat loss per meter of the uninsulated pipe?

8.37 A thick-walled, stainless steel (AISI 316) pipe of inside and outside diameters $D_i = 20$ mm and $D_o = 40$ mm is heated electrically to provide a uniform heat generation rate of $\dot{q} = 10^6$ W/m^3. The outer surface of the pipe is insulated, while water flows through the pipe at a rate of $\dot{m} = 0.1$ kg/s.

(a) If the water inlet temperature is $T_{m,i} = 20°C$ and the desired outlet temperature is $T_{m,o} = 40°C$, what is the required pipe length?

(b) What are the location and value of the maximum pipe temperature?

8.38 An air heater for an industrial application consists of an insulated, concentric tube annulus, for which air flows through a thin-walled inner tube. Saturated steam flows through the outer annulus, and condensation of the steam maintains a uniform temperature T_s on the tube surface.

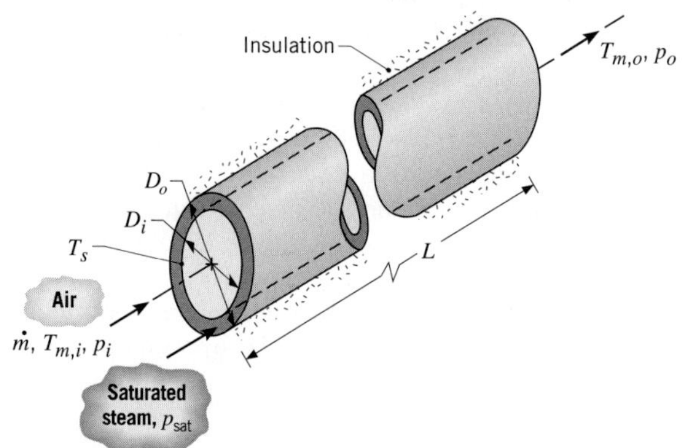

Consider conditions for which air enters a 50-mm-diameter tube at a pressure of 5 atm, a temperature of $T_{m,i} = 17°C$, and a flow rate of $\dot{m} = 0.03$ kg/s, while saturated steam at 2.455 bars condenses on the outer surface of the tube. If the length of the annulus is $L = 5$ m, what are the outlet temperature $T_{m,o}$ and pressure p_o of the air? What is the mass rate at which condensate leaves the annulus?

8.39 Consider fully developed conditions in a circular tube with constant surface temperature $T_s < T_m$. Determine whether a small- or large-diameter tube is more effective in minimizing heat loss from the flowing fluid characterized by a mass flow rate of $\dot{m}$. Consider both laminar and turbulent conditions.

8.40 Consider the encased pipe of Problem 4.29, but now allow for the difference between the mean temperature of the fluid, which changes along the pipe length, and that of the pipe.

(a) For the prescribed values of k, D, w, h, and T_∞ and a pipe of length $L = 100$ m, what is the outlet temperature $T_{m,o}$ of water that enters the pipe at a temperature of $T_{m,i} = 90°C$ and a flow rate of $\dot{m} = 2$ kg/s?

(b) What is the pressure drop of the water and the corresponding pump power requirement?

(c) Subject to the constraint that the width of the duct is fixed at $w = 0.30$ m, explore the effects of the flow rate and the pipe diameter on the outlet temperature.

8.41 Water flows through a thick-walled tube with an inner diameter of 12 mm and a length of 8 m. The tube is immersed in a well-stirred, hot reaction tank maintained at 85°C, and the conduction resistance of the tube wall (based on the inner surface area) is $R''_{\text{cd}} = 0.002\ \text{m}^2 \cdot \text{K/W}$. The inlet temperature of the process fluid is $T_{m,i} = 20°\text{C}$, and the flow rate is 33 kg/h.

(a) Estimate the outlet temperature of the process fluid, $T_{m,o}$. Assume, and then justify, fully developed flow and thermal conditions within the tube.

(b) Do you expect $T_{m,o}$ to increase or decrease if combined thermal and hydrodynamic entry conditions exist within the tube? Estimate the outlet temperature of the water for this condition.

8.42 Atmospheric air enters a 10-m-long, 150-mm-diameter uninsulated heating duct at 60°C and 0.04 kg/s. The duct surface temperature is approximately constant at $T_s = 15°\text{C}$.

(a) What are the outlet air temperature, the heat rate q, and pressure drop Δp for these conditions?

[(b)] To illustrate the tradeoff between heat transfer rate and pressure drop considerations, calculate q and Δp for diameters in the range from 0.1 to 0.2 m. In your analysis, maintain the total surface area, $A_s = \pi DL$, at the value computed for part (a). Plot q, Δp, and L as a function of the duct diameter.

8.43 NaK (45%/55%), which is an alloy of sodium and potassium, is used to cool fast neutron nuclear reactors. The NaK flows at a rate of $\dot{m} = 1$ kg/s through a $D = 50$-mm-diameter tube that has a surface temperature of $T_s = 450$ K. The NaK enters the tube at $T_{m,i} = 332$ K and exits at an outlet temperature of $T_{m,o} = 400$ K. Determine the tube length L and the local convective heat flux at the tube exit.

8.44 The products of combustion from a burner are routed to an industrial application through a thin-walled metallic duct of diameter $D_i = 1$ m and length $L = 100$ m. The gas enters the duct at atmospheric pressure and a mean temperature and velocity of $T_{m,i} = 1600$ K and $u_{m,i} = 10$ m/s, respectively. It must exit the duct at a temperature that is no less than $T_{m,o} = 1400$ K. What is the minimum thickness of an alumina-silica insulation ($k_{\text{ins}} = 0.125\ \text{W/m} \cdot \text{K}$) needed to meet the outlet requirement under worst case conditions for which the duct is exposed to ambient air at $T_\infty = 250$ K and a cross-flow velocity of $V = 15$ m/s? The properties of the gas may be approximated as those of air, and as a first estimate, the effect of the insulation thickness on the convection coefficient and thermal resistance associated with the cross flow may be neglected.

8.45 Liquid mercury at 0.5 kg/s is to be heated from 300 to 400 K by passing it through a 50-mm-diameter tube whose surface is maintained at 450 K. Calculate the required tube length by using an appropriate liquid metal convection heat transfer correlation. Compare your result with that which would have been obtained by using a correlation appropriate for $Pr \gtrsim 0.7$.

8.46 The surface of a 50-mm-diameter, thin-walled tube is maintained at 100°C. In one case air is in cross flow over the tube with a temperature of 25°C and a velocity of 30 m/s. In another case air is in fully developed flow through the tube with a temperature of 25°C and a mean velocity of 30 m/s. Compare the heat flux from the tube to the air for the two cases.

8.47 Consider a horizontal, thin-walled circular tube of diameter $D = 0.025$ m submerged in a container of n-octadecane (paraffin), which is used to store thermal energy. As hot water flows through the tube, heat is transferred to the paraffin, converting it from the solid to liquid state at the phase change temperature of $T_\infty = 27.4°\text{C}$. The latent heat of fusion and density of paraffin are $h_{sf} = 244$ kJ/kg and $\rho = 770\ \text{kg/m}^3$, respectively, and thermophysical properties of the water may be taken as $c_p = 4.185\ \text{kJ/kg} \cdot \text{K}$, $k = 0.653\ \text{W/m} \cdot \text{K}$, $\mu = 467 \times 10^{-6}\ \text{kg/s} \cdot \text{m}$, and $Pr = 2.99$.

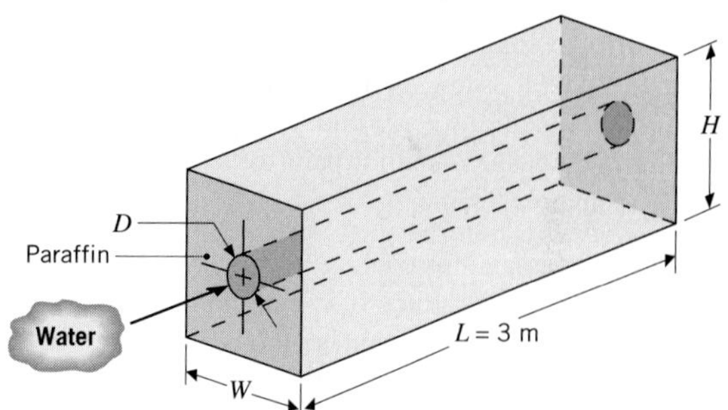

(a) Assuming the tube surface to have a uniform temperature corresponding to that of the phase change, determine the water outlet temperature and total heat transfer rate for a water flow rate of 0.1 kg/s and an inlet temperature of 60°C. If $H = W = 0.25$ m, how long would it take to completely liquefy the paraffin, from an initial state for which all the paraffin is solid and at 27.4°C?

[(b)] The liquefaction process can be accelerated by increasing the flow rate of the water. Compute and plot the heat rate and outlet temperature as a function of flow rate for $0.1 \leq \dot{m} \leq 0.5$ kg/s. How long would it take to melt the paraffin for $\dot{m} = 0.5$ kg/s?

8.48 Consider pressurized liquid water flowing at $\dot{m} = 0.1$ kg/s in a circular tube of diameter $D = 0.1$ m and length $L = 6$ m.

(a) If the water enters at $T_{m,i} = 500$ K and the surface temperature of the tube is $T_s = 510$ K, determine the water outlet temperature $T_{m,o}$.

(b) If the water enters at $T_{m,i} = 300$ K and the surface temperature of the tube is $T_s = 310$ K, determine the water outlet temperature $T_{m,o}$.

(c) If the water enters at $T_{m,i} = 300$ K and the surface temperature of the tube is $T_s = 647$ K, discuss whether the flow is laminar or turbulent.

8.49 Cooling water flows through the 25.4-mm-diameter thin-walled tubes of a steam condenser at 1 m/s, and a surface temperature of 350 K is maintained by the condensing steam. The water inlet temperature is 290 K, and the tubes are 5 m long.

(a) What is the water outlet temperature? Evaluate water properties at an assumed average mean temperature, $\overline{T}_m = 300$ K.

(b) Was the assumed value for $\overline{T}_m$ reasonable? If not, repeat the calculation using properties evaluated at a more appropriate temperature.

(c) A range of tube lengths from 4 to 7 m is available to the engineer designing this condenser. Generate a plot to show what coolant mean velocities are possible if the water outlet temperature is to remain at the value found for part (b). All other conditions remain the same.

8.50 The air passage for cooling a gas turbine vane can be approximated as a tube of 3-mm diameter and 75-mm length. The operating temperature of the vane is 650°C, and air enters the tube at 427°C.

(a) For an airflow rate of 0.18 kg/h, calculate the air outlet temperature and the heat removed from the vane.

(b) Generate a plot of the air outlet temperature as a function of flow rate for $0.1 \leq \dot{m} \leq 0.6$ kg/h. Compare this result with those for vanes having 2- and 4-mm-diameter tubes, with all other conditions remaining the same.

8.51 The core of a high-temperature, gas-cooled nuclear reactor has coolant tubes of 20-mm diameter and 780-mm length. Helium enters at 600 K and exits at 1000 K when the flow rate is 8×10^{-3} kg/s per tube.

(a) Determine the uniform tube wall surface temperature for these conditions.

(b) If the coolant gas is air, determine the required flow rate if the heat removal rate and tube wall surface temperature remain the same. What is the outlet temperature of the air?

8.52 Air at 200 kPa enters a 2-m-long, thin-walled tube of 25-mm diameter at 150°C and 6 m/s. Steam at 20 bars condenses on the outer surface.

(a) Determine the outlet temperature and pressure drop of the air, as well as the rate of heat transfer to the air.

(b) Calculate the parameters of part (a) if the pressure of the air is doubled.

8.53 Heated air required for a food-drying process is generated by passing ambient air at 20°C through long, circular tubes ($D = 50$ mm, $L = 5$ m) housed in a steam condenser. Saturated steam at atmospheric pressure condenses on the outer surface of the tubes, maintaining a uniform surface temperature of 100°C.

(a) If an airflow rate of 0.01 kg/s is maintained in each tube, determine the air outlet temperature $T_{m,o}$ and the total heat rate q for the tube.

(b) The air outlet temperature may be controlled by adjusting the tube mass flow rate. Compute and plot $T_{m,o}$ as a function of $\dot{m}$ for $0.005 \leq \dot{m} \leq 0.050$ kg/s. If a particular drying process requires approximately 1 kg/s of air at 75°C, what design and operating conditions should be prescribed for the air heater, subject to the constraint that the tube diameter and length be fixed at 50 mm and 5 m, respectively?

8.54 Consider laminar flow of a fluid with $Pr = 4$ that undergoes a combined entrance process within a constant surface temperature tube of length $L < x_{\text{fd},t}$ with a flow rate of $\dot{m}$. An engineer suggests that the total heat transfer rate can be improved if the tube is divided into N shorter tubes, each of length $L_N = L/N$ with a flow rate of $\dot{m}/N$. Determine an expression for the ratio of the heat transfer coefficient averaged over the N tubes, each experiencing a combined entrance process, to the heat transfer coefficient averaged over the single tube, $\overline{h}_{D,N}/\overline{h}_{D,1}$.

8.55 A common procedure for cooling a high-performance computer chip involves joining the chip to a heat sink within which circular microchannels are machined. During operation, the chip produces a uniform heat flux q''_c at its interface with the heat sink, while a liquid coolant (water) is routed through the channels. Consider a square chip and heat sink, each L on a side, with microchannels of diameter D and pitch $S = C_1 D$, where the constant C_1 is greater than unity. Water is supplied at an inlet temperature $T_{m,i}$ and a total mass flow rate $\dot{m}$ (for the entire heat sink).

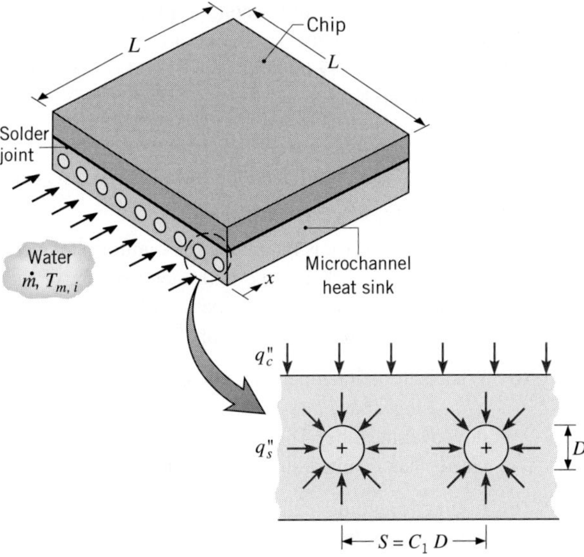

(a) Assuming that q_c'' is dispersed in the heat sink such that a uniform heat flux q_s'' is maintained at the surface of each channel, obtain expressions for the longitudinal distributions of the mean fluid, $T_m(x)$, and surface, $T_s(x)$, temperatures in each channel. Assume laminar, fully developed flow throughout each channel, and express your results in terms of $\dot{m}$, q_c'', C_1, D, and/or L, as well as appropriate thermophysical properties.

(b) For $L = 12$ mm, $D = 1$ mm, $C_1 = 2$, $q_c'' = 20$ W/cm^2, $\dot{m} = 0.010$ kg/s, and $T_{m,i} = 290$ K, compute and plot the temperature distributions $T_m(x)$ and $T_s(x)$.

(c) A common objective in designing such heat sinks is to maximize q_c'' while maintaining the heat sink at an acceptable temperature. Subject to prescribed values of $L = 12$ mm and $T_{m,i} = 290$ K and the constraint that $T_{s,\max} \leq 50°$C, explore the effect on q_c'' of variations in heat sink design and operating conditions.

8.56 One way to cool chips mounted on the circuit boards of a computer is to encapsulate the boards in metal frames that provide efficient pathways for conduction to supporting *cold plates*. Heat generated by the chips is then dissipated by transfer to water flowing through passages drilled in the plates. Because the plates are made from a metal of large thermal conductivity (typically aluminium or copper), they may be assumed to be at a temperature, $T_{s,\text{cp}}$.

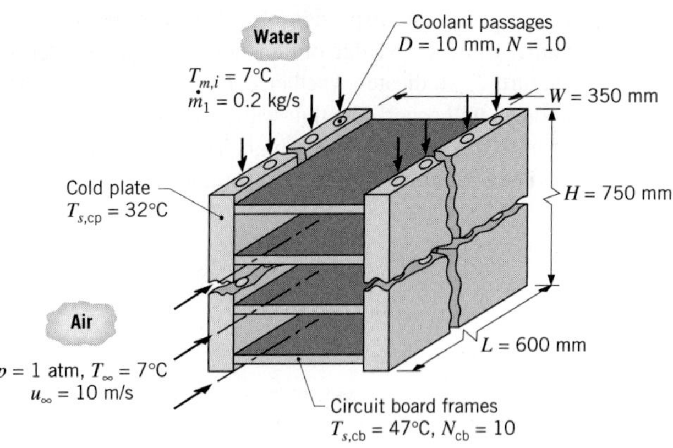

(a) Consider circuit boards attached to cold plates of height $H = 750$ mm and width $L = 600$ mm, each with $N = 10$ holes of diameter $D = 10$ mm. If operating conditions maintain plate temperatures of $T_{s,\text{cp}} = 32°$C with water flow at $\dot{m}_1 = 0.2$ kg/s per passage and $T_{m,i} = 7°$C, how much heat may be dissipated by the circuit boards?

(b) To enhance cooling, thereby allowing increased power generation without an attendant increase in system temperatures, a hybrid cooling scheme may be used. The scheme involves forced airflow over the encapsulated circuit boards, as well as water flow through the cold plates. Consider conditions for which $N_{\text{cb}} = 10$ circuit boards of width $W = 350$ mm are attached to the cold plates and their average surface temperature is $T_{s,\text{cb}} = 47°$C when $T_{s,\text{cp}} = 32°$C. If air is in parallel flow over the plates with $u_\infty = 10$ m/s and $T_\infty = 7°$C, how much of the heat generated by the circuit boards is transferred to the air?

8.57 Refrigerant-134a is being transported at 0.1 kg/s through a Teflon tube of inside diameter $D_i = 25$ mm and outside diameter $D_o = 28$ mm, while atmospheric air at $V = 25$ m/s and 300 K is in cross flow over the tube. What is the heat transfer per unit length of tube to Refrigerant-134a at 240 K?

8.58 Oil at 150°C flows *slowly* through a long, thin-walled pipe of 30-mm inner diameter. The pipe is suspended in a room for which the air temperature is 20°C and the convection coefficient at the outer tube surface is 11 W/m$^2\cdot$K. Estimate the heat loss per unit length of tube.

8.59 Exhaust gases from a wire processing oven are discharged into a tall stack, and the gas and stack surface temperatures at the outlet of the stack must be estimated. Knowledge of the outlet gas temperature $T_{m,o}$ is useful for predicting the dispersion of effluents in the thermal plume, while knowledge of the outlet stack surface temperature $T_{s,o}$ indicates whether condensation of the gas products will occur. The thin-walled, cylindrical stack is 0.5 m in diameter and 6.0 m high. The exhaust gas flow rate is 0.5 kg/s, and the inlet temperature is 600°C.

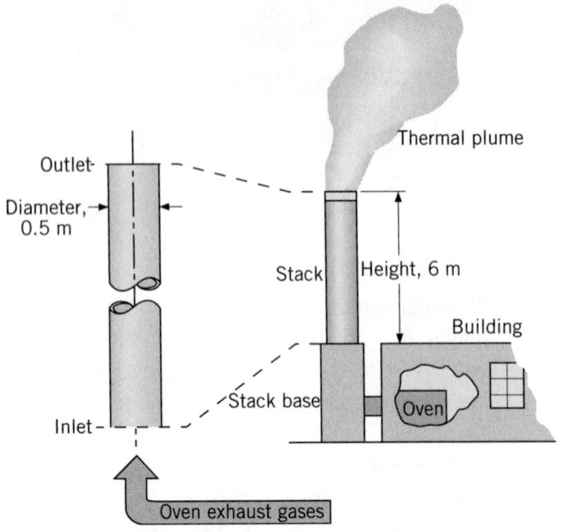

(a) Consider conditions for which the ambient air temperature and wind velocity are 4°C and 5 m/s, respectively. Approximating the thermophysical properties of the gas as those of atmospheric air, estimate the outlet gas and stack surface temperatures for the given conditions.

(b) The gas outlet temperature is sensitive to variations in the ambient air temperature and wind velocity. For $T_\infty = -25°C$, 5°C, and 35°C, compute and plot the gas outlet temperature as a function of wind velocity for $2 \le V \le 10$ m/s.

8.60 A hot fluid passes through a thin-walled tube of 10-mm diameter and 1-m length, and a coolant at $T_\infty = 25°C$ is in cross flow over the tube. When the flow rate is $\dot{m} = 18$ kg/h and the inlet temperature is $T_{m,i} = 85°C$, the outlet temperature is $T_{m,o} = 78°C$.

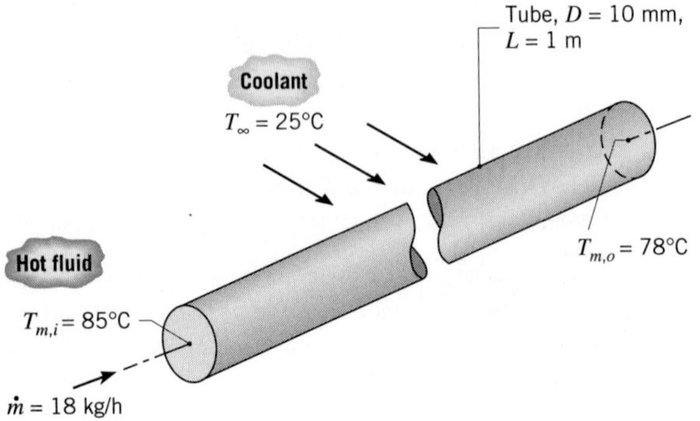

Assuming fully developed flow and thermal conditions in the tube, determine the outlet temperature, $T_{m,o}$, if the flow rate is increased by a factor of 2. That is, $\dot{m} = 36$ kg/h, with all other conditions the same. The thermophysical properties of the hot fluid are $\rho = 1079$ kg/m^3, $c_p = 2637$ J/kg·K, $\mu = 0.0034$ N·s/m^2, and $k = 0.261$ W/m·K.

8.61 Consider a thin-walled tube of 10-mm diameter and 2-m length. Water enters the tube from a large reservoir at $\dot{m} = 0.2$ kg/s and $T_{m,i} = 47°C$.

(a) If the tube surface is maintained at a uniform temperature of 27°C, what is the outlet temperature of the water, $T_{m,o}$? To obtain the properties of water, assume an average mean temperature of $\overline{T}_m = 300$ K.

(b) What is the exit temperature of the water if it is heated by passing air at $T_\infty = 100°C$ and $V = 10$ m/s in cross flow over the tube? The properties of air may be evaluated at an assumed film temperature of $T_f = 350$ K.

(c) In the foregoing calculations, were the assumed values of $\overline{T}_m$ and T_f appropriate? If not, use properly evaluated properties and recompute $T_{m,o}$ for the conditions of part (b).

8.62 Water at a flow rate of $\dot{m} = 0.215$ kg/s is cooled from 70°C to 30°C by passing it through a thin-walled tube of diameter $D = 50$ mm and maintaining a coolant at $T_\infty = 15°C$ in cross flow over the tube.

(a) What is the required tube length if the coolant is air and its velocity is $V = 20$ m/s?

(b) What is the tube length if the coolant is water and $V = 2$ m/s?

8.63 The problem of heat losses from a fluid moving through a buried pipeline has received considerable attention. Practical applications include the trans-Alaska pipeline, as well as power plant steam and water distribution lines. Consider a steel pipe of diameter D that is used to transport oil flowing at a rate $\dot{m}_o$ through a cold region. The pipe is covered with a layer of insulation of thickness t and thermal conductivity k_i and is buried in soil to a depth z (distance from the soil surface to the pipe centerline). Each section of pipe is of length L and extends between pumping stations in which the oil is heated to ensure low viscosity and hence low pump power requirements. The temperature of the oil entering the pipe from a pumping station and the temperature of the ground above the pipe are designated as $T_{m,i}$ and T_s, respectively, and are known.

Consider conditions for which the oil (o) properties may be approximated as $\rho_o = 900$ kg/m^3, $c_{p,o} = 2000$ J/kg·K, $\nu_o = 8.5 \times 10^{-4}$ m^2/s, $k_o = 0.140$ W/m·K,

$Pr_o = 10^4$; the oil flow rate is $\dot{m}_o = 500$ kg/s; and the pipe diameter is 1.2 m.

(a) Expressing your results in terms of D, L, z, t, $\dot{m}_o$, $T_{m,i}$, and T_s, as well as the appropriate oil (o), insulation (i), and soil (s) properties, obtain all the expressions needed to estimate the temperature $T_{m,o}$ of the oil leaving the pipe.

(b) If $T_s = -40°C$, $T_{m,i} = 120°C$, $t = 0.15$ m, $k_i = 0.05$ W/m·K, $k_s = 0.5$ W/m·K, $z = 3$ m, and $L = 100$ km, what is the value of $T_{m,o}$? What is the total rate of heat transfer q from a section of the pipeline?

(c) The operations manager wants to know the tradeoff between the burial depth of the pipe and insulation thickness on the heat loss from the pipe. Develop a graphical representation of this design information.

8.64 To maintain pump power requirements per unit flow rate below an acceptable level, operation of the oil pipeline of Problem 8.63 is subject to the constraint that the oil exit temperature $T_{m,o}$ exceed 110°C. For the values of $T_{m,i}$, T_s, D, t_i, z, L, and k_i prescribed in Problem 8.63, operating parameters that are variable and affect $T_{m,o}$ are the thermal conductivity of the soil and the flow rate of the oil. Depending on soil composition and moisture and the demand for oil, representative variations are $0.25 \leq k_s \leq 1.0$ W/m·K and $250 \leq \dot{m}_o \leq 500$ kg/s. Using the properties prescribed in Problem 8.63, determine the effect of the foregoing variations on $T_{m,o}$ and the total heat rate q. What is the worst case operating condition? If necessary, what adjustments could be made to ensure that $T_{m,o} \geq 110°C$ for the worst case conditions?

8.65 Consider a thin-walled, metallic tube of length $L = 1$ m and inside diameter $D_i = 3$ mm. Water enters the tube at $\dot{m} = 0.015$ kg/s and $T_{m,i} = 97°C$.

(a) What is the outlet temperature of the water if the tube surface temperature is maintained at 27°C?

(b) If a 0.5-mm-thick layer of insulation of $k = 0.05$ W/m·K is applied to the tube and its outer surface is maintained at 27°C, what is the outlet temperature of the water?

(c) If the outer surface of the insulation is no longer maintained at 27°C but is allowed to exchange heat by free convection with ambient air at 27°C, what is the outlet temperature of the water? The free convection heat transfer coefficient is 5 W/m^2·K.

8.66 A circular tube of diameter $D = 0.2$ mm and length $L = 100$ mm imposes a constant heat flux of $q'' = 20 \times 10^3$ W/m^2 on a fluid with a mass flow rate of $\dot{m} = 0.1$ g/s. For an inlet temperature of $T_{m,i} = 29°C$, determine the tube wall temperature at $x = L$ for pure water. Evaluate fluid properties at $\overline{T} = 300$ K. For the same conditions, determine the tube wall temperature at $x = L$ for the nanofluid of Example 2.2.

8.67 Repeat Problem 8.66 for a circular tube of diameter $D = 2$ mm, an applied heat flux of $q'' = 200{,}000$ W/m^2, and a mass flow rate of $\dot{m} = 10$ g/s.

8.68 Heat is to be removed from a reaction vessel operating at 75°C by supplying water at 27°C and 0.12 kg/s through a thin-walled tube of 15-mm diameter. The convection coefficient between the tube outer surface and the fluid in the vessel is 3000 W/m^2·K.

(a) If the outlet water temperature cannot exceed 47°C, what is the maximum rate of heat transfer from the vessel?

(b) What tube length is required to accomplish the heat transfer rate of part (a)?

8.69 A heating contractor must heat 0.2 kg/s of water from 15°C to 35°C using hot gases in cross flow over a thin-walled tube.

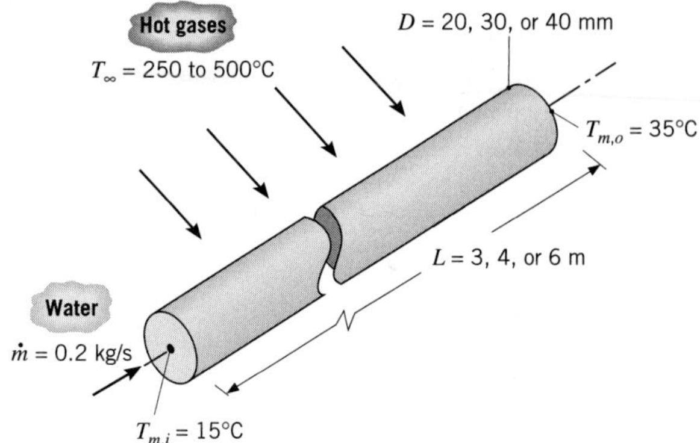

Your assignment is to develop a series of design graphs that can be used to demonstrate acceptable combinations of tube dimensions (D and L) and of hot gas conditions (T_∞ and V) that satisfy this requirement. In your analysis, consider the following parameter ranges: $D = 20$, 30, or 40 mm; $L = 3$, 4, or 6 m; $T_\infty = 250$, 375, or 500°C; and $20 \leq V \leq 40$ m/s.

8.70 A thin-walled tube with a diameter of 6 mm and length of 20 m is used to carry exhaust gas from a smoke stack to the laboratory in a nearby building for analysis. The gas enters the tube at 200°C and with a mass flow rate of 0.003 kg/s. Autumn winds at a temperature of 15°C blow directly across the tube at a velocity of 5 m/s. Assume the thermophysical properties of the exhaust gas are those of air.

(a) Estimate the average heat transfer coefficient for the exhaust gas flowing inside the tube.

(b) Estimate the heat transfer coefficient for the air flowing across the outside of the tube.

(c) Estimate the overall heat transfer coefficient U and the temperature of the exhaust gas when it reaches the laboratory.

8.71 A 50-mm-diameter, thin-walled metal pipe covered with a 25-mm-thick layer of insulation (0.085 W/m·K) and carrying superheated steam at atmospheric pressure is suspended from the ceiling of a large room. The steam temperature entering the pipe is 120°C, and the air temperature is 20°C. The convection heat transfer coefficient on the outer surface of the covered pipe is 10 W/m^2·K. If the velocity of the steam is 10 m/s, at what point along the pipe will the steam begin condensing?

8.72 A thin-walled, uninsulated 0.3-m-diameter duct is used to route chilled air at 0.05 kg/s through the attic of a large commercial building. The attic air is at 37°C, and natural circulation provides a convection coefficient of 2 W/m^2·K at the outer surface of the duct. If chilled air enters a 15-m-long duct at 7°C, what is its exit temperature and the rate of heat gain? Properties of the chilled air may be evaluated at an assumed average temperature of 300 K.

8.73 Pressurized water at $T_{m,i} = 200°C$ is pumped at $\dot{m} = 2$ kg/s from a power plant to a nearby industrial user through a thin-walled, round pipe of inside diameter $D = 1$ m. The pipe is covered with a layer of insulation of thickness $t = 0.15$ m and thermal conductivity $k = 0.05$ W/m·K. The pipe, which is of length $L = 500$ m, is exposed to a cross flow of air at $T_\infty = -10°C$ and $V = 4$ m/s. Obtain a differential equation that could be used to solve for the variation of the mixed mean temperature of the water $T_m(x)$ with the axial coordinate. As a first approximation, the internal flow may be assumed to be fully developed throughout the pipe. Express your results in terms of $\dot{m}$, V, T_∞, D, t, k, and appropriate water (w) and air (a) properties. Evaluate the heat loss per unit length of the pipe at the inlet. What is the mean temperature of the water at the outlet?

8.74 Water at 290 K and 0.2 kg/s flows through a Teflon tube ($k = 0.35$ W/m·K) of inner and outer radii equal to 10 and 13 mm, respectively. A thin electrical heating tape wrapped around the outer surface of the tube delivers a uniform surface heat flux of 2000 W/m^2, while a convection coefficient of 25 W/m^2·K is maintained on the outer surface of the tape by ambient air at 300 K. What is the fraction of the power dissipated by the tape, which is transferred to the water? What is the outer surface temperature of the Teflon tube?

8.75 The temperature of flue gases flowing through the large stack of a boiler is measured by means of a thermocouple enclosed within a cylindrical tube as shown. The tube axis is oriented normal to the gas flow, and the thermocouple senses a temperature T_t corresponding to that of the tube surface. The gas flow rate and temperature are designated as $\dot{m}_g$ and T_g, respectively, and the gas flow may be assumed to be fully developed. The stack is fabricated from sheet metal that is at a uniform temperature T_s and is exposed to ambient air at T_∞ and large surroundings at T_{sur}. The convection coefficient associated with the outer surface of the duct is designated as h_o, while those associated with the inner surface of the duct and the tube surface are designated as h_i and h_t, respectively. The tube and duct surface emissivities are designated as ε_t and ε_s, respectively.

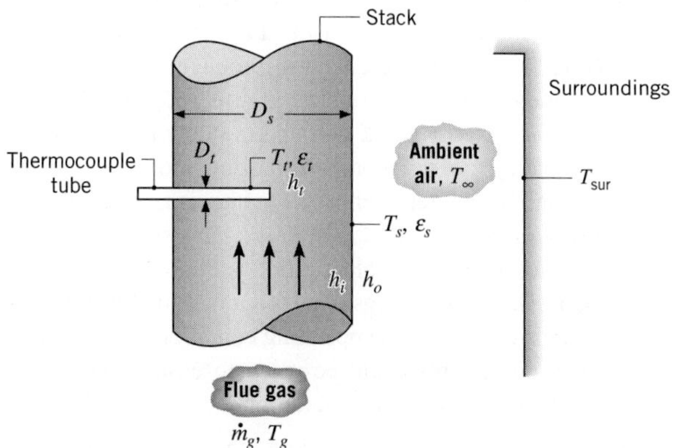

(a) Neglecting conduction losses along the thermocouple tube, develop an analysis that could be used to predict the error $(T_g - T_t)$ in the temperature measurement.

(b) Assuming the flue gas to have the properties of atmospheric air, evaluate the error for $T_t = 300°C$, $D_s = 0.6$ m, $D_t = 10$ mm, $\dot{m}_g = 1$ kg/s, $T_\infty = T_{sur} = 27°C$, $\varepsilon_t = \varepsilon_s = 0.8$, and $h_o = 25$ W/m^2·K.

8.76 In a biomedical supplies manufacturing process, a requirement exists for a large platen that is to be maintained at 45 ± 0.25°C. The proposed design features the attachment of heating tubes to the platen at a relative spacing S. The thick-walled, copper tubes have an inner diameter of $D_i = 8$ mm and are attached to the platen with a high thermal conductivity solder, which provides a contact width of $2D_i$. The heating fluid (ethylene glycol) flows through each tube at a fixed rate of $\dot{m} = 0.06$ kg/s. The platen has a thickness of $w = 25$ mm and is fabricated from a stainless steel with a thermal conductivity of 15 W/m·K.

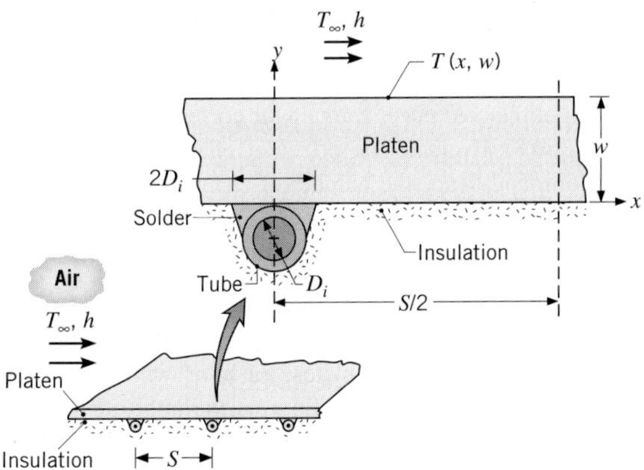

Considering the two-dimensional cross section of the platen shown in the inset, perform an analysis to determine the heating fluid temperature T_m and the tube spacing S required to maintain the surface temperature of the platen, $T(x, w)$, at $45 \pm 0.25°C$, when the ambient temperature is 25°C and the convection coefficient is 100 W/m^2·K.

8.77 Consider the ground source heat pump of Problem 5.100 under winter conditions for which the liquid is discharged from the heat pump into high-density polyethylene tubing of thickness $t = 8$ mm and thermal conductivity $k = 0.47$ W/m·K. The tubing is routed through soil that maintains a uniform temperature of approximately 10°C at the tube outer surface. The properties of the fluid may be approximated as those of water.

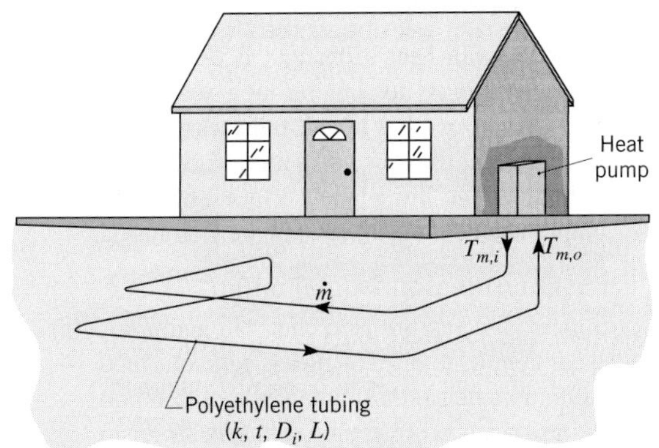

(a) For a tube inner diameter and flow rate of $D_i = 25$ mm and $\dot{m} = 0.03$ kg/s and a fluid inlet temperature of $T_{m,i} = 0°C$, determine the tube outlet temperature (heat pump inlet temperature), $T_{m,o}$, as a function of the tube length L for $10 \leq L \leq 50$ m.

(b) Recommend an appropriate length for the system. How would your recommendation be affected by variations in the liquid flow rate?

8.78 For a sharp-edged inlet and a combined entry region, the average Nusselt number may be computed from Equation 8.63, with $C = 24\, Re_D^{-0.23}$ and $m = 0.815 - 2.08 \times 10^{-6}\, Re_D$ [23]. Determine $\overline{Nu}_D/Nu_{D,\mathrm{fd}}$ at $x/D = 10$ and 60 for $Re_D = 10^4$ and 10^5.

8.79 Fluid enters a thin-walled tube of 5-mm diameter and 2-m length with a flow rate of 0.04 kg/s and a temperature of $T_{m,i} = 85°C$. The tube surface is maintained at a temperature of $T_s = 25°C$, and for this operating condition, the outlet temperature is $T_{m,o} = 31.1°C$. What is the outlet temperature if the flow rate is doubled? Fully developed, turbulent flow may be assumed to exist in both cases, and the fluid properties may be assumed to be independent of temperature.

Noncircular Ducts

8.80 Air at 3×10^{-4} kg/s and 27°C enters a rectangular duct that is 1 m long and 4 mm × 16 mm on a side. A uniform heat flux of 600 W/m^2 is imposed on the duct surface. What is the temperature of the air and of the duct surface at the outlet?

8.81 Air at 25°C flows at 30×10^{-6} kg/s within 100-mm-long channels used to cool a high thermal conductivity metal mold. Assume the flow is hydrodynamically and thermally fully developed.

(a) Determine the heat transferred to the air for a circular channel ($D = 10$ mm) when the mold temperature is 50°C (case A).

(b) Using new manufacturing methods (see Problem 8.105), channels of complex cross section can be readily fabricated within metal objects, such as molds. Consider air flowing under the same conditions as in case A, except now the channel is segmented into six smaller triangular sections. The flow area of case A is equal to the total flow area of case B. Determine the heat transferred to the air for the segmented channel.

(c) Compare the pressure drops for cases A and B.

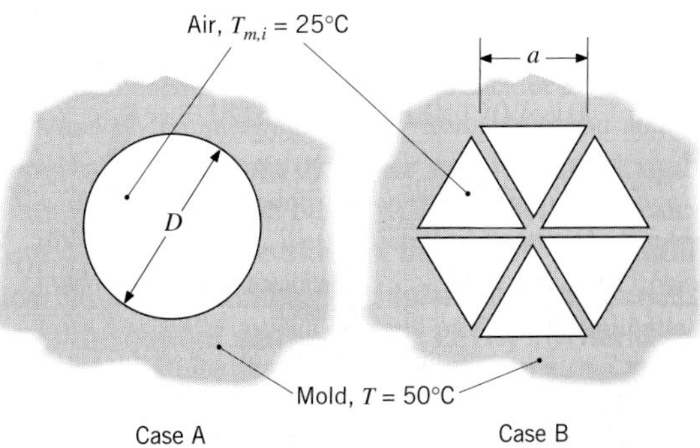

8.82 A *cold plate* is an active cooling device that is attached to a heat-generating system in order to dissipate the heat while maintaining the system at an acceptable temperature. It is typically fabricated from a material of high thermal conductivity, k_{cp}, within which channels are machined and a coolant is passed. Consider a copper cold plate of height H and width W on a side, within which water passes through square channels of width $w = h$. The transverse spacing between channels δ is twice the spacing between the sidewall of an outer channel and the sidewall of the cold plate.

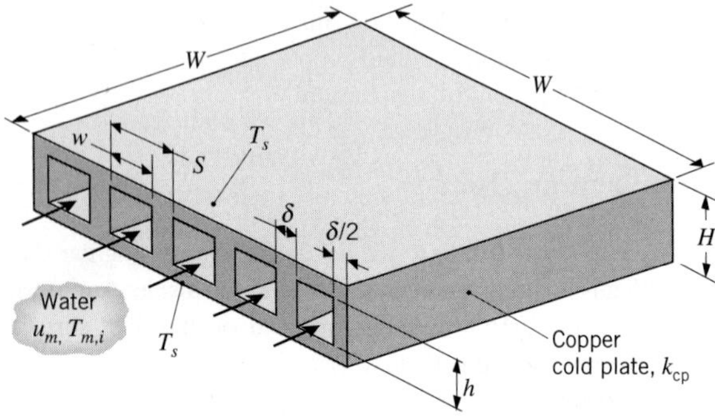

Consider conditions for which *equivalent* heat-generating systems are attached to the top and bottom of the cold plate, maintaining the corresponding surfaces at the same temperature T_s. The mean velocity and inlet temperature of the coolant are u_m and $T_{m,i}$, respectively.

(a) Assuming fully developed turbulent flow throughout each channel, obtain a system of equations that may be used to evaluate the total rate of heat transfer to the cold plate, q, and the outlet temperature of the water, $T_{m,o}$, in terms of the specified parameters.

(b) Consider a cold plate of width $W = 100$ mm and height $H = 10$ mm, with 10 square channels of width $w = 6$ mm and a spacing of $\delta = 4$ mm between channels. Water enters the channels at a temperature of $T_{m,i} = 300$ K and a velocity of $u_m = 2$ m/s. If the top and bottom cold plate surfaces are at $T_s = 360$ K, what is the outlet water temperature and the total rate of heat transfer to the cold plate? The thermal conductivity of the copper is 400 W/m·K, while average properties of the water may be taken to be $\rho = 984$ kg/m^3, $c_p = 4184$ J/kg·K, $\mu = 489 \times 10^{-6}$ N·s/m^2, $k = 0.65$ W/m·K, and $Pr = 3.15$. Is this a good cold plate design? How could its performance be improved?

8.83 The cold plate design of Problem 8.82 has not been optimized with respect to selection of the channel width, and we wish to explore conditions for which the rate of heat transfer may be enhanced. Assume that the width and height of the copper cold plate are fixed at $W = 100$ mm and $H = 10$ mm, while the channel height and spacing between channels are fixed at $h = 6$ mm and $\delta = 4$ mm. The mean velocity and inlet temperature of the water are maintained at $u_m = 2$ m/s and $T_{m,i} = 300$ K, while equivalent heat-generating systems attached to the top and bottom of the cold plate maintain the corresponding surfaces at 360 K. Evaluate the effect of changing the channel width, and hence the number of channels, on the rate of heat transfer to the cold plate. Include consideration of the limiting case for which $w = 96$ mm (one channel).

8.84 A device that recovers heat from high-temperature combustion products involves passing the combustion gas between parallel plates, each of which is maintained at 350 K by water flow on the opposite surface. The plate separation is 40 mm, and the gas flow is fully developed. The gas may be assumed to have the properties of atmospheric air, and its mean temperature and velocity are 1000 K and 60 m/s, respectively.

(a) What is the heat flux at the plate surface?

(b) If a third plate, 20 mm thick, is suspended midway between the original plates, what is the surface heat flux for the original plates? Assume the temperature and *flow rate* of the gas to be unchanged and radiation effects to be negligible.

8.85 Air at 1 atm and 285 K enters a 2-m-long rectangular duct with cross section 75 mm × 150 mm. The duct is maintained at a constant surface temperature of 400 K, and the air mass flow rate is 0.10 kg/s. Determine the heat transfer rate from the duct to the air and the air outlet temperature.

8.86 A double-wall heat exchanger is used to transfer heat between liquids flowing through semicircular copper tubes. Each tube has a wall thickness of $t = 3$ mm and an inner radius of $r_i = 20$ mm, and good contact is maintained at the plane surfaces by tightly wound straps. The tube outer surfaces are well insulated.

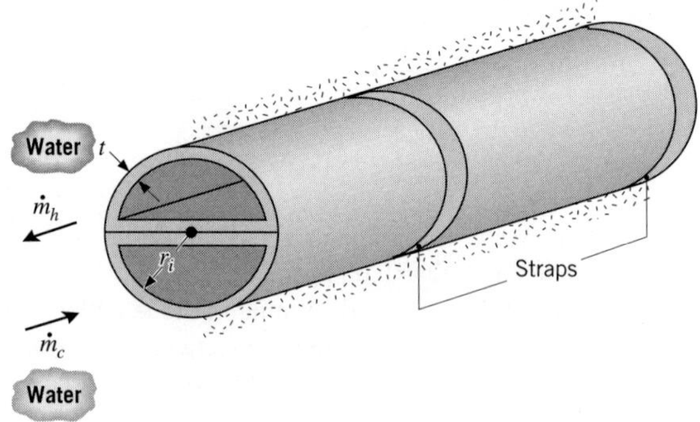

(a) If hot and cold water at mean temperatures of $T_{h,m} = 330$ K and $T_{c,m} = 290$ K flow through the adjoining tubes at $\dot{m}_h = \dot{m}_c = 0.2$ kg/s, what is the rate of heat transfer per unit length of tube? The wall contact resistance is 10^{-5} m²·K/W. Approximate the properties of both the hot and cold water as $\mu = 800 \times 10^{-6}$ kg/s·m, $k = 0.625$ W/m·K, and $Pr = 5.35$. *Hint*: Heat transfer is enhanced by conduction through the semicircular portions of the tube walls, and each portion may be subdivided into two straight fins with adiabatic tips.

(b) Using the thermal model developed for part (a), determine the heat transfer rate per unit length when the fluids are ethylene glycol. Also, what effect will fabricating the exchanger from an aluminum alloy have on the heat rate? Will increasing the thickness of the tube walls have a beneficial effect?

8.87 Consider laminar, fully developed flow in a channel of constant surface temperature T_s. For a given mass flow rate and channel length, determine which rectangular channel, $b/a = 1.0$, 1.43, or 2.0, will provide the highest heat transfer rate. Is this heat transfer rate greater than, equal to, or less than the heat transfer rate associated with a circular tube?

8.88 You have been asked to perform a feasibility study on the design of a blood warmer to be used during the transfusion of blood to a patient. This exchanger is to heat blood taken from the bank at 10°C to 37°C at a flow rate of 200 ml/min. The blood passes through a rectangular cross-section tube, 6.4 mm × 1.6 mm, which is sandwiched between two plates held at a constant temperature of 40°C.

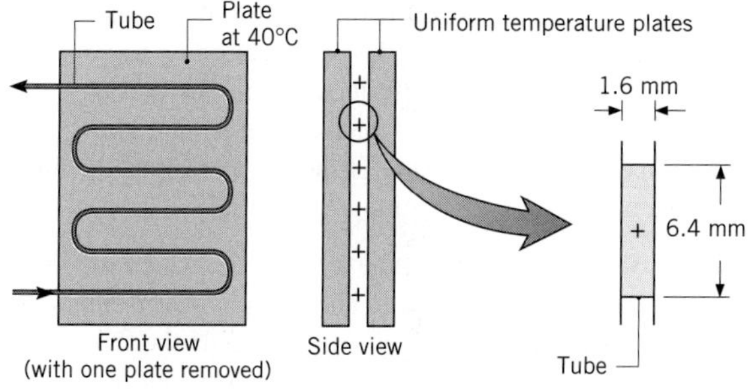

(a) Compute the length of the tubing required to achieve the desired outlet conditions at the specified flow rate. Assume the flow is fully developed and the blood has the same properties as water.

(b) Assess your assumptions and indicate whether your analysis over- or underestimates the necessary length.

8.89 A coolant flows through a rectangular channel (*gallery*) within the body of a mold used to form metal injection parts. The gallery dimensions are $a = 90$ mm and $b = 9.5$ mm, and the fluid flow rate is 1.3×10^{-3} m³/s. The coolant temperature is 15°C, and the mold wall is at an approximately uniform temperature of 140°C.

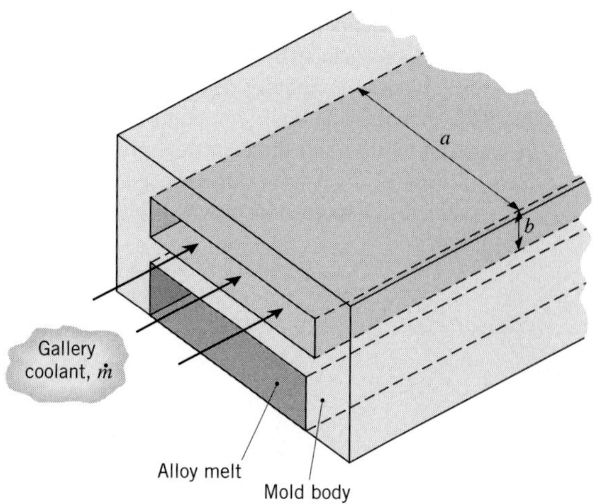

To minimize corrosion damage to the expensive mold, it is customary to use a heat transfer fluid such as ethylene glycol, rather than process water. Compare the convection coefficients of water and ethylene glycol for this application. What is the tradeoff between thermal performance and minimizing corrosion?

8.90 An electronic circuit board dissipating 50 W is sandwiched between two ducted, forced-air-cooled heat sinks. The sinks are 150 mm in length and have 20 rectangular passages 6 mm × 25 mm. Atmospheric air at a volumetric flow rate of 0.060 m³/s and 27°C is drawn through the sinks by a blower. Estimate the operating temperature of the board and the pressure drop across the sinks.

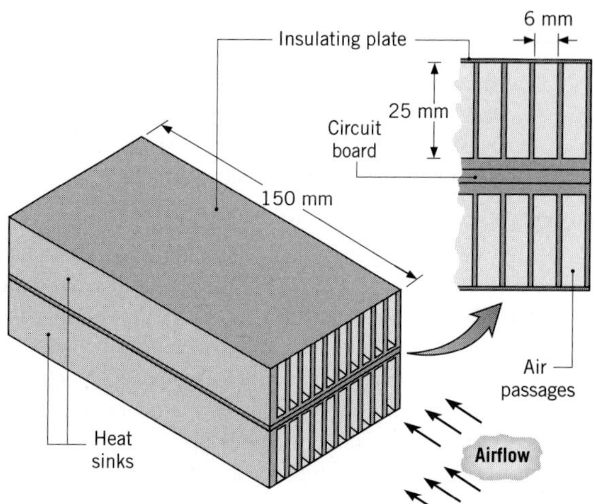

8.91 To slow down large prime movers like locomotives, a process termed dynamic electric braking is used to switch the traction motor to a generator mode in which mechanical power from the drive wheels is absorbed and used to generate electrical current. As shown in the schematic, the electric power is passed through a resistor grid (*a*), which consists of an array of metallic blades electrically connected in series (*b*). The blade material is a high-temperature, high electrical resistivity alloy, and the electrical power is dissipated as heat by internal volumetric generation. To cool the blades, a motor-fan moves high-velocity air through the grid.

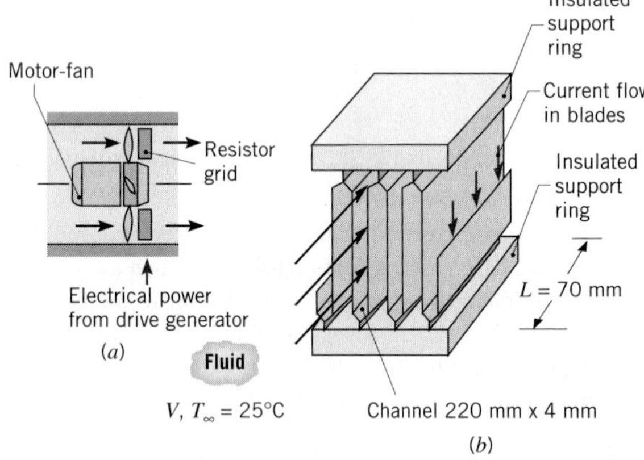

(a) Treating the space between the blades as a rectangular channel of 220-mm × 4-mm cross section and 70-mm length, estimate the heat removal rate per blade if the airstream has an inlet temperature and velocity of 25°C and 50 m/s, respectively, while the blade has an operating temperature of 600°C.

(b) On a locomotive pulling a 10-car train, there may be 2000 of these blades. Based on your result from part (a), how long will it take to slow a train whose total mass is 10^6 kg from a speed of 120 km/h to 50 km/h using dynamic electric braking?

8.92 A printed circuit board (PCB) is cooled by laminar, fully developed airflow in adjoining, parallel-plate channels of length L and separation distance a. The channels may be assumed to be of infinite extent in the transverse direction, and the upper and lower surfaces are insulated. The temperature T_s of the PCB board is uniform, and airflow with an inlet temperature of $T_{m,i}$ is driven by a pressure difference Δp.

Calculate the average heat removal rate per unit area (W/m^2) from the PCB.

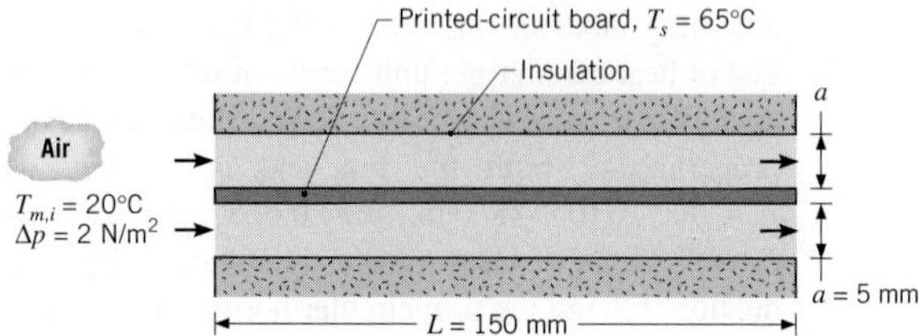

8.93 Water at $\dot{m} = 0.02$ kg/s and $T_{m,i} = 20°C$ enters an annular region formed by an inner tube of diameter $D_i = 25$ mm and an outer tube of diameter $D_o = 100$ mm. Saturated steam flows through the inner tube, maintaining its surface at a uniform temperature of $T_{s,i} = 100°C$, while the outer surface of the outer tube is well insulated. If fully developed conditions may be assumed throughout the annulus, how long must the system be to provide an outlet water temperature of 75°C? What is the heat flux from the inner tube at the outlet?

8.94 For the conditions of Problem 8.93, how long must the annulus be if the water flow rate is 0.30 kg/s instead of 0.02 kg/s?

8.95 Referring to Figure 8.11, consider conditions in an annulus having an outer surface that is insulated ($q''_o = 0$) and a uniform heat flux q''_i at the inner surface. Fully developed, laminar flow may be assumed to exist.

(a) Determine the velocity profile $u(r)$ in the annular region.

(b) Determine the temperature profile $T(r)$ and obtain an expression for the Nusselt number Nu_i associated with the inner surface.

8.96 Consider the air heater of Problem 8.38, but now with airflow through the annulus and steam flow through the inner tube. For the prescribed conditions and an outer tube diameter of $D_o = 65$ mm, determine the outlet temperature and pressure of the air, as well as the mass rate of steam condensation.

8.97 Consider a concentric tube annulus for which the inner and outer diameters are 25 and 50 mm. Water enters the annular region at 0.04 kg/s and 25°C. If the inner tube wall is heated electrically at a rate (per unit length) of $q' = 4000$ W/m, while the outer tube wall is insulated, how long must the tubes be for the water to achieve an outlet temperature of 85°C? What is the inner tube surface temperature at the outlet, where fully developed conditions may be assumed?

8.98 It is common practice to recover waste heat from an oil- or gas-fired furnace by using the exhaust gases to preheat the combustion air. A device commonly used for this purpose consists of a concentric pipe arrangement for which the exhaust gases are passed through the inner pipe, while the cooler combustion air flows through an annular passage around the pipe.

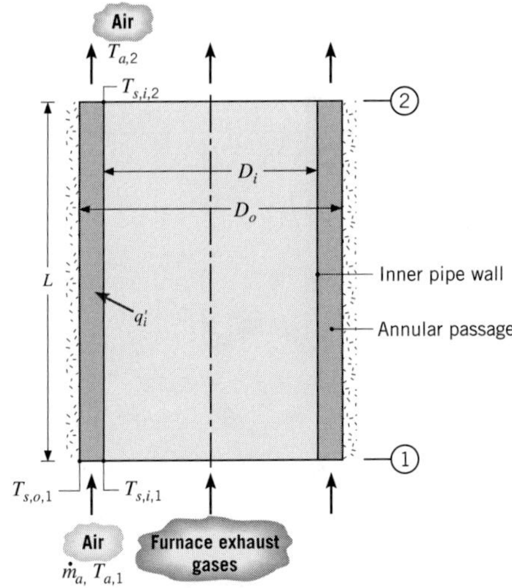

Consider conditions for which there is a uniform heat transfer rate per unit length, $q_i' = 1.25 \times 10^5$ W/m, from the exhaust gases to the pipe inner surface, while air flows through the annular passage at a rate of $\dot{m}_a = 2.1$ kg/s. The thin-walled inner pipe is of diameter $D_i = 2$ m, while the outer pipe, which is well insulated from the surroundings, is of diameter $D_o = 2.05$ m. The air properties may be taken to be $c_p = 1030$ J/kg·K, $\mu = 270 \times 10^{-7}$ N·s/m^2, $k = 0.041$ W/m·K, and $Pr = 0.68$.

(a) If air enters at $T_{a,1} = 300$ K and $L = 7$ m, what is the air outlet temperature $T_{a,2}$?

(b) If the airflow is fully developed throughout the annular region, what is the temperature of the inner pipe at the inlet ($T_{s,i,1}$) and outlet ($T_{s,i,2}$) sections of the device? What is the outer surface temperature $T_{s,o,1}$ at the inlet?

8.99 A concentric tube arrangement, for which the inner and outer diameters are 80 mm and 100 mm, respectively, is used to remove heat from a biochemical reaction occurring in a 1-m-long settling tank. Heat is generated uniformly within the tank at a rate of 10^5 W/m^3, and water is supplied to the annular region at a rate of 0.2 kg/s.

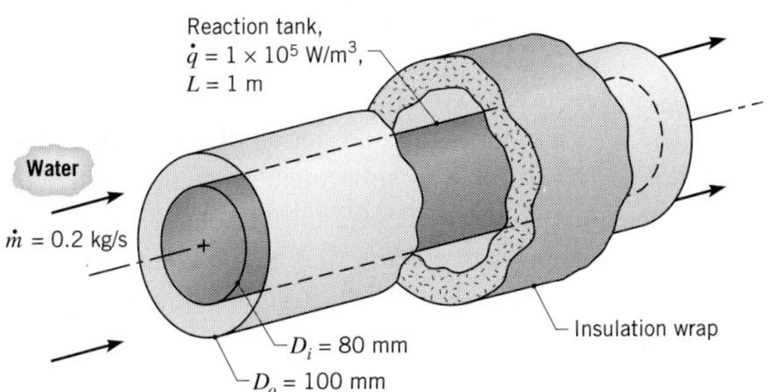

(a) Determine the inlet temperature of the supply water that will maintain an average tank surface temperature of 37°C. Assume fully developed flow and thermal conditions. Is this assumption reasonable?

(b) It is desired to have a slight, axial temperature gradient on the tank surface, since the rate of the biochemical reaction is highly temperature dependent. Sketch the axial variation of the water and surface temperatures along the flow direction for the following two cases: (i) the fully developed conditions of part (a), and (ii) conditions for which entrance effects are important. Comment on features of the temperature distributions. What change to the system or operating conditions would you make to reduce the surface temperature gradient?

Heat Transfer Enhancement

8.100 Consider the air cooling system and conditions of Problem 8.31, but with a prescribed pipe length of $L = 15$ m.

(a) What is the air outlet temperature, $T_{m,o}$? What is the fan power requirement?

(b) The convection coefficient associated with airflow in the pipe may be increased twofold by inserting a coiled spring along the length of the pipe to disrupt flow conditions near the inner surface. If such a heat transfer enhancement scheme is adopted, what is the attendant value of $T_{m,o}$? Use of the insert would not come without a corresponding increase in the fan power requirement. What is the power requirement if the friction factor is increased by 50%?

(c) After extended exposure to the water, a thin coating of organic matter forms on the outer surface of the pipe, and its thermal resistance (for a unit area of the outer surface) is $R''_{t,o} = 0.050$ m^2·K/W. What is the corresponding value of $T_{m,o}$ without the insert of part (b)?

8.101 Consider sterilization of the pharmaceutical product of Problem 8.27. To avoid any possibility of heating the product to an unacceptably high temperature, atmospheric steam is condensed on the exterior of the tube instead of using the resistance heater, providing a uniform surface temperature, $T_s = 100$°C.

(a) For the conditions of Problem 8.27, determine the required length of straight tube, L_s, that would be needed to increase the mean temperature of the pharmaceutical product from 25°C to 75°C.

(b) Consider replacing the straight tube with a coiled tube characterized by a coil diameter $C = 100$ mm and a coil pitch $S = 25$ mm. Determine the overall length of the coiled tube, L_{cl} (i.e., the product of the tube pitch and the number of coils), necessary to increase the mean temperature of the pharmaceutical to the desired value.

(c) Calculate the pressure drop through the straight tube and through the coiled tube.

(d) Calculate the steam condensation rate.

8.102 An engineer proposes to insert a solid rod of diameter D_i into a circular tube of diameter D_o to enhance heat transfer from the flowing fluid of temperature T_m to the outer tube wall of temperature $T_{s,o}$. Assuming laminar flow, calculate the ratio of the heat flux from the fluid to the outer tube wall with the rod to the heat flux without the rod, $q''_o/q''_{o,\text{wo}}$, for $D_i/D_o = 0, 0.10, 0.25$ and 0.50. The rod is placed concentrically within the tube.

8.103 An electrical power transformer of diameter 230 mm and height 500 mm dissipates 1000 W. It is desired to maintain its surface temperature at 47°C by supplying ethylene glycol at 24°C through thin-walled tubing of 20-mm diameter welded to the lateral surface of the transformer. All the heat dissipated by the transformer is assumed to be transferred to the ethylene glycol.

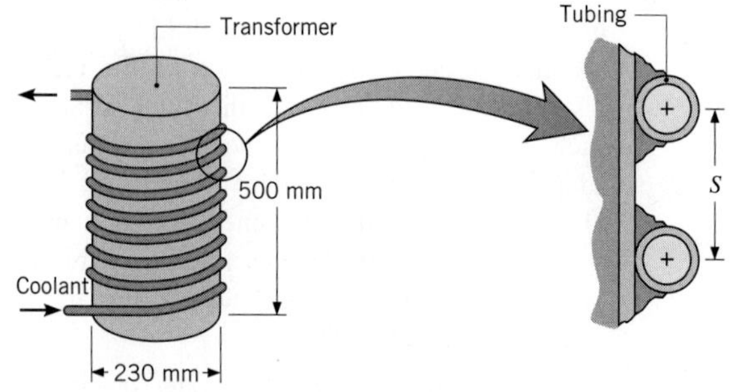

Assuming the maximum allowable temperature rise of the coolant to be 6°C, determine the required coolant flow rate, the total length of tubing, and the coil pitch S between turns of the tubing.

8.104 A *bayonet cooler* is used to reduce the temperature of a pharmaceutical fluid. The pharmaceutical fluid flows through the cooler, which is fabricated of 10-mm-diameter, thin-walled tubing with two 250-mm-long straight sections and a coil with six and a half turns and a coil diameter of 75 mm. A coolant flows outside the cooler, with a convection coefficient at the outside surface of $h_o = 500\ \text{W/m}^2 \cdot \text{K}$ and a coolant temperature of 20°C. Consider the situation where the pharmaceutical fluid enters at 90°C with a mass flow rate of 0.005 kg/s. The pharmaceutical has the following properties: $\rho = 1200\ \text{kg/m}^3$, $\mu = 4 \times 10^{-3}\ \text{N} \cdot \text{s/m}^2$, $c_p = 2000\ \text{J/kg} \cdot \text{K}$, and $k = 0.5\ \text{W/m} \cdot \text{K}$.

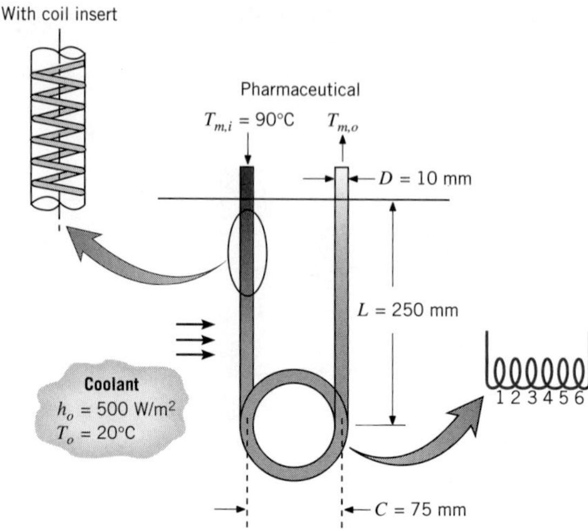

(a) Determine the outlet temperature of the pharmaceutical fluid.

(b) It is desired to further reduce the outlet temperature of the pharmaceutical. However, because the cooling process is just one part of an intricate processing operation, flow rates cannot be changed. A young engineer suggests that the outlet temperature might be reduced by inserting stainless steel coiled springs into the straight sections of the cooler with the notion that the springs will disturb the flow adjacent to the inner tube wall and, in turn, increase the heat transfer coefficient at the inner tube wall. A senior engineer asserts that insertion of the springs should double the heat transfer coefficient at the straight inner tube walls. Determine the outlet temperature of the pharmaceutical fluid with the springs inserted into the tubes, assuming the senior engineer is correct in his assertion.

(c) Would you expect the outlet temperature of the pharmaceutical to depend on whether the springs have a left-hand or right-hand spiral? Why?

8.105 The mold used in an injection molding process consists of a top half and a bottom half. Each half is 60 mm × 60 mm × 20 mm and is constructed of metal ($\rho = 7800\ \text{kg/m}^3$, $c = 450\ \text{J/kg} \cdot \text{K}$). The cold mold (100°C) is to be heated to 200°C with pressurized water (available at 275°C and a total flow rate of 0.02 kg/s) prior to injecting the thermoplastic material. The injection takes only a fraction of a second, and the hot mold (200°C) is subsequently cooled with cold water (available at 25°C and a total flow rate of 0.02 kg/s) prior to ejecting the molded part. After part ejection, which also takes a fraction of a second, the process is repeated.

(a) In conventional mold design, straight cooling (heating) passages are bored through the mold in a location where the passages will not interfere with the molded part. Determine the initial heating rate and the initial cooling rate of the mold when five 5-mm-diameter, 60-mm-long passages are bored in each half of the mold (10 passages total). The velocity distribution of the water is fully developed at the entrance of each passage in the hot (or cold) mold.

(b) New additive manufacturing processes, known as *selective freeform fabrication*, or *SFF*, are used to construct molds that are configured with *conformal cooling passages.* Consider the same mold as before, but now a 5-mm-diameter, coiled, conformal cooling passage is designed within each half of the SFF-manufactured mold. Each of the two coiled passages has $N = 2$ turns. The coiled passage does not interfere with the molded part. The conformal channels have a coil diameter $C = 50$ mm. The total water flow remains the same as in part (a) (0.01 kg/s per coil). Determine the initial heating rate and the initial cooling rate of the mold.

(c) Compare the surface areas of the conventional and conformal cooling passages. Compare the rate at which the mold temperature changes for molds configured with the conventional and conformal heating and cooling passages. Which cooling passage, conventional or conformal, will enable production of more parts per day? Neglect the presence of the thermoplastic material.

8.106 Consider the pharmaceutical product of Problem 8.27. Prior to finalizing the manufacturing process, test trials are performed to experimentally determine the dependence of the shelf life of the drug as a function of the sterilization temperature. Hence, the sterilization temperature must be carefully controlled in the trials. To promote good mixing of the pharmaceutical and, in turn, relatively uniform outlet temperatures across the exit tube area, experiments are performed using a device that is constructed of two interwoven coiled tubes, each of 10-mm diameter. The thin-walled tubing is welded to a solid high thermal conductivity rod of diameter $D_r = 40$ mm. One tube carries the pharmaceutical product at a mean velocity of $u_p = 0.1$ m/s and inlet temperature of 25°C, while the second tube carries pressurized liquid water at $u_w = 0.12$ m/s with an inlet temperature of 127°C. The tubes do not contact each other but are each welded to the solid metal rod, with each tube making 20 turns around the rod. The exterior of the apparatus is well insulated.

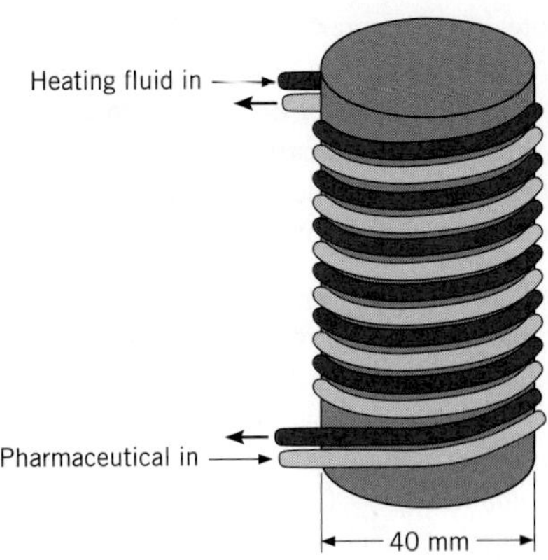

(a) Determine the outlet temperature of the pharmaceutical product. Evaluate the liquid water properties at 380 K.

(b) Investigate the sensitivity of the pharmaceutical's outlet temperature to the velocity of the pressurized water over the range $0.10 < u_w < 0.25$ m/s.

Flow in Small Channels

8.107 An extremely effective method of cooling high-power-density silicon chips involves etching microchannels in the back (noncircuit) surface of the chip. The channels are covered with a silicon cap, and cooling is maintained by passing water through the channels.

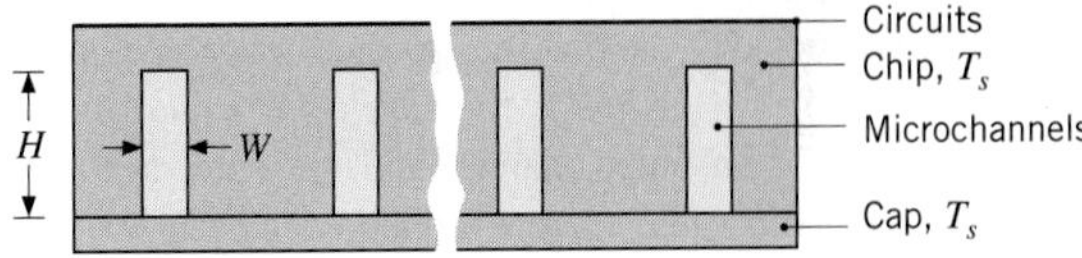

Consider a chip that is 10 mm × 10 mm on a side and in which fifty 10-mm-long rectangular microchannels, each of width $W = 50\ \mu$m and height $H = 200\ \mu$m, have been etched. Consider operating conditions for which water enters each microchannel at a temperature of 290 K and a flow rate of 10^{-4} kg/s, while the chip and cap are at a uniform temperature of 350 K. Assuming fully developed flow in the channel and that all the heat dissipated by the circuits is transferred to the water, determine the water outlet temperature and the chip power dissipation. Water properties may be evaluated at 300 K.

8.108 An ideal gas flows within a small diameter tube. Derive an expression for the transition density of the gas ρ_c below which microscale effects must be accounted for. Express your result in terms of the gas molecule diameter, universal gas constant, Boltzmann's constant, and the tube diameter. Evaluate the transition density for a $D = 10$-μm-diameter tube for hydrogen, air, and carbon dioxide. Compare the calculated transition densities with the gas density at atmospheric pressure and $T = 23°C$.

8.109 Consider the microchannel cooling arrangement of Problem 8.107. However, instead of assuming the entire chip and cap to be at a uniform temperature, adopt a more conservative (and realistic) approach that prescribes a temperature of $T_s = 350$ K at the base of the channels ($x = 0$) and allows for a decrease in temperature with increasing x along the side walls of each channel.

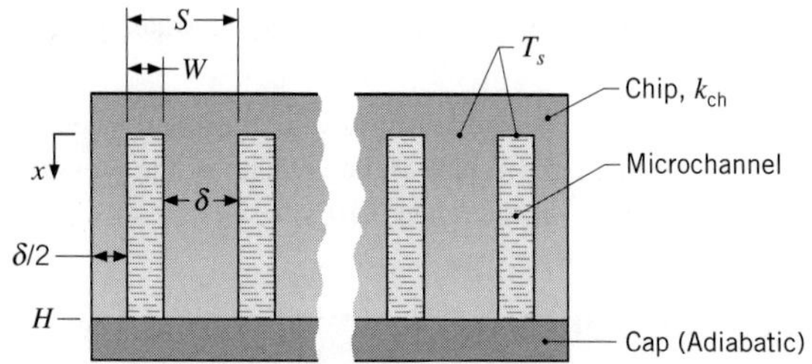

(a) For the operating conditions prescribed in Problem 8.107 and a chip thermal conductivity of $k_{ch} = 140$ W/m·K, determine the water outlet temperature and the chip power dissipation. Heat transfer from the sides of the chip to the surroundings and from the side walls of a channel to the cap may be neglected. Note that the spacing between channels, $\delta = S - W$, is twice the spacing between the side wall of an outer channel and the outer surface of the chip. The channel pitch is $S = L/N$, where $L = 10$ mm is the chip width and $N = 50$ is the number of channels.

(b) The channel geometry prescribed in Problem 8.107 and considered in part (a) is not optimized, and larger heat rates may be dissipated by adjusting related dimensions. Consider the effect of reducing the pitch to a value of $S = 100$ μm, while retaining a width of $W = 50$ μm and a flow rate per channel of $\dot{m}_1 = 10^{-4}$ kg/s.

8.110 The onset of turbulence in a gas flowing within a circular tube occurs at $Re_{D,c} \approx 2300$, while a transition from incompressible to compressible flow occurs at a critical Mach number of $Ma_c \approx 0.3$. Determine the critical tube diameter D_c, below which incompressible turbulent flow and heat transfer cannot exist for (i) air, (ii) CO_2, (iii) He. Evaluate properties at atmospheric pressure and a temperature of $T = 300$ K.

8.111 Due to its comparatively large thermal conductivity, water is a preferred fluid for convection cooling. However, in applications involving electronic devices, water must not come into contact with the devices, which would therefore have to be hermetically sealed. To circumvent related design and operational complexities and to ensure that the devices are not rendered inoperable by contact with the coolant, a dielectric fluid is commonly used in lieu of water. Many gases have excellent dielectric characteristics, and despite its poor heat transfer properties, air is the common choice for electronic cooling. However, there is an alternative, which involves a class of *perfluorinated liquids* that are excellent dielectrics and have heat transfer properties superior to those of gases.

Consider the microchannel chip cooling application of Problem 8.109 but now for a perfluorinated liquid with properties of $c_p = 1050$ J/kg·K, $k = 0.065$ W/m·K, $\mu = 0.0012$ N·s/m^2, and $Pr = 15$.

(a) For channel dimensions of $H = 200$ μm, $W = 50$ μm, and $S = 20$ μm, a chip thermal conductivity of $k_{ch} = 140$ W/m·K and width $L = 10$ mm, a channel base temperature ($x = 0$) of $T_s = 350$ K, a channel inlet temperature of $T_{m,i} = 290$ K, and a flow rate of $\dot{m}_1 = 10^{-4}$ kg/s per channel, determine the outlet temperature and the chip power dissipation for the dielectric liquid.

(b) Consider the foregoing conditions, but with air at a flow rate of $\dot{m}_1 = 10^{-6}$ kg/s used as the coolant. Using properties of $c_p = 1007$ J/kg·K, $k = 0.0263$ W/m·K, and $\mu = 185 \times 10^{-7}$ N·s/m^2, determine the air outlet temperature and the chip power dissipation.

8.112 Many of the solid surfaces for which values of the thermal and momentum accommodation coefficients have been measured are quite different from those used in micro- and nanodevices. Plot the Nusselt number Nu_D associated with fully developed laminar flow with constant surface heat flux versus tube diameter for $1\ \mu\text{m} \le D \le 1$ mm and (i) $\alpha_t = 1$, $\alpha_p = 1$, (ii) $\alpha_t = 0.1$, $\alpha_p = 0.1$, (iii) $\alpha_t = 1$, $\alpha_p = 0.1$, and (iv) $\alpha_t = 0.1$, $\alpha_p = 1$. For tubes of what diameter do the accommodation coefficients begin to influence convection heat transfer? For which combination of α_t and α_p does the Nusselt number exhibit the least sensitivity to changes in the diameter of the tube? Which combination results in Nusselt numbers greater than the conventional fully developed laminar value for constant heat flux conditions, $Nu_D = 4.36$? Which combination is associated with the smallest Nusselt numbers? What can you say about the ability to predict convection heat transfer coefficients in a small-scale device if the accommodation coefficients are not known for material from which the device is fabricated? Use properties of air at atmospheric pressure and $T = 300$ K.

8.113 A novel scheme for dissipating heat from the chips of a multichip array involves machining coolant channels in the ceramic substrate to which the chips are attached. The square chips ($L_c = 5$ mm) are aligned above each of the channels, with longitudinal and transverse pitches of $S_L = S_T = 20$ mm. Water flows through the square cross section ($W = 5$ mm) of each channel with a mean velocity of $u_m = 1$ m/s, and its properties may be approximated as $\rho = 1000$ kg/m^3, $c_p = 4180$ J/kg·K, $\mu = 855 \times 10^{-6}$ kg/s·m, $k = 0.610$ W/m·K, and $Pr = 5.8$. Symmetry in the transverse direction dictates the existence of equivalent conditions for each substrate section of length L_s and width S_T.

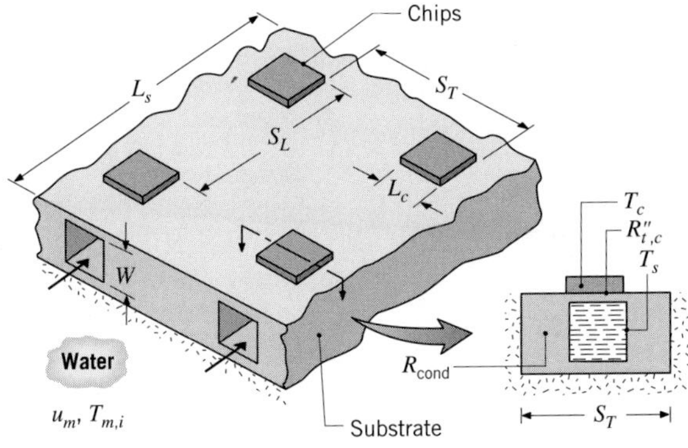

(a) Consider a substrate whose length in the flow direction is $L_s = 200$ mm, thereby providing a total of $N_L = 10$ chips attached in-line above each flow channel. To a good approximation, all the heat dissipated by the chips above a channel may be assumed to be transferred to the water flowing through the channel. If each chip dissipates 5 W, what is the temperature rise of the water passing through the channel?

(b) The chip-substrate contact resistance is $R''_{t,c} = 0.5 \times 10^{-4}$ m^2·K/W, and the three-dimensional conduction resistance for the $L_s \times S_T$ substrate section is $R_{\text{cond}} = 0.120$ K/W. If water enters the substrate at 25°C and is in fully developed flow, estimate the temperature T_c of the chips and the temperature T_s of the substrate channel surface.

8.114 Consider air flowing in a small-diameter steel tube. Graph the Nusselt number associated with fully developed laminar flow with constant surface heat flux for tube diameters ranging from 1 μm $\le D \le$ 1 mm. Evaluate air properties at $T = 350$ K and atmospheric pressure. The thermal and momentum accommodation coefficients are $\alpha_t = 0.92$ and $\alpha_p = 0.87$, respectively. Compare the Nusselt number you calculate to the value provided in Equation 8.53, $Nu_D = 4.36$.

8.115 An experiment is designed to study microscale forced convection. Water at $T_{m,i} = 300$ K is to be heated in a straight, circular glass tube with a 50-μm inner diameter and a wall thickness of 1 mm. Warm water at $T_\infty = 350$ K, $V = 2$ m/s is in cross flow over the exterior tube surface. The experiment is to be designed to cover the operating range $1 \le Re_D \le 2000$, where Re_D is the Reynolds number associated with the internal flow.

(a) Determine the tube length L that meets a design requirement that the tube be twice as long as the thermal entrance length associated with the highest Reynolds number of interest. Evaluate water properties at 305 K.

(b) Determine the water outlet temperature, $T_{m,o}$, that is expected to be associated with $Re_D = 2000$. Evaluate the heating water (water in cross flow over the tube) properties at 330 K.

(c) Calculate the pressure drop from the entrance to the exit of the tube for $Re_D = 2000$.

(d) Based on the calculated flow rate and pressure drop in the tube, estimate the height of a column of water (at 300 K) needed to supply the necessary pressure at the tube entrance and the time needed to collect 0.1 liter of water. Discuss how the outlet temperature of the water flowing from the tube, $T_{m,o}$, might be measured.

8.116 Determine the tube diameter that corresponds to a 10% reduction in the convection heat transfer coefficient for thermal and momentum accommodation coefficients of $\alpha_t = 0.92$ and $\alpha_p = 0.89$, respectively. Determine the channel spacing, a, that is associated with a 10% reduction in h using the same accommodation coefficients. The gas is air at $T = 350$ K and atmospheric pressure for both the tube and the parallel plate configurations. The flow is laminar and fully developed with constant surface heat flux.

8.117 An experiment is devised to measure liquid flow and convective heat transfer rates in microscale channels. The mass flow rate through a channel is determined by measuring the amount of liquid that has flowed through the channel and dividing by the duration of the experiment. The mean temperature of the outlet fluid is also measured. To minimize the time needed to perform the experiment (that is, to collect a significant amount of liquid so that its mass and temperature can be accurately measured), *arrays* of microchannels are typically used. Consider an array of microchannels of circular cross section, each with a nominal diameter of 50 μm, fabricated into a copper block. The channels are 20 mm long, and the block is held at 310 K. Water at an inlet temperature of 300 K is forced into the channels from a pressurized plenum, so that a pressure difference of 2.5×10^6 Pa exists from the entrance to the exit of each channel.

In many microscale systems, the characteristic dimensions are similar to the tolerances that can be controlled during the manufacture of the experimental apparatus. Hence, careful consideration of the effect of machining tolerances must be made when interpreting the experimental results.

(a) Consider the case in which three microchannels are machined in the copper block. The channel diameters exhibit some deviation due to manufacturing constraints and are of actual diameter 45 μm, 50 μm, and 55 μm, respectively. Calculate the mass flow rate through each of the three channels, along with the mean outlet temperature of each channel.

(b) If the water exiting each of the three channels is collected and mixed in a single container, calculate the average flow rate through each of the three channels and the average mixed temperature of the water that is collected from all three channels.

(c) The enthusiastic experimentalist uses the average flow rate and the average mixed outlet temperature to analyze the performance of the average (50 μm) diameter channel and concludes that flow rates and heat transfer coefficients are increased and decreased, respectively, by about 5% when forced convection occurs in microchannels. Comment on the validity of the experimentalist's conclusion.

CHAPTER

Free Convection

9

In preceding chapters we considered convection transfer in fluid flows that originate from an *external forcing* condition. For example, fluid motion may be induced by a fan or a pump, or it may result from propulsion of a solid through the fluid. In the presence of a temperature gradient, *forced convection* heat transfer will occur.

Now we consider situations for which there is no *forced* velocity, yet convection currents exist within the fluid. Such situations are referred to as *free* or *natural convection,* and they originate when a *body force* acts on a fluid in which there are *density gradients.* The net effect is a *buoyancy force,* which induces free convection currents. In the most common case, the density gradient is due to a temperature gradient, and the body force is due to the gravitational field.

Since free convection flow velocities are generally much smaller than those associated with forced convection, the corresponding convection transfer rates are also smaller. It is perhaps tempting to therefore attach less significance to free convection processes. This temptation should be resisted. In many systems involving multimode heat transfer effects, free convection provides the largest resistance to heat transfer and therefore plays an important role in the design or performance of the system. Moreover, when it is desirable to minimize heat transfer rates or to minimize operating cost, free convection is often preferred to forced convection.

There are, of course, many applications. Free convection strongly influences the operating temperatures of power generating and electronic devices. It plays a major role in a vast array of thermal manufacturing applications. Free convection is important in establishing temperature distributions within buildings and in determining heat losses or heat loads for heating, ventilating, and air conditioning systems. Free convection distributes the poisonous products of combustion during fires and is relevant to the environmental sciences, where it drives oceanic and atmospheric motions, as well as the related heat transfer processes.

In this chapter our objectives are to obtain an appreciation for the physical origins and nature of buoyancy-driven flows and to acquire tools for performing related heat transfer calculations.

9.1 *Physical Considerations*

In free convection fluid motion is due to buoyancy forces within the fluid, while in forced convection it is externally imposed. *Buoyancy is due to the combined presence of a fluid density gradient* and *a body force that is proportional to density.* In practice, the body force is usually *gravitational,* although it may be a centrifugal force in rotating fluid machinery or a Coriolis force in atmospheric and oceanic rotational motions. There are also several ways in which a mass density gradient may arise in a fluid, but for the most common situation it is due to the presence of a temperature gradient. We know that the density of gases and liquids depends on temperature, generally decreasing (due to fluid expansion) with increasing temperature ($\partial\rho/\partial T < 0$).

In this text we focus on free convection problems in which the density gradient is due to a temperature gradient and the body force is gravitational. However, the presence of a fluid density gradient in a gravitational field does not ensure the existence of free convection currents. Consider the conditions of Figure 9.1. A fluid is enclosed by two large, horizontal plates of different temperature ($T_1 \neq T_2$). In case *a* the temperature of the lower plate

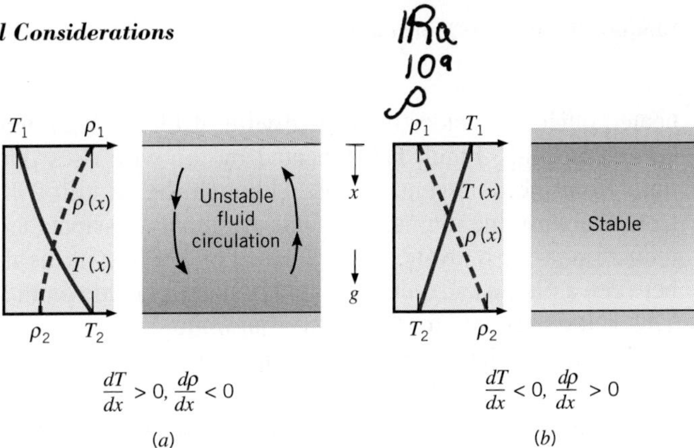

FIGURE 9.1 Conditions in a fluid between large horizontal plates at different temperatures: (*a*) Unstable temperature gradient. (*b*) Stable temperature gradient.

exceeds that of the upper plate, and the density decreases in the direction of the gravitational force. If the temperature difference exceeds a critical value, conditions are *unstable* and buoyancy forces are able to overcome the retarding influence of viscous forces. The gravitational force on the denser fluid in the upper layers exceeds that acting on the lighter fluid in the lower layers, and the designated circulation pattern will exist. The heavier fluid will descend, being warmed in the process, while the lighter fluid will rise, cooling as it moves. However, this condition does not characterize case *b*, for which $T_1 > T_2$ and the density no longer decreases in the direction of the gravitational force. Conditions are now *stable,* and there is no bulk fluid motion. In case *a* heat transfer occurs from the bottom to the top surface by free convection; for case *b* heat transfer (from top to bottom) occurs by conduction.

Free convection flows may be classified according to whether the flow is bounded by a surface. In the absence of an adjoining surface, *free boundary flows* may occur in the form of a *plume* or a *buoyant jet* (Figure 9.2). A plume is associated with fluid rising from a submerged

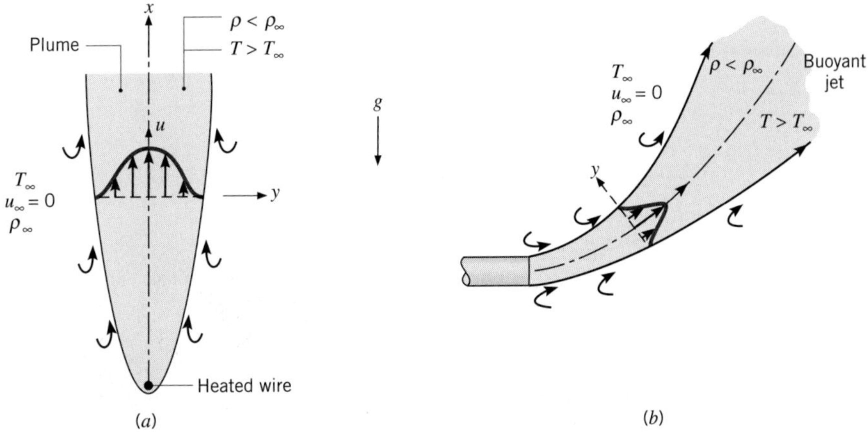

FIGURE 9.2 Buoyancy-driven free boundary layer flows in an extensive, quiescent medium. (*a*) Plume formation above a heated wire. (*b*) Buoyant jet associated with a heated discharge.

heated object. Consider the heated wire of Figure 9.2*a*, which is immersed in an *extensive, quiescent* fluid.[1] Fluid that is heated by the wire rises due to buoyancy forces, entraining fluid from the quiescent region. Although the width of the plume increases with distance from the wire, the plume itself will eventually dissipate as a result of viscous effects and a reduction in the buoyancy force caused by cooling of the fluid in the plume. The distinction between a plume and a buoyant jet is generally made on the basis of the *initial* fluid velocity. This velocity is zero for the plume, but finite for the buoyant jet. Figure 9.2*b* shows a heated fluid being discharged as a horizontal jet into a quiescent medium of lower temperature. The vertical motion that the jet begins to assume is due to the buoyancy force. Such a condition occurs when warm water from the condenser of a central power station is discharged into a reservoir of cooler water. Free boundary flows are discussed in considerable detail by Jaluria [1] and Gebhart et al. [2].

In this text we focus on free convection flows bounded by a surface, and a classic example relates to boundary layer development on a heated vertical plate (Figure 9.3). The plate is immersed in an extensive, quiescent fluid, and with $T_s > T_\infty$ the fluid close to the plate is less dense than fluid that is further removed. Buoyancy forces therefore induce a free convection boundary layer in which the heated fluid rises vertically, entraining fluid from the quiescent region. The resulting velocity distribution is unlike that associated with forced convection boundary layers. In particular, the velocity is zero as $y \rightarrow \infty$, as well as at $y = 0$. A free convection boundary layer also develops if $T_s < T_\infty$. In this case, however, fluid motion is downward.

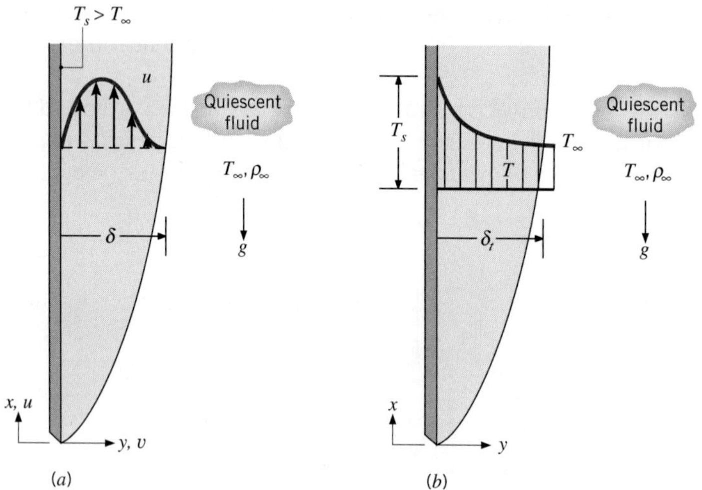

FIGURE 9.3 Boundary layer development on a heated vertical plate: (*a*) Velocity boundary layer. (*b*) Thermal boundary layer.

[1] An extensive medium is, in principle, an infinite medium. Since a quiescent fluid is one that is otherwise at rest, the velocity of fluid far from the heated wire is zero.

9.2 The Governing Equations for Laminar Boundary Layers

As for forced convection, the equations that describe momentum and energy transfer in free convection originate from the related conservation principles. Moreover, the specific processes are much like those that dominate in forced convection. Inertia and viscous forces remain important, as does energy transfer by advection and diffusion. The difference between the two flows is that, in free convection, a major role is played by buoyancy forces. Such forces, in fact, drive the flow.

Consider a laminar boundary layer flow (Figure 9.3) that is driven by buoyancy forces. Assume steady, two-dimensional, constant property conditions in which the gravity force acts in the negative x-direction. Also, with one exception, assume the fluid to be incompressible. The exception involves accounting for the effect of variable density only in the buoyancy force, since it is this variation that induces fluid motion. Finally, assume that the boundary layer approximations of Section 6.4.1 are valid.

With the foregoing simplifications the x-momentum equation (Equation E.2) reduces to the boundary layer equation (Equation 6.16), except that the body force term X is retained. If the only contribution to this force is made by gravity, the body force per unit volume is $X = -\rho g$, where g is the local acceleration due to gravity. The appropriate form of the x-momentum equation is then

$$u\frac{\partial u}{\partial x} + v\frac{\partial u}{\partial y} = -\frac{1}{\rho}\frac{dp_\infty}{dx} - g + \nu\frac{\partial^2 u}{\partial y^2} \tag{9.1}$$

where dp_∞/dx is the free stream pressure gradient in the quiescent region *outside* the boundary layer. In this region, $u = 0$ and Equation 9.1 reduces to

$$\frac{dp_\infty}{dx} = -\rho_\infty g \tag{9.2}$$

Substituting Equation 9.2 into 9.1, we obtain the following expression:

$$u\frac{\partial u}{\partial x} + v\frac{\partial u}{\partial y} = g(\Delta\rho/\rho) + \nu\frac{\partial^2 u}{\partial y^2} \tag{9.3}$$

where $\Delta\rho = \rho_\infty - \rho$. This expression must apply at every point in the free convection boundary layer.

The first term on the right-hand side of Equation 9.3 is the buoyancy force per unit mass, and flow originates because the density ρ is a variable. If density variations are due only to temperature variations, the term may be related to a fluid property known as the *volumetric thermal expansion coefficient*

$$\beta = -\frac{1}{\rho}\left(\frac{\partial \rho}{\partial T}\right)_p \tag{9.4}$$

This *thermodynamic* property of the fluid provides a measure of the amount by which the density changes in response to a change in temperature at constant pressure. If it is expressed in the following approximate form,

$$\beta \approx -\frac{1}{\rho}\frac{\Delta\rho}{\Delta T} = -\frac{1}{\rho}\frac{\rho_\infty - \rho}{T_\infty - T}$$

it follows that

$$(\rho_\infty - \rho) \approx \rho\beta(T - T_\infty)$$

This simplification is known as the *Boussinesq approximation*, and substituting into Equation 9.3, the x-momentum equation becomes

$$u\frac{\partial u}{\partial x} + v\frac{\partial u}{\partial y} = g\beta(T - T_\infty) + \nu\frac{\partial^2 u}{\partial y^2} \tag{9.5}$$

where it is now apparent how the buoyancy force, which drives the flow, is related to the temperature difference.

Since buoyancy effects are confined to the momentum equation, the mass and energy conservation equations are unchanged from forced convection. Equations 6.15 and 6.17 may then be used to complete the problem formulation. The set of governing equations is then

$$\frac{\partial u}{\partial x} + \frac{\partial v}{\partial y} = 0 \tag{9.6}$$

$$u\frac{\partial u}{\partial x} + v\frac{\partial u}{\partial y} = g\beta(T - T_\infty) + \nu\frac{\partial^2 u}{\partial y^2} \tag{9.7}$$

$$u\frac{\partial T}{\partial x} + v\frac{\partial T}{\partial y} = \alpha\frac{\partial^2 T}{\partial y^2} \tag{9.8}$$

Note that viscous dissipation has been neglected in the energy equation, (9.8), an assumption that is certainly reasonable for the small velocities associated with free convection. In the mathematical sense the appearance of the buoyancy term in Equation 9.7 complicates matters. No longer may the hydrodynamic problem, given by Equations 9.6 and 9.7, be uncoupled from and solved to the exclusion of the thermal problem, given by Equation 9.8. The solution to the momentum equation depends on knowledge of T, and hence on the solution to the energy equation. Equations 9.6 through 9.8 are therefore strongly coupled and must be solved simultaneously.

Free convection effects obviously depend on the expansion coefficient β. The manner in which β is obtained depends on the fluid. For an ideal gas, $\rho = p/RT$ and

$$\beta = -\frac{1}{\rho}\left(\frac{\partial \rho}{\partial T}\right)_p = \frac{1}{\rho}\frac{p}{RT^2} = \frac{1}{T} \tag{9.9}$$

where T is the *absolute* temperature. For liquids and nonideal gases, β must be obtained from appropriate property tables (Appendix A).

9.3 Similarity Considerations

Let us now consider the dimensionless parameters that govern free convective flow and heat transfer for the vertical plate. As for forced convection (Chapter 6), the parameters may be obtained by nondimensionalizing the governing equations. Introducing

$$x^* \equiv \frac{x}{L} \qquad y^* \equiv \frac{y}{L}$$

hydrodynamic
entry region

$$u^* \equiv \frac{u}{u_0} \qquad v^* \equiv \frac{v}{u_0} \qquad T^* \equiv \frac{T - T_\infty}{T_s - T_\infty}$$

where L is a characteristic length and u_0 is a reference velocity,[2] the x-momentum and energy equations (9.7 and 9.8) reduce to

$$u^* \frac{\partial u^*}{\partial x^*} + v^* \frac{\partial u^*}{\partial y^*} = \frac{g\beta(T_s - T_\infty)L}{u_0^2} T^* + \frac{1}{Re_L} \frac{\partial^2 u^*}{\partial y^{*2}} \tag{9.10}$$

$$u^* \frac{\partial T^*}{\partial x^*} + v^* \frac{\partial T^*}{\partial y^*} = \frac{1}{Re_L\, Pr} \frac{\partial^2 T^*}{\partial y^{*2}} \tag{9.11}$$

The dimensionless parameter in the first term on the right-hand side of Equation 9.10 is a direct consequence of the buoyancy force. The reference velocity u_0 can be specified to simplify the form of the equation. It is convenient to choose $u_0^2 = g\beta(T_s - T_\infty)L$, so that the term multiplying T^* becomes unity. Then, Re_L becomes $[g\beta(T_s - T_\infty)L^3/\nu^2]^{1/2}$. It is customary to define the *Grashof number* Gr_L as the square of this Reynolds number:

$$Gr_L \equiv \frac{g\beta(T_s - T_\infty)L^3}{\nu^2} \tag{9.12}$$

As a result, Re_L in Equations 9.10 and 9.11 is replaced by $Gr_L^{1/2}$, and we see that the Grashof number (or more precisely, $Gr_L^{1/2}$) plays the same role in free convection that the Reynolds number plays in forced convection. Based on the resulting form of Equations 9.10 and 9.11, we expect heat transfer correlations of the form $Nu_L = f(Gr_L, Pr)$ in free convection. Recall that the *Reynolds number* provides a measure of the *ratio of the inertial to viscous forces* acting on a fluid element. In contrast, the *Grashof number* is a measure of the *ratio of the buoyancy forces to the viscous forces* acting on the fluid.

When forced and free convection effects are comparable, the situation is more complex. For example, consider the boundary layer of Figure 9.3, but with a non-zero free stream velocity, u_∞. In this case, it is more convenient to choose the characteristic velocity as u_∞ (so that the free stream boundary condition for the dimensionless velocity, u^*, is simply $u^*(y^* \to \infty) \to 1$). Then the T^* term in Equation 9.10 will be multiplied by Gr_L/Re_L^2, and the resulting Nusselt number expressions will be of the form $Nu_L = f(Re_L, Gr_L, Pr)$. Generally, the combined effects of free and forced convection must be considered when $Gr_L/Re_L^2 \approx 1$. If the inequality $Gr_L/Re_L^2 \ll 1$ is satisfied, free convection effects may be neglected and $Nu_L = f(Re_L, Pr)$. Conversely, if $Gr_L/Re_L^2 \gg 1$, forced convection effects may be neglected and $Nu = f(Gr_L, Pr)$, as indicated in the preceding paragraph for pure free convection. Additional discussion of combined free and forced convection is provided in Section 9.9.

9.4 *Laminar Free Convection on a Vertical Surface*

Numerous solutions to the laminar free convection boundary layer equations have been obtained, and a special case that has received much attention involves free convection from

[2]Since free stream conditions are quiescent in free convection, there is no logical external reference velocity (V or u_∞), as in forced convection.

an isothermal vertical surface in an extensive quiescent medium (Figure 9.3). For this geometry Equations 9.6 through 9.8 must be solved subject to boundary conditions of the form[3]

$$y = 0: \qquad u = v = 0 \qquad T = T_s$$

$$y \to \infty: \qquad u \to 0 \qquad T \to T_\infty$$

A similarity solution to the foregoing problem has been obtained by Ostrach [3]. The solution involves transforming variables by introducing a similarity variable of the form

$$\eta \equiv \frac{y}{x}\left(\frac{Gr_x}{4}\right)^{1/4} \tag{9.13}$$

and representing the velocity components in terms of a stream function defined as

$$\psi(x, y) \equiv f(\eta)\left[4\nu\left(\frac{Gr_x}{4}\right)^{1/4}\right] \tag{9.14}$$

With the foregoing definition of the stream function, the x-velocity component may be expressed as

$$u = \frac{\partial \psi}{\partial y} = \frac{\partial \psi}{\partial \eta}\frac{\partial \eta}{\partial y} = 4\nu\left(\frac{Gr_x}{4}\right)^{1/4} f'(\eta)\frac{1}{x}\left(\frac{Gr_x}{4}\right)^{1/4}$$

$$= \frac{2\nu}{x} Gr_x^{1/2} f'(\eta) \tag{9.15}$$

where primed quantities indicate differentiation with respect to η. Hence $f'(\eta) \equiv df/d\eta$. Evaluating the y-velocity component $v = -\partial\psi/\partial x$ in a similar fashion and introducing the dimensionless temperature

$$T^* \equiv \frac{T - T_\infty}{T_s - T_\infty} \tag{9.16}$$

the three original partial differential equations (9.6 through 9.8) may then be reduced to two ordinary differential equations of the form

$$f''' + 3ff'' - 2(f')^2 + T^* = 0 \tag{9.17}$$

$$T^{*\prime\prime} + 3Pr f T^{*\prime} = 0 \tag{9.18}$$

where f and T^* are functions only of η and the double and triple primes, respectively, refer to second and third derivatives with respect to η. Note that f is the key dependent variable for the velocity boundary layer and that the continuity equation (9.6) is automatically satisfied through introduction of the stream function.

The transformed boundary conditions required to solve the momentum and energy equations (9.17 and 9.18) are of the form

$$\eta = 0: \qquad f = f' = 0 \qquad T^* = 1$$

$$\eta \to \infty \qquad f' \to 0 \qquad T^* \to 0$$

---

[3]The boundary layer approximations are assumed in using Equations 9.6 through 9.8. However, the approximations are only valid for $(Gr_x Pr) \gtrsim 10^4$. Below this value (close to the leading edge), the boundary layer thickness is too large relative to the characteristic length x to ensure the validity of the approximations.

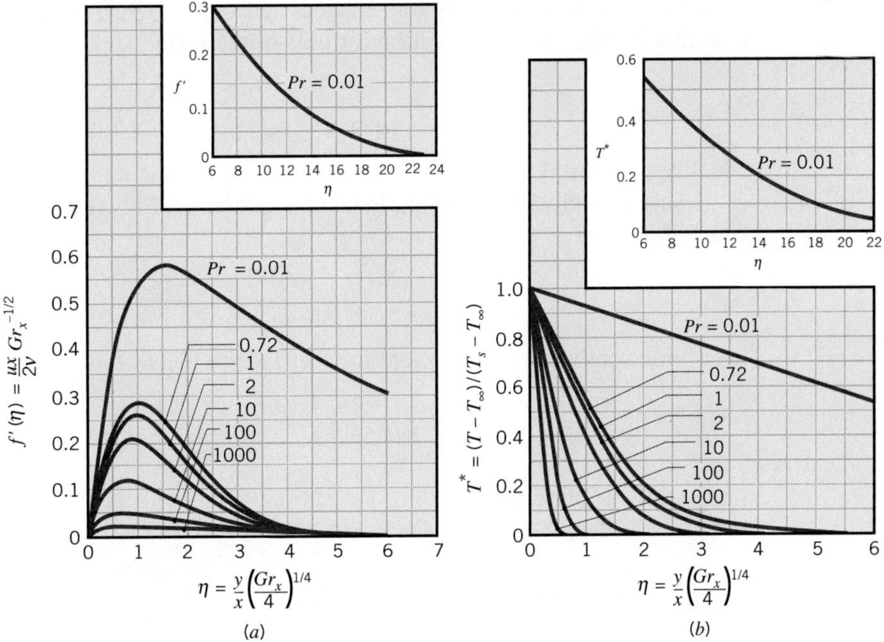

FIGURE 9.4 Laminar free convection boundary layer conditions on an isothermal, vertical surface. (*a*) Velocity profiles. (*b*) Temperature profiles [3].

A numerical solution has been obtained by Ostrach [3], and selected results are shown in Figure 9.4. Values of the x-velocity component u and the temperature T at any value of x and y may be obtained from Figure 9.4*a* and Figure 9.4*b*, respectively.

Figure 9.4*b* may also be used to infer the appropriate form of the heat transfer correlation. Using Newton's law of cooling for the local convection coefficient h, the local Nusselt number may be expressed as

$$Nu_x = \frac{hx}{k} = \frac{[q_s''/(T_s - T_\infty)]\,x}{k}$$

Using Fourier's law to obtain q_s'' and expressing the surface temperature gradient in terms of η, Equation 9.13, and T^*, Equation 9.16, it follows that

$$q_s'' = -k\frac{\partial T}{\partial y}\bigg|_{y=0} = -\frac{k}{x}(T_s - T_\infty)\left(\frac{Gr_x}{4}\right)^{1/4}\frac{dT^*}{d\eta}\bigg|_{\eta=0}$$

Hence

$$Nu_x = \frac{hx}{k} = -\left(\frac{Gr_x}{4}\right)^{1/4}\frac{dT^*}{d\eta}\bigg|_{\eta=0} = \left(\frac{Gr_x}{4}\right)^{1/4} g(Pr) \tag{9.19}$$

which acknowledges that the dimensionless temperature gradient at the surface is a function of the Prandtl number $g(Pr)$. This dependence is evident from Figure 9.4*b* and has been determined numerically for selected values of Pr [3]. The results have been correlated to within 0.5% by an interpolation formula of the form [4]

$$g(Pr) = \frac{0.75\, Pr^{1/2}}{(0.609 + 1.221\, Pr^{1/2} + 1.238\, Pr)^{1/4}} \tag{9.20}$$

which applies for $0 \leq Pr \leq \infty$.

Using Equation 9.19 for the local convection coefficient and substituting for the local Grashof number,

$$Gr_x = \frac{g\beta(T_s - T_\infty)x^3}{\nu^2}$$

the average convection coefficient for a surface of length L is then

$$\bar{h} = \frac{1}{L}\int_0^L h\,dx = \frac{k}{L}\left[\frac{g\beta(T_s - T_\infty)}{4\nu^2}\right]^{1/4} g(Pr)\int_0^L \frac{dx}{x^{1/4}}$$

Integrating, it follows that

$$\overline{Nu}_L = \frac{\bar{h}L}{k} = \frac{4}{3}\left(\frac{Gr_L}{4}\right)^{1/4} g(Pr) \tag{9.21}$$

or substituting from Equation 9.19, with $x = L$,

$$\overline{Nu}_L = \tfrac{4}{3} Nu_L \tag{9.22}$$

The foregoing results apply irrespective of whether $T_s > T_\infty$ or $T_s < T_\infty$. If $T_s < T_\infty$, conditions are inverted from those of Figure 9.3. The leading edge is at the top of the plate, and positive x is defined in the direction of the gravity force.

9.5 *The Effects of Turbulence*

It is important to note that free convection boundary layers are not restricted to laminar flow. As with forced convection, *hydrodynamic instabilities* may arise. That is, disturbances in the flow may be amplified, leading to transition from laminar to turbulent flow. This process is shown schematically in Figure 9.5 for a heated vertical plate.

Transition in a free convection boundary layer depends on the relative magnitude of the buoyancy and viscous forces in the fluid. It is customary to correlate its occurrence in

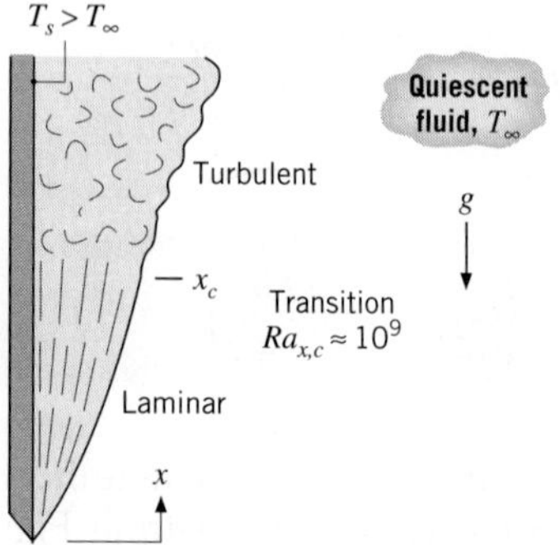

FIGURE 9.5 Free convection boundary layer transition on a vertical plate.

terms of the *Rayleigh number,* which is simply the product of the Grashof and Prandtl numbers. For vertical plates the critical Rayleigh number is

$$Ra_{x,c} = Gr_{x,c}\,Pr = \frac{g\beta(T_s - T_\infty)x^3}{\nu\alpha} \approx 10^9 \qquad (9.23)$$

An extensive discussion of stability and transition effects is provided by Gebhart et al. [2].

As in forced convection, transition to turbulence has a strong effect on heat transfer. Hence the results of the foregoing section apply only if $Ra_L \lesssim 10^9$. To obtain appropriate correlations for turbulent flow, emphasis is placed on experimental results.

EXAMPLE 9.1

Consider a 0.25-m-long vertical plate that is at 70°C. The plate is suspended in quiescent air that is at 25°C. Estimate the velocity boundary layer thickness and maximum upward velocity at the trailing edge of the plate. How does the boundary layer thickness compare with the thickness that would exist if the air were flowing over the plate at a free stream velocity of 5 m/s?

SOLUTION

Known: Vertical plate is in quiescent air at a lower temperature.

Find: Velocity boundary layer thickness and maximum upward velocity at trailing edge. Compare boundary layer thickness with value corresponding to an air speed of 5 m/s.

Schematic:

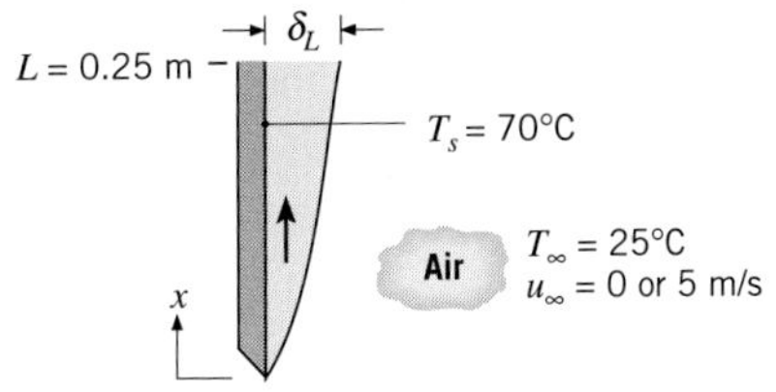

Assumptions:

1. Ideal gas.
2. Constant properties.
3. Buoyancy effects negligible when $u_\infty = 5$ m/s.

Properties: Table A.4, air ($T_f = 320.5$ K): $\nu = 17.95 \times 10^{-6}$ m²/s, $Pr = 0.7$, $\beta = T_f^{-1} = 3.12 \times 10^{-3}$ K^{-1}.

Analysis: For the quiescent air, Equation 9.12 gives

$$Gr_L = \frac{g\beta(T_s - T_\infty)L^3}{\nu^2}$$

$$= \frac{9.8\ \text{m/s}^2 \times (3.12 \times 10^{-3}\ \text{K}^{-1})(70 - 25)^\circ\text{C}(0.25\ \text{m})^3}{(17.95 \times 10^{-6}\ \text{m}^2/\text{s})^2} = 6.69 \times 10^7$$

Hence $Ra_L = Gr_L\, Pr = 4.68 \times 10^7$ and, from Equation 9.23, the free convection boundary layer is laminar. The analysis of Section 9.4 is therefore applicable. From the results of Figure 9.4*a*, it follows that, for $Pr = 0.7$, $\eta \approx 6.0$ at the edge of the boundary layer, that is, at $y \approx \delta$. Hence

$$\delta_L \approx \frac{6L}{(Gr_L/4)^{1/4}} = \frac{6(0.25\text{ m})}{(1.67 \times 10^7)^{1/4}} = 0.024\text{ m} \qquad \triangleleft$$

From Figure 9.4*a*, it can be seen that the maximum velocity corresponds to $f'(\eta) \approx 0.28$ and the velocity is

$$u = \frac{2\nu f'(\eta) Gr_L^{1/2}}{L} \approx \frac{2 \times 17.95 \times 10^{-6}\text{ m}^2/\text{s} \times 0.28 \times (6.69 \times 10^7)^{1/2}}{0.25\text{ m}} = 0.33\text{ m/s} \qquad \triangleleft$$

For airflow at $u_\infty = 5$ m/s

$$Re_L = \frac{u_\infty L}{\nu} = \frac{(5\text{ m/s}) \times 0.25\text{ m}}{17.95 \times 10^{-6}\text{ m}^2/\text{s}} = 6.97 \times 10^4$$

and the boundary layer is laminar. Hence, from Equation 7.17,

$$\delta_L \approx \frac{5L}{Re_L^{1/2}} = \frac{5(0.25\text{ m})}{(6.97 \times 10^4)^{1/2}} = 0.0047\text{ m} \qquad \triangleleft$$

Comments:

1. Free convection boundary layers typically have smaller velocities than in forced convection, which leads to thicker boundary layers. In turn, free convection boundary layers typically pose a larger resistance to heat transfer than forced convection boundary layers.
2. $(Gr_L/Re_L^2) = 0.014 \ll 1$, and the assumption of negligible buoyancy effects for $u_\infty = 5$ m/s is justified.

9.6 *Empirical Correlations: External Free Convection Flows*

In the preceding sections, we considered free convection associated with laminar boundary layer development adjacent to a heated vertical plate and transition of the laminar flow to a turbulent state. In doing so, we introduced two dimensionless parameters, the Grashof number *Gr* and the Rayleigh number *Ra*, which also appear in empirical correlations for free convection involving both laminar and turbulent flow conditions and in geometries other than a flat plate.

In this section we summarize empirical correlations that have been developed for common *immersed* (external flow) geometries. The correlations are suitable for many engineering calculations and are often of the form

$$\overline{Nu}_L = \frac{\bar{h}L}{k} = C\,Ra_L^n \tag{9.24}$$

where the Rayleigh number,

$$Ra_L = Gr_L\,Pr = \frac{g\beta(T_s - T_\infty)L^3}{\nu\alpha} \tag{9.25}$$

is based on the characteristic length L of the geometry. Typically, $n = \frac{1}{4}$ and $\frac{1}{3}$ for laminar and turbulent flows, respectively. For turbulent flow it then follows that $\bar{h}_L$ is independent of L. Note that all properties are evaluated at the film temperature, $T_f \equiv (T_s + T_\infty)/2$.

9.6.1 The Vertical Plate

Expressions of the form given by Equation 9.24 have been developed for the vertical plate [5–7]. For laminar flow ($10^4 \lesssim Ra_L \lesssim 10^9$), $C = 0.59$ and $n = 1/4$, and for turbulent flow ($10^9 \lesssim Ra_L \lesssim 10^{13}$), $C = 0.10$ and $n = 1/3$. A correlation that may be applied over the *entire* range of Ra_L has been recommended by Churchill and Chu [8] and is of the form

$$\overline{Nu}_L = \left\{0.825 + \frac{0.387\,Ra_L^{1/6}}{[1 + (0.492/Pr)^{9/16}]^{8/27}}\right\}^2 \tag{9.26}$$

Although Equation 9.26 is suitable for most engineering calculations, slightly better accuracy may be obtained for laminar flow by using [8]

$$\overline{Nu}_L = 0.68 + \frac{0.670\,Ra_L^{1/4}}{[1 + (0.492/Pr)^{9/16}]^{4/9}} \qquad Ra_L \lesssim 10^9 \tag{9.27}$$

When the Rayleigh number is moderately large, the second term on the right-hand side of Equations 9.26 and 9.27 dominates, and the correlations are the same form as Equation 9.24, except that the constant, C, is replaced by a function of Pr. Equation 9.27 is then in excellent quantitative agreement with the analytical solution given by Equations 9.21 and 9.20. In contrast, when the Rayleigh number is small, the first term on the right-hand side of Equations 9.26 and 9.27 dominates, and the equations yield the same behavior since $0.825^2 \approx 0.68$. The presence of leading constants in Equations 9.26 and 9.27 accounts for the fact that, for small Rayleigh number, the boundary layer assumptions become invalid and conduction parallel to the plate is important.

It is important to recognize that the foregoing results have been obtained for an isothermal plate (constant T_s). If the surface condition is, instead, one of uniform heat flux (constant q_s''), the temperature difference ($T_s - T_\infty$) will vary with x, increasing from the leading edge. An approximate procedure for determining this variation may be based on results [8, 9] showing that $\overline{Nu}_L$ correlations obtained for the isothermal plate may still be used to an excellent approximation, if $\overline{Nu}_L$ and Ra_L are defined in terms of the temperature difference at the midpoint of the plate, $\Delta T_{L/2} = T_s(L/2) - T_\infty$. Hence, with $\bar{h} \equiv q_s''/\Delta T_{L/2}$, a correlation such as Equation 9.27 could be used to determine $\Delta T_{L/2}$ (for example, using a trial-and-error technique), and hence the midpoint surface temperature $T_s(L/2)$. If it is assumed that $Nu_x \propto Ra_x^{1/4}$ over the entire plate, it follows that

$$\frac{q_s''x}{k\Delta T} \propto \Delta T^{1/4} x^{3/4}$$

or

$$\Delta T \propto x^{1/5}$$

Hence the temperature difference at any x is

$$\Delta T_x \approx \frac{x^{1/5}}{(L/2)^{1/5}} \Delta T_{L/2} = 1.15 \left(\frac{x}{L}\right)^{1/5} \Delta T_{L/2} \tag{9.28}$$

A more detailed discussion of constant heat flux results is provided by Churchill [10].

The foregoing results may also be applied to *vertical* cylinders of height L, if the boundary layer thickness δ is much less than the cylinder diameter D. This condition is known to be satisfied [11] when

$$\frac{D}{L} \gtrsim \frac{35}{Gr_L^{1/4}}$$

Cebeci [12] and Minkowycz and Sparrow [13] present results for slender, vertical cylinders not meeting this condition, where transverse curvature influences boundary layer development and enhances the rate of heat transfer.

EXAMPLE 9.2

A glass-door firescreen, used to reduce exfiltration of room air through a chimney, has a height of 0.71 m and a width of 1.02 m and reaches a temperature of 232°C. If the room temperature is 23°C, estimate the convection heat rate from the fireplace to the room.

SOLUTION

Known: Glass screen situated in fireplace opening.

Find: Heat transfer by convection between screen and room air.

Schematic:

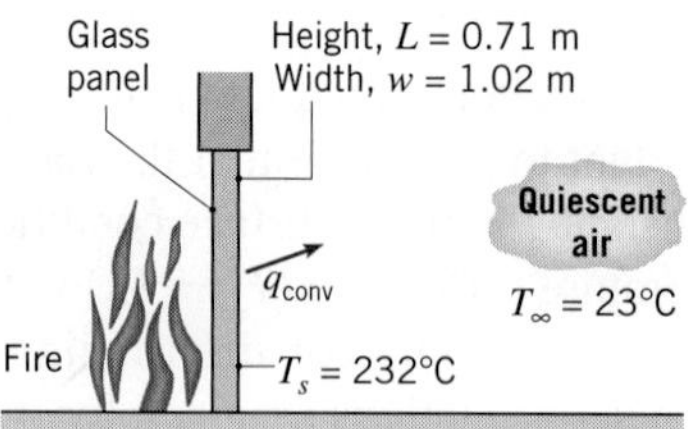

Assumptions:

1. Screen is at a uniform temperature T_s.
2. Room air is quiescent.
3. Ideal gas.
4. Constant properties.

Properties: Table A.4, air ($T_f = 400$ K): $k = 33.8 \times 10^{-3}$ W/m·K, $\nu = 26.4 \times 10^{-6}$ m²/s, $\alpha = 38.3 \times 10^{-6}$ m²/s, $Pr = 0.690$, $\beta = (1/T_f) = 0.0025\ \text{K}^{-1}$.

Analysis: The rate of heat transfer by free convection from the panel to the room is given by Newton's law of cooling

$$q = \bar{h}A_s(T_s - T_\infty)$$

where $\bar{h}$ may be obtained from knowledge of the Rayleigh number. Using Equation 9.25,

$$Ra_L = \frac{g\beta(T_s - T_\infty)L^3}{\alpha\nu}$$

$$= \frac{9.8 \text{ m/s}^2 \times 0.0025 \text{ K}^{-1} \times (232 - 23)°\text{C} \times (0.71 \text{ m})^3}{38.3 \times 10^{-6} \text{ m}^2/\text{s} \times 26.4 \times 10^{-6} \text{ m}^2/\text{s}} = 1.813 \times 10^9$$

and from Equation 9.23 it follows that transition to turbulence occurs on the panel. The appropriate correlation is then given by Equation 9.26

$$\overline{Nu}_L = \left\{0.825 + \frac{0.387\, Ra_L^{1/6}}{[1 + (0.492/Pr)^{9/16}]^{8/27}}\right\}^2$$

$$\overline{Nu}_L = \left\{0.825 + \frac{0.387(1.813 \times 10^9)^{1/6}}{[1 + (0.492/0.690)^{9/16}]^{8/27}}\right\}^2 = 147$$

Hence

$$\bar{h} = \frac{\overline{Nu}_L \cdot k}{L} = \frac{147 \times 33.8 \times 10^{-3} \text{ W/m} \cdot \text{K}}{0.71 \text{ m}} = 7.0 \text{ W/m}^2 \cdot \text{K}$$

and

$$q = 7.0 \text{ W/m}^2 \cdot \text{K} \, (1.02 \times 0.71) \text{ m}^2 \, (232 - 23)°\text{C} = 1060 \text{ W} \qquad \triangleleft$$

Comments:

1. Radiation heat transfer effects are often significant relative to free convection. Using Equation 1.7 and assuming $\varepsilon = 1.0$ for the glass surface and $T_{\text{sur}} = 23°\text{C}$, the net rate of radiation heat transfer between the glass and the surroundings is

$$q_{\text{rad}} = \varepsilon A_s \sigma (T_s^4 - T_{\text{sur}}^4)$$

$$q_{\text{rad}} = 1(1.02 \times 0.71)\text{m}^2 \times 5.67 \times 10^{-8} \text{ W/m}^2 \cdot \text{K}^4 \, (505^4 - 296^4) \text{ K}^4$$

$$q_{\text{rad}} = 2355 \text{ W}$$

   Hence in this case radiation heat transfer exceeds free convection heat transfer by more than a factor of 2.

2. The effects of radiation and free convection on heat transfer from the glass depend strongly on its temperature. With $q \propto T_s^4$ for radiation and $q \propto T_s^n$ for free convection, where $1.25 < n < 1.33$, we expect the relative influence of radiation to increase with increasing temperature. This behavior is revealed by computing and plotting the heat rates as a function of temperature for $50 \leq T_s \leq 250°\text{C}$.

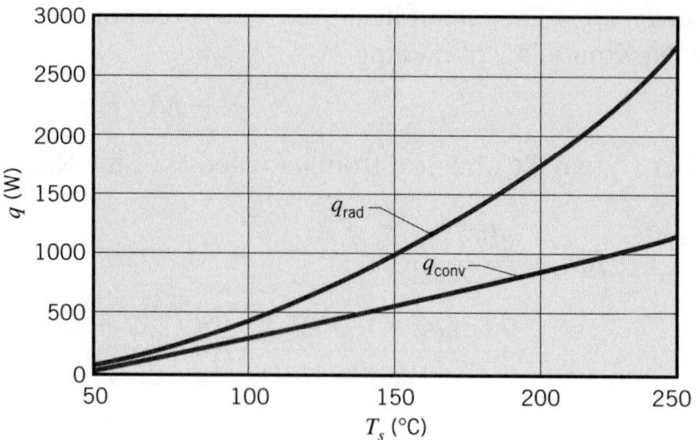

For each value of T_s used to generate the foregoing free convection results, air properties were determined at the corresponding value of T_f.

9.6.2 Inclined and Horizontal Plates

For a vertical plate that is hot (or cold) relative to an ambient fluid, the plate is aligned with the gravitational vector, and the buoyancy force acts exclusively to induce fluid motion in the upward (or downward) direction. However, if the plate is inclined with respect to gravity, the buoyancy force has a component normal, as well as parallel, to the plate surface. With a reduction in the buoyancy force parallel to the surface, there is a reduction in fluid velocities along the plate, and one might expect there to be an attendant reduction in convection heat transfer. Whether, in fact, there is such a reduction depends on whether one is interested in heat transfer from the top or bottom surface of the plate.

As shown in Figure 9.6*a*, if the plate is cold, the y-component of the buoyancy force, which is normal to the plate, acts to maintain the descending boundary layer flow in contact with the top surface of the plate. Since the x-component of the gravitational acceleration is reduced to $g \cos \theta$, fluid velocities along the plate are reduced and there is an attendant reduction in convection heat transfer to the top surface. However, at the bottom surface, the y-component of the buoyancy force acts to move fluid from the surface, and boundary layer development is interrupted by the discharge of parcels of cool fluid from the surface (Figure 9.6*a*). The resulting flow is three-dimensional, and, as shown by the spanwise (z-direction) variations of Figure 9.6*b*, the cool fluid discharged from the bottom surface is continuously replaced by the warmer ambient fluid. The displacement of cool boundary layer fluid by the warmer ambient and the attendant reduction in the thermal boundary layer thickness act to increase convection heat transfer to the bottom surface. In fact, heat transfer enhancement due to the three-dimensional flow typically exceeds the reduction associated with the reduced x-component of g, and the combined effect is to increase heat transfer to the bottom surface. Similar trends characterize a hot plate (Figure 9.6*c,d*), and the three-dimensional flow is now associated with the upper surface, from which parcels of warm fluid are discharged. Such flows have been observed by several investigators [14–16].

In an early study of heat transfer from inclined plates, Rich [17] suggested that convection coefficients could be determined from vertical plate correlations, if g is replaced by $g \cos \theta$ in computing the plate Rayleigh number. Since then, however, it has been determined that this approach is only satisfactory for the top and bottom surfaces of cold and hot plates, respectively. It is not appropriate for the top and bottom surfaces of hot and cold plates,

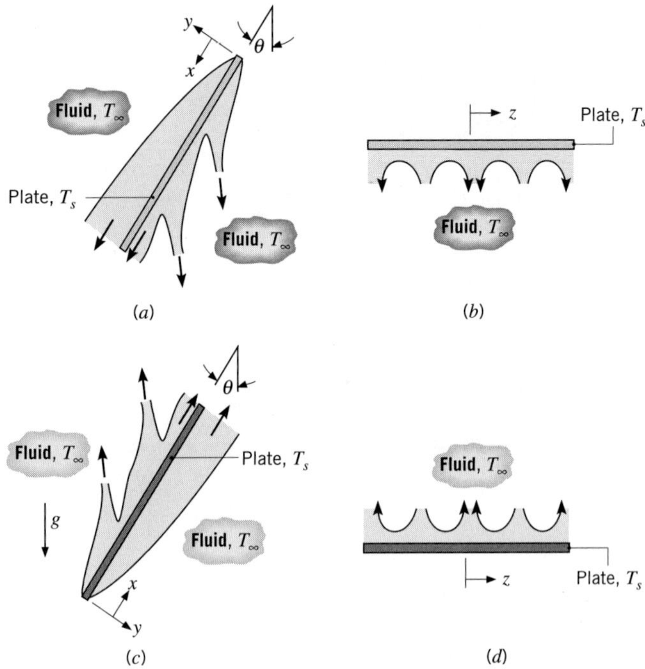

FIGURE 9.6 Buoyancy-driven flows on an inclined plate: (*a*) Side view of flows at top and bottom surfaces of a cold plate ($T_s < T_\infty$). (*b*) End view of flow at bottom surface of cold plate. (*c*) Side view of flows at top and bottom surfaces of a hot plate ($T_s > T_\infty$). (*d*) End view of flow at top surface of hot plate.

respectively, where the three-dimensionality of the flow has limited the ability to develop generalized correlations. At the top and bottom surfaces of cold and hot inclined plates, respectively, it is therefore recommended that, for $0 \leq \theta \lesssim 60°$, g be replaced by $g \cos \theta$ and that Equation 9.26 or 9.27 be used to compute the average Nusselt number. For the opposite surfaces, no recommendations are made, and the literature should be consulted [14–16].

If the plate is horizontal, the buoyancy force is exclusively normal to the surface. As for the inclined plate, flow patterns and heat transfer depend strongly on whether the surface is cold or hot and on whether it is facing upward or downward. For a cold surface facing upward (Figure 9.7*a*) and a hot surface facing downward (Figure 9.7*d*), the tendency of the fluid to descend and ascend, respectively, is impeded by the plate. The flow must move horizontally before it can descend or ascend from the edges of the plate, and convection heat transfer is somewhat ineffective. In contrast, for a cold surface facing downward (Figure 9.7*b*) and a hot surface facing upward (Figure 9.7*c*), flow is driven by descending and ascending parcels of fluid, respectively. Conservation of mass dictates that cold (warm) fluid descending (ascending) from a surface be replaced by ascending (descending) warmer (cooler) fluid from the ambient, and heat transfer is much more effective.

For horizontal plates of various shapes (for example, squares, rectangles, or circles), there is a need to define the characteristic length for use in the Nusselt and Rayleigh numbers. Experiments have shown [18, 19] that a single set of correlations can be used for a variety of different plate shapes when the characteristic length is defined as

$$L \equiv \frac{A_s}{P} \tag{9.29}$$

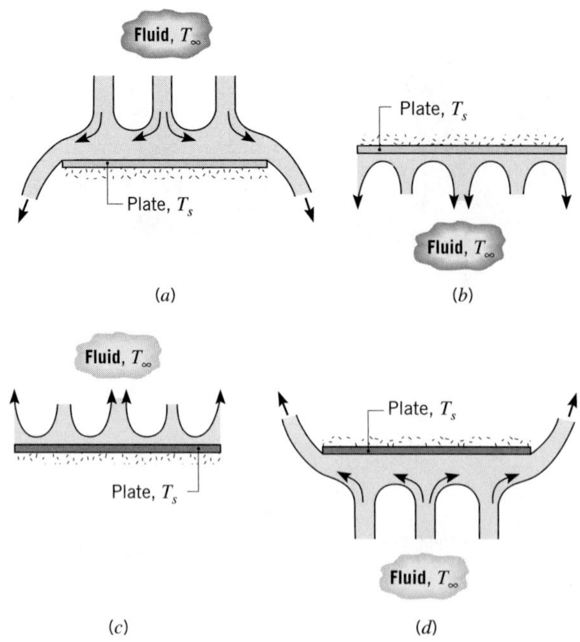

FIGURE 9.7 Buoyancy-driven flows on horizontal cold ($T_s < T_\infty$) and hot ($T_s > T_\infty$) plates: (*a*) Top surface of cold plate. (*b*) Bottom surface of cold plate. (*c*) Top surface of hot plate. (*d*) Bottom surface of hot plate.

where A_s and P are the plate surface area (one side) and perimeter, respectively. Using this characteristic length, the recommended correlations for the average Nusselt number are

Upper Surface of Hot Plate or Lower Surface of Cold Plate [19]:

$$\overline{Nu}_L = 0.54\, Ra_L^{1/4} \quad (10^4 \lesssim Ra_L \lesssim 10^7,\ Pr \gtrsim 0.7) \tag{9.30}$$

$$\overline{Nu}_L = 0.15\, Ra_L^{1/3} \quad (10^7 \lesssim Ra_L \lesssim 10^{11},\ \text{all } Pr) \tag{9.31}$$

Lower Surface of Hot Plate or Upper Surface of Cold Plate [20]:

$$\overline{Nu}_L = 0.52\, Ra_L^{1/5} \quad (10^4 \lesssim Ra_L \lesssim 10^9,\ Pr \gtrsim 0.7) \tag{9.32}$$

Additional correlations can be found in [21].

EXAMPLE 9.3

Airflow through a long rectangular heating duct that is 0.75 m wide and 0.3 m high maintains the outer duct surface at 45°C. If the duct is uninsulated and exposed to air at 15°C in the crawlspace beneath a home, what is the heat loss from the duct per meter of length?

SOLUTION

Known: Surface temperature of a long rectangular duct.

Find: Heat loss from duct per meter of length.

Schematic:

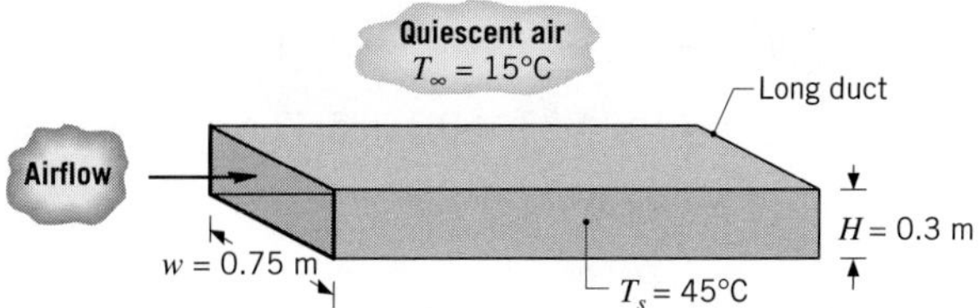

Assumptions:

1. Ambient air is quiescent.
2. Surface radiation effects are negligible.
3. Ideal gas.
4. Constant properties.

Properties: Table A.4, air ($T_f = 303$ K): $\nu = 16.2 \times 10^{-6}\ \text{m}^2/\text{s}$, $\alpha = 22.9 \times 10^{-6}\ \text{m}^2/\text{s}$, $k = 0.0265\ \text{W/m}\cdot\text{K}$, $\beta = 0.0033\ \text{K}^{-1}$, $Pr = 0.71$.

Analysis: Surface heat loss is by free convection from the vertical sides and the horizontal top and bottom. From Equation 9.25

$$Ra_L = \frac{g\beta(T_s - T_\infty)L^3}{\nu\alpha} = \frac{(9.8\ \text{m/s}^2)(0.0033\ \text{K}^{-1})(30\ \text{K})\ L^3\ (\text{m}^3)}{(16.2 \times 10^{-6}\ \text{m}^2/\text{s})(22.9 \times 10^{-6}\ \text{m}^2/\text{s})}$$

$$Ra_L = 2.62 \times 10^9 L^3$$

For the two sides, $L = H = 0.3$ m. Hence $Ra_L = 7.07 \times 10^7$. The free convection boundary layer is therefore laminar, and from Equation 9.27

$$\overline{Nu}_L = 0.68 + \frac{0.670\ Ra_L^{1/4}}{[1 + (0.492/Pr)^{9/16}]^{4/9}}$$

The convection coefficient associated with the sides is then

$$\bar{h}_s = \frac{k}{H}\overline{Nu}_L$$

$$\bar{h}_s = \frac{0.0265\ \text{W/m}\cdot\text{K}}{0.3\ \text{m}}\left\{0.68 + \frac{0.670(7.07 \times 10^7)^{1/4}}{[1 + (0.492/0.71)^{9/16}]^{4/9}}\right\} = 4.23\ \text{W/m}^2\cdot\text{K}$$

For the top and bottom, $L = (A_s/P) \approx (w/2) = 0.375$ m. Hence $Ra_L = 1.38 \times 10^8$, and from Equations 9.31 and 9.32, respectively,

$$\bar{h}_t = [k/(w/2)] \times 0.15\ Ra_L^{1/3} = \frac{0.0265\ \text{W/m}\cdot\text{K}}{0.375\ \text{m}} \times 0.15(1.38 \times 10^8)^{1/3}$$
$$= 5.47\ \text{W/m}^2\cdot\text{K}$$

$$\bar{h}_b = [k/(w/2)] \times 0.52\ Ra_L^{1/5} = \frac{0.0265\ \text{W/m}\cdot\text{K}}{0.375\ \text{m}} \times 0.52(1.38 \times 10^8)^{1/5}$$
$$= 1.56\ \text{W/m}^2\cdot\text{K}$$

The rate of heat loss per unit length of duct is then

$$q' = 2q'_s + q'_t + q'_b = (2\overline{h}_s \cdot H + \overline{h}_t \cdot w + \overline{h}_b \cdot w)(T_s - T_\infty)$$

$$q' = (2 \times 4.23 \times 0.3 + 5.47 \times 0.75 + 1.56 \times 0.75)(45 - 15)\ \text{W/m}$$

$$q' = 234\ \text{W/m} \qquad \triangleleft$$

Comments:

1. The heat loss may be reduced by insulating the duct. We consider this option for a 25-mm-thick layer of blanket insulation ($k = 0.035$ W/m·K) that is wrapped around the duct.

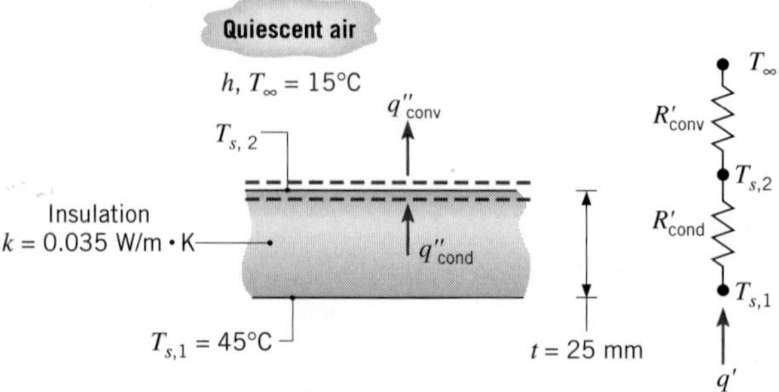

The heat loss at each surface may be expressed as

$$q' = \frac{T_{s,1} - T_\infty}{R'_{\text{cond}} + R'_{\text{conv}}}$$

where R'_{conv} is associated with free convection from the outer surface and hence depends on the unknown temperature $T_{s,2}$. This temperature may be determined by applying an energy balance to the outer surface, from which it follows that

$$q''_{\text{cond}} = q''_{\text{conv}}$$

or

$$\frac{(T_{s,1} - T_{s,2})}{(t/k)} = \frac{(T_{s,2} - T_\infty)}{(1/\overline{h})}$$

Since different convection coefficients are associated with the sides, top, and bottom ($\overline{h}_s$, $\overline{h}_t$, and $\overline{h}_b$), a separate solution to this equation must be obtained for each of the three surfaces. The solutions are iterative, since the properties of air and the convection coefficients depend on T_s. Performing the calculations, we obtain

Sides	$T_{s,2} = 24°\text{C}$,	$\overline{h}_s = 3.18\ \text{W/m}^2 \cdot \text{K}$
Top	$T_{s,2} = 23°\text{C}$,	$\overline{h}_t = 3.66\ \text{W/m}^2 \cdot \text{K}$
Bottom	$T_{s,2} = 30°\text{C}$,	$\overline{h}_b = 1.36\ \text{W/m}^2 \cdot \text{K}$

Neglecting heat loss through the corners of the insulation, the total heat rate per unit length of duct is then

$$q' = 2q'_s + q'_t + q'_b$$

$$q' = \frac{2H(T_{s,1} - T_\infty)}{(t/k) + (1/\overline{h}_s)} + \frac{w(T_{s,1} - T_\infty)}{(t/k) + (1/\overline{h}_t)} + \frac{w(T_{s,1} - T_\infty)}{(t/k) + (1/\overline{h}_b)}$$

which yields

$$q' = (17.5 + 22.8 + 15.5)\ \text{W/m} = 55.8\ \text{W/m}$$

The insulation therefore provides a 76% reduction in heat loss to the ambient air by natural convection.

2. Although they have been neglected, radiation losses may still be significant. From Equation 1.7 with ε assumed to be unity and $T_{sur} = 288$ K, $q'_{rad} = 398$ W/m for the uninsulated duct. Inclusion of radiation effects in the energy balance for the insulated duct would reduce the outer surface temperatures, thereby reducing the convection heat rates. With radiation, however, the total heat rate ($q'_{conv} + q'_{rad}$) would increase.

9.6.3 The Long Horizontal Cylinder

This important geometry has been studied extensively, and many existing correlations have been reviewed by Morgan [22]. For an isothermal cylinder, Morgan suggests an expression of the form

$$\overline{Nu}_D = \frac{\bar{h}D}{k} = C\,Ra_D^n \tag{9.33}$$

where C and n are given in Table 9.1 and Ra_D and $\overline{Nu}_D$ are based on the cylinder diameter. In contrast, Churchill and Chu [23] have recommended a single correlation for a wide Rayleigh number range:

$$\overline{Nu}_D = \left\{0.60 + \frac{0.387\,Ra_D^{1/6}}{[1 + (0.559/Pr)^{9/16}]^{8/27}}\right\}^2 \qquad Ra_D \lesssim 10^{12} \tag{9.34}$$

The foregoing correlations provide the average Nusselt number over the entire circumference of an isothermal cylinder. As shown in Figure 9.8 for a heated cylinder, local Nusselt numbers are influenced by boundary layer development, which begins at $\theta = 0$ and concludes at $\theta < \pi$ with formation of a plume ascending from the cylinder. If the flow remains laminar over the entire surface, the distribution of the local Nusselt number with θ is characterized by a maximum at $\theta = 0$ and a monotonic decay with increasing θ. This decay would be disrupted at Rayleigh numbers sufficiently large ($Ra_D \gtrsim 10^9$) to permit transition to turbulence within the boundary layer. If the cylinder is cooled relative to the ambient fluid, boundary layer development begins at $\theta = \pi$, the local Nusselt number is a maximum at this location, and the plume descends from the cylinder.

TABLE 9.1 Constants of Equation 9.33 for free convection on a horizontal circular cylinder [22]

Ra_D	C	n
10^{-10}–10^{-2}	0.675	0.058
10^{-2}–10^{2}	1.02	0.148
10^{2}–10^{4}	0.850	0.188
10^{4}–10^{7}	0.480	0.250
10^{7}–10^{12}	0.125	0.333

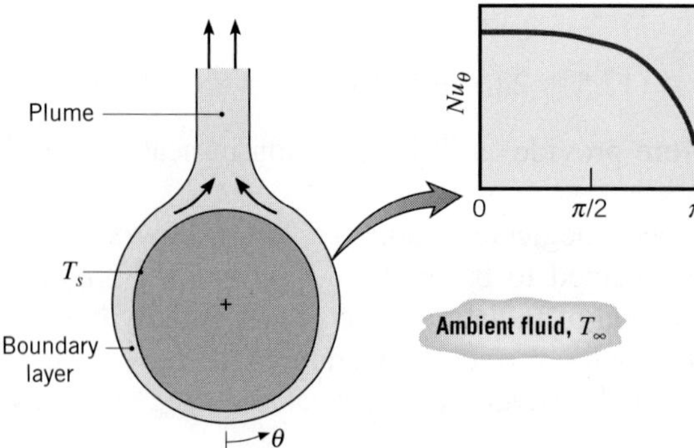

FIGURE 9.8 Boundary layer development and Nusselt number distribution on a heated horizontal cylinder.

EXAMPLE 9.4

The fluid of Example 2.2 is characterized by a thermal conductivity, density, specific heat, and dynamic viscosity of 0.705 W/m·K, 1146 kg/m^3, 3587 J/kg·K, and 962×10^{-6} N·s/m^2, respectively. An experiment is conducted in which a long aluminum rod of diameter $D = 20$ mm and initial temperature $T_i = 32°\text{C}$ is suddenly immersed horizontally into a large bath of the fluid at a temperature of $T_\infty = 22°\text{C}$. At $t = 65$ s, the measured temperature of the rod is $T_f = 23°\text{C}$. Determine the thermal expansion coefficient of the fluid β.

SOLUTION

Known: Initial and final temperatures of aluminum rod of known diameter. Temperature and properties of the fluid.

Find: Thermal expansion coefficient of the fluid.

Schematic:

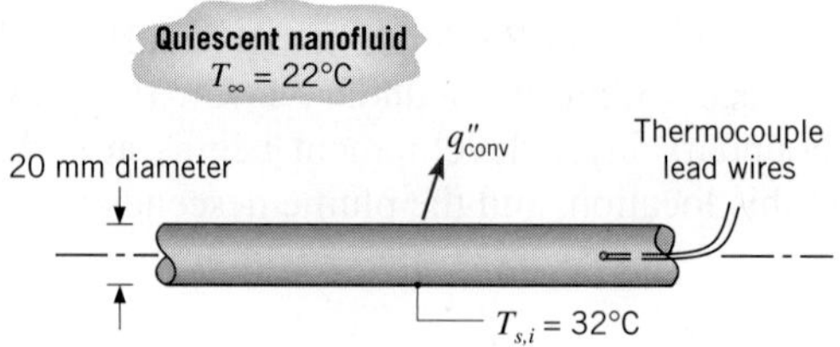

Assumptions:

1. Constant properties.
2. Spatially uniform rod temperature (applicability of the lumped capacitance approximation).

Properties: Table A.1, aluminum ($\bar{T} = 300$ K): $\rho_s = 2702$ kg/m^3, $c_{p,s} = 903$ J/kg·K, $k_s = 237$ W/m·K.

Analysis: Because the temperature difference between the rod and the fluid decreases with time, we expect the convection heat transfer coefficient to decrease as cooling proceeds. Since $\overline{h}$ depends on buoyancy forces established by temperature differences, the analysis of Section 5.3.3 may be applied. From Equations 5.28 and 5.26

$$\frac{\theta}{\theta_i} = \left[\frac{nC_1A_{s,c}\theta_i^n}{\rho_s V c_{p,s}} t + 1\right]^{-1/n} \tag{1}$$

where $\overline{h} = C_1(T_s - T_\infty)^n$ and $\theta = T_s - T_\infty$. From Equation 9.33, $\overline{Nu}_D = C\,Ra_D^n$. Substituting definitions of the Nusselt and Rayleigh numbers into Equation 9.33 yields

$$\overline{h} = C\frac{k_l}{D}\left[\frac{g\beta D^3}{\nu_l\alpha_l}\right]^n (T_s - T_\infty)^n \tag{2}$$

From a comparison of Equation 2 with the expression $\overline{h} = C_1(T_s - T_\infty)^n$, it is evident that

$$C_1 = C\frac{k_l}{D}\left[\frac{g\beta D^3}{\nu_l\alpha_l}\right]^n \tag{3}$$

Defining a final excess temperature as $\theta_f = T_{s,f} - T_\infty$ at $t = t_f$ and noting that $\nu_l = \mu_l/\rho_l$, $\alpha_l = k_l/\rho_l c_{p,l}$, and $A_{s,c}/V = 4/D$, Equation 3 may be substituted into Equation 1, yielding

$$\beta = \frac{k_l\mu_l}{c_{p,l}\rho_l^2 g D^3}\left\{\frac{\rho_s c_{p,s}D^2}{4k_l C n t_f \theta_i^n}\right\}\left[\left(\frac{\theta_f}{\theta_i}\right)^{-n} - 1\right]\Bigg\}^{1/n} \tag{4}$$

For now we will assume the Rayleigh number falls in the range $10^4 \leq Ra_D \leq 10^7$, for which $C = 0.480$ and $n = 0.250$ from Table 9.1. Hence the thermal expansion coefficient is

$$\beta = \frac{0.705\text{ W/m}\cdot\text{K} \times 962 \times 10^{-6}\text{ N}\cdot\text{s/m}^2}{3587\text{ J/kg}\cdot\text{K} \times (1146\text{ kg/m}^3)^2 \times 9.8\text{ m/s}^2 \times (20 \times 10^{-3}\text{ m})^3}$$

$$\times \left\{\frac{2702\text{ kg/m}^3 \times 903\text{ J/kg}\cdot\text{K} \times (20 \times 10^{-3}\text{ m})^2}{4 \times 0.705\text{ W/m}\cdot\text{K} \times 0.480 \times 0.25 \times 65\text{ s} \times (10\text{ K})^{0.25}}\left[\left(\frac{1.0}{10}\right)^{-0.25} - 1\right]\right\}^{1/0.25}$$

$$= 261 \times 10^{-6}\text{ K}^{-1}$$ ◁

Using this value of the thermal expansion coefficient, the Rayleigh number based on the initial temperature difference is

$$Ra_{D,\max} = \frac{g\beta\theta_i D^3}{\nu_l\alpha_l} = \frac{g\beta\rho^2 c_{p,l}\theta_i D^3}{\mu_l k_l}$$

$$= \frac{9.8\text{ m/s}^2 \times 261 \times 10^{-6}\text{ K}^{-1} \times (1145\text{ kg/m}^3)^2 \times 3587\text{ J/kg}\cdot\text{K} \times 10\text{ K} \times (20 \times 10^{-3}\text{ m})^3}{962 \times 10^{-6}\text{ N}\cdot\text{s/m}^2 \times 0.705\text{ W/m}\cdot\text{K}}$$

$$= 1.42 \times 10^6$$

Since $\theta_f = \theta_i/10$, the minimum value of the Rayleigh number during the cooling process is $Ra_{D,\min} = Ra_{D,\max}/10 = 1.42 \times 10^5$. Therefore, $10^4 < Ra_{D,\min} < Ra_{D,\max} < 10^7$, and the values of C and n selected from Table 9.1 are appropriate. Hence the foregoing value of the thermal expansion coefficient is correct.

Comments:

1. The manner in which the rod temperature decreases during the cooling process may be determined from Equation 1 with C_1 obtained from Equation 3. The rod temperature history is shown in the figure.

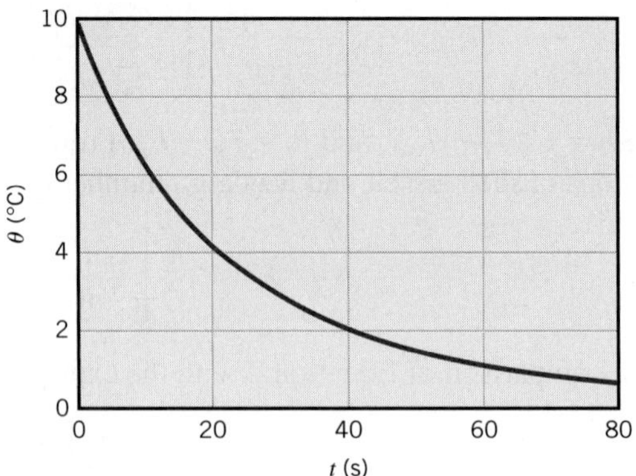

2. Since the temperature difference between the rod and the fluid decreases with time, the Rayleigh number also decreases as cooling proceeds. This leads to a gradual reduction in the convection heat transfer coefficient during cooling, as can be determined by solving Equation 2 once $\theta(t)$ is known.

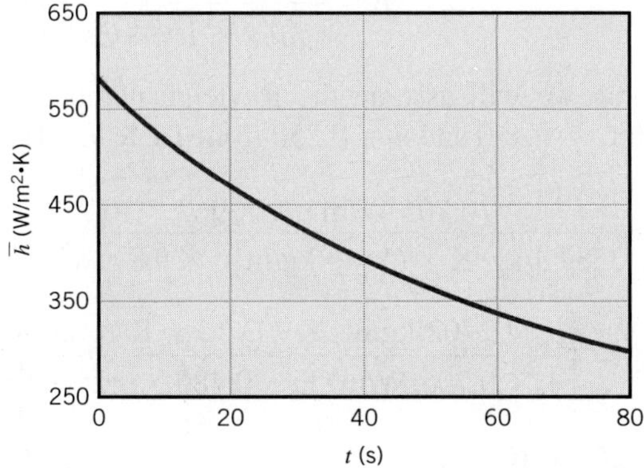

3. The maximum value of the convection heat transfer coefficient is $\overline{h}_{max} = 584\ \text{W/m}^2 \cdot \text{K}$. This corresponds to a maximum Biot number of $Bi_{max} = \overline{h}_{max}(D/2)/k_s = 584\ \text{W/m}^2 \cdot \text{K} \times (20 \times 10^{-3}\ \text{m}/2)/237\ \text{W/m} \cdot \text{K} = 0.025$ when the criterion of Equation 5.10 is applied in a conservative fashion. Since $Bi_{max} < 0.1$, we conclude that the lumped capacitance approximation is valid.

4. Because the rod temperature continually decreases, the buoyancy forces within the fluid decrease with time. Hence fluid velocities continually evolve as the temperature difference between the rod and the fluid slowly decays. Equation 9.33 is strictly applicable only for steady-state conditions. In applying the correlation here, we have implicitly assumed that the instantaneous heat transfer rate from the rod is the same as the steady-state heat transfer rate *if* the same temperature difference exists between the rod and the fluid. This assumption often yields predictions of acceptable accuracy and is referred to as the *quasi-steady approximation*.

9.6.4 Spheres

The following correlation due to Churchill [10] is recommended for spheres in fluids of $Pr \gtrsim 0.7$ and for $Ra_D \lesssim 10^{11}$.

$$\overline{Nu}_D = 2 + \frac{0.589\, Ra_D^{1/4}}{[1 + (0.469/Pr)^{9/16}]^{4/9}} \tag{9.35}$$

In the limit as $Ra_D \rightarrow 0$, Equation 9.35 reduces to $\overline{Nu}_D = 2$, which corresponds to heat transfer by conduction between a spherical surface and a stationary infinite medium, in a manner consistent with Equations 7.48 and 7.49.

Recommended correlations from this section are summarized in Table 9.2. Results for other immersed geometries and special conditions are presented in comprehensive reviews by Churchill [10] and Raithby and Hollands [21].

TABLE 9.2 Summary of free convection empirical correlations for immersed geometries

Geometry	Recommended Correlation	Restrictions
1. Vertical plates[a]	Equation 9.26	None
2. Inclined plates Cold surface up or hot surface down	Equation 9.26 $g \rightarrow g \cos\theta$	$0 \leq \theta \lesssim 60°$
3. Horizontal plates (a) Hot surface up or cold surface down	Equation 9.30 Equation 9.31	$10^4 \lesssim Ra_L \lesssim 10^7$, $Pr \gtrsim 0.7$ $10^7 \lesssim Ra_L \lesssim 10^{11}$
(b) Cold surface up or hot surface down	Equation 9.32	$10^4 \lesssim Ra_L \lesssim 10^9$, $Pr \gtrsim 0.7$

TABLE 9.2 *Continued*

Geometry	Recommended Correlation	Restrictions
4. Horizontal cylinder	Equation 9.34	$Ra_D \lesssim 10^{12}$
5. Sphere	Equation 9.35	$Ra_D \lesssim 10^{11}$ $Pr \gtrsim 0.7$

[a] The correlation may be applied to a vertical cylinder if $(D/L) \gtrsim (35/Gr_L^{1/4})$.

9.7 Free Convection Within Parallel Plate Channels

A common free convection geometry involves vertical (or inclined) parallel plate channels that are open to the ambient at opposite ends (Figure 9.9). The plates could constitute a fin array used to enhance free convection heat transfer from a base surface to which the fins are attached, or they could constitute an array of circuit boards with heat-dissipating electronic components. Surface thermal conditions may be idealized as being isothermal or isoflux and symmetrical ($T_{s,1} = T_{s,2}$; $q''_{s,1} = q''_{s,2}$) or asymmetrical ($T_{s,1} \neq T_{s,2}$; $q''_{s,1} \neq q''_{s,2}$).

For vertical channels ($\theta = 0$) buoyancy acts exclusively to induce motion in the streamwise (x) direction and, beginning at $x = 0$, boundary layers develop on each surface. For short channels and/or large spacings (small L/S), independent boundary layer development occurs at each surface and conditions correspond to those for an *isolated plate* in an infinite, quiescent medium. For large L/S, however, boundary layers developing on opposing surfaces eventually merge to yield a fully developed condition. If the channel is inclined, there is a component of the buoyancy force normal, as well as parallel,

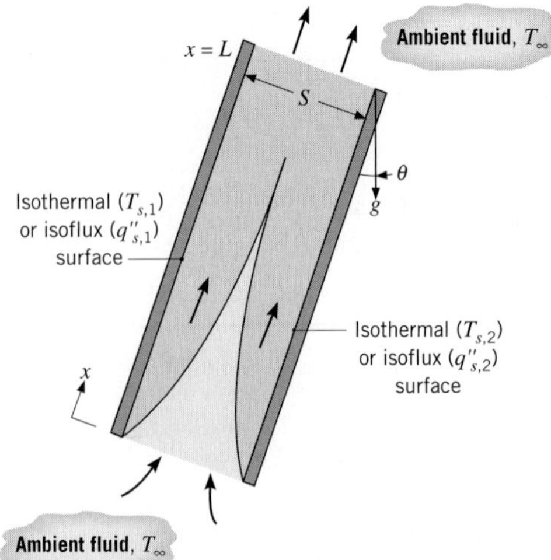

FIGURE 9.9 Free convection flow between heated parallel plates with opposite ends exposed to a quiescent fluid.

to the streamwise direction, and conditions may strongly be influenced by development of a three-dimensional, secondary flow.

9.7.1 Vertical Channels

Beginning with the benchmark paper by Elenbaas [24], the vertical orientation has been studied extensively for symmetrically and asymmetrically heated plates with isothermal or isoflux surface conditions. For *symmetrically heated, isothermal plates,* Elenbaas obtained the following semiempirical correlation:

$$\overline{Nu}_S = \frac{1}{24} Ra_S \left(\frac{S}{L}\right)\left\{1 - \exp\left[-\frac{35}{Ra_S(S/L)}\right]\right\}^{3/4} \tag{9.36}$$

where the average Nusselt and Rayleigh numbers are defined as

$$\overline{Nu}_S = \left(\frac{q/A}{T_s - T_\infty}\right)\frac{S}{k} \tag{9.37}$$

and

$$Ra_S = \frac{g\beta(T_s - T_\infty)S^3}{\alpha\nu} \tag{9.38}$$

Equation 9.36 was developed for air as the working fluid, and its range of applicability is

$$\left[10^{-1} \lesssim \frac{S}{L} Ra_S \lesssim 10^5\right]$$

Knowledge of the average Nusselt number for a plate therefore permits determination of the total heat rate for the plate. In the fully developed limit ($S/L \rightarrow 0$), Equation 9.36 reduces to

$$\overline{Nu}_{S(\mathrm{fd})} = \frac{Ra_S(S/L)}{24} \tag{9.39}$$

Retention of the L dependence results from defining $\overline{Nu}_S$ in terms of the fixed inlet (ambient) temperature and not in terms of the fluid mixed-mean temperature, which is not explicitly known. For the common condition corresponding to adjoining isothermal ($T_{s,1}$) and insulated ($q''_{s,2} = 0$) plates, the fully developed limit yields the following expression for the isothermal surface [25]:

$$\overline{Nu}_{S(\mathrm{fd})} = \frac{Ra_S(S/L)}{12} \tag{9.40}$$

For isoflux surfaces, it is more convenient to define a local Nusselt number as

$$Nu_{S,L} = \left(\frac{q''_s}{T_{s,L} - T_\infty}\right)\frac{S}{k} \tag{9.41}$$

and to correlate results in terms of a modified Rayleigh number defined as

$$Ra^*_S = \frac{g\beta q''_s S^4}{k\alpha\nu} \tag{9.42}$$

The subscript L refers to conditions at $x = L$, where the plate temperature is a maximum. For symmetric, isoflux plates the fully developed limit corresponds to [25]

$$Nu_{S,L(\text{fd})} = 0.144[Ra_S^*(S/L)]^{1/2} \tag{9.43}$$

and for asymmetric isoflux conditions with one surface insulated ($q''_{s,2} = 0$) the limit is

$$Nu_{S,L(\text{fd})} = 0.204[Ra_S^*(S/L)]^{1/2} \tag{9.44}$$

Combining the foregoing relations for the fully developed limit with available results for the isolated plate limit, Bar-Cohen and Rohsenow [25] obtained Nusselt number correlations applicable to the complete range of S/L. For isothermal and isoflux conditions, respectively, the correlations are of the form

$$\overline{Nu}_S = \left[\frac{C_1}{(Ra_S\,S/L)^2} + \frac{C_2}{(Ra_S\,S/L)^{1/2}}\right]^{-1/2} \tag{9.45}$$

$$Nu_{S,L} = \left[\frac{C_1}{Ra_S^*\,S/L} + \frac{C_2}{(Ra_S^*\,S/L)^{2/5}}\right]^{-1/2} \tag{9.46}$$

where the constants C_1 and C_2 are given in Table 9.3 for the different surface thermal conditions. In each case the fully developed and isolated plate limits correspond to Ra_S (or Ra_S^*)$S/L \lesssim 10$ and Ra_S (or Ra_S^*)$S/L \gtrsim 100$, respectively.

Bar-Cohen and Rohsenow [25] used the foregoing correlations to infer the optimum plate spacing S_{opt} for maximizing heat transfer from an array of isothermal plates, as well as the spacing S_{max} needed to maximize heat transfer from each plate in the array. Existence of an optimum for the array results from the fact that, although heat transfer from each plate decreases with decreasing S, the number of plates that may be placed in a prescribed volume increases. Hence S_{opt} maximizes heat transfer from the array by yielding a maximum for the product of $\bar{h}$ and the total plate surface area. In contrast, to maximize heat transfer from each plate, S_{max} must be large enough to preclude overlap of adjoining boundary layers, such that the isolated plate limit remains valid over the entire plate.

Consideration of the optimum plate spacing is particularly important for vertical parallel plates used as fins to enhance heat transfer by natural convection from a base surface of fixed width W. With the temperature of the fins exceeding that of the ambient fluid, flow between the fins is induced by buoyancy forces. However, resistance to the flow is associated with viscous forces imposed by the surface of the fins, and the rate of mass flow between adjoining fins is governed by a balance between buoyancy and viscous forces. Since viscous forces increase with decreasing S, there is an accompanying reduction in the flow rate, and hence $\bar{h}$. However, for fixed W, the attendant increase in the number of fins increases the total surface area A_s and yields a maximum in $\bar{h}A_s$ for $S = S_{\text{opt}}$. For $S < S_{\text{opt}}$,

TABLE 9.3 Heat transfer parameters for free convection between vertical parallel plates

Surface Condition	C_1	C_2	S_{opt}	$S_{\text{max}}/S_{\text{opt}}$
Symmetric isothermal plates ($T_{s,1} = T_{s,2}$)	576	2.87	$2.71(Ra_S/S^3L)^{-1/4}$	1.71
Symmetric isoflux plates ($q''_{s,1} = q''_{s,2}$)	48	2.51	$2.12(Ra_S^*/S^4L)^{-1/5}$	4.77
Isothermal/adiabatic plates ($T_{s,1}$, $q''_{s,2} = 0$)	144	2.87	$2.15(Ra_S/S^3L)^{-1/4}$	1.71
Isoflux/adiabatic plates ($q''_{s,1} = q''_{s,2} = 0$)	24	2.51	$1.69(Ra_S^*/S^4L)^{-1/5}$	4.77

the amount by which $\overline{h}$ is diminished by viscous effects exceeds the increase in A_s; for $S > S_{opt}$, the amount by which A_s is diminished exceeds the increase in $\overline{h}$.

For isoflux plates, the total volumetric heat rate simply increases with decreasing S. However, the need to maintain T_s below prescribed limits precludes reducing S to extremely small values. Hence S_{opt} may be defined as that value of S which yields the maximum volumetric heat dissipation per unit temperature difference, $T_s(L) - T_\infty$. The spacing S_{max} that yields the lowest possible surface temperature for a prescribed heat flux, without regard to volumetric considerations, is again the value of S that precludes boundary layer merger. Values of S_{opt} and S_{max}/S_{opt} are presented in Table 9.3 for plates of negligible thickness.

In using the foregoing correlations, fluid properties are evaluated at average temperatures of $\overline{T} = (T_s + T_\infty)/2$ for isothermal surfaces and $\overline{T} = (T_{s,L} + T_\infty)/2$ for isoflux surfaces.

9.7.2 Inclined Channels

Experiments have been performed by Azevedo and Sparrow [16] for inclined channels in water. Symmetric isothermal plates and isothermal-insulated plates were considered for $0 \leq \theta \leq 45°$ and conditions within the isolated plate limit, $Ra_S(S/L) > 200$. Although three-dimensional secondary flows were observed at the lower plate, when it was heated, data for all experimental conditions were correlated to within $\pm 10\%$ by

$$\overline{Nu}_S = 0.645[Ra_S(S/L)]^{1/4} \tag{9.47}$$

Departures of the data from the correlation were most pronounced at large tilt angles with bottom surface heating and were attributed to heat transfer enhancement by the three-dimensional secondary flow. Fluid properties are evaluated at $\overline{T} = (T_s + T_\infty)/2$.

9.8 *Empirical Correlations: Enclosures*

The foregoing results pertain to free convection between a surface and an extensive fluid medium. However, engineering applications frequently involve heat transfer between surfaces that are at different temperatures and are separated by an *enclosed* fluid. In this section we present correlations that are pertinent to the most common geometries.

9.8.1 Rectangular Cavities

The rectangular cavity (Figure 9.10) has been studied extensively, and comprehensive reviews of both experimental and theoretical results are available [26, 27]. Two of the opposing walls are maintained at different temperatures ($T_1 > T_2$), while the remaining

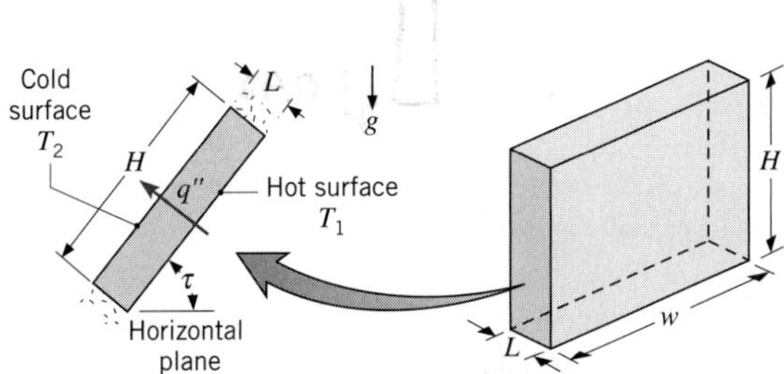

FIGURE 9.10 Free convection in a rectangular cavity.

walls are insulated from the surroundings. The tilt angle τ between the heated and cooled surfaces and the horizontal can vary from 0° (*horizontal cavity* with bottom heating) to 90° (*vertical cavity* with sidewall heating) to 180° (*horizontal cavity* with top heating). The heat flux across the cavity, which is expressed as

$$q'' = h(T_1 - T_2) \tag{9.48}$$

can depend strongly on the aspect ratio H/L, as well as the value of τ. For large values of the aspect ratio w/L, its dependence on w/L is small and may be neglected for the purposes of this text.

The horizontal cavity heated from below ($\tau = 0$) has been considered by many investigators. For H/L, $w/L \gg 1$, and Rayleigh numbers less than a critical value of $Ra_{L,c} = 1708$, buoyancy forces cannot overcome the resistance imposed by viscous forces and there is no advection within the cavity. Hence heat transfer from the bottom to the top surface occurs by conduction or, for a gas, by conduction and radiation. Since conditions correspond to one-dimensional conduction through a plane fluid layer, the convection coefficient is $h = k/L$ and $Nu_L = 1$. However, for

$$Ra_L \equiv \frac{g\beta(T_1 - T_2)L^3}{\alpha\nu} > 1708$$

conditions are thermally unstable and there is advection within the cavity. For Rayleigh numbers in the range $1708 < Ra_L \lesssim 5 \times 10^4$, fluid motion consists of regularly spaced roll cells (Figure 9.11), while for larger Rayleigh numbers, the cells break down and the fluid motion evolves through many different patterns before becoming turbulent.

As a first approximation, convection coefficients for the horizontal cavity heated from below may be obtained from the following correlation proposed by Globe and Dropkin [28]:

$$\overline{Nu}_L = \frac{\bar{h}L}{k} = 0.069\, Ra_L^{1/3}\, Pr^{0.074} \qquad 3 \times 10^5 \lesssim Ra_L \lesssim 7 \times 10^9 \tag{9.49}$$

where all properties are evaluated at the average temperature, $\bar{T} \equiv (T_1 + T_2)/2$. The correlation applies for values of L/H sufficiently small to ensure a negligible effect of the sidewalls. More detailed correlations, which apply over a wider range of Ra_L, have been proposed [29, 30]. In concluding the discussion of horizontal cavities, it is noted that in the absence of radiation, for heating from above ($\tau = 180°$), heat transfer from the top to the bottom surface is exclusively by conduction ($Nu_L = 1$), irrespective of the value of Ra_L.

In the vertical rectangular cavity ($\tau = 90°$), the vertical surfaces are heated and cooled, while the horizontal surfaces are adiabatic. As shown in Figure 9.12, fluid motion is characterized by a recirculating or cellular flow for which fluid ascends along the hot wall and

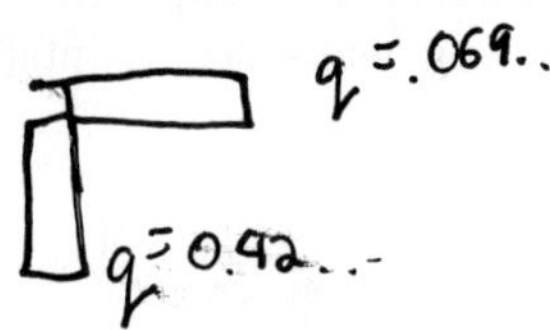

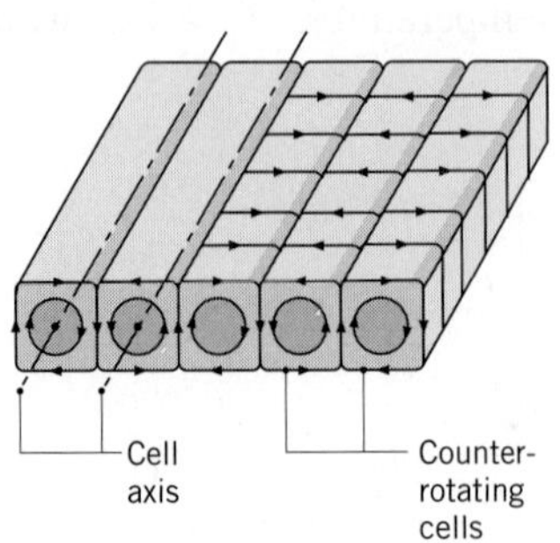

FIGURE 9.11 Longitudinal roll cells characteristic of advection in a horizontal fluid layer heated from below ($1708 < Ra_L \lesssim 5 \times 10^4$).

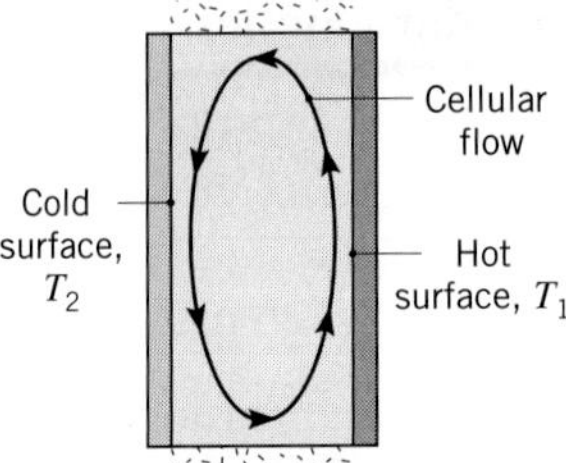

FIGURE 9.12 Cellular flow in a vertical cavity with different sidewall temperatures.

descends along the cold wall. For small Rayleigh numbers, $Ra_L \lesssim 10^3$, the buoyancy-driven flow is weak and, in the absence of radiation, heat transfer is primarily by conduction across the fluid. Hence, from Fourier's law, the Nusselt number is again $Nu_L = 1$. With increasing Rayleigh number, the cellular flow intensifies and becomes concentrated in thin boundary layers adjoining the sidewalls. The core becomes nearly stagnant, although additional cells can develop in the corners and the sidewall boundary layers eventually undergo transition to turbulence. For aspect ratios in the range $1 \lesssim (H/L) \lesssim 10$, the following correlations have been suggested [27]:

$$\overline{Nu}_L = 0.22\left(\frac{Pr}{0.2 + Pr}Ra_L\right)^{0.28}\left(\frac{H}{L}\right)^{-1/4} \tag{9.50}$$

$$\left[\begin{array}{l} 2 \lesssim \dfrac{H}{L} \lesssim 10 \\ Pr \lesssim 10^5 \\ 10^3 \lesssim Ra_L \lesssim 10^{10} \end{array}\right]$$

$$\overline{Nu}_L = 0.18\left(\frac{Pr}{0.2 + Pr}Ra_L\right)^{0.29} \tag{9.51}$$

$$\left[\begin{array}{l} 1 \lesssim \dfrac{H}{L} \lesssim 2 \\ 10^{-3} \lesssim Pr \lesssim 10^5 \\ 10^3 \lesssim \dfrac{Ra_L\, Pr}{0.2 + Pr} \end{array}\right]$$

while for larger aspect ratios, the following correlations have been proposed [31]:

$$\overline{Nu}_L = 0.42\, Ra_L^{1/4}\, Pr^{0.012}\left(\frac{H}{L}\right)^{-0.3} \quad \left[\begin{array}{l} 10 \lesssim \dfrac{H}{L} \lesssim 40 \\ 1 \lesssim Pr \lesssim 2 \times 10^4 \\ 10^4 \lesssim Ra_L \lesssim 10^7 \end{array}\right] \tag{9.52}$$

$$\overline{Nu}_L = 0.046\, Ra_L^{1/3} \quad \left[\begin{array}{l} 1 \lesssim \dfrac{H}{L} \lesssim 40 \\ 1 \lesssim Pr \lesssim 20 \\ 10^6 \lesssim Ra_L \lesssim 10^9 \end{array}\right] \tag{9.53}$$

TABLE 9.4 Critical angle for inclined rectangular cavities

(H/L)	1	3	6	12	>12
τ^*	25°	53°	60°	67°	70°

Convection coefficients computed from the foregoing expressions are to be used with Equation 9.48. Again, all properties are evaluated at the mean temperature, $(T_1 + T_2)/2$.

Studies of free convection in tilted cavities are often stimulated by applications involving flat-plate solar collectors [32–37]. For such cavities, the fluid motion consists of a combination of the roll structure of Figure 9.11 and the cellular structure of Figure 9.12. Typically, transition between the two types of fluid motion occurs at a critical tilt angle, τ^*, with a corresponding change in the value of $\overline{Nu}_L$. For large aspect ratios, $(H/L) \gtrsim 12$, and tilt angles less than the critical value τ^* given in Table 9.4, the following correlation due to Hollands et al. [37] is in excellent agreement with available data:

$$\overline{Nu}_L = 1 + 1.44\left[1 - \frac{1708}{Ra_L \cos\tau}\right]^{\cdot}\left[1 - \frac{1708(\sin 1.8\tau)^{1.6}}{Ra_L \cos\tau}\right] + \left[\left(\frac{Ra_L \cos\tau}{5830}\right)^{1/3} - 1\right]^{\cdot} \qquad \left[\begin{matrix}\frac{H}{L} \gtrsim 12 \\ 0 < \tau \lesssim \tau^*\end{matrix}\right] \tag{9.54}$$

The notation $[\ ]^{\cdot}$ implies that, if the quantity in brackets is negative, it must be set equal to zero. The implication is that, if the Rayleigh number is less than a critical value $Ra_{L,c} = 1708/\cos\tau$, there is no flow within the cavity. For small aspect ratios Catton [27] suggests that reasonable results may be obtained from a correlation of the form

$$\overline{Nu}_L = \overline{Nu}_L(\tau = 0)\left[\frac{\overline{Nu}_L(\tau = 90°)}{\overline{Nu}_L(\tau = 0)}\right]^{\tau/\tau^*}(\sin\tau^*)^{(\tau/4\tau^*)} \qquad \left[\begin{matrix}\frac{H}{L} \lesssim 12 \\ 0 < \tau \lesssim \tau^*\end{matrix}\right] \tag{9.55}$$

Beyond the critical tilt angle, the following correlations due to Ayyaswamy and Catton [32] and Arnold et al. [35], respectively, have been recommended [27] for all aspect ratios (H/L):

$$\overline{Nu}_L = \overline{Nu}_L(\tau = 90°)(\sin\tau)^{1/4} \qquad \tau^* \lesssim \tau < 90° \tag{9.56}$$

$$\overline{Nu}_L = 1 + [\overline{Nu}_L(\tau = 90°) - 1]\sin\tau \qquad 90° < \tau < 180° \tag{9.57}$$

9.8.2 Concentric Cylinders

Free convection heat transfer in the annular space between *long,* horizontal concentric cylinders (Figure 9.13) has been considered by Raithby and Hollands [38]. Flow in the annular region is characterized by two cells that are symmetric about the vertical midplane. If the inner cylinder is heated and the outer cylinder is cooled ($T_i > T_o$), fluid ascends and descends along the inner and outer cylinders, respectively. If $T_i < T_o$, the cellular flows are reversed. The heat transfer rate (W) between the two cylinders, each of length L, is

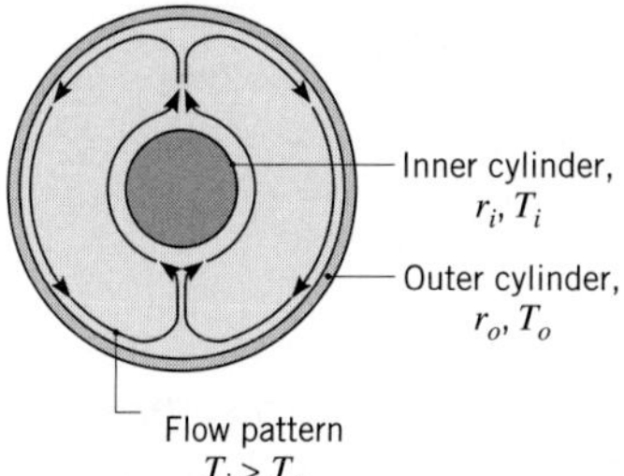

FIGURE 9.13 Free convection flow in the annular space between long, horizontal, concentric cylinders or concentric spheres of inner radius r_i and outer radius r_o.

expressed by Equation 3.32 (with an *effective thermal conductivity,* k_{eff}, replacing the molecular thermal conductivity, k) as

$$q = \frac{2\pi L k_{\text{eff}}(T_i - T_o)}{\ln(r_o/r_i)} \tag{9.58}$$

We see that the effective conductivity of a fictitious *stationary* fluid will transfer the same amount of heat as the actual *moving* fluid. The suggested correlation for k_{eff} is

$$\frac{k_{\text{eff}}}{k} = 0.386\left(\frac{Pr}{0.861 + Pr}\right)^{1/4} Ra_c^{1/4} \tag{9.59}$$

where the length scale in Ra_c is given by

$$L_c = \frac{2[\ln(r_o/r_i)]^{4/3}}{(r_i^{-3/5} + r_o^{-3/5})^{5/3}} \tag{9.60}$$

Equation 9.59 may be used for the range $0.7 \lesssim Pr \lesssim 6000$ and $Ra_c \lesssim 10^7$. Properties are evaluated at the mean temperature, $T_m = (T_i + T_o)/2$. Of course, the minimum heat transfer rate between the cylinders cannot fall below the conduction limit; therefore, $k_{\text{eff}} = k$ if the value of k_{eff}/k predicted by Equation 9.59 is less than unity. A more detailed correlation, which accounts for cylinder eccentricity effects, has been developed by Kuehn and Goldstein [39].

9.8.3 Concentric Spheres

Raithby and Hollands [38] have also considered free convection heat transfer between concentric spheres (Figure 9.13) and express the total heat transfer rate by Equation 3.40 (with an effective thermal conductivity, k_{eff}, replacing the molecular thermal conductivity, k) as

$$q = \frac{4\pi k_{\text{eff}}(T_i - T_o)}{(1/r_i) - (1/r_o)} \tag{9.61}$$

The effective thermal conductivity is

$$\frac{k_{\text{eff}}}{k} = 0.74\left(\frac{Pr}{0.861 + Pr}\right)^{1/4} Ra_s^{1/4} \tag{9.62}$$

where the length scale in Ra_s is given by

$$L_s = \frac{\left(\frac{1}{r_i} - \frac{1}{r_o}\right)^{4/3}}{2^{1/3}(r_i^{-7/5} + r_o^{-7/5})^{5/3}} \tag{9.63}$$

The result may be used to a reasonable approximation for $0.7 \lesssim Pr \lesssim 4000$ and $Ra_s \lesssim 10^4$. Properties are evaluated at $T_m = (T_i + T_o)/2$, and $k_{\text{eff}} = k$ if the value of k_{eff}/k predicted by Equation 9.62 is less than unity.

EXAMPLE 9.5

A long tube of 0.1-m diameter is maintained at 120°C by passing steam through its interior. A radiation shield is installed concentric to the tube with an air gap of 10 mm. If the shield is at 35°C, estimate the heat transfer by free convection from the tube per unit length. What is the heat loss if the space between the tube and the shield is filled with glass-fiber blanket insulation?

SOLUTION

Known: Temperatures and diameters of a steam tube and a concentric radiation shield.

Find:

1. Heat loss per unit length of tube.
2. Heat loss if air space is filled with glass-fiber blanket insulation.

Schematic:

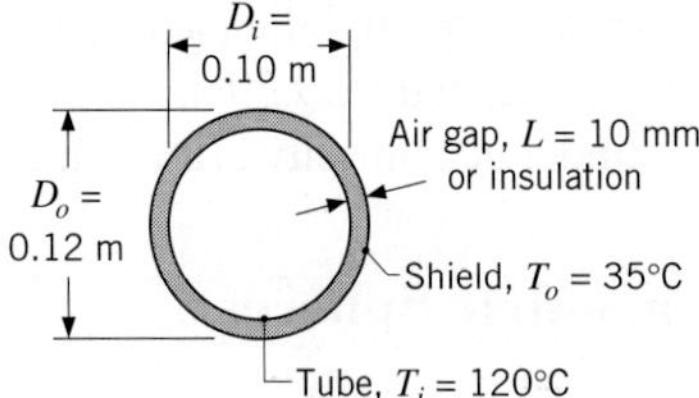

Assumptions:

1. Radiation heat transfer may be neglected.
2. Contact resistance with insulation is negligible.
3. Ideal gas.
4. Constant properties.

Properties: Table A.4, air [$T = (T_i + T_o)/2 = 350$ K]: $k = 0.030$ W/m·K, $\nu = 20.92 \times 10^{-6}$ m^2/s, $\alpha = 29.9 \times 10^{-6}$ m^2/s, $Pr = 0.70$, $\beta = 0.00285$ K^{-1}. Table A.3, insulation, glass-fiber ($T \approx 300$ K): $k = 0.038$ W/m·K.

Analysis:

1. From Equation 9.58, the heat loss per unit length by free convection is

$$q' = \frac{2\pi k_{\text{eff}}(T_i - T_o)}{\ln(r_o/r_i)}$$

where k_{eff} may be obtained from Equations 9.59 and 9.60. With

$$L_c = \frac{2[\ln(r_o/r_i)]^{4/3}}{(r_i^{-3/5} + r_o^{-3/5})^{5/3}} = \frac{2[\ln(0.06\text{ m}/0.05\text{ m})]^{4/3}}{(0.05^{-3/5} + 0.06^{-3/5})^{5/3}\text{ m}^{-1}} = 0.00117\text{ m}$$

we find

$$Ra_c = \frac{g\beta(T_i - T_o)L_c^3}{\nu\alpha} = \frac{9.8\text{ m/s}^2 \times 0.00285\text{ K}^{-1} \times (120 - 35)°\text{C} \times (0.00117\text{ m})^3}{20.92 \times 10^{-6}\text{ m}^2/\text{s} \times 29.9 \times 10^{-6}\text{ m}^2/\text{s}}$$

$$= 171$$

The effective thermal conductivity is then

$$k_{\text{eff}} = 0.386k\left(\frac{Pr}{0.861 + Pr}\right)^{1/4} Ra_c^{1/4}$$

$$= 0.386 \times 0.030\text{ W/m}\cdot\text{K}\left(\frac{0.70}{0.861 + 0.70}\right)^{1/4}(171)^{1/4} = 0.0343\text{ W/m}\cdot\text{K}$$

and the heat loss is

$$q' = \frac{2\pi k_{\text{eff}}(T_i - T_o)}{\ln(r_o/r_i)} = \frac{2\pi(0.0343\text{ W/m}\cdot\text{K})}{\ln(0.06\text{ m}/0.05\text{ m})}(120 - 35)°\text{C} = 100\text{ W/m} \quad \triangleleft$$

2. With insulation in the space between the tube and the shield, heat loss is by conduction; comparing Equation 3.32 and Equation 9.58,

$$q'_{\text{ins}} = q'\frac{k_{\text{ins}}}{k_{\text{eff}}} = 100\text{ W/m}\,\frac{0.038\text{ W/m}\cdot\text{K}}{0.0343\text{ W/m}\cdot\text{K}} = 111\text{ W/m} \quad \triangleleft$$

Comments: Although there is slightly more heat loss by conduction through the insulation than by free convection across the air space, the total heat loss across the air space may exceed that through the insulation because of the effects of radiation. The heat loss due to radiation may be minimized by using a radiation shield of low emissivity, and the means for calculating the loss will be developed in Chapter 13.

9.9 *Combined Free and Forced Convection*

In dealing with forced convection (Chapters 6 through 8), we ignored the effects of free convection. This was, of course, an assumption; for, as we now know, free convection is likely when there is an unstable temperature gradient. Similarly, in the preceding sections of this chapter, we assumed that forced convection was negligible. It is now time to acknowledge that situations may arise for which free and forced convection effects are comparable, in which case it is inappropriate to neglect either process. In Section 9.3 we indicated that free convection is negligible if $(Gr_L/Re_L^2) \ll 1$ and that forced convection is negligible if $(Gr_L/Re_L^2) \gg 1$. Hence the *combined free* and *forced* (or *mixed*) *convection* regime is generally one for which $(Gr_L/Re_L^2) \approx 1$.

The effect of buoyancy on heat transfer in a forced flow is strongly influenced by the direction of the buoyancy force relative to that of the flow. Three special cases that have been studied extensively correspond to buoyancy-induced and forced motions having the same direction (*assisting* flow), opposite directions (*opposing* flow), and perpendicular directions (*transverse* flow). Upward and downward forced motions over a hot vertical plate are examples of assisting and opposing flows, respectively. Examples of transverse flow include horizontal motion over a hot cylinder, sphere, or horizontal plate. In assisting and transverse flows, buoyancy acts to enhance the rate of heat transfer associated with pure forced convection; in opposing flows, it acts to decrease this rate.

It has become common practice to correlate mixed convection heat transfer results for external and internal flows by an expression of the form

$$Nu^n = Nu_F^n \pm Nu_N^n \tag{9.64}$$

For the specific geometry of interest, the Nusselt numbers Nu_F and Nu_N are determined from existing correlations for pure forced and natural (free) convection, respectively. The plus sign on the right-hand side of Equation 9.64 applies for assisting and transverse flows, while the minus sign applies for opposing flow. The best correlation of data is often obtained for $n = 3$, although values of 7/2 and 4 may be better suited for transverse flows involving horizontal plates and cylinders (or spheres), respectively.

Equation 9.64 should be viewed as a first approximation, and any serious treatment of a mixed convection problem should be accompanied by an examination of the open literature. Mixed convection flows received considerable attention in the late 1970s to middle 1980s, and comprehensive literature reviews are available [40–43]. The flows are endowed with a variety of rich and unusual features that can complicate heat transfer predictions. For example, in a horizontal, parallel-plate channel, three-dimensional flows in the form of longitudinal vortices are induced by bottom heating, and the longitudinal variation of the Nusselt number is characterized by a decaying oscillation [44, 45]. Moreover, in channel flows, significant asymmetries may be associated with convection heat transfer at top and bottom surfaces [46]. Finally, we note that, although buoyancy effects can significantly enhance heat transfer for laminar forced convection flows, enhancement is typically negligible if the forced flow is turbulent [47].

9.10 *Summary*

We have considered convective flows that originate in part or exclusively from buoyancy forces, and we have introduced the dimensionless parameters needed to characterize such flows. You should be able to discern when free convection effects are important and to quantify the associated heat transfer rates. An assortment of empirical correlations has been provided for this purpose.

To test your understanding of related concepts, consider the following questions.

- What is an *extensive, quiescent* fluid?
- What conditions are required for a buoyancy-driven flow?
- How does the velocity profile in the free convection boundary layer on a heated vertical plate differ from the velocity profile in the boundary layer associated with forced flow over a parallel plate?
- What is the general form of the buoyancy term in the x-momentum equation for a free convection boundary layer? How may it be approximated if the flow is due to temperature variations? What is the name of the approximation?
- What is the physical interpretation of the *Grashof number*? What is the *Rayleigh number*? How does each parameter depend on the characteristic length?
- For a hot horizontal plate in quiescent air, do you expect heat transfer to be larger for the top or bottom surface? Why? For a cold horizontal plate in quiescent air, do you expect heat transfer to be larger for the top or bottom surface? Why?
- For free convection within a vertical parallel plate channel, what kind of force balance governs the flow rate in the channel?
- For a vertical channel with isothermal plates, what is the physical basis for existence of an optimum spacing?
- What is the nature of flow in a cavity whose vertical surfaces are heated and cooled? What is the nature of flow in an annular space between concentric cylindrical surfaces that are heated and cooled?
- What is meant by the term *mixed convection*? How can one determine if mixed convection effects should be considered in a heat transfer analysis? Under what conditions is heat transfer enhanced by mixed convection? Under what conditions is it reduced?

References

1. Jaluria, Y., *Natural Convection Heat and Mass Transfer*, Pergamon Press, New York, 1980.
2. Gebhart, B., Y. Jaluria, R. L. Mahajan, and B. Sammakia, *Buoyancy-Induced Flows and Transport,* Hemisphere Publishing, Washington, DC, 1988.
3. Ostrach, S., "An Analysis of Laminar Free Convection Flow and Heat Transfer About a Flat Plate Parallel to the Direction of the Generating Body Force," National Advisory Committee for Aeronautics, Report 1111, 1953.
4. LeFevre, E. J., "Laminar Free Convection from a Vertical Plane Surface," *Proc. Ninth Int. Congr. Appl. Mech.*, Brussels, Vol. 4, 168, 1956.
5. McAdams, W. H., *Heat Transmission,* 3rd ed., McGraw-Hill, New York, 1954, Chap. 7.
6. Warner, C. Y., and V. S. Arpaci, *Int. J. Heat Mass Transfer,* **11,** 397, 1968.
7. Bayley, F. J., *Proc. Inst. Mech. Eng.,* **169,** 361, 1955.
8. Churchill, S. W., and H. H. S. Chu, *Int. J. Heat Mass Transfer,* **18,** 1323, 1975.
9. Sparrow, E. M., and J. L. Gregg, *Trans. ASME,* **78,** 435, 1956.
10. Churchill, S. W., "Free Convection Around Immersed Bodies," in G. F. Hewitt, Exec. Ed., *Heat Exchanger Design Handbook*, Section 2.5.7, Begell House, New York, 2002.
11. Sparrow, E. M., and J. L. Gregg, *Trans. ASME,* **78,** 1823, 1956.
12. Cebeci, T., "Laminar-Free-Convective Heat Transfer from the Outer Surface of a Vertical Slender Circular Cylinder," *Proc. Fifth Int. Heat Transfer Conf.*, Paper NC1.4, pp. 15–19, 1974.
13. Minkowycz, W. J., and E. M. Sparrow, *J. Heat Transfer,* **96,** 178, 1974.
14. Vliet, G. C., *Trans. ASME,* **91C,** 511, 1969.
15. Fujii, T., and H. Imura, *Int. J. Heat Mass Transfer,* **15,** 755, 1972.
16. Azevedo, L. F. A., and E. M. Sparrow, *J. Heat Transfer,* **107,** 893, 1985.
17. Rich, B. R., *Trans. ASME,* **75,** 489, 1953.

18. Goldstein, R. J., E. M. Sparrow, and D. C. Jones, *Int. J. Heat Mass Transfer,* **16,** 1025, 1973.
19. Lloyd, J. R., and W. R. Moran, *J. Heat Transfer,* **96,** 443, 1974.
20. Radziemska, E., and W. M. Lewandowski, *Applied Energy*, **68**, 347, 2001.
21. Raithby, G. D., and K. G. T. Hollands, in W. M. Rohsenow, J. P. Hartnett, and Y. I. Cho, Eds., *Handbook of Heat Transfer Fundamentals,* Chap. 4, McGraw-Hill, New York, 1998.
22. Morgan, V. T., "The Overall Convective Heat Transfer from Smooth Circular Cylinders," in T. F. Irvine and J. P. Hartnett, Eds., *Advances in Heat Transfer,* Vol. 11, Academic Press, New York, 1975, pp. 199–264.
23. Churchill, S. W., and H. H. S. Chu, *Int. J. Heat Mass Transfer,* **18,** 1049, 1975.
24. Elenbaas, W., *Physica,* **9,** 1, 1942.
25. Bar-Cohen, A., and W. M. Rohsenow, *J. Heat Transfer,* **106,** 116, 1984.
26. Ostrach, S., "Natural Convection in Enclosures," in J. P. Hartnett and T. F. Irvine, Eds., *Advances in Heat Transfer,* Vol. 8, Academic Press, New York, 1972, pp. 161–227.
27. Catton, I., "Natural Convection in Enclosures," *Proc. 6th Int. Heat Transfer Conf.*, Toronto, Canada, 1978, Vol. 6, pp. 13–31.
28. Globe, S., and D. Dropkin, *J. Heat Transfer,* **81C,** 24, 1959.
29. Hollands, K. G. T., G. D. Raithby, and L. Konicek, *Int. J. Heat Mass Transfer,* **18,** 879, 1975.
30. Churchill, S. W., "Free Convection in Layers and Enclosures," in G. F. Hewitt, Exec. Ed., *Heat Exchanger Design Handbook*, Section 2.5.8, Begell House, New York, 2002.
31. MacGregor, R. K., and A. P. Emery, *J. Heat Transfer,* **91,** 391, 1969.
32. Ayyaswamy, P. S., and I. Catton, *J. Heat Transfer,* **95,** 543, 1973.
33. Catton, I., P. S. Ayyaswamy, and R. M. Clever, *Int. J. Heat Mass Transfer,* **17,** 173, 1974.
34. Clever, R. M., *J. Heat Transfer,* **95,** 407, 1973.
35. Arnold, J. N., I. Catton, and D. K. Edwards, "Experimental Investigation of Natural Convection in Inclined Rectangular Regions of Differing Aspect Ratios," ASME Paper 75-HT-62, 1975.
36. Buchberg, H., I. Catton, and D. K. Edwards, *J. Heat Transfer,* **98,** 182, 1976.
37. Hollands, K. G. T., S. E. Unny, G. D. Raithby, and L. Konicek, *J. Heat Transfer,* **98,** 189, 1976.
38. Raithby, G. D., and K. G. T. Hollands, "A General Method of Obtaining Approximate Solutions to Laminar and Turbulent Free Convection Problems," in T. F. Irvine and J. P. Hartnett, Eds., *Advances in Heat Transfer,* Vol. 11, Academic Press, New York, 1975, pp. 265–315.
39. Kuehn, T. H., and R. J. Goldstein, *Int. J. Heat Mass Transfer,* **19,** 1127, 1976.
40. Churchill, S. W., "Combined Free and Forced Convection around Immersed Bodies," in G. F. Hewitt, Exec. Ed., *Heat Exchanger Design Handbook*, Section 2.5.9, Begell House, New York, 2002.
41. Churchill, S. W., "Combined Free and Forced Convection in Channels," in G. F. Hewitt, Exec. Ed., *Heat Exchanger Design Handbook*, Section 2.5.10, Begell House, New York, 2002.
42. Chen, T. S., and B. F. Armaly, in S. Kakac, R. K. Shah, and W. Aung, Eds., *Handbook of Single-Phase Convective Heat Transfer,* Chap. 14, Wiley-Interscience, New York, 1987.
43. Aung, W., in S. Kakac, R. K. Shah, and W. Aung, Eds., *Handbook of Single-Phase Convective Heat Transfer,* Chap. 15, Wiley-Interscience, New York, 1987.
44. Incropera, F. P., A. J. Knox, and J. R. Maughan, *J. Heat Transfer,* **109,** 434, 1987.
45. Maughan, J. R., and F. P. Incropera, *Int. J. Heat Mass Transfer,* **30,** 1307, 1987.
46. Osborne, D. G., and F. P. Incropera, *Int. J. Heat Mass Transfer,* **28,** 207, 1985.
47. Osborne, D. G., and F. P. Incropera, *Int. J. Heat Mass Transfer,* **28,** 1337, 1985.

Problems

Properties and General Considerations

9.1 The one-dimensional plane wall of Figure 3.1 is of thickness $L = 75$ mm and thermal conductivity $k = 5\ \text{W/m}\cdot\text{K}$. The fluid temperatures are $T_{\infty,1} = 200°\text{C}$ and $T_{\infty,2} = 100°\text{C}$, respectively. Using the minimum and maximum typical values of the convection heat transfer coefficients listed in Table 1.1, determine the minimum and maximum steady-state heat fluxes through the wall for (i) free convection in gases, (ii) free convection in liquids, (iii) forced convection in gases, (iv) forced convection in liquids, and (v) convection with phase change.

9.2 Using the values of density for water in Table A.6, calculate the volumetric thermal expansion coefficient at 300 K from its definition, Equation 9.4, and compare your result with the tabulated value.

9.3 Consider an object of characteristic length 0.01 m and a situation for which the temperature difference is 30°C. Evaluating thermophysical properties at the prescribed conditions, determine the Rayleigh number for the following fluids: air (1 atm, 400 K), helium (1 atm, 400 K), glycerin (285 K), and water (310 K).

9.4 To assess the efficacy of different liquids for cooling by natural convection, it is convenient to introduce a *figure of merit*, F_N, which combines the influence of all pertinent fluid properties on the convection coefficient. If the Nusselt number is governed by an expression of the form, $Nu_L \sim Ra^n$, obtain the corresponding relationship between F_N and the fluid properties. For a representative value of $n = 0.33$, calculate values of F_N for air ($k = 0.026$ W/m·K, $\beta = 0.0035\ \text{K}^{-1}$, $\nu = 1.5 \times 10^{-5}\ \text{m}^2/\text{s}$, $Pr = 0.70$), water ($k = 0.600$ W/m·K, $\beta = 2.7 \times 10^{-4}\ \text{K}^{-1}$, $\nu = 10^{-6}\ \text{m}^2/\text{s}$, $Pr = 5.0$), and a dielectric liquid ($k = 0.064$ W/m·K, $\beta = 0.0014\ \text{K}^{-1}$, $\nu = 10^{-6}\ \text{m}^2/\text{s}$, $Pr = 25$). What fluid is the most effective cooling agent?

9.5 In many cases, we are concerned with free convection involving gases that are contained within sealed enclosures. Consider air at 27°C and pressures of 1, 10, and 100 bars. Determine the *figure of merit* described in Problem 9.4 for each of these three pressures. Which air pressure will provide the most effective cooling? *Hint:* See Problem 6.22.

Vertical Plates

9.6 The heat transfer rate due to free convection from a vertical surface, 1 m high and 0.6 m wide, to quiescent air that is 20 K colder than the surface is known. What is the ratio of the heat transfer rate for that situation to the rate corresponding to a vertical surface, 0.6 m high and 1 m wide, when the quiescent air is 20 K warmer than the surface? Neglect heat transfer by radiation and any influence of temperature on the relevant thermophysical properties of air.

9.7 Consider a large vertical plate with a uniform surface temperature of 130°C suspended in quiescent air at 25°C and atmospheric pressure.

(a) Estimate the boundary layer thickness at a location 0.25 m measured from the lower edge.

(b) What is the maximum velocity in the boundary layer at this location and at what position in the boundary layer does the maximum occur?

(c) Using the similarity solution result, Equation 9.19, determine the heat transfer coefficient 0.25 m from the lower edge.

(d) At what location on the plate measured from the lower edge will the boundary layer become turbulent?

9.8 For laminar free convection flow on a vertical plate, the recommended values of C and n for use in the correlation of Equation 9.24 are 0.59 and 1/4, respectively. Derive the values of C from the similarity solution, Equation 9.21, for $Pr = 0.01$, 1, 10, and 100.

9.9 Consider an array of vertical rectangular fins, which is to be used to cool an electronic device mounted in quiescent, atmospheric air at $T_\infty = 27°\text{C}$. Each fin has $L = 20$ mm and $H = 150$ mm and operates at an approximately uniform temperature of $T_s = 77°\text{C}$.

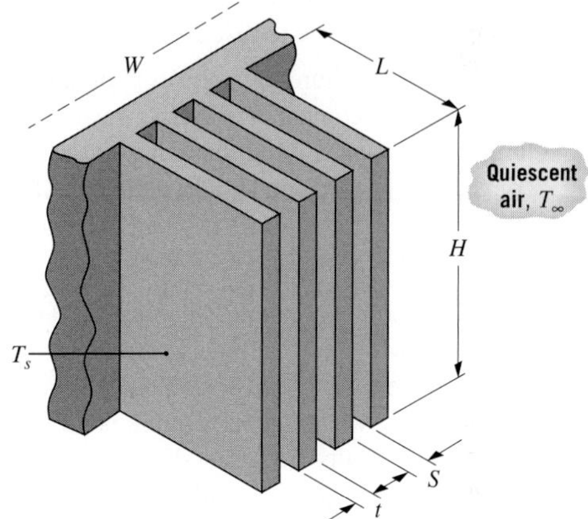

(a) Viewing each fin surface as a vertical plate in an infinite, quiescent medium, briefly describe why there exists an optimum fin spacing S. Using Figure 9.4, estimate the optimum value of S for the prescribed conditions.

(b) For the optimum value of S and a fin thickness of $t = 1.5$ mm, estimate the rate of heat transfer from the fins for an array of width $W = 355$ mm.

9.10 A number of thin plates are to be cooled by vertically suspending them in a water bath at a temperature of 20°C. If the plates are initially at 54°C and are 0.15 m long, what minimum spacing would prevent interference between their free convection boundary layers?

9.11 Beginning with the free convection correlation of the form given by Equation 9.24, show that for air at atmospheric pressure and a film temperature of 400 K, the average heat transfer coefficient for a vertical plate can be expressed as

$$\bar{h}_L = 1.40\left(\frac{\Delta T}{L}\right)^{1/4} \qquad 10^4 < Ra_L < 10^9$$

$$\bar{h}_L = 0.98\Delta T^{1/3} \qquad 10^9 < Ra_L < 10^{13}$$

9.12 A solid object is to be cooled by submerging it in a quiescent fluid, and the associated free convection coefficient is given by $\bar{h} = C\Delta T^{1/4}$, where C is a constant and $\Delta T = T - T_\infty$.

(a) Using the results of Section 5.3.3, obtain an expression for the time required for the object to cool from an initial temperature T_i to a final temperature T_f.

(b) Consider a highly polished, 150-mm square aluminum alloy (2024) plate of 5-mm thickness, initially at 225°C, and suspended vertically in ambient air at 25°C. Using the appropriate approximate correlation from Problem 9.11, determine the time required for the plate to reach 80°C.

(c) Plot the temperature–time history obtained from part (b) and compare with the results from a lumped capacitance analysis using a constant free convection coefficient, $\bar{h}_o$. Evaluate $\bar{h}_o$ from an appropriate correlation based on an average surface temperature of $\bar{T} = (T_i + T_f)/2$.

9.13 A square aluminum plate 5 mm thick and 200 mm on a side is heated while vertically suspended in quiescent air at 40°C. Determine the average heat transfer coefficient for the plate when its temperature is 15°C by two methods: using results from the similarity solution to the boundary layer equations, and using results from an empirical correlation.

9.14 An aluminum alloy (2024) plate, heated to a uniform temperature of 227°C, is allowed to cool while vertically suspended in a room where the ambient air and surroundings are at 27°C. The plate is 0.3 m square with a thickness of 15 mm and an emissivity of 0.25.

(a) Develop an expression for the time rate of change of the plate temperature, assuming the temperature to be uniform at any time.

(b) Determine the initial rate of cooling (K/s) when the plate temperature is 227°C.

(c) Justify the uniform plate temperature assumption.

(d) Compute and plot the temperature history of the plate from $t = 0$ to the time required to reach a temperature of 30°C. Compute and plot the corresponding variations in the convection and radiation heat transfer rates.

9.15 The plate described in Problem 9.14 has been used in an experiment to determine the free convection heat transfer coefficient. At an instant of time when the plate temperature was 127°C, the time rate of change of this temperature was observed to be −0.0465 K/s. What is the corresponding free convection heat transfer coefficient? Compare this result with an estimate based on a standard empirical correlation.

9.16 Determine the average convection heat transfer coefficient for the 2.5-m-high vertical walls of a home having respective interior air and wall surface temperatures of (a) 20 and 10°C and (b) 27 and 37°C.

9.17 Consider a vertical plate of dimension 0.25 m × 0.50 m that is at $T_s = 100°C$ in a quiescent environment at $T_\infty = 20°C$. In the interest of minimizing heat transfer from the plate, which orientation, (A) or (B), is preferred? What is the convection heat transfer from the front surface of the plate when it is in the preferred orientation?

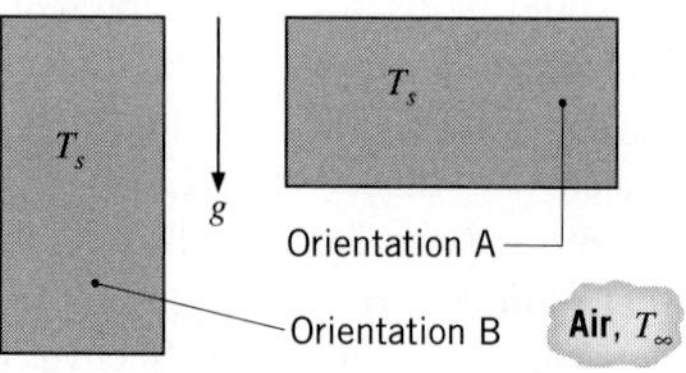

9.18 During a winter day, the window of a patio door with a height of 1.8 m and width of 1.0 m shows a frost line near its base. The room wall and air temperatures are 15°C.

(a) Explain why the window would show a frost layer at the base rather than at the top.

(b) Estimate the heat loss through the window due to free convection and radiation. Assume the window has a uniform temperature of 0°C and the emissivity of the glass surface is 0.94. If the room has electric baseboard heating, estimate the corresponding daily cost of the window heat loss for a utility rate of 0.18 \$/kW·h.

9.19 A vertical, thin pane of window glass that is 1 m on a side separates quiescent room air at $T_{\infty,i} = 20°C$ from quiescent ambient air at $T_{\infty,o} = -20°C$. The walls of the room and the external surroundings (landscape, buildings, etc.) are also at $T_{sur,i} = 20°C$ and $T_{sur,o} = -20°C$, respectively.

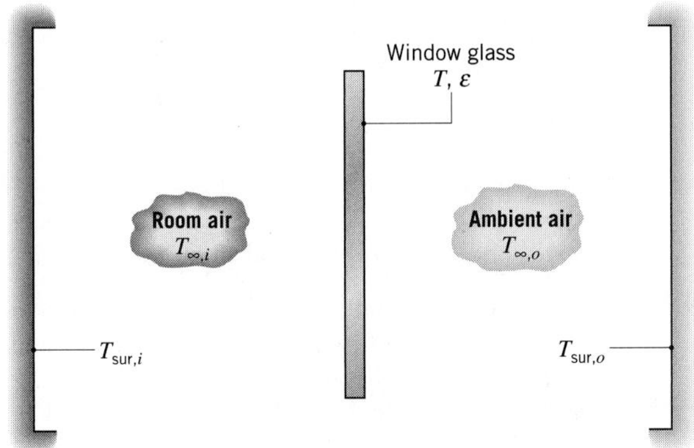

If the glass has an emissivity of $\varepsilon = 1$, what is its temperature T? What is the rate of heat loss through the glass?

9.20 Consider the conditions of Problem 9.19, but now allow for a difference between the inner and outer surface temperatures, $T_{s,i}$ and $T_{s,o}$, of the window. For a glass thickness and thermal conductivity of $t_g = 10$ mm and $k_g = 1.4$ W/m·K, respectively, evaluate $T_{s,i}$ and $T_{s,o}$. What is the heat loss through the window?

9.21 A household oven door of 0.5-m height and 0.7-m width reaches an average surface temperature of 32°C during operation. Estimate the heat loss to the room with ambient air at 22°C. If the door has an emissivity of 1.0 and the surroundings are also at 22°C, comment on the heat loss by free convection relative to that by radiation.

9.22 Consider a vertical, single-pane window of equivalent width and height ($W = L = 1$ m). The interior surface is exposed to the air and walls of a room, which are each at 18°C. Under cold ambient conditions for which a thin layer of frost has formed on the inner surface, what is the heat loss through the window? How would your analysis be affected by a frost layer whose thickness is not negligible? During incipience of frost formation, where would you expect the frost to begin to develop on the window? The frost may be assumed to have an emissivity of $\varepsilon = 0.90$.

9.23 Consider laminar flow about a vertical isothermal plate of length L, providing an average heat transfer coefficient of $\bar{h}_L$. If the plate is divided into N smaller plates, each of length, $L_N = L/N$, determine an expression for the ratio of the heat transfer coefficient averaged over the N plates to the heat transfer coefficient averaged over the single plate, $\bar{h}_{L,N}/\bar{h}_{L,1}$.

9.24 Consider the conveyor system described in Problem 7.24, but under conditions for which the conveyor is not moving and the air is quiescent. Radiation effects and interactions between boundary layers on adjoining surfaces may be neglected.

(a) For the prescribed plate dimensions and initial temperature, as well as the prescribed air temperature, what is the initial rate of heat transfer from one of the plates?

(b) How long does it take for a plate to cool from 300°C to 100°C? Comment on the assumption of negligible radiation.

9.25 A thin-walled container with a hot process fluid at 50°C is placed in a quiescent, cold water bath at 10°C. Heat transfer at the inner and outer surfaces of the container may be approximated by free convection from a vertical plate.

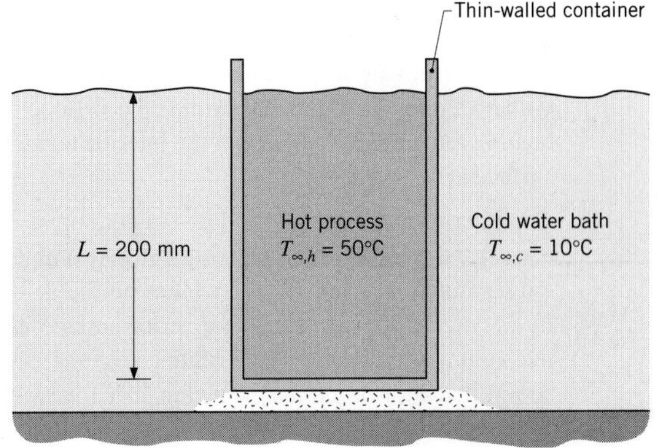

(a) Determine the overall heat transfer coefficient between the hot process fluid and the cold water bath. Assume the properties of the hot process fluid are those of water.

(b) Generate a plot of the overall heat transfer coefficient as a function of the hot process fluid temperature $T_{\infty,h}$ for the range 20 to 60°C, with all other conditions remaining the same.

9.26 Consider an experiment to investigate the transition to turbulent flow in a free convection boundary layer that develops along a vertical plate suspended in a large room. The plate is constructed of a thin heater that is sandwiched between two aluminum plates and may be assumed to be isothermal. The heated plate is 1 m high and 2 m wide. The quiescent air and the surroundings are both at 25°C.

(a) The exposed surfaces of the aluminum plate are painted with a very thin coating of high emissivity ($\varepsilon = 0.95$) paint. Determine the electrical power that must be supplied to the heater to sustain the plate at a temperature of $T_s = 35°C$. How much of the plate is exposed to turbulent conditions in the free convection boundary layer?

(b) The experimentalist speculates that the roughness of the paint is affecting the transition to turbulence in the boundary layer and decides to remove the paint and polish the aluminum surface ($\varepsilon = 0.05$). If the same power is supplied to the plate as in part (a), what is the steady-state plate temperature? How much of the plate is exposed to turbulent conditions in the free convection boundary layer?

9.27 The vertical rear window of an automobile is of thickness $L = 8$ mm and height $H = 0.5$ m and contains fine-meshed heating wires that can induce nearly uniform volumetric heating, $\dot{q}$(W/m^3).

(a) Consider steady-state conditions for which the interior surface of the window is exposed to quiescent air at 10°C, while the exterior surface is exposed to air at −10°C moving in parallel flow over the surface with a velocity of 20 m/s. Determine the volumetric heating rate needed to maintain the interior window surface at $T_{s,i} = 15°C$.

(b) The interior and exterior window temperatures, $T_{s,i}$ and $T_{s,o}$, depend on the compartment and ambient temperatures, $T_{\infty,i}$ and $T_{\infty,o}$, as well as on the velocity u_∞ of air flowing over the exterior surface and the volumetric heating rate $\dot{q}$. Subject to the constraint that $T_{s,i}$ is to be maintained at 15°C, we wish to develop guidelines for varying the heating rate in response to changes in $T_{\infty,i}$, $T_{\infty,o}$, and/or u_∞. If $T_{\infty,i}$ is maintained at 10°C, how will $\dot{q}$ and $T_{s,o}$ vary with $T_{\infty,o}$ for $-25 \le T_{\infty,o} \le 5°C$ and $u_\infty = 10$, 20, and 30 m/s? If a constant vehicle speed is maintained, such that $u_\infty = 30$ m/s, how will $\dot{q}$ and $T_{s,o}$ vary with $T_{\infty,i}$ for $5 \le T_{\infty,i} \le 20°C$ and $T_{\infty,o} = -25$, −10, and 5°C?

9.28 Determine the maximum allowable uniform heat flux that may be imposed at a wall heating panel 1 m high if the maximum temperature is not to exceed 37°C when the ambient air temperature is 25°C.

9.29 The components of a vertical circuit board, 150 mm on a side, dissipate 5 W. The back surface is well insulated and the front surface is exposed to quiescent air at 27°C.

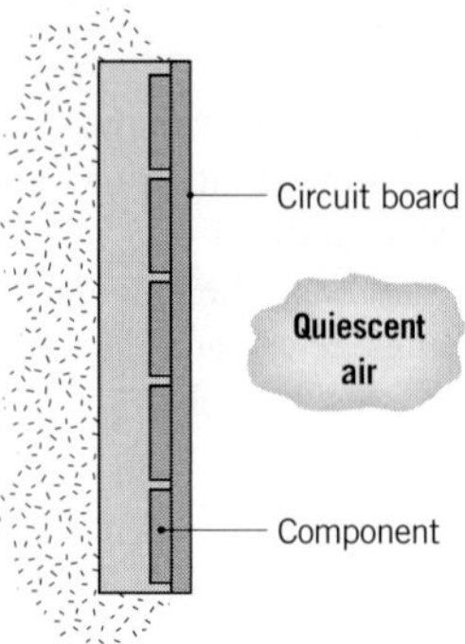

Assuming a uniform surface heat flux, what is the maximum temperature of the board? What is the temperature of the board for an isothermal surface condition?

9.30 Circuit boards are mounted to interior vertical surfaces of a rectangular duct of height $H = 400$ mm and length $L = 800$ mm. Although the boards are cooled by forced convection heat transfer to air flowing through the duct, not all of the heat dissipated by the electronic components is transferred to the flow. Some of the heat is instead transferred by conduction to the vertical walls of the duct and then by natural convection and radiation to the ambient (atmospheric) air and surroundings, which are at equivalent temperatures of $T_\infty = T_{sur} = 20°C$. The walls are metallic and, to a first approximation, may be assumed to be isothermal at a temperature T_s.

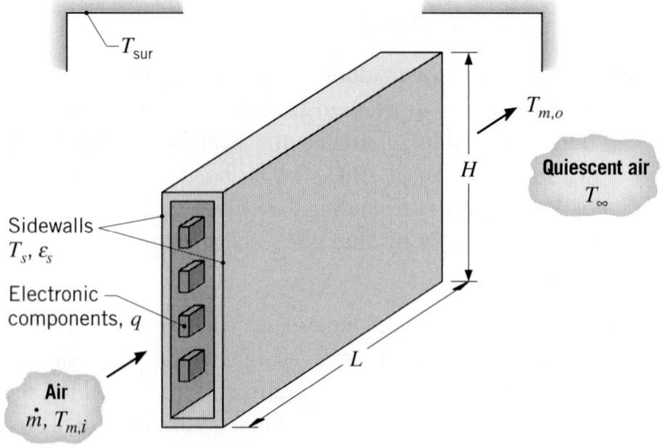

(a) Consider conditions for which the electronic components dissipate 200 W and air enters the duct at a flow rate of $\dot{m} = 0.015$ kg/s and a temperature of $T_{m,i} = 20°C$. If the emissivity of the side walls is $\varepsilon_s = 0.15$ and the outlet temperature of the air is $T_{m,o} = 30°C$, what is the surface temperature T_s?

(b) To reduce the temperature of the electronic components, it is desirable to enhance heat transfer from the side walls. Assuming no change in the airflow conditions, what is the effect on T_s of applying a high emissivity coating ($\varepsilon_s = 0.90$) to the side walls?

(c) If there is a loss of airflow while power continues to be dissipated, what are the resulting values of T_s for $\varepsilon_s = 0.15$ and $\varepsilon_s = 0.90$?

9.31 A refrigerator door has a height and width of $H = 1$ m and $W = 0.65$ m, respectively, and is situated in a large room for which the air and walls are at $T_\infty = T_{sur} = 25°C$. The door consists of a layer of polystyrene insulation ($k = 0.03$ W/m·K) sandwiched between thin sheets of steel ($\varepsilon = 0.6$) and polypropylene. Under normal operating conditions, the inner surface of the door is maintained at a fixed temperature of $T_{s,i} = 5°C$.

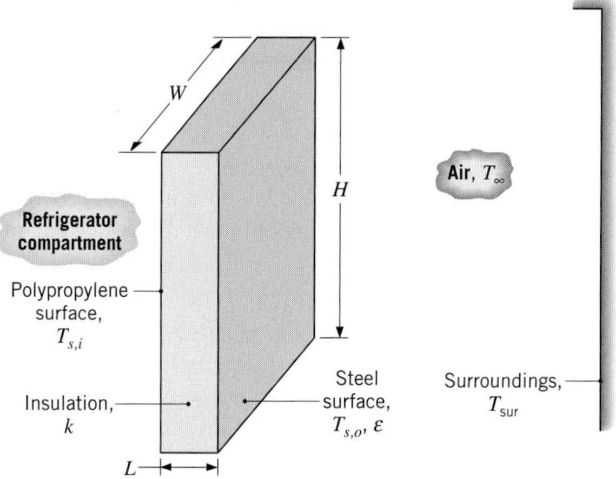

(a) Estimate the heat gain through the door for the worst case condition corresponding to no insulation ($L = 0$).

(b) Compute and plot the heat gain and the outer surface temperature $T_{s,o}$ as a function of insulation thickness for $0 \le L \le 25$ mm.

9.32 Air at 3 atm and 100°C is discharged from a compressor into a vertical receiver of 2.5-m height and 0.75-m diameter. Assume that the receiver wall has negligible thermal resistance, is at a uniform temperature, and that heat transfer at its inner and outer surfaces is by free convection from a vertical plate. Neglect radiation exchange and any losses from the top.

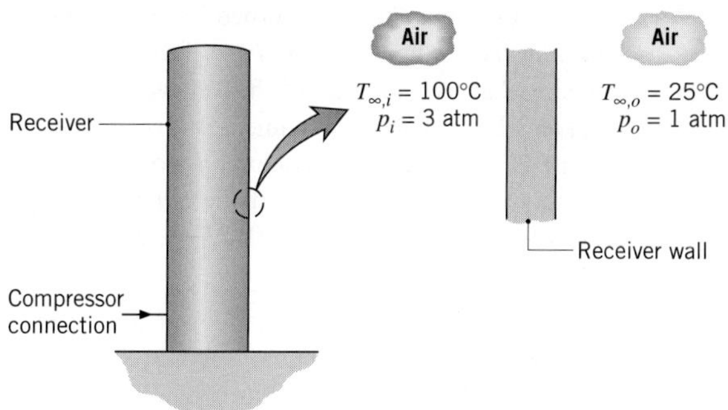

(a) Estimate the receiver wall temperature and the heat transfer to the ambient air at 25°C. To facilitate use of the free convection correlations with appropriate film temperatures, assume that the receiver wall temperature is 60°C.

(b) Were the assumed film temperatures of part (a) reasonable? If not, use an iteration procedure to find consistent values.

(c) Now consider two features of the receiver neglected in the previous analysis: (i) radiation exchange from the exterior surface of emissivity 0.85 to large surroundings, also at 25°C; and (ii) the thermal resistance of a 20-mm-thick wall with a thermal conductivity of 0.25 W/m·K. Represent the system by a thermal circuit and estimate the wall temperatures and the heat transfer rate.

9.33 In the *central receiver* concept of a solar power plant, many heliostats at ground level are used to direct a concentrated solar flux q''_s to the receiver, which is positioned at the top of a tower. However, even with absorption of all the solar flux by the outer surface of the receiver, losses due to free convection and radiation reduce the collection efficiency below the maximum possible value of 100%. Consider a cylindrical receiver of diameter $D = 7$ m, length $L = 12$ m, and emissivity $\varepsilon = 0.20$.

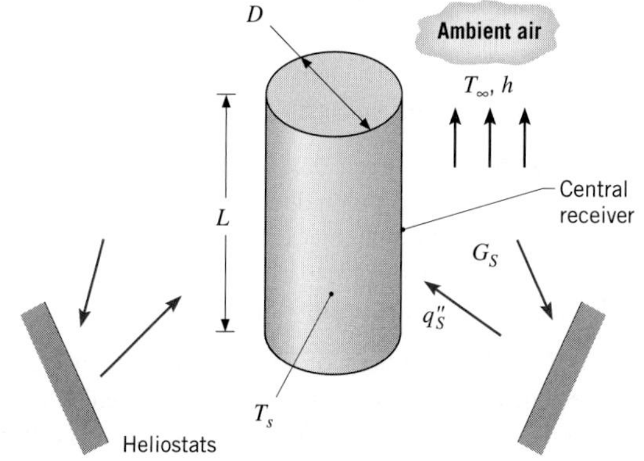

(a) If all of the solar flux is absorbed by the receiver and a surface temperature of $T_s = 800$ K is maintained, what is the rate of heat loss from the receiver? The ambient air is quiescent at a temperature of $T_\infty = 300$ K, and irradiation from the surroundings may be neglected. If the corresponding value of the solar flux is $q_S'' = 10^5$ W/m², what is the collector efficiency?

(b) The surface temperature of the receiver is affected by design and operating conditions within the power plant. Over the range from 600 to 1000 K, plot the variation of the convection, radiation, and total heat rates as a function of T_s. For a fixed value of $q_S'' = 10^5$ W/m², plot the corresponding variation of the receiver efficiency.

Horizontal and Inclined Plates

9.34 Consider the transformer of Problem 8.103, whose lateral surface is being maintained at 47°C by a forced convection coolant line removing 1000 W. It is desired to explore cooling of the transformer by free convection and radiation, assuming the surface to have an emissivity of 0.80.

(a) Determine how much power could be removed by free convection and radiation from the lateral *and* the upper horizontal surfaces when the ambient temperature and the surroundings are at 27°C.

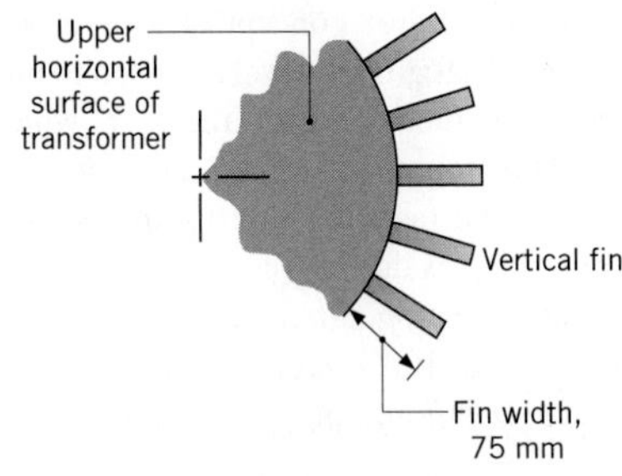

(b) Vertical fins, 5 mm thick, 75 mm wide, and 500 mm long, can easily be welded to the lateral surface. What is the heat removal rate by free convection if 30 such fins are attached?

9.35 Airflow through a long, 0.2-m-square air conditioning duct maintains the outer duct surface temperature at 10°C. If the horizontal duct is uninsulated and exposed to air at 35°C in the crawlspace beneath a home, what is the heat gain per unit length of the duct?

9.36 Consider the conditions of Example 9.3, including the effect of adding insulation of thickness t and thermal conductivity $k = 0.035$ W/m·K to the duct. We wish to now include the effect of radiation on the outer surface temperatures and the total heat loss per unit length of duct.

(a) If $T_{s,1} = 45$°C, $t = 25$ mm, $\varepsilon = 1$, and $T_{sur} = 288$ K, what are the temperatures of the side, top, and bottom surfaces? What are the corresponding heat losses per unit length of duct?

(b) For the top surface, compute and plot $T_{s,2}$ and q' as a function of insulation thickness for $0 \le t \le 50$ mm. The exposed duct surface ($t = 0$) may also be assumed to have an emissivity of $\varepsilon = 1$.

9.37 An electrical heater in the form of a horizontal disk of 400-mm diameter is used to heat the bottom of a tank filled with engine oil at a temperature of 5°C. Calculate the power required to maintain the heater surface temperature at 70°C.

9.38 Consider a horizontal 6-mm-thick, 100-mm-long straight fin fabricated from plain carbon steel ($k = 57$ W/m·K, $\varepsilon = 0.5$). The base of the fin is maintained at 150°C, while the quiescent ambient air and the surroundings are at 25°C. Assume the fin tip is adiabatic.

(a) Estimate the fin heat rate per unit width, q_f'. Use an average fin surface temperature of 125°C to estimate the free convection coefficient and the linearized radiation coefficient. How sensitive is your estimate to the choice of the average fin surface temperature?

(b) Generate a plot of q_f' as a function of the fin emissivity for $0.05 \le \varepsilon \le 0.95$. On the same coordinates, show the fraction of the total heat rate due to radiation exchange.

9.39 The thermal conductivity and surface emissivity of a material may be determined by heating its bottom surface and exposing its top surface to quiescent air and large surroundings of equivalent temperatures, $T_\infty = T_{sur} = 25$°C. The remaining surfaces of the sample/heater are well insulated.

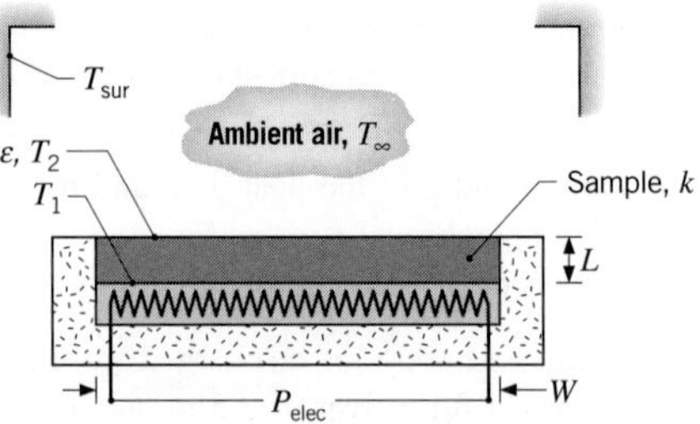

Consider a sample of thickness $L = 25$ mm and a square planform of width $W = 250$ mm. In an experiment performed under steady-state conditions, temperature measurements made at the lower and upper surface of the sample yield values of $T_1 = 150$°C and $T_2 = 100$°C, respectively, for a power input of $P_{\text{elec}} = 70$ W. What are the thermal conductivity and emissivity of the sample?

9.40 Convection heat transfer coefficients for a hot horizontal surface facing upward may be determined by a gage whose specific features depend on whether the temperature of the surroundings is known. For configuration A, a copper disk, which is electrically heated from below, is encased in an insulating material such that all of the heat is transferred by convection and radiation from the top surface. If the surface emissivity and the temperatures of the air and surroundings are known, the convection coefficient may be determined from measurement of the electrical power and the surface temperature of the disk. Configuration B is used in situations for which the temperature of the surroundings is not known. A thin, insulating strip separates semicircular disks with independent electrical heaters and different emissivities. If the emissivities and temperature of the air are known, the convection coefficient may be determined from measurement of the electrical power supplied to each of the disks in order to maintain them at a common temperature.

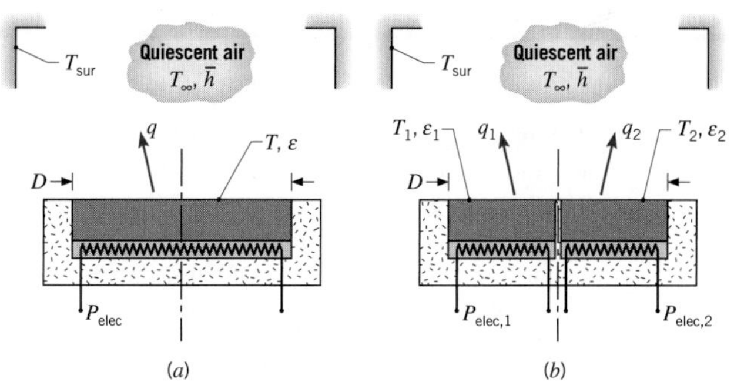

(a) In an application of configuration A to a disk of diameter $D = 160$ mm and emissivity $\varepsilon = 0.8$, values of $P_{\text{elec}} = 10.8$ W and $T = 67°\text{C}$ are measured for $T_\infty = T_{\text{sur}} = 27°\text{C}$. What is the corresponding value of the average convection coefficient? How does it compare with predictions based on a standard correlation?

(b) Now consider an application of configuration *B* for which $T_\infty = 17°\text{C}$ and T_{sur} is unknown. With $D = 160$ mm, $\varepsilon_1 = 0.8$, and $\varepsilon_2 = 0.1$, values of $P_{\text{elect},1} = 9.70$ W and $P_{\text{elec},2} = 5.67$ W are measured when $T_1 = T_2 = 77°\text{C}$. Determine the corresponding values of the convection coefficient and the temperature of the surroundings. How does the convection coefficient compare with predictions by an appropriate correlation?

9.41 A circular grill of diameter 0.25 m and emissivity 0.9 is maintained at a constant surface temperature of 130°C. What electrical power is required when the room air and surroundings are at 24°C?

9.42 Many laptop computers are equipped with thermal management systems that involve liquid cooling of the central processing unit (CPU), transfer of the heated liquid to the back of the laptop screen assembly, and dissipation of heat from the back of the screen assembly by way of a flat, isothermal heat spreader. The cooled liquid is recirculated to the CPU and the process continues. Consider an aluminum heat spreader that is of width $w = 275$ mm and height $L = 175$ mm. The screen assembly is oriented at an angle $\theta = 30°$ from the vertical direction, and the heat spreader is attached to the $t =$ 3-mm-thick plastic housing with a thermally conducting adhesive. The plastic housing has a thermal conductivity of $k = 0.21$ W/m·K and emissivity of $\varepsilon = 0.85$. The contact resistance associated with the heat spreader-housing interface is $R''_{t,c} = 2.0 \times 10^{-4}\ \text{m}^2\cdot\text{K/W}$. If the CPU generates, on average, 15 W of thermal energy, what is the temperature of the heat spreader when $T_\infty = T_{\text{sur}} = 23°\text{C}$? Which thermal resistance (contact, conduction, radiation, or free convection) is the largest?

9.43 Consider the roof of the refrigerated truck compartment described in Problem 7.20, but under conditions for which the truck is parked ($V = 0$). All other conditions remain unchanged. For $\alpha_S = \varepsilon = 0.5$, determine the outer surface temperature, $T_{s,o}$, and the heat load imposed on the refrigeration system. *Hint*: Assume $T_{s,o} > T_\infty$ and $Ra_L > 10^7$.

9.44 The 4 m × 4 m horizontal roof of an uninsulated aluminum melting furnace is comprised of a 0.08-m-thick fireclay brick refractory covered by a 5-mm-thick steel (AISI 1010) plate. The refractory surface exposed to the furnace gases is maintained at 1700 K during operation, while the outer surface of the steel is exposed to the air and walls of a large room at 25°C. The emissivity of the steel is $\varepsilon = 0.3$.

(a) What is the rate of heat loss from the roof?

(b) If a 20-mm-thick layer of alumina-silica insulation (64 kg/m^3) is placed between the refractory and the steel, what is the new rate of heat loss from the roof? What is the temperature at the inner surface of the insulation?

(c) One of the process engineers claims that the temperature at the inner surface of the insulation found in part (b) is too high for safe, long-term operation. What thickness of fireclay brick would reduce this temperature to 1350 K?

9.45 At the end of its manufacturing process, a silicon wafer of diameter $D = 150$ mm, thickness $\delta = 1$ mm, and emissivity $\varepsilon = 0.65$ is at an initial temperature of $T_i = 325°C$ and is allowed to cool in quiescent, ambient air and large surroundings for which $T_\infty = T_{sur} = 25°C$.

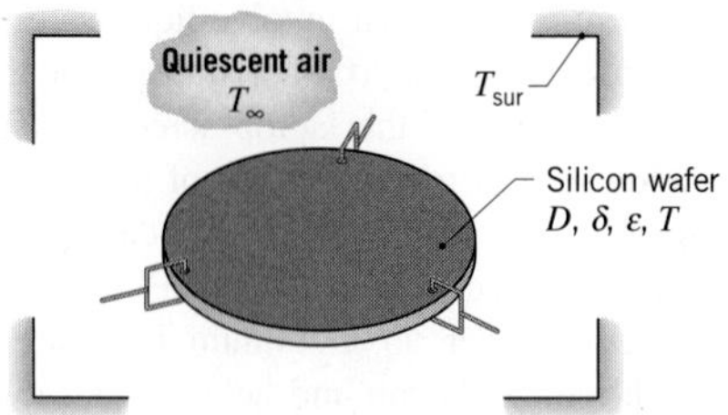

(a) What is the initial rate of cooling?

(b) How long does it take for the wafer to reach a temperature of 50°C? Comment on how the relative effects of convection and radiation vary with time during the cooling process.

9.46 A 200-mm-square, 10-mm-thick tile has the thermophysical properties of Pyrex ($\varepsilon = 0.80$) and emerges from a curing process at an initial temperature of $T_i = 140°C$. The backside of the tile is insulated while the upper surface is exposed to ambient air and surroundings at 25°C.

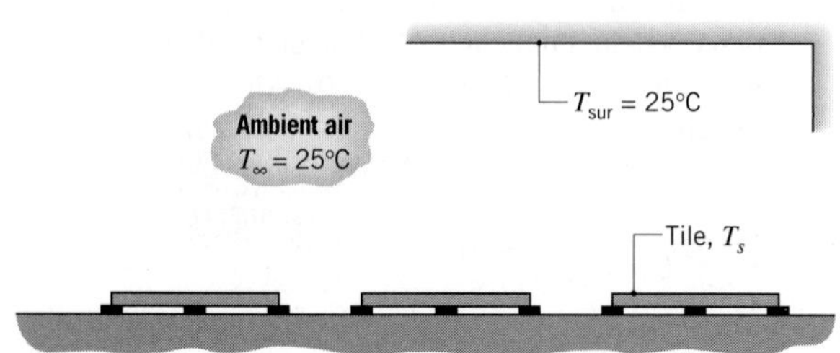

(a) Estimate the time required for the tile to cool to a final, safe-to-touch temperature of $T_f = 40°C$. Use an average tile surface temperature of $\overline{T} = (T_i + T_f)/2$ to estimate the average free convection coefficient and the linearized radiation coefficient. How sensitive is your estimate to the assumed value for $\overline{T}$?

(b) Estimate the required cooling time if ambient air is blown in parallel flow over the tile with a velocity of 10 m/s.

9.47 Integrated circuit (IC) boards are stacked within a duct and dissipate a total of 500 W. The duct has a square cross section with $w = H = 150$ mm and a length of 0.5 m. Air flows into the duct at 25°C and 1.2 m^3/min, and the convection coefficient between the air and the inner surfaces of the duct is $\overline{h}_i = 50$ W/m$^2 \cdot$K. The entire outer surface of the duct, which is anodized with an emissivity of 0.5, is exposed to ambient air and large surroundings at 25°C.

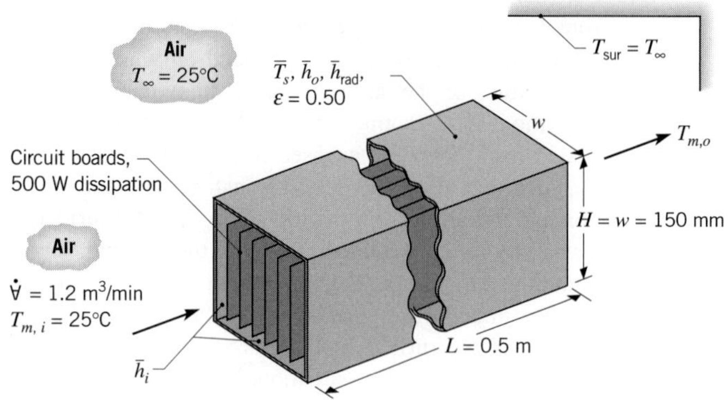

Your assignment is to develop a model to estimate the outlet temperature of the air, $T_{m,o}$, and the average surface temperature of the duct, $\overline{T}_s$.

(a) Assuming a surface temperature of 37°C, estimate the average free convection coefficient, $\overline{h}_o$, for the outer surface of the duct.

(b) Assuming a surface temperature of 37°C, estimate the average linearized radiation coefficient, $\overline{h}_{rad}$, for the outer surface of the duct.

(c) Perform an energy balance on the duct by considering the dissipation of electrical power in the ICs, the rate of change in the energy of air flowing through the duct, and the rate of heat transfer from the air in the duct to the surroundings. Express the last process in terms of thermal resistances between the mean temperature, $\overline{T}_m$, of the air in the duct and the temperature of the ambient air and the surroundings.

(d) Substitute numerical values into the expression of part (c) and calculate the air outlet temperature, $T_{m,o}$. Estimate the corresponding value of $\overline{T}_s$. Comment on your results and the assumptions inherent in your model.

9.48 A highly polished aluminum plate of length 0.5 m and width 0.2 m is subjected to an airstream at a temperature of 23°C and a velocity of 10 m/s. Because of upstream conditions, the flow is turbulent over the entire length of the plate. A series of segmented, independently controlled heaters is attached to the lower side of the plate to maintain approximately isothermal conditions over the entire plate. The electrical heater covering the section between the positions $x_1 = 0.2$ m and $x_2 = 0.3$ m is shown in the schematic.

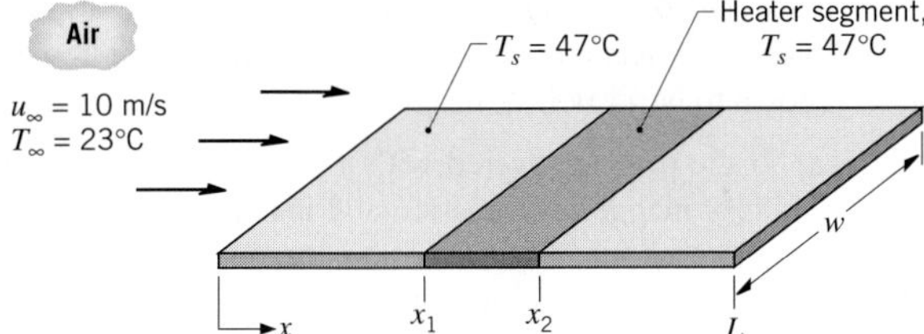

(a) Estimate the electrical power that must be supplied to the designated heater segment to maintain the plate surface temperature at $T_s = 47°C$.

(b) If the blower that maintains the airstream velocity over the plate malfunctions, but the power to the heaters remains constant, estimate the surface temperature of the designated segment. Assume that the ambient air is extensive, quiescent, and at 23°C.

9.49 The average free convection coefficient for the exterior surfaces of a long, horizontal rectangular duct exposed to a quiescent fluid can be estimated from the Hahn–Didion (H–D) correlation [ASHRAE Proceedings, Part 1, pp. 262–67, 1972]

$$\overline{Nu}_P = 0.55\, Ra_P^{1/4}\left(\frac{H}{P}\right)^{1/8} \qquad Ra_P \leq 10^7$$

where the characteristic length is the half-perimeter, $P = (w + H)$, and w and H are the horizontal width and vertical height, respectively, of the duct. The thermophysical properties are evaluated at the film temperature.

(a) Consider a horizontal 0.15-m-square duct with a surface temperature of 35°C in ambient air at 15°C. Calculate the average convection coefficient and the heat rate per unit length using the H–D correlation.

(b) Calculate the average convection coefficient and the heat rate per unit length considering the duct as formed by vertical plates (sides) and horizontal plates (top and bottom). Do you expect this estimate to be lower or higher than that obtained with the H–D correlation? Explain the difference, if any.

(c) Using an appropriate correlation, calculate the average convection coefficient and the heat rate per unit length for a duct of circular cross section having a perimeter equal to the wetted perimeter of the rectangular duct of part (a). Do you expect this estimate to be lower or higher than that obtained with the H–D correlation? Explain the difference, if any.

9.50 Certain wood stove designs rely exclusively on heat transfer by radiation and natural convection to the surroundings. Consider a stove that forms a cubical enclosure, $L_s = 1$ m on a side, in a large room. The exterior walls of the stove have an emissivity of $\varepsilon = 0.8$ and are at an operating temperature of $T_{s,s} = 500$ K.

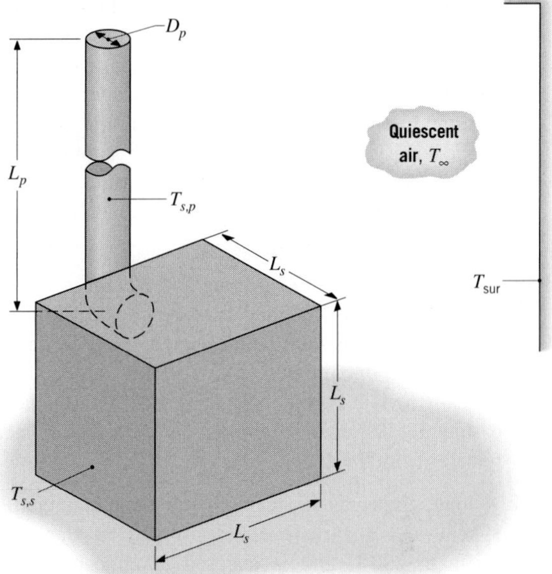

The stove pipe, which may be assumed to be isothermal at an operating temperature of $T_{s,p} = 400$ K, has a diameter of $D_p = 0.25$ m and a height of $L_p = 2$ m, extending from stove to ceiling. The stove is in a large room whose air and walls are at $T_\infty = T_{sur} = 300$ K. Neglecting heat transfer from the small horizontal section of the pipe and radiation exchange between the pipe and stove, estimate the rate at which heat is transferred from the stove and pipe to the surroundings.

9.51 A plate 1 m × 1 m, inclined at an angle of 45°, is exposed to a net radiation heat flux of 300 W/m^2 at its bottom surface. If the top surface of the plate is well insulated, estimate the temperature the plate reaches when the ambient air is quiescent and at a temperature of 0°C.

Horizontal Cylinders and Spheres

9.52 A horizontal rod 5 mm in diameter is immersed in water maintained at 18°C. If the rod surface temperature is 56°C, estimate the free convection heat transfer rate per unit length of the rod.

9.53 A horizontal uninsulated steam pipe passes through a large room whose walls and ambient air are at 300 K. The pipe of 150-mm diameter has an emissivity of 0.85 and an outer surface temperature of 400 K. Calculate the heat loss per unit length from the pipe.

9.54 As discussed in Section 5.2, the lumped capacitance approximation may be applied if $Bi < 0.1$, and, when implemented in a conservative fashion for a long cylinder, the characteristic length is the cylinder radius. After its extrusion, a long glass rod of diameter $D = 15$ mm is suspended horizontally in a room and cooled from its initial temperature by natural convection and radiation. At what rod temperatures may the lumped capacitance approximation be applied? The temperature of the quiescent air is the same as that of the surroundings, $T_\infty = T_{sur} = 27°C$, and the glass emissivity is $\varepsilon = 0.94$.

9.55 Beverage in cans 150 mm long and 60 mm in diameter is initially at 27°C and is to be cooled by placement in a refrigerator compartment at 4°C. In the interest of maximizing the cooling rate, should the cans be laid horizontally or vertically in the compartment? As a first approximation, neglect heat transfer from the ends.

9.56 A long, uninsulated steam line with a diameter of 89 mm and a surface emissivity of 0.8 transports steam at 200°C and is exposed to atmospheric air and large surroundings at an equivalent temperature of 20°C.

(a) Calculate the heat loss per unit length for a calm day.

(b) Calculate the heat loss on a breezy day when the wind speed is 8 m/s.

(c) For the conditions of part (a), calculate the heat loss with a 20-mm-thick layer of insulation ($k = 0.08$ W/m·K). Would the heat loss change significantly with an appreciable wind speed?

9.57 Consider Problem 8.47. A more realistic solution would account for the resistance to heat transfer due to free convection in the paraffin during melting. Assuming the tube surface to have a uniform temperature of 55°C and the paraffin to be an infinite, quiescent liquid, determine the convection coefficient associated with the outer surface. Using this result and recognizing that the tube surface temperature is not known, determine the water outlet temperature, the total heat transfer rate, and the time required to completely liquefy the paraffin, for the prescribed conditions. Thermophysical properties associated with the liquid state of the paraffin are $k = 0.15$ W/m·K, $\beta = 8 \times 10^{-4}\ K^{-1}$, $\rho = 770\ kg/m^3$, $\nu = 5 \times 10^{-6}\ m^2/s$, and $\alpha = 8.85 \times 10^{-8}\ m^2/s$.

9.58 A horizontal tube of 12.5-mm diameter with an outer surface temperature of 240°C is located in a room with an air temperature of 20°C. Estimate the heat transfer rate per unit length of the tube due to free convection.

9.59 Saturated steam at 4 bars absolute pressure with a mean velocity of 3 m/s flows through a horizontal pipe whose inner and outer diameters are 55 and 65 mm, respectively. The heat transfer coefficient for the steam flow is known to be 11,000 W/m^2·K.

(a) If the pipe is covered with a 25-mm-thick layer of 85% magnesia insulation and is exposed to atmospheric air at 25°C, determine the rate of heat transfer by free convection to the room per unit length of the pipe. If the steam is saturated at the inlet of the pipe, estimate its quality at the outlet of a pipe 30 m long.

(b) Net radiation to the surroundings also contributes to heat loss from the pipe. If the insulation has a surface emissivity of $\varepsilon = 0.8$ and the surroundings are at $T_{sur} = T_\infty = 25°C$, what is the rate of heat transfer to the room per unit length of pipe? What is the quality of the outlet flow?

(c) The heat loss may be reduced by increasing the insulation thickness and/or reducing its emissivity. What is the effect of increasing the insulation thickness to 50 mm if $\varepsilon = 0.8$? Of decreasing the emissivity to 0.2 if the insulation thickness is 25 mm? Of reducing the emissivity to 0.2 and increasing the insulation thickness to 50 mm?

9.60 A horizontal electrical cable of 25-mm diameter has a heat dissipation rate of 30 W/m. If the ambient air temperature is 27°C, estimate the surface temperature of the cable.

9.61 An electric immersion heater, 10 mm in diameter and 300 mm long, is rated at 550 W. If the heater is horizontally positioned in a large tank of water at 20°C, estimate its surface temperature. Estimate the surface temperature if the heater is accidentally operated in air at 20°C.

9.62 The maximum surface temperature of the 20-mm-diameter shaft of a motor operating in ambient air at 27°C should not exceed 87°C. Because of power dissipation within the motor housing, it is desirable to reject as much heat as possible through the shaft to the ambient air. In this problem, we will investigate several methods for heat removal.

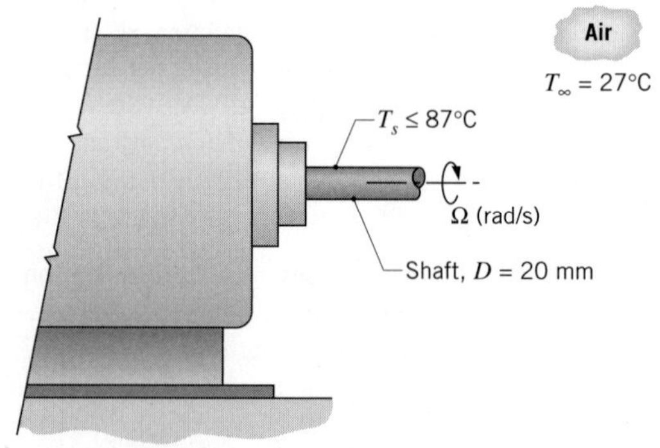

(a) For rotating cylinders, a suitable correlation for estimating the convection coefficient is of the form

$$\overline{Nu}_D = 0.133\, Re_D^{2/3}\, Pr^{1/3}$$
$$(Re_D < 4.3 \times 10^5, \quad 0.7 < Pr < 670)$$

where $Re_D \equiv \Omega D^2/\nu$ and Ω is the rotational velocity (rad/s). Determine the convection coefficient and the maximum heat rate per unit length as a function of rotational speed in the range from 5000 to 15,000 rpm.

(b) Estimate the free convection coefficient and the maximum heat rate per unit length for the stationary shaft. Mixed free and forced convection effects may become significant for $Re_D < 4.7(Gr_D^3/Pr)^{0.137}$. Are free convection effects important for the range of rotational speeds designated in part (a)?

(c) Assuming the emissivity of the shaft is 0.8 and the surroundings are at the ambient air temperature, is radiation exchange important?

(d) If ambient air is in cross flow over the shaft, what air velocities are required to remove the heat rates determined in part (a)?

9.63 Consider a horizontal pin fin of 6-mm diameter and 60-mm length fabricated from plain carbon steel ($k = 57$ W/m·K, $\varepsilon = 0.5$). The base of the fin is maintained at 150°C, while the quiescent ambient air and the surroundings are at 25°C. Assume the fin tip is adiabatic.

(a) Estimate the fin heat rate, q_f. Use an average fin surface temperature of 125°C in estimating the free convection coefficient and the linearized radiation coefficient. How sensitive is this estimate to your choice of the average fin surface temperature?

(b) Use the finite-difference method of solution to obtain q_f when the convection and radiation coefficients are based on local, rather than average, temperatures for the fin. How does your result compare with the analytical solution of part (a)?

9.64 Consider the hot water pipe of Problem 7.56, but under conditions for which the ambient air is not in cross flow over the pipe and is, instead, quiescent. Accounting for the effect of radiation with a pipe emissivity of $\varepsilon_p = 0.6$, what is the corresponding daily cost of heat loss per unit length of the uninsulated pipe?

9.65 Common practice in chemical processing plants is to clad pipe insulation with a durable, thick aluminum foil. The functions of the foil are to confine the batt insulation and to reduce heat transfer by radiation to the surroundings. Because of the presence of chlorine (at chlorine or seaside plants), the aluminum foil surface, which is initially bright, becomes etched with in-service time. Typically, the emissivity might change from 0.12 at installation to 0.36 with extended service. For a 300-mm-diameter foil-covered pipe whose surface temperature is 90°C, will this increase in emissivity due to degradation of the foil finish have a significant effect on heat loss from the pipe? Consider two cases with surroundings and ambient air at 25°C: (a) quiescent air and (b) a cross-wind velocity of 10 m/s.

9.66 Consider the electrical heater of Problem 7.49. If the blower were to malfunction, terminating airflow while the heater continued to operate at 1000 W/m, what temperature would the heater assume? How long would it take to come within 10°C of this temperature? Allow for radiation exchange between the heater ($\varepsilon = 0.8$) and the duct walls, which are also at 27°C.

9.67 A computer code is being developed to analyze a 12.5-mm-diameter, cylindrical sensor used to determine ambient air temperature. The sensor experiences free convection while positioned horizontally in quiescent air at $T_\infty = 27$°C. For the temperature range from 30 to 80°C, derive an expression for the convection coefficient as a function of only $\Delta T = T_s - T_\infty$, where T_s is the sensor temperature. Evaluate properties at an appropriate film temperature and show what effect this approximation has on the convection coefficient estimate.

9.68 A thin-walled tube of 20-mm diameter passes hot fluid at a mean temperature of 45°C in an experimental flow loop. The tube is mounted horizontally in quiescent air at a temperature of 15°C. To satisfy the stringent temperature control requirements of the experiment, it was decided to wind thin electrical heating tape on the outer surface of the tube to prevent heat loss from the hot fluid to the ambient air.

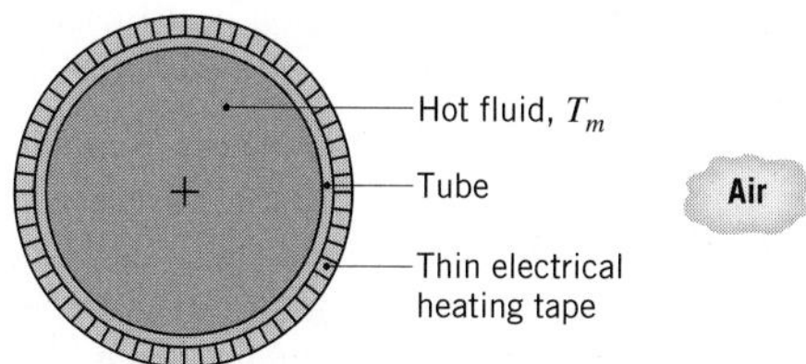

(a) Neglecting radiation heat loss, calculate the heat flux q''_e that must be supplied by the electrical tape to ensure a uniform fluid temperature.

(b) Assuming the emissivity of the tape is 0.95 and the surroundings are also at 15°C, calculate the required heat flux.

(c) The heat loss may be reduced by wrapping the heating tape in a layer of insulation. For 85% magnesia insulation ($k = 0.050$ W/m·K) having a surface emissivity of $\varepsilon = 0.60$, compute and plot the required heat flux q''_e as a function of insulation thickness in the range from 0 to 20 mm. For this range, compute and plot the convection and radiation heat rates per unit tube length as a function of insulation thickness.

9.69 A billet of stainless steel, AISI 316, with a diameter of 150 mm and a length of 500 mm emerges from a heat treatment process at 200°C and is placed in an unstirred oil bath maintained at 20°C.

(a) Determine whether it is advisable to position the billet in the bath with its centerline horizontal or vertical in order to decrease the cooling time.

(b) Estimate the time for the billet to cool to 30°C for the preferred arrangement.

9.70 Long stainless steel rods of 50-mm diameter are preheated to a uniform temperature of 1000 K before being suspended from an overhead conveyor for transport to a hot forming operation. The conveyor is in a large room whose walls and air are at 300 K.

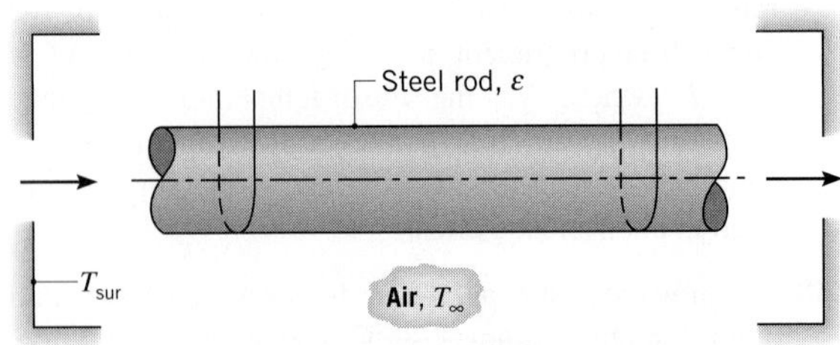

(a) Assuming the linear motion of the rod to have a negligible effect on convection heat transfer from its surface, determine the average convection coefficient at the start of the transport process.

(b) If the surface emissivity of the rod is $\varepsilon = 0.40$, what is the effective radiation heat transfer coefficient at the start of the transport process?

(c) Assuming a constant cumulative (radiation plus convection) heat transfer coefficient corresponding to the results of parts (a) and (b), what is the maximum allowable conveyor transit time, if the centerline temperature of the rod must exceed 900 K for the forming operation? Properties of the steel are $k = 25\ \text{W/m}\cdot\text{K}$ and $\alpha = 5.2 \times 10^{-6}\ \text{m}^2/\text{s}$.

(d) Heat transfer by convection and radiation are actually decreasing during the transfer operation. Accounting for this reduction, reconsider the conditions of part (c) and obtain a more accurate estimate of the maximum allowable conveyor transit time.

9.71 Hot air flows from a furnace through a 0.15-m-diameter, thin-walled steel duct with a velocity of 3 m/s. The duct passes through the crawlspace of a house, and its uninsulated exterior surface is exposed to quiescent air and surroundings at 0°C.

(a) At a location in the duct for which the mean air temperature is 70°C, determine the heat loss per unit duct length and the duct wall temperature. The duct outer surface has an emissivity of 0.5.

(b) If the duct is wrapped with a 25-mm-thick layer of 85% magnesia insulation ($k = 0.050\ \text{W/m}\cdot\text{K}$) having a surface emissivity of $\varepsilon = 0.60$, what are the duct wall temperature, the outer surface temperature, and the heat loss per unit length?

9.72 A biological fluid moves at a flow rate of $\dot{m} = 0.02$ kg/s through a coiled, thin-walled, 5-mm-diameter tube submerged in a large water bath maintained at 50°C. The fluid enters the tube at 25°C.

(a) Estimate the length of the tube and the number of coil turns required to provide an exit temperature of $T_{m,o} = 38°\text{C}$ for the biological fluid. Assume that the water bath is an extensive, quiescent medium, that the coiled tube approximates a horizontal tube, and that the biological fluid has the thermophysical properties of water.

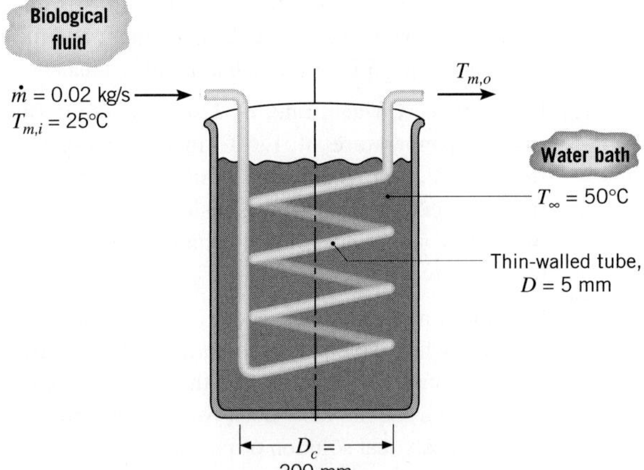

(b) The flow rate through the tube is controlled by a pump that experiences throughput variations of approximately ±10% at any one setting. This condition is of concern to the project engineer because the corresponding variation of the exit temperature of the biological fluid could influence the downstream process. What variation would you expect in $T_{m,o}$ for a ±10% change in $\dot{m}$?

9.73 Consider a batch process in which 200 L of a pharmaceutical are heated from 25°C to 70°C by saturated steam condensing at 2.455 bars as it flows through a coiled tube of 15-mm diameter and 15-m length. At any time during the process, the liquid may be approximated as an infinite, quiescent medium of uniform temperature and may be assumed to have constant properties of $\rho = 1100\ \text{kg/m}^3$, $c = 2000\ \text{J/kg}\cdot\text{K}$, $k = 0.25\ \text{W/m}\cdot\text{K}$, $\nu = 4.0 \times 10^{-6}\ \text{m}^2\text{/s}$, $Pr = 10$, and $\beta = 0.002\ \text{K}^{-1}$. The thermal resistances of the condensing steam and tube wall may be neglected.

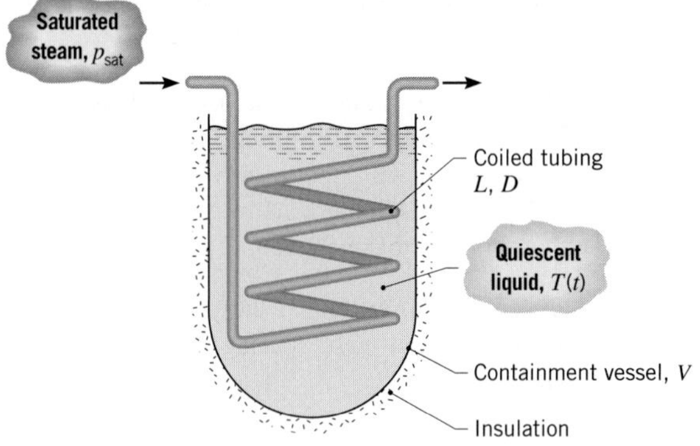

(a) What is the initial rate of heat transfer to the pharmaceutical?

(b) Neglecting heat transfer between the tank and its surroundings, how long does it take to heat the pharmaceutical to 70°C? Plot the corresponding variation with time of the fluid temperature and the convection coefficient at the outer surface of the tube. How much steam is condensed during the heating process?

9.74 In the analytical treatment of the fin with uniform cross-sectional area, it was assumed that the convection heat transfer coefficient is constant along the length of the fin. Consider an AISI 316 steel fin of 6-mm diameter and 50-mm length (with insulated tip) operating under conditions for which $T_b = 125°\text{C}$, $T_\infty = 27°\text{C}$, $T_{sur} = 27°\text{C}$, and $\varepsilon = 0.6$.

(a) Estimate average values of the fin heat transfer coefficients for free convection (h_c) and radiation exchange (h_r). Use these values to predict the tip temperature and fin effectiveness.

(b) Use a numerical method of solution to estimate the foregoing parameters when the convection and radiation coefficients are based on local, rather than average, values for the fin.

9.75 A hot fluid at 35°C is to be transported through a tube horizontally positioned in quiescent air at 25°C. Which of the tube shapes, each of equal cross-sectional area, would you use in order to minimize heat losses to the ambient air by free convection?

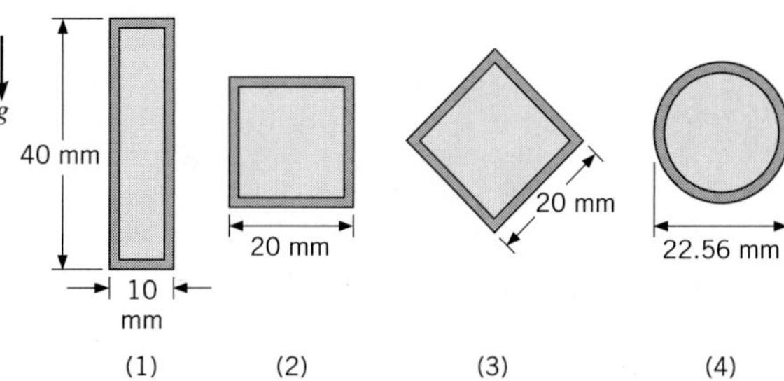

Use the following correlation of Lienhard [*Int. J. Heat Mass Transfer,* **16**, 2121, 1973] to approximate the laminar convection coefficient for an immersed body on which the boundary layer does not separate from the surface,

$$\overline{Nu}_l = 0.52\, Ra_l^{1/4}$$

The characteristic length l is the length of travel of the fluid in the boundary layer across the shape surface. Compare this correlation to that given for a sphere to test its utility.

9.76 Consider a 2-mm-diameter sphere immersed in a fluid at 300 K and 1 atm.

(a) If the fluid around the sphere is quiescent and extensive, show that the conduction limit of heat transfer from the sphere can be expressed as $Nu_{D,\text{cond}} = 2$. *Hint:* Begin with the expression for the thermal resistance of a hollow sphere, Equation 3.41, letting $r_2 \rightarrow \infty$, and then expressing the result in terms of the Nusselt number.

(b) Considering free convection, at what surface temperature will the Nusselt number be twice that for the conduction limit? Consider air and water as the fluids.

(c) Considering forced convection, at what velocity will the Nusselt number be twice that for the conduction limit? Consider air and water as the fluids.

9.77 A sphere of 25-mm diameter contains an embedded electrical heater. Calculate the power required to maintain the surface temperature at 94°C when the sphere is exposed to a quiescent medium at 20°C for: (a) air at atmospheric pressure, (b) water, and (c) ethylene glycol.

Parallel Plate Channels

9.78 Consider two long vertical plates maintained at uniform temperatures $T_{s,1} > T_{s,2}$. The plates are open at their ends and are separated by the distance $2L$.

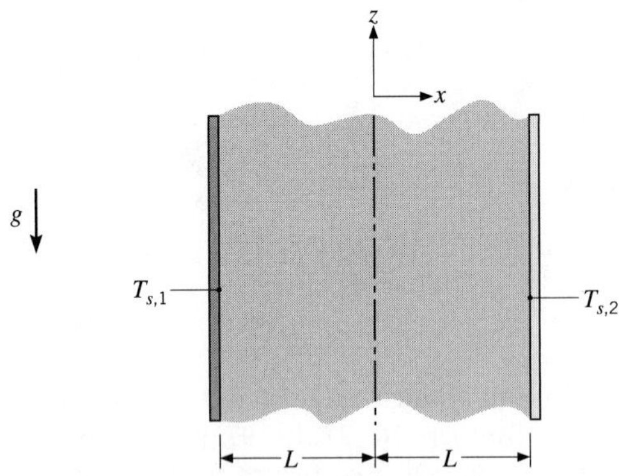

(a) Sketch the velocity distribution in the space between the plates.

(b) Write appropriate forms of the continuity, momentum, and energy equations for laminar flow between the plates.

(c) Evaluate the temperature distribution, and express your result in terms of the mean temperature, $T_m = (T_{s,1} + T_{s,2})/2$.

(d) Estimate the vertical pressure gradient by assuming the density to be a constant ρ_m corresponding to T_m. Substituting from the Boussinesq approximation, obtain the resulting form of the momentum equation.

(e) Determine the velocity distribution.

9.79 Consider the conditions of Problem 9.9, but now view the problem as one involving free convection between vertical, parallel plate channels. What is the optimum fin spacing S? For this spacing and the prescribed values of t and W, what is the rate of heat transfer from the fins?

9.80 A vertical array of circuit boards is immersed in quiescent ambient air at $T_\infty = 17°C$. Although the components protrude from their substrates, it is reasonable, as a first approximation, to assume *flat* plates with uniform surface heat flux q''_s. Consider boards of length and width $L = W = 0.4$ m and spacing $S = 25$ mm. If the maximum allowable board temperature is 77°C, what is the maximum allowable power dissipation per board?

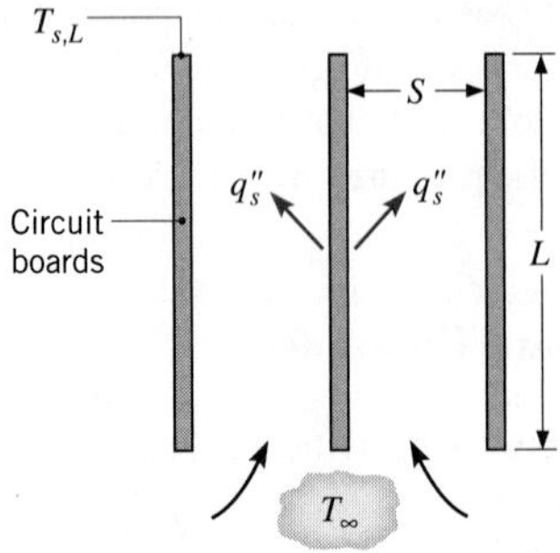

9.81 Determined to reduce the $7 per week cost associated with heat loss through their patio window by convection and radiation, the tenants of Problem 9.18 cover the inside of the window with a 50-mm-thick sheet of extruded insulation. Because they are not very handy around the house, the insulation is installed poorly, resulting in an $S = 5$-mm gap between the extruded insulation and the window pane, allowing the room air to infiltrate into the space between the pane and the insulation.

(a) Determine the window heat loss and associated weekly cost with the ill-fitting insulation in place. The insulation will significantly reduce the radiation losses through the window. Losses will be due almost entirely to convection.

(b) Plot the heat loss through the patio window as a function of the gap spacing for $1 \text{ mm} \leq S \leq 20 \text{ mm}$.

9.82 The front door of a dishwasher of width 580 mm has a vertical air vent that is 500 mm in height with a 20-mm spacing between the inner tub operating at 52°C and an outer plate that is thermally insulated.

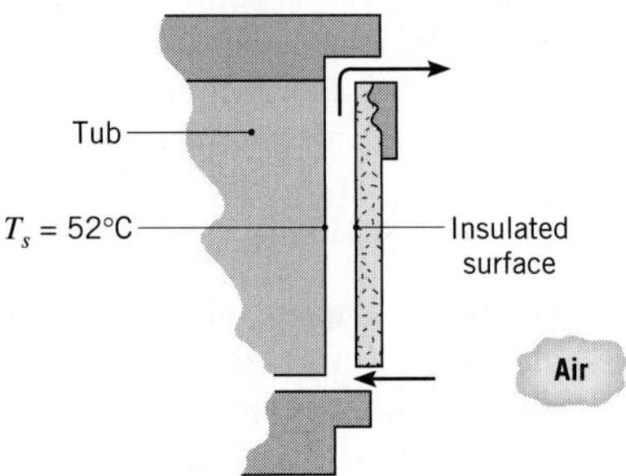

(a) Determine the heat loss from the tub surface when the ambient air is 27°C.

(b) A change in the design of the door provides the opportunity to increase or decrease the 20-mm spacing by 10 mm. What recommendations would you offer with regard to how the change in spacing will alter heat losses?

9.83 A natural convection air heater consists of an array of parallel, equally spaced vertical plates, which may be maintained at a fixed temperature T_s by embedded electrical heaters. The plates are of length and width $L = W = 300$ mm and are in quiescent, atmospheric air at $T_\infty = 20°C$. The total width of the array cannot exceed a value of $W_{ar} = 150$ mm.

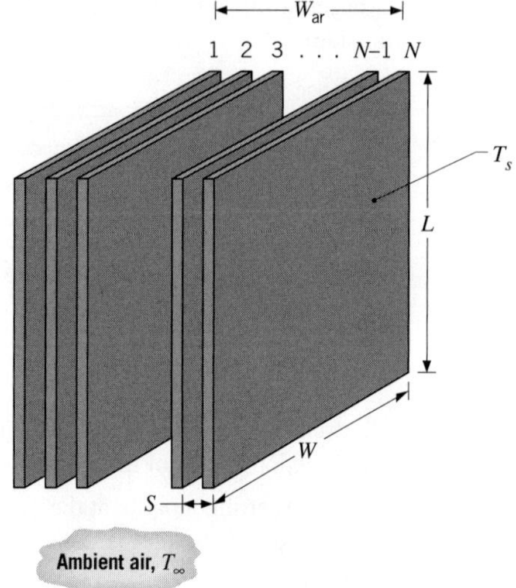

For $T_s = 75°C$, what is the plate spacing S that maximizes heat transfer from the array? For this spacing, how many plates comprise the array and what is the corresponding rate of heat transfer from the array?

9.84 A bank of drying ovens is mounted on a rack in a room with an ambient air temperature of 27°C. The cubical ovens are 500 mm to a side, and the spacing between the ovens is 15 mm.

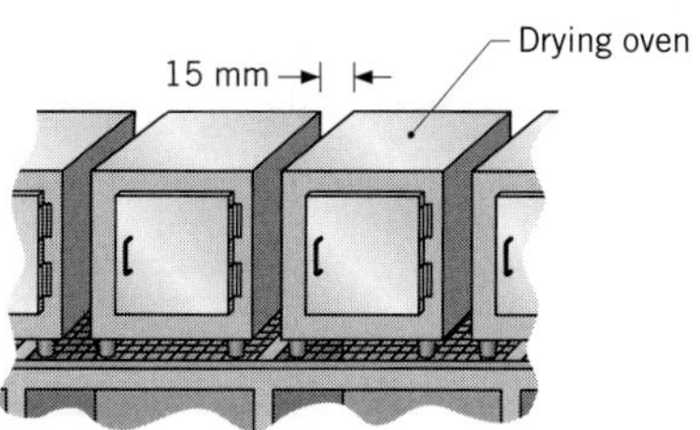

(a) Estimate the heat loss from a facing side of an oven when its surface temperature is 47°C.

(b) Explore the effect of the spacing on the heat loss. At what spacing is the heat loss a maximum? Describe the boundary layer behavior for this condition. Can this condition be analyzed by treating the side of an oven as an isolated vertical plate?

9.85 A solar collector consists of a parallel plate channel that is connected to a water storage plenum at the bottom and to a heat sink at the top. The channel is inclined $\theta = 30°$ from the vertical and has a transparent cover plate. Solar radiation transmitted through the cover plate and the water maintains the isothermal absorber plate at a temperature $T_s = 67°C$, while water returned to the reservoir from the heat sink is at $T_\infty = 27°C$. The system operates as a *thermosyphon,* for which water flow is driven exclusively by buoyancy forces. The plate spacing and length are $S = 15$ mm and $L = 1.5$ m.

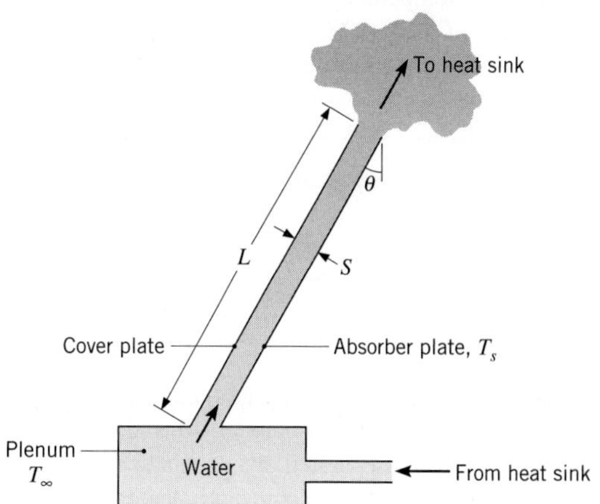

Assuming the cover plate to be adiabatic with respect to convection heat transfer to or from the water, estimate the rate of heat transfer per unit width normal to the flow direction (W/m) from the absorber plate to the water.

Rectangular Cavities

9.86 As is evident from the property data of Tables A.3 and A.4, the thermal conductivity of glass at room temperature is more than 50 times larger than that of air. It is therefore desirable to use windows of double-pane construction, for which the two panes of glass enclose an air space. If heat transfer across the air space is by conduction, the corresponding thermal resistance may be increased by increasing the thickness L of the space. However, there are limits to the efficacy of such a measure, since convection currents are induced if L exceeds a critical value, beyond which the thermal resistance decreases.

Consider atmospheric air enclosed by vertical panes at temperatures of $T_1 = 22°C$ and $T_2 = -20°C$. If the critical Rayleigh number for the onset of convection is $Ra_L \approx 2000$, what is the maximum allowable spacing for conduction across the air? How is this spacing affected by the temperatures of the panes? How is it affected by the pressure of the air, as, for example, by partial evacuation of the space?

9.87 A building window pane that is 1.2 m high and 0.8 m wide is separated from the ambient air by a storm window of the same height and width. The air space between the two windows is 0.06 m thick. If the building and storm windows are at 20 and $-10°C$, respectively, what is the rate of heat loss by free convection across the air space?

9.88 To reduce heat losses, a horizontal rectangular duct that is $W = 0.80$ m wide and $H = 0.3$ m high is encased in a metal radiation shield. The duct wall and shield are separated by an air gap of thickness $t = 0.06$ m. For a duct wall temperature of $T_d = 40°C$ and a shield temperature of $T_{sh} = 20°C$, determine the convection heat loss per unit length from the duct.

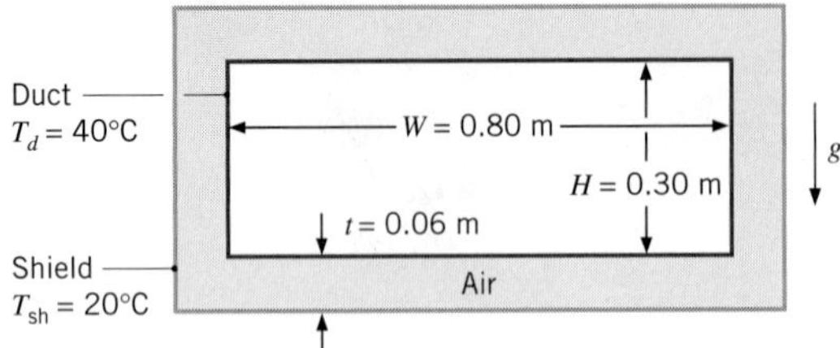

9.89 The absorber plate and the adjoining cover plate of a flat-plate solar collector are at 70 and 35°C, respectively, and are separated by an air space of 0.05 m. What is the rate of free convection heat transfer per unit surface area between the two plates if they are inclined at an angle of 60° from the horizontal?

9.90 Consider a thermal storage system in which the phase change material (paraffin) is housed in a large container whose bottom, horizontal surface is maintained at $T_s = 50°C$ by warm water delivered from a solar collector.

(a) Neglecting the change in sensible energy of the liquid phase, estimate the amount of paraffin that is melted over a five-hour period beginning with an initial liquid layer at the bottom of the container of thickness $s_i = 10$ mm. The paraffin of Problems 8.47 and 9.57 is used as the phase change material and is initially at the phase change temperature, $T_{mp} = 27.4°C$. The bottom surface area of the container is $A = 2.5\ m^2$.

(b) Compare the amount of energy needed to melt the paraffin to the amount of energy required to increase the temperature of the same amount of liquid from the phase change temperature to the average liquid temperature, $(T_s + T_{mp})/2$.

(c) Neglecting the change in sensible energy of the liquid phase, estimate the amount of paraffin that would melt over a five-hour time period if the hot plate is placed at the top of the container and $s_i = 10$ mm.

9.91 A rectangular cavity consists of two parallel, 0.5-m-square plates separated by a distance of 50 mm, with the lateral boundaries insulated. The heated plate is maintained at 325 K and the cooled plate at 275 K. Estimate the heat flux between the surfaces for three orientations of the cavity using the notation of Figure 9.6: vertical with $\tau = 90°$, horizontal with $\tau = 0°$, and horizontal with $\tau = 180°$.

9.92 Consider a horizontal flat roof section having the same dimensions as a vertical wall section. For both sections, the surfaces exposed to the air gap are at 18°C (inside) and −10°C (outside).

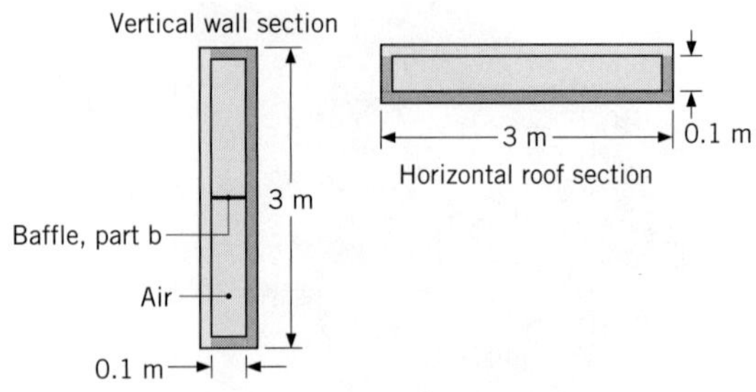

(a) Estimate the ratio of the convection heat rate for the horizontal section to that of the vertical section.

(b) What effect will inserting a baffle at the mid-height of the vertical section have on the convection heat rate for that section?

9.93 A 50-mm-thick air gap separates two horizontal metal plates that form the top surface of an industrial furnace. The bottom plate is at $T_h = 200°C$ and the top plate is at $T_c = 50°C$. The plant operator wishes to provide insulation between the plates to minimize heat loss. The relatively hot temperatures preclude use of foamed or felt insulation materials. Evacuated insulation materials cannot be used due to the harsh industrial environment and their expense. A young engineer suggests that equally spaced, thin horizontal sheets of aluminum foil may be inserted in the gap to eliminate natural convection and minimize heat loss through the air gap.

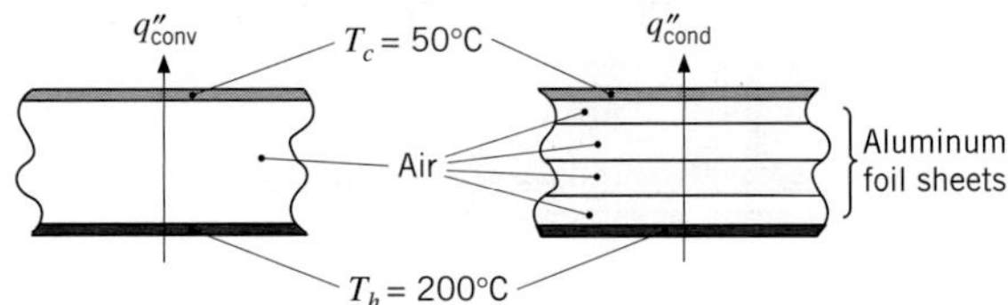

(a) Determine the convective heat flux across the gap when no insulation is in place.

(b) Determine the minimum number of foil sheets that must be inserted in the gap to eliminate free convection.

(c) Determine the conduction heat flux across the air gap with the foil sheets in place.

9.94 The space between the panes of a double-glazed window can be filled with either air or carbon dioxide at atmospheric pressure. The window is 1.5 m high and the spacing between the panes can be varied. Develop an analysis to predict the convection heat transfer rate across the window as a function of pane spacing and determine, under otherwise identical conditions, whether air or carbon dioxide will yield the smaller rate. Illustrate the results of your analysis for two surface-temperature conditions: winter (−10°C, 20°C) and summer (35°C, 25°C).

9.95 A vertical, double-pane window, which is 1 m on a side and has a 25-mm gap filled with atmospheric air, separates quiescent room air at $T_{\infty,i} = 20°C$ from quiescent ambient air at $T_{\infty,o} = -20°C$. Radiation exchange between the window panes, as well as between each pane and its surroundings, may be neglected.

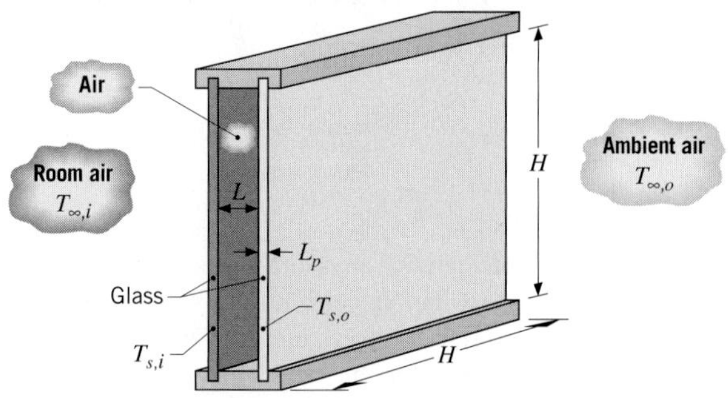

(a) Neglecting the thermal resistance associated with conduction heat transfer across each pane, determine the corresponding temperature of each pane and the rate of heat transfer through the window.

(b) Comment on the validity of neglecting the conduction resistance of the panes if each is of thickness $L_p = 6$ mm.

9.96 The top surface (0.5 m × 0.5 m) of an oven is 60°C for a particular operating condition when the room air is 23°C. To reduce heat loss from the oven and to minimize burn hazard, it is proposed to create a 50-mm air space by adding a cover plate.

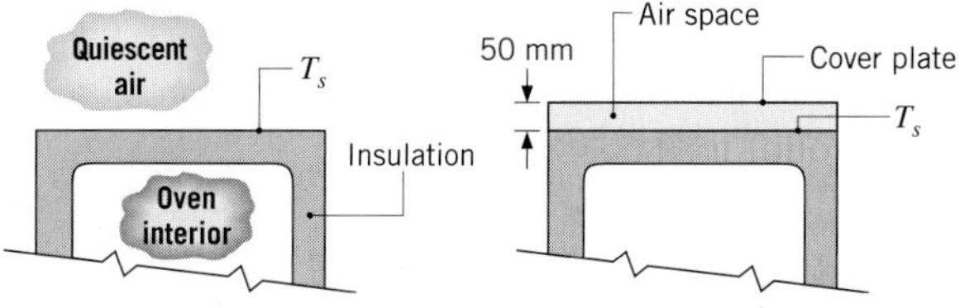

(a) Assuming the same oven surface temperature T_s for both situations, estimate the reduction in the convection heat loss resulting from installation of the cover plate. What is the temperature of the cover plate?

(b) Explore the effect of the cover plate spacing on the convection heat loss and the cover plate temperature for spacings in the range $5 \leq L \leq 50$ mm. Is there an optimum spacing?

9.97 Consider window blinds that are installed in the air space between the two panes of a vertical double-pane window. The window is $H = 0.5$ m high and $w = 0.5$ m wide, and includes $N = 19$ individual blinds that are each $L = 25$ mm wide. When the blinds are open, 20 smaller, square enclosures are formed along the height of the window. In the closed position, the blinds form a nearly continuous sheet with two $t = 12.5$ mm open gaps at the top and bottom of the enclosure. Determine the convection heat transfer rate between the inner pane, which is held at $T_{s,i} = 20°C$, and the outer pane, which is at $T_{s,o} = -20°C$, when the blinds are in the open and closed positions, respectively. Explain why the closed blinds have little effect on the convection heat transfer rate across the cavity.

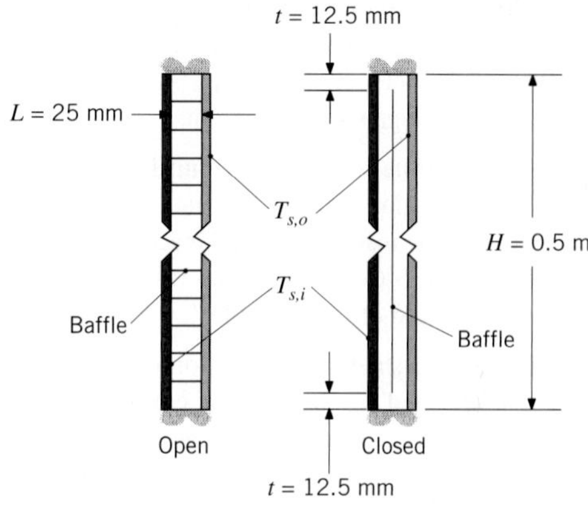

9.98 A solar water heater consists of a flat-plate collector that is coupled to a storage tank. The collector consists of a transparent cover plate and an absorber plate that are separated by an air gap.

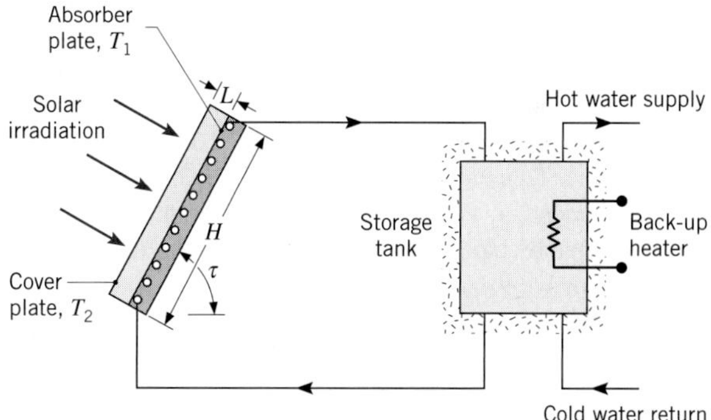

Although much of the solar energy collected by the absorber plate is transferred to a working fluid passing through a coiled tube brazed to the back of the absorber, some of the energy is lost by free convection and net radiation transfer across the air gap. In Chapter 13, we will evaluate the contribution of radiation exchange to this loss. For now, we restrict our attention to the free convection effect.

(a) Consider a collector that is inclined at an angle of $\tau = 60°$ and has dimensions of $H = w = 2$ m on a side, with an air gap of $L = 30$ mm. If the absorber and cover plates are at $T_1 = 70°C$ and $T_2 = 30°C$, respectively, what is the rate of heat transfer by free convection from the absorber plate?

(b) The heat loss by free convection depends on the spacing between the plates. Compute and plot the heat loss as a function of spacing for $5 \leq L \leq 50$ mm. Is there an optimum spacing?

Concentric Cylinders and Spheres

9.99 Consider the cylindrical, 0.12-m-diameter radiation shield of Example 9.5 that is installed concentric with a 0.10-m-diameter tube carrying steam. The spacing provides an air gap of $L = 10$ mm.

(a) Calculate the heat loss per unit length of the tube by convection when a second shield of 0.14-m diameter is installed, with the second shield maintained at 35°C. Compare the result to that for the single shield of the example.

(b) In the two-shield configuration of part (a), the air gaps formed by the annular concentric tubes are $L = 10$ mm. Calculate the heat loss per unit length if the gap dimension is $L = 15$ mm. Do you expect the heat loss to increase or decrease?

9.100 The effective thermal conductivity k_{eff} for concentric cylinders and concentric spheres is provided in Equations 9.59 and 9.62, respectively. Derive expressions for the critical Rayleigh numbers associated with the cylindrical and spherical geometries, $Ra_{c,\text{crit}}$ and $Ra_{s,\text{crit}}$, respectively, below which k_{eff} is minimized. Evaluate $Ra_{c,\text{crit}}$ and $Ra_{s,\text{crit}}$ for air, water, and glycerin at a mean temperature of 300 K. For specified inner and outer surface temperatures and inner cylinder or sphere radii, comment on the heat transfer rate for outer cylinder or sphere radii corresponding to $Ra_{c,\text{crit}}$ and $Ra_{s,\text{crit}}$, respectively.

9.101 A solar collector design consists of an inner tube enclosed concentrically in an outer tube that is transparent to solar radiation. The tubes are thin walled with inner and outer diameters of 0.10 and 0.15 m, respectively. The annular space between the tubes is completely enclosed and filled with air at atmospheric pressure. Under operating conditions for which the inner and outer tube surface temperatures are 70 and 30°C, respectively, what is the convective heat loss per meter of tube length across the air space?

9.102 A proposed method to reduce heat losses from a horizontal, isothermal cylinder placed within a large room is to encase it within a larger cylinder, as shown in the schematic, with all surfaces painted with a low emissivity coating.

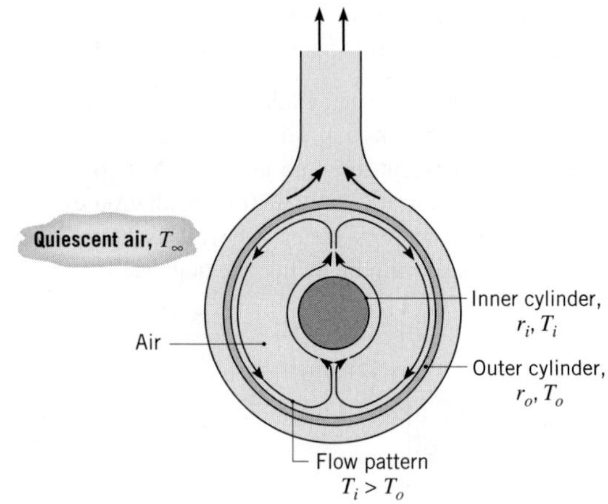

Air at atmospheric pressure exists within the annular region. For a concentrically located inner cylinder of surface temperature $T_i = 70$°C and radius $r_i = 20$ mm, and for an ambient temperature of $T_\infty = 30$°C, determine the optimal radius of the outer cylinder r_o that will minimize the convection heat loss. Compare the convection heat loss from the inner cylinder with the optimally sized outer cylinder in place to the heat loss without the outer cylinder. Is the approach effective?

9.103 It has been proposed to use large banks of rechargeable, lithium ion batteries to power electric vehicles. The cylindrical batteries, each of which is of radius $r_i = 9$ mm and length $L = 65$ mm, undergo exothermic electrochemical reactions while being discharged. Since excessively high temperatures damage the batteries, it is proposed to encase them in a phase change material that melts when the batteries discharge (and resolidifies when the batteries are charged; charging is associated with an endothermic electrochemical reaction). Consider the paraffin of Problems 8.47 and 9.57.

(a) At an instant in time during the discharge of a battery, liquid paraffin occupies an annular region of outer radius $r_o = 19$ mm around the battery, which is generating $\dot{E}_g = 1$ W of thermal energy. Determine the surface temperature of the battery.

(b) At the time of interest in part (a), what is the rate at which the liquid annulus radius is increasing?

(c) Plot the battery surface temperature versus the outer radius of the liquid-filled annulus. Explain the relative insensitivity of the battery surface temperature to the size of the annulus for $15 \text{ mm} \leq r_o \leq 30$ mm.

9.104 Free convection occurs between concentric spheres. The inner sphere is of diameter $D_i = 50$ mm and temperature $T_i = 50$°C, while the outer sphere is maintained at $T_o = 20$°C. Air is in the gap between the spheres. What outer sphere diameter is required so that the convection heat transfer from the inner sphere is the same as if it were placed in a large, quiescent environment with air at $T_\infty = 20°C$?

9.105 The surfaces of two long, horizontal, concentric thin-walled tubes having radii of 100 and 125 mm are maintained at 300 and 400 K, respectively. If the annular space is pressurized with nitrogen at 5 atm, estimate the convection heat transfer rate per unit length of the tubes.

9.106 Consider the phase change material (PCM) of Problems 8.47 and 9.57. The PCM is housed in a long, horizontal, and insulated cylindrical enclosure of diameter $D_e = 200$ mm, which in turn includes a concentric, heated inner cylinder of diameter $D_i = 30$ mm. Initially, the PCM is entirely solid and at its phase change temperature. The inner cylinder temperature is suddenly raised to $T_h = 50°C$. Assuming the PCM melts to form an expanding concentric liquid region about the heated tube such as the one shown in the schematic, determine how long it takes to melt half of the PCM.

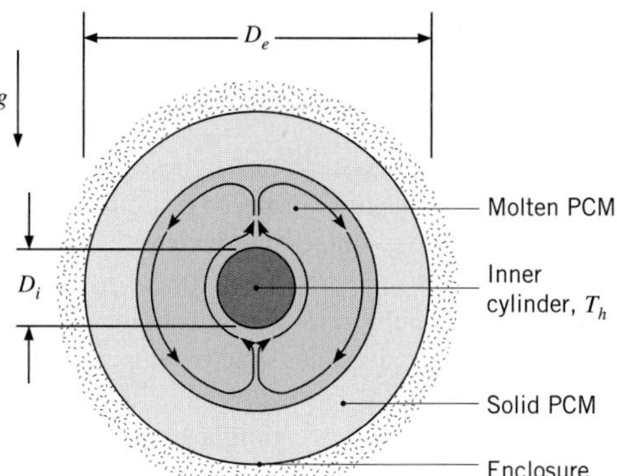

9.107 Liquid nitrogen is stored in a thin-walled spherical vessel of diameter $D_i = 1$ m. The vessel is positioned concentrically within a larger, thin-walled spherical container of diameter $D_o = 1.10$ m, and the intervening cavity is filled with atmospheric helium.

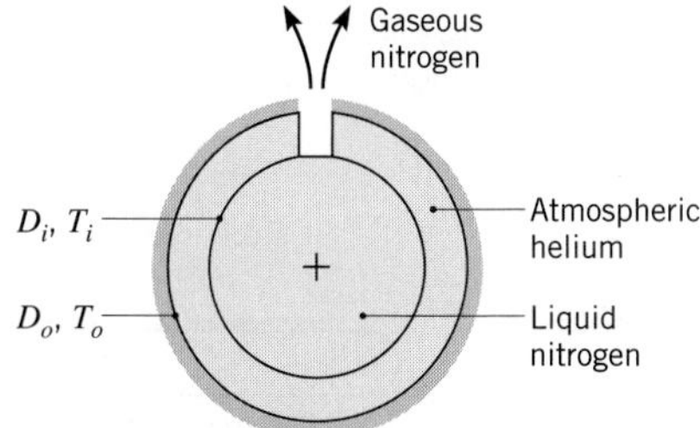

Under normal operating conditions, the inner and outer surface temperatures are $T_i = 77$ K and $T_o = 283$ K. If the latent heat of vaporization of nitrogen is 2×10^5 J/kg, what is the mass rate $\dot{m}$ (kg/s) at which gaseous nitrogen is vented from the system?

9.108 The human eye contains aqueous humor, which separates the external cornea and the internal iris–lens structure. It is hypothesized that, in some individuals, small flakes of pigment are intermittently liberated from the iris and migrate to, and subsequently damage, the cornea. Approximating the geometry of the enclosure formed by the cornea and iris–lens structure as a pair of concentric hemispheres of outer radius $r_o = 10$ mm and inner radius $r_i = 7$ mm, respectively, investigate whether free convection can occur in the aqueous humor by evaluating the effective thermal conductivity ratio, k_{eff}/k. If free convection can occur, it is possible that the damaging particles are advected from the iris to the cornea. The iris–lens structure is at the core temperature, $T_i = 37°C$, while the cornea temperature is measured to be $T_o = 34°C$. The properties of the aqueous humor are $\rho = 990\ \text{kg/m}^3$, $k = 0.58\ \text{W/m}\cdot\text{K}$, $c_p = 4.2 \times 10^3\ \text{J/kg}\cdot\text{K}$, $\mu = 7.1 \times 10^{-4}\ \text{N}\cdot\text{s/m}^2$, and $\beta = 3.2 \times 10^{-4}\ \text{K}^{-1}$.

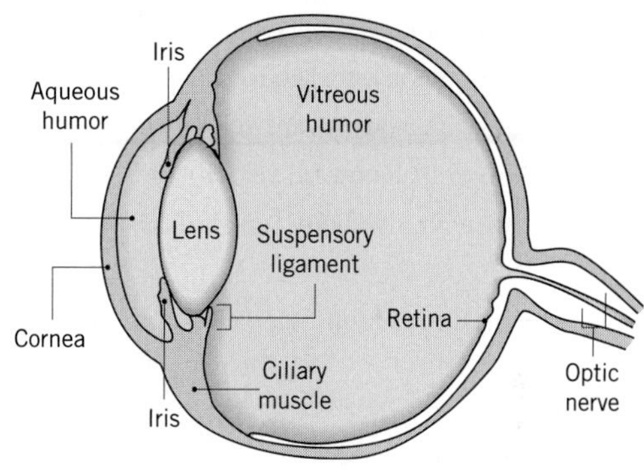

Mixed Convection

9.109 A horizontal, 25-mm diameter cylinder is maintained at a uniform surface temperature of 35°C. A fluid with a velocity of 0.05 m/s and temperature of 20°C is in cross flow over the cylinder. Determine whether heat transfer by free convection will be significant for (i) air, (ii) water, (iii) engine oil, and (iv) mercury.

9.110 According to experimental results for parallel airflow over a uniform temperature, heated vertical plate, the effect of free convection on the heat transfer convection coefficient will be 5% when $Gr_L/Re_L^2 = 0.08$. Consider a heated vertical plate 0.3 m long, maintained at a surface temperature of 60°C in atmospheric air at 25°C. What is the minimum vertical velocity required of the airflow such that free convection effects will be less than 5% of the heat transfer rate?

9.111 A vertical array of circuit boards of 150-mm height is to be air cooled such that the board temperature does not exceed 60°C when the ambient temperature is 25°C.

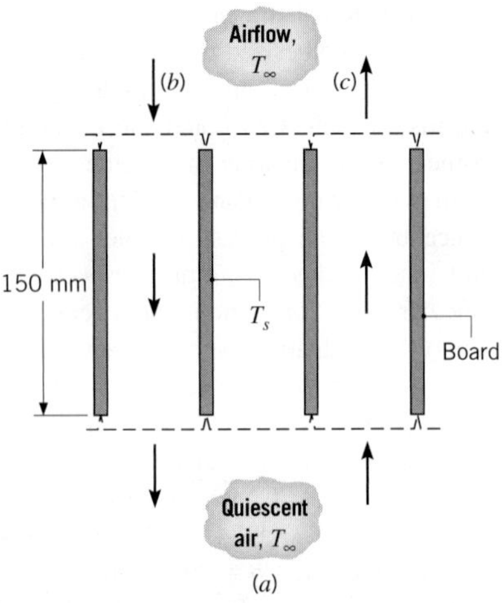

Assuming isothermal surface conditions, determine the allowable electrical power dissipation per board for the cooling arrangements:

(a) Free convection only (no forced airflow).

(b) Airflow with a downward velocity of 0.6 m/s.

(c) Airflow with an upward velocity of 0.3 m/s.

(d) Airflow with a velocity (upward or downward) of 5 m/s.

9.112 A probe, used to measure the velocity of air in a low-speed wind tunnel, is fabricated of an $L = 100$ mm long, $D = 8$-mm outside diameter horizontal aluminum tube. Power resistors are inserted into the stationary tube and dissipate $P = 1.5$ W. The surface temperature of the tube is determined experimentally by measuring the emitted radiation from the exterior of the tube. To maximize surface emission, the exterior of the tube is painted with flat black paint having an emissivity of $\varepsilon = 0.95$.

(a) For air at a temperature and cross flow velocity of $T_\infty = 25°C$, $V = 0.1$ m/s, respectively, determine the surface temperature of the tube. The surroundings temperature is $T_{sur} = 25°C$.

(b) For the conditions of part (a), plot the tube surface temperature versus the cross flow velocity over the range $0.05 \text{ m/s} \leq V \leq 1 \text{ m/s}$.

9.113 A horizontal 100-mm-diameter pipe passing hot oil is to be used in the design of an industrial water heater. Based on a typical water draw rate, the velocity over the pipe is 0.5 m/s. The hot oil maintains the outer surface temperature at 85°C and the water temperature is 37°C.

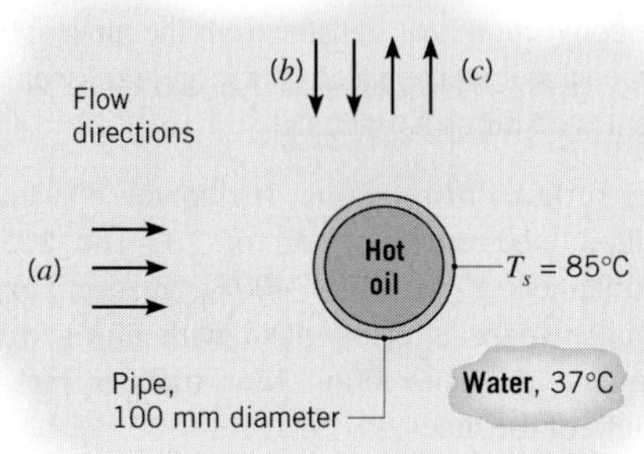

Investigate the effect of flow direction on the heat rate (W/m) for (a) horizontal, (b) downward, and (c) upward flow.

9.114 Determine the heat transfer rate from the steel plates of Problem 7.24 accounting for free convection from the plate surfaces. What is the corresponding rate of change of the plate temperature? Plot the heat transfer coefficient associated with free convection, forced convection, and mixed convection for air velocities ranging from $2 \leq u_\infty \leq 10$ m/s. The velocity of the plate is small compared to the air velocity.

9.115 An experiment involves heating a very small sphere that is suspended by a fine string in air with a laser beam in order to induce the highest sphere temperature possible. After inspecting Equation 9.64, a research assistant suggests inducing a uniform downward airflow to *exactly* offset free convection from the sphere, thereby minimizing heat losses and maximizing the steady-state sphere temperature. In the limiting case of a *very small* sphere, what is the minimum value of the convection heat transfer coefficient expressed in terms of the sphere diameter and thermal conductivity of the air?

9.116 Square panels (250 mm × 250 mm) with a decorative, highly reflective plastic finish are cured in an oven at 125°C and cooled in quiescent air at 29°C. Quality considerations dictate that the panels remain horizontal and that the cooling rate be controlled. To increase productivity in the plant, it is proposed to replace the batch cooling method with a conveyor system having a velocity of 0.5 m/s.

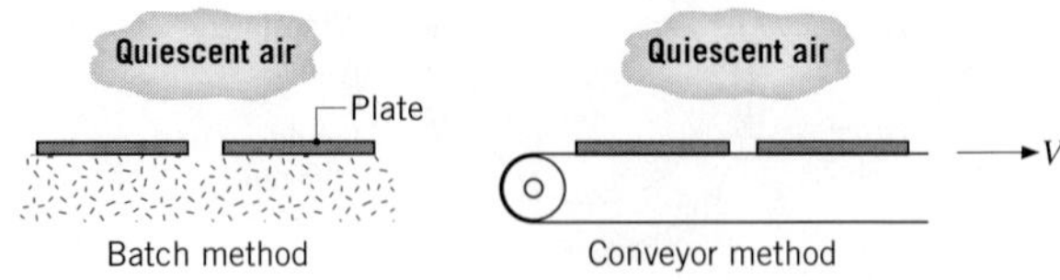

Compare the initial (immediately after leaving the oven) convection heat transfer rates for the two methods.

CHAPTER 10

Boiling and Condensation

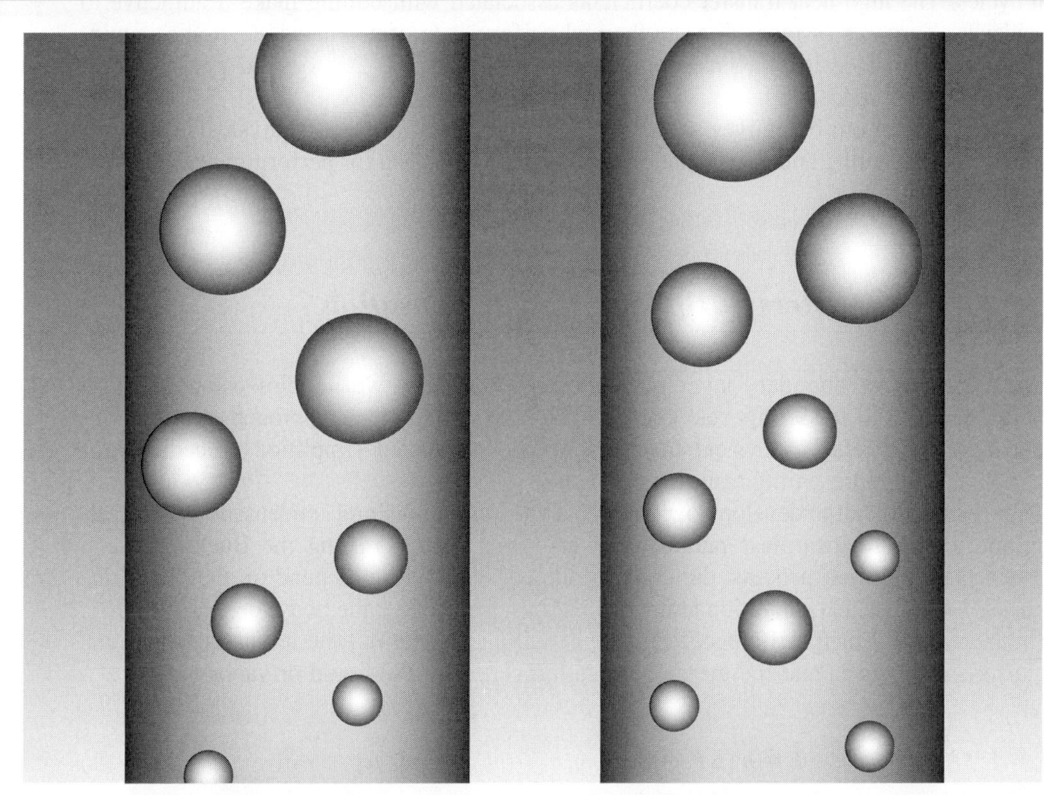

In this chapter we focus on convection processes associated with the change in phase of a fluid. In particular, we consider processes that can occur at a solid–liquid or solid–vapor interface, namely, *boiling* and *condensation*. For these cases *latent* heat effects associated with the phase change are significant. The change from the liquid to the vapor state due to boiling is sustained by heat transfer from the solid surface; conversely, condensation of a vapor to the liquid state results in heat transfer to the solid surface.

Since they involve fluid motion, boiling and condensation are classified as forms of the convection mode of heat transfer. However, they are characterized by unique features. Because there is a phase change, heat transfer to or from the fluid can occur without influencing the fluid temperature. In fact, through boiling or condensation, large heat transfer rates may be achieved with small temperature differences. In addition to the *latent heat* h_{fg}, two other parameters are important in characterizing the processes, namely, the *surface tension* σ at the liquid–vapor interface and the *density difference* between the two phases. This difference induces a *buoyancy force,* which is proportional to $g(\rho_l - \rho_v)$. Because of combined latent heat and buoyancy-driven flow effects, boiling and condensation heat transfer coefficients and rates are generally much larger than those characteristic of convection heat transfer without phase change.

Many engineering applications that are characterized by high heat fluxes involve boiling and condensation. In a closed-loop power cycle, pressurized liquid is converted to vapor in a *boiler.* After expansion in a turbine, the vapor is restored to its liquid state in a *condenser*, whereupon it is pumped to the boiler to repeat the cycle. *Evaporators*, in which the boiling process occurs, and condensers are also essential components in vapor-compression refrigeration cycles. The high heat transfer coefficients associated with boiling make it attractive to consider for purposes of managing the thermal performance of advanced electronics equipment. The rational design of such components dictates that the associated phase change processes be well understood.

In this chapter our objectives are to develop an appreciation for the physical conditions associated with boiling and condensation and to provide a basis for performing related heat transfer calculations.

10.1 Dimensionless Parameters in Boiling and Condensation

In our treatment of boundary layer phenomena (Chapter 6), we nondimensionalized the governing equations to identify relevant dimensionless groups. This approach enhanced our understanding of related physical mechanisms and suggested simplified procedures for generalizing and representing heat transfer results.

Since it is difficult to develop governing equations for boiling and condensation processes, the appropriate dimensionless parameters can be obtained by using the Buckingham pi theorem [1]. For either process, the convection coefficient could depend on the difference between the surface and saturation temperatures, $\Delta T = |T_s - T_{\text{sat}}|$, the body force arising from the liquid–vapor density difference, $g(\rho_l - \rho_v)$, the latent heat h_{fg}, the surface tension σ, a characteristic length L, and the thermophysical properties of the liquid or vapor: ρ, c_p, k, μ. That is,

$$h = h\,[\Delta T, g(\rho_l - \rho_v), h_{fg}, \sigma, L, \rho, c_p, k, \mu] \tag{10.1}$$

Since there are 10 variables in 5 dimensions (m, kg, s, J, K), there are (10 − 5) = 5 pi-groups, which can be expressed in the following forms:

$$\frac{hL}{k} = f\left[\frac{\rho g(\rho_l - \rho_v)L^3}{\mu^2}, \frac{c_p \Delta T}{h_{fg}}, \frac{\mu c_p}{k}, \frac{g(\rho_l - \rho_v)L^2}{\sigma}\right] \tag{10.2a}$$

or, defining the dimensionless groups,

$$Nu_L = f\left[\frac{\rho g\,(\rho_l - \rho_v)L^3}{\mu^2}, Ja, Pr, Bo\right] \tag{10.2b}$$

The Nusselt and Prandtl numbers are familiar from our earlier single-phase convection analyses. The new dimensionless parameters are the Jakob number *Ja*, the Bond number *Bo*, and a nameless parameter that bears a strong resemblance to the Grashof number (see Equation 9.12 and recall that $\beta \Delta T \approx \Delta\rho/\rho$). This unnamed parameter represents the effect of buoyancy-induced fluid motion on heat transfer. The Jakob number is the ratio of the maximum sensible energy absorbed by the liquid (vapor) to the latent energy absorbed by the liquid (vapor) during condensation (boiling). In many applications, the sensible energy is much less than the latent energy and *Ja* has a small numerical value. The Bond number is the ratio of the buoyancy force to the surface tension force. In subsequent sections, we will delineate the role of these parameters in boiling and condensation.

10.2 *Boiling Modes*

When evaporation occurs at a solid–liquid interface, it is termed *boiling*. The process occurs when the temperature of the surface T_s exceeds the saturation temperature T_{sat} corresponding to the liquid pressure. Heat is transferred from the solid surface to the liquid, and the appropriate form of Newton's law of cooling is

$$q''_s = h(T_s - T_{\text{sat}}) = h\,\Delta T_e \tag{10.3}$$

where $\Delta T_e \equiv T_s - T_{\text{sat}}$ is termed the *excess temperature.* The process is characterized by the formation of vapor bubbles, which grow and subsequently detach from the surface. Vapor bubble growth and dynamics depend, in a complicated manner, on the excess temperature, the nature of the surface, and thermophysical properties of the fluid, such as its surface tension. In turn, the dynamics of vapor bubble formation affect liquid motion near the surface and therefore strongly influence the heat transfer coefficient.

Boiling may occur under various conditions. For example, in *pool boiling* the liquid is quiescent and its motion near the surface is due to free convection and to mixing induced by bubble growth and detachment. In contrast, for *forced convection boiling,* fluid motion is induced by external means, as well as by free convection and bubble-induced mixing. Boiling may also be classified according to whether it is *subcooled* or *saturated.* In subcooled boiling, the temperature of most of the liquid is below the saturation temperature and bubbles formed at the surface may condense in the liquid. In contrast, the temperature of the liquid slightly exceeds the saturation temperature in *saturated boiling.* Bubbles formed at the surface are then propelled through the liquid by buoyancy forces, eventually escaping from a free surface.

10.3 Pool Boiling

Saturated pool boiling, as shown in Figure 10.1, has been studied extensively. Although there is a sharp increase in the liquid temperature close to the solid surface, the temperature through most of the liquid remains slightly above saturation. Bubbles generated at the liquid–solid interface rise to the liquid–vapor interface, where the vapor is ultimately transported across the interface. An appreciation for the underlying physical mechanisms may be obtained by examining the *boiling curve.*

10.3.1 The Boiling Curve

Nukiyama [2] was the first to identify different regimes of pool boiling using the apparatus of Figure 10.2. The heat flux from a horizontal *nichrome* wire to saturated water was determined by measuring the current flow I and potential drop E. The wire temperature was determined from knowledge of the manner in which its electrical resistance varied with temperature. This arrangement is termed *power-controlled* heating, wherein the wire temperature T_s (hence the excess temperature ΔT_e) is the dependent variable and the power setting (hence the heat flux q''_s) is the independent variable. Following the arrows of the *heating curve* of Figure 10.3, it is evident that as power is applied, the heat flux increases, at first slowly and then very rapidly, with excess temperature.

Nukiyama observed that boiling, as evidenced by the presence of bubbles, did not begin until $\Delta T_e \approx 5°\text{C}$. With further increase in power, the heat flux increased to very high levels until *suddenly,* for a value slightly larger than q''_{max}, the wire temperature jumped to the melting point and burnout occurred. However, repeating the experiment with a *platinum* wire having a higher melting point (2045 K vs. 1500 K), Nukiyama was able to maintain heat fluxes above q''_{max} without burnout. When he subsequently reduced the power, the variation of ΔT_e with q''_s followed the *cooling curve* of Figure 10.3. When the heat flux reached the minimum point q''_{min}, a further decrease in power caused the excess temperature to drop abruptly, and the process followed the original heating curve back to the saturation point.

Nukiyama believed that the hysteresis effect of Figure 10.3 was a consequence of the power-controlled method of heating, where ΔT_e is a dependent variable. He also believed that by using a heating process permitting the independent control of ΔT_e, the missing (dashed) portion of the curve could be obtained. His conjecture was subsequently confirmed by Drew and Mueller [3]. By condensing steam inside a tube at

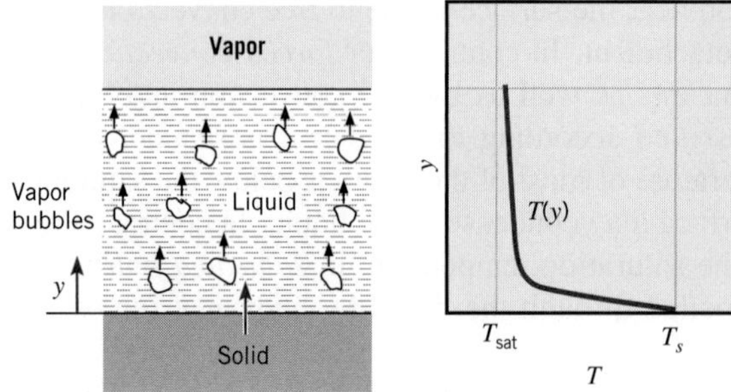

FIGURE 10.1 Temperature distribution in saturated pool boiling with a liquid–vapor interface.

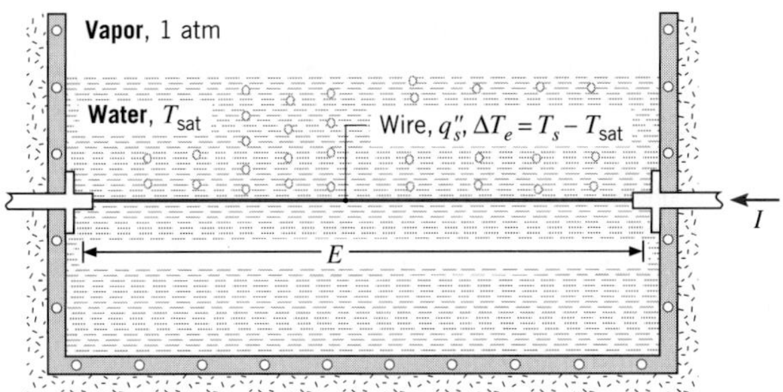

FIGURE 10.2 Nukiyama's power-controlled heating apparatus for demonstrating the boiling curve.

different pressures, they were able to control the value of ΔT_e for boiling of a low boiling point organic fluid at the tube outer surface and thereby obtain the missing portion of the boiling curve.

10.3.2 Modes of Pool Boiling

An appreciation for the underlying physical mechanisms may be obtained by examining the different modes, or regimes, of pool boiling. These regimes are identified in the boiling curve of Figure 10.4. The specific curve pertains to water at 1 atm, although similar trends characterize the behavior of other fluids. From Equation 10.3 we note that q_s'' depends on the convection coefficient h, as well as on the excess temperature ΔT_e. Different boiling regimes may be delineated according to the value of ΔT_e.

Free Convection Boiling Free convection boiling is said to exist if $\Delta T_e \leq \Delta T_{e,A}$, where $\Delta T_{e,A} \approx 5°C$. The surface temperature must be somewhat above the saturation temperature in order to sustain bubble formation. As the excess temperature is increased, bubble inception will eventually occur, but below point A (referred to as the *onset of nucleate boiling*, ONB), fluid motion is determined principally by free convection effects. According to

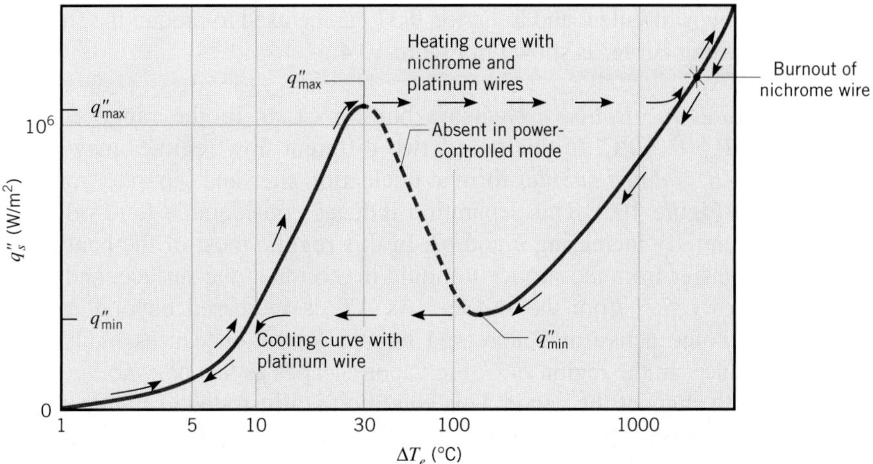

FIGURE 10.3 Nukiyama's boiling curve for saturated water at atmospheric pressure.

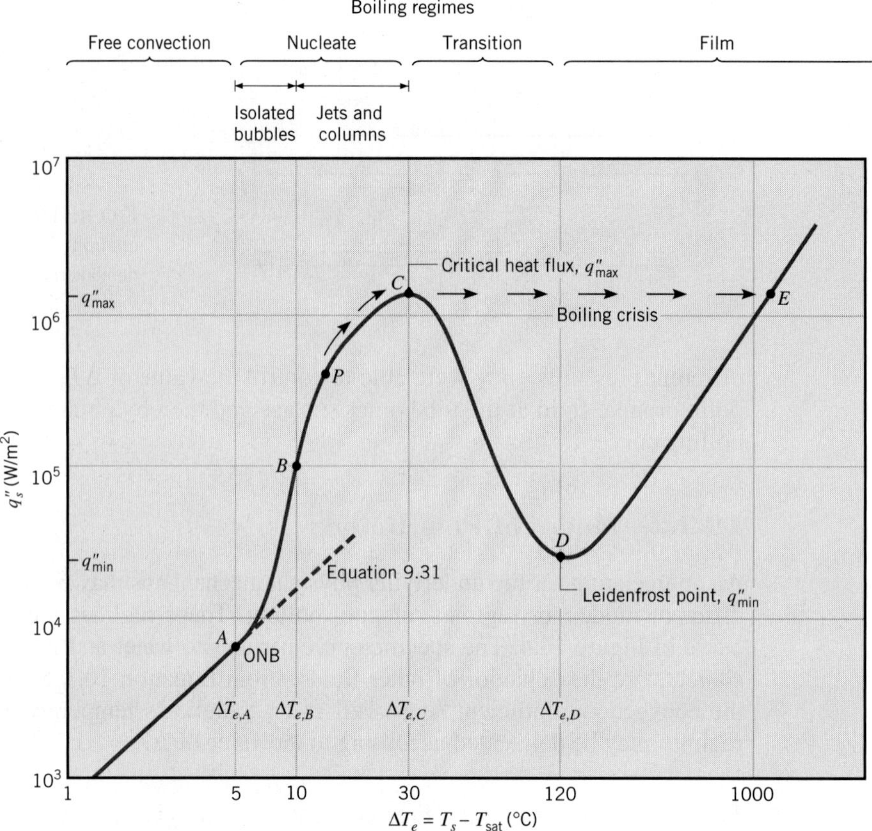

FIGURE 10.4 Typical boiling curve for water at 1 atm: surface heat flux q''_s as a function of excess temperature, $\Delta T_e \equiv T_s - T_{sat}$.

whether the flow is laminar or turbulent, h varies as ΔT_e to the $\frac{1}{4}$ or $\frac{1}{3}$ power, respectively, in which case q''_s varies as ΔT_e to the $\frac{5}{4}$ or $\frac{4}{3}$ power. For a large horizontal plate, the fluid flow is turbulent and Equation 9.31 can be used to predict the free convection portion of the boiling curve, as shown in Figure 10.4.

Nucleate Boiling Nucleate boiling exists in the range $\Delta T_{e,A} \leq \Delta T_e \leq \Delta T_{e,C}$, where $\Delta T_{e,C} \approx 30°\text{C}$. In this range, two different flow regimes may be distinguished. In region *A–B, isolated bubbles* form at nucleation sites and separate from the surface, as illustrated in Figure 10.2. This separation induces considerable fluid mixing near the surface, substantially increasing h and q''_s. In this regime most of the heat exchange is through direct transfer from the surface to liquid in motion at the surface, and not through the vapor bubbles rising from the surface. As ΔT_e is increased beyond $\Delta T_{e,B}$, more nucleation sites become active and increased bubble formation causes bubble interference and coalescence. In the region *B–C*, the vapor escapes as *jets* or *columns,* which subsequently merge into slugs of the vapor. This condition is illustrated in Figure 10.5*a*. Interference between the densely populated bubbles inhibits the motion of liquid near the surface. Point *P* of Figure 10.4 represents a change in the behavior of the boiling curve. Before point *P*, the

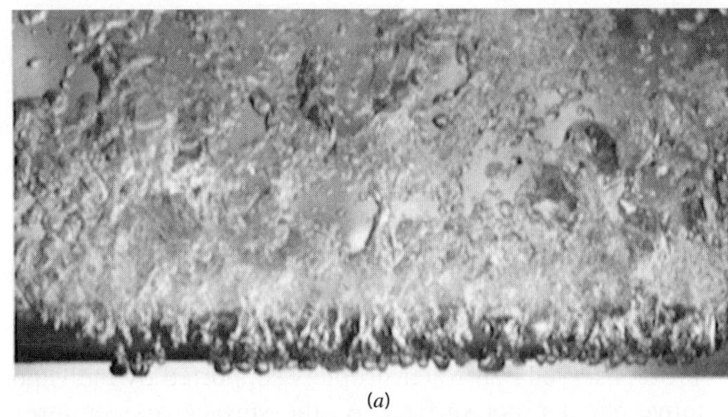

(*a*)

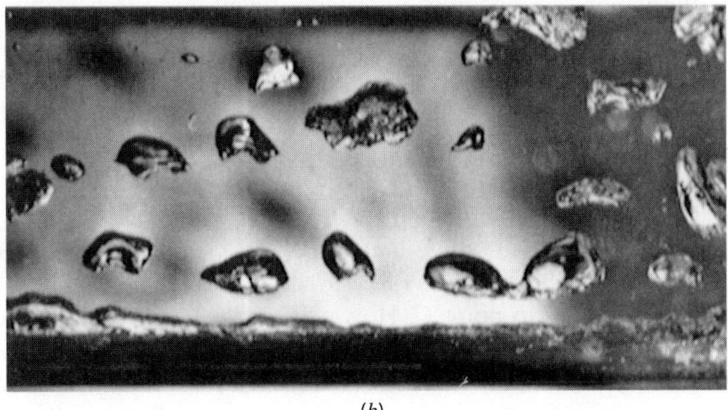

(*b*)

FIGURE 10.5 Boiling of methanol on a horizontal tube. (*a*) Nucleate boiling in the jets and columns regime. (*b*) Film boiling. (Photographs courtesy of Professor J. W. Westwater, University of Illinois at Champaign-Urbana.)

boiling curve can be approximated as a straight line on a log–log plot, meaning that $q''_s \propto \Delta T_e^n$. Beyond this point, the heat flux increases more slowly as ΔT_e is increased. At some point between *P* and *C*, the decaying increase of the heat flux leads to a reduction of the heat transfer coefficient $h = q''_s/\Delta T_e$. The maximum heat flux, $q''_{s,C} = q''_{\max}$, is usually termed the *critical heat flux,* and in water at atmospheric pressure it exceeds 1 MW/m^2. At the point of this maximum, considerable vapor is being formed, making it difficult for liquid to continuously wet the surface.

Because high heat transfer rates and convection coefficients are associated with small values of the excess temperature, it is desirable to operate many engineering devices in the nucleate boiling regime. The approximate magnitude of the convection coefficient may be inferred by using Equation 10.3 with the boiling curve of Figure 10.4. Dividing q''_s by ΔT_e, it is evident that convection coefficients in excess of 10^4 W/m$^2 \cdot$K are characteristic of this regime. These values are considerably larger than those normally corresponding to convection with no phase change.

Transition Boiling The region corresponding to $\Delta T_{e,C} \le \Delta T_e \le \Delta T_{e,D}$, where $\Delta T_{e,D} \approx$ 120°C, is termed *transition boiling, unstable film boiling,* or *partial film boiling.* Bubble formation is now so rapid that a vapor film or blanket begins to form on the surface. At any point on the surface, conditions may oscillate between film and nucleate boiling, but the fraction of the total surface covered by the film increases with increasing ΔT_e. Because the thermal conductivity of the vapor is much less than that of the liquid, h (and q''_s) must decrease with increasing ΔT_e.

Film Boiling Film boiling exists for $\Delta T_e \geq \Delta T_{e,D}$. At point D of the boiling curve, referred to as the *Leidenfrost point,* the heat flux is a minimum, $q''_{s,D} = q''_{\min}$, and the surface is completely covered by a *vapor blanket.* Heat transfer from the surface to the liquid occurs by conduction and radiation through the vapor. It was Leidenfrost who in 1756 observed that water droplets supported by the vapor film slowly boil away as they move about a hot surface. As the surface temperature is increased, radiation through the vapor film becomes more significant and the heat flux increases with increasing ΔT_e. Figure 10.5*b* illustrates the nature of the vapor formation and bubble dynamics associated with film boiling. The photographs of Figure 10.5 were obtained for the boiling of methanol on a horizontal tube.

Although the foregoing discussion of the boiling curve assumes that control may be maintained over T_s, it is important to remember the Nukiyama experiment and be mindful of the many applications that involve controlling q''_s (e.g., in a nuclear reactor or in an electric resistance heater) rather than ΔT_e. Consider starting at point P in Figure 10.4 and gradually increasing q''_s. The value of ΔT_e, and hence the value of T_s, will also increase, following the boiling curve to point C. However, any increase in q''_s beyond point C will induce a sharp increase from $\Delta T_{e,C}$ to $\Delta T_{e,E} \equiv T_{s,E} - T_{\text{sat}}$. Because $T_{s,E}$ may exceed the melting point of the solid, system failure may occur. For this reason point C is often termed the *burnout point* or the *boiling crisis,* and accurate knowledge of the critical heat flux (CHF), $q''_{s,C} \equiv q''_{\max}$, is important. Although we may want to operate a heat transfer surface close to the CHF, we would rarely want to exceed it.

10.4 *Pool Boiling Correlations*

From the shape of the boiling curve and the fact that various physical mechanisms characterize the different regimes, it is no surprise that a multiplicity of heat transfer correlations exist for the boiling process. For the region below $\Delta T_{e,A}$ of the boiling curve (Figure 10.4), appropriate free convection correlations from Chapter 9 can be used to estimate heat transfer coefficients and heat rates. In this section we review some of the more widely used correlations for nucleate and film boiling.

10.4.1 Nucleate Pool Boiling

The analysis of nucleate boiling requires prediction of the number of surface nucleation sites and the rate at which bubbles originate from each site. While mechanisms associated with this boiling regime have been studied extensively, complete and reliable mathematical models have yet to be developed. Yamagata et al. [4] were the first to show the influence of nucleation sites on the heat rate and to demonstrate that q''_s is approximately proportional to ΔT_e^3. It is desirable to develop correlations that reflect this relationship between the surface heat flux and the excess temperature.

In Section 10.3.2 we noted that within region *A-B* of Figure 10.4, most of the heat exchange is due to direct transfer from the heated surface to the liquid. Hence, the boiling phenomena in this region may be thought of as a type of liquid phase forced convection in which the fluid motion is induced by the rising bubbles. We have seen that forced convection correlations are generally of the form

$$\overline{Nu}_L = C_{fc}\, Re_L^{m_{fc}}\, Pr^{n_{fc}} \tag{7.1}$$

and Equation 7.1 may provide insight into how pool boiling data can be correlated, provided that a length scale and a characteristic velocity can be identified for inclusion in the Nusselt and Reynolds numbers. The subscript fc is added to the constants that appear in Equation 7.1 to remind us that they apply to this *forced convection* expression. As we saw in Chapter 7, these constants are determined experimentally for complicated flows. Because it is postulated that the rising bubbles mix the liquid, an appropriate length scale for relatively large heater surfaces is the bubble diameter, D_b. The diameter of the bubble upon its departure from the heated surface may be determined from a force balance in which the buoyancy force (which promotes bubble departure and is proportional to D_b^3) is equal to the surface tension force (which adheres the bubble to the surface and is proportional to D_b), resulting in the expression

$$D_b \propto \sqrt{\frac{\sigma}{g(\rho_l - \rho_v)}} \tag{10.4a}$$

The constant of proportionality depends on the angle of contact between the liquid, its vapor, and the solid surface; the contact angle depends on the particular liquid and solid surface that is considered. The subscripts l and v denote the saturated liquid and vapor states, respectively, and σ (N/m) is the surface tension.

A characteristic velocity for the agitation of the liquid may be found by dividing the distance the liquid travels to fill in behind a departing bubble (proportional to D_b) by the time between bubble departures, t_b. The time t_b is equal to the energy it takes to form a vapor bubble (proportional to D_b^3), divided by the rate at which heat is added over the solid–vapor contact area (proportional to D_b^2). Thus,

$$V \propto \frac{D_b}{t_b} \propto \frac{D_b}{\left(\dfrac{\rho_l h_{fg} D_b^3}{q_s'' D_b^2}\right)} \propto \frac{q_s''}{\rho_l h_{fg}} \tag{10.4b}$$

Substituting Equations 10.4a and 10.4b into Equation 7.1, absorbing the proportionalities into the constant C_{fc}, and substituting the resulting expression for h into Equation 10.3 provides the following expression, where the constants $C_{s,f}$ and n are newly introduced and the exponent m_{fc} in Equation 7.1 has an experimentally determined value of 2/3:

$$q_s'' = \mu_l h_{fg}\left[\frac{g(\rho_l - \rho_v)}{\sigma}\right]^{1/2}\left(\frac{c_{p,l}\,\Delta T_e}{C_{s,f}\, h_{fg}\, Pr_l^n}\right)^3 \tag{10.5}$$

Equation 10.5 was developed by Rohsenow [5] and is the first and most widely used correlation for nucleate boiling. All properties are for the liquid, except for ρ_v, and all should be evaluated at T_{sat}. The coefficient $C_{s,f}$ and the exponent n depend on the solid–fluid combination, and representative experimentally determined values are presented in Table 10.1. Values for other surface–fluid combinations may be obtained

TABLE 10.1 Values of $C_{s,f}$ for various surface–fluid combinations [5–7]

Surface–Fluid Combination	$C_{s,f}$	n
Water–copper		
Scored	0.0068	1.0
Polished	0.0128	1.0
Water–stainless steel		
Chemically etched	0.0133	1.0
Mechanically polished	0.0132	1.0
Ground and polished	0.0080	1.0
Water–brass	0.0060	1.0
Water–nickel	0.006	1.0
Water–platinum	0.0130	1.0
n-Pentane–copper		
Polished	0.0154	1.7
Lapped	0.0049	1.7
Benzene–chromium	0.0101	1.7
Ethyl alcohol–chromium	0.0027	1.7

from the literature [6–8]. Values of the surface tension and the latent heat of vaporization of water are presented in Table A.6 and for selected fluids in Table A.5. Values for additional fluids may be obtained from any recent edition of the *Handbook of Chemistry and Physics.* If Equation 10.5 is rewritten in terms of a Nusselt number based on an arbitrary length scale L, it will be in the form $Nu_L \propto Ja^2\, Pr^{1-3n}\, Bo^{1/2}$. Comparing with Equation 10.2b, we see that only the first dimensionless parameter does not appear. If the Nusselt number is based on the characteristic bubble diameter given in Equation 10.4a, the expression reduces to the simpler form $Nu_{D_b} \propto Ja^2\, Pr^{1-3n}$.

The Rohsenow correlation applies only for clean surfaces. When it is used to estimate the heat flux, errors can amount to $\pm 100\%$. However, since $\Delta T_e \propto (q_s'')^{1/3}$, this error is reduced by a factor of 3 when the expression is used to estimate ΔT_e from knowledge of q_s''. Also, since $q_s'' \propto h_{fg}^{-2}$ and h_{fg} decreases with increasing saturation pressure (temperature), the nucleate boiling heat flux will increase as the liquid is pressurized.

10.4.2 Critical Heat Flux for Nucleate Pool Boiling

We recognize that the critical heat flux, $q_{s,C}'' = q_{\max}''$, represents an important point on the boiling curve. We may wish to operate a boiling process close to this point, but we appreciate the danger of dissipating heat in excess of this amount. Kutateladze [9], through dimensional analysis, and Zuber [10], through a hydrodynamic stability analysis, obtained an expression which can be approximated as

$$q_{\max}'' = C h_{fg} \rho_v \left[\frac{\sigma g(\rho_l - \rho_v)}{\rho_v^2} \right]^{1/4} \tag{10.6}$$

which is independent of surface material and is weakly dependent upon the heated surface geometry through the leading constant, C. For large horizontal cylinders, for spheres, and for many large finite heated surfaces, use of a leading constant with the value $C = \pi/24 \approx 0.131$ (the Zuber constant) agrees with experimental data to within 16% [11]. For large horizontal plates, a value of $C = 0.149$ gives better agreement with experimental data. The properties in Equation 10.6 are evaluated at the saturation temperature. Equation 10.6 applies when the characteristic length of the heater surface, L, is large relative to the bubble diameter, D_b. However, when the heater is small, such that the Confinement number, $Co = \sqrt{\sigma/(g[\rho_l - \rho_v])}/L = Bo^{-1/2}$ [12], is greater than approximately 0.2, a correction factor must be applied to account for the small size of the heater. Lienhard [11] reports correction factors for various geometries, including horizontal plates, cylinders, spheres, and vertically and horizontally oriented ribbons.

It is important to note that the critical heat flux depends strongly on pressure, mainly through the pressure dependence of surface tension and the heat of vaporization. Cichelli and Bonilla [13] have experimentally demonstrated that the peak flux increases with pressure up to one-third of the critical pressure, after which it falls to zero at the critical pressure.

10.4.3 Minimum Heat Flux

The transition boiling regime is of little practical interest, as it may be obtained only by controlling the surface temperature. While no adequate theory has been developed for this regime, conditions can be characterized by periodic, *unstable* contact between the liquid and the heated surface. However, the upper limit of this regime is of interest because it corresponds to formation of a *stable* vapor blanket or film and to a minimum heat flux condition. If the heat flux drops below this minimum, the film will collapse, causing the surface to cool and nucleate boiling to be reestablished.

Zuber [10] used stability theory to derive the following expression for the minimum heat flux, $q''_{s,D} = q''_{\min}$, from a large horizontal plate.

$$q''_{\min} = C\rho_v h_{fg}\left[\frac{g\sigma(\rho_l - \rho_v)}{(\rho_l + \rho_v)^2}\right]^{1/4} \tag{10.7}$$

where the properties are evaluated at the saturation temperature. The constant, $C = 0.09$, has been experimentally determined by Berenson [14]. This result is accurate to approximately 50% for most fluids at moderate pressures but provides poorer estimates at higher pressures [15]. A similar result has been obtained for horizontal cylinders [16].

10.4.4 Film Pool Boiling

At excess temperatures beyond the Leidenfrost point, a continuous vapor film blankets the surface and there is no contact between the liquid phase and the surface. Because conditions in the stable vapor film bear a strong resemblance to those of laminar film condensation (Section 10.7), it is customary to base film boiling correlations on results obtained from condensation theory. One such result, which applies to film boiling on a cylinder or sphere of diameter D, is of the form

$$\overline{Nu}_D = \frac{\overline{h}_{\text{conv}} D}{k_v} = C\left[\frac{g(\rho_l - \rho_v)h'_{fg}D^3}{\nu_v k_v (T_s - T_{\text{sat}})}\right]^{1/4} \tag{10.8}$$

The correlation constant C is 0.62 for horizontal cylinders [17] and 0.67 for spheres [11]. The corrected latent heat h'_{fg} accounts for the sensible energy required to maintain temperatures within the vapor blanket above the saturation temperature. Although it may be approximated as $h'_{fg} = h_{fg} + 0.80c_{p,v}(T_s - T_{\text{sat}})$, it is known to depend weakly on the Prandtl number of the vapor [18]. Vapor properties are evaluated at the system pressure and the film temperature, $T_f = (T_s + T_{\text{sat}})/2$, whereas ρ_l and h_{fg} are evaluated at the saturation temperature.

At elevated surface temperatures ($T_s \gtrsim 300°\text{C}$), radiation heat transfer across the vapor film becomes significant. Since radiation acts to increase the film thickness, it is not reasonable to assume that the radiative and convective processes are simply additive. Bromley [17] investigated film boiling from the outer surface of horizontal tubes and suggested calculating the total heat transfer coefficient from a transcendental equation of the form

$$\bar{h}^{4/3} = \bar{h}_{\text{conv}}^{4/3} + \bar{h}_{\text{rad}}\bar{h}^{1/3} \tag{10.9}$$

If $\bar{h}_{\text{rad}} < \bar{h}_{\text{conv}}$, a simpler form may be used:

$$\bar{h} = \bar{h}_{\text{conv}} + \tfrac{3}{4}\bar{h}_{\text{rad}} \tag{10.10}$$

The effective radiation coefficient $\bar{h}_{\text{rad}}$ is expressed as

$$\bar{h}_{\text{rad}} = \frac{\varepsilon\sigma(T_s^4 - T_{\text{sat}}^4)}{T_s - T_{\text{sat}}} \tag{10.11}$$

where ε is the emissivity of the solid (Table A.11) and σ is the Stefan–Boltzmann constant.

Note that the analogy between film boiling and film condensation does not hold for small surfaces with high curvature because of the large disparity between vapor and liquid film thicknesses for the two processes. The analogy is also questionable for a vertical surface, although satisfactory predictions have been obtained for limited conditions.

10.4.5 Parametric Effects on Pool Boiling

In this section we briefly consider other parameters that can affect pool boiling, confining our attention to the gravitational field, liquid subcooling, and solid surface conditions.

The influence of the *gravitational field* on boiling must be considered in applications involving space travel and rotating machinery. This influence is evident from appearance of the gravitational acceleration g in the foregoing expressions. Siegel [19], in his review of low gravity effects, confirms that the $g^{1/4}$ dependence in Equations 10.6, 10.7, and 10.8 (for the maximum and minimum heat fluxes and for film boiling) is correct for values of g as low as $0.10\ \text{m/s}^2$. For nucleate boiling, however, evidence indicates that the heat flux is nearly independent of gravity, which is in contrast to the $g^{1/2}$ dependence of Equation 10.5. Above-normal gravitational forces show similar effects, although near the ONB, gravity can influence bubble-induced convection.

If liquid in a pool boiling system is maintained at a temperature that is less than the saturation temperature, the liquid is said to be *subcooled*, where $\Delta T_{\text{sub}} \equiv T_{\text{sat}} - T_l$. In the natural convection regime, the heat flux increases typically as $(T_s - T_l)^n$ or $(\Delta T_e + \Delta T_{\text{sub}})^n$, where $5/4 \leq n \leq 4/3$ depending on the geometry of the heated surface. In contrast, for nucleate boiling, the influence of subcooling is considered to be negligible, although the

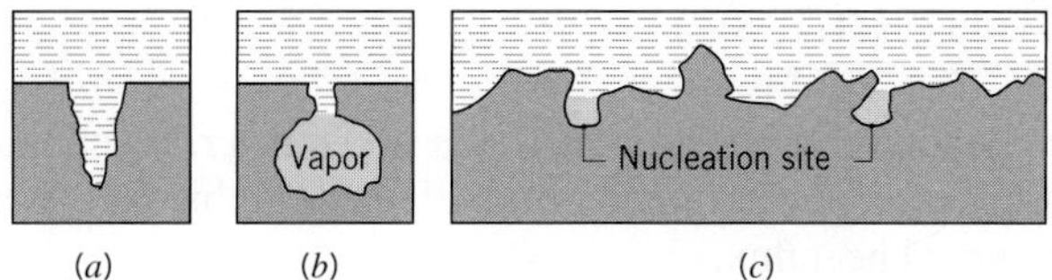

FIGURE 10.6 Formation of nucleation sites. (*a*) Wetted cavity with no trapped vapor. (*b*) Reentrant cavity with trapped vapor. (*c*) Enlarged profile of a roughened surface.

maximum and minimum heat fluxes, q''_{max} and q''_{min}, are known to increase linearly with ΔT_{sub}. For film boiling, the heat flux increases strongly with increasing ΔT_{sub}.

The influence of *surface roughness* (by machining, grooving, scoring, or sandblasting) on the maximum and minimum heat fluxes and on film boiling is negligible. However, as demonstrated by Berenson [20], increased surface roughness can cause a large increase in heat flux for the nucleate boiling regime. As Figure 10.6 illustrates, a roughened surface has numerous cavities that serve to trap vapor, providing more and larger sites for bubble growth. It follows that the nucleation site density for a rough surface can be substantially larger than that for a smooth surface. However, under prolonged boiling, the effects of surface roughness generally diminish, indicating that the new, large sites created by roughening are not stable sources of vapor entrapment.

Special surface arrangements that provide stable *augmentation* (*enhancement*) of nucleate boiling are available commercially and have been reviewed by Webb [21]. *Enhancement surfaces* are of two types: (1) coatings of very porous material formed by sintering, brazing, flame spraying, electrolytic deposition, or foaming, and (2) mechanically machined or formed double-reentrant cavities to ensure continuous vapor trapping (see Figure 10.7). Such surfaces provide for continuous renewal of vapor at the nucleation sites and heat transfer augmentation by more than an order of magnitude. Active augmentation techniques, such as surface wiping–rotation, surface vibration, fluid vibration, and electrostatic fields, have also been reviewed by Bergles [22, 23]. However, because such techniques complicate the boiling system and, in many instances, impair reliability, they have found little practical application.

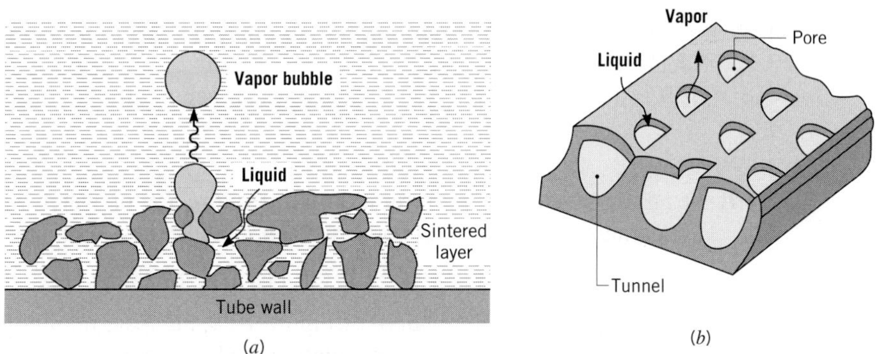

FIGURE 10.7 Typical structured enhancement surfaces for augmentation of nucleate boiling. (*a*) Sintered metallic coating. (*b*) Mechanically formed double-reentrant cavity.

EXAMPLE 10.1

The bottom of a copper pan, 0.3 m in diameter, is maintained at 118°C by an electric heater. Estimate the power required to boil water in this pan. What is the evaporation rate? Estimate the critical heat flux.

SOLUTION

Known: Water boiling in a copper pan of prescribed surface temperature.

Find:

1. Power required by electric heater to cause boiling.
2. Rate of water evaporation due to boiling.
3. Critical heat flux corresponding to the burnout point.

Schematic:

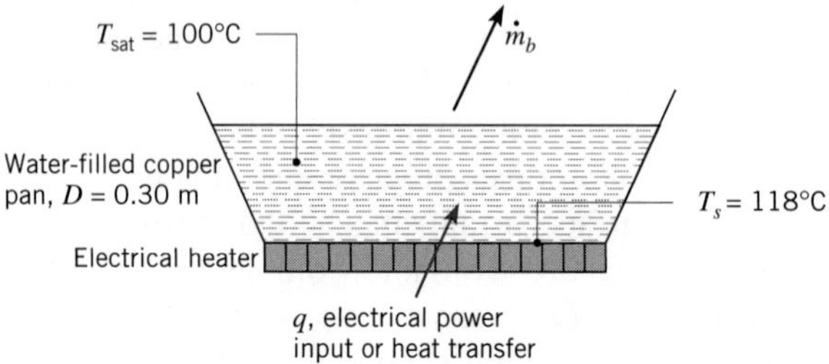

Assumptions:

1. Steady-state conditions.
2. Water exposed to standard atmospheric pressure, 1.01 bar.
3. Water at uniform temperature T_{sat} = 100°C.
4. Large pan bottom surface of polished copper.
5. Negligible losses from heater to surroundings.

Properties: Table A.6, saturated water, liquid (100°C): $\rho_l = 1/v_f = 957.9\ \text{kg/m}^3$, $c_{p,l} = c_{p,f} = 4.217\ \text{kJ/kg}\cdot\text{K}$, $\mu_l = \mu_f = 279 \times 10^{-6}\ \text{N}\cdot\text{s/m}^2$, $Pr_l = Pr_f = 1.76$, $h_{fg} = 2257\ \text{kJ/kg}$, $\sigma = 58.9 \times 10^{-3}\ \text{N/m}$. Table A.6, saturated water, vapor (100°C): $\rho_v = 1/v_g = 0.5956\ \text{kg/m}^3$.

Analysis:

1. From knowledge of the saturation temperature T_{sat} of water boiling at 1 atm and the temperature of the heated copper surface T_s, the excess temperature ΔT_e is

$$\Delta T_e \equiv T_s - T_{sat} = 118°\text{C} - 100°\text{C} = 18°\text{C}$$

According to the boiling curve of Figure 10.4, nucleate pool boiling will occur and the recommended correlation for estimating the heat transfer rate per unit area of plate surface is given by Equation 10.5.

$$q''_s = \mu_l h_{fg}\left[\frac{g(\rho_l - \rho_v)}{\sigma}\right]^{1/2}\left(\frac{c_{p,l}\,\Delta T_e}{C_{s,f}\,h_{fg}\,Pr_l^n}\right)^3$$

The values of $C_{s,f}$ and n corresponding to the polished copper surface–water combination are determined from the experimental results of Table 10.1, where $C_{s,f} = 0.0128$ and $n = 1.0$. Substituting numerical values, the boiling heat flux is

$$q''_s = 279 \times 10^{-6}\ \text{N}\cdot\text{s/m}^2 \times 2257 \times 10^3\ \text{J/kg}$$

$$\times \left[\frac{9.8\ \text{m/s}^2\,(957.9 - 0.5956)\ \text{kg/m}^3}{58.9 \times 10^{-3}\ \text{N/m}}\right]^{1/2}$$

$$\times \left(\frac{4.217 \times 10^3\ \text{J/kg}\cdot\text{K} \times 18^\circ\text{C}}{0.0128 \times 2257 \times 10^3\ \text{J/kg} \times 1.76}\right)^3 = 836\ \text{kW/m}^2$$

Hence the boiling heat transfer rate is

$$q_s = q''_s \times A = q''_s \times \frac{\pi D^2}{4}$$

$$q_s = 8.36 \times 10^5\ \text{W/m}^2 \times \frac{\pi(0.30\ \text{m})^2}{4} = 59.1\ \text{kW} \quad \triangleleft$$

2. Under steady-state conditions all heat addition to the pan will result in water evaporation from the pan. Hence

$$q_s = \dot{m}_b h_{fg}$$

where $\dot{m}_b$ is the rate at which water evaporates from the free surface to the room. It follows that

$$\dot{m}_b = \frac{q_s}{h_{fg}} = \frac{5.91 \times 10^4\ \text{W}}{2257 \times 10^3\ \text{J/kg}} = 0.0262\ \text{kg/s} = 94\ \text{kg/h} \quad \triangleleft$$

3. The critical heat flux for nucleate pool boiling can be estimated from Equation 10.6:

$$q''_{\max} = 0.149 h_{fg}\rho_v \left[\frac{\sigma g(\rho_l - \rho_v)}{\rho_v^2}\right]^{1/4}$$

Substituting the appropriate numerical values,

$$q''_{\max} = 0.149 \times 2257 \times 10^3\ \text{J/kg} \times 0.5956\ \text{kg/m}^3$$

$$\times \left[\frac{58.9 \times 10^{-3}\ \text{N/m} \times 9.8\ \text{m/s}^2\,(957.9 - 0.5956)\ \text{kg/m}^3}{(0.5956\ \text{kg/m}^3)^2}\right]^{1/4}$$

$$q''_{\max} = 1.26\ \text{MW/m}^2 \quad \triangleleft$$

Comments:

1. Note that the critical heat flux $q''_{\max} = 1.26\ \text{MW/m}^2$ represents the maximum heat flux for boiling water at normal atmospheric pressure. Operation of the heater at $q''_s = 0.836\ \text{MW/m}^2$ is therefore below the critical condition.
2. Using Equation 10.7, the minimum heat flux at the Leidenfrost point is $q''_{\min} = 18.9\ \text{kW/m}^2$. Note from Figure 10.4 that, for this condition, $\Delta T_e \approx 120^\circ\text{C}$.

EXAMPLE 10.2

A metal-clad heating element of 6-mm diameter and emissivity $\varepsilon = 1$ is horizontally immersed in a water bath. The surface temperature of the metal is 255°C under steady-state boiling conditions. Estimate the power dissipation per unit length of the heater.

SOLUTION

Known: Boiling from outer surface of horizontal cylinder in water.

Find: Power dissipation per unit length for the cylinder, q_s'.

Schematic:

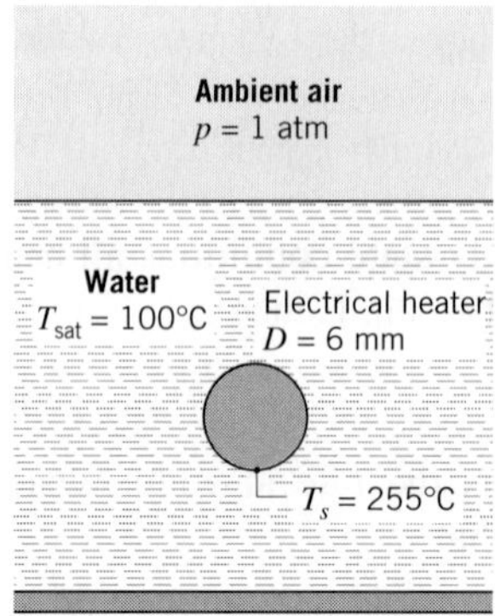

Assumptions:

1. Steady-state conditions.
2. Water exposed to standard atmospheric pressure and at uniform temperature T_{sat}.

Properties: Table A.6, saturated water, liquid (100°C): $\rho_l = 1/v_f = 957.9\ \text{kg/m}^3$, $h_{fg} = 2257\ \text{kJ/kg}$. Table A.4, water vapor at atmospheric pressure ($T_f \approx 450\ \text{K}$): $\rho_v = 0.4902\ \text{kg/m}^3$, $c_{p,v} = 1.980\ \text{kJ/kg} \cdot \text{K}$, $k_v = 0.0299\ \text{W/m} \cdot \text{K}$, $\mu_v = 15.25 \times 10^{-6}\ \text{N} \cdot \text{s/m}^2$.

Analysis: The excess temperature is

$$\Delta T_e = T_s - T_{\text{sat}} = 255°\text{C} - 100°\text{C} = 155°\text{C}$$

According to the boiling curve of Figure 10.4, film pool boiling conditions are achieved, in which case heat transfer is due to both convection and radiation. The heat transfer rate follows from Equation 10.3, written on a per unit length basis for a cylindrical surface of diameter D:

$$q_s' = q_s'' \pi D = \overline{h} \pi D \, \Delta T_e$$

The heat transfer coefficient $\overline{h}$ is calculated from Equation 10.9,

$$\overline{h}^{4/3} = \overline{h}_{\text{conv}}^{4/3} + \overline{h}_{\text{rad}} \overline{h}^{1/3}$$

where the convection and radiation heat transfer coefficients follow from Equations 10.8 and 10.11, respectively. For the convection coefficient:

$$\bar{h}_{\text{conv}} = 0.62\left[\frac{k_v^3 \rho_v(\rho_l - \rho_v)g(h_{fg} + 0.8c_{p,v}\,\Delta T_e)}{\mu_v D\,\Delta T_e}\right]^{1/4}$$

$$\bar{h}_{\text{conv}} = 0.62$$

$$\times\left[\frac{(0.0299\ \text{W/m}\cdot\text{K})^3 \times 0.4902\ \text{kg/m}^3\,(957.9 - 0.4902)\ \text{kg/m}^3 \times 9.8\ \text{m/s}^2}{1}\right.$$

$$\left.\times\frac{(2257 \times 10^3\ \text{J/kg} + 0.8 \times 1.98 \times 10^3\ \text{J/kg}\cdot\text{K} \times 155°\text{C})}{15.25 \times 10^{-6}\ \text{N}\cdot\text{s/m}^2 \times 6 \times 10^{-3}\ \text{m} \times 155°\text{C}}\right]^{1/4}$$

$$\bar{h}_{\text{conv}} = 238\ \text{W/m}^2\cdot\text{K}$$

For the radiation heat transfer coefficient:

$$\bar{h}_{\text{rad}} = \frac{\varepsilon\sigma(T_s^4 - T_{\text{sat}}^4)}{T_s - T_{\text{sat}}}$$

$$\bar{h}_{\text{rad}} = \frac{5.67 \times 10^{-8}\ \text{W/m}^2\cdot\text{K}^4\,(528^4 - 373^4)\text{K}^4}{(528 - 373)\ \text{K}} = 21.3\ \text{W/m}^2\cdot\text{K}$$

Solving Equation 10.9 by trial and error,

$$\bar{h}^{4/3} = 238^{4/3} + 21.3\bar{h}^{1/3}$$

it follows that

$$\bar{h} = 254.1\ \text{W/m}^2\cdot\text{K}$$

Hence the heat transfer rate per unit length of heater element is

$$q_s' = 254.1\ \text{W/m}^2\cdot\text{K} \times \pi \times 6 \times 10^{-3}\ \text{m} \times 155°\text{C} = 742\ \text{W/m} \qquad \triangleleft$$

Comments: Equation 10.10 is also appropriate for estimating $\bar{h}$; it provides a value of 254.0 W/m$^2\cdot$K.

10.5 *Forced Convection Boiling*

In *pool boiling*, fluid flow is due primarily to the buoyancy-driven motion of bubbles originating from the heated surface. In contrast, for *forced convection boiling,* flow is due to a directed (bulk) motion of the fluid, as well as to buoyancy effects. Conditions depend strongly on geometry, which may involve *external* flow over heated plates and cylinders or *internal* (duct) flow. Internal, forced convection boiling is commonly referred to as *two-phase flow* and is characterized by rapid changes from liquid to vapor in the flow direction.

10.5.1 External Forced Convection Boiling

For external flow over a heated plate, the heat flux can be estimated by standard forced convection correlations up to the inception of boiling. As the temperature of the heated plate is increased, nucleate boiling will occur, causing the heat flux to increase. If vapor generation is not extensive and the liquid is subcooled, Bergles and Rohsenow [24] suggest a method for estimating the total heat flux in terms of components associated with pure forced convection and pool boiling.

Both forced convection and subcooling are known to increase the critical heat flux $q''_{\max}$ for nucleate boiling. Experimental values as high as 35 MW/m^2 (compared with 1.3 MW/m^2 for pool boiling of water at 1 atm) have been reported [25]. For a liquid of velocity V moving in cross flow over a cylinder of diameter D, Lienhard and Eichhorn [26] have developed the following expressions for low- and high-velocity flows, where properties are evaluated at the saturation temperature.

Low Velocity:

$$\frac{q''_{\max}}{\rho_v h_{fg} V} = \frac{1}{\pi}\left[1 + \left(\frac{4}{We_D}\right)^{1/3}\right] \tag{10.12}$$

High Velocity:

$$\frac{q''_{\max}}{\rho_v h_{fg} V} = \frac{(\rho_l/\rho_v)^{3/4}}{169\pi} + \frac{(\rho_l/\rho_v)^{1/2}}{19.2\pi\, We_D^{1/3}} \tag{10.13}$$

The Weber number We_D is the ratio of inertia to surface tension forces and has the form

$$We_D \equiv \frac{\rho_v V^2 D}{\sigma} \tag{10.14}$$

The high- and low-velocity regions, respectively, are determined by whether the heat flux parameter $q''_{\max}/\rho_v h_{fg} V$ is less than or greater than $[(0.275/\pi)\,(\rho_l/\rho_v)^{1/2} + 1]$. In most cases, Equations 10.12 and 10.13 correlate $q''_{\max}$ data within 20%.

10.5.2 Two-Phase Flow

Internal forced convection boiling is associated with bubble formation at the inner surface of a heated tube through which a liquid is flowing. Bubble growth and separation are strongly influenced by the flow velocity, and hydrodynamic effects differ significantly from those corresponding to pool boiling. The process is accompanied by the existence of a variety of two-phase flow patterns.

Consider flow development in a vertical tube that is subjected to a constant surface heat flux, through which fluid is moving in the upward direction, as shown in Figure 10.8. Heat transfer to the subcooled liquid that enters the tube is initially by *single-phase forced convection* and may be predicted using the correlations of Chapter 8. Farther down the tube, the wall temperature exceeds the saturation temperature of the liquid, and vaporization is initiated in the *subcooled flow boiling region*. This region is characterized by large radial temperature gradients, with bubbles forming adjacent to the heated wall and subcooled liquid flowing near the center of the tube. The thickness of the bubble region increases

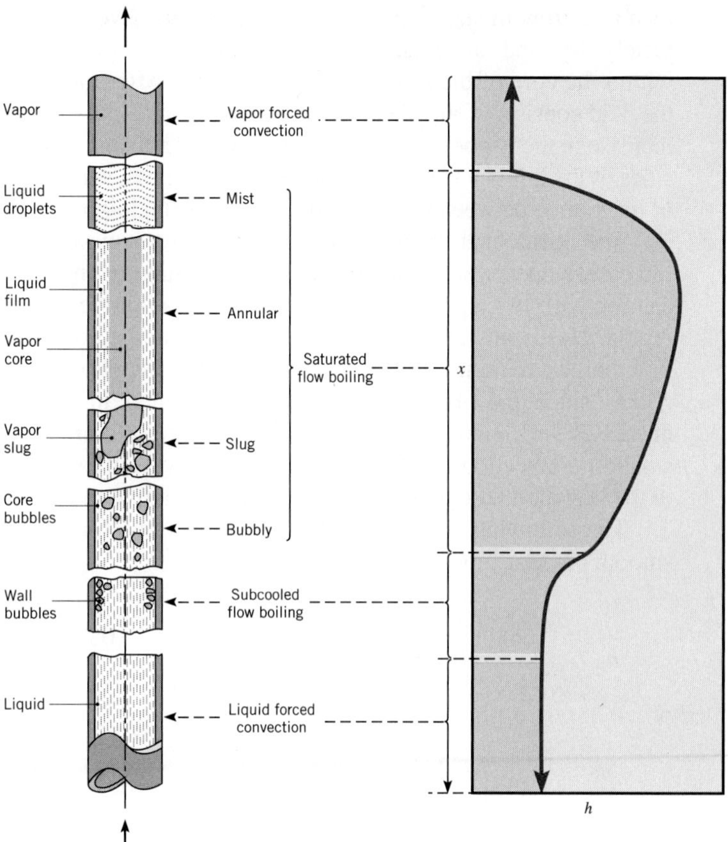

FIGURE 10.8 Flow regimes for forced convection boiling in a tube.

farther downstream, and eventually, the core of the liquid reaches the saturation temperature of the fluid. Bubbles can then exist at any radial location, and the time-averaged mass fraction of vapor in the fluid,[1] X, exceeds zero at any radial location. This marks the beginning of the *saturated flow boiling region*. Within the saturated flow boiling region, the mean vapor mass fraction defined as

$$\overline{X} \equiv \frac{\int_{A_c} \rho u(r,x) X dA_c}{\dot{m}}$$

increases and, due to the large density difference between the vapor and liquid phases, the mean velocity of the fluid, u_m, increases substantially.

The first stage of the saturated flow boiling region corresponds to the *bubbly flow regime*. As $\overline{X}$ increases further, individual bubbles coalesce to form slugs of vapor. This *slug-flow regime* is followed by an *annular-flow regime* in which the liquid forms a film on the tube wall. This film moves along the inner surface of the tube, while vapor moves at a larger velocity through the core of the tube. Dry spots eventually appear on the inner surface of the

[1]This term is often referred to as the *quality* of a two-phase fluid.

tube and grow in size within a *transition regime*. Eventually, the entire tube surface is completely dry, and all remaining liquid is in the form of droplets that travel at high velocity within the core of the tube in the *mist regime*. After the droplets are completely vaporized, the fluid consists of superheated vapor in a *second* single-phase forced convection region. The increase in the vapor fraction along the tube length, along with the large difference in the densities of the liquid and vapor phases, increases the mean velocity of the fluid by several orders of magnitude between the first and the second single-phase forced convection regions.

The local heat transfer coefficient varies significantly as $\overline{X}$ and u_m decrease and increase, respectively, along the length of the tube, x. In general, the heat transfer coefficient can increase by approximately an order of magnitude through the subcooled flow boiling region. Heat transfer coefficients are further increased in the early stages of the saturated flow boiling region. Conditions become more complex deeper in the saturated flow boiling region since the convection coefficient, defined in Equation 10.3, *either* increases or decreases with increasing $\overline{X}$, depending on the fluid and tube wall material. Typically, the smallest convection coefficients exist in the second (vapor) forced convection region owing to the low thermal conductivity of the vapor relative to that of the liquid.

The following correlation has been developed for the saturated flow boiling region in smooth circular tubes [27, 28]:

$$\frac{h}{h_{sp}} = 0.6683\left(\frac{\rho_l}{\rho_v}\right)^{0.1} \overline{X}^{0.16}(1-\overline{X})^{0.64} f(Fr) + 1058\left(\frac{q_s''}{\dot{m}''h_{fg}}\right)^{0.7}(1-\overline{X})^{0.8} G_{s,f} \tag{10.15a}$$

or

$$\frac{h}{h_{sp}} = 1.136\left(\frac{\rho_l}{\rho_v}\right)^{0.45} \overline{X}^{0.72}(1-\overline{X})^{0.08} f(Fr) + 667.2\left(\frac{q_s''}{\dot{m}''h_{fg}}\right)^{0.7}(1-\overline{X})^{0.8} G_{s,f} \tag{10.15b}$$

$$0 < \overline{X} \lesssim 0.8$$

where $\dot{m}'' = \dot{m}/A_c$ is the mass flow rate per unit cross-sectional area. In utilizing Equation 10.15, the larger value of the heat transfer coefficient, h, should be used. In this expression, the liquid phase *Froude number* is $Fr = (\dot{m}''/\rho_l)^2/gD$ and the coefficient $G_{s,f}$ depends on the surface–fluid combination, with representative values given in Table 10.2. Equation 10.15 applies for horizontal as well as vertical tubes, where the *stratification parameter, f(Fr)*, accounts for stratification of the liquid and vapor phases that may occur for horizontal tubes. Its value is unity for vertical tubes and for horizontal tubes with $Fr \gtrsim 0.04$. For horizontal tubes with $Fr \lesssim 0.04$, $f(Fr) = 2.63\, Fr^{0.3}$. All properties are evaluated at the saturation temperature, T_{sat}. The single-phase convection coefficient, h_{sp}, is associated with the liquid forced convection region of Figure 10.8 and is obtained from Equation 8.62 with properties evaluated at T_{sat}. Because Equation 8.62 is for turbulent flow, it is recommended that Equation 10.15 not be applied to situations where the liquid single-phase convection is laminar. Equation 10.15 is applicable when the channel dimension is large relative to the bubble diameter, that is, for Confinement numbers, $Co = \sqrt{\sigma/(g[\rho_l - \rho_v])}/D_h \lesssim 1/2$ [3].

In order to use Equation 10.15, the mean vapor mass fraction, $\overline{X}$, must be known. For negligible changes in the fluid's kinetic and potential energy as well as negligible work, Equation 1.12d may be rearranged to yield

$$\overline{X}(x) = \frac{q_s''\pi D x}{\dot{m} h_{fg}} \tag{10.16}$$

TABLE 10.2 Values of $G_{s,f}$ for various surface–fluid combinations [27, 28]

Fluid in Commercial Copper Tubing	$G_{s,f}$
Kerosene	0.488
Refrigerant R-134a	1.63
Refrigerant R-152a	1.10
Water	1.00
For stainless steel tubing, use $G_{s,f} = 1$.	

where the origin of the x-coordinate, $x = 0$, corresponds to the axial location where $\overline{X}$ begins to exceed zero, and the change in enthalpy, $u_t + pv$, is equal to the change in $\overline{X}$ multiplied by the enthalpy of vaporization, h_{fg}.

Correlations for the subcooled flow boiling region and annular as well as mist regimes are available in the literature [28]. For constant heat flux conditions, critical heat fluxes may occur in the subcooled flow boiling region, in the saturated flow boiling region where $\overline{X}$ is large, or in the vapor forced convection region. Critical heat flux conditions may lead to melting of the tube material in extreme conditions [29]. Additional discussions of flow boiling are available in the literature [7, 30–33]. Extensive databases consisting of thousands of experimentally measured values of the critical heat flux for wide ranges of operating conditions are also available [34, 35].

10.5.3 Two-Phase Flow in Microchannels

Two-phase microchannels feature forced convection boiling of a liquid through circular or noncircular tubes having hydraulic diameters ranging from 10 to 1000 μm, resulting in extremely high heat transfer rates [36, 37]. In these situations, the characteristic bubble size can occupy a significant fraction of the tube diameter and the Confinement number can become very large ($Co \gtrsim 1/2$). Hence, different types of flow regimes exist, including regimes where the bubbles occupy nearly the full diameter of the heated tube [38]. This can lead to a dramatic increase in the convection coefficient, h, corresponding to the peak in Figure 10.8. Thereafter, h decreases with increasing x as it does in Figure 10.8. Equation 10.15 cannot be used to predict correct values of the heat transfer coefficient and does not even predict correct trends for microchannel flow boiling cases. Recourse must be made to more sophisticated modeling [36, 39].

10.6 *Condensation: Physical Mechanisms*

Condensation occurs when the temperature of a vapor is reduced below its saturation temperature. In industrial equipment, the process commonly results from contact between the vapor and a cool *surface* (Figures 10.9*a*, *b*). The latent energy of the vapor is released, heat is transferred to the surface, and the condensate is formed. Other common modes are *homogeneous* condensation (Figure 10.9*c*), where vapor condenses out as droplets suspended in a gas phase to form a fog, and *direct contact* condensation (Figure 10.9*d*), which occurs when vapor is brought into contact with a cold liquid. In this chapter we will consider only surface condensation.

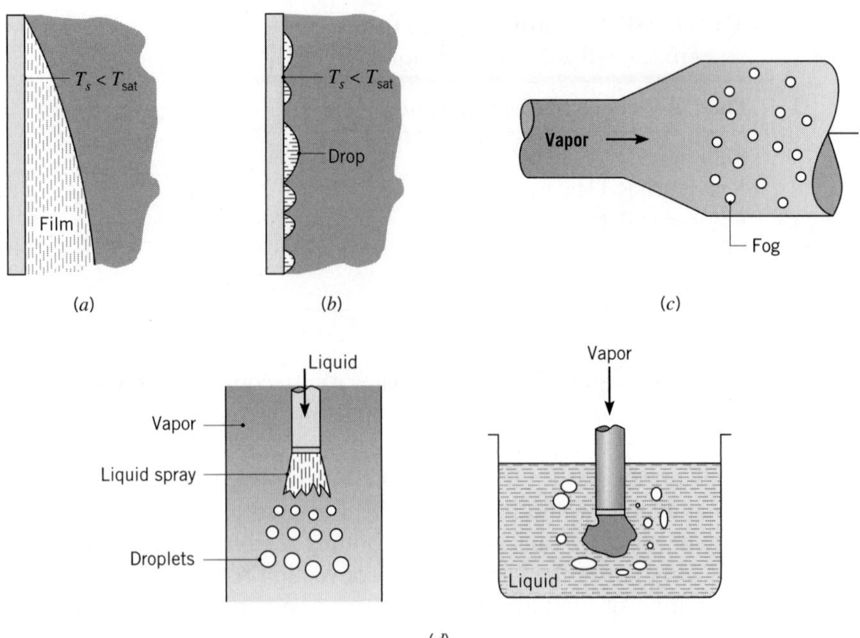

FIGURE 10.9 Modes of condensation. (*a*) Film. (*b*) Dropwise condensation on a surface. (*c*) Homogeneous condensation or fog formation resulting from increased pressure due to expansion. (*d*) Direct contact condensation.

As shown in Figures 10.9*a*, *b*, condensation may occur in one of two ways, depending on the condition of the surface. The dominant form of condensation is one in which a liquid film covers the entire condensing surface, and under the action of gravity the film flows continuously from the surface. *Film condensation* is generally characteristic of clean, uncontaminated surfaces. However, if the surface is coated with a substance that inhibits wetting, it is possible to maintain *dropwise condensation.* The drops form in cracks, pits, and cavities on the surface and may grow and coalesce through continued condensation. Typically, more than 90% of the surface is covered by drops, ranging from a few micrometers in diameter to agglomerations visible to the naked eye. The droplets flow from the surface due to the action of gravity. Film and dropwise condensation of steam on a vertical copper surface are shown in Figure 10.10. A thin coating of cupric oleate was applied to the left-hand portion of the surface to promote the dropwise condensation. A thermocouple probe of 1-mm diameter extends across the photograph.

Regardless of whether it is in the form of a film or droplets, the condensate provides a resistance to heat transfer between the vapor and the surface. Because this resistance increases with condensate thickness, which increases in the flow direction, it is desirable to use short vertical surfaces or horizontal cylinders in situations involving film condensation. Most condensers therefore consist of horizontal tube bundles through which a liquid coolant flows and around which the vapor to be condensed is circulated. In terms of maintaining high condensation and heat transfer rates, droplet formation is superior to film formation. In dropwise condensation most of the heat transfer is through drops of less than 100-μm diameter, and transfer rates that are more than an order of magnitude larger than those associated with film condensation may be achieved. It is therefore common practice to use surface coatings that inhibit wetting, and hence stimulate dropwise condensation. Silicones, Teflon, and an assortment of waxes and fatty acids are often used for this purpose.

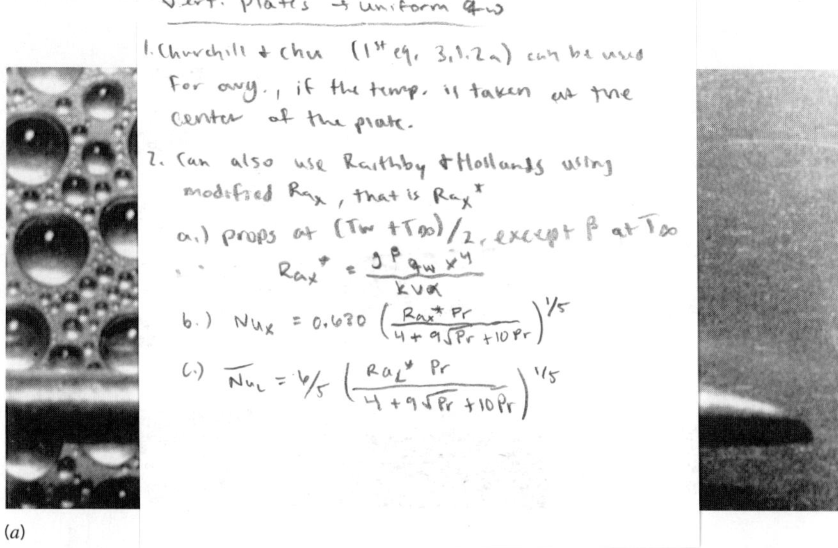

(*a*)

FIGURE 10.10 Condensation on a vertical surface. (*a*) Dropwise. (*b*) Film. (Photograph courtesy of Professor J. W. Westwater, University of Illinois at Champaign-Urbana.)

However, such coatings gradually lose their effectiveness due to oxidation, fouling, or outright removal, and film condensation eventually occurs.

Although it is desirable to achieve dropwise condensation in industrial applications, it is often difficult to maintain this condition. For this reason and because the convection coefficients for film condensation are smaller than those for the dropwise case, condenser design calculations are often based on the assumption of film condensation. In the remaining sections of this chapter, we focus on film condensation and mention only briefly results available for dropwise condensation.

10.7 *Laminar Film Condensation on a Vertical Plate*

As shown in Figure 10.11*a*, there may be several complicating features associated with film condensation. The film originates at the top of the plate and flows downward under the influence of gravity. The thickness δ and the condensate mass flow rate $\dot{m}$ increase with increasing x because of continuous condensation at the liquid–vapor interface, which is at T_{sat}. There is then heat transfer from this interface through the film to the surface, which is maintained at $T_s < T_{sat}$. In the most general case the vapor may be superheated ($T_{v,\infty} > T_{sat}$) and may be part of a mixture containing one or more noncondensable gases. Moreover, there exists a finite shear stress at the liquid–vapor interface, contributing to a velocity gradient in the vapor, as well as in the film [40, 41].

Despite the complexities associated with film condensation, useful results may be obtained by making assumptions that originated with an analysis by Nusselt [42].

1. Laminar flow and constant properties are assumed for the liquid film.
2. The gas is assumed to be a pure vapor and at a uniform temperature equal to T_{sat}. With no temperature gradient in the vapor, heat transfer to the liquid–vapor interface can occur only by condensation at the interface and not by conduction from the vapor.
3. The shear stress at the liquid–vapor interface is assumed to be negligible, in which case $\partial u/\partial y|_{y=\delta} = 0$. With this assumption and the foregoing assumption of a uniform vapor

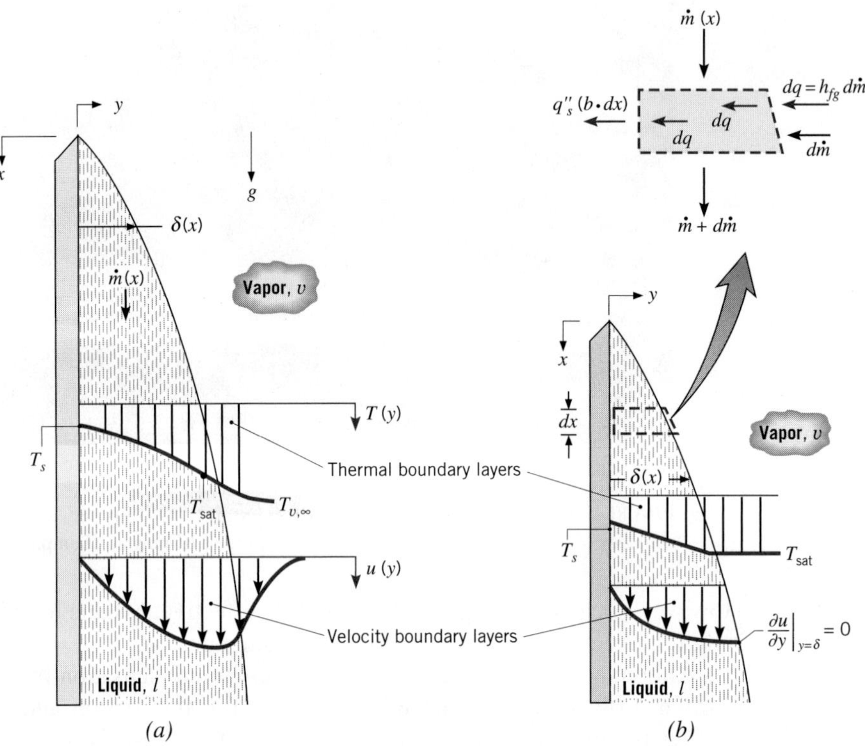

FIGURE 10.11 Boundary layer effects related to film condensation on a vertical surface. (*a*) Without approximation. (*b*) With assumptions associated with Nusselt's analysis, for a vertical plate of width *b*.

temperature, there is no need to consider the vapor velocity or thermal boundary layers shown in Figure 10.11*a*.

4. Momentum and energy transfer by advection in the condensate film are assumed to be negligible. This assumption is reasonable by virtue of the low velocities associated with the film. It follows that heat transfer across the film occurs only by conduction, in which case the liquid temperature distribution is linear.

Film conditions resulting from the assumptions are shown in Figure 10.11*b*.

The *x*-momentum equation for the film can be found from Equation 9.1, with $\rho = \rho_l$ and $\nu = \nu_l$ for the liquid, and with the sign of the gravity term changed since x is now in the direction of gravity. The pressure gradient is obtained from Equation 9.2 and is $dp_\infty/dx = +\rho_v g$, since the free stream density is the vapor density. From the fourth approximation, momentum advection terms may be neglected, and the *x*-momentum equation may be expressed as

$$\frac{\partial^2 u}{\partial y^2} = -\frac{g}{\mu_l}(\rho_l - \rho_v) \tag{10.17}$$

Integrating twice and applying boundary conditions of the form $u(0) = 0$ and $\partial u/\partial y|_{y=\delta} = 0$, the velocity profile in the film becomes

$$u(y) = \frac{g(\rho_l - \rho_v)\delta^2}{\mu_l}\left[\frac{y}{\delta} - \frac{1}{2}\left(\frac{y}{\delta}\right)^2\right] \tag{10.18}$$

From this result the condensate mass flow rate per unit width $\Gamma(x)$ may be obtained in terms of an integral involving the velocity profile:

$$\frac{\dot{m}(x)}{b} = \int_0^{\delta(x)} \rho_l u(y)\, dy \equiv \Gamma(x) \tag{10.19}$$

Substituting from Equation 10.18, it follows that

$$\Gamma(x) = \frac{g\rho_l(\rho_l - \rho_v)\delta^3}{3\mu_l} \tag{10.20}$$

The specific variation with x of δ, and hence of Γ, may be obtained by first applying the conservation of energy requirement to the differential element shown in Figure 10.11*b*. At a portion of the liquid–vapor interface of unit width and length dx, the rate of heat transfer into the film, dq, must equal the rate of energy release due to condensation at the interface. Hence

$$dq = h_{fg}\, d\dot{m} \tag{10.21}$$

Since advection is neglected, it also follows that the rate of heat transfer across the interface must equal the rate of heat transfer to the surface. Hence

$$dq = q_s''(b \cdot dx) \tag{10.22}$$

Since the liquid temperature distribution is linear, Fourier's law may be used to express the surface heat flux as

$$q_s'' = \frac{k_l(T_{\text{sat}} - T_s)}{\delta} \tag{10.23}$$

Combining Equations 10.19 and 10.21 through 10.23, we then obtain

$$\frac{d\Gamma}{dx} = \frac{k_l(T_{\text{sat}} - T_s)}{\delta h_{fg}} \tag{10.24}$$

Differentiating Equation 10.20, we also obtain

$$\frac{d\Gamma}{dx} = \frac{g\rho_l(\rho_l - \rho_v)\delta^2}{\mu_l}\frac{d\delta}{dx} \tag{10.25}$$

Combining Equations 10.24 and 10.25, it follows that

$$\delta^3 d\delta = \frac{k_l\mu_l(T_{\text{sat}} - T_s)}{g\rho_l(\rho_l - \rho_v)h_{fg}}\, dx$$

Integrating from $x = 0$, where $\delta = 0$, to any x-location of interest on the surface,

$$\delta(x) = \left[\frac{4k_l\mu_l(T_{\text{sat}} - T_s)x}{g\rho_l(\rho_l - \rho_v)h_{fg}}\right]^{1/4} \tag{10.26}$$

This result may then be substituted into Equation 10.20 to obtain $\Gamma(x)$.

An improvement to the foregoing result for $\delta(x)$ was made by Nusselt [42] and Rohsenow [43], who showed that, with the inclusion of thermal advection effects, a term is added to the latent heat of vaporization. In lieu of h_{fg}, Rohsenow recommended using a modified latent heat of the form $h'_{fg} = h_{fg} + 0.68c_{p,l}(T_{\text{sat}} - T_s)$, or in terms of the Jakob number,

$$h'_{fg} = h_{fg}(1 + 0.68\, Ja) \tag{10.27}$$

More recently, Sadasivan and Lienhard [18] have shown that the modified latent heat depends weakly on the Prandtl number of the liquid.

The surface heat flux may be expressed as

$$q''_s = h_x(T_{\text{sat}} - T_s) \tag{10.28}$$

Substituting from Equation 10.23, the local convection coefficient is then

$$h_x = \frac{k_l}{\delta} \tag{10.29}$$

or, from Equation 10.26, with h_{fg} replaced by h'_{fg},

$$h_x = \left[\frac{g\rho_l(\rho_l - \rho_v)k_l^3 h'_{fg}}{4\mu_l(T_{\text{sat}} - T_s)x}\right]^{1/4} \tag{10.30}$$

Since h_x depends on $x^{-1/4}$, it follows that the average convection coefficient for the entire plate is

$$\bar{h}_L = \frac{1}{L}\int_0^L h_x dx = \frac{4}{3}h_L$$

or

$$\bar{h}_L = 0.943\left[\frac{g\rho_l(\rho_l - \rho_v)k_l^3 h'_{fg}}{\mu_l(T_{\text{sat}} - T_s)L}\right]^{1/4} \tag{10.31}$$

The average Nusselt number then has the form

$$\overline{Nu}_L = \frac{\bar{h}_L L}{k_l} = 0.943\left[\frac{\rho_l g(\rho_l - \rho_v)h'_{fg}L^3}{\mu_l k_l(T_{\text{sat}} - T_s)}\right]^{1/4} \tag{10.32}$$

In using this equation in conjunction with Equation 10.27, all liquid properties should be evaluated at the film temperature $T_f = (T_{\text{sat}} + T_s)/2$. The vapor density ρ_v and latent heat of vaporization h_{fg} should be evaluated at T_{sat}.

A more detailed boundary layer analysis of film condensation on a vertical plate has been performed by Sparrow and Gregg [40]. Their results, confirmed by Chen [44], indicate that errors associated with using Equation 10.32 are less than 3% for $Ja \leq 0.1$ and $1 \leq Pr \leq 100$. Dhir and Lienhard [45] have also shown that Equation 10.32 may be used for inclined plates, if g is replaced by $g \cdot \cos\theta$, where θ is the angle between the vertical and the surface. However, it must be used with caution for large values of θ and does not apply if $\theta = \pi/2$. The expression may be used for condensation on the inner or outer surface of a vertical tube of radius R, if $R \gg \delta$.

The total heat transfer to the surface may be obtained by using Equation 10.31 with the following form of Newton's law of cooling:

$$q = \overline{h}_L A(T_{sat} - T_s) \tag{10.33}$$

The total condensation rate may then be determined from the relation

$$\dot{m} = \frac{q}{h'_{fg}} = \frac{\overline{h}_L A(T_{sat} - T_s)}{h'_{fg}} \tag{10.34}$$

Equations 10.33 and 10.34 are generally applicable to any surface geometry, although the form of $\overline{h}_L$ will vary according to geometry and flow conditions.

10.8 Turbulent Film Condensation

As for all previously discussed convection phenomena, turbulent flow conditions may exist in film condensation. Consider the vertical surface of Figure 10.12*a*. The transition criterion may be expressed in terms of a Reynolds number defined as

$$Re_\delta \equiv \frac{4\Gamma}{\mu_l} \tag{10.35}$$

With the condensate mass flow rate given by $\dot{m} = \rho_l u_m b\delta$, the Reynolds number may be expressed as

$$Re_\delta = \frac{4\dot{m}}{\mu_l b} = \frac{4\rho_l u_m \delta}{\mu_l} \tag{10.36}$$

where u_m is the average velocity in the film and δ, the film thickness, is the characteristic length. As in the case of single-phase boundary layers, the Reynolds number is an indicator of flow conditions. As shown in Figure 10.12*b*, for $Re_\delta \lesssim 30$, the film is laminar and wave free. For increased Re_δ, ripples or waves form on the condensate film, and at $Re_\delta \approx 1800$ the transition from laminar to turbulent flow is complete.

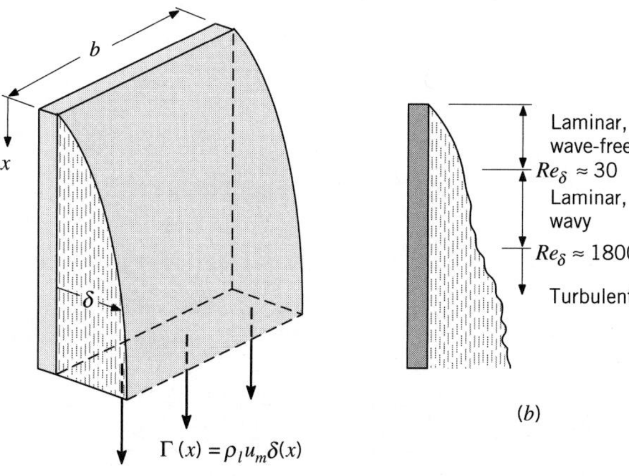

FIGURE 10.12 Film condensation on a vertical plate. (*a*) Condensate rate for plate of width *b*. (*b*) Flow regimes.

For the wave-free laminar regime ($Re_\delta \lesssim 30$), Equations 10.35 and 10.20 may be combined to yield

$$Re_\delta = \frac{4g\rho_l(\rho_l - \rho_v)\delta^3}{3\mu_l^2} \tag{10.37}$$

Assuming $\rho_l \gg \rho_v$, Equations 10.26, 10.31, and 10.37 may be combined to provide an expression for an average modified Nusselt number associated with condensation in the wave-free laminar regime:

$$\overline{Nu}_L = \frac{\overline{h}_L(\nu_l^2/g)^{1/3}}{k_l} = 1.47\,Re_\delta^{-1/3} \qquad Re_\delta \lesssim 30 \tag{10.38}$$

where the average heat transfer coefficient $\overline{h}_L$ is associated with condensation over the entire plate. When the flow at the bottom of the plate is in the laminar, wavy regime, Kutateladze [46] recommends a correlation of the form

$$\overline{Nu}_L = \frac{\overline{h}_L(\nu_l^2/g)^{1/3}}{k_l} = \frac{Re_\delta}{1.08\,Re_\delta^{1.22} - 5.2} \qquad 30 \lesssim Re_\delta \lesssim 1800 \tag{10.39}$$

and when the flow at the bottom of the plate is in the turbulent regime, Labuntsov [47] recommends

$$\overline{Nu}_L = \frac{\overline{h}_L(\nu_l^2/g)^{1/3}}{k_l} = \frac{Re_\delta}{8750 + 58\,Pr_l^{-0.5}(Re_\delta^{0.75} - 253)} \qquad Re_\delta \gtrsim 1800,\ Pr_l \gtrsim 1 \tag{10.40}$$

Graphical representation of the foregoing correlations is provided in Figure 10.13, and the trends have been verified experimentally by Gregorig et al. [48] for water over the range $1 < Re_\delta < 7200$. All properties are evaluated as for laminar film condensation, as explained beneath Equation 10.32.

The Reynolds number in Equations 10.38 through 10.40 is associated with the film thickness δ that exists at the bottom of the condensing surface, $x = L$. If δ is unknown, it is

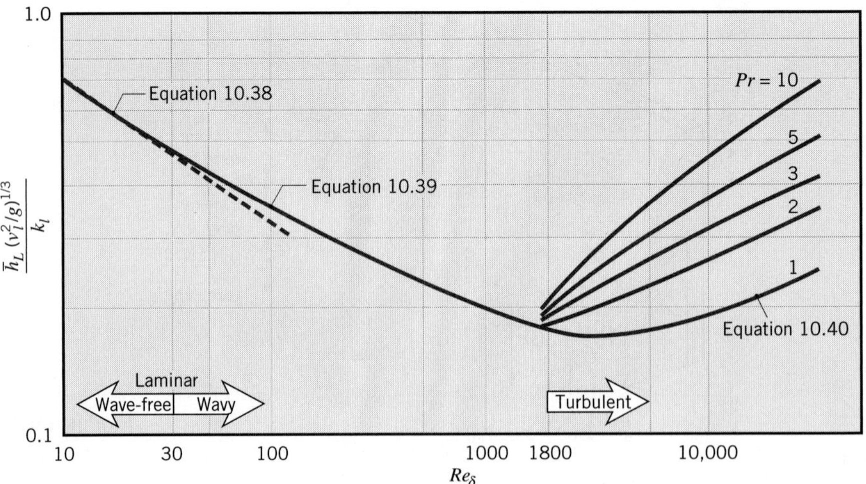

FIGURE 10.13 Modified Nusselt number for condensation on a vertical plate.

preferable to rewrite these equations in a form that eliminates Re_δ. To do so, Equations 10.34 and 10.36 may be combined with the definition of the average Nusselt number to provide

$$Re_\delta = 4P \frac{\bar{h}_L(\nu_l^2/g)^{1/3}}{k_l} = 4P\,\overline{Nu}_L \tag{10.41}$$

where the dimensionless parameter P is

$$P = \frac{k_l L(T_{sat} - T_s)}{\mu_l h'_{fg}(\nu_l^2/g)^{1/3}} \tag{10.42}$$

Substituting Equation 10.41 into Equations 10.38, 10.39, and 10.40, we can solve for the average Nusselt numbers in terms of P to yield

$$\overline{Nu}_L = \frac{\bar{h}_L(\nu_l^2/g)^{1/3}}{k_l} = 0.943\,P^{-1/4} \qquad P \lesssim 15.8 \tag{10.43}$$

$$\overline{Nu}_L = \frac{\bar{h}_L(\nu_l^2/g)^{1/3}}{k_l} = \frac{1}{P}(0.68\,P + 0.89)^{0.82} \qquad 15.8 \lesssim P \lesssim 2530 \tag{10.44}$$

$$\overline{Nu}_L = \frac{\bar{h}_L(\nu_l^2/g)^{1/3}}{k_l} = \frac{1}{P}[(0.024\,P - 53)Pr_l^{1/2} + 89]^{4/3} \qquad P \gtrsim 2530,\ Pr_l \geq 1 \tag{10.45}$$

Equation 10.43 is identical to Equation 10.32 with $\rho_l \gg \rho_v$.

For a particular problem, P may be determined from Equation 10.42, after which the average Nusselt number or average heat transfer coefficient may be found from Equation 10.43, 10.44, or 10.45.

EXAMPLE 10.3

The outer surface of a vertical tube, which is 1 m long and has an outer diameter of 80 mm, is exposed to saturated steam at atmospheric pressure and is maintained at 50°C by the flow of cool water through the tube. What is the rate of heat transfer to the coolant, and what is the rate at which steam is condensed at the surface?

SOLUTION

Known: Dimensions and temperature of a vertical tube experiencing condensation of steam at its outer surface.

Find: Heat transfer and condensation rates.

Schematic:

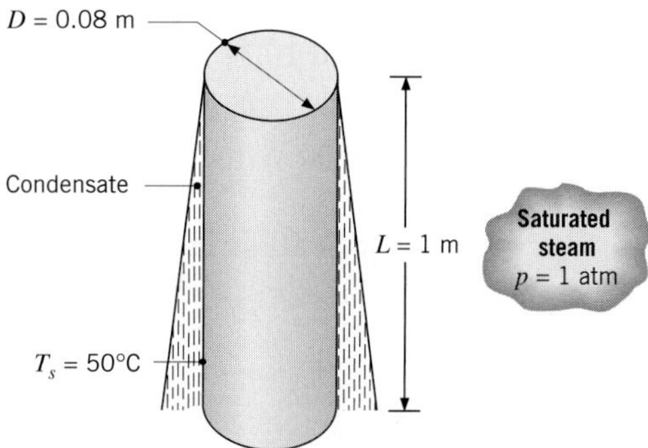

Assumptions:

1. The condensate film thickness is small relative to the cylinder diameter.
2. Negligible concentration of noncondensable gases in the steam.

Properties: Table A.6, saturated vapor (p = 1.0133 bars): $T_{sat} = 100°C$, $\rho_v = 1/v_g = 0.596\ kg/m^3$, $h_{fg} = 2257\ kJ/kg$. Table A.6, saturated liquid ($T_f = 75°C$): $\rho_l = 1/v_f = 975\ kg/m^3$, $\mu_l = 375 \times 10^{-6}\ N \cdot s/m^2$, $k_l = 0.668\ W/m \cdot K$, $c_{p,l} = 4193\ J/kg \cdot K$, $\nu_l = \mu_l/\rho_l = 385 \times 10^{-9}\ m^2/s$.

Analysis: Since we assume the film thickness is small relative to the cylinder diameter, we may use the correlations of Sections 10.7 and 10.8. With

$$Ja = \frac{c_{p,l}(T_{sat} - T_s)}{h_{fg}} = \frac{4193\ J/kg \cdot K(100 - 50)\ K}{2257 \times 10^3\ J/kg} = 0.0929$$

it follows that

$$h'_{fg} = h_{fg}(1 + 0.68\, Ja) = 2257\ kJ/kg\ (1.0632) = 2400\ kJ/kg$$

From Equation 10.42,

$$P = \frac{k_l L(T_{sat} - T_s)}{\mu_l h'_{fg}(\nu_l^2/g)^{1/3}}$$

$$= \frac{0.668\ W/m \cdot K \times 1\ m \times (100 - 50)\ K}{375 \times 10^{-6}\ N \cdot s/m^2 \times 2.4 \times 10^6\ J/kg \left[\dfrac{(385 \times 10^{-9}\ m^2/s)^2}{9.8\ m/s^2}\right]^{1/3}} = 1501$$

Therefore, Equation 10.44 applies:

$$\overline{Nu}_L = \frac{1}{P}(0.68\,P + 0.89)^{0.82} = \frac{1}{1501}(0.68 \times 1501 + 0.89)^{0.82} = 0.20$$

Then

$$\bar{h}_L = \frac{\overline{Nu}_L k_l}{(\nu_l^2/g)^{1/3}} = \frac{0.20 \times 0.668\ W/m \cdot K}{\left[\dfrac{(385 \times 10^{-9}\ m^2/s)^2}{9.8\ m/s^2}\right]^{1/3}} = 5300\ W/m^2 \cdot K$$

and from Equations 10.33 and 10.34

$$q = \bar{h}_L(\pi DL)(T_{sat} - T_s) = 5300\ W/m^2 \cdot K \times \pi \times 0.08\ m \times 1\ m\ (100 - 50)\ K = 66.6\ kW \quad \triangleleft$$

$$\dot{m} = \frac{q}{h'_{fg}} = \frac{66.6 \times 10^3\ W}{2.4 \times 10^6\ J/kg} = 0.0276\ kg/s \quad \triangleleft$$

Note that using Equation 10.26, with the corrected latent heat, the film thickness at the bottom of the tube $\delta(L)$ for the wave-free laminar assumption is

$$\delta(L) = \left[\frac{4k_l\mu_l(T_{\text{sat}} - T_s)L}{g\rho_l(\rho_l - \rho_v)h'_{fg}}\right]^{1/4}$$

$$\delta(L) = \left[\frac{4 \times 0.668\ \text{W/m}\cdot\text{K} \times 375 \times 10^{-6}\ \text{kg/s}\cdot\text{m}\ (100 - 50)\ \text{K} \times 1\ \text{m}}{9.8\ \text{m/s}^2 \times 975\ \text{kg/m}^3\ (975 - 0.596)\ \text{kg/m}^3 \times 2.4 \times 10^6\ \text{J/kg}}\right]^{1/4}$$

$$\delta(L) = 2.18 \times 10^{-4}\ \text{m} = 0.218\ \text{mm}$$

Hence $\delta(L) \ll (D/2)$, and use of the vertical plate correlation for a vertical cylinder is justified.

Comments:

1. The condensation heat and mass rates may be increased by increasing the length of the tube. For $1 \leq L \leq 2$ m, the calculations yield the variations shown below, for which $1000 \leq Re_\delta \leq 2330$ or $1500 \leq P \leq 3010$. The foregoing calculations were performed by using the wavy-laminar correlation, Equation 10.44, for $P \leq 2530$ ($L \leq 1.68$ m) and Equation 10.45, for $P > 2530$ ($L > 1.68$ m). Note, however, that the correlations do not provide equivalent results at $P = 2530$. In particular, Equation 10.45 is a function of Pr, whereas Equation 10.44 is not.

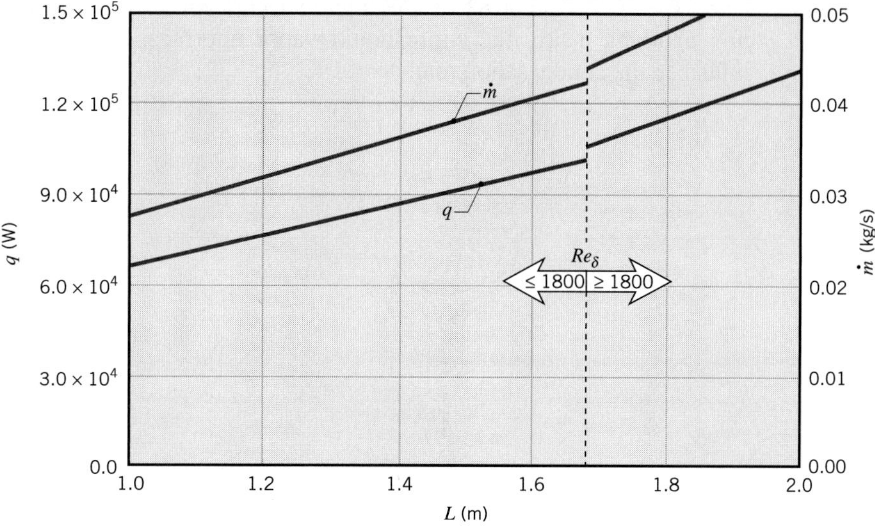

2. If a noncondensable gas such as air is mixed with the steam, heat transfer and condensation rates can be reduced significantly. This is due to multiple effects [36]. For example, q and $\dot{m}$ can drop by 65% if the steam contains only 1% air by weight. Steam condensers that operate at subatmospheric pressure, such as those utilized in Rankine cycles, must be meticulously designed to prevent infiltration of air.

10.9 Film Condensation on Radial Systems

The Nusselt analysis of Section 10.7 may be extended to laminar film condensation on the outer surface of a sphere or a horizontal tube (Figures 10.14*a,b*) and the average Nusselt number has the form

$$\overline{Nu}_D = \frac{\bar{h}_D D}{k_l} = C\left[\frac{\rho_l g(\rho_l - \rho_v)h'_{fg}D^3}{\mu_l k_l(T_{sat} - T_s)}\right]^{1/4} \quad (10.46)$$

where $C = 0.826$ for the sphere [49] and 0.729 for the tube [45]. The properties in this equation and in Equations 10.48 and 10.49 below are evaluated as explained beneath Equation 10.32.

When a liquid–vapor interface is curved, such as those of Figure 10.14, pressure differences are established across the interface by the effects of surface tension. This pressure difference is described by the *Young–Laplace* equation, which for a two-dimensional system may be written

$$\Delta p = p_v - p_l = \frac{\sigma}{r_c} \quad (10.47)$$

where r_c is the local radius of curvature of the liquid–vapor interface. If r_c varies along the interface (and the vapor pressure p_v is constant), the pressure on the liquid side of the interface is nonuniform, influencing the velocity distribution within the liquid and the heat transfer rate. For the unfinned tube of Figure 10.14*b*, the interface curvature is relatively large, $r_c \approx D/2$, except where the liquid sheet departs from the bottom of the tube. Hence $p_l \approx p_v$ along nearly the entire liquid–vapor interface, and the surface tension does not influence the condensation rate.

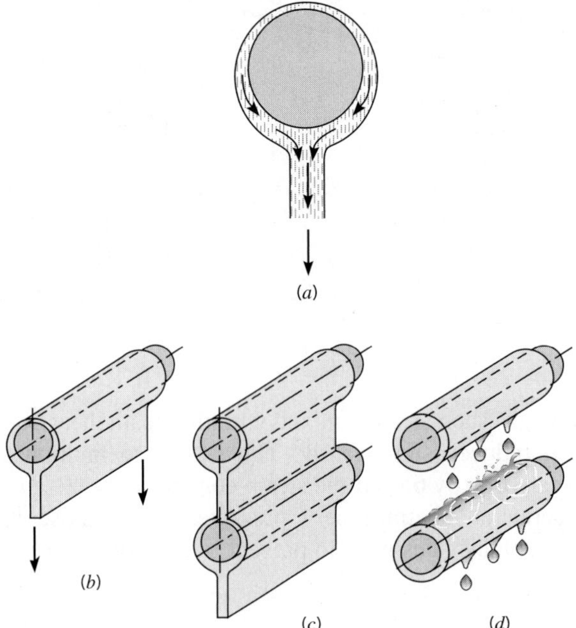

FIGURE 10.14 Film condensation on (*a*) a sphere, (*b*) a single horizontal tube, (*c*) a vertical tier of horizontal tubes with a continuous condensate sheet, and (*d*) with dripping condensate.

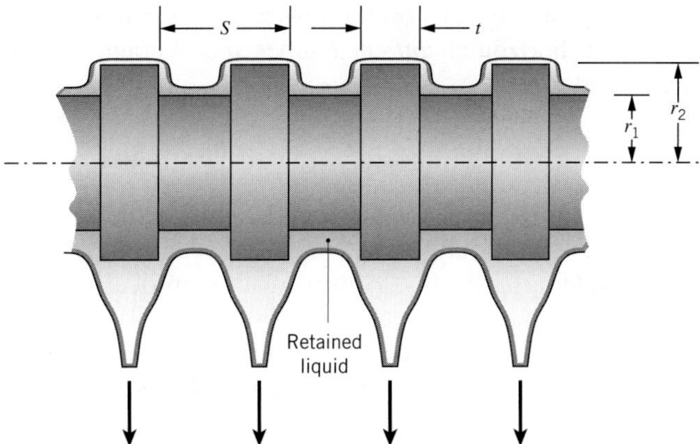

FIGURE 10.15 Condensation on a horizontal finned tube.

Condensation on a tube with annular fins is shown in Figure 10.15. In this case, the sharp corners of the finned tube lead to large variations in the liquid–vapor interface curvature, and surface tension effects can be important. For the finned tube, surface tension forces tend to increase heat transfer rates near the fin tips by reducing the film thickness and decrease heat transfer rates in the inter-fin region by retaining condensate. Just as the liquid layer is thicker on the bottom of a sphere (Figure 10.14*a*) or unfinned horizontal tube (Figure 10.14*b*), there is more retained condensate on the underside of the horizontal finned tube.

Heat transfer rates for the finned tube q_{ft} may be related to those for a corresponding unfinned tube q_{uft} by an *enhancement ratio*, $\varepsilon_{\text{ft}} = q_{\text{ft}}/q_{\text{uft}}$. The degree of enhancement depends primarily on the fluid, the ambient pressure, and the fin geometry, and is weakly dependent on the difference between the tube and ambient temperatures [50]. Small fins, relative to those commonly used for single-phase convection, promote a highly curved liquid surface and, in turn, can enhance heat transfer significantly. The small fins can be fabricated by, for example, removing material from a tube of radius r_2 as shown in Figure 10.15, thereby eliminating contact resistances at the tube–fin interface. Moreover, when manufactured from a metal of high thermal conductivity such as copper, it is often reasonable to assume the tube and small fins have the same, uniform temperature.

Heat transfer correlations for finned tubes tend to be cumbersome and have restricted ranges of application [51]. For design purposes, however, correlations derived by Rose [50] may be used to estimate the *minimum* enhancement associated with the use of a finned tube. This minimum enhancement occurs when condensate is retained in the *entire* inter-fin region, and is

$$\varepsilon_{\text{ft,min}} = \frac{q_{\text{ft,min}}}{q_{\text{uft}}} = \frac{tr_2}{Sr_1}\left[\frac{r_1}{r_2} + 1.02\,\frac{\sigma r_1}{(\rho_l - \rho_v)gt^3}\right]^{1/4} \tag{10.48}$$

where ρ_l and ρ_v are evaluated as described below Equation 10.32 and the surface tension σ is evaluated at T_{sat}. Actual enhancements exceed $\varepsilon_{\text{ft,min}}$ and have been reported to be in the range $2 \leq \varepsilon_{\text{ft}} \leq 4$ for water [50]. Procedures for estimating the heat transfer rates associated with nonisothermal fins are provided by Briggs and Rose [52].

For vertically aligned tubes with a continuous condensate sheet, as shown in Figure 10.14*c*, the heat transfer rate associated with the lower tubes is less than that of the top tube because the films on the lower tubes are thicker than on the top tube. For a vertical tier of N horizontal *unfinned* tubes the average coefficient (over all N tubes) may be expressed as

$$\bar{h}_{D,N} = \bar{h}_D N^n \tag{10.49}$$

where $\bar{h}_D$ is the heat transfer coefficient for the top tube given by Equation 10.46. The Nusselt analysis may be extended to account for the increasing tube-to-tube film thickness, yielding $n = -1/4$. However, an empirical value of $n = -1/6$ is often found to be more appropriate [53].

The discrepancy between the analytical and empirical values of n may be attributed to several effects. The analysis is based on the assumption of a continuous adiabatic sheet of condensate spanning the tubes, as illustrated in Figure 10.14*c*. However, heat transfer to the liquid sheet and its increase in momentum as it falls freely under gravity also increase overall heat transfer rates. Chen [54] accounted for these influences and reported their effects in terms of the Jakob number and the number of tubes in the tier, N. For $Ja < 0.1$, heat transfer was predicted to be enhanced by less than 15%. Larger measured values of $\bar{h}_{D,N}$ might also be attributed to condensate dripping, as illustrated in Figure 10.14*d*. As individual drops impinge on the lower tube, turbulence and waves propagate throughout the film, enhancing heat transfer. For tubes with annular fins, lateral propagation of condensate is hindered by the fins, directly exposing more of the lower tube surface to vapor and resulting in values of n in the range $-1/6 < n \lesssim 0$ [53].

If the length-to-diameter ratio of an unfinned tube exceeds $1.8 \tan\theta$ [55], Equations 10.46 and 10.49 may be applied to inclined tubes by replacing g with $g \cos\theta$, where the angle θ is measured from the horizontal position. For either finned or unfinned tubes, the presence of noncondensable gases will decrease the convection coefficients relative to values obtained from the foregoing correlations.

EXAMPLE 10.4

The tube bank of a steam condenser consists of a square array of 400 tubes, each of diameter $D = 2r_1 = 6$ mm.

1. If horizontal, unfinned tubes are exposed to saturated steam at a pressure of $p = 0.15$ bar and the tube surface is maintained at $T_s = 25°\text{C}$, what is the rate at which steam is condensed per unit length of the tube bank?
2. If annular fins of height $h = r_2 - r_1 = 1$ mm, thickness $t = 1$ mm, and pitch $S = 2$ mm are added, determine the minimum condensation rate per unit length of tubing.

SOLUTION

Known: Configuration and surface temperature of unfinned and finned condenser tubes exposed to saturated steam at 0.15 bar.

Find:

1. Condensation rate per unit length of unfinned tubing.
2. Minimum condensation rate per unit length of finned tubing.

Schematic:

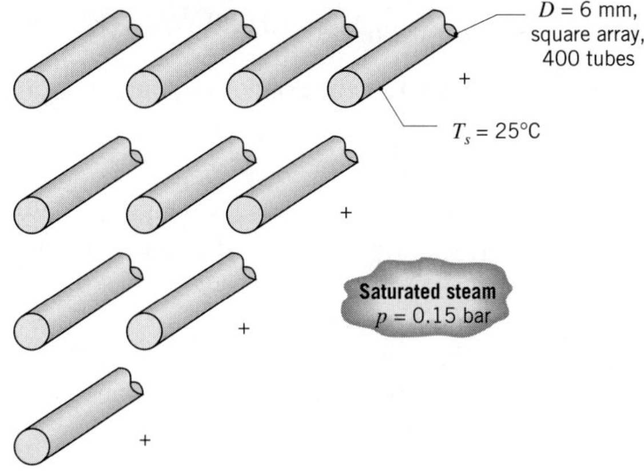

Assumptions:

1. Spatially uniform cylinder and fin temperature.
2. Average heat transfer coefficient varies with tube row with $n = -1/6$ in Equation 10.49.
3. Negligible concentration of noncondensable gases in the steam.

Properties: Table A.6, saturated vapor ($p = 0.15$ bar): $T_{\text{sat}} = 327\text{ K} = 54°\text{C}$, $\rho_v = 1/v_g = 0.098\text{ kg/m}^3$, $h_{fg} = 2373\text{ kJ/kg}$, $\sigma = 0.0671\text{ N/m}$. Table A.6, saturated water ($T_f = 312.5\text{ K}$): $\rho_l = 1/v_f = 992\text{ kg/m}^3$, $\mu_l = 663 \times 10^{-6}\text{ N}\cdot\text{s/m}^2$, $k_l = 0.631\text{ W/m}\cdot\text{K}$, $c_{p,l} = 4178\text{ J/kg}\cdot\text{K}$.

Analysis:

1. Equation 10.46 may be rearranged to yield an expression for the convection coefficient for the top, unfinned tube which is of the form

$$\overline{h}_D = C\left[\frac{\rho_l g(\rho_l - \rho_v)k_l^3 h'_{fg}}{\mu_l(T_{\text{sat}} - T_s)D}\right]^{1/4}$$

where $C = 0.729$ for a tube and

$$\begin{aligned} h'_{fg} &= h_{fg}(1 + 0.68\,Ja) = h_{fg} + 0.68c_{p,l}(T_{\text{sat}} - T_s) \\ &= 2373 \times 10^3\text{ J/kg} + 0.68 \times 4178\text{ J/kg}\cdot\text{K} \times (327 - 298)\text{ K} \\ &= 2455\text{ kJ/kg} \end{aligned}$$

Therefore

$$\overline{h}_D = 0.729\left[\frac{\begin{array}{c}992\text{ kg/m}^3 \times 9.8\text{ m/s}^2 \times (992 - 0.098)\text{ kg/m}^3 \\ \times (0.631\text{ W/m}\cdot\text{K})^3 \times 2455 \times 10^3\text{ J/kg}\end{array}}{663 \times 10^{-6}\text{ kg/s}\cdot\text{m} \times (327 - 298)\text{ K} \times 6 \times 10^{-3}\text{ m}}\right]^{1/4}$$

$$= 10{,}980\text{ W/m}^2\cdot\text{K}$$

From Equation 10.49 the array-averaged convection coefficient is

$$\bar{h}_{D,N} = \bar{h}_D N^n = 10{,}980 \text{ W/m}^2 \cdot \text{K} \times 20^{-1/6} = 6667 \text{ W/m}^2 \cdot \text{K}$$

From Equation 10.34 the condensation rate per unit length of tubing is

$$\dot{m}'_{\text{uft}} = N \times N \frac{\bar{h}_{D,N}(\pi D)(T_{\text{sat}} - T_s)}{h'_{fg}}$$

$$= 20 \times 20 \times 6667 \text{ W/m}^2 \cdot \text{K} \times \pi \times 6 \times 10^{-3} \text{ m} \times (327 - 298) \text{ K}/2455 \times 10^3 \text{ J/kg}$$

$$= 0.594 \text{ kg/s} \cdot \text{m} \qquad \triangleleft$$

2. From Equation 10.48, the minimum enhancement attributable to the annular fins is

$$\varepsilon_{\text{ft,min}} = \frac{q_{\text{ft,min}}}{q_{\text{uft}}} = \frac{\dot{m}'_{\text{ft,min}}}{\dot{m}'_{\text{uft}}} = \frac{tr_2}{Sr_1}\left[\frac{r_1}{r_2} + 1.02 \frac{\sigma r_1}{(\rho_l - \rho_v)gt^3}\right]^{1/4}$$

$$= \frac{1 \times 4}{2 \times 3}\left[\frac{3}{4} + 1.02 \frac{0.0671 \text{ N/m} \times 3 \times 10^{-3} \text{ m}}{(992 - 0.098) \text{ kg/m}^3 \times 9.8 \text{ m/s}^2 \times (1 \times 10^{-3} \text{ m})^3}\right]^{1/4}$$

$$= 1.44$$

Therefore, the minimum condensation rate for the finned tubes is

$$\dot{m}'_{\text{ft,min}} = \varepsilon_{\text{ft,min}} \dot{m}'_{\text{uft}} = 1.44 \times 0.594 \text{ kg/s} \cdot \text{m} = 0.855 \text{ kg/s} \cdot \text{m} \qquad \triangleleft$$

Comment: A value of $n = -1/6$ was used in Equation 10.49. However, for finned tubes the value of n is expected to be between zero and $-1/6$. For $n = 0$, the condensation rate per unit length of tubing would be

$$\dot{m}'_{\text{ft,min}} = \varepsilon_{\text{ft,min}} \times N \times N \frac{\bar{h}_D(\pi D)(T_{\text{sat}} - T_s)}{h'_{fg}}$$

$$= 1.44 \times 20 \times 20 \times 10{,}980 \text{ W/m}^2 \cdot \text{K} \times \pi \times 6 \times 10^{-3} \text{ m} \times (327 - 298) \text{ K}/2455 \times 10^3 \text{ J/kg}$$

$$= 1.41 \text{ kg/s} \cdot \text{m}$$

The preceding rate is for a *nonoptimized* condition where condensate fills the entire inter-fin region. Actual enhancements of $\varepsilon_{\text{ft,max}} \approx 4$ might be expected [50]. For $\varepsilon_{\text{ft,max}} = 4$ and $n = 0$, the condensation rate would be

$$\dot{m}'_{\text{ft}} = \varepsilon_{\text{ft,max}} \times N \times N \frac{\bar{h}_D(\pi D)(T_{\text{sat}} - T_s)}{h'_{\text{fg}}}$$

$$= 4 \times 20 \times 20 \times 10{,}980 \text{ W/m}^2 \cdot \text{K} \times \pi \times 6 \times 10^{-3} \text{ m} \times (327 - 298) \text{ K}/2455 \times 10^3 \text{ J/kg}$$

$$= 3.91 \text{ kg/s} \cdot \text{m}$$

Hence the condensation rate could potentially be increased by $100 \times (3.91 - 0.594)$ kg/s·m/ 0.594 kg/s·m = 559% by using finned tubes.

10.10 *Condensation in Horizontal Tubes*

Condensers used for refrigeration and air-conditioning systems generally involve vapor condensation inside horizontal or vertical tubes. Conditions within the tube depend strongly on the velocity of the vapor flowing through the tube, the mass fraction of vapor X, which decreases along the tube as condensation occurs, and the fluid properties. If the vapor velocity is small, condensation occurs in the manner depicted in Figure 10.16*a* for a horizontal tube. The fluid condenses in the upper regions of the tube wall and flows downward to a larger pool of liquid. In turn, the liquid pool is propelled down the length of the tube by shear forces imparted by the flowing vapor. For low vapor velocities such that

$$Re_{v,i} = \left(\frac{\rho_v u_{m,v} D}{\mu_v}\right)_i < 35{,}000 \tag{10.50}$$

where i refers to the tube inlet, heat transfer occurs predominantly through the falling condensate film. Dobson and Chato [56] recommend use of Equation 10.46 with $C = 0.555$ and $h'_{fg} = h_{fg} + 0.375c_{p,l}(T_{\text{sat}} - T_s)$. The value of C is less than that recommended for condensation on the outside of a cylinder ($C = 0.729$) because heat transfer associated with the condensate pool is small. Property evaluation is explained beneath Equation 10.32.

At high vapor velocities the two-phase flow becomes turbulent and annular (Figure 10.16*b*). The vapor occupies the core of the annulus, which diminishes in diameter as the thickness of the outer condensate layer increases in the flow direction. Dobson and Chato [56] recommend an empirical correlation for a local heat transfer coefficient h of the form

$$Nu_D = \frac{hD}{k_l} = 0.023\, Re_{D,l}^{0.8}\, Pr_l^{0.4}\left[1 + \frac{2.22}{X_{tt}^{0.89}}\right] \tag{10.51a}$$

where $Re_{D,l} = 4\dot{m}(1 - X)/(\pi D \mu_l)$, $X \equiv \dot{m}_v/\dot{m}$ is the mass fraction of vapor in the fluid, and X_{tt} is the *Martinelli parameter* corresponding to the existence of turbulent flow in both the liquid and vapor phases

$$X_{tt} = \left(\frac{1 - X}{X}\right)^{0.9}\left(\frac{\rho_v}{\rho_l}\right)^{0.5}\left(\frac{\mu_l}{\mu_v}\right)^{0.1} \tag{10.51b}$$

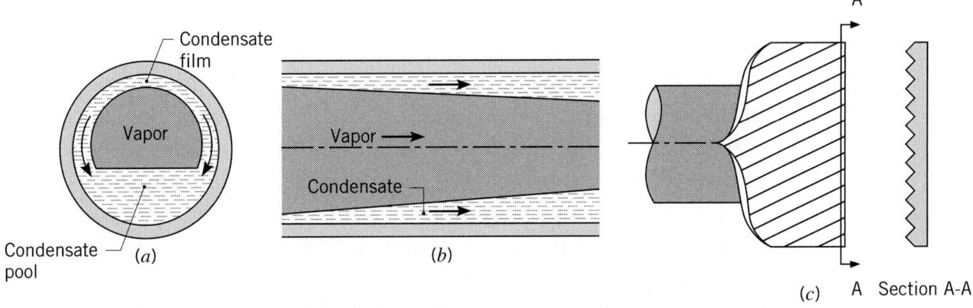

FIGURE 10.16 Film condensation in a horizontal tube. (*a*) Cross section of condensate flow for low vapor velocities. (*b*) Longitudinal section of condensate flow for large vapor velocities. (*c*) Microfins arranged in a helical pattern.

In generating Equations 10.51, Dobson and Chato evaluated all properties at the saturation temperature T_{sat}. The equations are recommended for use when the mass flow rate per unit cross-sectional tube area exceeds $500 \ \text{kg/s} \cdot \text{m}^2$ [56]. For single-phase liquid convection, $X \rightarrow 0$, $X_{tt} \rightarrow \infty$ and $Re_{D,l} \rightarrow Re_D$. In this case Equation 10.51a reduces to the Dittus–Boelter correlation, Equation 8.60, except for the exponent on the Prandtl number.

Condensation inside tubes at intermediate vapor velocities (or at low vapor mass fractions) is characterized by a variety of complex flow regimes. Heat transfer correlations have been developed for the individual regimes, and recommendations for their use are included in Dobson and Chato [56]. Condensation inside smaller tubes is influenced by surface tension effects and other considerations [36].

Condensation rates can be increased by adding small fins to the interior of the tube. Microfin tubes are typically made of copper with triangular or trapezoidal-shaped fins of height 0.1 to 0.25 mm as shown in Figure 10.16*c*. Heat transfer is increased due to the increase in the copper surface area, but also by turbulence induced by the fin structure and surface tension effects similar to those discussed for Figure 10.15. The fins are typically arranged in a helical or herringbone pattern down the tube length, with heat transfer rates enhanced by 50 to 180% [51].

10.11 *Dropwise Condensation*

Typically, heat transfer coefficients for dropwise condensation are an order of magnitude larger than those for film condensation. In fact, in heat exchanger applications for which dropwise condensation is promoted, other thermal resistances may be significantly larger than that due to condensation and, therefore, reliable correlations for the condensation process are not needed.

Of the many surface–fluid systems studied [57, 58], most of the data are for steam condensation on well-promoted copper surfaces—that is, surfaces for which wetting is inhibited—and are correlated by an expression of the form [59]

$$\bar{h}_{dc} = 51{,}104 + 2044\,T_{sat}(^\circ\text{C}) \qquad 22^\circ\text{C} \lesssim T_{sat} \lesssim 100^\circ\text{C} \tag{10.52}$$

$$\bar{h}_{dc} = 255{,}510 \qquad 100^\circ\text{C} \lesssim T_{sat} \tag{10.53}$$

where the heat transfer coefficient has units of $(\text{W/m}^2 \cdot \text{K})$. The heat transfer rate and condensation rate can be calculated from Equations 10.33 and 10.34, where h'_{fg} is given by Equation 10.27, and properties are evaluated as explained beneath Equation 10.32. The effect of subcooling, $T_{sat} - T_s$, on $\bar{h}_{dc}$ is small and may be neglected.

The effect of noncondensable vapors in the steam can be very important and has been studied by Shade and Mikic [60]. In addition, if the condensing surface material does not conduct as well as copper or silver, its thermal resistance becomes a factor. Since all the heat is transferred to the drops, which are very small and widely distributed over the surface, heat flow lines within the surface material near the active areas of condensation will *crowd*, inducing a *constriction* resistance. This effect has been studied by Hannemann and Mikic [61].

10.12 *Summary*

This chapter identifies the essential physical features of boiling and condensation processes and presents correlations suitable for approximate engineering calculations. However, a great deal of additional information is available, and much of it has been summarized in several extensive reviews of the subject [7, 15, 25, 30–33, 36, 51, 56, 58–59, 61–67].

You may test your understanding of heat transfer with phase change by addressing the following questions.

- What is *pool boiling*? *Forced convection boiling*? *Subcooled boiling*? *Saturated boiling*?
- How is the *excess temperature* defined?
- Sketch the *boiling curve* and identify key regimes and features. What is the *critical heat flux*? What is the *Leidenfrost point*? How does progression along the boiling curve occur if the surface heat flux is controlled? What is the nature of the hysteresis effect? How does progression along the boiling curve occur if the surface temperature is controlled?
- How does heat flux depend on the excess temperature in the *nucleate boiling* regime?
- What modes of heat transfer are associated with *film boiling*?
- How is the amount of liquid *subcooling* defined?
- To what extent is the boiling heat flux influenced by the magnitude of the gravitational field, liquid subcooling, and surface roughness?
- How do two-phase flow and heat transfer in microchannels differ from two-phase flow and heat transfer in larger tubes?
- How does *dropwise condensation* differ from *film condensation*? Which mode of condensation is characterized by larger heat transfer rates?
- For laminar film condensation on a vertical surface, how do the local and average convection coefficients vary with distance from the leading edge?
- How is the Reynolds number defined for film condensation on a vertical surface? What are the corresponding flow regimes?
- How does surface tension affect condensation on or in finned tubes?

References

1. Fox, R. W., A. T. McDonald, and P. J. Pritchard, *Introduction to Fluid Mechanics*, 6th ed. Wiley, Hoboken, NJ, 2003.
2. Nukiyama, S., *J. Japan Soc. Mech. Eng.*, **37**, 367, 1934 (Translation: *Int. J. Heat Mass Transfer*, **9**, 1419, 1966).
3. Drew, T. B., and C. Mueller, *Trans. AIChE*, **33**, 449, 1937.
4. Yamagata, K., F. Kirano, K. Nishiwaka, and H. Matsuoka, *Mem. Fac. Eng. Kyushu*, **15**, 98, 1955.
5. Rohsenow, W. M., *Trans. ASME*, **74**, 969, 1952.
6. Vachon, R. I., G. H. Nix, and G. E. Tanger, *J. Heat Transfer*, **90**, 239, 1968.
7. Collier, J. G., and J. R. Thome, *Convective Boiling and Condensation*, 3rd ed., Oxford University Press, New York, 1996.
8. Pioro I. L., *Int. J. Heat Mass Transfer*, **42**, 2003, 1999.
9. Kutateladze, S. S., *Kotloturbostroenie*, **3**, 10, 1948.
10. Zuber, N., *Trans. ASME*, **80**, 711, 1958.
11. Lienhard, J. H., *A Heat Transfer Textbook*, 2nd ed., Prentice-Hall, Englewood Cliffs, NJ, 1987.
12. Nakayama, W., A. Yabe, P. Kew, K. Cornwell, S. G. Kandlikar, and V. K. Dhir, in S. G. Kandlikar, M. Shoji, and V. K. Dhir, Eds., *Handbook of Phase Change: Boiling and Condensation*, Chap. 16, Taylor & Francis, New York, 1999.

13. Cichelli, M. T., and C. F. Bonilla, *Trans. AIChE*, **41**, 755, 1945.
14. Berenson, P. J., *J. Heat Transfer*, **83**, 351, 1961.
15. Hahne, E., and U. Grigull, *Heat Transfer in Boiling*, Hemisphere/Academic Press, New York, 1977.
16. Lienhard, J. H., and P. T. Y. Wong, *J. Heat Transfer*, **86**, 220, 1964.
17. Bromley, L. A., *Chem. Eng. Prog.*, **46**, 221, 1950.
18. Sadasivan, P., and J. H. Lienhard, *J. Heat Transfer*, **109**, 545, 1987.
19. Siegel, R., *Adv. Heat Transfer*, **4**, 143, 1967.
20. Berenson, P. J., *Int. J. Heat Mass Transfer*, **5**, 985, 1962.
21. Webb, R. L., *Heat Transfer Eng.*, **2**, 46, 1981, and *Heat Transfer Eng.*, **4**, 71, 1983.
22. Bergles, A. E., "Enhancement of Heat Transfer," *Heat Transfer 1978*, Vol. 6, pp. 89–108, Hemisphere Publishing, New York, 1978.
23. Bergles, A. E., in G. F. Hewitt, Exec. Ed., *Heat Exchanger Design Handbook*, Section 2.7.9, Begell House, New York, 2002.
24. Bergles, A. E., and W. H. Rohsenow, *J. Heat Transfer*, **86**, 365, 1964.
25. van Stralen, S., and R. Cole, *Boiling Phenomena*, McGraw-Hill/Hemisphere, New York, 1979.
26. Lienhard, J. H., and R. Eichhorn, *Int. J. Heat Mass Transfer*, **19**, 1135, 1976.
27. Kandlikar, S. G., *J. Heat Transfer*, **112**, 219, 1990.
28. Kandlikar, S. G., and H. Nariai, in S. G. Kandlikar, M. Shoji, and V. K. Dhir, Eds., *Handbook of Phase Change: Boiling and Condensation*, Chap. 15, Taylor & Francis, New York, 1999.
29. Celata, G. P., and A. Mariani, in S. G. Kandlikar, M. Shoji, and V. K. Dhir, Eds., *Handbook of Phase Change: Boiling and Condensation*, Chap. 17, Taylor & Francis, New York, 1999.
30. Tong, L. S., and Y. S. Tang, *Boiling Heat Transfer and Two Phase Flow*, 2nd ed., Taylor & Francis, New York, 1997.
31. Rohsenow, W. M., in W. M. Rohsenow and J. P. Hartnett, Eds., *Handbook of Heat Transfer*, Chap. 13, McGraw-Hill, New York, 1973.
32. Griffith, P., in W. M. Rohsenow and J. P. Hartnett, Eds., *Handbook of Heat Transfer*, Chap. 14, McGraw-Hill, New York, 1973.
33. Ginoux, J. N., *Two-Phase Flow and Heat Transfer*, McGraw-Hill/Hemisphere, New York, 1978.
34. Hall, D. D., and I. Mudawar, *Int. J. Heat Mass Transfer*, **43**, 2573, 2000.
35. Hall, D. D., and I. Mudawar, *Int. J. Heat Mass Transfer*, **43**, 2605, 2000.
36. Faghri, A., and Y. Zhang, *Transport Phenomena in Multiphase Systems*, Elsevier, Amsterdam, 2006.
37. Qu, W., and I. Mudawar, *Int. J. Heat Mass Transfer*, **46**, 2755, 2003.
38. Ghiaasiaan, S. M., and S. I. Abdel-Khalik, in J. P. Hartnett, T. F. Irvine, Y. I. Cho, and G. A. Greene, Eds., *Advances in Heat Transfer*, Vol. 34, Academic Press, New York, 2001.
39. Qu, W., and I. Mudawar, *Int. J. Heat Mass Transfer*, **46**, 2773, 2003.
40. Sparrow, E. M., and J. L. Gregg, *J. Heat Transfer*, **81**, 13, 1959.
41. Koh, J. C. Y., E. M. Sparrow, and J. P. Hartnett, *Int. J. Heat Mass Transfer*, **2**, 69, 1961.
42. Nusselt, W., *Z. Ver. Deut. Ing.*, **60**, 541, 1916.
43. Rohsenow, W. M., *Trans. ASME*, **78**, 1645, 1956.
44. Chen, M. M., *J. Heat Transfer*, **83**, 48, 1961.
45. Dhir, V. K., and J. H. Lienhard, *J. Heat Transfer*, **93**, 97, 1971.
46. Kutateladze, S. S., *Fundamentals of Heat Transfer*, Academic Press, New York, 1963.
47. Labuntsov, D. A., *Teploenergetika*, **4**, 72, 1957.
48. Gregorig, R., J. Kern, and K. Turek, *Wärme Stoffübertrag.*, **7**, 1, 1974.
49. Popiel, Cz. O., and L. Boguslawski, *Int. J. Heat Mass Transfer*, **18**, 1486, 1975.
50. Rose, J. W., *Int. J. Heat Mass Transfer*, **37**, 865, 1994.
51. Cavallini, A., G. Censi, D. Del Col, L. Doretti, G. A. Longo, L. Rossetto, and C. Zilio, *Int. J. Refrig.* **26**, 373, 2003.
52. Briggs, A., and J. W. Rose, *Int. J. Heat Mass Transfer*, **37**, 457, 1994.
53. Murase, T., H. S. Wang, and J. W. Rose, *Int. J. Heat Mass Transfer*, **49**, 3180, 2006.
54. Chen, M. M., *J. Heat Transfer*, **83**, 55, 1961.
55. Selin, G., "Heat Transfer by Condensing Pure Vapours Outside Inclined Tubes," *International Developments in Heat Transfer*, Part 2, International Heat Transfer Conference, University of Colorado, pp. 278–289, ASME, New York, 1961.
56. Dobson, M. K., and J. C. Chato, *J. Heat Transfer*, **120**, 193, 1998.
57. Tanner, D. W., D. Pope, C. J. Potter, and D. West, *Int. J. Heat Mass Transfer*, **11**, 181, 1968.
58. Rose, J. W., *Proc. Instn. Mech. Engrs. A: Power and Energy*, **216**, 115, 2001.
59. Griffith, P., in G. F. Hewitt, Exec. Ed., *Heat Exchanger Design Handbook*, Section 2.6.5, Hemisphere Publishing, New York, 1990.
60. Shade, R., and B. Mikic, "The Effects of Noncondensable Gases on Heat Transfer During Dropwise Condensation," Paper 67b presented at the 67th Annual

Meeting of the American Institute of Chemical Engineers, Washington, DC, 1974.

61. Hannemann, R., and B. Mikic, *Int. J. Heat Mass Transfer*, **19**, 1309, 1976.

62. Marto, P. J., in W. M. Rohsenow, J. P. Hartnett, and Y. I. Cho, Eds., *Handbook of Heat Transfer*, 3rd ed., Chap. 14, McGraw-Hill, New York, 1998.

63. Collier, J. G., and V. Wadekar, in G. F. Hewitt, Exec. Ed., *Heat Exchanger Design Handbook*, Section 2.7.2, Begell House, New York, 2002.

64. Butterworth, D., in D. Butterworth and G. F. Hewitt, Eds., *Two-Phase Flow and Heat Transfer*, Oxford University Press, London, 1977, pp. 426–462.

65. McNaught, J., and D. Butterworth, in G. F. Hewitt, Exec. Ed., *Heat Exchanger Design Handbook*, Section 2.6.2, Begell House, New York, 2002.

66. Rose, J. W., *Int. J. Heat Mass Transfer*, **24**, 191, 1981.

67. Pioro, L. S., and I. L. Pioro, *Industrial Two-Phase Thermosyphons*, Begell House, New York, 1997.

Problems

General Considerations

10.1 Show that, for water at 1-atm pressure with $T_s - T_{\text{sat}} = 10°\text{C}$, the Jakob number is much less than unity. What is the physical significance of this result? Verify that this conclusion applies to other fluids.

10.2 The surface of a horizontal, 7-mm-diameter cylinder is maintained at an excess temperature of 5°C in saturated water at 1 atm. Estimate the heat flux using an appropriate free convection correlation and compare your result to the boiling curve of Figure 10.4. Repeat the calculation for a horizontal, 7-μm-diameter wire at the same excess temperature. What can you say about the general applicability of Figure 10.4 to all situations involving boiling of water at 1 atm?

10.3 The role of surface tension in bubble formation can be demonstrated by considering a spherical bubble of pure saturated vapor in *mechanical* and *thermal* equilibrium with its superheated liquid.

(a) Beginning with an appropriate free-body diagram of the bubble, perform a force balance to obtain an expression of the bubble radius,

$$r_b = \frac{2\sigma}{p_{\text{sat}} - p_l}$$

where p_{sat} is the pressure of the saturated vapor and p_l is the pressure of the superheated liquid outside the bubble.

(b) On a p–v diagram, represent the bubble and liquid states. Discuss what changes in these conditions will cause the bubble to grow or collapse.

(c) Calculate the bubble size under equilibrium conditions for which the vapor is saturated at 101°C and the liquid pressure corresponds to a saturation temperature of 100°C.

10.4 Estimate the heat transfer coefficient, h, associated with Points A, B, C, D, and E in Figure 10.4. Which point is associated with the largest value of h? Which point corresponds to the smallest value of h? Determine the thickness of the vapor blanket at the Leidenfrost point, neglecting radiation heat transfer through the blanket. Assume the solid is a flat surface.

Nucleate Boiling and Critical Heat Flux

10.5 A long, 1-mm-diameter wire passes an electrical current dissipating 3150 W/m and reaches a surface temperature of 126°C when submerged in water at 1 atm. What is the boiling heat transfer coefficient? Estimate the value of the correlation coefficient $C_{s,f}$.

10.6 Estimate the nucleate pool boiling heat transfer coefficient for water boiling at atmospheric pressure on the outer surface of a platinum-plated 10-mm-diameter tube maintained 10°C above the saturation temperature.

10.7 Plot the nucleate boiling heat flux for saturated water at atmospheric pressure on a large, horizontal polished copper plate, over the excess temperature range $5°\text{C} \leq \Delta T_e \leq 30°\text{C}$. Compare your results with Figure 10.4. Also find the excess temperature corresponding to the critical heat flux.

10.8 A simple expression to account for the effect of pressure on the nucleate boiling convection coefficient in water $(\text{W/m}^2 \cdot \text{K})$ is

$$h = C(\Delta T_e)^n \left(\frac{p}{p_a}\right)^{0.4}$$

where p and p_a are the system pressure and standard atmospheric pressure, respectively. For a horizontal plate and the range $15 < q''_s < 235\ \text{kW/m}^2$, $C = 5.56$ and $n = 3$. Units of ΔT_e are kelvins. Compare predictions from this expression with the Rohsenow correlation ($C_{s,f} = 0.013$, $n = 1$) for pressures of 2 and 5 bars with $\Delta T_e = 10°\text{C}$.

10.9 In Example 10.1 we considered conditions for which vigorous boiling occurs in a pan of water, and we determined the electric power (heat rate) required to maintain a prescribed temperature for the bottom of the pan. However, the electric power is, in fact, the control (independent) variable, from which the temperature of the pan follows.

(a) For nucleate boiling in the copper pan of Example 10.1, compute and plot the temperature of the pan as a function of the heat rate for $1 \leq q \leq 100$ kW.

(b) If the water is initially at room temperature, it must, of course, be heated for a period of time before it will boil. Consider conditions shortly after heating is initiated and the water is at 20°C. Estimate the temperature of the pan bottom for a heat rate of 8 kW.

10.10 Calculate the critical heat flux on a large horizontal surface for the following fluids at 1 atm: mercury, ethanol, and refrigerant R-134a. Compare these results to the critical heat flux for water at 1 atm.

10.11 Water at atmospheric pressure boils on the surface of a large horizontal copper tube. The heat flux is 90% of the critical value. The tube surface is initially scored; however, over time the effects of scoring diminish and the boiling eventually exhibits behavior similar to that associated with a polished surface. Determine the tube surface temperature immediately after installation and after prolonged service.

10.12 The bottom of a copper pan, 150 mm in diameter, is maintained at 115°C by the heating element of an electric range. Estimate the power required to boil the water in this pan. Determine the evaporation rate. What is the ratio of the surface heat flux to the critical heat flux? What pan temperature is required to achieve the critical heat flux?

10.13 A nickel-coated heater element with a thickness of 15 mm and a thermal conductivity of $50\ \text{W/m} \cdot \text{K}$ is exposed to saturated water at atmospheric pressure. A thermocouple is attached to the back surface, which is well insulated. Measurements at a particular operating condition yield an electrical power dissipation in the heater element of $6.950 \times 10^7\ \text{W/m}^3$ and a temperature of $T_o = 266.4°\text{C}$.

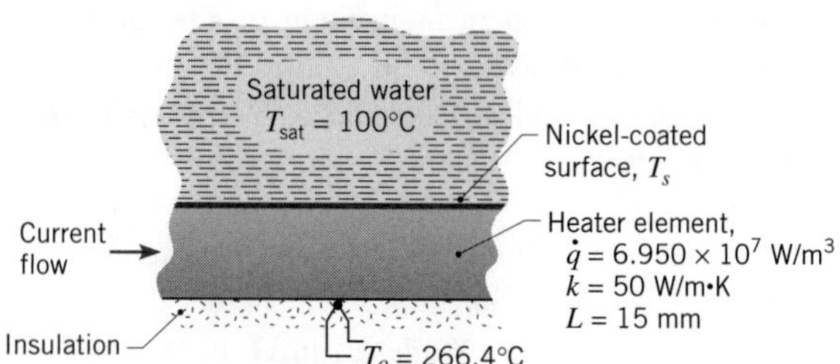

(a) From the foregoing data, calculate the surface temperature, T_s, and the heat flux at the exposed surface.

(b) Using the surface heat flux determined in part (a), estimate the surface temperature by applying an appropriate boiling correlation.

10.14 Advances in very large scale integration (VLSI) of electronic devices on a chip are often restricted by the ability to cool the chip. For mainframe computers, an array of several hundred chips, each of area 25 mm², may be mounted on a ceramic substrate. A method of cooling the array is by immersion in a low boiling point fluid such as refrigerant R-134a. At 1 atm and 247 K, properties of the saturated liquid are $\mu = 1.46 \times 10^{-4}\ \text{N} \cdot \text{s/m}^2$, $c_p = 1551\ \text{J/kg} \cdot \text{K}$, and $Pr = 3.2$. Assume values of $C_{s,f} = 0.004$ and $n = 1.7$.

(a) Estimate the power dissipated by a single chip if it is operating at 50% of the critical heat flux. What is the corresponding value of the chip temperature?

(b) Compute and plot the chip temperature as a function of surface heat flux for $0.25 \leq q''_s/q''_{max} \leq 0.90$.

10.15 Saturated ethylene glycol at 1 atm is heated by a horizontal chromium-plated surface which has a diameter of 200 mm and is maintained at 480 K. Estimate the heating power requirement and the rate of evaporation. What fraction is the power requirement of the maximum power associated with the critical heat flux? At 470 K, properties of the saturated liquid are $\mu = 0.38 \times 10^{-3}\ \text{N} \cdot \text{s/m}^2$, $c_p = 3280\ \text{J/kg} \cdot \text{K}$, and $Pr = 8.7$. The saturated vapor density is $\rho = 1.66\ \text{kg/m}^3$. Assume nucleate boiling constants of $C_{s,f} = 0.01$ and $n = 1.0$.

10.16 Copper tubes 25 mm in diameter and 0.75 m long are used to boil saturated water at 1 atm.

(a) If the tubes are operated at 75% of the critical heat flux, how many tubes are needed to provide a vapor production rate of 750 kg/h? What is the corresponding tube surface temperature?

(b) Compute and plot the tube surface temperature as a function of heat flux for $0.25 \leq q''_s/q''_{max} < 0.90$. On the same graph, plot the corresponding number of tubes needed to provide the prescribed vapor production rate.

10.17 Consider a gas-fired boiler in which five coiled, thin-walled, copper tubes of 25-mm diameter and 8-m length are submerged in pressurized water at 4.37 bars. The walls of the tubes are scored and may be assumed to be isothermal. Combustion gases enter each of the tubes at a temperature of $T_{m,i} = 700°C$ and a flow rate of $\dot{m} = 0.08$ kg/s, respectively.

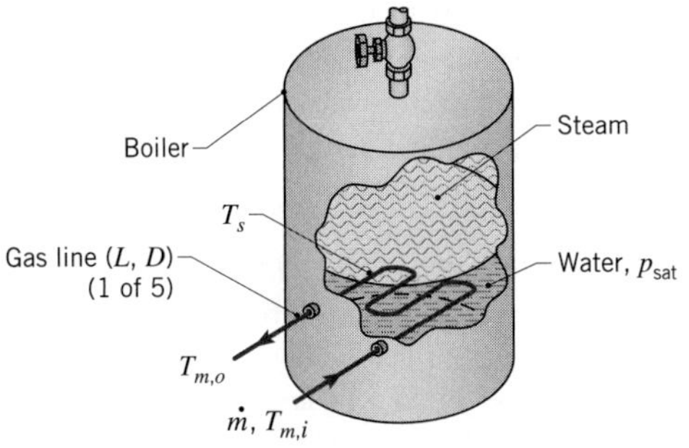

(a) Determine the tube wall temperature T_s and the gas outlet temperature $T_{m,o}$ for the prescribed conditions. As a first approximation, the properties of the combustion gases may be taken as those of air at 700 K.

(b) Over time the effects of scoring diminish, leading to behavior similar to that of a polished copper surface. Determine the wall temperature and gas outlet temperature for the aged condition.

10.18 Estimate the current at which a 1-mm-diameter nickel wire will burn out when submerged in water at atmospheric pressure. The electrical resistance of the wire is 0.129 Ω/m.

10.19 Estimate the power (W/m²) required to maintain a brass plate at $\Delta T_e = 15°C$ while boiling saturated water at 1 atm. What is the power requirement if the water is pressurized to 10 atm? At what fraction of the critical heat flux is the plate operating?

10.20 A dielectric fluid at atmospheric pressure is heated with a 0.5-mm-diameter, horizontal platinum wire. Determine the temperature of the wire when the wire is heated at 50% of the critical heat flux. The properties of the fluid are $c_{p,l} = 1300$ J/kg·K, $h_{fg} = 142$ kJ/kg, $k_l = 0.075$ W/m·K, $\nu_l = 0.32 \times 10^{-6}$ m²/s, $\rho_l = 1400$ kg/m³, $\rho_v = 7.2$ kg/m³, $\sigma = 12.4 \times 10^{-3}$ N/m, $T_{sat} = 34°C$. Assume the nucleate boiling constants are $C_{s,f} = 0.005$ and $n = 1.7$. For small horizontal cylinders, the critical heat flux is found by multiplying the value associated with large horizontal cylinders by a correction factor F, where $F = 0.89 + 2.27 \exp(-3.44\, Co^{-1/2})$. The Confinement number is based on the radius of the cylinder, and the range of applicability for the correction factor is $1.3 \lesssim Co \lesssim 6.7$ [11].

10.21 It has been demonstrated experimentally that the critical heat flux is highly dependent on pressure, primarily through the pressure dependence of the fluid surface tension and latent heat of vaporization. Using Equation 10.6, calculate values of q''_{max} for water on a large horizontal surface as a function of pressure. Demonstrate that the peak critical heat flux occurs at approximately one-third the critical pressure ($p_c = 221$ bars). Since all common fluids have this characteristic, suggest what coordinates should be used to plot critical heat flux–pressure values to obtain a universal curve.

10.22 In applying dimensional analysis, Kutateladze [9] postulated that the critical heat flux varies with the heat of vaporization, vapor density, surface tension, and the bubble diameter parameter given in Equation 10.4a. Verify that dimensional analysis would yield the following expression for the critical heat flux:

$$q''_{max} = C h_{fg} \rho_v^{1/2} D_b^{-1/2} \sigma^{1/2}$$

10.23 A silicon chip of thickness $L = 2.5$ mm and thermal conductivity $k_s = 135$ W/m·K is cooled by boiling a saturated fluorocarbon liquid ($T_{sat} = 57°C$) on its surface. The electronic circuits on the bottom of the chip produce a uniform heat flux of $q''_o = 5 \times 10^4$ W/m², while the sides of the chip are perfectly insulated.

Properties of the saturated fluorocarbon are $c_{p,l} = 1100$ J/kg·K, $h_{fg} = 84{,}400$ J/kg, $\rho_l = 1619.2$ kg/m³, $\rho_v = 13.4$ kg/m³, $\sigma = 8.1 \times 10^{-3}$ N/m, $\mu_l = 440 \times 10^{-6}$ kg/m·s, and $Pr_l = 9.01$. In addition, the nucleate boiling constants are $C_{s,f} = 0.005$ and $n = 1.7$.

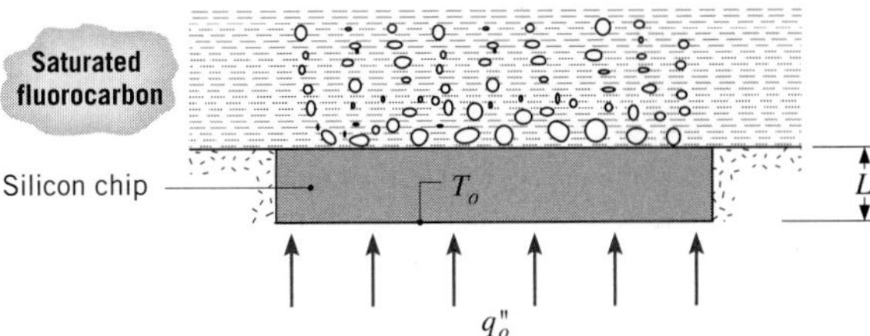

(a) What is the steady-state temperature T_o at the bottom of the chip? If, during testing of the chip, q''_o is increased to 90% of the critical heat flux, what is the new steady-state value of T_o?

(b) Compute and plot the chip surface temperatures (top and bottom) as a function of heat flux for $0.20 \leq q''_o/q''_{max} \leq 0.90$. If the maximum allowable chip temperature is 80°C, what is the maximum allowable value of q''_o?

10.24 What is the critical heat flux for boiling water at 1 atm on a large horizontal surface on the surface of the moon, where the gravitational acceleration is one-sixth that of the earth?

10.25 A heater for boiling a saturated liquid consists of two concentric stainless steel tubes packed with dense boron nitride powder. Electrical current is passed through the inner tube, creating uniform volumetric heating $\dot{q}$ (W/m^3). The exposed surface of the outer tube is in contact with the liquid and the boiling heat flux is given as

$$q''_s = C(T_s - T_{\text{sat}})^3$$

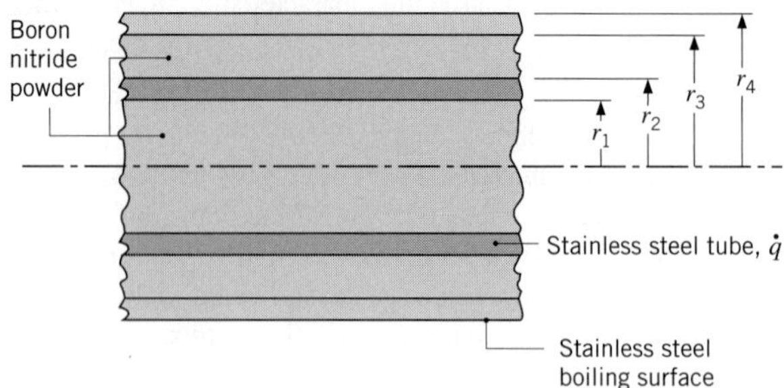

It is feared that under high-power operation the stainless steel tubes would severely oxidize if temperatures exceed $T_{\text{ss},x}$ or that the boron nitride would deteriorate if its temperature exceeds $T_{\text{bn},x}$. Presuming that the saturation temperature of the liquid (T_{sat}) and the boiling surface temperature (T_s) are prescribed, derive expressions for the maximum temperatures in the stainless steel (ss) tubes and in the boron nitride (bn). Express your results in terms of geometric parameters (r_1, r_2, r_3, r_4), thermal conductivities (k_{ss}, k_{bn}), and the boiling parameters (C, T_{sat}, T_s).

10.26 A device for performing boiling experiments consists of a copper bar ($k = 400$ W/m·K), which is exposed to a boiling liquid at one end, encapsulates an electrical heater at the other end, and is well insulated from its surroundings at all but the exposed surface. Thermocouples inserted in the bar are used to measure temperatures at distances of $x_1 = 10$ mm and $x_2 = 25$ mm from the surface.

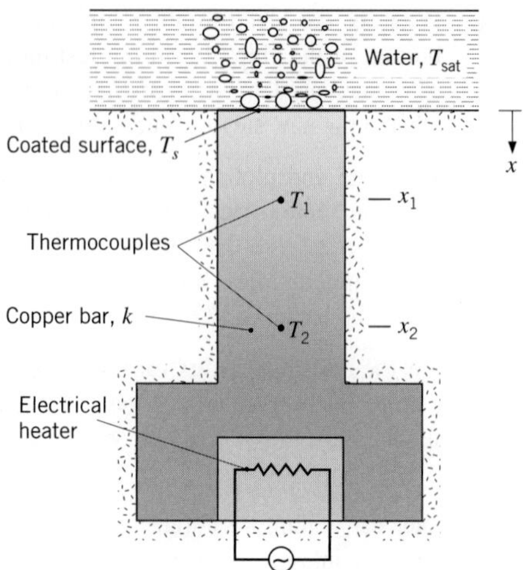

(a) An experiment is performed to determine the boiling characteristics of a special coating applied to the exposed surface. Under steady-state conditions, nucleate boiling is maintained in saturated water at atmospheric pressure, and values of $T_1 = 133.7°\text{C}$ and $T_2 = 158.6°\text{C}$ are recorded. If $n = 1$, what value of the coefficient $C_{s,f}$ is associated with the Rohsenow correlation?

(b) Assuming applicability of the Rohsenow correlation with the value of $C_{s,f}$ determined from part (a), compute and plot the excess temperature ΔT_e as a function of the boiling heat flux for $10^5 \leq q''_s \leq 10^6$ W/m^2. What are the corresponding values of T_1 and T_2 for $q''_s = 10^6$ W/m^2? If q''_s were increased to 1.5×10^6 W/m^2, could the foregoing results be extrapolated to infer the corresponding values of ΔT_e, T_1, and T_2?

Minimum Heat Flux and Film Boiling

10.27 A small copper sphere, initially at a uniform, elevated temperature $T(0) = T_i$, is suddenly immersed in a large fluid bath maintained at T_{sat}. The initial temperature of the sphere exceeds the Leidenfrost point corresponding to the temperature T_D of Figure 10.4.

(a) Sketch the variation of the average sphere temperature, $\overline{T}(t)$, with time during the quenching process. Indicate on this sketch the temperatures T_i, T_D, and T_{sat}, as well as the regimes of film, transition, and nucleate boiling and the regime of single-phase convection. Identify key features of the temperature history.

(b) At what time(s) in this cooling process do you expect the surface temperature of the sphere to deviate most from its center temperature? Explain your answer.

10.28 A sphere made of aluminum alloy 2024 with a diameter of 20 mm and a uniform temperature of 500°C is suddenly immersed in a saturated water bath maintained at atmospheric pressure. The surface of the sphere has an emissivity of 0.25.

(a) Calculate the total heat transfer coefficient for the initial condition. What fraction of the total coefficient is contributed by radiation?

(b) Estimate the temperature of the sphere 30 s after it is immersed in the bath.

10.29 A disk-shaped turbine rotor is heat-treated by quenching in water at $p = 1$ atm. Initially, the rotor is at a uniform temperature of $T_i = 1100°\text{C}$ and the water is at its boiling point as the rotor is lowered into the quenching bath by a harness.

(a) Assuming lumped-capacitance behavior and constant properties for the rotor, carefully plot the rotor temperature versus time, pointing out important features of your $T(t)$ curve. The rotor is in Orientation A.

(b) If the rotor is reoriented so that its large surfaces are horizontal (Orientation B), would the rotor temperature decrease more rapidly or less rapidly relative to Orientation A?

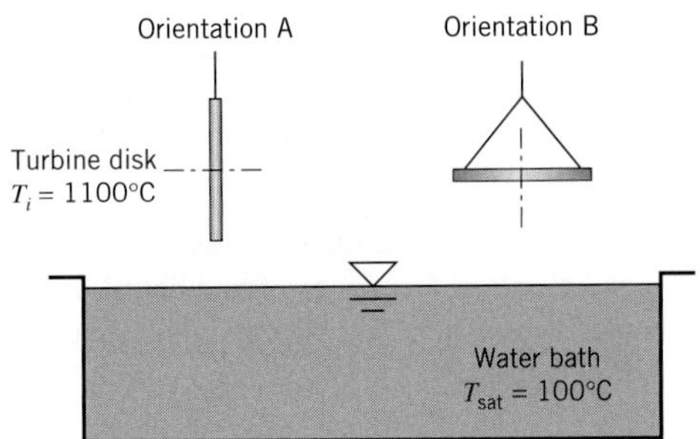

10.30 A steel bar, 20 mm in diameter and 200 mm long, with an emissivity of 0.9, is removed from a furnace at 455°C and suddenly submerged horizontally in a water bath under atmospheric pressure. Estimate the initial heat transfer rate from the bar.

10.31 Electrical current passes through a horizontal, 2-mm-diameter conductor of emissivity 0.5 when immersed in water under atmospheric pressure.

(a) Estimate the power dissipation per unit length of the conductor required to maintain the surface temperature at 555°C.

(b) For conductor diameters of 1.5, 2.0, and 2.5 mm, compute and plot the power dissipation per unit length as a function of surface temperature for $250 \leq T_s \leq 650°C$. On a separate figure, plot the percentage contribution of radiation as a function of T_s.

10.32 Consider a horizontal, D = 1-mm-diameter platinum wire suspended in saturated water at atmospheric pressure. The wire is heated by an electrical current. Determine the heat flux from the wire at the instant when the surface of the wire reaches its melting point. Determine the corresponding centerline temperature of the wire. Due to oxidation at very high temperature, the wire emissivity is $\varepsilon = 0.80$ when it burns out. The water vapor properties at the film temperature of 1209 K are $\rho_v = 0.189\ \text{kg/m}^3$, $c_{p,v} = 2404\ \text{J/kg}\cdot\text{K}$, $\nu_v = 231 \times 10^{-6}\ \text{m}^2\text{/s}$, $k_v = 0.113\ \text{W/m}\cdot\text{K}$.

10.33 A heater element of 5-mm diameter is maintained at a surface temperature of 350°C when immersed horizontally in water under atmospheric pressure. The element sheath is stainless steel with a mechanically polished finish having an emissivity of 0.25.

(a) Calculate the electrical power dissipation and the rate of vapor production per unit heater length.

(b) If the heater were operated at the same power dissipation rate in the nucleate boiling regime, what temperature would the surface achieve? Calculate the rate of vapor production per unit length for this operating condition.

(c) Sketch the boiling curve and represent the two operating conditions of parts (a) and (b). Compare the results of your analysis. If the heater element is operated in the power-controlled mode, explain how you would achieve these two operating conditions beginning with a cold element.

10.34 The thermal energy generated by a silicon chip increases in proportion to its clock speed. The silicon chip of Problem 10.23 is designed to operate in the nucleate boiling regime at approximately 30% of the critical heat flux. A sudden surge in the chip's clock speed triggers film boiling, after which the clock speed and power dissipation return to their design values.

(a) In which boiling regime does the chip operate after the power dissipation returns to its design value?

(b) To return to the nucleate boiling regime, how much must the clock speed be reduced relative to the design value?

10.35 A cylinder of 120-mm diameter at 1000 K is quenched in saturated water at 1 atm. Describe the quenching process and estimate the maximum heat removal rate per unit length during the process.

10.36 A 1-mm-diameter horizontal platinum wire of emissivity $\varepsilon = 0.25$ is operated in saturated water at 1-atm pressure.

(a) What is the surface heat flux if the surface temperature is $T_s = 800$ K?

(b) For emissivities of 0.1, 0.25, and 0.95, generate a log–log plot of the heat flux as a function of surface excess temperature, $\Delta T_e \equiv T_s - T_{sat}$, for $150 \leq \Delta T_e \leq 550$ K. Show the critical heat flux and the Leidenfrost point on your plot. Separately, plot the percentage contribution of radiation to the total heat flux for $150 \leq \Delta T_e \leq 550$ K.

10.37 As strip steel leaves the last set of rollers in a hot rolling mill, it is quenched by planar water jets before being coiled. Due to the large plate temperatures, film boiling is achieved shortly downstream of the jet impingement region.

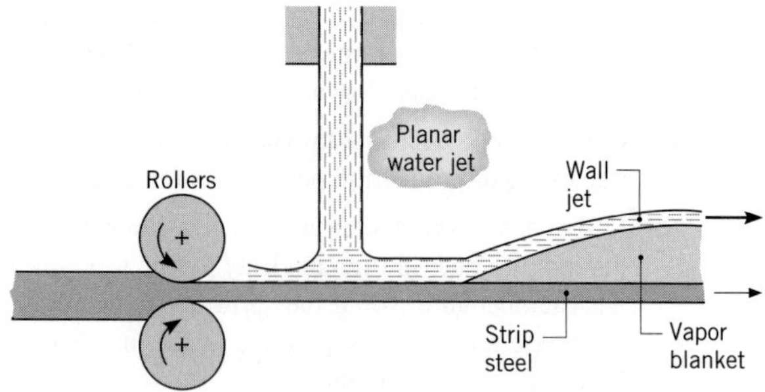

Consider conditions for which the strip steel beneath the vapor blanket is at a temperature of 907 K and has an emissivity of 0.35. Neglecting the effects of the strip and jet motions and assuming boiling within the film to be approximated by that associated with a large horizontal cylinder of 1-m diameter, estimate the rate of heat transfer per unit surface area from the strip to the wall jet.

10.38 A polished copper sphere of 10-mm diameter, initially at a prescribed elevated temperature T_i, is quenched in a saturated (1 atm) water bath. Using the lumped capacitance method of Section 5.3.3, estimate the time for the sphere to cool (a) from $T_i = 130°\text{C}$ to 110°C and (b) from $T_i = 550°\text{C}$ to 220°C. Make use of the average sphere temperatures in evaluating properties. Plot the temperature history for each quenching process.

Forced Convection Boiling

10.39 A tube of 2-mm diameter is used to heat saturated water at 1 atm, which is in cross flow over the tube. Calculate and plot the critical heat flux as a function of water velocity over the range 0 to 2 m/s. On your plot, identify the pool boiling region and the transition region between the low- and high-velocity ranges. *Hint:* Problem 10.20 contains relevant information for pool boiling on small-diameter cylinders.

10.40 Saturated water at 1 atm and velocity 2 m/s flows over a cylindrical heating element of diameter 5 mm. What is the maximum heating rate (W/m) for nucleate boiling?

10.41 A vertical steel tube carries water at a pressure of 10 bars. Saturated liquid water is pumped into the $D = 0.1$-m-diameter tube at its bottom end ($x = 0$) with a mean velocity of $u_m = 0.05$ m/s. The tube is exposed to combusting pulverized coal, providing a uniform heat flux of $q'' = 100{,}000 \text{ W/m}^2$.

(a) Determine the tube wall temperature and the quality of the flowing water at $x = 15$ m. Assume $G_{s,f} = 1$.

(b) Determine the tube wall temperature at a location beyond $x = 15$ m where single-phase flow of the vapor exists at a mean temperature of T_{sat}. Assume the vapor at this location is also at a pressure of 10 bars.

(c) Plot the tube wall temperature in the range $-5 \text{ m} \le x \le 30 \text{ m}$.

10.42 Consider refrigerant R-134a flowing in a smooth, horizontal, 10-mm-inner-diameter tube of wall thickness 2 mm. The refrigerant is at a saturation temperature of 15°C (for which $\rho_{v,\text{sat}} = 23.75 \text{ kg/m}^3$) and flows at a rate of 0.01 kg/s. Determine the maximum wall temperature associated with a heat flux of 10^5 W/m^2 at the inner wall at a location 0.4 m downstream from the onset of boiling for tubes fabricated of (a) pure copper and (b) AISI 316 stainless steel.

10.43 Determine the tube diameter associated with $p = 1$ atm and a critical Confinement number of 0.5 for ethanol, mercury, water, R-134a, and the dielectric fluid of Problem 10.23.

Film Condensation

10.44 Saturated steam at 0.1 bar condenses with a convection coefficient of $6800 \text{ W/m}^2\cdot\text{K}$ on the outside of a brass tube having inner and outer diameters of 16.5 and 19 mm, respectively. The convection coefficient for water flowing inside the tube is $5200 \text{ W/m}^2\cdot\text{K}$. Estimate the steam condensation rate per unit length of the tube when the mean water temperature is 30°C.

10.45 Consider a container exposed to a saturated vapor, T_{sat}, having a cold bottom surface, $T_s < T_{\text{sat}}$, and with insulated sidewalls.

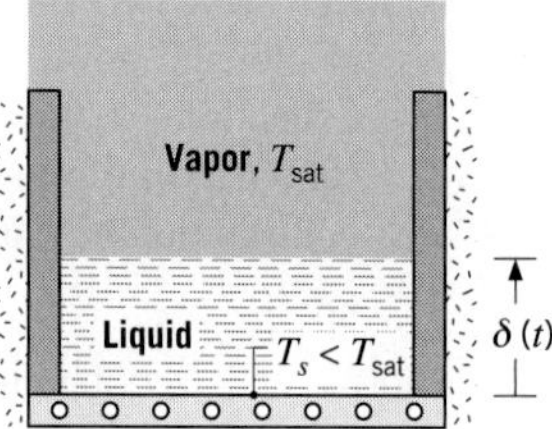

Assuming a linear temperature distribution for the liquid, perform a surface energy balance on the liquid–vapor interface to obtain the following expression for the growth rate of the liquid layer:

$$\delta(t) = \left[\frac{2k_l(T_{\text{sat}} - T_s)}{\rho_l h_{fg}} t\right]^{1/2}$$

Calculate the thickness of the liquid layer formed in 1 h for a 200-mm^2 bottom surface maintained at 80°C and exposed to saturated steam at 1 atm. Compare this result with the condensate formed by a *vertical* plate of the same dimensions for the same period of time.

10.46 Saturated steam at 1 atm condenses on the outer surface of a vertical, 100-mm-diameter pipe 1 m long, having a uniform surface temperature of 94°C. Estimate the total condensation rate and the heat transfer rate to the pipe.

10.47 Determine the total condensation rate and the heat transfer rate for Problem 10.46 when the steam is saturated at 1.5 bars.

10.48 Consider wave-free laminar condensation on a vertical isothermal plate of length L, providing an average heat transfer coefficient of $\bar{h}_L$. If the plate is divided into N smaller plates, each of length $L_N = L/N$, determine an expression for the ratio of the heat transfer coefficient averaged over the N plates to the heat transfer coefficient averaged over the single plate, $\bar{h}_{L,N}/\bar{h}_{L,1}$.

10.49 A vertical plate 500 mm high and 200 mm wide is to be used to condense saturated steam at 1 atm.

(a) At what surface temperature must the plate be maintained to achieve a condensation rate of $\dot{m} = 25$ kg/h?

(b) Compute and plot the surface temperature as a function of condensation rate for $15 \le \dot{m} \le 50$ kg/h.

(c) On the same graph and for the same range of $\dot{m}$, plot the surface temperature as a function of condensation rate if the plate is 200 mm high and 500 mm wide.

10.50 A 2 m × 2 m vertical plate is exposed on one side to saturated steam at atmospheric pressure and on the other side to cooling water that maintains a plate temperature of 50°C.

(a) What is the rate of heat transfer to the coolant? What is the rate at which steam condenses on the plate?

(b) For plates inclined at an angle θ from the vertical, the average convection coefficient for condensation on the upper surface, $\bar{h}_{L(\text{incl})}$, may be approximated by an expression of the form, $\bar{h}_{L(\text{incl})} \approx (\cos\theta)^{1/4} \cdot \bar{h}_{L(\text{vert})}$, where $\bar{h}_{L(\text{vert})}$ is the average coefficient for the vertical orientation. If the 2 m × 2 m plate is inclined 45° from the normal, what are the rates of heat transfer and condensation?

10.51 Saturated ethylene glycol vapor at 1 atm is exposed to a vertical plate 300 mm high and 100 mm wide having a uniform temperature of 420 K. Estimate the heat transfer rate to the plate and the condensation rate. Approximate the liquid properties as those corresponding to saturated conditions at 373 K (Table A.5).

10.52 A vertical plate 2.5 m high, maintained at a uniform temperature of 54°C, is exposed to saturated steam at atmospheric pressure.

(a) Estimate the condensation and heat transfer rates per unit width of the plate.

(b) If the plate height were halved, would the flow regime stay the same or change?

(c) For $54 \le T_s \le 90$°C, plot the condensation rate as a function of plate temperature for the two plate heights of parts (a) and (b).

10.53 Two configurations are being considered in the design of a condensing system for steam at 1 atm employing a vertical plate maintained at 90°C. The first configuration is a single vertical plate $L \times w$ and the second consists of two vertical plates $(L/2) \times w$, where L and w are the vertical and horizontal dimensions, respectively. Which configuration would you choose?

10.54 The condenser of a steam power plant consists of a square (in-line) array of 625 tubes, each of 25-mm diameter. Consider conditions for which saturated steam at 0.105 bars condenses on the outer surface of each tube, while a tube wall temperature of 17°C is maintained by the flow of cooling water through the tubes. What is the rate of heat transfer to the water per unit length of the tube array? What is the corresponding condensation rate?

10.55 The condenser of a steam power plant consists of AISI 302 stainless steel tubes ($k_s = 15$ W/m·K), each of outer and inner diameters $D_o = 30$ mm and $D_i = 26$ mm, respectively. Saturated steam at 0.135 bar condenses on the outer surface of a tube, while water at a mean temperature of $T_m = 290$ K is in fully developed flow through the tube.

(a) For a water flow rate of $\dot{m} = 0.25$ kg/s, what is the outer surface temperature $T_{s,o}$ of the tube and the rates of heat transfer and steam condensation per unit tube length? As a first estimate, you may evaluate the properties of the liquid film at the saturation temperature. If one wishes to increase the transfer rates, what is the limiting factor that should be addressed?

(b) Explore the effect of the water flow rate on $T_{s,o}$ and the rate of heat transfer per unit length.

10.56 Saturated vapor from a chemical process condenses at a slow rate on the inner surface of a vertical, thin-walled cylindrical container of length L and diameter D. The container wall is maintained at a uniform temperature T_s by flowing cold water across its outer surface.

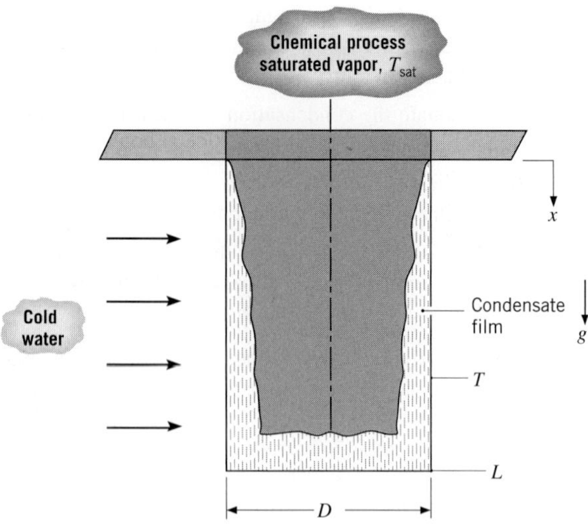

Derive an expression for the time, t_f, required to fill the container with condensate, assuming that the condensate film is laminar. Express your result in terms of D, L, $(T_{sat} - T_s)$, g, and appropriate fluid properties.

10.57 Determine the total condensation rate and heat transfer rate for the process of Problem 10.46 when the pipe is oriented at angles of $\theta = 0$, 30, 45, and 60° from the horizontal.

10.58 A horizontal tube of 50-mm outer diameter, with a surface temperature of 34°C, is exposed to steam at 0.2 bar. Estimate the condensation rate and heat transfer rate per unit length of the tube.

10.59 The tube of Problem 10.58 is modified by milling sharp-cornered grooves around its periphery, as in Figure 10.15. The 2-mm-deep grooves are each 2 mm wide with a pitch of $S = 4$ mm. Estimate the minimum condensation and heat transfer rates per unit length that would be expected for the modified tube. How much is the performance enhanced relative to the original tube of Problem 10.58?

10.60 A horizontal tube 1 m long with a surface temperature of 70°C is used to condense saturated steam at 1 atm.

(a) What diameter is required to achieve a condensation rate of 125 kg/h?

[(b)] Plot the condensation rate as a function of surface temperature for $70 \leq T_s \leq 90°C$ and tube diameters of 125, 150, 175 mm.

10.61 Saturated steam at a pressure of 0.1 bar is condensed over a square array of 100 tubes each of diameter 8 mm.

(a) If the tube surfaces are maintained at 27°C, estimate the condensation rate per unit tube length.

[(b)] Subject to the requirement that the total number of tubes and the tube diameter are fixed at 100 and 8 mm, respectively, what options are available for increasing the condensation rate? Assess these options quantitatively.

10.62 A horizontal, thin-walled concentric tube heat exchanger of 0.19-m length is to be used to heat deionized water from 40 to 60°C at a flow rate of 5 kg/s. The deionized water flows through the inner tube of 30-mm diameter while saturated steam at 1 atm is supplied to the annulus formed with the outer tube of 60-mm diameter. The thermophysical properties of the deionized water are $\rho = 982.3$ kg/m^3, $c_p = 4181$ J/kg·K, $k = 0.643$ W/m·K, $\mu = 548 \times 10^{-6}$ N·s/m^2, and $Pr = 3.56$. Estimate the convection coefficients for both sides of the tube and determine the inner tube wall outlet temperature. Does condensation provide a fairly uniform inner tube wall temperature equal approximately to the saturation temperature of the steam?

10.63 A technique for cooling a multichip module involves submerging the module in a saturated fluorocarbon liquid. Vapor generated due to boiling at the module surface is condensed on the outer surface of copper tubing suspended in the vapor space above the liquid. The thin-walled tubing is of diameter $D = 10$ mm and is coiled in a horizontal plane. It is cooled by water that enters at 285 K and leaves at 315 K. All the heat dissipated by the chips within the module is transferred from a 100-mm × 100-mm boiling surface, at which the flux is 10^5 W/m^2, to the fluorocarbon liquid, which is at $T_{sat} = 57°C$. Liquid properties are $k_l = 0.0537$ W/m·K, $c_{p,l} = 1100$ J/kg·K, $h'_{fg} \approx h_{fg} = 84{,}400$ J/kg, $\rho_l = 1619.2$ kg/m^3, $\rho_v = 13.4$ kg/m^3, $\sigma = 8.1 \times 10^{-3}$ N/m, $\mu_l = 440 \times 10^{-6}$ kg/m·s, and $Pr_l = 9$.

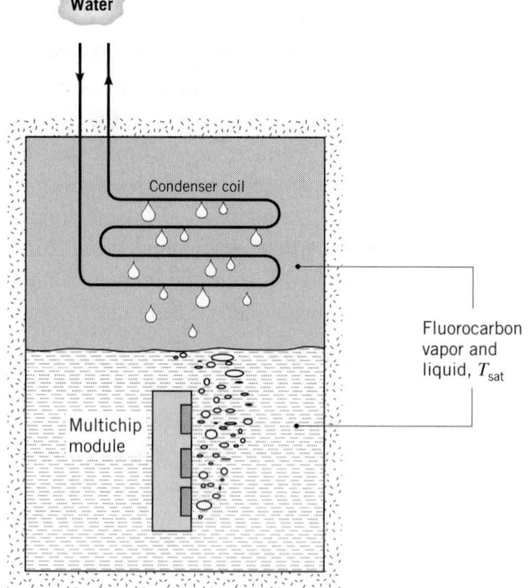

(a) For the prescribed heat dissipation, what is the required condensation rate (kg/s) and water flow rate (kg/s)?

(b) Assuming fully developed flow throughout the tube, determine the tube surface temperature at the coil inlet and outlet.

(c) Assuming a uniform tube surface temperature of $T_s = 53.0°C$, determine the required length of the coil.

10.64 Determine the rate of condensation on a 100-mm-diameter sphere with a surface temperature of 150°C in saturated ethylene glycol vapor at 1 atm. Approximate the liquid properties as those corresponding to saturated conditions at 373 K (Table A.5).

10.65 A 10-mm-diameter copper sphere, initially at a uniform temperature of 50°C, is placed in a large container filled with saturated steam at 1 atm. Using the lumped capacitance method, estimate the time required for the sphere to reach an equilibrium condition. How much condensate (kg) was formed during this period?

Condensation in Tubes

10.66 The Clean Air Act prohibited the production of chlorofluorocarbons (CFCs) in the United States as of 1996. One widely used CFC, refrigerant R-12, has been replaced by R-134a in many applications because of their similar properties, including a low boiling point at atmospheric pressure, $T_{sat} = 243$ K and 246.9 K for R-12 and R-134a, respectively. Compare the performance of these two refrigerants under the following conditions. The saturated refrigerant vapor at 310 K is condensed as it flows through a 30-mm-diameter, 0.8-m-long tube whose wall temperature is maintained at 290 K. If vapor enters the tube at a flow rate of 0.010 kg/s, what is the rate of condensation and the flow rate of vapor leaving the tube? The relevant properties of R-12 at $T_{sat} = 310$ K are $\rho_v = 50.1$ kg/m^3, $h_{fg} = 160$ kJ/kg, and $\mu_v = 150 \times 10^{-7}$ N·s/m^2 and those of liquid R-12 at $T_f = 300$ K are $\rho_l = 1306$ kg/m^3, $c_{p,l} = 978$ J/kg·K, $\mu_l = 2.54 \times 10^{-4}$ N·s/m^2, $k_l = 0.072$ W/m·K. The properties of the saturated R-134a vapor are $\rho_v = 46.1$ kg/m^3, $h_{fg} = 166$ kJ/kg, and $\mu_v = 136 \times 10^{-7}$ N·s/m^2.

10.67 Saturated steam at 1.5 bars condenses inside a horizontal, 75-mm-diameter pipe whose surface is maintained at 100°C. Assuming low vapor velocities and film condensation, estimate the heat transfer coefficient and the condensation rate per unit length of the pipe.

10.68 Consider the situation of Problem 10.67 at relatively high vapor velocities, with a fluid mass flow rate of $\dot{m} = 2.5$ kg/s.

(a) Determine the heat transfer coefficient and condensation rate per unit length of tube for a mass fraction of vapor of $X = 0.2$.

(b) Plot the heat transfer coefficient and the condensation rate for $0.1 \leq X \leq 0.3$.

10.69 Refrigerant R-22 with a mass flow rate of $\dot{m} = 8.75 \times 10^{-3}$ kg/s is condensed inside a 7-mm-diameter tube. Annular flow is observed. The saturation temperature of the pressurized refrigerant is $T_{sat} = 45°C$, and the wall temperature is $T_s = 40°C$. Vapor properties are $\rho_v = 77$ kg/m^3 and $\mu_v = 15 \times 10^{-6}$ N·s/m^2.

(a) Determine the heat transfer coefficient and the heat transfer and condensation rates per unit length at a quality of $X = 0.5$.

(b) Plot the condensation rate per unit length over the range $0.2 < X < 0.8$.

Dropwise Condensation

10.70 Consider Problem 10.44. In an effort to increase the condensation rate, an engineer proposes to apply an $L = 100$-μm-thick Teflon coating to the exterior surface of the brass tube to promote dropwise condensation. Estimate the new condensation convection coefficient and the steam condensation rate per unit length of the tube after application of the coating. Comment on the proposed scheme's effect on the condensation rate (the condensation rate per unit length in Problem 10.44 is approximately 1×10^{-3} kg/s).

10.71 Wetting of some metallic surfaces can be inhibited by means of ion implantation of the surface prior to its use, thereby promoting dropwise condensation. The degree of wetting inhibition and, in turn, the efficacy of the implantation process vary from metal to metal. Consider a vertical metal plate that is exposed to saturated steam at atmospheric pressure. The plate is $t = 1$ mm thick, and its vertical and horizontal dimensions are $L = 250$ mm and $b = 100$ mm, respectively. The temperature of the plate surface that is exposed to the steam is found to be $T_s = 90°C$ when the opposite surface of the metal plate is held at a cold temperature, T_c.

(a) Determine T_c for 2024-T6 aluminum. Assume the ion-implantation process does not promote dropwise condensation for this metal.

(b) Determine T_c for AISI 302 stainless steel, assuming the ion-implantation process is effective in promoting dropwise condensation.

Combined Boiling/Condensation

10.72 A passive technique for cooling heat-dissipating integrated circuits involves submerging the ICs in a low boiling point dielectric fluid. Vapor generated in cooling the circuits is condensed on vertical plates suspended in the vapor cavity above the liquid. The temperature of the plates is maintained below the saturation temperature, and during steady-state operation a balance is established between the rate of heat transfer to the condenser plates and the rate of heat dissipation by the ICs.

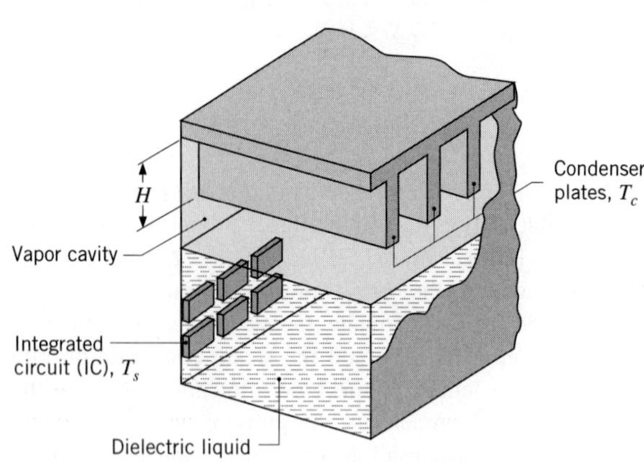

Consider conditions for which the 25-mm² surface area of each IC is submerged in a fluorocarbon liquid for which $T_{sat} = 50°C$, $\rho_l = 1700\ kg/m^3$, $c_{p,l} = 1005\ J/kg \cdot K$, $\mu_l = 6.80 \times 10^{-4}\ kg/s \cdot m$, $k_l = 0.062\ W/m \cdot K$, $Pr_l = 11.0$, $\sigma = 0.013\ N/m$, $h_{fg} = 1.05 \times 10^5\ J/kg$, $C_{s,f} = 0.004$, and $n = 1.7$. If the integrated circuits are operated at a surface temperature of $T_s = 75°C$, what is the rate at which heat is dissipated by each circuit? If the condenser plates are of height $H = 50$ mm and are maintained at a temperature of $T_c = 15°C$ by an internal coolant, how much condenser surface area must be provided to balance the heat generated by 500 integrated circuits?

10.73 A thermosyphon consists of a closed container that absorbs heat along its boiling section and rejects heat along its condensation section. Consider a thermosyphon made from a thin-walled mechanically polished stainless steel cylinder of diameter D. Heat supplied to the thermosyphon boils saturated water at atmospheric pressure on the surfaces of the lower boiling section of length L_b and is then rejected by condensing vapor into a thin film, which falls by gravity along the wall of the condensation section of length L_c back into the boiling section. The two sections are separated by an insulated section of length L_i. The top surface of the condensation section may be treated as being insulated. The thermosyphon dimensions are $D = 20$ mm, $L_b = 20$ mm, $L_c = 40$ mm, and $L_i = 40$ mm.

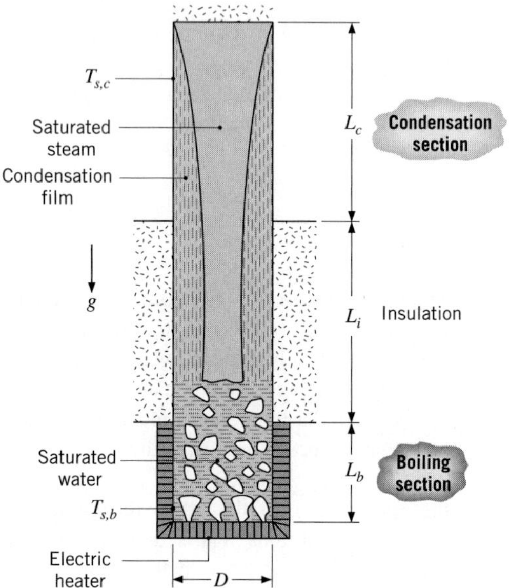

(a) Find the mean surface temperature, $T_{s,b}$, of the boiling surface if the nucleate boiling heat flux is to be maintained at 30% of the critical heat flux.

(b) Find the total condensation flow rate, $\dot{m}$, and the mean surface temperature of the condensation section, $T_{s,c}$.

10.74 A novel scheme for cooling computer chips uses a thermosyphon containing a saturated fluorocarbon. The chip is brazed to the bottom of a cuplike container, within which heat is dissipated by boiling and subsequently transferred to an external coolant (water) via condensation on the inner surface of a thin-walled tube.

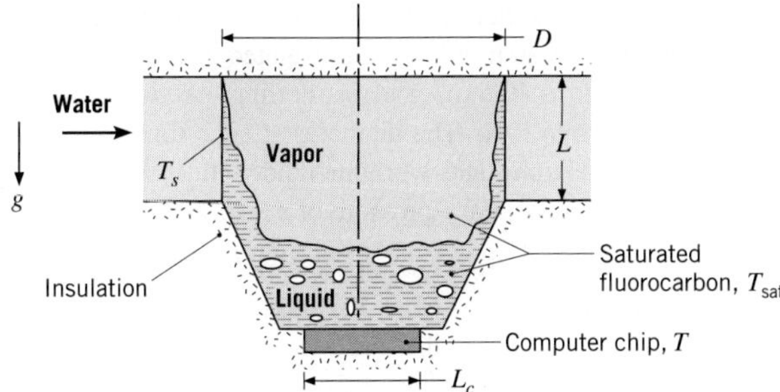

The nucleate boiling constants and the properties of the fluorocarbon are provided in Problem 10.23. In addition, $k_l = 0.054$ W/m·K.

(a) If the chip operates under steady-state conditions and its surface heat flux is maintained at 90% of the critical heat flux, what is its temperature T? What is the total power dissipation if the chip width is $L_c = 20$ mm on a side?

(b) If the tube diameter is $D = 30$ mm and its surface is maintained at $T_s = 25°C$ by the water, what tube length L is required to maintain the designated conditions?

10.75 A condenser–boiler section contains a 2-m × 2-m copper plate operating at a uniform temperature of $T_s = 100°C$ and separating saturated steam, which is condensing, from a saturated liquid-X, which experiences nucleate pool boiling. A portion of the boiling curve for liquid-X is shown as follows. Both saturated steam and saturated liquid-X are supplied to the system, while water condensate and vapor-X are removed by means not shown in the sketch. At a pressure of 1 bar, fluid-X has a saturation temperature and a latent heat of vaporization of $T_{sat} = 80°C$ and $h_{fg} = 700{,}000$ J/kg, respectively.

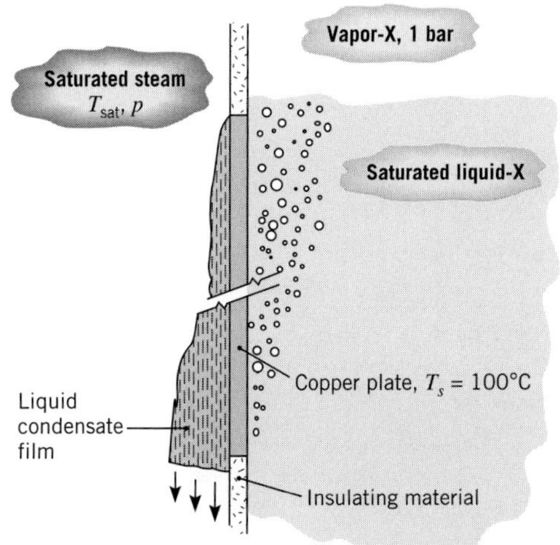

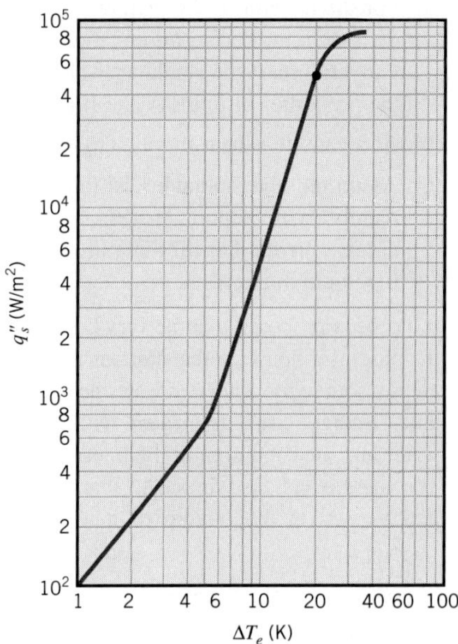

(a) Estimate the rates of evaporation and condensation (kg/s) for the two fluids.

(b) Determine the saturation temperature T_{sat} and pressure p for the steam, assuming that film condensation occurs.

10.76 A thin-walled cylindrical container of diameter D and height L is filled to a height y with a low boiling point liquid (A) at $T_{sat,A}$. The container is located in a large chamber filled with the vapor of a high boiling point fluid (B). Vapor-B condenses into a laminar film on the outer surface of the cylindrical container, extending from the location of the liquid-A free surface. The condensation process sustains nucleate boiling in liquid-A along the container wall according to the relation $q'' = C(T_s - T_{sat})^3$, where C is a known empirical constant.

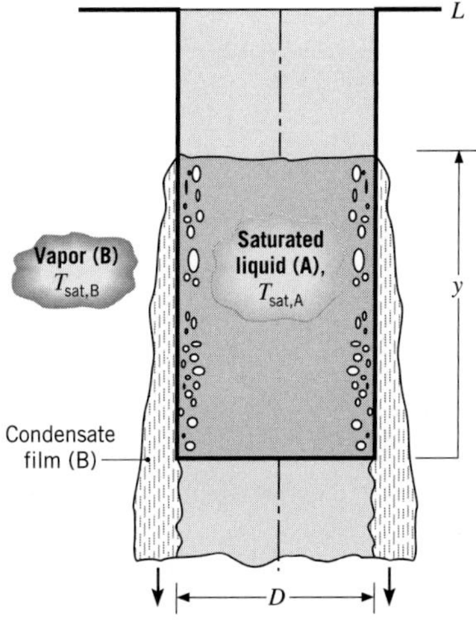

(a) For the portion of the wall covered with the condensate film, derive an equation for the average temperature of the container wall, T_s. Assume that the properties of fluids A and B are known.

(b) At what rate is heat supplied to liquid-A?

(c) Assuming the container is initially filled completely with liquid, that is, $y = L$, derive an expression for the time required to evaporate all the liquid in the container.

10.77 It has been proposed that the very hot air trapped inside the attic of a house in the summer may be used as the energy source for a *passive* water heater installed in the attic. Energy costs associated with heating the cool water and air conditioning the house are both reduced. Ten thermosyphons, similar to that of Problem 10.73, are inserted in the bottom of a well-insulated water heater. Each thermosyphon has a condensing section that is $L_c = 50$ mm long, an insulated section that is of length $L_i = 40$ mm, and a boiling section that is $L_b = 30$ mm long. The diameter of each thermosyphon is $D = 20$ mm. The working fluid within the thermosyphons is water at a pressure of $p = 0.047$ bars.

(a) Determine the heating rate delivered by the 10 thermosyphons when boiling occurs at 25% of the CHF. What are the mean temperatures of the boiling and condensing sections?

(b) At night the attic air temperature drops below the temperature of the water. Estimate the heat loss from the hot water tank to the cool attic, assuming losses through the tank insulation are negligible and the stainless steel tube wall thickness of each thermosyphon is very small.

CHAPTER

Heat Exchangers

11

The process of heat exchange between two fluids that are at different temperatures and separated by a solid wall occurs in many engineering applications. The device used to implement this exchange is termed a *heat exchanger*, and specific applications may be found in space heating and air-conditioning, power production, waste heat recovery, and chemical processing.

In this chapter our objectives are to introduce performance parameters for assessing the efficacy of a heat exchanger and to develop methodologies for designing a heat exchanger or for predicting the performance of an existing exchanger operating under prescribed conditions.

11.1 *Heat Exchanger Types*

Heat exchangers are typically classified according to *flow arrangement* and *type of construction.* The simplest heat exchanger is one for which the hot and cold fluids move in the same or opposite directions in a *concentric tube* (or *double-pipe*) construction. In the *parallel-flow* arrangement of Figure 11.1*a*, the hot and cold fluids enter at the same end, flow in the same direction, and leave at the same end. In the *counterflow* arrangement of Figure 11.1*b*, the fluids enter at opposite ends, flow in opposite directions, and leave at opposite ends.

Alternatively, the fluids may move in *cross flow* (perpendicular to each other), as shown by the *finned* and *unfinned* tubular heat exchangers of Figure 11.2. The two configurations

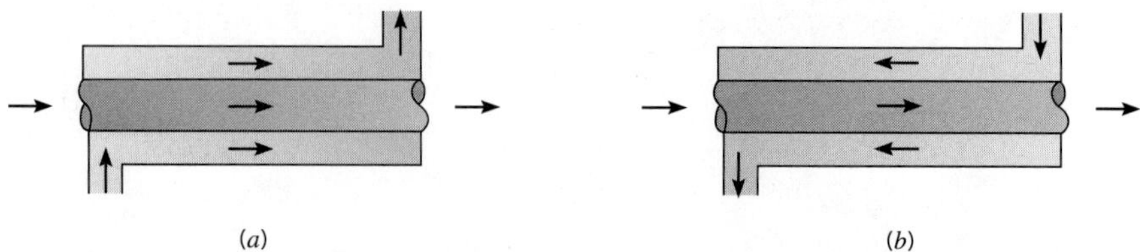

FIGURE 11.1 Concentric tube heat exchangers. (*a*) Parallel flow. (*b*) Counterflow.

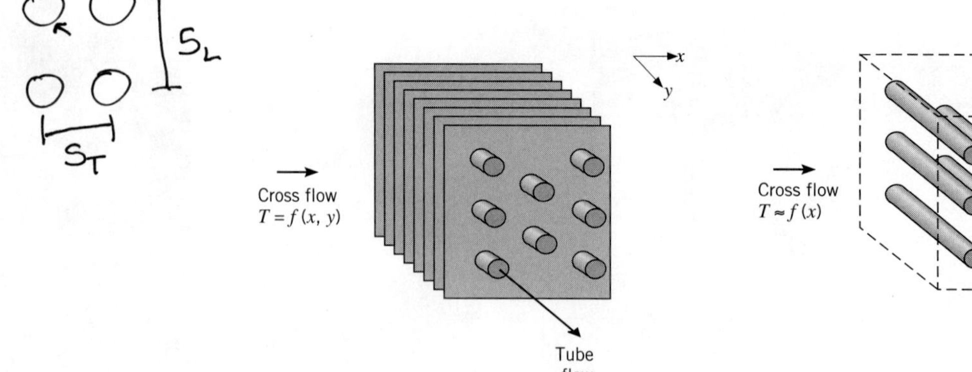

FIGURE 11.2 Cross-flow heat exchangers. (*a*) Finned with both fluids unmixed. (*b*) Unfinned with one fluid mixed and the other unmixed.

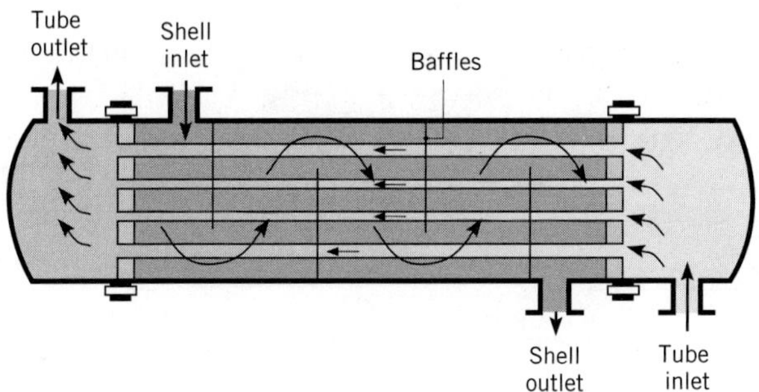

FIGURE 11.3 Shell-and-tube heat exchanger with one shell pass and one tube pass (cross-counterflow mode of operation).

are typically differentiated by an idealization that treats fluid motion over the tubes as *unmixed* or *mixed*. In Figure 11.2*a*, the cross-flowing fluid is said to be unmixed because the fins inhibit motion in a direction (y) that is transverse to the main-flow direction (x). In this case the cross-flowing fluid temperature varies with x and y. In contrast, for the unfinned tube bundle of Figure 11.2*b*, fluid motion, hence mixing, in the transverse direction is possible, and temperature variations are primarily in the main-flow direction. Since the tube flow is unmixed in either heat exchanger, both fluids are unmixed in the finned exchanger, while the cross-flowing fluid is mixed and the tube fluid is unmixed in the unfinned exchanger. The nature of the mixing condition influences heat exchanger performance.

Another common configuration is the *shell-and-tube* heat exchanger [1]. Specific forms differ according to the number of shell-and-tube passes, and the simplest form, which involves single tube and shell *passes*, is shown in Figure 11.3. Baffles are usually installed to increase the convection coefficient of the shell-side fluid by inducing turbulence and a cross-flow velocity component relative to the tubes. In addition, the baffles physically support the tubes, reducing flow-induced tube vibration. Baffled heat exchangers with one shell pass and two tube passes and with two shell passes and four tube passes are shown in Figures 11.4*a* and 11.4*b*, respectively.

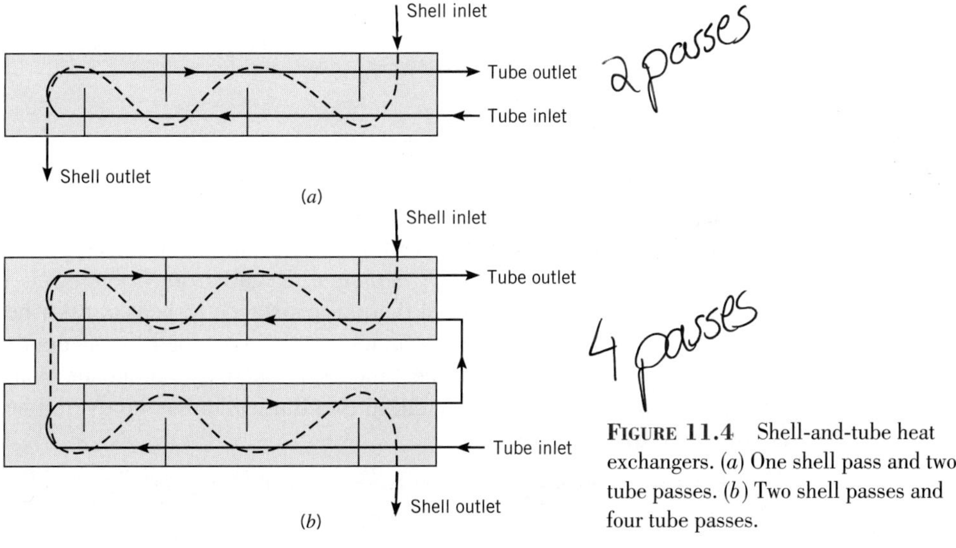

FIGURE 11.4 Shell-and-tube heat exchangers. (*a*) One shell pass and two tube passes. (*b*) Two shell passes and four tube passes.

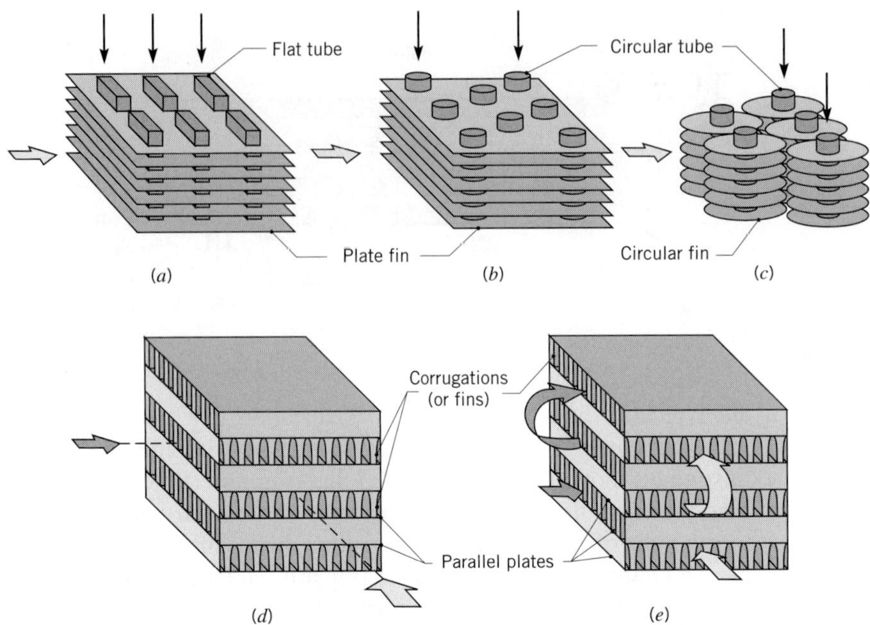

FIGURE 11.5 Compact heat exchanger cores. (*a*) Fin–tube (flat tubes, continuous plate fins). (*b*) Fin–tube (circular tubes, continuous plate fins). (*c*) Fin–tube (circular tubes, circular fins). (*d*) Plate–fin (single pass). (*e*) Plate–fin (multipass).

A special and important class of heat exchangers is used to achieve a very large ($\gtrsim 400\ m^2/m^3$ for liquids and $\gtrsim 700\ m^2/m^3$ for gases) heat transfer surface area per unit volume. Termed *compact heat exchangers*, these devices have dense arrays of finned tubes or plates and are typically used when at least one of the fluids is a gas, and is hence characterized by a small convection coefficient. The tubes may be *flat* or *circular*, as in Figures 11.5*a* and 11.5*b*, *c*, respectively, and the fins may be *plate* or *circular*, as in Figures 11.5*a*, *b*, and 11.5*c*, respectively. Parallel-plate heat exchangers may be finned or corrugated and may be used in single-pass (Figure 11.5*d*) or multipass (Figure 11.5*e*) modes of operation. Flow passages associated with compact heat exchangers are typically small ($D_h \lesssim 5$ mm), and the flow is usually laminar.

11.2 *The Overall Heat Transfer Coefficient*

An essential, and often the most uncertain, part of any heat exchanger analysis is determination of the *overall heat transfer coefficient.* Recall from Equation 3.19 that this coefficient is defined in terms of the total thermal resistance to heat transfer between two fluids. In Equations 3.18 and 3.36, the coefficient was determined by accounting for conduction and convection resistances between fluids separated by composite plane and cylindrical walls, respectively. For a wall separating two fluid streams, the overall heat transfer coefficient may be expressed as

$$\frac{1}{UA} = \frac{1}{U_cA_c} = \frac{1}{U_hA_h} = \frac{1}{(hA)_c} + R_w + \frac{1}{(hA)_h} \tag{11.1a}$$

where c and h refer to the cold and hot fluids, respectively. Note that calculation of the UA product does not require designation of the hot or cold side ($U_cA_c = U_hA_h$). However, calculation of an overall coefficient depends on whether it is based on the cold or hot side surface area, since $U_c \neq U_h$ if $A_c \neq A_h$. The conduction resistance R_w is obtained from Equation 3.6 for a plane wall or Equation 3.33 for a cylindrical wall.

It is important to acknowledge, however, that Equation 11.1a applies only to *clean, unfinned* surfaces. During normal heat exchanger operation, surfaces are often subject to fouling by fluid impurities, rust formation, or other reactions between the fluid and the wall material. The subsequent deposition of a film or scale on the surface can greatly increase the resistance to heat transfer between the fluids. This effect can be treated by introducing an additional thermal resistance in Equation 11.1a, termed the *fouling factor,* R_f. Its value depends on the operating temperature, fluid velocity, and length of service of the heat exchanger.

In addition, we know that fins are often added to surfaces exposed to either or both fluids and that, by increasing the surface area, they reduce the overall resistance to heat transfer. Accordingly, with inclusion of surface fouling and fin (extended surface) effects, the overall heat transfer coefficient is modified as follows:

$$\frac{1}{UA} = \frac{1}{(\eta_o hA)_c} + \frac{R''_{f,c}}{(\eta_o A)_c} + R_w + \frac{R''_{f,h}}{(\eta_o A)_h} + \frac{1}{(\eta_o hA)_h} \tag{11.1b}$$

Although representative fouling factors (R''_f) are listed in Table 11.1, the factor is a variable during heat exchanger operation (increasing from zero for a clean surface, as deposits accumulate on the surface). Comprehensive discussions of fouling are provided in References 2 through 4.

The quantity η_o in Equation 11.1b is termed the *overall surface efficiency* or *temperature effectiveness* of a finned surface. It is defined such that, for the hot or cold surface without fouling, the heat transfer rate is

$$q = \eta_o hA(T_b - T_\infty) \tag{11.2}$$

where T_b is the base surface temperature (Figure 3.21) and A is the total (fin plus exposed base) surface area. The quantity was introduced in Section 3.6.5, and the following expression was derived:

$$\eta_o = 1 - \frac{A_f}{A}(1 - \eta_f) \tag{11.3}$$

where A_f is the entire fin surface area and η_f is the efficiency of a single fin. To be consistent with the nomenclature commonly used in heat exchanger analysis, the ratio of fin surface area to the total surface area has been expressed as A_f/A. This representation differs

TABLE 11.1 Representative Fouling Factors [1]

Fluid	R''_f (m$^2 \cdot$ K/W)
Seawater and treated boiler feedwater (below 50°C)	0.0001
Seawater and treated boiler feedwater (above 50°C)	0.0002
River water (below 50°C)	0.0002–0.001
Fuel oil	0.0009
Refrigerating liquids	0.0002
Steam (nonoil bearing)	0.0001

from that of Section 3.6.5, where the ratio is expressed as NA_f/A_t, with A_f representing the area of a single fin and A_t the total surface area. If a straight or pin fin of length L (Figure 3.17) is used and an adiabatic tip is assumed, Equations 3.81 and 3.91 yield

$$\eta_f = \frac{\tanh(mL)}{mL} \tag{11.4}$$

where $m = (2h/kt)^{1/2}$ and t is the fin thickness. For several common fin shapes, the efficiency may be obtained from Table 3.5.

Note that, as written, Equation 11.2 corresponds to negligible fouling. However, if fouling is significant, the convection coefficient in Equation 11.2 must be replaced by a *partial* overall heat transfer coefficient of the form $U_p = h/(1 + hR''_f)$. In contrast to Equation 11.1b, which provides the overall heat transfer coefficient between the hot and cold fluids, U_p is termed a partial coefficient because it only includes the convection coefficient and fouling factor associated with one fluid and its adjoining surface. Partial coefficients for the hot and cold sides are then $U_{p,h} = h_h/(1 + h_h R''_{f,h})$ and $U_{p,c} = h_c/(1 + h_c R''_{f,c})$, respectively. Equation 11.3 may still be used to evaluate η_o for the hot and/or cold side, but U_p must be used in lieu of h to evaluate the corresponding fin efficiency. Moreover, it is readily shown that the second and fourth terms on the right-hand side of Equation 11.1b may be deleted if the convection coefficients in the first and fifth terms are replaced by $U_{p,c}$ and $U_{p,h}$, respectively.

The wall conduction term in Equation 11.1a or 11.1b may often be neglected, since a thin wall of large thermal conductivity is generally used. Also, one of the convection coefficients is often much smaller than the other and hence dominates determination of the overall coefficient. For example, if one of the fluids is a gas and the other is a liquid or a liquid–vapor mixture experiencing boiling or condensation, the gas-side convection coefficient is much smaller. It is in such situations that fins are used to enhance gas-side convection. Representative values of the overall coefficient are summarized in Table 11.2.

For the unfinned, tubular heat exchangers of Figures 11.1 through 11.4, Equation 11.1b reduces to

$$\begin{aligned} \frac{1}{UA} &= \frac{1}{U_i A_i} = \frac{1}{U_o A_o} \\ &= \frac{1}{h_i A_i} + \frac{R''_{f,i}}{A_i} + \frac{\ln(D_o/D_i)}{2\pi kL} + \frac{R''_{f,o}}{A_o} + \frac{1}{h_o A_o} \end{aligned} \tag{11.5}$$

where subscripts i and o refer to inner and outer tube surfaces ($A_i = \pi D_i L$, $A_o = \pi D_o L$), which may be exposed to either the hot or the cold fluid.

TABLE 11.2 Representative Values of the Overall Heat Transfer Coefficient

Fluid Combination	U (W/m$^2 \cdot$ K)
Water to water	850–1700
Water to oil	110–350
Steam condenser (water in tubes)	1000–6000
Ammonia condenser (water in tubes)	800–1400
Alcohol condenser (water in tubes)	250–700
Finned-tube heat exchanger (water in tubes, air in cross flow)	25–50

The overall heat transfer coefficient may be determined from knowledge of the hot and cold fluid convection coefficients and fouling factors and from appropriate geometric parameters. For unfinned surfaces, the convection coefficients may be estimated from correlations presented in Chapters 7 and 8. For standard fin configurations, the coefficients may be obtained from results compiled by Kays and London [5].

11.3 Heat Exchanger Analysis: Use of the Log Mean Temperature Difference

To design or to predict the performance of a heat exchanger, it is essential to relate the total heat transfer rate to quantities such as the inlet and outlet fluid temperatures, the overall heat transfer coefficient, and the total surface area for heat transfer. Two such relations may readily be obtained by applying overall energy balances to the hot and cold fluids, as shown in Figure 11.6. In particular, if q is the total rate of heat transfer between the hot and cold fluids and there is negligible heat transfer between the exchanger and its surroundings, as well as negligible potential and kinetic energy changes, application of the steady flow energy equation, Equation 1.12d, gives

$$q = \dot{m}_h(i_{h,i} - i_{h,o}) \tag{11.6a}$$

and

$$q = \dot{m}_c(i_{c,o} - i_{c,i}) \tag{11.7a}$$

where i is the fluid enthalpy. The subscripts h and c refer to the hot and cold fluids, whereas the subscripts i and o designate the fluid inlet and outlet conditions. If the fluids are not undergoing a phase change and constant specific heats are assumed, these expressions reduce to

$$q = \dot{m}_h c_{p,h}(T_{h,i} - T_{h,o}) \tag{11.6b}$$

and

$$q = \dot{m}_c c_{p,c}(T_{c,o} - T_{c,i}) \tag{11.7b}$$

where the temperatures appearing in the expressions refer to the *mean* fluid temperatures at the designated locations. Note that Equations 11.6 and 11.7 are independent of the flow arrangement and heat exchanger type.

Another useful expression may be obtained by relating the total heat transfer rate q to the temperature difference ΔT between the hot and cold fluids, where

$$\Delta T \equiv T_h - T_c \tag{11.8}$$

Such an expression would be an extension of Newton's law of cooling, with the overall heat transfer coefficient U used in place of the single convection coefficient h. However,

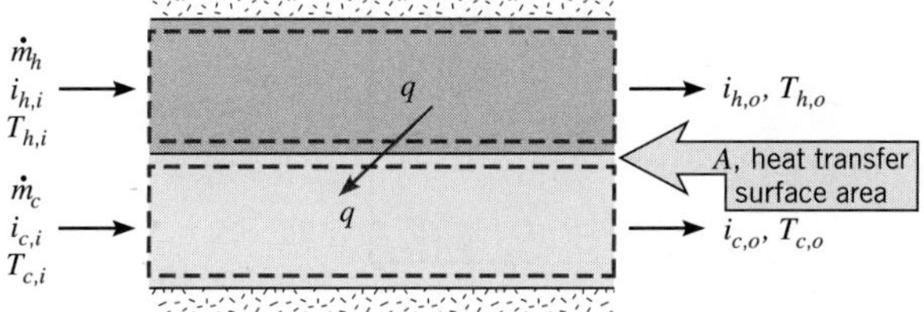

FIGURE 11.6 Overall energy balances for the hot and cold fluids of a two-fluid heat exchanger.

since ΔT varies with position in the heat exchanger, it is necessary to work with a rate equation of the form

$$q = UA\Delta T_m \tag{11.9}$$

where ΔT_m is an appropriate *mean* temperature difference. Equation 11.9 may be used with Equations 11.6 and 11.7 to perform a heat exchanger analysis. Before this can be done, however, the specific form of ΔT_m must be established.

11.3.1 The Parallel-Flow Heat Exchanger

The hot and cold mean fluid temperature distributions associated with a parallel-flow heat exchanger are shown in Figure 11.7. The temperature difference ΔT is initially large but decays with increasing x, approaching zero asymptotically. It is important to note that, for such an exchanger, the outlet temperature of the cold fluid never exceeds that of the hot fluid. In Figure 11.7 the subscripts 1 and 2 designate opposite ends of the heat exchanger. This convention is used for all types of heat exchangers considered. For parallel flow, it follows that $T_{h,i} = T_{h,1}$, $T_{h,o} = T_{h,2}$, $T_{c,i} = T_{c,1}$, and $T_{c,o} = T_{c,2}$.

The form of ΔT_m may be determined by applying an energy balance to differential elements in the hot and cold fluids. Each element is of length dx and heat transfer surface area dA, as shown in Figure 11.7. The energy balances and the subsequent analysis are subject to the following assumptions.

1. The heat exchanger is insulated from its surroundings, in which case the only heat exchange is between the hot and cold fluids.
2. Axial conduction along the tubes is negligible.
3. Potential and kinetic energy changes are negligible.
4. The fluid specific heats are constant.
5. The overall heat transfer coefficient is constant.

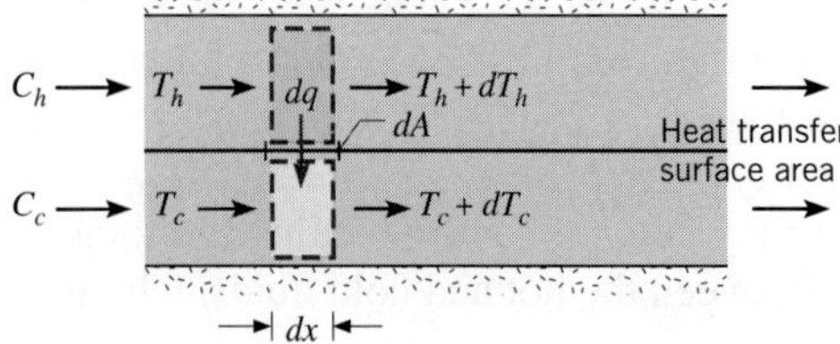

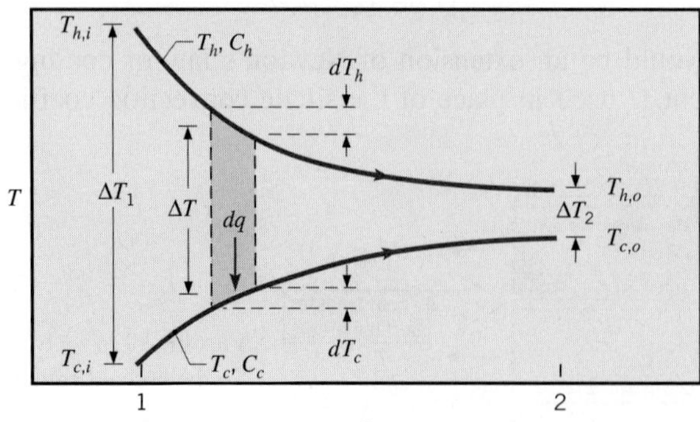

FIGURE 11.7 Temperature distributions for a parallel-flow heat exchanger.

The specific heats may of course change as a result of temperature variations, and the overall heat transfer coefficient may change because of variations in fluid properties and flow conditions. However, in many applications such variations are not significant, and it is reasonable to work with average values of $c_{p,c}$, $c_{p,h}$, and U for the heat exchanger.

Applying an energy balance to each of the differential elements of Figure 11.7, it follows that

$$dq = -\dot{m}_h c_{p,h}\, dT_h \equiv -C_h\, dT_h \tag{11.10}$$

and

$$dq = \dot{m}_c c_{p,c}\, dT_c \equiv C_c dT_c \tag{11.11}$$

where C_h and C_c are the hot and cold fluid *heat capacity rates*, respectively. These expressions may be integrated across the heat exchanger to obtain the overall energy balances given by Equations 11.6b and 11.7b. The heat transfer across the surface area dA may also be expressed as

$$dq = U\,\Delta T\, dA \tag{11.12}$$

where $\Delta T = T_h - T_c$ is the *local* temperature difference between the hot and cold fluids.

To determine the integrated form of Equation 11.12, we begin by substituting Equations 11.10 and 11.11 into the differential form of Equation 11.8

$$d(\Delta T) = dT_h - dT_c$$

to obtain

$$d(\Delta T) = -dq\left(\frac{1}{C_h} + \frac{1}{C_c}\right)$$

Substituting for dq from Equation 11.12 and integrating across the heat exchanger, we obtain

$$\int_1^2 \frac{d(\Delta T)}{\Delta T} = -U\left(\frac{1}{C_h} + \frac{1}{C_c}\right)\int_1^2 dA$$

or

$$\ln\left(\frac{\Delta T_2}{\Delta T_1}\right) = -UA\left(\frac{1}{C_h} + \frac{1}{C_c}\right) \tag{11.13}$$

Substituting for C_h and C_c from Equations 11.6b and 11.7b, respectively, it follows that

$$\ln\left(\frac{\Delta T_2}{\Delta T_1}\right) = -UA\left(\frac{T_{h,i} - T_{h,o}}{q} + \frac{T_{c,o} - T_{c,i}}{q}\right)$$

$$= -\frac{UA}{q}[(T_{h,i} - T_{c,i}) - (T_{h,o} - T_{c,o})]$$

Recognizing that, for the parallel-flow heat exchanger of Figure 11.7, $\Delta T_1 = (T_{h,i} - T_{c,i})$ and $\Delta T_2 = (T_{h,o} - T_{c,o})$, we then obtain

$$q = UA\frac{\Delta T_2 - \Delta T_1}{\ln(\Delta T_2/\Delta T_1)}$$

Comparing the above expression with Equation 11.9, we conclude that the appropriate average temperature difference is a *log mean temperature difference,* ΔT_{lm}. Accordingly, we may write

$$q = UA\,\Delta T_{\text{lm}} \tag{11.14}$$

where

$$\Delta T_{\text{lm}} = \frac{\Delta T_2 - \Delta T_1}{\ln(\Delta T_2/\Delta T_1)} = \frac{\Delta T_1 - \Delta T_2}{\ln(\Delta T_1/\Delta T_2)} \tag{11.15}$$

Remember that, for the *parallel-flow exchanger,*

$$\begin{bmatrix} \Delta T_1 \equiv T_{h,1} - T_{c,1} = T_{h,i} - T_{c,i} \\ \Delta T_2 \equiv T_{h,2} - T_{c,2} = T_{h,o} - T_{c,o} \end{bmatrix} \tag{11.16}$$

Referring back to Section 8.3.3, it can be seen that there is a strong similarity between the preceding analysis and the analysis of internal tube flow in which heat transfer occurs between the flowing fluid and either a surface at constant temperature or an external fluid at constant temperature. For this reason, internal tube flow is sometimes referred to as a *single stream heat exchanger*. Equations 8.43 and 8.44 or Equations 8.45a and 8.46a are analogous to Equations 11.14 and 11.15.

11.3.2 The Counterflow Heat Exchanger

The hot and cold fluid temperature distributions associated with a counterflow heat exchanger are shown in Figure 11.8. In contrast to the parallel-flow exchanger, this configuration provides for heat transfer between the hotter portions of the two fluids at one end, as well as between the colder portions at the other. For this reason, the change in the temperature difference, $\Delta T = T_h - T_c$, with respect to x is nowhere as large as it is for the inlet region of the parallel-flow exchanger. Note that the outlet temperature of the cold fluid may now exceed the outlet temperature of the hot fluid.

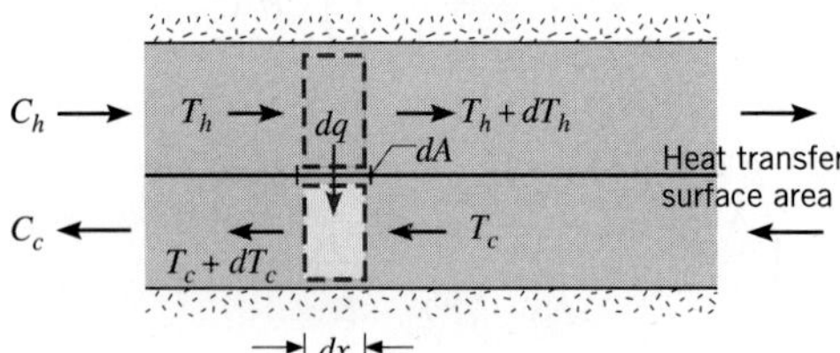

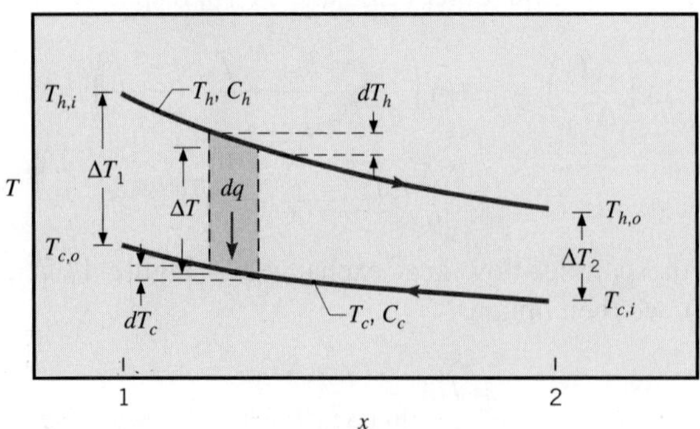

FIGURE 11.8 Temperature distributions for a counterflow heat exchanger.

Equations 11.6b and 11.7b apply to any heat exchanger and hence may be used for the counterflow arrangement. Moreover, from an analysis like that performed in Section 11.3.1, it may be shown that Equations 11.14 and 11.15 also apply. However, for the *counterflow exchanger* the endpoint temperature differences must now be defined as

$$\begin{bmatrix} \Delta T_1 \equiv T_{h,1} - T_{c,1} = T_{h,i} - T_{c,o} \\ \Delta T_2 \equiv T_{h,2} - T_{c,2} = T_{h,o} - T_{c,i} \end{bmatrix} \tag{11.17}$$

Note that, for the same inlet and outlet temperatures, the log mean temperature difference for counterflow exceeds that for parallel flow, $\Delta T_{\mathrm{lm,CF}} > \Delta T_{\mathrm{lm,PF}}$. Hence the surface area required to effect a prescribed heat transfer rate q is smaller for the counterflow than for the parallel-flow arrangement, assuming the same value of U. Also note that $T_{c,o}$ can exceed $T_{h,o}$ for counterflow but not for parallel flow.

11.3.3 Special Operating Conditions

It is useful to note certain special conditions under which heat exchangers may be operated. Figure 11.9*a* shows temperature distributions for a heat exchanger in which the hot fluid has a heat capacity rate, $C_h \equiv \dot{m}_h c_{p,h}$, which is much larger than that of the cold fluid, $C_c \equiv \dot{m}_c c_{p,c}$. For this case the temperature of the hot fluid remains approximately constant throughout the heat exchanger, while the temperature of the cold fluid increases. The same condition is achieved if the hot fluid is a condensing vapor. Condensation occurs at constant temperature, and, for all practical purposes, $C_h \rightarrow \infty$. Conversely, in an evaporator or a boiler (Figure 11.9*b*), it is the cold fluid that experiences a change in phase and remains at a nearly uniform temperature ($C_c \rightarrow \infty$). The same effect is achieved without phase change if $C_h \ll C_c$. Note that, with condensation or evaporation, the heat rate is given by Equation 11.6a or 11.7a, respectively. Conditions illustrated in Figure 11.9*a* or 11.9*b* also characterize an internal tube flow (or *single stream heat exchanger*) exchanging heat with a surface at constant temperature or an external fluid at constant temperature.

The third special case (Figure 11.9*c*) involves a counterflow heat exchanger for which the heat capacity rates are equal ($C_h = C_c$). The temperature difference ΔT must then be constant throughout the exchanger, in which case $\Delta T_1 = \Delta T_2 = \Delta T_{\mathrm{lm}}$.

Although flow conditions are more complicated in multipass and cross-flow heat exchangers, Equations 11.6, 11.7, 11.14, and 11.15 may still be used if modifications are made to the definition of the log mean temperature difference [6].

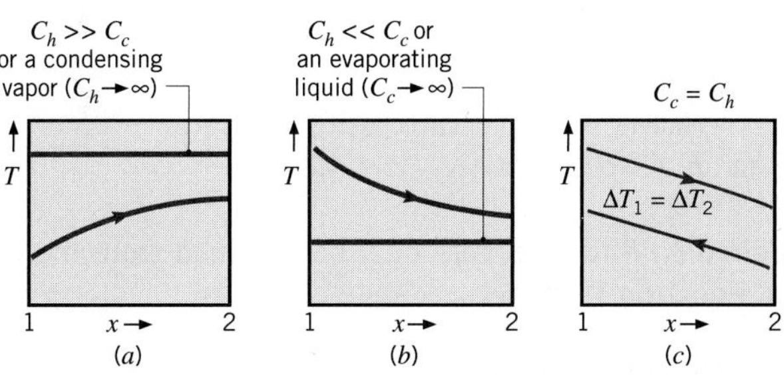

FIGURE 11.9 Special heat exchanger conditions. (*a*) $C_h \gg C_c$ or a condensing vapor. (*b*) An evaporating liquid or $C_h \ll C_c$. (*c*) A counterflow heat exchanger with equivalent fluid heat capacities ($C_h = C_c$).

Methodologies for using the LMTD method for other heat exchanger types are included in Section 11S.1.

EXAMPLE 11.1

A counterflow, concentric tube heat exchanger is used to cool the lubricating oil for a large industrial gas turbine engine. The flow rate of cooling water through the inner tube (D_i = 25 mm) is 0.2 kg/s, while the flow rate of oil through the outer annulus (D_o = 45 mm) is 0.1 kg/s. The oil and water enter at temperatures of 100 and 30°C, respectively. How long must the tube be made if the outlet temperature of the oil is to be 60°C?

SOLUTION

Known: Fluid flow rates and inlet temperatures for a counterflow, concentric tube heat exchanger of prescribed inner and outer diameter.

Find: Tube length to achieve a desired hot fluid outlet temperature.

Schematic:

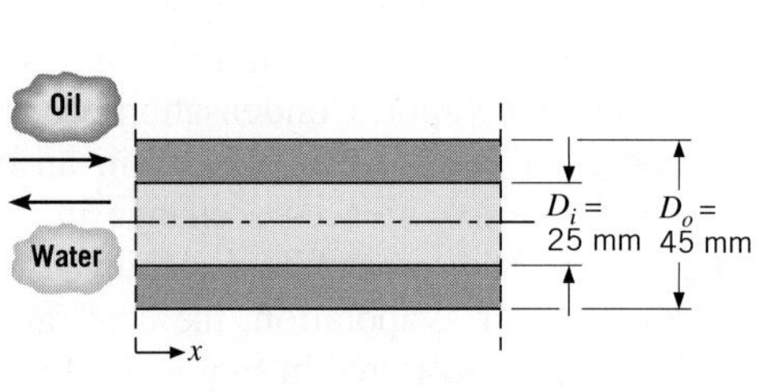

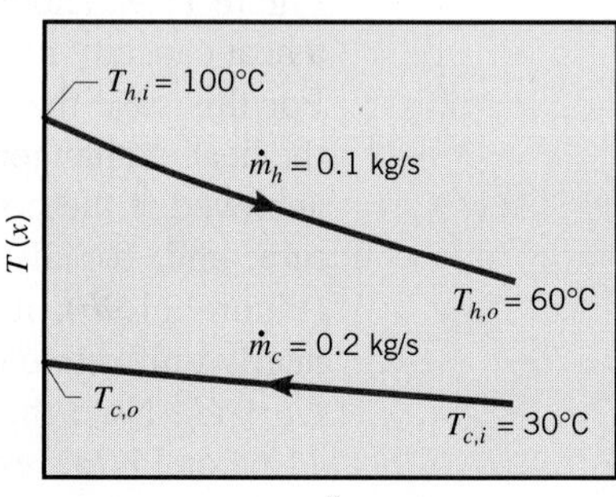

Assumptions:

1. Negligible heat loss to the surroundings.
2. Negligible kinetic and potential energy changes.
3. Constant properties.
4. Negligible tube wall thermal resistance and fouling factors.
5. Fully developed conditions for the water and oil (U independent of x).

Properties: Table A.5, unused engine oil ($\overline{T}_h$ = 80°C = 353 K): c_p = 2131 J/kg·K, μ = 3.25 × 10^{-2} N·s/m^2, k = 0.138 W/m·K. Table A.6, water ($\overline{T}_c \approx$ 35°C): c_p = 4178 J/kg·K, μ = 725 × 10^{-6} N·s/m^2, k = 0.625 W/m·K, Pr = 4.85.

Analysis: The required heat transfer rate may be obtained from the overall energy balance for the hot fluid, Equation 11.6b.

$$q = \dot{m}_h c_{p,h}(T_{h,i} - T_{h,o})$$

$$q = 0.1 \text{ kg/s} \times 2131 \text{ J/kg}\cdot\text{K}\,(100 - 60)^\circ\text{C} = 8524 \text{ W}$$

Applying Equation 11.7b, the water outlet temperature is

$$T_{c,o} = \frac{q}{\dot{m}_c c_{p,c}} + T_{c,i}$$

$$T_{c,o} = \frac{8524 \text{ W}}{0.2 \text{ kg/s} \times 4178 \text{ J/kg} \cdot \text{K}} + 30°\text{C} = 40.2°\text{C}$$

Accordingly, use of $\overline{T}_c = 35°\text{C}$ to evaluate the water properties was a good choice. The required heat exchanger length may now be obtained from Equation 11.14,

$$q = UA\,\Delta T_{\text{lm}}$$

where $A = \pi D_i L$ and from Equations 11.15 and 11.17

$$\Delta T_{\text{lm}} = \frac{(T_{h,i} - T_{c,o}) - (T_{h,o} - T_{c,i})}{\ln\,[(T_{h,i} - T_{c,o})/(T_{h,o} - T_{c,i})]} = \frac{59.8 - 30}{\ln\,(59.8/30)} = 43.2°\text{C}$$

From Equation 11.5 the overall heat transfer coefficient is

$$U = \frac{1}{(1/h_i) + (1/h_o)}$$

For water flow through the tube,

$$Re_D = \frac{4\dot{m}_c}{\pi D_i \mu} = \frac{4 \times 0.2 \text{ kg/s}}{\pi(0.025 \text{ m})725 \times 10^{-6} \text{ N} \cdot \text{s/m}^2} = 14{,}050$$

Accordingly, the flow is turbulent and the convection coefficient may be computed from Equation 8.60

$$Nu_D = 0.023\, Re_D^{4/5}\, Pr^{0.4}$$

$$Nu_D = 0.023\,(14{,}050)^{4/5}\,(4.85)^{0.4} = 90$$

Hence

$$h_i = Nu_D \frac{k}{D_i} = \frac{90 \times 0.625 \text{ W/m} \cdot \text{K}}{0.025 \text{ m}} = 2250 \text{ W/m}^2 \cdot \text{K}$$

For the flow of oil through the annulus, the hydraulic diameter is, from Equation 8.71, $D_h = D_o - D_i = 0.02$ m, and the Reynolds number is

$$Re_D = \frac{\rho u_m D_h}{\mu} = \frac{\rho(D_o - D_i)}{\mu} \times \frac{\dot{m}_h}{\rho\pi(D_o^2 - D_i^2)/4}$$

$$Re_D = \frac{4\dot{m}_h}{\pi\,(D_o + D_i)\,\mu} = \frac{4 \times 0.1 \text{ kg/s}}{\pi(0.045 + 0.025) \text{ m} \times 3.25 \times 10^{-2} \text{ kg/s} \cdot \text{m}} = 56.0$$

The annular flow is therefore laminar. Assuming uniform temperature along the inner surface of the annulus and a perfectly insulated outer surface, the convection coefficient at the inner surface may be obtained from Table 8.2. With $(D_i/D_o) = 0.56$, linear interpolation provides

$$Nu_i = \frac{h_o D_h}{k} = 5.63$$

and

$$h_o = 5.63 \frac{0.138 \text{ W/m} \cdot \text{K}}{0.020 \text{ m}} = 38.8 \text{ W/m}^2 \cdot \text{K}$$

The overall convection coefficient is then

$$U = \frac{1}{(1/2250 \text{ W/m}^2 \cdot \text{K}) + (1/38.8 \text{ W/m}^2 \cdot \text{K})} = 38.1 \text{ W/m}^2 \cdot \text{K}$$

and from the rate equation it follows that

$$L = \frac{q}{U\pi D_i \, \Delta T_{\text{lm}}} = \frac{8524 \text{ W}}{38.1 \text{ W/m}^2 \cdot \text{K} \; \pi \, (0.025 \text{ m}) \, (43.2°\text{C})} = 65.9 \text{ m} \qquad \triangleleft$$

Comments:

1. The hot side convection coefficient controls the rate of heat transfer between the two fluids, and the low value of h_o is responsible for the large value of L. Incorporation of heat transfer enhancement methods, such as described in Section 8.7, could be used to decrease the size of the heat exchanger.
2. Because $h_i \gg h_o$, the tube wall temperature will follow closely that of the coolant water. Accordingly, the assumption of uniform wall temperature, which is inherent in the use of Table 8.2 to obtain h_o, is reasonable.

EXAMPLE 11.2

The counterflow, concentric tube heat exchanger of Example 11.1 is replaced with a compact, plate-type heat exchanger that consists of a stack of thin metal sheets, separated by N gaps of width a. The oil and water flows are subdivided into $N/2$ individual flow streams, with the oil and water moving in opposite directions within alternating gaps. It is desirable for the stack to be of a cubical geometry, with a characteristic exterior dimension L. Determine the exterior dimensions of the heat exchanger as a function of the number of gaps if the flow rates, inlet temperatures, and desired oil outlet temperature are the same as in Example 11.1. Compare the pressure drops of the water and oil streams within the plate-type heat exchanger to the pressure drops of the flow streams in Example 11.1, if 60 gaps are specified.

SOLUTION

Known: Configuration of a plate-type heat exchanger. Fluid flow rates, inlet temperatures, and desired oil outlet temperature.

Find:

1. Exterior dimensions of the heat exchanger.
2. Pressure drops within the plate-type heat exchanger with $N = 60$ gaps, and the concentric tube heat exchanger of Example 11.1.

Schematic:

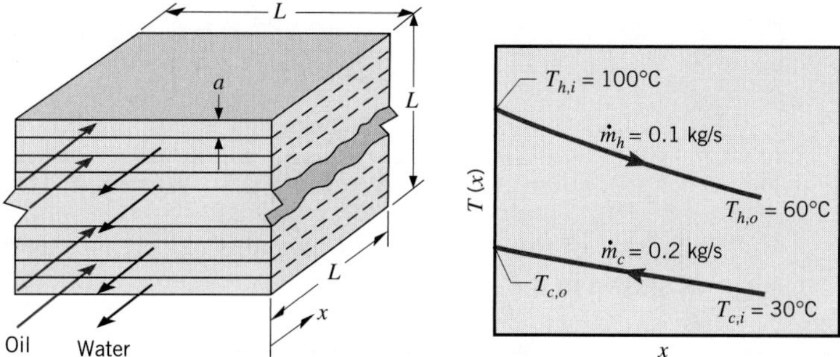

Assumptions:

1. Negligible heat loss to the surroundings.
2. Negligible kinetic and potential energy changes.
3. Constant properties.
4. Negligible plate thermal resistance and fouling factors.
5. Fully developed conditions for the water and oil.
6. Identical gap-to-gap heat transfer coefficients.
7. Heat exchanger exterior dimension is large compared to the gap width.

Properties: See Example 11.1. In addition, Table A.5, unused engine oil ($\overline{T}_h = 353$ K): $\rho = 852.1$ kg/m^3. Table A.6, water ($\overline{T}_c \approx 35°$C): $\rho = v_f^{-1} = 994$ kg/m^3.

Analysis:

1. The gap width may be related to the overall dimension of the heat exchanger by the expression $a = L/N$, and the total heat transfer area is $A = L^2(N-1)$. Assuming $a \ll L$ and the existence of laminar flow, the Nusselt number for each interior gap is provided in Table 8.1 and is

$$Nu_D = \frac{hD_h}{k} = 7.54$$

From Equation 8.66, the hydraulic diameter is $D_h = 2a$. Combining the preceding expressions yields for the water:

$$h_c = 7.54kN/2L = 7.54 \times 0.625 \text{ W/m} \cdot \text{K} \times N/2L = (2.36 \text{ W/m} \cdot \text{K})N/L$$

Likewise, for the oil:

$$h_h = 7.54kN/2L = 7.54 \times 0.138 \text{ W/m} \cdot \text{K} \times N/2L = (0.520 \text{ W/m} \cdot \text{K})N/L$$

and the overall convection coefficient is

$$U = \frac{1}{1/h_c + 1/h_h}$$

From Example 11.1, the required log mean temperature difference and heat transfer rate are $\Delta T_{\text{lm}} = 43.2°$C and $q = 8524$ W, respectively. From Equation 11.14 it follows that

$$UA = \frac{L^2(N-1)}{1/h_c + 1/h_h} = \frac{q}{\Delta T_{\text{lm}}}$$

which may be rearranged to yield

$$L = \frac{q}{\Delta T_{\text{lm}}(N-1)} \left[\frac{1}{h_c L} + \frac{1}{h_h L} \right]$$

$$= \frac{8524 \text{ W}}{43.2^\circ\text{C}(N-1)N} \left[\frac{1}{2.36 \text{ W/m}\cdot\text{K}} + \frac{1}{0.520 \text{ W/m}\cdot\text{K}} \right] = \frac{463 \text{ m}}{N(N-1)} \quad \triangleleft$$

The size of the compact heat exchanger decreases as the number of gaps is increased, as shown in the figure below.

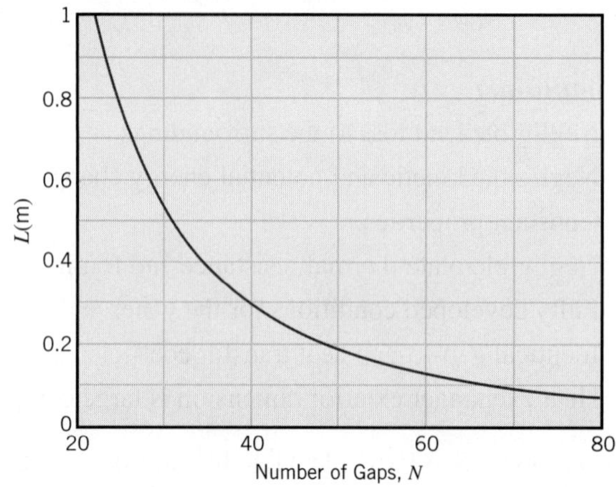

2. For $N = 60$ gaps, the stack dimension is $L = 0.131$ m from the results of part 1, and the gap width is $a = L/N = 0.131 \text{ m}/60 = 0.00218$ m.

The hydraulic diameter is $D_h = 0.00436$ m, and the mean velocity in each water-filled gap is

$$u_m = \frac{\dot{m}}{\rho L^2/2} = \frac{2 \times 0.2 \text{ kg/s}}{994 \text{ kg/m}^3 \times 0.131^2 \text{ m}^2} = 0.0235 \text{ m/s}$$

providing a Reynolds number of

$$Re_D = \frac{\rho u_m D_h}{\mu} = \frac{994 \text{ kg/m}^3 \times 0.0235 \text{ m/s} \times 0.00436 \text{ m}}{725 \times 10^{-6} \text{ N}\cdot\text{s/m}^2} = 141$$

For the oil-filled gaps

$$u_m = \frac{\dot{m}}{\rho L^2/2} = \frac{2 \times 0.1 \text{ kg/s}}{852.1 \text{ kg/m}^3 \times 0.131^2 \text{ m}^2} = 0.0137 \text{ m/s}$$

yielding a Reynolds number of

$$Re_D = \frac{\rho u_m D_h}{\mu} = \frac{852.1 \text{ kg/m}^3 \times 0.0137 \text{ m/s} \times 0.00436 \text{ m}}{3.25 \times 10^{-2} \text{ N}\cdot\text{s/m}^2} = 1.57$$

Therefore, the flow is laminar for both fluids, as assumed in part 1. Equations 8.19 and 8.22a may be used to calculate the pressure drop for the water:

$$\Delta p = \frac{64}{Re_D} \cdot \frac{\rho u_m^2}{2D_h} \cdot L = \frac{64}{141} \times \frac{994 \text{ kg/m}^3 \times 0.0235^2 \text{ m}^2/\text{s}^2}{2 \times 0.00436 \text{ m}} \times 0.131 \text{ m}$$

$$= 3.76 \text{ N/m}^2 \quad \triangleleft$$

Similarly, for the oil

$$\Delta p = \frac{64}{Re_D} \cdot \frac{\rho u_m^2}{2D_h} \cdot L = \frac{64}{1.57} \times \frac{852.1 \text{ kg/m}^3 \times 0.0137^2 \text{ m}^2\text{/s}^2}{2 \times 0.00436 \text{ m}} \times 0.131 \text{ m}$$

$$= 98.2 \text{ N/m}^2 \qquad \triangleleft$$

For Example 11.1, the friction factor associated with the water flow may be calculated using Equation 8.21, and for a smooth surface condition is $f = (0.790 \ln(14{,}050) - 1.64)^{-2} = 0.0287$. The mean velocity is $u_m = 4\dot{m}/(\rho\pi D_i^2) = 4 \times 0.2 \text{ kg/s}/(994 \text{ kg/m}^3 \times \pi \times 0.025^2 \text{ m}^2) = 0.410 \text{ m/s}$, and the pressure drop is

$$\Delta p = f \cdot \frac{\rho u_m^2}{2D_h} \cdot L = 0.0287 \times \frac{994 \text{ kg/m}^3 \times 0.410^2 \text{ m}^2\text{/s}^2}{2 \times 0.025 \text{ m}} \times 65.9 \text{ m}$$

$$= 6310 \text{ N/m}^2 \qquad \triangleleft$$

For the oil flowing in the annular region, the mean velocity is $u_m = 4\dot{m}/[\rho\pi (D_o^2 - D_i^2)] = 4 \times 0.1 \text{ kg/s}/[852.1 \text{ kg/m}^3 \times \pi \times (0.045^2 - 0.025^2)\text{m}^2] = 0.107 \text{ m/s}$, and the pressure drop is

$$\Delta p = \frac{64}{Re_D} \cdot \frac{\rho u_m^2}{2D_h} \cdot L = \frac{64}{56} \times \frac{852.1 \text{ kg/m}^3 \times 0.107^2 \text{ m}^2\text{/s}^2}{2 \times 0.020 \text{ m}} \times 65.9 \text{ m}$$

$$= 18{,}300 \text{ N/m}^2 \qquad \triangleleft$$

Comments:

1. Increasing the number of gaps increases the *UA* product by simultaneously providing more surface area and increasing the heat transfer coefficients associated with the flow of the fluids through smaller passages.
2. The area-to-volume ratio of the $N = 60$ heat exchanger is $L^2(N - 1)/L^3 = (N - 1)/L = (60 - 1)/0.131 \text{ m} = 451 \text{ m}^2\text{/m}^3$.
3. The volume occupied by the concentric tube heat exchanger is $V = \pi D_o^2 L/4 = \pi \times 0.045^2 \text{ m}^2 \times 65.9 \text{ m}/4 = 0.10 \text{ m}^3$, while the volume of the compact plate-type exchanger is $V = L^3 = 0.131^3 \text{ m}^3 = 0.0022 \text{ m}^3$. Use of the plate-type heat exchanger results in a 97.8% reduction in volume relative to the conventional, concentric tube heat exchanger.
4. The pressure drops associated with use of the compact heat exchanger are significantly less than for a conventional concentric tube configuration. Pressure drops are reduced by 99.9% and 99.5% for the water and oil flows, respectively.
5. Fouling of the heat transfer surfaces may result in a decrease in the gap width, as well as an associated reduction in heat transfer rate and increase in pressure drop.
6. Because $h_c > h_h$, the temperatures of the thin metal sheets will follow closely that of the water, and, as in Example 11.1, the assumption of uniform temperature conditions to obtain h_c and h_h is reasonable.
7. One method to fabricate such a heat exchanger is presented in C. F. McDonald, *Appl. Thermal Engin.*, **20**, 471, 2000.

11.4 Heat Exchanger Analysis: The Effectiveness–NTU Method

It is a simple matter to use the log mean temperature difference (LMTD) method of heat exchanger analysis when the fluid inlet temperatures are known and the outlet temperatures are specified or readily determined from the energy balance expressions, Equations 11.6b and 11.7b. The value of ΔT_{lm} for the exchanger may then be determined. However, if only the inlet temperatures are known, use of the LMTD method requires a cumbersome iterative procedure. It is therefore preferable to employ an alternative approach termed the *effectiveness*–NTU (or NTU) method.

11.4.1 Definitions

To define the *effectiveness of a heat exchanger*, we must first determine the *maximum possible heat transfer rate, q_{max},* for the exchanger. This heat transfer rate could, in principle, be achieved in a counterflow heat exchanger (Figure 11.8) of infinite length. In such an exchanger, one of the fluids would experience the maximum possible temperature difference, $T_{h,i} - T_{c,i}$. To illustrate this point, consider a situation for which $C_c < C_h$, in which case, from Equations 11.10 and 11.11, $|dT_c| > |dT_h|$. The cold fluid would then experience the larger temperature change, and since $L \rightarrow \infty$, it would be heated to the inlet temperature of the hot fluid ($T_{c,o} = T_{h,i}$). Accordingly, from Equation 11.7b,

$$C_c < C_h: \qquad q_{max} = C_c(T_{h,i} - T_{c,i})$$

Similarly, if $C_h < C_c$, the hot fluid would experience the larger temperature change and would be cooled to the inlet temperature of the cold fluid ($T_{h,o} = T_{c,i}$). From Equation 11.6b, we then obtain

$$C_h < C_c: \qquad q_{max} = C_h(T_{h,i} - T_{c,i})$$

From the foregoing results we are then prompted to write the general expression

$$q_{max} = C_{min}(T_{h,i} - T_{c,i}) \tag{11.18}$$

where C_{min} is equal to C_c or *C_h, whichever is smaller*. For prescribed hot and cold fluid inlet temperatures, Equation 11.18 provides the maximum heat transfer rate that could possibly be delivered by an exchanger. A quick mental exercise should convince the reader that the maximum possible heat transfer rate is *not* equal to $C_{max}(T_{h,i} - T_{c,i})$. If the fluid having the larger heat capacity rate were to experience the maximum possible temperature change, conservation of energy in the form $C_c(T_{c,o} - T_{c,i}) = C_h(T_{h,i} - T_{h,o})$ would require that the other fluid experience yet a larger temperature change. For example, if $C_{max} = C_c$ and one argues that it is possible for $T_{c,o}$ to be equal to $T_{h,i}$, it follows that $(T_{h,i} - T_{h,o}) = (C_c/C_h)(T_{h,i} - T_{c,i})$, in which case $(T_{h,i} - T_{h,o}) > (T_{h,i} - T_{c,i})$. Such a condition is clearly impossible.

It is now logical to define the *effectiveness, ε,* as the ratio of the actual heat transfer rate for a heat exchanger to the maximum possible heat transfer rate:

$$\varepsilon \equiv \frac{q}{q_{max}} \tag{11.19}$$

From Equations 11.6b, 11.7b, and 11.18, it follows that

$$\varepsilon = \frac{C_h(T_{h,i} - T_{h,o})}{C_{min}(T_{h,i} - T_{c,i})} \tag{11.20}$$

or

$$\varepsilon = \frac{C_c(T_{c,o} - T_{c,i})}{C_{\min}(T_{h,i} - T_{c,i})} \tag{11.21}$$

By definition the effectiveness, which is dimensionless, must be in the range $0 \le \varepsilon \le 1$. It is useful because, if ε, $T_{h,i}$, and $T_{c,i}$ are known, the actual heat transfer rate may readily be determined from the expression

$$q = \varepsilon C_{\min}(T_{h,i} - T_{c,i}) \tag{11.22}$$

For any heat exchanger it can be shown that [5]

$$\varepsilon = f\left(\text{NTU}, \frac{C_{\min}}{C_{\max}}\right) \tag{11.23}$$

where $C_{\min}/C_{\max}$ is equal to C_c/C_h or C_h/C_c, depending on the relative magnitudes of the hot and cold fluid heat capacity rates. The *number of transfer units* (NTU) is a dimensionless parameter that is widely used for heat exchanger analysis and is defined as

$$\text{NTU} \equiv \frac{UA}{C_{\min}} \tag{11.24}$$

11.4.2 Effectiveness–NTU Relations

To determine a specific form of the effectiveness–NTU relation, Equation 11.23, consider a *parallel-flow* heat exchanger for which $C_{\min} = C_h$. From Equation 11.20 we then obtain

$$\varepsilon = \frac{T_{h,i} - T_{h,o}}{T_{h,i} - T_{c,i}} \tag{11.25}$$

and from Equations 11.6b and 11.7b it follows that

$$\frac{C_{\min}}{C_{\max}} = \frac{\dot{m}_h c_{p,h}}{\dot{m}_c c_{p,c}} = \frac{T_{c,o} - T_{c,i}}{T_{h,i} - T_{h,o}} \tag{11.26}$$

Now consider Equation 11.13, which may be expressed as

$$\ln\left(\frac{T_{h,o} - T_{c,o}}{T_{h,i} - T_{c,i}}\right) = -\frac{UA}{C_{\min}}\left(1 + \frac{C_{\min}}{C_{\max}}\right)$$

or from Equation 11.24

$$\frac{T_{h,o} - T_{c,o}}{T_{h,i} - T_{c,i}} = \exp\left[-\text{NTU}\left(1 + \frac{C_{\min}}{C_{\max}}\right)\right] \tag{11.27}$$

Rearranging the left-hand side of this expression as

$$\frac{T_{h,o} - T_{c,o}}{T_{h,i} - T_{c,i}} = \frac{T_{h,o} - T_{h,i} + T_{h,i} - T_{c,o}}{T_{h,i} - T_{c,i}}$$

and substituting for $T_{c,o}$ from Equation 11.26, it follows that

$$\frac{T_{h,o} - T_{c,o}}{T_{h,i} - T_{c,i}} = \frac{(T_{h,o} - T_{h,i}) + (T_{h,i} - T_{c,i}) - (C_{\min}/C_{\max})(T_{h,i} - T_{h,o})}{T_{h,i} - T_{c,i}}$$

or from Equation 11.25

$$\frac{T_{h,o} - T_{c,o}}{T_{h,i} - T_{c,i}} = -\varepsilon + 1 - \left(\frac{C_{\min}}{C_{\max}}\right)\varepsilon = 1 - \varepsilon\left(1 + \frac{C_{\min}}{C_{\max}}\right)$$

Substituting the above expression into Equation 11.27 and solving for ε, we obtain for the *parallel-flow heat exchanger*

$$\varepsilon = \frac{1 - \exp\{-\text{NTU}[1 + (C_{\min}/C_{\max})]\}}{1 + (C_{\min}/C_{\max})} \tag{11.28a}$$

Since precisely the same result may be obtained for $C_{\min} = C_c$, Equation 11.28a applies for any parallel-flow heat exchanger, irrespective of whether the minimum heat capacity rate is associated with the hot or cold fluid.

Similar expressions have been developed for a variety of heat exchangers [5], and representative results are summarized in Table 11.3, where C_r is the *heat capacity ratio* $C_r \equiv C_{\min}/C_{\max}$. In deriving Equation 11.31a for a shell-and-tube heat exchanger with multiple shell passes, it is assumed that the total NTU is equally distributed between shell passes of the same arrangement, $\text{NTU} = n(\text{NTU})_1$. In order to determine ε, $(\text{NTU})_1$ would first be calculated using the heat transfer area for one shell, ε_1 would then be calculated from Equation 11.30a, and ε would finally be calculated from Equation 11.31a. Note that for $C_r = 0$, as in a boiler, condenser, or single stream heat exchanger, ε is given by Equation 11.35a for

TABLE 11.3 Heat Exchanger Effectiveness Relations [5]

Flow Arrangement	Relation		
Parallel flow	$\varepsilon = \dfrac{1 - \exp[-\text{NTU}(1 + C_r)]}{1 + C_r}$		(11.28a)
Counterflow	$\varepsilon = \dfrac{1 - \exp[-\text{NTU}(1 - C_r)]}{1 - C_r\exp[-\text{NTU}(1 - C_r)]}$	$(C_r < 1)$	
	$\varepsilon = \dfrac{\text{NTU}}{1 + \text{NTU}}$	$(C_r = 1)$	(11.29a)
Shell-and-tube			
One shell pass (2, 4, . . . tube passes)	$\varepsilon_1 = 2\left\{1 + C_r + (1 + C_r^2)^{1/2} \times \dfrac{1 + \exp[-(\text{NTU})_1(1 + C_r^2)^{1/2}]}{1 - \exp[-(\text{NTU})_1(1 + C_r^2)^{1/2}]}\right\}^{-1}$		(11.30a)
n shell passes ($2n$, $4n$, . . . tube passes)	$\varepsilon = \left[\left(\dfrac{1 - \varepsilon_1 C_r}{1 - \varepsilon_1}\right)^n - 1\right]\left[\left(\dfrac{1 - \varepsilon_1 C_r}{1 - \varepsilon_1}\right)^n - C_r\right]^{-1}$		(11.31a)
Cross-flow (single pass)			
Both fluids unmixed	$\varepsilon = 1 - \exp\left[\left(\dfrac{1}{C_r}\right)(\text{NTU})^{0.22}\{\exp[-C_r(\text{NTU})^{0.78}] - 1\}\right]$		(11.32)
$C_{\max}$ (mixed), $C_{\min}$ (unmixed)	$\varepsilon = \left(\dfrac{1}{C_r}\right)(1 - \exp\{-C_r[1 - \exp(-\text{NTU})]\})$		(11.33a)
$C_{\min}$ (mixed), $C_{\max}$ (unmixed)	$\varepsilon = 1 - \exp(-C_r^{-1}\{1 - \exp[-C_r(\text{NTU})]\})$		(11.34a)
All exchangers ($C_r = 0$)	$\varepsilon = 1 - \exp(-\text{NTU})$		(11.35a)

all flow arrangements. Hence, for this special case, it follows that heat exchanger behavior is independent of flow arrangement. For the cross-flow heat exchanger with both fluids unmixed, Equation 11.32 is exact only for $C_r = 1$. However, it may be used to a good approximation for all $0 < C_r \leq 1$. For $C_r = 0$, Equation 11.35a must be used.

In heat exchanger design calculations (Section 11.5), it is more convenient to work with ε–NTU relations of the form

$$\text{NTU} = f\left(\varepsilon, \frac{C_{\min}}{C_{\max}}\right)$$

Explicit relations for NTU as a function of ε and C_r are provided in Table 11.4. Note that Equation 11.32 may not be manipulated to yield a direct relationship for NTU as a function of ε and C_r. Note also that to determine the NTU for a shell-and-tube heat exchanger with multiple shell passes, ε would first be calculated for the entire heat exchanger. The variables F and ε_1 would then be calculated using Equations 11.31c and 11.31b, respectively. The parameter E would subsequently be determined from Equation 11.30c and substituted into Equation 11.30b to find $(\text{NTU})_1$. Finally, this result would be multiplied by n to obtain the NTU for the entire exchanger, as indicated in Equation 11.31d.

The foregoing expressions are represented graphically in Figures 11.10 through 11.15. When using Figure 11.13, the abscissa corresponds to the total number of transfer units,

TABLE 11.4 Heat Exchanger NTU Relations

Flow Arrangement	Relation		
Parallel flow	$\text{NTU} = -\dfrac{\ln[1 - \varepsilon(1 + C_r)]}{1 + C_r}$		(11.28b)
Counterflow	$\text{NTU} = \dfrac{1}{C_r - 1}\ln\left(\dfrac{\varepsilon - 1}{\varepsilon C_r - 1}\right)$	$(C_r < 1)$	
	$\text{NTU} = \dfrac{\varepsilon}{1 - \varepsilon}$	$(C_r = 1)$	(11.29b)
Shell-and-tube			
One shell pass (2, 4, . . . tube passes)	$(\text{NTU})_1 = -(1 + C_r^2)^{-1/2}\ln\left(\dfrac{E - 1}{E + 1}\right)$		(11.30b)
	$E = \dfrac{2/\varepsilon_1 - (1 + C_r)}{(1 + C_r^2)^{1/2}}$		(11.30c)
n shell passes ($2n$, $4n$, . . . tube passes)	Use Equations 11.30b and 11.30c with $\varepsilon_1 = \dfrac{F - 1}{F - C_r}$ $\quad F = \left(\dfrac{\varepsilon C_r - 1}{\varepsilon - 1}\right)^{1/n}$ $\quad \text{NTU} = n(\text{NTU})_1$		(11.31b, c, d)
Cross-flow (single pass)			
$C_{\max}$ (mixed), $C_{\min}$ (unmixed)	$\text{NTU} = -\ln\left[1 + \left(\dfrac{1}{C_r}\right)\ln(1 - \varepsilon C_r)\right]$		(11.33b)
$C_{\min}$ (mixed), $C_{\max}$ (unmixed)	$\text{NTU} = -\left(\dfrac{1}{C_r}\right)\ln[C_r\ln(1 - \varepsilon) + 1]$		(11.34b)
All exchangers ($C_r = 0$)	$\text{NTU} = -\ln(1 - \varepsilon)$		(11.35b)

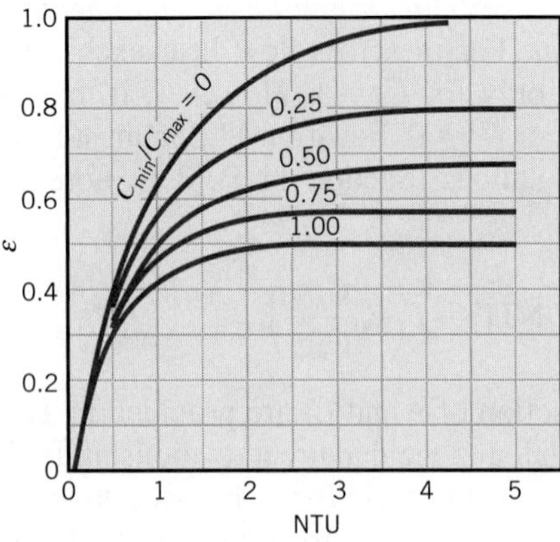

FIGURE 11.10 Effectiveness of a parallel-flow heat exchanger (Equation 11.28).

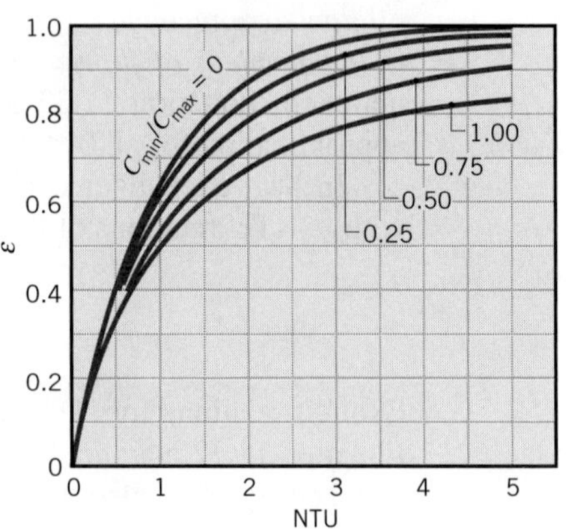

FIGURE 11.11 Effectiveness of a counterflow heat exchanger (Equation 11.29).

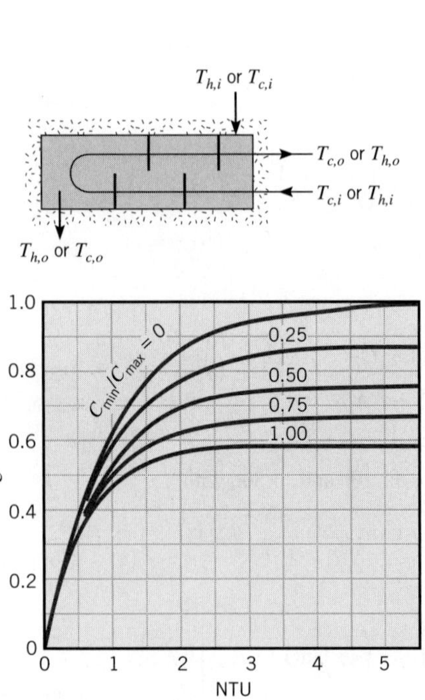

FIGURE 11.12 Effectiveness of a shell-and-tube heat exchanger with one shell and any multiple of two tube passes (two, four, etc., tube passes) (Equation 11.30).

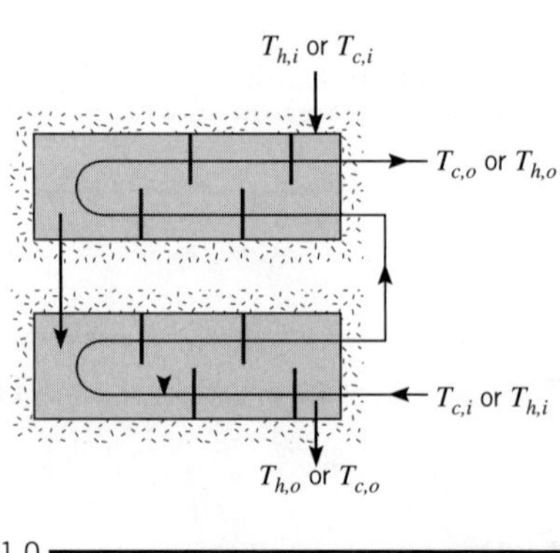

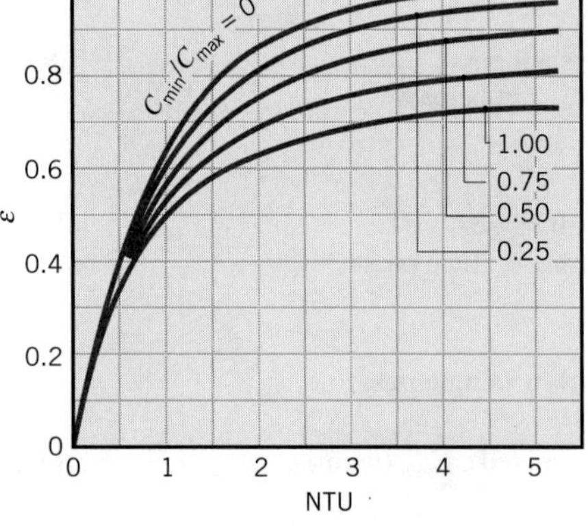

FIGURE 11.13 Effectiveness of a shell-and-tube heat exchanger with two shell passes and any multiple of four tube passes (four, eight, etc., tube passes) (Equation 11.31 with $n = 2$).

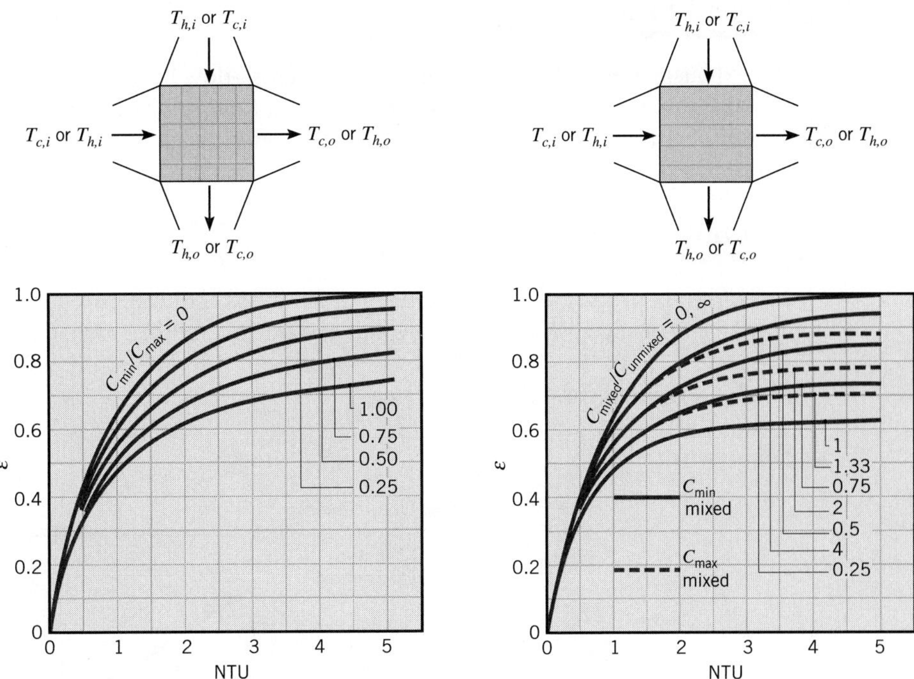

FIGURE 11.14 Effectiveness of a single-pass, cross-flow heat exchanger with both fluids unmixed (Equation 11.32).

FIGURE 11.15 Effectiveness of a single-pass, cross-flow heat exchanger with one fluid mixed and the other unmixed (Equations 11.33, 11.34).

NTU $= n(\text{NTU})_1$. For Figure 11.15 the solid curves correspond to C_{min} mixed and C_{max} unmixed, while the dashed curves correspond to C_{min} unmixed and C_{max} mixed. Note that for $C_r = 0$, all heat exchangers have the same effectiveness, which may be computed from Equation 11.35a. Moreover, if NTU $\lesssim 0.25$, all heat exchangers have approximately the same effectiveness, regardless of the value of C_r, and ε may again be computed from Equation 11.35a. More generally, for $C_r > 0$ and NTU $\gtrsim 0.25$, the counterflow exchanger is the most effective. For any exchanger, maximum and minimum values of the effectiveness are associated with $C_r = 0$ and $C_r = 1$, respectively.

As noted previously, in the context of cross-flow heat exchangers, the terms *mixed* and *unmixed* are idealizations representing limiting cases of actual flow conditions. That is, most flows are neither completely mixed nor unmixed, but exhibit partial degrees of mixing. This issue has been addressed by DiGiovanni and Webb [7], and algebraic expressions have been developed to determine the ε–NTU relationship for arbitrary values of partial mixing.

We also note that both the LMTD and ε–NTU methods approach heat exchanger analysis from a global perspective and provide no information concerning conditions within the exchanger. Although flow and temperature variations within a heat exchanger may be determined using commercial CFD (computational fluid dynamic) computer codes, simpler numerical procedures may be adopted. Such procedures have been applied by Ribando et al. to determine temperature variations in concentric tube and shell-and-tube heat exchangers [8].

EXAMPLE 11.3

Hot exhaust gases, which enter a finned-tube, cross-flow heat exchanger at 300°C and leave at 100°C, are used to heat pressurized water at a flow rate of 1 kg/s from 35 to 125°C. The overall heat transfer coefficient based on the gas-side surface area is $U_h = 100\ \text{W/m}^2\cdot\text{K}$. Determine the required gas-side surface area A_h using the NTU method.

SOLUTION

Known: Inlet and outlet temperatures of hot gases and water used in a finned-tube, cross-flow heat exchanger. Water flow rate and gas-side overall heat transfer coefficient.

Find: Required gas-side surface area.

Schematic:

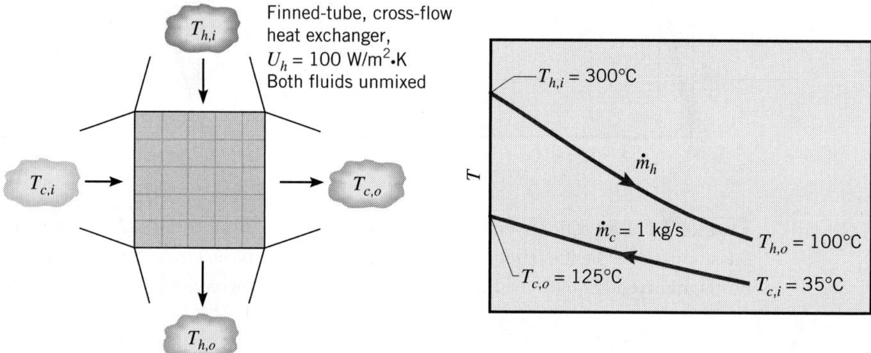

Assumptions:

1. Negligible heat loss to the surroundings and kinetic and potential energy changes.
2. Constant properties.

Properties: Table A.6, water ($\overline{T}_c = 80°\text{C}$): $c_{p,c} = 4197\ \text{J/kg}\cdot\text{K}$.

Analysis: The required surface area may be obtained from knowledge of the number of transfer units, which, in turn, may be obtained from knowledge of the ratio of heat capacity rates and the effectiveness. To determine the minimum heat capacity rate, we begin by computing

$$C_c = \dot{m}_c c_{p,c} = 1\ \text{kg/s} \times 4197\ \text{J/kg}\cdot\text{K} = 4197\ \text{W/K}$$

Since $\dot{m}_h$ is not specified, C_h is obtained by combining the overall energy balances, Equations 11.6b and 11.7b:

$$C_h = \dot{m}_h c_{p,h} = C_c \frac{T_{c,o} - T_{c,i}}{T_{h,i} - T_{h,o}} = 4197 \frac{125 - 35}{300 - 100} = 1889\ \text{W/K} = C_{\text{min}}$$

From Equation 11.18

$$q_{\text{max}} = C_{\text{min}}(T_{h,i} - T_{c,i}) = 1889\ \text{W/K}\ (300 - 35)°\text{C} = 5.00 \times 10^5\ \text{W}$$

From Equation 11.7b the actual heat transfer rate is

$$q = C_c(T_{c,o} - T_{c,i}) = 4197 \text{ W/K } (125 - 35)°\text{C}$$

$$q = 3.78 \times 10^5 \text{ W}$$

Hence from Equation 11.19 the effectiveness is

$$\varepsilon = \frac{q}{q_{\max}} = \frac{3.78 \times 10^5 \text{ W}}{5.00 \times 10^5 \text{ W}} = 0.755$$

With

$$\frac{C_{\min}}{C_{\max}} = \frac{1889}{4197} = 0.45$$

it follows from Figure 11.14 that

$$\text{NTU} = \frac{U_h A_h}{C_{\min}} \approx 2.0$$

or

$$A_h = \frac{2.0(1889 \text{ W/K})}{100 \text{ W/m}^2 \cdot \text{K}} = 37.8 \text{ m}^2 \qquad \triangleleft$$

Comments:

1. Equation 11.32 may be solved iteratively or by trial and error to yield NTU = 2.0, which is in excellent agreement with the estimate obtained from the charts.
2. With the heat exchanger sized ($A_h = 37.8 \text{ m}^2$) and placed into operation, its actual performance is subject to uncontrolled variations in the exhaust gas inlet temperature ($200 \leq T_{h,i} \leq 400°\text{C}$) and to gradual degradation of the heat exchanger surfaces due to fouling (U_h decreasing from 100 to 60 W/m$^2 \cdot$K). For a fixed value of $C_{\min} = C_h =$ 1889 W/K, the reduction in U_h corresponds to a reduction in the NTU (to NTU $\approx$ 1.20) and hence to a reduction in the heat exchanger effectiveness, which can be computed from Equation 11.32. The effect of the variations on the water outlet temperature has been computed and is plotted as follows:

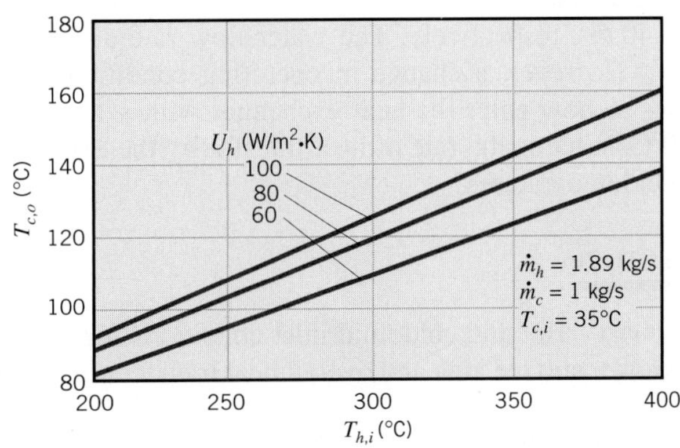

If the intent is to maintain a fixed water outlet temperature of $T_{c,o} = 125°C$, adjustments in the flow rates, $\dot{m}_c$ and $\dot{m}_h$, could be made to compensate for the variations. The model equations could be used to determine the adjustments and hence as a basis for designing the requisite *controller.*

11.5 *Heat Exchanger Design and Performance Calculations*

Two general types of heat exchanger problems are commonly encountered by the practicing engineer.

In the *heat exchanger design problem*, the fluid inlet temperatures and flow rates, as well as a desired hot or cold fluid outlet temperature, are prescribed. The design problem is then one of specifying a specific heat exchanger type and determining its size—that is, the heat transfer surface area A—required to achieve the desired outlet temperature. The design problem is commonly encountered when a heat exchanger is to be custom-built for a specific application. Alternatively, in a *heat exchanger performance calculation*, an existing heat exchanger is analyzed to determine the heat transfer rate and the fluid outlet temperatures for prescribed flow rates and inlet temperatures. The performance calculation is commonly associated with the use of off-the-shelf heat exchanger types and sizes available from a vendor.

For heat exchanger design problems, the NTU method may be used by first calculating ε and (C_{min}/C_{max}). The appropriate equation (or chart) may then be used to obtain the NTU value, which in turn may be used to determine A. For a performance calculation, the NTU and (C_{min}/C_{max}) values may be computed and ε may then be determined from the appropriate equation (or chart) for a particular exchanger type. Since q_{max} may also be computed from Equation 11.18, it is a simple matter to determine the actual heat transfer rate from the requirement that $q = \varepsilon q_{max}$. Both fluid outlet temperatures may then be determined from Equations 11.6b and 11.7b.

EXAMPLE 11.4

Consider the heat exchanger design of Example 11.3, that is, a finned-tube, cross-flow heat exchanger with a gas-side overall heat transfer coefficient and area of $100\ W/m^2 \cdot K$ and $40\ m^2$, respectively. The water flow rate and inlet temperature remain at 1 kg/s and 35°C. However, a change in operating conditions for the hot gas generator causes the gases to now enter the heat exchanger with a flow rate of 1.5 kg/s and a temperature of 250°C. What is the rate of heat transfer by the exchanger, and what are the gas and water outlet temperatures?

SOLUTION

Known: Hot and cold fluid inlet conditions for a finned-tube, cross-flow heat exchanger of known surface area and overall heat transfer coefficient.

Find: Heat transfer rate and fluid outlet temperatures.

Schematic:

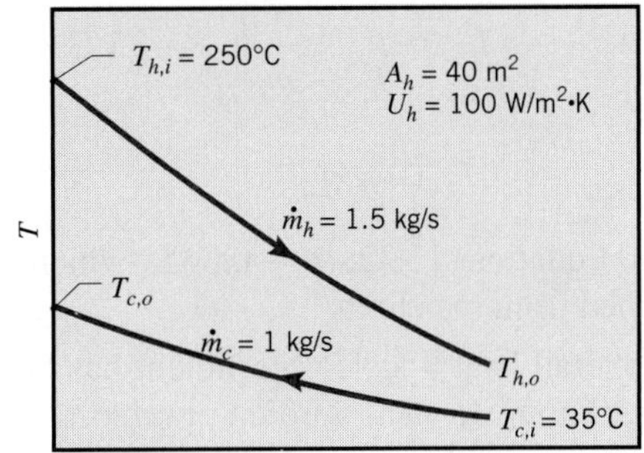

Assumptions:

1. Negligible heat loss to surroundings and kinetic and potential energy changes.
2. Constant properties (unchanged from Example 11.3).

Analysis: The problem may be classified as one requiring a heat exchanger *performance calculation.* The heat capacity rates are

$$C_c = \dot{m}_c c_{p,c} = 1 \text{ kg/s} \times 4197 \text{ J/kg} \cdot \text{K} = 4197 \text{ W/K}$$

$$C_h = \dot{m}_h c_{p,h} = 1.5 \text{ kg/s} \times 1000 \text{ J/kg} \cdot \text{K} = 1500 \text{ W/K} = C_{\min}$$

in which case

$$\frac{C_{\min}}{C_{\max}} = \frac{1500}{4197} = 0.357$$

The number of transfer units is

$$\text{NTU} = \frac{U_h A_h}{C_{\min}} = \frac{100 \text{ W/m}^2 \cdot \text{K} \times 40 \text{ m}^2}{1500 \text{ W/K}} = 2.67$$

From Figure 11.14 the heat exchanger effectiveness is then $\varepsilon \approx 0.82$, and from Equation 11.18 the maximum possible heat transfer rate is

$$q_{\max} = C_{\min}(T_{h,i} - T_{c,i}) = 1500 \text{ W/K } (250 - 35)\text{°C} = 3.23 \times 10^5 \text{ W}$$

Accordingly, from the definition of ε, Equation 11.19, the actual heat transfer rate is

$$q = \varepsilon q_{\max} = 0.82 \times 3.23 \times 10^5 \text{ W} = 2.65 \times 10^5 \text{ W} \qquad \triangleleft$$

It is now a simple matter to determine the outlet temperatures from the overall energy balances. From Equation 11.6b

$$T_{h,o} = T_{h,i} - \frac{q}{\dot{m}_h c_{p,h}} = 250\text{°C} - \frac{2.65 \times 10^5 \text{ W}}{1500 \text{ W/K}} = 73.3\text{°C} \qquad \triangleleft$$

and from Equation 11.7b

$$T_{c,o} = T_{c,i} + \frac{q}{\dot{m}_c c_{p,c}} = 35°\text{C} + \frac{2.65 \times 10^5\ \text{W}}{4197\ \text{W/K}} = 98.1°\text{C} \quad \triangleleft$$

Comments:

1. From Equation 11.32, $\varepsilon = 0.845$, which is in good agreement with the estimate obtained from the charts.
2. The overall heat transfer coefficient has tacitly been assumed to be unaffected by the change in $\dot{m}_h$. In fact, with an approximately 20% reduction in $\dot{m}_h$, there would be a significant, albeit smaller percentage, reduction in U_h.
3. As discussed in the Comment of Example 11.3, flow rate adjustments could be made to maintain a fixed water outlet temperature. If, for example, the outlet temperature must be maintained at $T_{c,o} = 125°\text{C}$, the water flow rate could be reduced to an amount prescribed by Equation 11.7b. That is,

$$\dot{m}_c = \frac{q}{c_{p,c}(T_{c,o} - T_{c,i})} = \frac{2.65 \times 10^5\ \text{W}}{4197\ \text{J/kg}\cdot\text{K}\ (125 - 35)°\text{C}} = 0.702\ \text{kg/s}$$

The change in flow rate has again been presumed to have a negligible effect on U_h. In this case the assumption is good, since the dominant contribution to U_h is made by the gas-side, and not the water-side, convection coefficient.

EXAMPLE 11.5

The condenser of a large steam power plant is a heat exchanger in which steam is condensed to liquid water. Assume the condenser to be a *shell-and-tube* heat exchanger consisting of a single shell and 30,000 tubes, each executing two passes. The tubes are of thin wall construction with $D = 25$ mm, and steam condenses on their outer surface with an associated convection coefficient of $h_o = 11{,}000\ \text{W/m}^2\cdot\text{K}$. The heat transfer rate that must be effected by the exchanger is $q = 2 \times 10^9$ W, and this is accomplished by passing cooling water through the tubes at a rate of 3×10^4 kg/s (the flow rate per tube is therefore 1 kg/s). The water enters at 20°C, while the steam condenses at 50°C. What is the temperature of the cooling water emerging from the condenser? What is the required tube length L per pass?

SOLUTION

Known: Heat exchanger consisting of single shell and 30,000 tubes with two passes each.

Find:

1. Outlet temperature of the cooling water.
2. Tube length per pass to achieve required heat transfer.

Schematic:

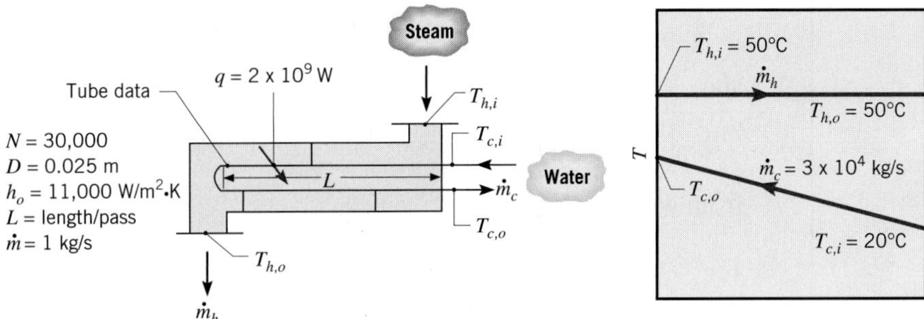

Assumptions:

1. Negligible heat transfer between exchanger and surroundings and negligible kinetic and potential energy changes.
2. Tube internal flow and thermal conditions fully developed.
3. Negligible thermal resistance of tube material and fouling effects.
4. Constant properties.

Properties: Table A.6, water (assume $\overline{T}_c \approx 27°\text{C} = 300$ K): $\rho = 997$ kg/m^3, $c_p = 4179$ J/kg·K, $\mu = 855 \times 10^{-6}$ N·s/m^2, $k = 0.613$ W/m·K, $Pr = 5.83$.

Analysis:

1. The cooling water outlet temperature may be obtained from the overall energy balance, Equation 11.7b. Accordingly,

$$T_{c,o} = T_{c,i} + \frac{q}{\dot{m}_c c_{p,c}} = 20°\text{C} + \frac{2 \times 10^9 \text{ W}}{3 \times 10^4 \text{ kg/s} \times 4179 \text{ J/kg} \cdot \text{K}}$$

$$T_{c,o} = 36.0°\text{C} \qquad \triangleleft$$

2. The problem may be classified as one requiring a *heat exchanger design calculation*. First, we determine the overall heat transfer coefficient for use in the NTU method.

From Equation 11.5

$$U = \frac{1}{(1/h_i) + (1/h_o)}$$

where h_i may be estimated from an internal flow correlation. With

$$Re_D = \frac{4\dot{m}}{\pi D \mu} = \frac{4 \times 1 \text{ kg/s}}{\pi(0.025 \text{ m})855 \times 10^{-6} \text{ N} \cdot \text{s/m}^2} = 59{,}567$$

the flow is turbulent and from Equation 8.60

$$Nu_D = 0.023\, Re_D^{4/5}\, Pr^{0.4} = 0.023(59{,}567)^{0.8}(5.83)^{0.4} = 308$$

Hence

$$h_i = Nu_D \frac{k}{D} = 308\, \frac{0.613 \text{ W/m} \cdot \text{K}}{0.025 \text{ m}} = 7543 \text{ W/m}^2 \cdot \text{K}$$

$$U = \frac{1}{[(1/7543) + (1/11{,}000)] \text{ m}^2 \cdot \text{K/W}} = 4474 \text{ W/m}^2 \cdot \text{K}$$

Using the design calculation methodology, we note that

$$C_h = C_{max} = \infty$$

and

$$C_{min} = \dot{m}_c c_{p,c} = 3 \times 10^4 \text{ kg/s} \times 4179 \text{ J/kg} \cdot \text{K} = 1.25 \times 10^8 \text{ W/K}$$

from which

$$\frac{C_{min}}{C_{max}} = C_r = 0$$

The maximum possible heat transfer rate is

$$q_{max} = C_{min}(T_{h,i} - T_{c,i}) = 1.25 \times 10^8 \text{ W/K} \times (50 - 20) \text{ K} = 3.76 \times 10^9 \text{ W}$$

from which

$$\varepsilon = \frac{q}{q_{max}} = \frac{2 \times 10^9 \text{ W}}{3.76 \times 10^9 \text{ W}} = 0.532$$

From Equation 11.35b or Figure 11.12, we find NTU = 0.759. From Equation 11.24, it follows that the tube length per pass is

$$L = \frac{\text{NTU} \cdot C_{min}}{U(N2\pi D)} = \frac{0.759 \times 1.25 \times 10^8 \text{ W/K}}{4474 \text{ W/m}^2 \cdot \text{K} (30{,}000 \times 2 \times \pi \times 0.025 \text{ m})} = 4.51 \text{ m} \quad \triangleleft$$

Comments:

1. Recognize that L is the tube length per pass, in which case the total length per tube is 9.0 m. The entire length of tubing in the condenser is $N \times L \times 2 = 30{,}000 \times 4.51 \text{ m} \times 2 = 271{,}000$ m or 271 km.
2. Over time, the performance of the heat exchanger would be degraded by fouling on both the inner and outer tube surfaces. A representative maintenance schedule would call for taking the heat exchanger off-line and cleaning the tubes when fouling factors reached values of $R''_{f,i} = R''_{f,o} = 10^{-4} \text{ m}^2 \cdot \text{K/W}$. To determine the effect of fouling on performance, the ε–NTU method may be used to calculate the total heat rate as a function of the fouling factor, with $R''_{f,o}$ assumed to equal $R''_{f,i}$. The following results are obtained:

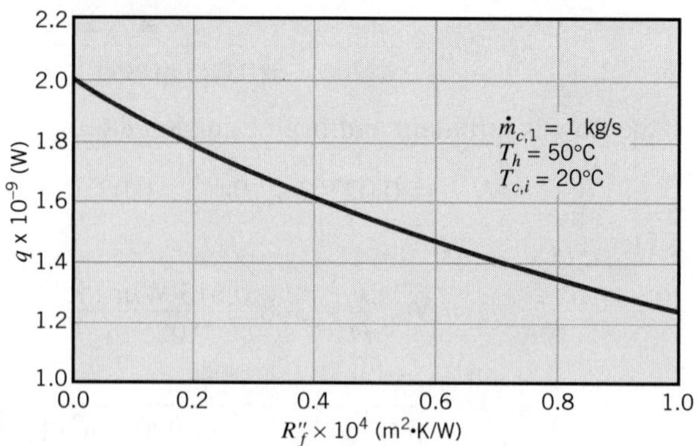

To maintain the requirement of $q = 2 \times 10^9$ W with the maximum allowable fouling and the restriction of $\dot{m}_{c,1} = 1$ kg/s, the tube length or the number of tubes would have to be increased. Keeping the length per pass at $L = 4.51$ m, $N = 48{,}300$ tubes would be needed to transfer 2×10^9 W for $R''_{f,i} = R''_{f,o} = 10^{-4}$ m$^2 \cdot$K/W. The corresponding increase in the total flow rate to $\dot{m}_c = N\dot{m}_{c,1} = 48{,}300$ kg/s would have the beneficial effect of reducing the water outlet temperature to $T_{c,o} = 29.9$°C, thereby ameliorating potentially harmful effects associated with discharge into the environment. The additional tube length associated with increasing the number of tubes to $N = 48{,}300$ is 165 km, which would result in a significant increase in the capital cost of the condenser.

3. The steam plant generates 1250 MW of electricity with a wholesale value of \$0.05 per kW·h. If the plant is shut down for 48 hours to clean the condenser tubes, the loss in revenue for the plant's owner is 48 h $\times$ 1250 $\times$ 10^6 W $\times$ \$0.05/(1 $\times$ 10^3 W·h) = \$3 million.

4. Assuming a smooth surface condition within each tube, the friction factor may be determined from Equation 8.21, $f = (0.790 \ln(59{,}567) - 1.64)^{-2} = 0.020$. The pressure drop within one tube of length $L = 9$ m may be determined from Equation 8.22a, where $u_m = 4\dot{m}/(\rho\pi D^2) = (4 \times 1 \text{ kg/s})/(997 \text{ kg/m}^3 \times \pi \times 0.025^2 \text{ m}^2) = 2.04$ m/s.

$$\Delta p = f\frac{\rho u_m^2}{2D}L = 0.020\frac{997 \text{ kg/m}^3(2.04 \text{ m/s})^2}{2(0.025 \text{ m})}9.0 \text{ m} = 15{,}300 \text{ N/m}^2$$

Therefore, the power required to pump the cooling water through the 48,300 tubes may be found by using Equation 8.22b and is

$$P = \frac{\Delta p \dot{m}}{\rho} = \frac{15{,}300 \text{ N/m}^2 \times 48{,}300 \text{ kg/s}}{997 \text{ kg/m}^3} = 742{,}000 \text{ W} = 0.742 \text{ MW}$$

The cooling water pump is driven by an electric motor. If the combined efficiency of the pump and motor is 87%, the annual cost to overcome friction losses in the condenser tubes is 24 h/day $\times$ 365 days/yr $\times$ 0.742 $\times$ 10^6 W $\times$ \$0.05/1 $\times$ 10^3 W·h/0.87 = \$374,000.

5. Optimal condenser designs are based on the desired thermal performance and environmental considerations as well as on the capital cost, operating cost, and maintenance cost associated with the device.

EXAMPLE 11.6

A geothermal power plant utilizes pressurized, deep groundwater at $T_G = 147$°C as the heat source for an *organic Rankine cycle*, the operation of which is described further in Comment 2. An evaporator, consisting of a vertically oriented shell-and-tube heat exchanger with one shell pass and one tube pass, transfers heat between the tube side groundwater and the counterflowing shell-side organic fluid of the power cycle. The organic fluid enters the shell side of the evaporator as a subcooled liquid at $T_{c,i} = 27$°C, and exits the evaporator as a saturated vapor of *quality* $X_{R,o} = 1$ and temperature $T_{c,o} = T_{\text{sat}} = 122$°C. Within the evaporator, heat transfer occurs between liquid groundwater and the organic fluid in Stage A with $U_{\text{A}} = 900$ W/m$^2 \cdot$K, and between liquid groundwater and boiling organic fluid in Stage B with $U_B = 1200\ \text{W/m}^2 \cdot \text{K}$. For groundwater and organic fluid flow rates of $\dot{m}_G = 10\ \text{kg/s}$ and $\dot{m}_R = 5.2\ \text{kg/s}$, respectively, determine the required evaporator heat transfer surface area. The specific heat of the liquid organic fluid of the Rankine cycle is $c_{p,R} = 1300\ \text{J/kg} \cdot \text{K}$, and its latent heat of vaporization is $h_{fg} = 110\ \text{kJ/kg}$.

SOLUTION

Known: Mass flow rates of groundwater and Rankine cycle organic fluids. Inlet and outlet temperatures and qualities of organic fluid. Inlet temperature of groundwater. Overall heat transfer coefficients for top and bottom stages of evaporator.

Find: Required evaporator heat transfer surface area.

Schematic:

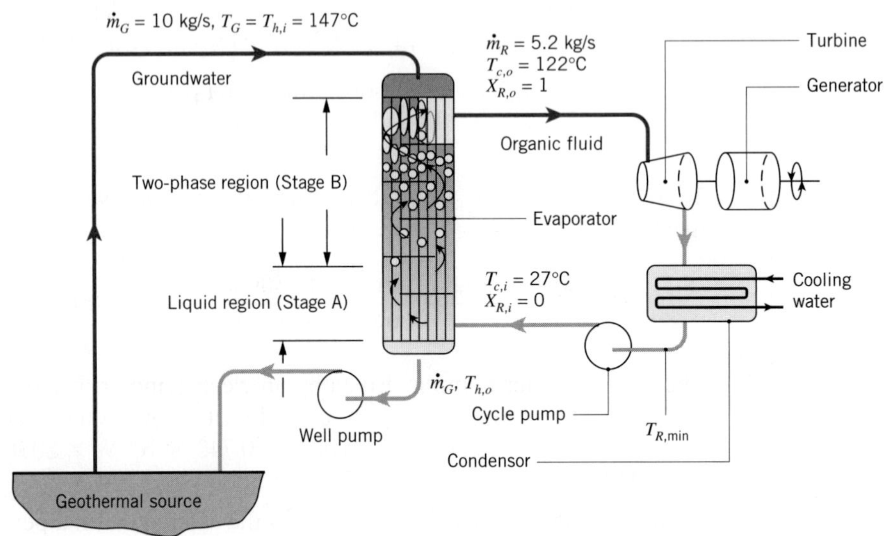

Assumptions:

1. Steady-state conditions.
2. Constant properties.
3. Negligible heat loss to surroundings and kinetic and potential energy changes.

Properties: Table A.6, water (assume $\overline{T} \approx 405\ \text{K}$): $c_{p,G} = 4267\ \text{J/kg} \cdot \text{K}$.

Analysis: Applying the conservation of energy principle to the organic fluid within the evaporator consisting of Stages A and B yields

$$q = q_A + q_B = \dot{m}_R[c_{p,R}(T_{\text{sat}} - T_{c,i}) + h_{fg}]$$

$$= 5.2\ \text{kg/s}[1300\ \text{J/kg} \cdot \text{K}\ (122 - 27)^\circ\text{C} + 110 \times 10^3\ \text{J/kg}]$$

$$= 642 \times 10^3\ \text{W} + 572 \times 10^3\ \text{W} = 1.214 \times 10^6\ \text{W} = 1.214\ \text{MW}$$

The groundwater temperature exiting the evaporator may be determined from an energy balance on the hot stream

$$T_{h,o} = T_{h,i} - \frac{q}{\dot{m}_G c_{p,G}} = 147°\text{C} - \frac{1.214 \times 10^6 \text{ W}}{10 \text{ kg/s} \times 4267 \text{ J/kg} \cdot \text{K}} = 118.5°\text{C}$$

The inlet and outlet temperatures for the cold stream are

$$T_{c,i,\text{A}} = T_{c,i} = 27°\text{C}; \; T_{c,o,\text{A}} = 122°\text{C}; \; T_{c,i,\text{B}} = T_{c,o,\text{A}} = 122°\text{C}; \; T_{c,o,\text{B}} = T_{c,o} = 122°\text{C}$$

while for the hot stream

$$T_{h,i,\text{B}} = T_{h,i} = 147°\text{C}; \; T_{h,o,\text{B}} = T_{h,i,\text{B}} - \frac{q_\text{B}}{\dot{m}_G c_{p,G}} = 147°\text{C} - \frac{572 \times 10^3 \text{ W}}{10 \text{ kg/s} \times 4267 \text{ J/kg} \cdot \text{K}} = 133.6°\text{C}$$

$$T_{h,i,\text{A}} = T_{h,o,\text{B}} = 133.6°\text{C}; \; T_{h,o,\text{A}} = T_{h,o} = 118.5°\text{C}$$

The heat capacity rates in the bottom stage (A) of the evaporator are

$$C_h = \dot{m} c_{p,h} = \dot{m}_G c_{p,G} = 10 \text{ kg/s} \times 4267 \text{ J/kg} \cdot \text{K} = 42{,}670 \text{ W/K}$$

$$C_c = \dot{m} c_{p,c} = \dot{m}_R c_{p,R} = 5.2 \text{ kg/s} \times 1300 \text{ J/kg} \cdot \text{K} = 6760 \text{ W/K}$$

$$C_{r,\text{A}} = \frac{C_{\min,\text{A}}}{C_{\max,\text{A}}} = \frac{6760}{42{,}670} = 0.158$$

Therefore, the effectiveness associated with the bottom stage of the evaporator is

$$\varepsilon_\text{A} = \frac{q_\text{A}}{C_{\min,\text{A}}(T_{h,i,\text{A}} - T_{c,i,\text{A}})} = \frac{642 \times 10^3 \text{ W}}{6760 \text{ W/K} \times (133.6 - 27)°\text{C}} = 0.891$$

The NTU can be calculated from the relation for a counterflow heat exchanger, Equation 11.29b, as

$$\text{NTU}_\text{A} = \frac{1}{C_{r,\text{A}} - 1} \ln\left(\frac{\varepsilon_\text{A} - 1}{\varepsilon_\text{A} C_{r,\text{A}} - 1}\right) = \frac{1}{0.158 - 1} \ln\left(\frac{0.891 - 1}{0.891 \times 0.158 - 1}\right) = 2.45$$

Hence the required heat transfer area for Stage A is

$$A_\text{A} = \frac{\text{NTU}_\text{A} C_{\min,\text{A}}}{U_\text{A}} = \frac{2.45 \times 6760 \text{ W/K}}{900 \text{ W/m}^2 \cdot \text{K}} = 18.4 \text{ m}^2$$

Phase change occurs in the organic fluid in the top (B) stage. Therefore $C_{r,\text{B}} = 0$ and $C_{\min,\text{B}} =$ 42,670 W/K. The effectiveness for Stage B is

$$\varepsilon_\text{B} = \frac{q_\text{B}}{C_{\min,\text{B}}(T_{h,i,\text{B}} - T_{c,i,\text{B}})} = \frac{572 \times 10^3 \text{ W}}{42{,}670 \text{ W/K} \times (147 - 122)°\text{C}} = 0.536$$

From Equation 11.35b

$$\text{NTU}_\text{B} = -\ln(1 - \varepsilon_\text{B}) = -\ln(1 - 0.536) = 0.768$$

and

$$A_\text{B} = \frac{\text{NTU}_\text{B} C_{\min,\text{B}}}{U_\text{B}} = \frac{0.768 \times 42{,}670 \text{ W/K}}{1200 \text{ W/m}^2 \cdot \text{K}} = 27.3 \text{ m}^2$$

Therefore, the entire heat transfer area is

$$A = A_A + A_B = 18.4 \text{ m}^2 + 27.3 \text{ m}^2 = 45.7 \text{ m}^2 \quad \triangleleft$$

Comments:

1. Although a baffled, shell-and-tube heat exchanger is used, there is only one tube pass, and it is appropriate to assume counterflow conditions.
2. Thermodynamic cycles can be described in terms of their temperature-entropy, or $T - s$ diagrams [9]. A Rankine cycle with water as the working fluid is shown in the diagram on the left below. Also included is the *vapor dome* of water, under which there exists a two-phase mixture of liquid and vapor. Superheated water vapor exists to the right of the vapor dome. Note that within the turbine the water quality is $X_t > 0$, and saturated liquid droplets are mixed with the saturated vapor. The droplets can impinge on the turbine blades, causing failure of the turbine. As such, most water-based Rankine cycles require the addition of an expensive *superheater* to ensure that condensation does not occur within the turbine.

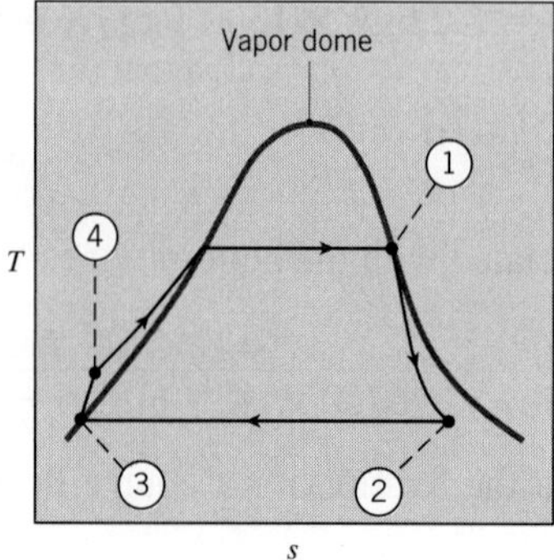

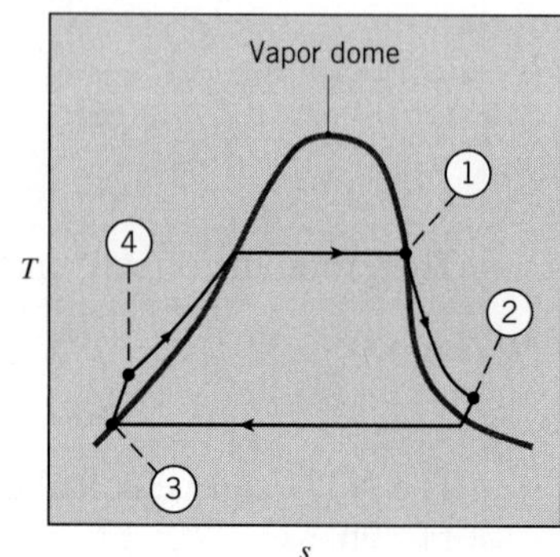

(1) Evaporator outlet, turbine inlet
(2) Turbine outlet, condenser inlet
(3) Condenser outlet, pump inlet
(4) Pump outlet, evaporator inlet

Many organic fluids are characterized by a vapor dome such as that shown in the diagram on the right. Note that, in contrast to using water as the working fluid, condensation cannot occur in the turbine. Hence, a superheater is not needed, making organic Rankine cycles attractive for a broad range of applications such as geothermal energy generation, conversion of waste heat from large turbine or diesel engines to electricity, and concentrating solar power applications as in Example 8.5 where cost reduction is critical [10].

3. The temperatures and mass flow rates in this problem correspond to an electric power generation of 250 kW using pentafluorpropane (R245fa) as the organic working fluid with high and low cycle pressures of $p = 20$ and 1.2 bars, respectively.

11.6 *Additional Considerations*

Because there are many important applications, heat exchanger research and development has had a long history. Such activity is by no means complete, however, as many talented workers continue to seek ways of improving design and performance. In fact, with heightened concern for energy conservation, there has been a steady and substantial increase in activity. A focal point for this work has been *heat transfer enhancement,* which includes the search for special heat exchanger surfaces through which increased heat transfer rates may be achieved. Also, as discussed in Section 11.1 and illustrated in Example 11.2, *compact heat exchangers* are typically used when enhancement is desired and at least one of the fluids is a gas. Many different tubular and plate configurations have been considered, where differences are due primarily to fin design and arrangement.

In addition to application to heat exchanger analysis, the LMTD and NTU methods are powerful tools that may also be applied to similar thermal systems, as illustrated in the following two examples.

EXAMPLE 11.7

A small copper heat sink with dimensions $W_1 = W_2 = 40$ mm, $L_b = 1.0$ mm, $S = 1.6$ mm, $t = 0.8$ mm, and $L_f = 5$ mm has a uniform maximum temperature of $T_h = 50°\text{C}$ on its bottom surface. An insulating cap is placed on the top of the heat sink. Water is used as the coolant, entering the heat sink at $T_{m,i} = 30°\text{C}$ and $u_m = 1.75$ m/s, providing an average heat transfer coefficient of $\bar{h} = 7590\ \text{W/m}^2\cdot\text{K}$. Determine the heat transfer rate from the hot surface to the water.

SOLUTION

Known: Dimensions of copper heat sink, maximum heat sink temperature and water inlet temperature, mean velocity, and average heat transfer coefficient.

Find: Heat transfer rate.

Schematic:

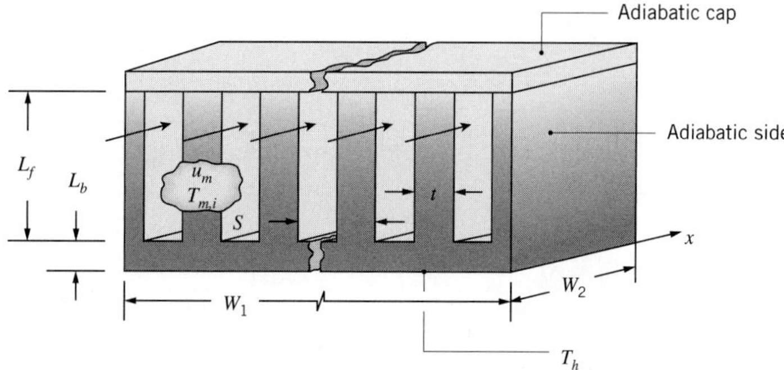

A discussion of compact heat exchanger analysis is presented in Section 11S.2.

Assumptions:

1. Steady-state conditions.
2. Adiabatic tips on heat sink fins.
3. Adiabatic heat sink sides, front and back surfaces.
4. Isothermal bottom surface temperature T_h.
5. Constant properties.
6. Negligible axial conduction in the heat sink.

Properties: Table A.1, copper ($T = 300$ K): $k_{Cu} = 401$ W/m·K. Table A.6, water (assume $\overline{T} \approx 310$ K): $\rho = 993$ kg/m^3, $c_p = 4178$ J/kg·K.

Analysis: Since the heat sink's bottom surface temperature is spatially uniform, and axial conduction is neglected, the heat sink's thermal behavior corresponds to a single stream heat exchanger as shown in Figure 11.9*a*. Specifically, the bottom surface temperature does not vary in the *x*-direction, but the water temperature increases as it flows through the heat sink. Hence, we may use Equation 11.22 to determine the heat transfer rate,

$$q = \varepsilon C_{\min}(T_{h,i} - T_{c,i}) \tag{1}$$

where $C_{\min} = C_c = \dot{m}_c c_{p,c}$ and $C_r \to 0$. From Section 11.2 and the discussion surrounding Equations 8.45b and 8.46b, we note that the term $1/UA$ used in the definition of NTU corresponds to the overall thermal resistance between the two fluid streams of a heat exchanger. In this example, $UA = 1/R_{\text{tot}}$ where R_{tot} is the total thermal resistance between the bottom of the heat sink and the fluid. Therefore, Equation 11.35a may be written as

$$\varepsilon = 1 - \exp(-\text{NTU}) = 1 - \exp\left(-\frac{UA}{C_{\min}}\right) = 1 - \exp\left(-\frac{1}{R_{\text{tot}}C_{\min}}\right) \tag{2}$$

Once $C_{\min}$ and R_{tot} are evaluated, the effectiveness can be found from Equation 2, and the heat rate may be determined from Equation 1.

The number of fins is equal to the number of channels and is $N = W_1/S = 40\text{ mm}/1.6\text{ mm} = 25$. The minimum heat capacity rate is

$$\begin{aligned} C_{\min} &= \dot{m}_C c_{p,c} = N\rho u_m L_f (S - t) c_p \\ &= 25 \times 993\text{ kg/m}^3 \times 1.75\text{ m/s} \times 0.005\text{ m} \times (0.0016\text{ m} - 0.0008\text{ m}) \times 4178\text{ J/kg}\cdot\text{K} \\ &= 726\text{ W/K} \end{aligned}$$

The total thermal resistance is calculated in Comment 4 and is $R_{\text{tot}} = 17.8 \times 10^{-3}$ K/W. From Equation 2,

$$\varepsilon = 1 - \exp\left(-\frac{1}{R_{\text{tot}}C_{\min}}\right) = 1 - \exp\left(-\frac{1}{17.8 \times 10^{-3}\text{ K/W} \times 726\text{ W/K}}\right) = 0.0745$$

and from Equation 1,

$$q = \varepsilon C_{\min}(T_h - T_{c,i}) = 0.0745 \times 726\text{ W/K} \times (50^\circ\text{C} - 30^\circ\text{C}) = 1080\text{ W} \quad \triangleleft$$

Comments:

1. If the water temperature is assumed to be constant as it flows through the heat sink, the heat rate is $q = (T_h - T_{c,i})/R_{\text{tot}} = (50°\text{C} - 30°\text{C})/17.8 \times 10^{-3}\ \text{K/W} = 1120\ \text{W}$. The assumption of constant water temperature leads to an overestimate of the actual heat rate.
2. The outlet temperature of the water is $T_{c,o} = T_{c,i} + q/C_{\text{min}} = 30°\text{C} + (1080\ \text{W})/(726\ \text{W/K}) = 31.5°\text{C}$.
3. From Equations 11.15 and 11.16, $\Delta T_{\text{lm}} = [(T_b - T_{c,i}) - (T_b - T_{c,o})]/\ln[(T_b - T_{c,i})/(T_b - T_{c,o})] = [31.5°\text{C} - 30°\text{C}]/\ln[(50°\text{C} - 30°\text{C})/(50°\text{C} - 31.5°\text{C})] = 19.2°\text{C}$ and $q = \Delta T_{\text{lm}}/R_{\text{tot}} = 19.2°\text{C}/17.8 \times 10^{-3}\ \text{K/W} = 1080\ \text{W}$. Therefore, the appropriate mean temperature difference shown in the following thermal circuit, ΔT_m, is the *log mean temperature difference* [11]. As such, this problem could have been solved using an LMTD approach, but an iterative solution would have been required.
4. The total thermal resistance corresponds to the following thermal circuit.

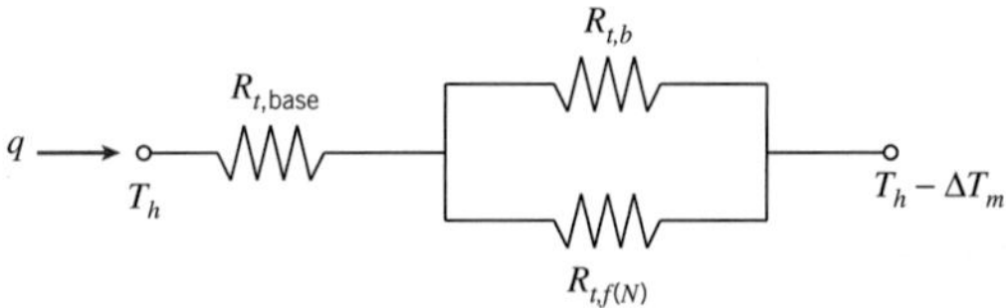

where ΔT_m is the appropriate mean temperature difference between the bottom of the heat sink base and the fluid. The thermal resistance of the base is

$$R_{t,\text{base}} = L_b/(k_{\text{Cu}} W_1 W_2) = (0.001\ \text{m})/(401\ \text{W/m} \cdot \text{K} \times 0.040\ \text{m} \times 0.040\ \text{m})$$
$$= 1.56 \times 10^{-3}\ \text{K/W}$$

The parallel resistances in the thermal circuit represent the fins and unfinned portion of the base. The combination of these two resistances is the overall thermal resistance of the fin array, as given by Equation 3.108 with Equation 11.3:

$$R_{t,o} = \frac{1}{\eta_o \bar{h} A} = \frac{1}{\bar{h}[A - A_f(1 - \eta_f)]}$$

In this expression, A_f is the surface area of all the fins and $A = A_f + A_b$, where A_b is the area of the unfinned portion of the base. Thus

$$A_f = 2L_f W_2 N = 2 \times 0.005\ \text{m} \times 0.040\ \text{m} \times 25 = 0.01\ \text{m}^2$$

and

$$A = A_f + (W_1 - Nt)W_2 = 0.01\ \text{m}^2 + (0.040\ \text{m} - 25 \times 0.0008\ \text{m}) \times 0.040\ \text{m}$$
$$= 0.0108\ \text{m}^2$$

The quantity η_f is the efficiency of a single fin, given by Equation 11.4. We first calculate

$$mL_f = \sqrt{2\bar{h}/k_{\text{Cu}} t}\, L_f$$
$$= \sqrt{2 \times 7590\ \text{W/m}^2 \cdot \text{K}/(401\ \text{W/m} \cdot \text{K} \times 0.0008\ \text{m})} \times 0.005\ \text{m} = 1.09$$

Then

$$\eta_f = \frac{\tanh(mL_f)}{mL_f} = \frac{\tanh(1.09)}{1.09} = 0.732$$

$$R_{t,o} = \frac{1}{\bar{h}[A - A_f(1-\eta_f)]}$$

$$= \frac{1}{7590 \text{ W/m}^2\cdot\text{K}[0.0108 \text{ m}^2 - 0.01 \text{ m}^2(1-0.732)]} = 0.0162 \text{ K/W}$$

Therefore, the total thermal resistance is

$$R_{\text{tot}} = R_{t,\text{base}} + R_{t,o} = 1.56 \times 10^{-3} \text{ K/W} + 0.0162 \text{ K/W} = 17.8 \times 10^{-3} \text{ K/W}$$

EXAMPLE 11.8

Spherical steel balls of diameter $D = 10$ mm are cooled from an initial temperature of $T_{h,i} = 1000$ K by submersing them in an insulated oil bath initially at $T_{c,i} = 300$ K. The total mass of the balls is $m_h = 200$ kg, while the mass of oil is $m_c = 500$ kg. The convection coefficient associated with the spheres and the oil is $h = 40 \text{ W/m}^2\cdot\text{K}$, and the steel properties are $k_h = 40 \text{ W/m}\cdot\text{K}$, $\rho_h = 7800 \text{ kg/m}^3$, and $c_h = 600 \text{ J/kg}\cdot\text{K}$. Determine the steady-state ball and oil temperatures, and the time needed for the balls to reach a temperature of $T_{h,f} = 500$ K.

SOLUTION

Known: Mass, diameter, properties, and initial temperature of steel spheres. Mass and initial temperature of oil bath.

Find: Steady-state ball and sphere temperatures, time to cool balls to $T_{h,f} = 500$ K.

Schematic:

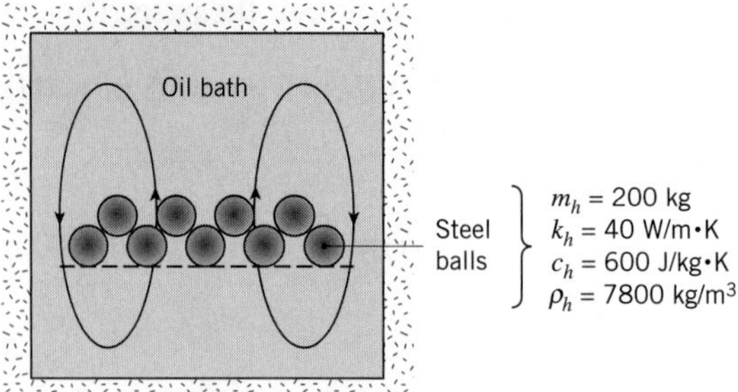

Assumptions:

1. Constant properties.
2. Negligible heat loss from oil bath.

Properties: Table A.5, engine oil (assume $\overline{T} \approx 350$ K): $c_c = 2118$ J/kg·K.

Analysis: We begin by examining whether the lumped capacitance analysis of Chapter 5 may be applied to the balls. With $L_c = r_o/3$, it follows from Equation 5.10 that

$$Bi = \frac{h(r_o/3)}{k_h} = \frac{40 \text{ W/m}^2\cdot\text{K} \times (0.005 \text{ m}/3)}{40 \text{ W/m}\cdot\text{K}} = 0.0017$$

Accordingly, Equation 5.10 is satisfied, and the spheres are nearly isothermal at any instant of time. Treating the steel balls collectively, the sphere and average oil temperatures, T_h and T_c, respectively, may be determined from an energy balance of the form

$$\Delta E_c = -\Delta E_h = \Delta E = C_{t,c}(T_c - T_{c,i}) = C_{t,h}(T_{h,i} - T_h) \tag{1}$$

where $C_{t,c} = m_c c_c = 500 \text{ kg} \times 2118 \text{ J/kg}\cdot\text{K} = 1.06 \times 10^6$ J/K and $C_{t,h} = m_h c_h = 200 \text{ kg} \times 600 \text{ J/kg}\cdot\text{K} = 120 \times 10^3$ J/K are the thermal capacitances of the oil and balls, respectively, as defined in Equation 5.7. The steady-state temperature is achieved when $T_c = T_h = T_{ss}$, and is,

$$T_{ss} = \frac{C_{t,h}T_{h,i} + C_{t,c}T_{c,i}}{C_{t,h} + C_{t,c}} = \frac{120 \times 10^3 \text{ J/K} \times 1000 \text{ K} + 1.06 \times 10^6 \text{ J/K} \times 300 \text{ K}}{120 \times 10^3 \text{ J/K} + 1.06 \times 10^6 \text{ J/K}} = 371 \text{ K} \quad \triangleleft$$

Since the oil temperature increases with time, the lumped capacitance analysis of Chapter 5, which presumes a constant ambient temperature T_∞ is not valid. Rather, heat transfer follows the process described in the schematic below.

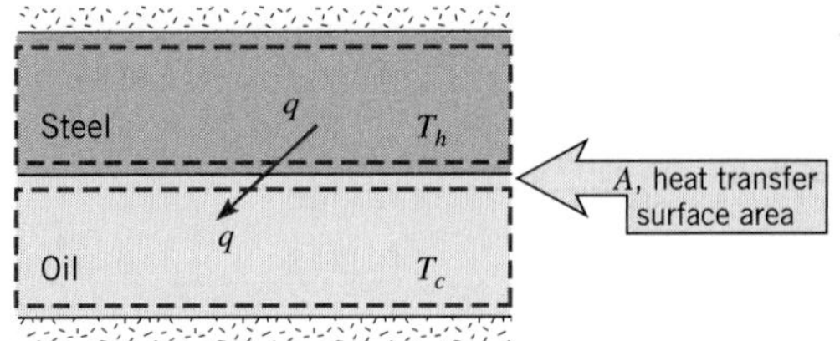

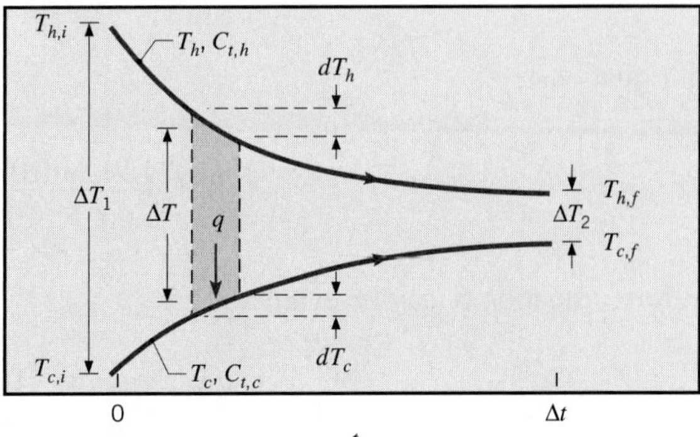

This process is analogous to that of the parallel-flow heat exchanger shown in Figure 11.7, where the total heat transfer area for the N spheres is

$$A = N4\pi r_o^2 = \frac{m_h}{\rho_h(4/3)\pi r_o^3} 4\pi r_o^2 = \frac{3m_h}{\rho_h r_o} = \frac{3 \times 200 \text{ kg}}{7800 \text{ kg/m}^3 \times 0.005 \text{ m}} = 15.5 \text{ m}^2$$

Applying an energy balance to each of the differential elements shown schematically above yields

$$dE = qdt = -C_{t,h}\,dT_h \quad \text{and} \quad dE = qdt = C_{t,c}\,dT_c \tag{2a, b}$$

where

$$q = hA(T_h - T_c) = UA\Delta T = UA(T_h - T_c) \tag{3}$$

Substituting expressions for dT_h and dT_c from Equations 2a,b into the expression $d(\Delta T) = dT_h - dT_c$ gives

$$d(\Delta T) = -\frac{qdt}{C_{t,h}} - \frac{qdt}{C_{t,c}} \tag{4}$$

Combining Equations 3 and 4 yields the relationship

$$d(\Delta T) = -\frac{UA\Delta T}{C_{t,h}}\,dt - \frac{UA\Delta T}{C_{t,c}}\,dt$$

Separating variables and integrating,

$$\int_1^2 \frac{d(\Delta T)}{\Delta T} = -UA\left(\frac{1}{C_{t,h}} + \frac{1}{C_{t,c}}\right)\int_{t_1}^{t_2} dt$$

or

$$\ln\left(\frac{\Delta T_2}{\Delta T_1}\right) = -UA\left(\frac{1}{C_{t,h}} + \frac{1}{C_{t,c}}\right)\Delta t \tag{5}$$

which may be rearranged to provide

$$\Delta t = -\frac{1}{UA}\ln\left[\frac{T_{h,f} - T_{c,f}}{T_{h,i} - T_{c,i}}\right] \Bigg/ \left[\frac{1}{C_{t,h}} + \frac{1}{C_{t,c}}\right] \tag{6}$$

From Equation 1,

$$T_{c,f} = T_{c,i} + \frac{C_{t,h}}{C_{t,c}}(T_{h,i} - T_{h,f}) = 300\text{ K} + \frac{120 \times 10^3\text{ J/K}}{1.06 \times 10^6\text{ J/K}}(1000\text{ K} - 500\text{ K}) = 357\text{ K}$$

Therefore, Equation 6 may be evaluated as

$$\Delta t = -\frac{1}{40\text{ W/m}^2\cdot\text{K} \times 15.5\text{ m}^2}\ln\left[\frac{500\text{ K} - 357\text{ K}}{1000\text{ K} - 300\text{ K}}\right] \Bigg/ \left[\frac{1}{120 \times 10^3\text{ K/J}} + \frac{1}{1.06 \times 10^6\text{ K/J}}\right]$$

$$= 278\text{ s}$$ ◁

Comments:

1. The sphere and oil temperature histories are plotted below. Note the asymptotic approach of both temperatures to the steady-state value, $T_{ss} = 371$ K.

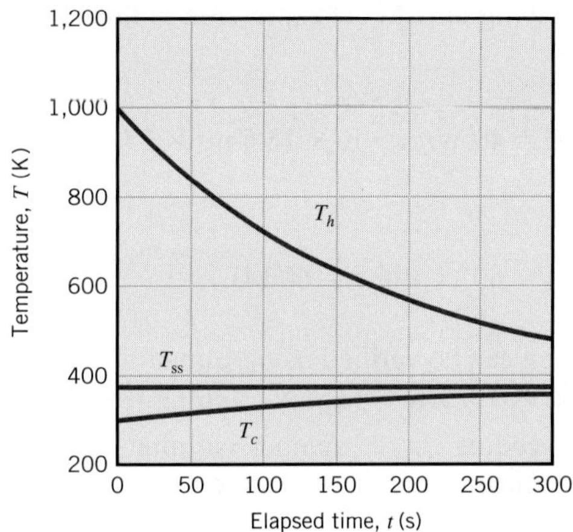

2. Equation 5 may be rewritten as

$$\frac{T_{h,f} - T_{c,f}}{T_{h,i} - T_{c,i}} = \exp\left[-UA\left(\frac{1}{C_{t,h}} + \frac{1}{C_{t,c}}\right)\Delta t\right]$$

and for an *infinitely large* oil bath, $C_{t,c} \rightarrow \infty$, and $T_{c,f} = T_{c,i} = T_\infty$. Hence,

$$\frac{T_{h,f} - T_\infty}{T_{h,i} - T_\infty} = \exp\left[-\left(\frac{UA}{\Sigma_h m_h}\right)\Delta t\right]$$

which is equivalent to Equation 5.6 of the lumped capacitance analysis.

For the infinitely large bath, the preceding equation may be rearranged to yield

$$\Delta t = -\frac{1}{40\ \text{W/m}^2\cdot\text{K} \times 15.5\ \text{m}^2}\ln\left[\frac{500\ \text{K} - 300\ \text{K}}{1000\ \text{K} - 300\ \text{K}}\right] \Bigg/ \left[\frac{1}{120 \times 10^3\ \text{K/J}}\right] = 244\ \text{s}$$

As such, the spheres cool more rapidly in the large bath, as expected since the oil temperature remains at 300 K throughout the process.

3. From Equation 1 we note that $C_{t,c} = \Delta E/(T_{c,f} - T_{c,i})$ and $C_{t,h} = \Delta E/(T_{h,i} - T_{h,f})$. Substituting these two expressions into Equation 6 yields

$$\ln\left(\frac{\Delta T_2}{\Delta T_1}\right) = -UA\Delta t\left(\frac{T_{h,i} - T_{h,f}}{\Delta E} + \frac{T_{c,f} - T_{c,i}}{\Delta E}\right) = -\frac{UA\Delta t}{\Delta E}(\Delta T_1 - \Delta T_2)$$

which can be rearranged to provide an expression involving a *log mean temperature difference* that is of the same form as Equation 11.15

$$\Delta E = UA\Delta t\frac{\Delta T_2 - \Delta T_1}{\ln(\Delta T_2/\Delta T_1)} = UA\Delta t\Delta T_{\text{lm}}$$

Applying the LMTD expression to this problem yields

$$\Delta E = 40\ \text{W/m}^2\cdot\text{K} \times 15.5\ \text{m}^2 \times 278\ \text{s}\left[\frac{(500\ \text{K} - 357\ \text{K}) - (1000\ \text{K} - 300\ \text{K})}{\ln\left(\dfrac{500\ \text{K} - 357\ \text{K}}{1000\ \text{K} - 300\ \text{K}}\right)}\right]$$

$$= 60 \times 10^6\ \text{J} = 60\ \text{MJ}$$

which can be verified using Equation 1, with $\Delta E = C_{t,c}(T_{c,f} - T_{c,i}) = C_{t,h}(T_{h,i} - T_{h,f}) = 1.06 \times 10^6\ \text{J/K} \times (357\ \text{K} - 300\ \text{K}) = 120 \times 10^3\ \text{J/K} \times (1000\ \text{K} - 500\ \text{K}) = 60\ \text{MJ}$.

4. Proceeding in a manner similar to that for the parallel-flow, two-fluid heat exchanger, note that the maximum possible change in thermal energy of the spheres or the oil is

$$\Delta E_{\max} \equiv C_{t,\min}(T_{h,i} - T_{c,i})$$

Likewise, a modified effectiveness and NTU may be defined as

$$\varepsilon^* = \frac{\Delta E}{\Delta E_{\max}} \quad \text{and} \quad \text{NTU}^* = \frac{UA\Delta t}{C_{t,\min}}$$

It may be shown that, with $C_{t,r} = C_{t,\min}/C_{t,\max}$,

$$\varepsilon^* = \frac{1 - \exp[-\text{NTU}^*(1 + C_{t,r})]}{1 + C_{t,r}} \quad \text{and} \quad \text{NTU}^* = -\frac{\ln[1 - \varepsilon^*(1 + C_{t,r})]}{1 + C_{t,r}}$$

which are of the same form as Equations 11.28a and 11.28b, respectively. For this problem, $C_{t,r} = 120 \times 10^3\ \text{J/K}/1.06 \times 10^6\ \text{J/K} = 0.113$ and $\varepsilon^* = \Delta E/\Delta E_{\max} = 60 \times 10^6\ \text{J/K}/[120 \times 10^3\ \text{J/K} \times (1000\ \text{K} - 300\ \text{K})] = 0.714$. Therefore,

$$\text{NTU}^* = -\frac{\ln[1 - 0.714(1 + 0.113)]}{1 + 0.113} = 1.42$$

and

$$\Delta t = \text{NTU}^* C_{t,\min}/UA = 1.42 \times 120 \times 10^3\ \text{J/K}/[40\ \text{W/m}^2\cdot\text{K} \times 15.5\ \text{m}^2] = 278\ \text{s}$$

which is in agreement with the problem solution.

5. This problem illustrates the value of recognizing *analogous behavior* that characterizes various thermal systems. In general, the LMTD and ε–NTU heat exchanger analyses can be used to determine the transient thermal responses of two materials between which heat is exchanged, if each material can be characterized by a unique temperature at any time.

11.7 Summary

In this chapter we have developed tools that will allow you to perform approximate heat exchanger calculations. More detailed considerations of the subject are available in the literature, including treatment of the uncertainties associated with heat exchanger analysis [3, 4, 7, 12–18].

Although we have restricted attention to heat exchangers involving separation of hot and cold fluids by a stationary wall, there are other important options. For example, *evaporative* heat exchangers enable *direct contact* between a liquid and a gas (there is no separating wall), and because of latent energy effects, large heat transfer rates per unit volume are possible. Also, for gas-to-gas heat exchange, use is often made of *regenerators* in which the same space is alternately occupied by the hot and cold gases. In a fixed regenerator such as a packed bed, the hot and cold gases alternately enter a stationary, porous solid. In a rotary regenerator, the porous solid is a rotating wheel, which alternately exposes its surfaces to the continuously flowing hot and cold gases. Detailed descriptions of such heat exchangers are available in the literature [3, 4, 12, 15, 19–22].

You should test your understanding of fundamental issues by addressing the following questions.

- What are the two possible arrangements for a *concentric tube heat exchanger*? For each arrangement, what restrictions are associated with the fluid outlet temperatures?
- As applied to a *cross-flow heat exchanger*, what is meant by the terms *mixed* and *unmixed*? In what sense are they idealizations of actual conditions?
- Why are baffles used in a *shell-and-tube heat exchanger*?
- What is the principal distinguishing feature of a *compact heat exchanger*?
- What effect does *fouling* have on the overall heat transfer coefficient and hence the performance of a heat exchanger?
- What effect do *finned surfaces* have on the overall heat transfer coefficient and hence the performance of a heat exchanger? When is the use of fins most appropriate?
- When can the overall heat transfer coefficient be expressed as $U = (h_i^{-1} + h_o^{-1})^{-1}$?
- What is the appropriate form of the mean temperature difference for the two fluids of a parallel or counterflow heat exchanger?
- What can be said about the change in temperature of a saturated fluid undergoing evaporation or condensation in a heat exchanger?
- Will the fluid having the minimum or the maximum heat capacity rate experience the largest temperature change in a heat exchanger?
- Why is the maximum possible heat rate for a heat exchanger *not* equal to $C_{\max}(T_{h,i} - T_{c,i})$? Can the outlet temperature of the cold fluid ever exceed the inlet temperature of the hot fluid?
- What is the *effectiveness* of a heat exchanger? What is its range of possible values? What is the *number of transfer units*? What is its range of possible values?
- Generally, how does the effectiveness change if the size (surface area) of a heat exchanger is increased? If the overall heat transfer coefficient is increased? If the ratio of heat capacity rates is decreased? As manifested by the number of transfer units, are there limitations to the foregoing trends? What penalty is associated with increasing the size of a heat exchanger? With increasing the overall heat transfer coefficient?

References

1. *Standards of the Tubular Exchange Manufacturers Association*, 6th ed., Tubular Exchanger Manufacturers Association, New York, 1978.
2. Chenoweth, J. M., and M. Impagliazzo, Eds., *Fouling in Heat Exchange Equipment*, American Society of Mechanical Engineers Symposium Volume HTD-17, ASME, New York, 1981.
3. Kakac, S., A. E. Bergles, and F. Mayinger, Eds., *Heat Exchangers*, Hemisphere Publishing, New York, 1981.
4. Kakac, S., R. K. Shah, and A. E. Bergles, Eds., *Low Reynolds Number Flow Heat Exchangers*, Hemisphere Publishing, New York, 1983.
5. Kays, W. M., and A. L. London, *Compact Heat Exchangers*, 3rd ed., McGraw-Hill, New York, 1984.
6. Bowman, R. A., A. C. Mueller, and W. M. Nagle, *Trans. ASME*, **62**, 283, 1940.
7. DiGiovanni, M. A., and R. L. Webb, *Heat Transfer Eng.*, **10**, 61, 1989.
8. Ribando, R. J., G. W. O'Leary, and S. Carlson-Skalak, *Comp. Appl. Eng. Educ.*, **5**, 231, 1997.
9. Moran, M. J., and H. N. Shapiro, *Engineering Thermodynamics*, 6th ed., Wiley, Hoboken, NJ, 2008.
10. Dai, Y., J. Wang, and L. Gao, *Energy Conv. Management*, **50**, 576, 2009.
11. Webb, R. L., *Trans. ASME, J. Heat Transfer*, **129**, 899, 2007.
12. Shah, R. K., C. F. McDonald, and C. P. Howard, Eds., *Compact Heat Exchangers*, American Society of Mechanical Engineers Symposium Volume HTD-10, ASME, New York, 1980.
13. Webb, R. L., in G. F. Hewitt, Exec. Ed., *Heat Exchanger Design Handbook*, Section 3.9, Begell House, New York, 2002.
14. Marner, W. J., A. E. Bergles, and J. M. Chenoweth, *Trans. ASME, J. Heat Transfer*, **105**, 358, 1983.
15. G. F. Hewitt, Exec. Ed., *Heat Exchanger Design Handbook*, Vols. 1–5, Begell House, New York, 2002.
16. Webb, R. L., and N.-H. Kim, *Principles of Enhanced Heat Transfer*, 2nd ed., Taylor & Francis, New York, 2005.
17. Andrews, M. J., and L. S. Fletcher, *ASME/JSME Thermal Eng. Conf.*, **4**, 359, 1995.
18. James, C. A., R. P. Taylor, and B. K. Hodge, *ASME/JSME Thermal Eng. Conf.*, **4**, 337, 1995.
19. Coppage, J. E., and A. L. London, *Trans. ASME*, **75**, 779, 1953.
20. Treybal, R. E., *Mass-Transfer Operations*, 3rd ed., McGraw-Hill, New York, 1980.
21. Sherwood, T. K., R. L. Pigford, and C. R. Wilkie, *Mass Transfer*, McGraw-Hill, New York, 1975.
22. Schmidt, F. W., and A. J. Willmott, *Thermal Energy Storage and Regeneration*, Hemisphere Publishing, New York, 1981.

Problems

Overall Heat Transfer Coefficient

11.1 In a fire-tube boiler, hot products of combustion flowing through an array of thin-walled tubes are used to boil water flowing over the tubes. At the time of installation, the overall heat transfer coefficient was 400 W/m$^2\cdot$K. After 1 year of use, the inner and outer tube surfaces are fouled, with corresponding fouling factors of $R''_{f,i} = 0.0015$ and $R''_{f,o} = 0.0005\ \text{m}^2\cdot\text{K/W}$, respectively. Should the boiler be scheduled for cleaning of the tube surfaces?

11.2 A type-302 stainless steel tube of inner and outer diameters $D_i = 22$ mm and $D_o = 27$ mm, respectively, is used in a cross-flow heat exchanger. The fouling factors, R''_f, for the inner and outer surfaces are estimated to be 0.0004 and 0.0002 m$^2\cdot$K/W, respectively.

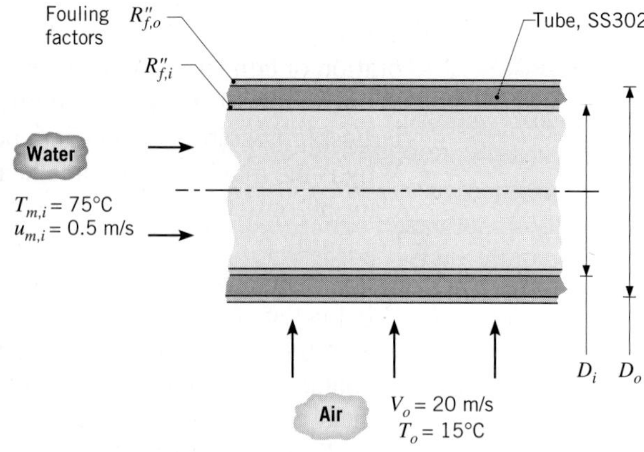

(a) Determine the overall heat transfer coefficient based on the outside area of the tube, U_o. Compare the thermal resistances due to convection, tube wall conduction, and fouling.

(b) Instead of air flowing over the tube, consider a situation for which the cross-flow fluid is water at 15°C with a velocity of $V_o = 1$ m/s. Determine the overall heat transfer coefficient based on the outside area of the tube, U_o. Compare the thermal resistances due to convection, tube wall conduction, and fouling.

(c) For the water–air conditions of part (a) and mean velocities, $u_{m,i}$, of 0.2, 0.5, and 1.0 m/s, plot the overall heat transfer coefficient as a function of the cross-flow velocity for $5 \leq V_o \leq 30$ m/s.

(d) For the water–water conditions of part (b) and cross-flow velocities, V_o, of 1, 3, and 8 m/s, plot the overall heat transfer coefficient as a function of the mean velocity for $0.5 \leq u_{m,i} \leq 2.5$ m/s.

11.3 A shell-and-tube heat exchanger is to heat an acidic liquid that flows in unfinned tubes of inside and outside diameters $D_i = 10$ mm and $D_o = 11$ mm, respectively. A hot gas flows on the shell side. To avoid corrosion of the tube material, the engineer may specify either a Ni-Cr-Mo corrosion-resistant metal alloy ($\rho_m = 8900$ kg/m^3, $k_m = 8$ W/m·K) or a polyvinylidene fluoride (PVDF) plastic ($\rho_p = 1780$ kg/m^3, $k_p = 0.17$ W/m·K). The inner and outer heat transfer coefficients are $h_i = 1500$ W/m^2·K and $h_o = 200$ W/m^2·K, respectively.

(a) Determine the ratio of plastic to metal tube surface areas needed to transfer the same amount of heat.

(b) Determine the ratio of plastic to metal mass associated with the two heat exchanger designs.

(c) The cost of the metal alloy per unit mass is three times that of the plastic. Determine which tube material should be specified on the basis of cost.

11.4 A steel tube ($k = 50$ W/m·K) of inner and outer diameters $D_i = 20$ mm and $D_o = 26$ mm, respectively, is used to transfer heat from hot gases flowing over the tube ($h_h = 200$ W/m^2·K) to cold water flowing through the tube ($h_c = 8000$ W/m^2·K). What is the cold-side overall heat transfer coefficient U_c? To enhance heat transfer, 16 straight fins of rectangular profile are installed longitudinally along the outer surface of the tube. The fins are equally spaced around the circumference of the tube, each having a thickness of 2 mm and a length of 15 mm. What is the corresponding overall heat transfer coefficient U_c?

11.5 A heat recovery device involves transferring energy from the hot flue gases passing through an annular region to pressurized water flowing through the inner tube of the annulus. The inner tube has inner and outer diameters of 24 and 30 mm and is connected by eight struts to an insulated outer tube of 60-mm diameter. Each strut is 3 mm thick and is integrally fabricated with the inner tube from carbon steel ($k = 50$ W/m·K).

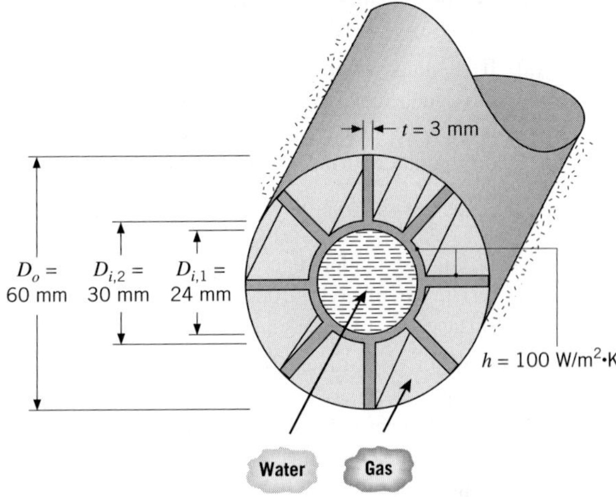

Consider conditions for which water at 300 K flows through the inner tube at 0.161 kg/s while flue gases at 800 K flow through the annulus, maintaining a convection coefficient of 100 W/m^2·K on both the struts and the outer surface of the inner tube. What is the rate of heat transfer per unit length of tube from gas to the water?

11.6 A novel design for a condenser consists of a tube of thermal conductivity 200 W/m·K with longitudinal fins snugly fitted into a larger tube. Condensing refrigerant at 45°C flows axially through the inner tube, while water at a flow rate of 0.012 kg/s passes through the six channels around the inner tube. The pertinent diameters are $D_1 = 10$ mm, $D_2 = 14$ mm, and $D_3 = 50$ mm, while the fin thickness is $t = 2$ mm. Assume that the convection coefficient associated with the condensing refrigerant is extremely large.

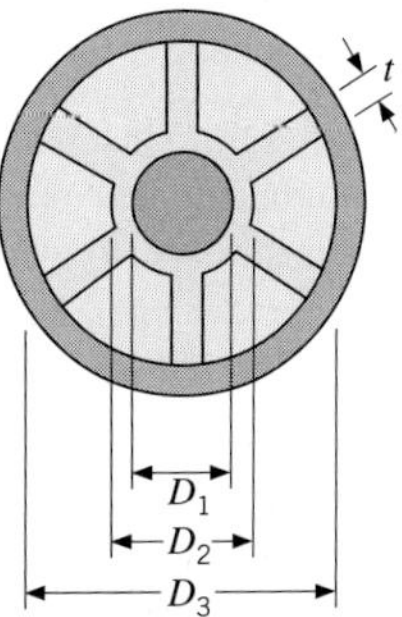

Determine the heat removal rate per unit tube length in a section of the tube for which the water is at 15°C.

11.7 The condenser of a steam power plant contains $N = 1000$ brass tubes ($k_t = 110$ W/m·K), each of inner and outer diameters, $D_i = 25$ mm and $D_o = 28$ mm, respectively. Steam condensation on the outer surfaces of the tubes is characterized by a convection coefficient of $h_o = 10{,}000$ W/m^2·K.

(a) If cooling water from a large lake is pumped through the condenser tubes at $\dot{m}_c = 400$ kg/s, what is the overall heat transfer coefficient U_o based on the outer surface area of a tube? Properties of the water may be approximated as $\mu = 9.60 \times 10^{-4}$ N·s/m^2, $k = 0.60$ W/m·K, and Pr = 6.6.

(b) If, after extended operation, fouling provides a resistance of $R''_{f,i} = 10^{-4}$ m^2·K/W, at the inner surface, what is the value of U_o?

(c) If water is extracted from the lake at 15°C and 10 kg/s of steam at 0.0622 bars are to be condensed, what is the corresponding temperature of the water leaving the condenser? The specific heat of the water is 4180 J/kg·K.

11.8 Thin-walled aluminum tubes of diameter $D = 10$ mm are used in the condenser of an air conditioner. Under normal operating conditions, a convection coefficient of $h_i = 5000$ W/m^2·K is associated with condensation on the inner surface of the tubes, while a coefficient of $h_o = 100$ W/m^2·K is maintained by airflow over the tubes.

(a) What is the overall heat transfer coefficient if the tubes are unfinned?

(b) What is the overall heat transfer coefficient based on the inner surface, U_i, if aluminum annular fins of thickness $t = 1.5$ mm, outer diameter $D_o = 20$ mm, and pitch $S = 3.5$ mm are added to the outer surface? Base your calculations on a 1-m-long section of tube. Subject to the requirements that $t \geq 1$ mm and $(S - t) \geq 1.5$ mm, explore the effect of variations in t and S on U_i. What combination of t and S would yield the best heat transfer performance?

11.9 A finned-tube, cross-flow heat exchanger is to use the exhaust of a gas turbine to heat pressurized water. Laboratory measurements are performed on a prototype version of the exchanger, which has a surface area of 10 m^2, to determine the overall heat transfer coefficient as a function of operating conditions. Measurements made under particular conditions, for which $\dot{m}_h = 2$ kg/s, $T_{h,i} = 325$°C, $\dot{m}_c = 0.5$ kg/s, and $T_{c,i} = 25$°C, reveal a water outlet temperature of $T_{c,o} = 150$°C. What is the overall heat transfer coefficient of the exchanger?

11.10 Water at a rate of 45,500 kg/h is heated from 80 to 150°C in a heat exchanger having two shell passes and eight tube passes with a total surface area of 925 m^2. Hot exhaust gases having approximately the same thermophysical properties as air enter at 350°C and exit at 175°C. Determine the overall heat transfer coefficient.

11.11 A novel heat exchanger concept consists of a large number of extruded polypropylene sheets ($k = 0.17$ W/m·K), each having a fin-like geometry, that are subsequently stacked and melted together to form the heat exchanger core. Besides being inexpensive, the heat exchanger can be easily recycled at the end of its life. Carbon dioxide at a mean temperature of 10°C and pressure of 2 atm flows in the cool channels at a mean velocity of $u_m = 0.1$ m/s. Air at 30°C and 2 atm flows at 0.2 m/s in the warm channels. Neglecting the thermal contact resistance at the welded interface, determine the product of the overall heat transfer coefficient and heat transfer area, UA, for a heat exchanger core consisting of 200 cool channels and 200 warm channels.

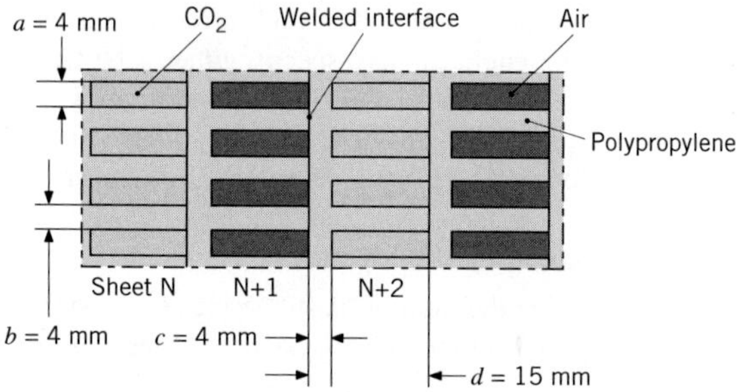

Design and Performance Calculations

11.12 The properties and flow rates for the hot and cold fluids of a heat exchanger are shown in the following table. Which fluid limits the heat transfer rate of the exchanger? Explain your choice.

	Hot fluid	Cold fluid
Density, kg/m^3	997	1247
Specific heat, J/kg·K	4179	2564
Thermal conductivity, W/m·K	0.613	0.287
Viscosity, N·s/m^2	8.55×10^{-4}	1.68×10^{-4}
Flow rate, m^3/h	14	16

11.13 A process fluid having a specific heat of 3500 J/kg·K and flowing at 2 kg/s is to be cooled from 80°C to 50°C with chilled water, which is supplied at a temperature of 15°C and a flow rate of 2.5 kg/s. Assuming an overall heat transfer coefficient of 2000 W/m^2·K, calculate the required heat transfer areas for the following exchanger configurations: (a) parallel flow, (b) counterflow, (c) shell-and-tube, one shell pass and two tube passes, and (d) cross-flow, single pass, both fluids unmixed. Compare the results of your analysis. Your work can be reduced by using *IHT*.

11.14 A shell-and-tube exchanger (two shells, four tube passes) is used to heat 10,000 kg/h of pressurized water from 35 to 120°C with 5000 kg/h pressurized water entering the exchanger at 300°C. If the overall heat transfer coefficient is 1500 W/m$^2\cdot$K, determine the required heat exchanger area.

11.15 Consider the heat exchanger of Problem 11.14. After several years of operation, it is observed that the outlet temperature of the cold water reaches only 95°C rather than the desired 120°C for the same flow rates and inlet temperatures of the fluids. Determine the cumulative (inner and outer surface) fouling factor that is the cause of the poorer performance.

11.16 The hot and cold inlet temperatures to a concentric tube heat exchanger are $T_{h,i} = 200$°C, $T_{c,i} = 100$°C, respectively. The outlet temperatures are $T_{h,o} = 110$°C and $T_{c,o} = 125$°C. Is the heat exchanger operating in a parallel flow or in a counterflow configuration? What is the heat exchanger effectiveness? What is the NTU? Phase change does not occur in either fluid.

11.17 A concentric tube heat exchanger of length $L = 2$ m is used to thermally process a pharmaceutical product flowing at a mean velocity of $u_{m,c} = 0.1$ m/s with an inlet temperature of $T_{c,i} = 20$°C. The inner tube of diameter $D_i = 10$ mm is thin walled, and the exterior of the outer tube ($D_o = 20$ mm) is well insulated. Water flows in the annular region between the tubes at a mean velocity of $u_{m,h} = 0.2$ m/s with an inlet temperature of $T_{h,i} = 60$°C. Properties of the pharmaceutical product are $\nu = 10 \times 10^{-6}$ m^2/s, $k = 0.25$ W/m$\cdot$K, $\rho = 1100$ kg/m^3, and $c_p = 2460$ J/kg$\cdot$K. Evaluate water properties at $\overline{T}_h = 50$°C.

(a) Determine the value of the overall heat transfer coefficient U.

(b) Determine the mean outlet temperature of the pharmaceutical product when the exchanger operates in the counterflow mode.

(c) Determine the mean outlet temperature of the pharmaceutical product when the exchanger operates in the parallel-flow mode.

11.18 A counterflow, concentric tube heat exchanger is designed to heat water from 20 to 80°C using hot oil, which is supplied to the annulus at 160°C and discharged at 140°C. The thin-walled inner tube has a diameter of $D_i = 20$ mm, and the overall heat transfer coefficient is 500 W/m$^2\cdot$K. The design condition calls for a total heat transfer rate of 3000 W.

(a) What is the length of the heat exchanger?

(b) After 3 years of operation, performance is degraded by fouling on the water side of the exchanger, and the water outlet temperature is only 65°C for the same fluid flow rates and inlet temperatures. What are the corresponding values of the heat transfer rate, the outlet temperature of the oil, the overall heat transfer coefficient, and the water-side fouling factor, $R''_{f,c}$?

11.19 Consider the counterflow, concentric tube heat exchanger of Example 11.1. The designer wishes to consider the effect of the cooling water flow rate on the tube length. All other conditions, including the outlet oil temperature of 60°C, remain the same.

(a) From the analysis of Example 11.1, we saw that the overall coefficient U is dominated by the hot-side convection coefficient. Assuming the water properties are independent of temperature, calculate U as a function of the water flow rate. Justify a constant value of U in the calculations of part (b).

(b) Calculate and plot the required exchanger tube length L and the water outlet temperature $T_{c,o}$ as a function of the cooling water flow rate for $0.15 \leq \dot{m}_c \leq 0.30$ kg/s.

11.20 Consider a concentric tube heat exchanger with an area of 50 m^2 operating under the following conditions:

	Hot fluid	**Cold fluid**
Heat capacity rate, kW/K	6	3
Inlet temperature, °C	60	30
Outlet temperature, °C	—	54

(a) Determine the outlet temperature of the hot fluid.

(b) Is the heat exchanger operating in counterflow or parallel flow, or can't you tell from the available information?

(c) Calculate the overall heat transfer coefficient.

(d) Calculate the effectiveness of this exchanger.

(e) What would be the effectiveness of this exchanger if its length were made very large?

11.21 As part of a senior project, a student was given the assignment to design a heat exchanger that meets the following specifications:

	$\dot{m}$ **(kg/s)**	$T_{m,i}$ **(°C)**	$T_{m,o}$ **(°C)**
Hot water	28	90	—
Cold water	27	34	60

Like many real-world situations, the customer hasn't revealed, or doesn't know, additional requirements that would allow you to proceed directly to a final configuration. At the outset, it is helpful to make a first-cut design based upon simplifying assumptions, which can be evaluated to determine what additional requirements and trade-offs should be considered by the customer.

(a) Design a heat exchanger to meet the foregoing specifications. List and explain your assumptions. *Hint:* Begin by finding the required value for UA and using representative values of U to determine A.

(b) Evaluate your design by identifying what features and configurations could be explored with your customer in order to develop more complete specifications.

11.22 A shell-and-tube heat exchanger must be designed to heat 2.5 kg/s of water from 15 to 85°C. The heating is to be accomplished by passing hot engine oil, which is available at 160°C, through the shell side of the exchanger. The oil is known to provide an average convection coefficient of $h_o = 400\ \text{W/m}^2\cdot\text{K}$ on the outside of the tubes. Ten tubes pass the water through the shell. Each tube is thin walled, of diameter $D = 25$ mm, and makes eight passes through the shell. If the oil leaves the exchanger at 100°C, what is its flow rate? How long must the tubes be to accomplish the desired heating?

11.23 A concentric tube heat exchanger for cooling lubricating oil is comprised of a thin-walled inner tube of 25-mm diameter carrying water and an outer tube of 45-mm diameter carrying the oil. The exchanger operates in counterflow with an overall heat transfer coefficient of 60 W/m² · K and the tabulated average properties.

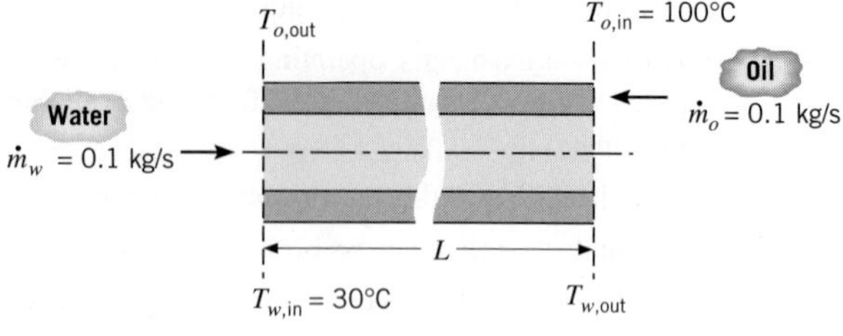

Properties	Water	Oil
ρ (kg/m³)	1000	800
c_p (J/kg·K)	4200	1900
ν (m²/s)	7×10^{-7}	1×10^{-5}
k (W/m·K)	0.64	0.134
Pr	4.7	140

(a) If the outlet temperature of the oil is 60°C, determine the total heat transfer and the outlet temperature of the water.

(b) Determine the length required for the heat exchanger.

11.24 A counterflow, concentric tube heat exchanger used for engine cooling has been in service for an extended period of time. The heat transfer surface area of the exchanger is 5 m², and the *design value* of the overall convection coefficient is 38 W/m²·K. During a test run, engine oil flowing at 0.1 kg/s is cooled from 110°C to 66°C by water supplied at a temperature of 25°C and a flow rate of 0.2 kg/s. Determine whether fouling has occurred during the service period. If so, calculate the fouling factor, R''_f(m² · K/W).

11.25 An automobile radiator may be viewed as a cross-flow heat exchanger with both fluids unmixed. Water, which has a flow rate of 0.05 kg/s, enters the radiator at 400 K and is to leave at 330 K. The water is cooled by air that enters at 0.75 kg/s and 300 K.

(a) If the overall heat transfer coefficient is 200 W/m²·K, what is the required heat transfer surface area?

(b) A manufacturing engineer claims ridges can be stamped on the finned surface of the exchanger, which could greatly increase the overall heat transfer coefficient. With all other conditions remaining the same and the heat transfer surface area determined from part (a), generate a plot of the air and water outlet temperatures as a function of U for $200 \leq U \leq 400\ \text{W/m}^2\cdot\text{K}$. What benefits result from increasing the overall convection coefficient for this application?

11.26 Hot air for a large-scale drying operation is to be produced by routing the air over a tube bank (unmixed), while products of combustion are routed through the tubes. The surface area of the cross-flow heat exchanger is $A = 25$ m², and for the proposed operating conditions, the manufacturer specifies an overall heat transfer coefficient of $U = 35\ \text{W/m}^2\cdot\text{K}$. The air and the combustion gases may each be assumed to have a specific heat of $c_p = 1040\ \text{J/kg}\cdot\text{K}$. Consider conditions for which combustion gases flowing at 1 kg/s enter the heat exchanger at 800 K, while air at 5 kg/s has an inlet temperature of 300 K.

(a) What are the air and gas outlet temperatures?

(b) After extended operation, deposits on the inner tube surfaces are expected to provide a fouling resistance of $R''_f = 0.004\ \text{m}^2\cdot\text{K/W}$. Should operation be suspended in order to clean the tubes?

(c) The heat exchanger performance may be improved by increasing the surface area and/or the overall heat transfer coefficient. Explore the effect of such changes on the air outlet temperature for $500 \le UA \le 2500$ W/K.

11.27 In a dairy operation, milk at a flow rate of 250 L/h and a *cow-body* temperature of 38.6°C must be chilled to a safe-to-store temperature of 13°C or less. Ground water at 10°C is available at a flow rate of 0.72 m^3/h. The density and specific heat of milk are 1030 kg/m^3 and 3860 J/kg·K, respectively.

(a) Determine the *UA* product of a counterflow heat exchanger required for the chilling process. Determine the length of the exchanger if the inner pipe has a 50-mm diameter and the overall heat transfer coefficient is $U = 1000$ W/m^2·K.

(b) Determine the outlet temperature of the water.

(c) Using the value of *UA* found in part (a), determine the milk outlet temperature if the water flow rate is doubled. What is the outlet temperature if the flow rate is halved?

11.28 A shell-and-tube heat exchanger with one shell pass and two tube passes is used as a *regenerator,* to preheat milk before it is pasteurized. Cold milk enters the regenerator at $T_{c,i} = 5$°C, while hot milk, which has completed the pasteurization process, enters at $T_{h,i} =$ 70°C. After leaving the regenerator, the heated milk enters a second heat exchanger, which raises its temperature from $T_{c,o}$ to 70°C.

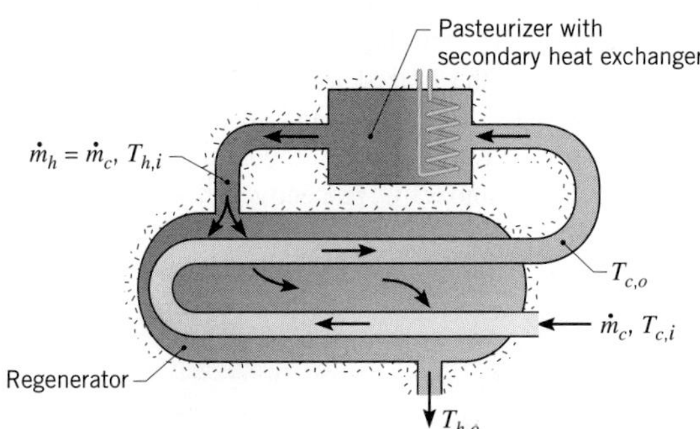

(a) A regenerator is to be used in a pasteurization process for which the flow rate of the milk is $\dot{m}_c = \dot{m}_h = 5$ kg/s. For this flow rate, the manufacturer of the regenerator specifies an overall heat transfer coefficient of 2000 W/m^2·K. If the desired effectiveness of the regenerator is 0.5, what is the requisite heat transfer area? What are the corresponding rate of heat recovery and the fluid outlet temperatures? Refer to Problem 11.27 for the properties of milk.

(b) If the hot fluid in the secondary heat exchanger derives its energy from the combustion of natural gas and the burner has an efficiency of 90%, what would be the annual savings in energy and fuel costs associated with installation of the regenerator? The facility operates continuously throughout the year, and the cost of natural gas is $C_{ng} = \$0.02$/MJ.

11.29 A twin-tube, counterflow heat exchanger operates with balanced flow rates of 0.003 kg/s for the hot and cold airstreams. The cold stream enters at 280 K and must be heated to 340 K using hot air at 360 K. The average pressure of the airstreams is 1 atm and the maximum allowable pressure drop for the cold air is 10 kPa. The tube walls may be assumed to act as fins, each with an efficiency of 100%.

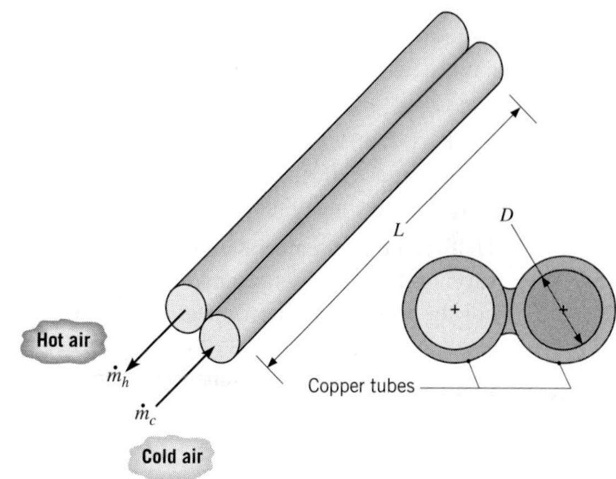

(a) Determine the tube diameter D and length L that satisfy the prescribed heat transfer and pressure drop requirements.

[(b)] For the diameter D and length L found in part (a), generate plots of the cold stream outlet temperature, the heat transfer rate, and pressure drop as a function of balanced flow rates in the range from 0.002 to 0.004 kg/s. Comment on your results.

11.30 A 5-m-long, twin-tube, counterflow heat exchanger, such as that illustrated in Problem 11.29, is used to heat air for a drying operation. Each tube is made from plain carbon steel ($k = 60$ W/m·K) and has an inner diameter and wall thickness of 50 mm and 4 mm, respectively. The thermal resistance per unit length of the brazed joint connecting the tubes is $R_t' = 0.01$ m·K/W. Consider conditions for which air enters one tube at a pressure of 5 atm, a temperature of 17°C, and flow rate of 0.030 kg/s, while saturated steam at 2.455 bar condenses in the other tube. The convection coefficient for condensation may be approximated as 5000 W/m^2·K. What is the air outlet temperature? What is the mass rate at which condensate leaves the system? *Hint*: Account for the effects of circumferential conduction in the tubes by treating them as extended surfaces.

11.31 Hot water for an industrial washing operation is produced by recovering heat from the flue gases of a furnace. A cross-flow heat exchanger is used, with the gases passing over the tubes and the water making a single pass through the tubes. The steel tubes ($k = 60\ \text{W/m}\cdot\text{K}$) have inner and outer diameters of $D_i = 15$ mm and $D_o = 20$ mm, while the staggered tube array has longitudinal and transverse pitches of $S_T = S_L = 40$ mm. The plenum in which the array is installed has a width (corresponding to the tube length) of $W = 2$ m and a height (normal to the tube axis) of $H = 1.2$ m. The number of tubes in the transverse plane is therefore $N_T \approx H/S_T = 30$. The gas properties may be approximated as those of atmospheric air, and the convection coefficient associated with water flow in the tubes may be approximated as $3000\ \text{W/m}^2\cdot\text{K}$.

(a) If 50 kg/s of water are to be heated from 290 to 350 K by 40 kg/s of flue gases entering the exchanger at 700 K, what is the gas outlet temperature and how many tube rows N_L are required?

(b) The water outlet temperature may be controlled by varying the gas flow rate and/or inlet temperature. For the value of N_L determined in part (a) and the prescribed values of H, W, S_T, $\dot{m}_c$, and $T_{c,i}$, compute and plot $T_{c,o}$ as a function of $\dot{m}_h$ over the range $20 \leq \dot{m}_h \leq 40$ kg/s for values of $T_{h,i} = 500$, 600, and 700 K. Also plot the corresponding variations of $T_{h,o}$. If $T_{h,o}$ must not drop below 400 K to prevent condensation of corrosive vapors on the heat exchanger surfaces, are there any constraints on $\dot{m}_h$ and $T_{h,i}$?

11.32 A single-pass, cross-flow heat exchanger uses hot exhaust gases (mixed) to heat water (unmixed) from 30 to 80°C at a rate of 3 kg/s. The exhaust gases, having thermophysical properties similar to air, enter and exit the exchanger at 225 and 100°C, respectively. If the overall heat transfer coefficient is $200\ \text{W/m}^2\cdot\text{K}$, estimate the required surface area.

11.33 Consider the fluid conditions and overall heat transfer coefficient of Problem 11.32 for a concentric tube heat exchanger operating in parallel flow. The thin-walled separator tube has a diameter of 100 mm.

(a) Determine the required length for the exchanger.

(b) Assuming water flow inside the separator tube to be fully developed, estimate the convection heat transfer coefficient.

(c) Using the overall coefficient and the inlet temperatures from Problem 11.32, plot the heat transfer rate and fluid outlet temperatures as a function of the tube length for $60 \leq L \leq 400$ m and the parallel-flow configuration.

(d) If the exchanger were operated in counterflow with the same overall coefficient and inlet temperatures, what would be the reduction in the required length relative to the value found in part (a)?

(e) For the counterflow configuration, plot the effectiveness and fluid outlet temperatures as a function of the tube length for $60 \leq L \leq 400$ m.

11.34 The compartment heater of an automobile exchanges heat between warm radiator fluid and cooler outside air. The flow rate of water is large compared to the air, and the effectiveness, ε, of the heater is known to depend on the flow rate of air according to the relation, $\varepsilon \sim \dot{m}_{\text{air}}^{-0.2}$.

(a) If the fan is switched to high and $\dot{m}_{\text{air}}$ is doubled, determine the percentage increase in the heat added to the car, if fluid inlet temperatures remain the same.

(b) For the low-speed fan condition, the heater warms outdoor air from 0 to 30°C. When the fan is turned to medium, the airflow rate increases 50% and the heat transfer increases 20%. Find the new outlet temperature.

11.35 A counterflow, twin-tube heat exchanger is made by brazing two circular nickel tubes, each 40 m long, together as shown below. Hot water flows through the smaller tube of 10-mm diameter and air at atmospheric pressure flows through the larger tube of 30-mm diameter. Both tubes have a wall thickness of 2 mm. The thermal contact conductance per unit length of the brazed joint is 100 W/m·K. The mass flow rates of the water and air are 0.04 and 0.12 kg/s, respectively. The inlet temperatures of the water and air are 85 and 23°C, respectively.

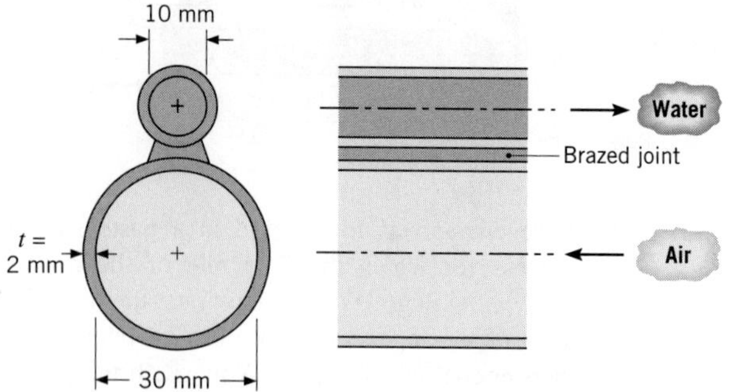

Employ the ε–NTU method to determine the outlet temperature of the air. *Hint:* Account for the effects of circumferential conduction in the walls of the tubes by treating them as extended surfaces.

11.36 Consider a *coupled* shell-in-tube heat exchange device consisting of two identical heat exchangers A and B.

Air flows on the shell side of heat exchanger A, entering at $T_{h,i,A} = 520$ K and $\dot{m}_{h,A} = 10$ kg/s. Ammonia flows in the shell of heat exchanger B, entering at $T_{c,i,B} = 280$ K, $\dot{m}_{c,B} = 5$ kg/s. The tube-side flow is common to both heat exchangers and consists of water at a flow rate $\dot{m}_{c,A} = \dot{m}_{h,B}$ with two tube passes. The UA product increases with water flow rate for heat exchanger A as expressed by the relation $UA_A = a + b\dot{m}_{c,A}$ where $a = 6000$ W/K and $b = 100$ J/kg·K. For heat exchanger B, $UA_B = 1.2UA_A$.

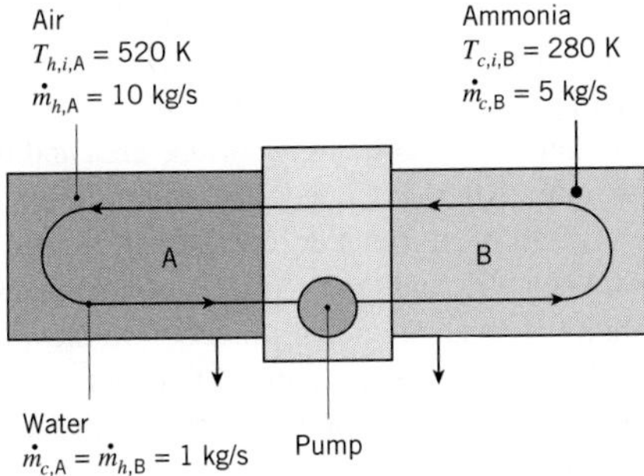

(a) For $\dot{m}_{c,A} = \dot{m}_{h,B} = 1$ kg/s, determine the outlet air and ammonia temperatures, as well as the heat transfer rate.

(b) The plant engineer wishes to fine-tune the heat exchanger performance by installing a variable-speed pump to allow adjustment of the water flow rate. Plot the outlet air and outlet ammonia temperatures versus the water flow rate over the range $0 \text{ kg/s} \leq \dot{m}_{c,A} = \dot{m}_{h,B} \leq 2$ kg/s.

11.37 Consider Problem 11.36.

(a) For $\dot{m}_{c,A} = \dot{m}_{h,B} = 10$ kg/s, determine the outlet air and ammonia temperatures, as well as the heat transfer rate.

(b) Plot the outlet air and outlet ammonia temperatures versus the water flow rate over the range $5 \text{ kg/s} \leq \dot{m}_{c,A} = \dot{m}_{h,B} \leq 50$ kg/s.

11.38 For health reasons, public spaces require the continuous exchange of a specified mass of stale indoor air with fresh outdoor air. To conserve energy during the heating season, it is expedient to recover the thermal energy in the exhausted, warm indoor air and transfer it to the incoming, cold fresh air. A *coupled* single-pass, cross-flow heat exchanger with both fluids unmixed is installed in the intake and return ducts of a heating system as shown in the schematic. Water containing an anti-freeze agent is used as the working fluid in the coupled heat exchange device, which is composed of individual heat exchangers A and B. Hence, heat is transferred from the warm stale air to the cold fresh air by way of the pumped water.

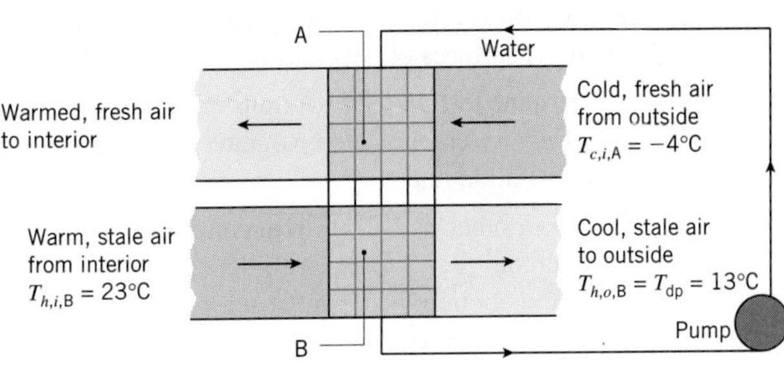

Consider a specified air mass flow rate (in each duct) of $\dot{m} = 1.50$ kg/s, an overall heat transfer coefficient–area product of $UA = 2500$ W/K (for each heat exchanger), an outdoor temperature of $T_{c,i,A} = -4°C$ and an indoor temperature of $T_{h,i,B} = 23°C$. Since the warm air has been humidified, excessive heat transfer can result in unwanted condensation in the ductwork. What water flow rate is necessary to maximize heat transfer while ensuring the outlet temperature associated with heat exchanger B does not fall below the dew point temperature, $T_{h,o,B} = T_{dp} = 13°C$? *Hint*: Assume the maximum heat capacity rate is associated with the air.

11.39 A cross-flow heat exchanger used in a cardiopulmonary bypass procedure cools blood flowing at 5 L/min from a body temperature of 37°C to 25°C in order to induce body hypothermia, which reduces metabolic and oxygen requirements. The coolant is ice water at 0°C, and its flow rate is adjusted to provide an outlet temperature of 15°C. The heat exchanger operates with both fluids unmixed, and the overall heat transfer coefficient is 750 W/m^2·K. The density and specific heat of the blood are 1050 kg/m^3 and 3740 J/kg·K, respectively.

(a) Determine the heat transfer rate for the exchanger.

(b) Calculate the water flow rate.

(c) What is the surface area of the heat exchanger?

(d) Calculate and plot the blood and water outlet temperatures as a function of the water flow rate for the range 2 to 4 L/min, assuming all other parameters remain unchanged. Comment on how the changes in the outlet temperatures are affected by changes in the water flow rate. Explain this behavior and why it is an advantage for this application.

11.40 Saturated steam at 0.14 bar is condensed in a shell-and-tube heat exchanger with one shell pass and two tube passes consisting of 130 brass tubes, each with a length per pass of 2 m. The tubes have inner and outer diameters of 13.4 and 15.9 mm, respectively. Cooling water enters the tubes at 20°C with a mean velocity of

1.25 m/s. The heat transfer coefficient for condensation on the outer surfaces of the tubes is 13,500 W/m$^2 \cdot$ K.

(a) Determine the overall heat transfer coefficient, the cooling water outlet temperature, and the steam condensation rate.

(b) With all other conditions remaining the same, but accounting for changes in the overall coefficient, plot the cooling water outlet temperature and the steam condensation rate as a function of the water flow rate for $10 \leq \dot{m}_c \leq 30$ kg/s.

11.41 A feedwater heater that supplies a boiler consists of a shell-and-tube heat exchanger with one shell pass and two tube passes. One hundred thin-walled tubes each have a diameter of 20 mm and a length (per pass) of 2 m. Under normal operating conditions water enters the tubes at 10 kg/s and 290 K and is heated by condensing saturated steam at 1 atm on the outer surface of the tubes. The convection coefficient of the saturated steam is 10,000 W/m$^2 \cdot$ K.

(a) Determine the water outlet temperature.

(b) With all other conditions remaining the same, but accounting for changes in the overall heat transfer coefficient, plot the water outlet temperature as a function of the water flow rate for $5 \leq \dot{m}_c \leq 20$ kg/s.

(c) On the plot of part (b), generate two additional curves for the water outlet temperature as a function of flow rate for fouling factors of $R''_f = 0.0002$ and 0.0005 m$^2 \cdot$ K/W.

11.42 Saturated steam at 110°C is condensed in a shell-and-tube heat exchanger (1 shell pass; 2, 4, · · · tube passes) with a *UA* value of 2.5 kW/K. Cooling water enters at 40°C.

(a) Calculate the cooling water flow rate required to maintain a heat rate of 150 kW.

(b) Assuming that *UA* is independent of flow rate, calculate and plot the water flow rate required to provide heat rates over the range from 130 to 160 kW. Comment on the validity of your assumption.

11.43 A shell-and-tube heat exchanger (1 shell pass, 2 tube passes) is to be used to condense 2.73 kg/s of saturated steam at 340 K. Condensation occurs on the outer tube surfaces, and the corresponding convection coefficient is 10,000 W/m$^2 \cdot$ K. The temperature of the cooling water entering the tubes is 15°C, and the exit temperature is not to exceed 30°C. Thin-walled tubes of 19-mm diameter are specified, and the mean velocity of water flow through the tubes is to be maintained at 0.5 m/s.

(a) What is the minimum number of tubes that should be used, and what is the corresponding tube length per pass?

(b) To reduce the size of the heat exchanger, it is proposed to increase the water-side convection coefficient by inserting a wire mesh in the tubes. If the mesh increases the convection coefficient by a factor of two, what is the required tube length per pass?

11.44 Saturated water vapor leaves a steam turbine at a flow rate of 1.5 kg/s and a pressure of 0.51 bar. The vapor is to be completely condensed to saturated liquid in a shell-and-tube heat exchanger that uses city water as the cold fluid. The water enters the thin-walled tubes at 17°C and is to leave at 57°C. Assuming an overall heat transfer coefficient of 2000 W/m$^2 \cdot$ K, determine the required heat exchanger surface area and the water flow rate. After extended operation, fouling causes the overall heat transfer coefficient to decrease to 1000 W/m$^2 \cdot$ K, and to completely condense the vapor, there must be an attendant reduction in the vapor flow rate. For the same water inlet temperature and flow rate, what is the new vapor flow rate required for complete condensation?

11.45 A two-fluid heat exchanger has inlet and outlet temperatures of 65 and 40°C for the hot fluid and 15 and 30°C for the cold fluid. Can you tell whether this exchanger is operating under counterflow or parallel-flow conditions? Determine the effectiveness of the heat exchanger.

11.46 The human brain is especially sensitive to elevated temperatures. The cool blood in the veins leaving the face and neck and returning to the heart may contribute to thermal regulation of the brain by cooling the arterial blood flowing to the brain. Consider a vein and artery running between the chest and the base of the skull for a distance $L = 250$ mm, with mass flow rates of 3×10^{-3} kg/s in opposite directions in the two vessels. The vessels are of diameter $D = 5$ mm and are separated by a distance $w = 7$ mm. The thermal conductivity of the surrounding tissue is $k_t = 0.5$ W/m $\cdot$ K. If the arterial blood enters at 37°C and the venous blood enters at 27°C, at what temperature will the arterial blood exit? If the arterial blood becomes overheated, and the body responds by halving the blood flow rate, how much hotter can the entering arterial blood be and still maintain its exit temperature below 37°C? *Hint:* If we assume that all the heat leaving the artery enters the vein, then heat transfer between the two vessels can be modeled using a relationship found in Table 4.1. Approximate the blood properties as those of water.

11.47 Consider a *very long,* concentric tube heat exchanger having hot and cold water inlet temperatures of 85 and 15°C. The flow rate of the hot water is twice that of the cold water. Assuming equivalent hot and cold water specific heats, determine the hot water outlet temperature for the following modes of operation: (a) counterflow and (b) parallel flow.

11.48 A plate-fin heat exchanger is used to condense a saturated refrigerant vapor in an air-conditioning system. The vapor has a saturation temperature of 45°C, and a condensation rate of 0.015 kg/s is dictated by system performance requirements. The frontal area of the condenser is fixed at $A_{fr} = 0.25\ m^2$ by installation requirements, and a value of $h_{fg} = 135$ kJ/kg may be assumed for the refrigerant.

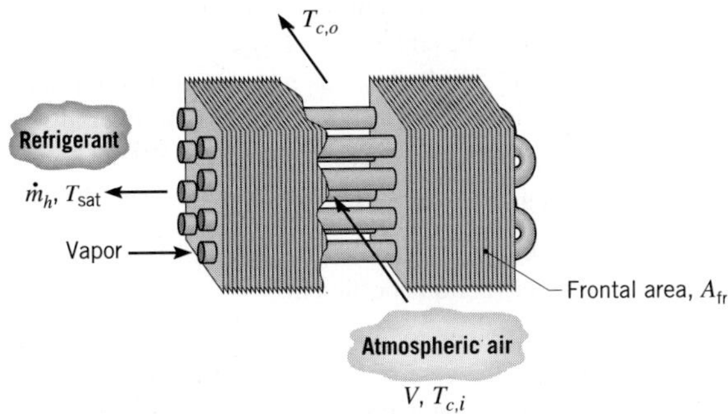

(a) The condenser design is to be based on a nominal air inlet temperature of $T_{c,i} = 30°C$ and nominal air inlet velocity of $V = 2$ m/s for which the manufacturer of the heat exchanger core indicates an overall coefficient of $U = 50\ W/m^2 \cdot K$. What is the corresponding value of the heat transfer surface area required to achieve the prescribed condensation rate? What is the air outlet temperature?

(b) From the manufacturer of the heat exchanger core, it is also known that $U \propto V^{0.7}$. During daily operation the air inlet temperature is not controllable and may vary from 27 to 38°C. If the heat exchanger area is fixed by the result of part (a), what is the range of air velocities needed to maintain the prescribed condensation rate? Plot the velocity as a function of the air inlet temperature.

11.49 A shell-and-tube heat exchanger is to heat 10,000 kg/h of water from 16 to 84°C by hot engine oil flowing through the shell. The oil makes a single shell pass, entering at 160°C and leaving at 94°C, with an average heat transfer coefficient of $400\ W/m^2 \cdot K$. The water flows through 11 brass tubes of 22.9-mm inside diameter and 25.4-mm outside diameter, with each tube making four passes through the shell.

(a) Assuming fully developed flow for the water, determine the required tube length per pass.

(b) For the tube length found in part (a), plot the effectiveness, fluid outlet temperatures, and water-side convection coefficient as a function of the water flow rate for $5000 \leq \dot{m}_c \leq 15{,}000$ kg/h, with all other conditions remaining the same.

11.50 In a supercomputer, signal propagation delays are reduced by resorting to high-density circuit arrangements which are cooled by immersing them in a special dielectric liquid. The fluid is pumped in a closed loop through the computer and an adjoining shell-and-tube heat exchanger having one shell and two tube passes.

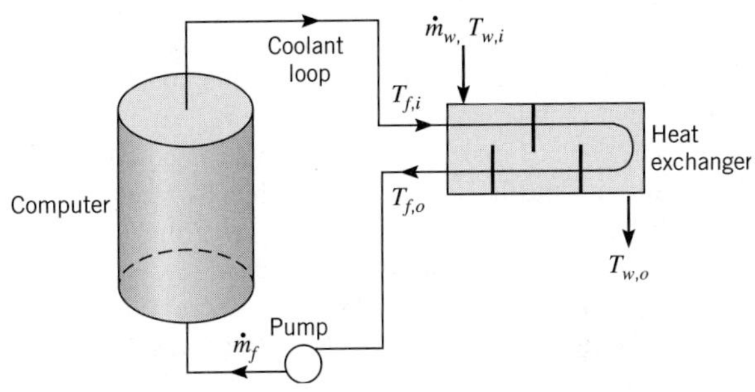

During normal operation, heat generated within the computer is transferred to the dielectric fluid passing through the computer at a flow rate of $\dot{m}_f = 4.81$ kg/s. In turn, the fluid passes through the tubes of the heat exchanger and the heat is transferred to water passing over the tubes. The dielectric fluid may be assumed to have constant properties of $c_p = 1040\ J/kg \cdot K$, $\mu = 7.65 \times 10^{-4}\ kg/s \cdot m$, $k = 0.058\ W/m \cdot K$, and $Pr = 14$. During normal operation, chilled water at a flow rate of $\dot{m}_w = 2.5$ kg/s and an inlet temperature of $T_{w,i} = 5°C$ passes over the tubes. The water has a specific heat of $4200\ J/kg \cdot K$ and provides an average convection coefficient of $10{,}000\ W/m^2 \cdot K$ over the outer surface of the tubes.

(a) If the heat exchanger consists of 72 thin-walled tubes, each of 10-mm diameter, and fully developed flow is assumed to exist within the tubes, what is the convection coefficient associated with flow through the tubes?

(b) If the dielectric fluid enters the heat exchanger at $T_{f,i} = 25°C$ and is to leave at $T_{f,o} = 15°C$, what is the required tube length per pass?

(c) For the exchanger with the tube length per pass determined in part (b), plot the outlet temperature of the dielectric fluid as a function of its flow rate for $4 \leq \dot{m}_f \leq 6$ kg/s. Account for corresponding changes in the overall heat transfer coefficient, but assume all other conditions to remain the same.

(d) The site specialist for the computer facilities is concerned about changes in the performance of the water chiller supplying the cold water ($\dot{m}_w$, $T_{w,i}$) and their effect on the outlet temperature $T_{f,o}$ of the dielectric fluid. With all other conditions remaining the same, determine the effect of a ±10% change in the cold water flow rate on $T_{f,o}$.

(e) Repeat the performance analysis of part (d) to determine the effect of a ±3°C change in the water inlet temperature on $T_{f,o}$, with all other conditions remaining the same.

11.51 Untapped geothermal sites in the United States have the estimated potential to deliver 100,000 MW (electric) of new, clean energy. The key component in a geothermal power plant is a heat exchanger that transfers thermal energy from hot, geothermal brine to a second fluid that is evaporated in the heat exchanger. The cooled brine is reinjected into the geothermal well after it exits the heat exchange, while the vapor exiting the heat exchanger serves as the working fluid of a Rankine cycle. Consider a geothermal power plant designed to deliver $P = 25$ MW (electric) operating at a thermal efficiency of $\eta = 0.20$. Pressurized hot brine at $T_{h,i} = 200°C$ is sent to the tube side of a shell-and-tube heat exchanger, while the Rankine cycle's working fluid enters the shell side at $T_{c,i} = 45°C$. The brine is reinjected into the well at $T_{h,o} = 80°C$.

(a) Assuming the brine has the properties of water, determine the required brine flow rate, the required effectiveness of the heat exchanger, and the required heat transfer surface area. The overall heat transfer coefficient is $U = 4000\ W/m^2$.

(b) Over time, the brine fouls the heat transfer surfaces, resulting in $U = 2000\ W/m^2$. For the operating conditions of part (a), determine the electric power generated by the geothermal plant under fouled heat exchanger conditions.

11.52 An energy storage system is proposed to absorb thermal energy collected during the day with a solar collector and release thermal energy at night to heat a building. The key component of the system is a shell-and-tube heat exchanger with the shell side filled with *n*-octadecane (see Problem 8.47).

(a) Warm water from the solar collector is delivered to the heat exchanger at $T_{h,i} = 40°C$ and $\dot{m} = 2$ kg/s through the tube bundle consisting of 50 tubes, two tube passes, and a tube length per pass of $L_t = 2$ m. The thin-walled, metal tubes are of diameter $D = 25$ mm. Free convection exists within the molten *n*-octadecane, providing an average heat transfer coefficient of $h_o = 25\ W/m^2 \cdot K$ on the outside of each tube. Determine the volume of *n*-octadecane that is melted over a 12-h period. If the total volume of *n*-octadecane is to be 50% greater than the volume melted over 12 h, determine the diameter of the $L_s = 2.2$-m-long shell.

(b) At night, water at $T_{c,i} = 15°C$ is supplied to the heat exchanger, increasing the water temperature and solidifying the *n*-octadecane. Do you expect the heat transfer rate to be the same, greater than, or less than the heat transfer rate in part (a)? Explain your reasoning.

11.53 A shell-and-tube heat exchanger consists of 135 thin-walled tubes in a double-pass arrangement, each of 12.5-mm diameter with a total surface area of 47.5 m^2. Water (the tube-side fluid) enters the heat exchanger at 15°C and 6.5 kg/s and is heated by exhaust gas entering at 200°C and 5 kg/s. The gas may be assumed to have the properties of atmospheric air, and the overall heat transfer coefficient is approximately 200 $W/m^2 \cdot K$.

(a) What are the gas and water outlet temperatures?

(b) Assuming fully developed flow, what is the tube-side convection coefficient?

(c) With all other conditions remaining the same, plot the effectiveness and fluid outlet temperatures as a function of the water flow rate over the range from 6 to 12 kg/s.

(d) What gas inlet temperature is required for the exchanger to supply 10 kg/s of hot water at an outlet temperature of 42°C, all other conditions remaining the same? What is the effectiveness for this operating condition?

11.54 An ocean thermal energy conversion system is being proposed for electric power generation. Such a system is based on the standard power cycle for which the working fluid is evaporated, passed through a turbine, and subsequently condensed. The system is to be used in very special locations for which the oceanic water temperature near the surface is approximately 300 K, while the temperature at reasonable depths is approximately 280 K. The warmer water is used as a heat source to evaporate the working fluid, while the colder water is used as a heat sink for condensation of the fluid. Consider a power plant that is to generate 2 MW of electricity at an efficiency (electric power output per heat input) of 3%. The evaporator is a heat exchanger consisting of a single shell with many tubes executing two passes. If the working fluid is evaporated at its phase change temperature of 290 K, with ocean water entering at 300 K and leaving at 292 K, what is the heat exchanger area required for the evaporator? What flow rate must be maintained for the water passing through the evaporator? The overall heat transfer coefficient may be approximated as 1200 $W/m^2 \cdot K$.

11.55 A single-pass, cross-flow heat exchanger with both fluids unmixed is being used to heat water ($\dot{m}_c = 2$ kg/s, $c_p = 4200$ J/kg·K) from 20°C to 100°C with hot exhaust gases ($c_p = 1200$ J/kg·K) entering at 320°C. What mass flow rate of exhaust gases is required? Assume that UA is equal to its design value of 4700 W/K, independent of the gas mass flow rate.

11.56 Saturated process steam at 1 atm is condensed in a shell-and-tube heat exchanger (one shell, two tube passes). Cooling water enters the tubes at 15°C with an average velocity of 3.5 m/s. The tubes are thin walled and made of copper with a diameter of 14 mm and length of 0.5 m. The convective heat transfer coefficient for condensation on the outer surface of the tubes is 21,800 W/m^2·K.

(a) Find the number of tubes/pass required to condense 2.3 kg/s of steam.

(b) Find the outlet water temperature.

(c) Find the maximum possible condensation rate that could be achieved with this heat exchanger using the same water flow rate and inlet temperature.

(d) Using the heat transfer surface area found in part (a), plot the water outlet temperature and steam condensation rate for water mean velocities in the range from 1 to 5 m/s. Assume that the shell-side convection coefficient remains unchanged.

11.57 The chief engineer at a university that is constructing a large number of new student dormitories decides to install a counterflow concentric tube heat exchanger on each of the dormitory shower drains. The thin-walled copper drains are of diameter $D_i = 50$ mm. Wastewater from the shower enters the heat exchanger at $T_{h,i} = 38$°C while fresh water enters the dormitory at $T_{c,i} = 10$°C. The wastewater flows down the vertical wall of the drain in a thin, *falling film*, providing $h_h = 10{,}000$ W/m^2·K.

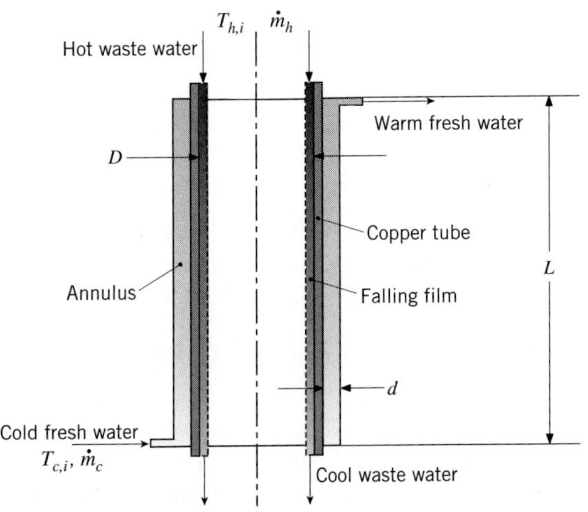

(a) If the annular gap is $d = 10$ mm, the heat exchanger length is $L = 1$ m, and the water flow rate is $\dot{m} = 10$ kg/min, determine the heat transfer rate and the outlet temperature of the warmed fresh water.

(b) If a helical spring is installed in the annular gap so the fresh water is forced to follow a spiral path from the inlet to the fresh water outlet, resulting in $h_c = 9050$ W/m^2·K, determine the heat transfer rate and the outlet temperature of the fresh water.

(c) Based on the result for part (b), calculate the daily savings if 15,000 students each take a 10-minute shower per day and the cost of water heating is \$0.07/kW·h.

11.58 A shell-and-tube heat exchanger with one shell pass and 20 tube passes uses hot water on the tube side to heat oil on the shell side. The single copper tube has inner and outer diameters of 20 and 24 mm and a length per pass of 3 m. The water enters at 87°C and 0.2 kg/s and leaves at 27°C. Inlet and outlet temperatures of the oil are 7 and 37°C. What is the average convection coefficient for the tube outer surface?

11.59 The oil in an engine is cooled by air in a cross-flow heat exchanger where both fluids are unmixed. Atmospheric air enters at 30°C and 0.53 kg/s. Oil at 0.026 kg/s enters at 75°C and flows through a tube of 10-mm diameter. Assuming fully developed flow and constant wall heat flux, estimate the oil-side heat transfer coefficient. If the overall convection coefficient is 53 W/m^2·K and the total heat transfer area is 1 m^2, determine the effectiveness. What is the exit temperature of the oil?

11.60 A recuperator is a heat exchanger that heats the air used in a combustion process by extracting energy from the products of combustion (the flue gas). Consider using a single-pass, cross-flow heat exchanger as a recuperator.

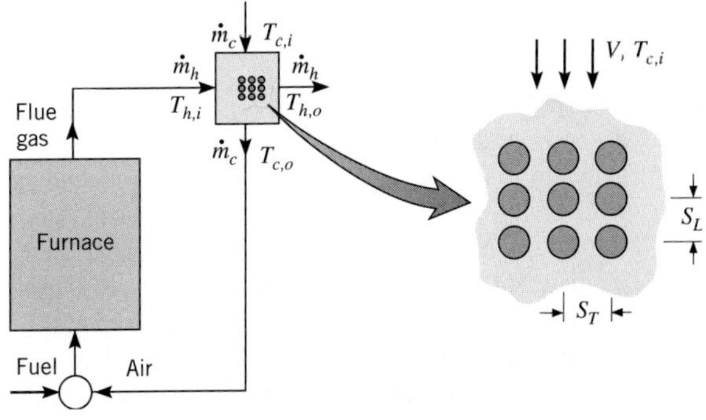

Eighty (80) silicon carbide ceramic tubes ($k = 20$ W/m·K) of inner and outer diameters equal to 55 and

80 mm, respectively, and of length $L = 1.4$ m are arranged as an aligned tube bank of longitudinal and transverse pitches $S_L = 100$ mm and $S_T = 120$ mm, respectively. Cold air is in cross flow over the tube bank with upstream conditions of $V = 1$ m/s and $T_{c,i} = 300$ K, while hot flue gases of inlet temperature $T_{h,i} = 1400$ K pass through the tubes. The tube outer surface is clean, while the inner surface is characterized by a fouling factor of $R''_f = 2 \times 10^{-4}$ m$^2 \cdot$K/W. The air and flue gas flow rates are $\dot{m}_c = 1.0$ kg/s and $\dot{m}_h = 1.05$ kg/s, respectively. As first approximations, (1) evaluate all required air properties at 1 atm and 300 K, (2) assume the flue gas to have the properties of air at 1 atm and 1400 K, and (3) assume the tube wall temperature to be at 800 K for the purpose of treating the effect of variable properties on convection heat transfer.

(a) If there is a 1% fuel savings associated with each 10°C increase in the temperature of the combustion air ($T_{c,o}$) above 300 K, what is the percentage fuel savings for the prescribed conditions?

(b) The performance of the recuperator is strongly influenced by the product of the overall heat transfer coefficient and the total surface area, UA. Compute and plot $T_{c,o}$ and the percentage fuel savings as a function of UA for $300 \le UA \le 600$ W/K. Without changing the flow rates, what measures may be taken to increase UA?

11.61 Consider operation of the furnace–recuperator combination of Problem 11.60 under conditions for which chemical energy is converted to thermal energy in the combustor at a rate of $q_{\text{comb}} = 2.0 \times 10^6$ W and energy is transferred from the combustion gases to the load in the furnace at a rate of $q_{\text{load}} = 1.4 \times 10^6$ W. Assuming equivalent flow rates ($\dot{m}_c = \dot{m}_h = 1.0$ kg/s) and specific heats ($c_{p,c} = c_{p,h} = 1200$ J/kg·K) for the cold air and flue gases in the recuperator, determine $T_{h,i}$, $T_{h,o}$, and $T_{c,o}$ when $T_{c,i} = 300$ K and the recuperator has an effectiveness of $\varepsilon = 0.30$. What value of the effectiveness would be needed to achieve a combustor air inlet temperature of 800 K?

11.62 It is proposed that the exhaust gas from a natural gas–powered electric generation plant be used to generate steam in a shell-and-tube heat exchanger with one shell and one tube pass. The steel tubes have a thermal conductivity of 40 W/m·K, an inner diameter of 50 mm, and a wall thickness of 4 mm. The exhaust gas, whose flow rate is 2 kg/s, enters the heat exchanger at 400°C and must leave at 215°C. To limit the pressure drop within the tubes, the tube gas velocity should not exceed 25 m/s. If saturated water at 11.7 bar is supplied to the shell side of the exchanger, determine the required number of tubes and their length. Assume that the properties of the exhaust gas can be approximated as those of atmospheric air and that the water-side thermal resistance is negligible. However, account for fouling on the gas side of the tubes and use a fouling resistance of 0.0015 m$^2 \cdot$K/W.

11.63 A recuperator is a heat exchanger that heats air used in a combustion process by extracting energy from the products of combustion. It can be used to increase the efficiency of a gas turbine by increasing the temperature of air entering the combustor.

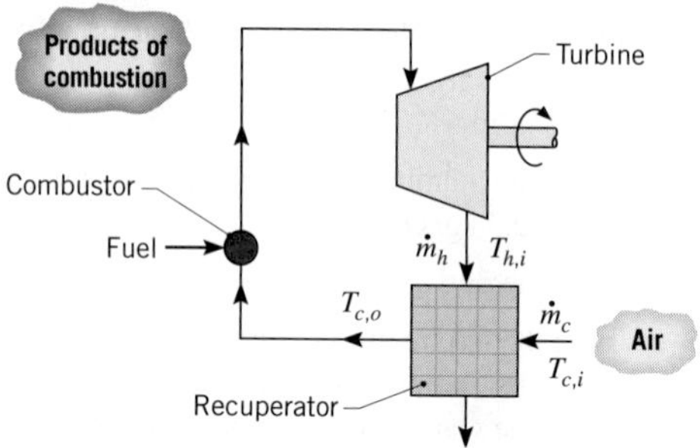

Consider a system for which the recuperator is a cross-flow heat exchanger with both fluids unmixed and the flow rates associated with the turbine exhaust and the air are $\dot{m}_h = 6.5$ kg/s and $\dot{m}_c = 6.2$ kg/s, respectively. The corresponding value of the overall heat transfer coefficient is $U = 100$ W/m$^2 \cdot$K.

(a) If the gas and air inlet temperatures are $T_{h,i} = 700$ K and $T_{c,i} = 300$ K, respectively, what heat transfer surface area is needed to provide an air outlet temperature of $T_{c,o} = 500$ K? Both the air and the products of combustion may be assumed to have a specific heat of 1040 J/kg·K.

(b) For the prescribed conditions, compute and plot the air outlet temperature as a function of the heat transfer surface area.

11.64 A concentric tube heat exchanger uses water, which is available at 15°C, to cool ethylene glycol from 100 to 60°C. The water and glycol flow rates are each 0.5 kg/s. What are the maximum possible heat transfer rate and effectiveness of the exchanger? Which is preferred, a parallel-flow or counterflow mode of operation?

11.65 Water is used for both fluids (unmixed) flowing through a single-pass, cross-flow heat exchanger. The hot water enters at 90°C and 10,000 kg/h, while the cold water enters at 10°C and 20,000 kg/h. If the effectiveness of the exchanger is 60%, determine the cold water exit temperature.

11.66 A cross-flow heat exchanger consists of a bundle of 32 tubes in a 0.6-m^2 duct. Hot water at 150°C and a mean velocity of 0.5 m/s enters the tubes having inner and outer diameters of 10.2 and 12.5 mm. Atmospheric air at 10°C enters the exchanger with a volumetric flow rate of 1.0 m^3/s. The convection heat transfer coefficient on the tube outer surfaces is 400 W/$m^2 \cdot$ K. Estimate the fluid outlet temperatures.

11.67 Exhaust gas from a furnace is used to preheat the combustion air supplied to the furnace burners. The gas, which has a flow rate of 15 kg/s and an inlet temperature of 1100 K, passes through a bundle of tubes, while the air, which has a flow rate of 10 kg/s and an inlet temperature of 300 K, is in cross flow over the tubes. The tubes are unfinned, and the overall heat transfer coefficient is 100 W/$m^2 \cdot$K. Determine the total tube surface area required to achieve an air outlet temperature of 850 K. The exhaust gas and the air may each be assumed to have a specific heat of 1075 J/kg $\cdot$ K.

11.68 Derive Equation 11.35a. *Hint*: See Section 8.3.3.

11.69 A liquefied natural gas (LNG) regasification facility utilizes a vertical heat exchanger or *vaporizer* that consists of a shell with a single-pass tube bundle used to convert the fuel to its vapor form for subsequent delivery through a land-based pipeline. Pressurized LNG is off-loaded from an oceangoing tanker to the bottom of the vaporizer at $T_{c,i} = -155°C$ and $\dot{m}_{LNG} = 150$ kg/s and flows through the shell. The pressurized LNG has a vaporization temperature of $T_f = -75°C$ and specific heat $c_{p,l} = 4200$ J/kg$\cdot$K. The specific heat of the vaporized natural gas is $c_{p,v} = 2210$ J/kg$\cdot$K while the gas has a latent heat of vaporization of $h_{fg} = 575$ kJ/kg. The LNG is heated with seawater flowing through the tubes, also introduced at the bottom of the vaporizer, that is available at $T_{h,i} = 20°C$ with a specific heat of $c_{p,SW} = 3985$ J/kg$\cdot$K. If the gas is to leave the vaporizer at $T_{c,o} = 8°C$ and the seawater is to exit the device at $T_{h,o} = 10°C$, determine the required vaporizer heat transfer area. *Hint*: Divide the vaporizer into three sections, as shown in the schematic, with $U_A = 150$ W/$m^2\cdot$K, $U_B = 260$ W/$m^2\cdot$K, and $U_C = 40$ W/$m^2\cdot$K.

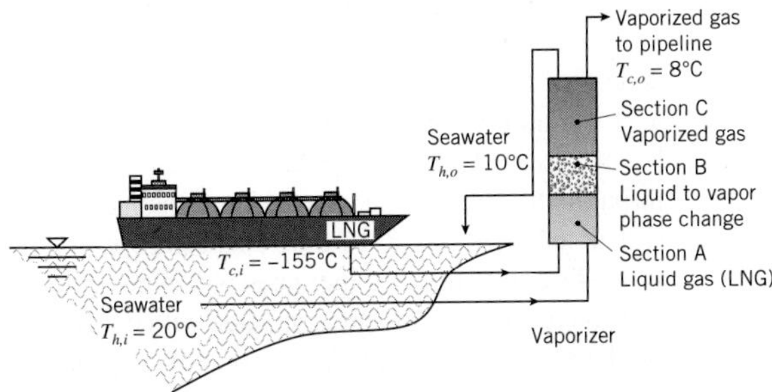

11.70 Work Problem 11.69 for the situation where the seawater is introduced to the top of the vaporizer, resulting in counterflowing natural gas and seawater.

11.71 Cooling of outdoor electronic equipment such as in telecommunications towers is difficult due to seasonal and diurnal variations of the air temperature, and potential fouling of heat exchange surfaces due to dust accumulation or insect nesting. A concept to provide a nearly constant sink temperature in a hermetically sealed environment is shown below. The cool surface is maintained at nearly constant groundwater temperature ($T_1 = 5°C$) while the hot surface is subjected to a constant heat load from the electronic equipment ($q_2 = 50$ W, T_2). Connecting the surfaces is a concentric tube of length $L = 10$ m with $D_i = 100$ mm and $D_o = 150$ mm. A fan moves air at a mass flow rate of $\dot{m} = 0.0325$ kg/s and dissipates $P = 10$ W of thermal energy. Heat transfer to the cool surface is described by $q_1'' = \bar{h}_1(T_{h,o} - T_1)$ while heat transfer from the hot surface is described by $q_2'' = \bar{h}_2(T_2 - T_{f,o})$ where $T_{f,o}$ is the fan outlet temperature. The values of $\bar{h}_1$ and $\bar{h}_2$ are 40 and 60 W/$m^2\cdot$K, respectively. To isolate the electronics from ambient temperature variations, the entire device is insulated at its outer surfaces. The design engineer is concerned that conduction through the wall of the inner tube may adversely affect the device performance. Determine the value of T_2 for the limiting cases of (i) no conduction resistance in the inner tube wall and (ii) infinite conduction resistance in the inner tube wall. Does the proposed device maintain maximum temperatures below 80°C?

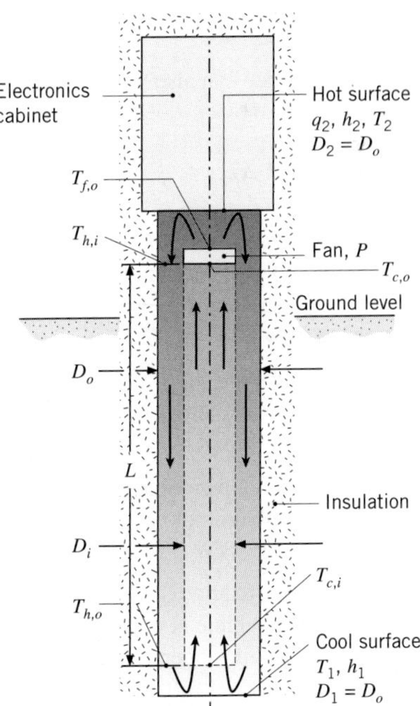

11.72 A shell-and-tube heat exchanger consisting of one shell pass and two tube passes is used to transfer heat from an ethylene glycol–water solution (shell side)

supplied from a rooftop solar collector to pure water (tube side) used for household purposes. The tubes are of inner and outer diameters $D_i = 3.6$ mm and $D_o = 3.8$ mm, respectively. Each of the 100 tubes is 0.8 m long (0.4 m per pass), and the heat transfer coefficient associated with the ethylene glycol–water mixture is $h_o = 11{,}000$ W/m$^2\cdot$K.

(a) For pure copper tubes, calculate the heat transfer rate from the ethylene glycol–water solution ($\dot{m} = 2.5$ kg/s, $T_{h,i} = 80$°C) to the pure water ($\dot{m} =$ 2.5 kg/s, $T_{c,i} = 20$°C). Determine the outlet temperatures of both streams of fluid. The density and specific heat of the ethylene glycol–water mixture are 1040 kg/m^3 and 3660 J/kg$\cdot$K, respectively.

(b) It is proposed to replace the copper tube bundle with a bundle composed of high-temperature nylon tubes of the same diameter and tube wall thickness. The nylon is characterized by a thermal conductivity of $k_n = 0.31$ W/m$\cdot$K. Determine the tube length required to transfer the same amount of energy as in part (a).

11.73 In analyzing thermodynamic cycles involving heat exchangers, it is useful to express the heat rate in terms of an overall thermal resistance R_t and the inlet temperatures of the hot and cold fluids,

$$q = \frac{(T_{h,i} - T_{c,i})}{R_t}$$

The heat transfer rate can also be expressed in terms of the rate equations,

$$q = UA\,\Delta T_{\text{lm}} = \frac{1}{R_{\text{lm}}}\Delta T_{\text{lm}}$$

(a) Derive a relation for R_{lm}/R_t for a *parallel-flow* heat exchanger in terms of a single dimensionless parameter B, which does not involve any fluid temperatures but only U, A, C_h, C_c (or $C_{\min}, C_{\max}$).

(b) Calculate and plot R_{lm}/R_t for values of $B = 0.1$, 1.0, and 5.0. What conclusions can be drawn from the plot?

11.74 The power needed to overcome wind and friction drag associated with an automobile traveling at a constant velocity of 25 m/s is 9 kW.

(a) Determine the required heat transfer area of the radiator if the vehicle is equipped with an internal combustion engine operating at an efficiency of 21%. (Assume 79% of the energy generated by the engine is in the form of waste heat removed by the radiator.) The inlet and outlet mean temperatures of the water with respect to the radiator are $T_{m,i} = 400$ K and $T_{m,o} = 330$ K, respectively. Cooling air is available at 3 kg/s and 300 K. The radiator may be analyzed as a cross-flow heat exchanger with both fluids unmixed with an overall heat transfer coefficient of 400 W/m$^2\cdot$K.

(b) Determine the required water mass flow rate and heat transfer area of the radiator if the vehicle is equipped with a fuel cell operating at 50% efficiency. The fuel cell operating temperature is limited to approximately 85°C, so the inlet and outlet mean temperatures of the water with respect to the radiator are $T_{m,i} = 355$ K and $T_{m,o} = 330$ K, respectively. The air inlet temperature is as in part (a). Assume the flow rate of air is proportional to the surface area of the radiator. *Hint:* Iteration is required.

(c) Determine the required heat transfer area of the radiator and the outlet mean temperature of the water for the fuel cell–equipped vehicle if the mass flow rate of the water is the same as in part (a).

11.75 An air conditioner operating between indoor and outdoor temperatures of 23 and 43°C, respectively, removes 5 kW from a building. The air conditioner can be modeled as a reversed Carnot heat engine with refrigerant as the working fluid. The efficiency of the motor for the compressor and fan is 80%, and 0.2 kW is required to operate the fan.

(a) Assuming negligible thermal resistances (Problem 11.73) between the refrigerant in the condenser and the outside air and between the refrigerant in the evaporator and the inside air, calculate the power required by the motor.

(b) If the thermal resistances between the refrigerant and the air in the evaporator and condenser sections are the same, 3×10^{-3} K/W, determine the temperature required by the refrigerant in each section. Calculate the power required by the motor.

11.76 In a Rankine power system, 1.5 kg/s of steam leaves the turbine as saturated vapor at 0.51 bar. The steam is condensed to saturated liquid by passing it over the tubes of a shell-and-tube heat exchanger, while liquid water, having an inlet temperature of $T_{c,i} = 280$ K, is passed through the tubes. The condenser contains 100 thin-walled tubes, each of 10-mm diameter, and the total water flow rate through the tubes is 15 kg/s. The average convection coefficient associated with condensation on the outer surface of the tubes may be approximated as $\bar{h}_o = 5000$ W/m$^2\cdot$K. Appropriate property values for the liquid water are $c_p = 4178$ J/kg$\cdot$K, $\mu = 700 \times 10^{-6}$ kg/s$\cdot$m, $k = 0.628$ W/m$\cdot$K, and $Pr = 4.6$.

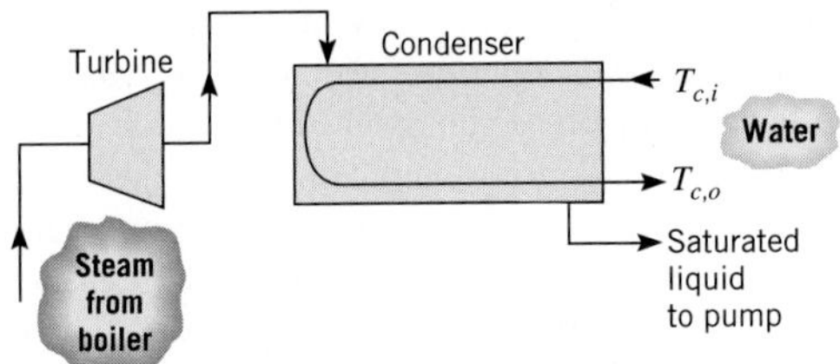

(a) What is the water outlet temperature?

(b) What is the required tube length (per tube)?

(c) After extended use, deposits accumulating on the inner and outer tube surfaces provide a cumulative fouling factor of 0.0003 $m^2 \cdot K/W$. For the prescribed inlet conditions and the computed tube length, what mass fraction of the vapor is condensed?

(d) For the tube length computed in part (b) and the fouling factor prescribed in part (c), explore the extent to which the water flow rate and inlet temperature may be varied (within physically plausible ranges) to improve the condenser performance. Represent your results graphically, and draw appropriate conclusions.

11.77 Consider a Rankine cycle with saturated steam leaving the boiler at a pressure of 2 MPa and a condenser pressure of 10 kPa.

(a) Calculate the thermal efficiency of the ideal Rankine cycle for these operating conditions.

(b) If the net reversible work for the cycle is 0.5 MW, calculate the required flow rate of cooling water supplied to the condenser at 15°C with an allowable temperature rise of 10°C.

(c) Design a shell-and-tube heat exchanger (one-shell, multiple-tube passes) that will meet the heat rate and temperature conditions required of the condenser. Your design should specify the number of tubes and their diameter and length.

11.78 Consider the Rankine cycle of Problem 11.77, which rejects 2.3 MW to the condenser, which is supplied with a cooling water flow rate of 70 kg/s at 15°C.

(a) Calculate UA, a parameter that is indicative of the size of the condenser required for this operating condition.

(b) Consider now the situation where the overall heat transfer coefficient for the condenser, U, is reduced by 10% because of fouling. Determine the reduction in the thermal efficiency of the cycle caused by fouling, assuming that the cooling water flow rate and water temperature remain the same and that the condenser is operated at the same steam pressure.

11.79 Consider a concentric tube heat exchanger characterized by a uniform overall heat transfer coefficient and operating under the following conditions:

	$\dot{m}$ (kg/s)	c_p (J/kg·K)	T_i (°C)	T_o (°C)
Cold fluid	0.125	4200	40	95
Hot fluid	0.125	2100	210	—

What is the maximum possible heat transfer rate? What is the heat exchanger effectiveness? Should the heat exchanger be operated in parallel flow or in counterflow? What is the ratio of the required areas for these two flow conditions?

11.80 The floor space of any facility that houses shell-and-tube heat exchangers must be sufficiently large so the tube bundle can be serviced easily. A rule of thumb is that the floor space must be at least 2.5 times the length of the tube bundle so that the bundle can be completely removed from the shell (hence the absolute minimum floor space is twice the tube bundle length) and subsequently cleaned, repaired, or replaced easily (associated with the extra half bundle length floor space). The room in which the heat exchanger of Problem 11.22 is to be installed is 8 m long and, therefore, the 4.7-m-long heat exchanger is too large for the facility. Will a shell-and-tube heat exchanger with two shells, one above the other, be sufficiently small to fit into the facility? Each shell has 10 tubes and 8 tube passes.

11.81 Consider the influence of a finite sheet thickness in Example 11.2, when there are 40 gaps.

(a) Determine the exterior dimension, L, of the heat exchanger core for a sheet thickness of $t = 0.8$ mm for pure aluminum ($k_{al} = 237$ W/m·K) and polyvinylidene fluoride (PVDF, $k_{pv} = 0.17$ W/m·K) sheets. Neglect the thickness of the top and bottom exterior plates.

(b) Plot the heat exchanger core dimension as a function of the sheet thickness for aluminum and PVDF over the range $0 \leq t \leq 1$ mm.

11.82 Hot exhaust gases are used in a shell-and-tube exchanger to heat 2.5 kg/s of water from 35 to 85°C. The gases, assumed to have the properties of air, enter at 200°C and leave at 93°C. The overall heat transfer coefficient is 180 $W/m^2 \cdot K$. Using the effectiveness–NTU method, calculate the area of the heat exchanger.

11.83 In open heart surgery under hypothermic conditions, the patient's blood is cooled before the surgery and rewarmed afterward. It is proposed that a concentric tube, counterflow heat exchanger of length 0.5 m be used for this purpose, with the thin-walled inner tube having a diameter of 55 mm. The specific heat of the blood is 3500 J/kg·K.

(a) If water at $T_{h,i} = 60°C$ and $\dot{m}_h = 0.10$ kg/s is used to heat blood entering the exchanger at $T_{c,i} = 18°C$ and $\dot{m}_c = 0.05$ kg/s, what is the temperature of the blood leaving the exchanger? The overall heat transfer coefficient is 500 W/m^2·K.

(b) The surgeon may wish to control the heat rate q and the outlet temperature $T_{c,o}$ of the blood by altering the flow rate and/or inlet temperature of the water during the rewarming process. To assist in the development of an appropriate controller for the prescribed values of $\dot{m}_c$ and $T_{c,i}$, compute and plot q and $T_{c,o}$ as a function of $\dot{m}_h$ for $0.05 \leq \dot{m}_h \leq 0.20$ kg/s and values of $T_{h,i} = 50$, 60, and 70°C. Since the dominant influence on the overall heat transfer coefficient is associated with the blood flow conditions, the value of U may be assumed to remain at 500 W/m^2·K. Should certain operating conditions be excluded?

11.84 Ethylene glycol and water, at 60 and 10°C, respectively, enter a shell-and-tube heat exchanger for which the total heat transfer area is 15 m^2. With ethylene glycol and water flow rates of 2 and 5 kg/s, respectively, the overall heat transfer coefficient is 800 W/m^2·K.

(a) Determine the rate of heat transfer and the fluid outlet temperatures.

(b) Assuming all other conditions to remain the same, plot the effectiveness and fluid outlet temperatures as a function of the flow rate of ethylene glycol for $0.5 \leq \dot{m}_h \leq 5$ kg/s.

11.85 A boiler used to generate saturated steam is in the form of an unfinned, cross-flow heat exchanger, with water flowing through the tubes and a high-temperature gas in cross flow over the tubes. The gas, which has a specific heat of 1120 J/kg·K and a mass flow rate of 10 kg/s, enters the heat exchanger at 1400 K. The water, which has a flow rate of 3 kg/s, enters as saturated liquid at 450 K and leaves as saturated vapor at the same temperature. If the overall heat transfer coefficient is 50 W/m^2·K and there are 500 tubes, each of 0.025-m diameter, what is the required tube length?

11.86 Waste heat from the exhaust gas of an industrial furnace is recovered by mounting a bank of unfinned tubes in the furnace stack. Pressurized water at a flow rate of 0.025 kg/s makes a single pass through *each* of the tubes, while the exhaust gas, which has an upstream velocity of 5.0 m/s, moves in cross flow over the tubes at 2.25 kg/s. The tube bank consists of a square array of 100 thin-walled tubes (10 × 10), each 25 mm in diameter and 4 m long. The tubes are aligned with a transverse pitch of 50 mm. The inlet temperatures of the water and the exhaust gas are 300 and 800 K, respectively. The water flow is fully developed, and the gas properties may be assumed to be those of atmospheric air.

(a) What is the overall heat transfer coefficient?

(b) What are the fluid outlet temperatures?

(c) Operation of the heat exchanger may vary according to the demand for hot water. For the prescribed heat exchanger design and inlet conditions, compute and plot the rate of heat recovery and the fluid outlet temperatures as a function of water flow rate per tube for $0.02 \leq \dot{m}_{c,1} \leq 0.20$ kg/s.

11.87 A heat exchanger consists of a bank of 1200 thin-walled tubes with air in cross flow over the tubes. The tubes are arranged in-line, with 40 longitudinal rows (along the direction of airflow) and 30 transverse rows. The tubes are 0.07 m in diameter and 2 m long, with transverse and longitudinal pitches of 0.14 m. The hot fluid flowing through the tubes consists of saturated steam condensing at 400 K. The convection coefficient of the condensing steam is much larger than that of the air.

(a) If air enters the heat exchanger at $\dot{m}_c = 12$ kg/s, 300 K, and 1 atm, what is its outlet temperature?

(b) The condensation rate may be controlled by varying the airflow rate. Compute and plot the air outlet temperature, the heat rate, and the condensation rate as a function of flow rate for $10 \leq \dot{m}_c \leq 50$ kg/s.

Additional Considerations

11.88 Derive the expression for the modified effectiveness ε^*, given in Comment 4 of Example 11.8.

11.89 Consider Problem 3.144a.

(a) Using an appropriate correlation from Chapter 8, determine the air inlet velocity for each channel in the heat sink. Assume laminar flow and evaluate air properties at $T = 300$ K.

(b) Accounting for the increase in air temperature as it flows through the heat sink, determine the chip power q_c and the outlet temperature of the air exiting each channel. Assume the airflow along the outer surfaces provides a similar cooling effect as airflow in the channels.

(c) If the air velocity is reduced by half, determine the chip power and the air outlet temperature.

11.90 Work Problem 7.29, taking into account the increase in temperature of the water as it flows through the heat sink. Properties of water are listed in Problem 7.29, along with $\rho = 995\ \text{kg/m}^3$ and $c_p = 4178\ \text{J/kg}\cdot\text{K}$. *Hint*: Assume the water does not escape through the upper surface of the heat sink and that the boundary layers on each fin surface do not merge, allowing evaluation of the heat transfer coefficient using a correlation from Chapter 7. Also, see Problem 11.68.

11.91 The heat sink of Problem 7.29 is considered for an application in which the power dissipation is only 70 W, and the engineer proposes to use air at $T_\infty = 20°\text{C}$ for cooling. Taking into account the increase in temperature of the air as it flows through the heat sink, plot the allowable power dissipation and the air exit temperature as a function of the air velocity over the range $1\ \text{m/s} \leq u_\infty \leq 5\ \text{m/s}$, with the constraint that the base temperature not exceed $T_b = 70°\text{C}$. Properties of the air may be approximated as $k = 0.027\ \text{W/m}\cdot\text{K}$, $\nu = 16.4 \times 10^{-6}\ \text{m}^2/\text{s}$, $Pr = 0.706$, $\rho = 1.145\ \text{kg/m}^3$, and $c_p = 1007\ \text{J/kg}\cdot\text{K}$. *Hint*: Assume the air does not escape through the upper surface of the heat sink, use a correlation for internal flow, and see Problem 11.68.

11.92 Solve Problem 8.109a using the effectiveness–NTU method.

11.93 Consider Problem 7.113. Estimate the heat transfer rate to the air, accounting for both the increase in the air temperature as it flows through the foam and the thermal resistance associated with conduction in the foam in the x-direction. Do you expect the actual heat transfer rate to the air to be equal to, less than, or greater than the value you have calculated?

11.94 The metallic foam of Problem 7.113 is brazed to the surface of a silicon chip of width $W = 25$ mm on a side. The foam heat sink is $L = 10$ mm tall. Air at $T_i = 27°\text{C}$, $V = 5$ m/s impinges on the foam heat sink while the chip surface is maintained at 70°C. Determine the heat transfer rate from the chip. To calculate a conservative estimate of the heat transfer rate, neglect convection and radiation from the top and sides of the heat sink.

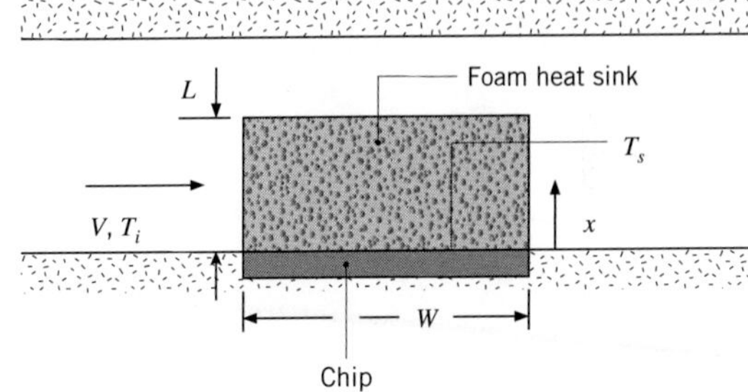

CHAPTER 12

Radiation: Processes and Properties

We have come to recognize that heat transfer by conduction and convection requires the presence of a temperature gradient in some form of matter. In contrast, heat transfer by *thermal radiation* requires no matter. It is an extremely important process, and in the physical sense it is perhaps the most interesting of the heat transfer modes. It is relevant to many industrial heating, cooling, and drying processes, as well as to energy conversion methods that involve fossil fuel combustion and solar radiation.

In this chapter our objective is to consider the means by which thermal radiation is generated, the specific nature of the radiation, and the manner in which it interacts with matter. We give particular attention to radiative interactions at a surface and to the properties that must be introduced to describe these interactions. In Chapter 13 we focus on means for computing radiative exchange between two or more surfaces.

12.1 *Fundamental Concepts*

Consider a solid that is initially at a higher temperature T_s than that of its surroundings T_{sur}, but around which there exists a vacuum (Figure 12.1). The presence of the vacuum precludes energy loss from the surface of the solid by conduction or convection. However, our intuition tells us that the solid will cool and eventually achieve thermal equilibrium with its surroundings. This cooling is associated with a reduction in the internal energy stored by the solid and is a direct consequence of the *emission* of thermal radiation from the surface. In turn, the surface will intercept and absorb radiation originating from the surroundings. However, if $T_s > T_{sur}$ the *net* heat transfer rate by radiation $q_{rad,net}$ is *from* the surface, and the surface will cool until T_s reaches T_{sur}.

We associate thermal radiation with the rate at which energy is emitted by matter as a result of its finite temperature. At this moment thermal radiation is being emitted by all the matter that surrounds you: by the furniture and walls of the room, if you are indoors, or by the ground, buildings, and the atmosphere and sun if you are outdoors. The mechanism of emission is related to energy released as a result of oscillations or transitions of the many electrons that constitute matter. These oscillations are, in turn, sustained by the internal energy, and therefore the temperature, of the matter. Hence we associate the emission of thermal radiation with thermally excited conditions within the matter.

All forms of matter emit radiation. For gases and for semitransparent solids, such as glass and salt crystals at elevated temperatures, emission is a *volumetric phenomenon,* as

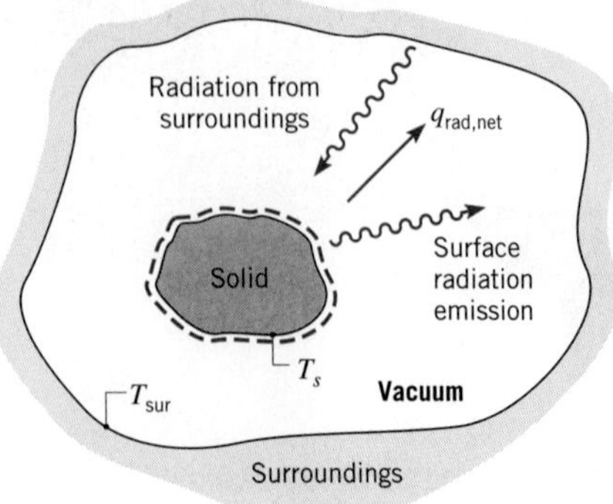

FIGURE 12.1 Radiation cooling of a hot solid.

illustrated in Figure 12.2. That is, radiation emerging from a finite volume of matter is the integrated effect of local emission throughout the volume. However, in this text we concentrate on situations for which radiation can be treated as a *surface phenomenon.* In most solids and liquids, radiation emitted from interior molecules is strongly absorbed by adjoining molecules. Accordingly, radiation that is emitted from a solid or a liquid originates from molecules that are within a distance of approximately 1 μm from the exposed surface. It is for this reason that emission from a solid or a liquid into an adjoining gas or a vacuum can be viewed as a surface phenomenon, except in situations involving nanoscale or microscale devices.

We know that radiation originates due to emission by matter and that its subsequent transport does not require the presence of any matter. But what is the nature of this transport? One theory views radiation as the propagation of a collection of particles termed *photons* or *quanta.* Alternatively, radiation may be viewed as the propagation of *electromagnetic waves.* In any case we wish to attribute to radiation the standard wave properties of frequency ν and wavelength λ. For radiation propagating in a particular medium, the two properties are related by

$$\lambda = \frac{c}{\nu} \tag{12.1}$$

where c is the speed of light in the medium. For propagation in a vacuum, $c_o = 2.998 \times 10^8$ m/s. The unit of wavelength is commonly the micrometer (μm), where 1 μm $= 10^{-6}$ m.

The complete electromagnetic spectrum is delineated in Figure 12.3. The short wavelength gamma rays, X rays, and ultraviolet (UV) radiation are primarily of interest to the high-energy physicist and the nuclear engineer, while the long wavelength microwaves and radio waves ($\lambda > 10^5$ μm) are of concern to the electrical engineer. It is the intermediate portion of the spectrum, which extends from approximately 0.1 to 100 μm and includes a

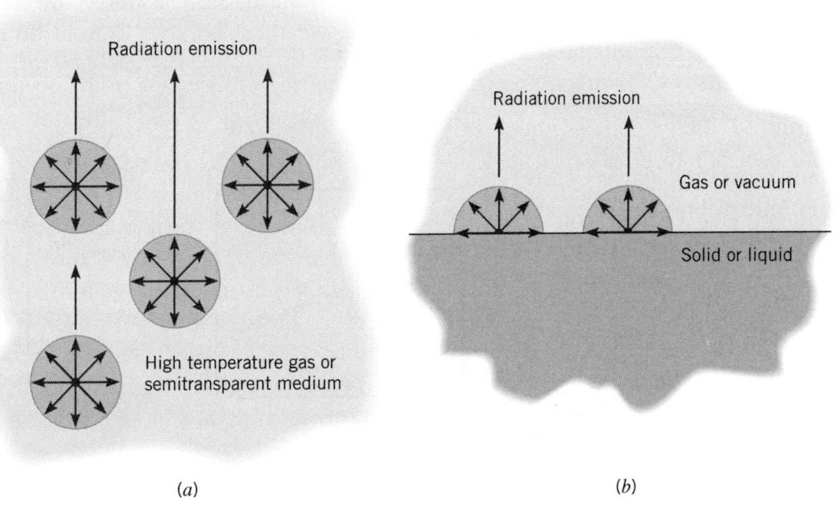

FIGURE 12.2 The emission process. (*a*) As a volumetric phenomenon. (*b*) As a surface phenomenon.

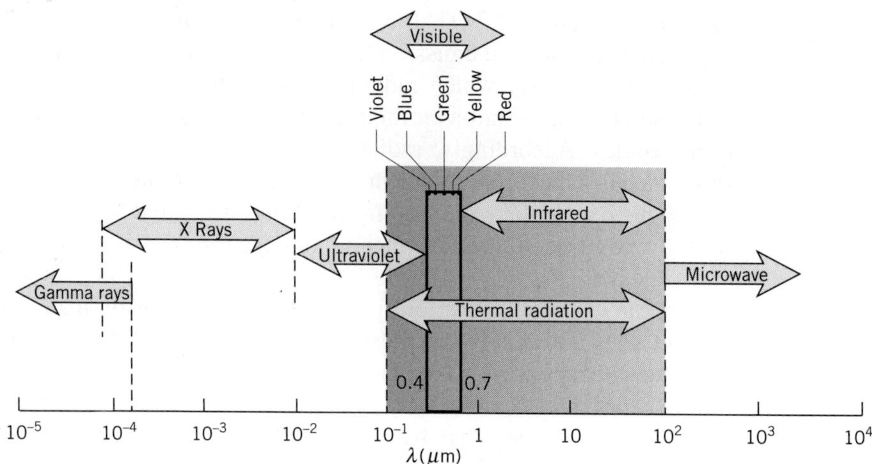

FIGURE 12.3 Spectrum of electromagnetic radiation.

portion of the UV and all of the visible and infrared (IR), that is termed *thermal radiation* because it is both caused by and affects the thermal state or temperature of matter. For this reason, thermal radiation is pertinent to heat transfer.

Thermal radiation emitted by a surface encompasses a range of wavelengths. As shown in Figure 12.4*a*, the magnitude of the radiation varies with wavelength, and the term *spectral* is used to refer to the nature of this dependence. As we will find, both the magnitude of the radiation at any wavelength and the *spectral distribution* vary with the nature and temperature of the emitting surface.

The spectral nature of thermal radiation is one of two features that complicates its description. The second feature relates to its *directionality.* As shown in Figure 12.4*b*, a surface may emit preferentially in certain directions, creating a *directional distribution* of the emitted radiation. To quantify the emission, absorption, reflection, and transmission concepts introduced in Chapter 1, we must be able to treat both spectral and directional effects.

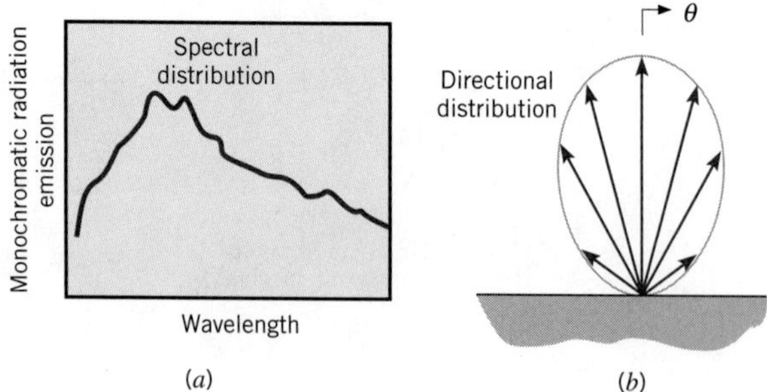

FIGURE 12.4 Radiation emitted by a surface. (*a*) Spectral distribution. (*b*) Directional distribution.

12.2 Radiation Heat Fluxes

Various types of heat fluxes are pertinent to the analysis of radiation heat transfer. Table 12.1 lists four distinct radiation fluxes that can be defined at a surface such as the one in Figure 12.2*b*. The *emissive power*, E (W/m^2), is the rate at which radiation is emitted from a surface per unit surface area, over all wavelengths and in all directions. In Chapter 1, this emissive power was related to the behavior of a *blackbody* through the relation $E = \varepsilon\sigma T_s^4$ (Equation 1.5), where ε is a surface property known as the *emissivity*.

Radiation from the surroundings, which may consist of multiple surfaces at various temperatures, is incident upon the surface. The surface might also be irradiated by the sun or by a laser. In any case, we define the *irradiation, G* (W/m^2), as the rate at which radiation is incident upon the surface per unit surface area, over all wavelengths and from all directions. The two remaining heat fluxes of Table 12.1 are readily described once we consider the fate of the irradiation arriving at the surface.

When radiation is incident upon a *semitransparent medium*, portions of the irradiation may be reflected, absorbed, and transmitted, as discussed in Section 1.2.3 and illustrated in Figure 12.5*a*. Transmission refers to radiation passing through the medium, as

TABLE 12.1 Radiative fluxes (over all wavelengths and in all directions)

Flux (W/m^2)	Description	Comment
Emissive power, E	Rate at which radiation is emitted from a surface per unit area	$E = \varepsilon\sigma T_s^4$
Irradiation, G	Rate at which radiation is incident upon a surface per unit area	Irradiation can be reflected, absorbed, or transmitted
Radiosity, J	Rate at which radiation leaves a surface per unit area	For an opaque surface $J = E + \rho G$
Net radiative flux, $q''_{rad} = J - G$	Net rate of radiation leaving a surface per unit area	For an opaque surface $q''_{rad} = \varepsilon\sigma T_s^4 - \alpha G$

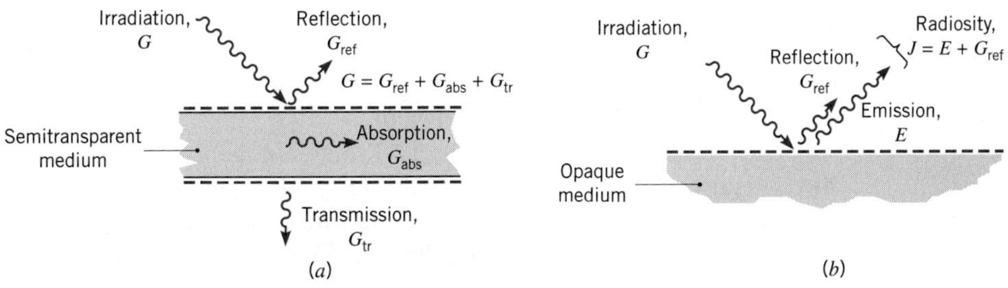

FIGURE 12.5 Radiation at a surface. (*a*) Reflection, absorption, and transmission of irradiation for a semitransparent medium. (*b*) The radiosity for an opaque medium.

occurs when a layer of water or a glass plate is irradiated by the sun or artificial lighting. Absorption occurs when radiation interacts with the medium, causing an increase in the internal thermal energy of the medium. Reflection is the process of incident radiation being redirected away from the surface, with no effect on the medium. We define reflectivity ρ as the fraction of the irradiation that is reflected, absorptivity α as the fraction of the irradiation that is absorbed, and transmissivity τ as the fraction of the irradiation that is transmitted. Because all of the irradiation must be reflected, absorbed, or transmitted, it follows that

$$\rho + \alpha + \tau = 1 \tag{12.2}$$

A medium that experiences no transmission ($\tau = 0$) is *opaque,* in which case

$$\rho + \alpha = 1 \tag{12.3}$$

With this understanding of the partitioning of the irradiation into reflected, absorbed, and transmitted components, two additional and useful radiation fluxes can be defined. The *radiosity,* J (W/m^2), of a surface accounts for *all* the radiant energy leaving the surface. For an opaque surface, it includes emission and the reflected portion of the irradiation, as illustrated in Figure 12.5*b*. It is therefore expressed as

$$J = E + G_{\text{ref}} = E + \rho G \tag{12.4}$$

Radiosity can also be defined at a surface of a semitransparent medium. In that case, the radiosity leaving the top surface of Figure 12.5*a* (not shown) would include radiation transmitted through the medium from below.

Finally, the *net* radiative flux *from* a surface, q''_{rad} (W/m^2), is the difference between the outgoing and incoming radiation

$$q''_{\text{rad}} = J - G \tag{12.5}$$

Combining Equations 12.5, 12.4, 12.3, and 1.4, the net flux for an opaque surface is

$$q''_{\text{rad}} = E + \rho G - G = \varepsilon\sigma T_s^4 - \alpha G \tag{12.6}$$

A similar expression may be written for a semitransparent surface involving the transmissivity. Because it affects the temperature distribution within the system, the net radiative flux (or net radiation heat transfer rate, $q_{\text{rad}} = q''_{\text{rad}}A$), is an important quantity in heat transfer analysis. As will become evident, the quantities E, G, and J are typically used to determine q''_{rad}, but they are also intrinsically important in applications involving *radiation detection* and *temperature measurement.*

The various fluxes in Table 12.1 may, in general, be quantified only when the directional and spectral nature of the radiation is known. Directional effects are considered by introducing the concept of *radiation intensity* in Section 12.3, while spectral effects are treated by introducing the concept of blackbody radiation in Section 12.4. The emissive power of a real surface will be related to that of the blackbody through the definition of *emissivity* in Section 12.5. The spectral and directional characteristics of the emissivity, absorptivity, reflectivity, and transmissivity of real surfaces are included in Sections 12.5 and 12.6.

12.3 Radiation Intensity

Radiation that leaves a surface can propagate in all possible directions (Figure 12.4*b*), and we are often interested in knowing its directional distribution. Also, radiation incident upon a surface may come from different directions, and the manner in which the surface responds to this radiation depends on the direction. Such directional effects can be of primary importance in determining the net radiative heat transfer rate and may be treated by introducing the concept of *radiation intensity*.

12.3.1 Mathematical Definitions

Due to its nature, mathematical treatment of radiation heat transfer involves the extensive use of the spherical coordinate system. From Figure 12.6*a*,we recall that the differential plane angle $d\alpha$ is defined by a region between the rays of a circle and is measured as the ratio of the arc length dl on the circle to the radius r of the circle. Similarly, from Figure 12.6*b*, the differential solid angle $d\omega$ is defined by a region between the rays of a sphere and is measured as the ratio of the area dA_n on the sphere to the sphere's radius squared. Accordingly,

$$d\omega \equiv \frac{dA_n}{r^2} \tag{12.7}$$

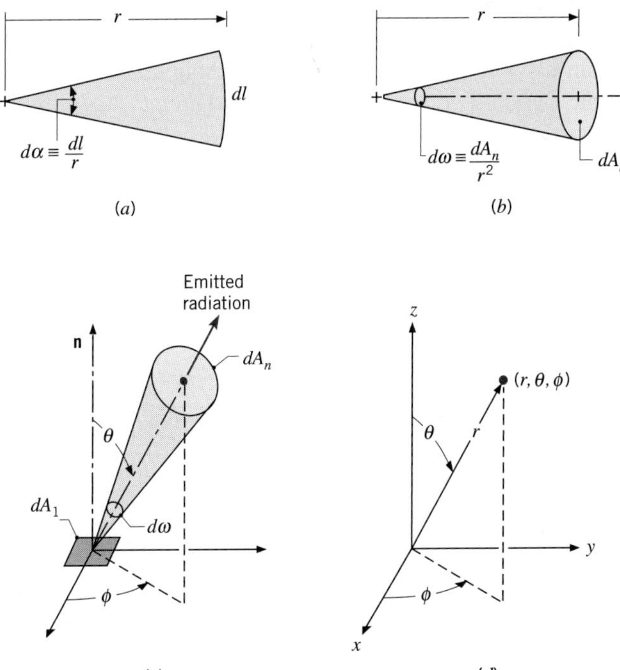

FIGURE 12.6 Mathematical definitions. (*a*) Plane angle. (*b*) Solid angle. (*c*) Emission of radiation from a differential area dA_1 into a solid angle $d\omega$ subtended by dA_n at a point on dA_1. (*d*) The spherical coordinate system.

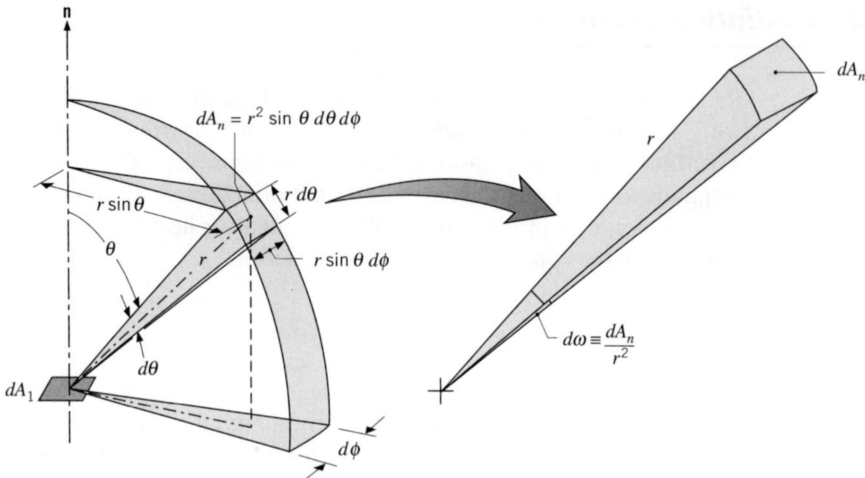

FIGURE 12.7 The solid angle subtended by dA_n at a point on dA_1 in the spherical coordinate system.

Consider emission in a particular direction from an element of surface area dA_1, as shown in Figure 12.6*c*. The direction may be specified in terms of the zenith and azimuthal angles, θ and ϕ, respectively, of a spherical coordinate system (Figure 12.6*d*). The area dA_n, through which the radiation passes, subtends a differential solid angle $d\omega$ when viewed from a point on dA_1. As shown in Figure 12.7, the area dA_n is a rectangle of dimension $r\, d\theta \times r \sin\theta\, d\phi$; thus, $dA_n = r^2 \sin\theta\, d\theta\, d\phi$. Accordingly,

$$d\omega = \sin\theta\, d\theta\, d\phi \tag{12.8}$$

When viewed from a point on an *opaque* surface area element dA_1, radiation may be emitted into any direction defined by a hypothetical hemisphere above the surface. The solid angle associated with the entire hemisphere may be obtained by integrating Equation 12.8 over the limits $\phi = 0$ to $\phi = 2\pi$ and $\theta = 0$ to $\theta = \pi/2$. Hence,

$$\int_h d\omega = \int_0^{2\pi}\int_0^{\pi/2} \sin\theta\, d\theta\, d\phi = 2\pi \int_0^{\pi/2} \sin\theta\, d\theta = 2\pi \text{ sr} \tag{12.9}$$

where the subscript h refers to integration over the hemisphere. Note that the unit of the solid angle is the steradian (sr), analogous to radians for plane angles.

12.3.2 Radiation Intensity and Its Relation to Emission

Returning to Figure 12.6*c*, we now consider the rate at which emission from dA_1 passes through dA_n. This quantity may be expressed in terms of the *spectral intensity* $I_{\lambda,e}$ of the emitted radiation. We formally define $I_{\lambda,e}$ as the *rate at which radiant energy is emitted at the wavelength* λ *in the* (θ, ϕ) *direction, per unit area of the emitting surface normal to this direction, per unit solid angle about this direction, and per unit wavelength interval*

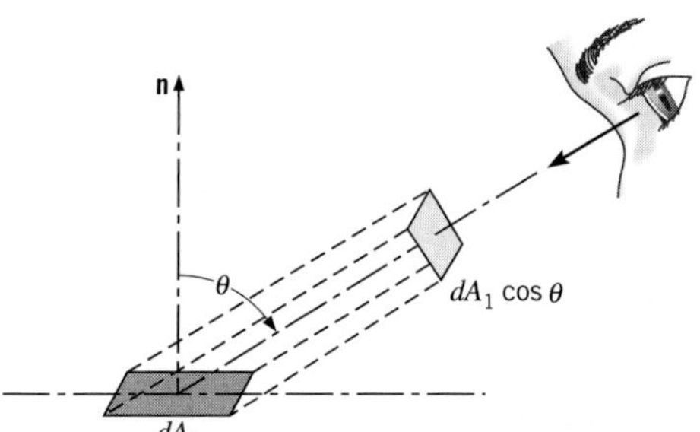

FIGURE 12.8 The projection of dA_1 normal to the direction of radiation.

$d\lambda$ *about* λ. Note that the area used to define the intensity is the component of dA_1 perpendicular to the direction of the radiation. From Figure 12.8, we see that this projected area is equal to $dA_1 \cos\theta$. In effect it is how dA_1 would appear to an observer situated on dA_n. The spectral intensity, which has units of $\text{W/m}^2 \cdot \text{sr} \cdot \mu\text{m}$, is then

$$I_{\lambda,e}(\lambda, \theta, \phi) \equiv \frac{dq}{dA_1 \cos\theta \cdot d\omega \cdot d\lambda} \tag{12.10}$$

where $(dq/d\lambda) \equiv dq_\lambda$ is the rate at which radiation of wavelength λ leaves dA_1 and passes through dA_n. Rearranging Equation 12.10, it follows that

$$dq_\lambda = I_{\lambda,e}(\lambda, \theta, \phi) dA_1 \cos\theta \, d\omega \tag{12.11}$$

where dq_λ has the units of $\text{W}/\mu\text{m}$. This important expression allows us to compute the rate at which radiation emitted by a surface propagates into the region of space defined by the solid angle $d\omega$ about the (θ, ϕ) direction. However, to compute this rate, the spectral intensity $I_{\lambda,e}$ of the emitted radiation must be known. The manner in which this quantity may be determined is discussed later, in Sections 12.4 and 12.5. Expressing Equation 12.11 per unit area of the emitting surface and substituting from Equation 12.8, the spectral radiation *flux* associated with dA_1 is

$$dq''_\lambda = I_{\lambda,e}(\lambda, \theta, \phi) \cos\theta \sin\theta \, d\theta \, d\phi \tag{12.12}$$

If the spectral and directional distributions of $I_{\lambda,e}$ are known, that is, $I_{\lambda,e}$ (λ, θ, ϕ) is known, the heat flux associated with emission into any finite solid angle or over any finite wavelength interval may be determined by integrating Equation 12.12. For example, we define the *spectral, hemispherical emissive power* E_λ $(\text{W/m}^2 \cdot \mu\text{m})$ as the rate at which radiation of wavelength λ is emitted in *all directions* from a surface per unit wavelength interval $d\lambda$ about λ and per unit surface area. Thus, E_λ is the spectral heat flux associated with emission into a hypothetical hemisphere above dA_1, as shown in Figure 12.9, or

$$E_\lambda(\lambda) = q''_\lambda(\lambda) = \int_0^{2\pi} \int_0^{\pi/2} I_{\lambda,e}(\lambda, \theta, \phi) \cos\theta \sin\theta \, d\theta \, d\phi \tag{12.13}$$

Note that E_λ is a flux based on the *actual* surface area, whereas $I_{\lambda,e}$ is based on the *projected* area. The $\cos\theta$ term appearing in the integrand is a consequence of this difference.

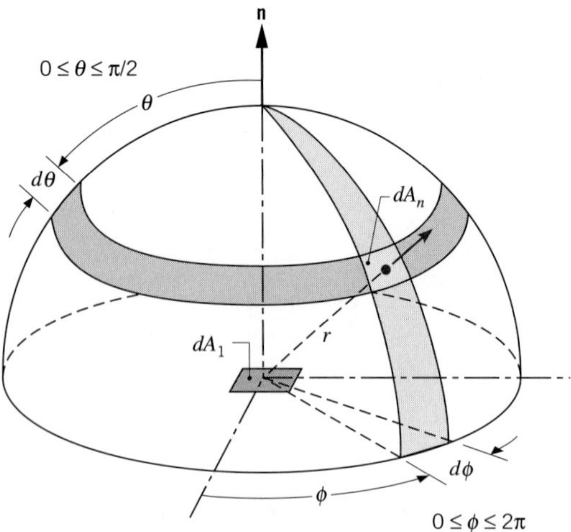

FIGURE 12.9 Emission from a differential element of area dA_1 into a hypothetical hemisphere centered at a point on dA_1.

The *total, hemispherical emissive power, E* (W/m^2), is the rate at which radiation is emitted per unit area at all possible wavelengths and in all possible directions. Accordingly,

$$E = \int_0^\infty E_\lambda(\lambda)\, d\lambda \tag{12.14}$$

or from Equation 12.13

$$E = \int_0^\infty \int_0^{2\pi} \int_0^{\pi/2} I_{\lambda,e}(\lambda, \theta, \phi) \cos\theta \sin\theta \, d\theta \, d\phi \, d\lambda \tag{12.15}$$

Since the term "emissive power" implies emission in all directions, the adjective "hemispherical" is redundant and is often dropped. One then speaks of the *spectral emissive power* E_λ, or the *total emissive power E* that was first introduced in Equation 1.5 and again in Table 12.1.

Although the directional distribution of surface emission varies according to the nature of the surface, there is a special case that provides a reasonable approximation for many surfaces. We speak of a *diffuse emitter* as a surface for which the intensity of the emitted radiation is independent of direction, in which case $I_{\lambda,e}(\lambda, \theta, \phi) = I_{\lambda,e}(\lambda)$. Removing $I_{\lambda,e}$ from the integrand of Equation 12.13 and performing the integration, it follows that

$$E_\lambda(\lambda) = \pi I_{\lambda,e}(\lambda) \tag{12.16}$$

Similarly, from Equation 12.15

$$E = \pi I_e \tag{12.17}$$

where I_e is the *total intensity* of the emitted radiation. Note that the constant appearing in the above expressions is π, not 2π, and has the unit steradians.

EXAMPLE 12.1

A small surface of area $A_1 = 10^{-3}$ m^2 is known to emit diffusely, and from measurements the total intensity associated with emission in the normal direction is $I_n = 7000$ W/m$^2 \cdot$ sr.

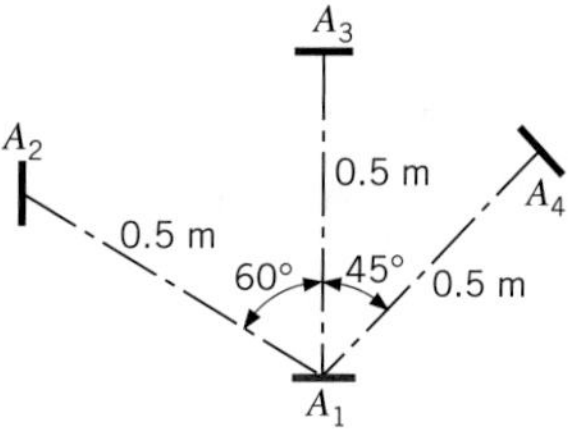

Radiation emitted from the surface is intercepted by three other surfaces of area $A_2 = A_3 = A_4 = 10^{-3}$ m^2, which are 0.5 m from A_1 and are oriented as shown. What is the intensity associated with emission in each of the three directions? What are the solid angles subtended by the three surfaces when viewed from A_1? What is the rate at which radiation emitted by A_1 is intercepted by the three surfaces?

SOLUTION

Known: Normal intensity of diffuse emitter of area A_1 and orientation of three surfaces relative to A_1.

Find:

1. Intensity of emission in each of the three directions.
2. Solid angles subtended by the three surfaces.
3. Rate at which radiation is intercepted by the three surfaces.

Schematic:

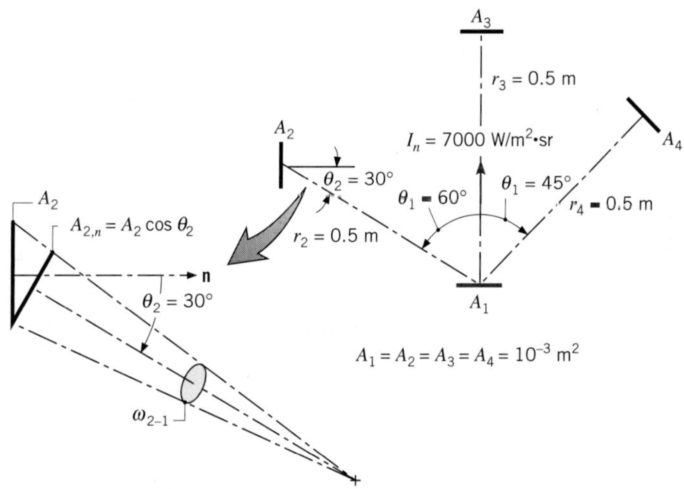

Assumptions:

1. Surface A_1 emits diffusely.
2. A_1, A_2, A_3, and A_4 may be approximated as differential surfaces, $(A_j/r_j^2) \ll 1$.

Analysis:

1. From the definition of a diffuse emitter, we know that the intensity of the emitted radiation is independent of direction. Hence

$$I = 7000\,\text{W/m}^2 \cdot \text{sr} \qquad \triangleleft$$

for each of the three directions.

2. Treating A_2, A_3, and A_4 as differential surface areas, the solid angles may be computed from Equation 12.7

$$d\omega \equiv \frac{dA_n}{r^2}$$

where dA_n is the projection of the surface normal to the direction of the radiation. Since surfaces A_3 and A_4 are normal to the direction of radiation, the solid angles subtended by these surfaces can be directly found from this equation as

$$\omega_{3-1} = \omega_{4-1} = \frac{A_3}{r^2} = \frac{10^{-3}\ \text{m}^2}{(0.5\ \text{m})^2} = 4.00 \times 10^{-3}\ \text{sr} \qquad \triangleleft$$

Since surface A_2 is not normal to the direction of radiation, we use $dA_{n,2} = dA_2 \cos\theta_2$, where θ_2 is the angle between the surface normal and the direction of the radiation. Thus

$$\omega_{2-1} = \frac{A_2 \cos\theta_2}{r^2} = \frac{10^{-3}\ \text{m}^2 \times \cos 30°}{(0.5\,\text{m})^2} = 3.46 \times 10^{-3}\ \text{sr} \qquad \triangleleft$$

3. Approximating A_1 as a differential surface, the rate at which radiation is intercepted by each of the three surfaces may be found from Equation 12.11, which, for the total radiation, may be expressed as

$$q_{1-j} = I \times A_1 \cos\theta_1 \times \omega_{j-1}$$

where θ_1 is the angle between the normal to surface 1 and the direction of the radiation. Hence

$$q_{1-2} = 7000\ \text{W/m}^2 \cdot \text{sr}\ (10^{-3}\ \text{m}^2 \times \cos 60°)\ 3.46 \times 10^{-3}\ \text{sr}$$
$$= 12.1 \times 10^{-3}\ \text{W} \qquad \triangleleft$$

$$q_{1-3} = 7000\ \text{W/m}^2 \cdot \text{sr}\ (10^{-3}\ \text{m}^2 \times \cos 0°)\ 4.00 \times 10^{-3}\ \text{sr}$$
$$= 28.0 \times 10^{-3}\ \text{W} \qquad \triangleleft$$

$$q_{1-4} = 7000\ \text{W/m}^2 \cdot \text{sr}\ (10^{-3}\ \text{m}^2 \times \cos 45°)\ 4.00 \times 10^{-3}\ \text{sr}$$
$$= 19.8 \times 10^{-3}\,\text{W} \qquad \triangleleft$$

Comments:

1. Note the different values of θ_1 for the emitting surface and the values of θ_2, θ_3, and θ_4 for the receiving surfaces.
2. If the surfaces were not small relative to the square of the separation distance, the solid angles and radiation heat transfer rates would have to be obtained by integrating Equations 12.8 and 12.11, respectively, over the appropriate surface areas.

3. Any spectral component of the radiation rate can also be obtained using these procedures, if the spectral intensity I_λ is known.
4. Even though the intensity of the emitted radiation is independent of direction, the rate at which radiation is intercepted by the three surfaces differs significantly due to differences in the solid angles and projected areas. For example, consider moving surface A_4 to various θ_1 positions, keeping A_4 normal to the direction of radiation and r_4 constant at 0.5 m as shown in Figure (*a*) below.

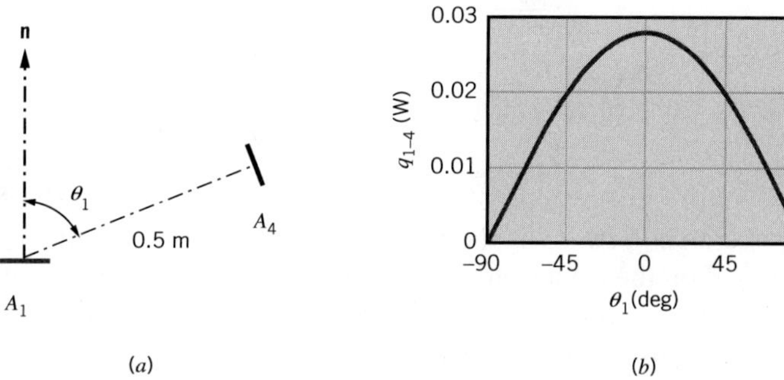

Under these conditions $\omega_{4-1} = 4.00 \times 10^{-3}$ sr is constant and

$$q_{1-4} = I \times A_1 \omega_{4-1} \cos\theta_1 = [7000 \text{ W/m}^2 \cdot \text{sr} \times 10^{-3} \text{ m}^2 \times 4.00 \times 10^{-3} \text{ sr}] \times \cos\theta_1$$

The energy that is emitted from A_1 and subsequently intercepted by A_4 is plotted in Figure (*b*) above. Consistent with our intuition, the energy intercepted by surface A_4 is maximum at $\theta_1 = 0°$ ($q_{1-4} = 28.0 \times 10^{-3}$ W) since an observer on A_4 would see the largest projected area of A_1. Also consistent with our intuition, the energy intercepted by A_4 is zero at $\theta_1 = \pm 90°$, even though the intensity of the radiation emitted from A_1 is independent of θ. At $\theta_1 = \pm 90°$, an observer on A_4 would be unable to see A_1 and hence would intercept none of the energy emitted from A_1. Many real surfaces emit radiation in a manner that is approximately diffuse.

12.3.3 Relation to Irradiation

The foregoing concepts may be extended to *incident* radiation (Figure 12.10). Such radiation may originate from emission and reflection occurring at other surfaces and will have spectral and directional distributions determined by the spectral intensity $I_{\lambda,i}(\lambda, \theta, \phi)$. This quantity is defined as the rate at which radiant energy of wavelength λ is incident from the (θ, ϕ) direction, per unit area of the *intercepting surface* normal to this direction, per unit solid angle about this direction, and per unit wavelength interval $d\lambda$ about λ.

The intensity of the incident radiation may be related to the irradiation, which encompasses radiation incident *from all directions.* The *spectral irradiation* G_λ(W/m$^2 \cdot \mu$m) is defined as the rate at which radiation of wavelength λ is incident on a surface, per unit area of the surface and per unit wavelength interval $d\lambda$ about λ. Accordingly,

$$G_\lambda(\lambda) = \int_0^{2\pi}\int_0^{\pi/2} I_{\lambda,i}(\lambda, \theta, \phi) \cos\theta \sin\theta \, d\theta \, d\phi \qquad (12.18)$$

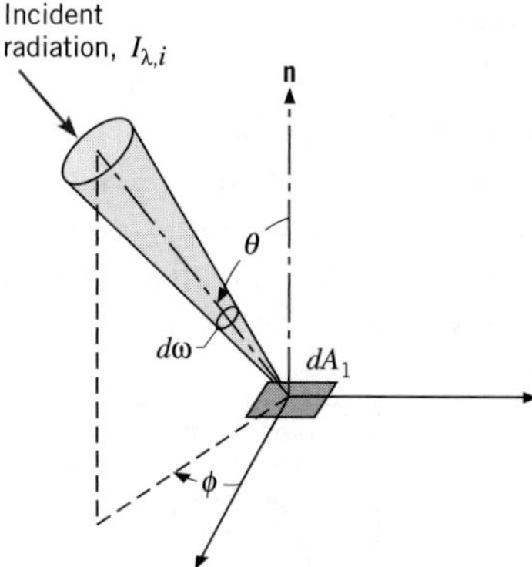

FIGURE 12.10 Directional nature of incident radiation.

where $\sin\theta\, d\theta\, d\phi$ is the unit solid angle. The $\cos\theta$ factor originates because G_λ is a flux based on the actual surface area, whereas $I_{\lambda,i}$ is defined in terms of the projected area. If the *total irradiation* G (W/m^2) represents the rate at which radiation is incident per unit area from all directions and at all wavelengths, it follows that

$$G = \int_0^\infty G_\lambda(\lambda)\, d\lambda \tag{12.19}$$

or from Equation 12.18

$$G = \int_0^\infty \int_0^{2\pi} \int_0^{\pi/2} I_{\lambda,i}(\lambda, \theta, \phi) \cos\theta \sin\theta\, d\theta\, d\phi\, d\lambda \tag{12.20}$$

The total irradiation was first introduced in Section 1.2.3 and again in Table 12.1. If the incident radiation is *diffuse*, $I_{\lambda,i}$ is independent of θ and ϕ and it follows that

$$G_\lambda(\lambda) = \pi I_{\lambda,i}(\lambda) \tag{12.21}$$

and

$$G = \pi I_i \tag{12.22}$$

EXAMPLE 12.2

The spectral distribution of surface irradiation is as follows:

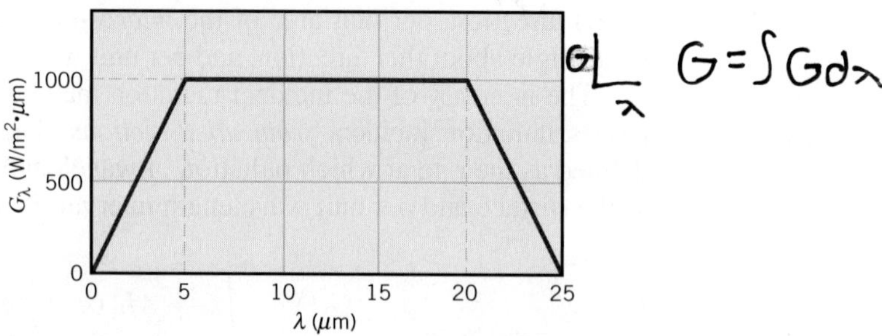

What is the total irradiation?

SOLUTION

Known: Spectral distribution of surface irradiation.

Find: Total irradiation.

Analysis: The total irradiation may be obtained from Equation 12.19.

$$G = \int_0^\infty G_\lambda \, d\lambda$$

The integral is readily evaluated by breaking it into parts. That is,

$$G = \int_0^{5\,\mu\text{m}} G_\lambda \, d\lambda + \int_{5\,\mu\text{m}}^{20\,\mu\text{m}} G_\lambda \, d\lambda + \int_{20\,\mu\text{m}}^{25\,\mu\text{m}} G_\lambda \, d\lambda + \int_{25\,\mu\text{m}}^{\infty} G_\lambda \, d\lambda$$

Hence

$$\begin{aligned} G &= \tfrac{1}{2}(1000\ \text{W/m}^2 \cdot \mu\text{m})(5 - 0)\ \mu\text{m} + (1000\ \text{W/m}^2 \cdot \mu\text{m})(20 - 5)\ \mu\text{m} \\ &\quad + \tfrac{1}{2}(1000\ \text{W/m}^2 \cdot \mu\text{m})(25 - 20)\ \mu\text{m} + 0 \\ &= (2500 + 15{,}000 + 2500)\ \text{W/m}^2 \\ G &= 20{,}000\ \text{W/m}^2 \end{aligned}$$

◁

Comments: Generally, radiation sources do not provide such a regular spectral distribution for the irradiation. However, the procedure of computing the total irradiation from knowledge of the spectral distribution remains the same, although evaluation of the integral is likely to involve more detail.

12.3.4 Relation to Radiosity for an Opaque Surface

As discussed in Section 12.2, the radiosity accounts for *all* the radiant energy leaving a surface. Since this radiation includes the *reflected* portion of the irradiation, as well as direct emission (Figure 12.5*b*), the radiosity is generally different from the emissive power. The *spectral radiosity* J_λ (W/m$^2 \cdot \mu$m) represents the rate at which radiation of wavelength λ leaves a unit area of the surface, per unit wavelength interval $d\lambda$ about λ. Since it accounts for radiation leaving in all directions, it is related to the intensity associated with emission and reflection, $I_{\lambda,e+r}(\lambda, \theta, \phi)$, by the expression

$$J_\lambda(\lambda) = \int_0^{2\pi} \int_0^{\pi/2} I_{\lambda,e+r}(\lambda, \theta, \phi) \cos\theta \sin\theta \, d\theta \, d\phi \tag{12.23}$$

Hence the *total radiosity* J (W/m^2) associated with the entire spectrum is

$$J = \int_0^\infty J_\lambda(\lambda) \, d\lambda \tag{12.24}$$

or

$$J = \int_0^\infty \int_0^{2\pi} \int_0^{\pi/2} I_{\lambda,e+r}(\lambda, \theta, \phi) \cos\theta \sin\theta \, d\theta \, d\phi \, d\lambda \tag{12.25}$$

This quantity is included in Table 12.1. If the surface is both a *diffuse reflector* and a *diffuse emitter,* $I_{\lambda,e+r}$ is independent of θ and ϕ, and it follows that

$$J_\lambda(\lambda) = \pi I_{\lambda,e+r} \tag{12.26}$$

and

$$J = \pi I_{e+r} \tag{12.27}$$

Again, note that the radiation flux, in this case the radiosity, is based on the actual surface area, while the intensity is based on the projected area.

12.3.5 Relation to the Net Radiative Flux for an Opaque Surface

As expressed in Equation 12.5, the net radiative flux from an opaque surface is equal to the difference between the outgoing radiosity J and the incoming irradiation G. From Equations 12.20 and 12.25, Equation 12.5 may be written in terms of the intensities associated with emission, reflection, and irradiation as

$$\begin{aligned} q''_{\text{rad}} = &\int_0^\infty \int_0^{2\pi} \int_0^{\pi/2} I_{\lambda,e+r}(\lambda, \theta, \phi) \cos\theta \sin\theta \, d\theta \, d\phi \, d\lambda \\ &- \int_0^\infty \int_0^{2\pi} \int_0^{\pi/2} I_{\lambda,i}(\lambda, \theta, \phi) \cos\theta \sin\theta \, d\theta \, d\phi \, d\lambda \end{aligned} \tag{12.28}$$

Hence the net radiative heat flux can be determined if the various intensities are known. The formal integration of Equation 12.28 is sometimes carried out in practice but will not be performed here. Rather, as will become evident in Sections 12.4 through 12.7, evaluation of the net radiative flux can be simplified by expressing the various intensities in terms of the intensity associated with a perfectly emitting and absorbing surface, the blackbody, and the emissivity, absorptivity, and reflectivity of the surface.

12.4 *Blackbody Radiation*

To evaluate the emissive power, irradiation, radiosity, or net radiative heat flux of a real opaque surface, we must quantify the spectral intensities used in Equations 12.13, 12.18, 12.23, and 12.28. To do so, it is useful to first introduce the concept of a *blackbody*.

1. *A blackbody absorbs all incident radiation, regardless of wavelength and direction.*
2. *For a prescribed temperature and wavelength, no surface can emit more energy than a blackbody.*
3. *Although the radiation emitted by a blackbody is a function of wavelength and temperature, it is independent of direction. That is, the blackbody is a diffuse emitter.*

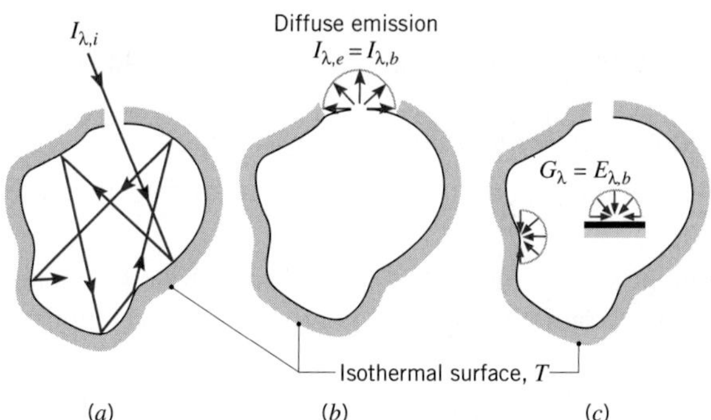

FIGURE 12.11 Characteristics of an isothermal blackbody cavity. (*a*) Complete absorption. (*b*) Diffuse emission from an aperture. (*c*) Diffuse irradiation of interior surfaces.

As the perfect absorber and emitter, the blackbody serves as a *standard* against which the radiative properties of actual surfaces may be compared.

Although closely approximated by some surfaces, it is important to note that no surface has precisely the properties of a blackbody. The closest approximation is achieved by a *cavity* whose inner surface is at a uniform temperature. If radiation enters the cavity through a small aperture (Figure 12.11*a*), it is likely to experience many reflections before reemergence. Since some radiation is absorbed by the inner surface upon each reflection, it is eventually almost entirely absorbed by the cavity, and blackbody behavior is approximated. From thermodynamic principles it may then be argued that radiation leaving the aperture depends only on the surface temperature and corresponds to blackbody emission (Figure 12.11*b*). Since blackbody emission is diffuse, the spectral intensity $I_{\lambda,b}$ of radiation leaving the cavity is independent of direction. Moreover, since the radiation field in the cavity, which is the cumulative effect of emission and reflection from the cavity surface, must be of the same form as the radiation emerging from the aperture, it also follows that a blackbody radiation field exists within the cavity. Accordingly, any small surface in the cavity (Figure 12.11*c*) experiences irradiation for which $G_\lambda = E_{\lambda,b}(\lambda, T)$. This surface is diffusely irradiated, regardless of its orientation. *Blackbody radiation exists within the cavity irrespective of whether the cavity surface is highly reflecting or absorbing.*

12.4.1 The Planck Distribution

The blackbody spectral intensity is well known, having first been determined by Planck [1]. It is

$$I_{\lambda,b}(\lambda,T) = \frac{2hc_o^2}{\lambda^5[\exp(hc_o/\lambda k_B T) - 1]} \tag{12.29}$$

where $h = 6.626 \times 10^{-34}$ J·s and $k_B = 1.381 \times 10^{-23}$ J/K are the universal Planck and Boltzmann constants, respectively, $c_o = 2.998 \times 10^8$ m/s is the speed of light in vacuum, and T is the *absolute* temperature of the blackbody (K). Since the blackbody is a diffuse emitter, it follows from Equation 12.16 that its spectral emissive power is

$$E_{\lambda,b}(\lambda,T) = \pi I_{\lambda,b}(\lambda,T) = \frac{C_1}{\lambda^5[\exp(C_2/\lambda T) - 1]} \tag{12.30}$$

where the first and second radiation constants are $C_1 = 2\pi hc_o^2 = 3.742 \times 10^8\ \text{W}\cdot\mu\text{m}^4/\text{m}^2$ and $C_2 = (hc_o/k_B) = 1.439 \times 10^4\ \mu\text{m}\cdot\text{K}$.

Equation 12.30, known as the *Planck distribution,* or *Planck's law*, is plotted in Figure 12.12 for selected temperatures. Several important features should be noted.

1. The emitted radiation varies *continuously* with wavelength.[1]
2. At any wavelength the magnitude of the emitted radiation increases with increasing temperature.
3. The spectral region in which the radiation is concentrated depends on temperature, with *comparatively* more radiation appearing at shorter wavelengths as the temperature increases.
4. A significant fraction of the radiation emitted by the sun, which may be approximated as a blackbody at 5800 K, is in the visible region of the spectrum. In contrast, for $T \lesssim 800$ K, emission is predominantly in the infrared region of the spectrum and is not visible to the eye.

12.4.2 Wien's Displacement Law

From Figure 12.12 we see that the blackbody spectral distribution has a maximum and that the corresponding wavelength λ_{max} depends on temperature. The nature of this dependence may be obtained by differentiating Equation 12.30 with respect to λ and setting the result equal to zero. In so doing, we obtain

$$\lambda_{max} T = C_3 \tag{12.31}$$

where the third radiation constant is $C_3 = 2898\ \mu\text{m}\cdot\text{K}$.

Equation 12.31 is known as *Wien's displacement law,* and the locus of points described by the law is plotted as the dashed line of Figure 12.12. According to this result, the maximum spectral emissive power is displaced to shorter wavelengths with increasing temperature. This emission is in the middle of the visible spectrum ($\lambda \approx 0.50\ \mu$m) for solar radiation, since the sun emits approximately as a blackbody at 5800 K. For a blackbody at 1000 K, peak emission occurs at 2.90 μm, with some of the emitted radiation appearing visible as red light. With increasing temperature, shorter wavelengths become more prominent, until eventually significant emission occurs over the entire visible spectrum. For example, a tungsten filament lamp operating at 2900 K ($\lambda_{max} = 1\ \mu$m) emits white light, although most of the emission remains in the IR region.

12.4.3 The Stefan–Boltzmann Law

Substituting the Planck distribution, Equation 12.30, into Equation 12.14, the total emissive power of a blackbody E_b may be expressed as

$$E_b = \int_0^\infty \frac{C_1}{\lambda^5[\exp(C_2/\lambda T) - 1]}\, d\lambda$$

---

[1]The *continuous* nature of blackbody emission can be determined *only* by considering the *discontinuous* energy states of atomic matter. Planck's derivation of the blackbody intensity distribution is one of the most important discoveries in quantum physics [2].

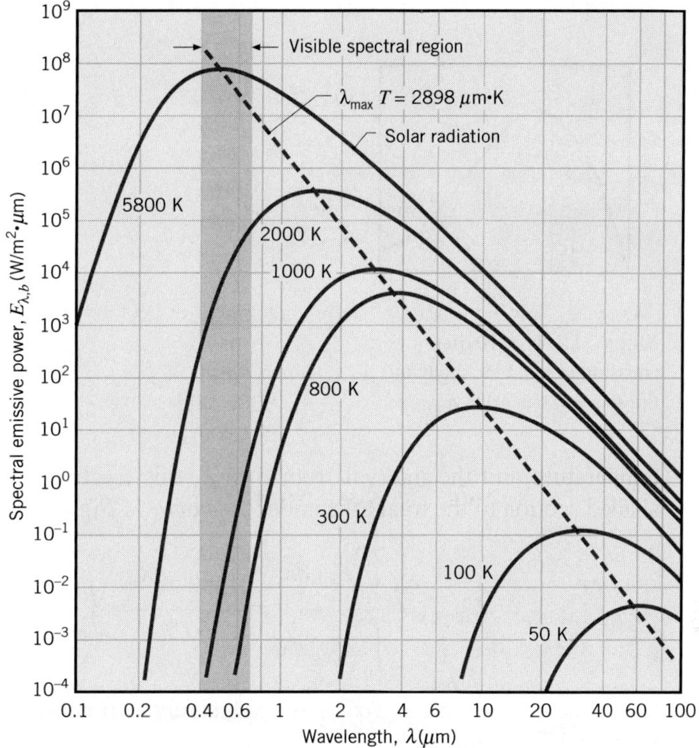

FIGURE 12.12 Spectral blackbody emissive power.

Performing the integration, it may be shown that

$$E_b = \sigma T^4 \tag{12.32}$$

where the *Stefan–Boltzmann* constant, which depends on C_1 and C_2, has the numerical value

$$\sigma = 5.670 \times 10^{-8}\ \text{W/m}^2 \cdot \text{K}^4$$

This simple, yet important, result is termed the *Stefan–Boltzmann law.* It enables calculation of the amount of radiation emitted in all directions and over all wavelengths simply from knowledge of the temperature of the blackbody. Because this emission is diffuse, it follows from Equation 12.17 that the total intensity associated with blackbody emission is

$$I_b = \frac{E_b}{\pi} \tag{12.33}$$

12.4.4 Band Emission

To account for spectral effects, it is often necessary to know the fraction of the total emission from a blackbody that is in a certain wavelength interval or *band.* For a prescribed

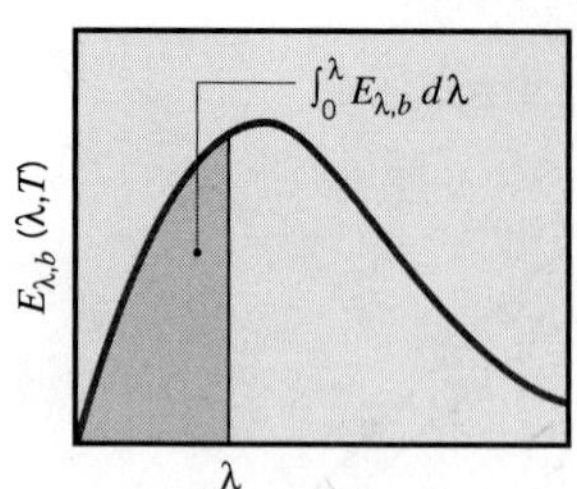

FIGURE 12.13 Radiation emission from a blackbody in the spectral band 0 to λ.

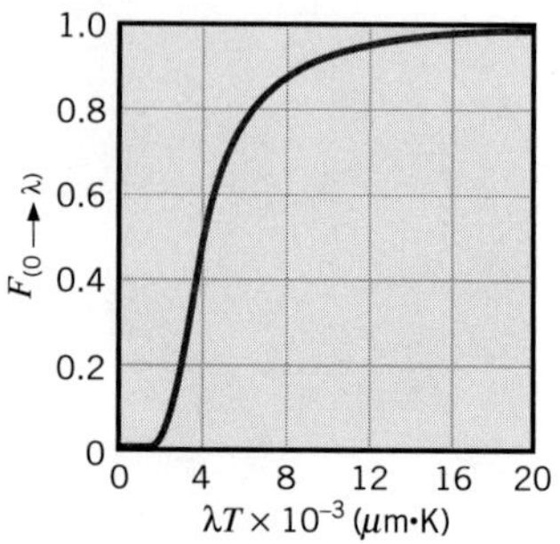

FIGURE 12.14 Fraction of the total blackbody emission in the spectral band from 0 to λ as a function of λT.

temperature and the interval from 0 to λ, this fraction is determined by the ratio of the shaded section to the total area under the curve of Figure 12.13. Hence

$$F_{(0\to\lambda)} \equiv \frac{\int_0^\lambda E_{\lambda,b}\, d\lambda}{\int_0^\infty E_{\lambda,b}\, d\lambda} = \frac{\int_0^\lambda E_{\lambda,b}\, d\lambda}{\sigma T^4} = \int_0^{\lambda T} \frac{E_{\lambda,b}}{\sigma T^5}\, d(\lambda T) = f(\lambda T) \tag{12.34}$$

Since the integrand $(E_{\lambda,b}/\sigma T^5)$ is exclusively a function of the wavelength–temperature product λT, the integral of Equation 12.34 may be evaluated to obtain $F_{(0\to\lambda)}$ as a function of only λT. The results are presented in Table 12.2 and Figure 12.14. They may also be used to obtain the fraction of the radiation between any two wavelengths λ_1 and λ_2, since

$$F_{(\lambda_1\to\lambda_2)} = \frac{\int_0^{\lambda_2} E_{\lambda,b}\, d\lambda - \int_0^{\lambda_1} E_{\lambda,b}\, d\lambda}{\sigma T^4} = F_{(0\to\lambda_2)} - F_{(0\to\lambda_1)} \tag{12.35}$$

TABLE 12.2 Blackbody Radiation Functions

λT (μm·K)	$F_{(0\to\lambda)}$	$I_{\lambda,b}(\lambda,T)/\sigma T^5$ (μm·K·sr)$^{-1}$	$\dfrac{I_{\lambda,b}(\lambda,T)}{I_{\lambda,b}(\lambda_{max},T)}$
200	0.000000	0.375034×10^{-27}	0.000000
400	0.000000	0.490335×10^{-13}	0.000000
600	0.000000	0.104046×10^{-8}	0.000014
800	0.000016	0.991126×10^{-7}	0.001372
1,000	0.000321	0.118505×10^{-5}	0.016406
1,200	0.002134	0.523927×10^{-5}	0.072534
1,400	0.007790	0.134411×10^{-4}	0.186082
1,600	0.019718	0.249130	0.344904
1,800	0.039341	0.375568	0.519949
2,000	0.066728	0.493432	0.683123
2,200	0.100888	0.589649×10^{-4}	0.816329
2,400	0.140256	0.658866	0.912155
2,600	0.183120	0.701292	0.970891
2,800	0.227897	0.720239	0.997123
2,898	0.250108	0.722318×10^{-4}	1.000000

TABLE 12.2 *Continued*

λT ($\mu m \cdot K$)	$F_{(0 \to \lambda)}$	$I_{\lambda,b}(\lambda, T)/\sigma T^5$ $(\mu m \cdot K \cdot sr)^{-1}$	$\frac{I_{\lambda,b}(\lambda, T)}{I_{\lambda,b}(\lambda_{max}, T)}$
3,000	0.273232	0.720254×10^{-4}	0.997143
3,200	0.318102	0.705974	0.977373
3,400	0.361735	0.681544	0.943551
3,600	0.403607	0.650396	0.900429
3,800	0.443382	0.615225×10^{-4}	0.851737
4,000	0.480877	0.578064	0.800291
4,200	0.516014	0.540394	0.748139
4,400	0.548796	0.503253	0.696720
4,600	0.579280	0.467343	0.647004
4,800	0.607559	0.433109	0.599610
5,000	0.633747	0.400813	0.554898
5,200	0.658970	0.370580×10^{-4}	0.513043
5,400	0.680360	0.342445	0.474092
5,600	0.701046	0.316376	0.438002
5,800	0.720158	0.292301	0.404671
6,000	0.737818	0.270121	0.373965
6,200	0.754140	0.249723×10^{-4}	0.345724
6,400	0.769234	0.230985	0.319783
6,600	0.783199	0.213786	0.295973
6,800	0.796129	0.198008	0.274128
7,000	0.808109	0.183534	0.254090
7,200	0.819217	0.170256×10^{-4}	0.235708
7,400	0.829527	0.158073	0.218842
7,600	0.839102	0.146891	0.203360
7,800	0.848005	0.136621	0.189143
8,000	0.856288	0.127185	0.176079
8,500	0.874608	0.106772×10^{-4}	0.147819
9,000	0.890029	0.901463×10^{-5}	0.124801
9,500	0.903085	0.765338	0.105956
10,000	0.914199	0.653279×10^{-5}	0.090442
10,500	0.923710	0.560522	0.077600
11,000	0.931890	0.483321	0.066913
11,500	0.939959	0.418725	0.057970
12,000	0.945098	0.364394×10^{-5}	0.050448
13,000	0.955139	0.279457	0.038689
14,000	0.962898	0.217641	0.030131
15,000	0.969981	0.171866×10^{-5}	0.023794
16,000	0.973814	0.137429	0.019026
18,000	0.980860	0.908240×10^{-6}	0.012574
20,000	0.985602	0.623310	0.008629
25,000	0.992215	0.276474	0.003828
30,000	0.995340	0.140469×10^{-6}	0.001945
40,000	0.997967	0.473891×10^{-7}	0.000656
50,000	0.998953	0.201605	0.000279
75,000	0.999713	0.418597×10^{-8}	0.000058
100,000	0.999905	0.135752	0.000019

Additional blackbody functions are listed in the third and fourth columns of Table 12.2. The third column facilitates calculation of the spectral intensity for a prescribed wavelength and temperature. In lieu of computing this quantity from Equation 12.29, it may be obtained by simply multiplying the tabulated value of $I_{\lambda,b}/\sigma T^5$ by σT^5. The fourth column is used to obtain a quick estimate of the ratio of the spectral intensity at any wavelength to that at λ_{max}.

EXAMPLE 12.3

Determine an expression for the net radiative heat flux at the surface of the small solid object of Figure 12.1 in terms of the surface and surroundings temperatures and the Stefan–Boltzmann constant. The small object is a blackbody.

SOLUTION

Known: Surface temperature of a small blackbody, T_s, and the surroundings temperature, T_{sur}.

Find: Expression for the net radiative flux at the surface of the small object, q''_{rad}.

Assumptions: Small object experiences blackbody irradiation.

Schematic:

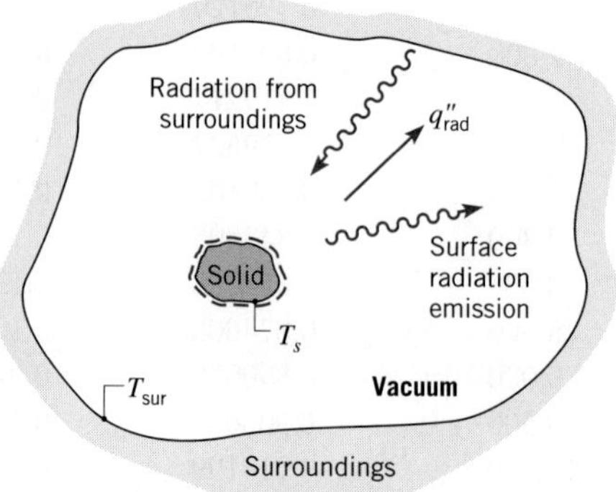

Analysis: Since none of the irradiation is reflected from the small object, Equation 12.28 may be written as

$$q''_{rad} = \int_0^\infty \int_0^{2\pi} \int_0^{\pi/2} I_{\lambda,e}(\lambda, \theta, \phi) \cos\theta \sin\theta \, d\theta \, d\phi \, d\lambda - \int_0^\infty \int_0^{2\pi} \int_0^{\pi/2} I_{\lambda,i}(\lambda, \theta, \phi) \cos\theta \sin\theta \, d\theta \, d\phi \, d\lambda \tag{1}$$

The intensity emitted by the small object corresponds to that of a blackbody. Hence

$$I_{\lambda,e}(\lambda, \theta, \phi) = I_{\lambda,b}(\lambda, T_s) \tag{2}$$

The intensity corresponding to the irradiation is also black. Therefore

$$I_{\lambda,i}(\lambda, \theta, \phi) = I_{\lambda,b}(\lambda, T_{sur}) \tag{3}$$

Since the blackbody intensity is diffuse, it is independent of angles θ and ϕ. Therefore, substituting Equations 2 and 3 into Equation 1 yields

$$q''_{\text{rad}} = \int_0^{2\pi}\int_0^{\pi/2} \cos\theta \sin\theta \, d\theta \, d\phi \times \int_0^{\infty} I_{\lambda,b}(\lambda, T_{\text{sur}})\, d\lambda$$

$$-\int_0^{2\pi}\int_0^{\pi/2} \cos\theta \sin\theta \, d\theta \, d\phi \times \int_0^{\infty} I_{\lambda,b}(\lambda, T_s)\, d\lambda$$

$$= \pi\left[\int_0^{\infty} I_{\lambda,b}(\lambda, T_{\text{sur}})\, d\lambda - \int_0^{\infty} I_{\lambda,b}(\lambda, T_s)\, d\lambda\right]$$

Substituting from Equations 12.32 and 12.33 yields

$$q''_{\text{rad}} = \sigma(T_s^4 - T_{\text{sur}}^4) \quad \triangleleft$$

which is identical to Equation 1.7 with $\varepsilon = 1$.

EXAMPLE 12.4

Consider a large isothermal enclosure that is maintained at a uniform temperature of 2000 K. Calculate the emissive power of the radiation that emerges from a small aperture on the enclosure surface. What is the wavelength λ_1 below which 10% of the emission is concentrated? What is the wavelength λ_2 above which 10% of the emission is concentrated? Determine the maximum spectral emissive power and the wavelength at which this emission occurs. What is the irradiation incident on a small object placed inside the enclosure?

SOLUTION

Known: Large isothermal enclosure at uniform temperature.

Find:

1. Emissive power of a small aperture on the enclosure.
2. Wavelengths below which and above which 10% of the radiation is concentrated.
3. Spectral emissive power and wavelength associated with maximum emission.
4. Irradiation on a small object inside the enclosure.

Schematic:

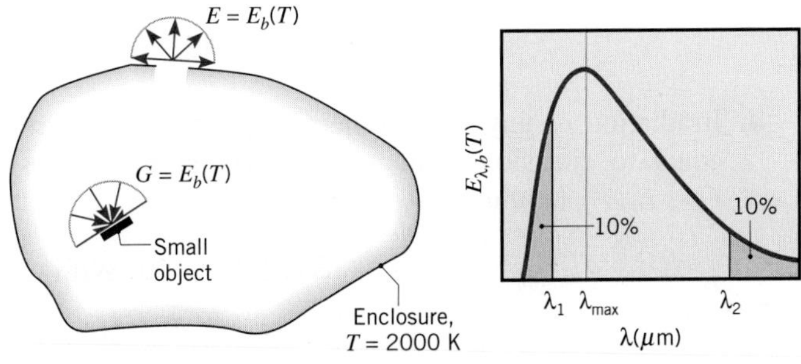

Assumptions: Areas of aperture and object are very small relative to enclosure surface.

Analysis:

1. Emission from the aperture of any isothermal enclosure will have the characteristics of blackbody radiation. Hence, from Equation 12.32,

$$E = E_b(T) = \sigma T^4 = 5.670 \times 10^{-8}\ \text{W/m}^2 \cdot \text{K}^4 (2000\ \text{K})^4$$
$$E = 9.07 \times 10^5\ \text{W/m}^2 \qquad \triangleleft$$

2. The wavelength λ_1 corresponds to the upper limit of the spectral band $(0 \rightarrow \lambda_1)$ containing 10% of the emitted radiation. With $F_{(0\rightarrow\lambda_1)} = 0.10$ it follows from Table 12.2 that $\lambda_1 T = 2195\ \mu\text{m} \cdot \text{K}$. Hence

$$\lambda_1 = 1.1\ \mu\text{m} \qquad \triangleleft$$

The wavelength λ_2 corresponds to the lower limit of the spectral band $(\lambda_2 \rightarrow \infty)$ containing 10% of the emitted radiation. With

$$F_{(\lambda_2\rightarrow\infty)} = 1 - F_{(0\rightarrow\lambda_2)} = 0.1$$
$$F_{(0\rightarrow\lambda_2)} = 0.9$$

it follows from Table 12.2 that $\lambda_2 T = 9382\ \mu\text{m} \cdot \text{K}$. Hence

$$\lambda_2 = 4.69\ \mu\text{m} \qquad \triangleleft$$

3. From Wien's displacement law, Equation 12.31, $\lambda_{\text{max}} T = 2898\ \mu\text{m} \cdot \text{K}$. Hence

$$\lambda_{\text{max}} = 1.45\ \mu\text{m} \qquad \triangleleft$$

The spectral emissive power associated with this wavelength may be computed from Equation 12.30 or from the third column of Table 12.2. For $\lambda_{\text{max}} T = 2898\ \mu\text{m} \cdot \text{K}$ it follows from Table 12.2 that

$$I_{\lambda,b}(1.45\ \mu\text{m}, T) = 0.722 \times 10^{-4} \sigma T^5$$

Hence

$$I_{\lambda,b}(1.45\ \mu\text{m}, 2000\ \text{K}) = 0.722 \times 10^{-4}\ (\mu\text{m} \cdot \text{K} \cdot \text{sr})^{-1} \times 5.67 \times 10^{-8}\ \text{W/m}^2 \cdot \text{K}^4\ (2000\ \text{K})^5$$
$$I_{\lambda,b}(1.45\ \mu\text{m}, 2000\ \text{K}) = 1.31 \times 10^5\ \text{W/m}^2 \cdot \text{sr} \cdot \mu\text{m}$$

Since the emission is diffuse, it follows from Equation 12.16 that

$$E_{\lambda,b} = \pi I_{\lambda,b} = 4.12 \times 10^5\ \text{W/m}^2 \cdot \mu\text{m} \qquad \triangleleft$$

4. Irradiation of any small object inside the enclosure may be approximated as being equal to emission from a blackbody at the enclosure surface temperature. Hence $G = E_b(T)$, in which case

$$G = 9.07 \times 10^5\ \text{W/m}^2 \qquad \triangleleft$$

EXAMPLE 12.5

A surface emits as a blackbody at 1500 K. What is the rate per unit area (W/m^2) at which it emits radiation over all directions corresponding to $0° \leq \theta \leq 60°$ and over the wavelength interval $2\ \mu m \leq \lambda \leq 4\ \mu m$?

SOLUTION

Known: Temperature of a surface that emits as a blackbody.

Find: Rate of emission per unit area over all directions between $\theta = 0°$ and $60°$ and over all wavelengths between $\lambda = 2$ and $4\ \mu m$.

Schematic:

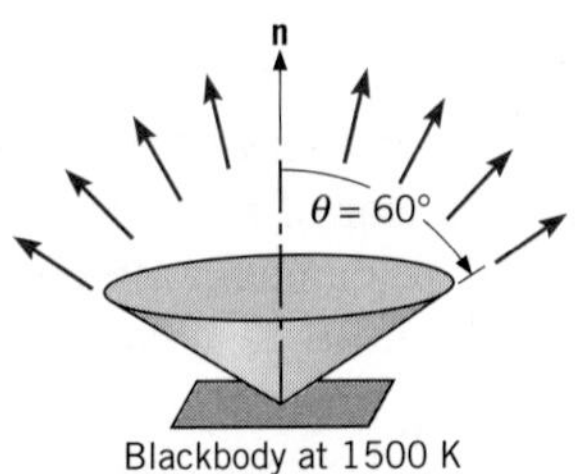

Assumptions: Surface emits as a blackbody.

Analysis: The desired emission may be inferred from Equation 12.15, with the limits of integration restricted as follows:

$$\Delta E = \int_2^4 \int_0^{2\pi} \int_0^{\pi/3} I_{\lambda,b} \cos\theta \sin\theta \, d\theta \, d\phi \, d\lambda$$

or, since a blackbody emits diffusely,

$$\Delta E = \int_2^4 I_{\lambda,b} \left(\int_0^{2\pi} \int_0^{\pi/3} \cos\theta \sin\theta \, d\theta \, d\phi \right) d\lambda$$

$$\Delta E = \int_2^4 I_{\lambda,b} \left(2\pi \frac{\sin^2\theta}{2} \bigg|_0^{\pi/3} \right) d\lambda = 0.75 \int_2^4 \pi I_{\lambda,b} \, d\lambda$$

Substituting from Equation 12.16 and multiplying and dividing by E_b, this result may be put in a form that allows for use of Table 12.2 in evaluating the spectral integration. In particular,

$$\Delta E = 0.75 E_b \int_2^4 \frac{E_{\lambda,b}}{E_b} \, d\lambda = 0.75 E_b \, [F_{(0\rightarrow 4)} - F_{(0\rightarrow 2)}]$$

where from Table 12.2

$$\lambda_1 T = 2\ \mu m \times 1500\ K = 3000\ \mu m \cdot K: \qquad F_{(0\rightarrow 2)} = 0.273$$

$$\lambda_2 T = 4\ \mu m \times 1500\ K = 6000\ \mu m \cdot K: \qquad F_{(0\rightarrow 4)} = 0.738$$

Hence

$$\Delta E = 0.75(0.738 - 0.273)E_b = 0.75(0.465)E_b$$

From Equation 12.31, it then follows that

$$\Delta E = 0.75(0.465)5.67 \times 10^{-8}\ \text{W/m}^2 \cdot \text{K}^4\ (1500\ \text{K})^4 = 10^5\ \text{W/m}^2 \quad \triangleleft$$

Comments: The total, hemispherical emissive power is reduced by 25% and 53.5% due to the directional and spectral restrictions, respectively.

12.5 *Emission from Real Surfaces*

Having developed the notion of a blackbody to describe *ideal* surface behavior, we may now consider the behavior of *real* surfaces. Recall that the blackbody is an ideal emitter in the sense that no surface can emit more radiation than a blackbody at the same temperature. It is therefore convenient to designate the blackbody as a reference in describing emission from a real surface. A surface radiative property known as the *emissivity*[2] may then be defined as the *ratio* of the radiation emitted by the surface to the radiation emitted by a blackbody at the same temperature.

It is important to acknowledge that, in general, the spectral radiation emitted by a real surface differs from the Planck distribution (Figure 12.15*a*). Moreover, the directional distribution (Figure 12.15*b*) may be other than diffuse. Hence the emissivity may assume different values according to whether one is interested in emission at a given wavelength or in a given direction, or in integrated averages over wavelength and direction.

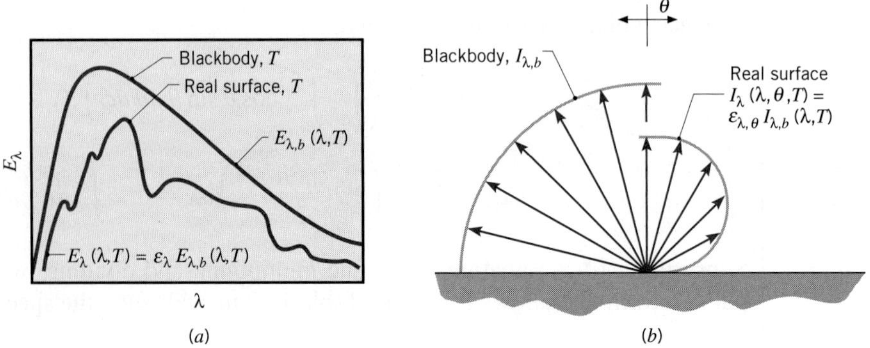

FIGURE 12.15 Comparison of blackbody and real surface emission. (*a*) Spectral distribution. (*b*) Directional distribution.

[2]In this text we use the *-ivity,* rather than the *-ance,* ending for material radiative properties (e.g., "emissivity" rather than "emittance"). Although efforts are being made to reserve the *-ivity* ending for optically smooth, uncontaminated surfaces, no such distinction is made in much of the literature, and none is made in this text.

The emissivity that accounts for emission over all wavelengths and in all directions is the *total, hemispherical* emissivity, which is the ratio of the total emissive power of a real surface, $E(T)$, to the total emissive power of a blackbody at the same temperature, $E_b(T)$. That is,

$$\varepsilon(T) \equiv \frac{E(T)}{E_b(T)} \tag{12.36}$$

If the total, hemispherical emissivity of a surface is known, it is a simple matter to express its emissive power in terms of the emissive power of a blackbody by combining Equation 12.36 with Equation 12.32, namely

$$E(T) = \varepsilon(T)E_b(T) = \varepsilon(T)\sigma T^4 \tag{12.37}$$

While Equation 12.37 is simple in form, its simplicity is deceptive in that $\varepsilon(T)$ itself depends on the spectral and directional characteristics of the surface emission. To develop an appropriate understanding of Equation 12.37, we define the *spectral, directional emissivity* $\varepsilon_{\lambda,\theta}(\lambda, \theta, \phi, T)$ of a surface at the temperature T as the ratio of the intensity of the radiation emitted at the wavelength λ and in the direction of θ and ϕ to the intensity of the radiation emitted by a blackbody at the same values of T and λ. Hence

$$\varepsilon_{\lambda,\theta}(\lambda, \theta, \phi, T) \equiv \frac{I_{\lambda,e}(\lambda, \theta, \phi, T)}{I_{\lambda,b}(\lambda, T)} \tag{12.38}$$

Note how the subscripts λ and θ designate interest in a specific wavelength and direction for the emissivity. In contrast, terms appearing within parentheses designate functional dependence on wavelength, direction, and/or temperature. The absence of directional variables in the parentheses of the denominator in Equation 12.38 implies that the intensity is independent of direction, which is, of course, a characteristic of blackbody emission. In like manner a *total, directional emissivity* ε_θ, which represents a spectral average of $\varepsilon_{\lambda,\theta}$, may be defined as

$$\varepsilon_\theta(\theta, \phi, T) \equiv \frac{I_e(\theta, \phi, T)}{I_b(T)} \tag{12.39}$$

For most engineering calculations, it is desirable to work with surface properties that represent directional averages. A *spectral, hemispherical emissivity* is therefore defined as

$$\varepsilon_\lambda(\lambda, T) \equiv \frac{E_\lambda(\lambda, T)}{E_{\lambda,b}(\lambda, T)} \tag{12.40}$$

It may be related to the directional emissivity $\varepsilon_{\lambda,\theta}$ by substituting the expression for the spectral emissive power, Equation 12.13, to obtain

$$\varepsilon_\lambda(\lambda, T) = \frac{\int_0^{2\pi}\int_0^{\pi/2} I_{\lambda,e}(\lambda, \theta, \phi, T)\cos\theta \sin\theta \, d\theta \, d\phi}{\int_0^{2\pi}\int_0^{\pi/2} I_{\lambda,b}(\lambda, T)\cos\theta \sin\theta \, d\theta \, d\phi}$$

In contrast to Equation 12.13, the temperature dependence of emission is now acknowledged. From Equation 12.38 and the fact that $I_{\lambda,b}$ is independent of θ and ϕ, it follows that

$$\varepsilon_\lambda(\lambda, T) = \frac{\int_0^{2\pi}\int_0^{\pi/2} \varepsilon_{\lambda,\theta}(\lambda, \theta, \phi, T) \cos\theta \sin\theta \, d\theta \, d\phi}{\int_0^{2\pi}\int_0^{\pi/2} \cos\theta \sin\theta \, d\theta \, d\phi} \tag{12.41}$$

Assuming $\varepsilon_{\lambda,\theta}$ to be independent of ϕ, which is a reasonable assumption for most surfaces, and evaluating the denominator, we obtain

$$\varepsilon_\lambda(\lambda, T) = 2\int_0^{\pi/2} \varepsilon_{\lambda,\theta}(\lambda, \theta, T) \cos\theta \sin\theta \, d\theta \tag{12.42}$$

The *total, hemispherical emissivity,* which represents an average over all possible directions and wavelengths, is defined in Equation 12.36. Substituting Equations 12.14 and 12.40 into Equation 12.36, it follows that

$$\varepsilon(T) = \frac{\int_0^{\infty} \varepsilon_\lambda(\lambda, T) E_{\lambda,b}(\lambda, T) \, d\lambda}{E_b(T)} \tag{12.43}$$

If the emissivities of a surface are known, it is a simple matter to compute its emission characteristics. For example, if $\varepsilon_\lambda(\lambda, T)$ is known, it may be used with Equations 12.30 and 12.40 to compute the spectral emissive power of the surface at any wavelength and temperature,

$$E_\lambda(\lambda, T) = \varepsilon_\lambda(\lambda, T) E_{\lambda,b}(\lambda, T) = \frac{C_1 \varepsilon_\lambda(\lambda, T)}{\lambda^5[\exp(C_2/\lambda T) - 1]} \tag{12.44}$$

As noted previously, if $\varepsilon(T)$ is known, it may be used to compute the emissive power of the surface at any temperature, as in Equation 12.37. Measurements have been performed to determine these properties for many different materials and surface coatings.

The directional emissivity of a *diffuse emitter* is a constant, independent of direction. However, although this condition is often a reasonable *approximation,* all surfaces exhibit some departure from diffuse behavior. Representative variations of ε_θ with θ are shown schematically in Figure 12.16 for conducting and nonconducting materials. For conductors

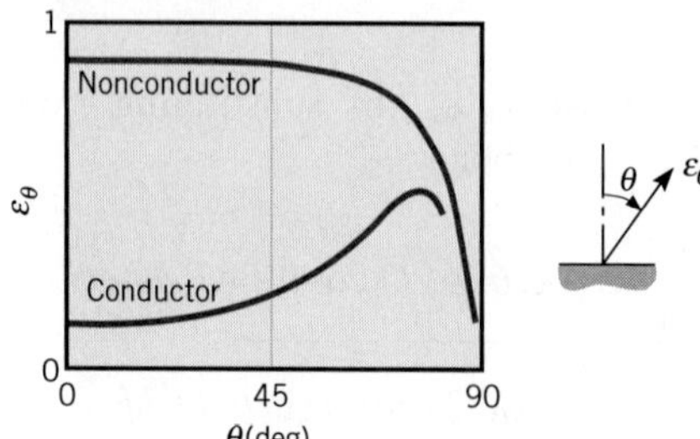

FIGURE 12.16 Representative directional distributions of the total, directional emissivity.

ε_θ is approximately constant over the range $\theta \lesssim 40°$, after which it increases with increasing θ but ultimately decays to zero. In contrast, for nonconductors ε_θ is approximately constant for $\theta \lesssim 70°$, beyond which it decreases sharply with increasing θ. One implication of these variations is that, although there are preferential directions for emission, the hemispherical emissivity ε will not differ markedly from the normal emissivity ε_n, corresponding to $\theta = 0$. In fact the ratio rarely falls outside the range $1.0 \leq (\varepsilon/\varepsilon_n) \leq 1.3$ for conductors and the range of $0.95 \leq (\varepsilon/\varepsilon_n) \leq 1.0$ for nonconductors. Hence to a reasonable approximation,

$$\varepsilon \approx \varepsilon_n \tag{12.45}$$

Note that, although the foregoing statements were made for the total emissivity, they also apply to spectral components.

Since the spectral distribution of emission from real surfaces departs from the Planck distribution (Figure 12.15*a*), we do not expect the value of the spectral emissivity ε_λ to be independent of wavelength. Representative spectral distributions of ε_λ are shown in Figure 12.17. The manner in which ε_λ varies with λ depends on whether the solid is a conductor or nonconductor, as well as on the nature of the surface coating.

Representative values of the total, normal emissivity ε_n are plotted in Figures 12.18 and 12.19 and listed in Table A.11. Several generalizations may be made.

1. The emissivity of metallic surfaces is generally small, achieving values as low as 0.02 for highly polished gold and silver.
2. The presence of oxide layers may significantly increase the emissivity of metallic surfaces. From Figure 12.18, contrast the values of 0.3 and 0.7 for stainless steel at 900 K, depending on whether it is polished or heavily oxidized.
3. The emissivity of nonconductors is comparatively large, generally exceeding 0.6.
4. The emissivity of conductors increases with increasing temperature; however, depending on the specific material, the emissivity of nonconductors may either increase or decrease with increasing temperature. Note that the variations of ε_n with T shown in

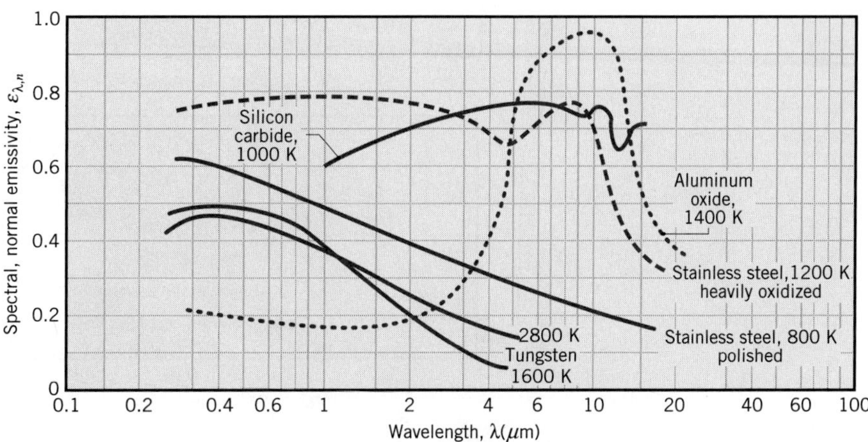

FIGURE 12.17 Spectral dependence of the spectral, normal emissivity $\varepsilon_{\lambda,n}$ of selected materials.

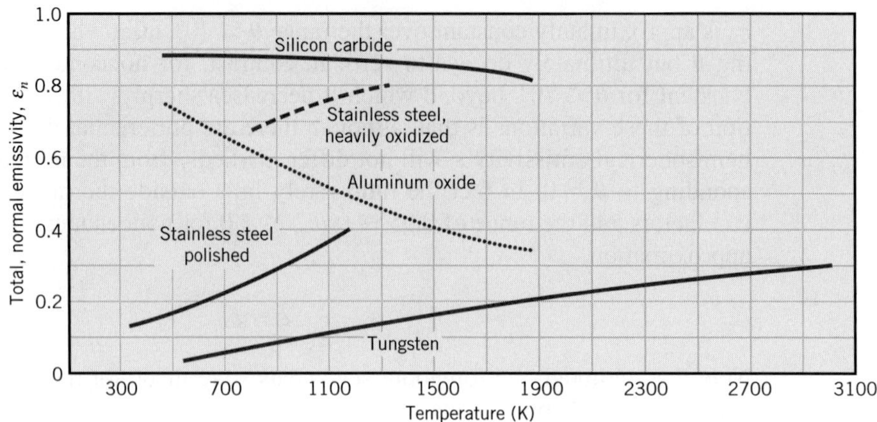

FIGURE 12.18 Temperature dependence of the total, normal emissivity ε_n of selected materials.

Figure 12.18 are consistent with the spectral distributions of $\varepsilon_{\lambda,n}$ shown in Figure 12.17. These trends follow from Equation 12.43. Although the spectral distribution of $\varepsilon_{\lambda,n}$ is approximately independent of temperature, there is proportionately more emission at lower wavelengths with increasing temperature. Hence, if $\varepsilon_{\lambda,n}$ increases with decreasing wavelength for a particular material, ε_n will increase with increasing temperature for that material.

It should be recognized that the emissivity depends strongly on the nature of the surface, which can be influenced by the method of fabrication, thermal cycling, and chemical reaction with its environment. More comprehensive surface emissivity compilations are available in the literature [3–6].

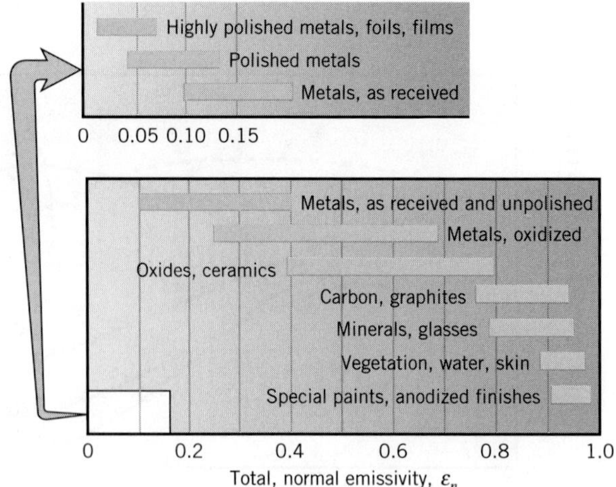

FIGURE 12.19 Representative values of the total, normal emissivity ε_n.

EXAMPLE 12.6

A diffuse surface at 1600 K has the spectral, hemispherical emissivity shown as follows.

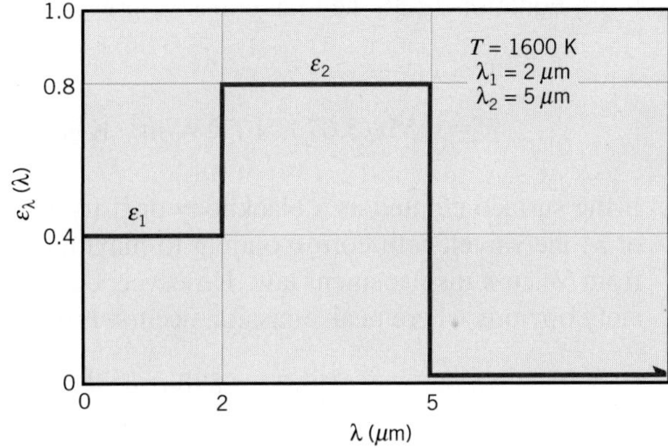

Determine the total, hemispherical emissivity and the total emissive power. At what wavelength will the spectral emissive power be a maximum?

SOLUTION

Known: Spectral, hemispherical emissivity of a diffuse surface at 1600 K.

Find:

1. Total, hemispherical emissivity.
2. Total emissive power.
3. Wavelength at which spectral emissive power will be a maximum.

Assumptions: Surface is a diffuse emitter.

Analysis:

1. The total, hemispherical emissivity is given by Equation 12.43, where the integration may be performed in parts as follows:

$$\varepsilon = \frac{\int_0^\infty \varepsilon_\lambda E_{\lambda,b}\,d\lambda}{E_b} = \frac{\varepsilon_1 \int_0^2 E_{\lambda,b}\,d\lambda}{E_b} + \frac{\varepsilon_2 \int_2^5 E_{\lambda,b}\,d\lambda}{E_b}$$

or

$$\varepsilon = \varepsilon_1 F_{(0\rightarrow 2\,\mu\text{m})} + \varepsilon_2[F_{(0\rightarrow 5\,\mu\text{m})} - F_{(0\rightarrow 2\,\mu\text{m})}]$$

From Table 12.2 we obtain

$$\lambda_1 T = 2\,\mu\text{m} \times 1600\,\text{K} = 3200\,\mu\text{m}\cdot\text{K}: \qquad F_{(0\rightarrow 2\,\mu\text{m})} = 0.318$$

$$\lambda_2 T = 5\,\mu\text{m} \times 1600\,\text{K} = 8000\,\mu\text{m}\cdot\text{K}: \qquad F_{(0\rightarrow 5\,\mu\text{m})} = 0.856$$

Hence

$$\varepsilon = 0.4 \times 0.318 + 0.8[0.856 - 0.318] = 0.558 \quad \triangleleft$$

2. From Equation 12.36 the total emissive power is

$$E = \varepsilon E_b = \varepsilon \sigma T^4$$

$$E = 0.558(5.67 \times 10^{-8}\ \text{W/m}^2 \cdot \text{K}^4)(1600\ \text{K})^4 = 207\ \text{kW/m}^2 \quad \triangleleft$$

3. If the surface emitted as a blackbody or if its emissivity were a constant, independent of λ, the wavelength corresponding to maximum spectral emission could be obtained from Wien's displacement law. However, because ε_λ varies with λ, it is not immediately obvious where peak emission occurs. From Equation 12.31 we know that

$$\lambda_{\text{max}} = \frac{2898\ \mu\text{m} \cdot \text{K}}{1600\ \text{K}} = 1.81\ \mu\text{m}$$

The spectral emissive power at this wavelength may be obtained by using Equation 12.40 with Table 12.2. That is,

$$E_\lambda(\lambda_{\text{max}}, T) = \varepsilon_\lambda(\lambda_{\text{max}}) E_{\lambda,b}(\lambda_{\text{max}}, T)$$

or, since the surface is a diffuse emitter,

$$E_\lambda(\lambda_{\text{max}}, T) = \pi \varepsilon_\lambda(\lambda_{\text{max}}) I_{\lambda,b}(\lambda_{\text{max}}, T)$$

$$= \pi \varepsilon_\lambda(\lambda_{\text{max}}) \frac{I_{\lambda,b}(\lambda_{\text{max}}, T)}{\sigma T^5} \times \sigma T^5$$

$$E_\lambda(1.81\ \mu\text{m}, 1600\ \text{K}) = \pi \times 0.4 \times 0.722 \times 10^{-4}\ (\mu\text{m} \cdot \text{K} \cdot \text{sr})^{-1} \times 5.67 \times 10^{-8}\ \text{W/m}^2 \cdot \text{K}^4 \times (1600\ \text{K})^5 = 54\ \text{kW/m}^2 \cdot \mu\text{m}$$

Since $\varepsilon_\lambda = 0.4$ from $\lambda = 0$ to $\lambda = 2\ \mu$m, the foregoing result provides the maximum spectral emissive power for the region $\lambda < 2\ \mu$m. However, with the change in ε_λ that occurs at $\lambda = 2\ \mu$m, the value of E_λ at $\lambda = 2\ \mu$m may be larger than that for $\lambda = 1.81\ \mu$m. To determine whether this is, in fact, the case, we compute

$$E_\lambda(\lambda_1, T) = \pi \varepsilon_\lambda(\lambda_1) \frac{I_{\lambda,b}(\lambda_1, T)}{\sigma T^5} \times \sigma T^5$$

where, for $\lambda_1 T = 3200\ \mu\text{m} \cdot \text{K}$, $[I_{\lambda,b}(\lambda_1, T)/\sigma T^5] = 0.706 \times 10^{-4}\ (\mu\text{m} \cdot \text{K} \cdot \text{sr})^{-1}$. Hence

$$E_\lambda(2\ \mu\text{m}, 1600\ \text{K}) = \pi \times 0.80 \times 0.706 \times 10^{-4}\ (\mu\text{m} \cdot \text{K} \cdot \text{sr})^{-1} \times 5.67 \times 10^{-8}\ \text{W/m}^2 \cdot \text{K}^4\ (1600\ \text{K})^5$$

$$E_\lambda(2\ \mu\text{m}, 1600\ \text{K}) = 105.5\ \text{kW/m}^2 \cdot \mu\text{m} > E_\lambda(1.81\ \mu\text{m}, 1600\ \text{K})$$

and peak emission occurs at

$$\lambda = \lambda_1 = 2\ \mu\text{m} \quad \triangleleft$$

Comments: For the prescribed spectral distribution of ε_λ, the spectral emissive power will vary with wavelength as shown.

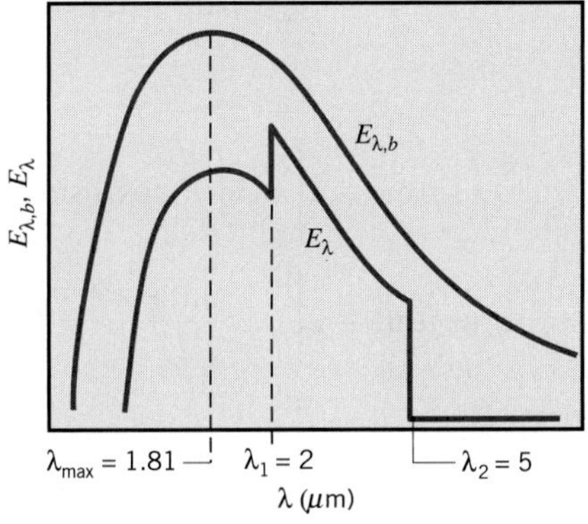

EXAMPLE 12.7

Measurements of the spectral, directional emissivity of a metallic surface at $T = 2000$ K and $\lambda = 1.0\ \mu$m yield a directional distribution that may be approximated as follows:

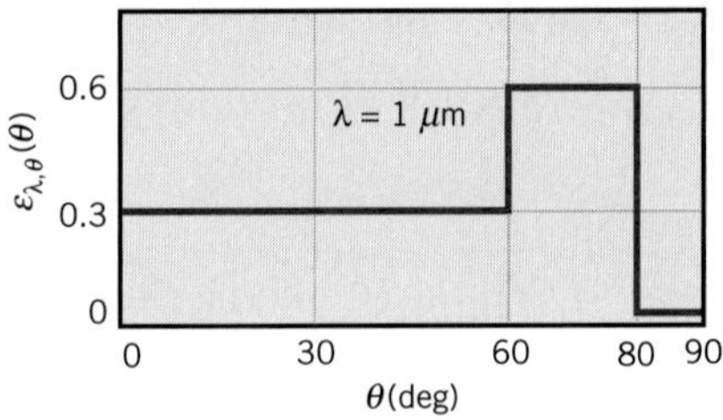

Determine corresponding values of the spectral, normal emissivity; the spectral, hemispherical emissivity; the spectral intensity of radiation emitted in the normal direction; and the spectral emissive power.

SOLUTION

Known: Directional distribution of $\varepsilon_{\lambda,\theta}$ at $\lambda = 1\ \mu$m for a metallic surface at 2000 K.

Find:

1. Spectral, normal emissivity $\varepsilon_{\lambda,n}$ and spectral, hemispherical emissivity ε_λ.
2. Spectral, normal intensity $I_{\lambda,n}$ and spectral emissive power E_λ.

Analysis:

1. From the measurement of $\varepsilon_{\lambda,\theta}$ at $\lambda = 1\ \mu$m, we see that

$$\varepsilon_{\lambda,n} = \varepsilon_{\lambda,\theta}(1\ \mu\text{m}, 0°) = 0.3 \qquad \triangleleft$$

From Equation 12.42, the spectral, hemispherical emissivity is

$$\varepsilon_\lambda(1\ \mu\text{m}) = 2\int_0^{\pi/2} \varepsilon_{\lambda,\theta} \cos\theta \sin\theta \, d\theta$$

or

$$\varepsilon_\lambda(1\ \mu\text{m}) = 2\left[0.3\int_0^{\pi/3} \cos\theta \sin\theta \, d\theta + 0.6\int_{\pi/3}^{4\pi/9} \cos\theta \sin\theta \, d\theta\right]$$

$$\varepsilon_\lambda(1\ \mu\text{m}) = 2\left[0.3 \left.\frac{\sin^2\theta}{2}\right|_0^{\pi/3} + 0.6 \left.\frac{\sin^2\theta}{2}\right|_{\pi/3}^{4\pi/9}\right]$$

$$= 2\left[\frac{0.3}{2}(0.75) + \frac{0.6}{2}(0.97 - 0.75)\right]$$

$$\varepsilon_\lambda(1\ \mu\text{m}) = 0.36 \qquad \triangleleft$$

2. From Equation 12.38 the spectral intensity of radiation emitted at $\lambda = 1\ \mu$m in the normal direction is

$$I_{\lambda,n}(1\ \mu\text{m}, 0°, 2000\ \text{K}) = \varepsilon_{\lambda,\theta}(1\ \mu\text{m}, 0°) I_{\lambda,b}(1\ \mu\text{m}, 2000\ \text{K})$$

where $\varepsilon_{\lambda,\theta}(1\ \mu\text{m}, 0°) = 0.3$ and $I_{\lambda,b}(1\ \mu\text{m}, 2000\ \text{K})$ may be obtained from Table 12.2. For $\lambda T = 2000\ \mu\text{m}\cdot\text{K}$, $(I_{\lambda,b}/\sigma T^5) = 0.493 \times 10^{-4}\ (\mu\text{m}\cdot\text{K}\cdot\text{sr})^{-1}$ and

$$I_{\lambda,b} = 0.493 \times 10^{-4}\ (\mu\text{m}\cdot\text{K}\cdot\text{sr})^{-1} \times 5.67 \times 10^{-8}\ \text{W/m}^2\cdot\text{K}^4\ (2000\ \text{K})^5$$

$$I_{\lambda,b} = 8.95 \times 10^4\ \text{W/m}^2\cdot\mu\text{m}\cdot\text{sr}$$

Hence

$$I_{\lambda,n}(1\ \mu\text{m}, 0°, 2000\ \text{K}) = 0.3 \times 8.95 \times 10^4\ \text{W/m}^2\cdot\mu\text{m}\cdot\text{sr}$$

$$I_{\lambda,n}(1\ \mu\text{m}, 0°, 2000\ \text{K}) = 2.69 \times 10^4\ \text{W/m}^2\cdot\mu\text{m}\cdot\text{sr} \qquad \triangleleft$$

From Equation 12.40 the spectral emissive power for $\lambda = 1\ \mu$m and $T = 2000$ K is

$$E_\lambda(1\ \mu\text{m}, 2000\ \text{K}) = \varepsilon_\lambda(1\ \mu\text{m}) E_{\lambda,b}(1\ \mu\text{m}, 2000\ \text{K})$$

where

$$E_{\lambda,b}(1\ \mu\text{m}, 2000\ \text{K}) = \pi I_{\lambda,b}(1\ \mu\text{m}, 2000\ \text{K})$$

$$E_{\lambda,b}(1\ \mu\text{m}, 2000\ \text{K}) = \pi\ \text{sr} \times 8.95 \times 10^4\ \text{W/m}^2\cdot\mu\text{m}\cdot\text{sr}$$

$$= 2.81 \times 10^5\ \text{W/m}^2\cdot\mu\text{m}$$

Hence

$$E_\lambda(1\ \mu\text{m}, 2000\ \text{K}) = 0.36 \times 2.81 \times 10^5\ \text{W/m}^2\cdot\mu\text{m}$$

or

$$E_\lambda(1\ \mu\text{m}, 2000\ \text{K}) = 1.01 \times 10^5\ \text{W/m}^2\cdot\mu\text{m} \qquad \triangleleft$$

12.6 Absorption, Reflection, and Transmission by Real Surfaces

In the preceding section, we learned that emission from a real surface is associated with a surface property termed the emissivity ε. To determine the net radiative heat flux from the surface, it is also necessary to consider properties that determine the absorption, reflection, and transmission of the irradiation. In Section 12.3.3 we defined the *spectral irradiation* G_λ (W/m$^2 \cdot \mu$m) as the rate at which radiation of wavelength λ is incident on a surface per unit area of the surface and per unit wavelength interval $d\lambda$ about λ. It may be incident from all possible directions, and it may originate from several different sources. The *total irradiation* G (W/m^2) encompasses all spectral contributions and may be evaluated from Equation 12.19.

In the most general situation the irradiation interacts with a *semitransparent medium,* such as a layer of water or a glass plate. As shown in Figure 12.20 for a spectral component of the irradiation, portions of this radiation may be *reflected, absorbed,* and *transmitted.* From a radiation balance on the medium, it follows that

$$G_\lambda = G_{\lambda,\text{ref}} + G_{\lambda,\text{abs}} + G_{\lambda,\text{tr}} \tag{12.46}$$

In general, determination of these components is complex, depending on the upper and lower surface conditions, the wavelength of the radiation, and the composition and thickness of the medium. Moreover, conditions may be strongly influenced by *volumetric* effects occurring within the medium.

In a simpler situation, which pertains to most engineering applications, the medium is *opaque* to the incident radiation. In this case, $G_{\lambda,\text{tr}} = 0$ and the remaining absorption and reflection processes may be treated as *surface phenomena.* That is, they are controlled by processes occurring within a fraction of a micrometer from the irradiated surface. It is therefore appropriate to speak of irradiation being absorbed and reflected *by the surface,* with the relative magnitudes of $G_{\lambda,\text{abs}}$ and $G_{\lambda,\text{ref}}$ depending on λ and the nature of the surface material. There is no net effect of the reflection process on the medium, while absorption has the effect of increasing the internal thermal energy of the medium.

It is interesting to note that surface absorption and reflection are responsible for our perception of *color.* Unless the surface is at a high temperature ($T_s \gtrsim 1000$ K), such that it is *incandescent,* color is in no way due to emission, which is concentrated in the IR region, and is hence imperceptible to the eye. Color is instead due to selective reflection and absorption of the visible portion of the irradiation that is incident from the sun or an artificial source of light. A shirt is "red" because it contains a pigment that preferentially absorbs the blue, green, and yellow components of the incident light. Hence the relative

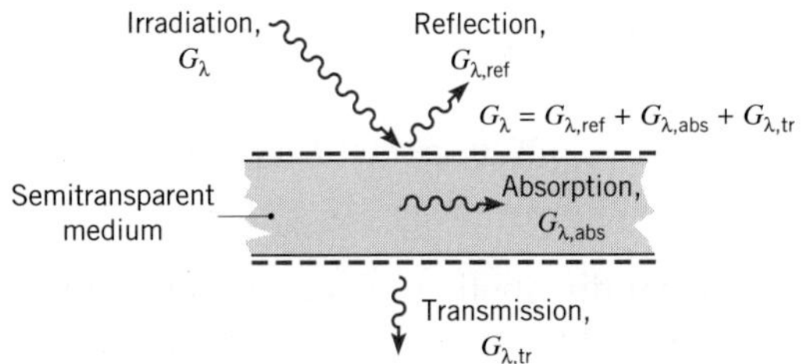

FIGURE 12.20 Spectral absorption, reflection, and transmission processes associated with a semitransparent medium.

contributions of these components to the reflected light, which is seen, is diminished, and the red component is dominant. Similarly, a leaf is "green" because its cells contain chlorophyll, a pigment that shows strong absorption in the blue and the red and preferential reflection in the green. A surface appears "black" if it absorbs all incident visible radiation, and it is "white" if it reflects this radiation. However, we must be careful how we interpret such *visual* effects. For a prescribed irradiation, the "color" of a surface may not indicate its overall capacity as an absorber or reflector, since much of the irradiation may be in the IR region. A "white" surface such as snow, for example, is highly reflective to visible radiation but strongly absorbs IR radiation, thereby approximating blackbody behavior at long wavelengths.

In Section 12.2, we introduced properties to characterize the absorption, reflection, and transmission processes. In general these properties depend on surface material and finish, surface temperature, and the wavelength and direction of the incident radiation. These properties are considered in the following subsections.

12.6.1 Absorptivity

The absorptivity is a property that determines the fraction of the irradiation absorbed by a surface. Determination of the property is complicated by the fact that, like emission, it may be characterized by both a directional and spectral dependence. The *spectral, directional absorptivity,* $\alpha_{\lambda,\theta}(\lambda, \theta, \phi)$, of a surface is defined as the fraction of the spectral intensity incident in the direction of θ and ϕ that is absorbed by the surface. Hence

$$\alpha_{\lambda,\theta}(\lambda, \theta, \phi) \equiv \frac{I_{\lambda,i,\text{abs}}(\lambda, \theta, \phi)}{I_{\lambda,i}(\lambda, \theta, \phi)} \tag{12.47}$$

In this expression, we have neglected any dependence of the absorptivity on the surface temperature. Such a dependence is small for most spectral radiative properties.

It is implicit in the foregoing result that surfaces may exhibit selective absorption with respect to the wavelength and direction of the incident radiation. For most engineering calculations, however, it is desirable to work with surface properties that represent directional averages. We therefore define a *spectral, hemispherical absorptivity* $\alpha_\lambda(\lambda)$ as

$$\alpha_\lambda(\lambda) \equiv \frac{G_{\lambda,\text{abs}}(\lambda)}{G_\lambda(\lambda)} \tag{12.48}$$

which, from Equations 12.18 and 12.47, may be expressed as

$$\alpha_\lambda(\lambda) = \frac{\int_0^{2\pi}\int_0^{\pi/2} \alpha_{\lambda,\theta}(\lambda, \theta, \phi) I_{\lambda,i}(\lambda, \theta, \phi) \cos\theta \sin\theta \, d\theta \, d\phi}{\int_0^{2\pi}\int_0^{\pi/2} I_{\lambda,i}(\lambda, \theta, \phi) \cos\theta \sin\theta \, d\theta \, d\phi} \tag{12.49}$$

Hence α_λ depends on the directional distribution of the incident radiation, as well as on the wavelength of the radiation and the nature of the absorbing surface. Note that, if the incident radiation is diffusely distributed and $\alpha_{\lambda,\theta}$ is independent of ϕ, Equation 12.49 reduces to

$$\alpha_\lambda(\lambda) = 2\int_0^{\pi/2} \alpha_{\lambda,\theta}(\lambda, \theta) \cos\theta \sin\theta \, d\theta \tag{12.50}$$

The *total, hemispherical absorptivity,* α, represents an integrated average over both direction and wavelength. It is defined as the fraction of the total irradiation absorbed by a surface

$$\alpha \equiv \frac{G_{\text{abs}}}{G} \tag{12.51}$$

and, from Equations 12.19 and 12.48, it may be expressed as

$$\alpha = \frac{\int_0^\infty \alpha_\lambda(\lambda) G_\lambda(\lambda)\, d\lambda}{\int_0^\infty G_\lambda(\lambda)\, d\lambda} \tag{12.52}$$

Accordingly, α depends on the spectral distribution of the incident radiation, as well as on its directional distribution and the nature of the absorbing surface. Note that, although α is approximately independent of surface temperature, the same may not be said for the total, hemispherical emissivity, ε. From Equation 12.43 it is evident that this property is strongly temperature dependent.

Because α depends on the spectral distribution of the irradiation, its value for a surface exposed to solar radiation may differ appreciably from its value for the same surface exposed to longer wavelength radiation originating from a source of lower temperature. Since the spectral distribution of solar radiation is nearly proportional to that of emission from a blackbody at 5800 K, it follows from Equation 12.52 that the total absorptivity to solar radiation α_S may be approximated as

$$\alpha_S \approx \frac{\int_0^\infty \alpha_\lambda(\lambda) E_{\lambda,b}(\lambda, 5800\text{ K})\, d\lambda}{\int_0^\infty E_{\lambda,b}(\lambda, 5800\text{ K})\, d\lambda} \tag{12.53}$$

The integrals appearing in this equation may be evaluated by using the blackbody radiation function $F_{(0\rightarrow\lambda)}$ of Table 12.2.

12.6.2 Reflectivity

The reflectivity is a property that determines the fraction of the incident radiation reflected by a surface. However, its specific definition may take several different forms, because the property is inherently *bidirectional* [7]. That is, in addition to depending on the direction of the incident radiation, it also depends on the direction of the reflected radiation. We shall avoid this complication by working exclusively with a reflectivity that represents an integrated average over the hemisphere associated with the reflected radiation and therefore provides no information concerning the directional distribution of this radiation. Accordingly, the *spectral, directional reflectivity,* $\rho_{\lambda,\theta}(\lambda, \theta, \phi)$, of a surface is defined as the fraction of the spectral intensity incident in the direction of θ and ϕ, which is reflected by the surface. Hence

$$\rho_{\lambda,\theta}(\lambda, \theta, \phi) \equiv \frac{I_{\lambda,i,\text{ref}}(\lambda, \theta, \phi)}{I_{\lambda,i}(\lambda, \theta, \phi)} \tag{12.54}$$

The *spectral, hemispherical reflectivity* $\rho_\lambda(\lambda)$ is then defined as the fraction of the spectral irradiation that is reflected by the surface. Accordingly,

$$\rho_\lambda(\lambda) \equiv \frac{G_{\lambda,\text{ref}}(\lambda)}{G_\lambda(\lambda)} \tag{12.55}$$

which is equivalent to

$$\rho_\lambda(\lambda) = \frac{\int_0^{2\pi}\int_0^{\pi/2} \rho_{\lambda,\theta}(\lambda, \theta, \phi) I_{\lambda,i}(\lambda, \theta, \phi) \cos\theta \sin\theta \, d\theta \, d\phi}{\int_0^{2\pi}\int_0^{\pi/2} I_{\lambda,i}(\lambda, \theta, \phi) \cos\theta \sin\theta \, d\theta \, d\phi} \tag{12.56}$$

The *total, hemispherical reflectivity* ρ is then defined as

$$\rho \equiv \frac{G_{\text{ref}}}{G} \tag{12.57}$$

in which case

$$\rho = \frac{\int_0^\infty \rho_\lambda(\lambda) G_\lambda(\lambda) \, d\lambda}{\int_0^\infty G_\lambda(\lambda) \, d\lambda} \tag{12.58}$$

Surfaces may be idealized as *diffuse* or *specular,* according to the manner in which they reflect radiation (Figure 12.21). Diffuse reflection occurs if, regardless of the direction of the incident radiation, the intensity of the reflected radiation is independent of the reflection angle. In contrast, if all the reflection is in the direction of θ_2, which equals the incident angle θ_1, specular reflection is said to occur. Although no surface is perfectly diffuse or specular, the latter condition is more closely approximated by polished, mirror-like surfaces and the former condition by rough surfaces. The assumption of diffuse reflection is reasonable for most engineering applications.

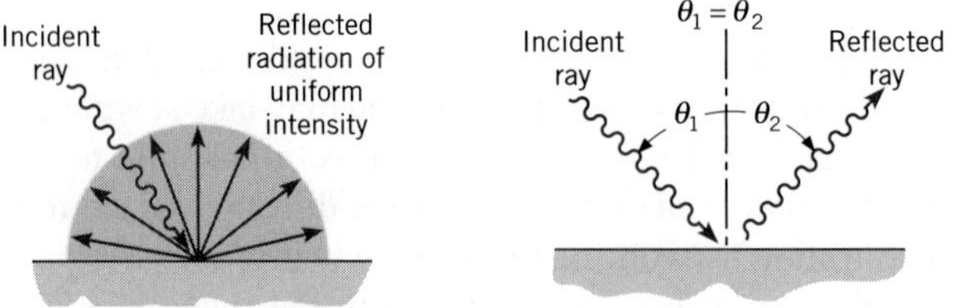

FIGURE 12.21 Diffuse and specular reflection.

12.6.3 Transmissivity

Although treatment of the response of a semitransparent material to incident radiation is a complicated problem [7], reasonable results may often be obtained through the use of hemispherical transmissivities defined as

$$\tau_\lambda = \frac{G_{\lambda,\text{tr}}(\lambda)}{G_\lambda(\lambda)} \tag{12.59}$$

and

$$\tau = \frac{G_{\text{tr}}}{G} \tag{12.60}$$

The total transmissivity τ is related to the spectral component τ_λ by

$$\tau = \frac{\int_0^\infty G_{\lambda,\text{tr}}(\lambda)\,d\lambda}{\int_0^\infty G_\lambda(\lambda)\,d\lambda} = \frac{\int_0^\infty \tau_\lambda(\lambda)G_\lambda(\lambda)\,d\lambda}{\int_0^\infty G_\lambda(\lambda)\,d\lambda} \tag{12.61}$$

12.6.4 Special Considerations

From the radiation balance of Equation 12.46 and the foregoing definitions,

$$\rho_\lambda + \alpha_\lambda + \tau_\lambda = 1 \tag{12.62}$$

for a *semitransparent* medium. This is analogous to Equation 12.2 but on a spectral basis. Of course, if the medium is *opaque*, there is no transmission, and absorption and reflection are surface processes for which

$$\alpha_\lambda + \rho_\lambda = 1 \tag{12.63}$$

which is analogous to Equation 12.3. Hence knowledge of one property implies knowledge of the other.

Spectral distributions of the normal reflectivity and absorptivity are plotted in Figure 12.22 for selected *opaque* surfaces. A material such as glass or water, which is semitransparent at short wavelengths, becomes opaque at longer wavelengths. This behavior is shown in Figure 12.23, which presents the spectral transmissivity of several common *semitransparent* materials. Note that the transmissivity of glass is affected by its iron content and that the transmissivity of plastics, such as Tedlar, is greater than that of glass in the IR region. These factors have an important bearing on the selection of cover plate materials for solar collector applications, design and selection of windows for energy conservation, and specification of materials for fabrication of the optical components of infrared imaging systems. Values for the total transmissivity to solar radiation of common collector cover plate materials are presented in Table A.12, along with surface solar absorptivities and low-temperature emissivities.

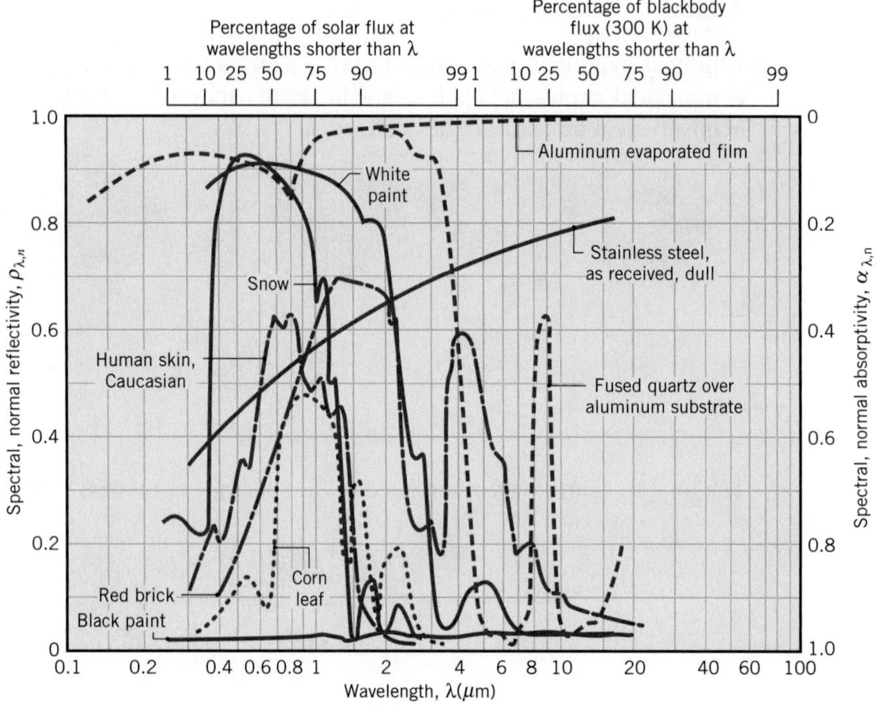

FIGURE 12.22 Spectral dependence of the spectral, normal absorptivity $\alpha_{\lambda,n}$ and reflectivity $\rho_{\lambda,n}$ of selected opaque materials.

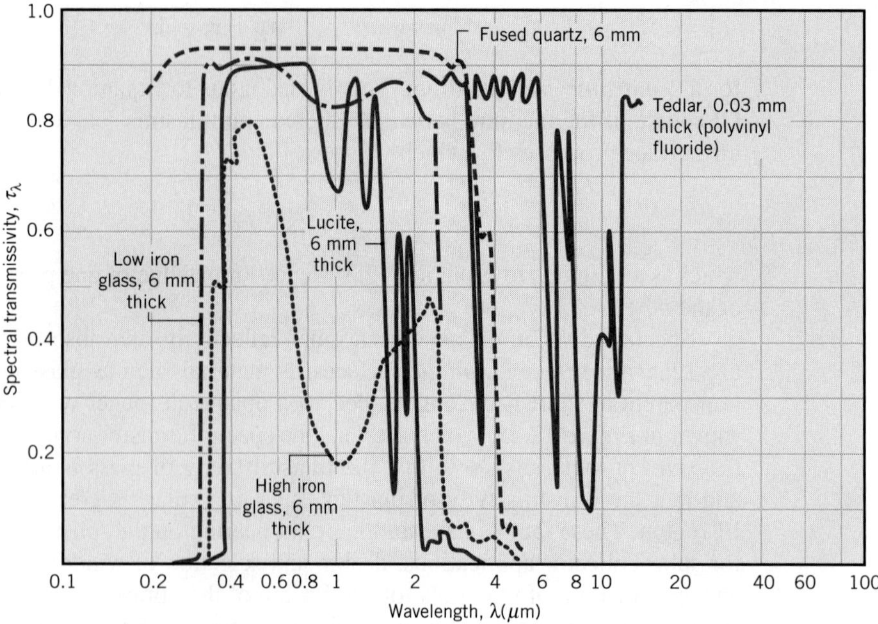

FIGURE 12.23 Spectral dependence of the spectral transmissivities τ_λ of selected semitransparent materials.

EXAMPLE 12.8

The spectral, hemispherical absorptivity of an opaque surface and the spectral irradiation at the surface are as shown.

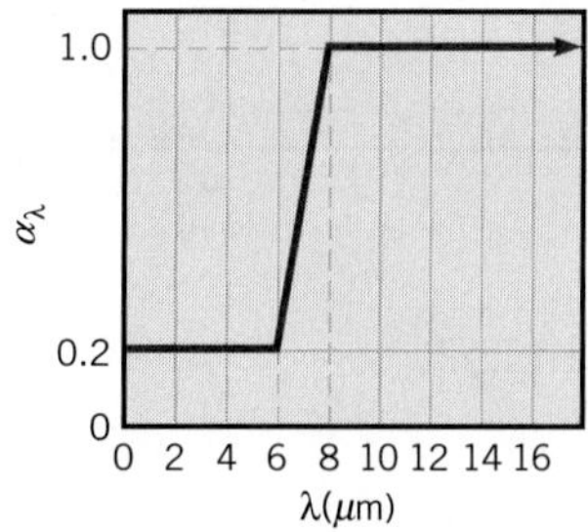

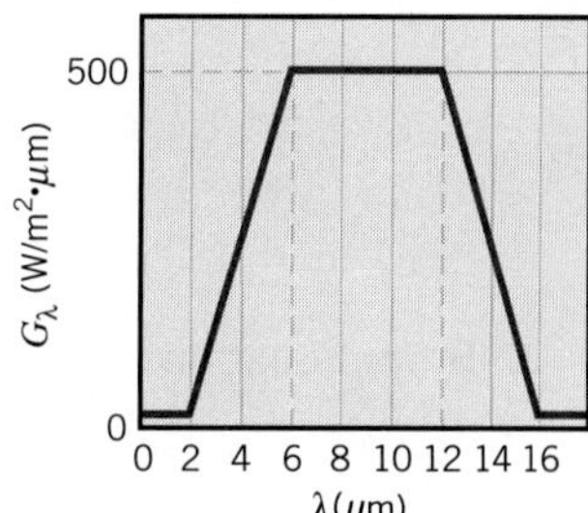

How does the spectral, hemispherical reflectivity vary with wavelength? What is the total, hemispherical absorptivity of the surface? If the surface is initially at 500 K and has a total, hemispherical emissivity of 0.8, how will its temperature change upon exposure to the irradiation?

SOLUTION

Known: Spectral, hemispherical absorptivity and irradiation of a surface. Surface temperature (500 K) and total, hemispherical emissivity (0.8).

Find:

1. Spectral distribution of reflectivity.
2. Total, hemispherical absorptivity.
3. Nature of surface temperature change.

Schematic:

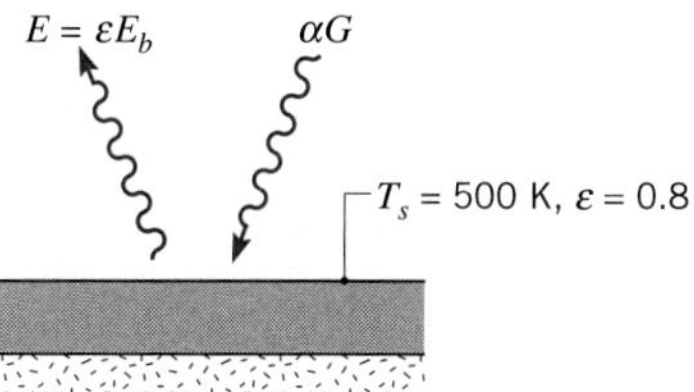

Assumptions:

1. Surface is opaque.
2. Surface convection effects are negligible.
3. Back surface is insulated.

Analysis:

1. From Equation 12.63, $\rho_\lambda = 1 - \alpha_\lambda$. Hence from knowledge of $\alpha_\lambda(\lambda)$, the spectral distribution of ρ_λ is as shown.

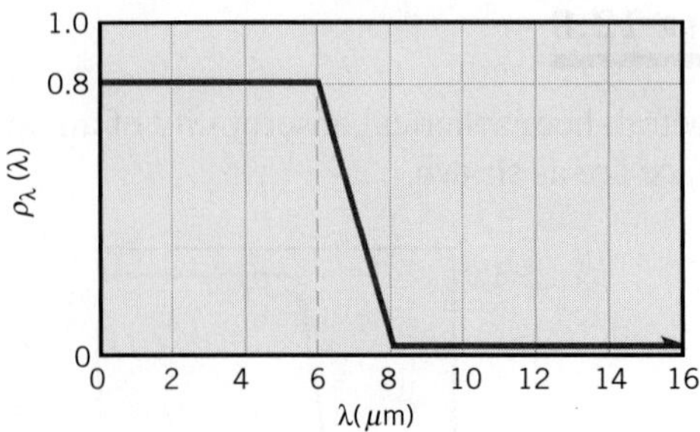

2. From Equations 12.51 and 12.52,

$$\alpha = \frac{G_{\text{abs}}}{G} = \frac{\int_0^\infty \alpha_\lambda G_\lambda \, d\lambda}{\int_0^\infty G_\lambda \, d\lambda}$$

or, subdividing the integral into parts,

$$\alpha = \frac{0.2\int_2^6 G_\lambda \, d\lambda + 500\int_6^8 \alpha_\lambda \, d\lambda + 1.0\int_8^{16} G_\lambda \, d\lambda}{\int_2^6 G_\lambda \, d\lambda + \int_6^{12} G_\lambda \, d\lambda + \int_{12}^{16} G_\lambda \, d\lambda}$$

$$\begin{aligned}\alpha = \{&0.2(\tfrac{1}{2})500 \text{ W/m}^2\cdot\mu\text{m}\,(6-2)\,\mu\text{m} \\ &+ 500 \text{ W/m}^2\cdot\mu\text{m}\,[0.2(8-6)\,\mu\text{m} + (1-0.2)(\tfrac{1}{2})(8-6)\,\mu\text{m}] \\ &+ [1 \times 500 \text{ W/m}^2\cdot\mu\text{m}\,(12-8)\,\mu\text{m} \\ &+ 1(\tfrac{1}{2})500 \text{ W/m}^2\cdot\mu\text{m}\,(16-12)\,\mu\text{m}]\} \\ &\div [(\tfrac{1}{2})500 \text{ W/m}^2\cdot\mu\text{m}\,(6-2)\,\mu\text{m} + 500 \text{ W/m}^2\cdot\mu\text{m}\,(12-6)\,\mu\text{m} \\ &+ (\tfrac{1}{2})500 \text{ W/m}^2\cdot\mu\text{m}\,(16-12)\,\mu\text{m}]\end{aligned}$$

Hence

$$\alpha = \frac{G_{\text{abs}}}{G} = \frac{(200 + 600 + 3000) \text{ W/m}^2}{(1000 + 3000 + 1000) \text{ W/m}^2} = \frac{3800 \text{ W/m}^2}{5000 \text{ W/m}^2} = 0.76 \qquad \triangleleft$$

3. Neglecting convection effects, the net heat flux *to* the surface is

$$q''_{\text{net}} = \alpha G - E = \alpha G - \varepsilon\sigma T^4$$

Hence

$$q''_{\text{net}} = 0.76(5000 \text{ W/m}^2) - 0.8 \times 5.67 \times 10^{-8} \text{ W/m}^2\cdot\text{K}^4(500 \text{ K})^4$$

$$q''_{\text{net}} = 3800 - 2835 = 965 \text{ W/m}^2$$

Since $q''_{\text{net}} > 0$, the surface temperature will *increase* with time. ◁

Example 12.9

The cover glass on a flat-plate solar collector has a low iron content, and its spectral transmissivity may be approximated by the following distribution.

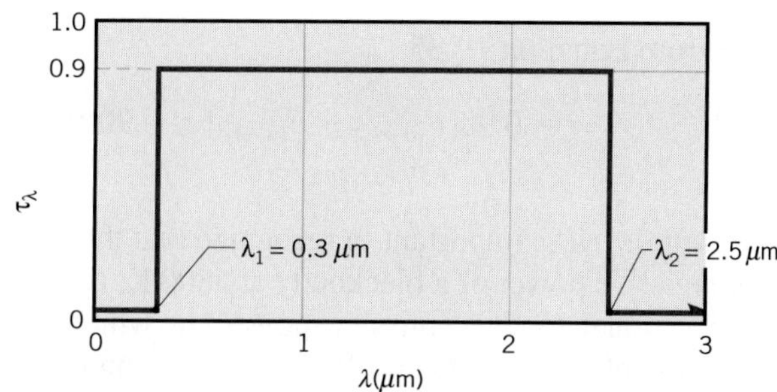

What is the total transmissivity of the cover glass to solar radiation?

Solution

Known: Spectral transmissivity of solar collector cover glass.

Find: Total transmissivity of cover glass to solar radiation.

Assumptions: Spectral distribution of solar irradiation is proportional to that of blackbody emission at 5800 K.

Analysis: From Equation 12.61 the total transmissivity of the cover is

$$\tau = \frac{\int_0^\infty \tau_\lambda G_\lambda \, d\lambda}{\int_0^\infty G_\lambda \, d\lambda}$$

where the irradiation G_λ is due to solar emission. Having assumed that the sun emits as a blackbody at 5800 K, it follows that

$$G_\lambda(\lambda) \propto E_{\lambda,b}(5800 \text{ K})$$

With the proportionality constant canceling from the numerator and denominator of the expression for τ, we obtain

$$\tau = \frac{\int_0^\infty \tau_\lambda E_{\lambda,b}(5800 \text{ K}) \, d\lambda}{\int_0^\infty E_{\lambda,b}(5800 \text{ K}) \, d\lambda}$$

or, for the prescribed spectral distribution of $\tau_\lambda(\lambda)$,

$$\tau = 0.90 \frac{\int_{0.3}^{2.5} E_{\lambda,b}(5800 \text{ K}) \, d\lambda}{E_b(5800 \text{ K})}$$

From Table 12.2

$$\lambda_1 = 0.3\ \mu\text{m}, T = 5800\ \text{K}: \qquad \lambda_1 T = 1740\ \mu\text{m}\cdot\text{K}, F_{(0\to\lambda_1)} = 0.0335$$

$$\lambda_2 = 2.5\ \mu\text{m}, T = 5800\ \text{K}: \qquad \lambda_2 T = 14{,}500\ \mu\text{m}\cdot\text{K}, F_{(0\to\lambda_2)} = 0.9664$$

Hence from Equation 12.35

$$\tau = 0.90[F_{(0\to\lambda_2)} - F_{(0\to\lambda_1)}] = 0.90(0.9664 - 0.0335) = 0.84 \qquad \triangleleft$$

Comments: It is important to recognize that the irradiation at the cover plate is not equal to the emissive power of a blackbody at 5800 K, $G_\lambda \neq E_{\lambda,b}$ (5800 K). It is simply assumed to be proportional to this emissive power, in which case it is assumed to have a spectral distribution of the same form. With G_λ appearing in both the numerator and denominator of the expression for τ, it is then possible to replace G_λ by $E_{\lambda,b}$.

12.7 *Kirchhoff's Law*

In the foregoing sections we separately considered real surface properties associated with emission and irradiation. In Sections 12.7 and 12.8 we consider conditions for which the emissivity and absorptivity are equal.

Consider a *large, isothermal enclosure* of surface temperature T_s, within which several small bodies are confined (Figure 12.24). Since these bodies are small relative to the enclosure, they have a negligible influence on the radiation field, which is due to the cumulative effect of emission and reflection by the enclosure surface. Recall that, regardless of its radiative properties, such a surface forms a *blackbody cavity.* Accordingly, regardless of its orientation, the irradiation experienced by any body in the cavity is diffuse and equal to emission from a blackbody at T_s.

$$G = E_b(T_s) \tag{12.64}$$

Under steady-state conditions, *thermal equilibrium* must exist between the bodies and the enclosure. Hence $T_1 = T_2 = \cdots = T_s$, and the net rate of energy transfer to each

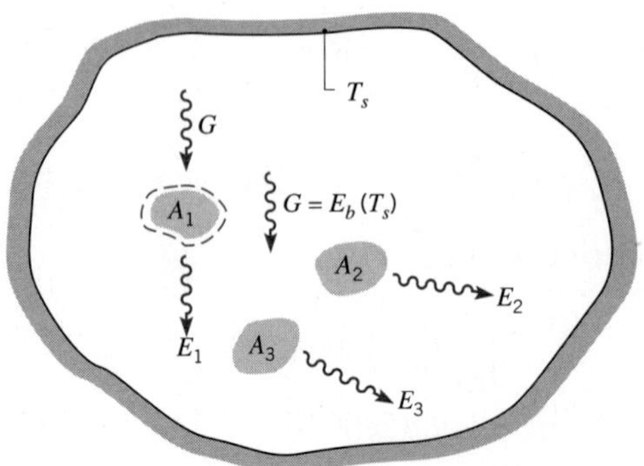

FIGURE 12.24 Radiative exchange in an isothermal enclosure.

surface must be zero. Applying an energy balance to a control surface about body 1, it follows that

$$\alpha_1 G A_1 - E_1(T_s) A_1 = 0$$

or, from Equation 12.64,

$$\frac{E_1(T_s)}{\alpha_1} = E_b(T_s)$$

Since this result must apply to each of the confined bodies, we then obtain

$$\frac{E_1(T_s)}{\alpha_1} = \frac{E_2(T_s)}{\alpha_2} = \cdots = E_b(T_s) \tag{12.65}$$

This relation is known as *Kirchhoff's law.* A major implication is that, since $\alpha \leq 1$, $E(T_s) \leq E_b(T_s)$. Hence *no real surface can have an emissive power exceeding that of a black surface at the same temperature,* and the notion of the blackbody as an ideal emitter is confirmed.

From the definition of the total, hemispherical emissivity, Equation 12.36, an alternative form of Kirchhoff's law, is

$$\frac{\varepsilon_1}{\alpha_1} = \frac{\varepsilon_2}{\alpha_2} = \cdots = 1$$

Hence, for any surface in the enclosure,

$$\varepsilon = \alpha \tag{12.66}$$

That is, the total, hemispherical emissivity of the surface is equal to its total, hemispherical absorptivity *if* isothermal conditions exist *and* no net radiation heat transfer occurs at any of the surfaces.

We will later find that calculations of radiative exchange between surfaces are greatly simplified *if* Equation 12.66 may be applied to each of the surfaces. However, the restrictive conditions inherent in its derivation should be remembered. In particular, the surface irradiation has been assumed to correspond to emission from a blackbody at the same temperature as the surface. In Section 12.8 we consider other, less restrictive, conditions for which Equation 12.66 is applicable.

The preceding derivation may be repeated for spectral conditions, and for any surface in the enclosure it follows that

$$\varepsilon_\lambda = \alpha_\lambda \tag{12.67}$$

Conditions associated with the use of Equation 12.67 are less restrictive than those associated with Equation 12.66. In particular, it will be shown that Equation 12.67 is applicable if the irradiation is diffuse *or* if the surface is diffuse. A form of Kirchhoff's law for which there are *no restrictions* involves the spectral, directional properties.

$$\varepsilon_{\lambda,\theta} = \alpha_{\lambda,\theta} \tag{12.68}$$

This equality is *always* applicable because $\varepsilon_{\lambda,\theta}$ and $\alpha_{\lambda,\theta}$ are *inherent* surface properties. That is, respectively, they are independent of the spectral and directional distributions of the emitted and incident radiation.

More detailed developments of Kirchhoff's law are provided by Planck [1] and Howell et al. [7].

12.8 The Gray Surface

In Chapter 13 we will find that the problem of predicting radiant energy exchange between surfaces is greatly simplified if Equation 12.66 may be assumed to apply for each individual surface. It is therefore important to examine whether this equality may be applied to conditions other than those for which it was derived, namely, irradiation due to emission from a blackbody at the same temperature as the surface.

Accepting the fact that the spectral, directional emissivity and absorptivity are equal under any conditions, Equation 12.68, we begin by considering the conditions associated with using Equation 12.67. According to the definitions of the spectral, hemispherical properties, Equations 12.41 and 12.49, we are really asking under what conditions, if any, the following equality will hold:

$$\varepsilon_\lambda = \frac{\int_0^{2\pi}\int_0^{\pi/2} \varepsilon_{\lambda,\theta} \cos\theta \sin\theta \, d\theta \, d\phi}{\int_0^{2\pi}\int_0^{\pi/2} \cos\theta \sin\theta \, d\theta \, d\phi} \stackrel{?}{=} \frac{\int_0^{2\pi}\int_0^{\pi/2} \alpha_{\lambda,\theta} I_{\lambda,i} \cos\theta \sin\theta \, d\theta \, d\phi}{\int_0^{2\pi}\int_0^{\pi/2} I_{\lambda,i} \cos\theta \sin\theta \, d\theta \, d\phi} = \alpha_\lambda \tag{12.69}$$

Since $\varepsilon_{\lambda,\theta} = \alpha_{\lambda,\theta}$, it follows by inspection of Equation 12.69 that Equation 12.67 is applicable if *either* of the following conditions is satisfied:

1. The *irradiation* is *diffuse* ($I_{\lambda,i}$ is independent of θ and ϕ).
2. The *surface* is *diffuse* ($\varepsilon_{\lambda,\theta}$ and $\alpha_{\lambda,\theta}$ are independent of θ and ϕ).

The first condition is a reasonable approximation for many engineering calculations; the second condition is reasonable for many surfaces, particularly for electrically nonconducting materials (Figure 12.16).

Assuming the existence of either diffuse irradiation or a diffuse surface, we now consider what *additional* conditions must be satisfied for Equation 12.66 to be valid. From Equations 12.43 and 12.52, the equality applies if

$$\varepsilon = \frac{\int_0^\infty \varepsilon_\lambda E_{\lambda,b}(\lambda, T) \, d\lambda}{E_b(T)} \stackrel{?}{=} \frac{\int_0^\infty \alpha_\lambda G_\lambda(\lambda) \, d\lambda}{G} = \alpha \tag{12.70}$$

Since $\varepsilon_\lambda = \alpha_\lambda$, it follows by inspection of Equation 12.70 that Equation 12.66 applies if *either* of the following conditions is satisfied:

1. The irradiation corresponds to emission from a blackbody at the surface temperature T, in which case $G_\lambda(\lambda) = E_{\lambda,b}(\lambda, T)$ and $G = E_b(T)$.
2. The *surface* is *gray* (α_λ and ε_λ are independent of λ).

Note that the first condition corresponds to the major assumption required for the derivation of Kirchhoff's law (Section 12.7).

Because the total absorptivity of a surface depends on the spectral distribution of the irradiation, it cannot be stated unequivocally that $\alpha = \varepsilon$. For example, a particular surface may be highly absorbing to radiation in one spectral region and virtually nonabsorbing in another region (Figure 12.25*a*). Accordingly, for the two possible irradiation fields $G_{\lambda,1}(\lambda)$

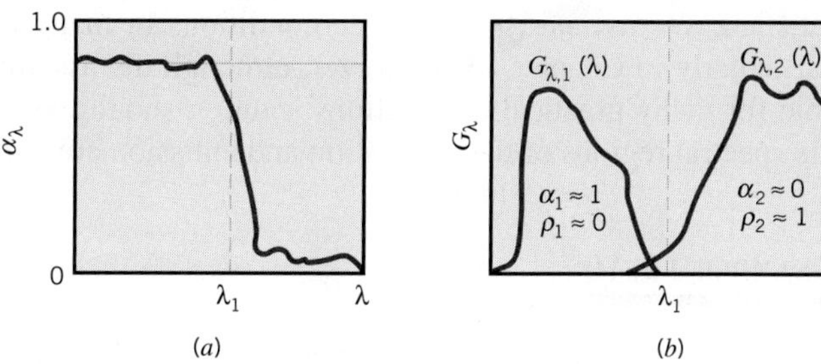

FIGURE 12.25 Spectral distribution of (*a*) the spectral absorptivity of a surface and (*b*) the spectral irradiation at the surface.

and $G_{\lambda,2}(\lambda)$ of Figure 12.25*b*, the values of α will differ drastically. In contrast, the value of ε is independent of the irradiation. Hence there is *no* basis for stating that α is *always* equal to ε.

To assume gray surface behavior, hence the validity of Equation 12.66, it is not necessary for α_λ and ε_λ to be independent of λ over the entire spectrum. Practically speaking, a *gray surface* may be defined as *one for which α_λ and ε_λ are independent of λ over the spectral regions of the irradiation and the surface emission.* For example, from Equation 12.70 it is readily shown that gray surface behavior may be assumed for the conditions of Figure 12.26. That is, the irradiation and surface emission are concentrated in a region for which the spectral properties of the surface are approximately constant. Accordingly,

$$\varepsilon = \frac{\varepsilon_{\lambda,o}\int_{\lambda_1}^{\lambda_2} E_{\lambda,b}(\lambda,T)\,d\lambda}{E_b(T)} = \varepsilon_{\lambda,o} \qquad \text{and} \qquad \alpha = \frac{\alpha_{\lambda,o}\int_{\lambda_3}^{\lambda_4} G_\lambda(\lambda)\,d\lambda}{G} = \alpha_{\lambda,o}$$

in which case $\alpha = \varepsilon = \varepsilon_{\lambda,o}$. However, if the irradiation were in a spectral region corresponding to $\lambda < \lambda_1$ or $\lambda > \lambda_4$, gray surface behavior could not be assumed.

A surface for which $\alpha_{\lambda,\theta}$ and $\varepsilon_{\lambda,\theta}$ are independent of θ and λ is termed a *diffuse, gray surface* (diffuse because of the directional independence and gray because of the wavelength independence). It is a surface for which both Equations 12.66 and 12.67 are

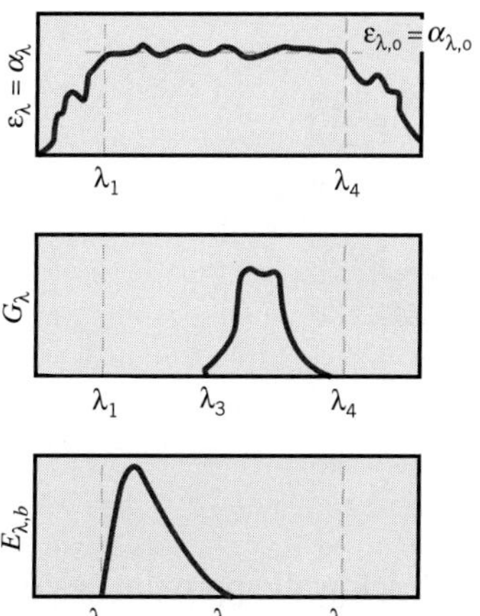

FIGURE 12.26 A set of conditions for which gray surface behavior may be assumed.

satisfied. We *assume* such surface conditions in many of our subsequent considerations, particularly in Chapter 13. However, although the assumption of a gray surface is reasonable for many practical applications, caution should be exercised in its use, particularly if the spectral regions of the irradiation and emission are widely separated.

EXAMPLE 12.10

A diffuse, fire brick wall of temperature $T_s = 500$ K has the spectral emissivity shown and is exposed to a bed of coals at 2000 K.

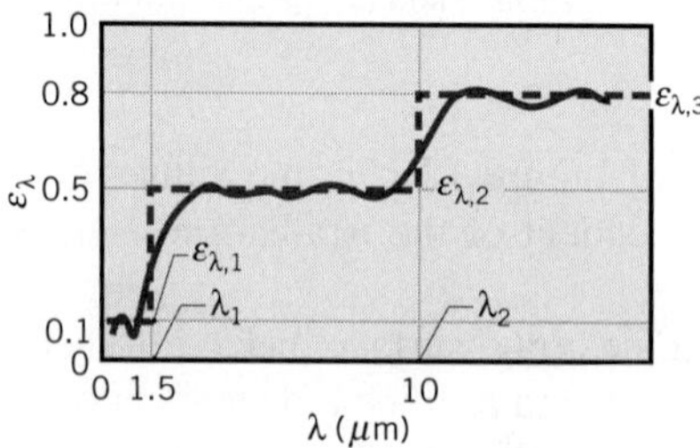

Determine the total, hemispherical emissivity and emissive power of the fire brick wall. What is the total absorptivity of the wall to irradiation resulting from emission by the coals?

SOLUTION

Known: Brick wall of surface temperature $T_s = 500$ K and prescribed $\varepsilon_\lambda(\lambda)$ is exposed to coals at $T_c = 2000$ K.

Find:

1. Total, hemispherical emissivity of the fire brick wall.
2. Total emissive power of the brick wall.
3. Absorptivity of the wall to irradiation from the coals.

Schematic:

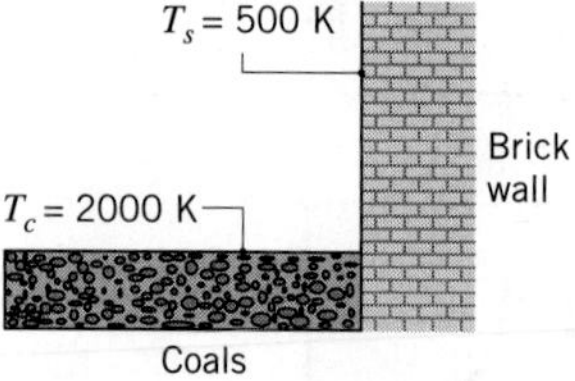

Assumptions:

1. Brick wall is opaque and diffuse.
2. Spectral distribution of irradiation at the brick wall approximates that due to emission from a blackbody at 2000 K.

Analysis:

1. The total, hemispherical emissivity follows from Equation 12.43.

$$\varepsilon(T_s) = \frac{\int_0^\infty \varepsilon_\lambda(\lambda) E_{\lambda,b}(\lambda, T_s)\, d\lambda}{E_b(T_s)}$$

Breaking the integral into parts,

$$\varepsilon(T_s) = \varepsilon_{\lambda,1} \frac{\int_0^{\lambda_1} E_{\lambda,b}\, d\lambda}{E_b} + \varepsilon_{\lambda,2} \frac{\int_{\lambda_1}^{\lambda_2} E_{\lambda,b}\, d\lambda}{E_b} + \varepsilon_{\lambda,3} \frac{\int_{\lambda_2}^{\infty} E_{\lambda,b}\, d\lambda}{E_b}$$

and introducing the blackbody functions, it follows that

$$\varepsilon(T_s) = \varepsilon_{\lambda,1} F_{(0\rightarrow\lambda_1)} + \varepsilon_{\lambda,2}[F_{(0\rightarrow\lambda_2)} - F_{(0\rightarrow\lambda_1)}] + \varepsilon_{\lambda,3}[1 - F_{(0\rightarrow\lambda_2)}]$$

From Table 12.2

$$\lambda_1 T_s = 1.5\ \mu\text{m} \times 500\ \text{K} = 750\ \mu\text{m}\cdot\text{K}: \qquad F_{(0\rightarrow\lambda_1)} = 0.000$$

$$\lambda_2 T_s = 10\ \mu\text{m} \times 500\ \text{K} = 5000\ \mu\text{m}\cdot\text{K}: \qquad F_{(0\rightarrow\lambda_2)} = 0.634$$

Hence

$$\varepsilon(T_s) = 0.1 \times 0 + 0.5 \times 0.634 + 0.8\,(1 - 0.634) = 0.610 \qquad \triangleleft$$

2. From Equations 12.32 and 12.36, the total emissive power is

$$E(T_s) = \varepsilon(T_s) E_b(T_s) = \varepsilon(T_s) \sigma T_s^4$$

$$E(T_s) = 0.61 \times 5.67 \times 10^{-8}\ \text{W/m}^2\cdot\text{K}^4 (500\ \text{K})^4 = 2162\ \text{W/m}^2 \qquad \triangleleft$$

3. From Equation 12.52, the total absorptivity of the wall to radiation from the coals is

$$\alpha = \frac{\int_0^\infty \alpha_\lambda(\lambda) G_\lambda(\lambda)\, d\lambda}{\int_0^\infty G_\lambda(\lambda)\, d\lambda}$$

Since the surface is diffuse, $\alpha_\lambda(\lambda) = \varepsilon_\lambda(\lambda)$. Moreover, since the spectral distribution of the irradiation approximates that due to emission from a blackbody at 2000 K, $G_\lambda(\lambda) \propto E_{\lambda,b}(\lambda, T_c)$. It follows that

$$\alpha = \frac{\int_0^\infty \varepsilon_\lambda(\lambda) E_{\lambda,b}(\lambda, T_c)\, d\lambda}{\int_0^\infty E_{\lambda,b}(\lambda, T_c)\, d\lambda}$$

Breaking the integral into parts and introducing the blackbody functions, we then obtain

$$\alpha = \varepsilon_{\lambda,1} F_{(0\rightarrow\lambda_1)} + \varepsilon_{\lambda,2}[F_{(0\rightarrow\lambda_2)} - F_{(0\rightarrow\lambda_1)}] + \varepsilon_{\lambda,3}[1 - F_{(0\rightarrow\lambda_2)}]$$

From Table 12.2

$$\lambda_1 T_c = 1.5\ \mu\text{m} \times 2000\ \text{K} = 3000\ \mu\text{m}\cdot\text{K}: \qquad F_{(0\rightarrow\lambda_1)} = 0.273$$

$$\lambda_2 T_c = 10\ \mu\text{m} \times 2000\ \text{K} = 20{,}000\ \mu\text{m}\cdot\text{K}: \qquad F_{(0\rightarrow\lambda_2)} = 0.986$$

Hence

$$\alpha = 0.1 \times 0.273 + 0.5(0.986 - 0.273) + 0.8(1 - 0.986) = 0.395 \qquad \triangleleft$$

Comments:

1. The emissivity depends on the surface temperature T_s, while the absorptivity depends on the spectral distribution of the irradiation, which depends on the temperature of the source T_c.
2. The surface is not gray, $\alpha \neq \varepsilon$. This result is to be expected. Since emission is associated with $T_s = 500$ K, its spectral maximum occurs at $\lambda_{max} \approx 6\ \mu$m. In contrast, since irradiation is associated with emission from a source at $T_c = 2000$ K, its spectral maximum occurs at $\lambda_{max} \approx 1.5\ \mu$m. Even though ε_λ and α_λ are equal, because they are not constant over the spectral ranges of both emission and irradiation, $\alpha \neq \varepsilon$. For the prescribed spectral distribution of $\varepsilon_\lambda = \alpha_\lambda$, ε and α decrease with increasing T_s and T_c, respectively, and it is only for $T_s = T_c$ that $\varepsilon = \alpha$. The foregoing expressions for ε and α may be used to determine their equivalent variation with T_s and T_c, and the following result is obtained:

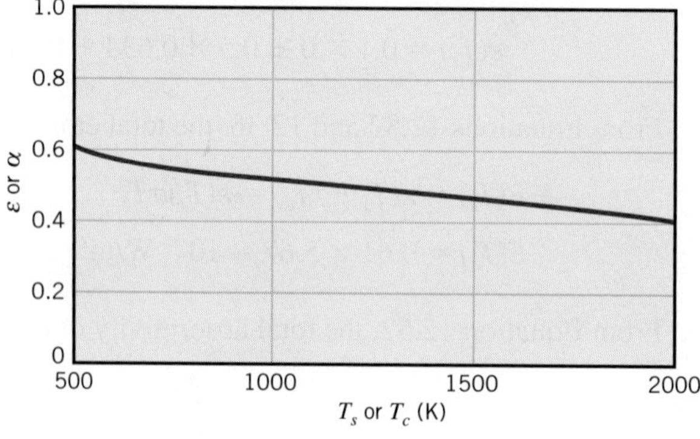

IHT

EXAMPLE 12.11

A *small,* solid metallic sphere has an opaque, diffuse coating for which $\alpha_\lambda = 0.8$ for $\lambda \leq 5\ \mu$m and $\alpha_\lambda = 0.1$ for $\lambda > 5\ \mu$m. The sphere, which is initially at a uniform temperature of 300 K, is inserted into a *large* furnace whose walls are at 1200 K. Determine the total, hemispherical absorptivity and emissivity of the coating for the initial condition and for the final, steady-state condition.

SOLUTION

Known: Small metallic sphere with spectrally selective absorptivity, initially at $T_s = 300$ K, is inserted into a large furnace at $T_f = 1200$ K.

Find:

1. Total, hemispherical absorptivity and emissivity of sphere coating for the initial condition.
2. Values of α and ε after sphere has been in furnace a long time.

Schematic:

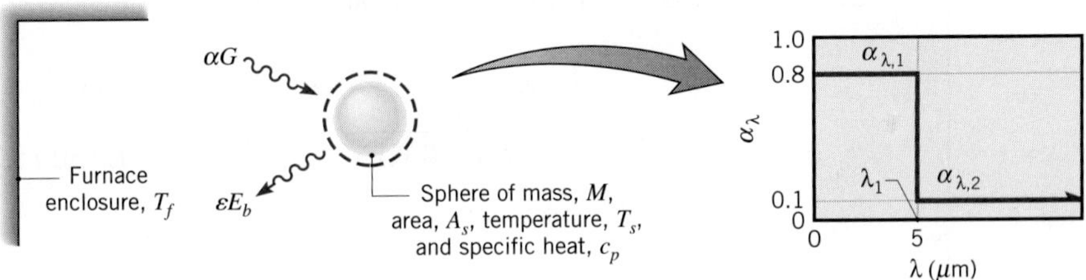

Assumptions:

1. Coating is opaque and diffuse.
2. Since furnace surface is much larger than that of sphere, irradiation approximates emission from a blackbody at T_f.

Analysis:

1. From Equation 12.52 the total, hemispherical absorptivity is

$$\alpha = \frac{\int_0^\infty \alpha_\lambda(\lambda) G_\lambda(\lambda)\, d\lambda}{\int_0^\infty G_\lambda(\lambda)\, d\lambda}$$

or, with $G_\lambda = E_{\lambda,b}(T_f) = E_{\lambda,b}(\lambda, 1200\text{ K})$,

$$\alpha = \frac{\int_0^\infty \alpha_\lambda(\lambda) E_{\lambda,b}(\lambda, 1200\text{ K})\, d\lambda}{E_b(1200\text{ K})}$$

Hence

$$\alpha = \alpha_{\lambda,1} \frac{\int_0^{\lambda_1} E_{\lambda,b}(\lambda, 1200\text{ K})\, d\lambda}{E_b(1200\text{ K})} + \alpha_{\lambda,2} \frac{\int_{\lambda_1}^\infty E_{\lambda,b}(\lambda, 1200\text{ K})\, d\lambda}{E_b(1200\text{ K})}$$

or

$$\alpha = \alpha_{\lambda,1} F_{(0\to\lambda_1)} + \alpha_{\lambda,2}[1 - F_{(0\to\lambda_1)}]$$

From Table 12.2,

$$\lambda_1 T_f = 5\ \mu\text{m} \times 1200\text{ K} = 6000\ \mu\text{m}\cdot\text{K}: \qquad F_{(0\to\lambda_1)} = 0.738$$

Hence

$$\alpha = 0.8 \times 0.738 + 0.1\,(1 - 0.738) = 0.62$$

The total, hemispherical emissivity follows from Equation 12.43.

$$\varepsilon(T_s) = \frac{\int_0^\infty \varepsilon_\lambda E_{\lambda,b}(\lambda, T_s)\, d\lambda}{E_b(T_s)}$$

Since the surface is diffuse, $\varepsilon_\lambda = \alpha_\lambda$ and it follows that

$$\varepsilon = \alpha_{\lambda,1} \frac{\int_0^{\lambda_1} E_{\lambda,b}(\lambda, 300\text{ K})\, d\lambda}{E_b(300\text{ K})} + \alpha_{\lambda,2} \frac{\int_{\lambda_1}^\infty E_{\lambda,b}(\lambda, 300\text{ K})\, d\lambda}{E_b(300\text{ K})}$$

or,

$$\varepsilon = \alpha_{\lambda,1}\, F_{(0\to\lambda_1)} + \alpha_{\lambda,2}[1 - F_{(0\to\lambda_1)}]$$

From Table 12.2,

$$\lambda_1 T_s = 5\ \mu\text{m} \times 300\text{ K} = 1500\ \mu\text{m}\cdot\text{K}: \qquad F_{(0\to\lambda_1)} = 0.014$$

Hence

$$\varepsilon = 0.8 \times 0.014 + 0.1(1 - 0.014) = 0.11 \qquad \triangleleft$$

2. Because the spectral characteristics of the coating and the furnace temperature remain fixed, there is no change in the value of α with increasing time. However, as T_s increases with time, the value of ε will change. After a sufficiently long time, $T_s = T_f$, and $\varepsilon = \alpha$ ($\varepsilon = 0.62$).

Comments:

1. The equilibrium condition that eventually exists ($T_s = T_f$) corresponds precisely to the condition for which Kirchhoff's law was derived. Hence α must equal ε.
2. Approximating the sphere as a lumped capacitance and neglecting convection, an energy balance for a control volume about the sphere yields

$$\dot{E}_{\text{in}} - \dot{E}_{\text{out}} = \dot{E}_{\text{st}}$$

$$(\alpha G)A_s - (\varepsilon\sigma T_s^4)A_s = Mc_p \frac{dT_s}{dt}$$

The differential equation could be solved to determine $T(t)$ for $t > 0$, and the variation in ε that occurs with increasing time would have to be included in the solution.

12.9 *Environmental Radiation*

Solar radiation is essential to all life on earth. Through the process of photosynthesis, solar radiation satisfies the human need for food, fiber, and fuel. Utilizing thermal and photovoltaic processes, it also has the potential to satisfy considerable demand for heat and electricity.

Together, solar radiation and radiation emitted by surfaces of the earth's land and oceans comprise what is commonly termed *environmental radiation*. It is the interaction of environmental radiation with the earth's atmosphere that determines the temperature of our planet.

12.9.1 Solar Radiation

The sun is a nearly spherical radiation source that is 1.39×10^9 m in diameter and is located 1.50×10^{11} m from the earth. As noted previously, the sun emits approximately as a blackbody at 5800 K. As the radiation emitted by the sun travels through space, the radiation flux decreases because of the greater (spherical) area through which it passes. At the outer edge of the earth's atmosphere, the flux of solar energy has decreased by a factor of $(r_s/r_d)^2$, where r_s is the radius of the sun and r_d is the distance from the sun to the earth. The *solar constant,*[3] S_c, is defined as the flux of solar energy incident on a surface oriented normal to the sun's rays, at the outer edge of the earth's atmosphere, when the earth is at its mean distance from the sun (Figure 12.27). It has a value of 1368 ± 0.65 W/m^2. For a *horizontal* surface (that is, parallel to the earth's surface), solar radiation appears as a beam of nearly *parallel rays* that form an angle θ, the *zenith angle*, relative to the surface normal. The *extraterrestrial solar irradiation*, $G_{S,o}$, defined for a horizontal surface, depends on the geographic latitude, as well as the time of day and year. It may be determined from an expression of the form

$$G_{S,o} = S_c \cdot f \cdot \cos\theta \tag{12.71}$$

The quantity f is a correction factor to account for the eccentricity of the earth's orbit about the sun ($0.97 \lesssim f \lesssim 1.03$). On a time- and surface area-averaged basis, the earth receives $S_c \times \pi r_e^2/4\pi r_e^2 = S_c/4 = 342$ W/m^2 of solar irradiation. The diameter of the earth is $d_e = 2r_e = 1.27 \times 10^7$ m.

As shown in Figure 12.28*a*, the spectral distribution of extraterrestrial solar irradiation *approximates* that of a blackbody at 5800 K. The radiation is concentrated in the short

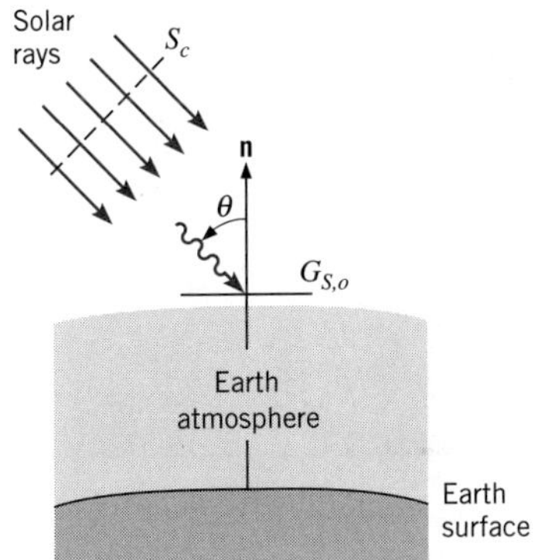

FIGURE 12.27 Directional nature of solar radiation outside the earth's atmosphere.

[3]The term *solar constant* is a misnomer in that its value varies with time in a predictable manner. Radiation emitted by the sun undergoes an 11-year cycle with peak emission (+0.65 W/m^2) corresponding to periods of high sunspot activity [8].

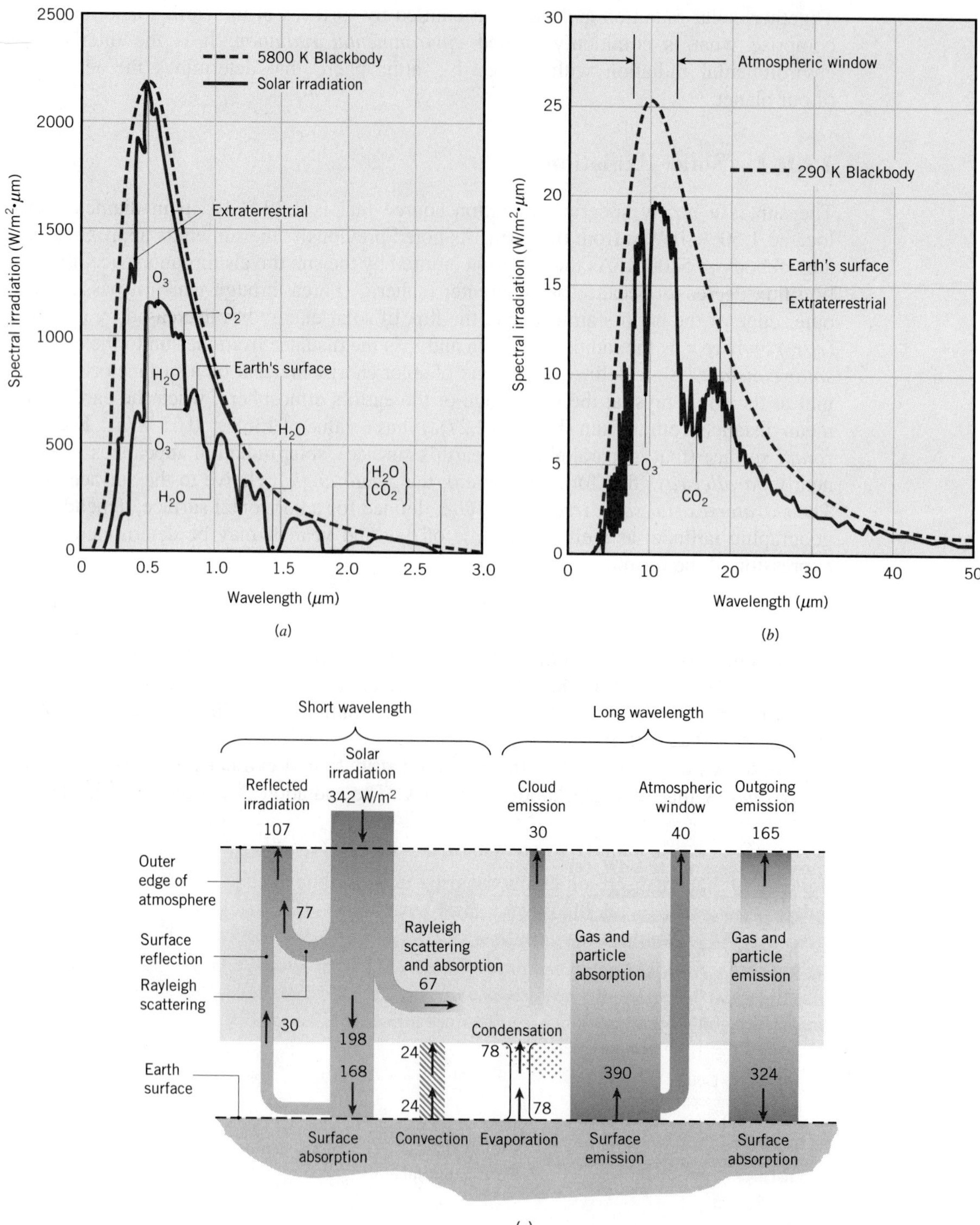

FIGURE 12.28 Solar and environmental radiation. (*a*) Spectral distribution of downward-propagating short wavelength solar radiation. (*b*) Spectral distribution of upward-propagating long wavelength environmental radiation. (*c*) Energy balance on the atmosphere for moderate temperature and cloudy conditions [9].

wavelength region ($0.2 \lesssim \lambda \lesssim 3\ \mu m$) of the spectrum with the peak occurring at approximately 0.50 μm. However, as the solar radiation propagates through the earth's atmosphere, its magnitude and both its spectral and directional distributions experience substantial modification. This change is due to *absorption* and *scattering* of the radiation by atmospheric constituents. The effect of absorption by the atmospheric gases O_3 (ozone), H_2O, O_2, and CO_2 is shown on the lower curve of Figure 12.28*a* corresponding to solar irradiation at the earth's surface, after it has passed through the atmosphere. Absorption by ozone is strong in the UV region, providing considerable attenuation below 0.4 μm and complete attenuation below 0.3 μm. In the visible region there is some absorption by O_3 and O_2, and in the near and far IR regions absorption is dominated by water vapor. Throughout the solar spectrum, there is also continuous absorption of short wavelength radiation by the dust and aerosol content of the atmosphere, including the products of fossil fuel combustion such as soot.

Atmospheric scattering provides for *redirection* of the sun's rays and therefore also affects solar radiation reaching the earth's surface. Two types of scattering are shown in Figure 12.29. *Rayleigh* (or *molecular*) *scattering* is caused by very small gas molecules. It occurs when the ratio of the effective molecule diameter to the wavelength of the radiation, $\pi D/\lambda$, is much less than unity and provides for nearly uniform scattering of the radiation in all directions. In contrast, *Mie scattering* by larger dust or soot particles occurs when $\pi D/\lambda$ is approximately unity and is concentrated in the direction of the incident rays. Hence virtually all Mie-scattered radiation strikes the earth's surface in directions close to that of the sun's rays.

12.9.2 The Atmospheric Radiation Balance

In addition to solar irradiation from above, the atmosphere is irradiated from below by the earth's surface. Since the average temperature of the earth is approximately 290 K, this upward-propagating radiation is concentrated at longer wavelengths, as shown in Figure 12.28*b*. The spectral distribution of the earth's emission varies smoothly relative to the distribution of the extraterrestrial solar irradiation of Figure 12.28*a*; the smooth variation is also characteristic of many engineered surfaces. However, like the downward-propagating solar irradiation of Figure 12.28*a*, the terrestrial emission is modified by absorption and

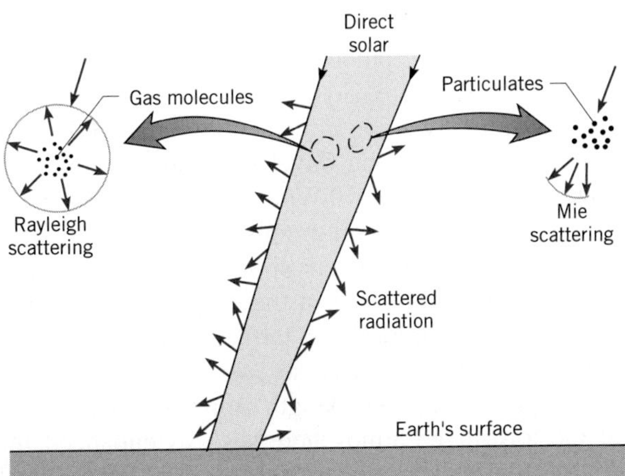

FIGURE 12.29 Scattering of solar radiation in the earth's atmosphere.

scattering as it propagates upward through the atmosphere. Absorption by water vapor occurs throughout the spectrum. Strong ozone absorption is noted in the wavelength region around 9 μm and robust CO_2 absorption spans the wavelength region $13 \lesssim \lambda \lesssim 16$ μm. The majority of terrestrial emission in the $8 \lesssim \lambda \lesssim 13$ μm *atmospheric window* propagates to the outer edge of the atmosphere, except in the spectral range associated with strong ozone absorption. Rayleigh and Mie scattering of the long wavelength terrestrial emission is triggered by the presence of various particles and aerosols in the atmosphere.

The modification of both the downward-propagating extraterrestrial solar irradiation and the upward-propagating terrestrial emission due to absorption and scattering have a profound influence on the energy balance of the atmosphere. For *either* the upward- or downward-propagating radiation, the net effect is the heating of the atmosphere since the energy content of the radiation leaving the atmosphere is less than that of the corresponding incoming radiation. However, this heating is balanced by cooling due to radiation emitted by atmospheric constituents.

A representative *equilibrium* energy balance (Figure 12.28*c*) shows the *partitioning* of the short wavelength solar irradiation and the long wavelength terrestrial emission [9]. Of the surface- and time-averaged value of 342 W/m^2 for the solar irradiation at the outer edge of earth's atmosphere, 77 W/m^2 is reflected back to space, primarily by Rayleigh scattering, while 67 W/m^2 heats the atmosphere through the effects of absorption, including absorption by soot, dust, and clouds. The remaining portion of the solar irradiation (198 W/m^2) reaches the earth's surface, where 30 W/m^2 reflects back to space and 168 W/m^2 is absorbed.

The partitioning of the long wavelength radiation associated with emission from the earth's surface is more complex. The time-averaged surface emission (390 W/m^2) is mainly absorbed by the atmosphere, except for 40 W/m^2 corresponding to the atmospheric window. The remaining 350 W/m^2 component of the surface emission combines with the short wavelength radiation absorption (67 W/m^2), convection (24 W/m^2) from the earth's surface, and condensation in the form of precipitation in the lowest regions of the atmosphere (78 W/m^2) to heat the bulk of the atmosphere. In turn, the heated atmospheric gases emit radiation at long wavelengths resulting in a radiation flux of 165 W/m^2 at the outer edge of the atmosphere, and a corresponding downward radiation flux of 324 W/m^2 at the earth's surface. Emission from clouds accounts for a radiation flux of 30 W/m^2. Since conditions are *assumed* to be in equilibrium, the *net* heat transfer at the outer edge of the atmosphere and at the earth's surface are both zero.

In reality, conditions are not in equilibrium, since the absorption and scattering of both short- and long-wavelength radiation evolve in response to the changing chemical and particulate content of our atmosphere. Anthropogenic activity that influences the makeup of the atmosphere is primarily related to the combustion of fossil fuels, leading to increases in the CO_2 and aerosol content of the atmosphere. Hence gas absorption and aerosol-induced scattering (and absorption) are continually being affected by human activity. In general, as the CO_2 content of the atmosphere increases, long wavelength radiation emitted by the surface of the earth and absorbed by the atmosphere (350 W/m^2) will be absorbed closer to the earth's surface, resulting in a decrease in the outgoing emission (165 W/m^2) of long-wavelength radiation at the top of the atmosphere and a corresponding increase in radiative flux to the earth's surface (324 W/m^2). With a reduction in the outgoing long wavelength emission, the *net* heat transfer at the top of the atmosphere, termed *radiative forcing*, is to the atmosphere, and atmospheric temperatures must increase.

The reduction in the outgoing long-wavelength emission could be offset by increases in other components at the top of the atmosphere such as enhanced reflection of short-wavelength solar irradiation due to Rayleigh scattering (107 W/m^2) [10] or reflection of short-wavelength radiation from the earth's surface (30 W/m^2) [11]. However, as engineers improve the efficiency and cleanliness of combustion processes, for example, the concentrations of aerosols and particles that are responsible for scattering are correspondingly reduced. Decreasing the production of pollutants by improving combustion technology has important health benefits for humankind, but ironically may lead to a reduction in the beneficial short-wavelength reflection (107 W/m^2), increasing the temperature of the atmosphere and also potentially changing the convection and precipitation that occurs in the lower atmosphere, influencing and modifying weather patterns [12–14]. Clearly, fossil fuel combustion and the associated environmental radiation heat transfer effects are complex, and, while they are not completely understood, they may have a profound impact at the global scale.

12.9.3 Terrestrial Solar Irradiation

Solar irradiation at the earth's surface may be utilized in a wide variety of engineering applications, including but not limited to heating and electricity generation. Increased usage of solar irradiation for these purposes reduces our dependence on fossil fuels and, in turn, can mitigate the potential for atmospheric warming. As such, knowledge of the nature of the solar irradiation at the earth's surface is crucial. Detailed treatment of solar energy technologies is left to the literature [15–19].

The cumulative effect of the scattering processes on the *directional distribution* of solar radiation striking the earth's surface is shown in Figure 12.30*a*. That portion of the radiation that has penetrated the atmosphere without having been scattered (or absorbed) is in the direction of the zenith angle and is termed the *direct radiation*. The scattered radiation is incident from all directions, although its intensity is largest for directions close to that of the direct radiation. The scattered radiation may vary from approximately 10% of the total solar radiation on a clear day to nearly 100% on a very overcast day. The scattered component of the solar radiation is often *approximated* as being independent of direction (Figure 12.30*b*), or *diffuse*.

As evident in Figure 12.28*c*, the long-wavelength forms of the environmental radiation include emission from the earth's surface, as well as gas emission from certain atmospheric constituents. The emissive power from the earth's surface may be computed in the conventional manner. That is,

$$E = \varepsilon \sigma T^4 \tag{12.72}$$

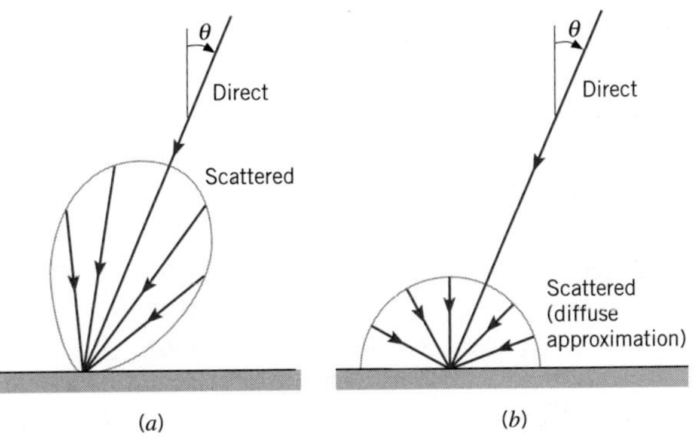

FIGURE 12.30 Directional distribution of solar radiation at the earth's surface. (*a*) Actual distribution. (*b*) Diffuse approximation.

where ε and T are the surface emissivity and temperature, respectively. As implied in Figure 12.28*b*, emissivities are generally close to unity; that of water, for example, is approximately 0.97. Using $\varepsilon = 0.97$ and $\overline{E} = 390 \text{ W/m}^2$ from Figure 12.28*c*, the effective radiation temperature of the earth is $\overline{T} = 291$ K. Emission is concentrated in the spectral region from approximately 4 to 40 μm, with the peak occurring at approximately 10 μm, as evident in Figure 12.28*b*.

Downward-propagating atmospheric emission that is incident upon the earth's surface is largely due to the CO_2 and H_2O content of the atmosphere and is concentrated in the spectral regions from 5 to 8 μm and above 13 μm. Although the spectral distribution of atmospheric emission does not correspond to that of a blackbody, its contribution to irradiation of the earth's surface can be estimated using Equation 12.32. In particular, irradiation at the earth's surface due to atmospheric emission may be expressed in the form

$$G_{\text{atm}} = \sigma T_{\text{sky}}^4 \tag{12.73}$$

where T_{sky} is termed the *effective sky temperature*. Its value depends on atmospheric conditions, and for the moderate temperature, cloudy conditions of Figure 12.28*c*, $G_{\text{atm}} = 324 \text{ W/m}^2$ and $T_{\text{sky}} = 275$ K. Actual values range from a low of 230 K under a cold, clear sky to a high of approximately 285 K under warm, cloudy conditions. When its value is small, as on a cool, clear night, an exposed pool of water may freeze even though the air temperature exceeds 273 K.

We close by recalling that values of the spectral properties of a surface at short wavelengths may be appreciably different from values at long wavelengths (Figures 12.17 and 12.22). Since solar radiation is concentrated in the short wavelength region of the spectrum and surface emission is at much longer wavelengths, it follows that many surfaces may not be approximated as gray in their response to solar irradiation. In other words, the solar absorptivity of a surface α_S may differ from its emissivity ε. Values of α_S and the emissivity at moderate temperatures are presented in Table 12.3 for representative surfaces. Note that the ratio α_S/ε is an important engineering parameter. Small values are desired if the surface is intended to reject heat; large values are required if the surface is intended to collect solar energy.

TABLE 12.3 Solar Absorptivity α_S and Emissivity ε of Surfaces Having Spectral Absorptivity Given in Figure 12.22

Surface	α_S	ε (300 K)	α_S/ε
Evaporated aluminum film	0.09	0.03	3.0
Fused quartz on aluminum film	0.19	0.81	0.24
White paint on metallic substrate	0.21	0.96	0.22
Black paint on metallic substrate	0.97	0.97	1.0
Stainless steel, as received, dull	0.50	0.21	2.4
Red brick	0.63	0.93	0.68
Human skin (Caucasian)	0.62	0.97	0.64
Snow	0.28	0.97	0.29
Corn leaf	0.76	0.97	0.78

Example 12.12

A flat-plate solar collector with no cover plate has a selective absorber surface of emissivity 0.1 and solar absorptivity 0.95. At a given time of day the absorber surface temperature T_s is 120°C when the solar irradiation is 750 W/m^2, the effective sky temperature is -10°C, and the ambient air temperature T_∞ is 30°C. Assume that the heat transfer convection coefficient for the calm day conditions can be estimated from

$$\overline{h} = 0.22(T_s - T_\infty)^{1/3}\ \text{W/m}^2\cdot\text{K}$$

Calculate the useful heat removal rate (W/m^2) from the collector for these conditions. What is the corresponding efficiency of the collector?

Solution

Known: Operating conditions for a flat-plate solar collector.

Find:

1. Useful heat removal rate per unit area, q''_u (W/m^2).
2. Efficiency η of the collector.

Schematic:

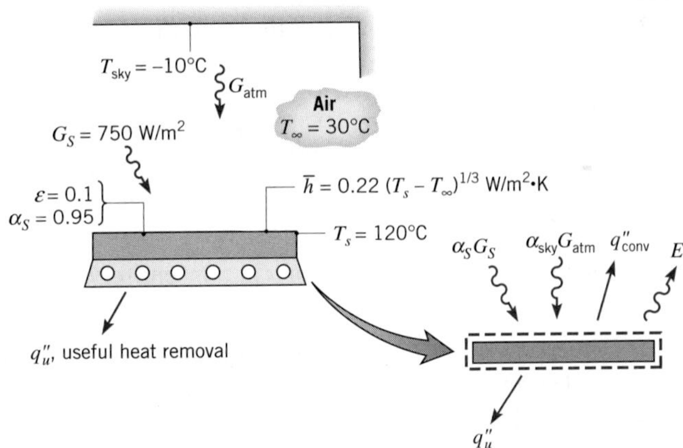

Assumptions:

1. Steady-state conditions.
2. Bottom of collector well insulated.
3. Absorber surface diffuse.

Analysis:

1. Performing an energy balance on the absorber,

$$\dot{E}_{in} - \dot{E}_{out} = 0$$

or, per unit surface area,

$$\alpha_S G_S + \alpha_{sky} G_{atm} - q''_{conv} - E - q''_u = 0$$

From Equation 12.73,

$$G_{atm} = \sigma T_{sky}^4$$

Since the atmospheric irradiation is concentrated in approximately the same spectral region as that of surface emission, it is reasonable to assume that

$$\alpha_{sky} \approx \varepsilon = 0.1$$

With

$$q''_{conv} = \bar{h}(T_s - T_\infty) = 0.22(T_s - T_\infty)^{4/3} \qquad \text{and} \qquad E = \varepsilon\sigma T_s^4$$

it follows that

$$q''_u = \alpha_S G_S + \varepsilon\sigma T_{sky}^4 - 0.22(T_s - T_\infty)^{4/3} - \varepsilon\sigma T_s^4$$

$$q''_u = \alpha_S G_S - 0.22(T_s - T_\infty)^{4/3} - \varepsilon\sigma(T_s^4 - T_{sky}^4)$$

$$q''_u = 0.95 \times 750\ \text{W/m}^2 - 0.22(120 - 30)^{4/3}\ \text{W/m}^2$$
$$-0.1 \times 5.67 \times 10^{-8}\ \text{W/m}^2 \cdot \text{K}^4(393^4 - 263^4)\ \text{K}^4$$

$$q''_u = (712.5 - 88.7 - 108.1)\ \text{W/m}^2 = 516\ \text{W/m}^2 \qquad \triangleleft$$

2. The collector efficiency, defined as the fraction of the solar irradiation extracted as useful energy, is then

$$\eta = \frac{q''_u}{G_S} = \frac{516\ \text{W/m}^2}{750\ \text{W/m}^2} = 0.69 \qquad \triangleleft$$

Comments:

1. Since the spectral range of G_{atm} is entirely different from that of G_S, it would be *incorrect* to assume that $\alpha_{sky} = \alpha_S$.
2. The convection heat transfer coefficient is extremely low ($\bar{h} \approx 1\ \text{W/m}^2 \cdot \text{K}$). With a modest increase to $\bar{h} = 5\ \text{W/m}^2 \cdot \text{K}$, the useful heat flux and the efficiency are reduced to $q''_u = 154\ \text{W/m}^2$ and $\eta = 0.21$. A cover plate can contribute significantly to reducing convection (and radiation) heat loss from the absorber plate.

12.10 *Summary*

Many new and important ideas have been introduced in this chapter, and at this stage you may well be confused, particularly by the terminology. However, the subject matter has been developed in a systematic fashion, and a careful rereading of the material should make you more comfortable with its application. A glossary has been provided in Table 12.4 to assist you in assimilating the terminology.

TABLE 12.4 Glossary of Radiative Terms

Term	Definition
Absorption	The process of converting radiation intercepted by matter to internal thermal energy.
Absorptivity	Fraction of the incident radiation absorbed by matter. Equations 12.47, 12.48, and 12.51. Modifiers: *directional, hemispherical, spectral, total.*
Blackbody	The ideal emitter and absorber. Modifier referring to ideal behavior. Denoted by the subscript b.
Diffuse	Modifier referring to the directional independence of the intensity associated with emitted, reflected, or incident radiation.
Directional	Modifier referring to a particular direction. Denoted by the subscript θ.
Directional distribution	Refers to variation with direction.
Emission	The process of radiation production by matter at a finite temperature. Modifiers: *diffuse, blackbody, spectral.*
Emissive power	Rate of radiant energy emitted by a surface in all directions per unit area of the surface, E (W/m^2). Modifiers: *spectral, total, blackbody.*
Emissivity	Ratio of the radiation emitted by a surface to the radiation emitted by a blackbody at the same temperature. Equations 12.36, 12.38, 12.39, and 12.40. Modifiers: *directional, hemispherical, spectral, total.*
Gray surface	A surface for which the spectral absorptivity and the emissivity are independent of wavelength over the spectral regions of surface irradiation and emission.
Hemispherical	Modifier referring to all directions in the space above a surface.
Intensity	Rate of radiant energy propagation in a particular direction, per unit area normal to the direction, per unit solid angle about the direction, I ($W/m^2 \cdot sr$). Modifier: *spectral.*
Irradiation	Rate at which radiation is incident on a surface from all directions per unit area of the surface, G (W/m^2). Modifiers: *spectral, total, diffuse.*
Kirchhoff's law	Relation between emission and absorption properties for surfaces irradiated by a blackbody at the same temperature. Equations 12.65, 12.66, 12.67, and 12.68.
Planck's law	Spectral distribution of emission from a blackbody. Equation 12.30.
Radiosity	Rate at which radiation leaves a surface due to emission and reflection in all directions per unit area of the surface, J (W/m^2). Modifiers: *spectral, total.*
Reflection	The process of redirection of radiation incident on a surface. Modifiers: *diffuse, specular.*
Reflectivity	Fraction of the incident radiation reflected by matter. Equations 12.54, 12.55, and 12.57. Modifiers: *directional, hemispherical, spectral, total.*
Semitransparent	Refers to a medium in which radiation absorption is a volumetric process.
Solid angle	Region subtended by an element of area on the surface of a sphere with respect to the center of the sphere, ω (sr). Equations 12.7 and 12.8.
Spectral	Modifier referring to a single-wavelength (monochromatic) component. Denoted by the subscript λ.
Spectral distribution	Refers to variation with wavelength.

TABLE 12.4 *Continued*

Term	Definition
Specular	Refers to a surface for which the angle of reflected radiation is equal to the angle of incident radiation.
Stefan–Boltzmann law	Emissive power of a blackbody. Equation 12.32.
Thermal radiation	Electromagnetic energy emitted by matter at a finite temperature and concentrated in the spectral region from approximately 0.1 to 100 μm.
Total	Modifier referring to all wavelengths.
Transmission	The process of thermal radiation passing through matter.
Transmissivity	Fraction of the incident radiation transmitted by matter. Equations 12.59 and 12.60. Modifiers: *hemispherical, spectral, total.*
Wien's displacement law	Locus of the wavelength corresponding to peak emission by a blackbody. Equation 12.31.

Test your understanding of the terms and concepts introduced in this chapter by addressing the following questions.

- What is the nature of radiation? What two important features characterize radiation?
- What is the physical origin of radiation *emission* from a surface? How does emission affect the thermal energy of a material?
- In what region of the electromagnetic spectrum is *thermal radiation* concentrated?
- What is the *spectral intensity* of radiation emitted by a surface? On what variables does it depend? How may knowledge of this dependence be used to determine the rate at which matter loses thermal energy due to emission from its surface?
- What is a *steradian*? How many steradians are associated with a hemisphere?
- What is the distinction between *spectral* and *total* radiation? Between *directional* and *hemispherical* radiation?
- What is *total emissive power*? What role does it play in a surface energy balance?
- What is a *diffuse emitter*? For such an emitter, how is the intensity related to the total emissive power?
- What is *irradiation*? How is it related to the intensity of incident radiation, if the radiation is diffuse?
- What is *radiosity*? What role do the total radiosity and the total irradiation play in a surface energy balance?
- What are the characteristics of a *blackbody*? Does such a thing actually exist in nature? What is the principal role of blackbody behavior in radiation analysis?
- What is the *Planck distribution*? What is *Wien's displacement law*?
- From memory, sketch the spectral distribution of radiation emission from a blackbody at three temperatures, $T_1 < T_2 < T_3$. Identify salient features of the distributions.
- In what region of the electromagnetic spectrum is radiation emission from a surface at room temperature concentrated? What is the spectral region of concentration for a surface at 1000°C? For the surface of the sun?

- What is the *Stefan–Boltzmann law*? How would you determine the total intensity of radiation emitted by a blackbody at a prescribed temperature?
- How would you approximate the total irradiation of a small surface in a large isothermal enclosure?
- In the term *total, hemispherical emissivity*, to what do the adjectives *total* and *hemispherical* refer?
- How does the directional emissivity of a material change as the zenith angle associated with emission approaches 90°?
- If the spectral emissivity of a material increases with increasing wavelength, how does its total emissivity vary with temperature?
- What is larger, the emissivity of a polished metal or an oxidized metal? A refractory brick or ice?
- What processes accompany irradiation of a *semitransparent* material? An *opaque* material?
- Are glass and water semitransparent or opaque materials?
- How is the perceived color of a material determined by its response to irradiation in the infrared portion of the spectrum? How is its color affected by its temperature?
- Can snow be thought of as a good absorber or reflector of incident infrared radiation?
- How is the thermal energy of a material affected by the absorption of incident radiation? By the reflection of incident radiation?
- Can the total absorptivity of an opaque surface at a fixed temperature differ according to whether irradiation is from a source at room temperature or from the sun? Can its reflectivity differ? Its emissivity?
- What is a *diffuse reflector*? A *specular reflector*? How does the roughness of a surface affect the nature of reflection for the surface?
- Under what conditions is there equivalence between the spectral, directional emissivity of a surface and the spectral, directional absorptivity? The spectral, hemispherical emissivity and the spectral, hemispherical absorptivity? The total, hemispherical emissivity and the total, hemispherical absorptivity?
- What is a *gray surface*?
- How does the gas and aerosol content of the atmosphere modify the spectral variation of downward-propagating solar radiation? How does the content of the atmosphere modify the spectral variation of upward-propagating terrestrial emission?
- What is meant by radiative forcing, and what is the impact of such forcing on the temperature of earth's atmosphere?
- How can the effective radiation temperature of the earth be calculated from the atmospheric radiation balance? What is the effective sky temperature, and how can it be determined from the atmospheric radiation balance?
- What is the directional nature of solar radiation outside the earth's atmosphere? At the surface of the earth?
- What is the primary difference between Rayleigh and Mie scattering? In the context of solar and environmental radiation, how do these scattering phenomena affect the temperature of earth's atmosphere? How might anthropogenic activity affect radiation scattering, absorption, and emission in the atmosphere?

References

1. Planck, M., *The Theory of Heat Radiation,* Dover Publications, New York, 1959.
2. Zetteli, N., *Quantum Mechanics Concepts and Applications*, Wiley, Chichester, 2001.
3. Gubareff, G. G., J. E. Janssen, and R. H. Torberg, *Thermal Radiation Properties Survey,* 2nd ed., Honeywell Research Center, Minneapolis, 1960.
4. Wood, W. D., H. W. Deem, and C. F. Lucks, *Thermal Radiative Properties,* Plenum Press, New York, 1964.
5. Touloukian, Y. S., *Thermophysical Properties of High Temperature Solid Materials,* Macmillan, New York, 1967.
6. Touloukian, Y. S., and D. P. DeWitt, *Thermal Radiative Properties,* Vols. 7, 8, and 9, from *Thermophysical Properties of Matter,* TPRC Data Series, Y.S. Touloukian and C. Y. Ho, Eds., IFI Plenum, New York, 1970–1972.
7. Howell, J. R., R. Siegel, and M. P. Menguc, *Thermal Radiation Heat Transfer*, 5th ed., Taylor & Francis, New York, 2010.
8. National Academy of Sciences, *Solar Influences on Global Change*, National Academy Press, Washington, D.C. 2004.
9. Kiehl, J. T., and K. E. Trenberth, *Bull. Am. Met. Soc.* **78**, 197, 1997.
10. Myhre, G., *Science*, **325**, 187, 2009.
11. Akbari, H., S. Menon, and A. Rosenfeld, *Climate Change*, **94**, 275, 2009.
12. Arneth, A., N. Unger, M. Kulmala, and M. O Andreae, *Science*, **326**, 672, 2009.
13. Shindell, D. T., G. Faluvegi, D. M. Koch, G. A. Schmidt, N. Unger, and S. E. Bauer, *Science*, **326**, 716, 2009.
14. Ramanathan, V., P. J. Crutzen, J. T. Kiehl, and D. Rosenfeld, *Science*, **294**, 2119, 2001.
15. Duffie, J. A., and W. A. Beckman, *Solar Engineering of Thermal Processes*, 3rd ed., Wiley, Hoboken, NJ, 2006.
16. Goswami, D. Y., F. Kreith, and J. F. Kreider, *Principles of Solar Energy*, 2nd ed., Taylor & Francis, New York, 2002.
17. Howell, J. R., R. B. Bannerot, and G.C. Vliet, *Solar-Thermal Energy Systems, Analysis and Design,* McGraw-Hill, New York, 1982.
18. Kalogirou, S. A., *Prog. Energy Comb. Sci.*, **30**, 231, 2004.
19. Kalogirou, S. A., *Solar Energy Engineering: Processes and Systems*, Elsevier, Oxford, 2009.

Problems

Radiation Heat Fluxes

12.1 Consider an opaque horizontal plate that is well insulated on its back side. The irradiation on the plate is 2500 W/m^2, of which 500 W/m^2 is reflected. The plate is at 227°C and has an emissive power of 1200 W/m^2. Air at 127°C flows over the plate with a heat transfer convection coefficient of 15 W/m$^2 \cdot$K. Determine the emissivity, absorptivity, and radiosity of the plate. What is the net heat transfer rate per unit area?

12.2 A horizontal, opaque surface at a steady-state temperature of 77°C is exposed to an airflow having a free stream temperature of 27°C with a convection heat transfer coefficient of 28 W/m$^2 \cdot$K. The emissive power of the surface is 628 W/m^2, the irradiation is 1380 W/m^2, and the reflectivity is 0.40. Determine the absorptivity of the surface. Determine the net radiation heat transfer rate for this surface. Is this heat transfer to the surface or from the surface? Determine the combined heat transfer rate for the surface. Is this heat transfer to the surface or from the surface?

12.3 The top surface of an $L = 5$-mm-thick anodized aluminum plate is irradiated with $G = 1000$ W/m^2 while being simultaneously exposed to convection conditions characterized by $h = 40$ W/m$^2 \cdot$K and $T_\infty = 30$°C. The bottom surface of the plate is insulated. For a plate temperature of 400 K as well as $\alpha = 0.14$ and $\varepsilon = 0.76$, determine the radiosity at the top plate surface, the net radiation heat flux at the top surface, and the rate at which the temperature of the plate is changing with time.

12.4 A horizontal semitransparent plate is uniformly irradiated from above and below, while air at $T_\infty = 300$ K flows over the top and bottom surfaces, providing a uniform convection heat transfer coefficient of $h = 40$ W/m$^2 \cdot$K. The absorptivity of the plate to the irradiation is 0.40. Under steady-state conditions measurements made with a radiation detector above the top surface indicate a radiosity (which includes transmission, as well as reflection and emission) of $J = 5000$ W/m^2, while the plate is at a uniform temperature of $T = 350$ K.

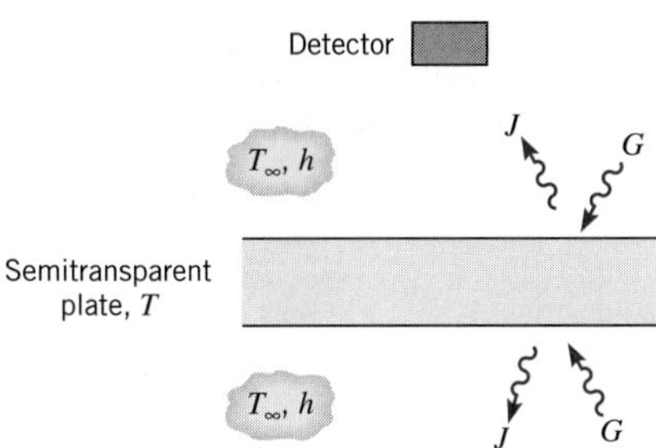

Determine the irradiation G and the emissivity of the plate. Is the plate gray ($\varepsilon = \alpha$) for the prescribed conditions?

Intensity, Emissive Power, and Irradiation

12.5 What is the irradiation at surfaces A_2, A_3, and A_4 of Example 12.1 due to emission from A_1?

12.6 Consider a small surface of area $A_1 = 10^{-4}\ \text{m}^2$, which emits diffusely with a total, hemispherical emissive power of $E_1 = 5 \times 10^4\ \text{W/m}^2$.

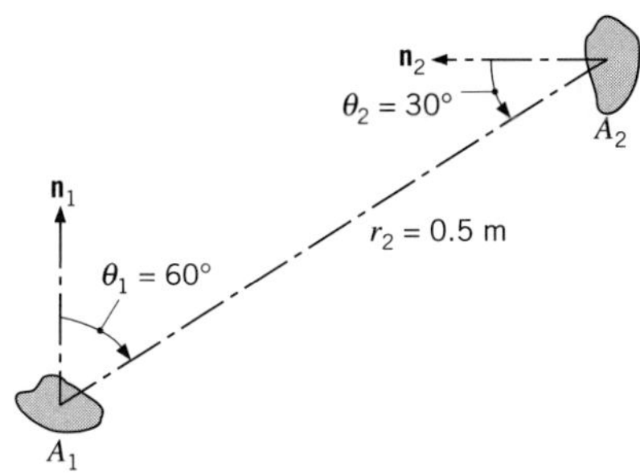

(a) At what rate is this emission intercepted by a small surface of area $A_2 = 5 \times 10^{-4}\ \text{m}^2$, which is oriented as shown?

(b) What is the irradiation G_2 on A_2?

(c) For zenith angles of $\theta_2 = 0$, 30, and 60°, plot G_2 as a function of the separation distance for $0.25 \leq r_2 \leq 1.0$ m.

12.7 A furnace with an aperture of 20-mm diameter and emissive power of $3.72 \times 10^5\ \text{W/m}^2$ is used to calibrate a heat flux gage having a sensitive area of $1.6 \times 10^{-5}\ \text{m}^2$.

(a) At what distance, measured along a normal from the aperture, should the gage be positioned to receive irradiation of 1000 W/m²?

(b) If the gage is tilted off normal by 20°, what will be its irradiation?

(c) For tilt angles of 0, 20, and 60°, plot the gage irradiation as a function of the separation distance for values ranging from 100 to 300 mm.

12.8 A small radiant source A_1 emits diffusely with an intensity $I_1 = 1.2 \times 10^5\ \text{W/m}^2 \cdot \text{sr}$. The radiation detector A_2 is aligned normal to the source at a distance of $L_o = 0.2$ m. An opaque screen is positioned midway between A_1 and A_2 to prevent radiation from the source reaching the detector. The small surface A_m is a perfectly diffuse mirror that permits radiation emitted from the source to be reflected into the detector.

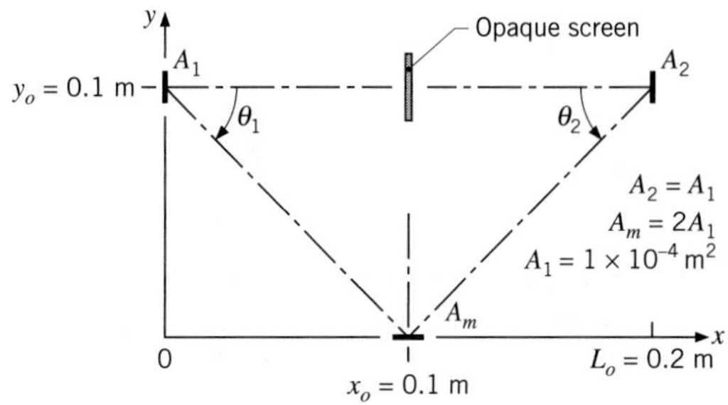

(a) Calculate the radiant power incident on A_m due to emission from the source A_1, $q_{1\to m}$(W).

(b) Assuming that the radiant power, $q_{1\to m}$, is perfectly and diffusely reflected, calculate the intensity leaving A_m, I_m (W/m² · sr).

(c) Calculate the radiant power incident on A_2 due to the reflected radiation leaving A_m, $q_{m\to 2}(\mu\text{W})$.

(d) Plot the radiant power $q_{m\to 2}$ as a function of the lateral separation distance y_o for the range $0 \leq y_o \leq 0.2$ m. Explain the features of the resulting curve.

12.9 According to its directional distribution, solar radiation incident on the earth's surface may be divided into two components. The *direct* component consists of parallel rays incident at a fixed zenith angle θ, while the *diffuse* component consists of radiation that may be approximated as being diffusely distributed with θ.

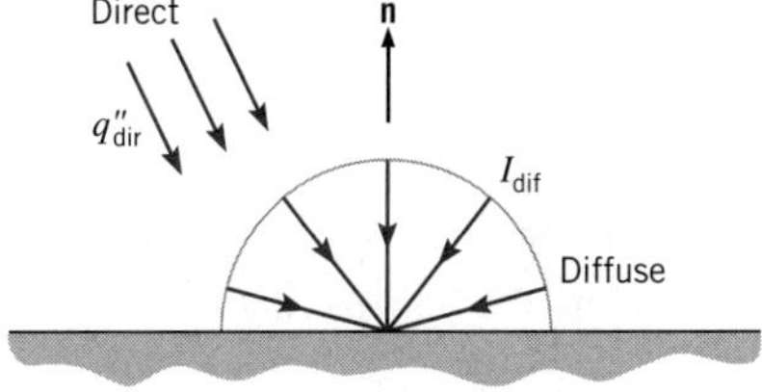

Consider clear sky conditions for which the direct radiation is incident at $\theta = 30°$, with a total flux (based on an area that is normal to the rays) of $q''_{\text{dir}} = 1000\ \text{W/m}^2$, and the total intensity of the diffuse radiation is $I_{\text{dif}} = 70\ \text{W/m}^2 \cdot \text{sr}$. What is the total solar irradiation at the earth's surface?

12.10 Solar radiation incident on the earth's surface may be divided into the direct and diffuse components described in Problem 12.9. Consider conditions for a day in which the intensity of the direct solar radiation is $I_{dir} = 210 \times 10^7$ W/m²·sr in the solid angle subtended by the sun with respect to the earth, $\Delta\omega_s = 6.74 \times 10^{-5}$ sr. The intensity of the diffuse radiation is $I_{dif} = 70$ W/m²·sr.

(a) What is the total solar irradiation at the earth's surface when the direct radiation is incident at $\theta = 30°$?

(b) Verify the prescribed value for $\Delta\omega_s$, recognizing that the diameter of the sun is 1.39×10^9 m and the distance between the sun and the earth is 1.496×10^{11} m (1 astronomical unit).

12.11 On an overcast day the directional distribution of the solar radiation incident on the earth's surface may be approximated by an expression of the form $I_i = I_n \cos\theta$, where $I_n = 80$ W/m²·sr is the total intensity of radiation directed normal to the surface and θ is the zenith angle. What is the solar irradiation at the earth's surface?

12.12 During radiant heat treatment of a thin-film material, its shape, which may be hemispherical (a) or spherical (b), is maintained by a relatively low air pressure (as in the case of a rubber balloon). Irradiation on the film is due to emission from a radiant heater of area $A_h = 0.0052$ m², which emits diffusely with an intensity of $I_{e,h} = 169{,}000$ W/m²·sr.

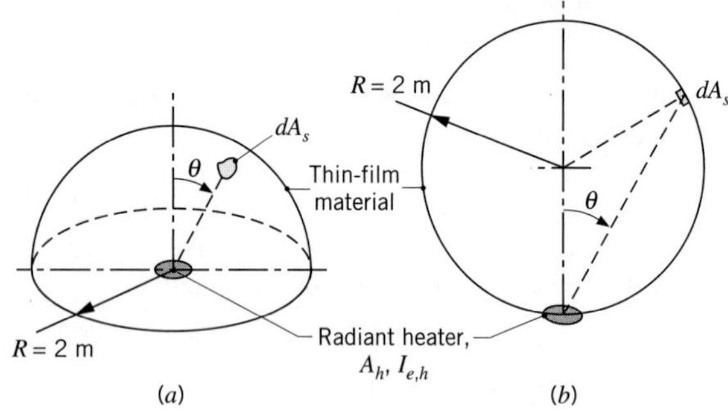

(a) Obtain an expression for the irradiation on the film as a function of the zenith angle θ.

(b) Based on the expressions derived in part (a), which shape provides the more uniform irradiation G and hence provides better quality control for the treatment process?

12.13 To initiate a process operation, an infrared motion sensor (radiation detector) is employed to determine the approach of a hot part on a conveyor system. To set the sensor's amplifier discriminator, the engineer needs a relationship between the sensor output signal, S, and the position of the part on the conveyor. The sensor output signal is proportional to the rate at which radiation is incident on the sensor.

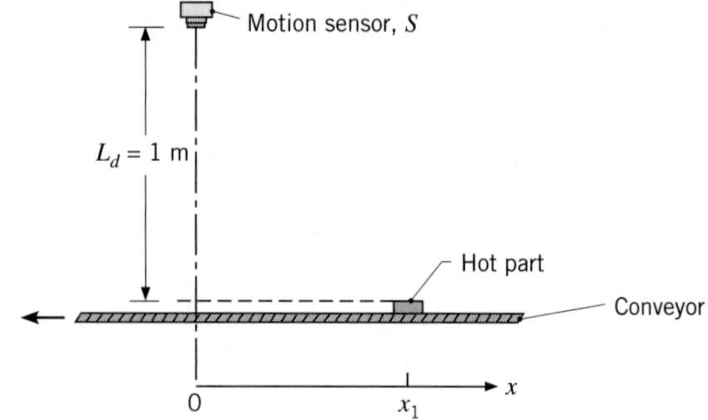

(a) For $L_d = 1$ m, at what location x_1 will the sensor signal S_1 be 75% of the signal corresponding to the position directly beneath the sensor, S_o $(x = 0)$?

(b) For values of $L_d = 0.8$, 1.0, and 1.2 m, plot the signal ratio, S/S_o, versus part position, x, for signal ratios in the range from 0.2 to 1.0. Compare the x-locations for which $S/S_o = 0.75$.

12.14 A small radiant heat source of area $A_1 = 2 \times 10^{-4}$ m² emits diffusely with an intensity $I_1 = 1000$ W/m²·sr. A second small area, $A_2 = 1 \times 10^{-4}$ m², is located as shown in the sketch.

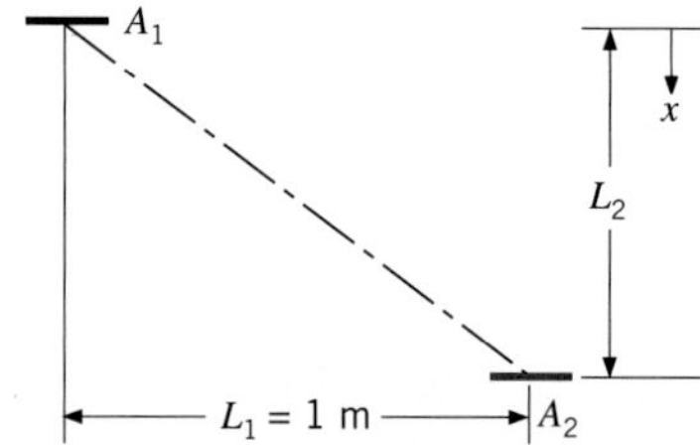

(a) Determine the irradiation of A_2 for $L_2 = 0.5$ m.

(b) Plot the irradiation of A_2 over the range $0 \leq L_2 \leq 10$ m.

12.15 Determine the fraction of the total, hemispherical emissive power that leaves a diffuse surface in the directions $\pi/4 \leq \theta \leq \pi/2$ and $0 \leq \phi \leq \pi$.

12.16 The spectral distribution of the radiation emitted by a diffuse surface may be approximated as follows.

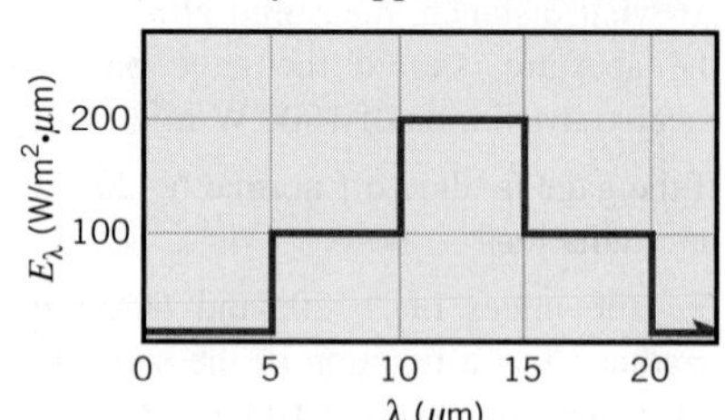

(a) What is the total emissive power?

(b) What is the total intensity of the radiation emitted in the normal direction and at an angle of 30° from the normal?

(c) Determine the fraction of the emissive power leaving the surface in the directions $\pi/4 \leq \theta \leq \pi/2$.

12.17 Consider a 5-mm-square, diffuse surface ΔA_o having a total emissive power of $E_o = 4000\ \text{W/m}^2$. The radiation field due to emission into the hemispherical space above the surface is diffuse, thereby providing a uniform intensity $I(\theta, \phi)$. Moreover, if the space is a nonparticipating medium (nonabsorbing, nonscattering, and nonemitting), the intensity is independent of radius for any (θ, ϕ) direction. Hence intensities at any points P_1 and P_2 would be equal.

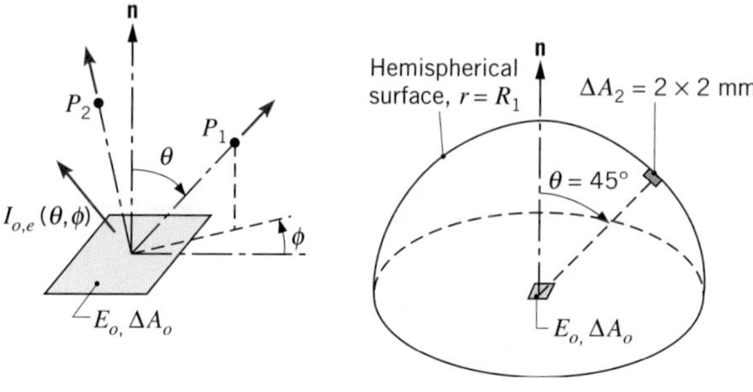

(a) What is the rate at which radiant energy is emitted by ΔA_o, q_{emit}?

(b) What is the intensity $I_{o,e}$ of the radiation field emitted from the surface ΔA_o?

(c) Beginning with Equation 12.13 and presuming knowledge of the intensity $I_{o,e}$, obtain an expression for q_{emit}.

(d) Consider the hemispherical surface located at $r = R_1 = 0.5$ m. Using the conservation of energy requirement, determine the rate at which radiant energy is incident on this surface due to emission from ΔA_o.

(e) Using Equation 12.10, determine the rate at which radiant energy leaving ΔA_o is intercepted by the small area ΔA_2 located in the direction $(45°, \phi)$ on the hemispherical surface. What is the irradiation on ΔA_2?

(f) Repeat part (e) for the location $(0°, \phi)$. Are the irradiations at the two locations equal?

(g) Using Equation 12.18, determine the irradiation G_1 on the hemispherical surface at $r = R_1$.

Blackbody Radiation

12.18 Assuming blackbody behavior, determine the temperature of, and the energy emitted by, areas A_1 in Example 12.1 and Problems 12.8 and 12.14, as well as area A_h in Problem 12.12.

12.19 The dark surface of a ceramic stove top may be approximated as a blackbody. The "burners," which are integral with the stove top, are heated from below by electric resistance heaters.

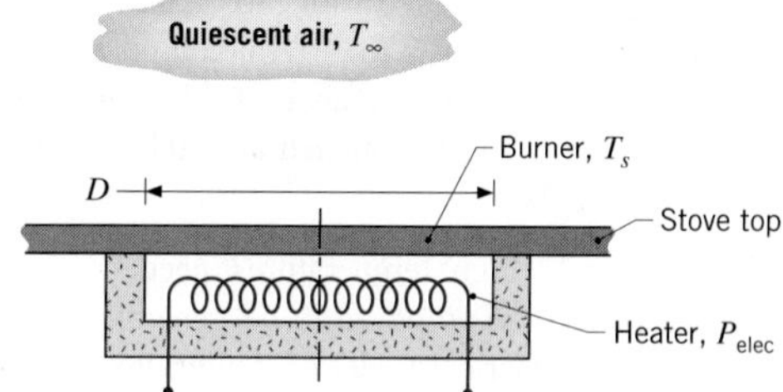

(a) Consider a burner of diameter $D = 200$ mm operating at a uniform surface temperature of $T_s = 250°\text{C}$ in ambient air at $T_\infty = 20°\text{C}$. Without a pot or pan on the burner, what are the rates of heat loss by radiation and convection from the burner? If the efficiency associated with energy transfer from the heaters to the burners is 90%, what is the electric power requirement? At what wavelength is the spectral emission a maximum?

(b) Compute and plot the effect of the burner temperature on the heat rates for $100 \leq T_s \leq 350°\text{C}$.

12.20 The energy flux associated with solar radiation incident on the outer surface of the earth's atmosphere has been accurately measured and is known to be $1368\ \text{W/m}^2$. The diameters of the sun and earth are 1.39×10^9 and 1.27×10^7 m, respectively, and the distance between the sun and the earth is 1.5×10^{11} m.

(a) What is the emissive power of the sun?

(b) Approximating the sun's surface as black, what is its temperature?

(c) At what wavelength is the spectral emissive power of the sun a maximum?

(d) Assuming the earth's surface to be black and the sun to be the only source of energy for the earth, estimate the earth's surface temperature.

12.21 A small flat plate is positioned just beyond the earth's atmosphere and is oriented such that the normal to the plate passes through the center of the sun. Refer to Problem 12.20 for pertinent earth–sun dimensions.

(a) What is the solid angle subtended by the sun about a point on the surface of the plate?

(b) Determine the incident intensity, I_i, on the plate using the known value of the solar irradiation above the earth's atmosphere ($G_S = 1368$ W/m^2).

(c) Sketch the incident intensity I_i as a function of the zenith angle θ, where θ is measured from the normal to the plate.

12.22 A spherical aluminum shell of inside diameter $D = 2$ m is evacuated and is used as a radiation test chamber. If the inner surface is coated with carbon black and maintained at 600 K, what is the irradiation on a small test surface placed in the chamber? If the inner surface were not coated and maintained at 600 K, what would the irradiation be?

12.23 The extremely high temperatures needed to trigger nuclear fusion are proposed to be generated by laser-irradiating a spherical pellet of deuterium and tritium fuel of diameter $D_p = 1.8$ mm.

(a) Determine the maximum fuel temperature that can be achieved by irradiating the pellet with 200 lasers, each producing a power of $P = 500$ W. The pellet has an absorptivity $\alpha = 0.3$ and emissivity $\varepsilon = 0.8$.

(b) The pellet is placed inside a cylindrical enclosure. Two laser entrance holes are located at either end of the enclosure and have a diameter of $D_{LEH} = 2$ mm. Determine the maximum temperature that can be generated within the enclosure.

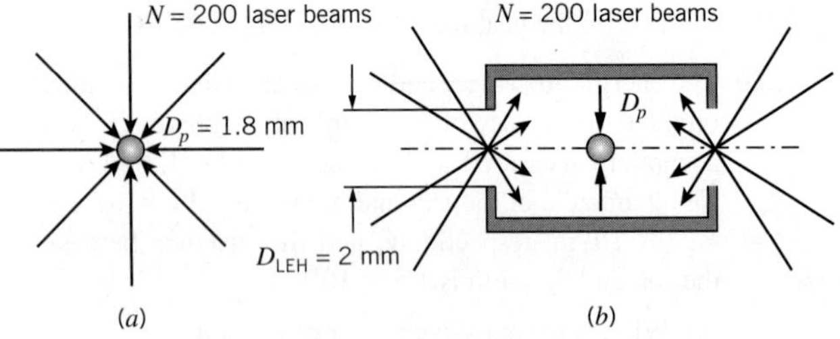

12.24 An enclosure has an inside area of 100 m^2, and its inside surface is black and is maintained at a constant temperature. A small opening in the enclosure has an area of 0.02 m^2. The radiant power emitted from this opening is 70 W. What is the temperature of the interior enclosure wall? If the interior surface is maintained at this temperature, but is now polished, what will be the value of the radiant power emitted from the opening?

12.25 Assuming the earth's surface is black, estimate its temperature if the sun has an equivalent blackbody temperature of 5800 K. The diameters of the sun and earth are 1.39×10^9 and 1.27×10^7 m, respectively, and the distance between the sun and earth is 1.5×10^{11} m.

12.26 A proposed method for generating electricity from solar irradiation is to concentrate the irradiation into a cavity that is placed within a large container of a salt with a high melting temperature. If all heat losses are neglected, part of the solar irradiation entering the cavity is used to melt the salt while the remainder is used to power a Rankine cycle. (The salt is melted during the day and is resolidified at night in order to generate electricity around the clock.)

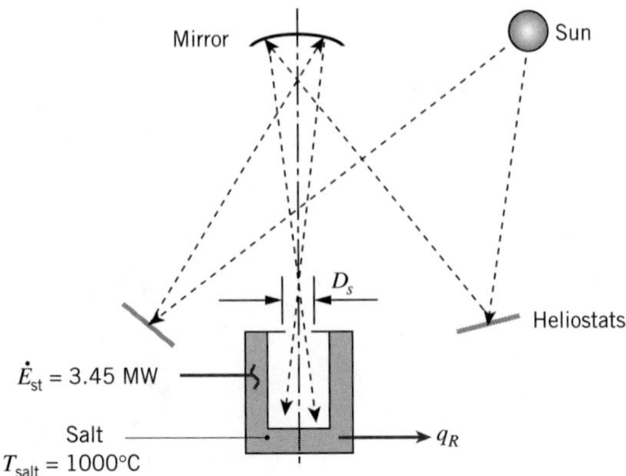

Consider conditions for which the solar power entering the cavity is $q_{sol} = 7.50$ MW and the time rate of change of energy stored in the salt is $\dot{E}_{st} = 3.45$ MW. For a cavity opening of diameter $D_s = 1$ m, determine the heat transfer to the Rankine cycle, q_R. The temperature of the salt is maintained at its melting point, $T_{salt} = T_m = 1000°$C. Neglect heat loss by convection and irradiation from the surroundings.

12.27 Approximations to Planck's law for the spectral emissive power are the Wien and Rayleigh-Jeans spectral distributions, which are useful for the extreme low and high limits of the product λT, respectively.

(a) Show that the Planck distribution will have the form

$$E_{\lambda,b}(\lambda, T) \approx \frac{C_1}{\lambda^5} \exp\left(-\frac{C_2}{\lambda T}\right)$$

when $C_2/\lambda T \gg 1$ and determine the error (compared to the exact distribution) for the condition $\lambda T = 2898$ μm·K. This form is known as Wien's law.

(b) Show that the Planck distribution will have the form

$$E_{\lambda,b}(\lambda, T) \approx \frac{C_1}{C_2}\frac{T}{\lambda^4}$$

when $C_2/\lambda T \ll 1$ and determine the error (compared to the exact distribution) for the condition $\lambda T = 100{,}000\ \mu\text{m}\cdot\text{K}$. This form is known as the Rayleigh-Jeans law.

12.28 Estimate the wavelength corresponding to maximum emission from each of the following surfaces: the sun, a tungsten filament at 2500 K, a heated metal at 1500 K, human skin at 305 K, and a cryogenically cooled metal surface at 60 K. Estimate the fraction of the solar emission that is in the following spectral regions: the ultraviolet, the visible, and the infrared.

12.29 Thermal imagers have radiation detectors that are sensitive to a spectral region and provide white-black or color images with shading to indicate relative temperature differences in the scene. The imagers, which have appearances similar to a video camcorder, have numerous applications, such as for equipment maintenance to identify overheated motors or electrical transformers and for fire-fighting service to determine the direction of fire spread and to aid search and rescue for victims. The most common operating spectral regions are 3 to 5 μm and 8 to 14 μm. The selection of a particular region typically depends on the temperature of the scene, although the atmospheric conditions (water vapor, smoke, etc.) may also be important.

(a) Determine the band emission fractions for each of the spectral regions, 3 to 5 μm and 8 to 14 μm, for temperatures of 300 and 900 K.

(b) Using the *Tools/Radiation/Band Emission Factor* feature within *IHT*, calculate and plot the band emission factors for each of the spectral regions for the temperature range 300 to 1000 K. Identify the temperatures at which the fractions are a maximum. What conclusions can you draw from this graph concerning the choice of an imager for an application?

(c) The noise-equivalent temperature (NET) is a specification of the imager that indicates the minimum temperature change that can be resolved in the image scene. Consider imagers operating at the maximum-fraction temperatures identified in part (b). For each of these conditions, determine the sensitivity (%) required of the radiation detector in order to provide a NET of 5°C. Explain the significance of your results. *Note:* The sensitivity (% units) can be defined as the difference in the band emission fractions for two temperatures differing by the NET, divided by the band emission fraction at one of the temperatures.

12.30 A furnace with a long, isothermal, graphite tube of diameter $D = 12.5$ mm is maintained at $T_f = 2000$ K and is used as a blackbody source to calibrate heat flux gages. Traditional heat flux gages are constructed as blackened thin films with thermopiles to indicate the temperature change caused by absorption of the incident radiant power over the entire spectrum. The traditional gage of interest has a sensitive area of 5 mm^2 and is mounted coaxial with the furnace centerline, but positioned at a distance of $L = 60$ mm from the beginning of the heated section. The cool extension tube serves to shield the gage from extraneous radiation sources and to contain the inert gas required to prevent rapid oxidation of the graphite tube.

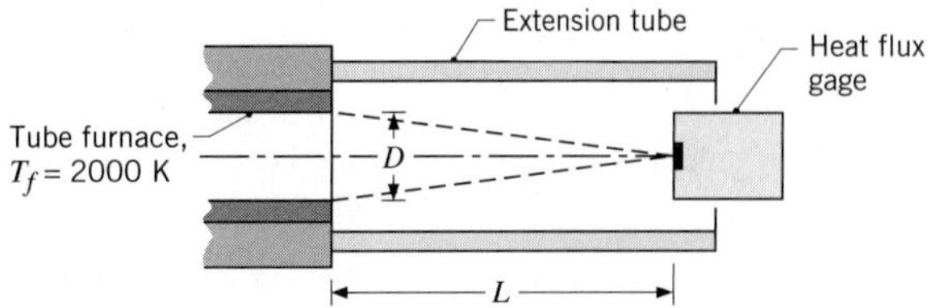

(a) Calculate the heat flux (W/m^2) on the traditional gage for this condition, assuming that the extension tube is cold relative to the furnace.

(b) The traditional gage is replaced by a solid-state (photoconductive) heat flux gage of the same area, but sensitive only to the spectral region between 0.4 and 2.5 μm. Calculate the radiant heat flux incident on the solid-state gage within the prescribed spectral region.

(c) Calculate and plot the total heat flux and the heat flux in the prescribed spectral region for the solid-state gage as a function of furnace temperature for the range $2000 \leq T_f \leq 3000$ K. Which gage will have an output signal that is more sensitive to changes in the furnace temperature?

12.31 Photovoltaic materials convert sunlight directly to electric power. Some of the photons that are incident upon the material displace electrons that are in turn collected to create an electric current. The overall efficiency of a photovoltaic panel, η, is the ratio of electrical energy produced to the energy content of the incident radiation. The efficiency depends primarily on two properties of the photovoltaic material, (i) the *band gap*, which identifies the energy states of photons having the potential to be converted to electric current, and (ii) the interband gap conversion efficiency, η_{bg}, which is the fraction of the total energy of photons within the band gap that is converted to electricity. Therefore, $\eta = \eta_{bg}F_{bg}$ where F_{bg} is the fraction of the photon energy incident on the surface within the band gap. Photons that are either outside the material's band gap or within the band gap but not converted to electrical energy are either reflected from the panel or absorbed and converted to thermal energy.

Consider a photovoltaic material with a band gap of $1.1 \leq B \leq 1.8$ eV, where B is the energy state of a photon. The wavelength is related to the energy state of a photon by the relationship $\lambda = 1240$ eV·nm/B. The incident solar irradiation approximates that of a blackbody at 5800 K and $G_S = 1000$ W/m^2.

(a) Determine the wavelength range of solar irradiation corresponding to the band gap.

(b) Determine the overall efficiency of the photovoltaic material if the interband gap efficiency is $\eta_{bg} = 0.50$.

(c) If half of the incident photons that are not converted to electricity are absorbed and converted to thermal energy, determine the heat absorption per unit surface area of the panel.

12.32 An electrically powered, ring-shaped radiant heating element is maintained at a temperature of $T_h = 3000$ K and is used in a manufacturing process to heat a small part having a surface area of $A_p = 0.007$ m^2. The surface of the heating element may be assumed to be black.

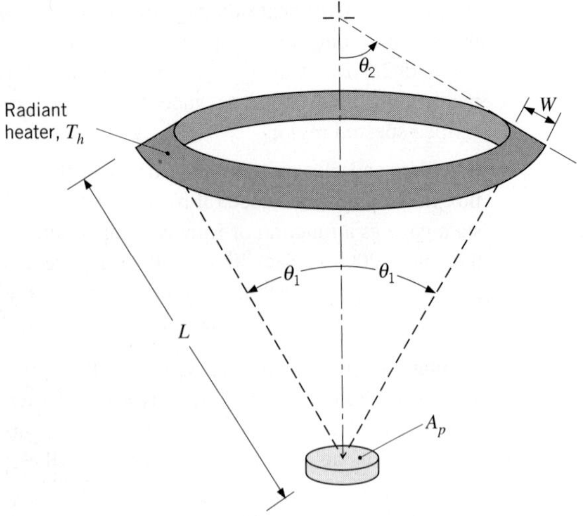

For $\theta_1 = 30°$, $\theta_2 = 60°$, $L = 3$ m, and $W = 30$ mm, what is the rate at which radiant energy emitted by the heater is incident on the part?

12.33 Isothermal furnaces with small apertures approximating a blackbody are frequently used to calibrate heat flux gages, radiation thermometers, and other radiometric devices. In such applications, it is necessary to control power to the furnace such that the variation of temperature and the spectral intensity of the aperture are within desired limits.

(a) By considering the Planck spectral distribution, Equation 12.30, show that the ratio of the fractional change in the spectral intensity to the fractional change in the temperature of the furnace has the form

$$\frac{dI_\lambda/I_\lambda}{dT/T} = \frac{C_2}{\lambda T} \frac{1}{1 - \exp(-C_2/\lambda T)}$$

(b) Using this relation, determine the allowable variation in temperature of the furnace operating at 2000 K to ensure that the spectral intensity at 0.65 μm will not vary by more than 0.5%. What is the allowable variation at 10 μm?

Properties: Emissivity

12.34 For materials A and B, whose spectral hemispherical emissivities vary with wavelength as shown below, how does the total, hemispherical emissivity vary with temperature? Explain briefly.

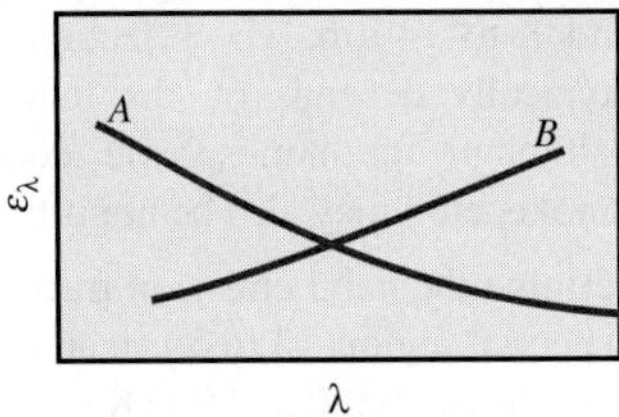

12.35 A small metal object, initially at $T_i = 1000$ K, is cooled by radiation in a low-temperature vacuum chamber. One of two thin coatings can be applied to the object so that spectral hemispherical emissivities vary with wavelength as shown. For which coating will the object most rapidly reach a temperature of $T_f = 500$ K?

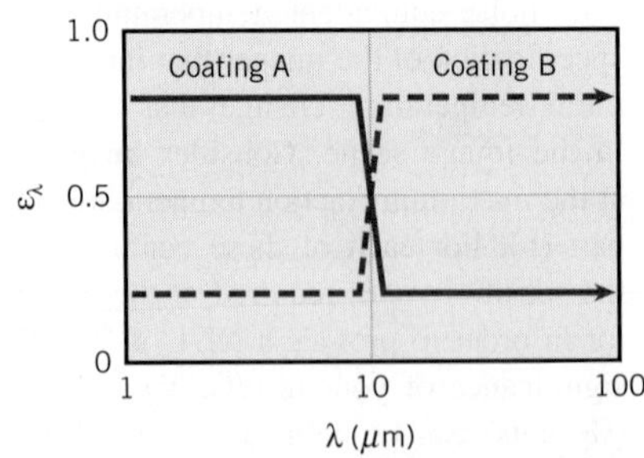

12.36 The directional total emissivity of nonmetallic materials may be approximated as $\varepsilon_\theta = \varepsilon_n \cos\theta$, where ε_n is the normal emissivity. Show that the total hemispherical emissivity for such materials is 2/3 of the normal emissivity.

12.37 Consider the metallic surface of Example 12.7. Additional measurements of the spectral, hemispherical emissivity yield a spectral distribution which may be approximated as follows:

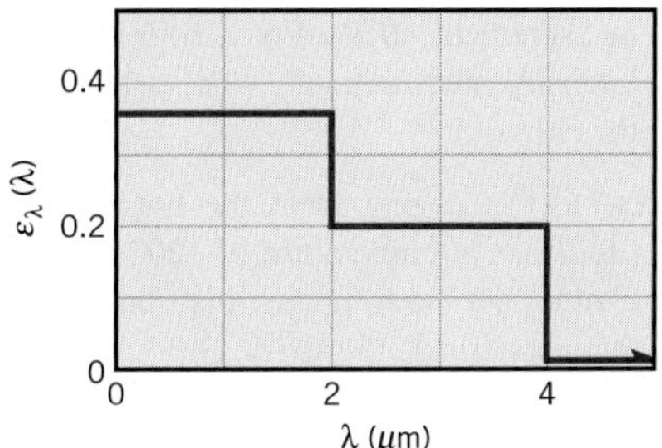

(a) Determine corresponding values of the total, hemispherical emissivity ε and the total emissive power E at 2000 K.

(b) Plot the emissivity as a function of temperature for $500 \le T \le 3000$ K. Explain the variation.

12.38 The spectral emissivity of unoxidized titanium at room temperature is well described by the expression $\varepsilon_\lambda = 0.52\lambda^{-0.5}$ for 0.3 μm $\le \lambda \le 30$ μm.

(a) Determine the emissive power associated with an unoxidized titanium surface at $T = 300$ K. Assume the spectral emissivity is $\varepsilon_\lambda = 0.1$ for $\lambda > 30$ μm.

(b) Determine the value of λ_{max} for the emissive power of the surface in part (a).

12.39 The spectral, directional emissivity of a diffuse material at 2000 K has the following distribution:

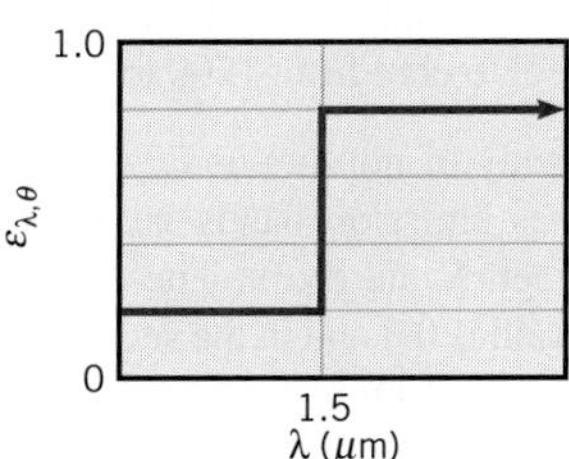

Determine the total, hemispherical emissivity at 2000 K. Determine the emissive power over the spectral range 0.8 to 2.5 μm *and* for the directions $0 \le \theta \le 30°$.

12.40 A diffuse surface is characterized by the spectral hemispherical emissivity distribution shown. Considering surface temperatures over the range $300 \le T_s \le 1000$ K, at what temperature will the emissive power be minimized?

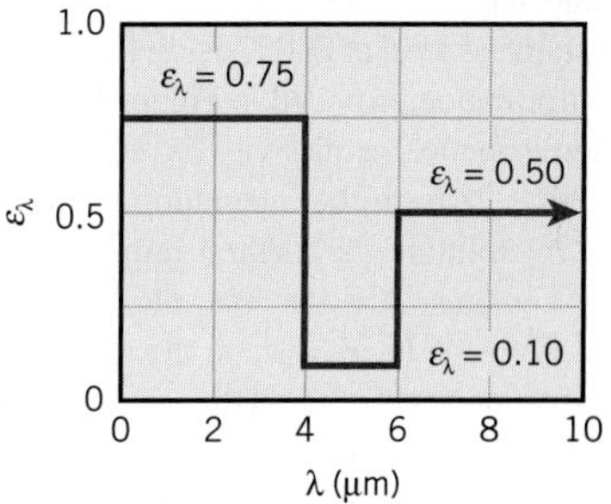

12.41 Consider the directionally selective surface having the directional emissivity ε_θ, as shown,

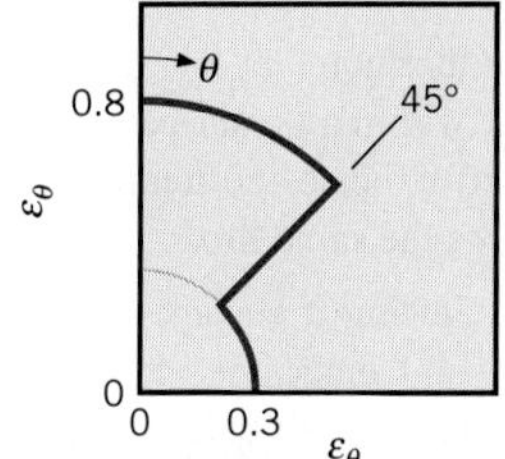

Assuming that the surface is isotropic in the ϕ direction, calculate the ratio of the normal emissivity ε_n to the hemispherical emissivity ε_h.

12.42 A sphere is suspended in air in a dark room and maintained at a uniform incandescent temperature. When first viewed with the naked eye, the sphere appears to be brighter around the rim. After several hours, however, it appears to be brighter in the center. Of what type material would you reason the sphere is made? Give plausible reasons for the nonuniformity of brightness of the sphere and for the changing appearance with time.

12.43 A proposed *proximity meter* is based on the physical arrangement of Problem 12.14. The sensing area of a meter that is installed on a vehicle, A_2, is irradiated by a stationary warm object, A_1. The sensor's electrical output signal is proportional to its irradiation.

(a) The object temperature and emissivity are 200°C and $\varepsilon = 0.85$, respectively. Determine the distance, $L_{2,crit}$, associated with the maximum sensor output signal. Assume the object is a diffuse emitter.

(b) If the object emits as a nonmetallic material, the total directional emissivity may be approximated as $\varepsilon_\theta = \varepsilon_n \cos\theta$, where ε_n is the normal emissivity (Problem 12.36). Determine the distance $L_{2,crit}$ associated with the maximum sensor output signal.

(c) Calculate and plot the irradiation of A_2 over the range $0 \leq L_2 \leq 10$ m.

12.44 Estimate the total, hemispherical emissivity ε for polished stainless steel at 800 K using Equation 12.43 along with information provided in Figure 12.17. Assume that the hemispherical emissivity is equal to the normal emissivity. Perform the integration using a *band calculation*, by splitting the integral into five bands, each of which contains 20% of the blackbody emission at 800 K. For each band, assume the average emissivity is that associated with the median wavelength within the band λ_m, for which half of the blackbody radiation within the band is above λ_m (and half is below λ_m). For example, the first band runs from $\lambda = 0$ to λ_1, such that $F_{(0 \to \lambda_1)} = 0.2$, and the median wavelength for the first band is chosen such that $F_{(0 \to \lambda_m)} = 0.1$. Also determine the surface emissive power.

12.45 A radiation thermometer is a device that responds to a radiant flux within a prescribed spectral interval and is calibrated to indicate the temperature of a blackbody that produces the same flux.

(a) When viewing a surface at an elevated temperature T_s and emissivity less than unity, the thermometer will indicate an apparent temperature referred to as the brightness or spectral radiance temperature T_λ. Will T_λ be greater than, less than, or equal to T_s?

(b) Write an expression for the spectral emissive power of the surface in terms of Wien's spectral distribution (see Problem 12.27) and the spectral emissivity of the surface. Write the equivalent expression using the spectral radiance temperature of the surface and show that

$$\frac{1}{T_s} = \frac{1}{T_\lambda} + \frac{\lambda}{C_2} \ln \varepsilon_\lambda$$

where λ represents the wavelength at which the thermometer operates.

(c) Consider a radiation thermometer that responds to a spectral flux centered about the wavelength 0.65 μm. What temperature will the thermometer indicate when viewing a surface with $\varepsilon_\lambda(0.65\ \mu\text{m}) = 0.9$ and $T_s = 1000$ K? Verify that Wien's spectral distribution is a reasonable approximation to Planck's law for this situation.

12.46 For a prescribed wavelength λ, measurement of the spectral intensity $I_{\lambda,e}(\lambda, T) = \varepsilon_\lambda I_{\lambda,b}$ of radiation emitted by a diffuse surface may be used to determine the surface temperature, if the spectral emissivity ε_λ is known, or the spectral emissivity, if the temperature is known.

(a) Defining the uncertainty of the temperature determination as dT/T, obtain an expression relating this uncertainty to that associated with the intensity measurement, dI_λ/I_λ. For a 10% uncertainty in the intensity measurement at $\lambda = 10\ \mu$m, what is the uncertainty in the temperature for $T = 500$ K? For $T = 1000$ K?

(b) Defining the uncertainty of the emissivity determination as $d\varepsilon_\lambda/\varepsilon_\lambda$, obtain an expression relating this uncertainty to that associated with the intensity measurement, dI_λ/I_λ. For a 10% uncertainty in the intensity measurement, what is the uncertainty in the emissivity?

12.47 Sheet steel emerging from the hot roll section of a steel mill has a temperature of 1200 K, a thickness of $\delta = 3$ mm, and the following distribution for the spectral, hemispherical emissivity.

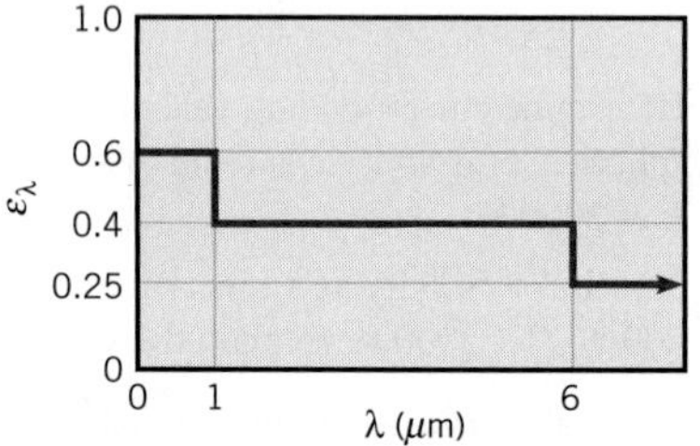

The density and specific heat of the steel are 7900 kg/m^3 and 640 J/kg·K, respectively. What is the total, hemispherical emissivity? Accounting for emission from both sides of the sheet and neglecting conduction, convection, and radiation from the surroundings, determine the initial time rate of change of the sheet temperature $(dT/dt)_i$. As the steel cools, it oxidizes and its total, hemispherical emissivity increases. If this increase may be correlated by an expression of the form $\varepsilon = \varepsilon_{1200}[1200\text{ K}/T\text{ (K)}]$, how long will it take for the steel to cool from 1200 to 600 K?

12.48 A large body of nonluminous gas at a temperature of 1200 K has emission bands between 2.5 and 3.5 μm and between 5 and 8 μm. The effective emissivity in the first band is 0.8 and in the second 0.6. Determine the emissive power of this gas.

Absorptivity, Reflectivity, and Transmissivity

12.49 An opaque surface with the prescribed spectral, hemispherical reflectivity distribution is subjected to the spectral irradiation shown.

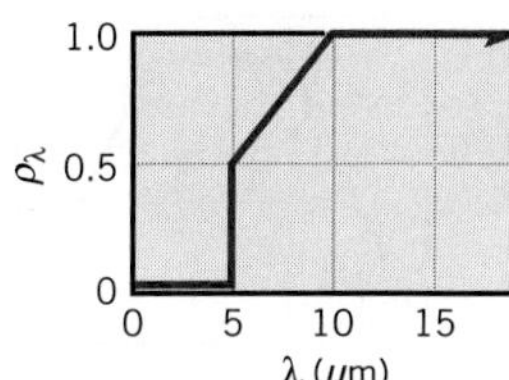

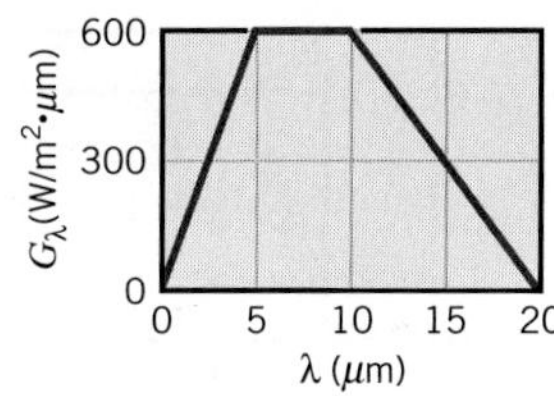

(a) Sketch the spectral, hemispherical absorptivity distribution.

(b) Determine the total irradiation on the surface.

(c) Determine the radiant flux that is absorbed by the surface.

(d) What is the total, hemispherical absorptivity of this surface?

12.50 A small, opaque, diffuse object at $T_s = 400$ K is suspended in a large furnace whose interior walls are at $T_f = 2000$ K. The walls are diffuse and gray and have an emissivity of 0.20. The spectral, hemispherical emissivity for the surface of the small object is given below.

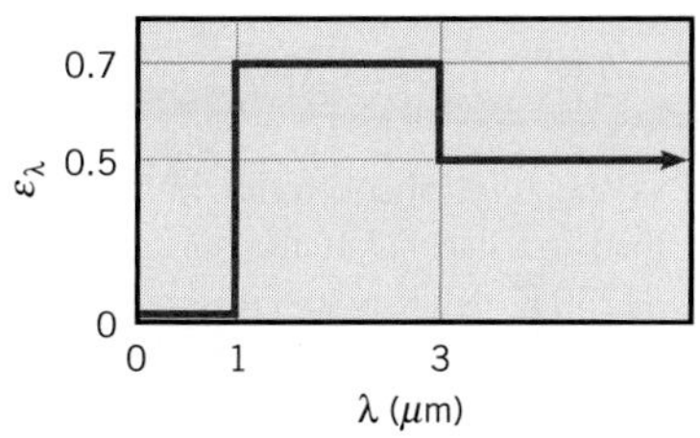

(a) Determine the total emissivity and absorptivity of the surface.

(b) Evaluate the reflected radiant flux and the net radiative flux *to* the surface.

(c) What is the spectral emissive power at $\lambda = 2\ \mu$m?

(d) What is the wavelength $\lambda_{1/2}$ for which one-half of the total radiation emitted by the surface is in the spectral region $\lambda \geq \lambda_{1/2}$?

12.51 The spectral reflectivity distribution for white paint (Figure 12.22) can be approximated by the following stair-step function:

λ (μm)	<0.4	0.4–3.0	>3.0
α_λ	0.75	0.15	0.96

A small flat plate coated with this paint is suspended inside a large enclosure, and its temperature is maintained at 400 K. The surface of the enclosure is maintained at 3000 K and the spectral distribution of its emissivity has the following characteristics:

λ (μm)	<2.0	>2.0
ε_λ	0.2	0.9

(a) Determine the total emissivity, ε, of the enclosure surface.

(b) Determine the total emissivity, ε, and absorptivity, α, of the plate.

12.52 An opaque surface, 2 m × 2 m, is maintained at 400 K and is simultaneously exposed to solar irradiation with $G_S = 1200$ W/m². The surface is diffuse and its spectral absorptivity is $\alpha_\lambda = 0$, 0.8, 0, and 0.9 for $0 \leq \lambda \leq 0.5\ \mu$m, $0.5\ \mu\text{m} < \lambda \leq 1\ \mu$m, $1\ \mu\text{m} < \lambda \leq 2\ \mu$m, and $\lambda > 2\ \mu$m, respectively. Determine the absorbed irradiation, emissive power, radiosity, and net radiation heat transfer from the surface.

12.53 Consider Problem 4.51.

(a) The students are each given a flat, *first-surface* silver mirror with which they collectively irradiate the wooden ship at location B. The reflection from the mirror is specular, and the silver's reflectivity is 0.98. The solar irradiation of each mirror, perpendicular to the direction of the sun's rays, is $G_S = 1000$ W/m². How many students are needed to conduct the experiment if the solar absorptivity of the wood is $\alpha_w = 0.80$ and the mirror is oriented at an angle of 45° from the direction of G_S?

(b) If the students are given *second-surface* mirrors that consist of a sheet of plain glass that has polished silver on its back side, how many students are needed to conduct the experiment? *Hint:* See Problem 12.62.

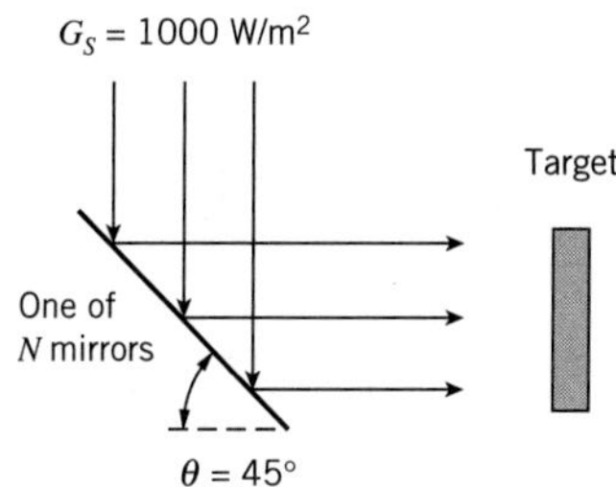

12.54 A diffuse, opaque surface at 700 K has spectral emissivities of $\varepsilon_\lambda = 0$ for $0 \leq \lambda \leq 3\ \mu$m, $\varepsilon_\lambda = 0.5$ for $3\ \mu\text{m} < \lambda \leq 10\ \mu$m, and $\varepsilon_\lambda = 0.9$ for $10\ \mu\text{m} < \lambda < \infty$. A radiant flux of 1000 W/m², which is uniformly distributed between 1 and 6 μm, is incident on the surface at an angle of 30° relative to the surface normal.

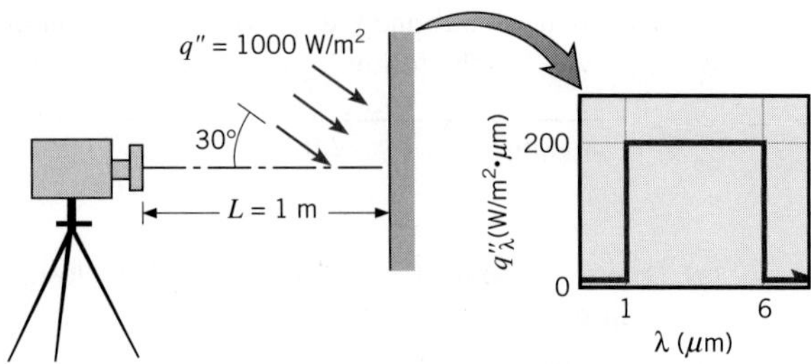

Calculate the total radiant power from a 10^{-4} m² area of the surface that reaches a radiation detector positioned along the normal to the area. The aperture of the detector is 10^{-5} m², and its distance from the surface is 1 m.

12.55 A small disk 5 mm in diameter is positioned at the center of an isothermal, hemispherical enclosure. The disk is diffuse and gray with an emissivity of 0.7 and is maintained at 900 K. The hemispherical enclosure, maintained at 300 K, has a radius of 100 mm and an emissivity of 0.85.

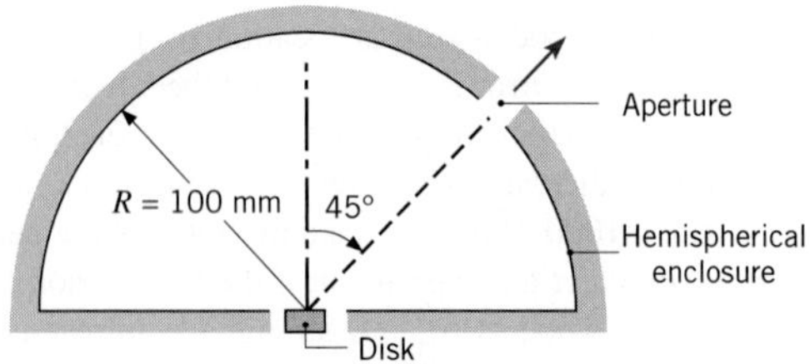

Calculate the radiant power leaving an aperture of diameter 2 mm located on the enclosure as shown.

12.56 The spectral, hemispherical absorptivity of an opaque surface is as shown.

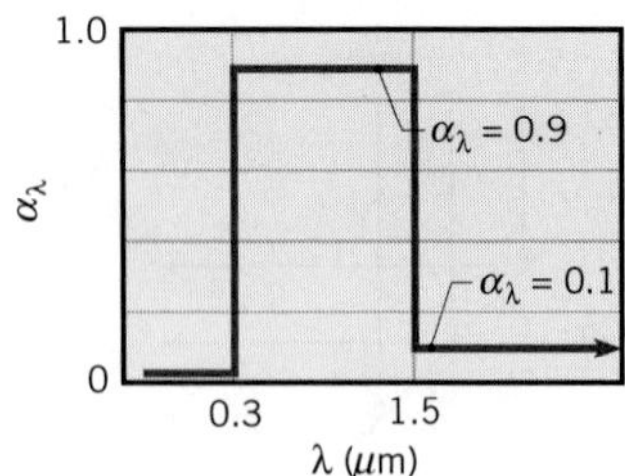

What is the solar absorptivity, α_S? If it is assumed that $\varepsilon_\lambda = \alpha_\lambda$ and that the surface is at a temperature of 340 K, what is its total, hemispherical emissivity?

12.57 The spectral, hemispherical absorptivity of an opaque surface and the spectral distribution of radiation incident on the surface are as shown.

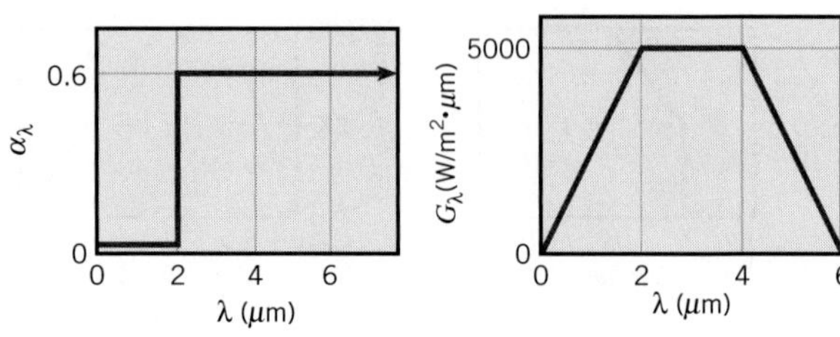

What is the total, hemispherical absorptivity of the surface? If it is assumed that $\varepsilon_\lambda = \alpha_\lambda$ and that the surface is at 1000 K, what is its total, hemispherical emissivity? What is the net radiant heat flux to the surface?

12.58 Consider an opaque, diffuse surface for which the spectral absorptivity and irradiation are as follows:

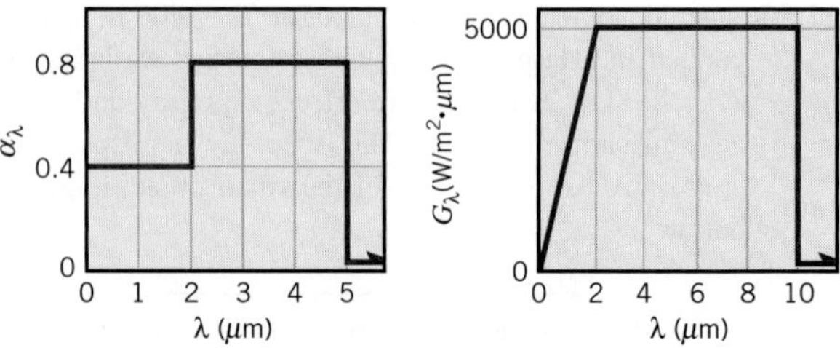

What is the total absorptivity of the surface for the prescribed irradiation? If the surface is at a temperature of 1250 K, what is its emissive power? How will the surface temperature vary with time, for the prescribed conditions?

12.59 The spectral emissivity of an opaque, diffuse surface is as shown.

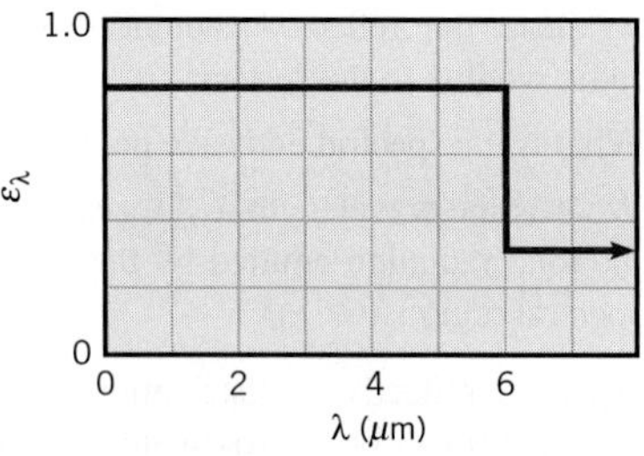

(a) If the surface is maintained at 1000 K, what is the total, hemispherical emissivity?

(b) What is the total, hemispherical absorptivity of the surface when irradiated by large surroundings of emissivity 0.8 and temperature 1500 K?

(c) What is the radiosity of the surface when it is maintained at 1000 K and subjected to the irradiation prescribed in part (b)?

(d) Determine the net radiation flux into the surface for the conditions of part (c).

(e) Plot each of the parameters featured in parts (a)–(d) as a function of the surface temperature for $750 \leq T \leq 2000$ K.

12.60 Radiation leaves a furnace of inside surface temperature 1500 K through an aperture 20 mm in diameter. A portion of the radiation is intercepted by a detector that is 1 m from the aperture, has a surface area of 10^{-5} m^2, and is oriented as shown.

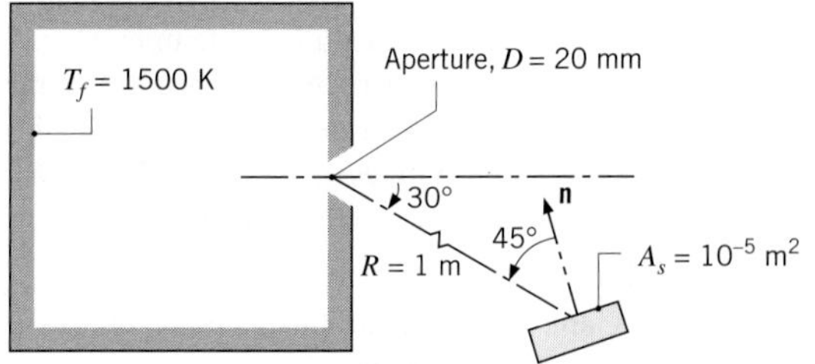

If the aperture is open, what is the rate at which radiation leaving the furnace is intercepted by the detector? If the aperture is covered with a diffuse, semitransparent material of spectral transmissivity $\tau_\lambda = 0.8$ for $\lambda \leq 2\ \mu$m and $\tau_\lambda = 0$ for $\lambda > 2\ \mu$m, what is the rate at which radiation leaving the furnace is intercepted by the detector?

12.61 The spectral transmissivity of a 1-mm-thick layer of liquid water can be approximated as follows:

$$\tau_{\lambda 1} = 0.99 \quad 0 \leq \lambda \leq 1.2\ \mu\text{m}$$
$$\tau_{\lambda 2} = 0.54 \quad 1.2\ \mu\text{m} < \lambda \leq 1.8\ \mu\text{m}$$
$$\tau_{\lambda 3} = 0 \quad 1.8\ \mu\text{m} < \lambda$$

(a) Liquid water can exist only below its critical temperature, $T_c = 647.3$ K. Determine the maximum possible total transmissivity of a 1-mm-thick layer of liquid water when the water is housed in an opaque container and boiling does not occur. Assume the irradiation is that of a blackbody.

(b) Determine the transmissivity of a 1-mm-thick layer of liquid water associated with melting the platinum wire used in Nukiyama's boiling experiment, as described in Section 10.3.1.

(c) Determine the total transmissivity of a 1-mm-thick layer of liquid water exposed to solar irradiation. Assume the sun emits as a blackbody at T_s = 5800 K.

12.62 The spectral transmissivity of plain and tinted glass can be approximated as follows:

Plain glass:	$\tau_\lambda = 0.9$	$0.3 \leq \lambda \leq 2.5\ \mu$m
Tinted glass:	$\tau_\lambda = 0.9$	$0.5 \leq \lambda \leq 1.5\ \mu$m

Outside the specified wavelength ranges, the spectral transmissivity is zero for both glasses. Compare the solar energy that could be transmitted through the glasses. With solar irradiation on the glasses, compare the visible radiant energy that could be transmitted.

12.63 Referring to the distribution of the spectral transmissivity of low iron glass (Figure 12.23), describe briefly what is meant by the "greenhouse effect." That is, how does the glass influence energy transfer to and from the contents of a greenhouse?

12.64 The spectral absorptivity α_λ and spectral reflectivity ρ_λ for a spectrally selective, diffuse material are as shown.

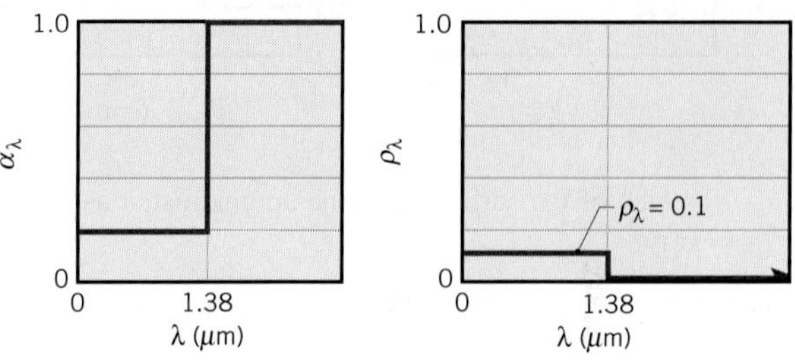

(a) Sketch the spectral transmissivity τ_λ.

(b) If solar irradiation with G_S = 750 W/m^2 and the spectral distribution of a blackbody at 5800 K is incident on this material, determine the fractions of the irradiation that are transmitted, reflected, and absorbed by the material.

(c) If the temperature of this material is 350 K, determine the emissivity ε.

(d) Determine the net heat flux by radiation to the material.

12.65 Consider a large furnace with opaque, diffuse, gray walls at 3000 K having an emissivity of 0.85. A small, diffuse, spectrally selective object in the furnace is maintained at 300 K.

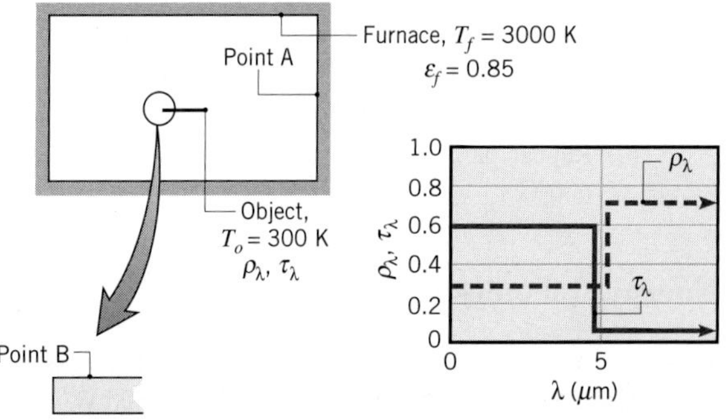

For the specified points on the furnace wall (A) and the object (B), indicate values for ε, α, E, G, and J.

12.66 Four diffuse surfaces having the spectral characteristics shown are at 300 K and are exposed to solar radiation.

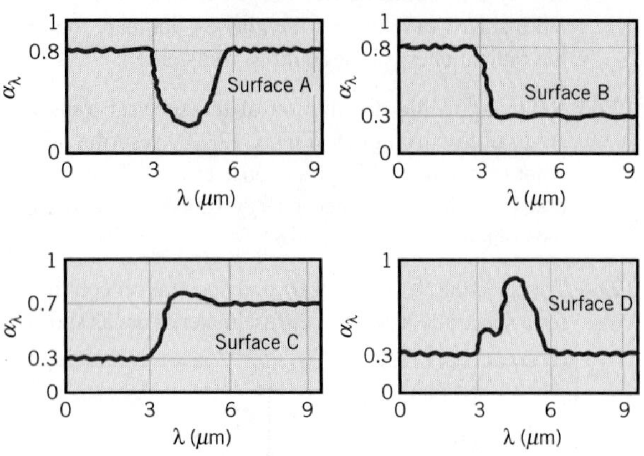

Which of the surfaces may be approximated as being gray?

12.67 Consider a material that is gray, but directionally selective with $\alpha_\theta(\theta, \phi) = 0.5(1 - \cos\phi)$. Determine the hemispherical absorptivity α when collimated solar flux irradiates the surface of the material in the direction $\theta = 45°$ and $\phi = 0°$. Determine the hemispherical emissivity ε of the material.

12.68 The spectral transmissivity of a 50-μm-thick polymer film is measured over the wavelength range $2.5\ \mu m \leq \lambda \leq 15\ \mu m$. The spectral distribution may be approximated as $\tau_\lambda = 0.80$ for $2.5\ \mu m \leq \lambda \leq 7\ \mu m$, $\tau_\lambda = 0.05$ for $7\ \mu m < \lambda \leq 13\ \mu m$, and $\tau_\lambda = 0.55$ for $13\ \mu m < \lambda \leq 15\ \mu m$. Transmissivity data outside the range cannot be acquired due to limitations associated with the instrumentation. An engineer wishes to determine the total transmissivity of the film.

(a) Estimate the *maximum possible* total transmissivity of the film associated with irradiation from a blackbody at $T = 30°C$.

(b) Estimate the *minimum possible* total transmissivity of the film associated with irradiation from a blackbody at $T = 30°C$.

(c) Repeat parts (a) and (b) for a blackbody at $T = 600°C$.

Energy Balances and Properties

12.69 An opaque, horizontal plate has a thickness of $L = 21$ mm and thermal conductivity $k = 25$ W/m·K. Water flows adjacent to the bottom of the plate and is at a temperature of $T_{\infty,w} = 25°C$. Air flows above the plate at $T_{\infty,a} = 260°C$ with $h_a = 40\ W/m^2 \cdot K$. The top of the plate is diffuse and is irradiated with $G = 1450\ W/m^2$, of which $435\ W/m^2$ is reflected. The steady-state top and bottom plate temperatures are $T_t = 43°C$ and $T_b = 35°C$, respectively. Determine the transmissivity, reflectivity, absorptivity, and emissivity of the plate. Is the plate gray? What is the radiosity associated with the top of the plate? What is the convection heat transfer coefficient associated with the water flow?

12.70 Two small surfaces, A and B, are placed inside an isothermal enclosure at a uniform temperature. The enclosure provides an irradiation of $6300\ W/m^2$ to each of the surfaces, and surfaces A and B absorb incident radiation at rates of 5600 and $630\ W/m^2$, respectively. Consider conditions after a long time has elapsed.

(a) What are the net heat fluxes for each surface? What are their temperatures?

(b) Determine the absorptivity of each surface.

(c) What are the emissive powers of each surface?

(d) Determine the emissivity of each surface.

12.71 A diffuse surface having the following spectral characteristics is maintained at 500 K when situated in a large furnace enclosure whose walls are maintained at 1500 K:

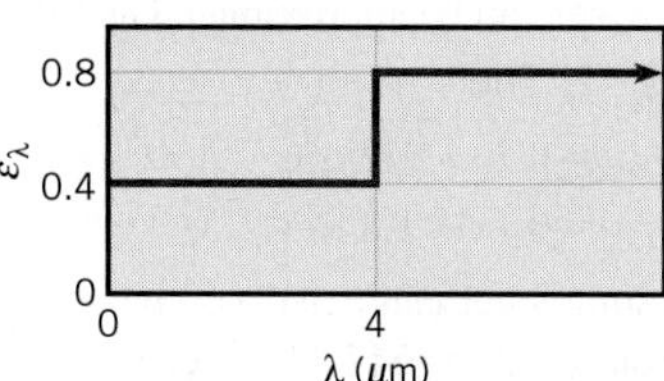

(a) Sketch the spectral distribution of the surface emissive power E_λ and the emissive power $E_{\lambda,b}$ that the surface would have if it were a blackbody.

(b) Neglecting convection effects, what is the net heat flux to the surface for the prescribed conditions?

(c) Plot the net heat flux as a function of the surface temperature for $500 \leq T \leq 1000$ K. On the same coordinates, plot the heat flux for a diffuse, gray surface with total emissivities of 0.4 and 0.8.

(d) For the prescribed spectral distribution of ε_λ, how do the total emissivity and absorptivity of the surface vary with temperature in the range $500 \leq T \leq 1000$ K?

12.72 Consider an opaque, diffuse surface whose spectral reflectivity varies with wavelength as shown. The surface is at 750 K, and irradiation on one side varies with wavelength as shown. The other side of the surface is insulated.

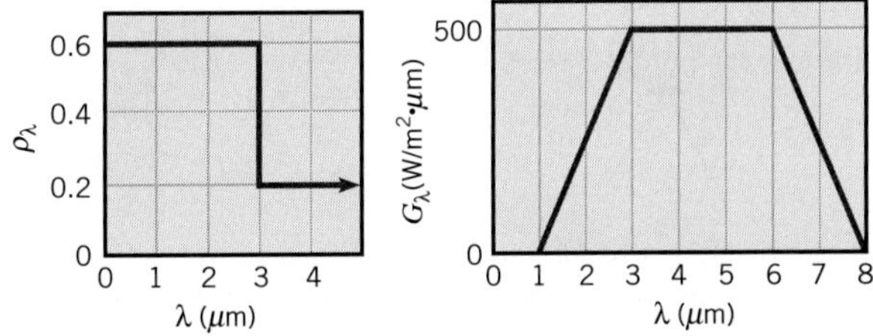

What are the total absorptivity and emissivity of the surface? What is the net radiative heat flux to the surface?

12.73 A special diffuse glass with prescribed spectral radiative properties is heated in a large oven. The walls of the oven are lined with a diffuse, gray refractory brick having an emissivity of 0.75 and are maintained at $T_w = 1800$ K. Consider conditions for which the glass temperature is $T_g = 750$ K.

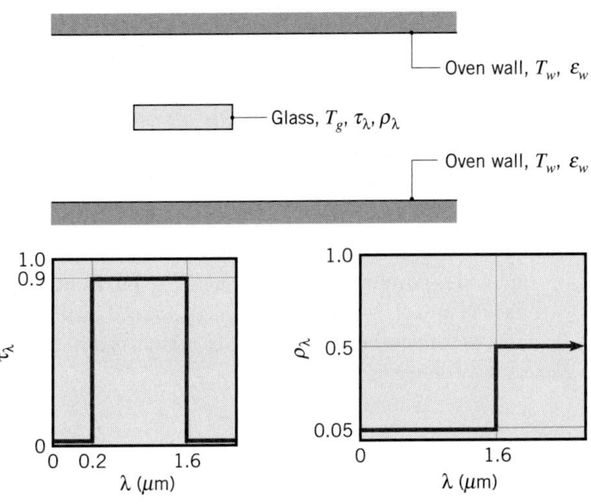

(a) What are the total transmissivity τ, the total reflectivity ρ, and the total emissivity ε of the glass?

(b) What is the net radiative heat flux, $q''_{\text{net,in}}$ (W/m²), to the glass?

(c) For oven wall temperatures of 1500, 1800, and 2000 K, plot $q''_{\text{net,in}}$ as a function of glass temperature for $500 \leq T_g \leq 800$ K.

12.74 The 50-mm peephole of a large furnace operating at 450°C is covered with a material having $\tau = 0.8$ and $\rho = 0$ for irradiation originating from the furnace. The material has an emissivity of 0.8 and is opaque to irradiation from a source at room temperature. The outer surface of the cover is exposed to surroundings and ambient air at 27°C with a convection heat transfer coefficient of 50 W/m²·K. Assuming that convection effects on the inner surface of the cover are negligible, calculate the heat loss by the furnace and the temperature of the cover.

12.75 The window of a large vacuum chamber is fabricated from a material of prescribed spectral characteristics. A collimated beam of radiant energy from a solar simulator is incident on the window and has a flux of 3000 W/m². The inside walls of the chamber, which are large compared to the window area, are maintained at 77 K. The outer surface of the window is subjected to surroundings and room air at 25°C, with a convection heat transfer coefficient of 15 W/m²·K.

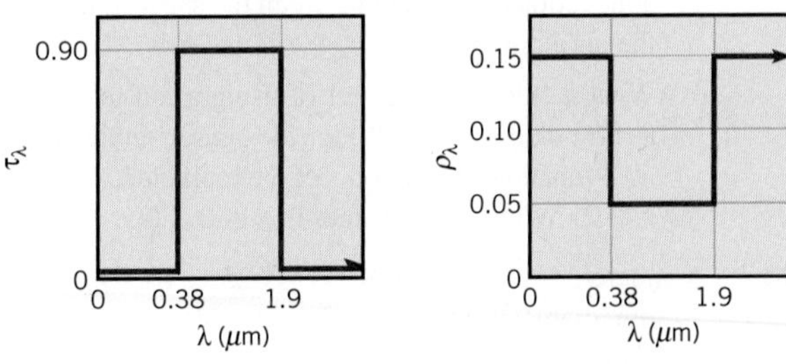

(a) Determine the transmissivity of the window material to radiation from the solar simulator, which approximates the solar spectral distribution.

(b) Assuming that the window is insulated from its chamber mounting arrangement, what steady-state temperature does the window reach?

(c) Calculate the net radiation transfer per unit area of the window to the vacuum chamber wall, excluding the transmitted simulated solar flux.

12.76 A thermocouple whose surface is diffuse and gray with an emissivity of 0.6 indicates a temperature of 180°C when used to measure the temperature of a gas flowing through a large duct whose walls have an emissivity of 0.85 and a uniform temperature of 450°C.

(a) If the convection heat transfer coefficient between the thermocouple and the gas stream is $h = 125$ W/m²·K and there are negligible conduction losses from the thermocouple, determine the temperature of the gas.

(b) Consider a gas temperature of 125°C. Compute and plot the thermocouple *measurement error* as a function of the convection coefficient for $10 \leq h \leq 1000$ W/m²·K. What are the implications of your results?

12.77 A thermocouple inserted in a 4-mm-diameter stainless steel tube having a diffuse, gray surface with an emissivity of 0.4 is positioned horizontally in a large air-conditioned room whose walls and air temperature are 30 and 20°C, respectively.

(a) What temperature will the thermocouple indicate if the air is quiescent?

(b) Compute and plot the thermocouple *measurement error* as a function of the surface emissivity for $0.1 \le \varepsilon \le 1.0$.

12.78 A temperature sensor embedded in the tip of a small tube having a diffuse, gray surface with an emissivity of 0.8 is centrally positioned within a large air-conditioned room whose walls and air temperature are 30 and 20°C, respectively.

(a) What temperature will the sensor indicate if the convection coefficient between the sensor tube and the air is 5 W/m²·K?

(b) What would be the effect of using a fan to induce airflow over the tube? Plot the sensor temperature as a function of the convection coefficient for $2 \le h \le 25$ W/m²·K and values of $\varepsilon = 0.2$, 0.5, and 0.8.

12.79 A sphere ($k = 185$ W/m·K, $\alpha = 7.25 \times 10^{-5}$ m²/s) of 30-mm diameter whose surface is diffuse and gray with an emissivity of 0.8 is placed in a large oven whose walls are of uniform temperature at 600 K. The temperature of the air in the oven is 400 K, and the convection heat transfer coefficient between the sphere and the oven air is 15 W/m²·K.

(a) Determine the net heat transfer to the sphere when its temperature is 300 K.

(b) What will be the steady-state temperature of the sphere?

(c) How long will it take for the sphere, initially at 300 K, to come within 20 K of the steady-state temperature?

(d) For emissivities of 0.2, 0.4, and 0.8, plot the elapsed time of part (c) as a function of the convection coefficient for $10 \le h \le 25$ W/m²·K.

Radiation Detection

12.80 A thermograph is a device responding to the radiant power from the scene, which reaches its radiation detector within the spectral region 9–12 μm. The thermograph provides an image of the scene, such as the side of a furnace, from which the surface temperature can be determined.

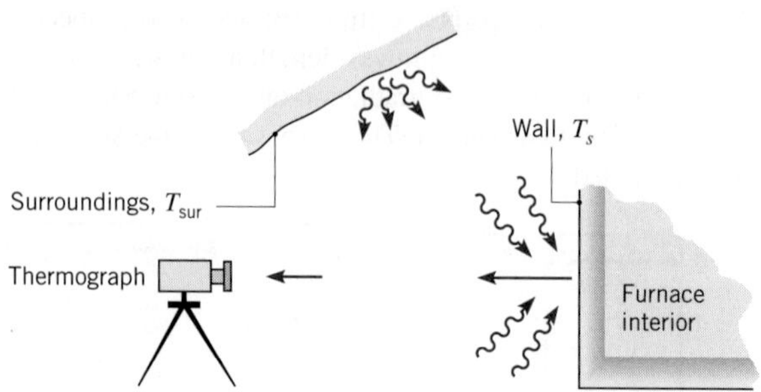

(a) For a black surface at 60°C, determine the emissive power for the spectral region 9–12 μm.

(b) Calculate the radiant power (W) received by the thermograph in the same range (9–12 μm) when viewing, in a normal direction, a small black wall area, 200 mm², at $T_s = 60$°C. The solid angle ω subtended by the aperture of the thermograph when viewed from the target is 0.001 sr.

(c) Determine the radiant power (W) received by the thermograph for the same wall area (200 mm²) and solid angle (0.001 sr) when the wall is a gray, opaque, diffuse material at $T_s = 60$°C with emissivity 0.7 and the surroundings are black at $T_{sur} = 23$°C.

12.81 A radiation thermometer is a radiometer calibrated to indicate the temperature of a blackbody. A steel billet having a diffuse, gray surface of emissivity 0.8 is heated in a furnace whose walls are at 1500 K. Estimate the temperature of the billet when the radiation thermometer viewing the billet through a small hole in the furnace indicates 1160 K.

12.82 A radiation detector has an aperture of area $A_d = 10^{-6}$ m² and is positioned at a distance of $r = 1$ m from a surface of area $A_s = 10^{-4}$ m². The angle formed by the normal to the detector and the surface normal is $\theta = 30°$.

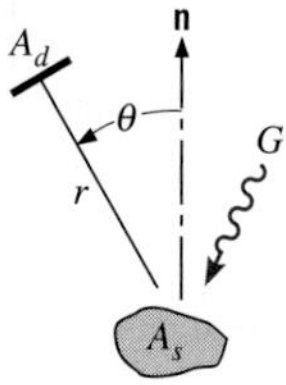

The surface is at 500 K and is opaque, diffuse, and gray with an emissivity of 0.7. If the surface irradiation is 1500 W/m², what is the rate at which the detector intercepts radiation from the surface?

12.83 A small anodized aluminum block at 35°C is heated in a large oven whose walls are diffuse and gray with $\varepsilon = 0.85$ and maintained at a uniform temperature of

175°C. The anodized coating is also diffuse and gray with $\varepsilon = 0.92$. A radiation detector views the block through a small opening in the oven and receives the radiant energy from a small area, referred to as the target, A_t, on the block. The target has a diameter of 3 mm, and the detector receives radiation within a solid angle 0.001 sr centered about the normal from the block.

(a) If the radiation detector views a small, but deep, hole drilled into the block, what is the total power (W) received by the detector?

(b) If the radiation detector now views an area on the block surface, what is the total power (W) received by the detector?

12.84 Consider the diffuse, gray opaque disk A_1, which has a diameter of 10 mm, an emissivity of 0.3, and is at a temperature of 400 K. Coaxial to the disk A_1, there is a black, ring-shaped disk A_2 at 1000 K having the dimensions shown in the sketch. The backside of A_2 is insulated and does not directly irradiate the cryogenically cooled detector disk A_3, which is of diameter 10 mm and is located 2 m from A_1.

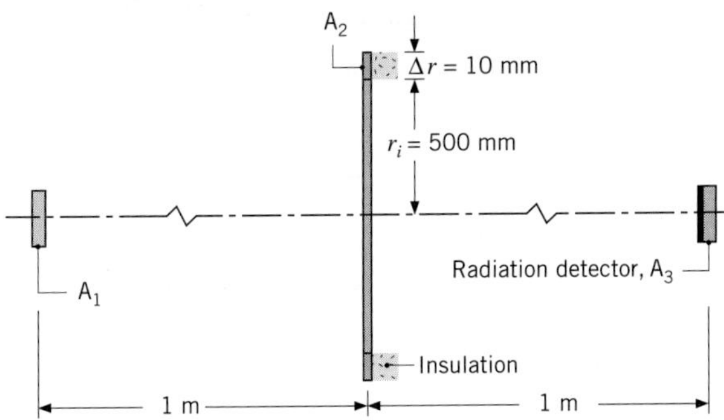

Calculate the rate at which radiation is incident on A_3 due to emission and reflection from A_1.

12.85 An infrared (IR) thermograph is a radiometer that provides an image of the target scene, indicating the apparent temperature of elements in the scene by a black–white brightness or blue–red color scale. Radiation originating from an element in the target scene is incident on the radiation detector, which provides a signal proportional to the incident radiant power. The signal sets the image brightness or color scale for the image pixel associated with that element. A scheme is proposed for field calibration of an infrared thermograph having a radiation detector with a 3- to 5-μm spectral bandpass. A heated metal plate, which is maintained at 327°C and has four diffuse, gray coatings with different emissivities, is viewed by the IR thermograph in surroundings for which $T_{sur} = 87$°C.

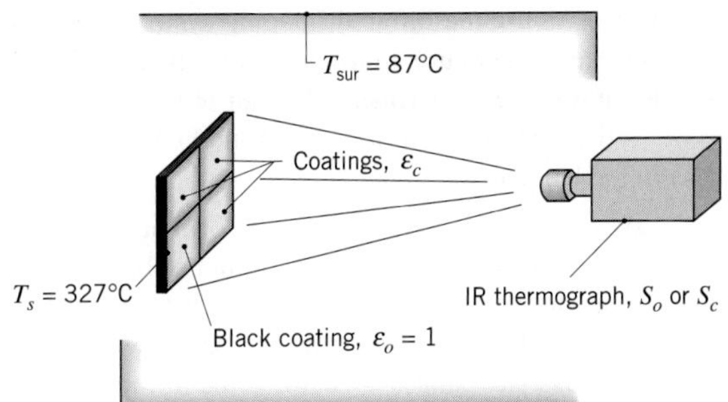

(a) Consider the thermograph output when viewing the black coating, $\varepsilon_o = 1$. The radiation reaching the detector is proportional to the product of the blackbody emissive power (or emitted intensity) at the temperature of the surface and the band emission fraction corresponding to the IR thermograph spectral bandpass. The proportionality constant is referred to as the responsivity, $R(\mu V \cdot m^2/W)$. Write an expression for the thermograph output signal, S_o, in terms of R, the coating blackbody emissive power, and the appropriate band emission fraction. Assuming $R = 1\ \mu V \cdot m^2/W$, evaluate S_o (μV).

(b) Consider the thermograph output when viewing one of the coatings for which the emissivity ε_c is less than unity. Radiation from the coating reaches the detector due to emission and the reflection of irradiation from the surroundings. Write an expression for the signal, S_c, in terms of R, the coating blackbody emissive power, the blackbody emissive power of the surroundings, the coating emissivity, and the appropriate band emission fractions. For the diffuse, gray coatings, the reflectivity is $\rho_c = 1 - \varepsilon_c$.

(c) Assuming $R = 1\ \mu V \cdot m^2/W$, evaluate the thermograph signals, S_c (μV), when viewing panels with emissivities of 0.8, 0.5, and 0.2.

(d) The thermograph is calibrated so that the signal S_o (with the black coating) will give a correct scale indication of $T_s = 327$°C. The signals from the other three coatings, S_c, are less than S_o. Hence the thermograph will indicate an apparent (blackbody) temperature less than T_s. Estimate the temperatures indicated by the thermograph for the three panels of part (c).

12.86 A charge-coupled device (CCD) infrared imaging system (see Problem 12.85) operates in a manner similar to a digital video camera. Instead of being sensitive to irradiation in the visible part of the spectrum, however, each small sensor in the infrared system's CCD array is sensitive in the spectral region 9–12 μm. Note that the system is designed to only view radiation coming from directly in front of it. An experimenter wishes to use the infrared imaging system to map the surface temperature distribution of a heated object in a wind tunnel experiment. The air temperature in the wind tunnel, as well as the surroundings temperature in the laboratory, is 23°C.

(a) In a preliminary test of the concept, the experimenter views a small aluminum billet located in the wind tunnel that is at a billet temperature of 50°C. The aluminum is coated with a high-emissivity paint, $\varepsilon = 0.96$. If the infrared imaging system is calibrated to indicate the temperature of a blackbody, what temperature will be indicated by the infrared imaging system as it is used to view the aluminum billet through a 6-mm-thick fused quartz window?

(b) In a subsequent experiment, the experimenter replaces the quartz window with a thin (130-μm-thick) household polyethylene film with $\tau \approx 0.78$ within the spectral range of the imaging system. What temperature will be indicated by the infrared imaging system when it is used to view the aluminum billet through the polyethylene film?

12.87 A diffuse, spherical object of diameter and temperature 9 mm and 600 K, respectively, has an emissivity of 0.95. Two very sensitive radiation detectors, each with an aperture area of 300×10^{-6} m², detect the object as it passes over at high velocity from left to right as shown in the schematic. The detectors capture hemispherical irradiation and are equipped with filters characterized by $\tau_\lambda = 0.9$ for $\lambda < 2.5$ μm and $\tau_\lambda = 0$ for $\lambda \geq 2.5$ μm. At time $t_1 = 0$, detectors A and B indicate irradiations of $G_{A,1} = 5.060$ mW/m² and $G_{B,1} = 5.000$ mW/m², respectively. At time $t_2 = 4$ ms, detectors A and B indicate irradiations of $G_{A,2} = 5.010$ mW/m² and $G_{B,2} = 5.050$ mW/m², respectively. The environment is at 300 K. Determine the velocity components of the particle, v_x and v_y. Determine when and where the particle will strike a horizontal plane located at $y = 0$. *Hint:* The object is located at an elevation above $y = 2$ m when it is detected. Assume the object's trajectory is a straight line in the plane of the page. Recall that the projected area of a sphere is a circle.

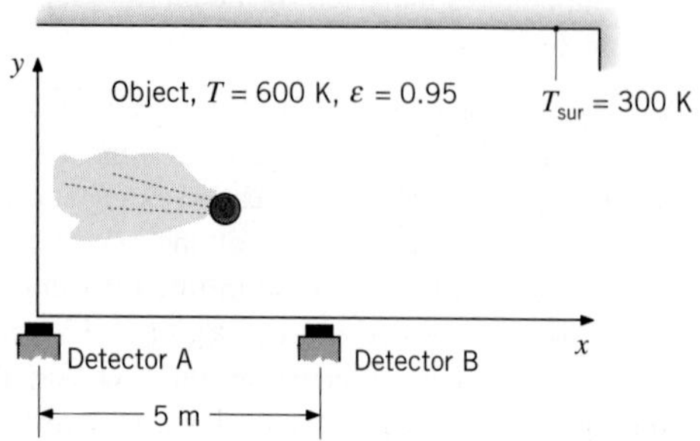

12.88 A radiation detector having a sensitive area of $A_d = 4 \times 10^{-6}$ m² is configured to receive radiation from a target area of diameter $D_t = 40$ mm when located a distance of $L_t = 1$ m from the target. For the experimental apparatus shown in the sketch, we wish to determine the emitted radiation from a hot sample of diameter $D_s = 20$ mm. The temperature of the aluminum sample is $T_s = 700$ K and its emissivity is $\varepsilon_s = 0.1$. A ring-shaped cold shield is provided to minimize the effect of radiation from outside the sample area, but within the target area. The sample and the shield are diffuse emitters.

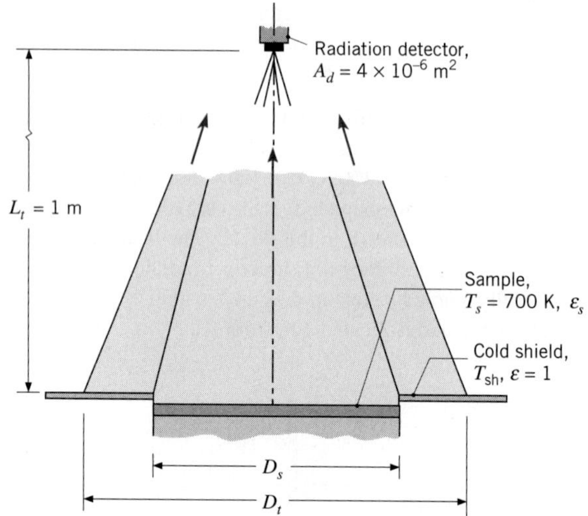

(a) Assuming the shield is black, at what temperature, T_{sh}, should the shield be maintained so that its emitted radiation is 1% of the total radiant power received by the detector?

(b) Subject to the parametric constraint that radiation emitted from the cold shield is 0.05, 1, or 1.5% of the total radiation received by the detector, plot the required cold shield temperature, T_{sh}, as a function of the sample emissivity for $0.05 \leq \varepsilon_s \leq 0.35$.

12.89 A *two-color pyrometer* is a device that is used to measure the temperature of a diffuse surface, T_s. The device measures the spectral, directional intensity emitted by the surface at two distinct wavelengths separated by $\Delta\lambda$. Calculate and plot the ratio of the intensities $I_{\lambda+\Delta\lambda,e}(\lambda + \Delta\lambda, \theta, \phi, T_s)$ and $I_{\lambda,e}(\lambda, \theta, \phi, T_s)$ as a function of the surface temperature over the range $500\text{ K} \leq T_s \leq 1000\text{ K}$ for $\lambda = 5$ μm and $\Delta\lambda = 0.1, 0.5$, and 1 μm. Comment on the sensitivity to temperature and on whether the ratio depends on the emissivity of the surface. Discuss the tradeoffs associated with specification of the various values of $\Delta\lambda$. *Hint*: The change in the emissivity over small wavelength intervals is modest for most solids, as evident in Figure 12.17.

12.90 Consider a two-color pyrometer such as in Problem 12.89 that operates at $\lambda_1 = 0.65\ \mu m$ and $\lambda_2 = 0.63\ \mu m$. Using Wien's law (see Problem 12.27) determine the temperature of a sheet of stainless steel if the ratio of radiation detected is $I_{\lambda_1}/I_{\lambda_2} = 2.15$.

Applications

12.91 Square plates freshly sprayed with an epoxy paint must be cured at 140°C for an extended period of time. The plates are located in a large enclosure and heated by a bank of infrared lamps. The top surface of each plate has an emissivity of $\varepsilon = 0.8$ and experiences convection with a ventilation airstream that is at $T_\infty = 27°C$ and provides a convection coefficient of $h = 20\ W/m^2 \cdot K$. The irradiation from the enclosure walls is estimated to be $G_{wall} = 450\ W/m^2$, for which the plate absorptivity is $\alpha_{wall} = 0.7$.

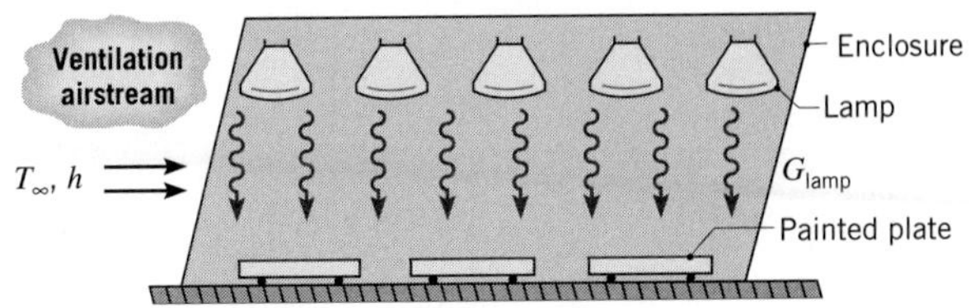

(a) Determine the irradiation that must be provided by the lamps, G_{lamp}. The absorptivity of the plate surface for this irradiation is $\alpha_{lamp} = 0.6$.

(b) For convection coefficients of $h = 15$, 20, and $30\ W/m^2 \cdot K$, plot the lamp irradiation, G_{lamp}, as a function of the plate temperature, T_s, for $100 \le T_s \le 300°C$.

(c) For convection coefficients in the range from 10 to $30\ W/m^2 \cdot K$ and a lamp irradiation of $G_{lamp} = 3000\ W/m^2$, plot the airstream temperature T_∞ required to maintain the plate at $T_s = 140°C$.

12.92 An apparatus commonly used for measuring the reflectivity of materials is shown below. A water-cooled sample, of 30-mm diameter and temperature $T_s = 300\ K$, is mounted flush with the inner surface of a large enclosure. The walls of the enclosure are gray and diffuse with an emissivity of 0.8 and a uniform temperature $T_f = 1000\ K$. A small aperture is located at the bottom of the enclosure to permit sighting of the sample or the enclosure wall. The spectral reflectivity ρ_λ of an opaque, diffuse sample material is as shown. The heat transfer coefficient for convection between the sample and the air within the cavity, which is also at 1000 K, is $h = 10\ W/m^2 \cdot K$.

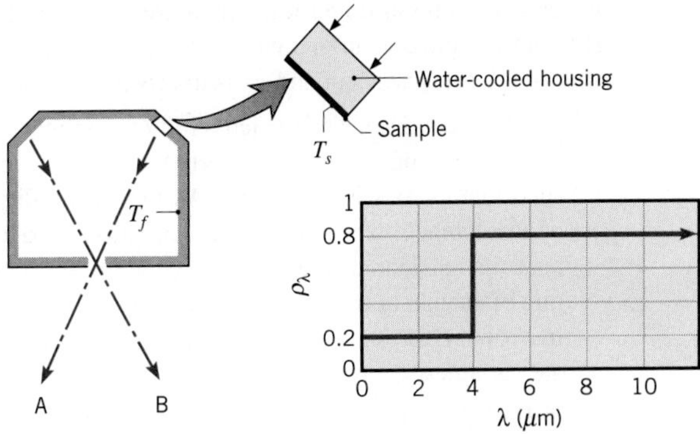

(a) Calculate the absorptivity of the sample.

(b) Calculate the emissivity of the sample.

(c) Determine the heat removal rate (W) by the coolant.

(d) The ratio of the radiation in the A direction to that in the B direction will give the reflectivity of the sample. Briefly explain why this is so.

12.93 A very small sample of an opaque surface is initially at 1200 K and has the spectral, hemispherical absorptivity shown.

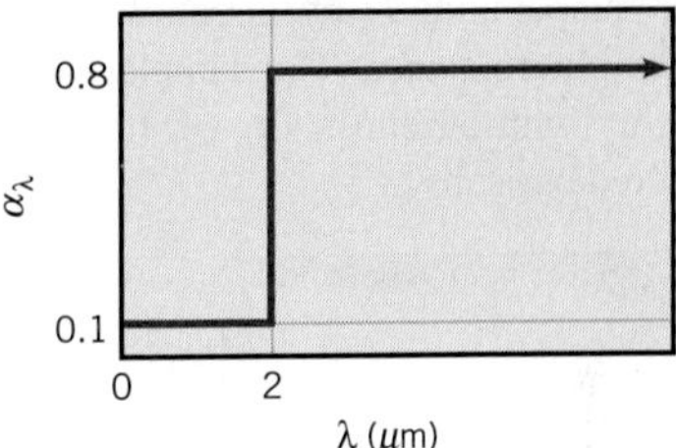

The sample is placed inside a very large enclosure whose walls have an emissivity of 0.2 and are maintained at 2400 K.

(a) What is the total, hemispherical absorptivity of the sample surface?

(b) What is its total, hemispherical emissivity?

(c) What are the values of the absorptivity and emissivity after the sample has been in the enclosure a long time?

(d) For a 10-mm-diameter spherical sample in an evacuated enclosure, compute and plot the variation of the sample temperature with time, as it is heated from its initial temperature of 1200 K.

12.94 A manufacturing process involves heating long copper rods, which are coated with a thin film, in a large furnace whose walls are maintained at an elevated temperature T_w. The furnace contains quiescent nitrogen gas at 1-atm pressure and a temperature of $T_\infty = T_w$. The film is a diffuse surface with a spectral emissivity of $\varepsilon_\lambda = 0.9$ for $\lambda \leq 2\ \mu\text{m}$ and $\varepsilon_\lambda = 0.4$ for $\lambda > 2\ \mu\text{m}$.

(a) Consider conditions for which a rod of diameter D and initial temperature T_i is inserted in the furnace, such that its axis is horizontal. Assuming validity of the lumped capacitance approximation, derive an equation that could be used to determine the rate of change of the rod temperature at the time of insertion. Express your result in terms of appropriate variables.

(b) If $T_w = T_\infty = 1500$ K, $T_i = 300$ K, and $D = 10$ mm, what is the initial rate of change of the rod temperature? Confirm the validity of the lumped capacitance approximation.

(c) Compute and plot the variation of the rod temperature with time during the heating process.

12.95 A procedure for measuring the thermal conductivity of solids at elevated temperatures involves placement of a sample at the bottom of a large furnace. The sample is of thickness L and is placed in a square container of width W on a side. The sides are well insulated. The walls of the cavity are maintained at T_w, while the bottom surface of the sample is maintained at a much lower temperature T_c by circulating coolant through the sample container. The sample surface is diffuse and gray with an emissivity ε_s. Its temperature T_s is measured optically.

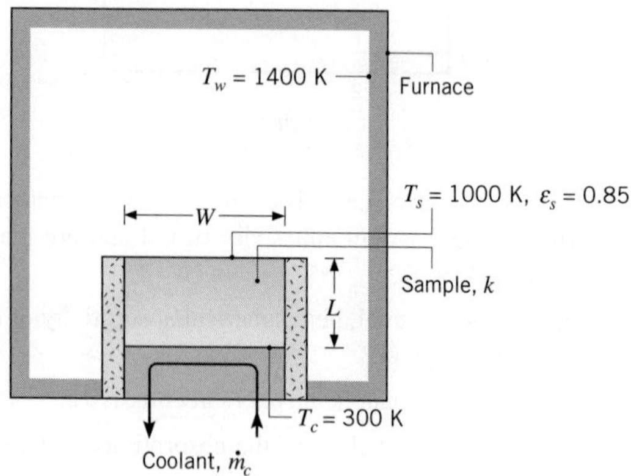

(a) Neglecting convection effects, obtain an expression from which the sample thermal conductivity may be evaluated in terms of measured and known quantities (T_w, T_s, T_c, ε_s, L). The measurements are made under steady-state conditions. If $T_w = 1400$ K, $T_s = 1000$ K, $\varepsilon_s = 0.85$, $L = 0.015$ m, and $T_c = 300$ K, what is the sample thermal conductivity?

(b) If $W = 0.10$ m and the coolant is water with a flow rate of $\dot{m}_c = 0.1$ kg/s, is it reasonable to assume a uniform bottom surface temperature T_c?

12.96 One scheme for extending the operation of gas turbine blades to higher temperatures involves applying a ceramic coating to the surfaces of blades fabricated from a superalloy such as inconel. To assess the reliability of such coatings, an apparatus has been developed for testing samples under laboratory conditions. The sample is placed at the bottom of a large vacuum chamber whose walls are cryogenically cooled and which is equipped with a radiation detector at the top surface. The detector has a surface area of $A_d = 10^{-5}$ m^2, is located at a distance of $L_{s\text{-}d} = 1$ m from the sample, and views radiation originating from a portion of the ceramic surface having an area of $\Delta A_c = 10^{-4}$ m^2. An electric heater attached to the bottom of the sample dissipates a uniform heat flux, q''_h, which is transferred upward through the sample. The bottom of the heater and sides of the sample are well insulated.

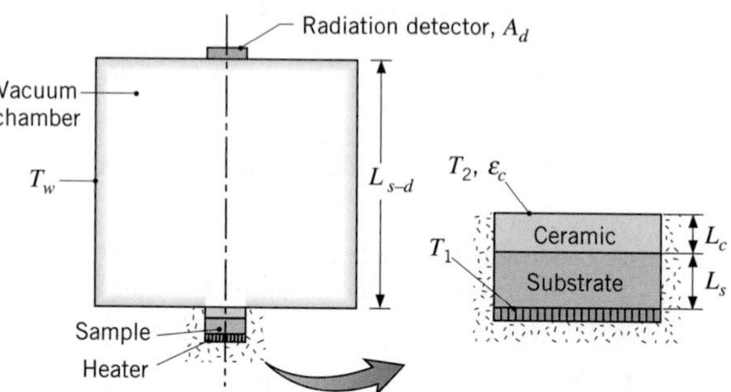

Consider conditions for which a ceramic coating of thickness $L_c = 0.5$ mm and thermal conductivity $k_c = 6$ W/m·K has been sprayed on a metal substrate of thickness $L_s = 8$ mm and thermal conductivity $k_s = 25$ W/m·K. The opaque surface of the ceramic may be approximated as diffuse and gray, with a total, hemispherical emissivity of $\varepsilon_c = 0.8$.

(a) Consider steady-state conditions for which the bottom surface of the substrate is maintained at $T_1 = 1500$ K, while the chamber walls (including the surface of the radiation detector) are maintained at $T_w = 90$ K. Assuming negligible thermal contact resistance at the ceramic–substrate interface, determine the ceramic top surface temperature T_2 and the heat flux q''_h.

(b) For the prescribed conditions, what is the rate at which radiation emitted by the ceramic is intercepted by the detector?

(c) After repeated experiments, numerous cracks develop at the ceramic–substrate interface, creating an interfacial thermal contact resistance. If T_w and q''_h are maintained at the conditions associated with part (a), will T_1 increase, decrease, or remain the same? Similarly, will T_2 increase, decrease, or remain the same? In each case, justify your answer.

12.97 The equipment for heating a wafer during a semiconductor manufacturing process is shown schematically. The wafer is heated by an ion beam source (not shown) to a uniform, steady-state temperature. The large chamber contains the process gas, and its walls are at a uniform temperature of $T_{ch} = 400$ K. A 5 mm × 5 mm target area on the wafer is viewed by a radiometer, whose objective lens has a diameter of 25 mm and is located 500 mm from the wafer. The line-of-sight of the radiometer is 30° off the wafer normal.

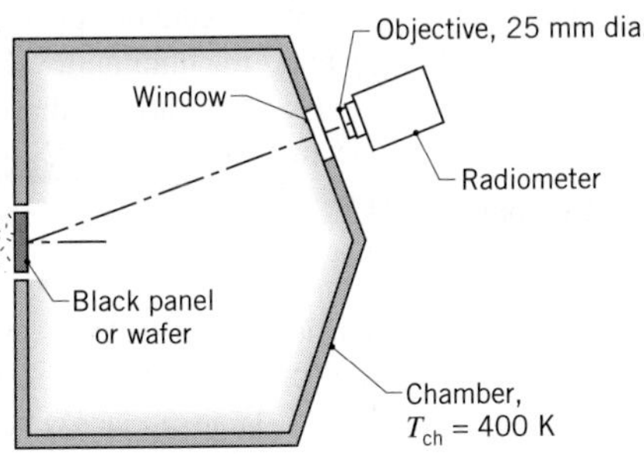

(a) In a preproduction test of the equipment, a black panel ($\varepsilon \approx 1.0$) is mounted in place of the wafer. Calculate the radiant power (W) received by the radiometer if the temperature of the panel is 800 K.

(b) The wafer, which is opaque, diffuse-gray with an emissivity of 0.7, is now placed in the equipment, and the ion beam is adjusted so that the power received by the radiometer is the same as that found for part (a). Calculate the temperature of the wafer for this heating condition.

12.98 The fire brick of Example 12.10 is used to construct the walls of a brick oven. The irradiation on the interior surface of the wall is $G = 50{,}000$ W/m^2 and has a spectral distribution proportional to that of a blackbody at 2000 K. The temperature of the gases adjacent to the inner wall of the oven is 500 K, and the convection heat transfer coefficient is 25 W/m^2. Find the wall surface temperature if the heat loss through the wall is negligible. If the brick wall is 0.1 m thick and of thermal conductivity $k_b = 1.0$ W/m·K, and is insulated with a 0.1-m-thick layer of thermal conductivity $k_i = 0.05$ W/m·K, what is the steady-state interior wall surface temperature if the temperature of the external surface of the insulation is 300 K?

12.99 A laser-materials-processing apparatus encloses a sample in the form of a disk of diameter $D = 25$ mm and thickness $w = 1$ mm. The sample has a diffuse surface for which the spectral distribution of the emissivity, $\varepsilon_\lambda(\lambda)$, is prescribed. To reduce oxidation, an inert gas stream of temperature $T_\infty = 500$ K and convection coefficient $h = 50$ W/m$^2\cdot$K flows over the sample upper and lower surfaces. The apparatus enclosure is large, with isothermal walls at $T_{enc} = 300$ K. To maintain the sample at a suitable operating temperature of $T_s = 2000$ K, a collimated laser beam with an operating wavelength of $\lambda = 0.5$ μm irradiates its upper surface.

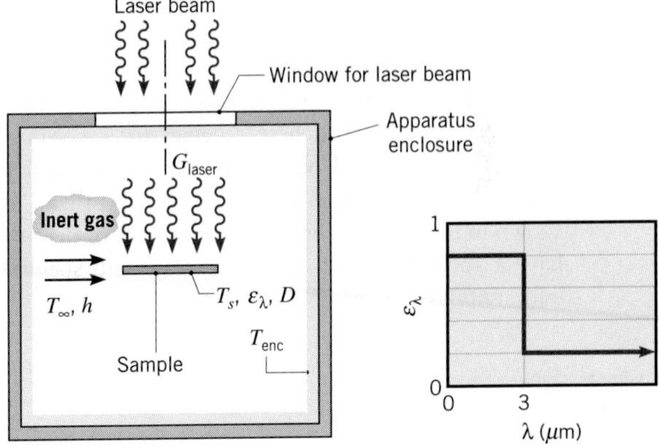

(a) Determine the total emissivity ε of the sample.

(b) Determine the total absorptivity α of the sample for irradiation from the enclosure walls.

(c) Perform an energy balance on the sample and determine the laser irradiation, G_{laser}, required to maintain the sample at $T_s = 2000$ K.

(d) Consider a *cool-down* process, when the laser and the inert gas flow are deactivated. Sketch the total emissivity as a function of the sample temperature, $T_s(t)$, during the process. Identify key features, including the emissivity for the final condition ($t \rightarrow \infty$).

(e) Estimate the time to cool a sample from its operating condition at $T_s(0) = 2000$ K to a *safe-to-touch* temperature of $T_s(t) = 40°$C. Use the lumped capacitance method and include the effect of convection to the inert gas with $h = 50$ W/m$^2\cdot$K and $T_\infty = T_{enc} = 300$ K. The thermophysical properties of the sample material are $\rho = 3900$ kg/m^3, $c_p = 760$ J/kg·K, and $k = 45$ W/m·K.

12.100 A cylinder of 30-mm diameter and 150-mm length is heated in a large furnace having walls at 1000 K, while air at 400 K is circulating at 3 m/s. Estimate the steady-state cylinder temperature under the following specified conditions.

(a) The cylinder is in cross flow, and its surface is diffuse and gray with an emissivity of 0.5.

(b) The cylinder is in cross flow, but its surface is spectrally selective with $\alpha_\lambda = 0.1$ for $\lambda \leq 3\ \mu m$ and $\alpha_\lambda = 0.5$ for $\lambda > 3\ \mu m$.

(c) The cylinder surface is positioned such that the airflow is longitudinal and its surface is diffuse and gray.

(d) For the conditions of part (a), compute and plot the cylinder temperature as a function of the air velocity for $1 \leq V \leq 20$ m/s.

12.101 An instrumentation transmitter pod is a box containing electronic circuitry and a power supply for sending sensor signals to a base receiver for recording. Such a pod is placed on a conveyor system, which passes through a large vacuum brazing furnace as shown in the sketch. The exposed surfaces of the pod have a special diffuse, opaque coating with spectral emissivity as shown.

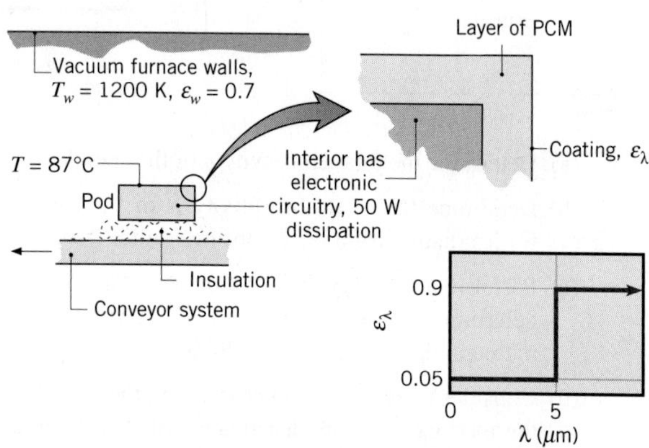

To stabilize the temperature of the pod and prevent overheating of the electronics, the inner surface of the pod is surrounded by a layer of a phase-change material (PCM) having a fusion temperature of 87°C and a heat of fusion of 25 kJ/kg. The pod has an exposed surface area of 0.040 m^2 and the mass of the PCM is 1.6 kg. Furthermore, it is known that the power dissipated by the electronics is 50 W. Consider the situation when the pod enters the furnace at a uniform temperature of 87°C and all the PCM is in the solid state. How long will it take before all the PCM changes to the liquid state?

12.102 A thin-walled plate separates the interior of a large furnace from surroundings at 300 K. The plate is fabricated from a ceramic material for which diffuse surface behavior may be assumed and the exterior surface is air cooled. With the furnace operating at 2400 K, convection at the interior surface may be neglected.

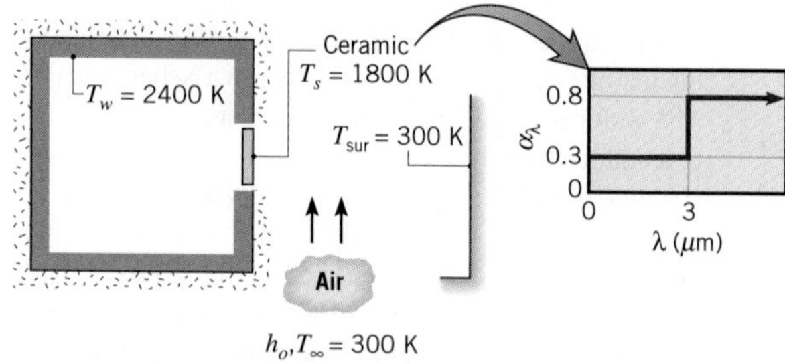

(a) If the temperature of the ceramic plate is not to exceed 1800 K, what is the minimum value of the outside convection coefficient, h_o, that must be maintained by the air-cooling system?

(b) Compute and plot the plate temperature as a function of h_o for $50 \leq h_o \leq 250$ W/m$^2\cdot$K.

12.103 A thin coating, which is applied to long, cylindrical copper rods of 10-mm diameter, is cured by placing the rods horizontally in a large furnace whose walls are maintained at 1300 K. The furnace is filled with nitrogen gas, which is also at 1300 K and at a pressure of 1 atm. The coating is diffuse, and its spectral emissivity has the distribution shown.

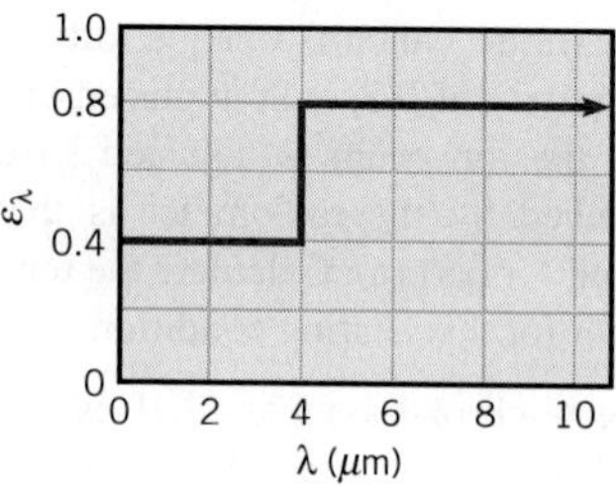

(a) What are the emissivity and absorptivity of the coated rods when their temperature is 300 K?

(b) What is the initial rate of change of their temperature?

(c) What are the emissivity and absorptivity of the coated rods when they reach a steady-state temperature?

(d) Estimate the time required for the rods to reach 1000 K.

12.104 A large combination convection–radiation oven is used to heat-treat a small cylindrical product of diameter 25 mm and length 0.2 m. The oven walls are at a uniform temperature of 1000 K, and hot air at 750 K is in cross flow over the cylinder with a velocity of 5 m/s. The cylinder surface is opaque and diffuse with the spectral emissivity shown.

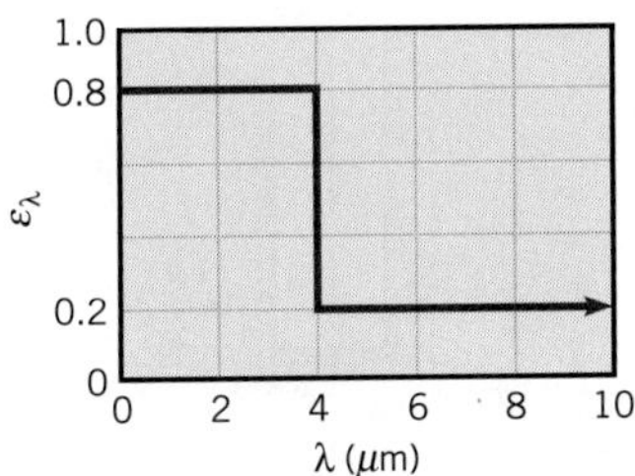

(a) Determine the rate of heat transfer to the cylinder when it is first placed in the oven at 300 K.

(b) What is the steady-state temperature of the cylinder?

(c) How long will it take for the cylinder to reach a temperature that is within 50°C of its steady-state value?

12.105 A 10-mm-thick workpiece, initially at 25°C, is to be annealed at a temperature above 725°C for a period of at least 5 minutes and then cooled. The workpiece is opaque and diffuse, and the spectral distribution of its emissivity is shown schematically. Heating is effected in a large furnace with walls and circulating air at 750°C and a convection coefficient of 100 W/m$^2\cdot$K. The thermophysical properties of the workpiece are $\rho = 2700$ kg/m^3, $c = 885$ J/kg$\cdot$K, and $k = 165$ W/m$\cdot$K.

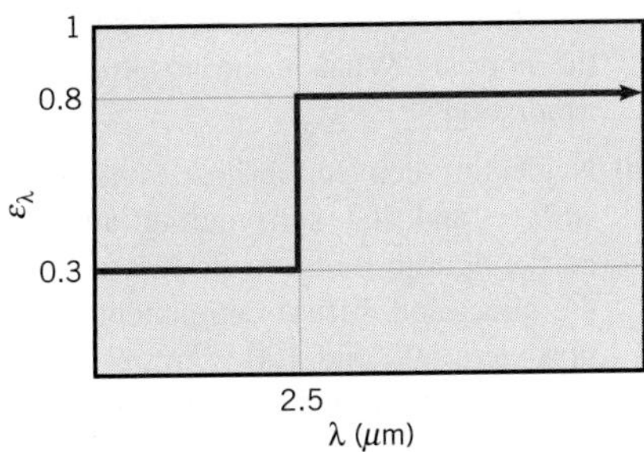

(a) Calculate the emissivity and the absorptivity of the workpiece when it is placed in the furnace at its initial temperature of 25°C.

(b) Determine the net heat flux into the workpiece for this initial condition. What is the corresponding rate of change in temperature, dT/dt, for the workpiece?

(c) Calculate the time for the workpiece to cool from 750°C to a safe-to-touch temperature of 40°C, if the surroundings and cooling air temperature are 25°C and the convection coefficient is 100 W/m$^2\cdot$K.

12.106 After being cut from a large single-crystal boule and polished, silicon wafers undergo a high-temperature annealing process. One technique for heating the wafer is to irradiate its top surface using high-intensity, tungsten-halogen lamps having a spectral distribution approximating that of a blackbody at 2800 K. To determine the lamp power and the rate at which radiation is absorbed by the wafer, the equipment designer needs to know its absorptivity as a function of temperature. Silicon is a semiconductor material that exhibits a characteristic *band edge*, and its spectral absorptivity may be idealized as shown schematically. At low and moderate temperatures, silicon is semitransparent at wavelengths larger than that of the band edge, but becomes nearly opaque above 600°C.

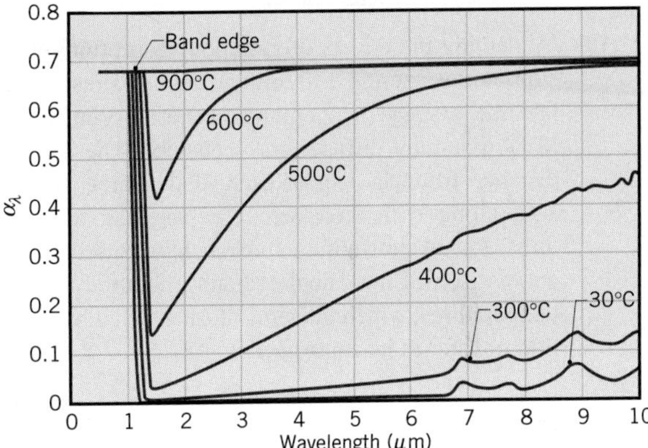

(a) What are the 1% limits of the spectral band that includes 98% of the blackbody radiation corresponding to the spectral distribution of the lamps? Over what spectral region do you need to know the spectral absorptivity?

(b) How do you expect the total absorptivity of silicon to vary as a function of its temperature? Sketch the variation and explain its key features.

(c) Calculate the total absorptivity of the silicon wafer for the lamp irradiation and each of the five temperatures shown schematically. From the data, calculate the emissivity of the wafer at 600 and 900°C. Explain your results and why the emissivity changes with temperature. *Hint*: Within *IHT*, create a *look-up* table to specify values of the spectral properties and the LOOKUPVAL and INTEGRAL functions to perform the necessary integrations.

(d) If the wafer is in a vacuum and radiation exchange only occurs at one face, what is the irradiation needed to maintain a wafer temperature of 600°C?

Environmental Radiation

12.107 Solar irradiation of 1100 W/m^2 is incident on a large, flat, horizontal metal roof on a day when the wind blowing over the roof causes a convection heat transfer coefficient of 25 W/m$^2 \cdot$K. The outside air temperature is 27°C, the metal surface absorptivity for incident solar radiation is 0.60, the metal surface emissivity is 0.20, and the roof is well insulated from below.

(a) Estimate the roof temperature under steady-state conditions.

(b) Explore the effect of changes in the absorptivity, emissivity, and convection coefficient on the steady-state temperature.

12.108 Neglecting the effects of radiation absorption, emission, and scattering within their atmospheres, calculate the average temperature of Earth, Venus, and Mars assuming diffuse, gray behavior. The average distance from the sun of each of the three planets, L_{s-p}, along with their *measured* average temperatures, $\overline{T}_p$, are shown in the table below. Based upon a comparison of the calculated and measured average temperatures, which planet is most affected by radiation transfer in its atmosphere?

Planet	L_{s-p} (m)	$\overline{T}_p$ (K)
Venus	1.08×10^{11}	735
Earth	1.50×10^{11}	287
Mars	2.30×10^{11}	227

12.109 A deep cavity of 50-mm diameter approximates a blackbody and is maintained at 250°C while exposed to solar irradiation of 800 W/m^2 and surroundings and ambient air at 25°C. A thin window of spectral transmissivity and reflectivity 0.9 and 0, respectively, for the spectral range 0.2 to 4 μm is placed over the cavity opening. In the spectral range beyond 4 μm, the window behaves as an opaque, diffuse, gray body of emissivity 0.95. Assuming that the convection coefficient on the upper surface of the window is 10 W/m$^2 \cdot$K, determine the temperature of the window and the power required to maintain the cavity at 250°C.

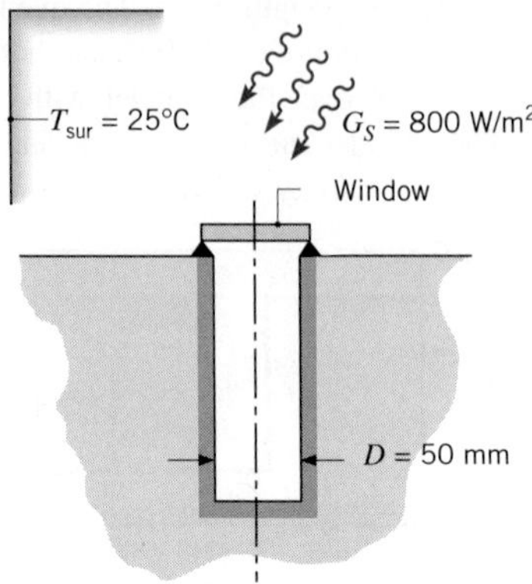

12.110 Consider the evacuated tube solar collector described in part (d) of Problem 1.87 of Chapter 1. In the interest of maximizing collector efficiency, what spectral radiative characteristics are desired for the outer tube and for the inner tube?

12.111 Solar flux of 900 W/m^2 is incident on the top side of a plate whose surface has a solar absorptivity of 0.9 and an emissivity of 0.1. The air and surroundings are at 17°C and the convection heat transfer coefficient between the plate and air is 20 W/m$^2 \cdot$K. Assuming that the bottom side of the plate is insulated, determine the steady-state temperature of the plate.

12.112 Consider an opaque, gray surface whose directional absorptivity is 0.8 for $0 \leq \theta \leq 60°$ and 0.1 for $\theta > 60°$. The surface is horizontal and exposed to solar irradiation comprised of direct and diffuse components.

(a) What is the surface absorptivity to direct solar radiation that is incident at an angle of 45° from the normal? What is the absorptivity to diffuse irradiation?

(b) Neglecting convection heat transfer between the surface and the surrounding air, what would be the equilibrium temperature of the surface if the direct and diffuse components of the irradiation were 600 and 100 W/m^2, respectively? The back side of the surface is insulated.

12.113 The absorber plate of a solar collector may be coated with an opaque material for which the spectral, directional absorptivity is characterized by relations of the form

$$\alpha_{\lambda,\theta}(\lambda, \theta) = \alpha_1 \cos\theta \qquad \lambda < \lambda_c$$

$$\alpha_{\lambda,\theta}(\lambda, \theta) = \alpha_2 \qquad \lambda > \lambda_c$$

The zenith angle θ is formed by the sun's rays and the plate normal, and α_1 and α_2 are constants.

(a) Obtain an expression for the total, hemispherical absorptivity, α_S, of the plate to solar radiation incident at $\theta = 45°$. Evaluate α_S for $\alpha_1 = 0.93$, $\alpha_2 = 0.25$, and a cut-off wavelength of $\lambda_c = 2\ \mu m$.

(b) Obtain an expression for the total, hemispherical emissivity ε of the plate. Evaluate ε for a plate temperature of $T_p = 60°C$ and the prescribed values of α_1, α_2, and λ_c.

(c) For a solar flux of $q''_S = 1000\ W/m^2$ incident at $\theta = 45°$ and the prescribed values of α_1, α_2, λ_c, and T_p, what is the net radiant heat flux, q''_{net}, to the plate?

(d) Using the prescribed conditions and the *Radiation/Band Emission Factor* option in the *Tools* section of *IHT* to evaluate $F_{(0\rightarrow\lambda_c)}$, explore the effect of λ_c on α_S, ε, and q''_{net} for the wavelength range $0.7 \le \lambda_c \le 5\ \mu m$.

12.114 A contractor must select a roof covering material from the two diffuse, opaque coatings with $\alpha_\lambda(\lambda)$ as shown. Which of the two coatings would result in a lower roof temperature? Which is preferred for summer use? For winter use? Sketch the spectral distribution of α_λ that would be ideal for summer use. For winter use.

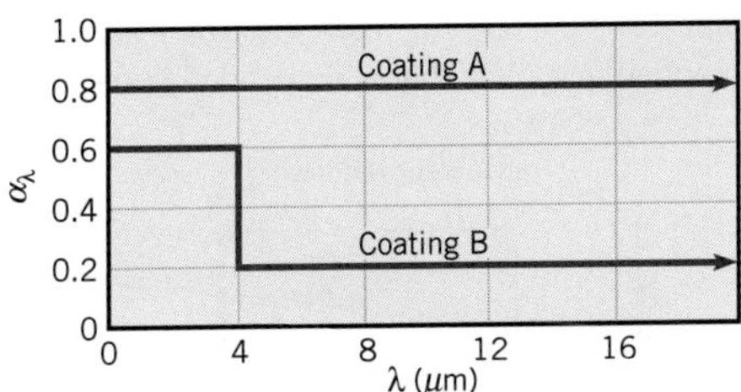

12.115 It is not uncommon for the night sky temperature in desert regions to drop to $-40°C$. If the ambient air temperature is 20°C and the convection coefficient for still air conditions is approximately $5\ W/m^2 \cdot K$, can a shallow pan of water freeze?

12.116 Plant leaves possess small channels that connect the interior moist region of the leaf to the environment. The channels, called *stomata*, pose the primary resistance to moisture transport through the entire plant, and the diameter of an individual stoma is sensitive to the level of CO_2 in the atmosphere. Consider a leaf of corn (maize) whose top surface is exposed to solar irradiation of $G_S = 600\ W/m^2$ and an effective sky temperature of $T_{sky} = 0°C$. The bottom side of the leaf is irradiated from the ground which is at a temperature of $T_g = 20°C$. Both the top and bottom of the leaf are subjected to convective conditions characterized by $h = 35\ W/m^2 \cdot K$, $T_\infty = 25°C$ and also experience evaporation through the stomata. Assuming the evaporative flux of water vapor is $50 \times 10^{-6}\ kg/m^2 \cdot s$ under rural atmospheric CO_2 concentrations and is reduced to $5 \times 10^{-6}\ kg/m^2 \cdot s$ when ambient CO_2 concentrations are doubled near an urban area, calculate the leaf temperature in the rural and urban locations. The heat of vaporization of water is $h_{fg} = 2400\ kJ/kg$ and assume $\alpha = \varepsilon = 0.97$ for radiation exchange with the sky and the ground, and $\alpha_S = 0.76$ for solar irradiation.

12.117 In the central receiver concept of solar energy collection, a large number of heliostats (reflectors) provide a concentrated solar flux of $q''_S = 80{,}000\ W/m^2$ to the receiver, which is positioned at the top of a tower.

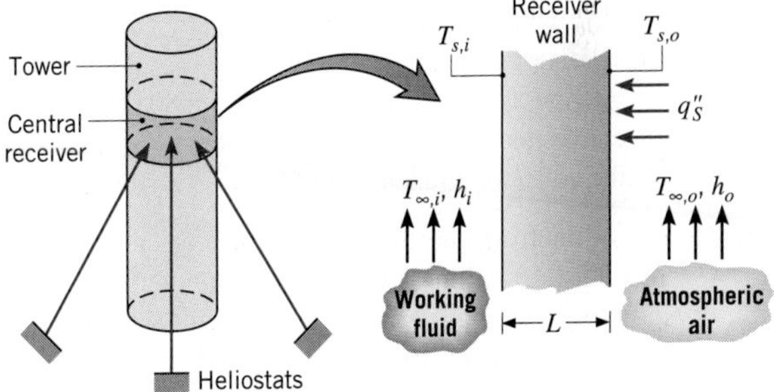

The receiver wall is exposed to the solar flux at its outer surface and to atmospheric air for which $T_{\infty,o} = 300\ K$ and $h_o = 25\ W/m^2 \cdot K$. The outer surface is opaque and diffuse, with a spectral absorptivity of $\alpha_\lambda = 0.9$ for $\lambda < 3\ \mu m$ and $\alpha_\lambda = 0.2$ for $\lambda > 3\ \mu m$. The inner surface is exposed to a working fluid (a pressurized liquid) for which $T_{\infty,i} = 700\ K$ and $h_i = 1000\ W/m^2 \cdot K$. The outer surface is also exposed to surroundings for which $T_{sur} = 300\ K$. If the wall is fabricated from a high-temperature material for which $k = 15\ W/m \cdot K$, what is the minimum thickness L needed to ensure that the outer surface temperature does not exceed $T_{s,o} = 1000\ K$? What is the collection efficiency associated with this thickness?

12.118 Consider the central receiver of Problem 12.117 to be a cylindrical shell of outer diameter $D = 7\ m$ and length $L = 12\ m$. The outer surface is opaque and diffuse, with a spectral absorptivity of $\alpha_\lambda = 0.9$ for $\lambda < 3\ \mu m$ and $\alpha_\lambda = 0.2$ for $\lambda > 3\ \mu m$. The surface is exposed to *quiescent* ambient air for which $T_\infty = 300\ K$.

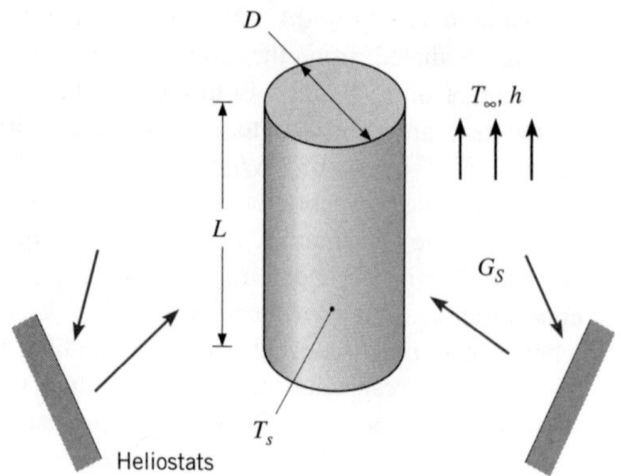

(a) Consider representative operating conditions for which solar irradiation at $G_S = 80{,}000\ \text{W/m}^2$ is uniformly distributed over the receiver surface and the surface temperature is $T_s = 800$ K. Determine the rate at which energy is collected by the receiver and the corresponding collector efficiency.

(b) The surface temperature is affected by conditions internal to the receiver. For $G_S = 80{,}000\ \text{W/m}^2$, compute and plot the rate of energy collection and the collector efficiency for $600 \le T_s \le 1000$ K.

12.119 Radiation from the atmosphere or sky can be estimated as a fraction of the blackbody radiation corresponding to the air temperature near the ground, T_{air}. That is, irradiation from the sky can be expressed as $G_{\text{atm}} = \varepsilon_{\text{sky}}\sigma T_{\text{air}}^4$ and for a clear night sky, the emissivity is correlated by an expression of the form $\varepsilon_{\text{sky}} = 0.741 + 0.0062T_{\text{dp}}$, where T_{dp} is the dew point temperature (°C). Consider a flat plate exposed to the night sky and in ambient air at 15°C with a relative humidity of 70%. Assume the back side of the plate is insulated, and that the convection coefficient on the front side can be estimated by the correlation $h(\text{W/m}^2 \cdot \text{K}) = 1.25\Delta T^{1/3}$, where ΔT is the absolute value of the plate-to-air temperature difference. Will dew form on the plate if the surface is (a) clean and metallic with $\varepsilon = 0.23$, and (b) painted with $\varepsilon = 0.85$?

12.120 A thin sheet of glass is used on the roof of a greenhouse and is irradiated as shown.

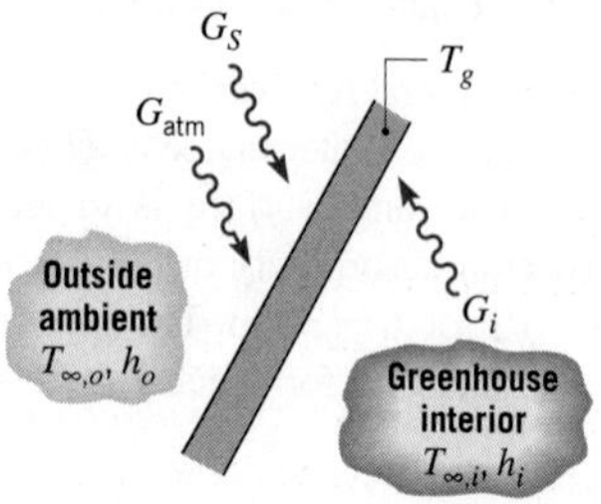

The irradiation comprises the total solar flux G_S, the flux G_{atm} due to atmospheric emission (sky radiation), and the flux G_i due to emission from interior surfaces. The fluxes G_{atm} and G_i are concentrated in the far IR region ($\lambda \gtrsim 8\ \mu\text{m}$). The glass may also exchange energy by convection with the outside and inside atmospheres. The glass may be assumed to be totally transparent for $\lambda < 1\ \mu\text{m}$ ($\tau_\lambda = 1.0$ for $\lambda < 1\ \mu\text{m}$) and opaque, with $\alpha_\lambda = 1.0$ for $\lambda \ge 1\ \mu\text{m}$.

(a) Assuming steady-state conditions, with all radiative fluxes uniformly distributed over the surfaces and the glass characterized by a uniform temperature T_g, write an appropriate energy balance for a unit area of the glass.

(b) For $T_g = 27°\text{C}$, $h_i = 10\ \text{W/m}^2 \cdot \text{K}$, $G_S = 1100\ \text{W/m}^2$, $T_{\infty,o} = 24°\text{C}$, $h_o = 55\ \text{W/m}^2 \cdot \text{K}$, $G_{\text{atm}} = 250\ \text{W/m}^2$, and $G_i = 440\ \text{W/m}^2$, calculate the temperature of the greenhouse ambient air, $T_{\infty,i}$.

12.121 A solar furnace consists of an evacuated chamber with transparent windows, through which concentrated solar radiation is passed. Concentration may be achieved by mounting the furnace at the focal point of a large curved reflector that tracks radiation incident directly from the sun. The furnace may be used to evaluate the behavior of materials at elevated temperatures, and we wish to design an experiment to assess the durability of a diffuse, spectrally selective coating for which $\alpha_\lambda = 0.95$ in the range $\lambda \le 4.5\ \mu\text{m}$ and $\alpha_\lambda = 0.03$ for $\lambda > 4.5\ \mu\text{m}$. The coating is applied to a plate that is suspended in the furnace.

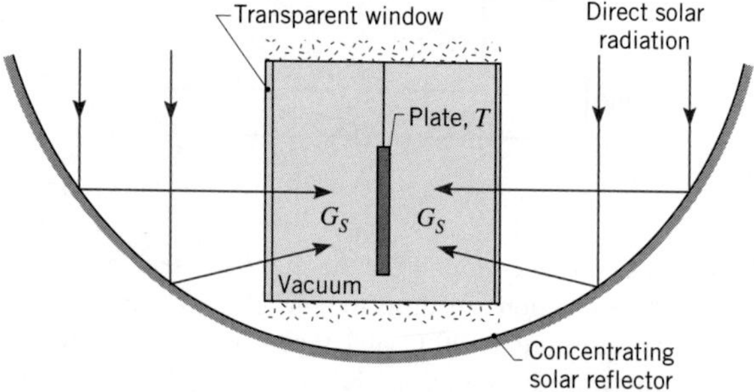

(a) If the experiment is to be operated at a steady-state plate temperature of $T = 2000$ K, how much solar irradiation G_S must be provided? The irradiation may be assumed to be uniformly distributed over the plate surface, and other sources of incident radiation may be neglected.

(b) The solar irradiation may be *tuned* to allow operation over a range of plate temperatures. Compute and plot G_S as a function of temperature for $500 \le T \le 3000$ K. Plot the corresponding values of α and ε as a function of T for the designated range.

12.122 The flat roof on the refrigeration compartment of a food delivery truck is of length $L = 5$ m and width $W = 2$ m. It is fabricated from thin sheet metal to which a fiberboard insulating material of thickness $t = 25$ mm and thermal conductivity $k = 0.05$ W/m·K is bonded. During normal operation, the truck moves at a velocity of $V = 30$ m/s in air at $T_\infty = 27°C$, with a rooftop solar irradiation of $G_S = 900$ W/m^2 and with the interior surface temperature maintained at $T_{s,i} = -13°C$.

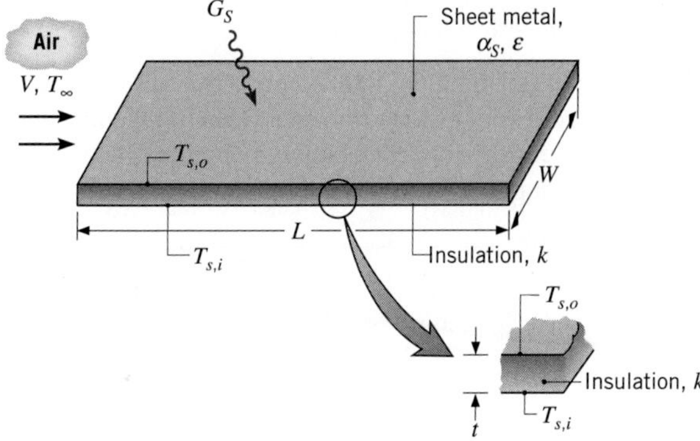

(a) The owner has the option of selecting a roof coating from one of the three paints listed in Table A.12 (Parsons Black, Acrylic White, or Zinc Oxide White). Which should be chosen and why?

(b) For the preferred paint of part (a), determine the steady-state value of the outer surface temperature $T_{s,o}$. The boundary layer is tripped at the leading edge of the roof, and turbulent flow may be assumed to exist over the entire roof. Properties of the air may be taken to be $\nu = 15 \times 10^{-6}$ m^2/s, $k = 0.026$ W/m·K, and $Pr = 0.71$.

(c) What is the load (W) imposed on the refrigeration system by heat transfer through the roof?

(d) Explore the effect of the truck velocity on the outer surface temperature and the heat load.

12.123 Growers use giant fans to prevent grapes from freezing when the effective sky temperature is low. The grape, which may be viewed as a thin skin of negligible thermal resistance enclosing a volume of sugar water, is exposed to ambient air and is irradiated from the sky above and ground below. Assume the grape to be an isothermal sphere of 15-mm diameter, and assume uniform blackbody irradiation over its top and bottom hemispheres due to emission from the sky and the earth, respectively.

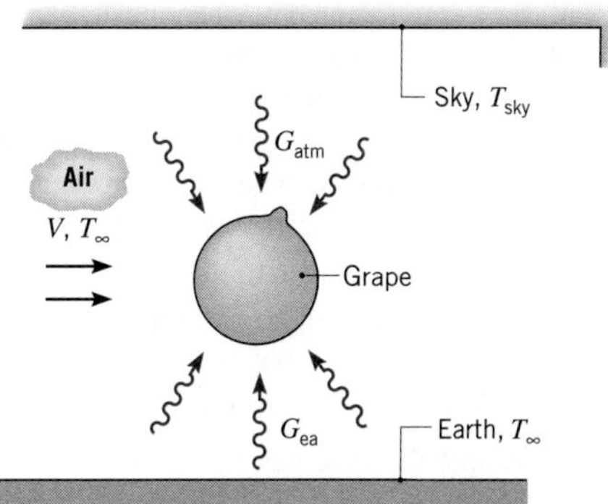

(a) Derive an expression for the rate of change of the grape temperature. Express your result in terms of a convection coefficient and appropriate temperatures and radiative quantities.

(b) Under conditions for which $T_{sky} = 235$ K, $T_\infty = 273$ K, and the fan is off ($V = 0$), determine whether the grapes will freeze. To a good approximation, the skin emissivity is 1 and the grape thermophysical properties are those of sugarless water. However, because of the sugar content, the grape freezes at $-5°C$.

(c) With all conditions remaining the same, except that the fans are now operating with $V = 1$ m/s, will the grapes freeze?

12.124 A circular metal disk having a diameter of 0.4 m is placed firmly against the ground in a barren horizontal region where the earth is at a temperature of 280 K. The effective sky temperature is also 280 K. The disk is exposed to quiescent ambient air at 300 K and direct solar irradiation of 745 W/m^2. The surface of the disk is diffuse with $\varepsilon_\lambda = 0.9$ for $0 < \lambda < 1$ μm and $\varepsilon_\lambda = 0.2$ for $\lambda > 1$ μm. After some time has elapsed, the disk achieves a uniform, steady-state temperature. The thermal conductivity of the soil is 0.52 W/m·K.

(a) Determine the fraction of the incident solar irradiation that is absorbed.

(b) What is the emissivity of the disk surface?

(c) For a steady-state disk temperature of 340 K, employ a suitable correlation to determine the average free convection heat transfer coefficient at the upper surface of the disk.

(d) Show that a disk temperature of 340 K does indeed yield a steady-state condition for the disk.

12.125 The neighborhood cat likes to sleep on the roof of our shed in the backyard. The roofing surface is weathered galvanized sheet metal ($\varepsilon = 0.65$, $\alpha_S = 0.8$). Consider a cool spring day when the ambient air temperature is 10°C and the convection coefficient can be estimated from an empirical correlation of the form $\bar{h} = 1.0\Delta T^{1/3}$, where ΔT is the difference between the surface and ambient temperatures. Assume the sky temperature is -40°C.

(a) Assuming the backside of the roof is well insulated, calculate the roof temperature when the solar irradiation is 600 W/m^2. Will the cat enjoy sleeping under these conditions?

(b) Consider the case when the backside of the roof is not insulated, but is exposed to ambient air with the same convection coefficient relation and experiences radiation exchange with the ground, also at the ambient air temperature. Calculate the roof temperature and comment on whether the roof will be a comfortable place for the cat to snooze.

12.126 The exposed surface of a power amplifier for an earth satellite receiver of area 130 mm × 130 mm has a diffuse, gray, opaque coating with an emissivity of 0.5. For typical amplifier operating conditions, the surface temperature is 58°C under the following environmental conditions: air temperature, $T_\infty = 27$°C; sky temperature, $T_{sky} = -20$°C; convection coefficient, $h =$ 15 W/m$^2 \cdot$K; and solar irradiation, $G_S = 800$ W/m^2.

(a) For the above conditions, determine the electrical power being generated within the amplifier.

(b) It is desired to reduce the surface temperature by applying one of the diffuse coatings (A, B, C) shown as follows.

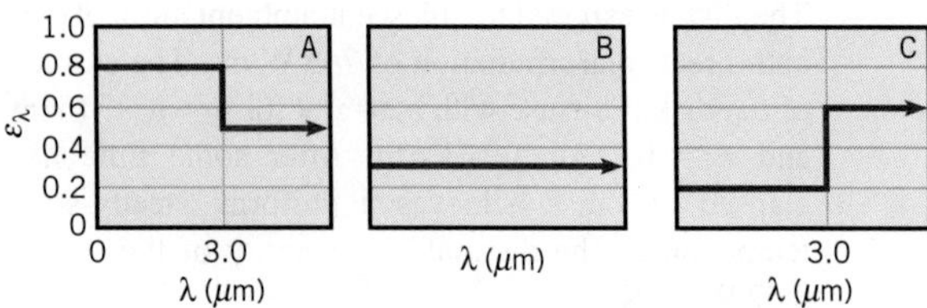

Which coating will result in the coolest surface temperature for the same amplifier operating and environmental conditions?

12.127 Consider a thin opaque, horizontal plate with an electrical heater on its backside. The front side is exposed to ambient air that is at 20°C and provides a convection heat transfer coefficient of 10 W/m$^2 \cdot$K, solar irradiation of 600 W/m^2, and an effective sky temperature of -40°C.

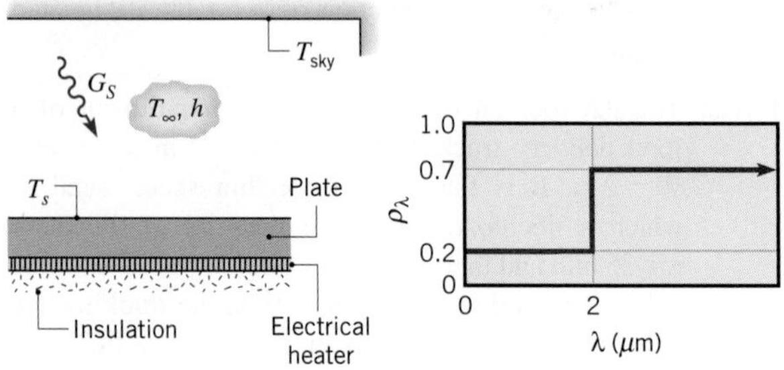

What is the electrical power (W/m^2) required to maintain the plate surface temperature at $T_s = 60$°C if the plate is diffuse and has the designated spectral, hemispherical reflectivity?

12.128 The oxidized-aluminum wing of an aircraft has a chord length of $L_c = 4$ m and a spectral, hemispherical emissivity characterized by the following distribution.

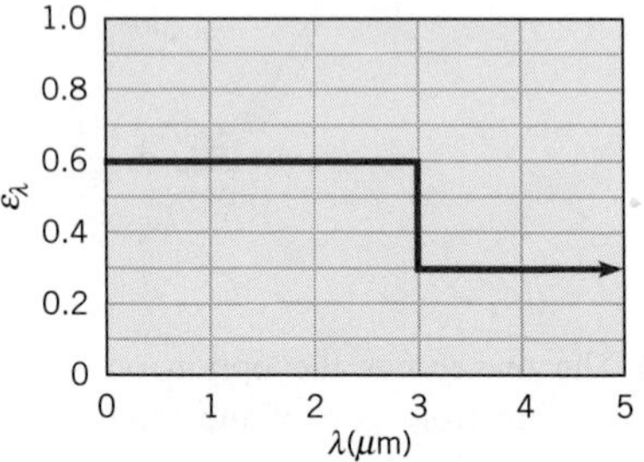

(a) Consider conditions for which the plane is on the ground where the air temperature is 27°C, the solar irradiation is 800 W/m^2, and the effective sky temperature is 270 K. If the air is quiescent, what is the temperature of the top surface of the wing? The wing may be approximated as a horizontal, flat plate.

(b) When the aircraft is flying at an elevation of approximately 9000 m and a speed of 200 m/s, the air temperature, solar irradiation, and effective sky temperature are -40°C, 1100 W/m^2, and 235 K, respectively. What is the temperature of the wing's top surface? The properties of the air may be approximated as $\rho = 0.470$ kg/m^3, $\mu = 1.50 \times 10^{-5}$ N$\cdot$s/m^2, $k = 0.021$ W/m$\cdot$K, and $Pr = 0.72$.

Space Radiation

12.129 Two plates, one with a black painted surface and the other with a special coating (chemically oxidized copper) are in earth orbit and are exposed to solar radiation. The solar rays make an angle of 30° with the normal to the plate. Estimate the equilibrium temperature of each plate assuming they are diffuse and that the solar flux is 1368 W/m^2. The spectral absorptivity of the black painted surface can be approximated by $\alpha_\lambda = 0.95$ for $0 \leq \lambda \leq \infty$ and that of the special coating by $\alpha_\lambda = 0.95$ for $0 \leq \lambda < 3\ \mu$m and $\alpha_\lambda = 0.05$ for $\lambda \geq 3\ \mu$m.

12.130 A spherical satellite of diameter D is in orbit about the earth and is coated with a diffuse material for which the spectral absorptivity is $\alpha_\lambda = 0.6$ for $\lambda \leq 3\ \mu$m and $\alpha_\lambda = 0.3$ for $\lambda > 3\ \mu$m. When it is on the "dark" side of the earth, the satellite sees irradiation from the earth's surface only. The irradiation may be assumed to be incident as parallel rays, and its magnitude is $G_E = 340$ W/m^2. On the "bright" side of the earth the satellite sees the earth irradiation G_E plus the solar irradiation $G_S = 1368$ W/m^2. The spectral distribution of radiation from the earth may be approximated as that of a blackbody at 280 K, and the temperature of the satellite may be assumed to remain below 500 K.

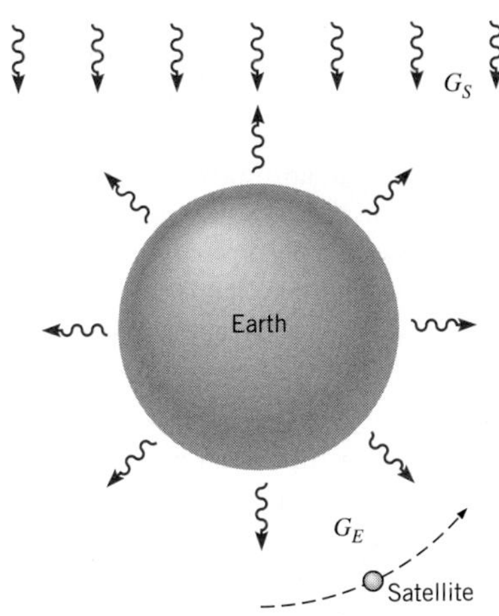

What is the steady-state temperature of the satellite when it is on the dark side of the earth and when it is on the bright side?

12.131 A radiator on a proposed satellite solar power station must dissipate heat being generated within the satellite by radiating it into space. The radiator surface has a solar absorptivity of 0.5 and an emissivity of 0.95. What is the equilibrium surface temperature when the solar irradiation is 1000 W/m^2 and the required heat dissipation is 1500 W/m^2?

12.132 A spherical satellite in near-earth orbit is exposed to solar irradiation of 1368 W/m^2. To maintain a desired operating temperature, the thermal control engineer intends to use a checker pattern for which a fraction F of the satellite surface is coated with an evaporated aluminum film ($\varepsilon = 0.03$, $\alpha_S = 0.09$), and the fraction $(1-F)$ is coated with a white, zinc-oxide paint ($\varepsilon = 0.85$, $\alpha_S = 0.22$). Assume the satellite is isothermal and has no internal power dissipation. Determine the fraction F of the checker pattern required to maintain the satellite at 300 K.

12.133 An annular fin of thickness t is used as a radiator to dissipate heat for a space power system. The fin is insulated on the bottom and may be exposed to solar irradiation G_S. The fin is coated with a diffuse, spectrally selective material whose spectral reflectivity is specified.

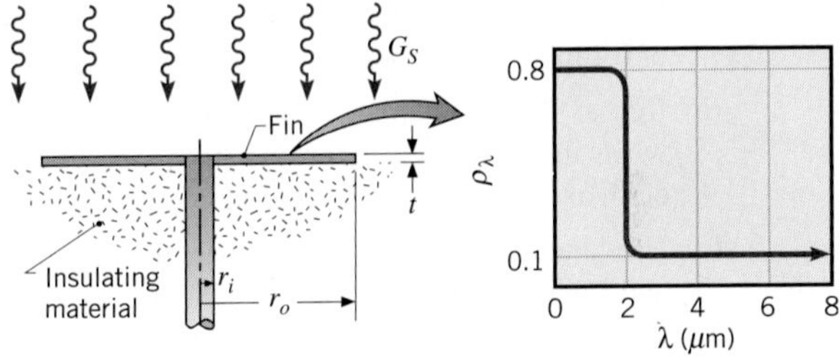

Heat is conducted to the fin through a solid rod of radius r_i, and the exposed upper surface of the fin radiates to free space, which is essentially at absolute zero temperature.

(a) If conduction through the rod maintains a fin base temperature of $T(r_i) = T_b = 400$ K and the fin efficiency is 100%, what is the rate of heat dissipation for a fin of radius $r_o = 0.5$ m? Consider two cases, one for which the radiator is exposed to the sun with $G_S = 1000$ W/m^2 and the other with no exposure ($G_S = 0$).

(b) In practice, the fin efficiency will be less than 100% and its temperature will decrease with increasing radius. Beginning with an appropriate control volume, derive the differential equation that determines the steady-state, radial temperature distribution in the fin. Specify appropriate boundary conditions.

12.134 A rectangular plate of thickness t, length L, and width W is proposed for use as a radiator in a spacecraft application. The plate material has a thermal conductivity of 300 W/m·K, a solar absorptivity of 0.45, and an emissivity of 0.9. The radiator is exposed to solar radiation only on its top surface, while both surfaces are exposed to deep space at a temperature of 4 K.

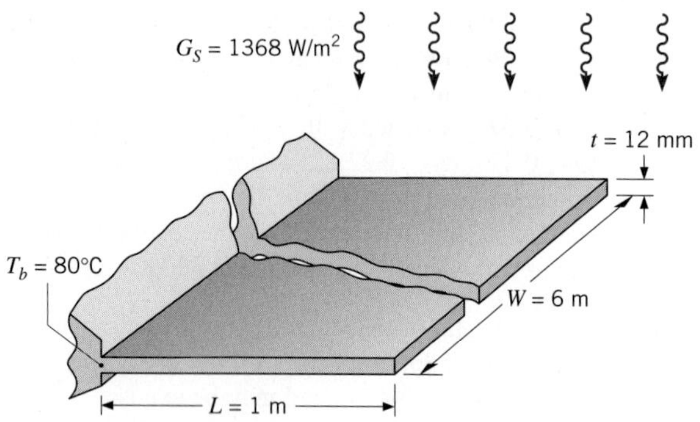

(a) If the base of the radiator is maintained at $T_b = 80°C$, what is its tip temperature and the rate of heat rejection? Use a computer-based, finite-difference method with a space increment of 0.1 m to obtain your solution.

(b) Repeat the calculation of part (a) for the case when the space ship is on the dark side of the earth and is not exposed to the sun.

(c) Use your computer code to calculate the heat rate and tip temperature for $G_S = 0$ and an extremely large value of the thermal conductivity. Compare your results to those obtained from a hand calculation that assumes the radiator to be at a uniform temperature T_b. What other approach might you use to validate your code?

12.135 The directional absorptivity of a gray surface varies with θ as follows.

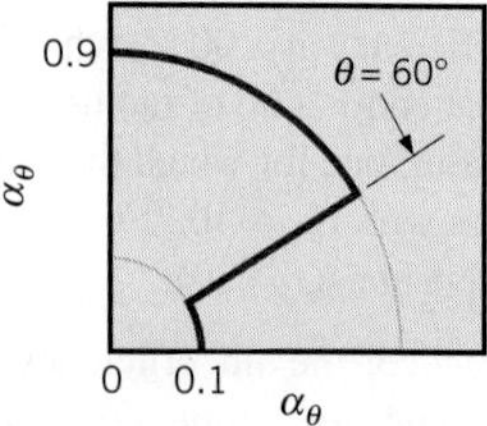

(a) What is the ratio of the normal absorptivity α_n to the hemispherical emissivity of the surface?

(b) Consider a plate with these surface characteristics on both sides in earth orbit. If the solar flux incident on one side of the plate is $q''_S = 1368 \text{ W/m}^2$, what equilibrium temperature will the plate assume if it is oriented normal to the sun's rays? What temperature will it assume if it is oriented at 75° to the sun's rays?

12.136 Two special coatings are available for application to an absorber plate installed below the cover glass described in Example 12.9. Each coating is diffuse and is characterized by the spectral distributions shown.

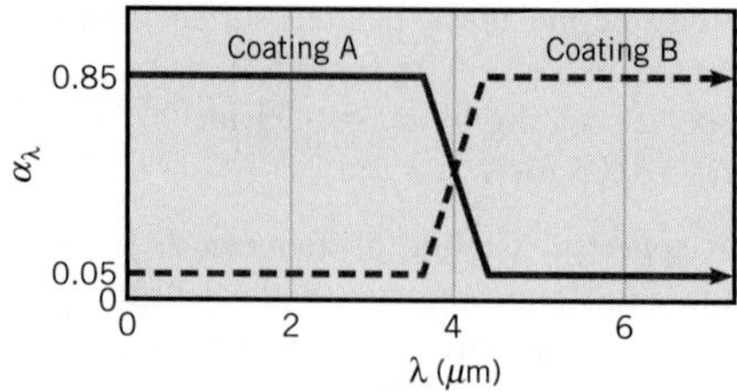

Which coating would you select for the absorber plate? Explain briefly. For the selected coating, what is the rate at which radiation is absorbed per unit area of the absorber plate if the total solar irradiation at the cover glass is $G_S = 1000 \text{ W/m}^2$?

12.137 Consider the spherical satellite of Problem 12.130. Instead of the entire satellite being coated with a material that is spectrally selective, half of the satellite is covered with a diffuse gray coating characterized by $\alpha_1 = 0.6$. The other half of the satellite is coated with a diffuse gray material with $\alpha_2 = 0.3$.

(a) Determine the steady-state satellite temperature when the satellite is on the bright side of the earth with the high-absorptivity coating facing the sun. Determine the steady-state satellite temperature when the low-absorptivity coating faces the sun. *Hint*: Assume one hemisphere of the satellite is irradiated by the sun and the opposite hemisphere is irradiated by the earth.

(b) Determine the steady-state satellite temperature when the satellite is on the dark side of the earth with the high-absorptivity coating facing the earth. Determine the steady-state satellite temperature when the low-absorptivity coating faces the earth.

(c) Identify a scheme to minimize the temperature variations of the satellite as it travels between the bright and dark sides of the earth.

12.138 A spherical capsule of 3-m radius is fired from a space platform in earth orbit, such that it travels toward the center of the sun at 16,000 km/s. Assume that the capsule is a lumped capacitance body with a density–specific heat product of $4 \times 10^6 \text{ J/m}^3 \cdot \text{K}$ and that its surface is black.

(a) Derive a differential equation for predicting the capsule temperature as a function of time. Solve this equation to obtain the temperature as a function of time in terms of capsule parameters and its initial temperature T_i.

(b) If the capsule begins its journey at 20°C, predict the position of the capsule relative to the sun at which its destruction temperature, 150°C, is reached.

12.139 The spectral absorptivity of aluminum coated with a thin layer of silicon dioxide may be approximated as $\alpha_{\lambda,1} = 0.98$ for $\lambda < \lambda_c$ and $\alpha_{\lambda,2} = 0.05$ for $\lambda \geq \lambda_c$ where the *cutoff wavelength* is $\lambda_c = 0.15\ \mu\text{m}$ under normal circumstances.

(a) Determine the equilibrium temperature of a flat piece of the coated aluminum that is exposed to solar irradiation, $G_S = 1368\ \text{W/m}^2$ on its upper surface. The opposite surface is insulated.

(b) The cutoff wavelength can be modified by varying the coating thickness. Determine the value of λ_c that will maximize the equilibrium temperature of the surface.

12.140 Consider the spherical satellite of Problem 12.130. By changing the thickness of the diffuse material used for the coating, engineers can control the *cutoff wavelength* that marks the boundary between $\alpha_\lambda = 0.6$ and $\alpha_\lambda = 0.3$.

(a) What cutoff wavelength will minimize the steady-state temperature of the satellite when it is on the bright side of the earth? Using this coating, what will the steady-state temperature on the dark side of the earth be?

(b) What cutoff wavelength will maximize the steady-state temperature of the satellite when it is on the dark side of the earth? What will the corresponding steady-state temperature be on the bright side?

12.141 A solar panel mounted on a spacecraft has an area of 1 m^2 and a solar-to-electrical power conversion efficiency of 12%. The side of the panel with the photovoltaic array has an emissivity of 0.8 and a solar absorptivity of 0.8. The back side of the panel has an emissivity of 0.7. The array is oriented normal to solar irradiation of 1500 W/m^2.

(a) Determine the steady-state temperature of the panel and the electrical power (W) produced for the prescribed conditions.

(b) If the panel were a thin plate without the solar cells, but with the same radiative properties, determine the temperature of the plate for the prescribed conditions. Compare this result with that from part (a). Are they the same or different? Explain why.

(c) Determine the temperature of the solar panel 1500 s after the spacecraft is eclipsed by a planet. The thermal capacity of the panel per unit area is 9000 J/m$^2 \cdot$ K.

CHAPTER 13

Radiation Exchange Between Surfaces

Having thus far restricted our attention to radiative processes that occur at a *single surface*, we now consider the problem of radiative exchange between two or more surfaces. This exchange depends strongly on the surface geometries and orientations, as well as on their radiative properties and temperatures. Initially, we assume that the surfaces are separated by a *nonparticipating medium.* Since such a medium neither emits, absorbs, nor scatters, it has no effect on the transfer of radiation between surfaces. A vacuum meets these requirements exactly, and most gases meet them to an excellent approximation.

Our first objective is to establish geometrical features of the radiation exchange problem by developing the notion of a *view factor*. Our second objective is to develop procedures for predicting radiative exchange between surfaces that form an *enclosure*. We will limit our attention to surfaces that are assumed to be opaque, diffuse, and gray. We conclude our consideration of radiation exchange between surfaces by considering the effects of a *participating medium*, namely, an intervening gas that emits and absorbs radiation.

13.1 *The View Factor*

To compute radiation exchange between any two surfaces, we must first introduce the concept of a *view factor* (also called a *configuration* or *shape factor*).

13.1.1 The View Factor Integral

The view factor F_{ij} is defined as the *fraction of the radiation leaving surface i that is intercepted by surface j.* To develop a general expression for F_{ij}, we consider the arbitrarily oriented surfaces A_i and A_j of Figure 13.1. Elemental areas on each surface, dA_i and dA_j, are connected by a line of length R, which forms the polar angles θ_i and θ_j, respectively, with the surface normals $\mathbf{n}_i$ and $\mathbf{n}_j$. The values of R, θ_i, and θ_j vary with the position of the elemental areas on A_i and A_j.

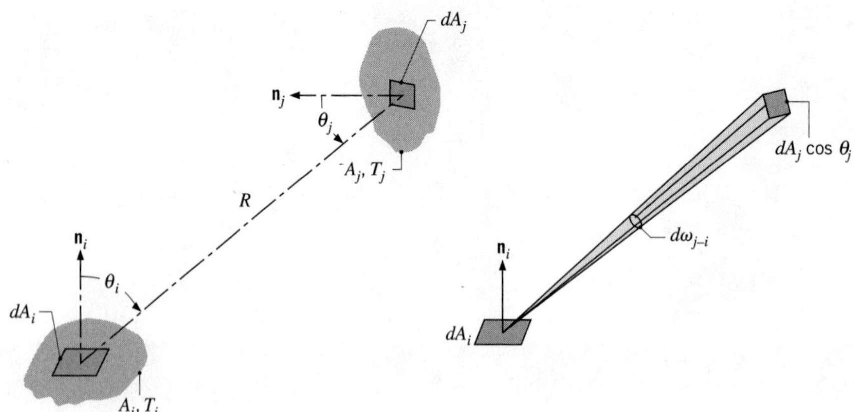

FIGURE 13.1 View factor associated with radiation exchange between elemental surfaces of area dA_i and dA_j.

From the definition of the radiation intensity, Section 12.3.2, and Equation 12.11, the rate at which radiation *leaves* dA_i and is *intercepted* by dA_j may be expressed as

$$dq_{i\to j} = I_{e+r,i} \cos\theta_i \, dA_i \, d\omega_{j-i}$$

where $I_{e+r,i}$ is the intensity of radiation leaving surface i by emission and reflection and $d\omega_{j-i}$ is the solid angle subtended by dA_j when viewed from dA_i. With $d\omega_{j-i} = (\cos\theta_j \, dA_j)/R^2$ from Equation 12.7, it follows that

$$dq_{i\to j} = I_{e+r,i} \frac{\cos\theta_i \cos\theta_j}{R^2} dA_i \, dA_j$$

Assuming that surface i *emits* and *reflects diffusely* and substituting from Equation 12.27, we then obtain

$$dq_{i\to j} = J_i \frac{\cos\theta_i \cos\theta_j}{\pi R^2} dA_i \, dA_j$$

The total rate at which radiation leaves surface i and is intercepted by j may then be obtained by integrating over the two surfaces. That is,

$$q_{i\to j} = J_i \int_{A_i} \int_{A_j} \frac{\cos\theta_i \cos\theta_j}{\pi R^2} dA_i \, dA_j$$

where it is assumed that the radiosity J_i is uniform over the surface A_i. From the definition of the view factor as the fraction of the radiation that leaves A_i and is intercepted by A_j,

$$F_{ij} = \frac{q_{i\to j}}{A_i J_i}$$

it follows that

$$F_{ij} = \frac{1}{A_i} \int_{A_i} \int_{A_j} \frac{\cos\theta_i \cos\theta_j}{\pi R^2} dA_i \, dA_j \qquad (13.1)$$

Similarly, the view factor F_{ji} is defined as the fraction of the radiation that leaves A_j and is intercepted by A_i. The same development then yields

$$F_{ji} = \frac{1}{A_j} \int_{A_i} \int_{A_j} \frac{\cos\theta_i \cos\theta_j}{\pi R^2} dA_i \, dA_j \qquad (13.2)$$

Either Equation 13.1 or 13.2 may be used to determine the view factor associated with any two surfaces that are *diffuse emitters* and *reflectors* and have *uniform radiosity.*

13.1.2 View Factor Relations

An important view factor relation is suggested by Equations 13.1 and 13.2. In particular, equating the integrals appearing in these equations, it follows that

$$A_i F_{ij} = A_j F_{ji} \qquad (13.3)$$

This expression, termed the *reciprocity relation*, is useful in determining one view factor from knowledge of the other.

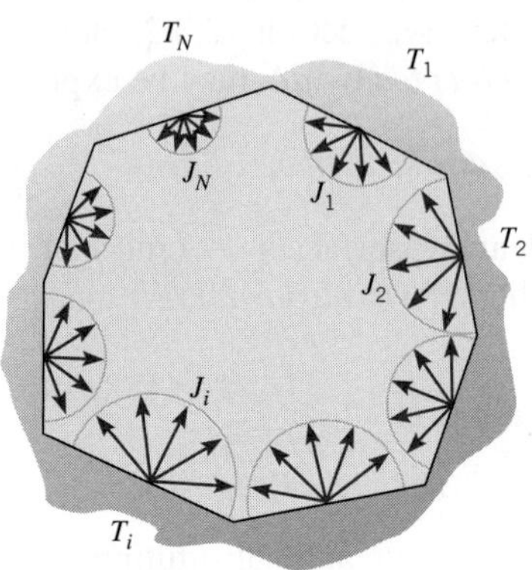

FIGURE 13.2 Radiation exchange in an enclosure.

Another important view factor relation pertains to the surfaces of an *enclosure* (Figure 13.2). From the definition of the view factor, the *summation rule*

$$\sum_{j=1}^{N} F_{ij} = 1 \tag{13.4}$$

may be applied to each of the N surfaces in the enclosure. This rule follows from the conservation requirement that all radiation leaving surface i must be intercepted by the enclosure surfaces. The term F_{ii} appearing in this summation represents the fraction of the radiation that leaves surface i and is directly intercepted by i. If the surface is concave, it *sees itself* and F_{ii} is nonzero. However, for a plane or convex surface, $F_{ii} = 0$.

To calculate radiation exchange in an enclosure of N surfaces, a total of N^2 view factors is needed. This requirement becomes evident when the view factors are arranged in the matrix form:

$$\begin{bmatrix} F_{11} & F_{12} & \cdots & F_{1N} \\ F_{21} & F_{22} & \cdots & F_{2N} \\ \vdots & \vdots & & \vdots \\ F_{N1} & F_{N2} & \cdots & F_{NN} \end{bmatrix}$$

However, all the view factors need not be calculated *directly.* A total of N view factors may be obtained from the N equations associated with application of the summation rule, Equation 13.4, to each of the surfaces in the enclosure. In addition, $N(N-1)/2$ view factors may be obtained from the $N(N-1)/2$ applications of the reciprocity relation, Equation 13.3, which are possible for the enclosure. Accordingly, only $[N^2 - N - N(N-1)/2] = N(N-1)/2$ view factors need be determined directly. For example, in a three-surface enclosure this requirement corresponds to only $3(3-1)/2 = 3$ view factors. The remaining six view factors may be obtained by solving the six equations that result from use of Equations 13.3 and 13.4.

To illustrate the foregoing procedure, consider a simple, two-surface enclosure involving the spherical surfaces of Figure 13.3. Although the enclosure is characterized by $N^2 = 4$ view factors (F_{11}, F_{12}, F_{21}, F_{22}), only $N(N-1)/2 = 1$ view factor need be determined directly. In this case such a determination may be made by *inspection.* In particular, since all radiation leaving the inner surface must reach the outer surface, it follows that $F_{12} = 1$. The same may not be said of radiation leaving the outer surface, since this surface sees itself. However, from the reciprocity relation, Equation 13.3, we obtain

$$F_{21} = \left(\frac{A_1}{A_2}\right) F_{12} = \left(\frac{A_1}{A_2}\right)$$

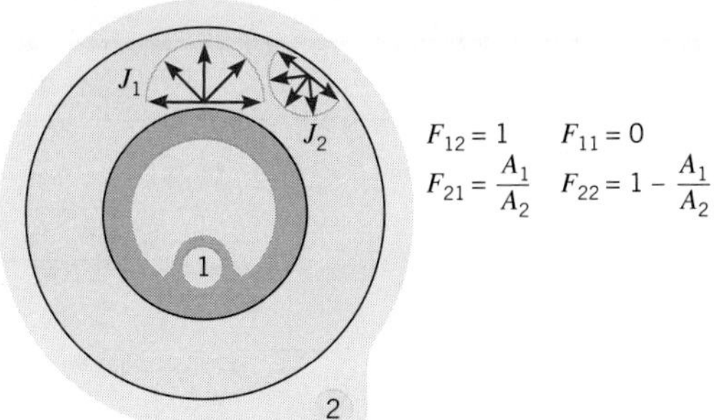

FIGURE 13.3 View factors for the enclosure formed by two spheres.

From the summation rule, we also obtain

$$F_{11} + F_{12} = 1$$

in which case $F_{11} = 0$, and

$$F_{21} + F_{22} = 1$$

in which case

$$F_{22} = 1 - \left(\frac{A_1}{A_2}\right)$$

For more complicated geometries, the view factor may be determined by solving the double integral of Equation 13.1. Such solutions have been obtained for many different surface arrangements and are available in equation, graphical, and tabular form [1–4]. Results for several common geometries are presented in Tables 13.1 and 13.2 and Figures 13.4 through 13.6. The configurations of Table 13.1 are assumed to be infinitely long (in a

TABLE 13.1 View Factors for Two-Dimensional Geometries [4]

Geometry	Relation
Parallel Plates with Midlines Connected by Perpendicular 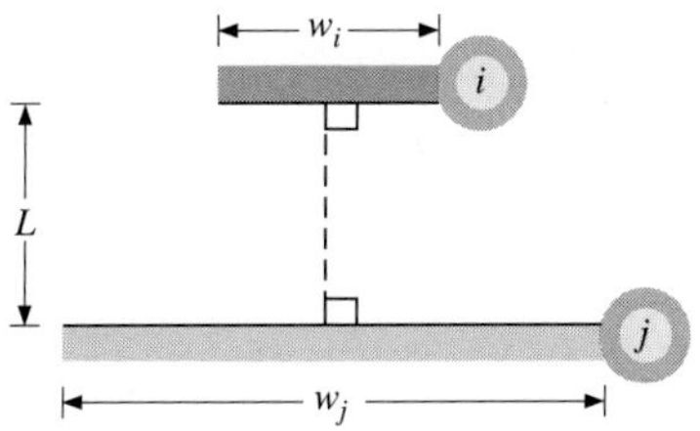	$F_{ij} = \dfrac{[(W_i + W_j)^2 + 4]^{1/2} - [(W_j - W_i)^2 + 4]^{1/2}}{2W_i}$ $W_i = w_i/L,\ W_j = w_j/L$
Inclined Parallel Plates of Equal Width and a Common Edge 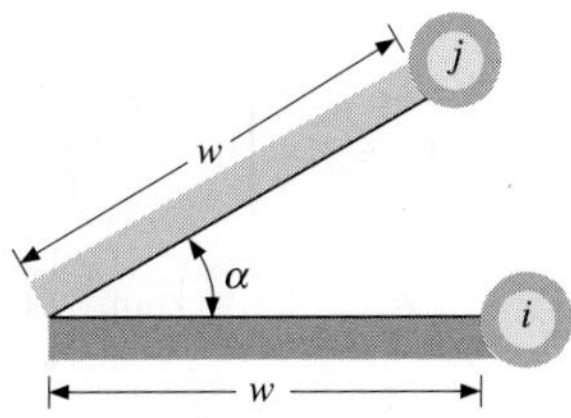	$F_{ij} = 1 - \sin\left(\dfrac{\alpha}{2}\right)$

(continues)

TABLE 13.1 Continued

Geometry	Relation
Perpendicular Plates with a Common Edge	
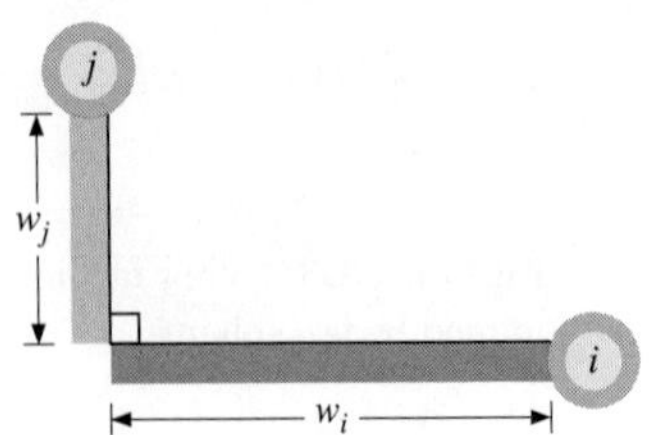	$F_{ij} = \dfrac{1 + (w_j/w_i) - [1 + (w_j/w_i)^2]^{1/2}}{2}$
Three-Sided Enclosure	
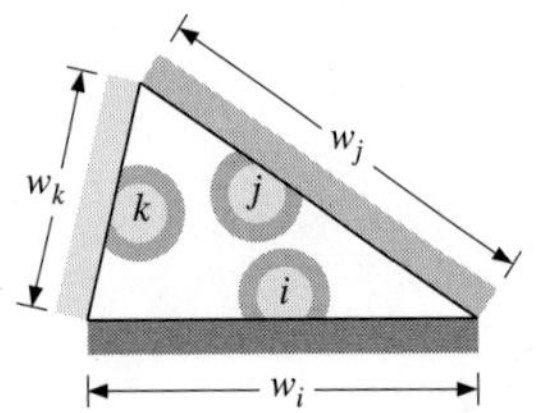	$F_{ij} = \dfrac{w_i + w_j - w_k}{2w_i}$
Parallel Cylinders of Different Radii	
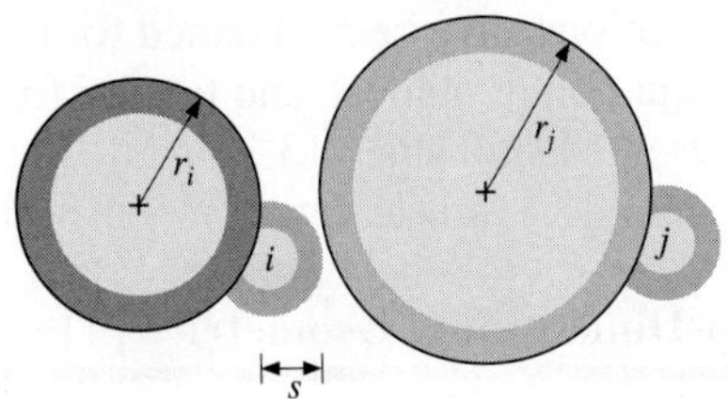	$F_{ij} = \dfrac{1}{2\pi}\Big\{\pi + [C^2 - (R+1)^2]^{1/2} - [C^2 - (R-1)^2]^{1/2} + (R-1)\cos^{-1}\Big[\Big(\dfrac{R}{C}\Big) - \Big(\dfrac{1}{C}\Big)\Big] - (R+1)\cos^{-1}\Big[\Big(\dfrac{R}{C}\Big) + \Big(\dfrac{1}{C}\Big)\Big]\Big\}$ $R = r_j/r_i,\ S = s/r_i$ $C = 1 + R + S$
Cylinder and Parallel Rectangle	
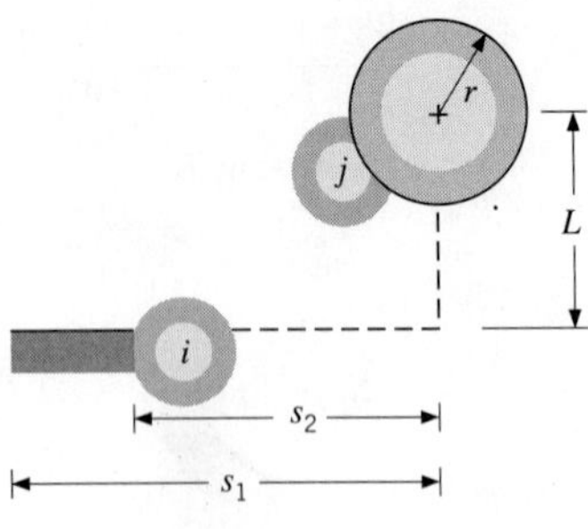	$F_{ij} = \dfrac{r}{s_1 - s_2}\Big[\tan^{-1}\dfrac{s_1}{L} - \tan^{-1}\dfrac{s_2}{L}\Big]$
Infinite Plane and Row of Cylinders	
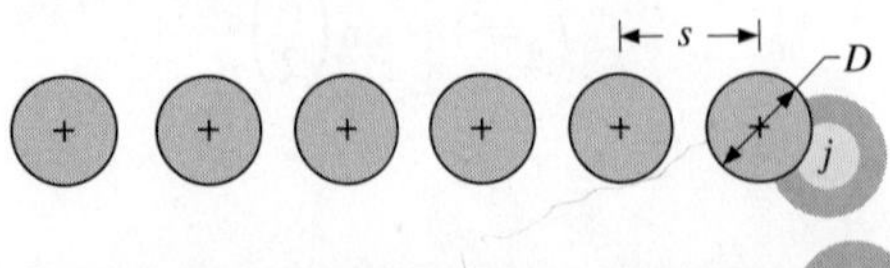	$F_{ij} = 1 - \Big[1 - \Big(\dfrac{D}{s}\Big)^2\Big]^{1/2} + \Big(\dfrac{D}{s}\Big)\tan^{-1}\Big[\Big(\dfrac{s^2 - D^2}{D^2}\Big)^{1/2}\Big]$

TABLE 13.2 View Factors for Three-Dimensional Geometries [4]

Geometry	Relation
Aligned Parallel Rectangles (Figure 13.4) 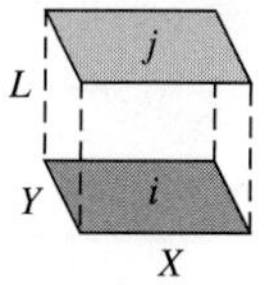	$\bar{X} = X/L,\ \bar{Y} = Y/L$ $F_{ij} = \dfrac{2}{\pi \bar{X}\bar{Y}}\left\{\ln\left[\dfrac{(1+\bar{X}^2)(1+\bar{Y}^2)}{1+\bar{X}^2+\bar{Y}^2}\right]^{1/2} + \bar{X}(1+\bar{Y}^2)^{1/2}\tan^{-1}\dfrac{\bar{X}}{(1+\bar{Y}^2)^{1/2}} + \bar{Y}(1+\bar{X}^2)^{1/2}\tan^{-1}\dfrac{\bar{Y}}{(1+\bar{X}^2)^{1/2}} - \bar{X}\tan^{-1}\bar{X} - \bar{Y}\tan^{-1}\bar{Y}\right\}$
Coaxial Parallel Disks (Figure 13.5) 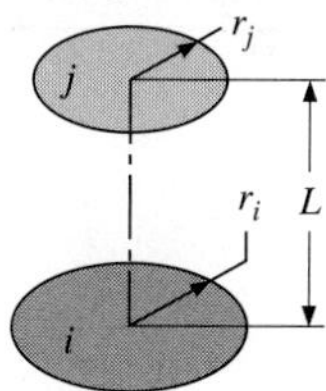	$R_i = r_i/L,\ R_j = r_j/L$ $S = 1 + \dfrac{1+R_j^2}{R_i^2}$ $F_{ij} = \dfrac{1}{2}\{S - [S^2 - 4(r_j/r_i)^2]^{1/2}\}$
Perpendicular Rectangles with a Common Edge (Figure 13.6) 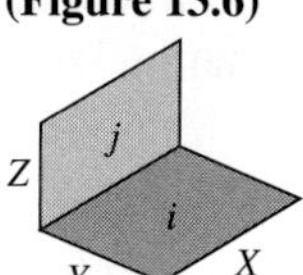	$H = Z/X,\ W = Y/X$ $F_{ij} = \dfrac{1}{\pi W}\left(W\tan^{-1}\dfrac{1}{W} + H\tan^{-1}\dfrac{1}{H} - (H^2+W^2)^{1/2}\tan^{-1}\dfrac{1}{(H^2+W^2)^{1/2}} + \dfrac{1}{4}\ln\left\{\dfrac{(1+W^2)(1+H^2)}{1+W^2+H^2}\left[\dfrac{W^2(1+W^2+H^2)}{(1+W^2)(W^2+H^2)}\right]^{W^2} \times \left[\dfrac{H^2(1+H^2+W^2)}{(1+H^2)(H^2+W^2)}\right]^{H^2}\right\}\right)$

direction perpendicular to the page) and are hence two-dimensional. The configurations of Table 13.2 and Figures 13.4 through 13.6 are three-dimensional.

It is useful to note that the results of Figures 13.4 through 13.6 may be used to determine other view factors. For example, the view factor for an end surface of a cylinder (or a truncated cone) relative to the lateral surface may be obtained by using the results of Figure 13.5 with the summation rule, Equation 13.4. Moreover, Figures 13.4 and 13.6 may be used to obtain other useful results if two additional view factor relations are developed.

The first relation concerns the additive nature of the view factor for a subdivided surface and may be inferred from Figure 13.7. Considering radiation from surface i to surface j, which is divided into n components, it is evident that

$$F_{i(j)} = \sum_{k=1}^{n} F_{ik} \tag{13.5}$$

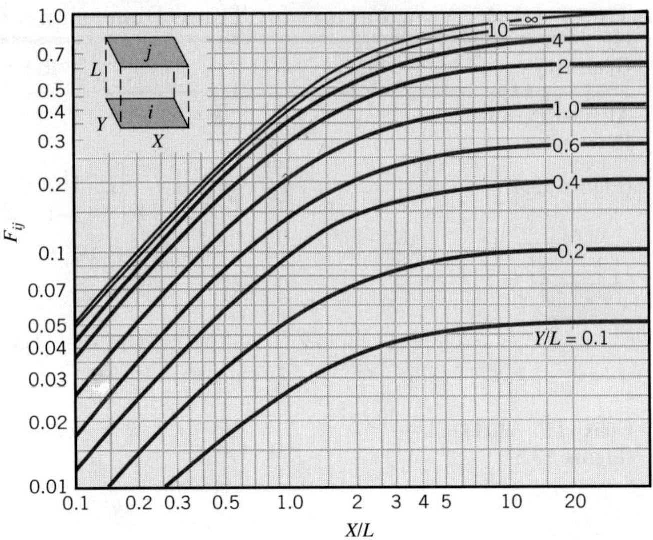

FIGURE 13.4 View factor for aligned parallel rectangles.

where the parentheses around a subscript indicate that it is a composite surface, in which case (j) is equivalent to $(1, 2, \ldots, k, \ldots, n)$. This expression simply states that radiation reaching a composite surface is the sum of the radiation reaching its parts. Although it pertains to subdivision of the receiving surface, it may also be used to obtain the second view factor relation, which pertains to subdivision of the originating surface. Multiplying

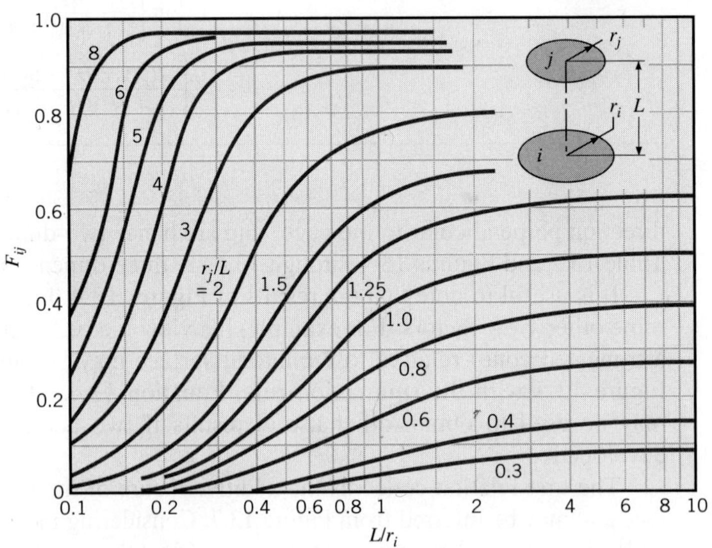

FIGURE 13.5 View factor for coaxial parallel disks.

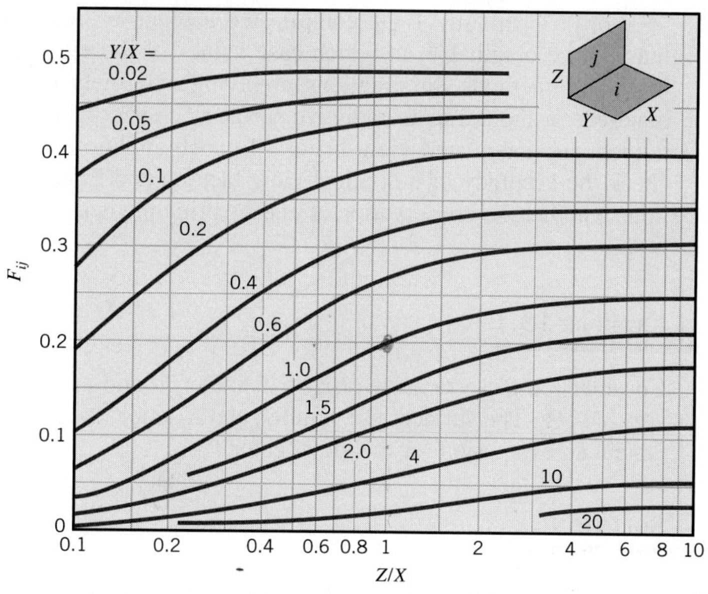

FIGURE 13.6 View factor for perpendicular rectangles with a common edge.

Equation 13.5 by A_i and applying the reciprocity relation, Equation 13.3, to each of the resulting terms, it follows that

$$A_j F_{(j)i} = \sum_{k=1}^{n} A_k F_{ki} \tag{13.6}$$

or

$$F_{(j)i} = \frac{\sum_{k=1}^{n} A_k F_{ki}}{\sum_{k=1}^{n} A_k} \tag{13.7}$$

Equations 13.6 and 13.7 may be applied when the originating surface is composed of several parts.

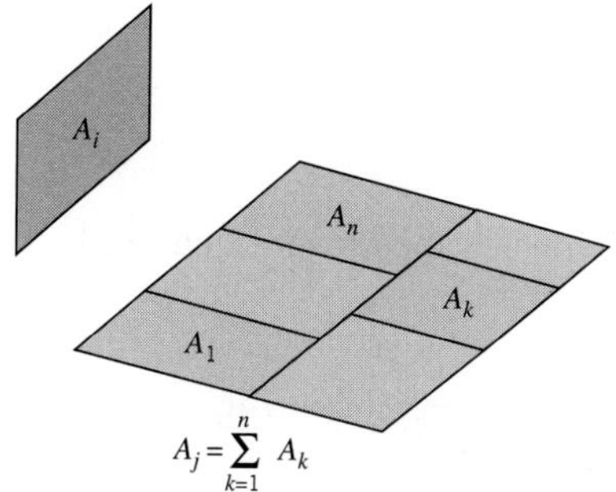

FIGURE 13.7 Areas used to illustrate view factor relations.

For problems involving complicated geometries, analytical solutions to Equation 13.1 may not be obtainable, in which case values of the view factors must be estimated using numerical methods. In situations involving extremely complex structures that may have hundreds or thousands of radiative surfaces, considerable error may be associated with the numerically calculated view factors. In such situations, Equation 13.3 should be used to check the accuracy of individual view factors, and Equation 13.4 should be used to determine whether the conservation of energy principle is satisfied [5].

EXAMPLE 13.1

Consider a diffuse circular disk of diameter D and area A_j and a plane diffuse surface of area $A_i \ll A_j$. The surfaces are parallel, and A_i is located at a distance L from the center of A_j. Obtain an expression for the view factor F_{ij}.

SOLUTION

Known: Orientation of small surface relative to large circular disk.

Find: View factor of small surface with respect to disk, F_{ij}.

Schematic:

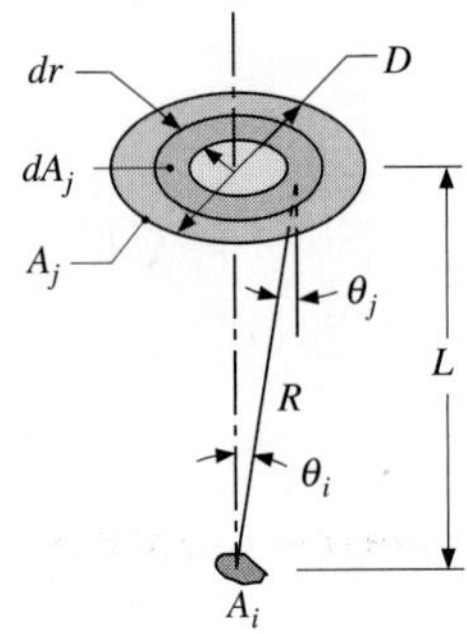

Assumptions:

1. Diffuse surfaces.
2. $A_i \ll A_j$.
3. Uniform radiosity on surface A_i.

Analysis: The desired view factor may be obtained from Equation 13.1.

$$F_{ij} = \frac{1}{A_i} \int_{A_i} \int_{A_j} \frac{\cos\theta_i \cos\theta_j}{\pi R^2} dA_i \, dA_j$$

Recognizing that θ_i, θ_j, and R are approximately independent of position on A_i, this expression reduces to

$$F_{ij} = \int_{A_j} \frac{\cos\theta_i \cos\theta_j}{\pi R^2} dA_j$$

or, with $\theta_i = \theta_j \equiv \theta$,

$$F_{ij} = \int_{A_j} \frac{\cos^2\theta}{\pi R^2} dA_j$$

With $R^2 = r^2 + L^2$, $\cos\theta = (L/R)$, and $dA_j = 2\pi r\, dr$, it follows that

$$F_{ij} = 2L^2 \int_0^{D/2} \frac{r\, dr}{(r^2 + L^2)^2} = \frac{D^2}{D^2 + 4L^2} \qquad \triangleleft \quad (13.8)$$

Comments:

1. Equation 13.8 may be used to quantify the asymptotic behavior of the curves in Figure 13.5 as the radius of the lower circle, r_i, approaches zero.
2. The preceding geometry is one of the simplest cases for which the view factor may be obtained from Equation 13.1. Geometries involving more detailed integrations are considered in the literature [1, 3].

EXAMPLE 13.2

Determine the view factors F_{12} and F_{21} for the following geometries:

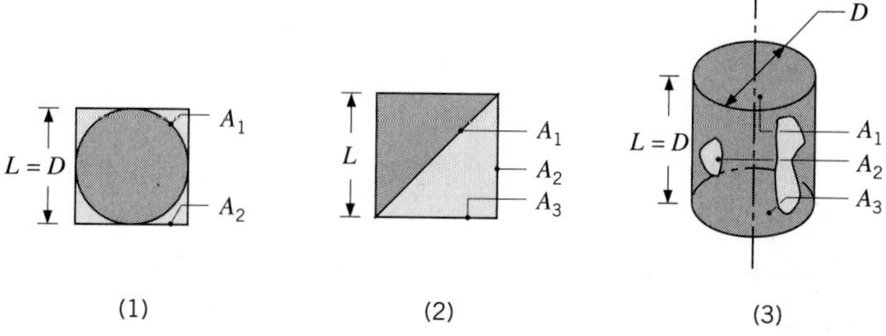

1. Sphere of diameter D inside a cubical box of length $L = D$.
2. One side of a diagonal partition within a long square duct.
3. End and side of a circular tube of equal length and diameter.

SOLUTION

Known: Surface geometries.

Find: View factors.

Assumptions: Diffuse surfaces with uniform radiosities.

Analysis: The desired view factors may be obtained from inspection, the reciprocity rule, the summation rule, and/or use of the charts.

1. Sphere within a cube:
   By inspection, $F_{12} = 1$ ◁

   By reciprocity, $F_{21} = \dfrac{A_1}{A_2} F_{12} = \dfrac{\pi D^2}{6L^2} \times 1 = \dfrac{\pi}{6}$ ◁

2. Partition within a square duct:
   From summation rule, $F_{11} + F_{12} + F_{13} = 1$
   where $F_{11} = 0$
   By symmetry, $F_{12} = F_{13}$
   Hence $F_{12} = 0.50$ ◁

   By reciprocity, $F_{21} = \dfrac{A_1}{A_2} F_{12} = \dfrac{\sqrt{2}L}{L} \times 0.5 = 0.71$ ◁

3. Circular tube:
   From Table 13.2 or Figure 13.5, with $(r_3/L) = 0.5$ and $(L/r_1) = 2$, $F_{13} = 0.172$
   From summation rule, $F_{11} + F_{12} + F_{13} = 1$
   or, with $F_{11} = 0$, $F_{12} = 1 - F_{13} = 0.828$ ◁

   From reciprocity, $F_{21} = \dfrac{A_1}{A_2} F_{12} = \dfrac{\pi D^2/4}{\pi DL} \times 0.828 = 0.207$ ◁

Comment: The geometric surfaces may, in reality, not be characterized by uniform radiosities. The consequences of nonuniform radiosity are discussed in Example 13.3.

13.2 *Blackbody Radiation Exchange*

In general, radiation may leave a surface due to both reflection and emission, and on reaching a second surface, experience reflection as well as absorption. However, matters are simplified for surfaces that may be approximated as blackbodies, since there is no reflection. Hence energy leaves only as a result of emission, and all incident radiation is absorbed.

Consider radiation exchange between two black surfaces of arbitrary shape (Figure 13.8). Defining $q_{i \to j}$ as the rate at which radiation *leaves* surface i and is *intercepted* by surface j, it follows that

$$q_{i \to j} = (A_i J_i) F_{ij} \tag{13.9}$$

or, since radiosity equals emissive power for a black surface ($J_i = E_{bi}$),

$$q_{i \to j} = A_i F_{ij} E_{bi} \tag{13.10}$$

Similarly,

$$q_{j \to i} = A_j F_{ji} E_{bj} \tag{13.11}$$

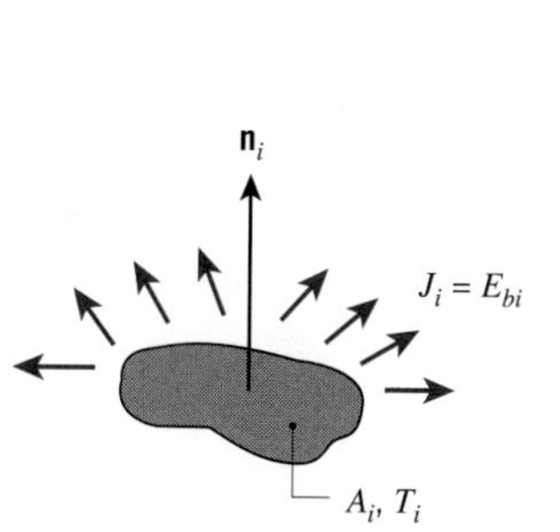

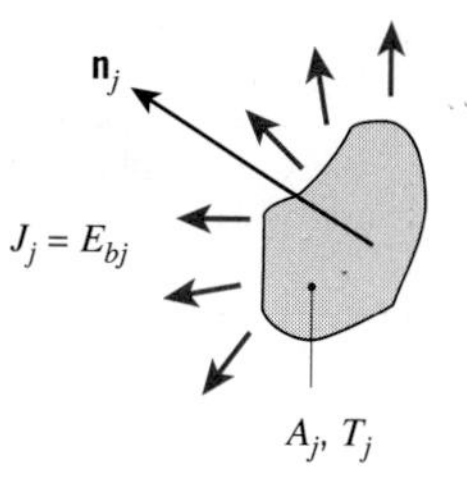

FIGURE 13.8 Radiation transfer between two surfaces that may be approximated as blackbodies.

The *net radiative exchange* between the two surfaces may then be defined as

$$q_{ij} = q_{i \to j} - q_{j \to i} \tag{13.12}$$

from which it follows that

$$q_{ij} = A_i F_{ij} E_{bi} - A_j F_{ji} E_{bj}$$

or, from Equations 12.32 and 13.3

$$q_{ij} = A_i F_{ij} \sigma (T_i^4 - T_j^4) \tag{13.13}$$

Equation 13.13 provides the *net* rate at which radiation *leaves* surface i as a result of its interaction with j, which is equal to the *net* rate at which j *gains* radiation due to its interaction with i.

The foregoing result may also be used to evaluate the net radiation transfer from any surface in an *enclosure* of black surfaces. With N surfaces maintained at different temperatures, the net transfer of radiation from surface i is due to exchange with the remaining surfaces and may be expressed as

$$q_i = \sum_{j=1}^{N} A_i F_{ij} \sigma (T_i^4 - T_j^4) \tag{13.14}$$

The net radiative heat flux, $q_i'' = q_i / A_i$, was denoted as q''_{rad} in Chapters 1 and 12. The subscript *rad* has been dropped here for convenience.

EXAMPLE 13.3

A furnace cavity, which is in the form of a cylinder of 50-mm diameter and 150-mm length, is open at one end to large surroundings that are at 27°C. The bottom of the cavity is heated independently, as are three annular sections that comprise the sides of the cavity. All interior surfaces of the cavity may be approximated as blackbodies and are maintained at 1650°C. What is the required electrical power input to the bottom surface of the cavity? What is the electrical power to the top, middle, and bottom sections of the cavity sides? The backs of the electrically heated surfaces are well insulated.

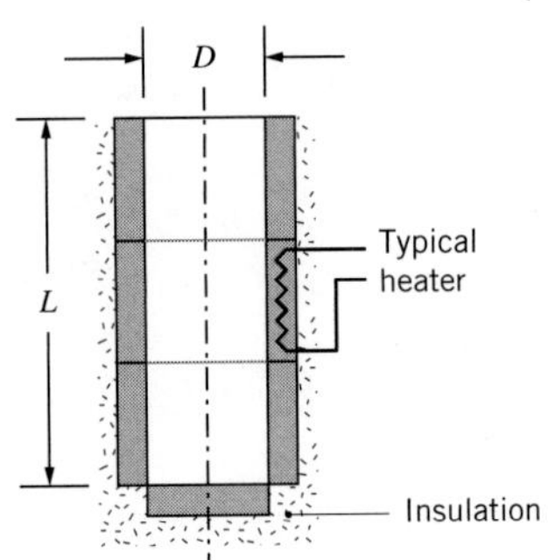

SOLUTION

Known: Temperature of furnace surfaces and surroundings.

Find: Electrical power required to maintain four sections of the furnace at the prescribed temperature.

Schematic:

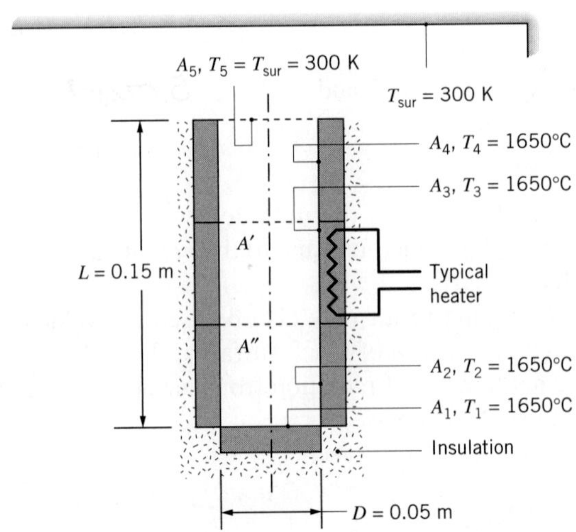

Assumptions:

1. Interior surfaces behave as blackbodies with uniform radiosity and irradiation.
2. Heat transfer by convection is negligible.
3. Backs of electrically heated surfaces are adiabatic.

Analysis: Subject to the foregoing assumptions, the only heat loss from the furnace is by radiation through the opening. Because the surroundings are large, the irradiation from the surroundings is equal to emission from a blackbody at T_{sur}, as discussed in Section 12.7. Furthermore, since radiation heat transfer between the furnace and the surroundings must pass through the opening, the radiation exchange may be analyzed as if it were between the furnace and a hypothetical black surface 5 at the opening, with $T_5 = T_{sur}$. This approach is discussed in detail in Example 13.4. The electrical power delivered to each surface balances the corresponding radiation loss, which may be obtained from Equation 13.14. After employing Equation 13.3, we may write the following equations for surfaces 1 through 4.

Surface 1: $$q_1 = A_1F_{15}\sigma(T_1^4 - T_5^4) = A_5F_{51}\sigma(T_1^4 - T_5^4) \quad (1)$$

Surface 2: $$q_2 = A_2F_{25}\sigma(T_2^4 - T_5^4) = A_5F_{52}\sigma(T_2^4 - T_5^4) \quad (2)$$

Surface 3: $$q_3 = A_3F_{35}\sigma(T_3^4 - T_5^4) = A_5F_{53}\sigma(T_3^4 - T_5^4) \quad (3)$$

Surface 4: $$q_4 = A_4F_{45}\sigma(T_4^4 - T_5^4) = A_5F_{54}\sigma(T_4^4 - T_5^4) \quad (4)$$

We will determine the view factors by first defining two hypothetical surfaces A' and A'' as shown in the schematic. From Table 13.2 with $(r_i/L) = (r_j/L) = (0.025\text{ m}/0.150\text{ m}) = 0.167$, $F_{51} = 0.0263$. With $(r_i/L) = (r_j/L) = (0.025\text{ m}/0.100\text{ m}) = 0.25$, $F_{5A''} = 0.0557$ so that $F_{52} = F_{5A''} - F_{51} = 0.0557 - 0.0263 = 0.0294$. Likewise, with $(r_i/L) = (r_j/L) = (0.025\text{ m}/0.050\text{ m}) = 0.5$, $F_{5A'} = 0.172$ so that $F_{53} = F_{5A'} - F_{5A''} = 0.172 - 0.0557 = 0.1163$. Finally, $F_{54} = 1 - F_{5A'} = 1 - 0.172 = 0.828$. The electrical power delivered to each of the four furnace surfaces can now be determined by solving Equations 1 through 4 for the radiation loss from each surface with $A_5 = \pi D^2/4 = \pi \times (0.05\text{ m})^2/4 = 0.00196\text{ m}^2$.

$$q_1 = 0.00196\text{ m}^2 \times 0.0263 \times 5.67 \times 10^{-8}\text{ W/m}^2\cdot\text{K}^4 \times (1923\text{ K}^4 - 300\text{ K}^4) = 39.9\text{ W} \quad \triangleleft$$

$$q_2 = 0.00196\text{ m}^2 \times 0.0294 \times 5.67 \times 10^{-8}\text{ W/m}^2\cdot\text{K}^4 \times (1923\text{ K}^4 - 300\text{ K}^4) = 44.7\text{ W} \quad \triangleleft$$

$$q_3 = 0.00196\text{ m}^2 \times 0.1163 \times 5.67 \times 10^{-8}\text{ W/m}^2\cdot\text{K}^4 \times (1923\text{ K}^4 - 300\text{ K}^4) = 177\text{ W} \quad \triangleleft$$

$$q_4 = 0.00196\text{ m}^2 \times 0.828 \times 5.67 \times 10^{-8}\text{ W/m}^2\cdot\text{K}^4 \times (1923\text{ K}^4 - 300\text{ K}^4) = 1260\text{ W} \quad \triangleleft$$

Comments:

1. Adding the view factors corresponding to surface 5 yields

$$F_{51} + F_{52} + F_{53} + F_{54} + F_{55} = 0.0263 + 0.0294 + 0.1163 + 0.828 + 0 = 1$$

   Hence the enclosure rule, Equation 13.4, is satisfied, indicating that the view factors have been calculated correctly. Alternatively, the enclosure rule could have been utilized to determine one of the view factors used in the problem solution.

2. The radiation heat loss from the furnace is $q_{\text{tot}} = q_1 + q_2 + q_3 + q_4 = 1522\text{ W} = 1.522\text{ kW}$. If the furnace were to have been treated as a single surface f, the heat loss could be quickly calculated as $q_{\text{tot}} = A_5F_{5f}\sigma(T_f^4 - T_{\text{sur}}^4) = 0.00196\text{ m}^2 \times 1 \times 5.67 \times 10^{-8}\text{ W/m}^2\cdot\text{K} \times (1923\text{ K}^4 - 300\text{ K}^4) = 1.522\text{ kW}$. The answer is the same as determined in the problem solution since $F_{5f} = 1 = F_{51} + F_{52} + F_{53} + F_{54} + F_{55}$.

3. We have assumed that each surface i is *isothermal* and characterized by *uniform radiosity*, J_i, as well as *uniform irradiation*, G_i. Because the isothermal furnace walls are treated as blackbodies, $J_i = E_{bi}$ and the assumption of uniform radiosity is valid. However, the irradiation distribution of the furnace surfaces is *not* uniform since, for example, irradiation from the cool surroundings influences the upper region of an annular surface more than the lower region. To quantify this effect, the *local* radiation heat flux along the vertical furnace wall may be determined by considering a ring element of differential area $dA = \pi Ddx$ as shown. Both sides of Equation 13.13 may be

divided by dA yielding $q''(x) = F_{dA-A_5}\sigma(T_f^4 - T_{sur}^4)$ where the view factor from the differential ring element to the opening, area A_5 is [4]

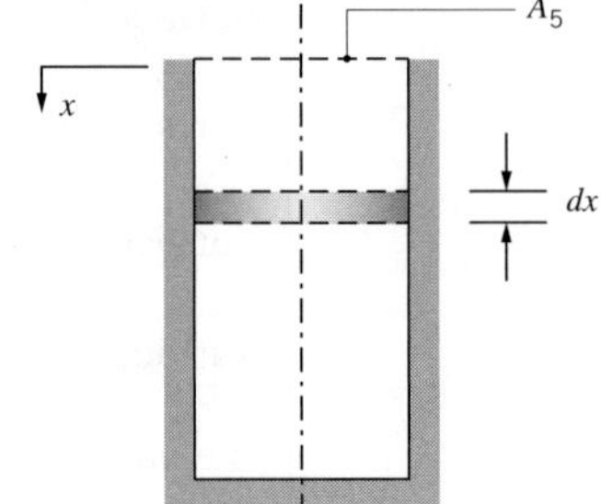

$$F_{dA-A_5} = \frac{(x/D)^2 + 1/2}{\sqrt{1 + (x/D)^2}} - x/D$$

Substituting the expression for F_{dA-A_5} into the equation for the heat flux $q''(x)$ yields the heat flux distribution shown below. Also shown are the *average* heat fluxes associated with the three furnace segments, $\overline{q_2''} = q_2/(\pi DL/3) = 44.7\ \text{W}/(\pi \times 0.05\ \text{m} \times 0.15\ \text{m}/3) = 5690\ \text{W/m}^2 = 5.69\ \text{kW/m}^2$, $\overline{q_3''} = q_3/(\pi DL/3) = 22.5\ \text{kW/m}^2$, and $\overline{q_4''} = q_4/(\pi DL/3) = 160\ \text{kW/m}^2$.

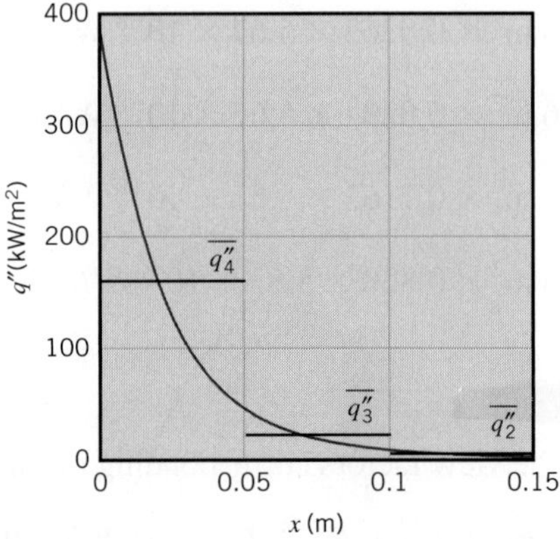

Because of the nonuniformity of the irradiation along the sidewalls of the cavity, the local heat flux is highly nonuniform with the largest value occurring adjacent to the furnace opening. If local temperatures or heat fluxes are of interest, it is necessary to subdivide the various *geometric surfaces* into smaller *radiative surfaces.* This may be done either analytically, as demonstrated here, or numerically. Computational determination of local temperatures or heat fluxes may involve hundreds or perhaps thousands of radiative surfaces, even for simple geometries such as in this example.

13.3 *Radiation Exchange Between Opaque, Diffuse, Gray Surfaces in an Enclosure*

In general, radiation may leave an opaque surface due to both reflection and emission, and on reaching a second opaque surface, experience reflection as well as absorption. In an enclosure, such as that of Figure 13.9*a*, radiation may experience multiple reflections off all surfaces, with partial absorption occurring at each.

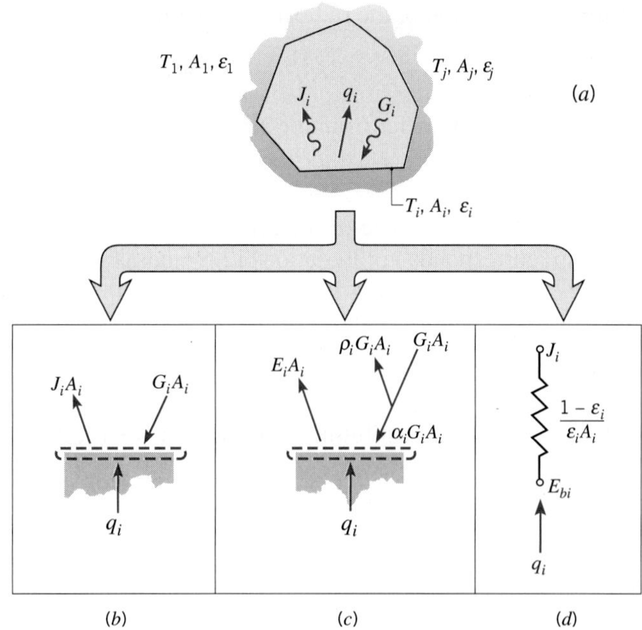

FIGURE 13.9 Radiation exchange in an enclosure of diffuse, gray surfaces with a nonparticipating medium. (*a*) Schematic of the enclosure. (*b*) Radiative balance according to Equation 13.15. (*c*) Radiative balance according to Equation 13.17. (*d*) Resistance representing net radiation transfer from a surface.

Analyzing radiation exchange in an enclosure may be simplified by making certain assumptions. Each surface of the enclosure is assumed to be *isothermal* and to be characterized by a *uniform radiosity* and a *uniform irradiation.* The surfaces are also assumed to be *opaque* ($\tau = 0$) and to have emissivities, absorptivities, and reflectivities that are independent of direction (the surfaces are *diffuse*) *and* independent of wavelength (the surfaces are *gray*). It was shown in Section 12.8 that under these conditions the emissivity is equal to the absorptivity, $\varepsilon = \alpha$ (a form of Kirchhoff's law). Finally, the medium within the enclosure is taken to be *nonparticipating*. The problem is generally one in which either the temperature T_i or the net radiative heat flux q''_i associated with each of the surfaces is known. The objective is to use this information to determine the unknown radiative heat fluxes and temperatures associated with each of the surfaces.

13.3.1 Net Radiation Exchange at a Surface

The term q_i, which is the *net* rate at which radiation *leaves* surface i, represents the net effect of radiative interactions occurring at the surface (Figure 13.9*b*). It is the rate at which energy would have to be transferred to the surface by other means to maintain it at a constant temperature. It is equal to the difference between the surface radiosity and irradiation and from Equation 12.5 may be expressed as

$$q_i = A_i(J_i - G_i) \tag{13.15}$$

Using the definition of the radiosity, Equation 12.4,

$$J_i \equiv E_i + \rho_i G_i \tag{13.16}$$

the net radiative transfer from the surface may be expressed as

$$q_i = A_i(E_i - \alpha_i G_i) \tag{13.17}$$

where use has been made of the relationship $\alpha_i = 1 - \rho_i$ for an opaque surface. This relationship corresponds to Equation 12.6 and is illustrated in Figure 13.9*c*. Noting that $E_i = \varepsilon_i E_{bi}$ and recognizing that $\rho_i = 1 - \alpha_i = 1 - \varepsilon_i$ for an opaque, diffuse, gray surface, the radiosity may also be expressed as

$$J_i = \varepsilon_i E_{bi} + (1 - \varepsilon_i)G_i \tag{13.18}$$

Solving for G_i and substituting into Equation 13.15, it follows that

$$q_i = A_i\left(J_i - \frac{J_i - \varepsilon_i E_{bi}}{1 - \varepsilon_i}\right)$$

or

$$q_i = \frac{E_{bi} - J_i}{(1 - \varepsilon_i)/\varepsilon_i A_i} \tag{13.19}$$

Equation 13.19 provides a convenient representation for the net radiative heat transfer rate from a surface. This transfer, which is represented in Figure 13.9*d*, is associated with the driving potential $(E_{bi} - J_i)$ and a *surface radiative resistance* of the form $(1 - \varepsilon_i)/\varepsilon_i A_i$. Hence, if the emissive power that the surface would have if it were black exceeds its radiosity, there is net radiation heat transfer from the surface; if the inverse is true, the net transfer is to the surface.

It is sometimes the case that one of the surfaces is very large relative to the other surfaces under consideration. For example, the system might consist of multiple small surfaces in a large room. In this case, the area of the large surface is effectively infinite $(A_i \rightarrow \infty)$, and we see that its surface radiative resistance, $(1 - \varepsilon_i)/\varepsilon_i A_i$, is effectively zero, just as it would be for a black surface $(\varepsilon_i = 1)$. Hence, $J_i = E_{bi}$, and *a surface which is large relative to all other surfaces under consideration can be treated as if it were a blackbody*. This important conclusion was reached in Section 12.7 and utilized in Example 13.3, where it was based on a physical argument and has now been confirmed from our treatment of gray surface radiation exchange. Again, the physical explanation is that, even though the large surface may reflect some of the irradiation incident upon it, it is so big that there is a high probability that the reflected radiation reaches another point on the same large surface. After many such reflections, all the radiation that was originally incident on the large surface is absorbed by the large surface, and none ever reaches any of the smaller surfaces.

13.3.2 Radiation Exchange Between Surfaces

To use Equation 13.19, the surface radiosity J_i must be known. To determine this quantity, it is necessary to consider radiation exchange between the surfaces of the enclosure.

The irradiation of surface i can be evaluated from the radiosities of all the surfaces in the enclosure. In particular, from the definition of the view factor, it follows that the total rate at which radiation reaches surface i from all surfaces, including i, is

$$A_i G_i = \sum_{j=1}^{N} F_{ji} A_j J_j$$

or from the reciprocity relation, Equation 13.3,

$$A_i G_i = \sum_{j=1}^{N} A_i F_{ij} J_j$$

Canceling the area A_i and substituting into Equation 13.15 for G_i,

$$q_i = A_i \left(J_i - \sum_{j=1}^{N} F_{ij} J_j \right)$$

or, from the summation rule, Equation 13.4,

$$q_i = A_i \left(\sum_{j=1}^{N} F_{ij} J_i - \sum_{j=1}^{N} F_{ij} J_j \right)$$

Hence

$$q_i = \sum_{j=1}^{N} A_i F_{ij} (J_i - J_j) = \sum_{j=1}^{N} q_{ij} \tag{13.20}$$

Radiation Network Approach Equation 13.20 equates the net rate of radiation transfer from surface i, q_i, to the sum of components q_{ij} related to radiative exchange with the other surfaces. Each component may be represented by a *network element* for which $(J_i - J_j)$ is the driving potential and $(A_i F_{ij})^{-1}$ is a *space* or *geometrical resistance* (Figure 13.10).

Combining Equations 13.19 and 13.20, we then obtain

$$\frac{E_{bi} - J_i}{(1 - \varepsilon_i)/\varepsilon_i A_i} = \sum_{j=1}^{N} \frac{J_i - J_j}{(A_i F_{ij})^{-1}} \tag{13.21}$$

As shown in Figure 13.10, this expression represents a radiation balance for the radiosity *node* associated with surface i. The rate of radiation transfer (current flow) to i through its surface resistance must equal the net rate of radiation transfer (current flows) from i to all other surfaces through the corresponding geometrical resistances.

Note that Equation 13.21 is especially useful when the surface temperature T_i (hence E_{bi}) is known. Although this situation is typical, it does not always apply. In particular, situations may arise for which the net radiation transfer rate at the surface q_i, rather than the temperature T_i, is known. In such cases the preferred form of the radiation balance is Equation 13.20, rearranged as

$$q_i = \sum_{j=1}^{N} \frac{J_i - J_j}{(A_i F_{ij})^{-1}} \tag{13.22}$$

Use of network representations was first suggested by Oppenheim [6]. The network is built by first identifying nodes associated with the radiosities of each of the N surfaces of the enclosure. The method provides a useful tool for visualizing radiation exchange in the

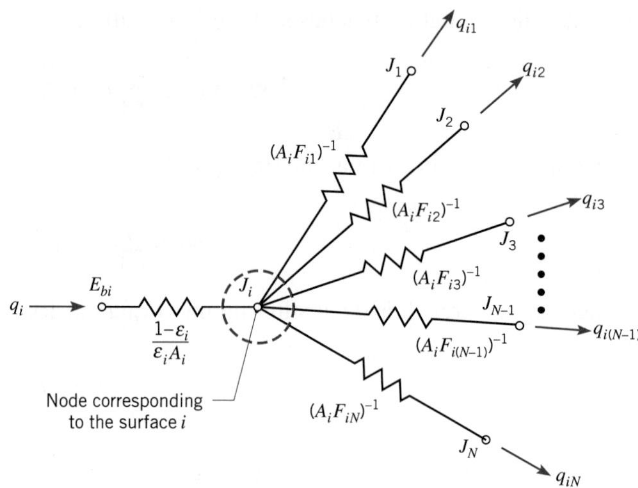

FIGURE 13.10 Network representation of radiative exchange between surface i and the remaining surfaces of an enclosure.

enclosure and, at least for simple enclosures, may be used as the basis for predicting this exchange.

Direct Approach An alternative *direct approach* to solving radiation enclosure problems involves writing Equation 13.21 for each surface at which T_i is known, and writing Equation 13.22 for each surface at which q_i is known. The resulting set of N linear, algebraic equations is solved for $J_1, J_2, \ldots, J_N$. With knowledge of the J_i, Equation 13.19 may then be used to determine the net radiation heat transfer rate q_i at each surface of known T_i or the value of T_i at each surface of known q_i. For any number N of surfaces in the enclosure, the foregoing problem may readily be solved by the iteration or matrix inversion methods of Chapter 4 and Appendix D.

EXAMPLE 13.4

In manufacturing, the special coating on a curved solar absorber surface of area $A_2 = 15\ \text{m}^2$ is cured by exposing it to an infrared heater of width $W = 1$ m. The absorber and heater are each of length $L = 10$ m and are separated by a distance of $H = 1$ m. The upper surface of the absorber and the lower surface of the heater are insulated.

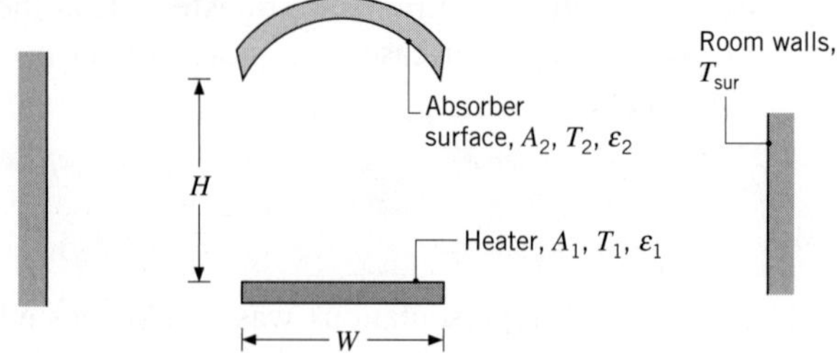

The heater is at $T_1 = 1000$ K and has an emissivity of $\varepsilon_1 = 0.9$, while the absorber is at $T_2 = 600$ K and has an emissivity of $\varepsilon_2 = 0.5$. The system is in a large room whose walls are at 300 K. What is the net rate of heat transfer to the absorber surface?

SOLUTION

Known: A curved, solar absorber surface with a special coating is being cured by use of an infrared heater in a large room.

Find: Net rate of heat transfer to the absorber surface.

Schematic:

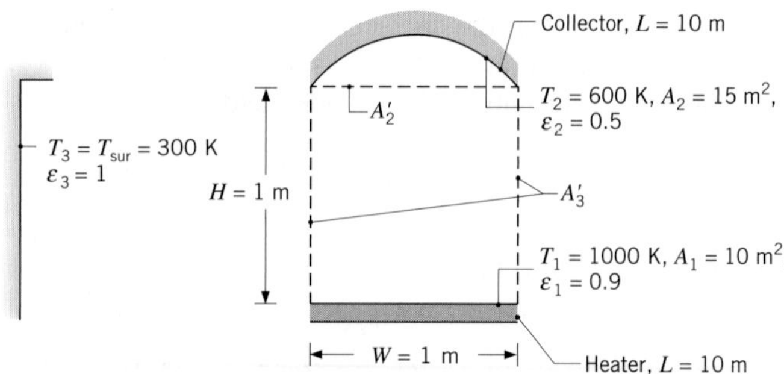

Assumptions:

1. Steady-state conditions exist.
2. Convection effects are negligible.
3. Absorber and heater surfaces are diffuse and gray and are characterized by uniform irradiation and radiosity.
4. The surrounding room is large and therefore behaves as a blackbody.

Analysis: The system may be viewed as a three-surface enclosure, with the third surface being the large surrounding room, which behaves as a blackbody. We are interested in obtaining the net rate of radiation transfer to surface 2. We solve the problem using both the radiation network and direct approaches.

Radiation Network Approach The radiation network is constructed by first identifying nodes associated with the radiosities of each surface, as shown in step 1 in the following schematic. Then each radiosity node is connected to each of the other radiosity nodes through the appropriate space resistance, as shown in step 2. We will treat the surroundings as having a large but unspecified area, which introduces difficulty in expressing the space resistances $(A_3F_{31})^{-1}$ and $(A_3F_{32})^{-1}$. Fortunately, from the reciprocity relation (Equation 13.3), we can replace A_3F_{31} with A_1F_{13} and A_3F_{32} with A_2F_{23}, which are more readily obtained. The final step is to connect the blackbody emissive powers associated with the temperature of each surface to the radiosity nodes, using the appropriate form of the surface resistance.

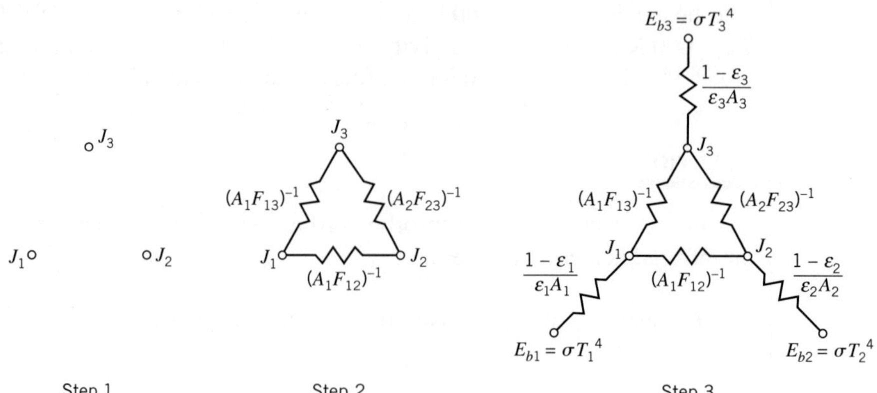

In this problem, the surface resistance associated with surface 3 is zero according to assumption 4; therefore, $J_3 = E_{b3} = \sigma T_3^4 = 459\ \text{W/m}^2$.

Summing currents at the J_1 node yields

$$\frac{\sigma T_1^4 - J_1}{(1-\varepsilon_1)/\varepsilon_1 A_1} = \frac{J_1 - J_2}{1/A_1F_{12}} + \frac{J_1 - \sigma T_3^4}{1/A_1F_{13}} \tag{1}$$

while summing the currents at the J_2 node results in

$$\frac{\sigma T_2^4 - J_2}{(1-\varepsilon_2)/\varepsilon_2 A_2} = \frac{J_2 - J_1}{1/A_1F_{12}} + \frac{J_2 - \sigma T_3^4}{1/A_2F_{23}} \tag{2}$$

The view factor F_{12} may be obtained by recognizing that $F_{12} = F_{12'}$, where A_2' is shown in the schematic as the rectangular base of the absorber surface. Then, from Figure 13.4 or Table 13.2, with $Y/L = 10/1 = 10$ and $X/L = 1/1 = 1$,

$$F_{12} = 0.39$$

From the summation rule, and recognizing that $F_{11} = 0$, it also follows that

$$F_{13} = 1 - F_{12} = 1 - 0.39 = 0.61$$

The last needed view factor is F_{23}. We recognize that, since radiation propagating from surface 2 to surface 3 must pass through the hypothetical surface A_2',

$$A_2F_{23} = A_2'F_{2'3}$$

and from symmetry $F_{2'3} = F_{13}$. Thus

$$F_{23} = \frac{A_2'}{A_2}F_{13} = \frac{10\ \text{m}^2}{15\ \text{m}^2} \times 0.61 = 0.41$$

We may now solve Equations 1 and 2 for J_1 and J_2. Recognizing that $E_{b1} = \sigma T_1^4 = 56{,}700\ \text{W/m}^2$ and canceling the area A_1, we can express Equation 1 as

$$\frac{56{,}700 - J_1}{(1-0.9)/0.9} = \frac{J_1 - J_2}{1/0.39} + \frac{J_1 - 459}{1/0.61}$$

or

$$-10J_1 + 0.39J_2 = -510{,}582 \tag{3}$$

Noting that $E_{b2} = \sigma T_2^4 = 7348\ \text{W/m}^2$ and dividing by the area A_2, we can express Equation 2 as

$$\frac{7348 - J_2}{(1 - 0.5)/0.5} = \frac{J_2 - J_1}{15\ \text{m}^2/(10\ \text{m}^2 \times 0.39)} + \frac{J_2 - 459}{1/0.41}$$

or

$$0.26J_1 - 1.67J_2 = -7536 \tag{4}$$

Solving Equations 3 and 4 simultaneously yields $J_2 = 12{,}487\ \text{W/m}^2$.

An expression for the net rate of heat transfer *from* the absorber surface, q_2, may be written upon inspection of the radiation network and is

$$q_2 = \frac{\sigma T_2^4 - J_2}{(1 - \varepsilon_2)/\varepsilon_2 A_2}$$

resulting in

$$q_2 = \frac{(7348 - 12{,}487)\text{W/m}^2}{(1 - 0.5)/(0.5 \times 15\ \text{m}^2)} = -77.1\ \text{kW}$$

Hence, the net heat transfer rate *to* the absorber is $q_{\text{net}} = -q_2 = 77.1\ \text{kW}$. ◁

Direct Approach Using the direct approach, we write Equation 13.21 for each of the three surfaces. We use reciprocity to rewrite the space resistances in terms of the known view factors from above and to eliminate A_3.

Surface 1

$$\frac{\sigma T_1^4 - J_1}{(1 - \varepsilon_1)/\varepsilon_1 A_1} = \frac{J_1 - J_2}{1/A_1 F_{12}} + \frac{J_1 - J_3}{1/A_1 F_{13}} \tag{5}$$

Surface 2

$$\frac{\sigma T_2^4 - J_2}{(1 - \varepsilon_2)/\varepsilon_2 A_2} = \frac{J_2 - J_1}{1/A_2 F_{21}} + \frac{J_2 - J_3}{1/A_2 F_{23}} = \frac{J_2 - J_1}{1/A_1 F_{12}} + \frac{J_2 - J_3}{1/A_2 F_{23}} \tag{6}$$

Surface 3

$$\frac{\sigma T_3^4 - J_3}{(1 - \varepsilon_3)/\varepsilon_3 A_3} = \frac{J_3 - J_1}{1/A_3 F_{31}} + \frac{J_3 - J_2}{1/A_3 F_{32}} = \frac{J_3 - J_1}{1/A_1 F_{13}} + \frac{J_3 - J_2}{1/A_2 F_{23}} \tag{7}$$

Substituting values of the areas, temperatures, emissivities, and view factors into Equations 5 through 7 and solving them simultaneously, we obtain $J_1 = 51{,}541\ \text{W/m}^2$, $J_2 = 12{,}487\ \text{W/m}^2$, and $J_3 = 459\ \text{W/m}^2$. Equation 13.19 may then be written for surface 2 as

$$q_2 = \frac{\sigma T_2^4 - J_2}{(1 - \varepsilon_2)/\varepsilon_2 A_2}$$

This expression is identical to the expression that was developed using the radiation network. Hence, $q_2 = -77.1\ \text{kW}$. ◁

Comments:

1. In order to solve Equations 5 through 7 simultaneously, we must first multiply both sides of Equation 7 by $(1 - \varepsilon_3)/\varepsilon_3 A_3 = 0$ to avoid division by zero, resulting in the simplified form of Equation 7, which is $J_3 = \sigma T_3^4$.
2. If we substitute $J_3 = \sigma T_3^4$ into Equations 5 and 6, it is evident that Equations 5 and 6 are identical to Equations 1 and 2, respectively.
3. The direct approach is recommended for problems involving $N \geq 4$ surfaces, since radiation networks become quite complex as the number of surfaces increases.
4. As will be seen in Section 13.4, the radiation network approach is particularly useful when thermal energy is transferred to or from surfaces by additional means, that is, by conduction and/or convection. In these *multimode* heat transfer situations, the additional energy delivered to or taken from the surface can be represented by additional current into or out of a node.
5. Recognize the utility of using a hypothetical surface (A_2') to simplify the evaluation of view factors.
6. We could have approached the solution in a slightly different manner. Radiation leaving surface 1 must pass through the openings (hypothetical surface 3′) in order to reach the surroundings. Thus, we can write.

$$F_{13} = F_{13'}$$
$$A_1 F_{13} = A_1 F_{13'} = A_3' F_{3'1}$$

   A similar relationship can be written for exchange between surface 2 and the surroundings, that is, $A_2 F_{23} = A_3' F_{3'2}$. Thus, the space resistances which connect to radiosity node 3 in the radiation network above can be replaced by space resistances pertaining to surface 3′. The resistance network would be unchanged, and the space resistances would have the same values as those determined in the foregoing solution. However, it may be more convenient to calculate the view factors by utilizing the hypothetical surfaces 3′. With the surface resistance for surface 3 equal to zero, we see that *openings of enclosures that exchange radiation with large surroundings may be treated as hypothetical, nonreflecting black surfaces* ($\varepsilon_3 = 1$) *whose temperature is equal to that of the surroundings* ($T_3 = T_{sur}$).
7. The heater and absorber surfaces would not be characterized by uniform irradiation or radiosity. The calculated heat rate could be checked by dividing the heater and absorber into subsurfaces, and repeating the analysis.

13.3.3 The Two-Surface Enclosure

The simplest example of an enclosure is one involving two surfaces that exchange radiation only with each other. Such a two-surface enclosure is shown schematically in Figure 13.11*a*. Since there are only two surfaces, the net rate of radiation transfer *from* surface 1, q_1, must equal the net rate of radiation transfer *to* surface 2, $-q_2$, and both quantities must equal the net rate at which radiation is exchanged between 1 and 2. Accordingly,

$$q_1 = -q_2 = q_{12}$$

The radiation transfer rate may be determined by applying Equation 13.21 to surfaces 1 and 2 and solving the resulting two equations for J_1 and J_2. The results could then be used with Equation 13.19 to determine q_1 (or q_2). However, in this case the desired result is more readily obtained by working with the network representation of the enclosure shown in Figure 13.11*b*.

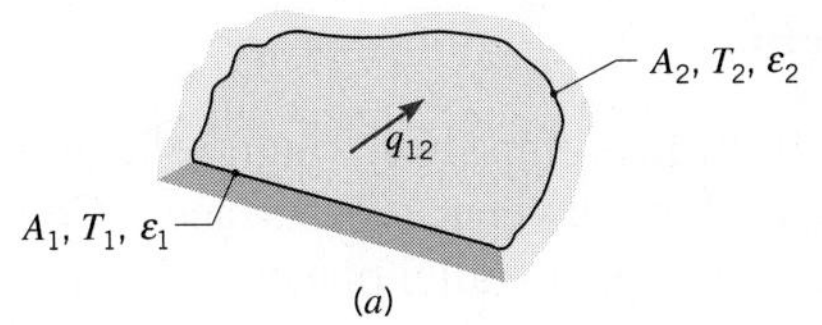

(*a*)

$$q_1 \rightarrow E_{b1} \quad \frac{1-\varepsilon_1}{\varepsilon_1 A_1} \quad J_1 \quad \frac{1}{A_1 F_{12}} \quad J_2 \quad \frac{1-\varepsilon_2}{\varepsilon_2 A_2} \quad E_{b2} \rightarrow -q_2$$

$$q_1 = \frac{E_{b1}-J_1}{(1-\varepsilon_1)/\varepsilon_1 A_1} \qquad q_{12} \qquad -q_2 = \frac{J_2 - E_{b2}}{(1-\varepsilon_2)/\varepsilon_2 A_2}$$

(*b*)

FIGURE 13.11 The two-surface enclosure. (*a*) Schematic. (*b*) Network representation.

From Figure 13.11*b* we see that the total resistance to radiation exchange between surfaces 1 and 2 is comprised of the two surface resistances and the geometrical resistance. Hence, substituting from Equation 12.32, the net radiation exchange between surfaces may be expressed as

$$q_{12} = q_1 = -q_2 = \frac{\sigma(T_1^4 - T_2^4)}{\dfrac{1-\varepsilon_1}{\varepsilon_1 A_1} + \dfrac{1}{A_1 F_{12}} + \dfrac{1-\varepsilon_2}{\varepsilon_2 A_2}} \tag{13.23}$$

The foregoing result may be used for any two isothermal diffuse, gray surfaces that *form an enclosure* and *are each characterized by uniform radiosity and irradiation.* Important special cases are summarized in Table 13.3.

TABLE 13.3 Special Diffuse, Gray, Two-Surface Enclosures

Large (Infinite) Parallel Planes

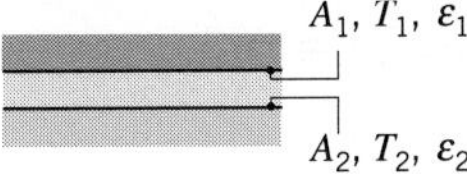

$$A_1 = A_2 = A \qquad F_{12} = 1 \qquad q_{12} = \frac{A\sigma(T_1^4 - T_2^4)}{\dfrac{1}{\varepsilon_1} + \dfrac{1}{\varepsilon_2} - 1} \tag{13.24}$$

Long (Infinite) Concentric Cylinders

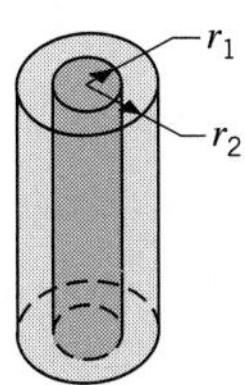

$$\frac{A_1}{A_2} = \frac{r_1}{r_2} \qquad F_{12} = 1 \qquad q_{12} = \frac{\sigma A_1(T_1^4 - T_2^4)}{\dfrac{1}{\varepsilon_1} + \dfrac{1-\varepsilon_2}{\varepsilon_2}\left(\dfrac{r_1}{r_2}\right)} \tag{13.25}$$

Concentric Spheres

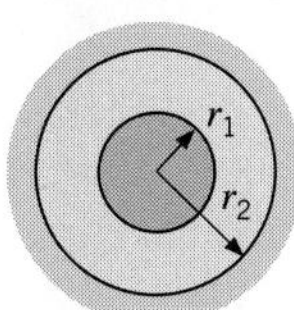

$$\frac{A_1}{A_2} = \frac{r_1^2}{r_2^2} \qquad F_{12} = 1 \qquad q_{12} = \frac{\sigma A_1(T_1^4 - T_2^4)}{\dfrac{1}{\varepsilon_1} + \dfrac{1-\varepsilon_2}{\varepsilon_2}\left(\dfrac{r_1}{r_2}\right)^2} \tag{13.26}$$

Small Convex Object in a Large Cavity

$$\frac{A_1}{A_2} \approx 0 \qquad F_{12} = 1 \qquad q_{12} = \sigma A_1 \varepsilon_1 (T_1^4 - T_2^4) \tag{13.27}$$

13.3.4 Radiation Shields

Radiation shields constructed from low emissivity (high reflectivity) materials can be used to reduce the net radiation transfer between two surfaces. Consider placing a radiation shield, surface 3, between the two large, parallel planes of Figure 13.12*a*. Without the radiation shield, the net rate of radiation transfer between surfaces 1 and 2 is given by Equation 13.24. However, with the radiation shield, additional resistances are present, as shown in Figure 13.12*b*, and the heat transfer rate is reduced. Note that the emissivity associated with one side of the shield ($\varepsilon_{3,1}$) may differ from that associated with the opposite side ($\varepsilon_{3,2}$) and the radiosities will always differ. Summing the resistances and recognizing that $F_{13} = F_{32} = 1$, it follows that

$$q_{12} = \frac{A_1\sigma(T_1^4 - T_2^4)}{\dfrac{1}{\varepsilon_1} + \dfrac{1}{\varepsilon_2} + \dfrac{1-\varepsilon_{3,1}}{\varepsilon_{3,1}} + \dfrac{1-\varepsilon_{3,2}}{\varepsilon_{3,2}}} \tag{13.28}$$

Note that the resistances associated with the radiation shield become very large when the emissivities $\varepsilon_{3,1}$ and $\varepsilon_{3,2}$ are very small.

Equation 13.28 may be used to determine the net heat transfer rate if T_1 and T_2 are known. From knowledge of q_{12} and the fact that $q_{12} = q_{13} = q_{32}$, the value of T_3 may then be determined by expressing Equation 13.24 for q_{13} or q_{32}.

The foregoing procedure may readily be extended to problems involving multiple radiation shields. In the special case for which all the emissivities are equal, it may be shown that, with N shields,

$$(q_{12})_N = \frac{1}{N+1}(q_{12})_0 \tag{13.29}$$

where $(q_{12})_0$ is the radiation transfer rate with no shields ($N = 0$).

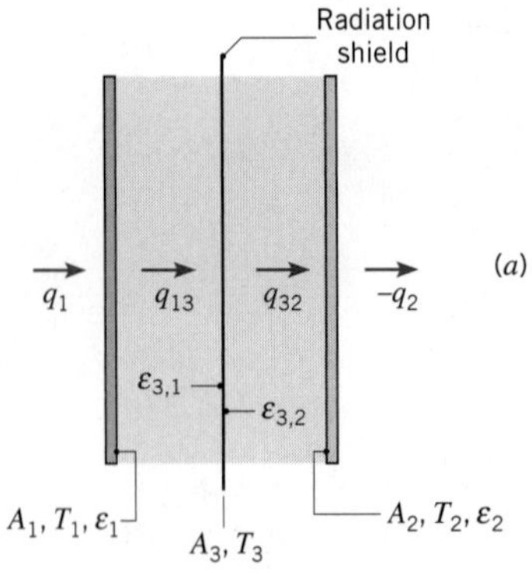

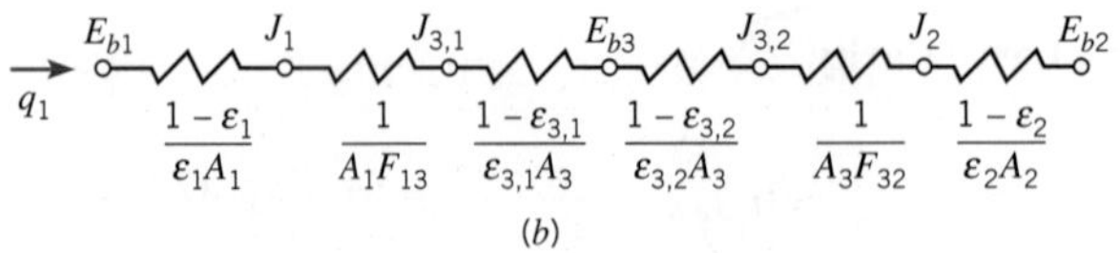

FIGURE 13.12 Radiation exchange between large parallel planes with a radiation shield. (*a*) Schematic. (*b*) Network representation.

EXAMPLE 13.5

A cryogenic fluid flows through a long tube of 20-mm diameter, the outer surface of which is diffuse and gray with $\varepsilon_1 = 0.02$ and $T_1 = 77$ K. This tube is concentric with a larger tube of 50-mm diameter, the inner surface of which is diffuse and gray with $\varepsilon_2 = 0.05$ and $T_2 = 300$ K. The space between the surfaces is evacuated. Calculate the heat gain by the cryogenic fluid per unit length of tubes. If a thin radiation shield of 35-mm diameter and $\varepsilon_3 = 0.02$ (both sides) is inserted midway between the inner and outer surfaces, calculate the change (percentage) in heat gain per unit length of the tubes.

SOLUTION

Known: Concentric tube arrangement with diffuse, gray surfaces of different emissivities and temperatures.

Find:

1. Heat gain by the cryogenic fluid passing through the inner tube.
2. Percentage change in heat gain with radiation shield inserted midway between inner and outer tubes.

Schematic:

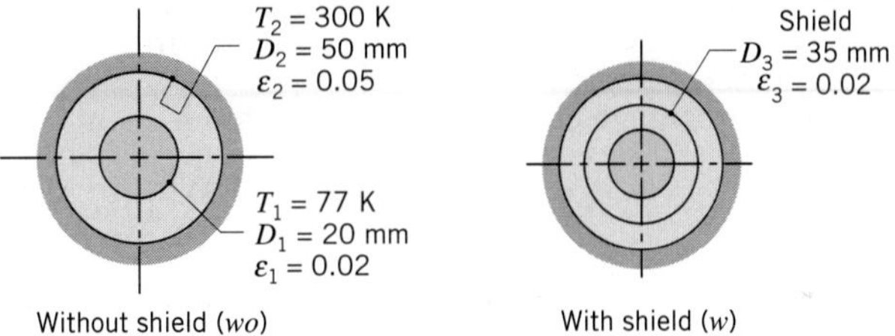

Assumptions:

1. Surfaces are diffuse and gray and characterized by uniform irradiation and radiosity.
2. Space between tubes is evacuated.
3. Conduction resistance for radiation shield is negligible.
4. Concentric tubes form a two-surface enclosure (end effects are negligible).

Analysis:

1. The network representation of the system without the shield is shown in Figure 13.11, and the heat rate may be obtained from Equation 13.25, where

$$q = \frac{\sigma(\pi D_1 L)(T_1^4 - T_2^4)}{\dfrac{1}{\varepsilon_1} + \dfrac{1-\varepsilon_2}{\varepsilon_2}\left(\dfrac{D_1}{D_2}\right)}$$

Hence

$$q' = \frac{q}{L} = \frac{5.67 \times 10^{-8}\,\text{W/m}^2\cdot\text{K}^4\,(\pi \times 0.02\,\text{m})[(77\,\text{K})^4 - (300\,\text{K})^4]}{\dfrac{1}{0.02} + \dfrac{1-0.05}{0.05}\left(\dfrac{0.02\,\text{m}}{0.05\,\text{m}}\right)}$$

$$q' = -0.50\,\text{W/m} \quad \triangleleft$$

2. The network representation of the system with the shield is shown in Figure 13.12, and the heat rate is now

$$q = \frac{E_{b1} - E_{b2}}{R_{\text{tot}}} = \frac{\sigma(T_1^4 - T_2^4)}{R_{\text{tot}}}$$

where

$$R_{\text{tot}} = \frac{1-\varepsilon_1}{\varepsilon_1(\pi D_1 L)} + \frac{1}{(\pi D_1 L)F_{13}} + 2\left[\frac{1-\varepsilon_3}{\varepsilon_3(\pi D_3 L)}\right] + \frac{1}{(\pi D_3 L)F_{32}} + \frac{1-\varepsilon_2}{\varepsilon_2(\pi D_2 L)}$$

or

$$R_{\text{tot}} = \frac{1}{L}\left\{\frac{1-0.02}{0.02(\pi \times 0.02\text{ m})} + \frac{1}{(\pi \times 0.02\text{ m})1} + 2\left[\frac{1-0.02}{0.02(\pi \times 0.035\text{ m})}\right] + \frac{1}{(\pi \times 0.035\text{ m})1} + \frac{1-0.05}{0.05(\pi \times 0.05\text{ m})}\right\}$$

$$R_{\text{tot}} = \frac{1}{L}(779.9 + 15.9 + 891.3 + 9.1 + 121.0) = \frac{1817}{L}\left(\frac{1}{\text{m}^2}\right)$$

Hence

$$q' = \frac{q}{L} = \frac{5.67 \times 10^{-8}\text{ W/m}^2\cdot\text{K}^4\,[(77\text{ K})^4 - (300\text{ K})^4]}{1817\text{ (1/m)}} = -0.25\text{ W/m} \quad \triangleleft$$

The percentage change in the heat gain is then

$$\frac{q'_w - q'_{wo}}{q'_{wo}} \times 100 = \frac{(-0.25\text{ W/m}) - (-0.50\text{ W/m})}{-0.50\text{ W/m}} \times 100 = -50\% \quad \triangleleft$$

Comment: Because the geometries are concentric and the specified emissivities and prescribed surface temperatures are spatially uniform, each surface is characterized by uniform irradiation and radiosity distributions. Hence the calculated heat transfer rates would not change if the cylindrical surfaces were to be subdivided into smaller radiative surfaces.

13.3.5 The Reradiating Surface

The assumption of a *reradiating surface* is common to many industrial applications. This idealized surface is characterized by *zero* net radiation transfer ($q_i = 0$). It is closely approached by real surfaces that are well insulated on one side *and* for which convection effects may be neglected on the opposite (radiating) side. With $q_i = 0$, it follows from Equations 13.15 and 13.19 that $G_i = J_i = E_{bi}$. Hence, if the radiosity of a reradiating surface is known, its temperature is readily determined. In an enclosure, the equilibrium temperature of a reradiating surface is determined by its interaction with the other surfaces, and it is *independent of the emissivity of the reradiating surface.*

A three-surface enclosure, for which the third surface, surface R, is reradiating, is shown in Figure 13.13*a*, and the corresponding network is shown in Figure 13.13*b*. Surface R is presumed to be well insulated, and convection effects are assumed to be negligible. Hence, with $q_R = 0$, the net radiation *transfer* from surface 1 must equal the net radiation

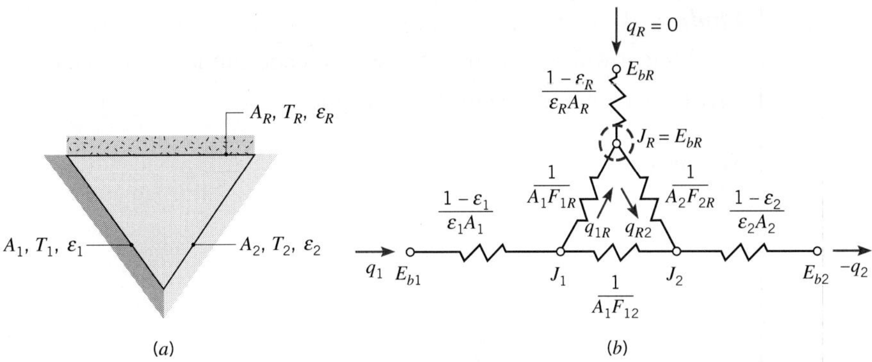

FIGURE 13.13 A three-surface enclosure with one surface reradiating. (*a*) Schematic. (*b*) Network representation.

transfer to surface 2. The network is a simple series–parallel arrangement, and from its analysis it is readily shown that

$$q_1 = -q_2 = \frac{E_{b1} - E_{b2}}{\dfrac{1-\varepsilon_1}{\varepsilon_1 A_1} + \dfrac{1}{A_1F_{12} + [(1/A_1F_{1R}) + (1/A_2F_{2R})]^{-1}} + \dfrac{1-\varepsilon_2}{\varepsilon_2 A_2}} \tag{13.30}$$

Knowing $q_1 = -q_2$, Equation 13.19 may be applied to surfaces 1 and 2 to determine their radiosities J_1 and J_2. Knowing J_1, J_2, and the geometrical resistances, the radiosity of the reradiating surface J_R may be determined from the radiation balance

$$\frac{J_1 - J_R}{(1/A_1F_{1R})} - \frac{J_R - J_2}{(1/A_2F_{2R})} = 0 \tag{13.31}$$

The temperature of the reradiating surface may then be determined from the requirement that $\sigma T_R^4 = J_R$.

Note that the general procedure described in Section 13.3.2 may be applied to enclosures with reradiating surfaces. For each such surface, it is appropriate to use Equation 13.22 with $q_i = 0$.

EXAMPLE 13.6

A paint baking oven consists of a long, triangular duct in which a heated surface is maintained at 1200 K and another surface is insulated. Painted panels, which are maintained at 500 K, occupy the third surface. The triangle is of width $W = 1$ m on a side, and the heated and insulated surfaces have an emissivity of 0.8. The emissivity of the panels is 0.4. During steady-state operation, at what rate must energy be supplied to the heated side per unit length of the duct to maintain its temperature at 1200 K? What is the temperature of the insulated surface?

SOLUTION

Known: Surface properties of a long triangular duct that is insulated on one side and heated and cooled on the other sides.

Find:

1. Rate at which heat must be supplied per unit length of duct.
2. Temperature of the insulated surface.

Schematic:

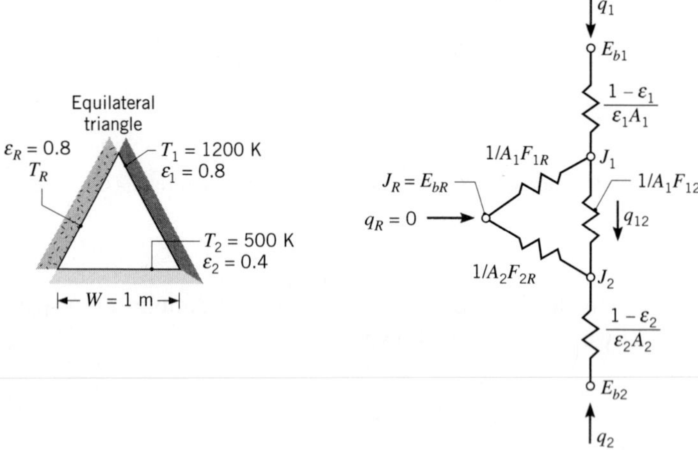

Assumptions:

1. Steady-state conditions exist.
2. All surfaces are opaque, diffuse, gray, and of uniform radiosity and irradiation.
3. Convection effects are negligible.
4. Surface R is reradiating.
5. End effects are negligible.

Analysis:

1. The system may be modeled as a three-surface enclosure with one surface reradiating. The rate at which energy must be supplied to the heated surface may then be obtained from Equation 13.30:

$$q_1 = \frac{E_{b1} - E_{b2}}{\dfrac{1-\varepsilon_1}{\varepsilon_1 A_1} + \dfrac{1}{A_1F_{12} + [(1/A_1F_{1R}) + (1/A_2F_{2R})]^{-1}} + \dfrac{1-\varepsilon_2}{\varepsilon_2 A_2}}$$

From symmetry, $F_{12} = F_{1R} = F_{2R} = 0.5$. Also, $A_1 = A_2 = W \cdot L$, where L is the duct length. Hence

$$q_1' = \frac{q_1}{L} = \frac{5.67 \times 10^{-8}\,\text{W/m}^2\cdot\text{K}^4\,(1200^4 - 500^4)\,\text{K}^4}{\dfrac{1-0.8}{0.8 \times 1\,\text{m}} + \dfrac{1}{1\,\text{m} \times 0.5 + (2+2)^{-1}\,\text{m}} + \dfrac{1-0.4}{0.4 \times 1\,\text{m}}}$$

or

$$q_1' = 37\,\text{kW/m} = -q_2' \quad \triangleleft$$

2. The temperature of the insulated surface may be obtained from the requirement that $J_R = E_{bR}$, where J_R may be obtained from Equation 13.31. However, to use this expression J_1 and J_2 must be known. Applying the surface energy balance, Equation 13.19, to surfaces 1 and 2, it follows that

$$J_1 = E_{b1} - \frac{1-\varepsilon_1}{\varepsilon_1 W} q_1' = 5.67 \times 10^{-8}\ \text{W/m}^2 \cdot \text{K}^4\ (1200\ \text{K})^4$$

$$- \frac{1 - 0.8}{0.8 \times 1\ \text{m}} \times 37{,}000\ \text{W/m} = 108{,}323\ \text{W/m}^2$$

$$J_2 = E_{b2} - \frac{1-\varepsilon_2}{\varepsilon_2 W} q_2' = 5.67 \times 10^{-8}\ \text{W/m}^2 \cdot \text{K}^4\ (500\ \text{K})^4$$

$$- \frac{1 - 0.4}{0.4 \times 1\ \text{m}} (-37{,}000\ \text{W/m}) = 59{,}043\ \text{W/m}^2$$

From the energy balance for the reradiating surface, Equation 13.31, it follows that

$$\frac{108{,}323 - J_R}{\dfrac{1}{W \times L \times 0.5}} - \frac{J_R - 59{,}043}{\dfrac{1}{W \times L \times 0.5}} = 0$$

Hence

$$J_R = 83{,}683\ \text{W/m}^2 = E_{bR} = \sigma T_R^4$$

$$T_R = \left(\frac{83{,}683\ \text{W/m}^2}{5.67 \times 10^{-8}\ \text{W/m}^2 \cdot \text{K}^4}\right)^{1/4} = 1102\ \text{K} \qquad \triangleleft$$

Comments:

1. We would expect the temperature of the reradiating surface to be higher in regions adjacent to surface 1 and lower in regions closer to surface 2. Our intuition corresponds to the fact that the surface irradiation and radiosity distributions are not uniform, calling into question the validity of Assumption 2. The temperature distribution of the reradiating surface could be determined by use of an analytical or numerical approach, as described in Comment 3 of Example 13.3. If each geometric surface were to be subdivided into 10 smaller elements, however, we would need $(3 \times 10)^2 = 900$ view factors. Precise prediction of radiation heat transfer rates in enclosures whose geometric surfaces are not characterized by uniform radiosity or irradiation distributions involves a trade-off between accuracy and computational effort.
2. The results are independent of the value of ε_R.
3. This problem may also be solved using the direct approach. The solution involves first determining the three unknown radiosities J_1, J_2, and J_R. The governing equations are

obtained by writing Equation 13.21 for the two surfaces of known temperature, 1 and 2, and Equation 13.22 for surface R. The three equations are

$$\frac{E_{b1} - J_1}{(1 - \varepsilon_1)/\varepsilon_1 A_1} = \frac{J_1 - J_2}{(A_1 F_{12})^{-1}} + \frac{J_1 - J_R}{(A_1 F_{1R})^{-1}}$$

$$\frac{E_{b2} - J_2}{(1 - \varepsilon_2)/\varepsilon_2 A_2} = \frac{J_2 - J_1}{(A_2 F_{21})^{-1}} + \frac{J_2 - J_R}{(A_2 F_{2R})^{-1}}$$

$$0 = \frac{J_R - J_1}{(A_R F_{R1})^{-1}} + \frac{J_R - J_2}{(A_R F_{R2})^{-1}}$$

Canceling the area A_1, the first equation reduces to

$$\frac{117{,}573 - J_1}{0.25} = \frac{J_1 - J_2}{2} + \frac{J_1 - J_R}{2}$$

or

$$10J_1 - J_2 - J_R = 940{,}584 \qquad (1)$$

Similarly, for surface 2,

$$\frac{3544 - J_2}{1.50} = \frac{J_2 - J_1}{2} + \frac{J_2 - J_R}{2}$$

or

$$-J_1 + 3.33J_2 - J_R = 4725 \qquad (2)$$

and for the reradiating surface

$$0 = \frac{J_R - J_1}{2} + \frac{J_R - J_2}{2}$$

or

$$-J_1 - J_2 + 2J_R = 0 \qquad (3)$$

Solving Equations 1, 2, and 3 simultaneously yields

$$J_1 = 108{,}328 \text{ W/m}^2 \quad J_2 = 59{,}018 \text{ W/m}^2 \quad \text{and} \quad J_R = 83{,}673 \text{ W/m}^2$$

Recognizing that $J_R = \sigma T_R^4$, it follows that

$$T_R = \left(\frac{J_R}{\sigma}\right)^{1/4} = \left(\frac{83{,}673 \text{ W/m}^2}{5.67 \times 10^{-8} \text{ W/m}^2 \cdot \text{K}^4}\right)^{1/4} = 1102 \text{ K}$$

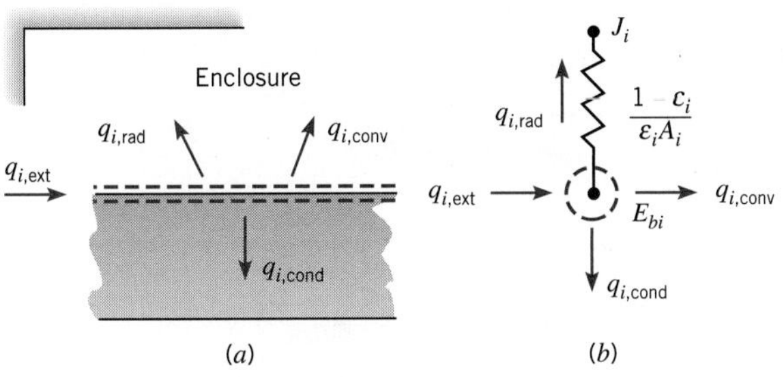

FIGURE 13.14 Multimode heat transfer from a surface in an enclosure. (*a*) Surface energy balance. (*b*) Circuit representation.

13.4 *Multimode Heat Transfer*

Thus far, radiation exchange in an enclosure has been considered under conditions for which conduction and convection could be neglected. However, in many applications, convection and/or conduction are comparable to radiation and must be considered in the heat transfer analysis.

Consider the general surface condition of Figure 13.14*a*. In addition to exchanging energy by radiation with other surfaces of the enclosure, there may be external heat addition to the surface, as, for example, by electric heating, and heat transfer from the surface by convection and conduction. From a surface energy balance, it follows that

$$q_{i,\text{ext}} = q_{i,\text{rad}} + q_{i,\text{conv}} + q_{i,\text{cond}} \tag{13.32}$$

where $q_{i,\text{rad}}$, the net rate of radiation transfer from the surface, is determined by standard procedures for an enclosure. Hence, in general, $q_{i,\text{rad}}$ may be determined from Equation 13.19 or 13.20, while for special cases such as a two-surface enclosure and a three-surface enclosure with one reradiating surface, it may be determined from Equations 13.23 and 13.30, respectively. The surface network element of the radiation circuit is modified according to Figure 13.14*b*, where $q_{i,\text{ext}}$, $q_{i,\text{cond}}$, and $q_{i,\text{conv}}$ represent current flows to or from the surface node. Note, however, that while $q_{i,\text{cond}}$ and $q_{i,\text{conv}}$ are proportional to temperature differences, $q_{i,\text{rad}}$ is proportional to the difference between temperatures raised to the fourth power. Conditions are simplified if the back of the surface is insulated, in which case $q_{i,\text{cond}} = 0$. Moreover, if there is no external heating and convection is negligible, the surface is reradiating.

EXAMPLE 13.7

Consider an air heater consisting of a semicircular tube for which the plane surface is maintained at 1000 K and the other surface is well insulated. The tube radius is 20 mm, and both surfaces have an emissivity of 0.8. If atmospheric air flows through the tube at 0.01 kg/s and $T_m = 400$ K, what is the rate at which heat must be supplied per unit length to maintain the plane surface at 1000 K? What is the temperature of the insulated surface?

SOLUTION

Known: Airflow conditions in tubular heater and heater surface conditions.

Find: Rate at which heat must be supplied and temperature of insulated surface.

Schematic:

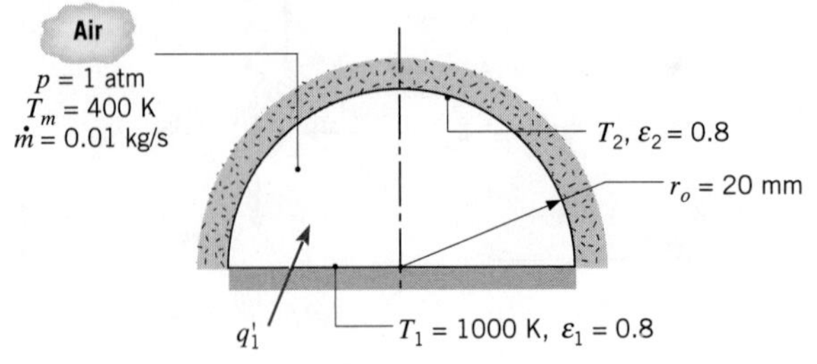

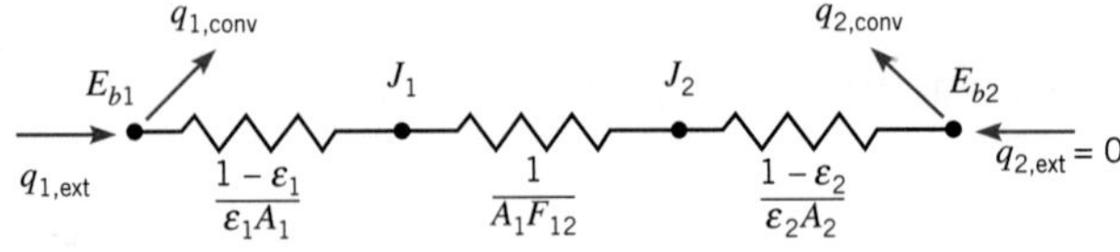

Assumptions:

1. Steady-state conditions.
2. Diffuse, gray surfaces experiencing uniform irradiation and radiosity.
3. Negligible tube end effects and axial variations in gas temperature.
4. Fully developed flow.

Properties: Table A.4, air (1 atm, 400 K): $k = 0.0338$ W/m·K, $\mu = 230 \times 10^{-7}$ kg/s·m, $c_p = 1014$ J/kg·K, $Pr = 0.69$.

Analysis: Since the semicircular surface is well insulated and there is no external heat addition, a surface energy balance yields

$$-q_{2,\text{rad}} = q_{2,\text{conv}}$$

Since the tube constitutes a two-surface enclosure, the net radiation transfer to surface 2 may be evaluated from Equation 13.23. Hence

$$\frac{\sigma(T_1^4 - T_2^4)}{\dfrac{1-\varepsilon_1}{\varepsilon_1 A_1} + \dfrac{1}{A_1 F_{12}} + \dfrac{1-\varepsilon_2}{\varepsilon_2 A_2}} = hA_2(T_2 - T_m)$$

where the view factor is $F_{12} = 1$ and, per unit length, the surface areas are $A_1 = 2r_o$ and $A_2 = \pi r_o$. With

$$Re_D = \frac{\rho u_m D_h}{\mu} = \frac{\dot{m} D_h}{A_c \mu} = \frac{\dot{m} D_h}{(\pi r_o^2/2)\mu}$$

the hydraulic diameter is

$$D_h = \frac{4A_c}{P} = \frac{2\pi r_o}{\pi + 2} = \frac{0.04\pi \text{ m}}{\pi + 2} = 0.0244 \text{ m}$$

Hence

$$Re_D = \frac{0.01 \text{ kg/s} \times 0.0244 \text{ m}}{(\pi/2)(0.02 \text{ m})^2 \times 230 \times 10^{-7} \text{ kg/s} \cdot \text{m}} = 16{,}900$$

From the Dittus–Boelter equation,

$$Nu_D = 0.023\, Re_D^{4/5}\, Pr^{0.4}$$

$$Nu_D = 0.023(16{,}900)^{4/5}(0.69)^{0.4} = 47.8$$

$$h = \frac{k}{D_h} Nu_D = \frac{0.0338 \text{ W/m} \cdot \text{K}}{0.0244 \text{ m}} 47.8 = 66.2 \text{ W/m}^2 \cdot \text{K}$$

Dividing both sides of the energy balance by A_1, it follows that

$$\frac{5.67 \times 10^{-8} \text{ W/m}^2 \cdot \text{K}^4 [(1000)^4 - T_2^4]\text{K}^4}{\dfrac{1-0.8}{0.8} + 1 + \dfrac{1-0.8}{0.8}\dfrac{2}{\pi}} = 66.2\frac{\pi}{2}(T_2 - 400) \text{ W/m}^2$$

or

$$5.67 \times 10^{-8}\, T_2^4 + 146.5\, T_2 - 115{,}313 = 0$$

A trial-and-error solution yields

$$T_2 = 696 \text{ K}$$ ◁

From an energy balance at the heated surface,

$$q_{1,\text{ext}} = q_{1,\text{rad}} + q_{1,\text{conv}} = q_{2,\text{conv}} + q_{1,\text{conv}}$$

Hence, on a unit length basis,

$$q'_{1,\text{ext}} = h\pi r_o(T_2 - T_m) + h2r_o(T_1 - T_m)$$

$$q'_{1,\text{ext}} = 66.2 \times 0.02[\pi(696 - 400) + 2(1000 - 400)] \text{ W/m}$$

$$q'_{1,\text{ext}} = (1231 + 1589) \text{ W/m} = 2820 \text{ W/m}$$ ◁

Comments:

1. The irradiation, radiosity, and convection heat flux distributions along the surfaces would not be uniform. As a consequence, we would expect the temperature of the insulated surface to be higher in the corner regions adjacent to surface 1 and lower in the crown of the enclosure. Determination of the irradiation, radiosity, temperature, and convection heat flux distributions along the various surfaces would require a more complex analysis incorporating many radiative surfaces.
2. Applying an energy balance to a differential control volume about the air, it follows that

$$\frac{dT_m}{dx} = \frac{q'_1}{\dot{m}c_p} = \frac{2820 \text{ W/m}}{0.01 \text{ kg/s}\,(1014 \text{ J/kg} \cdot \text{K})} = 278 \text{ K/m}$$

Hence the air temperature change is significant, and a more representative analysis would subdivide the tube into axial zones and would allow for variations in air and insulated surface temperatures between zones. Moreover, a two-surface analysis of radiation exchange would no longer be appropriate.

13.5 Implications of the Simplifying Assumptions

Although we have developed a means for predicting radiation exchange between surfaces, it is important to be cognizant of the inherent limitations of our analyses. Recall that we have analyzed *idealized surfaces* that are isothermal, opaque, and gray that emit and reflect diffusely and are characterized by uniform radiosity and irradiation distributions. Moreover, the enclosures that we have analyzed have been assumed to house media that are nonparticipating; that is, they neither absorb nor scatter the surface radiation, and emit no radiation.

The analysis techniques presented in this chapter may often be used to obtain first estimates and, in many cases, sufficiently accurate results for radiation heat transfer involving multiple surfaces that form an enclosure. In some cases, however, the assumptions are inappropriate, and more refined prediction methods are needed. Although beyond the scope of this text, the methods are discussed in more advanced treatments of radiation transfer [3, 7–12].

13.6 Radiation Exchange with Participating Media

Except for our discussion of environmental radiation (Section 12.9), we have said little about gaseous radiation, having confined our attention to radiation exchange at the surface of an opaque solid or liquid. For *nonpolar* gases, such as O_2 or N_2, such neglect is justified, since the gases do not emit radiation and are essentially transparent to incident thermal radiation. However, the same may not be said for polar molecules, such as CO_2, H_2O (vapor), NH_3, and hydrocarbon gases, which emit and absorb over a wide temperature range. For such gases matters are complicated by the fact that, unlike radiation from a solid or a liquid, which is distributed continuously with wavelength, gaseous radiation is concentrated in specific *wavelength intervals* (called bands). Moreover, gaseous radiation is not a surface phenomenon, but is instead a *volumetric* phenomenon.

13.6.1 Volumetric Absorption

Spectral radiation absorption in a gas (or in a semitransparent liquid or solid) is a function of the absorption coefficient κ_λ (1/m) and the thickness L of the medium (Figure 13.15). If a monochromatic beam of intensity $I_{\lambda,0}$ is incident on the medium, the intensity is reduced due to absorption, and the reduction occurring in an infinitesimal layer of thickness dx may be expressed as

$$dI_\lambda(x) = -\kappa_\lambda I_\lambda(x)\, dx \qquad (13.33)$$

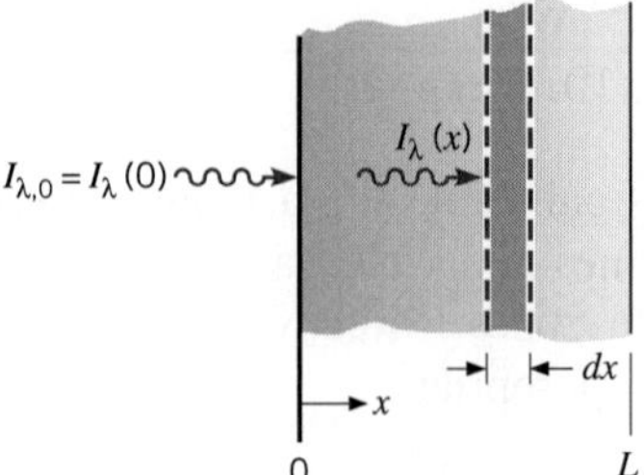

FIGURE 13.15 Absorption in a gas or liquid layer.

Separating variables and integrating over the entire layer, we obtain

$$\int_{I_{\lambda,0}}^{I_{\lambda,L}} \frac{dI_\lambda(x)}{I_\lambda(x)} = -\kappa_\lambda \int_0^L dx$$

where κ_λ is assumed to be independent of x. It follows that

$$\frac{I_{\lambda,L}}{I_{\lambda,0}} = e^{-\kappa_\lambda L} \tag{13.34}$$

This exponential decay, termed *Beer's law*, is a useful tool in approximate radiation analysis. It may, for example, be used to infer the overall spectral absorptivity of the medium. In particular, with the transmissivity defined as

$$\tau_\lambda = \frac{I_{\lambda,L}}{I_{\lambda,0}} = e^{-\kappa_\lambda L} \tag{13.35}$$

the absorptivity is

$$\alpha_\lambda = 1 - \tau_\lambda = 1 - e^{-\kappa_\lambda L} \tag{13.36}$$

If Kirchhoff's law is assumed to be valid, $\alpha_\lambda = \varepsilon_\lambda$, Equation 13.36 also provides the spectral emissivity of the medium.

13.6.2 Gaseous Emission and Absorption

A common engineering calculation is one that requires determination of the radiant heat flux from a gas to an adjoining surface. Despite the complicated spectral and directional effects inherent in such calculations, a simplified procedure may be used. The method was developed by Hottel [13] and involves determining radiation emission from a hemispherical gas mass of temperature T_g to a surface element dA_1, which is located at the center of the hemisphere's base. Emission from the gas per unit area of the surface is expressed as

$$E_g = \varepsilon_g \sigma T_g^4 \tag{13.37}$$

where the gas emissivity ε_g was determined by correlating available data. In particular, ε_g was correlated in terms of the temperature T_g and total pressure p of the gas, the partial pressure p_g of the radiating species, and the radius L of the hemisphere.

Results for the emissivity of water vapor are plotted in Figure 13.16 as a function of the gas temperature, for a total pressure of 1 atm, and for different values of the product of the vapor partial pressure and the hemisphere radius. To evaluate the emissivity for total pressures other than 1 atm, the emissivity from Figure 13.16 must be multiplied by the correction factor C_w from Figure 13.17. Similar results were obtained for carbon dioxide and are presented in Figures 13.18 and 13.19.

The foregoing results apply when water vapor or carbon dioxide appear *separately* in a mixture with other species that are nonradiating. However, the results may readily be extended to situations in which water vapor and carbon dioxide appear *together* in a mixture with other nonradiating gases. In particular, the total gas emissivity may be expressed as

$$\varepsilon_g = \varepsilon_w + \varepsilon_c - \Delta\varepsilon \tag{13.38}$$

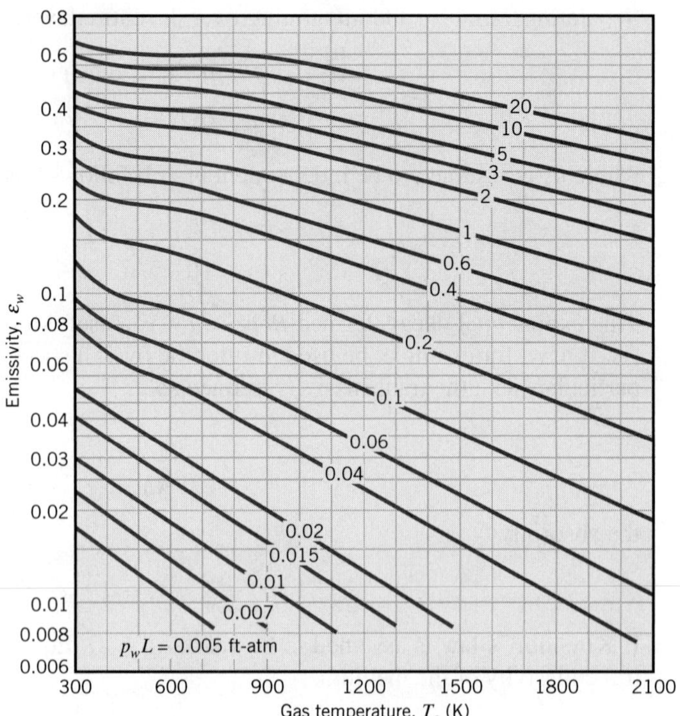

FIGURE 13.16 Emissivity of water vapor in a mixture with nonradiating gases at 1-atm total pressure and of hemispherical shape [13]. Used with permission.

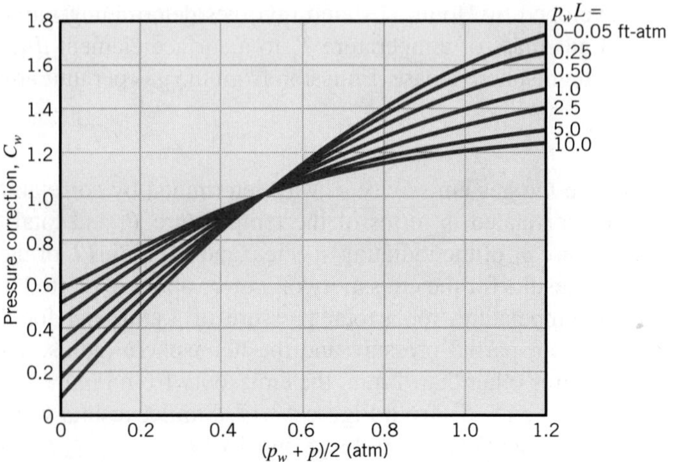

FIGURE 13.17 Correction factor for obtaining water vapor emissivities at pressures other than ($\varepsilon_{w,p\neq 1\text{atm}} = C_w\varepsilon_{w,p=1\text{atm}}$) [13]. Used with permission.

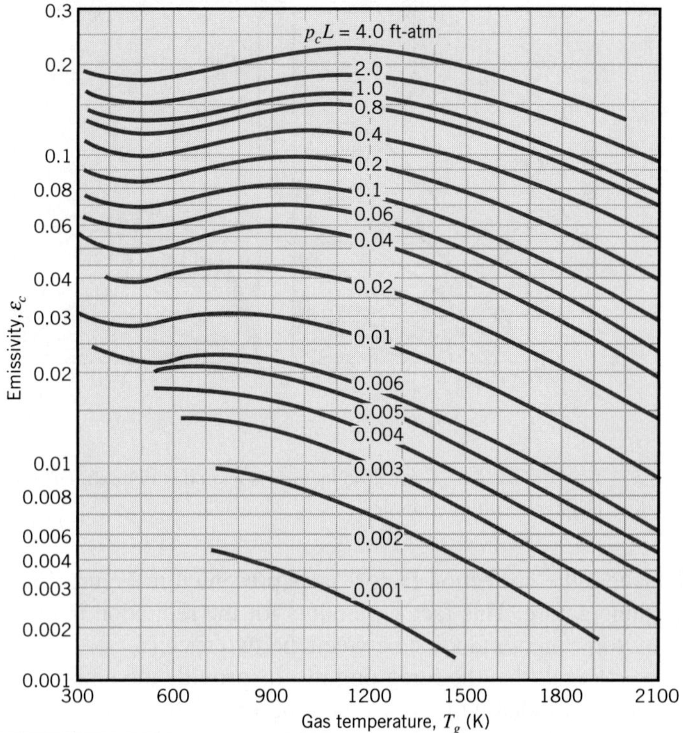

FIGURE 13.18 Emissivity of carbon dioxide in a mixture with nonradiating gases at 1-atm total pressure and of hemispherical shape [13]. Used with permission.

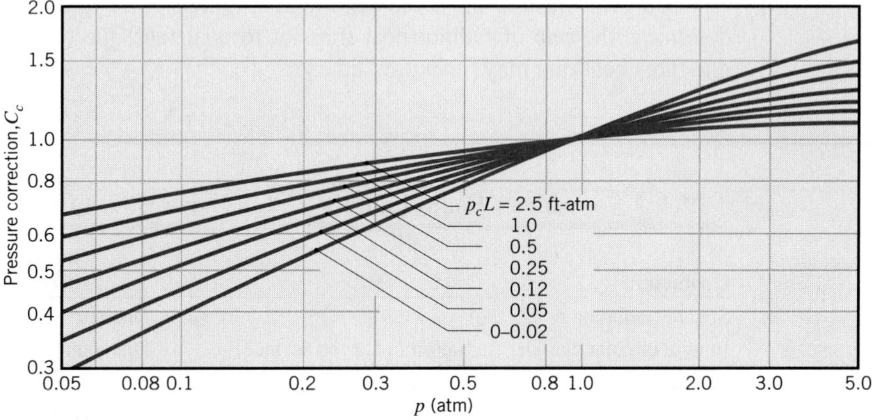

FIGURE 13.19 Correction factor for obtaining carbon dioxide emissivities at pressures other than 1 atm ($\varepsilon_{c,p\neq 1\text{atm}} = C_c\varepsilon_{c,p=1\text{atm}}$) [13]. Used with permission.

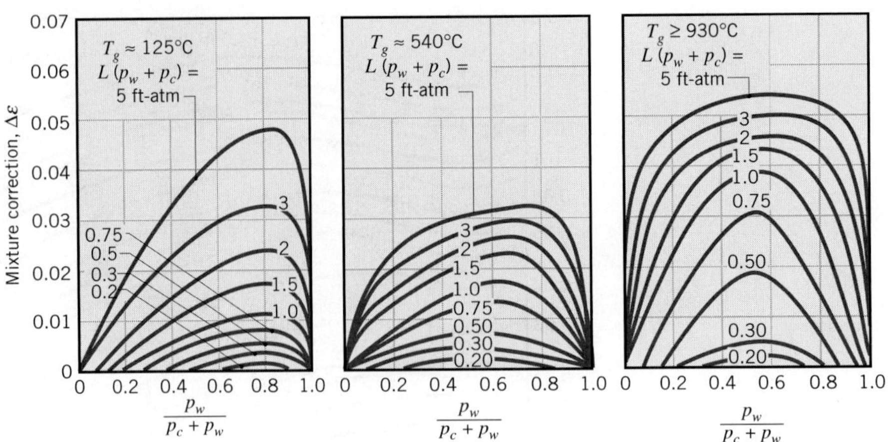

FIGURE 13.20 Correction factor associated with mixtures of water vapor and carbon dioxide [13]. Used with permission.

where the correction factor $\Delta\varepsilon$ is presented in Figure 13.20 for different values of the gas temperature. This factor accounts for the reduction in emission associated with the mutual absorption of radiation between the two species.

Recall that the foregoing results provide the emissivity of a hemispherical gas mass of radius L radiating to an element of area at the center of its base. However, the results may be extended to other gas geometries by introducing the concept of a *mean beam length*, L_e. The quantity was introduced to correlate, in terms of a single parameter, the dependence of gas emissivity on both the size and the shape of the gas geometry. It may be interpreted as the radius of a hemispherical gas mass whose emissivity is equivalent to that for the geometry of interest. Its value has been determined for numerous gas shapes [13], and representative results are listed in Table 13.4. Replacing L by L_e in Figures 13.16 through 13.20, the emissivity associated with the geometry of interest may then be determined.

Using the results of Table 13.4 with Figures 13.16 through 13.20, it is possible to determine the rate of radiant heat transfer to a surface due to emission from an adjoining gas. This heat rate may be expressed as

$$q = \varepsilon_g A_s \sigma T_g^4 \tag{13.39}$$

TABLE 13.4 Mean Beam Lengths L_e for Various Gas Geometries

Geometry	Characteristic Length	L_e
Sphere (radiation to surface)	Diameter (D)	$0.65D$
Infinite circular cylinder (radiation to curved surface)	Diameter (D)	$0.95D$
Semi-infinite circular cylinder (radiation to base)	Diameter (D)	$0.65D$
Circular cylinder of equal height and diameter (radiation to entire surface)	Diameter (D)	$0.60D$
Infinite parallel planes (radiation to planes)	Spacing between planes (L)	$1.80L$
Cube (radiation to any surface)	Side (L)	$0.66L$
Arbitrary shape of volume V (radiation to surface of area A)	Volume to area ratio (V/A)	$3.6V/A$

where A_s is the surface area. If the surface is black, it will, of course, absorb all this radiation. A black surface will also emit radiation, and the net rate at which radiation is exchanged between the surface at T_s and the gas at T_g is

$$q_{\text{net}} = A_s\sigma(\varepsilon_g T_g^4 - \alpha_g T_s^4) \tag{13.40}$$

For water vapor and carbon dioxide the required gas absorptivity α_g may be evaluated from the emissivity by expressions of the form [13]

Water:

$$\alpha_w = C_w \left(\frac{T_g}{T_s}\right)^{0.45} \times \varepsilon_w \left(T_s, p_w L_e \frac{T_s}{T_g}\right) \tag{13.41}$$

Carbon dioxide:

$$\alpha_c = C_c \left(\frac{T_g}{T_s}\right)^{0.65} \times \varepsilon_c \left(T_s, p_c L_e \frac{T_s}{T_g}\right) \tag{13.42}$$

where ε_w and ε_c are evaluated from Figures 13.16 and 13.18, respectively, and C_w and C_c are evaluated from Figures 13.17 and 13.19, respectively. Note, however, that in using Figures 13.16 and 13.18, T_g is replaced by T_s and $p_w L_e$ or $p_c L_e$ is replaced by $p_w L_e(T_s/T_g)$ or $p_c L_e(T_s/T_g)$, respectively. Note also that, in the presence of both water vapor and carbon dioxide, the total gas absorptivity may be expressed as

$$\alpha_g = \alpha_w + \alpha_c - \Delta\alpha \tag{13.43}$$

where $\Delta\alpha = \Delta\varepsilon$ is obtained from Figure 13.20.

13.7 *Summary*

In this chapter we focused on the analysis of radiation exchange between the surfaces of an enclosure, and in treating this exchange we introduced the concept of a *view factor*. Because knowledge of this geometrical quantity is essential to determining radiation exchange between any two diffuse surfaces, you should be familiar with the means by which it may be determined. You should also be adept at performing radiation calculations for an enclosure of *isothermal*, *opaque*, *diffuse*, and *gray* surfaces of *uniform radiosity* and *irradiation.* Moreover, you should be familiar with the results that apply to simple cases such as an enclosure of black surfaces, a two-surface enclosure, or a three-surface enclosure with a reradiating surface. Finally, be cognizant that in situations where the radiosity and irradiation distributions are nonuniform, it may be necessary to perform a radiation exchange analysis involving many surfaces in order to determine local surface temperatures or local heat fluxes.

Test your understanding of pertinent concepts by addressing the following questions.

- What is a *view factor*? What assumptions are typically associated with computing the view factor between two surfaces?
- What is the *reciprocity relation* for view factors? What is the *summation rule*?

- Can the view factor of a surface with respect to itself be nonzero? If so, what kind of surface exhibits such behavior?
- What is a *nonparticipating medium*?
- What assumptions are inherent in treating radiation exchange between surfaces of an enclosure that may not be approximated as blackbodies? Comment on the validity of the assumptions and when or where they are most likely to break down.
- How is the radiative resistance of a surface in an enclosure defined? What is the driving potential that relates this resistance to the net rate of radiation transfer from the surface? What is the resistance if the surface may be approximated as a blackbody?
- How is the geometrical resistance associated with radiative exchange between two surfaces of an enclosure defined? What is the driving potential that relates this resistance to the net rate of radiation transfer between the surfaces?
- What is a *radiation shield* and how is net radiation transfer between two surfaces affected by an intervening shield? Is it advantageous for a shield to have a large surface absorptivity or reflectivity?
- What is a *reradiating surface*? Under what conditions may a surface be approximated as reradiating? What is the relation between radiosity, blackbody emissive power, and irradiation for a reradiating surface? Does the temperature of such a surface depend on its radiative properties?
- What can be said about a surface of an enclosure for which net radiation heat transfer to the surface is balanced by convection heat transfer from the surface to a gas in the enclosure? Is the surface reradiating? Is the back side of the surface adiabatic?
- Consider a surface in an enclosure for which net radiation transfer from the surface exceeds convection heat transfer from a gas in the enclosure. What other process or processes must occur at the surface, either independently or collectively?
- What molecular features render a gas nonemitting and nonabsorbing? What features allow for emission and absorption of radiation by a gas?
- What features distinguish emission and absorption of radiation by a gas from that of an opaque solid?
- How does the intensity of radiation propagating through a semitransparent medium vary with distance in the medium? What can be said about this variation if the absorption coefficient is very large? If the coefficient is very small?

References

1. Hamilton, D. C., and W. R. Morgan, "Radiant Interchange Configuration Factors," National Advisory Committee for Aeronautics, Technical Note 2836, 1952.
2. Eckert, E. R. G., "Radiation: Relations and Properties," in W. M. Rohsenow and J. P. Hartnett, Eds., *Handbook of Heat Transfer*, 2nd ed., McGraw-Hill, New York, 1973.
3. Howell, J. R., R. Siegel, and M. P. Menguc, *Thermal Radiation Heat Transfer*, 5th ed., Taylor & Francis, New York, 2010.
4. Howell, J. R., *A Catalog of Radiation Configuration Factors*, McGraw-Hill, New York, 1982.
5. Emery, A. F., O. Johansson, M. Lobo, and A. Abrous, *J. Heat Transfer*, **113**, 413, 1991.
6. Oppenheim, A. K., *Trans. ASME*, **65,** 725, 1956.
7. Hottel, H. C., and A. F. Sarofim, *Radiative Transfer*, McGraw-Hill, New York, 1967.
8. Tien, C. L., "Thermal Radiation Properties of Gases," in J. P. Hartnett and T. F. Irvine, Eds., *Advances in Heat Transfer*, Vol. 5, Academic Press, New York, 1968.
9. Sparrow, E. M., "Radiant Interchange Between Surfaces Separated by Nonabsorbing and Nonemitting Media," in W. M. Rohsenow and J. P. Hartnett, Eds., *Handbook of Heat Transfer*, McGraw-Hill, New York, 1973.
10. Dunkle, R. V., "Radiation Exchange in an Enclosure with a Participating Gas," in W. M. Rohsenow and J. P. Hartnett, Eds., *Handbook of Heat Transfer*, McGraw-Hill, New York, 1973.

11. Sparrow, E. M., and R. D. Cess, *Radiation Heat Transfer*, Hemisphere Publishing, New York, 1978.

12. Edwards, D. K., *Radiation Heat Transfer Notes*, Hemisphere Publishing, New York, 1981.

13. Hottel, H. C., and R. B. Egbert, *AIChE J.*, **38**, 531, 1942.

Problems

View Factors

13.1 Determine F_{12} and F_{21} for the following configurations using the reciprocity theorem and other basic shape factor relations. Do not use tables or charts.

(a) Long duct

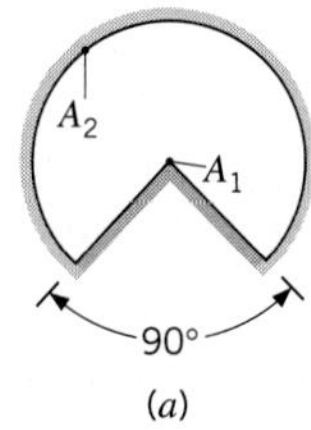

(*a*)

(b) Small sphere of area A_1 under a concentric hemisphere of area $A_2 = 2A_1$

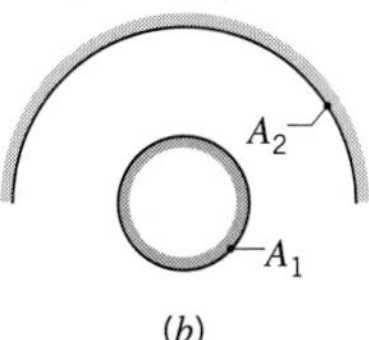

(*b*)

(c) Long duct. What is F_{22} for this case?

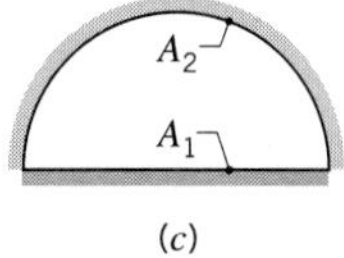

(*c*)

(d) Long inclined plates (point *B* is directly above the center of A_1)

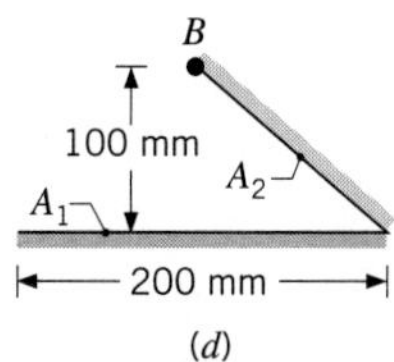

(*d*)

(e) Sphere lying on infinite plane

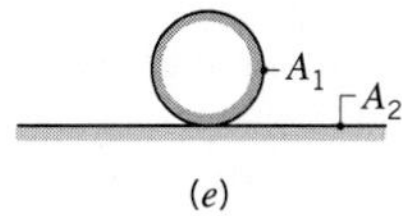

(*e*)

(f) Hemisphere–disk arrangement

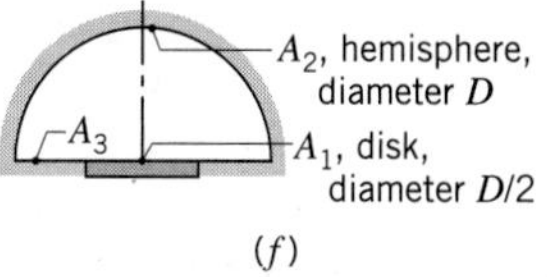

(*f*)

(g) Long, open channel

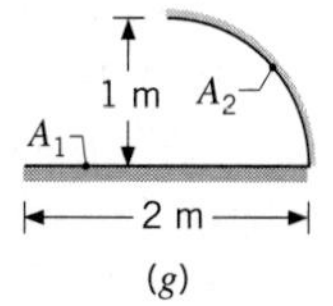

(*g*)

(h) Long concentric cylinders

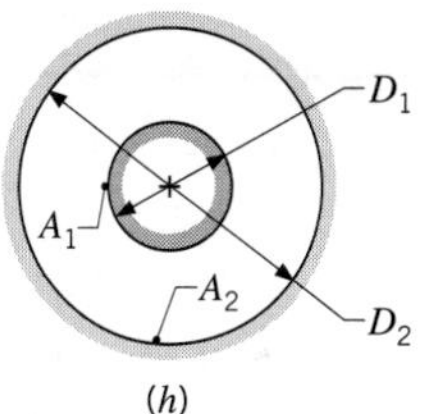

(*h*)

13.2 Consider the following grooves, each of width *W*, that have been machined from a solid block of material.

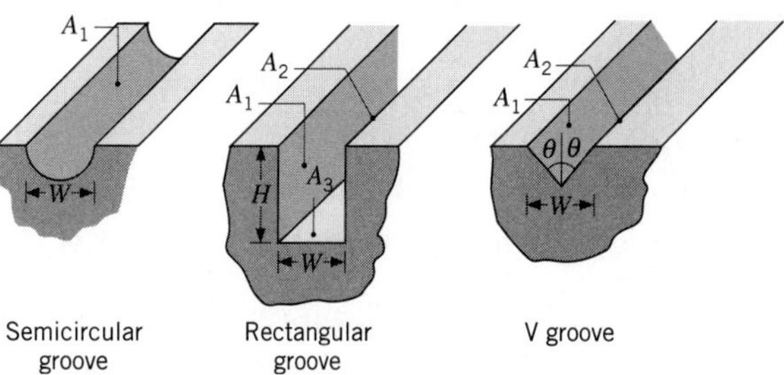

(a) For each case obtain an expression for the view factor of the groove with respect to the surroundings outside the groove.

(b) For the *V* groove, obtain an expression for the view factor F_{12}, where A_1 and A_2 are opposite surfaces.

(c) If $H = 2W$ in the rectangular groove, what is the view factor F_{12}?

13.3 Derive expressions for the view factor F_{12} associated with the following arrangements. Express your results in terms of A_1, A_2, and any appropriate hypothetical surface area, as well as the view factor for coaxial parallel disks (Table 13.2, Figure 13.5).

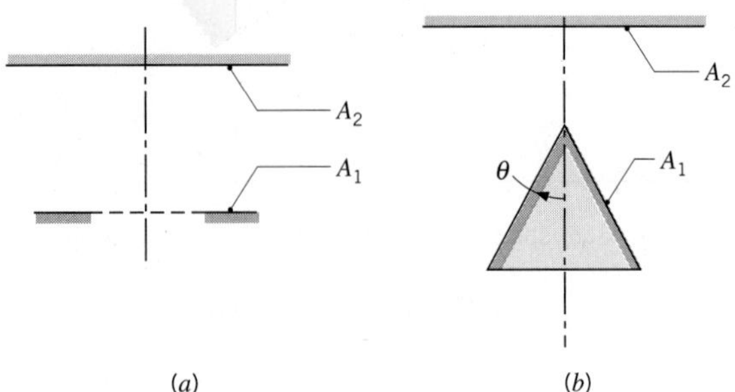

(a) A circular disk and a coaxial, ring-shaped disk.

(b) A circular disk and a coaxial, right-circular cone. Sketch the variation of F_{12} with θ for $0 \leq \theta \leq \pi/2$, and explain the key features.

13.4 A right-circular cone and a right-circular cylinder of the same diameter and length (A_2) are positioned coaxially at a distance L_o from the circular disk (A_1) shown schematically. The inner base and lateral surfaces of the cylinder may be treated as a single surface, A_2. The hypothetical area corresponding to the opening of the cone and cylinder is identified as A_3.

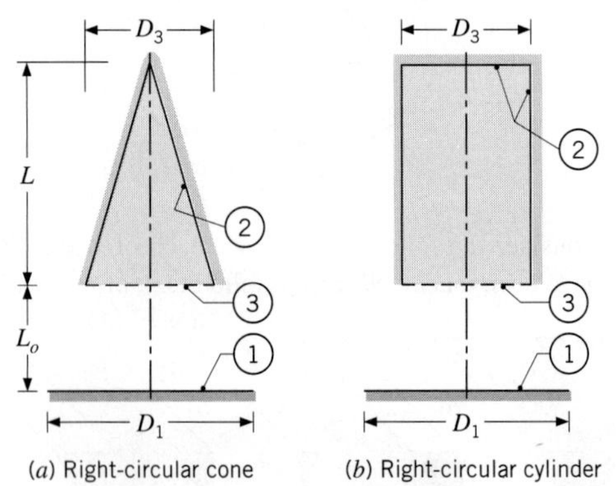

(a) Right-circular cone (b) Right-circular cylinder

(a) Show that, for both arrangements, $F_{21} = (A_1/A_2)F_{13}$ and $F_{22} = 1 - (A_3/A_2)$, where F_{13} is the view factor between two coaxial, parallel disks (Table 13.2).

(b) For $L = L_o = 50$ mm and $D_1 = D_3 = 50$ mm, calculate F_{21} and F_{22} for the conical and cylindrical configurations and compare their relative magnitudes. Explain any similarities and differences.

(c) Do the relative magnitudes of F_{21} and F_{22} change for the conical and cylindrical configurations as L increases and all other parameters remain fixed? In the limit of very large L, what do you expect will happen? Sketch the variations of F_{21} and F_{22} with L, and explain the key features.

13.5 Consider the two parallel, coaxial, ring-shaped disks shown schematically.

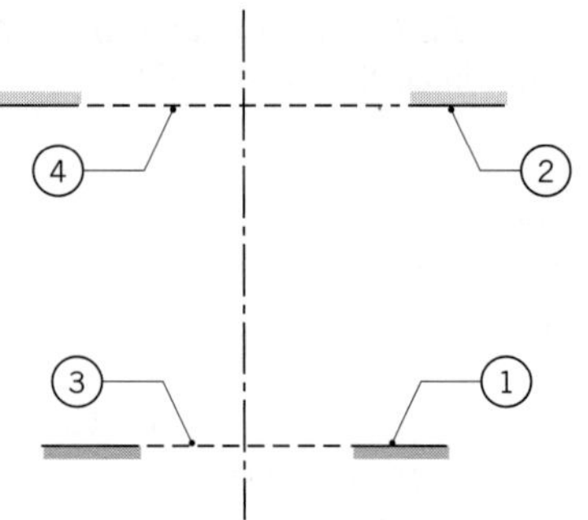

Show that F_{12} can be expressed as

$$F_{12} = \frac{1}{A_1}\{A_{(1,3)}F_{(1,3)(2,4)} - A_3F_{3(2,4)} - A_4(F_{4(1,3)} - F_{43})\}$$

where all view factors on the right-hand side of the equation can be evaluated from Figure 13.5 or Table 13.2 for coaxial parallel disks.

13.6 The "crossed-strings" method of Hottel [7] provides a simple means to calculate view factors between surfaces that are of infinite extent in one direction. For two such surfaces (a) with unobstructed views of one another, the view factor is of the form

$$F_{12} = \frac{1}{2w_1}[(ac + bd) - (ad + bc)]$$

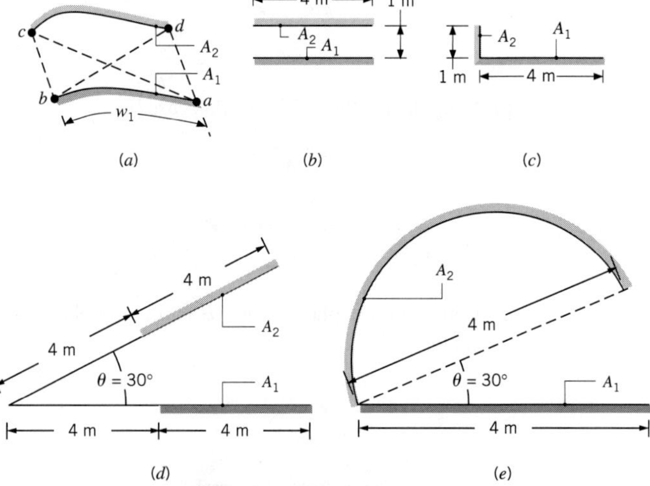

Use this method to evaluate the view factors F_{12} for sketches (b) through (e). Compare your results with those from using the appropriate graphs, analytical expressions, and view factor relations.

13.7 Consider the right-circular cylinder of diameter D, length L, and the areas A_1, A_2, and A_3 representing the base, inner, and top surfaces, respectively.

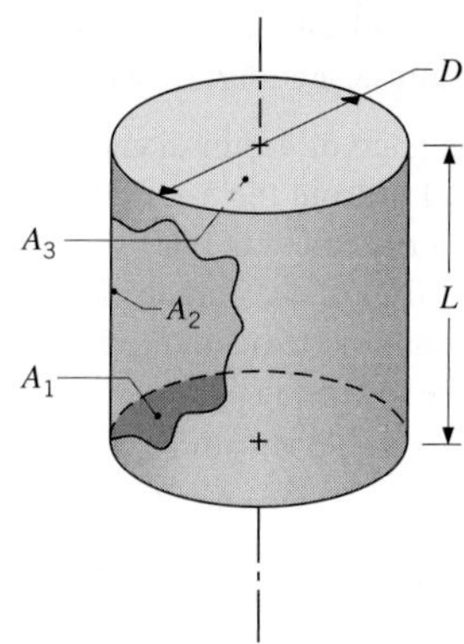

(a) Show that the view factor between the base of the cylinder and the inner surface has the form $F_{12} = 2H[(1 + H^2)^{1/2} - H]$, where $H = L/D$.

(b) Show that the view factor for the inner surface to itself has the form $F_{22} = 1 + H - (1 + H^2)^{1/2}$.

13.8 Consider the parallel rectangles shown schematically.

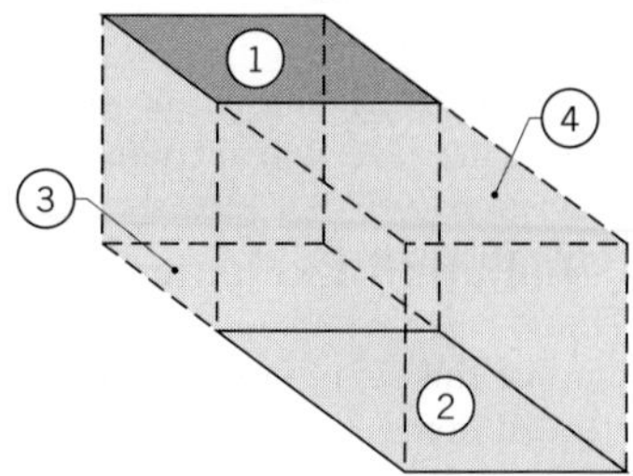

Show that the view factor F_{12} can be expressed as

$$F_{12} = \frac{1}{2A_1}[A_{(1,4)}F_{(1,4)(2,3)} - A_1F_{13} - A_4F_{42}]$$

where all view factors on the right-hand side of the equation can be evaluated from Figure 13.4 (see Table 13.2) for aligned parallel rectangles.

13.9 Consider the perpendicular rectangles shown schematically.

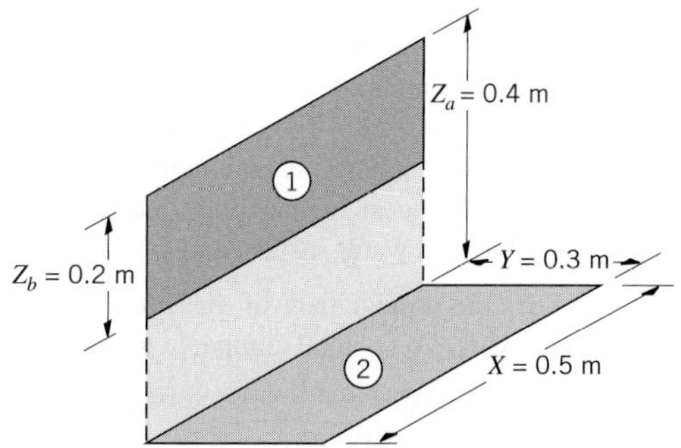

(a) Determine the shape factor F_{12}.

(b) For rectangle widths of $X = 0.5$, 1.5, and 5 m, plot F_{12} as a function of Z_b for $0.05 \leq Z_b \leq 0.4$ m.

Co pare your results with the view factor obtained fro the two-dimensional relation for perpendicular lates with a common edge (Table 13.1).

13.10 The reciprocity relation, the summation rule, and Equations 13.5 to 13.7 can be used to develop view factor relations that allow for applications of Figure 13.4 and/or 13.6 to more complex configurations. Consider the view factor F_{14} for surfaces 1 and 4 of the following geometry. These surfaces are perpendicular but do not share a common edge.

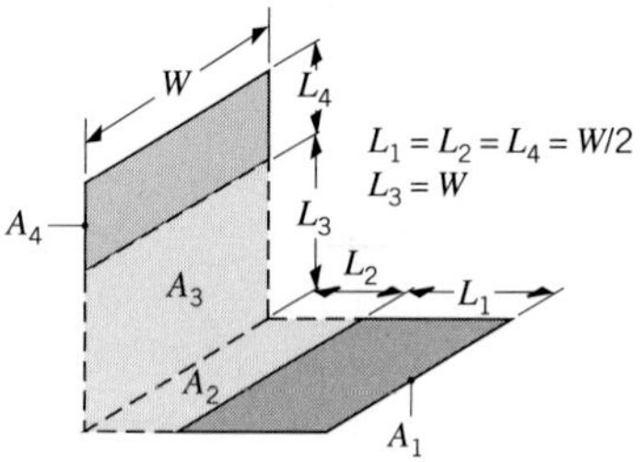

(a) Obtain the following expression for the view factor F_{14}:

$$F_{14} = \frac{1}{A_1}[(A_1 + A_2)F_{(1,2)(3,4)} + A_2F_{23} - (A_1 + A_2)F_{(1,2)3} - A_2F_{2(3,4)}]$$

(b) If $L_1 = L_2 = L_4 = (W/2)$ and $L_3 = W$, what is the value of F_{14}?

13.11 Determine the shape factor, F_{12}, for the rectangles shown.

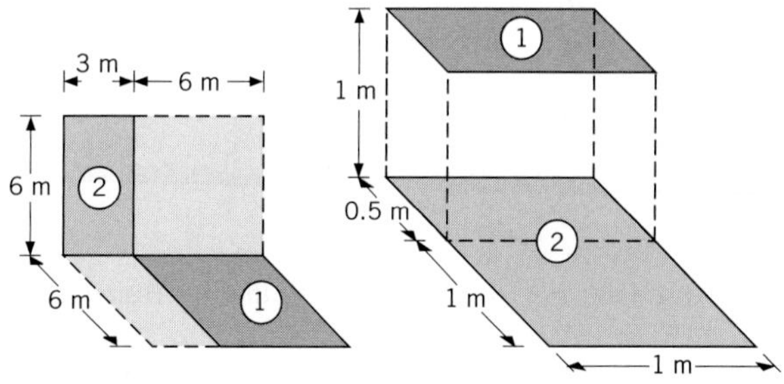

(a) Perpendicular rectangles without a common edge.

(b) Parallel rectangles of unequal areas.

13.12 Consider parallel planes of infinite extent normal to the page as shown in the sketch.

(a) Determine F_{12} using the results of Figure 13.4.

(b) Determine F_{12} using the first case of Table 13.1.

(c) Determi using Hottel's crossed-string method describe lem 13.6.

(d) D g the second case of Table 13.1.

(e) Determine F_{12} using the results of Figure 13.4 if the dimensions are increased to $w = L = 2$ m.

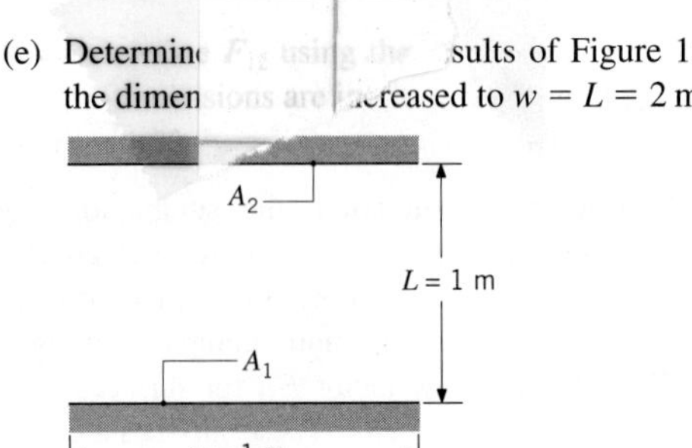

13.13 Consider the parallel planes of infinite extent normal to the page having opposite edges aligned as shown in the sketch.

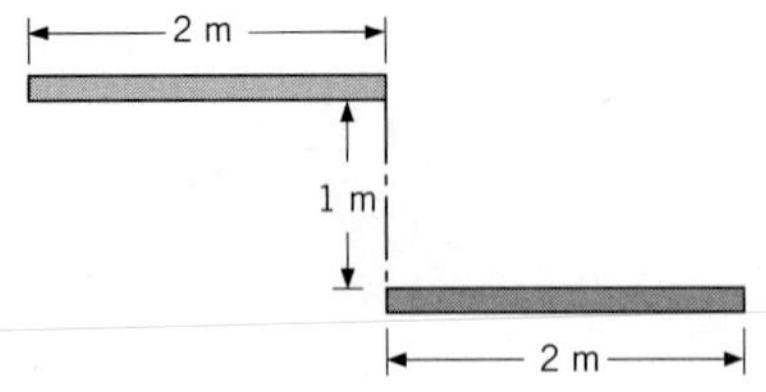

(a) Using appropriate view factor relations and the results for opposing parallel planes, develop an expression for the view factor F_{12}.

(b) Use Hottel's crossed-string method described in Problem 13.6 to determine the view factor.

13.14 Consider two diffuse surfaces A_1 and A_2 on the inside of a spherical enclosure of radius R. Using the following methods, derive an expression for the view factor F_{12} in terms of A_2 and R.

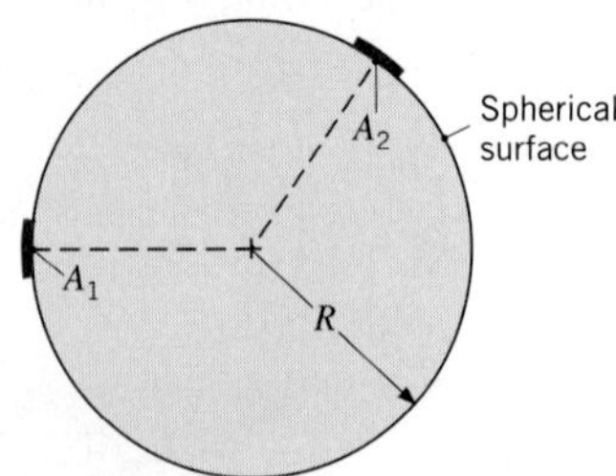

(a) Find F_{12} by beginning with the expression $F_{ij} = q_{i \to j}/A_i J_i$.

(b) Find F_{12} using the view factor integral, Equation 13.1.

13.15 As shown in the sketch, consider the disk A_1 located coaxially 1 m distant, but tilted 30° off the normal, from the ring-shaped disk A_2.

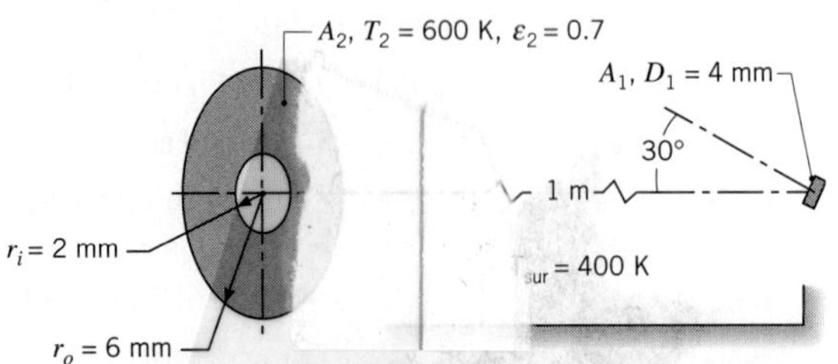

What is the irradiation on A_1 due to radiation from A_2, which is a diffuse, gray surface with an emissivity of 0.7?

13.16 A heat flux gage of 4-mm diameter is positioned normal to and 1 m from the 5-mm-diameter aperture of a blackbody furnace at 1000 K. The diffuse, gray cover shield ($\varepsilon = 0.2$) of the furnace has an outer diameter of 100 mm and its temperature is 350 K. The furnace and gage are located in a large room whose walls have an emissivity of 0.8 and are at 300 K.

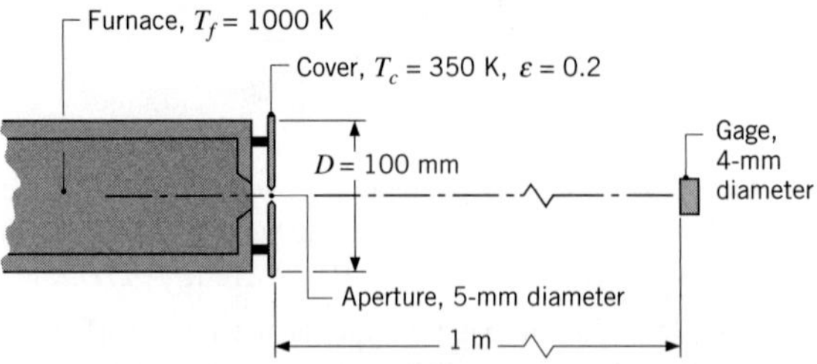

(a) What is the irradiation on the gage, G_g (W/m^2), considering only emission from the aperture of the furnace?

(b) What is the irradiation on the gage due to radiation from the cover and aperture?

Blackbody Radiation Exchange

13.17 A circular ice rink 25 m in diameter is enclosed by a hemispherical dome 35 m in diameter. If the ice and dome surfaces may be approximated as blackbodies and are at 0 and 15°C, respectively, what is the net rate of radiative transfer from the dome to the rink?

13.18 A drying oven consists of a long semicircular duct of diameter $D = 1$ m.

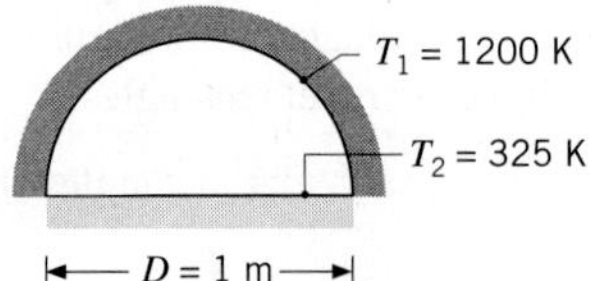

Materials to be dried cover the base of the oven while the wall is maintained at 1200 K. What is the drying rate per unit length of the oven (kg/s·m) if a water-coated layer of material is maintained at 325 K during the drying process? Blackbody behavior may be assumed for the water surface and the oven wall.

13.19 Consider the arrangement of the three black surfaces shown, where A_1 is small compared to A_2 or A_3.

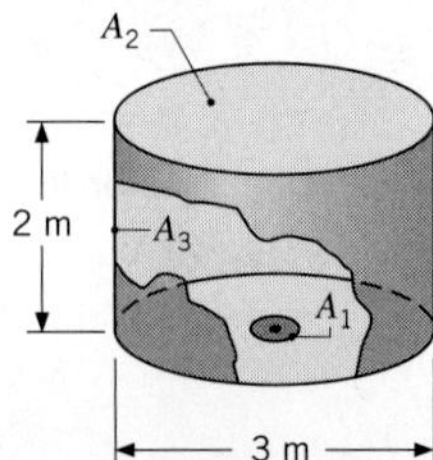

Determine the value of F_{13}. Calculate the net radiation heat transfer from A_1 to A_3 if $A_1 = 0.05\ \text{m}^2$, $T_1 = 1000$ K, and $T_3 = 500$ K.

13.20 A long, V-shaped part is heat treated by suspending it in a tubular furnace with a diameter of 2 m and a wall temperature of 1000 K. The "vee" is 1-m long on a side and has an angle of 60°.

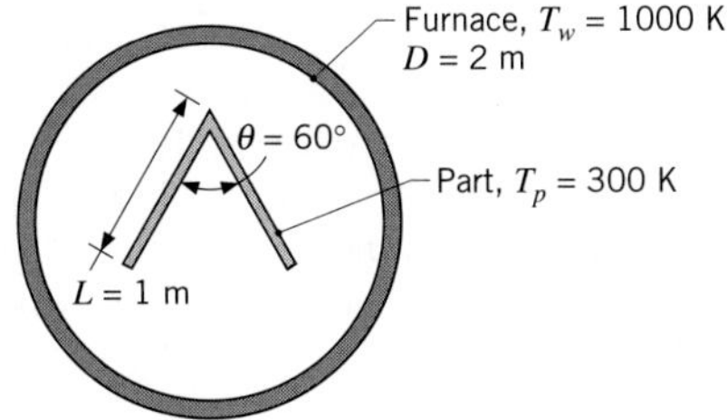

If the wall of the furnace and the surfaces of the part may be approximated as blackbodies and the part is at an initial temperature of 300 K, what is the net rate of radiation heat transfer per unit length to the part?

13.21 Consider coaxial, parallel, black disks separated a distance of 0.20 m. The lower disk of diameter 0.40 m is maintained at 500 K and the surroundings are at 300 K.

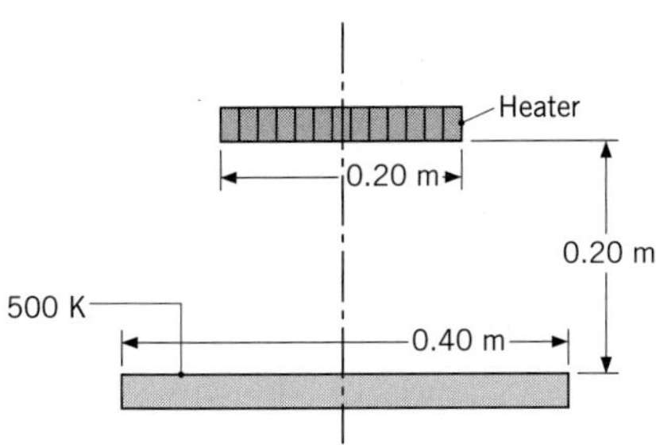

What temperature will the upper disk of diameter 0.20 m achieve if electrical power of 17.5 W is supplied to the heater on the back side of the disk?

13.22 Rework Problem 13.21 but with both disks having diameter $D = 0.20$ m. Next, consider the coaxial disks to be replaced by aligned squares of length X on a side so that $X^2 = \pi D^2/4$. Using a software package such as *IHT*, graph the upper square and disk temperatures over the range $0 < L \leq 1$ m, where L is the separation distance between the surfaces.

13.23 A tubular heater with a black inner surface of uniform temperature $T_s = 1000$ K irradiates a coaxial disk.

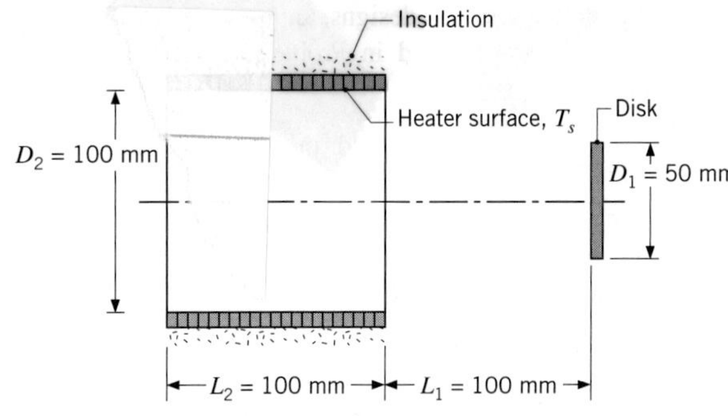

(a) Determine the radiant power from the heater which is incident on the disk, $q_{s \to 1}$. What is the irradiation on the disk, G_1?

(b) For disk diameters of $D_1 = 25$, 50, and 100 mm, plot $q_{s \to 1}$ and G_1 as a function of the separation distance L_1 for $0 \leq L_1 \leq 200$ mm.

13.24 A circular plate of 500-mm diameter is maintained at $T_1 = 600$ K and is positioned coaxial to a conical shape. The back side of the cone is well insulated. The plate and the cone, whose surfaces are black, are located in a large, evacuated enclosure whose walls are at 300 K.

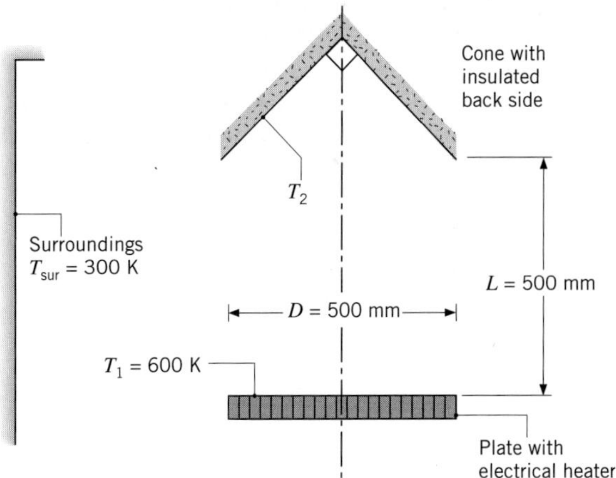

(a) What is the temperature of the conical surface, T_2?

(b) What is the electrical power that would be required to maintain the circular plate at 600 K?

13.25 Two furnace designs, as illustrated and dimensioned in Problem 13.4, are used to heat a disk-shaped work piece (A_1). The power supplied to the conical and cylindrical furnaces (A_2) is 50 W. The work piece is located in a large room at a temperature of 300 K, and its back side is well insulated. Assuming all surfaces are black, determine the temperature of the work piece, T_1, and the temperature of the furnace inner surface,

T_2, for each of the designs. In your analysis, use the expressions provided in Problem 13.4 for the view factors, F_{21} and F_{23}.

13.26 A furnace is constructed in three sections, which include insulated circular (2) and cylindrical (3) sections, as well as an intermediate cylindrical section (1) with embedded electrical resistance heaters. The overall length and diameter are 200 mm and 100 mm, respectively, and the cylindrical sections are of equal length. The surroundings are at 0 K.

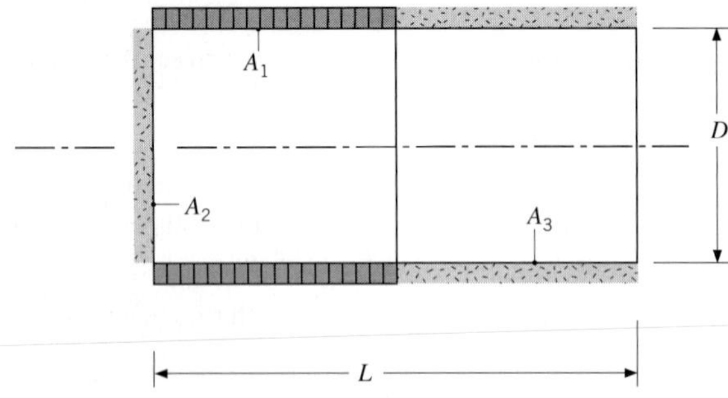

(a) If all the surfaces are black, determine the electrical power, q_1, required to maintain the heated section at 1000 K.

(b) What are the temperatures of the insulated sections, T_2 and T_3?

(c) For $D = 100$ mm, generate a plot of q_1, T_2, and T_3 as functions of the length-to-diameter ratio, with $1 \leq L/D \leq 5$.

13.27 To enhance heat rejection from a spacecraft, an engineer proposes to attach an array of rectangular fins to the outer surface of the spacecraft and to coat all surfaces with a material that approximates blackbody behavior.

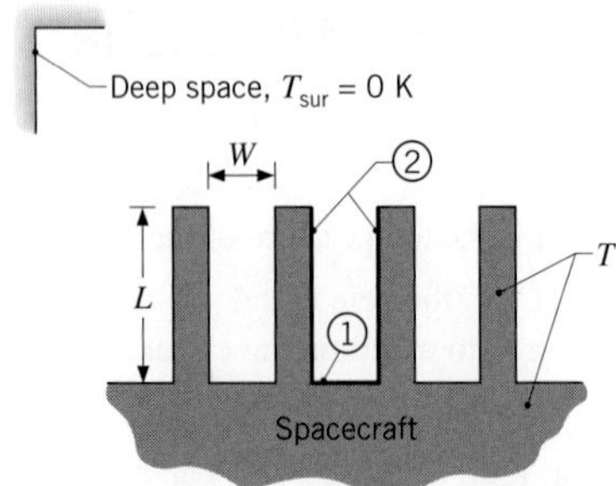

Consider the U-shaped region between adjoining fins and subdivide the surface into components associated with the base (1) and the side (2). Obtain an expression for the rate per unit length at which radiation is transferred from the surface to deep space, which may be approximated as a blackbody at absolute zero temperature. The fins and the base may be assumed to be isothermal at a temperature T. Comment on your result. Does the engineer's proposal have merit?

13.28 Determine the temperatures of surfaces 1 through 4 of the furnace cavity of Example 13.3 if the entire furnace experiences a uniform heat flux corresponding to a total power of 1522 W.

13.29 A cylindrical cavity of diameter D and depth L is machined in a metal block, and conditions are such that the base and side surfaces of the cavity are maintained at $T_1 = 1000$ K and $T_2 = 700$ K, respectively. Approximating the surfaces as black, determine the emissive power of the cavity if $L = 20$ mm and $D = 10$ mm.

13.30 In the arrangement shown, the lower disk has a diameter of 30 mm and a temperature of 500 K. The upper surface, which is at 1000 K, is a ring-shaped disk with inner and outer diameters of 0.15 m and 0.2 m. This upper surface is aligned with and parallel to the lower disk and is separated by a distance of 1 m.

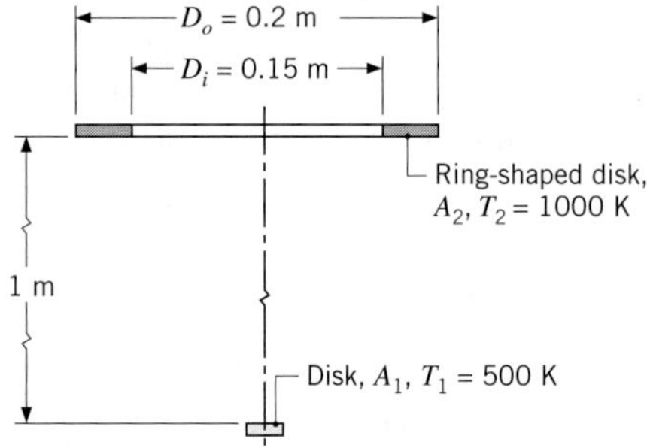

Assuming both surfaces to be blackbodies, calculate their net radiative heat exchange.

13.31 Two plane coaxial disks are separated by a distance $L = 0.20$ m. The lower disk (A_1) is solid with a diameter $D_o = 0.80$ m and a temperature $T_1 = 300$ K. The upper disk (A_2), at temperature $T_2 = 1000$ K, has the same outer diameter but is ring-shaped with an inner diameter $D_i = 0.40$ m.

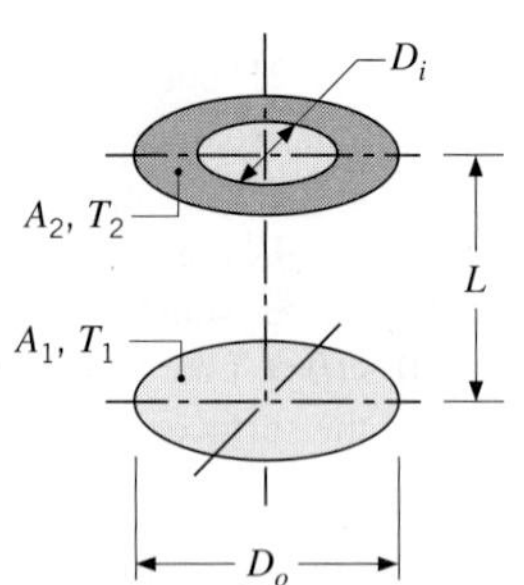

Assuming the disks to be blackbodies, calculate the net radiative heat exchange between them.

13.32 A radiometer views a small target (1) that is being heated by a ring-shaped disk heater (2). The target has an area of $A_1 = 0.0004\ \text{m}^2$, a temperature of $T_1 = 500\ \text{K}$, and a diffuse, gray emissivity of $\varepsilon_1 = 0.8$. The heater operates at $T_2 = 1000\ \text{K}$ and has a black surface. The radiometer views the entire sample area with a solid angle of $\omega = 0.0008$ sr.

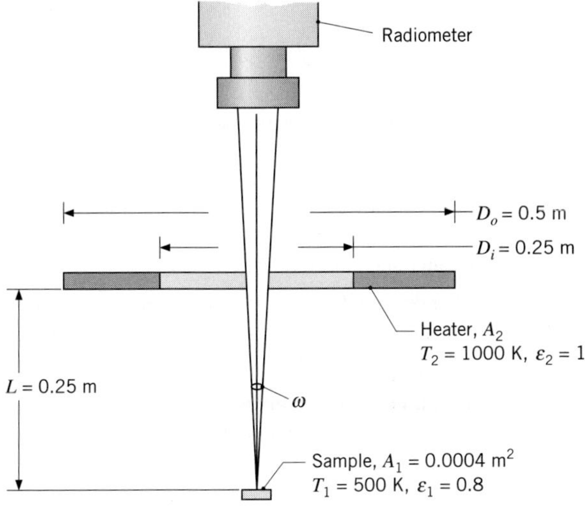

(a) Write an expression for the radiant power leaving the target which is collected by the radiometer, in terms of the target radiosity J_1 and relevant geometric parameters. Leave in symbolic form.

(b) Write an expression for the target radiosity J_1 in terms of its irradiation, emissive power, and appropriate radiative properties. Leave in symbolic form.

(c) Write an expression for the irradiation on the target, G_1, due to emission from the heater in terms of the heater emissive power, the heater area, and an appropriate view factor. Use this expression to numerically evaluate G_1.

(d) Use the foregoing expressions and results to determine the radiant power collected by the radiometer.

13.33 A meter to measure the power of a laser beam is constructed with a thin-walled, black conical cavity that is well insulated from its housing. The cavity has an opening of $D = 10$ mm and a depth of $L = 12$ mm. The meter housing and surroundings are at a temperature of 25.0°C.

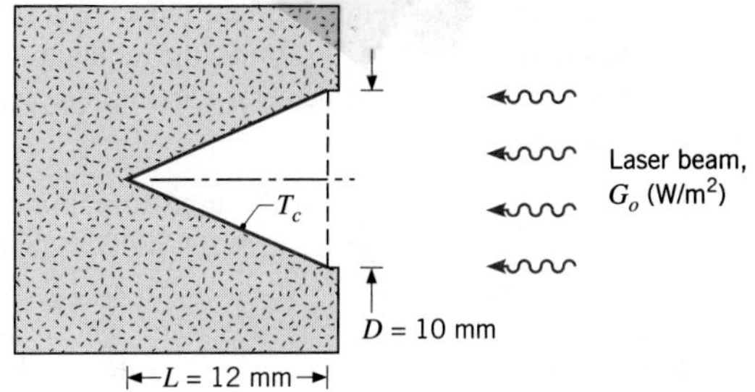

A fine-wire thermocouple attached to the surface indicates a temperature rise of 10.1°C when a laser beam is incident upon the meter. What is the radiant flux of the laser beam, G_o (W/m^2)?

13.34 An electrically heated sample is maintained at a surface temperature of $T_s = 500\ \text{K}$. The sample coating is diffuse but spectrally selective, with the spectral emissivity distribution shown schematically. The sample is irradiated by a furnace located coaxially at a distance of $L_{sf} = 750$ mm. The furnace has isothermal walls with an emissivity of $\varepsilon_f = 0.7$ and a uniform temperature of $T_f = 3000\ \text{K}$. A radiation detector of area $A_d = 8 \times 10^{-5}\ \text{m}^2$ is positioned at a distance of $L_{sd} = 1.0$ m from the sample along a direction that is 45° from the sample normal. The detector is sensitive to spectral radiant power only in the spectral region from 3 to 5 μm. The sample surface experiences convection with a gas for which $T_\infty = 300\ \text{K}$ and $h = 20\ \text{W/m}^2\cdot\text{K}$. The surroundings of the sample mount are large and at a uniform temperature of $T_{sur} = 300\ \text{K}$.

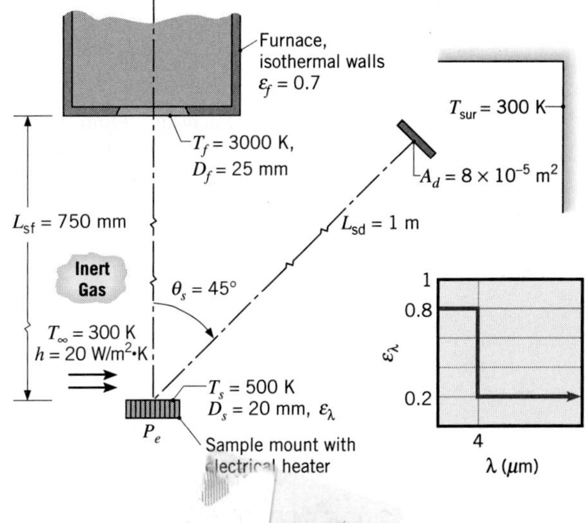

(a) Determine the electrical power, P_e, required to maintain the sample at $T_s = 500\ \text{K}$.

(b) Considering emission and reflected irradiation from the sample, determine the radiant power that is incident on the detector within the spectral region from 3 to 5 μm.

13.35 The arrangement shown is to be used to calibrate a heat flux gage. The gage has a black surface that is 10 mm in diameter and is maintained at 17°C by means of a water-cooled backing plate. The heater, 200 mm in diameter, has a black surface that is maintained at 800 K and is located 0.5 m from the gage. The surroundings and the air are at 27°C and the convection heat transfer coefficient between the gage and the air is 15 W/m²·K.

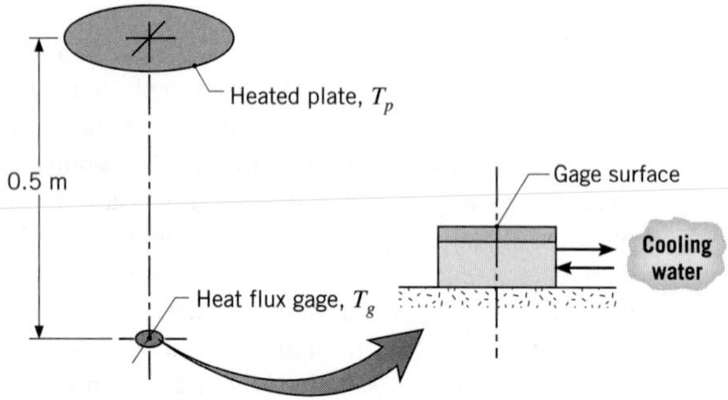

(a) Determine the net radiation exchange between the heater and the gage.

(b) Determine the net transfer of radiation to the gage per unit area of the gage.

(c) What is the net heat transfer rate to the gage per unit area of the gage?

(d) If the gage is constructed according to the description of Problem 3.107, what heat flux will it indicate?

13.36 A long, cylindrical heating element of 20-mm diameter operating at 700 K in vacuum is located 40 mm from an insulated wall of low thermal conductivity.

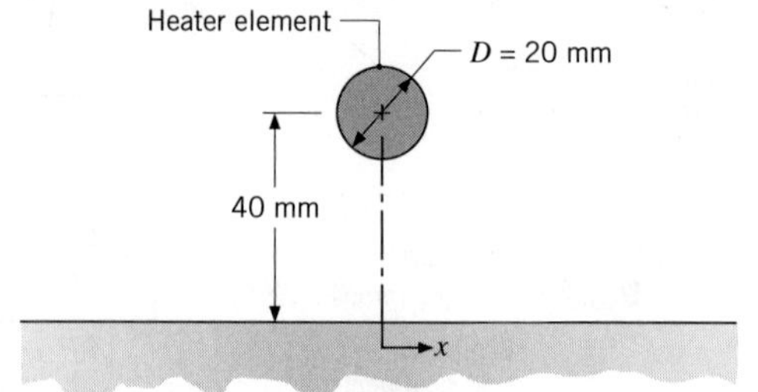

(a) Assuming both the element and the wall are black, estimate the maximum temperature reached by the wall when the surroundings are at 300 K.

(b) Calculate and plot the steady-state wall temperature distribution over the range $-100\text{ mm} \leq x \leq 100\text{ mm}$.

13.37 Water flowing through a large number of long, circular, thin-walled tubes is heated by means of hot parallel plates above and below the tube array. The space between the plates is evacuated, and the plate and tube surfaces may be approximated as blackbodies.

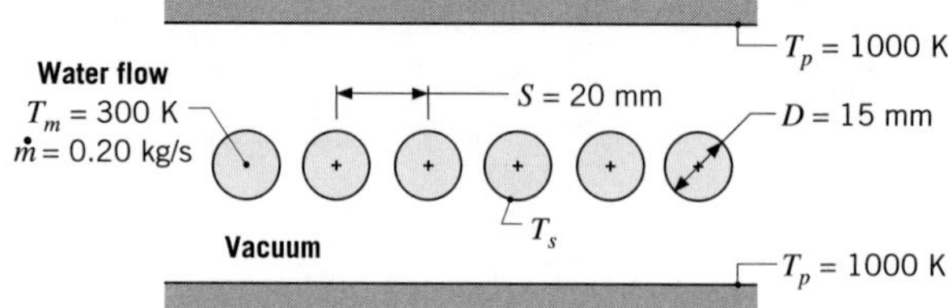

(a) Neglecting axial variations, determine the tube surface temperature, T_s, if water flows through each tube at a mass rate of $\dot{m} = 0.20$ kg/s and a mean temperature of $T_m = 300$ K.

(b) Compute and plot the surface temperature as a function of flow rate for $0.05 \leq \dot{m} \leq 0.25$ kg/s.

13.38 A row of regularly spaced, cylindrical heating elements is used to maintain an insulated furnace wall at 500 K. The opposite wall is at a uniform temperature of 300 K.

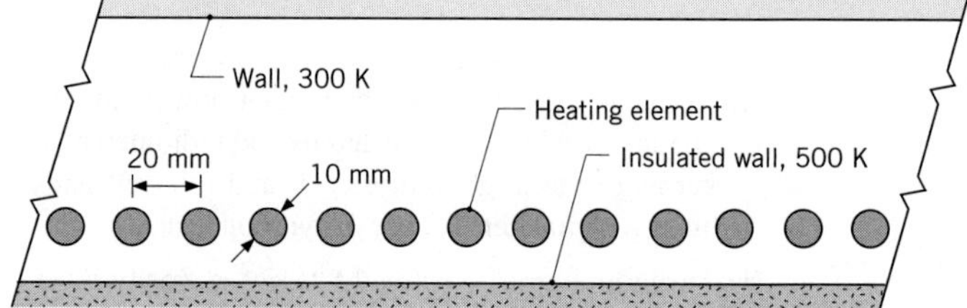

The insulated wall experiences convection with air at 450 K and a convection coefficient of 200 W/m²·K. Assuming the walls and elements are black, estimate the required operating temperature for the elements.

13.39 A manufacturing process calls for heating long copper rods, which are coated with a thin film having $\varepsilon = 1$, by placing them in a large evacuated oven whose surface is maintained at 1650 K. The rods are of 10-mm diameter and are placed in the oven with an initial temperature of 300 K.

(a) What is the initial rate of change of the rod temperature?

(b) How long must the rods remain in the oven to achieve a temperature of 1000 K?

(c) The heating process may be accelerated by routing combustion gases, also at 1650 K, through the oven. For convection coefficients of 10, 100, and 500 $W/m^2 \cdot K$, determine the time required for the rods to reach 1000 K.

13.40 Consider the very long, inclined black surfaces (A_1, A_2) maintained at uniform temperatures of $T_1 = 1000$ K and $T_2 = 800$ K.

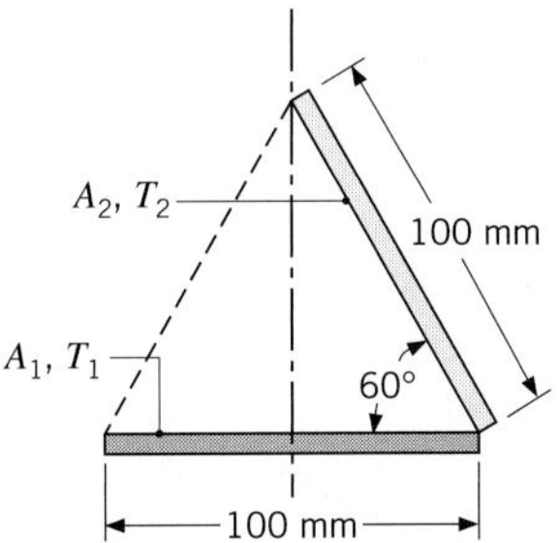

Determine the net radiation exchange between the surfaces per unit length of the surfaces. Consider the configuration when a black surface (A_3), whose back side is insulated, is positioned along the dashed line shown. Calculate the net radiation transfer to surface A_2 per unit length of the surface and determine the temperature of the insulated surface A_3.

13.41 Many products are processed in a manner that requires a specified product temperature as a function of time. Consider a product in the shape of long, 10-mm-diameter cylinders that are conveyed slowly through a processing oven as shown in the schematic. The product exhibits near-black behavior and is attached to the conveying apparatus at the product's ends. The surroundings are at 300 K, while the radiation panel heaters are each isothermal at 500 K and have surfaces that exhibit near-black behavior. An engineer proposes a novel oven design with a tilting top surface so as to be able to quickly change the thermal response of the product.

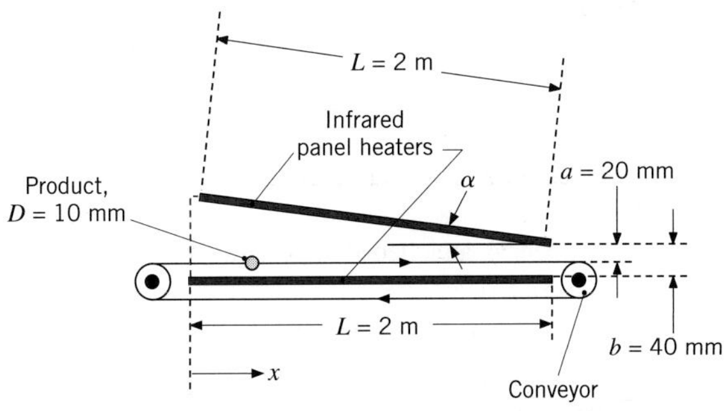

(a) Determine the radiation per unit length incident upon the product at $x = 0.5$ m and $x = 1$ m for $\alpha = 0$.

(b) Determine the radiation per unit length incident upon the product at $x = 0.5$ m and $x = 1$ m for $\alpha = \pi/15$. *Hint:* The view factor from the cylinder to the left-hand side surroundings can be found by summing the view factors from the cylinder to the two surfaces shown as red dashed lines in the schematic.

Two-Surface Enclosures

13.42 Consider two very large parallel plates with diffuse, gray surfaces.

Determine the irradiation and radiosity for the upper plate. What is the radiosity for the lower plate? What is the net radiation exchange between the plates per unit area of the plates?

13.43 A flat-bottomed hole 6 mm in diameter is bored to a depth of 24 mm in a diffuse, gray material having an emissivity of 0.8 and a uniform temperature of 1000 K.

(a) Determine the radiant power leaving the opening of the cavity.

(b) The effective emissivity ε_e of a cavity is defined as the ratio of the radiant power leaving the cavity to that from a blackbody having the area of the cavity opening and a temperature of the inner surfaces of the cavity. Calculate the effective emissivity of the cavity described above.

(c) If the depth of the hole were increased, would ε_e increase or decrease? What is the limit of ε_e as the depth increases?

13.44 Consider a double-pane window. The glass surface may be treated with a low-emissivity coating to reduce its emissivity from $\varepsilon = 0.95$ to $\varepsilon = 0.05$. Determine the radiation heat flux between two glass sheets for case 1: $\varepsilon_1 = \varepsilon_2 = 0.95$, case 2: $\varepsilon_1 = \varepsilon_2 = 0.05$, and case 3: $\varepsilon_1 = 0.05$, $\varepsilon_2 = 0.95$. The glass temperatures are $T_1 = 20°C$ and $T_2 = 0°C$, respectively.

13.45 Consider a long V groove 10 mm deep machined on a block that is maintained at 1000 K.

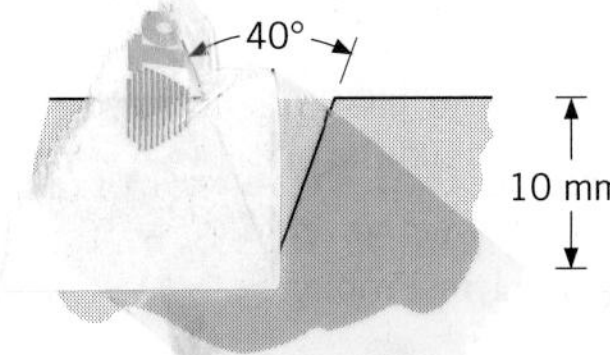

If the groove surfaces are diffuse and gray with an emissivity of 0.6, determine the radiant flux leaving the groove to its surroundings. Also determine the effective emissivity of the cavity, as defined in Problem

13.46 In Problems 12.20 and 12.25, we estimated the earth's surface temperature, assuming the earth is black. Most of the earth's surface is water, which has a hemispherical emissivity of $\varepsilon = 0.96$. In reality, the water surface is not flat but has waves and ripples.

(a) Assuming the wave geometry can be closely approximated as two-dimensional and as shown in the schematic, determine the effective emissivity of the water surface, as defined in Problem 13.43, for $\alpha = 3\pi/4$.

[(b)] Calculate and plot the effective emissivity of the water surface, normalized by the hemispherical emissivity of water ($\varepsilon_{\text{eff}}/\varepsilon$) over the range $\pi/2 \leq \alpha \leq \pi$.

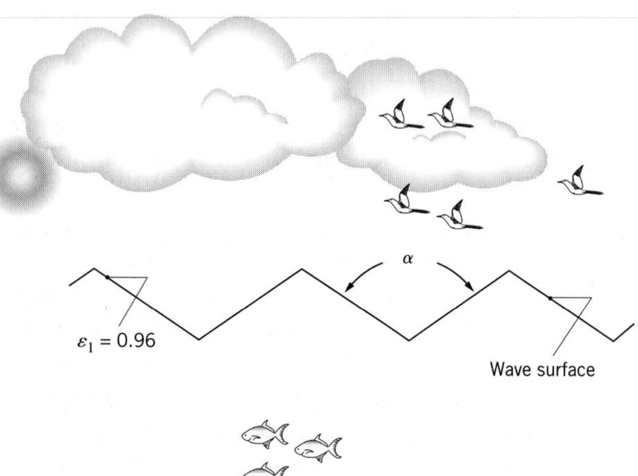

13.47 Consider the cavities formed by a cone, cylinder, and sphere having the same opening size (d) and major dimension (L), as shown in the diagram.

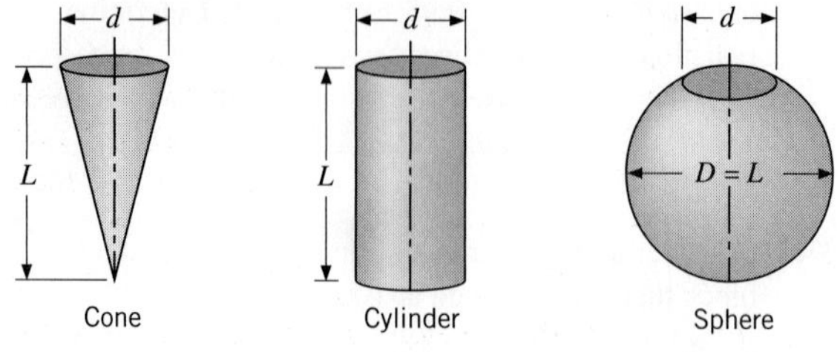

(a) Find the view factor between the inner surface of each cavity and the opening of the cavity.

(b) Find the effective emissivity of each cavity, ε_e, as defined in Problem 13.43, assuming the inner walls are diffuse and gray with an emissivity of ε_w.

[(c)] For each cavity and wall emissivities of $\varepsilon_w = 0.5$, 0.7, and 0.9, plot ε_e as a function of the major dimension-to-opening size ratio, L/d, over a range from 1 to 10.

13.48 Consider the attic of a home located in a hot climate. The floor of the attic is characterized by a width of $L_1 = 10$ m while the roof makes an angle of $\theta = 30°$ from the horizontal direction, as shown in the schematic. The homeowner wishes to reduce the heat load to the home by adhering bright aluminum foil ($\varepsilon_f = 0.07$) onto the surfaces of the attic space. Prior to installation of the foil, the surfaces are of emissivity $\varepsilon_o = 0.85$.

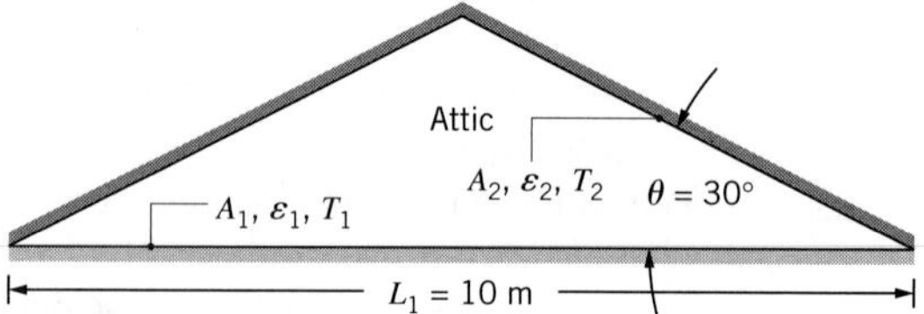

(a) Consider installation on the bottom of the attic roof only. Determine the ratio of the radiation heat transfer after to before the installation of the foil.

(b) Determine the ratio of the radiation heat transfer after to before installation if the foil is installed only on the top of the attic floor.

(c) Determine the ratio of the radiation heat transfer if the foil is installed on both the roof bottom and the floor top.

13.49 A long, thin-walled horizontal tube 100 mm in diameter is maintained at 120°C by the passage of steam through its interior. A radiation shield is installed around the tube, providing an air gap of 10 mm between the tube and the shield, and reaches a surface temperature of 35°C. The tube and shield are diffuse, gray surfaces with emissivities of 0.80 and 0.10, respectively. What is the radiant heat transfer from the tube per unit length?

13.50 A $t = 5$-mm-thick sheet of anodized aluminum is used to reject heat in a space power application. The edge of the sheet is attached to a hot source, and the sheet is maintained at nearly isothermal conditions at $T = 300$ K. The sheet is not subjected to irradiation.

(a) Determine the net radiation heat transfer from both sides of the 200 mm × 200 mm sheet to deep space.

(b) An engineer suggests boring 3-mm-diameter holes through the sheet. The holes are spaced

5 mm apart. The interior surfaces of the holes are anodized after they are bored. Determine the net radiation heat transfer from both sides of the sheet to deep space.

(c) As an alternative design, the 3-mm-diameter flat-bottomed holes are not bored completely through the sheet but are bored to depths of 2 mm on each side, leaving a 1-mm-thick web of aluminum separating the bottoms of the holes located on opposite sides of the sheet. Determine the net radiation heat transfer from both sides of the sheet to deep space.

(d) Compare the ratio of the net radiation heat transfer to the mass of the sheet for the three designs.

13.51 Consider the spacecraft heat rejection scheme of Problem 13.27, but under conditions for which surfaces 1 and 2 may not be approximated as blackbodies.

(a) For isothermal surfaces of temperature $T = 325$ K and emissivity $\varepsilon = 0.7$ and a U-section of width $W = 25$ mm and length $L = 125$ mm, determine the rate per unit length (normal to the page) at which radiation is transferred from a section to deep space.

(b) Explore the effect of the emissivity on the rate of heat rejection, and contrast your results with those for emission exclusively from the base of the section.

13.52 An electronic device dissipating 50 W is attached to the inner surface of an isothermal cubical container that is 120 mm on a side. The container is located in the much larger service bay of the space shuttle, which is evacuated and whose walls are at 150 K. If the outer surface of the container has an emissivity of 0.8 and the thermal resistance between the surface and the device is 0.1 K/W, what are the temperatures of the surface and the device? All surfaces of the container may be assumed to exchange radiation with the service bay, and heat transfer through the container restraint may be neglected.

13.53 A very long electrical conductor 10 mm in diameter is concentric with a cooled cylindrical tube 50 mm in diameter whose surface is diffuse and gray with an emissivity of 0.9 and temperature of 27°C. The electrical conductor has a diffuse, gray surface with an emissivity of 0.6 and is dissipating 6.0 W per meter of length. Assuming that the space between the two surfaces is evacuated, calculate the surface temperature of the conductor.

13.54 Liquid oxygen is stored in a thin-walled, spherical container 0.8 m in diameter, which is enclosed within a second thin-walled, spherical container 1.2 m in diameter. The opaque, diffuse, gray container surfaces have an emissivity of [illegible] and are separated by an evacuated space. If the outer surface is at 280 K and the inner surface is at 95 K, what is the mass rate of oxygen lost due to evaporation? (The latent heat of vaporization of oxygen is 2.13×10^5 J/kg.)

13.55 Two concentric spheres of diameter $D_1 = 0.8$ m and $D_2 = 1.2$ m are separated by an air space and have surface temperatures of $T_1 = 400$ K and $T_2 = 300$ K.

(a) If the surfaces are black, what is the net rate of radiation exchange between the spheres?

(b) What is the net rate of radiation exchange between the surfaces if they are diffuse and gray with $\varepsilon_1 = 0.5$ and $\varepsilon_2 = 0.05$?

(c) What is the net rate of radiation exchange if D_2 is increased to 20 m, with $\varepsilon_2 = 0.05$, $\varepsilon_1 = 0.5$, and $D_1 = 0.8$ m? What error would be introduced by assuming blackbody behavior for the outer surface ($\varepsilon_2 = 1$), with all other conditions remaining the same?

(d) For $D_2 = 1.2$ m and emissivities of $\varepsilon_1 = 0.1$, 0.5, and 1.0, compute and plot the net rate of radiation exchange as a function of ε_2 for $0.05 \leq \varepsilon_2 \leq 1.0$.

13.56 Determine the steady-state temperatures of two radiation shields placed in the evacuated space between two infinite planes at temperatures of 600 and 325 K. All the surfaces are diffuse and gray with emissivities of 0.7.

13.57 Consider two large (infinite) parallel planes that are diffuse-gray with temperatures and emissivities of T_1, ε_1 and T_2, ε_2. Show that the ratio of the radiation transfer rate with multiple shields, N, of emissivity ε_s to that with no shields, $N = 0$, is

$$\frac{q_{12,N}}{q_{12,0}} = \frac{[1/\varepsilon_1 + 1/\varepsilon_2 - 1]}{[1/\varepsilon_1 + 1/\varepsilon_2 - 1] + N[2/\varepsilon_s - 1]}$$

where $q_{12,N}$ and $q_{12,0}$ represent the radiation heat transfer rates for N shields and no shields, respectively.

13.58 Consider two large, diffuse, gray, parallel surfaces separated by a small distance. If the surface emissivities are 0.8, what emissivity should a thin radiation shield have to reduce the radiation heat transfer rate between the two surfaces by a factor of 10?

13.59 Heat transfer by radiation occurs between two large parallel plates, which are maintained at temperatures T_1 and T_2, with $T_1 > T_2$. To reduce the rate of heat transfer between the plates, it is proposed that they be separated by a thin shield tha[illegible] different emissivities on opposite surfaces. In p[illegible] one surface has the emissivity $\varepsilon_s < 0.5$, while the opposite surface has an emissivity of $2\varepsilon_s$.

(a) How should the shield be oriented to provide the larger reduction in heat transfer between the plates? That is, should the surface of emissivity ε_s or that of emissivity $2\varepsilon_s$ be oriented toward the plate at T_1?

(b) What orientation will result in the larger value of the shield temperature T_s?

13.60 The end of a cylindrical liquid cryogenic propellant tank in free space is to be protected from external (solar) radiation by placing a thin metallic shield in front of the tank. Assume the view factor F_{ts} between the tank and the shield is unity; all surfaces are diffuse and gray, and the surroundings are at 0 K.

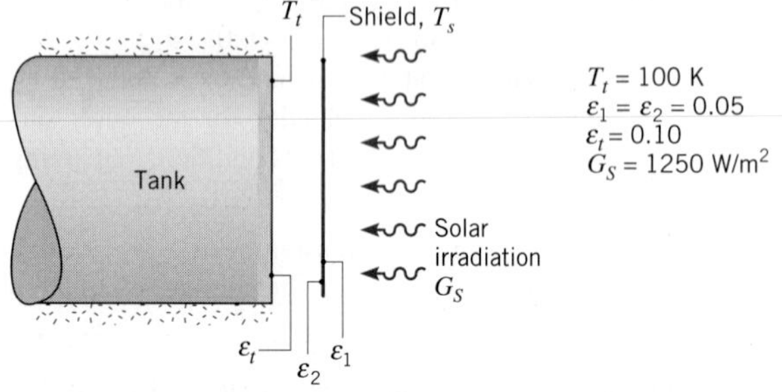

Find the temperature of the shield T_s and the heat flux (W/m²) to the end of the tank.

13.61 At the bottom of a very large vacuum chamber whose walls are at 300 K, a black panel 0.1 m in diameter is maintained at 77 K. To reduce the heat gain to this panel, a radiation shield of the same diameter D and an emissivity of 0.05 is placed very close to the panel. Calculate the net heat gain to the panel.

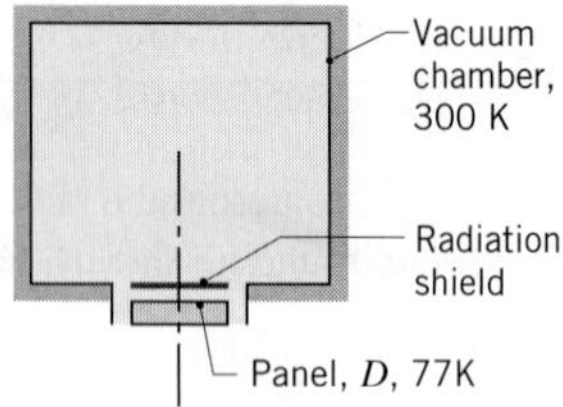

13.62 A furnace is located next to a dense array of cryogenic fluid piping. The ice-covered piping approximates a plane surface with an average temperature of $T_p = 0°C$ and an emissivity of $\varepsilon_p = 0.6$. The furnace wall has a temperature of $T_f = 200°C$ and an emissivity of $\varepsilon_f = 0.9$. To protect the refrigeration equipment and piping from excessive heat loading, reflective aluminum radiation shielding with an emissivity of $\varepsilon_s = 0.1$ is placed between the piping and the furnace wall, as shown in the schematic. Assume all surfaces are diffuse-gray.

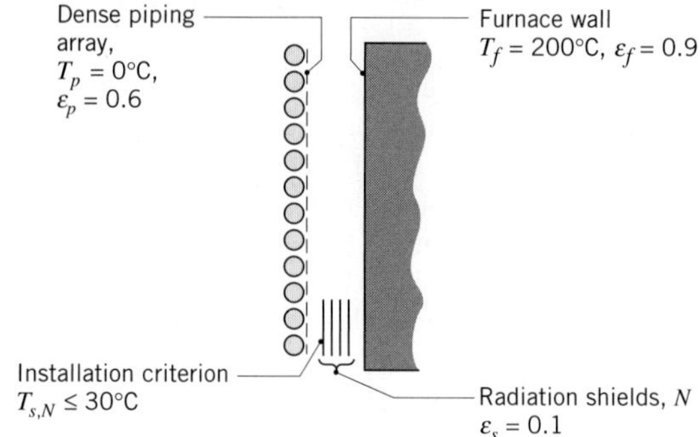

If the temperature of the shield closest to the piping $T_{s,N}$ must be less than 30°C, how many radiation shields, N, must be installed between the piping and the furnace wall?

13.63 A cryogenic fluid flows through a tube 20 mm in diameter, the outer surface of which is diffuse and gray with an emissivity of 0.02 and temperature of 77 K. This tube is concentric with a larger tube of 50-mm diameter, the inner surface of which is diffuse and gray with an emissivity of 0.05 and temperature of 300 K. The space between the surfaces is evacuated. Determine the heat gain by the cryogenic fluid per unit length of the inner tube. If a thin-walled radiation shield that is diffuse and gray with an emissivity of 0.02 (both sides) is inserted midway between the inner and outer surfaces, calculate the change (percentage) in heat gain per unit length of the inner tube.

13.64 A diffuse, gray radiation shield of 60-mm diameter and emissivities of $\varepsilon_{2,i} = 0.01$ and $\varepsilon_{2,o} = 0.1$ on the inner and outer surfaces, respectively, is concentric with a long tube transporting a hot process fluid. The tube surface is black with a diameter of 20 mm. The region interior to the shield is evacuated. The exterior surface of the shield is exposed to a large room whose walls are at 17°C and experiences convection with air at 27°C and a convection heat transfer coefficient of 10 W/m²·K.

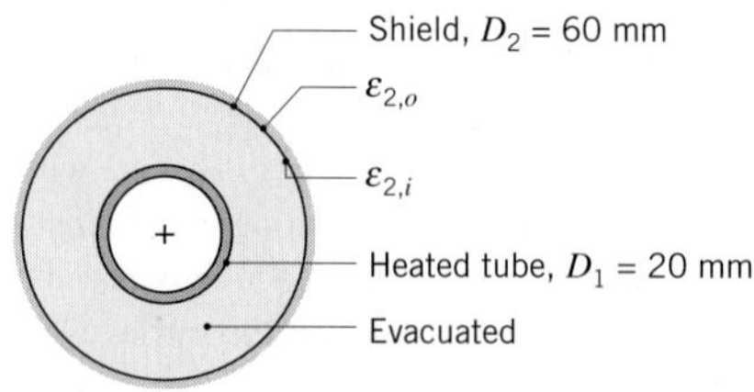

Determine the operating temperature for the inner tube if the shield temperature is maintained at 42°C.

Enclosures with a Reradiating Surface

13.65 Consider the three-surface enclosure shown. The lower plate (A_1) is a black disk of 200-mm diameter and is supplied with a heat rate of 10,000 W. The upper plate (A_2), a disk coaxial to A_1, is a diffuse, gray surface with $\varepsilon_2 = 0.8$ and is maintained at $T_2 = 473$ K. The diffuse, gray sides between the plates are perfectly insulated. Assume convection heat transfer is negligible.

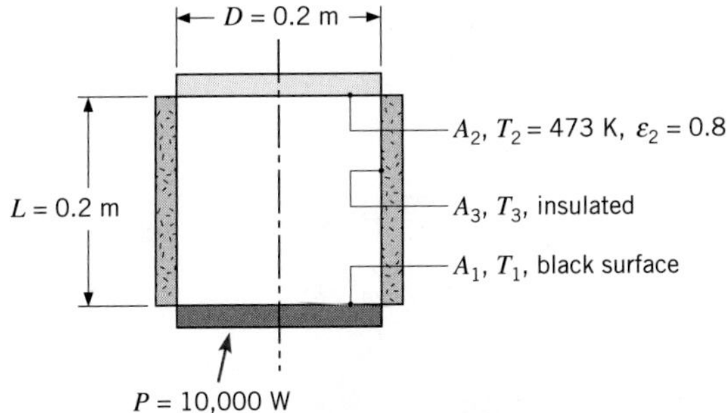

Determine the operating temperature of the lower plate T_1 and the temperature of the insulated side T_3.

13.66 A furnace has the form of a truncated conical section, as shown in the schematic. The floor of the furnace has an emissivity of $\varepsilon_1 = 0.7$ and is maintained at 1000 K with a heat flux of 2200 W/m^2. The lateral wall is perfectly insulated with an emissivity of $\varepsilon_3 = 0.3$. Assume that all the surfaces are diffuse-gray.

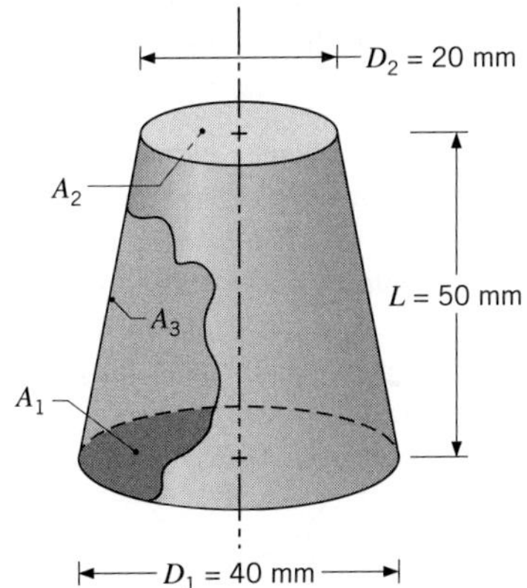

(a) Determine the temperature of the upper surface, T_2, if its emissivity is $\varepsilon_2 = 0.5$. What is the temperature of the lateral wall, T_3?

(b) If conditions for the furnace floor remain unchanged, but all surfaces are black instead of diffuse-gray, determine T_2 and T_3. From the diffuse and blackbody enclosure analyses, what can you say about the influence of ε_2 on your results?

13.67 Two paralle[illegible] 0.4 m in diameter and separated by [illegible]d in a large room whose walls are ma[illegible] K. One of the disks is maintained at [illegible]erature of 500 K with an emissivity of 0.[illegible]ack side of the second disk is well insulate[illegible]ks are diffuse, gray surfaces, determine the t[illegible]e of the insulated disk.

13.68 Coatings applied to long [illegible]ic strips are cured by installing the strips along the walls of a long oven of square cross section.

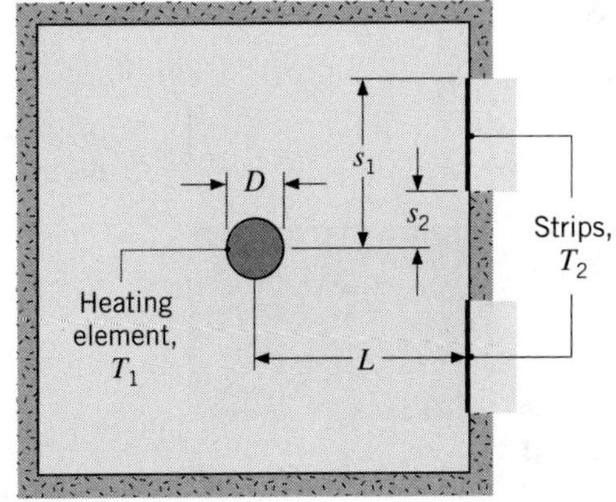

Thermal conditions in the oven are maintained by a long silicon carbide rod (heating element), which is of diameter $D = 20$ mm and is operated at $T_1 = 1700$ K. Each of two strips on one side wall has the same orientation relative to the rod ($s_1 = 60$ mm, $s_2 = 20$ mm, $L = 80$ mm) and is operated at $T_2 = 600$ K. All surfaces are diffuse and gray with $\varepsilon_1 = 0.9$ and $\varepsilon_2 = 0.4$. Assuming the oven to be well insulated at all but the strip surfaces and neglecting convection effects, determine the heater power requirement per unit length (W/m).

13.69 Long, cylindrical bars are heat-treated in an infrared oven. The bars, of diameter $D = 50$ mm, are placed on an insulated tray and are heated with an overhead infrared panel maintained at temperature $T_p = 800$ K with $\varepsilon_p = 0.85$. The bars are at $T_b = 300$ K and have an emissivity of $\varepsilon_b = 0.92$.

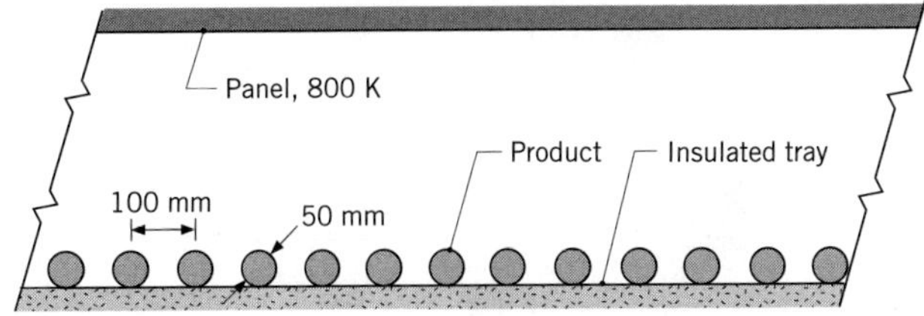

(a) For a product spacing of $s = 100$ mm and a product length of $L = 1$ m, determine the radiation heat flux delivered to the product. Determine the heat flux at the surface of the pane[illegible]ter.

(b) Plot the radiation heat flux experienced by the product and the panel heater radiation heat flux over the range 50 mm $\leq s \leq$ 250 mm.

13.70 A molten aluminum alloy at 900 K is poured into a cylindrical container that is well insulated from large surroundings at 300 K. The inner diameter of the container is 250 mm, and the distance from the surface of the melt to the top of the container is 100 mm.

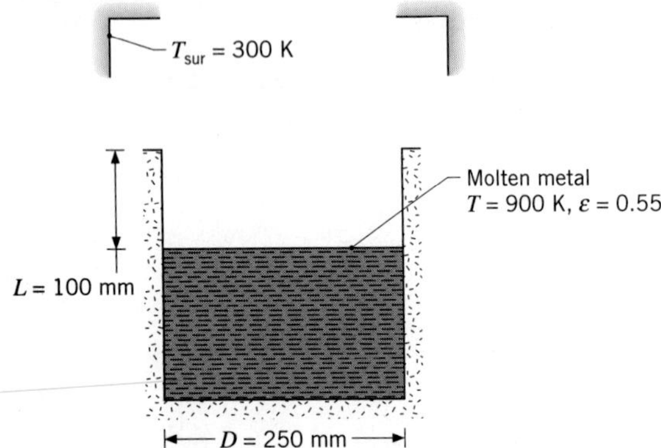

If the oxidized aluminum at the surface of the melt has an emissivity of 0.55, what is the net rate of radiation heat transfer from the melt?

13.71 A long, hemicylindrical (1-m radius) shaped furnace used to heat treat sheet metal products is comprised of three zones. The heating zone (1) is constructed from a ceramic plate of emissivity 0.85 and is operated at 1600 K by gas burners. The load zone (2) consists of sheet metal products, assumed to be black surfaces, that are to be maintained at 500 K. The refractory zone (3) is fabricated from insulating bricks having an emissivity of 0.6. Assume steady-state conditions, diffuse, gray surfaces, and negligible convection.

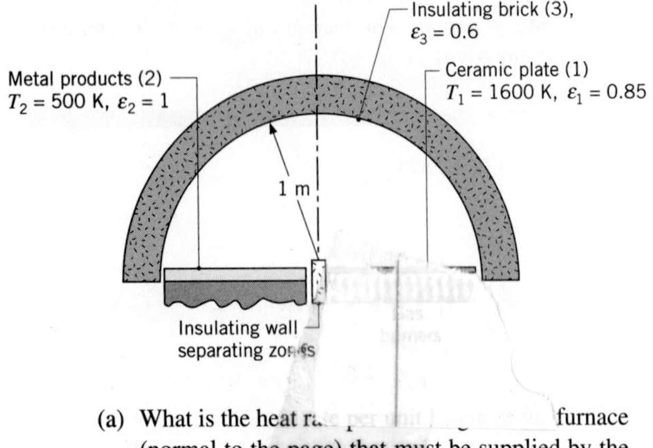

(a) What is the heat rate per unit length of the furnace (normal to the page) that must be supplied by the gas burners for the prescribed conditions?

(b) What is the temperature of the insulating brick surface for the prescribed conditions?

13.72 The bottom of a steam-producing still of 200-mm diameter is heated by radiation. The heater, maintained at 1000°C and separated 100 mm from the still, has the same diameter as the still bottom. The still bottom and heater surfaces are black.

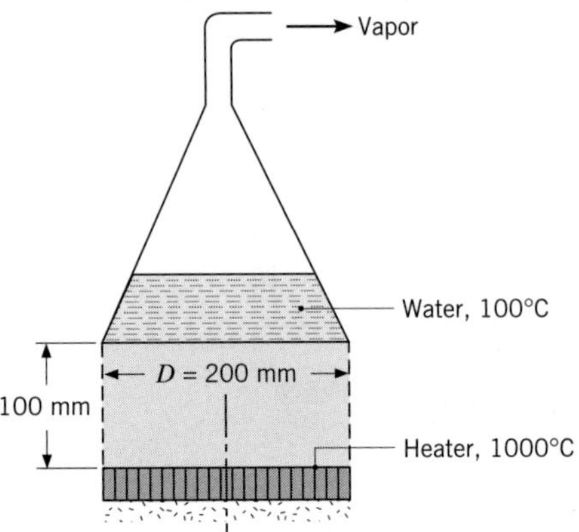

(a) By what factor could the vapor production rate be increased if the cylindrical sides (dashed surface) were insulated rather than open to the surroundings maintained at 27°C?

(b) For heater temperatures of 600, 800, and 1000°C, plot the net rate of radiation heat transfer to the still as a function of the separation distance over the range from 25 to 100 mm. Consider the cylindrical sides to be insulated and all other conditions to remain the same.

13.73 A long cylindrical heater element of diameter D = 10 mm, temperature T_1 = 1500 K, and emissivity ε_1 = 1 is used in a furnace. The bottom area A_2 is a diffuse, gray surface with ε_2 = 0.6 and is maintained at T_2 = 500 K. The side and top walls are fabricated from an insulating, refractory brick that is diffuse and gray with ε = 0.9. The length of the furnace normal to the page is very large compared to the width w and height h.

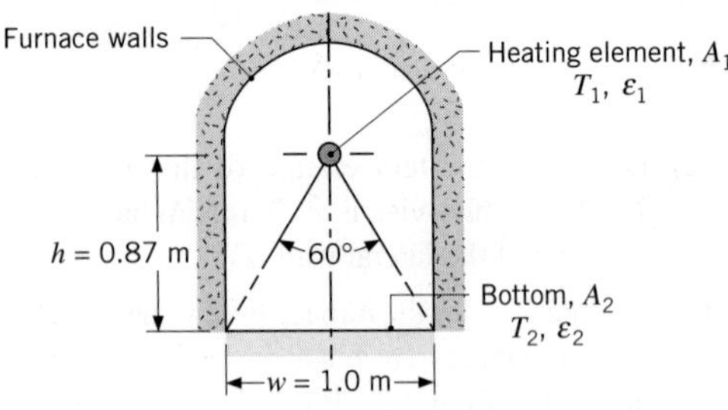

Neglecting convection and treating the furnace walls as isothermal, determine the power per unit length that must be provided to the heating element to maintain steady-state conditions. Calculate the temperature of the furnace wall.

13.74 A radiative heater consists of a bank of ceramic tubes with internal heating elements. The tubes are of diameter $D = 20$ mm and are separated by a distance $s = 50$ mm. A reradiating surface is positioned behind the heating tubes as shown in the schematic. Determine the net radiative heat flux to the heated material when the heating tubes ($\varepsilon_h = 0.87$) are maintained at 1000 K. The heated material ($\varepsilon_m = 0.26$) is at a temperature of 500 K.

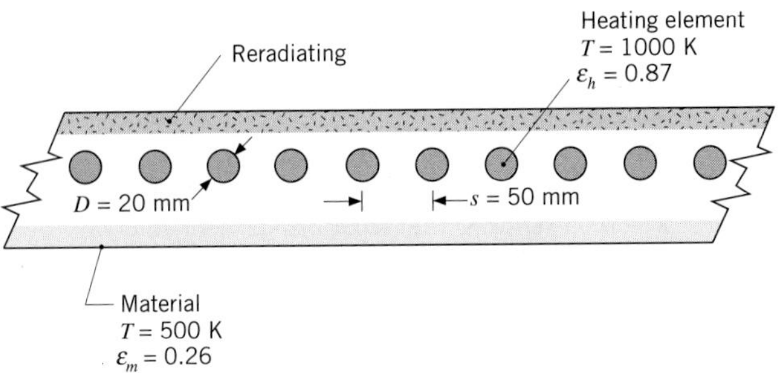

13.75 Consider a long duct constructed with diffuse, gray walls 1 m wide.

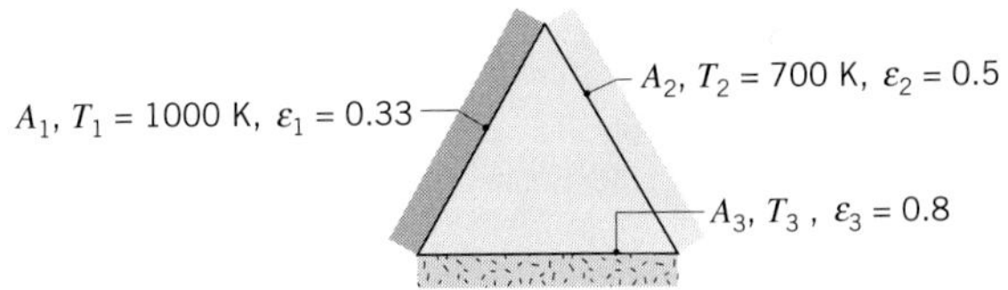

(a) Determine the net radiation transfer from surface A_1 per unit length of the duct.

(b) Determine the temperature of the insulated surface A_3.

(c) What effect would changing the value of ε_3 have on your result? After considering your assumptions, comment on whether you expect your results to be exact.

13.76 The coating on a 1 m × 2 m metallic surface is cured by placing it 0.5 m below an electrically heated surface of equivalent dimensions, and the assembly is exposed to large surroundings at 300 K. The heater is well insulated on its top side and is aligned with the coated surface. Both the heater and coated surfaces may be approximated as blackbodies. During the curing procedure, the heater and coated surfaces are maintained at 700 and 400 K, respectively. Convection effects may be neglected.

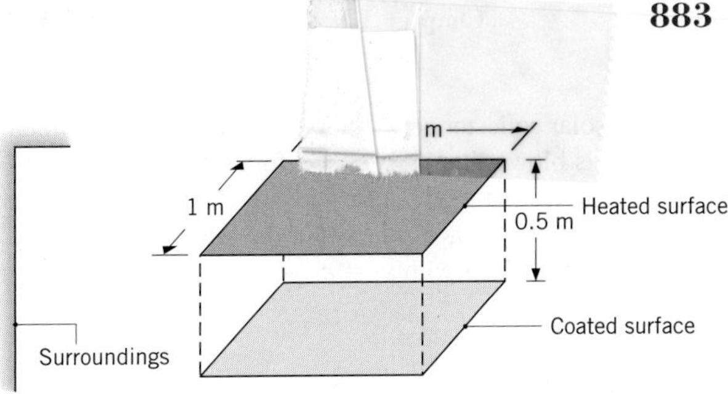

(a) What is the electrical power, q_{elec}, required to operate the heater?

(b) What is q_{elec} if the surfaces are connected with reradiating sidewalls?

(c) The facility may be used to cure different surface coatings. With the heater still approximated as a blackbody, compute and plot q_{elec} as a function of the coating's emissivity for values ranging from 0.1 to 1.0. Perform the calculations for both open and reradiating sides.

13.77 A cubical furnace 2 m on a side is used for heat treating steel plate. The top surface of the furnace consists of electrical radiant heaters that have an emissivity of 0.8 and a power input of 1.5×10^5 W. The sidewalls consist of a well-insulated refractory material, while the bottom consists of the steel plate, which has an emissivity of 0.4. Assume diffuse, gray surface behavior for the heater and the plate, and consider conditions for which the plate is at 300 K. What are the corresponding temperatures of the heater surface and the sidewalls?

13.78 An electric furnace consisting of two heater sections, top and bottom, is used to heat treat a coating that is applied to both surfaces of a thin metal plate inserted midway between the heaters.

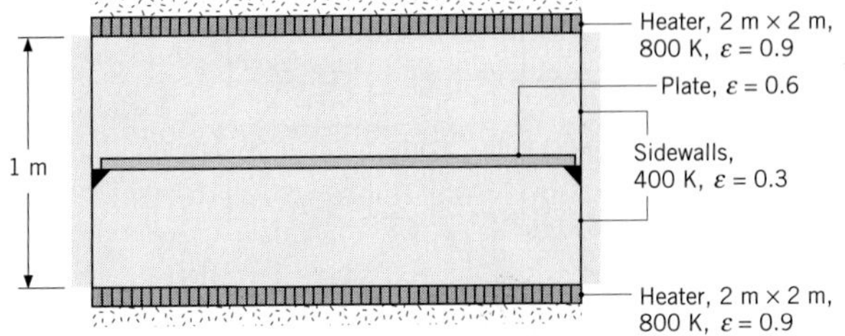

The heaters and the plate are 2 m × 2 m on a side, and each heater is separated from the plate by a distance of 0.5 m. Each heater is well insulated on its back side and has an emissivity of 0.9 at its exposed surface. The plate and the sidewalls have emissivities of 0.6 and 0.3, respectively. Sketch the equivalent radiation network for the system and label all pertinent resistances and potentials. For the prescribed conditions, obtain the required electrical power and the plate temperature.

13.79 A solar collector consists of a long duct through which air is blown; its cross section forms an equilateral triangle 1 m on a side. One side consists of a glass cover of emissivity $\varepsilon_1 = 0.9$, while the other two sides are absorber plates with $\varepsilon_2 = \varepsilon_3 = 1.0$.

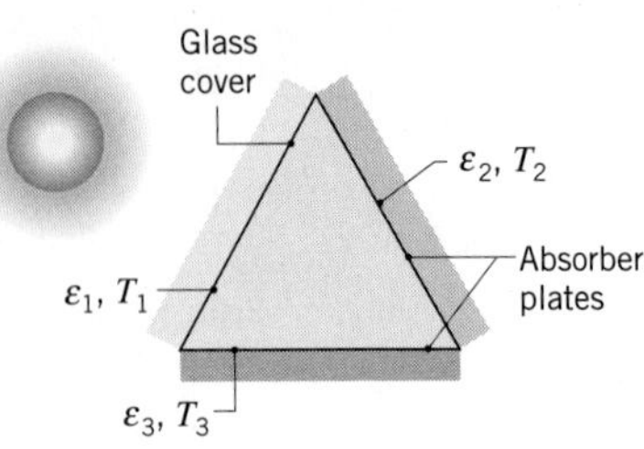

During operation the surface temperatures are known to be $T_1 = 25°C$, $T_2 = 60°C$, and $T_3 = 70°C$. What is the net rate at which radiation is transferred to the cover due to exchange with the absorber plates?

13.80 A frictionless piston of diameter $D_p = 50$ mm, thickness $\delta = 10$ mm, and density $\rho = 8000\ \text{kg/m}^3$ is placed in a vertical cylinder of total length $L = 100$ mm. The cylinder is bored into a material of very low thermal conductivity, and a mass of air $M_a = 75 \times 10^{-6}$ kg is enclosed in the volume beneath the cylinder. The surroundings are at $T_{sur} = 300$ K while the bottom surface of the cylinder is maintained at T_1. The emissivity of all the surfaces is $\varepsilon = 0.3$.

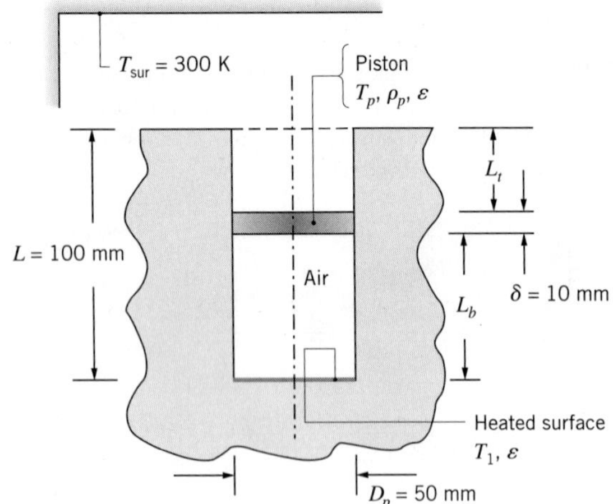

What is the distance between the bottom of the piston and the bottom of the cylinder, L_b, for $T_1 = 300$, 450, and 600 K? What is the piston temperature? *Hint*: Neglect convection heat transfer and assume that the average air temperature is $\overline{T}_a = (T_p + T_1)/2$.

13.81 The cylindrical peephole in a furnace wall of thickness $L = 250$ mm has a diameter of $D = 125$ mm. The furnace interior has a temperature of 1300 K, and the surroundings outside the furnace have a temperature of 300 K.

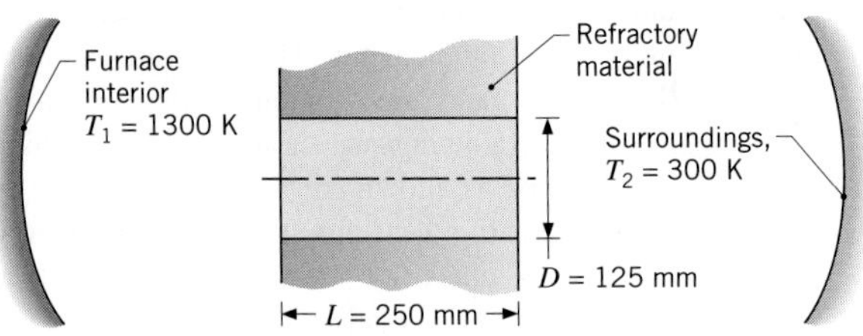

Determine the heat loss by radiation through the peephole.

13.82 A composite wall is comprised of two large plates separated by sheets of refractory insulation, as shown in the schematic. In the installation process, the sheets of thickness $L = 50$ mm and thermal conductivity $k = 0.05\ \text{W/m}\cdot\text{K}$ are separated at 1-m intervals by gaps of width $w = 10$ mm. The hot and cold plates have temperatures and emissivities of $T_1 = 400°C$, $\varepsilon_1 = 0.85$ and $T_2 = 35°C$, $\varepsilon_2 = 0.5$, respectively. Assume that the plates and insulation are diffuse-gray surfaces.

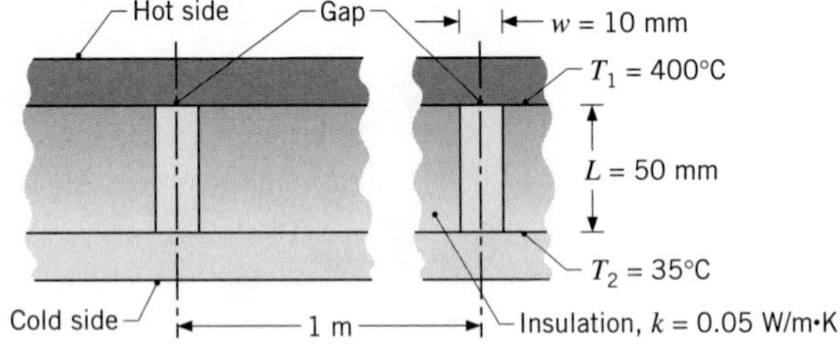

(a) Determine the heat loss by radiation through the gap per unit length of the composite wall (normal to the page).

(b) Recognizing that the gaps are located on a 1-m spacing, determine what fraction of the total heat loss through the composite wall is due to transfer by radiation through the insulation gap.

13.83 A small disk of diameter $D_1 = 50$ mm and emissivity $\varepsilon_1 = 0.6$ is maintained at a temperature of $T_1 = 900$ K. The disk is covered with a hemispherical radiation shield of the same diameter and an emissivity of $\varepsilon_2 = 0.02$ (both sides). The disk and cap are located at the bottom of a large evacuated refractory container ($\varepsilon_4 = 0.85$), facing another disk of diameter $D_3 = D_1$, emissivity $\varepsilon_3 = 0.4$, and temperature $T_3 = 400$ K. The view factor F_{23} of the shield with respect to the upper disk is 0.3.

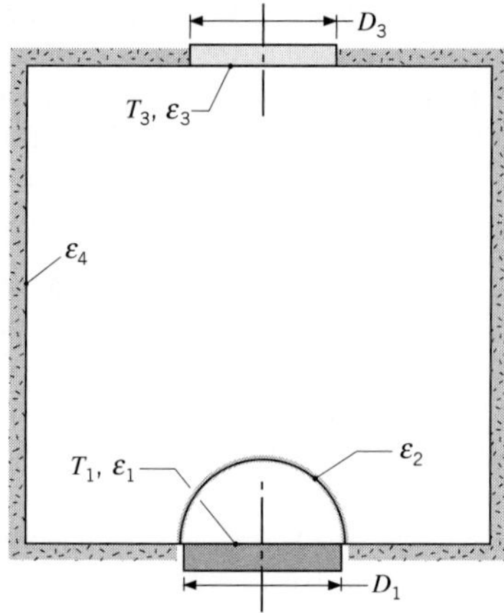

(a) Construct an equivalent thermal circuit for the above system. Label all nodes, resistances, and currents.

(b) Find the net rate of heat transfer between the hot disk and the rest of the system.

Enclosures: Three or More Surfaces

13.84 Consider a cylindrical cavity of diameter $D = 100$ mm and depth $L = 50$ mm whose sidewall and bottom are diffuse and gray with an emissivity of 0.6 and are at a uniform temperature of 1500 K. The top of the cavity is open and exposed to surroundings that are large and at 300 K.

(a) Calculate the net radiation heat transfer from the cavity, treating the bottom and sidewall of the cavity as one surface (q_A).

(b) Calculate the net radiation heat transfer from the cavity, treating the bottom and sidewall of the cavity as two separate surfaces (q_B).

(c) Plot the percentage difference between q_A and q_B as a function of L over the range 5 mm $\leq L \leq$ 100 mm.

13.85 Consider a circular furnace that is 0.3 m long and 0.3 m in diameter. The two ends have diffuse, gray surfaces that are maintained at 400 and 500 K with emissivities of 0.4 and 0.5, respectively. The lateral surface is also diffuse and gray with an emissivity of 0.8 and a temperature of 800 K. Determine the net radiative heat transfer from each of the surfaces.

13.86 Consider two very large metal parallel plates. The top plate is at a temperature $T_t = 400$ K while the bottom plate is at $T_b = 300$ K. The desired net radiation heat flux between the two plates is $q'' = 330$ W/m^2.

(a) If the two surfaces have the same radiative properties, show that the required surface emissivity is $\varepsilon = 0.5$.

(b) Metal surfaces at relatively low temperatures tend to have emissivities much less than 0.5 (see Table A.11). An engineer proposes to apply a checker pattern, similar to that of Problem 12.132, onto each of the metal surfaces so that half of each surface is characterized by the low emissivity of the bare metal and the other half is covered with the high-emissivity paint. If the average of the high and low emissivities is 0.5, will the net radiative heat flux between the surfaces be the desired value?

13.87 Two convex objects are inside a large vacuum enclosure whose walls are maintained at $T_3 = 300$ K. The objects have the same area, 0.2 m^2, and the same emissivity, 0.2. The view factor from object 1 to object 2 is $F_{12} = 0.3$. Embedded in object 2 is a heater that generates 400 W. The temperature of object 1 is maintained at $T_1 = 200$ K by circulating a fluid through channels machined into it. At what rate must heat be supplied (or removed) by the fluid to maintain the desired temperature of object 1? What is the temperature of object 2?

13.88 Consider the diffuse, gray, four-surface enclosure with all sides equal as shown. The temperatures of three surfaces are specified, while the fourth surface is well insulated and can be treated as a reradiating surface. Determine the temperature of the reradiating surface (4).

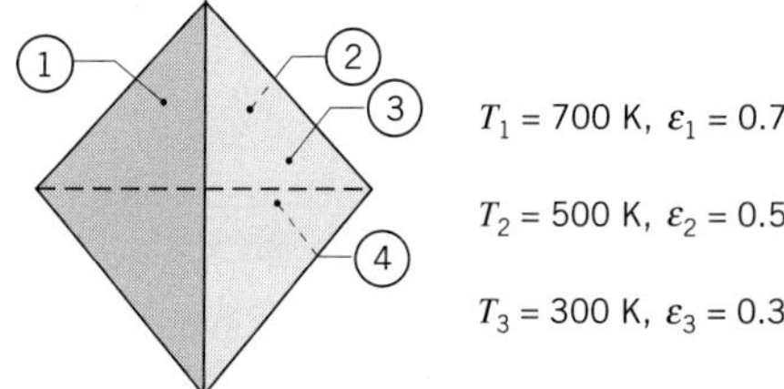

13.89 A room is represented by the following enclosure, where the ceiling (1) has an emissivity of 0.8 and is maintained at 40°C by embedded electrical heating elements. Heaters are also used to maintain the floor (2) of emissivity 0.9 at 50°C. The right wall (3) of emissivity 0.7 reaches a temperature of 15°C on a cold winter day. The left wall (4) and end walls (5A) and (5B) are very well insulated. To simplify the analysis, treat the two end walls as a single surface (5).

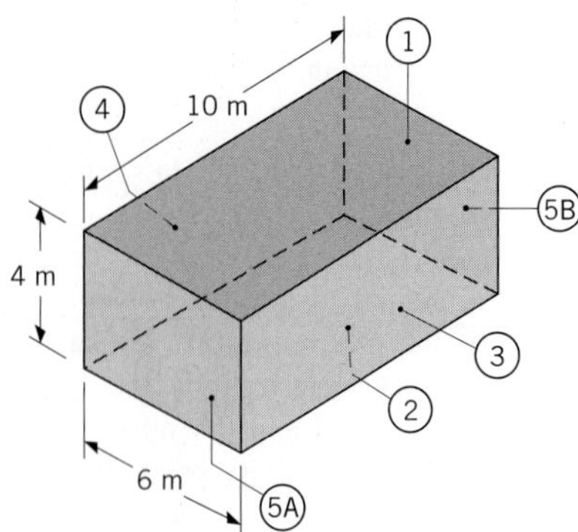

Assuming the surfaces are diffuse and gray, find the net radiation heat transfer from each surface.

13.90 A cylindrical furnace for heat-treating materials in a spacecraft environment has a 90-mm diameter and an overall length of 180 mm. Heating elements in the 135-mm-long section (1) maintain a refractory lining of $\varepsilon_1 = 0.8$ at 800°C. The linings for the bottom (2) and upper (3) sections are made of the same refractory material, but are insulated.

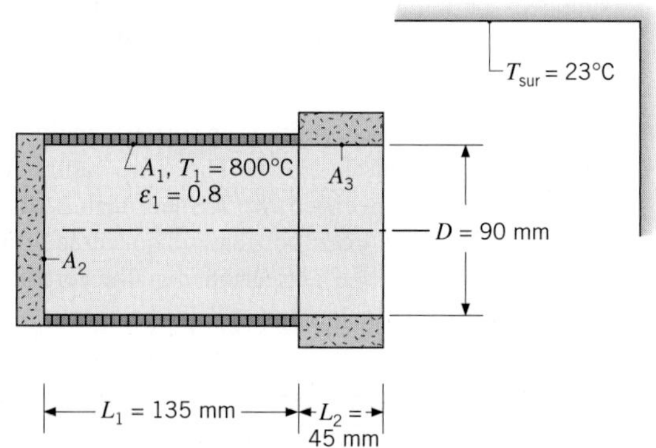

Determine the power required to maintain the furnace operating conditions with the surroundings at 23°C.

13.91 A laboratory oven has a cubical interior chamber 1 m on a side with interior surfaces that are of emissivity $\varepsilon = 0.85$.

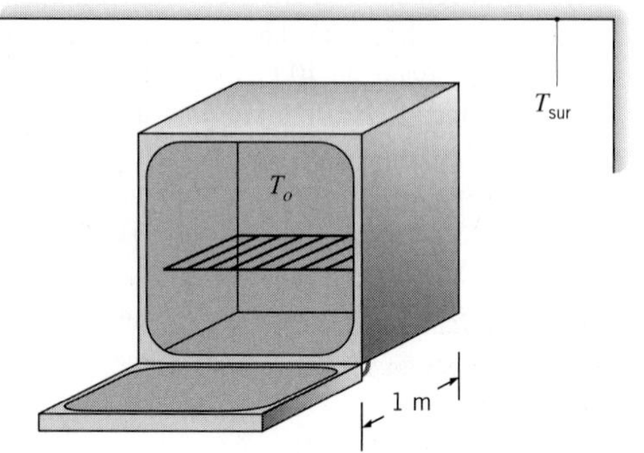

Determine the initial rate of radiation heat transfer to the laboratory in which the oven is placed when the oven door is opened. The oven and surroundings temperatures are $T_o = 375°C$ and $T_{sur} = 20°C$, respectively. Treat the oven door as one surface and the remaining five interior furnace walls as another.

13.92 An observation cabin is located in a hot-strip mill directly over the line, as shown in the schematic. The cabin floor is exposed to a portion of the hot strip that is at a temperature of $T_{ss} = 920°C$ and has an emissivity of $\varepsilon_{ss} = 0.85$, as well as to *surroundings* within the mill (not shown) at a temperature of $T_{sur} = 80°C$. To protect the operating personnel in the cabin, the floor must be maintained at $T_f = 50°C$.

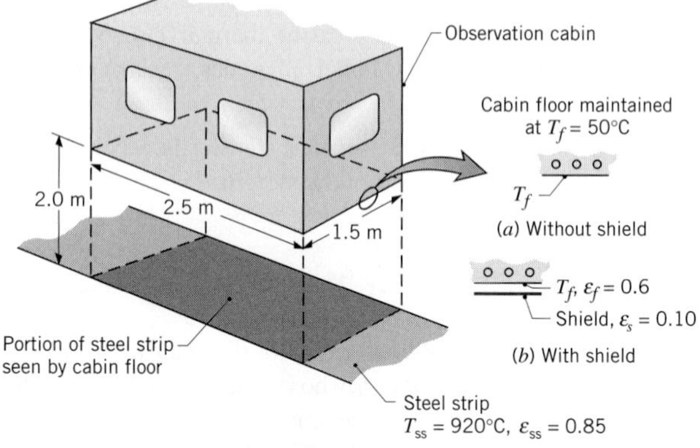

Determine the coolant system heat removal rate required to maintain the cabin floor at 50°C for the following conditions: (a) when the floor, with an emissivity of $\varepsilon_f = 0.6$, is directly exposed to the strip; and (b) when a radiation shield of emissivity $\varepsilon_s = 0.1$ is installed between the floor and the strip.

13.93 A small oven consists of a cubical box of dimension $L = 0.1$ m, as shown. The floor of the box consists of a heater that supplies $P = 400$ W. The remaining walls lose heat to the surroundings outside the oven, which maintains their temperatures at $T_3 = 400$ K. A spherical object of diameter $D = 30$ mm is placed at the center of the oven. Sometime after the sphere is placed in the oven, its temperature is $T_1 = 420$ K. All surfaces have emissivities of 0.4.

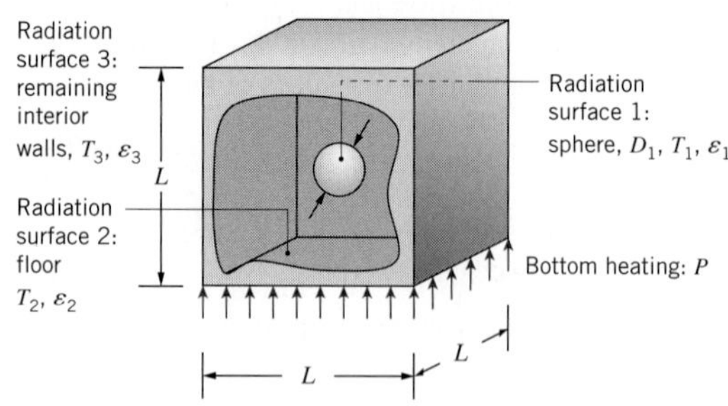

(a) Find the following view factors: F_{12}, F_{13}, F_{21}, F_{31}, F_{23}, F_{32}, F_{33}.

(b) Determine the temperature of the floor and the net rate of heat transfer leaving the sphere due to radiation. Is the sphere under steady-state conditions?

Multimode Heat Transfer: Introductory

13.94 An opaque, diffuse, gray (200 mm × 200 mm) plate with an emissivity of 0.8 is placed over the opening of a furnace and is known to be at 400 K at a certain instant. The bottom of the furnace, having the same dimensions as the plate, is black and operates at 1000 K. The sidewalls of the furnace are well insulated. The top of the plate is exposed to ambient air with a convection coefficient of 25 W/m²·K and to large surroundings. The air and surroundings are each at 300 K.

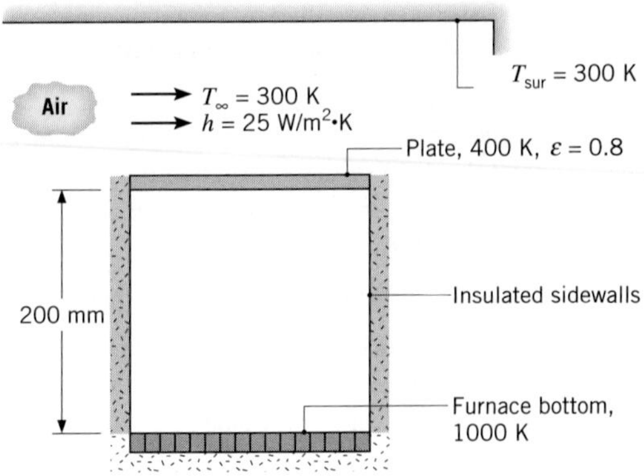

(a) Evaluate the net radiative heat transfer to the bottom surface of the plate.

(b) If the plate has mass and specific heat of 2 kg and 900 J/kg·K, respectively, what will be the change in temperature of the plate with time, dT_p/dt? Assume convection to the bottom surface of the plate to be negligible.

(c) Extending the analysis of part (b), generate a plot of the change in temperature of the plate with time, dT_p/dt, as a function of the plate temperature for $350 \leq T_p \leq 900$ K and all other conditions remaining the same. What is the steady-state temperature of the plate?

13.95 A tool for processing silicon wafers is housed within a vacuum chamber whose walls are black and maintained by a coolant at $T_{vc} = 300$ K. The thin silicon wafer is mounted close to, but not touching, a chuck, which is electrically heated and maintained at the temperature T_c. The surface of the chuck facing the wafer is black. The wafer temperature is $T_w = 700$ K, and its surface is diffuse and gray with an emissivity of $\varepsilon_w = 0.6$. The function of the grid, a thin metallic foil positioned coaxial with the wafer and of the same diameter, is to control the power of the ion beam reaching the wafer. The grid surface is black with a temperature of $T_g = 500$ K. The effect of the ion beam striking the wafer is to apply a uniform heat flux of $q''_{ib} = 600$ W/m². The top surface of the wafer is subjected to the flow of a process gas for which $T_\infty = 500$ K and $h = 10$ W/m²·K. Since the gap between the wafer and chuck, δ, is very small, flow of the process gas in this region may be neglected.

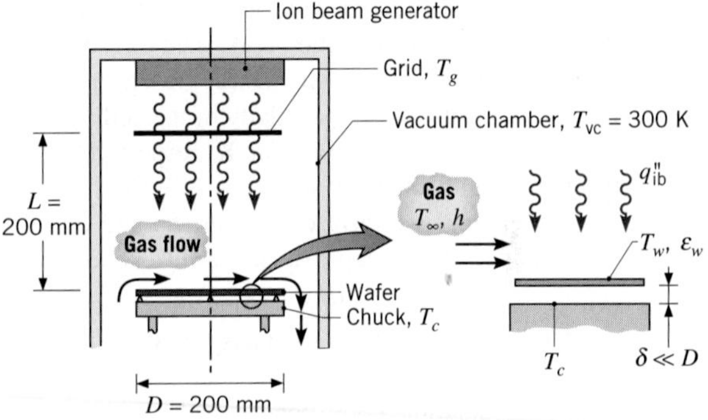

(a) Represent the wafer schematically, showing a control surface and all relevant thermal processes.

(b) Perform an energy balance on the wafer and determine the chuck temperature T_c.

13.96 Consider Problem 6.17. The stationary plate, ambient air, and surroundings are at $T_\infty = T_{sur} = 20°$C. If the rotating disk temperature is $T_s = 80°$C, what is the total power dissipated from the disk's top surface for $g = 2$ mm, $\Omega = 150$ rad/s for the case when both the stationary plate and disk are painted with Parsons black paint? Over time, the paint on the rotating disk is worn off by dust in the air, exposing the base metal, which has an emissivity of $\varepsilon = 0.10$. Determine the total power dissipated from the disk's worn top surface.

13.97 Most architects know that the ceiling of an ice-skating rink must have a high reflectivity. Otherwise, condensation may occur on the ceiling, and water may drip onto the ice, causing bumps on the skating surface. Condensation will occur on the ceiling when its surface temperature drops below the dew point of the rink air. Your assignment is to perform an analysis to determine the effect of the ceiling emissivity on the ceiling temperature, and hence the propensity for condensation.

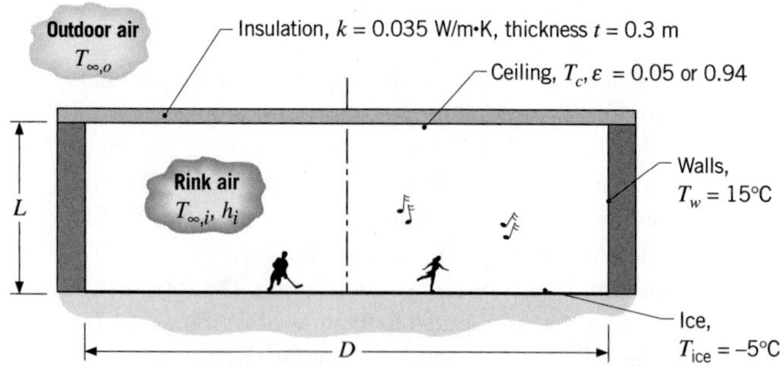

The rink has a diameter of $D = 50$ m and a height of $L = 10$ m, and the temperatures of the ice and walls are $-5°C$ and 15°C, respectively. The rink air temperature is 15°C, and a convection coefficient of 5 W/m$^2 \cdot$K characterizes conditions on the ceiling surface. The thickness and thermal conductivity of the ceiling insulation are 0.3 m and 0.035 W/m·K, respectively, and the temperature of the outdoor air is $-5°C$. Assume that the ceiling is a diffuse-gray surface and that the walls and ice may be approximated as blackbodies.

(a) Consider a flat ceiling having an emissivity of 0.05 (highly reflective panels) or 0.94 (painted panels). Perform an energy balance on the ceiling to calculate the corresponding values of the ceiling temperature. If the relative humidity of the rink air is 70%, will condensation occur for either or both of the emissivities?

(b) For each of the emissivities, calculate and plot the ceiling temperature as a function of the insulation thickness for $0.1 \leq t \leq 1$ m. Identify conditions for which condensation will occur on the ceiling.

13.98 Boiler tubes exposed to the products of coal combustion in a power plant are subject to fouling by the ash (mineral) content of the combustion gas. The ash forms a solid deposit on the tube outer surface, which reduces heat transfer to a pressurized water/steam mixture flowing through the tubes. Consider a thin-walled boiler tube ($D_t = 0.05$ m) whose surface is maintained at $T_t = 600$ K by the boiling process. Combustion gases flowing over the tube at $T_\infty = 1800$ K provide a convection coefficient of $\bar{h} = 100$ W/m$^2 \cdot$K, while radiation from the gas and boiler walls to the tube may be approximated as that originating from large surroundings at $T_{sur} = 1500$ K.

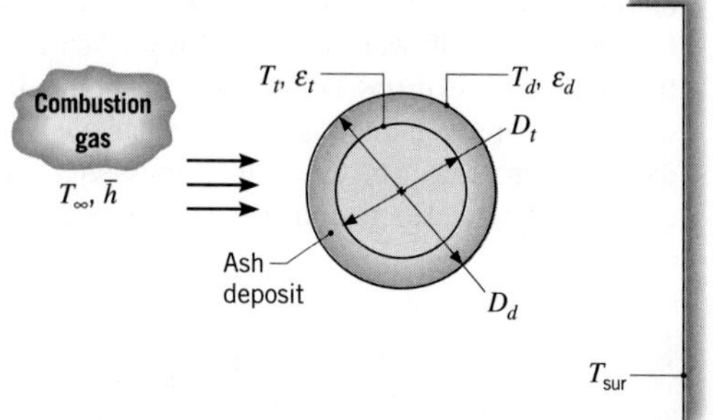

(a) If the tube surface is diffuse and gray, with $\varepsilon_t = 0.8$, and there is *no* ash deposit layer, what is the rate of heat transfer per unit length, q', to the boiler tube?

(b) If a deposit layer of diameter $D_d = 0.06$ m and thermal conductivity $k = 1$ W/m·K forms on the tube, what is the deposit surface temperature, T_d? The deposit is diffuse and gray, with $\varepsilon_d = 0.9$, and T_t, T_∞, $\bar{h}$, and T_{sur} remain unchanged. What is the net rate of heat transfer per unit length, q', to the boiler tube?

(c) Explore the effect of variations in D_d and $\bar{h}$ on q', as well as on relative contributions of convection and radiation to the net heat transfer rate. Represent your results graphically.

13.99 Consider two very large parallel plates. The bottom plate is warmer than the top plate, which is held at a constant temperature of $T_1 = 330$ K. The plates are separated by $L = 0.1$ m, and the gap between the two surfaces is filled with air at atmospheric pressure. The heat flux from the bottom plate is $q'' = 250$ W/m^2.

(a) Determine the temperature of the bottom plate and the ratio of the convective to radiative heat fluxes for $\varepsilon_1 = \varepsilon_2 = 0.5$. Evaluate air properties at $T = 350$ K.

(b) Repeat part (a) for $\varepsilon_1 = \varepsilon_2 = 0.25$ and 0.75.

13.100 Coated metallic disks are cured by placing them at the top of a cylindrical furnace whose bottom surface is electrically heated and whose sidewall may be approximated as a reradiating surface. Curing is accomplished by maintaining a disk at $T_2 = 400$ K for an extended period. The electrically heated surface is maintained at $T_1 = 800$ K and is mounted on a ceramic base material of thermal conductivity $k = 20$ W/m·K. The bottom of the base material, as well as the ambient air and large surroundings above the disk, are maintained at a temperature of 300 K. Emissivities of the heater and the disk inner and outer surfaces are $\varepsilon_1 = 0.9$, $\varepsilon_{2,i} = 0.5$, and $\varepsilon_{2,o} = 0.9$, respectively.

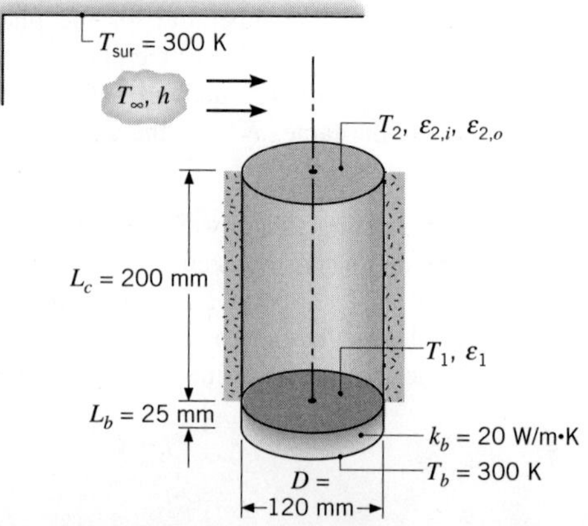

Assuming steady-state operation and neglecting convection within the cylindrical cavity, determine the electrical power that must be supplied to the heater and the convection coefficient h that must be maintained at the outer surface of the disk in order to satisfy the prescribed conditions.

13.101 A double-glazed window consists of two panes of glass, each of thickness $t = 6$ mm. The inside room temperature is $T_i = 20°C$ with $h_i = 7.7\ W/m^2 \cdot K$, while the outside temperature is $T_o = -10°C$ with $h_o = 25\ W/m^2 \cdot K$. The gap between the glass sheets is of thickness $L = 5$ mm and is filled with a gas. The glass surfaces may be treated with a low-emissivity coating to reduce their emissivity from $\varepsilon = 0.95$ to $\varepsilon = 0.05$. Determine the heat flux through the window for case 1: $\varepsilon_1 = \varepsilon_2 = 0.95$, case 2: $\varepsilon_1 = \varepsilon_2 = 0.05$, and case 3: $\varepsilon_1 = 0.05$, $\varepsilon_2 = 0.95$. Consider either air or argon of thermal conductivity $k_{Ar} = 17.7 \times 10^{-3}$ W/m·K to be within the gap. Radiation heat transfer occurring at the external surfaces of the two glass sheets is negligible, as is free convection between the glass sheets.

13.102 Electrical conductors, in the form of parallel plates of length $L = 40$ mm, have one edge mounted to a ceramic insulated base and are spaced a distance $w = 10$ mm apart. The plates are exposed to large isothermal surroundings at $T_{sur} = 300$ K. The conductor (1) and ceramic (2) surfaces are diffuse and gray with emissivities of $\varepsilon_1 = 0.8$ and $\varepsilon_2 = 0.6$, respectively. For a prescribed operating current in the conductors, their temperature is $T_1 = 500$ K.

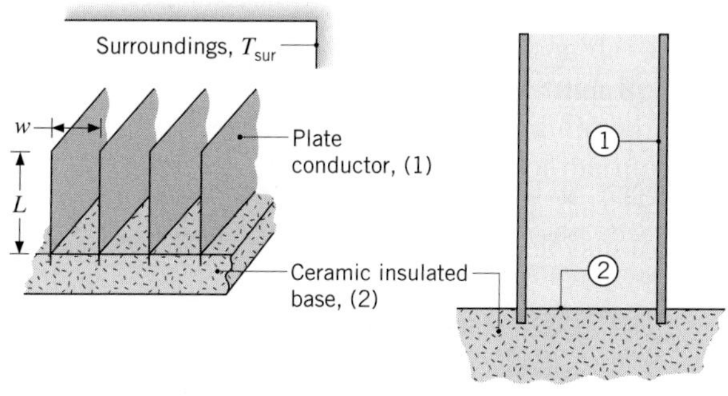

(a) Determine the electrical power dissipated in a conductor plate per unit length, q_1', considering only radiation exchange. What is the temperature of the insulated base, T_2?

(b) Determine q_1' and T_2 when the surfaces experience convection with an airstream at 300 K and a convection coefficient of $h = 25\ W/m^2 \cdot K$.

13.103 The spectral absorptivity of a *large* diffuse surface is $\alpha_\lambda = 0.9$ for $\lambda < 1\ \mu m$ and $\alpha_\lambda = 0.3$ for $\lambda \geq 1\ \mu m$. The bottom of the surface is well insulated, while the top may be exposed to one of two different conditions.

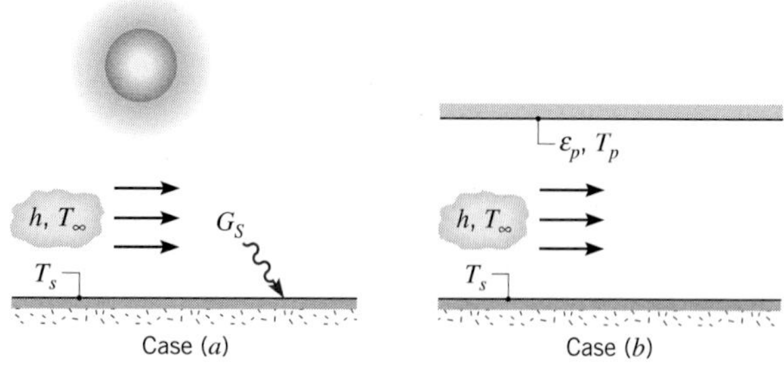

(a) In case (*a*) the surface is exposed to the sun, which provides an irradiation of $G_S = 1200\ W/m^2$, and to an airflow for which $T_\infty = 300$ K. If the surface temperature is $T_s = 320$ K, what is the convection coefficient associated with the airflow?

(b) In case (*b*) the surface is shielded from the sun by a large plate and an airflow is maintained between the plate and the surface. The plate is diffuse and gray with an emissivity of $\varepsilon_p = 0.8$. If $T_\infty = 300$ K and the convection coefficient is equivalent to the result obtained in part (a), what is the plate temperature T_p that is needed to maintain the surface at $T_s = 320$ K?

13.104 A long uniform rod of 50-mm diameter with a thermal conductivity of 15 W/m·K is heated internally by volumetric energy generation of 20 kW/m³. The rod is positioned coaxially within a larger circular tube of 60-mm diameter whose surface is maintained at 500°C. The annular region between the rod and the tube is evacuated, and their surfaces are diffuse and gray with an emissivity of 0.2.

(a) Determine the center and surface temperatures of the rod.

(b) Determine the center and surface temperatures of the rod if atmospheric air occupies the annular space.

(c) For tube diameters of 60, 100, and 1000 mm and for both the evacuated and atmospheric conditions, compute and plot the center and surface temperatures as a function of equivalent surface emissivities in the range from 0.1 to 1.0.

13.105 The cross section of a long circular tube, which is divided into two semicylindrical ducts by a thin wall, is shown.

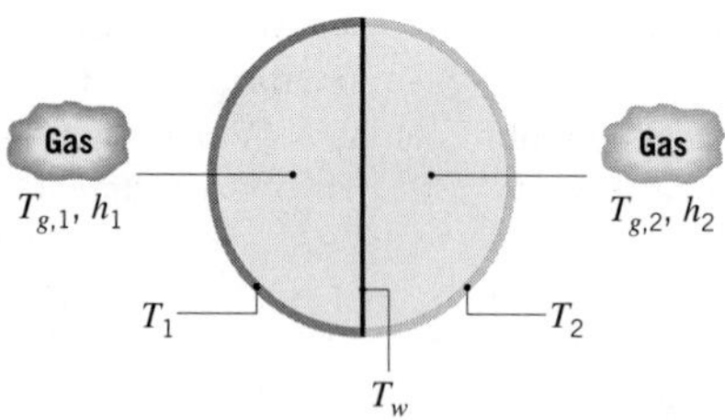

Sides 1 and 2 are maintained at temperatures of $T_1 = 600$ K and $T_2 = 400$ K, respectively, while the mean temperatures of gas flows through ducts 1 and 2 are $T_{g,1} = 571$ K and $T_{g,2} = 449$ K, respectively. The foregoing temperatures are invariant in the axial direction. The gases provide surface convection coefficients of $h_1 = h_2 = 5$ W/m$^2 \cdot$K, while all duct surfaces may be approximated as blackbodies ($\varepsilon_1 = \varepsilon_2 = \varepsilon_w = 1$). What is the duct wall temperature, T_w? By performing an energy balance on the gas in side 1, verify that $T_{g,1}$ is, in fact, equal to 571 K.

13.106 Cylindrical pillars similar to those of Problem 4.22 are positioned between the glass sheets with a pillar-to-pillar spacing of W. The inside surface of one glass sheet is treated with a low-emissivity coating characterized by $\varepsilon_1 = 0.05$. The second inner surface has an emissivity of $\varepsilon_2 = 0.95$. Determine the ratio of conduction to radiation heat transfer through a square unit area of dimension $W \times W$ for $W = 10$, 20, and 30 mm. The stainless steel pillar is located at the center of the unit area and has a length of $L = 0.4$ mm and diameter of $D = 0.15$ mm. The contact resistances and glass temperatures are the same as in Problem 4.22.

13.107 A row of regularly spaced, cylindrical heating elements (1) is used to cure a surface coating that is applied to a large panel (2) positioned below the elements. A second large panel (3), whose top surface is well insulated, is positioned above the elements. The elements are black and maintained at $T_1 = 600$ K, while the panel has an emissivity of $\varepsilon_2 = 0.5$ and is maintained at $T_2 = 400$ K. The cavity is filled with a nonparticipating gas and convection heat transfer occurs at surfaces 1 and 2, with $\bar{h}_1 = 10$ W/m$^2 \cdot$K and $\bar{h}_2 = 2$ W/m$^2 \cdot$K. (Convection at the insulated panel may be neglected.)

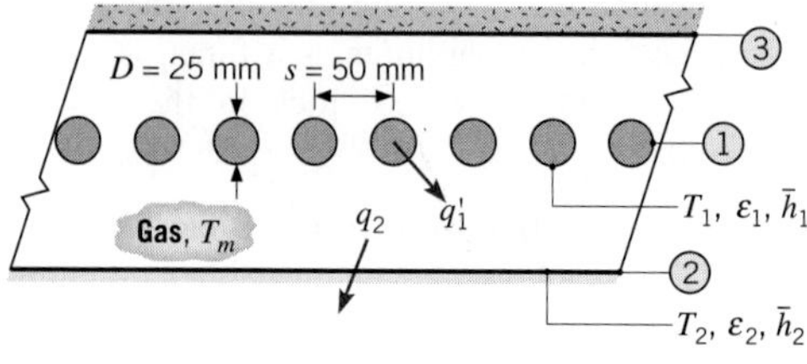

(a) Evaluate the mean gas temperature, T_m.

(b) What is the rate per unit axial length at which electrical energy must be supplied to each element to maintain its prescribed temperature?

(c) What is the rate of heat transfer to a portion of the coated panel that is 1 m wide by 1 m long?

13.108 Applying high-emissivity paints to radiating surfaces is a common technique used to enhance heat transfer by radiation.

(a) For large parallel plates, determine the radiation heat flux across the gap when the surfaces are at $T_1 = 350$ K, $T_2 = 300$ K, $\varepsilon_1 = \varepsilon_2 = \varepsilon_s = 0.85$.

(b) Determine the radiation heat flux when a very thin layer of high-emissivity paint, $\varepsilon_p = 0.98$, is applied to both surfaces.

(c) Determine the radiation heat flux when the paint layers are each $L = 2$ mm thick and the thermal conductivity of the paint is $k = 0.21$ W/m$\cdot$K.

(d) Plot the heat flux across the gap for the bare surface as a function of ε_s, with $0.05 \leq \varepsilon_s \leq 0.95$. Show on the same plot the heat flux for the painted surface with very thin paint layers and the painted surface with $L = 2$-mm-thick paint layers.

Multimode Heat Transfer: Advanced

13.109 Options for thermally shielding the top ceiling of a large furnace include the use of an insulating material of thickness L and thermal conductivity k, case (a), or an air space of equivalent thickness formed by installing a steel sheet above the ceiling, case (b).

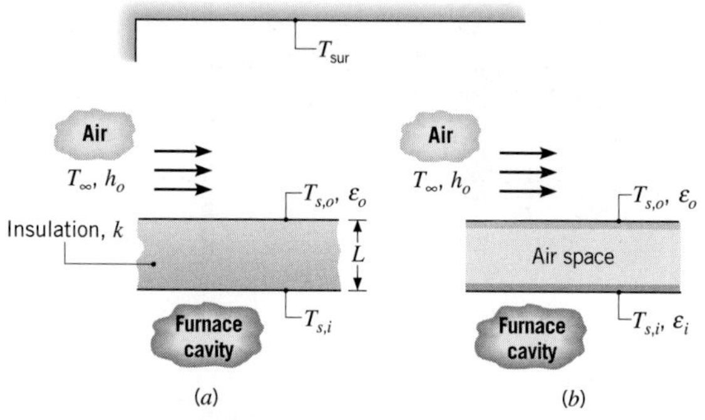

(a) Develop mathematical models that could be used to assess which of the two approaches is better. In both cases the interior surface is maintained at the same temperature, $T_{s,i}$, and the ambient air and surroundings are at equivalent temperatures ($T_\infty = T_{sur}$).

(b) If $k = 0.090\ \text{W/m}\cdot\text{K}$, $L = 25\ \text{mm}$, $h_o = 25\ \text{W/m}^2\cdot\text{K}$, the surfaces are diffuse and gray with $\varepsilon_i = \varepsilon_o = 0.50$, $T_{s,i} = 900\ \text{K}$, and $T_\infty = T_{sur} = 300\ \text{K}$, what is the outer surface temperature $T_{s,o}$ and the heat loss per unit surface area associated with each option?

(c) For each case, assess the effect of surface radiative properties on the outer surface temperature and the heat loss per unit area for values of $\varepsilon_i = \varepsilon_o$ ranging from 0.1 to 0.9. Plot your results.

13.110 The composite insulation shown, which was described in Chapter 1 (Problem 1.86e), is being considered as a ceiling material.

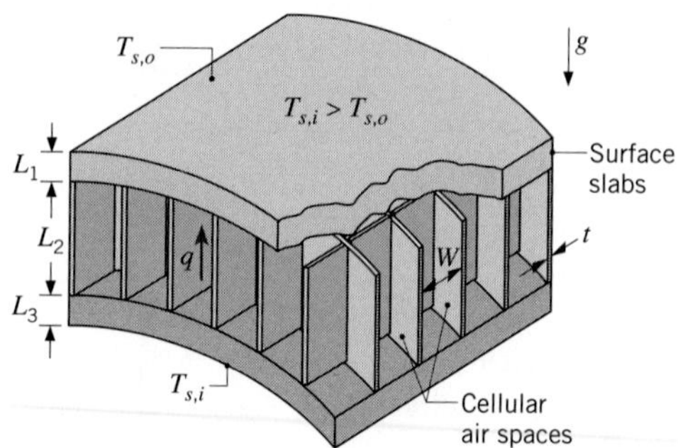

It is proposed that the outer and inner slabs be made from low-density particle board of thicknesses $L_1 = L_3 = 12.5\ \text{mm}$ and that the honeycomb core be constructed from a high-density particle board. The square cells of the core are to have length $L_2 = 50\ \text{mm}$, width $W = 10\ \text{mm}$, and wall thickness $t = 2\ \text{mm}$. The emissivity of both particle boards is approximately 0.85, and the honeycomb cells are filled with air at 1-atm pressure. To assess the effectiveness of the insulation, its total thermal resistance must be evaluated under representative operating conditions for which the bottom (inner) surface temperature is $T_{s,i} = 25°\text{C}$ and the top (outer) surface temperature is $T_{s,o} = -10°\text{C}$. To assess the effect of free convection in the air space, assume a cell temperature difference of 20°C and evaluate air properties at 7.5°C. To assess the effect of radiation across the air space, assume inner surface temperatures of the outer and inner slabs to be −5 and 15°C, respectively.

13.111 Hot coffee is contained in a cylindrical thermos bottle that is of length $L = 0.3\ \text{m}$ and is lying on its side (horizontally).

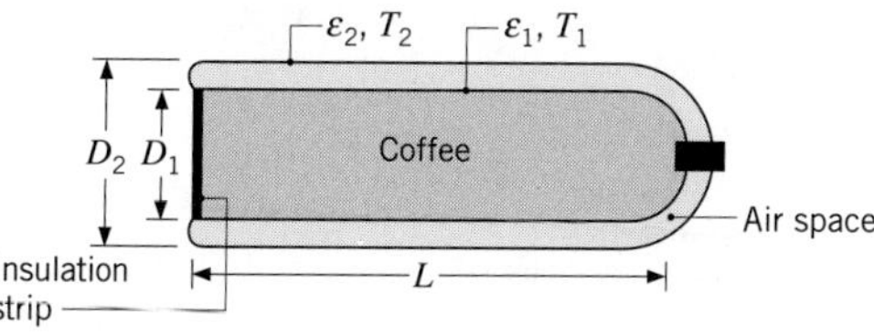

The coffee container consists of a glass flask of diameter $D_1 = 0.07\ \text{m}$, separated from an aluminum housing of diameter $D_2 = 0.08\ \text{m}$ by air at atmospheric pressure. The outer surface of the flask and the inner surface of the housing are silver coated to provide emissivities of $\varepsilon_1 = \varepsilon_2 = 0.25$. If these surface temperatures are $T_1 = 75°\text{C}$ and $T_2 = 35°\text{C}$, what is the heat loss from the coffee?

13.112 Consider a vertical, double-pane window for the conditions prescribed in Problem 9.94. That is, vertical panes at temperatures of $T_1 = 22°\text{C}$ and $T_2 = -20°\text{C}$ are separated by atmospheric air, and the critical Rayleigh number for the onset of convection is $Ra_{L,c} = 2000$.

(a) What is the conduction heat flux across the air gap for the optimal spacing L_{op} between the panes?

(b) If the glass has an emissivity of $\varepsilon_g = 0.90$, what is the total heat flux across the gap?

(c) What is the total heat flux if a special, low-emissivity coating ($\varepsilon_c = 0.10$) is applied to one of the panes at its air-glass interface? What is the total heat flux if both panes are coated?

13.113 Consider the double-pane window of Problem 9.95, for which 1 m × 1 m panes are separated by a 25-mm gap of atmospheric air. The window panes are approximately isothermal and separate quiescent room air at $T_{\infty,i} = 20°\text{C}$ from quiescent ambient air at $T_{\infty,o} = -20°\text{C}$.

(a) For glass panes of emissivity $\varepsilon_g = 0.90$, determine the temperature of each pane and the rate of heat transfer through the window.

(b) Quantify the improvements in energy conservation that may be effected if the space between the panes is evacuated and/or a low emissivity coating ($\varepsilon_c = 0.1$) is applied to the surface of each pane adjoining the gap.

13.114 A flat-plate solar collector, consisting of an absorber plate and single cover plate, is inclined at an angle of $\tau = 60°$ relative to the horizontal.

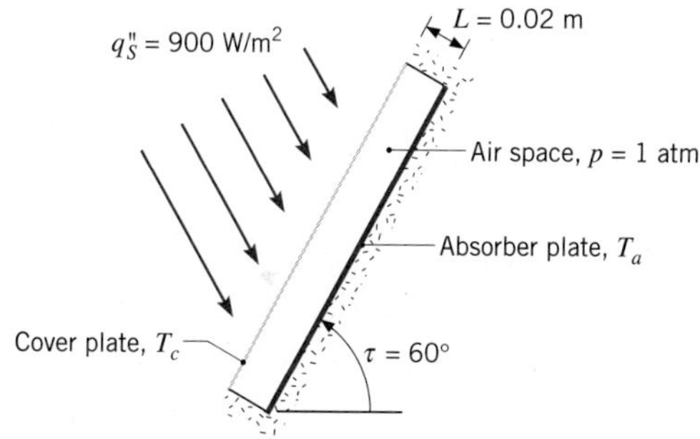

Consider conditions for which the incident solar radiation is collimated at an angle of 60° relative to the horizontal and the solar flux is 900 W/m^2. The cover plate is perfectly transparent to solar radiation ($\lambda \leq 3$ μm) and is opaque to radiation of larger wavelengths. The cover and absorber plates are diffuse surfaces having the spectral absorptivities shown.

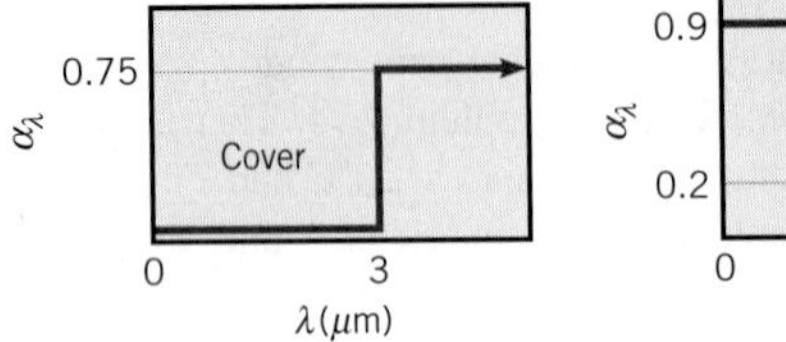

The length and width of the absorber and cover plates are much larger than the plate spacing L. What is the rate at which solar radiation is absorbed per unit area of the absorber plate? With the absorber plate well insulated from below and absorber and cover plate temperatures T_a and T_c of 70°C and 27°C, respectively, what is the heat loss per unit area of the absorber plate?

13.115 Consider the tube and radiation shield of Problem 13.49, but now account for free convection in the gap between the tube and the shield.

(a) What is the total rate of heat transfer per unit length between the tube and the shield?

[(b)] Explore the effect of variations in the shield diameter on the total heat rate, as well as on the contributions due to convection and radiation.

13.116 Consider the tube and radiation shield of Problem 13.49, but now account for free convection in the gap between the tube and the shield, as well as for the fact that the temperature of the shield may not be arbitrarily prescribed but, in fact, depends on the nature of the surroundings. If the radiation shield is exposed to quiescent ambient air and large surroundings, each at a temperature of 20°C, what is the temperature of the shield? What is the rate of heat loss from the tube per unit length?

13.117 Consider the flat-plate solar collector of Problem 9.98. The absorber plate has a coating for which $\varepsilon_1 = 0.96$, and the cover plate has an emissivity of $\varepsilon_2 = 0.92$. With respect to radiation exchange, both plates may be approximated as diffuse, gray surfaces.

(a) For the conditions of Problem 9.98, what is the rate of heat transfer by free convection from the absorber plate and the net rate of radiation exchange between the plates?

[(b)] The temperature of the absorber plate varies according to the flow rate of the working fluid routed through the coiled tube. With all other parameters remaining as prescribed, compute and plot the free convection and radiant heat rates as a function of the absorber plate temperature for $50 \leq T_1 \leq 100$°C.

13.118 The lower side of a 400-mm-diameter disk is heated by an electric furnace, while the upper side is exposed to quiescent, ambient air and surroundings at 300 K. The radiant furnace (negligible convection) is of circular construction with the bottom surface ($\varepsilon_1 = 0.6$) and cylindrical side surface ($\varepsilon_2 = 1.0$) maintained at $T_1 = T_2 = 500$ K. The surface of the disk facing the radiant furnace is black ($\varepsilon_{d,1} = 1.0$), while the upper surface has an emissivity of $\varepsilon_{d,2} = 0.8$. Assume the plate and furnace surfaces to be diffuse and gray.

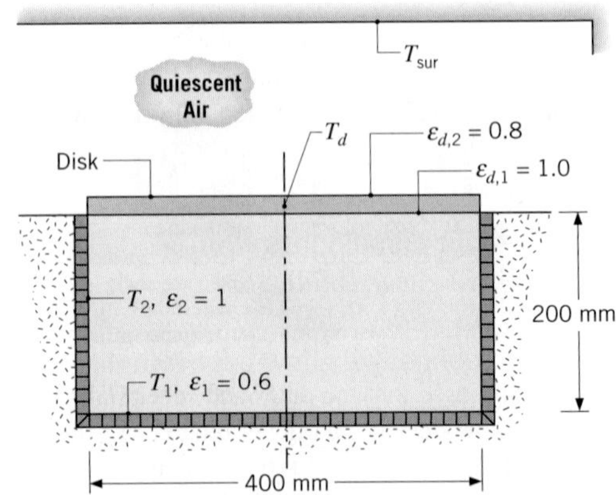

(a) Determine the net heat transfer rate to the disk, $q_{net,d}$, when $T_d = 400$ K.

[(b)] Plot $q_{net,d}$ as a function of the disk temperature for $300 \leq T_d \leq 500$ K, with all other conditions remaining the same. What is the steady-state temperature of the disk?

13.119 The surface of a radiation shield facing a black hot wall at 400 K has a reflectivity of 0.95. Attached to the back side of the shield is a 25-mm-thick sheet of insulating material having a thermal conductivity of 0.016 W/m·K. The overall heat transfer coefficient (convection and radiation) at the surface exposed to the ambient air and surroundings at 300 K is 10 W/m^2·K.

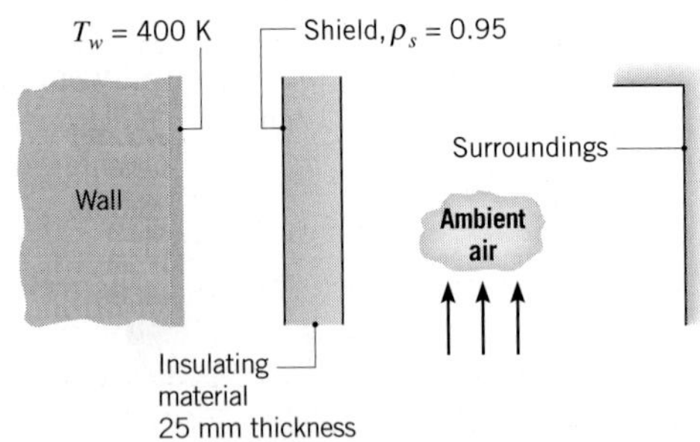

(a) Assuming negligible convection in the region between the wall and the shield, estimate the heat loss per unit area from the hot wall.

(b) Perform a parameter sensitivity analysis on the insulation system, considering the effects of shield reflectivity, ρ_s, and insulation thermal conductivity, k. What influence do these parameters have on the heat loss from the hot wall? What is the effect of an increased overall coefficient on the heat loss? Show the results of your analysis in a graphical format.

13.120 The fire tube of a hot water heater consists of a long circular duct of diameter $D = 0.07$ m and temperature $T_s = 385$ K, through which combustion gases flow at a temperature of $T_{m,g} = 900$ K. To enhance heat transfer from the gas to the tube, a thin partition is inserted along the midplane of the tube. The gases may be assumed to have the thermophysical properties of air and to be radiatively nonparticipating.

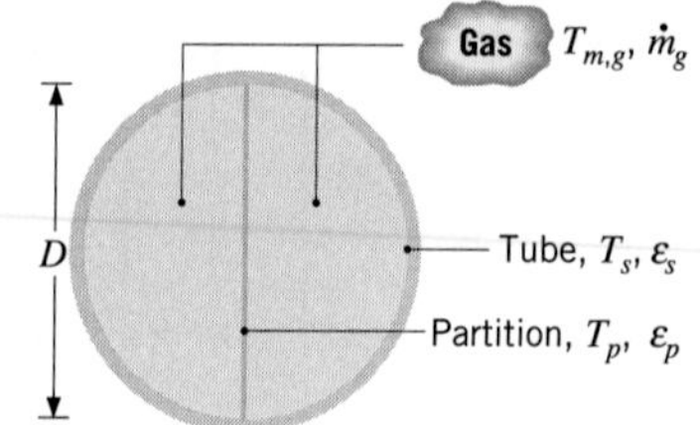

(a) With no partition and a gas flow rate of $\dot{m}_g = 0.05$ kg/s, what is the rate of heat transfer per unit length, q', to the tube?

(b) For a gas flow rate of $\dot{m}_g = 0.05$ kg/s and emissivities of $\varepsilon_s = \varepsilon_p = 0.5$, determine the partition temperature T_p and the total rate of heat transfer q' to the tube.

(c) For $\dot{m}_g = 0.02$, 0.05, and 0.08 kg/s and equivalent emissivities $\varepsilon_p = \varepsilon_s \equiv \varepsilon$, compute and plot T_p and q' as a function of ε for $0.1 \leq \varepsilon \leq 1.0$. For $\dot{m}_g = 0.05$ kg/s and equivalent emissivities, plot the convective and radiative contributions to q' as a function of ε.

13.121 Consider Problem 9.93 with $N = 4$ sheets of thin aluminum foil ($\varepsilon_f = 0.07$), equally spaced throughout the 50-mm gap so as to form five individual air gaps, each 10 mm thick. The hot and cold surfaces of the enclosure are characterized by $\varepsilon = 0.85$.

(a) Neglecting conduction or convection in the air, determine the heat flux through the system.

(b) Accounting for conduction but neglecting radiation, determine the heat flux through the system. The effect of variable properties is important. Calculate the air properties for each gap independently, based on the average gap temperature.

(c) Accounting for both conduction and radiation, determine the heat flux through the system. Calculate the air properties for each gap independently.

(d) Is natural convection negligible in part (c)? Explain why or why not.

13.122 Consider the conditions of Problem 9.107. Accounting for radiation, as well as convection, across the helium-filled cavity, determine the mass rate at which gaseous nitrogen is vented from the system. The cavity surfaces are diffuse and gray with emissivities of $\varepsilon_i = \varepsilon_o = 0.3$. If the cavity is evacuated, how may surface conditions be altered to further reduce the evaporation? Support your recommendation with appropriate calculations.

13.123 A special surface coating on a square panel that is 5 m × 5 m on a side is cured by placing the panel directly under a radiant heat source having the same dimensions. The heat source is diffuse and gray and operates with a power input of 75 kW. The top surface of the heater, as well as the bottom surface of the panel, may be assumed to be well insulated, and the arrangement exists in a large room with air and wall temperatures of 25°C. The surface coating is diffuse and gray, with an emissivity of 0.30 and an upper temperature limit of 400 K. Neglecting convection effects, what is the minimum spacing that may be maintained between the heater and the panel to ensure that the panel temperature will not exceed 400 K? Allowing for convection effects at the coated surface of the panel, what is the minimum spacing?

13.124 A long rod heater of diameter $D_1 = 10$ mm and emissivity $\varepsilon_1 = 1.0$ is coaxial with a well-insulated, semi-cylindrical reflector of diameter $D_2 = 1$ m. A long panel of width $W = 1$ m is aligned with the reflector and is separated from the heater by a distance of $H = 1$ m. The panel is coated with a special paint ($\varepsilon_3 = 0.7$), which is cured by maintaining it at 400 K. The panel is well insulated on its back side, and the entire system is located in a large room where the walls and the atmospheric, quiescent air are at 300 K. Heat transfer by convection may be neglected for the reflector surface.

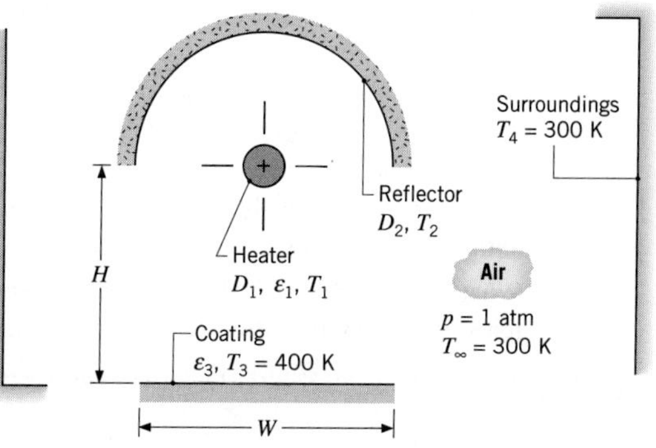

(a) Sketch the equivalent thermal circuit for the system and label all pertinent resistances and potentials.

(b) Expressing your results in terms of appropriate variables, write the system of equations needed to determine the heater and reflector temperatures, T_1 and T_2, respectively. Determine these temperatures for the prescribed conditions.

(c) Determine the rate at which electrical power must be supplied per unit length of the rod heater.

13.125 A radiant heater, which is used for surface treatment processes, consists of a long cylindrical heating element of diameter $D_1 = 0.005$ m and emissivity $\varepsilon_1 = 0.80$. The heater is partially enveloped by a long, thin parabolic reflector whose inner and outer surface emissivities are $\varepsilon_{2i} = 0.10$ and $\varepsilon_{2o} = 0.80$, respectively. Inner and outer surface areas per unit length of the reflector are each $A'_{2i} = A'_{2o} = 0.20$ m, and the average convection coefficient for the combined inner and outer surfaces is $\bar{h}_{2(i,o)} = 2$ W/m$^2 \cdot$K. The system may be assumed to be in an infinite, quiescent medium of atmospheric air at $T_\infty = 300$ K and to be exposed to large surroundings at $T_{sur} = 300$ K.

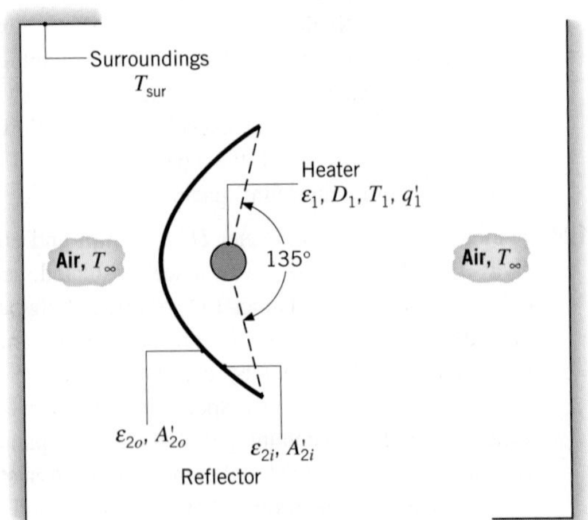

(a) Sketch the appropriate radiation circuit, and write expressions for each of the network resistances.

(b) If, under steady-state conditions, electrical power is dissipated in the heater at $P'_1 = 1500$ W/m and the heater surface temperature is $T_1 = 1200$ K, what is the *net* rate at which *radiant* energy is transferred from the heater?

(c) What is the net rate at which radiant energy is transferred from the heater to the surroundings?

(d) What is the temperature, T_2, of the reflector?

13.126 A steam generator consists of an in-line array of tubes, each of outer diameter $D = 10$ mm and length $L = 1$ m. The longitudinal and transverse pitches are each $S_L = S_T = 20$ mm, while the numbers of longitudinal and transverse rows are $N_L = 20$ and $N_T = 5$. Saturated water (liquid) enters the tubes at a pressure of 2.5 bars, and its flow rate is adjusted to ensure that it leaves the tubes as saturated vapor. Boiling that occurs in the tubes maintains a uniform tube wall temperature of 400 K.

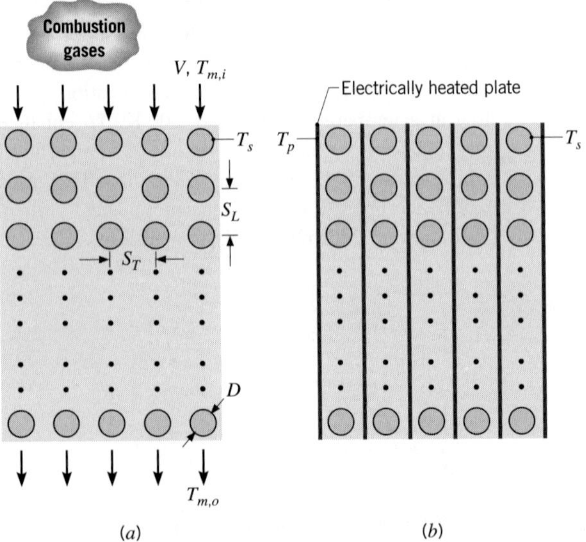

(a) Consider case (*a*) for which products of combustion *enter* the tube bank with velocity and temperature of $V = 10$ m/s and $T_{m,i} = 1200$ K, respectively. Determine the average gas-side convection coefficient, the gas outlet temperature, and the rate of steam production in kg/s. Properties of the gas may be approximated to be those of atmospheric air at an average temperature of 900 K.

(b) An alternative steam generator design, case (*b*), consists of the same tube arrangement, but the gas flow is replaced by an evacuated space with electrically heated plates inserted between each line of tubes. If the plates are maintained at a uniform temperature of $T_p = 1200$ K, what is the rate of steam production? The plate and tube surfaces may be approximated as blackbodies.

(c) Consider conditions for which the plates are installed, as in case (*b*), and the high-temperature products of combustion flow over the tubes, as in case (*a*). The plates are no longer electrically heated, but their thermal conductivity is sufficiently large to ensure a uniform plate temperature. Comment on factors that influence the plate temperature and the gas temperature distribution. Contrast (qualitatively) the gas outlet temperature and the steam generation rate with the results of case (*a*).

13.127 A wall-mounted natural gas heater uses combustion on a porous catalytic pad to maintain a ceramic plate of emissivity $\varepsilon_c = 0.95$ at a uniform temperature of $T_c = 1000$ K. The ceramic plate is separated from a glass plate by an air gap of thickness $L = 50$ mm. The surface of the glass is diffuse, and its spectral transmissivity and absorptivity may be approximated as $\tau_\lambda = 0$ and $\alpha_\lambda = 1$ for $0 \leq \lambda \leq 0.4\ \mu$m, $\tau_\lambda = 1$ and $\alpha_\lambda = 0$ for $0.4 < \lambda \leq 1.6\ \mu$m, and $\tau_\lambda = 0$ and $\alpha_\lambda = 0.9$ for $\lambda > 1.6\ \mu$m. The exterior surface of the glass is exposed to quiescent ambient air and large surroundings for which $T_\infty = T_{sur} = 300$ K. The height and width of the heater are $H = W = 2$ m.

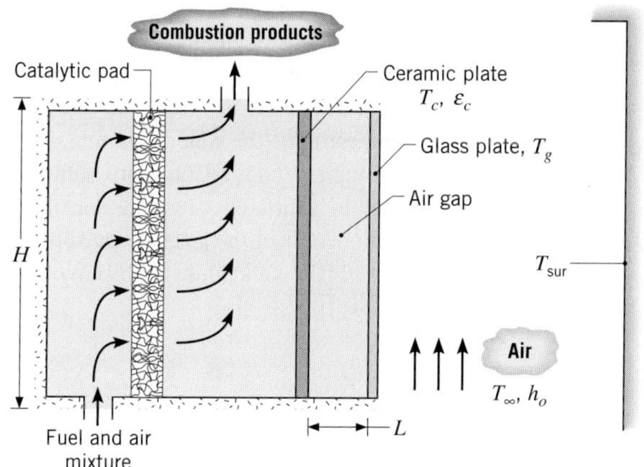

(a) What is the total transmissivity of the glass to irradiation from the ceramic plate? Can the glass be approximated as opaque and gray?

(b) For the prescribed conditions, evaluate the glass temperature, T_g, and the rate of heat transfer from the heater, q_h.

(c) A fan may be used to control the convection coefficient h_o at the exterior surface of the glass. Compute and plot T_g and q_h as a function of h_o for $10 \leq h_o \leq 100$ W/m$^2 \cdot$K.

Participating Media

13.128 Pure, solid silicon is produced from a melt, as, for example, in Problem 1.42. Solid silicon is semitransparent, and the spectral absorption coefficient distribution for pure silicon may be approximated as

$$\kappa_{\lambda,1} = 10^8\ \text{m}^{-1} \qquad (\lambda \leq 0.4\ \mu\text{m})$$
$$\kappa_{\lambda,2} = 0\ \text{m}^{-1} \qquad (0.4\ \mu\text{m} < \lambda \leq 8\ \mu\text{m})$$
$$\kappa_{\lambda,3} = 10^2\ \text{m}^{-1} \qquad (8\ \mu\text{m} < \lambda \leq 25\ \mu\text{m})$$
$$\kappa_{\lambda,4} = 0\ \text{m}^{-1} \qquad (25\ \mu\text{m} < 1)$$

(a) Determine the total absorption coefficient defined as

$$\kappa = \frac{\int_0^\infty \kappa_\lambda G_\lambda(\lambda) d\lambda}{G}$$

for irradiation from surroundings at a temperature equal to the melting point of silicon.

(b) Estimate the total transmissivity, total absorptivity, and total emissivity of an $L = 150$-μm-thick sheet of solid silicon at its melting point.

13.129 A furnace having a spherical cavity of 0.5-m diameter contains a gas mixture at 1 atm and 1400 K. The mixture consists of CO_2 with a partial pressure of 0.25 atm and nitrogen with a partial pressure of 0.75 atm. If the cavity wall is black, what is the cooling rate needed to maintain its temperature at 500 K?

13.130 A gas turbine combustion chamber may be approximated as a long tube of 0.4-m diameter. The combustion gas is at a pressure and temperature of 1 atm and 1000°C, respectively, while the chamber surface temperature is 500°C. If the combustion gas contains CO_2 and water vapor, each with a mole fraction of 0.15, what is the net radiative heat flux between the gas and the chamber surface, which may be approximated as a blackbody?

13.131 A flue gas at 1-atm total pressure and a temperature of 1400 K contains CO_2 and water vapor at partial pressures of 0.05 and 0.10 atm, respectively. If the gas flows through a long flue of 1-m diameter and 400 K surface temperature, determine the net radiative heat flux from the gas to the surface. Blackbody behavior may be assumed for the surface.

13.132 A furnace consists of two large parallel plates separated by 0.75 m. A gas mixture comprised of O_2, N_2, CO_2, and water vapor, with mole fractions of 0.20, 0.50, 0.15, and 0.15, respectively, flows between the plates at a total pressure of 2 atm and a temperature of 1300 K. If the plates may be approximated as blackbodies and are maintained at 500 K, what is the net radiative heat flux to the plates?

13.133 In an industrial process, products of combustion at a temperature and pressure of 2000 K and 1 atm, respectively, flow through a long, 0.25-m-diameter pipe whose inner surface is black. The combustion gas contains CO_2 and water vapor, each at a partial pressure of 0.10 atm. The gas may be assumed to have the thermophysical properties of atmospheric air and to be in fully developed flow with $\dot{m} = 0.25$ kg/s. The pipe is cooled by passing water in cross flow over its outer surface. The upstream velocity and temperature of the water are 0.30 m/s and 300 K, respectively. Determine the pipe wall temperature and heat flux. *Hint:* Emission from the pipe wall may be neglected.

13.134 Noting from Figure 13.16 that gas emissivity can be increased by adding water vapor, it is proposed to enhance heat transfer by injecting saturated steam at 100°C at the entrance of the pipe of Problem 13.133. The mass flow rate of injected steam is 50% of the mass flow rate of water vapor in the combustion products. Determine the gas radiation to the pipe wall without and with steam injection. *Hint*: The temperature of the hot gas decreases, and the partial pressures of the water vapor and CO_2 change when relatively cool steam is injected. Use your knowledge of gas mixtures from thermodynamics to calculate the partial pressures and mass flow rates. Assume the molecular weight of the original mixture is that of air.

13.135 Waste heat recovery from the exhaust (flue) gas of a melting furnace is accomplished by passing the gas through a vertical metallic tube and introducing saturated water (liquid) at the bottom of an annular region around the tube.

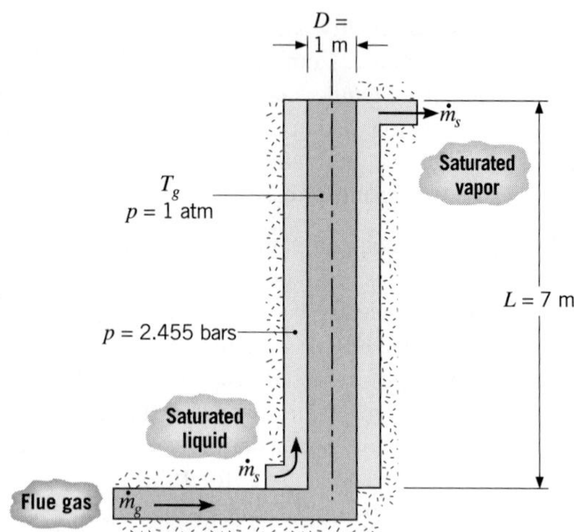

The tube length and inside diameter are 7 and 1 m, respectively, and the tube inner surface is black. The gas in the tube is at atmospheric pressure, with CO_2 and $H_2O(v)$ partial pressures of 0.1 and 0.2 atm, respectively, and its mean temperature may be approximated as $T_g = 1400$ K. The gas flow rate is $\dot{m} = 2$ kg/s. If saturated water is introduced at a pressure of 2.455 bars, estimate the water flow rate $\dot{m}_s$ for which there is complete conversion from saturated liquid at the inlet to saturated vapor at the outlet. Thermophysical properties of the gas may be approximated as $\mu = 530 \times 10^{-7}$ kg/s·m, $k = 0.091$ W/m·K, and $Pr = 0.70$.

APPENDIX A

Thermophysical Properties of Matter[1]

Table		*Page*

[1]The convention used to present numerical values of the properties is illustrated by this example:

T (K)	$\nu \cdot 10^7$ (m^2/s)	$k \cdot 10^3$ (W/m·K)
300	0.349	521

where $\nu = 0.349 \times 10^{-7}$ m^2/s and $k = 521 \times 10^{-3} = 0.521$ W/m·K at 300 K.

TABLE A.1 Thermophysical Properties of Selected Metallic Solids[a]

Composition	Melting Point (K)	Properties at 300 K: ρ (kg/m^3)	c_p (J/kg · K)	k (W/m · K)	$\alpha \cdot 10^6$ (m^2/s)	Properties at Various Temperatures (K), k (W/m · K)/c_p (J/kg · K): 100	200	400	600	800	1000	1200	1500	2000	2500
Aluminum															
Pure	933	2702	903	237	97.1	302	237	240	231	218					
						482	798	949	1033	1146					
Alloy 2024-T6 (4.5% Cu, 1.5% Mg, 0.6% Mn)	775	2770	875	177	73.0	65	163	186	186						
						473	787	925	1042						
Alloy 195, Cast (4.5% Cu)		2790	883	168	68.2			174	185						
								—	—						
Beryllium	1550	1850	1825	200	59.2	990	301	161	126	106	90.8	78.7			
						203	1114	2191	2604	2823	3018	3227	3519		
Bismuth	545	9780	122	7.86	6.59	16.5	9.69	7.04							
						112	120	127							
Boron	2573	2500	1107	27.0	9.76	190	55.5	16.8	10.6	9.60	9.85				
						128	600	1463	1892	2160	2338				
Cadmium	594	8650	231	96.8	48.4	203	99.3	94.7							
						198	222	242							
Chromium	2118	7160	449	93.7	29.1	159	111	90.9	80.7	71.3	65.4	61.9	57.2	49.4	
						192	384	484	542	581	616	682	779	937	
Cobalt	1769	8862	421	99.2	26.6	167	122	85.4	67.4	58.2	52.1	49.3	42.5		
						236	379	450	503	550	628	733	674		
Copper															
Pure	1358	8933	385	401	117	482	413	393	379	366	352	339			
						252	356	397	417	433	451	480			
Commercial bronze (90% Cu, 10% Al)	1293	8800	420	52	14		42	52	59						
							785	460	545						
Phosphor gear bronze (89% Cu, 11% Sn)	1104	8780	355	54	17		41	65	74						
							—	—	—						
Cartridge brass (70% Cu, 30% Zn)	1188	8530	380	110	33.9	75	95	137	149						
							360	395	425						
Constantan (55% Cu, 45% Ni)	1493	8920	384	23	6.71	17	19								
						237	362								
Germanium	1211	5360	322	59.9	34.7	232	96.8	43.2	27.3	19.8	17.4	17.4			
						190	290	337	348	357	375	395			

Table A.1 *Continued*

Composition	Melting Point (K)	Properties at 300 K ρ (kg/m^3)	c_p (J/kg · K)	k (W/m · K)	$\alpha \cdot 10^6$ (m^2/s)	Properties at Various Temperatures (K) k (W/m · K)/c_p (J/kg · K) 100	200	400	600	800	1000	1200	1500	2000	2500
Gold	1336	19300	129	317	127	327	323	311	298	284	270	255			
						109	124	131	135	140	145	155			
Iridium	2720	22500	130	147	50.3	172	153	144	138	132	126	120	111		
						90	122	133	138	144	153	161	172		
Iron															
Pure	1810	7870	447	80.2	23.1	134	94.0	69.5	54.7	43.3	32.8	28.3	32.1		
						216	384	490	574	680	975	609	654		
Armco (99.75% pure)		7870	447	72.7	20.7	95.6	80.6	65.7	53.1	42.2	32.3	28.7	31.4		
						215	384	490	574	680	975	609	654		
Carbon steels															
Plain carbon (Mn ≤ 1%, Si ≤ 0.1%)		7854	434	60.5	17.7			56.7	48.0	39.2	30.0				
								487	559	685	1169				
AISI 1010		7832	434	63.9	18.8			58.7	48.8	39.2	31.3				
								487	559	685	1168				
Carbon–silicon (Mn ≤ 1%, 0.1% < Si ≤ 0.6%)		7817	446	51.9	14.9			49.8	44.0	37.4	29.3				
								501	582	699	971				
Carbon–manganese–silicon (1% < Mn ≤ 1.65%, 0.1% < Si ≤ 0.6%)		8131	434	41.0	11.6			42.2	39.7	35.0	27.6				
								487	559	685	1090				
Chromium (low) steels															
$\frac{1}{2}$Cr–$\frac{1}{4}$Mo–Si (0.18% C, 0.65% Cr, 0.23% Mo, 0.6% Si)		7822	444	37.7	10.9			38.2	36.7	33.3	26.9				
								492	575	688	969				
1 Cr–$\frac{1}{2}$Mo (0.16% C, 1% Cr, 0.54% Mo, 0.39% Si)		7858	442	42.3	12.2			42.0	39.1	34.5	27.4				
								492	575	688	969				
1 Cr–V (0.2% C, 1.02% Cr, 0.15% V)		7836	443	48.9	14.1			46.8	42.1	36.3	28.2				
								492	575	688	969				

Stainless steels															
AISI 302		8055	480	15.1	3.91			17.3 512	20.0 559	22.8 585	25.4 606				
AISI 304	1670	7900	477	14.9	3.95	9.2 272	12.6 402	16.6 515	19.8 557	22.6 582	25.4 611	28.0 640	31.7 682		
AISI 316		8238	468	13.4	3.48			15.2 504	18.3 550	21.3 576	24.2 602				
AISI 347		7978	480	14.2	3.71			15.8 513	18.9 559	21.9 585	24.7 606				
Lead	601	11340	129	35.3	24.1	39.7 118	36.7 125	34.0 132	31.4 142						
Magnesium	923	1740	1024	156	87.6	169 649	159 934	153 1074	149 1170	146 1267					
Molybdenum	2894	10240	251	138	53.7	179 141	143 224	134 261	126 275	118 285	112 295	105 308	98 330	90 380	86 459
Nickel															
Pure	1728	8900	444	90.7	23.0	164 232	107 383	80.2 485	65.6 592	67.6 530	71.8 562	76.2 594	82.6 616		
Nichrome (80% Ni, 20% Cr)	1672	8400	420	12	3.4			14 480	16 525	21 545					
Inconel X-750 (73% Ni, 15% Cr, 6.7% Fe)	1665	8510	439	11.7	3.1	8.7 —	10.3 372	13.5 473	17.0 510	20.5 546	24.0 626	27.6 —	33.0 —		
Niobium	2741	8570	265	53.7	23.6	55.2 188	52.6 249	55.2 274	58.2 283	61.3 292	64.4 301	67.5 310	72.1 324	79.1 347	
Palladium	1827	12020	244	71.8	24.5	76.5 168	71.6 227	73.6 251	79.7 261	86.9 271	94.2 281	102 291	110 307		
Platinum															
Pure	2045	21450	133	71.6	25.1	77.5 100	72.6 125	71.8 136	73.2 141	75.6 146	78.7 152	82.6 157	89.5 165	99.4 179	
Alloy 60Pt–40Rh (60% Pt, 40% Rh)	1800	16630	162	47	17.4			52 —	59 —	65 —	69 —	73 —	76 —		
Rhenium	3453	21100	136	47.9	16.7	58.9 97	51.0 127	46.1 139	44.2 145	44.1 151	44.6 156	45.7 162	47.8 171	51.9 186	
Rhodium	2236	12450	243	150	49.6	186 147	154 220	146 253	136 274	127 293	121 311	116 327	110 349	112 376	
Silicon	1685	2330	712	148	89.2	884 259	264 556	98.9 790	61.9 867	42.2 913	31.2 946	25.7 967	22.7 992		
Silver	1235	10500	235	429	174	444 187	430 225	425 239	412 250	396 262	379 277	361 292			
Tantalum	3269	16600	140	57.5	24.7	59.2 110	57.5 133	57.8 144	58.6 146	59.4 149	60.2 152	61.0 155	62.2 160	64.1 172	65.6 189
Thorium	2023	11700	118	54.0	39.1	59.8 99	54.6 112	54.5 124	55.8 134	56.9 145	56.9 156	58.7 167			
Tin	505	7310	227	66.6	40.1	85.2 188	73.3 215	62.2 243							

TABLE A.1 *Continued*

Composition	Melting Point (K)	Properties at 300 K				Properties at Various Temperatures (K) k (W/m · K)/c_p (J/kg · K)									
		ρ (kg/m³)	c_p (J/kg · K)	k (W/m · K)	$\alpha \cdot 10^6$ (m²/s)	100	200	400	600	800	1000	1200	1500	2000	2500
Titanium	1953	4500	522	21.9	9.32	30.5 300	24.5 465	20.4 551	19.4 591	19.7 633	20.7 675	22.0 620	24.5 686		
Tungsten	3660	19300	132	174	68.3	208 87	186 122	159 137	137 142	125 145	118 148	113 152	107 157	100 167	95 176
Uranium	1406	19070	116	27.6	12.5	21.7 94	25.1 108	29.6 125	34.0 146	38.8 176	43.9 180	49.0 161			
Vanadium	2192	6100	489	30.7	10.3	35.8 258	31.3 430	31.3 515	33.3 540	35.7 563	38.2 597	40.8 645	44.6 714	50.9 867	
Zinc	693	7140	389	116	41.8	117 297	118 367	111 402	103 436						
Zirconium	2125	6570	278	22.7	12.4	33.2 205	25.2 264	21.6 300	20.7 322	21.6 342	23.7 362	26.0 344	28.8 344	33.0 344	

[a]Adapted from References 1–7.

TABLE A.2 Thermophysical Properties of Selected Nonmetallic Solids[a]

Composition	Melting Point (K)	Properties at 300 K: ρ (kg/m^3)	c_p (J/kg · K)	k (W/m · K)	$\alpha \cdot 10^6$ (m^2/s)	Properties at Various Temperatures (K), k (W/m · K)/c_p (J/kg · K): 100	200	400	600	800	1000	1200	1500	2000	2500
Aluminum oxide,	2323	3970	765	46	15.1	450	82	32.4	18.9	13.0	10.5				
sapphire						—	—	940	1110	1180	1225				
Aluminum oxide,	2323	3970	765	36.0	11.9	133	55	26.4	15.8	10.4	7.85	6.55	5.66	6.00	
polycrystalline						—	—	940	1110	1180	1225	—	—	—	
Beryllium oxide	2725	3000	1030	272	88.0			196	111	70	47	33	21.5	15	
								1350	1690	1865	1975	2055	2145	2750	
Boron	2573	2500	1105	27.6	9.99	190	52.5	18.7	11.3	8.1	6.3	5.2			
						—	—	1490	1880	2135	2350	2555			
Boron fiber epoxy (30% vol) composite	590	2080													
k, ∥ to fibers				2.29		2.10	2.23	2.28							
k, ⊥ to fibers				0.59		0.37	0.49	0.60							
c_p			1122			364	757	1431							
Carbon															
Amorphous	1500	1950	—	1.60	—	0.67	1.18	1.89	2.19	2.37	2.53	2.84	3.48		
						—	—	—	—	—	—	—	—		
Diamond, type IIa	—	3500	509	2300	—	10,000	4000	1540							
insulator						21	194	853							
Graphite, pyrolytic	2273	2210													
k, ∥ to layers				1950		4970	3230	1390	892	667	534	448	357	262	
k, ⊥ to layers				5.70		16.8	9.23	4.09	2.68	2.01	1.60	1.34	1.08	0.81	
c_p			709			136	411	992	1406	1650	1793	1890	1974	2043	
Graphite fiber epoxy (25% vol) composite	450	1400													
k, heat flow ∥ to fibers				11.1		5.7	8.7	13.0							
k, heat flow ⊥ to fibers				0.87		0.46	0.68	1.1							
c_p			935			337	642	1216							
Pyroceram,	1623	2600	808	3.98	1.89	5.25	4.78	3.64	3.28	3.08	2.96	2.87	2.79		
Corning 9606						—	—	908	1038	1122	1197	1264	1498		

TABLE A.2 *Continued*

Composition	Melting Point (K)	Properties at 300 K				Properties at Various Temperatures (K) k (W/m · K)/c_p (J/kg · K)									
		ρ (kg/m^3)	c_p (J/kg · K)	k (W/m · K)	$\alpha \cdot 10^6$ (m^2/s)	100	200	400	600	800	1000	1200	1500	2000	2500
Silicon carbide	3100	3160	675	490	230			— 880	— 1050	— 1135	87 1195	58 1243	30 1310		
Silicon dioxide, crystalline (quartz)	1883	2650													
k, ∥ to c axis				10.4		39	16.4	7.6	5.0	4.2					
k, ⊥ to c axis				6.21		20.8	9.5	4.70	3.4	3.1					
c_p			745			—	—	885	1075	1250					
Silicon dioxide, polycrystalline (fused silica)	1883	2220	745	1.38	0.834	0.69 —	1.14 —	1.51 905	1.75 1040	2.17 1105	2.87 1155	4.00 1195			
Silicon nitride	2173	2400	691	16.0	9.65	— —	— 578	13.9 778	11.3 937	9.88 1063	8.76 1155	8.00 1226	7.16 1306	6.20 1377	
Sulfur	392	2070	708	0.206	0.141	0.165 403	0.185 606								
Thorium dioxide	3573	9110	235	13	6.1			10.2 255	6.6 274	4.7 285	3.68 295	3.12 303	2.73 315	2.5 330	
Titanium dioxide, polycrystalline	2133	4157	710	8.4	2.8			7.01 805	5.02 880	3.94 910	3.46 930	3.28 945			

[a]Adapted from References 1, 2, 3 and 6.

Table A.3 Thermophysical Properties of Common Materials[a]

Structural Building Materials

	Typical Properties at 300 K		
Description/Composition	**Density, ρ (kg/m^3)**	**Thermal Conductivity, k (W/m·K)**	**Specific Heat, c_p (J/kg·K)**
Building Boards			
Asbestos–cement board	1920	0.58	—
Gypsum or plaster board	800	0.17	—
Plywood	545	0.12	1215
Sheathing, regular density	290	0.055	1300
Acoustic tile	290	0.058	1340
Hardboard, siding	640	0.094	1170
Hardboard, high density	1010	0.15	1380
Particle board, low density	590	0.078	1300
Particle board, high density	1000	0.170	1300
Woods			
Hardwoods (oak, maple)	720	0.16	1255
Softwoods (fir, pine)	510	0.12	1380
Masonry Materials			
Cement mortar	1860	0.72	780
Brick, common	1920	0.72	835
Brick, face	2083	1.3	—
Clay tile, hollow			
1 cell deep, 10 cm thick	—	0.52	—
3 cells deep, 30 cm thick	—	0.69	—
Concrete block, 3 oval cores			
Sand/gravel, 20 cm thick	—	1.0	—
Cinder aggregate, 20 cm thick	—	0.67	—
Concrete block, rectangular core			
2 cores, 20 cm thick, 16 kg	—	1.1	—
Same with filled cores	—	0.60	—
Plastering Materials			
Cement plaster, sand aggregate	1860	0.72	—
Gypsum plaster, sand aggregate	1680	0.22	1085
Gypsum plaster, vermiculite aggregate	720	0.25	—

TABLE A.3 *Continued*

Insulating Materials and Systems

Description/Composition	Density, ρ (kg/m^3)	Thermal Conductivity, k (W/m·K)	Specific Heat, c_p (J/kg·K)
	Typical Properties at 300 K		
Blanket and Batt			
Glass fiber, paper faced	16	0.046	—
	28	0.038	—
	40	0.035	—
Glass fiber, coated; duct liner	32	0.038	835
Board and Slab			
Cellular glass	145	0.058	1000
Glass fiber, organic bonded	105	0.036	795
Polystyrene, expanded			
Extruded (R-12)	55	0.027	1210
Molded beads	16	0.040	1210
Mineral fiberboard; roofing material	265	0.049	—
Wood, shredded/cemented	350	0.087	1590
Cork	120	0.039	1800
Loose Fill			
Cork, granulated	160	0.045	—
Diatomaceous silica, coarse	350	0.069	—
Powder	400	0.091	—
Diatomaceous silica, fine powder	200	0.052	—
	275	0.061	—
Glass fiber, poured or blown	16	0.043	835
Vermiculite, flakes	80	0.068	835
	160	0.063	1000
Formed/Foamed-in-Place			
Mineral wool granules with asbestos/inorganic binders, sprayed	190	0.046	—
Polyvinyl acetate cork mastic; sprayed or troweled	—	0.100	—
Urethane, two-part mixture; rigid foam	70	0.026	1045
Reflective			
Aluminum foil separating fluffy glass mats; 10–12 layers, evacuated; for cryogenic applications (150 K)	40	0.00016	—
Aluminum foil and glass paper laminate; 75–150 layers; evacuated; for cryogenic application (150 K)	120	0.000017	—
Typical silica powder, evacuated	160	0.0017	—

TABLE A.3 *Continued*

Industrial Insulation

Description/ Composition	Maximum Service Temperature (K)	Typical Density (kg/m³)	Typical Thermal Conductivity, *k* (W/m · K), at Various Temperatures (K)													
			200	215	230	240	255	270	285	300	310	365	420	530	645	750
Blankets																
Blanket, mineral fiber,	920	96–192									0.038	0.046	0.056	0.078		
metal reinforced	815	40–96									0.035	0.045	0.058	0.088		
Blanket, mineral fiber,	450	10				0.036	0.038	0.040	0.043	0.048	0.052	0.076				
glass; fine fiber,																
organic bonded		12				0.035	0.036	0.039	0.042	0.046	0.049	0.069				
		16				0.033	0.035	0.036	0.039	0.042	0.046	0.062				
		24				0.030	0.032	0.033	0.036	0.039	0.040	0.053				
		32				0.029	0.030	0.032	0.033	0.036	0.038	0.048				
		48				0.027	0.029	0.030	0.032	0.033	0.035	0.045				
Blanket, alumina–																
silica fiber	1530	48												0.071	0.105	0.150
		64												0.059	0.087	0.125
		96												0.052	0.076	0.100
		128												0.049	0.068	0.091
Felt, semirigid;	480	50–125						0.035	0.036	0.038	0.039	0.051	0.063			
organic bonded	730	50	0.023	0.025	0.026	0.027	0.029	0.030	0.032	0.033	0.035	0.051	0.079			
Felt, laminated;																
no binder	920	120											0.051	0.065	0.087	
Blocks, Boards, and Pipe Insulations																
Asbestos paper, laminated and corrugated																
4-ply	420	190								0.078	0.082	0.098				
6-ply	420	255								0.071	0.074	0.085				
8-ply	420	300								0.068	0.071	0.082				
Magnesia, 85%	590	185									0.051	0.055	0.061			
Calcium silicate	920	190									0.055	0.059	0.063	0.075	0.089	0.104

Table A.3 *Continued*

Industrial Insulation (Continued)

Description/ Composition	Maximum Service Temperature (K)	Typical Density (kg/m^3)	Typical Thermal Conductivity, k (W/m · K), at Various Temperatures (K)													
			200	215	230	240	255	270	285	300	310	365	420	530	645	750
Cellular glass	700	145			0.046	0.048	0.051	0.052	0.055	0.058	0.062	0.069	0.079			
Diatomaceous	1145	345												0.092	0.098	0.104
silica	1310	385												0.101	0.100	0.115
Polystyrene, rigid																
Extruded (R-12)	350	56	0.023	0.023	0.022	0.023	0.023	0.025	0.026	0.027	0.029					
Extruded (R-12)	350	35	0.023	0.023	0.023	0.025	0.025	0.026	0.027	0.029						
Molded beads	350	16	0.026	0.029	0.030	0.033	0.035	0.036	0.038	0.040						
Rubber, rigid foamed	340	70						0.029	0.030	0.032	0.033					
Insulating Cement																
Mineral fiber (rock, slag or glass)																
With clay binder	1255	430									0.071	0.079	0.088	0.105	0.123	
With hydraulic setting binder	922	560									0.108	0.115	0.123	0.137		
Loose Fill																
Cellulose, wood or paper pulp	—	45							0.038	0.039	0.042					
Perlite, expanded	—	105	0.036	0.039	0.042	0.043	0.046	0.049	0.051	0.053	0.056					
Vermiculite, expanded	—	122			0.056	0.058	0.061	0.063	0.065	0.068	0.071					
		80			0.049	0.051	0.055	0.058	0.061	0.063	0.066					

TABLE A.3 *Continued*

Other Materials

Description/ Composition	Temperature (K)	Density, ρ (kg/m^3)	Thermal Conductivity, k (W/m·K)	Specific Heat, c_p (J/kg·K)
Asphalt	300	2115	0.062	920
Bakelite	300	1300	1.4	1465
Brick, refractory				
Carborundum	872	—	18.5	—
	1672	—	11.0	—
Chrome brick	473	3010	2.3	835
	823		2.5	
	1173		2.0	
Diatomaceous	478	—	0.25	—
silica, fired	1145	—	0.30	
Fireclay, burnt 1600 K	773	2050	1.0	960
	1073	—	1.1	
	1373	—	1.1	
Fireclay, burnt 1725 K	773	2325	1.3	960
	1073		1.4	
	1373		1.4	
Fireclay brick	478	2645	1.0	960
	922		1.5	
	1478		1.8	
Magnesite	478	—	3.8	1130
	922	—	2.8	
	1478		1.9	
Clay	300	1460	1.3	880
Coal, anthracite	300	1350	0.26	1260
Concrete (stone mix)	300	2300	1.4	880
Cotton	300	80	0.06	1300
Foodstuffs				
Banana (75.7% water content)	300	980	0.481	3350
Apple, red (75% water content)	300	840	0.513	3600
Cake, batter	300	720	0.223	—
Cake, fully baked	300	280	0.121	—
Chicken meat, white	198	—	1.60	—
(74.4% water content)	233	—	1.49	
	253		1.35	
	263		1.20	
	273		0.476	
	283		0.480	
	293		0.489	
Glass				
Plate (soda lime)	300	2500	1.4	750
Pyrex	300	2225	1.4	835

TABLE A.3 *Continued*

Other Materials (Continued)

Description/ Composition	Temperature (K)	Density, ρ (kg/m^3)	Thermal Conductivity, k (W/m·K)	Specific Heat, c_p (J/kg·K)
Ice	273	920	1.88	2040
	253	—	2.03	1945
Leather (sole)	300	998	0.159	—
Paper	300	930	0.180	1340
Paraffin	300	900	0.240	2890
Rock				
Granite, Barre	300	2630	2.79	775
Limestone, Salem	300	2320	2.15	810
Marble, Halston	300	2680	2.80	830
Quartzite, Sioux	300	2640	5.38	1105
Sandstone, Berea	300	2150	2.90	745
Rubber, vulcanized				
Soft	300	1100	0.13	2010
Hard	300	1190	0.16	—
Sand	300	1515	0.27	800
Soil	300	2050	0.52	1840
Snow	273	110	0.049	—
		500	0.190	—
Teflon	300	2200	0.35	—
	400		0.45	—
Tissue, human				
Skin	300	—	0.37	—
Fat layer (adipose)	300	—	0.2	—
Muscle	300	—	0.5	—
Wood, cross grain				
Balsa	300	140	0.055	—
Cypress	300	465	0.097	—
Fir	300	415	0.11	2720
Oak	300	545	0.17	2385
Yellow pine	300	640	0.15	2805
White pine	300	435	0.11	—
Wood, radial				
Oak	300	545	0.19	2385
Fir	300	420	0.14	2720

[a]Adapted from References 1 and 8–13.

TABLE A.4 Thermophysical Properties of Gases at Atmospheric Pressure[a]

T (K)	ρ (kg/m^3)	c_p (kJ/kg · K)	$\mu \cdot 10^7$ (N · s/m^2)	$\nu \cdot 10^6$ (m^2/s)	$k \cdot 10^3$ (W/m · K)	$\alpha \cdot 10^6$ (m^2/s)	Pr
Air, $\mathcal{M}$ = 28.97 kg/kmol							
100	3.5562	1.032	71.1	2.00	9.34	2.54	0.786
150	2.3364	1.012	103.4	4.426	13.8	5.84	0.758
200	1.7458	1.007	132.5	7.590	18.1	10.3	0.737
250	1.3947	1.006	159.6	11.44	22.3	15.9	0.720
300	1.1614	1.007	184.6	15.89	26.3	22.5	0.707
350	0.9950	1.009	208.2	20.92	30.0	29.9	0.700
400	0.8711	1.014	230.1	26.41	33.8	38.3	0.690
450	0.7740	1.021	250.7	32.39	37.3	47.2	0.686
500	0.6964	1.030	270.1	38.79	40.7	56.7	0.684
550	0.6329	1.040	288.4	45.57	43.9	66.7	0.683
600	0.5804	1.051	305.8	52.69	46.9	76.9	0.685
650	0.5356	1.063	322.5	60.21	49.7	87.3	0.690
700	0.4975	1.075	338.8	68.10	52.4	98.0	0.695
750	0.4643	1.087	354.6	76.37	54.9	109	0.702
800	0.4354	1.099	369.8	84.93	57.3	120	0.709
850	0.4097	1.110	384.3	93.80	59.6	131	0.716
900	0.3868	1.121	398.1	102.9	62.0	143	0.720
950	0.3666	1.131	411.3	112.2	64.3	155	0.723
1000	0.3482	1.141	424.4	121.9	66.7	168	0.726
1100	0.3166	1.159	449.0	141.8	71.5	195	0.728
1200	0.2902	1.175	473.0	162.9	76.3	224	0.728
1300	0.2679	1.189	496.0	185.1	82	257	0.719
1400	0.2488	1.207	530	213	91	303	0.703
1500	0.2322	1.230	557	240	100	350	0.685
1600	0.2177	1.248	584	268	106	390	0.688
1700	0.2049	1.267	611	298	113	435	0.685
1800	0.1935	1.286	637	329	120	482	0.683
1900	0.1833	1.307	663	362	128	534	0.677
2000	0.1741	1.337	689	396	137	589	0.672
2100	0.1658	1.372	715	431	147	646	0.667
2200	0.1582	1.417	740	468	160	714	0.655
2300	0.1513	1.478	766	506	175	783	0.647
2400	0.1448	1.558	792	547	196	869	0.630
2500	0.1389	1.665	818	589	222	960	0.613
3000	0.1135	2.726	955	841	486	1570	0.536
Ammonia (NH_3), $\mathcal{M}$ = 17.03 kg/kmol							
300	0.6894	2.158	101.5	14.7	24.7	16.6	0.887
320	0.6448	2.170	109	16.9	27.2	19.4	0.870
340	0.6059	2.192	116.5	19.2	29.3	22.1	0.872
360	0.5716	2.221	124	21.7	31.6	24.9	0.872
380	0.5410	2.254	131	24.2	34.0	27.9	0.869

TABLE A.4 *Continued*

T (K)	ρ (kg/m^3)	c_p (kJ/kg $\cdot$ K)	$\mu \cdot 10^7$ (N $\cdot$ s/m^2)	$\nu \cdot 10^6$ (m^2/s)	$k \cdot 10^3$ (W/m $\cdot$ K)	$\alpha \cdot 10^6$ (m^2/s)	Pr
Ammonia (NH_3) *(continued)*							
400	0.5136	2.287	138	26.9	37.0	31.5	0.853
420	0.4888	2.322	145	29.7	40.4	35.6	0.833
440	0.4664	2.357	152.5	32.7	43.5	39.6	0.826
460	0.4460	2.393	159	35.7	46.3	43.4	0.822
480	0.4273	2.430	166.5	39.0	49.2	47.4	0.822
500	0.4101	2.467	173	42.2	52.5	51.9	0.813
520	0.3942	2.504	180	45.7	54.5	55.2	0.827
540	0.3795	2.540	186.5	49.1	57.5	59.7	0.824
560	0.3708	2.577	193	52.0	60.6	63.4	0.827
580	0.3533	2.613	199.5	56.5	63.8	69.1	0.817
Carbon Dioxide (CO_2), $\mathcal{M}$ = 44.01 kg/kmol							
280	1.9022	0.830	140	7.36	15.20	9.63	0.765
300	1.7730	0.851	149	8.40	16.55	11.0	0.766
320	1.6609	0.872	156	9.39	18.05	12.5	0.754
340	1.5618	0.891	165	10.6	19.70	14.2	0.746
360	1.4743	0.908	173	11.7	21.2	15.8	0.741
380	1.3961	0.926	181	13.0	22.75	17.6	0.737
400	1.3257	0.942	190	14.3	24.3	19.5	0.737
450	1.1782	0.981	210	17.8	28.3	24.5	0.728
500	1.0594	1.02	231	21.8	32.5	30.1	0.725
550	0.9625	1.05	251	26.1	36.6	36.2	0.721
600	0.8826	1.08	270	30.6	40.7	42.7	0.717
650	0.8143	1.10	288	35.4	44.5	49.7	0.712
700	0.7564	1.13	305	40.3	48.1	56.3	0.717
750	0.7057	1.15	321	45.5	51.7	63.7	0.714
800	0.6614	1.17	337	51.0	55.1	71.2	0.716
Carbon Monoxide (CO), $\mathcal{M}$ = 28.01 kg/kmol							
200	1.6888	1.045	127	7.52	17.0	9.63	0.781
220	1.5341	1.044	137	8.93	19.0	11.9	0.753
240	1.4055	1.043	147	10.5	20.6	14.1	0.744
260	1.2967	1.043	157	12.1	22.1	16.3	0.741
280	1.2038	1.042	166	13.8	23.6	18.8	0.733
300	1.1233	1.043	175	15.6	25.0	21.3	0.730
320	1.0529	1.043	184	17.5	26.3	23.9	0.730
340	0.9909	1.044	193	19.5	27.8	26.9	0.725
360	0.9357	1.045	202	21.6	29.1	29.8	0.725
380	0.8864	1.047	210	23.7	30.5	32.9	0.729
400	0.8421	1.049	218	25.9	31.8	36.0	0.719
450	0.7483	1.055	237	31.7	35.0	44.3	0.714
500	0.67352	1.065	254	37.7	38.1	53.1	0.710
550	0.61226	1.076	271	44.3	41.1	62.4	0.710
600	0.56126	1.088	286	51.0	44.0	72.1	0.707

TABLE A.4 *Continued*

T (K)	ρ (kg/m^3)	c_p (kJ/kg·K)	$\mu \cdot 10^7$ (N·s/m^2)	$\nu \cdot 10^6$ (m^2/s)	$k \cdot 10^3$ (W/m·K)	$\alpha \cdot 10^6$ (m^2/s)	Pr
Carbon Monoxide (CO) *(continued)*							
650	0.51806	1.101	301	58.1	47.0	82.4	0.705
700	0.48102	1.114	315	65.5	50.0	93.3	0.702
750	0.44899	1.127	329	73.3	52.8	104	0.702
800	0.42095	1.140	343	81.5	55.5	116	0.705
Helium (He), $\mathcal{M}$ = 4.003 kg/kmol							
100	0.4871	5.193	96.3	19.8	73.0	28.9	0.686
120	0.4060	5.193	107	26.4	81.9	38.8	0.679
140	0.3481	5.193	118	33.9	90.7	50.2	0.676
160	—	5.193	129	—	99.2	—	—
180	0.2708	5.193	139	51.3	107.2	76.2	0.673
200	—	5.193	150	—	115.1	—	—
220	0.2216	5.193	160	72.2	123.1	107	0.675
240	—	5.193	170	—	130	—	—
260	0.1875	5.193	180	96.0	137	141	0.682
280	—	5.193	190	—	145	—	—
300	0.1625	5.193	199	122	152	180	0.680
350	—	5.193	221	—	170	—	—
400	0.1219	5.193	243	199	187	295	0.675
450	—	5.193	263	—	204	—	—
500	0.09754	5.193	283	290	220	434	0.668
550	—	5.193	—	—	—	—	—
600	—	5.193	320	—	252	—	—
650	—	5.193	332	—	264	—	—
700	0.06969	5.193	350	502	278	768	0.654
750	—	5.193	364	—	291	—	—
800	—	5.193	382	—	304	—	—
900	—	5.193	414	—	330	—	—
1000	0.04879	5.193	446	914	354	1400	0.654
Hydrogen (H_2), $\mathcal{M}$ = 2.016 kg/kmol							
100	0.24255	11.23	42.1	17.4	67.0	24.6	0.707
150	0.16156	12.60	56.0	34.7	101	49.6	0.699
200	0.12115	13.54	68.1	56.2	131	79.9	0.704
250	0.09693	14.06	78.9	81.4	157	115	0.707
300	0.08078	14.31	89.6	111	183	158	0.701
350	0.06924	14.43	98.8	143	204	204	0.700
400	0.06059	14.48	108.2	179	226	258	0.695
450	0.05386	14.50	117.2	218	247	316	0.689
500	0.04848	14.52	126.4	261	266	378	0.691
550	0.04407	14.53	134.3	305	285	445	0.685

TABLE A.4 *Continued*

T (K)	ρ (kg/m^3)	c_p (kJ/kg·K)	$\mu \cdot 10^7$ (N·s/m^2)	$\nu \cdot 10^6$ (m^2/s)	$k \cdot 10^3$ (W/m·K)	$\alpha \cdot 10^6$ (m^2/s)	Pr
Hydrogen (H_2) *(continued)*							
600	0.04040	14.55	142.4	352	305	519	0.678
700	0.03463	14.61	157.8	456	342	676	0.675
800	0.03030	14.70	172.4	569	378	849	0.670
900	0.02694	14.83	186.5	692	412	1030	0.671
1000	0.02424	14.99	201.3	830	448	1230	0.673
1100	0.02204	15.17	213.0	966	488	1460	0.662
1200	0.02020	15.37	226.2	1120	528	1700	0.659
1300	0.01865	15.59	238.5	1279	568	1955	0.655
1400	0.01732	15.81	250.7	1447	610	2230	0.650
1500	0.01616	16.02	262.7	1626	655	2530	0.643
1600	0.0152	16.28	273.7	1801	697	2815	0.639
1700	0.0143	16.58	284.9	1992	742	3130	0.637
1800	0.0135	16.96	296.1	2193	786	3435	0.639
1900	0.0128	17.49	307.2	2400	835	3730	0.643
2000	0.0121	18.25	318.2	2630	878	3975	0.661
Nitrogen (N_2), $\mathcal{M}$ = 28.01 kg/kmol							
100	3.4388	1.070	68.8	2.00	9.58	2.60	0.768
150	2.2594	1.050	100.6	4.45	13.9	5.86	0.759
200	1.6883	1.043	129.2	7.65	18.3	10.4	0.736
250	1.3488	1.042	154.9	11.48	22.2	15.8	0.727
300	1.1233	1.041	178.2	15.86	25.9	22.1	0.716
350	0.9625	1.042	200.0	20.78	29.3	29.2	0.711
400	0.8425	1.045	220.4	26.16	32.7	37.1	0.704
450	0.7485	1.050	239.6	32.01	35.8	45.6	0.703
500	0.6739	1.056	257.7	38.24	38.9	54.7	0.700
550	0.6124	1.065	274.7	44.86	41.7	63.9	0.702
600	0.5615	1.075	290.8	51.79	44.6	73.9	0.701
700	0.4812	1.098	321.0	66.71	49.9	94.4	0.706
800	0.4211	1.122	349.1	82.90	54.8	116	0.715
900	0.3743	1.146	375.3	100.3	59.7	139	0.721
1000	0.3368	1.167	399.9	118.7	64.7	165	0.721
1100	0.3062	1.187	423.2	138.2	70.0	193	0.718
1200	0.2807	1.204	445.3	158.6	75.8	224	0.707
1300	0.2591	1.219	466.2	179.9	81.0	256	0.701
Oxygen (O_2), $\mathcal{M}$ = 32.00 kg/kmol							
100	3.945	0.962	76.4	1.94	9.25	2.44	0.796
150	2.585	0.921	114.8	4.44	13.8	5.80	0.766
200	1.930	0.915	147.5	7.64	18.3	10.4	0.737
250	1.542	0.915	178.6	11.58	22.6	16.0	0.723
300	1.284	0.920	207.2	16.14	26.8	22.7	0.711

TABLE A.4 *Continued*

T (K)	ρ (kg/m^3)	c_p (kJ/kg·K)	$\mu \cdot 10^7$ ($N \cdot s/m^2$)	$\nu \cdot 10^6$ (m^2/s)	$k \cdot 10^3$ (W/m·K)	$\alpha \cdot 10^6$ (m^2/s)	Pr
Oxygen (O_2) *(continued)*							
350	1.100	0.929	233.5	21.23	29.6	29.0	0.733
400	0.9620	0.942	258.2	26.84	33.0	36.4	0.737
450	0.8554	0.956	281.4	32.90	36.3	44.4	0.741
500	0.7698	0.972	303.3	39.40	41.2	55.1	0.716
550	0.6998	0.988	324.0	46.30	44.1	63.8	0.726
600	0.6414	1.003	343.7	53.59	47.3	73.5	0.729
700	0.5498	1.031	380.8	69.26	52.8	93.1	0.744
800	0.4810	1.054	415.2	86.32	58.9	116	0.743
900	0.4275	1.074	447.2	104.6	64.9	141	0.740
1000	0.3848	1.090	477.0	124.0	71.0	169	0.733
1100	0.3498	1.103	505.5	144.5	75.8	196	0.736
1200	0.3206	1.115	532.5	166.1	81.9	229	0.725
1300	0.2960	1.125	588.4	188.6	87.1	262	0.721
Water Vapor (Steam), $\mathcal{M}$ = 18.02 kg/kmol							
380	0.5863	2.060	127.1	21.68	24.6	20.4	1.06
400	0.5542	2.014	134.4	24.25	26.1	23.4	1.04
450	0.4902	1.980	152.5	31.11	29.9	30.8	1.01
500	0.4405	1.985	170.4	38.68	33.9	38.8	0.998
550	0.4005	1.997	188.4	47.04	37.9	47.4	0.993
600	0.3652	2.026	206.7	56.60	42.2	57.0	0.993
650	0.3380	2.056	224.7	66.48	46.4	66.8	0.996
700	0.3140	2.085	242.6	77.26	50.5	77.1	1.00
750	0.2931	2.119	260.4	88.84	54.9	88.4	1.00
800	0.2739	2.152	278.6	101.7	59.2	100	1.01
850	0.2579	2.186	296.9	115.1	63.7	113	1.02

[a]Adapted from References 8, 14, and 15.

TABLE A.5 Thermophysical Properties of Saturated Fluids[a]

Saturated Liquids

T (K)	ρ (kg/m^3)	c_p (kJ/kg · K)	$\mu \cdot 10^2$ (N · s/m^2)	$\nu \cdot 10^6$ (m^2/s)	$k \cdot 10^3$ (W/m · K)	$\alpha \cdot 10^7$ (m^2/s)	Pr	$\beta \cdot 10^3$ (K^{-1})
Engine Oil (Unused)								
273	899.1	1.796	385	4280	147	0.910	47,000	0.70
280	895.3	1.827	217	2430	144	0.880	27,500	0.70
290	890.0	1.868	99.9	1120	145	0.872	12,900	0.70
300	884.1	1.909	48.6	550	145	0.859	6400	0.70
310	877.9	1.951	25.3	288	145	0.847	3400	0.70
320	871.8	1.993	14.1	161	143	0.823	1965	0.70
330	865.8	2.035	8.36	96.6	141	0.800	1205	0.70
340	859.9	2.076	5.31	61.7	139	0.779	793	0.70
350	853.9	2.118	3.56	41.7	138	0.763	546	0.70
360	847.8	2.161	2.52	29.7	138	0.753	395	0.70
370	841.8	2.206	1.86	22.0	137	0.738	300	0.70
380	836.0	2.250	1.41	16.9	136	0.723	233	0.70
390	830.6	2.294	1.10	13.3	135	0.709	187	0.70
400	825.1	2.337	0.874	10.6	134	0.695	152	0.70
410	818.9	2.381	0.698	8.52	133	0.682	125	0.70
420	812.1	2.427	0.564	6.94	133	0.675	103	0.70
430	806.5	2.471	0.470	5.83	132	0.662	88	0.70
Ethylene Glycol [$C_2H_4(OH)_2$]								
273	1130.8	2.294	6.51	57.6	242	0.933	617	0.65
280	1125.8	2.323	4.20	37.3	244	0.933	400	0.65
290	1118.8	2.368	2.47	22.1	248	0.936	236	0.65
300	1114.4	2.415	1.57	14.1	252	0.939	151	0.65
310	1103.7	2.460	1.07	9.65	255	0.939	103	0.65
320	1096.2	2.505	0.757	6.91	258	0.940	73.5	0.65
330	1089.5	2.549	0.561	5.15	260	0.936	55.0	0.65
340	1083.8	2.592	0.431	3.98	261	0.929	42.8	0.65
350	1079.0	2.637	0.342	3.17	261	0.917	34.6	0.65
360	1074.0	2.682	0.278	2.59	261	0.906	28.6	0.65
370	1066.7	2.728	0.228	2.14	262	0.900	23.7	0.65
373	1058.5	2.742	0.215	2.03	263	0.906	22.4	0.65
Glycerin [$C_3H_5(OH)_3$]								
273	1276.0	2.261	1060	8310	282	0.977	85,000	0.47
280	1271.9	2.298	534	4200	284	0.972	43,200	0.47
290	1265.8	2.367	185	1460	286	0.955	15,300	0.48
300	1259.9	2.427	79.9	634	286	0.935	6780	0.48
310	1253.9	2.490	35.2	281	286	0.916	3060	0.49
320	1247.2	2.564	21.0	168	287	0.897	1870	0.50

TABLE A.5 *Continued*

Saturated Liquids (Continued)

T (K)	ρ (kg/m^3)	c_p (kJ/kg·K)	$\mu \cdot 10^2$ (N·s/m^2)	$\nu \cdot 10^6$ (m^2/s)	$k \cdot 10^3$ (W/m·K)	$\alpha \cdot 10^7$ (m^2/s)	Pr	$\beta \cdot 10^3$ (K^{-1})
Refrigerant-134a ($C_2H_2F_4$)								
230	1426.8	1.249	0.04912	0.3443	112.1	0.629	5.5	2.02
240	1397.7	1.267	0.04202	0.3006	107.3	0.606	5.0	2.11
250	1367.9	1.287	0.03633	0.2656	102.5	0.583	4.6	2.23
260	1337.1	1.308	0.03166	0.2368	97.9	0.560	4.2	2.36
270	1305.1	1.333	0.02775	0.2127	93.4	0.537	4.0	2.53
280	1271.8	1.361	0.02443	0.1921	89.0	0.514	3.7	2.73
290	1236.8	1.393	0.02156	0.1744	84.6	0.491	3.5	2.98
300	1199.7	1.432	0.01905	0.1588	80.3	0.468	3.4	3.30
310	1159.9	1.481	0.01680	0.1449	76.1	0.443	3.3	3.73
320	1116.8	1.543	0.01478	0.1323	71.8	0.417	3.2	4.33
330	1069.1	1.627	0.01292	0.1209	67.5	0.388	3.1	5.19
340	1015.0	1.751	0.01118	0.1102	63.1	0.355	3.1	6.57
350	951.3	1.961	0.00951	0.1000	58.6	0.314	3.2	9.10
360	870.1	2.437	0.00781	0.0898	54.1	0.255	3.5	15.39
370	740.3	5.105	0.00580	0.0783	51.8	0.137	5.7	55.24
Refrigerant-22 ($CHClF_2$)								
230	1416.0	1.087	0.03558	0.2513	114.5	0.744	3.4	2.05
240	1386.6	1.100	0.03145	0.2268	109.8	0.720	3.2	2.16
250	1356.3	1.117	0.02796	0.2062	105.2	0.695	3.0	2.29
260	1324.9	1.137	0.02497	0.1884	100.7	0.668	2.8	2.45
270	1292.1	1.161	0.02235	0.1730	96.2	0.641	2.7	2.63
280	1257.9	1.189	0.02005	0.1594	91.7	0.613	2.6	2.86
290	1221.7	1.223	0.01798	0.1472	87.2	0.583	2.5	3.15
300	1183.4	1.265	0.01610	0.1361	82.6	0.552	2.5	3.51
310	1142.2	1.319	0.01438	0.1259	78.1	0.518	2.4	4.00
320	1097.4	1.391	0.01278	0.1165	73.4	0.481	2.4	4.69
330	1047.5	1.495	0.01127	0.1075	68.6	0.438	2.5	5.75
340	990.1	1.665	0.00980	0.0989	63.6	0.386	2.6	7.56
350	920.1	1.997	0.00831	0.0904	58.3	0.317	2.8	11.35
360	823.4	3.001	0.00668	0.0811	53.1	0.215	3.8	23.88
Mercury (Hg)								
273	13,595	0.1404	0.1688	0.1240	8180	42.85	0.0290	0.181
300	13,529	0.1393	0.1523	0.1125	8540	45.30	0.0248	0.181
350	13,407	0.1377	0.1309	0.0976	9180	49.75	0.0196	0.181
400	13,287	0.1365	0.1171	0.0882	9800	54.05	0.0163	0.181
450	13,167	0.1357	0.1075	0.0816	10,400	58.10	0.0140	0.181
500	13,048	0.1353	0.1007	0.0771	10,950	61.90	0.0125	0.182
550	12,929	0.1352	0.0953	0.0737	11,450	65.55	0.0112	0.184
600	12,809	0.1355	0.0911	0.0711	11,950	68.80	0.0103	0.187

TABLE A.5 *Continued*

Saturated Liquid–Vapor, 1 atm[b]

Fluid	T_{sat} (K)	h_{fg} (kJ/kg)	ρ_f (kg/m^3)	ρ_g (kg/m^3)	$\sigma \cdot 10^3$ (N/m)
Ethanol	351	846	757	1.44	17.7
Ethylene glycol	470	812	1111[c]	—	32.7
Glycerin	563	974	1260[c]	—	63.0[c]
Mercury	630	301	12,740	3.90	417
Refrigerant R-134a	247	217	1377	5.26	15.4
Refrigerant R-22	232	234	1409	4.70	18.1

[a]Adapted from References 15–19.
[b]Adapted from References 8, 20, and 21.
[c]Property value corresponding to 300 K.

TABLE A.6 Thermophysical Properties of Saturated Water[a]

Temperature, T (K)	Pressure, p (bars)[b]	Specific Volume (m³/kg) $v_f \cdot 10^3$	Specific Volume (m³/kg) v_g	Heat of Vaporization, h_{fg} (kJ/kg)	Specific Heat (kJ/kg · K) $c_{p,f}$	Specific Heat (kJ/kg · K) $c_{p,g}$	Viscosity (N · s/m²) $\mu_f \cdot 10^6$	Viscosity (N · s/m²) $\mu_g \cdot 10^6$	Thermal Conductivity (W/m · K) $k_f \cdot 10^3$	Thermal Conductivity (W/m · K) $k_g \cdot 10^3$	Prandtl Number Pr_f	Prandtl Number Pr_g	Surface Tension, $\sigma_f \cdot 10^3$ (N/m)	Expansion Coefficient, $\beta_f \cdot 10^6$ (K⁻¹)	Temperature, T (K)
273.15	0.00611	1.000	206.3	2502	4.217	1.854	1750	8.02	569	18.2	12.99	0.815	75.5	−68.05	273.15
275	0.00697	1.000	181.7	2497	4.211	1.855	1652	8.09	574	18.3	12.22	0.817	75.3	−32.74	275
280	0.00990	1.000	130.4	2485	4.198	1.858	1422	8.29	582	18.6	10.26	0.825	74.8	46.04	280
285	0.01387	1.000	99.4	2473	4.189	1.861	1225	8.49	590	18.9	8.81	0.833	74.3	114.1	285
290	0.01917	1.001	69.7	2461	4.184	1.864	1080	8.69	598	19.3	7.56	0.841	73.7	174.0	290
295	0.02617	1.002	51.94	2449	4.181	1.868	959	8.89	606	19.5	6.62	0.849	72.7	227.5	295
300	0.03531	1.003	39.13	2438	4.179	1.872	855	9.09	613	19.6	5.83	0.857	71.7	276.1	300
305	0.04712	1.005	29.74	2426	4.178	1.877	769	9.29	620	20.1	5.20	0.865	70.9	320.6	305
310	0.06221	1.007	22.93	2414	4.178	1.882	695	9.49	628	20.4	4.62	0.873	70.0	361.9	310
315	0.08132	1.009	17.82	2402	4.179	1.888	631	9.69	634	20.7	4.16	0.883	69.2	400.4	315
320	0.1053	1.011	13.98	2390	4.180	1.895	577	9.89	640	21.0	3.77	0.894	68.3	436.7	320
325	0.1351	1.013	11.06	2378	4.182	1.903	528	10.09	645	21.3	3.42	0.901	67.5	471.2	325
330	0.1719	1.016	8.82	2366	4.184	1.911	489	10.29	650	21.7	3.15	0.908	66.6	504.0	330
335	0.2167	1.018	7.09	2354	4.186	1.920	453	10.49	656	22.0	2.88	0.916	65.8	535.5	335
340	0.2713	1.021	5.74	2342	4.188	1.930	420	10.69	660	22.3	2.66	0.925	64.9	566.0	340
345	0.3372	1.024	4.683	2329	4.191	1.941	389	10.89	664	22.6	2.45	0.933	64.1	595.4	345
350	0.4163	1.027	3.846	2317	4.195	1.954	365	11.09	668	23.0	2.29	0.942	63.2	624.2	350
355	0.5100	1.030	3.180	2304	4.199	1.968	343	11.29	671	23.3	2.14	0.951	62.3	652.3	355
360	0.6209	1.034	2.645	2291	4.203	1.983	324	11.49	674	23.7	2.02	0.960	61.4	697.9	360
365	0.7514	1.038	2.212	2278	4.209	1.999	306	11.69	677	24.1	1.91	0.969	60.5	707.1	365
370	0.9040	1.041	1.861	2265	4.214	2.017	289	11.89	679	24.5	1.80	0.978	59.5	728.7	370
373.15	1.0133	1.044	1.679	2257	4.217	2.029	279	12.02	680	24.8	1.76	0.984	58.9	750.1	373.15
375	1.0815	1.045	1.574	2252	4.220	2.036	274	12.09	681	24.9	1.70	0.987	58.6	761	375
380	1.2869	1.049	1.337	2239	4.226	2.057	260	12.29	683	25.4	1.61	0.999	57.6	788	380
385	1.5233	1.053	1.142	2225	4.232	2.080	248	12.49	685	25.8	1.53	1.004	56.6	814	385
390	1.794	1.058	0.980	2212	4.239	2.104	237	12.69	686	26.3	1.47	1.013	55.6	841	390
400	2.455	1.067	0.731	2183	4.256	2.158	217	13.05	688	27.2	1.34	1.033	53.6	896	400
410	3.302	1.077	0.553	2153	4.278	2.221	200	13.42	688	28.2	1.24	1.054	51.5	952	410
420	4.370	1.088	0.425	2123	4.302	2.291	185	13.79	688	29.8	1.16	1.075	49.4	1010	420
430	5.699	1.099	0.331	2091	4.331	2.369	173	14.14	685	30.4	1.09	1.10	47.2		430

TABLE A.6 *Continued*

Temperature, T (K)	Pressure, p (bars)[b]	Specific Volume (m³/kg)		Heat of Vaporization, h_{fg} (kJ/kg)	Specific Heat (kJ/kg · K)		Viscosity (N · s/m²)		Thermal Conductivity (W/m · K)		Prandtl Number		Surface Tension, $\sigma_f \cdot 10^3$ (N/m)	Expansion Coefficient, $\beta_f \cdot 10^6$ (K⁻¹)	Temperature, T (K)
		$v_f \cdot 10^3$	v_g		$c_{p,f}$	$c_{p,g}$	$\mu_f \cdot 10^6$	$\mu_g \cdot 10^6$	$k_f \cdot 10^3$	$k_g \cdot 10^3$	Pr_f	Pr_g			
440	7.333	1.110	0.261	2059	4.36	2.46	162	14.50	682	31.7	1.04	1.12	45.1		440
450	9.319	1.123	0.208	2024	4.40	2.56	152	14.85	678	33.1	0.99	1.14	42.9		450
460	11.71	1.137	0.167	1989	4.44	2.68	143	15.19	673	34.6	0.95	1.17	40.7		460
470	14.55	1.152	0.136	1951	4.48	2.79	136	15.54	667	36.3	0.92	1.20	38.5		470
480	17.90	1.167	0.111	1912	4.53	2.94	129	15.88	660	38.1	0.89	1.23	36.2		480
490	21.83	1.184	0.0922	1870	4.59	3.10	124	16.23	651	40.1	0.87	1.25	33.9	—	490
500	26.40	1.203	0.0766	1825	4.66	3.27	118	16.59	642	42.3	0.86	1.28	31.6	—	500
510	31.66	1.222	0.0631	1779	4.74	3.47	113	16.95	631	44.7	0.85	1.31	29.3	—	510
520	37.70	1.244	0.0525	1730	4.84	3.70	108	17.33	621	47.5	0.84	1.35	26.9	—	520
530	44.58	1.268	0.0445	1679	4.95	3.96	104	17.72	608	50.6	0.85	1.39	24.5	—	530
540	52.38	1.294	0.0375	1622	5.08	4.27	101	18.1	594	54.0	0.86	1.43	22.1	—	540
550	61.19	1.323	0.0317	1564	5.24	4.64	97	18.6	580	58.3	0.87	1.47	19.7	—	550
560	71.08	1.355	0.0269	1499	5.43	5.09	94	19.1	563	63.7	0.90	1.52	17.3	—	560
570	82.16	1.392	0.0228	1429	5.68	5.67	91	19.7	548	76.7	0.94	1.59	15.0	—	570
580	94.51	1.433	0.0193	1353	6.00	6.40	88	20.4	528	76.7	0.99	1.68	12.8	—	580
590	108.3	1.482	0.0163	1274	6.41	7.35	84	21.5	513	84.1	1.05	1.84	10.5	—	590
600	123.5	1.541	0.0137	1176	7.00	8.75	81	22.7	497	92.9	1.14	2.15	8.4	—	600
610	137.3	1.612	0.0115	1068	7.85	11.1	77	24.1	467	103	1.30	2.60	6.3	—	610
620	159.1	1.705	0.0094	941	9.35	15.4	72	25.9	444	114	1.52	3.46	4.5	—	620
625	169.1	1.778	0.0085	858	10.6	18.3	70	27.0	430	121	1.65	4.20	3.5	—	625
630	179.7	1.856	0.0075	781	12.6	22.1	67	28.0	412	130	2.0	4.8	2.6	—	630
635	190.9	1.935	0.0066	683	16.4	27.6	64	30.0	392	141	2.7	6.0	1.5	—	635
640	202.7	2.075	0.0057	560	26	42	59	32.0	367	155	4.2	9.6	0.8	—	640
645	215.2	2.351	0.0045	361	90	—	54	37.0	331	178	12	26	0.1	—	645
647.3[c]	221.2	3.170	0.0032	0	∞	∞	45	45.0	238	238	∞	∞	0.0	—	647.3[c]

[a]Adapted from Reference 22.
[b]1 bar = 10^5 N/m².
[c]Critical temperature.

TABLE A.7 Thermophysical Properties of Liquid Metals[a]

Composition	Melting Point (K)	T (K)	ρ (kg/m^3)	c_p (kJ/kg · K)	$\nu \cdot 10^7$ (m^2/s)	k (W/m · K)	$\alpha \cdot 10^5$ (m^2/s)	Pr
Bismuth	544	589	10,011	0.1444	1.617	16.4	1.138	0.0142
		811	9739	0.1545	1.133	15.6	1.035	0.0110
		1033	9467	0.1645	0.8343	15.6	1.001	0.0083
Lead	600	644	10,540	0.159	2.276	16.1	1.084	0.024
		755	10,412	0.155	1.849	15.6	1.223	0.017
		977	10,140	—	1.347	14.9	—	—
Potassium	337	422	807.3	0.80	4.608	45.0	6.99	0.0066
		700	741.7	0.75	2.397	39.5	7.07	0.0034
		977	674.4	0.75	1.905	33.1	6.55	0.0029
Sodium	371	366	929.1	1.38	7.516	86.2	6.71	0.011
		644	860.2	1.30	3.270	72.3	6.48	0.0051
		977	778.5	1.26	2.285	59.7	6.12	0.0037
NaK, (45%/55%)	292	366	887.4	1.130	6.522	25.6	2.552	0.026
		644	821.7	1.055	2.871	27.5	3.17	0.0091
		977	740.1	1.043	2.174	28.9	3.74	0.0058
NaK, (22%/78%)	262	366	849.0	0.946	5.797	24.4	3.05	0.019
		672	775.3	0.879	2.666	26.7	3.92	0.0068
		1033	690.4	0.883	2.118	—	—	—
PbBi, (44.5%/55.5%)	398	422	10,524	0.147	—	9.05	0.586	—
		644	10,236	0.147	1.496	11.86	0.790	0.189
		922	9835	—	1.171	—	—	—
Mercury	234			See Table A.5				

[a]Adapted from Reference 23.

TABLE A.8 Total, Normal (n) or Hemispherical (h) Emissivity of Selected Surfaces

Metallic Solids and Their Oxides[a]

Description/Composition		Emissivity, ε_n or ε_h, at Various Temperatures (K)										
		100	200	300	400	600	800	1000	1200	1500	2000	2500
Aluminum												
Highly polished, film	(h)	0.02	0.03	0.04	0.05	0.06						
Foil, bright	(h)	0.06	0.06	0.07								
Anodized	(h)			0.82	0.76							
Chromium												
Polished or plated	(n)	0.05	0.07	0.10	0.12	0.14						
Copper												
Highly polished	(h)			0.03	0.03	0.04	0.04	0.04				
Stably oxidized	(h)					0.50	0.58	0.80				
Gold												
Highly polished or film	(h)	0.01	0.02	0.03	0.03	0.04	0.05	0.06				
Foil, bright	(h)	0.06	0.07	0.07								
Molybdenum												
Polished	(h)					0.06	0.08	0.10	0.12	0.15	0.21	0.26
Shot-blasted, rough	(h)					0.25	0.28	0.31	0.35	0.42		
Stably oxidized	(h)					0.80	0.82					
Nickel												
Polished	(h)					0.09	0.11	0.14	0.17			
Stably oxidized	(h)					0.40	0.49	0.57				
Platinum												
Polished	(h)						0.10	0.13	0.15	0.18		
Silver												
Polished	(h)			0.02	0.02	0.03	0.05	0.08				
Stainless steels												
Typical, polished	(n)			0.17	0.17	0.19	0.23	0.30				
Typical, cleaned	(n)			0.22	0.22	0.24	0.28	0.35				
Typical, lightly oxidized	(n)						0.33	0.40				
Typical, highly oxidized	(n)						0.67	0.70	0.76			
AISI 347, stably oxidized	(n)					0.87	0.88	0.89	0.90			
Tantalum												
Polished	(h)								0.11	0.17	0.23	0.28
Tungsten												
Polished	(h)							0.10	0.13	0.18	0.25	0.29

TABLE A.8 *Continued*

Nonmetallic Substances[b]

Description/Composition		Temperature (K)	Emissivity ε
Aluminum oxide	(*n*)	600	0.69
		1000	0.55
		1500	0.41
Asphalt pavement	(*h*)	300	0.85–0.93
Building materials			
Asbestos sheet	(*h*)	300	0.93–0.96
Brick, red	(*h*)	300	0.93–0.96
Gypsum or plaster board	(*h*)	300	0.90–0.92
Wood	(*h*)	300	0.82–0.92
Cloth	(*h*)	300	0.75–0.90
Concrete	(*h*)	300	0.88–0.93
Glass, window	(*h*)	300	0.90–0.95
Ice	(*h*)	273	0.95–0.98
Paints			
Black (Parsons)	(*h*)	300	0.98
White, acrylic	(*h*)	300	0.90
White, zinc oxide	(*h*)	300	0.92
Paper, white	(*h*)	300	0.92–0.97
Pyrex	(*n*)	300	0.82
		600	0.80
		1000	0.71
		1200	0.62
Pyroceram	(*n*)	300	0.85
		600	0.78
		1000	0.69
		1500	0.57
Refractories (furnace liners)			
Alumina brick	(*n*)	800	0.40
		1000	0.33
		1400	0.28
		1600	0.33
Magnesia brick	(*n*)	800	0.45
		1000	0.36
		1400	0.31
		1600	0.40
Kaolin insulating brick	(*n*)	800	0.70
		1200	0.57
		1400	0.47
		1600	0.53
Sand	(*h*)	300	0.90
Silicon carbide	(*n*)	600	0.87
		1000	0.87
		1500	0.85
Skin	(*h*)	300	0.95
Snow	(*h*)	273	0.82–0.90

TABLE A.8 *Continued*

Nonmetallic Substances[b]

Description/Composition		Temperature (K)	Emissivity ε
Soil	(*h*)	300	0.93–0.96
Rocks	(*h*)	300	0.88–0.95
Teflon	(*h*)	300	0.85
		400	0.87
		500	0.92
Vegetation	(*h*)	300	0.92–0.96
Water	(*h*)	300	0.96

[a]Adapted from Reference 1.
[b]Adapted from References 1, 9, 24, and 25.

TABLE A.9 **Solar Radiative Properties for Selected Materials**[a]

Description/Composition	α_S	ε^b	α_S/ε	τ_S
Aluminum				
Polished	0.09	0.03	3.0	
Anodized	0.14	0.84	0.17	
Quartz overcoated	0.11	0.37	0.30	
Foil	0.15	0.05	3.0	
Brick, red (Purdue)	0.63	0.93	0.68	
Concrete	0.60	0.88	0.68	
Galvanized sheet metal				
Clean, new	0.65	0.13	5.0	
Oxidized, weathered	0.80	0.28	2.9	
Glass, 3.2-mm thickness				
Float or tempered				0.79
Low iron oxide type				0.88
Metal, plated				
Black sulfide	0.92	0.10	9.2	
Black cobalt oxide	0.93	0.30	3.1	
Black nickel oxide	0.92	0.08	11	
Black chrome	0.87	0.09	9.7	
Mylar, 0.13-mm thickness				0.87
Paints				
Black (Parsons)	0.98	0.98	1.0	
White, acrylic	0.26	0.90	0.29	
White, zinc oxide	0.16	0.93	0.17	
Plexiglas, 3.2-mm thickness				0.90
Snow				
Fine particles, fresh	0.13	0.82	0.16	
Ice granules	0.33	0.89	0.37	
Tedlar, 0.10-mm thickness				0.92
Teflon, 0.13-mm thickness				0.92

[a]Adapted with permission from Reference 25.
[b]The emissivity values in this table correspond to a surface temperature of approximately 300 K.

References

1. Touloukian, Y. S., and C. Y. Ho, Eds., *Thermophysical Properties of Matter,* Vol. 1, *Thermal Conductivity of Metallic Solids;* Vol. 2, *Thermal Conductivity of Nonmetallic Solids;* Vol. 4, *Specific Heat of Metallic Solids;* Vol. 5, *Specific Heat of Nonmetallic Solids;* Vol. 7, *Thermal Radiative Properties of Metallic Solids;* Vol. 8, *Thermal Radiative Properties of Nonmetallic Solids;* Vol. 9, *Thermal Radiative Properties of Coatings,* Plenum Press, New York, 1972.
2. Touloukian, Y. S., and C. Y. Ho, Eds., *Thermophysical Properties of Selected Aerospace Materials,* Part I: Thermal Radiative Properties; Part II: Thermophysical Properties of Seven Materials. Thermophysical and Electronic Properties Information Analysis Center, CINDAS, Purdue University, West Lafayette, IN, 1976.
3. Ho, C. Y., R. W. Powell, and P. E. Liley, *J. Phys. Chem. Ref. Data,* **3,** Supplement 1, 1974.
4. Desai, P. D., T. K. Chu, R. H. Bogaard, M. W. Ackermann, and C. Y. Ho, Part I: Thermophysical Properties of Carbon Steels; Part II: Thermophysical Properties of Low Chromium Steels; Part III: Thermophysical Properties of Nickel Steels; Part IV: Thermophysical Properties of Stainless Steels. CINDAS Special Report, Purdue University, West Lafayette, IN, September 1976.
5. American Society for Metals, *Metals Handbook,* Vol. 1, *Properties and Selection of Metals,* 8th ed., ASM, Metals Park, OH, 1961.
6. Hultgren, R., P. D. Desai, D. T. Hawkins, M. Gleiser, K. K. Kelley, and D. D. Wagman, *Selected Values of the Thermodynamic Properties of the Elements,* American Society of Metals, Metals Park, OH, 1973.
7. Hultgren, R., P. D. Desai, D. T. Hawkins, M. Gleiser, and K. K. Kelley, *Selected Values of the Thermodynamic Properties of Binary Alloys,* American Society of Metals, Metals Park, OH, 1973.
8. American Society of Heating, Refrigerating and Air Conditioning Engineers, *ASHRAE Handbook of Fundamentals,* ASHRAE, New York, 1981.
9. Mallory, J. F., *Thermal Insulation,* Van Nostrand Reinhold, New York, 1969.
10. Hanley, E. J., D. P. DeWitt, and R. E. Taylor, "The Thermal Transport Properties at Normal and Elevated Temperature of Eight Representative Rocks," *Proceedings of the Seventh Symposium on Thermophysical Properties,* American Society of Mechanical Engineers, New York, 1977.
11. Sweat, V. E., "A Miniature Thermal Conductivity Probe for Foods," American Society of Mechanical Engineers, Paper 76-HT-60, August 1976.
12. Kothandaraman, C. P., and S. Subramanyan, *Heat and Mass Transfer Data Book,* Halsted Press/Wiley, Hoboken, NJ, 1975.
13. Chapman, A. J., *Heat Transfer,* 4th ed., Macmillan, New York, 1984.
14. Vargaftik, N. B., *Tables of Thermophysical Properties of Liquids and Gases,* 2nd ed., Hemisphere Publishing, New York, 1975.
15. Eckert, E. R. G., and R. M. Drake, *Analysis of Heat and Mass Transfer,* McGraw-Hill, New York, 1972.
16. Vukalovich, M. P., A. I. Ivanov, L. R. Fokin, and A. T. Yakovelev, *Thermophysical Properties of Mercury,* State Committee on Standards, State Service for Standards and Handbook Data, Monograph Series No. 9, Izd. Standartov, Moscow, 1971.
17. Tillner-Roth, R., and H. D. Baehr, *J. Phys. Chem. Ref. Data*, **23**, 657, 1994.
18. Kamei, A., S. W. Beyerlein, and R. T. Jacobsen, *Int. J. Thermophysics*, **16**, 1155, 1995.
19. Lemmon, E. W., M. O. McLinden, and M. L. Huber, *NIST Standard Reference Database* 23: Reference Fluid Thermodynamic and Transport Properties-REFPROP, Version 7.0 National Institute of Standards and Technology, Standard Reference Data Program, Gaithersburg, 2002.
20. Bolz, R. E., and G. L. Tuve, Eds., *CRC Handbook of Tables for Applied Engineering Science*, 2nd ed., CRC Press, Boca Raton, FL, 1979.
21. Liley, P. E., private communication, School of Mechanical Engineering, Purdue University, West Lafayette, IN, May 1984.
22. Liley, P. E., Steam Tables in SI Units, private communication, School of Mechanical Engineering, Purdue University, West Lafayette, IN, March 1984.
23. *Liquid Materials Handbook*, 23rd ed., The Atomic Energy Commission, Department of the Navy, Washington, DC, 1952.
24. Gubareff, G. G., J. E. Janssen, and R. H. Torborg, *Thermal Radiation Properties Survey,* Minneapolis-Honeywell Regulator Company, Minneapolis, MN, 1960.
25. Kreith, F., and J. F. Kreider, *Principles of Solar Energy,* Hemisphere Publishing, New York, 1978.

APPENDIX B

Mathematical Relations and Functions

Section		*Page*

B.1 Hyperbolic Functions[1]

x	sinh x	cosh x	tanh x	x	sinh x	cosh x	tanh x
0.00	0.0000	1.0000	0.00000	2.00	3.6269	3.7622	0.96403
0.10	0.1002	1.0050	0.09967	2.10	4.0219	4.1443	0.97045
0.20	0.2013	1.0201	0.19738	2.20	4.4571	4.5679	0.97574
0.30	0.3045	1.0453	0.29131	2.30	4.9370	5.0372	0.98010
0.40	0.4108	1.0811	0.37995	2.40	5.4662	5.5569	0.98367
0.50	0.5211	1.1276	0.46212	2.50	6.0502	6.1323	0.98661
0.60	0.6367	1.1855	0.53705	2.60	6.6947	6.7690	0.98903
0.70	0.7586	1.2552	0.60437	2.70	7.4063	7.4735	0.99101
0.80	0.8881	1.3374	0.66404	2.80	8.1919	8.2527	0.99263
0.90	1.0265	1.4331	0.71630	2.90	9.0596	9.1146	0.99396
1.00	1.1752	1.5431	0.76159	3.00	10.018	10.068	0.99505
1.10	1.3356	1.6685	0.80050	3.50	16.543	16.573	0.99818
1.20	1.5095	1.8107	0.83365	4.00	27.290	27.308	0.99933
1.30	1.6984	1.9709	0.86172	4.50	45.003	45.014	0.99975
1.40	1.9043	2.1509	0.88535	5.00	74.203	74.210	0.99991
1.50	2.1293	2.3524	0.90515	6.00	201.71	201.72	0.99999
1.60	2.3756	2.5775	0.92167	7.00	548.32	548.32	1.0000
1.70	2.6456	2.8283	0.93541	8.00	1490.5	1490.5	1.0000
1.80	2.9422	3.1075	0.94681	9.00	4051.5	4051.5	1.0000
1.90	3.2682	3.4177	0.95624	10.000	11013	11013	1.0000

[1]The hyperbolic functions are defined as

$$\sinh x = \tfrac{1}{2}(e^x - e^{-x}) \qquad \cosh x = \tfrac{1}{2}(e^x + e^{-x}) \qquad \tanh x = \frac{e^x - e^{-x}}{e^x + e^{-x}} = \frac{\sinh x}{\cosh x}$$

The derivatives of the hyperbolic functions of the variable u are given as

$$\frac{d}{dx}(\sinh u) = (\cosh u)\frac{du}{dx} \qquad \frac{d}{dx}(\cosh u) = (\sinh u)\frac{du}{dx} \qquad \frac{d}{dx}(\tanh u) = \left(\frac{1}{\cosh^2 u}\right)\frac{du}{dx}$$

B.2 *Gaussian Error Function*[1]

w	erf w	w	erf w	w	erf w
0.00	0.00000	0.36	0.38933	1.04	0.85865
0.02	0.02256	0.38	0.40901	1.08	0.87333
0.04	0.04511	0.40	0.42839	1.12	0.88679
0.06	0.06762	0.44	0.46622	1.16	0.89910
0.08	0.09008	0.48	0.50275	1.20	0.91031
0.10	0.11246	0.52	0.53790	1.30	0.93401
0.12	0.13476	0.56	0.57162	1.40	0.95228
0.14	0.15695	0.60	0.60386	1.50	0.96611
0.16	0.17901	0.64	0.63459	1.60	0.97635
0.18	0.20094	0.68	0.66378	1.70	0.98379
0.20	0.22270	0.72	0.69143	1.80	0.98909
0.22	0.24430	0.76	0.71754	1.90	0.99279
0.24	0.26570	0.80	0.74210	2.00	0.99532
0.26	0.28690	0.84	0.76514	2.20	0.99814
0.28	0.30788	0.88	0.78669	2.40	0.99931
0.30	0.32863	0.92	0.80677	2.60	0.99976
0.32	0.34913	0.96	0.82542	2.80	0.99992
0.34	0.36936	1.00	0.84270	3.00	0.99998

[1]The Gaussian error function is defined as

$$\text{erf}\, w = \frac{2}{\sqrt{\pi}} \int_0^w e^{-v^2}\, dv$$

The complementary error function is defined as

$$\text{erfc}\, w \equiv 1 - \text{erf}\, w$$

B.3 *The First Four Roots of the Transcendental Equation, $\xi_n \tan \xi_n = Bi$, for Transient Conduction in a Plane Wall*

$Bi = \frac{hL}{k}$	ξ_1	ξ_2	ξ_3	ξ_4
0	0	3.1416	6.2832	9.4248
0.001	0.0316	3.1419	6.2833	9.4249
0.002	0.0447	3.1422	6.2835	9.4250
0.004	0.0632	3.1429	6.2838	9.4252
0.006	0.0774	3.1435	6.2841	9.4254
0.008	0.0893	3.1441	6.2845	9.4256
0.01	0.0998	3.1448	6.2848	9.4258
0.02	0.1410	3.1479	6.2864	9.4269
0.04	0.1987	3.1543	6.2895	9.4290
0.06	0.2425	3.1606	6.2927	9.4311
0.08	0.2791	3.1668	6.2959	9.4333
0.1	0.3111	3.1731	6.2991	9.4354
0.2	0.4328	3.2039	6.3148	9.4459
0.3	0.5218	3.2341	6.3305	9.4565
0.4	0.5932	3.2636	6.3461	9.4670
0.5	0.6533	3.2923	6.3616	9.4775
0.6	0.7051	3.3204	6.3770	9.4879
0.7	0.7506	3.3477	6.3923	9.4983
0.8	0.7910	3.3744	6.4074	9.5087
0.9	0.8274	3.4003	6.4224	9.5190
1.0	0.8603	3.4256	6.4373	9.5293
1.5	0.9882	3.5422	6.5097	9.5801
2.0	1.0769	3.6436	6.5783	9.6296
3.0	1.1925	3.8088	6.7040	9.7240
4.0	1.2646	3.9352	6.8140	9.8119
5.0	1.3138	4.0336	6.9096	9.8928
6.0	1.3496	4.1116	6.9924	9.9667
7.0	1.3766	4.1746	7.0640	10.0339
8.0	1.3978	4.2264	7.1263	10.0949
9.0	1.4149	4.2694	7.1806	10.1502
10.0	1.4289	4.3058	7.2281	10.2003
15.0	1.4729	4.4255	7.3959	10.3898
20.0	1.4961	4.4915	7.4954	10.5117
30.0	1.5202	4.5615	7.6057	10.6543
40.0	1.5325	4.5979	7.6647	10.7334
50.0	1.5400	4.6202	7.7012	10.7832
60.0	1.5451	4.6353	7.7259	10.8172
80.0	1.5514	4.6543	7.7573	10.8606
100.0	1.5552	4.6658	7.7764	10.8871
∞	1.5708	4.7124	7.8540	10.9956

B.4 *Bessel Functions of the First Kind*

x	$J_0(x)$	$J_1(x)$
0.0	1.0000	0.0000
0.1	0.9975	0.0499
0.2	0.9900	0.0995
0.3	0.9776	0.1483
0.4	0.9604	0.1960
0.5	0.9385	0.2423
0.6	0.9120	0.2867
0.7	0.8812	0.3290
0.8	0.8463	0.3688
0.9	0.8075	0.4059
1.0	0.7652	0.4400
1.1	0.7196	0.4709
1.2	0.6711	0.4983
1.3	0.6201	0.5220
1.4	0.5669	0.5419
1.5	0.5118	0.5579
1.6	0.4554	0.5699
1.7	0.3980	0.5778
1.8	0.3400	0.5815
1.9	0.2818	0.5812
2.0	0.2239	0.5767
2.1	0.1666	0.5683
2.2	0.1104	0.5560
2.3	0.0555	0.5399
2.4	0.0025	0.5202

B.5 *Modified Bessel Functions*[1] *of the First and Second Kinds*

x	$e^{-x}I_0(x)$	$e^{-x}I_1(x)$	$e^{x}K_0(x)$	$e^{x}K_1(x)$
0.0	1.0000	0.0000	∞	∞
0.2	0.8269	0.0823	2.1407	5.8334
0.4	0.6974	0.1368	1.6627	3.2587
0.6	0.5993	0.1722	1.4167	2.3739
0.8	0.5241	0.1945	1.2582	1.9179
1.0	0.4657	0.2079	1.1445	1.6361
1.2	0.4198	0.2152	1.0575	1.4429
1.4	0.3831	0.2185	0.9881	1.3010
1.6	0.3533	0.2190	0.9309	1.1919
1.8	0.3289	0.2177	0.8828	1.1048
2.0	0.3085	0.2153	0.8416	1.0335
2.2	0.2913	0.2121	0.8056	0.9738
2.4	0.2766	0.2085	0.7740	0.9229
2.6	0.2639	0.2046	0.7459	0.8790
2.8	0.2528	0.2007	0.7206	0.8405
3.0	0.2430	0.1968	0.6978	0.8066
3.2	0.2343	0.1930	0.6770	0.7763
3.4	0.2264	0.1892	0.6579	0.7491
3.6	0.2193	0.1856	0.6404	0.7245
3.8	0.2129	0.1821	0.6243	0.7021
4.0	0.2070	0.1787	0.6093	0.6816
4.2	0.2016	0.1755	0.5953	0.6627
4.4	0.1966	0.1724	0.5823	0.6453
4.6	0.1919	0.1695	0.5701	0.6292
4.8	0.1876	0.1667	0.5586	0.6142
5.0	0.1835	0.1640	0.5478	0.6003
5.2	0.1797	0.1614	0.5376	0.5872
5.4	0.1762	0.1589	0.5279	0.5749
5.6	0.1728	0.1565	0.5188	0.5633
5.8	0.1696	0.1542	0.5101	0.5525
6.0	0.1666	0.1520	0.5019	0.5422
6.4	0.1611	0.1479	0.4865	0.5232
6.8	0.1561	0.1441	0.4724	0.5060
7.2	0.1515	0.1405	0.4595	0.4905
7.6	0.1473	0.1372	0.4476	0.4762
8.0	0.1434	0.1341	0.4366	0.4631
8.4	0.1398	0.1312	0.4264	0.4511
8.8	0.1365	0.1285	0.4168	0.4399
9.2	0.1334	0.1260	0.4079	0.4295
9.6	0.1305	0.1235	0.3995	0.4198
10.0	0.1278	0.1213	0.3916	0.4108

[1] $I_{n+1}(x) = I_{n-1}(x) - (2n/x)I_n(x)$

APPENDIX C

Thermal Conditions Associated with Uniform Energy Generation in One-Dimensional, Steady-State Systems

In Section 3.5 the problem of conduction with thermal energy generation is considered for one-dimensional, steady-state conditions. The form of the heat equation differs, according to whether the system is a plane wall, a cylindrical shell, or a spherical shell (Figure C.1). In each case, there are several options for the boundary condition at each surface, and hence a greater number of possibilities for specific forms of the temperature distribution and heat rate (or heat flux).

An alternative to solving the heat equation for each possible combination of boundary conditions involves obtaining a solution by prescribing *boundary conditions of the first kind,* Equation 2.31, at both surfaces and then applying an energy balance to each surface at which the temperature is unknown. For the geometries of Figure C.1, with uniform temperatures $T_{s,1}$ and $T_{s,2}$ prescribed at each surface, solutions to appropriate forms of the heat equation are readily obtained and are summarized in Table C.1. The temperature distributions may be used with Fourier's law to obtain corresponding distributions for the heat flux and heat rate. If $T_{s,1}$ and $T_{s,2}$ are both known for a particular problem, the expressions of Table C.1 provide all that is needed to completely determine related thermal conditions. If $T_{s,1}$ and/or $T_{s,2}$ are not known, the results may still be used with surface energy balances to determine the desired thermal conditions.

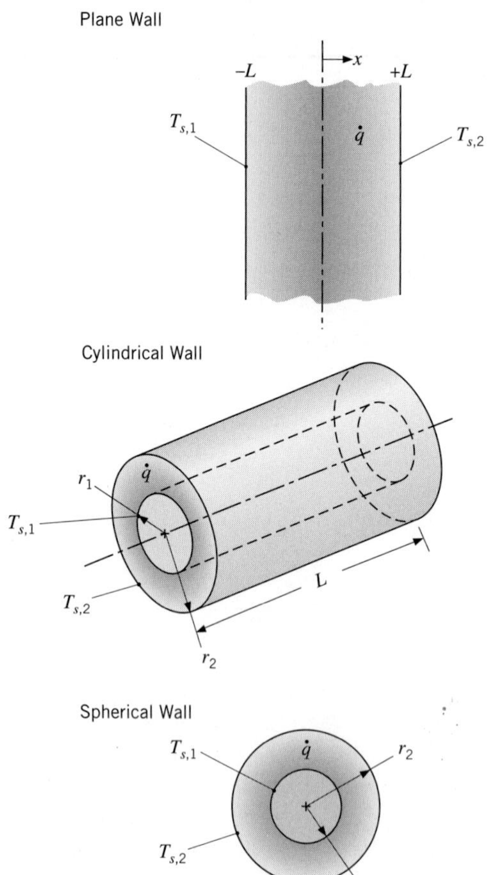

FIGURE C.1 One-dimensional conduction systems with uniform thermal energy generation: a plane wall with asymmetric surface conditions, a cylindrical shell, and a spherical shell.

TABLE C.1 One-Dimensional, Steady-State Solutions to the Heat Equation for Plane, Cylindrical, and Spherical Walls with Uniform Generation and Asymmetrical Surface Conditions

Temperature Distribution

Plane Wall

$$T(x) = \frac{\dot{q}L^2}{2k}\left(1 - \frac{x^2}{L^2}\right) + \frac{T_{s,2} - T_{s,1}}{2}\frac{x}{L} + \frac{T_{s,1} + T_{s,2}}{2} \quad (C.1)$$

Cylindrical Wall

$$T(r) = T_{s,2} + \frac{\dot{q}r_2^2}{4k}\left(1 - \frac{r^2}{r_2^2}\right) - \left[\frac{\dot{q}r_2^2}{4k}\left(1 - \frac{r_1^2}{r_2^2}\right) + (T_{s,2} - T_{s,1})\right]\frac{\ln(r_2/r)}{\ln(r_2/r_1)} \quad (C.2)$$

Spherical Wall

$$T(r) = T_{s,2} + \frac{\dot{q}r_2^2}{6k}\left(1 - \frac{r^2}{r_2^2}\right) - \left[\frac{\dot{q}r_2^2}{6k}\left(1 - \frac{r_1^2}{r_2^2}\right) + (T_{s,2} - T_{s,1})\right]\frac{(1/r) - (1/r_2)}{(1/r_1) - (1/r_2)} \quad (C.3)$$

Heat Flux

Plane Wall

$$q''(x) = \dot{q}x - \frac{k}{2L}(T_{s,2} - T_{s,1}) \quad (C.4)$$

Cylindrical Wall

$$q''(r) = \frac{\dot{q}r}{2} - \frac{k\left[\frac{\dot{q}r_2^2}{4k}\left(1 - \frac{r_1^2}{r_2^2}\right) + (T_{s,2} - T_{s,1})\right]}{r\ln(r_2/r_1)} \quad (C.5)$$

Spherical Wall

$$q''(r) = \frac{\dot{q}r}{3} - \frac{k\left[\frac{\dot{q}r_2^2}{6k}\left(1 - \frac{r_1^2}{r_2^2}\right) + (T_{s,2} - T_{s,1})\right]}{r^2[(1/r_1) - (1/r_2)]} \quad (C.6)$$

Heat Rate

Plane Wall

$$q(x) = \left[\dot{q}x - \frac{k}{2L}(T_{s,2} - T_{s,1})\right]A_x \quad (C.7)$$

Cylindrical Wall

$$q(r) = \dot{q}\pi L r^2 - \frac{2\pi L k}{\ln(r_2/r_1)} \cdot \left[\frac{\dot{q}r_2^2}{4k}\left(1 - \frac{r_1^2}{r_2^2}\right) + (T_{s,2} - T_{s,1})\right] \quad (C.8)$$

Spherical Wall

$$q(r) = \frac{\dot{q}4\pi r^3}{3} - \frac{4\pi k\left[\frac{\dot{q}r_2^2}{6k}\left(1 - \frac{r_1^2}{r_2^2}\right) + (T_{s,2} - T_{s,1})\right]}{(1/r_1) - (1/r_2)} \quad (C.9)$$

Alternative surface conditions could involve specification of a uniform surface heat flux (*boundary condition of the second kind,* Equation 2.32 or 2.33) or a convection condition (*boundary condition of the third kind,* Equation 2.34). In each case, the surface temperature would not be known but could be determined by applying a surface energy balance. The forms that such balances may take are summarized in Table C.2. Note that, to accommodate situations for which a surface of interest may adjoin a composite wall in which there is no generation, the boundary condition of the third kind has been applied by using the overall heat transfer coefficient U in lieu of the convection coefficient h.

TABLE C.2 Alternative Surface Conditions and Energy Balances for One-Dimensional, Steady-State Solutions to the Heat Equation for Plane, Cylindrical, and Spherical Walls with Uniform Generation

Plane Wall

Uniform Surface Heat Flux

$$x = -L: \quad q''_{s,1} = -\dot{q}L - \frac{k}{2L}(T_{s,2} - T_{s,1}) \tag{C.10}$$

$$x = +L: \quad q''_{s,2} = \dot{q}L - \frac{k}{2L}(T_{s,2} - T_{s,1}) \tag{C.11}$$

Prescribed Transport Coefficient and Ambient Temperature

$$x = -L: \quad U_1(T_{\infty,1} - T_{s,1}) = -\dot{q}L - \frac{k}{2L}(T_{s,2} - T_{s,1}) \tag{C.12}$$

$$x = +L: \quad U_2(T_{s,2} - T_{\infty,2}) = \dot{q}L - \frac{k}{2L}(T_{s,2} - T_{s,1}) \tag{C.13}$$

Cylindrical Wall

Uniform Surface Heat Flux

$$r = r_1: \quad q''_{s,1} = \frac{\dot{q}r_1}{2} - \frac{k\left[\frac{\dot{q}r_2^2}{4k}\left(1 - \frac{r_1^2}{r_2^2}\right) + (T_{s,2} - T_{s,1})\right]}{r_1 \ln(r_2/r_1)} \tag{C.14}$$

$$r = r_2: \quad q''_{s,2} = \frac{\dot{q}r_2}{2} - \frac{k\left[\frac{\dot{q}r_2^2}{4k}\left(1 - \frac{r_1^2}{r_2^2}\right) + (T_{s,2} - T_{s,1})\right]}{r_2 \ln(r_2/r_1)} \tag{C.15}$$

Prescribed Transport Coefficient and Ambient Temperature

$$r = r_1: \quad U_1(T_{\infty,1} - T_{s,1}) = \frac{\dot{q}r_1}{2} - \frac{k\left[\frac{\dot{q}r_2^2}{4k}\left(1 - \frac{r_1^2}{r_2^2}\right) + (T_{s,2} - T_{s,1})\right]}{r_1 \ln(r_2/r_1)} \tag{C.16}$$

$$r = r_2: \quad U_2(T_{s,2} - T_{\infty,2}) = \frac{\dot{q}r_2}{2} - \frac{k\left[\frac{\dot{q}r_2^2}{4k}\left(1 - \frac{r_1^2}{r_2^2}\right) + (T_{s,2} - T_{s,1})\right]}{r_2 \ln(r_2/r_1)} \tag{C.17}$$

Spherical Wall

Uniform Surface Heat Flux

$$r = r_1: \quad q''_{s,1} = \frac{\dot{q}r_1}{3} - \frac{k\left[\frac{\dot{q}r_2^2}{6k}\left(1 - \frac{r_1^2}{r_2^2}\right) + (T_{s,2} - T_{s,1})\right]}{r_1^2[(1/r_1) - (1/r_2)]} \tag{C.18}$$

$$r = r_2: \quad q''_{s,2} = \frac{\dot{q}r_2}{3} - \frac{k\left[\frac{\dot{q}r_2^2}{6k}\left(1 - \frac{r_1^2}{r_2^2}\right) + (T_{s,2} - T_{s,1})\right]}{r_2^2[(1/r_1) - (1/r_2)]} \tag{C.19}$$

TABLE C.2 *Continued*

Prescribed Transport Coefficient and Ambient Temperature

$$r = r_1: \quad U_1(T_{\infty,1} - T_{s,1}) = \frac{\dot{q}r_1}{3} - \frac{k\left[\dfrac{\dot{q}r_2^2}{6k}\left(1 - \dfrac{r_1^2}{r_2^2}\right) + (T_{s,2} - T_{s,1})\right]}{r_1^2[(1/r_1) - (1/r_2)]} \tag{C.20}$$

$$r = r_2: \quad U_2(T_{s,2} - T_{\infty,2}) = \frac{\dot{q}r_2}{3} - \frac{k\left[\dfrac{\dot{q}r_2^2}{6k}\left(1 - \dfrac{r_1^2}{r_2^2}\right) + (T_{s,2} - T_{s,1})\right]}{r_2^2[(1/r_1) - (1/r_2)]} \tag{C.21}$$

As an example, consider a plane wall for which a uniform (known) surface temperature $T_{s,1}$ is prescribed at $x = -L$ and a uniform heat flux $q''_{s,2}$ is prescribed at $x = +L$. Equation C.11 may be used to evaluate $T_{s,2}$, and Equations C.1, C.4, and C.7 may then be used to determine the temperature, heat flux, and heat rate distributions, respectively.

Special cases of the foregoing configurations involve a plane wall with one adiabatic surface, a solid cylinder (a circular rod), and a sphere (Figure C.2). Subject to the requirements that $dT/dx|_{x=0} = 0$ and $dT/dr|_{r=0} = 0$, the corresponding forms of the heat equation may be solved to obtain Equations C.22 through C.24 of Table C.3. The solutions are based on prescribing a

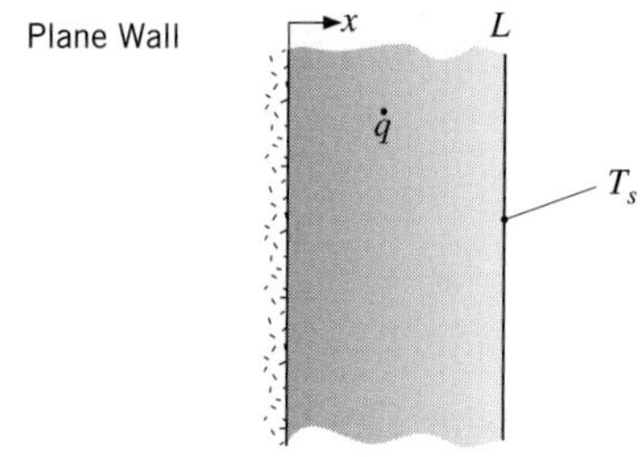

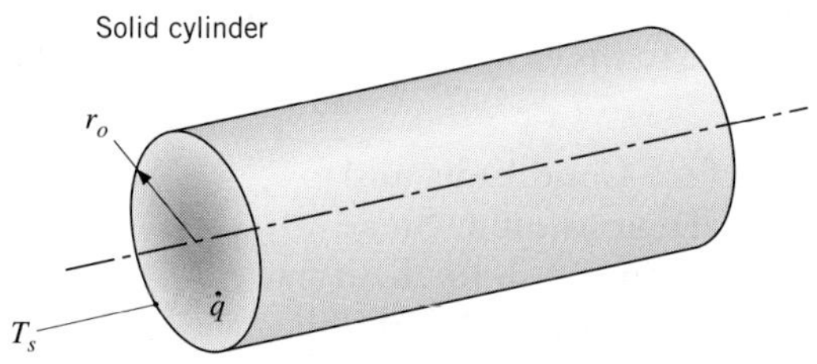

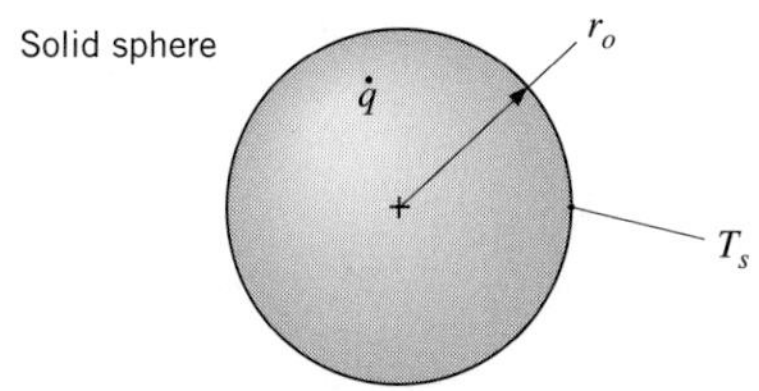

FIGURE C.2 One-dimensional conduction systems with uniform thermal energy generation: a plane wall with one adiabatic surface, a cylindrical rod, and a sphere.

Table C.3 One-Dimensional, Steady-State Solutions to the Heat Equation for Uniform Generation in a Plane Wall with One Adiabatic Surface, a Solid Cylinder, and a Solid Sphere

	Temperature Distribution	
Plane Wall	$T(x) = \frac{\dot{q}L^2}{2k}\left(1 - \frac{x^2}{L^2}\right) + T_s$	(C.22)
Circular Rod	$T(r) = \frac{\dot{q}r_o^2}{4k}\left(1 - \frac{r^2}{r_o^2}\right) + T_s$	(C.23)
Sphere	$T(r) = \frac{\dot{q}r_o^2}{6k}\left(1 - \frac{r^2}{r_o^2}\right) + T_s$	(C.24)
	Heat Flux	
Plane Wall	$q''(x) = \dot{q}x$	(C.25)
Circular Rod	$q''(r) = \frac{\dot{q}r}{2}$	(C.26)
Sphere	$q''(r) = \frac{\dot{q}r}{3}$	(C.27)
	Heat Rate	
Plane Wall	$q(x) = \dot{q}xA_x$	(C.28)
Circular Rod	$q(r) = \dot{q}\pi Lr^2$	(C.29)
Sphere	$q(r) = \frac{\dot{q}4\pi r^3}{3}$	(C.30)

uniform temperature T_s at $x = L$ and $r = r_o$. Using Fourier's law with the temperature distributions, the heat flux (Equations C.25 through C.27) and heat rate (Equations C.28 through C.30) distributions may also be obtained. If T_s is not known, it may be determined by applying a surface energy balance, appropriate forms of which are summarized in Table C.4.

Table C.4 Alternative Surface Conditions and Energy Balances for One-Dimensional, Steady-State Solutions to the Heat Equation for Uniform Generation in a Plane Wall with One Adiabatic Surface, a Solid Cylinder, and a Solid Sphere

Prescribed Transport Coefficient and Ambient Temperature

Plane Wall

$x = L$: $\quad \dot{q}L = U(T_s - T_\infty)$ (C.31)

Circular Rod

$r = r_o$: $\quad \frac{\dot{q}r_o}{2} = U(T_s - T_\infty)$ (C.32)

Sphere

$r = r_o$: $\quad \frac{\dot{q}r_o}{3} = U(T_s - T_\infty)$ (C.33)

APPENDIX D

The Gauss–Seidel Method

The Gauss–Seidel method is an example of an iterative approach for solving systems of linear algebraic equations, such as that represented by Equation 4.47, reproduced below.

$$\begin{aligned} a_{11}T_1 + a_{12}T_2 + a_{13}T_3 + \cdots + a_{1N}T_N &= C_1 \\ a_{21}T_1 + a_{22}T_2 + a_{23}T_3 + \cdots + a_{2N}T_N &= C_2 \\ \vdots \quad\quad \vdots \quad\quad \vdots \quad\quad \vdots \quad\quad \vdots \quad &\quad \vdots \\ a_{N1}T_1 + a_{N2}T_2 + a_{N3}T_3 + \cdots + a_{NN}T_N &= C_N \end{aligned} \tag{4.47}$$

For small numbers of equations, Gauss–Seidel iteration can be performed by hand. Application of the Gauss–Seidel method to the system of equations represented by Equation 4.47 is facilitated by the following procedure.

1. To whatever extent possible, the equations should be reordered to provide diagonal elements whose magnitudes are larger than those of other elements in the same row. That is, it is desirable to sequence the equations such that $|a_{11}| > |a_{12}|, |a_{13}|, \ldots, |a_{1N}|$; $|a_{22}| > |a_{21}|, |a_{23}|, \ldots, |a_{2N}|$; and so on.
2. After reordering, each of the N equations should be written in explicit form for the temperature associated with its diagonal element. Each temperature in the solution vector would then be of the form

$$T_i^{(k)} = \frac{C_i}{a_{ii}} - \sum_{j=1}^{i-1} \frac{a_{ij}}{a_{ii}} T_j^{(k)} - \sum_{j=i+1}^{N} \frac{a_{ij}}{a_{ii}} T_j^{(k-1)} \tag{D.1}$$

   where $i = 1, 2, \ldots, N$. The superscript k refers to the level of the iteration.
3. An initial ($k = 0$) value is assumed for each temperature T_i. Subsequent computations may be reduced by selecting values based on rational estimates of the actual solution.
4. Setting $k = 1$ in Equation D.1, values of $T_i^{(1)}$ are then calculated by substituting assumed (second summation, $k - 1 = 0$) or new (first summation, $k = 1$) values of T_j into the right-hand side. This step is the first ($k = 1$) iteration.
5. Using Equation D.1, the iteration procedure is continued by calculating new values of $T_i^{(k)}$ from the $T_j^{(k)}$ values of the current iteration, where $1 \le j \le i - 1$, and the $T_j^{(k-1)}$ values of the previous iteration, where $i + 1 \le j \le N$.
6. The iteration is terminated when a prescribed *convergence criterion* is satisfied. The criterion may be expressed as

$$\left|T_i^{(k)} - T_i^{(k-1)}\right| \le \varepsilon \tag{D.2}$$

   where ε represents an error in the temperature that is considered to be acceptable.

If step 1 can be accomplished for each equation, the resulting system is said to be *diagonally dominant,* and the rate of convergence is maximized (the number of required iterations is minimized). However, convergence may also be achieved in many situations for which diagonal dominance cannot be obtained, although the rate of convergence is slowed. The manner in which new values of T_i are computed (steps 4 and 5) should also be noted. Because the T_i for a particular iteration are calculated sequentially, each value can be computed by using the *most recent estimates* of the other T_i. This feature is implicit in Equation D.1, where the value of each unknown is updated as soon as possible, that is, for $1 \le j \le i - 1$.

An example problem that utilizes the Gauss–Seidel method is included in Section 4S.2.

APPENDIX E

The Convection Transfer Equations

In Chapter 2 we considered a stationary substance in which heat is transferred by conduction and developed means for determining the temperature distribution within the substance. We did so by applying *conservation of energy* to a differential control volume (Figure 2.11) and deriving a differential equation that was termed the *heat equation*. For a prescribed geometry and boundary conditions, the equation may be solved to determine the corresponding temperature distribution.

If the substance is not stationary, conditions become more complex. For example, if conservation of energy is applied to a differential control volume in a moving fluid, the effects of fluid motion (*advection*) on energy transfer across the surfaces of the control volume must be considered, along with those of conduction. The resulting differential equation, which provides the basis for predicting the temperature distribution, now requires knowledge of the velocity equations derived by applying *conservation of mass* and *Newton's second law of motion* to a differential control volume.

In this appendix we consider conditions involving flow of a *viscous fluid* in which there is concurrent *heat transfer*. We restrict our attention to the *steady, two-dimensional flow* of an *incompressible fluid* with *constant properties* in the x- and y-directions of a Cartesian coordinate system, and present the differential equations that may be used to predict velocity and temperature fields within the fluid. These equations can be derived by applying Newton's second law of motion and conservation of mass and energy to a differential control volume in the fluid.

E.1 Conservation of Mass

One conservation law that is pertinent to the flow of a viscous fluid is that matter can be neither created nor destroyed. For steady flow, this law requires that *the net rate at which mass enters a control volume* (inflow − outflow) *must equal zero*. Applying this law to a differential control volume in the flow yields

$$\frac{\partial u}{\partial x} + \frac{\partial v}{\partial y} = 0 \tag{E.1}$$

where u and v are the x- and y-components of the *mass average velocity*.

Equation E.1, the *continuity equation*, is a general expression of the *overall mass* conservation requirement, and it must be satisfied at every point in the fluid, provided that the fluid can be approximated as *incompressible*, that is, constant density.

E.2 Newton's Second Law of Motion

The second fundamental law that is pertinent to the flow of a viscous fluid is *Newton's second law of motion*. For a differential control volume in the fluid, under steady conditions, this requirement states that *the sum of all forces acting on the control volume must equal the net rate at which momentum leaves the control volume* (outflow − inflow).

Two kinds of forces may act on the fluid: *body forces*, which are proportional to the volume, and *surface forces*, which are proportional to area. Gravitational, centrifugal, magnetic, and/or electric fields may contribute to the total body force, and we designate the x- and

---

These equations are derived in Section 6S.1.

y-components of this force per unit volume of fluid as X and Y, respectively. The surface forces are due to the fluid static pressure as well as to *viscous stresses*.

Applying Newton's second law of motion (in the x- and y-directions) to a differential control volume in the fluid, accounting for body and surface forces, yields

$$\rho\left(u\frac{\partial u}{\partial x} + v\frac{\partial u}{\partial y}\right) = -\frac{\partial p}{\partial x} + \mu\left(\frac{\partial^2 u}{\partial x^2} + \frac{\partial^2 u}{\partial y^2}\right) + X \tag{E.2}$$

$$\rho\left(u\frac{\partial v}{\partial x} + v\frac{\partial v}{\partial y}\right) = -\frac{\partial p}{\partial y} + \mu\left(\frac{\partial^2 v}{\partial x^2} + \frac{\partial^2 v}{\partial y^2}\right) + Y \tag{E.3}$$

where p is the pressure and μ is the fluid viscosity.

We should not lose sight of the physics represented by Equations E.2 and E.3. The two terms on the left-hand side of each equation represent the *net* rate of momentum flow from the control volume. The terms on the right-hand side, taken in order, account for the net pressure force, the net viscous forces, and the body force. These equations must be satisfied at each point in the fluid, and with Equation E.1 they may be solved for the velocity field.

E.3 *Conservation of Energy*

As mentioned at the beginning of this Appendix, in Chapter 2 we considered a stationary substance in which heat is transferred by conduction and applied conservation of energy to a differential control volume (Figure 2.11) to derive the heat equation. When conservation of energy is applied to a differential control volume *in a moving fluid* under steady conditions, it expresses that the net rate at which energy enters the control volume, plus the rate at which heat is added, minus the rate at which work is done by the fluid in the control volume, is equal to zero. After much manipulation, the result can be rewritten as a *thermal energy equation.* For steady, two-dimensional flow of an incompressible fluid with constant properties, the resulting differential equation is

$$\rho c_p\left(u\frac{\partial T}{\partial x} + v\frac{\partial T}{\partial y}\right) = k\left(\frac{\partial^2 T}{\partial x^2} + \frac{\partial^2 T}{\partial y^2}\right) + \mu\Phi + \dot{q} \tag{E.4}$$

where T is the temperature, c_p is the specific heat at constant pressure, k is the thermal conductivity, $\dot{q}$ is the volumetric rate of thermal energy generation, and $\mu\Phi$, the *viscous dissipation*, is defined as

$$\mu\Phi \equiv \mu\left\{\left(\frac{\partial u}{\partial y} + \frac{\partial v}{\partial x}\right)^2 + 2\left[\left(\frac{\partial u}{\partial x}\right)^2 + \left(\frac{\partial v}{\partial y}\right)^2\right]\right\} \tag{E.5}$$

The same form of the thermal energy equation, Equation E.4, also applies to an ideal gas with negligible pressure variation.

In Equation E.4, the terms on the left-hand side account for the net rate at which thermal energy leaves the control volume due to bulk fluid motion (advection), while the terms on the right-hand side account for net inflow of energy due to conduction, viscous dissipation, and generation. Viscous dissipation represents the net rate at which mechanical work is irreversibly converted to thermal energy due to viscous effects in the fluid. The generation term characterizes conversion from other forms of energy (such as chemical, electrical, electromagnetic, or nuclear) to thermal energy.

APPENDIX F

Boundary Layer Equations for Turbulent Flow

It has been noted in Section 6.3 that turbulent flow is inherently *unsteady*. This behavior is shown in Figure F.1, where the variation of an arbitrary flow property P is plotted as a function of time at some location in a turbulent boundary layer. The property P could be a velocity component or the fluid temperature, and at any instant it may be represented as the sum of a *time-mean* value $\overline{P}$ and a fluctuating component P'. The average is taken over a time that is large compared with the period of a typical fluctuation, and if $\overline{P}$ is independent of time, the time-mean flow is said to be *steady*.

Since engineers are typically concerned with the time-mean properties, $\overline{P}$, the difficulty of solving the time-dependent governing equations is often eliminated by averaging the equations over time. For steady (in the mean), incompressible, constant property, boundary layer flow with negligible viscous dissipation, using well-established time-averaging procedures [1], the following forms of the continuity, x-momentum, and energy conservation equations may be obtained:

$$\frac{\partial \overline{u}}{\partial x} + \frac{\partial \overline{v}}{\partial y} = 0 \tag{F.1}$$

$$\overline{u}\frac{\partial \overline{u}}{\partial x} + \overline{v}\frac{\partial \overline{u}}{\partial y} = -\frac{1}{\rho}\frac{d\overline{p}_\infty}{dx} + \frac{1}{\rho}\frac{\partial}{\partial y}\left(\mu\frac{\partial \overline{u}}{\partial y} - \rho\overline{u'v'}\right) \tag{F.2}$$

$$\overline{u}\frac{\partial \overline{T}}{\partial x} + \overline{v}\frac{\partial \overline{T}}{\partial y} = \frac{1}{\rho c_p}\frac{\partial}{\partial y}\left(k\frac{\partial \overline{T}}{\partial y} - \rho c_p\overline{v'T'}\right) \tag{F.3}$$

The equations are like those for the laminar boundary layer, Equations 6.15 through 6.17 (after neglecting viscous dissipation), except for the presence of additional terms of the form $\overline{a'b'}$. These terms account for the effect of the turbulent fluctuations on momentum and energy transport.

On the basis of the foregoing results, it is customary to speak of a *total* shear stress and *total* heat flux, which are defined as

$$\tau_{\text{tot}} = \left(\mu\frac{\partial \overline{u}}{\partial y} - \rho\overline{u'v'}\right) \tag{F.4}$$

$$q''_{\text{tot}} = -\left(k\frac{\partial \overline{T}}{\partial y} - \rho c_p\overline{v'T'}\right) \tag{F.5}$$

and consist of contributions due to molecular diffusion and turbulent mixing. From the form of these equations we see how momentum and energy transfer rates are enhanced by the existence of turbulence. The term $-\rho\overline{u'v'}$ appearing in Equation F.4 represents the momentum

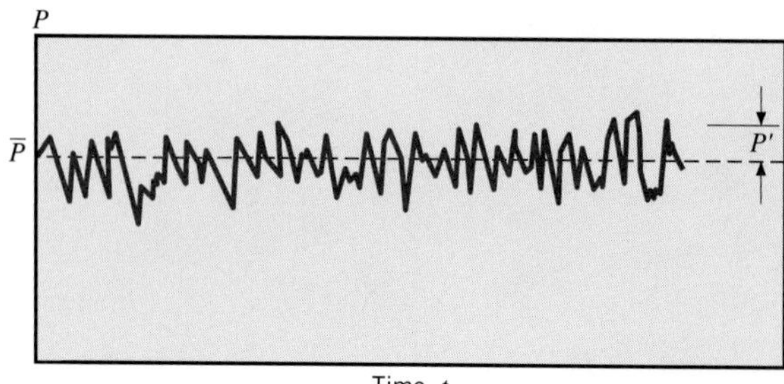

FIGURE F.1 Property variation with time at some point in a turbulent boundary layer.

flux due to the turbulent fluctuations, and it is often termed the *Reynolds stress*. The term $\rho c_p \overline{v'T'}$ in Equation F.5 represents the heat flux due to the turbulent fluctuations. Unfortunately, these new terms introduced by the time-averaging process are additional unknowns, so that the number of unknowns exceeds the number of equations. Resolving this problem is the subject of the field of *turbulence modeling* [2].

References

1. Kays, W. M., M. E. Crawford, and B. Weigand, *Convective Heat and Mass Transfer*, 4th ed., McGraw-Hill Higher Education, Boston, 2005.
2. Wilcox, D. C., *Turbulence Modeling for CFD*, 2nd ed., DCW Industries, La Cañada, 1998.

APPENDIX G

An Integral Laminar Boundary Layer Solution for Parallel Flow over a Flat Plate

An alternative approach to solving the boundary layer equations involves the use of an approximate *integral* method. The approach was originally proposed by von Kárman [1] in 1921 and first applied by Pohlhausen [2]. It is without the mathematical complications inherent in the *exact* (*similarity*) method of Section 7.2.1; yet it can be used to obtain reasonably accurate results for the key boundary layer parameters (δ, δ_t, C_f, and h). Although the method has been used with some success for a variety of flow conditions, we restrict our attention to parallel flow over a flat plate, subject to the same restrictions enumerated in Section 7.2.1, that is, *incompressible laminar flow* with *constant fluid properties* and *negligible viscous dissipation.*

To use the method, the boundary layer equations, Equations 7.3 through 7.5, must be cast in integral form. These forms are obtained by integrating the equations in the y-direction across the boundary layer. For example, integrating Equation 7.3, we obtain

$$\int_0^\delta \frac{\partial u}{\partial x}\,dy + \int_0^\delta \frac{\partial v}{\partial y}\,dy = 0 \tag{G.1}$$

or, since $v = 0$ at $y = 0$,

$$v(y = \delta) = -\int_0^\delta \frac{\partial u}{\partial x}\,dy \tag{G.2}$$

Similarly, from Equation 7.4, we obtain

$$\int_0^\delta u\frac{\partial u}{\partial x}\,dy + \int_0^\delta v\frac{\partial u}{\partial y}\,dy = \nu\int_0^\delta \frac{\partial}{\partial y}\left(\frac{\partial u}{\partial y}\right)dy$$

or, integrating the second term on the left-hand side by parts,

$$\int_0^\delta u\frac{\partial u}{\partial x}\,dy + uv\Big|_0^\delta - \int_0^\delta u\frac{\partial v}{\partial y}\,dy = \nu\frac{\partial u}{\partial y}\Big|_0^\delta$$

Substituting from Equations 7.3 and G.2, we obtain

$$\int_0^\delta u\frac{\partial u}{\partial x}\,dy - u_\infty\int_0^\delta \frac{\partial u}{\partial x}\,dy + \int_0^\delta u\frac{\partial u}{\partial x}\,dy = -\nu\frac{\partial u}{\partial y}\Big|_{y=0}$$

or

$$u_\infty\int_0^\delta \frac{\partial u}{\partial x}\,dy - \int_0^\delta 2u\frac{\partial u}{\partial x}\,dy = \nu\frac{\partial u}{\partial y}\Big|_{y=0}$$

Therefore

$$\int_0^\delta \frac{\partial}{\partial x}(u_\infty \cdot u - u \cdot u)\,dy = \nu\frac{\partial u}{\partial y}\Big|_{y=0}$$

Rearranging, we then obtain

$$\frac{d}{dx}\left[\int_0^\delta (u_\infty - u)u\,dy\right] = \nu\frac{\partial u}{\partial y}\Big|_{y=0} \tag{G.3}$$

Equation G.3 is the integral form of the boundary layer momentum equation. In a similar fashion, the following integral form of the boundary layer energy equation may be obtained:

$$\frac{d}{dx}\left[\int_0^{\delta_t} (T_\infty - T)u\,dy\right] = \alpha\frac{\partial T}{\partial y}\Big|_{y=0} \tag{G.4}$$

Equations G.3 through G.4 satisfy the x-momentum and energy requirements in an *integral* (or *average*) fashion over the entire boundary layer. In contrast, the original conservation equations, (7.4) and (7.5), satisfy the conservation requirements *locally,* that is, at each point in the boundary layer.

The integral equations can be used to obtain *approximate* boundary layer solutions. The procedure involves first *assuming* reasonable functional forms for the unknowns u and T in terms of the corresponding (*unknown*) boundary layer thicknesses. The assumed forms must satisfy appropriate boundary conditions. Substituting these forms into the integral equations, expressions for the boundary layer thicknesses may be determined and the assumed functional forms may then be completely specified. Although this method is approximate, it frequently leads to accurate results for the surface parameters.

Consider the hydrodynamic boundary layer, for which appropriate boundary conditions are

$$u(y=0) = \left.\frac{\partial u}{\partial y}\right|_{y=\delta} = 0 \qquad \text{and} \qquad u(y=\delta) = u_\infty$$

From Equation 7.4 it also follows that, since $u = v = 0$ at $y = 0$,

$$\left.\frac{\partial^2 u}{\partial y^2}\right|_{y=0} = 0$$

With the foregoing conditions, we could approximate the velocity profile as a third-degree polynomial of the form

$$\frac{u}{u_\infty} = a_1 + a_2\left(\frac{y}{\delta}\right) + a_3\left(\frac{y}{\delta}\right)^2 + a_4\left(\frac{y}{\delta}\right)^3$$

and apply the conditions to determine the coefficients a_1 to a_4. It is easily verified that $a_1 = a_3 = 0$, $a_2 = \frac{3}{2}$ and $a_4 = -\frac{1}{2}$, in which case

$$\frac{u}{u_\infty} = \frac{3}{2}\frac{y}{\delta} - \frac{1}{2}\left(\frac{y}{\delta}\right)^3 \tag{G.5}$$

The velocity profile is then specified in terms of the unknown boundary layer thickness δ. This unknown may be determined by substituting Equation G.5 into G.3 and integrating over y to obtain

$$\frac{d}{dx}\left(\frac{39}{280}u_\infty^2\,\delta\right) = \frac{3}{2}\frac{\nu u_\infty}{\delta}$$

Separating variables and integrating over x, we obtain

$$\frac{\delta^2}{2} = \frac{140}{13}\frac{\nu x}{u_\infty} + \text{constant}$$

However, since $\delta = 0$ at the leading edge of the plate ($x = 0$), the integration constant must be zero and

$$\delta = 4.64\left(\frac{\nu x}{u_\infty}\right)^{1/2} = \frac{4.64x}{Re_x^{1/2}} \tag{G.6}$$

Substituting Equation G.6 into Equation G.5 and evaluating $\tau_s = \mu(\partial u/\partial y)_s$, we also obtain

$$C_{f,x} = \frac{\tau_s}{\rho u_\infty^2/2} = \frac{0.646}{Re_x^{1/2}} \tag{G.7}$$

Despite the approximate nature of the foregoing procedure, Equations G.6 and G.7 compare quite well with results obtained from the exact solution, Equations 7.17 and 7.18.

In a similar fashion one could assume a temperature profile of the form

$$T^* = \frac{T - T_s}{T_\infty - T_s} = b_1 + b_2\left(\frac{y}{\delta_t}\right) + b_3\left(\frac{y}{\delta_t}\right)^2 + b_4\left(\frac{y}{\delta_t}\right)^3$$

and determine the coefficients from the conditions

$$T^*(y = 0) = \left.\frac{\partial T^*}{\partial y}\right|_{y=\delta_t} = 0$$

$$T^*(y = \delta_t) = 1$$

as well as

$$\left.\frac{\partial^2 T^*}{\partial y^2}\right|_{y=0} = 0$$

which is inferred from the energy equation (7.5). We then obtain

$$T^* = \frac{3}{2}\frac{y}{\delta_t} - \frac{1}{2}\left(\frac{y}{\delta_t}\right)^3 \tag{G.8}$$

Substituting Equations G.5 and G.8 into Equation G.4, we obtain, after some manipulation and assuming $Pr \gtrsim 1$,

$$\frac{\delta_t}{\delta} = \frac{Pr^{-1/3}}{1.026} \tag{G.9}$$

This result is in good agreement with that obtained from the exact solution, Equation 7.22. Moreover, the heat transfer coefficient may then be computed from

$$h = \frac{-k\,\partial T/\partial y|_{y=0}}{T_s - T_\infty} = \frac{3}{2}\frac{k}{\delta_t}$$

Substituting from Equations G.6 and G.9, we obtain

$$Nu_x = \frac{hx}{k} = 0.332\, Re_x^{1/2}\, Pr^{1/3} \tag{G.10}$$

This result agrees precisely with that obtained from the exact solution, Equation 7.21.

References

1. von Kármán, T., *Z. Angew. Math. Mech.*, **1,** 232, 1921.

2. Pohlhausen, K., *Z. Angew. Math. Mech.*, **1,** 252, 1921.

Index

NOTE: Page references preceded by a "W" refer to pages that are located on the Web site www.wiley.com/college/incropera. Page numbers followed by "n" refer to footnotes on the page.

D

E

Conversion Factors

Acceleration	1 m/s^2	= 4.2520×10^7 ft/h^2
Area	1 m^2	= 1550.0 $in.^2$
		= 10.764 ft^2
Density	1 kg/m^3	= 0.06243 lb_m/ft^3
Energy	1 J (0.2388 cal)	= 9.4782×10^{-4} Btu
Force	1 N	= 0.22481 lb_f
Heat transfer rate	1 W	= 3.4121 Btu/h
Heat flux	1 W/m^2	= 0.3170 $Btu/h \cdot ft^2$
Heat generation rate	1 W/m^3	= 0.09662 $Btu/h \cdot ft^3$
Heat transfer coefficient	1 $W/m^2 \cdot K$	= 0.17611 $Btu/h \cdot ft^2 \cdot °F$
Kinematic viscosity and diffusivities	1 m^2/s	= 3.875×10^4 ft^2/h
Latent heat	1 J/kg	= 4.2992×10^{-4} Btu/lb_m
Length	1 m	= 39.370 in.
		= 3.2808 ft
	1 km	= 0.62137 mile
Mass	1 kg	= 2.2046 lb_m
Mass density	1 kg/m^3	= 0.06243 lb_m/ft^3
Mass flow rate	1 kg/s	= 7936.6 lb_m/h
Power	1 kW	= 3412.1 Btu/h
		= 1.341 hp
Pressure and stress[1]	1 N/m^2 (1 Pa)	= 0.020885 lb_f/ft^2
		= 1.4504×10^{-4} $lb_f/in.^2$
		= 4.015×10^{-3} in. water
		= 2.953×10^{-4} in. Hg
	1.0133×10^5 N/m^2	= 1 standard atmosphere
	1×10^5 N/m^2	= 1 bar
Specific heat	1 $kJ/kg \cdot K$	= 0.2388 $Btu/lb_m \cdot °F$
Temperature	K	= (5/9)°R
		= (5/9)(°F + 459.67)
		= °C + 273.15
Temperature difference	1 K	= 1°C
		= (9/5)°R = (9/5)°F
Thermal conductivity	1 $W/m \cdot K$	= 0.57779 $Btu/h \cdot ft \cdot °F$
Thermal resistance	1 K/W	= 0.52753 $°F/h \cdot Btu$
Viscosity (dynamic)[2]	1 $N \cdot s/m^2$	= 2419.1 $lb_m/ft \cdot h$
		= 5.8015×10^{-6} $lb_f \cdot h/ft^2$
Volume	1 m^3	= 6.1023×10^4 $in.^3$
		= 35.315 ft^3
		= 264.17 gal (U.S.)
Volume flow rate	1 m^3/s	= 1.2713×10^5 ft^3/h
		= 2.1189×10^3 ft^3/min
		= 1.5850×10^4 gal/min

[1] The SI name for the quantity pressure is pascal (Pa) having units N/m^2 or $kg/m \cdot s^2$.
[2] Also expressed in equivalent units of $kg/s \cdot m$.